Features and Benefits

Mathematics Connections

Integrated and Applied

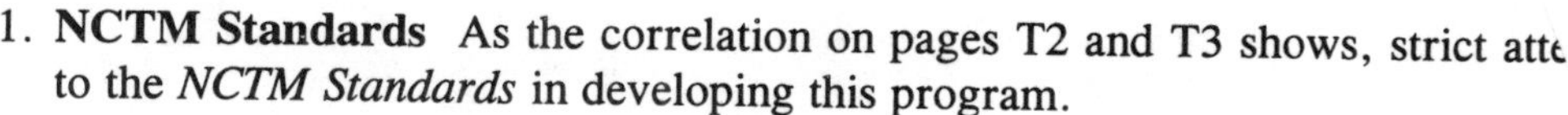

1. **NCTM Standards** As the correlation on pages T2 and T3 shows, strict att to the *NCTM Standards* in developing this program.

2. **Connections** Connections to interdisciplinary areas (page 101), consumerism 282-283), geometry (pages 58-59), and algebra (pages 26-27) enhance learning (page 85), and to various areas of mathematics, such as probability (pages while increasing student interest.

3. **Integrated** Algebra is introduced in Chapter 1 (pages 24-31) and integrated throughout the text (pages 286-287). Geometry (pages 92-93), statistics (page 117), consumer math (page 157), business math (page 141), and probability (pages 282-283) are other mathematical concepts integrated throughout the program.

4. **Applications** Because the ability to grasp concepts and skills is greatly enhanced when they are motivated by real-world applications, each lesson opens with an application. This is strengthened by including *Applications* in every exercise set (Exercises 30-33 on page 77). In addition, there are lessons that focus on specific applications (pages 74-75 and 84).

5. **Problem Solving** Each Problem-Solving Strategy lesson relates a strategy to a real-world situation (pages 136-137). Problem solving is a focus of many lessons, not just strategy lessons (pages 244-245).

6. **Technology** The calculator as a problem-solving tool is integrated throughout the program (Exercises 26 and 32-37 on page 49). Computer-related lessons are also included (page 222).

7. **Tech-Prep** The Tech Prep Handbook contains career information, educational prerequisites, mathematical exercises, and career-oriented projects on seven different careers in the tech-prep field (pages 516-518).

8. **Mixed Review** To help students maintain prior-taught skills and concepts, a *Mixed Review* is included in many of the lessons. Each problem in the mixed review is referenced to the related lesson (page 185).

9. **Extra Practice** With the exception of application and problem-solving lessons, extra practice is provided for all lessons (pages 448-484).

10. **Portfolios/Journals** *Portfolio* suggestions ask students to select items from their work that represents different aspects of their mathematical knowledge (page 97). *Journal Entries* give students the opportunity to keep a log of their thoughts and ideas about the mathematics they are studying (page 225).

11. **Teacher Support** The unique *Teacher's Wraparound Edition* includes all the elements that comprise a complete lesson plan for each lesson (pages 150-151).

12. **Teacher's Classroom Resources** This package contains an extensive set of supplementary materials that are designed to meet the complete spectrum of student needs and interests. See the brochure pages that follow.

Glencoe Presents ...

THE MATHEMATICS SKILLS YOUR STUDENTS NEED FOR REAL-WORLD SUCCESS

MATHEMATICS CONNECTIONS

INCLUDES A TECH PREP HANDBOOK

INCLUDES A TECH PREP HANDBOOK

GLENCOE

Meets NCTM Standards

FINALLY! A HIGH SCHOOL PROGRAM THAT MAKES MATHEMATICS UNDERSTANDABLE AND RELEVANT

Mathematics Connections: Integrated and Applied **brings an exciting new dimension to mathematics teaching and learning. Designed for students who've had little success with mathematics in the past, *Mathematics Connections* offers a new emphasis on tech prep applications to prepare your students for life and work in the real world.**

Focusing on seven career possibilities, this program helps your students see the importance of mathematics in daily life and motivates them to learn mathematical principles and skills.

A variety of engaging hands-on activities and applications helps students think critically and communicate mathematically.

New assessment strategies show you ways to gauge your students' mastery of concepts and skills.

An algebra strand is integrated throughout the text.

ACTIVITY: Similar Figures

Suppose you don't draw very well but you want to enlarge a picture of your favorite cartoon. There is a way you can draw it and make it look like the original.

Materials: pencil, half-inch grid paper, ruler

$\frac{1}{4}$-inch grid

$\frac{1}{2}$-inch grid

Procedures

1. A quarter-inch grid has been placed over the cartoon. How many squares long and how many squares wide is the grid?
2. Make sure the half-inch grid has the same number of squares.
3. Study the drawings above and explain how the grids are used to enlarge the cartoon.

On Your Own

4. Now trace the larger grid and all the lines within each square. Then complete the drawing.

Cooperative Groups

Materials: a favorite cartoon, pencil, half-inch grid paper

5. Make an enlargement of your favorite cartoon. Decide what you need to do to get started.
6. Does your cartoon look like the original? If so, congratulations!

Communicate Your Ideas

7. What is the same about your drawing and the original? How is your drawing different from the original?

SIMILAR FIGURES 279

- **Hands-on Activities** keep students interested in learning. Especially effective in groups, these activities encourage students to make observations, draw conclusions, and explore new ways of approaching mathematical problems.

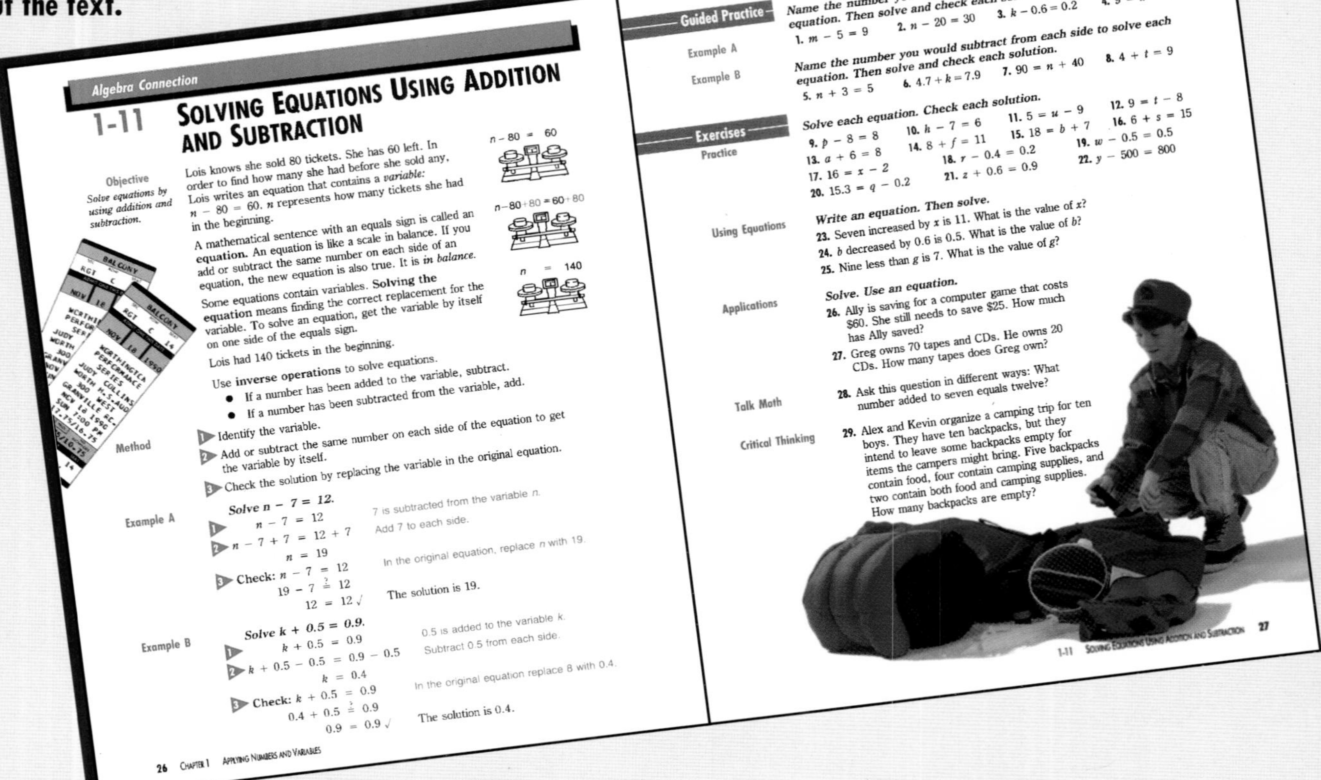

Algebra Connection

1-11 SOLVING EQUATIONS USING ADDITION AND SUBTRACTION

Objective
Solve equations by using addition and subtraction.

Lois knows she sold 80 tickets. She has 60 left. In order to find how many she had before she sold any, Lois writes an equation that contains a *variable:* $n - 80 = 60$. n represents how many tickets she had in the beginning.

A mathematical sentence with an equals sign is called an **equation.** An equation is like a scale in balance. If you add or subtract the same number on each side of an equation, the new equation is also true. It is *in balance.*

Some equations contain variables. **Solving the equation** means finding the correct replacement for the variable. To solve an equation, get the variable by itself on one side of the equals sign.

Lois had 140 tickets in the beginning.

$n - 80 = 60$

$n - 80 + 80 = 60 + 80$

$n = 140$

Use **inverse operations** to solve equations.
- If a number has been added to the variable, subtract.
- If a number has been subtracted from the variable, add.

Method

1. Identify the variable.
2. Add or subtract the same number on each side of the equation to get the variable by itself.
3. Check the solution by replacing the variable in the original equation.

Example A

Solve $n - 7 = 12$.

1. $n - 7 = 12$ — 7 is subtracted from the variable n.
2. $n - 7 + 7 = 12 + 7$ — Add 7 to each side.
   $n = 19$
3. Check: $n - 7 = 12$ — In the original equation, replace n with 19.
   $19 - 7 \stackrel{?}{=} 12$
   $12 = 12$ ✓ — The solution is 19.

Example B

Solve $k + 0.5 = 0.9$.

1. $k + 0.5 = 0.9$ — 0.5 is added to the variable k.
2. $k + 0.5 - 0.5 = 0.9 - 0.5$ — Subtract 0.5 from each side.
   $k = 0.4$
3. Check: $k + 0.5 = 0.9$ — In the original equation replace k with 0.4.
   $0.4 + 0.5 \stackrel{?}{=} 0.9$
   $0.9 = 0.9$ ✓ — The solution is 0.4.

26 CHAPTER 1 APPLYING NUMBERS AND VARIABLES

Guided Practice

Example A: *Name the number you would add to each side to solve each equation. Then solve and check each solution.*

1. $m - 5 = 9$ 2. $n - 20 = 30$ 3. $k - 0.6 = 0.2$ 4. $9 = x - 9$

Example B: *Name the number you would subtract from each side to solve each equation. Then solve and check each solution.*

5. $n + 3 = 5$ 6. $4.7 + k = 7.9$ 7. $90 = n + 40$ 8. $4 + t = 9$

Exercises

Practice: *Solve each equation. Check each solution.*

9. $p - 8 = 8$ 10. $h - 7 = 6$ 11. $5 = u - 9$ 12. $9 = t - 8$
13. $a + 6 = 8$ 14. $8 + f = 11$ 15. $18 = b + 7$ 16. $6 + s = 15$
17. $16 = x - 2$ 18. $r - 0.4 = 0.2$ 19. $w - 0.5 = 0.5$
20. $15.3 = q - 0.2$ 21. $z + 0.6 = 0.9$ 22. $y - 500 = 800$

Using Equations: *Write an equation. Then solve.*

23. Seven increased by x is 11. What is the value of x?
24. b decreased by 0.6 is 0.5. What is the value of b?
25. Nine less than g is 7. What is the value of g?

Applications: *Solve. Use an equation.*

26. Ally is saving for a computer game that costs $60. She still needs to save $25. How much has Ally saved?
27. Greg owns 70 tapes and CDs. He owns 20 CDs. How many tapes does Greg own?

Talk Math

28. Ask this question in different ways: What number added to seven equals twelve?

Critical Thinking

29. Alex and Kevin organize a camping trip for ten boys. They have ten backpacks, but they intend to leave some backpacks empty for items the campers might bring. Five backpacks contain food, four contain camping supplies, and two contain both food and camping supplies. How many backpacks are empty?

1-11 SOLVING EQUATIONS USING ADDITION AND SUBTRACTION 27

AN EMPHASIS ON CAREERS BRINGS MATHEMATICS TO LIFE

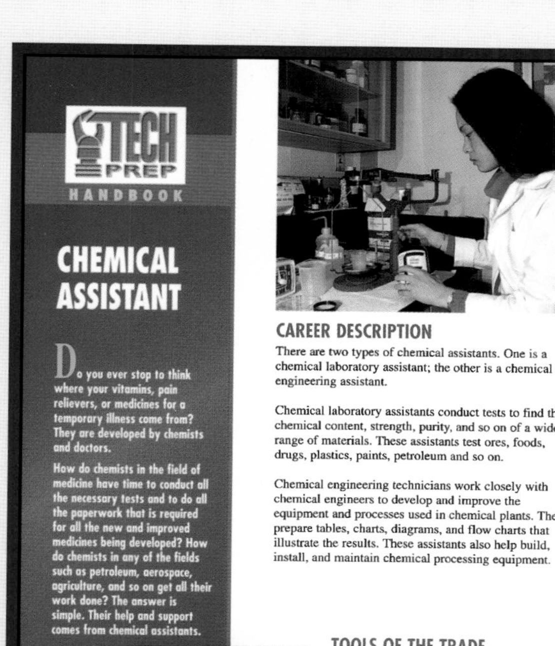

TECH PREP HANDBOOK

CHEMICAL ASSISTANT

Do you ever stop to think where your vitamins, pain relievers, or medicines for a temporary illness come from? They are developed by chemists and doctors.

How do chemists in the field of medicine have time to conduct all the necessary tests and to do all the paperwork that is required for all the new and improved medicines being developed? How do chemists in any of the fields such as petroleum, aerospace, agriculture, and so on get all their work done? The answer is simple. Their help and support comes from chemical assistants.

CAREER DESCRIPTION

There are two types of chemical assistants. One is a chemical laboratory assistant; the other is a chemical engineering assistant.

Chemical laboratory assistants conduct tests to find the chemical content, strength, purity, and so on of a wide range of materials. These assistants test ores, foods, drugs, plastics, paints, petroleum and so on.

Chemical engineering technicians work closely with chemical engineers to develop and improve the equipment and processes used in chemical plants. They prepare tables, charts, diagrams, and flow charts that illustrate the results. These assistants also help build, install, and maintain chemical processing equipment.

TOOLS OF THE TRADE

Some of the tools of the trade are gamma counters that test the amount of protein in cancer cells, laboratory equipment such as Bunsen burners, flasks and tubes, and large industrial machines. The tools are too numerous to mention all of them.

534 Tech Prep Handbook

- The **Tech Prep Handbook** in the Student Edition contains information on seven different careers in tech prep fields—**Laser Technician, Agribusiness Specialist, Graphic Arts Professional, Robotics Technician, Automotive Mechanic, Biomedical Equipment Specialist,** and **Chemical Assistant.**

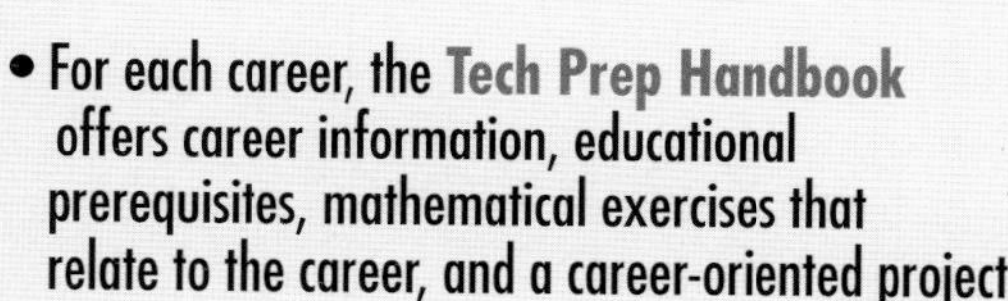

- For each career, the **Tech Prep Handbook** offers career information, educational prerequisites, mathematical exercises that relate to the career, and a career-oriented project.

On pages 531–533, you can learn how biomedical equipment specialists use mathematics in their jobs.

- Margin features in appropriate chapters direct students to the **Tech Prep Handbook.**

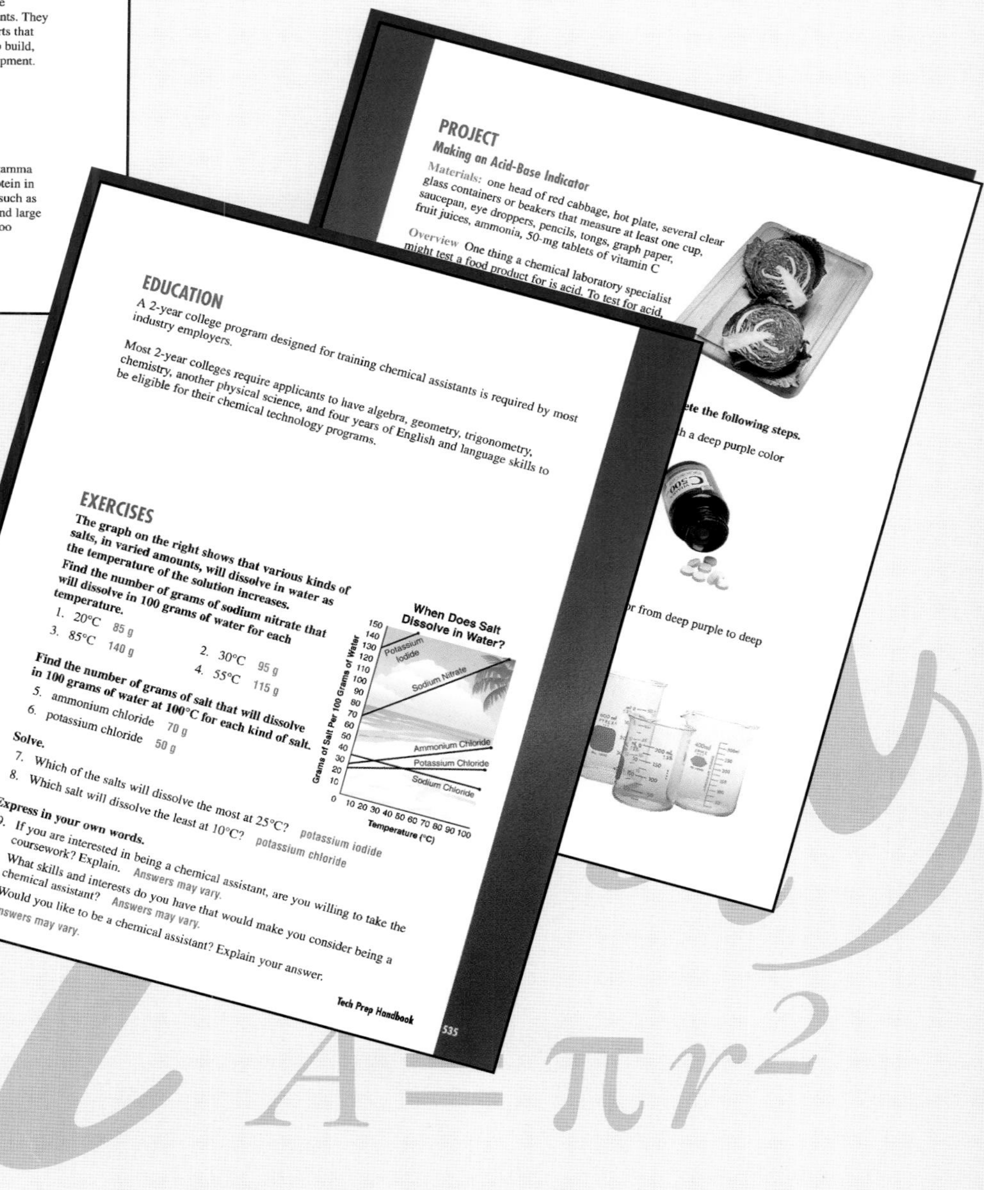

PROJECT

Making an Acid-Base Indicator

Materials: one head of red cabbage, hot plate, several clear glass containers or beakers that measure at least one cup, saucepan, eye droppers, pencils, tongs, graph paper, fruit juices, ammonia, 50-mg tablets of vitamin C

Overview One thing a chemical laboratory specialist might test a food product for is acid. To test for acid,

...ete the following steps.

...h a deep purple color

...r from deep purple to deep

EDUCATION

A 2-year college program designed for training chemical assistants is required by most industry employers.

Most 2-year colleges require applicants to have algebra, geometry, trigonometry, chemistry, another physical science, and four years of English and language skills to be eligible for their chemical technology programs.

EXERCISES

The graph on the right shows that various kinds of salts, in varied amounts, will dissolve in water as the temperature of the solution increases.

Find the number of grams of sodium nitrate that will dissolve in 100 grams of water for each temperature.

1. 20°C 85 g
2. 30°C 95 g
3. 85°C 140 g
4. 55°C 115 g

Find the number of grams of salt that will dissolve in 100 grams of water at 100°C for each kind of salt.

5. ammonium chloride 70 g
6. potassium chloride 50 g

Solve.

7. Which of the salts will dissolve the most at 25°C? potassium iodide
8. Which salt will dissolve the least at 10°C? potassium chloride

Express in your own words.

9. If you are interested in being a chemical assistant, are you willing to take the coursework? Explain. Answers may vary.
10. What skills and interests do you have that would make you consider being a chemical assistant? Answers may vary.
11. Would you like to be a chemical assistant? Explain your answer. Answers may vary.

Tech Prep Handbook 535

$A = \pi r^2$

APPLICATIONS AND COMMUNICATIONS FEATURES MOTIVATE AND ENRICH LEARNING

6. Select an assignment from this chapter that shows photos or sketches of physical models you made.

- **Portfolio** and **Journal Entry** suggestions strengthen your students' ability to think and communicate mathematically. Every chapter contains icons to identify **Journal** and **Portfolio** suggestions that enhance students writing and communication skills.

20. Name a situation where you would use a graph to prove a point to a committee. Tell what kind of graph you would use.

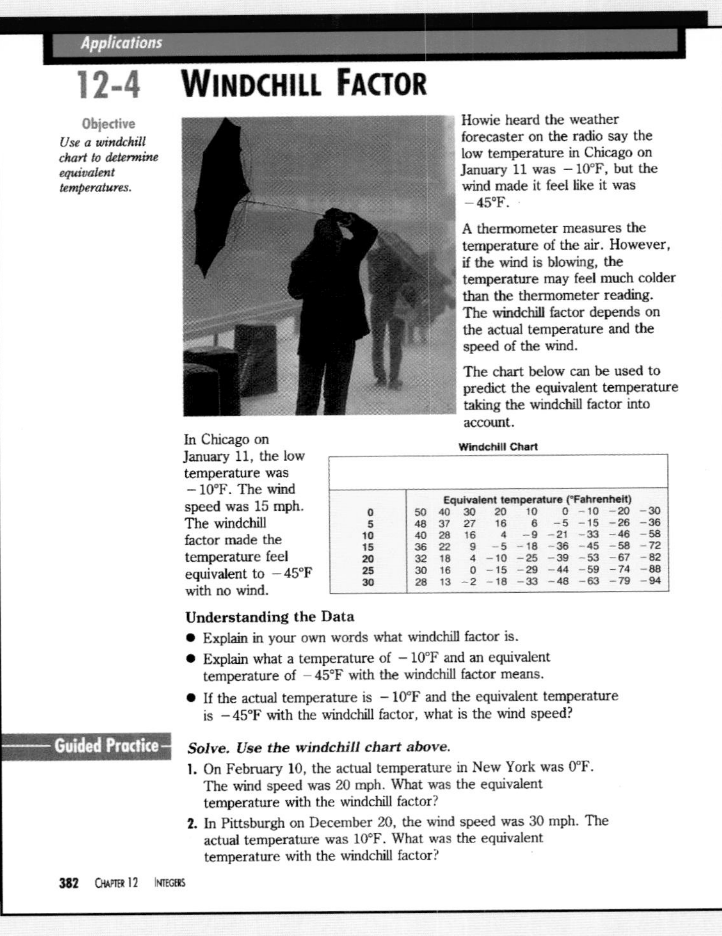

Applications

12-4 WINDCHILL FACTOR

Objective
Use a windchill chart to determine equivalent temperatures.

Howie heard the weather forecaster on the radio say the low temperature in Chicago on January 11 was −10°F, but the wind made it feel like it was −45°F.

A thermometer measures the temperature of the air. However, if the wind is blowing, the temperature may feel much colder than the thermometer reading. The windchill factor depends on the actual temperature and the speed of the wind.

The chart below can be used to predict the equivalent temperature taking the windchill factor into account.

In Chicago on January 11, the low temperature was −10°F. The wind speed was 15 mph. The windchill factor made the temperature feel equivalent to −45°F with no wind.

Windchill Chart

	Equivalent temperature (°Fahrenheit)								
0	50	40	30	20	10	0	−10	−20	−30
5	48	37	27	16	6	−5	−15	−26	−36
10	40	28	16	4	−9	−21	−33	−46	−58
15	36	22	9	−5	−18	−36	−45	−58	−72
20	32	18	4	−10	−25	−39	−53	−67	−82
25	30	16	0	−15	−29	−44	−59	−74	−88
30	28	13	−2	−18	−33	−48	−63	−79	−94

Understanding the Data

- Explain in your own words what windchill factor is.
- Explain what a temperature of −10°F and an equivalent temperature of −45°F with the windchill factor means.
- If the actual temperature is −10°F and the equivalent temperature is −45°F with the windchill factor, what is the wind speed?

Guided Practice

Solve. Use the windchill chart above.

1. On February 10, the actual temperature in New York was 0°F. The wind speed was 20 mph. What was the equivalent temperature with the windchill factor?
2. In Pittsburgh on December 20, the wind speed was 30 mph. The actual temperature was 10°F. What was the equivalent temperature with the windchill factor?

382 CHAPTER 12 INTEGERS

- **Applications** are integrated throughout the text to help your students apply what they've learned. A two-page **Application Lesson** in each chapter challenges students to apply mathematics principles to real-life situations and develops their capacity for problem solving.

UNIQUE FEATURES EXTEND CHAPTER CONTENT AND MAKE IT EASY FOR STUDENTS TO RETAIN WHAT THEY'VE LEARNED

- **Computer Applications** show students how to use this technology to solve problems.

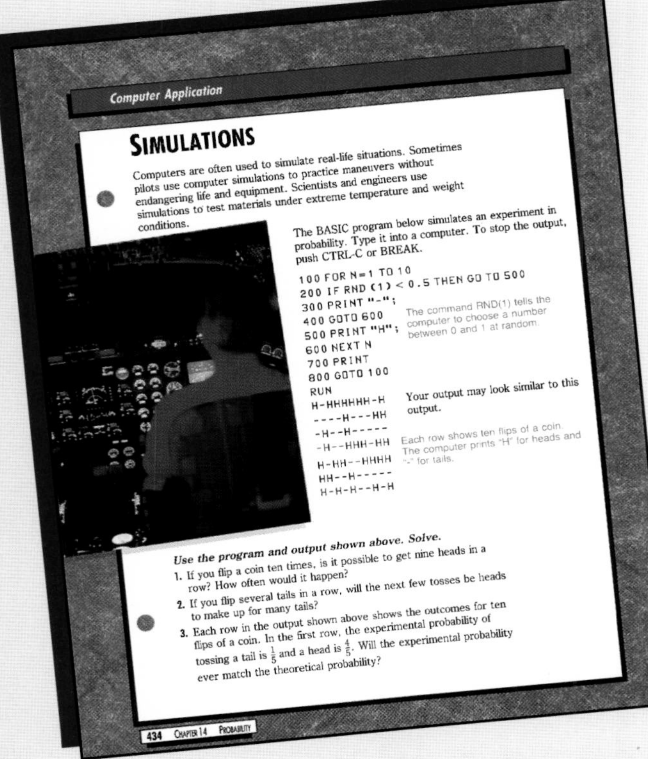

Computer Application

SIMULATIONS

Computers are often used to simulate real-life situations. Sometimes pilots use computer simulations to practice maneuvers without endangering life and equipment. Scientists and engineers use simulations to test materials under extreme temperature and weight conditions.

The BASIC program below simulates an experiment in probability. Type it into a computer. To stop the output, push CTRL-C or BREAK.

```
100 FOR N=1 TO 10
200 IF RND(1) < 0.5 THEN GO TO 500
300 PRINT "-";
400 GOTO 600
500 PRINT "H";
600 NEXT N
700 PRINT
800 GOTO 100
RUN
```

The command RND(1) tells the computer to choose a number between 0 and 1 at random.

Your output may look similar to this output.

```
H-HHHHHH-H
----H---HH
-H--H-----
-H--HHH-HH
H-HH--HHHH
HH--H-----
H-H-H--H-H
```

Each row shows ten flips of a coin. The computer prints "H" for heads and "-" for tails.

Use the program and output shown above. Solve.

1. If you flip a coin ten times, is it possible to get nine heads in a row? How often would it happen?
2. If you flip several tails in a row, will the next few tosses be heads to make up for many tails?
3. Each row in the output shown above shows the outcomes for ten flips of a coin. In the first row, the experimental probability of tossing a tail is $\frac{1}{5}$ and a head is $\frac{4}{5}$. Will the experimental probability ever match the theoretical probability?

434 CHAPTER 14 PROBABILITY

Find each percentage. Use a fraction for the percent.

17. 25% of 40 **18.** 80% of 200 **19.** 50% of 86

21. 80% of 915 **22.** $37\frac{1}{2}$% of 320 **23.** $87\frac{1}{2}$% of 8

- **Calculator Exercises** give students the opportunity to explore concepts.

Mixed Review			
Lesson 2-8	*Find the perimeter.* **40.** square; side, 13.3 cm		**41.** rectangle; length, 4 in. width, 3.2 in.
Lesson 5-7	*Divide.* **42.** $3\frac{1}{4} \div 2\frac{1}{2}$	**43.** $8\frac{2}{3} \div 2\frac{3}{4}$	**44.** $10\frac{3}{5} \div 3\frac{1}{3}$

- **Mixed Review** exercises reinforce learning by presenting a mixture of previously taught concepts. The reviews offer a lesson reference before each problem, making it easy for students to restudy important ideas and skills.

- **Chapter Reviews** and **Chapter Tests** contain free- response questions. **Cumulative Reviews/Tests** contain both free-response and multiple choice questions.

For information about **Performance Assessment**, see page 11.

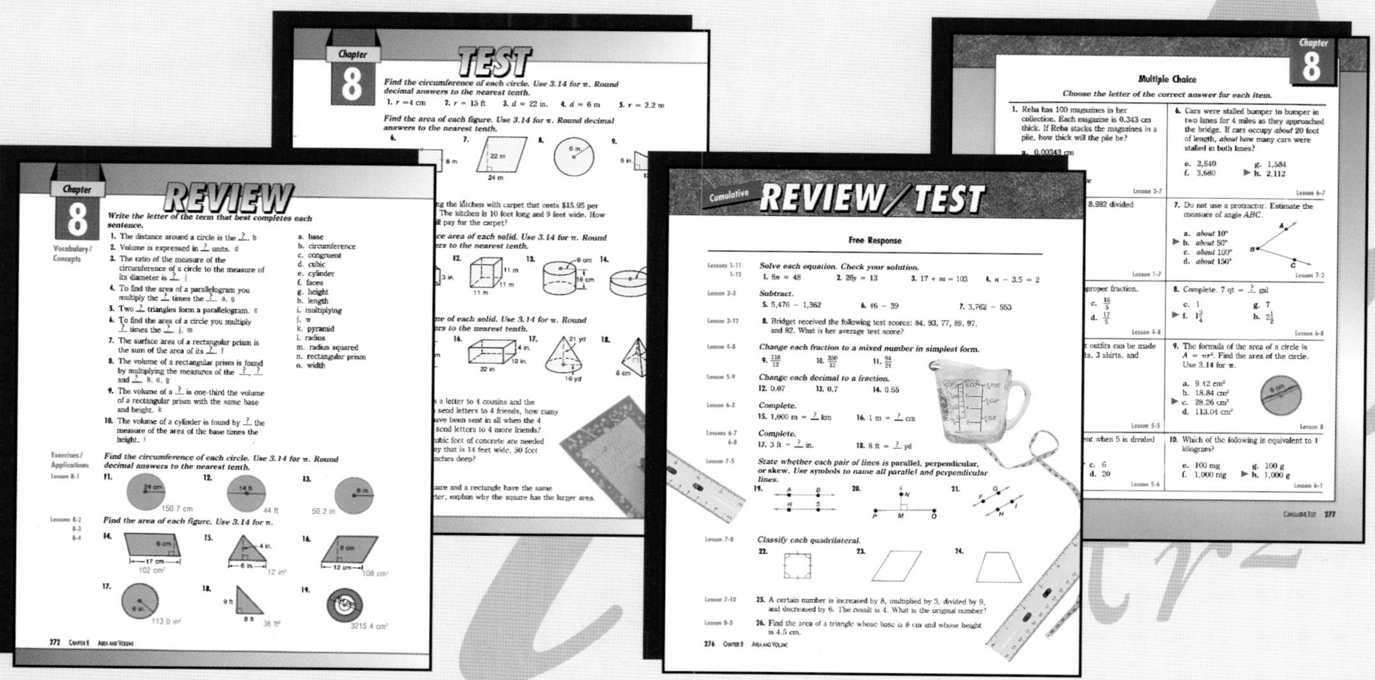

Chapter 8 TEST

Chapter 8 REVIEW

Cumulative REVIEW/TEST

Free Response

Chapter 8 Multiple Choice

CONNECTIONS AMONG MATHEMATICS DISCIPLINES GIVE YOUR STUDENTS A BETTER UNDERSTANDING OF MATHEMATICS

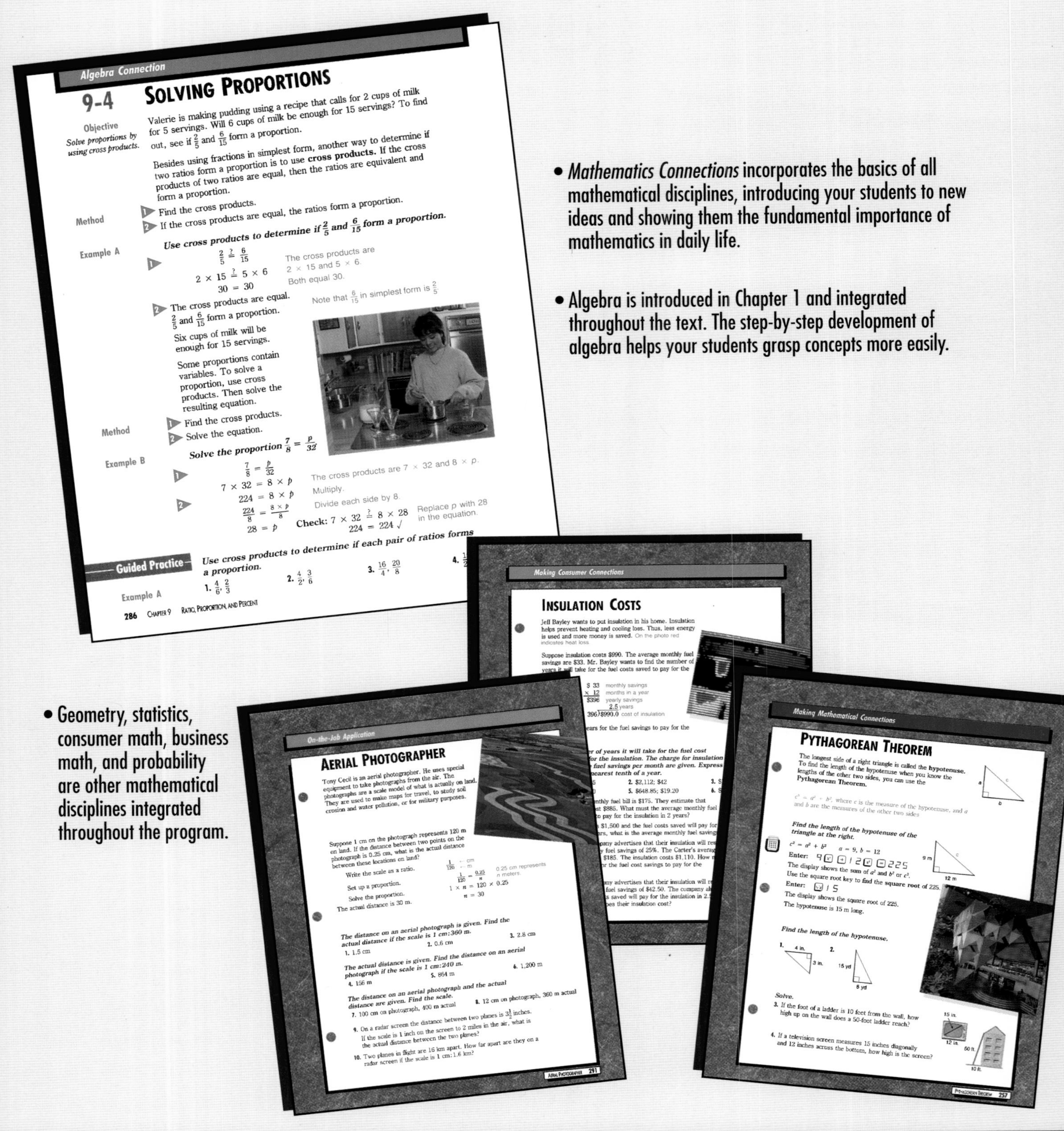
Algebra Connection

9-4 SOLVING PROPORTIONS

Objective
Solve proportions by using cross products.

Valerie is making pudding using a recipe that calls for 2 cups of milk for 5 servings. Will 6 cups of milk be enough for 15 servings? To find out, see if $\frac{2}{5}$ and $\frac{6}{15}$ form a proportion.

Besides using fractions in simplest form, another way to determine if two ratios form a proportion is to use **cross products.** If the cross products of two ratios are equal, then the ratios are equivalent and form a proportion.

Method
1. Find the cross products.
2. If the cross products are equal, the ratios form a proportion.

Example A
Use cross products to determine if $\frac{2}{5}$ and $\frac{6}{15}$ form a proportion.

1. $\frac{2}{5} \stackrel{?}{=} \frac{6}{15}$ — The cross products are 2×15 and 5×6.
$2 \times 15 \stackrel{?}{=} 5 \times 6$
$30 = 30$ — Both equal 30.

2. The cross products are equal. $\frac{2}{5}$ and $\frac{6}{15}$ form a proportion. — Note that $\frac{6}{15}$ in simplest form is $\frac{2}{5}$.
Six cups of milk will be enough for 15 servings.

Some proportions contain variables. To solve a proportion, use cross products. Then solve the resulting equation.

Method
1. Find the cross products.
2. Solve the equation.

Example B
Solve the proportion $\frac{7}{8} = \frac{p}{32}$.

1. $\frac{7}{8} = \frac{p}{32}$
$7 \times 32 = 8 \times p$ — The cross products are 7×32 and $8 \times p$.
2. $224 = 8 \times p$ — Multiply.
$\frac{224}{8} = \frac{8 \times p}{8}$ — Divide each side by 8.
$28 = p$ — **Check:** $7 \times 32 \stackrel{?}{=} 8 \times 28$, $224 = 224$ ✓ — Replace p with 28 in the equation.

Guided Practice
Use cross products to determine if each pair of ratios forms a proportion.

Example A
1. $\frac{4}{6}, \frac{2}{3}$ 2. $\frac{4}{2}, \frac{3}{6}$ 3. $\frac{16}{4}, \frac{20}{8}$ 4.

286 CHAPTER 9 RATIO, PROPORTION, AND PERCENT

Making Consumer Connections

INSULATION COSTS

Jeff Bayley wants to put insulation in his home. Insulation helps prevent heating and cooling loss. Thus, less energy is used and more money is saved.

Suppose insulation costs $990. The average monthly fuel savings are $33. Mr. Bayley wants to find the number of years it will take for the fuel costs saved to pay for the

On-the-Job Application

AERIAL PHOTOGRAPHER

Tony Cecil is an aerial photographer. He uses special equipment to take photographs from the air. The photographs are a scale model of what is actually on land. They are used to make maps for travel, to study soil erosion and water pollution, or for military purposes.

291

Making Mathematical Connections

PYTHAGOREAN THEOREM

The longest side of a right triangle is called the **hypotenuse.** To find the length of the hypotenuse when you know the lengths of the other two sides, you can use the Pythagorean Theorem.

257

- *Mathematics Connections* incorporates the basics of all mathematical disciplines, introducing your students to new ideas and showing them the fundamental importance of mathematics in daily life.
- Algebra is introduced in Chapter 1 and integrated throughout the text. The step-by-step development of algebra helps your students grasp concepts more easily.
- Geometry, statistics, consumer math, business math, and probability are other mathematical disciplines integrated throughout the program.

MATHEMATICS CONNECTIONS DEVELOPS CRITICAL-THINKING SKILLS AND PROBLEM-SOLVING ABILITIES

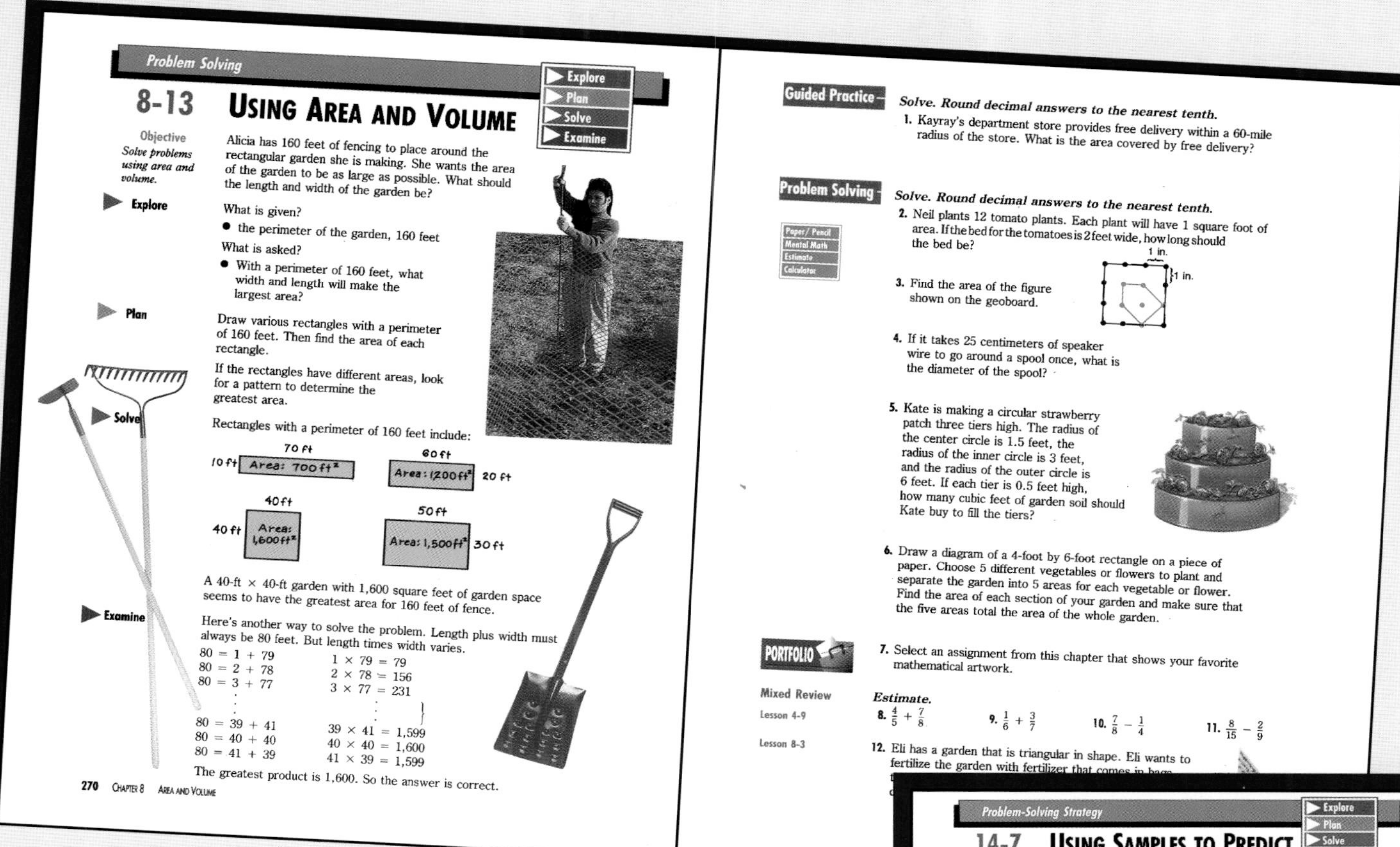

Problem Solving

8-13 USING AREA AND VOLUME

Explore
Plan
Solve
Examine

Objective
Solve problems using area and volume.

Alicia has 160 feet of fencing to place around the rectangular garden she is making. She wants the area of the garden to be as large as possible. What should the length and width of the garden be?

Explore

What is given?
- the perimeter of the garden, 160 feet

What is asked?
- With a perimeter of 160 feet, what width and length will make the largest area?

Plan

Draw various rectangles with a perimeter of 160 feet. Then find the area of each rectangle.

If the rectangles have different areas, look for a pattern to determine the greatest area.

Solve

Rectangles with a perimeter of 160 feet include:

A 40-ft × 40-ft garden with 1,600 square feet of garden space seems to have the greatest area for 160 feet of fence.

Examine

Here's another way to solve the problem. Length plus width must always be 80 feet. But length times width varies.

80 = 1 + 79	1 × 79 = 79
80 = 2 + 78	2 × 78 = 156
80 = 3 + 77	3 × 77 = 231
⋮	⋮
80 = 39 + 41	39 × 41 = 1,599
80 = 40 + 40	40 × 40 = 1,600
80 = 41 + 39	41 × 39 = 1,599

The greatest product is 1,600. So the answer is correct.

270 CHAPTER 8 AREA AND VOLUME

Guided Practice

Solve. Round decimal answers to the nearest tenth.

1. Kayray's department store provides free delivery within a 60-mile radius of the store. What is the area covered by free delivery?

Problem Solving

Paper/Pencil
Mental Math
Estimate
Calculator

Solve. Round decimal answers to the nearest tenth.

2. Neil plants 12 tomato plants. Each plant will have 1 square foot of area. If the bed for the tomatoes is 2 feet wide, how long should the bed be?
3. Find the area of the figure shown on the geoboard.

4. If it takes 25 centimeters of speaker wire to go around a spool once, what is the diameter of the spool?
5. Kate is making a circular strawberry patch three tiers high. The radius of the center circle is 1.5 feet, the radius of the inner circle is 3 feet, and the radius of the outer circle is 6 feet. If each tier is 0.5 feet high, how many cubic feet of garden soil should Kate buy to fill the tiers?
6. Draw a diagram of a 4-foot by 6-foot rectangle on a piece of paper. Choose 5 different vegetables or flowers to plant and separate the garden into 5 areas for each vegetable or flower. Find the area of each section of your garden and make sure that the five areas total the area of the whole garden.

PORTFOLIO

7. Select an assignment from this chapter that shows your favorite mathematical artwork.

Mixed Review

Estimate.

Lesson 4-9

8. $\frac{4}{5} + \frac{7}{8}$
9. $\frac{1}{6} + \frac{3}{7}$
10. $\frac{7}{8} - \frac{1}{4}$
11. $\frac{8}{15} - \frac{2}{9}$

Lesson 8-3

12. Eli has a garden that is triangular in shape. Eli wants to fertilize the garden with fertilizer that comes in bags

Problem-Solving Strategy

14-7 USING SAMPLES TO PREDICT

Explore
Plan
Solve
Examine

Objective
Solve verbal problems using samples to make predictions.

Johnson's Department Store conducts a survey of 75 shoppers in the mall and asks if they have a Johnson's credit card. Thirty of the 75 shoppers say that they have one. If about 3,200 people shop in the mall each day, predict how many of them probably have a Johnson's credit card.

Predictions about a large group of people based on the choices of a sample population are *estimates*. The best estimates are made when the sample population is representative of the larger group. For example, if you stand at the door of Johnson's Department Store to conduct the survey described above, the sample results would probably not be representative.

Explore

What is given?
- There are about 3,200 shoppers each day.
- In the sample of 75 shoppers, 30 have a credit card.

What is asked?
- How many shoppers in the mall probably have a Johnson's credit card?

Plan

Use the sample of 75 shoppers. Let n represent the total number of shoppers who probably have a Johnson's credit card. The ratio of n to 3,200 should be equivalent to the ratio of 30 to 75. Set up a proportion and solve for n.

Solve

$\frac{n}{3{,}200} = \frac{30}{75}$

3200 [×] 30 [÷] 75 [=] 1280

The number of shoppers that probably have a credit card is *about* 1,280.

Examine

30 out of 75 or less than half of the shoppers in the sample say they have a Johnson's credit card. Since half of 3,200 is 1,600, it seems reasonable to predict that about 1,280 shoppers have a Johnson's credit card.

Guided Practice

1. A sample of 100 seniors at Madison High School showed that 72 expect to continue their education after graduation. If there are 700 students in the class, predict the number of students who will continue their education.

440 CHAPTER 14 PROBABILITY

- **Problem-Solving** lessons, found throughout the text, give students opportunities to learn and use a four-step problem-solving plan—**Explore, Plan, Solve,** and **Examine**. The lessons show students that this method can be useful in solving real-life problems, as well as those found in the textbook.

- Each chapter contains a **Problem-Solving Strategy**, such as guess and check, work backwards, make a drawing, and make a list. These valuable techniques show students the many ways they can work through problems within the text and in other mathematics disciplines.

- **Critical thinking** exercises challenge students to develop and apply higher-order thinking skills.

Critical Thinking	**10.** Arrange eight congruent equilateral triangles to form a parallelogram.

THE TEACHER'S WRAPAROUND EDITION OFFERS UNIQUE TEACHING STRATEGIES TO HELP YOU REACH AND MOTIVATE YOUR STUDENTS

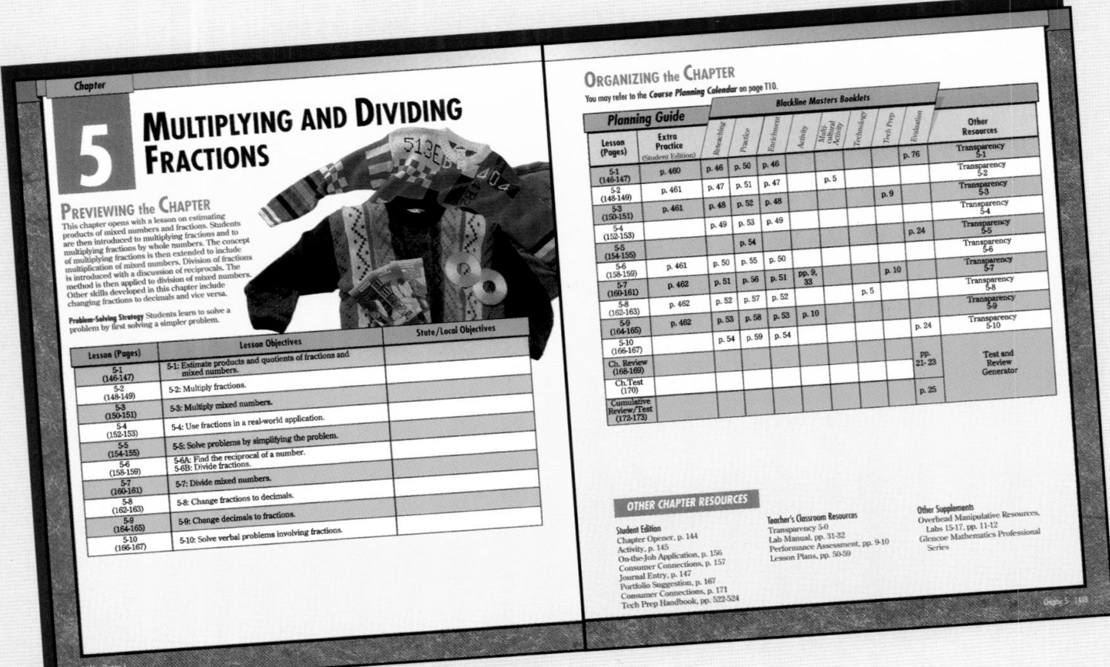

Chapter

5 MULTIPLYING AND DIVIDING FRACTIONS

PREVIEWING the CHAPTER

This chapter opens with a lesson on estimating products of mixed numbers and fractions. Students are then introduced to multiplying fractions and to multiplying fractions by whole numbers. The concept of multiplying fractions is then extended to include multiplication of mixed numbers. Division of fractions is introduced with a discussion of reciprocals. The method is then applied to division of mixed numbers. Other skills developed in this chapter include changing fractions to decimals and vice versa.

Problem-Solving Strategy Students learn to solve a problem by first solving a simpler problem.

Lesson (Pages)	Lesson Objectives	State/Local Objectives
5-1 (146-147)	5-1: Estimate products and quotients of fractions and mixed numbers.	
5-2 (148-149)	5-2: Multiply fractions.	
5-3 (150-151)	5-3: Multiply mixed numbers.	
5-4 (152-153)	5-4: Use fractions in a real-world application.	
5-5 (154-155)	5-5: Solve problems by simplifying the problem.	
5-6 (158-159)	5-6A: Find the reciprocal of a number. 5-6B: Divide fractions.	
5-7 (160-161)	5-7: Divide mixed numbers.	
5-8 (162-163)	5-8: Change fractions to decimals.	
5-9 (164-165)	5-9: Change decimals to fractions.	
5-10 (166-167)	5-10: Solve verbal problems involving fractions.	

ORGANIZING the CHAPTER

You may refer to the *Course Planning Calendar* on page T10.

Planning Guide

Lesson (Pages)	Extra Practice (Student Edition)	Blackline Masters Booklets: Reteaching	Practice	Enrichment	Activity	Multicultural Activity	Technology	Tech Prep	Evaluation	Other Resources
5-1 (146-147)	p. 460	p. 46	p. 50	p. 46					p. 76	Transparency 5-1
5-2 (148-149)	p. 461	p. 47	p. 51	p. 47		p. 5				Transparency 5-2
5-3 (150-151)	p. 461	p. 48	p. 52	p. 48				p. 9		Transparency 5-3
5-4 (152-153)		p. 49	p. 53	p. 49						Transparency 5-4
5-5 (154-155)			p. 54						p. 24	Transparency 5-5
5-6 (158-159)	p. 461	p. 50	p. 55	p. 50						Transparency 5-6
5-7 (160-161)	p. 462	p. 51	p. 56	p. 51	pp. 9, 33			p. 10		Transparency 5-7
5-8 (162-163)	p. 462	p. 52	p. 57	p. 52			p. 5			Transparency 5-8
5-9 (164-165)	p. 462	p. 53	p. 58	p. 53	p. 10					Transparency 5-9
5-10 (166-167)		p. 54	p. 59	p. 54					p. 24	Transparency 5-10
Ch. Review (168-169)									pp. 21-23	Test and Review Generator
Ch. Test (170)									p. 25	
Cumulative Review/Test (172-173)										

OTHER CHAPTER RESOURCES

Student Edition
Chapter Opener, p. 144
Activity, p. 145
On-the-Job Application, p. 156
Consumer Connections, p. 157
Journal Entry, p. 147
Portfolio Suggestion, p. 167
Consumer Connections, p. 171
Tech Prep Handbook, pp. 522-524

Teacher's Classroom Resources
Transparency 5-0
Lab Manual, pp. 31-32
Performance Assessment, pp. 9-10
Lesson Plans, pp. 50-59

Other Supplements
Overhead Manipulative Resources, Labs 15-17, pp. 11-12
Glencoe Mathematics Professional Series

- Four **Interleaf Pages** preceding each chapter preview the chapter, organize the content using a planning chart, and enhance chapter content.

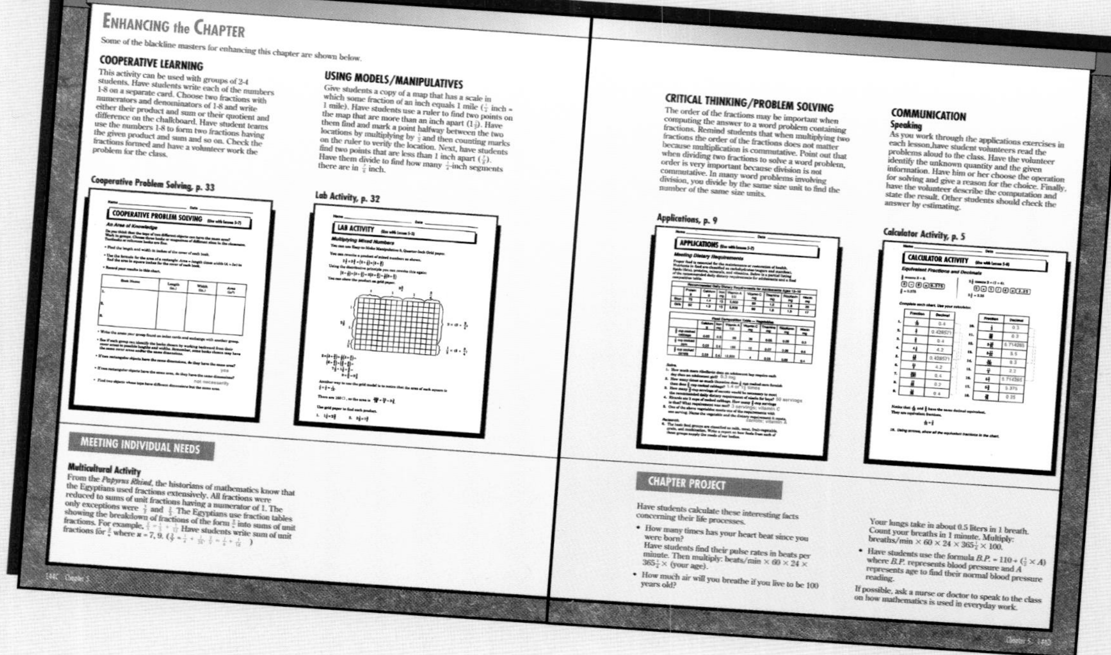

ENHANCING the CHAPTER

Some of the blackline masters for enhancing this chapter are shown below.

COOPERATIVE LEARNING

This activity can be used with groups of 2-4 students. Have students write each of the numbers 1-8 on a separate card. Choose two fractions with numerators and denominators of 1-8 and write either their product and sum or their quotient and difference on the chalkboard. Have student teams use the numbers 1-8 to form two fractions having the given product and sum and so on. Check the fractions formed and have a volunteer work the problem for the class.

USING MODELS/MANIPULATIVES

Give students a copy of a map that has a scale in which some fraction of an inch equals 1 mile ([illegible] inch = 1 mile). Have students use a ruler to find two points on the map that are more than an inch apart ([illegible]). Have them find and mark a point halfway between the two locations by multiplying by $\frac{1}{2}$ and then counting marks on the ruler to verify the location. Next, have students find two points that are less than 1 inch apart ([illegible]). Have them divide to find how many [illegible]-inch segments there are in [illegible] inch.

Cooperative Problem Solving, p. 33

COOPERATIVE PROBLEM SOLVING
An Area of Knowledge

Lab Activity, p. 32

LAB ACTIVITY
Multiplying Mixed Numbers

CRITICAL THINKING/PROBLEM SOLVING

The order of the fractions may be important when computing the answer to a word problem containing fractions. Remind students that when multiplying two fractions the order of the fractions does not matter because multiplication is commutative. Point out that when dividing two fractions to solve a word problem, order is very important because division is not commutative. In many word problems involving division, you divide by the same size unit to find the number of the same size units.

COMMUNICATION

Speaking

As you work through the applications exercises in each lesson,have student volunteers read the problems aloud to the class. Have the volunteer identify the unknown quantity and the given information. Have him or her choose the operation for solving and give a reason for the choice. Finally, have the volunteer describe the computation and state the result. Other students should check the answer by estimating.

Applications, p. 9

APPLICATIONS
Meeting Dietary Requirements

Calculator Activity, p. 5

CALCULATOR ACTIVITY
Equivalent Fractions and Decimals

MEETING INDIVIDUAL NEEDS

Multicultural Activity

From the *Papyrus Rhind*, the historians of mathematics know that the Egyptians used fractions extensively. All fractions were reduced to sums of unit fractions having a numerator of 1. The only exceptions were [illegible] and [illegible]. The Egyptians use fraction tables showing the breakdown of fractions of the form $\frac{2}{n}$ into sums of unit fractions. For example, [illegible]. Have students write sum of unit fractions for $\frac{2}{n}$ where $n = 7, 9$. ([illegible])

CHAPTER PROJECT

Have students calculate these interesting facts concerning their life processes.

- How many times has your heart beat since you were born?
  Have students find their pulse rates in beats per minute. Then multiply: beats/min $\times$ 60 $\times$ 24 $\times$ $365\frac{1}{4}$ $\times$ (your age).
- How much air will you breathe if you live to be 100 years old?
  Your lungs take in about 0.5 liters in 1 breath. Count your breaths in 1 minute. Multiply: breaths/min $\times$ 60 $\times$ 24 $\times$ $365\frac{1}{4}$ $\times$ 100.
- Have students use the formula $B.P. = 110 + (\frac{1}{2} \times A)$ where *B.P.* represents blood pressure and *A* represents age to find their normal blood pressure reading.

If possible, ask a nurse or doctor to speak to the class on how mathematics is used in everyday work.

- **Teaching Strategies** include Problem-Solving Activities, Lab Projects, Cooperative Learning, Multicultural Activities, Technology Features, Applications, and Connections Activities.

A SIX-STEP TEACHING PLAN MAKES IT EASY TO PLAN LESSONS AND MEET YOUR TEACHING OBJECTIVES

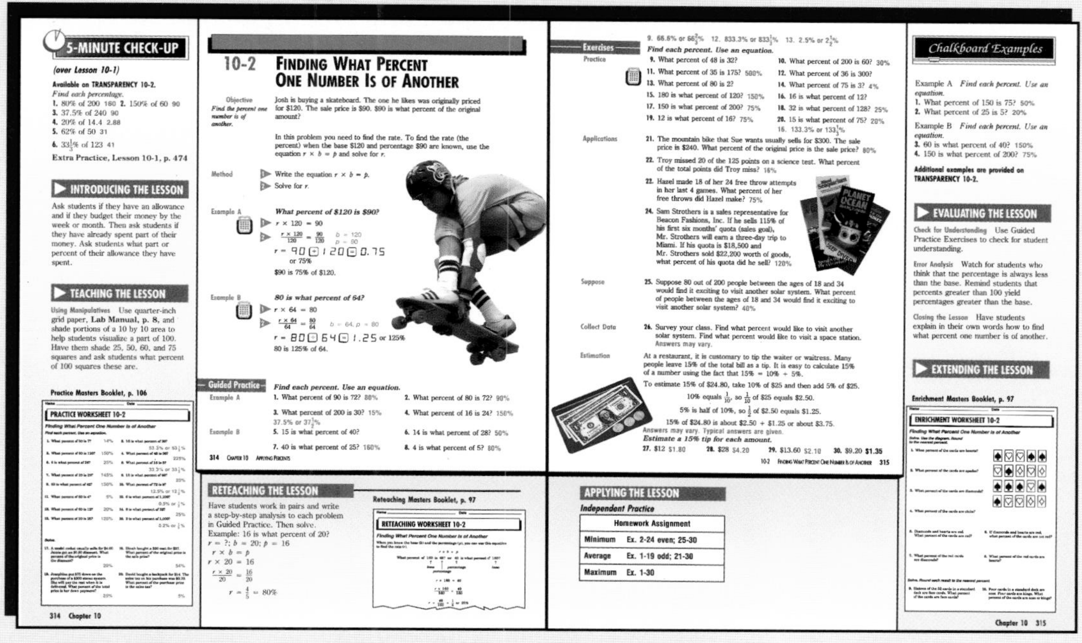

- The six-step plan is adaptable to all state teaching requirements and includes **Introducing, Teaching, Reteaching, Applying, Evaluating,** and **Extending the Lesson.**
- Side margins contain additional time-saving activities, including a **5-Minute Check** for bridging lesson content with previous lessons; **Chalkboard Examples; Error Analysis**, to monitor and adjust student learning; and **Enrichment** suggestions.
- Practice and Enrichment masters, found in the **Teacher's Classroom Resources**, are reduced and conveniently positioned within the side margins. The answers are highlighted in red, so you can evaluate worksheets quickly and easily.

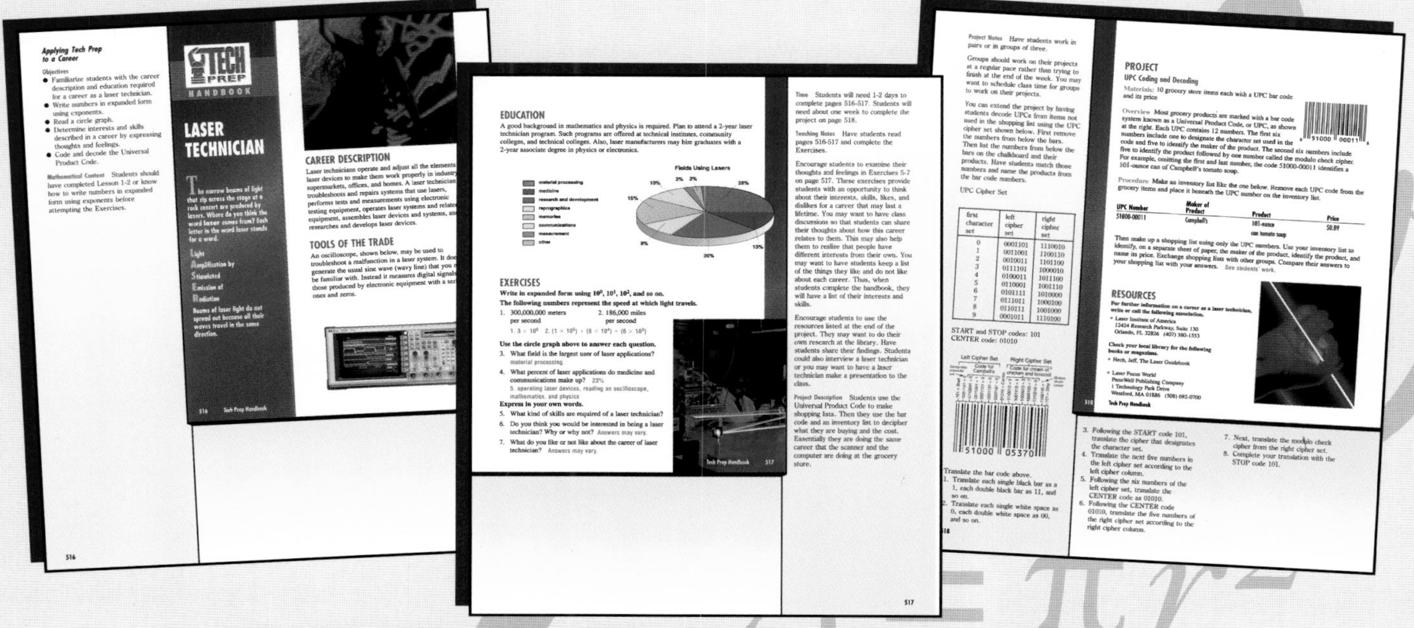

- Extensive information on tech prep careers is included. **Margin Notes** that accompany **Tech Prep Handbook** pages include objectives, teaching notes, project notes, and answers.

AN UNPARALLELED SELECTION OF TEACHE
MEANS YOU CAN TEACH MORE EFFECTIVEL

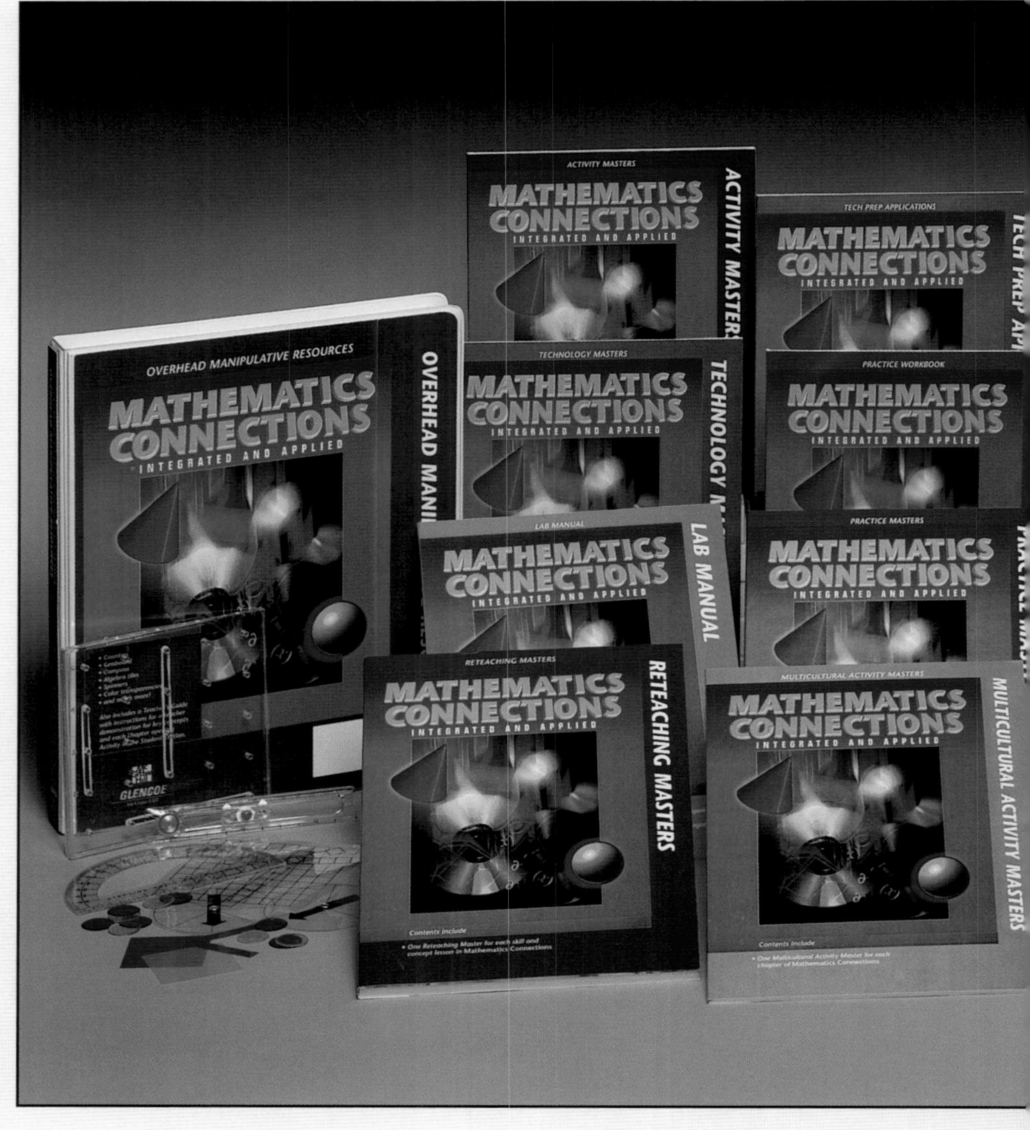

Applications

Activity Masters contain Application activities to connect content to real life and Cooperative Problem-Solving activities that help students learn mathematics concepts while they develop important interpersonal skills.

Multicultural Activity Masters present a unique cross-cultural array of historical situations that invite students to solve a problem based on chapter content.

New!

Tech Prep Applications contain information on the tech prep curriculum and include two tech prep applications for each chapter.

Technology Masters offer calculator and computer activities.

New!

Manipulatives

Overhead Manipulative Resources contain translucent manipulatives and transparencies for use on the overhead projector. The Teacher's Guide includes instructions for demonstrations of key concepts and suggestions for each chapter-opening activity in the student text.

The **Lab Manual** contains 22 pages of easy-to-make manipulatives and two activity pages for each chapter. One page is used as a recording sheet or model for each chapter-opening activity, and the other is an independent activity.

Meeting Individual Needs

Reteaching Masters explore lesson content using alternative methods, such as modeling and graphing.

Practice Masters contain exercises that give students opportunities to reinforce important mathematics skills. (Also available as a workbook).

RESOURCES IN LESS TIME

The **Test and Review Generator**, available in Apple, IBM, and Macintosh versions, allows you to create your own tests and quizzes.

New!

Performance Assessment booklet contains open-ended assessment items and a scoring rubic for each chapter.

Alternative Assessment in the Mathematics Classroom includes suggestions and ideas for projects, portfolios, questioning, and other assessment strategies.

And more!

Lesson Plans help you choose resources and learning materials for each lesson.

Involving Parents and the Community in the Mathematics Classroom presents suggestions on how parents and community members can be active participants in supporting mathematics instruction.

Cooperative Learning in the Mathematics Classroom includes ways to implement successful cooperative learning groups.

The **Transparency Package** contains more than 160 full-color transparencies that include captivating photographs, graphs, completely worked-out examples, and a repeat of the 5-Minute Check from the Teacher's Wraparound Edition.

Enrichment Masters contain stimulating puzzles and games for extending and enriching key concepts.

Spanish Resources provide a Spanish translation of the glossary in the Student Edition.

Assessment

Evaluation Masters include a thorough representation of chapter content through a variety of multiple choice and free-response tests, free-response quizzes, and cumulative reviews.

$A = \pi r^2$

MATHEMATICS CONNECTIONS: INTEGRATED AND APPLIED

Teaching New Topics... Developing Real-World Skills... Getting Students Ready for Life!

This unique program, with its emphasis on tech prep applications, gets students excited about math. They begin to discover the relevance of mathematics for daily living, and they are motivated to learn more about technical careers. As they become more confident problem solvers, your students take those critical first steps on the road to future success.

Student Edition	0-02-824795-7
Teacher's Wraparound Edition	0-02-824796-5
Teacher's Classroom Resources	0-02-824797-3
*Transparency Package	0-02-824799-X
*Reateaching Masters	0-02-824801-5
*Practice Masters	0-02-824802-3
*Practice Workbook	0-02-824803-1
*Enrichment Masters	0-02-824804-X
*Evaluation Masters	0-02-824805-8
*Technology Masters	0-02-824806-6
*Activity Masters	0-02-824807-4
*Multicultural Activity Masters	0-02-824808-2
*Lab Manual	0-02-824809-0
*Lesson Plans	0-02-824810-4
*Tech Prep Applications	0-02-824811-2
*Performance Assessment	0-02-824812-0
Overhead Manipulative Resources	0-02-824813-9
Spanish Resources	0-02-824798-1
Test and Review Generator (Apple)	0-02-824814-7
Test and Review Generatior (IBM)	0-02-824815-5
Test and Review Generator (Macintosh)	0-02-824816-3

*Included in Teacher's Classroom Resources

For more information contact your nearest regional office or call 1-800-334-7344

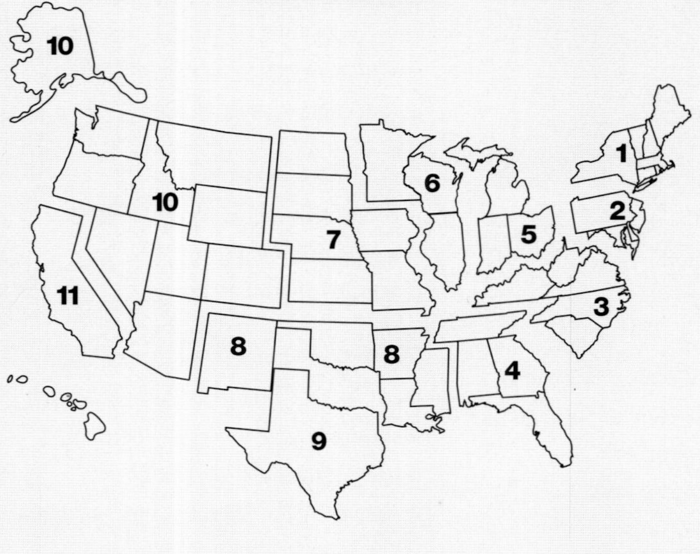

MA 91049-8
4-95

1. Northeast Region
Glencoe/McGraw-Hill
15 Trafalgar Square #201
Nashua, NH 03063-1968
Phone: 603-880-4701
Phone: 800-424-3451
Fax: 603-595-0204

2. Mid-Atlantic Region
Glencoe/McGraw-Hill
P.O. Box 458
Hightstown, NJ 08520-0458
Phone: 609-426-5560
Phone: 800-553-7515
Fax: 609-426-7063

3. Atlantic-Southeast Region
Glencoe/McGraw-Hill
Brookside Park
One Harbison Way, Suite 101
Columbia, SC 29212
Phone: 803-732-2365
Phone: 800-731-2365
Fax: 803-732-4582

4. Southeast Region
Glencoe/McGraw-Hill
6510 Jimmy Carter Boulevard
Norcross, GA 30071
Phone: 404-446-7493
Phone: 800-982-3992
Fax: 404-446-2356

5. Mid-America Region
Glencoe/McGraw-Hill
936 Eastwind Drive
Westerville, OH 43081
Phone: 614-890-1111
Phone: 800-848-1567
Fax: 614-899-4905

6. Great Lakes Region
Glencoe/McGraw-Hill
846 East Algonquin Road
Schaumburg, IL 60173
Phone: 708-397-8448
Phone: 800-762-4876
Fax: 708-397-9472

7. Mid-Continent Region
Glencoe/McGraw-Hill
846 East Algonquin Road
Schaumburg, IL 60173
Phone: 708-397-8448
Phone: 800-762-4876
Fax: 708-397-9472

8. Southwest Region
Glencoe/McGraw-Hill
320 Westway Place, Suite 550
Arlington, TX 76018
Phone: 817-784-2113
Phone: 800-828-5096
Fax: 817-784-2116

9. Texas Region
Glencoe/McGraw-Hill
320 Westway Place, Suite 550
Arlington, TX 76018
Phone: 817-784-2100
Phone: 800-828-5096
Fax: 817-784-2116

10. Western Region
Glencoe/McGraw-Hill
709 E. Riverpark Lane, Suite 150
Boise, ID 83706
Phone: 208-368-0300
Phone: 800-452-6126
Fax: 208-368-0303
Includes Alaska

11. California Region
Glencoe/McGraw-Hill
15319 Chatsworth Street
P. O. Box 9609
Mission Hills, CA 91346
Phone: 818-898-1391
Phone: 800-423-9534
Fax: 818-365-5489
Includes Hawaii

Glencoe Catholic School Region
Glencoe/McGraw-Hill
25 Crescent Street, 1st Floor
Stamford, CT 06906
Phone: 203-964-9109
Phone: 800-551-8766
Fax: 203-967-3108

Canada
McGraw-Hill Ryerson Ltd.
300 Water Street
Whitby, Ontario
Canada L1N 9B6
Phone: 905-430-5000
Phone: 800-565-5758
Fax: 905-430-5020

International
McGraw-Hill, Inc. International Group
Princeton-Hightstown Road
Hightstown, NJ 08520
Phone: 609-426-5934
Fax: 609-426-7917

DoDDS and Pacific Territories
McGraw-Hill School Publishing Company
1221 Avenue of the Americas
13th Floor
New York, NY 10020
Phone: 212-512-6128
Fax: 212-512-6050

GLENCOE
McGraw-Hill
P.O. BOX 508
COLUMBUS, OHIO 43216

TEACHER'S WRAPAROUND EDITION

MATHEMATICS CONNECTIONS

INTEGRATED AND APPLIED

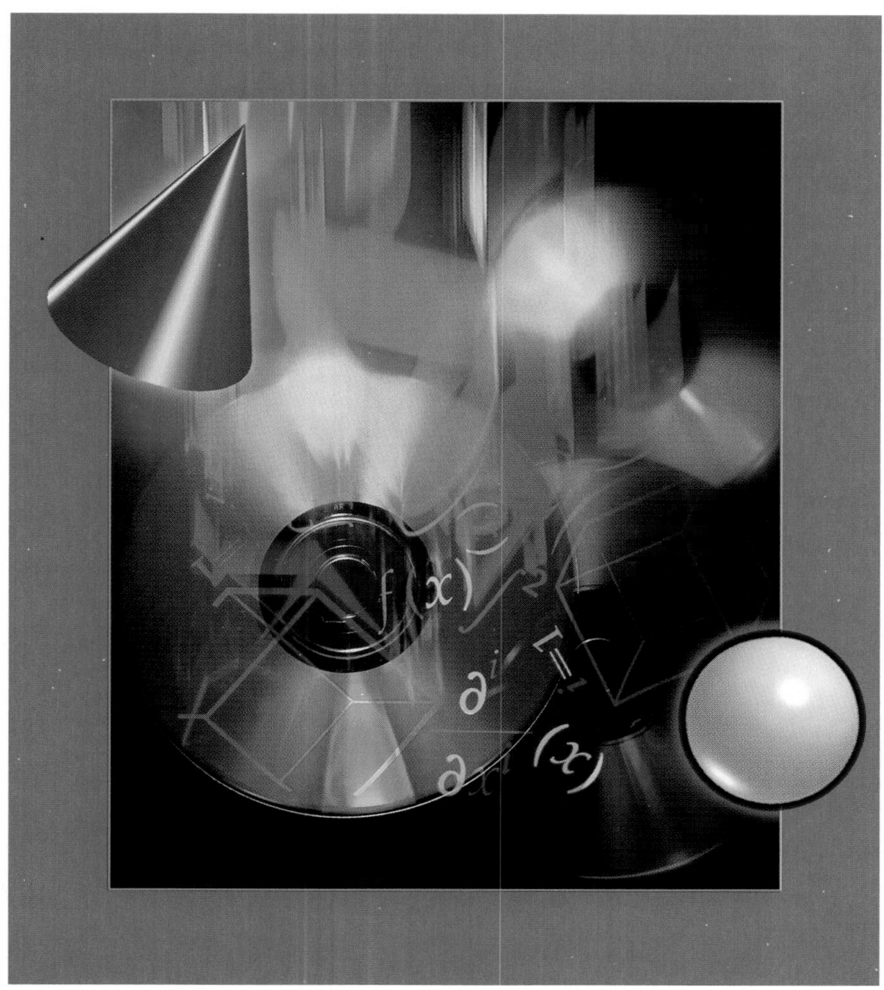

GLENCOE

McGraw-Hill

New York, New York Columbus, Ohio Mission Hills, California Peoria, Illinois

Consultant

Lee Yunker
Mathematics Department Head
West Chicago Community High School
West Chicago, Illinois

Reviewers

Wanda Sue Fowler
Mathematics Coordinator
Lee County Schools
Fort Myers, Florida

Sheryl G. Keith
Mathematics Teacher
Central High School
Evansville, Indiana

Patsy T. Malin
Secondary Mathematics Coordinator
El Dorado School District
El Dorado, Arkansas

Carl J. Minzenberger
Coordinator of Mathematics
Erie City School District
Erie, Pennsylvania

Dr. Charleen Mitchell DeRidder
Mathematics Supervisor
Knox County Schools
Knoxville, Tennessee

Roger L. O'Brien
Mathematics Supervisor
Polk County Schools
Bartow, Florida

Donna Marie Strickland
Mathematics Teacher
McGavock Comprehensive High School
Nashville, Tennessee

Susan C. Weaver
Mathematics Teacher
Hugo Junior High School
Hugo, Oklahoma

Printed in the United States of America.

Send all inquiries to:
Glencoe/McGraw-Hill
936 Eastwind Drive
Westerville, OH 43081

ISBN: 0-02-824795-7

2 3 4 5 6 7 8 9 10 RRW/LP 03 02 01 00 99 98 97 96 95

Authors

Dr. Robert B. Ashlock is a Professor of Education at Covenant College in Lookout Mountain, Georgia. He authored the book *Error Patterns in Computation,* and is widely recognized as an authority on common mathematical errors. He also served as a consultant on *Merrill Mathematics.*

Dr. Mary M. Hatfield is an Associate Professor of Mathematics Education at Arizona State University, Tempe, Arizona. She is active in professional mathematics organizations. She speaks frequently at the National and Regional NCTM conferences. She recently served on the Board of Directors of the National Council of Teachers of Mathematics, as President of the Arizona Association of Teachers of Mathematics, and is developing a CD-ROM to enhance mathematics teaching.

Dr. Howard L. Hausher is chairperson of the Mathematics and Computer Science Department at the California University of Pennsylvania in California, Pennsylvania. He holds certification as math supervisor of grades K-12. He does inservice sessions for public and private secondary schools on implementing the *NCTM Standards.* He evaluates high school math departments.

Mr. John H. Stoeckinger has recently retired from Carmel High School in Carmel, Indiana, where he was a mathematics teacher and department head. During his career, he was actively involved in teaching all levels of mathematics for grades seven through twelve. He also served as an author for *Merrill Mathematics.*

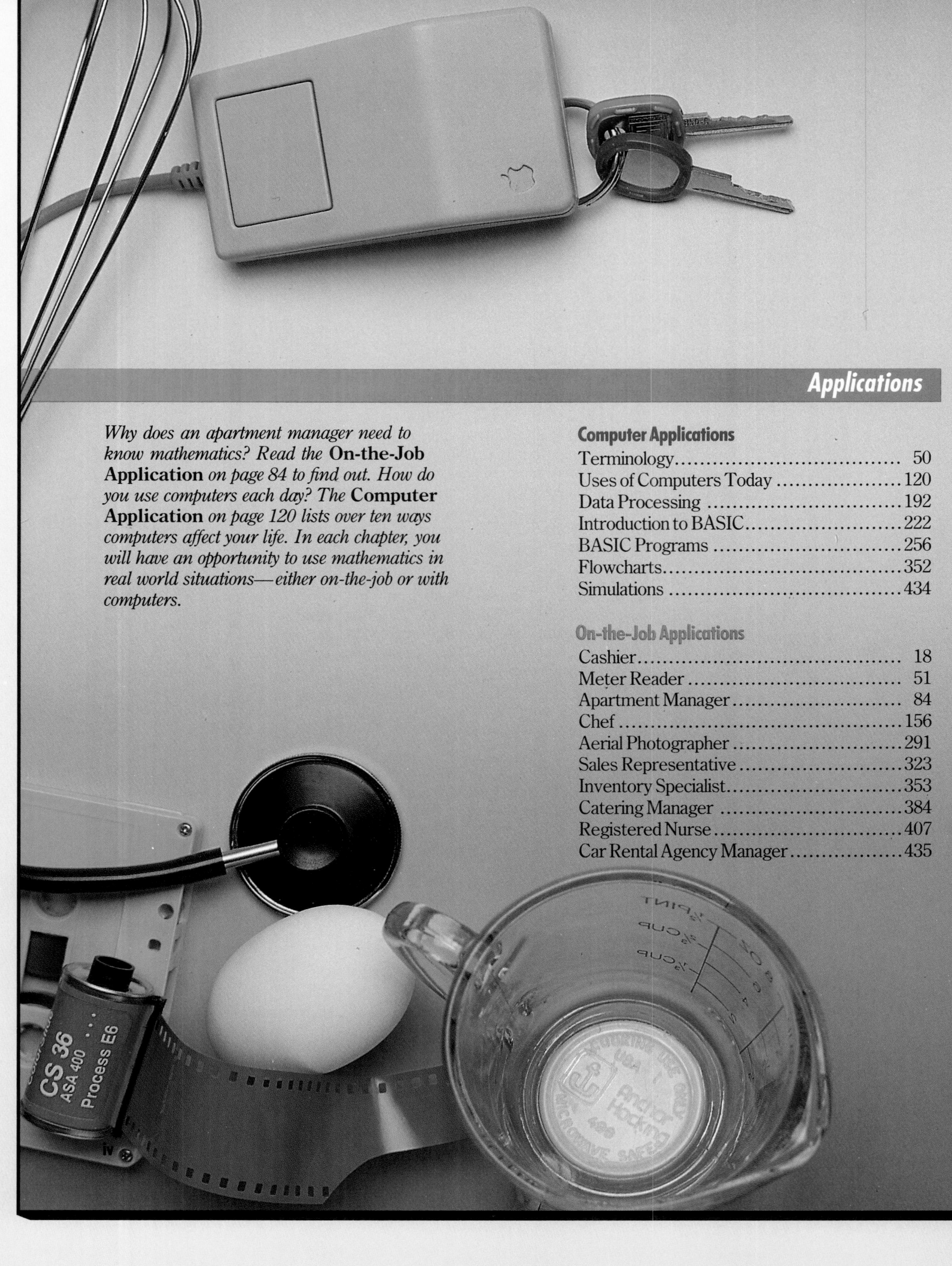

Applications

Why does an apartment manager need to know mathematics? Read the **On-the-Job Application** *on page 84 to find out. How do you use computers each day? The* **Computer Application** *on page 120 lists over ten ways computers affect your life. In each chapter, you will have an opportunity to use mathematics in real world situations—either on-the-job or with computers.*

Computer Applications

On-the-Job Applications

TABLE OF CONTENTS

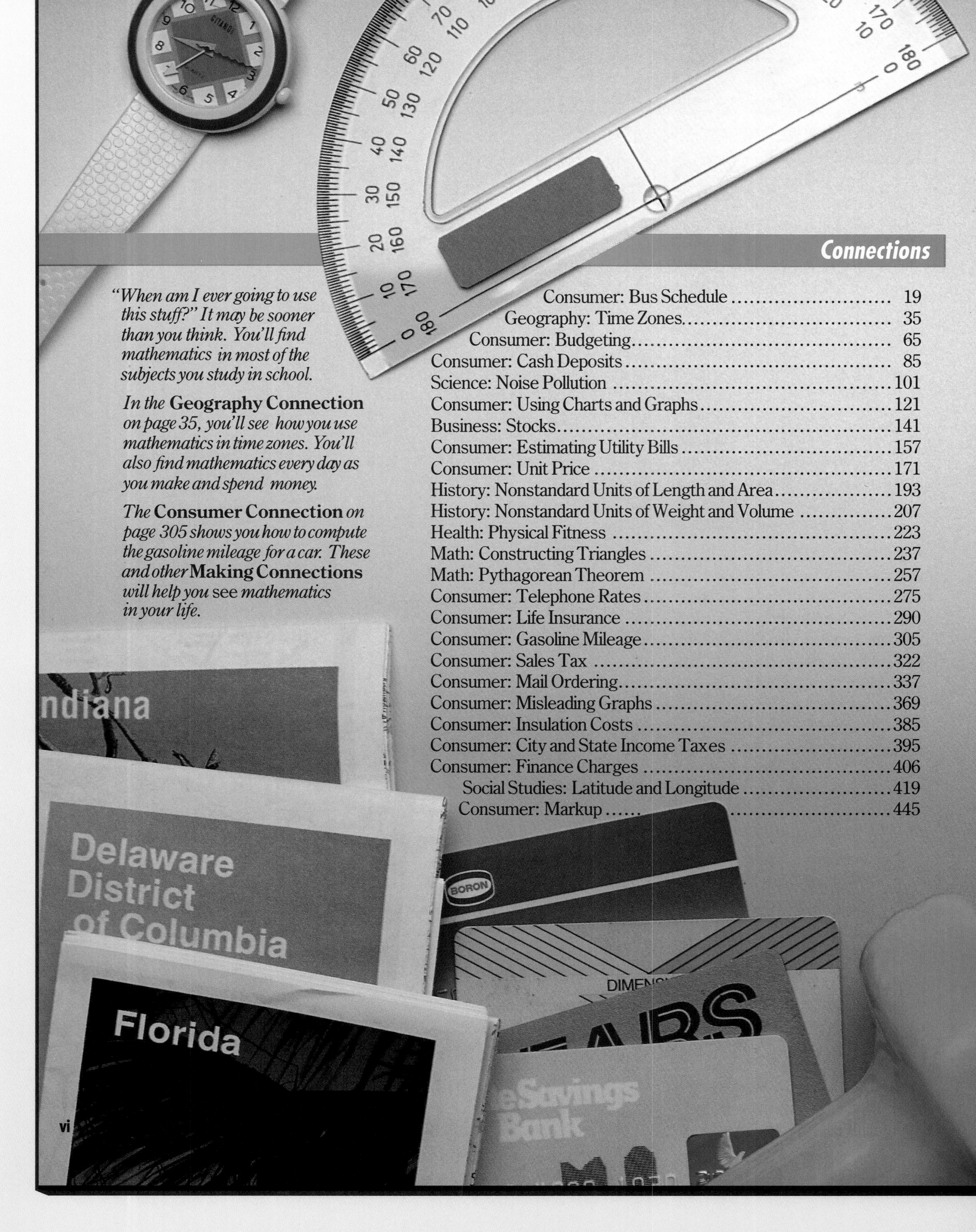

Connections

"When am I ever going to use this stuff?" It may be sooner than you think. You'll find mathematics in most of the subjects you study in school.

In the **Geography Connection** *on page 35, you'll see how you use mathematics in time zones. You'll also find mathematics every day as you make and spend money.*

The **Consumer Connection** *on page 305 shows you how to compute the gasoline mileage for a car. These and other* **Making Connections** *will help you see mathematics in your life.*

Chapter 3

CONNECTIONS

Chapter 4

Have you thought about what career you want to pursue after you graduate from high school? Do you want to go on to a 2-year college, attend a vocational or technical school, or go to a university? There are many options open to you.

With technology on the move, more and more careers are being created that weren't here 10 years, or 5 years, or even 1 year ago. Most of these careers are in high demand for qualified people. The careers that are in demand require education beyond high school. A few of these careers are listed in the table of contents shown below.

Are you aware of what your skills and interests are? You will find questions in the **Tech Prep Handbook** *on pages 515-536 that will help you think about what you like to do and what you already know how to do. One goal of Tech Prep is to get you to think about your future, to make plans for it, and to prepare for it while you're in high school.*

Once you have read about the careers in the **Tech Prep Handbook**, *do some research on your own at the public or school library or talk to your guidance counselor. Maybe your high school offers a Tech-Prep program, that you can consider enrolling in.*

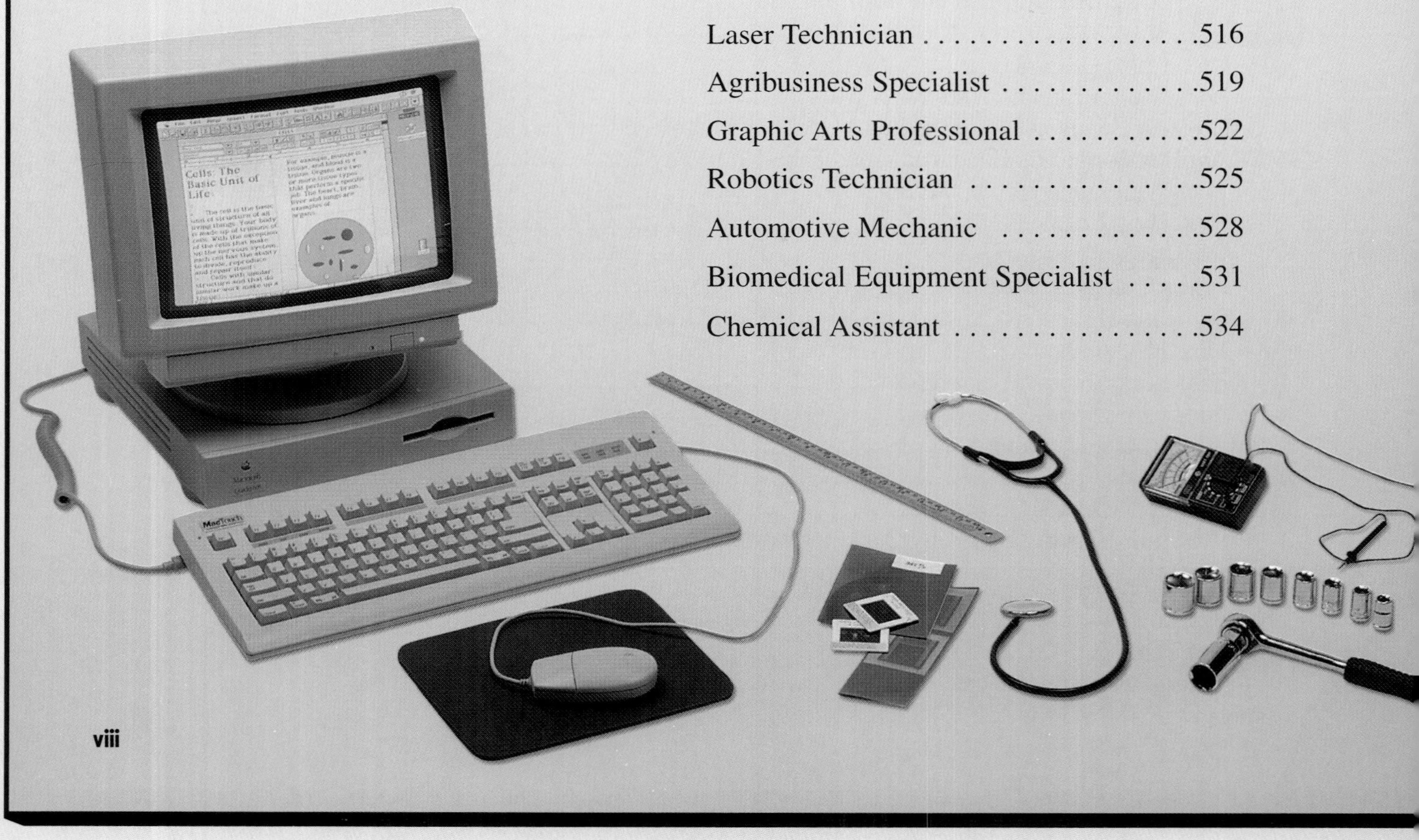

Chapter 5

CONNECTIONS

TOPIC	PAGES
Algebra	147,149,159, 165,191,195
Statistics	152,153,155, 163,201
Technology	162,192,197
Consumer Math	152-153,157, 171
Business Math	156,200-201
History	193,207
Science	183

Chapter 6

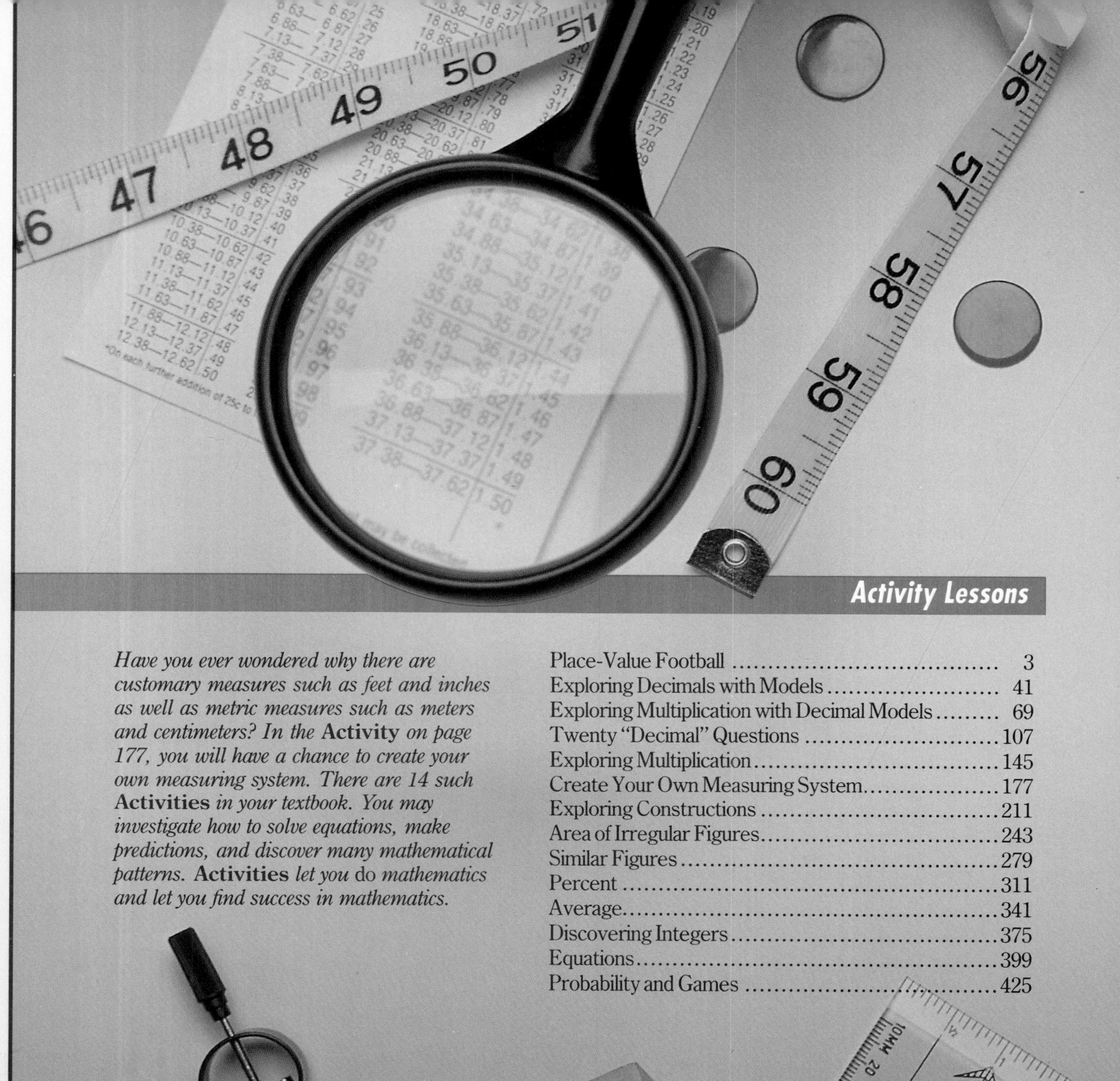

Activity Lessons

Have you ever wondered why there are customary measures such as feet and inches as well as metric measures such as meters and centimeters? In the **Activity** *on page 177, you will have a chance to create your own measuring system. There are 14 such* **Activities** *in your textbook. You may investigate how to solve equations, make predictions, and discover many mathematical patterns.* **Activities** *let you* do *mathematics and let you find success in mathematics.*

Chapter 7

Chapter 8

CONNECTIONS

Chapter 9

Chapter 10

Chapter 11

Review and Testing

CONNECTIONS

Chapter 12

Review and Testing

Chapter 13

Review and Testing

CONNECTIONS

TOPIC	PAGES
Algebra	405,413,415, 441
Geometry	401,413,441
Statistics	401,405,407, 411,415,433, 436,437,441
Technology	406,407,434, 435,440,445
Consumer Math	406,412-413, 445
Business Math	407,435, 436-437
Social Studies	419

Chapter 14

Review and Testing

TO THE TEACHER

The purpose of this 10-page Teacher's Guide is to provide an introduction to the format and philosophy of the *Mathematics Connections Teacher's Wraparound Edition*. It is also intended to relate some background information on several contemporary issues facing mathematics educators in the 1990s, such as tech prep, technology, cooperative learning, and the use of manipulatives in your classroom. Suggested time schedules for six-week and nine-week grading periods, as well as a course planning calendar are also included. A correlation of the NCTM Standards to *Mathematics Connections* are included for ease of use by teachers.

TEACHER'S GUIDE TABLE OF CONTENTS

How Merrill Mathematics Connections Meets the NCTM Standards

Standard	Student Edition Pages
Standard 1: Mathematics as Problem Solving	
In grades 9-12, the mathematics curriculum should include the refinement and extension of methods of mathematical problem solving.	17, 30,31, 45, 56, 57, 60, 61, 75, 82, 83, 96, 97, 127, 136, 137, 149, 154, 155, 161, 166, 167, 183, 188, 189, 199, 202, 203, 232, 233, 254, 255, 270, 271, 287, 300, 301, 318, 319, 321, 332, 333, 364, 365, 390, 391, 414, 415, 440, 441
Standard 2: Mathematics as Communication	
In grades 9-12, the mathematics curriculum should include the continued development of language and symbolism to communicate mathematical ideas.	3, 4, 5, 6, 7, 13, 25, 27, 29, 41, 45, 49, 59, 69, 73, 79, 85, 86, 91, 93, 101, 107, 111, 113, 115, 116, 117, 118, 119, 120, 125, 129, 131, 151, 155, 159, 163, 165, 177, 178, 179, 184, 186, 190, 191, 195, 203, 213, 215, 217, 219, 221, 225, 227, 229, 231, 240, 243, 247, 259, 265, 267, 269, 279, 281, 289, 299, 313, 327, 341, 347, 357, 372, 376, 379, 387, 399, 403, 405, 409, 422, 425, 427, 429, 433, 441
Standard 3: Mathematics as Reasoning	
In grades 9-12, the mathematics curriculum should include numerous and varied experiences that reinforce and extend logical reasoning skills.	15, 20, 21, 23, 41, 69, 111, 163, 217, 227, 229, 231, 232, 233, 247, 283, 357, 425, 433
Standard 4: Mathematics Connections	
In grades 9-12, the mathematics curriculum should include investigation of the connections and interplay among various mathematical topics and their applications.	24, 25, 26, 27, 28, 29, 35, 54, 55, 58, 59, 86, 87, 88, 89, 90, 91, 94, 95, 101, 116, 117, 118, 119, 141, 183, 187, 193, 207, 223, 237, 257, 280, 281, 282, 283, 284, 285, 286, 287, 288, 289, 292, 293, 296, 297, 298, 299
Standard 5: Algebra	
In grades 9-12, the mathematics curriculum should include the continued study of algebraic concepts and methods.	6, 7, 8, 24, 25, 26, 27, 28, 29, 30, 31, 45, 59, 86, 90, 91, 116, 117, 118, 119, 121, 141, 159, 165, 184, 186, 190, 376, 387, 399, 400, 401, 402, 403, 404, 405, 406, 407, 410, 411, 412, 413, 414, 415

STANDARD	STUDENT EDITION PAGES
STANDARD 6: FUNCTIONS	
In grades 9-12, the mathematics curriculum should include the continued study of functions.	77, 117, 121, 141, 153, 297, 317, 333, 348, 349, 350, 351, 354, 355, 356, 357, 362, 363, 367, 369, 383, 401, 408, 409, 410, 411, 412, 413, 414, 415, 416, 417, 418, 419, 420, 421, 422, 423, 424, 425, 426, 427, 428, 429, 430, 431, 432, 433, 434, 435, 436, 437, 438, 439, 440, 441
STANDARD 7: GEOMETRY FROM A SYNTHETIC PERSPECTIVE	
In grades 9-12, the mathematics curriculum should include the continued study of the geometry of two and three dimensions.	230, 231, 232, 233, 254, 255, 258, 259, 260, 261, 262, 263, 264, 265, 266, 267, 268, 269, 270, 271, 292, 293
STANDARD 8: GEOMETRY FROM AN ALGEBRAIC PERSPECTIVE	
In grades 9-12, the mathematics curriculum should include the study of the geometry of two and three dimensions from an algebraic point of view.	408, 409, 410, 411, 412, 413
STANDARD 10: STATISTICS	
In grades 9-12, the mathematics curriculum should include the continued study of data analysis and statistics.	341, 342, 343, 344, 345, 346, 347, 348, 349, 350, 351, 352, 353, 354, 355, 356, 357, 358, 359, 360, 361, 362, 363, 437, 441
STANDARD 11: PROBABILITY	
In grades 9-12, the mathematics curriculum should include the continued study of probability.	282, 283, 424, 425, 426, 427, 428, 429, 430, 431, 432, 433, 434, 435, 436, 437, 438, 439, 440, 441
STANDARD 12: DISCRETE MATHEMATICS	
In grades 9-12, the mathematics curriculum should include topics from discrete mathematics.	54, 55, 88, 89, 282, 283, 341, 342, 343, 344, 345, 346, 347, 348, 349, 350, 351, 352, 353, 354, 355, 356, 357, 358, 359, 360, 361, 362, 363, 424, 425, 426, 427, 428, 429, 430, 431, 432, 433, 434, 435, 436, 437, 438, 439, 440, 441
STANDARD 14: MATHEMATICS STRUCTURE	
In grades 9-12, the mathematics curriculum should include the study of mathematical structure.	4, 5, 6, 7, 8, 9, 10, 11, 14, 16, 26, 27, 28, 29, 30, 31, 42, 43, 46, 47, 70, 71, 76, 77, 374, 375, 376, 377, 378, 379, 380, 381, 382, 383, 386, 387, 388, 389, 399, 400, 401, 402, 403, 404, 405, 414, 415

COOPERATIVE LEARNING

Cooperative learning groups learn things together, not just do things together. Studies show that cooperative learning promotes more learning than competitive or individual learning experiences regardless of student age, subject matter, or learning activity. More difficult learning tasks, such as problem solving, critical thinking, and conceptual learning, come out far ahead when cooperative strategies are used. Studies also show that in classroom settings, adolescents learn more from each other about subject matter than from the teacher.

The basic elements of a cooperative learning group are as follows:

(1) Students must perceive that they "sink or swim together."

(2) Students are responsible for everyone else in the group, as well as them selves, in learning the assigned material.

(3) Students must see that they all have the same goal, that they need to divide up the tasks and share the responsibility equally, and that they will be given one evaluation or reward that will apply to all members of the group.

PREPARATION AND SOCIAL SKILLS

1. Arrange your room. Students in groups should face each other as they work together. It is helpful to number the tables so you can refer to groups by number, or have groups choose a name.

2. Decide on the size of the group. Groups work best when teams are composed of two to five students. The materials available might dictate the size of the group.

3. Assign students to groups. Each group should be mixed socially, racially, ethnically, sexually, and by learning abilities. Occasionally a student may insist on working alone. This student usually changes his or her mind when seeing that everyone else's grades are better, and the groups are having more fun.

4. Prepare students for cooperation. This is a critical step. Tell students about the rationale, procedures, and expected outcomes of this method of instruction. Students need to know that you are not forcing them to be friends, but asking them to work together as they will later on in life with people who come together for a specific purpose.

5. Explain group tasks. For more activities, each group will need someone to take notes, someone to summarize as the group progresses, and someone to make sure everyone is involved. Classes that have less experience using these methods may need these roles assigned to group members at the beginning of each activity. These jobs should be done by different students each time, so that one student does not feel burdened doing the same job all the time.

MORE DIFFICULT LEARNING TASKS, SUCH AS PROBLEM SOLVING AND CRITICAL THINKING, COME OUT FAR AHEAD WHEN COOPERATIVE STRATEGIES ARE USED.

6. Explain the day's lesson. On the chalkboard or overhead, write the following headings:

➤ **Form Groups**: list the number of students in each group.

Topic of the Day: general academic topic.

➤ **Task**: title of activity. At this time, go over instructions and relate the work to previous learning.

➤ **Goal**: indicate whether students will all do individual work of which you will select one person's paper or product to grade, or whether they will produce only one product per group. Student signatures on all the work indicates that they will accept the collected work for their grade.

➤ **Cooperative Skills**: After students have some experience working in cooperative groups, you can expect group members to exhibit some or all of the following higher level cooperative skills.

a. Express support and acceptance, both verbal and nonverbal, through eye contact, enthusiasm, and praise.

b. Ask for help or clarification about what is happening.

c. Suggest new ideas.

d. Use appropriate humor that stays on task.

e. Describe feelings using "I messages," such as, "I like the way your praised my idea."

f. Summarize and elaborate on what others have contributed.

Using Manipulatives in Your Mathematics Classroom

Most teachers agree that the use of manipulative materials helps students build a solid understanding of mathematical concepts and enhances students' achievement in mathematics. The universal anxiety today about achievement in mathematics should prompt educators in this field to heed results from recent studies which indicate that by using manipulative materials at every level, test scores are dramatically improved.

"Manipulative materials are the key to understanding operations and algorithms."[1] Because abstractions are an integral part of mathematics and because students derive their abstract ideas primarily from their experience, it is imperative they they experiment with a variety of manipulative materials on the concrete level to develop an understanding of algebraic symbols and concepts.

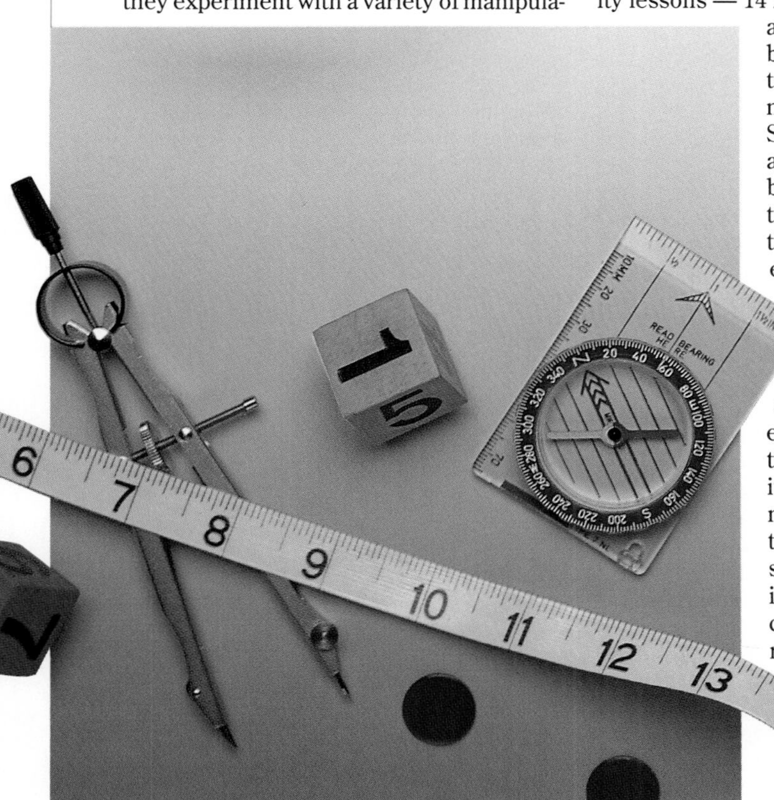

"The purpose of using manipulatives is to assist students in bridging the gap from their own concrete environment to the abstract level."[2] Affording students the opportunity for meaningful practice through modeling algorithms with manipulatives not only results in greater understanding of the concepts and skills in question but also provides a fun alternative to everyday, routine problem-solving exercises as well.

THE PURPOSE OF USING MANIPULATIVES IS TO ASSIST STUDENTS IN BRIDGING THE GAP FROM CONCRETE TO ABSTRACT.

Merrill Mathematics Connections contains many opportunities for students to use manipulatives to discover and explore mathematical concepts. The Activity lessons — 14 in all — guide students to actually do mathematics by using counters, base-ten blocks, and decimal models just to name a few. Students are also encouraged to make conjectures based on their observations. In lessons following the Activities, teachers are encouraged to allow the continued use of the manipulatives as necessary.

To make sure that students understand what is expected of each activity, the teacher must discuss its goal and model how the manipulative is to be used to achieve it. Encouraging students to suggest ways in which manipulatives can be used helps them to relate concepts and to develop mathematical insights.

The important thing to keep in mind is that most students do not automatically abstract the concepts they explore with materials: they must be led to see how the concepts relate to traditional algorithms. Summarizing and recording group activity results helps to focus on such relationships and algebraic concepts. Used in this way, manipulative materials are a justifiable means to a desirable end.

No presentation advocating the use of manipulative materials in the mathematics classroom would be totally complete without interjecting a note of caution. Despite the fact that manipulative materials are highly touted, and rightly so, they can, if used incorrectly, undo much of the good for which they are intended. through careless or erroneous use of manipulatives, students might conclude that there are two distinct mathematical worlds - one of manipulatives and another of symbols - and that each has its own rules.

Teachers must, therefore, direct students to see the need for precise and exact connection between the symbols and the manipulatives; otherwise they cannot possibly develop proper mathematical understanding. The point that must be stressed repeatedly in teaching with manipulatives in the mathematics classroom is that "symbols and manipulatives must always reflect the same concept."[3]

If mathematics educators sincerely want to challenge their students in the mathematics classroom, perhaps they must first endorse the fact that the use of manipulative materials holds the promise of increasing students' understanding of and achievement in mathematics, and then translate that belief into classroom practice.

1 Beattie, Ian D., "Modeling Operations and Algorithms," Arithmetic Teacher, February 1986, p. 23.
2 Hynes, Michael C., "Selection Criteria," Arithmetic Teacher, February 1986, p 11.
3 Bright, George W., "One Point of View; Using Manipulatives," Arithmetic Teacher, February 1986, p. 4.

Integrating Technology Into Your Mathematics Classroom

Perhaps the biggest fear, and certainly one that is most often expressed, is that of the technology taking the place of the teacher. This will never happen. There is no way that a computer can do what a teacher does, and in particular, sense what a teacher can. For example, no computer can sense uncertainty of an answer in a student's tone of voice. Few students will be able to turn on a calculator or computer and teach themselves with it. They need teachers to explain the technology and to help make connections.

> **Probably the biggest advantage of technology is the amount of time it can save you**

Appropriate Use of Technology

There are many different ideas and definitions of "appropriate use of technology," and rightly so. Every class and every teacher is different. However, there are some classifications and generalizations that can be made for different grade levels. Keep in mind that many of your students are just learning the technology. You should refrain from getting "high-tech" with them. Give them opportunities for success early so that they can begin to feel comfortable with technology.

Have your students complete an easy calculation for their first encounter. For the students new to technology, this will give them a sense of accomplishment because they can see something they have done; for the more technologically experienced students, this will give them a sense of confidence because they already know what is being done.

At any level of teaching, try to get students to make generalizations about what they see or do with technology.

Using Technology in Your Classroom

Technology opens up many new ideas and opportunities. It is up to you to decide how to present and use technology in your classroom. Many of the standard teaching techniques are appropriate, such as using cooperative groups, or having students work in pairs, but there are also new ways of teaching that are appropriate when using technology in your classroom. For example, you can assign lab partners in each class. Each pair of students would then work together whenever technology is used. This can also be used if you have a limited supply of equipment.

Technology in Mathematics Connections

The TI Challenger is the calculator upon which the calculator topics in this textbook are based. The TI Challenger offers a wide range of functions and operations for the mathematic's student, such as order of operations in Lesson 1-9 and fraction operations as in Chapter 4.

Changes in the Classroom

There will be some changes in your classroom with the onset of technology. One will be your role. It most likely will change from leader to guide. Students will begin to do things on their own and it will be up to you to keep them headed in the right direction. Students will also begin to ask more questions, including more higher-order questions, than before.

Advantages of Technology

Probably the biggest advantage of technology is the amount of time it can save you. The ease of editing errors and the speed of calculating make technology a major plus for the mathematics classroom.

There are numerous other advantages. One is that students will have a deeper understanding of the concepts being taught. as we develop toward a more pictorial society, students are becoming visual learners. Technology teaches them in a way that piques their interest and leads to questions that show a desire to learn.

Another advantage of technology is the cooperative efforts that develop among students. Since students will most likely need to work with others when they become members of the work force, it is best for them to learn to work with others while they are in school. With technology, you can teach your students to work cooperatively which will benefit them in school and in their jobs.

Computers and calculators are excellent tools for teaching mathematics concepts. Though there are still some who argue their use in the classroom, none can argue their educational advantage. Technology can teach students in ways that were never before possible. In the past, teachers could only dream of being able to do some of the things that they can now do with ease, thanks to technology.

MEETING INDIVIDUAL NEEDS

"WE CANNOT AFFORD TO HAVE THE MAJORITY OF OUR POPULATION MATHEMATICALLY ILLITERATE: EQUITY HAS BECOME AN ECONOMIC NECESSITY."

NCTM STANDARDS (1989, P.4)

MULTICULTURAL PERSPECTIVE

There is no doubt that the United States is a multicultural society. Changing demographics, and changing economic and social orders are having a tremendous impact on the schools in this country. But, the term multicultural represents more than just many cultures. There is a multicultural basis to all knowledge, even mathematics.

From the ancient Egyptians, who used the "Pythagorean" theorem fifteen hundred years before Pythagoras, to the ancient Chinese, who calculated the value of π to ten places twelve hundred years before the Europeans, mathematics as we know it today has been shaped by many cultures. Even the term algebra was contributed by an Arabian mathematician, Al-Khowarizmi.

What is the role of the mathematics educator in all this? Students should have the opportunity to learn about persons from all cultures who have been successful in their respective careers. To this end, the *Merrill Mathematics Connections Activities Masters Booklet* contains one multicultural worksheet for each chapter. The *Merrill Mathematics Connections Teacher's Wraparound Edition* also contains suggestions for meeting the needs of all students in the interleaf pages that precede the chapters.

As educators, we also must prepare all students for the new jobs of the future that will require mathematical literacy. The mathematics teacher must be in the vanguard of those who demand that all students be given the opportunity to study the more advanced forms of mathematics. It is our responsibility as educators to make every effort to prepare the students of today to participate in the complex world of tomorrow.

LIMITED ENGLISH PROFICIENCY NEEDS

One of the greatest factors contributing to the under-achievement in mathematics education for the language-minority student is his or her failure to understand the language of instruction. There are, however, strategies which the mathematics teacher can employ to help overcome the obstacles that beset language-minority students.

Ideally, these students would be afforded the opportunity of having new concepts and skills reinforced by discussing them in their native tongue. It may be that a bilingual teacher could meet this need, or perhaps a tutor proficient in the language could be procured to provide such service.

If the student is proficient in his or her native language, perhaps materials written in that language could be provided to supplement classroom instruction.

If the student is not especially literate in his or her own language, perhaps an oral approach using pictorial materials and/or manipulative devices would be feasible.

When a student does not respond to the prescribed expectations of the school, the teacher needs to substitute developmentally equivalent tasks to shape development. If, for example, the student does not participate in classroom discussion, the wise teacher will observe his or her verbal interaction with other students in informal, less structured environments.

Because students coming from diverse cultural backgrounds lack common educational experiences, the teacher soon recognizes that the only way to establish a basis for communication is to begin instruction with content that is familiar to one and all.

"The challenge is to find personally interesting and culturally relevant ways of creating new contexts for children, contexts in which the mastery of school skills can be meaningful and rewarding."[1]

By integrating the development of language in such new contexts, we may be able to open the door to the challenging and exciting world of mathematics for the language-minority student.

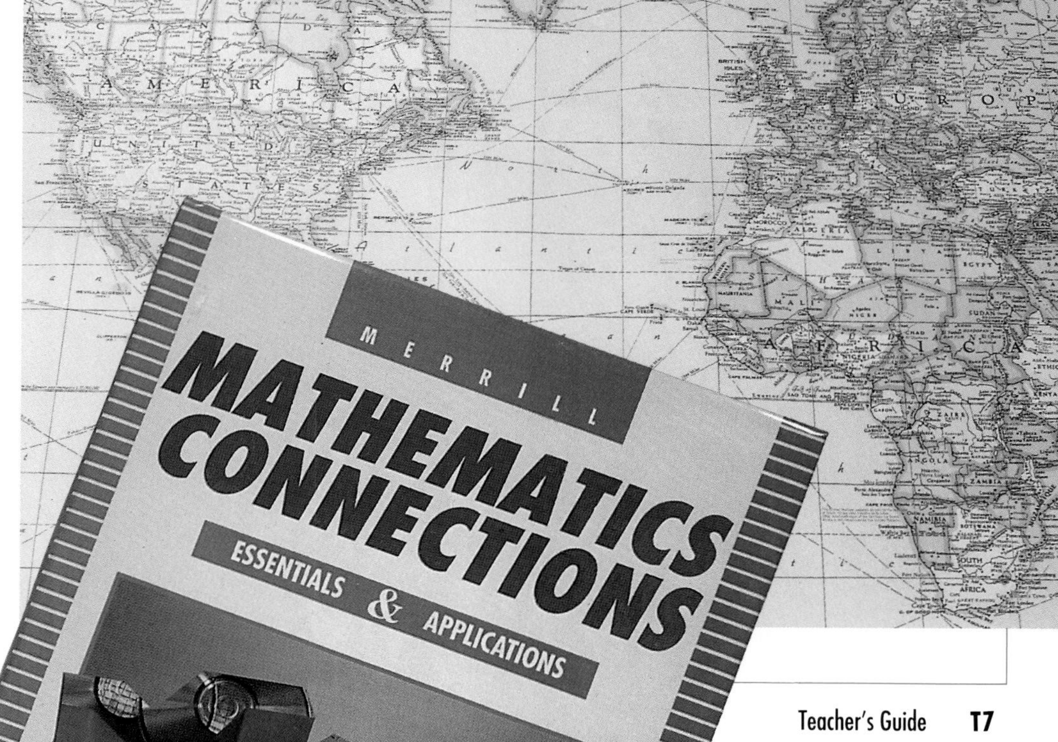

1 Bowman, Barbara T., "Educating Language-Minority Children: Challenges and Opportunities," Phi Delta Kappan, October 1989, p. 120.

Integrating Tech Prep Into Your Mathematics Classroom

What is Tech Prep?

Tech prep is a course of study whose goal is to create a United States work force that is technically literate. To accomplish this, a partnership among high schools, postsecondary educational institutions, business, industry, and labor must be developed. The reasoning underlying tech prep is that students are more willing to learn if they have a career goal and a support system enabling them to reach that goal.

A good program will also address the prevention of students dropping out of school and the issue of students reentering school.

Is Tech Prep for All Students?

Tech prep provides an alternative to the college-prep course of study that leads to a career goal. Because of this, tech prep is available to all students whether they are taking college prep, vocational, or a general course of study. It is especially beneficial to the student in the general course of study.

Why is Tech Prep Needed?

With the advancement of today's technology, employers need people with technical skills, who exhibit high performance, and have the ability to change with the rapidly changing technology. "More than half our young people leave school without the knowledge or foundation required to find and hold a job." according to a 1991 report from the U.S. Department of Labor.

More than half our young people leave school without the knowledge or foundation required to find and hold a job.

What Are the Goals of Tech Prep?

The goal of tech prep is to have students exit high school or the community/technical college with:

- marketable skills
- the academic credentials to pursue higher education; or
- both, so that the student can earn a living during a pattern of lifelong learning.

Another, but separate, goal of tech prep is to make the United States more competitive in the world economy.

What Qualifies as a Tech-Prep Program?

The following guidelines must be met to receive Federal grants under Title III.

- Tech prep must have a 2 + 2 design or a 4-year sequence with math, science, communication, and technology as the common core. It starts in the junior year of high school and continues through two years of postsecondary occupational education that ends with either a certificate or an associate degree. The education students obtain from the tech-prep program could also lead them to bachelor's and advanced degrees.

2
↓
last two years of high school
↓
yields a high school diploma
+
2
↓
two years of occupational postsecondary education or apprenticeship
↓
yields a two-year associate degree or certificate

- The second guideline emphasizes an agreement among all the educational systems involved to provide continuity in the program, continual progress and achievement, and ease of movement from one level to another. Communication between institutions is essential to the success of the program.
- The technical occupations must be current to the job market and skills and coursework are applicable to employer needs.
- Inservice training for both secondary and postsecondary teachers and counselors is necessary.
- The tech-prep program must be available to all students.
- Preparatory services such as recruiting, personal and career counseling, and occupational assessment should be available to students not in the program.

Also under Title III, tech-prep programs should receive priority if they:

- offer job placement
- offer transfers to 4-year baccalaureate programs
- are developed by the school, business, industry, and labor
- emphasize dropout prevention and reentry, and the needs of special populations

As a result of the tech-prep program, students will enter the work force as qualified technicians.

What is Included in Tech Prep?

The tech-prep course of study includes a common core of math, science, communications, and technology. The program provides preparation in at least one field of:

- engineering technology
- applied science
- mechanical, industrial, or practical art or trade
- agriculture
- health
- business

What Other Kinds of Opportunities Make a Successful Tech-Prep Program?

Career Exploration: Throughout elementary, middle/junior high, secondary, and post secondary education, students need to be made aware of current careers, how to explore career options, what skills and education are required to obtain that job, what the working conditions are like, where the jobs are located if not universal, and their own likes, dislikes, skills, and aptitudes.

"The program tolerates no "watered down" courses but maintains the same academic integrity as the college-prep curriculum expanding occupational education to include academic development."

The Merrill Mathematics Connections Tech Prep Handbook located on pages 515-536 of the Student Edition contains descriptions of several technical occupations, resources for additional job information, education and skills required for those jobs, and a description of their tools of the trade. Exercises are included that allow students an opportunity to explore their feelings about the job and evaluate the skills they possess.

Employer Participation: When business, government, and the education system work together, the program will be successful. If employers take an active role in tech prep, they can help students look at the work place realistically, help them form practical ideas about the work place, motivate them with expectations for that career, answer immediately any questions, qualms, or false impressions about a career, and predict the future of that job.

Work-based Learning Experiences: Business and educators can work together to provides students with work-based experiences such as shadowing, having a mentor, internships, apprenticeships, and so on.

Tech Prep Is Not!

Tech prep is not:

- remedial;
- a tracking system;
- a way to prevent students from entering a four-year college program;
- only for certain occupational programs; or
- unique to big schools.

"The program tolerates no 'watered down' courses but maintains the same academic integrity as the college-prep curriculum expanding occupational education to include academic development." as stated in the November, 1992 issue of the *National TechPrep Network.*

How Can You Learn More About Tech Prep?

To learn more about tech prep, contact the following organizations.

National Center for Research in Vocational Education
University of California at Berkeley
2150 Shattuck Avenue
Suite 1250
Berkeley, CA 94720-1674
800-762-4093

National TechPrep Network Center for Occupational
Research and Development
P.O. Box 21689
Waco, TX 76702-1689
800-231-3015

National Network for Curriculum Coordination in
Vocation Technical Education
National Tech Prep Clearing House
Room K-80
Sangamon State University
Springfield, IL 62794-9243
800-252-4822

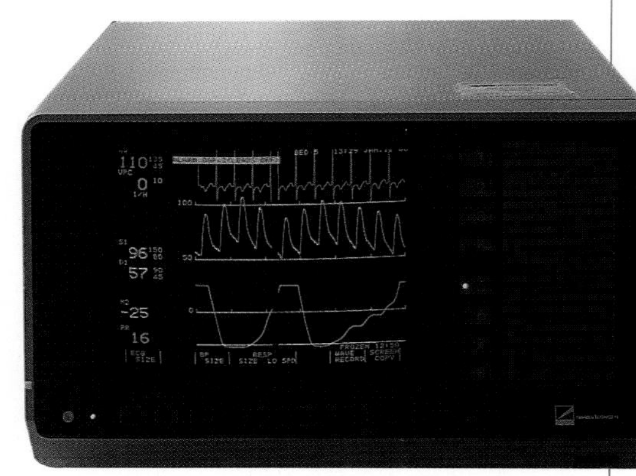

COURSE PLANNING CALENDAR

The charts on this page give suggested time schedules for two types of courses: minimum and standard and two types of grading periods: 9-weeks and 6-weeks.

The minimum course covers Chapters 1-12. It allows for extra time for longer sessions and for reteaching and review. The standard course covers Chapters 1-14. Generally, one day is allotted for each lesson, the Chapter Review, and the Chapter Test.

NINE WEEKS GRADING PERIOD	TYPE OF COURSE			
	MINIMUM		STANDARD	
	CHAPTER	DAYS	CHAPTER	DAYS
1	1 2 3 (Lessons 3-1 to 3-11)	15 12 11	1 2 3	14 10 14
2	3 (Lessons 3-12 to end) 4 5 6 (Lessons 6-1 to 6-6)	5 17 13 6	4 5 6	15 11 13
3	6 (Lessons 6-7 to end) 7 8 9 (lessons 9-6 to 9-5)	8 12 15 5	7 8 9 10 (Lessons 10-1 to 10-6)	11 14 11 6
4	9 (Lessons 9-6 to end) 10 11 12	7 12 13 9	10 (Lessons 10-7 to end) 11 12 13 14	5 12 8 8 8
TOTAL DAYS		160		160

SIX WEEKS GRADING PERIOD	TYPE OF COURSE			
	MINIMUM		STANDARD	
	CHAPTER	DAYS	CHAPTER	DAYS
1	1 2	15 12	1 2 3 (Lessons 3-1 to 3-3)	14 10 3
2	3 4 (Lessons 4-1 to 4-11)	16 11	3 (Lessons 3-4 to end) 4	11 15
3	4 (Lessons 4-12 to end) 5 6 (Lessons 6-1 to 6-8)	6 13 8	5 6 7 (Lessons 7-1 to 7-5)	11 13 5
4	6 (Lessons 6-9 to end) 7 8 (Lessons 8-1 to 8-9)	6 12 9	7 (Lessons 7-6 to end) 8 9 (Lessons 9-1 to 9-7)	6 14 7
5	8 (Lessons 8-10 to end) 9 10 (Lessons 10-1 to 10-8)	6 12 8	9 (Lessons 9-8 to end) 10 11	4 11 12
6	10 (Lessons 10-9 to end) 11 12	4 13 9	12 13 14	8 8 8
TOTAL DAYS		160		160

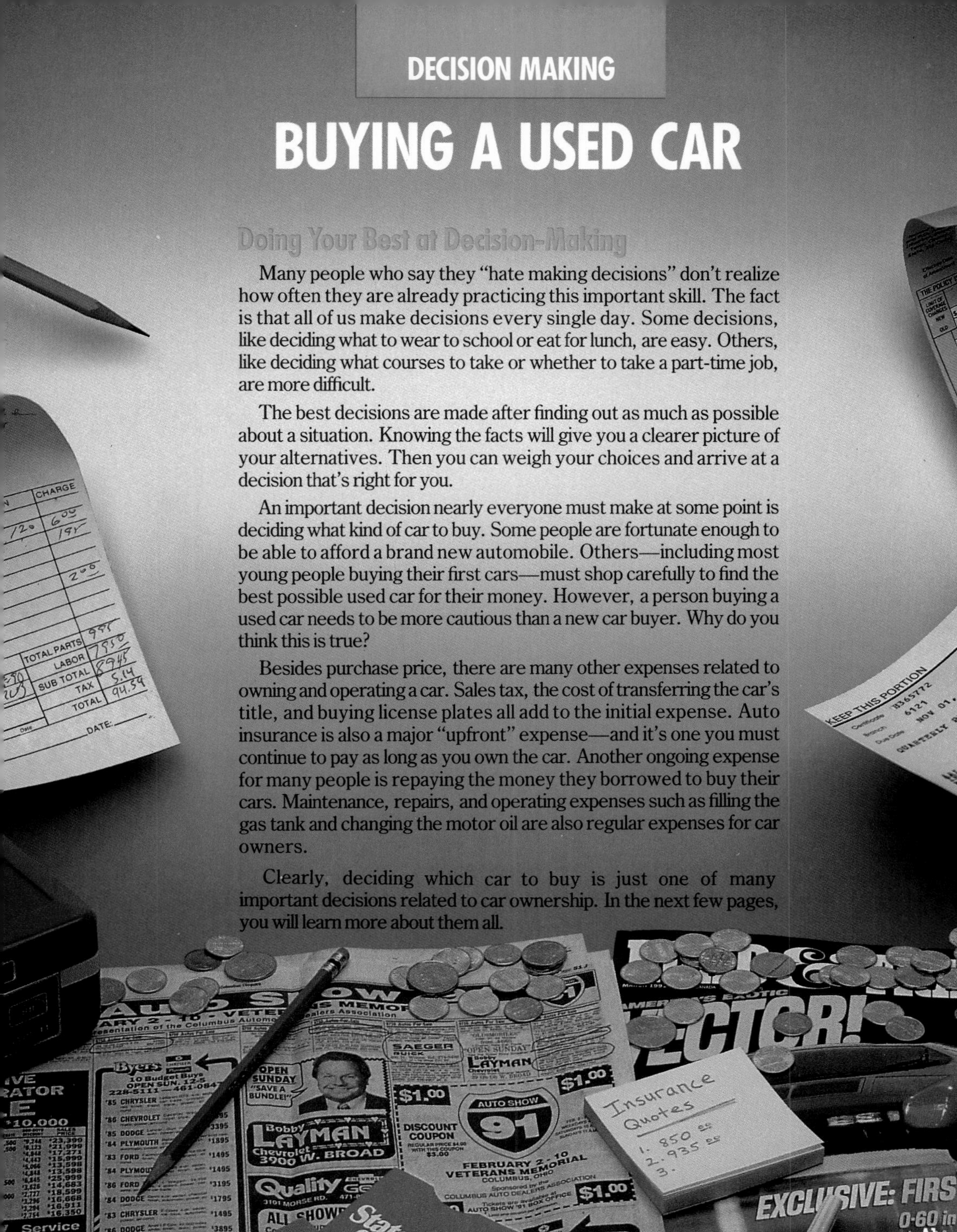
DECISION MAKING

BUYING A USED CAR

Doing Your Best at Decision-Making

Many people who say they "hate making decisions" don't realize how often they are already practicing this important skill. The fact is that all of us make decisions every single day. Some decisions, like deciding what to wear to school or eat for lunch, are easy. Others, like deciding what courses to take or whether to take a part-time job, are more difficult.

The best decisions are made after finding out as much as possible about a situation. Knowing the facts will give you a clearer picture of your alternatives. Then you can weigh your choices and arrive at a decision that's right for you.

An important decision nearly everyone must make at some point is deciding what kind of car to buy. Some people are fortunate enough to be able to afford a brand new automobile. Others—including most young people buying their first cars—must shop carefully to find the best possible used car for their money. However, a person buying a used car needs to be more cautious than a new car buyer. Why do you think this is true?

Besides purchase price, there are many other expenses related to owning and operating a car. Sales tax, the cost of transferring the car's title, and buying license plates all add to the initial expense. Auto insurance is also a major "upfront" expense—and it's one you must continue to pay as long as you own the car. Another ongoing expense for many people is repaying the money they borrowed to buy their cars. Maintenance, repairs, and operating expenses such as filling the gas tank and changing the motor oil are also regular expenses for car owners.

Clearly, deciding which car to buy is just one of many important decisions related to car ownership. In the next few pages, you will learn more about them all.

A1

Making a Decision Buying a Used Car

Making a Decision

The purpose of this 11-page section is to provide students with a foundation for the decision-making process. The problem presented is buying a used car. This section also helps ease students into the text prior to introducing the mathematical concepts of the book.

Page A1 points out that the best decisions are made after finding out as much as possible about a situation and weighing the facts before arriving at a decision that is right for the individual.

On pages A2 and A3 the common sense approach is stressed. The choices are narrowed, and the student is walked through the decision-making process that is used to evaluate several used cars before a final decision is made.

The process of actually purchasing a used car is discussed on pages A4 and A5. This includes method of payment, sales tax, title transfer and registration.

Financing is the focus of pages A6 and A7. On these pages, loan applications and monthly repayment charges are discussed.

The students will learn about the cost of car insurance on pages A8 and A9. The factors that affect the insurance rates and the various types of coverage available are presented here.

Pages A10 and A11 deal with the cost of routine care and maintenance of the car.

Test Sample Answer

Most new cars carry a warranty that guarantees the dealer will pay for certain repairs caused by faulty parts if they occur during a specified period. A standard warranty offered by many dealers covers the first year or first 30,000 miles of ownership. Most used cars sold on lots carry very limited 30-day warranties. Used cars sold by private owners rarely carry any warranty—which is why buying a used car is always riskier than buying a new one.

Objective
Apply the decision-making process to buying a used car.

▶ INTRODUCING THE LESSON

Have students tell what decisions have to be made by a group of four students who want to order pizza(s).

▶ TEACHING THE LESSON

Using Questioning

- Why it is necessary to have something particular in mind before shopping for a car?
- Are there any other costs in addition to the purchase price?
- What other things should you think about before making a final decision?

Using Discussion Discuss narrowing choices, gathering facts and information on the choices, and making the final decision.

Text Sample Answer

What kind of car do I need? How much can I afford to spend? How will I finance the car? Should I buy a used car through a private owner or a used car dealer?

How to Shop for a Used Car

When you shop for new shoes, chances are you have something quite particular in mind. For example, if you need a pair of athletic shoes you arrive at the store knowing you prefer a certain style of shoe in a particular price range. You limit the types of shoes you choose from based on your preferences and what you can afford to pay.

This same common sense approach holds true for buying a used car. A new red sports car may be everyone's dream, but reality for most of us is usually something much less expensive and much more practical! Based on common sense, can you think of some important questions you need to ask yourself before proceeding to search for a used car?

Let's assume that before you begin shopping, you narrow your choices to buying a large, 4-door vehicle manufactured in 1985 or later that costs $5,500 or less, including tax, title, registration, and insurance. To save money, you decide to buy from a private owner rather than a dealer who's likely to have a higher markup. Carefully read the following ads to find cars that meet your requirements. Refer to the abbreviation chart if necessary.

Abbreviation Chart

AC—air conditioning	**int**—interior
auto—automatic transmission	**HT**—hardtop
EC—excellent condition	**PB**—power brakes
dr—door	**PS**—power steering
mpg—miles per gallon	**PW**—power windows
gd—good	**spd**—speed
ext—exterior	**cyl**—cylinder

'85 JEEP CHEROKEE
Red. Great shape! Sunroof, alum. mags, cloth int, A/C, cass./stereo, cruise control. $4,500. 241-0112 after 5 P.M.

'85 CHEVY CAVALIER
Liftback. 5 speed with air. AM/FM, 60,000 miles, 25-30 mpg, 1 owner, $2,250. 325-0012 after 6 P.M.

'86 OLDS CUTLASS CIERA WAGON
60,000 miles. Excellent condition. One owner. Loaded with extras. $4,300. 771-1030. Ask for Mark.

'87 CHEVROLET SPRINT
Great on gas! 2-door, 5 speed, air, AM/FM/cassette. 41,000 miles. $4,000 or best offer. 772-0860 anytime.

'87 HONDA ACCORD
2 dr. low mileage, 5 spd., cruise, A/C. Must sell. $7,400/best offer. 224-0967 eves.

1. What criteria does the Honda Accord fail to meet? Too expensive and it's a two-door car.
2. Which is the only ad that mentions specific gas mileage? '85 Chevy Cavalier
3. Why do you think the price of the Chevy Cavalier is so much lower than that of the other cars? It's a small, inexpensive make of car that has been driven many miles.
4. You regularly need to haul supplies for your job, as well as pick up four friends on the way to school. To which vehicles would you narrow your choices? Cutlass Ciera wagon and the Jeep Cherokee

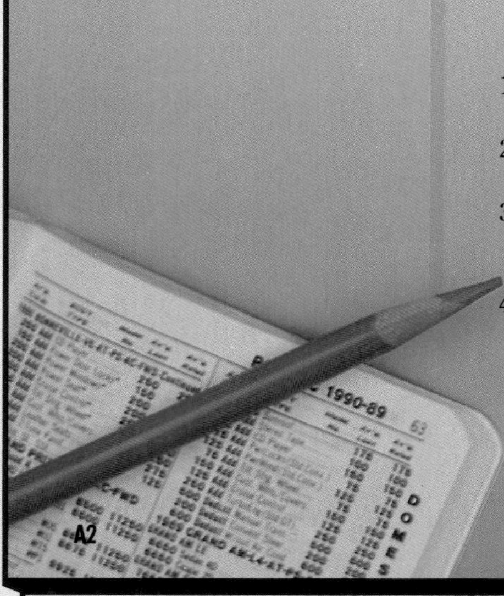

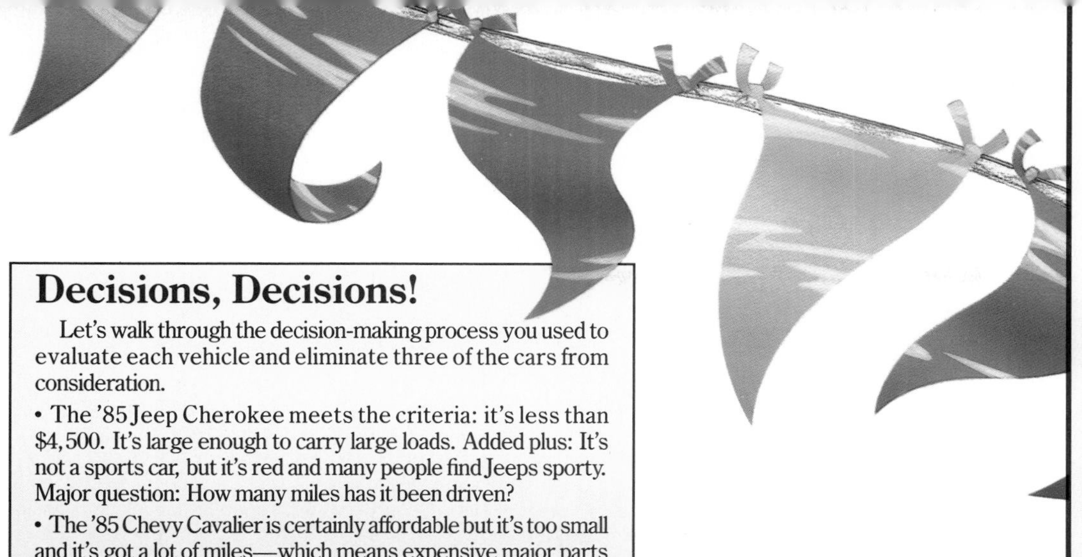

Decisions, Decisions!

Let's walk through the decision-making process you used to evaluate each vehicle and eliminate three of the cars from consideration.

- The '85 Jeep Cherokee meets the criteria: it's less than $4,500. It's large enough to carry large loads. Added plus: It's not a sports car, but it's red and many people find Jeeps sporty. Major question: How many miles has it been driven?
- The '85 Chevy Cavalier is certainly affordable but it's too small and it's got a lot of miles—which means expensive major parts may be wearing out.
- The '86 Cutlass Ciera wagon is within the price range and can carry a large load. It has as many miles as the Cavalier, but has had only one owner which implies the owner took good care of it. Major question: Does the owner have maintenance records to prove this?
- Despite the good mileage and price, the '87 Chevrolet Sprint won't do because it has only two doors—not practical for getting people and cargo in and out quickly.
- The Honda costs too much and has only two doors. Too bad! It's a nice car and a popular model.

Making Your Final Decision

Just as you try on new shoes before buying them, you must "try out" a car before making a purchase. Arrange with the seller to drive the car in the city and on the freeway. Ask yourself these questions: Does the car ride smoothly? Do the seat belts fasten properly in the front and rear seats? Do the doors and windows shut and lock properly? What other things should you check?

Even if everything looks and sounds right, never buy a used car without having it first checked out by a mechanic. The cost for this service varies, but count on paying at least $50. If the mechanic finds the car is basically sound but needs some minor repairs, use them as bargaining points to get the seller to lower the price. Unless the ad specifically states a firm price, most private sellers are quite willing—in fact they fully expect—to negotiate on their asking prices.

EVALUATING THE LESSON

Have students tell in their own words what steps they would follow when deciding to buy a used car.

EXTENDING THE LESSON

Have students make a list of five items they would look for when buying a used car. Then have them cut ads out of the newspaper for cars that meet their requirements. Have students examine the ads and make a decision on which car they would buy. Have students give reasons for their choice.

Text Sample Answer

JEEP: Does the sun roof work? Does the air conditioning work? Do the cassette player and stereo work? CIERA WAGON: Specifically, what are the "extras" the Wagon is loaded with? Ask the owner to point out and demonstrate each. BOTH VEHICLES: Are the tires in good shape? Does the glove compartment lock? Does the radio antenna go up and down? Is there rust under the floor mats?

Objective
Introduce students to the tax, title transfer, and registration of a car.

▶ INTRODUCING THE LESSON

Discuss the paperwork involved when you buy a car-new or used.
You must have proof of ownership and cars must be registered in the state you live in.

▶ TEACHING THE LESSON

Using Discussion Discuss the following procedures.

1. payment of sales tax to appropriate agencies.
2. title transfer
3. car registration and license plates

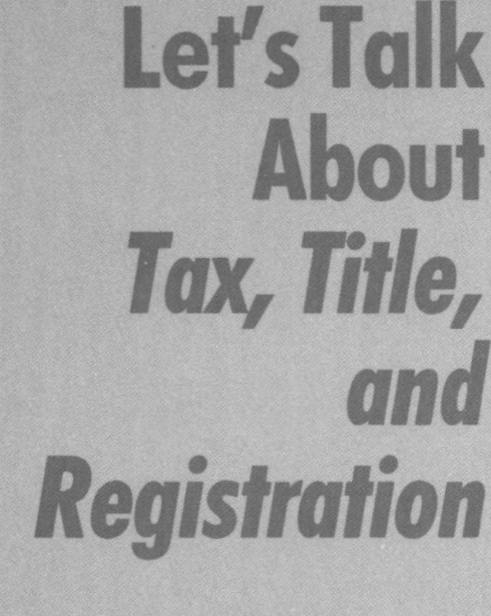

Let's Talk About *Tax, Title, and Registration*

The amount of paperwork involved in buying a car comes as a surprise to many first-time buyers. Some people are also surprised to learn that they must immediately pay a sales tax on their new or used cars, unless they live in a state where sales tax is not charged.

Here's how the process works: First, you must pay the seller "up front" for the full amount you, the buyer, and the seller have agreed upon. As previously mentioned, this may range from a couple of hundred to several hundred dollars below the seller's original asking price.

Used car dealers and new car dealers can help you arrange financing, a method for borrowing money to pay for the car. However, individual sellers—those private owners of cars who place ads in the newspapers—expect you to provide a check for the amount of the sale. Some buyers want a "certified" check, which means that your bank guarantees you have funds equal to the amount written on the check. Other sellers don't require a certified check, but will not release the car to you until your check "clears" the bank. When a check "clears," the amount of money specified on the check has been taken from the buyer's bank account and transferred to the seller's account.

Title Transfer

After receiving payment, the seller must have his or her signature notarized on the car title. The title is a paper that provides legal proof of ownership. A notarized signature is one witnessed by a "notary public," usually a Clerk of Courts in the county where the seller lives. The notary stamps the signature with an official seal. The seller gives or "transfers" the title to the buyer who then has his or her signature notarized to show that he or she is now owner of the car. The cost for transferring a title varies from county to county, but is usually around $5.

Sales Tax

When the title to the car is transferred to the new owner, the Clerk of Courts writes the price of the car on the back of the title. In some counties, the seller must produce a bill of sale showing what the buyer paid for the car. Sales tax, which usually ranges from 5 to 6 percent, is then figured on the cost of the transaction. The buyer must pay the sales tax to the Clerk of Courts at this time.

Let's assume that after having the '85 Jeep checked out by a mechanic, you and the seller agree to a sale price of $4,300—$200 lower than the asking price. Here is how you would figure a sales tax of 5 percent.

Multiple $4,300 by 0.05 because 5% = 0.05.

4300 [×] 0.05 [=] 215

The sales tax on the Jeep would be $215.

Registration

Registration of cars in most states is handled through another county agency called the Office of the Deputy Registrar. Registration simply means getting license plates for the car. To do so, you show the title with your name on it to prove you are now the owner, and fill out a new registration card. The fee for this is usually about $5. If the county requires you as the new owner to buy license plates rather than transferring the plates used by the seller into your name, be prepared to pay another $45 or so.

1. Why do you think the seller insists on a certified check or waits until the buyer's check clears his or her bank before giving the buyer the car? To make sure the check doesn't bounce
2. What is a certified check? One that the issuing bank guarantees is valid
3. Besides selling price, list the other charges related to buying the '85 Jeep Cherokee. Mechanic's fee: $50, Title transfer charge: $5, Sales tax: $215, Registration fee: $5, New license plates: $45
4. What is the total you have spent so far on the '85 Jeep Cherokee?
   a. $4,420 b. $4,520 c. $4,620 d. $4,720
5. What is the purpose of a title for a car? It provides legal proof of ownership

▶ EVALUATING THE LESSON

Have students explain about the tax, title transfer and registration of a car.

▶ EXTENDING THE LESSON

Have students research the amount of sales tax, title transfer fee, and registration fee that would have to be paid on a $4,300 vehicle in your area.

Objective
Examine the financing of a used car.

▶ INTRODUCING THE LESSON

Ask students why people finance a car. Ask them if they have ever had to borrow money from their parents or friends.

▶ TEACHING THE LESSON

Using Questioning

- What types of financial institutions offer auto loans?
- How old must you be to take out an auto loan?
- Are all auto loans the same? Explain.
- What kind of information must be provided on a loan application?

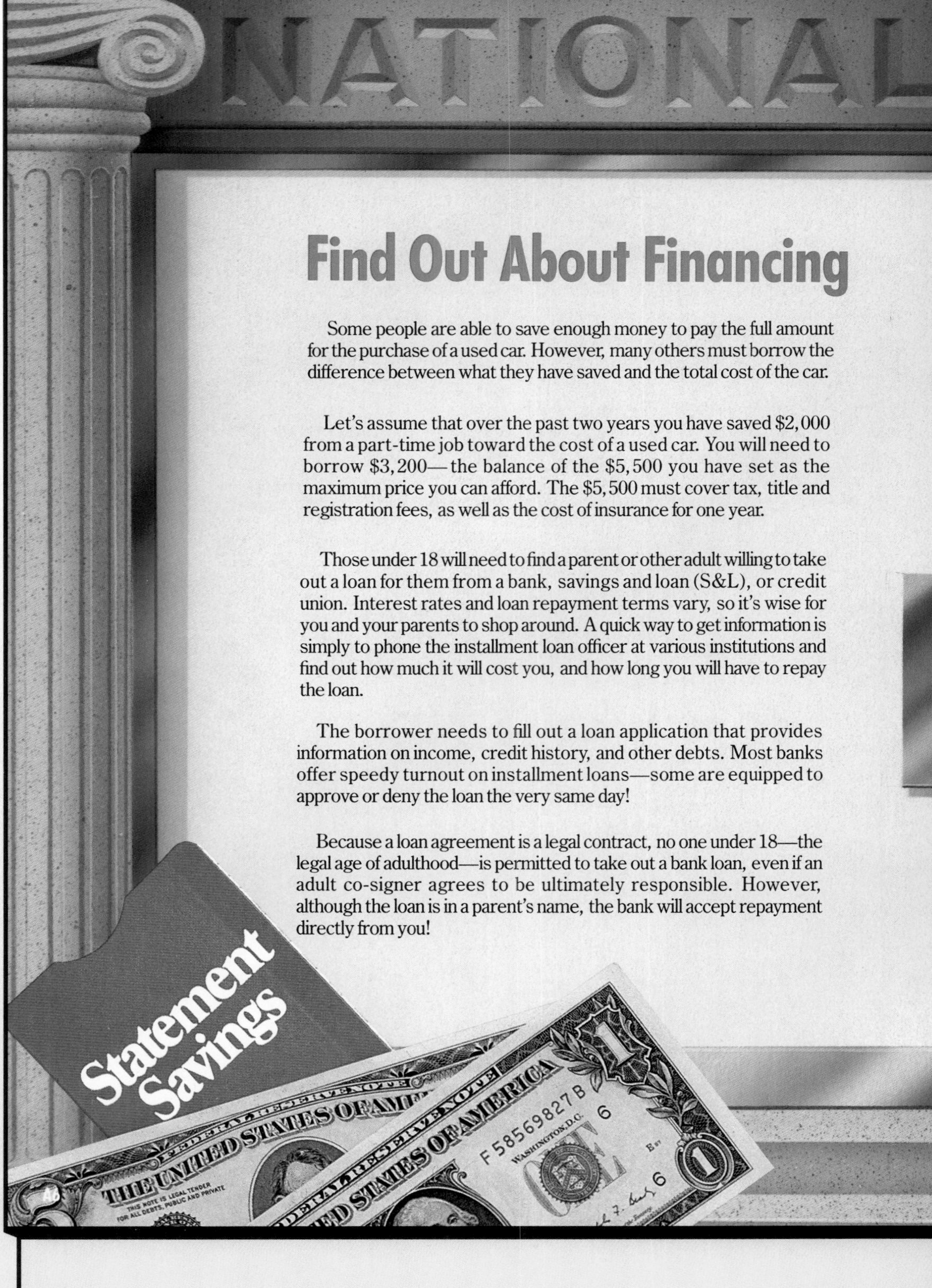

Find Out About Financing

Some people are able to save enough money to pay the full amount for the purchase of a used car. However, many others must borrow the difference between what they have saved and the total cost of the car.

Let's assume that over the past two years you have saved $2,000 from a part-time job toward the cost of a used car. You will need to borrow $3,200— the balance of the $5,500 you have set as the maximum price you can afford. The $5,500 must cover tax, title and registration fees, as well as the cost of insurance for one year.

Those under 18 will need to find a parent or other adult willing to take out a loan for them from a bank, savings and loan (S&L), or credit union. Interest rates and loan repayment terms vary, so it's wise for you and your parents to shop around. A quick way to get information is simply to phone the installment loan officer at various institutions and find out how much it will cost you, and how long you will have to repay the loan.

The borrower needs to fill out a loan application that provides information on income, credit history, and other debts. Most banks offer speedy turnout on installment loans—some are equipped to approve or deny the loan the very same day!

Because a loan agreement is a legal contract, no one under 18—the legal age of adulthood—is permitted to take out a bank loan, even if an adult co-signer agrees to be ultimately responsible. However, although the loan is in a parent's name, the bank will accept repayment directly from you!

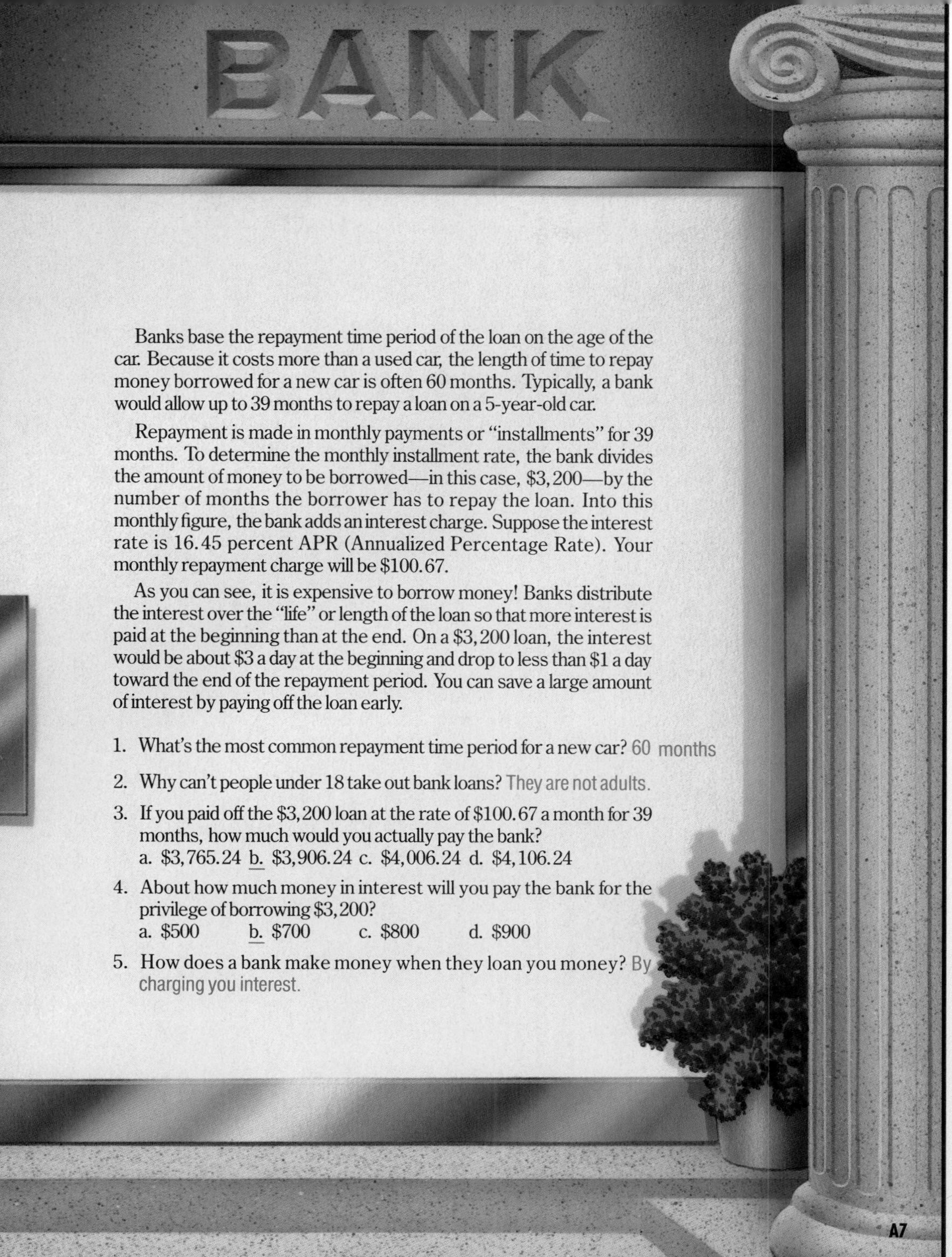

Banks base the repayment time period of the loan on the age of the car. Because it costs more than a used car, the length of time to repay money borrowed for a new car is often 60 months. Typically, a bank would allow up to 39 months to repay a loan on a 5-year-old car.

Repayment is made in monthly payments or "installments" for 39 months. To determine the monthly installment rate, the bank divides the amount of money to be borrowed—in this case, $3,200—by the number of months the borrower has to repay the loan. Into this monthly figure, the bank adds an interest charge. Suppose the interest rate is 16.45 percent APR (Annualized Percentage Rate). Your monthly repayment charge will be $100.67.

As you can see, it is expensive to borrow money! Banks distribute the interest over the "life" or length of the loan so that more interest is paid at the beginning than at the end. On a $3,200 loan, the interest would be about $3 a day at the beginning and drop to less than $1 a day toward the end of the repayment period. You can save a large amount of interest by paying off the loan early.

1. What's the most common repayment time period for a new car? 60 months
2. Why can't people under 18 take out bank loans? They are not adults.
3. If you paid off the $3,200 loan at the rate of $100.67 a month for 39 months, how much would you actually pay the bank?
   a. $3,765.24 b. $3,906.24 c. $4,006.24 d. $4,106.24
4. About how much money in interest will you pay the bank for the privilege of borrowing $3,200?
   a. $500 b. $700 c. $800 d. $900
5. How does a bank make money when they loan you money? By charging you interest.

A7

EVALUATING THE LESSON

Have students explain in their own words about financing a car.

EXTENDING THE LESSON

Have students gather information from several local lending institutions about interest rates and loan repayment terms for an '85 Jeep if they want to borrow $3,200. Compare information.

Objective
Examine auto insurance rates and types of coverage.

▶ INTRODUCING THE LESSON

Ask students why it is necessary to buy car insurance.

▶ TEACHING THE LESSON

Using Discussion Discuss the factors that affect the cost of car insurance and how young drivers can save on auto insurance. Also, discuss the basic types of insurance coverage.

Discuss why you must prove you have insurance in some states before you can register your car.

The Inside Story on Insurance

The high cost of insuring young drivers often comes as an unpleasant surprise to young people and their parents. However, insurance companies set higher rates for this age group for good reason: According to the Insurance Information Institute, more than 40 percent of the deaths among 16 to 22-year-olds result from auto accidents.

Rates for car insurance vary widely. Because they are involved in more accidents and get more traffic tickets, young unmarried males pay higher rates than any other group. Besides age and gender, these factors also affect the cost of car insurance:

- *How often you drive.* The more miles you drive each year, the higher the rate.
- *How well you drive.* Drivers with no accidents or citations have lower rates than others in their age group.
- *Where you live.* Because more auto accidents occur in cities, people who live in metropolitan areas generally pay more for insurance than those with similar driving records who live in the country.
- *What kind of car you drive.* Because they are more likely to be stolen—and to be driven faster and are therefore more likely to be involved in accidents—sports cars are much more expensive to insure than other types of automobiles.

How Young Drivers Can Save on Auto Insurance

Although young drivers as a group are charged higher auto insurance rates, individual drivers often qualify for discounts that lower the basic high cost for their age group. Here's how:

- *Qualify for a good student discount.* Offered by many insurance companies, good student rates provide a price break for students in the top 20% of their class, those with a "B" average, or those on the honor roll. Discounts vary and you must show proof that you qualify.
- *Drive the same car as your parents.* It's cheaper to "share the risk" with parents than to be the sole driver of a car in your own name.
- *Take a drivers' education course approved by the insurance company.* You will be asked to show proof you have passed the approved course. Most school-based courses qualify.
- *Maintain a good driving record.* Your rates will drop if you have no citations or violations three years or more after receiving your driver's license.

AUTO POLICY

What's Covered by Auto Insurance?

In most states, all drivers are required to carry some type of insurance. It's important to talk with a qualified agent who will take time to explain all the types of coverage, from least to most expensive.

Briefly, here are some basic types of insurance coverage:

• **Collision** This pays for repairing your car if an accident was your fault. Usually you have to pay the first $100 or $200 and the insurance company picks up the rest.

• **Uninsured Motorist** If a person without insurance causes an accident that damages your car, this type of insurance will pay for the damages.

• **Bodily Injury** If the other driver or any passengers are injured in an accident that was your fault, this insurance will help pay their medical bills or lawsuits they file against you for damage to their bodies or minds resulting from the accident.

• **Comprehensive** This covers collision and accident-related claims as well as such noncollision situations, such as vandalism.

Which type of coverage is indicated in the following situations?

1. Someone breaks into your car and steals a tape deck. Comprehensive
2. A passenger in your car is hurt when you run into a tree. Bodily injury
3. The driver whose car backs into yours has no insurance. Uninsured motorist
4. No one is hurt, but you crumple a fender when you back into a dumpster. Collision

▶ EVALUATING THE LESSON

Have students describe in their own words basic auto insurance.

▶ EXTENDING THE LESSON

Have students find out what it would cost to insure an '85 Jeep with a driver between the ages of 16-18 in your area.

Objective
Become familiar with the cost of routine care and maintenance of a car.

▶ INTRODUCING THE LESSON

Point out that any type of machine must be properly cared for in order to operate efficiently. Have students give examples.

▶ TEACHING THE LESSON

Using Questioning

- Why is car maintenance necessary?
- What types of things must be done regularly to a car? Periodically?
- What types of things could you or a family member do?
- What types of things will you need a mechanic to do?

Taking Care of Your Car

Like a bicycle, lawnmower, or any other machine, a car needs regular maintenance to operate efficiently. Maintenance includes simple but important tasks you can do yourself, such as regularly checking the oil, wiper fluid, transmission fluid and antifreeze coolant levels, and replacing or filling them when necessary. You will probably need to pay a mechanic to do more complicated tasks such as changing the shock absorbers or fan belt and performing a general tune-up.

Car maintenance is expensive but necessary. Mechanics charge more than $30 an hour for labor, not including the cost of parts. Many first-time car owners are shocked at the prices they must pay to keep their cars in good condition.

The following is a list of car parts and their prices.

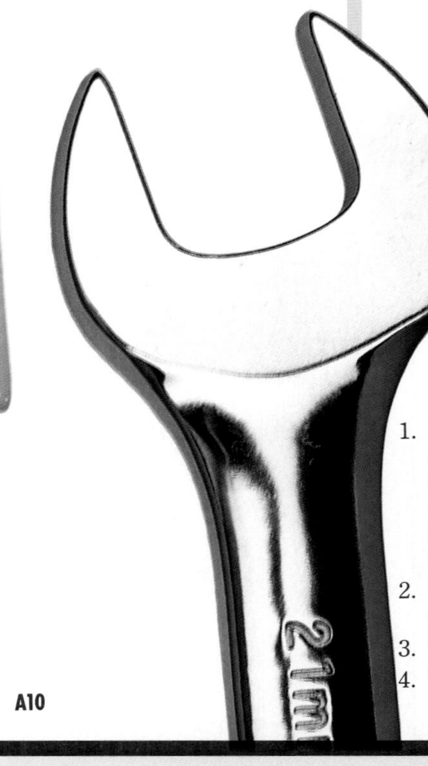

Car Parts/Fluids	Price
air filter	$4.44
alternator	47.53
antifreeze (gallon)	4.88
brake fluid (pint)	1.38
fan belt	2.99
fuel filter	1.73
headlight	4.89
oil (quart)	1.01
oil filter	3.44
PCV valve	1.68
shock absorber	10.88
spark plug	1.06
starter	31.24
transmission fluid (quart)	0.89
tune-up kit	6.29
water pump	17.44
wiper blade refills (pair)	3.08

1. You will need to change the oil and the oil filter in your car every 4,000 miles. To do so, you'll need to buy 5 quarts of oil at $1.01 a quart. The oil filter costs $3.44. Multiply the cost of oil ($1.01) × the number of quarts (5). How much will the oil cost (not including sales tax)?
   a. $5.00 b. $5.05 c. $5.10 d. $5.15
2. What is the combined cost of the oil and the oil filter?
   a. $8.29 b. $8.39 c. $8.49 d. $8.59
3. What is the most expensive item on the chart? Alternator
4. What is the least expensive item on the chart? Transmission fluid

Don't Forget About Gas!

When budgeting for a car, be prepared not only for maintenance and repair, but for the cost of regularly buying gasoline. Clearly, buying a car is just the beginning! The costs associated with insuring, operating and maintaining a car are considerable.

It's wise to plan ahead and budget for these costs—a car won't be of any use to you if you can't afford to drive it! The majority of cars get gas mileage of between 13 and 18 miles per gallon of gas in the city and 21 to 28 miles per gallon on the highway. If you drive 60 miles a week to school and work and can drive 15 miles on each gallon of gas, you'll need to buy 4 gallons of gas to meet your needs.

60 miles divided by 15 miles for each gallon = 4 gallons

5. If gas costs $1.30 a gallon, what is the bare minimum you will need to spend each week on gas?
   a. $4.20 b. $5.20 c. $6.20 d. $7.20
6. About how much would your bare minimum monthly gas budget be? There are about 4½ weeks in a month.
   a. $10 b. $15 c. $20 d. $25

   Either c or d is correct.

▶ EVALUATING THE LESSON

Have students tell about routine care and maintenance of a car.

▶ EXTENDING THE LESSON

Have students research the cost of an oil change from several car service centers. Which offers the best deal? Do any offer reduced rates or coupons at different times of the year?

Chapter

1

APPLYING NUMBERS AND VARIABLES

PREVIEWING the CHAPTER

This chapter begins with a review of those topics such as place value, exponents, comparing numbers, rounding, and estimation that are the foundation of all computation and problem solving to come. Charts are used to help illustrate practical applications of numbers and computation. The second half of the chapter introduces basic skills of algebra beginning with order of operations and evaluating expressions. The chapter concludes with students using inverse operations to solve one-step equations.

Problem-Solving Strategy Students use clues and matrix logic to deduce the answers to problems involving logical reasoning.

Lesson (Pages)	Lesson Objectives	State/Local Objectives
1-1 (4-5)	1-1: Use place value with whole numbers and decimals.	
1-2 (6-7)	1-2: Write numbers in expanded form using exponents.	
1-3 (8-9)	1-3: Compare whole numbers and decimals.	
1-4 (10-11)	1-4: Round whole numbers and decimals.	
1-5 (14-15)	1-5: Use a tax table to determine amount of income tax.	
1-6 (16-17)	1-6: Estimate sums and differences.	
1-7 (18-19)	1-7: Estimate products and quotients.	
1-8 (20-21)	1-8: Solve problems using matrix logic.	
1-9 (22-23)	1-9: Use the order of operations to find the value of an expression.	
1-10 (24-25)	1-10: Write and evaluate expressions using variables.	
1-11 (26-27)	1-11: Solve equations by using addition and subtraction.	
1-12 (28-29)	1-12: Solve equations by using multiplication and division.	
1-13 (30-31)	1-13: Solve verbal problems using an equation.	

ORGANIZING the CHAPTER

You may refer to the ***Course Planning Calendar*** on page T10.

You may refer to the Course Planning Calendar on page T10.

Planning Guide

Lesson (Pages)	Extra Practice (Student Edition)	Blackline Masters Booklets: Reteaching	Practice	Enrichment	Activity	Multi-cultural Activity	Technology	Tech Prep	Evaluation	Other Resources
1-1 (4-5)	p. 448	p. 1	p. 1	p. 1					p. 72	Transparency 1-1
1-2 (6-7)	p. 448	p. 2	p. 2	p. 2						Transparency 1-2
1-3 (8-9)	p. 448	p. 3	p. 3	p. 3	pp. 1, 29					Transparency 1-3
1-4 (10-11)	p. 449	p. 4	p. 4	p. 4	p. 2					Transparency 1-4
1-5 (12-13)		p. 5	p. 5	p. 5						Transparency 1-5
1-6 (14-15)	p. 449	p. 6	p. 6	p. 6			p. 1			Transparency 1-6
1-7 (16-17)	p. 449	p. 7	p. 7	p. 7				p. 1	p. 4	Transparency 1-7
1-8 (20-21)			p. 8							Transparency 1-8
1-9 (22-23)	p. 450	p. 8	p. 9	p. 8						Transparency 1-9
1-10 (24-25)	p. 450	p. 9	p. 10	p. 9		p. 1		p. 2		Transparency 1-10
1-11 (26-27)	p. 450	p. 10	p. 11	p. 10						Transparency 1-11
1-12 (28-29)	p. 451	p. 11	p. 12	p. 11						Transparency 1-12
1-13 (30-31)		p. 12	p. 13	p. 12					p. 4	Transparency 1-13
Ch. Review (32-33)									pp. 1-3	Test and Review Generator
Ch. Test (34)										
Cumulative Review/Test (36-37)									p. 5	

OTHER CHAPTER RESOURCES

Student Edition

Chapter Opener, p. 2
Activity, p. 3
On-the-Job Application, p. 18
Consumer Connections, p. 19
Journal Entry, p. 13
Portfolio Suggestion, p. 31
Geography Connections, p. 35
Tech Prep Handbook, pp. 516-518

Teacher's Classroom Resources

Transparency 1-0
Lab Manual, pp. 23-24
Performance Assesment, pp. 1-2
Lesson Plans, pp. 1-13

Other Supplements

Overhead Manipulative Resources, Labs 1-4, pp. 1-2
Glencoe Mathematics Professional Series

ENHANCING the CHAPTER

Some of the blackline masters for enhancing this chapter are shown below.

COOPERATIVE LEARNING

Have two pairs of students play this game of place value and comparison. Have each student pair draw a place-value chart extending from thousands to thousandths. A student from each pair rolls a die in turn. Each pair writes the digit on their respective chart. Each pair's goal is to try to make the greater number with seven rolls of the number cube. Since a digit cannot be moved once it has been written in the chart, each pair should discuss their strategy beforehand.

USING MODELS/MANIPULATIVES

Have students use algebra tiles representing a variable and ones to model one-step addition and multiplication equations. To solve an addition equation, first model it. Then remove the number of ones necessary to get x alone from each side of the equation. The ones remaining on the right side are the solution. To solve an equation like $3x = 9$, use three x pieces and nine ones to model the equation. Find how many ones there are for each x. Remove an x piece and a group of ones until one x and a group of three ones are left. The solution is 3.

Cooperative Problem Solving, p. 29

Name ______________________ Date ______________

COOPERATIVE PROBLEM SOLVING (Use with Lesson 1-3)

Comparison Shopping

As a class, decide on 10 items you might buy at a grocery or other type of store. Be sure to specify a measure and number for each.

List
1 lb Granny Smith apples
1 16-oz box of corn flakes
$\frac{1}{2}$ gal orange juice

Work in small groups. Each group should visit, call or find newspaper ads for a different store to find out the prices for items on the list.

1. As a class find and list the store with the best prices for each item.

Item 1 __________ 6 __________
2 __________ 7 __________
3 __________ 8 __________
4 __________ 9 __________
5 __________ 10 __________

2. Each group should find the total for all 10 items at their chosen store. Total: __________
3. At which store would it cost the least amount of money to buy all 10 items? __________
4. What are some reasons you might want to shop at the store with the best total price? __________
5. What are some reasons you might *not* want to shop at the store with the best total price? __________
6. Did all the stores carry the same brand of each item? Does the brand make a difference to you? __________
7. Work with your group to find the price of 2 different items at 3 stores. Decide which store has the best buy. Explain.

Lab Activity, p. 24

Name ______________________ Date ______________

LAB ACTIVITY (Use with Lesson 1-10)

Modeling Equations

You can cut out these models for variables and numbers and use them to model and solve equations.

Variables Numbers

Let ▯ stand for a variable and □ stand for 1.

$x + 3 = 5$

Subtract 3 from each side.

$x = 2$

$x - 2 = 4$

Add 2 to each side. $-2 + 2 = 0$

$x = 6$

$2x = 6$

Divide each side by 2.

$x = 3$

$\frac{x}{3} = 4$

Multiply each side by 3.

$x = 12$

Use models to solve.

1. $x + 11 = 18$ 2. $3x = 15$ 3. $x - 13 = 7$ 4. $\frac{x}{2} = 8$

MEETING INDIVIDUAL NEEDS

Students at risk

Special students often have difficulty dealing with numbers and with operations on numbers in the abstract. Use concrete objects when possible to introduce new topics or to express a problem. Base-ten blocks can be used to show the value of a number, to compare two numbers, to round a number, to estimate simple sums and differences, and to evaluate many expressions.

CRITICAL THINKING/PROBLEM SOLVING

An important key to unlocking the solution to a problem involving critical thinking is to realize that the information given may eliminate some possible answers. Eliminating possibilities saves the student valuable time in solving a problem. Even more important is the focusing of attention on the most promising solutions; thus offering a greater chance of success. Such problems often involve a pattern of unknown numbers or guessing a particular number.

COMMUNICATION

Speaking

Have the students in your class give an oral presentation of the content of the chapter by having them speak on selected topics. You may wish to have several students speak on each of the following topics: the relationship between place value and the expanded form of a number, alternative methods for comparing whole numbers and decimals, the appropriate circumstances for estimating, the necessity of rules for order of operations, and the use of equations to represent real-world problems.

Applications, p. 1

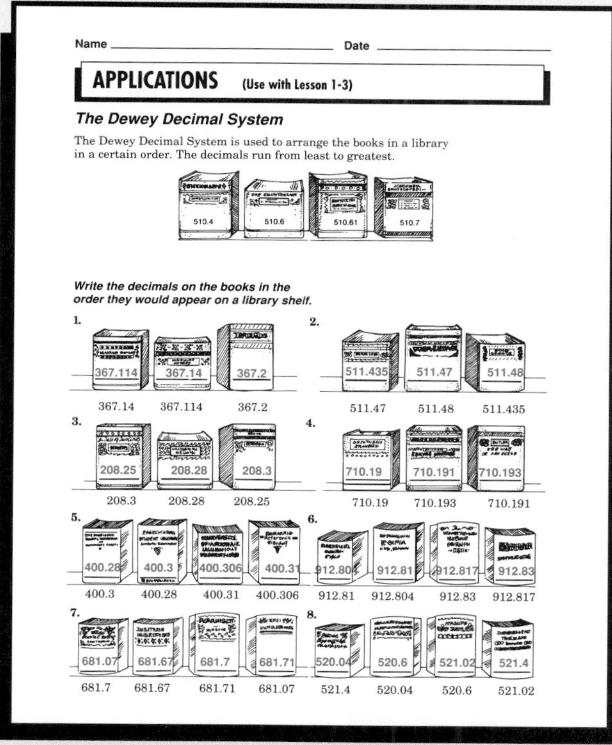

Name ______________ Date ______________

APPLICATIONS (Use with Lesson 1-3)

The Dewey Decimal System

The Dewey Decimal System is used to arrange the books in a library in a certain order. The decimals run from least to greatest.

Write the decimals on the books in the order they would appear on a library shelf.

1. 367.14 367.114 367.2 — 367.114, 367.14, 367.2
2. 511.47 511.48 511.435 — 511.435, 511.47, 511.48
3. 208.3 208.28 208.25 — 208.25, 208.28, 208.3
4. 710.19 710.193 710.191 — 710.19, 710.191, 710.193
5. 400.3 400.28 400.31 400.306 — 400.28, 400.3, 400.306, 400.31
6. 912.81 912.804 912.83 912.817 — 912.804, 912.81, 912.817, 912.83
7. 681.7 681.67 681.71 681.07 — 681.07, 681.67, 681.7, 681.71
8. 521.4 520.04 520.6 521.02 — 520.04, 520.6, 521.02, 521.4

Calculator Activity, p. 1

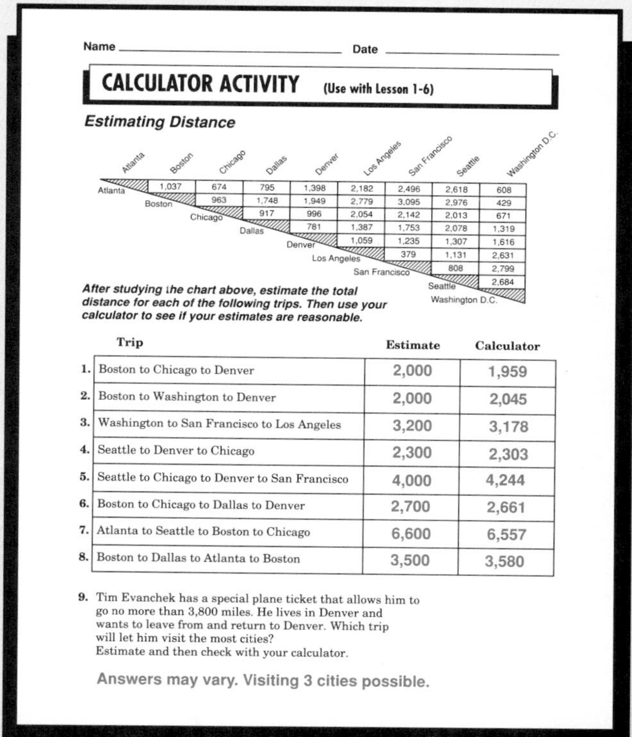

Name ______________ Date ______________

CALCULATOR ACTIVITY (Use with Lesson 1-6)

Estimating Distance

	Atlanta	Boston	Chicago	Dallas	Denver	Los Angeles	San Francisco	Seattle	Washington D.C.
Atlanta		1,037	674	795	1,398	2,182	2,496	2,618	608
Boston			963	1,748	1,949	2,779	3,095	2,976	429
Chicago				917	996	2,054	2,142	2,013	671
Dallas					781	1,387	1,753	2,078	1,319
Denver						1,059	1,235	1,307	1,616
Los Angeles							379	1,131	2,631
San Francisco								808	2,799
Seattle									2,684
Washington D.C.									

After studying the chart above, estimate the total distance for each of the following trips. Then use your calculator to see if your estimates are reasonable.

	Trip	Estimate	Calculator
1.	Boston to Chicago to Denver	2,000	1,959
2.	Boston to Washington to Denver	2,000	2,045
3.	Washington to San Francisco to Los Angeles	3,200	3,178
4.	Seattle to Denver to Chicago	2,300	2,303
5.	Seattle to Chicago to Denver to San Francisco	4,000	4,244
6.	Boston to Chicago to Dallas to Denver	2,700	2,661
7.	Atlanta to Seattle to Boston to Chicago	6,600	6,557
8.	Boston to Dallas to Atlanta to Boston	3,500	3,580

9. Tim Evanchek has a special plane ticket that allows him to go no more than 3,800 miles. He lives in Denver and wants to leave from and return to Denver. Which trip will let him visit the most cities? Estimate and then check with your calculator.

Answers may vary. Visiting 3 cities possible.

CHAPTER PROJECT

Have students use an almanac or other resource to find the populations of each state for the two most recent censuses. Have students show the following information in a chart like the one below: list states in order of their most recent populations from least to greatest; give populations for the two most recent censuses; compute and record the amounts of increase or decrease in population; write equations that represent the change in the population, using a variable for the amount of change.

State	1980	1990	Equation	+/−
Indiana	5,490,260	5,550,000	$5{,}490{,}260 + x = 5{,}550{,}000$	59,740 (+)
Ohio	10,797,624	10,791,000	$10{,}797{,}624 - x = 10{,}791{,}000$	6,624 (−)

Using the Chapter Opener

Transparency 1-0 is available in the Transparency Package. It provides a full-color visual and motivational activity that you can use to engage students in the mathematical content of the chapter.

Using Discussion

- Have students read the opening paragraph. Then ask, "What is the farthest distance you have ever ridden your bicycle?"
- Allow students the opportunity to speculate on the farthest distance a bicycle has ever been ridden. Encourage students to research this information. **The greatest outdoor distance covered in 24 hours is 548.9 miles. The greatest distance covered in 60 minutes unpaced is 31.3 miles.**
- Continue on with the last two questions in the opening paragraph. Encourage everyone to participate in the discussion.

The second paragraph is a motivational problem. Have students estimate the answer and discuss ways in which they arrived at their answer.

You may want to point out that estimating is useful when an exact answer is not needed.

Chapter 1

APPLYING NUMBERS AND VARIABLES

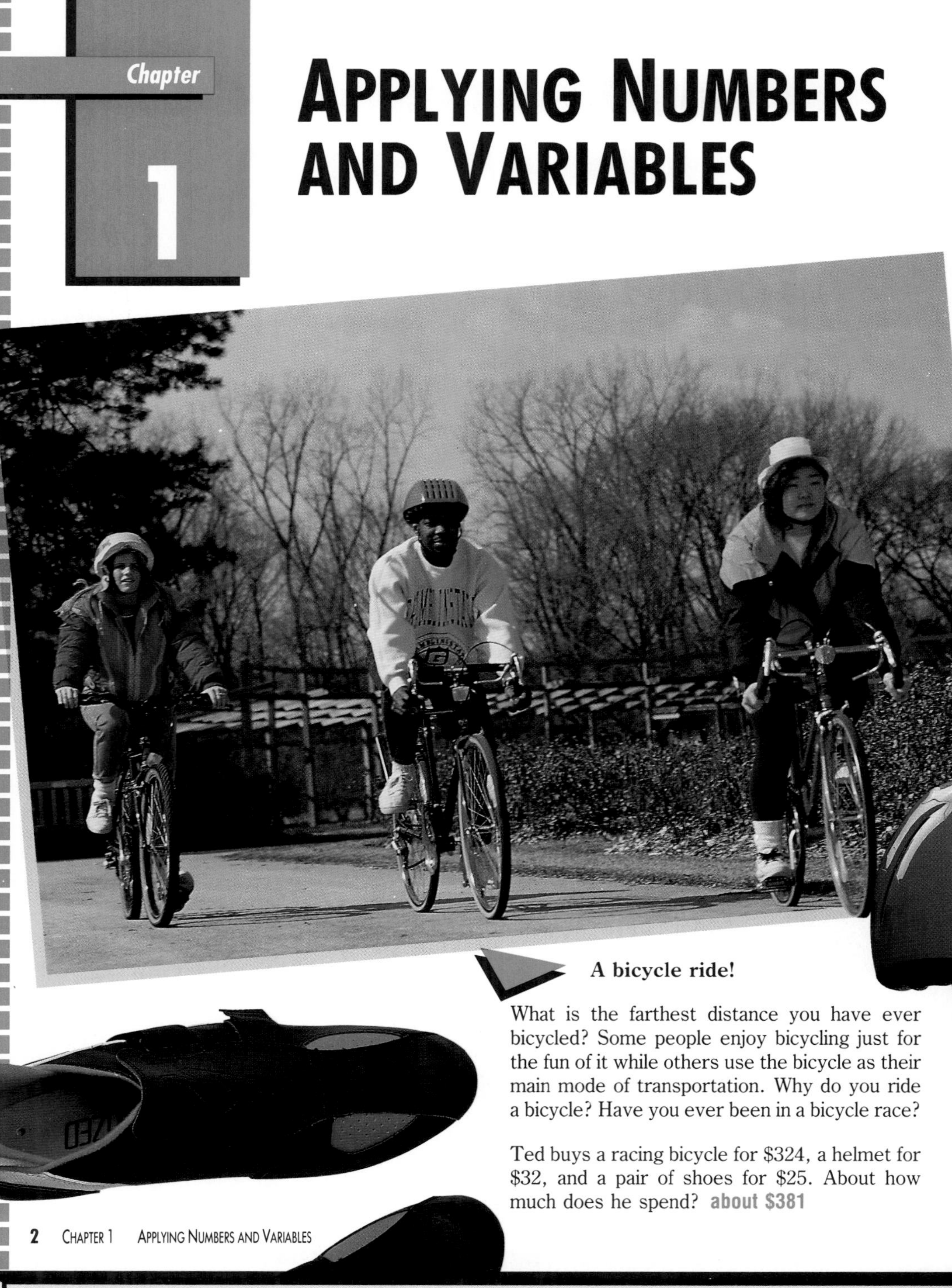

A bicycle ride!

What is the farthest distance you have ever bicycled? Some people enjoy bicycling just for the fun of it while others use the bicycle as their main mode of transportation. Why do you ride a bicycle? Have you ever been in a bicycle race?

Ted buys a racing bicycle for $324, a helmet for $32, and a pair of shoes for $25. About how much does he spend? **about $381**

ACTIVITY: Place-Value Football

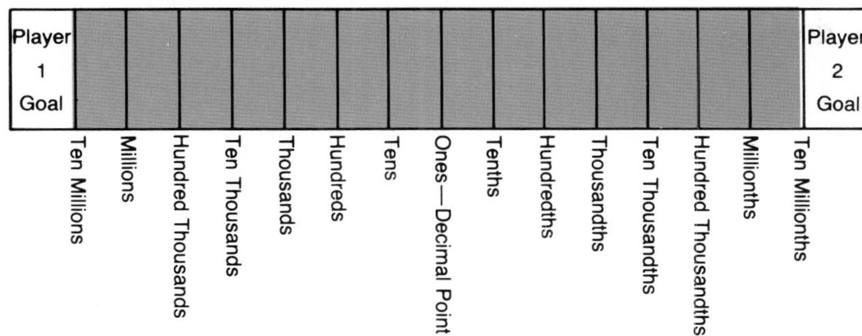

Materials: game markers, 15 index cards, pencil, paper or cardboard
Number of Players: 2

Making the Game Pieces

1. Write 5 as the first digit in each number from 50 million to 5 ten millionths on 15 index cards. For example: 50,000,000, 5,000,000, 500,000, . . . , 0.0000005.
2. On a piece of paper or cardboard, copy the gameboard shown above.

Playing the Game

3. Shuffle the index cards and place them face down.
4. Each player places a marker on the ones—decimal point yard line.
5. Player 1 turns over the first index card. Player 1 must identify the place-value position of the digit 5 on the card.
6. If Player 1 is correct, he or she moves one place-value yard line towards his or her goal. If Player 1 is incorrect, he or she moves one place-value yard line away from his or her goal.
7. Player 1 returns the index card to the bottom of the pile, and Player 2 takes a turn.
8. The first player to move into his or her goal area wins.

Variations of the Game

9. Draw a card and identify the place value that is two places to the right of the digit 5.
10. Draw two cards and name a place value that is between the two place values. If there is no place value between the two, name the larger place value.

Communicate Your Ideas

11. Create your own variation of the game. Make sure to adjust the rules to accommodate your changes.

INTRODUCING THE LESSON

Objective Recognize place-value positions.

TEACHING THE LESSON

Using Cooperative Groups Divide the students into groups of two. Have students begin by making the gameboard and game markers, one per group.

Have students read the instructions under Playing the Game.

EVALUATING THE LESSON

Have students play the game several times. Once they feel confident with the game, have them play the Variations of the Game.

Communicate Your Ideas You may want to offer a prize to the group who has developed the most original game.

EXTENDING THE LESSON

Closing the Lesson Encourage students to tell what they learned about place value from playing this game.

Activity Worksheet, Lab Manual, p. 23

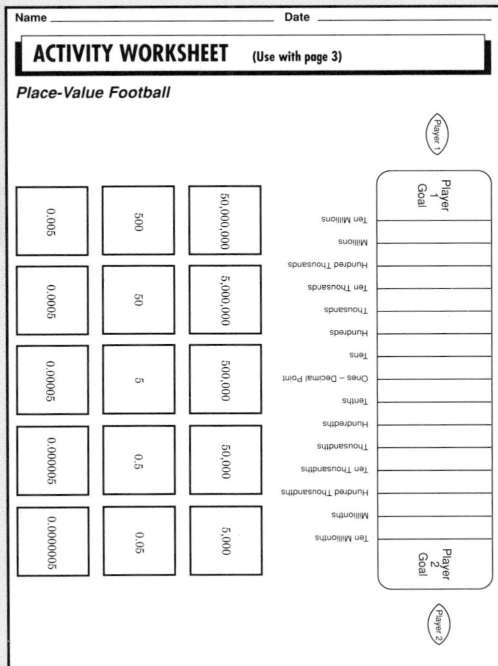
Name ______ Date ______

ACTIVITY WORKSHEET (Use with page 3)

Place-Value Football

5-MINUTE CHECK-UP

Available on TRANSPARENCY 1-1.
Write in standard form.
1. three hundred forty 340
2. three hundred four 304
3. three and four tenths 3.4
4. three and twenty-four hundredths 3.24

INTRODUCING THE LESSON

Show students an example of a very large amount of money written in the newspaper. Ask them how to read it.

TEACHING THE LESSON

Using Manipulatives Have students place base-ten blocks from **Lab Manual, p. 1,** in order (one of each size) and discuss patterns left-to-right and right-to-left. Assign a number to different blocks and identify the number.

Practice Masters Booklet, p. 1

Name ______ Date ______

PRACTICE WORKSHEET 1-1

Place Value

Name the place-value position for each digit in 106.3972.
1. the 0 tens 2. the 9 hundredths 3. the 3 tenths
4. the 6 ones 5. the 1 hundreds 6. the 2 ten thousandths

Complete the word name for each number.
7. 547,000 five hundred forty-seven thousand
8. 0.086 eighty-six thousandths
9. 0.159 one hundred fifty-nine thousandths
10. 3.07 three and seven hundredths

Write the number named by the 4 in each number.
11. 14,690 4,000
12. 480,000 400,000
13. 7.004 0.004
14. 18.345 0.04

Place the decimal point in each number to show the number named in words.
15. thirty-five hundredths 035 0.35
16. five hundred twenty thousandths 0520 0.520
17. two hundred nine and four tenths 2094 209.4
18. forty-four and four hundredths 4404 44.04

Write in standard form.
19. seventy-two thousand, three 72,003
20. three million, five hundred seven thousand, ninety 3,507,090
21. forty-three and seven tenths 43.7
22. eight and forty-seven thousandths 8.047

1-1 PLACE VALUE

Objective
Use place value with whole numbers and decimals.

Mr. Robinson is a NASA engineer who plans communications for space travel. He needs to know how long it will take a radio signal to travel 6,943.857 kilometers. This number can be read as *six thousand, nine hundred forty-three and eight hundred fifty-seven thousandths.* **can also be read as *six thousand, nine hundred forty-three point eight five seven***

Notice that the decimal point is read as *and*.

billions	hundred millions	ten millions	millions	hundred thousands	ten thousands	thousands	hundreds	tens	ones	tenths ($\frac{1}{10}$)	hundredths ($\frac{1}{100}$)	thousandths ($\frac{1}{1000}$)
						6	9	4	3	8	5	7
						6	0	0	0			
							9	0	0			
								4	0			
									3			
									0	8	0	0
									0	0	5	0
									0	0	0	7

A digit and its place-value position name a number. For example, in 6,943.857 the 9 in the hundreds place names the number 900. The 5 in the hundredths place names the number 0.05.

Example A ***Name the place-value position for the digits 4 and 7 in 6,943.857.***

4 tens 7 thousandths

Example B ***Write each number in words.***

Standard Form	Words
1,065	one thousand, sixty-five
10.430	ten and four hundred thirty thousandths

Example C ***Zeros can be annexed to the right of the decimal point without changing the value.***

4 = 4.0 = 4.00 5.1 = 5.10 = 5.100

Example D ***Numbers in standard form can be renamed by using place value.***

200 = 2 hundreds = 20 tens = 200 ones 0.030 = 30 thousandths = 3 hundredths

3. ten thousands

Guided Practice ***Name the place-value position for each digit in 10,472.365.***

Example A

1. 7 tens 2. 0 thousands 3. 1 4. 3 tenths
5. 6 hundredths 6. 5 thousandths 7. 4 hundreds 8. 2 ones

RETEACHING THE LESSON

Look for numbers used daily. Write the value of each digit.
Example: The price of a radio is $39.95.
39.95: 3 tens, 9 ones, 9 tenths, 5 hundredths

Do the same for the following numbers: the current year, your telephone number (replace the hyphen with a decimal), and the price of a car. **For example, $12,435.78.**

Reteaching Masters Booklet, p. 1.

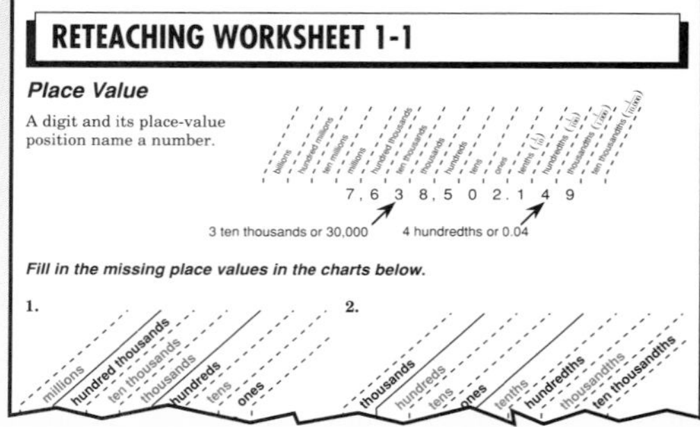

Name ______ Date ______

RETEACHING WORKSHEET 1-1

Place Value

A digit and its place-value position name a number.

7, 6 3 8, 5 0 2. 1 4 9

3 ten thousands or 30,000 4 hundredths or 0.04

Fill in the missing place values in the charts below.

1. millions, hundred thousands, ten thousands, thousands, hundreds, tens, ones
2. thousands, hundreds, tens, ones, tenths, hundredths, thousandths, ten thousandths

11. two thousand, seven and eight tenths

Example B *Write each number in words.* 9. six hundred three

9. 603 **10.** 164.05 **11.** 2,007.8

10. one hundred sixty-four and five hundredths

Example C *Annex zeros to write each number without changing their value.*

12. 0.7 0.70 **13.** 0.42 0.420 **14.** 900 900.0 **15.** 0.04 0.040

Example D *Replace each ■ with a number to make a true sentence.*

16. 0.050 = ■ thousandths 50 **17.** 5,000 = ■ hundreds 50

18. 600 = ■ tens 60 **19.** 0.080 = ■ hundredths 8

Exercises

Practice *Name the digit in each place-value position in 18,209.637.*

20. tenths 6 **21.** tens 0 **22.** hundredths 3 **23.** hundreds 2

24. ones 9 **25.** thousandths 7 **26.** thousands 8 **27.** ten thousands 1

Write each number in words. See margin.

28. 6,010 **29.** 100.05 **30.** 0.112 **31.** 303.03

Write in standard form.

32. 4 thousand 4,000 **33.** 30 hundreds 3,000 **34.** 90 tens 900

35. 48 hundredths 0.48 **36.** 5 thousandths 0.005 **37.** 25 thousandths 0.025

Write Math **38.** Write nine and six thousandths in standard form. 9.006

Applications **39.** Rank the cities from 1 through 5, with 1 having the greatest population.

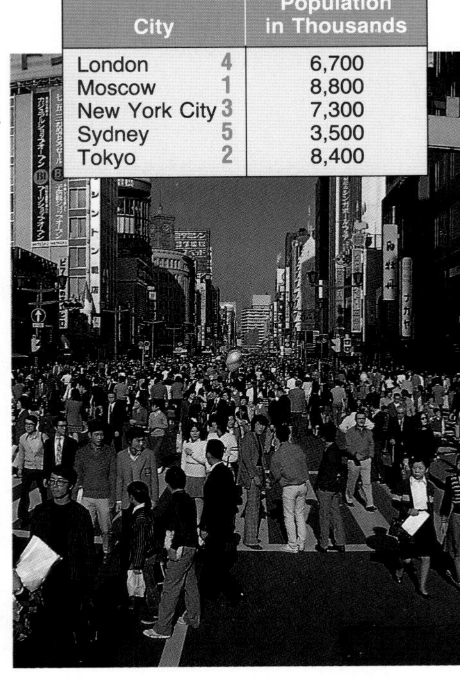

City		Population in Thousands
London	4	6,700
Moscow	1	8,800
New York City	3	7,300
Sydney	5	3,500
Tokyo	2	8,400

40. The chart shows the population of New York City as 7,300 thousand. What is the standard number for this number? 7,300,000

41. In Great Britain, 1,000,000,000 is read one thousand million. Read 73,000,000,000 as it is read in Great Britain and in the U.S.

seventy-three billion, seventy-three thousand million

Calculator 

42. Enter 456,036 in your calculator. Change it to 456,836 by using addition. What number did you add? 800

43. Enter 24.539 in your calculator. Change it to 24.509 by using subtraction. What number did you subtract? 0.030

Make Up a Problem **44.** Make up an addition problem using the digit 5 in two different place-value positions. See students' work.

Chalkboard Examples

Example A *Name the place-value position for each digit in 143,659.087.*

1. 0 **2.** 3 **3.** 5 tens **4.** 7

See margin Examples A-B.

Example B *Write each number in words.*

5. 26,304 **6.** 470.059

Example C *Annex zeros to write each number another way.*

7. 6 6.0 **8.** 3.2 3.20

Example D *Replace each ▩ with a number to make a true statement.*

9. 3,000 = ▩ hundreds 30

Additional examples are provided on TRANSPARENCY 1-1.

▶ EVALUATING THE LESSON

Check for Understanding Use Guided Practice Exercises to check for student understanding.

Error Analysis Watch for students who think that numbers with greater digits have greater value.

Closing the Lesson Ask students how to determine the standard form number for an expression like 40 tens.

▶ EXTENDING THE LESSON

Enrichment Masters Booklet, p. 1

Name ________________ Date ________________

ENRICHMENT WORKSHEET 1-1

Secret Codes

Sometimes messages need to be transmitted without others reading them. Many coding systems have been developed to keep secret messages safe.

One of the most common codes is the simple substitution code. The alphabet code below is one example of how one letter can be substituted for another.

A B C D E F G H I J K L M N O P Q R S T U V W X Y Z
L M N O P Q R S T U V W X Y Z A B C D E F G H I J K

MEET ME AT SCHOOL in this code would read XPPE XP LE DNSZZW. A secret message such as XLES EPDE ZY XZYOLJ would be decoded as MATH TEST ON MONDAY. In this code, each letter has been shifted eleven places.

Another code is called an arrangement code. This coding idea keeps the letters as they are, but arranges the message in scrambled form. To set up an arrangement code, the message must first be put into a rectangular arrangement like the one shown at the right.

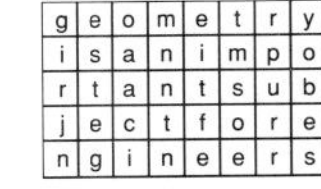

Since there are forty letters in the message, an 8 × 5 rectangle can be used. Then, rewrite the message by using top-to-bottom order instead of left-to-right order. The code above would read as follows.

GIRJN ESTEG OAACI MNNTN EITFE TMSOE RPURR YOBES

Solve the following messages using the substitution code above.

1. SLDEP XLVPD HLDEP HASTE MAKES WASTE

2. ALETPYNP TD L GTCEFP PATIENCE IS A VIRTUE

Use an arrangement grid to decode the following message.

3. YTOET OHUAI UARNT CVCDT AEAEO NYKAO YOU CAN'T HAVE YOUR CAKE AND EAT IT TOO

APPLYING THE LESSON

Independent Practice

Homework Assignment	
Minimum	Ex. 1-37 odd; 38-44 even
Average	Ex. 2-36 even; 38-44
Maximum	Ex. 1-44

Additional Answers

28. six thousand, ten
29. one hundred and five hundredths
30. one hundred twelve thousandths
31. three hundred three and three hundredths

Examples A-B Answers

1. tenths 2. thousands
4. thousandths
5. twenty-six thousand, three hundred four
6. four hundred seventy and fifty-nine thousandths

5-MINUTE CHECK-UP

(over Lesson 1-1)

Available on TRANSPARENCY 1-2.

Write in standard form.

1. three and twenty-five hundredths 3.25
2. ten and two hundredths 10.02
3. 60 hundreds 6,000

Extra Practice, Lesson 1-1, p. 448

▶ INTRODUCING THE LESSON

Ask students if anyone knows what a googol is? A googol is a 1 followed by 100 zeros. Yet it can be written with five digits as 10^{100}.

▶ TEACHING THE LESSON

Using Critical Thinking *In Example C, the number given has five digits. Why are there only four terms in the expanded form?* one number is zero

Practice Masters Booklet, p. 2

Name ______________________ Date ______________

PRACTICE WORKSHEET 1-2

Exponents

Write using exponents.

1. $5 \times 5 \times 5$ 5^3
2. $6 \times 6 \times 6 \times 6 \times 6$ 6^5
3. $9 \times 9 \times 9 \times 9 \times 9 \times 9$ 9^6
4. $7 \times 7 \times 7 \times 7 \times 7 \times 7 \times 7 \times 7$ 7^8
5. $2 \times 2 \times 2 \times 2 \times 2 \times 2 \times 2$ 2^7
6. $1 \times 1 \times 1 \times 1 \times 1$ 1^5
7. 3 squared 3^2
8. 2 to the fourth power 2^4
9. 8 cubed 8^3

Write in standard form.

10. $200 + 30 + 9$ 239
11. $(7 \times 1{,}000) + (3 \times 10)$ 7,030
12. $(6 \times 10^3) + (5 \times 10^2) + (3 \times 10^0)$ 6,503

Write in expanded form using 10^0, 10^1, 10^2 and so on.

13. 328 $(3 \times 10^2) + (2 \times 10^1) + (8 \times 10^0)$
14. 5,402 $(5 \times 10^3) + (4 \times 10^2) + (2 \times 10^0)$
15. 87,001 $(8 \times 10^4) + (7 \times 10^3) + (1 \times 10^0)$

Write as a product and then find the number named.

16. 5^3 $5 \times 5 \times 5 = 125$
17. 2^6 $2 \times 2 \times 2 \times 2 \times 2 \times 2 = 64$
18. 1^4 $1 \times 1 \times 1 \times 1 = 1$
19. 2^5 $2 \times 2 \times 2 \times 2 \times 2 = 32$
20. 6^3 $6 \times 6 \times 6 = 216$
21. 7^2 $7 \times 7 = 49$
22. 11^2 $11 \times 11 = 121$
23. 5^2 $5 \times 5 = 25$
24. 3^2 $3 \times 3 = 9$
25. 6^2 $6 \times 6 = 36$
26. 10^3 $10 \times 10 \times 10 = 1{,}000$
27. 8^2 $8 \times 8 = 64$

Solve.

28. Feisty Farlow made a deal to work for a grain company. He worked the first day for one grain of wheat. The second day he worked for two grains of wheat. The third day he worked for four grains of wheat. Each day the amount doubled. Express the number of grains of wheat that Feisty earned on the 10th day as a power.

1-2 Exponents

Objective
Write numbers in expanded form using exponents.

Marie Alexander is a scientist. She often works with numbers in the billions and greater. To save time and space, she uses *exponents* to write numbers.

The product 10×10 can be written as 10^2.

$10 \times 10 = 10^2$ ← exponent
2 factors base

The *exponent* is the number of times the *base* is used as a factor. Expressions written with exponents are called *powers.*

You read 10^2 as *10 to the second power* or *10 squared.*
You read 10^3 as *10 to the third power* or *10 cubed.*

Example A ***Write 8^3 as a product and then find the number named.***

$8^3 = 8 \times 8 \times 8$ 8 is used as a factor 3 times

$8 \times 8 \times 8 = 64 \times 8 = 512$

Example B ***Write $3 \times 3 \times 3 \times 3$ using exponents.***

$3 \times 3 \times 3 \times 3 = 3^4$ 3 is a factor 4 times.

The chart below shows how powers of 10 are related to place values. Each place value is 10 times the place value to its right.

1,000,000	100,000	10,000	1,000	100	10	1
10^6	10^5	10^4	10^3	10^2	10^1	10^0

Any nonzero number to the zero power is 1.

The number 26,097 is in **standard form.** The **expanded form** for a number can be written in several different ways.

Example C ***Write 26,097 in expanded form using each of three ways.***

- Use multiples of 1, 10, 100,. . . : $20{,}000 + 6{,}000 + 90 + 7$
- Use 1, 10, 100, . . . : $(2 \times 10{,}000) + (6 \times 1{,}000) + (9 \times 10) + (7 \times 1)$
- Use 10^0, 10^1, 10^2, . . . : $(2 \times 10^4) + (6 \times 10^3) + (9 \times 10^1) + (7 \times 10^0)$

Guided Practice

1. $10 \times 10 \times 10$; 1,000 2. 5×5; 25 3. $2 \times 2 \times 2 \times 2$; 16

Example A ***Write as a product and then find the number named.***

1. 10^3 **2.** 5^2 **3.** 2^4 **4.** 12 squared **5.** 20 cubed

4. 12×12; 144 5. $20 \times 20 \times 20$; 8,000

Example B ***Write using exponents.***

6. $7 \times 7 \times 7$ 7^3 **7.** 25×25 25^2 **8.** $5 \times 5 \times 5 \times 5$ 5^4

Example C ***Write in expanded form using 10^0, 10^1, 10^2, and so on.***

9. 156 **10.** 2,934 **11.** 708 **12.** 4,076 **13.** 96,000

See margin.

RETEACHING THE LESSON

Have students complete the chart.

Exponential Form	Product of Factors	Standard Form
2^4	$2 \times 2 \times 2 \times 2$	16
9^3	$9 \times 9 \times 9$	729
3^2	3×3	9
11^2	11×11	121
7^4	$7 \times 7 \times 7 \times 7$	2,401
6^2	6×6	36
5^3	$5 \times 5 \times 5$	125

Reteaching Masters Booklet, p. 2

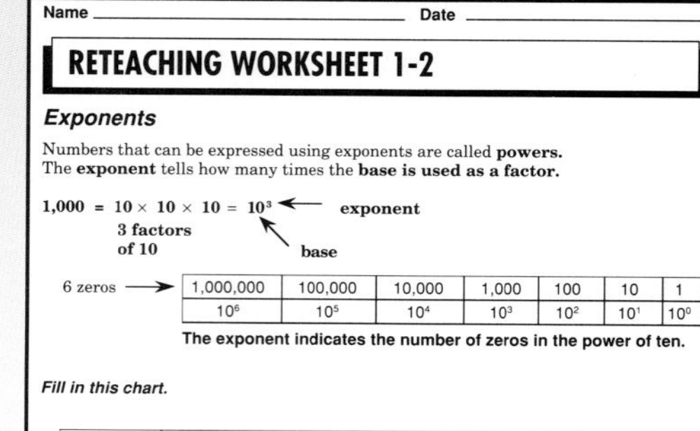

Name ______________________ Date ______________

RETEACHING WORKSHEET 1-2

Exponents

Numbers that can be expressed using exponents are called **powers.**
The **exponent** tells how many times the **base is used as a factor.**

1,000 = $10 \times 10 \times 10 = 10^3$ ← **exponent**
3 factors of 10 **base**

6 zeros →

1,000,000	100,000	10,000	1,000	100	10	1
10^6	10^5	10^4	10^3	10^2	10^1	10^0

The exponent indicates the number of zeros in the power of ten.

Fill in this chart.

14. 10 × 10 × 10 × 10; 10,000
16. 2 × 2 × 2 × 2 × 2; 32

Exercises

Practice

Write as a product and then find the number named.

14. 10^4 **15.** 3 cubed 3 × 3 × 3 **16.** 2 to the fifth power

Write using exponents.

17. 8 × 8 8^2 **18.** 7 × 7 × 7 7^3 **19.** 3 × 3 × 3 × 3 3^4

20. 250 × 250 250^2 **21.** 11 squared 11^2 **22.** 10 to the sixth power 10^6

Write in standard form.

23. 400 + 70 + 5 475 **24.** $(5 \times 10^3) + (6 \times 10^0)$ 5,006

Write in expanded form using 10^0, 10^1, 10^2, and so on. See margin.

25. 345 **26.** 8,246 **27.** 3,705 **28.** 90,240

Calculator

You can use the square key, x^2, on a calculator to square a number. Simply enter the number to be squared and then press the square key.

31. 662,596 32. 876,096

Use a calculator to find the number named.

29. 56^2 3,136 **30.** 70^2 4,900 **31.** 814^2 **32.** 936^2

33. 426^2 181,476 **34.** $2,011^2$ 4,044,121 **35.** $7,531^2$ 56,715,961 **36.** $9,999^2$ 99,980,001

FUN with MATH

Learn to send messages with your calculator.
See page 38.

Applications

37. One of the patterns of tiles in Josh's mosaic design has 5^2 tiles. How many tiles are in that pattern? 25 tiles

38. Sally is nearly 15 years old. She has been alive 362^2 hours. How many hours old is Sally? 131,044 hours

Critical Thinking

39. Use the chart on page 6 to study how powers of ten are related to place values. Find the pattern for exponents. What power of ten would you write for 0.1? 10^{-1}

Mixed Review

Lesson 1-1

Name the digit in each place-value position in 23,481.95.

40. thousands 3 **41.** ones 1 **42.** tenths 9 **43.** tens 8

45. 1,600

Write in standard form.

44. 5 hundred 500 **45.** 16 hundreds **46.** 5 tens 50

47. Maurice is offered a job with a computer company that will pay him 28 thousand dollars a year. What is the standard form of this number? $28,000

Chalkboard Examples

Example A *Write as a product and then find the number named.*

1. 3^4 81 **2.** 5 squared 25 **3.** 2 cubed 8

Example B *Write using exponents.*

4. 4 × 4 × 4 × 4 4^4

5. 100 × 100 100^2

Example C *Write in expanded form using 10^0, 10^1, 10^2, and so on.*

6. 40,760 $(4 \times 10^4) + (7 \times 10^2) + (6 \times 10^1)$

Additional examples are provided on TRANSPARENCY 1-2.

EVALUATING THE LESSON

Check for Understanding Use Guided Practice Exercises to check for student understanding.

Closing the Lesson Tell students that they will use exponents in algebra and scientific notation.

EXTENDING THE LESSON

Enrichment Work in small groups. Have students use calculators to find the greatest number that can be expressed with two digits. Use one digit as the base and one digit as the exponent.

Enrichment Masters Booklet, p. 2

Name ______ Date ______

ENRICHMENT WORKSHEET 1-2

Difference of Squares

Expressions such as $9^2 - 7^2$ and $56^2 - 49^2$ are called **the difference of squares.** It is easy to compute $9^2 - 7^2$.

$9^2 - 7^2 = 81 - 49 = 32$

However, $56^2 - 49^2$ is a little more difficult unless you have a calculator.

$56^2 - 49^2 = 3,136 - 2,401 = 735$

Try this method.

$9^2 - 7^2 = (9 + 7) \times (9 - 7) = 16 \times 2$ or 32

$56^2 - 49^2 = (56 + 49) \times (56 - 49) = 105 \times 7$ or 735

Use the method above to find the value of each expression.

1. $12^2 - 10^2$ 44 2. $16^2 - 11^2$ 135

3. $81^2 - 79^2$ 320 4. $95^2 - 90^2$ 925

5. $64^2 - 36^2$ 2,800 6. $77^2 - 23^2$ 5,400

7. $999^2 - 998^2$ 1,997 8. $649^2 - 549^2$ 119,800

9. $230^2 - 220^2$ 4,500 10. $3,568^2 - 3,567^2$ 7,135

APPLYING THE LESSON

Independent Practice

Homework Assignment	
Minimum	Ex. 2-46 even
Average	Ex. 1-35 odd; 37-47
Maximum	Ex. 1-47

Additional Answers

9. $(1 \times 10^2) + (5 \times 10^1) + (6 \times 10^0)$
10. $(2 \times 10^3) + (9 \times 10^2) + (3 \times 10^1) + (4 \times 10^0)$
11. $(7 \times 10^2) + (8 \times 10^0)$
12. $(4 \times 10^3) + (7 \times 10^1) + (6 \times 10^0)$
13. $(9 \times 10^4) + (6 \times 10^3)$
25. $(3 \times 10^2) + (4 \times 10^1) + (5 \times 10^0)$
26. $(8 \times 10^3) + (2 \times 10^2) + (4 \times 10^1) + (6 \times 10^0)$
27. $(3 \times 10^3) + (7 \times 10^2) + (5 \times 10^0)$
28. $(9 \times 10^4) + (2 \times 10^2) + (4 \times 10^1)$

5-MINUTE CHECK-UP

(over Lesson 1-2)

Available on TRANSPARENCY 1-3.

Write as a product and then find the number named.

1. 6^2 6 × 6, 36
2. 5^4 5 × 5 × 5 × 5 625
3. 10 squared 10 × 10, 100 4. 9 cubed 9 × 9 × 9 729

Extra Practice, Lesson 1-2, p. 448

INTRODUCING THE LESSON

From the newspaper, list the percent won for each team in a major league division. List the teams alphabetically. Ask students to name another way to list the teams. rank

TEACHING THE LESSON

Using Critical Thinking *How is it possible for the greater of two numbers to have fewer digits? How is it possible for the greater number to use only digits of lesser value?* placement of decimal point

Practice Masters Booklet, p. 3

Name ____________ Date ____________

PRACTICE WORKSHEET 1-3

Comparing Whole Numbers and Decimals

Fill in the blank with >, <, or = to make a true sentence.

1. 6 < 60		2. 0.80 = 0.8
3. 140 > 104		4. 0.72 < 0.8
5. 1.5 = 1.50		6. 3,200 > 3,020
7. 5.23 < 5.3		8. 8.9 > 8.88
9. 393 > 339		10. 989 < 998
11. 0.20 > 0.02		12. 1.56 > 1.55
13. 0.99 < 1.00		14. 13.4 > 1.43

Order the numbers from least to greatest.

15. 2,041; 2,001; 2,341; 2,011
2,001; 2,011; 2,041; 2,341

16. 3,342; 3,234; 4,332; 3,432; 2,432
2,432; 3,234; 3,342; 3,432; 4,3

17. 0.7, 0.67, 0.51, 0.03, 0.07
0.03, 0.07, 0.51, 0.67, 0.7

18. 0.8, 1.08, 1.8, 0.08, 1.88
0.08, 0.8, 1.08, 1.8, 1.88

19. 754,004; 75,403; 754,011; 745,004
75,403; 745,004; 754,004; 754,011

20. 2.12, 1.2, 1.112, 2.121
1.112, 1.2, 2.12, 2.121

21. 8.91, 8.919, 8.98, 8.989, 8.198
8.198, 8.91, 8.919, 8.98, 8.989

22. 1.3, 1.03, 1.303, 1.313, 1.033
1.03, 1.033, 1.3, 1.303, 1.313

1-3 COMPARING WHOLE NUMBERS AND DECIMALS

Objective
Compare whole numbers and decimals.

George has \$5.28 in his pocket. Jan has \$5.92 in her purse. Who has more money?

The whole numbers 5 and 6 are shown on the number line. The interval between 5 and 6 is separated into tenths and hundredths.

5.28 5.92
5 5.1 5.2 5.3 5.4 5.5 5.6 5.7 5.8 5.9 6

Numbers to the right are greater than numbers to the left.

5.28 *is less than* 5.92. 5.28 < 5.92

5.92 *is greater than* 5.28. 5.92 > 5.28

Decimals can also be compared by looking at the digits.

Method

1. Annex zeros, if necessary, so that each decimal has the same number of decimal places.
2. Starting at the left, compare digits in the same place-value position.
3. Use <, >, or =.

Examples

A ***Compare 8.2 and 8.17. Use <, >, or =.***

1. Annex a zero so 8.2 has the same number of decimal places as 8.17.
2. 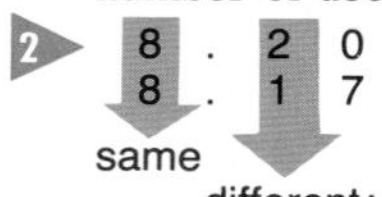

3. So 8.2 > 8.17.

B ***Compare 386 and 394. Use <, >, or =.***

1. Since there are no decimals, do not annex zeros.
2. 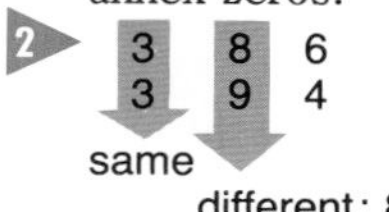

3. So 386 < 394.

C ***Order the following numbers from least to greatest.***

4.4, 2, 4.04, 6.775, 4.175, 4

1. Annex zeros so each number has three decimal places.
4.400
2.000
4.040
6.775
4.175
4.000
2. Order from least to greatest.
2.000
4.000
4.040
4.175
4.400
6.775
3. The order from least to greatest is 2, 4, 4.04, 4.175, 4.4, and 6.775.

RETEACHING THE LESSON

Compare whole numbers using the procedure of writing one number under the other and systematically comparing digits in the same place moving from the greatest to the least. Extend to one place decimals, two place decimals, etc. and then to decimals with an unequal number of decimal places.

Reteaching Masters Booklet, p. 3

Name ____________ Date ____________

RETEACHING WORKSHEET 1-3

Comparing Whole Numbers and Decimals

When comparing two numbers, begin with the greatest place-value position. Continue with the next greatest place-value position, and so on.

27 26	3.7 3.8
The greatest place-value position is tens. 2 = 2	The greatest place-value position is ones. 3 = 3
The next greatest place-value position is ones. 7 > 6	The next greatest place-value position is tenths. 0.7 < 0.8
So 27 > 26	So 3.7 < 3.8

2. = 3. > 4. <

Guided Practice

Replace each ● with <, >, or = to make a true sentence.

Example A
1. 0.2 ● 0.8 < **2.** 0.23 ● 0.230 **3.** 0.10 ● 0.01 **4.** 2.258 ● 20

Example B
5. 36 ● 89 < **6.** 450 ● 405 > **7.** 1,689 ● 1,689 =

Example C
Order from least to greatest.

8. 5.05, 5.51, 5.105, 5, 5.15
5, 5.05, 5.105, 5.15, 5.51

9. 372, 237, 723, 273, 327, 732
237, 273, 327, 372, 723, 732

Exercises

Practice

Replace each ● with <, >, or = to make a true sentence.

10. 0.201 ● 0.20 > **11.** 0.03 ● 0.030 = **12.** 4.923 ● 4.932 <

13. 7.017 ● 7.005 > **14.** 13.7 ● 13.07 > **15.** 0.612 ● 0.8 <

16. 0.002, 0.02, 0.021, 2.01, 2.1 17. 0.128, 0.821, 1.28, 1.82

Order from least to greatest.

16. 0.02, 0.021, 0.002, 2.01, 2.1

17. 0.128, 1.28, 1.82, 0.821,

18. 52, 86, 34, 20, 63, 27, 38
20, 27, 34, 38, 52, 63, 86

19. 1,008, 1,035, 1,002, 1,156
1,002, 1,008, 1,035, 1,156

Applications

20. The chart at the right lists winning times in the Olympic women's 100-meter hurdles. Rank the runners in order from 1 through 5. The fastest runner gets a rank of 1.

	Runner	Time
3	Annelie Ehrhardt	12.59 s
4	Johanna Schaller	12.77 s
2	Vera Komisova	12.56 s
5	Patoulidou Paraskevi	12.64 s
1	Jordanka Donkova	12.38 s

21. Which runners had times faster than 12.6 seconds? Ehrhardt, Komisova, Donkova

Number Sense

22. Which whole numbers are greater than 23.01 and less than 25.9? 24, 25

Show Math

23. Draw a number line that shows that 3.7 is greater than 3.2. See students' work.

Critical Thinking

24. Karen and Bob are sister and brother. Karen has twice as many brothers as she has sisters. Bob has as many brothers as he has sisters. How many girls are in the family? How many are boys? 3 girls, 4 boys

Collect Data

25. Find the heights of five tall buildings in your city or a city near you. Rank the height of the buildings in order from 1 through 5. The tallest building gets a rank of 1. Answers may vary.

Mixed Review

Lesson 1-1

Name the place-value position for each digit in 41,687.25.

26. 8 tens **27.** 5 hundredths **28.** 1 thousands **29.** 6 hundreds

30. 4^4

Lesson 1-2

Write each expression using exponents.

30. 4 × 4 × 4 × 4 **31.** 15 cubed 15^3 **32.** 5 × 5 × 5 5^3

33. Vince earns $3 an hour. In 4 years his hourly wage will be equal to the square of his hourly wage now. What will his hourly wage be in 4 years? $9

APPLYING THE LESSON

Independent Practice

Homework Assignment	
Minimum	Ex. 1-19 odd; 2-25; 26-33
Average	Ex. 2-18 even; 22-33
Maximum	Ex. 1-33

Chalkboard Examples

Example A, B *Replace each ● with <, >, or = to make a true sentence.*

1. 0.30 ● 0.03 > **2.** 123 ● 132 <

Example C Order from least to greatest.

3. 3.31, 3.03, 3.13, 3.103
3.03, 3.103, 3.13, 3.31

4. 63, 93, 60, 90, 89, 98
60, 63, 89, 90, 93, 98

Additional examples are provided on TRANSPARENCY 1-3.

EVALUATING THE LESSON

Check for Understanding Use Guided Practice Exercises to check for student understanding.

Error Analysis Be alert for students who compare greater digits before comparing place-value positions.

Closing the Lesson Ask students to tell in their own words how to determine which of two numbers is greater.

EXTENDING THE LESSON

Enrichment Masters Booklet, p. 3

Name ____________________ Date ____________

ENRICHMENT WORKSHEET 1-3

Base Ten–Base Five

The place values for base-five numerals are computed from powers of 5. The table below shows some powers of 5.

5^5	5^4	5^3	5^2	5^1	5^0
3,125	625	125	25	5	1

Translate 324_{five} to base-ten notation.

$$\begin{aligned} 324_{five} &= (3 \times 5^2) + (2 \times 5^1) + (4 \times 5^0) \\ &= (3 \times 25) + (2 \times 5) + (4 \times 1) \\ &= 75 + 10 + 4 \\ &= 89_{ten} \end{aligned}$$

Therefore, $324_{five} = 89_{ten}$.

Translate 316_{ten} to base-five notation.

Think of the greatest power of 5 equal to or less than 316. It is 5^3 or 125. Now, 316 ÷ 125 = 2 R66. The general form of the base-five numeral for 316_{ten} is 2 □ □ □. Taking the remainder 66 from the first step, find the groups of 5^2 (25), then groups of 5^1 (5), then groups of 5^0 (1) as shown below.

$$\begin{array}{r l} 316 & \\ \underline{-250} & \leftarrow 2 \times 5^3 \\ 66 & \\ \underline{-\ 50} & \leftarrow 2 \times 5^2 \\ 16 & \\ \underline{-\ 15} & \leftarrow 3 \times 5^1 \\ 1 & \leftarrow 1 \times 5^0 \end{array}$$

66 ÷ 25 = 2 R16 16 ÷ 5 = 3 R1 1 ÷ 1 = 1

Therefore, $316_{ten} = 2231_{five}$.

Translate each base-five numeral to a base-ten numeral and each base-ten numeral to a base-five numeral.

1. 101_{five} 26_{ten} **2.** 33_{five} 18_{ten} **3.** 4000_{five} 500_{ten} **4.** 24101_{five} 1776_{ten}

5. 79_{ten} 304_{five} **6.** 176_{ten} 1201_{five} **7.** 1976_{ten} 30401_{five} **8.** 5^8_{ten} 100000000_{five}

5-MINUTE CHECK-UP

(over Lesson 1-3)

Available on TRANSPARENCY 1-4.

Replace each ● with <, >, or = to make a true sentence.

1. 1.05 ● 0.87 > **2.** 0.300 ● 0.3 =
3. 6.45 ● 6.54 < **4.** 0.17 ● 0.017 >

Extra Practice, Lesson 1-3, p. 448

INTRODUCING THE LESSON

Write the current U.S. population on the board. Ask students if this is an exact or rounded number. Ask students if it is important to know the exact number. Why? rounded, no, population changes constantly

TEACHING THE LESSON

Using Cooperative Groups Give each group a list of numbers. Ask them to determine the whole numbers that would round to each of these numbers. For example, the numbers that round to 300 to the nearest hundred are 250-349.

Practice Masters Booklet, p. 4

Name ____________ Date ____________

PRACTICE WORKSHEET 1-4

Rounding Whole Numbers and Decimals

Round each number to the nearest whole number.

1. 8.3 — 8	**2.** 7.42 — 7	**3.** 321.68 — 322	**4.** 0.007 — 0

Round each number to the nearest thousand.

5. 2,703 — 3,000	**6.** 27,439 — 27,000	**7.** 882 — 1,000	**8.** 6,499 — 6,000

Round each number to the nearest tenth.

9. 4.71 — 4.7	**10.** 32.28 — 32.3	**11.** 178.69 — 178.7	**12.** 39.04 — 39.0

Round each number to the nearest hundredth.

13. 2.376 — 2.38	**14.** 0.039 — 0.04	**15.** 7.222 — 7.22	**16.** 3.4083 — 3.41

Round each number to the underlined place-value position.

17. 648 — 650	**18.** 93.174 — 93.17	**19.** 7.635 — 7.64	**20.** 5,298 — 5,000
21. 57.799 — 57.80	**22.** 17,862 — 17,900	**23.** 0.0081 — 0.008	**24.** 60.519 — 61
25. 0.735 — 0.74	**26.** 972 — 1,000	**27.** 201.899 — 200	**28.** 0.9999 — 1.00

Round each number to the nearest hundredth, tenth, and whole number.

29. 7.893 — 7.89, 7.9	**30.** 149.0769 — 149.08, 149.1	**31.** 9.598 — 9.60, 9.6	**32.** 0.0895 — 0.09, 0.1

1-4 ROUNDING WHOLE NUMBERS AND DECIMALS

Objective
Round whole numbers and decimals.

Roger Kingdom set a new Olympic record in 1988 when he ran the 110-meter hurdles in 12.98 seconds. How fast was his time to the nearest second?

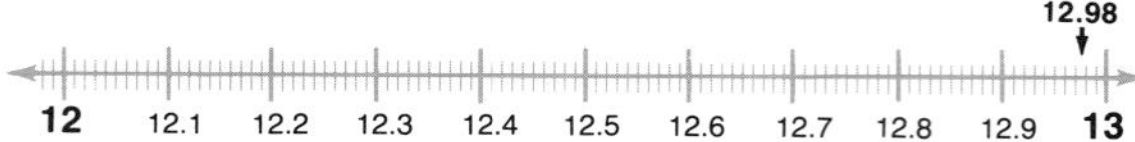

On a number line, numbers to the right are greater than numbers to the left. Note that 12.98 is closer to 13 than to 12. Therefore, 12.98 s rounded to the nearest second is 13 seconds.

Method

1. Look at the digit to the right of the place being rounded.
2. The digit remains the same if the digit to the right is 0, 1, 2, 3, or 4. Round up if the digit to the right is 5, 6, 7, 8, or 9.

Example A ***Round 91,681 to the nearest thousand.***

1. 91,681 — The digit to the right of the place being rounded is 6.
2. 91,681 → 92,000 — 91,681 rounded to the nearest thousand is 92,000.

Example B ***Round 3.409 to the nearest whole number.***

1. 3.409 — The digit to the right of the place being rounded is a 4.
2. 3.409 → 3 — 3.409 rounded to the nearest whole number is 3.

Example C ***Round 28.395 to the nearest hundredth.***

1. 28.395 — The digit to the right of the place being rounded is 5.
2. 28.395 → 28.40 — To the nearest hundredth, 28.395 rounds to 28.40.

RETEACHING THE LESSON

To help students who are having difficulty with rounding, have them underline the place-value being rounded to and circle the first digit to the right of underlined place value. Have students practice "talking aloud," asking in each case if the circled number is 5 or greater or 4 or less. If 5 or greater, increase the underlined place by 1, and if 4 or less, leave the underlined place the same.

Reteaching Masters Booklet, p. 4

Name ____________ Date ____________

RETEACHING WORKSHEET 1-4

Rounding Whole Numbers and Decimals

- Round to the nearest:

hundred	hundredth
5,344	41.376

- Look at the digit to the *right* of the place being rounded.

5,344 tens	41.376 thousandths

- The digit remains the same if the digit to the right is 0, 1, 2, 3, or 4.
- Round up if the digit to the right is 5, 6, 7, 8, or 9.

5,300 [4, same]	41.38 [6, up]

Round each number to the nearest whole number.

Guided Practice

Example A

Round each number to the nearest thousand.

1. 3,840 4,000 **2.** 1,434 1,000 **3.** 587 1,000 **4.** 9,520 10,000 **5.** 44,623 45,000

Example B

Round each number to the nearest whole number.

6. 1.255 1 **7.** 100.5 101 **8.** 49.72 50 **9.** 0.2 0 **10.** 9.5 10

Example C

Round each number to the nearest hundredth.

11. 0.782 0.78 **12.** 12.861 12.86 **13.** 0.097 0.10 **14.** $8.552 $8.55 **15.** $0.395 $0.40

Exercises

Practice 21. 1,000
22. 5,800
23. 6,730
24. 10,000
25. 80,000

Round each number to the underlined place-value position.

16. 16 20 **17.** 64 60 **18.** 736 700 **19.** 405 410 **20.** 230 200

21. 973 **22.** 5,848 **23.** 6,729 **24.** 9,915 **25.** 75,400

26. 83,912 **27.** 92,720,001 **28.** 69.5 **29.** $35.34 **30.** 0.752

31. 0.3 **32.** 399.6 **33.** 0.2749 **34.** 0.093 **35.** 0.995

26. 84,000 27. 93,000,000 28. 70.0 29. $40.00 30. 1.000

Write Math

Write each number in words. See margin.

36. 820 **37.** 6,035 **38.** 1.37 **39.** 0.025 **40.** 10.401

31. 0.0 32. 400.0 33. 0.2700 34. 0.100 35. 1.000

Applications

41. Clay bought a jacket for $35.40. To the nearest dollar, how much did Clay pay for the jacket? $35

FUN with MATH

Where and when was the mini skirt created?
See page 39.

42. Pamela has only $10 bills. How many $10 bills must she give the sales clerk when she buys a skirt for $23.29? 3 bills

43. A mutual fund reported a dividend of 6.954 shares to Mrs. McNair. To the nearest whole number, how many shares of stock did Mrs. McNair receive? 7 shares

44. The Dow Jones stock index closed at 2,878.57. To the nearest tenth, what was the closing figure? 2,878.6

Round each distance to the nearest ten miles.

45. Baltimore to Memphis 910

46. Nashville to Baltimore 702

47. Atlanta to Charlotte 240

48. The least whole number that rounds to 40 is 35. What is the greatest whole number that rounds to 40? 44

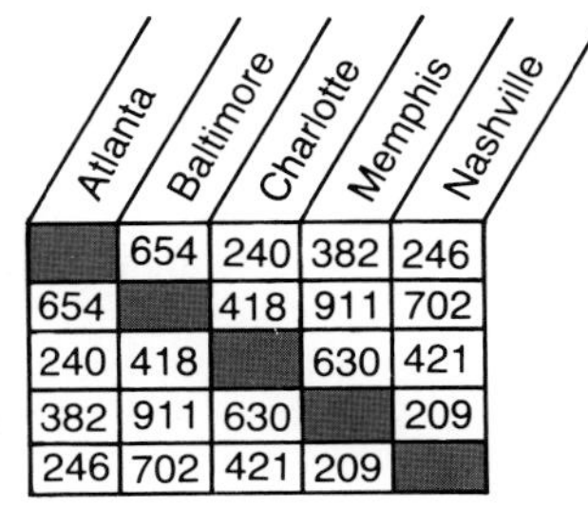

Mileage Chart

	Atlanta	Baltimore	Charlotte	Memphis	Nashville
Atlanta		654	240	382	246
Baltimore	654		418	911	702
Charlotte	240	418		630	421
Memphis	382	911	630		209
Nashville	246	702	421	209	

Number Sense

Paper/ Pencil
Mental Math
Estimate
Calculator

Many times it is not necessary to use exact numbers. Rounded numbers provide an idea of how great a number is. Write whether each statement contains a rounded or an exact number.

49. The next flight for Orlando leaves at 9:49 A.M. exact

50. Nearly 76% of Americans live in urban areas. rounded

51. Ty Cobb has a career record of 2,245 runs scored. exact

Chalkboard Examples

Example A *Round each number to the nearest thousand.*

1. 83,612 84,000 **2.** 102,487 102,000

Example B *Round each number to the nearest whole number.*

3. 37.49 37 **4.** 19.612 20

Example C *Round each number to the nearest hundredth.*

5. 70.242 70.24 **6.** 8.105 8.11

Additional examples are provided on TRANSPARENCY 1-4.

APPLYING THE LESSON

Independent Practice

Homework Assignment	
Minimum	Ex. 2-40 even; 41; 45; 48-51
Average	Ex. 1-51 odd
Maximum	Ex. 1-51

Additional Answers

36. eight hundred twenty
37. six thousand, thirty-five
38. one and thirty-seven hundredths
39. twenty-five thousandths
40. ten and four hundred one thousandths

EVALUATING THE LESSON

Check for Understanding Use Guided Practice Exercises to check for student understanding.

Error Analysis Be alert for students who round up a 9 by crowding the place-value position with 10.

Closing the Lesson Have students tell about times when they could use rounded numbers instead of exact numbers.

EXTENDING THE LESSON

Enrichment Masters Booklet, p. 4

ENRICHMENT WORKSHEET 1-4

Sequences

Sequences are lists of numbers, words, or pictures which have a predictable pattern or order to them. You have probably seen sequence problems on some tests that you have taken in school.

What comes next in these sequences?

A. 2, 4, 6, 8, 10, 12

B. April, August, December, February, January, July–alphabetical

C. 7 sides

You are familiar with the numbers, words, and shapes above, and you probably could predict what came next without too much trouble. But how can you detect a pattern when you have sequences which are not as familiar?

One method is to examine the differences between the numbers of the sequence.

Example: 6, 13, 20, 27, 34

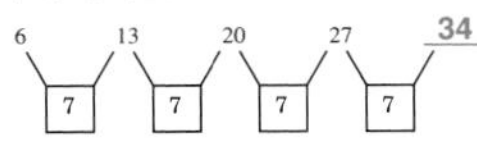

The differences between the pairs of numbers in the sequence are all equal to 7. Assume the next difference to be 7 also. So, the answer for the next number in this sequence is 27 + 7, or 34.

For each sequence, find the difference between each pair of numbers. Then find the next number in the sequence.

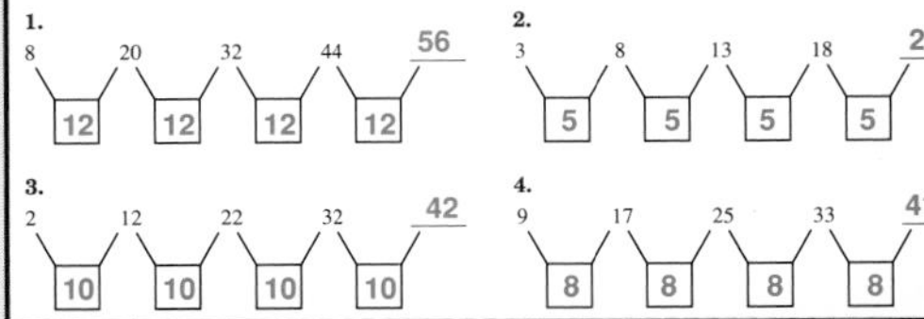

5-MINUTE CHECK-UP

(over Lesson 1-4)

Available on TRANSPARENCY 1-5.

Round each number to the indicated place-value position.

1. 708 710 **2.** 0.234 0.2
3. 2,649 2,600 **4.** 2.256 2.26

Write the name for each number in words.

5. 10.307 **6.** 1,450

5. ten and three hundred seven thousandths
6. one thousand four hundred fifty

Extra Practice, Lesson 1-4, p. 449

▶ INTRODUCING THE LESSON

Ask students how the federal, state and city governments raise money.

▶ TEACHING THE LESSON

Using Discussion Point out that generally city income tax is a flat percentage rate (that is a percentage of your gross income). State and federal income tax rates are graduated and use a tax table (that is a percentage dependent upon your taxable income - the higher your taxable income the higher percentage taxes you pay).

Practice Masters Booklet, p. 5

PRACTICE WORKSHEET 1-5

Applications: Using Tax Tables

Ten people work for the Xenon Gas Station. The table below shows their taxable income, filing status, federal withholding tax, and whether additional taxes are owed or a refund is due.

Use the tax table at the right to complete the table below.

If line 37 (taxable income) is—		And you are—			
At least	But less than	Single	Married filing jointly	Married filing separately	Head of a household
		Your tax is—			
10,000					
10,000	10,050	1,058	799	1,212	989
10,050	10,100	1,066	806	1,221	997
10,100	10,150	1,074	813	1,230	1,006
10,150	10,200	1,082	820	1,239	1,014
10,200	10,250	1,090	827	1,248	1,023
10,250	10,300	1,098	834	1,257	1,031
10,300	10,350	1,106	841	1,266	1,040
10,350	10,400	1,114	848	1,275	1,048
10,400	10,450	1,122	855	1,284	1,057
10,450	10,500	1,130	862	1,293	1,065
10,500	10,550	1,138	869	1,302	1,074
10,550	10,600	1,146	876	1,313	1,082
10,600	10,650	1,154	883	1,324	1,091
10,650	10,700	1,162	890	1,335	1,099
10,700	10,750	1,170	897	1,346	1,108
10,750	10,800	1,178	904	1,357	1,116

Name	Taxable Income	Filing Status	Federal Tax Withheld	Additional Tax Owed	Amount of Refund
Raul	$10,112	hh	$998	1. $8	none
Paula	$10,762	s	2. $908.95	$269.05	none
Anoki	$10,433	mj	$910	none	3. $55
Lewis	$10,525	4. s	$930	$208.00	none
Shamu	$10,340	ms	5. $1,310	none	$44
Christin	$10,613	6. mj	$900	none	7. $17
Pamela	$10,701	hh	8. $1,056	$52	none
Dan	$10,155	ms	$1,424	9. none	10. $185
Kerry	$10,205	s	11. $1,243	12. none	$153
Mel	$10,111	13. s	$999	$75	14. none

Applications

1-5 USING TAX TABLES

Objective
Use a tax table to determine amount of income tax.

Anyone who makes over a certain amount of money in a year must file a federal tax return. Instruction booklets provide tax tables to use in computing yearly federal income tax.

Gina Rivas earned $19,254 last year. She can find the tax she owes by locating her earnings on the table.

1. Look at the left side of the table. Locate the row that contains Gina's earnings.

   $19,254 is at least $19,250 but less than $19,300.

2. Look at the top of the table. Locate the column that indicates her filing status.

   Gina is single.

3. The amount of tax is at the intersection of the row and the column.

   Gina's federal income tax is $2,986.

If line 37 (taxable income) is—		And you are—			
At least	But less than	Single	Married filing jointly	Married filing separately	Head of a household
		Your tax is—			
19,000					
19,000	19,050	2,916	2,854	3,315	2,854
19,050	19,100	2,930	2,861	3,329	2,861
19,100	19,150	2,944	2,869	3,343	2,869
19,150	19,200	2,958	2,876	3,357	2,876
19,200	19,250	2,972	2,884	3,371	2,884
19,250	19,300	2,986	2,891	3,385	2,891
19,300	19,350	3,000	2,899	3,399	2,899
19,350	19,400	3,014	2,906	3,413	2,906
19,400	19,450	3,028	2,914	3,427	2,914
19,450	19,500	3,042	2,921	3,441	2,921
19,500	19,550	3,056	2,929	3,455	2,929
19,550	19,600	3,070	2,936	3,469	2,936
19,600	19,650	3,084	2,944	3,483	2,944
19,650	19,700	3,098	2,951	3,497	2,951
19,700	19,750	3,112	2,959	3,511	2,959
19,750	19,800	3,126	2,966	3,525	2,966
19,800	19,850	3,140	2,974	3,539	2,974
19,850	19,900	3,154	2,981	3,553	2,981
19,900	19,950	3,168	2,989	3,567	2,989
19,950	20,000	3,182	2,996	3,581	2,996

Understanding the Data

- Under what four categories do income tax filers fall? **single, married filing jointly, married filing separately, head of household**
- What are the least and greatest amounts in the tax table shown above? **$19,000-$19,999**
- In which category did you file if you paid $2,884 in taxes? **married filing jointly or head of household**

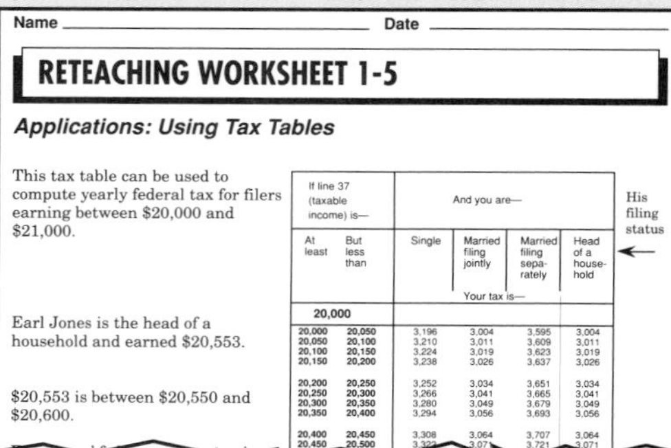

Guided Practice

Find the tax for each income from the table. The filing status is single.

1. $19,525 $3,056 **2.** $19,098 $2,930
3. $19,147 $2,944 **4.** $19,782 $3,126
5. $19,240 $2,972 **6.** $19,625 $3,084
7. $19,987 $3,182 **8.** $19,550 $3,070

RETEACHING THE LESSON

Use the tax table on page 12 to find the tax for each income.

	Income	Filing status
1.	$19,600	single $3,084
2.	$19,750	married filing jointly $2,966
3.	$19,444	head of household $2,914
4.	$19,350	married filing separately $3,413
5.	$19,825	single $3,140

Reteaching Masters Booklet, p. 5

Name ________________ Date ________________

RETEACHING WORKSHEET 1-5

Applications: Using Tax Tables

This tax table can be used to compute yearly federal tax for filers earning between $20,000 and $21,000.

Earl Jones is the head of a household and earned $20,553.

$20,553 is between $20,550 and $20,600.

If line 37 (taxable income) is—		And you are—			
At least	But less than	Single	Married filing jointly	Married filing separately	Head of a household ← His filing status
		Your tax is—			
20,000					
20,000	20,050	3,196	3,004	3,595	3,004
20,050	20,100	3,210	3,011	3,609	3,011
20,100	20,150	3,224	3,019	3,623	3,019
20,150	20,200	3,238	3,026	3,637	3,026
20,200	20,250	3,252	3,034	3,651	3,034
20,250	20,300	3,266	3,041	3,665	3,041
20,300	20,350	3,280	3,049	3,679	3,049
20,350	20,400	3,294	3,056	3,693	3,056
20,400	20,450	3,308	3,064	3,707	3,064
20,450	20,500	[illegible]	3,071	3,721	3,071

Exercises

Practice

Find the tax for each income and filing status.

9. $19,418; married filing separately **$3,427**

10. $19,284; head of a household **$2,891**

11. $19,922; married filing jointly **$2,989**

14. On a tax of $2,929, the income has to be less than $19,550. So the most Karen could make is $19,549.

12. Jim Sanders earned $19,827 as a drafter. He is single. What is his income tax?

13. Karen Johnson's income tax is $2,929. She files as head of a household. What is the most she could have earned?

12. $3,140 13. $19,549

14. For problem 13, explain how you determined the most Karen could earn.

JOURNAL ENTRY

15. Write an explanation to another student about how to use a tax table to determine income tax. **See students' work.**

Cooperative Groups

16. Choose one person in your group to contact your local IRS office or your local library for a federal tax booklet 1040 EZ. Make up problems about yearly incomes. Then figure the amount of tax that would have to be paid. **See students' work.**

Applications

17. Eric and Gladys Sneed file their taxes as married filing jointly. Eric earns $11,280 and Gladys earns $8,415. How much do they pay in taxes? **$2,951**

18. Lindsay Kulp earns $19,575 and files single. Ben and Marie Valdez earn $19,782 and file married filing jointly. Who pays more income taxes? **Lindsay**

19. Max Turner earns $19,457 and files as head of household. Louise and Kevin Early earn $19,457 and file as married filing jointly. Who pays more income taxes? **They pay the same amount.**

Critical Thinking

20. Use these clues to find three gameboard scores. The scores are all perfect squares and the scores total 590. The first two scores have the same three digits. **169, 196, 225**

Mixed Review

Lesson 1-3

Replace each ● with <, >, or = to make a true sentence.

21. 0.51 ● 0.42 > **22.** 0.494 ● 0.497 < **23.** 2.416 ● 2.410 >

26. 50,000 27. 0.880

Lesson 1-4

Round each number to the underlined place-value position.

24. 23 **20** **25.** 342 **300** **26.** 49,981 **27.** 0.876

28. What are the least and greatest whole numbers that round to 120? **115, 124**

APPLYING THE LESSON

Independent Practice

Homework Assignment	
Minimum	Ex. 1-13 odd; 14-28 even
Average	Ex. 2-28 even
Maximum	Ex. 1-28

Chalkboard Examples

Find the tax for a person earning $19,064, whose filing status is head of a household. **$2,861**

Find the tax for a person earning $19,950, whose filing status is married filing jointly. **$2,996**

Additional examples are provided on TRANSPARENCY 1-5.

EVALUATING THE LESSON

Check for Understanding Use Guided Practice Exercises to check for student understanding.

Error Analysis Check to make sure students are using the tax table correctly. Stress the meaning of the headings—At least; But less than.

Closing the Lesson Ask students to summarize what they have learned about using tax tables.

EXTENDING THE LESSON

Enrichment Masters Booklet, p. 5

Name ______________________ Date ____________

ENRICHMENT WORKSHEET 1-5

Percent and Tax

Money is deducted from most workers' pay for federal income tax.

Example: From a part-time job, John earned a gross pay of $89.75 a week. If he had only one exemption for himself, what would be his pay after tax?

$89.75 Gross pay
− 6.60 Federal tax
$83.15 Pay after tax

Find the pay after tax for each person's part-time job on the chart. Assume each has only one exemption.

Amount of Weekly Wages		Number of Withholding Exemptions Claimed			
At Least	But Less Than	0	1	2	3
		The Amount of Income Tax to be Withheld:			
$ 50.00	$ 51.00	$ 3.50	$ .60	$ .00	$ .00
51.00	52.00	3.60	.70	.00	.00
52.00	53.00	3.80	.90	.00	.00
53.00	54.00	3.90	1.00	.00	.00
54.00	55.00	4.10	1.20	.00	.00
55.00	56.00	4.20	1.30	.00	.00
56.00	57.00	4.40	1.50	.00	.00
57.00	58.00	4.50	1.60	.00	.00
58.00	59.00	4.70	1.80	.00	.00
59.00	60.00	4.80	1.90	.00	.00
60.00	62.00	5.10	2.20	.00	.00
62.00	64.00	5.40	2.50	.00	.00
64.00	66.00	5.70	2.80	.00	.00
66.00	68.00	6.10	3.10	.20	.00
68.00	70.00	6.40	3.40	.50	.00
70.00	72.00	6.80	3.70	.80	.00
72.00	74.00	7.10	4.00	1.10	.00
74.00	76.00	7.50	4.30	1.40	.00
76.00	78.00	7.90	4.60	1.70	.00
78.00	80.00	8.20	4.90	2.00	.00
80.00	82.00	8.60	5.20	2.30	.00
82.00	84.00	8.90	5.50	2.60	.00
84.00	86.00	9.30	5.80	2.90	.00
86.00	88.00	9.70	6.20	3.20	.30
88.00	90.00	10.00	6.60	3.50	.60

Person	Gross pay	Tax	Pay after tax
Sally	$75.00	$4.30	$70.70
Katie	$62.50	$2.50	$60.00
Daniel	$57.95	$1.60	$56.35
Patrick	$73.25	$4.00	$69.25
Peter	$67.50	$3.10	$64.40
Tricia	$87.75	$6.20	$81.55
Jacob	$77.00	$4.60	$72.40
Laura	$82.05	$5.50	$76.55
Sam	$50.50	$ .60	$49.90
Sarah	$66.01	$3.10	$62.91

(over Lesson 1-5)

Available on TRANSPARENCY 1-6.
Use the tax table on page 12 to find the tax for each income. The filing status is single.

1. $19,386 $3,014 **2.** $19,228 $2,972
3. $19,008 $2,916 **4.** $19,942 $3,168

INTRODUCING THE LESSON

Ask students if anyone has ever used estimation other than for school. Have students name situations where estimation comes in handy.

TEACHING THE LESSON

Using Discussion Discuss with students when an estimated answer is more appropriate than an exact answer.

Practice Masters Booklet, p. 6

Name ______________ Date ______________

PRACTICE WORKSHEET 1-6

Estimating Sums and Differences

Estimate. Answers may vary. Typical answers given.

1. 482 + 890 = 1,400
2. $4.37 + 0.93 = 5.30
3. 34.7 − 0.84 = 33.9
4. 837 − 359 = 400
5. 438 + 568 + 1,389 = 2,400
6. 7.33 + 40.27 + 0.6 = 48.2
7. 3,823 + 477 + 5,789 = 10,100
8. 0.27 + 16.8 + 0.34 = 17.4
9. 872 − 586 = 300
10. $86.06 − 28.49 = $60
11. 0.63 − 0.38 = 0.2
12. 72,685 − 4,499 = 69,000

13. 0.89 + 2.3 3.2 **14.** 81 − 18 60 **15.** 7.04 − 0.8 6.2

Solve. Round.

16. There are 345 white pages and 783 yellow pages in the local telephone directory. About how many more yellow pages than white pages does the directory contain?
500 pages

17. A truck weighing 6,675 pounds is carrying a load of apples weighing 953 pounds. About how much do the truck and the apples weigh?
7,700 pounds

Solve. Round to the nearest dollar.

18. Lenny bought a shirt for $17.89. About how much change will he receive from a $20 bill? $2

19. Judy has $30. Can she afford to buy a belt for $5.67, a skirt for $14.29, and tights for $13.95? no

1-6 ESTIMATING SUMS AND DIFFERENCES

Objective
Estimate sums and differences.

Pat is collecting paper to sell to Recycling Central. During the first month she collected 578 pounds of paper, and during the second month she collected 412 pounds. *About* how many pounds of paper did she collect in the two months?

In this case, an exact answer is not needed. You can *estimate* to find the sum.

Method

1. Round each number to the same place-value position.
2. Add or subtract.

Examples

A ***Estimate the sum of 578 and 412.***

1. Round to the nearest hundred.

578 → 600
+ 412 → + 400

2. Add → 1,000

Pat collected *about* 1,000 pounds of paper.

B ***Estimate the difference of 62,341 and 5,908.***

1. Round to the nearest thousand.

62,341 → 62,000
− 5,908 → − 6,000

2. Subtract → 56,000

The difference of 62,341 and 5,908 is *about* 56,000.

C ***Estimate the sum of 5.23, 17.52, and 6.491.***

1. Round to the nearest whole number.

5.23 → 5
17.52 → 18
+ 6.491 → + 6

2. Add. → 29

The sum of 5.23, 17.52, and 6.491 is *about* 29.

D ***Estimate the difference of 0.504 and 0.17.***

1. Round to the nearest tenth.

0.504 → 0.5
− 0.17 → − 0.2

2. Subtract. → 0.3

The difference of 0.504 and 0.17 is *about* 0.3.

RECYCLING EXCHANGE
BUYERS OF:
• ALUMINUM CANS
• GLASS
• NEWSPAPER
• NON-FERROUS METAL

Guided Practice
Examples A, C

Estimate each sum. Answers may vary. Typical answers are given.

1. 132 + 361 = 500
2. 9,136 + 958 = 10,000
3. 4,356 + 237 = 4,600
4. 0.035 + 0.24 + 0.471 = 0.7
5. 7.09 + 3.807 + 0.48 = 11.4

6. $13.27 + $9.58 $23
7. 86.12 + 57.9 + 4 148

RETEACHING THE LESSON

Round and estimate each sum.

1. 0.53 + 0.38 → 0.5 + 0.4 = 0.9
2. 3.78 + 2.64 → 4 + 3 = 7
3. 12.83 + 9.19 → 13 + 9 = 22

Reteaching Masters Booklet, p. 6

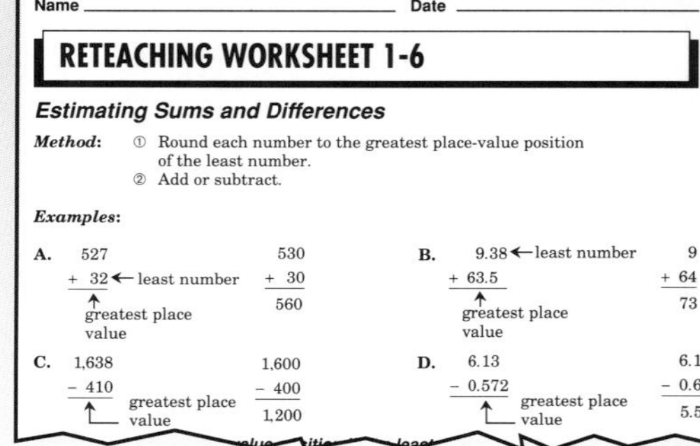

Name ______________ Date ______________

RETEACHING WORKSHEET 1-6

Estimating Sums and Differences

Method: ① Round each number to the greatest place-value position of the least number.
② Add or subtract.

Examples:

A. 527 + 32 (← least number; ↑ greatest place value) → 530 + 30 = 560

B. 9.38 (← least number) + 63.5 (↑ greatest place value) → 9 + 64 = 73

C. 1,638 − 410 (↑ greatest place value) → 1,600 − 400 = 1,200

D. 6.13 − 0.572 (↑ greatest place value) → 6.1 − 0.6 = 5.5

Examples B, D

Estimate each difference. **Answers may vary. Typical answers are given.**

8. 697 − 406 = **300**	**9.** 6,409 − 180 = **6,200**	**10.** 10.651 − 5.524 = **5.000**	**11.** 0.86 − 0.77 = **0.1**	**12.** \$18.05 − 9.64 = **\$ 8.00**

13. 65,495 − 46,485 **20,000** **14.** \$101.99 − \$24 **\$80**

Exercises

Practice

Estimate. **Answers may vary. Typical answers are given.**

15. 629 + 293 = **900**	**16.** 126 − 90 = **40**	**17.** 2,506 − 496 = **2,000**	**18.** 6.8 + 9.2 = **16.0**	**19.** 79.5 + 15.13 = **100.00**
20. 71.31 − 24.528 = **50.000**	**21.** 17.8 + 3.14 + 6.752 = **28.000**	**22.** 0.5 + 0.84 + 0.174 = **1.500**	**23.** 9.41 − 6.97 = **2.00**	**24.** 58.3 − 7.95 = **50.00**

25. 803 + 3,297 **4,100** **26.** 421,172 − 90,900 **330,000** **27.** \$3.49 + \$0.18 **\$3.67**

28. \$20 − \$5.77 **\$14** **29.** 87.6 − 57 **31** **30.** 3 − 1.63 **1**

31. Carolyn buys a cassette tape for \$7.89. Estimate her change from a \$10 bill. **\$2**

Applications

32. Lee has \$15. Estimate to determine if he can afford to buy tube socks for \$2.10, tennis balls for \$1.88, car wax for \$3.94, and 35mm film for \$4.88. **yes**

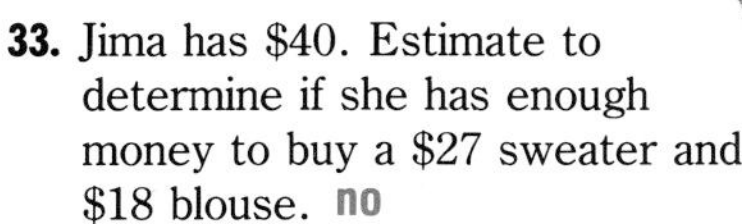

33. Jima has \$40. Estimate to determine if she has enough money to buy a \$27 sweater and \$18 blouse. **no**

Mental Math

34. Phyllis left for work at 7:35 A.M. If it takes her 27 minutes to drive to work, at what time will she arrive? **8:02 A.M.**

35. Buses stop at the corner of Fifth Street and Broad Street at 3:58 P.M., 4:10 P.M., 4:22 P.M. and so on until 8:00 P.M. Floyd leaves his office at 5:15 P.M. How much time does he have to get to the bus stop to catch the next bus? **7 minutes**

Suppose

36. Suppose a number rounds to 9,500 to the nearest hundred and it rounds to 9,460 when rounded to the nearest ten. What is the number if 5 is in the ones place? **9,455**

Determining Reasonable Answers

37. Peggy bought a jacket for \$19.99 and a blouse for \$34. The clerk charged her \$67.56. Do you think the total is correct? Why? **no**
The estimate is about \$50. Even with tax \$67.56 is too much.

APPLYING THE LESSON

Independent Practice

Homework Assignment	
Minimum	Ex. 2-30 even; 31; 35-37
Average	Ex. 1-29 odd; 31-37
Maximum	Ex. 1-37

Chalkboard Examples

Examples A, C *Estimate each sum.* **Answers may vary.**

1. 538 + 1,912 **2,400**

2. 0.0319 + 0.047 **0.08**

Examples B, D *Estimate each difference.* **Answers may vary.**

3. 7,546 − 4,390 **4,000**

4. 4.28 − 0.937 **3**

Additional examples are provided on TRANSPARENCY 1-6.

EVALUATING THE LESSON

Check for Understanding Use Guided Practice Exercises to check for student understanding.

Closing the Lesson Have students explain why it is important to know how to estimate.

EXTENDING THE LESSON

Enrichment Work in small groups. Have students roll a number cube 4 times and record digits in order as a 4-digit number. Repeat for a second number. Agree on estimates for their sum and their difference. Repeat the process for other numbers.

Enrichment Masters Booklet, p. 6

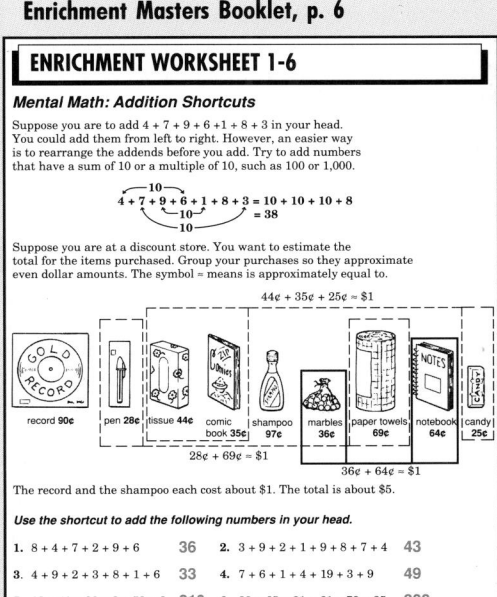

ENRICHMENT WORKSHEET 1-6

Mental Math: Addition Shortcuts

Suppose you are to add 4 + 7 + 9 + 6 + 1 + 8 + 3 in your head. You could add them from left to right. However, an easier way is to rearrange the addends before you add. Try to add numbers that have a sum of 10 or a multiple of 10, such as 100 or 1,000.

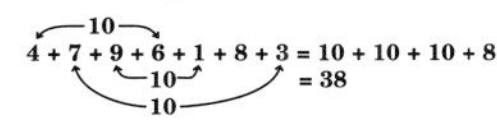

4 + 7 + 9 + 6 + 1 + 8 + 3 = 10 + 10 + 10 + 8 = 38

Suppose you are at a discount store. You want to estimate the total for the items purchased. Group your purchases so they approximate even dollar amounts. The symbol ≈ means is approximately equal to.

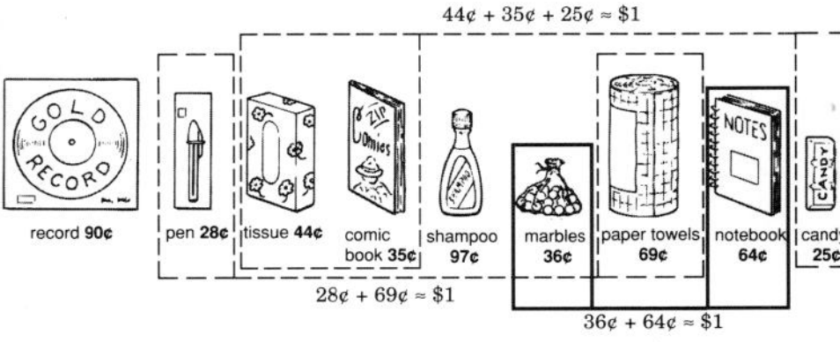

The record and the shampoo each cost about \$1. The total is about \$5.

Use the shortcut to add the following numbers in your head.

1. 8 + 4 + 7 + 2 + 9 + 6 **36** **2.** 3 + 9 + 2 + 1 + 9 + 8 + 7 + 4 **43**

3. 4 + 9 + 2 + 3 + 8 + 1 + 6 **33** **4.** 7 + 6 + 1 + 4 + 19 + 3 + 9 **49**

5. 12 + 41 + 88 + 2 + 59 + 8 **210** **6.** 39 + 65 + 21 + 61 + 79 + 35 **300**

Estimate the sum to the nearest dollar.

7. 46¢ + 87¢ + 32¢ + 13¢ + 98¢ + 54¢ + 66¢ ≈ **\$4**

8. 21¢ + 33¢ + 78¢ + 94¢ + 81¢ + 35¢ + 32¢ + 19¢ + 92¢ ≈ **\$5**

5-MINUTE CHECK-UP

(over Lesson 1-6)

Available on TRANSPARENCY 1-7.

Estimate each sum. **Answers may vary.**

1. 236 + 472 **2.** 68.13 + 1.681
700 70

Estimate each difference. **Answers may vary.**

3. 68.9 − 48 **4.** 7 − 2.61
20 4

Extra Practice, Lesson 1-6, p. 449

INTRODUCING THE LESSON

Suppose a bicycle costs $240. If you could save $18 each week, about how many weeks would it take you to save enough money to buy it? How would you solve this problem? Explain why an exact answer is not needed.

TEACHING THE LESSON

Using Data Have students go through grocery ads and make a grocery list. Have them buy large quantities of several canned goods and estimate their cost.

Practice Masters Booklet, p. 7

Name ______________ Date ______________

PRACTICE WORKSHEET 1-7

Estimating Products and Quotients.

Estimate.

1. 5.03 × 9.37 45	2. 112 × 694 70,000	3. 87.7 × 18.3 1,800
4. 0.803 × 483.5 400	5. 96 × 58 6,000	6. 375.5 × 28.1 12,000
7. 490 × 7 3,500	8. 32.9 × 68 2,100	9. 7,288 × 315 2,100,000
10. 5,399 × 0.79 4,000	11. 7,831 × 4.8 40,000	12. 3,900 × 49 200,000
13. 358 ÷ 91 4	14. 735.4 ÷ 76.9 9	15. 7,175 ÷ 9 800
16. 1,680 ÷ 38 40	17. 5,139 ÷ 194 25	18. 2,000 ÷ 6.9 300
19. 8)395 50	20. 7.2)50.2 7	21. 47)97 2
22. 79)55,794 700	23. 41)$15.85 $0.40	24. 8.5)$25.63 $3.00
25. 58)5,360 90	26. 29)2,000 70	27. 4.8)345.9 70

1-7 ESTIMATING PRODUCTS AND QUOTIENTS

Objective
Estimate products and quotients.

Harry's Grocery sells a 15.7 pound turkey for $0.98 a pound. About how much does the turkey cost?

Sometimes an estimate, not an exact answer, is all that's needed when you multiply. You only need to know *about* how much the turkey costs.

Method

1. Round each factor to its greatest place-value position. Do *not* change 1-digit factors.
2. Multiply.

Examples

A ***Estimate 15.7 × $0.98.***

1. 15.7 → 20
   0.98 → 1
2. 20 × 1 = 20
   The turkey costs *about* $20.

B ***Estimate 472 × 5.***

1. 472 → 500
   5 → 5
2. 500 × 5 = 2,500.
   The product of 472 and 5 is *about* 2,500.

Sometimes, you also need to estimate quotients.

Method

1. Round the divisor to its greatest place-value position. Do *not* change 1-digit divisors.
2. Round the dividend so it is a multiple of the divisor.
3. Divide.

Examples

C ***Estimate 357 ÷ 62.***

1. 62 → 60
2. Round 357 to 360 because 360 is a multiple of 60.
3. 60)360 = 6 ← estimate

357 ÷ 62 is *about* 6.

D ***Estimate 8,472.1 ÷ 26.5.***

1. 26.5 → 30
2. Round 8,472.1 to 9,000 because 9,000 is a multiple of 30.
3. 30)9,000 = 300 ← estimate

8,472.1 ÷ 26.5 is *about* 300.

Guided Practice

1. 1,800 2. 180,000 3. 0

Examples A, B

Estimate each product. **Answers may vary. Typical answers given.**

1. 637 × 3 **2.** 56 × 3,179 **3.** 3,450 × 0.15

4. 1,292 × 567 **5.** $10.87 × 6.7 **6.** 15.7 × 0.98

600,000 **$77** **16**

RETEACHING THE LESSON

Prepare a flow chart for students to follow for multiplication and/or division.

1. Round each factor to its greatest place-value position. 2. Multiply and/or divide. Adapt steps to the needs of the students. Encourage students to organize problems like examples. Have students work in pairs in order to help each other work through the problems using each step.

Reteaching Masters Booklet, p. 7

Name ______________ Date ______________

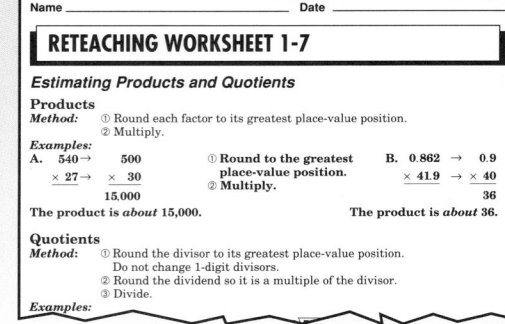

RETEACHING WORKSHEET 1-7

Estimating Products and Quotients

Products

Method: ① Round each factor to its greatest place-value position.
② Multiply.

Examples:

A. 540 → 500, × 27 → × 30; 15,000

① **Round to the greatest place-value position.**
② **Multiply.**

B. 0.862 → 0.9, × 41.9 → × 40; 36

The product is *about* 15,000. **The product is *about* 36.**

Quotients

Method: ① Round the divisor to its greatest place-value position. Do not change 1-digit divisors.
② Round the dividend so it is a multiple of the divisor.
③ Divide.

Examples:

9. 4 10. 60

Examples C, D

Estimate each quotient.

7. 123 ÷ 58 2 **8.** 735 ÷ 91 8 **9.** 11.9 ÷ 3.2 **10.** 369.7 ÷ 5.8

11. 32)6,402 200 **12.** 3.2)9.07 3 **13.** 23.8)459.75 20 **14.** 1.17)248.9 240

Exercises

16. 3,500 17. 21,000 18. 12 19. 68 20. 35,000 22. 100

Practice

23. 55
24. 10
25. $5
26 .05

Estimate. Answers may vary. Typical answers are given.

15. 59 × 2 120 **16.** 5 × 747 **17.** 738 × 29 **18.** 12 × 0.8

19. 34.15 × 1.78 **20.** 7,123 × 4.6 **21.** 602 ÷ 86 7 **22.** 4,763 ÷ 51

23. 55,114 ÷ 958 **24.** 41.6 ÷ 3.9 **25.** $31.85 ÷ 6 **26.** $1.19 ÷ 24

27. 528 × 36 20,000 **28.** 8,275 × 66 560,000 **29.** $23.79 × 31 $600 **30.** 0.03 × 0.02 0

31. 31)1,031 30 **32.** 691)3,532 5 **33.** 23.1)654.32 33 **34.** 1.23)457.8 400

35. Rachel estimates 4,254 × 32 to be about 1,200. What did Rachel do incorrectly? Did not multiply 4,000 × 30 correctly. Four zeros are required.

Applications

36. Gregory uses 0.5 pounds of hamburger for each sandwich. About how many sandwiches can he make with 6.38 pounds of hamburger? 12 sandwiches

37. Marcia drives 231.3 miles on 11.4 gallons of gasoline. About how many miles per gallon is this? 23 mpg

Estimation

38. One factor of 4,763 is 51. Estimate the other factor. 100

Choose the most reasonable estimate for each situation.

39. number of seats on a school bus
a. 60 **b.** 120 **c.** 180

40. population of the United States
a. 225 billion **b.** 225 million **c.** 225 thousand

41. Make up a problem that involves an estimate of a product. Use at least one decimal. Answers may vary.

APPLYING THE LESSON

Independent Practice

Homework Assignment	
Minimum	Ex. 1-37 odd; 38-41
Average	Ex. 2-34 even; 35-41
Maximum	Ex. 1-41

Chapter 1, Quiz A (Lessons 1-1 through 1-7) is available in the Evaluation Masters Booklet, p. 4.

Chalkboard Examples

Examples A, B *Estimate each product.* Answers may vary.

1. 367 × 7 2,800
2. 25 × 4,275 120,000
3. 1,905 × 3.7 8,000
4. 0.73 × 0.921 1

Examples C, D *Estimate each quotient.* Answers may vary.

5. 315 ÷ 43 8 **6.** 1,972 ÷ 48 40
7. 28.97 ÷ 5.8 5
8. 7,269 ÷ 0.941 7,000

Additional examples are provided on TRANSPARENCY 1-7.

EVALUATING THE LESSON

Check for Understanding Use Guided Practice Exercises to check for student understanding.

Error Analysis Be alert for students who round each number to the greatest place-value position of the *least* number.

Closing the Lesson Ask students to state in their own words how to round the dividend when estimating a quotient.

EXTENDING THE LESSON

Enrichment Masters Booklet, p. 7

ENRICHMENT WORKSHEET 1-7

Mental Math: Multiplying Shortcuts

Multiplying by 5. **5 × 84 = 420**	• Multiply by 10. • Take half the answer.	840 420
Multiplying by 15. **15 × 26 = 390**	• Multiply by 10. • Take half the answer. • Add the two products.	260 130 390
Multiplying by 9. **9 × 284 = 2,556**	• Multiply by 10. • Subtract the number.	2,840 2,556
Multiplying by 11. **11 × 315 = 3,465**	• Multiply by 10. • Add the number.	3,150 3,465
Multiplying by 99. **99 × 637 = 63,063**	• Multiply by 100. • Subtract the number.	63,700 63,063

Use the shortcuts to find each of the following products.

1. 5 × 392 1,960 **2.** 15 × 456 6,840 **3.** 9 × 726 6,534

4. 11 × 471 5,181 **5.** 5 × 638 3,190 **6.** 99 × 315 31,185

7. 5 × 2,471 12,355 **8.** 15 × 782 11,730 **9.** 9 × 980 8,820

10. 11 × 238 2,618 **11.** 99 × 148 14,652 **12.** 15 × 746 11,190

13. 9 × 3,684 33,156 **14.** 11 × 4,279 47,069 **15.** 99 × 1,472 145,728

16. Make up shortcuts for multiplying by 101 and 999.
Multiply by 100. Add the number.
Multiply by 1,000. Subtract the number.

Applying Mathematics to the Real World

This optional page shows how mathematics is used in the real world and also provides a change of pace.

Objective Count change from amount of purchase to the amount given.

Using Discussion Discuss why it is important to know the correct change. **A job like cashier may depend on it. As a consumer, you need to make sure you receive the correct change.**

Using Critical Thinking *Why can estimation be used to help check the amount of change?* **Estimation will give an idea about how much change there should be.**

Activity

Using Games Have students make a flash card of an item they would like to buy and its price. Give students various denominations of play money that they will hand to the cashier. Then show the flash card. The first student to say the correct amount of change they will receive gets one point.

On-the-Job Application

CASHIER

Brian Hunter is a cashier for Paper Supply Outlet. In his job, he needs to know how to count change to the customer by counting from the amount of purchase to the amount given.

Cindy buys paper and supplies for $1.29. She gives Brian a $5 bill. Brian counts from $1.29 to $5.

The change is 1 penny, 2 dimes, 2 quarters, and 3 one-dollar bills, or $3.71. The art below shows how the cashier would count back the change.

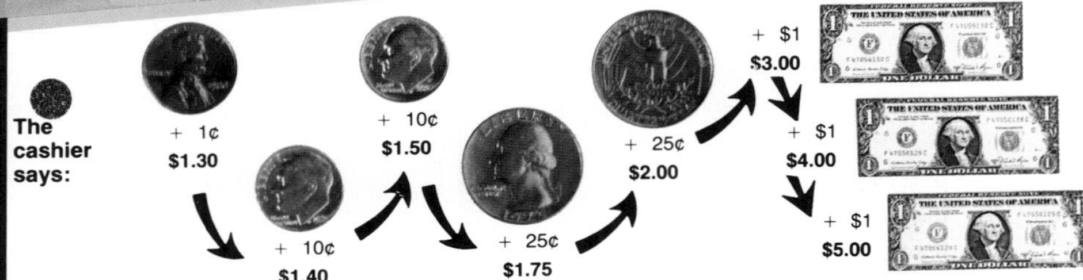

Copy and complete the chart.

	Amount of Purchase	Amount Given	Change							Total Change
			1¢	5¢	10¢	25¢	$1	$5	$10	
	$1.29	$5.00	1		2	2	3			$3.71
1.	$2.53	$5.00	2		2	1	2			$2.47
2.	$1.74	$2.00	1			1				$0.26
3.	$0.83	$10.00	2	1	1		4	1		$9.17
4.	$7.34	$20.00	1	1	1	2	2		1	$12.66

5. Suppose your bill is $4.62 and you gave Brian $10.02. How much change should you receive? **$5.40**

6. In problem 9, why would you give Brian $10.02 instead of just giving him $10?

7. Suppose Susan's bill is $8.17. Why would she give Brian $10.17 rather than just giving him $10? **to avoid receiving coins**

6. to avoid receiving pennies; to get fewer coins as change

Bus Schedule

The timetable below shows the departure (leave) and arrival (due) times for a bus route. The table gives times between 8:00 A.M. and 5:00 P.M.

MON THROUGH FRI—WEST						MON THROUGH FRI—EAST					
LEAVE	LEAVE	DUE	LEAVE	LEAVE	DUE	LEAVE	LEAVE	DUE	LEAVE	LEAVE	DUE
COUNTRY CLUB AND MAIN	COLLEGE AND LIVINGSTON	BROAD AND HIGH	COUNTRY CLUB AND MAIN	COLLEGE AND LIVINGSTON	BROAD AND HIGH	BROAD AND HIGH	OHIO AND LIVINGSTON	COUNTRY CLUB AND MAIN	BROAD AND HIGH	OHIO AND LIVINGSTON	COUNTRY CLUB AND MAIN
8:22	8:39	9:01	12:22	12:39	1:01	8:12	8:23	8:52	12:12	12:23	12:52
8:42	8:59	9:21	12:42	12:59	1:21	8:32	8:43	9:12	12:32	12:43	1:12
9:02	9:19	9:41	1:02	1:19	1:41	8:52	9:03	9:32	12:52	1:03	1:32
9:22	9:39	10:01	1:22	1:39	2:01	9:12	9:23	9:52	1:12	1:23	1:52
9:42	9:59	10:21	1:42	1:59	2:21	9:32	9:43	10:12	1:32	1:43	2:12
10:02	10:19	10:41	2:02	2:19	2:41	9:52	10:03	10:32	1:52	2:03	2:32
10:22	10:39	11:01	2:22	2:39	3:01	10:12	10:23	10:52	2:12	2:23	2:52
10:42	10:59	11:21	2:42	2:59	3:21	10:32	10:43	11:12	2:32	2:43	3:12
11:02	11:19	11:41	3:02	3:19	3:41	10:52	11:03	11:32	2:52	3:03	3:32
11:22	11:39	12:01	3:22	3:39	4:01	11:12	11:23	11:52	3:12	3:23	3:52
11:42	11:59	12:21	3:42	3:59	4:21	11:32	11:43	12:12	3:32	3:43	4:12
12:02	12:19	12:41	4:02	4:19	4:41	11:52	12:03	12:32	3:52	4:03	4:32

Ms. McKenzie boards the bus at Broad and High at 11:32 A.M. When will she arrive at Country Club and Main?

Find the 11:32 A.M. departure at Broad and High and read across to the arrival time at Country Club and Main.

Ms. McKenzie will arrive at 12:12 P.M.

1. Suppose you board at Country Club and Main at 10:42 A.M. When will you arrive at Broad and High? **11:21 A.M.**

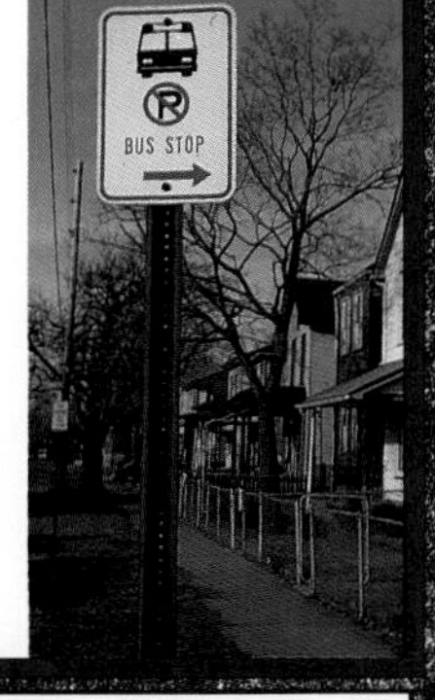

2. Mr. Juarez has a business appointment at Ohio and Livingston at 3:00 P.M. What is the latest he can board at Broad and High? **2:32 P.M.**
3. Suppose a bus broke down at Broad and High at 8:32 A.M. and it took one-half hour to repair it. How long would the bus have to wait after it was repaired to leave at the next scheduled time? **10 minutes**
4. Suppose you left Broad and High at 10:52 A.M. and arrived at Country Club and Main at 11:32 A.M. Then you realized you forgot your billfold and decided to return to Broad and High immediately. Can you be back in 40 minutes? Why or why not? **No, it's 10 minutes until the next bus leaves and 39 minutes for the bus ride.**

Applying Mathematics to the Real World

This optional page shows how mathematics is used in the real world and also provides a change of pace.

Objective Read a bus schedule.

Using Discussion Ask students to explain how they would read this schedule. Point out the importance of reading the heads—Leave and Due—at the top of each column.

Using Questioning

- What hours does this schedule represent? **8 A.M - 5 P.M.**
- Can this schedule be used to plan a round-trip? How do you know? **yes, goes from west to east and east to west**

Activity

Using Cooperative Groups Have students create a mystery game using certain times from the bus schedule as clues to where they have been. Have groups present the mysteries to the class to solve.

Problem-Solving Strategy

Explore
Plan
Solve
Examine

1-8 MATRIX LOGIC

Objective
Solve problems using matrix logic.

Steven, Diana, Janice, and Michael attend the same school. Each participates in a different school sport—basketball, football, track, or tennis. Diana does not like track or basketball. Steven does not participate in football or tennis. Janice prefers an indoor winter sport. Michael scored four touchdowns in the final game of the season. Name the sport in which each student participates.

What is given?

- Diana does not like track or basketball.
- Steven does not participate in football or tennis.
- Janice prefers an indoor winter sport. (basketball)
- Michael scored four touchdowns. (football)

Find each student's sport by listing the names and sports in a table. Then read each clue and write yes or no in the appropriate boxes.

Matrix logic is a logic or reasoning problem. It uses a rectangular array with the appropriate number of columns and rows in which you record what you have learned from clues. Once all the rows and columns are filled, you can deduce the answer.

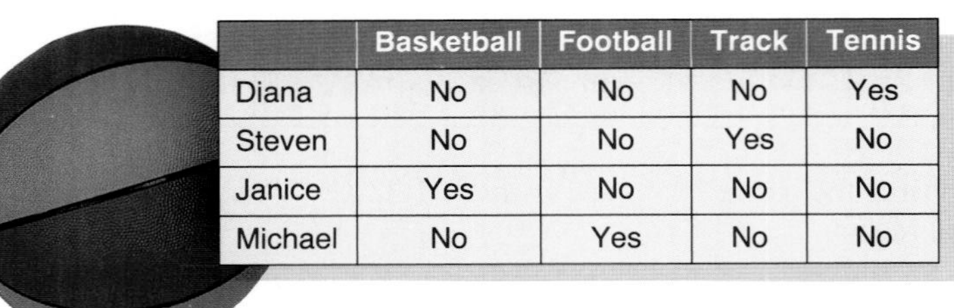

	Basketball	Football	Track	Tennis
Diana	No	No	No	Yes
Steven	No	No	Yes	No
Janice	Yes	No	No	No
Michael	No	Yes	No	No

Write no beside Diana's name for track and basketball. Write no beside Steven's name for football and tennis. Write yes beside Janice's name for basketball. Since only one student plays basketball, write no in all the remaining boxes in that row and column. Steven's sport has to be track. All remaining boxes in that row and column are no. Since Michael plays football, Diana's sport is tennis.

Guided Practice

Solve. Use matrix logic. 1. John-1; Aaron-2; Lyn-3; Sue-4; Al-5; Diane-6

1. Six friends live in consecutive houses on the same side of the street (#1-6). Lyn's house is #3. Diane's house is just beyond Al's. Lyn's closest neighbors are Aaron and Sue. Al's house is not #1-4. Aaron's house is before Sue's. If John lives in one of the houses, who lives in which house?
2. Dan, William, and Laurie have chosen careers as a teacher, a pediatrician, and a lawyer. If William does not like to be around children and Dan cannot stand the sight of blood, who has chosen which career? Dan-teacher; William-lawyer; Laurie-pediatrician

5-MINUTE CHECK-UP

(over Lesson 1-7)

Available on TRANSPARENCY 1-8.

Estimate. Answers may vary. Typical answers are given.

1. 534 × 7 3,500
2. 789 × 32 24,000
3. 248 ÷ 25 10 4. 7,926 ÷ 18 400

Extra Practice, Lesson 1-7, p. 449

INTRODUCING THE LESSON

Ask students how they would match a grey car and a red car with Jim and Zack knowing that Jim dislikes bright colors. Have them explain how they made this decision.

TEACHING THE LESSON

Using Discussion Discuss the importance of the process of elimination when using matrix logic. A yes in one box eliminates all other possibilities in that row and column.

Practice Masters Booklet, p. 8

PRACTICE WORKSHEET 1-8

Problem Solving Strategy: Matrix Logic

Use the clues to complete a matrix table for each problem below.

1. George, Gene, and Gina each have a favorite vegetable. None of the vegetables is the same. The three vegetables are broccoli, carrots, and peas. Use the table and the clues below to find out who likes which vegetable.
   a. The boys like green vegetables.
   b. Gene's vegetable is not round.

	broc	carrots	peas
George	N	N	Y
Gene	Y	N	N
Gina	N	Y	N

2. Michael, Mandy, Mario, and Maude all like different flavors of ice cream. Use a table to find out which person likes each flavor of ice cream best. The flavors are vanilla, chocolate, butterscotch, and strawberry.
   a. Michael does not like fruit-flavored ice cream.
   b. Mandy likes butterscotch.
   c. Mario does not like vanilla or chocolate.
   d. Maude does not like vanilla.

	van.	choc.	butter	straw.
Michael	Y	N	N	N
Mandy	N	N	Y	N
Mario	N	N	N	Y
Maude	N	Y	N	N

3. Cathy, Carlos, Charles, and Carrie all have favorite numbers that are different from one another. The numbers are 2, 3, 5, and 7, but are not necessarily in that order. Use a table to find out which person has which favorite number.
   a. Carrie's favorite number is less than 6 and greater than 2.
   b. Cathy likes the number that divides into 10 evenly.
   c. Carlos likes the greatest number.
   d. Charles likes the even number.

	2	3	5	7
Cathy	N	N	Y	N
Carrie	N	Y	N	N
Carlos	N	N	N	Y
Charles	Y	N	N	N

4. Four friends, Fred, Flora, Frank and Fran each have different phone area codes. The area codes they have are 201, 203, 212, and 213. Use the table to find each friend's area code.
   a. The sum of the digits in Fred's area code is not 5.
   b. No girl has an area code the sum of whose digits is divisible by 3.
   c. There is no 0 in Fred's or Fran's area code.

	201	203	212	213
Fred	N	N	N	Y
Flora	N	Y	N	N
Frank	Y	N	N	N
Fran	N	N	Y	N

RETEACHING THE LESSON

Work through each step of the following logic problem with the students.

The Green Tomatoes rock group has four members: Bob, Denny, Gil, and Pete. They play the bass, drums, guitar and piano. No one plays an instrument that begins with the same letter as his first name. Bob hates the piano. Pete and the drummer are best friends. Gil plays the bass which he borrowed from Bob. Which instrument does each person play? Bob, drums; Denny, piano; Gill, bass; Pete, guitar

Problem Solving

3. Paul-pepperoni and anchovies; David-sausage; Ben-pepperoni and sausage

Solve. Use any strategy.

3. Paul, David, and Ben order different individual pizzas for dinner. The pizzas delivered are: pepperoni and sausage; pepperoni and anchovies; and sausage. If David does not like fish, Paul does not like sausage, and Ben likes pepperoni, who ordered which pizza?

4. Four college graduates decide to have a reunion at a large hotel. Each person has a room on a different floor. Susan must ride the elevator down four floors to visit Corey. Greg is one floor below Kim. Corey has a room on the tenth floor. Kim must ride the elevator up six floors to visit Susan. Who is staying on which floor? **Greg-7; Kim-8; Corey-10; Susan-14**

5. Tracy had cake left over from her graduation party. She gave her grandparents half of the remaining pieces and half of that amount to her friend, Jerrod. If Jerrod took seven pieces home, how many pieces were left over before she gave any away? **28 pieces**

6. An odd number is greater than 7×3 and less than 9×4. Find the number if the sum of its digits is 11. **29**

7. Jack-rock; Kevin-heavy metal; Jason-jazz

7. Three friends like three different kinds of music. If Jack does not like jazz, Kevin does not like rock, and Jason does not like heavy metal or rock, who likes which kind of music?

FUN with MATH

How many grooves are on the edge of a dime?

See page 39.

8. Jay has coins in his pocket consisting of nickels, dimes, and quarters. He has three fewer nickels than dimes and two more dimes than quarters. How much money does he have if he has three quarters? **$1.35**

9. If you step into an elevator on the fifth floor and ride up thirteen floors, down six floors, and up nine floors, on which floor will you be? **21st floor**

10. How many squares can you find in this rectangle? (Hint: there are more than 18.) **32 squares**

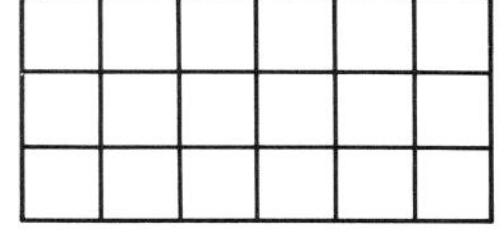

Round each number to the nearest whole number.

11. 3.266 **3** **12.** 46.019 **46** **13.** 9.048 **9** **14.** 63.51 **64**

Estimate.

15. $815 + 429$ **1,200**

16. $9.68 - 4.25$ **6**

17. $4,323 + 891$ **5,200**

18. $634 - 55$ **570**

19. Ryan fills his tank with 12.62 gallons of gasoline. His car gets 21.2 miles per gallon. About how far can Ryan travel on one tank of gasoline? **200 miles**

APPLYING THE LESSON

Independent Practice

Homework Assignment	
Minimum	Ex. 2-10 even; 11-19 odd
Average	Ex. 1-10; 11-19 odd
Maximum	Ex. 1-19

Chalkboard Examples

1. Four friends each have different pets. Joyce is afraid of dogs and horses. Ty has to buy litter for his pet. Julie does not have a cat or fish. Brian trains horses. Which person owns which pet? **Joyce-fish; Ty-cat; Julie-dog; Brian-horse**

Additional examples are provided on TRANSPARENCY 1-8.

EVALUATING THE LESSON

Check for Understanding Use Guided Practice Exercises to check for student understanding.

Closing the Lesson Ask students to name the steps in solving problems using matrix logic.

EXTENDING THE LESSON

Enrichment Have students use matrix logic to find the career and age of each person. John is not a teacher and is not 22. Mary is an artist but is not 17. One person is a student. Susan is 30. **John-student-17; Susan-teacher-30; Mary-artist-22**

5-MINUTE CHECK-UP

(over Lesson 1-8)

Available on TRANSPARENCY 1-9.

1. Gary, Angie, Bob, and Sue all own a favorite jacket. None of them own the same kind of jacket. Sue does not like tour or school jackets. Gary wears leather but does not wear jean jackets. Bob is very supportive of his school. Angie likes to attend rock tour concerts. Name the type of jacket each student owns.
Sue-jean; Gary-leather; Angie-tour; Bob-school

INTRODUCING THE LESSON

Write $23 + 82 \times 47$, and have students find the value with calculators. Ask how many students got 4,935 as their answer or 3,877 as their answer. After the lesson we will find out who is correct.

TEACHING THE LESSON

Using Critical Thinking Have students explain why different orders of operations produce different values. Consider $3 + 2 \times 5$.

Practice Masters Booklet, p. 9

Name ______ Date ______

PRACTICE WORKSHEET 1-9

Order of Operations

Name the operation that should be done first.

1. $7 + 6 \div 2$	division	**2.** $8 \times 3^2 \div 3$	find square
3. $2 \times (12 - 4)$	subtraction	**4.** $6 + 8 - 3$	addition
5. $16 \div 4 \times 2^2$	find square	**6.** $63 \div (4 + 3)$	addition
7. $[(2 + 3) \times 7] - 4^2$	addition	**8.** $52 \times [6 + (14 \div 2)]$	division

Find the value of each expression.

9. $63 \div (4 + 5)$	7	**10.** $12 + 7 \times 3^2$	75	**11.** $4 \times 5 - 3$	17
12. $24 \div 6 \times 2^2$	16	**13.** $24 \div (6 \times 2)$	2	**14.** $48 \div 6 \div 2$	4
15. $3^2 + 9 \times 5$	54	**16.** $7 - 3 + 9$	13	**17.** $52 \div 4 + 9$	22
18. $8 + 12 \div 4$	11	**19.** $21 - 6 - 4$	11	**20.** $(3^2 + 5) \times 7$	98
21. $(8 + 12) \div 4$	5	**22.** $21 - (6 - 4)$	19	**23.** $(3 + 5) \times 7$	56
24. $48 + 8 \div 4$	50	**25.** $(26 - 6) - 4^2$	4	**26.** $3 + 5 \times 7$	38

27. $27 \div (2 \times 4 + 1)$	3	**28.** $10^2 + (16 + 9 \times 7)$	179
29. $[36 \div (5 + 8 \div 2) + 1]$	5	**30.** $18 + [15 - (8 - 1)] \times 3$	42
31. $[(6 + 3) \times 4] \div [7 \times 2 - 2]$	3	**32.** $[2 + (3 - 1)] \times [(8 - 6)^2 \times 3]$	48

Write an expression and solve.

33. Jane packed 6 hams in a shipping crate. Each ham weighs 5 pounds. The crate weighs 2 pounds. How much does the total package weigh?
(6 x 5) + 2; 32 pounds

34. Martha wrote checks to each of her 7 grandchildren. The older three received \$15 each. The others received \$10. What was the total of Martha's checks?
(\$15 x 3) + (\$10 x 4); \$85

35. Stella correctly answered 19 problems on a math test. Six of the problems were worth 5 points each. The others were worth 4 points each. What was Stella's score?

36. Manuel sold 72 daily newspapers at 35¢ each and 75 Sunday newspapers at \$1.25 each. What was his total income?

1-9 ORDER OF OPERATIONS

Objective
Use the order of operations to find the value of an expression.

Cyd mows lawns during the summer. Monday, he mowed one lawn for \$6 and four other lawns for \$5 each. The expression $6 + 4 \times 5$ gives the total earned.

To be sure you find the correct value for an expression such as $6 + 4 \times 5$, use the following *order of operations.*

Incorrect	Correct
$6 + 4 \times 5$ → 10×5 → 50	$6 + 4 \times 5$ → $6 + 20$ → 26

Method

1. Find the value of all powers.
2. Multiply and/or divide from left to right.
3. Add and/or subtract from left to right.

Cyd earned \$26 mowing lawns.

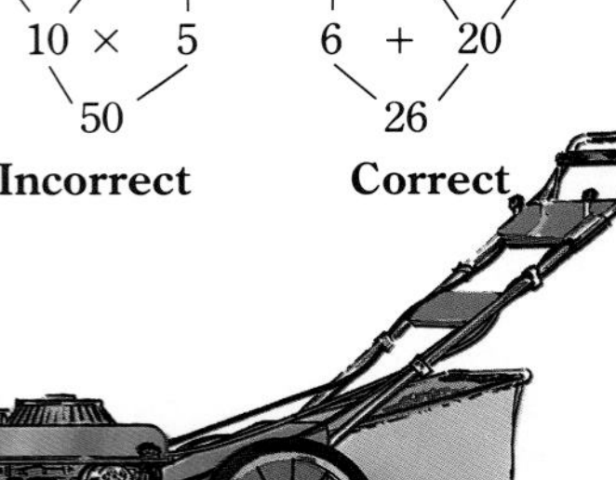

Example A ***Find the value of $18 - 4 \times 3^2 \div 6$.***

Step			
1	$18 - 4 \times 3^2 \div 6$	$= 18 - 4 \times 9 \div 6.$	Find the value of 3^2.
2		$= 18 - 36 \div 6.$	Multiply 4 by 9.
		$= 18 - 6$	Divide 36 by 6.
3		$= 12$	The value is 12.

When a different order of operations is needed, parentheses are used. First do operations inside the parentheses. Then do other operations.

Example B ***Find the value of $(0.7 + 0.3) \times (13 - 5)$.***

Step			
1	$(0.7 + 0.3) \times (13 - 5)$	$= 1 \times (13 - 5)$	$0.7 + 0.3 = 1$
2		$= 1 \times 8$	$13 - 5 = 8$
3		$= 8$	The value is 8.

Guided Practice

1. 50 2. 18 3. 48.3

Find the value of each expression.

Example A

1. $10 \times 4 + 5 \times 2$ **2.** $7 + 8 + 12 \div 4$ **3.** $54.3 - 42 \div 7$

4. $21 \div 7 + 4 \times 8$ **35** **5.** $2^3 \div 4 \times 1.2$ **2.4** **6.** $30 \div 6 - 8^2 \times 0$ **0**

Example B

7. $26 - (20 + 4)$ **2** **8.** $4 \times (0.7 + 1.3)$ **9.** $(25 + 15) \div 4$

10. $6 \times (9.3 - 0.3)$ **54** **11.** $(7 + 1) \times (3^2 - 1)$ **64** **12.** $(7 - 3)^2$ **16**

8. 8 9. 10

RETEACHING THE LESSON

Name the operation that you should do first to evaluate each expression.

1. $7 \times 4 - 6$ **multiply**
2. $100 - 6^2$ **evaluate power**
3. $3 + 5 \times 8$ **multiply**
4. $50 \div 2 \times 6$ **divide**
5. $16 + 7 - 4$ **add**
6. $7 - 2 \times 1$ **multiply**
7. $17 + 3^3$ **evaluate power**
8. $27 - 8 \div 4 + 3$ **divide**
9. $68 - 12 + 5$ **subtract**

Reteaching Masters Booklet, p. 8

Name ______ Date ______

RETEACHING WORKSHEET 1-9

Order of Operations

To evaluate: $6 + 2 \times 4^2 - 3 \div 3$

- First, evaluate all powers.
- Then, do all multiplication and division from left to right.
- Then, do all addition and subtraction from left to right.

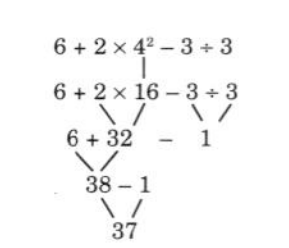

$2^3 + 3 \div 3 \times 6$
$8 + 3 \div 3 \times 6$
$8 + 1 \times 6$
$8 + 6$
14

If there are parentheses or brackets in an expression, do the

Exercises

13. 14 14. 39.6 15. 12 16. 1 19. 20

Practice

Find the value of each expression.

13. $12 - 5 + 9 - 2$ **14.** $6 \times 6 + 3.6$ **15.** $12 + 20 \div 4 - 5$

16. $6 \times 3 \div 9 - 1$ **17.** $(4^2 + 2^3) \times 5$ 120 **18.** $24 \div 8 - 2$ 1

19. $3 \times (4 + 5) - 7$ **20.** $4.3 + (24 \div 6)$ **21.** $(5^2 + 2) \div 3$ 9

22. $(40 \times 2) - (6 \times 10)$ **23.** $2 \times [5 \times (4 + 6) - 10]$

20. 8.3 22. 20 23. 80

Mental Math

Find the value of each expression mentally.

24. $2 + 24 \div 6$ 6 **25.** $2 \times (5^2 - 20)$ 10 **26.** $24 \div (10 + 2)$ 2

27. $10^2 - 2^2 \times 10$ 60 **28.** $(20 + 80) \times (64 - 8^2)$ 0

Applications

29. Bob buys six pairs of socks at $3.00 each. How much change does he receive from $20? Write an expression and solve. $2

30. Mrs. Anderson works 40 hours a week and makes $8.00 an hour. What is her take-home pay if $70.00 is withheld for taxes and insurance? Write an expression and solve. $250

Determining Reasonable Answers

31. Elliot worked this problem on his calculator. Without actually finding the correct answer, tell how you know that the answer he found is wrong.

A box of albums weighs 24 pounds. A box of tapes weighs 18 pounds. What would a shipment of 10 boxes of albums and 20 boxes of tapes weigh?

Elliot's answer: The shipment would weigh 360 pounds.

The tapes alone would weigh 360 pounds.

Critical Thinking

32. A clock strikes the number of hours each hour. How many times will the clock strike in a 24-hour day? 156 times

33. Find the value of $2 + 5 \times 3$ on your calculator.

33. Answers may vary.

Is the result 17? If your result is not 17, can you find a way to get a result of 17?

Use a calculator to find the value of each expression.

34. $5 + 3 \times 7$ 26 **35.** $12 \div 3 + 12$ 16 **36.** $36 \div 4 \times 3$ 27

Mixed Review

Estimate.

Lesson 1-6

37. $\begin{array}{r} 42.35 \\ - 16.48 \\ \hline 20 \end{array}$ **38.** $\begin{array}{r} 0.6 \\ + 0.94 \\ \hline 1.6 \end{array}$ **39.** $\begin{array}{r} 123.45 \\ - 11.82 \\ \hline 110 \end{array}$ **40.** $\begin{array}{r} \$3.62 \\ + 9.89 \\ \hline \$14 \end{array}$

Lesson 1-7

41. 16.4×2.8 48 **42.** $65.41 - 15.37$ 50 **43.** 763×8 6,400

Lesson 1-8

44. Jenny, Betty, and Ashley are traveling to Daytona, Miami, and Ft. Myers for vacation. Jenny is not going to Miami, Betty is not going to Ft. Myers, and Ashley is traveling the farthest distance south. Who is traveling to which city? Jenny-Ft. Myers, Betty-Daytona, Ashley-Miami

Chalkboard Examples

Example A *Find the value of each expression.*

1. $49 \div 7 + 3$ 10 **2.** $15 - 3 \times 4$ 3

3. $0.5 + 56 \div 7$ 8.5

4. $8^2 - 3 \times 8$ 40

Example B *Find the value of each expression.*

5. $2 \times (4 + 6)$ 20

6. $(5^2 + 5) \times (4^2 - 6)$ 300

7. $(0.8 + 0.2) \times (37 - 6^2) + 1^3$ 2

Additional examples are provided on TRANSPARENCY 1-9.

EVALUATING THE LESSON

Check for Understanding Use Guided Practice Exercises to check for student understanding.

Error Analysis Be alert for students who try to perform the operations in the order in which they appear especially when there are no parentheses.

Closing the Lesson Ask students to summarize what they have learned during this lesson.

EXTENDING THE LESSON

Enrichment Masters Booklet, p. 8

Name ______ Date ______

ENRICHMENT WORKSHEET 1-9

Two-Step Problems

Solve.

1. Anne and Lisa were looking for articles for their science notebook. Anne found 14 articles and Lisa found 11. Then, Lisa spilled a glass of milk and ruined six of the articles. How many were left? 19 articles

2. To get to his friend's home, Ben must take the bus for 50¢ and then the subway for 60¢. Today, Ben has a subway pass, which will save him 20¢. How much will it cost Ben to get to his friend's home? 90¢

3. In September, when Joe got his new apartment, he bought some new furniture. He charged $400 worth of furniture in June. In July, he paid $150 on his bill but later bought and charged an $80 coffee table. Now how much does Joe owe the furniture store? $330.00

4. Dominica listens to a radio talk show every Saturday morning for one hour. During the show, there are 5 two-minute commercials. How long is the "talk" part of the show? 50 minutes

5. On his vacation, Eric took four rolls of 24-exposure film. When he had them developed, a total of nine photos did not turn out. How many photos did Eric have of his vacation? 87 photos

6. There are four trays of muffins on the shelf. Each tray holds 72 muffins. A student comes in to pick up 20 dozen of them for an after-school party. Are there enough muffins available? yes

7. There were 483 guests registered at the Otani Hotel. On Friday, 318 checked out and 215 checked in. Now how many guests are registered at the hotel? 380 guests

8. Sally buys milk for $1.89, cheese for $2.39, juice for 99¢, and apples for $1.12. How much change does Sally receive from a $20 bill? $13.61

APPLYING THE LESSON

Independent Practice

Homework Assignment	
Minimum	Ex. 1-31 odd; 32; 33-43 odd
Average	Ex. 2-28 even; 29-44
Maximum	Ex. 1-44

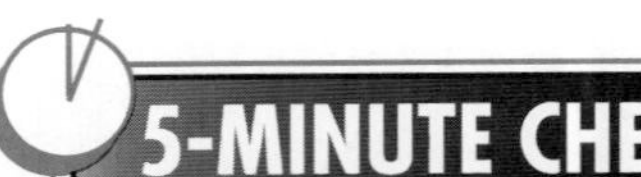

5-MINUTE CHECK-UP

(over Lesson 1-9)

Available on TRANSPARENCY 1-10.

Find the value of each expression.

1. $13 - 2 \times 3$ 7
2. $4 \times 2 + 30 \div 6$ 13
3. $5^2 + 4 \times 5$ 45
4. $6 + 8 + 15 \div 5$ 17
5. $8 - (3 + 4)$ 1
6. $2 \times (5 - 1) + 3^2$ 17

Extra Practice, Lesson 1-9, p. 450

INTRODUCING THE LESSON

I am thinking of a number. Multiply it by 4 then divide by 8. Write down what I just said . . . . We are going to learn a short way to write the same idea.

TEACHING THE LESSON

Using Critical Thinking *How many values for variables must you know to evaluate an expression?* You must know the value of each variable in the expression.

Practice Masters Booklet, p. 10

PRACTICE WORKSHEET 1-10

Evaluating Expressions

Find the value of each expression.

1. $16 - 4 \times 3 + 2$ 6
2. $15 + 3^2$ 24
3. $15 \div 3 + 4 \times 8$ 37
4. $4 \times (9 + 3)$ 48
5. $4 \times 2^2 - 2 \times 2$ 12
6. $5^2 - 3(5) - 8$ 2

Find the value of each expression if n = 3.

7. $6n$ 18
8. $4(n + 5)$ 32
9. $4n - 12$ 0
10. $(6 - n)^2 + 13$ 22
11. $\frac{n^2 + 6}{3}$ 5
12. $8(16 - n)$ 104

Find the value of each expression if t = 3, u = 2, and v = 5.

13. $5t - v$ 10
14. $5t - u^2$ 11
15. $t + 3u - v$ 4
16. $(v - u)^2 + t$ 12
17. $3tv + 8u$ 61
18. $t(v - u)$ 9

Find the value of each expression if a = 5, b = 3, and c = 7.

19. $a^2 + c^2$ 74
20. $6b^2$ 54
21. $a^2 - b$ 22
22. $a^2 - b^2$ 16
23. $(a + c)^2$ 144
24. $5(a - b)$ 10

Solve.

25. A science lab is in the shape of a rectangle. The lab is 32 feet long by 25 feet wide. Use $P = 2(l + w)$ to find the perimeter of the lab.
    114 feet
26. A courtyard at a high school is in the shape of a square. The length of each side, s, is 12 meters. Use $A = s^2$ to find the area of the courtyard.
    144 square meters
27. The math club uses a phone tree to get messages out. The president calls two members, who each call two others, who each call two others, and so on. How many are called at the fourth level?
    16
28. Write a formula to find the perimeter of your math classroom. Then find the perimeter.
    Answers will vary. Should include the lengths of all sides.

Algebra Connection

1-10 EVALUATE EXPRESSIONS

Objective
Write and evaluate expressions using variables.

Just for Feet Shoe Store adds \$5 to the wholesale cost of each pair of shoes to make a profit. What is the retail price of a pair of shoes?

You can let c represent the wholesale cost of any pair of shoes. Then $c + 5$ represents the retail price of the shoes.

A **mathematical expression** consists of numbers, operations like addition or multiplication, and sometimes variables. A **variable** is a symbol, usually a letter, used to represent a number. Any letter can be used.

Translate words into mathematical expressions.

	Words	Mathematical Expression
Example A	five more than a number	$d + 5$
Example B	six less than a number	$m - 6$
Example C	three times a number	$3 \times y$ or $3y$
Example D	a number divided by three	$n \div 3$ or $\frac{n}{3}$

The value of an expression depends on the number that replaces each variable.

Method

1. Replace each variable with the given number.
2. Perform the indicated operations using the order of operations.

Example E ***Find the value of 3n − 4 if n = 5.***

1. $3n - 4 = 3 \times 5 - 4$ Replace n with 5.
2. $= 15 - 4$ Multiply first.
   $= 11$ Subtract. The value is 11.

Example F ***Find the value of $a^2 + \frac{b}{3}$ if a = 4 and b = 2.***

1. $a^2 + \frac{b}{3} = 4^2 + \frac{2}{3}$ Replace a with 4 and b with 2.
2. $= 16 + \frac{2}{3}$ Find 4^2.
   $= 16\frac{2}{3}$ Add. The value is $16\frac{2}{3}$.

RETEACHING THE LESSON

Match.

1. 6 more than n b
2. the quotient of b and 4 d
3. 3 times some number f
4. 12 plus some number a
5. 9 less than c c
6. m decreased by 7 e

a. $12 + a$
b. $n + 6$
c. $c - 9$
d. $b \div 4$
e. $m - 7$
f. $3p$

Reteaching Masters Booklet, p. 9

Name _______________ Date _______________

RETEACHING WORKSHEET 1-10

Evaluating Expressions

To evaluate an expression, replace the variable with a number. Then, find the value of the expression.

Expression	$n - 7$	
Replace n with 10.	$10 - 7$	
Simplify.	3	**The value of the expression $n - 7$ is 3 when $n = 10$.**
Expression	$3n + 2$	
Replace n with 6.	$3 \times 6 + 2$	
Simplify.	20	**The value of the expression $3n + 2$ is 20 when $n = 6$.**

$1.\ 7 + n$ $2.\ n - 5$ $3.\ 16 \div n, \frac{16}{n}$

Guided Practice

Examples A-D

Write an expression for each phrase. Use n to represent the number.

1. the sum of 7 and a number
2. 5 less than a number
3. 16 divided by a number
4. the product of 8 and a number $8 \times n, 8n$

Examples E, F

Find the value of each expression if $n = 3$ and $p = 5$.

5. $42 - n$ 39
6. $p^2 + 3$ 28
7. $2np$ 30
8. $\frac{2 + n}{p}$ 1

Exercises

Practice

12. $n \div 2$ or $\frac{n}{2}$
13. $7n$
14. $n + 3$

Write an expression for each phrase. Use n for the number.

9. 8 more than a number $n + 8$
10. 17 less than a number $n - 17$
11. n less than 3 $3 - n$
12. a number divided by 2
13. the product of 7 and a number
14. a number increased by 3
15. a number decreased by 3 $n - 3$
16. 4 times a number $4 \times n$ or $4n$

Find the value of each expression if $n = 9$.

17. $8 + n$ 17
18. $15.7 - n$ 6.7
19. $n^2 + 7$ 88
20. $3n - 4$ 23
21. $\frac{2n}{3}$ 6

Find the value of each expression if $a = 2$, $b = 3$, and $c = 5$.

22. $ab + c$ 11
23. $\frac{a + b}{c}$ 1
24. $a^2 + b$ 7
25. a^2b 12
26. $(b + c)^2$ 64

Write Math

Write a verbal phrase for each expression. See margin.

27. $a + 7$
28. $26 - y$
29. $n \div 9$
30. $18m$
31. $y - 26$

Suppose

32. Suppose six is squared, then the product of four and nine is subtracted. What number results? 0

Applications

Copy and complete each table.

33.

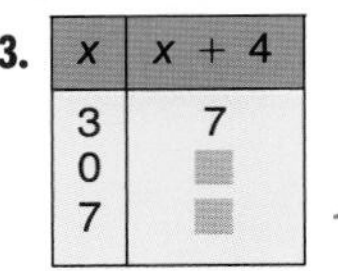

x	$x + 4$
3	7
0	▒
7	▒

4, 11

34.

m	$m + 11$
2	13
22	▒
33	▒

33, 44

35.

b	$b - 3$
8	5
▒	1
▒	14

4, 17

Calculator 

Most calculators have a memory key such as [STO], [M+], or [SUM]. Once a number has been put into memory, it can be recalled using a key such as [RCL], [MR], or [EXC].

The memory is useful when the same variable is used more than once in an expression. For example, find the value of $p^2 + 4p$ if $p = 3.1$.

Enter: 3.1 [STO] This puts 3.1 into memory and clears the display.

Enter: [x^2] [+] 4 [×] [RCL] [=] 22.01 The value is 22.01.

36. 45 37. 5.61 38. 80.64 39. 22.4196 40. 63.4041

Find the value of $p^2 + 4p$ for each value of p.

36. $p = 5$
37. $p = 1.1$
38. $p = 7.2$
39. $p = 3.14$
40. $p = 6.21$

APPLYING THE LESSON

Independent Practice

Homework Assignment	
Minimum	Ex. 1-40 odd
Average	Ex. 1-16; 18-40 even
Maximum	Ex. 1-40

Additional Answers

Answers may vary. Typical answers are given.

27. seven more than a number
28. y less than 26
29. a number divided by 9
30. 18 times some number
31. a number decreased by 26

Chalkboard Examples

Examples A-D *Write a mathematical expression for each phrase.*

1. five more than a number $n + 5$
2. seven less than a number $n - 7$
3. eight divided by a number $8 \div n$ or $\frac{8}{n}$

Examples E, F *Find the value of each expression if $a = 3$ and $b = 4$.*

4. $17 - a$ 14
5. $a^2 + b$ 13
6. $\frac{b + 2}{a}$ 2
7. $\frac{4a}{b}$ 3

Additional examples are provided on TRANSPARENCY 1-10.

EVALUATING THE LESSON

Check for Understanding Use Guided Practice Exercises to check for student understanding.

Error Analysis Be alert for students who write numbers in the order they appear in a phrase.

Closing the Lesson Ask students to state in their own words how to find the value of an expression if they know the value of each variable.

EXTENDING THE LESSON

Enrichment Masters Booklet, p. 9

Name ______ Date ______

ENRICHMENT WORKSHEET 1-10

Four Fours

In mathematics, there are many ways to express numbers. For instance, the number five can be expressed as 5, V, 4 + 1, 10 ÷ 2, 2 + 3, 8 − 3, 20 ÷ 4, and so on.

Operations can be combined to produce a desired result. For example, the number one can be expressed using four fours as shown below.

$\frac{4+4}{4+4} = 1$ $\frac{4 \times 4}{4 \times 4} = 1$ $\frac{4}{4} \times \frac{4}{4} = 1$ $\frac{44}{44} = 1$

Use four fours to express each result.

Answers may vary.

1. 2 $\frac{4}{4} + \frac{4}{4}$
2. 10 $\frac{44 - 4}{4}$
3. 7 $4 + 4 - \frac{4}{4}$
4. 6 $4 + \frac{4 + 4}{4}$
5. 9 $4 + 4 + \frac{4}{4}$
6. 3 $\frac{4 + 4 + 4}{4}$
7. 8 $4 \times \left(\frac{4 + 4}{4}\right)$
8. 11 $\frac{44}{\sqrt{4 \times 4}}$

5-MINUTE CHECK-UP

(over Lesson 1-10)

Available on TRANSPARENCY 1-11.

Write an expression for each phrase.

1. a number decreased by 4 $n - 4$

2. the product of 6 and m $6 \times m$ or $6m$

Find the value of each expression if $r = 2$ and $s = 4$.

3. $(s - r) \times 3$ 6 **4.** $\frac{2r}{s}$ 1

Extra Practice, Lesson 1-10, p. 450.

INTRODUCING THE LESSON

Ask students if they have ever seen candy weighed using a balance scale. If so, have students explain to the class how it worked. The point is to make the scale balance, both sides need to weigh the same.

TEACHING THE LESSON

Using Vocabulary Point out that addition and subtraction are inverse operations; one undoes the other. Can students name another pair of inverse operations? **heating and cooling, multiplication and division**

Practice Masters Booklet, p. 11

Name ______________________ Date ______________

PRACTICE WORKSHEET 1-11

Solving Equations Using Addition and Subtraction

Solve each equation. Check each solution.

1. $9 + s = 14$ $s = 5$	**2.** $4 + t = 5$ $t = 1$	**3.** $22 - r = 15$ $r = 7$
4. $18 - a = 6$ $a = 12$	**5.** $22 + 4 = b$ $b = 26$	**6.** $x + 9 = 9$ $x = 0$
7. $(6 + 2) = y$ $y = 8$	**8.** $9 + r = 26$ $r = 17$	**9.** $36 - 0 = a$ $a = 36$
10. $w - 30 = 2$ $w = 32$	**11.** $196 + 14 = b$ $b = 210$	**12.** $24 - p = 19$ $p = 5$
13. $20.8 + g = 21.8$ $g = 1$	**14.** $w = 4.2 + 1.55$ $w = 5.75$	**15.** $7.68 + l = 32$ $l = 24.32$
16. $t + 4.5 = 9$ $t = 4.5$	**17.** $4.019 = t - 6.11$ $t = 10.129$	**18.** $8.06 - d = 6.543$ $d = 1.517$

Write an equation. Then solve.

19. Ten increased by x is 17. What is the value of x?
$10 + x = 17$, $x = 7$

20. Thirteen decreased by r is 9. What is the value of r?
$13 - r = 9$, $r = 4$

21. Twelve less than t is 13. What is the value of t?
$t - 12 = 13$, $t = 25$

22. Five more than p is 7.3. What is the value of p?
$p + 5 = 7.3$, $p = 2.3$

Solve. Use an equation.

23. When a number is added to 12, the result is 27. Find the number.
$12 + n = 27$, 15

24. The difference of 6 and a number is 72. Find the number.
$n - 6 = 72$, 78

25. When a number is subtracted from 12, the result is 9. Find the number.
$12 - n = 9$, 3

26. When 15 is subtracted from a number, the result is 7.5. Find the number.
$n - 15 = 7.5$, 22.5

Algebra Connection

1-11 Solving Equations Using Addition and Subtraction

Objective

Solve equations by using addition and subtraction.

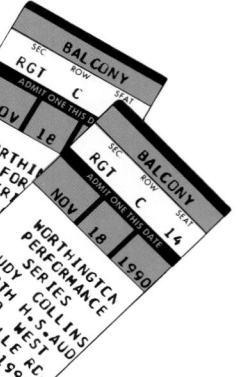

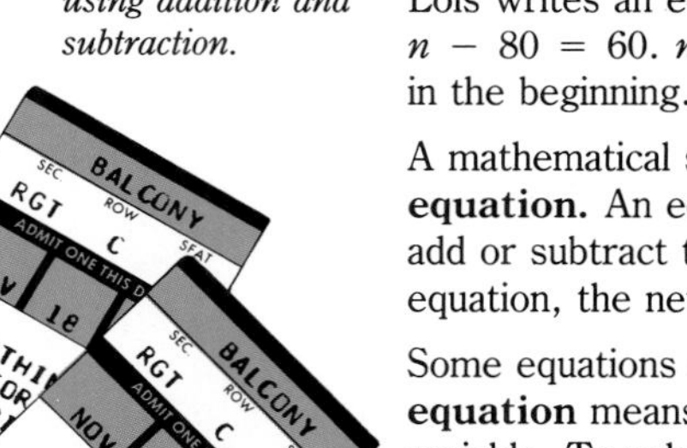

Lois knows she sold 80 tickets. She has 60 left. In order to find how many she had before she sold any, Lois writes an equation that contains a *variable:* $n - 80 = 60$. n represents how many tickets she had in the beginning.

$n - 80 = 60$

A mathematical sentence with an equals sign is called an **equation.** An equation is like a scale in balance. If you add or subtract the same number on each side of an equation, the new equation is also true. It is *in balance.*

$n - 80 + 80 = 60 + 80$

Some equations contain variables. **Solving the equation** means finding the correct replacement for the variable. To solve an equation, get the variable by itself on one side of the equals sign.

$n = 140$

Lois had 140 tickets in the beginning.

Use **inverse operations** to solve equations.

- If a number has been added to the variable, subtract.
- If a number has been subtracted from the variable, add.

Method

1. Identify the variable.
2. Add or subtract the same number on each side of the equation to get the variable by itself.
3. Check the solution by replacing the variable in the original equation.

Example A

Solve $n - 7 = 12$.

1. $n - 7 = 12$ — 7 is subtracted from the variable n.
2. $n - 7 + 7 = 12 + 7$ — Add 7 to each side.
   $n = 19$
3. **Check:** $n - 7 = 12$ — In the original equation, replace n with 19.
   $19 - 7 \stackrel{?}{=} 12$
   $12 = 12$ ✓ — The solution is 19.

Example B

Solve $k + 0.5 = 0.9$.

1. $k + 0.5 = 0.9$ — 0.5 is added to the variable k.
2. $k + 0.5 - 0.5 = 0.9 - 0.5$ — Subtract 0.5 from each side.
   $k = 0.4$
3. **Check:** $k + 0.5 = 0.9$ — In the original equation replace 8 with 0.4.
   $0.4 + 0.5 \stackrel{?}{=} 0.9$
   $0.9 = 0.9$ ✓ — The solution is 0.4.

RETEACHING THE LESSON

You may wish to use actual scales as an analogy to equations. Have students add and subtract equal weights on each side. Point out that the scales remain in balance as long as the total weight on each side is equal.

Reteaching Masters Booklet, p. 10

Name ______________________ Date ______________

RETEACHING WORKSHEET 1-11

Solving Equations Using Addition and Subtraction

When solving an equation, you must find an exact replacement for the variable.

An equation like $x + 6 = 13$ can be solved using subtraction.

$x + 6 - 6 = 13$

$x + 6 - 6 = 13 - 6$ **Subtract 6 from each side.**

$x = 7$

To check the solution, replace x with 7.

An equation like $s - 9 = 16$ can be solved using addition.

$s - 9 = 16$

$s - 9 + 9 = 16 + 9$ **Add 9 to each side.**

$s = 25$

To check the solution, replace s with 25.

Guided Practice

Example A

Name the number you would add to each side to solve each equation. Then solve and check each solution.

1. $m - 5 = 9$ 5; 14　**2.** $n - 20 = 30$ 20; 50　**3.** $k - 0.6 = 0.2$ 0.6; 0.8　**4.** $9 = x - 9$ 9; 18

Example B

Name the number you would subtract from each side to solve each equation. Then solve and check each solution.

5. $n + 3 = 5$ 3; 2　**6.** $4.7 + k = 7.9$ 4.7; 3.2　**7.** $90 = n + 40$ 40; 50　**8.** $4 + t = 9$ 4; 5

Exercises

Practice

9. 16　10. 13
11. 14　12. 17
13. 2　14. 3
15. 11　16. 9
17. 18　18. 0.6

Solve each equation. Check each solution.

9. $p - 8 = 8$　**10.** $h - 7 = 6$　**11.** $5 = u - 9$　**12.** $9 = t - 8$

13. $a + 6 = 8$　**14.** $8 + f = 11$　**15.** $18 = b + 7$　**16.** $6 + s = 15$

17. $16 = x - 2$　**18.** $r - 0.4 = 0.2$　**19.** $w - 0.5 = 0.5$

20. $15.3 = q - 0.2$　**21.** $z + 0.6 = 0.9$　**22.** $y - 500 = 800$

19. 1　20. 15.5　21. 0.3　22. 1,300

Using Equations

Write an equation. Then solve.

23. Seven increased by x is 11. What is the value of x? 4

24. b decreased by 0.6 is 0.5. What is the value of b? 1.1

25. Nine less than g is 7. What is the value of g? 16

Applications

Solve. Use an equation.

26. Ally is saving for a computer game that costs $60. She still needs to save $25. How much has Ally saved? $35

27. Greg owns 70 tapes and CDs. He owns 20 CDs. How many tapes does Greg own? 50 tapes

Talk Math

28. Ask this question in different ways: What number added to seven equals twelve?

Critical Thinking

29. Alex and Kevin organize a camping trip for ten boys. They have ten backpacks, but they intend to leave some backpacks empty for items the campers might bring. Five backpacks contain food, four contain camping supplies, and two contain both food and camping supplies. How many backpacks are empty?

28. Answers may vary. A sample answer is 12 is 7 more than what number?
29. 3 empty backpacks

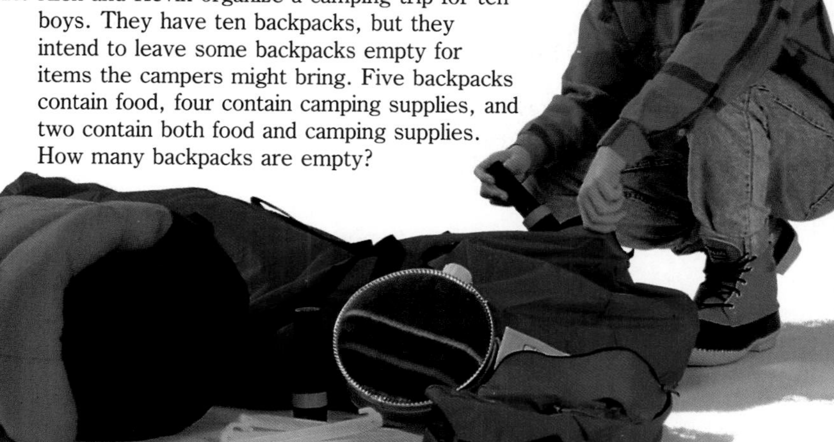

APPLYING THE LESSON

Independent Practice

Homework Assignment	
Minimum	Ex. 2-28 even; 29
Average	Ex. 1-25 odd; 26-29
Maximum	Ex. 1-29

Chalkboard Examples

Example A *Solve each equation. Check your solution.*

1. $r - 8 = 4$ 12　**2.** $6 = t - 7$ 13

3. $0.2 = k - 0.5$ 0.7

Example B *Solve each equation. Check your solution.*

4. $p + 7 = 15$ 8　**5.** $13 = q + 8$ 5

6. $0.5 + y = 0.9$ 0.4

Additional examples are provided on TRANSPARENCY 1-11.

EVALUATING THE LESSON

Check for Understanding Use Guided Practice Exercises to check for student understanding.

Closing the Lesson Ask students to summarize what they have learned about solving equations.

EXTENDING THE LESSON

Enrichment Work in small groups. Have students roll a number cube 2 times, then write an equation using both numbers, a variable, and addition or subtraction. Then have them solve the equation.

Enrichment Masters Booklet, p. 10

Name ______________________ Date ______________

ENRICHMENT WORKSHEET 1-11

Mind Reading

Make four cards having the following numbers and letters.

Ask someone to think of a number from 1 to 15. Ask the person which cards the number is on.

A: 1 3 5 7 9 11 13 15
B: 2 3 6 7 10 11 14 15

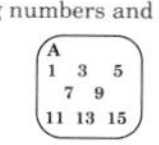

C: 4 5 6 7 12 13 14 15
D: 8 9 10 11 12 13 14 15

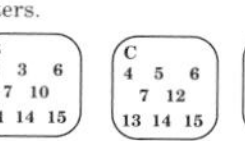

For example, suppose the number appears on cards **A, B,** and **C**. Add the least number on each card. This sum is **1 + 2 + 4 or 7**. The person's number is **7**.

Answer these questions about the cards.

1. What is the pattern of the numbers on card A?
odds; +2; one number, skip one
2. The least number on card B is 2. Note that two numbers (2 and 3) are on the card. Then two numbers are skipped (4 and 5). This pattern is repeated up to 15. Now explain the patterns on cards C and D.
C - four numbers, skip four; D - eight numbers, skip eight
3. The least numbers on card A through card D are 1, 2, 4, and 8. How are these numbers related to the base-two numeration system?
$1 = 2^0, 2 = 2^1, 4 = 2^2, 8 = 2^3$
4. Suppose you wanted to do this trick with five cards, A through E. What would be the least number on card E? How many numbers would be on each card?
$2^4 = 16$; 16

Tell the number chosen when it appears on the following cards, A through D.

5. C and D 12　6. A, B, and D 11　7. A, B, C, and D 15

Each of the following numbers is chosen. Write the corresponding 4-digit base-two numeral. Arrange the cards in base-two place-value position, D (eights), C (fours), B (twos), A (ones). Interpret 1 as on the card and 0 as not on the card. Use the base-two numeral to tell on which cards the given numbers appear.

8. 6 0110　9. 14 1110　10. 10 1010

Suppose you want to do this trick with five cards, A through E. Which numbers should appear on each of the following cards?

11. card A 1, 3, 5, 7, 9, 11, 13, 15, 17, 19, 21, 23, 25, 27, 29, 31
12. card B 2, 3, 6, 7, 10, 11, 14, 15, 18, 19, 22, 23, 26, 27, 30, 31
13. card C 4, 5, 6, 7, 12, 1 14, 15, 20, 21, 23, 28, 29, 30,
14. card D 8, 9, 10, 11, 12, 13, 14, 15, 24, 25, 26, 27, 28, 29, 30, 31
15. card E 16, 17, 18, 19, 20, 21, 22, 23 24, 25, 26, 27, 28, 29, 30, 31

5-MINUTE CHECK-UP

(over Lesson 1-11)

Available on TRANSPARENCY 1-12.

Solve each equation. Check your solution.

1. $n - 60 = 20$ 80
2. $0.7 = r - 0.1$ 0.8
3. $13 = m + 5$ 8
4. $200 + k = 600$ 400

Extra Practice, Lesson 1-11, p. 450

INTRODUCING THE LESSON

Show students these equations: $7 \times n = 35$, $7 = 35 \div n$, and $7 \times 35 = n$.

If you were solving these, which do you think you could solve by using the operation indicated with the sign?

TEACHING THE LESSON

Using Cooperative Groups Have each group write five equations that include a variable and multiplication or division. The five solutions must be 0, 1, 2, 3, and 4. Groups should exchange equations and solve.

Practice Masters Booklet, p. 12

Name ______ Date ______

PRACTICE WORKSHEET 1-12

Solving Equations Using Multiplication and Division

Solve each equation. Check each solution.

1. $9 \times s = 54$ $s = 6$	2. $4t = 20$ $t = 5$	3. $\frac{12}{r} = 6$ $r = 2$
4. $\frac{18}{a} = 6$ $a = 3$	5. $2.5n = 10$ $n = 4$	6. $\frac{x}{9} = 9$ $x = 81$
7. $0.3p = 3$ $p = 10$	8. $9 \times r = 27$ $r = 3$	9. $\frac{36}{n} = 4$ $n = 9$
10. $5q = 0.25$ $q = 0.05$	11. $\frac{196}{14} = b$ $b = 14$	12. $24 \times p = 19.2$ $p = 0.8$
13. $\frac{20.8}{g} = 1.3$ $g = 16$	14. $3m = 0.09$ $m = 0.03$	15. $0.5 = \frac{p}{2}$ $p = 1$
16. $10t = 9$ $t = 0.9$	17. $0.3r = 0.9$ $r = 3$	18. $0.1 = \frac{x}{7}$ $x = 0.7$

19. Eleven multiplied by r is 99. What is the value of r? 9
20. A number divided by 4 is 10. What is the number? 40
21. The product of n and 0.7 is 1.4. What is the value of n? 2
22. Twelve divided by 2.4 is p. What is the value of p? 5

Write an equation. Then solve.

23. When a number is multiplied by 12, the result is 60. Find the number. $12n = 60$, $n = 5$
24. The product of 6 and a number is 72. Find the number. $6n = 72$, $n = 12$
25. When a number is divided by 12, the result is 9. Find the number. $\frac{n}{12} = 9$, $n = 108$
26. When 15 is divided by a number, the result is 0.3. Find the number. $\frac{15}{n} = 0.3$, $n = 50$

Algebra Connection

1-12 SOLVING EQUATIONS USING MULTIPLICATION AND DIVISION

Objective
Solve equations by using multiplication and division.

The parking garage charges \$3.00 per hour. How many hours can Oliver park if he has only \$15? An equation can be written to solve this problem. The equation is in Example A.

Multiplication and division are *inverse operations.*

- If a variable is multiplied by a number, divide each side by that number.
- If a variable is divided by a number, multiply each side by that number.

Method

1. Identify the variable.
2. Multiply or divide each side of the equation by the same nonzero number to get the variable itself.
3. Check the solution by replacing the variable in the original equation with the solution.

Example A

Solve 3n = 15. $3n$ means $3 \times n$.

1. $3n = 15$ 3 is multiplied by the variable n.
2. $\frac{3n}{3} = \frac{15}{3}$ Divide each side by 3.

   $n = 5$
3. **Check:** $3n = 15$

   $3 \times 5 \stackrel{?}{=} 15$ Replace n with 5.

   $15 = 15\checkmark$ Oliver can park for 5 hours.

Example B

Solve $\frac{t}{4} = 9$.

1. $\frac{t}{4} = 9$ The variable, t, is divided by 4.
2. $\frac{t}{4} \times 4 = 9 \times 4$ Multiply each side by 4.

   $t = 36$
3. **Check:** $\frac{t}{4} = 9$

   $\frac{36}{4} \stackrel{?}{=} 9$ Replace t with 36

   $\mathbf{9 = 9}\checkmark$ The solution is 36.

RETEACHING THE LESSON

Students multiply or divide each side of each sentence below by the same number.

1. $13 \cdot 2 = 26$

   $\frac{13 \cdot 2}{13} = \frac{26}{13}$

   $2 = 2$
2. $\frac{14}{2} = 7$

   $\frac{14}{2} \times 2 = 7 \times 2$

   $14 = 14$

Reteaching Masters Booklet, p. 11

Name ______ Date ______

RETEACHING WORKSHEET 1-12

Solving Equations Using Multiplication and Division

An equation like $7c = 42$ can be solved using division.

$7c = 42$ **Divide each side by 7.**

$\frac{7c}{7} = \frac{42}{7}$

$c = 6$

To check the solution, replace c with 6.

$7c = 42$

An equation like $\frac{s}{5} = 8$ can be solved using multiplication.

$\frac{s}{5} = 8$ **Multiply each side by 5.**

$\frac{s}{5} \times 5 = 8 \times 5$

$s = 40$

To check the solution, replace s with 40.

$\frac{s}{5} = 8$

1. 3; 4 2. 7;3 3. 20;4 4. 3; 0.2

Guided Practice

Name the number you would divide each side by to solve each equation. Then solve and check each solution.

Example A

1. $3 \times r = 12$ **2.** $7m = 21$ **3.** $80 = 20y$ **4.** $0.6 = 3n$

Example B

Name the number you would multiply each side by to solve each equation. Then solve and check each equation.

5. $\frac{n}{5} = 7$ 5; 35 **6.** $\frac{f}{10} = 30$ 10; 300 **7.** $0.3 = \frac{d}{2}$ 2; 0.6 **8.** $10 = \frac{r}{25}$ 25; 250

Exercises

Practice

10. 100 13. 0.04
14. 60 21. 400
22. 0.8 23. 400

Solve each equation. Check each solution.

9. $9k = 72$ 8 **10.** $10y = 1{,}000$ **11.** $72y = 0$ 0 **12.** $25r = 25$ 1

13. $2m = 0.08$ **14.** $180 = 3n$ **15.** $\frac{q}{8} = 6$ 48 **16.** $\frac{m}{9} = 9$ 81

17. $\frac{s}{6} = 7$ 42 **18.** $9 = \frac{t}{6}$ 54 **19.** $0.4 = \frac{r}{2}$ 0.8 **20.** $30 = \frac{k}{3}$ 90

21. $800 = 2b$ **22.** $\frac{s}{4} = 0.2$ **23.** $3n = 1{,}200$ **24.** $0.1 = \frac{p}{5}$ 0.5

Using Algebra

Write an equation. Then solve.

25. The product of 6 and n is 72. What is the value of n? 12

26. Some number divided by 4 is 20. What is that number? 80

27. The sum of 200 and a number is 600. What is that number? 400

Applications

Solve. Use an equation.

28. Roger weighs 5 times what his brother weighs. Roger weighs 150 pounds. How much does his brother weigh? 30 pounds

29. Five friends share the cost of a lunch equally. Each person pays $4.00. What is the total cost of the lunch? $20

Research

30. Find out what the total operating budget for your school district is. Also find out how many students are enrolled in your school district. Determine the cost per student. Answers may vary.

Number Sense

31. The graph shows that $400.9 billion was spent for construction in 1991. Write this amount of money in standard form. $400,900,000,000

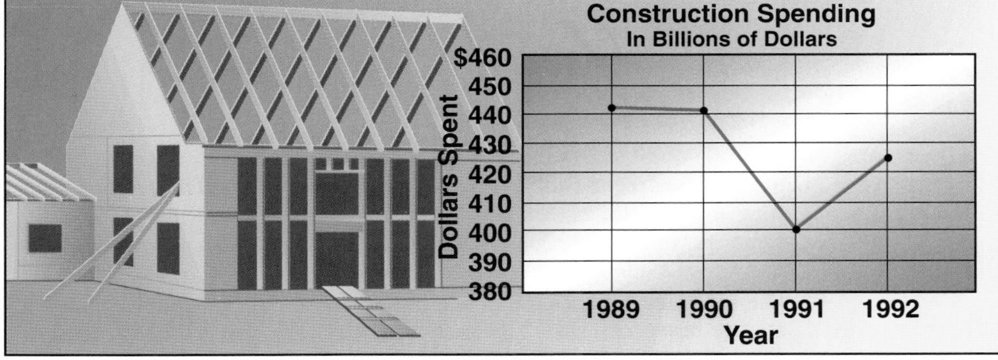

APPLYING THE LESSON

Independent Practice

Homework Assignment	
Minimum	Ex. 1-29 odd; 30-31
Average	Ex. 2-28 even; 30-31
Maximum	Ex. 1-31

Chalkboard Examples

Example A *Solve each equation. Check your solution.*

1. $4 \times t = 16$ 4 **2.** $5b = 35$ 7
3. $42 = 6n$ 7 **4.** $0.8 = 2p$ 0.4

Example B *Solve each equation. Check your solution.*

5. $\frac{m}{5} = 9$ 45 **6.** $\frac{r}{3} = 8$ 24

7. $200 = \frac{d}{4}$ 800 **8.** $0.20 = \frac{z}{2}$ 0.40

Additional examples are provided on TRANSPARENCY 1-12.

EVALUATING THE LESSON

Check for Understanding Use Guided Practice Exercises to check for student understanding.

Closing the Lesson Ask students to state in their own words how to solve an equation.

EXTENDING THE LESSON

Enrichment Have students make a drawing of a balance scale showing how the solution of equations using addition and subtraction is similar to solving equations using multiplication and division.

Enrichment Masters Booklet, p. 11

Name ______________________ Date ______

ENRICHMENT WORKSHEET 1-12

Magic Squares

A magic square is an arrangement of numbers where the sum of the numbers in each row, column, and diagonal is the same.

See if you can complete this magic square. Its magic number is 15. Every row, column, and diagonal must total 15.

8	3	4
1	5	9
6	7	2

Try to place the rest of the numbers 1–9 in the blank spaces so that the sum is 15 across each row, down each column, and along each diagonal.

Add 7

15	10	11
8	12	16
13	14	9

You can make one magic square from another one by adding some number to each element or by multiplying each element by a constant. ***Complete these and check.***

Multiply by 3

24	9	12
3	15	27
18	21	6

Magic squares have been of interest and a recreation to many people over the years. Benjamin Franklin created one that is 8 × 8 and contains the numbers from 1 to 64. It contains many smaller squares with interesting relationships.

14	1	8	11
15	5	4	10
2	16	9	7
3	12	13	6

Complete the following magic square.

The magic sum will be the sum of all 16 numbers divided by four.

$1 + 2 + 3 + 4 + ... + 13 + 14 + 15 + 16 = 136 \quad \frac{136}{4} = 34$

5-MINUTE CHECK-UP

(over Lesson 1-12)

Available on TRANSPARENCY 1-13.

Solve each equation. Check your solution.

1. $32 = 8n$ **4** **2.** $4t = 0.8$ **0.2**

3. $\frac{c}{6} = 9$ **54** **4.** $50 = \frac{400}{m}$ **8**

Extra Practice, Lesson 1-12, p. 451

▶ INTRODUCING THE LESSON

Have two students count their pocket change. Then have all students suggest an equation that would show the total amount of pocket change for the two students.

▶ TEACHING THE LESSON

Using Questioning

- What are the four steps named in the four-step plan to problem solving? **explore, plan, solve, examine**
- In the four-step plan, what is your definition for read? decide? solve? examine?

Practice Masters Booklet, p. 13

Name ______________________ Date ______________

PRACTICE WORKSHEET 1-13

Problem Solving: Write an Equation

Choose the correct equation to solve each problem. Then solve.

1. On Tuesday at Casa Burrito 136 orders of chicken fajitas were sold and 154 orders of shrimp fajitas were sold. How many orders of fajitas were sold? **290 fajitas**
   a. $154 - 136 = y$
   (b.) $154 + 136 = y$ **$y = 290$**
2. Tortillas are packed 50 to a box. How many tortillas are in 10 boxes? **500 tortillas**
   (a.) $50 \times 10 = z$
   b. $50 \div 10 = z$ **$z = 500$**

Write an equation to solve each problem. Then solve.

3. An order of chicken fajitas costs \$6.75. How much do 13 orders cost?
   13 x \$6.75 = n, n = \$87.75
4. To make the day's guacamolé José needs 39 more avocados. He has 42 avocados. How many avocados does the recipe call for?
   39 + 42 = n, n = 81
5. Mr. Serra spent \$15.75 for lunch for himself and his four children. If everyone had the same lunch, how much did each lunch cost?
   \$15.75 ÷ 5 = n, n = \$3.15
6. Mr. Serra gave the waiter a \$20 bill for the lunches. How much should he receive in change?
   \$20 - \$15.75 = n, n = \$4.25
7. Mr. Serra left a tip of \$3.00. How much did lunch and tip cost in all?
   \$15.75 + \$3 = n, n = \$18.75
8. If Mr. Serra takes his children out to lunch once a month and spends \$15.75 plus a \$3 tip each time, how much does he spend in a year?
   (\$15.75 + \$3) x 12 = n, n = \$225
9. A combination platter at Casa Burrito has 738 calories. If Carlos has a combination platter and wants to limit himself to 1,500 calories a day, how many more calories can he consume that day?
   1,500 - 738 = n, n = 762
10. A glass of fruit juice contains 154 calories. If Carlos has a glass with his combination platter how many calories will he have had at Casa Burrito?
   738 + 154 = n, n = 892

Problem Solving

1-13 WRITE AN EQUATION

▶ Explore
▶ Plan
▶ Solve
▶ Examine

Objective
Solve verbal problems using an equation.

Marcy has delivered mail to 56 homes. She has to deliver mail to 32 more homes. How many homes does Marcy have on her route?

Method

You can use the four-step plan to solve any verbal problem.

▶ Explore

What facts are given?

- The part already delivered is 56.
- The part yet to be delivered is 32.

What fact do you need to find?

- To how many homes does Marcy deliver mail?

▶ Plan

Write an equation.

Let n be the total number of homes on Marcy's route.

$$\underbrace{\text{part}}_{56} + \underbrace{\text{part}}_{32} = \underbrace{\text{total}}_{n}$$

▶ Solve

Estimate. $60 + 30 = 90$ Add. $56 + 32 = 88$

Marcy has 88 homes on her route.

▶ Examine

The answer is reasonable because 88 is close to the estimate of 90.

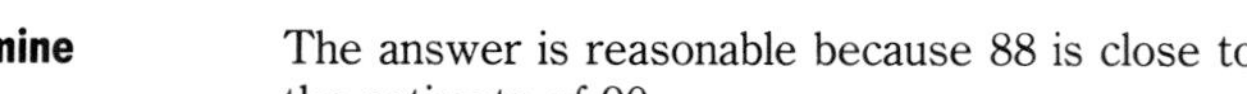

Guided Practice

Choose the correct equation to solve each problem. Then solve.

1. After it began raining, 346 people left the game. However, 632 people stayed. How many people attended the game?
   a. $632 + 346 = y$
   b. $632 - 346 = y$
2. Elin separates her audio tapes into five storage boxes. She puts 20 in each box. How many audio tapes does she have?
   a. $20 \times 5 = z$
   b. $20 \div 5 = z$

RETEACHING THE LESSON

Fill in the blanks in each equation, and then solve it.

1. Scott Pierce bought 5 pounds of dog food. Altogether he paid \$4.25. Find the price per pound.
   $\underline{5} \times p = \underline{\$4.25}$ **\$0.85**
2. Juanita paid \$3.70 less than the regular price for a pair of slacks that were on sale. She paid \$14.80. What was the regular price?
   $r - \underline{3.70} = \underline{14.80}$ **18.50**

Reteaching Masters Booklet, p. 12

Name ______________________ Date ______________

RETEACHING WORKSHEET 1-13

Problem Solving: Write an Equation

A disc jockey plays a record 4 minutes long and an album 37 minutes long. How many minutes of music are played?

Explore.	What is given? What do you need to find?	4 minutes; 37 minutes total minutes of music
Plan.	Write an equation. Let n be the total number of minutes of music. $4 + 37 = n$	
Solve.	Add 4 + 37. **4 + 37 = 41** $n = 41$	
Examine.	41 minutes of music are played. 41 is a reasonable answer.	

Problem Solving

Write an equation to solve each problem. Then solve.

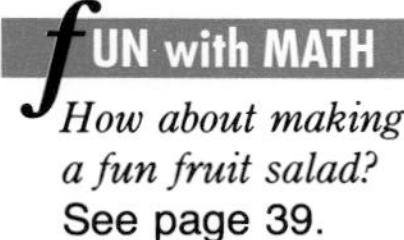

How about making a fun fruit salad? See page 39.

3. A television costs \$530. The stand costs \$60. What is the total cost of the television with the stand? **\$590**
4. Carlos spent \$50 today and has \$30 left in his wallet. How much money did Carlos have to start? **\$80**
5. Mr. David is allowed 1,600 calories a day in his diet. If he has already eaten 900 calories, how many more can he eat and stay on his diet? **700 calories**
6. Michele looks at three cars. The first costs \$10,200, the second costs \$10,900, and the third \$10,400. What is the difference in cost between the most and least expensive car? **\$700**

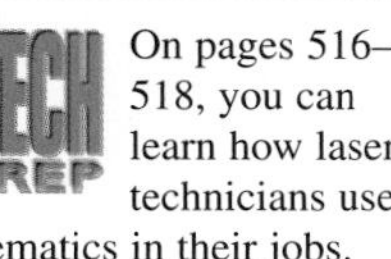

On pages 516–518, you can learn how laser technicians use mathematics in their jobs.

7. A stadium has 70 rows of seats with 30 seats in each row. How many seats are in the stadium? **2,100 seats**
8. The profit from a sale was divided equally among four people. Each received \$200. What was the total profit? **\$800**

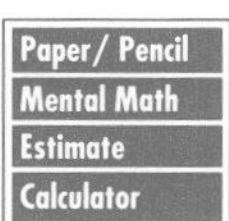

Paper/ Pencil
Mental Math
Estimate
Calculator

9. Jill puts \$10 into her savings account each week. How many weeks will it take her to save \$260? **26 weeks**
10. Mr. Sampsel bought 100 shares of Ajax stock and paid \$3,400. What was the cost per share? **\$34**

Applications

11. Mrs. Stigers paid \$3,000 down when she ordered the car. She will pay the remaining \$8,125 when the car is delivered. What is the cost of the car? **\$11,125**

12. A portfolio represents samples of your work, collected over a period of time. Begin your portfolio by choosing an assignment that you feel shows your best work. Explain why you selected it. Date each item as you place it in your portfolio. **See students' work.**

Critical Thinking

13. Draw the figure at the right without lifting up your pencil from the paper or crossing a line.

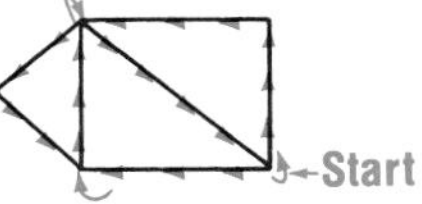

Mixed Review

Lesson 1-2

Write as a product then find the number named.

14. 6^3 **216** 15. 4 cubed **64** 16. 12 squared **144** 17. 9^4 **6,561**

Lesson 1-10

19. $11 - n$

Write an expression for each phrase. Use n for the number.

18. 6 plus a number $6 + n$
19. 11 decreased by a number
20. 12 times *n* $12n$
21. 23 increased by *n* $23 + n$

Lesson 1-11

Solve. Use an equation.

22. In 14 years Frank will be 30 years old. How old is Frank now?
n = Frank's age now; $n + 14 = 30$; $n = 16$

APPLYING THE LESSON

Independent Practice

Homework Assignment	
Minimum	Ex. 1-11 odd; 14-22 even; 23
Average	Ex. 1-11 odd; 13-22
Maximum	Ex. 1-22

Chapter 1, Quiz B (Lessons 1-8 through 1-13) is available in the Evaluation Masters Booklet, p. 4.

Chalkboard Examples

Write an equation to solve the problem. Then solve.

1. The Civic Club added 132 new members last year. Now they have 895 members. How many members did they have before last year? $895 - 132 = n$, 763 members

Additional examples are provided on TRANSPARENCY 1-13.

EVALUATING THE LESSON

Check for Understanding Use Guided Practice Exercises to check for student understanding.

Closing the Lesson Ask students to state in their own words how to decide what equation can be used to solve a problem.

EXTENDING THE LESSON

Enrichment Have students list as many different equations as they can for both classroom examples. Each equation must solve the problem.

1. $895 - 132 = n$, $895 - n = 132$, $n + 132 = 895$, $132 + n = 895$
2. $20 \times m = 60$, $m \times 20 = 60$, $60 \div 20 = m$, $60 \div m = 20$

Enrichment Masters Booklet, p. 12

ENRICHMENT WORKSHEET 1-13

Draw a Diagram

Diagrams may help you decide what operation to use when solving word problems about groups that are being combined or groups that are being separated.

Combining is done by
- *addition* if the groups are *unequal or*,
- *multiplication* if the groups are *equal*.

Yoko scored 18 points in the first game and 5 points in the second game. How many points did she score in all?

18	5
?	

2 unequal groups Add.

Ramo buys 6 tapes at 5 dollars each. How many dollars do the tapes cost?

5	5	5	5	5	5
?					

6 equal groups Multiply.

Separating is done by
- *subtraction* if the groups are *unequal or*,
- *division* if the groups are *equal*.

Willy has 250 baseball cards. Sandy has 300. How many more cards does Sandy have than Willy?

300	
250	?

Subtract. 2 unequal groups

Aubrey plans to drive 550 miles in 11 hours. How many miles will he have to drive each hour?

550										
?	?	?	?	?	?	?	?	?	?	?

Divide. 11 equal grou

Draw a diagram. Then, name the operation you would use to solve each problem.

1. The scout troop wants to sell 144 cheese balls. How many must each of the 8 members sell? **Division**

144							
?	?	?	?	?	?	?	?

2. Fred sees 8 old comic books at the antique show for \$1.75 each. How much will all 8 cost? **Multiplication**

\$1.75	\$1.75	\$1.75	\$1.75	\$1.75	\$1.75	\$1.75	\$1.75
?							

3. Mary baby-sat 28 hours. Kevin baby-sat 31 hours. How many more hours must Mary sit to be equal to Kevin? **Subtraction**

31	
28	?

4. Huey has to read 2 books. One book has 415 pages, and the other has 278 pages. How many pages does Huey have to read? **Addition**

415	278
?	

5. Jan ran 8.5 miles on Tuesday and 7.6 miles on Thursday. How many miles did Jan run? **Addition**

8.5	7.6
?	

6. For a party 5 people make favors. If 400 favors are needed, how many must each make? **Division**

400				
?	?	?	?	?

Using the Chapter Review

The Chapter Review is a comprehensive review of the concepts presented in this chapter. This review may be used to prepare students for the Chapter Test.

Chapter 1, Quizzes A and B, are provided in the Evaluation Masters Booklet as shown below.

Quiz A should be given after students have completed Lessons 1-1 through 1-7. Quiz B should be given after students have completed Lessons 1-8 through 1-13.

These quizzes can be used to obtain a quiz score or as a check of the concepts students need to review.

Evaluation Masters Booklet, p. 4

CHAPTER 1, QUIZ A (Lessons 1-1 through 1-7)

Name the digit in each place-value position in 78,032.45.

1. ten-thousands
2. tenths

Write in expanded form using 10^0, 10^1, and 10^2, and so on.

3. 548

Replace each ○ with >, <, or = to make a true sentence.

4. 4.876 ○ 48.76
5. 73.01 ○ 73.010

Round to the underlined place-value position.

6. 84,732
7. 0.947

Estimate.

8. 6.8 + 3.4 + 2.9
9. $8.7\overline{)53.8}$

10. Mr. Ortez is single and earned $19,110. How much tax does he pay?

at least	but less than	single
$19,050	$19,100	$2,930
$19,100	$19,150	$2,944

1. 7
2. 4
3. $(5 \times 10^2) + (4 \times 10^1) + (8 \times 10^0)$
4. <
5. =
6. 85,000
7. 0.9
8. 13
9. 6
10. $2,944

Name ______________________ Date ______________________

CHAPTER 1, QUIZ B (Lessons 1-8 through 1-13)

Find the value of each expression.

1. $3 \times (4 - 2) + 4$
2. $(3^2 - 1) \div 4$

Find the value of each expression if n = 6.

3. $8 - \frac{n}{2}$
4. $n^2 + 3.2$

Solve each equation.

5. $9 - b = 3$
6. $h + 4 = 12.3$
7. $\frac{s}{8} = 0.3$
8. $5r = 125$

Write an equation. Then solve.

9. Stan saves $60 a month. How much does he save in 12 months?

Solve. Use matrix logic.

10. Jon, Jan, and Jen each went on vaction. One person went by car, another by train, and the third by plane. Jan gets dizzy on planes and in cars. Jen did not pass through a security check before departing. Who used which kind of transportation?

1. 10
2. 2
3. 5
4. 39.2
5. 6
6. 8.3
7. 2.4
8. 25
9. 12 × $60 = n; n = $720
10. Jan, train; Jen, car; Jon, plane

Chapter 1 REVIEW

Vocabulary / Concepts

1. $(6 \times 10^2) + (4 \times 10^0)$ 6. dividend

Choose the word or number from the list at the right to complete each sentence.

1. In expanded form the number 604 is ___?___.
2. Both 5^2 and 5^3 are ___?___ of five. powers
3. When we say a sum is *about* 600, we are ___?___ the sum. estimating
4. Three less than a number can be written ___?___. $n - 3$
5. Eight and thirty thousandths written in standard form is ___?___. 8.030
6. When estimating a quotient, we round the ___?___ so it is a multiple of the divisor.
7. A mathematical sentence with an equals sign is called a(n) ___?___. equation
8. Multiplication and division are ___?___ operations. inverse
9. To solve an equation, we find the correct replacement for a(n) ___?___. variable

dividend
estimating
equation
exponents
inverse
powers
variable
0.830
$3 - n$
$6 \times 100 + 4 \times 10$
$(6 \times 10^2) + (4 \times 10^0)$
8.030
$n - 3$

Exercises / Applications

Lesson 1-1 ***Write in standard form.***

10. 15 thousands 15,000 11. 36 tens 360 12. 6 tenths 0.6 13. 50 ones 50

Lesson 1-2 ***Write in expanded form using 10^0, 10^1, 10^2, and so on.*** See margin.

14. 500 15. 24 16. 1,821 17. 2,650 18. 20,340 19. 7,005

20. 6, 6.02, 6.021, 6.2 21. 10.708, 10.78, 10.87, 10.871

Lesson 1-3 ***Order from least to greatest.***

20. 6.2, 6.02, 6.021, 6 21. 10.87, 10.708, 10.871, 10.78

Lesson 1-4 ***Round each number to the underlined place-value position.***

22. 829 800 23. 9,703 10,000 24. 36.098 36 25. 10.654 10.7 26. 19.0148 19.01

Lesson 1-5 ***Use the tax table on page 14 to find the tax for each income and filing status.***

27. $19,505; single $3,056
28. $19,650; head of household $2,951
29. $19,167; married filing jointly $2,876
30. $19,400; married filing separately $3,427

Lessons 1-6, 1-7 ***Estimate.*** Answers may vary. Typical answers given.

31. 409 + 2,803 3,200 32. 313,250 − 81,695 230,000 33. $2.39 + $3.64 $6.03

34. 9,000

34. 2,643 × 3 35. 629 × 38 24,000 36. 408 ÷ 52 8 37. 386 ÷ 3.704 100

Additional Answers

14. (5×10^2)
15. $(2 \times 10^1) + (4 \times 10^0)$
16. $(1 \times 10^3) + (8 \times 10^2) + (2 \times 10^1) + (1 \times 10^0)$
17. $(2 \times 10^3) + (6 \times 10^2) + (5 \times 10^1)$
18. $(2 \times 10^4) + (3 \times 10^2) + (4 \times 10^1)$
19. $(7 \times 10^3) + (5 \times 10^0)$

Lesson 1-8

Solve. Use matrix logic.

38. Spot, Lady, Fido, and Buffy each wear a collar. Spot's collar is not red. Lady's collar is not black or tan. Fido's fur is black and he wears a collar to match. Buffy's owner likes red but dislikes brown. What is the color of each dog's collar? **Spot-tan; Lady-brown; Fido-black; Buffy-red**

39. 90 40. 37 41. 24.6 44. 70

Lesson 1-9

Find the value of each expression.

39. $20 \times 4 + 2 \times 5$ **40.** $42 \div 6 + 6 \times 5$ **41.** $30.6 - 48 \div 8$

42. $(3 \times 7) - 1$ **20** **43.** $15 - (64 \div 8)$ **7** **44.** $(5 \times 8) + (60 \div 2)$

45. 11 46. 1

Lesson 1-10

Find the value of each expression if a = 4, b = 5, and c = 10.

45. $a^2 - b$ **46.** $c - (a + b)$ **47.** $\frac{ab}{c}$ **2** **48.** $\frac{5a + c}{6}$ **5** **49.** $(a + b)^2$ **81**

53. $k = 7$ 54. $a = 54$ 55. $p = 70$ 56. $t = 3$

Lessons 1-11, 1-12

Solve each equation. Check each solution.

50. $m - 7 = 7$ **$m = 14$** **51.** $y + 8 = 13$ **$y = 5$** **52.** $0.6 = n - 0.4$ **$n = 1$**

53. $8k = 56$ **54.** $6 = \frac{a}{9}$ **55.** $420 = 6p$ **56.** $\frac{1.8}{t} = 0.6$

Lessons 1-11, 1-12

Write an equation. Then solve.

57. Eight increased by m is 15. What is the value of m? **$m = 7$**

58. Some number decreased by 5 is 15. What is that number? **20**

59. Six multiplied by y is 48. What is the value of y? **$y = 8$**

60. Some number divided by 3 is 600. What is that number? **1,800**

64. $y = 900$

Lesson 1-13

Write an equation to solve each problem. Then solve.

61. Beth spent $40 for a pair of shoes. She has $25 remaining in her purse. How much money did she have before buying the shoes? **$65**

62. Kevin places nine baseball cards in each page of his album. All 30 pages of his album are full. How many baseball cards are in the album? **270 cards**

63. Becky has $55. Does she have enough money to buy a $28.50 skirt and a $22.95 blouse? **yes**

64. The product of 8 and y is 7,200. What is the value of y? **$y = 900$**

65. Mr. Painter works 40 hours a week and makes $7 an hour. What is his take-home pay if $60 is withheld for taxes? **$220**

66. Sixty students moved into the Logan school district. Now there are 420 students attending the school. How many students were attending before the students moved in? **360 students**

APPLYING THE LESSON

Independent Practice

Homework Assignment	
Minimum	Ex. 1-65 odd
Average	Ex. 1-9; 10-56 even; 58-66
Maximum	Ex. 1-66

Alternate Assessment Strategy

To provide a brief in-class review, you may wish to read the following questions to the class and require a verbal response.

1. What is the tens digit in 135? **3**
2. What is the standard form of 200 plus 70 plus 1? **271**
3. Which is greater, 0.25 or 0.31? **0.31**
4. Round 2.67 to the nearest whole number. **3**
5. Estimate the sum of 4,382 and 321. **4,700**
6. About how much is 14.8 times 10.4? **150**
7. In the expression $4 + 6 \times 8 - 3$, what operation would you perform first? **multiply 6 and 8**
8. Give an expression for the phrase 10 more than a number. **$n + 10$**
9. 5 added to a number is 24. What is the number? **19**
10. 3 multiplied by some number y is equal to 75. What is the value of y? **25**

Using the Chapter Test

This page may be used as a test or as an additional page of review if necessary. Two forms of a Chapter Test are provided in the Evaluation Masters Booklet. Form 2 (free response) is shown below, and Form 1 (multiple choice) is shown on page 37.

The **Tech Prep Applications Booklet** provides students with an opportunity to familiarize themselves with various types of technical vocations. The Tech Prep applications for this chapter can be found on pages 1–2.

Evaluation Masters Booklet, p. 3

CHAPTER 1 TEST, FORM 2

Write the number named by the 7 in each number.

1. 678,341.2 2. 6.007

Write in standard form.

3. $(5 \times 1{,}000) + (9 \times 10) + 3$
4. $(8 \times 10^4) + (3 \times 10^3) + (7 \times 10^2) + (4 \times 10^1) + (2 \times 10^0)$

Replace each ○ with >, <, or = to make a true sentence.

5. 487 ○ 4,780 6. 3.007 ○ 3.070

Round to the nearest hundredth.

7. 5.4462

Estimate.

8. $43.6 + 8.9$ 9. $4.26 - 0.382$ 10. $2.8\overline{)149.2}$ 11. 0.37×4.2

12. Find the value of $3 \times (19 - 17.5) \div 5$
13. Find the value of $n + 3.6$ if $n = 0.3$.

Solve each equation.

14. $\frac{3}{4} + m = \frac{7}{8}$ 15. $b - 3.7 = 5$

16. $\frac{x}{0.3} = 0.5$ 17. $3.1s = 93$

Write an equation. Then solve.

18. Mark's annual income is $25,000. He pays $3,609 in federal taxes. What is his pay after federal tax?

19. Solve.Use matrix logic. The four choices at a banquet were chicken, duck, eggs, and fish. Len, Mai, Nick, and Opal each chose a different meal. Opal does not eat meat. Len ordered his meal scrambled. Nick refused to eat an animal that quacked. What did Mai choose?

20. Solve. If it is 1,437 miles from Here to There, how far would it be round-trip?

Bonus Tell whether the product of 0.32×0.7 will be *greater* or *less than* either factor without doing the computation. Explain.

Answers: 1. 70,000 2. 0.007 3. 5,093 4. 83,742 5. < 6. < 7. 5.45 *8. 53 *9. 3.9 *10. 50 *11. 1.6 12. 0.9 13. 3.9 14. $m = \frac{1}{8}$ 15. $b = 8.7$ 16. $x = 0.15$ 17. $s = 30$ 18. $25,000 − $3,609 = ; n = $21,39 19. duck 20. 2,874 mi Bonus less, both factors are between 0 and 1

* Accept any reasonable estimate

Chapter 1

1. 700 2. 0.30 3. 370

Write in standard form.

1. 70 tens **2.** 30 hundredths **3.** $(3 \times 10^2) + (7 \times 10^1)$

4. 1,296 $6 \times 6 \times 6 \times 6$ 5. 27 $3 \times 3 \times 3$ 6. 25 5×5

Write as a product and then find the number named.

4. 6^4 **5.** 3^3 **6.** 5^2 **7.** 2^6

7. 64 $2 \times 2 \times 2 \times 2 \times 2 \times 2$

Replace each ● with <, >, or = to make a true sentence.

8. 36 ● 48 < **9.** 3.8 ● 3.80 = **10.** 0.6 ● 0.06 > **11.** 78 ● 79 <

Round each number to the underlined place-value position.

12. 876 900 **13.** 87.5 88 **14.** 1.098 1.100 **15.** 12,394 12,000

Solve.

16. John is single and earns $19,800. The tax table shows at least $19,750 but less than $19,800 is $3,126 or at least $19,800 but less than $19,850 is $3,140. How much tax does John owe? $3,140

Estimate. Answers may vary. Typical answers given.

17. $75 − $2.80 $72 **18.** 3,168 + 425 3,600 **19.** 33×780 24,000 **20.** $245 \div 48$ 5

Solve. Use matrix logic.

21. Joan is baby-sitting Nancy, Bobby, Susie, and Timmy. For a snack, she gives them fruit. Each child gets a different fruit. Nancy only likes red fruit. Timmy does not like bananas or pears. Susie can choose either a pear or a banana. Bobby likes bananas but does not like apples or oranges. Who gets which fruit?

Find the value of each expression.

22. $8 + 8 \times 4$ 40 **23.** $(40 + 60) \div 2$ 50 **24.** $(9 \times 4) - 6^2$ 0

21. Nancy-apple; Bobby-banana; Susie-pear; Timmy-orange

Find the value of each expression if $r = 2$, $s = 5$, and $t = 10$.

25. $t \times (s - r)$ 30 **26.** $2t \div 4$ 5 **27.** $\frac{st}{r}$ 25 **28.** rst 100

Solve each equation. Check each solution. 29. 1,200

29. $m + 9 = 15$ 6 **30.** $400 = n - 800$ **31.** $5{,}600 = 8q$ 700

Write an equation to solve each problem. Then solve.

32. At Vick's Pizza, you can get a free pizza with 26 coupons. Sue has 20 coupons. How many more must she save to have enough for a free pizza? 6 coupons

33. Nick walks five times as many blocks to school as George walks. George walks three blocks. How many blocks does Nick walk?

33. 15 blocks a. Any number will satisfy this equation. b. The solution is −2 c. There is no solution.

BONUS: ***Describe the solution of each equation.***

a. $y = y$ **b.** $x + 4 = 2$ **c.** $x + 2 = x + 3$

The **Performance Assessment Booklet** provides an alternative assessment for evaluating student progress. An assessment for this chapter can be found on pages 1–2.

A **Test and Review Generator** is provided in Apple, IBM, and Macintosh versions. You may use this software to create your own tests or worksheets, based on the needs of your students.

TIME ZONES

Mr. Gunther owns a construction company. He bids for jobs all over the country. On Monday, Mr. Gunther makes a call from Denver at 10:30 A.M. to New York City. What time is it in New York? Use the map showing time zones.

Mr. Gunther's call is made from the west to the east. The time is one hour later for each time zone crossed.

10:30 A.M. in Denver ↔ 12:30 P.M. in New York City

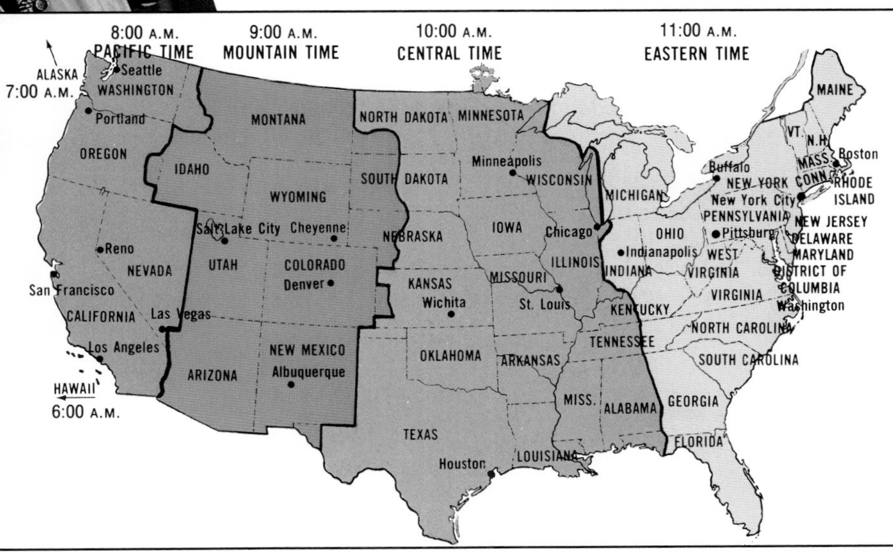

Suppose a call is made from the east to the west. The time is one hour earlier for each time zone crossed.

2:00 P.M. in East Florida ↔ 11:00 A.M. in California

For each city and time, give the time in the second city.

1. Boston 1:30 P.M. ↔ Wichita 12:30 P.M.
2. Portland 11:00 P.M. ↔ Chicago 1:00 A.M.
3. Cheyenne 8:35 P.M. ↔ Honolulu 5:35 P.M.
4. Jessica makes a call from Indianapolis at 4:30 P.M. to Stockholm, Sweden. Stockholm, Sweden, is six time zones from Indianapolis. What time is her call received in Stockholm? 10:30 P.M.

Applying Mathematics to the Real World

This optional page shows how mathematics is used in the real world and also provides a change of pace.

Objective Identify time zones.

Using Critical Thinking Ask students why it is not the same time all over the world. Time depends on the location of the sun. If it were the same time all over the world, sunrise might occur at noon in certain locations.

Using Discussion Discuss why it is important to know about the time zones.

Using Research Have students or groups of students research the history of how time zones were established. Have them explain the significance of the international date line.

Using the Cumulative Review

This page provides an aid for maintaining skills and concepts presented thus far in the text.

A Cumulative Review is also provided in the Evaluation Masters Booklet as shown below.

Evaluation Masters Booklet, p. 5

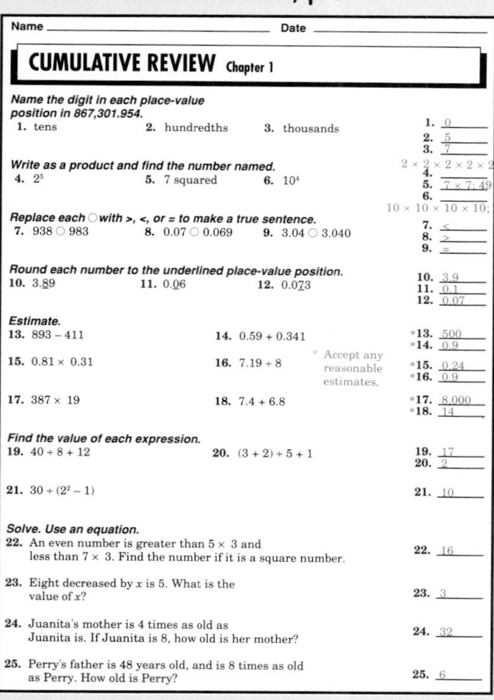

Name ______ Date ______

CUMULATIVE REVIEW Chapter 1

Name the digit in each place-value position in 867,301.954.

1. tens   2. hundredths   3. thousands

1. 0
2. 5
3. 7

Write as a product and find the number named.

4. 2^5 5. 7 squared   6. 10^4

4. 2 × 2 × 2 × 2 × 2
5. 7 × 7; 49
6. 10 × 10 × 10 × 10;

Replace each ○ with >, <, or = to make a true sentence.

7. 938 ○ 983   8. 0.07 ○ 0.069   9. 3.04 ○ 3.040

7. <
8. >
9. =

Round each number to the underlined place-value position.

10. 3.89   11. 0.06   12. 0.073

10. 3.9
11. 0.1
12. 0.07

Estimate.

13. 893 − 411   14. 0.59 + 0.341   15. 0.81 × 0.31   16. 7.19 ÷ 8   17. 387 × 19   18. 7.4 + 6.8

* Accept any reasonable estimates.

*13. 500
*14. 0.9
*15. 0.24
*16. 0.9
*17. 8,000
*18. 14

Find the value of each expression.

19. 40 ÷ 8 + 12   20. (3 + 2) ÷ 5 + 1   21. 30 ÷ (2² − 1)

19. 17
20. 2
21. 10

Solve. Use an equation.

22. An even number is greater than 5 × 3 and less than 7 × 3. Find the number if it is a square number.   22. 16

23. Eight decreased by *x* is 5. What is the value of *x*?   23. 3

24. Juanita's mother is 4 times as old as Juanita is. If Juanita is 8, how old is her mother?   24. 32

25. Perry's father is 48 years old, and is 8 times as old as Perry. How old is Perry?   25. 6

Cumulative REVIEW/TEST

Free Response

Lesson 1-1

Replace each ■ with a number to make a true sentence.

1. 40 = ■ ones **40**   2. 300 = ■ hundreds **3**   2,000 = ■ tens **200**

Name the digit in each place-value position in 9,105,832.467.

4. tens **3**   5. thousandths **7**   6. ten thousands **0**
7. ones **2**   8. tenths **4**   9. hundredths **6**

Write in standard form.

10. 6 hundred **600**   11. 20 tens **200**   12. 40 hundreds **4,000**

Lesson 1-2

Write as a product and then find the number named. **See margin.**

13. 3^4 14. 15 squared   15. 7 cubed   16. 2^5 17. 10^6

Write in expanded form using 10^0, 10^1 10^2, and so on. **See margin.**

18. 562   19. 4,381   20. 10,601   21. 33,332

Lesson 1-3

22. Terri would like to purchase a compact disc that costs $12.95 or a cassette that costs $8.95. She has $10.52. Which can she afford to buy? **cassette**
23. Which whole numbers are greater than 42.4 and less than 45.8? **43, 44, 45**
24. Which cyclist had a time greater than 10.273 s? **Jens Fiedler, Ken Carpenter**

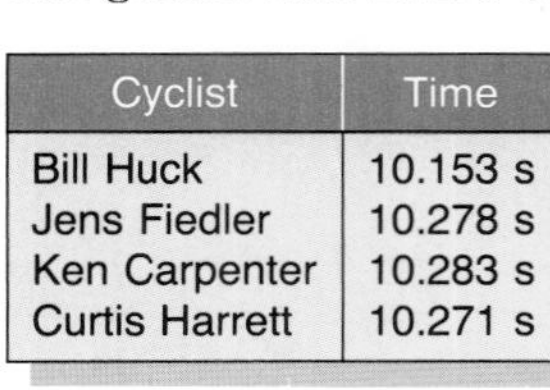
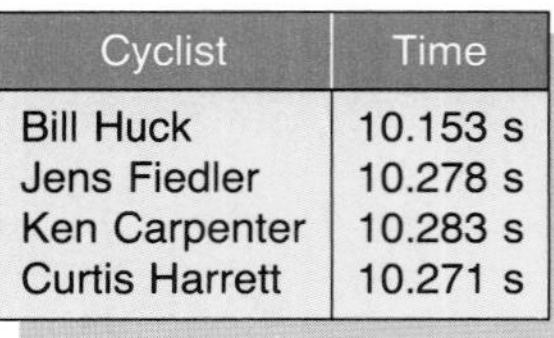

Cyclist	Time
Bill Huck	10.153 s
Jens Fiedler	10.278 s
Ken Carpenter	10.283 s
Curtis Harrett	10.271 s

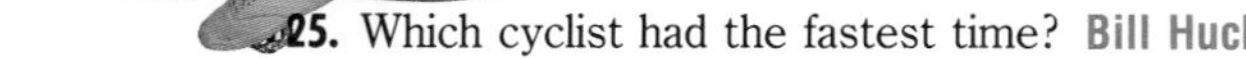

25. Which cyclist had the fastest time? **Bill Huck**

Lesson 1-4

Round each number to the underlined place-value position.

26. 3.021 **3.000**   27. 10.463 **10.500**   28. 0.642 **0.6**   29. 25.122 **25.12**
30. 87.5 **88.0**   31. 0.081 **0.1**   32. 28.195 **28.20**   33. 7.695 **7.70**
34. Janet's final grade average is 0.8273. To the nearest hundredth, what is her average? **0.83**

Lesson 1-6, 1-7

Estimate.

35. 6,189 + 523 **6,700**   36. 906 − 32 **880**   37. 0.715 + 0.328 **1**
38. 43.17 − 6.55 **36**   39. 0.85 − 0.2 **0.7**   40. 32 × 19 **600**
41. 126 × 5 **500**   42. 360 ÷ 24 **18**   43. 144 ÷ 18 **7**
44. Denise is having a party with 86 invited guests. She would like to have enough cans of soda for each guest to have 2 cans. *About* how many cans of soda should she buy? **180 cans**

APPLYING THE LESSON

Independent Practice

Homework Assignment	
Minimum	Ex. 2-44 even
Average	Ex. 1-44
Maximum	Ex. 1-44

Additional Answers

13. 3 × 3 × 3 × 3 = 81
14. 15 × 15 = 225
15. 7 × 7 × 7 = 343
16. 2 × 2 × 2 × 2 × 2 = 32
17. 10 × 10 × 10 × 10 × 10 × 10 = 1,000,000
18. $(5 \times 10^2) + (6 \times 10^1) + (2 \times 10^0)$
19. $(4 \times 10^3) + (3 \times 10^2) + (8 \times 10^1) + (1 \times 10^0)$
20. $(1 \times 10^4) + (6 \times 10^2) + (1 \times 10^0)$
21. $(3 \times 10^4) + (3 \times 10^3) + (3 \times 10^2) + (3 \times 10^1) + (2 \times 10^0)$

Chapter 1

Multiple Choice

Choose the letter of the correct answer for each item.

1. What is the place-value position of the 8 in 184,972?
   - a. tens
   - b. hundreds
   - c. thousands
   - ▶ d. ten thousands

   Lesson 1-1

2. What is the numeral for 91 hundreds?
   - e. 91
   - f. 910
   - ▶ g. 9,100
   - h. 91,000

   Lesson 1-1

3. What is 93,740 rounded to the nearest hundred?
   - a. 90,000
   - ▶ b. 93,700
   - c. 93,800
   - d. 94,000

   Lesson 1-4

4. Estimate the sum of 625 and 8,825.
   - e. 1,400
   - ▶ f. 9,400
   - g. 14,000
   - h. 15,000

   Lesson 1-6

5. Estimate the difference of 81,467 and 63,903.
   - a. 2,000
   - b. 8,000
   - ▶ c. 20,000
   - d. 140,000

   Lesson 1-6

6. What is the standard form of $(6 \times 10^4) + (5 \times 10^3) + (7 \times 10^0)$?
   - e. 657
   - ▶ f. 65,007
   - g. 65,700
   - h. 650,007

   Lesson 1-2

7. Find the value of $30 - 9 \times (2 + 1)$.
   - ▶ a. 3
   - b. 13
   - c. 21
   - d. 63

   Lesson 1-9

8. Bobby gives 18 baseball cards away and he has 32 left. How many baseball cards did he have before he gave any away?
   - e. 14
   - f. 40
   - ▶ g. 50
   - h. 576

   Lesson 1-11

9. An odd number is greater than 6×5 and less than 6×7. What is the number if the sum of its digits is 8?
   - a. 31
   - ▶ b. 35
   - c. 40
   - d. 39

   Lesson 1-8

10. What facts are given in this problem?
    A concert hall has 100 rows of seats with 40 seats in each row. Each seat sells for $15.95. How much money is made if the concert is a sell-out?
    - e. Who is giving the concert.
    - ▶ f. The total number of seats.
    - g. The number in attendance.
    - h. The cost of parking.

    Lesson 1-13

Using the Standardized Test

This test serves to familiarize students with standardized format while testing skills and concepts presented up to this point.

APPLYING THE LESSON

Independent Practice

Homework Assignment	
Minimum	Ex. 1-10
Average	Ex. 1-10
Maximum	Ex. 1-10

Evaluation Masters Booklet, pp. 1-2

Name ______________ Date ______________

CHAPTER 1 TEST, FORM 1

Write the letter of the correct answer on the blank at the right of the page.

1. Name the place-value position for the digit 8 in 6,821,354. — 1. B
   A. hundreds B. hundred-thousands C. thousands D. millions
2. Name the digit in the hundredths place in 438,217.065. — 2. D
   A. 2 B. 5 C. 4 D. 6
3. Write $(5 \times 10^4) + (3 \times 10^2) + (2 \times 10^1) + (7 \times 10^0)$ in standard form. — 3. C
   A. 5,327 B. 53,270 C. 50,327 D. 503,270
4. Write 3,708 in expanded form. — 4. C
   A. $(3 + 10^3) \times (7 + 10^2) \times (8 + 10^1)$
   B. $(3 \times 10^3) + (7 \times 10^2) + (8 \times 10^2)$
   C. $(3 \times 10^3) + (7 \times 10^2) + (8 \times 10^0)$
   D. $(3 \times 100) + (7 \times 10) + 8$
5. Order from least to greatest. 32, 0.32, 320, 3.2, 0.032 — 5. A
   A. 0.032, 0.32, 3.2, 32, 320 B. 0.32, 0.032, 32, 320, 3.2 C. 320, 32, 3.2, 0.32, 0.032 D. 3.2, 32, 0.32, 320, 0.032
6. Round 4̲7.983 to the underlined place-value position. — 6. D
   A. 48 B. 47 C. 47.98 D. 50
7. Round 0.0̲93 to the underlined place-value position. — 7. C
   A. 0.09 B. 1.0 C. 0.1 D. 0.08
8. Estimate. $43.9 + 25.4$ — 8. A
   A. 70 B. 60 C. 20 D. 80
9. Estimate. $9.47 - 0.31$ — 9. B
   A. 916 B. 9.2 C. 9.7 D. 9.8
10. Estimate. 0.73×4 — 10. B
    A. 28 B. 2.8 C. 0.28 D. 4
11. Estimate. $6.29 \div 7$ — 11. D
    A. 9 B. 9.00 C. 0.09 D. 0.9
12. Find the value of $5 \times (3 + 4.2) - 6.2$ — 12. C
    A. 13.2 B. 30 C. 29.8 D. 30.2
13. Find the value of $12 - 3r$ if $r = 2.1$. — 13. D
    A. $9r$ B. 9 C. 6.3 D. 5.7

Using Mathematical Connections

Objective Understand, appreciate, and enjoy the connections between mathematics and real-life phenomena and other disciplines.

The following data is background information for each event on the time-line.

History First math problems found recorded on papyrus in 1650 B.C.
In the nineteenth century a Scottish Egyptologist, Rhind, discovered what is often referred to as the Rhind papyrus. The papyrus dates back to 1650 B.C. and is thought to have been copied from a much older piece of work. It is a collection of problem solving, probably used in a school for scribes.

History First inflatable swimming aid in 880 B.C.
The first inflatable swimming aid was developed by the Assyrian army to assist soldiers in crossing turbulent rivers. This area is today within the nation of Iraq. Prepared skins of full-grown sheep and goats were filled with air by blowing through a single opening and tied securely. Soldiers swam across a river while holding one of these floats beneath them.

History First use of plus (+) and minus (−) signs in 1489
Johann Widman in Leipzig, Germany was the first person to communicate arithmetic using the signs + and −. These symbols had previously been used in warehouses as marks indicating excesses (+) and deficiencies (−) of inventory. The + sign is the result of joining together the two letters of the Latin word et meaning *and.* The − sign may be a reduction of the character m used in previous work to indicate minus.

History Binary arithmetic in 1703
Gottfried Wilhelm Leibnitz of Leipzig, Germany, is credited with formulating binary arithmetic that is the foundation of calculators and computers today. The binary numbering system is based on twos (2s) rather than tens (tens) as we use in decimals. Each element in a binary is represented by a digit value of either zero (0) or one (1) and is known as a bit.

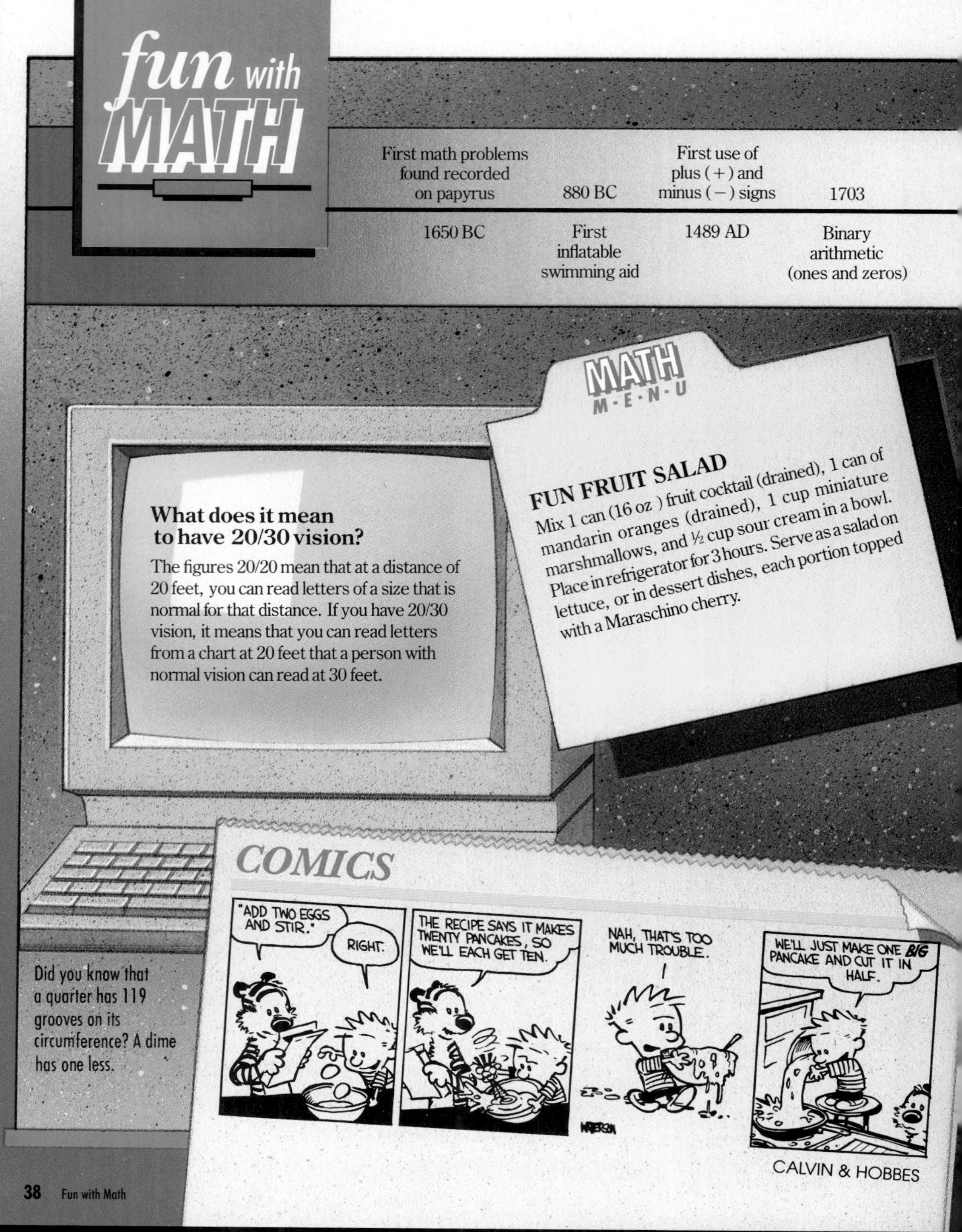

fun with MATH
First math problems found recorded on papyrus
1650 BC
880 BC
First inflatable swimming aid
First use of plus (+) and minus (−) signs
1489 AD
1703
Binary arithmetic (ones and zeros)
MATH M·E·N·U
FUN FRUIT SALAD
Mix 1 can (16 oz) fruit cocktail (drained), 1 can of mandarin oranges (drained), 1 cup miniature marshmallows, and ½ cup sour cream in a bowl. Place in refrigerator for 3 hours. Serve as a salad on lettuce, or in dessert dishes, each portion topped with a Maraschino cherry.
What does it mean to have 20/30 vision?
The figures 20/20 mean that at a distance of 20 feet, you can read letters of a size that is normal for that distance. If you have 20/30 vision, it means that you can read letters from a chart at 20 feet that a person with normal vision can read at 30 feet.
Did you know that a quarter has 119 grooves on its circumference? A dime has one less.
COMICS
"ADD TWO EGGS AND STIR."
RIGHT.
THE RECIPE SAYS IT MAKES TWENTY PANCAKES, SO WE'LL EACH GET TEN.
NAH, THAT'S TOO MUCH TROUBLE.
WE'LL JUST MAKE ONE BIG PANCAKE AND CUT IT IN HALF.
CALVIN & HOBBES
38 Fun with Math

Baseball Abner Doubleday Cooperstown, NY	1899	First mini skirt London, England	1980
1839	First tape recorder	1965	First compact disc (CD) digital audio system

JOKE!

Q: If two's company and three's a crowd, what are four and five?

A: Nine.

RIDDLE

Q. When can ten students stand under an umbrella without getting wet?

A. When it's not raining

How does a CD player work? During recording, music is converted to a binary code, made up of ones and zeros, by a computer. A laser beam, carrying the code of ones and zeros, burns a pit along each track of a CD for each zero in the code. It leaves a non-pit, or 'land' for each one in the code. A laser beam in your CD player reads the pits and lands of a CD. The built-in microcomputer converts the binary number code back to music.

TEASER

The Mayas used a number system of only two symbols—a dot and a line—to build numbers from 1 to 19. Complete the following.

• 1, •• 2, (?) 3, •••• 4, — 5
•/— 6, ••/— 7, (?) 8, (?) 9, ═ 10
•/═ 11, (?) 12, (?) 13, (?) 14, ≡ 15
•/≡ 16, (?) 17, (?) 18, ••••/≡ 19

An oval below a number multiplied it by 20:
••/⌓ = 2 × 20 = 40.

Write Mayan numbers for 60, 80, 100, and so on to 380.

QUIZ TIME

Substitute words for the numbers in the following story by doing the math on your calculator, then turning it upside down to read each word. (632 × 497) + 3433 and (2 × 5 × 7 × 110) + 18 went to the (677 + 271 − 946) ÷ 100. They walked up a (59,982 ÷ 13) (69,426 ÷ 9), (11,283 + 17,584) × 11 broke the (193 × 38) of her (1367 − 758) × 5 when (69 × 15) ÷ 3 stepped in a (49 × 56) + (12 × 80). (69 × 5) had to (66,666 + 77,777 + 88,888 + 99,999 + 45,474) around and her (91 × 7) hurt. (227 × 34) almost stepped on a (315,054 ÷ 9) (3 × 13 × 17).

Fun with Math 39

Physical Education Baseball, Abner Doubleday, Cooperstown, NY, 1839
Abner Doubleday is credited with having invented baseball in Cooperstown, New York, in 1839. However, publications were written about baseball in 1835 when Abner was only 16. Further, he attended West Point from 1838 to 1842 and was not allowed to leave campus until his later years in school. He could not have even visited Cooperstown before 1841.

Science First tape recorder in 1898
The first workable magnetic recording device was patented in 1898 by Valdemar Poulsen, a Danish electrical engineer. He called it a *telegraphone* and utilized a process of magnetizing steel wire.

Home Economics First mini-skirt in 1965
Mary Quant, a fashion designer, first presented the mini-skirt to the world at King's Road, Chelsea, London, England.

Science First Compact Disc (CD) digital audio system, in 1980
The first laser recorded compact disc, or CD, digital audio system was introduced in 1980 by Sony of Tokyo, Japan, with Philips of Eindhoven, Netherlands. CDs became commercially available in 1983 and quickly became popular. In 1988, manufacturers introduced CD-Vs, or Compact Disc Videos, which carry pictures along with music.

Language Arts Quiz
Answers to Quiz: **Leslie, Bill, zoo, high, hill, Leslie, heel, shoe, she, hole, she, hobble, leg, Bill, goose, egg**

Activity Have students write mathematical messages to one another using their calculators and the following: 0 = 0, 1 = i or I, 2 = Z, 3 = E, 4 = h, 5 = S, 6 = g, 7 = L, 8 = B, and 9 = b.

Teaser Answer

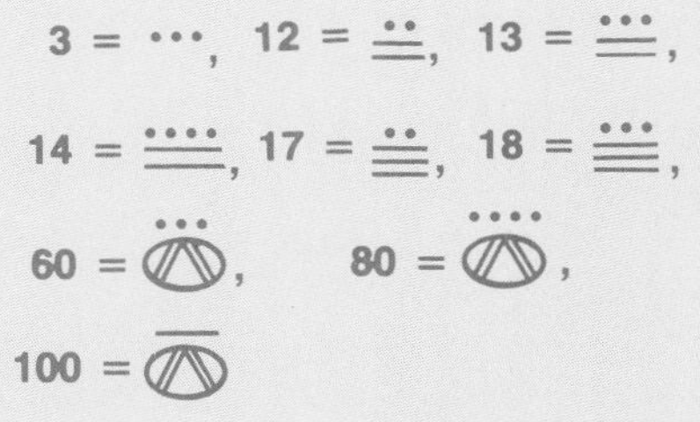

3 = •••, 12 = ••/═, 13 = •••/═,

14 = ••••/═, 17 = ••/≡, 18 = •••/≡,

60 = •••/⌓, 80 = ••••/⌓,

100 = —/⌓

Chapter

2 PATTERNS: ADDING AND SUBTRACTING

PREVIEWING the CHAPTER

In this chapter, addition and subtraction of whole numbers and of decimals are presented. Students are encouraged to use their estimating skills to determine whether the answer is reasonable. These skills are then applied to the consumer topics of checking accounts and budgeting, the algebraic topic of number sequences, and the geometric topic of perimeter.

Problem-Solving Strategy Students learn to choose the method of computation.

Lesson (Pages)	Lesson Objectives	State/Local Objectives
2-1 (42-43)	2-1: Add whole numbers.	
2-2 (44-45)	2-2: Add decimals.	
2-3 (46-47)	2-3: Subtract whole numbers.	
2-4 (48-49)	2-4: Subtract decimals.	
2-5 (52-53)	2-5: Write a check and use a check register.	
2-6 (54-55)	2-6: Identify and write arithmetic sequences.	
2-7 (56-57)	2-7: Choose an appropriate method of computation.	
2-8 (58-59)	2-8: Find the perimeter of a polygon.	
2-9 (60-61)	2-9: Identify missing and extra facts.	

ORGANIZING the CHAPTER

You may refer to the ***Course Planning Calendar*** on page T10.

Planning Guide

Lesson (Pages)	Extra Practice (Student Edition)	Blackline Masters Booklets								Other Resources
		Reteaching	Practice	Enrichment	Activity	Multi-cultural Activity	Technology	Tech Prep	Evaluation	
2-1 (42-43)	p. 451	p. 13	p. 14	p. 13			p.2		p.73	Transparency 2-1
2-2 (44-45)	p. 451	p. 14	p. 15	p. 14	p. 3					Transparency 2-2
2-3 (46-47)	p. 452	p. 15	p. 16	p. 15		p.2				Transparency 2-3
2-4 (48-49)	p.452	p. 16	p. 17	p. 16	p. 4			p. 3		Transparency 2-4
2-5 (52-53)		p. 17	p. 18	p. 17						Transparency 2-5
2-6 (54-55)	p. 452	p. 18	p. 19	p. 18				p. 4		Transparency 2-6
2-7 (56-57)			p. 20							Transparency 2-7
2-8 (58-59)	p. 453	p. 19	p. 21	p. 19	p. 30					Transparency 2-8
2-9 (60-61)		p. 20	p. 22	p. 20					9	Transparency 2-9
Ch. Review (62-63)									6-8	Test and Review Generator
Ch. Test (64)										
Cumulative Review/Test (66-67)									10	

OTHER CHAPTER RESOURCES

Student Edition
Chapter Opener, p. 40
Activity, p. 41
Computer Application, p. 50
On-the-Job Application, p. 51
Journal Entry, p. 59
Portfolio Suggestion, p. 61
Consumer Connections, p. 65

Teacher's Classroom Resources
Transparency 2-0
Lab Manual, pp. 25-26
Performance Assesment, pp. 3-4
Lesson Plans, pp. 14-22

Other Supplements
Overhead Manipulative Resources, Labs 5-6, p. 3
Glencoe Mathematics Professional Series

ENHANCING the CHAPTER

Some of the blackline masters for enhancing this chapter are shown below.

COOPERATIVE LEARNING

Provide each student with a 10 x 10 grid. Have them make and color six designs within the grid. Suggest the students' designs should result from a number sequence. Have them record the decimal value of each design and the pattern used to generate each design. Have students exchange drawings and find each others' decimal values and number pattern.

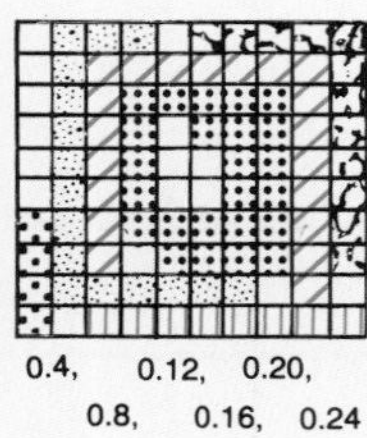

USING MODELS/MANIPULATIVES

Have students use play money in bills and coins to model the living costs and economy of the real world. Let one student be an employer who pays a salary to several student workers. Let others represent a landlord, a utility company, a phone company, a bank holding auto loans, and so on. Discuss reasonable salaries and costs to provide a valid simulation. Have students work with decimal amounts by adding, subtracting, and budgeting to simulate real-world transactions. If you have insufficient play bills, you may choose to reduce all amounts by a factor of ten.

Cooperative Problem Solving, p. 30

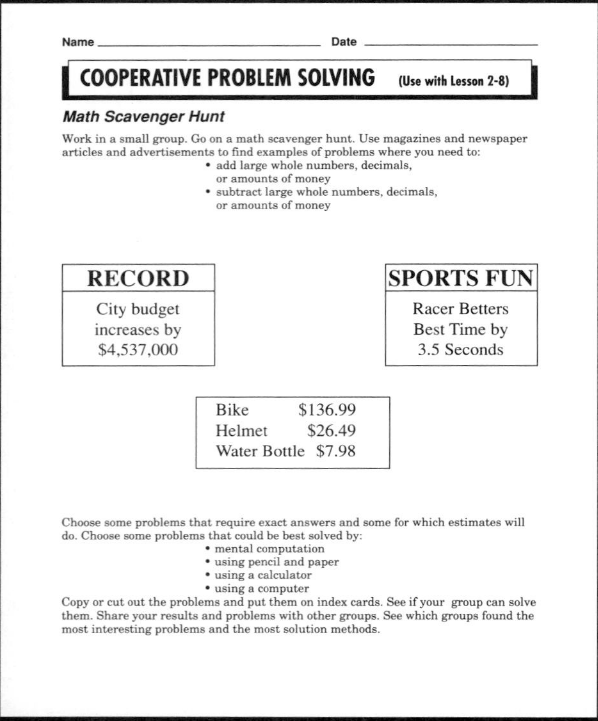

Name ______________ Date ______________

COOPERATIVE PROBLEM SOLVING (Use with Lesson 2-8)

Math Scavenger Hunt

Work in a small group. Go on a math scavenger hunt. Use magazines and newspaper articles and advertisements to find examples of problems where you need to:

- add large whole numbers, decimals, or amounts of money
- subtract large whole numbers, decimals, or amounts of money

RECORD
City budget increases by $4,537,000

SPORTS FUN
Racer Betters Best Time by 3.5 Seconds

Bike	$136.99
Helmet	$26.49
Water Bottle	$7.98

Choose some problems that require exact answers and some for which estimates will do. Choose some problems that could be best solved by:

- mental computation
- using pencil and paper
- using a calculator
- using a computer

Copy or cut out the problems and put them on index cards. See if your group can solve them. Share your results and problems with other groups. See which groups found the most interesting problems and the most solution methods.

Multicultural Activities, p. 2

Name ______________ Date ______________

MULTICULTURAL ACTIVITY (Use with Lesson 2-3)

Calendar Math

The calendar we use today is known as the Gregorian calendar. Although calendars from many different countries influenced the Gregorian calendar, Babylonian, Chinese, Egyptian, Athenian, and Roman calendars had the greatest influence on our present-day calendar.

Early Roman calendars contained so many errors that by Julius Caesar's time, the calendar was 80 days out of its astronomical place. So, he decreed that the year 46 B.C. should have 445 days! Thereafter, he decreed, all years should have 365 days, with a leap year every 4 years. However, each year in Caesar's calendar was actually about 11 minutes and 14 seconds too short. In 1752, this problem was corrected. In that year, the day following September 1 was declared to be September 13!

Our calendar will require no adjusting for thousands of years. Use it to solve the following problems.

1. If four days from tomorrow is Wednesday, what day was it the day before yesterday? Wednesday
2. If the first Monday in the month of May is the first prime number, what is the date of the last Monday in May? May 30
3. If tomorrow's date is the third odd number, what will the date be a week from today? the 11th
4. If three days before yesterday was Sunday, what day will it be a week from tomorrow? Friday
5. How many months have only 28 days? How many have only 30 days? How many have 31 days? some Februarys; 4; 7
6. Work with a partner. Make up calendar problems similar to those on this page for each other to solve. Answers may vary.

Research

7. Other countries use different calendars today. Find a different calendar from a country of your choice and report to the class on its similarities and differences.

MEETING INDIVIDUAL NEEDS

Mainstreaming

Learning-disabled students may have trouble adding and subtracting with decimals because the decimal point and the fractional parts introduce new potentials for misaligning the digits in the problem. For each addition and subtraction problem, have students draw a grid into which they can write the digits of the numbers and decimal points correctly. Make sure that the decimal points are aligned in the same column.

CRITICAL THINKING/PROBLEM SOLVING

When a student is confronted with a problem and says, "I don't know what to do," this usually indicates that the student does not understand the problem. Here are several suggestions for helping students to understand a problem.

- Ask leading questions that focus attention on the facts.
- Remind students that they may have solved a similar problem.
- Have students explain the problem in their own words.
- Draw a picture to model the action in a problem.

COMMUNICATION

Writing

Have students review the exercises from each lesson that they have completed in this chapter. Have them choose from the lessons exercises that are best solved by estimation, mental computation, a calculator, paper and pencil, and a computer. Record the exercises in a journal, identify the preferred method of computation, and give a reason for your choices.

Applications, p.3

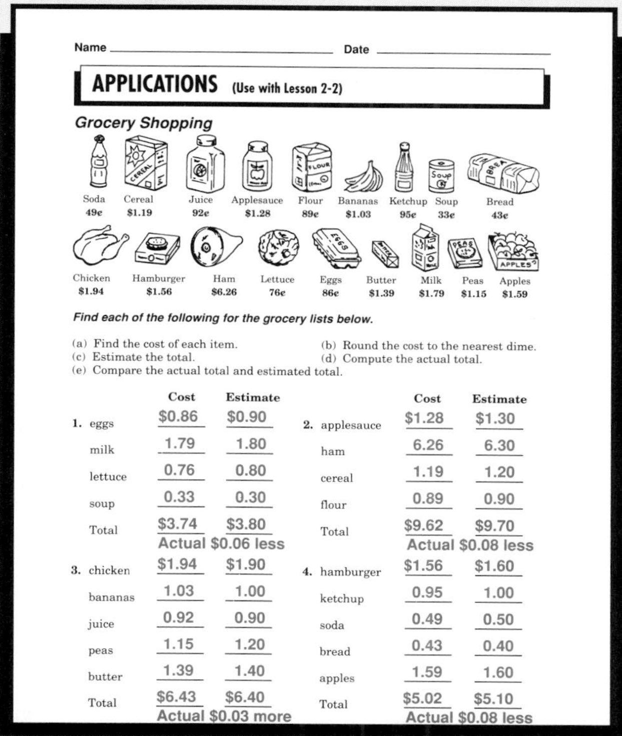

Name ______________ Date ______________

APPLICATIONS (Use with Lesson 2-2)

Grocery Shopping

Soda	Cereal	Juice	Applesauce	Flour	Bananas	Ketchup	Soup	Bread
49¢	$1.19	92¢	$1.28	89¢	$1.03	95¢	33¢	43¢

Chicken	Hamburger	Ham	Lettuce	Eggs	Butter	Milk	Peas	Apples
$1.94	$1.56	$6.26	76¢	86¢	$1.39	$1.79	$1.15	$1.59

Find each of the following for the grocery lists below.

(a) Find the cost of each item.
(b) Round the cost to the nearest dime.
(c) Estimate the total.
(d) Compute the actual total.
(e) Compare the actual total and estimated total.

	Cost	Estimate		Cost	Estimate
1. eggs	$0.86	$0.90	2. applesauce	$1.28	$1.30
milk	1.79	1.80	ham	6.26	6.30
lettuce	0.76	0.80	cereal	1.19	1.20
soup	0.33	0.30	flour	0.89	0.90
Total	$3.74	$3.80	Total	$9.62	$9.70
	Actual $0.06 less			Actual $0.08 less	
3. chicken	$1.94	$1.90	4. hamburger	$1.56	$1.60
bananas	1.03	1.00	ketchup	0.95	1.00
juice	0.92	0.90	soda	0.49	0.50
peas	1.15	1.20	bread	0.43	0.40
butter	1.39	1.40	apples	1.59	1.60
Total	$6.43	$6.40	Total	$5.02	$5.10
	Actual $0.03 more			Actual $0.08 less	

Computer Activity, p.16

Name ______________ Date ______________

COMPUTER ACTIVITY (Use with Lesson 2-7)

Using a Spreadsheet

A **spreadsheet** program can help to make tables. The spreadsheet allows the operator to enter information into **cells** that are organized in rows and columns. Most spreadsheets name columns with letters of the alphabet (A, B, C, ...) and rows with numbers (1, 2, 3, ...). An individual cell can be located by naming its column and then its row. For example, C12 names the cell in column 3, row 12. Z99 names the cell in column 26, row 99.

The diagram below shows the upper-left corner of a spreadsheet.

	A	B	C
1	Tickets	Price	Income
2			
3	160	$4.50	$720.00
4	170	4.50	765.00
5	1800	4.50	8,100.00
6	190	4.50	855.00
7	200	4.50	900.00
8	210	4.50	945.00
9	220	4.50	990.00
10	230	4.50	1,035.00
11	240	4.50	1,080.00
12	250	4.50	1,125.00

The operator has made a table that shows how much income a movie theater will take in when a certain number of tickets are sold.

The operator enters the price of the tickets and the number of tickets sold. The program then computes the total amount based on the formula that the operator enters.

Use the spreadsheet to answer.

1. Which number in column A seems to be an input error?
   1800
2. Which row shows the result for 220 paying customers?
   row 9
3. Which number in column C seems to be a formula error? (Computers cannot be expected to give correct information if the operator enters poor data or incorrect formulas.)
   $8,100.00
4. What is the entry in cell A3?
   160
5. Why do you think the operator stopped at 250 people?
   The theater's capacity is 250.
6. How much income do you think the theater would take in if 205 people bought tickets?
   $922.50

CHAPTER PROJECT

Have the class plan a budget for a school carnival. Allow the students to make a list of attractions and necessary items, including food and drink, and assign a cost to each item. Have them design a layout for the carnival given a shape and the total perimeter of the grounds. Students should estimate attendance and revenues from ticket sales, attractions, and food sales.

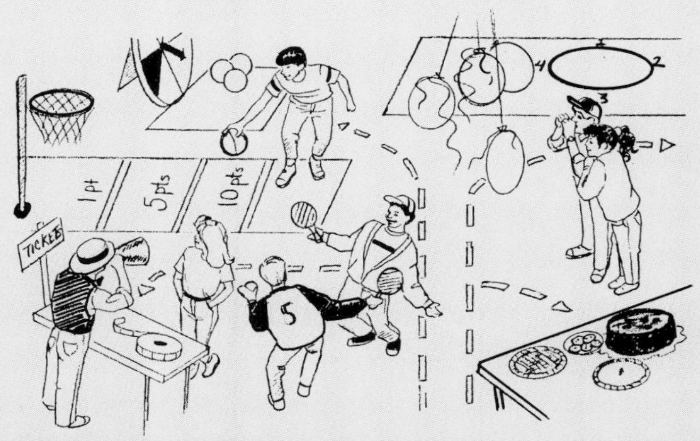

Using the Chapter Opener

Transparency 2-0 is available in the Transparency Package. It provides a full-color visual and motivational activity that you can use to engage students in the mathematical content of the chapter.

Using Questioning Have students read the opening paragraph.

- What kinds of transportation are available in this city (town)?
- How many of you have lived in other cities or towns before this one?
- What kinds of transportation were available where you last lived?
- If you had a choice, what kind of transportation would you choose to get to school or work? Why?

The second paragraph is a motivational problem. Have students try to solve the problem and discuss how they arrived at the answer.

The problem can be extended by asking the students to find the round trip distance.

You may want to include a discussion of the many areas in daily life that involve decimals.

Chapter 2

PATTERNS: ADDING AND SUBTRACTING

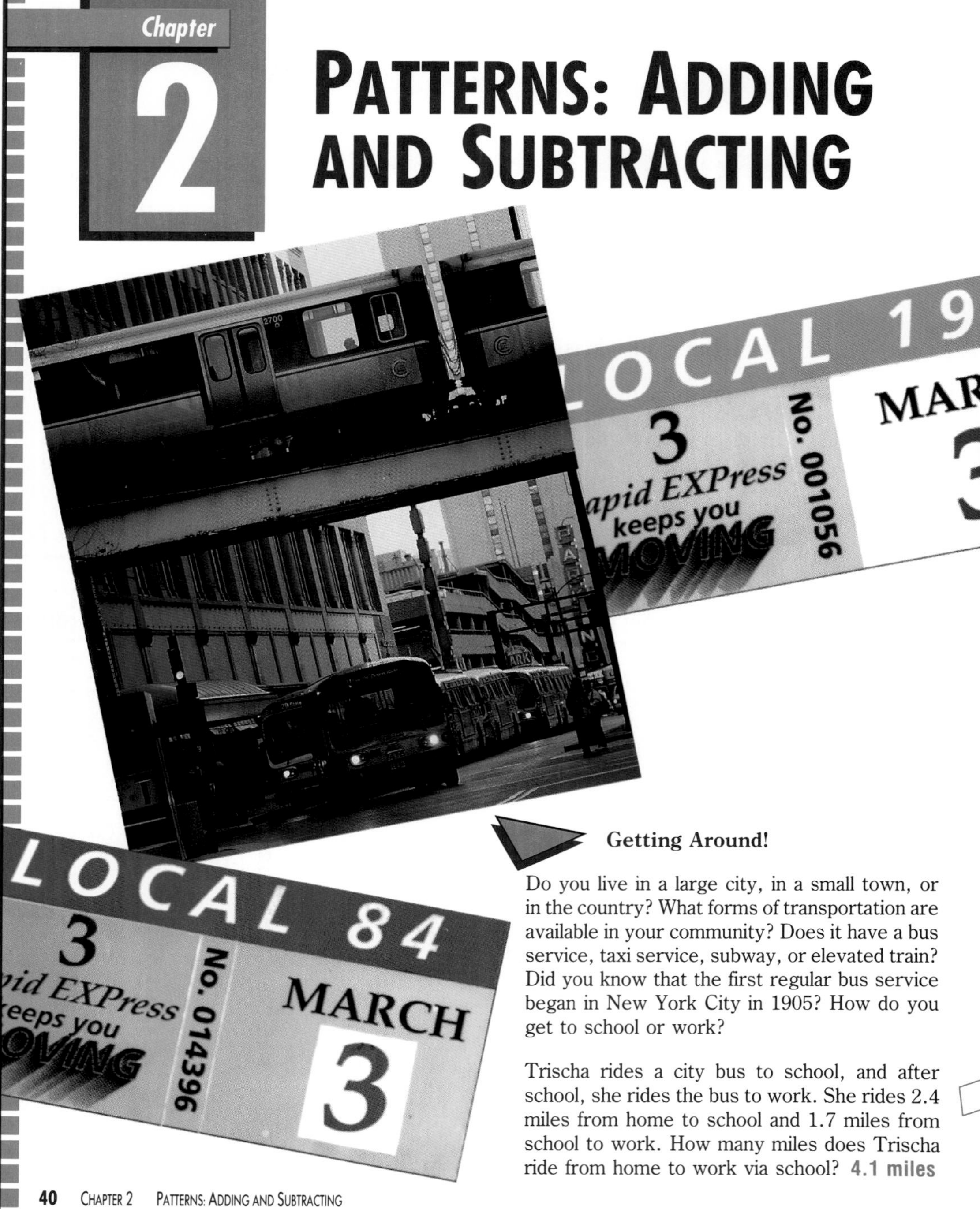

Getting Around!

Do you live in a large city, in a small town, or in the country? What forms of transportation are available in your community? Does it have a bus service, taxi service, subway, or elevated train? Did you know that the first regular bus service began in New York City in 1905? How do you get to school or work?

Trischa rides a city bus to school, and after school, she rides the bus to work. She rides 2.4 miles from home to school and 1.7 miles from school to work. How many miles does Trischa ride from home to work via school? **4.1 miles**

ACTIVITY: Exploring Decimals with Models

You can use decimal models to find the sum of decimals. Suppose you want to find the sum of 0.7 and 0.6.

Materials: decimal models (tenths and hundredths), pencil

1. Use two decimal models that are separated into tenths.
2. On the first decimal model, shade 7 tenths as shown.
3. On the second decimal model, shade 6 tenths as shown.

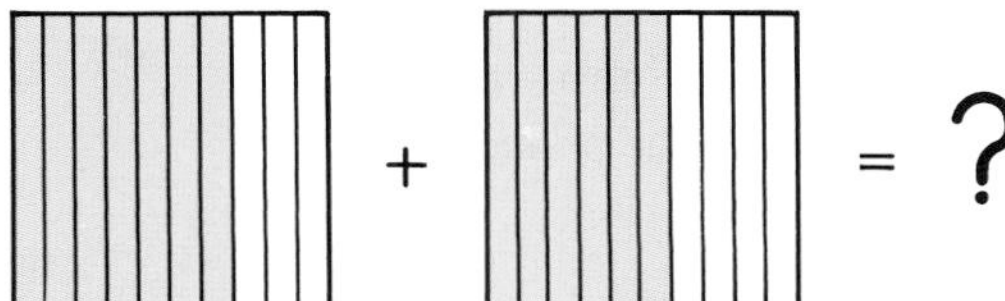

Cooperative Groups

Work together in groups of two or three.

4. Determine a way to use decimal models and shading to show the sum of 0.7 and 0.6. What is the sum? **See margin.**
5. Use decimal models to show the sum of 0.63 and 0.45. Discuss with your group how to represent the problem and the sum on decimal models. Explain how you arrived at the sum using decimal models. **See margin.**
6. Using decimal models, how would you find the sum of 0.3 and 0.09? **See margin.**
7. How could you show subtraction of decimals using decimal models? Use 0.7 − 0.4 as an example. **See margin.**

Communicate Your Ideas

8. Work with your group to draw a conclusion about how to add decimals without decimal models. Describe what happens to the decimal point. **See margin.**
9. What determines the type of decimal model, tenths, hundredths, and so on, you need to model a problem? **See margin.**

You will study more about adding and subtracting decimals in this chapter.

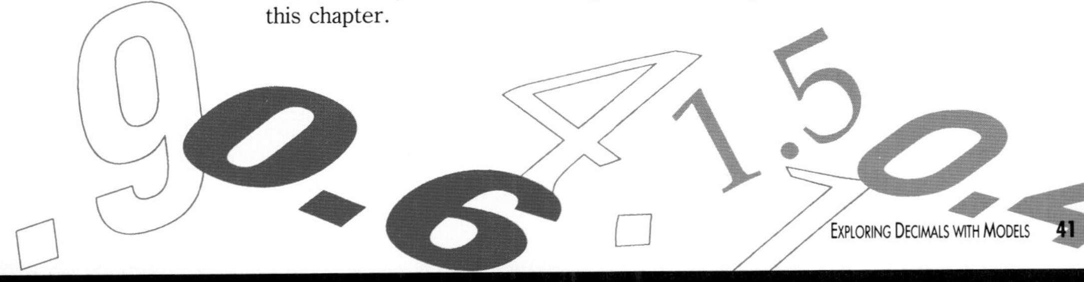

INTRODUCING THE LESSON

Objective Add and subtract decimals using models.

TEACHING THE LESSON

Using Models You may want to explain what the sections of a decimal model represent. Copy the decimal models in the **Lab Manual, p. 2.** Encourage the students to shade sections accurately and neatly to prevent errors.

EVALUATING THE LESSON

Communicate Your Ideas Have students make up an addition problem using two decimal addends and show the solution using decimal models.

EXTENDING THE LESSON

Closing the Lesson Allow groups time to discuss their answers with other groups. Students will study more about adding and subtracting decimals in this chapter.

Activity Worksheet, Lab Manual, p. 25

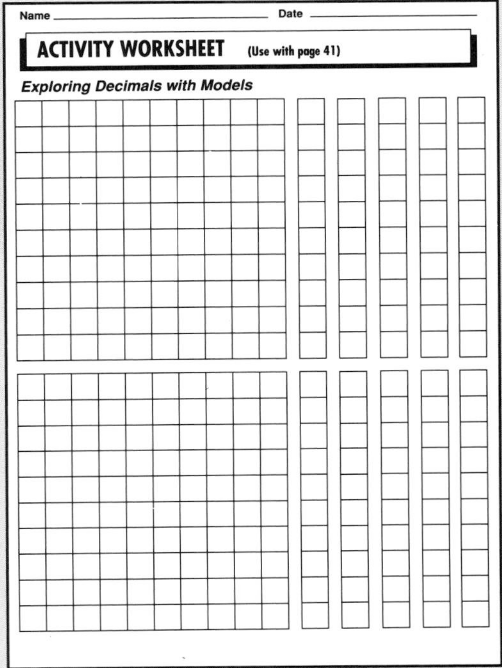
Name ______ Date ______

ACTIVITY WORKSHEET (Use with page 41)

Exploring Decimals with Models

Additional Answers

4. Count the number of tenths that are shaded on both decimal models. Since 13 are shaded, shade 10 tenths and 3 tenths on two decimal models.; 1.3
5. Use hundredths decimal models. Shade 63 on one model and 45 on the other. Since 108 hundredths are shaded, shade 100 hundredths and 8 hundredths on the second decimal models.
6. Since 3 tenths is the same as 30 hundredths, use hundredths decimal models. The sum is 39 hundredths.
7. Draw a box around 7 tenths on one decimal model. Shade 4 tenths in the box. The number of tenths left is the difference. 0.3
8. The decimal point in the answer aligns with the decimal point in the addends.
9. The largest number of decimal places determines the type of model to use.

2-1 ADDING WHOLE NUMBERS

Objective *Add whole numbers.*

At Rushmore High School, 33 students have signed up in the first week for volleyball intramurals. If there is room for 35 more students, how many will be able to play volleyball intramurals? To find the total number of students, add.

To add two or more whole numbers, you should begin at the right and add the numbers in each place-value position.

Method

1. Align the numbers by place-value position.
2. Add in each place-value position from right to left.

Example A ***33 + 35*** Estimate: 30 + 40 = 70

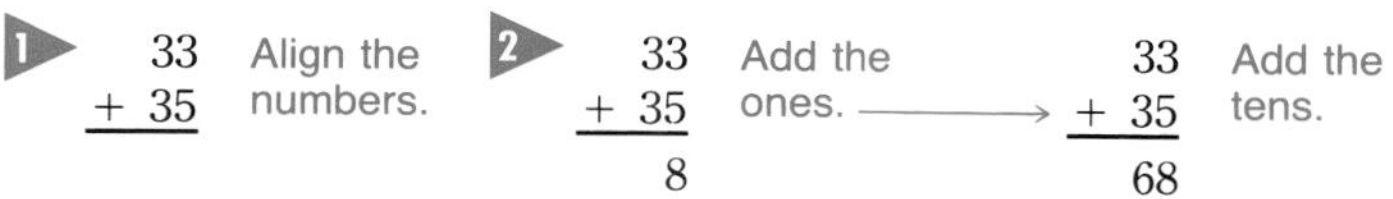

Compared to the estimate, is the answer reasonable?

There can be 68 students in Rushmore intramurals.

Example B ***249 + 28*** Estimate: 250 + 30 = 280

1: 249 + 28

2: 249 + 28 = 7 (carry 1) Write 17 ones as 1 ten and 7 ones. → 249 + 28 = 77 → 249 + 28 = 277

Is the answer reasonable?

Example C ***7,453 + 2,865*** Estimate: 7,000 + 3,000 = 10,000

1: 7,453 + 2,865

2: 7,453 + 2,865 = 10,318 (carry 1 1)

Based on the estimate, explain why the answer is reasonable.

Guided Practice

Add. **Encourage students to estimate each answer first.**

Example A

1. 43 + 36 = **79**
2. 20 + 18 = **38**
3. 81 + 17 = **98**
4. 45 + 83 = **128**
5. 62 + 75 = **137**

Example B

6. 56 + 863 **919**
7. 94 + 786 **880**
8. 598 + 75 **673**
9. 545 + 57 **602**

Example C

10. 4,726 + 28,538 **33,264**
11. 270 + 46,090 **46,360**
12. 76,118 + 768 **76,886**

5-MINUTE CHECK-UP

(over Lesson 1-13)

Available on TRANSPARENCY 2-1.

Write an equation and solve.

1. After climbing 110 ft, a hiker climbed 15 ft per half-hour for 3 hours. How many feet did he climb in all? **200 feet**

INTRODUCING THE LESSON

Ask students how they would describe adding 23 + 4 to a younger child. Ask them to describe adding 45 + 8.

TEACHING THE LESSON

Using Discussion Have students reverse the order of the addends in Example B and add. Discuss the use of the Commutative Property as a method of checking sums.

Practice Masters Booklet, p. 14

PRACTICE WORKSHEET 2-1

Adding Whole Numbers

Add.

1. 48 + 6 = **54**
2. 24 + 69 = **93**
3. 487 + 215 = **702**
4. 396 + 77 = **473**
5. 543 + 8 = **551**
6. 936 + 4,173 = **5,109**
7. 8,318 + 1,236 = **9,554**
8. 3,807 + 4,524 = **8,331**
9. 7,565 + 936 = **8,501**
10. 48 + 6,154 = **6,202**
11. 1,346 + 56,672 = **58,018**
12. 38,416 + 2,618 = **41,034**
13. 73,281 + 15,765 = **89,046**
14. 797 + 39,044 = **39,841**
15. 43,167 + 68 = **43,235**
16. 5,783 + 1,764 **7,547**
17. 5,383 + 279 **5,662**
18. 3,088 + 25,921 **29,009**
19. 52,675 + 3,678 **56,353**
20. 493 + 38,267 **38,760**
21. 72,786 + 45,439 **118,225**

Solve. Use the chart.

22. How many immigrants came to the U.S. during 1921–1940? **4,635,640 immigrants**
23. How many immigrants came to the U.S. during 1941–1960? **3,550,518 immigrants**
24. How many immigrants came to the U.S. during 1961–1980? **7,815,091 immigrants**

Immigrants to the U.S. from all countries	
1921–1930	4,107,209
1931–1940	528,431
1941–1950	1,035,039
1951–1960	2,515,479
1961–1970	3,321,777
1971–1980	4,493,314

RETEACHING THE LESSON

Use base-10 blocks with students having difficulty. Work with these exercises.

1. 275 + 187 = **462**
2. 226 + 82 = **308**
3. 706 + 945 = **1,651**
4. 161 + 159 = **320**

Reteaching Masters Booklet, p. 13

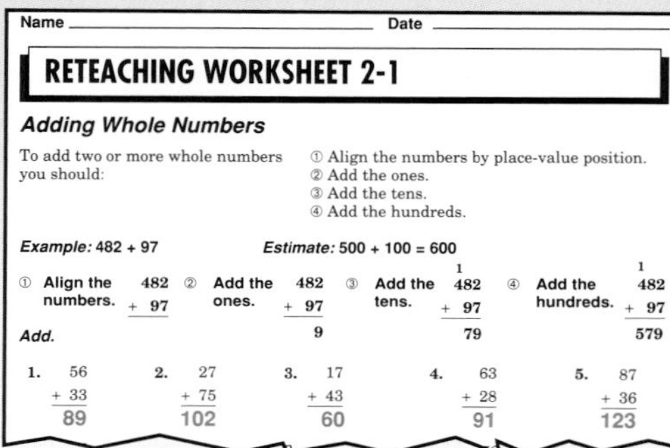

Name ____________ Date ____________

RETEACHING WORKSHEET 2-1

Adding Whole Numbers

To add two or more whole numbers you should:

① Align the numbers by place-value position.
② Add the ones.
③ Add the tens.
④ Add the hundreds.

Example: 482 + 97 *Estimate:* 500 + 100 = 600

① Align the numbers. 482 + 97
② Add the ones. 482 + 97 = 9
③ Add the tens. 482 + 97 = 79 (carry 1)
④ Add the hundreds. 482 + 97 = 579 (carry 1)

Add.

1. 56 + 33 = **89**
2. 27 + 75 = **102**
3. 17 + 43 = **60**
4. 63 + 28 = **91**
5. 87 + 36 = **123**

Exercises

Practice

Add. **Encourage students to estimate each answer first.**

13. 25 + 71 **96**
14. 32 + 45 **77**
15. 53 + 30 **83**
16. 792 + 25 **817**
17. 847 + 990 **1,837**

18. 2,914 + 3,087 **6,001**
19. 4,803 + 297 **5,100**
20. 3,408 + 5,697 **9,105**
21. 62,924 + 6,085 **69,009**
22. 2,816 + 13,847 **16,663**

23. 408 + 694 **1,102**
24. 23,088 + 257
25. 94,670 + 69,033

26. Add 23 and 257. **280**

24. 23,345 **25. 163,703**

27. Find the sum of 4,715 and 14,369. **19,084**

Applications

28. The annual budget for the city of Circleville is $985,000. The voters approved a $53,000 increase in the budget. What is the amount of the increased budget? **$1,038,000**

29. John drove 238 kilometers from London, England, to Liverpool, England. Then he drove 52 kilometers from Liverpool to Manchester. How many kilometers did he drive in all? **290 km**

30. Simsbury has three communities. Tariffville has 3,509 people, Weatogue has 2,878 people, and West Simsbury has 23,644 people. What is the total population of Simsbury? **30,031 people**

31. Lin Ye has $420 in her savings account. She deposits $35. What is her new balance? **$455**

Critical Thinking

32. Fill in the appropriate whole numbers in the magic square so that each row, column, and diagonal add up to 39.

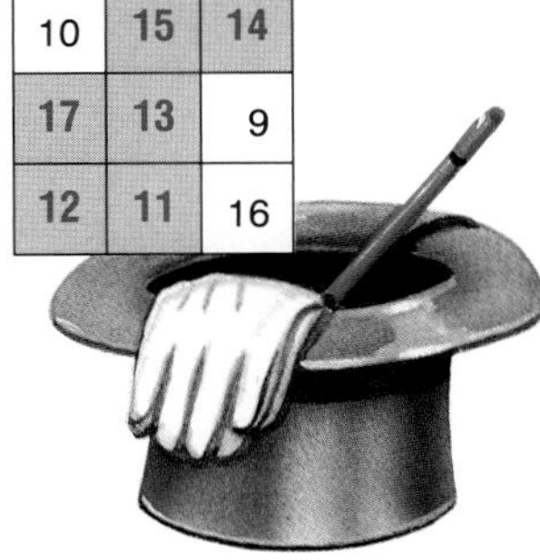

Cooperative Groups

33. In groups of two or three, make up five addition problems. Each problem must have a sum of 1,250. At least one of the problems must be a word problem. Exchange your problems with other groups to check that the problems are correct. **Answers may vary.**

Mixed Review

Lesson 1-9

Find the value of each expression.

34. $5 + 6 - 3 \times 2$ **5**
35. $16 \div 4 + 8 \times 3$ **28**
36. $6 \times (4 \div 2) + 8$ **20**
37. $3 + 7 \times 2 - 9$ **8**

Lesson 1-11

Solve each equation. Check your solution.

38. $x + 7 = 41$ **34**
39. $13 + b = 22$ **9**
40. $52 = m - 30$ **82**

Lesson 1-13

41. Angelo found the price of pizzas at three competing pizza shops. Shop #1 has a one-item pizza for $7.95. Shop #2 has the same pizza for $8.50, and shop #3 has the pizza for $10.25. What is the difference in price between the most expensive and least expensive? **$2.30**

APPLYING THE LESSON

Independent Practice

Homework Assignment	
Minimum	Ex. 1-41 odd
Average	Ex. 2-26 even; 28-41
Maximum	Ex. 1-41

Chalkboard Examples

Example A *Add.*

1. 42 + 26 **68** **2.** 15 + 61 **76**

Example B *Add.*

3. 136 + 28 **164** **4.** 527 + 46 **573**

Example C *Add.*

5. 6,822 + 1,393 **8,215**

6. 4,961 + 3,578 **8,539**

Additional examples are provided on TRANSPARENCY 2-1.

EVALUATING THE LESSON

Check for Understanding Use Guided Practice Exercises to check for student understanding.

Error Analysis Watch for students who align addends on the left rather than on the right. Remind students to match place-value positions.

Closing the Lesson Have students describe in their own words how to align a four-digit number and a two-digit number for addition.

EXTENDING THE LESSON

Enrichment Masters Booklet, p. 13

Name ______________________ Date ______________

ENRICHMENT WORKSHEET 2-1

Having Fun with Addition

The table below has a special arrangement of numbers that could let you amaze your friends and relatives. To make the table work, complete the following steps.

A "Patriotic" Puzzle

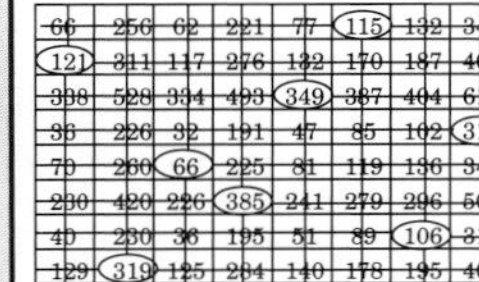

66	256	62	221	77	115	132	345
121	311	117	276	132	170	187	400
338	528	334	493	349	387	404	617
36	226	32	191	47	85	102	315
70	260	66	225	81	119	136	349
230	420	226	385	241	279	296	509
40	230	36	195	51	89	106	319
129	319	125	284	140	178	195	408

Step 1 Circle any number in the table. Cross out all of the other numbers that are in the same row (↔) and column (↕).

Step 2 Circle any other number that is not crossed out. Again cross out all of the numbers that are in the same row and column.

Step 3 Repeat until 8 numbers have been chosen. Add the circled numbers. The total should be 1,776.

115 + 121 + 349 + 315 + 66 + 385 + 106 + 319 = 1,776

No matter what 8 numbers are chosen from this table, the total will be the same.

Find the total of the table below. Can you find the "secret" of the table?

A "Futuristic" Puzzle

90	280	86	250	106	144	161	374
160	350	156	320	176	214	231	444
371	561	367	531	387	425	442	655
65	255	61	225	81	119	136	349
93	283	89	253	109	147	164	377
264	454	260	424	280	318	335	548
52	242	48	212	68	106	123	336
135	325	131	295	151	189	206	419

Total: 2,001

To construct a table like these:

1. Choose a total.
2. Find 16 numbers whose sum is that total.
3. Set up a table. 8 numbers above the first row and next to the first column.
4. Add the numbers as you would in an addition table.
5. Once you complete the table delete the original row and column.

5-MINUTE CHECK-UP

(over Lesson 2-1)

Available on TRANSPARENCY 2-2.

Add.

1. 34 + 48 **82**
2. 124 + 85 **209**
3. 3,461 + 673 **4,134**
4. 4,789 + 3,228 **8,017**
5. 236 + 457 + 2,915 **3,608**

Extra Practice, Lesson 2-1, p. 451

INTRODUCING THE LESSON

In a newspaper or magazine, have students find instances where decimals appear. Ask students to suggest situations that involve decimals.

TEACHING THE LESSON

Using Questioning

- In a whole number, where is the decimal point placed? **after the ones digit**

Practice Masters Booklet, p. 15

Name ________________ Date ________

PRACTICE WORKSHEET 2-2

Adding Decimals

Add.

1. 0.3 + 0.6 = **0.9**	2. 0.4 + 0.7 = **1.1**	3. 0.7 + 0.2 = **0.9**	4. 0.9 + 0.8 = **1.7**
5. 7.3 + 1.8 = **9.1**	6. 3.4 + 2.5 = **5.9**	7. 5.6 + 1.2 = **6.8**	8. 7.4 + 4.7 = **12.1**
9. 0.62 + 0.4 = **1.02**	10. 0.13 + 0.9 = **1.03**	11. $0.78 + 0.43 = **$1.21**	12. $0.98 + 0.12 = **$1.10**
13. 2.076 + 1.34 = **3.416**	14. 3.7 + 1.609 = **5.309**	15. 5.8 + 14.312 = **20.112**	16. 12.006 + 4.3 = **16.306**
17. 3.6 + 4.9 + 21.4 = **29.9**	18. $4.78 + 6.20 + 5.16 = **$16.14**	19. 0.1 + 4.08 + 19.164 = **23.344**	20. 9.1 + 22.006 + 40.07 = **71.176**

21. $16.49 + $26 **$42.49** 22. 7.439 + 0.88 **8.319** 23. 0.564 + 19.7 **20.264**

24. 0.976 + 23.4 **24.376** 25. $58 + 36¢ **$58.36** 26. 6.325 + 29 **35.325**

Solve.

27. Nam's speed-reading times are 2.1 minutes, 2.06 minutes, 1.98 minutes, and 1.9 minutes. What is his total reading time? **8.04 minutes**

28. Tina orders a hamburger for $1.59, salad for $0.89, and juice for $0.69. What is her total bill? **$3.17**

2-2 ADDING DECIMALS

Objective
Add decimals.

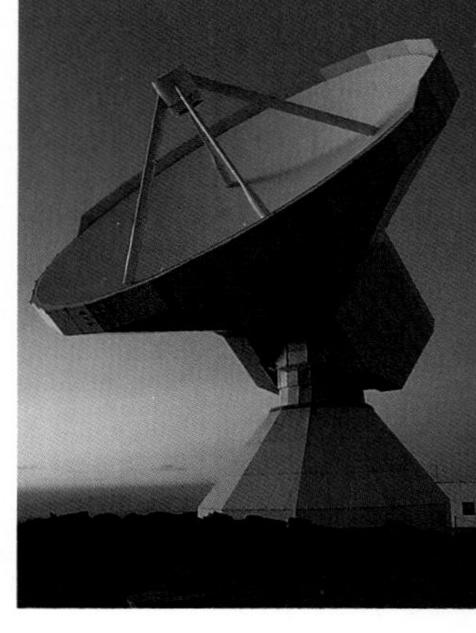

George Thompkin monitors transmitting equipment at Station 3 for Satellite Systems, Inc. Data beamed from Station 3 to a satellite takes 0.27 seconds. Data beamed back to Station 4 from the satellite takes 0.41 seconds. What is the total time it takes for data to travel from Station 3 to Station 4 through the satellite?

To add two or more decimals, begin at the right and add the numbers in each place-value position.

Method

1. Align the decimal points.
2. Add in each place-value position from right to left.

Example A

0.41 + 0.27 Estimate: 0.4 + 0.3 = 0.7

1. 0.41 + 0.27 — Align the decimal points.
2. 0.41 + 0.27 = 8 — Add the hundredths.
   0.41 + 0.27 = 0.68 — Add the tenths.

Based on the estimate, 0.7, explain why the answer is reasonable.

The data takes 0.68 seconds to travel from Station 3 to Station 4.

Example B

9.38 + 15 + 4.701 Estimate: 9 + 15 + 5 = 29

1. 9.380 + 15.000 + 4.701 — Align the decimal points. Annex zeros if you need to so each addend has the same number of decimal places.
2. 9.380 + 15.000 + 4.701 = 29.081

Is the answer reasonable?

Guided Practice

Add. **Encourage students to estimate the answer first.**

Example A

1. 0.85 + 0.12 = **0.97** **2.** 7.946 + 0.032 = **7.978** **3.** 26.2 + 3.9 = **30.1** **4.** 15.24 + 9.18 + 6.37 = **30.79**

Example B

5. 48.6 + 9.31 = **57.91** **6.** 76.7 + 15.678 = **92.378** **7.** 0.33 + 21.7 = **22.03** **8.** 126.2 + 71.08 = **197.28**

9. 1.56 + 7.2 + 3.73 **12.49** **10.** 5.98 + 12.4 + 3 **21.38**
11. 16.78 + 924 + 7.01 **947.79** **12.** 21,195 + 18.2 + 46 **21,259.2**

RETEACHING THE LESSON

Write the following whole numbers as decimals.

1. 8 **8.0** **2.** 10 **10.0** **3.** 23 **23.0**

Have students write whole numbers as a decimal, align decimal points, annex zeros, then add.

4. 9 + 1.7 **10.7** **5.** 15.1 + 6.34 **21.44**
6. 46 + 0.493 **46.493**
7. 23 + 0.09 + 6.8 **29.89**
8. 2.06 + 7.8 + 24 **33.86**
9. 0.66 + 17 + 8.7 **26.36**

Reteaching Masters Booklet, p. 14

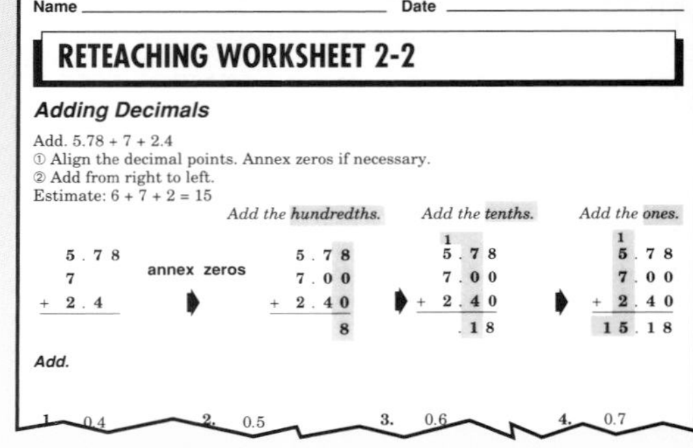
Name ________________ Date ________

RETEACHING WORKSHEET 2-2

Adding Decimals

Add. 5.78 + 7 + 2.4
① Align the decimal points. Annex zeros if necessary.
② Add from right to left.
Estimate: 6 + 7 + 2 = 15

5.78 + 7 + 2.4 → annex zeros → 5.78 + 7.00 + 2.40
Add the hundredths. 8
Add the tenths. .18
Add the ones. 15.18

Add.

1. 0.4 2. 0.5 3. 0.6 4. 0.7

Exercises

Practice

Add. Encourage students to estimate the answer first.

13. 0.23 + 0.18 **0.41**
14. 0.94 + 1.23 **2.17**
15. 0.81 + 1.9 **2.71**
16. 65.3 + 4.91 **70.21**
17. \$0.43 + 0.97 + 0.60 **\$2.00**
18. \$0.38 + 1.42 + 0.66 **\$2.46**
19. \$45.63 + 5.96 **\$51.59**
20. 0.07 + 0.19 + 1.26 **1.52**
21. 0.275 + 1.38 + 6.91 **8.565**
22. 2.05 + 0.156 + 0.071 **2.277**
23. \$355.92 + 105.17 + 481.37 **\$942.46**

24. 5.206
25. \$18.13
26. 8.33

24. 2.05 + 0.156 + 3
25. \$3.51 + \$14.62
26. 0.23 + 0.1 + 8
27. Find the sum of 0.036, 0.15, and 1.7. **1.886**
28. What is the total of 490, 7.3, and 26.946? **524.246**

Applications

29. The four members of a 400-meter relay team had times of 11.26 seconds, 11.07 seconds, 11.03 seconds, and 10.43 seconds. What was the total time for the relay team? **43.79 seconds**
30. Pat bought a tennis racket for \$84.98, a pair of shorts for \$12.67, a pair of socks for \$2.49, and a visor for \$5.39. What was Pat's total bill not including sales tax? **\$105.53**
31. Sarah's bank pays 5.1% interest on savings accounts. Since Sarah's balance is greater than \$1,500, she receives an extra 0.3% interest. What is the total percent of interest Sarah receives? **5.4%**

Using Variables

Write an expression for each phrase.

32. the sum of a number and 43 **$a + 43$**
33. 0.12 more than a number **$t + 0.12$**

Make Up a Problem

34. Make up a problem using decimals for which the sum is 23.46. Use three addends. **Answers may vary.**
35. Make up a problem using four decimal addends that has a sum of 50. **Answers may vary.**

Estimation

Deciding whether a computation will be over or under a certain value is called *reference point* estimation. For example, can you buy a 77¢ drink and a 26¢ pack of gum for \$1.00? You know \$0.75 plus \$0.25 is \$1.00. Since the items are more than \$0.75 and \$0.25, their sum will be more than \$1.00. Therefore, you cannot buy both for \$1.00. The reference point is \$1.00.

36. 25 + 25 = 50, so 26 + 27 is __?__ than 50. (greater, less) **greater**
37. 100 − 40 = 60, so 100 − 45 is __?__ than 60. (greater, less) **less**
38. 3 × 50 = 150, so 3 × 55 is __?__ than 150. (greater, less) **greater**

APPLYING THE LESSON

Independent Practice

Homework Assignment	
Minimum	Ex. 1-37 odd
Average	Ex. 2-28 even; 29-38
Maximum	Ex. 1-38

Chalkboard Examples

Example A *Add.*

1. 0.32 + 0.47 **0.79**
2. 1.34 + 3.22 **4.56**
3. 1.2 + 4.5 + 2.1 **7.8**

Example B *Add.*

4. 40.2 + 9.9 **50.1**
5. 3.423 + 5.778 **9.201**
6. 12.32 + 58 + 3.294 **73.614**

Additional examples are provided on TRANSPARENCY 2-2.

EVALUATING THE LESSON

Check for Understanding Use Guided Practice Exercises to check for student understanding.

Error Analysis Watch for students who do not align addends at the decimal point. Remind students to annex zeros so that each number has the same number of decimal places.

Closing the Lesson Have students explain in their own words how to align and add decimals.

EXTENDING THE LESSON

Enrichment Masters Booklet, p. 14

ENRICHMENT WORKSHEET 2-2

Casting out Nines

In the ninth century, casting out nines was first used to check computations. To cast out nines from a number, add the digits. If the sum is greater than 9, continue adding the digits. For example, 34 → 3 + 4 = 7. But 3,879 → 3 + 8 + 7 + 9 = 27 and 27 → 2 + 7 = 9. When you cast out 9 from 9, you get 0.

Casting out nines can be used to check multiplication problems.

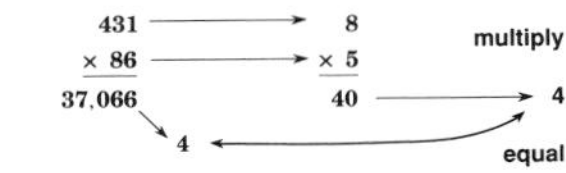

Since the check digits are equal, the answer 37,066 is *probably* correct. If the check digits were *not* equal, the answer would be incorrect.

In a similar way, casting out nines can be used to check addition problems.

814 → 4
+ 974 → + 2 (add)
1,788 → 6 (equal) 6

Cast out nines from the following numbers.

1. 40 **4**
2. 431 **8**
3. 86 **5**
4. 143 **8**
5. 860 **5**
6. 37,066 **4**
7. 9,754 **7**
8. 230,948 **8**

Check each answer by casting out nines.

9. 203 × 745 = 151,235 — **5 × 7 = 35 → 8; 8**
10. 48 + 75 = 123 — **3 + 3 = 6; 6**
11. 197 + 408 + 1,299 + 5,284 = 7,188 — **8 + 3 + 3 + 1 = 15 → 6; 6**
12. 5,481 × 3,794 = 20,794,914 — **0 × 5 = 0; 0**

13. Give an example of a problem with an incorrect answer that checks using casting out nines. **Answers will vary. Possible answer: 77 + 21 = 89; 5 + 3 ↔ 8**

5-MINUTE CHECK-UP

(over Lesson 2-2)

Available on TRANSPARENCY 2-3.

Add.

1. 2.1 + 3.6 + 4.2 **9.9**
2. 21.43 + 34.237 **55.667**
3. 34 + 1.2 + 3.57 **38.77**
4. \$1.29 + \$98.31 **\$99.60**
5. 12.45 + 2.521 + 8.2 + 2 **25.171**

Extra Practice, Lesson 2-2, p. 451

INTRODUCING THE LESSON

Write these items on the chalkboard: CD player \$250, video game system \$150, VCR \$200, television \$300. Ask students to imagine they have \$1,000 to spend. Ask students how much they will have left after buying the CD player, the VCR, and so on. You will study more about subtracting in this lesson.

TEACHING THE LESSON

Using Calculators Have students use calculators for the exercises in this lesson. Have students write an estimate before computing to determine if the answer is reasonable.

Practice Masters Booklet, p. 16

PRACTICE WORKSHEET 2-3

Subtracting Whole Numbers

Subtract.

1. 58 − 32 = **26**	2. 86 − 41 = **45**	3. 600 − 407 = **193**	4. 703 − 76 = **627**	5. 822 − 243 = **579**
6. 790 − 135 = **655**	7. 492 − 359 = **133**	8. 308 − 126 = **182**	9. 853 − 247 = **606**	10. 178 − 159 = **19**
11. 326 − 89 = **237**	12. 271 − 225 = **46**	13. 496 − 351 = **145**	14. 827 − 604 = **223**	15. 965 − 745 = **220**
16. 3,451 − 1,524 = **1,927**	17. 1,460 − 188 = **1,272**	18. 7,285 − 3,175 = **4,110**		
19. 24,002 − 16,126 = **7,876**	20. 5,100 − 3,405 = **1,695**	21. 47,316 − 27,213 = **20,103**		
22. 5,208 − 105 **5,103**	23. 3,642 − 2,234 **1,408**	24. 4,622 − 3,784 **838**		

Solve.

25. Maureen needs \$600 for the down payment on a computer. She has saved \$336 already. How much does she still need to save for the down payment? **\$264**

26. Yuri took a \$20 bill with him to the ball game. He paid \$7.50 at the gate for his ticket. He bought snacks for \$5.65 and a program for \$2.25. How much did he have left? **\$4.60**

2-3 SUBTRACTING WHOLE NUMBERS

Objective *Subtract whole numbers.*

Wilson Golf Course has a fleet of 94 golf carts. By 10:30 A.M., Helen's list shows that 37 carts have been rented. She subtracts to find out how many carts she has left to rent.

To subtract whole numbers, you should begin at the right and subtract the numbers in each place-value position. Sometimes you must rename before subtracting.

Method

1. Align the numbers by place-value position.
2. Subtract in each place-value position from right to left. Rename first if necessary.

Example A ***94 − 37*** Estimate: 90 − 40 = 50

1. 94 − 37 Align the numbers.
2. $\overset{8\,14}{94}$ − 37 = 7 Rename 9 tens as 8 tens and 10 ones. Subtract the ones. → 94 − 37 = 57 Subtract the tens.

Compared to the estimate, is the answer reasonable?

Helen has 57 carts left to rent.

Subtracting is the opposite of adding.

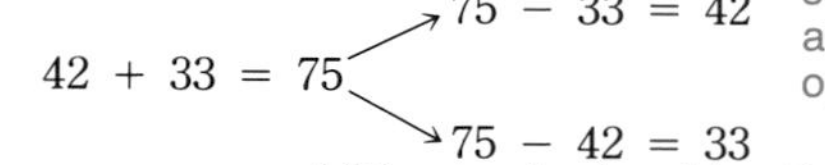

Addition and subtraction are inverse operations.

So you can use addition to check subtraction.

Example B ***601 − 378*** Estimate: 600 − 400 = 200

1. 601 − 378
2. $\overset{5\,9\,11}{601}$ − 378 = 3 Rename 60 tens as 59 tens and 10 ones. → 601 − 378 = 223

Check by using addition. 223 + 378 = 601 ✓

Based on the estimate, explain why the answer is reasonable.

Example C ***6,014 − 878*** Estimate: 6,000 − 900 = 5,100

1. 6,014 − 878
2. $\overset{5\,9\,10\,14}{6,014}$ − 878 = 5,136

Check by adding. 5,136 + 878 = 6,014 ✓ Is the answer reasonable?

Guided Practice *Examples A, B*

Subtract. **Encourage students to estimate each answer first.**

1. 72 − 47 **25**
2. 76 − 53 **23**
3. 67 − 28 **39**
4. 985 − 871 **114**
5. 563 − 455 **108**

RETEACHING THE LESSON

Using Cooperative Groups For students experiencing difficulty with regrouping, have manipulatives available, such as tongue depressors or base-ten blocks. Have students work easier problems using manipulatives to regroup and check their regrouping with other students.

Reteaching Masters Booklet, p. 15

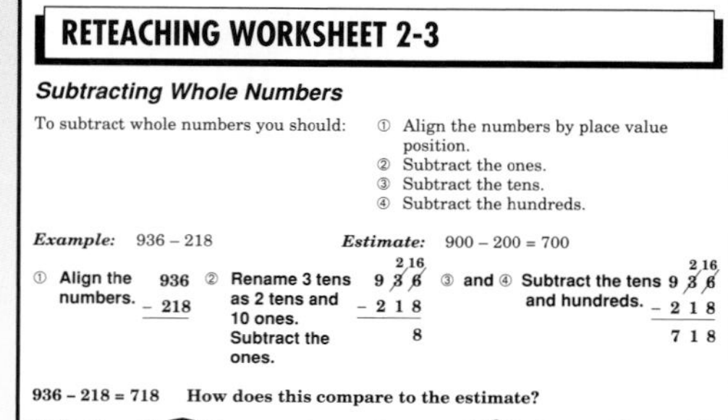

Name ______ Date ______

RETEACHING WORKSHEET 2-3

Subtracting Whole Numbers

To subtract whole numbers you should:

① Align the numbers by place value position.
② Subtract the ones.
③ Subtract the tens.
④ Subtract the hundreds.

Example: 936 − 218 *Estimate:* 900 − 200 = 700

① Align the numbers. 936 − 218

② Rename 3 tens as 2 tens and 10 ones. Subtract the ones. $9\overset{2}{3}\overset{16}{6}$ − 218 = 8

③ and ④ Subtract the tens and hundreds. $9\overset{2}{3}\overset{16}{6}$ − 218 = 718

936 − 218 = 718 **How does this compare to the estimate?**

Example C

6. 3,723 − 262 = **3,461**
7. 5,423 − 718 = **4,705**
8. 4,540 − 475 = **4,065**
9. 7,275 − 3,866 = **3,409**
10. 3,838 − 2,496 = **1,342**

Exercises

Practice

Subtract. **Encourage students to estimate each answer first.**

11. 55 − 23 = **32**
12. 95 − 43 = **52**
13. 88 − 65 = **23**
14. 800 − 506 = **294**
15. 503 − 85 = **418**
16. 29,006 − 19,137 = **9,869**
17. 4,100 − 3,308 = **792**
18. 6,209 − 108 = **6,101**
19. 47,603 − 37,501 = **10,102**
20. 2,506 − 496 = **2,010**

21. 3,402 − 2,324 **21. 1,078**
22. 3,016 − 2,305 **22. 711**
23. 4,557 − 3,593 **23. 964**
24. Subtract 1,661 from 23,208. **21,547**
25. Find the difference of 44,213 and 3,502. **40,711**

Applications

26. Kathy needs $800 for the down payment on a car. She has saved $476. How much does she still need to save? **$324**
27. The elevation at the top of a ski slope is 2,123 feet. The elevation at the bottom of the slope is 941 feet. What is the vertical height of the slope? **1,182 feet**
28. Lake Huron occupies about 23,010 square miles, Lake Erie about 9,930 square miles. *About* how much larger is Lake Huron than Lake Erie? **Answers may vary. A sample answer is 10,000 square miles.**
29. When Kyle took Elise to the movies, he had a $20 bill in his wallet. He paid $5 each for two tickets, $1.00 each for two drinks, and $2.50 for the large bucket of popcorn. How much money did Kyle have left from the $20 bill? **$5.50**

FUN with MATH
Why worry about the rain forests?
See page 104.

Research

30. Find out how many unmanned lunar probes have been launched by the United States or by the Soviet Union. Estimate how many total space miles have been covered by the probes. **U.S.S.R., 8 probes, about 1,840,000 miles; U.S., 7 probes, about 1,650,000 miles**

Critical Thinking

31. Robyn needs to thin 4 seedlings from a bed of 12 so that no more than 3 and no less than 1 remain in each row or column. Which seedlings does Robyn need to remove? **Answers may vary. A sample answer is given.**

Mixed Review

Solve each equation. Check your solution.

Lesson 1-12

32. $14k = 42$ **3**
33. $60 = 5y$ **12**
34. $36 = t \div 3$ **108**

Lesson 2-1

Add.

35. 68 + 19 = **87**
36. 163 + 21 = **184**
37. 821 + 601 = **1,422**
38. 32,181 + 1,427 = **33,608**

Lesson 2-1

39. What is the least sum of two consecutive numbers that are greater than 499? **1,001**

Chalkboard Examples

Examples A, B *Subtract.*
1. 78 − 24 **54**
2. 63 − 47 **16**
3. 503 − 236 **267**
4. 900 − 254 **646**

Example C *Subtract.*
5. 4,783 − 2,594 **2,189**
6. 7,032 − 1,473 **5,559**

Additional examples are provided on TRANSPARENCY 2-3.

EVALUATING THE LESSON

Check for Understanding Use Guided Practice Exercises to check for student understanding.

Closing the Lesson Have students explain in their own words when to rename before subtracting.

EXTENDING THE LESSON

Enrichment Have students make up five subtraction problems; one that requires no renaming and four that require renaming of 1, 2, 3, and 4 digits each. Have students compute their own answers, then trade with classmates to check.

Enrichment Masters Booklet, p. 15

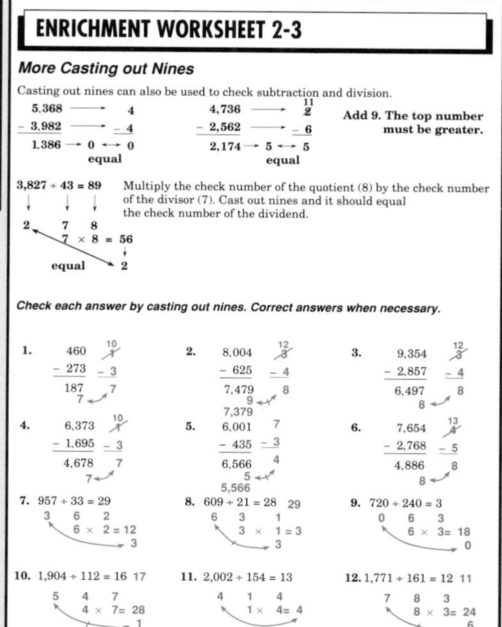

Name ______________ Date ______________

ENRICHMENT WORKSHEET 2-3

More Casting out Nines

Casting out nines can also be used to check subtraction and division.

5,368 → 4
− 3,982 → − 4
1,386 → 0 ↔ 0 equal

4,736 → 2 (11)
− 2,562 → − 6
2,174 → 5 ↔ 5 equal

Add 9. The top number must be greater.

3,827 ÷ 43 = 89 → 2, 7, 8
7 × 8 = 56 → 2
equal

Multiply the check number of the quotient (8) by the check number of the divisor (7). Cast out nines and it should equal the check number of the dividend.

Check each answer by casting out nines. Correct answers when necessary.

1. 460 − 273 = 187 (1 → 10, − 3, 7; 7)
2. 8,004 − 625 = 7,479 (3 → 12, − 4, 8; 9) 7,379
3. 9,354 − 2,857 = 6,497 (3 → 12, − 4, 8; 8)
4. 6,373 − 1,695 = 4,678 (1 → 10, − 3, 7; 7)
5. 6,001 − 435 = 6,566 (7, − 3, 4; 5) 5,566
6. 7,654 − 2,768 = 4,886 (4 → 13, − 5, 8; 8)
7. 957 ÷ 33 = 29 (3, 6, 2; 6 × 2 = 12; 3)
8. 609 ÷ 21 = 28 29 (6, 3, 1; 3 × 1 = 3; 3)
9. 720 ÷ 240 = 3 (0, 6, 3; 6 × 3 = 18; 0)
10. 1,904 ÷ 112 = 16 17 (5, 4, 7; 4 × 7 = 28; 1)
11. 2,002 ÷ 154 = 13 (4, 1, 4; 1 × 4 = 4)
12. 1,771 ÷ 161 = 12 11 (7, 8, 3; 8 × 3 = 24; 6)

APPLYING THE LESSON

Independent Practice

Homework Assignment	
Minimum	Ex. 2-30 even; 31-39
Average	Ex. 1-29 odd; 30-39
Maximum	Ex. 1-39

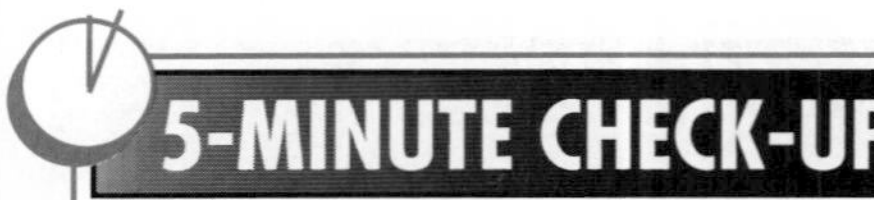

5-MINUTE CHECK-UP

(over Lesson 2-3)

Available on TRANSPARENCY 2-4.

Subtract.

1. 342 − 23 **319**
2. 121 − 49 **72**
3. 5,427 − 399 **5,028**
4. 4,654 − 2,764 **1,890**
5. 77,215 − 2,967 **74,248**

Extra Practice, Lesson 2-3, p. 452

▶ INTRODUCING THE LESSON

Ask students to tell how much change they have. Use two amounts that are under $1. Ask students if they know how to find the difference between the two amounts.

▶ TEACHING THE LESSON

Using Cooperative Groups In small groups, have students write a letter to a fifth grader explaining how to subtract decimals. Students should include an example that involves renaming.

Practice Masters Booklet, p. 17

Name ______________________ Date ______________

PRACTICE WORKSHEET 2-4

Subtracting Decimals

Subtract.

1. 8.4 − 3.3 = **5.1**	**2.** 7.3 − 5.4 = **1.9**	**3.** 8 − 3.1 = **4.9**	**4.** 0.6 − 0.21 = **0.39**
5. 0.41 − 0.17 = **0.24**	**6.** $0.86 − 0.17 = **$0.69**	**7.** $10.85 − 2.17 = **$8.68**	**8.** 0.13 − 0.07 = **0.06**
9. 0.7 − 0.36 = **0.34**	**10.** 0.727 − 0.451 = **0.276**	**11.** 1.61 − 0.9 = **0.71**	**12.** $10.02 − 0.88 = **$9.14**
13. 0.43 − 0.39 = **0.04**	**14.** 6.03 − 0.13 = **5.9**	**15.** $6.70 − 1.42 = **$5.28**	**16.** $5.00 − 2.76 = **$2.24**
17. 0.310 − 0.042 = **0.268**	**18.** $75.86 − 1.09 = **$74.77**	**19.** 63,508.76 − 51,429.08 = **12,079.68**	**20.** 2.0003 − 0.08 = **1.9203**

21. 11.680 − 4.23 **7.45** **22.** 8.42 − 5.526 **2.894**

23. Subtract $22.48 from $56.35. **$33.87**

24. Find the difference of 296.03 and 84.007. **212.023**

Solve.

25. The class collects $19.32 for a holiday project. They need $23.50. How much more must they collect? **$4.18**

26. Connie has 20 milliliters of sulfuric acid. Her experiment calls for 1.6 milliliters. How many milliliters will Connie have left? **18.4 mL**

2-4 SUBTRACTING DECIMALS

Objective
Subtract decimals.

Cora is the costume designer for the Orleans Summer Playhouse. She has 3.7 yards of felt in storage and needs 2.1 yards for King Henry's hat and shoes. After cutting the hat and shoes, how much felt does Cora have left?

To subtract decimals, begin at the right and subtract the numbers in each place-value position. It may be necessary to annex zeros.

Method

1. Align the decimal points. If necessary, annex zeros so both numbers have the same number of decimal places.
2. Subtract in each place-value position from right to left.

Example A

3.7 − 2.1 Estimate: 4 − 2 = 2

▶1 3.7 − 2.1 Align the decimal points.

▶2 3.7 − 2.1 = 1.6 Subtract tenths, then ones.

Compared to the estimate, is the answer reasonable? Check: 2.1 + 1.6 = 3.7 ✓

Cora has 1.6 yards of felt left.

Example B

6.391 − 4.62 Estimate: 6 − 5 = 1

▶1 6.391 − 4.620 Align the decimal points. Annex a zero.

▶2 (5 13) 6.391 − 4.620 = 1.771 Subtract the digits in each place-value position.

Is the answer reasonable? Check by adding.

Example C

9.3 − 3.712 Estimate: 9 − 4 = 5

▶1 9.300 − 3.712 Align the decimal points. Annex zeros so both numbers have three decimal places.

▶2 (8 12 9 10) 9.300 − 3.712 = 5.588 Subtract.

Is the answer reasonable? Check by adding.

Guided Practice

Subtract. **Encourage students to estimate each answer first.**

Examples A,B

1. 6.1 − 4.3 = **1.8** **2.** 17.35 − 9.17 = **8.18** **3.** 100.09 − 22.67 = **77.42** **4.** 0.497 − 0.16 = **0.337** **5.** 2.56 − 1.4 = **1.16**

Example C

6. 4.8 − 1.85 = **2.95** **7.** 7 − 1.63 = **5.37** **8.** 252.7 − 38.95 = **213.75** **9.** 38.7 − 9.82 = **28.88** **10.** 3,000 − 789.5 = **2,210.5**

RETEACHING THE LESSON

Write the following whole numbers as decimals.

1. 5 **5.0** **2.** 18 **18.0** **3.** 34 **34.0**

Have students write whole numbers as a decimal, align decimal points, annex zeros, then subtract.

4. 5 − 1.5 **3.5**
5. 18 − 6.872 **11.128**
6. 34 − 21.8 **12.2**
7. 112.8 − 81.93 **30.87**
8. 9.407 − 0.22 **9.187**

Reteaching Masters Booklet, p. 16

Name ______________________ Date ______________

RETEACHING WORKSHEET 2-4

Subtracting Decimals

To subtract 5.76 from 8, first align the decimal points. Then annex zeros if necessary.

Subtract from right to left.

Estimate: 8 − 6 = 2

8 − 5.76 ▶ *Subtract the hundredths.* (7 9 10) 8.00 − 5.76 = 4 ▶ *Subtract the tenths.* (7 9 10) 8.00 − 5.76 = .24 ▶ *Subtract the ones.* (7 9 10) 8.00 − 5.76 = 2.24

Subtract.

Exercises

Practice

Subtract. Encourage students to estimate each answer first.

11.	**12.**	**13.**	**14.**	**15.**
4.6 − 2.2 = **2.4**	0.64 − 0.21 = **0.43**	0.894 − 0.172 = **0.722**	2.51 − 0.8 = **1.71**	1.3 − 0.48 = **0.82**

16.	**17.**	**18.**	**19.**	**20.**
83.79 − 51.16 = **32.63**	4.968 − 2.3 = **2.668**	5.284 − 3.197 = **2.087**	9.06 − 7.752 = **1.308**	8.9 − 1.45 = **7.45**

21. 0.6 − 0.49 **0.11** **22.** 21 − 4.09 **16.91** **23.** 0.416 − 0.27 **0.146**

24. Subtract $23.25 from $68.93. **$45.68**

25. Find the difference of 568.335 and 34.008. **534.327**

Applications

26. The Reisser Corporation posted a profit of $15.7 million for the second quarter. This was a $0.9 million gain over the first quarter. What was the profit for the first quarter? What was the total profit for the first two quarters. **$14.8 million; $30.5 million**

27. The German mark closed at 1.645 on Wednesday. Six months ago the mark closed at 1.69. How much loss occurred over the six months? **0.045 Note: 1.645 means that one U.S. dollar is worth 1.645 marks.**

Interpreting Data

28. How much is spent annually for the pet pest prevention and electronic bug killers? **$309 million**

How long ago were numbers first used? See page 39.

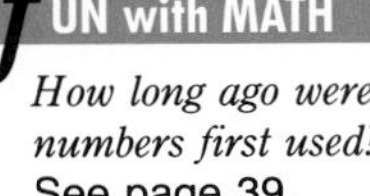

29. How much more is spent annually on do-it-yourself products than on electronic bug killers? **$341 million**

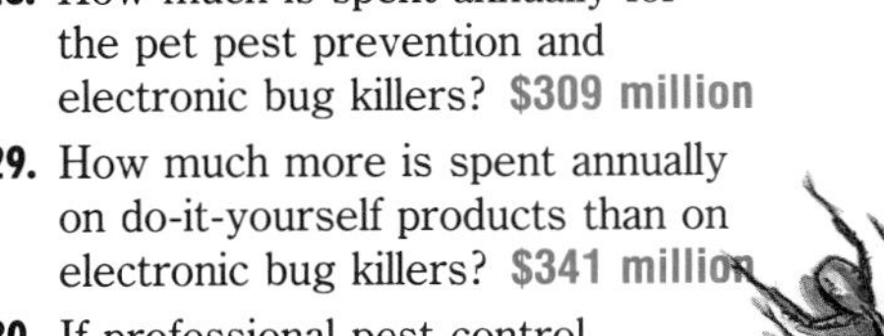

30. If professional pest control companies make 10 million calls a year, what is the average cost of one professional call? **$300**

Number Sense

31. Write 1.111 as the sum of four different decimals. **Answers may vary**

Calculator

Place value and addition can be used to "build" a decimal number.

5000 [+] 300 [+] 20 [+] 8 [+] 0.1 [+] 0.02 [+] 0.004 [=] 5328.124

Add with a calculator to "build" each number. **See students' work.**

32. 4,351.897 **33.** 946.12 **34.** 1,407.95 **35.** 9,720.822

Enter the first number in a calculator. Then add or subtract to get the remaining numbers in order. **See students' work.**

36. 134.56 → 134.59 → 134.69 → 135.69

37. 8,345.22 → 8,345.28 → 8,345.88 → 8,345.888

APPLYING THE LESSON

Independent Practice

Homework Assignment	
Minimum	Ex. 1-37 odd
Average	Ex. 2-24 even; 26-37
Maximum	Ex. 1-37

Chalkboard Examples

Examples A, B *Subtract.*

1. 5.93 − 2.71 **3.22**

2. 5.654 − 3.91 **1.744**

3. 8.337 − 2.56 **5.777**

Example C *Subtract.*

4. 4.4 − 2.861 **1.539**

5. 7.3 − 4.527 **2.773**

6. 9.1 − 3.644 **5.456**

Additional examples are provided on TRANSPARENCY 2-4.

EVALUATING THE LESSON

Check for Understanding Use Guided Practice Exercises to check for student understanding.

Closing the Lesson Have students explain in their own words when it is necessary to annex zeros in order to subtract.

EXTENDING THE LESSON

Enrichment Use the financial pages of a newspaper to have students find the loss or gain in currency values of 3 different countries over the last 6 months.

Enrichment Masters Booklet, p. 16

ENRICHMENT WORKSHEET 2-4

Using Charts and Graphs

Solve. Use the chart.

1. Juan borrows $500 for a dirt bike. What is the difference in monthly payments if he borrows the money for three years rather than two years? **$6.93 per month**
2. Elena's family borrows $1,800 for home repairs. What is the difference in monthly payments if they borrow the money for one year rather than two years? **$75.18**
3. If you borrow $1,000 for 3 years, what is the total amount you will pay in monthly payments? **$1,195.20**

LOAN	12 MO	24 MO	36 MO
$ 100	$ 8.88	$ 4.70	$ 3.32
200	17.76	9.41	6.64
300	26.65	14.12	9.96
400	35.53	18.82	13.28
500	44.41	23.53	16.60
600	53.30	28.24	19.92
700	62.18	32.94	23.24
800	71.06	37.65	26.56
900	79.95	42.36	29.88
1,000	88.83	47.06	33.20
1,100	97.72	51.77	36.53
1,200	106.60	56.48	39.85
1,300	115.48	61.18	43.17
1,400	124.37	65.89	46.49
1,500	133.25	70.60	49.81
1,600	142.13	75.30	53.13
1,700	151.02	80.01	56.45
1,800	159.90	84.72	59.77
1,900	168.79	89.42	63.09
2,000	177.67	94.13	66.41

Solve. Use the chart.

4.

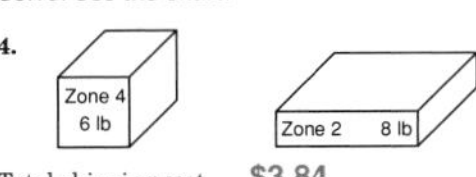

Total shipping cost **$3.84**

5.

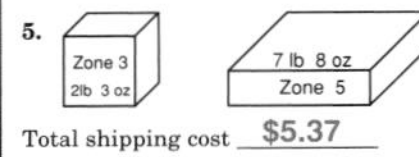

Zone 2 1 lb 1 oz

Total shipping cost **$5.37**

WEIGHT NOT TO EXCEED	POSTAGE RATES ZONES			
	2	3	4	5
1 lb	$ 1.21	$ 1.24	$ 1.28	$ 1.32
2 lb	1.31	1.35	1.42	1.50
3 lb	1.39	1.46	1.57	1.68
4 lb	1.48	1.57	1.72	1.87
5 lb	1.56	1.68	1.86	2.05
6 lb	1.65	1.80	2.01	2.24
7 lb	1.75	1.91	2.15	2.42
8 lb	1.83	2.02	2.31	2.60
9 lb	1.92	2.13	2.45	2.79
10 lb	2.00	2.25	2.59	2.97

Solve. Use the graph.

6. How many inches did Jason grow from age 3 to age 6? **5 in.**
7. At what age was Jason twice as tall as he was at age 1? **11 yr**
8. Between which ages did Jason grow the greatest number of inches? **9 and 10**

Jason's Height

Inches: 0, 25, 30, 35, 40, 45, 50

Years of Age: 1 2 3 4 5 6 7 8 9 10 11

Applying Mathematics to Technology

This optional page provides students with an experience in computer literacy and also provides a change of pace.

Objective Use computer terminology.

Using Questioning

- What tasks do you imagine a computer can do? **Answers should include items such as repeated or large calculations, word processing, tracking and sorting data, comparing data.**

Using Critical Thinking Ask students why programs for computers and instructions for a computer user need to be very precise. **Computers respond to instruction. They do not initiate any instructions and will perform only the task they have been instructed to do.**

Collect Data Have students collect data about the cost of hardware and software from newspapers, magazines, or computer stores. Have students collect the cost of their dream system.

Computer Application

TERMINOLOGY

A **computer** is an electronic device that can store, retrieve, and process data. The most widely used computer for educational purposes is the **microcomputer** or **personal computer (PC).**

A computer must be given a set of instructions called a **program.** Computer programs are also referred to as **software.** The computer itself is called the **hardware.**

The heart of the computer is the **central processing unit (CPU).** The CPU contains an **arithmetic/logic unit** and a **control unit.** The arithmetic/logic unit performs standard arithmetic operations and comparisons. The control unit directs the overall actions of the computer.

Information that is entered into a computer is called **input.** A keyboard is the most common input device. Other input devices are light pens, disk drives, and cassette tapes.

The CPU handles a limited amount of information at one time. So programs and data are stored in computer **memory** and are called for by the CPU as they are needed.

The results of computer processing are called **output.** Two output devices are the cathode-ray tube (CRT) and the printer. The CRT, along with its controls, is called a **monitor.** A printer produces **hard copy** of the computer output.

See margin.

1. What does PC stand for? **personal computer**
2. What is the set of instructions for a computer called? **a program**
3. Name and define the parts of a CPU.
4. Explain the difference between hardware and software.
5. Name the four basic parts of a computer system.
6. Name four input and two output devices.
7. What is a monitor? What does it display?
8. What is hard copy of computer output? How is it produced?

Additional Answers

2. a program
3. arithmetic/logic unit, performs arithmetic operations, performs comparisons; control unit, directs action
4. Hardware refers to the machinery; software refers to the instructions that operate the hardware.
5. CPU, memory unit, input device, output device
6. keyboard, light pen, disk drive, cassette tape; monitor, printer
7. a CRT unit along with its controls; displays both input and output
8. Hard copy is a printout of computer output produced by a printer

On-the-job Application

Meter Reader

Dawn Westrick is a meter reader for Southern Electric Company. On her route, Dawn reads electric meters and enters the readings in a hand-held computer. She also makes a quick check of the condition of the meters and connections.

Electric meters measure the number of kilowatt hours used. A **kilowatt hour (kWh)** is equal to 1 kilowatt of electricity used for 1 hour.

The electric meter shows 5 dials. Use the following steps to find the meter reading.

1. Choose the number that was just passed as the arrow rotates from 0 through 9. Note that the numbers on some dials run clockwise and some run counterclockwise.
2. Each dial is in its place-value position. This meter reads 26,453 kilowatt hours.

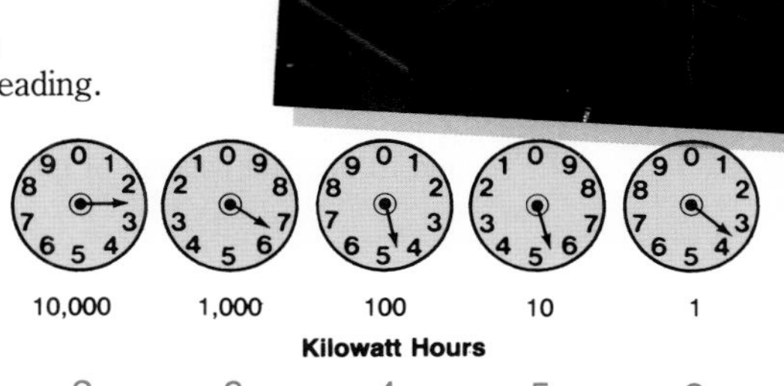

How many kilowatt hours are shown on each electric meter?

1.  **2.**  **3.**  **4.**  **5.**  **6.**

42,387 kWh 65,710 kWh 16,201 kWh 18,420 kWh
65,552 kWh 26,812 kWh

7. If electricity costs $0.028 per kWh, what is the monthly bill for a usage of 4,065 kilowatt hours? At this same rate of usage per month, what is the yearly cost for electricity? **$113.82; $1,365.84**

Applying Mathematics to the Work World

This optional page shows how mathematics is used in the work world and also provides a change of pace.

Objective Read an electric meter.

Using Data Have students check their electric bill at home and compute the charges for electricity.

Collect Data Have students contact the local electric company to find the types of appliances and hours of use that produce the highest charges for electric power. Have them recommend conservation practices suggested by their findings.

5-MINUTE CHECK-UP

(over Lesson 2-4)

Available on TRANSPARENCY 2-5.

Subtract.

1. 24,513 − 1,745 22,768
2. 14 − 2.364 11.636
3. 2.43 − 1.98 0.45
4. $129.25 − $30.90 $98.35
5. 596 − 261.493 334.507

Extra Practice, Lesson 2-4, p. 452

▶ INTRODUCING THE LESSON

Discuss with students how they would pay bills *without* a checking account. Discuss the advantages of using a checking account.

▶ TEACHING THE LESSON

Using Models Copy the blank checks and check register in the **Lab Manual, pp. 5-6.** Have students practice writing checks and filling in the register.

Practice Masters Booklet, p. 18

Name ______________________ Date ______________

PRACTICE WORKSHEET 2-5

Applications: Checking Accounts

Complete the balance column to find the amount of money Jennifer Chase has left in her checking account on April 30.

	NUMBER	DATE	DESCRIPTION OF TRANSACTION	PAYMENT/DEBIT (−)	DEPOSIT/CREDIT (+)	BALANCE	
						$1231	07
1.	201	4/1	Groden's Grocery	103.82		1127	25
2.	201	4/4	Big Al's Autos	111.26		1015	99
3.	203	4/7	Mickey's Cuisine	72.25		943	74
4.	204	4/8	Wasco Telephone Company	201.44		742	30
5.		4/9	deposit	—	473.53	1215	83
6.	205	4/10	A & T Insurance	386.89		828	94
7.	206	4/11	Wasco Gas	75.53		753	41
8.	207	4/11	Electric Company	195.40		558	01
9.	208	4/12	Mortgage	521.57		36	44
10.		4/12	deposit	—	713.98	750	42
11.	209	4/19	Loan	92.75		657	67
12.	210	4/21	Darlene's Department Store	118.65		539	02
13.	211	4/21	Super Sundry	5.00		534	02
14.		4/21	deposit	—	428.60	962	62
15.	212	4/23	Hirt's Furniture	219.74		742	88
16.	213	4/23	City of Charlesville	123.90		618	98
17.	214	4/24	Shoe Outlet	78.65		540	33
18.		4/24	deposit	—	628.35	1168	68
19.	215	4/25	Frank's Finer Foods	77.78		1090	90
20.	216	4/25	Pamela Beauty Salon	35.40		1055	50
21.	217	4/28	Buck's Garage	187.65		867	85
22.	218	4/28	Galordi Airlines	213.14		654	71
23.	219	4/30	Kelley's Kennel	97.80		556	91

Applications

2-5 CHECKING ACCOUNTS

Objective
Write a check and use a check register.

Rajiv Ramur opened a checking account after he began working part time. He uses checks to pay bills like car insurance with the money in his account. He uses deposit slips to add money to his account.

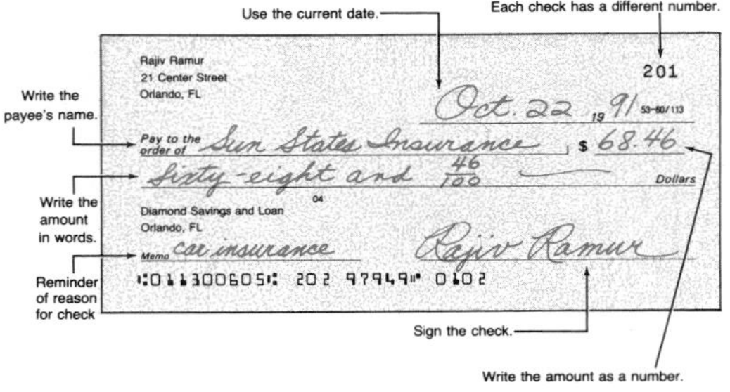

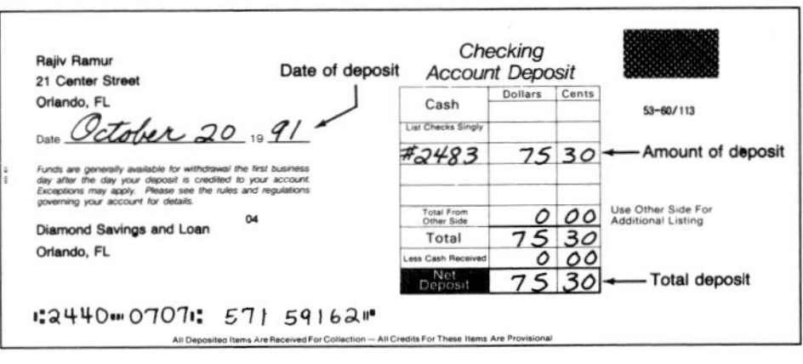

Rajiv keeps a check register to record checks and deposit amounts and to find the balance of his account.

Balance from previous page

CHECK NUMBER	DATE	DESCRIPTION OF TRANSACTION	PAYMENT/DEBIT (−)		✓ TAX	DEPOSIT/CREDIT (+)		BALANCE FWD.	
								$267	25
D	10/20	Deposit	$			$75	30	75	30
								342	55
201	10/22	Sun States Insurance	68	46				68	46
		car insurance						274	09
202	10/22	Tommy's Electronics	82	10				82	10
		CD repair						191	99
203	10/25	Terri's birthday	21	98				21	98
		gift						170	01
D	10/27	Deposit						63	00
								233	01
204	10/29	Best, Inc.-frame	86	20				86	20
								146	81

Payee's name
Reminder of reason for check
Balance after each transaction

Understanding the Data

- What is the name of the payee on the check? Sun States Insurance
- What is the amount of the deposit made on October 20? $75.30
- What is Rajiv's balance from the check register on October 29? $146.81

To find the balance in a check register, add each deposit and subtract the amount of each check.

Check for Understanding

1. Why is the amount of the check in numbers written close to the dollar sign? Numbers cannot be altered by someone else.

2. Why is the amount of the check in words written as far to the left as possible? Words cannot be altered by someone else.

3. What word replaces the decimal point in the amount when it is written in words? and

RETEACHING THE LESSON

Work with students in completing the balance column of the check register.

PAYMENT/DEBIT (−)		FEE (−)	✓ TAX	DEPOSIT/CREDIT (+)		BALANCE FWD.	
						$208	12
$128	56			$		79	56
				567	28	646	84
261	72					385	12
76	98					308	14
89	35					218	79
				7	10	225	89

Reteaching Masters Booklet, p. 17

Name ______________________ Date ______________

RETEACHING WORKSHEET 2-5

Applications: Checking Accounts

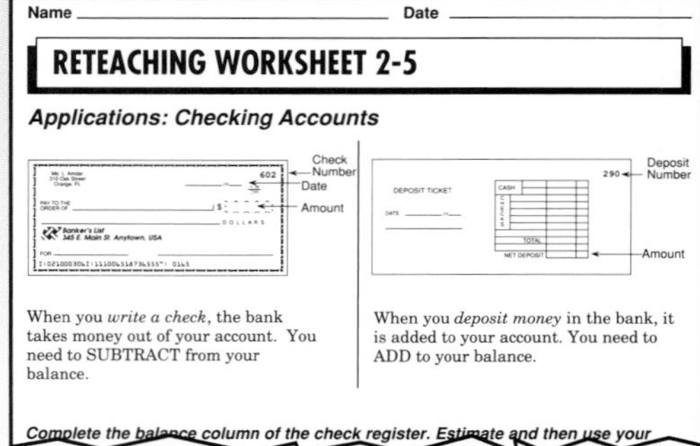

When you *write a check*, the bank takes money out of your account. You need to SUBTRACT from your balance.

When you *deposit money* in the bank, it is added to your account. You need to ADD to your balance.

Complete the balance column of the check register. Estimate and then use your

Exercises

Practice

Write each amount in words as you would on a check. **See margin.**

4. \$13.85 **5.** \$29.00 **6.** \$272.05 **7.** \$108.97

8. Copy and complete the balance column of the register. **See margin.**

9. What is the balance on June 21? **\$1,209.25**

10. What is the balance on June 30? **\$451.17**

CHECK NUMBER	DATE	DESCRIPTION OF TRANSACTION	PAYMENT/DEBIT (−)		✓ TAX	DEPOSIT/CREDIT (+)		BALANCE FWD.	
								\$564	78
328	6/15	Golden Eagle	\$62	12		\$		62	12
		groceries						502	66
329	6/17	Rosa's	13	76				13	76
		gift						488	90
D	6/20	Deposit				720	35	720	35
		paycheck						1,209	25
330	6/25	Asherton Corp.	515	00				515	15
		rent						694	25
331	6/25	Diamond National Bank	206	89				206	89
		car payment						487	36
332	6/25	Haven Fire + Casualty	63	50				63	50
		insurance						423	86
D	6/30	Deposit				27	31	27	31
		dividend						451	17

Suppose

11. Suppose check number 329 was recorded incorrectly. The amount should have been \$13.67. What should the corrected balance be on June 30? **\$451.26**

Research

12. Find out about checking account service charges and charges for using automated teller machines from literature available in banks or from the newspaper. **See students' work.**

Using Variables

Write an equation and solve.

13. The service charge, *s*, at First National Bank is \$0.10 for each check. Let *n* equal the number of checks. Write an equation to find the service charge for writing *n* checks. **$0.10 \times n = s$**

14. If Karen writes 23 checks on her checking account at First National Bank, how much is the service charge? Use the equation in Exercise 13. **\$2.30**

Critical Thinking

15. In two weeks, Katie savied \$12.60. Each day she saved \$0.10 more than the day before. How much did Katie save on the first day? **\$0.25**

Mixed Review

Lesson 1-10

Write an expression for each phrase.

16. the product of 13 and *z* **$13z$**

17. 5 less than *x* **$x - 5$**

18. 18 divided by a number **$18 \div n; \frac{18}{n}$**

19. a number increased by 14 **$n + 14$**

Lesson 2-2

Add.

20. $4.13 + 6.85 = $ **10.98**

21. $18.627 + 9.651 = $ **28.278**

22. $84.699 + 24.276 = $ **108.975**

23. $0.392 + 18.67 = $ **19.062**

Lesson 2-3

24. At the beginning of Pat's trip his odometer read 25,481. When he had completed his trip the odometer read 26,715. How many total miles did Pat travel? **1,234 miles**

APPLYING THE LESSON

Independent Practice

Homework Assignment	
Minimum	Ex. 1-24
Average	Ex. 1-24
Maximum	Ex. 1-24

Chapter 2, Quiz A (Lessons 2-1 through 2-5) is available in the Evaluation Masters Booklet, p. 9

Additional Answers

4. Thirteen and $\frac{85}{100}$
5. Twenty-nine and $\frac{00}{100}$
6. Two hundred seventy-two and $\frac{05}{100}$
7. One hundred eight and $\frac{97}{100}$
8. 62.12; 502.66; 13.76; 488.90; 720.35; 1,209.25; 515.00; 694.25; 206.89; 487.36; 63.50; 423.86; 27.31; 451.17

Chalkboard Examples

Write each amount in words as you would on a check.

1. \$578.50 **2.** \$25.99

1. five hundred seventy-eight and $\frac{50}{100}$
2. twenty-five and $\frac{99}{100}$

Additional examples are provided on TRANSPARENCY 2-5.

EVALUATING THE LESSON

Check for Understanding Use Guided Practice Exercises to check for student understanding.

Closing the Lesson Ask students to summarize what they have learned about writing a check and using a check register.

EXTENDING THE LESSON

Enrichment Solve.

1. Jackie has a balance of \$378.54 in her checking account. She deposited \$159.68 and wrote a check for \$79.95. What is her new balance? **\$458.27**

Enrichment Masters Booklet, p. 17

Name ____________ Date ____________

ENRICHMENT WORKSHEET 2-5

Using a Mileage Chart

Solve. Use the mileage chart.

Destinations From/To	Number of Miles		
	By Car	By Plane	By Train
Cincinnati to Boston	876	740	938
Los Angeles to Atlanta	2,197	1,936	2,285
Buffalo to Seattle	2,590	2,117	2,265
San Francisco to Miami	3,075	2,594	3,355

1. How many miles longer is a trip from Cincinnati to Boston by car than by plane? **136 mi**
2. How much shorter is a trip from Los Angeles to Atlanta by plane than by train? **349 mi**
3. How many miles longer is a trip from Buffalo to Seattle by car than by train? **325 mi**
4. How much shorter is a trip from San Francisco to Miami by plane than by train? **761 mi**
5. Jesse drives from Cincinnati to Boston. Esther drives from San Francisco to Miami. How much farther does Esther drive? **2,199 mi**
6. Loretta takes a plane from Los Angeles to Atlanta. Luis takes a plane from Buffalo to Seattle. How many more miles does Luis travel? **181 mi**
7. Find the distance of a round-trip between Los Angeles and Atlanta by train. **4,570 mi**
8. Find the distance of a round-trip between San Francisco and Miami by plane. **5,188 mi**
9. Natalie drives from Buffalo to Seattle. On her return trip, Natalie drives 2,078 miles to Dallas. Then she drives 1,346 miles to Buffalo. How much farther was Natalie's return trip to Buffalo than her trip to Seattle? **834 mi**
10. André takes a train from San Francisco to Miami. Then he flies 1,270 miles from Miami to New York. He flies 2,497 miles back home to San Francisco. What is the total number of miles André traveled? **7,122 mi**

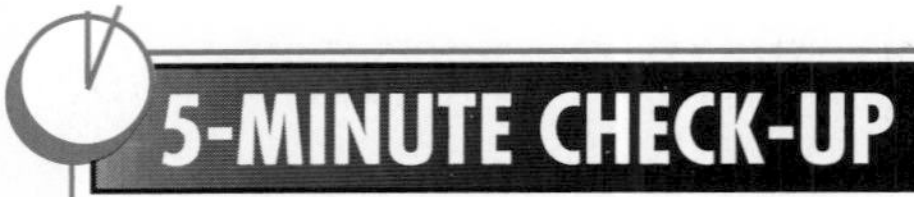

(over Lesson 2-5)

Available on TRANSPARENCY 2-6.

Solve.

1. Marie wrote checks for $22.45 and $65.23. She deposited $36.31. If her balance before the transactions was $259.34, find her new balance. **$207.97**

INTRODUCING THE LESSON

List the first ten multiplication facts for 2 and 3. Have students describe the patterns they see in the ordered lists.

TEACHING THE LESSON

Using Discussion Discuss with students the need to check more than one difference between consecutive terms. The difference between numbers must be the same for a sequence to be an arithmetic sequence.

Practice Masters Booklet, p. 19

Name ______ Date ______

PRACTICE WORKSHEET 2-6

Arithmetic Sequences

Write the next three numbers in each sequence.

1. 18, 20, 22, 24, 26, 28
2. 54, 48, 42, 36, 30, 24
3. 0.4, 2.4, 4.4, 6.4, 8.4, 10.4
4. 45, 57, 69, 81, 93, 105
5. 25.3, 22, 18.7, 15.4, 12.1, 8.8
6. 8.9, 8.5, 8.1, 7.7, 7.3, 6.9
7. 5, 11, 17, 23, 29, 35, 41
8. 11, 15, 19, 23, 27, 31, 35
9. 6, 9, 12, 15, 18, 21, 24
10. 2.1, 2.3, 2.5, 2.7, 2.9, 3.1, 3.3
11. 3.6, 3.1, 2.6, 2.1, 1.6, 1.1, 0.6
12. 8, 27, 46, 65, 84, 103, 122
13. 50, 42, 34, 26, 18, 10, 2
14. 20.6, 21.7, 22.8, 23.9, 25.0, 26.1, 27.2
15. 3.5, 7.0, 10.5, 14, 17.5, 21, 24.5
16. 65, 56, 47, 38, 29, 20, 11

State whether each sequence is arithmetic. If it is, write the next three terms. If is is not, write no.

17. 3, 9, 27, 81,... no
18. 4, 34, 64, 94,... 124, 154, 184
19. 21.3, 24.3, 30.3, 33.3,... no
20. 4.5, 9, 13.5, 18,... 22.5, 27, 31.5, 36

Copy and complete each arithmetic sequence.

21. 12, 20, 28, 36, 44, 52
22. 20.5, 23, 25.5, 28, 30.5, 33

Patterns Connection

2-6 ARITHMETIC SEQUENCES

Objective
Identify and write arithmetic sequences.

Paula is mailing photographs to her pen pal in New Zealand. She notices that the costs for mailing are $0.45, $0.62, and $0.79 for the first, second, and third ounce respectively. If this pattern continues, how much will it cost Paula to mail a six-ounce package?

A list of numbers that follows a certain pattern is called a **sequence.** If the difference between consecutive numbers in a sequence is the same, the sequence is called an **arithmetic sequence.**

5, 10, 15, 20, ... Each number in the sequence is 5 *more than* the number before it.

60, 45, 30, 15, ... Each number in the sequence is 15 *less than* the number before it.

You can extend an arithmetic sequence by using the following steps.

Method

1. List the numbers of the sequence in order.
2. Find the difference between any two consecutive numbers.
3. Add or subtract the difference to find the next number.

Example A

Find the next three numbers in the sequence 0.45, 0.62, 0.79, ... to find the cost of a six-ounce package.

1. List the numbers in order. 0.45 0.62 0.79
2. Find the difference between consecutive numbers. 0.62 − 0.45 = 0.17 0.79 − 0.62 = 0.17
   The difference is 0.17. If the difference between consecutive numbers is *not* the same, it is *not* an arithmetic sequence.
3. Add the difference to the last number. 0.79 + 0.17 = 0.96 0.96 + 0.17 = 1.13 1.13 + 0.17 + 1.30

It will cost Paula $1.30 to mail a six-ounce package.

Example B

Find the next three numbers in the sequence 85, 75, 65, 55, ...

1. 85 75 65 55
2. 85 − 75 = 10 75 − 65 = 10 65 − 55 = 10
   The difference is 10.
3. 55 − 10 = 45 45 − 10 = 35 35 − 10 = 25 Subtract since the sequence is decreasing.

The sequence is 85, 75, 65, 55, 45, 35, and 25.

1. 2; 14, 16, 18 3.0; 12.6, 15.6, 18.6

Guided Practice

Find the difference and the next three numbers for each sequence.

Example A
1. 6, 8, 10, 12, ?, ?, ?
2. 0.6, 3.6, 6.6, 9.6, ?, ?, ?

Example B
3. 9, 8, 7, 6, ?, ?, ? **1; 5, 4, 3**
4. 52, 49, 46, 43, ?, ?, ? **3; 40, 37, 34**

RETEACHING THE LESSON

Work with the students in completing the following sequences.

Find the next three numbers in each sequence.

1. 1, 2, 3, 4, 5, 6, 7
2. 2, 4, 6, 8, 10, 12, 14
3. 1, 3, 5, 7, 9, 11, 13
4. 7, 11, 15, 19, 23, 27, 31
5. 2, 3, 5, 7, 11, 13, 17, 19
6. Which of the above sequences are arithmetic sequences? **1, 2, 3, 4**

Reteaching Masters Booklet, p. 18

Name ______ Date ______

RETEACHING WORKSHEET 2-6

Arithmetic Sequences

4, 11, 18, 25, 32, 39, ...

The list of numbers above is called an arithmetic sequence. Each number is 7 more than the previous number. The common difference is 7.

11 − 4 = **7**, 18 − 11 = **7**, 25 − 18 = **7**, ...

Find the common difference for each arithmetic sequence.

1. 5, 14, 23, ... 9
2. 24, 39, 54, 69, ... 15
3. 6.8, 6.3, 5.8, 5.3, ... 0.5
4. 0.50, 0.54, 0.58, ... 0.04

11. 19.6, 21.3, 23 12. 100.1, 125.6, 151.1 8. 500, 600, 700

Exercises

Practice

Write the next three numbers in each sequence.

5. 100, 92, 84, 76, **68, 60, 52**
6. 10, 8.2, 6.4, ... **4.6, 2.8, 1**
7. 4, 14, 24, 34, ... **44, 54, 64**
8. 200, 300, 400, ...
9. 250, 225, 200, ... **175, 150, 125**
10. 89, 78, 67, ... **56, 45, 34**
11. 14.5, 16.2, 17.9, ...
12. 23.6, 49.1, 74.6, ...

***State whether each sequence is arithmetic. If it is, write the next three terms. If it is not, write* no.**

13. 6, 30, 150, 750, ... **no**
14. 200, 100, 50, 25, ... **no**
15. 6.1, 11.3, 16.5, ... **yes; 21.7, 26.9, 32.1**
16. 5.5, 5.2, 4.9, 4.6, ... **yes; 4.3, 4.0, 3.7**

Copy and complete each arithmetic sequence. **See margin.**

17. 9, 18, _?_, 36, _?_, _?_
18. 81, _?_, 91, 96, _?_, _?_, 111
19. 10.4, 26.2, _?_, _?_, 73.6, _?_
20. 26.9, _?_, 40.5, _?_, 54.1, _?_
21. _?_, 6.3, _?_, 4.9, _?_, 3.5
22. 108.4, 83.4, _?_, _?_, 8.4

Applications

23. In the first three weeks of typing class Keith's speed increased from 28 words per minute to 32 words per minute to 36 words per minute. If Keith continues to improve at this rate, how many more weeks will it take him to type 48 words per minute? **3 weeks**
24. Martin falls 16 feet in the first second, 48 feet in the second second, and 80 feet in the third second during a free-fall. At this rate, how many feet will he fall in the sixth second? **176 feet**

Number Sense

25. Copy and complete the table for the sequence. **36, 42; 3, 4, 5**

First	Second	Third	Fourth	Fifth	Sixth
12	18	24	30		
12 + 0×6	12 + 1×6	12 + 2×6	12 + _?_×6	12 + _?_×6	12 + _?_×6

26. In Exercise 25 how would you write an expression for the seventh number? the eighth number? **12 + 6 × 6; 12 + 7 × 6**

Using Variables

27. How would you write an expression for the *n*th number of the sequence in Exercise 25? **12 + (*n* − 1) × 6**

Estimation

You can estimate differences by changing the numbers in the problem to numbers that are easy to subtract mentally. Any one of the problems given below could be used to estimate difference of 97,584 and 33,751.

$$\begin{array}{r} 97{,}584 \\ -\ 33{,}751 \end{array} \longrightarrow \begin{array}{r} 95{,}000 \\ -\ 35{,}000 \\ \hline 60{,}000 \end{array} \quad \text{or} \quad \begin{array}{r} 95{,}000 \\ -\ 30{,}000 \\ \hline 65{,}000 \end{array} \quad \text{or} \quad \begin{array}{r} 100{,}000 \\ -\ 30{,}000 \\ \hline 70{,}000 \end{array}$$

Estimate. **Answers may vary. Typical answers are given. 31. 1,000**

28. $\begin{array}{r} 697 \\ -\ 83 \\ \hline \end{array}$ **620**
29. $\begin{array}{r} 355 \\ -\ 103 \\ \hline \end{array}$ **250**
30. $\begin{array}{r} 727 \\ -\ 538 \\ \hline \end{array}$ **200**
31. $\begin{array}{r} 6{,}243 \\ -\ 4{,}564 \\ \hline \end{array}$

Chalkboard Examples

Examples A, B *Find the next three numbers for each sequence.*

1. 7,12,17, . . . **22, 27, 32**
2. 2.4, 3.7, 5, . . . **6.3, 7.6, 8.9**
3. 26, 22, 18, . . . **14, 10, 6**
4. 9.1, 8.7, 8.3, . . . **7.9, 7.5, 7.1**

Additional examples are provided on TRANSPARENCY 2-6.

EVALUATING THE LESSON

Check for Understanding Use Guided Practice Exercises to check for student understanding.

Closing the Lesson Ask students to state in their own words how to determine whether a sequence is arithmetic.

EXTENDING THE LESSON

Enrichment Write the sequence 1, 1, 2, 3, 5, 8, . . . on the chalkboard. Ask students how the differences between consecutive terms are related. Then have them find the next three terms. **Succeeding terms are the sum of the previous two terms.; 13, 21, 34**

Enrichment Masters Booklet, p. 18

ENRICHMENT WORKSHEET 2-6

Finding Patterns in Sequences

Number sequences have a rule or pattern which is used to determine what comes next. You have probably seen problems like these on tests you have taken in school.

What number comes next in each sequence?

A. 5, 7, 9, 11, 13
B. 74, 64, 54, 44, 34
C. 68, 78, 88, 98, 108
D. 2, 7, 12, 17, 22, 27
E. 2, 3, 5, 8, 12, 17
F. 84, 79, 74, 69, 64, 59

How can you detect the pattern or rule to use in extending sequences like these? One method is to examine the differences between the numbers of the sequence.

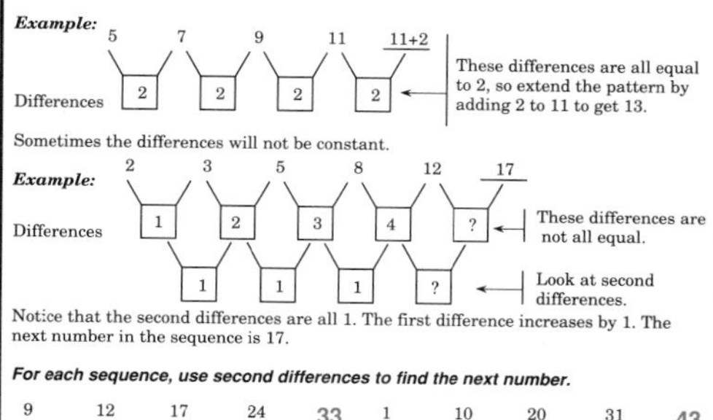

Example:
5 7 9 11 11+2
Differences
2 2 2 2
These differences are all equal to 2, so extend the pattern by adding 2 to 11 to get 13.
Sometimes the differences will not be constant.
Example:
2 3 5 8 12 17
Differences
1 2 3 4 ?
These differences are not all equal.
1 1 1 ?
Look at second differences.

Notice that the second differences are all 1. The first difference increases by 1. The next number in the sequence is 17.

For each sequence, use second differences to find the next number.

1. 9, 12, 17, 24, 33 (differences 3, 5, 7, 9; second differences 2, 2, 2)
2. 1, 10, 20, 31, 43 (differences 9, 10, 11, 12; second differences 1, 1, 1)

APPLYING THE LESSON

Independent Practice

Homework Assignment	
Minimum	Ex. 2-24 even; 25-31
Average	Ex. 1-23 odd; 25-31
Maximum	Ex. 1-31

Additional Answers

17. 27, 45, 54
18. 86, 101, 106
19. 42, 57.8, 89.4
20. 33.7, 47.3, 60.9
21. 7, 5.6, 4.2
22. 58.4, 33.4

5-MINUTE CHECK-UP

(over Lesson 2-6)

Available on TRANSPARENCY 2-7.

Write the next three numbers for each arithmetic sequence.

1. 24, 27.6, 31.2, . . . 34.8, 38.4, 42

2. 85, 72, 59, . . . 46, 33, 20

Extra Practice, Lesson 2-6, p. 452

INTRODUCING THE LESSON

Ask students to name the longest or most boring task they do or can imagine. Select one and ask students for different ways to accomplish the task more easily or more quickly. Then have the class select the most appropriate method.

TEACHING THE LESSON

Using Models Have students copy the flowchart on page 56 into their notebooks for easy reference. Have them include sample problems they have solved.

Practice Masters Booklet, p. 20

Name ____________________ Date ____________________

PRACTICE WORKSHEET 2-7

Problem-Solving Strategy: Choosing the Method of Computation

Solve each problem. State the method of computation used.

1. Lisa is ordering sporting goods equipment from a catalog. She orders gym shoes for $68.95, two wrist bands for $3.95 each, a volleyball for $18.49, and a jersey for $25.95. What is the total amount of Lisa's order? $121.29; calculator
2. Warren is planning a trip to Funworld with his friends. Warren knows that the entrance fee is $5 and that he will go on about 10 rides. The rides cost between $2 and $3 each. He will buy lunch, some snacks, and maybe a souvenir, too. How much money should Warren take to Funworld? Possible answer: about $50; estimate
3. Two classes at Lincoln School are raising money for new science equipment for the school. Twenty-five students in one class collected, on the average, $20 each. Twenty students in another class collected an average of $25 each. How much money was raised by the two classes? $1,000; mental math
4. On his last bank statement, Mark's checking account showed a balance of $324.88. Since then, he has made withdrawals of $30, $50, and $30, and deposits of $134.62 and $208.03. What is the current balance in Mark's account? $557.53; calculator
5. If Becky saves $400, she can buy a mountain bike. She has already saved $252. She figures that if she saves $37.50 a week from her baby-sitting job for the next 5 weeks, she will have enough money. Is Becky right? Explain. yes; possible answer: estimation and mental math
6. John wants a motorcycle that costs $2,195. He has saved $468.42 thus far. If he can save $47 a week, how many weeks will it take him to save the money he needs? 37 weeks; calculator
7. Use each of the digits 2, 3, 4, and 5 only once. Find two 2-digit numbers that will give the least possible answer when multiplied. 24 x 35; calculator or
8. A bicycle shop contains 48 bicycles and tricycles. If there is a total of 108 wheels, how many bicycles and how tricycles are in the store? 36 bicycles, 12 tricycles; estimation,

Explore
Plan
Solve
Examine

2-7 CHOOSING THE METHOD OF COMPUTATION

Objective
Choose an appropriate method of computation.

Sherri needs five pairs of socks for running camp. She saw an ad in the newspaper for a clearance sale at Sports Plus. If socks are on sale for $2.89 a pair, how much money should she take to the store?

For some problems, an estimate is enough to solve the problem.

Example A Since Sherri doesn't need to know the exact amount, she estimates.

$2.89 is about $3 $\quad$ $\$3 \times 5 = \$15$

Sherri decides to take $15 to the store.

Example B The clerk in the store needs to know exactly how much five pairs of socks will cost. The cash register finds the cost of five pairs of socks.

$\$2.89 \times 5 = \$14.45$

Five pairs of socks cost $14.45.

The following flowchart can be used to choose the method of computation.

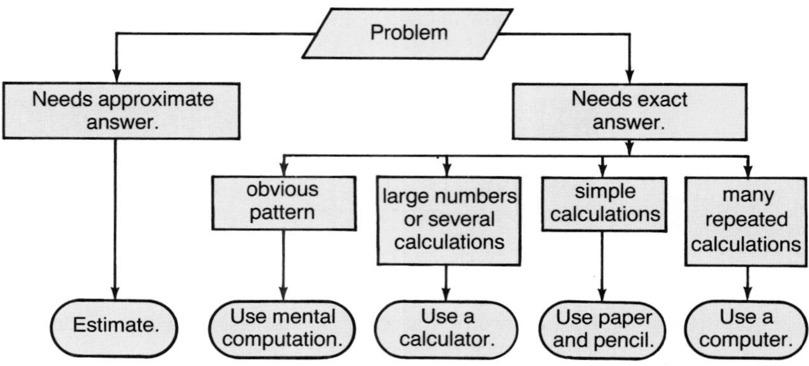

Guided Practice

Solve. State the method of computation used.

1. John is ordering parts from a catalog to fix his motorbike. The total bill must be prepaid with a check or money order. John needs a transmission for $97.95, a seat cover for $43.90, and two side baskets for $23.95 each. How much money should John send with his order? **$189.75; calculator**
2. Curtis is going to a movie with his friends. He wants to get popcorn and a drink before the movie starts. If a ticket costs $5.50, popcorn is $1.89, and a drink is $0.99, how much money should he take? **$9.00; estimate**

RETEACHING THE LESSON

Have students make-up problems to solve as a group using the situation described in the Skills Warm-Up. For example, I have placed the following items in my grocery cart: a loaf of bread costing $0.75, a bottle of soda costing $0.89 and a 99¢ carton of milk. I have $3. Do I have enough money? **estimate; yes**

Have students discuss choice of operation using flow chart on page 56.

Problem Solving

Paper/ Pencil
Mental Math
Estimate
Calculator

Solve. Use any strategy. **3. $603; mental computation**

3. The senior class of Westfield High School wants to donate money to plant trees at the school entrance. There are 201 students donating $3.00 each. Determine how much money the class is donating and state the method of computation used.

4. Sean works part-time while he is in school. If he earns at least $2,000, he will be able to buy a car. He works four hours a day, three days a week, for $3.85 per hour. If he works 35 weeks this year, will he be able to buy the car? **no**

5. A square, a circle, and a triangle are each a different color. The figure that is round is not green, one of the figures is red, and the blue figure has four equal sides. Find the color of each figure. **square, blue; circle, red; triangle, green**

FUN with MATH
Try making Pretzel Cookies.
See page 104.

6. Darlene chooses a number between 1 and 10. If she adds 9, subtracts 8, then adds 6; the result is 12. What number did Darlene choose? **5**

7. Susan's car uses approximately 9 gallons of gasoline each week. If she pays $1.29 a gallon, how much money should she plan to spend on gasoline each week? State the method of computation used. **$13; estimate**

8. Use each of the digits 1, 2, 3, and 4 only once. Find two 2-digit numbers that will give the greatest possible answer when multiplied. **41, 32**

Critical Thinking

9. In a farmyard containing chickens and pigs, there are 54 heads and 144 feet. How many chickens are in the enclosure? how many pigs? **36 chickens, 18 pigs**

Mixed Review

Lesson 2-1 ***Add.***

10.	**11.**	**12.**	**13.**
91 + 18 = **109**	862 + 714 = **1,576**	9,152 + 3,601 = **12,753**	54,918 + 16,875 = **71,793**

Lesson 2-3 **Subtract.**

14.	**15.**	**16.**	**17.**
67 − 58 = **9**	876 − 421 = **455**	953 − 72 = **881**	7,658 − 4,329 = **3,329**

Lesson 2-4

18. On her last bank statement, Sarah's savings account had a total of $394.82. Since then, Sarah has made withdrawals of $48.00 and $27.50. What is the current balance of her savings account? **$319.32**

APPLYING THE LESSON

Independent Practice

Homework Assignment	
Minimum	Ex. 2-18 even
Average	Ex. 1-18
Maximum	Ex. 1-18

Chalkboard Examples

Solve each problem and write the method of computation used.

1. How much does Angie need for art supplies for school? She needs paper that costs $4.75, three brushes that cost $1.39 each, and oil pastels priced at $4.35. **$13.27; Calculator**

2. Complete the sequence. 1, 4, 2, 5, 3, __6__, __4__, __7__, __5__, **pattern +3, -2, mental computation**

Additional examples are provided on TRANSPARENCY 2-7.

EVALUATING THE LESSON

Check for Understanding Use Guided Practice Exercises to check for student understanding.

Error Analysis Watch for students that have difficulty deciding when to use paper and pencil or a calculator. Stress that simple problems can often be solved more quickly using paper and pencil.

Closing the Lesson Tell students that they will use logical reasoning throughout life to make the most efficient choices that occur everyday.

5-MINUTE CHECK-UP

(over Lesson 2-7)

Available on TRANSPARENCY 2-8.

Solve. State the method of computation used.

1. Gary is filling out his company expense report. He is reimbursed 25¢ a mile for every mile driven on business. If Gary drove 1,238 miles on business this month, how much will he be reimbursed? **$309.50; calculator**

▶ INTRODUCING THE LESSON

Ask students to point out and name geometric shapes, triangles, squares, rectangles, and so on, in the classroom or in magazines. Ask them how to find the distance around each shape. What is this distance called?

▶ TEACHING THE LESSON

Using Manipulatives Have students find the perimeter of their desks, notebooks, or room, using rulers.

Practice Masters Booklet, p. 21

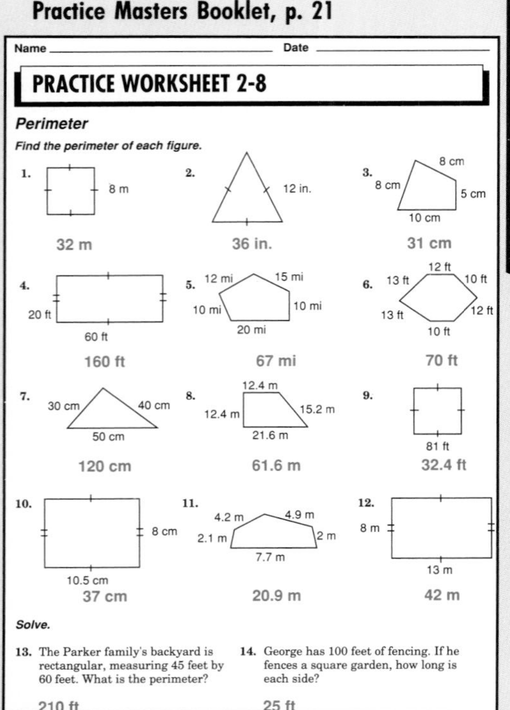

Name ______ Date ______

PRACTICE WORKSHEET 2-8

Perimeter

Find the perimeter of each figure.

1. 8 m — 32 m
2. 12 in. — 36 in.
3. 8 cm, 8 cm, 5 cm, 10 cm — 31 cm
4. 20 ft, 60 ft — 160 ft
5. 12 mi, 15 mi, 10 mi, 10 mi, 20 mi — 67 mi
6. 13 ft, 12 ft, 10 ft, 13 ft, 12 ft, 10 ft — 70 ft
7. 30 cm, 40 cm, 50 cm — 120 cm
8. 12.4 m, 12.4 m, 15.2 m, 21.6 m — 61.6 m
9. 81 ft — 32.4 ft
10. 8 cm, 10.5 cm — 37 cm
11. 4.2 m, 4.9 m, 2.1 m, 2 m, 7.7 m — 20.9 m
12. 8 m, 13 m — 42 m

Solve.

13. The Parker family's backyard is rectangular, measuring 45 feet by 60 feet. What is the perimeter? 210 ft
14. George has 100 feet of fencing. If he fences a square garden, how long is each side? 25 ft

Geometry Connection

2-8 PERIMETER

Objective
Find the perimeter of a polygon.

Luke Adams is building a fence around his backyard to screen the swimming pool. He needs to know the perimeter of the yard to order fencing materials.

The **perimeter (*P*)** of any figure is the distance around the figure. It is found by adding the measures of the sides.

Example A Find the perimeter of Luke's backyard shown at the right.

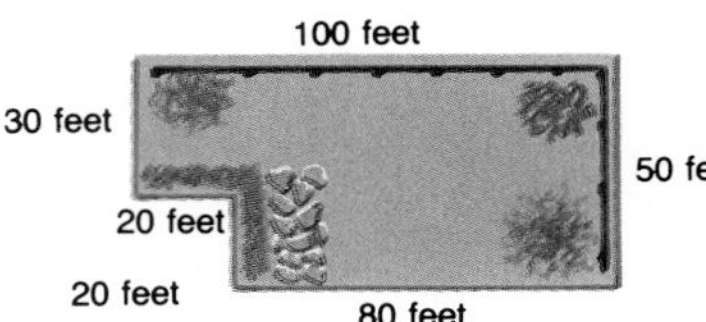

$P = 100 + 50 + 80 + 20 + 20 + 30$

$P = 300$

The perimeter is 300 feet.

Luke orders 300 feet of fencing.

You can also find the perimeter of a rectangle by multiplying the length by 2 and the width by 2, and adding the products.

Perimeter = 2 × length + 2 × width or $P = 2\ell + 2w$

Example B Find the perimeter of the rectangle at the right.

43 cm

29 cm

$P = 2\ell + 2w$ $\quad \ell = 43, w = 29$

$P = 2 \times 43 + 2 \times 29$

$P = 86 + 58$

$P = 144$ You can check the answer by adding the lengths of the sides.

Marks on the sides of a figure represent sides of equal measure.

The perimeter is 144 centimeters.

To find the perimeter of a square, multiply the length of one side by 4. Why?

Perimeter = 4 × length of one side or $P = 4s$

Example C Find the perimeter of a square with a side that measures 1.3 meters.

$P = 4s$

$P = 4 \times 1.3$

$P = 5.2$ or 5.2 meters. Check by adding.

Guided Practice *Find the perimeter of each figure.*

Examples A, B

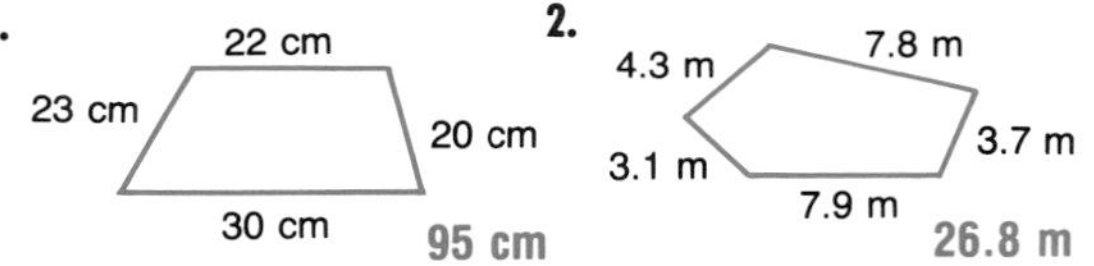

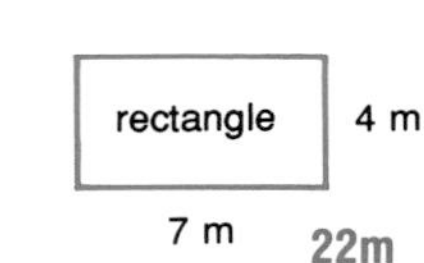

1. **95 cm** 2. **26.8 m** 3. **22m**

Example C *Find the perimeter of each square.*

4. side, 35 mm **140 mm** **5.** side, 12 cm **48 cm** **6.** side, 22 ft **88 ft** **7.** side, 14.1 cm **56.4 cm**

RETEACHING THE LESSON

Have students find the perimeter of objects found in the classroom. For example, their books, desk, notebooks, the room, etc. Suggest that they start at one vertex and add the lengths as they move around the figure. Record the measurements and then order them from greatest to least.

Reteaching Masters Booklet, p. 19

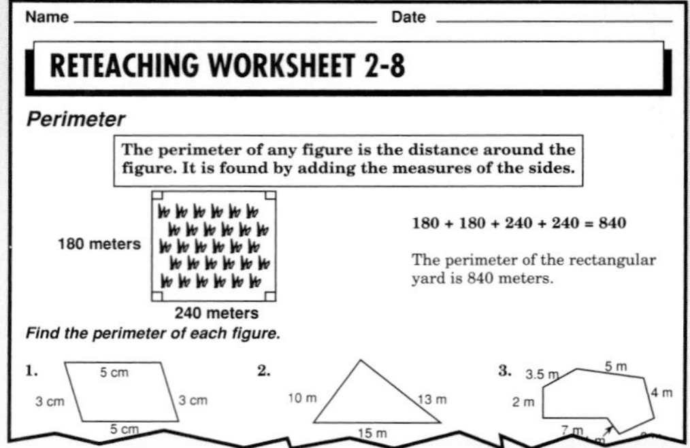

Name ______ Date ______

RETEACHING WORKSHEET 2-8

Perimeter

The perimeter of any figure is the distance around the figure. It is found by adding the measures of the sides.

180 meters

240 meters

180 + 180 + 240 + 240 = 840

The perimeter of the rectangular yard is 840 meters.

Find the perimeter of each figure.

1. 5 cm, 3 cm, 3 cm, 5 cm
2. 10 m, 13 m, 15 m
3. 3.5 m, 5 m, 4 m, 2 m, 7 m

Exercises

Practice

Find the perimeter of each figure.

8. 8 in.; 6 in.; 8 in.; 12 in. **34 in.**

9. 

24 ft

10. 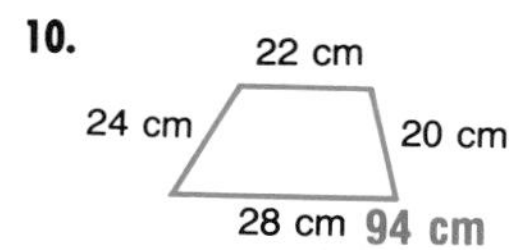

94 cm

11. square; side, 14 ft **56 ft**

12. rectangle; ℓ, 12 m, w, 3 m **30 m**

Applications

13. Pat is applying weatherstripping to a window on the north side of his house. If the window is 42 in. wide and 67 in. high, how many inches of weatherstripping does Pat need? **218 inches**

14. Miss Standish is outlining the bulletin board with ribbon using the school colors. The width is 8 feet and will be blue. The height is 4 feet and will be gold. How many feet of each color ribbon does she need? **16 ft of blue, 8 ft of gold**

Problem Solving

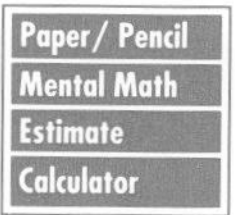

15. Juan is planting flowers around 4 sides of his patio. He is planning for 3 plants in every foot. How many plants does Juan need to fill four sides? **114 plants**

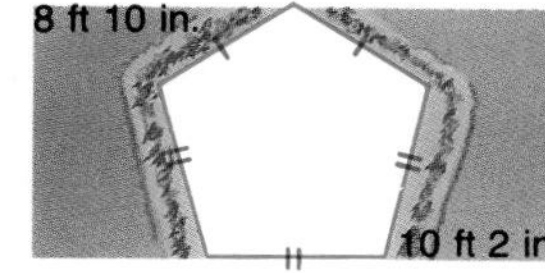

16. Erica wants to fence a triangular area in the backyard for a garden. The three sides measure 45 ft, 57 ft, and 47 ft. How many 25-foot rolls of fence will she need? **6 rolls**

17. A rectangular parking lot has a perimeter of 1,060 feet. Two of the sides measure 180 feet each. How long are the other two sides? **350 feet**

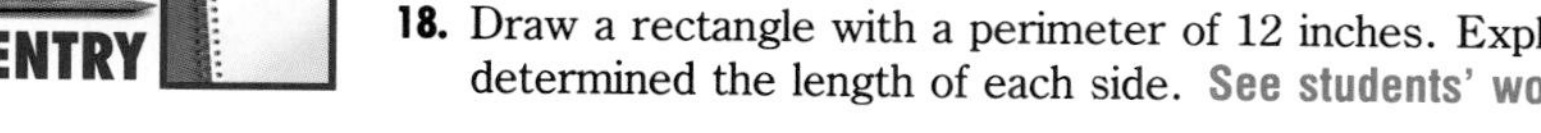

18. Draw a rectangle with a perimeter of 12 inches. Explain how you determined the length of each side. **See students' work.**

Using Algebra

Write an equation and solve.

19. A triangular park has a perimeter of 936 m. If the length of the three sides are the same, what is the length of one side? **$3s$ = 936, 312 m**

20. A rectangular stadium has a perimeter of 3,270 m. One side measures 545 m. What are the lengths of the other sides?

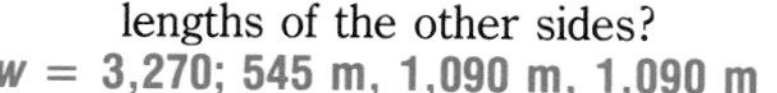

1,090 + $2w$ = 3,270; 545 m, 1,090 m, 1,090 m

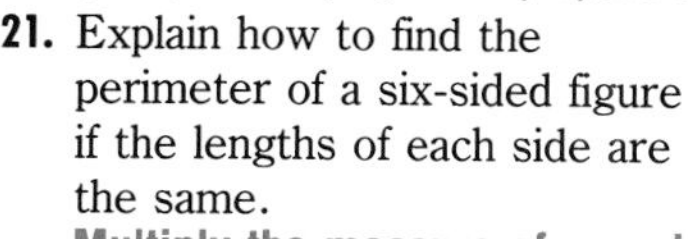

Write Math

21. Explain how to find the perimeter of a six-sided figure if the lengths of each side are the same.
Multiply the measure of one side by 6.

APPLYING THE LESSON

Independent Practice

Homework Assignment	
Minimum	**Ex. 2-20 even; 21**
Average	**Ex. 1-11 odd; 13-21**
Maximum	**Ex. 1-21**

Chalkboard Examples

Examples A, B *Find the perimeter.*

1. **41 cm** **2.** **20 m**

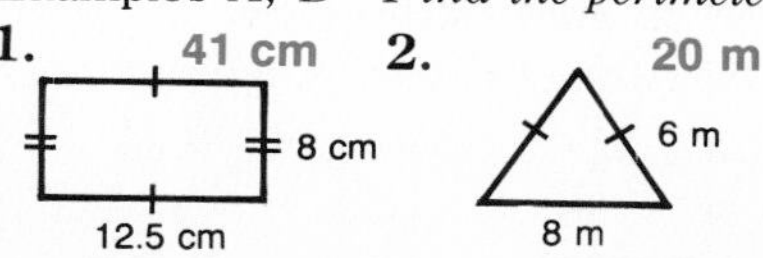

Example C *Find the perimeter of each square.*

3. side, 40 cm **160 cm**

4. side, 30.2 ft **120.8 ft**

Additional examples are provided on TRANSPARENCY 2-8.

▶ EVALUATING THE LESSON

Check for Understanding Use Guided Practice Exercises to check for student understanding.

Error Analysis Watch for students who add only the labeled sides of a figure. Remind students that perimeter includes every side of a figure.

Closing the Lesson Have students explain how to find the perimeter of a figure with four unequal sides and a figure with four equal sides.

▶ EXTENDING THE LESSON

Enrichment Masters Booklet, p. 19

Name ______________________ Date __________

ENRICHMENT WORKSHEET 2-8

Finding Sums of Arithmetic Series

The numbers 7, 11, 15, 19, 23, 31, 35 form an arithmetic sequence. It might take a long time to find the sum of the **series** 7 + 11 + 15 + ... + 35 even if you had a computer or a calculator. Formulas have been developed to make finding these sums much easier.

An **arithmetic series** is the sum of the first n terms of an arithmetic sequence, where n is the number of terms in the series. To find the sum of an arithmetic series using a formula, you need to know
- the first term,
- the common difference,
- and the number of terms.

Example: 7 + 11 + 15 + 19 + 23 + 27 + 31 + 35
The first term (a) is 7.
The common difference (d) is 4.
The number of terms (n) is 8.

Use the formula $S_n = \frac{n}{2}[2a + (n-1)d]$ and substitute to find the sum.

$S_8 = \frac{8}{2}[2(7) + (8-1)4]$

$S_8 = 4[14 + (7)4]$

$S_8 = 4[42]$

$S_8 = 168$

S_8 means the sum of the first eight terms in the series.

168 is the sum of 7 + 11 + 15 + 19 + 23 + 27 + 31 + 35.

Use the formula to find the sum of each arithmetic series.

1. 7 + 11 + 15 + 19 + 23 — 75
2. 12 + 17 + 22 + 27 + 32 + 37 — 147
3. 0 + 6 + 12 + 18 + ... + 36 — 126
4. 50 + 57 + 64 + ... + 92 — 497
5. 5 + 10 + 15 + 20 + ... + 65 — 455
6. 9 + 18 + 27 + 36 + ... + 72 — 324
7. 9 + 20 + 31 + 42 + ... + 86 — 380
8. 21 + 29 + 37 + 45 + 53 + ... + 93 — 570

5-MINUTE CHECK-UP

(over Lesson 2-8)

Available on TRANSPARENCY 2-9.

Find the perimeter of each figure.

1. 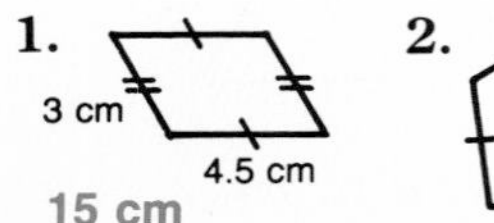

15 cm

2. 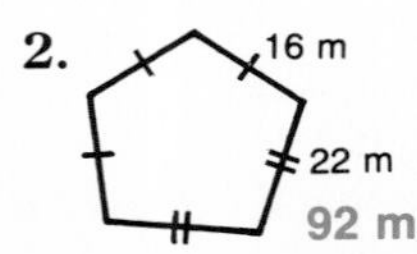

92 m

3. square; side, 6 ft 24 ft

Extra Practice, Lesson 2-8, p. 453

INTRODUCING THE LESSON

Have students tell about times when advertising is missing needed facts. For example, an advertised sale item is only available at the sale price with an additional purchase.

TEACHING THE LESSON

Using Data Encourage students to use paper and pencil to take notes when reading a problem. Gathering data as they read will help them to identify missing or extra facts.

Practice Masters Booklet, p. 22

PRACTICE WORKSHEET 2-9

Problem Solving: Identifying the Necessary Facts

If the problem has the necessary facts, solve. State any missing or extra facts.

1. Marcia earns $8.10 per hour. Each paycheck she puts $25 toward the purchase of U.S. Savings Bonds and $20 into her savings account. If Marcia gets paid every two weeks, how much will she save in a year?
   $1,170; earns $8.10 an hour — extra
2. Brad has a checking account balance of $832.76. He just wrote check #235 to his landlord for $290 to pay the rent for March and check #236 to the electric company for $32.15. What is Brad's new checking account balance?
   $510.61; check #235, 236 — extra
3. Charlie had the following payroll deductions last week: State Tax $52.18, Federal Tax $75.13, Health Insurance $5.76, Union Dues $8.50, FICA Tax $32.43, and City Tax $9.12. What were Charlie's total deductions for taxes?
   $168.86; health insurance, union dues — extra
4. The Lopez family has a monthly income of $2,450. Each month they pay $490 for rent, $550 for food, $42 for electricity, $90 for entertainment, and $70 for transportation. The remaining money goes to savings and travel. How much do they spend on travel? Missing: amount saved each month
5. Ralph's paycheck was $276.82 after $137.28 in deductions. He deposited the check in his checking account. He also withdrew $25 in cash and wrote a $43.87 check for groceries. How much money was left in Ralph's account?
   Missing: beginning checking balance
6. Sherrie's bank statement shows a balance of $628.97 in her checking account. She wrote 15 checks, but 3 checks totaling $153 are not included on this statement. What is Sherrie's correct checking account balance?
   $475.97; 15 checks, 3 checks — extra
7. The Foster family pays $520 rent each month. They average $130 per month for utilities (gas, electric, water, and telephone). They have lived in this house for 3 years. At this rate, how much rent do they pay in a year?
   $6,240; $130 utilities, 3 yr — extra
8. Gail and George budget their monthly household income as follows: $650 for housing, $700 for food, 10% for savings, $50 for transportation, and the rest for miscellaneous expenses. How much money do they save in a month?
   Missing: monthly income
9. Carmen puts hair care, cleaning, and new clothes in her monthly budget for clothing. Her total monthly budget is $850, of which $120 goes for clothing. Permanents cost $26 and she bought a $50 dress last week. What is Carmen's monthly budget for clothing?
   $120; total budget of $850, permanent $26, dress $50 — extra
10. Pam is paid weekly at the bank where she works. She gets a gross income of $21,580 per year and has weekly deductions of $135.27. What are Pam's total deductions for a year?
    $7,034.04; gross income $21,580 — extra.

Problem Solving

▶ Explore ▶ Plan ▶ Solve ▶ Examine

2-9 IDENTIFYING THE NECESSARY FACTS

Objective
Identify missing and extra facts.

At Rock Records Aaron sold 185 heavy metal tapes, 126 easy listening tapes, and 142 classical tapes on Thursday. How many heavy metal and easy listening tapes did he sell?

Always study a verbal problem carefully to determine what facts are given and what facts are needed. Be alert for extra facts you do not need to solve the problem.

▶Explore What is given?

- Aaron sold 185 heavy metal tapes, 126 easy listening tapes, and 142 classical tapes.

What is asked?

- How many heavy metal and easy listening tapes did Aaron sell?

The number of classical tapes sold is not needed.

▶Plan Add the sales for heavy metal and easy listening tapes to find the total sales.

▶Solve

$$\begin{array}{r} 185 \\ +\ 126 \\ \hline 311 \end{array}$$

Estimate: 200 + 100 = 300

Aaron sold 311 tapes.

▶Examine The answer is close to the estimate, so 311 is a reasonable solution.

Sometimes facts that you need are missing from the problem.

Nick is saving money to buy a guitar. He has saved $120. He borrows $50. How much more money does Nick need?

▶Explore What is given?

- Nick has saved $120 and borrowed $50.
  Nick has a total of $170.

What is asked?

- How much more money does Nick need to buy the guitar?

▶Plan Find the difference between the cost of the guitar and the money Nick has saved.

▶Solve You do not know the cost of the guitar. The problem does not have all the facts you need to solve it.

RETEACHING THE LESSON

State any facts not needed to solve the problem.

1. The post office opens at 8:00 A.M. It costs 29¢ per ounce to mail first-class packages. How much would it cost to mail a 4-ounce package first class? Post office opens at 8:00 AM.
2. A gift costs $7.05. The clerk was given a $20 bill. There is $1.20 tax on $20. What is the change from the $20? There is $1.20 tax on $20.

Reteaching Masters Booklet, p. 20

Name ____________ Date ____________

RETEACHING WORKSHEET 2-9

Problem Solving: Identifying the Necessary Facts

Mr. Johnston bought 2.6 pounds of apples and 2.2 pounds of peaches. If he divides the fruit among his grandchildren, how many pounds of fruit will each child receive?

Explore What is given? Mr. Johnston bought 2.6 pounds of apples and 2.2 pounds of peaches to divide among his grandchildren.
What is asked? How many pounds of fruit will each grandchild get?

Plan Add to find the total number of pounds of fruit.
Divide the sum by the number of grandchildren.

Solve
Step 1
total pounds of fruit = 2.6 + 2.2
= 4.8

Step 2
4.8 ÷ number of grandchildren = amount per child

You need to know how many

Guided Practice

If the problem has the necessary facts, solve. State any missing or extra facts.

1. Gasoline costs $1.28 a gallon and oil costs $1.69 a quart. What is the cost of two quarts of oil? **$3.38; The cost of the gasoline is not needed.**
2. The Photography Club has bake sales to help pay for darkroom supplies. This week they made $34.80, and last week they made $29.40. How much more do they need to purchase darkroom supplies? **missing the cost of darkroom supplies**

Problem Solving

Paper/Pencil
Mental Math
Estimate
Calculator

If the problem has the necessary facts, solve. State any missing or extra facts. **3. missing original amount of flour**

3. Frank wants to triple a recipe that calls for 2 cups of milk and 2 eggs. How many cups of flour should he use?
4. Claudia raises gerbils to sell to Petite Pets. If she sells 50 gerbils at $4 each, how much money will she earn? **$200**

FUN with MATH

When and where were skis first used? See page 104.

5. It will cost the Ski Club $250 to hire a disc jockey, $125 to rent decorations, and $80 to buy refreshments for their dance. How much money does the club need to sponsor the dance? **$455**
6. The Choice Ticket agency sold 150 tickets for the better seats at $28 each. General admission tickets were sold out at $25 each. What was the income from the ticket sales?

6. missing the number of general admission seats

Critical Thinking

7. You have 3 buckets, one holds 7 L and is filled with water. The other two buckets hold 5 L and 2 L and are empty. How can you pour

7. Pour 5 L into the 5 L bucket. There are 2 L left. Fill the 2-L bucket from the 5-L bucket. Add the 2 L to the 2 L in the large bucket.

8. Select an assignment from this chapter that shows your creativity and place it in your portfolio. **See students' work.**

Mixed Review

Lesson 1-12

Solve.

9. $4m = 0.008$ **0.002**
10. $500 = 2.5y$ **200**
11. $40 = \frac{g}{5}$ **200**
12. $0.9 = \frac{n}{2}$ **1.8**

Lesson 2-7

Solve. State the method of computation used.

13. Grant is going to order 3 magazine subscriptions for the next year. The magazines cost $38.95, $14.80, and $24.25. How much money does Grant need to pay for the magazines? **$78.00, calculator or paper/pencil**

Chalkboard Examples

Solve. State any missing or extra facts.

1. Lisa gave 2 tapes to Mary, 3 to Jesse, and 4 to Bob. If she had 32 tapes and 6 CDs before giving any away, how many tapes does Lisa have left? **23, 6 CDs extra**
2. Before boarding the bus for home, Jan left 2 books in his locker, 1 in the gym, and 1 in the band room. How many books did Jan bring home? **missing original number of books**

Additional examples are provided on TRANSPARENCY 2-9.

EVALUATING THE LESSON

Check for Understanding Use Guided Practice Exercises to check for student understanding.

Error Analysis Watch for students who use extra facts in computing an answer.

Closing the Lesson Have students summarize the 4-step problem-solving plan in their own words.

EXTENDING THE LESSON

Enrichment Masters Booklet, p. 20

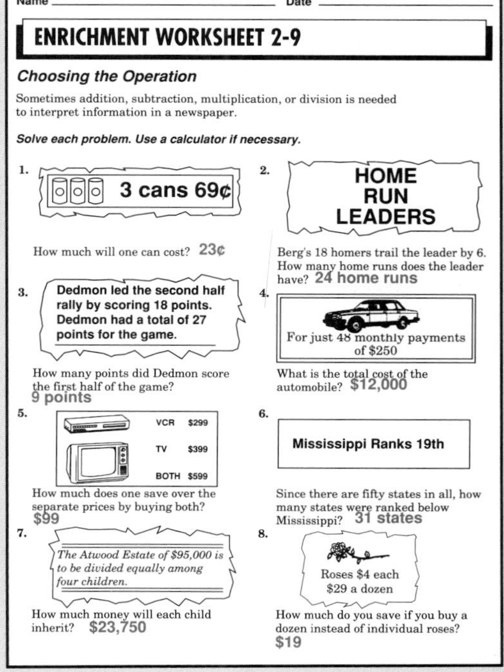

Name ______ Date ______

ENRICHMENT WORKSHEET 2-9

Choosing the Operation

Sometimes addition, subtraction, multiplication, or division is needed to interpret information in a newspaper.

Solve each problem. Use a calculator if necessary.

1. 3 cans 69¢
How much will one can cost? **23¢**

2. HOME RUN LEADERS
Berg's 18 homers trail the leader by 6. How many home runs does the leader have? **24 home runs**

3. Dedmon led the second half rally by scoring 18 points. Dedmon had a total of 27 points for the game.
How many points did Dedmon score the first half of the game? **9 points**

4. For just 48 monthly payments of $250
What is the total cost of the automobile? **$12,000**

5. VCR $299 TV $399 BOTH $599
How much does one save over the separate prices by buying both? **$99**

6. Mississippi Ranks 19th
Since there are fifty states in all, how many states were ranked below Mississippi? **31 states**

7. The Atwood Estate of $95,000 is to be divided equally among four children.
How much money will each child inherit? **$23,750**

8. Roses $4 each $29 a dozen
How much do you save if you buy a dozen instead of individual roses? **$19**

APPLYING THE LESSON

Independent Practice

Homework Assignment	
Minimum	Ex. 1-13
Average	Ex. 1-13
Maximum	Ex. 1-13

Chapter 2, Quiz B (Lessons 2-6 through 2-9) is available in the Evaluation Masters Booklet, p. 9.

Using the Chapter Review

The Chapter Review is a comprehensive review of the concepts presented in this chapter. This review may be used to prepare students for the Chapter Test.

Chapter 2, Quizzes A and B, are provided in the Evaluation Masters Booklet as shown below.

Quiz A should be given after students have completed Lessons 2-1 through 2-5. Quiz B should be given after students have completed Lessons 2-6 through 2-9.

These quizzes can be used to obtain a quiz score or as a check of the concepts students need to review.

Evaluation Masters Booklet, p. 9

CHAPTER 2, QUIZ A (Lessons 2-1 through 2-5)

Add.

1. 872 + 694 2. 7,648 + 13,582 3. 0.08 + 1.9 + 4.736

4. 39,109 + 16,883 5. 39.6 + 4.88

Subtract.

6. 9,378 – 4,556 7. 8,406 – 3,936

8. 39.6 – 4.88 9. 37.49 – 9.7

10. Kerry had a checkbook balance of $129.45. She wrote a check for $54.56. What was her new balance?

1. 1,566
2. 21,230
3. 6.716
4. 55,992
5. 44.48
6. 4,822
7. 4,470
8. 34.72
9. 27.79
10. $74.89

Name ______________ Date ______________

CHAPTER 2, QUIZ B (Lessons 2-6 through 2-9)

Write the next three numbers in each sequence.

1. 80, 74, 68, 62, ... 2. 22.5, 28, 33.5, ...

State whether each sequence is arithmetic. If it is, write the next three terms. If not, write no.

3. 3, 12, 36, ... 4. 12, 13.5, 15, ...

Find the perimeter of each figure.

5. 8.5 m, 12.2 m 6. 6 cm, 4 cm, 4 cm, 2.5 cm, 9.5 cm

Solve. State the method of computation you used.

7. A ticket to a game costs about $5, milk is $2.50 each, and sandwiches cost from $3 to $6. If Larry plans to eat his lunch at the game, about how much money should he take?

Solve if possible. State any missing or extra facts.

8. A rectangular lot has a perimeter of 290 feet. Two of the sides measure 60 feet. How long are the other two sides?
9. Eric gained 1.2 pounds in April and 2.5 pounds in May. In June he lost 4 pounds. How much does he weigh now?
10. On a test, Albert got a score of 77, José got a score of 81, Louise got a 72, and Inez got a 90. How many points higher than José's score was Inez's score?

1. 56, 50, 44
2. 39, 44.5, 5
3. no
4. 16.5, 18, 1
5. 41.4 m
6. 26 cm
7. about $12
8. 85 ft missing: weight on
9. 9 points; score and
10. score extr

Chapter 2

REVIEW

Choose the letter of the word at the right that best matches each description.

Vocabulary / Concepts

1. You can check a __?__ problem by using addition. **e**
2. When adding or subtracting decimal numbers, the __?__ should be aligned. **b**
3. When subtracting decimals, it is sometimes necessary to annex __?__. **g**
4. A list of numbers that follows a certain pattern is called a __?__. **d**
5. A check register is used to record __?__. **f**
6. The perimeter of any figure is found by __?__ the measures of the sides. **a**

a. adding
b. decimal points
c. perimeter
d. sequence
e. subtraction
f. transactions
g. zeros
h. multiplying

Exercises / Applications

Lessons 2-1, 2-2, 2-3, 2-4

Add or subtract.

7. 47 + 37 **84**
8. 419 + 54 **473**
9. 916 + 520 **1,436**
10. 0.93 + 0.84 **1.77**
11. 0.132 + 0.378 **0.510**
12. 0.132 − 0.017 **0.115**
13. 60.7 − 42.3 **18.4**
14. 375.3 − 190.4 **184.9**
15. 5,343 − 602 **4,741**
16. 5,047 − 2,085 **2,962**
17. 4.06 + 0.782 **4.842**
18. 4.9 − 1.807 **3.093**
19. 827 − 53 **774**
20. 0.78 + 3.461 **4.241**
21. 6.1 − 4.96 **1.14**
22. 5,126 + 2,899 **8,025**
23. 0.685 + 0.59 + 1.011 **2.286**
24. $200 − $133.62 **$66.38**
25. 38,876 + 3,603 **42,479**
26. 9.84 + 0.27 + 3.6 **13.71**
27. 43,040 − 5,503 **37,537**
28. 58,316 − 29,957 **28,359**

Lesson 2-5

Solve.

29. Edward has a balance of $86.09 in his checking account. If he deposits $356 and writes checks for $54.60 and $23.98, what is his new balance? **$363.51**

Copy and complete the balance column.

30. What is the balance on May 13? **$699.54**
31. What is the balance on May 16? **$794.37**
32. What is the balance on May 22? **$1,239.66**

CHECK NUMBER	DATE	DESCRIPTION OF TRANSACTION	PAYMENT/DEBIT (−)	✓ TAX	DEPOSIT/CREDIT (+)	BALANCE FWD.
						$786.55
607	5/12	Supermart groceries	$57.31		$	
608	5/13	Gina's Boutique gift	29.70			
D	5/15	Deposit dividend			110.50	
609	5/16	Fraley's Hardware hardware	15.62			
D	5/20	Deposit paycheck			471.87	
610	5/22	Classic Rock CD's	26.63			

Lesson 2-6

***State whether each sequence is arithmetic. If it is, write the next three numbers. If it is not, write* no.**

33. 11.3, 13.5, 15.7, ... **34.** 110, 135, 160, ...

35. 85.4, 75.3, 65.2, ... **36.** 54, 45, 36, ... yes; 27, 18, 9

33. yes; 17.9, 20.1, 22.3 34. yes; 185, 210, 235 35. yes; 55.1, 45, 34.9

Lesson 2-7

Solve. State the method of computation used.

Paper/ Pencil
Mental Math
Estimate
Calculator

37. A sports car gets 32.5 miles per gallon of gasoline. *About* how far can it go on 12 gallons of gasoline? 360 miles, estimate

38. Matt worked for 28 hours each week for the last two weeks and earned a total of $196 a week. How much was he paid per hour? $7/h; paper and pencil or calculator

Lesson 2-8

Find the perimeter of each figure.

39. 12 in. 5 in. 34 in.

40. 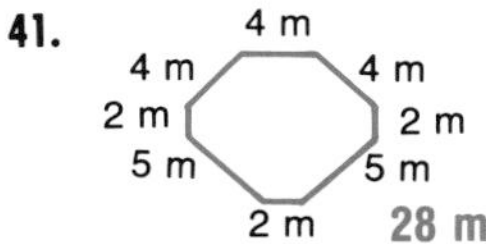

49 ft

41. 4 m, 4 m, 4 m, 2 m, 2 m, 5 m, 5 m, 2 m 28 m

42. square; side, 4.6 m 18.4 m

43. rectangle; ℓ, 32 ft; w, 12 ft 88 ft

Lesson 2-9

If the problem has the necessary information, solve. State any missing or extra facts.

44. In June, Troy could bench press 55.5 kg. By October, he could bench press an additional 5.75 kg. How much could Troy bench press in October? 61.25 kg

45. missing initial weight

45. Suzanne lost 4.6 pounds in February, 5.2 pounds in March, and 4.7 pounds in April. How much does she weigh now?

46. Benito's horse weighs 61 pounds at birth. The horse should weigh 1,100 pounds before he can carry Benito easily. How much weight does the horse need to gain? 1,039 pounds

47. The cheerleaders bought 350 pennants and 200 pompoms to sell. They received 282 orders for pennants and sold 35 pennants more during the football game. How many pennants do they have left to sell? 33; do not need number of pompoms

Alternate Assessment Strategy

To provide a brief in-class review, you may wish to read the following questions to the class and require a verbal response.

1. What is the sum of 15 and 35? 50
2. What is the sum of 4.6 and 8.2? 12.8
3. Find the difference of 82 and 63. 19
4. What is 14 minus 4.6? 9.4
5. What is the result when 12.56 is subtracted from 23.88? 11.32
6. What are the next 3 terms in the sequence 8, 14, 20, . . . ? 26, 32, 38
7. What is the difference in the sequence 4, 15, 26, 37, 48? 11
8. Find the perimeter of a square that has a side with length 4 cm. 16 cm
9. What is the perimeter of a triangle that has sides 11 m, 14 m, and 20 m? 45 m
10. Find the sum of 123.456 and 433.001. 556.457

APPLYING THE LESSON

Independent Practice

Homework Assignment	
Minimum	Ex. 1-6; 7-47 odd
Average	Ex. 1-6; 8-42 even; 44-47
Maximum	Ex. 1-47

Using the Chapter Test

This page may be used as a test or as an additional page of review if necessary. Two forms of a Chapter Test are provided in the Evaluation Masters Booklet. Form 2 (free response) is shown below, and Form 1 (multiple choice) is shown on page 67.

The **Tech Prep Applications Booklet** provides students with an opportunity to familiarize themselves with various types of technical vocations. The Tech Prep applications for this chapter can be found on pages 3–4.

The **Performance Assessment Booklet** provides an alternative assessment for evaluating student progress. An assessment for this chapter can be found on pages 3–4.

Evaluation Masters Booklet, p. 8

CHAPTER 2 TEST, FORM 2

Add.

1. 396 + 685 — 1. 1,081
2. 6,382 + 29,586 — 2. 35,968
3. 9,246 + 22,651 + 104,537 — 3. 136,434
4. 6.67 + 69.4 — 4. 76.07
5. 8.6 + 15 + 0.73 — 5. 24.33

Subtract.

6. 584 – 89 — 6. 495
7. 4,043 – 2,706 — 7. 1,337
8. 80,305 – 48,668 — 8. 31,637
9. 0.6 – 0.54 — 9. 0.06
10. 49.6 – 37.37 — 10. 12.23
11. 64 – 8.25 — 11. 55.75
12. 7.341 – 4.9 — 12. 2.441
13. What is the next number in this arithmetic sequence? 24, 31, 38, 45, ___ — 13. 52
14. What is the sixth number in this sequence? 30.7, 33, 35.3, ... — 14. 42.2

Solve.

15. Danielle had a checking balance of $505.26. She wrote a check for $85.95. What was her new balance? — 15. $419.31
16. What method of computation could you use? Dee bought a CD player for $399, compact disks for $58.95, and a case for $39.29. Is $500 enough to pay? — 16. Estimation
17. Find the perimeter of a rectangular plot that is 32 feet by 40 feet. — 17. 144 ft
18. Li is fencing her flower garden. The garden has 6 sides, each 9 meters in length. How long will the fence be? — 18. 54 m

Solve or state any missing or extra facts.

19. Mr. Reid bought a roll of film for $1.89 and tuna for $0.79. How much change did he receive from $5? — 19. $2.32
20. A gallon of orange juice costs $3.99, a gallon of milk costs $2.89, and a gallon of apple juice costs $2.29. What is the cost of a gallon of milk and a gallon of apple juice? — 20. gallon of orange juice costs $3.99

Bonus The fifth and sixth terms of an arithmetic sequence are 39 and 47. What is the common difference and first term of the sequence? — Bonus 8; 7

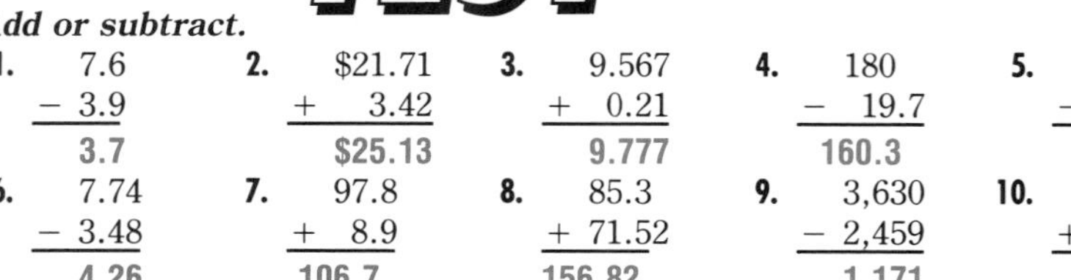

Add or subtract.

1. 7.6 − 3.9 = **3.7**
2. $21.71 + 3.42 = **$25.13**
3. 9.567 + 0.21 = **9.777**
4. 180 − 19.7 = **160.3**
5. 308 − 43.7 = **264.3**
6. 7.74 − 3.48 = **4.26**
7. 97.8 + 8.9 = **106.7**
8. 85.3 + 71.52 = **156.82**
9. 3,630 − 2,459 = **1,171**
10. 356 + 2,522 = **2,878**
11. 846 − 334 **512**
12. 256 + 16 **272**
13. 614 + 257 **871**
14. 630 − 225
15. 300 − 199 **101**
16. 359 + 2,330 **2,589**
17. 2,973 − 168
18. 680 + 3,945 **4,625**
19. 83.8 + 2.071 **85.871**
20. 78.4 − 9.47 **68.93**

14. 405
17. 2,805

***State whether each sequence is arithmetic. If it is, write the next three numbers. If it is not, write* no.**

21. 3.1, 4.3, 5.5, ... **6.7, 7.9, 9.1**
22. 50, 42, 34, ... **26, 18, 10**
23. 3, 15, 75, ... **no**
24. 243, 81, 27, ... **no**

Find the perimeter of each figure.

25. 4 m, 7 m **22 m**

26. 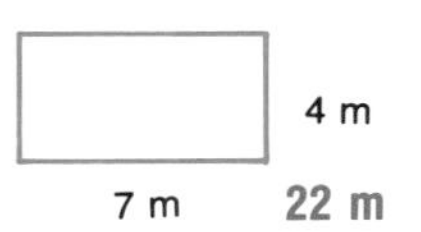

21 cm

27. 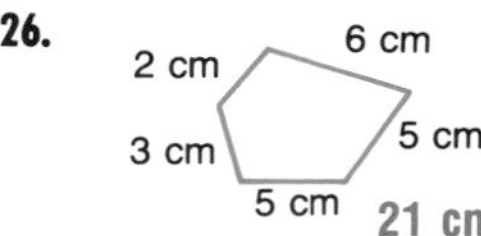

66 in.

28. square; side, 6.8 cm **27.2 cm**
29. square; side, 8 ft **32 ft**

If the problem has the necessary facts, solve. State any missing or extra facts.

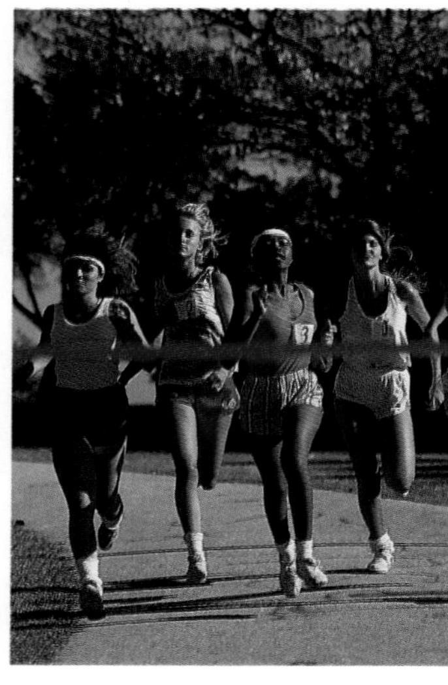

30. Randy deposited $25.67 to his checking account. He wrote checks for $12.35 and $36.11. His balance before these transactions was $632.75. What is his new balance? **$609.96**
31. If a city bus pass costs $2.10 a week, *about* how much will bus passes cost for one year? State the method of computation used. **estimation, approximately $104**
32. Charles earns $6.50 an hour. How much does he earn in a week?
33. Ella finished the mile in 6.54 minutes. Rose finished in 6.32 minutes. Who won and by how many seconds? **Rose; 13.2 seconds**

32. missing number of hours worked per week

BONUS: How do you know that $P = 4 \times s$ will not always give the same answer as $P = 2\ell + 2w$, but $P = 2\ell + 2w$ will always give the same answer as $P = 4 \times s$? **See margin.**

Additional Answers

BONUS: If ℓ and w equal s, 2ℓ plus $2w$ equals 4ℓ or $4w$ or $4s$.

A **Test and Review Generator** is provided in Apple, IBM, and Macintosh versions. You may use this software to create your own tests or worksheets, based on the needs of your students.

BUDGETING

Marie Hiroto is a student at Central State College. She wants to buy a computer and has saved $900 for the down payment. The monthly payment on the computer is $68 for 24 months.

Marie makes $310 each month working in one of the dining halls on campus. She prepares and follows a **budget** so that she can afford the computer.

ITEM	MONTHLY AMOUNT
Food	$ 82
Computer	68
Recreation	40
Clothes	35
Savings	40
Miscellaneous	20
Transportation	25
TOTAL	$310

Find the average amount spent in one week for each item. Assume that there are 4 weeks in a month.

1. Food $20.50
2. Recreation $10
3. Clothing $8.75
4. Transportation $6.25

From which item of the budget should each expense be taken?

5. blouse clothing
6. movie recreation
7. birthday gift misc.
8. dry cleaning clothing
9. lunch food
10. concert recreation

11. If Marie takes $5 a month from her clothes allowance and $8 a month from her recreation allowance and adds both amounts to savings, how much will she save in 6 months? $318

12. At the end of the month, Marie finds she has $12 left over from her food allowance, $6 left over from her recreation allowance, and $5 left over from her miscellaneous allowance. Suppose Marie always has these extra amounts. How would you adjust her budget using these extra amounts? Answers may vary.

13. Why is it a good practice to put money into savings on a regular basis? See margin.

14. Make up an income or allowance, and then plan a budget. See students' work

Applying Mathematics to the Real World

This optional page shows how mathematics is used in the real world and also provides a change of pace.

Objective Use a budget.

Using Critical Thinking After Exercise 13, ask students to hypothesize about why people who follow budgets are successful in handling their money. Answers should include reasons such as planning for an outcome helps to insure an outcome, organization helps prevent waste.

Ask students what other areas can be budgeted and why. time, space

Activity

Using Research Have students research the National Budget in the library.

Additional Answers

13. It is easy to spend the money on something else. Savings can be used for unexpected expenses or for expensive items to avoid buying on credit.

Using the Cumulative Review

This page provides an aid for maintaining skills and concepts presented thus far in the text.

A Cumulative Review is also provided in the Evaluation Masters Booklet as shown below.

Evaluation Masters Booklet, p. 10

Name ______________________ Date ______________

CUMULATIVE REVIEW 2 Chapters 1–2

Name the digit in each place-value position in 8,327,519.

1. hundreds 2. ten thousands 3. millions

1. 5
2. 2
3. 8

Estimate.

4. 72,761 + 5,992 5. 48.7 − 14.2 6. 276)14,998

*4. 79,000
*5. 35
*6. 50

* Accept any reasonable estimate.

Write using exponents.

7. 4 × 4 × 4 × 4 8. 7 × 7 × 7 9. 6 × 6 × 6 × 6 × 6

7. 4^4
8. 7^3
9. 6^5

Order from least to greatest.

10. 16.7, 16.17, 16.1, 16.71 11. 0.810, 0.081, 0.180, 0.801

10. 16.1, 16.1… 16.7, 16.7…
11. 0.081, 0.1… 0.801, 0.8…

Add or subtract.

12. 98 + 327 13. 6,604 − 3,992 14. 56.88 − 3.91 15. 2.7 + 0.55

16. 15.4 − 8.47 17. 55,706 + 32,993 18. 0.7 − 0.22

12. 425
13. 2,612
14. 52.97
15. 3.25
16. 6.93
17. 88,699
18. 0.48

Find the value of each expression.

19. 3 × 7 + 4 20. 24 − 2 × 9 21. [36 + 3 × (2 + 3)]

19. 25
20. 6
21. 51

Write the next three numbers in each sequence.

22. 14.7, 19.7, 24.7, ... 23. 88, 75, 62, ...

22. 29.7, 34.7…
23. 49, 36, 2…

Solve. State any missing or extra facts.

24. Sam had a balance of $420.25 in his checking account. He made a deposit of $75 and wrote a check for $107.50. What was his new balance?

24. $387.75

25. Anna's yard is in the shape of a rectangle. It is 30 feet wide. What is the perimeter of her yard?

25. missing: … of yard

Cumulative REVIEW/TEST

Free Response

Lesson 1-1

Round each number to the underlined place-value position.

1. 749 **750** 2. 3,279 **3,300** 3. 633 **600** 4. 60,280
5. 4,803 **4,800** 6. 90,550 7. 393,488 8. 2,683 **3,000**

4. 60,000 6. 90,600 7. 390,000

Lesson 1-6

Estimate.

9. 697 + 77 = **780** 10. 280 + 390 = **700** 11. 655 + 4,848 = **5,500** 12. 2,684 + 3,724 = **7,000** 13. 96,278 + 3,843 = **100,000**

Lesson 1-8

14. Mike, Steve, and Laura each have a baseball card collection. They have rookie cards of Willie Mays, Mickey Mantle, and Brooks Robinson. Mike does not have the Willie Mays card. Steve does not have the Mickey Mantle card and Laura does not have Mickey Mantle or Brooks Robinsons. Which boy has which card? **Mike-Mickey Mantle; Steve-Brooks Robinson; Laura-Willie Mays**

Lesson 1-9

Find the value of each expression. **16. 36**

15. (64 − 8) ÷ 4 **14** 16. 144 ÷ 12 × (4 − 1) 17. 64 − 8 ÷ 4 **62**
18. 36 ÷ (6 × 3) − 2 **0** 19. 9 × (21 + 7) **252** 20. 9 × 21 + 7 **196**

Lesson 2-1

Add.

21. 12 + 67 = **79** 22. 52 + 159 = **211** 23. 342 + 4,181 = **4,523** 24. 2,827 + 1,354 = **4,181** 25. 6,184 + 56,424 = **62,608**

26. 576 + 91 **667** 27. 8,897 + 659 **9,556** 28. 679 + 347 **1,026**

29. During a half hour, 28 mini-vans, 97 sedans, and 36 sports cars passed Erika's house. How many passed in all? **161**

Lesson 2-3

Subtract.

30. 53 − 19 = **34** 31. 991 − 88 = **903** 32. 1,471 − 927 = **544** 33. 2,617 − 2,419 = **198** 34. 10,451 − 8,367 = **2,084**

35. 223 − 92 **131** 36. 928 − 233 **695** 37. 2,415 − 754 **1,661**

38. Alice earned $45 last week. She bought new shoes that cost $27. How much money does she have left? **$18**

Lesson 2-6

39. Jude plans to collect CDs. Each month he plans to purchase 3 CDs. He begins his collection with 4 CDs. How many CDs will he have in 7 months? **25 CDs**

APPLYING THE LESSON

Independent Practice

Homework Assignment	
Minimum	Ex. 1-39 odd
Average	Ex. 1-39
Maximum	Ex. 1-39

Chapter 2

Multiple Choice

Choose the letter of the correct answer for each item.

1. The first week he began his savings account, Caleb saved $7. The second week his total amount was $21. The third week he had a total of $35. How much will he have saved in 7 weeks?
   a. $77 ► c. $91
   b. $105 d. $133

   Lesson 2-6

6. If Arnold spends $3,992 of $8,825 that he has in his savings account, how much does he have left?
   ► e. $4,833
   f. $4,933
   g. $5,173
   h. $5,833

   Lesson 2-3

2. Find the value of $14 + 6 \times 3$.
   e. 60 g. 18
   ► f. 32 h. 46

   Lesson 1-9

7. What is the difference of 37,153 and 1,279?
   a. 24,363 c. 35,884
   ► b. 35,874 d. 35,974

   Lesson 2-3

3. Estimate the product of 68 and 5,428.
   a. 35,000
   ► b. 350,000
   c. 3,500,000
   d. 35,000,000

   Lesson 1-7

8. What number, when multiplied by 3, equals 42?
   ► e. 14
   f. 39
   g. 45
   h. 126

   Lesson 1-12

4. Margaret adds 150 to 874. What is her result?
   e. 724
   f. 824
   g. 924
   ► h. 1,024

   Lesson 2-1

9. What is the sum of 1,436 and 93,806?
   a. 94,242
   b. 95,232
   ► c. 95,242
   d. 108,166

   Lesson 2-1

5. Which equation would you use to solve this problem?
   Tim used $15 to purchase gas and oil for his car. If the oil cost $3.52, how much did he pay for gas?
   a. $15 + 3.52 = y$
   ► b. $3.52 + y = 15$
   c. $x - 3.52 = 15$
   d. none of these

   Lesson 1-11

10. Jonathan wants to purchase a computer for $899, a computer desk for $90, a printer for $350, and a box of computer paper for $32.50. Can Jonathon afford the computer system? What fact is missing?
    e. the amount of set-up time
    f. the cost of the computer with printer
    ► g. the amount Jonathan has to spend
    h. the size of the computer

    Lesson 2-9

Using the Standardized Test

This test serves to familiarize students with standardized format while testing skills and concepts presented up to this point.

APPLYING THE LESSON

Independent Practice

Homework Assignment	
Minimum	Ex. 1-10
Average	Ex. 1-10
Maximum	Ex. 1-10

Evaluation Masters Booklet, pp. 6-7

Name ______________ Date ______________

CHAPTER 2 TEST, FORM 1

Write the letter of the correct answer on the blank at the right of the page.

1. Add. 763 + 875 — A. 15,138 B. 1,638 C. 1,538 D. 1,528 — 1. B
2. Add. 96 + 685 — A. 1,645 B. 681 C. 791 D. 781 — 2. D
3. Add. 7,482 + 38,546 — A. 46,028 B. 36,028 C. 36,928 D. not given — 3. A
4. Add. 270 + 35,060 — A. 37,760 B. 35,330 C. 36,330 D. 35,087 — 4. B
5. Add. 9,846 + 19,651 + 206,532 — A. 236,929 B. 236,030 C. 226,030 D. 236,029 — 5. D
6. Add. 5.67 + 68.4 — A. 73.71 B. 74.07 C. 74.11 D. 12.51 — 6. B
7. Add. 38 + 24.17 — A. 23.55 B. 62.07 C. 24.55 D. not given — 7. D
8. Add. 4.6 + 17 + 0.83 — A. 1.46 B. 21.33 C. 14.6 D. 22.43 — 8. D
9. Subtract. 536 – 79 — A. 457 B. 543 C. 557 D. not given — 9. A
10. Subtract. 907 – 238 — A. 769 B. 669 C. 679 D. 771 — 10. B
11. Subtract. 5,062 – 3,406 — A. 2,456 B. 1,656 C. 2,464 D. 1,556 — 11. B
12. Subtract. 90,205 – 38,648 — A. 62,443 B. 52,657 C. 51,667 D. not given — 12. D
13. Subtract. 0.7 – 0.33 — A. 10.03 B. 3.7 C. 0.37 D. 0.47 — 13. C
14. Subtract. 0.568 – 0.393 — A. 0.175 B. 0.285 C. 0.235 D. 0.165 — 14. A
15. Subtract. 59.8 – 37.47 — A. 22.43 B. 22.33 C. 22.47 D. 21.43 — 15. B
16. Subtract. 84 – 6.15 — A. 7.785 B. 77.85 C. 7.875 D. not given — 16. B

6

Chapter

3 PATTERNS: MULTIPLYING AND DIVIDING

PREVIEWING the CHAPTER

In this chapter, multiplication and division of whole numbers and decimals is introduced. Students are reminded to estimate the answer before performing the computation. Both multiplication and division with decimals are introduced by multiplying or dividing by whole numbers. Before students divide by a decimal they have learned to multiply by powers of ten. These skills are applied to consumer applications on credit and deposits, the algebraic topic of formulas, and the geometric topic of area.

Problem-Solving Strategy Students learn to solve problems by looking for a pattern.

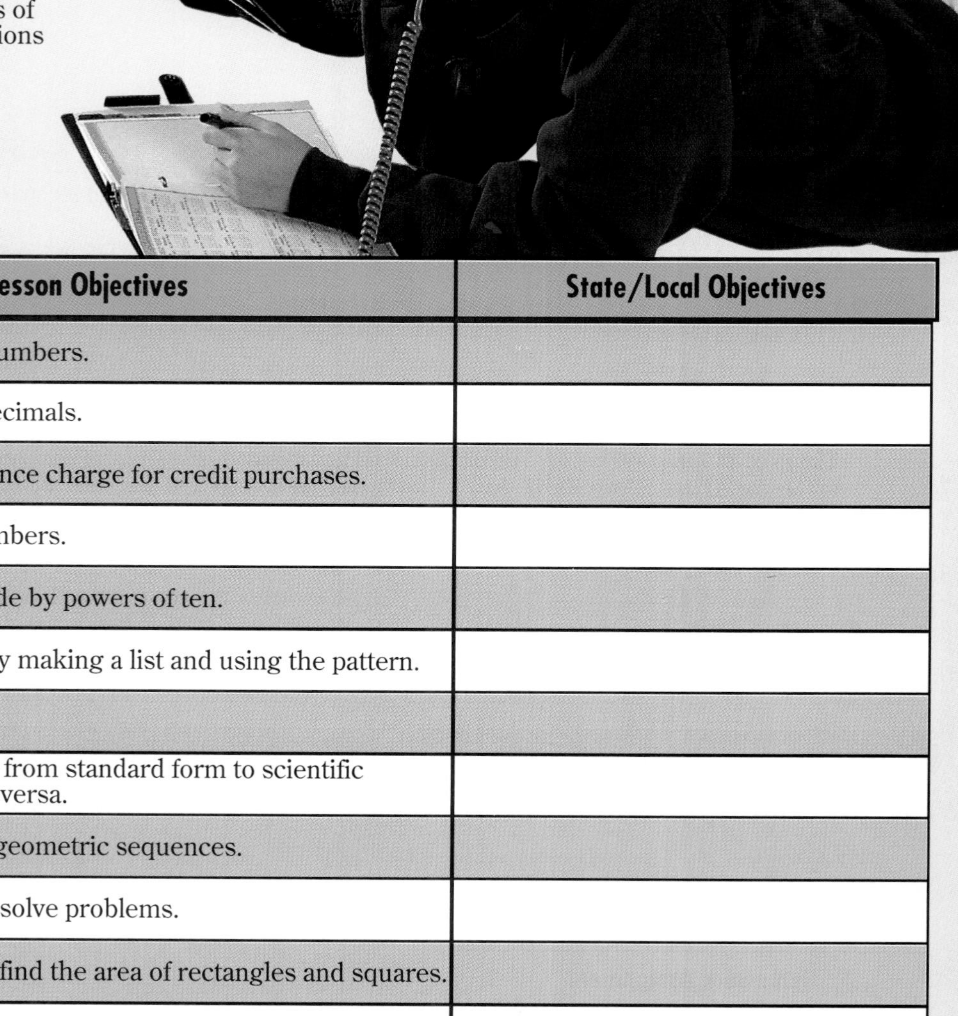

Lesson (Pages)	Lesson Objectives	State/Local Objectives
3-1 (70-71)	3-1: Multiply whole numbers.	
3-2 (72-73)	3-2: Multiply using decimals.	
3-3 (74-75)	3-3: Compute the finance charge for credit purchases.	
3-4 (76-77)	3-4: Divide whole numbers.	
3-5 (78-79)	3-5: Multiply and divide by powers of ten.	
3-6 (80-81)	3-6: Solve problems by making a list and using the pattern.	
3-7 (82-83)	3-7: Divide decimals.	
3-8 (86-87)	3-8: Change numbers from standard form to scientific notation and vice versa.	
3-9 (88-89)	3-9: Identify and find geometric sequences.	
3-10 (90-91)	3-10: Use formulas to solve problems.	
3-11 (92-93)	3-11: Use formulas to find the area of rectangles and squares.	
3-12 (94-95)	3-12: Find the mean of a set of data.	
3-13 (96-97)	3-13: Solve multi-step problems.	

ORGANIZING the CHAPTER

You may refer to the ***Course Planning Calendar*** on page T10.

Planning Guide

Lesson (Pages)	Extra Practice (Student Edition)	Blackline Masters Booklets: Reteaching	Practice	Enrichment	Activity	Multi-cultural Activity	Technology	Tech Prep	Evaluation	Other Resources
3-1 (70-71)	p. 453	p. 21	p. 23	p. 21		p. 3			p. 74	Transparency 3-1
3-2 (72-73)	p. 453	p. 22	p. 24	p. 22			p. 3			Transparency 3-2
3-3 (74-75)	p. 454	p. 23	p. 25	p. 23						Transparency 3-3
3-4 (76-77)	p. 454	p. 24	p. 26	p. 24	p. 5					Transparency 3-4
3-5 (78-79)	p. 454	p. 25	p. 27	p. 25						Transparency 3-5
3-6 (80-81)			p. 28							Transparency 3-6
3-7 (82-83)	p. 454	p. 26	p. 29	p. 26	p. 31				p. 14	Transparency 3-7
3-8 (86-87)	p. 455	p. 27	p. 30	p. 27						Transparency 3-8
3-9 (88-89)	p. 455	p. 28	p. 31	p. 28				p. 5		Transparency 3-9
3-10 (90-91)	p. 455	p. 29	p. 32	p. 29						Transparency 3-10
3-11 (92-93)	p. 456	p. 30	p. 33	p. 30			p. 17			Transparency 3-11
3-12 (94-95)	p. 456	p. 31	p. 34	p. 31						Transparency 3-12
3-13 (96-97)		p. 32	p. 35	p. 32	p. 6			p. 6	p. 14	Transparency 3-13
Ch. Review (98-99)									pp. 11-13	Test and Review Generator
Ch. Test (100)										
Cumulative Review/Test (102-103)									p. 15	

OTHER CHAPTER RESOURCES

Student Edition
Chapter Opener, p. 68
Activity, p. 69
On-the-Job Application, p. 84
Consumer Connections, p. 85
Journal Entry, p. 87
Portfolio Suggestion, p. 97
Science Connections, p. 101
Tech Prep Handbook, pp. 519-521

Teacher's Classroom Resources
Transparency 3-0
Lab Manual, pp. 27-28
Performance Assessment, pp. 5-6
Lesson Plans, pp. 23-35

Other Supplements
Overhead Manipulative Resources, Labs 8-10, pp. 4-7
Glencoe Mathematics Professional Series

Enhancing the Chapter

Some of the blackline masters for enhancing this chapter are shown below.

COOPERATIVE LEARNING

Use base-ten blocks to introduce the concepts of multiplication as repeated addition and division as separating into equal groups. Model a problem such as 324 x 3 and show students that the product is the same as 324 + 324 + 324. Model a problem such as 120 ÷ 3 and show students that 120 can be separated into 3 groups of 40. Show students that the same concepts hold true for multiplication and division involving decimals. Use the blocks as wholes, tenths, and hundredths to model 1.2 x 3 and 4.68 ÷ 2.

USING MODELS/MANIPULATIVES

Have students work in pairs to play the following game. Make a game board by filling a 5 x 5 grid with different whole numbers and decimals. Players in turn toss two coins or markers onto the game board and then divide the greater number by the lesser number as indicated by the markers. A correct answer is worth 2 points. If an answer is challenged by the opponent, both players use multiplication to determine the correct answer. A successful challenge is worth 1 point to the challenger.

Cooperative Problem Solving, p. 31

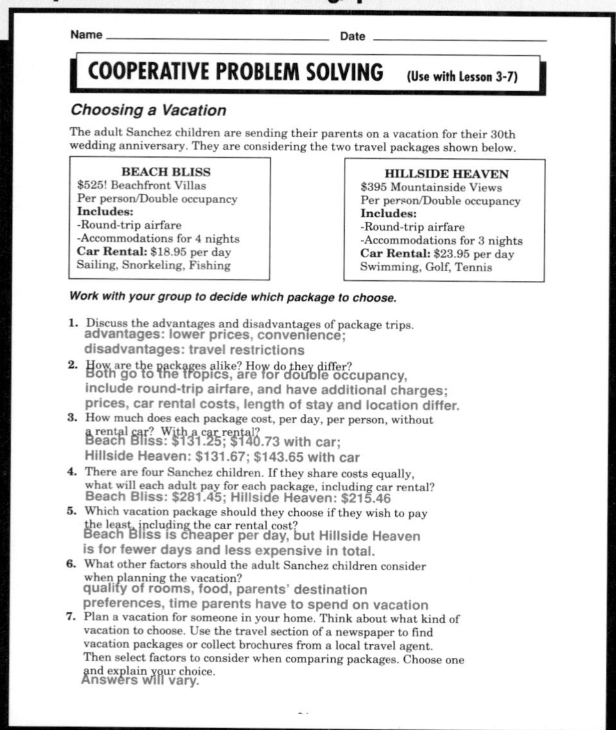

Name ______ Date ______

COOPERATIVE PROBLEM SOLVING (Use with Lesson 3-7)

Choosing a Vacation

The adult Sanchez children are sending their parents on a vacation for their 30th wedding anniversary. They are considering the two travel packages shown below.

BEACH BLISS	HILLSIDE HEAVEN
$525! Beachfront Villas	$395 Mountainside Views
Per person/Double occupancy	Per person/Double occupancy
Includes:	**Includes:**
-Round-trip airfare	-Round-trip airfare
-Accommodations for 4 nights	-Accommodations for 3 nights
Car Rental: $18.95 per day	**Car Rental:** $23.95 per day
Sailing, Snorkeling, Fishing	Swimming, Golf, Tennis

Work with your group to decide which package to choose.

1. Discuss the advantages and disadvantages of package trips.
   advantages: lower prices, convenience; disadvantages: travel restrictions
2. How are the packages alike? How do they differ?
   Both go to the tropics, are for double occupancy, include round-trip airfare, and have additional charges; prices, car rental costs, length of stay and location differ.
3. How much does each package cost, per day, per person, without a rental car? With a car rental?
   Beach Bliss: $131.25; $140.73 with car; Hillside Heaven: $131.67; $143.65 with car
4. There are four Sanchez children. If they share costs equally, what will each adult pay for each package, including car rental?
   Beach Bliss: $281.45; Hillside Heaven: $215.46
5. Which vacation package should they choose if they wish to pay the least, including the car rental cost?
   Beach Bliss is cheaper per day, but Hillside Heaven is for fewer days and less expensive in total.
6. What other factors should the adult Sanchez children consider when planning the vacation?
   quality of rooms, food, parents' destination preferences, time parents have to spend on vacation
7. Plan a vacation for someone in your home. Think about what kind of vacation to choose. Use the travel section of a newspaper to find vacation packages or collect brochures from a local travel agent. Then select factors to consider when comparing packages. Choose one and explain your choice.
   Answers will vary.

Lab Activity, p. 28

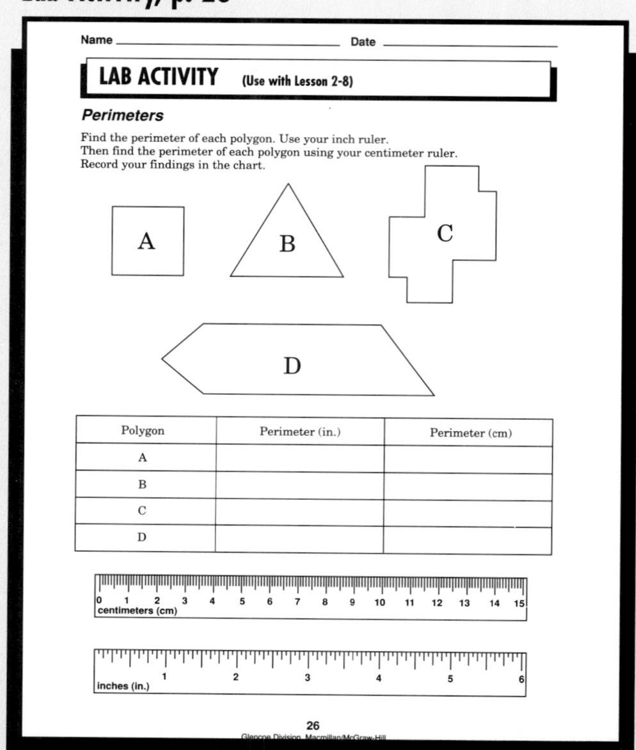

Name ______ Date ______

LAB ACTIVITY (Use with Lesson 2-8)

Perimeters

Find the perimeter of each polygon. Use your inch ruler.
Then find the perimeter of each polygon using your centimeter ruler.
Record your findings in the chart.

Polygon	Perimeter (in.)	Perimeter (cm)
A		
B		
C		
D		

26

Glencoe Division, Macmillan/McGraw-Hill

MEETING INDIVIDUAL NEEDS

Limited English Proficiency Students

Before beginning the chapter, scan the pages for the words factor, product, divisor, dividend, and quotient. Note where these words are used and pay special attention to them when teaching LEP students. Identify numbers in multiplication and division problems by pointing out their position in the problem and the appropriate name. Carefully explain the relationships among these numbers. In the applications lessons you may need to explain the meaning and use of credit and deposit. Also explain the situations used to introduce these terms.

CRITICAL THINKING/PROBLEM SOLVING

The ability to see patterns and form conclusions is an important thinking skill in this chapter. Form a 3 x 3 magic square using the following numbers listed in order by row: 36, 31, 38; 37, 35, 33; 32, 39, 34. Have some students confirm that the sums of the rows, the columns, and the diagonals each equal 105. Have some students divide each number by 10, add the decimals, and report the results. Have others divide each number by 100, and report the results. Have the class formulate a conclusion. They should discover the magic sum reflects the operation.

COMMUNICATION

Speaking

Have students give an oral presentation with examples on the following topics.

- Placing the decimal point in a product of decimals
- How to divide by a decimal
- Multiplying and dividing by a power of ten and writing a number in scientific notation
- Finding area both geometrically and algebraically

Applications, p. 5

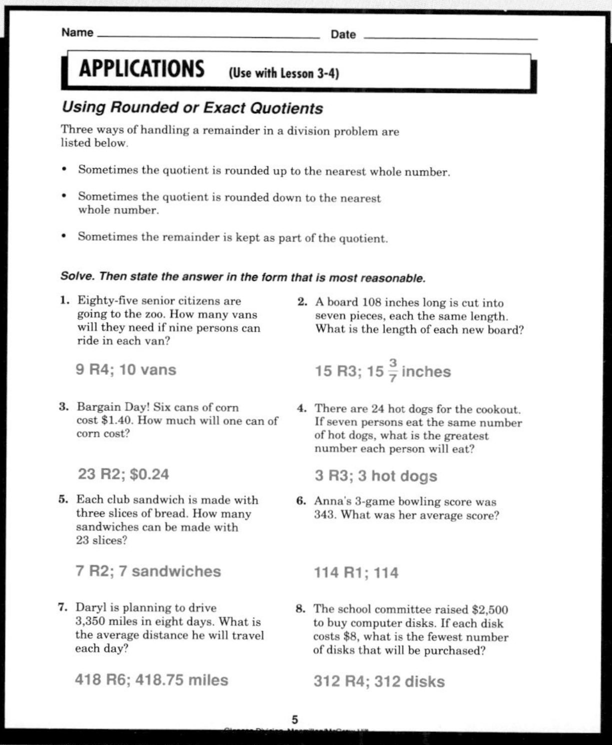

Name ____________ Date ____________

APPLICATIONS (Use with Lesson 3-4)

Using Rounded or Exact Quotients

Three ways of handling a remainder in a division problem are listed below.

- Sometimes the quotient is rounded up to the nearest whole number.
- Sometimes the quotient is rounded down to the nearest whole number.
- Sometimes the remainder is kept as part of the quotient.

Solve. Then state the answer in the form that is most reasonable.

1. Eighty-five senior citizens are going to the zoo. How many vans will they need if nine persons can ride in each van?
   9 R4; 10 vans
2. A board 108 inches long is cut into seven pieces, each the same length. What is the length of each new board?
   15 R3; $15\frac{3}{7}$ inches
3. Bargain Day! Six cans of corn cost $1.40. How much will one can of corn cost?
   23 R2; $0.24
4. There are 24 hot dogs for the cookout. If seven persons eat the same number of hot dogs, what is the greatest number each person will eat?
   3 R3; 3 hot dogs
5. Each club sandwich is made with three slices of bread. How many sandwiches can be made with 23 slices?
   7 R2; 7 sandwiches
6. Anna's 3-game bowling score was 343. What was her average score?
   114 R1; 114
7. Daryl is planning to drive 3,350 miles in eight days. What is the average distance he will travel each day?
   418 R6; 418.75 miles
8. The school committee raised $2,500 to buy computer disks. If each disk costs $8, what is the fewest number of disks that will be purchased?
   312 R4; 312 disks

5

Calculator Activity, p. 3

Name ____________ Date ____________

CALCULATOR ACTIVITY (Use with Lesson 3-2)

Greater? Less? Equal?

When you multiply a number by another number, is the product greater than, less than, or equal to the original number?

Try to discover when:

- a product is greater than (>) the original number
- a product is less than (<) the original number
- a product is equal to (=) the original number

Use your calculator to complete the table.

	Original Number	Multiplier	Is the multiplier >, <, or = 1?	Product	Is the product >, <, or = the original number?
1.	16	4	>	64	>
2.	24	0.75	<	18	<
3.	2.5	14.3	>	35.75	>
4.	14	0.5	<	7	<
5.	2.93	0.07	<	0.2051	<
6.	0.932	2.75	>	2.563	>
7.	0.83	0.62	<	0.5146	<
8.	16.32	1.00	=	16.32	=
9.	144.77	0.001	<	0.14477	<

10. If the multiplier is > 1, the product is greater than the original number.
11. If the multiplier is < 1, the product is less than the original number.
12. If the multiplier is 1, the product is the same as the original number.

3

Glencoe Division, Macmillan/McGraw-Hill

CHAPTER PROJECT

Discover the asteroids using Bode's Rule:

* List the planets in a vertical column in order from the sun. Place a ? between Mars and Jupiter.
* Draw five columns to the right of the list. Place the number 4 in the first column next to each planet, including the ? row.
* In the next column, place a 0 in Mercury's row, 3 in Venus's row, and 6 in Earth's row. Continue down the column, placing a number in each row that is twice the number above it.
* Add the numbers in the first column. Place the sum in the third column.
* Divide each sum by ten. Place the quotient in the fourth column. Label this column "Distance by Bode's Rule."
* Label the fifth column "Actual Distance." Research each planet's distance from the sun in astronomical units (au). Place in the column.
* The last two columns should correspond closely. According to Bode's Rule, there should be a planet between Mars and Jupiter. This is how Bode "discovered" the asteroids. Which two planets do not fit Bode's Rule? Neptune and Pluto

Using the Chapter Opener

Transparency 3-0 is available in the Transparency Package. It provides a full-color visual and motivational activity that you can use to engage students in the mathematical content of the chapter.

Have students read the opening paragraph.

Using Questioning

- What time of day are long distance calls most economical? **Check the telephone directory for the time and rates in your area.**
- If you live on the East Coast and want to call the West Coast during business hours, what time of day will it be least expensive to make the call? **after 5 P.M. East Coast time**
- If you live on the West Coast and want to call the East Coast during business hours, what time of day will it be least expensive to call? **before 8 A.M.**

The second paragraph is a real-world application of finding the cost of a long distance call. Have students list the steps they need to solve the problem and then solve.

Remind students that long distance calls are subject to tax. When checking their own telephone bill, the amount of tax needs to be considered.

Allow students the opportunity to compare rates for long-distance calls for day, evening, nights, person-to-person, and station-to-station calls by showing them the rates out of the local telephone directory.

Chapter

3 PATTERNS: MULTIPLYING AND DIVIDING

Pick up the phone!

Did you know the cost of a long-distance telephone call depends on the time of day, the location of the caller, the city called, and the duration of the call? During business hours, rates are usually higher. The time of day for the *caller* is the time of day on which the rates are based.

During business hours, the cost of a call from Indianapolis to Columbus is $0.21 for the first minute and $0.22 for each additional minute. From 5:00 P.M. to 11:00 P.M. the cost of a call is $0.14 for the first minute and $0.15 for each additional minute. How much less is a 10-minute call from Indianapolis to Columbus after 5:00 P.M. than during business hours? **$2.19, $1.49, $0.70 difference**

ACTIVITY: Multiplication with Decimal Models

In this chapter you will multiply whole numbers and decimals.

As with addition, you can use decimal models to multiply. Suppose you want to find the product of 0.3 and 0.5.

Materials: decimal models, pencil

1. Use a decimal model that is divided into hundredths.

2. Shade 3 tenths. **3.** Shade 5 tenths. This model represents the product.

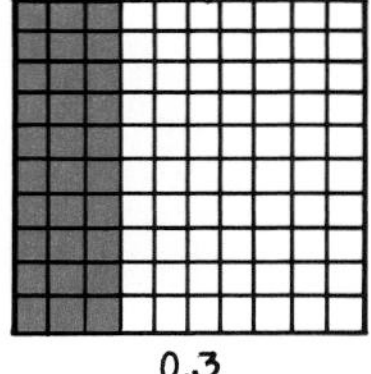

0.3

×

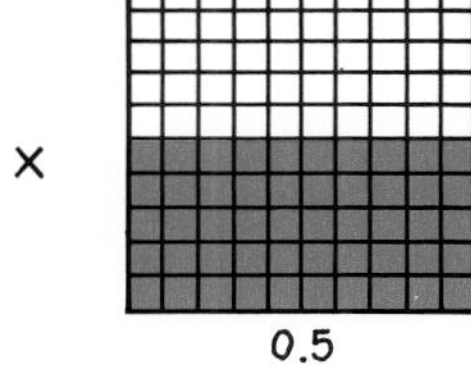

0.5

=

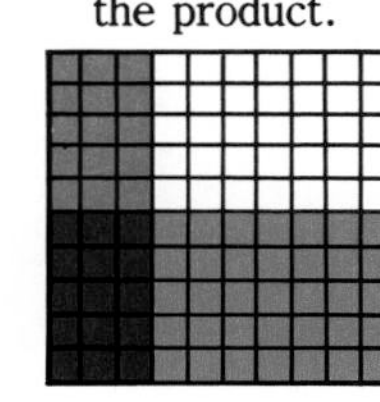

Cooperative Groups

Work together in groups of three or four.

4. Explain how the shaded models show the product of 0.3 and 0.5. What is the product? How did you find the product? **See margin.**

5. Discuss with your group how to show 0.7 × 0.4 with decimal models. Then find the product of 0.7 and 0.4 using models. **See margin.**

6. Discuss how to show the product of 3 and 0.2 using decimal models. Which model should you use? What is the product? **See margin.**

7. In Exercises 4, 5, and 6, how many decimal places were there in each factor? How many decimal places were there in each product? **4; 1 and 1, 2 5; 1 and 1, 2 6; 0 and 1, 1**

Communicate Your Ideas

8. With your group, draw a conclusion about how the number of decimal places in a product is related to the number of decimal places in the factors. **See margin.**

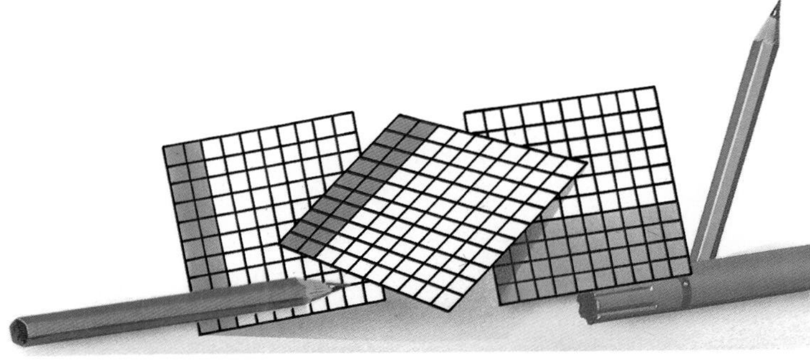

▶ INTRODUCING THE LESSON

Objective Find products of decimals using decimal models.

▶ TEACHING THE LESSON

Materials: hundredths decimal models, **Lab Manual, p. 2,** 2 different colored pencils

You may want to discuss what the sections of a decimal model represent. Encourage students to work neatly and accurately to prevent errors.

▶ EVALUATING THE LESSON

Using Cooperative Groups Have students work in small groups of three or four on Exercises 4-8.

Communicate Your Ideas Groups may discuss with other groups their conclusions about placing the decimal point in a product.

▶ EXTENDING THE LESSON

Closing the Lesson Allow groups time to compare answers with other groups.

Activity Worksheet, Lab Manual, p. 27

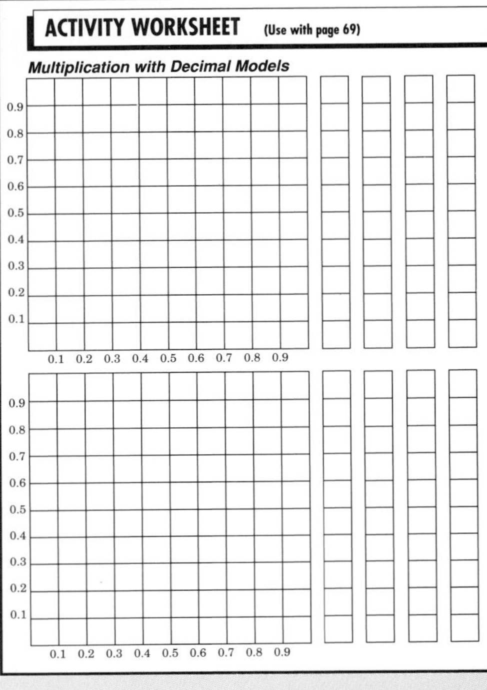

ACTIVITY WORKSHEET (Use with page 69)

Multiplication with Decimal Models

Additional Answers

4. 15 hundredths, Count the number of hundredths that are shaded twice if one model is placed on top of the other.
5. Shade 7 columns. Shade 4 rows. Count the number of hundredths that are shaded twice. 0.28
8. A good answer should refer to adding the number of decimal places in the factors.

5-MINUTE CHECK-UP

(over Lesson 2-9)

Available on TRANSPARENCY 3-1.

Solve. State any missing or extra facts.

1. Paul has checks for $21.43, $30.31, and $45.29 to deposit. What is his new balance after depositing these checks? missing balance before deposit
2. Jill ran a lap in 41.6 seconds. It took Susan 3.6 more seconds. How long did it take Susan? 45.2 seconds

▶ INTRODUCING THE LESSON

Have students jog in place for 2 minutes, then find a pulse point and count the beats for 15 seconds. Multiply this by 4 to find beats/minute.

▶ TEACHING THE LESSON

Using Cooperative Learning For the group activity on p. 71, make number cubes by covering dots on dice with press-on circles. This activity can be extended by using the least possible product to determine the winner or by using several rounds before determining a winner.

Practice Masters Booklet, p. 23

Name ________________ Date ________

PRACTICE WORKSHEET 3-1

Multiplying Whole Numbers

Multiply.

1. 49 × 8 = 392
2. $64 × 7 = $448
3. 97 × 6 = 582
4. 108 × 5 = 540
5. $231 × 9 = $2,079
6. $735 × 6 = $4,410
7. 2,431 × 5 = 12,155
8. 50 × 12 = 600
9. $87 × 23 = $2,001
10. 92 × 37 = 3,404
11. 465 × 30 = 13,950
12. $721 × 20 = $14,420
13. $409 × 31 = $12,679
14. 657 × 26 = 17,082
15. 582 × 34 = 19,788
16. 6,432 × 30 = 192,960
17. $2,759 × 50 = $137,950
18. 3,506 × 24 = 84,144
19. $5,720 × 32 = $183,040
20. $9,874 × 46 = $454,204
21. 732 × 400 = 292,800
22. $694 × 500 = $347,000
23. 503 × 307 = 154,421
24. $620 × 805 = $499,100
25. 756 × 327 = 247,212
26. 506 × 481 = 243,386
27. 2,407 × 303 = 729,321
28. $6,409 × 342 = $2,191,878

Solve.

29. If a car travels at a constant speed of 55 miles per hour, how far will it travel in 6 hours? 330 mi
30. Juan is reading a book that has 16 chapters. Each chapter has 23 pages. How many pages are in the book? 368 pages

3-1 MULTIPLYING WHOLE NUMBERS

Objective *Multiply whole numbers.*

Aaron likes to do three aerobic workouts each week. During one workout, he raises his heartbeat to 98 beats per minute. If he maintains this rate for 20 minutes, how many times will his heart beat? To find out, multiply 98 by 20.

Method

1. Align the numbers on the right.
2. Multiply by the number in each place-value position from right to left.
3. Add the products obtained in Step 2.

Example A ***20 × 98*** Estimate: 20 × 100 = 2,000

- Step 1: 98 × 20
- Step 2: 98 × 20 → 0   0 × 98 = 0
- Result: 98 × 20 = 1,960 (carry 1)

When multiplying by multiples of 10, there is no need to write another row.

Compared to the estimate, is the answer reasonable?

Aaron's heart will beat 1,960 times in twenty minutes.

Example B ***27 × 53*** Estimate: 30 × 50 = 1,500

- Step 1: 53 × 27
- Step 2: 53 × 27 → 371 (7 × 53), 106 (2 × 53)
- Step 3: 371 + 1 06 = 1,431   Is the answer reasonable?

Example C ***65 × 238*** Estimate: 70 × 200 = 14,000

- Step 1: 238 × 65
- Step 2: 238 × 65 → 1190 (5 × 238), 1428 (6 × 238)
- Step 3: 1 190 + 14 28 = 15,470

Is the answer reasonable?

Guided Practice

Multiply. **Encourage students to estimate each answer first.**

Examples A, B

1. 63 × 20 = 1,260
2. 31 × 30 = 930
3. 75 × 19 = 1,425
4. 73 × 32 = 2,336
5. 26 × 58 = 1,508

Example C

6. 142 × 55 = 7,810
7. 427 × 14 = 5,978
8. 580 × 58 = 33,640
9. 232 × 10 = 2,320
10. 742 × 30 = 22,260

RETEACHING THE LESSON

Some students may be more successful using lattice multiplication than the standard multiplication algorithm.

36 × 57 = 2,052

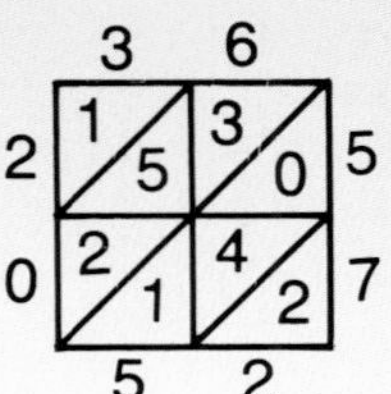

Reteaching Masters Booklet, p. 21

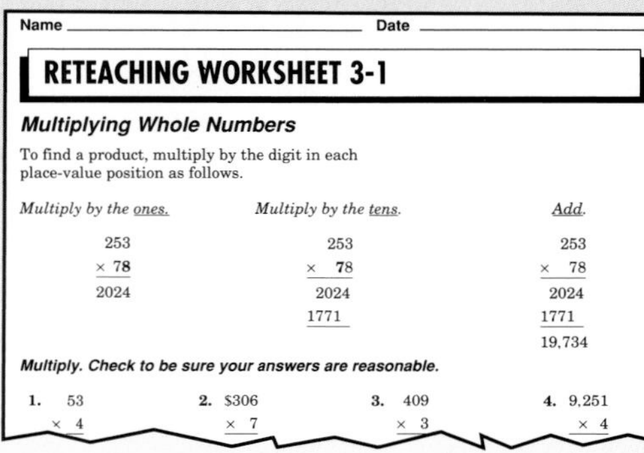
Name ________________ Date ________

RETEACHING WORKSHEET 3-1

Multiplying Whole Numbers

To find a product, multiply by the digit in each place-value position as follows.

Multiply by the ones. 253 × 78 → 2024

Multiply by the tens. 253 × 78 → 2024, 1771

Add. 253 × 78 → 2024 + 1771 = 19,734

Multiply. Check to be sure your answers are reasonable.

1. 53 × 4
2. $306 × 7
3. 409 × 3
4. 9,251 × 4

Exercises

Practice

Multiply. Encourage students to estimate each answer first.

11. 76 × 43 = 3,268 **12.** 88 × 35 = 3,080 **13.** 12 × 80 = 960 **14.** 96 × 60 = 5,760 **15.** 82 × 30 = 2,460

16. 948 × 90 = 85,320 **17.** 741 × 42 = 31,122 **18.** 290 × 83 = 24,070 **19.** 214 × 31 = 6,634 **20.** 803 × 70 = 56,210

21. 3,803 × 52 = 197,756 **22.** 2,341 × 77 = 180,257 **23.** 4,861 × 30 = 145,830 **24.** 3,219 × 61 = 196,359

25. 4,728 × 83 392,424 **6.** 70 × 14,966 1,047,620 **27.** 48 × 26,091 1,252,368

28. Find the product of 38 and 75. 2,850

Applications

29. Chuck is reading two books this month. One has 16 chapters. The other has 12 chapters. If the book with 16 chapters has 28 pages in each chapter, how many pages does the 16-chapter book have? 448 pages

30. Pierre is spraying the yard for insects. He mixes 12 tablespoons of insect repellant with one gallon of water. He needs to make 15 gallons of spray. How many tablespoons of repellant will he need? 180 tablespoons

Using Equations

Solve each equation.

31. $4 \times y = 60$ 15 **32.** $20 \times 12 = q$ 240 **33.** $25c = 175$ 7

34. $z \times 6 = 48$ 8 **35.** $11 \times 12 = n$ 132 **36.** $9 \times 8 = y$ 72

Cooperative Groups

37. Use four number cubes, with two cubes marked with the digits 0 through 5 and two cubes marked with the digits 4 through 9. Form groups of three or four. Each player takes a turn and rolls all four cubes. Arrange the digits rolled to form 2 two-digit numbers or a one-digit and a three-digit number that have the greatest product possible. The player with the greatest product is the winner.

FUN with MATH

When was Rubik's cube invented?

See page 105.

Critical Thinking

38. Use the numbers 7, 1, 5, and 8 only once to write a whole number multiplication problem with a product as close as possible to 1,000. 58 × 17 = 986

Mental Math

Multiply \$15 by 8. Think: \$15 = \$10 + \$5

8 × \$15: 8 × \$10 = \$80; 8 × \$5 = \$40; \$120

Multiply. Write only the answers.

39. 7 × 18 126 **40.** 13 × 9 117 **41.** 12 × 15 180 **42.** 78 × 4 312

Chalkboard Examples

Examples A, B *Multiply.*

1. 30 × 15 450

2. 23 × 46 1,058

Example C *Multiply.*

3. 31 × 709 21,979

4. 74 × 480 35,520

Additional examples are provided on TRANSPARENCY 3-1.

EVALUATING THE LESSON

Check for Understanding Use Guided Practice Exercises to check for student understanding.

Error Analysis Stress careful alignment of digits. Grid paper may be helpful. When multiplying by the tens digit, emphasize that the product is begun in the tens column under the tens digit. Annex a zero in the ones place.

Closing the Lesson Ask students to summarize what they have learned from the cooperative group activity on p. 71.

APPLYING THE LESSON

Independent Practice

Homework Assignment	
Minimum	Ex. 2-36 even; 37-42
Average	Ex. 1-27 odd; 28-42
Maximum	Ex. 1-42

EXTENDING THE LESSON

Enrichment Masters Booklet, p. 21

Name ______ Date ______

ENRICHMENT WORKSHEET 3-1

Pictographs

Solve. Use the pictograph.

Canoes Rented in One Day	
Blackfork	4½ symbols
Wilderness	4 symbols
Sugar Creek	3 symbols
Logan Valley	5 symbols
Hawkeye	3½ symbols
Each symbol represents 8 canoes.	

1. How many canoes were rented in a day at Wilderness? 32 canoes
2. How many more canoes were rented at Logan Valley than at Sugar Creek? 16 canoes
3. At \$7 for each canoe, how much money is collected at Hawkeye? \$196
4. Blackfork charges \$10 for each canoe rental. How much money is collected? \$360
5. Wilderness charges \$7 per canoe and Sugar Creek charges \$8. How much more money is collected at Wilderness? \$32
6. Find the total number of canoes rented in one day. 160 canoes

Solve.

7. Kyle pays \$195 rent each month for his one-bedroom apartment. How much does Kyle pay in one year? \$2,340
8. Swenson's Bakery makes 225 dozen cookies every day. How many cookies is this? How many cookies are made in one week? 2,700 cookies per day; 18,900 cookies per week
9. Juanita drives 55 miles per hour on the highway. How far does she drive in seven hours? 385 miles
10. Kirby reads 66 words in one minute. How many words does Kirby read in 50 minutes? 3,300 words

5-MINUTE CHECK-UP

(over Lesson 3-1)

Available on TRANSPARENCY 3-2.

Multiply.

1. 45×22 990
2. 50×12 600
3. 628×14 8,792
4. 690×201 138,690

Extra Practice, Lesson 3-1, p. 453

▶ INTRODUCING THE LESSON

Ask students if they have made purchases of more than one of the same item. Ask them how they computed the cost and why they need to know the total cost.

▶ TEACHING THE LESSON

Using Manipulatives Hand out one 1×8 inch strip of paper and have students tear off about one-tenth of the strip. Ask students to tear off one-tenth of the one-tenth strip. Remind students that one tenth of one tenth is the same as 0.1×0.1. Ask students if the product of two decimals, each less than one, is lesser or greater than either of the decimal factors.

Practice Masters Booklet, p. 24

Name ______________ Date ______________

PRACTICE WORKSHEET 3-2

Multiplying Decimals

Multiply.

1. 0.7 × 3 = 2.1	2. 2.1 × 9 = 18.9	3. 7.87 × 6 = 47.22	4. 0.46 × 7 = 3.22
5. 0.3 × 0.4 = 0.12	6. 0.8 × 0.4 = 0.32	7. 24 × 0.3 = 7.2	8. 0.71 × 0.2 = 0.142
9. 0.5 × 0.7 = 0.35	10. 0.9 × 0.6 = 0.54	11. 0.74 × 0.3 = 0.222	12. 0.36 × 0.8 = 0.288
13. 1.44 × 0.6 = 0.864	14. 5.27 × 0.7 = 3.689	15. 2.86 × 0.04 = 0.1144	16. 0.329 × 0.35 = 0.11515
17. 5.06 × 1.2 = 6.072	18. 71.4 × 2.9 = 207.06	19. 7.64 × 0.29 = 2.2156	20. 5.28 × 0.52 = 2.7456
21. 7.24 × 5.9 = 42.716	22. 0.114 × 0.89 = 0.10146	23. 6.75 × 9.7 = 65.475	24. 0.837 × 0.56 = 0.46872

Solve.

25. Marie travels 5.6 miles per hour on her bike. If she rides for 1.25 hours, how far does she ride? 7 miles

26. Almonds are $1.78 per pound. How much change does Boris receive if he pays for 1.5 pounds with a $5 bill? $2.33

3-2 MULTIPLYING DECIMALS

Objective *Multiply using decimals.*

Allison makes $4.25 an hour working at the snack bar in Fenway Park. She worked 6 hours on Tuesday and from 9:00 A.M. to 12:30 P.M. on Wednesday. To find out how much she earned on Tuesday, multiply $4.25 by 6.

When multiplying decimals, multiply as with whole numbers. The number of decimal places in the product is the same as the sum of the number of decimal places in the factors.

Method

1. Multiply as with whole numbers.
2. Find the sum of the number of decimal places in the factors.
3. To place the decimal point in the product, start at the right and count the number of decimal places needed. If more decimal places are needed, annex zeros on the left. Then insert the decimal point.

Example A ***6 × $4.25*** Estimate: $6 \times 4 = 24$

1. Align the numbers on the right and multiply.

$$\begin{array}{r} 4.25 \\ \times \quad 6 \\ \hline 25.50 \end{array}$$

2. 2 decimal places
3. Starting at the right, count 2 decimal places.

Check by adding.

$$\begin{array}{r} 4.25 \\ 4.25 \\ 4.25 \\ 4.25 \\ 4.25 \\ +\ 4.25 \\ \hline 25.50 \end{array}$$

How much did she earn on Wednesday? $14.88

Allison earned $25.50 on Tuesday. Based on the estimate, explain why the answer is reasonable.

Example B ***0.02 × 1.35*** Estimate: $0 \times 1 = 0$

1.
$$\begin{array}{r} 1.35 \\ \times\ 0.02 \\ \hline .0270 \end{array}$$

2. 4 decimal places
3. To make 4 decimal places in the product, annex a zero on the left.

$0.02 \times 1.35 = 0.0270$ How does this compare with the estimate?

Guided Practice **Example A**

Multiply. **Encourage students to estimate the product first.**

1. 9.2×6 55.2
2. 8.4×7 58.8
3. 3.2×19 60.8
4. 0.95×5 4.75
5. 2.31×12 27.72

RETEACHING THE LESSON

Have students practice decimal multiplication by playing a game of Low Score Roll. You will need two dice: one numbered 0.1, 0.2, 0.3, etc., and one numbered 0.01, 0.02, 0.03, etc. Each player, in turn, rolls the dice and multiplies the numbers. This is his score for the round. Play continues for four rounds and all scores are added. The player with the lowest score wins.

Reteaching Masters Booklet, p. 22

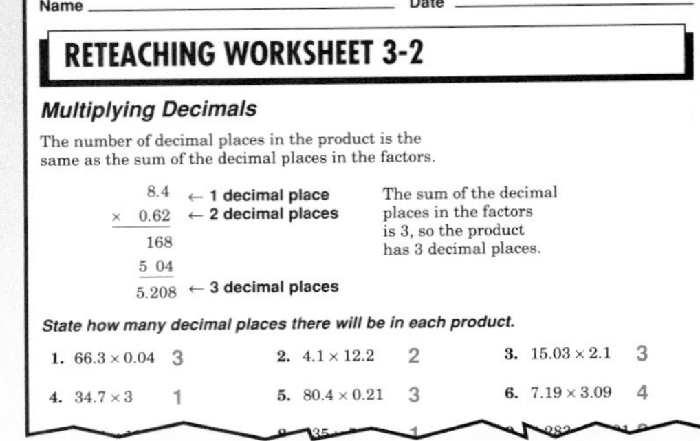

Name ______________ Date ______________

RETEACHING WORKSHEET 3-2

Multiplying Decimals

The number of decimal places in the product is the same as the sum of the decimal places in the factors.

$$\begin{array}{r} 8.4 \\ \times\ 0.62 \\ \hline 168 \\ 5\ 04 \\ \hline 5.208 \end{array}$$

8.4 ← 1 decimal place
0.62 ← 2 decimal places
5.208 ← 3 decimal places

The sum of the decimal places in the factors is 3, so the product has 3 decimal places.

State how many decimal places there will be in each product.

1. 66.3 × 0.04 3	2. 4.1 × 12.2 2	3. 15.03 × 2.1 3
4. 34.7 × 3 1	5. 80.4 × 0.21 3	6. 7.19 × 3.09 4

Example B

6. 0.3 × 0.2 = **0.06**
7. 0.47 × 0.2 = **0.094**
8. 0.03 × 1.5 = **0.045**
9. 0.002 × 0.7 = **0.0014**
10. 0.012 × 0.1 = **0.0012**

Exercises

Practice

Multiply. **Encourage students to estimate the product first.**

11. 0.5 × 7 **3.5**
12. 0.4 × 0.9 **0.36**
13. 0.05 × 8 **0.40**
14. 4.5 × 0.06 **0.27**
15. 0.023 × 51 **1.173**
16. 2.6 × 4.7 **12.22**
17. 0.07 × 0.3 **0.021**
18. 0.881 × 0.5 **0.4405**
19. 0.83 × 36 **29.88**
20. $8.45 × 27 **$228.15**

21. 8.09 × 0.5 **4.045**
22. 332.5 × 7.3 **2,427.25**
23. 4.53 × 0.01 **0.0453**
24. 239.75 × 0.1 **23.975**
25. 6.82 × 0.05 **0.3410**
26. 147.5 × 0.3 **44.25**
27. 3.14 × 0.15
28. 64.2 × 0.021
29. Find the product of 0.3 and 52.
30. Multiply 0.368 and 0.6.

27. 0.4710 28. 1.3482 29. 15.6 30. 0.2208

Application

31. Boy Scout Troop #672 is going to the baseball game. Each boy is allowed to buy a hot dog for $2.25, a drink for $1.25, and peanuts for $0.75. If 23 boys are going to the game, how much money is needed to buy all three items for the troop? **$97.75**

Use the chart for Exercises 32-33.

32. Suppose the owners of the White Sox decided to increase the seating capacity in Comisky Park by the number of seats in the Kingdome. If the expansion costs $257 per seat, how much will the total expansion cost? **$15,343,414**

Tickets, please!

About 50,000 fans can be seated in each Major League baseball stadium.

Seating Capacity

Kingdome 59,702 (Seattle Mariners)

Comisky Park 44,702 (Chicago White Sox)

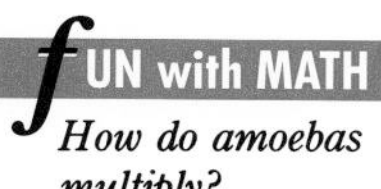

FUN with MATH
How do amoebas multiply?
See page 104.

33. How many more seats are in the Kingdome than in Comisky Park? Suppose the owners of the Mariners claim that they have more than $1\frac{1}{3}$ the seating capacity of Comisky Park. Is this a correct statement? Explain your answer. **See margin.**
34. Use decimal models to show the product for Exercise 12. **See students' work.**

Estimation

Estimate each product and decide if the decimal point is in the correct place. If not, write the correct product.

35. 2 × 0.7 = 14 **1.4**
36. 2 × 4.1 = 8.2 **correct**
37. $1.23 × 8 = $9.84 **correct**
38. 1.2 × 71 = 852 **85.2**

Chalkboard Examples

Example A *Multiply.*
1. 65 × 2.4 **156**
2. 7 × 4.662 **32.634**

Example B *Multiply.*
3. 0.1 × 0.4 **0.04**
4. 0.043 × 0.5 **0.0215**

Additional examples are provided on TRANSPARENCY 3-2.

EVALUATING THE LESSON

Check for Understanding Use Guided Practice Exercises to check for student understanding.

Error Analysis Watch for students who fail to annex zeros on the left when needed. Remind students to sum the decimal places from *each* factor before placing the decimal point.

Closing the Lesson Have students state in their own words how to decide how many decimal places are needed in the product of two decimals.

EXTENDING THE LESSON

Enrichment Masters Booklet, p. 22

ENRICHMENT WORKSHEET 3-2

Gasoline Pumps

Service station gasoline pumps use decimal fractions to indicate price per gallon, number of gallons, and total price. The pump automatically multiplies the price per gallon by the number of gallons to give the total price of the purchase.

REGULAR: Total $ 19.74; Gallons 15.8; Price per Gallon $ 1.249

Example:
① 15.8 × $1.249 = $19.7342 — The total price is $19.74 rounded up to the next cent.
② $19.74 ÷ $1.249 = 15.8 — Divide the total price by the price per gallon to find the number of gallons.

Fill in the correct total price on each pump.

1. UNLEADED: Gallons 10.2; Price per Gallon $ 1.369 — **$13.96**
2. DIESEL: Gallons 12.4; Price per Gallon $ 1.659 — **$20.57**
3. REGULAR: Gallons 3.5; Price per Gallon $ 1.499 — **$5.25**

Check each pump to see if the total price is correct. Write right or wrong.

4. DIESEL: Total $ 24.63; Gallons 15.4; Price per Gallon $ 1.599 — **right**
5. REGULAR: Total $ 15.00; Gallons 12.3; Price per Gallon $ 1.219 — **right**
6. UNLEADED: Total $ 9.75; Gallons 8.3; Price per Gallon $ 1.159 — **wrong**

Fill in the number of gallons rounded to the nearest tenth.

7. REGULAR: Total $ 20.00; Price per Gallon $ 1.249 — **16.0**
8. UNLEADED: Total $ 15.00; Price per Gallon $ 1.049 — **14.3**
9. DIESEL: Total $ 10.00; Price per Gallon $ 1.129 — **8.9**

APPLYING THE LESSON

Independent Practice

Homework Assignment	
Minimum	Ex. 1-31 odd; 32-38
Average	Ex. 2-30 even; 31-38
Maximum	Ex. 1-38

Additional Answers

33. 15,000 seats; yes; $1\frac{1}{3}$ of 44,702 is 59,454, so Kingdome has more than $1\frac{1}{3}$ the seating

5-MINUTE CHECK-UP

(over Lesson 3-2)

Available on TRANSPARENCY 3-3.

Multiply.

1. 2.6 × 4.04 10.504
2. 3.92 × 45 176.4
3. 0.28 × 1.112 0.31136
4. 0.03 × 0.06 0.0018

Extra Practice, Lesson 3-2, p. 453

▶ INTRODUCING THE LESSON

Ask students to give examples of some very expensive items they would like to buy but do not have the money to pay for the item all at once. Ask students when it is advisable to buy on credit.

▶ TEACHING THE LESSON

Using Questioning Which two numbers from the computation of the finance charge can be compared in order to judge whether or not the finance charge is reasonable? finance charge ÷ cash price; can be changed to a decimal or fraction in lowest terms and compared to an interest rate.

Practice Masters Booklet, p. 25

Name ______________ Date ______________

PRACTICE WORKSHEET 3-3

Applications: Buying on Credit

Complete the table.

Cash Price	Down Payment	Monthly Payment	Number of Months	Total Cost	Finance Charge
$366	$10	$35	12	1. $430	2. $64
$900	$100	$75	12	3. $1,000	4. $100
$1,800	$250	$50	36	5. $2,050	6. $250
$6,500	$1,250	$164	36	7. $7,154	8. $654
$2,995	$875	$115	24	9. $3,635	10. $640
$8,229	$1,500	$226	36	11. $9,636	12. $1,407
$4,789	$750	$218	24	13. $5,982	14. $1,193

Solve.

15. With a down payment of $1,000, Kim can buy a $12,000 car by making 36 monthly payments of $336. What is her total cost? What is the finance charge?
$13,096; $1,096
16. With a down payment of $50, Carlos can buy a $489 camera by making 12 monthly payments of $41. What is the total cost of the camera? What is the finance charge?
$542; $53
17. A stereo system costs $1,450. Jake can buy it on credit by putting $300 down and making 24 monthly payments of $64. What is the finance charge?
$386
18. Ramona bought furniture on credit by paying $375 and agreeing to pay the rest in monthly payments of $140 for two years. After 8 payments, she wants to pay the remaining cost. What is the remaining cost?
$2,240
19. Refer to Problem 18. If the furniture store gives Ramona $250 off the total price for paying early, what is the total amount she will pay for the furniture?
$3,485
20. A $4,000 motorcycle is on sale for $600 less. Anna can buy the bike with a down payment of $500 and monthly payments of $148 for two years. How much will she pay for the motorcycle? What is the finance charge?
$4,052; $652

Applications

3-3 BUYING ON CREDIT

Objective

Compute the finance charge for credit purchases.

Doug Halpern is buying the video camera described in the ad at the right on credit. After a down payment of $200, he will pay $50.50 a month for 24 months. What is the finance charge Doug will pay?

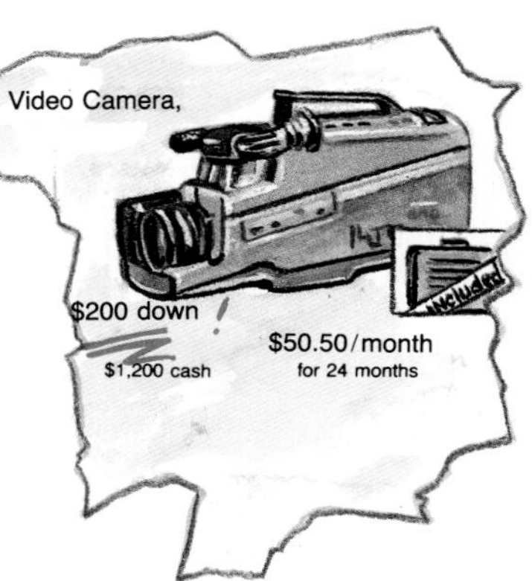

Many people buy expensive items on credit. They pay for the item over a period of time. Some purchases require a small portion of the credit price, called a **down payment**, to be paid at the time of purchase. The buyer makes a monthly payment for a given amount of time. The credit price of the item will be greater than the cash price. The difference between the credit price and the cash price is the **finance charge.**

Find the finance charge Doug will pay as follows. Multiply the amount of the monthly payment by the number of months.

50.50 [×] 24 [=] 1212

Add the down payment, if there is one, to find the credit price.

1212 [+] 200 [=] 1412

To find the finance charge, subtract the cash price from the credit price.

1412 [−] 1200 [=] 212

Understanding the Data

- What is the cash price? $1,200
- What is the amount of the down payment? $200
- What is the amount of the monthly payment? $50.50
- What is the finance charge? $212

Guided Practice

Copy and complete the table.

	Cash Price	Down Payment	Monthly Payment	Number of Months	Credit Price	Finance Charge
1.	$384	$10	$36	12	$442	$58
2.	$1,900	$250	$56.99	36	$2,301.64	$401.64
3.	$5,996	$1,496	$156	36	$7,112	$1,116
4.	$3,050	$600	$130.75	24	$3,738	$688
5.	$1,200	$100	$83	18	$1,594	$394

RETEACHING THE LESSON

Have students solve each problem.

1. A color television is advertised for $560. You can also buy this television for $25 down and $55 a month for 12 months. What is the amount of finance charge you will pay? $125
2. Jason bought a personal computer for $200 down and $60 a month for 2 years. The advertised price of the computer was $1149. How much finance charge did Jason pay? $491

Reteaching Masters Booklet, p. 23

Name ______________ Date ______________

RETEACHING WORKSHEET 3-3

Applications: Buying on Credit

Elena is buying the computer shown in the ad at the right *on credit*. She makes a *down payment* of $100. Then she will pay $60 a month for 12 months.

Computer
$700 Cash
or $60 per month for 12 months and downpayment of $100

When you buy on credit you pay more than the cash price.

A **down payment** is a small part of the total price.
The **total** cost of an item = down payment + total of monthly payments.
The **finance charge** = total cost − cash price.

What is the amount of the finance charge Elena will pay?

1. First find the total cost.
Multiply to find the total of the monthly payments. $12 × $60 = $720

Applications

6. What is the credit price of this car? **$11,620**

7. How much can you save by paying cash for the camera? **$24**

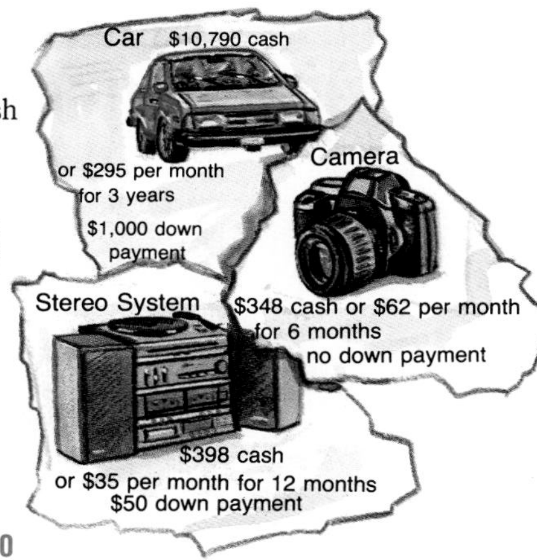

8. How much is the finance charge if you purchase the stereo system on credit? **$72**

9. Jane Peters bought living room furniture for $400 down and $100 a month for 2 years. After making 6 payments, she wants to pay off the remaining cost so that she won't have monthly payments. What is the remaining cost of the furniture? **$1,800**

10. Refer to Exercise 9. If the store gives Jane $100 off the total price for paying off the remaining cost early, what is the total price Jane will pay for the furniture? **$2,700**

Collect Data

11. From newspaper ads, find the credit price of a car, a major appliance, or a stereo system. Find the finance charge if the cash price is not paid. **Answers may vary.**

Make Up a Problem

12. Make up a problem in which the finance charge is $240 and there is no down payment. **See students' work.**

Number Sense

13. The price of a car, in dollars, rounds to 9,460 when rounded to the nearest ten. What is the price of the car if 5 is in the ones place? **$9,455**

Mixed Review

Lesson 1-9

Find the value of each expression.

14. $14 \div 2 + 6 \times 3$ **25**

15. $16 \times 2 \div 4 + 20$ **28**

16. $18 \div 9 \times [3 \times (20 - 11)]$ **54**

Lesson 2-8

Find the perimeter.

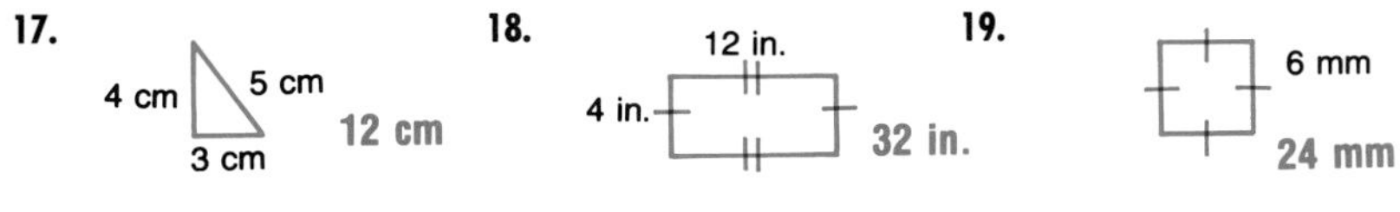

Lesson 3-1

20. Josiah began jogging and plans to jog 15 miles a week. How many miles will he jog in one year? **780 miles**

APPLYING THE LESSON

Independent Practice

Homework Assignment	
Minimum	Ex. 1-20
Average	Ex. 1-20
Maximum	Ex. 1-20

Chalkboard Examples

1. The cash price of a stereo is $1,100. If Jenny makes a $100 down payment and pays $55 a month for 24 months, what is the total cost of the stereo? What is the finance charge? **$1,420; $320**

Additional examples are provided on TRANSPARENCY 3-3.

EVALUATING THE LESSON

Check for Understanding Use Guided Practice Exercises to check for student understanding.

Closing the Lesson Ask students to summarize what they have learned about buying on credit.

EXTENDING THE LESSON

Enrichment Have students copy the following chart and complete.

Item	Down Payment	Monthly Payment	No. of Payments	Total Paid
Motor Bike	$100	$42.50	24	**$1,120**
Stereo	$200	**$50**	36	$2,000
T.V.	$50	**$85**	12	$1,070

Enrichment Masters Booklet, p. 23

Name ______ Date ______

ENRICHMENT WORKSHEET 3-3

Using a Slide Rule

Before the widespread use of electronic calculators, slide rules were used to do many calculations. A slide rule cannot be used to add or subtract. However, it can be used to multiply, divide, raise numbers to powers, and perform many other operations. A slide rule has three parts.

1. the body 2. the slide 3. the cursor with hairline

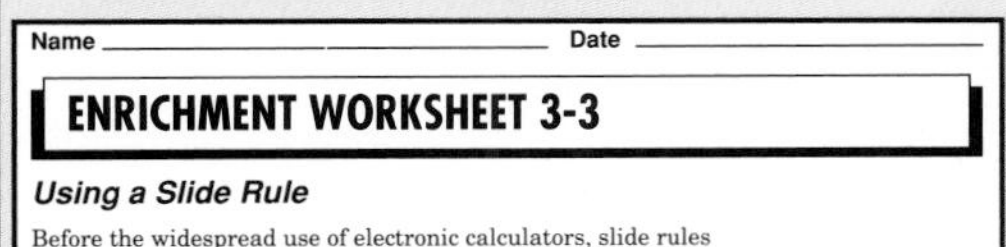

Graduated scales are marked on both the body and the slide. The scales are labeled with different letters. The C and D scales are used to multiply. The C and D scales are identical. Unlike a ruler, the distances between any two consecutive integers are not the same.

The scale does not use decimal points. The numbers 21,000; 21; 2.1; and 0.0021 are all located by the arrow.

Mark the following numbers on either scale on the diagram of the slide rule.

1. W, 30 2. X, 4.5 3. Y, 18 4. Z, 1.1

On a slide rule, you multiply two numbers by adding the corresponding lengths on the C and D scales. Multiply 1.5 times 4. See the slide rule pictured above. Place the 1 of the C scale opposite 1.5 on the D scale. Slide the cursor so the hairline is over 4 on the D scale. Read the product, 6, on the D scale.

Using the diagram above, multiply 1.5 by the following numbers.

5. 2.0 **3** 6. 400 **600** 7. 1.4 **2.1** 8. 100 **150**

5-MINUTE CHECK-UP

(over Lesson 3-3)

Available on TRANSPARENCY 3-4.

1. Terry is buying a kayak on credit. He makes a down payment of $200 and payments of $32 a month for 2 years. If the cash price of the kayak is $875, what is the finance charge? **$93**

▶ INTRODUCING THE LESSON

Suppose the class had 7 dozen T-shirts printed. The cost of the shirts came to $1,512. How much does one T-shirt cost? **$18**

Have students discuss in groups how the cost can be found.

▶ TEACHING THE LESSON

Using Critical Thinking Ask students what happens to the quotient if the divisor is halved, if the divisor is doubled. **The quotient doubles; the quotient is halved.**

Practice Masters Booklet, p. 26

Name ______________ Date ______________

PRACTICE WORKSHEET 3-4

Dividing Whole Numbers

Divide. Check with multiplication.

1. 9)927 **103**	2. 8)648 **81**	3. 6)624 **104**	4. 3)930 **310**
5. 7)494 **70 R4**	6. 2)626 **313**	7. 4)972 **243**	8. 9)592 **65 R7**
9. 20)400 **20**	10. 32)440 **13 R24**	11. 14)68 **4 R12**	12. 31)63 **2 R1**
13. 42)135 **3 R9**	14. 40)98 **2 R18**	15. 28)776 **27 R20**	16. 62)354 **5 R44**
17. 67)8,143 **121 R36**	18. 36)7,288 **202 R16**	19. 41)3,895 **95**	20. 52)3,652 **70 R12**

Solve.

21. If Jenny drives an average of 45 miles an hour, how long does it take her to drive 585 miles? **13 h**

22. In 13 days, 2,834 people visit the exhibit. On the average, how many people visit each day? **218 people**

3-4 DIVIDING WHOLE NUMBERS

Objective
Divide whole numbers.

Marcella told her parents that she had 1,108 days left until she graduates from high school. How many weeks does she have left until graduation? To solve this problem, you must divide 1,108 by 7.

Method

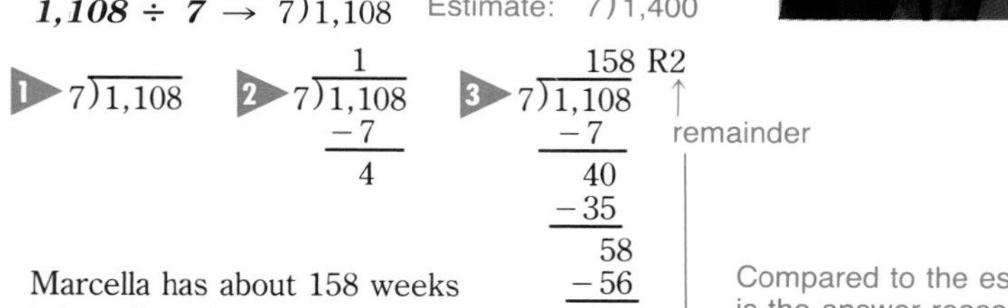

1. Divide the greatest place-value position possible.
2. Multiply and subtract.
3. Repeat 1 and 2 as necessary.

Example A

1,108 ÷ 7 → 7)1,108   Estimate: 7)1,400 = 200

1. 7)1,108
2. 7)1,108 → 1; −7; 4
3. 7)1,108 → 158 R2; −7; 40; −35; 58; −56; 2 ← remainder

Marcella has about 158 weeks left until graduation.

Compared to the estimate, is the answer reasonable?

Multiplication and division are inverse operations, so you can use multiplication to check division.

Check by multiplying. Add the remainder.

$$\begin{array}{rl} 158 & \text{quotient} \\ \times\ 7 & \text{divisor} \\ \hline 1{,}106 & \\ +\ 2 & \text{remainder} \\ \hline 1{,}108\ \checkmark & \text{dividend} \end{array}$$

The answer is correct.

A *quotient* is the result of dividing one number, the *dividend,* by another, the *divisor.*

Use estimation to help determine the digits in the quotient.

Method

1. Round the divisor to the greatest place-value position.
2. Divide in each place-value position from greatest to least. Estimate each digit of the quotient using the rounded divisor. Revise the estimate if necessary. Then divide.

Example B

3,405 ÷ 36

1. 36 → 40
2. 36)3,405 → 8; −2 88; 52

Note that 52 is greater than 36. So increase the estimate.

40)3,200 = 80

36)3,405 → 94 R21; −3 24; 165; −144; 21

40)160 = 4

Check: 36 × 94 = 144 + 3 24 = 3,384; + 21 = 3,405

Is the answer reasonable?

RETEACHING THE LESSON

Have students answer these two questions before dividing.

- *Where should the first digit be placed?*
- *How many digits should there be in the quotient?*

Have students mark where the first digit should be placed, write how many digits there should be in the quotient, then divide.

1. 4)84 **21**
2. 22)198 **9**

Reteaching Masters Booklet, p. 24

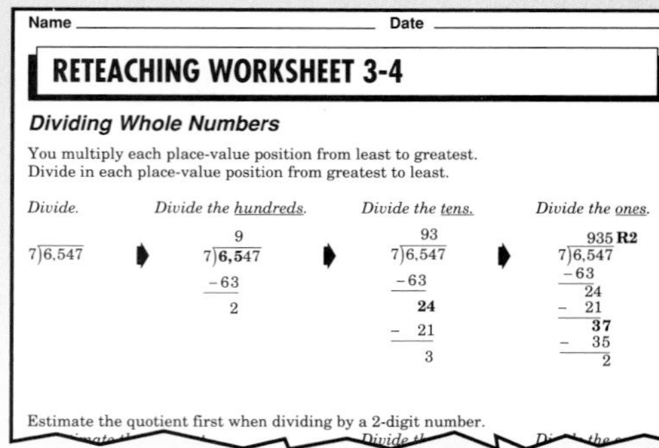

Name ______________ Date ______________

RETEACHING WORKSHEET 3-4

Dividing Whole Numbers

You multiply each place-value position from least to greatest. Divide in each place-value position from greatest to least.

Divide. 7)6,547 ▶ *Divide the hundreds.* 7)**6**,547 → 9; −63; 2 ▶ *Divide the tens.* 7)6,547 → 93; −63; **24**; −21; 3 ▶ *Divide the ones.* 7)6,547 → 935 **R2**; −63; 24; −21; **37**; −35; 2

Estimate the quotient first when dividing by a 2-digit number.

Guided Practice

Example A — ***Divide. Check using multiplication.***

1. $7\overline{)189}$ **27**
2. $8\overline{)768}$ **96**
3. $6\overline{)980}$ **163 R2**
4. $2\overline{)457}$ **228 R1**
5. $5\overline{)4,625}$ **925**

Example B — ***Use estimation to write the number of digits in each quotient. Then divide.***

6. $29\overline{)89}$ **3 R2; 1 digit**
7. $53\overline{)162}$ **3 R3; 1 digit**
8. $47\overline{)310}$ **6 R28; 1 digit**
9. $58\overline{)725}$ **12 R29; 2 digits**
10. $23\overline{)165}$ **7 R4; 1 digit**

Exercises

Write the number of digits in the quotient.

11. 2,450 ÷ 540 **1 digit**
12. 645 ÷ 27 **2 digits**
13. 248 ÷ 19 **2 digits**
14. 820 ÷ 4 **3 digits**

Practice — ***Divide. Check using multiplication.***

15. $5\overline{)270}$ **54**
16. $7\overline{)5,040}$ **720**
17. $8\overline{)512}$ **64**
18. $6\overline{)403}$ **67 R1**
19. $39\overline{)176}$ **4 R20**
20. $13\overline{)117}$ **9**
21. $19\overline{)115}$ **6 R1**
22. $33\overline{)250}$ **7 R19**
23. $810\overline{)3,258}$ **4 R18**
24. $709\overline{)5,000}$ **7 R37**
25. $450\overline{)3,750}$ **8 R150**
26. 1,300 ÷ 28 **46 R12**
27. 2,700 ÷ 68 **39 R48**
28. 1,225 ÷ 13 **94 R3**
29. What is the quotient if 733 is divided by 62? **11 R51**

Applications

30. 137 adults and 79 players are guests at the sports award banquet. You decide to seat the guests in groups of 8 at each table. How many tables will be needed? **27 tables**
31. Jolanda wants to save $2,500 to buy a used car. She earns $96 a week from her paper route and babysitting. If she saves half of her earnings each week, how many weeks will it take her to save the money? **approximately 52 weeks**

32. 300 people

Using Data — ***Complete. Use the graph at the right.***

32. In the survey of 1,000 people, how many said they do not recycle their trash because it took too much time? (30% = 0.30)
33. Find the number of people in each category. The total number of people should be 1,000. (19% = 0.19; 12 % = 0.12; 8% = 0.08; 23% = 0.23) **19%, 190; 12%, 120; 8%, 80; 23%, 230**

Collect Data

34. Conduct a survey in your class, school, or neighborhood to find the reasons people do not recycle. Compare your results to this survey. **See students' work.**

APPLYING THE LESSON

Independent Practice

Homework Assignment	
Minimum	Ex. 2-34 even
Average	Ex. 1-29 odd; 30-34
Maximum	Ex. 1-34

Chalkboard Examples

Example A *Divide.*

1. 746 ÷ 8 **93 R2**
2. 2,145 ÷ 7 **306 R3**

Example B *Divide.*

3. 567 ÷ 27 **21**
4. 1,620 ÷ 34 **47 R22**

Additional examples are provided on TRANSPARENCY 3-4.

EVALUATING THE LESSON

Check for Understanding Use Guided Practice Exercises to check for student understanding.

Error Analysis Watch for students who fail to add the remainder when checking. Remind students that not all quotients will be whole numbers or will divide evenly.

Closing the Lesson Ask students to state in their own words why multiplication can be used to check division.

EXTENDING THE LESSON

Enrichment Masters Booklet, p. 24

Name ______________________ Date ______________

ENRICHMENT WORKSHEET 3-4

Currency and Foreign Exchange

If you are traveling to a foreign country, you can change United States dollars to the currency of the foreign country. The rates for a recent day at which banks will exchange United States currency for foreign currency are given below.

Country	Monetary Unit	Currency Rates: Bank will buy at:	Bank will sell at:
Canada	dollar	0.6860	0.7440
France	franc	0.1370	0.1490
Great Britain	pound	1.3980	1.5160
Germany	deutsche mark	0.4215	0.4580

Examples:

A. To change $100 U.S. for Canadian dollars: $100 U.S. ÷ 0.7440 = $134.41 Canadian
↑ **selling rate of exchange**

B. To change 100 francs to U.S. dollars: 100 francs × 0.1370 = $13.70 U.S.
↑ **buying rate of exchange**

Solve.

1. Mr. Fong has moved to the United States from Canada. If he opens a checking account with $2,000 in Canadian currency, what is the value of his checking account in U.S. dollars? **$1,372**
2. Mrs. Carr is planning a vacation in England. She wants to buy $4,000 (U.S.) worth of British pounds. How many pounds will she have for her trip? **2,638.52 pounds**
3. Mrs. Carr had to cancel her trip to Great Britain. How much in U.S. dollars did she receive when she sold her British pounds? **$3,688.65**
4. Mr. Otto sold a property in Germany for 15,000 deutsche marks. How much money will he receive when he exchanges the marks for U.S. dollars? **$6,322.50**
5. Suppose that next year the selling rate of francs increases to 0.5214. Will it be more or less expensive for an American to travel in France? **more expensive**
6. Suppose that next year the buying rate of the pound increases. Will it be more or less expensive for a person from England to travel in the United States? **less expensive**

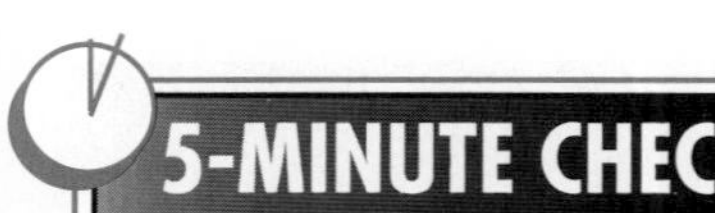

5-MINUTE CHECK-UP

(over Lesson 3-4)

Available on TRANSPARENCY 3-5.

Divide.

1. 198 ÷ 9 **22**
2. 2,609 ÷ 62 **42 R5**
3. 2,576 ÷ 56 **46**
4. 6,400 ÷ 78 **82 R4**

Extra Practice, Lesson 3-4, p. 454

▶ INTRODUCING THE LESSON

Have students name practical situations in which powers of ten are used. Will banking and stock market areas use powers of ten? **yes**

▶ TEACHING THE LESSON

Using Connections Put a metric equivalency table on the chalkboard or overhead. Have students make up problems that involve metric measures.

Practice Masters Booklet, p. 27

PRACTICE WORKSHEET 3-5

Multiplying and Dividing by Powers of 10

Multiply.

1. 14 × 10 **140** **2.** 27 × 100 **2,700** **3.** 58 × 1,000 **58,000**
4. 256 × 10^1 **2,560** **5.** 361 × 10^2 **36,100** **6.** 495 × 1,000 **495,000**
7. 7.6 × 10 **76** **8.** 5.2 × 100 **520** **9.** 8.9 × 10^3 **8,900**
10. 4.21 × 10 **42.1** **11.** 3.73 × 100 **373** **12.** 6.85 × 1,000 **6,850**

Divide.

13. 8,024 ÷ 10 **802.4** **14.** 6,371 ÷ 10^2 **63.71** **15.** 5,406 ÷ 1,000 **5.406**
16. 436 ÷ 10^1 **43.6** **17.** 218 ÷ 100 **2.18** **18.** 153 ÷ 1,000 **0.153**
19. 1.5 ÷ 10 **0.15** **20.** 3.8 ÷ 100 **0.038** **21.** 7.1 ÷ 10^3 **0.0071**
22. 0.25 ÷ 10 **0.025** **23.** 0.93 ÷ 100 **0.0093** **24.** 0.84 ÷ 1,000 **0.00084**

Replace each variable with a power of 10 to make a true sentence.

25. $57 \div y = 0.057$ $y = 10^3$ **26.** $3.522 \times p = 352.2$ $p = 10^2$
27. $93 \div x = 9.3$ $x = 10^1$ **28.** $64.027 \times q = 640.27$ $q = 10^1$
29. $0.083 \div w = 0.000083$ $w = 10^3$ **30.** $502.3 \times s = 5{,}023$ $s = 10^1$
31. $99.94 \times z = 99{,}940$ $z = 10^3$ **32.** $0.049 \times t = 4.9$ $t = 10^2$
33. $2 \times r = 200$ $r = 10^2$ **34.** $7.324 \div y = 0.07324$ $y = 10^2$

3-5 MULTIPLYING AND DIVIDING BY POWERS OF 10

Objective
Multiply and divide by powers of 10.

The automated packing system at Machine Builders Supply packs number 8 common nails in boxes weighing 3.7 pounds. If the machine packs 10 boxes in five minutes, how many pounds of nails does it pack in 10 boxes? How many pounds of nails does it pack in 100 boxes? in 1,000 boxes?

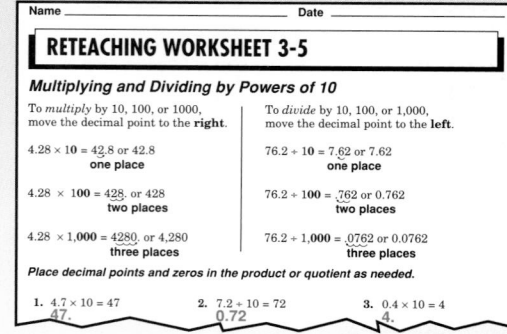

To multiply a number by 10, 100, or 1,000, move the decimal point to the right.

Method

1. Move the decimal point to the right one place for 10, two places for 100, or three places for 1,000.
2. Annex zeros on the right, if necessary, to move the correct number of decimal places.

Examples

A ***Multiply.***

$3.7 \times 10^1 = 37.$
$3.7 \times 10^2 = 370.$
$3.7 \times 10^3 = 3{,}700.$

B ***Multiply.***

$0.195 \times 10 = 1.95$
$0.195 \times 100 = 19.5$
$0.195 \times 1{,}000 = 195.$

The automated system packs 37 pounds of nails in 10 boxes, 370 pounds in 100 boxes, and 3,700 pounds in 1,000 boxes.

To divide a number by 10, 100, or 1,000, move the decimal point to the left.

Method

1. Move the decimal point to the left one place for 10, two places for 100, or three places for 1,000.
2. Insert zeros on left, if necessary, to move the correct number of decimal places.

Examples

C ***Divide.***

$5{,}234.8 \div 10 = 523.48$
$5{,}234.8 \div 100 = 52.348$
$5{,}234.8 \div 1{,}000 = 5.2348$

D ***Divide.***

$49 \div 10^1 = 4.9$
$49 \div 10^2 = 0.49$
$49 \div 10^3 = 0.049$

Guided Practice

Multiply.

Example A
1. 6.54×10^1 **65.4** **2.** 8.204×10^3 **8,204** **3.** 45×10^2 **4,500** **4.** 340×10^3 **340,000**

Example B
5. 4.9 × 10 **49** **6.** 5.12 × 100 **512** **7.** 62.9 × 1,000 **62,900** **8.** 3.7 × 100 **370**

RETEACHING THE LESSON

Using discussion helps students understand and learn the relationship between the exponent and the number of zeros.

Exponential Form	Standard Number
10^0	1
10^1	10
10^2	100
...	...
10^{200}	1 followed by 200 zeros

Reteaching Masters Booklet, p. 25

Name ____________ Date ____________

RETEACHING WORKSHEET 3-5

Multiplying and Dividing by Powers of 10

To *multiply* by 10, 100, or 1000, move the decimal point to the **right**.

4.28 × **10** = 42.8 or 42.8
one place

4.28 × **100** = 428. or 428
two places

4.28 × **1,000** = 4280. or 4,280
three places

To *divide* by 10, 100, or 1,000, move the decimal point to the **left**.

76.2 ÷ **10** = 7.62 or 7.62
one place

76.2 ÷ **100** = .762 or 0.762
two places

76.2 ÷ **1,000** = .0762 or 0.0762
three places

Place decimal points and zeros in the product or quotient as needed.

1. 4.7 × 10 = 47 **47.** **2.** 7.2 ÷ 10 = 72 **0.72** **3.** 0.4 × 10 = 4 **4.**

Example C *Divide.*

9. 4.48 ÷ 1,000 **0.00448**
10. 94.1 ÷ 10 **9.41**
11. 14.5 ÷ 100 **0.145**
12. 341.2 ÷ 100 **3.412**

Example D

13. 63.7 ÷ 10^2 **0.637**
14. 23.9 ÷ 10^3 **0.0239**
15. 39 ÷ 10^1 **3.9**
16. 48 ÷ 10^3 **0.048**

Exercises

Practice *Multiply.*

17. 17.5 × 100
18. 4.59 × 10
19. 3.72 × 100
20. 9.1 × 10^1
21. 0.178 × 1,000
22. 5.15 × 10^3
23. 14.7 × 1,000
24. 50.38 × 10
25. 23.04 × 10^1
26. 72.9 × 10^2
27. 28.4 × 10^3
28. 0.58 × 10

17. 1,750 18. 45.9 19. 372 20. 91 21. 178 22. 5,150 23. 14,700 24. 503.8 25. 230.4 26. 7,290 27. 28,400 28. 5.8

Divide.

29. 0.93 ÷ 10
30. 36.2 ÷ 10^2
31. 5.29 ÷ 10
32. 46.1 ÷ 100
33. 281.7 ÷ 1,000
34. 0.3 ÷ 10
35. 11.8 ÷ 10^3
36. 84.7 ÷ 100
37. 6.05 ÷ 10^3
38. 9.4 ÷ 10^2
39. 50.7 ÷ 10^1
40. 0.03 ÷ 10

29. 0.093 30. 0.362 31. 0.529 32. 0.461 33. 0.2817 34. 0.03 35. 0.0118 36. 0.847 37. 0.00605 38. 0.094 39. 5.07 40. 0.003

Write Math

41. Explain why the decimal point moves right when multiplying by powers of ten and why the decimal point moves left when dividing by powers of ten. **See margin.**

Applications

42. A safe contains $1,000 in dimes. How many dimes are in the safe? **10,000 dimes**
43. The safe contains $10,000 in $10 bills. How many $10 bills are in the safe? **1,000 $10 bills**
44. The diameter of Earth at its equator is 7,926 miles. The diameter of Jupiter at its equator is 88,700 miles. Jupiter's diameter is *about* how many times as large as Earth's? **about 11 times as large**

FUN with MATH
When and where was the decimal system first used?
See page 104.

Using Equations ***Solve each equation.***

45. $10x = 3.45$ **0.345**
46. $100y = 980.5$ **9.805**
47. $6.84q = 0.00684$ **0.001**
48. $42.23s = 4{,}223$ **100**

Mixed Review *Subtract.*

Lesson 2-4

49. 4.83 − 0.12 = **4.71**
50. 68.19 − 53.02 = **15.17**
51. 98.008 − 17.012 = **80.996**
52. 16.019 − 9.887 = **6.132**

Lesson 3-4

53. At the local library the librarian had 7,700 books in the children's department to arrange on 220 shelves. How many books did he put on each shelf if he divided them equally among the shelves? **35 books**

Chalkboard Examples

Examples A, B *Multiply.*

1. 14.32 × 10^2 **1,432**
2. 7,215 × 10 **72,150**
3. 0.56 × 10,000 **5,600**

Examples C, D *Divide.*

4. 9.3 ÷ 1,000 **0.0093**
5. 645.2 ÷ 10^2 **6.452**
6. 6.17 ÷ 10^4 **0.000617**

Additional examples are provided on TRANSPARENCY 3-5.

EVALUATING THE LESSON

Check for Understanding Use Guided Practice Exercises to check for student understanding.

Closing the Lesson Have students describe how to multiply and divide by powers of ten.

EXTENDING THE LESSON

Enrichment Have students research money facts, weight of a $1 bill, length of a $1 bill, and so on. Then apply powers of ten to the facts and find the weight of 100 $1 bills, 1,000 $1 bills, and so on.

Enrichment Masters Booklet, p. 25

ENRICHMENT WORKSHEET 3-5

Multiplying and Dividing by 10, 100, and 1,000

Find the products and quotients to complete the crossnumber puzzle.

ACROSS

1. 2.6 × 8.5
4. 4 × 16.5
6. 1.875 × 2
7. 4.31 + 0.82 + 1.05 + 5.08
9. 100 and 59 thousandths
11. 45.3 ÷ 10
13. 0.494 × 1,000
14. 21.03 + 4.839 + 201.431
15. 720 ÷ 10
16. 2.5 × 2.8
17. 0.005 × 1,000
18. 14.75 − 1.75
19. 10.4 + 26 + 0.39 + 5.98
23. 2,890 ÷ 100
25. 264 ÷ 100
26. 37.3 × 23.78
29. 45.9 − 6.38
31. 464 × 0.5
32. 0.022 × 1,000
33. 438 ÷ 100

		1. 2	2. 2	3. .1		4. 6	5. 6		
	6. 3	.7	5		7. 1	1	.2	8. 6	
9. 1	0	0	.0	10. 5	9		11. 4	.5	12. 3
	13. 4	9	4			14. 2	2	7	3
15. 7	2								16. .7
17. 5								18. 1	3
19. 4	20. 2	21. .7	22. 7			23. 2	24. 8	.9	
25. 2	.6	4		26. 8	27. 8	6	.9	9	28. 4
	29. 3	9	30. .5	2		31. 2	3	2	
		32. 2	2		33. 4	.3	8		

DOWN

1. 2.709 × 10
2. 250 + 0.4
3. 0.12 ÷ 1.2
4. 31.865 + 29.135
5. 1560.5 × 0.04
6. 6,084 × 0.5
7. 1,900 ÷ 100
8. 3.6 × 18.25
9. 0.8 × 1.25
10. 500 ÷ 100
12. 33 + 0.73
14. 0.8 × 2.5
15. 75,420 ÷ 10
18. 19 and 92 hundredths
20. 263 ÷ 10
21. 7,492 ÷ 1,000
22. 0.35 ÷ 0.05
23. 2.623 × 1,000
24. 89 + 0.38
26. 8,200 ÷ 100
27. 0.08 × 100
28. 2.72 + 1.28
30. 52 ÷ 10
33. 1.6 + 2.4

APPLYING THE LESSON

Independent Practice

Homework Assignment	
Minimum	Ex. 1-53 odd
Average	Ex. 2-40 even; 41-53
Maximum	Ex. 1-53

Additonal Answers

41. Answers may vary. Answers should include multiplying by a number greater than 1 results in a greater number and dividing by a number greater than 1 results in a lesser number.

5-MINUTE CHECK-UP

(over Lesson 3-5)

Available on TRANSPARENCY 3-6.

Multiply or divide.

1. 7.43×10^2 743

2. $513.4 \div 10^3$ 0.5134

3. 0.953×10^4 9,530

4. $209.04 \div 10^3$ 0.20904

Extra Practice, Lesson 3-5, p. 454

INTRODUCING THE LESSON

Ask students to list activities or situations in which a list is helpful, such as shopping lists, homework assignments, research materials. Ask students why lists are helpful.

TEACHING THE LESSON

Using Discussion In the Example on page 80, show students how to rewrite the amounts under Amount Saved as follows: \$20, $\$4 \times 1 + \$20$, $\$4 \times 2 + \$20$, $\$4 \times 3 + \$20$, and so on. Rewriting the Amount of Increase as $\$4 \times 0$, $\$4 \times 1$, $\$4 \times 2$, and so on, will help reveal the pattern.

Practice Masters Booklet, p. 28

Name ______________________ Date ______________________

PRACTICE WORKSHEET 3-6

Problem Solving Strategy: Look for a Pattern

Solve.

1. Ralph and Ella are playing a game called "Guess My Rule." Ralph has kept track of his guesses and Ella's responses in this table.

Ralph	0	1	2	3	4	5	6
Ella	10	9	8	7	6	5	

Look for a pattern and predict Ella's response for the number 6. Describe this pattern. 4; subtract the guessed number from 10

2. Mollie is using the following chart to help her figure prices for tickets.

Tickets	1	2	3	4
Price	\$7.50	\$12.50	\$17.50	\$22.50

A customer came in and ordered 10 tickets. How much should Mollie charge for this order? \$52.50

3. Brad needs to set up a coding system for files in the library using combinations of letters. He has begun this table.

Letters	1	2	3	4	5
Combinations	1	4	9	16	

1 LETTER	2 LETTERS	3 LETTERS	4 LETTERS
AA	AB BA BB AA	AA BC AB CA AC CB BA CC BB	AA BC DA AB BD DB AC CA DC AD CB DD BA CC BB CD

How many files can Brad code using the letters A, B, C, D, and E? 25 files

4. If the library has 400 items to code, how many letters will the librarian need to use? 20 letters

5. Billie needs to make a tower of soup cans as a display in a grocery store. Each layer of the tower will be in the shape of a rectangle. The length and the width of each layer will be one less than the layer below it.

top layer

second layer

third layer

How many cans will be needed for the fifth layer of the tower? 42 cans

How many cans will be needed for a 10-layer tower? 570 cans

▶ Explore ▶ Plan ▶ Solve ▶ Examine

3-6 LOOKING FOR A PATTERN

Objective

Solve problems by making a list and using the pattern.

Rita saves \$20 during the first month at her new job. She plans to increase the amount she saves by an additional \$4 a month for 10 months. What is the amount Rita will save in the tenth month?

You can make a list to find the pattern.

▶ Explore

What is given?

- Rita saves \$20 the first month.
- Rita increases the amount an additional \$4 each month.

What is asked?

- What is the amount Rita saves in the tenth month?

▶ Plan

Make a list and look for the pattern. Then apply it to month 10.

▶ Solve

Month	Amount Saved	Amount of Increase in Savings
1	\$20	\$ 0
2	\$24	\$ 4
3	\$28	\$ 8
4	\$32	\$12
5	\$36	\$16

Notice that the amount of increase from the first month is one *less* than the number of the month multiplied by 4. For each month add the increase to the first \$20 to find the amount saved.

Applying the pattern to month 10, the expression $(10 - 1) \times 4 + 20$ is the amount saved.

$$(10 - 1) \times 4 + 20 = 9 \times 4 + 20$$
$$= 36 + 20 \text{ or } 56$$

▶ Examine

Rita saves \$56 during month 10. You can check the solution by extending the list to month 10. The solution is correct.

Guided Practice

Solve. Make a list.

1. Jack sets patio blocks in a triangular area of his garden as shown. If he sets 10 rows, how many patio blocks will he use? 55 blocks

2. Find the sum of the first ten whole numbers that are multiples of three. Remember that zero is a whole number. 135

RETEACHING THE LESSON

Have students continue the following patterns two more steps, then find the total number of dots.

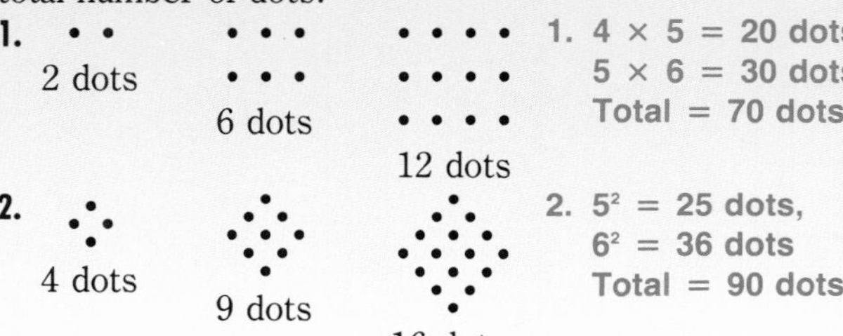

1. $4 \times 5 = 20$ dots, $5 \times 6 = 30$ dots, Total $= 70$ dots

2. $5^2 = 25$ dots, $6^2 = 36$ dots, Total $= 90$ dots

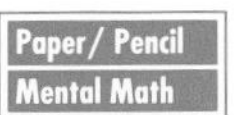

Problem Solving

Paper/ Pencil
Mental Math
Estimate
Calculator

Solve. Use any strategy.

3. Cans of soup are stacked in a triangular arrangement in a supermarket display. There is one can in the top row, three in the next, and five in the third. How many cans are in a display that has eight rows? **64 cans**

4. Determine if the sequence 56, 48, 40 is arithmetic. If it is, find the next three numbers. **32, 24, 16**

5. Sam plans to put one-half of his paycheck into his savings account and one-fourth of his paycheck into his checking account. Sam plans to spend one-eighth of his paycheck on cassette tapes and one-eighth of his paycheck on skateboard accessories. How much will Sam spend on cassette tapes if his paycheck is $372? **$46.50**

6. Glen, Doug, and Mike date three girls with eyes that are blue, green, and brown. Suppose Glen dates a girl with eyes the color of grass and Mike's date does not have brown eyes. Figure out the eye color of each boy's date.

6. Mike, blue eyes; Doug, brown eyes; Glen, green eyes

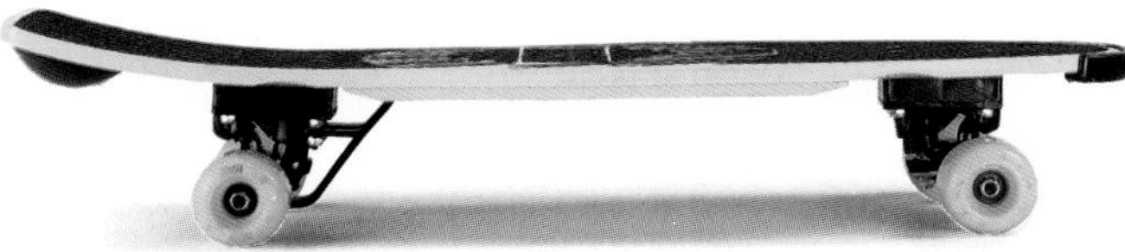

TECH PREP On pages 519–521, you can learn how agribusiness specialists use mathematics in their jobs.

7. Andrea drove 4 hours at an average speed of 52 miles per hour. Then she drove 3 hours at an average speed of 43 miles per hour. How far did Andrea drive altogether? **337 miles**

8. Harry's Sporting Goods advertises socks at 2 pair for $5, running shorts for $9, and T-shirts for $14. If Berni needs 4 pairs of socks and 2 T-shirts, *about* how much money should she take to the store? **about $40**

9. Gina says she ran a mile 0.5 seconds faster than Suzi and 0.7 seconds faster than Nicole. If Suzi ran the mile in 7 minutes 49.5 seconds, how fast did Nicole run the mile? **7 minutes 49.7 seconds**

Critical Thinking

10. Find the 1-digit numbers # and * represent that will make the statement true. #*.# × #* = #*#.* **# = 1, * = 0; 10.1 × 10 = 101.0**

Mixed Review

Lesson 1-4

Round each number to the nearest hundredth.

11. 0.841 **0.84** **12.** 11.976 **11.98** **13.** 0.4901 **0.49** **14.** 1.0097 **1.01**

Lesson 2-2

Add.

15. $68.45 + 1.79 = $ **70.24** **16.** $9.881 + 6.728 = $ **16.609** **17.** $0.381 + 4.2 = $ **4.581** **18.** $146.2 + 91.093 = $ **237.293**

Lesson 2-7

19. Darren chooses a number between 1 and 10. Starting with his number, he subtracts 3, adds 8, subtracts 12, and then adds 11. The result is 14. Which number did Darren choose? **10**

Chalkboard Examples

Solve by looking for a pattern.

1. Every fifth day, Jim increases the number of laps he does in the pool by 5. If Jim starts with 7 laps on the first day, how many laps will he do on the thirtieth day? **pattern: 7 + number of the day less one times 5, day 30 → 7 + 6 × 5; 37**

Additional examples are provided on TRANSPARENCY 3-6.

EVALUATING THE LESSON

Check for Understanding Use Guided Practice Exercises to check for student understanding.

Error Analysis Watch for students who do not list enough examples before looking for the pattern. Remind students that the pattern will be more apparent if they list four or five examples instead of two or three.

Closing the Lesson Have students explain how to make a list to look for a pattern.

APPLYING THE LESSON

Independent Practice

Homework Assignment	
Minimum	Ex. 2-18 even
Average	Ex. 1-19
Maximum	Ex. 1-19

5-MINUTE CHECK-UP

(over Lesson 3-6)

Available on TRANSPARENCY 3-7.

Solve by looking for a pattern.

1. In the first row Frank planted 4 seedlings. He increases the number of seedlings in each successive row by 3. How many seedlings will he plant in the 14th row? **row number × 3 + 1; 43 seedlings**

INTRODUCING THE LESSON

The Jennings counted extra pocket change they have at home. They had $3.55 in nickels, $10.80 in dimes, and $8.75 in quarters. How many of each coin do they have? **n, 71; d, 108; q, 35**

TEACHING THE LESSON

Using Critical Thinking Have students explain why moving the decimal point the same number of places in the dividend and the divisor does not change the value of the quotient. It is the same as multiplying the numerator and denominator of a fraction by the same power of 10.

Practice Masters Booklet, p. 29

Name ________________ Date ________________

PRACTICE WORKSHEET 3-7

Dividing Decimals

Divide.

1. 6)1.38 **0.23**	2. 7)7.21 **1.03**	3. 9)0.36 **0.04**	4. 3)9.18 **3.06**
5. 0.8)6.4 **8**	6. 0.7)0.63 **0.9**	7. 0.3)27 **90**	8. 8)2 **0.25**
9. 14)0.56 **0.04**	10. 8)32.2 **4.025**	11. 5.3)0.3869 **0.073**	12. 0.8)3.76 **4.7**
13. 2.3)1.61 **0.7**	14. 7.8)0.39 **0.05**	15. 4.4)16.72 **3.8**	16. 7.9)$41.08 **$5.20**
17. 7.1)0.0426 **0.006**	18. 6.5)$769.34 **$118.36**	19. 0.83)0.1909 **0.23**	20. 7.9)0.1185 **0.015**
21. $0.84 ÷ 12 **$0.07**	22. 0.414 ÷ 3 **0.138**	23. 25 ÷ 0.5 **50**	24. 139.4 ÷ 4.1 **34**

Solve.

25. How many nickels can Jeb get for $7.35? **147 nickels**

26. Kara can park her car for 1 hour for 75¢. If she spends $3.00 on parking, how long does she park? **4 hours**

3-7 DIVIDING DECIMALS

Objective *Divide decimals.*

Jeff and three friends bought Darryl a video game for his birthday. How much will the gift cost each person if they share the total price of $46.28 equally? To solve this problem, divide $46.28 by 4.

Method

1. Write the decimal point in the quotient directly above the decimal point in the dividend.
2. Divide as with whole numbers.

Example A ***4)46.28*** Estimate: 4)40 = 10

4)46.28 — 1 Write the decimal point in the quotient.

$$\begin{array}{r} 11.57 \\ 4\overline{)46.28} \\ -4 \\ \hline 6 \\ -4 \\ \hline 2\,2 \\ -2\,0 \\ \hline 28 \\ -28 \\ \hline 0 \end{array}$$

2 Divide as with whole numbers.

The gift will cost each person $11.57.

Based on the estimate, explain why the answer is reasonable.

When the divisor is a decimal, multiply the divisor by 10, 100, 1,000, and so on, until the divisor is a whole number. Multiply the dividend by the same number.

Method

1. Move the decimal point in the divisor to the right until a whole number is obtained.
2. Move the decimal point in the dividend the same number of places. Then write the decimal point in the quotient directly above the adjusted decimal point.
3. Divide.

Example B ***9.8 ÷ 0.002***

0.002.)9.800. (1, 2)

Annex two zeros in the dividend so the decimal point can be moved.

3

$$\begin{array}{r} 4{,}900 \\ 2\overline{)9{,}800} \\ -8 \\ \hline 1\,8 \\ -1\,8 \\ \hline 0 \end{array}$$

Guided Practice

Divide. **Encourage students to estimate the quotient first.**

Example A

1. 8)40.8 **5.1**
2. 2)8.6 **4.3**
3. 3)56.7 **18.9**
4. 13)39.39 **3.03**
5. 45)0.585 **0.013**

RETEACHING THE LESSON

Have the students solve the following problems. Monitor closely as they work and reteach as needed.

1. 0.8)2.4 **3**
2. 0.3)0.201 **0.67**
3. 0.04)0.092 **2.3**
4. 0.06)12 **200**

Emphasize that the division problem is changed unless the same thing is done to both sides.

Reteaching Masters Booklet, p. 26

Name ________________ Date ________________

RETEACHING WORKSHEET 3-7

Dividing Decimals

To divide a decimal by a whole number, place the decimal point for the quotient directly over the decimal point in the dividend. 8)3.76 **Then divide as you would with whole numbers.** 8)3.76 = 0.47; − 3 2; 56; 56

Sometimes, you need to annex zeros to the dividend. Divide until the remainder is zero. 4)4.2 ▸ 4)4.20 = 1.05; − 4; 20; 20

To divide by a decimal, multiply both the divisor and dividend by the same power of 10 so the divisor is a whole number. Then divide as with whole numbers.

Example B 6. $0.4\overline{)7.6}$ 19 7. $0.3\overline{)0.84}$ 2.8 8. $4.8\overline{)1.2}$ 0.25 9. $1.8\overline{)43.74}$ 24.3 10. $0.1\overline{)6}$ 60

Exercises

Divide. Encourage students to estimate the quotient first.

Practice

11. $7\overline{)7.84}$ 1.12
12. $4\overline{)9.2}$ 2.3
13. $8\overline{)2.08}$ 0.26
14. $8\overline{)0.024}$ 0.003
15. $0.25\overline{)5}$ 20
16. $0.4\overline{)5}$ 12.5
17. $0.09\overline{)61.2}$ 680
18. $1.6\overline{)44.8}$ 28
19. $1.2\overline{)3.12}$ 2.6
20. $1.25\overline{)200}$ 160
21. $0.3\overline{)6.12}$ 20.4
22. $0.5\overline{)32.15}$ 64.3
23. 52 ÷ 1.3 40
24. 3.755 ÷ 5
25. 9.6 ÷ 12
26. 16 ÷ 0.64
27. 15 ÷ 0.5 30
28. 0.312 ÷ 6
29. 220.1 ÷ 3.1
30. $0.72 ÷ 12

24. 0.751 25. 0.8 26. 25 28. 0.052 29. 71 30. $0.06

Number Sense

Place the missing decimal point in the answer to each problem.

31. 5.6 ÷ 8 = 7 0.7
32. 5.01 × 6 = 3006
33. 30.24 ÷ 27 = 112
34. 73.2 × 6 = 4392
35. 0.4 ÷ 0.2 = 2
36. 0.980 ÷ 4 = 245
37. 2.9 × 1.8 = 522
38. 16 ÷ 0.64 = 25
39. 6 ÷ 15 = 4 0.4

32. 30.06 33. 1.12 34. 439.2 35. 2.0 36. 0.245 37. 5.22 38. 25

Applications

40. A 3-ounce can of chili powder costs $1.29. How much does one ounce cost? At that price, how much do twenty ounces cost?

40. $0.43; $8.60

41. Shelley buys a piece of ribbon 21.6 yards long. She cuts it into 15 pieces. How long is each piece to the nearest tenth of a yard? 1.44 yards

Research

42. Find the names of the five longest rivers in the world and list them in order from longest to shortest. See margin.

Mental Math

To add 0.9, you can add 1 and then subtract 0.1 because 0.9 = 1 − 0.1.

2.57 + 0.9 = 2.57 + 1 − 0.1 = 3.57 − 0.1 = 3.47

Using a shortcut like the one shown above, find each sum mentally.

43. 1.76 + 0.8 2.56
44. 7.2 + 1.9 9.1
45. 0.85 + 0.9 1.75
46. 4.2 + 1.9 6.1
47. 3.4 + 0.9 4.3
48. 14.5 + 0.8 15.3

APPLYING THE LESSON

Independent Practice

Homework Assignment	
Minimum	Ex. 2-48 even
Average	Ex. 1-39 odd; 40-48
Maximum	Ex. 1-48

Chapter 3, Quiz A (Lessons 3-1 through 3-7) is available in the Evaluation Masters Booklet, p. 14.

Additional Answers

42. Answers may vary according to sources used. Nile, 4,145 mi; Amazon, 3,915 mi; Chang Jiang, 3,900 mi; Mississippi-Missouri-Red Rock, 3,741 mi; Ob, 3,362 mi.

Chalkboard Examples

Example A *Divide.*

1. 11.28 ÷ 12 0.94
2. 0.161 ÷ 7 0.023

Example B *Divide.*

3. 7 ÷ 0.02 350
4. 4.2 ÷ 1.2 3.5
5. 2.12 ÷ 0.02 106

Additional examples are provided on TRANSPARENCY 3-7.

EVALUATING THE LESSON

Check for Understanding Use Guided Practice Exercises to check for student understanding.

Error Analysis Watch for students who fail to record a zero in the quotient. Have students use lined paper turned sideways or grid paper to align columns when dividing.

Closing the Lesson Ask students to summarize how to divide decimals and when to move the decimal point in the divisor and dividend.

EXTENDING THE LESSON

Enrichment Masters Booklet, p. 26

Name ______ Date ______

ENRICHMENT WORKSHEET 3-7

Using Decimals

Solve.

1. A 64-ounce box of detergent sells for $3.39. What is the cost per ounce? Round to the nearest tenth of a cent. $0.053 per ounce
2. Find the area of a rectangle whose length is 3.8 cm and whose width is 2.3 cm. 8.74 cm²
3. Ski equipment is rented for $3.50 an hour. John has $14.00. For how many hours can he rent ski equipment? 4 hours
4. Harold drove his car 228.7 miles on 10.8 gallons of gasoline. To the nearest tenth of a mile, how many miles did he drive per gallon? 21.2 mpg
5. Millie is saving quarters to buy a new stereo. How many quarters will she have when she has saved $97.75? 391 quarters
6. Mei had 15 hits for 43 times at bat. What was her batting average? (batting average = hits ÷ times at bat) .349 (rounded to the nearest thousandth)
7. Luke bought 4.35 pounds of ground beef for $1.89 per pound. What was the total cost of the ground beef? $8.23
8. The ABC Store needs 1,000 packing boxes now and will need 500 more in two months. The boxes cost 8.34¢ each in lots of fewer than 1,000. They cost 8.11¢ each in lots of 1,000 or more. How much will be saved by buying all 1,500 boxes now? $1.15

Applying Mathematics to the Work World

This optional page shows how mathematics is used in the work world and also provides a change of pace.

Objective Find the cost of renting an apartment and research expenses associated with renting.

Using Questioning

- List other ways an apartment manager might use mathematics on-the-job.

Using Discussion Discuss with students the different deposits that landlords may require.

You may wish to discuss the different occupations that permit a person time to manage an apartment complex, such as writer, consultant, and so on. There are also different options offered by landlords that include a salary, as well as free rent.

Activity

Using Cooperative Groups Assign each group a different yearly salary and a set of typical monthly bills except rent. Have them create a budget that includes all the bills. Then have students determine how much rent they could afford.

On-the-Job Application

APARTMENT MANAGER

Carlen Jackson manages the Whispering Wood Apartments. He has an agreement with the landlord to live in an apartment, rent free, in return for his duties as a manager.

Mr. Jackson's duties include showing and renting apartments, performing routine maintenance, and collecting the rent each month.

During November, Mr. Jackson has 8 apartments rented for \$540 each and 12 other apartments rented for \$595 each. How much rent should Mr. Jackson collect for November?

Rent	×	Number of Rented Apartments	=	Amount of Rent He Should Collect
\$540		8		\$ 4,320
\$595		12		+ \$ 7,140
			total rent	\$11,460

Mr. Jackson should collect \$11,460 for the month.

1. Ms. Evans manages a town house complex. In one month, she has six town houses rented for \$600 each and nine rented for \$655 each. How much rent should she collect for the month? **\$9,495**

2. For a one-bedroom apartment, the rent is \$475, the security deposit is the same as one month's rent, and the utility deposit is \$150. What is the total deposit required? The total deposit and the first month's rent are collected at the same time. How much should the manager collect the first month? **\$625; \$1,100**

3. Research the reasons that landlords require a security deposit and utility deposit. Find out if a security deposit is always equal to one month's rent. Find out if these are the only deposits that landlords require. **Answers may vary.**

4. Managing Whispering Wood apartments is Mr. Jackson's parttime job. If the apartment Mr. Jackson lives in would normally rent for \$595 a month, how much is he receiving annually as income in the form of free rent? **\$7,140**

Cash Deposits

People in many occupations or jobs, such as store managers, cashiers, or newspaper carriers, may handle large numbers of bills and coins.

Armen delivers newspapers in the morning to earn extra money. He collects from his customers on Thursday and pays his route manager on Friday. This week he collected the following bills and coins.

11–$5 bills
64–$1 bills
75–quarters
42–dimes
37–nickels

Find the total amount of money Armen collected.

1. Multiply the number of each type of bill or coin by its value.

11 × $5 = $55.00
64 × $1 = $64.00
75 × $0.25 = $18.75
42 × $0.10 = $ 4.20
37 × $0.05 = $ 1.85

Armen collected $143.80.

2. Add the total values.

$55.00
64.00
18.75
4.20
+ 1.85
$143.80

1. The owner of Qwick Vending Machine Company deposits 1,643 quarters, 319 dimes and 740 nickels. What is the amount of money deposited? **$479.65**
2. A food store manager deposited 416 $20 bills, 329 $10 bills, 60 $5 bills, and 6,784 $1 bills. She also deposited 194 quarters, 137 dimes, 116 nickels, and 208 pennies. What was the amount of the deposit? **$18,764.08**
3. At noon, the manager counted $2,640.71. She deposited everything but 4 $10 bills, 20 $5 bills, 100 $1 bills, and 200 each of quarters, dimes, nickels, and pennies. What was the amount of the manager's deposit? **$2,318.71**
4. Name some types of businesses that have to handle large amounts of cash. Describe some of the problems involved with handling large amounts of cash. **See margin.**

Applying Mathematics to the Real World

This optional page shows how mathematics is used in the real world and also provides a change of pace.

Objective Find the total of varying amounts of bills and coins.

Using Discussion Ask students to list and discuss all the occupations that handle large amounts of money. **cashiers, bankers, restaurant managers, and so on.**

Discuss why people who handle large amounts of money separate it into groups of the same denomination. **The money is easier to count, more efficient use of time, more accurate results.**

Activity

Using Cooperative Groups Have students pretend they are having a T-shirt sale and need $50 in coins and one-dollar bills. Have students make different combinations of pennies, nickels, dimes, quarters, and one-dollar bills that add up to $50. **Answers may vary.**

Additional Answers

4. Answers may vary. Typical answers are fast food establishments, drive-thrus, car wash, and cleaners. Problems involved include accuracy of counting, depositing cash after banking hours, employee and non-employee theft, and so on.

5-MINUTE CHECK-UP

(over Lesson 3-7)

Available on TRANSPARENCY 3-8.

Divide.

1. 128.8 ÷ 23 5.6
2. 37.2 ÷ 1.5 24.8
3. 97.188 ÷ 2.1 46.28
4. 40.02 ÷ .03 1,334

Extra Practice, Lesson 3-7, p. 454

INTRODUCING THE LESSON

Write several numbers on the chalkboard that are in millions and billions. Ask students what it would be like to work with large numbers all the time. Discuss with students that large numbers can be written in a shorter, easier to handle form.

TEACHING THE LESSON

Using Calculators Have students explore how their calculators show numbers that exceed the display. Discuss how they can still use the calculator with large numbers using scientific notation.

3-8 Scientific Notation

Objective
Change numbers from standard form to scientific notation and vice versa.

Mr. Andretti's computer holds 42 megabytes of data. One megabyte is one million bytes, so this computer holds 42 × 1,000,000 or 42,000,000 bytes.

Scientific notation is a way to write very large numbers, such as 42,000,000, or very small numbers.

To write a number using scientific notation, write the number as a product. One factor is a number that is at least 1 but less than 10. The other factor is a power of 10.

Method

1. Move the decimal point to find a number that is at least 1 but less than 10.
2. Count the number of places the decimal point was moved. This is the exponent for the power of ten.

Example A ***Write 42,000,000 in scientific notation.***

1. 42,000,000 4.2 is at least 1 but less than 10.
2. The decimal point was moved 7 places.

$42,000,000 = 4.2 \times 10^7$ The computer holds 4.2×10^7 bytes.

Example B ***Write 7,968 in scientific notation.***

1. 7,968 7.968 is at least 1 but less than 10.
2. The decimal point was moved 3 places.

$7,968 = 7.968 \times 10^3$

To write a number in standard form, remember the shortcut for multiplying by powers of ten.

Example C ***Write 6.82×10^4 in standard form.***

1. 6.8200
2. Move the decimal point 4 places to the right. Annex 2 zeros.

$6.82 \times 10^4 = 68,200$

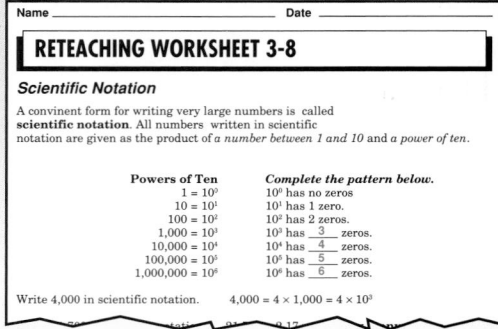

Guided Practice

1. 4.68×10^2 2. 1.4×10^1 3. 1.543×10^4 4. 2.66×10^5 5. 5×10^4 6. 1.4×10^2 7. 2.1×10^7 8. 3.5×10^6 9. 2.07×10^3 10. 5.4006×10^5

Examples A, B ***Write in scientific notation.***

1. 468	**2.** 14	**3.** 15,430	**4.** 266,000	**5.** 50,000
6. 140	**7.** 21,000,000	**8.** 3,500,000	**9.** 2,070	**10.** 540,060

Example C ***Write in standard form.*** 11. 575 12. 3,100 13. 271,000 14. 76,800 15. 38,000,000

11. 5.75×10^2	**12.** 3.1×10^3	**13.** 2.71×10^5	**14.** 7.68×10^4
15. 3.8×10^7	**16.** 1.09×10^6 1,090,000	**17.** 4.002×10^3 4,002	**18.** 1×10^5 100,000

RETEACHING THE LESSON

Use the following steps to correct errors when changing to scientific notation.

1. Write the number with one digit to the left of the decimal point.
2. Write the multiplication sign.
3. Write the second factor as 10.
4. Determine the exponent by counting the number of places the decimal was moved to the left.

Practice Masters Booklet, p. 30

Name ______________ Date ______________

PRACTICE WORKSHEET 3-8

Scientific Notation

Write in scientific notation.

1. 860 8.6×10^2	**2.** 2,000 2×10^3	**3.** 7,200 7.2×10
4. 840,000 8.4×10^5	**5.** 163,000 1.63×10^5	**6.** 87,400 8.74×10
7. 2,340 2.34×10^3	**8.** 3 million 3×10^6	**9.** 595,000 5.95×10
10. 480 4.8×10^2	**11.** 14,380 1.438×10^4	**12.** 6 thousand 6×10^3
13. 2,540 2.54×10^3	**14.** 13,800 1.38×10^4	**15.** 352,000 3.52×10^5
16. 156 1.56×10^2	**17.** 4,230,000 4.23×10^6	**18.** 37,700 3.77×10^4
19. 5,220,000 5.22×10^6	**20.** 455,000 4.55×10^5	**21.** 25,200,000 2.52×10^7
22. 43,200 4.32×10^4	**23.** 20 billion 2.0×10^{10}	**24.** 4 million 4×10^6

Write in standard form.

25. 1.9×10^5 190,000	**26.** 3.7×10^3 3,700	**27.** 6.82×10^6 6,820,000
28. 1.67×10^3 1,670	**29.** 5.38×10^5 538,000	**30.** 9.73×10^4 97,300
31. 4.4×10^9 4,400,000,000	**32.** 2.3×10^8 230,000,000	**33.** 7.85×10^7 78,500,000
34. 7.65×10^5 765,000	**35.** 8.79×10^6 8,790,000	**36.** 4.92×10^3 4,920
37. 3.19×10^5 319,000	**38.** 1.5×10^9 1,500,000,000	**39.** 7.92×10^3 7,920
40. 2.77×10^8 277,000,000	**41.** 2×10^6 2,000,000	**42.** 6.01×10^7 60,100,000

Reteaching Masters Booklet, p. 27

Name ______________ Date ______________

RETEACHING WORKSHEET 3-8

Scientific Notation

A convinent form for writing very large numbers is called **scientific notation**. All numbers written in scientific notation are given as the product of *a number between 1 and 10* and *a power of ten.*

Powers of Ten	***Complete the pattern below.***
$1 = 10^0$	10^0 has no zeros
$10 = 10^1$	10^1 has 1 zero.
$100 = 10^2$	10^2 has 2 zeros.
$1,000 = 10^3$	10^3 has 3 zeros.
$10,000 = 10^4$	10^4 has 4 zeros.
$100,000 = 10^5$	10^5 has 5 zeros.
$1,000,000 = 10^6$	10^6 has 6 zeros.

Write 4,000 in scientific notation. $4,000 = 4 \times 1,000 = 4 \times 10^3$

19. 3.5×10^5 20. 1.982×10^3 21. 9.2372×10^4 22. 5.7×10^7
23. 6.7×10^6 24. 6.791×10^3 25. 1.5×10^1 26. 9.36×10^4

Exercises

Practice

30. 690 31. 431
32. 9,070
33. 53,000
34. 8,270
35. 217,000
36. 3,430,000
37. 750,000

Write in scientific notation.

19. 350,000	**20.** 1,982	**21.** 92,372	**22.** 57,000,000
23. 6,700,000	**24.** 6,791	**25.** 15	**26.** 93,600
27. 6 million 6×10^6	**28.** 149 million 1.49×10^8		**29.** 417 billion 4.17×10^{11}

Write in standard form.

30. 6.9×10^2	**31.** 4.31×10^2	**32.** 9.07×10^3	**33.** 5.3×10^4
34. 8.27×10^3	**35.** 2.17×10^5	**36.** 3.43×10^6	**37.** 7.5×10^5
38. 3.694×10^5 369,400	**39.** 6.38×10^9 6,380,000,000	**40.** 4.6×10^8 460,000,000	**41.** 5.35×10^6 5,350,000

Applications

42. 907,000 days; 9.07×10^5 days

43. 25,730,000 miles; 2.573×10^7 miles

For Exercises 42-44, write your answer in standard form and in scientific notation.

42. It takes *about* 90,700 days for Pluto to complete one revolution around the Sun. There are 9.07×10^4 days in one Pluto year. How many Earth days are there in 10 Pluto years?

43. Earth's average distance from the sun is 9.296×10^7 miles. Venus is an average of 6.723×10^7 miles from the Sun. On the average, how much farther is Earth from the Sun than Venus?

44. How many seconds are there in a year (365 days)? 31,536,000 seconds; 3.1536×10^7

Critical Thinking

45. Have you lived a million seconds? Have you lived a billion seconds? Use a calculator to find out. Answers may vary.

46. Complete this sentence. "The one thing I did not understand in this lesson was ___." See students' work.

Calculator

Most calculators can display only eight digits at one time. So numbers with more than eight digits are usually shown in scientific notation.

Add 35,000,000 and 77,000,000 on a calculator.

Enter: 35000000 [+] 77000000 [=] 1.12 08

The readout 1.12 08 means 1.12×10^8.

The sum is 1.12×10^8 or 112,000,000.

Write each sum or product in scientific notation and then in standard form. See margin.

47. 68,000,000 + 74,000,000	**48.** 99,999,999 + 1
49. 80,000,000 + 80,000,000	**50.** 99,999,000 × 1,000
51. 1,000,000 × 100,000	**52.** 18 × 365 × 24 × 60 × 60

Chalkboard Examples

Write the numbers of national park visitors in 1989 in scientific notation. 1. 1.89×10^6 2. 3.4×10^6

1. Shenandoah 1.89 million
2. Mt. Ranier 3.4 million
3. Crater Lake 450,000 4.5×10^5
4. Everglades 967,900 9.679×10^5

Additional examples are provided on TRANSPARENCY 3-8.

EVALUATING THE LESSON

Check for Understanding Use Guided Practice Exercises to check for student understanding.

Closing the Lesson Tell students that they will use this lesson when they read about government spending and financial news.

EXTENDING THE LESSON

Enrichment Have students research newspapers for examples of numbers written in compact form such as $4.56 billion. Students may read about the problems caused by such practices in John Paulos Allen's book, *Innumeracy*.

Enrichment Masters Booklet, p. 27

Name ______________ Date ______________

ENRICHMENT WORKSHEET 3-8

Scientific Notation

Multiplication and division can be performed on numbers written in scientific notation.

$(3 \times 10^2) \times (6 \times 10^3)$	$(8 \times 10^4) \div (2 \times 10^2)$
$(3 \times 6) \times (10^2 \times 10^3)$	$(8 \div 2) \times (10^4 \div 10^2)$
$18 \times (10 \times 10 \times 10 \times 10 \times 10)$	$\frac{8}{2} \times \frac{10 \times 10 \times 10 \times 10}{10 \times 10}$
18×10^5 or 1,800,000	4×10^2 or 400

Multiply or divide.

1. $(3 \times 10^2) \times (4 \times 10^2)$ 12×10^4 or 1.2×10^5
2. $(5 \times 10\ ) \times (6 \times 10^4)$ 30×10^5 or 3×10^6
3. $(7 \times 10^5) \times (8 \times 10^3)$ 56×10^8 or 5.6×10^9
4. $(2.3 \times 10^2) \times (4.5 \times 10^3)$ 10.35×10^5 or 1.035×10^6
5. $(9 \times 10^4) \div (3 \times 10^3)$ 3×10^1
6. $(7 \times 10^3) \div (7 \times 10)$ 1×10^2
7. $(8 \times 10^5) \div (4 \times 10^5)$ 2×10^0 or 2
8. $(5.2 \times 10^4) \div (2.6 \times 10^2)$ 2×10^2
9. Explain how addition is used in multiplying powers of 10. The exponent of the product is the sum of the exponents of the factors.
10. Explain how subtraction is used in dividing powers of 10. The divisor's exponent is subtracted from the dividend's exponent to get the exponent of the quotient.

APPLYING THE LESSON

Independent Practice

Homework Assignment	
Minimum	Ex. 1-45 odd; 46-52
Average	Ex. 2-40 even; 42-52
Maximum	Ex. 1-52

Additional Answers

47. 1.42×10^8; 142,000,000
48. 1×10^8; 100,000,000
49. 1.6×10^8; 160,000,000
50. 9.9999×10^{10}; 99,999,000,000
51. 1×10^{11}; 100,000,000,000
52. 5.67648×10^8; 567,648,000

5-MINUTE CHECK-UP

(over Lesson 3-8)

Available on TRANSPARENCY 3-9.

Write in scientific notation.

1. 934,000 9.34×10^5

2. 43.2 million 4.32×10^7

Write in standard form.

3. 6.018×10^6 6,018,000

4. 9.074×10^4 90,740

5. 8.4×10^9 8,400,000,000

Extra Practice, Lesson 3-8, p. 455

▶ INTRODUCING THE LESSON

Ask students to list situations in mathematics and other areas in which patterns are used or can be found.

You may wish to provide examples, such as quilting, nature-Fibonnaci sequences, traffic control, and so on.

▶ TEACHING THE LESSON

Using Discussion Emphasize to students that if the common ratio is not the same for any two consecutive terms in a sequence, then the sequence is not geometric.

Practice Masters Booklet, p. 31

PRACTICE LESSON 3-9

Geometric Sequences

Find the common ratio and write the next three terms in each geometric sequence.

1. 6, 18, 54, 162, □, □, □ 3
486; 1,458; 4,374

2. 14, 28, 56, 112, □, □, □ 2
224; 448; 896

3. 405, 135, 45, 15, □, □, □ $\frac{1}{3}$ or 0.333
5, 1.6, 0.5

4. 48, 24, 12, 6, □, □, □ $\frac{1}{2}$ or 0.5
3, 1.5, 0.75

5. 3, 15, 75, 375, □, □, □ 5
1,875; 9,375; 46,875

6. 0.5, 5, 50, 500, □, □, □ 10
5,000; 50,000; 500,000

7. 3, 6, 12, 24, □, □, □ 2
48; 96; 192

8. 4, 20, 100, 500, □, □, □ 5
2,500; 12,500; 62,500

9. 3, 12, 48, 192, □, □, □ 4
768; 3,072; 12,288

10. 1.4, 2.8, 5.6, 11.2, □, □, □ 2
22.4; 44.8; 89.6

11. 1.3, 2.6, 5.2, 10.4, □, □, □ 2
20.8; 41.6; 83.2

12. 800, 200, 50, 12.5, □, □, □ $\frac{1}{4}$ or 0.25
3.125; 0.78125; 0.1953125

13. 0.12, 0.48, 1.92, 7.68, □, □, □ 4
30.72; 122.88; 491.52

14. 5, 5.5, 6.05, 6.655, □, □, □ 1.1
7.3205; 8.05255; 8.857805

15. 2, 12, 72, 432, □, □, □ 6
2,592; 15,552; 93,312

16. 2.7, 8.1, 24.3, 72.9, □, □, □ 3
218.7; 656.1; 1,968.3

17. 2, 3, 4.5, 6.75, □, □, □ 1.5
10.125; 15.1875; 22.78125

18. 8.8, 4.4, 2.2, 1.1, □, □, □ 0.5
0.55; 0.275; 0.1375

19. 80.6, 40.3, 20.15, 10.075, □, □, □ 0.5
5.0375; 2.51875; 1.259375

20. 1.5, 12, 96, 768, □, □, □ 8
6,144; 49,152; 393,216

21. 0.8, 2, 5, 12.5, □, □, □ 2.5
31.25; 78.125; 195.3125

22. 900, 450, 225, 112.5, □, □, □ 0.5
56.25; 28.125; 14.0625

Solve.

23. Inez is playing a video game. The longer she keeps her token active, the higher the point values become, doubling every minute. If the point value at the start was 20, what will the value be if she keeps her token active for 10 minutes? 20,480 points

24. Water is being lost through a leak in a tank. The hole causing the leak is widening and the water loss has been tripled each day. If 2 gallons were lost on the first day, how many gallons would be lost on the seventh day? 1,458 gallons

Patterns Connection

3-9 GEOMETRIC SEQUENCES

Objective

Identify and find geometric sequences.

Imagine that the Starlight Theater's ticket sales totaled $32,000 in 1992, $64,000 in 1993, and $128,000 in 1994. If ticket sales continue to grow at this rate, what amount will they total in 1997?

You know a list of numbers that follows a certain pattern is called a sequence. In a **geometric sequence,** consecutive numbers are found by multiplying the term before it by the same number. This number is called the **common ratio.**

1, 5, 25, 125, ... Each number in the sequence is 5 times the number before it.

96, 48, 24, 12, ... Each number in the sequence is 0.5 times the number before it.

The yearly ticket sales, $32,000, $64,000 and $128,000 form a geometric sequence.

To continue the pattern of numbers, find the common ratio. Then multiply the last number by the common ratio to find the next number.

Method

1. To find the common ratio, divide any number by the one before it.
2. Multiply the last number in the sequence by the common ratio to find the next number.

Example A

Find the common ratio and write the next three terms of the sequence 32,000; 64,000; 128,000; ...

1. Divide to find the common ratio.
   $64{,}000 \div 32{,}000 = 2$ $128{,}000 \div 64{,}000 = 2$
   The common ratio is 2.
2. Multiply the last number in the sequence by the common ratio to find the next number.
   $128{,}000 \times 2 = 256{,}000$ $256{,}000 \times 2 = 512{,}000$
   $512{,}000 \times 2 = 1{,}024{,}000$

The sequence is 32,000; 64,000; 128,000; 256,000; 512,000; 1,024,000.

The ticket sales in 1997 will be $1,024,000.

Example B

Find the common ratio and write the next three numbers in the sequence 9,375; 1,875; 375; ...

1. Divide a number by the one before it to find the common ratio.
   $1{,}875 \div 9{,}375 = 0.2$ $375 \div 1{,}875 = 0.2$
   The common ratio is 0.2
2. $375 \times 0.2 = 75$ $75 \times 0.2 = 15$ $15 \times 0.2 = 3$

The sequence is 9,375; 1,875; 375; 75; 15; 3.

RETEACHING THE LESSON

Have students find the next four numbers in each sequence. The pattern for numbers are given in parentheses.

1. 4, 16, 64, 256, 1,024, 4,096, 16,384 (multiply by 4)

2. 64, 32, 16, 8, 4, 2, 1 (divide by 2)

3. 9, 27, 81, 243, 729, 2,187, 6,561 (multiply by 3)

4. 46,656, 7,776, 1,296, 216, 36, 6, 1 (divide by 6)

Reteaching Masters Booklet, p. 28

Name ______ Date ______

RETEACHING WORKSHEET 3-9

Geometric Sequences

A sequence in which each term after the first is the product of the same number and the preceding term is called a **geometric sequence.**

3, 6, 12, 24, 48,...

Each term after the first is found by multiplying the preceding term by 2.
The 2 is called a **common ratio.**

State whether each sequence is geometric. If so, name the common ratio.

1. 5, 15, 45, 135, ... yes; 3

2. 7, 35, 175, 875,... yes; 5

1. 2; 34.4, 68.8; 137.6 2. 10; 56,000; 560,000; 5,600,000
3. 5; 937.5; 4,687.5; 23,437.5 4. 0.1; 0.091, 0.0091, 0.00091

Guided Practice

Find the common ratio and write the next three terms in each sequence.

Example A

1. 4.3, 8.6, 17.2, ... **2.** 56, 560, 5,600, ... **3.** 7.5, 37.5, 187.5, ...

Example B

4. 91, 9.1, 0.91, ... **5.** 200, 80, 32, ... **6.** 20, 10, 5, ...

5. 0.4; 12.8, 5.12, 2.048 6. 0.5; 2.5, 1.25, 0.625

Exercises

Find the common ratio and write the next three numbers in each sequence. See margin.

Practice

7. 23, 92, 368, ... **8.** 25, 75, 225, ... **9.** 2, 10, 50, ...

10. 50, 10, 2, ... **11.** 600, 240, 96, ... **12.** 240, 144, 86.4, ...

13. 1.2, 1.44, 1.728, ... **14.** 3, 12, 48, ... **15.** 2, 6, 18, ...

16. 3,000, 900, 270, ... **17.** 64, 32, 16, ... **18.** 100, 90, 81, ...

Applications

19. The pitch of a musical note depends on the number of vibrations per second. A note one octave higher than a given note vibrates twice as many times as the given note. If the number of vibrations per second for middle C is 256, what is the number of vibrations per second for a note three octaves above middle C? 2,048 vibrations

20. What is the number of vibrations per second for a note four octaves below middle C? Refer to Exercise 19. 16 vibrations

21. Tom and Cheryl Jackson opened a savings account with $5 for their son Robert on his first birthday. On his second birthday, they put $15 in the account. Each year they deposit 3 times the amount of the previous year's deposit. On which of Robert's birthdays will they make the first four-digit deposit? 6th birthday

22. *about* 0.1 24. *about* 10 27. *about* 10

Number Sense

Is the answer* about *1,* about *10, or* about *0.1?

22. 0.03 × 6 **23.** 3.1 × 0.3 *about* 1 **24.** 3.1 × 3.1

25. 6 ÷ 40 *about* 0.1 **26.** 5 ÷ 0.5 *about* 10 **27.** 1.1 ÷ 0.1

Critical Thinking

28. Julie has a collection of 12 math books. Seven of the books are about algebra and five of the books are about geometry. If she told you to choose any combination of books, one on each subject, how many possible choices would you have? 35 choices

Mental Math

Find each product or quotient mentally. 33. 12,309 34. 1,230,900

29. 0.778 × 100 77.8 **30.** 0.778 × 1,000 778 **31.** 0.778 × 10,000 7,780

33. 12,309

32. 12.309 × 10 123.09 **33.** 12.309 × 1,000 **34.** 12.309 × 100,000

34. 1,230,900

35. 42.53 ÷ 10 4.253 **36.** 42.53 ÷ 100 0.4253 **37.** 42.53 ÷ 1,000 0.04253

Chalkboard Examples

Example A *Write the next 3 numbers.*

1. 9, 18, 36, . . . 72, 144, 288

2. 48, 192, 768, . . . 3,072, 12,288, 49,152

Example B *Write the next 3 numbers.*

3. 1600, 400, 100, . . . 25, 6.25, 1.56

4. 131.25, 26.25, 5.25, . . . 1.05, 0.21, 0.042

Additional examples are provided on TRANSPARENCY 3-9.

EVALUATING THE LESSON

Check for Understanding Use Guided Practice Exercises to check for student understanding.

Closing the Lesson Have students tell why understanding number patterns is a helpful skill.

EXTENDING THE LESSON

Enrichment Have students research other patterns in nature that follow a pattern, a swinging pendulum, a bouncing ball, and so on. Physics textbooks are one source of this information.

Enrichment Masters Booklet, p. 28

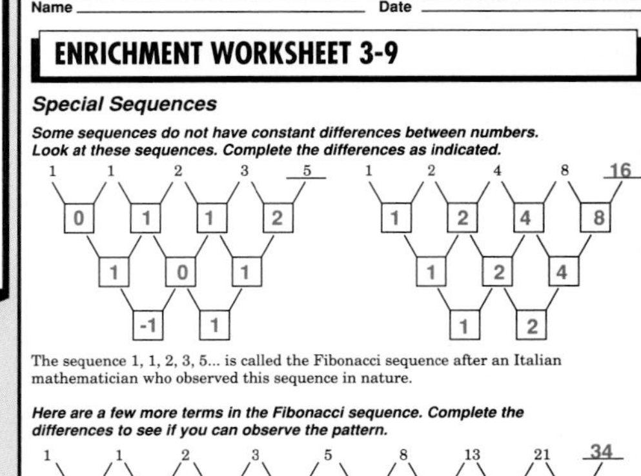
Name ______ Date ______

ENRICHMENT WORKSHEET 3-9

Special Sequences

Some sequences do not have constant differences between numbers. Look at these sequences. Complete the differences as indicated.

1 1 2 3 5 — 0 1 1 2 — 1 0 1 — -1 1

1 2 4 8 16 — 1 2 4 8 — 1 2 4 — 1 2

The sequence 1, 1, 2, 3, 5... is called the Fibonacci sequence after an Italian mathematician who observed this sequence in nature.

Here are a few more terms in the Fibonacci sequence. Complete the differences to see if you can observe the pattern.

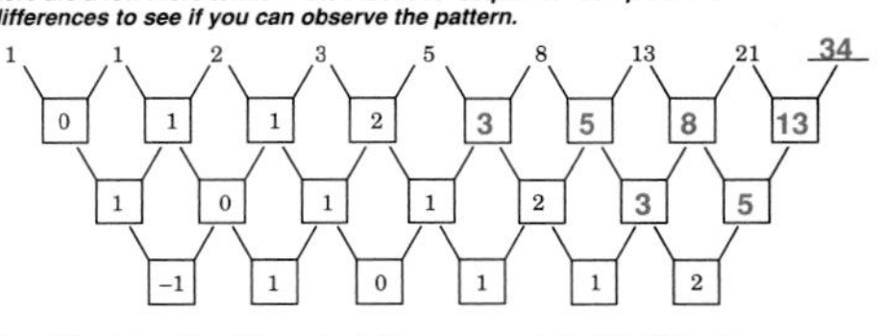
1 1 2 3 5 8 13 21 34

0 1 1 2 3 5 8 13

1 0 1 1 2 3 5

-1 1 0 1 1 2

One of the interesting things about this sequence is that its differences contain the sequence itself. Can you see how to extend it?

The "rule" for Fibonacci-type sequences is that each term is the sum of the previous two terms. Apply that rule to extend the following sequences.

1. 5, 9, 14, 23, 37, 60, 97, 157 2. 9, 1, 10, 11, 21, 32, 53, 85

3. 3, 9, 12, 21, 33, 54, 87, 141 4. 8, 7, 15, 22, 37, 59, 96, 155

5. 3, 3, 6, 9, 15, 24, 39, 63 6. 4, 3, 7, 10, 17, 27, 44, 71

APPLYING THE LESSON

Independent Practice

Homework Assignment	
Minimum	Ex. 2-36 even
Average	Ex. 1-17 odd; 19-37
Maximum	Ex. 1-37

Additional Answers

7. 4; 1,472, 5,888, 23,552
8. 3; 675, 2,025, 6,075
9. 5; 250, 1,250, 6,250
10. 0.2; 0.4, 0.08, 0.016
11. 0.4; 38.4, 15.36, 6.144
12. 0.6; 51.84, 31.104, 18.6624
13. 1.2; 2.0736, 2.48832, 2.985984
14. 4; 192, 768, 3,072
15. 3; 54, 162, 486
16. 0.3; 81, 24.3, 7.29
17. 0.5; 8, 4, 2
18. 0.9; 72.9, 65.61, 59.049

5-MINUTE CHECK-UP

(over Lesson 3-9)

Available on TRANSPARENCY 3-10.

Write the next three terms in each sequence.

1. 20, 30, 45, . . . **67.5, 101.25, 151.875**
2. 1, 6, 36, . . . **216, 1,296, 7,776**
3. 20, 50, 125, . . . **312.5, 781.25, 1953.125**
4. 1,000, 200, 40, . . . **8, 1.6, 0.32**

▶ INTRODUCING THE LESSON

Discuss with students what they think a formula is. One useful analogy is that formulas are like recipes.

Have students name some situations that use formulas to solve problems, such as gas mileage, area of geometric figures, loan payments, and so on.

▶ TEACHING THE LESSON

Using Connections Ask students to collect formulas used in other classes. Emphasize the need in everyday life for using formulas to solve problems.

Practice Masters Booklet, p. 32

Name ____________________ Date ____________________

PRACTICE WORKSHEET 3-10

Formulas

A formula for finding the amount of work needed is W = Fd.
Find the amount of work needed.

1. $F = 12$ lb; $d = 20$ ft **240 ft-lb**
2. $F = 28$ lb; $d = 150$ ft **4,200 ft-lb**
3. $F = 90$ lb; $d = 25$ ft **2,250 ft-lb**
4. $F = 175$ lb; $d = 20.5$ ft **3,587.5 ft-lb**
5. $F = 99$ lb; $d = 5.5$ ft **544.5 ft-lb**
6. $F = 150$ lb; $d = 50$ ft **7,500 ft-lb**

Find the circumference of each circle. Use 3.14 for π.

7. $d = 33$ cm **103.62 cm**
8. $r = 26$ ft **163.28 ft**
9. $r = 4.2$ in. **26.376 in.**
10. $d = 7.5$ m **23.55 m**

A formula for determining a normal blood pressure reading (B.P.) *is* B.P. $= 110 + \frac{A}{2}$
The A *stands for age in years. Find the normal blood presssure for each age.*

15. 16 **118**
16. 78 **149**
17. 21 **120.5**
18. 35 **127.5**
19. 56 **138**
20. 65 **142.5**
21. 37 **128.5**
22. 75 **147.5**

Write a formula. Then solve.

23. The air distance from Boston, Massachusetts to San Francisco, California is about 2,700 mi. If a flight from Boston to San Francisco takes 6 h, at what rate does the plane fly? **$r = 2,700 \div 6$, 450 mph**
24. The road distance from Boston to San Francisco is about 3,200 mi. To the nearest hundredth, how many hours of driving time will it take Ms. Costello if her rate is 55 mph? **$h = 3,200 \div 55$, 58.18 h**
25. Ms. Costello's car averages 28 miles per gallon of gasoline. To the nearest hundredth, how many gallons will be used for the drive from Boston to San Francisco? **$g = 3,200 \div 28$, 114.29 gal**
26. If gasoline on average costs $1.35 per gallon, to the nearest cent how much will Ms. Costello spend on gasoline for the drive from Boston to San Francisco? **c = $1.35 x 114.29, $154.29**

Algebra Connection

3-10 FORMULAS

Objective
Use formulas to solve problems.

Mr. and Mrs. Perez travel 630 miles to the Grand Canyon. If they drive at an average rate of 60 miles per hour, how long will it take?

A **formula** shows how certain quantities are related. You can use a formula to find an unknown quantity if you know the other quantities.

Method

1. Write the appropriate formula.
2. Replace each variable with the appropriate number.
3. Solve.

A formula that relates distance, rate (speed), and time is $d = rt$. The variable d represents distance, r represents rate or speed, and t represents time.

Example A ***Find the time it will take Mr. and Mrs. Perez to drive to the Grand Canyon using d = rt.***

1. $d = rt$ Write the formula.
2. $630 = 60t$ Replace d with 630, r with 60.
3. $\frac{630}{60} = \frac{60}{60}t$ Solve.

$10.5 = t$ It will take 10.5 hours to drive to the Grand Canyon.

Determine the rushing average for Emmitt Smith of the Dallas Cowboys. If Smith carried the ball 373 times for 1,713 yards, what was his rushing average for the year?

The formula for a player's rushing average (r) with a total of yards rushed (y) in n carries of the ball is $r = y \div n$.

Example B ***Find Smith's rushing average using r = y ÷ n.***

1. $r = y \div n$ **$y = 1,713$, $n = 373$**
2. $r = 1,713 \div 373$
3. $r = 4.592$

Smith's rushing average was 4.592 yards per carry.

Guided Practice

Example A

Solve for d, r, or t using the formula d = rt.

1. 264 miles, 55 mph **4.8 hr**
2. 2.5 mph, 2 hours **5 miles**
3. 30 miles, 12 mph **2.5 mph**
4. 650 mph, 2.2 hours **1,430 miles**

RETEACHING THE LESSON

Solve.

1. Use $D = 7w$, where D represents the number of days in w weeks, to determine the number of days in 52 weeks. **364 days**
2. Use $C = 0.29 \times n$, where C represents the cost in dollars of n 29¢ postage stamps, to determine the cost of fifty 29¢ stamps. **$14.50**

Reteaching Masters Booklet, p. 29

Name ____________________ Date ____________________

RETEACHING WORKSHEET 3-10

Formulas

A **formula** shows how certain quantities are related.
The formula for the distance traveled by a moving object is
d = rt. In the formula, d represents distance in kilometers (km), r represents the rate in kilometers per hour (km/h), and t represents the time in hours (h).

Suppose r is 40 kilometers per hour and t is 3 hours.
Find the distance traveled (d).

$d = rt$

$d = 40 \times 3$ **Replace *r* with 40 and *t* with 3.**

$d = 120$ The distance traveled is 120 kilometers.

Solve for d *using the formula* d = rt.

Example B ***Solve for r, y, or n using the formula r = y ÷ n.*** See margin.

5. 165 yards, 15 carries
6. 10.5 yards per carry, 28 carries
7. 171 yards, 18 carries
8. 7.6 yards per carry, 190 yards

Exercises

Practice ***Solve for d, r, or t using the formula d = rt.***

9. 4 miles, 1 hour **4 mph**
10. 4.25 hours, 55 mph **233.75 miles**
11. 9 miles, 6 mph **1.5 hours**
12. 880 yards, 20 min **44 yds/min**

Solve for y, n, or r using the formula r = y ÷ n.

13. 216 yards, 7.2 yards per carry **30 carries**
14. 160 yards, 25 carries **6.4 yards per carry**
15. 99 yards, 15 carries **6.6 yards per carry**
16. 14 carries, 9 yards per carry **126 yards rushed**

A formula for finding the gas mileage of a car is $s = \frac{m}{g}$, where s is gas mileage in miles per gallon (mpg), m is the number of miles driven, and g is the number of gallons of gas used. Find the gas mileage.

17. 220 miles, 10 gallons **22 mpg**
18. 240 miles, 15 gallons **16 mpg**
19. 120.5 miles, 5 gallons **24.1 mpg**
20. 137.2 miles, 8 gallons **17.15 mpg**
21. 374 miles, 8.5 gallons **44 mpg**
22. 400 miles, 12.5 gallons **32 mpg**

23. $s = 216 \div 8$, 27 mpg

Applications ***Write a formula. Then solve.***

23. At the start of Guido's trip, the odometer read 5,123.7. At the end of the trip, the odometer read 5,339.7. If Guido's car used 8 gallons of gasoline, how many miles per gallon is this?
24. Bo Jackson ran for 168 yards in 24 carries during a football game in Oakland. What was his rushing average for the game? **$r = 168 \div 24$; 7 yards per carry**

Connections

25. An important formula in physics is Ohm's law. This formula relates voltage (volts, V), current (amperes, I), and resistance (ohms, R). Ohm's law states that voltage equals amperes of current multiplied by resistance in ohms, or $V = I \times R$. If a toaster draws 4 amperes of current when connected to a 120-volt circuit, what is the resistance? **30 ohms**
26. If a 12-volt car battery has a resistance of 4 ohms with the engine running, what is the current? Use the formula in Exercise 25. **3 amps**

Mixed Review ***Solve each equation. Check your solutions.*** **28. 71.2 29. 5.47**

Lesson 1-11

27. $19 = y + 7$ **12**
28. $t - 60 = 11.2$
29. $m + 1.53 = 7$

Lesson 1-12 ***Solve each equation. Check your solutions.*** **30. 20 33. 10**

30. $900 = 45b$
31. $\frac{n}{5} = 14$ **70**
32. $\frac{c}{6} = 1.5$ **9**
33. $18g = 180$

Lesson 3-5

34. The computer club purchased 450 floppy discs for a total of $648. What was the cost of each disc? **$1.44**

APPLYING THE LESSON

Independent Practice

Homework Assignment	
Minimum	Ex. 2-34 even
Average	Ex. 1-21 odd; 23-34
Maximum	Ex. 1-34

Additional Answers

5. 11 yards per carry
6. 294 yards
7. 9.5 yards per carry
8. 25 carries

Chalkboard Examples

Find the time using d = rt.

1. $d = 180$ km, $r = 60$ km/hr, $t = ?$ **3 hours**

Find the yards rushed using $r = y \div n$.

2. $r = 7.2$ yds/carry, $n = 13$ carries, $y = ?$ **93.6 yards**

Additional examples are provided on TRANSPARENCY 3-10.

EVALUATING THE LESSON

Check for Understanding Use Guided Practice Exercises to check for student understanding.

Closing the Lesson Tell students that they will use formulas to find area and volume of other figures in Chapter 8.

EXTENDING THE LESSON

Enrichment Give students the formula for changing Fahrenheit temperature to Celsius, $C = \frac{5}{9}(F - 32)$. From a daily newspaper, have students change the Fahrenheit temperature to Celsius for five countries around the world.

Enrichment Masters Booklet, p. 29

Name ______ Date ______

ENRICHMENT WORKSHEET 3-10

Patterns and sequences

Find the next two terms for each sequence. You may want to use a calculator.

1. 3, 7, 11, 15, 19
2. 25, 22, 19, 16, 13
3.
4.
5. 5, 7, 9, 11, 13
6. 125, 116, 107, 98, 89
7.
8. 3, 12, 48, 192, 768
9. 48, 24, 12, 6, 3
10.
11.
12.
13. 256, 64, 16, 4, 1
14. 3, 11, 19, 27, 35
15.
16. 121, 100, 81, 64, 49
17. 5, 15, 45, 135, 405
18. 2, 4, 8, 16, 32, 64
19. 1, 4, 9, 16, 25, 36
20. 6, 13, 27, 48, 76, 111
21. 1, 1, 2, 4, 7, 11, 16
22. 1, 5, 2, 5, 3, 5, 4

5-MINUTE CHECK-UP

(over Lesson 3-10)

Available on TRANSPARENCY 3-11.

Solve using the formula given.

1. $m = 272.8$ miles, $g = 12.4$ gal, $s = ?$
   $s = m \div g$ **22 mpg**
2. If you drive 55 miles per hour for 4.5 hours, how far will you drive? $d = rt$ **247.5 miles**

Extra Practice, Lesson 3-10, p. 455

▶ INTRODUCING THE LESSON

Have students draw a rectangle with an area of 12 square centimeters. Make a transparency of **Lab Manual, p. 8.** Have students record and label their rectangles on the transparency.

▶ TEACHING THE LESSON

Using Cooperative Groups Change the Introduction problem to an area of 11 cm². In groups have students discuss the difference in finding sides of a rectangle when the area is a number that has many whole number factors or a number that has only one and itself as whole number factors.

Practice Masters Booklet, p. 33

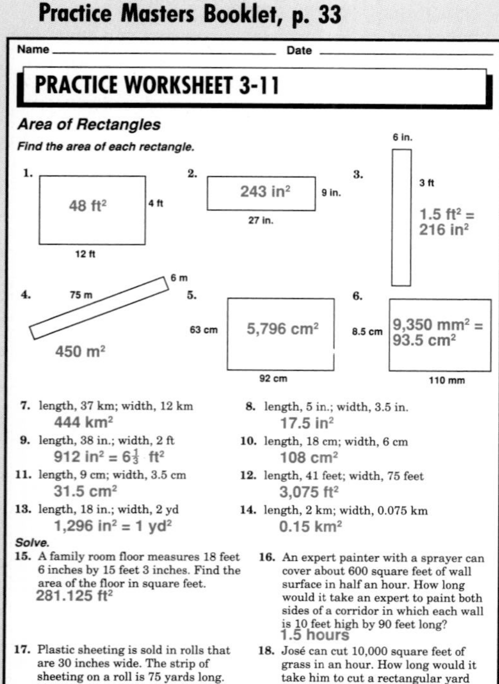
Name ______ Date ______

PRACTICE WORKSHEET 3-11

Area of Rectangles

Find the area of each rectangle.

1. 12 ft, 4 ft — 48 ft²
2. 27 in., 9 in. — 243 in²
3. 6 in., 3 ft — 1.5 ft² = 216 in²
4. 75 m, 6 m — 450 m²
5. 92 cm, 63 cm — 5,796 cm²
6. 110 mm, 8.5 cm — 9,350 mm² = 93.5 cm²
7. length, 37 km; width, 12 km — 444 km²
8. length, 5 in.; width, 3.5 in. — 17.5 in²
9. length, 38 in.; width, 2 ft — 912 in² = $6\frac{1}{3}$ ft²
10. length, 18 cm; width, 6 cm — 108 cm²
11. length, 9 cm; width, 3.5 cm — 31.5 cm²
12. length, 41 feet; width, 75 feet — 3,075 ft²
13. length, 18 in.; width, 2 yd — 1,296 in² = 1 yd²
14. length, 2 km; width, 0.075 km — 0.15 km²

Solve.

15. A family room floor measures 18 feet 6 inches by 15 feet 3 inches. Find the area of the floor in square feet. 281.125 ft²
16. An expert painter with a sprayer can cover about 600 square feet of wall surface in half an hour. How long would it take an expert to paint both sides of a corridor in which each wall is 10 feet high by 90 feet long? 1.5 hours
17. Plastic sheeting is sold in rolls that are 30 inches wide. The strip of sheeting on a roll is 75 yards long. How many square feet of sheeting are on a roll?
18. José can cut 10,000 square feet of grass in an hour. How long would it take him to cut a rectangular yard that measures 50 yards by 100 yards? 4.5 hours

Geometry Connection

3-11 AREA OF RECTANGLES

Objective
Use formulas to find the area of rectangles and squares.

Mark fenced in a small rectangular pen for his puppy. He made the pen 5 feet long and 3 feet wide. How many square feet of area does the pen enclose?

Area is the number of square units that covers a surface.

1 ft
1 ft
1 square foot
1 ft²

A drawing of the rectangular pen is shown below.

Example A

Find the area of the rectangle by counting the squares.

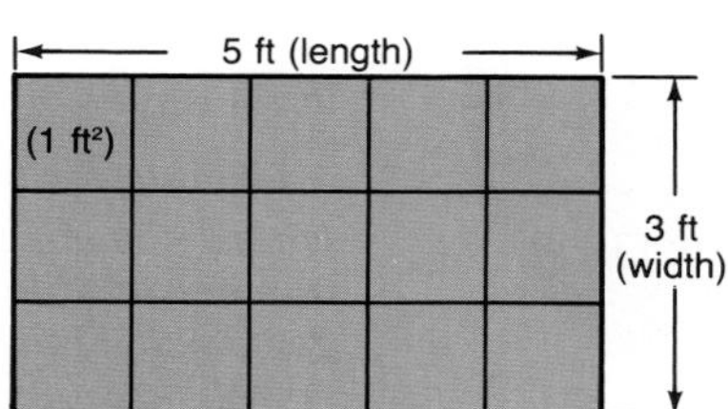

There are 15 square feet.

The area is 15 ft².

Mark's puppy's pen encloses an area of 15 square feet.

Example B

Find the area of the puppy pen. Use the formula $A = \ell w$.

A = area ℓ = length w = width

$A = \ell w$ length = 5, width = 3
$A = 5 \times 3$
$A = 15$ The area is 15 ft².

This is the same area found by counting squares.

A square is a rectangle in which all sides have the same length. Since the length and width are the same, use s to stand for both. So the formula for the area of a square is $A = s \times s$ or $A = s^2$.

Example C

Find the area of a square with sides 9 inches long.

$A = s^2$ $s = 9$
$A = 9^2$
$A = 81$ The area is 81 in².

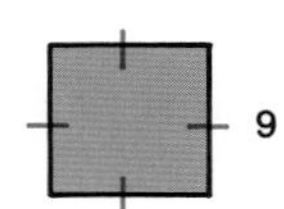

Guided Practice
Example A

Find the area of each rectangle by counting the squares.

1. 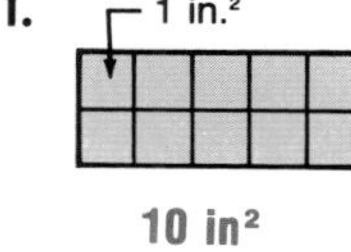

10 in²

2. 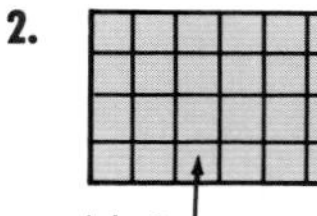

32 in²

3. 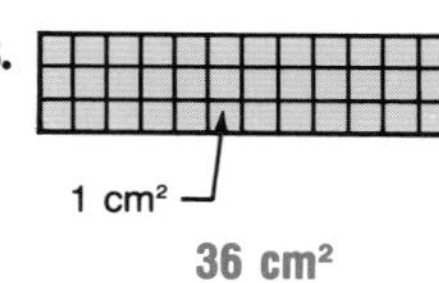

36 cm²

RETEACHING THE LESSON

Give students rectangles and parallelograms drawn on grid paper. For each figure, have them calculate the area and then count squares to verify their answers.

Reteaching Masters Booklet, p. 30

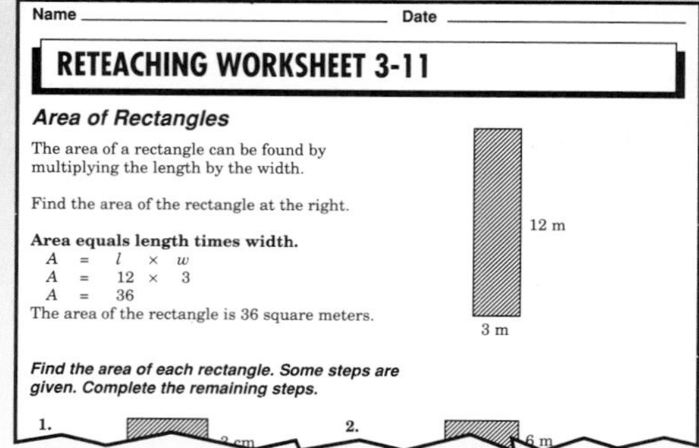
Name ______ Date ______

RETEACHING WORKSHEET 3-11

Area of Rectangles

The area of a rectangle can be found by multiplying the length by the width.

Find the area of the rectangle at the right.

Area equals length times width.

$A = l \times w$
$A = 12 \times 3$
$A = 36$

The area of the rectangle is 36 square meters.

12 m
3 m

Find the area of each rectangle. Some steps are given. Complete the remaining steps.

1. 3 cm
2. 6 m

Examples B, C

Find the area of each rectangle.

4. 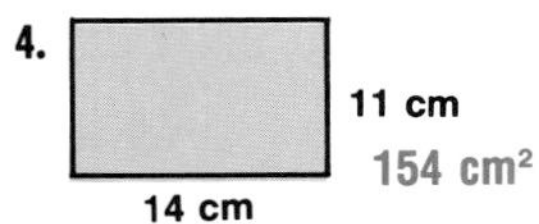

154 cm²

5. 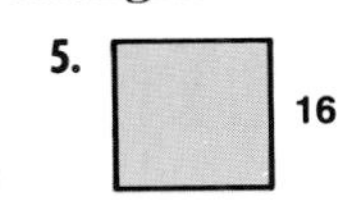

256 m²

6. 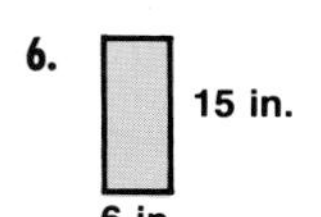

90 in²

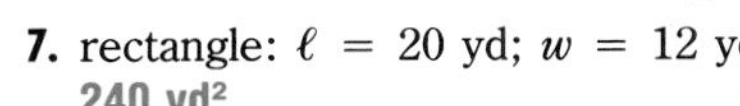

7. rectangle: $\ell = 20$ yd; $w = 12$ yd 240 yd²

8. square: $s = 40$ cm 1,600 cm²

Exercises

Practice

Find the area of each rectangle.

9. 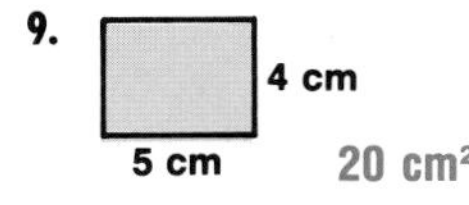

20 cm²

10. 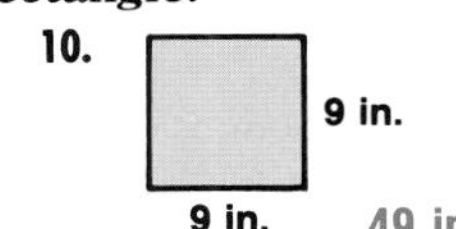

49 in²

11. 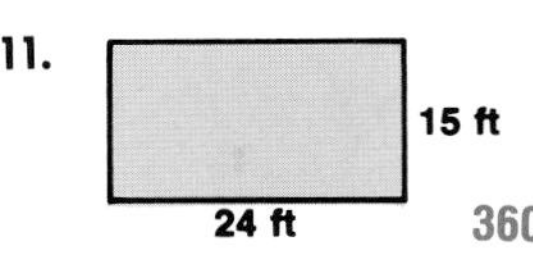

360 f

fUN with MATH

Challenge a friend to the game of Decimal Maze.
See page 105.

12. 13. 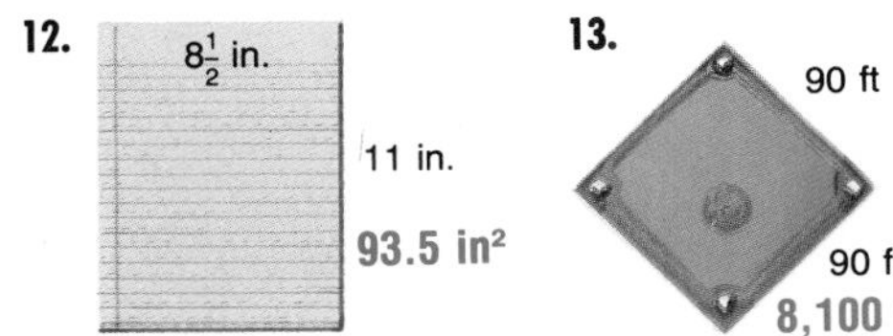

93.5 in²
8,100 ft²

14. 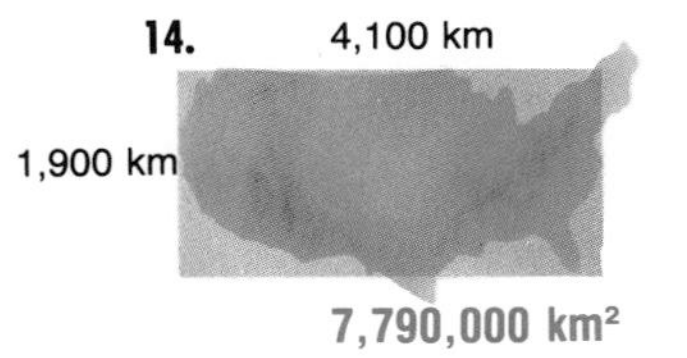

7,790,000 km²

Applications

15. Kevin Lewis needs to replace sod for a section of lawn. The section is 10.5 feet long and 6 feet wide. If sod costs \$2 a square foot, how much will it cost Kevin to replace the sod? \$126

16. Eileen is making name tags out of poster board 22 inches by 28 inches. She would like each name tag to be 2 inches by 4 inches. How many name tags can she cut from one poster board? 77 tags

Suppose

17. Suppose a square has a total area of 64 square feet. What is the length of each side? 8 feet

18. Suppose a rectangle with an area of 884 square meters has a length of 34 meters. What is the width of the rectangle? 26 m

Show Math

19. Draw a square with an area of 16 square inches. See students' work.

Critical Thinking

20. The area of the figure shown at the right is 384 square inches. The figure is made up of 6 squares. What is the perimeter of the figure? 112 in

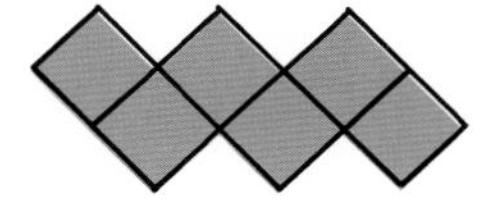

Calculator

You can use the 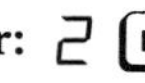 EE key on a calculator to perform operations on numbers written in scientific notation.

Add 2×10^9 and 6×10^9. Enter: 2 EE 9  + 6  EE 9  = 8 09

The sum is $8 \times 10^9 = 8{,}000{,}000{,}000$ or 8 billion.

Write each answer in scientific notation and then in standard form. See margin.

21. $(3.51 \times 10^{15}) + (2.47 \times 10^{15})$

22. $(3.95 \times 10^{11}) - (2.31 \times 10^{11})$

Chalkboard Examples

Example A *Find the area of each rectangle by counting the squares.*

1. 2. 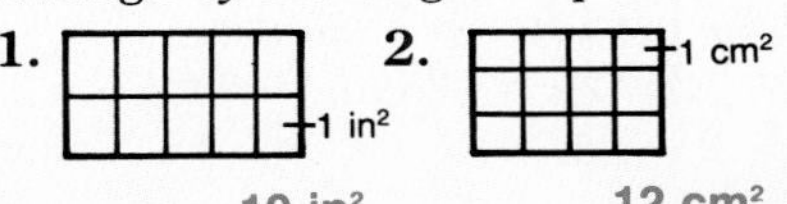

10 in² 12 cm²

Examples B, C *Find the area of each rectangle.*

3. $l = 36$ in., $w = 24$ in. 864 in²

4. $s = 11$ feet 121 ft²

Additional examples are provided on TRANSPARENCY 3-11.

EVALUATING THE LESSON

Check for Understanding Use Guided Practice Exercises to check for student understanding.

Error Analysis Watch for students who confuse area and perimeter. Remind students of the definition of each.

Closing the Lesson Ask students to summarize what they have learned about area of rectangles.

EXTENDING THE LESSON

Enrichment Masters Booklet, p. 30

Name ____________ Date ____________

ENRICHMENT WORKSHEET 3-11

Using Decimals

Solve.

1. A diver received scores of 43.6, 38.29, 40, and 37.7. How many points did she score on her four dives? 159.59 points

2. Gary bought 12.4 gallons of gasoline at the cash only island at \$1.129 per gallon. If he had used his credit card it would have cost \$1.219 per gallon. How much did he save by paying cash? \$1.12

3. Wynona used 415 kilowatt hours (kWh) of electricity last month. The cost is 5.313¢ per kWh. What was the total amount of her electric bill? \$22.05

4. At the start of the trip the odometer reading was 34,256.7 miles. On returning home, the reading was 36,581.1 miles. How many miles were driven on the trip? 2,324.4 miles

5. Mark bought a loaf of bread for 95¢, a dozen eggs for \$1.35, two cans of beans at 3 for \$1.00, and a newspaper for 45¢. How much change did he receive from \$10? \$6.58

6. Shawn ran the mile in 4 minutes 38.7 seconds. Tim ran the same race in 4 minutes 52.5 seconds. By how many seconds did Shawn beat Tim? 13.8 seconds

7. A rectangular metal plate in a piece of machinery is 5.62 cm wide and 6.03 cm long. To the nearest hundredth, what is the area of the plate? 33.89 cm²

8. When Ralph works overtime, he gets paid 1.5 times as much per hour as he gets regularly. The regular hourly rate is \$4.30. How much does he make for a regular week of 35 hours plus 4 hours overtime? \$176.30

APPLYING THE LESSON

Independent Practice

Homework Assignment	
Minimum	Ex. 1-19 odd
Average	Ex. 2-12 even; 13-20
Maximum	Ex. 1-20

Additional Answers

19. 5.98×10^{15}; 5,980,000,000,000,000

20. 1.64×10^{11}; 164,000,000,000

5-MINUTE CHECK-UP

(over Lesson 3-11)

Available on TRANSPARENCY 3-12.

Find the area of each rectangle

1. $l = 12$ m, $w = 25$ m, 300 m²
2. $l = 2.4$ cm, $w = 4.7$ cm. 11.28 cm²
3. A napkin is 14 in. on each side. What is the area of the napkin? **196 in²**
4. A 3 in. × 5 in. index card contains how many square inches? 15 in²

Extra Practice, Lesson 3-11, p. 456

INTRODUCING THE LESSON

List each student's name and the distance he or she lives from school. Have students determine the distance that best represents the average distance each student lives from school.

TEACHING THE LESSON

Using Discussion Discuss with students the meaning of the word *average*. Make sure they realize that an average is a reference point in a set of numbers. Point out that the average they are finding is the mean, but that there are other averages (median, mode).

Practice Masters Booklet, p. 34

Name ______________________ Date ______________

PRACTICE WORKSHEET 3-12

Statistics: Average (Mean)

Find the mean for each set of data. Round to the nearest tenth or cent.

1. 10, 10, 15, 10, 4, 1, 15
   9.3
2. 59, 69, 73, 74, 61, 67, 59, 58
   65
3. \$175, \$176, \$172, \$177, \$177, \$175, \$175, \$176, \$176, \$177, \$173
   \$175.36
4. 75, 80, 73, 74, 80, 80, 76, 74, 67, 70
   74.9
5. 1.4, 1.5, 1.4, 1.44, 1.39, 1.48, 1.47, 1.49, 1.49, 1.42, 1.42, 1.40
   1.4
6. 144, 143, 143, 138, 137, 146, 135, 141, 135, 147, 138, 134
   140.1
7. 0.65, 1.62, 0.63, 1.66, 0.62, 0.65, 1.66, 0.64, 0.62, 1.62, 0.65
   1.0
8. \$1.60, \$1.56, \$1.60, \$1.63, \$1.59, \$1.60, \$1.63, \$1.61, \$1.58, \$1.60
   \$1.60

Solve.

9. In the first five basketball games of the season, Jojo scored 12, 15, 22, 9, and 18 points. What was Jojo's average for these five games?
   15.2
10. Millen went bowling with her friends. Her scores were 182, 151, 127, and 167. What was Millen's average for these four games?
   156.75

Statistics Connection

3-12 AVERAGE (MEAN)

Objective

Find the mean of a set of data.

Vicki Riegle wants to buy a used car that gets high gasoline mileage. After looking at five cars, she makes a chart comparing their mileages to show her parents.

	Miles per Gallon
Sentra	33
Escort	33
Camaro	20
LeMans	28
Turcel	26

One average of a set of numbers is called the **mean.** Mean is the most common type of average.

Method

1. Find the sum of all the numbers.
2. Divide the sum by the number of addends.

Example A

Find the average for the mileages in the chart.

1. Find the sum of all the numbers.

   $33 + 33 + 20 + 28 + 26 = 140$

2. There are 5 addends. Divide the sum by 5.

   $\frac{140}{5} = 28$

The average of the mileages in the chart is 28 miles per gallon.

Example B

Find the average for these numbers: 3.4, 1.8, 2.6, 1.8, 2.3, and 3.2. Round to the nearest tenth.

1. $3.4 + 1.8 + 2.6 + 1.8 + 2.3 + 3.2 = 15.1$
2. $\frac{15.1}{6} \approx 2.51$ ≈ stands for "is approximately equal to."

The average, rounded to the nearest tenth, is 2.5.

Guided Practice

Example A

Find the average for each set of numbers. 3. 6 5. 38

1. 2, 3, 7, 8, 10 6
2. 1, 3, 5, 7 4
3. 1, 3, 4, 4, 7, 9, 14
4. 17, 18, 20, 13 17
5. 10, 40, 60, 30, 50
6. 23, 37, 20, 16 24

Example B

Find the average for each set of numbers. Round to the nearest tenth. 8. 10.4 9. 119.7 12. 4.1

7. 58, 61, 60 59.7
8. 14, 13, 10, 9, 6
9. 129, 130, 100
10. 0.8, 0.5, 0.6 0.6
11. 4.8, 6.4, 7.2 6.1
12. 4.5, 2.3, 6.0, 3.5

RETEACHING THE LESSON

Work with students to find the average for each set of numbers.

1. 18, 20, 15, 23 19
2. 98, 60, 95, 98, 94, 92 89.5
3. 10, 200, 10, 10, 100 66
4. 120, 135, 116, 121, 150 128.4

Reteaching Masters Booklet, p. 31

Name ______________________ Date ______________

RETEACHING WORKSHEET 3-12

Statistics: Average (Mean)

Joni went bowling. Her scores for five games are shown at the right. She wants to find her average score. Remember, *average* is another name for *mean*.

To find the **average**, add the numbers. Then divide the sum by the number of addends.

Joni's average for 5 games is 100.

Scores
98
93
115
89
+ 105
500

$\frac{100}{5)500}$

Find the mean for each set of data.

1. 2, 3, 4, 11
   5
2. 1, 5, 8, 10
   6

Exercises

Practice
15. 175 16. 0.9
17. 1.2 18. 2.6
23. 3.6 24. 58.3

Find the average for each set of numbers. Round to the nearest tenth.

13. 84, 111, 150 **115** **14.** 63, 47, 130 **80** **15.** 145, 230, 100, 225
16. 0.98, 0.85, 0.75 **17.** 1.1, 1.25, 1.34 **18.** 2.46, 3.7, 1.6, 2.55
19. 5, 8, 2, 7, 7, 9, 6, 4 **6** **20.** 92, 87, 97, 80, 95 **90.2**
21. 476, 833, 729, 548 **646.5** **22.** 801, 897, 843, 865, 854 **852**
23. 2.5, 4.73, 3.86, 3.39, 2.98, 4.24 **24.** 113.4, 6.6, 85, 16.16, 70.12
25. $50,000, $37,500, $43,900, $76,900, $46,000, $48,580 **$50,480**

Write Math

26. Name a situation when it would be better to find the average as a rounded answer rather than as an exact answer. **Answers may vary.**

27. Explain how you would use a calculator to find the average. **Answers may vary.**

Applications

28. In Princeton, the total rainfall for each month in a recent year was as follows:
1.2 in., 1.8 in., 2.4 in., 3.9 in., 4.7 in., 5.5 in., 4.5 in., 4.7 in., 7.2 in., 6.8 in., 3.0 in., 1.1 in.
What is the average? **3.9 in.**

29. Two basketballs are bought for $17.99 each. Another costs $32.99. What is the average cost of all three? **$22.99**

Critical Thinking

30. Emmy Explorer walked one mile north, then one mile west, and then one mile south. She arrived back where she started. Where did Emmy start? **the South Pole**

Estimation

Sometimes the clustering strategy can be used to estimate the mean. For example, the numbers in the table cluster around 10,000,000. So 10,000,000 is a good estimate of the average number of vehicles sold each year in the United States between 1983 and 1988.

U.S. Motor Vehicle Retail Sales (rounded to the nearest thousand)	
1983	9,182,000
1984	10,390,000
1985	11,038,000
1986	11,460,000
1987	10,278,000
1988	11,225,000

Answers may vary. Typical answers are given.

Use the clustering strategy to estimate each average.

31. 82, 77, 82, 76, 79, 78, 81 **80** **32.** 396, 391, 411, 407, 389 **400**
33. 6,637, 5,952, 5,848, 6,207 **6,000** **34.** 3.6, 2.75, 3.8, 3, 1.6, 2.34 **3**

APPLYING THE LESSON

Independent Practice

Homework Assignment	
Minimum	Ex. 2-34 even
Average	Ex. 1-25 odd; 26-34
Maximum	Ex. 1-34

Chalkboard Examples

Examples A, B *Find the average for each set of numbers. Round to the nearest tenth if necessary.*

1. 34, 63, 47, 52, 55, 44 **49.2**
2. 95, 124, 108, 110 **109.3**
3. 46, 83, 57, 64 **62.5**
4. 1.88, 0.93, 1.46 **1.4**

Additional examples are provided on TRANSPARENCY 3-12.

▶ EVALUATING THE LESSON

Check for Understanding Use Guided Practice Exercises to check for student understanding.

Error Analysis Remind students to align columns carefully when adding. Make sure they do not count the renamed number when counting addends to determine the divisor.

Closing the Lesson Have students describe in their own words how to find the mean for a set of numbers.

▶ EXTENDING THE LESSON

Enrichment Masters Booklet, p. 31

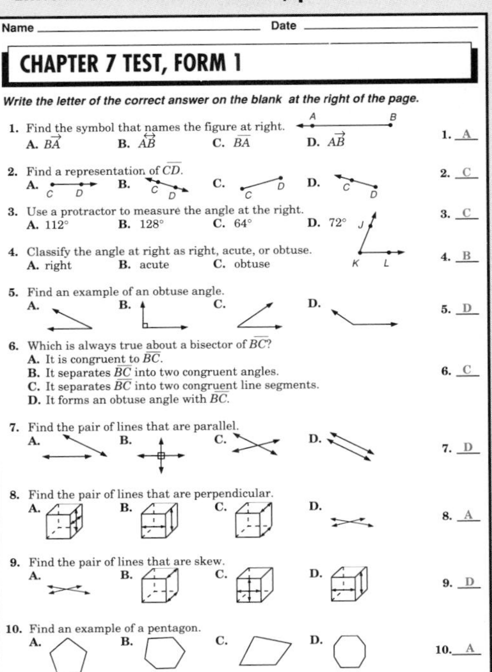

Name ______ Date ______

CHAPTER 7 TEST, FORM 1

Write the letter of the correct answer on the blank at the right of the page.

1. Find the symbol that names the figure at right. 1. A
A. $\overrightarrow{BA}$ B. $\overleftrightarrow{AB}$ C. $\overline{BA}$ D. $\overrightarrow{AB}$

2. Find a representation of $\overline{CD}$. 2. C
A. B. C. D.

3. Use a protractor to measure the angle at the right. 3. C
A. 112° B. 128° C. 64° D. 72°

4. Classify the angle at right as right, acute, or obtuse. 4. B
A. right B. acute C. obtuse

5. Find an example of an obtuse angle. 5. D
A. B. C. D.

6. Which is always true about a bisector of $\overline{BC}$? 6. C
A. It is congruent to $\overline{BC}$.
B. It separates $\overline{BC}$ into two congruent angles.
C. It separates $\overline{BC}$ into two congruent line segments.
D. It forms an obtuse angle with $\overline{BC}$.

7. Find the pair of lines that are parallel. 7. D
A. B. C. D.

8. Find the pair of lines that are perpendicular. 8. A
A. B. C. D.

9. Find the pair of lines that are skew. 9. D
A. B. C. D.

10. Find an example of a pentagon. 10. A
A. B. C. D.

5-MINUTE CHECK-UP

(over Lesson 3-12)

Available on TRANSPARENCY 3-13.

Find the average for each set of numbers. Round to the nearest tenth.

1. 65, 82, 72 **73**
2. 0.7, 0.9, 0.3, 0.5 **0.6**
3. 119, 90, 120, 95, 107 **106.2**
4. 3.58, 2.69, 4.47, 3.16 **3.5**

Extra Practice, Lesson 3-12, p. 456

INTRODUCING THE LESSON

Have students list the necessary steps to arrange vacation travel. Compare students' lists and count the number of steps. Point out that everyday problems are most often solved using multiple steps.

TEACHING THE LESSON

Using Discussion Remind students that the numbers used in the magic square on p. 99 are *not* the products but the number of decimal places in the product. Review magic squares.

Practice Masters Booklet, p. 35

Name ______________________ Date ______________

PRACTICE WORKSHEET 3-13

Problem Solving: Multi-Step Problems

Solve.

1. Roy receives $45 for his birthday. He buys a new book for $7.98, a shirt for $13.99, and a puzzle game for $4.59. How much does he have left to put in his savings account? **$18.44**
2. Connie earns money painting house numbers on curbs. In three weeks, she earns $175, $145, and $130. She pays a supply bill of $81.30. How much does she have left? **$368.70**
3. Earl, Steve, and Don walk dogs for their summer spending money. They walk 5 dogs for 8 weeks and are paid $6 per week for each dog. What is each boy's share if they divide the money evenly? **$80.00**
4. Pat earned $22.75. Her parents gave her $20. How much more does she need to buy a sweater that costs $50? **$7.25**
5. Luis earns $45 each week. He saves $12.75 each week. In how many weeks will he have enough money to buy a CD player that costs $135? **11 weeks**
6. Duwayne's lunch cost $5.35. He had a sandwich for $2.95, a drink for $0.89, and dessert. How much did dessert cost? **$1.51**
7. Mara works at a restaurant. She makes $3.75 an hour. She worked 20 hours and made $104.50 in tips. How much did she earn? **$179.50**
8. Crystal bought 2 shirts for $19.95 each and 3 pairs of socks at $3.89 a pair. Tax is $3.49. She gives the clerk three twenty-dollar bills. What is her change? **$4.94**

Use the passbook to find the balance on each of the following dates.

THE CENTRAL BANK — ACCOUNT NUMBER 07-1692-03 — IN ACCOUNT WITH Robin Jeffries

DATE	INTEREST	WITHDRAWALS	DEPOSITS	BALANCE

9. Sept. 15 **$118.10**
10. Oct. 31 **$139.67**
11. How much must Robin save to have $150? **$10.33**

96 Chapter 3

Problem Solving

Explore · Plan · Solve · Examine

3-13 MULTI-STEP PROBLEMS

Objective
Solve multi-step problems.

Monica works at Monterey's Mexican restaurant as a waitress. During the first week of October, she worked 12 hours and made $52 in tips. If Monica earns $4.25 an hour, how much did she make that week?

You may need to use more than one operation to solve a problem.

Explore

What is given?

- the hourly wage
- the number of hours worked
- the amount of tips earned

What is asked?

- Monica's total earnings during the first week of October

Plan

Hourly wages are found by multiplying the number of hours worked by the hourly rate. Total earnings are found by adding hourly wages and tips.

Solve

Estimate: $10 \times 4 = 40$ $\quad 40 + 50 = 90$

$12 \times 4.25 = 51$ Find the wages earned.

$51 + 52 = 103$ Add the tips.

Monica earned $103 during the first week of October.

Examine

$103 is close to the estimate. The answer is reasonable.

Guided Practice

1. Brett saws seven 0.2-meter pieces, from a 2.5-meter board. How long is the remaining piece? **1.1 m**

2. In his first three games, Larry scored 7, 15, and 21 points for Walnut Ridge. Geoff scored 22, 3, and 14 points in his first three games for DeSales. Which player has a better scoring average after three games? **Larry**

3. Susan earns $28.70 working 7 hours on Saturdays at Laura's Gifts. In December, Susan worked an extra 6 hours a week. At this rate, what does Susan earn in one week during December? **$53.30**

RETEACHING THE LESSON

Use steps A and B to solve the problem below.

In each of five weeks, Darla saved $46.10, $76.05, $25.00, $93.45, and $108.25. What was the average amount that she saved each week?

A. What was the total amount she saved? **$348.85**

B. What was the average amount Darla saved each week? **$69.77**

Reteaching Masters Booklet, p. 32

Name ______________________ Date ______________

RETEACHING WORKSHEET 3-13

Problem Solving: Multi-Step Problems

The monthly rainfall from May to August was 3.26 inches, 0.837 inches, 4.1 inches, and 1.003 inches. What was the average monthly rainfall during this period?

Explore What is given? 3.26 in., 0.837 in., 4.1 in., and 1.003 in. of rainfall
What is asked? average monthly rainfall

Plan To find the average rainfall, first add to find the total rainfall. Then divide the sum by the number of months.

Solve Add 3.26, 0.837, 4.1 … 4 addends Divide $4\overline{)9.2}$ = 2.3 …

Problem Solving

4. Keith's test scores are 76, 89, 85, and 93. What is his average score? 85.75
5. Hal's lunch cost $4.85. He bought a hamburger for $2.50 and fries for $1.35. He also bought a large iced tea. What was the cost of the iced tea? $1.00
6. Refer to Exercise 5. If Hal leaves a tip of $0.75 and the tax on his lunch is $0.30, how much does he have left from $10? $4.10

7. Select one of the assignments from this chapter that you found particularly challenging. Place it in your portfolio. See students' work.

Number Sense

8. Find the number of decimal places in each product. Use the answers to make a magic square. Each row, column, and diagonal should make the same sum.

0.06×0.05 4
2.4×5 1
7.420×8.0064 7
0.124×0.65 5
0.00067×0.812 8
5.5×3.2 2
0.29×1.4 3
2.1×3.71272 6
3.30245×8.9632 9

6	1	8	15
7	5	3	15
2	9	4	15
15	15	15	15

9. When you buy a car, you have many additional expenses; one is automobile insurance. Make up a problem using the table. Exchange with a classmate to solve. Discuss your solution. See students' work.

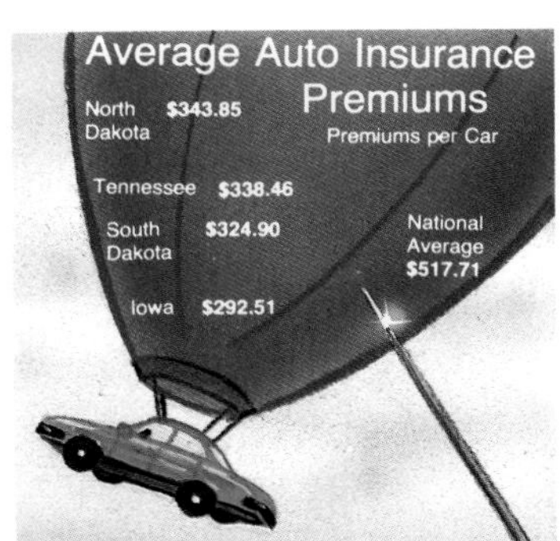

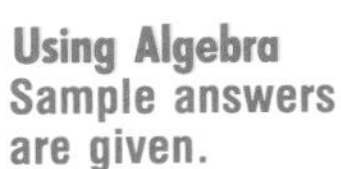

Using Algebra
Sample answers are given.

Write a mathematical expression for each verbal expression.

10. twice a number and 6 more $2n + 6$
11. a number divided by four $x \div 4$
12. the sum of a number and two, all times four $4(n + 2)$
13. the sum of a number divided by three, and 10 more $(t \div 3) + 10$

Mixed Review

Lesson 3-1

Multiply.

14. 32×4 128
15. 481×12 5,772
16. 631×23 14,513
17. $7{,}801 \times 18$ 140,418

Lesson 3-4

Divide. Check using multiplication.

18. $15\overline{)229}$ 15 R4
19. $91\overline{)819}$ 9
20. $42\overline{)4{,}325}$ 102 R41
21. $647\overline{)8{,}741}$ 13 R330

Lesson 3-7

22. Adrianne Hunter borrowed $10,000 from First Bank to buy a car. She will pay back a total of $13,728 that includes interest. Adrianne makes 48 equal payments. What is the amount of each payment? $286

Chalkboard Examples

1. Football tickets cost $12.75 each. Baseball tickets cost $7 each. If you buy 3 football and 6 baseball tickets, how much will it cost? $80.25
2. You buy 2 hamburgers at $2.50 each, 2 fries at $1.35 each, and a drink for $1.75. What is the change from ten dollars? 55¢

Additional examples are provided on TRANSPARENCY 3-13.

EVALUATING THE LESSON

Check for Understanding Use Guided Practice Exercises to check for student understanding.

Closing the Lesson Ask students to state in their own words how to solve multi-step problems.

EXTENDING THE LESSON

Enrichment You have 30 coins in dimes and nickels. What are the greatest and least possible values of the coins if you must have at least two nickels and two dimes? $2.90, $1.60

Enrichment Masters Booklet, p. 32

Name ______________ Date ______________

ENRICHMENT WORKSHEET 3-13

Multi-Step Problems

Tyrone bought 4 shirts and 4 ties. Each shirt cost $8.95 and each tie cost $3.95. How much change did Tyrone get from 3 twenty-dollar bills?

There is often more than one way to solve a multi-step problem.

Method 1:
a) Find the cost of the shirts.
b) Find the cost of the ties.
c) Find the total cost.
d) Subtract the total cost from $60.

Method 2:
a) Add the cost of 1 shirt and 1 tie.
b) Multiply this cost by 4.
c) Subtract the total cost from $60.

Study each problem carefully to see if there is more than one way to solve it. If so, solve it both ways and compare your answers.

1. Diana had exactly $20 when she left to go shopping. Her bill at the drug store was $8.15. At the grocery store, her bill was $7.93. How much money does Diana have left? $3.92 (two ways)

   $20.00 − 8.15 = $11.85; $11.85 − 7.93 = $3.92 or $8.15 + 7.93 = $16.08; $20.00 − 16.08 = $3.92

2. Mike's tips for Monday through Thursday were $5.50, $4.70, $3.35, and $5.25. What was the average amount Mike received in tips for the 4 days? $4.70 (one way)

   $5.50 + 4.70 + 3.35 + 5.25 = $18.80
   $18.80 ÷ 4 = $4.70

3. Manuela bought 4 tubes of lipstick for $1.35 each and 4 bottles of matching nail polish for $1.19 each. How much change did Manuela receive from $15? $4.84 (two ways)

   $1.35 × 4 = $5.40; $1.19 × 4 = $4.76; $5.40 + 4.76 = $10.16; $15.00 − 10.16 = $4.84
   or
   $1.35 + 1.19 = $2.54; $2.54 × 4 = $10.16; $15.00 − 10.16 = $4.84

4. Tony has 2 baseball card albums. One album has 6 cards on each of 33 pages. The other album has 4 cards on each of 25 pages. How many baseball cards does Tony have? 298 cards (one way)

   33 × 6 = 198; 25 × 4 = 100; 198 + 100 = 298

APPLYING THE LESSON

Independent Practice

Homework Assignment	
Minimum	Ex. 1-9; 10-22 even
Average	Ex. 1-22
Maximum	Ex. 1-22

Chapter 3, Quiz B (Lessons 3-8 through 3-13) is available in the Evaluation Masters Booklet, p. 14.

Using the Chapter Review

The Chapter Review is a comprehensive review of the concepts presented in this chapter. This review may be used to prepare students for the Chapter Test.

Chapter 3, Quizzes A and B, are provided in the Evaluation Masters Booklet as shown below.

Quiz A should be given after students have completed Lessons 3-1 through 3-7. Quiz B should be given after students have completed Lessons 3-8 through 3-13.

These quizzes can be used to obtain a quiz score or as a check of the concepts students need to review.

Evaluation Masters Booklet, p. 14

CHAPTER 3, QUIZ A (Lessons 3-1 through 3-7)

Multiply.

1. 53×287
2. 34×0.07
3. $2.39 \times 1,000$
4. 0.045×10^2

Divide.

5. 73)4,492
6. 0.6)26.28
7. $87.22 \div 100$
8. $0.17 \div 10^3$
9. The cash price of a rug is $1,200. Walter buys the rug on credit. He makes a down payment of $200 and makes 12 monthly payments of $95. What is the finance charge?
10. The cash price of a car is $13,000. Anita buys it on credit. She puts $2,500 down and makes monthly payments of $325 for 3 years. What is the finance charge?

1. 15,211
2. 2.38
3. 2,390
4. 4.5
5. 61 R39
6. 43.8
7. 0.8722
8. 0.00017
9. $140
10. $1,200

Name ____________ Date ____________

CHAPTER 3, QUIZ B (Lessons 3-8 through 3-13)

Write in scientific notation.

1. 2,600
2. 347,000

Write in standard form.

3. 4.67×10^3
4. 2.03×10^5
5. Find the common ratio. Write the next three terms in the geometric sequence: 14, 42, 126, . . .
6. Solve for d, r, or t using the formula $d = rt$. $r = 45$ mph, $t = 3$ hours, $d = ?$

Find the area of each rectangle.

7. 3.5 m, 4 m
8. 6.2 in., 6.2 in.
9. Find the mean for the set of data. Round to the nearest tenth. 11.7, 8.0, 7.9, 6.6
10. Daniel bought gloves for $12.95 and shoelaces for $1.99. He gave the clerk twenty dollars. How much change should he receive?

1. 2.6×10^3
2. 3.47×10^5
3. 4,670
4. 203,000
5. 3; 378; 1,134; 3,402
6. $d = 135$ mi
7. 14 m^2
8. 38.44 in^2
9. 8.6
10. $5.06

Chapter 3 REVIEW

Vocabulary / Concepts

Choose a word or number from the list at the right that best completes each sentence.

1. To divide by 100, move the decimal point __?__ places to the __?__. **two, left**
2. The area of a rectangle is found by multiplying the __?__ by the width. **length**
3. If you move the decimal point two places in the divisor, you must move the decimal point two places in the __?__. **dividend**
4. To multiply by 1,000, move the decimal point __?__ places to the __?__. **three, right**
5. To find the common ratio in a geometric sequence, divide any number by the one __?__ it. **before**
6. In standard form, 1.4×10^2 equals __?__. **140**
7. 9,520,000,000 in scientific notation is __?__. **7. 9.52×10^9**
8. The result of multiplying two numbers is the __?__. **8. product**
9. One average of a set of data is the __?__. **mean**

1.4
140
1,400
9.52×10^9
952×10^9
after
before
dividend
divisor
left
mean
product
quotient
right
three
two

Exercises / Applications

Lessons 3-1, 3-2

Multiply. Estimate the products first.

10. 58×3 **174**
11. 29×18 **522**
12. 716×30 **21,480**
13. 98×27 **2,646**
14. $4,881 \times 79$ **385,599**
15. 3.21×7 **22.47**
16. $\$79.52 \times 8.5$ **$675.92**
17. $\$53.07 \times 13$ **$689.91**
18. 1.93×7.8 **15.054**
19. 0.58×2.4 **1.392**

Lesson 3-3

Solve.

20. A car is advertised at a cash price of $9,850. To purchase the car on credit, the down payment is $250 and the monthly payment is $282 for 36 months. What is the finance charge if the car is purchased on credit? **$552**

Lesson 3-4

Divide. Check using multiplication.

21. 7)494 **70 R4**
22. 7)642 **91 R5**
23. 5)7.285 **1.457**
24. 16)3,841 **240 R1**
25. 98)588 **6**

Lesson 3-5

Multiply or divide.

26. $2.39 \div 10$ **0.239**
27. 65.7×10^2 **6,570**
28. 4.382×10 **43.82**
29. $14.8 \div 1,000$ **0.0148**

Lesson 3-6

Solve.

30. Suppose Mr. Gomez typed 1 page on Monday. Then on each successive day he types 2 additional pages. How many pages will he type on Friday? **9 pages**

Lesson 3-7

Divide.

31. $0.24\overline{)0.15}$ 0.625
32. $1.3\overline{)0.91}$ 0.7
33. $6.7\overline{)23.45}$ 3.5
34. $0.5\overline{)4.8}$ 9.6
35. $0.3\overline{)2.04}$ 6.8
36. $0.6\overline{)0.507}$ 0.845
37. $0.8\overline{)0.38}$ 0.475
38. $0.9\overline{)3.078}$ 3.42
39. $9.6\overline{)5.4}$ 0.5625
40. $2.9\overline{)9.28}$ 3.2

Lesson 3-8

Write in scientific notation. 41. 7.99×10^4 42. 2.66×10^5

41. 79,900
42. 266,000
43. 86 million 8.6×10^7
44. 6,300,000,000 6.3×10^9

Write in standard form.

45. 4.9×10^2 490
46. 3.862×10^3 3,862
47. 9.7×10^3 9,700
48. 1.496×10^4 14,960

Lesson 3-9

Write the next three numbers in the sequence.

49. 800, 240, 72, . . . 21.6, 6.48, 1.944
50. 24, 12, 6, . . . 3, 1.5, 0.75
51. 26, 39, 58.5, . . . 87.75, 131.625, 197.4375

Lesson 3-10

Solve for d, r, or t using $d = rt$.

52. $r = 45$ mph, $t = 3$ h $d = 135$ mi
53. $r = 55$ mph, $d = 165$ mi $t = 3$h

Lesson 3-11

Find the area of each rectangle.

54. 12 in. by 5 in. 60 in.^{2}
55. 16 m by 10 m 160 m^2
56. 8 cm 64 cm^2

Lesson 3-12

58. 20.4

Find the average for each set of data. Round to the nearest tenth.

57. 1, 5, 6, 4, 3, 8, 9, 10, 7, 2 5.5
58. 21, 25, 16, 18, 26, 17, 19, 20, 22
59. \$4.00, \$5.60, \$7.80, \$9.35, \$8.65, \$9.10, \$4.10 \$6.90

Lesson 3-13

60. Carole Jefferson pays 12 equal additional charges on cable TV service that total \$27 per year. How much is each service charge? \$2.25
61. A long-distance telephone call costs \$0.18 for the first minute and \$0.15 for each additional minute. What is the cost of a 12-minute call? \$1.83
62. Darren wants to buy a radio that costs \$129. He saves \$6.50 each week. So far he has saved \$52. How many more weeks will it take to save enough money to buy the radio? 12 weeks
63. Bryce Canyon National Park in Utah had 1.1 million visitors in 1989. The same year, the Canyonlands National Park in Utah had 260,000 visitors. How many more people visited Bryce Canyon than the Canyonlands? 840,000 people

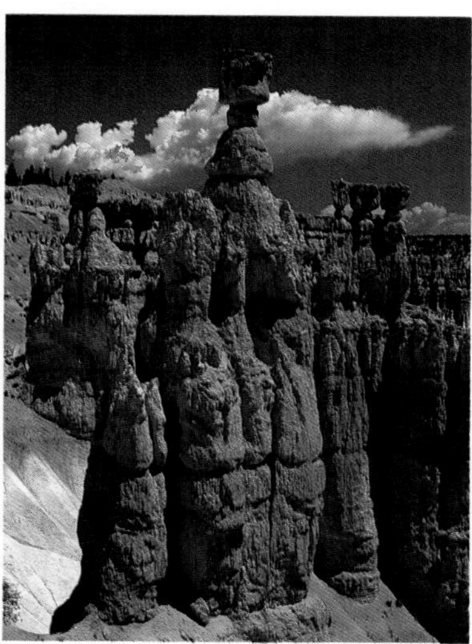

APPLYING THE LESSON

Independent Practice

Homework Assignment	
Minimum	Ex. 1-9; 10-62 even
Average	Ex. 1-9; 11-63 odd
Maximum	Ex. 1-63

Alternate Assessment Strategy

To provide a brief in-class review, you may wish to read the following questions to the class and require a verbal response.

1. What is the product of 15 and 12? 180
2. Multiply 8.2 and 2.1. What is the result? 17.22
3. Find the quotient when 135 is divided by 5. 27
4. What is the result when 3.6 is divided by 0.4? 9
5. What is the product when 5.643 is multiplied by 100? 564.3
6. Divide 189.45 by 10. 18.945
7. What is the scientific notation for 3,408,000? 3.408×10^6
8. What is the common ratio in the sequence 2, 3, 4.5, 6.75? 1.5
9. Using the formula $d = rt$, find the distance if $r = 60$ mph and $t = 3.5$ hours. $d = 210$ miles
10. Find the area of a rectangle if the length is 12 in and the width is 14 in. 168 in^2

Using the Chapter Test

This page may be used as a test or as an additional page of review if necessary. Two forms of a Chapter Test are provided in the Evaluation Masters Booklet. Form 2 (free response) is shown below, and Form 1 (multiple choice) is shown on page 103.

The **Tech Prep Applications Booklet** provides students with an opportunity to familiarize themselves with various types of technical vocations. The Tech Prep applications for this chapter can be found on pages 5–6.

Chapter 3 TEST

Multiply.

1. 72 × 49 **3,528**
2. 6,244 × 53 **330,932**
3. 4.345×10^3 **4,345**
4. 7.923×10^6 **7,923,000**

Divide.

5. $7\overline{)642}$ **8.91 R5**
6. $49\overline{)816}$ **16 R32**
7. $45.03 \div 10^4$ **0.004503**
8. Gretyl Haenes will pay $32.50 a month for 24 months for a computer system. If she made a $100 down payment, what is the credit price of the system? **$880**

Write in scientific notation.

9. 5,280 **5.280×10^3**
10. 392,000 **3.92×10^5**

Solve for d, r, or t using the formula d = rt. Round to the nearest tenth.

11. 34 kilometers, 2.6 hours **13.1 kph**
12. 450 miles per hour, 3.8 hours **12. 1,710 miles**

Find the area of each rectangle.

13. 2 m, 3.5 m **7 m²**

14. 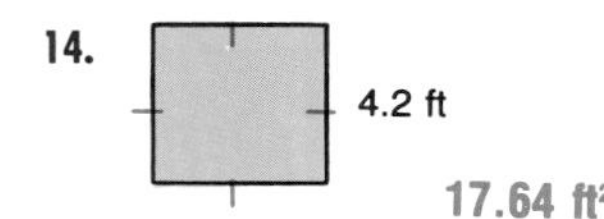

17.64 ft²

Write the next three numbers in each sequence.

15. 3, 15, 75, . . . **375, 1,875, 9,375**
16. 600, 300, 150 . . . **75, 37.5, 18.75**

Find the average for each set of numbers.

17. 9, 8, 4, 7, 6, 11 **7.5**
18. 3.2, 4.6, 1.2, 3.5 **3.125**

19. Mr. Schultz buys numerals to put on the doors of each apartment in a 99-unit apartment building. The apartments are numbered 1 through 99. How many of each digit 0, 1, 2, 3, 4, 5, 6, 7, 8, 9 should Mr. Schultz buy? **19. 9 zeros and 20 of each other digit**
20. Mr. Jamison bought a compact disc system on credit for a $65 down payment and a monthly payment of $38.60 for 12 months. What is the total cost of the system? **$528.20**

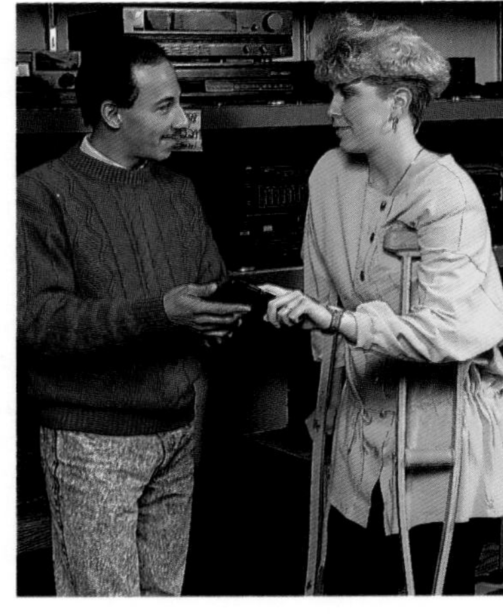

BONUS: These are . . .
4 × 12, 6 × 8, 1 × 48
These are not . . .
3 × 8, 4 × 7, 2 × 13
Which of these are?
<u>2 × 24</u>, <u>3 × 16</u>, 4 × 14, 5 × 16

The **Performance Assessment Booklet** provides an alternative assessment for evaluating student progress. An assessment for this chapter can be found on pages 5–6.

A **Test and Review Generator** is provided in Apple, IBM, and Macintosh versions. You may use this software to create your own tests or worksheets, based on the needs of your students.

Evaluation Masters Booklet, p. 13

Name ______________ Date ______________

CHAPTER 3 TEST, FORM 2

Multiply.

1. 54 × 30
2. 73 × 0.6
3. 51 × 0.92
4. 5.8 × 10
5. 8.32×10^2

Divide.

6. $29\overline{)375}$
7. 8,274 ÷ 34
8. $64 \div 10^2$
9. $9\overline{)3.06}$
10. $0.56\overline{)28}$
11. Write 3,170 in scientific notation.
12. Write 2.4×10^5 in standard form.
13. Find the common ratio of the sequence 4, 48, 576,... Write the term that comes next in the sequence.

Find the area of each rectangle.

14. 4 in., 6.5 in.
15. 1.1 in., 1.1 in.

Find the mean of each set of data. Round to the nearest tenth.

16. 9.6, 6.8, 12.4
17. 15, 19, 11, 8, 14

Solve.

18. Alonzo bought 4.7 pounds of fish at $3.80 a pound. He gave the clerk a $20. How much change should he receive?
19. Ty bought a car on credit. He made a down payment of $1,200 and will pay $300 per month for 36 months. What is the finance charge if the cash price of the car is $10,500?
20. Solve for d, r, or t using the formula $d = rt$. $d = 40$ miles, $t = 4$ hours, $r = ?$

Bonus Elise is putting snapshots that measure 3.5 in. by 5.1 in. into a 9.2 in. by 11.3 in. album. How many snapshots will fit on a page?

Answers: 1. 1,620 2. 43.8 3. 46.92 4. 58 5. 832 6. 12 R27 7. 243 R12 8. 0.64 9. 0.34 10. 50 11. 3.17×10^3 12. 240,000 13. 12; 6,912 14. 26 in² 15. 1.21 in² 16. 9.6 17. 13.4 18. $2.14 19. $1,500 20. 10 mph Bonus 4 snapshots

NOISE POLLUTION

Heidi works in a noisy downtown office. She notices that the office seems noisier than yesterday. Then she remembers that traffic is being rerouted past the office because of construction. How much more noise does a passing car make than Heidi's office?

Sound intensity is measured in decibels.

The chart below lists the intensity of various sounds and the meaning of each measurement.

Sound	Decibels	Mathematical Language
barely heard	0	10^0
breathing	10	10^1
rustling leaves	20	10^2
talking	40	10^4
noisy office	60	10^6
vacuum cleaner	70	10^7
car	80	10^8
subway train	100	10^{10}
motorcycle	110	10^{11}
jet airplane	130	10^{13}

How many times more intense is the sound of a passing car than the sound of a noisy office?

car → 80 decibels → 10^8

noisy office → 60 decibels → 10^6

The sound of a passing car is 10^2 or 100 times more intense than a noisy office.

How many times more intense is the first sound than the second?

1. subway train, noisy office **10,000 times**

2. breathing, barely heard sound **10 times**

3. car, vacuum cleaner **10 times**

4. jet airplane, breathing **10^{12} or 1,000,000,000,000**

5. At the Summerville rock concert, the sound intensity in row 20 was measured at 110 decibels. How much more intense was the sound at the concert than sound of normal talking? **10^7 or 10,000,000**

Applying Mathematics to the Real World

This optional page shows how mathematics is used in the real world and also provides a change of pace.

Objective Determine how much more intense one sound is than another.

Using Models Remind students that when comparing two numbers they are using a ratio. On the chalkboard write the ratio 10^5 to 10^2 as a fraction as shown.

$$\frac{10^5}{10^2} = \frac{10 \times 10 \times 10 \times 10 \times 10}{10 \times 10} = 10^3$$

Then divide through the common factors in the numerator and denominator. Emphasize to students that the result of dividing the common factors is the same as subtracting the exponents since the base of the exponent is the same in the numerator and the denominator.

Activity

Using Data Have students research the damage noise can do, what levels cause damage to the human ear, and what levels are distracting to on-the-job employees.

Using the Cumulative Review

This page provides an aid for maintaining skills and concepts presented thus far in the text.

A Cumulative Review is also provided in the Evaluation Masters Booklet as shown below.

Evaluation Masters Booklet, p. 15

Name ______________________ Date ______________

CUMULATIVE REVIEW Chapters 1–3

Name the digit in each place-value position in 6,527,314.

1. tens 2. thousands 3. millions

1. 1
2. 7
3. 6

Estimate.

4. 71,761 + 29,992 5. 56.7 + 34.8 6. 73)27,998

*4. 100,000
*5. 92
*6. 400

* Accept any reasonable estimate.

Write using exponents.

7. $3 \times 3 \times 3 \times 3$ 8. $9 \times 9 \times 9 \times 9 \times 9$

7. 3^4
8. 9^5

Order from least to greatest.

9. 26.6, 26.66, 27.6, 26.61 10. 0.401, 0.410, 0.040, 0.004

9. 26.6, 26.61, 26.66, 27.6
10. 0.004, 0.040, 0.401, 0.410

Add or subtract.

11. 78 + 3.9 12. 466.04 − 399.2 13. 56.88 + 16.2 + 0.55

11. 81.9
12. 66.84
13. 73.63

Multiply or divide.

14. 15.6 ÷ 52 15. 7.06×3.9 16. 4.7×10^3

14. 0.3
15. 275.34
16. 4,700

Find the value of each expression.

17. $3 \times 9 - 4$ 18. $32 \div 4 + 9$ 19. $36 + 3 \times 7$

17. 23
18. 17
19. 57

Write in scientific notation.

20. 388 21. 43,200 22. 385,000

20. 3.88×10^2
21. 4.32×10^4
22. 3.85×10^5

23. Find the common ratio. Write the next three numbers in the geometric sequence: 11, 44, 176, . . .

23. 4; 704; 2,816; 11,264

Solve. State any missing or extra facts.

24. Jared made a deposit of $85 into his checking account and then wrote a check for $207.50. What is his new balance?

24. missing beginning balance

25. Fiona earns $4 an hour at Rachel's Diner. She got $274 in tips last week. If she worked for 35 hours, how much did she earn in all?

25. $414

Cumulative REVIEW/TEST

Free Response

Lesson 1-10 ***Write an expression for each phrase.***

1. 7 more than a number $n + 7$
2. 8 less than y $y - 8$
3. the product of 6 and m $6m$
4. the quotient of 4 and w $4 \div w$
5. the sum of g and 32 $g + 32$
6. 9 times a number $9n$

Lessons 1-11, 1-12 ***Solve each equation. Check your solution.*** 9. 15 12. $36\frac{1}{3}$

7. $14y = 294$ 21
8. $76 = 4x$ 19
9. $17 + y = 32$
10. $13y = 52$ 4
11. $8x = 4$ $\frac{1}{2}$
12. $3\frac{2}{3} + y = 40$

Lesson 1-13 ***Write an equation to solve this problem. Then solve.***

13. A train travels at a constant rate of 80 miles per hour. How far will it travel in 15 hours? 1,200 miles

Lessons 2-2, 2-4 ***Add or subtract.***

18. 6.52
19. 6.09
20. 21.61

14. $17.4 + 19.2$ 28.6
15. $1.85 + 1.2$ 3.05
16. $6.71 - 2.7$ 4.01
17. $32.4 - 6.27$ 26.13
18. $7.82 - 1.3$
19. $6.45 - 0.36$
20. $20.6 + 1.01$

21. A book that regularly sells for $13.45 is discounted by $1.50. What is the sale price of the book? $11.95

Lesson 3-1 ***Multiply.***

26. 9,205
27. 19,296
28. 132,330

22. 90×70 6,300
23. 280×0.9 252
24. 24×1.1 26.4
25. 42×22 924
26. 263×35
27. 603×32
28. $4{,}010 \times 33$

Lesson 3-2

29. Ben paid $16.50 for 3 classic Elvis singles at a garage sale. He sold 2 of the records to Sheila at the same price he paid for them. How much did Sheila pay for the 2 singles? $11

Lessons 3-2, 3-7 ***Multiply or divide.***

30. 4.9×7 34.3
31. 3.82×5.1 19.482
32. 2.061×8.8 18.1368
33. 15)5.4 0.36
34. 2.7)16.74 6.2
35. 7.8×6.42 50.076
36. $1.08 \div 1.8$ 0.6
37. $210 \div 280$ 0.75

APPLYING THE LESSON

Independent Practice

Homework Assignment	
Minimum	Ex. 1-37 odd
Average	Ex. 1-37
Maximum	Ex. 1-37

Chapter 3

Multiple Choice

Choose the letter of the correct answer for each item.

1. The average attendance at 77 Red Sox home games was 26,026. What was the total attendance for 77 games?
a. 338
b. 8,796,788
c. 563,024
▶ **d.** 2,004,002

Lesson 3-1

2. Subtract 7,996 from 56,489.
▶ **e.** 48,493
f. 49,493
g. 49,593
h. 51,513

Lesson 2-3

3. What is 101,740 rounded to the nearest thousand?
a. 100,000
b. 101,000
c. 101,700
▶ **d.** 102,000

Lesson 1-4

4. Estimate the difference of 28,683 and 7,405.
e. 2,000
f. 12,000
▶ **g.** 22,000
h. 32,000

Lesson 2-3

5. What is the quotient if you divide 21,264 by 12?
a. 17,720
▶ **b.** 1,772
c. 255,168
d. 1,936

Lesson 3-4

6. The total ticket sales for the All-Country concert are $200,824. 38 ticket outlets sold a total of 7,724 tickets. What was the price of each ticket?
e. $31
f. $20
▶ **g.** $26
h. $36

Lesson 3-4

7. What is the sum of 6,360 and 9,863?
a. 15,123
b. 15,223
c. 16,123
▶ **d.** 16,223

Lesson 2-1

8. If Sue Thomas earns $15,184 a year, what are her weekly earnings?
e. $298
f. $12,560
▶ **g.** $292
h. $1,265

Lesson 2-3

9. Which facts are given? Jody purchased a life insurance policy worth $50,000. Her monthly payment is $24.84.
a. name of insurance company
▶ **b.** payment per month
c. length of time for payments
d. how her payments will be made

Lesson 2-9

10. Mars is an average 1.4171×10^8 miles from the sun. About how many miles is Mars from the sun?
e. 141,000
▶ **f.** 141,000,000
g. 141,000,000,000
h. none of the above

Lesson 3-8

Using the Standardized Test

This test familiarizes students with a standardized format while testing skills and concepts presented up to this point.

APPLYING THE LESSON

Independent Practice

Homework Assignment	
Minimum	Ex. 1-10
Average	Ex. 1-10
Maximum	Ex. 1-10

Evaluation Masters Booklet, pp. 11-12

Name ______________ Date ______________

CHAPTER 3 TEST , FORM 1

Write the letter of the correct answer on the blank at the right of the page.

1. Multiply. 20×74
A. 1,580 **B.** 1,480 **C.** 148 **D.** 158 — 1. B

2. Multiply. 63×862
A. 54,306 **B.** 5,436 **C.** 53,306 **D.** 5,486 — 2. A

3. Multiply. 68×0.4
A. 2,720 **B.** 2.72 **C.** 27.2 **D.** 0.272 — 3. C

4. Multiply. 0.94×48
A. 0.4512 **B.** 451.2 **C.** 45.12 **D.** 4.512 — 4. C

5. Multiply. 0.67×0.9
A. 0.603 **B.** 6.03 **C.** 603 **D.** 60.3 — 5. A

6. Ed buys a car on credit. He makes a down payment of $1,500 and pays $250 per month for 36 months. What is the finance charge if the cash price of the car is $9,500?
A. $250 **B.** $1,000 **C.** $10,500 **D.** $0 — 6. B

7. A stereo system sells for $850. Mario makes a down payment of $200 and pays $65 per month for 12 months. What is the finance charge?
A. $780 **B.** $880 **C.** $65 **D.** $130 — 7. D

8. Divide. $196 \div 7$
A. 209 **B.** 28 **C.** 29 **D.** 27 — 8. B

9. Divide. $29\overline{)357}$
A. 13 R7 **B.** 12 **C.** 13 **D.** 12 R9 — 9. D

10. $9,264 \div 42$
A. 22 R24 **B.** 221 **C.** 221 R4 **D.** 220 R4 — A

11. Multiply. 5.4×10
A. 0.54 **B.** 54 **C.** 540 **D.** 5.4 — 11. B

12. Multiply. 3.82×10^2
A. 38.2 **B.** 382 **C.** 3,820 **D.** 0.0382 — 12. B

13. Divide. $43.6 \div 1,000$
A. 0.0436 **B.** 0.436 **C.** 436,000 **D.** 43,600 — 13. A

Using Mathematical Connections

Objective Understand, appreciate, and enjoy the connections between mathematics and real-life phenomena and other disciplines.

The following data is background information for each event on the time line.

History Earliest use of numbers, Congo (Zaire), Africa, 6000 BC
A find in the Congo (Zaire) in Africa has markings on a bone tool handle 8,000 years old. The bone has notches arranged in definite patterns. The discoverer, Dr. J. de Heinzelm, claims it is a numeration system.

Physical Education First evidence of the use of snow skis, Northern Norway, 2000 BC
The earliest use of snow skis can be traced to 2000 BC during the Bronze Age. Primitive skis over 4,000 years old have been found in marshes in Scandinavia.

Activity Have students use the encyclopedia, ski books, and magazines to make a list of ski terms, their meanings, and country of origin.

Game Rubik's cube, 1975
The idea of the magic cube was developed independently in both Hungary and Japan. In 1975, architect Erno Rubik of Budapest, Hungary, was the first to patent the simple and ingenious mechanism that enables small cubes to be rotated. What makes the cube a unique puzzle are the games of restoring patterns on each face. A twelve-year old Hungarian boy restored the correct patterns in a record time of twelve seconds.

Activity Have students disassemble a Rubik's cube, study its mechanism for rotating the cubes, and reassemble it correctly.

Science Calculating machine, 1843
In 1822, Charles Babbage, a young Englishman, designed a calculating machine that could calculate and print out results. The machine was designed to include some 50,000 cog teeth and gear wheels. The machine was not built in Babbage's lifetime because the machinists of his time did not have the capability to build such extremely technical parts.

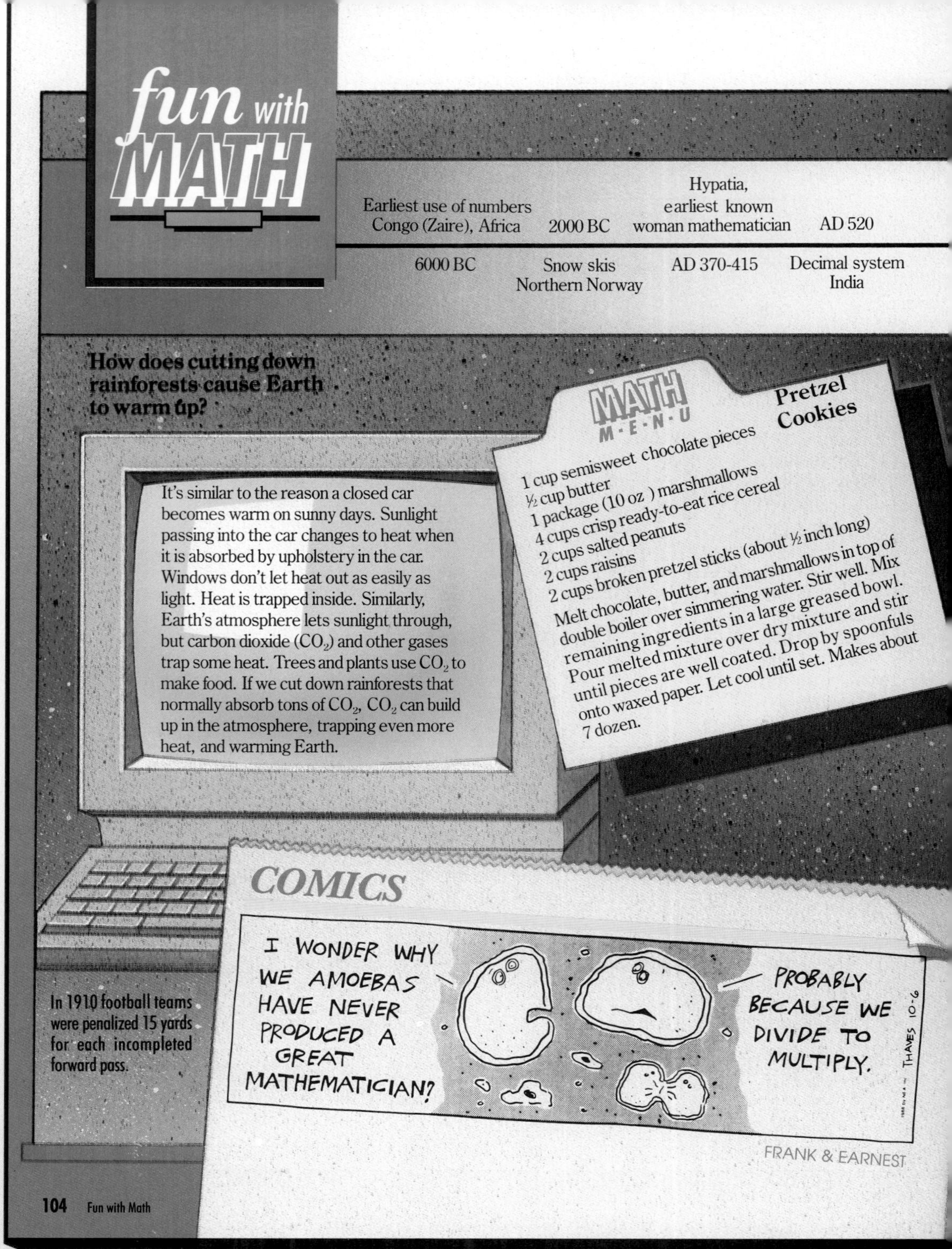

fun with MATH

Earliest use of numbers Congo (Zaire), Africa — 6000 BC

Snow skis Northern Norway — 2000 BC

Hypatia, earliest known woman mathematician — AD 370-415

Decimal system India — AD 520

How does cutting down rainforests cause Earth to warm up?

It's similar to the reason a closed car becomes warm on sunny days. Sunlight passing into the car changes to heat when it is absorbed by upholstery in the car. Windows don't let heat out as easily as light. Heat is trapped inside. Similarly, Earth's atmosphere lets sunlight through, but carbon dioxide (CO_2) and other gases trap some heat. Trees and plants use CO_2 to make food. If we cut down rainforests that normally absorb tons of CO_2, CO_2 can build up in the atmosphere, trapping even more heat, and warming Earth.

MATH M-E-N-U

Pretzel Cookies

1 cup semisweet chocolate pieces
½ cup butter
1 package (10 oz) marshmallows
4 cups crisp ready-to-eat rice cereal
2 cups salted peanuts
2 cups raisins
2 cups broken pretzel sticks (about ½ inch long)

Melt chocolate, butter, and marshmallows in top of double boiler over simmering water. Stir well. Mix remaining ingredients in a large greased bowl. Pour melted mixture over dry mixture and stir until pieces are well coated. Drop by spoonfuls onto waxed paper. Let cool until set. Makes about 7 dozen.

In 1910 football teams were penalized 15 yards for each incompleted forward pass.

COMICS

I WONDER WHY WE AMOEBAS HAVE NEVER PRODUCED A GREAT MATHEMATICIAN?
PROBABLY BECAUSE WE DIVIDE TO MULTIPLY.
THAVES 10-6

FRANK & EARNEST

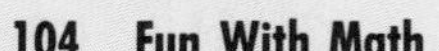

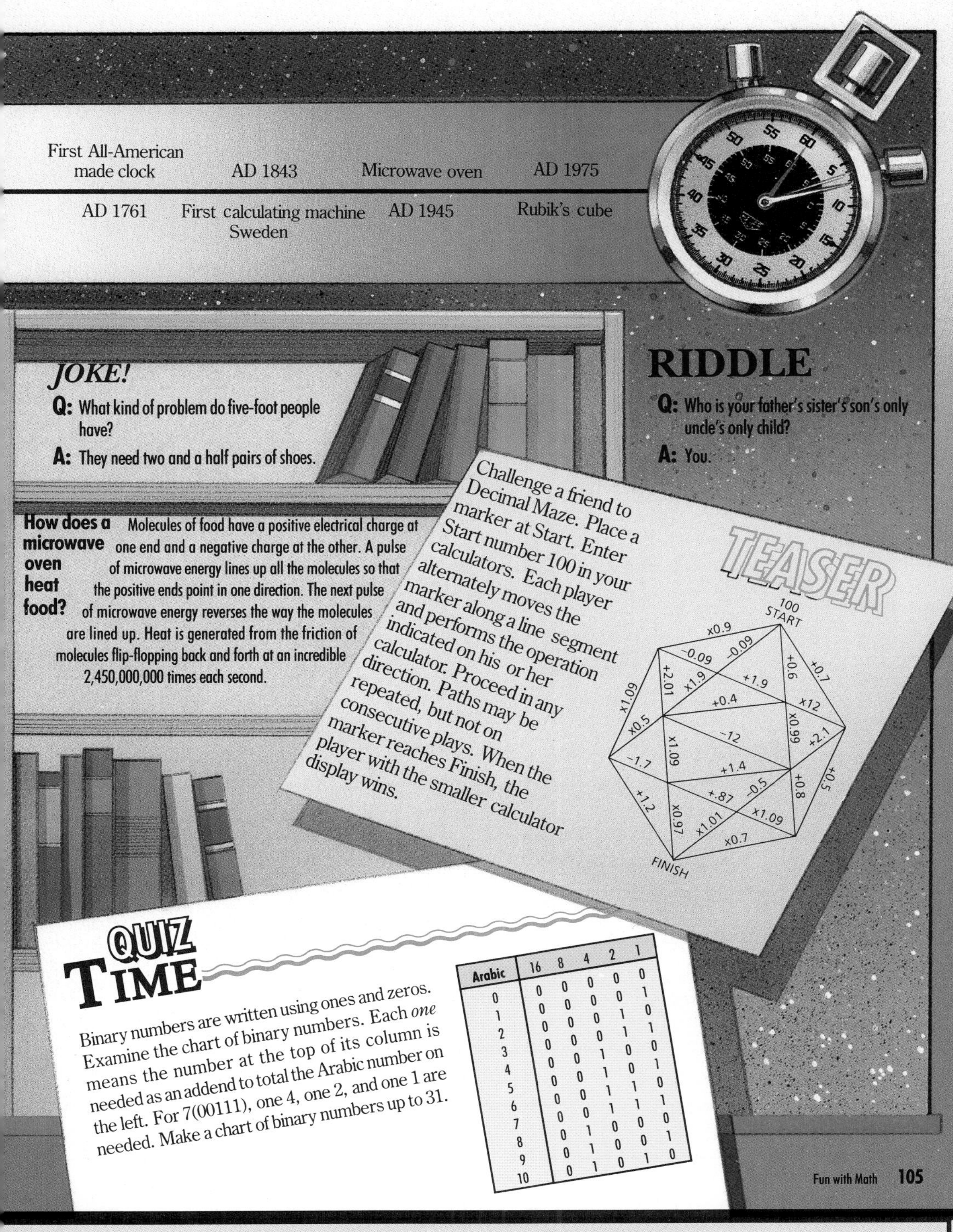

First All-American made clock — AD 1761

AD 1843 — First calculating machine Sweden

Microwave oven — AD 1945

AD 1975 — Rubik's cube

RIDDLE

Q: Who is your father's sister's son's only uncle's only child?

A: You.

JOKE!

Q: What kind of problem do five-foot people have?

A: They need two and a half pairs of shoes.

How does a microwave oven heat food? Molecules of food have a positive electrical charge at one end and a negative charge at the other. A pulse of microwave energy lines up all the molecules so that the positive ends point in one direction. The next pulse of microwave energy reverses the way the molecules are lined up. Heat is generated from the friction of molecules flip-flopping back and forth at an incredible 2,450,000,000 times each second.

TEASER

Challenge a friend to Decimal Maze. Place a marker at Start. Enter Start number 100 in your calculators. Each player alternately moves the marker along a line segment and performs the operation indicated on his or her calculator. Proceed in any direction. Paths may be repeated, but not on consecutive plays. When the marker reaches Finish, the player with the smaller calculator display wins.

QUIZ TIME

Binary numbers are written using ones and zeros. Examine the chart of binary numbers. Each *one* means the number at the top of its column is needed as an addend to total the Arabic number on the left. For 7(00111), one 4, one 2, and one 1 are needed. Make a chart of binary numbers up to 31.

Arabic	16	8	4	2	1
0	0	0	0	0	0
1	0	0	0	0	1
2	0	0	0	1	0
3	0	0	0	1	1
4	0	0	1	0	0
5	0	0	1	0	1
6	0	0	1	1	0
7	0	0	1	1	1
8	0	1	0	0	0
9	0	1	0	0	1
10	0	1	0	1	0

Fun with Math 105

Home Economics First microwave oven, 1945

Microwaves are very high frequency radio waves with very short wavelengths; hence, the name "microwave." Microwaves are generated by a high frequency radar tube called a *magnetron.* The magnetron was invented in England in the late 1930s. It wasn't until 1963 that a solid-state source of microwave power was discovered by J. B. Gunn of the I.B.M. Watson Research Center.

Activity Have students explain why microwaveable dishes do not also become heated.

History Hypatia lived from AD 370-415 and is the earliest known woman mathematician.

Hypatia was the first woman in mathematics of whom there is considerable knowledge. She was born in AD 370 in Alexandria, Greece. Hypatia was the author of several works in mathematics that have been recorded as hers since the 10th century.

History Decimal system, India, AD 520

A prominent Hindu astronomer and mathematician, Aryabhata, was the first known to use the decimal system. Until AD 520, Hindu astronomers used fractions based on the number 60, sexagesimal fractions. Aryabhata transferred the system of sexagesimal fractions to a system of integers based on the number ten, or a decimal system.

Industrial Art First All-American made clock in 1761

Benjamin Bannecker, a free-born African American, was an essayist, inventor, mathematician, and astronomer. He lived in Ellicot, Maryland, and taught himself mathematics and astronomy. In 1761, Benjamin completed the first clock wholly made in America. It was constructed entirely out of wood and kept accurate time until his death in 1806.

Bannecker became the first black presidential appointee in the U.S. as assistant surveyor under Major Pierre Charles L'Enfant, architect of our national capitol, Washington, D.C.

Quiz

Have students work in groups of 3 or 4 to complete the suggested binary chart of numbers from 1 to 31. Help them to understand that to build each binary number they are to place a one in enough columns so that when the Arabic numbers heading all columns with a one in them are added up, they equal the Arabic number desired in the column on the left.

Answers to chart

(0) 00000, (1) 00001, (2) 00010, (3) 00011, (4) 00100, (5) 00101, (6) 00110, (7) 00111, (8) 01000, (9) 01001, (10) 01010, (11) 01011, (12) 01100, (13) 01101, (14) 01110, (15) 01111, (16) 10000, (17) 10001, (18) 10010, (19) 10011, (20) 10100, (21) 10101, (22) 10110, (23) 10111, (24) 11000, (25) 11001, (26) 11010, (27) 11011, (28) 11100, (29) 11101, (30) 11110, (31) 11111

Chapter

4 ADDING AND SUBTRACTING FRACTIONS

PREVIEWING the CHAPTER

The prerequisite skills for operations with fractions are developed in this chapter. The concepts of factors and divisibility lead to finding prime factorizations, greatest common factors, least common multiples, and equivalent fractions. Greatest common factors are used to simplify fractions. Least common multiples are used to find the least common denominator for two or more unlike fractions. The least common denominator is used in comparing, adding, and subtracting unlike fractions and mixed numbers.

Problem-Solving Strategy Students learn to solve problems by making a drawing.

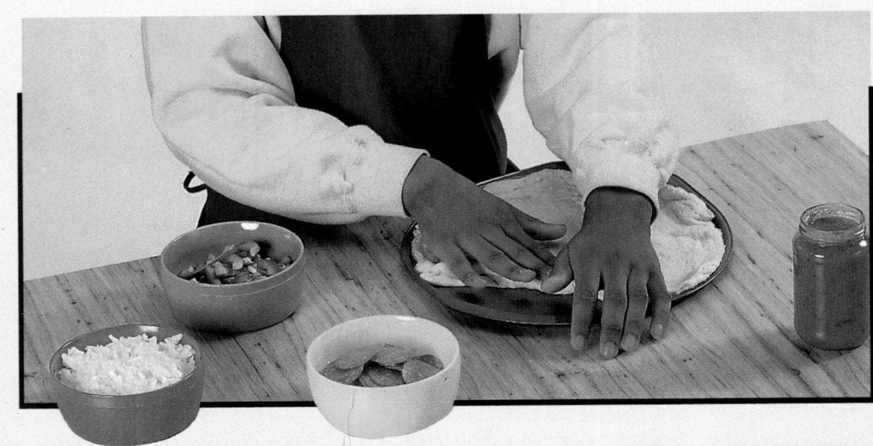

Lesson (Pages)	Lesson Objectives	State/Local Objectives
4-1 (108-109)	4-1: Identify factors of a whole number.	
4-2 (110-111)	4-2: Write the prime factorization of a whole number.	
4-3 (112-113)	4-3: Find the greatest common factor and the least common multiple of two or more numbers.	
4-4 (114-115)	4-4: Read an earnings statement.	
4-5 (116-117)	4-5: Find fractions equivalent to given fractions.	
4-6 (118-119)	4-6: Write fractions in simplest form.	
4-7 (122-123)	4-7: Compare and order fractions.	
4-8 (124-125)	4-8: Change improper fractions to mixed numbers and vice versa.	
4-9 (126-127)	4-9: Estimate the sum and difference of fractions.	
4-10 (128-129)	4-10: Add proper fractions.	
4-11 (130-131)	4-11: Add mixed numbers.	
4-12 (132-133)	4-12: Subtract proper fractions.	
4-13 (134-135)	4-13: Subtract mixed numbers.	
4-14 (136-137)	4-14: Solve verbal problems by making a drawing.	

ORGANIZING the CHAPTER

You may refer to the ***Course Planning Calendar*** on page T10.

Planning Guide

Lesson (Pages)	Extra Practice (Student Edition)	Blackline Masters Booklets: Reteaching	Practice	Enrichment	Activity	Multi-cultural Activity	Technology	Tech Prep	Evaluation	Other Resources
4-1 (108-109)	p. 456	p. 33	p. 36	p. 33			p. 18		p. 75	Transparency 4-1
4-2 (110-111)	p. 457	p. 34	p. 37	p. 34						Transparency 4-2
4-3 (112-113)	p. 457	p. 35	p. 38	p. 35						Transparency 4-3
4-4 (114-115)		p. 36	p. 39	p. 36						Transparency 4-4
4-5 (116-117)	p. 457	p. 37	p. 40	p. 37				p. 7		Transparency 4-5
4-6 (118-119)	p. 458	p. 38	p. 41	p. 38						Transparency 4-6
4-7 (122-123)	p. 458	p. 39	p. 42	p. 39	p. 32		p. 4		p. 19	Transparency 4-7
4-8 (124-125)	p. 458	p. 40	p. 43	p. 40						Transparency 4-8
4-9 (126-127)	p. 459	p. 41	p. 44	p. 41						Transparency 4-9
4-10 (128-129)	p. 459	p. 42	p. 45	p. 42		p. 4				Transparency 4-10
4-11 (130-131)	p. 459	p. 43	p. 46	p. 43						Transparency 4-11
4-12 (132-133)	p. 460	p. 44	p. 47	p. 44						Transparency 4-12
4-13 (134-135)	p. 460	p. 45	p. 48	p. 45	pp. 7-8			p. 8		Transparency 4-13
4-14 (136-137)			p. 49						p. 19	Transparency 4-14
Ch. Review (138-139)									pp. 16-18	Test and Review Generator
Ch. Test (140)										
Cumulative Review/Test (141-142)									p. 20	

OTHER CHAPTER RESOURCES

Student Edition
Chapter Opener, p. 106
Activity, p. 107
Computer Application, p. 120
Consumer Connections, p. 121
Journal Entry, p. 129
Portfolio Suggestion, p. 137
Business Connections, p. 141

Teacher's Classroom Resources
Transparency 4-0
Lab Manual, pp. 29-30
Performance Assessment, pp. 7-8
Lesson Plans, pp. 36-49

Other Supplements
Overhead Manipulative Resources, Labs 11-14, pp. 8-10
Glencoe Mathematics Professional Series

ENHANCING the CHAPTER

Some of the blackline masters for enhancing this chapter are shown below.

COOPERATIVE LEARNING

This game can be played by 2-4 students. Have students make a game board as follows. Make a large 6 x 6 square grid. In each space, write either two unlike fractions, two mixed numbers, or a combination of the two. To begin play, each player in turn tosses a penny onto the grid to determine which fractions to use. Toss another coin to determine the operation, addition (heads), or subtraction (tails). Perform the computation. If the answer is correct, place a colored marker (different for each player) on the space. If the answer is incorrect, the turn passes. The first player to mark four adjacent spaces wins.

USING MODELS/MANIPULATIVES

Have students write the fractions $\frac{2}{3}$ and $\frac{1}{2}$ on a separate piece of paper. Place corresponding fraction strips below each fraction. Below each model place equivalent fraction strips ($\frac{2}{3}$: $\frac{4}{6}$, $\frac{6}{9}$, $\frac{8}{12}$; $\frac{1}{2}$: $\frac{2}{4}$, $\frac{3}{6}$, $\frac{4}{8}$, $\frac{6}{12}$). Have students explain the concept of equivalent fractions. Add $\frac{2}{3}$ and $\frac{1}{2}$ by choosing corresponding equivalent strips having the same total parts. Have students relate common multiples and common denominators describing their choice. Use the fraction strips to find the sum. Then simplify.

Cooperative Problem Solving, p. 32

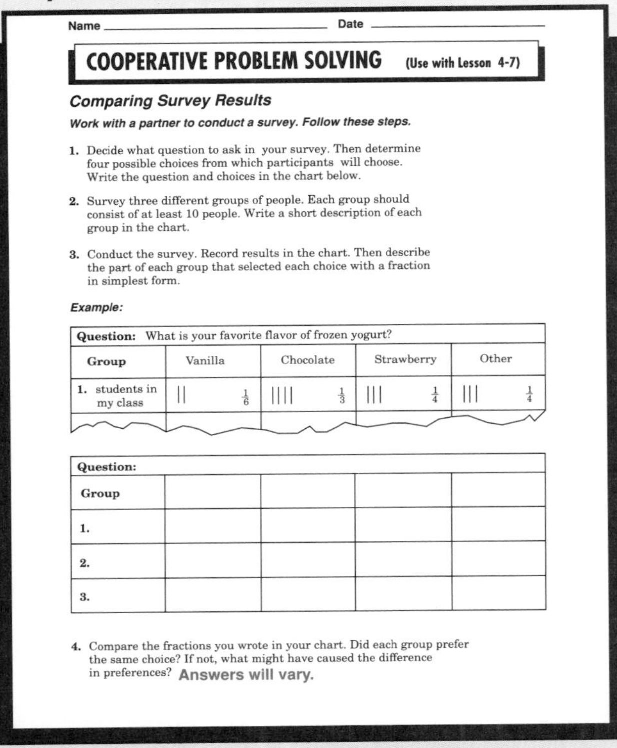

Name ______________________ Date ______________

COOPERATIVE PROBLEM SOLVING (Use with Lesson 4-7)

Comparing Survey Results

Work with a partner to conduct a survey. Follow these steps.

1. Decide what question to ask in your survey. Then determine four possible choices from which participants will choose. Write the question and choices in the chart below.
2. Survey three different groups of people. Each group should consist of at least 10 people. Write a short description of each group in the chart.
3. Conduct the survey. Record results in the chart. Then describe the part of each group that selected each choice with a fraction in simplest form.

Example:

Question: What is your favorite flavor of frozen yogurt?				
Group	Vanilla	Chocolate	Strawberry	Other
1. students in my class	|| $\frac{1}{6}$	|||| $\frac{1}{3}$	||| $\frac{1}{4}$	||| $\frac{1}{4}$

Question:				
Group				
1.				
2.				
3.				

4. Compare the fractions you wrote in your chart. Did each group prefer the same choice? If not, what might have caused the difference in preferences? Answers will vary.

Multicultural Activities, p. 4

Name ______________________ Date ______________

MULTICULTURAL ACTIVITY (Use with Lesson 4-10)

Egyptian Fractions

In approximately 1650 B.C., the Egyptian scribe A'h-mose, copied an earlier mathematical text onto what is known as the Rhind papyrus. (It was discovered by the Scottish archaeologist, A. H. Rhind, in 1858.) Because of their mathematical system, early Egyptians needed to use only unit fractions (fractions with a numerator of 1, such as $\frac{1}{3}$) and fractions with a number of 2.

Unit fractions were written with a special symbol above the denominator.

$\frac{1}{2}$ was written [symbol] ← symbol for 2 $\frac{1}{11}$ was written [symbol] ← symbol for 11

There was a special symbol for $\frac{2}{3}$: [symbol].

All other fractions having a numerator of 2 were written as the sum of two or more unit fractions.

Examples: $\frac{2}{5} = \frac{1}{3} + \frac{1}{15}$ $\frac{2}{101} = \frac{1}{101} + \frac{1}{202} + \frac{1}{303} + \frac{1}{606}$

Write each fraction as the sum of two unit fractions.
(Hint: The denominator of one of the unit fractions will be a multiple of the denominator of the given fraction.)

1. $\frac{2}{7}$ $\frac{1}{4} + \frac{1}{28}$
2. $\frac{2}{9}$ $\frac{1}{6} + \frac{1}{18}$
3. $\frac{2}{11}$ $\frac{1}{6} + \frac{1}{66}$
4. $\frac{2}{15}$ $\frac{1}{10} + \frac{1}{30}$
5. $\frac{2}{21}$ $\frac{1}{14} + \frac{1}{42}$
6. $\frac{2}{25}$ $\frac{1}{15} + \frac{1}{75}$
7. $\frac{2}{39}$ $\frac{1}{26} + \frac{1}{78}$
8. $\frac{2}{45}$ $\frac{1}{27} + \frac{1}{135}$
9. The Rhind papyrus includes a table showing how to write all fractions with a numerator of 2 and an odd number denominator from 5 to 101 as the sum of unit fractions. Why wasn't it necessary to include fractions with an even number denominator?
These can be simplified to unit fractions by dividing numerator and denominator by 2.

MEETING INDIVIDUAL NEEDS

Students at Risk

Stress the importance of understanding the use of the terms factor and multiple. Point out that factor may be used as a noun or a verb. For example, "2 and 3 are factors of 6" and "When you factor, you get 6 and 1, and 3 and 2." Multiple is used as a noun. Prepare a bulletin board display of these terms. Have students use these terms accurately and frequently.

CRITICAL THINKING/PROBLEM SOLVING

The solution to this non-routine problem uses the concepts of divisibility and least common multiple.

A coach wanted to separate his players so that the same number of players were on each team. When he separated the players by two, by threes, by fours, by fives, or by sixes, there was always one player left over. What is the least number of players the coach could have had? **The smallest number divisible by 2, 3, 4, 5, or 6 is 60. Thus, when 61 is divided by 2, 3, 4, 5, or 6, the remainder is 1. The coach had 61 players.**

COMMUNICATION

Writing

Have students use a specific addition or subtraction problem involving unlike fractions or mixed numbers to explain in writing how the skills of divisibility, finding the greatest common factor, finding the least common multiple, writing equivalent fractions, and comparing fractions are used to find the sum or difference.

Applications, p. 7

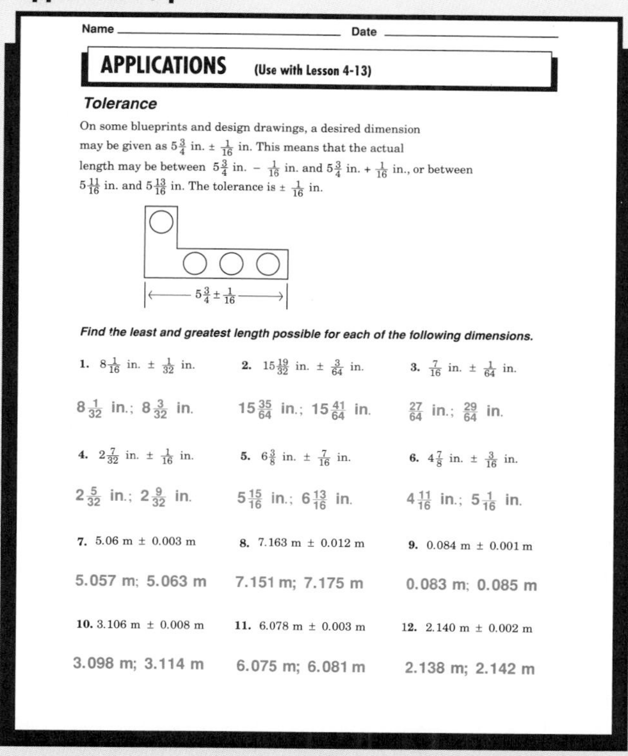

Name ______________ Date ______________

APPLICATIONS (Use with Lesson 4-13)

Tolerance

On some blueprints and design drawings, a desired dimension may be given as $5\frac{3}{4}$ in. ± $\frac{1}{16}$ in. This means that the actual length may be between $5\frac{3}{4}$ in. − $\frac{1}{16}$ in. and $5\frac{3}{4}$ in. + $\frac{1}{16}$ in., or between $5\frac{11}{16}$ in. and $5\frac{13}{16}$ in. The tolerance is ± $\frac{1}{16}$ in.

Find the least and greatest length possible for each of the following dimensions.

1. $8\frac{1}{16}$ in. ± $\frac{1}{32}$ in. — $8\frac{1}{32}$ in.; $8\frac{3}{32}$ in.
2. $15\frac{19}{32}$ in. ± $\frac{3}{64}$ in. — $15\frac{35}{64}$ in.; $15\frac{41}{64}$ in.
3. $\frac{7}{16}$ in. ± $\frac{1}{64}$ in. — $\frac{27}{64}$ in.; $\frac{29}{64}$ in.
4. $2\frac{7}{32}$ in. ± $\frac{1}{16}$ in. — $2\frac{5}{32}$ in.; $2\frac{9}{32}$ in.
5. $6\frac{3}{8}$ in. ± $\frac{7}{16}$ in. — $5\frac{15}{16}$ in.; $6\frac{13}{16}$ in.
6. $4\frac{7}{8}$ in. ± $\frac{3}{16}$ in. — $4\frac{11}{16}$ in.; $5\frac{1}{16}$ in.
7. 5.06 m ± 0.003 m — 5.057 m; 5.063 m
8. 7.163 m ± 0.012 m — 7.151 m; 7.175 m
9. 0.084 m ± 0.001 m — 0.083 m; 0.085 m
10. 3.106 m ± 0.008 m — 3.098 m; 3.114 m
11. 6.078 m ± 0.003 m — 6.075 m; 6.081 m
12. 2.140 m ± 0.002 m — 2.138 m; 2.142 m

Computer Activity, p. 18

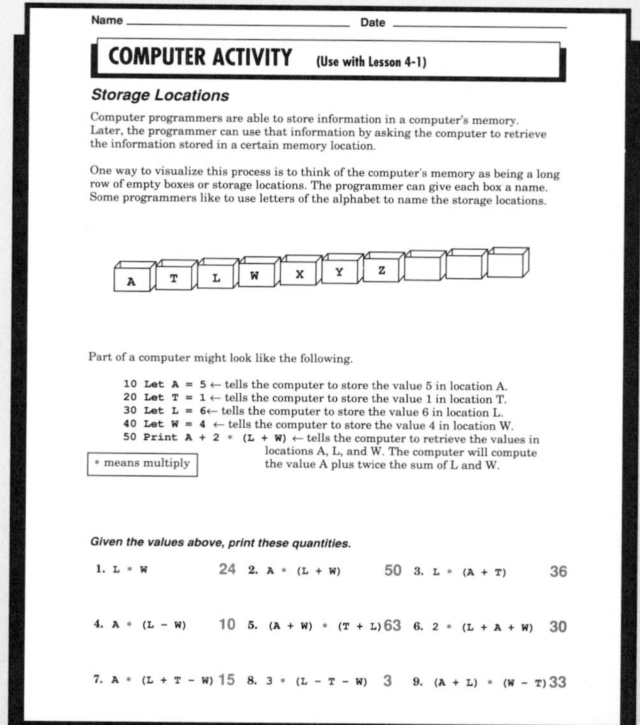

Name ______________ Date ______________

COMPUTER ACTIVITY (Use with Lesson 4-1)

Storage Locations

Computer programmers are able to store information in a computer's memory. Later, the programmer can use that information by asking the computer to retrieve the information stored in a certain memory location.

One way to visualize this process is to think of the computer's memory as being a long row of empty boxes or storage locations. The programmer can give each box a name. Some programmers like to use letters of the alphabet to name the storage locations.

Part of a computer might look like the following.

```
10 Let A = 5 ← tells the computer to store the value 5 in location A.
20 Let T = 1 ← tells the computer to store the value 1 in location T.
30 Let L = 6← tells the computer to store the value 6 in location L.
40 Let W = 4 ← tells the computer to store the value 4 in location W.
50 Print A + 2 * (L + W) ← tells the computer to retrieve the values in
                          locations A, L, and W. The computer will compute
                          the value A plus twice the sum of L and W.
```

* means multiply

Given the values above, print these quantities.

1. L * W — 24
2. A * (L + W) — 50
3. L * (A + T) — 36
4. A * (L − W) — 10
5. (A + W) * (T + L) — 63
6. 2 * (L + A + W) — 30
7. A * (L + T − W) — 15
8. 3 * (L − T − W) — 3
9. (A + L) * (W − T) — 33

CHAPTER PROJECT

Have each student bring to class a recipe containing fractional amounts for a favorite food. Then have students determine the amount of ingredients necessary to double and then to triple the number of servings. Finally, have them determine the minimum capacity of a container to hold the ingredients by finding the total amount of contents of the original recipe.

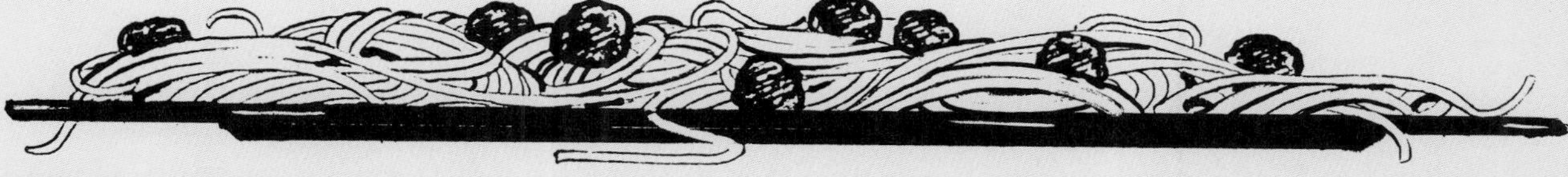

Using the Chapter Opener

Transparency 4-0 is available in the Transparency Package. It provides a full-color visual and motivational activity that you can use to engage students in the mathematical content of the chapter.

Using Discussion

- Have students read the opening paragraph. Ask students how many of them have ever made a pizza.
- Allow students the opportunity to share their experience in baking or cooking. You may suggest that students try to make a pizza at home tonight and see how well it turns out.
- Encourage everyone to participate in the discussion of the last three questions in the opening paragraph.

The second paragraph is a motivational problem. Have students answer the question and discuss how they arrived at their answer.

Chapter 4

FRACTIONS: ADDING AND SUBTRACTING

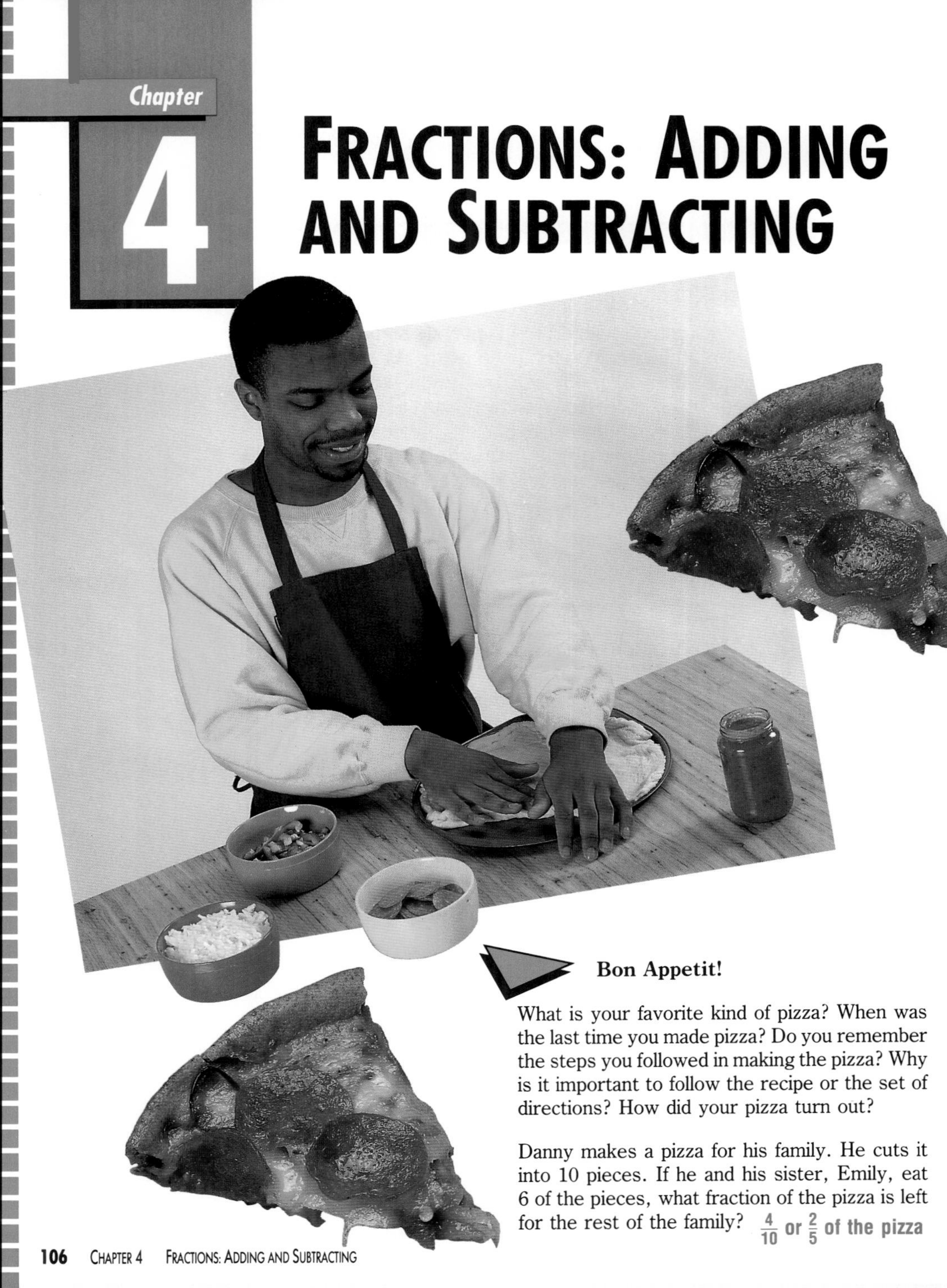

Bon Appetit!

What is your favorite kind of pizza? When was the last time you made pizza? Do you remember the steps you followed in making the pizza? Why is it important to follow the recipe or the set of directions? How did your pizza turn out?

Danny makes a pizza for his family. He cuts it into 10 pieces. If he and his sister, Emily, eat 6 of the pieces, what fraction of the pizza is left for the rest of the family? $\frac{4}{10}$ or $\frac{2}{5}$ of the pizza

ACTIVITY: Twenty "Decimal" Questions

The object of the game is to find two factors and their product by asking twenty questions. You may only ask questions that can be answered *yes* or *no.*

Materials: three dice or three number cubes with the digits 1, 2, 3, 4, 5, 6; paper and pencil; calculators are optional

Divide the class into two teams, Team 1 and Team 2.

Playing the Game

1. Team 2 rolls three number cubes and makes one 1-digit and one 2-digit number from the digits rolled.
2. Team 2 may place a decimal point in one or both of their numbers. For example, a 2-digit number from the digits 4 and 6 may be 0.46, 4.6, or 46. 0.046 is not allowed.
3. Team 2 multiplies the numbers and records the product.
4. Team 1 may ask up to 20 questions in order to find the numbers and their product. Some sample questions are listed below.
   - Is one number a multiple of 5?
   - Is there a 5 in the ones place?
   - Are there two decimal places in the product?
   - Is the product less than 1?
5. Keep a record of the number of questions Team 1 asks to find the answers.
6. Team 1 then takes a turn to roll the cubes and makes two numbers and finds the product.
7. The team that uses fewer questions to find the answer is the winner.

Variations of the Game

8. a. Cover the faces on one die or number cube and change the digits to: 7, 8, 9, 0, 0, 0.
   b. Roll four number cubes or four dice and make two 2-digit numbers.

Communicate Your Ideas

9. Create your own variations of the game. Adjust the rules to accommodate your changes.

INTRODUCING THE LESSON

Objective Identify factors that result in whole numbers and decimal products.

TEACHING THE LESSON

Using Cooperative Groups Divide your class into two groups. Have students begin by reading "Playing the Game."

EVALUATING THE LESSON

Have students play the game several times. Once they feel confident with the game, have students try Variations of the Game.

Communicate Your Ideas Students may want to share their games with each other.

EXTENDING THE LESSON

Closing the Lesson Encourage students to tell what they learned about finding factors from playing this game.

Activity Worksheet, Lab Manual, p. 29

Name ______ Date ______

ACTIVITY WORKSHEET (Use with page 107)

Twenty "Decimal" Questions

Team ______	Team ______
Digits ☐ ☐ ☐	Digits ☐ ☐ ☐
Numbers ___ ___	Numbers ___ ___
Product ______	Product ______
Questions	Questions
1 2 3 4 5	1 2 3 4 5
6 7 8 9 10	6 7 8 9 10
11 12 13 14 15	11 12 13 14 15
16 17 18 19 20	16 17 18 19 20

Keep a record of the questions your team asks.

1. ______	6. ______	11. ______	16. ______
2. ______	7. ______	12. ______	17. ______
3. ______	8. ______	13. ______	18. ______
4. ______	9. ______	14. ______	19. ______
5. ______	10. ______	15. ______	20. ______

Tab 3 Tab Tab 6 5 1 2 Tab 4 Tab Tab Tab

Tab 3 Tab Tab 6 5 1 2 Tab 4 Tab Tab Tab

5-MINUTE CHECK-UP

(over Lesson 3-13)

Available on TRANSPARENCY 4-1.

1. The Adams' orchard produced 2,775 bushels of apples. They spent $4,350 for harvesting and the apples sold at $5.25 a bushel. How much was the profit? **$10,218.75**

▶ INTRODUCING THE LESSON

Use graph paper to show the factors of a number such as 8. Then have students draw the possible arrays of 18 square units to show its factors.

▶ TEACHING THE LESSON

Using Models Have students draw all possible rectangular arrays of 36, 54, and so on, square units.

Show students how factors can be paired (unless the number is a perfect square). For example, show 27.

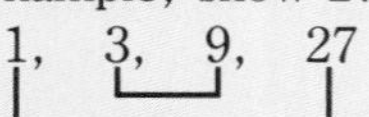

1, 3, 9, 27

Practice Masters Booklet, p. 36

Name ______ Date ______

PRACTICE WORKSHEET 4-1

Factors and Divisibility

Find all the factors of each number.

1. 8 1, 2, 4, 8
2. 15 1, 3, 5, 15
3. 21 1, 3, 7, 21
4. 27 1, 3, 9, 27
5. 33 1, 3, 11, 33
6. 36 1, 2, 3, 4, 6, 9, 12, 18, 36
7. 42 1, 2, 3, 6, 7, 14, 21, 42
8. 48 1, 2, 3, 4, 6, 8, 12, 16, 24, 48
9. 50 1, 2, 5, 10, 25, 50
10. 54 1, 2, 3, 6, 9, 18, 27, 54

State whether each number is divisible by 2, 3, 5, 9, or 10.

11. 110 2, 5, 10
12. 225 3, 5, 9
13. 315 3, 5, 9
14. 405 3, 5, 9
15. 918 2, 3, 9
16. 243 3, 9
17. 630 2, 3, 5, 9, 10
18. 735 3, 5
19. 1,233 3, 9
20. 2,460 2, 3, 5, 10
21. 5,103 3, 9
22. 8,001 3, 9
23. 9,270 2, 3, 5, 9, 10
24. 44,127 3, 9
25. 117,930 2, 3, 5, 10

4-1 FACTORS AND DIVISIBILITY

Objective
Identify factors of a whole number.

Jamie has 6 different patches to sew on his jacket. He can arrange them in a rectangular pattern. Two rectangular displays are shown at the right.

Factors of a number divide that number so that the remainder is zero. You can find the factors of a number by dividing. If the remainder is zero, the divisor and quotient are factors of the number.

What are the factors of 6? 1, 2, 3, 6

Two other rectangular displays are 6 × 1 and 3 × 2.

Examples

A ***Find the factors of 12.***

12 ÷ 1 = 12
12 ÷ 2 = 6
12 ÷ 3 = 4
12 ÷ 4

Stop dividing since you know 4 is a factor.

The factors of 12 are 1, 2, 3, 4, 6, and 12.

B ***Find the factors of 27.***

27 ÷ 1 = 27
27 ÷ 2 = 13 R1
27 ÷ 3 = 9
27 ÷ 4 = 6 R3
27 ÷ 5 = 5 R2
27 ÷ 6 = 4 R3
27 ÷ 7 = 3 R6
27 ÷ 8 = 3 R3

The factors of 27 are 1, 3, 9, and 27.

Stop dividing. Why?

A whole number is *divisible* by its factors. You can use these divisibility rules for 2, 3, 5, 9, and 10.

2: A number is divisible by 2 if its ones digit is 0, 2, 4, 6, or 8.
5: A number is divisible by 5 if its ones digit is 0 or 5.
10: A number is divisible by 10 if its ones digit is 0.
3: A number is divisible by 3 if the sum of its digits is divisible by 3.
9: A number is divisible by 9 if the sum of its digits is divisible by 9.

Determine whether each number is divisible by 2, 3, 5, 9, or 10.

		Ones Digit	By 2?	By 5?	By 10?	Sum of Digits	By 3?	By 9?
Example C	160	0	yes	yes	yes	1 + 6 + 0 = 7	no	no
Example D	255	5	no	yes	no	2 + 5 + 5 = 12	yes	no
Example E	3,357	7	no	no	no	3 + 3 + 5 + 7 = 18	yes	yes

RETEACHING THE LESSON

Review divisibility rules with exercises such as these.

Which of the following numbers: 12, 15, 18, 20, 24, 30, 40, 45, 54, 56, 150

1. is divisible by 2? **12, 18, 20, 24, 30, 40, 54, 56, 150**
2. is divisible by 3? **12, 15, 18, 24, 30, 45, 54, 150**
3. is divisible by 5? **15, 20, 30, 40, 45, 150**
4. is divisible by 9? **18, 45, 54**
5. is divisible by 10? **20, 30, 40, 150**

Reteaching Masters Booklet, p. 33

Name ______ Date ______

RETEACHING WORKSHEET 4-1

Factors and Divisibility

Find the factors of 18.

18 ÷ 1 = 18
18 ÷ 2 = 9
18 ÷ 3 = 6
18 ÷ 4 = 4 R2
18 ÷ 5 = 3 R3
18 ÷ 6 = 3

The factors of 18 are 1, 2, 3, 6, 9, and 18.

Find the factors of 25.

25 ÷ 1 = 25
25 ÷ 2 = 12 R1
25 ÷ 3 = 8 R1
25 ÷ 4 = 6 R1
25 ÷ 5 = 5

The factors of 25 are 1, 5, and 25.

5 is multiplied by itself to equal 25, so it cannot be paired with another factor.

Guided Practice

1. 1,2,3,6 2. 1,2,5,10 3. 1,2,3,5,6,10,15,30 4. 1,5,11,55

Examples A, B

Find all the factors of each number.

1. 6 **2.** 10 **3.** 30 **4.** 55 **5.** 120

5. 1,2,3,4,5,6,8,10,12,15,20,24,30,40,60,120

Examples C-E

State whether each number is divisible by 2, 3, 5, 9, or 10.

6. 72 2,3,9 **7.** 236 2 **8.** 102 2,3 **9.** 957 3 **10.** 3,485 5

Exercises

Practice

Find all the factors of each number. See margin.

11. 9 **12.** 16 **13.** 18 **14.** 20 **15.** 21

16. 48 **17.** 36 **18.** 25 **19.** 42 **20.** 65

State whether each number is divisible by 2, 3, 5, 9, or 10.

26. 2,3,5,10
30. 2,3,9

21. 89 none **22.** 64 2 **23.** 125 5 **24.** 156 2,3 **25.** 216 2,3,9

26. 330 **27.** 225 3,5,9 **28.** 524 2 **29.** 1,986 2,3 **30.** 2,052

Number Sense

31. Answers may vary. Sample answers are 1,050 and 1,260.

31. Write a 4-digit number that is divisible by 2, 6, 7, and 10.

32. What is the greatest 3-digit number that is divisible by 2, 3, and 5? 990

33. Name the first year of the twenty-first century that is divisible by only 1 and itself. 2003

34. What number is a factor of every number? 1

Applications

35. Janet and two of her friends went fishing. They caught 57 fish. Can they divide the number of fish evenly among themselves? yes If so, how many fish would each girl receive? 19 fish

36. How many trips must a six-passenger airplane make to fly 159 passengers to a summer wilderness camp? 27 trips

37. Todd made 81 cookies. How many ways could he package the cookies so that there would be the same number in each package with no cookies left over? in packages of 1, 3, 9, or 81

Critical Thinking

38. Suppose you had to choose 3 more players out of 6 to be on your basketball team. Your choices are Jana, Ted, Ken, Melissa, Angie, and Lee. How many different groups of three could you choose? List all possible groups. 20 See margin.

Using Variables

39. If the product of 17 and n is 51, what is the value of n? 3

40. If the product of 8 and b is 72, what is the value of b? 9

APPLYING THE LESSON

Independent Practice

Homework Assignment	
Minimum	Ex. 2-40 even
Average	Ex. 1-30 odd; 31-40
Maximum	Ex. 1-40

Additional Answers

11. 1, 3, 9
12. 1, 2, 4, 8, 16
13. 1, 2, 3, 6, 9, 18
14. 1, 2, 4, 5, 10, 20
15. 1, 3, 7, 21
16. 1, 2, 3, 4, 6, 8, 12, 16, 24, 48
17. 1, 2, 3, 4, 6, 9, 12, 18, 36
18. 1, 5, 25
19. 1, 2, 3, 6, 7, 14, 21, 42
20. 1, 5, 13, 65
38. Jana, Ted, Ken
Jana, Ted, Melissa
Jana, Ted, Angie
Jana, Ted, Lee
Jana, Ken, Melissa
Jana, Ken, Angie
Jana, Ken, Lee
Jana, Melissa, Angie
Jana, Melissa, Lee
Jana, Angie, Lee
Ted, Ken, Melissa
Ted, Ken, Angie
Ted, Ken, Lee
Ken, Melissa, Angie
Ken, Melissa, Lee
Ken, Angie, Lee
Melissa, Angie, Lee

Chalkboard Examples

Examples A, B *Find all the factors of each number.* 1. 1, 2, 7, 14

1. 14 **2.** 28

2. 1, 2, 4, 7, 14, 28

Examples C-E *State whether each number is divisible by 2, 3, 5, 9, or 10.*

3. 84 2, 3 **4.** 44 2 **5.** 235 5

Additional examples are provided on TRANSPARENCY 4-1.

EVALUATING THE LESSON

Check for Understanding Use Guided Practice Exercises to check for student understanding.

Closing the Lesson Have students explain how the rules for divisibility may be utilized when finding the prime factorization of a number.

EXTENDING THE LESSON

Enrichment Have students explain how to determine the greatest factor of a number other than itself. Use a number such as 18,275. Divide by increasing primes.

Enrichment Masters Booklet, p. 33

Name ______ Date ______

ENRICHMENT WORKSHEET 4-1

Divisibility

Divisibility Rule for 7

Determine whether 4,032 is divisible by 7.

- 4,032 Cross out the ones digit.
 −4 Double the value of the digit in the ones place.
 399 Subtract the product from the rest of the number.
 −18 If the difference is divisible by 7, stop. If not, repeat.
 21 **Since 21 is divisible by 7, then 4,032 is divisible by 7.**

Divisibility Rules for 11

Determine whether 5,159 is divisible by 11.

- 5,159 Cross out the ones digit.
 −9 Subtract the value of the ones digit from the rest of the number.
 506 If the difference is divisible by 11, stop. If not, repeat.
 −6
 21 (sic) **Since 44 is divisible by 11, then 5,159 is divisible by 11.**

- 5,159
 5 + 5 = 10 Find the sum of the values of the odd-numbered digits.
 1 + 9 = −10 Find the sum of the values of the even-numbered digits.
 0 Subtract the sums. If the difference is divisible by 11, the number is divisible by 11.

Since 0 is divisible by 11, then 5,159 is divisible by 11.

State whether each number is divisible by 7 or 11.

1. 266 7	2. 4,312 7, 11	3. 8,976 11
4. 936 none	5. 13,293 7	6. 7,085 none
7. 2,957 none	8. 3,124 11	9. 6,545 7, 11

5-MINUTE CHECK-UP

(over Lesson 4-1)

Available on TRANSPARENCY 4-2

Find all the factors of each number.

1. 15 **2.** 68 **3.** 87 **4.** 43 **5.** 132

1. 1, 3, 5, 15 2. 1, 2, 4, 17, 34, 68
3. 1, 3, 29, 87 4. 1, 43
5. 1, 2, 3, 4, 6, 11, 12, 22, 33, 44, 66, 132

Extra Practice, Lesson 4-1, p 456

INTRODUCING THE LESSON

Prime numbers can be found by using the sieve of Erathosthenes. A ten-by-ten array of numbers, 1 through 100, can be made by each student. Have the students circle the number 2 and cross out all numbers that are divisible by 2, circle the number 3 and cross out all numbers that are divisible by 3, and so on. Note that all the circled numbers are prime.

TEACHING THE LESSON

Using Discussion Have the class suggest different number pairs that have a product of 48 (2 × 24, 3 × 16, 4 × 12, 6 × 8) and discuss with them how any pair can be used as the first step in the factor tree.

Practice Masters Booklet, p. 37

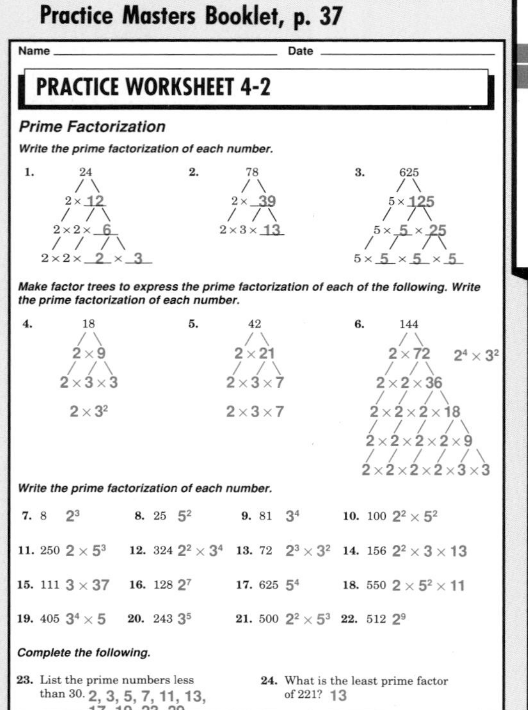

Name ______ Date ______

PRACTICE WORKSHEET 4-2

Prime Factorization

Write the prime factorization of each number.

1. 24 → 2 × 12 → 2 × 2 × 6 → 2 × 2 × 2 × 3
2. 78 → 2 × 39 → 2 × 3 × 13
3. 625 → 5 × 125 → 5 × 5 × 25 → 5 × 5 × 5 × 5

Make factor trees to express the prime factorization of each of the following. Write the prime factorization of each number.

4. 18 → 2 × 9 → 2 × 3 × 3; 2×3^2
5. 42 → 2 × 21 → 2 × 3 × 7; 2 × 3 × 7
6. 144 → 2 × 72 → 2 × 2 × 36 → 2 × 2 × 2 × 18 → 2 × 2 × 2 × 2 × 9 → 2 × 2 × 2 × 2 × 3 × 3; $2^4 \times 3^2$

Write the prime factorization of each number.

7. 8 2^3 **8.** 25 5^2 **9.** 81 3^4 **10.** 100 $2^2 \times 5^2$

11. 250 2×5^3 **12.** 324 $2^2 \times 3^4$ **13.** 72 $2^3 \times 3^2$ **14.** 156 $2^2 \times 3 \times 13$

15. 111 3×37 **16.** 128 2^7 **17.** 625 5^4 **18.** 550 $2 \times 5^2 \times 11$

19. 405 $3^4 \times 5$ **20.** 243 3^5 **21.** 500 $2^2 \times 5^3$ **22.** 512 2^9

Complete the following.

23. List the prime numbers less than 30. 2, 3, 5, 7, 11, 13, 17, 19, 23, 29

24. What is the least prime factor of 221? 13

4-2 PRIME FACTORIZATION

Objective
Write the prime factorization of a whole number.

The front row bleacher seats for the cheering section at Allen High School has 5 seats. This is 1 row of 5 students or 1 × 5 students.

A **prime number** has exactly two factors, 1 and the number itself. 5 is a prime number.

5 = 1 × 5 The only factors of 5 are 1 and 5.

A **composite** number has more than two factors.

16 = 1 × 16 The factors of 16
16 = 2 × 8 are 1, 2, 4, 8,
16 = 4 × 4 and 16.

The numbers 0 and 1 are neither prime nor composite.

This is known as the Fundamental Theorem of Arithmetic.

Every composite number can be expressed as the product of prime numbers. This is called the **prime factorization** of a number. You can use a **factor tree** to find the prime factorization of a number.

Method

1. Find any factor pair of the number.
2. Continue finding factors of each factor until all factors are prime.

Point out to students that factor trees for a given number need not be identical.

Examples

Write the prime factorization for each number.

A 54

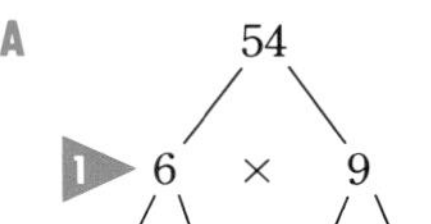

The prime factorization of 54 is 2 × 3 × 3 × 3 or 2×3^3.

B 175

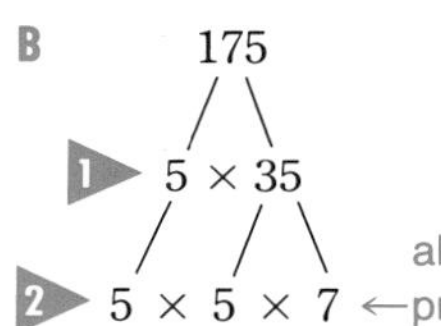

The prime factorization of 175 is 5 × 5 × 7 or $5^2 \times 7$.

Guided Practice

State whether each number is prime or composite.

1. 7 P **2.** 19 P **3.** 15 C **4.** 38 C **5.** 419 P **6.** 231 C

7. $2^3 \times 3$ 8. 2 × 5 × 7 9. 2 × 3 × 17 10. 11^2 11. $2^2 \times 41$ 12. $5^2 \times 3^2$

Examples A, B

Write the prime factorization of each number.

7. 24 **8.** 70 **9.** 102 **10.** 121 **11.** 164 **12.** 225

RETEACHING THE LESSON

Have students try a modified form of short division. Just write the usual 72 ÷ 2 as 2)72. The answer is then placed below the horizontal branch. All of the divisors and the final quotient form the prime factorization of the original dividend.

Example:
2)72 The prime
2)36 factorization of 72 is
2)18 2 × 2 × 2 × 3 × 3
3)9 or $2^3 \times 3^2$.
3

Reteaching Masters Booklet, p. 34

Name ______ Date ______

RETEACHING WORKSHEET 4-2

Prime Factorization

Factor trees help to find the prime factorization of a number.

88 → 2 × 44 → 2 × 2 × 22 → 2 × 2 × 2 × 11

The prime factorization of 88 is 2 × 2 × 2 × 11, or $2^3 \times 11$.

Complete each factor tree.

1. 24 → 2 × 12 → 2 × 2 × 6
2. 78 → 2 × 39 → 2 × 3 × 13
3. 625 → 5 × 125 → 5 × 5 × 25

31. 5×31 32. 2×5^2 33. $2^5 \times 3$ 34. 3×67 35. 2×5^4 36. $2^3 \times 331$

Exercises

Practice

State whether each number is prime or composite.

13. 2 P	**14.** 9 C	**15.** 11 P	**16.** 24 C	**17.** 29 P	**18.** 35 C
19. 73 P	**20.** 153 C	**21.** 61 P	**22.** 291 C	**23.** 671 C	**24.** 501 C

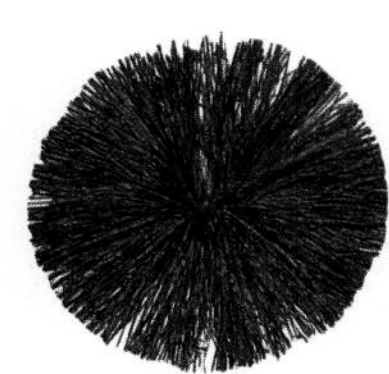

25. $2^2 \times 5$ 26. 5×13 27. $2^2 \times 13$ 28. $2 \times 3 \times 5$ 29. $2^2 \times 7$

Write the prime factorization of each number. 30. $2^3 \times 3^2$

25. 20	**26.** 65	**27.** 52	**28.** 30	**29.** 28	**30.** 72
31. 155	**32.** 50	**33.** 96	**34.** 201	**35.** 1,250	**36.** 2,648

Number Sense

37. List the first ten prime numbers.

38. List the first ten composite numbers.

39. Find the least prime number that is greater than 50. 53

40. Find the greatest prime number that is less than 2,000. 1,999

37. 2,3,5,7,11,13,17,19,23,29

38. 4,6,8,9,10,12,14,15,16

Write Math

41. Explain why an even number greater than 2 cannot be prime. All even numbers are divisible by 2.

Cooperative Groups

42. Prime numbers that differ by 2 are called *twin primes.* One such pair is 3 and 5. Discuss with your group how you would find all pairs of twin primes less than 100. Then find all the pairs.
3,5; 5,7; 11,13; 17,19; 29,31; 41,43; 59,61; 71,73

Calculator

Use a calculator to find each quotient. Then answer each question.

43. $468 \div 2 = ?$ 234
$468 \div 3 = ?$ 156
$468 \div 13 = ?$ 36
Is 468 divisible by $2 \times 3 \times 13$?
yes

44. $3{,}150 \div 3 = ?$ 1,050
$3{,}150 \div 5 = ?$ 630
$3{,}150 \div 8 = ?$ 393.75
Is 3,150 divisible by $3 \times 5 \times 8$?
no

Write a Conclusion

45. Study the results in Exercises 43 and 44. What can you say about a number that is divisible by several relatively prime numbers?

45. The number is divisible by the product of any combination of the relatively prime numbers.

Mixed Review

Lesson 2-3

Subtract.

46. $63 - 14$ = 49

47. $819 - 68$ = 751

48. $988 - 699$ = 289

49. $4{,}382 - 695$ = 3,687

50. $62{,}103 - 40{,}674$ = 21,429

Lesson 3-4

51. $18\overline{)723}$ 40 R3

52. $22\overline{)2{,}210}$ 100 R10

53. $47\overline{)6{,}584}$ 140 R4

54. $432\overline{)6{,}879}$ 15 R399

Lesson 3-5

55. Alex bought a 24-can case of soda at store A for \$5.29. At store B he bought two 12 packs of soda for \$2.49 each. How much did he pay per can at each store? Which store had the better buy?
A-\$0.22, B-\$0.21, Store B

Chalkboard Examples

Example A, B *Write the prime factorization of each number.*

1. 32 **2.** 86 **3.** 120

1. $2 \times 2 \times 2 \times 2 \times 2$ **2.** 2×43

3. $2 \times 2 \times 2 \times 3 \times 5$

4. 72 **5.** 68 **6.** 240

4. $2 \times 2 \times 2 \times 3 \times 3$

5. $2 \times 2 \times 17$

6. $2 \times 2 \times 2 \times 2 \times 3 \times 5$

Additional examples are provided on TRANSPARENCY 4-2.

EVALUATING THE LESSON

Check for Understanding Use Guided Practice Exercises to check for student understanding.

Error Analysis Be alert for students who confuse prime factorization and listing all factors of a number.

Closing the Lesson Ask the students to state in their own words how they would find the prime factorization of a number.

EXTENDING THE LESSON

Enrichment Masters Booklet, p. 34

ENRICHMENT WORKSHEET 4-2

Sieve of Eratosthenes

Eratosthenes was born about 275 B.C. He developed a systematic method, called the sieve of Eratosthenes, which sifts out all numbers except the prime numbers.

Complete the sieve of Eratosthenes for the 200 numbers listed. Circle all prime numbers. Cross out all nonprime numbers. Follow these steps.

1. One (1) is neither prime nor composite, so cross it out.
2. Two (2) is prime, so circle it. Then, cross out every second number after 2. How many even numbers are prime? 1
3. Three (3) is prime so circle it. Then, cross out every third number after 3. (Some will have been crossed out already.)
4. Circle the next greater number that is not crossed out. (It is 5.) Then, cross out every fifth number after 5.
5. Repeat the process described in step 4 until all numbers are either circled or crossed out.
6. How many primes less than 100 are there? 25 Less than 200? 46

1	2	3	4	5	6	7	8	9	10	11	12	13	14	15
16	17	18	19	20	21	22	23	24	25	26	27	28	29	30
31	32	33	34	35	36	37	38	39	40	41	42	43	44	45
46	47	48	49	50	51	52	53	54	55	56	57	58	59	60
61	62	63	64	65	66	67	68	69	70	71	72	73	74	75
76	77	78	79	80	81	82	83	84	85	86	87	88	89	90
91	92	93	94	95	96	97	98	99	100	101	102	103	104	105
106	107	108	109	110	111	112	113	114	115	116	117	118	119	120
121	122	123	124	125	126	127	128	129	130	131	132	133	134	135
136	137	138	139	140	141	142	143	144	145	146	147	148	149	150
151	152	153	154	155	156	157	158	159	160	161	162	163	164	165
166	167	168	169	170	171	172	173	174	175	176	177	178	179	180
181	182	183	184	185	186	187	188	189	190	191	192	193	194	195
196	197	198	199	200										

APPLYING THE LESSON

Independent Practice

Homework Assignment	
Minimum	Ex. 1-55 odd
Average	Ex. 2-40 even; 41-55
Maximum	Ex. 1-55

(over Lesson 4-2)

Available on TRANSPARENCY 4-3.

Write the prime factorization of each number.

1. 9 3×3
2. 28 $2 \times 2 \times 7$
3. 42 $2 \times 3 \times 7$
4. 102 $2 \times 3 \times 17$

Extra Practice, Lesson 4-2, p. 457

▶ INTRODUCING THE LESSON

Ask students to name traits they have in common with a brother or sister, mother or father, or a cousin.

▶ TEACHING THE LESSON

Using Critical Thinking Ask students how the method of finding the GCF and the LCM are alike and different.

Practice Masters Booklet, p. 38

Name ________________ Date ________________

PRACTICE WORKSHEET 4-3

Greatest Common Factor and Least Common Multiple

Find the GCF of each group of numbers.

1. 12 1, 2, 3, 4, 6, 12, or $2 \times 2 \times 3$
27 1, 3, 9, 27, or $3 \times 3 \times 3$
GCF: 3

2. 25 1, 5, 25, or 5×5
30 1, 2, 3, 5, 6, 10, 15, 30 or $2 \times 3 \times 5$
GCF: 5

3. 16 1, 2, 4, 8, 16 or $2 \times 2 \times 2 \times 2$
24 1, 2, 3, 4, 6, 8, 12, 24 or $2 \times 2 \times 2 \times 3$
GCF: 8

4. 48 1, 2, 3, 4, 6, 8, 12, 16, 24, 48 or $2 \times 2 \times 2 \times 2 \times 3$
60 1, 2, 3, 4, 5, 6, 10, 12, 15, 20, 30, 60 or $2 \times 2 \times 3 \times 5$
GCF: 12

5. 60 1, 2, 3, 4, 5, 6, 10, 12, 15, 20, 30, 60 or $2 \times 2 \times 3 \times 5$
75 1, 3, 5, 15, 25, 75 or $3 \times 5 \times 5$
GCF: 15

6. 54 1, 2, 3, 6, 9, 18, 27, 54 or $2 \times 3 \times 3 \times 3$
72 1, 2, 3, 4, 6, 8, 9, 12, 18, 24, 36, 72 or $2 \times 2 \times 2 \times 3 \times 3$
GCF: 18

Find the LCM of each group of numbers.

7. 6 0, 6, 12, 18, 24, 30, . . . or 2×3
30 0, 30, 60, . . . or $2 \times 3 \times 5$
LCM: 30

8. 30 0, 30, 60, . . . or $2 \times 3 \times 5$
10 0, 10, 20, 30, 40, . . . or 2×5
LCM: 30

9. 12 0, 12, 24, 36, 48, 60, 72, 84, . . . or $2 \times 2 \times 3$
42 0, 42, 84, . . . or $2 \times 3 \times 7$
LCM: 84

10. 12 0, 12, 24, 36, 48, . . . or $2 \times 2 \times 3$
18 0, 18, 36, 54, . . . or $2 \times 3 \times 3$
LCM: 36

11. 8 0, 8, 16, 24, 32, 40, . . . or $2 \times 2 \times 2$
10 0, 10, 20, 30, 40, 50, . . . or 2×5
LCM: 40

12. 15 0, 15, 30, 45, 60, 75, . . . or 3×5
75 0, 75, 150, . . . or $3 \times 5 \times 5$
LCM: 75

4-3 GREATEST COMMON FACTOR AND LEAST COMMON MULTIPLE

Objective

Find the greatest common factor and the least common multiple of two or more numbers.

Fraternal twins may have common features such as color of eyes, color of hair, or they may be the same sex. Likewise, two or more numbers have common factors and common multiples. That is, different numbers have factors or multiples that are the same.

The **greatest common factor (GCF)** of two or more numbers is the greatest number that is a factor of both numbers.

Here's one way to find the GCF of 12 and 18. List the factors of 12 and 18.

factors of 12: 1, 2, 3, 4, 6, 12
factors of 18: 1, 2, 3, 6, 9, 18

The common factors of 12 and 18, shown in red, are 1, 2, 3, and 6. The greatest common factor (GCF) of 12 and 18 is 6.

Another way to find the GCF of two or more numbers is given below.

Method

1. Write the prime factorization for each number.
2. Circle the common prime factors.
3. Find the product of the common prime factors.

Example A

Find the GCF of 16 and 20.

1. $16 = 2 \times 2 \times 2 \times 2$
   $20 = 2 \times 2 \times 5$
2. $16 = (2) \times (2) \times 2 \times 2$
   $20 = (2) \times (2) \times 5$
3. GCF: $2 \times 2 = 4$

The GCF of 16 and 20 is 4.

When you multiply a number by the whole numbers 0, 1, 2, 3, and so on, you get multiples of the number.

The **least common multiple (LCM)** of two or more numbers is the least nonzero number that is a multiple of each number.

multiples of 6: 0, 6, 12, 18, 24, 30, 36, 42, . . .
multiples of 9: 0, 9, 18, 27, 36, 45, . . .

Two common multiples of 6 and 9 are shown in red. The least nonzero common multiple (LCM) of 6 and 9 is 18.

You can also use prime factors to find the LCM of two or more numbers.

Method

1. Write the prime factorization for each number.
2. Circle the common prime factors and each remaining factor.
3. Find the product of the prime factors using each common prime factor only once.

Example B

Find the LCM of 6 and 10.

1. $6 = 2 \times 3$
   $10 = 2 \times 5$
2. $6 = (2) \times (3)$
   $10 = (2) \times (5)$
3. 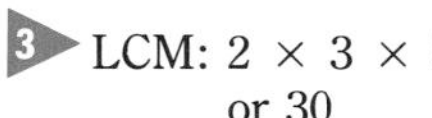
   LCM: $2 \times 3 \times 5$ or 30

The LCM of 6 and 10 is 30.

RETEACHING THE LESSON

Finding the LCM may be confusing for some students. Have them align factors as shown below, then find the LCM.

Find the LCM of 12 and 30.

$12 = 2 \times 2 \times 3$
$30 = \quad\ \ 2 \times 3 \times 5$
LCM $= 2 \times 2 \times 3 \times 5 = 60$

Find the LCM for each group of numbers using the method shown above.

1. 20, 25 100
2. 18, 24 72
3. 3, 5, 8 120

Reteaching Masters Booklet, p. 35

Name ________________ Date ________________

RETEACHING WORKSHEET 4-3

Greatest Common Factor and Least Common Multiple

You can find the GCF of 8 and 12 by listing their factors.
You can find the LCM of 8 and 12 by listing their multiples.

You can also find the GCF and LCM of 8 and 12 by using their prime factorizations.

$8 = 2 \times (2) \times (2)$
$12 = (2) \times (2) \times 3$
GCF: $2 \times 2 = 4$

$8 = (2) \times (2) \times (2)$
$12 = (2) \times (2) \times (3)$
LCM: $2 \times 2 \times 2 \times 3 = 24$

Find the GCF of each group of numbers. List all the factors of each number or use prime factorization.

Guided Practice

Example A

Find the GCF of each group of numbers.

1. 9, 12 **3**
2. 20, 30 **10**
3. 15, 56 **1**
4. 81, 108 **27**
5. 6, 12, 18 **6**

Example B

Find the LCM of each group of numbers.

6. 3, 4 **12**
7. 12, 21 **84**
8. 10, 25 **50**
9. 12, 35 **420**
10. 8, 12, 24 **24**

Exercises

Practice

Find the GCF of each group of numbers.

11. 8, 18 **2**
12. 6, 9 **3**
13. 4, 12 **4**
14. 18, 24 **6**
15. 8, 24 **8**
16. 17, 51 **17**
17. 65, 95 **5**
18. 42, 48 **6**
19. 64, 32 **32**
20. 72, 144 **72**

Find the LCM of each group of numbers.

21. 5, 6 **30**
22. 9, 27 **27**
23. 12, 15 **60**
24. 8, 12 **24**
25. 5, 15 **15**
26. 13, 39 **39**
27. 16, 24 **48**
28. 18, 20 **180**
29. 21, 14 **42**
30. 25, 30 **150**

Applications

31. Mrs. Jacobs is paid $10.25 an hour as a practical nurse. To the nearest cent, how much is she paid for a 7.5-hour day? **$76.88**
32. Barney saws an 18-foot log into 2-foot pieces. How many cuts will he make? **8 cuts**
33. Sweat suits usually sell for $49.99. They are on sale for $35. How much is saved if three sweat suits are bought? **$44.97**

Using Variables

34. If n is a prime number, what is the GCF of $12n$ and $15n$? **$3n$**
35. Find the prime factorization of $5c + 6d$ if $c = 4$ and $d = 5$. **$2 \times 5 \times 5$**
36. If p is a prime number, what is the LCM of $5p$ and $15p$? **$15p$**

Suppose

37. Suppose your real estate taxes, car insurance, and water bill all came due in June. The real estate taxes are due every six months. The car insurance is due every four months and the water bill is due every three months. Name the next month that all three bills will come due at the same time. **June**

39. Christian Goldbach suggested that every even whole number greater than or equal to 4 is the sum of two primes.

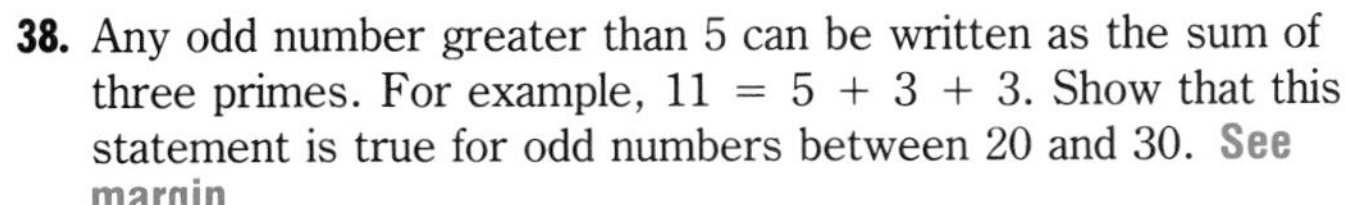

38. Any odd number greater than 5 can be written as the sum of three primes. For example, $11 = 5 + 3 + 3$. Show that this statement is true for odd numbers between 20 and 30. **See margin.**

Research

39. Write a report about Goldbach's Conjecture.

Chalkboard Examples

Example A *Find the GCF of each group of numbers.*

1. 30, 45 **15**
2. 12, 42 **6**

Example B *Find the LCM of each group of numbers.*

3. 28, 42 **84**
4. 7, 13 **91**

Additional examples are provided on TRANSPARENCY 4-3.

▶ EVALUATING THE LESSON

Check for Understanding Use Guided Practice Exercises to check for student understanding.

Closing the Lesson Ask students to state in their own words how they would find the GCF and the LCM of a pair of numbers.

▶ EXTENDING THE LESSON

Enrichment *When you fold a sheet of paper in half, you get 2 parts of equal size. Suppose you fold the paper in half again. How many parts of the same size do you have? Continue folding the paper in half 3 more times. How many parts do you have?* **32 parts** *How are the parts related to the number of folds?* **Each time a fold is made the parts double.**

Enrichment Masters Booklet, p. 35

Name ______________________ Date ____________

ENRICHMENT WORKSHEET 4-3

Prime Factorization: Ladder Method

Write the prime factorization of 96.

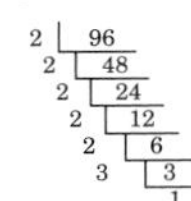

- Choose any prime number (for example, 2) that divides evenly into your number.
- Continue dividing the quotient by primes until you reach a quotient of 1.
- The numbers that you used for divisors make up the prime factorization.

$96 = 2 \times 2 \times 2 \times 2 \times 2 \times 3 = 2^5 \times 3$

Use the ladder method to write the prime factorization of each number.

1. 42 — 2 | 42, 3 | 21, 7 | 7, 1 — $2 \times 3 \times 7$
2. 116 — 2 | 116, 2 | 58, 29 | 29, 1 — $2^2 \times 29$
3. 100 — 2 | 100, 2 | 50, 5 | 25, 5 | 5, 1 — $2^2 \times 5^2$
4. 140 — 2 | 140, 2 | 70, 5 | 35, 7 | 7, 1 — $2^2 \times 5 \times 7$
5. 175 — 5 | 175, 5 | 35, 7 | 7, 1 — $5^2 \times 7$
6. 432 — 2 | 432, 2 | 216, 2 | 108, 2 | 54, 3 | 27, 3 | 9, 3 | 3, 1 — $2^4 \times 3^3$
7. 550 — 2 | 550, 5 | 275, 5 | 55, 11 | 11, 1 — $2 \times 5^2 \times 11$
8. 620 — 2 | 620, 2 | 310, 5 | 155, 31 | 31, 1 — $2^2 \times 5 \times 31$
9. 1,260 — 2 | 1,260, 2 | 630, 3 | 315, 3 | 105, 5 | 35, 7 | 7

APPLYING THE LESSON

Independent Practice

Homework Assignment	
Minimum	Ex. 1-39 odd
Average	Ex. 2-30 even; 31-39
Maximum	Ex. 1-39

Additional Answers

38. $21 = 7 + 7 + 7$
$23 = 17 + 3 + 3$
$25 = 19 + 3 + 3$
$27 = 13 + 11 + 3$
$29 = 23 + 3 + 3$

5-MINUTE CHECK-UP

(over Lesson 4-3)

Available on TRANSPARENCY 4-4.

Find the GCF of each group of numbers.

1. 30, 40 **10**
2. 65, 91 **13**

Find the LCM of each group of numbers.

3. 15, 25 **75**
4. 17, 51 **51**

Extra Practice, Lesson 4-3, p. 457

INTRODUCING THE LESSON

Suppose you work at a fast food restaurant. You make $4.25 per hour and you worked 20 hours last week. You made $85. When you get your paycheck, will it be for $85?

TEACHING THE LESSON

Using Discussion Point out that everyone must pay federal, state and local taxes and FICA. There are also other optional deductions. Have students name these optional deductions.

Practice Masters Booklet, p. 39

PRACTICE WORKSHEET 4-4

Applications: Total Deductions and Take-Home Pay

Complete the chart.

	Name	Gross Pay	Total Tax Deduction	Total Personal Deduction	Take-Home Pay
1.	S. Cook	$247.80	$37.17	$35.00	**$175.63**
2.	R. Choi	$215.62	$29.48	$8.75	**$177.39**
3.	T. Brady	$195.75	$27.16	$20.00	**$148.59**
4.	L. Sanchez	**$228.30**	$32.53	$9.95	$185.82
5.	F. Hyde	**$239.44**	$35.46	$15.50	$188.48
6.	P. Morgan	**$175.25**	$23.92	$7.25	$144.08
7.	N. Hill	$188.50	**$24.18**	$9.40	$154.92
8.	D. Tallchief	$203.22	$25.81	**$18.75**	$158.66
9.	R. Sanders	$217.95	$29.76	**$24.00**	$164.19
10.	J. Ortega	$224.50	**$31.27**	$17.50	$175.73
11.	L. Horton	$206.74	$27.44	$9.36	**$169.94**
12.	A. Cheng	$185.75	**$22.19**	$5.00	$158.56
13.	T. McVay	$200.50	$24.52	$8.25	**$167.73**
14.	S. Shelton	**$236.40**	$38.73	$25.00	$172.67
15.	M. Jordan	$242.83	$36.84	**$18.50**	$187.49

Applications

4-4 TOTAL DEDUCTIONS AND TAKE-HOME PAY

Objective
Read an earnings statement.

Sandy Graves is an office manager for Carpenter's Lumber Company. Sandy checks her earnings statement to make sure it is correct. This earnings statement was attached to her last paycheck.

Carpenter's Lumber Company			Weekly Earnings Statement
Sandy Graves			
Pay Period Ending	Pay Date	Check Number	
8/16	8/16	43075	

Hours	Rate	Earnings	Type	Deduction	Type
40.00	8.19	327.60	Reg	1.25	Dental
				8.54	Health
				10.17	401 K
				.54	AccIns
				10.00	CrUnion
				2.00	UnitedFd

This Pay	Gross Pay	Federal	F.I.C.A.	State	Local	Net Pay
	327.60	52.42	24.57	9.83	3.28	205.00

F.I.C.A. means Federal Insurance Contribution Act or Social Security.

Sandy adds to check her total tax deductions.

Federal Tax	$52.42
Social Security (F.I.C.A.)	24.57
State Tax	9.83
Local Tax	+ 3.28
	$90.10

Sandy also requests her employer to make these personal deductions.

Dental Insurance	$ 1.25
Health Insurance	8.54
401 K (Retirement)	10.17
Accident Insurance	0.54
Credit Union	10.00
United Way	+ 2.00
	$32.50

The *net pay* or take-home pay is found by subtracting the total tax and personal deductions from the *gross pay,* her weekly wage.

$90.10	$327.60	gross pay
+ 32.50	− 122.60	deductions
$122.60	$205.00	net pay

Understanding the Data

- How many hours a week does Sandy work? **40 hours**
- How much does she make an hour? **$8.19**
- Explain how you would check to see if her gross pay is correct. Then check to see if it is correct. **Multiply hours worked times hourly pay. It is correct.**

Guided Practice

1. Mr. Garcia's gross pay is $569.19. His total tax deductions are $156.54 and his personal deductions are $68.49. What are his total deductions and his net pay? **$225.03, $344.16**

RETEACHING THE LESSON

Find the net pay for each gross pay, tax deduction, and personal deduction.

1. $354; $68.44; $35.40 **$250.16**
2. $170; $20.40; $8.50 **$141.10**
3. $750; $146.50; $69.38 **$534.12**
4. $298; $44.70; $32.78 **$220.52**

Reteaching Masters Booklet, p. 36

Name ____________ Date ____________

RETEACHING WORKSHEET 4-4

Applications: Total Deductions and Take-Home Pay

This is a part of the earning statement attached to Michael Green's paycheck.

Hours	Rate	Gross Pay	Total Tax Deduction	Total Personal Deduction	Take-Home Pay
40	$6.75	$270.00	$40.50	$12.15	$217.35

To find **gross pay**, multiply hours worked times rate of pay.

40 × $6.75 = $270.00

To find **take-home pay**, first add the deductions.

$40.50 + $12.15 = $52.65

Then subtract the total deductions from gross pay.

$270.00 − $52.65 = $217.35

Exercises

Complete the chart.

Practice
2. $147.92
3. $141.83
4. $166.88
5. $201.84
6. $134.79

Name	Gross Pay	Total Tax Deduction	Total Personal Deduction	Take-Home Pay
L. Adams	$187.60	$25.43	$14.25	2. ■
D. Block	$193.35	$34.82	$16.70	3. ■
M. Federico	$215.70	$38.64	$10.18	4. ■
J. Mason	$243.92	$29.58	$12.50	5. ■
B. Klenk	$174.51	$31.05	$ 8.67	6. ■

Applications

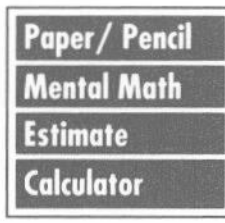

7. Tom Poling has $36.20 deducted weekly from his paycheck for federal tax, $4.52 for state tax, and $2.05 for city tax. How much total tax does he pay yearly? **$2,224.04**
8. Jenny Spalding's gross pay is $472.36. Her total tax deductions are $129.91 and her personal deductions are $53.49. What is her net pay? **$288.96**
9. Todd Benson's gross pay is $394.93, and his net pay is $289.27. What are his total deductions? **$105.66**
10. Refer to Sandy's earnings statement on page 114. Sandy uses her credit union for a savings account. How much savings will Sandy have in one year? **$520**

Interpreting Data

11. In your own words, explain the difference between gross pay and net pay. **gross pay–total earnings; net pay–take-home pay**
12. Name some ways you could make your net pay higher without getting a raise. **See margin.**

Suppose

13. Suppose your net pay is less than or greater than normal and you know of no changes. What would you do? **See margin.**

Research

14. Contact the nearest Social Security (F.I.C.A.) office to ask for a booklet that explains the laws, benefits, and current rate taken out of earnings. Determine what benefits are available. If you have a job, check the rate against your earnings statement.

Mixed Review

Lesson 3-1

Multiply.

15. 82×5 = **410**
16. 46×15 = **690**
17. $1{,}432 \times 91$ = **130,312**
18. $6{,}423 \times 108$ = **693,684**

Lesson 3-10

Solve for d, r, or t using the formula $d = rt$.

19. $d = 240$ miles, $t = 6$ hours, $r =$ __?__ **40 mph**
20. $r = 58$ mph, $t = 8.25$ hours, $d =$ __?__ **478.5 miles**
21. $d = 48$ miles, $r = 16$ mph, $t =$ __?__ **3 hours**

Lesson 3-11

22. Erin is going to carpet her bedroom floor. Her bedroom is rectangular in shape and measures 12 feet by 9 feet. How many square yards of carpet does she need to cover her floor? **12 yd²**

APPLYING THE LESSON

Independent Practice

Homework Assignment	
Minimum	Ex. 2-22 even
Average	Ex. 1-22
Maximum	Ex. 1-22

Additional Answers

12. reduce personal deductions, increase number of tax exemptions within the law, work more hours
13. compare it to previous statements, find the error, contact your employer or payroll department immediately

Chalkboard Examples

1. Janet works 40 hours a week and makes $392.40. Her total tax deductions are $98.10 and her personal deductions are $31.29. What are her total deductions and net pay? **$129.39; $263.01**

Additional examples are provided on TRANSPARENCY 4-4.

EVALUATING THE LESSON

Check for Understanding Use Guided Practice Exercises to check for student understanding.

Closing the Lesson Ask students what they have learned about total deductions and take home pay.

EXTENDING THE LESSON

Enrichment Have students use the local newspaper to find a job they would like and for which the hourly rate or salary is listed. Then have students determine their weekly salary or gross pay. Using a tax booklet, have them estimate their weekly taxes. Students could make up personal deductions and find their net pay.

Enrichment Masters Booklet, p. 36

Name ______ Date ______

ENRICHMENT WORKSHEET 4-4

Perfect Numbers

A natural number that equals the sum of all its divisors except itself is called a perfect number.

Divisors of 6: 1, 2, 3, 6

6 = 1 + 2 + 3

So 6 is a perfect number.

State whether each number is a perfect number.

1. 1 — no; only divisor is itself
2. 4 — no; 4 ≠ 1 + 2
3. 12 — no; 12 ≠ 1 + 2 + 3 + 4 + 6
4. 16 — no; 16 ≠ 1 + 2 + 4 + 8
5. 28 — yes; 28 = 1 + 2 + 4 + 7 + 14
6. 48 — no; 48 ≠ 1 + 2 + 3 + 4 + 6 + 8 + 12 + 16 + 24
7. 100 — no; 100 ≠ 1 + 2 + 4 + 5 + 10 + 20 + 25 + 50
8. 150 — no; 150 ≠ 1 + 2 + 3 + 5 + 6 + 10 + 15 + 25 + 30 + 50 + 75
9. 496 — yes; 496 = 1 + 2 + 4 + 8 + 16 + 31 + 62 + 124 + 248
10. 500 — no; 500 ≠ 1 + 2 + 4 + 5 + 10 + 20 + 25 + 50 + 100 + 125 + 250
11. 1,000 — no; 1,000 ≠ 1 + 2 + 4 + 5 + 8 + 10 + 20 + 25 + 40 + 50 + 100 + 125 + 200 + 250 + 500
12. 1,111 — no; 1,111 ≠ 1 + 11 + 101
13. 8,128 — yes; 8,128 = 1 + 2 + 4 + 8 + 16 + 32 + 64 + 127 + 254 + 508 + 1,016 + 2,032 + 4,064

(over Lesson 4-4)

Available on TRANSPARENCY 4-5.

1. Janet Hope's gross pay is \$215.01. Her total tax deductions are \$36.20 and her total personal deductions are \$22.57. What is her take-home pay? **\$156.24**

▶ INTRODUCING THE LESSON

Have students cut two pizza shapes out of paper. Then ask them to determine different ways of cutting the pizzas so that four people may take a different number of pieces but still get the same amount of pizza.

▶ TEACHING THE LESSON

Using Discussion Discuss with students the fact that the fraction $\frac{3}{4}$ also indicates 3 divided by 4 ($3 \div 4$ or $4\overline{)3}$). Also discuss with students that they are really multiplying or dividing by 1 when finding equivalent fractions.

Practice Masters Booklet, p. 40

Name ______________________ Date ______________

PRACTICE WORKSHEET 4-5

Equivalent Fractions

Replace each ☐ with a number so that the fractions are equivalent.

1. $\frac{2}{3} = \frac{\boxed{4}}{6}$	2. $\frac{3}{16} = \frac{\boxed{9}}{48}$	3. $\frac{1}{3} = \frac{\boxed{4}}{12}$	4. $\frac{6}{7} = \frac{\boxed{42}}{49}$
5. $\frac{15}{20} = \frac{\boxed{3}}{4}$	6. $\frac{4}{12} = \frac{\boxed{1}}{3}$	7. $\frac{4}{16} = \frac{\boxed{1}}{4}$	8. $\frac{5}{10} = \frac{\boxed{1}}{2}$
9. $\frac{3}{4} = \frac{\boxed{9}}{12}$	10. $\frac{3}{4} = \frac{\boxed{6}}{8}$	11. $\frac{1}{3} = \frac{\boxed{2}}{6}$	12. $\frac{1}{2} = \frac{\boxed{3}}{6}$
13. $\frac{7}{7} = \frac{\boxed{1}}{1}$	14. $\frac{3}{7} = \frac{\boxed{18}}{42}$	15. $\frac{3}{15} = \frac{\boxed{1}}{5}$	16. $\frac{4}{32} = \frac{\boxed{1}}{8}$
17. $\frac{1}{25} = \frac{\boxed{4}}{100}$	18. $\frac{7}{9} = \frac{\boxed{14}}{18}$	19. $\frac{0}{3} = \frac{\boxed{0}}{36}$	20. $\frac{6}{18} = \frac{\boxed{1}}{3}$
21. $\frac{1}{3} = \frac{\boxed{5}}{15}$	22. $\frac{3}{7} = \frac{\boxed{9}}{21}$	23. $\frac{2}{5} = \frac{\boxed{10}}{25}$	24. $\frac{3}{10} = \frac{\boxed{30}}{100}$

Name a fraction equivalent to each fraction.

25. $\frac{1}{3}$ $\frac{2}{6}$ 26. $\frac{1}{2}$ $\frac{5}{10}$ 27. $\frac{1}{4}$ $\frac{2}{8}$ 28. $\frac{3}{6}$ $\frac{1}{2}$ 29. $\frac{5}{8}$ $\frac{10}{16}$ 30. $\frac{44}{48}$ $\frac{11}{12}$

Answers may vary. Sample answers are given.

31. Write three fractions that are equivalent to $\frac{6}{10}$. $\frac{12}{20}$, $\frac{18}{30}$, $\frac{24}{40}$

32. Name three fractions that are equivalent to $\frac{35}{100}$. $\frac{7}{20}$, $\frac{14}{40}$, $\frac{21}{60}$

Answers may vary. Sample answers are given.

4-5 EQUIVALENT FRACTIONS

Objective
Find fractions equivalent to given fractions.

Katie tries to get 8 hours of sleep each night. If she does this the rest of her life, she will spend $\frac{8}{24}$ or $\frac{1}{3}$ of her life sleeping.

A fraction, such as $\frac{1}{3}$, may be used to name part of a whole or a group.

$\frac{1}{3}$ ← numerator
← denominator

The *denominator* tells the part of objects or equal-sized parts. The *numerator* tells the number of objects or parts being considered.

The figures at the right show that $\frac{1}{3}$ and $\frac{8}{24}$ name the same number. They are called **equivalent fractions.**

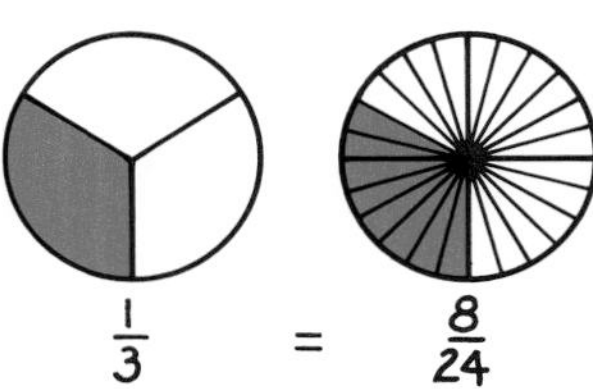

You can multiply or divide the numerator and denominator of a fraction by the same nonzero number to find an equivalent fraction.

Examples

Replace each ■ with a number so that the fractions are equivalent.

A $\frac{3}{7} = \frac{■}{21}$

Multiply the numerator and the denominator by 3. This is like multiplying by what number? **1**

$\frac{3}{7} = \frac{■}{21} \rightarrow \frac{3}{7} = \frac{9}{21}$ (× 3 above and below)

B $\frac{12}{36} = \frac{■}{9}$

Divide the numerator and the denominator by 4. This is like dividing by what number? **1**

$\frac{12}{36} = \frac{■}{9} \rightarrow \frac{12}{36} = \frac{3}{9}$ (÷ 4 above and below)

Name another fraction equivalent to $\frac{3}{9}$.

A sample answer is $\frac{1}{3}$.

Guided Practice

Replace each ■ with a number so that the fractions are equivalent.

Example A

1. $\frac{6}{7} = \frac{■}{28}$ **24** 2. $\frac{1}{2} = \frac{■}{14}$ **7** 3. $\frac{2}{3} = \frac{■}{6}$ **4** 4. $\frac{3}{4} = \frac{■}{8}$ **6** 5. $\frac{7}{9} = \frac{■}{18}$ **14**

RETEACHING THE LESSON

Complete with a name for 1 that makes the statement true.

1. $\frac{1}{3} \times \frac{3}{3} = \frac{3}{9}$ 2. $\frac{3}{4} \times \frac{5}{5} = \frac{15}{20}$

Write a number in the numerator or denominator that will make the fractions equivalent.

3. $\frac{4}{5} = \frac{16}{20}$ 4. $\frac{3}{7} = \frac{9}{21}$

5. $\frac{15}{18} = \frac{5}{6}$ 6. $\frac{33}{55} = \frac{3}{5}$

Reteaching Masters Booklet, p. 37

Name ______________________ Date ______________

RETEACHING WORKSHEET 4-5

Equivalent Fractions

Equivalent fractions name the same number.

Multiplying or dividing the numerator and the denominator by the same nonzero number gives an equivalent fraction.

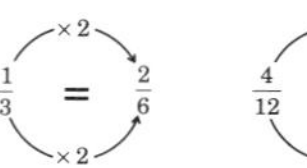

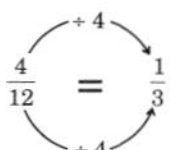

Replace each ☐ with a number so that the fractions are equivalent.

Example B *Replace each ■ with a number so that the fractions are equivalent.*

6. $\frac{6}{12} = \frac{\blacksquare}{2}$ 1 **7.** $\frac{4}{20} = \frac{\blacksquare}{10}$ 2 **8.** $\frac{4}{16} = \frac{\blacksquare}{4}$ 1 **9.** $\frac{6}{15} = \frac{\blacksquare}{5}$ 2 **10.** $\frac{6}{18} = \frac{\blacksquare}{3}$ 1

Exercises

Practice

11\. 15 12. 21
14\. 12 25. 18
26\. 4 29. 18

Replace each ■ with a number so that the fractions are equivalent.

11. $\frac{5}{9} = \frac{\blacksquare}{27}$ **12.** $\frac{7}{12} = \frac{\blacksquare}{36}$ **13.** $\frac{6}{1} = \frac{\blacksquare}{4}$ 24 **14.** $\frac{2}{2} = \frac{\blacksquare}{12}$ **15.** $\frac{3}{16} = \frac{\blacksquare}{48}$ 9

16. $\frac{6}{16} = \frac{\blacksquare}{8}$ 3 **17.** $\frac{5}{5} = \frac{\blacksquare}{1}$ 1 **18.** $\frac{10}{25} = \frac{\blacksquare}{5}$ 2 **19.** $\frac{15}{9} = \frac{\blacksquare}{3}$ 5 **20.** $\frac{12}{18} = \frac{\blacksquare}{9}$ 6

21. $\frac{3}{4} = \frac{\blacksquare}{12}$ 9 **22.** $\frac{1}{3} = \frac{\blacksquare}{6}$ 2 **23.** $\frac{3}{15} = \frac{\blacksquare}{5}$ 1 **24.** $\frac{12}{16} = \frac{\blacksquare}{4}$ 3 **25.** $\frac{3}{7} = \frac{\blacksquare}{42}$

26. $\frac{1}{25} = \frac{\blacksquare}{100}$ **27.** $\frac{0}{36} = \frac{\blacksquare}{3}$ 0 **28.** $\frac{6}{32} = \frac{\blacksquare}{16}$ 3 **29.** $\frac{9}{32} = \frac{\blacksquare}{64}$ **30.** $\frac{10}{5} = \frac{\blacksquare}{1}$ 2

Answers may vary. Sample answers given.

Name a fraction equivalent to each fraction.

31. $\frac{2}{3}$ $\frac{4}{6}$ **32.** $\frac{5}{8}$ $\frac{10}{16}$ **33.** $\frac{1}{4}$ $\frac{2}{8}$ **34.** $\frac{2}{6}$ $\frac{1}{3}$ **35.** $\frac{5}{10}$ $\frac{1}{2}$ **36.** $\frac{10}{16}$ $\frac{5}{8}$

Applications

37\. two fractions equivalent to $\frac{6}{20}$
38\. two fractions equivalent to $\frac{5}{20}$
39\. two fractions equivalent to $\frac{9}{20}$
40\. two fractions equivalent to $\frac{11}{20}$

Write two equivalent fractions for each of the following. Use the circle graph.

U.S. Population by Age

60 and over
0–19 years
35–59 years
20–34 years

37. What fraction of the population is from 0-19 years old?

38. What fraction of the population is from 20-34 years old?

39. What fraction of the population is 35 years old or older?

40. What fraction of the population is 34 years old or younger?

41. The trans-Alaska pipeline delivers about 2 million barrels of oil per day. The pipeline is 800 miles long. About how many barrels are delivered in 7.5 days? **15 million barrels**

Write Math

42. Write four fractions that are equivalent to three-fourths. Write each fraction in words. **See margin.**

Suppose

43. Suppose the denominator of a fraction is increased, what happens to the value of the fraction? **decreases**

Critical Thinking

44. There is a gallon of punch in a punch bowl and a gallon of lemonade in another punch bowl. Mickey puts a ladle of punch in the lemonade. Then Mickey takes a ladle of lemonade and puts it in the punch. Now, is there more punch in the lemonade or more lemonade in the punch? **See margin.**

Chalkboard Examples

Example A *Replace each ■ with a number so that the fractions are equivalent.*

1. $\frac{1}{4} = \frac{\blacksquare}{8}$ 2 **2.** $\frac{2}{3} = \frac{\blacksquare}{15}$ 10

Example B *Replace each ■ with a number so that the fractions are equivalent.*

3. $\frac{12}{24} = \frac{\blacksquare}{2}$ 1 **4.** $\frac{10}{25} = \frac{\blacksquare}{5}$ 2

Additional examples are provided on TRANSPARENCY 4-5.

EVALUATING THE LESSON

Check for Understanding Use Guided Practice Exercises to check for student understanding.

Error Analysis Watch for students that add or subtract the same number to the numerator and denominator. Stress that they multiply or divide by 1.

Closing the Lesson Tell students that they will use equivalent fractions when they add or subtract fractions.

EXTENDING THE LESSON

Enrichment Masters Booklet, p. 37

ENRICHMENT WORKSHEET 4-5

Equivalent Fractions Drawings may vary.

Solve. Make a drawing to illustrate your answer. Typical drawings are shown

1. The pizza was cut into six slices. Jon wanted $\frac{1}{2}$ of the pizza so Nancy gave him 3 of the 6 slices. Did he get what he wanted? yes

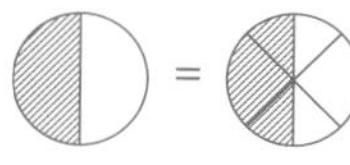

2. Wayne had eight silver dollars. Four of them were dated 1899. He said that $\frac{1}{3}$ were dated 1899. Was he correct?

no, more than $\frac{1}{3}$ were.

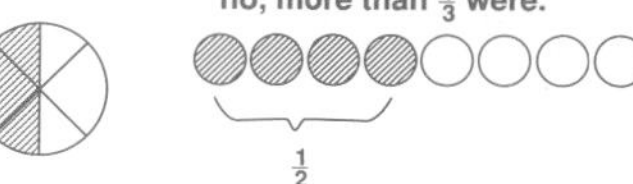

3. Ten scouts were camping. Three scouts had compasses. Toya told her father that $\frac{1}{2}$ of the scouts had compasses. Was she correct?

no

ⓧ ⓧ ⓧ X X
X X X X X

4. Sixteen students belonged to the math club. One-fourth of them received special awards. The school newspaper headline read "Four Math Club Members Get Awards." Was the headline correct? yes

X X X X
X X X X
X X X X
X X X X

5. Betsy is an amateur photographer. On her last roll of 24 photos, $\frac{3}{8}$ did not turn out. The photo lab gave her credit for 6 photos. Was that correct?

X X X
X X X
X X X } $\frac{3}{8}$ no
X X X
X X X
X X X
X X X
X X X

6. Six friends went out for lunch. Two-thirds of them ordered tacos. The waiter wrote down 3 tacos. Was that correct? no

X X
X X } $\frac{2}{3}$
X X

APPLYING THE LESSON

Independent Practice

Homework Assignment	
Minimum	Ex. 2-44 even
Average	Ex. 1-35 odd; 37-44
Maximum	Ex. 1-44

Additional Answers

42\. six-eighths, nine-twelfths, twelve-sixteenths, fifteen-twentieths Answers may vary.

44\. Neither. There is as much punch in the lemonade as there is lemonade in the punch. No matter how many round trips are made with the ladle, whatever amount of punch is taken is exactly the amount of lemonade that replaced it, and vice versa.

5-MINUTE CHECK-UP

(over Lesson 4-5)

Available on TRANSPARENCY 4-6.

Replace each ▩ with a number so that the fractions are equivalent.

1. $\frac{1}{4} = \frac{▩}{12}$ 3 **2.** $\frac{2}{5} = \frac{▩}{15}$ 6

3. $\frac{2}{8} = \frac{▩}{4}$ 1 **4.** $\frac{10}{30} = \frac{▩}{3}$ 1

Extra Practice, Lesson 4-5, p. 457

▶ INTRODUCING THE LESSON

Have students name situations in which they had to share something equally with a family member and how they went about making the two parts equal. How did they know the two parts were equal?

▶ TEACHING THE LESSON

Using Critical Thinking Have students explain the advantage of using the GCF to simplify fractions.

Practice Masters Booklet, p. 41

PRACTICE WORKSHEET 4-6

Simplifying Fractions

Simplify.

1. $\frac{5}{10}$ $\frac{1}{2}$	**2.** $\frac{4}{10}$ $\frac{2}{5}$	**3.** $\frac{9}{12}$ $\frac{3}{4}$	**4.** $\frac{18}{24}$ $\frac{3}{4}$
5. $\frac{9}{24}$ $\frac{3}{8}$	**6.** $\frac{16}{20}$ $\frac{4}{5}$	**7.** $\frac{13}{39}$ $\frac{1}{3}$	**8.** $\frac{16}{48}$ $\frac{1}{3}$
9. $\frac{8}{16}$ $\frac{1}{2}$	**10.** $\frac{9}{12}$ $\frac{3}{4}$	**11.** $\frac{15}{18}$ $\frac{5}{6}$	**12.** $\frac{5}{35}$ $\frac{1}{7}$
13. $\frac{20}{24}$ $\frac{5}{6}$	**14.** $\frac{14}{16}$ $\frac{7}{8}$	**15.** $\frac{16}{18}$ $\frac{8}{9}$	**16.** $\frac{9}{15}$ $\frac{3}{5}$
17. $\frac{7}{21}$ $\frac{1}{3}$	**18.** $\frac{6}{16}$ $\frac{3}{8}$	**19.** $\frac{20}{36}$ $\frac{5}{9}$	**20.** $\frac{33}{39}$ $\frac{11}{13}$
21. $\frac{18}{24}$ $\frac{3}{4}$	**22.** $\frac{24}{36}$ $\frac{2}{3}$	**23.** $\frac{32}{48}$ $\frac{2}{3}$	**24.** $\frac{66}{121}$ $\frac{6}{11}$
25. $\frac{17}{34}$ $\frac{1}{2}$	**26.** $\frac{4}{40}$ $\frac{1}{10}$	**27.** $\frac{26}{39}$ $\frac{2}{3}$	**28.** $\frac{20}{50}$ $\frac{2}{5}$
29. $\frac{25}{75}$ $\frac{1}{3}$	**30.** $\frac{32}{80}$ $\frac{2}{5}$	**31.** $\frac{60}{72}$ $\frac{5}{6}$	**32.** $\frac{30}{45}$ $\frac{2}{3}$

Member	Tickets sold
Ryko	73
Paul	80
Amy	75
Miguel	50
Yvonne	92
Jerome	70

Solve. Use the chart.

33. Rank the members from 1 through 6. The one who sold the most tickets gets a rank of 1.

34. Each member had 100 tickets to sell. Write each member's sale as a fraction in simplest form.

1. Yvonne - $\frac{23}{25}$
2. Paul - $\frac{4}{5}$
3. Amy - $\frac{3}{4}$
4. Ryko - $\frac{73}{100}$
5. Jerome - $\frac{7}{10}$
6. Miguel - $\frac{1}{2}$

4-6 SIMPLIFYING FRACTIONS

Objective
Write fractions in simplest form.

In the 1988 presidential election, George Bush carried 40 out of 50 states. You could write the fraction of states that he carried using the fractions $\frac{40}{50}$, $\frac{8}{10}$, and $\frac{4}{5}$.

The fraction $\frac{4}{5}$ is in simplest form.

A fraction is in *simplest form* when the GCF of the numerator and denominator is 1. To write a fraction in simplest form, divide both the numerator and the denominator by common factors until their GCF is 1.

Method

1. Divide the numerator and the denominator by a common factor.
2. Continue dividing by common factors until the GCF is 1.

Examples

Simplify each fraction.

A $\frac{24}{30}$

$$\frac{24}{30} \overset{\div 2}{=} \frac{12}{15} \overset{\div 3}{=} \frac{4}{5}$$

The GCF of 4 and 5 is 1.

B $\frac{16}{24}$

$$\frac{16}{24} \overset{\div 2}{=} \frac{8}{12} \overset{\div 2}{=} \frac{4}{6} \overset{\div 2}{=} \frac{2}{3}$$

How could the GCF be used to simplify $\frac{16}{24}$?

C $\frac{5}{15}$

$$\frac{5}{15} \overset{\div 5}{=} \frac{1}{3}$$

D $\frac{7}{9}$

The GCF of 7 and 9 is 1 so the fraction is already in simplest form.

Guided Practice

Simplify each fraction.

Example A **1.** $\frac{12}{16}$ $\frac{3}{4}$ **2.** $\frac{28}{32}$ $\frac{7}{8}$ **3.** $\frac{75}{100}$ $\frac{3}{4}$ **4.** $\frac{8}{16}$ $\frac{1}{2}$ **5.** $\frac{6}{18}$ $\frac{1}{3}$ **6.** $\frac{27}{36}$ $\frac{3}{4}$

Example B **7.** $\frac{16}{64}$ $\frac{1}{4}$ **8.** $\frac{8}{16}$ $\frac{1}{2}$ **9.** $\frac{50}{100}$ $\frac{1}{2}$ **10.** $\frac{24}{40}$ $\frac{3}{5}$ **11.** $\frac{32}{80}$ $\frac{2}{5}$ **12.** $\frac{8}{24}$ $\frac{1}{3}$

RETEACHING THE LESSON

Complete.

1. $\frac{12 \div 3}{15 \div 3} = \frac{4}{5}$ **2.** $\frac{30 \div 5}{35 \div 5} = \frac{6}{7}$

3. $\frac{28 \div 4}{32 \div 4} = \frac{7}{8}$ **4.** $\frac{21 \div 7}{28 \div 7} = \frac{3}{4}$

Reteaching Masters Booklet, p. 38

Name ____________________ Date ____________

RETEACHING WORKSHEET 4-6

Simplifying Fractions

A fraction is in simplest form when the GCF of the numerator and denominator is 1.
To simplify a fraction, divide the numerator and denominator by common factors until the GCF is 1.
This may take more than one step.

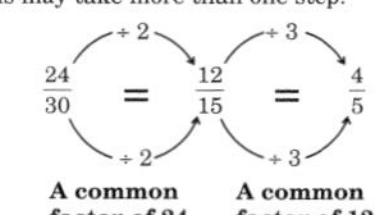

A common factor of 24 and 30 is 2. **A common factor of 12 and 15 is 3.**

To simplify a fraction in just one step, divide both numerator and denominator by their GCF.

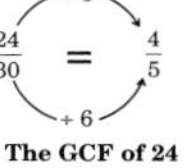

The GCF of 24 and 30 is 60.

Complete.

Examples C, D

13. $\frac{20}{25}$ $\frac{4}{5}$ 14. $\frac{4}{10}$ $\frac{2}{5}$ 15. $\frac{3}{5}$ $\frac{3}{5}$ 16. $\frac{14}{19}$ $\frac{14}{19}$ 17. $\frac{9}{12}$ $\frac{3}{4}$ 18. $\frac{6}{8}$ $\frac{3}{4}$

19. $\frac{15}{18}$ $\frac{5}{6}$ 20. $\frac{9}{20}$ $\frac{9}{20}$ 21. $\frac{8}{21}$ $\frac{8}{21}$ 22. $\frac{10}{15}$ $\frac{2}{3}$ 23. $\frac{9}{24}$ $\frac{3}{8}$ 24. $\frac{6}{31}$ $\frac{6}{31}$

Exercises

Practice

Simplify each fraction.

25. $\frac{18}{32}$ $\frac{9}{16}$ 26. $\frac{15}{36}$ $\frac{5}{12}$ 27. $\frac{27}{72}$ $\frac{3}{8}$ 28. $\frac{23}{69}$ $\frac{1}{3}$ 29. $\frac{36}{48}$ $\frac{3}{4}$ 30. $\frac{6}{30}$ $\frac{1}{5}$

31. $\frac{15}{27}$ $\frac{5}{9}$ 32. $\frac{20}{24}$ $\frac{5}{6}$ 33. $\frac{6}{16}$ $\frac{3}{8}$ 34. $\frac{4}{6}$ $\frac{2}{3}$ 35. $\frac{7}{21}$ $\frac{1}{3}$ 36. $\frac{18}{27}$ $\frac{2}{3}$

37. $\frac{24}{30}$ $\frac{4}{5}$ 38. $\frac{8}{10}$ $\frac{4}{5}$ 39. $\frac{16}{18}$ $\frac{8}{9}$ 40. $\frac{12}{36}$ $\frac{1}{3}$ 41. $\frac{30}{50}$ $\frac{3}{5}$ 42. $\frac{14}{21}$ $\frac{2}{3}$

43. Write $\frac{6}{9}$ in simplest form. $\frac{2}{3}$

44. What is the simplest form of $\frac{6}{12}$? $\frac{1}{2}$

Applications

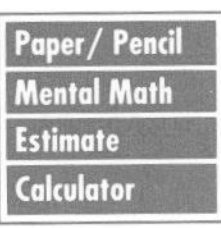

45. In the 1980 presidential election, Ronald Reagan carried 44 of the 50 states. What fraction, in simplest form, of the states did he carry? $\frac{22}{25}$

46. On the average, 12 of the 30 days in June in Pittsburgh are rainy. What fraction, in simplest form, of the days in June are not rainy? $\frac{3}{5}$

Mental Math

How long has the equal sign been around? See page 175.

Place operation signs (+, −, ×, or ÷) and parentheses, if needed, in each equation to make a true statement.

47. 5 __?__ 5 __?__ 5 = 50

48. 5 __?__ 5 __?__ 5 = 6 ÷, +

49. 5 __?__ 5 __?__ 5 = $\frac{1}{2}$

50. 5 __?__ 5 __?__ 5 = 30 ×, +

47. (5 + 5) × 5 49. 5 ÷ (5 + 5)

Show Math

51. Draw a rectangle and divide it into equal-sized parts in such a way that shows three-fourths is equal to six-eighths. See margin.

Critical Thinking

52. A pile of dimes is counted by 2s, 3s, 4s, and 5s but there is always one remaining. When counted by 7s, none are left over. What is the total value of the coins in the pile? $30.10

Mixed Review

Lesson 1-6

Estimate.

53. 65 + 81 = 150

54. 983 − 426 = 600

55. 6,423 + 718 = 7,100

56. 7,773 − 4,281 = 4,000

Lesson 3-2

Multiply.

57. 6.4 × .02 = 0.128

58. 9.23 × 1.42 = 13.1066

59. 72.6 × 8.3 = 602.58

60. 12.001 × 3.19 = 38.28319

Lesson 3-13

61. Marcus and his two brothers were given 8 boxes of 12 pencils each. They shared the pencils equally. How many pencils did each receive? 32 pencils

APPLYING THE LESSON

Independent Practice

Homework Assignment	
Minimum	Ex. 1-61 odd; 52
Average	Ex. 2-44 even; 45-61
Maximum	Ex. 1-61

Additional Answers

51.

Chalkboard Examples

Example A *Simplify each fraction.*

1. $\frac{12}{18}$ $\frac{2}{3}$ 2. $\frac{16}{20}$ $\frac{4}{5}$ 3. $\frac{6}{12}$ $\frac{1}{2}$

Example B *Simplify each fraction.*

4. $\frac{16}{32}$ $\frac{1}{2}$ 5. $\frac{24}{64}$ $\frac{3}{8}$ 6. $\frac{12}{36}$ $\frac{1}{3}$

Example C, D *Simplify each fraction.*

7. $\frac{3}{12}$ $\frac{1}{4}$ 8. $\frac{7}{8}$ $\frac{7}{8}$ 9. $\frac{25}{40}$ $\frac{5}{8}$

Additional examples are provided on TRANSPARENCY 4-6.

EVALUATING THE LESSON

Check for Understanding Use Guided Practice Exercises to check for student understanding.

Closing the Lesson Ask students to summarize what they have learned about simplifying fractions.

EXTENDING THE LESSON

Enrichment Have students research the Euclidean Algorithm. Then use the Euclidean Algorithm to find the GCF for 168 and 216, 180 and 216, and 735 and 1,925. 24, 36, 35

Enrichment Masters Booklet, p. 38

ENRICHMENT WORKSHEET 4-6

Density of Rational Numbers

Shown below is a portion of the number line containing all points between 0 and $\frac{1}{4}$. The coordinate of point F is $\frac{1}{8}$, which is half of $\frac{1}{4}$.

A B C D E F G H I J
0 $\frac{1}{16}$ $\frac{3}{32}$ $\frac{1}{8}$ $\frac{3}{16}$ $\frac{1}{4}$

The coordinate of point C is $\frac{1}{16}$, which is half of $\frac{1}{8}$.
Between any two rational numbers, there are an unlimited number of rational numbers. This property is called the **density property** of rational numbers.

M N O P Q
$\frac{3}{5}$ $\frac{13}{20}$ $\frac{4}{5}$

$\frac{13}{20} > \frac{3}{5}$ and $\frac{13}{20} < \frac{4}{5}$

The easiest point to locate is the point halfway between the two points. This is half the sum of the two coordinates.

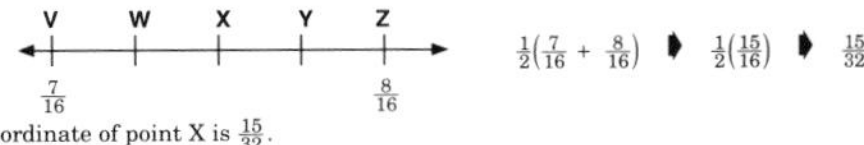

$\frac{1}{2}\left(\frac{7}{16} + \frac{8}{16}\right)$ ➧ $\frac{1}{2}\left(\frac{15}{16}\right)$ ➧ $\frac{15}{32}$

The coordinate of point X is $\frac{15}{32}$.

State the coordinates of the following points.

1. B $\frac{1}{32}$ 2. D $\frac{5}{64}$ 3. G $\frac{5}{32}$ 4. I $\frac{7}{32}$

5. O $\frac{7}{10}$ 6. P $\frac{3}{4}$ 7. W $\frac{29}{64}$ 8. Y $\frac{31}{64}$

Name the rational number that is halfway between the two given numbers on a number line.

9. $\frac{1}{2}$ and $\frac{3}{4}$ $\frac{5}{8}$ 10. 0 and 5 $\frac{5}{2}$ 11. 4 and $5\frac{1}{4}$ $\frac{37}{8}$ or $4\frac{5}{8}$

12. $1\frac{1}{2}$ and $\frac{5}{8}$ $\frac{17}{16}$ or $1\frac{1}{16}$ 13. $\frac{1}{2}$ and $\frac{2}{3}$ $\frac{7}{12}$ 14. $\frac{7}{8}$ and $\frac{8}{9}$ $\frac{129}{144}$

15. On the number line below, locate and label $\frac{2}{3}$, $\frac{3}{4}$, $\frac{3}{5}$, $\frac{1}{3}$, $\frac{1}{2}$, $\frac{1}{5}$, $1\frac{3}{4}$, $\frac{7}{10}$, and $1\frac{1}{20}$.

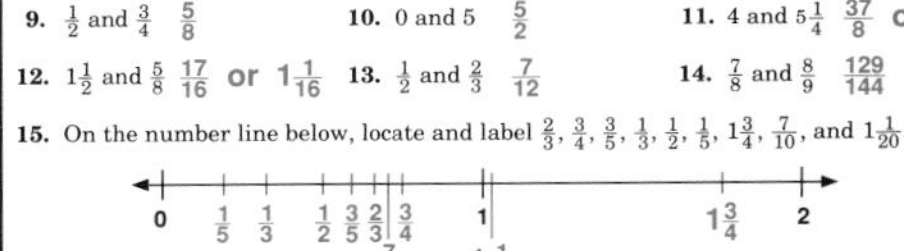

Applying Mathematics to Technology

This optional page shows how mathematics is used in the real world and also provides a change of pace.

Objective Familiarize students with uses of computers.

Using Discussion Ask students how computers could be used in each of the following areas:

- Education: tutoring
- Law Enforcement: aid in dispatching police, information on prisoners, sketching suspects
- Scientific Research: organize and analyze information, help develop theories

Activity

Using Research Have students pick an occupation which interests them and research how computers are used in that occupation. Suggest that students talk to someone in that occupation. Then have students give a short presentation to the class.

USES OF COMPUTERS TODAY

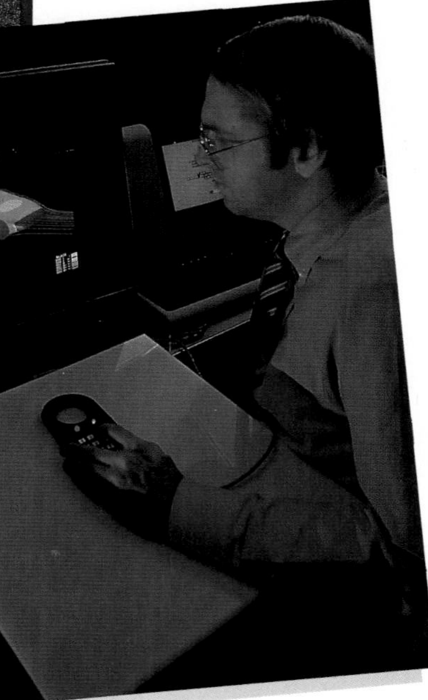

Computers are widely used in today's society. In the medical field, a computer can help a doctor make faster and more accurate diagnoses. It can analyze a patient's medical history and the results of laboratory tests. A computer can help medical researchers study the effects of drugs and procedures before they are used on patients.

In businesses, computers are largely used to help with paperwork, record keeping, and information retrieval. Word-processing software makes it possible for computers to be used in place of typewriters.

Many department stores and supermarkets are using *point-of-sale (POS) terminals* in place of cash registers. Part of a POS terminal is an optical scanning device. This device is used to read a *Universal Product Code (UPC)* that is printed on each item.

Many banks provide *automated teller machines.* These machines can be used to transfer funds from one account to another, deposit funds, withdraw funds, pay utility bills, and get a balance statement. The machines are connected directly to the bank's computer.

Drawings, music, sculpture, and poetry are being generated by computers. Computers are also being used to produce special effects in movies and animated films.

See margin.

1. How are computers being used in the medical field?
2. How are businesses using computers?
3. What do POS and UPC stand for? How are they used by department stores and supermarkets?
4. **Research** How are computers being used in air and space travel?

Additional Answers

1. diagnosis, analyze histories and laboratory results, effects of drugs and procedures.
2. paperwork, record keeping, information retrieval, and word processing.
3. point of sale, Universal Product Code, in place of cash registers.
4. Computers help control air traffic. In space travel, computers are used to simulate space flights and the conditions encountered on such flights. They also help control the actual flight.

USING CHARTS AND GRAPHS

Stacey Mills works in the advertising department of a recordings manufacturer. Her job is to increase the number of customers the company has. She puts together a report that lists the wholesale value of recordings for 1991.

To conserve space and make the numbers easy to read, Stacey writes the numbers in shortened form.

For example, the wholesale value of CD's was $4,337,700,000. Written in shortened form, 4,337,700,000 is 4,337.7 million and read *four billion, three hundred thirty-seven million, seven hundred thousand.* Many newspapers and magazines write large amounts in this way.

Manufacturers' Value of Recordings in 1991 (in millions)	
Singles	63.9
LPs	29.4
CDs	4,337.7
Cassettes	3,019.6
Cassette singles	230.4

LP–Long Play
CD–Compact Disk

Use the appropriate table. Write your answers in standard form.

1. What was the wholesale value of singles in 1991? **$63,900,000**

2. What was the total wholesale value of recordings in the five categories in 1991? **$7,681,000,000**

3. How much more was the wholesale value of cassettes than CDs in 1991? (do not include cassette singles) **$1,318,100,000**

4. In 1970, Joe Namath earned $150,000. How much was that salary worth in 1990? **$502,191**

5. How much more did Joe Montana earn in 1990 than Sammy Baugh earned in 1940? **$3,988,000**

6. What is the difference in salary from what Walter Payton earned in 1980 and what he would earn in 1990? **$273,817**

7. Comparing the salary earned and the 1990 salary, what conclusion can you come to about the value of the dollar? **It has probably decreased in value.**

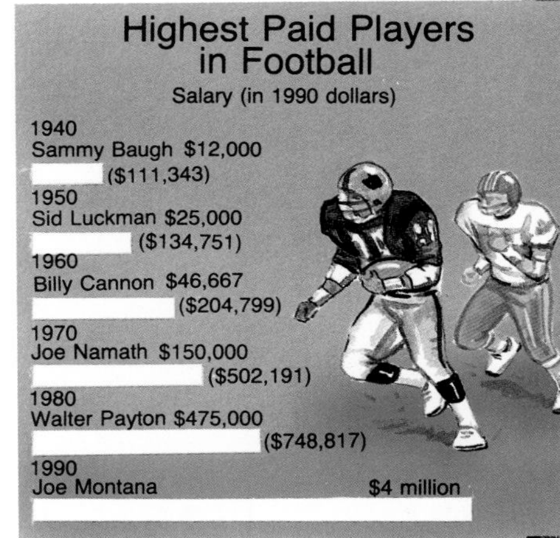

Applying Mathematics to the Real World

This optional page shows how mathematics is used in the real world and also provides a change of pace.

Objective Use charts and graphs to solve problems.

Using Critical Thinking Ask students why someone would use a chart or graph to present information. **The chart or graph presents the information in an organized manner that makes it easier to read.**

Using Questioning What are some types of graphs that can be used to present information? **Bar, line, circle, pictographs, histograms**

Activity

Using Data Have students conduct an opinion survey on an agreed upon topic. Then have each group make and present a chart or graph that represents their results.

5-MINUTE CHECK-UP

(over Lesson 4-6)

Available on TRANSPARENCY 4-7.
Simplify each fraction.

1. $\frac{14}{16}$ $\frac{7}{8}$ **2.** $\frac{6}{6}$ 1 **3.** $\frac{5}{15}$ $\frac{1}{3}$ **4.** $\frac{24}{32}$ $\frac{3}{4}$

Extra Practice, Lesson 4-6, p. 458

▶ INTRODUCING THE LESSON

Ask students to name situations when they compared fractions. (Hint: Who received the larger piece of the candy bar or the cake?)

▶ TEACHING THE LESSON

Using Models Have the students draw a number line to show that one-fourth is less than three-fourths and to show that one-half is greater than one-fourth, but less than three-fourths.

Practice Masters Booklet, p. 42

PRACTICE WORKSHEET 4-7

Comparing Fractions

Write >, <, or = in each circle to make a true sentence.

1. $\frac{1}{3}$ (<) $\frac{3}{4}$	2. $\frac{3}{8}$ (>) $\frac{1}{8}$	3. $\frac{5}{6}$ (>) $\frac{3}{7}$
4. $\frac{1}{2}$ (>) $\frac{2}{7}$	5. $\frac{16}{20}$ (=) $\frac{4}{5}$	6. $\frac{3}{4}$ (>) $\frac{2}{7}$
7. $\frac{3}{5}$ (<) $\frac{4}{5}$	8. $\frac{2}{3}$ (>) $\frac{2}{5}$	9. $\frac{1}{3}$ (<) $\frac{5}{7}$
10. $\frac{2}{7}$ (<) $\frac{4}{7}$	11. $\frac{3}{16}$ (>) $\frac{1}{8}$	12. $\frac{1}{4}$ (<) $\frac{3}{4}$
13. $\frac{1}{6}$ (>) $\frac{1}{7}$	14. $\frac{5}{9}$ (<) $\frac{2}{3}$	15. $\frac{2}{11}$ (=) $\frac{6}{33}$
16. $\frac{3}{5}$ (>) $\frac{3}{8}$	17. $\frac{2}{7}$ (<) $\frac{2}{5}$	18. $\frac{3}{9}$ (=) $\frac{15}{45}$
19. $\frac{3}{5}$ (<) $\frac{2}{3}$	20. $\frac{4}{9}$ (>) $\frac{2}{5}$	21. $\frac{4}{5}$ (>) $\frac{7}{10}$
22. $\frac{3}{4}$ (>) $\frac{3}{5}$	23. $\frac{1}{3}$ (>) $\frac{1}{4}$	24. $\frac{1}{4}$ (>) $\frac{1}{5}$
25. $\frac{5}{6}$ (<) $\frac{6}{7}$	26. $\frac{5}{8}$ (>) $\frac{4}{7}$	27. $\frac{7}{8}$ (<) $\frac{9}{10}$
28. $\frac{2}{3}$ (<) $\frac{5}{6}$	29. $\frac{9}{10}$ (<) $\frac{11}{12}$	30. $\frac{3}{10}$ (<) $\frac{1}{3}$

Order the following fractions from least to greatest.

31. $\frac{2}{3}, \frac{3}{4}, \frac{1}{2}, \frac{1}{8}$ $\frac{1}{8}, \frac{1}{2}, \frac{2}{3}, \frac{3}{4}$ 32. $\frac{1}{3}, \frac{1}{6}, \frac{1}{9}, \frac{1}{2}$ $\frac{1}{9}, \frac{1}{6}, \frac{1}{3}, \frac{1}{2}$

33. $\frac{5}{6}, \frac{2}{3}, \frac{3}{4}, \frac{7}{12}$ $\frac{7}{12}, \frac{2}{3}, \frac{3}{4}, \frac{5}{6}$ 34. $\frac{2}{5}, \frac{3}{10}, \frac{1}{4}, \frac{3}{20}$ $\frac{3}{20}, \frac{1}{4}, \frac{3}{10}, \frac{2}{5}$

4-7 COMPARING FRACTIONS

Objective
Compare and order fractions.

Four-fifths of Earth's surface is covered by water. One-fifth of Earth's surface is covered by land. Is Earth's surface covered mostly by land or water?

To compare fractions with the same denominator, compare the numerators.

Example A ***Compare $\frac{4}{5}$ and $\frac{1}{5}$.***

You can use a number line to compare fractions.

Compare the numerators. $4 > 1$

Since $4 > 1$, it follows that $\frac{4}{5} > \frac{1}{5}$.

Earth's surface is covered mostly by water.

To compare fractions with different denominators, find equivalent fractions that have the same denominator. To do this, find their **least common denominator,** the LCM of the denominators. Then compare the numerators.

Method

1. Find the least common denominator (LCD).
2. Find equivalent fractions using the LCD as the denominator.
3. Compare the numerators.

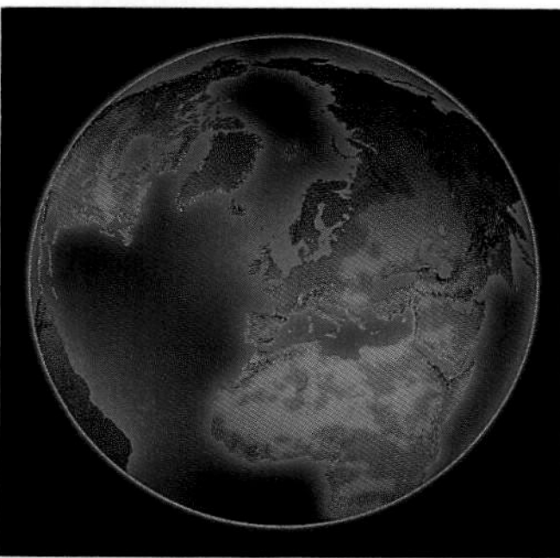

Example B ***Compare $\frac{4}{5}$ and $\frac{5}{6}$.***

1. Find the LCD of 5 and 6. $5 = (5)$
$6 = (2) \times (3)$
LCD: $5 \times 2 \times 3$ or 30

2. $\frac{4}{5} = \frac{24}{30}$ (× 6) $\frac{5}{6} = \frac{25}{30}$ (× 5)

3. Since $24 < 25$, it follows that $\frac{24}{30} < \frac{25}{30}$ or $\frac{4}{5} < \frac{5}{6}$.

Guided Practice ***Replace each ● with <, >, or = to make a true sentence.***

Example A

1. $\frac{2}{3}$ ● $\frac{1}{3}$ > **2.** $\frac{6}{7}$ ● $\frac{4}{7}$ > **3.** $\frac{3}{8}$ ● $\frac{7}{8}$ < **4.** $\frac{11}{16}$ ● $\frac{15}{16}$ < **5.** $\frac{9}{11}$ ● $\frac{3}{11}$ >

Example B

6. $\frac{7}{8}$ ● $\frac{3}{4}$ > **7.** $\frac{1}{3}$ ● $\frac{1}{5}$ > **8.** $\frac{3}{7}$ ● $\frac{4}{5}$ < **9.** $\frac{4}{16}$ ● $\frac{1}{4}$ = **10.** $\frac{5}{12}$ ● $\frac{2}{3}$ <

RETEACHING THE LESSON

Some students may have difficulty renaming fractions using the LCD.

Find the LCD for each pair of fractions. Then rename each fraction using the LCD.

1. $\frac{1}{3}, \frac{3}{4}$ $\frac{4}{12}, \frac{9}{12}$

2. $\frac{5}{6}, \frac{3}{7}$ $\frac{35}{42}, \frac{18}{42}$

3. $\frac{4}{5}, \frac{7}{10}$ $\frac{8}{10}, \frac{7}{10}$

Reteaching Masters Booklet, p. 39

Name ______________________ Date ______________

RETEACHING WORKSHEET 4-7

Comparing Fractions

To compare fractions with different denominators, find equivalent fractions that have the same denominator.

The least common denominator is the easiest to use. It is the LCM of the original denominators.

Compare $\frac{2}{3}$ and $\frac{1}{2}$.
The LCM of 3 and 2 is 6.

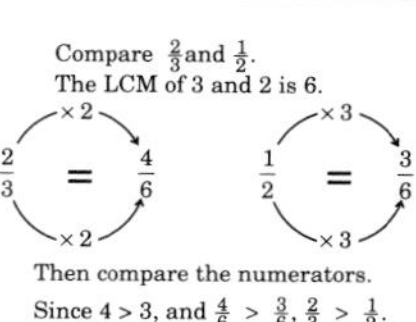

Then compare the numerators.
Since $4 > 3$, and $\frac{4}{6} > \frac{3}{6}$, $\frac{2}{3} > \frac{1}{2}$.

Find the least common denominator for each pair of fractions. Then rename each pair using the LCM.

1. $\frac{1}{3}, \frac{1}{6}$ $\frac{2}{6}, \frac{1}{6}$ 2. $\frac{1}{4}, \frac{1}{8}$ $\frac{2}{8}, \frac{1}{8}$ 3. $\frac{1}{3}, \frac{1}{8}$ $\frac{8}{24}, \frac{3}{24}$ 4. $\frac{7}{8}, \frac{7}{12}$ $\frac{21}{24}, \frac{14}{24}$

Exercises

Replace each ● with <, >, or = to make a true sentence.

Practice

11. $\frac{0}{5}$ ● $\frac{3}{5}$ $<$ **12.** $\frac{8}{9}$ ● $\frac{6}{9}$ $>$ **13.** $\frac{5}{6}$ ● $\frac{2}{6}$ $>$ **14.** $\frac{3}{8}$ ● $\frac{5}{8}$ $<$ **15.** $\frac{7}{10}$ ● $\frac{3}{10}$ $>$

16. $\frac{2}{9}$ ● $\frac{3}{9}$ $<$ **17.** $\frac{1}{12}$ ● $\frac{5}{12}$ $<$ **18.** $\frac{6}{7}$ ● $\frac{5}{7}$ $>$ **19.** $\frac{2}{7}$ ● $\frac{4}{14}$ $=$ **20.** $\frac{3}{4}$ ● $\frac{3}{8}$ $>$

21. $\frac{2}{3}$ ● $\frac{6}{9}$ $=$ **22.** $\frac{7}{10}$ ● $\frac{3}{5}$ $>$ **23.** $\frac{7}{11}$ ● $\frac{3}{11}$ $>$ **24.** $\frac{6}{16}$ ● $\frac{3}{8}$ $=$ **25.** $\frac{5}{7}$ ● $\frac{2}{3}$ $>$

26. $\frac{4}{9}$ ● $\frac{1}{2}$ $<$ **27.** $\frac{3}{8}$ ● $\frac{5}{6}$ $<$ **28.** $\frac{2}{5}$ ● $\frac{2}{3}$ $<$ **29.** $\frac{9}{15}$ ● $\frac{11}{15}$ $<$ **30.** $\frac{3}{5}$ ● $\frac{7}{8}$ $<$

Order the following fractions from least to greatest. **See margin.**

31. $\frac{1}{2}, \frac{1}{3}, \frac{1}{5}, \frac{1}{4}$ **32.** $\frac{3}{8}, \frac{1}{4}, \frac{7}{8}, \frac{3}{4}$ **33.** $\frac{1}{3}, \frac{5}{6}, \frac{3}{8}, \frac{7}{8}$

34. $\frac{5}{6}, \frac{2}{3}, \frac{3}{4}, \frac{4}{5}$ **35.** $\frac{5}{16}, \frac{1}{4}, \frac{1}{2}, \frac{11}{16}$ **36.** $\frac{3}{7}, \frac{2}{5}, \frac{4}{9}, \frac{5}{11}$

Applications

37. Africa

37. About $\frac{1}{12}$ of the world's population lives in North America. About $\frac{1}{8}$ of the population lives in Africa. Which continent has a greater population?

38. In a recent year, $\frac{3}{4}$ of the domestic motor vehicles sold in the U.S. were cars and $\frac{5}{24}$ were trucks. Were more cars or trucks sold in the U.S.? **cars**

39. Bananas are on sale for $0.39 a pound. About how many pounds can Mrs. Barry buy for $2.00? **about 5 pounds**

Critical Thinking

40. Notice the page numbers in the open book. The product of these page numbers is 7,482. Suppose you open the book to two other page numbers whose product is 30,450. To which pages do you have the book open? **174, 175**

Mental Math

Justin Robinson's answers to a math quiz on adding and subtracting decimals are shown below.

***Without using pencil and paper to calculate, determine whether each answer is reasonable. Write* yes *or* no.**

41. 12.1 + 0.65 = <u>14.75</u> **no** **42.** 52.1 − 0.45 = <u>51.65</u> **yes**

43. 8.9 + 13 + 9.5 = <u>19.7</u> **no** **44.** 5 + 5.1 + 23.41 = <u>23.97</u> **no**

45. 53.7 − 36 = <u>17.7</u> **yes** **46.** 35.7 − 24.83 = <u>10.87</u> **yes**

47. 82.6 + 15.25 = <u>97.85</u> **yes** **48.** 9.36 − 2.4 = <u>9.12</u> **no**

49. 10.17 + 9.16 + 21.3 = <u>25.43</u> **no** **50.** 113.52 − 91.65 = <u>21.87</u> **yes**

Chalkboard Examples

Example A *Replace each ● with <, >, or = to make a true sentence.*

1. $\frac{2}{5}$ ● $\frac{3}{5}$ $<$ **2.** $\frac{7}{9}$ ● $\frac{2}{9}$ $>$

Example B *Replace each ● with <, >, or = to make a true sentence.*

3. $\frac{2}{3}$ ● $\frac{3}{5}$ $>$ **4.** $\frac{5}{6}$ ● $\frac{5}{7}$ $>$

Additional examples are provided on TRANSPARENCY 4-7.

EVALUATING THE LESSON

Check for Understanding Use Guided Practice Exercises to check for student understanding.

Error Analysis Watch for students that compare only the numerators or only the denominators when ordering fractions. Stress the need to make the denominators the same before attempting to order the fractions.

Closing the Lesson Ask the students to state in their own words how they would compare two fractions that have different denominators.

EXTENDING THE LESSON

Enrichment Masters Booklet, p. 39

Name ______________________ Date ______________

ENRICHMENT WORKSHEET 4-7

Problem Solving

Solve.

1. About $\frac{2}{25}$ of the world's population lives in North America. About $\frac{1}{10}$ lives in Africa. Which continent has the greater number of people? **Africa**

2. Colorado has about $\frac{17}{50}$ of its land in forests. About $\frac{2}{5}$ of the land in Hawaii is forests. In which state is the greater part of land forests? **Hawaii**

3. Two building lots were sold. One contained $1\frac{3}{5}$ acres and the other $1\frac{7}{10}$ acres. Which lot was larger? **$1\frac{7}{10}$ acres**

4. Mary's best time in the 1,000-meter race is 4 min $40\frac{1}{5}$ seconds. Karen's best time is 4 min $40\frac{3}{10}$ seconds. Who had the better time? **Mary**

Solve these problems from the early 1900s.

5. A boy worked for $\frac{2}{3}$ of a day for Mr. Adams and $\frac{3}{4}$ of a day for Mr. May. For whom did he work longer? **Mr. May**

6. A girl paid $\$1\frac{3}{4}$ for a sled, $\$\frac{1}{2}$ for a book, and $\$1\frac{4}{5}$ for a doll. Which purchase cost the most? **doll**

7. A dressmaker deposited in a savings book $\$3\frac{2}{5}$ on Monday, $\$3\frac{1}{2}$ on Tuesday, and $\$3\frac{7}{10}$ on Wednesday. On which day did she deposit the most money? **Wednesday**

8. The expenses of a party were $\$3\frac{4}{5}$ for railroad tickets, $\$3\frac{3}{4}$ for carriages, and $\$3\frac{7}{10}$ for provisions. Which expense was the greatest? **railroad tickets**

9. A grocer sold $1\frac{1}{2}$ dozen eggs to person A, $1\frac{2}{3}$ dozen to person B, and $1\frac{5}{6}$ dozen to person C. To which person did he sell most eggs? **to person C**

10. A man planted $\frac{5}{6}$ of an acre in corn, $\frac{2}{3}$ of an acre in potatoes, and $\frac{4}{5}$ of an acre in wheat. Of which crop did he plant the most? **corn**

APPLYING THE LESSON

Independent Practice

Homework Assignment	
Minimum	**Ex. 2-50 even**
Average	**Ex. 2-36 even; 37-50**
Maximum	**Ex. 1-50**

Chapter 4, Quiz A (Lesson 4-1 through 4-7) is available in the Evaluation Masters Booklet, p. 19.

Additional Answers

31. $\frac{1}{5}, \frac{1}{4}, \frac{1}{3}, \frac{1}{2}$

32. $\frac{1}{4}, \frac{3}{8}, \frac{3}{4}, \frac{7}{8}$

33. $\frac{1}{3}, \frac{3}{8}, \frac{5}{6}, \frac{7}{8}$

34. $\frac{2}{3}, \frac{3}{4}, \frac{4}{5}, \frac{5}{6}$

35. $\frac{1}{4}, \frac{5}{16}, \frac{1}{2}, \frac{11}{16}$

36. $\frac{2}{5}, \frac{3}{7}, \frac{4}{9}, \frac{5}{11}$

5-MINUTE CHECK-UP

(over Lesson 4-7)

Available on TRANSPARENCY 4-8.
Replace each ● with <, >, or = to make a true sentence.

1. $\frac{3}{4}$ ● $\frac{5}{6}$ < **2.** $\frac{4}{16}$ ● $\frac{1}{4}$ =

3. $\frac{2}{3}$ ● $\frac{5}{16}$ > **4.** $\frac{11}{16}$ ● $\frac{5}{8}$ >

Extra Practice, Lesson 4-7, p. 458

▶ INTRODUCING THE LESSON

Ask students to name different ways that they tell time or how they state how much time has passed from one event to another. (a quarter past two means $2\frac{1}{4}$ hours after twelve.

▶ TEACHING THE LESSON

Using Manipulatives Give each student 9 wedges from the **Lab Manual, fraction models, page 4.** Each wedge represents $\frac{1}{6}$ of a circle. Have students use the wedges to form a circle, using extra wedges to form part of another circle. Ask students to name the improper fraction.

Practice Masters Booklet, p. 43

Name ______________________ Date ______________

PRACTICE WORKSHEET 4-8

Mixed Numbers

Change each fraction to a mixed number in simplest form.

1. $\frac{9}{5}$ $1\frac{4}{5}$	**2.** $\frac{7}{6}$ $1\frac{1}{6}$	**3.** $\frac{19}{12}$ $1\frac{7}{12}$	**4.** $\frac{14}{10}$ $1\frac{2}{5}$
5. $\frac{15}{9}$ $1\frac{2}{3}$	**6.** $\frac{7}{2}$ $3\frac{1}{2}$	**7.** $\frac{3}{2}$ $1\frac{1}{2}$	**8.** $\frac{5}{4}$ $1\frac{1}{4}$
9. $\frac{9}{7}$ $1\frac{2}{7}$	**10.** $\frac{9}{2}$ $4\frac{1}{2}$	**11.** $\frac{11}{4}$ $2\frac{3}{4}$	**12.** $\frac{12}{5}$ $2\frac{2}{5}$
13. $\frac{6}{4}$ $1\frac{1}{2}$	**14.** $\frac{9}{6}$ $1\frac{1}{2}$	**15.** $\frac{21}{15}$ $1\frac{2}{5}$	**16.** $\frac{55}{16}$ $3\frac{7}{16}$
17. $\frac{16}{3}$ $5\frac{1}{3}$	**18.** $\frac{14}{5}$ $2\frac{4}{5}$	**19.** $\frac{24}{20}$ $1\frac{1}{5}$	**20.** $\frac{22}{6}$ $3\frac{2}{3}$

Change each mixed number to an improper fraction.

21. $3\frac{1}{16}$ $\frac{49}{16}$	**22.** $2\frac{3}{4}$ $\frac{11}{4}$	**23.** $1\frac{3}{8}$ $\frac{11}{8}$	**24.** $1\frac{5}{12}$ $\frac{17}{12}$
25. $7\frac{3}{5}$ $\frac{38}{5}$	**26.** $6\frac{5}{8}$ $\frac{53}{8}$	**27.** $3\frac{1}{3}$ $\frac{10}{3}$	**28.** $1\frac{7}{9}$ $\frac{16}{9}$
29. $2\frac{3}{16}$ $\frac{35}{16}$	**30.** $1\frac{2}{3}$ $\frac{5}{3}$	**31.** $3\frac{3}{10}$ $\frac{33}{10}$	**32.** $4\frac{3}{25}$ $\frac{103}{25}$
33. $4\frac{2}{5}$ $\frac{22}{5}$	**34.** $6\frac{1}{2}$ $\frac{13}{2}$	**35.** $4\frac{5}{6}$ $\frac{29}{6}$	**36.** $1\frac{1}{100}$ $\frac{101}{100}$
37. $2\frac{5}{8}$ $\frac{21}{8}$	**38.** $3\frac{1}{6}$ $\frac{19}{6}$	**39.** $4\frac{3}{5}$ $\frac{23}{5}$	**40.** $1\frac{49}{50}$ $\frac{99}{50}$

Solve.

41. How many pounds of margarine does Joel have if he has seven quarter-pound sticks?
$1\frac{3}{4}$ pounds

42. How many quarters did Nancy play if she played in $2\frac{3}{4}$ basketball games?
11 quarters

4-8 MIXED NUMBERS

Objective
Change improper fractions to mixed numbers and vice versa.

A survey showed that two-thirds of the people attending an event at the Fort Wayne Coliseum were Indianapolis Colts fans.

A fraction, like $\frac{2}{3}$, that has a numerator that is less than the denominator is a **proper fraction.**

A fraction, like $\frac{7}{4}$ or $\frac{5}{5}$, that has a numerator that is greater than or equal to the denominator is an **improper fraction.**

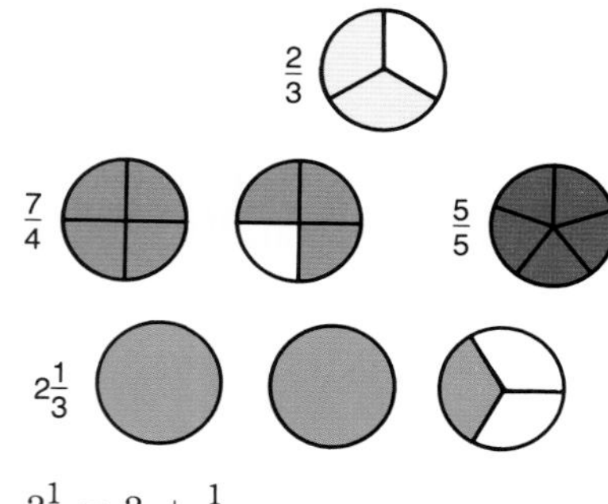

A **mixed number**, like $2\frac{1}{3}$, indicates the sum of a whole number and a fraction.

$2\frac{1}{3} = 2 + \frac{1}{3}$

An improper fraction may be changed to a mixed number by dividing the numerator by the denominator.

Remember that the bar separating the numerator and the denominator indicates division.

Examples ***Change each fraction to a mixed number in simplest form.***

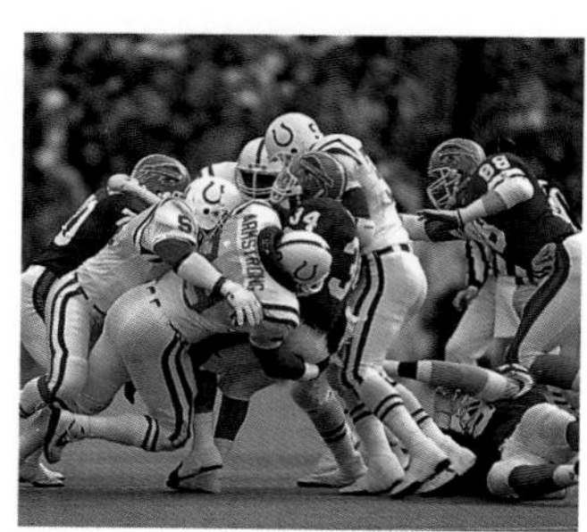

A $\frac{18}{4}$

$$\begin{array}{r} 4 \text{ R2} \\ 4\overline{)18} \\ -\ 16 \\ \hline 2 \end{array} \longrightarrow 4\frac{2}{4} = 4\frac{1}{2}$$

So $\frac{18}{4} = 4\frac{1}{2}$.

B $\frac{30}{6}$

$$\begin{array}{r} 5 \\ 6\overline{)30} \\ -\ 30 \\ \hline 0 \end{array} \longrightarrow 5$$

The remainder is 0.

So $\frac{30}{6} = 5$.

You can use the following method to change a mixed number to an improper fraction.

Method

1. Multiply the whole number by the denominator and add the numerator.
2. Write the sum over the denominator.

Example C ***Change $3\frac{2}{5}$ to an improper fraction.***

$$3\frac{2}{5} \xrightarrow{\ 1\ } \frac{(3 \times 5) + 2}{5} \xrightarrow{\ 2\ } \frac{17}{5}$$

Explain how this diagram shows that $3\frac{2}{5} = \frac{17}{5}$.

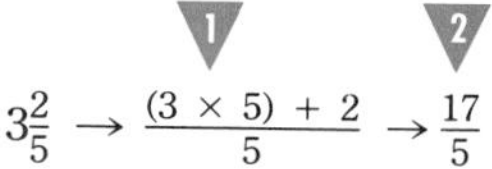

RETEACHING THE LESSON

Use number lines to change an improper fraction to a mixed number and to change a mixed number to an improper fraction.

Change $\frac{9}{4}$ to a mixed number.

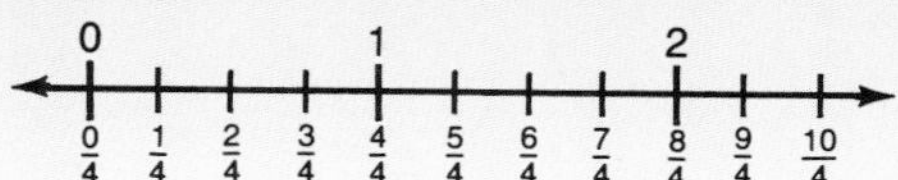

Now, count by whole units, $\frac{9}{4} = 2\frac{1}{4}$.

Reteaching Masters Booklet, p. 40

Name ______________________ Date ______________

RETEACHING WORKSHEET 4-8

Mixed Numbers

A fraction that has a numerator that is less than the denominator is called a **proper fraction.**

$\frac{2}{3}$ $\frac{5}{8}$ $\frac{3}{10}$

A fraction that has a numerator that is greater than or equal to the denominator is an **improper fraction.**

$\frac{5}{5}$ $\frac{6}{2}$

A **mixed number** indicates the sum of a whole number and a fraction.

$3\frac{1}{3}$

$3\frac{1}{3} = 3 + \frac{1}{3}$

Divide to change an improper fraction such as $\frac{14}{4}$ to a mixed number.

Multiply and add to change a mixed number such as $2\frac{3}{5}$ to an improper fraction.

Guided Practice

Change each fraction to a mixed number in simplest form.

Example A
1. $\frac{5}{3}$ $1\frac{2}{3}$ 2. $\frac{7}{6}$ $1\frac{1}{6}$ 3. $\frac{9}{6}$ $1\frac{1}{2}$ 4. $\frac{42}{8}$ $5\frac{1}{4}$ 5. $\frac{24}{9}$ $2\frac{2}{3}$ 6. $\frac{14}{5}$ $2\frac{4}{5}$

Example B
7. $\frac{6}{6}$ 1 8. $\frac{12}{4}$ 3 9. $\frac{10}{5}$ 2 10. $\frac{36}{9}$ 4 11. $\frac{42}{7}$ 6 12. $\frac{81}{9}$ 9

Example C
Change each mixed number to an improper fraction.

13. $1\frac{3}{8}$ $\frac{11}{8}$ 14. $1\frac{7}{12}$ $\frac{19}{12}$ 15. $3\frac{2}{3}$ $\frac{11}{3}$ 16. $4\frac{2}{5}$ $\frac{22}{5}$ 17. $3\frac{5}{7}$ $\frac{26}{7}$ 18. $11\frac{2}{3}$ $\frac{35}{3}$

Exercises

Practice
Change each fraction to a mixed number in simplest form.

19. $\frac{8}{7}$ $1\frac{1}{7}$ 20. $\frac{9}{5}$ $1\frac{4}{5}$ 21. $\frac{21}{7}$ 3 22. $\frac{7}{4}$ $1\frac{3}{4}$ 23. $\frac{9}{7}$ $1\frac{2}{7}$ 24. $\frac{19}{8}$ $2\frac{3}{8}$

25. $\frac{11}{5}$ $2\frac{1}{5}$ 26. $\frac{12}{8}$ $1\frac{1}{2}$ 27. $\frac{16}{6}$ $2\frac{2}{3}$ 28. $\frac{33}{9}$ $3\frac{2}{3}$ 29. $\frac{76}{9}$ $8\frac{4}{9}$ 30. $\frac{40}{24}$ $1\frac{2}{3}$

Change each mixed number to an improper fraction.

31. $1\frac{2}{5}$ $\frac{7}{5}$ 32. $2\frac{3}{4}$ $\frac{11}{4}$ 33. $1\frac{7}{8}$ $\frac{15}{8}$ 34. $2\frac{1}{2}$ $\frac{5}{2}$ 35. $2\frac{9}{10}$ $\frac{29}{10}$ 36. $3\frac{5}{8}$ $\frac{29}{8}$

Write Math
Write the name of each fraction in words.

37. $\frac{1}{2}$ one-half 38. $1\frac{2}{3}$ one and two-thirds 39. $2\frac{7}{8}$ two and seven-eighths 40. $\frac{9}{5}$ nine-fifths

Using Expressions
41. If $2 + \frac{1}{3}$ can be written as the improper fraction $\frac{7}{3}$, how would you write $n + \frac{1}{3}$ as an improper fraction? $\frac{3n + 1}{3}$

Critical Thinking
42. Without bending or breaking the sticks, how would you place six ice cream sticks so that each of the six sticks touches each of the others? See margin.

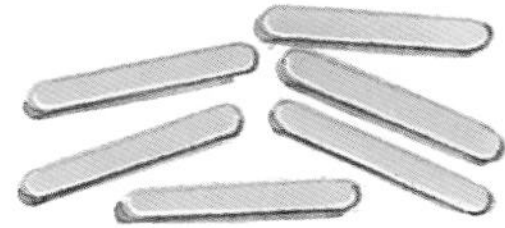

Applications
43. Two lots for home building were sold. One lot contained $1\frac{3}{5}$ acres and the other $1\frac{7}{10}$ acres. Which lot was larger? $1\frac{7}{10}$ acres

44. A dress pattern calls for $3\frac{5}{8}$ yards of material. Melissa bought $3\frac{3}{4}$ yards. Did she buy enough material? yes, $3\frac{3}{4} > 3\frac{5}{8}$

45. Bart plans to set up a lawn mowing business for his summer job. He can buy a used lawn mower for $50. If he saves $4.75 a week, how long will it take him to save for the mower? 11 weeks

Mixed Review
Lesson 1-9
Find the value of each expression.

46. $16 \times 2 \div 8 + 20$ 24 47. $9 \div 3 + 14 \div 7 + 18$ 23

Lesson 3-9
Find the common ratio. Write the next 3 terms in each sequence.

48. 5.6, 11.2, 22.4 44.8, 89.6, 179.2 49. 729, 243, 81 27, 9, 3

Lesson 4-6
50. Four months out of the year, Mrs. Joel lives in Florida. The rest of the year Mrs. Joel lives in Minnesota. What fraction, in simplest form, of the year does Mrs. Joel live in Florida? $\frac{1}{3}$ year

APPLYING THE LESSON

Independent Practice

Homework Assignment	
Minimum	Ex. 2-50 even
Average	Ex. 1-35 odd; 37-50
Maximum	Ex. 1-50

Additional Answers

42.

Chalkboard Examples

Example A *Change each fraction to a mixed number in simplest form.*

1. $\frac{24}{5}$ $4\frac{4}{5}$ 2. $\frac{20}{6}$ $3\frac{1}{3}$ 3. $\frac{14}{3}$ $4\frac{2}{3}$

Example B *Change each fraction to a mixed number in simplest form.*

4. $\frac{6}{6}$ 1 5. $\frac{24}{8}$ 3 6. $\frac{72}{9}$ 8

Example C *Change each mixed number to an improper fraction.*

7. $4\frac{2}{3}$ $\frac{14}{3}$ 8. $5\frac{3}{8}$ $\frac{43}{8}$ 9. $2\frac{7}{9}$ $\frac{25}{9}$

Additional examples are provided on TRANSPARENCY 4-8.

EVALUATING THE LESSON

Check for Understanding Use Guided Practice Exercises to check for student understanding.

Closing the Lesson Ask students to summarize how to change improper fractions to mixed numbers.

EXTENDING THE LESSON

Enrichment Have students write each of the following as a mixed number.

1. 6 days and 30 hours = ___?___ days $7\frac{1}{4}$
2. 1 year and 4 months = ___?___ years $1\frac{1}{3}$

Enrichment Masters Booklet, p. 40

Name ______________ Date ______________

ENRICHMENT WORKSHEET 4-8

Mixed Numbers

When comparing mixed numbers, first look at the whole numbers. If they are different, there is no need to compare the fractions.

Example: $6\frac{2}{3} \square 5\frac{1}{2}$

$6 > 5$, so $6\frac{2}{3} > 5\frac{1}{2}$

If they are the same, then compare only the fractions using the LCD and equivalent fractions.

Example: $5\frac{1}{3} \square 5\frac{6}{17}$

$\frac{1}{3} \square \frac{6}{17} \rightarrow \frac{17}{51} < \frac{18}{51}$

$\frac{17}{51} < \frac{18}{51}$ so $5\frac{1}{3} < 5\frac{6}{17}$

Fill in each blank with >, <, or = to make a true sentence.

1. $3\frac{3}{5}$ _>_ $3\frac{1}{5}$ 2. $2\frac{3}{4}$ _<_ $3\frac{1}{7}$ 3. $6\frac{1}{5}$ _<_ $6\frac{2}{7}$

4. $11\frac{2}{3}$ _>_ $11\frac{5}{8}$ 5. $4\frac{3}{5}$ _=_ $4\frac{6}{10}$ 6. $18\frac{7}{8}$ _<_ $19\frac{5}{6}$

7. $35\frac{1}{3}$ _<_ $35\frac{2}{5}$ 8. $1\frac{6}{9}$ _>_ $1\frac{6}{12}$ 9. $42\frac{7}{9}$ _>_ $42\frac{5}{7}$

10. $2\frac{3}{4}$ _>_ $1\frac{5}{6}$ 11. $64\frac{2}{3}$ _<_ $64\frac{3}{4}$ 12. $3\frac{2}{7}$ _>_ $3\frac{1}{5}$

5-MINUTE CHECK-UP

(over Lesson 4-8)

Available on TRANSPARENCY 4-9.

Change each fraction to a mixed number in simplest form.

1. $\frac{12}{5}$ $2\frac{2}{5}$ **2.** $\frac{27}{12}$ $2\frac{1}{4}$ **3.** $\frac{35}{4}$ $8\frac{3}{4}$

Change each mixed number to an improper fraction.

4. $4\frac{2}{3}$ $\frac{14}{3}$ **5.** $8\frac{1}{2}$ $\frac{17}{2}$ **6.** $7\frac{4}{5}$ $\frac{39}{5}$

Extra Practice, Lesson 4-8, p. 458

▶ INTRODUCING THE LESSON

Ask students if they have ever baked cookies that called for nuts or raisins. Since these items cannot be purchased by the cup or parts of a cup, ask how they decided how much they needed.

▶ TEACHING THE LESSON

Using Cooperative Groups Have the students work in groups of three and discuss the rounding process while working the Guided Practice exercises.

Practice Masters Booklet, p. 44

Name ______________________ Date ______________

PRACTICE WORKSHEET 4-9

Estimating Sums and Differences

Estimate.

Answers may vary. Typical answers are given.

1.	$\frac{7}{16}+\frac{2}{9}$	$\frac{1}{2}$	**2.**	$\frac{5}{6}+\frac{1}{3}$	1
3.	$\frac{1}{4}-\frac{1}{3}$	0	**4.**	$\frac{2}{15}+2\frac{1}{20}$	2
5.	$\frac{24}{25}-\frac{1}{2}$	$\frac{1}{2}$	**6.**	$1\frac{1}{3}+\frac{2}{5}$	$1\frac{1}{2}$
7.	$\frac{4}{5}-\frac{1}{8}$	1	**8.**	$\frac{5}{8}+\frac{4}{9}$	1
9.	$3\frac{2}{5}-1\frac{1}{4}$	$2\frac{1}{2}$	**10.**	$\frac{11}{12}-\frac{1}{3}$	1
11.	$4\frac{2}{5}+\frac{5}{6}$	$5\frac{1}{2}$	**12.**	$6\frac{7}{8}-\frac{2}{3}$	6
13.	$4\frac{7}{12}-1\frac{3}{4}$	$2\frac{1}{2}$	**14.**	$9\frac{7}{10}+\frac{4}{5}$	$10\frac{1}{2}$
15.	$4\frac{2}{3}+10\frac{3}{8}$	$15\frac{1}{2}$	**16.**	$18\frac{1}{4}-12\frac{3}{5}$	$5\frac{1}{2}$
17.	$7\frac{7}{15}-3\frac{1}{12}$	$4\frac{1}{2}$	**18.**	$12\frac{5}{9}+8\frac{5}{8}$	21

4-9 ESTIMATING SUMS AND DIFFERENCES

Objective
Estimate the sum and difference of fractions.

Michael volunteers a couple of hours each Saturday at the Village Retirement Center. One Saturday he and Mr. Svenson walked $\frac{9}{10}$ mile to attend a festival in the park. On the return trip, they took a shortcut that was a $\frac{4}{5}$-mile walk. *About* how far did they walk to the park and back? (The problem is solved in Example F.)

To round fractions before estimating their sum or difference, you can use the following guidelines.

- If the numerator is much less than the denominator, round the fraction to 0.
- If the numerator is about half of the denominator, round the fraction to $\frac{1}{2}$.
- If the numerator is a little less than the denominator, round the fraction to 1.

Examples

State whether each fraction is close to 0, $\frac{1}{2}$, or 1. You can see that $\frac{1}{5}$ is close to 0, $\frac{9}{16}$ is close to $\frac{1}{2}$, and $\frac{4}{5}$ is close to 1.

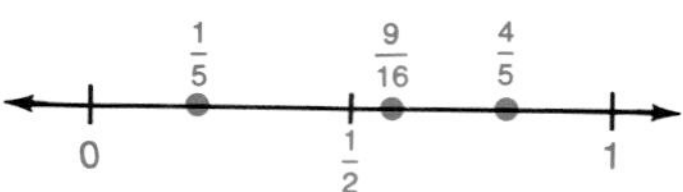

A $\frac{4}{5} \rightarrow 1$ 4 is a little less than 5.

B $\frac{9}{16} \rightarrow \frac{1}{2}$ 9 is about half of 16.

C $\frac{1}{5} \rightarrow 0$ 1 is much less than 5.

Mixed numbers are usually rounded to the nearest whole number.

Examples

Round each mixed number to the nearest whole number.

D $1\frac{5}{6} \rightarrow 2$

E $14\frac{5}{16} \rightarrow 14$

You can estimate sums and differences of fractions and mixed numbers.

Method

1 Round each fraction or mixed number.

2 Add or subtract.

Examples

F ***To find out about how far Michael and Mr. Svenson walked, estimate $\frac{9}{10}+\frac{4}{5}$.***

1 $\frac{9}{10} \rightarrow 1$ $\frac{4}{5} \rightarrow 1$

2 $1 + 1 = 2$

Michael and Mr. Svenson walked about 2 miles.

G ***Estimate $2\frac{3}{4}-\frac{3}{8}$.***

1 $2\frac{3}{4} \rightarrow 3$ $\frac{3}{8} \rightarrow \frac{1}{2}$

2 $3 - \frac{1}{2} = 2\frac{1}{2}$

$2\frac{3}{4} - \frac{3}{8}$ is about $2\frac{1}{2}$.

RETEACHING THE LESSON

1. State whether each fraction is closer to 0, to $\frac{1}{2}$ or to 1.

$\frac{4}{5}, \frac{7}{9}, \frac{3}{8}, \frac{2}{3}, \frac{1}{6}, \frac{7}{8}, \frac{3}{10}, \frac{5}{8}, \frac{7}{10}, \frac{5}{6}$

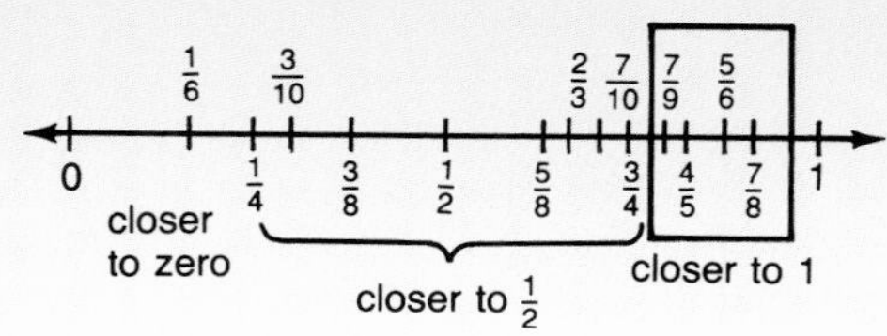

Reteaching Masters Booklet, p. 41

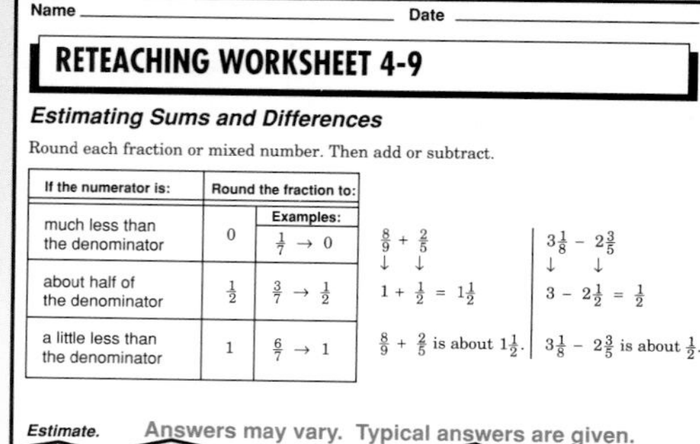

Name ______________________ Date ______________

RETEACHING WORKSHEET 4-9

Estimating Sums and Differences

Round each fraction or mixed number. Then add or subtract.

If the numerator is:	Round the fraction to:	Examples:
much less than the denominator	0	$\frac{1}{7} \rightarrow 0$
about half of the denominator	$\frac{1}{2}$	$\frac{3}{7} \rightarrow \frac{1}{2}$
a little less than the denominator	1	$\frac{6}{7} \rightarrow 1$

$\frac{8}{9} + \frac{2}{5}$
↓ ↓
$1 + \frac{1}{2} = 1\frac{1}{2}$

$\frac{8}{9} + \frac{2}{5}$ is about $1\frac{1}{2}$.

$3\frac{1}{8} - 2\frac{3}{5}$
↓ ↓
$3 - 2\frac{1}{2} = \frac{1}{2}$

$3\frac{1}{8} - 2\frac{3}{5}$ is about $\frac{1}{2}$.

Estimate. Answers may vary. Typical answers are given.

Guided Practice

Examples A-C

State whether each fraction is close to 0, $\frac{1}{2}$, or 1. Use a number line to help.

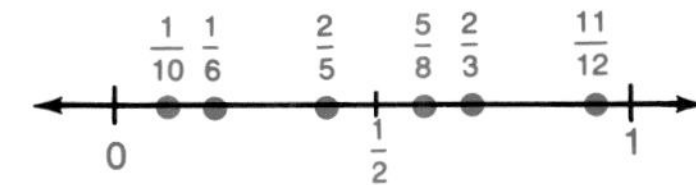

1. $\frac{11}{12}$ 1 **2.** $\frac{5}{8}$ $\frac{1}{2}$ **3.** $\frac{2}{5}$ $\frac{1}{2}$ **4.** $\frac{1}{10}$ 0 **5.** $\frac{1}{6}$ 0 **6.** $\frac{2}{3}$ 1

Examples D, E

Round each mixed number to the nearest whole number.

7. $6\frac{2}{3}$ 7 **8.** $18\frac{1}{4}$ 18 **9.** $27\frac{5}{8}$ 28 **10.** $32\frac{2}{5}$ 32 **11.** $12\frac{7}{9}$ 13 **12.** $20\frac{1}{3}$ 20

Estimate. Answers may vary. Typical answers are given.

Example F

13. $\frac{9}{16} + \frac{13}{24}$ 1 **14.** $\frac{5}{6} + \frac{4}{9}$ $1\frac{1}{2}$ **15.** $3\frac{1}{4} + \frac{2}{3}$ 4 **16.** $5\frac{1}{2} + 6\frac{7}{12}$ 12

Example G

17. $\frac{5}{12} - \frac{1}{10}$ $\frac{1}{2}$ **18.** $\frac{4}{5} - \frac{1}{3}$ 1 **19.** $3\frac{2}{5} - \frac{7}{8}$ $2\frac{1}{2}$ **20.** $3\frac{5}{11} - 2\frac{1}{2}$ 1

Exercises

Practice

Estimate. Answers may vary. Typical answers are given.

21. $\frac{2}{5} + \frac{1}{3}$ $\frac{1}{2}$ **22.** $\frac{1}{5} + \frac{4}{10}$ $\frac{1}{2}$ **23.** $\frac{1}{6} + \frac{4}{9}$ $\frac{1}{2}$ **24.** $\frac{3}{4} + \frac{7}{16}$ $1\frac{1}{2}$

25. $\frac{5}{6} - \frac{7}{12}$ $\frac{1}{2}$ **26.** $\frac{5}{8} - \frac{1}{10}$ $\frac{1}{2}$ **27.** $12\frac{3}{4} - \frac{3}{8}$ $12\frac{1}{2}$ **28.** $7\frac{2}{3} - \frac{3}{5}$ 7

29. $14\frac{1}{10} - 6\frac{4}{5}$ 7 **30.** $8\frac{1}{3} + 2\frac{1}{6}$ 10 **31.** $4\frac{7}{8} + 7\frac{3}{4}$ 13 **32.** $11\frac{11}{12} - 5\frac{1}{4}$ 7

Make Up a Problem

33. Make up a problem that has two addends with an estimated sum of 1. Use fractions and/or mixed numbers. Answers may vary.

Applications

34. Carlos runs $1\frac{1}{2}$ miles on Monday, $2\frac{1}{3}$ miles on Tuesday, and $1\frac{5}{6}$ miles on Wednesday. About how many miles did he run in all?

34. about $5\frac{1}{2}$ miles

35. Which race is longer, a $1\frac{7}{16}$-mile race or a $1\frac{3}{8}$-mile race?

35. $1\frac{7}{16}$-mile race

36. A pane of glass costs 99¢. What does it cost to buy glass to replace six windows that have four panes each? $23.76

FUN with MATH

Ready for a game of Korean tic tac toe? See page 175.

Mixed Review

Write as a product and then find the number named.

Lesson 1-2

37. 4 cubed **38.** 2^6 **39.** 5 squared **40.** 8^2

37. 4 × 4 × 4 = 64
38. 2 × 2 × 2 × 2 × 2 × 2 = 64
39. 5 × 5 = 25
40. 8 × 8 = 64

Lesson 1-7

Estimate.

41. 16 × 8 = 200 **42.** 0.6)12 = 12 **43.** 481 × 322 = 150,000 **44.** 22.1)437.2 = 22

Lesson 2-8

45. Marcia walks 5 miles each day. She walks 0.5 mile on Elm St., 0.25 mile on Oak St., 0.75 mile on Main St., and 1.00 mile on Town St. How many times does she walk this route each day? 2 times

Chalkboard Examples

Examples A-C *State whether each fraction is closer to 0, $\frac{1}{2}$, or 1.*

1. $\frac{9}{10}$ 1 **2.** $\frac{1}{8}$ 0 **3.** $\frac{4}{9}$ $\frac{1}{2}$

Examples D-E *Round each mixed number to the nearest whole number.*

4. $8\frac{1}{2}$ 9 **5.** $14\frac{1}{9}$ 14 **6.** $22\frac{7}{8}$ 23

Example F-G *Estimate.*

7. $\frac{2}{3} + \frac{1}{10}$ $\frac{2}{3}$ **8.** $8\frac{2}{3} - 5\frac{1}{2}$ 3

Additional examples are provided on TRANSPARENCY 4-9.

EVALUATING THE LESSON

Check for Understanding Use Guided Practice Exercises to check for student understanding.

Closing the Lesson Tell the students that they will estimate sums and differences of fractions to check the reasonableness of computed sums and differences.

EXTENDING THE LESSON

Enrichment Estimate the sum of the fractions in this series.

$\frac{1}{3} + \frac{2}{3} + \frac{3}{3} + , . . . , + \frac{9}{3}$ 15

Enrichment Masters Booklet, p. 41

Name ________ Date ________

ENRICHMENT WORKSHEET 4-9

Estimating with Fractions

Estimate whether the answer is closer to 0, $\frac{1}{2}$ or 1. Check with your calculator.

Answers may vary. Typical answers are given. Checks are rounded to the nearest tenth.

Example: $\frac{3}{7} + \frac{4}{5}$ = (3 ÷ 7) + (4 ÷ 5) = 1.23 (rounded)

1. $\frac{6}{17} + \frac{2}{9}$	$\frac{1}{2}$; 0.58	**2.** $\frac{5}{7} + \frac{1}{3}$	1; 1.05
3. $\frac{1}{4} + \frac{1}{3}$	$\frac{1}{2}$; 0.58	**4.** $\frac{2}{11} + \frac{1}{19}$	0; 0.23
5. $\frac{6}{25} + \frac{3}{4}$	1; 0.99	**6.** $\frac{9}{13} + \frac{2}{5}$	1; 1.09
7. $\frac{4}{5} + \frac{1}{8}$	1; 0.93	**8.** $\frac{6}{19} + \frac{4}{11}$	$\frac{1}{2}$; 0.68
9. $\frac{1}{21} + \frac{7}{62} + \frac{1}{51}$	0; 0.18	**10.** $\frac{9}{38} + \frac{1}{16} + \frac{1}{3}$	$\frac{1}{2}$; 0.63
11. $\frac{2}{15} + \frac{3}{11} + \frac{6}{77}$	$\frac{1}{2}$; 0.48	**12.** $\frac{2}{17} + \frac{3}{28} + \frac{5}{11}$	$\frac{1}{2}$; 0.68
13. $\frac{6}{53} + \frac{2}{29} + \frac{1}{97}$	0; 0.19	**14.** $\frac{6}{103} + \frac{4}{75} + \frac{6}{85}$	0; 0.18
15. $\frac{19}{37} + \frac{2}{33} + \frac{3}{7}$	1; 1.00	**16.** $\frac{16}{47} + \frac{25}{74} + \frac{3}{13}$	1; 0.91
17. $\frac{8}{107} + \frac{7}{93} + \frac{8}{91} + \frac{2}{101}$	$\frac{1}{2}$; 0.26	**18.** $\frac{4}{37} + \frac{5}{81} + \frac{1}{84} + \frac{7}{85}$	$\frac{1}{2}$; 0.26

APPLYING THE LESSON

Independent Practice

Homework Assignment	
Minimum	Ex. 1-45 odd
Average	Ex. 2-32 even; 33-45
Maximum	Ex. 1-45

5-MINUTE CHECK-UP

(over Lesson 4-9)

Available on TRANSPARENCY 4-10.

Estimate.

1. $\frac{1}{4} + \frac{2}{3}$ 1
2. $\frac{15}{16} - \frac{1}{8}$ 1
3. $2\frac{1}{3} + 3\frac{5}{6}$ 6
4. $1\frac{3}{7} - \frac{1}{9}$ $1\frac{1}{2}$

Extra Practice, Lesson 4-9, p. 459

INTRODUCING THE LESSON

Ask students to tell about experiences they have had with sewing or woodworking. Ask if anyone has ever tried to double a recipe.

TEACHING THE LESSON

Using Models Have students use number lines from **Lab Manual, page 7.** Have students label them in fourths and whole numbers. Ask students to show how they would represent $\frac{1}{4} + \frac{1}{4} = \frac{2}{4}$ or $\frac{1}{2}$.

Practice Masters Booklet, p. 45

Name ____________ Date ____________

PRACTICE WORKSHEET 4-10

Adding Fractions

Add.

1. $\frac{4}{7} + \frac{2}{7}$	$\frac{6}{7}$	2. $\frac{6}{9} + \frac{3}{9}$	1	3. $\frac{11}{15} + \frac{2}{15}$	$\frac{13}{15}$
4. $\frac{11}{15} + \frac{7}{15}$	$1\frac{1}{5}$	5. $\frac{15}{20} + \frac{7}{20}$	$1\frac{1}{10}$	6. $\frac{9}{11} + \frac{8}{11}$	$1\frac{6}{11}$
7. $\frac{12}{20} + \frac{7}{20}$	$\frac{19}{20}$	8. $\frac{1}{5} + \frac{1}{4}$	$\frac{9}{20}$	9. $\frac{6}{7} + \frac{6}{7}$	$1\frac{5}{7}$
10. $\frac{7}{10} + \frac{3}{5}$	$1\frac{3}{10}$	11. $\frac{5}{8} + \frac{1}{4}$	$\frac{7}{8}$	12. $\frac{8}{15} + \frac{2}{3}$	$1\frac{1}{5}$
13. $\frac{5}{12} + \frac{1}{3}$	$\frac{3}{4}$	14. $\frac{2}{3} + \frac{1}{6}$	$\frac{5}{6}$	15. $\frac{5}{6} + \frac{5}{18}$	$1\frac{1}{9}$
16. $\frac{3}{5} + \frac{2}{3}$	$1\frac{4}{15}$	17. $\frac{1}{2} + \frac{3}{7}$	$\frac{13}{14}$	18. $\frac{1}{2} + \frac{3}{8}$	$\frac{7}{8}$
19. $\frac{1}{6} + \frac{3}{5}$	$\frac{23}{30}$	20. $\frac{7}{8} + \frac{5}{6}$	$1\frac{17}{24}$	21. $\frac{1}{12} + \frac{2}{5}$	$\frac{29}{60}$

22. $\frac{1}{8} + \frac{5}{6} + \frac{7}{12}$ $1\frac{13}{24}$
23. $\frac{1}{2} + \frac{2}{3} + \frac{1}{6}$ $1\frac{1}{3}$

Solve.

24. Paul earns $48.24 each week. He saves $\frac{1}{4}$ of his earnings and spends $\frac{2}{3}$ on clothes and school supplies. What fraction of his earnings does he set aside for savings, clothes, and school supplies? $\frac{11}{12}$
25. In the 8th grade, $\frac{11}{25}$ of the students wear watches. In the 9th grade, $\frac{3}{5}$ of the students wear watches. In which grade do more students wear watches? 9th grade

4-10 ADDING FRACTIONS

Objective
Add proper fractions.

Barbara is making a jacket. She needs $\frac{5}{8}$ yard of backing for the front of the jacket and $\frac{2}{8}$ yard of backing for the collar. How much backing does she need to buy?

To add fractions with like denominators, add the numerators and write the sum over the denominator. Simplify if necessary.

Examples ***Add.***

A $\frac{5}{8} + \frac{2}{8}$ Estimate: $\frac{1}{2} + \frac{1}{2} = 1$

$\frac{5}{8} + \frac{2}{8} = \frac{7}{8}$ Compared to the estimate, is the answer reasonable?

Barbara needs $\frac{7}{8}$ yard of backing.

B $\frac{3}{18} + \frac{17}{18}$ Estimate: $0 + 1 = 1$

$\frac{3}{18} + \frac{17}{18} = \frac{20}{18} \rightarrow 18\overline{)20}$ 1 R2

$= 1\frac{2}{18}$ or $1\frac{1}{9}$

Is the answer reasonable?

Add fractions with unlike denominators as follows.

Method

1. Find the least common denominator (LCD).
2. Rename the fractions using the LCD.
3. Add. Simplify if necessary.

Example C ***Add*** $\frac{7}{8} + \frac{5}{12}$.

1. $8 = 2 \times 2 \times 2$
   $12 = 2 \times 2 \times 3$
   LCD: $2 \times 2 \times 2 \times 3$ or 24

Estimate: $1 + \frac{1}{2} = 1\frac{1}{2}$

2. $\frac{7}{8} + \frac{5}{12} \rightarrow \frac{21}{24} + \frac{10}{24}$

3. $\frac{21}{24} + \frac{10}{24} = \frac{31}{24}$ Change to a mixed number.

$\frac{31}{24} \rightarrow 24\overline{)31}$ 1 R7 $\rightarrow 1\frac{7}{24}$

Is the answer reasonable?

RETEACHING THE LESSON

Add. Write each sum as a fraction or mixed number in simplest form.

1. $\frac{1}{7} + \frac{5}{7} = \frac{6}{7}$
2. $\frac{1}{8} + \frac{3}{8} = \frac{1}{2}$
3. $\frac{9}{11} + \frac{6}{11} = 1\frac{4}{11}$
4. $\frac{3}{10} + \frac{9}{10} = 1\frac{1}{5}$
5. $\frac{1}{8} + \frac{3}{8} + \frac{3}{8} = \frac{7}{8}$

Add. Write each sum in simplest form.

6. $\frac{1}{2} + \frac{1}{3} = \frac{5}{6}$
7. $\frac{1}{4} + \frac{1}{3} = \frac{7}{12}$

Reteaching Masters Booklet, p. 42

Name ____________ Date ____________

RETEACHING WORKSHEET 4-10

Adding Fractions

To add fractions with like denominators, add the numerators. Write the sum over the denominator. Simplify if necessary.

$\frac{4}{12} + \frac{10}{12} = \frac{14}{12}$ ▸ $12\overline{)14}$ 1 R2 ▸ $1\frac{2}{12}$ or $1\frac{1}{6}$

To add fractions with unlike denominators, first rename the fractions using the LCD. Then add. Simplify if necessary.

	Find the LCD.	*Rename each fraction using the LCD.*	*Add and simplify if necessary.*
$\frac{1}{9} + \frac{2}{6}$	$9 = 3 \times 3$; $6 = 2 \times 3$; LCD: $2 \times 3 \times 3$ or 18	$\frac{1}{9} = \frac{2}{18}$; $\frac{2}{6} = \frac{6}{18}$	$\frac{2}{18} + \frac{6}{18} = \frac{8}{18} = \frac{4}{9}$

Guided Practice

Add.

Example A

1. $\frac{1}{4} + \frac{2}{4}$ $\frac{3}{4}$
2. $\frac{2}{5} + \frac{2}{5}$ $\frac{4}{5}$
3. $\frac{13}{32} + \frac{15}{32}$ $\frac{7}{8}$
4. $\frac{3}{9} + \frac{3}{9}$ $\frac{2}{3}$

Example B

5. $\frac{3}{8} + \frac{7}{8}$ $1\frac{1}{4}$
6. $\frac{3}{4} + \frac{2}{4}$ $1\frac{1}{4}$
7. $\frac{5}{6} + \frac{4}{6}$ $1\frac{1}{2}$
8. $\frac{7}{10} + \frac{9}{10}$ $1\frac{3}{5}$

Example C

9. $\frac{3}{4} + \frac{1}{8}$ $\frac{7}{8}$
10. $\frac{1}{8} + \frac{1}{6}$ $\frac{7}{24}$
11. $\frac{3}{5} + \frac{2}{7}$ $\frac{31}{35}$
12. $\frac{1}{3} + \frac{6}{7}$ $1\frac{4}{21}$

Exercises

Practice

16. $1\frac{1}{3}$ 17. $1\frac{1}{4}$

25. $1\frac{5}{6}$

Add.

13. $\frac{1}{5} + \frac{3}{5}$ $\frac{4}{5}$
14. $\frac{2}{7} + \frac{2}{7}$ $\frac{4}{7}$
15. $\frac{1}{9} + \frac{5}{9}$ $\frac{2}{3}$
16. $\frac{7}{12} + \frac{9}{12}$
17. $\frac{11}{16} +$
18. $\frac{2}{3} + \frac{1}{6}$ $\frac{5}{6}$
19. $\frac{3}{4} + \frac{1}{3}$ $1\frac{1}{12}$
20. $\frac{2}{3} + \frac{1}{8}$ $\frac{19}{24}$
21. $\frac{7}{10} + \frac{1}{2}$ $1\frac{1}{5}$
22. $\frac{1}{4} + \frac{3}{8}$ $\frac{5}{8}$
23. $\frac{7}{8} + \frac{5}{6}$ $1\frac{17}{24}$
24. $\frac{3}{4} + \frac{1}{5}$ $\frac{19}{20}$
25. $\frac{9}{10} + \frac{14}{15}$
26. $\frac{2}{9} + \frac{1}{6}$ $\frac{7}{18}$
27. $\frac{2}{3} + \frac{7}{9}$ $1\frac{4}{9}$
28. Find the sum of $\frac{2}{5}$, $\frac{1}{9}$, and $\frac{4}{15}$. $\frac{7}{9}$
29. Add $\frac{1}{4}$, $\frac{1}{3}$, and $\frac{7}{9}$. $1\frac{13}{36}$

Applications

30. Hans worked on his science homework for $\frac{7}{8}$ hour and his math homework for $\frac{2}{3}$ hour. How long did he work on science and math? $1\frac{13}{24}$ hours

31. $\frac{7}{12}$ can

31. Bev mixes $\frac{1}{3}$ can of yellow paint with $\frac{1}{4}$ can of blue paint to make green paint. How many cans of green paint does she make?

Using Algebra

35. $1\frac{5}{8}$

Find the value for each expression if $a = \frac{1}{2}$, $b = \frac{3}{4}$, and $c = \frac{3}{8}$.

32. $a + b$ $1\frac{1}{4}$
33. $b + c$ $1\frac{1}{8}$
34. $a + c$ $\frac{7}{8}$
35. $a + b + c$

36. Explain how you could use prime factorization to add fractions with unlike denominators. See students' work.

Critical Thinking

37. A bookworm chewed his way from Volume 1 page 1 to Volume 2, last page. The books are on a bookshelf standing next to each other. If the pages in each volume are 3 inches thick and the covers are $\frac{1}{4}$ inch thick each, how many inches did the bookworm chew? $\frac{1}{2}$ inch

Applying the Lesson

Independent Practice

Homework Assignment	
Minimum	**Ex. 1-37 odd; 38-45**
Average	**Ex. 2-32 even; 34-45**
Maximum	**Ex. 1-45**

Chalkboard Examples

Example A *Add.*

1. $\frac{1}{9} + \frac{3}{9}$ $\frac{4}{9}$
2. $\frac{5}{24} + \frac{11}{24}$ $\frac{2}{3}$

Example B *Add.*

3. $\frac{7}{9} + \frac{4}{9}$ $1\frac{2}{9}$
4. $\frac{2}{5} + \frac{4}{5}$ $1\frac{1}{5}$

Example C *Add.*

5. $\frac{1}{9} + \frac{2}{3}$ $\frac{7}{9}$
6. $\frac{2}{3} + \frac{5}{6}$ $1\frac{1}{2}$

Additional examples are provided on TRANSPARENCY 4-10.

Evaluating the Lesson

Check for Understanding Use Guided Practice Exercises to check for student understanding.

Error Analysis Watch for students who use the original numerators with the new denominator.

Closing the Lesson Ask students to state in their own words how they would add fractions with unlike denominators.

Extending the Lesson

Enrichment Masters Booklet, p. 42

ENRICHMENT WORKSHEET 4-10

Addition Tables

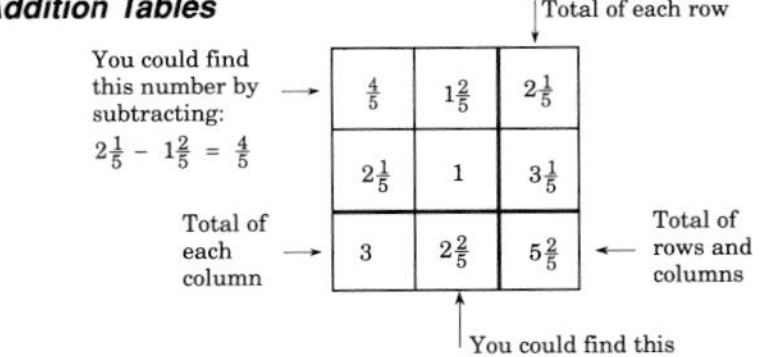

Complete each addition table.

1.

$1\frac{1}{8}$	$\frac{5}{8}$	$1\frac{3}{4}$
$\frac{1}{8}$	1	$1\frac{1}{8}$
$1\frac{1}{4}$	$1\frac{5}{8}$	$2\frac{7}{8}$

2.

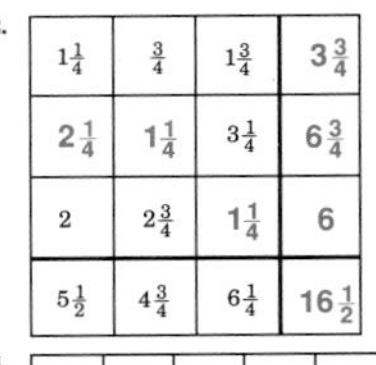

$1\frac{1}{4}$	$\frac{3}{4}$	$1\frac{3}{4}$	$3\frac{3}{4}$
$2\frac{1}{4}$	$1\frac{1}{4}$	$3\frac{1}{4}$	$6\frac{3}{4}$
2	$2\frac{3}{4}$	$1\frac{1}{4}$	6
$5\frac{1}{2}$	$4\frac{3}{4}$	$6\frac{1}{4}$	$16\frac{1}{2}$

3.

$\frac{1}{4}$	$\frac{5}{6}$	$\frac{7}{12}$	$1\frac{2}{3}$
$1\frac{5}{12}$	$\frac{1}{3}$	$\frac{3}{4}$	$2\frac{1}{2}$
$\frac{1}{2}$	$\frac{3}{4}$	1	$2\frac{1}{4}$
$2\frac{1}{6}$	$1\frac{11}{12}$	$2\frac{1}{3}$	$6\frac{5}{12}$

4.

$3\frac{2}{3}$	1	$1\frac{2}{3}$	$1\frac{1}{3}$	3	$10\frac{2}{3}$
$1\frac{2}{3}$	$2\frac{2}{3}$	4	3	2	$13\frac{1}{3}$
$2\frac{2}{3}$	$1\frac{2}{3}$	2	$\frac{2}{3}$	$4\frac{2}{3}$	$11\frac{2}{3}$
$2\frac{2}{3}$	$2\frac{1}{3}$	3	$1\frac{1}{3}$	$\frac{1}{3}$	$9\frac{2}{3}$
$\frac{1}{3}$	$4\frac{1}{3}$	$2\frac{1}{3}$	$\frac{1}{3}$	$2\frac{1}{3}$	$9\frac{2}{3}$
11	12	13	$6\frac{2}{3}$	$12\frac{1}{3}$	55

5-MINUTE CHECK-UP

(over Lesson 4-10)

Available on TRANSPARENCY 4-11.

Add.

1. $\frac{3}{7} + \frac{2}{7}$ $\frac{5}{7}$ **2.** $\frac{8}{9} + \frac{7}{9}$ $1\frac{2}{3}$

3. $\frac{1}{4} + \frac{1}{5}$ $\frac{9}{20}$ **4.** $\frac{7}{8} + \frac{3}{4}$ $1\frac{5}{8}$

Extra Practice, Lesson 4-10, p. 459

▶ INTRODUCING THE LESSON

Give students copies of the fraction bars from the **Lab Manual, page 3.** Ask students to use the bars to find the sum of $1\frac{2}{6}$ and $2\frac{3}{6}$. $3\frac{5}{6}$

Have students discuss how they found their answers.

▶ TEACHING THE LESSON

Using Models Use a number line from the **Lab Manual, page 7.** Have students add $1\frac{1}{4}$ and $1\frac{1}{2}$. Discuss with students how they arrived at their answer.

Practice Masters Booklet, p. 46

Name ______________ Date ______________

PRACTICE WORKSHEET 4-11

Adding Mixed Numbers

Add.

1. $2\frac{1}{8} + 3\frac{3}{8}$ $5\frac{1}{2}$
2. $5\frac{5}{6} + 8\frac{1}{6}$ 14
3. $7\frac{3}{4} + 1\frac{2}{4}$ $9\frac{1}{4}$
4. $8\frac{3}{8} + 2\frac{7}{8}$ $11\frac{1}{4}$
5. $5\frac{3}{8} + 4\frac{1}{4}$ $9\frac{5}{8}$
6. $7\frac{3}{5} + 2\frac{1}{3}$ $9\frac{14}{15}$
7. $4\frac{1}{2} + 10\frac{3}{7}$ $14\frac{13}{14}$
8. $9\frac{5}{6} + 7\frac{2}{9}$ $17\frac{1}{18}$
9. $6\frac{3}{5} + 3\frac{1}{4}$ $9\frac{17}{20}$
10. $8\frac{2}{3} + 4\frac{5}{7}$ $13\frac{8}{21}$
11. $5\frac{1}{6} + 6\frac{6}{7}$ $12\frac{1}{42}$
12. $3\frac{5}{8} + 7\frac{1}{3}$ $10\frac{23}{24}$
13. $3\frac{1}{6} + 5\frac{5}{6}$ 9
14. $12\frac{5}{9} + 7\frac{5}{9}$ $20\frac{1}{9}$
15. $2\frac{1}{8} + 1\frac{3}{5}$ $3\frac{29}{40}$
16. $6\frac{3}{5} + 3\frac{1}{6}$ $9\frac{23}{30}$
17. $3\frac{4}{5} + 1\frac{9}{10}$ $5\frac{7}{10}$
18. $12\frac{11}{12} + 9\frac{2}{3}$ $22\frac{7}{12}$
19. $9\frac{7}{8} + 8\frac{5}{6}$ $18\frac{17}{24}$
20. $14\frac{7}{20} + 6\frac{4}{5}$ $21\frac{3}{20}$
21. $2\frac{1}{4} + 1\frac{1}{2}$ $3\frac{3}{4}$
22. $22\frac{7}{8} + 7\frac{1}{6}$ $30\frac{1}{24}$
23. $15\frac{1}{2} + 4\frac{3}{4}$ $20\frac{1}{4}$
24. $12\frac{4}{5} + 8\frac{5}{6}$ $21\frac{19}{30}$
25. $4\frac{1}{12} + 3\frac{5}{12} + 5\frac{7}{12}$ $13\frac{1}{12}$
26. $2\frac{1}{2} + 1\frac{1}{4} + \frac{1}{8}$ $3\frac{7}{8}$
27. $2\frac{4}{9} + 3\frac{2}{3} + 1\frac{1}{6}$ $7\frac{5}{18}$
28. $4\frac{5}{12} + 1\frac{3}{8} + 1\frac{5}{6}$ $7\frac{5}{8}$

4-11 ADDING MIXED NUMBERS

Objective
Add mixed numbers.

Ling Wu weighed $6\frac{3}{4}$ pounds at birth. She gained $13\frac{3}{4}$ pounds in her first year. How much did she weigh on her first birthday?

You can use the following steps to add mixed numbers.

Method

1. Add the fractions. Rename first if necessary.
2. Add the whole numbers.
3. Rename and simplify if necessary.

Example A ***Add $13\frac{3}{4} + 6\frac{3}{4}$.*** Estimate: 14 + 7 = 21

Step 1: $13\frac{3}{4} + 6\frac{3}{4}$ → $\frac{6}{4}$

Step 2: $13\frac{3}{4} + 6\frac{3}{4}$ → $19\frac{6}{4}$

Step 3: $\frac{6}{4} = 1\frac{2}{4}$ or $1\frac{1}{2}$

$19 + 1\frac{1}{2} = 20\frac{1}{2}$

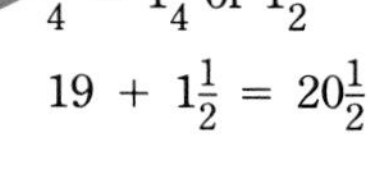

Ling Wu weighed $20\frac{1}{2}$ pounds on her first birthday.

Based on the estimate, explain why the answer, $20\frac{1}{2}$, is reasonable.

Example B ***Add $16\frac{1}{2} + 5\frac{5}{8}$.*** Estimate: 17 + 6 = 23

Step 1: $16\frac{1}{2} + 5\frac{5}{8}$ → $16\frac{4}{8} + 5\frac{5}{8}$ → $\frac{9}{8}$

Step 2: $16\frac{4}{8} + 5\frac{5}{8}$ → $21\frac{9}{8}$

Step 3: $\frac{9}{8} = 1\frac{1}{8}$

$21 + 1\frac{1}{8} = 22\frac{1}{8}$

2. $13\frac{8}{11}$ 5. $15\frac{1}{2}$ 8. $12\frac{2}{3}$

Guided Practice *Add.*

Example A

1. $4\frac{2}{5} + 7\frac{1}{5}$ $11\frac{3}{5}$
2. $9\frac{3}{11} + 4\frac{5}{11}$
3. $9\frac{1}{8} + 6\frac{3}{8}$ $15\frac{1}{2}$
4. $4\frac{3}{8} + 5\frac{3}{8}$ $9\frac{3}{4}$

Example B

5. $8\frac{1}{5} + 7\frac{3}{10}$
6. $2\frac{1}{8} + 5\frac{3}{4}$ $7\frac{7}{8}$
7. $5\frac{1}{2} + 3\frac{1}{3}$ $8\frac{5}{6}$
8. $4\frac{1}{4} + 8\frac{5}{12}$

RETEACHING THE LESSON

Rename each mixed number.

1. $3\frac{6}{4}$ $4\frac{1}{2}$
2. $10\frac{5}{2}$ $12\frac{1}{2}$
3. $15\frac{21}{7}$ 18

Write the sum in simplest form.

4. $4\frac{2}{3} + 5\frac{1}{4}$ $9\frac{11}{12}$
5. $9\frac{4}{5} + 3\frac{7}{10}$ $13\frac{1}{2}$
6. $26\frac{3}{8} + 24\frac{3}{4} = 51\frac{1}{8}$

Reteaching Masters Booklet, p. 43

Name ______________ Date ______________

RETEACHING WORKSHEET 4-11

Adding Mixed Numbers

To add mixed numbers, add the fractions and then the whole numbers.

Add $1\frac{2}{3}$ and $4\frac{3}{4}$.

$1\frac{2}{3} + 4\frac{3}{4}$

Find the LCD. 3 = ③ 4 = ② × ② LCD: 2 × 2 × 3 or 12

Rename each fraction using the LCD. $1\frac{2}{3} = 1\frac{8}{12}$, $+ 4\frac{3}{4} = 4\frac{9}{12}$

Add and simplify if necessary. $1\frac{2}{3} = 1\frac{8}{12}$, $+ 4\frac{3}{4} = 4\frac{9}{12}$, $5\frac{17}{12} = 6\frac{5}{12}$

Add.

1. $3\frac{1}{8} + 2\frac{3}{4}$
2. $7\frac{1}{10} + 3\frac{3}{5}$
3. $13\frac{4}{7} + 12\frac{1}{14}$
4. $9\frac{1}{6}$ + ...
5. $17\frac{3}{5} + 5\frac{2}{5}$

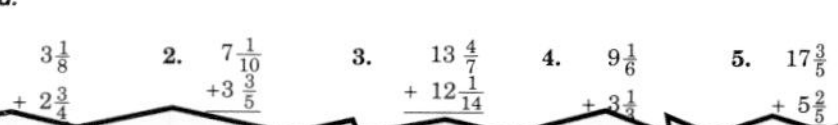

Exercises

Add. 10. $11\frac{3}{5}$ 12. $17\frac{1}{2}$ 13. $16\frac{19}{24}$ 14. $19\frac{11}{18}$ 15. $11\frac{3}{4}$ 16. $25\frac{19}{20}$ 19. $11\frac{1}{2}$

Practice
20. $12\frac{5}{8}$ 22. $20\frac{22}{45}$
23. $24\frac{1}{12}$
24. $13\frac{3}{14}$
26. $19\frac{7}{18}$

9. $1\frac{3}{7} + 8\frac{2}{7}$ $9\frac{5}{7}$ **10.** $4\frac{3}{10} + 7\frac{3}{10}$ **11.** $4\frac{3}{8} + 3\frac{7}{8}$ $8\frac{1}{4}$ **12.** $5\frac{3}{4} + 11\frac{3}{4}$

13. $7\frac{1}{6} + 9\frac{5}{8}$ **14.** $4\frac{4}{9} + 15\frac{1}{6}$ **15.** $6\frac{7}{12} + 5\frac{1}{6}$ **16.** $18\frac{3}{4} + 7\frac{1}{5}$

17. $4\frac{7}{8} + 5\frac{3}{8}$ $10\frac{1}{4}$ **18.** $3\frac{3}{4} + 8\frac{7}{8}$ $12\frac{5}{8}$ **19.** $4\frac{4}{5} + 6\frac{7}{10}$ **20.** $9\frac{11}{16} + 2\frac{15}{16}$

21. $8\frac{2}{3} + 6\frac{1}{6}$ $14\frac{5}{6}$ **22.** $9\frac{8}{9} + 10\frac{3}{5}$ **23.** $17\frac{1}{3} + 6\frac{3}{4}$ **24.** $4\frac{5}{7} + 8\frac{1}{2}$

25. Add $1\frac{1}{12}$, $3\frac{1}{2}$, and $4\frac{1}{6}$. $8\frac{3}{4}$ **26.** Find the sum of $7\frac{8}{9}$, $2\frac{5}{6}$, and $8\frac{2}{3}$.

Using Algebra

Solve each equation.

27. $a - 2 = 3\frac{1}{4}$ $5\frac{1}{4}$ **28.** $b - \frac{4}{4} = 8\frac{1}{5}$ $9\frac{1}{5}$ **29.** $x - 8\frac{3}{8} = 17\frac{1}{2}$ $25\frac{7}{8}$

33. $13\frac{7}{8}$

Evaluate each expression if $a = \frac{3}{4}$, $b = 5\frac{1}{2}$, and $c = 7\frac{5}{8}$.

30. $a + b$ $6\frac{1}{4}$ **31.** $c + a$ $8\frac{3}{8}$ **32.** $b + c$ $13\frac{1}{8}$ **33.** $a + b + c$

Applications
34. $23\frac{1}{2}$ feet

34. Paul has two sections of rope that he will tie together. One section is $6\frac{3}{4}$ feet long and the other is $17\frac{1}{2}$ feet long. If he loses $\frac{3}{4}$ of a foot when he ties them together, what is the total length?

35. Mrs. Baker had $5\frac{1}{2}$ yards of curtain material. She needs another $2\frac{7}{8}$ yards. How much material did she need in all? $8\frac{3}{8}$ yards

36. Mr. James can drive his car for about 22¢ a mile. The trip home is 415 miles. Will $80 be enough to pay for the car expenses? Explain your answer. no; $0.22 \times 415 = 91.30$; > $80

Write Math

37. Why would it not be necessary to use the parentheses in the expression $a + (b + c)$? Since all operations are addition, the sum is the same without or with parentheses.

Critical Thinking

38. Use all the digits 1, 2, 3, 4, and 5 and addition (in the denominator only) to write a fraction that is equal to one-half.

38. $\frac{5}{1 + 2 + 3 + 4}$

Mixed Review

Add.

Lesson 2-2

39.	40.	41.	42.
143.6	$82.07	72.6	5.012
+ 18.2	+ $16.95	+ 1.03	+ 71.8
161.8	$99.02	73.63	76.812

Lesson 2-6

Complete each sequence.

43. 18, <u>21</u>, 24, <u>27</u>, <u>30</u>, 33 **44.** <u>1.3</u>, 4.6, 7.9, <u>11.2</u>, <u>14.5</u>, 17.8

Lesson 2-9

If the problem has the necessary facts, solve. State any missing or extra facts.

NORTHEAST—SKYWAE

Spacious two BR townhomes, 1½ bath, fin. bsmt., fully equipped eat-in kitchen, patio. Close to busline, shopping, dining & freeways. ONLY $410. Open Sat. & Sun. 9–6. Weekdays by appt. 222-3333, 444-0000. Located in Strawberry Farms. ASK ABOUT FREE RENT SPECIAL. Bette Varca & Co., Realtors.

45. A newspaper ad costs $5 for the first 2 lines and $1.50 for each additional line. Bold print costs $0.50 extra per line. How much does a regular ad cost if it has 10 lines with no bold print? $17; do not need bold type cost

Chalkboard Examples

Example A *Add.*

1. $7\frac{9}{16} + 4\frac{5}{16}$ $11\frac{7}{8}$ **2.** $7\frac{4}{7} + 8\frac{4}{7}$ $16\frac{1}{7}$

Example B *Add.*

3. $3\frac{1}{5} + 6\frac{1}{2}$ $9\frac{7}{10}$ **4.** $7\frac{3}{4} + 3\frac{5}{12}$ $11\frac{1}{6}$

Additional examples are provided on TRANSPARENCY 4-11.

EVALUATING THE LESSON

Check for Understanding Use Guided Practice Exercises to check for student understanding.

Closing the Lesson Tell students that they will use addition of fractions in cooking, sewing, woodworking, and higher levels of mathematics and so on.

EXTENDING THE LESSON

Enrichment Have students make up a problem using fraction models that represent addition of mixed numbers. Have them write an equation and solve it. Then ask them to give the problem to a classmate to write an equation and solve. Have them check to see if both equations and answers match.

Enrichment Masters Booklet, p. 43

ENRICHMENT WORKSHEET 4-11

Magic Squares

In a magic square, each row, column, and diagonal has the same sum.

Complete each magic square. Name the magic sum.

1.

1	$3\frac{1}{2}$	3
$4\frac{1}{2}$	$2\frac{1}{2}$	$\frac{1}{2}$
2	$1\frac{1}{2}$	4

magic sum $7\frac{1}{2}$

2.

2	$\frac{1}{4}$	$1\frac{1}{2}$
$\frac{3}{4}$	$1\frac{1}{4}$	$1\frac{3}{4}$
1	$2\frac{1}{4}$	$\frac{1}{2}$

magic sum $3\frac{3}{4}$

3.

1	$1\frac{5}{8}$	$\frac{3}{4}$
$\frac{7}{8}$	$1\frac{1}{8}$	$1\frac{3}{8}$
$1\frac{1}{2}$	$\frac{5}{8}$	$1\frac{1}{4}$

magic sum $3\frac{3}{8}$

4.

$\frac{5}{6}$	$2\frac{11}{12}$	$2\frac{1}{2}$
$3\frac{3}{4}$	$2\frac{1}{12}$	$\frac{5}{12}$
$1\frac{2}{3}$	$1\frac{1}{4}$	$3\frac{1}{3}$

magic sum $6\frac{1}{4}$

5.

$3\frac{1}{5}$	$7\frac{1}{5}$	$1\frac{3}{5}$
$2\frac{2}{5}$	4	$5\frac{3}{5}$
$6\frac{2}{5}$	$\frac{4}{5}$	$4\frac{4}{5}$

magic sum 12

6.

$4\frac{1}{2}$	$\frac{5}{12}$	$3\frac{1}{3}$
$1\frac{7}{12}$	$2\frac{3}{4}$	$3\frac{11}{12}$
$2\frac{1}{6}$	$5\frac{1}{12}$	1

magic sum $8\frac{1}{4}$

7.

$9\frac{1}{6}$	$\frac{23}{30}$	$1\frac{11}{30}$	$7\frac{11}{30}$
$2\frac{17}{30}$	$6\frac{1}{6}$	$5\frac{17}{30}$	$4\frac{11}{30}$
$4\frac{29}{30}$	$3\frac{23}{30}$	$3\frac{1}{6}$	$6\frac{23}{30}$
$1\frac{29}{30}$	$7\frac{29}{30}$	$8\frac{17}{30}$	$\frac{1}{6}$

magic sum $18\frac{2}{3}$

8.

$8\frac{1}{6}$	$\frac{7}{12}$	$1\frac{1}{8}$	$6\frac{13}{24}$
$2\frac{5}{24}$	$5\frac{11}{24}$	$4\frac{11}{12}$	$3\frac{5}{6}$
$4\frac{3}{8}$	$3\frac{7}{24}$	$2\frac{3}{4}$	6
$1\frac{2}{3}$	$7\frac{1}{12}$	$7\frac{5}{8}$	$\frac{1}{24}$

magic sum $16\frac{5}{12}$

APPLYING THE LESSON

Independent Practice

Homework Assignment	
Minimum	Ex. 1-37 odd
Average	Ex. 2-28 even; 30-37
Maximum	Ex. 1-37

5-MINUTE CHECK-UP

(over Lesson 4-11)

Available on TRANSPARENCY 4-12.

Add.

1. $3\frac{1}{8} + 1\frac{3}{8}$ $4\frac{1}{2}$ **2.** $4\frac{1}{3} + 5\frac{1}{5}$ $9\frac{8}{15}$

3. $3\frac{1}{3} + 6\frac{1}{6}$ $9\frac{1}{2}$ **4.** $7\frac{1}{4} + 3\frac{2}{5}$ $10\frac{13}{20}$

Extra Practice, Lesson 4-11, p. 459

INTRODUCING THE LESSON

Give students copies of a ruler from the **Lab Manual, page 10.** Ask them to identify $\frac{1}{8}$, $\frac{1}{4}$, $\frac{1}{2}$, and the whole numbers. Ask if anyone can determine how to subtract $\frac{1}{4}$ from $\frac{3}{4}$.

TEACHING THE LESSON

Critical Thinking Have students describe the relationship between $\frac{3}{4} - \frac{1}{2} - \frac{1}{8}$ and $\frac{3}{4} - (\frac{1}{2} + \frac{1}{8})$. They represent the same problem and equal $\frac{1}{8}$.

Practice Masters Booklet, p. 47

Name ________________ Date ________________

PRACTICE WORKSHEET 4-12

Subtracting Fractions

Subtract.

1. $\frac{13}{20} - \frac{7}{20}$ $\frac{3}{10}$ **2.** $\frac{7}{12} - \frac{5}{12}$ $\frac{1}{6}$ **3.** $\frac{9}{10} - \frac{7}{10}$ $\frac{1}{5}$

4. $\frac{1}{2} - \frac{1}{3}$ $\frac{1}{6}$ **5.** $\frac{7}{8} - \frac{5}{6}$ $\frac{1}{24}$ **6.** $\frac{13}{16} - \frac{7}{12}$ $\frac{11}{48}$

7. $\frac{7}{8} - \frac{7}{12}$ $\frac{7}{24}$ **8.** $\frac{7}{9} - \frac{3}{5}$ $\frac{8}{45}$ **9.** $\frac{3}{4} - \frac{13}{20}$ $\frac{1}{10}$

10. $\frac{7}{8} - \frac{2}{3}$ $\frac{5}{24}$ **11.** $\frac{3}{4} - \frac{2}{3}$ $\frac{1}{12}$ **12.** $\frac{5}{6} - \frac{5}{8}$ $\frac{5}{24}$

13. $\frac{3}{4} - \frac{1}{2}$ $\frac{1}{4}$ **14.** $\frac{5}{8} - \frac{9}{16}$ $\frac{1}{16}$ **15.** $\frac{11}{12} - \frac{1}{8}$ $\frac{19}{24}$

16. $\frac{1}{2} - \frac{1}{8}$ $\frac{3}{8}$ **17.** $\frac{11}{12} - \frac{1}{4}$ $\frac{2}{3}$ **18.** $\frac{3}{4} - \frac{1}{8}$ $\frac{5}{8}$

19. $\frac{8}{15} - \frac{1}{6}$ $\frac{11}{30}$ **20.** $\frac{3}{10} - \frac{1}{5}$ $\frac{1}{10}$ **21.** $\frac{5}{6} - \frac{7}{9}$ $\frac{1}{18}$

22. $\frac{19}{27} - \frac{10}{27}$ $\frac{1}{3}$ **23.** $\frac{11}{20} - \frac{2}{5}$ $\frac{3}{20}$ **24.** $\frac{3}{4} - \frac{5}{12}$ $\frac{1}{3}$

Solve.

25. Mrs. Smith has $\frac{3}{4}$ gallon of milk. She uses $\frac{1}{2}$ gallon. How much does she have left? $\frac{1}{4}$ gallon

26. Mr. Morales has 11 eggs. He uses $\frac{1}{4}$ dozen. How many dozen eggs does he have left? $\frac{2}{3}$ dozen

4-12 SUBTRACTING FRACTIONS

Objective
Subtract proper fractions.

Maxine is building a bird feeder as part of her science project. A peg that helps hold the feeder together needs to be $\frac{5}{8}$ inch long. The peg she has is $\frac{7}{8}$ inch long. How much does Maxine need to saw off the peg?

To subtract fractions with like denominators, subtract the numerators and write the difference over the denominator. Simplify if necessary.

Example A ***Subtract $\frac{7}{8} - \frac{5}{8}$.***

$\frac{7}{8} - \frac{5}{8} = \frac{2}{8}$ Estimate: $1 - \frac{1}{2} = \frac{1}{2}$

$= \frac{1}{4}$ Compared to the estimate, is the answer reasonable?

Maxine needs to saw off $\frac{1}{4}$ inch of the peg.

Subtract fractions with unlike denominators as follows.

Method

1. Find the least common denominator.
2. Rename the fractions using the LCD.
3. Subtract. Simplify if necessary.

Example B ***Subtract $\frac{5}{6} - \frac{1}{4}$.*** Estimate: $1 - \frac{1}{2} = \frac{1}{2}$

1. $6 = 2 \times 3$, $4 = 2 \times 2$ — LCD: $2 \times 3 \times 2$ or 12

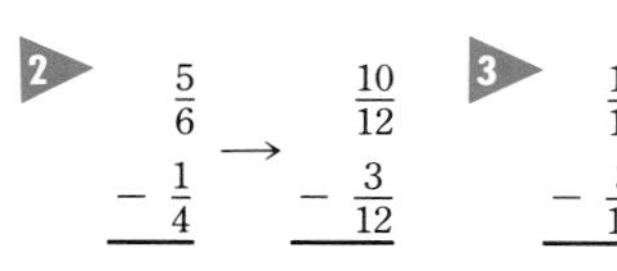

2. $\frac{5}{6} - \frac{1}{4} \rightarrow \frac{10}{12} - \frac{3}{12}$

3. $\frac{10}{12} - \frac{3}{12} = \frac{7}{12}$

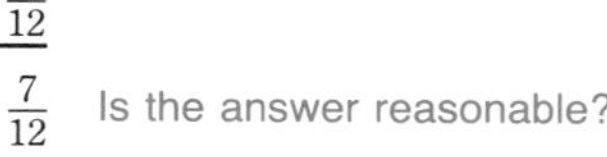

Is the answer reasonable?

Guided Practice *Subtract.*

Example A

1. $\frac{6}{9} - \frac{3}{9}$ $\frac{1}{3}$ **2.** $\frac{7}{18} - \frac{5}{18}$ $\frac{1}{9}$ **3.** $\frac{20}{21} - \frac{8}{21}$ $\frac{4}{7}$ **4.** $\frac{5}{11} - \frac{1}{11}$ $\frac{4}{11}$ **5.** $\frac{16}{20} - \frac{12}{20}$ $\frac{1}{5}$

Example B

6. $\frac{1}{2} - \frac{1}{3}$ $\frac{1}{6}$ **7.** $\frac{2}{3} - \frac{1}{4}$ $\frac{5}{12}$ **8.** $\frac{7}{10} - \frac{4}{9}$ $\frac{23}{90}$ **9.** $\frac{5}{6} - \frac{3}{4}$ $\frac{1}{12}$ **10.** $\frac{13}{24} - \frac{3}{8}$ $\frac{1}{6}$

RETEACHING THE LESSON

Subtract. Write each difference in simplest form.

1. $\frac{8}{9} - \frac{5}{9}$ $\frac{1}{3}$ **2.** $\frac{13}{18} - \frac{5}{18}$ $\frac{4}{9}$ **3.** $\frac{5}{4} - \frac{3}{8}$ $\frac{7}{8}$

4. $\frac{7}{10} - \frac{3}{5} = \frac{1}{10}$ **5.** $\frac{15}{24} - \frac{1}{4} = \frac{3}{8}$

6. $\frac{27}{25} - \frac{8}{15} = \frac{41}{75}$ **7.** $\frac{9}{16} - \frac{3}{10} = \frac{21}{80}$

Reteaching Masters Booklet, p. 44

Name ________________ Date ________________

RETEACHING WORKSHEET 4-12

Subtracting Fractions

To subtract fractions with unlike denominators, first rename each fraction using the LCD.
Then subtract and simplify if necessary.

Find the LCD.	*Rename each fraction using the LCD.*	*Subtract.*
$\frac{2}{5} - \frac{1}{10}$; $5 = 5$, $10 = 2 \times 5$; LCD: 2×5 or 10	$\frac{2}{5} = \frac{4}{10}$, $-\frac{1}{10} = \frac{1}{10}$	$\frac{4}{10} - \frac{1}{10} = \frac{3}{10}$ Answer is already in simplest form.

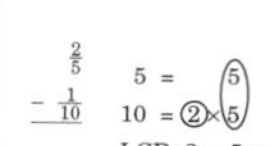

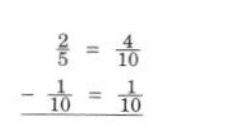

Rename each pair of fractions using the given denominator. Then subtract. Write each difference in simplest form.

Exercises

Practice

13. $\frac{7}{20}$

Subtract.

11. $\frac{7}{8} - \frac{3}{8}$ $\frac{1}{2}$ **12.** $\frac{5}{6} - \frac{4}{6}$ $\frac{1}{6}$ **13.** $\frac{15}{20} - \frac{8}{20}$ **14.** $\frac{5}{7} - \frac{2}{7}$ $\frac{3}{7}$ **15.** $\frac{17}{18} - \frac{5}{18}$ $\frac{2}{3}$

16. $\frac{4}{15} - \frac{1}{15}$ $\frac{1}{5}$ **17.** $\frac{1}{4} - \frac{1}{5}$ $\frac{1}{20}$ **18.** $\frac{2}{5} - \frac{1}{3}$ $\frac{1}{15}$ **19.** $\frac{1}{2} - \frac{1}{2}$ 0 **20.** $\frac{5}{8} - \frac{1}{6}$ $\frac{11}{24}$

21. $\frac{7}{8} - \frac{5}{6}$ $\frac{1}{24}$ **22.** $\frac{3}{4} - \frac{2}{3}$ $\frac{1}{12}$ **23.** $\frac{23}{28} - \frac{9}{28}$ $\frac{1}{2}$ **24.** $\frac{4}{7} - \frac{1}{3}$ $\frac{5}{21}$ **25.** $\frac{5}{6} - \frac{4}{9}$ $\frac{7}{18}$

26. Subtract $\frac{1}{3}$ from $\frac{6}{7}$. $\frac{11}{21}$ **27.** Find the difference of $\frac{7}{8}$ and $\frac{5}{12}$. $\frac{11}{24}$

Using Equations

Solve each equation.

28. $a + \frac{3}{8} = \frac{5}{8}$ $\frac{1}{4}$ **29.** $x - \frac{1}{2} = \frac{1}{6}$ $\frac{2}{3}$ **30.** $c + \frac{3}{5} = \frac{4}{5}$ $\frac{1}{5}$

Applications

31. $\frac{1}{2}$ cup

31. Jamie was making brownies using a recipe that called for $\frac{1}{4}$ cup of oil. She was talking on the phone at the same time and accidently poured $\frac{3}{4}$ cup of oil into the mixing bowl. How much did she need to remove from the bowl?

32. $\frac{3}{4}$ cup

32. Max is making a taco salad. He needs $\frac{1}{4}$ cup of cheddar cheese. He already has $\frac{1}{2}$ cup of mozzerella in the measuring cup. If he puts the cheddar cheese in the cup on top of the mozzerella, how much cheese should be in the cup?

Critical Thinking

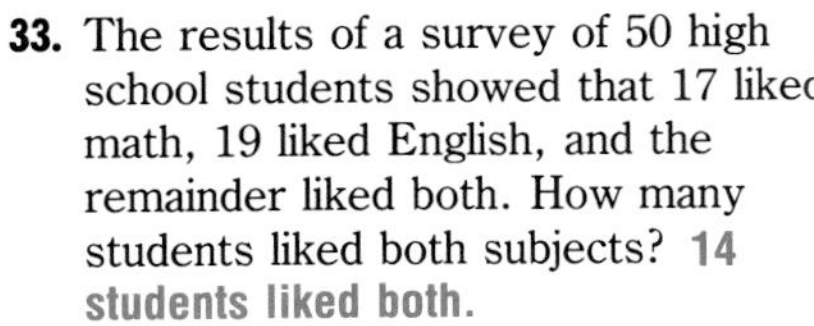

33. The results of a survey of 50 high school students showed that 17 liked math, 19 liked English, and the remainder liked both. How many students liked both subjects? **14 students liked both.**

Estimation

Front-end estimation is one way to find the sum of mixed numbers. Consider $2\frac{7}{8} + 1\frac{1}{10} + 4\frac{5}{6} + 3$.

First add only the whole numbers, the "front end" of each addend. $2 + 1 + 4 + 3 = 10$

Then mentally estimate the sum of the fractions as a whole number. $\frac{7}{8} + \frac{1}{10} + \frac{5}{6}$ is about 2.

So, $2\frac{7}{8} + 1\frac{1}{10} + 4\frac{5}{6} + 3$ is about $10 + 2$ or 12.

Estimate. **Answers may vary. Typical answers are given.**

34. $6\frac{1}{2} + 10\frac{1}{5} + 4\frac{3}{5}$ 21 **35.** $7\frac{1}{2} + 11\frac{5}{8} + \frac{1}{16}$ 19 **36.** $3\frac{5}{6} + 4\frac{1}{3} + 4\frac{1}{3} + 4$ 16

APPLYING THE LESSON

Independent Practice

Homework Assignment	
Minimum	Ex. 1-35 odd
Average	Ex. 2-30 even; 31-36
Maximum	Ex. 1-36

Chalkboard Examples

Example A *Subtract.*

1. $\frac{5}{7} - \frac{1}{7}$ $\frac{4}{7}$ **2.** $\frac{7}{8} - \frac{3}{8}$ $\frac{1}{2}$

Example B *Subtract.*

3. $\frac{5}{6} - \frac{3}{8}$ $\frac{11}{24}$ **4.** $\frac{4}{5} - \frac{11}{20}$ $\frac{1}{4}$

Additional examples are provided on TRANSPARENCY 4-12.

EVALUATING THE LESSON

Check for Understanding Use Guided Practice Exercises to check for student understanding.

Error Analysis Watch for students who subtract both the numerators and the denominators of the fraction. Suggest that they think of the denominators as labels and labels cannot be added or subtracted.

Closing the Lesson Have students tell about times when they could use subtraction of fractions.

EXTENDING THE LESSON

Enrichment Masters Booklet, p. 44

Name ______ Date ______

ENRICHMENT WORKSHEET 4-12

Subtracting Fractions

Complete each subtraction table.

1.

$\frac{7}{8}$	$\frac{1}{4}$	$\frac{5}{8}$
$\frac{3}{8}$	$\frac{1}{8}$	$\frac{1}{4}$
$\frac{1}{2}$	$\frac{1}{8}$	$\frac{3}{8}$

2.

$\frac{2}{3}$	$\frac{3}{8}$	$\frac{7}{24}$
$\frac{1}{2}$	$\frac{1}{4}$	$\frac{1}{4}$
$\frac{1}{6}$	$\frac{1}{8}$	$\frac{1}{24}$

3.

$\frac{3}{4}$	$\frac{3}{8}$	$\frac{3}{8}$
$\frac{1}{6}$	$\frac{1}{8}$	$\frac{1}{24}$
$\frac{7}{12}$	$\frac{1}{4}$	$\frac{1}{3}$

4.

$\frac{4}{5}$	$\frac{1}{2}$	$\frac{3}{10}$
$\frac{1}{4}$	$\frac{1}{5}$	$\frac{1}{20}$
$\frac{11}{20}$	$\frac{3}{10}$	$\frac{1}{4}$

5.

$\frac{11}{12}$	$\frac{1}{3}$	$\frac{7}{12}$
$\frac{3}{4}$	$\frac{1}{6}$	$\frac{7}{12}$
$\frac{1}{6}$	$\frac{1}{6}$	0

6.

$\frac{13}{16}$	$\frac{1}{4}$	$\frac{9}{16}$
$\frac{3}{8}$	$\frac{3}{16}$	$\frac{3}{16}$
$\frac{7}{16}$	$\frac{1}{16}$	$\frac{3}{8}$

5-MINUTE CHECK-UP

(over Lesson 4-12)

Available on TRANSPARENCY 4-13.

Subtract.

1. $\frac{6}{7} - \frac{3}{7}$ **$\frac{3}{7}$**
2. $\frac{8}{9} - \frac{2}{9}$ **$\frac{2}{3}$**
3. $\frac{5}{6} - \frac{3}{8}$ **$\frac{11}{24}$**
4. $\frac{2}{3} - \frac{3}{8}$ **$\frac{7}{24}$**

Extra Practice, Lesson 4-12, p. 460

▶ INTRODUCING THE LESSON

Ask students to tell about times when they have had to subtract mixed numbers. **sewing, cooking, woodworking**

▶ TEACHING THE LESSON

Using Critical Thinking Ask students to explain how to rename the whole number and fraction when the subtrahend is a fraction and less than the minuend. For example, $8\frac{1}{3} - 4\frac{4}{5}$.

Practice Masters Booklet, p. 48

Name ______________________ Date ______________

PRACTICE WORKSHEET 4-13

Subtracting Mixed Numbers

Subtract.

1. $7\frac{3}{4} - 6\frac{1}{4}$ **$1\frac{1}{2}$**	2. $9\frac{7}{8} - 2\frac{3}{8}$ **$7\frac{1}{2}$**	3. $8 - 6\frac{1}{5}$ **$1\frac{4}{5}$**	4. $10\frac{7}{12} - 6$ **$4\frac{7}{12}$**
5. $2\frac{1}{8} - 1\frac{1}{16}$ **$1\frac{1}{16}$**	6. $12\frac{4}{5} - 9\frac{3}{10}$ **$3\frac{1}{2}$**	7. $6\frac{2}{3} - 4\frac{1}{4}$ **$2\frac{5}{12}$**	8. $7\frac{5}{6} - 1\frac{3}{4}$ **$6\frac{1}{12}$**
9. $7\frac{1}{2} - 3\frac{3}{4}$ **$3\frac{3}{4}$**	10. $4\frac{7}{12} - 2\frac{2}{3}$ **$1\frac{11}{12}$**	11. $6\frac{2}{5} - 3\frac{1}{2}$ **$2\frac{9}{10}$**	12. $11\frac{1}{5} - 7\frac{1}{3}$ **$3\frac{13}{15}$**

13. $5\frac{11}{12} - 2\frac{1}{6}$ **$3\frac{3}{4}$** 14. $8 - 7\frac{4}{15}$ **$\frac{11}{15}$** 15. $5\frac{6}{7} - 5\frac{1}{2}$ **$\frac{5}{14}$**

16. $18\frac{1}{6} - 7\frac{7}{9}$ **$10\frac{7}{18}$** 17. $16\frac{9}{10} - 8\frac{3}{10}$ **$8\frac{3}{5}$** 18. $14\frac{1}{7} - 8\frac{4}{7}$ **$5\frac{4}{7}$**

Solve.

19. After the stock rose $2\frac{3}{8}$ points, it closed at $28\frac{1}{4}$. What was the opening price? **$25\frac{7}{8}$**

20. Sue and Jerry are making muffins for breakfast. They need $2\frac{1}{2}$ cups of flour, but they have only $1\frac{3}{4}$ cups. How much more flour do they need? **$\frac{3}{4}$ cup**

4-13 SUBTRACTING MIXED NUMBERS

Objective
Subtract mixed numbers.

Barney Lopez has to send his computer printer back to the manufacturer for repairs. Since he must pay the postage, he wants to keep the weight low without allowing damage to the printer. The printer weighs $8\frac{1}{4}$ pounds. When packaged for shipping it weighs $13\frac{3}{4}$ pounds. What is the weight of the packing materials?

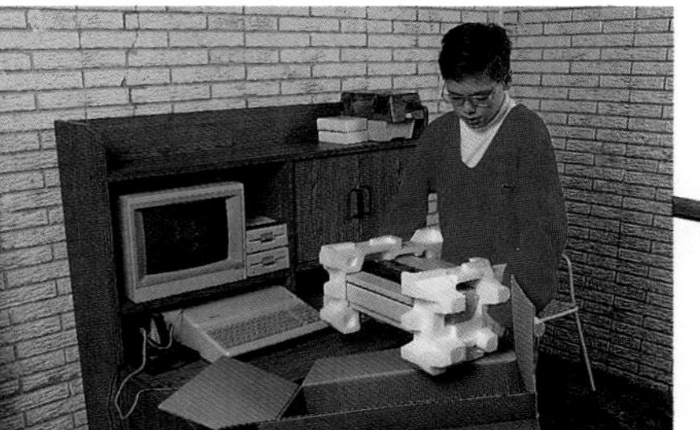

Subtracting mixed numbers is similiar to adding mixed numbers.

Method

1. Subtract the fractions. Rename first if necessary.
2. Subtract the whole numbers.
3. Rename and simplify if necessary.

Examples

A $13\frac{3}{4} - 8\frac{1}{4}$ Estimate: $14 - 8 = 6$

▶1 $13\frac{3}{4} - 8\frac{1}{4}$: $\frac{2}{4}$ ▶2 $13\frac{3}{4} - 8\frac{1}{4} = 5\frac{2}{4}$

▶3 $5\frac{2}{4} = 5\frac{1}{2}$

The weight of the packing materials is $5\frac{1}{2}$ pounds.

B $12\frac{1}{4} - 8\frac{1}{6}$ Estimate: $12 - 8 = 5$

▶1 $12\frac{1}{4} - 8\frac{1}{6} \longrightarrow 12\frac{3}{12} - 8\frac{2}{12}$: $\frac{1}{12}$ ▶2 $12\frac{3}{12} - 8\frac{2}{12} = 4\frac{1}{12}$

Is the answer reasonable?

Example C $9\frac{1}{4} - 6\frac{5}{8}$ Estimate: $9 - 7 = 2$

▶1 $9\frac{1}{4} - 6\frac{5}{8} \longrightarrow 9\frac{2}{8} - 6\frac{5}{8} \longrightarrow 8\frac{10}{8} - 6\frac{5}{8}$: $\frac{5}{8}$ Since $\frac{2}{8} < \frac{5}{8}$, rename $9\frac{2}{8}$ as $8\frac{10}{8}$.

▶2 $8\frac{10}{8} - 6\frac{5}{8} = 2\frac{5}{8}$

Based on the estimate, explain why the answer, $2\frac{5}{8}$, is reasonable.

RETEACHING THE LESSON

Complete.

1. $7\frac{2}{5} = 6\frac{7}{5}$ 2. $1\frac{4}{8} = \frac{12}{8}$ 3. $8 = 7\frac{5}{5}$

Subtract. Write each difference in simplest form.

4. $3\frac{1}{3} - 1\frac{1}{6}$ **$2\frac{1}{6}$** 5. $10\frac{1}{4} - 3\frac{2}{3}$ **$6\frac{7}{12}$**

Reteaching Masters Booklet, p. 45

Name ______________________ Date ______________

RETEACHING WORKSHEET 4-13

Subtracting Mixed Numbers

To subtract mixed numbers, you may need to rename more than once.

Subtract $2\frac{3}{4}$ from $4\frac{1}{3}$.

	Find the LCD.	*Rename each fraction using the LCD.*	*Rename again so the fractions can be subtracted.*	*Subtract and simplify if necessary.*
$4\frac{1}{3} - 2\frac{3}{4}$	3 = ③; 4 = ② × ②; LCD: 2 × 2 × 3 or 12	$4\frac{4}{12} - 2\frac{9}{12}$	$3\frac{16}{12} - 2\frac{9}{12}$	$3\frac{16}{12} - 2\frac{9}{12} = 1\frac{7}{12}$

Subtract. Write each difference in simplest form.

Example D $10 - 3\frac{2}{7}$ Estimate: 10 − 3 = 7

1 $\begin{array}{r} 10 \\ -\ 3\frac{2}{7} \\ \hline \end{array} \rightarrow \begin{array}{r} 9\frac{7}{7} \\ -\ 3\frac{2}{7} \\ \hline \frac{5}{7} \end{array}$ Rename 10 as $9\frac{7}{7}$.

2 $\begin{array}{r} 9\frac{7}{7} \\ -\ 3\frac{2}{7} \\ \hline 6\frac{5}{7} \end{array}$ Is the answer reasonable?

Guided Practice

Example A, Example B, Example C, Example D

Subtract. 7. $11\frac{1}{8}$ 8. $6\frac{4}{9}$ 10. $6\frac{17}{20}$ 12. $4\frac{23}{24}$ 16. $13\frac{3}{7}$

1. $3\frac{8}{9} - 1\frac{2}{9}$ $2\frac{2}{3}$ **2.** $5\frac{3}{4} - 4\frac{1}{4}$ $1\frac{1}{2}$ **3.** $7\frac{5}{8} - 2\frac{3}{8}$ $5\frac{1}{4}$ **4.** $9\frac{4}{5} - 6\frac{1}{5}$ $3\frac{3}{5}$

5. $8\frac{4}{5} - 5\frac{3}{4}$ $3\frac{1}{20}$ **6.** $7\frac{1}{3} - 3\frac{1}{5}$ $4\frac{2}{15}$ **7.** $24\frac{3}{8} - 13\frac{1}{4}$ **8.** $17\frac{5}{6} - 11\frac{7}{18}$

9. $5\frac{1}{3} - 1\frac{1}{2}$ $3\frac{5}{6}$ **10.** $15\frac{3}{5} - 8\frac{3}{4}$ **11.** $7\frac{1}{6} - 2\frac{7}{8}$ $4\frac{7}{24}$ **12.** $15\frac{5}{8} - 10\frac{2}{3}$

13. $6 - 2\frac{2}{3}$ $3\frac{1}{3}$ **14.** $7 - 2\frac{1}{4}$ $4\frac{3}{4}$ **15.** $10 - \frac{5}{8}$ $9\frac{3}{8}$ **16.** $25 - 11\frac{4}{7}$

Exercises

Practice

28. $2\frac{19}{24}$ 30. $15\frac{3}{5}$ 31. $1\frac{3}{8}$ 34. $19\frac{1}{2}$ 35. $10\frac{5}{8}$

Subtract. 17. $10\frac{3}{8}$ 19. $3\frac{2}{3}$ 20. $8\frac{11}{24}$ 21. $8\frac{1}{3}$ 23. $1\frac{33}{40}$

17. $16\frac{11}{16} - 6\frac{5}{16}$ **18.** $9\frac{7}{9} - 6\frac{4}{9}$ $3\frac{1}{3}$ **19.** $10\frac{11}{12} - 7\frac{1}{4}$ **20.** $17\frac{5}{6} - 9\frac{3}{8}$

21. $17\frac{1}{6} - 8\frac{5}{6}$ **22.** $4\frac{1}{7} - 2\frac{3}{4}$ $1\frac{11}{28}$ **23.** $3\frac{5}{8} - 1\frac{4}{5}$ **24.** $9\frac{5}{8} - 7\frac{5}{6}$ $1\frac{19}{24}$

25. $6\frac{1}{2} - 1\frac{2}{3}$ $4\frac{5}{6}$ **26.** $9\frac{3}{5} - \frac{4}{5}$ $8\frac{4}{5}$ **27.** $9\frac{2}{5} - 6\frac{9}{10}$ $2\frac{1}{2}$ **28.** $5\frac{3}{8} - 2\frac{7}{12}$

29. $14 - \frac{2}{7}$ $13\frac{5}{7}$ **30.** $50 - 34\frac{2}{5}$ **31.** $36 - 34\frac{5}{8}$ **32.** $10 - 1\frac{1}{9}$ $8\frac{8}{9}$

33. $(3\frac{3}{4} + 4\frac{1}{4}) - 2\frac{1}{2}$ $5\frac{1}{2}$ **34.** $(16 + 14\frac{7}{8}) - 11\frac{3}{8}$ **35.** $(8\frac{7}{8} + 5\frac{1}{4}) - 3\frac{1}{2}$

36. Subtract $2\frac{3}{8}$ from 20. $17\frac{5}{8}$

37. Find the difference of $9\frac{5}{8}$ and $7\frac{5}{6}$. $1\frac{19}{24}$

Applications

FUN with MATH
Which country invented the first wheelbarrow?
See page 174.

38. $\frac{13}{100}$ 39. $\frac{22}{75}$ 40. $\frac{17}{24}$ 41. $\frac{5}{12}$ 42. Middle East-Africa and Soviet Union

38. How much more energy does Europe use than it produces?

39. How much more energy do the Americas produce than China?

40. How much more energy do the Americas use than the Soviet Union?

41. How much energy is used by the Soviet Union, and Europe?

42. Which two regions produce about the same amount of energy?

43. List the regions from greatest to least based on energy use.

World's Energy Production and Usage

Region	Produced	Used
The Americas 1	$\frac{1}{3}$	$\frac{7}{8}$
Middle East-Africa 5	$\frac{3}{10}$	$\frac{1}{20}$
Soviet Union 3	$\frac{1}{5}$	$\frac{1}{6}$
Western Europe 2	$\frac{3}{25}$	$\frac{1}{4}$
China 4	$\frac{1}{25}$	$\frac{7}{100}$

Collect Data

44. Take a survey to find what fractional part of each day students study. Then make a table to compare study time of boys to girls according to grade level by arranging the fractions from greatest to least. **See student's work.**

APPLYING THE LESSON

Independent Practice

Homework Assignment	
Minimum	Ex. 2-44 even
Average	Ex. 1-37 odd; 38-44
Maximum	Ex. 1-44

Chalkboard Examples

Example A *Subtract.*

1. $8\frac{3}{5} - 5\frac{1}{5}$ $3\frac{2}{5}$ **2.** $12\frac{9}{10} - 6\frac{3}{10}$ $6\frac{3}{5}$

Example B *Subtract.*

3. $6\frac{3}{7} - 2\frac{3}{14}$ $4\frac{3}{14}$ **4.** $6\frac{2}{3} - 1\frac{5}{9}$ $5\frac{1}{9}$

Example C *Subtract.*

5. $11\frac{1}{5} - 8\frac{2}{5}$ $2\frac{4}{5}$ **6.** $13\frac{1}{3} - 8\frac{2}{5}$ $4\frac{14}{15}$

Example D *Subtract.*

7. $12\frac{4}{7} - 9$ $3\frac{4}{7}$ **8.** $20 - 2\frac{1}{2}$ $17\frac{1}{2}$

Additional examples are provided on TRANSPARENCY 4-13.

EVALUATING THE LESSON

Check for Understanding Use Guided Practice Exercises to check for student understanding.

Error Analysis Watch for students who are having trouble renaming before subtracting.

Closing the Lesson Have students state in their own words how to subtract mixed numbers.

EXTENDING THE LESSON

Enrichment Masters Booklet, p. 45

Name ______ Date ______

ENRICHMENT WORKSHEET 4-13

Subtraction with Mixed Numbers

Solve. Use the chart. Write your answers in simplest form.

1. How much more fabric does it take to make the jacket than the shirt? $1\frac{5}{8}$ yd

2. How many yards of fabric does it take to make both the shirt and the pants? $4\frac{1}{8}$ yd

3. Helene has $5\frac{3}{8}$ yards of fabric. Does she have enough to make both the pants and the jacket? no

Sew-Easy Pattern $2.50

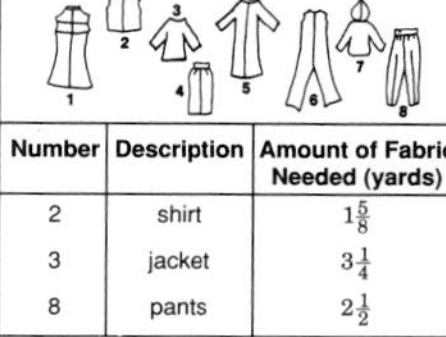

Number	Description	Amount of Fabric Needed (yards)
2	shirt	$1\frac{5}{8}$
3	jacket	$3\frac{1}{4}$
8	pants	$2\frac{1}{2}$

4. Lois has $3\frac{5}{12}$ yards of fabric. She cuts out the pattern for the shirt. How much fabric does she have left over? $1\frac{19}{24}$ yd

5. Pat has $7\frac{3}{16}$ yards of fabric. She cuts out the shirt and the jacket. How much extra fabric does she have? $2\frac{5}{16}$ yd

6. Michael makes two shirts and one pair of pants. How much fabric does he need? $5\frac{3}{4}$ yd

7. What is the total amount of fabric needed to make the shirt, pants, and jacket? $7\frac{3}{8}$ yd

Solve. Write your answer in simplest form.

8. Nick buys $\frac{3}{4}$ pound of grapes and $1\frac{2}{3}$ pounds of cherries. He buys enough plums to make a total of 4 pounds of fruit. How many pounds of plums does he buy? $1\frac{7}{12}$ pounds

5-MINUTE CHECK-UP

(over Lesson 4-13)

Available on TRANSPARENCY 4-14.

Subtract.

1. $12\frac{5}{7} - 9\frac{3}{7}$ $3\frac{2}{7}$
2. $9\frac{1}{4} - 7\frac{1}{6}$ $2\frac{1}{12}$
3. $14\frac{2}{9} - 12\frac{2}{3}$ $1\frac{5}{9}$
4. $16 - 8\frac{3}{5}$ $7\frac{2}{5}$

Extra Practice, Lesson 4-13, p. 460

INTRODUCING THE LESSON

Ask students what the easiest way is to give directions. to draw a map Have students draw a map from their house to school. Then while holding the map up, ask students to give verbal directions.

TEACHING THE LESSON

Using Cooperative Groups Divide students into small groups and have each group write a problem similar to the ones on pages 136-137. Once their group has solved it they should pass it to another group to solve.

Practice Masters Booklet, p. 49

PRACTICE WORKSHEET 4-14

Problem-Solving Strategy: Make a Drawing

Solve. Drawings may vary.

1. A loaf of bread is 10 inches long. How many cuts are necessary to cut it into 12 equal slices?

11 cuts

2. A floor is $6\frac{2}{3}$ feet long and $5\frac{3}{4}$ feet wide. How many 1-foot square tiles are needed to cover the floor?

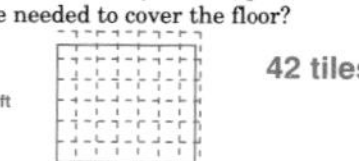

42 tiles

3. Kima runs 5 yards every time Randy runs 4 yards. If Randy runs 24 yards, how far does Kima run?

4 4 4 4 4 4
5 5 5 5 5 5

30 yards

4. Each bead on an add-a-bead necklace costs $2. How much do 6 beads cost?

$2 $2 $2 $2 $2 $2

$12

5. In how many ways can you make two right triangles from an equilateral triangle with a single straight cut?

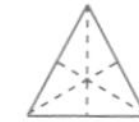

3 ways

6. Every time Jennifer earns a dollar she saves $0.35. If she has earned $7, how much has she saved?

$1 $1 $1 $1 $1 $1 $1
savings $0.35

$2.45

7. The Cameron family is seated at a circular table for a holiday dinner. If the third person is directly across from the ninth person, and each person is equidistant from the adjacent persons, how many people are seated at the table?

12 people

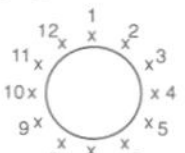

8. Mr. Gardner is building a 100-foot fence across the back of his property. If he places the fence posts 10 feet apart, how many posts will he need?

11 posts

x–x–x–x–x–x–x–x–x–x–x
10 ft

Problem-Solving Strategy

Explore
Plan
Solve
Examine

4-14 MAKE A DRAWING

Objective

Solve verbal problems by making a drawing.

Paul left his home and jogged north 2 blocks, then east 3 blocks, and then south 2 blocks. How many blocks east of his home was he when he stopped jogging?

Sometimes it helps to make a drawing when you are deciding how to solve a problem.

Explore What facts are given?

- The jogging path is:
  north 2 blocks
  east 3 blocks
  south 2 blocks

What fact do you need to find?

- How many blocks east of his home does Paul stop jogging?

Plan Draw a map that shows Paul's house (the starting point), the jogging path, and the stopping point.

Solve Draw a map of Paul's jogging path.

From the map, Paul stopped jogging 3 blocks east of his home.

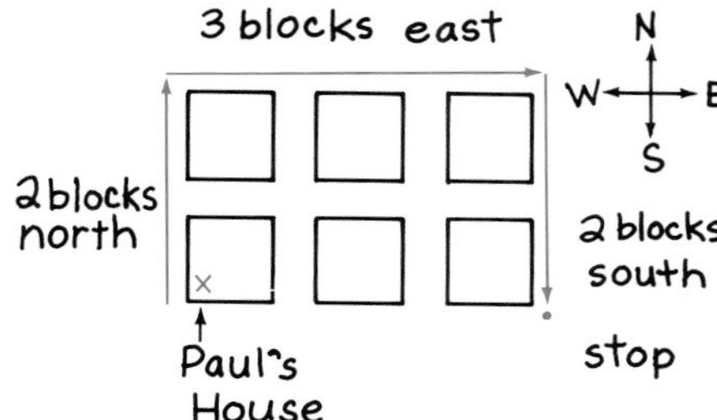

Examine Look at the drawing of your map. Did you draw 2 blocks north, 3 blocks east, and 2 blocks south? Yes. So you can see Paul is 3 blocks east of his home.

Guided Practice

Solve. Make a drawing.

1. Dario paints 3 sections of fence in the time it takes Marty to paint 2 sections of fence. Together they paint 40 sections of fence. How many sections does Marty paint? 16 sections
2. Each page of a photo album is $8\frac{1}{2}$ inches by 11 inches. At most how many 4-inch by 5-inch pictures will fit on one page? 4 pictures

RETEACHING THE LESSON

Have students work in small groups to draw a picture or diagram for each problem on page 137. Help students draw a scale model by giving them grid paper and hints as to what scale to use. For example, in problem 3, let one square equal 1 inch.

Problem Solving

Solve. Use any strategy.

3. Erica plans to sew two squares of cloth together along one side to make a placemat. Then she wants to sew a braid around the outside. Each square of cloth is $8\frac{1}{2}$ inches by $8\frac{1}{2}$ inches. How many inches of braid does Ashley need? **51 inches**
4. It cost $0.10 to cut and weld a link. How much does it cost to make a chain out of 10 links? **$0.90**
5. A lizard climbs a 30-foot pole. Each day it climbs up 7 feet and each night he slips back 4 feet. How many days will it take the lizard to reach the top? **9 days**
6. Luke can rake a lawn that is 60 meters by 60 meters in 4 hours. At the same rate, how long will it take him to rake a lawn that is 30 meters by 30 meters? **1 hour**
7. Mrs. Fern's dance class is standing evenly spaced in a circle. If the sixth person is directly opposite the sixteenth person, how many persons are in the circle? **20 persons**
8. Three spiders are on a 9-foot wall. The gray spider is 4 feet from the top. The brown spider is 7 feet from the bottom. The black spider is 3 feet below the brown spider. Which spider is the highest? **brown spider**

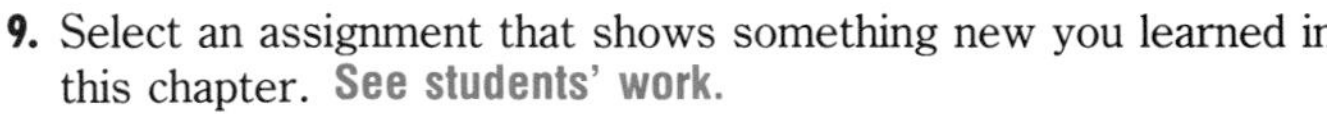

9. Select an assignment that shows something new you learned in this chapter. **See students' work.**

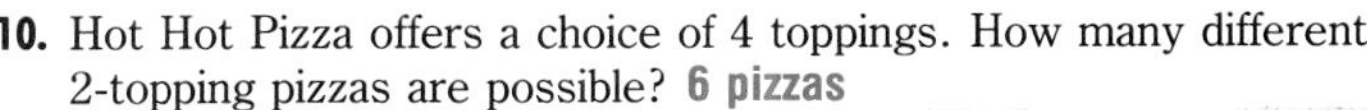

10. Hot Hot Pizza offers a choice of 4 toppings. How many different 2-topping pizzas are possible? **6 pizzas**
11. How many different ways can you buy 4 attached postage stamps? Two possible ways are shown at the right. **See answers on p. T521.**

PORTFOLIO

9. Heather has 17 cassettes and Tai has 9.

Mixed Review

Lesson 2-4

Subtract.

12. 49.6 − 18.4 = **31.2**
13. 90.1 − 48.6 = **41.5**
14. 800 − 34.5 = **765.5**
15. 7.773 − 4.2 = **3.573**

Lesson 3-7

Multiply or divide.

16. 18.4 × 100
17. 9.6 × 10 **96**
18. 37.2 × 100
19. 62.8 × 10
20. 0.84 ÷ 10
21. 62.4 ÷ 100
22. 15.9 ÷ 1,000
23. 46 ÷ 10 **4.6**

16. 1,840 18. 3,720 19. 628 20. 0.084 21. 0.624 22. 0.0159

Lesson 4-9

24. Antonio does $\frac{1}{8}$ of all his homework during study hall. He does $\frac{1}{4}$ of all his homework when he gets home from school, and he does $\frac{1}{2}$ of all his homework before he goes to bed. Does Antonio complete all of his homework? If not, how much does he still need to complete?

24. no, $\frac{1}{8}$

Chalkboard Examples

1. A receiver on the football team catches the kickoff at the twenty-yard line, gains ten yards, runs right three yards, then runs back 2 yards to avoid a defensive player, and is tackled. What yard line is he on at the end of the play, and how many yards does he gain? **28-yard line, 8 yards**

Additional examples are provided on TRANSPARENCY 4-14.

EXTENDING THE LESSON

Check for Understanding Use Guided Practice Exercises to check for student understanding.

Error Analysis Watch for students who cannot picture a problem in their minds. Encourage them to draw a step-by-step diagram as they read each problem.

Closing the Lesson Tell students that they will use this strategy when they use a road map to plan the best route for a vacation, when making plans for a deck, room addition, and so on, and whenever they need to see the problem visually.

APPLYING THE LESSON

Independent Practice

Homework Assignment	
Minimum	**Ex. 2-24 even**
Average	**Ex. 1-24**
Maximum	**Ex. 1-24**

Chapter 4, Quiz B (Lessons 4-8 through 4-14) is available in the Evaluation Masters Booklet, p. 19.

Using the Chapter Review

The Chapter Review is a comprehensive review of the concepts presented in this chapter. This review may be used to prepare students for the Chapter Test.

Chapter 4, Quizzes A and B, are provided in the Evaluation Masters Booklet as shown below.

Quiz A should be given after students have completed Lessons 4-1 through 4-7. Quiz B should be given after students have completed Lessons 4-8 through 4-14.

These quizzes can be used to obtain a quiz score or as a check of the concepts students need to review.

Evaluation Masters Booklet, p. 19

Name ____________ Date ____________

CHAPTER 4, QUIZ A (Lessons 4-1 through 4-7)

Write the prime factorization of the following number.

1. 90 — 1. $2 \times 3^2 \times 5$

Find the GCF or LCM of each group of numbers.

2. 30, 42 (GCF) 3. 6, 28 (LCM) — 2. 6 3. 84

Solve.

4. Tai's gross pay is $348.75. His total tax deductions are $52.92 and his total personal deductions are $8.50. What is his net pay? — 4. $287.33

Replace each □ with a number so that the fractions are equivalent.

5. $\frac{3}{8} = \frac{\square}{24}$ 6. $\frac{12}{20} = \frac{\square}{5}$ — 5. 9 6. 3

Simplify each fraction.

7. $\frac{18}{27}$ 8. $\frac{20}{24}$ — 7. $\frac{2}{3}$ 8. $\frac{5}{6}$

Replace each ○ with <, >, or = to make a true sentence.

9. $\frac{5}{9}$ ○ $\frac{7}{9}$ 10. $\frac{2}{3}$ ○ $\frac{3}{5}$ — 9. < 10. >

Name ____________ Date ____________

CHAPTER 4, QUIZ B (Lessons 4-8 through 4-14)

Change each fraction to a mixed number in the simplest form.

1. $\frac{34}{8}$ 2. $\frac{40}{7}$ — 1. $4\frac{1}{4}$ 2. $5\frac{5}{7}$

Estimate.

3. $5\frac{1}{10} + 3\frac{4}{9}$ — *3. $8\frac{1}{2}$ (* Accept any reasonable estimate.)

Add or subtract.

4. $\frac{1}{2} + \frac{3}{4}$ 5. $3\frac{4}{5} + 7\frac{4}{5}$ — 4. $1\frac{1}{4}$ 5. $11\frac{3}{5}$

6. $15\frac{3}{4} + 9\frac{5}{6}$ 7. $\frac{15}{16} - \frac{9}{16}$ — 6. $25\frac{7}{12}$ 7. $\frac{3}{8}$

8. $11\frac{2}{3} - 7\frac{1}{4}$ 9. $18\frac{1}{5} - 9\frac{2}{3}$ — 8. $4\frac{5}{12}$ 9. $8\frac{8}{15}$

Solve.

10. If a groove is to be made across a sidewalk every 4 feet, how many grooves will there be in a 24-foot-long sidewalk? — 10. 5 grooves

Chapter 4 REVIEW

Vocabulary / Concepts

Choose the word or number from the list at the right to complete each sentence.

Word list: $\frac{2}{3}$, $\frac{3}{4}$, 5, 10, 23, 30, composite, greatest common factor (GCF), least common denominator (LCD), mixed number, prime

1. The least common multiple of 10 and 15 is __?__. **30**
2. The number 27 is an example of a __?__ number. **composite**
3. __?__ is the simplest form of $\frac{16}{24}$. $\frac{2}{3}$
4. The fraction $\frac{12}{16}$ is equivalent to the fraction __?__. 4. $\frac{3}{4}$
5. 14 is the __?__ of 28 and 42. **GCF**
6. __?__ is a factor of 675 **5**
7. A __?__ indicates the sum of a whole number and a fraction. **mixed number**
8. In the fraction $\frac{23}{30}$, the numerator is __?__. **23**

Exercises / Applications

Lesson 4-1 ***Find all the factors of each number.***

9. 12 10. 18 11. 29 **1,29** 12. 38 13. 42

9. 1,2,3,4,6,12 10. 1,2,3,6,9,18 12. 1,2,19,38 13. 1,2,3,6,7,14,21,42

Lesson 4-1 ***State whether each number is divisible by 2, 3, 5, 9, or 10.***

14. 16 **2** 15. 21 **3** 16. 41 **none** 17. 54 **2,3,9** 18. 108 **2,3,9**

Lesson 4-2 ***Write the prime factorization for each number.***

19. 98 20. 78 21. 90 22. 475 23. 156

19. 2 × 7 × 7 20. 2 × 3 × 13 21. 2 × 3 × 3 × 5 22. 5 × 5 × 19 23. 2 × 2 × 3 × 13

Lesson 4-3 ***Find the GCF of each group of numbers.***

24. 6, 8 **2** 25. 32, 48 **16** 26. 18, 24 **6** 27. 54, 200 **2** 28. 12, 38, 62 **2**

Lesson 4-3 ***Find the LCM of each group of numbers.***

29. 7, 4 **28** 30. 16, 40 **80** 31. 8, 12 **24** 32. 6, 14 **42** 33. 2, 6, 10 **30**

Lesson 4-4 ***Solve.***

34. Jorge Ortega's gross pay is $523.78 a week. His total tax deductions are $142.09 and his personal deductions are $45.89. What are his total deductions and net pay? **$187.98; $335.80**
35. Ms. Chan has $89.42 deducted weekly from her paycheck for federal tax, $13.24 for state tax, and $6.99 for city tax. How much total tax does she pay yearly? **$5,701.80**

Lesson 4-5 — 40. 18 43. 21 44. 22

Replace each ■ with a number so the fractions are equivalent.

36. $\frac{4}{5} = \frac{■}{25}$ 20 **37.** $\frac{3}{8} = \frac{■}{32}$ 12 **38.** $\frac{21}{24} = \frac{■}{8}$ 7 **39.** $\frac{8}{36} = \frac{■}{9}$ 2 **40.** $\frac{9}{16} = \frac{■}{32}$

41. $\frac{2}{15} = \frac{■}{45}$ 6 **42.** $\frac{27}{45} = \frac{■}{5}$ 3 **43.** $\frac{42}{50} = \frac{■}{25}$ **44.** $\frac{11}{30} = \frac{■}{60}$ **45.** $\frac{36}{90} = \frac{■}{10}$ 4

Lesson 4-6

Simplify each fraction.

46. $\frac{4}{8}$ $\frac{1}{2}$ **47.** $\frac{12}{16}$ $\frac{3}{4}$ **48.** $\frac{21}{24}$ $\frac{7}{8}$ **49.** $\frac{6}{42}$ $\frac{1}{7}$ **50.** $\frac{32}{40}$ $\frac{4}{5}$ **51.** $\frac{38}{72}$ $\frac{19}{36}$

Lesson 4-7

Replace each ● with <, >, or = to make a true sentence.

52. $\frac{3}{5}$ ● $\frac{12}{20}$ = **53.** $\frac{5}{6}$ ● $\frac{7}{18}$ > **54.** $\frac{4}{15}$ ● $\frac{9}{16}$ < **55.** $\frac{4}{13}$ ● $\frac{4}{9}$ < **56.** $\frac{7}{8}$ ● $\frac{5}{6}$ >

Lesson 4-8

Change each fraction to a mixed number in simplest form.

57. $\frac{22}{3}$ $7\frac{1}{3}$ **58.** $\frac{37}{5}$ $7\frac{2}{5}$ **59.** $\frac{27}{4}$ $6\frac{3}{4}$ **60.** $\frac{94}{8}$ $11\frac{3}{4}$ **61.** $\frac{39}{5}$ $7\frac{4}{5}$ **62.** $\frac{83}{7}$ $11\frac{6}{7}$

Lesson 4-8

Change each mixed number to an improper fraction.

63. $4\frac{5}{8}$ $\frac{37}{8}$ **64.** $3\frac{6}{7}$ $\frac{27}{7}$ **65.** $6\frac{2}{3}$ $\frac{20}{3}$ **66.** $5\frac{4}{5}$ $\frac{29}{5}$ **67.** $7\frac{8}{9}$ $\frac{71}{9}$ **68.** $8\frac{9}{10}$ $\frac{89}{10}$

Lesson 4-9

Estimate. Answers may vary. Typical answers are given.

69. $\frac{2}{3} + \frac{1}{6}$ 1 **70.** $\frac{9}{20} + \frac{11}{12}$ **71.** $1\frac{3}{8} + 7\frac{1}{9}$ **72.** $\frac{7}{8} - \frac{1}{2}$ $\frac{1}{2}$ **73.** $14\frac{2}{9} - 7\frac{5}{6}$

Lessons 4-10, 4-11

Add. 70. $1\frac{1}{2}$ 71. $8\frac{1}{2}$ 73. 6

74. $\frac{3}{5} + \frac{1}{5}$ $\frac{4}{5}$ **75.** $\frac{5}{8} + \frac{5}{8}$ $1\frac{1}{4}$ **76.** $\frac{1}{3} + \frac{3}{10}$ $\frac{19}{30}$ **77.** $\frac{4}{5} + \frac{1}{4}$ **78.** $\frac{5}{6} + \frac{1}{10}$ $\frac{14}{15}$

79. $2\frac{4}{9} + 3\frac{7}{9}$ **80.** $1\frac{5}{8} + 4\frac{7}{8}$ **81.** $3\frac{5}{11} + 6\frac{4}{11}$ **82.** $14\frac{2}{3} + 4\frac{5}{12}$ **83.** $7\frac{2}{3} + 5\frac{1}{4}$

Lessons 4-12, 4-13

Subtract. 77. $1\frac{1}{20}$ 79. $6\frac{2}{9}$ 80. $6\frac{1}{2}$ 81. $9\frac{9}{11}$ 82. $19\frac{1}{12}$ 83. $12\frac{11}{12}$

84. $\frac{7}{12} - \frac{1}{12}$ $\frac{1}{2}$ **85.** $\frac{9}{10} - \frac{3}{10}$ $\frac{3}{5}$ **86.** $\frac{14}{15} - \frac{8}{15}$ $\frac{2}{5}$ **87.** $\frac{5}{6} - \frac{2}{9}$ $\frac{11}{18}$ **88.** $\frac{1}{2} - \frac{1}{14}$ $\frac{3}{7}$

89. $4\frac{4}{5} - 2\frac{2}{5}$ **90.** $7\frac{1}{8} - 5\frac{7}{8}$ **91.** $6\frac{3}{4} - 3\frac{1}{3}$ **92.** $12\frac{4}{7} - 9$ **93.** $20 - 2\frac{1}{2}$

89. $2\frac{2}{5}$
90. $1\frac{1}{4}$
91. $3\frac{5}{12}$
92. $3\frac{4}{7}$
93. $17\frac{1}{2}$ 94. $1\frac{3}{8}$ lb

94. Janice and Julie are identical twins. At birth, Janice weighed $6\frac{5}{8}$ lb and Julie weighed 8 lb. What was the difference in their weight?

Lesson 4-14

Solve. Use any strategy.

95. After a bake sale, there was one cake left. Anna cut the cake into thirds to share it with two co-workers. She then remembered Janice in the kitchen. How could Anna make one more cut to have four equal parts? See margin.

Alternate Assessment Strategy

To provide a brief in-class review, you may wish to read the following questions to the class and require a verbal response.

1. What are all the factors of 35? 1, 5, 7, 35
2. State the prime factorization of the number 28. $2^2 \times 7$
3. Find the GCF of 22 and 33. 11
4. Name a fraction equivalent to $\frac{2}{7}$. Answers may vary.
5. Simplify the fraction $\frac{18}{27}$. $\frac{2}{3}$
6. Compare the fractions $\frac{5}{7}$ and $\frac{3}{4}$. Which is greater? $\frac{3}{4}$
7. Change $\frac{18}{4}$ to a mixed number. $4\frac{1}{2}$
8. Estimate the sum of $\frac{3}{4}$ and $\frac{2}{3}$. 2
9. Find the sum of $\frac{3}{5}$ and $\frac{7}{9}$. $1\frac{17}{45}$
10. What is the result when $\frac{4}{9}$ is subtracted from $\frac{8}{9}$? $\frac{4}{9}$

APPLYING THE LESSON

Independent Practice

Homework Assignment	
Minimum	**Ex. 1-8; 9-95 odd**
Average	**Ex. 1-8; 2-92 even; 94-95**
Maximum	**Ex. 1-95**

Additional Answers

95. [diagram: rectangle divided into thirds by vertical lines, with dashed lines and one horizontal cut]

Using the Chapter Test

This page may be used as a test or as an additional page of review if necessary. Two forms of a Chapter Test are provided in the Evaluation Masters Booklet. Form 2 (free response) is shown below, and Form 1 (multiple choice) is shown on page 143.

The **Tech Prep Applications Booklet** provides students with an opportunity to familiarize themselves with various types of technical vocations. The Tech Prep applications for this chapter can be found on pages 7–8.

Evaluation Masters Booklet, p. 18

CHAPTER 4 TEST, FORM 2

State whether the following number is divisible by 2, 3, 5, 9, or 10.
1. 30 — 1. 2, 3, 5, 10

Write the prime factorization of the following number.
2. 140 — 2. $2^2 \times 5 \times 7$

Find the GCF of the following group of numbers.
3. 60,150 — 3. 30

Find the LCM of the following group of numbers.
4. 12, 30 — 4. 60

Solve.
5. Gina's gross pay is $295.50. Her total tax deductions are $48.17, and her total personal deductions are $12.00. What is her net pay? — 5. $235.33

Name a fraction equivalent to the following fraction.
6. $\frac{3}{4}$ — 6. $\frac{6}{8}, \frac{9}{12}, \frac{12}{16}$,

Simplify the following fraction.
7. $\frac{24}{30}$ — 7. $\frac{4}{5}$

Replace ○ with <, >, or = to make a true sentence.
8. $\frac{5}{9}$ ○ $\frac{2}{3}$ — 8. <

Change the following fraction to a mixed number in simplest form.
9. $\frac{30}{9}$ — 9. $3\frac{1}{3}$

Estimate. * Accept any reasonable estimate.
10. $5\frac{6}{7}+\frac{3}{8}$ 11. $9\frac{4}{7}-3\frac{1}{9}$

Add or subtract.
12. $\frac{1}{8}+\frac{3}{8}$ 13. $\frac{3}{5}+\frac{2}{3}$
14. $4\frac{5}{9}+8\frac{7}{9}$ 15. $6\frac{3}{4}+8\frac{1}{8}$
16. $\frac{13}{15}-\frac{4}{15}$ 17. $\frac{9}{10}-\frac{5}{6}$
18. $5\frac{4}{5}-1\frac{7}{10}$ 19. $14\frac{2}{3}-9\frac{3}{4}$

*10. $6\frac{1}{2}$ *11. $6\frac{1}{2}$ 12. $\frac{1}{2}$ 13. $1\frac{4}{15}$ 14. $13\frac{1}{3}$ 15. $14\frac{7}{8}$ 16. $\frac{3}{5}$ 17. $\frac{1}{15}$ 18. $4\frac{1}{10}$ 19. $4\frac{11}{12}$

Solve.
20. Four points are on a circle. How many line segments are needed to connect each point with each of the other points? — 20. 6 segments

Bonus The sum of three positive mixed numbers is always greater than what whole number? — Bonus 3

Chapter 4 TEST

Write the prime factorization for each number.

1. 24 $2 \times 2 \times 2 \times 3$
2. 54 $2 \times 3 \times 3 \times 3$
3. 81 $3 \times 3 \times 3 \times 3$

Replace each ■ with a number so that the fractions are equivalent.

4. $\frac{1}{3} = \frac{■}{18}$ 6
5. $\frac{21}{21} = \frac{■}{9}$ 9
6. $\frac{12}{20} = \frac{■}{5}$ 3

Solve.

7. Tanya Smith's gross pay is $425.81. Her total tax deductions are $119.39 and her personal deductions are $58.07. What is her net pay? **$248.35**

Simplify each fraction.

8. $\frac{9}{12}$ $\frac{3}{4}$
9. $\frac{8}{16}$ $\frac{1}{2}$
10. $\frac{15}{20}$ $\frac{3}{4}$

Replace each ● with <, >, or = to make a true sentence.

11. $\frac{8}{9}$ ● $\frac{7}{9}$ >
12. $\frac{2}{3}$ ● $\frac{3}{4}$ <
13. $\frac{5}{6}$ ● $\frac{10}{12}$ =

Change each mixed number to an improper fraction.

14. $2\frac{2}{3}$ $\frac{8}{3}$
15. $6\frac{4}{9}$ $\frac{58}{9}$
16. $8\frac{3}{10}$ $\frac{83}{10}$

Estimate. **Answers may vary. Typical answers are given.**

17. $\frac{9}{10} + \frac{1}{8}$ 1
18. $2\frac{1}{3} + 7\frac{1}{4}$ $9\frac{1}{2}$
19. $\frac{7}{8} - \frac{1}{5}$ 1

Add or subtract.

20. $\frac{1}{5} + \frac{2}{5}$ $\frac{3}{5}$
21. $\frac{2}{5} + \frac{1}{4}$ $\frac{13}{20}$
22. $\frac{5}{12} + \frac{7}{8}$ $1\frac{7}{24}$
23. $3\frac{1}{5} + 5\frac{4}{5}$ 9
24. $7\frac{1}{6} + 2\frac{2}{3}$ $9\frac{5}{6}$
25. $13\frac{5}{7} + 6\frac{1}{4}$ $19\frac{27}{28}$
26. $\frac{5}{12} - \frac{1}{12}$ $\frac{1}{3}$
27. $\frac{7}{9} - \frac{1}{6}$ $\frac{11}{18}$
28. $\frac{5}{6} - \frac{2}{3}$ $\frac{1}{6}$
29. $7\frac{5}{6} - 5\frac{1}{6}$ $2\frac{2}{3}$
30. $7\frac{3}{8} - 2\frac{1}{4}$ $5\frac{1}{8}$
31. $8\frac{1}{9} - 1\frac{4}{9}$ $6\frac{2}{3}$

Solve. Use any strategy.

32. Andy ran on three different trails. The first was $1\frac{1}{2}$ miles long, the second $2\frac{1}{4}$, and the third $1\frac{3}{4}$. How far did he run? **$5\frac{1}{2}$ miles.**
33. The chess club has 6 members. If every member shakes hands with every other member, how many handshakes are there? **15 handshakes**

BONUS: Explain when the statement $\frac{a}{b} + \frac{c}{b} = 1$ is true. **When $a + c = b$**

The **Performance Assessment Booklet** provides an alternative assessment for evaluating student progress. An assessment for this chapter can be found on pages 7–8.

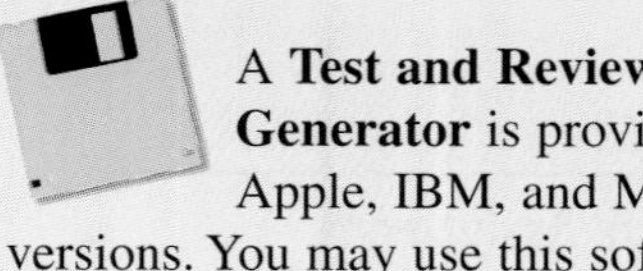

A **Test and Review Generator** is provided in Apple, IBM, and Macintosh versions. You may use this software to create your own tests or worksheets, based on the needs of your students.

Stock		High	Low	Last	Chg.
Herculs	2.24	$32\frac{1}{4}$	$28\frac{7}{8}$	30	$-2\frac{1}{4}$
Hrshey	.90	$35\frac{1}{8}$	$32\frac{3}{4}$	$34\frac{1}{4}$	$-\frac{3}{8}$
HewlPk	.42	$35\frac{1}{8}$	31	$33\frac{3}{4}$	$-\frac{7}{8}$
Hexcel	.44	$10\frac{3}{4}$	$9\frac{3}{4}$	10	$-\frac{3}{4}$
Hibern	1.00	$14\frac{7}{8}$	$12\frac{3}{4}$	$12\frac{7}{8}$	-2
HiShear	.22	11	$7\frac{7}{8}$	$7\frac{7}{8}$	-3
Hilnco	.90	$4\frac{7}{8}$	$4\frac{3}{8}$	$4\frac{1}{2}$	$-\frac{3}{8}$
Hlncll	1.08	$5\frac{1}{2}$	$4\frac{5}{8}$	$4\frac{5}{8}$	$-\frac{7}{8}$
Hilnlll	1.14	$6\frac{1}{4}$	$5\frac{7}{8}$	$5\frac{7}{8}$	$-\frac{3}{8}$

STOCKS

Tim and Nadine Jackson invest some of their money in stocks. Investing in stocks can be very risky. So the Jacksons study the stock market carefully. The list at the right shows some stock trades on the New York Stock Exchange. Some company names are abbreviated.

Hrshey → Hershey

HomeSh → Home Shopping Club

The profit or loss is found by comparing the quote of the share when it is bought with the quote when it is sold. What profit or loss would Tim and Nadine have if they buy 100 shares of Hercules at $28\frac{7}{8}$ and sell it for $32\frac{1}{4}$?

The price of each share is quoted in dollars to the nearest $\frac{1}{8}$ of a dollar.

selling price → $32\frac{1}{4}$ → $32\frac{1}{4}$

buying price → $28\frac{7}{8}$ → $-\ 28\frac{7}{8}$

$$32\frac{1}{4} - 28\frac{7}{8} = 3\frac{3}{8}$$

Tim and Nadine would make a profit of $3\frac{3}{8}$ or \$3.375 on each share. The profit on 100 shares is \$3.375 × 100 or \$337.50.

1. What is the highest quote for HiShear stock this year? **11**
2. What profit or loss would there be if 100 shares of Hercules were bought at $28\frac{1}{8}$ and sold at the last quote? **profit of \$187.50**

Suppose the graph at the right shows the net change in the HewlPk stock over one week.

3. What was the closing price Wednesday? **35**
4. What was the change between Thursday and Friday? **−2**
5. What was the overall change during the week shown? **+1**

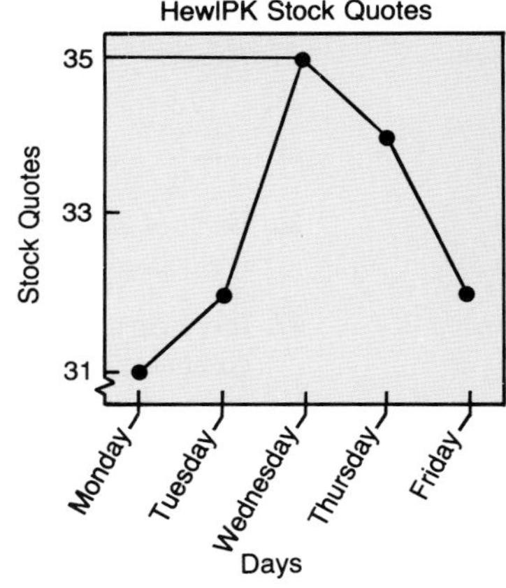

Applying Mathematics to the Real World

This optional page shows how mathematics is used in the real world and also provides a change of pace.

Objective Determine amount of profit or loss on stock.

Using Discussion Discuss the partial listing of the New York Stock Exchange. How is it read? What do the numbers represent?

Point out that a stockbroker is the person who buys the stock for a consumer. The stockbroker is then paid a commission on every stock transaction.

Using Critical Thinking What would be the reason anyone would graph what happens to certain stocks over a period of time? **People try to predict what will happen to decide if they should buy stocks.**

Activity

Using Data Tell students they have each inherited \$1,000 to invest in stocks. Students must choose a stock and plot its activity on a line graph over the next two weeks. At the end, students can find whether they made a profit or loss in their stock.

Using the Cumulative Review

This page provides an aid for maintaining skills and concepts presented thus far in the text.

A Cumulative Review is also provided in the Evaluation Masters Booklet as shown below.

Evaluation Masters Booklet, p. 20

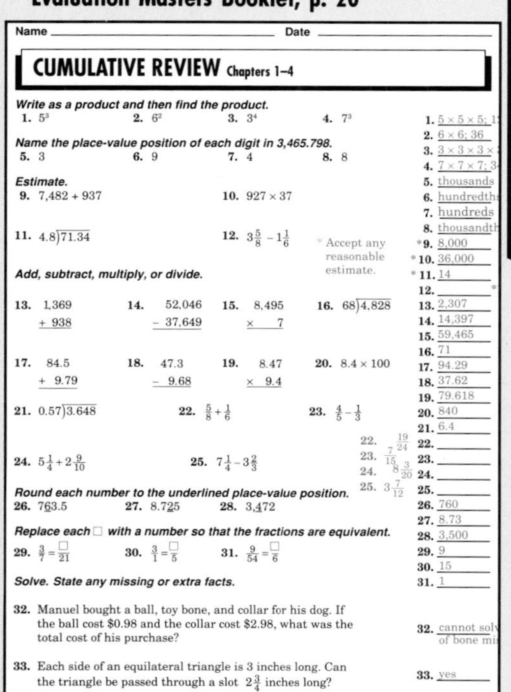

Name ____________ Date ____________

CUMULATIVE REVIEW Chapters 1–4

Write as a product and then find the product.

1. 5^3 2. 6^2 3. 3^4 4. 7^3

Name the place-value position of each digit in 3,465.798.

5. 3 6. 9 7. 4 8. 8

Estimate.

9. 7,482 + 937 10. 927 × 37

11. 4.8)71.34 12. $3\frac{5}{8} - 1\frac{1}{6}$

*Accept any reasonable estimate.

Add, subtract, multiply, or divide.

13. 1,369 + 938 14. 52,046 − 37,649 15. 8,495 × 7 16. 68)4,828

17. 84.5 + 9.79 18. 47.3 − 9.68 19. 8.47 × 9.4 20. 8.4 × 100

21. 0.57)3.648 22. $\frac{5}{8} + \frac{1}{6}$ 23. $\frac{4}{5} - \frac{1}{3}$

24. $5\frac{1}{4} + 2\frac{9}{10}$ 25. $7\frac{1}{4} - 3\frac{2}{3}$

22. $\frac{19}{24}$ 23. $\frac{7}{15}$ 24. $8\frac{3}{20}$ 25. $3\frac{7}{12}$

Round each number to the underlined place-value position.

26. 763.5 27. 8.725 28. 3,472

Replace each □ with a number so that the fractions are equivalent.

29. $\frac{3}{7} = \frac{\square}{21}$ 30. $\frac{3}{1} = \frac{\square}{5}$ 31. $\frac{9}{54} = \frac{\square}{6}$

Solve. State any missing or extra facts.

32. Manuel bought a ball, toy bone, and collar for his dog. If the ball cost $0.98 and the collar cost $2.98, what was the total cost of his purchase?

33. Each side of an equilateral triangle is 3 inches long. Can the triangle be passed through a slot $2\frac{3}{4}$ inches long?

1. 5 × 5 × 5; 1
2. 6 × 6; 36
3. 3 × 3 × 3 ×
4. 7 × 7 × 7; 3
5. thousands
6. hundredths
7. hundreds
8. thousandth
*9. 8,000
*10. 36,000
*11. 14
12. ____ *
13. 2,307
14. 14,397
15. 59,465
16. 71
17. 94.29
18. 37.62
19. 79.618
20. 840
21. 6.4
22. ____
23. ____
24. ____
25. ____
26. 760
27. 8.73
28. 3,500
29. 9
30. 15
31. 1
32. cannot sol of bone mi
33. yes

Cumulative REVIEW/TEST

Free Response

Lesson 1-1

Round each number to the underlined place-value position.

1. 628 **600** **2.** 5,354 **5,350** **3.** 3,890 **4,000** **4.** 31,617 **30,000** **5.** 850,638 **850,600**

Lessons 1-6, 1-7

Estimate.

6. 351 + 490 **900** **7.** 6,082 − 693 **5,400** **8.** 495 × 57 **30,000** **9.** 55)436 **7** **10.** 8,774 − 4,876 **4,000**

11. 7,762 + 856 **8,700** **12.** 6,493 ÷ 74 **90** **13.** 325 × 608 **180,000**

14. 5,000 − 3,837 **1,000** **15.** 639 × 471 **300,000** **16.** 48,470 ÷ 583 **80**

Lesson 1-13

Write an equation to solve this problem. Then solve.

17. Alison has a collection of 483 stamps. How many stamps does she need to add to her collection to raise the total to 625?
483 + n = 625; 142

Lessons 2-1, 2-3

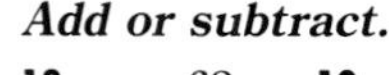

Add or subtract.

28. 10,600

18. 62 + 853 **915** **19.** 540 + 389 **929** **20.** 316 − 178 **138** **21.** 506 − 38 **468** **22.** 160 + 825 **985** **23.** 600 − 352 **248**

24. 8,706 + 537 **9,243** **25.** 480 − 309 **171** **26.** 7,800 − 279 **7,521**

27. 512 + 4,246 **4,758** **28.** 5,123 + 5,477 **29.** 6,231 − 2,345 **3,886**

30. A car dealer sold 1,774 cars one year and 1,967 cars the next year. How many cars were sold in the two years? **3,741 cars**

Lesson 3-1

Multiply.

31. 73 × 2 **146** **32.** 646 × 5 **3,230** **33.** 2,374 × 4 **9,496** **34.** 126 × 58 **7,308** **35.** 710 × 14 **9,940**

36. Each of 231 persons donates $21 to the Heart Fund. How much is donated in all? **$4,851**

Lesson 3-13

37. Jerry earns $20,800 a year. He would like to purchase a computer that costs $1,200. How long will it take him to save enough money to buy the computer if he saves $200 every two weeks? **12 weeks**

Lesson 4-2

Write the prime factorization for each number. **See margin.**

38. 36 **39.** 136 **40.** 720 **41.** 111 **42.** 1,800

Lesson 4-5

43. 70

Replace each ■ with a number so the fractions are equivalent.

43. $\frac{7}{10} = \frac{■}{100}$ **44.** $\frac{15}{20} = \frac{■}{4}$ **3** **45.** $\frac{2}{7} = \frac{■}{14}$ **4** **46.** $\frac{3}{6} = \frac{■}{2}$ **1** **47.** $\frac{8}{8} = \frac{■}{6}$ **6**

APPLYING THE LESSON

Independent Practice

Homework Assignment	
Minimum	Ex. 1-47 odd
Average	Ex. 1-47
Maximum	Ex. 1-47

Additional Answers

38. 2 × 2 × 3 × 3
39. 2 × 2 × 2 × 17
40. 2 × 2 × 2 × 2 × 3 × 3 × 5
41. 3 × 37
42. 2 × 2 × 2 × 3 × 3 × 5 × 5

Chapter 4

Multiple Choice

Choose the letter of the correct answer for each item.

1. How can 6 to the third power be written?
 a. 3×6
 b. 6×3
 ▶ c. $6 \times 6 \times 6$
 d. $6 + 6 + 6$
 Lesson 1-2

2. About how much is 17×53?
 e. 350
 ▶ f. 1,000
 g. 10,000
 h. *none of the above*
 Lesson 1-7

3. What is the result when 6.71 is subtracted from 3,028.5?
 a. 2,961.4
 ▶ b. 3,021.79
 c. 3,021.89
 d. *none of the above*
 Lesson 2-4

4. A cheetah can run 32 meters per second. How far can the cheetah run in 2.5 seconds?
 e. 8 meters
 f. 12.8 meters
 ▶ g. 80 meters
 h. 800 meters
 Lesson 3-2

5. Denny received 90, 87, 89, and 93 on his last four history exams. He needs a total of 455 points to get an A for the quarter. What must he score on the last exam?
 a. 90
 b. 95
 ▶ c. 96
 d. 93
 Lesson 2-1

6. Write 17,500,000 in scientific notation.
 e. 1.75×10^5
 ▶ f. 1.75×10^7
 g. 17.5×10^6
 h. 17.5×10^7
 Lesson 3-8

7. Find the sum of 28.5 and 6.
 a. 29.1
 ▶ b. 34.5
 c. 44.5
 d. 88.5
 Lesson 2-2

8. Herb divides 1,320 marbles among 27 people. What is the result?
 e. 4 R27
 ▶ f. 48 R24
 g. 48 R34
 h. 49
 Lesson 3-4

9. A book carton weighs 1 pound. What is the weight when a book carton is filled with ten books? State the missing fact.
 a. weight of address labels
 b. shipping costs
 ▶ c. weight of each book
 d. *none of the above*
 Lesson 2-9

10. Janet cuts two pieces from a ribbon that is 1 foot long. One piece is $\frac{1}{2}$ foot long and the other is $\frac{1}{3}$ foot long. How long is the remaining ribbon?
 ▶ e. $\frac{1}{6}$ foot
 f. $\frac{3}{5}$ foot
 g. $\frac{4}{5}$ foot
 h. $\frac{5}{6}$ foot
 Lesson 4-12

APPLYING THE LESSON

Independent Practice

Homework Assignment	
Minimum	Ex. 1-10
Average	Ex. 1-10
Maximum	Ex. 1-10

Using the Standardized Test

This test serves to familiarize students with standardized format while testing skills and concepts presented up to this point.

Evaluation Masters Booklet, p. 16-17

CHAPTER 4 TEST, FORM 1

Write the letter of the correct answer on the blank at the right of the page.

1. Find all the factors of 30.
 A. 1, 2, 3, 4, 5, 6, 10, 15 B. 1, 2, 3, 4, 5, 6, 10
 C. 1, 2, 3, 5, 6, 10, 15, 30 D. 1, 2, 3, 6, 10, 15 — 1. C

2. Find the number divisible by 9.
 A. 573 B. 3,470 C. 875 D. 2,448 — 2. D

3. Find the prime factorization of 120.
 A. $3 \times 4 \times 10$ B. $2^3 \times 3 \times 5$
 C. $2 \times 3 \times 4 \times 5$ D. $2 \times 3 \times 5^2$ — 3. B

4. Find the GCF of 60 and 84.
 A. 12 B. 6 C. 15 D. 24 — 4. A

5. Find the LCM of 50 and 60.
 A. 1,500 B. 300 C. 600 D. 3,000 — 5. B

6. Marc's gross pay is \$246.38. His total tax deductions are \$43.22, and his personal deductions are \$15.00. What is his net pay?
 A. \$304.60 B. \$203.16 C. \$58.22 D. \$188.16 — 6. D

7. Find a number to replace □ in $\frac{18}{27} = \frac{\square}{3}$ so that the fractions are equivalent.
 A. 1 B. 4 C. 9 D. 2 — 7. D

8. Find a fraction equivalent to $\frac{3}{8}$.
 A. $\frac{15}{40}$ B. $\frac{3}{4}$ C. $\frac{6}{24}$ D. $\frac{12}{16}$ — 8. A

9. Simplify the fraction $\frac{36}{48}$.
 A. $\frac{6}{8}$ B. $\frac{9}{12}$ C. $\frac{3}{4}$ D. $\frac{1}{2}$ — 9. C

10. Simplify the fraction $\frac{27}{54}$.
 A. $\frac{1}{2}$ B. $\frac{1}{3}$ C. $\frac{9}{18}$ D. $\frac{13}{27}$ — 10. A

11. Find the true sentence.
 A. $\frac{3}{7} < \frac{2}{7}$ B. $\frac{3}{7} > \frac{2}{7}$ C. $\frac{2}{7} > \frac{3}{7}$ D. $\frac{2}{7} = \frac{3}{7}$ — 11. B

12. Find the true sentence.
 A. $\frac{8}{12} < \frac{12}{18}$ B. $\frac{8}{12} > \frac{12}{18}$ C. $\frac{12}{18} = \frac{8}{12}$ D. $\frac{12}{18} < \frac{8}{12}$ — 12. C

13. Find a mixed number equivalent to $\frac{28}{8}$.
 A. $3\frac{1}{2}$ B. 4 C. $2\frac{1}{2}$ D. $3\frac{3}{8}$ — 13. A

Chapter

5 MULTIPLYING AND DIVIDING FRACTIONS

PREVIEWING the CHAPTER

This chapter opens with a lesson on estimating products of mixed numbers and fractions. Students are then introduced to multiplying fractions and to multiplying fractions by whole numbers. The concept of multiplying fractions is then extended to include multiplication of mixed numbers. Division of fractions is introduced with a discussion of reciprocals. The method is then applied to division of mixed numbers. Other skills developed in this chapter include changing fractions to decimals and vice versa.

Problem-Solving Strategy Students learn to solve a problem by first solving a simpler problem.

Lesson (Pages)	Lesson Objectives	State/Local Objectives
5-1 (146-147)	5-1: Estimate products and quotients of fractions and mixed numbers.	
5-2 (148-149)	5-2: Multiply fractions.	
5-3 (150-151)	5-3: Multiply mixed numbers.	
5-4 (152-153)	5-4: Use fractions in a real-world application.	
5-5 (154-155)	5-5: Solve problems by simplifying the problem.	
5-6 (158-159)	5-6A: Find the reciprocal of a number. 5-6B: Divide fractions.	
5-7 (160-161)	5-7: Divide mixed numbers.	
5-8 (162-163)	5-8: Change fractions to decimals.	
5-9 (164-165)	5-9: Change decimals to fractions.	
5-10 (166-167)	5-10: Solve verbal problems involving fractions.	

ORGANIZING the CHAPTER

You may refer to the ***Course Planning Calendar*** on page T10.

Planning Guide		Blackline Masters Booklets								
Lesson (Pages)	**Extra Practice** (Student Edition)	*Reteaching*	*Practice*	*Enrichment*	*Activity*	*Multi-cultural Activity*	*Technology*	*Tech Prep*	*Evaluation*	**Other Resources**
5-1 (146-147)	p. 460	p. 46	p. 50	p. 46					p. 76	Transparency 5-1
5-2 (148-149)	p. 461	p. 47	p. 51	p. 47		p. 5				Transparency 5-2
5-3 (150-151)	p. 461	p. 48	p. 52	p. 48				p. 9		Transparency 5-3
5-4 (152-153)		p. 49	p. 53	p. 49						Transparency 5-4
5-5 (154-155)			p. 54						p. 24	Transparency 5-5
5-6 (158-159)	p. 461	p. 50	p. 55	p. 50						Transparency 5-6
5-7 (160-161)	p. 462	p. 51	p. 56	p. 51	pp. 9, 33			p. 10		Transparency 5-7
5-8 (162-163)	p. 462	p. 52	p. 57	p. 52			p. 5			Transparency 5-8
5-9 (164-165)	p. 462	p. 53	p. 58	p. 53	p. 10					Transparency 5-9
5-10 (166-167)		p. 54	p. 59	p. 54					p. 24	Transparency 5-10
Ch. Review (168-169)									pp. 21- 23	Test and Review Generator
Ch.Test (170)										
Cumulative Review/Test (172-173)									p. 25	

OTHER CHAPTER RESOURCES

Student Edition

Chapter Opener, p. 144
Activity, p. 145
On-the-Job Application, p. 156
Consumer Connections, p. 157
Journal Entry, p. 147
Portfolio Suggestion, p. 167
Consumer Connections, p. 171
Tech Prep Handbook, pp. 522-524

Teacher's Classroom Resources

Transparency 5-0
Lab Manual, pp. 31-32
Performance Assessment, pp. 9-10
Lesson Plans, pp. 50-59

Other Supplements

Overhead Manipulative Resources, Labs 15-17, pp. 11-12
Glencoe Mathematics Professional Series

ENHANCING the CHAPTER

Some of the blackline masters for enhancing this chapter are shown below.

COOPERATIVE LEARNING

This activity can be used with groups of 2-4 students. Have students write each of the numbers 1-8 on a separate card. Choose two fractions with numerators and denominators of 1-8 and write either their product and sum or their quotient and difference on the chalkboard. Have student teams use the numbers 1-8 to form two fractions having the given product and sum and so on. Check the fractions formed and have a volunteer work the problem for the class.

USING MODELS/MANIPULATIVES

Give students a copy of a map that has a scale in which some fraction of an inch equals 1 mile ($\frac{1}{4}$ inch = 1 mile). Have students use a ruler to find two points on the map that are more than an inch apart ($1\frac{1}{2}$). Have them find and mark a point halfway between the two locations by multiplying by $\frac{1}{2}$ and then counting marks on the ruler to verify the location. Next, have students find two points that are less than 1 inch apart ($\frac{7}{8}$). Have them divide to find how many $\frac{1}{4}$-inch segments there are in $\frac{7}{8}$ inch.

Cooperative Problem Solving, p. 33

Name ______________________ Date ______________________

COOPERATIVE PROBLEM SOLVING (Use with Lesson 5-7)

An Area of Knowledge

Do you think that the tops of two different objects can have the same area? Work in groups. Choose three books or magazines of different sizes in the classroom. Textbooks or reference books are fine.

- Find the length and width in inches of the cover of each book.
- Use the formula for the area of a rectangle: Area = *length* times *width* ($A = lw$) to find the area in square inches for the cover of each book.
- Record your results in this chart.

Book Name	Length (in.)	Width (in.)	Area (in²)
1.			
2.			
3.			

- Write the areas your group found on index cards and exchange with another group.
- See if each group can identify the books chosen by working backward from their cover areas to possible lengths and widths. Remember, some books chosen may have the same cover areas and/or the same dimensions.
- If two rectangular objects have the same dimensions, do they have the same area? yes
- If two rectangular objects have the same area, do they have the same dimensions? not necessarily
- Find two objects whose tops have different dimensions but the same area.

Lab Activity, p. 32

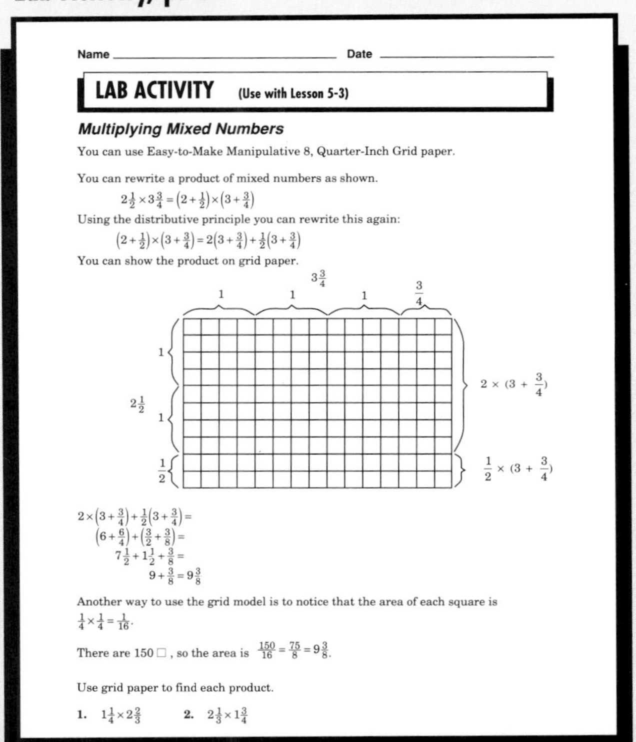

Name ______________________ Date ______________________

LAB ACTIVITY (Use with Lesson 5-3)

Multiplying Mixed Numbers

You can use Easy-to-Make Manipulative 8, Quarter-Inch Grid paper.

You can rewrite a product of mixed numbers as shown.

$$2\frac{1}{2} \times 3\frac{3}{4} = \left(2+\frac{1}{2}\right)\times\left(3+\frac{3}{4}\right)$$

Using the distributive principle you can rewrite this again:

$$\left(2+\frac{1}{2}\right)\times\left(3+\frac{3}{4}\right) = 2\left(3+\frac{3}{4}\right)+\frac{1}{2}\left(3+\frac{3}{4}\right)$$

You can show the product on grid paper.

$$2\times\left(3+\frac{3}{4}\right)+\frac{1}{2}\left(3+\frac{3}{4}\right)=$$
$$\left(6+\frac{6}{4}\right)+\left(\frac{3}{2}+\frac{3}{8}\right)=$$
$$7\frac{1}{2}+1\frac{1}{2}+\frac{3}{8}=$$
$$9+\frac{3}{8}=9\frac{3}{8}$$

Another way to use the grid model is to notice that the area of each square is $\frac{1}{4}\times\frac{1}{4}=\frac{1}{16}$.

There are 150 □, so the area is $\frac{150}{16}=\frac{75}{8}=9\frac{3}{8}$.

Use grid paper to find each product.

1. $1\frac{1}{4}\times 2\frac{2}{3}$ 2. $2\frac{1}{3}\times 1\frac{3}{4}$

MEETING INDIVIDUAL NEEDS

Multicultural Activity

From the *Papyrus Rhind,* the historians of mathematics know that the Egyptians used fractions extensively. All fractions were reduced to sums of unit fractions having a numerator of 1. The only exceptions were $\frac{1}{2}$ and $\frac{2}{3}$. The Egyptians use fraction tables showing the breakdown of fractions of the form $\frac{2}{n}$ into sums of unit fractions. For example, $\frac{2}{5} = \frac{1}{3} + \frac{1}{15}$ Have students write sum of unit fractions for $\frac{2}{n}$ where $n = 7, 9$. ($\frac{2}{7} = \frac{1}{4} + \frac{1}{28}$, $\frac{2}{9} = \frac{1}{6} + \frac{1}{18}$)

CRITICAL THINKING/PROBLEM SOLVING

The order of the fractions may be important when computing the answer to a word problem containing fractions. Remind students that when multiplying two fractions the order of the fractions does not matter because multiplication is commutative. Point out that when dividing two fractions to solve a word problem, order is very important because division is not commutative. In many word problems involving division, you divide by the same size unit to find the number of the same size units.

COMMUNICATION

Speaking

As you work through the applications exercises in each lesson,have student volunteers read the problems aloud to the class. Have the volunteer identify the unknown quantity and the given information. Have him or her choose the operation for solving and give a reason for the choice. Finally, have the volunteer describe the computation and state the result. Other students should check the answer by estimating.

Applications, p. 9

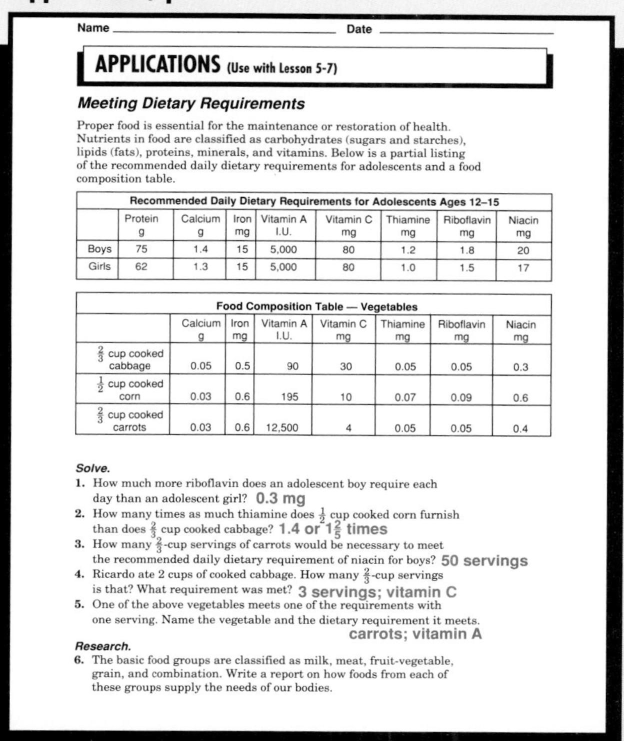

Name ______________ Date ______________

APPLICATIONS (Use with Lesson 5-7)

Meeting Dietary Requirements

Proper food is essential for the maintenance or restoration of health. Nutrients in food are classified as carbohydrates (sugars and starches), lipids (fats), proteins, minerals, and vitamins. Below is a partial listing of the recommended daily dietary requirements for adolescents and a food composition table.

Recommended Daily Dietary Requirements for Adolescents Ages 12–15

	Protein g	Calcium g	Iron mg	Vitamin A I.U.	Vitamin C mg	Thiamine mg	Riboflavin mg	Niacin mg
Boys	75	1.4	15	5,000	80	1.2	1.8	20
Girls	62	1.3	15	5,000	80	1.0	1.5	17

Food Composition Table — Vegetables

	Calcium g	Iron mg	Vitamin A I.U.	Vitamin C mg	Thiamine mg	Riboflavin mg	Niacin mg
$\frac{2}{3}$ cup cooked cabbage	0.05	0.5	90	30	0.05	0.05	0.3
$\frac{1}{2}$ cup cooked corn	0.03	0.6	195	10	0.07	0.09	0.6
$\frac{2}{3}$ cup cooked carrots	0.03	0.6	12,500	4	0.05	0.05	0.4

Solve.

1. How much more riboflavin does an adolescent boy require each day than an adolescent girl? **0.3 mg**
2. How many times as much thiamine does $\frac{1}{2}$ cup cooked corn furnish than does $\frac{2}{3}$ cup cooked cabbage? **1.4 or $1\frac{2}{5}$ times**
3. How many $\frac{2}{3}$-cup servings of carrots would be necessary to meet the recommended daily dietary requirement of niacin for boys? **50 servings**
4. Ricardo ate 2 cups of cooked cabbage. How many $\frac{2}{3}$-cup servings is that? What requirement was met? **3 servings; vitamin C**
5. One of the above vegetables meets one of the requirements with one serving. Name the vegetable and the dietary requirement it meets. **carrots; vitamin A**

Research.

6. The basic food groups are classified as milk, meat, fruit-vegetable, grain, and combination. Write a report on how foods from each of these groups supply the needs of our bodies.

Calculator Activity, p. 5

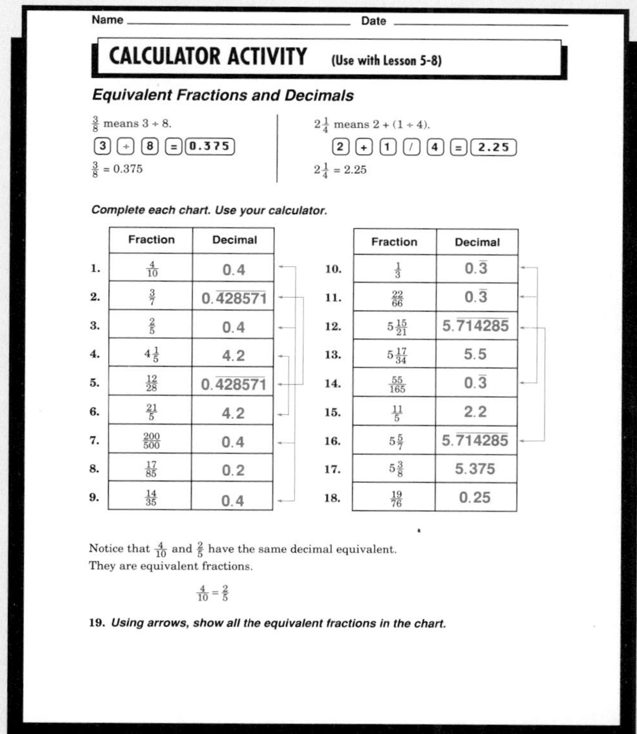

Name ______________ Date ______________

CALCULATOR ACTIVITY (Use with Lesson 5-8)

Equivalent Fractions and Decimals

$\frac{3}{8}$ means $3 \div 8$.

[3] [÷] [8] [=] [0.375]

$\frac{3}{8} = 0.375$

$2\frac{1}{4}$ means $2 + (1 \div 4)$.

[2] [+] [1] [/] [4] [=] [2.25]

$2\frac{1}{4} = 2.25$

Complete each chart. Use your calculator.

	Fraction	Decimal
1.	$\frac{4}{10}$	0.4
2.	$\frac{3}{7}$	$0.\overline{428571}$
3.	$\frac{2}{5}$	0.4
4.	$4\frac{1}{5}$	4.2
5.	$\frac{12}{28}$	$0.\overline{428571}$
6.	$\frac{21}{5}$	4.2
7.	$\frac{200}{500}$	0.4
8.	$\frac{17}{85}$	0.2
9.	$\frac{14}{35}$	0.4

	Fraction	Decimal
10.	$\frac{1}{3}$	$0.\overline{3}$
11.	$\frac{22}{66}$	$0.\overline{3}$
12.	$5\frac{15}{21}$	$5.\overline{714285}$
13.	$5\frac{17}{34}$	5.5
14.	$\frac{55}{165}$	$0.\overline{3}$
15.	$\frac{11}{5}$	2.2
16.	$5\frac{5}{7}$	$5.\overline{714285}$
17.	$5\frac{3}{8}$	5.375
18.	$\frac{19}{76}$	0.25

Notice that $\frac{4}{10}$ and $\frac{2}{5}$ have the same decimal equivalent. They are equivalent fractions.

$$\frac{4}{10} = \frac{2}{5}$$

19. ***Using arrows, show all the equivalent fractions in the chart.***

CHAPTER PROJECT

Have students calculate these interesting facts concerning their life processes.

- How many times has your heart beat since you were born?
  Have students find their pulse rates in beats per minute. Then multiply: beats/min $\times$ 60 $\times$ 24 $\times$ $365\frac{1}{4}$ $\times$ (your age).
- How much air will you breathe if you live to be 100 years old?
  Your lungs take in about 0.5 liters in 1 breath. Count your breaths in 1 minute. Multiply: breaths/min $\times$ 60 $\times$ 24 $\times$ $365\frac{1}{4}$ $\times$ 100.
- Have students use the formula $B.P. = 110 + (\frac{1}{2} \times A)$ where $B.P.$ represents blood pressure and A represents age to find their normal blood pressure reading.

If possible, ask a nurse or doctor to speak to the class on how mathematics is used in everyday work.

Using the Chapter Opener

Transparency 5-0 is available in the Transparency Package. It provides a full-color visual and motivational activity that you can use to engage students in the mathematical content of the chapter.

- Have students read the opening paragraph and allow time for students to exchange ideas about job opportunities.
- Suggest sources of information about successes other young people have had running their own businesses. The library is one source, as well as the small business groups in your area. One book you may wish to suggest is Allan Smith's *Teenage MONEYMAKING Guide.*

The second paragraph is a motivational problem. Have students find the solution and discuss the ways they obtained their answers.

Chapter 5

FRACTIONS: MULTIPLYING AND DIVIDING

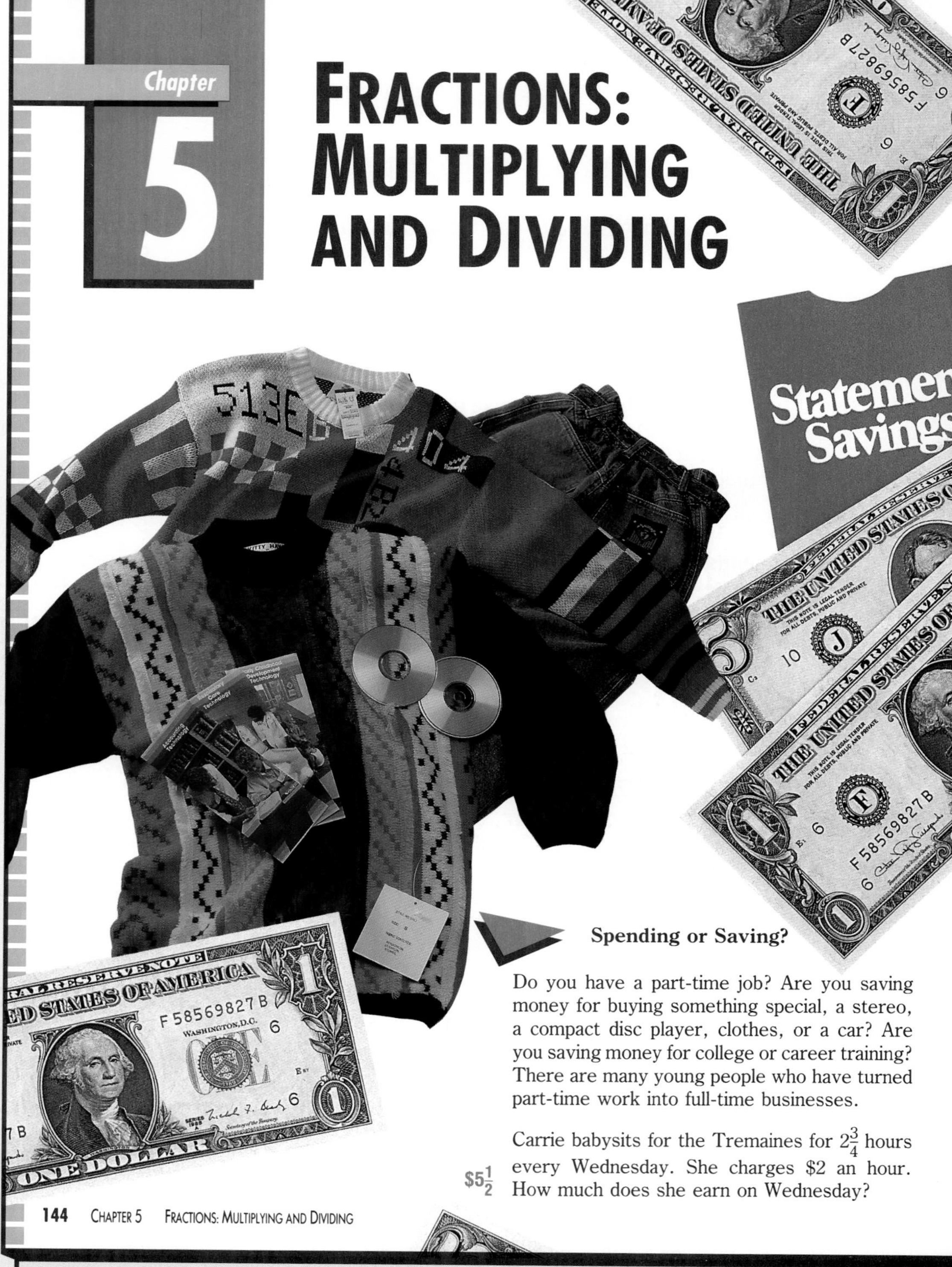

Spending or Saving?

Do you have a part-time job? Are you saving money for buying something special, a stereo, a compact disc player, clothes, or a car? Are you saving money for college or career training? There are many young people who have turned part-time work into full-time businesses.

Carrie babysits for the Tremaines for $2\frac{3}{4}$ hours every Wednesday. She charges \$2 an hour. How much does she earn on Wednesday? $\$5\frac{1}{2}$

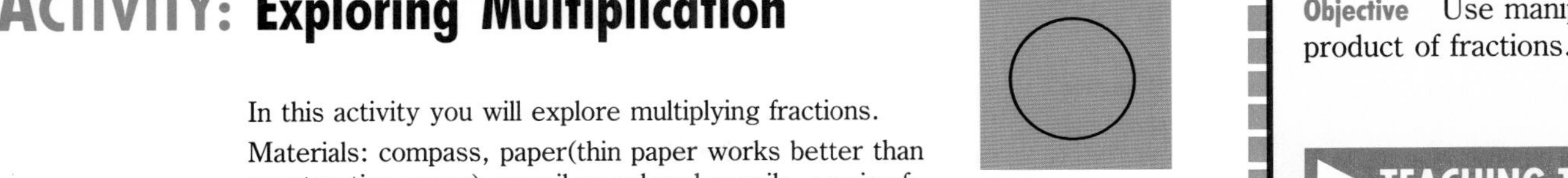

ACTIVITY: Exploring Multiplication

In this activity you will explore multiplying fractions.

Materials: compass, paper(thin paper works better than construction paper), pencil or colored pencils, a pair of scissors

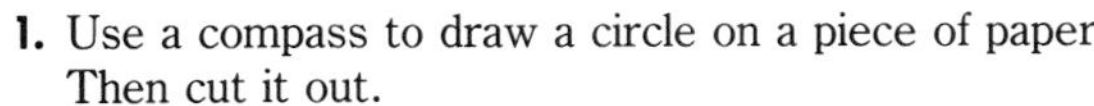

1. Use a compass to draw a circle on a piece of paper. Then cut it out.
2. Fold the circle in half, in half again, and then in half again, making sharp creases.
3. Unfold the circle.

See answers on p. T521.

Cooperative Groups ***Work together in small groups.***

4. Into how many parts do the creases separate the circle? What fraction represents one part of the circle?
5. To find the product of 5 and $\frac{1}{8}$, decide how many parts of the circle you should shade. What is the product of 5 and $\frac{1}{8}$?
6. Write a sentence describing how to find the product of any whole number and a fraction with a numerator of 1.

Extend the Concept

7. Cut a piece of paper into a 2-inch by 8-inch strip. Discuss with your group how to find $\frac{1}{2}$ of $\frac{1}{4}$, or $\frac{1}{2} \times \frac{1}{4}$ using the paper. Then find the product.
8. Cut a second strip of paper the same size as the one in Exercise 7. Discuss with your group how to show the product of $\frac{2}{3}$ and $\frac{3}{4}$. Then find the product.

Communicate Your Ideas

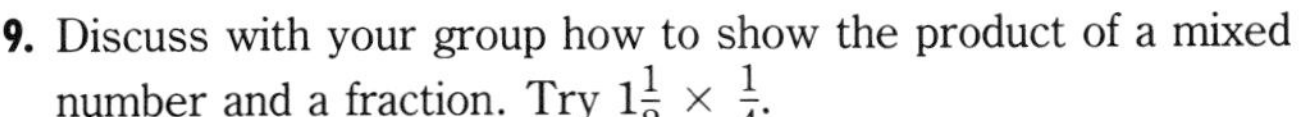

9. Discuss with your group how to show the product of a mixed number and a fraction. Try $1\frac{1}{2} \times \frac{1}{4}$.
10. With your group, discuss the results of Exercises 4-8 and write some ideas about the product of fractions that you have discovered.

INTRODUCING THE LESSON

Objective Use manipulatives to find the product of fractions.

TEACHING THE LESSON

Using Manipulatives In Exercises 7-9, encourage students to color the creases with one color for the first fold and a second color for the second set of folds. This will help them to separate and count the sections of the strip.

EVALUATING THE LESSON

Communicate Your Ideas Encourage students to write their solution to Exercise 9 and ideas from Exercise 10 in complete sentences. An alternative to complete sentences is outlining the solution steps.

EXTENDING THE LESSON

Closing the Lesson Encourage groups to share their ideas about products of fractions with other groups.

Activity Worksheet, Lab Manual, p. 31

ACTIVITY WORKSHEET (Use with page 145)

Exploring Multiplication

Use for Exercise 7.

Use for Exercise 8.

Use for Exercise 9.

5-MINUTE CHECK-UP

(over Lesson 4-14)

Available on TRANSPARENCY 5-1.

Solve.

1. From his home, Mark walked six blocks west, two blocks north, three blocks west, and two blocks south. How many blocks away from home is Mark? **9 blocks west**

▶ INTRODUCING THE LESSON

Have students estimate the area of the classroom. Then form a unit fraction using 1 for the numerator and the number of students in the class for the denominator. Have students try to estimate how much area of the classroom would be his or hers if each had an equal share.

▶ TEACHING THE LESSON

Using Cooperative Groups In groups of two or three, have students use the newspaper or a book of world records to make up 6 problems, 3 using proper fractions and 3 using mixed numbers. Then have groups exchange problems.

Practice Masters Booklet, p. 50

Name ________________ Date ________________

PRACTICE WORKSHEET 5-1

Estimating Products and Quotients

Estimate. **Answers may vary. Typical answers are given.**

1. $4\frac{1}{3} \times 3$ 12	2. $2\frac{1}{3} \times 1\frac{7}{8}$ 4	3. $5\frac{7}{8} \times 3\frac{2}{3}$ 24	4. $2\frac{1}{7} \times 3\frac{1}{6}$ 6
5. $7\frac{2}{7} \div 1\frac{1}{3}$ 7	6. $4\frac{2}{3} \div 1\frac{1}{6}$ 5	7. $14\frac{5}{6} \div 3\frac{1}{3}$ 5	8. $5\frac{1}{2} \div 2\frac{6}{7}$ 2
9. $4\frac{2}{5} \times 3\frac{1}{7}$ 12	10. $3\frac{3}{5} \times 1\frac{2}{7}$ 4	11. $2\frac{2}{5} \div 1\frac{4}{5}$ 1	12. $27\frac{1}{3} \div 9\frac{1}{3}$ 3
13. $6\frac{7}{8} \times \frac{7}{8}$ 7	14. $14\frac{1}{6} \div 2\frac{3}{4}$ 5	15. $1\frac{7}{8} \div 1\frac{5}{6}$ 1	16. $8 \times 5\frac{5}{9}$ 48
17. $23\frac{1}{2} \div 3\frac{5}{8}$ 6	18. $6\frac{3}{10} \times 2\frac{5}{8}$ 18	19. $2\frac{1}{4} \times 5\frac{3}{8}$ 10	20. $33\frac{1}{3} \div 4\frac{2}{5}$ 8
21. $16\frac{5}{6} \div 3\frac{9}{10}$ 4	22. $2\frac{7}{12} \times 1\frac{5}{12}$ 3	23. $9\frac{7}{8} \times 1\frac{9}{16}$ 20	24. $28\frac{2}{7} \div 9\frac{1}{4}$ 3

25. Is $\frac{5}{6}$ of 3 less than or greater than 3? (less than)

26. Is $\frac{1}{3}$ of $2\frac{1}{2}$ less than, greater than, or equal to $2\frac{1}{2}$? (less than)

27. Is $\frac{3}{8}$ of $\frac{1}{2}$ less than or greater than 0? (greater than)

28. Is $\frac{4}{5}$ of $1\frac{3}{4}$ less than or greater than $1\frac{3}{4}$? (less than)

29. Is $2\frac{1}{2} \div 4$ less than or greater than $2\frac{1}{2}$? (less than)

30. Is $3\frac{5}{6} \div \frac{2}{3}$ less than or greater than $3\frac{5}{6}$? (greater than)

31. Is $4\frac{1}{5} \div 3\frac{1}{5}$ less than or greater than $4\frac{1}{5}$? (less than)

32. Is $8 \div 4\frac{2}{3}$ less than or greater than 8? (less than)

5-1 ESTIMATING PRODUCTS AND QUOTIENTS

Objective

Estimate products and quotients of fractions and mixed numbers.

Mr. Sykes's science class discussed a newspaper article on energy consumption. The article stated that an average of $2\frac{5}{6}$ quadrillion BTUs of energy were used by each person in each household in the United States during 1985. Mr. Sykes asked the class *about* how many quadrillion BTUs of energy were used by each household if each household, on average, has about $3\frac{1}{5}$ people.

As with whole numbers and decimals, round mixed numbers before estimating products.

When estimating quotients, round the divisor to a whole number. Round the dividend to the nearest multiple of the divisor.

Examples

A ***Estimate $2\frac{5}{6} \times 3\frac{1}{5}$.***

$2\frac{5}{6}$ rounds to 3.

$3\frac{1}{5}$ rounds to 3.

$3 \times 3 = 9$

An average household used about 9 quadrillion BTUs.

B ***Estimate $14\frac{5}{8} \div 2\frac{1}{4}$.***

$2\frac{1}{4}$ rounds to 2.

Round $14\frac{5}{8}$ to 14 since 14 is the nearest multiple of 2.

Since $14 \div 2$ is 7, the estimate is 7.

To estimate the product of fractions or a fraction and a mixed number, note that multiplying by a proper fraction produces a product *less* than the other factor.

When you divide by a proper fraction, note that the quotient is always *greater* than the dividend.

Examples

C ***Is $\frac{1}{4}$ of $\frac{2}{3}$ less than or greater than $\frac{2}{3}$?***

Because you are finding $\frac{1}{4}$ of $\frac{2}{3}$, or only a *part* of $\frac{2}{3}$, the result will be less than $\frac{2}{3}$.

D ***Is $4 \div \frac{1}{3}$ less than or greater than 4?***

Dividing 4 by 1 gives 4 parts or a quotient of 4. Dividing 4 by a number less than 1 such as $\frac{1}{3}$ gives *more* than 4 parts or a quotient greater than 4.

Guided Practice

Example A

Estimate each product. **Answers may vary. Typical answers are given.**

1. $2\frac{3}{4} \times 7$ **21** **2.** $1\frac{1}{4} \times 1\frac{3}{5}$ **2** **3.** $3\frac{3}{4} \times 2\frac{2}{3}$ **12** **4.** $3\frac{1}{8} \times 3\frac{2}{5}$ **9**

RETEACHING THE LESSON

Review rounding fractions.

The numerator is:	The fraction is close to:
• much less than the denominator	0
• about half the denominator	$\frac{1}{2}$
• more than half the denominator	1

Reteaching Masters Booklet, p. 46

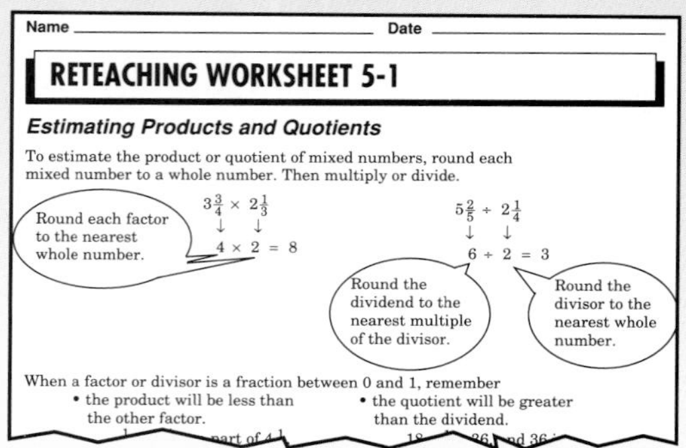
Name ________________ Date ________________

RETEACHING WORKSHEET 5-1

Estimating Products and Quotients

To estimate the product or quotient of mixed numbers, round each mixed number to a whole number. Then multiply or divide.

$3\frac{3}{4} \times 2\frac{1}{3}$ → $4 \times 2 = 8$

Round each factor to the nearest whole number.

$5\frac{2}{5} \div 2\frac{1}{4}$ → $6 \div 2 = 3$

Round the dividend to the nearest multiple of the divisor.

Round the divisor to the nearest whole number.

When a factor or divisor is a fraction between 0 and 1, remember

- the product will be less than the other factor.
- the quotient will be greater than the dividend.

Estimate each quotient. Answers may vary. Typical answers are given.

Example B
5. $5 \div 3\frac{1}{3}$ 2 **6.** $10\frac{1}{2} \div 2\frac{1}{4}$ 5 **7.** $4\frac{2}{3} \div 2$ 2 **8.** $8\frac{1}{3} \div 1\frac{5}{6}$ 4

Examples C, D
9. Is $\frac{3}{4}$ of 2 less than or greater than 2? **10.** Is $15 \div \frac{1}{4}$ less than or greater than 15?

Exercises

Estimate. Answers may vary. Typical answers are given.

Practice
11. $4\frac{1}{2} \times 2\frac{3}{4}$ 15 **12.** $5\frac{1}{3} \times 4\frac{1}{2}$ 25 **13.** $9\frac{3}{5} \div 4\frac{1}{5}$ 2 **14.** $8\frac{2}{3} \div 1\frac{3}{5}$ 4

15. $20\frac{5}{8} \div 3\frac{2}{5}$ 7 **16.** $17\frac{7}{8} \div 2\frac{3}{16}$ 9 **17.** $2\frac{5}{6} \times 1\frac{7}{8}$ 6 **18.** $1\frac{9}{16} \times 4\frac{4}{5}$ 10

19. Is $\frac{3}{4}$ of 2 less than or greater than 2? **20.** Is $15 \div \frac{1}{4}$ less than or greater than 15?

21. Is $\frac{1}{2}$ of $\frac{1}{2}$ less than, greater than, or equal to $\frac{1}{2}$?

22. Is $6\frac{2}{3} \div \frac{3}{5}$ less than or greater than 6?

Applications
23. Alec's car used 7 gallons of gasoline last week. If he paid $1.29 a gallon for gasoline, *about* how much did he spend on gasoline for the week? *about* $9

24. A $3\frac{1}{2}$-ounce ice cream bar contains 5 grams of fat per ounce. *About* how many grams of fat are contained in the whole bar? 20 grams

25. Linda breaks a seed-treat stick into 3 pieces for her guinea pig. If she feeds the guinea pig $1\frac{2}{3}$ sticks per week, *about* how long will a supply of $8\frac{1}{3}$ sticks last? *about* 4 weeks

26. If the Smith's fuel bill for last year was $1,180, *about* how much was their average monthly fuel bill? *about* $100

FUN with MATH
What does your refrigerator do to the ozone layer?
See page 174.

Using Algebra
Solve each equation.
27. $x + 6 = 11$ 5 **28.** $h - 9 = 4$ 13 **29.** $\frac{t}{4} = 9$ 36

JOURNAL ENTRY

30. Describe a situation in your life where you need to use estimation. See students' work.

Mixed Review
Divide. Round to the nearest hundredth.

Lesson 3-7
31. $6.3\overline{)43.7}$ 6.94 **32.** $1.8\overline{)9.76}$ 5.42 **33.** $22.4\overline{)484.3}$ 21.62

Lesson 3-8
Write in standard form.
34. 4.8×10^4 48,000 **35.** 6.23×10^3 6,230 **36.** 9.01×10^5 901,000

Lesson 4-10
37. Lindsey uses $\frac{3}{4}$ cup of sugar in her cookies. She uses $\frac{2}{3}$ cup of sugar for her cake. How much sugar does she use for both? $1\frac{5}{12}$ cups

Chalkboard Examples

Examples A, B *Estimate.*
Answers may vary. Typical answers are given.

1. $2\frac{1}{8} \times 1\frac{1}{3}$ 2 **2.** $9\frac{3}{4} \div 1\frac{5}{8}$ 5

Examples C, D

3. Is $\frac{1}{2}$ of $\frac{1}{4}$ less than or greater than $\frac{1}{4}$? less than

4. Is $6\frac{2}{3} \div \frac{3}{5}$ less than or greater than 6? greater than

Additional examples are provided on TRANSPARENCY 5-1.

▶ EVALUATING THE LESSON

Check for Understanding Use Guided Practice Exercises to check for student understanding.

Closing the Lesson Ask students to explain in their own words how to multiply and divide mixed numbers.

▶ EXTENDING THE LESSON

Enrichment Masters Booklet, p. 46

Name ______ Date ______

ENRICHMENT WORKSHEET 5-1

Estimation

Estimation is helpful in determining whether an answer is reasonable. For the following, choose the better estimate.

$2\frac{3}{7} + 3\frac{2}{5}$ **greater than 6** **less than 6**

To estimate, compare the fractional part of each mixed number to $\frac{1}{2}$.
Note that $\frac{3}{7} < \frac{1}{2}$ and $\frac{2}{5} < \frac{1}{2}$. So $\frac{3}{7} + \frac{2}{5} < 1$.
Since $2 + 3 = 5$ and $\frac{3}{7} + \frac{2}{5} < 1$, it follows that $2\frac{3}{7} + 3\frac{2}{5}$ is less than 6.

Underline the better estimate.

1.	$8\frac{4}{5} + 7\frac{8}{9}$	greater than 16	less than 16
2.	$6\frac{4}{17} + 3\frac{3}{4}$	greater than 10	less than 10
3.	$16\frac{3}{4} - 7\frac{1}{4}$	greater than 9	less than 9
4.	$36\frac{2}{3} - 12\frac{3}{4}$	greater than 24	less than 24
5.	$7\frac{1}{7} \times 3\frac{7}{8}$	greater than 24	less than 24
6.	$\frac{7}{12} \times 25$	greater than 16	less than 16
7.	$5\frac{9}{17} \div 7\frac{5}{6}$	greater than 1	less than 1
8.	$\frac{7}{12} \div \frac{9}{10}$	greater than 1	less than 1

Estimate. Give your answer as a whole number. Do not compute.
Answers may vary. Typical answers are given.

9. $\frac{2}{3} + \frac{4}{9}$ 1 **10.** $\frac{6}{7} + 1\frac{5}{8}$ 3 **11.** $\frac{1}{12} + 2\frac{2}{17}$ 2 **12.** $10\frac{6}{11} + 9\frac{7}{13}$ 20

13. $5\frac{5}{7} - \frac{18}{19}$ 5 **14.** $7\frac{1}{5} - 1\frac{3}{4}$ 5 **15.** $8\frac{1}{2} - 5\frac{2}{3}$ 3 **16.** $27\frac{8}{9} - 19\frac{1}{8}$ 9

17. $20 \times \frac{1}{9}$ 2 **18.** $\frac{7}{8} \times 7$ 7 **19.** $1\frac{5}{6} \times \frac{1}{2}$ 1 **20.** $\frac{1}{3} \times 3\frac{1}{2}$ 1

21. $\frac{3}{7} \div \frac{4}{9}$ 1 **22.** $6\frac{1}{7} \div \frac{1}{2}$ 12 **23.** $4\frac{2}{7} \div 2\frac{1}{3}$ 2 **24.** $8\frac{8}{9} \div 3\frac{1}{6}$ 3

APPLYING THE LESSON

Independent Practice

Homework Assignment	
Minimum	Ex. 1-37 odd
Average	Ex. 2-24 even; 25-37
Maximum	Ex. 1-37

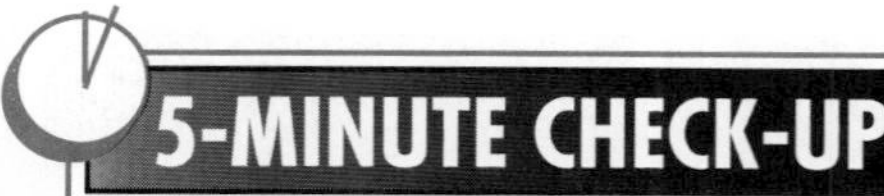

5-MINUTE CHECK-UP

(over Lesson 5-1)

Available on TRANSPARENCY 5-2.

Estimate.

1. $1\frac{4}{5} \times 6$ 12
2. $3\frac{7}{8} \div 2$ 2
3. $4\frac{2}{3} \times 2\frac{1}{8}$ 10
4. $8\frac{9}{10} \div 2\frac{6}{7}$ 3

Extra Practice, Lesson 5-1, p. 460

INTRODUCING THE LESSON

Ask students to pretend they're having a party with 4 pizzas, 8 slices in each, four 6-packs of soda pop, and 3 gallons of ice cream. Have students find equal portions for each member of the class.

TEACHING THE LESSON

Using Cooperative Groups In small groups, have students make up four multiplication problems. Two must be problems in which the numerator and denominator can be divided by a common factor. Two must be problems in which the numerator and denominator have no common factors. Have groups exchange problems and solve.

Practice Masters Booklet, p. 51

PRACTICE WORKSHEET 5-2

Multiplying Fractions

Multiply.

1. $\frac{1}{2} \times \frac{1}{6}$	$\frac{1}{12}$	2. $\frac{1}{3} \times \frac{1}{2}$	$\frac{1}{6}$	3. $\frac{1}{3} \times \frac{4}{9}$	$\frac{4}{27}$
4. $\frac{3}{4} \times \frac{5}{7}$	$\frac{15}{28}$	5. $\frac{2}{3} \times \frac{7}{9}$	$\frac{14}{27}$	6. $\frac{3}{5} \times \frac{4}{7}$	$\frac{12}{35}$
7. $\frac{1}{3} \times 5$	$1\frac{2}{3}$	8. $6 \times \frac{1}{2}$	3	9. $\frac{2}{3} \times 9$	6
10. $\frac{7}{8} \times \frac{2}{5}$	$\frac{7}{20}$	11. $\frac{2}{3} \times \frac{3}{4}$	$\frac{1}{2}$	12. $\frac{5}{8} \times \frac{2}{5}$	$\frac{1}{4}$
13. $\frac{3}{8} \times \frac{5}{6}$	$\frac{5}{16}$	14. $\frac{3}{8} \times \frac{4}{9}$	$\frac{1}{6}$	15. $\frac{8}{9} \times \frac{3}{16}$	$\frac{1}{6}$
16. $\frac{5}{9} \times \frac{3}{5}$	$\frac{1}{3}$	17. $\frac{7}{8} \times \frac{4}{7}$	$\frac{1}{2}$	18. $\frac{15}{16} \times \frac{4}{5}$	$\frac{3}{4}$
19. $\frac{5}{6} \times \frac{3}{4}$	$\frac{5}{8}$	20. $\frac{2}{3} \times \frac{9}{10}$	$\frac{3}{5}$	21. $\frac{7}{10} \times \frac{5}{6}$	$\frac{7}{12}$
22. $\frac{4}{5} \times \frac{5}{16}$	$\frac{1}{4}$	23. $\frac{3}{10} \times \frac{6}{7}$	$\frac{9}{35}$	24. $\frac{7}{12} \times \frac{3}{14}$	$\frac{1}{8}$
25. $\frac{5}{8} \times \frac{9}{10}$	$\frac{9}{16}$	26. $\frac{7}{8} \times \frac{4}{9}$	$\frac{7}{18}$	27. $\frac{7}{10} \times \frac{15}{16}$	$\frac{21}{32}$
28. $\frac{1}{2} \times \frac{2}{3} \times \frac{3}{4}$	$\frac{1}{4}$	29. $\frac{1}{2} \times \frac{1}{3} \times \frac{4}{7}$	$\frac{2}{21}$	30. $\frac{9}{16} \times \frac{2}{3} \times \frac{5}{6}$	$\frac{5}{16}$
31. $\frac{5}{12} \times \frac{4}{5} \times \frac{3}{4}$	$\frac{1}{4}$	32. $\frac{7}{8} \times \frac{3}{14} \times \frac{7}{9}$	$\frac{7}{48}$	33. $\frac{13}{16} \times \frac{8}{9} \times \frac{2}{3}$	$\frac{13}{27}$

Solve.

34. What is $\frac{1}{2}$ of $\frac{2}{3}$ cup of sugar? $\frac{1}{3}$ cup

35. What is $\frac{3}{4}$ of a 3-pound box of raisins? $2\frac{1}{4}$ pounds

5-2 MULTIPLYING FRACTIONS

Objective
Multiply fractions.

Lisa is a baker for Mm-Mm Good Caterers. Her recipe for filling 100 cream puffs calls for $\frac{1}{2}$ cup of vanilla flavoring. If Lisa needs to make $\frac{3}{4}$ of a recipe, how much vanilla flavoring does she need? Lisa needs to find $\frac{3}{4}$ of $\frac{1}{2}$.

$\frac{1}{2}$ cup of vanilla

You are making $\frac{3}{4}$ of the recipe.

$\frac{3}{4}$ of the vanilla is shaded blue.

You can see that $\frac{3}{4}$ of $\frac{1}{2}$ is $\frac{3}{8}$.

Method

1. Multiply the numerators. Multiply the denominators.
2. Write the product in simplest form.

Example A ***Multiply $\frac{3}{4} \times \frac{1}{2}$.***

1. $\frac{3}{4} \times \frac{1}{2} = \frac{3 \times 1}{4 \times 2}$
2. $= \frac{3}{8}$

Lisa needs $\frac{3}{8}$ cup of vanilla flavoring.

Example B ***Multiply $8 \times \frac{2}{3}$.*** Explain why the product should be a little more than 4.

1. Write 8 as a fraction. $8 = \frac{8}{1}$ $\quad \frac{8}{1} \times \frac{2}{3} = \frac{8 \times 2}{1 \times 3}$
2. $= \frac{16}{3}$ or $5\frac{1}{3}$

You can use the following shortcut to simplify a product.

$$\frac{3}{5} \times \frac{10}{21} = \frac{3 \times 10}{5 \times 21}$$

$$= \frac{\overset{1}{\cancel{3}} \times \overset{2}{\cancel{10}}}{\underset{1}{\cancel{5}} \times \underset{7}{\cancel{21}}}$$

Divide both the numerator and the denominator by 3 and by 5.
3 is the GCF of 3 and 21.
5 is the GCF of 10 and 5.

$$= \frac{2}{7}$$ ← simplest form

Guided Practice ***Multiply.***

Example A

1. $\frac{3}{4} \times \frac{1}{4}$ $\frac{3}{16}$
2. $\frac{2}{3} \times \frac{5}{6}$ $\frac{5}{9}$
3. $\frac{2}{3} \times \frac{4}{5}$ $\frac{8}{15}$
4. $\frac{3}{8} \times \frac{4}{5}$ $\frac{3}{10}$
5. $\frac{4}{9} \times \frac{3}{4}$ $\frac{1}{3}$

RETEACHING THE LESSON

Multiply. Write each answer in simplest form.

1. $\frac{5}{6} \times \frac{1}{3} = \frac{5 \times 1}{6 \times 3} = \frac{5}{18}$
2. $\frac{3}{7} \times \frac{2}{3} = \frac{\overset{1}{\cancel{3}} \times 2}{7 \times \underset{1}{\cancel{3}}} = \frac{2}{7}$
3. $\frac{1}{4} \times \frac{3}{5} = \frac{3}{20}$
4. $9 \times \frac{4}{7} = 5\frac{1}{7}$
5. $\frac{5}{8} \times \frac{8}{9} = \frac{5}{9}$
6. $\frac{3}{5} \times \frac{25}{27} = \frac{5}{9}$

Reteaching Masters Booklet, p. 47

Name ______________________ Date ______________

RETEACHING WORKSHEET 5-2

Multiplying Fractions

To multiply fractions, first multiply the numerators. Then multiply the denominators. Write the product in simplest form.

Multiply numerators. Multiply denominators. $\frac{2}{5} \times \frac{3}{4} = \frac{2 \times 3}{5 \times 4}$

Divide the numerator and denominator by their GCF (2) to simplify. $= \frac{\overset{3}{\cancel{6}}}{\underset{10}{\cancel{20}}}$ $= \frac{3}{10}$

You can use a shortcut. First simplify. Then multiply. $\frac{2}{5} \times \frac{3}{4} = \frac{\overset{1}{\cancel{2}} \times 3}{5 \times \underset{2}{\cancel{4}}}$ $= \frac{3}{10}$

Divide a factor in the numerator (2) and a factor in the denominator (4) by their GCF (2).

Multiply.

1. $\frac{1}{2} \times \frac{7}{9}$ $\frac{7}{18}$
2. $\frac{4}{9} \times \frac{3}{8}$ $\frac{1}{6}$
3. $\frac{1}{8} \times \frac{2}{3}$ $\frac{1}{12}$

Example B

6. $\frac{1}{7} \times 3$ $\frac{3}{7}$ 7. $3 \times \frac{7}{16}$ $1\frac{5}{16}$ 8. $\frac{5}{14} \times 42$ 15 9. $7 \times \frac{5}{8}$ $4\frac{3}{8}$ 10. $\frac{2}{3} \times 9$ 6

Exercises

Practice

Multiply.

11. $\frac{2}{9} \times \frac{2}{7}$ $\frac{4}{63}$
12. $\frac{2}{3} \times \frac{1}{4}$ $\frac{1}{6}$
13. $\frac{5}{7} \times \frac{8}{9}$ $\frac{40}{63}$
14. $\frac{4}{5} \times \frac{1}{2}$ $\frac{2}{5}$
15. $\frac{1}{2} \times \frac{8}{9}$ $\frac{4}{9}$
16. $\frac{4}{5} \times \frac{1}{9}$ $\frac{4}{45}$
17. $\frac{7}{8} \times \frac{8}{9}$ $\frac{7}{9}$
18. $\frac{7}{9} \times \frac{2}{21}$ $\frac{2}{27}$
19. $\frac{3}{4} \times 5$ $3\frac{3}{4}$
20. $\frac{2}{3} \times 5$ $3\frac{1}{3}$
21. $16 \times \frac{5}{8}$ 10
22. $64 \times \frac{5}{16}$ 20
23. $\frac{3}{4} \times \frac{3}{5} \times \frac{5}{6}$ $\frac{3}{8}$
24. $\frac{6}{7} \times \frac{1}{3} \times \frac{1}{5}$ $\frac{2}{35}$
25. $\frac{7}{8} \times \frac{2}{7} \times \frac{9}{10}$ $\frac{9}{40}$
26. Find the product of $\frac{3}{7}$ and $\frac{21}{33}$. $\frac{3}{11}$
27. *True* or *false:* $\frac{2}{3} \times \frac{12}{30} = \frac{4}{15}$. true

Applications

28. Bev wants to wallpaper one-half of her bedroom in a pastel print and the other half in the coordinating stripe. If each roll is the same size and it takes $5\frac{1}{2}$ rolls of wallpaper to do the whole bedroom, how many rolls of each pattern does Bev need?

28. $2\frac{3}{4}$ rolls

29. A rod and reel usually sell for $65. They are on sale for three-fourths of the regular price. What is the sale price? $48.75

Critical Thinking

30. Move one straight line segment to make this true statement different, but true, also.

77 + 1 = 78 71 + 7 = 78

Using Variables

Evaluate each expression if a = 4, b = $\frac{3}{4}$, and c = $\frac{2}{3}$.

31. $b + c$ $1\frac{5}{12}$ 32. $a - (b + c)$ $2\frac{7}{12}$ 33. ac $2\frac{2}{3}$

Make Up a Problem

34. Make up a problem using two proper fractions. The answer must be $3\frac{3}{4}$. Answers may vary.

Mixed Review

Lesson 2-2

Add.

35. 48.5 + 16.7 = 65.2
36. 6.23 + 10.8 = 17.03
37. 1.07 + 2.98 = 4.05

Lesson 3-1

Multiply.

38. 13 × 14 = 182
39. 45 × 8 = 360
40. 51 × 63 = 3,213

Lesson 3-13

41. Kisha is building a fence around her rectangular garden. The fencing she wants to use costs $0.75 per foot. The garden measures 12 feet by 10 feet. How much will fencing for her garden cost? $33

Chalkboard Examples

Examples A, B *Multiply.*

1. $\frac{2}{3} \times \frac{3}{4}$ $\frac{1}{2}$
2. $\frac{2}{7} \times \frac{7}{8}$ $\frac{1}{4}$
3. $5 \times \frac{3}{8}$ $\frac{15}{8}$ or $1\frac{7}{8}$
4. $7 \times \frac{9}{14}$ $4\frac{1}{2}$

Additional examples are provided on TRANSPARENCY 5-2.

EVALUATING THE LESSON

Check for Understanding Use Guided Practice Exercises to check for student understanding.

Closing the Lesson Have students explain in their own words when to divide the numerator and denominator of fractions by a common factor when multiplying.

EXTENDING THE LESSON

Enrichment Have students use the following fact, or research similar facts in the library to find the amount of oil used by a 3-person household.

- One person uses the equivalent of 391 gallons of oil a year. 1,173 gal

Enrichment Masters Booklet, p. 47

Name ______________ Date ______________

ENRICHMENT WORKSHEET 5-2

Fractions and Exponents

$2^3 = 2 \times 2 \times 2 = 8$ and $5^3 = 5 \times 5 \times 5 = 125$; so $\left(\frac{2}{5}\right)^3 = \frac{2}{5} \times \frac{2}{5} \times \frac{2}{5} = \frac{8}{125}$.

Evaluate the following powers.

1. $\left(\frac{1}{4}\right)^2$ $\frac{1}{16}$
2. $\left(\frac{2}{3}\right)^2$ $\frac{4}{9}$
3. $\left(\frac{3}{4}\right)^2$ $\frac{9}{16}$
4. $\left(\frac{3}{10}\right)^2$ $\frac{9}{100}$
5. $\left(\frac{1}{6}\right)^2$ $\frac{1}{36}$
6. $\left(\frac{4}{5}\right)^2$ $\frac{16}{25}$
7. $\left(\frac{1}{8}\right)^2$ $\frac{1}{64}$
8. $\left(\frac{9}{10}\right)^2$ $\frac{81}{100}$
9. $\left(\frac{7}{8}\right)^2$ $\frac{49}{64}$
10. $\left(\frac{1}{3}\right)^3$ $\frac{1}{27}$
11. $\left(\frac{1}{2}\right)^4$ $\frac{1}{16}$
12. $\left(\frac{3}{4}\right)^3$ $\frac{27}{64}$
13. $\left(\frac{2}{3}\right)^4$ $\frac{16}{81}$
14. $\left(\frac{2}{9}\right)^3$ $\frac{8}{729}$
15. $\left(\frac{1}{10}\right)^5$ $\frac{1}{100{,}000}$
16. $\left(\frac{4}{7}\right)^3$ $\frac{64}{343}$
17. $\left(\frac{1}{3}\right)^5$ $\frac{1}{243}$
18. $\left(\frac{5}{9}\right)^3$ $\frac{125}{729}$
19. $\left(\frac{3}{10}\right)^4$ $\frac{81}{10{,}000}$
20. $\left(\frac{1}{2}\right)^6$ $\frac{1}{64}$
21. $\left(\frac{1}{4}\right)^5$ $\frac{1}{1{,}024}$

Find the missing exponent to make each a true sentence.

22. $\left(\frac{1}{2}\right)^{\_} = \frac{1}{256}$ 8
23. $\left(\frac{3}{5}\right)^{\_} = \frac{81}{625}$ 4
24. $\left(\frac{2}{3}\right)^{\_} = \frac{8}{27}$ 3
25. $\left(\frac{7}{10}\right)^{\_} = \frac{16{,}807}{100{,}000}$ 5

APPLYING THE LESSON

Independent Practice

Homework Assignment	
Minimum	Ex. 1-37 odd; 38-41
Average	Ex. 2-26 even; 28-41
Maximum	Ex. 1-41

5-MINUTE CHECK-UP

(over Lesson 5-2)

Available on TRANSPARENCY 5-3.

Multiply.

1. $\frac{3}{5} \times \frac{4}{7}$ $\frac{12}{35}$ **2.** $\frac{5}{8} \times \frac{3}{4}$ $\frac{15}{32}$

3. $16 \times \frac{1}{4}$ 4 **4.** $\frac{5}{8} \times 12$ $7\frac{1}{2}$

Extra Practice, Lesson 5-2, p. 461

INTRODUCING THE LESSON

Use a financial page of the newspaper and have students choose a stock whose price includes a fraction. Then have students imagine that the value of their stock has increased 2 times. Ask students to try to find the new value of their stock.

TEACHING THE LESSON

Using Critical Thinking Ask students how to multiply $6 \times 3\frac{1}{2}$ without changing $3\frac{1}{2}$ to an improper fraction. **Use the distributive property, $6 \times 3 + 6 \times \frac{1}{2}$.**

5-3 MULTIPLYING MIXED NUMBERS

Objective
Multiply mixed numbers.

Jeff made eight sections of a silver necklace and decorated them with hand-hammered designs and turquoise stones. Each section is $2\frac{3}{4}$ inches long. How long will the necklace be when all eight sections are joined? Assume that no length is lost when the sections are joined.

You need to multiply $2\frac{3}{4}$ by 8. Multiply mixed numbers as follows.

Method

1. Rename each mixed number as an improper fraction.
2. Multiply the fractions.
3. Write the product in simplest form.

You may want to review renaming mixed numbers as improper fractions.

Example A

Multiply $8 \times 2\frac{3}{4}$. Estimate: $8 \times 3 = 24$

1. $8 = \frac{8}{1}$

   $2\frac{3}{4} = \frac{(2 \times 4) + 3}{4}$ or $\frac{11}{4}$

2. $8 \times 2\frac{3}{4} = \frac{8}{1} \times \frac{11}{4}$

   $= \frac{\overset{2}{\cancel{8}} \times 11}{1 \times \underset{1}{\cancel{4}}}$

3. $= \frac{22}{1}$ or 22

Compared to the estimate, is the answer reasonable?

The necklace will be 22 inches long.

Example B

Multiply $2\frac{5}{8} \times 1\frac{3}{7}$. Estimate: $3 \times 1 = 3$

1. $2\frac{5}{8} = \frac{(2 \times 8) + 5}{8}$ or $\frac{21}{8}$

   $1\frac{3}{7} = \frac{(1 \times 7) + 3}{7}$ or $\frac{10}{7}$

2. $2\frac{5}{8} \times 1\frac{3}{7} = \frac{21}{8} \times \frac{10}{7}$

   $= \frac{\overset{3}{\cancel{21}} \times \overset{5}{\cancel{10}}}{\underset{4}{\cancel{8}} \times \underset{1}{\cancel{7}}}$

3. $= \frac{15}{4}$ or $3\frac{3}{4}$

Is the answer reasonable?

Guided Practice

Multiply. **Encourage students to estimate each product first.**

Example A

1. $3 \times 1\frac{3}{8}$ $4\frac{1}{8}$ **2.** $2\frac{1}{6} \times 4$ $8\frac{2}{3}$

3. $2\frac{5}{8} \times 3$ $7\frac{7}{8}$ **4.** $7 \times 2\frac{2}{3}$ $18\frac{2}{3}$

5. $12 \times 3\frac{1}{4}$ 39 **6.** $5\frac{1}{5} \times 25$ 130

Example B

7. $1\frac{1}{6} \times 1\frac{2}{5}$ $1\frac{19}{30}$ **8.** $2\frac{2}{9} \times 1\frac{4}{7}$ $3\frac{31}{63}$

9. $2\frac{2}{3} \times \frac{2}{3}$ $1\frac{7}{9}$ **10.** $9\frac{3}{8} \times 3\frac{1}{5}$ 30

11. $3\frac{1}{3} \times 3\frac{4}{5}$ $12\frac{2}{3}$ **12.** $2\frac{4}{5} \times 2\frac{6}{7}$ 8

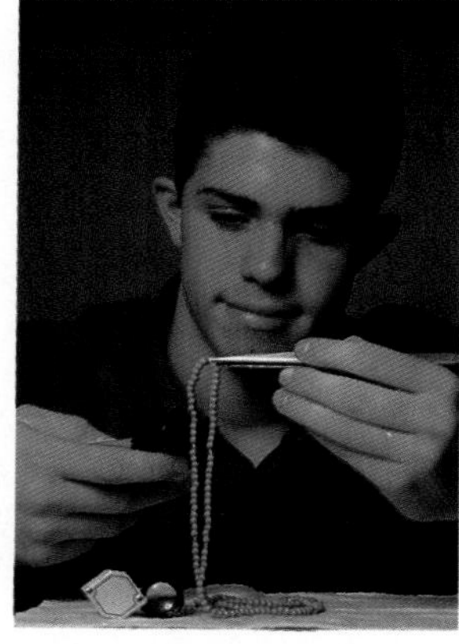

Practice Masters Booklet, p. 52

Name ______ Date ______

PRACTICE WORKSHEET 5-3

Multiplying Mixed Numbers

Multiply.

1. $2\frac{1}{8} \times 4\frac{4}{5}$	$10\frac{1}{5}$	**2.** $2\frac{3}{4} \times 1\frac{2}{3}$	$4\frac{7}{12}$	**3.** $1\frac{5}{8} \times 1\frac{1}{2}$	$2\frac{7}{16}$
4. $6\frac{3}{10} \times 3\frac{1}{3}$	21	**5.** $3\frac{4}{5} \times 2\frac{2}{3}$	$10\frac{2}{15}$	**6.** $2\frac{1}{10} \times 2\frac{1}{7}$	$4\frac{1}{2}$
7. $1\frac{1}{5} \times 3\frac{1}{8}$	$3\frac{3}{4}$	**8.** $5\frac{1}{2} \times 2\frac{2}{3}$	$14\frac{2}{3}$	**9.** $6\frac{2}{3} \times 2\frac{1}{10}$	14
10. $2\frac{2}{7} \times 4\frac{3}{8}$	10	**11.** $2\frac{1}{2} \times 1\frac{3}{5}$	4	**12.** $3\frac{1}{2} \times 1\frac{3}{5}$	$5\frac{3}{5}$
13. $4\frac{2}{3} \times 1\frac{7}{8}$	$8\frac{3}{4}$	**14.** $3\frac{3}{4} \times 2\frac{4}{5}$	$10\frac{1}{2}$	**15.** $7\frac{1}{3} \times 2\frac{5}{11}$	18
16. $3\frac{1}{2} \times 3\frac{1}{2}$	$12\frac{1}{4}$	**17.** $1\frac{1}{5} \times 2\frac{3}{4}$	$3\frac{3}{10}$	**18.** $1\frac{3}{5} \times 3\frac{5}{6}$	$6\frac{2}{15}$
19. $3\frac{6}{10} \times 8\frac{1}{12}$	$29\frac{1}{10}$	**20.** $4\frac{3}{7} \times 1\frac{4}{5}$	$7\frac{34}{35}$	**21.** $3\frac{2}{3} \times 5\frac{1}{4}$	$19\frac{1}{4}$
22. $6\frac{3}{7} \times 2\frac{5}{8}$	$16\frac{7}{8}$	**23.** $3\frac{1}{5} \times 2\frac{5}{16}$	$7\frac{2}{5}$	**24.** $4\frac{3}{5} \times 3\frac{1}{3}$	$15\frac{1}{3}$
25. $2\frac{3}{4} \times \frac{2}{5} \times 8$	$8\frac{4}{5}$	**26.** $3\frac{1}{3} \times \frac{3}{8} \times 2\frac{1}{6}$	$2\frac{17}{24}$	**27.** $4\frac{1}{2} \times 2\frac{1}{4} \times 2\frac{2}{3}$	27

Solve.

28. A hose fills $\frac{5}{16}$ of a bucket in one minute. How much of a bucket will it fill in 3 minutes? $\frac{15}{16}$ bucket

29. Gae proofread $5\frac{2}{3}$ pages in one hour. How many pages can she proofread in $7\frac{1}{2}$ hours? $42\frac{1}{2}$

RETEACHING THE LESSON

Multiply. Write each product in simplest form.

1. $3\frac{1}{2} \times 1\frac{1}{3} = \frac{14}{3}$ or $4\frac{2}{3}$

2. $1\frac{1}{3} \times 4\frac{7}{8} = \frac{13}{2}$ or $6\frac{1}{2}$

3. $2\frac{1}{7} \times 1\frac{24}{25} = \frac{21}{5}$ or $4\frac{1}{5}$

4. $1\frac{1}{2} \times 6\frac{2}{3} = 10$ **5.** $1\frac{7}{8} \times 1\frac{3}{5} = 3$

6. $1\frac{1}{2} \times 2\frac{1}{6} \times 1\frac{1}{3} = \frac{13}{3}$ or $4\frac{1}{3}$

Reteaching Masters Booklet, p. 48

Name ______ Date ______

RETEACHING WORKSHEET 5-3

Multiplying Mixed Numbers

To multiply mixed numbers, first rename the mixed numbers as fractions. Then multiply the fractions.

First rename the mixed number. Then multiply.

$5\frac{3}{4} \times \frac{4}{7}$

$5\frac{3}{4} = \frac{(5 \times 4) + 3}{4} = \frac{23}{4}$

Write the product in simplest form.

$5\frac{3}{4} \times \frac{4}{7} = \frac{23 \times \overset{1}{\cancel{4}}}{\underset{1}{\cancel{4}} \times 7} = \frac{23}{7}$

$= 3\frac{2}{7}$

Multiply.

1. $1\frac{2}{3} \times \frac{3}{4}$ $1\frac{1}{4}$ **2.** $5\frac{1}{8} \times \frac{2}{3}$ $3\frac{5}{12}$ **3.** $2\frac{1}{4} \times \frac{2}{5}$ $\frac{9}{10}$

Exercises

Practice

Multiply. **Encourage students to estimate each product first.**

13. $2\frac{1}{4} \times 2\frac{1}{7}$ $4\frac{23}{28}$
14. $2\frac{2}{3} \times 5\frac{1}{3}$ $14\frac{2}{9}$
15. $3\frac{1}{5} \times 1\frac{1}{8}$ $3\frac{3}{5}$
16. $1\frac{1}{6} \times \frac{7}{8}$ $1\frac{1}{48}$
17. $7\frac{1}{2} \times 3\frac{1}{3}$ 25
18. $2\frac{6}{7} \times 2\frac{4}{5}$ 8
19. $1\frac{5}{6} \times 2\frac{1}{2}$ $4\frac{7}{12}$
20. $8\frac{1}{3} \times 3\frac{2}{5}$ $28\frac{1}{3}$
21. $\frac{3}{8} \times 2\frac{2}{3}$ 1
22. $3\frac{1}{2} \times 1\frac{1}{2}$ $5\frac{1}{4}$
23. $6\frac{1}{4} \times 8\frac{1}{5}$ $51\frac{1}{4}$
24. $1\frac{1}{6} \times 1\frac{3}{5}$ $1\frac{13}{15}$
25. $7\frac{1}{2} \times \frac{2}{5} \times 15$ 45
26. $5\frac{1}{2} \times \frac{3}{5} \times 8\frac{1}{3}$ $27\frac{1}{2}$
27. $3\frac{4}{5} \times 3\frac{1}{3} \times 1\frac{1}{5}$ $15\frac{1}{5}$
28. Multiply $2\frac{1}{2}$ and $4\frac{1}{3}$. $10\frac{5}{6}$
29. What is the product of 6 and $7\frac{1}{24}$? $42\frac{1}{4}$

Applications

30. If it takes 180 gallons of water to fill a waterbed and water weighs $8\frac{1}{3}$ pounds per gallon, how much does the water in the bed weigh? **1,500 pounds**
31. A manager spends $\frac{1}{3}$ of her $7\frac{1}{2}$-hour day in meetings and the rest in directing her staff. How many hours a day does she spend directing her staff? **5 hours**
32. Carla's car travels $22\frac{1}{2}$ miles on one gallon of gasoline. How far can it travel on $4\frac{1}{2}$ gallons of gasoline? **$101\frac{1}{4}$ mile**

FUN with MATH

Suppose you were to cross a horse with a kangaroo?

See page 174.

Write Math

33. Write a letter to a sixth-grade student explaining how to multiply mixed numbers. Be sure to use an example that shows the shortcut on page 148. **Answers may vary.**

Critical Thinking

34. A horse, rider, and saddle weigh 1,210 pounds. The horse weighs 9 times as much as the rider. The rider weighs 12 times as much as the saddle. How much does the horse weigh? **1,080 pounds**

Suppose

35. Suppose you are driving 800 miles from Ohio to Massachusetts. Your car gets $23\frac{1}{2}$ miles to a gallon of gasoline and has a 12-gallon gas tank. Can you make the trip with 3 tankfuls? **yes**
36. Suppose, in Exercise 35, you had a 15-gallon gas tank. Could you make the trip in 2 tankfuls? **no**

Mental Math

A **unit fraction** is a fraction that has a numerator of 1. Multiplying by a unit fraction is the same as dividing by the denominator of the unit fraction.

$$\frac{1}{4} \text{ of } 360 \rightarrow 360 \div 4 \text{ or } 90$$

Suppose you need to find $\frac{3}{4}$ of 360. Note that $\frac{3}{4}$ of 360 is 3 times $\frac{1}{4}$ of 360. So $\frac{3}{4}$ of 360 is 3×90 or 270.

Multiply. Write only the answer.

37. $\frac{1}{2} \times 70$ 35
38. $\frac{1}{3} \times 240$ 80
39. $\frac{1}{4} \times 160$ 40
40. $\frac{1}{5} \times 250$ 50

APPLYING THE LESSON

Independent Practice

Homework Assignment	
Minimum	Ex. 1-33 odd; 34-40 even
Average	Ex. 2-28 even; 30-40
Maximum	Ex. 1-40

Chalkboard Examples

Examples A, B *Multiply.*

1. $1\frac{1}{6} \times 13$ $15\frac{1}{6}$
2. $9 \times 2\frac{2}{3}$ 24
3. $1\frac{3}{4} \times 1\frac{1}{2}$ $2\frac{5}{8}$
4. $2\frac{1}{2} \times 1\frac{1}{4}$ $3\frac{1}{8}$

Additional examples are provided on TRANSPARENCY 5-3.

EVALUATING THE LESSON

Check for Understanding Use Guided Practice Exercises to check for student understanding.

Error Analysis Watch for students who do not change whole numbers to fraction form, and then multiply both numerator and denominator of the other factor by the whole number. Remind students that the method for multiplying fractions applies to numbers in fraction form.

Closing the Lesson Ask students to explain how to change a mixed number to an improper fraction.

EXTENDING THE LESSON

Enrichment Masters Booklet, p. 48

Name ______ Date ______

ENRICHMENT WORKSHEET 5-3

Using a Circle Graph

The circle graph shows the result of a poll of the students at Wells School on a certain day. There are 720 students at Wells. How many students ride to school on bicycles?

Wells School Transportation

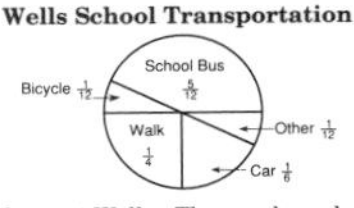

Explore There are 720 students at Wells. The number who ride bicycles is $\frac{1}{12}$ of the whole.

Plan Find $\frac{1}{12}$ of 720.

Solve $\frac{1}{12} \times 720 = \frac{1}{12} \times \frac{720}{1}$

$$= \frac{1 \times \overset{60}{\cancel{720}}}{\underset{1}{\cancel{12}} \times 1}$$

$$= 60$$

Examine Since $\frac{1}{10}$ of 720 is 72, $\frac{1}{12}$ of 720 should be less than 72. The answer 60 seems reasonable.

Solve. Use the circle graph.

1. How many students at Wells School ride the bus? **300 students**
2. How many students do not walk to school? **540 students**
3. How many students either ride in cars or walk to school? **300 students**
4. How many students use another form of transportation? **60 students**
5. Suppose 210 students walk to school and the results of the poll are the same. How many students are at Wells? **840 students**
6. Suppose there are 480 students at Wells and the results are the same. How many students ride in cars? **80 students**

5-MINUTE CHECK-UP

(over Lesson 5-3)

Availble on TRANPARENCY 5-4.

Multiply.

1. $18 \times 2\frac{1}{3}$ 42 **2.** $9 \times 3\frac{3}{4}$ $33\frac{3}{4}$

3. $1\frac{1}{8} \times \frac{2}{3}$ $\frac{3}{4}$ **4.** $2\frac{2}{9} \times 5\frac{5}{8}$ $12\frac{1}{2}$

Extra Practice, Lesson 5-3, p. 461

▶ INTRODUCING THE LESSON

Ask students to name any energy conservation methods that they know, such as turning off lights when not in use, turning down the heat a few degrees at night, and so on.

▶ TEACHING THE LESSON

Using Discussion Discuss with students the types of appliances that use more electricity than others, those that heat and cool. Have students discuss methods for efficient use of this kind of appliance.

Practice Masters Booklet, p. 53

Name ______________________ Date ______________

PRACTICE WORKSHEET 5-4

Applications: Energy and Coal

The chart gives the energy equivalents for one year of appliance operation.

Use the chart to find the energy equivalent in tons of coal for using each appliance.

Appliance	Energy Equivalent in Tons of Coal
Range	$\frac{1}{2}$
Microwave oven	$\frac{1}{10}$
Water heater	2
Lighting a 6-room house	$\frac{1}{3}$
Refrigerator	1
Radio	$\frac{1}{20}$
Dishwasher	$\frac{1}{5}$
Color TV	$\frac{1}{4}$

1. lighting a 6-room house for 3 years
   1 ton
2. a range for $\frac{1}{2}$ year
   $\frac{1}{4}$ ton
3. a color TV for 3 years
   $\frac{3}{4}$ ton
4. a water heater for $\frac{1}{4}$ year
   $\frac{1}{2}$ ton
5. a microwave oven for $\frac{3}{4}$ year
   $\frac{3}{40}$ ton
6. a radio for 4 years
   $\frac{1}{5}$ ton
7. a range for 1 month
   $\frac{1}{24}$ ton
8. lighting a 6-room house for 7 months
   $\frac{7}{36}$ ton
9. a microwave oven and a dishwasher for 1 year
   $\frac{3}{10}$ ton
10. a water heater and a range for 2 years
    5 tons
11. a radio and a color TV for $\frac{1}{2}$ year
    3 ton

5-4 ENERGY AND COAL

Objective

Use fractions in a real world application.

Energy and its uses are an important concern for everyone. The manufacturers of appliances put labels on their products to show the consumer the savings that can be expected by using their energy efficient appliances.

Coal is often used to produce energy in the form of electricity. How much coal does it take to produce electricity to run a color television for one year? The chart at the right lists the energy equivalents in tons of coal for one year of operation for several appliances. The electricity needed to run a color television for one year is produced by $\frac{1}{4}$ ton of coal.

	Appliance	Energy Equivalent In Tons of Coal
3	Range	$\frac{1}{2}$
7	Microwave oven	$\frac{1}{10}$
1	Water heater	2
4	Lighting a 6-room house	$\frac{1}{3}$
2	Refrigerator	1
8	Radio	$\frac{1}{20}$
6	Dishwasher	$\frac{1}{5}$
5	Color TV	$\frac{1}{4}$

Understanding the Data

- Which appliance, a dishwasher or a color TV, takes more electricity to operate for one year? **color TV**
- What types of appliances, ones that heat and cool or ones that produce light and sound, take more electricity to operate? **ones that heat and cool**

Exercises

Use the chart to find the energy equivalent in tons of coal for using each appliance. 1. $\frac{11}{20}$ ton 2. $\frac{1}{5}$ ton 6. $\frac{1}{4}$ ton

1. a range and a radio for 1 year
2. a microwave oven for 2 years
3. a color TV for $\frac{1}{2}$ year $\frac{1}{8}$ ton
4. a dishwasher for 5 years 1 ton
5. a refrigerator for $\frac{1}{4}$ year $\frac{1}{4}$ ton
6. lighting a 6-room house for $\frac{3}{4}$ year
7. a color TV for 2 years $\frac{1}{2}$ ton
8. a water heater for 1 month $\frac{1}{6}$ ton
9. a radio for 6 months $\frac{1}{40}$ ton
10. List the appliances in the chart according to the amount of energy they use. List in order from greatest to least energy use. **See the chart.**

RETEACHING THE LESSON

Have students work in small groups and discuss operations necessary to solve the exercises on page 153. Then, have students solve the problems independently.

Reteaching Masters Booklet, p. 49

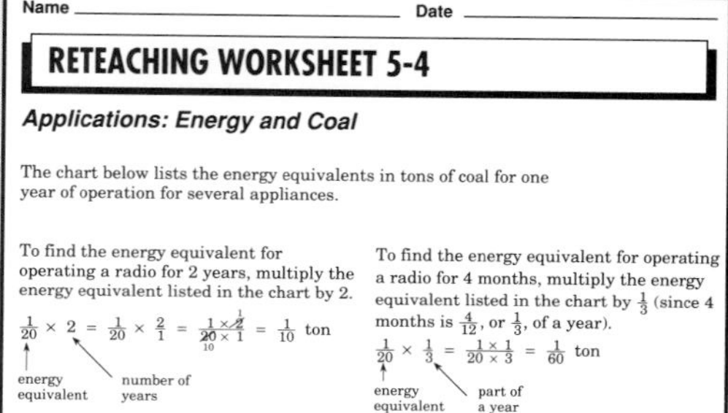

Name ______________________ Date ______________

RETEACHING WORKSHEET 5-4

Applications: Energy and Coal

The chart below lists the energy equivalents in tons of coal for one year of operation for several appliances.

To find the energy equivalent for operating a radio for 2 years, multiply the energy equivalent listed in the chart by 2.

$\frac{1}{20} \times 2 = \frac{1}{20} \times \frac{2}{1} = \frac{1 \times \cancel{2}^{1}}{\cancel{20}_{10} \times 1} = \frac{1}{10}$ ton

energy equivalent — number of years

To find the energy equivalent for operating a radio for 4 months, multiply the energy equivalent listed in the chart by $\frac{1}{3}$ (since 4 months is $\frac{4}{12}$, or $\frac{1}{3}$, of a year).

$\frac{1}{20} \times \frac{1}{3} = \frac{1 \times 1}{20 \times 3} = \frac{1}{60}$ ton

energy equivalent — part of a year

Use the chart to find the energy equivalent in … of coal … each appliance

Problem Solving

Coal is found in many areas of the United States. There are coal mines in 26 states. The circle graph at the right shows areas where the coal deposits are located.

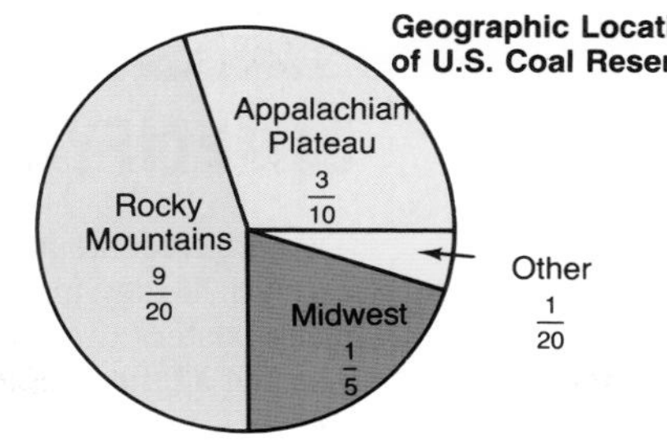

11. What part of the coal deposits are found in the Midwest and the Rocky Mountains? **$\frac{13}{20}$ of the total deposits**
12. Which area has the most coal deposits? **Rocky Mountains**
13. Which area has the greater coal deposits, the Appalachian Plateau or the Midwest?
14. About $\frac{4}{5}$ of the coal in the Appalachian Plateau is mined underground. What part of the total U.S. coal deposits is this?
15. About $\frac{2}{5}$ of the coal in the Rocky Mountains can be mined on the surface. What part of the total U.S. coal reserves is this?
16. Recently, there were about 470 billion tons of coal in deposits. About how much of this coal was in the Midwest? **94 billion tons**
17. Chin's family has a monthly income of $1,940. They spend about $\frac{1}{30}$ of their income for electricity. About how much do they spend each month for electricity? ***about* $65**

14. $\frac{6}{25}$ of total coal deposits

15. $\frac{9}{50}$ of the total coal deposits

Suppose

18. Suppose Mrs. Crewe installed a solar heating system in her home for $5,720. If she saves $40 a month on her heating bill, in *about* how many years will the savings equal the cost of the solar heating system? ***about* 12 years**
19. The U.S. produces about 740 million tons of coal in one year. If Canada produces $\frac{1}{25}$ of that amount, *about* how many tons of coal does Canada produce? ***about* 30 million tons**

Mixed Review

Write each number as a power of another number in two different ways.

Lesson 1-2

20. 64 **4^3, 8^2**
21. 81 **3^4, 9^2**
22. 1 **1^2, 1^3**
23. 10,000 **10^4, 100^2**

Lesson 4-1

Find all the factors of each number. **See margin.**

24. 32
25. 45
26. 24
27. 54

Lesson 5-2

28. One winter day only $\frac{5}{8}$ of the entire student body of Central High School was in school. How many students were in school that day if the total number of students who attend Central is 1,120? **700 students**

Chalkboard Examples

Have students use the chart on page 152 to find the energy equivalent in tons of coal for using each appliance.

1. a color TV and radio for 2 years **$\frac{3}{5}$**
2. How many years does it take for a microwave to use the energy equivalent to 1 ton of coal? **10 years**

Additional examples are provided on TRANSPARENCY 5-4.

EVALUATING THE LESSON

Check for Understanding Use Guided Practice Exercises to check for student understanding.

Closing the Lesson Tell students they will use their knowledge of types of appliances and their energy usage in choosing appliances for their own use.

EXTENDING THE LESSON

Enrichment Have students research the energy savings labels found on new appliances. The labels show monthly savings. Have students find yearly savings for these appliances.

Enrichment Masters Booklet, p. 49

ENRICHMENT WORKSHEET 5-4

Using a Recipe

Solve. Use the recipe card. Write each answer in simplest form.

Hot Cocoa
Makes 10 Servings (about $\frac{2}{3}$ cup each)
$\frac{1}{2}$ cup sugar
$\frac{1}{3}$ cup cocoa
$\frac{1}{8}$ teaspoon salt
$1\frac{1}{2}$ cups water
$4\frac{1}{2}$ cups milk
$\frac{1}{4}$ teaspoon vanilla
Mix well. Heat slowly in large saucepan on stovetop or heat individually in cups in microwave oven.

1. Phillip makes one recipe (10 servings) of hot cocoa. How many cups of hot cocoa does he make? $6\frac{2}{3}$ cups
2. Nikki makes fifteen $\frac{2}{3}$-cup servings of hot cocoa. How many cups of cocoa does she make altogether? 10 cups
3. Bruce doubles the recipe. How many cups of cocoa are needed? $\frac{2}{3}$ cup
4. Rita makes $2\frac{1}{2}$ recipes of hot cocoa. How much vanilla is needed? $\frac{5}{8}$ teaspoon
5. Charlotte has 14 cups of milk. Does she have enough to triple the recipe? yes
6. Tim has 2 quarts of milk. Does he have enough milk to double the recipe? no
7. Mrs. Lee needs to make 50 servings of hot cocoa. How much vanilla will Mrs. Lee need? $1\frac{1}{4}$ teaspoon
8. Dan needs to make 70 servings of hot cocoa. How much water will he need? $10\frac{1}{2}$ cups
9. Jerry is making one recipe of hot cocoa. Will all the ingredients fit in a 2-quart saucepan? yes
10. Hai makes makes forty-two $\frac{2}{3}$-cup servings of hot cocoa. How many cups of hot cocoa does he make altogether? 28 cups

APPLYING THE LESSON

Independent Practice

Homework Assignment	
Minimum	Ex. 2-28 even
Average	Ex. 1-28
Maximum	Ex. 1-28

Additional Answers

24. 1, 2, 4, 8, 16, 32
25. 1, 3, 5, 9, 15, 45
26. 1, 2, 3, 4, 6, 8, 12, 24
27. 1, 2, 3, 6, 9, 18, 27, 54

5-MINUTE CHECK-UP

(over Lesson 5-4)

Available on TRANSPARENCY 5-5.

1. If a color TV uses the energy equivalent of $\frac{1}{4}$ ton of coal a year, how much coal would it take to operate the TV for the average life of a television, $7\frac{1}{2}$ years? $1\frac{7}{8}$ tons

INTRODUCING THE LESSON

Show students the amount of the national deficit from a newspaper. Ask them to determine about how much the deficit would be if it were reduced by one-half. Encourage them to simplify the problem.

TEACHING THE LESSON

Using Questioning

- In the first step of the solution in the example on page 154, why do you find how long it takes for *one* student to make a poster? The key to simplifying a problem is to find the facts for *one*. Then it is simpler to find the facts for any multiple of one.

Practice Masters Booklet, p. 54

Name ______ Date ______

PRACTICE WORKSHEET 5-5

Problem-Solving Strategy: Simplifying the Problem

Solve.

On Monday, or day number 1, Sarah, Rich, and Claire heard a funny joke on the radio. These people each told the joke to 3 more people on day number 2, who each told the joke to 3 more people on day number 3. The pattern continued.

1. How many days passed before 100 people heard the joke? 4 days
2. How many people heard the joke on day 6? By the end of day 6, how many people altogether had heard the joke? 729 people; 1,092 people
3. The summer camp has 7 buildings arranged in a circle. Paths must be constructed joining every building to every other building. How many paths are needed? 21 paths
4. In a basketball tournament, each team plays until it loses a game, then it is out of the tournament. If 64 teams are in the tournament at the start, how many games must be played to determine the tournament winner? 63 games
5. Four people can unpack 24 crates in 2 hours. How many crates can 5 people unpack in 3 hours? 45 crates
6. A loaf of raisin bread needs to be cut into 20 slices. How many cuts are necessary? 19 cuts

Problem-Solving Strategy

Explore / Plan / Solve / Examine

5-5 SIMPLIFYING THE PROBLEM

Objective
Solve problems by simplifying the problem.

Two art students are making posters for the school carnival. If they make two posters in two hours, how many posters can six students make in ten hours? Assume all the students work at the same rate.

Explore

What is given?

- Two students make two posters in two hours.

What is asked?

- How many posters can six students make in ten hours?

Plan

Simplify the problem by finding how long it takes each student to make one poster. Then find how many posters each can make in ten hours and multiply by the number of students, or six.

Solve

$2 \div 2 = 1$ Each student makes 1 poster in 2 hours.

How many posters can one student make in 10 hours?

Divide 10 by 2 since each poster takes 2 hours.

$10 \div 2 = 5$ Each student can make 5 posters in 10 hours.

How many posters can 6 students make in 10 hours? $6 \times 5 = 30$

Six students can make 30 posters in ten hours.

Examine

Check your solution by thinking another way. If two students can make two posters in two hours, six students can make three times as many posters in two hours or 6 posters in two hours. In ten hours the students can make 5 times as many or 30 posters. The solution is 30 posters.

2. Since the last cut yields two pieces, only 15 cuts are necessary.

Guided Practice

Solve. Use simplifying the problem.

1. Two boys can mow four lawns in two hours. How many lawns can six boys mow in eight hours? Assume all the boys work at the same rate. 48 lawns
2. Sixteen people are sharing a super-sized submarine sandwich. How many cuts must be made to divide the sandwich equally among the sixteen people? Explain your answer.

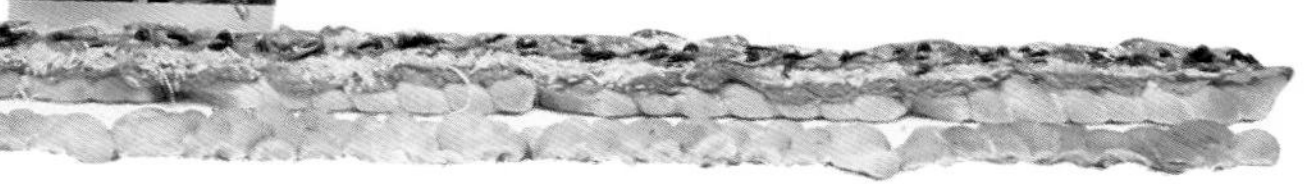

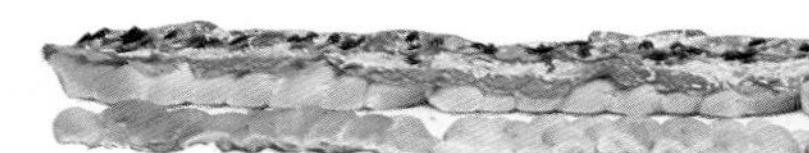

RETEACHING THE LESSON

Ask the following questions about Exercise 1.

- How many lawns can one boy mow in two hours? 2
- How many lawns can one boy mow in four hours? 4
- How many lawns can one boy mow in six hours? 6
- How many lawns can one boy mow in eight hours? 8
- How many lawns can six boys mow in eight hours? 48

Problem Solving

Paper/ Pencil
Mental Math
Estimate
Calculator

Solve. Use any strategy.

3. A plane flying at an altitude of 33,700 feet is directly above a submarine at a depth of 400 feet. Find the vertical distance between them. **(337 + 4) × 100 = 34,100 feet**

4. A rectangular field is fenced on two adjacent sides by a stone wall. The length of the field is 63 meters. The area of the field is 1,323 square meters. Determine how much fencing is needed to enclose the two sides not fenced in by stone walls. **84 meters**

5. Find the number of seconds in the month of July. Write the method of computation used: estimation, mental math, calculator, or paper and pencil. **2,678,400 seconds, calculator**

6. Chris, Peter, Maria, Danielle, Sarah, and Mike have tickets to a rock concert. Their tickets are seats 1-6 in row L. The lowest numbered seat is on the aisle. Chris is sitting between Maria and Sarah. Sarah is sitting on the aisle. Peter is in seat 4 and is between two girls. Find the number of each person's seat. **Sarah-1; Chris-2; Maria-3; Peter-4; Danielle-5; Mike-6**

7. Five different colored pairs of socks are mixed up in a drawer. If socks are pulled out at random, what is the greatest number of socks that could be chosen before you get a matching pair? **6 socks**

8. A jet can seat 525 passengers. Find the number of passengers on the plane if four seats are occupied for every empty seat. **420 passengers**

Research

9. In the library, find information about the history of calculators. See the Fun Pages, 104 and 105, in Chapter 3 to get started. **See students' work.**

Critical Thinking

10. How many different routes can Peter take from Room 101 to Room 205 if the hallway from Room 104 to Room 203 does not intersect with the hallway from Room 103 to Room 204? **4 routes**

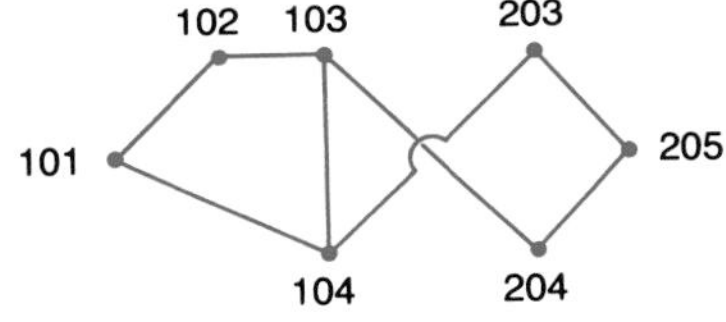

Mixed Review

Find the average for each set of numbers. Round to nearest tenth.

Lesson 3-12

11. 18, 24, 25 **22.3**
12. 4.5, 6.2, 5.8 **5.5**
13. 0.83, 0.74, 0.78 **0.8**

Lesson 4-2

Write the prime factorization of each number.

14. 18 **2×3^2**
15. 20 **$2^2 \times 5$**
16. 28 **$2^2 \times 7$**
17. 42 **$2 \times 3 \times 7$**
18. 102 **$2 \times 3 \times 17$**

Lesson 4-11

19. Miranda works for $3\frac{1}{2}$ hours on Saturday and $4\frac{1}{3}$ hours on Sunday. How many hours does she work on both days? **$7\frac{5}{6}$ hours**

APPLYING THE LESSON

Independent Practice

Homework Assignment	
Minimum	Ex. 1-19
Average	Ex. 1-19
Maximum	Ex. 1-19

Chapter 5, Quiz A (Lessons 5-1 through 5-5) is available in the Evaluation Masters Booklet, p. 24.

Chalkboard Examples

Solve by simplifying the problem.

1. Fred and Al work together and mow lawns at the same rate. They can mow three lawns in two hours. How long will it take Al to mow $4\frac{1}{2}$ lawns alone? **6 hours**
2. How many lawns can Fred and Al mow in 8 hours? **12 lawns**

Additional examples are provided on TRANSPARENCY 5-5.

EVALUATING THE LESSON

Check for Understanding Use Guided Practice Exercises to check for student understanding.

Error Analysis Watch for students who have trouble determining how much one person can accomplish when given how much several can accomplish working together. Emphasize that this first computation determines the answer and careful reading now will yield the correct answer.

Closing the Lesson Tell students they will use this simplifying procedure when deciding how many workers are needed on a committee to complete a project, such as making dance decorations.

Applying Mathematics to the Real World

This optional page shows how mathematics is used in the real world and also provides a change of pace.

Objective Find the total cost of ordering food for a group.

Using Discussion Discuss with students the uses a chef may have for mathematics; for example, increasing and decreasing recipes for different numbers of people, computing and ordering food supplies, computing and ordering kitchen supplies, computing cost per serving, and so on.

Activity

Using Data Have students collect menus from different restaurants or from advertisements in the food section of the newspaper. Have students work in groups of three or four. Each student places an *order*. Have students exchange orders and find the total cost of the order.

Making Career Connections

CHEF

Diane Seriani is the head chef in a hotel restaurant. Chef Seriani coordinates the work of the kitchen staff and directs preparation of special dishes. One of the house specialties is Grilled Lemon Chicken.

Grilled Lemon Chicken

6 pieces chicken (leg, thigh, or split breast)
$\frac{1}{4}$ pound corn oil margarine
$\frac{1}{2}$ cup lemon juice
1 tbsp poultry seasoning
1 tsp dried parsley, crumbled
$1\frac{1}{2}$ cups long grain and wild rice, uncooked
$\frac{1}{2}$ tsp salt

Cook rice according to package directions. In a small saucepan, combine lemon juice, poultry seasoning, margarine, and parsley. Bring mixture to a boil and remove from heat. Baste chicken with lemon mixture. Broil 4 inches from heat, placing breasts with rib side facing heat, 20 to 25 minutes. Turn chicken and baste again. Broil an additional 10 minutes until brown. Serve on the rice. Pour remaining lemon mixture on chicken. Serves 4.

Suppose Chef Seriani doubles the recipe. How much rice does she need?

cups of rice needed for one recipe ⟶ $\frac{3}{2} \times 2 = \frac{6}{2}$ or 3 ⟵ cups of rice needed for double recipe

(double ↓ the 2; change $1\frac{1}{2}$ to an improper fraction)

How much of each item is needed to double the recipe?

1. margarine $\frac{1}{2}$ **pound** **2.** lemon juice $\frac{1}{4}$ **cup** **3.** poultry seasoning **2 tbsp**

How many ounces of each item is needed for $\frac{3}{4}$ of the recipe?

4. salt $\frac{3}{8}$ **tsp** **5.** parsley $\frac{3}{4}$ **tsp** **6.** rice $1\frac{1}{8}$ **cups**

7. How many pieces of chicken are needed to serve 16 people? **24 pieces**

8. If 2 cups of uncooked rice is equivalent to one pound, how many pounds of rice are needed to serve 16 people? **3 lb**

9. Tomorrow Chef Seriani needs 2 extra cooks. Each will work 4 extra hours to prepare for a banquet. The rate for cooks is \$10.25. How much will this overtime help cost? **\$82**

Estimating Utility Bills

Pam and Mike Spiker live in an apartment. They have to pay three utility bills. Each month they receive an electric bill and a gas bill. Every three months—four times a year—they receive a water bill.

The amount of electricity, natural gas, and water they use is measured using the following units. The cost per unit is also given below.

Utility	Electricity	Natural Gas	Water
Unit	Kilowatt hour (kWh)	100 cubic feet (CCF)	cubic feet (ft^3)
Cost per unit	5.6¢	56.5¢	7.85¢

Mrs. Spiker estimates each utility bill to see if the amount charged is reasonable.

Estimate the cost of 906 kWh of electricity at 5.6¢ per kWh.

906		900	Round 906 to 900.
× 5.6¢	→	× 6¢	Round 5.6¢ to 6¢.
		5400¢	Cost in cents

Write 5400¢ as $54. Mrs. Spiker estimates the electric bill to be $54.

Estimate the amount of each electric bill. The number of kilowatt-hours (kWh) and the cost per kWh are given.

1. 563 kWh, 8.23¢ **$48**
2. 795 kWh, 7.6¢ **$64**
3. 241 kWh, 5.5¢ **$12**

Estimate the amount of each natural gas bill. The amount of gas used (CCF) and the cost per CCF are given.

4. 64 CCF, 56.6¢ **$36**
5. 125 CCF, 48.9¢ **$50**
6. 187 CCF, 59¢ **$120**

7. Last month the Jones family used 945 kWh of electricity at 7.6¢ per kWh and 57 CCF of gas at 58.2¢ per CCF. Estimate the total of these two utility bills. **$72, $36; $108**

8. Estimate each utility bill. Then estimate the total. 894 kWh of electricity at 8.9¢ per kWh, 94 CCF of gas at 49.9¢ per CCF, and 726 ft^3 of water at 5.7¢ per ft^3. **$81, $50, $42; $170 to $180**

Applying Mathematics to the Real World

This optional page shows how mathematics is used in the real world and also provides a change of pace.

Objective Estimate the cost of a utility bill.

Using Discussion Discuss with students some of the reasons for different rates for electricity in different parts of the United States. **The source of power for producing electricity will affect its price. The cost of supplying the electricity will also affect the price so proximity to the source may be a factor.**

Discuss with students that the cost per unit for utilities is not the only charge that appears on utility bills. Depending on the area of the country, households may pay surcharges, tax, service charges, and so on, so that an estimate based on unit cost will not match the amount on a utility statement.

Activity

Using Research Have students research the local utility rates. Have students find the number of kWh of electricity, the CCF of natural gas (if using gas), and ft^3 of water their household uses in a month and estimate the bills.

5-MINUTE CHECK-UP

(over Lesson 5-5)

Available on TRANSPARENCY 5-6.

Solve by simplifying the problem.

1. Last year Mr. Terrel filled his pool in 18 hours using two identical hoses. This year he bought 4 more hoses. How long will it take to fill the pool with 6 hoses? **6 hr**

INTRODUCING THE LESSON

Using Manipulatives Have students use two identical pieces of paper. Mark one sheet with the number 1. Cut the other sheet into fourths. Label each piece $\frac{1}{4}$. Ask students how many times a fourth fits onto 1 (like puzzle pieces). **The answer, 4, is the result of $1 \div \frac{1}{4}$.**

TEACHING THE LESSON

Using Critical Thinking Ask students how they would prove the answer in Example D on page 158 is correct. **A good answer should refer to the inverse operations of multiplication and division.**

Practice Masters Booklet, p. 55

Name ______________________ Date ______________

PRACTICE WORKSHEET 5-6

Dividing Fractions

Divide.

1.	$4 \div \frac{1}{2}$	8	2.	$\frac{4}{9} \div \frac{2}{3}$	$\frac{2}{3}$
3.	$\frac{5}{6} \div 5$	$\frac{1}{6}$	4.	$\frac{9}{10} \div \frac{3}{5}$	$1\frac{1}{2}$
5.	$\frac{8}{13} \div \frac{4}{7}$	$1\frac{1}{13}$	6.	$\frac{5}{12} \div \frac{10}{11}$	$\frac{11}{24}$
7.	$\frac{6}{7} \div 2$	$\frac{3}{7}$	8.	$\frac{2}{15} \div \frac{2}{5}$	$\frac{1}{3}$
9.	$\frac{4}{7} \div \frac{8}{11}$	$\frac{11}{14}$	10.	$8 \div \frac{2}{3}$	12
11.	$\frac{3}{5} \div \frac{6}{7}$	$\frac{7}{10}$	12.	$\frac{1}{9} \div \frac{5}{6}$	$\frac{2}{15}$
13.	$20 \div \frac{4}{5}$	25	14.	$\frac{2}{9} \div \frac{7}{18}$	$\frac{4}{7}$
15.	$\frac{7}{11} \div \frac{21}{22}$	$\frac{2}{3}$	16.	$\frac{8}{15} \div \frac{12}{25}$	$1\frac{1}{9}$
17.	$\frac{16}{19} \div \frac{4}{5}$	$1\frac{1}{19}$	18.	$\frac{5}{12} \div \frac{5}{8}$	$\frac{2}{3}$
19.	$\frac{14}{17} \div \frac{21}{23}$	$\frac{46}{51}$	20.	$\frac{9}{11} \div \frac{3}{4}$	$1\frac{1}{11}$
21.	$18 \div \frac{21}{25}$	$21\frac{3}{7}$	22.	$\frac{6}{13} \div 2$	$\frac{3}{13}$
23.	$\frac{13}{16} \div \frac{5}{32}$	$5\frac{1}{5}$	24.	$\frac{11}{12} \div \frac{7}{24}$	$3\frac{1}{7}$

Solve.

25. Rosa's birthday cake was cut into pieces such that each piece was $\frac{1}{32}$ of the entire cake. If $\frac{3}{4}$ of the cake was eaten, how many pieces were eaten? **24 pieces**

5-6 DIVIDING FRACTIONS

Objective

Find the reciprocal of a number.
Divide fractions.

The Student Council at Jefferson High is planning a cookout. Harry found the best price on hamburger was $16.50 for a 10-pound package. How many $\frac{1}{4}$-pound patties can Harry make from 10 pounds of hamburger? Harry divided 10 by $\frac{1}{4}$ to find the answer. He made this drawing to show 10 pounds of hamburger.

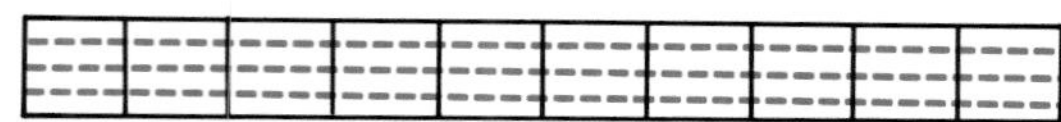

The dashed lines show that each pound makes four $\frac{1}{4}$-pound hamburger patties. Ten pounds makes 10 times 4 or 40 hamburger patties. Example C shows another way to solve this problem.

Two numbers whose product is 1, such as 4 and $\frac{1}{4}$, are **reciprocals**.

Examples

A ***What is the reciprocal of $\frac{5}{12}$?***

Since $\frac{5}{12} \times \frac{12}{5} = 1$, the reciprocal of $\frac{5}{12}$ is $\frac{12}{5}$.

B ***What is the reciprocal of 3?***

Since $3 \times \frac{1}{3} = 1$, the reciprocal of 3 is $\frac{1}{3}$.

To divide by a fraction, multiply by its reciprocal.

Method

1. Rewrite the problem as a product of the dividend and the reciprocal of the divisor.
2. Multiply.
3. Write the result in simplest form if necessary.

Examples

C $10 \div \frac{1}{4}$ The quotient will be greater than 10.

1. The reciprocal of $\frac{1}{4}$ is 4.
$10 \div \frac{1}{4} = \frac{10}{1} \times \frac{4}{1}$
2. $= \frac{40}{1}$
3. $= 40$

Harry can make 40 hamburger patties from 10 pounds of hamburger.

D $\frac{5}{9} \div \frac{5}{12}$

1. The reciprocal of $\frac{5}{12}$ is $\frac{12}{5}$.
$\frac{5}{9} \div \frac{5}{12} = \frac{5}{9} \times \frac{12}{5}$
2. $= \frac{\overset{1}{\cancel{5}}}{\underset{3}{\cancel{9}}} \times \frac{\overset{4}{\cancel{12}}}{\underset{1}{\cancel{5}}}$

3. $= \frac{4}{3}$ or $1\frac{1}{3}$

Example E $\frac{14}{15} \div 7$ Will the quotient be less than or greater than 7? **less than**

1. The reciprocal of 7 is $\frac{1}{7}$.
$\frac{14}{15} \div 7 = \frac{14}{15} \times \frac{1}{7}$
2. $= \frac{\overset{2}{\cancel{14}}}{15} \times \frac{1}{\underset{1}{\cancel{7}}}$ or $\frac{2}{15}$

RETEACHING THE LESSON

Find quotients for the following exercises. Then check answers by multiplying the quotient by the divisor.

1. $\frac{1}{4} \div \frac{1}{3} =$ **$\frac{3}{4}$**
2. $\frac{1}{2} \div \frac{1}{5} =$ **$2\frac{1}{2}$**
3. $\frac{1}{5} \div \frac{3}{5} =$ **$\frac{1}{3}$**
4. $\frac{5}{12} \div \frac{5}{6} =$ **$\frac{1}{2}$**
5. $\frac{1}{2} \div \frac{7}{16} =$ **$1\frac{1}{7}$**
6. $\frac{2}{3} \div 4 =$ **$\frac{1}{6}$**
7. $\frac{3}{5} \div 8 =$ **$\frac{3}{40}$**
8. $6 \div \frac{1}{3} =$ **18**
9. $18 \div \frac{9}{10} =$ **20**

Reteaching Masters Booklet, p. 50

Name ______________________ Date ______________

RETEACHING WORKSHEET 5-6

Dividing Fractions

Two numbers whose product is 1 are **reciprocals**.

3 and $\frac{1}{3}$ are reciprocals. $\frac{3}{1} \times \frac{1}{3} = 1$

$\frac{4}{5}$ and $\frac{5}{4}$ are reciprocals. $\frac{4}{5} \times \frac{5}{4} = 1$

To divide by a fraction, multiply by its reciprocal.

$7 \div \frac{3}{4}$

$7 \div \frac{3}{4} = \frac{7}{1} \times \frac{4}{3}$ **The reciprocal of $\frac{3}{4}$ is $\frac{4}{3}$.**

$= \frac{7 \times 4}{3 \times 1}$

$= \frac{28}{3}$ or $9\frac{1}{3}$ **Simplify.**

$\frac{5}{8} \div \frac{11}{16}$

$\frac{5}{8} \div \frac{11}{16} = \frac{5}{8} \times \frac{16}{11}$ **The reciprocal of $\frac{11}{16}$ is $\frac{16}{11}$.**

$= \frac{5 \times \overset{2}{\cancel{16}}}{\underset{1}{\cancel{8}} \times 11}$

$= \frac{10}{11}$ **Simplify.**

Guided Practice

Examples A, B

Name the reciprocal of each number.

1. $\frac{8}{9}$ $\frac{9}{8}$　**2.** $\frac{2}{3}$ $\frac{3}{2}$　**3.** $\frac{3}{20}$ $\frac{20}{3}$　**4.** $\frac{5}{7}$ $\frac{7}{5}$　**5.** 5 $\frac{1}{5}$　**6.** 12 $\frac{1}{12}$

Divide.

Example C

7. $4 \div \frac{3}{4}$ $5\frac{1}{3}$　**8.** $10 \div \frac{5}{8}$ 16　**9.** $14 \div \frac{7}{9}$ 18　**10.** $6 \div \frac{3}{7}$ 14

Examples D, E

11. $\frac{2}{3} \div \frac{1}{4}$ $2\frac{2}{3}$　**12.** $\frac{7}{15} \div \frac{5}{6}$ $\frac{14}{25}$　**13.** $\frac{3}{8} \div 5$ $\frac{3}{40}$　**14.** $\frac{11}{12} \div 11$ $\frac{1}{12}$

Exercises

Practice

Name the reciprocal of each number.

15. $\frac{19}{20}$ $\frac{20}{19}$　**16.** 24 $\frac{1}{24}$　**17.** $\frac{11}{16}$ $\frac{16}{11}$　**18.** $\frac{1}{6}$ 6

Divide.

19. $\frac{3}{4} \div \frac{2}{5}$ $1\frac{7}{8}$　**20.** $\frac{7}{8} \div \frac{3}{5}$ $1\frac{11}{24}$　**21.** $\frac{5}{12} \div \frac{2}{3}$ $\frac{5}{8}$　**22.** $\frac{2}{3} \div \frac{2}{3}$ 1

23. $6 \div \frac{3}{8}$ 16　**24.** $3 \div \frac{2}{3}$ $4\frac{1}{2}$　**25.** $2 \div \frac{1}{2}$ 4　**26.** $8 \div \frac{2}{3}$ 12

27. $\frac{1}{4} \div 9$ $\frac{1}{36}$　**28.** $\frac{7}{8} \div 3$ $\frac{7}{24}$　**29.** $\frac{3}{5} \div 12$ $\frac{1}{20}$　**30.** $\frac{3}{8} \div 3$ $\frac{1}{8}$

31. Divide $\frac{7}{8}$ by $\frac{1}{2}$. $1\frac{3}{4}$

32. What is the quotient if $\frac{3}{10}$ is divided by 2? $\frac{3}{20}$

Applications

33. Angie has $\frac{7}{8}$ yard of material left over after making chair pads. If it takes $\frac{1}{6}$ yard to make a napkin ring, does Angie have enough material left to make 6 matching napkin rings?
no; $\frac{7}{8} \div \frac{1}{6} = 5\frac{1}{4}$

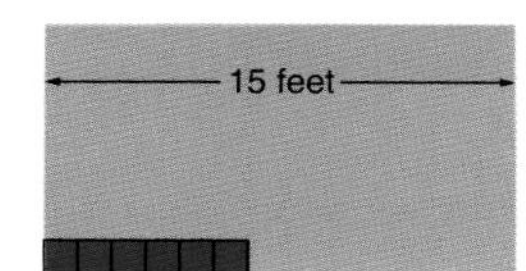

34. How many $\frac{5}{6}$-foot vinyl tiles will fit along the edge of a 15-foot room? Make no allowance for seams.
18 tiles

Talk Math

35. Tell how this drawing shows $6 \div \frac{1}{3} = 18$. See margin

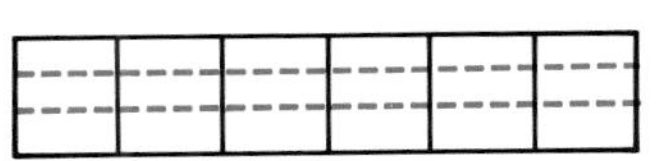

Using Variables

Evaluate each expression if $p = \frac{1}{2}$, $q = \frac{3}{4}$, and $r = \frac{1}{3}$.

36. $q \div r$ $2\frac{1}{4}$　**37.** $p \div q$ $\frac{2}{3}$　**38.** $p \times q$ $\frac{3}{8}$　**39.** $q \times r$ $\frac{1}{4}$

Critical Thinking

40. A knitting pattern for the body of a sweater is given below.

Row 1	Knit 1, Purl 1, . . .	Row 2	Knit 1, Purl 1, . . .
Row 3	Purl 1, Knit 1, . . .	Row 4	Purl 1, Knit 1, . . .
Row 5	Knit 1, Purl 1, . . .	Row 6	Knit 1, Purl 1, . . .
Row 7	Purl 1, Knit 1, . . .	and so on.	

Does Row 35 begin with *Knit 1* or *Purl 1*? Purl 1

APPLYING THE LESSON

Independent Practice

Homework Assignment	
Minimum	Ex. 1-39 odd; 40
Average	Ex. 2-30 even; 31-40
Maximum	Ex. 1-40

Additional Answers

35. When the six rectangles are divided into thirds, there are 18 parts.

Chalkboard Examples

Examples A, B *Name the reciprocal of each number.*

1. $\frac{5}{12}$ $\frac{12}{5}$　**2.** $\frac{5}{6}$ $\frac{6}{5}$　**3.** 101 $\frac{1}{101}$

Examples C, D, E *Divide.*

4. $34 \div \frac{1}{2}$ 68　**5.** $8 \div \frac{4}{5}$ 10

6. $\frac{4}{5} \div \frac{1}{5}$ 4　**7.** $\frac{4}{7} \div 8$ $\frac{1}{14}$

Additional examples are provided on TRANSPARENCY 5-6

EVALUATING THE LESSON

Check for Understanding Use Guided Practice Exercises to check for student understanding.

Error Analysis Watch for students who cancel common factors before using the reciprocal. Emphasize that the shortcut applies only to multiplying.

Closing the Lesson Have the students list the steps needed to divide a fraction by a fraction.

EXTENDING THE LESSON

Enrichment Masters Booklet, p. 50

Name ______________ Date ______________

ENRICHMENT WORKSHEET 5-6

Complex Fractions

A **complex fraction** is a fraction whose numerator or denominator, or both, contains a fraction.

There are two methods to evaluate complex fractions.

Method One: Multiply both the numerator and the denominator by the LCM of the denominators within them.

Method Two: Write the complex fraction as a division problem and divide.

Complex Fraction	Method One	Method Two
$\frac{\frac{2}{3}}{6}$	$\frac{\frac{2}{3} \times 3}{6 \times 3} = \frac{2}{18}$ or $\frac{1}{9}$	$\frac{2}{3} \div 6 = \frac{2}{3} \times \frac{1}{6}$ or $\frac{1}{9}$
$\frac{\frac{4}{9}}{\frac{5}{6}}$	$\frac{\frac{4}{9} \times 18}{\frac{5}{6} \times 18} = \frac{8}{15}$	$\frac{4}{9} \div \frac{5}{6} = \frac{4}{9} \times \frac{6}{5}$ or $\frac{8}{15}$

Evaluate each complex fraction using Method One. Then use Method Two and compare your answers. They should be the same, of course.

1. $\frac{1}{\frac{3}{7}}$ $2\frac{1}{3}$　**2.** $\frac{\frac{9}{10}}{\frac{3}{5}}$ $1\frac{1}{2}$　**3.** $\frac{\frac{5}{18}}{\frac{2}{9}}$ $1\frac{1}{4}$　**4.** $\frac{\frac{7}{24}}{\frac{21}{32}}$ $\frac{4}{9}$

5. $\frac{9}{\frac{3}{4}}$ 12　**6.** $\frac{4}{\frac{3}{5}}$ $6\frac{2}{3}$　**7.** $\frac{8}{\frac{4}{5}}$ 10　**8.** $\frac{12}{\frac{5}{6}}$ $14\frac{2}{5}$

9. $\frac{\frac{6}{7}}{3}$ $\frac{2}{7}$　**10.** $\frac{\frac{8}{55}}{16}$ $\frac{1}{110}$　**11.** $\frac{\frac{5}{8}}{\frac{3}{4}}$ $\frac{5}{6}$　**12.** $\frac{\frac{18}{25}}{90}$ $\frac{1}{125}$

5-MINUTE CHECK-UP

(over Lesson 5-6)

Available on TRANSPARENCY 5-7.

Name the reciprocal of each number.

1. $\frac{3}{7}$ $\frac{7}{3}$ **2.** 7 $\frac{1}{7}$ **3.** $\frac{8}{5}$ $\frac{5}{8}$

Divide.

4. $\frac{1}{2} \div \frac{7}{8}$ $\frac{4}{7}$ **5.** $\frac{5}{6} \div 3$ $\frac{5}{18}$ **6.** $8 \div \frac{2}{3}$ 12

Extra Practice, Lesson 5-6, p. 461

INTRODUCING THE LESSON

Use pages of the newspaper, a science textbook, or a social studies textbook to have students find examples of mixed numbers.

TEACHING THE LESSON

Using Questioning

- Is the quotient greater than or less than the dividend when the divisor is a proper fraction? greater than
- Explain in your own words how to change a mixed number to an improper fraction. Multiply the whole number by the denominator and add the numerator. Place the result over the original denominator.

Practice Masters Booklet, p. 56

PRACTICE WORKSHEET 5-7

Dividing Mixed Numbers

Divide.

1. $3 \div 1\frac{1}{3}$ $2\frac{1}{4}$	**2.** $14 \div 1\frac{3}{4}$ 8	**3.** $\frac{5}{7} \div 2\frac{1}{7}$ $\frac{1}{3}$
4. $1\frac{4}{5} \div \frac{3}{5}$ 3	**5.** $4\frac{5}{6} \div 3\frac{8}{9}$ $1\frac{17}{70}$	**6.** $2\frac{1}{4} \div 6$ $\frac{3}{8}$
7. $2\frac{4}{7} \div 1\frac{1}{4}$ $2\frac{2}{35}$	**8.** $3\frac{1}{8} \div 4\frac{7}{8}$ $\frac{25}{39}$	**9.** $10\frac{2}{7} \div 6\frac{1}{7}$ $1\frac{29}{43}$
10. $3\frac{5}{7} \div \frac{6}{22}$ $13\frac{13}{21}$	**11.** $\frac{7}{15} \div 1\frac{2}{15}$ $\frac{7}{17}$	**12.** $3\frac{8}{9} \div 1\frac{13}{15}$ $2\frac{1}{12}$
13. $12 \div 3\frac{3}{7}$ $3\frac{1}{2}$	**14.** $2\frac{8}{9} \div \frac{3}{9}$ $8\frac{2}{3}$	**15.** $\frac{4}{5} \div 1\frac{7}{10}$ $\frac{8}{17}$
16. $\frac{4}{5} \div 3\frac{1}{2}$ $\frac{8}{35}$	**17.** $2\frac{1}{3} \div 8$ $\frac{7}{24}$	**18.** $1\frac{1}{5} \div 2\frac{5}{8}$ $\frac{16}{35}$
19. $3\frac{3}{4} \div \frac{7}{12}$ $6\frac{3}{7}$	**20.** $5\frac{1}{8} \div \frac{1}{6}$ $30\frac{3}{4}$	**21.** $7\frac{2}{3} \div 6\frac{1}{2}$ $1\frac{7}{39}$
22. $3\frac{1}{9} \div 6\frac{1}{9}$ $\frac{28}{55}$	**23.** $2\frac{6}{7} \div 6\frac{2}{5}$ $\frac{25}{56}$	**24.** $4\frac{1}{6} \div 2\frac{1}{2}$ $1\frac{2}{3}$
25. $4\frac{4}{5} \div 4\frac{8}{9}$ $\frac{54}{55}$	**26.** $11\frac{1}{5} \div 3\frac{3}{5}$ $3\frac{1}{9}$	**27.** $3\frac{3}{5} \div 9$ $\frac{2}{5}$
28. $4\frac{9}{10} \div 1\frac{1}{20}$ $4\frac{2}{3}$	**29.** $1\frac{7}{20} \div 4\frac{3}{5}$ $\frac{27}{92}$	**30.** $7\frac{1}{2} \div 1\frac{5}{19}$ $5\frac{15}{16}$

Solve.

31. Strips $\frac{2}{9}$ of a yard wide must be cut from $4\frac{1}{2}$ yards of fabric. How many strips can be cut? 20 strips

32. In 7 ounces of fertilizer, there are $1\frac{3}{5}$ ounces phosphorus. What part of the fertilizer is phosphorus? $\frac{8}{35}$

5-7 DIVIDING MIXED NUMBERS

Objective
Divide mixed numbers.

Justin jogs $12\frac{1}{2}$ miles a day when training for a wrestling event. If the course he jogs is $2\frac{1}{2}$ miles long, how many times does he jog the course each day?

Method

1. Rename each mixed number as an improper fraction.
2. Divide the fractions.
3. Write the result in simplest form, if necessary.

Example A $12\frac{1}{2} \div 2\frac{1}{2}$ Estimate: $12 \div 3 = 4$

1. $12\frac{1}{2} = \frac{(12 \times 2) + 1}{2}$ or $\frac{25}{2}$

   $2\frac{1}{2} = \frac{(2 \times 2) + 1}{2}$ or $\frac{5}{2}$

2. $12\frac{1}{2} \div 2\frac{1}{2} = \frac{25}{2} \div \frac{5}{2}$

   $= \frac{\overset{5}{\cancel{25}}}{\underset{1}{\cancel{2}}} \times \frac{\overset{1}{\cancel{2}}}{\underset{1}{\cancel{5}}}$ The reciprocal of $\frac{5}{2}$ is $\frac{2}{5}$.

   $= \frac{5}{1}$

3. $= 5$

Subtract $2\frac{1}{2}$ five times to check the answer.

$12\frac{1}{2} - 2\frac{1}{2} - 2\frac{1}{2} - 2\frac{1}{2} - 2\frac{1}{2} - 2\frac{1}{2} = 0$

Based on the estimate, explain why the answer is reasonable.

Justin jogs around the course 5 times.

Example B $2\frac{2}{3} \div 4$ Estimate: $2 \div 4 = \frac{2}{4}$ or $\frac{1}{2}$

1. $2\frac{2}{3} = \frac{(2 \times 3) + 2}{3}$ or $\frac{8}{3}$

2. $2\frac{2}{3} \div 4 = \frac{8}{3} \div 4$ The reciprocal of 4 is $\frac{1}{4}$.

   $= \frac{\overset{2}{\cancel{8}}}{3} \times \frac{1}{\underset{1}{\cancel{4}}}$

   $= \frac{2}{3}$ ← simplest form

   Is the answer reasonable?

Guided Practice

Divide. Encourage students to estimate each quotient first.

Example A

1. $1\frac{2}{3} \div 1\frac{1}{5}$ $1\frac{7}{18}$ **2.** $8\frac{1}{4} \div 5\frac{1}{6}$ $1\frac{37}{62}$ **3.** $2\frac{2}{5} \div \frac{2}{5}$ 6 **4.** $\frac{1}{3} \div 1\frac{5}{6}$ $\frac{2}{11}$

5. $\frac{2}{5} \div 5\frac{4}{5}$ $\frac{2}{29}$ **6.** $3\frac{3}{8} \div 2\frac{1}{5}$ $1\frac{47}{88}$ **7.** $5\frac{1}{4} \div 8\frac{1}{6}$ $\frac{9}{14}$ **8.** $1\frac{1}{8} \div \frac{9}{10}$ $1\frac{1}{4}$

Example B

9. $3\frac{1}{3} \div 4$ $\frac{5}{6}$ **10.** $1\frac{3}{4} \div 2$ $\frac{7}{8}$ **11.** $1\frac{1}{4} \div 15$ $\frac{1}{12}$ **12.** $9 \div 1\frac{4}{5}$ 5

13. $6 \div 3\frac{1}{5}$ $1\frac{7}{8}$ **14.** $12 \div 1\frac{3}{5}$ $7\frac{1}{2}$ **15.** $10 \div 13\frac{3}{4}$ $\frac{8}{11}$ **16.** $10\frac{1}{2} \div 9$ $1\frac{1}{6}$

RETEACHING THE LESSON

Have students write a step-by-step analysis of each problem. For example:

$8\frac{2}{5} \div 3\frac{1}{2} = \frac{42}{5} \div \frac{7}{2}$ Rename mixed numbers.

$= \frac{42}{5} \times \frac{2}{7}$ Multiply by the reciprocal.

$= \frac{\overset{6}{\cancel{42}}}{5} \times \frac{2}{\underset{1}{\cancel{7}}}$ The GCF of 7 and 42 is 7. Multiply.

$= \frac{12}{5} = 2\frac{2}{5}$ Rename improper fractions.

Reteaching Masters Booklet, p. 51

Name ______________ Date ______________

RETEACHING WORKSHEET 5-7

Dividing Mixed Numbers

To divide with mixed numbers, first rename the mixed numbers as fractions. Then divide the fractions.

$5\frac{1}{2} \div 2\frac{1}{4}$

First rename the mixed numbers. $5\frac{1}{2} = \frac{(5 \times 2) + 1}{2} = \frac{11}{2}$ $2\frac{1}{4} = \frac{(2 \times 4) + 1}{4} = \frac{9}{4}$

$5\frac{1}{2} \div 2\frac{1}{4} = \frac{11}{2} \div \frac{9}{4}$

Multiply by the reciprocal of $\frac{9}{4}$. $= \frac{11}{2} \times \frac{4}{9}$

Simplify. $= \frac{11 \times \overset{2}{\cancel{4}}}{\underset{1}{\cancel{2}} \times 9}$

Write the quotient in simplest form. $= \frac{22}{9}$ or $2\frac{4}{9}$

Rename each mixed or whole number as an improper fraction. Then write its reciprocal.

Exercises

Practice

Divide. Encourage students to estimate each quotient first.

17. $1\frac{1}{8} \div 3$ $\frac{3}{8}$ **18.** $2\frac{1}{2} \div 3$ $\frac{5}{6}$ **19.** $6 \div 2\frac{2}{5}$ $2\frac{1}{2}$ **20.** $6 \div 9\frac{3}{5}$ $\frac{5}{8}$

21. $14 \div 2\frac{1}{10}$ $6\frac{2}{3}$ **22.** $6 \div 3\frac{1}{2}$ $1\frac{5}{7}$ **23.** $3\frac{3}{5} \div 12$ $\frac{3}{10}$ **24.** $13\frac{3}{4} \div 5$ $2\frac{3}{4}$

25. $1\frac{3}{4} \div 4\frac{3}{8}$ $\frac{2}{5}$ **26.** $8\frac{2}{5} \div 1\frac{2}{5}$ 6 **27.** $6\frac{3}{5} \div 6$ $1\frac{1}{10}$ **28.** $1\frac{2}{3} \div \frac{2}{3}$ $2\frac{1}{2}$

29. $5 \div 2\frac{1}{2}$ 2 **30.** $\frac{1}{4} \div 2\frac{1}{3}$ $\frac{3}{28}$ **31.** $12\frac{3}{4} \div 4\frac{7}{8}$ $2\frac{8}{13}$ **32.** $1\frac{7}{8} \div 5$ $\frac{3}{8}$

33. Divide $3\frac{1}{8}$ by $\frac{5}{8}$. 5

34. What is the quotient if $\frac{1}{2}$ is divided by $1\frac{1}{2}$? $\frac{1}{3}$

Applications

35. A $3\frac{3}{4}$-foot board is cut into pieces that are $\frac{1}{8}$ of a foot long. How many pieces are cut from the board? 30 pieces

36. Don has two stacks of $\frac{3}{8}$-inch plywood sheets. One stack is 30 inches high. The other is 24 inches high. How many sheets of plywood are there in both stacks? 144 sheets

37. How many boards of 6-inch siding does it take to cover the side of a house if each board covers $5\frac{1}{4}$ inches and the height of the wall is 105 inches? 20 boards

Critical Thinking

38. A football team can score 6 points for a touchdown, 1 point for a point after a touchdown, 2 points for a safety, and 3 points for a field goal. Write 5 different ways a team can score 34 points. See margin.

Cooperative Groups

39. In groups of three or four, use a newspaper ad for a building supply company and make up three problems. In one problem you must use multiplication of fractions to solve it. In another problem you must use division of fractions to solve it. In the third problem you must use mixed numbers to write the problem. Answers may vary.

Problem Solving

40. The product of two fractions is $15\frac{1}{8}$. One of the fractions is $2\frac{3}{4}$. What is the other fraction? $5\frac{1}{2}$

Estimation

Most department stores have sales several times a year. You can estimate the savings mentally to be sure you are charged the correct price.

CLEARANCE SALE!
All Shoes $\frac{1}{4}$ off
All Sweaters $\frac{1}{3}$ off
All Shirts $\frac{1}{3}$ off

About how much are a pair of shoes that regularly cost $32.99?

Think: $\frac{1}{4}$ of \$32.99 → $\frac{1}{4}$ of \$32 or \$8 ← estimated savings
\$32.99 − \$8 → \$33 − \$8 or \$25 ← estimated sale price

Estimate the sale price of each item. Write only your answer. 43. $29

41. a $27.50 sweater $18 **42.** a $14.99 shirt $12 **43.** a $39 pair of shoes

44. an $18.95 shirt $15 **45.** a $45 pair of shoes $34 **46.** a $38.99 sweater $26

Chalkboard Examples

Example A *Divide.*

1. $7\frac{1}{2} \div 2\frac{1}{4}$ $3\frac{1}{3}$ **2.** $8\frac{3}{4} \div 1\frac{2}{5}$ $6\frac{1}{4}$

Example B *Divide.*

3. $5\frac{1}{4} \div$ $\frac{3}{4}$ **4.** $16 \div 1\frac{2}{5}$ $11\frac{3}{7}$

Additional examples are provided on TRANSPARENCY 5-7.

▶ EVALUATING THE LESSON

Check for Understanding Use Guided Practice Exercises to check for student understanding.

Closing the Lesson Ask students to summarize the steps to divide a mixed number by a proper fraction.

▶ EXTENDING THE LESSON

Enrichment Find $\frac{3}{4} \times \frac{1}{2}$ on a calculator.

3 [÷] 4 × 1 [÷] 2 [=] 0.375

Find $\frac{5}{7} \times \frac{7}{5}$ on a calculator and explain the answer. Since calculators compute in a decimal mode and $\frac{5}{7}$ is a repeating decimal, $0.\overline{9}$ will appear on the calculator.

Enrichment Masters Booklet, p. 51

Name ______________________ Date ______________

ENRICHMENT WORKSHEET 5-7

Dividing with Mixed Numbers

Complete.

1. $5\frac{2}{3} \div 5\frac{2}{3} = \underline{1}$ **2.** $8\frac{4}{9} \div 1\frac{1}{3} = \underline{6\frac{1}{3}}$ **3.** $\underline{9\frac{1}{2}} \div \frac{19}{20} = 10$

4. $8\frac{3}{4} \div \underline{2\frac{1}{2}} = 3\frac{1}{2}$ **5.** $5\frac{3}{7} \div 3\frac{1}{6} = \underline{1\frac{5}{7}}$ **6.** $4\frac{2}{3} \div \underline{2\frac{1}{6}} = 2\frac{2}{13}$

7. $\underline{15} \div 3\frac{3}{4} = 4$ **8.** $18 \div 4\frac{1}{5} = \underline{4\frac{2}{7}}$ **9.** $20 \div \underline{3\frac{3}{4}} = 5\frac{1}{3}$

Divide $\frac{3}{5}$ of $\frac{5}{13}$ of $\frac{4}{9}$ of $3\frac{1}{4}$ by $\frac{5}{8}$ of $\frac{12}{25}$ of $\frac{4}{9}$ of $\frac{2}{3}$ of 6. "Of" means to multiply.

$$\left(\frac{3}{5} \times \frac{5}{13} \times \frac{4}{9} \times \frac{13}{4}\right) \div \left(\frac{5}{8} \times \frac{12}{25} \times \frac{4}{9} \times \frac{2}{3} \times 6\right)$$

$$\frac{3}{5} \times \frac{5}{13} \times \frac{4}{9} \times \frac{13}{4} \times \frac{8}{5} \times \frac{25}{12} \times \frac{9}{4} \times \frac{3}{2} \times \frac{1}{6} = \frac{5}{8}$$

10. Divide $\frac{7}{8}$ of $\frac{13}{38}$ of $\frac{16}{27}$ of $6\frac{3}{4}$ by $\frac{13}{19}$ of $\frac{14}{29}$ of $14\frac{1}{2}$. $\frac{1}{4}$

11. Divide $\frac{9}{14}$ of $2\frac{1}{3}$ of $\frac{16}{47}$ of $7\frac{5}{6}$ by $1\frac{5}{9}$ of $\frac{15}{28}$ of $\frac{16}{45}$. $13\frac{1}{2}$

12. Divide $\frac{19}{42}$ of $\frac{28}{33}$ of $\frac{11}{14}$ of $7\frac{1}{9}$ by $\frac{23}{35}$ of $\frac{5}{8}$ of $\frac{16}{23}$ of $\frac{8}{27}$ of $1\frac{7}{12}$. 16

APPLYING THE LESSON

Independent Practice

Homework Assignment	
Minimum	Ex. 2-46 even
Average	Ex. 1-33 odd; 35-46
Maximum	Ex. 1-46

Additional Answers

38. Answers may vary. Sample answer is given.
6 + 6 + 6 + 6 + 6 + 3 + 1
6 + 6 + 6 + 6 + 3 + 3 + 3 + 1
6 + 6 + 6 + 6 + 3 + 3 + 2 + 1 + 1
6 + 6 + 6 + 6 + 6 + 2 + 1 + 1
6 + 6 + 6 + 3 + 3 + 3 + 3 + 2 + 1 + 1

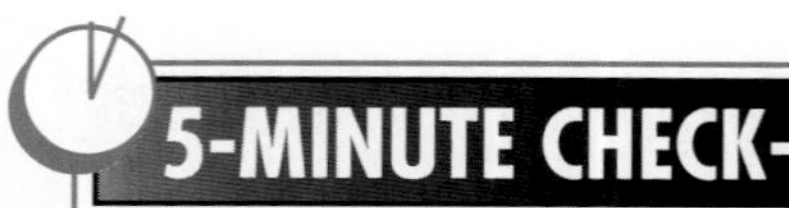

5-MINUTE CHECK-UP

(over Lesson 5-7)

Available on TRANSPARENCY 5-8.

Divide.

1. $7\frac{1}{2} \div 2\frac{1}{2}$ 3 **2.** $16 \div 4\frac{2}{3}$ $3\frac{3}{7}$

3. $1\frac{3}{4} \div 4\frac{3}{8}$ $\frac{2}{5}$ **4.** $6\frac{3}{5} \div 6$ $1\frac{1}{10}$

Extra Practice, Lesson 5-7, p. 462

▶ INTRODUCING THE LESSON

Ask students why they do or do not use a book bag to carry their books, or why they use the front door or the back door to enter the school. Point out that there are many things we do every day because it is *convenient.* Decimals are often more convenient to use than fractions.

▶ TEACHING THE LESSON

Using Critical Thinking Ask students how to change fractions with a denominator that is a power of ten to a decimal without dividing. The numerator is placed in the place-value position corresponding to the power of the denominator.

Practice Masters Booklet, p. 57

Name ______________________ Date ______________

PRACTICE WORKSHEET 5-8

Changing Fractions to Decimals

Change each fraction to a decimal. Use bar notation to show a repeating decimal.

1. $\frac{3}{4}$ 0.75	**2.** $\frac{2}{5}$ 0.4	**3.** $\frac{7}{8}$ 0.875
4. $\frac{1}{3}$ $0.\overline{3}$	**5.** $\frac{4}{9}$ $0.\overline{4}$	**6.** $\frac{3}{11}$ $0.\overline{27}$
7. $\frac{17}{20}$ 0.85	**8.** $\frac{5}{6}$ $0.8\overline{3}$	**9.** $\frac{3}{16}$ 0.1875
10. $\frac{8}{33}$ $0.\overline{24}$	**11.** $\frac{7}{12}$ $0.58\overline{3}$	**12.** $\frac{14}{25}$ 0.56
13. $\frac{7}{10}$ 0.7	**14.** $\frac{5}{8}$ 0.625	**15.** $\frac{11}{15}$ $0.7\overline{3}$
16. $\frac{8}{9}$ $0.\overline{8}$	**17.** $\frac{15}{16}$ 0.9375	**18.** $\frac{1}{12}$ $0.08\overline{3}$
19. $\frac{7}{20}$ 0.35	**20.** $\frac{5}{18}$ $0.2\overline{7}$	**21.** $\frac{3}{10}$ 0.3
22. $\frac{2}{15}$ $0.1\overline{3}$	**23.** $\frac{23}{50}$ 0.46	**24.** $\frac{21}{25}$ 0.84

Change each fraction to a mixed decimal.

25. $\frac{2}{3}$ $0.66\frac{2}{3}$	**26.** $\frac{5}{7}$ $0.71\frac{3}{7}$	**27.** $\frac{4}{11}$ $0.36\frac{4}{11}$
28. $\frac{7}{9}$ $0.77\frac{7}{9}$	**29.** $\frac{1}{6}$ $0.16\frac{2}{3}$	**30.** $\frac{5}{12}$ $0.41\frac{2}{3}$
31. $\frac{11}{30}$ $0.36\frac{2}{3}$	**32.** $\frac{2}{9}$ $0.22\frac{2}{9}$	**33.** $\frac{4}{15}$ $0.26\frac{2}{3}$

5-8 CHANGING FRACTIONS TO DECIMALS

Objective
Change fractions to decimals.

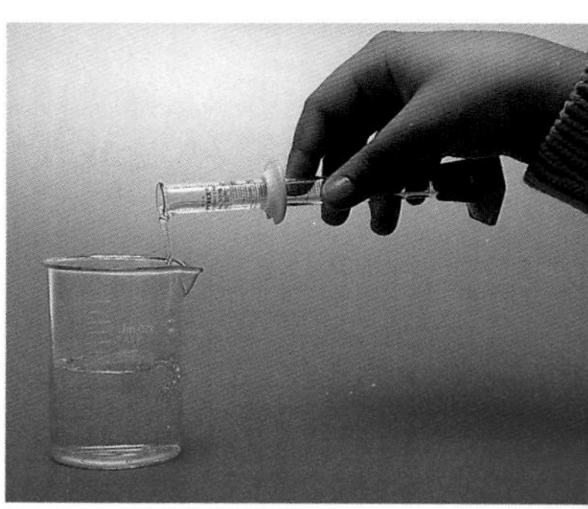

Joy needs to add $\frac{1}{4}$ liter of a salt solution to the mixture in her beaker. The graduated cylinder is marked in decimal parts of a liter, so Joy needs to change the fraction to a decimal. She divides 1 by 4 to change $\frac{1}{4}$ to a decimal.

To change a fraction to a decimal, divide the numerator by the denominator. If the division ends or terminates with a remainder of zero, the quotient is called a **terminating decimal.** If a remainder of zero cannot be obtained, the digits in the quotient repeat. The quotient is called a **repeating decimal.**

Method

1. Divide the numerator by the denominator.
2. Stop when a remainder of zero is obtained or a pattern develops in the quotient.

Examples

A ***Change $\frac{1}{4}$ to a decimal.***

Joy measures to the line marked 0.25 liter.

B ***Change $\frac{2}{3}$ to a decimal.***

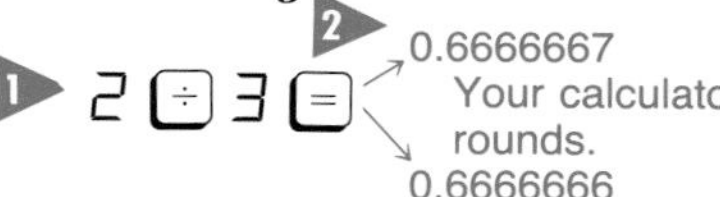

0.6666667 Your calculator rounds.
0.6666666 Your calculator carries another digit in memory.

$\frac{2}{3} = 0.\overline{6}$ The bar is used to show digits that repeat.

Fractions that are equivalent to repeating decimals may also be written as mixed decimals.

Example C ***Change $\frac{4}{7}$ to a mixed decimal.***

$$\begin{array}{r} 0.57\frac{1}{7} \\ 7\overline{)4.00} \\ -3\,5 \\ \hline 50 \\ -\ 49 \\ \hline 1 \end{array}$$

$\frac{4}{7} = 0.57\frac{1}{7}$.

Guided Practice — Example A *Change each fraction to a decimal.*

1. $\frac{2}{5}$ 0.4 **2.** $\frac{3}{8}$ 0.375 **3.** $\frac{3}{4}$ 0.75 **4.** $\frac{5}{16}$ 0.3125

RETEACHING THE LESSON

State whether each of the following is true or false. Correct each false statement by changing the decimal form.

1. $\frac{1}{8} = 0.12\overline{5}$ false $= 0.125$

2. $\frac{5}{4} = 1.25$ true

3. $\frac{5}{12} = 0.\overline{41}$ false $= 0.41\overline{6}$

Reteaching Masters Booklet, p. 52

Name ______________________ Date ______________

RETEACHING WORKSHEET 5-8

Changing Fractions to Decimals

Method: ① Divide the numerator by the denominator.
② Stop when a remainder of zero is obtained or a pattern develops in the quotient.

Change $\frac{13}{20}$ to a decimal.

$$\begin{array}{r} 0.65 \\ 20\overline{)13.00} \\ -12\,0 \\ \hline 100 \\ -\ 100 \\ \hline 0 \end{array}$$

① (division) ② → 0

Change $\frac{1}{6}$ to a decimal.

$$\begin{array}{r} 0.166 \\ 6\overline{)1.00} \\ -\ 6 \\ \hline 40 \\ -\ 36 \\ \hline 40 \\ -36 \\ \hline 4 \end{array}$$

$\frac{1}{6} = 0.166...$
$= 0.1\overline{6}$

The bar shows which digit(s) repeats. ②

Repeating decimals may also be written as mixed decimals.

$$\begin{array}{r} 0.16\frac{4}{6} \\ 6\overline{)1.0} \\ -\ 6 \\ \hline 40 \\ -\ 36 \\ \hline 4 \end{array}$$

Example B *Change each fraction to a decimal. Use bar notation.*

5. $\frac{1}{9}$ $0.\overline{1}$ 6. $\frac{5}{33}$ $0.\overline{15}$ 7. $\frac{5}{6}$ $0.8\overline{3}$ 8. $\frac{7}{12}$ $0.58\overline{3}$

Example C *Change each fraction to a mixed decimal.*

9. $\frac{2}{3}$ $0.66\frac{2}{3}$ 10. $\frac{7}{13}$ $0.53\frac{11}{13}$ 11. $\frac{8}{9}$ $0.88\frac{8}{9}$ 12. $\frac{13}{24}$ $0.54\frac{1}{6}$

Exercises

Practice *Change each fraction to a decimal. Use bar notation to show a repeating decimal.*

13. $\frac{7}{10}$ 0.7 14. $\frac{12}{25}$ 0.48 15. $\frac{11}{20}$ 0.55 16. $\frac{5}{8}$ 0.625 17. $\frac{3}{11}$ $0.\overline{27}$

18. $\frac{7}{16}$ 0.4375 19. $\frac{2}{3}$ $0.\overline{6}$ 20. $\frac{5}{9}$ $0.\overline{5}$ 21. $\frac{7}{9}$ $0.\overline{7}$ 22. $\frac{10}{33}$ $0.\overline{30}$

23. $\frac{28}{45}$ $0.6\overline{2}$ 24. $\frac{9}{16}$ 0.5625 25. $\frac{3}{16}$ 0.1875 26. $\frac{31}{40}$ 0.775 27. $\frac{11}{12}$ $0.91\overline{6}$

Change each fraction to a mixed decimal.

28. $\frac{2}{7}$ $0.28\frac{4}{7}$ 29. $\frac{19}{30}$ $0.63\frac{1}{3}$ 30. $\frac{6}{7}$ $0.85\frac{5}{7}$ 31. $\frac{1}{3}$ $0.33\frac{1}{3}$ 32. $\frac{13}{15}$ $0.86\frac{2}{3}$

33. Write $\frac{1}{125}$ as a decimal. 0.008

34. Write $\frac{5}{11}$ as a decimal. $0.\overline{45}$

Applications

35. The length of one bolt is $\frac{5}{6}$ inch. The length of another is 0.875 inch. Which bolt is longer?

FUN with MATH *Ever thought about how a VCR works?* See page 175.

36. A box of electrical connectors weighs $6\frac{1}{4}$ pounds. If the box weighs $\frac{1}{4}$ pound and each connector weighs $\frac{1}{16}$ pound, how many connectors are in the box? 96 connectors

Cooperative Groups

37. In groups of three or four, use a calculator to list examples of fractions that can be expressed as repeating decimals. Examine your list to discover what kind of numbers in the denominator result in a repeating decimal. Answers may vary. See margin.

Critical Thinking

38. If [sunflower divided into sixths] represents $3\frac{1}{2}$, then [half sunflower] represents __?__. $1\frac{3}{4}$

Using Data *Solve. Use the chart.*

39. If 1,181 students were surveyed, how many considered themselves competent in computer games? 89% = 0.89 1,051 students

Collect Data See student's work.

40. Conduct your own survey of the students in your class or in your school. Compare your results to the results of the survey.

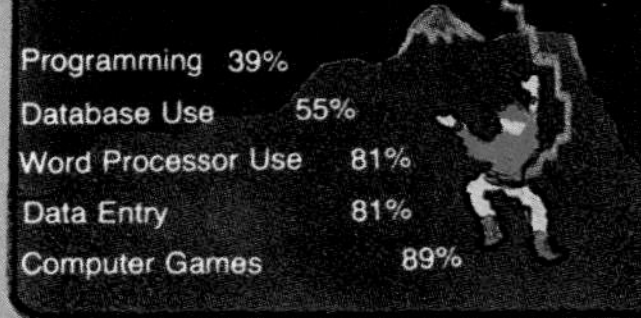

Chalkboard Examples

Example A *Change each fraction to a decimal.*

1. $\frac{7}{25}$ 0.28 2. $\frac{3}{16}$ 0.1875

Example B *Change each fraction to a decimal. Use bar notation to show a repeating decimal.*

3. $\frac{2}{11}$ $0.\overline{18}$ 4. $\frac{7}{12}$ $0.58\overline{3}$

Example C *Change each fraction to a mixed decimal.*

5. $\frac{4}{9}$ $0.44\frac{4}{9}$ 6. $\frac{5}{6}$ $0.83\frac{1}{3}$

Additional examples are provided on TRANSPARENCY 5-8.

EVALUATING THE LESSON

Check for Understanding Use Guided Practice Exercises to check for student understanding.

Closing the Lesson Ask students to summarize how to change a fraction to a decimal.

EXTENDING THE LESSON

Enrichment Masters Booklet, p. 52

ENRICHMENT WORKSHEET 5-8

Repeating Decimals

Write each of the elevenths as a decimal. Continue the division until the pattern repeats twice.

1. $\frac{1}{11}$ 0.0909. . . 2. $\frac{2}{11}$ 0.1818. . . 3. $\frac{3}{11}$ 0.2727. . .

4. $\frac{4}{11}$ 0.3636. . . 5. $\frac{5}{11}$ 0.4545. . . 6. $\frac{6}{11}$ 0.5454. . .

7. $\frac{7}{11}$ 0.6363. . . 8. $\frac{8}{11}$ 0.7272. . .

9. $\frac{9}{11}$ 0.8181. . . 10. $\frac{10}{11}$ 0.9090. . .

11. How many different patterns are there? Be careful. 10

12. How many digits are there in each pattern? 2

Write each of the thirteenths as a decimal. Continue the division until the decimal repeats twice.

13. $\frac{1}{13}$ 0.076923076923. . . 14. $\frac{2}{13}$ 0.153846153846. . . 15. $\frac{3}{13}$ 0.230769230769. . .

16. $\frac{4}{13}$ 0.307692307692. . . 17. $\frac{5}{13}$ 0.384615384615. . . 18. $\frac{6}{13}$ 0.461538461538. . .

19. $\frac{7}{13}$ 0.538461538461. . . 20. $\frac{8}{13}$ 0.615384615384. . . 21. $\frac{9}{13}$ 0.692307692307. . .

22. $\frac{10}{13}$ 0.769230769230. . . 23. $\frac{11}{13}$ 0.846153846153. . . 24. $\frac{12}{13}$ 0.923076923076. . .

25. How many different patterns are there? 12

26. How many digits repeat in each pattern? 6

27. Try to discover the relationship between the prime number denominators, N, the number of digits in the pattern, D, and the number of different patterns, P. $P = N - 1$ $D = \frac{P}{2}$

APPLYING THE LESSON

Independent Practice

Homework Assignment	
Minimum	Ex. 2-36 even; 37-40
Average	Ex. 1-35 odd; 36-40
Maximum	Ex. 1-40

Additional Answers

37. Answers may vary. The denominators are numbers with factors other than 2 or 5.

5-MINUTE CHECK-UP

(over Lesson 5-8)

Available on TRANSPARENCY 5-9.

Change each fraction to a decimal. Use bar notation to show a repeating decimal.

1. $\frac{5}{8}$ 0.625 **2.** $\frac{3}{5}$ 0.6

3. $\frac{2}{3}$ $0.\overline{6}$ **4.** $\frac{5}{6}$ $0.8\overline{3}$

Change each fraction to a mixed decimal.

5. $\frac{6}{7}$ $0.85\frac{5}{7}$ **6.** $\frac{2}{3}$ $0.66\frac{2}{3}$

Extra Practice, Lesson 5-8, p. 462

INTRODUCING THE LESSON

Ask students if they think it is easier to enter decimals or fractions into calculators and computers.

TEACHING THE LESSON

Using Questioning How do you determine the denominator of the fraction when changing a decimal to a fraction? the least place-value position of the decimal, or the number of zeros matches the number of digits in the decimal

Practice Masters Booklet, p. 58

Name ______________ Date ______________

PRACTICE WORKSHEET 5-9

Changing Decimals to Fractions

Change each decimal to a fraction.

1. 0.54 $\frac{27}{50}$	**2.** 0.06 $\frac{3}{50}$	**3.** 0.75 $\frac{3}{4}$
4. 0.48 $\frac{12}{25}$	**5.** 0.9 $\frac{9}{10}$	**6.** 0.005 $\frac{1}{200}$
7. 0.25 $\frac{1}{4}$	**8.** 0.625 $\frac{5}{8}$	**9.** 0.375 $\frac{3}{8}$
10. 0.4 $\frac{2}{5}$	**11.** 0.45 $\frac{9}{20}$	**12.** 0.62 $\frac{31}{50}$
13. 0.096 $\frac{12}{125}$	**14.** 0.357 $\frac{357}{1,000}$	**15.** 0.225 $\frac{9}{40}$
16. 0.79 $\frac{79}{100}$	**17.** 0.256 $\frac{32}{125}$	**18.** 0.08 $\frac{2}{25}$
19. 0.006 $\frac{3}{500}$	**20.** 0.126 $\frac{63}{500}$	**21.** 0.875 $\frac{7}{8}$

Change each mixed decimal to a fraction.

22. $0.55\frac{5}{9}$ $\frac{5}{9}$	**23.** $0.66\frac{2}{3}$ $\frac{2}{3}$	**24.** $0.27\frac{3}{11}$ $\frac{3}{11}$
25. $0.16\frac{2}{3}$ $\frac{1}{6}$	**26.** $0.57\frac{1}{7}$ $\frac{4}{7}$	**27.** $0.41\frac{2}{3}$ $\frac{5}{12}$

Write <, >, or = in each ◯ to make a true sentence.

28. $0.\overline{7} > \frac{7}{10}$	**29.** $\frac{3}{7} > 0.428$	**30.** $0.\overline{6} = \frac{2}{3}$
31. $0.2 < \frac{2}{9}$	**32.** $0.8 = \frac{4}{5}$	**33.** $0.83 < \frac{5}{6}$

5-9 CHANGING DECIMALS TO FRACTIONS

Objective
Change decimals to fractions.

Missy is shopping for pizza supplies. She asks the clerk in the meat department for $\frac{7}{8}$ pound of ground sausage. The package the clerk handed to Missy was labeled 0.875 pound. Missy changed 0.875 to a fraction to make sure she had $\frac{7}{8}$ pound of ground sausage.

To change a terminating decimal to a fraction, write the decimal as a fraction and simplify.

Method

1. Write the digits of the decimal as the numerator. Use the appropriate power of ten (10, 100, 1,000, and so on) as the denominator.
2. Simplify.

Examples

A ***Change 0.875 to a fraction.***

1. $0.875 = \frac{\overset{7}{\cancel{875}}}{\underset{8}{\cancel{1,000}}}$ The GCF is 125.
2. $= \frac{7}{8}$

Missy found she had $\frac{7}{8}$ pound of sausage.

B ***Change 0.32 to a fraction.***

1. $0.32 = \frac{\overset{8}{\cancel{32}}}{\underset{25}{\cancel{100}}}$ The GCF is 4.
2. $= \frac{8}{25}$

To change a mixed decimal to a fraction, remember that a fraction indicates division.

Example C ***Change $0.16\frac{2}{3}$ to a fraction.***

1. $0.16\frac{2}{3} = \frac{16\frac{2}{3}}{100}$
2. $\frac{16\frac{2}{3}}{100} = 16\frac{2}{3} \div 100$ Write the fraction as a division problem.

$= \frac{50}{3} \div 100$ Write $16\frac{2}{3}$ as an improper fraction: $\frac{(16 \times 3) + 2}{3}$

$= \frac{\overset{1}{\cancel{50}}}{3} \times \frac{1}{\underset{2}{\cancel{100}}}$

$= \frac{1}{6}$

Guided Practice

Change each decimal to a fraction.

Example A **1.** 0.006 $\frac{3}{500}$ **2.** 0.084 $\frac{21}{250}$ **3.** 0.125 $\frac{1}{8}$ **4.** 0.650 $\frac{13}{20}$ **5.** 0.408 $\frac{51}{125}$

Example B **6.** 0.98 $\frac{49}{50}$ **7.** 0.53 $\frac{53}{100}$ **8.** 0.31 $\frac{31}{100}$ **9.** 0.64 $\frac{16}{25}$ **10.** 0.05 $\frac{1}{20}$

Example C *Change each mixed decimal to a fraction.*

11. $0.14\frac{2}{7}$ $\frac{1}{7}$ **12.** $0.66\frac{2}{3}$ $\frac{2}{3}$ **13.** $0.22\frac{2}{9}$ $\frac{2}{9}$ **14.** $0.85\frac{5}{7}$ $\frac{6}{7}$ **15.** $0.71\frac{3}{7}$ $\frac{5}{7}$

RETEACHING THE LESSON

Have students follow the steps outlined below to change a decimal to a fraction.

	Example
1. Read the decimal aloud.	0.25 is read 25 hundredths
2. Write the decimal as a fraction.	25 hundredths $= \frac{25}{100}$
3. Simplify.	$\frac{25}{100} = \frac{1}{4}$

Reteaching Masters Booklet, p. 53

Name ______________ Date ______________

RETEACHING WORKSHEET 5-9

Changing Decimals to Fractions

Method: ① Write the digits of the decimal as the numerator. Use the appropriate power of ten (10, 100, 1000, and so on) as the denominator.
② Simplify.

Change 0.75 to a fraction.

① $0.75 = \frac{75}{100}$
② $= \frac{3}{4}$
So, $0.75 = \frac{3}{4}$.

Change 0.125 to a fraction.

① $0.125 = \frac{125}{1,000}$
② $= \frac{1}{8}$
So, $0.125 = \frac{1}{8}$.

Change $0.37\frac{1}{2}$ to a fraction.

① $0.37\frac{1}{2} = \frac{37\frac{1}{2}}{100}$
$= 37\frac{1}{2} \div 100$
$= \frac{75}{2} \div 100$
② $= \frac{\overset{3}{\cancel{75}}}{2} \times \frac{1}{\underset{4}{\cancel{100}}} = \frac{3}{8}$
So, $0.37\frac{1}{2} = \frac{3}{8}$.

Change each decimal to a fraction.

Exercises

Practice

Change each decimal to a fraction. 23. $\frac{101}{1,000}$ 27. $\frac{56}{125}$

16. 0.5 $\frac{1}{2}$	**17.** 0.8 $\frac{4}{5}$	**18.** 0.32 $\frac{8}{25}$	**19.** 0.75 $\frac{3}{4}$
20. 0.54 $\frac{27}{50}$	**21.** 0.38 $\frac{19}{50}$	**22.** 0.744 $\frac{93}{125}$	**23.** 0.101
24. 0.303 $\frac{303}{1,000}$	**25.** 0.486 $\frac{243}{500}$	**26.** 0.626 $\frac{313}{500}$	**27.** 0.448
28. 0.074 $\frac{37}{500}$	**29.** 0.008 $\frac{1}{125}$	**30.** 9.36 $9\frac{9}{25}$	**31.** 10.18 $10\frac{9}{50}$

Change each mixed decimal to a fraction.

32. $0.11\frac{1}{9}$ $\frac{1}{9}$	**33.** $0.33\frac{1}{3}$ $\frac{1}{3}$	**34.** $0.83\frac{1}{3}$ $\frac{5}{6}$	**35.** $0.09\frac{1}{11}$ $\frac{1}{11}$

Replace each ● with =, <, or > to make a true sentence.

36. 0.25 ● $\frac{1}{4}$ =	**37.** $0.\overline{1}$ ● $\frac{1}{9}$ =	**38.** $\frac{2}{7}$ ● 0.286 <	**39.** $\frac{9}{10}$ ● $0.\overline{9}$ <

Applications

40. yes, 1.75 < 1.89

40. Scott's meatloaf recipe calls for $1\frac{3}{4}$ pounds of ground beef. The price label on a ground beef package indicates that it contains 1.89 pounds of ground beef. Does the package contain enough ground beef to make the meatloaf?

FUN with MATH

What, besides pizza, originated in Italy?
See page 175.

41. A 10-pound spool of solder for making a stained glass window contains 9.375 feet of solder. How many $\frac{5}{8}$-foot pieces of solder can be cut from the spool? 15 pieces

42. Suppose David Robinson's field-goal average was 0.501 in 1993. We say he makes 5 out of 10 field goals. Suppose your field-goal average is 0.40. You can say you make _?_ out of _?_ field goals. 2,5; 4,10; and so on.

Number Sense

A method for finding a fraction that is halfway between a given pair of fractions is to find their average or one-half their sum.

43. Find a fraction halfway between $\frac{7}{9}$ and $\frac{8}{9}$. $\frac{15}{18}$

44. Find a fraction halfway between $\frac{6}{11}$ and $\frac{7}{11}$. $\frac{13}{22}$

45. Find a decimal that is halfway between 0.67 and 0.68. 0.675

Using Algebra

Evaluate each expression if a = 0.25, b = $\frac{3}{8}$, and c = 1.375. 49. 0.515625 or $\frac{33}{64}$

46. $a + b$ 0.625 or $\frac{5}{8}$	**47.** $b - a$ 0.125 or $\frac{1}{8}$	**48.** ac 0.34375	**49.** bc

APPLYING THE LESSON

Independent Practice

Homework Assignment	
Minimum	Ex. 2-48 even
Average	Ex. 1-39 odd; 40-49
Maximum	Ex. 1-49

Chalkboard Examples

Examples A, B *Change each decimal to a fraction.*

1. 0.375 $\frac{3}{8}$ **2.** 0.08 $\frac{2}{25}$

Example C *Change each mixed decimal to a fraction.*

3. $0.28\frac{4}{7}$ $\frac{2}{7}$ **4.** $0.06\frac{2}{3}$ $\frac{1}{15}$

Additional examples are provided on TRANSPARENCY 5-9.

EVALUATING THE LESSON

Check for Understanding Use Guided Practice Exercises to check for student understanding.

Closing the Lesson Have students state in their own words how to change a decimal to a fraction.

EXTENDING THE LESSON

Enrichment Find the fractional equivalent for $0.\overline{63}$. Let $n = 0.\overline{63}$.

$100n = 63.\overline{63}$	Multiply by 100.
$-\ n = 0.\overline{63}$	Subtract.
$99n = 63$	Solve for n.
$n = \frac{63}{99}$ or $\frac{7}{11}$	

Enrichment Masters Booklet, p. 53

Name ________________ Date ________________

ENRICHMENT WORKSHEET 5-9

Decimals to Fractions

Every fraction can be expressed as a **terminating** or a **repeating** decimal.

$\frac{3}{4} = 0.75$ $\frac{2}{3} = 0.\overline{6}$

Every terminating decimal can be expressed as a fraction.

$0.65 \rightarrow \frac{65}{100} \rightarrow \frac{13}{20}$

Repeating decimals, such as $0.\overline{48}$, can be changed to fractions as follows.

- Some number is equal to $0.\overline{48}$. Call the number n.
- One hundred times n or $100 \times n$ is equal to $48.\overline{48}$.

$$100 \times n = 48.\overline{48}$$
$$-\ n = 0.\overline{48}$$
$$99 \times n = 48$$
$$n = \frac{48}{99} \text{ or } \frac{16}{33}$$

Write each decimal as a fraction.

1. $0.\overline{1}$ $10 \times n =$ □ 1.1; $-\ n =$ □ 0.1; □ $\times n =$ □ 9, 1; $n =$ □ $\frac{1}{9}$

2. $0.\overline{36}$ $\frac{4}{11}$

3. $0.\overline{54}$ $\frac{27}{50}$

4. $0.\overline{8}$ $\frac{8}{9}$

5. 0.828 $\frac{207}{250}$

6. $0.\overline{714285}$ $\frac{5}{7}$

5-MINUTE CHECK-UP

(over Lesson 5-9)

Available on TRANSPARENCY 5-10.

Change each decimal to a fraction.

1. 0.28 $\frac{7}{25}$
2. 0.375 $\frac{3}{8}$
3. 0.680 $\frac{17}{25}$
4. $0.06\frac{1}{4}$ $\frac{1}{16}$
5. 0.0125 $\frac{1}{80}$
6. $0.66\frac{2}{3}$ $\frac{2}{3}$

Extra Practice, Lesson 5-9, p. 462

INTRODUCING THE LESSON

Ask students how many times they answered a question or solved a problem since they awoke this morning. Ask them how many problems they expect to solve before the day is over. Point out how much of daily life is spent in problem solving.

TEACHING THE LESSON

Using Questioning

- Which phrases in a problem indicate the operation needed to solve the problem? How many more, how many left, how many in all, and so on.

Practice Masters Booklet, p. 59

Name ____________ Date ____________

PRACTICE WORKSHEET 5-10

Problem Solving: Using Fractions

Solve. Write each fraction in simplest form.

1. XYZ stock increased $\frac{5}{8}$ one week, $\frac{3}{4}$ the second week, and $\frac{1}{8}$ the third week. What was the stock's average weekly price increase? $\frac{1}{2}$
2. The regular price of a tennis racket is $24. What is the sale price during the "$\frac{1}{3}$-off" sale? $16
3. A machine is switched off and repaired because $\frac{7}{8}$ of the bolts it produced were defective. Out of 64 bolts, how many were defective? 56 bolts
4. There are only $2\frac{1}{4}$ yards of ribbon left on a roll. Sheila needs $\frac{1}{3}$ yard for a bow for each of 7 dolls she is making. Is there enough ribbon? If not, how much more is needed? no; $\frac{1}{12}$ yard
5. Tim has four $8\frac{1}{2}$-foot boards that he could use for shelves. He decided to make just 3 shelves from each board so that none would be wasted. How long will each shelf be? $2\frac{5}{6}$ feet
6. Four $12\frac{7}{8}$-inch pieces of molding are needed to frame a picture. Since the color may differ a little in different molding strips, Jane wants to cut the 4 pieces from the same strip. How long must the strip be? $51\frac{1}{2}$ inches
7. Adult shoe sizes start at size 1, which has an inside length of $8\frac{7}{12}$ inches. There is a $\frac{1}{3}$-inch difference in full sizes. How long is an adult size 5 shoe? $9\frac{11}{12}$ inches
8. A recipe for 12 dozen muffins calls for $4\frac{1}{2}$ cups of oatmeal. Steve wants to make only six dozen muffins. How much oatmeal should he use? $2\frac{1}{4}$ cups

Explore / Plan / Solve / Examine

5-10 USING FRACTIONS

Objective

Solve verbal problems involving fractions.

Janie bought an aquarium. She uses a $2\frac{1}{4}$-gallon bucket to fill it with $22\frac{1}{2}$ gallons of water. How many buckets of water does it take?

What facts are known?

- Janie fills the aquarium with $22\frac{1}{2}$ gallons of water.
- She uses a bucket that holds $2\frac{1}{4}$ gallons of water.

What do you need to find?

- the number of buckets of water it will take to fill the aquarium

Divide the total number of gallons by the number of gallons in one bucket.

Estimate 22 ÷ 2 = 11, 21 ÷ 3 = 7. It takes between 7 and 11 buckets to fill the aquarium.

$$22\frac{1}{2} \div 2\frac{1}{4} = \frac{45}{2} \div \frac{9}{4} = \frac{\overset{5}{\cancel{45}}}{\underset{1}{\cancel{2}}} \times \frac{\overset{2}{\cancel{4}}}{\underset{1}{\cancel{9}}} = 10$$

It takes 10 buckets of water to fill the aquarium.

Compared to the estimate the answer is reasonable. You can use repeated subtraction to show the answer is correct.

$$22\frac{1}{2} - \left(2\frac{1}{4} + 2\frac{1}{4} + 2\frac{1}{4} + 2\frac{1}{4} + 2\frac{1}{4} + 2\frac{1}{4} + 2\frac{1}{4} + 2\frac{1}{4} + 2\frac{1}{4} + 2\frac{1}{4}\right) = 0$$

Guided Practice

1. Boyd buys $8\frac{3}{4}$ yards of blue fabric and $6\frac{7}{8}$ yard of plaid fabric to make a sleeping bag. How many yards of fabric does he buy in all? $15\frac{5}{8}$ yards
2. To hang some of her paintings, Heather needs six pieces of wire each $1\frac{3}{4}$ feet long. At least how many feet of wire should she buy? $10\frac{1}{2}$ feet
3. A survey indicated that of the 660 students at Harry Johnson High School, $\frac{2}{5}$ participate in the sports programs. How many students participate in sports? 264 students

RETEACHING THE LESSON

Have students discuss the operations necessary to solve these verbal problems. Then solve.

1. Three-fifths of 180 students are taking Spanish. How many are taking Spanish? multiply; 108
2. Eduardo needs $2\frac{1}{2}$ cups of sugar. A measuring scoop holds $\frac{1}{4}$ cup. How many scoops of sugar does he need? divide; 10

Reteaching Masters Booklet, p. 54

Name ____________ Date ____________

RETEACHING WORKSHEET 5-10

Problem Solving: Using Fractions

Kari ate $\frac{3}{8}$ of a pizza. Her brother ate $\frac{2}{3}$ as much. How much pizza did he eat?

Explore	Kari ate $\frac{3}{8}$ of a pizza. Her brother ate $\frac{2}{3}$ as much.
Plan	Find $\frac{2}{3}$ of $\frac{3}{8}$.
Solve	$\frac{2}{3} \times \frac{3}{8} = \frac{\overset{1}{\cancel{2}} \times \overset{1}{\cancel{3}}}{\underset{1}{\cancel{3}} \times \underset{4}{\cancel{8}}} = \frac{1}{4}$ Kari's brother ate $\frac{1}{4}$ of the pizza.
Examine	Since $\frac{2}{3} < 1$, $\frac{2}{3}$ of $\frac{3}{8}$ should be less than $\frac{3}{8}$. $\frac{1}{4} < \frac{3}{8}$, so the answer is reasonable.

Solve.

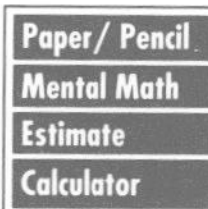

Problem Solving

Paper/ Pencil
Mental Math
Estimate
Calculator

4. Rita and three friends divide $\frac{2}{3}$ of a pizza evenly among themselves. How much pizza does each person get? $\frac{1}{6}$ **of the pizza**

5. Horace walks one block in $1\frac{3}{4}$ minutes. At that rate, about how many minutes does it take him to walk $7\frac{1}{2}$ blocks to school? **about 14 minutes**

6. Ben works $5\frac{1}{4}$ hours overtime on Saturday and $3\frac{1}{2}$ hours overtime on Sunday. How many hours of overtime does he work on the weekend?

6. $8\frac{3}{4}$ **hours**

7. The Mississippi River is about $2\frac{1}{3}$ times as long as the Ohio River. If the Mississippi River is about 2,350 miles long, about how long is the Ohio River? **about 1,000 miles long**

8. On a recent trip Tom averaged $48\frac{1}{2}$ miles per hour. How far did he travel in $3\frac{3}{4}$ hours? $181\frac{7}{8}$ **miles**

9. Joan drove $62\frac{1}{2}$ miles in $1\frac{1}{3}$ hours on Thursday. How many miles per hour did she average? $46\frac{7}{8}$ **miles**

10. There are $5\frac{1}{2}$ cups of sugar in a canister. Clare uses $2\frac{3}{4}$ cups for one recipe and $1\frac{3}{4}$ cups for another recipe. How much sugar is left in the canister? **1 cup**

TECH PREP On pages 522–524, you can learn how graphic arts [pro]fessionals use [mat]hematics in their jobs.

11. Mr. Acker has $8\frac{1}{2}$ bushels of apples. He sells $\frac{3}{4}$ of them for $0.79 a pound. How many bushels of apples are left? $2\frac{1}{8}$ **bushels**

12. Juan buys $2\frac{1}{2}$ pounds of ground beef at $1.90 a pound and half as much ground pork at $2.80 a pound. He gives the cashier $10. How much change does he receive? **$1.75**

Critical Thinking

13. Four monkeys can eat 4 sacks of peanuts in 3 minutes. How many monkeys, eating at the same rate, will it take to eat 100 sacks of peanuts in one hour? **5 monkeys**

14. Review the items in your portfolio. Make a table of contents of the items, noting why each item was chosen. Replace any items that are no longer appropriate. **See students' work.**

Mixed Review

Lesson 2-3

Subtract.

15. $82 - 16 =$ **66**
16. $95 - 73 =$ **22**
17. $436 - 58 =$ **378**
18. $204 - 196 =$ **8**

Lesson 4-3

Find the GCF of each group of numbers.

19. 21, 35 **7**
20. 15, 45 **15**
21. 81, 99 **9**
22. 34, 102 **17**

Lesson 4-8

23. Albert ran $4\frac{7}{10}$ miles and Leonard ran $4\frac{5}{8}$ miles. Which runner ran farther? **Albert**

APPLYING THE LESSON

Independent Practice

Homework Assignment	
Minimum	Ex. 1-23 odd
Average	Ex. 1-23
Maximum	Ex. 1-23

Chapter 5, Quiz B (Lessons 5-6 through 5-10) is available in the Evaluation Masters Booklet, p. 24.

Chalkboard Examples

1. A roll of TV cable contained $48\frac{1}{2}$ feet of cable. Jason uses $\frac{2}{3}$ of the cable on a job. How many feet of cable remain on the roll? $16\frac{1}{6}$ **feet**

Additional examples are provided on TRANSPARENCY 5-10.

EVALUATING THE LESSON

Check for Understanding Use Guided Practice Exercises to check for student understanding.

Error Analysis Watch for students who use a scanning approach to reading a problem. Emphasize to students that reading mathematics needs a different technique than reading a novel.

Closing the Lesson Have students give examples of real-life problems they have solved that involved using fractions.

EXTENDING THE LESSON

Enrichment Masters Booklet, p. 54

Name ______ Date ______

ENRICHMENT WORKSHEET 5-10

Problem Solving

Solve. Write your fractions in simplest form.

1. Martha's pumpkin pie recipe calls for four cups of pumpkin pie filling. Each can of filling contains $\frac{3}{4}$ cup. How many cans of filling does Martha need? **6 cans**

2. Martha's recipe also calls for two cups of evaporated milk. Each can of evaporated milk contains $\frac{2}{3}$ cup. How may cans does she need? **3 cans**

3. Mitch found $\frac{5}{8}$ of a gallon of frozen yogurt in the freezer. He ate $\frac{1}{5}$ of it. How much frozen yogurt did Mitch eat? $\frac{1}{8}$ **gallon**

4. A chili recipe calls for $\frac{3}{4}$ cup of green pepper. Steve makes $\frac{1}{2}$ of the recipe. How much green pepper does he use? $\frac{3}{8}$ **cup**

5. The Petrie family uses about $\frac{3}{4}$ of a box of cereal each week. How long will 8 boxes of cereal last? $10\frac{2}{3}$ **weeks**

6. One-half of the students in Toyoko's class are boys. One-eighth of the boys have red hair. What fraction of the class are boys with red hair? $\frac{1}{16}$ **of the class**

7. A $\frac{9}{16}$-acre lot is to be divided into three equal sections. How large will each section be? $\frac{3}{16}$ **acre**

8. A book has 240 pages. So far, Sean has read $\frac{3}{4}$ of it. How many more pages does Sean have to read? **60 pages**

Using the Chapter Review

The Chapter Review is a comprehensive review of the concepts presented in this chapter. This review may be used to prepare students for the Chapter Test.

Chapter 5, Quizzes A and B, are provided in the Evaluation Masters Booklet as shown below.

Quiz A should be given after students have completed Lessons 5-1 through 5-5. Quiz B should be given after students have completed Lessons 5-6 through 5-10.

These quizzes can be used to obtain a quiz score or as a check of the concepts students need to review.

Evaluation Masters Booklet, p. 24

Name ______________ Date ______________

CHAPTER 5, QUIZ A (Lessons 5-1 through 5-5)

Estimate. * Accept any reasonable estimate.

1. $3\frac{1}{5} \times 5\frac{4}{7}$ 2. $9\frac{3}{5} \div 5\frac{1}{8}$ *1. 18 *2. 2

Multiply. 3. ______

3. $\frac{4}{5} \times \frac{1}{2}$ 4. $\frac{5}{6} \times 3$ 5. $\frac{7}{9} \times \frac{3}{14}$ 4. $\frac{2}{5}$ 4. ______

5. $2\frac{1}{2}$ 5. ______

6. $1\frac{1}{3} \times 3\frac{1}{2}$ 7. $4 \times 3\frac{5}{6}$ 8. $6\frac{3}{4} \times 8\frac{2}{3}$ 6. $\frac{1}{6}$ $4\frac{2}{3}$ 6. ______

7. $15\frac{1}{3}$ 7. ______

9. A range uses the energy equivalent of $\frac{1}{2}$ ton of coal per year. Find the energy equivalent for operating a range for $\frac{1}{4}$ year. 8. $58\frac{1}{2}$ 8. ______ 9. $\frac{1}{8}$ ton 9. ______

10. How many cuts are needed to slice a loaf of bread into 24 pieces? 10. 23 cuts

Name ______________ Date ______________

CHAPTER 5, QUIZ B (Lessons 5-6 through 5-10)

1. Name the reciprocal of 7. 1. $\frac{1}{7}$

Divide. 2. $\frac{7}{9}$

2. $\frac{5}{9} \div \frac{5}{7}$ 3. $5\frac{1}{3} \div 4$ 4. $2\frac{4}{5} \div 2\frac{1}{10}$

5. Change $\frac{7}{15}$ to a decimal. Use bar notation to show a repeating decimal. 3. $1\frac{1}{3}$ 3. ______ 4. $1\frac{1}{3}$ 4. ______

6. Change $\frac{9}{11}$ to a mixed decimal. 5. $0.4\overline{6}$

Change the folowing decimal and mixed decimal to fractions. 6. $0.81\frac{9}{11}$ 6. ______

7. 0.32 8. $0.43\frac{1}{4}$ 7. $\frac{8}{25}$ 7. ______ 8. $\frac{173}{400}$ 8. ______

Solve.

9. A carpenter nails three 2-by-4's together to make a post. If a 2-by-4 is $1\frac{3}{8}$ inches thick, how thick is the post? 9. $4\frac{1}{8}$ inches

10. It took Ms. Carbone $1\frac{1}{2}$ hours to paint the kitchen and $2\frac{3}{4}$ hours to paint the den. How long did it take her to paint both rooms? 10. $4\frac{1}{4}$ hours

Chapter 5 REVIEW

Vocabulary/Concepts

Choose the letter of the number, word, or phrase at the right that best matches each description.

1. a decimal whose digits end, or terminate, like 0.25 i
2. an estimate of $16\frac{1}{2} \div 1\frac{3}{4}$ c
3. a number that consists of a decimal and a fraction, like $0.83\frac{1}{3}$ e
4. two numbers whose product is 1 g
5. an estimate of $2\frac{1}{4} \times 5\frac{2}{3}$ d
6. a decimal whose digits repeat in groups of one or more, like 0.525252 ... h

a. $\frac{1}{2}$
b. 2
c. 8
d. 12
e. mixed decimal
f. mixed numeral
g. reciprocals
h. repeating decimal
i. terminating decimal

Exercises/Applications

Lesson 5-1

Estimate. Answers may vary. Typical answers are given.

7. $4\frac{2}{3} \times 9$ 45
8. $4\frac{1}{2} \times 2\frac{1}{8}$ 10
9. $7\frac{1}{8} \div 3\frac{4}{5}$ 2
10. $12\frac{2}{5} \div 3\frac{3}{4}$ 3

11. Is $\frac{2}{3}$ of 5 less than or greater than 5? (less than)

12. Is $\frac{5}{6} \div \frac{3}{4}$ less than or greater than $\frac{5}{6}$? (greater than)

Lesson 5-2

Multiply.

13. $\frac{4}{9} \times \frac{1}{4}$ $\frac{1}{9}$
14. $\frac{3}{5} \times \frac{5}{8}$ $\frac{3}{8}$
15. $\frac{5}{6} \times 8$ $6\frac{2}{3}$
16. $\frac{7}{12} \times \frac{4}{5}$ $\frac{7}{15}$
17. $\frac{2}{5} \times \frac{9}{10}$ $\frac{9}{25}$
18. $7 \times \frac{3}{14}$ $1\frac{1}{2}$
19. $\frac{4}{5} \times \frac{5}{9}$ $\frac{4}{9}$
20. $\frac{4}{7} \times \frac{7}{8}$ $\frac{1}{2}$
21. $\frac{3}{4} \times \frac{8}{9}$ $\frac{2}{3}$
22. $\frac{3}{5} \times 25$ 15

Lesson 5-3

Multiply.

23. $\frac{3}{4} \times 1\frac{2}{5}$ $1\frac{1}{20}$
24. $7\frac{1}{2} \times 3\frac{3}{5}$ 27
25. $6 \times 4\frac{3}{5}$ $27\frac{3}{5}$
26. $2\frac{1}{2} \times \frac{2}{3}$ $1\frac{2}{3}$
27. $6\frac{1}{4} \times 2\frac{2}{3}$ $16\frac{2}{3}$
28. $9\frac{1}{4} \times 12$ 111
29. $\frac{3}{11} \times 3\frac{2}{3}$ 1
30. $2\frac{2}{9} \times 2\frac{4}{7}$ $5\frac{5}{7}$

Lesson 5-4

Use the table on page 152 to find energy equivalents for each appliance.

31. a dishwasher for 6 months $\frac{1}{10}$ ton
32. a color TV for 8 months $\frac{1}{6}$ ton
33. a radio and TV for 2 years $\frac{3}{5}$ ton
34. a refrigerator for 4 years 4 tons

Lesson 5-5

Solve. Use simplifying the problem.

35. Three students can make 6 floral centerpieces for the prom in 1 hour. How many centerpieces can the whole prom committee make in 2 hours if there are 9 people on the committee and they all work at the same rate? 36 centerpieces

Lesson 5-6 **Name the reciprocal of each number.**

36. 7 $\frac{1}{7}$ **37.** $\frac{3}{4}$ $\frac{4}{3}$ **38.** $\frac{7}{12}$ $\frac{12}{7}$ **39.** 6 $\frac{1}{6}$ **40.** $\frac{7}{5}$ $\frac{5}{7}$ **41.** $\frac{9}{2}$ $\frac{2}{9}$

Lesson 5-6 **Divide.**

42. $\frac{3}{8} \div \frac{1}{2}$ $\frac{3}{4}$ **43.** $\frac{7}{10} \div \frac{2}{5}$ $1\frac{3}{4}$ **44.** $\frac{8}{9} \div 4$ $\frac{2}{9}$ **45.** $\frac{5}{8} \div \frac{5}{6}$ $\frac{3}{4}$ **46.** $9 \div \frac{3}{8}$ 24

47. $\frac{9}{10} \div 6$ $\frac{3}{20}$ **48.** $\frac{5}{9} \div \frac{9}{10}$ $\frac{50}{81}$ **49.** $\frac{11}{12} \div \frac{3}{4}$ $1\frac{2}{9}$ **50.** $15 \div \frac{5}{7}$ 21 **51.** $\frac{5}{12} \div \frac{5}{7}$ $\frac{7}{12}$

Lesson 5-7 **Divide.**

52. $10 \div 3\frac{3}{4}$ $2\frac{2}{3}$ **53.** $1\frac{2}{5} \div 2\frac{2}{3}$ $\frac{21}{40}$ **54.** $3\frac{1}{5} \div 1\frac{1}{3}$ $2\frac{2}{5}$ **55.** $2\frac{2}{5} \div \frac{3}{10}$ 8

56. $3\frac{1}{2} \div 5\frac{1}{2}$ $\frac{7}{11}$ **57.** $\frac{8}{9} \div 2\frac{2}{5}$ $\frac{10}{27}$ **58.** $1\frac{5}{8} \div 2$ $\frac{13}{16}$ **59.** $5\frac{1}{2} \div 2\frac{3}{4}$ 2

Lesson 5-8 **Change each fraction to a decimal. Use bar notation to show a repeating decimal.**

60. $\frac{2}{3}$ $0.\overline{6}$ **61.** $\frac{1}{10}$ 0.1 **62.** $\frac{11}{15}$ $0.7\overline{3}$ **63.** $\frac{13}{25}$ 0.52 **64.** $\frac{7}{22}$ $0.3\overline{18}$ **65.** $\frac{1}{99}$ $0.\overline{01}$

Lesson 5-9 **Change each decimal to a fraction.**

66. 0.20 $\frac{1}{5}$ **67.** 0.034 $\frac{17}{500}$ **68.** 0.36 $\frac{9}{25}$ **69.** 0.25 $\frac{1}{4}$ **70.** 0.425 $\frac{17}{40}$

Lesson 5-10 **Solve.**

71. A sugar scoop holds $\frac{2}{3}$ cup. How many scoops will equal 2 cups of sugar? 3 scoops

72. Leonard has $5\frac{1}{2}$ rows of green beans in his garden. He has $\frac{2}{3}$ as many rows of peas as green beans. How many rows of peas does he have? $3\frac{2}{3}$ rows

73. Alice saves $\frac{1}{5}$ of her salary and spends $\frac{1}{6}$ of her salary for transportation. What part of her salary does she have left?

73. $\frac{19}{20}$ of her salary

74. Mike walked $12\frac{1}{4}$ miles in $2\frac{1}{2}$ hours to raise money for Central City's Fund for the Homeless. What was the average distance he walked each hour? $4\frac{9}{10}$ miles

75. If $2\frac{3}{5}$ is multiplied by a number, the product is $17\frac{1}{3}$. What is the number? $6\frac{2}{3}$

APPLYING THE LESSON

Independent Practice

Homework Assignment	
Minimum	Ex. 1-6; 7-75 odd
Average	Ex. 1-6; 8-68 even; 71-75
Maximum	Ex. 1-75

Alternate Assessment Strategy

To provide a brief in-class review, you may wish to read the following questions to the class and require a verbal response.

1. *About* how much is $3\frac{1}{3}$ multiplied by $4\frac{2}{3}$? 15
2. What is the product of $\frac{4}{5}$ and $\frac{3}{8}$? $\frac{3}{10}$
3. Multiply $3\frac{1}{3}$ and $4\frac{1}{5}$. What is you result? 14
4. What is the reciprocal of $\frac{7}{9}$? $\frac{9}{7}$
5. Divide $\frac{2}{7}$ by $\frac{4}{21}$. $1\frac{1}{2}$
6. What is the quotient when $4\frac{1}{3}$ is divided by 2? $2\frac{1}{6}$
7. Change $\frac{4}{5}$ to a decimal. 0.8
8. Change 0.45 to a fraction. $\frac{9}{20}$
9. Name the reciprocal of $2\frac{3}{5}$. $\frac{5}{13}$
10. Find the product of $\frac{2}{9}$ and $\frac{3}{10}$. $\frac{1}{15}$

Using the Chapter Test

This page may be used as a test or as an additional page of review if necessary. Two forms of a Chapter Test are provided in the Evaluation Masters Booklet. Form 2 (free response) is shown below, and Form 1 (multiple choice) is shown on page 175.

The **Tech Prep Applications Booklet** provides students with an opportunity to familiarize themselves with various types of technical vocations. The Tech Prep applications for this chapter can be found on pages 9–10.

Evaluation Masters Booklet, p. 23

Name ______________________ Date ______________

CHAPTER 5 TEST, FORM 2

Estimate.

1. $3\frac{6}{7} \times 5\frac{1}{8}$
2. $2\frac{1}{2} \times 4\frac{2}{9}$
3. $5\frac{2}{3} \div 1\frac{3}{4}$
4. $11\frac{2}{3} \div 3\frac{5}{8}$

* Accept any reasonable estimate.

*1. 20 *2. 12 *3. 3 *4. 3

Multiply.

5. $\frac{1}{3} \times \frac{2}{5}$
6. $\frac{3}{4} \times \frac{4}{7}$
7. $4 \times \frac{5}{16}$
8. $1\frac{2}{3} \times 2\frac{2}{5}$
9. $3\frac{3}{4} \times 20$
10. $5\frac{1}{6} \times \frac{6}{7}$

5. $\frac{2}{15}$ 6. $\frac{3}{7}$ 7. $1\frac{1}{4}$ 8. 4 9. 75 10. $4\frac{3}{7}$

Divide.

11. $\frac{5}{6} \div \frac{1}{2}$
12. $\frac{9}{20} \div \frac{3}{4}$
13. $2\frac{7}{9} \div 1\frac{2}{3}$
14. $3\frac{4}{7} \div 5$
15. Change $\frac{7}{12}$ to a decimal. Use bar notation to show a repeating decimal.

11. $1\frac{2}{3}$ 12. $\frac{3}{5}$ 13. $1\frac{2}{3}$ 14. $\frac{5}{7}$ 15. $0.58\overline{3}$

Change the following decimal and mixed decimal to fractions.

16. 0.62
17. $0.08\frac{1}{3}$
18. A color TV uses the energy equivalent of $\frac{1}{4}$ ton of coal per year. Find the energy equivalent for operating a color TV for $\frac{1}{2}$ year.
19. Four people can assemble six plant hangers in five hours. How many hangers can 12 people assemble in 10 hours?
20. A recipe calls for $2\frac{1}{2}$ cans of pineapple. If a can contains $10\frac{1}{2}$ ounces, how many ounces of pineapple are needed?

16. $\frac{31}{50}$ 17. $\frac{1}{12}$ 18. $\frac{1}{8}$ ton 19. 36 hangers 20. $26\frac{1}{4}$ ounces

Bonus If you divide a mixed number by a mixed number your quotient must be greater than which whole number?

Bonus 0

Chapter 5 TEST

Estimate. **Answers may vary. Typical answers are given.**

1. $3\frac{1}{7} \times 1\frac{7}{9}$ **6** **2.** $8\frac{1}{3} \times 4\frac{4}{5}$ **40** **3.** $2\frac{1}{3} \div 1\frac{1}{3}$ **2** **4.** $5\frac{1}{2} \div 1\frac{3}{4}$ **3**

5. Is $10 \div \frac{1}{5}$ less than or greater than 10? (greater than underlined)

6. Is $\frac{1}{5}$ of 5 less than, equal to, or greater than 5? (less than underlined)

Multiply.

7. $2 \times \frac{2}{3}$ **$1\frac{1}{3}$** **8.** $\frac{5}{8} \times 6$ **$3\frac{3}{4}$** **9.** $\frac{1}{3} \times \frac{3}{4}$ **$\frac{1}{4}$** **10.** $\frac{7}{18} \times \frac{9}{14}$ **$\frac{1}{4}$**

11. $1\frac{1}{4} \times 2\frac{1}{4}$ **$2\frac{13}{16}$** **12.** $1\frac{5}{6} \times 8\frac{2}{11}$ **15** **13.** $2\frac{2}{3} \times 5\frac{1}{7}$ **$13\frac{5}{7}$** **14.** $1\frac{5}{16} \times 3\frac{3}{7}$ **$4\frac{1}{2}$**

Divide.

15. $6 \div \frac{1}{3}$ **18** **16.** $\frac{5}{8} \div 5$ **$\frac{1}{8}$** **17.** $\frac{7}{8} \div \frac{3}{4}$ **$1\frac{1}{6}$** **18.** $\frac{3}{7} \div \frac{4}{5}$ **$\frac{15}{28}$**

19. $5\frac{3}{4} \div 1\frac{1}{3}$ **$4\frac{5}{16}$** **20.** $4\frac{1}{2} \div 1\frac{1}{2}$ **3** **21.** $26 \div 1\frac{3}{10}$ **20** **22.** $4\frac{1}{8} \div 1\frac{3}{4}$ **$2\frac{5}{14}$**

23. Gary and Suki can make 3 dozen cupcakes for the bakesale in a half-hour. How many dozens of cupcakes can they make in 1 hour if Peter and Jillian help? Assume Peter and Jillian work at the same rate. **12 dozens**

Change each fraction to a decimal. Use bar notation to show a repeating decimal.

24. $\frac{2}{5}$ **0.4** **25.** $\frac{5}{16}$ **0.3125** **26.** $\frac{5}{11}$ **$0.\overline{45}$** **27.** $\frac{7}{12}$ **$0.58\overline{3}$**

Change each decimal to a fraction.

28. 0.6 **$\frac{3}{5}$** **29.** 0.35 **$\frac{7}{20}$** **30.** 0.875 **$\frac{7}{8}$** **31.** 0.92 **$\frac{23}{25}$**

Solve. **32. $146\frac{1}{4}$ miles**

32. Janet's car will travel $22\frac{1}{2}$ miles on one gallon of gas. At this rate, how far will it travel on $6\frac{1}{2}$ gallons of gas?

33. Betty has $6\frac{1}{4}$ bags of plant food. She uses $1\frac{1}{4}$ bags each month. How many months will her supply of plant food last? **5 months**

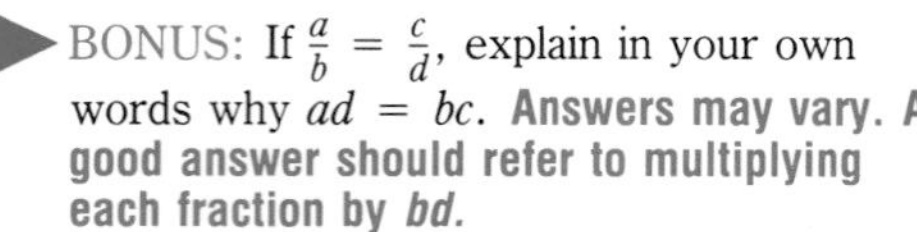

BONUS: If $\frac{a}{b} = \frac{c}{d}$, explain in your own words why $ad = bc$. **Answers may vary. A good answer should refer to multiplying each fraction by *bd*.**

The **Performance Assessment Booklet** provides an alternative assessment for evaluating student progress. An assessment for this chapter can be found on pages 9–10.

A **Test and Review Generator** is provided in Apple, IBM, and Macintosh versions. You may use this software to create your own tests or worksheets, based on the needs of your students.

UNIT PRICE

Careful shopping can lead to considerable savings. This is especially true for food shopping. Many grocery stores now display the unit price so that comparison shopping is easier. It is a good idea to compare different brands for quality and price. To tell which is the better buy of two items equal in quality but different in size, find the cost of one unit of the product. This is the **unit price.**

Which is the better buy, $1.59 for 12 oz of peanut butter, or $2.94 for 28 oz of peanut butter?

Small Jar of Peanut Butter	**Large Jar of Peanut Butter**
$1.59 for 12 ounces	$2.94 for 28 ounces
Find the cost for 1 ounce.	Find the cost for 1 ounce.
1.59 ÷ 12 = 0.1325	2.94 ÷ 28 = 0.105

Since 10.5¢ < 13.25¢, the large jar is the better buy. $0.1325 = 13.25¢
$0.105 = 10.5¢

Assume the quality in each case is the same. Determine which size is the better buy for each item.

1.	cereal	10 oz for $1.79	15 oz for $2.39
2.	ketchup	14 oz for $0.89	28 oz for $1.29
3.	peaches	16 oz for $0.77	29 oz for $1.52
4.	bread	16 oz for $0.85	24 oz for $1.39

5. Which is the better buy, a 6.5-ounce bag of potato chips for $1.69 or a 14.5-ounce bag for $2.99?

6. Brittany buys a 10-ounce bottle of shampoo for $2.79. She has a coupon for 20¢ off the regular price. What is the cost per ounce of the shampoo with the coupon? **$0.259**

7. At the Corner Market, one quart of milk costs 63¢, one-half gallon costs $1.25, and one gallon costs $2.53. Which is the best buy? **half gallon**

8. George is in charge of buying paper plates, napkins, and cups for the sophomore dance. Third Street Market sells 100 plates for $1.39. Joy's Supermarket sells 50 plates for $0.74. Which store has the better buy? **Third Street Market**

Applying Mathematics to the Real World

This optional page shows how mathematics is used in the real world and also provides a change of pace.

Objective Find and use the unit price to comparison shop.

Using Discussion Discuss with students the term *unit price*. You may wish to discuss some of the different units in which different items are sold, such as material by the yard, lumber by the foot, and so on.

Activity

Using Calculators Have students use calculators to do Exercises 1-7 on p. 171. Round answers to the nearest tenth of a cent.

Using the Cumulative Review

This page provides an aid for maintaining skills and concepts presented thus far in the text.

A Cumulative Review is also provided in the Evaluation Masters Booklet as shown below.

Additional Answers

1. 3 × 3 × 3 × 3 = 81
2. 7 × 7 × 7 = 343
3. 10 × 10 × 10 × 10 × 10 = 100,000
4. 2 × 2 × 2 × 2 = 16
5. 1 × 1 × 1 × 1 × 1 × 1 × 1 × 1 × 1 × 1 = 1
6. 6 × 6 = 36
7. 8 × 8 = 64
8. 5 × 5 × 5 = 125

32. 1, 2, 3, 4, 6, 9, 12, 18, 36
33. 1, 89
34. 1, 3, 5, 9, 15, 45
35. 1, 3, 13, 39
36. 1, 2, 3, 4, 6, 8, 9, 12, 16, 18, 24, 36, 48, 72, 144

Evaluation Masters Booklet, p. 25

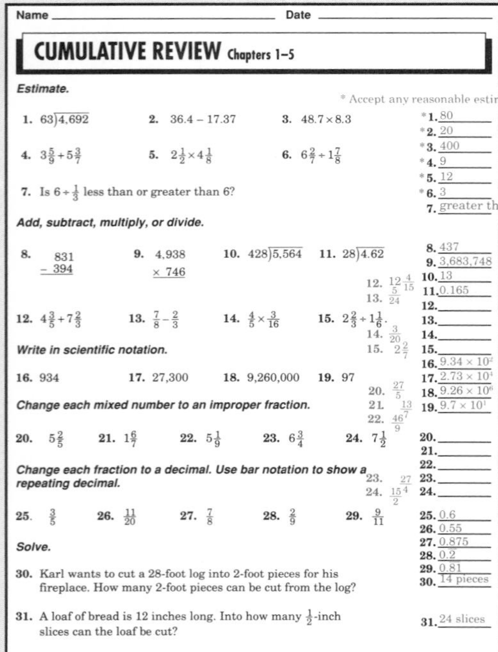

Name ____________ Date ____________

CUMULATIVE REVIEW Chapters 1–5

Estimate.

* Accept any reasonable estimate.

1. $63\overline{)4,692}$ 2. 36.4 − 17.37 3. 48.7 × 8.3

4. $3\frac{5}{9} + 5\frac{3}{7}$ 5. $2\frac{1}{2} \times 4\frac{1}{8}$ 6. $6\frac{2}{7} \div 1\frac{7}{8}$

7. Is $6 \div \frac{1}{3}$ less than or greater than 6?

*1. 80
*2. 20
*3. 400
*4. 9
*5. 12
*6. 3
7. greater than 6

Add, subtract, multiply, or divide.

8. 831 − 394 9. 4,938 × 746 10. $428\overline{)5,564}$ 11. $28\overline{)4.62}$

12. $4\frac{3}{5} + 7\frac{2}{3}$ 13. $\frac{7}{8} - \frac{2}{3}$ 14. $\frac{4}{5} \times \frac{3}{16}$ 15. $2\frac{2}{3} \div 1\frac{1}{6}$.

8. 437
9. 3,683,748
10. 13
11. 0.165
12. $12\frac{4}{15}$
13. $\frac{5}{24}$
14. $\frac{3}{20}$
15. $2\frac{2}{7}$

Write in scientific notation.

16. 934 17. 27,300 18. 9,260,000 19. 97

16. 9.34×10^2
17. 2.73×10^4
18. 9.26×10^6
19. 9.7×10^1

Change each mixed number to an improper fraction.

20. $5\frac{2}{5}$ 21. $1\frac{6}{7}$ 22. $5\frac{1}{9}$ 23. $6\frac{3}{4}$ 24. $7\frac{1}{2}$

20. $\frac{27}{5}$
21. $\frac{13}{7}$
22. $\frac{46}{9}$
23. $\frac{27}{4}$
24. $\frac{15}{2}$

Change each fraction to a decimal. Use bar notation to show a repeating decimal.

25. $\frac{3}{5}$ 26. $\frac{11}{20}$ 27. $\frac{7}{8}$ 28. $\frac{2}{9}$ 29. $\frac{9}{11}$

25. 0.6
26. 0.55
27. 0.875
28. $0.\overline{2}$
29. $0.\overline{81}$

Solve.

30. Karl wants to cut a 28-foot log into 2-foot pieces for his fireplace. How many 2-foot pieces can be cut from the log?

31. A loaf of bread is 12 inches long. Into how many $\frac{1}{2}$-inch slices can the loaf be cut?

30. 14 pieces
31. 24 slices

Cumulative REVIEW/TEST

Free Response

Lesson 1-2 ***Write as a product and then find the product.*** **See margin.**

1. 3^4 2. 7^3 3. 10^5 4. 2^4
5. 1^{10} 6. 6^2 7. 8^2 8. 5^3

Lesson 1-9 ***Find the value of each expression.*** **14. 16.4**

9. 6 + 18 ÷ 3 **12**
10. 30 − 5 × 4 **10**
11. (9 + 21) ÷ 3 **10**
12. 4.6 × 2 ÷ 4 **2.3**
13. 7.8 × (15 + 5) **156**
14. 37.8 − 25 + 3.6
15. 16.4 − (8.5 + 0.1) **7.8**
16. $3 \times 10^3 + 2 \times 10^3$ **5,000**
17. $1.6 \times 10^2 + 2.7$ **162.7**

Lesson 2-3

18. The Big Time Amusement Park had an average daily attendance of 23,760 during July. This was 5,807 more than during June. Find the average daily attendance in June. **17,953 people**

Lesson 3-1 ***Multiply.*** **22. 404.974**

19. 54 × 72 **3,888**
20. 87 × 37 **3,219**
21. 20 × 384 **7,680**
22. 86 × 4.709
23. 3,000 × 21 **63,000**
24. 50 × 2,000 **100,000**

Lesson 3-4 ***Divide.***

25. $10\overline{)40}$ **4**
26. $60\overline{)300}$ **5**
27. $23\overline{)92}$ **4**
28. $60\overline{)11,895}$ **198 R15**
29. $30\overline{)12,000}$ **400**
30. $48\overline{)600}$ **12 R24**

Lesson 3-13

31. A hotel room costs $29 a day plus $2.24 tax. What is the charge on a room for four days? **$124.96**

Lesson 4-1 ***Find all the factors of each number.*** **See margin.**

32. 36 33. 89 34. 45 35. 39 36. 144

Lesson 4-3 ***Find the GCF and the LCM of each group of numbers.***

37. 36, 40 **4; 360**
38. 15, 50 **5; 150**
39. 9, 25 **1; 225**
40. 4, 28 **4; 28**

Lesson 4-12

41. A recipe calls for $2\frac{3}{4}$ cups of flour. Mark has measured 1 cup. How much more flour does he need? **$1\frac{3}{4}$ cups**

Lessons 5-2, 5-3 ***Multiply.***

42. $\frac{3}{4} \times \frac{5}{8}$ **$\frac{15}{32}$**
43. $\frac{2}{7} \times \frac{1}{6}$ **$\frac{1}{21}$**
44. $\frac{2}{3} \times \frac{9}{10}$ **$\frac{3}{5}$**
45. $\frac{7}{9} \times 20$ **$15\frac{5}{9}$**
46. $1\frac{3}{4} \times \frac{4}{5}$ **$1\frac{2}{5}$**
47. $\frac{11}{18} \times 2\frac{2}{3}$ **$1\frac{17}{27}$**
48. $7\frac{5}{6} \times 18$ **141**
49. $3\frac{3}{5} \times 10\frac{1}{9}$ **$36\frac{2}{5}$**

APPLYING THE LESSON

Independent Practice

Homework Assignment	
Minimum	Ex. 1-49 odd
Average	Ex. 1-49
Maximum	Ex. 1-49

Chapter 5

Multiple Choice

Choose the letter of the correct answer for each item.

1. What is 238.675 rounded to the nearest hundredth?
a. 200 ▶ **c.** 238.68
b. 238.67 **d.** 238.7
Lesson 1-4

2. Elia spends $4.50 at the record store and $36.79 at the clothing store. She arrives home with $12.61. Which question can you answer from the information above?
e. At what store did she buy clothes?
f. How many miles between stores?
▶ **g.** How much money did she have to begin with?
h. How much change she received at the record store?
Lesson 2-2

3. John withdraws $25.68 from his savings of $810.50. What is the amount left in his account?
a. $553.70 **c.** $784.98
▶ **b.** $784.82 **d.** $794.98
Lesson 2-4

4. What is the product of 0.05 and 1,000?
e. 0.005 ▶ **g.** 50
f. 5 **h.** 5,000
Lesson 3-2

5. In one game, Pat kicks field goals from 43, 38, and 41 yards. To the nearest tenth of a yard, what is the average length of the field goals?
a. 40.6 yards
▶ **b.** 40.7 yards
c. 46.6 yards
d. 46.7 yards
Lesson 3-12

6. Linda Fada earns $16,500 per year. What is her weekly wage?
▶ **e.** $317.31 **g.** $3,173
f. $317.37 **h.** $3,173.07
Lesson 3-4

7. Order from least to greatest.
$\frac{1}{2}, \frac{5}{8}, \frac{6}{5}, \frac{7}{16}$
▶ **a.** $\frac{7}{16}, \frac{1}{2}, \frac{5}{8}, \frac{6}{5}$
b. $\frac{1}{2}, \frac{5}{8}, \frac{6}{5}, \frac{7}{16}$
c. $\frac{1}{2}, \frac{7}{16}, \frac{5}{8}, \frac{6}{5}$
d. $\frac{7}{16}, \frac{5}{8}, \frac{6}{5}, \frac{1}{2}$
Lesson 4-7

8. If you combine $5\frac{3}{4}$ cups of flour with $4\frac{7}{8}$ cups of flour, how much flour will you have?
e. $9\frac{1}{8}$ cups **g.** $9\frac{5}{6}$ cups
▶ **f.** $10\frac{5}{8}$ cups **h.** $9\frac{7}{8}$ cups
Lesson 4-11

9. *About* how much is 1,180 divided 375?
a. 1 **c.** 30
▶ **b.** 3 **d.** 40
Lesson 1-7

10. If $4\frac{1}{2}$ dozen doughnuts are divided between 2 tenth-grade classes, how many doughnuts does each person receive? Which fact is missing?
e. the number of classes
f. the price of the doughnuts
▶ **g.** the number of people in each class
h. none of the above
Lesson 5-7

APPLYING THE LESSON

Independent Practice

Homework Assignment	
Minimum	Ex. 1-10
Average	Ex. 1-10
Maximum	Ex. 1-10

Using the Cumulative Test

This test familiarizes students with a standardized format while testing skills and concepts presented up to this point.

Evaluation Masters Booklet, pp. 21-22

Name ______ Date ______

CHAPTER 5 TEST, FORM 1

Write the letter of the correct answer on the blank at the right of the page.

1. Estimate. $4\frac{1}{3}\times2\frac{4}{5}$
A. 8 **B.** 15 **C.** 9 **D.** 12 — 1. D
2. Estimate. $5\frac{9}{10}\times2\frac{2}{3}$
A. 10 **B.** 15 **C.** 18 **D.** 12 — 2. C
3. Multiply. $\frac{1}{4}\times\frac{1}{2}$
A. $\frac{1}{3}$ **B.** $\frac{1}{8}$ **C.** $\frac{1}{6}$ **D.** $\frac{1}{16}$ — 3. B
4. Multiply. $\frac{3}{4}\times\frac{8}{9}$
A. $\frac{2}{3}$ **B.** $\frac{3}{4}$ **C.** $\frac{3}{5}$ **D.** $\frac{4}{5}$ — 4. A
5. Multiply. $15\times\frac{2}{3}$
A. 5 **B.** 10 **C.** 8 **D.** 9 — 5. B
6. Multiply. $2\frac{1}{3}\times3\frac{1}{2}$
A. $2\frac{1}{3}$ **B.** $2\frac{4}{5}$ **C.** $7\frac{5}{6}$ **D.** $8\frac{1}{6}$ — 6. D
7. Multiply. $4\frac{2}{5}\times3$
A. $12\frac{1}{5}$ **B.** $4\frac{1}{6}$ **C.** $13\frac{1}{5}$ **D.** 5 — 7. C
8. Multiply. $5\frac{5}{6}\times2\frac{4}{5}$
A. $17\frac{2}{5}$ **B.** $4\frac{2}{3}$ **C.** $3\frac{1}{2}$ **D.** $16\frac{1}{3}$ — 8. D
9. Estimate. $9\frac{5}{6}\div2\frac{1}{8}$
A. $4\frac{1}{2}$ **B.** 3 **C.** 5 **D.** 6 — 9. C
10. Estimate. $6\div\frac{1}{2}$
A. 3 **B.** 12 **C.** 2 **D.** 4 — 10. B
11. Name the reciprocal of $\frac{2}{5}$.
A. 10 **B.** 3 **C.** 7 **D.** $\frac{5}{2}$ — 11. D
12. Divide. $\frac{2}{3}\div\frac{3}{5}$
A. $1\frac{1}{9}$ **B.** $\frac{2}{5}$ **C.** $\frac{5}{8}$ **D.** $\frac{9}{10}$ — 12. A
13. Divide. $\frac{7}{12}\div\frac{2}{3}$
A. $\frac{3}{4}$ **B.** $\frac{3}{5}$ **C.** $\frac{7}{8}$ **D.** $\frac{7}{18}$ — 13. C

Using Mathematical Connections

Objective Understand, appreciate, and enjoy the connections between mathematics and real-life phenomena and other disciplines.

The following data is background information for each event on the time line.

History Pictographic accounting system, Ur, Sumer (Babylon), 3500 BC
In 3500 BC scribes in the city of Ur in the ancient Sumerian empire kept pictographic accounts of their temple receipts and disbursements on clay tables. Pictographs were the first of three forms of writing in history.

History First Fire Depatment, Rome 32 BC
What has been called the first fire department can be traced to about 32 BC in imperial Rome. Caesar Augustus organized the *Vigiles*, which consisted of seven squads of 100 to 1,000 men each. Their "fire engines" consisted of huge water-filled syringes used to squirt a stream of water on a fire.

Science / Technical The first steel mill, Tanzania, Africa, AD 0-500
Africans living on the western shores of Lake Victoria in Tanzania, produced carbon steel 1,500 to 2,000 years ago. They used pre-heated forced-draft furnaces, a method that was technologically more sophisticated than any developed in Europe until the mid-19th century. The temperature achieved in the blast furnace of the African steel-smelting machine was 1,800° C.

Technical The Wheelbarrow, China AD 230
It is said that the wheelbarrow, which the Chinese called the "wooden ox," was invented by an official named Chuke Liang about AD 230. However, carvings on Chinese tombs show wheelbarrows being used at least two centuries earlier.

History The equal sign, England, 1545
The symbol = for equivalence was introduced by an Englishman Robert Recorde in 1545. He wrote several mathematics books. He used parallel lines = as a symbol for equality, because "no two things could be more equal."

fun with MATH

First accounting system Ur, Sumer (Babylonia)	3500 BC	First fire department Rome	32 BC	First steel mill Tanzania, Africa	AD 0-500	The wheelbarrow China	AD 230

CFCs could outlive you!

If you have a refrigerator in your home or an air conditioner in your car, its cooling fluid probably contains CFCs, or chlorofluorocarbons. Ozone is a form of oxygen with three oxygen atoms, O_3. The ozone layer protects us from dangerous ultraviolet radiation from the sun. CFCs rise to the upper atmosphere, releasing chlorine atoms that combine with ozone molecules, in effect, slowly destroying the ozone layer. One CFC molecule can remain in the upper atmosphere for more than 100 years.

MATH MENU

Carrot Bars

4 eggs (beaten) 2 cups sugar
2 tsp baking soda 2 tsp cinnamon
2 tsp salt 1½ cups vegetable oil
2½ cups flour ½ cup nuts (large pieces)
3 small jars strained carrots (baby food)
Combine and mix ingredients in order given. Bake 30-40 minutes at 350° on greased 10″ × 15″ jelly-roll pan or cookie sheet.
Frosting: 1/2 tsp vanilla 3 1/2 cups powdered sugar
1/2 cup margarine 1 8-oz pkg of cream cheese
Combine ingredients and beat until smooth. Spread on cooled bars.

When playing golf, Mary Queen of Scots called the boys who fetched her golf balls "cadets," pronounced "cadday"; hence "caddie." Today we use the term caddy."

COMICS

174 Fun with Math

The equal sign
England
AD 1718-1799
The "Real McCoy"
Ypsilanti, Michigan
AD 1956
AD 1545
The "Witch" of Agnesi
Bologna, Italy
AD 1872
The VTR
California
JOKE!
Q: What professional football team has a name which is the same as 7²?
A: San Francisco Forty-Niners.
RIDDLE
Q: To what number can you add a letter and make it three less?
A: IX (9), prefixed by an 'S' makes it SIX.
How does a VCR work?
When recording, the VCR converts incoming TV signals into electric current. The current travels to the head, an electromagnet. The head magnetizes the particles on the tape into patterns that represent video signals, audio signals, and control signals. When playing the tape back, the head reads the magnetic patterns on the tape and creates an electric current. The current is sent to the TV and transformed into sounds and pictures.
Challenge a friend to a Korean form of tic-tac-toe, called ko-no. You will need two red and two blue playing pieces, or two types of coins. Place the red pieces on the circles marked A and B, the blue pieces on C and D. The player on A and B goes first. The player attempts to move either playing piece to the empty circle along a connecting segment. Jumping a piece is not allowed. The object of the game is to block your opponent from moving his/her playing pieces.
TEASER
A
B
C
D
QUIZ TIME
Why is a Black Hole black?
1/2 + 2/4 = ____ of
3/6 − 1/3 = ____ a
1/5 × 2/4 = ____ is
7/12 − 1/3 = ____ even
6/16 − 3/8 = ____ escape
5/6 × 2/3 = ____ gravity
3/4 − 1/8 = ____ Black
5/6 − 1/3 = ____ so
4/9 × 7/8 = ____ light
1/2 − 6/10 = ____ the
1/9 + 5/9 = ____ in
11/15 − 2/5 = ____ Hole
1/4 × 6/12 = ____ great
15/16 − 3/4 = ____ can't
3/5 + 2/6 = ____ force
5/6
5/9
6
1/2
7/18
14/15
2/3
1/3
1/8
5/4
1
1/6
1/10
1/4
0
ANSWER
Fun with Math 175

History / Language The "Witch" of Agnesi, Bologna, Italy, 1718-1799
One of the most extraordinary women scholars of all time was Maria Gaetana Agnesi. Agnesi's greatest accomplishment was her two-volume work called *Analytical Institutions.* Maria presented what is known as a *versed* sine curve or *versiera.* However, *versiera* was also an abbreviation for *avversiera*, or "wife of the devil." Later translations of Maria's text used *versiera* as "witch." As a result of these mistranslations, the curve presented by Maria came to be known as the "Witch of Agnesi."

Technical The "Real McCoy," Ypsilanti, Michigan, 1872
Elijah J. McCoy was born in Canada, the son of runaway slaves. In 1870, McCoy developed a cup with a tiny stop-cock that regulated the flow of oil onto moving parts of machines. This was the first lubricator; and since then, millions of machines have been equipped with some version of his invention including the vehicles in our space program. No one would consider purchasing a machine unless they were guaranteed that it was equipped with the "Real McCoy," and the slang term was born.

Science The VTR, California 1956
Development of videotape recorders (VTRs) began during the 1940s. However, the first VTR that was capable of recording a television picture of broadcasting quality on magnetic tape was not invented until 1956 by the Ampex Corporation of California. Color VTRs were introduced by Ampex and RCA in 1959. In 1969 the Sony Corporaton of Japan introduced a VCR that used a 20 mm cassette tape. After 1975 and the introduction of the Betamax format by Sony and the VHS format by Matsushita Corporation, prices fell below $1,000, and the market for VCRs for the home took off.

Environment CFCs could outlive you! In 1988, the use of CFCs worldwide reached 2 1/2 billion pounds. The Montreal Protocol of 1986 is an international agreement that mandates a freeze on CFC production with a total phase-out by the year 2000. This has stimulated research for ozone-benign CFC replacements.

Quiz Answer

1, 1/6, 1/10, 1/4, 0, 5/9, 6, 1/2, 7/18, 5/6, 2/3, 1/3, 1/8, 5/4, 14/15; The force of gravity in a Black Hole is so great even light can't escape.

Chapter

6

MEASUREMENT

PREVIEWING the CHAPTER

In this chapter, both the metric system of measurement and the customary system of measurement are discussed. After a lesson on measuring length in the metric system, students are taught to change form one metric unit of measure to another. Students then review the customary units of length, weight, and capacity. The relationship between the temperature units degrees Celsius and degrees Fahrenheit is presented using formulas. Finally, units of time are discussed.

Problem-Solving Strategy Students learn to solve a problem by acting it out.

Lesson (Pages)	Lesson Objectives	State/Local Objectives
6-1 (178-179)	6-1: Identify and use metric prefixes,	
6-2 (180-181)	6-2: Estimate and measure metric units of length.	
6-3 (182-183)	6-3: Change metric units of length.	
6-4 (184-185)	6-4: Estimate and use metric units of mass.	
6-5 (186-187)	6-5: Estimate and use metric units of capacity.	
6-6 (188-189)	6-6: Solve problems by acting out a solution.	
6-7 (190-191)	6-7: Change from one customary unit of length to another.	
6-8 (194-195)	6-8: Change customary units of weight and capacity.	
6-9 (196-197)	6-9: Estimate temperature and use formulas to change Celsius to Fahrenheit and vice versa.	
6-10 (198-199)	6-10: Change from one unit of time to another.	
6-11 (200-201)	6-11: Compute the number of hours worked using 24-hour notation.	
6-12 (202-203)	6-12: Solve problems involving measurements.	

ORGANIZING the CHAPTER

You may refer to the ***Course Planning Calendar*** on page T10.

Planning Guide

Lesson (Pages)	Extra Practice (Student Edition)	Blackline Masters Booklets: Reteaching	Practice	Enrichment	Activity	Multi-cultural Activity	Technology	Tech Prep	Evaluation	Other Resources
6-1 (178-179)	p. 463	p. 55	p. 60	p. 55			p. 20			Transparency 6-1
6-2 (180-181)	p. 463	p. 56	p. 61	p. 56						Transparency 6-2
6-3 (182-183)	p. 463	p. 57	p. 62	p. 57						Transparency 6-3
6-4 (184-185)	p. 464	p. 58	p. 63	p. 58		p. 6				Transparency 6-4
6-5 (186-187)	p. 464	p. 59	p. 64	p. 59	p. 34					Transparency 6-5
6-6 (188-189)			p. 65						p. 29	Transparency 6-6
6-7 (190-191)	p. 464	p. 60	p. 66	p. 60						Transparency 6-7
6-8 (194-195)	p 465	p. 61	p. 67	p. 61	p. 11			p. 11		Transparency 6-8
6-9 (196-197)	p. 465	p. 62	p. 68	p. 62				p. 12		Transparency 6-9
6-10 (198-199)	p 465	p. 63	p. 69	p. 63	p. 12		p. 6			Transparency 6-10
6-11 (200-201)		p. 64	p. 70	p. 64						Transparency 6-11
6-12 (202-203)		p. 65	p. 71	p. 65					p. 29	Transparency 6-12
Ch. Review (204-205)									pp. 26-28	Test and Review Generator
Ch. Test (206)										
Cumulative Review/Test (208-209)									p. 30	

OTHER CHAPTER RESOURCES

Student Edition
Chapter Opener, p. 176
Activity, p. 177
Computer Application, p. 192
History Connections, p. 193
Journal Entry, p. 183
Portfolio Suggestion, p. 203
History Connections, p. 207

Teacher's Classroom Resources
Transparency 6-0
Lab Manual, pp. 33-34
Performance Assessment, pp. 11-12
Lesson Plans, pp. 60-71

Other Supplements
Overhead Manipulative Resources, Labs 18-19, pp. 13-14
Glencoe Mathematics Professional Series

ENHANCING the CHAPTER

Some of the blackline masters for enhancing this chapter are shown below.

COOPERATIVE LEARNING

Have cooperative learning groups of two students research the following customary measures: rod, furlong, land league, peck, bushel, and long ton. Have students identify each unit as a unit of length, weight, or capacity. Have them compare each unit with more common customary units and with each other (1 bu = 4 pk). Have students exchange information and note equivalencies among the units.

USING MODELS/MANIPULATIVES

Have students create a display of objects or pictures of objects that approximate each of the common metric and customary units of length, mass or weight, and capacity. Units represented should include meter, decimeter, centimeter, millimeter, liter, kilogram, gram, yard, foot, inch, gallon, quart, pint, pound, and ounce. Be sure that students clearly label each display.

Cooperative Problem Solving, p. 33

Name ____________ Date ____________

COOPERATIVE PROBLEM SOLVING (Use with Lesson 5-7)

An Area of Knowledge

Do you think that the tops of two different objects can have the same area? Work in groups. Choose three books or magazines of different sizes in the classroom. Textbooks or reference books are fine.

- Find the length and width in inches of the cover of each book.
- Use the formula for the area of a rectangle: Area = *l*ength times *w*idth ($A = lw$) to find the area in square inches for the cover of each book.
- Record your results in this chart.

Book Name	Length (in.)	Width (in.)	Area (in²)
1.			
2.			
3.			

- Write the areas your group found on index cards and exchange with another group.
- See if each group can identify the books chosen by working backward from their cover areas to possible lengths and widths. Remember, some books chosen may have the same cover areas and/or the same dimensions.
- If two rectangular objects have the same dimensions, do they have the same area? **yes**
- If two rectangular objects have the same area, do they have the same dimensions? **not necessarily**
- Find two objects whose tops have different dimensions but the same area.

Multicultural Activities, p. 6

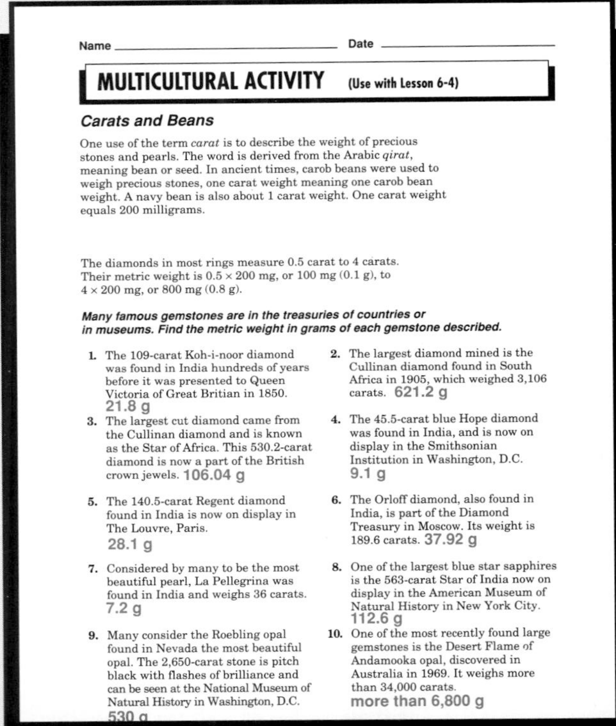

Name ____________ Date ____________

MULTICULTURAL ACTIVITY (Use with Lesson 6-4)

Carats and Beans

One use of the term *carat* is to describe the weight of precious stones and pearls. The word is derived from the Arabic *qirat*, meaning bean or seed. In ancient times, carob beans were used to weigh precious stones, one carat weight meaning one carob bean weight. A navy bean is also about 1 carat weight. One carat weight equals 200 milligrams.

The diamonds in most rings measure 0.5 carat to 4 carats. Their metric weight is 0.5 × 200 mg, or 100 mg (0.1 g), to 4 × 200 mg, or 800 mg (0.8 g).

Many famous gemstones are in the treasuries of countries or in museums. Find the metric weight in grams of each gemstone described.

1. The 109-carat Koh-i-noor diamond was found in India hundreds of years before it was presented to Queen Victoria of Great Britian in 1850. **21.8 g**
2. The largest diamond mined is the Cullinan diamond found in South Africa in 1905, which weighed 3,106 carats. **621.2 g**
3. The largest cut diamond came from the Cullinan diamond and is known as the Star of Africa. This 530.2-carat diamond is now a part of the British crown jewels. **106.04 g**
4. The 45.5-carat blue Hope diamond was found in India, and is now on display in the Smithsonian Institution in Washington, D.C. **9.1 g**
5. The 140.5-carat Regent diamond found in India is now on display in The Louvre, Paris. **28.1 g**
6. The Orloff diamond, also found in India, is part of the Diamond Treasury in Moscow. Its weight is 189.6 carats. **37.92 g**
7. Considered by many to be the most beautiful pearl, La Pellegrina was found in India and weighs 36 carats. **7.2 g**
8. One of the largest blue star sapphires is the 563-carat Star of India now on display in the American Museum of Natural History in New York City. **112.6 g**
9. Many consider the Roebling opal found in Nevada the most beautiful opal. The 2,650-carat stone is pitch black with flashes of brilliance and can be seen at the National Museum of Natural History in Washington, D.C. **530 g**
10. One of the most recently found large gemstones is the Desert Flame of Andamooka opal, discovered in Australia in 1969. It weighs more than 34,000 carats. **more than 6,800 g**

MEETING INDIVIDUAL NEEDS

Limited English Proficiency Students

If you have LEP students in your classroom from countries in which the metric system is commonly used, encourage them to share their knowledge of the prefixes with other students. If they have lived in an environment in which the metric system is commonly used, they can make a positive contribution to the class discussion. Since we still use the customary system in this country, have English-speaking students help those who are unfamiliar with these measures with pronunciation and meaning of the units.

CRITICAL THINKING/PROBLEM SOLVING

Sometimes you have to use the information you have in a new way to solve a problem.

Julian needs to draw a line that is 6 inches long. He does not have a ruler. He does have some sheets of $8\frac{1}{2}$ in. x 11 in. notebook paper. Describe how Julian can use the paper to measure 6 inches. (Add the measure of 2 paper widths, 17 inches. Then subtract the measure of 1 paper length, 11 inches. The difference is 6 inches.)

COMMUNICATION

Writing

Have students write conversion charts for metric measures and for customary measures showing the relative values of metric measures of length, mass, and capacity; and the relative values of customary measures of length, weight, and capacity. Students' charts should show that metric units differ by factors of ten. Charts for customary measures can show equivalencies between units. Students will use charts for reference.

Applications, p. 11

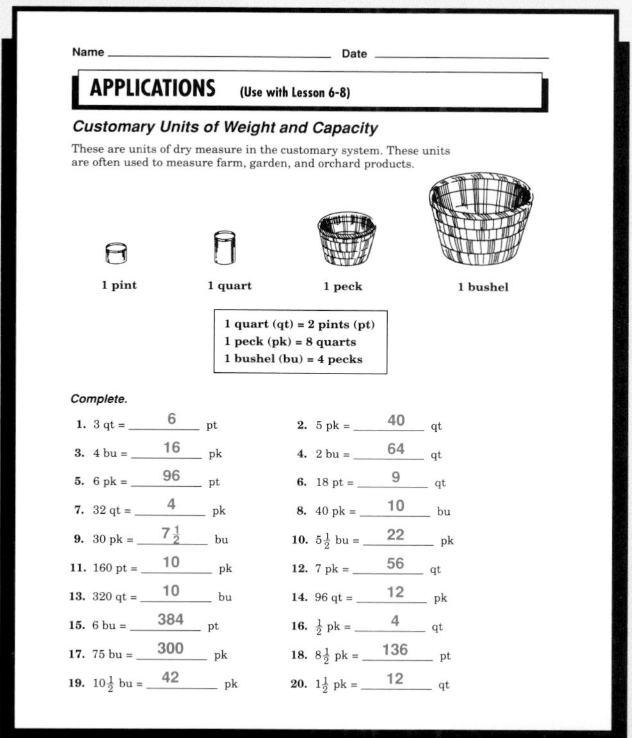

Name ______________ Date ______________

APPLICATIONS (Use with Lesson 6-8)

Customary Units of Weight and Capacity

These are units of dry measure in the customary system. These units are often used to measure farm, garden, and orchard products.

1 pint　1 quart　1 peck　1 bushel

1 quart (qt) = 2 pints (pt)
1 peck (pk) = 8 quarts
1 bushel (bu) = 4 pecks

Complete.

1. 3 qt = 6 pt
2. 5 pk = 40 qt
3. 4 bu = 16 pk
4. 2 bu = 64 qt
5. 6 pk = 96 pt
6. 18 pt = 9 qt
7. 32 qt = 4 pk
8. 40 pk = 10 bu
9. 30 pk = $7\frac{1}{2}$ bu
10. $5\frac{1}{2}$ bu = 22 pk
11. 160 pt = 10 pk
12. 7 pk = 56 qt
13. 320 qt = 10 bu
14. 96 qt = 12 pk
15. 6 bu = 384 pt
16. $\frac{1}{2}$ pk = 4 qt
17. 75 bu = 300 pk
18. $8\frac{1}{2}$ pk = 136 pt
19. $10\frac{1}{2}$ bu = 42 pk
20. $1\frac{1}{2}$ pk = 12 qt

Computer Activity, p. 20

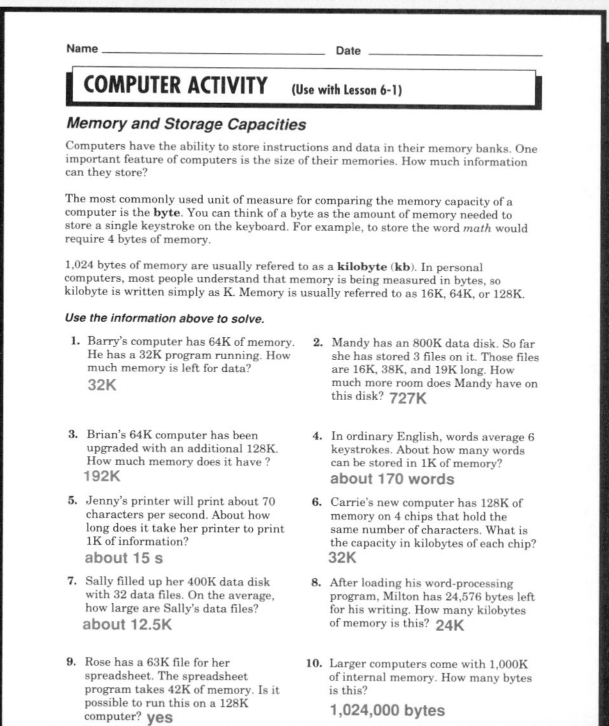

Name ______________ Date ______________

COMPUTER ACTIVITY (Use with Lesson 6-1)

Memory and Storage Capacities

Computers have the ability to store instructions and data in their memory banks. One important feature of computers is the size of their memories. How much information can they store?

The most commonly used unit of measure for comparing the memory capacity of a computer is the **byte**. You can think of a byte as the amount of memory needed to store a single keystroke on the keyboard. For example, to store the word *math* would require 4 bytes of memory.

1,024 bytes of memory are usually refered to as a **kilobyte** (**kb**). In personal computers, most people understand that memory is being measured in bytes, so kilobyte is written simply as K. Memory is usually referred to as 16K, 64K, or 128K.

Use the information above to solve.

1. Barry's computer has 64K of memory. He has a 32K program running. How much memory is left for data? 32K
2. Mandy has an 800K data disk. So far she has stored 3 files on it. Those files are 16K, 38K, and 19K long. How much more room does Mandy have on this disk? 727K
3. Brian's 64K computer has been upgraded with an additional 128K. How much memory does it have? 192K
4. In ordinary English, words average 6 keystrokes. About how many words can be stored in 1K of memory? about 170 words
5. Jenny's printer will print about 70 characters per second. About how long does it take her printer to print 1K of information? about 15 s
6. Carrie's new computer has 128K of memory on 4 chips that hold the same number of characters. What is the capacity in kilobytes of each chip? 32K
7. Sally filled up her 400K data disk with 32 data files. On the average, how large are Sally's data files? about 12.5K
8. After loading his word-processing program, Milton has 24,576 bytes left for his writing. How many kilobytes of memory is this? 24K
9. Rose has a 63K file for her spreadsheet. The spreadsheet program takes 42K of memory. Is it possible to run this on a 128K computer? yes
10. Larger computers come with 1,000K of internal memory. How many bytes is this? 1,024,000 bytes

CHAPTER PROJECT

Have students perform experiments using meter sticks, yard sticks, gallon containers, liter containers, and quart containers to make a conversion chart between some metric measures and customary measures and vice versa. Have students find the following approximations.
1 cm ≈ $\frac{3}{8}$ in., 1 m ≈ $3\frac{1}{4}$ ft, 1 m ≈ $39\frac{5}{16}$ in., 1 L ≈ $\frac{1}{4}$ gal, 1 L ≈ $1\frac{1}{2}$ qt, 1 in. ≈ 2.5 cm, 1 ft ≈ 30.5 cm, 1 gal ≈ 3.5 L

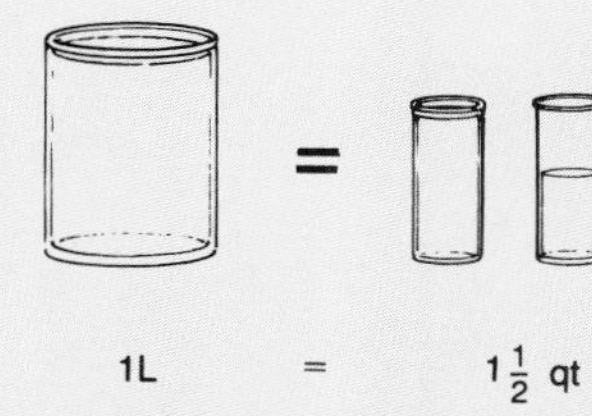

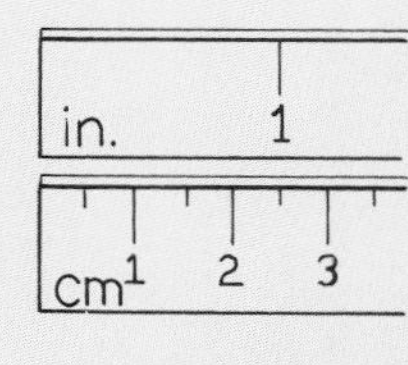

Using the Chapter Opener

Transparency 6-0 is available in the Transparency Package. It provides a full-color visual and motivational activity that you can use to engage students in the mathematical content of the chapter.

Using Discussion

- Have students read the opening paragraph. Ask students how many have pets and what kinds of pets they have.
- Allow time for students to discuss the use of pets in nursing homes. Then ask if students know any other ways in which pets help people. Some may volunteer the use of dogs as watchdogs or cats that keep areas free of rodents or vermin. Some may talk about the heroic rescues some dogs have performed and the protection a dog may offer.
- Students may wish to discuss the use of dogs and monkeys as aids to people who are blind, deaf, or physically handicapped. These animals are not only pets, but working animals.

The second paragraph is a motivational problem. Have students solve the problem and discuss the ways they arrived at the answer.

Chapter 6

MEASUREMENT

How much is that doggie in the window?

Do you have a pet? What kind of pet do you have and what makes your pet special to you? If you don't have a pet, what kind would you like to have? Pets are a responsibility and need feeding, care, and attention every day.

Julio's German shepherd puppy eats 400 grams of puppy food a day. How many days will a 6-kilogram bag of puppy food last? **15 days**

ACTIVITY: Create Your Own Measuring System

In this activity, you will make your own measuring device and create your own system for measuring capacity. Take some time to read the activity with your group before preparing any materials and planning your system.

This chapter is about the standard units of measure that everyone uses. You can have some fun creating your own system.

Materials: large glass (straight sides are the easiest to use); spoon, tablespoon, or coffee measure; small container of water; masking tape; felt-tip pen; other containers like paper cups, coffee cans, small bottles, and so on.
Have students work on a level surface.

Cooperative Groups

1. Prepare the large glass by putting a piece of tape vertically on the outside from top to bottom.

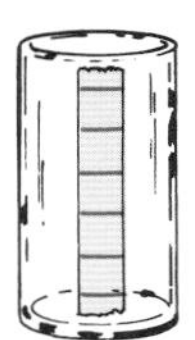

2. Decide with your group how many spoonfuls you will use as the next smallest amount after one spoonful. Use the spoon to add the number of spoonfuls of water that you have chosen and mark the tape with a line. Write the total number of spoonfuls next to the line.
3. Now add the same number of spoonfuls as you did in Step 2, and mark and number the tape again. Continue to add and mark until you have the number of spoonfuls that will equal **one** of the largest unit in your system.
4. Think up a name that will represent half of the largest unit and also a name for the whole unit.
5. Use your measuring device to find the capacity of other containers, some larger than your device and some smaller, such as paper cups, small bottles, and so on.
6. Set up a table or chart for changing from spoonfuls to the unit for one half, from spoonfuls to the whole unit, and from the half unit to the whole unit.

Communicate Your Ideas

7. Is your system easy to use? Is it easy to change from one unit to another? **See margin.**
8. Does your system have units that are small enough to measure a small capacity and units that are large enough to measure a larger capacity? **See margin.**
9. Write about a system of length or weight that you would like to create. Include the names of the units you would use and a chart for changing from one unit to another. **Answers may vary.**

Additional Answers

7. Answers may vary. A good system will resemble the metric system.
8. Answers may vary. A good system will range from a spoonful to about a cupful.

▶ INTRODUCING THE LESSON

Objective Make and use a system of capacity.

You may want to have students read the entire activity before beginning so that they know a system will be developed.

▶ TEACHING THE LESSON

Using Vocabulary Discuss with students the need to have simple, easily remembered names for their units.

Using Cooperative Groups In groups of three or four, have groups assign tasks to each member. For example, one should measure and mark the glass. One should keep records and one should make the table for changing units.

▶ EVALUATING THE LESSON

Communicate Your Ideas Allow students time to discuss the attributes of other systems that were created.

▶ EXTENDING THE LESSON

Closing the Lesson Ask students to summarize the attributes of a good system of measurement.

Activity Worksheet, Lab Manual, p. 33

Name ______________ Date ______________

ACTIVITY WORKSHEET (Use with page 177)

Create Your Own Measuring System

Here is how one group of students did this activity.

Number of spoonfuls added: 2
Number of spoonfuls in largest unit: 10
Name of largest unit: NET
Number of spoonfuls in half unit: 5
Name of half unit: EVIF

Container	unit: NET	half unit: EVIF
Large vase	4	8
Bowl	2	4
Coffee can	16	32
Cup	$\frac{1}{2}$	1

Record your data here.

Number of spoonfuls you added: ________
Number of spoonfuls in your largest unit: ________
Name of your largest unit: ________
Number of spoonfuls in your half unit: ________
Name of your half unit: ________

Container	unit: ________	half unit: ________

5-MINUTE CHECK-UP

(over Lesson 5-10)

Available on TRANSPARENCY 6-1.
Solve.

1. Tami works $3\frac{3}{4}$ hours a day. How many hours does she work in 5 days? **$18\frac{3}{4}$ hr**

2. Lenny bought an 8-cup container of cleaning fluid. He used $1\frac{1}{2}$ cups to clean his brushes and $1\frac{2}{3}$ cups to clean his car mats. How much was left? **$4\frac{5}{6}$ cups**

INTRODUCING THE LESSON

Ask students to name items that are labeled with metric units, or if they know which sports or Olympic records use metric measurement.

TEACHING THE LESSON

Using Vocabulary Have students associate words they know with the prefixes for metric units in order to remember their meanings. For example, a century is 100 years, so centi- means hundredth. Millenium means one thousand years, so milli- means thousandth.

Practice Masters Booklet, p. 60

Name ____________ Date ____________

PRACTICE WORKSHEET 6-1

The Metric System

Name the place value related to each prefix.

1. kilo thousands	2. deci tenths
3. centi hundredths	4. milli thousandths
5. hecto hundreds	6. deka tens

Complete. Use the place-value chart.

7. 1 kilometer = 1,000 meter(s)
8. 1 centimeter = 0.01 meter(s)
9. 1 millimeter = 0.001 meter(s)
10. 1 dekagram = 10 gram(s)
11. 1 decigram = 0.1 gram(s)
12. 1 hectoliter = 100 liters(s)

Name the metric unit for each measurement.

13. 0.001 liter milliliter	16. 100 meters hectometer
14. 1,000 grams kilogram	17. 10 meters dekameter
15. 0.01 gram centigram	18. 0.1 meter decimeter

Underline the larger unit.

19. 1 deciliter or centiliter	20. 1 centigram or 1 kilogram
21. 1 hectometer or 1 millimeter	22. 1 dekaliter or 1 deciliter

6-1 THE METRIC SYSTEM

Objective
Identify and use metric prefixes.

Tom wanted to know why his teacher said the metric system is easier system to use and remember than the customary system. He found there are fewer terms to remember. Also, changing from one unit to another is very similar to changing money; ten pennies equal one dime, ten dimes equal one dollar.

The basic units in the metric system are the **meter** (length), **gram** (mass), and **liter** (capacity).

Prefixes and **basic units** are used to name units larger and smaller than the basic units. The chart at the right shows how the prefixes relate to decimal place values and powers of ten.

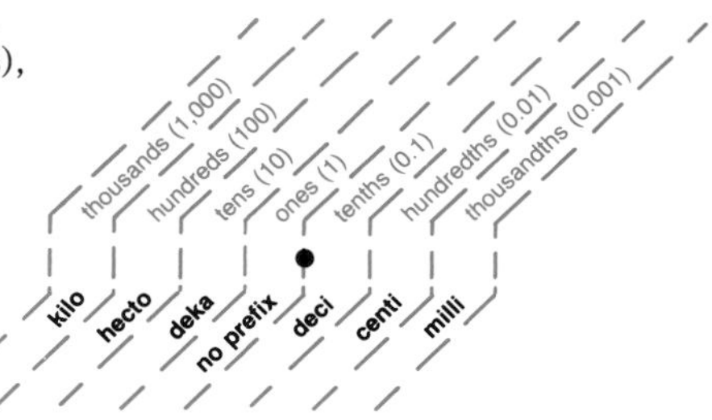

Some examples of metric units and their meanings are given below.

prefix basic unit	prefix basic unit	prefix basic unit
centimeter	kilogram	milliliter
0.01 meter	1,000 grams	0.001 liter

Look at the chart at the right for the basic unit of a meter. Each unit is equal to ten times the unit to its right. For example, 1 kilometer equals 10 hectometers. Each unit is equal to one-tenth of the unit to its left. For example, 1 meter equals 0.1 dekameter.

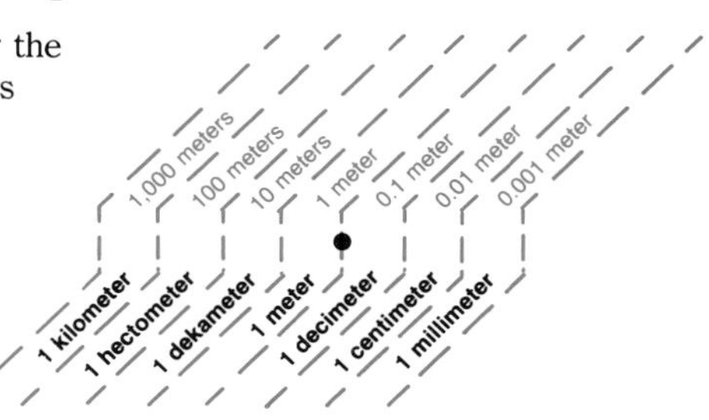

The system is the same for changing units of mass and capacity.

Guided Practice

Name the place value related to each prefix.

1. milli **thousandths**	**2.** kilo **thousands**	**3.** centi **hundredths**
4. deka **tens**	**5.** deci **tenths**	**6.** hecto **hundreds**

Complete. Use the place-value chart.

7. 1 kilogram = _?_ hectogram(s) **10**	**8.** 1 dekagram = _?_ gram(s) **10**
9. 1 liter = _?_ milliliter(s) **1,000**	**10.** 1 kiloliter = _?_ liter(s) **1,000**

Name the metric unit for each measurement.

11. 10 grams **dekagram**	**12.** 0.01 gram **centigram**
13. 0.001 meter **millimeter**	**14.** 0.1 meter **decimeter**

RETEACHING THE LESSON

Have students make a place-value chart based on powers of ten. Then have them put in metric prefixes that correspond. Their charts should be similar to the one on page 178, with the powers of ten listed first since those are familiar.

Reteaching Masters Booklet, p. 55

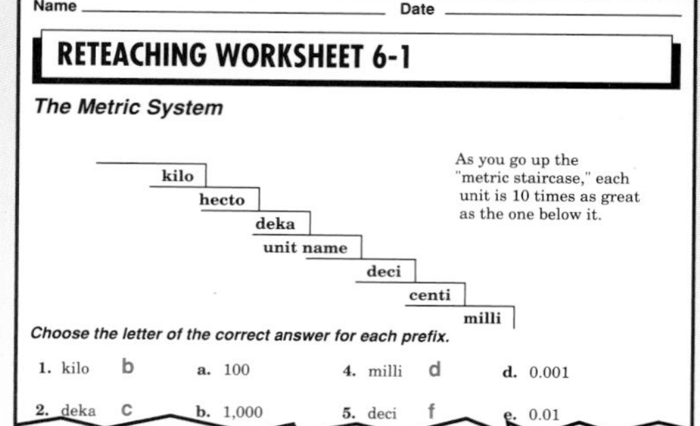
Name ____________ Date ____________

RETEACHING WORKSHEET 6-1

The Metric System

As you go up the "metric staircase," each unit is 10 times as great as the one below it.

Choose the letter of the correct answer for each prefix.

1. kilo b	a. 100	4. milli d	d. 0.001
2. deka c	b. 1,000	5. deci f	e. 0.01

Exercises

Practice

Complete. Use the place-value chart.

15. 1 kilometer = ? meter(s) 1,000 **16.** 1 deciliter = ? liter(s) 0.1

17. 1 milligram = ? gram(s) 0.001 **18.** 1 dekameter = ? meter(s) 10

19. 1 centiliter = ? liter(s) 0.01 **20.** 1 hectogram = ? gram(s) 100

21. millimeter 22. kiloliter 23. hectogram 24. decimeter 26. centigram

Name the metric unit for each measurement.

21. 0.001 meter **22.** 1,000 liters **23.** 100 grams

24. 0.1 meter **25.** 10 liters dekaliter **26.** 0.01 gram

FUN with MATH

Where and when did the metric system originate? See page 241.

Name the larger unit. 28. centigram 30. kilometer

27. 1 meter or 1 kilometer kilometer **28.** 1 centigram or 1 milligram

29. 1 milliliter or 1 liter liter **30.** 1 kilometer or 1 centimeter

31. Change 1,000 centimeters to dekameters. 1 dekameter

32. Is 100 millimeters equivalent to 1 meter or 1 decimeter?

Applications

33. 7 spools

33. Eric is the school play sound technician. He needs 100 meters of cable to extend the ceiling microphones. How many 15-meter spools of cable should he buy?

34. 10 packets

34. In science class Sharyn needs to add 1 dekagram of silver nitrate to the water in her beaker. How many 1-gram packets should she add?

Write Math See margin.

35. If the U.S. adopts the metric system, what are some of the problems you expect to develop?

36. Name three areas involving measure that would be improved by using the metric system. Answers may vary.

Research

37. Who developed the metric system? What is the policy in the U.S. for using metric measure? See margin.

38. In a grocery or hardware store, find three items you could buy that are labeled in metric units. See margin.

Mixed Review

Lesson 3-7

Multiply.

39. 4.63×100 463 **40.** 5.82×10 58.2 **41.** $16.4 \times 1{,}000$ 16,400 **42.** 6.97×10^2 697

Lesson 4-13

Subtract.

43. $8\frac{5}{6} - 3\frac{1}{6}$ $5\frac{2}{3}$ **44.** $16\frac{4}{7} - 8\frac{3}{7}$ $8\frac{1}{7}$ **45.** $6\frac{4}{9} - 3\frac{1}{6}$ $3\frac{5}{18}$ **46.** $5\frac{3}{5} - 2\frac{1}{2}$ $3\frac{1}{10}$

Lesson 5-8

47. Kevin purchased a $6\frac{3}{8}$ inch length of pipe to replace a pipe that was to be 6.425 inches long. Did Kevin purchase the right length pipe? If not, is the pipe he purchased too short or too long? no; too short

Chalkboard Examples

Name the place value related to each prefix.

1. no prefix ones **2.** hecto hundreths

Complete. Use the place value chart.

3. 10 dekameters = ? hectometers 1

4. 10 hectograms = ? kilograms 1

Name the metric unit for each measurement.

5. 100 meters 1 hectometer

6. 0.01 liter 1 centiliter

Additional examples are provided on TRANSPARENCY 6-1

EVALUATING THE LESSON

Check for Understanding Use Guided Practice Exercises to check for student understanding.

Closing the Lesson Ask students to list the six prefixes of the metric system and their meanings.

EXTENDING THE LESSON

Enrichment Have students research other metric prefixes and their meanings. This information is available in world almanac books and encyclopedias.

Enrichment Masters Booklet, p. 55

Name ____________ Date ____________

ENRICHMENT WORKSHEET 6-1

Metric Prefixes

Write the metric prefixes in descending order with each root word below. Give the meaning of the new unit named.

second (a unit of time)		watt (a unit of electric power)	
kilosecond	= 1,000 seconds	6. kilowatt	= 1,000 watts
1. hectosecond	= 100 seconds	7. hectowatt	= 100 watts
2. dekasecond	= 10 seconds	8. dekawatt	= 10 watts
3. decisecond	= 0.1 second	9. deciwatt	= 0.1 watt
4. centisecond	= 0.01 second	10. centiwatt	= 0.01 watt
5. millisecond	= 0.001 second	11. milliwatt	= 0.001 watt

joule (a unit of work or energy)		hertz (a unit of frequency)	
12. kilojoule	= 1,000 joules	18. kilohertz	= 1,000 hertz
13. hectojoule	= 100 joules	19. hectohertz	= 100 hertz
14. dekajoule	= 10 joules	20. dekahertz	= 10 hertz
15. decijoule	= 0.1 joule	21. decihertz	= 0.1 hertz
16. centijoule	= 0.01 joule	22. centihertz	= 0.01 hertz
17. millijoule	= 0.001 joule	23. millihertz	= 0.001 hertz

Answer each of the following.

24. Whose picture is on the "kilodollar"? Grover Cleveland
25. Whose picture is on the "hectodollar"? Benjamin Franklin
26. Whose picture is on the "dekadollar"? Alexander Hamilton
27. Whose picture is on the "decidollar"? Franklin Roosevelt
28. Whose picture is on the "centidollar"? Abraham Lincoln

APPLYING THE LESSON

Independent Practice

Homework Assignment	
Minimum	Ex. 1-47 odd
Average	Ex. 2-32 even; 33-47
Maximum	Ex. 1-47

Additional Answers

35. changes for road signs, records for sports events, and so on

37. The current metric system was developed in 1795 by the French Academy of Sciences and was expanded in 1960. In the United States the Metric Conversion Act of 1975 specified a national policy of *voluntary* use of the metric system and established the U.S. Metric Board to help ease the change.

38. soft drinks, nuts and bolts, boxes of crackers, sets of tools, measuring tapes, and so on

5-MINUTE CHECK-UP

(over Lesson 6-1)

Available on TRANSPARENCY 6-2.
Name the metric unit for each measurement.

1. 0.01 meter **1 centimeter**
2. 1,000 meters **1 kilometer**
3. 10 grams **1 dekagram**
4. 100 liters **1 hectoliter**
5. 0.1 liter **1 deciliter**

Extra Practice, Lesson 6-1, p. 463

INTRODUCING THE LESSON

Have students pretend they are taking a 3-day bike tour 100 km long. Have them plan where they would stop if they could get lodging at the following points; 20 km, 30 km, 45 km, 55 km, 65 km, 70 km, 85 km, and 100 km.

TEACHING THE LESSON

Using Discussion Discuss with students what it means to "think metric." An analogy can be made to speaking another language. When it is automatic to estimate in metric units instead of the customary system, then a person is *thinking* metric. For example, many people now think of soft drinks as being available in 2-liter bottles.

Practice Masters Booklet, p. 61

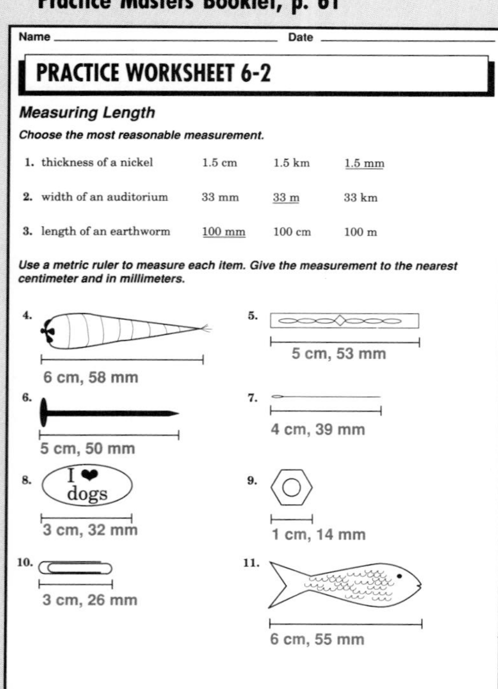

Name ____________ Date ____________

PRACTICE WORKSHEET 6-2

Measuring Length

Choose the most reasonable measurement.

1. thickness of a nickel	1.5 cm	1.5 km	<u>1.5 mm</u>
2. width of an auditorium	33 mm	<u>33 m</u>	33 km
3. length of an earthworm	<u>100 mm</u>	100 cm	100 m

Use a metric ruler to measure each item. Give the measurement to the nearest centimeter and in millimeters.

4. 6 cm, 58 mm
5. 5 cm, 53 mm
6. 5 cm, 50 mm
7. 4 cm, 39 mm
8. I ❤ dogs — 3 cm, 32 mm
9. 1 cm, 14 mm
10. 3 cm, 26 mm
11. 6 cm, 55 mm

6-2 MEASURING LENGTH

Objective
Estimate and measure metric units of length.

In the 1992 Summer Olympics, Yang Wenyi swam the 50-meter freestyle and won in 24.76 seconds. She set a winning record time for the gold medal.

The most commonly used metric units of length are the **meter (m), centimeter (cm), millimeter (mm),** and **kilometer (km).**

One meter (1 m) is about the height of a car fender.

One centimeter (1 cm) is about the width of a large paper clip.

One millimeter (1 mm) is about the thickness of a dime.

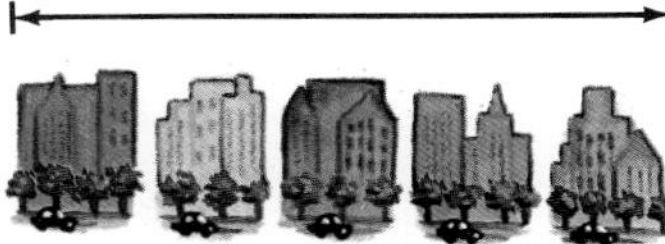

One kilometer (1 km) is about the length of 5 city blocks.

You can use a metric ruler to measure objects in centimeters or millimeters.

Example A ***Use the portion of a metric ruler shown below to find the length of the safety pin in centimeters and millimeters.***

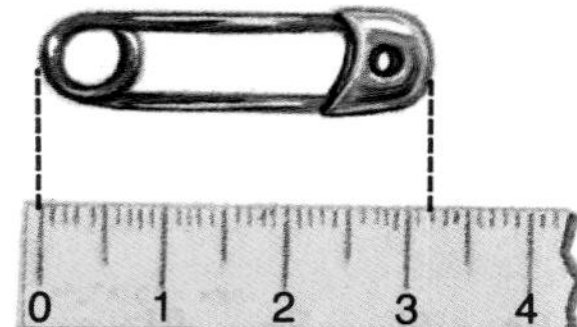

The length of the safety pin is between 3 and 4 centimeters or 30 and 40 millimeters. Since each small unit represents 1 millimeter, the safety pin is about 32 millimeters long. This can also be read as 3.2 centimeters.

Guided Practice

***Name the most reasonable unit for measuring each item. Use* kilometer, meter, centimeter, *or* millimeter.** **3. millimeter**

1. height of a tree **meter**
2. length of a pencil **centimeter**
3. thickness of an electrical wire
4. width of a classroom **meter**
5. distance from Dallas to Tampa **kilometer**
6. width of a staple **millimeter**

RETEACHING THE LESSON

Some students may have difficulty choosing the most reasonable unit for measuring. Have students use a metric ruler or tape measure to find each measurement.

1. the width of their math book
2. the width of their thumbnail
3. the width of their shoe
4. the height of their desk
5. the length of a pencil

Reteaching Masters Booklet, p. 56

Name ____________ Date ____________

RETEACHING WORKSHEET 6-2

Measuring Length

The ruler below can be used to measure items in centimeters and millimeters.

0 1 2 3 4 5 6 7 8 9 10 11 12 13
centimeters (cm)

Centimeters: Each numbered space stands for 1 centimeter.

Millimeters: Each of the 10 smaller spaces between a pair of numbers stands for 1 millimeter. 10 millimeters equal 1 centimeter.

Measure the length of each segment, from 0 to each letter, in centimeters.

A D F B E C H G

Exercises

Practice

Choose the most reasonable measurement.

7. diameter of a bicycle wheel	<u>66 cm</u>	66 m	66 km
8. length of a bolt	21 cm	21 m	<u>21 mm</u>
9. width of notebook paper	22 mm	<u>22 cm</u>	22 km
10. width of school hallway	5 mm	5 cm	<u>5 m</u>
11. width of a door	120 mm	120 km	<u>120 cm</u>

Use a metric ruler to measure each item. Give the measurement in centimeters and also in millimeters.

12.

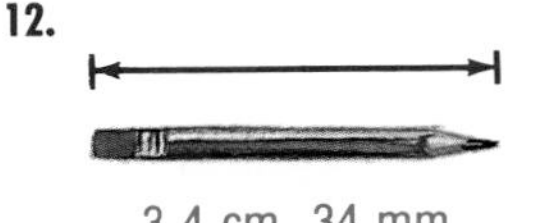

3.4 cm, 34 mm

13.

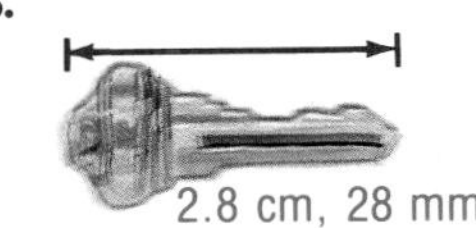

2.8 cm, 28 mm

14.

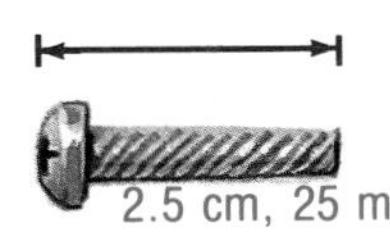

2.5 cm, 25 mm

Applications

15. Sing Lu makes flower arrangements for Caufield's Flowers. If it takes 1.2 m to make a bow for an arrangement, how many bows can Sing Lu make if he has 30 m of ribbon? **25 bows**

16. A pencil, measured to the nearest centimeter, is 13 centimeters long. Between what two measurements does the pencil have to be? **$\geq$ 12.5 cm and $<$ 13.5 cm**

17. If the air distance from San Francisco to New York is 4,139 kilometers and the air distance from San Francisco to Chicago is 2,990 kilometers, how much farther is it from San Francisco to New York than to Chicago? **1,149 km**

Critical Thinking

18. The Big K ranch brought six horses to the fair for the calf-roping contest. They have six adjacent stalls for the horses. Your job is to assign each horse to a stall. Where will you put them if Lola cannot be next to Lady, Trudy cannot be next to King, Lucky cannot be next to Jim, and Lady behaves well when she is next to Lucky. **See margin.**

Number Sense

19. Locate three items in your classroom that are about 1 centimeter, about 1 millimeter, and about 1 meter in length or height. **Answers may vary.**

Estimation

You can estimate the length or height of an object by comparing it to an object with which you are familiar.

20. Noelle is about half as tall as her father. If she is 92 cm tall, about how tall is her father?

21. A classroom wall is 24 blocks high. If each block is about 20 centimeters high and there is about 1 centimeter of mortar between blocks, about how many meters high is the wall?

20. about 180 cm **21. about 5 m**

APPLYING THE LESSON

Independent Practice

Homework Assignment	
Minimum	Ex. 2-20 even
Average	Ex. 1-13 odd; 15-21
Maximum	Ex. 1-21

Additional Answers

18. Answers may vary. A sample answer is given. Trudy, Jim, Lady, Lucky, King, Lola

Chalkboard Examples

Example A *Name the most reasonable unit for measuring each item.*

1. height of a candle **cm**
2. length of the gymnasium **m**
3. width of a window **m**
4. distance to the grocery store **km**

Additional examples are provided on TRANSPARENCY 6-2

EVALUATING THE LESSON

Check for Understanding Use the Guided Practice Exercises to check for student understanding.

Closing the Lesson Have students give examples of items and the metric unit they would most likely use to measure it.

EXTENDING THE LESSON

Enrichment In small groups, have students measure the length of several large items, such as the school, the school property, and the parking lot. Then have them measure several small items, such as a dollar bill, a pencil, a paper clip, and so on. Have them record their measurements and compare with other groups.

Enrichment Masters Booklet, p. 56

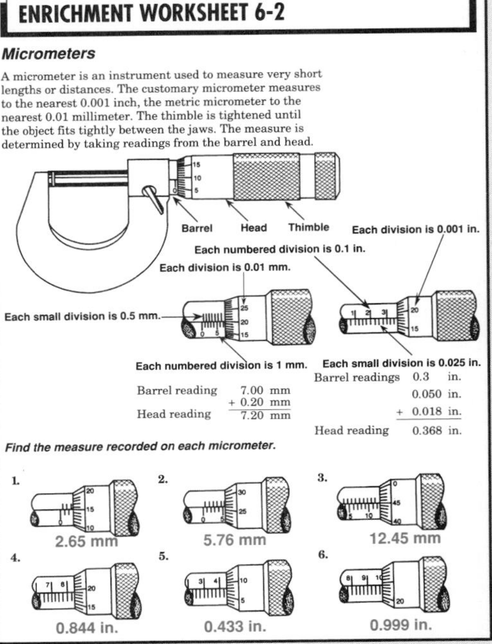

Name ______________________ Date ______________

ENRICHMENT WORKSHEET 6-2

Micrometers

A micrometer is an instrument used to measure very short lengths or distances. The customary micrometer measures to the nearest 0.001 inch, the metric micrometer to the nearest 0.01 millimeter. The thimble is tightened until the object fits tightly between the jaws. The measure is determined by taking readings from the barrel and head.

Barrel reading	7.00 mm
	+ 0.20 mm
Head reading	7.20 mm

Barrel readings	0.3 in.
	0.050 in.
	+ 0.018 in.
Head reading	0.368 in.

Find the measure recorded on each micrometer.

1. 2.65 mm
2. 5.76 mm
3. 12.45 mm
4. 0.844 in.
5. 0.433 in.
6. 0.999 in.

5-MINUTE CHECK-UP

(over Lesson 6-2)

Available on TRANSPARENCY 6-3.

Choose the most reasonable measurement.

1. height of your desk
   68 cm   68 mm   68 m
2. width of a dollar bill
   66 cm   66 mm   66 m
3. from your desk to school door
   18 m   18 cm   18 km
4. from Los Angeles to Seattle
   1,620 km   1,620 m   1,620 cm

Extra Practice, Lesson 6-2, p. 463

▶ INTRODUCING THE LESSON

Ask students to list the ingredients of their favorite recipe and the amounts needed. Ask students if they can buy the exact amount needed; for example, 2 c flour, 1 c sugar. Ask students to suggest other items that cannot be bought in exact amounts.

▶ TEACHING THE LESSON

Using Cooperative Groups In groups of three or four, have students make up three word problems that involve changing metric length to a different unit of length. Have groups exchange and solve. Have the original group check the solutions.

Practice Masters Booklet, p. 62

Name ______________ Date ______________

PRACTICE WORKSHEET 6-3

Changing Metric Units

Complete.

1. 3 km = 3,000 m	2. 4 m = 4,000 mm
3. 5 mm = 0.5 cm	4. 20 km = 2,000,000 cm
5. 30 cm = 0.3 m	6. 820 mm = 0.82 m
7. 9 cm = 90 mm	8. 50 m = 5,000 cm
9. 99 m = 0.099 km	10. 5 m = 5,000 mm
11. 12 mm = 1.2 cm	12. 0.2 km = 20,000 cm
13. 39 cm = 0.39 m	14. 82 mm = 0.082 m
15. 92 cm = 920 mm	16. 1.3 m = 130 cm
17. 905 m = 0.905 km	18. 3.1 km = 3,100 m
19. 3.7 m = 3,700 mm	20. 50 mm = 5 cm
21. 0.9 km = 90,000 cm	22. 39.5 cm = 0.395 m
23. 132 cm = 1.32 m	24. 21 cm = 210 mm
25. 0.34 m = 34 cm	26. 57 m = 0.057 km

Solve.

27. The handlebars of Jackie's racing bike are 97 centimeters above the ground. How many millimeters is this?
    970 mm

28. The distance from Oak Street to Main Street is 0.75 kilometers. How many meters is this?
    750 m

6-3 CHANGING METRIC UNITS

Objective
Change metric units of length.

Mr. Santos wants to run at least 4,000 meters a day while training for the 5,000-meter run at Boyce Park. His office is 4.3 kilometers from his home. If Mr. Santos runs to his office in the morning instead of taking the bus, will he run at least 4,000 meters?

To change metric units, multiply or divide by powers of ten. When changing a larger unit to a smaller unit, you should *multiply*. When changing a smaller unit to a larger unit, you should *divide*.

Method

1. Determine if you are changing from smaller to larger units or vice versa.
2. Divide to change from smaller to larger units. Multiply to change from larger to smaller units.
3. Determine the multiple of ten by which to multiply or divide. Use this diagram.

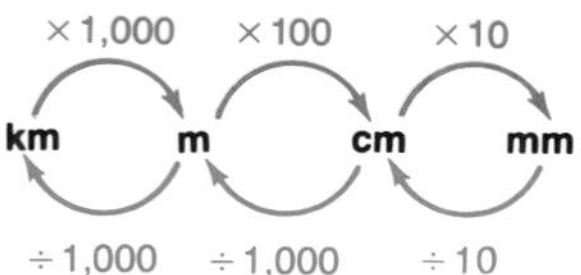

Example A ***Change 4.3 kilometers to meters.***

1. kilometers → meters
   You are changing from larger to smaller units.
2. There will be more units after the change. Multiply.
3. There are 1,000 meters in a kilometer. Multiply by 1,000.   4.3 km = ? m   × 1,000
   4.3 km = 4,300 m

Mr. Santos will run 4,300 meters.

Example B ***Change 375 centimeters to meters.***

1. centimeters → meters
   You are changing from smaller to larger units.
2. Divide.
3. There are 100 centimeters in a meter. Divide by 100.   375 cm = ? m   ÷ 100
   375 cm = 3.75 m

Guided Practice

Example A

Complete.

1. 3 m = ? cm **300**
2. 8 km = ? m **8,000**
3. 2 cm = ? mm **20**
4. 1.2 m = ? cm **120**
5. 0.4 cm = ? mm **4**
6. 0.96 km = ? m **960**

RETEACHING THE LESSON

Complete each statement.

1. To change millimeters to centimeters, divide by ___?___. **10**
2. To change meters to millimeters, multiply by ___?___. **1,000**
3. To change meters to centimeters, ___?___ by 100. **multiply**
4. To change meters to kilometers, ___?___ by 1,000. **divide**
5. To change centimeters to meters, divide by ___?___. **100**

Reteaching Masters Booklet, p. 57

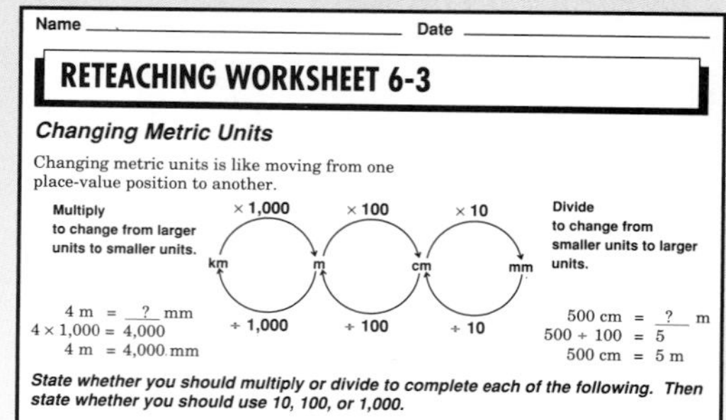

Name ______________ Date ______________

RETEACHING WORKSHEET 6-3

Changing Metric Units

Changing metric units is like moving from one place-value position to another.

Multiply to change from larger units to smaller units.

× 1,000 × 100 × 10
km m cm mm
÷ 1,000 ÷ 100 ÷ 10

Divide to change from smaller units to larger units.

4 m = ? mm
4 × 1,000 = 4,000
4 m = 4,000 mm

500 cm = ? m
500 ÷ 100 = 5
500 cm = 5 m

State whether you should multiply or divide to complete each of the following. Then state whether you should use 10, 100, or 1,000.

1. 9,000 mm = ? m
2. 7.3 km = ? m × 1,000
3. 55 m = ? cm × 100

Example B *Complete.*

7. 20 mm = _?_ cm **2** 8. 400 cm = _?_ m **4** 9. 4.6 m = _?_ km **0.0046**
10. 27.5 cm = _?_ m **0.275** 11. 9.78 m = _?_ km **0.00978** 12. 5,670 mm = _?_ cm **567**

Exercises

Practice *Complete.* **25. 0.0087**

13. 700 mm = _?_ cm **70** 14. 0.8 m = _?_ cm **80** 15. 350 cm = _?_ m **3.5**
16. 3 km = _?_ m **3,000** 17. 0.4 km = _?_ m **400** 18. 40 mm = _?_ cm **4**
19. 5,000 m = _?_ km **5** 20. 0.9 cm = _?_ mm **9** 21. 3 mm = _?_ cm **0.3**
22. 2,600 m = _?_ km **2.6** 23. 4.2 m = _?_ cm **420** 24. 40 cm = _?_ m **0.4**
25. 8.7 m = _?_ km 26. 15 mm = _?_ cm **1.5** 27. 3.6 cm = _?_ mm **36**
28. How many meters are in 38.2 kilometers? **38,200 m**

Applications

29. In air, sound at a particular frequency travels at about 343 meters per second. In water, sound at the same frequency travels at about 1.435 kilometers per second. How many meters per second faster does this sound wave travel in water than in air? **1,092 mps**

FUN with MATH

Why do we only see one side of the moon?
See page 240.

30. Kim walks every day for 1 hour as part of her aerobic program. If she walks at a rate of 1 kilometer in 15 minutes, how many meters will she walk in 1 hour? **4,000 m**

31. Suppose Dr. Seuss' book *Green Eggs and Ham* has sold 3.8 million copies. Imagine that the books are stacked on top of each other. If each book is 13 mm thick, how many kilometers high would the stack be? **49.4 km**

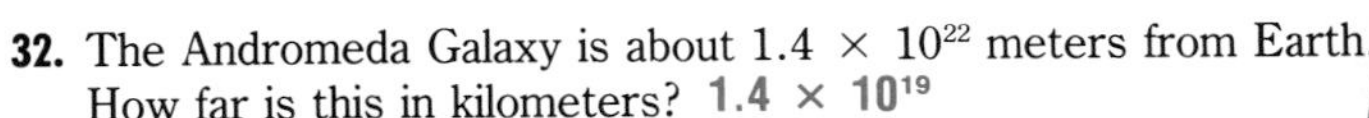

32. The Andromeda Galaxy is about 1.4×10^{22} meters from Earth. How far is this in kilometers? **1.4×10^{19}**

JOURNAL ENTRY

33. Write a few sentences about something new you learned in this lesson. Give examples. **See students' work.**

Critical Thinking

34. Nick has 4 pictures to put in a grouping frame. The largest picture will only fit in the top left-hand corner. The other three pictures will fit in any of the other corners. How many different ways can Nick put the pictures in the frame? **6 ways**

Mixed Review *Write in scientific notation.*

Lesson 3-8

35. 7,010,000 **7.01×10^6** 36. 830,000 **8.3×10^5** 37. 48,000,000 **4.8×10^7** 38. 53,700 **5.37×10^4**

Lesson 4-1 *State whether each number is divisible by 2, 3, 5, 9, or 10.*

39. 14 **2** 40. 24 **2, 3** 41. 33 **3** 42. **3, 5, 9** 43. 90 **2, 3, 5, 9, 10**

Lesson 4-7

44. At the school cafeteria $\frac{8}{15}$ of the students ate pizza and $\frac{9}{20}$ of the students ate hamburgers. Which food did more students eat? **pizza**

APPLYING THE LESSON

Independent Practice

Homework Assignment	
Minimum	Ex. 2-30 even; 31-44
Average	Ex. 1-27 odd; 28-44
Maximum	Ex. 1-44

Chalkboard Examples

Example A *Complete.*
1. 6 cm = _?_ mm **60**
2. 4 m = _?_ cm **400**
3. 9 km = _?_ m **9,000**

Example B *Complete.*
4. 82 mm = _?_ cm **8.2**
5. 943 cm = _?_ m **9.43**
6. 93 m = _?_ km **0.093**

Additional examples are provided on TRANSPARENCY 6-3.

EVALUATING THE LESSON

Check for Understanding Use the Guided Practice Exercises to check for student understanding.

Error Analysis Watch for students who multiply when they should divide and vice versa. Remind students to check equivalent measurements using this expression: smaller number larger unit = larger number smaller unit

Closing the Lesson Ask students to state in their own words the procedure for changing from smaller to larger units and from larger to smaller units.

EXTENDING THE LESSON

Enrichment Masters Booklet, p. 57

Name ______________ Date ______________

ENRICHMENT WORKSHEET 6-3

Comparing Systems of Measurement

Fill each ◯ with <, >, or ≈ to make a true sentence.

1. 10 cm (≈) 4 in. 2. 25 ft (<) 9 m 3. 55 mi (>) 80 km
4. 100 m (>) 300 yd 5. 16 km (≈) 10 mi 6. 150 cm (>) 4 ft
7. 30 mi (>) 40 km 8. 10 m (>) 20 ft 9. 2 m (≈) 6.6 ft

Solve.

10. Patty ran the 880-yard event in 2 minutes and 52 seconds. At this rate, would it take her more or less time to run the 800-meter event? **less**
11. The swimming pool at school is 25 yards long. About how many laps does it take to make a mile? **about 70**
12. On a Canadian road the speed limit is 50 kilometers per hour. Mike's car shows miles per hour. About how many miles per hour is this Canadian speed limit? **about 30 mph**
13. The swimming pool in which Rita practices is 50 meters long. About how many laps does Rita have to swim in order to swim one mile? **about 32 laps**
14. Sarah bought a 50-meter roll of packing tape. It takes about 25 centimeters of tape to seal one package. About how many packages can Sarah seal? **about 200 packages**
15. Tom bought 100 meters of clothesline for $1.89. His brother bought 100 yards of comparable quality clothesline for the same price. Who got the better buy? **Tom**
16. A U.S. basketball team averages 6 feet 8 inches in height. A Soviet team averages 2.1 meters in height. Which team has the taller average height? **the Soviet team**
17. Freeway speeds in Canada are posted at 100 kilometers per hour. U.S. speed limits are now 65 miles per hour. Which speed will let you drive faster? **65 mph**
18. Al bought a 10-foot-long strip of molding for $0.54. Ross bought a 3-yard strip of the same molding for $0.55. Who got the better buy? **Al**
19. Harold long-jumped 17 feet 7 inches on his first try and 6 yards 1 inch on his second try. How much longer was Harold's second jump than his first? **6 in.**

5-MINUTE CHECK-UP

(over Lesson 6-3)

Available on TRANSPARENCY 6-4.

Complete.

1. 490 cm = __?__ m **4.9**
2. 81 m = __?__ mm **81,000**
3. 2.5 km = __?__ m **2,500**
4. 708 cm = __?__ mm **7,080**
5. 35 m = __?__ km **0.035**
6. 37 mm = __?__ m **0.037**

Extra Practice, Lesson 6-3, p. 463

▶ INTRODUCING THE LESSON

Ask students to imagine being weightless and ask if they would like to spend some time in weightlessness. Remind students that weight changes depending on the pull of gravity, but mass remains the same.

▶ TEACHING THE LESSON

Using Manipulatives Have students construct a simple balance from a hexagonal-shaped pencil and a ruler. Have them balance objects such as elastic bands, erasers, and chalk against a paper clip or several paper clips.

Practice Masters Booklet, p. 63

Name ____________ Date ____________

PRACTICE WORKSHEET 6-4

Measuring Mass

Complete.

1. 54 g = 54,000 mg	2. 7 kg = 7,000 g
3. 8.5 kg = 8,500 g	4. 2.7 g = 2,700 mg
5. 458 g = 0.458 kg	6. 3 mg = 0.003 g
7. 78 kg = 78,000 g	8. 3,952 g = 3.952 kg
9. 16 g = 0.016 kg	10. 41.7 g = 41,700 mg
11. 4,351 mg = 4.351 g	12. 2 g = 0.002 kg
13. 906 g = 0.906 kg	14. 38 mg = 0.038 g
15. 7.5 g = 0.0075 kg	16. 6,000 g = 6 kg
17. 520 mg = 0.520 g	18. 25,480 g = 25.48 kg

Solve.

19. The King High School yearbook has a mass of 520 grams. If 430 students ordered the yearbook, what is the mass in kilograms of all these yearbooks combined? 223.6 kg

20. Sandy weighed the newspapers that she delivered last Sunday. They weighed 75 kilograms in all. If Sandy delivered 62 Sunday papers, about how much did one paper weigh? 1.2 kg

21. The total weight of 3 kittens is 8.7 kilograms. What is the average weight of each kitten in grams? 2,900 g

22. One egg weighs about 35 grams. About how much do a dozen eggs weigh in kilograms? about 0.42 kg

6-4 MEASURING MASS

Objective
Estimate and use metric units of mass.

In Chang's chemistry experiment, the water dissolved 523 milligrams of sodium chloride (table salt). How many grams of sodium chloride did the water dissolve?

The most commonly used metric units of mass are the **gram (g), kilogram (kg),** and **milligram (mg).**

The mass of a large paper clip is about one gram (1 g).

The mass of a baseball bat is about one kilogram (1 kg).

The mass of a grain of salt is about one milligram (1 mg)

Changing metric units of mass is similar to changing metric units of length.

×1,000 ×1,000
kg g mg
÷1,000 ÷1,000

Method

1. Determine if you are changing from smaller to larger units or vice versa.
2. Divide to change from smaller to larger units. Multiply to change from larger to smaller units.
3. Determine the multiple of ten by which to multiply or divide.

Examples

A ***Change 523 milligrams to grams.***

1. milligrams → grams
   smaller → larger
2. Divide.
3. There are 1,000 milligrams in a gram.

523 mg = __?__ g ÷ 1,000

523 mg = 0.523 g

The water dissolved 0.523 g of sodium chloride.

B ***Change 28 kilograms to grams.***

1. kilograms → grams
   larger → smaller
2. Multiply.
3. There are 1,000 grams in a kilogram.

28 kg = __?__ g × 1,000

28 kg = 28,000 g

Guided Practice

Choose the better estimate for the mass of each item.

1. 1 vitamin tablet, 1 g or 1 kg
2. a feather, 17 mg or 17 g
3. automobile, 1,000 g or 1,000 kg
4. pencil, 10 g or 10 mg
5. grocery bag, 80 g or 80 mg
6. bowling pin, 1.4 g or 1.4 kg

Complete.

Example A 7. 4.9 mg = __?__ g **0.0049** 8. 5,862 g = __?__ kg **5.862** 9. 8,754 mg = __?__ g **8.754**

Example B 10. 67 g = __?__ mg **67,000** 11. 73 kg = __?__ g **73,000** 12. 9.7 g = __?__ mg **9,700**

RETEACHING THE LESSON

Have students use the diagram on page 184 to complete the following.

1. To change milligrams to grams, __divide__ by 1000.
2. To change kilograms to grams, __multiply__ by 1000.
3. To change grams to milligrams, __multiply__ by 1000.
4. To change grams to kilograms, __divide__ by 1000.
5. 3,000 mg = __3__ g

Reteaching Masters Booklet, p. 58

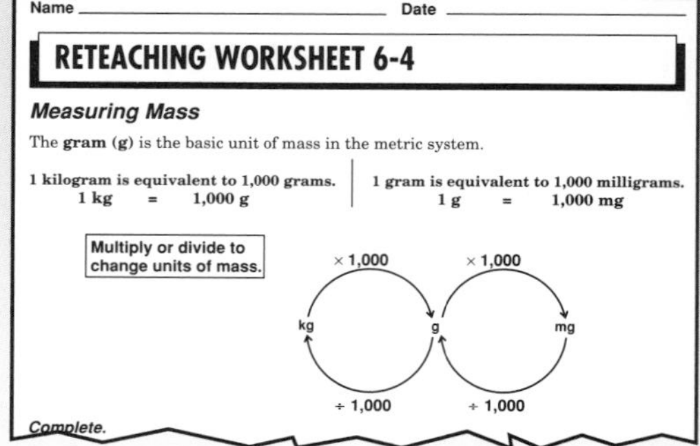

Name ____________ Date ____________

RETEACHING WORKSHEET 6-4

Measuring Mass

The **gram (g)** is the basic unit of mass in the metric system.

1 kilogram is equivalent to 1,000 grams.
1 kg = 1,000 g

1 gram is equivalent to 1,000 milligrams.
1 g = 1,000 mg

Multiply or divide to change units of mass.

×1,000 ×1,000
kg g mg
÷1,000 ÷1,000

Complete.

Exercises

Practice

Complete. 16. 2.197 20. 1.775 22. 17,800

13. 8 kg = ? g **8,000** **14.** 62 g = ? mg **62,000** **15.** 752 mg = ? g **0.752**

16. 2,197 g = ? kg **17.** 9.4 g = ? mg **9,400** **18.** 6.8 kg = ? g **6,800**

19. 912 g = ? kg **0.912** **20.** 1,775 mg = ? g **21.** 24 kg = ? g **24,000**

22. 17.8 g = ? mg **23.** 5 mg = ? g **0.005** **24.** 40 g = ? kg **0.040**

25. Change 25.8 kilograms to grams. **25,800 grams**

26. How many kilograms are in 260,000 milligrams? **0.26 kilograms**

Applications

27. Ace Athletic Club has 4 different sizes of hand weights in the workout room. The lightest weight is 0.5 kg. If each size increases by 500 g, how many kilograms does the heaviest size weigh? **2 kg**

28. Kurt has a jar full of nickels and a jar full of pennies. He wants to know how much money he has in nickels. Kurt knows that a nickel has a mass of 5 grams. The nickels from his jar weigh 2.5 kilograms. How many nickels does Kurt have? **500 nickels**

29. A case of cans of vegetable soup weighs 4 kilograms. If the weight of the packaging is 40 grams and 1 can contains 330 grams of soup, how many cans of soup are there in the case? **12 cans**

Number Sense
11 mg-11 g; 1 m-1 cm; 200 g-200 mg; 1 m-1 cm

30. The following paragraph has four errors in units of mass or length. Find and correct the errors.

The mass of a hummingbird is about 11 mg and is light compared to the mass of an orange, about 330 g. The bill of a hummingbird is about 1 m long, so it can get nectar from flowers. Another insect that gets nectar is the honeybee that has a total mass of about 200 g. A honeybee is about 1 cm long with a wingspan of about 1 m so it can fly easily.

Mixed Review

Add.

Lesson 2-1

31. 42 + 87 = **129** **32.** 631 + 825 = **1,456** **33.** 7,340 + 283 = **7,623** **34.** 4,522 + 8,766 = **13,288** **35.** 10,487 + 68,313 = **78,800**

Lesson 3-4

36. 16)483 **30 R3** **37.** 42)723 **17 R9** **38.** 123)6,458 **52 R62** **39.** 381)19,482 **51 R51**

Lesson 5-10

40. Janelle noted that over the last 45 days, $\frac{4}{5}$ of the days were sunny. How many days were sunny? **36 days**

APPLYING THE LESSON

Independent Practice

Homework Assignment	
Minimum	Ex. 2-40 even
Average	Ex. 1-25 odd; 27-40
Maximum	Ex. 1-40

Chalkboard Examples

Example A *Complete.*
1. 8 g = ? kg **0.008**
2. 67 mg = ? g **0.067**
3. 500 mg = ? g **0.5**

Example B *Complete.*
4. 75 kg = ? g **75,000**
5. 19.4 g = ? mg **19,400**

Additional examples are provided on TRANSPARENCY 6-4.

EVALUATING THE LESSON

Check for Understanding Use the Guided Practice Exercises to check for student understanding.

Error Analysis Encourage students having difficulty changing from one unit of mass to another to use the diagram.

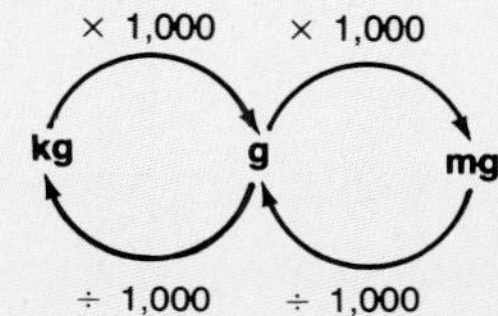

Closing the Lesson Have students explain in their own words how to change grams to kilograms.

EXTENDING THE LESSON

Enrichment Masters Booklet, p. 58

Name ____ Date ____

ENRICHMENT WORKSHEET 6-4

Precision

A measurement is *never* exact. It becomes more and more precise as the unit of measure used becomes smaller.

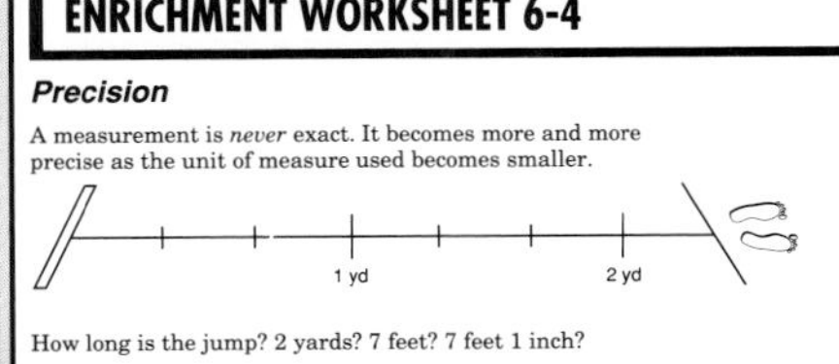

How long is the jump? 2 yards? 7 feet? 7 feet 1 inch?

Each answer is correct. 7 feet is more precise than 2 yards. 7 feet 1 inch is more precise than 7 feet.

The unit to be used depends on how precise you need to be.

Choose the unit necessary to give you the required precision.

1. baking a cake	hour	minutes	seconds
2. running the 100-m dash	hour	minutes	seconds
3. distance to school	miles	feet	inches
4. your height	meters	centimeters	millimeters

Choose the more precise measurement.

5.	2L	1,900 mL	6.	6 gal	25 qt
7.	4,200 g	4 kg	8.	45 oz	3 lb
9.	5 km	5,010 m	10.	15 ft	175 in.
11.	31 fl oz	4 cups	12.	7 g	6,750 mg

5-MINUTE CHECK-UP

(over Lesson 6-4)

Available on TRANSPARENCY 6-5.

Complete.

1. 517 g = ? kg **0.517**
2. 2.34 kg = ? g **2,340**
3. 559 mg = ? g **0.559**
4. 250 g = ? kg **0.25**
5. 9.3 g = ? mg **9,300**
6. 3.9 kg = ? g **3,900**

Extra Practice, Lesson 6-4, p. 464

INTRODUCING THE LESSON

Ask students if they think a 3-liter engine will outperform a 2.5 liter engine in the same car and why they think so. Ask them to name other everyday items that people choose based on their capacity. **Students should be familiar with many grocery items.**

TEACHING THE LESSON

Using Manipulatives Have students estimate the capacity of common items, such as a coffee cup, drinking glass, cereal/soup bowl, cartons used for milk or juice, and so on. Then measure with a liter container and metric measuring cups.

Practice Masters Booklet, p. 64

Name ________________ Date ________________

PRACTICE WORKSHEET 6-5

Measuring Capacity

Complete.

1. 2 kL = 2,000 L	2. 5 L = 5,000 mL
3. 125 mL = 0.125 L	4. 0.52 kL = 520 L
5. 12.4 kL = 12,400 L	6. 15 L = 0.015 kL
7. 75 mL = 0.075 L	8. 309 mL = 0.309 L
9. 900 L = 0.9 kL	10. 0.06 L = 60 mL
11. 85 kL = 85,000 L	12. 3,500 L = 3.5 kL
13. 425 L = 0.425 kL	14. 29.3 L = 29,300 mL
15. 0.25 L = 250 mL	16. 0.4 kL = 400 L
17. 32 L = 32,000 mL	18. 51.3 L = 0.0513 kL

Solve.

19. The social committee figures it needs one 180-milliliter serving of punch for each of the 140 guests at the reception. How many liters of punch do they need in all? **25.2 L**

20. The Adams family car has a gas tank with a capacity of 50 liters. After a trip they filled the tank with 42 liters of gas. How much gas was in the tank before they filled it? **8 L**

21. Seventeen people attended the Forbes family picnic. They had 10 liters of lemonade at the start, and it was all gone at the end of the day. If each person had an equal amount of lemonade, about how many milliliters did each drink? **About 588 mL**

22. The daily dosage of medicine for Joe's allergy is 2 milliliters. The bottle contained 0.24 liter when it was full. How many days will this medicine last? **120 days**

6-5 MEASURING CAPACITY

Objective
Estimate and use metric units of capacity.

Mary and Jake each spend $0.75 a day for a drink at lunch. Jake buys 1 carton of fruit juice from the vending machine in the cafeteria for $0.75. The carton contains 300 milliliters of juice. Mary brings a 0.5-liter carton of juice that she buys in the grocery store for $0.75. Who gets more juice for the same amount of money? Mary changes 0.5 liter to milliliters to find out.

The most commonly used metric units of capacity are the **liter (L), kiloliter (kL),** and **milliliter (mL).**

The capacity of the pitcher is about one liter (1 L).

The capacity of the average above-ground swimming pool is 15 kiloliters (15 kL).

The capacity of the eyedropper is about one milliliter (1 mL)

Changing metric units of capacity is similar to changing metric units of length and mass.

Method

1. Determine if you are changing from smaller units to larger units or vice versa.
2. Divide to change from smaller to larger units. Multiply to change from larger to smaller units.
3. Determine the multiple of ten by which to multiply or divide.

×1,000 ×1,000
kL L mL
÷1,000 ÷1,000

Examples

A ***Change 0.5 liters to milliliters.***

1. liters → milliliters
larger → smaller
2. Multiply.
3. There are 1,000 milliliters in a liter.
0.5 L = ? mL ×1,000
0.5 L = 500 mL
Mary gets more juice than Jake for the same amount of money.

B ***Change 4,560 liters to kiloliters.***

1. liters → kiloliters
smaller → larger
2. Divide.
3. There are 1,000 liters in a kiloliter.
4,560 L = ? kL ÷1,000
4,560 L = 4.560 kL

Guided Practice

***Name the most reasonable unit for measuring the capacity of each item. Use* kiloliter, liter, *or* milliliter.**

1. aquarium **liter**
2. large coffee pot **liter**
3. jar of mayonnaise **liter**
4. bottle of perfume **milliliter**

RETEACHING THE LESSON

Complete.

1. To change liters to milliliters, **multiply** by 1,000.
2. To change kiloliters to liters, **multiply** by **1,000**.
3. To change milliliters to liters, **divide** by **1,000**.
4. To change liters to kiloliters, **divide** by **1,000**.
5. 4L = **4,000** mL
6. 2 kL = **2,000** L

Reteaching Masters Booklet, p. 59

Name ________________ Date ________________

RETEACHING WORKSHEET 6-5

Measuring Capacity

The **liter (L)** is the basic unit of capacity in the metric system.

1 kiloliter is equivalent to 1,000 liters.
1 kL = 1,000 L

1 liter is equivalent to 1,000 milliliters.
1 L = 1,000 mL

Multiply or divide to change units of capacity.

×1,000 ×1,000
kL L mL
÷1,000 ÷1,000

Complete.

1. 8,620 mL = 8.62 L
2. 340.5 kL = 340,500 L

Example A — ***Complete.***

5. 42 L = ? mL **42,000** **6.** 0.9 L = ? mL **900** **7.** 36 kL = ? L **36,000**

Example B

8. 791 L = ? kL **0.791** **9.** 235.6 mL = ? L **0.2356** **10.** 4 L = ? kL **0.004**

Exercises

Practice

Complete. **15. 0.234 21. 0.262 24. 34,200**

11. 6 L = ? mL **6,000** **12.** 3 kL = ? L **3,000** **13.** 7.5 kL = ? L **7,500**

14. 0.03 L = ? mL **30** **15.** 234 mL = ? L **16.** 58 L = ? kL **0.058**

17. 127 L = ? kL **0.127** **18.** 42 mL = ? L **0.042** **19.** 28 L = ? mL **28,000**

20. 0.22 kL = ? L **220** **21.** 262 mL = ? L **22.** 500 L = ? kL **0.5**

23. 17 kL = ? L **17,000** **24.** 34.2 L = ? mL **25.** 96.8 L = ? kL **0.0968**

26. Change 34.2 L to milliliters. **34,200 mL**

27. How many milliliters are in 0.78 kiloliters? **780,000 mL**

Applications

28. When it was Charlene's turn to bring the drink for the photography club meeting, she brought three 2-liter bottles of punch. How many milliliters did each person have to drink if there were fifteen people at the meeting and they shared equally? **400 mL**

29. A pharmacist fills a prescription for pediatric antibiotic liquid. The dosage prescribed by the doctor is 2 mL 3 times a day for 5 days. How many milliliters of antibiotic should the pharmacist prepare? **30 mL**

FUN with MATH

Lasers are used in medicine. Find out what a laser is.
See page 241.

30. The Boosters Club buys cases of lemonade to sell at the track meets for $4.25. There are 12 cans of lemonade in a case. Each can contains 180 mL of lemonade. If the club wants to make a profit of $0.20 a can, how much should it charge for a can of lemonade? **$0.55**

Suppose

31. Suppose Benney's Sweet Shoppe has a drink dispenser that holds 75 L. Mr. Benney blended a punch that he wants to sell in a 300-mL cup. How many cups can he sell from a full dispenser? **250 cups**

Research

32. Interview a pharmacist, nurse, doctor, or lab technician about the use of the metric system in medicine. Write a paragraph about your interview. **See students' work.**

Mixed Review

Lesson 2-4

Subtract.

33. 18.5 − 6.2 = **12.3** **34.** 9.03 − 4.67 = **4.36** **35.** 25.4 − 6.73 = **18.67**

Lesson 4-3

Find the LCM of each group of numbers.

36. 5,7 **35** **37.** 9, 12 **36** **38.** 17, 51 **51**

Lesson 5-2

39. A leather jacket sells for $185. It is on sale for $\frac{1}{5}$ off the regular price. How much money is saved by buying the jacket on sale? **$37**

APPLYING THE LESSON

Independent Practice

Homework Assignment	
Minimum	Ex. 1-31 odd; 32-39
Average	Ex. 2-26 even; 28-39
Maximum	Ex. 1-39

Chalkboard Examples

Example A *Complete.*

1. 6.5 L = ? mL **6,500**
2. 98 L = ? mL **98,000**
3. 8.2 kL = ? L **8,200**

Example B *Complete.*

4. 1,565 mL = ? L **1.565**
5. 45 L = ? kL **0.045**
6. 93.1 mL = ? L **0.0931**

Additional examples are provided on TRANSPARENCY 6-5.

EVALUATING THE LESSON

Check for Understanding Use Guided Practice Exercises to check for student understanding.

Closing the Lesson Have students list items they know that are sold with metric units of capacity.

EXTENDING THE LESSON

Enrichment The average family of 4 uses about 1,600 liters of water in a month. How many liters will they use in a year at this rate? Then have students research other water facts and make up a problem about their findings.
19,200 L/year

Enrichment Masters Booklet, p. 59

Name ______ Date ______

ENRICHMENT WORKSHEET 6-5

Mass, Capacity, Volume

In the metric system of measurement, units of mass, volume, and capacity are interrelated. Volume and capacity are related as follows.

1 milliliter (mL)	=	1 cubic centimeter (cm^3)
1 liter (L)	=	1,000 cubic centimeters
1 liter	=	1 cubic decimeter (dm^3)
1 cubic decimeter	=	1,000 milliliters

The mass of water is related to volume and capacity in the following way. (The temperature of the water has to be 4°C at an atmospheric pressure of 760 mm of mercury.)
1 mL or 1 cm^3 of water has a mass of 1 gram (g).
1 L or 1 dm^3 of water has a mass of 1 kilogram (kg).

Complete.

1. 10 cm^3 = **10** mL **2.** 2 L = **2,000** cm^3 **3.** 4 dm^3 = **4** L

4. 7.5 L = **7,500** mL **5.** 500 mL = **0.5** dm^3 **6.** 0.2 L = **200** cm^3

7. 0.016 dm^3 = **16** mL **8.** 185 cm^3 = **0.185** dm^3 **9.** 7 cm^3 = **0.007** L

10. 0.008 L = **8** cm^3 **11.** 7.4 dm^3 = **7.4** L **12.** 1.04 dm^3 = **1,040** mL

Solve.

13. Find the mass, in grams, of 28 cm^3 of water. **28 g**

14. Find the mass, in grams, of 0.34 dm^3 of water. **340 g**

15. Find the mass, in kilograms, of 4,783 mL of water. **4.783 kg**

16. Find the volume, in cubic decimeters, of 75 g of water. **0.075 dm^3**

17. Find the capacity, in milliliters, of 0.471 kg of water. **471 mL**

18. How many cubic millimeters are in a cubic centimeter? In a cubic decimeter? In a cubic meter? **1,000; 1,000,000; 1,000,000,000**

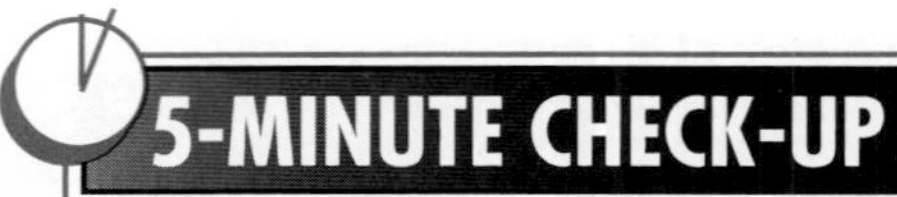

5-MINUTE CHECK-UP

(over Lesson 6-5)

Available on TRANSPARENCY 6-6.

Complete.

1. 4 kL = ? L 4,000
2. 543 mL = ? L 0.543
3. 25 mL = ? kL 0.000025
4. 139 L = ? kL 0.139
5. 472 L = ? mL 472,000

Extra Practice, Lesson 6-5, p. 464

INTRODUCING THE LESSON

Ask students if they have ever lost something and had to retrace their steps and actions to locate the missing article. Discuss students' answers.

TEACHING THE LESSON

Using Manipulatives Have students work in pairs and use play money to act out the following problem.
If you buy a rare stamp for $25, sell it for $30, buy it back for $35, and then sell it again for $45, how much money will you gain or lose? **Gain $15**

Practice Masters Booklet, p. 65

Name ______ Date ______

PRACTICE WORKSHEET 6-6

Problem-Solving Strategy: Acting It Out

Solve.

1. Twenty votes were cast for club president. You won by 4 votes over your opponent. How many votes did you receive? 12 votes
2. For every $3 your friend earned, you earned $5. How much did you earn if your friend earned $12? $20
3. You have 4 different-colored bangle bracelets. How many different combinations of 2 or 3 bracelets can you wear? 10
4. Sue is shorter than Tom and taller than Bob. Tom is taller than Bob and shorter than Amy. Put the people in order from tallest to shortest. Amy, Tom, Sue, Bob
5. Sam has more money than Sue. Allen has less money than Sue. Does Sam have more money than Allen? yes
6. Four people are going out to dinner. How many different ways can the people be seated at a square table if one person only sits on each side and one person never moves? 6
7. There are 5 people. How many different committees of 3 people each can be formed? 10 committees
8. Joan lives next door to Bob. Bob lives next door to Joe. Does Joan live next door to Joe? no

Problem-Solving Strategy

Explore / Plan / Solve / Examine

6-6 ACTING IT OUT

Objective
Solve problems by acting out a solution.

George collects baseball cards and posters. He decides to sell the poster collection so he can buy some special baseball cards. On Saturday, George sells every third poster. The next Saturday, he sells every third poster that he has left, and so on, for four Saturdays. How many posters does George have left if he started with twenty posters?

Explore What is given?
- George had twenty posters.
- George sells every third poster, four times.

What is asked?
- How many posters are left after George sells every third poster four times?

Plan The problem could be solved by having twenty students act out the sale of the posters.

Solve Twenty students stand in a line and count off by ones to twenty. Each student whose number is a multiple of three sits down. This indicates that the poster they represent has been sold.

Students count off again by ones. Then every student whose number is a multiple of three sits down. Students repeat the procedure two more times.

There are five students left standing, so the solution is five posters.

Examine Look at the diagram below. The numbers represent the students. The blue circles represent the first group to sit. The green circles represent the second group to sit. The red circles represent the third group, and the yellow circles, the fourth group.

1	2	(3)	(4)	(5)	(6)	(7)
(8)	(9)	10	(11)	(12)	(13)	14
(15)	16	(17)	(18)	(19)	20	

There are five numbers left so the solution is correct.

Guided Practice

Solve. Use acting it out.

1. At Savory Snacks, pretzels are moved from the ovens to packaging on a conveyor belt in a line of 25. Every fifth pretzel is tested for weight. Every third pretzel is tested for size. How many of the 25 pretzels are not checked or tested? **13 pretzels**
2. Dominique bought a classic rock album for $5. She sold it for $8.00. Then she bought it back for $10.50, and sold it again for $13.00. How much money did Dominique lose or gain in buying and selling the album? **$5.50 gain**

RETEACHING THE LESSON

Work with students to solve these problems.

1. A bicycle wheel has 40 spokes. How many spaces does it have between spokes? **40 spaces**
2. How many times does a person saw through a log to make 4 equal lengths? **3 times**
3. How many cuts to make 4 equal pieces of pie? **2 cuts**

Problem Solving

- Paper/ Pencil
- Mental Math
- Estimate
- Calculator

Solve. Use any strategy.

3. Leah initially ordered 250 flowers to sell at the homecoming game. After checking people's preferences, she changes her order for 50 carnations to 20 roses, changes 40 roses to 50 orchids, and changes 10 orchids to 20 mums. How many flowers will Leah have for sale? **240 flowers**
4. Ms. Dorman is planning to drive her daughter to college, a 500-mile trip. If her car gets 32 miles per gallon and has an 18-gallon gas tank, can she make the trip without stopping for more gas? **yes, 576-gal range**
5. Main Street runs north and south, and Third Avenue runs east and west. From the intersection of Main and Third, John jogs 5 blocks south, 6 blocks west, 3 blocks north, and 2 blocks east. How many blocks west of Third Avenue is John? **4 blocks west**
6. Jeanne, Mary, and some friends order a pizza that is cut into 8 pieces. Mary takes the first piece and passes the pizza around the table. If each friend has one piece and there is one piece left when the pizza returns to Mary, how many friends joined Jeanne and Mary? **5 friends**
7. What whole number between 50 and 70 is a multiple of 3, 4, and 5? **60**
8. Jody starts baking her cherry pie at 450° F for $\frac{1}{4}$ of an hour. Then she bakes the pie at 350° for $\frac{5}{6}$ of an hour. How long is the pie in the oven altogether? **$1\frac{1}{12}$ hours**
9. Juan and Susan are painting opposite sides of the fence. If Susan paints one-fourth of her side in an hour and Juan paints one-half of his side in the same time, how much longer will it take Susan than Juan to paint her side of the fence? **2 hours longer**

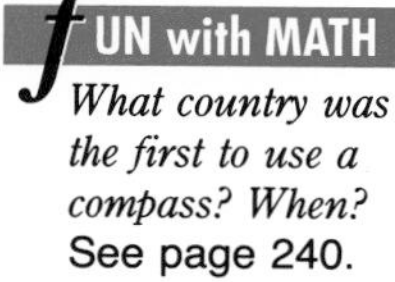

FUN with MATH

What country was the first to use a compass? When?
See page 240.

Mixed Review

Lesson 5-1 — *Estimate.*

10. $6\frac{1}{4} \times 3\frac{4}{5}$ **24**
11. $8\frac{5}{8} \times 6\frac{3}{4}$ **63**
12. $10\frac{1}{6} \div 1\frac{7}{8}$ **5**
13. $16\frac{3}{8} - 4\frac{1}{7}$ **4**

Lesson 5-6 — *Divide.*

14. $\frac{6}{7} \div \frac{3}{8}$ **$2\frac{2}{7}$**
15. $\frac{5}{9} \div \frac{10}{9}$ **$\frac{1}{2}$**
16. $\frac{6}{11} \div 3$ **$\frac{2}{11}$**
17. $\frac{7}{12} \div \frac{2}{3}$ **$\frac{7}{8}$**

Lesson 6-2

18. Mario bought 50 meters of rope to make a rope ladder to use in a tree house. If each rung of the ladder uses $2\frac{1}{2}$ meters, how many rungs will there be? **20 rungs**

APPLYING THE LESSON

Independent Practice

Homework Assignment	
Minimum	Ex. 1-18
Average	Ex. 1-18
Maximum	Ex. 1-18

Chapter 6, Quiz A (Lessons 6-1 through 6-6) is available in the Evaluation Masters Booklet, p. 29.

Chalkboard Examples

Solve. Use acting it out.

1. Quality control at Jones' Manufacturing finds that every eleventh bolt is defective in length and every fifteenth bolt is defective in threading. Out of every 60 bolts, how many are not defective? **51 bolts**

Additional examples are provided on TRANSPARENCY 6-6.

EVALUATING THE LESSON

Check for Understanding Use Guided Practice Exercises to check for student understanding.

Error Analysis Watch for students who do not follow the facts in the problem when acting it out. Have students write the facts in list form and number them. Then have students act out each step from their list.

Closing the Lesson Ask students to explain in their own words the kinds of problems that are best to act out.

5-MINUTE CHECK-UP

(over Lesson 6-6)

Available on TRANSPARENCY 6-7.

Solve. Use acting it out.

1. Jim has a dog, a cat, and a bird to move from the shuttle to the space station. He can't leave the dog with the cat or the bird with the cat if he is not there. He can only carry 2 animals at a time through the airlock. How should Jim carry the animals to the station? **Take the dog and cat over and leave the dog. Take the cat back and pick up the bird. Take both back.**

INTRODUCING THE LESSON

Tell students to imagine they are prehistoric people and have no standard measures. Ask students how they would ensure fair trading of items like animal pelts, dried herbs, and so on.

TEACHING THE LESSON

Using Manipulatives Have students measure objects in the classroom using rulers or yardsticks. Record four measurements for each object, an exact measurement, and to the nearest quarter inch, nearest half inch, and nearest inch.

Practice Masters Booklet, p. 66

Name ______________ Date ______________

PRACTICE WORKSHEET 6-7

Customary Units of Length

Complete.

1. 3 mi = 15,840 ft	**2.** 4 yd = 12 ft
3. 21 ft = 7 yd	**4.** 72 in. = 6 ft
5. 9 mi = 15,840 yd	**6.** 840 in. = 70 ft
7. 1 mi = 63,360 in.	**8.** 84 in. = $2\frac{1}{3}$ yd
9. 90 ft = 1,080 in.	**10.** 2 ft = 24 in.
11. 5 yd = 180 in.	**12.** 12 mi = 63,360 ft
13. 13 mi = 22,880 yd	**14.** 39 yd = 117 ft
15. 72 in. = 2 yd	**16.** 96 in. = 8 ft
17. $1\frac{1}{3}$ ft = 16 in.	**18.** 95 mi = 167,200 yd
19. 7,040 yd = 4 mi	**20.** $3\frac{1}{2}$ mi = 18,480 ft
21. 54 in. = $4\frac{1}{2}$ ft	**22.** $\frac{1}{2}$ mi = 2,640 ft
23. 42 ft = 14 yd	**24.** 132 yd = 44 ft
25. 26,436 ft = 5 mi 36 ft or 12 yd	**26.** 4,368 in. = 121 yd 1 ft

Solve.

27. The track around the field is 440 yards long. How many laps must be run to cover one mile?

28. Marco's favorite shot in basketball is 15 feet from the basket. How many inches is that?

6-7 CUSTOMARY UNITS OF LENGTH

Objective

Charge from one customary unit of length to another.

Henry James High School is voting for student council members on Thursday. Molly bought 8 yards each of red, white, and blue bunting to decorate the tables. How many 2-foot lengths of each color can Molly cut? She needs to change the number of yards to feet and divide.

The most commonly used customary units of length are the **inch (in. or ″), foot (ft or ′), yard (yd),** and **mile (mi).**

To change customary units, multiply or divide by the appropriate number.

1 foot	=	12 inches
1 yard	=	3 feet or 36 inches
1 mile	=	5,280 feet or 1,760 yards

Method

1. Determine if you are changing from smaller to larger units or vice versa.
2. Divide to change from smaller to larger units. Multiply to change from larger to smaller units.
3. Determine the number by which to multiply or divide.

Example A ***How many 2-foot lengths can Molly cut from 8 yards of bunting.***

1. yards → feet
You are changing from larger to smaller units.
2. There will be more units after the change.
Multiply. 8 yards = _?_ feet
3. There are 3 feet in one yard.
Multiply by 3. 8 yards = 24 feet

Molly divides 24 feet by 2 to find the number of 2-foot lengths in 8 yards. 24 ÷ 2 = 12

Molly can cut 12 2-foot lengths from 8 yards of bunting.

Example B ***Change 50 inches to feet and inches.***

1. inches → feet smaller → larger
2. Divide. 50 in. = _?_ ft __ in.
3. Divide by 12. ÷ 12

50 in. = 4 ft 2 in.

$$\begin{array}{r} 4 \text{ R2} \\ 12\overline{)50} \\ -48 \\ \hline 2 \end{array}$$

Guided Practice

Complete.

Example A

1. 3 ft = _?_ in. **36** **2.** 2 yd = _?_ ft **6** **3.** 2 mi = _?_ yd **3,520**

4. 3 yd = _?_ in. **108** **5.** 2 mi = _?_ ft **10,560** **6.** 15 yd = _?_ ft **45**

Example B

7. 24 in. = _?_ ft **2** **8.** 72 in. = _?_ yd **2** **9.** 15 ft = _?_ yd **5**

10. 30 in. = _?_ ft _?_ in. **2,6** **11.** 23 ft = _?_ yd _?_ ft **7,2** **12.** 76 in. = _?_ ft _?_ in. **6,4**

RETEACHING THE LESSON

Work with students to complete the following.

1. 36 in. = 3 ft
2. 6 ft = 72 in.
3. 1 mi = 5,280 ft
4. 108 in. = 3 yd
5. 16 ft = 5 yd 1 ft
6. 13 yd 2 ft = 41 ft

Reteaching Masters Booklet, p. 60

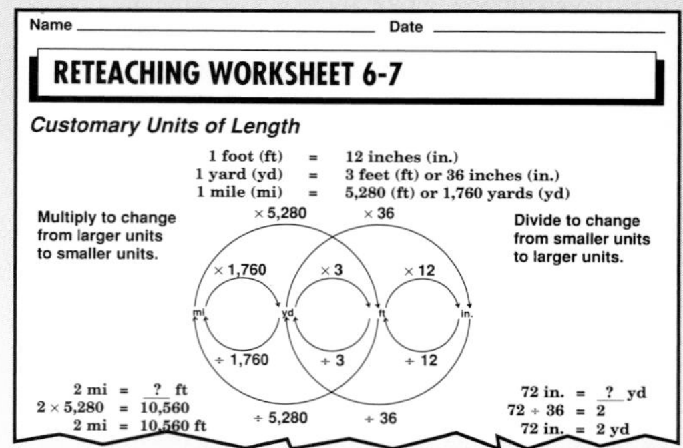
Name ______________ Date ______________

RETEACHING WORKSHEET 6-7

Customary Units of Length

1 foot (ft) = 12 inches (in.)
1 yard (yd) = 3 feet (ft) or 36 inches (in.)
1 mile (mi) = 5,280 (ft) or 1,760 yards (yd)

Multiply to change from larger units to smaller units.

Divide to change from smaller units to larger units.

2 mi = _?_ ft
2 × 5,280 = 10,560
2 mi = 10,560 ft

72 in. = _?_ yd
72 ÷ 36 = 2
72 in. = 2 yd

Exercises

Practice

Complete. 16. 5, 7 17. 7, 2 25. 2, 2

13. 5 ft = ? in. 60
14. 4 yd = ? ft 12
15. 2 yd = ? in. 72
16. 67 in. = ? ft ? in.
17. 86 in. = ? ft ? in.
18. 36 in. = ? ft 3
19. 27 ft = ? yd 9
20. 10 ft = ? in. 120
21. 180 in. = ? yd 5
22. $\frac{1}{2}$ mi = ? yd 880
23. $1\frac{1}{2}$ ft = ? in. 18
24. $2\frac{1}{4}$ yd = ? in. 81
25. 96 in. = ? yd ? ft
26. 18 in. = ? ft $1\frac{1}{2}$
27. 54 in. = ? yd $1\frac{1}{2}$

Applications

28. Romona walks $\frac{1}{2}$ mile to school on Mondays, Wednesdays, and Fridays. On Tuesdays and Thursdays she walks $2\frac{1}{4}$ miles to the athletic club. How much farther is the walk to the athletic club? $1\frac{3}{4}$ miles
29. In the high jump, Allan cleared 5 ft 3 in. David jumped 13 inches higher. How high was David's jump? 6 ft 4 in.

30. Nan walked an average of $2\frac{1}{2}$ miles a day for three days. The first day she walked 3 miles, the second day she walked $1\frac{3}{4}$ miles. How far did she walk the third day? $2\frac{3}{4}$ miles

Cooperative Groups

31. In groups of three or four, measure each person's height with a yardstick and a meterstick. Find the average height of the group in customary units and in metric units. Which system is easier to use? See margin.

Talk Math

32. In your own words, explain how to find the average of 5 ft 2 in., 4 ft 10 in., 5 ft 3 in., 4 ft 5 in., and 3 ft 11 in. See margin.

Using Equations

33. The formula that relates shoe size s and foot length f (in inches) for men is $s = 3f - 24$. What is the shoe size for a man whose foot is 11 inches long? size 9

Estimation

34. While shopping, Celia found a bin of different-sized candles on sale. She knows a dollar bill is about 6 inches long, so she used a bill to measure the unmarked candles. If the white candles measure about $2\frac{1}{3}$ dollar bills long and the red candles measure about $1\frac{1}{2}$ dollar bills long, *about* how long is each candle? white, 14 inches; red, 9 inches

Chalkboard Examples

Example A *Complete.*

1. 4 ft = ? in. 48
2. 12 yd = ? ft 36
3. 3.5 ft = ? in. 42
4. 2 yd = ? ft 6

Example B *Complete.*

5. 18 in. = ? ft ? in.
6. 48 in. = ? yd ? ft
7. 32 ft = ? yd ? ft
8. 60 in. = ? ft

5. 1, 6 6. 1, 1 7. 10, 2 8. 5

Additional examples are provided on TRANSPARENCY 6-7.

EVALUATING THE LESSON

Check for Understanding Use the Guided Practice Exercises to check for student understanding.

Error Analysis Watch for students who do not change a mixed measurement to one unit before multiplying or dividing.

Closing the Lesson Have students tell how often measurements are used in everyday life.

APPLYING THE LESSON

Independent Practice

Homework Assignment	
Minimum	Ex. 2-30 even; 31-34
Average	Ex. 1-27 odd; 28-34
Maximum	Ex. 1-34

Additional Answers

31. Answers may vary. Students may respond to the ease of computing in the metric system or the familiarity of using the customary system.
32. Answers may vary. A sample response is to change each measurement to inches, add them together, divide by 5, and change the result to feet and inches.

EXTENDING THE LESSON

Enrichment Masters Booklet, p. 60

Name ______________ Date ______________

ENRICHMENT WORKSHEET 6-7

Customary Units of Length

Name the best unit for measuring each of the following. Use inch, foot, yard, or mile.

1. length of your thumb inch
2. distance from town to town mile
3. height of a television tower foot or yard
4. thickness of a tile floor inch
5. width of a notepad inch
6. depth of a light snowfall inch
7. length of a kite string foot or yard
8. distance of an automobile race mile
9. height of a satellite mile
10. width of a camera inch
11. distance around a tree inch
12. distance walked in 2 minutes foot or yard

Choose the best estimate.

13. height of a 5-year-old child a. 48 in. b. 48 ft c. 48 yd
14. width of a bunk bed a. 30 yd b. 30 ft c. 30 in.
15. diameter of a pizza a. 1 in. b. 1 ft c. 1 yd
16. length of an ocean liner a. 150 mi b. 150 yd c. 150 ft
17. diameter of a volleyball a. 8 in. b. 8 ft c. 8 yd
18. length of a telephone cord a. 1 mi b. 1 yd c. 1 ft
19. thickness of a hamburger bun a. 2 in. b. 2 ft c. 2 yd
20. length of a shoelace a. 1 mi b. 1 yd c. 1 ft

Use a customary ruler to measure each line segment to the nearest $\frac{1}{8}$ inch.

21. ______ $2\frac{1}{2}$ in.
22. ______ $2\frac{7}{8}$ in.
23. ______ 1 in.
24. ______ $\frac{3}{8}$ in.
25. ______ 2 in.
26. ______ $2\frac{1}{8}$ in.
27. ______ 3 in.
28. ______ $1\frac{3}{8}$ in.

Solve.

29. In 7 minutes Pam runs 1 mile and Simi runs 2,000 yards. Who runs faster? Simi
30. Cal jumped 6 yards. Tami jumped 15 feet. How much farther did Cal jump than Tami? 3 ft or 1 yd

Applying Mathematics to the Work World

This optional page shows how mathematics is used in the work world and also provides a change of pace.

Objective Find typing rates per minute and typing rates per hour.

Using Discussion Discuss with students the various types of businesses and industries that employ data entry clerks.

Discuss the type of education or training that is necessary.

Activity

Using Computers Give students five minutes to type a paragraph and have them calculate how many words per minute they can type.

Also have students calculate how many words they can type in 10 minutes, 3 minutes, and 1 hour.

Computer Application

DATA PROCESSING

Peter Soanes works for the credit card division of a bank. He enters data for individual accounts into the computer system. He must type quickly and accurately.

Peter had to take a typing test to get his job in data processing. In five minutes he typed 325 words and made 5 errors. How many words did he type per minute? The formula used to determine typing speed is $\frac{\text{number of words}}{\text{number of minutes}}$ − number of errors.

number of minutes → $5\overline{)325}$ (quotient 65) ← number of words

Subtract one for each error made.

65 − 5 = 60

Peter typed 60 words per minute. At this rate, how many words can he type in one hour?

60 × 60 = 3,600

Peter can type 3,600 words in one hour.

Find the number of words typed per minute.

1. 300 words in 5 minutes with 6 errors **54 words per minute**
2. 210 words in 3 minutes with 4 errors **66 words per minute**

Find the number of words typed in one hour.

3. 48 words per minute **2,880 words per hour**
4. 63 words per minute **3,780 words per hour**
5. Richard plans to type his 8-page term paper on Wednesday. He can fit about 250 words on each page. If Richard types 50 words per minute, about how long will it take him to type 8 pages? **40 minutes**
6. Anne can type 50 words per minute. If she works for 4 hours and takes one 15-minute break, how many words can she type? **11,250 words**
7. Kevin can type 63 words per minute with 2 errors. If he works for 7 hours and takes two 20-minute breaks, how many words can he type? **23,180**

History Connection

Nonstandard Units of Length and Area

Linear measure probably developed around 10,000 to 8,000 BC before the development of measures of weight and capacity. Measurements were made by comparing distances with available units. People used their fingers, hands, arms, and feet as measuring tools. Since these differ from person to person, the units were not standard.

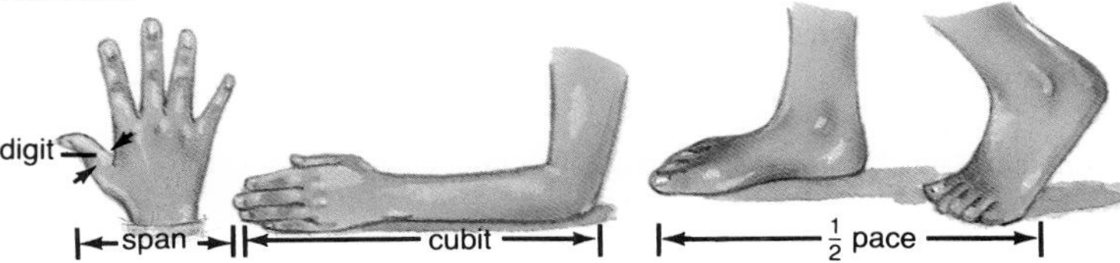

In early England, an acre of land was measured in different (nonstandard) ways. An acre might be the amount of land that could be plowed in one year, or it might be several lengths of a stick or rod. One old English system that may have been used through the Middle Ages is shown at the right.

1 hide = 120 acres
1 virgate = 30 acres or $\frac{1}{4}$ hide

Name the most reasonable unit for measuring each item. Use digit, span, cubit, or pace. **2. digit 4. span or digit**

1. length of a car **pace or cubit**
2. width of a paper clip
3. width of a skateboard **span**
4. height of a candle

Choose the best estimate for each of the following.

5. height of a desk — <u>a. 2 cubits</u> — b. 2 paces
6. width of a book — a. 1 cubit — <u>b. 1 span</u>
7. length of a chalkboard — <u>a. 4 paces</u> — b. 4 digits

Complete.

8. 1 hide = __?__ virgate(s) **4**
9. $\frac{1}{2}$ hide = __?__ acres **60**
10. 8 virgates = __?__ hide(s) **2**
11. Could one person's span equal another person's cubit? **possibly, large adult and small child**

Applying Mathematics to the Real World

This optional page shows the historical development of mathematics and also provides a change of pace.

Objective Familiarize students with the historical development of measures of length and area.

Using Discussion Discuss with students the measures that remain today whose units retain a name based on parts of the body, such as feet. Horses are measured in hands.

Activity

Using Manipulatives Have students measure the classroom, desks, pencils, paper, etc. using these nonstandard units of measure. Compare measurements. Discuss the differences in the answers and the reasons for these differences.

5-MINUTE CHECK-UP

(over Lesson 6-7)

Available on TRANSPARENCY 6-8.

Complete.

1. 56 yd = __?__ ft 168
2. 16 ft = __?__ in. 192
3. 48 in. = __?__ ft 4
4. 68 in. = __?__ ft __?__ in 5, 8
5. $3\frac{3}{4}$ ft = __?__ in. 45
6. 27 ft = __?__ yd 9

Extra Practice, Lesson 6-7, p. 464

INTRODUCING THE LESSON

Ask students to pretend they are employees of a moving company and they will pack the entire contents of the classroom to move to another building. Have them suggest the size and number of boxes needed.

TEACHING THE LESSON

Using Questioning

- Is it worthwhile to memorize the equivalencies of the customary system of weight and capacity? Explain. Yes, it is easier to buy what you need quickly and accurately when you can change amounts mentally.

Practice Masters Booklet, p. 67

Name ______________ Date ______________

PRACTICE WORKSHEET 6-8

Customary Units of Weight and Capacity

Complete.

1. 2 T = __4,000__ lb
2. 16,000 lb = __8__ T
3. 48 oz = __3__ lb
4. 1.5 lb = __24__ oz
5. 6 lb = __96__ oz
6. 3 T = __6,000__ lb
7. 3 lb 4 oz = __52__ oz
8. $2\frac{1}{2}$ T = __5,000__ lb
9. $\frac{3}{4}$ lb = __12__ oz
10. 8 qt = __2__ gal
11. 3 pt = __6__ c
12. 12 c = __6__ pt
13. 5 pt = __$2\frac{1}{2}$__ qt
14. 4 gal = __16__ qt
15. 18 qt = __72__ c
16. 64 fl oz = __2__ qt
17. $2\frac{3}{4}$ gal = __22__ pt
18. $\frac{1}{2}$ c = __4__ fl oz

Solve.

19. Darwin is buying fruit juice for a party. A 32-fluid-ounce bottle costs $1.19. How much will it cost for 20 one-cup servings? $5.95
20. An 8-ounce package of ground beef costs $1.12. How much will it cost for a pound and a half of ground beef? $3.36

6-8 CUSTOMARY UNITS OF WEIGHT AND CAPACITY

Objective

Change customary units of weight and capacity.

Melinda and Jeff are making ice cream. The recipe calls for 8 cups of heavy cream. At the store, heavy cream is available in pints. They change cups to pints to find the amount to buy.

The most commonly used customary units of weight are the **ounce (oz), pound (lb),** and **ton (T).**

1 pound = 16 ounces
1 ton = 2,000 pounds

The most commonly used customary units of capacity are the **fluid ounce (fl oz), cup (c), pint (pt), quart (qt),** and **gallon (gal).**

1 cup = 8 fluid ounces
1 pint = 2 cups
1 quart = 2 pints
1 gallon = 4 quarts

Changing customary units of weight and capacity is similar to changing customary units of length.

Method

1. Determine if you are changing from smaller to larger units or vice versa.
2. Divide to change from smaller to larger units. Multiply to change from larger to smaller units.
3. Determine the number by which to multiply or divide.

Examples

A ***Change 8 cups to pints.***

1. cups → pints
   smaller → larger
2. Divide.
3. There are 2 cups in a pint. So divide by 2.

8 c = __?__ pt ÷ 2

8 c = 4 pt

Melinda and Jeff buy 4 pints of heavy cream.

B ***Change 2 pounds to ounces.***

1. pounds → ounces
   larger → smaller
2. Multiply.
3. There are 16 ounces in a pound. So multiply by 16.

2 lb = __?__ oz × 16

2 lb = 32 oz

Guided Practice

Complete. 2. 8 4. 3 6. $1\frac{1}{2}$

Example A

1. 2 c = __?__ fl oz 16
2. 2 gal = __?__ qt
3. 40 fl oz = __?__ c 5
4. 12 qt = __?__ gal
5. 10 c = __?__ pt 5
6. 3 pt = __?__ qt

Example B

7. 2 T = __?__ lb 4,000
8. 48 oz = __?__ lb
9. 96 oz = __?__ lb 6
10. 5 lb = __?__ oz
11. 3 lb = __?__ oz 48
12. 128 oz = __?__ lb

8. 3 10. 80 12. 8

RETEACHING THE LESSON

For each of the following, tell whether you should multiply or divide and by what number.

1. Change pounds to ounces. multiply by 16
2. Change ounces to pounds. divide by 16
3. Change quarts to gallons. divide by 4
4. Change pounds to tons. divide by 2000

Reteaching Masters Booklet, p. 61

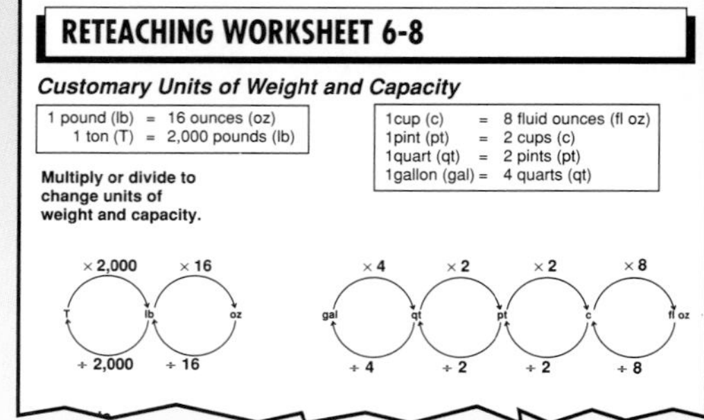

Name ______________ Date ______________

RETEACHING WORKSHEET 6-8

Customary Units of Weight and Capacity

1 pound (lb) = 16 ounces (oz)
1 ton (T) = 2,000 pounds (lb)

1cup (c) = 8 fluid ounces (fl oz)
1pint (pt) = 2 cups (c)
1quart (qt) = 2 pints (pt)
1gallon (gal) = 4 quarts (qt)

Multiply or divide to change units of weight and capacity.

Exercises

Practice

Complete.

13. 8 gal = ? qt 32 **14.** 14 pt = ? qt 7 **15.** 4 c = ? pt 2

16. 8 qt = ? gal 2 **17.** 5 pt = ? c 10 **18.** $2\frac{1}{2}$ pt = ? c 5

19. $5\frac{1}{2}$ qt = ? pt 11 **20.** $\frac{1}{2}$ lb = ? oz 8 **21.** 96 fl oz = ? c 12

22. 3 T = ? lb 6,000 **23.** 1 gal = ? pt 8 **24.** 1 qt = ? fl oz 32

25. 1 gal = ? c 16 **26.** 500 lb = ? T $\frac{1}{4}$ **27.** 1 pt = ? fl oz 16

28. 4 gal = ? c 64 **29.** 3 qt = ? fl oz 96 **30.** 6 gal = ? pt 48

31. How many pounds are in 112 ounces? 7 pounds

32. Change 2.5 quarts to cups. 10 cups

Number Sense

***Give the most appropriate unit of capacity. Use* ounces, cups, pints, quarts, *and* gallons.**

33. bath tub gallon **34.** soup spoon ounce

35. coffee cup cup, ounce **36.** apple cider quart, gallon

Applications

37. Bernie jogs 3 hours a week. If she loses an average of 8 ounces a week by jogging, how many weeks will it take her to lose 10 pounds? 20 weeks

38. Rich swims 6 hours a week in order to lose 10 ounces each week. How many weeks will it take him to lose $5\frac{1}{2}$ pounds? 8.8 weeks

39. about 59 pounds

39. A newborn hooded seal pup weighs about 45 pounds at birth. In the first 4 days of life the pup gains about 3.5 pounds a day. How much will the average pup weigh in 4 days?

40. Ken's recipe for chili calls for $2\frac{1}{2}$ cups of tomato sauce. Ken doubles the recipe for his party Friday night. If Ken buys three 14-ounce cans, will he have enough tomato sauce? yes, 3 × 14 = 42

Using Variables

Evaluate each expression if c = 8, q = 32, and p = 16.

41. $4c$ 32 **42.** $q \div c$ 4 **43.** $p \div c$ 4 **44.** $q \div p$ 2

Talk Math

See margin.

45. Explain in your own words how to change 8 gallons to the equivalent number of ounces.

Critical Thinking

46. You are arranging the games for the volleyball intramural finals. There are 8 teams ranked 1 to 8 in the tournament. Draw a schedule tree like the one at the right and assign the teams so that the two top ranked teams might play the final match against each other.

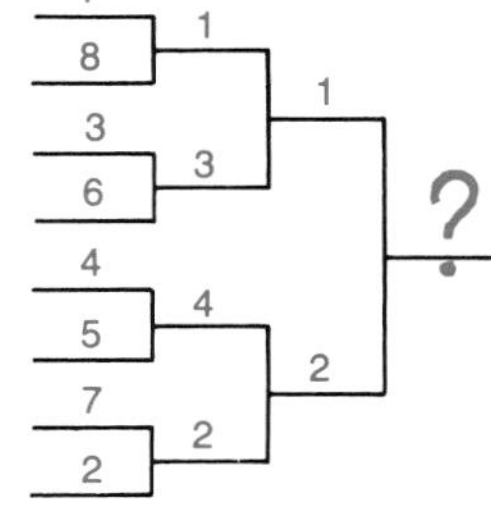

Chalkboard Examples

Example A *Complete.*

1. 48 oz = ? 3 lb

2. 2 c = ? q 0.5 or $\frac{1}{2}$ qt

Example B *Complete.*

3. 2.5 lb = ? oz 40 oz

4. 0.5 gal = ? qt 2 qt

Additional examples are provided on TRANSPARENCY 6-8.

EVALUATING THE LESSON

Check for Understanding Use the Guided Practice Exercises to check for student understanding.

Closing the Lesson Have students explain in their own words how to change from one customary unit of capacity to another.

EXTENDING THE LESSON

Enrichment Have students find the relative pull of gravity on Jupiter. Then have students find their own weight on Jupiter, the weight of their pet, the weight of their textbook, and so on. weight on Jupiter = 2.64 × weight on Earth

Enrichment Masters Booklet, p. 61

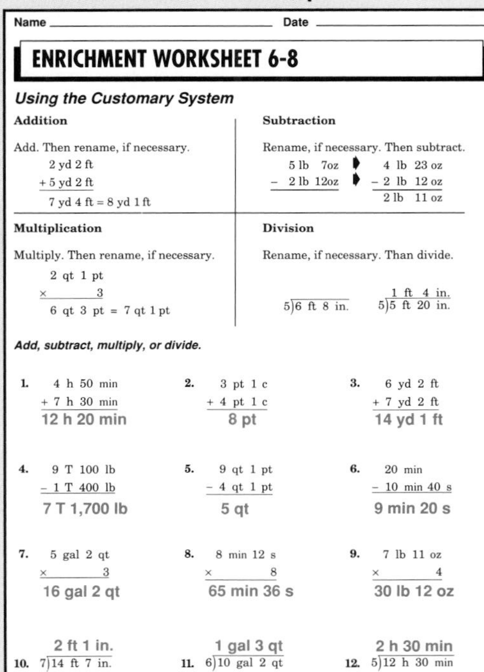

Name ______ Date ______

ENRICHMENT WORKSHEET 6-8

Using the Customary System

Addition

Add. Then rename, if necessary.

2 yd 2 ft
+ 5 yd 2 ft
7 yd 4 ft = 8 yd 1 ft

Subtraction

Rename, if necessary. Then subtract.

5 lb 7oz → 4 lb 23 oz
− 2 lb 12oz → − 2 lb 12 oz
2 lb 11 oz

Multiplication

Multiply. Then rename, if necessary.

2 qt 1 pt
× 3
6 qt 3 pt = 7 qt 1 pt

Division

Rename, if necessary. Than divide.

5)6 ft 8 in. → 1 ft 4 in. / 5)5 ft 20 in.

Add, subtract, multiply, or divide.

1. 4 h 50 min + 7 h 30 min = 12 h 20 min
2. 3 pt 1 c + 4 pt 1 c = 8 pt
3. 6 yd 2 ft + 7 yd 2 ft = 14 yd 1 ft
4. 9 T 100 lb − 1 T 400 lb = 7 T 1,700 lb
5. 9 qt 1 pt − 4 qt 1 pt = 5 qt
6. 20 min − 10 min 40 s = 9 min 20 s
7. 5 gal 2 qt × 3 = 16 gal 2 qt
8. 8 min 12 s × 8 = 65 min 36 s
9. 7 lb 11 oz × 4 = 30 lb 12 oz
10. 7)14 ft 7 in. = 2 ft 1 in.
11. 6)10 gal 2 qt = 1 gal 3 qt
12. 5)12 h 30 min = 2 h 30 min

APPLYING THE LESSON

Independent Practice

Homework Assignment	
Minimum	Ex. 2-44 even; 45-46
Average	Ex. 1-35 odd; 37-46
Maximum	Ex. 1-46

Additional Answers

45. Find the number of fluid ounces in one gallon, 128, and multiply by 8.

5-MINUTE CHECK-UP

(over Lesson 6-8)

Available on TRANSPARENCY 6-9.

Complete.

1. 160 oz = ? lb 10
2. 20 oz = ? lb ? oz 1, 4
3. 5 pt = ? c 10
4. 16 qt = ? gal 4
5. 4 lb = ? oz 64
6. 2 gal = ? pt 16
7. 4 c = ? qt 1
8. 8 c = ? pt 4

Extra Practice, Lesson 6-8, p. 465

INTRODUCING THE LESSON

Discuss the effects of temperature change on the environment and the theory of global warming, that is, air pollution and changes in the flora of the globe contribute to higher average temperatures.

TEACHING THE LESSON

Using Calculators Have students use calculators for Exercises 1-34 on pp. 196 and 197. Use the example in the Calculator Exercises on p. 197 to demonstrate the use of the calculator.

Practice Masters Booklet, p. 68

Name ______________ Date ______________

PRACTICE WORKSHEET 6-9

Formulas and Temperature

Choose the better temperature.

1. drinking water, 5°C or 45°C
2. a summer day, 25°C or 75°C
3. hot chocolate, 50°F or 105°F
4. a fall day, –12°C or 12° C
5. snow, ⁻42°F or 30°F
6. a comfortable room, 68°F or 20°F
7. a baking oven, 205°C or 500°C
8. a car's engine at high speed, 90°C or 32°C

Find the equivalent temperature to the nearest degree.

9. 7°C 45°F
10. 35°C 95°F
11. 272°C 522°F
12. 58°C 136°F
13. 163°C 325°F
14. 118°C 244°F
15. 54°C 129°F
16. 67°C 153°F
17. 12°C 54°F

Find the equivalent temperature to the nearest degree.

18. 59°F 15°C
19. 93°F 34°C
20. 446°F 230°C
21. 26°F –3°C
22. 84°F 29°C
23. 41°F 5°C
24. 72°F 22°C
25. 122°F 50°C
26. 107°F 42°C

Solve.

27. Would you need to wear a sweater if the room temperature was 25°C? no
28. Would you need to turn on the heater in the car if the temperature was 5°C? yes
29. The noon temperature was reported at 83°F. The 6:00 P.M. temperature was 57°F. What was the drop in temperature? 26°F
30. A Canadian recipe called for an oven temperature of 230°C. About what temperature would this be in degrees Fahrenheit? about 450°F

6-9 FORMULAS AND TEMPERATURE

Objective

Estimate temperature and use formulas to change Celsius to Fahrenheit and vice versa.

Jonine received a letter from her pen pal Bryan in Australia. He told her that he had been on vacation in Sidney, and the temperature had been about 25°C every day. Jonine changed the Celsius temperature to Fahrenheit.

Temperature is commonly measured in degrees Celsius (°C) or degrees Fahrenheit (°F). A Celsius thermometer and a Fahrenheit thermometer are shown at the right.

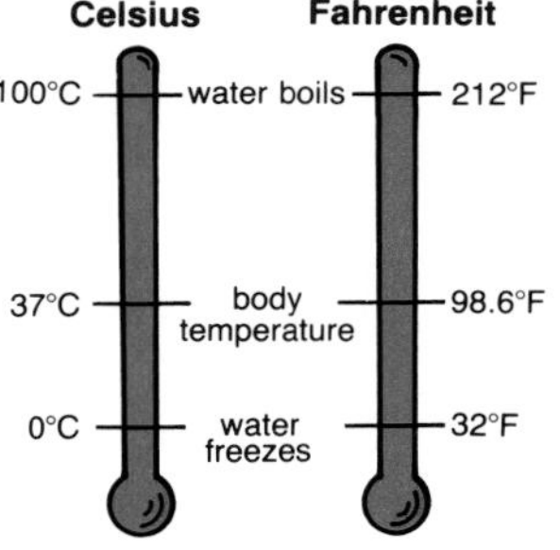

The formulas for changing temperature scales are given below.

Celsius to Fahrenheit: $F = \frac{9}{5} \times C + 32$

Fahrenheit to Celsius: $C = \frac{5}{9} \times (F - 32)$

Method

1. Determine if you are changing from Celsius to Fahrenheit units or vice versa.
2. Choose the correct formula.
3. Perform the required calculations.

Example A

Change 25°C to degrees Fahrenheit.

1. Celsius → Fahrenheit
2. Use $F = \frac{9}{5} \times C + 32$.
3. $F = \frac{9}{5} \times 25 + 32$ Replace C with 25.

$F = 45 + 32$ $\frac{9}{5} \times 25 = 45$

$F = 77$

The temperature in Sidney was about 77°F every day.

Example B

Change 68°F to degrees Celsius.

1. Fahrenheit → Celsius
2. Use $C = \frac{5}{9} \times (F - 32)$.
3. $C = \frac{5}{9} \times (68 - 32)$ Replace F with 68.

$C = \frac{5}{9} \times (36)$ $\frac{5}{\cancel{9}_1} \times \frac{\cancel{36}^4}{1} = 20$

$C = 20$ 68°F is the same as 20°C.

Guided Practice

Choose the better temperature for each activity.

1. ice hockey, 120°F or 20°F
2. swimming, 30°C or 10°C
3. snow skiing, 59°F or 23°F
4. jogging, 70°F or 10°F

Example A ***Find the equivalent Fahrenheit temperature to the nearest degree.***

5. 60°C 140°F 6. 15°C 59°F 7. 50°C 122°F 8. 85°C 185°F 9. 30°C 86°F

Example B ***Find the equivalent Celsius temperature to the nearest degree.***

10. 41°F 5°C 11. 59°F 15°C 12. 149°F 65°C 13. 239°F 115°C 14. 194°F 90°C

RETEACHING THE LESSON

Have the students work in pairs and decide which formula should be used to solve each problem. Then, have students solve the problems.

Change each to degrees Fahrenheit.

1. 25°C 77°F 2. 50°C 122°F

Change each to degrees Celsius.

3. 41°F 5°C 4. 95°F 35°C

Reteaching Masters Booklet, p. 62

Name ______________ Date ______________

RETEACHING WORKSHEET 6-9

Formulas and Temperature

Most countries that use the metric system use the Celsius thermometer to measure temperature.

It is named after Anders Celsius, a Swedish astronomer, who invented it in 1742.

To change from degrees Celsius to degrees Fahrenheit, multiply degrees Celsius by $\frac{9}{5}$ and add 32.

°C 100° ← water boils → 212° °F

0° ← water freezes → 32°

Some countries still use the Fahrenheit thermometer to measure temperature.

It is named after Daniel Fahrenheit, a German instrument maker, who invented it in 1714.

To change from degrees Fahrenheit to degrees Celsius, subtract 32 and multiply the difference by $\frac{5}{9}$.

Change 15°C to degrees Fahrenheit.

Change 86°F to degrees Celsius.

Exercises

15. 212°F

Practice

Find the equivalent Fahrenheit temperature to the nearest degree.

15. 100°C **16.** 5°C **41°F** **17.** 53°C **127°F** **18.** 47°C **117°F** **19.** 122°C **252°F**
20. 27°C **81°F** **21.** 13°C **55°F** **22.** 18°C **64°F** **23.** 63°C **145°F** **24.** 33°C **91°F**

Find the equivalent Celsius temperature to the nearest degree.

25. 95°F **35°C** **26.** 104°F **40°C** **27.** 100°F **38°C** **28.** 77°F **25°C** **29.** 212°F **100°C**
30. 32°F **0°C** **31.** 90°F **32°C** **32.** 80°F **27°C** **33.** 34°F **1°C** **34.** 87°F **31°C**

Applications

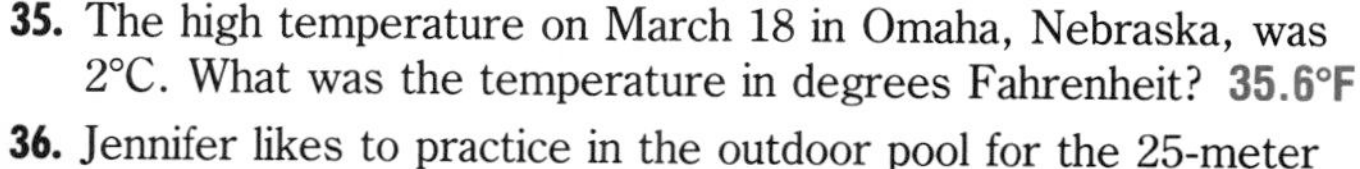

35. The high temperature on March 18 in Omaha, Nebraska, was 2°C. What was the temperature in degrees Fahrenheit? **35.6°F**

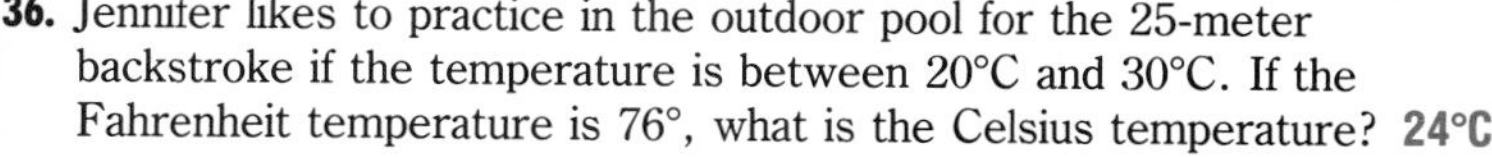

36. Jennifer likes to practice in the outdoor pool for the 25-meter backstroke if the temperature is between 20°C and 30°C. If the Fahrenheit temperature is 76°, what is the Celsius temperature? **24°C**

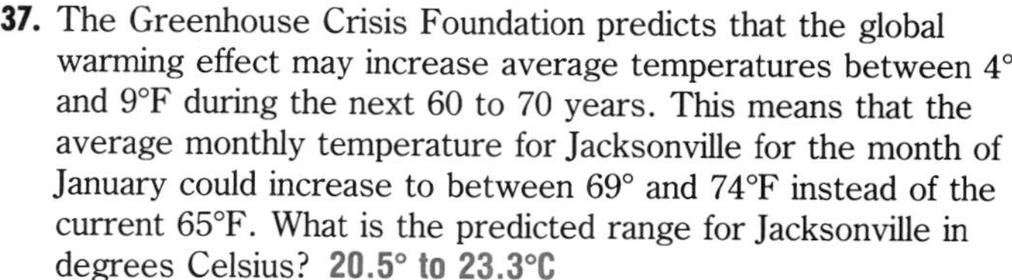

37. The Greenhouse Crisis Foundation predicts that the global warming effect may increase average temperatures between 4° and 9°F during the next 60 to 70 years. This means that the average monthly temperature for Jacksonville for the month of January could increase to between 69° and 74°F instead of the current 65°F. What is the predicted range for Jacksonville in degrees Celsius? **20.5° to 23.3°C**

Cooperative Groups

38. In groups of three or four, collect pictures from magazines of the group's favorite activities or sports and place each picture on a separate piece of paper. On index cards, write appropriate Celsius temperatures for each picture. See if other groups can match the temperature with the activity. **See students' work.**

Critical Thinking

39. Hector was leading his flock of geese south for the winter. There were 5 geese behind him in each wing of the V-formation. Hector dropped back after 1 hour to the tail of the right-hand section. Alternating one leader from each side of the formation, each new leader heads the flock for one hour. In how many hours will Hector have to lead again? **10 hours**

Calculator

Find the equivalent Fahrenheit temperature for 60°C.

9 [÷] 5 [×] 60 [+] 32 [=] 140

60°C is equivalent to 140°F.

Find the equivalent Celsius temperature for 140°F.

140 [−] 32 [=] 108 [×] 5 [÷] 9 [=] 60

140°F is equivalent to 60°C.

Complete. Use a calculator.

40. 32°C = ? F **90°** **41.** 70°C = ? F **158°**
42. 99°F = ? C **37°** **43.** 212°F = ? C **100°**

Chalkboard Examples

Example A *Find the equivalent Fahrenheit temperature to the nearest degree.*

1. 53°C **127°F** **2.** 5°C **41°F**
3. 90°C **194°F** **4.** 200°C **392°F**

Example B *Find the equivalent Celsius temperature to the nearest degree.*

5. 77°F **25°C** **6.** 100°F **38°C**
7. 23°F **−5°C** **8.** 10°F **−12°C**

Additional examples are provided on TRANSPARENCY 6-9.

EVALUATING THE LESSON

Check for Understanding Use the Guided Practice Exercises to check for student understanding.

Error Analysis Watch for students who ignore the parentheses in the formula for Fahrenheit temperature. Remind students to write the formula with variables before replacing with numbers.

Closing the Lesson Have students suggest different activities that can be performed at 0°, 15°, and 30° C.

EXTENDING THE LESSON

Enrichment Masters Booklet, p. 62

Name ______ Date ______

ENRICHMENT WORKSHEET 6-9

Converting Temperatures Using a Graph

Use the graph to help you complete the following. Give your answers to the nearest degree.

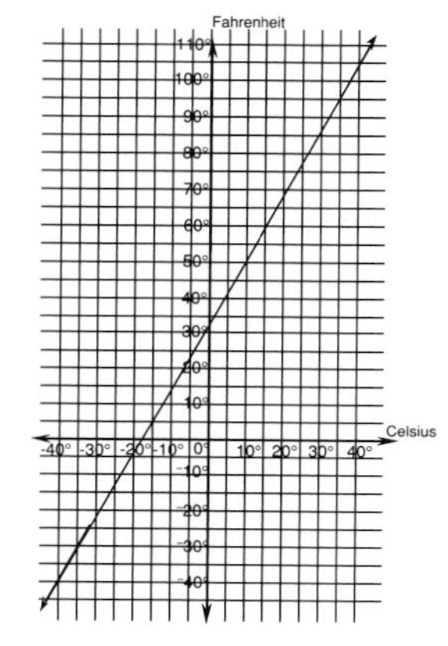

1. 40°C = 105 °F
2. −30°C = −22 °F
3. 40°F = 4 °C
4. 100°F = 37 °C
5. 20°C = 69 °F
6. 32°F = 0 °C
7. −20°C = −28 °F
8. 10°C = 51 °F
9. 80°F = 26 °C
10. −5°C = 24 °F
11. 60°F = 15 °C
12. 35°C = 97 °F
13. 0°F = −18 °C
14. 25°C = 78 °F
15. 18°F = −7 °C
16. −35°F = −37 °C
17. 105°F = 40 °C
18. 70°F = 20 °C
19. −40°C = −40 °F
20. −40°F = −40 °C

APPLYING THE LESSON

Independent Practice

Homework Assignment	
Minimum	Ex. 1-37 odd; 38-43
Average	Ex. 2-34 even; 35-43
Maximum	Ex. 1-43

5-MINUTE CHECK-UP

(over Lesson 6-9)

Available on TRANSPARENCY 6-10.
Find the equivalent Fahrenheit temperature to the nearest degree.

1. 47°C **117°F** **2.** 30°C **86°F**

Find the equivalent Celsius temperature to the nearest degree.

3. 212°F **100°C** **4.** 100°F **38°C**

Extra Practice, Lesson 6-9, p. 465

▶ INTRODUCING THE LESSON

Have students work in groups of three or four and find an average for the number of blinks of the eye for 15 seconds. Have students find the number of blinks in one class period.

▶ TEACHING THE LESSON

Using Vocabulary Discuss with students the meanings of the terms quarter-past, half-past, and quarter 'til. Ask students if they think digital clocks have affected the way we talk about time.

Practice Masters Booklet, p. 69

Name ________________ Date ________________

PRACTICE WORKSHEET 6-10

Measuring Time

Complete.

1. 45 min = 2,700 s	2. 4 d = 5,760 min
3. 72 h = 3 d	4. 3 h = 180 min
5. 8 h = 28,800 s	6. 7 d = 10,080 min
7. 75 min = 1 h 15 min	8. 8 h 20 min = 500 min
9. 15 h = 900 min	10. 8 d = 192 h
11. 3,000 min = 2 d 2 h	12. 10,080 s = 2 h 48 min

Find the elapsed time.

13. from 7:15 A.M. to 10:36 A.M. 3 h 21 min	14. from 2:48 P.M. to 9:16 P.M. 6 h 28 min
15. from 3:15 P.M. to noon 20 h 45 min	16. from midnight to 5:57 A.M. 5 h 57 min
17. from 8:40 A.M. to 5:30 P.M. 8 h 50 min	18. from 9:45 P.M. to 7:50 A.M. 10 h 5 min

Solve.

19. Bill punched the time clock at the factory at 7:55 A.M. He worked for 8 hours and 25 minutes and had a 45-minute lunch break. What time did he punch out? 5:05 P.M.

20. Lorie worked six days last week. She worked 7 hours and 15 minutes each day. She gets paid overtime for time worked over 40 hours in a week. How much overtime did she work? $3\frac{1}{2}$ h

6-10 MEASURING TIME

Objective
Change from one unit of time to another.

As part of her weight reducing program, Pat plays tennis for one hour three times a week. In three weeks Pat can lose about one pound. How many minutes a week does Pat play tennis?

Units of time are the same in the customary and metric systems. Common units of time are shown at the right.

1 day (d) = 24 hours (h)
1 hour (h) = 60 minutes (min)
1 minute (min) = 60 seconds (s)

Method

1. Determine if you are changing from smaller to larger units or vice versa.
2. Divide to change from smaller to larger units. Multiply to change from larger to smaller units.
3. Determine the number by which to multiply or divide.

Example A ***Change 3 hours to minutes.***

1. hours → minutes
2. larger → smaller
3. Multiply by 60.

3 h = ? min
3 × 60 = 180
3 h = 180 min

Pat plays tennis for 180 minutes a week.

Example B ***Change 325 seconds to minutes.***

1. seconds → minutes
2. smaller → larger
3. Divide by 60.

325 s = ? min ÷ 60
325 s = 5 min 25 s

$$\begin{array}{r} 5 \text{ R25} \\ 60\overline{)325} \\ -300 \\ \hline 25 \end{array}$$

Elapsed time is the amount of time that has passed from one time to another. Remember to watch for A.M. and P.M. changes when you determine elapsed time.

Example C ***Find the elapsed time from 11:00 A.M. to 4:30 P.M.***

Beginning
11:00 A.M.

11:00 A.M. to 12:00 noon is 1 hour

12:00 noon to 4:30 P.M. is $4\frac{1}{2}$ hours.

Ending
4:30 P.M.

The elapsed time is $1 + 4\frac{1}{2}$ or $5\frac{1}{2}$ hours.

Guided Practice ***Name the larger measurement.***

1. 60 min or 1 h 5 min **2.** 3 days or 60 h **3.** 6 h or 350 min
4. 8 min or 500 s **5.** 30 h or 2 days **6.** 4 h 10 min or 300 min

RETEACHING THE LESSON

Give the students models of clock faces showing a certain time. Ask students what time it will be after a certain amount of time has elapsed.

What time will it be in 2 hours 25 minutes? **1:30**

Reteaching Masters Booklet, p. 63

Name ________________ Date ________________

RETEACHING WORKSHEET 6-10

Measuring Time

1 day (d) = 24 hours (h)
1 hour (h) = 60 minutes (min)
1 minute (min) = 60 seconds (s)

Multiply or divide units of time.
d → h: × 24, h → min: × 60, min → s: × 60
s → min: ÷ 60, min → h: ÷ 60, h → d: ÷ 24

Elapsed time is the amount of time that passes from one time to another.

If both times are A.M. or both are P.M., you can find elapsed time by subtracting the earlier time from the later time.

If one time is A.M. and one is P.M., you need to add the elapsed

Complete.

Example A
7. 5 min = ? s **300** **8.** 8 h = ? min **480** **9.** 4 d = ? h **96**

Example B
10. 48 h = ? d **2** **11.** 360 s = ? min **6** **12.** 1,440 min = ? d **1**

Find the elapsed time.

Example C
13. from 4:10 P.M. to 7:35 P.M. **3 h 25 min** **14.** from 11:25 P.M. to 1:48 A.M. **2 h 23 min**

Exercises

Practice

Complete.

15. 10 d = ? h **240** **16.** 345,600 s = ? d **4** **17.** 72 h = ? d **3**

18. $\frac{1}{2}$ h = ? min **30** **19.** 45 min = ? h $\frac{3}{4}$ **20.** $\frac{1}{3}$ h = ? min **20**

21. 1 day = ? min **1,440** **22.** 1 h 30 min = ? min **90** **23.** $4\frac{1}{2}$ min = ? s **270**

Find the elapsed time. **24. 2 h 42 min** **25. 13 h 30 min**

24. from 3:30 A.M. to 6:12 A.M. **25.** from 10:30 A.M. to midnight

Applications

26. Nina wants to get to the airport in plenty of time so she won't miss her plane. If it takes about 35 minutes to drive to the airport, about 20 minutes to park the car and ride the shuttle, and she plans to check in 45 minutes early for her 8:20 A.M. flight, what time should she leave home for the airport? **about 6:40 A.M.**

27. Janet Evans won the 400-meter freestyle in the 1988 Summer Olympics with a time of 4 minutes 3.85 seconds. If you disregard the hundredths of a second, how many seconds is this? **243 sec**

28. Brandon likes to sleep about 7 hours a night. If he went to bed at 10:45 P.M. and got up at 6:15 A.M., how much more or less than 7 hours did he sleep? **30 min more**

Make Up a Problem

29. Using your school schedule, make up two problems that relate to time. One problem must use elapsed time. One problem must involve changing units of time. **See students' work.**

Critical Thinking

30. On a digital clock during a 12-hour period, which numeral appears more often, *1* or *0*? **numeral 0**

31. If a standard chiming clock became a 24-hour clock, it would strike 24 times at midnight. How many times would the clock strike during one whole day? **300 times**

Mental Math

Solve Exercise 32 by counting. Count the hours, then count the minutes. For example, Jamie checked out a video that has a running time of 97 minutes. If she begins watching it at 8:30 P.M., what time will the video be over?

Think: 97 minutes is 7 minutes more than an hour and a half. An hour and a half from 8:30 P.M. is 10:00 P.M., so 7 more minutes will be 10:07 P.M.

32. Larry begins watching a video movie at 7:20 P.M. His brother wants to know if he can watch his favorite sitcom on the same T.V. If the video is 128 minutes long, when will it be over? **9:28 P.M.**

Chalkboard Examples

Example A *Complete.*

1. 6 h = ? min **360**

2. 3 d = ? min **4,320**

Example B *Complete.*

3. 720 s = ? min **12**

4. 5,400 s = ? h **1.5**

Example C *Find the elapsed time.*

5. From 11:20 A.M. to 5:15 P.M. **5 hr 55 min**

Additional examples are provided on TRANSPARENCY 6-10.

EVALUATING THE LESSON

Check for Understanding Use the Guided Practice Exercises to check for student understanding.

Closing the Lesson Ask students to state in their own words how to find elapsed time.

EXTENDING THE LESSON

Enrichment Have students use their class schedule to find the elapsed time from the beginning of their first class to lunch time. Then find the elapsed time from the beginning of their first class after lunch to the end of their last class.

Enrichment Masters Booklet, p. 63

Name ______ Date ______

ENRICHMENT WORKSHEET 6-10

Other Units of Time

In addition to the units of time used in everyday life, there are others that are very interesting. Some are units to measure very, very small time intervals and others, of course, are units to measure very large time intervals. For example,

1 chronon = 1 billionth of a trillionth of a second
1 quartan = 4 days
1 bimestrial = 2 months

Use the metric prefixes on pages 178 of your textbook and a dictionary to find definitions for each unit of time.

1. picosecond — 1 trillionth of a second
2. nanosecond — 1 billionth of a second
3. microsecond — 1 millionth of a second
4. millisecond — 1 thousandth of a second
5. fortnight — 2 weeks
6. decade — 10 years
7. century — 100 years
8. bicentenary — 200 years
9. tercentenary — 300 years
10. millennium — 1,000 years

APPLYING THE LESSON

Independent Practice

Homework Assignment	
Minimum	Ex. 1-31 odd; 32
Average	Ex. 2-24 even; 26-32
Maximum	Ex. 1-32

5-MINUTE CHECK-UP

(over Lesson 6-10)

Available on TRANSPARENCY 6-1.

Complete.

1. 22 min = _?_ s **1,320**

2. 12 h = _?_ min **720**

Find the elapsed time.

3. from 10:30 A.M. to 2:15 P.M. **3 h 45 min**

Extra Practice, Lesson 6-10, p. 465

▶ INTRODUCING THE LESSON

Ask students if they have "punched a clock." Students may have experience with time cards, or they may have a parent, relative, or friend who uses a time card. Allow students time to discuss their experiences.

▶ TEACHING THE LESSON

Materials: index cards (optional)

Using Cooperative Groups In groups of three or four, have students create a one-week time card on an index card. The time card must have:

- hours that total 8 hours a day
- hours that total a 40-hour week for 5 working days
- 2 IN and 2 OUT times per day

Have groups exchange time cards and check the number of hours.

Practice Masters Booklet, p. 70

PRACTICE WORKSHEET 6-11

Applications: Time Cards

Compute the working hours for each day.

1.

IN	OUT
7:00	11:00
12:00	16:00

8 h

2.

IN	OUT
9:30	12:30
13:00	16:30

6 h 30 min

3.

IN	OUT
8:45	12:15
13:15	16:15

6 h 30 min

4.

IN	OUT
9:00	11:50
12:30	17:00

7 h 20 min

5.

IN	OUT
7:50	11:30
12:15	17:05

8 h 30 min

6.

IN	OUT
10:35	14:00
14:30	19:15

8 h 10 min

7.

IN	OUT
7:45	11:45
12:15	16:45

8 h 30 min

8.

IN	OUT
13:30	16:45
17:15	21:20

7 h 20 min

9.

IN	OUT
8:25	12:00
12:45	17:15

8 h 5 min

10.

IN	OUT
12:20	16:15
16:45	21:30

8 h 40 min

11.

IN	OUT
7:15	11:30
12:15	16:00

8 h

12.

IN	OUT
11:50	17:05
17:45	20:15

7 h 45 min

Solve.

13. Michael's time card showed IN times of 8:30 and 12:45 and OUT times of 12:05 and 16:55 for Tuesday. If Michael earns $5.25 an hour, how much did he earn on Tuesday? **$40.69**

14. Stacy gets paid overtime for any hours worked over 8 hours in a day. On Friday Stacy's time card showed IN times of 7:55 and 13:05 and OUT times of 12:00 and 17:30. Did Stacy work overtime on Friday? If, so how much overtime? **yes; 30 min**

Applications

6-11 TIME CARDS

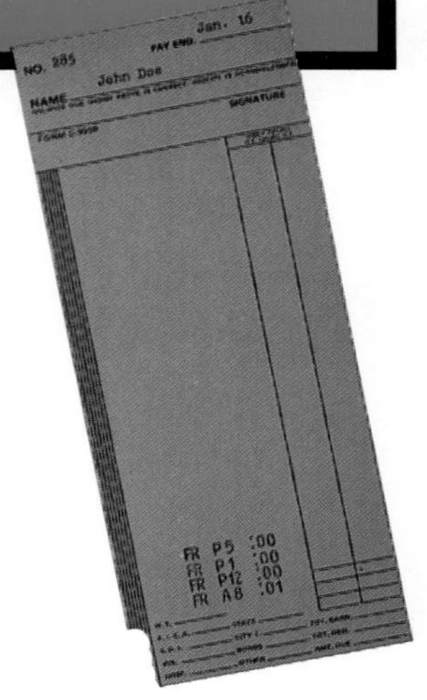

Objective

Compute the number of hours worked using 24-hour notation.

April punches a time clock when she begins and ends her work shift. On Wednesday her time card showed IN times of 7:45 and 12:30 and OUT times of 12:00 and 14:30. How many hours did she work on Wednesday?

Finding the number of hours worked from time cards is similar to finding elapsed time. Many time clocks print the time in 24-hour notation. It is easier to find elapsed time using 24-hour notation because you do not have to adjust for A.M. to P.M. changes in the same 24-hour day.

The A.M. times are 00:01 to 12:00. The P.M. times are 12:01 to 24:00. For example, 14:30 means the same as 2:30 P.M.

April subtracts to find the amount of time she worked on Wednesday.

Before Lunch

OUT	12:00 (1 60)
IN	− 7:45
	4:15

Rename 1 h as 60 min.

After Lunch

OUT	14:30
IN	− 12:30
	2:00

Total Time

4 h 15 min	before lunch
2 h 00 min	after lunch
6 h 15 min	

April worked 6 hours and 15 minutes on Wednesday.

Understanding the Data

- What time did April punch in Wednesday morning? **7:45 A.M.**
- What P.M. time corresponds to 17:00 in 24-hour notation? **5:00 P.M.**
- What P.M. time corresponds to 21:30 in 24-hour notation? **9:30 P.M.**

Check for Understanding

Change each 12-hour time to 24-hour notation.

1. 1:35 P.M. **13:35** **2.** 7:20 P.M. **19:20** **3.** 2:10 A.M. **02:10**

4. 6:40 A.M. **06:40** **5.** 4:50 P.M. **16:50** **6.** 12:10 A.M. **00:10**

Compute the working hours for each day.

7.

IN	OUT
8:00	12:00
13:00	17:00

8 h

8.

IN	OUT
8:30	12:00
13:00	17:30

8 h

9.

IN	OUT
7:45	11:45
13:15	17:15

8 h

RETEACHING THE LESSON

Compute the working hours for each day.

1.

IN	OUT
7:00	11:30
12:45	16:00

$7\frac{3}{4}$ hrs

2.

IN	OUT
6:45	11:30
12:15	15:00

$7\frac{1}{2}$ hrs

Reteaching Masters Booklet, p. 64

RETEACHING WORKSHEET 6-11

Applications: Time Cards

Mario's time card from Monday shows that he started work at 8:30, left for lunch at 11:45, returned from lunch at 12:30, and left work for the day at 17:15.

IN	OUT
8:30	11:45
12:30	17:15

17:15 is an example of time written in 24-hour notation.

12-hour time	24-hour time
7:00 A.M.	07:00
12:00 noon	12:00
12:30 P.M.	12:30
5:15 P.M.	17:15

12 hours before noon plus 5 hours after noon

To find the hours Mario worked on Monday.

subtract, then add and rename if necessary.

Before Lunch After Lunch Total Working Hours

Exercises

Compute the working hours for each day.

Practice

10.

IN	OUT
7:15	12:00
13:30	18:00

9 h 15 min

11.

IN	OUT
9:30	13:00
13:30	16:30

6 h 30 min

12.

IN	OUT
9:15	12:15
13:45	18:15

7 h 30 min

13.

IN	OUT
13:30	18:30
19:00	22:10

8 h 10 min

14.

IN	OUT
10:50	14:15
15:00	19:20

7 h 45 min

15.

IN	OUT
8:45	11:10
11:40	16:25

7 h 10 min

Applications

16. Sherri's time card showed IN times of 09:00 and 13:15 and OUT times of 12:15 and 16:30 for Thursday. If she makes $4.75 an hour, how much will Sherri earn Thursday? **$30.88**

17. Howard was scheduled to work from 2:00 P.M. on Saturday until 7:00 P.M. If his time card shows he punched in at 14:20, was Howard on time? If not, how late was he? **no; 20 minutes late**

Critical Thinking

18. A jeweler has 6 lengths of gold chain. Each length has 4 links. To join any 2 lengths, a link must first be cut, looped through a closed link, and soldered together. It costs $5 to cut a link and $10 to solder it together. What is the least expensive way to rejoin the six lengths into a 24-link length? How much will it cost? **See margin.**

Use the table for Exercises 19–21.

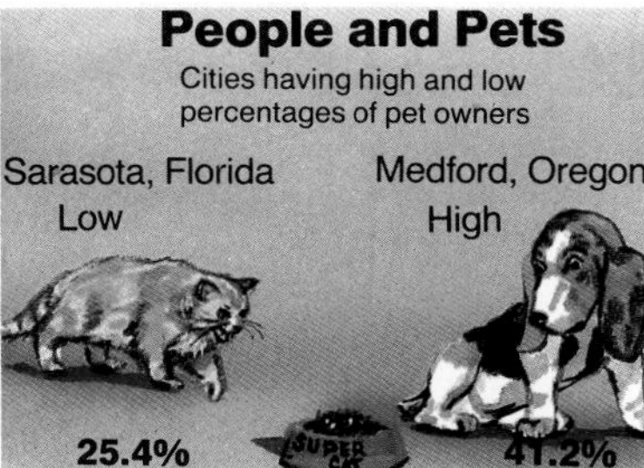

Suppose

19. Suppose the number of households in Medford, Oregon, is 675. *About* how many households own pets? (41.2% = 0.412) **about 278 households**

20. Suppose *about* one-sixth of the households that own pets in Medford own more than one pet. How many households is this? **about 46 households**

Collect Data

21. Survey the members of your class to find how many own pets. Find the number of pet owners in the class compared to the total number in the class. Is the portion of the class that owns pets closer to the portion of Medford, Oregon (0.412), or Sarasota, Florida (0.254)? **Answers may vary. See students' work.**

Mixed Review

Lesson 4-11

Add. **See margin.**

22. $6\frac{1}{8} + 4\frac{2}{8}$ **23.** $15\frac{1}{4} + 6\frac{2}{3}$ **24.** $5\frac{2}{5} + 3\frac{8}{9}$ **25.** $9\frac{2}{7} + 8\frac{2}{3}$

Lesson 5-3

Multiply. **See margin.**

26. $4\frac{1}{5} \times 3\frac{2}{3}$ **27.** $6\frac{2}{5} \times 5$ **28.** $2\frac{1}{4} \times 3\frac{1}{3}$ **29.** $1\frac{1}{8} \times 3\frac{2}{3}$

Lesson 6-3

30. At the print shop they print posters that are 1.2 meters long. If 300 posters are printed end to end, what is the length of paper, in km, needed to print the posters? **0.36 km**

Chalkboard Examples

Compute the working hours for each day.

1.

IN	OUT
8:30	12:15
15:30	19:00

7 h 15 min

2.

IN	OUT
22:30	2:45
3:15	7:00

8 h

Additional examples are provided on TRANSPARENCY 6-11.

EVALUATING THE LESSON

Check for Understanding Use the Guided Practice Exercises to check for student understanding.

Error Analysis Watch for students who rename 1 hour as 10 minutes as if they were subtracting whole numbers instead of minutes. Remind students to borrow 1 hour or *60 minutes.*

Closing the Lesson Have students explain in their own words how a clock is marked in 24-hour notation.

EXTENDING THE LESSON

Enrichment Masters Booklet, p. 64

ENRICHMENT WORKSHEET 6-11

Bus Schedules

The timetable shows the times that the Number 10 bus leaves (LV) from three downtown locations and the times that it is scheduled to arrive (AR) at four westside locations on Saturday. Heavy type indicates P.M. times. A blank indicates that the bus does not stop at this location.

Bus Number 10
WESTBOUND Saturdays

LV Jackson & Wisconsin	LV 12th & Wisconsin	LV 35th & Wisconsin	AR 67th & Fairview	AR County Hospital	AR 124th Street	AR Brookfield Square
AM						
4:45	5:07	5:15	5:22			
5:18	5:26	5:34		5:46	5:54	
5:38	5:47	5:55		6:07	6:15	
5:47	5:56	6:04	6:13			
5:59	6:08	6:16		6:28	6:36	
6:17	6:26	6:34	6:43			
6:18	6:27	6:35		6:47	6:55	
6:39	6:48	6:56		7:08	7:16	
6:45	6:54	7:02	7:11			
7:00	7:09	7:17		7:29	7:37	
7:14	7:23	7:31	7:40			
7:21	7:30	7:38		7:50	7:58	
7:42	7:51	7:59		8:11	8:19	
7:44	7:53	8:01	8:10			
8:03	8:12	8:20		8:32	8:40	8:47
8:14	8:23	8:31	8:40			
8:24	8:33	8:41		8:53	9:01	9:08
8:43	8:52	9:00	9:09			
8:45	8:54	9:02		9:14		
9:07	9:18	9:27		9:40	9:48	9:56
9:14	9:26	9:35	9:45			
9:28	9:39	9:48		10:01		
9:47	9:59	10:08		10:20	10:28	10:36
9:49	10:00	10:09	10:20			
10:06	10:17	10:26		10:39		
10:21	10:32	10:41	10:52			
10:24	10:36	10:45		10:57	11:06	11:13
10:43	10:54	11:04		11:16	11:25	11:33
10:54	11:06	11:15	11:26			
11:01	11:13	11:22		11:35		
11:21	11:32	11:42		11:54	**12:03**	**12:11**
11:27	11:39	11:48	11:59			
11:39	11:50	**12:00**		**12:12**		
11:58	**12:10**	**12:19**		**12:32**	**12:41**	**12:49**
PM						
12:01	**12:12**	**12:22**	**12:32**			
12:14	**12:26**	**12:35**		**12:48**		
12:30	**12:42**	**12:51**		**1:04**	**1:13**	**1:21**
12:35	**12:46**	**12:56**	**1:06**			
12:46	**12:58**	**1:07**		**1:20**		
1:02	**1:14**	**1:23**		**1:36**	**1:45**	**1:53**
1:09	**1:21**	**1:30**	**1:41**			

Complete. Use the timetable.

1. Suppose you need to get to 124th Street before 8:00 A.M. What is the latest time you could be at Jackson and Wisconsin? **7:21 A.M.**
2. If you are at the bus stop on 12th and Wisconsin at 7:10 A.M., how long will you have to wait for the next Number 10 bus? **13 minutes**
3. José boards the Number 10 bus on 35th and Wisconsin at 8:41 A.M. How long is his ride to Brookfield Square? **27 minutes**
4. Irma has an 11:15 A.M. appointment at County Hospital. How late can she be at Jackson and Wisconsin to catch the Number 10 bus? **10:24 A.M.**
5. How long does it take the bus leaving County Hospital at 5:46 A.M. to get to 124th Street? **8 minutes**
6. Which is the first bus that takes longer than 8 minutes to go from County Hospital to 124th Street? Can you think of a reason for this? **10:57 A.M.; more traffic**
7. How many minutes does it take the 8:03 A.M. bus to make the entire trip from Jackson and Wisconsin to Brookfield Square? **44 minutes**
8. Refer to Exercise 7. How many minutes longer does it take the 11:58 A.M. bus to make the trip? **7 minutes**

APPLYING THE LESSON

Independent Practice

Homework Assignment	
Minimum	**Ex. 2-20 even; 21-30**
Average	**Ex. 1-15 odd; 16-30**
Maximum	**Ex. 1-30**

Additional Answers

18. Cut each link on one four-link length. Use each link between the other five lengths. $60

22. $10\frac{3}{8}$ 23. $21\frac{11}{12}$ 24. $9\frac{13}{45}$ 25. $17\frac{20}{21}$

26. $15\frac{2}{5}$ 27. 32 28. $7\frac{1}{2}$ 29. $4\frac{1}{8}$

5-MINUTE CHECK-UP

(over Lesson 6-11)

Available on TRANSPARENCY 6-12.

Change each 12-hour time to 24-hour notation.

1. 6:35 A.M. 06:35
2. 7:50 P.M. 19:50
3. 12:15 A.M. 00:15
4. 6:12 P.M. 18:12

Compute the working hours for the day.

5.

IN	OUT	
8:10	11:40	6 h 55 min
12:20	15:45	

INTRODUCING THE LESSON

Ask students to suggest everyday situations that involve measurement. Survey the class to find out how many times a week they need to estimate or use exact measurement.

TEACHING THE LESSON

Using Critical Thinking Have students explain why problem solving involving measurements may require additional information. For example, light travels about 186,000 mps. How far does it travel in a minute, an hour? **You need to know the number of seconds in a minute or minutes in an hour.**

Practice Masters Booklet, p. 71

Name ______________________ Date ______________

PRACTICE WORKSHEET 6-12

Problem Solving: Using Measurements

Solve.

1. Marshall's flight is scheduled to leave Dallas at 8:46 A.M. and arrive at Chicago at 12:30 P.M. How long is the flight? 3 h 44 min
2. Six tablecloths, each 150 centimeters long, are needed to cover a row of picnic tables that have been placed end-to-end. About how many meters long is the row of tables? about 9 m
3. On a trip, Fred bought gasoline three times — 18 gallons, 14.5 gallons, and 16.5 gallons. What was the total cost at $1.12 per gallon? $54.88
4. If each banana has a mass of about 225 g, how much will you have to pay for 6 bananas at 64¢ per kilogram? $0.86
5. On the first four days of her vacation, Carrie drove 215 miles, 360 miles, 280 miles, and 155 miles. What was the average daily distance? 252.5 mi
6. The weather forecast is for a low overnight of 7°C and a high tomorrow of 16°C. How many degrees will the temperature have to rise? 9°C
7. You have a bad cough for which the doctor has given you medicine. The prescribed dosage is 5 milliliters, 4 times a day. How many milliliters must be in the bottle if there is enough for 1 week? 140 mL
8. You are driving a truck loaded with bales of straw over a country road. The bridge you have crossed has posted a limit of 5 tons. Your truck weighs about 5,000 pounds and each of the 80 bales of straw weighs about 50 pounds. Is it safe for you to cross the bridge? What is the greatest number of bales you can safely haul across the bridge? yes; 100 bales

Explore
Plan
Solve
Examine

6-12 USING MEASUREMENTS

Objective
Solve problems involving measurements.

State marching band competition will be held in 1 week and José wants 8 hours of extra practice. If José practices the trumpet for 35 minutes every day before school and on Saturday and Sunday, will he practice for 8 hours?

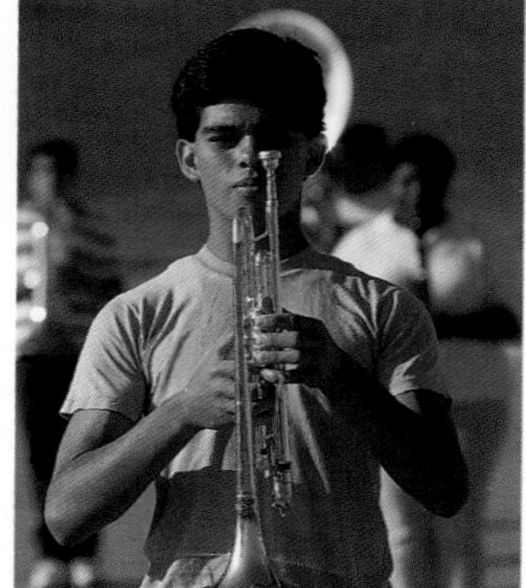

Explore

- What is given?
  José practices 35 minutes for 7 days a week.
- What is asked?
  At 35 minutes a day, will José practice for 8 hours in a week?

Plan Find the number of *minutes* José practices each week by multiplying 35 minutes by 7. Change the number of minutes to hours by dividing by 60.

Solve Estimate: José practices a little more than $\frac{1}{2}$ hour each day. In 7 days, he practices a little more than 7 times $\frac{1}{2}$ or $3\frac{1}{2}$ hours.

$$\begin{array}{r} 35 \text{ minutes} \\ \times\ 7 \text{ days} \\ \hline 245 \text{ minutes} \end{array} \qquad \begin{array}{r} 4 \text{ R5} \\ 60\overline{)245} \\ -240 \\ \hline 5 \end{array}$$

José practices 4 hours 5 minutes. He will not practice an extra 8 hours.

Examine To solve another way, think how many days are in one week. There are 7. If José practices one whole hour a day for 7 days, he will practice an extra 7 hours. Since one whole hour a day is not enough, 35 minutes a day is not enough.

Guided Practice

1. Kyle has 2 boards for shelving. One is 18 inches longer than the other. If the shorter board is 6 feet 3 inches long, how long is the other board? **7 ft 9 in.**
2. Karen brought 3 quarts of orange juice in her thermal jug for the break at hockey practice. How many 4-ounce glasses can she serve to the team? **24 glasses**
3. Kurt added walking 5 days a week to his reducing program. If he walks 40 minutes a day for 5 days, how many hours will he walk? **$3\frac{1}{3}$ h**
4. A rectangular garden measures 55 feet wide by 83 feet long. Jed has 100 yards of fencing to enclose the garden. How many yards of fencing will Jed have left after enclosing the garden? **8 yards**

RETEACHING THE LESSON

Have students use the steps outlined on page 202 to solve each problem.
Solve.

1. Terry is in training for a marathon. On Monday he ran 15 miles, on Tuesday 18 miles, on Wednesday 20 miles, on Thursday 15 miles, and on Friday 15 miles. What was the average number of miles he ran each day? **16.6 miles**

Reteaching Masters Booklet, p. 65

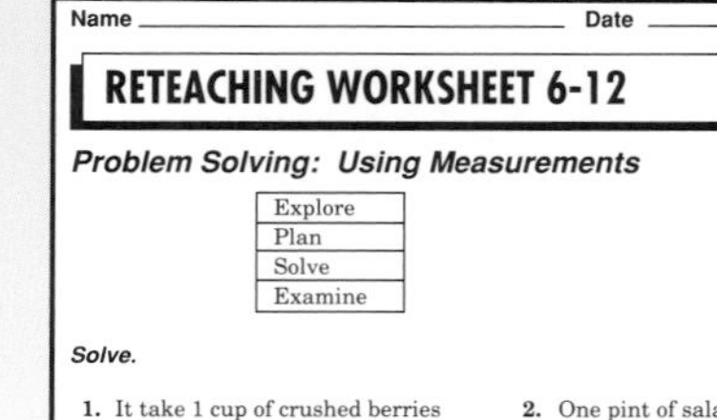

Name ______________________ Date ______________

RETEACHING WORKSHEET 6-12

Problem Solving: Using Measurements

Explore
Plan
Solve
Examine

Solve.

1. It take 1 cup of crushed berries to make 2 quarts of punch. How many cups of crushed berries are needed to make 6 gallons of punch? 12 c
2. One pint of salad dressing has 10 fl oz of oil. If one serving of salad dressing is 2 fl oz, how much oil is in a serving? $1\frac{1}{4}$ fl oz

Problem Solving

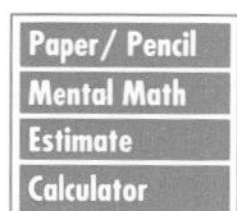

1 mm too large

5. Linda's blouse pattern calls for buttons that are 9 millimeters in diameter. Linda chose buttons that are 1 centimeter in diameter. By how many millimeters are the buttons too large or too small?

6. Kent wants to measure the top of his desk for a new blotter, but he didn't want to walk downstairs for the yardstick. Kent knows that a sheet of typing paper is 8.5 inches wide and 11 inches long. He measures the top of his desk with a sheet of typing paper using only the side that measures 11 inches. If the desk measures $2\frac{1}{2}$ sheets by 4 sheets, what are the approximate dimensions of the desk? **27 or 28 inches by 44 inches**

7. A dime has a mass of 2.27 grams. What is the mass of a roll of dimes that is worth $5? **113.5 grams**

8. The Wangler's dog Gracie had three puppies. The first one weighed 1 pound 3 ounces, the next one 1 pound 1 ounce, and the last one 1 pound 5 ounces. What was the average weight of the puppies? **1 lb 3 oz**

9. Select your favorite word problem from this chapter. Attach a note explaining why it is your favorite. **See students' work.**

Mixed Review

Lesson 1-6

Estimate.

10. 586 + 381 = **900**

11. 4,621 + 314 = **4,900**

12. 8.12 + 3.98 + 4.76 = **17.00**

13. 3,512 − 2,621 = **1,000**

Lesson 3-1

Multiply.

14. 62 × 3 = **186**

15. 143 × 12 = **1,716**

16. 280 × 45 = **12,600**

17. 651 × 231 = **150,381**

Lesson 6-7

18. Max lives 8 blocks from the school. On his way home from drama club, he alternately walks two blocks, then jogs two blocks. If each block is 210 yards long, how many blocks does Max jog? **840 yards**

APPLYING THE LESSON

Independent Practice

Homework Assignment	
Minimum	Ex. 1-18
Average	Ex. 1-18
Maximum	Ex. 1-18

Chapter 6, Quiz B (Lessons 6-7 through 6-12) is available in the Evaluation Masters Booklet, p. 29.

Chalkboard Examples

Solve.

1. Some mountain climbers on Mt Rainier climbed from Camp Muir at 10,188 ft to Columbia Crest at 14,410 ft. How far did they climb? **4,222 ft**
2. Climbers left Camp Muir at 2 A.M. and arrived at the crest at 8:45 A.M.. How long was the climb? **6 hr 45 min**

Additional examples are provided on TRANSPARENCY 6-12.

EVALUATING THE LESSON

Check for Understanding Use the Guided Practice Exercises to check for student understanding.

Closing the Lesson Have students tell about times when they need to solve problems involving measurements.

EXTENDING THE LESSON

Enrichment Have students make up three word problems using information from newspaper or magazine articles about sports, cars, and so on. Exchange problems and have classmates solve.

Enrichment Masters Booklet, p. 65

ENRICHMENT WORKSHEET 6-12

Significant Digits

Significant digits are the digits in the numeral of a measurement that are needed to name the number of units of measure. How many significant digits are in the measurements 47 m, 5<u>0</u> cm, 50 mm, 1.25 km, and 0.080 cm?

47 m The unit of measure is 1 m. The number of units named is 47. Thus, there are 2 significant digits.

5<u>0</u> cm The unit of measure is 1 cm. The number of units named is 50. Thus, there are 2 significant digits.

50 mm The unit of measure is 10 mm. The number of units named is 5. Thus, there is 1 significant digit.

1.25 km The unit of measure is 0.01 km. The number of units named is 125. Thus, there are 3 significant digits.

0.080 cm The unit of measure is 0.001 cm. The number of units named is 80. Thus, there are 2 significant digits.

Note that underlined zeros (<u>0</u>) are significant. Zeros between significant zeros are significant. Also, final zeros to the right of the decimal point are significant. However, zeros that serve only to locate the decimal point are not significant.

Complete.

	Measurement	Unit of Measure	Number of Units	Number of Significant Digits
1.	498 millimeters	1 mm	498	3
2.	1,5<u>0</u>0 kilometers	10 km	150	3
3.	2,540 kilograms	10 kg	254	3
4.	10.8 seconds	0.1 s	108	3
5.	0.480 meters	0.001 m	480	3
6.	0.095 liters	0.001 L	95	2
7.	109 grams	1 g	109	3
8.	16.3 centimeters	0.1 cm	163	3
9.	4,001 minutes	1 min	4,001	4
10.	10,001 hours	1 h	10,001	5

Using the Chapter Review

The Chapter Review is a comprehensive review of the concepts presented in this chapter. This review may be used to prepare students for the Chapter Test.

Chapter 6, Quizzes A and B, are provided in the Evaluation Masters Booklet as shown below.

Quiz A should be given after students have completed Lessons 6-1 through 6-6. Quiz B should be given after students have completed Lessons 6-7 through 6-12.

These quizzes can be used to obtain a quiz score or as a check of the concepts students need to review.

Evaluation Masters Booklet, p. 29

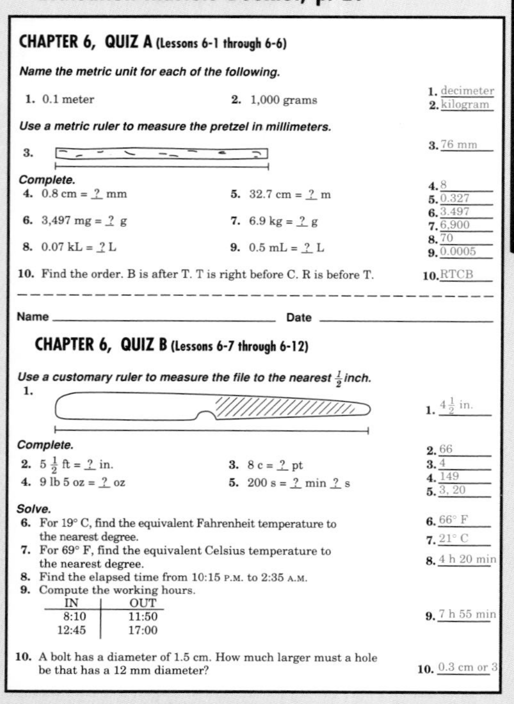

CHAPTER 6, QUIZ A (Lessons 6-1 through 6-6)

Name the metric unit for each of the following.

1. 0.1 meter 2. 1,000 grams 1. decimeter 2. kilogram

Use a metric ruler to measure the pretzel in millimeters.

3. 3. 76 mm

Complete.

4. 0.8 cm = ? mm 5. 32.7 cm = ? m 4. 8 5. 0.327

6. 3,497 mg = ? g 7. 6.9 kg = ? g 6. 3.497 7. 6,900

8. 0.07 kL = ? L 9. 0.5 mL = ? L 8. 70 9. 0.0005

10. Find the order. B is after T. T is right before C. R is before T. 10. RTCB

Name ______ Date ______

CHAPTER 6, QUIZ B (Lessons 6-7 through 6-12)

Use a customary ruler to measure the file to the nearest $\frac{1}{2}$ inch.

1. 1. $4\frac{1}{2}$ in.

Complete.

2. $5\frac{1}{2}$ ft = ? in. 3. 8 c = ? pt 2. 66 3. 4

4. 9 lb 5 oz = ? oz 5. 200 s = ? min ? s 4. 149 5. 3, 20

Solve.

6. For 19° C, find the equivalent Fahrenheit temperature to the nearest degree. 6. 66° F
7. For 69° F, find the equivalent Celsius temperature to the nearest degree. 7. 21° C
8. Find the elapsed time from 10:15 P.M. to 2:35 A.M. 8. 4 h 20 min
9. Compute the working hours. 9. 7 h 55 min

IN	OUT
8:10	11:50
12:45	17:00

10. A bolt has a diameter of 1.5 cm. How much larger must a hole be that has a 12 mm diameter? 10. 0.3 cm or 3

Chapter 6 REVIEW

Vocabulary / Concepts

Choose a word or numeral from the list at the right to complete each sentence.

0.001, 4, 100, 1,000, Celsius, customary, Fahrenheit, foot, gram, kilo, liter, metric, time

1. There are ___?___ centimeters in one meter. 100
2. The prefix ___?___ means 1,000. kilo
3. The basic unit for capacity in the metric system is the ___?___. liter
4. The prefix milli means ___?___. 0.001
5. Water freezes at 0 degrees ___?___. Celsius
6. The basic unit for mass in the metric system is the ___?___. gram
7. Units of ___?___ are the same in the customary and metric systems. time
8. There are ___?___ grams in one kilogram. 1,000
9. The most commonly used ___?___ units of weight are the ounce, pound, and ton. customary
10. There are ___?___ quarts in one gallon. 4
11. There are 12 inches in one ___?___. foot

Exercises / Applications

13. centimeter 14. meter 16. kilometer

Name the metric unit for each measurement.

Lesson 6-1

12. 1,000 grams kilogram 13. 0.01 meter 14. 100 centimeters
15. 0.001 liter milliliter 16. 1,000 meters 17. 0.001 gram milligram

Lesson 6-2

Use a metric ruler to measure each segment. Give the measurement in centimeters and also in millimeters.

18. ________ 1.9 cm; 19 mm
19. ____________________ 4.2 cm; 42 mm
20. __________________ 3.7 cm; 37 mm
21. ______________________ 5.1 cm; 51 mm

Lesson 6-3

Complete.

22. 400 mm = ? cm 40 23. 0.5 m = ? cm 50 24. 5 km = ? m 5,000
25. 100 cm = ? m 1 26. 2,372 m = ? km 27. 4 mm = ? cm 0.4

Lesson 6-4, 6-5

Complete. 26. 2.372 29. 34,000 30. 0.346 31. 4.536 32. 4,900

28. 4 kg = ? g 4,000 29. 34 g = ? mg 30. 346 mg = ? g
31. 4,536 g = ? kg 32. 4.9 g = ? mg 33. 8.5 kg = ? g 8,500

Lesson 6-6

34. The Mustangs were on their own 22-yard line. The quarterback for the Mustangs threw the football 22 yards forward from 6 yards behind the line of scrimmage. The wide receiver caught the ball but decided he couldn't advance so he threw a lateral (no forward or backward distance) to another receiver. The second receiver ran 11 yards forward before he was tackled. Where did the referee spot the ball? 49-yard line of Mustangs

38. 34 40. 122

Lessons 6-7, 6-8

Complete.

35. 3 ft = ? in. 36 **36.** 7 yd = ? ft 21 **37.** 120 in. = ? ft 10

38. 2 lb 2 oz = ? oz **39.** 36 ft = ? yd 12 **40.** 7 lb 10 oz = ? oz

41. $2\frac{1}{2}$ ft = ? in. 30 **42.** 3 lb = ? oz 48 **43.** 2 T = ? lb 4,000

44. 5 c = ? fl oz 40 **45.** $2\frac{1}{2}$ lb = ? oz 40 **46.** 64 fl oz = ? c 8

Lesson 6-9

Find the equivalent Fahrenheit temperature or Celsius temperature to the nearest degree.

47. 30°C 86°F **48.** 120°C 248°F **49.** 100°F 38°C **50.** 36°F 2°C

Choose the best temperature.

51. a summer day	100°C	42°C	42°F
52. an ice cube	26°F	62°F	26°C
53. winter day	2°C	75°C	75°F
54. swimming outdoors	25°C	0°C	25°F

Lesson 6-10

Complete. 58. 270 59. 302

55. 3 min = ? s 180 **56.** 72 h = ? d 3 **57.** 7 d = ? h 168

58. 4 h 30 min = ? min **59.** 5 min 2 s = ? s **60.** $\frac{1}{2}$ h = ? min 30

Lesson 6-11

Change each 12-hour time to 24-hour notation.

61. 2:33 A.M. 02:33 **62.** 5:56 P.M. 17:56 **63.** 11:29 P.M. 23:29

Compute the working hours for each day. 66. 7 h 30 min

64. 5 h 38 min
65. 6 h 59 min

64. IN	OUT	**65.** IN	OUT	**66.** IN	OUT
9:32	11:50	8:21	12:06	6:45	11:00
12:20	3:40	12:56	4:10	11:30	2:45

Lesson 6-12
67. 9,600 kg

67. For a Friday evening flight to Boston from Newark, a Boeing 737 carried 120 passengers. If the average weight of a passenger was 64 kg and the average weight of each passenger's luggage was 16 kg, what was the total weight of the passengers and their luggage?

68. 2 cans

68. Melinda needs 3 cups of apricot nectar for a gelatin salad. If each can of nectar contains 14 ounces, how many cans should Melinda buy?

69. The longest Wimbledon tennis match was in 1982 between Jimmy Connors and John McEnroe. It lasted 4 hours 16 minutes. How many minutes did the match last? 256 min

Alternate Assessment Strategy
To provide a brief in-class review, you may wish to read the following questions to the class and require a verbal response.

1. How many kilometers are equal to 400 meters? 0.4 km
2. Is the length of a pencil closer to 8 cm or 8 mm? 8 cm
3. 5,000 mm equal how many centimeters? 500 cm
4. How many grams equal 3 kilograms? 3,000 g
5. How many liters are there in 240 mL? 0.24 L
6. How many inches are in 2.5 feet? 30 in.
7. What are the number of ounces in 2 gallons? 256 oz
8. Find the equivalent Celsius temperature of 41°F. 5°C
9. What is the temperature, in Fahrenheit, of 85°C? 185°F
10. How many seconds are in 15 minutes? 900 s

APPLYING THE LESSON

Independent Practice

Homework Assignment	
Minimum	Ex. 1-11; 12-68 even
Average	Ex. 1-11; 13-65 odd; 67-69
Maximum	Ex. 1-69

Using the Chapter Test

This page may be used as a test or as an additional review if necessary. Two forms of a Chapter Test are provided in the Evaluation Masters Booklet. Form 2 (free response) is shown below, and Form 1 (multiple choice) is shown on page 209.

The **Tech Prep Applications Booklet** provides students with an opportunity to familiarize themselves with various types of technical vocations. The Tech Prep applications for this chapter can be found on pages 11–12.

Evaluation Masters Booklet, p. 28

Name ______________ Date ______________

CHAPTER 6 TEST, FORM 2

Name the metric unit for each of the following.

1. 1,000 grams 2. 0.001 liter — 1. kg 2. mL

Use a metric ruler to measure the wire in centimeters.

3. — 3. 9 cm

Complete.

4. 38 mm = ? cm 5. 7.9 km = ? m 6. 24 g = ? mg
7. 64.1 g = ? kg 8. 12.5 mL = ? L 9. 17.3 kL = ? L

4. 3.8 5. 7,900 6. 24,000 7. 0.0641 8. 0.0125 9. 17,300

Use a customary ruler to measure the needle to the nearest inch.

10. — 10. 2 in.

Complete.

11. 3 mi. = ? ft 12. $3\frac{1}{2}$ gal = ? qt
13. 52 oz = ? lb ? oz 14. 6 min 15 s = ? s

11. 15,840 12. 14 13. 3, 4 14. 375

Solve.

15. Find the order. Q is before C, but not first. C is after G, but before W. — 15. GQCW
16. For 30°C, find the equivalent Fahrenheit temperature. — 16. 86° F
17. For 98°F, find the equivalent Celsius temperature to the nearest degree. — 17. 37°C
18. Find the elapsed time from 7:20 P.M. to 1:30 A.M. — 18. 6 h 10 min
19. Compute the working hours. — 19. 7h 40 min

IN	OUT
8:10	12:15
13:55	17:30

20. Five boards 3 ft 4 in. long are nailed end to end. What is the total length of the boards? — 20. 16 ft 8 in.

Bonus Two gas tanks will hold 55.5 L and 85 L. If gas costs \$0.35 per liter, how much more will it cost to fill the larger tank? **Bonus** \$10.33

Chapter 6 TEST

Complete. 9. 0.0036 12. 45,500 18. 3, 12

1. 400 cm = ? mm 4,000 **2.** 0.9 m = ? cm 90 **3.** 2 km = ? m 2,000
4. 260 cm = ? m 2.6 **5.** 9,000 m = ? km 9 **6.** 17 mm = ? cm 1.7
7. 30 g = ? kg 0.03 **8.** 3.42 g = ? mg 3,420 **9.** 3.6 g = ? kg
10. 10 L = ? mL 10,000 **11.** 705 mL = ? L 0.705 **12.** 45.5 L = ? mL
13. 9 ft = ? yd 3 **14.** 32 oz = ? lb 2 **15.** 7 ft = ? in. 84
16. $\frac{1}{2}$ gal = ? qt 2 **17.** 10 pt = ? c 20 **18.** 60 oz = ? lb ? oz
19. 3 pt = ? c 6 **20.** 381 min = ? h ? min 6, 21 **21.** 2 h 45 min = ? min 165

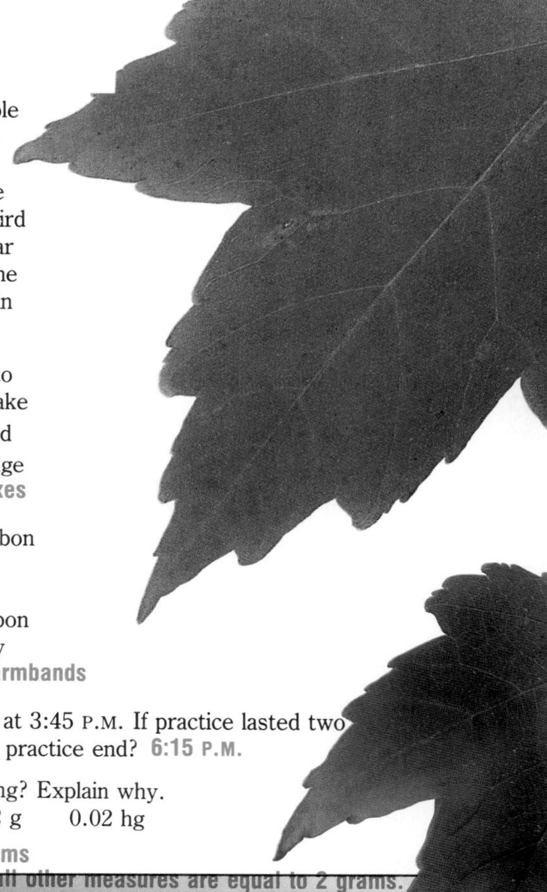

Solve.

22. Smythe's tree farm has 30 maple trees planted in each row. The first year Smythe sold every fourth tree from each row. The second year they sold every third tree in a row, and the third year they sold every other tree in the row. How many trees are left in each row? **8 trees**

23. Rosa made 5 pounds of fudge to sell at the Photography Club bake sale. She packaged it in $\frac{1}{4}$-pound boxes. How many boxes of fudge does Rosa have to sell? **20 boxes**

24. Coach Thompson uses blue ribbon armbands in physical education class to separate players into teams. She cuts 8 yards of ribbon into 1.5-foot pieces. How many armbands does she have? **16 armbands**

25. Kevin went to football practice at 3:45 P.M. If practice lasted two and a half hours, what time did practice end? **6:15 P.M.**

BONUS: Which one does not belong? Explain why.
2,000 mg 20 cg 2 g 0.02 hg

Bonus: Twenty centigrams does not belong since all other measures are equal to 2 grams.

The **Performance Assessment Booklet** provides an alternative assessment for evaluating student progress. An assessment for this chapter can be found on pages 11–12.

A **Test and Review Generator** is provided in Apple, IBM, and Macintosh versions. You may use this software to create your own tests or worksheets, based on the needs of your students.

History Connection

Nonstandard Units of Weight and Volume

In ancient Egypt the Qedet system of weights was based on a unit of grain. The type of a grain was probably barley. The city of Heliopolis used a system that was based on 140 grains that equaled one qedet. Since the value of a qedet varied between cities, this unit was *not* standard.

Cities kept their own standards for weights in their temples. These were often decorated shapes of stone marked with the weight they represented.

The animal figure is a 10-qedet weight carved from malachite.

The Qedet System

70 grains = $\frac{1}{2}$ qedet (or 1 drachm)
140 grains = 1 qedet
10 qedet = 1 deben
10 deben = 1 sep

The ancient Romans used a system of volume based on containers of a certain size. They kept standards of each in their temples.

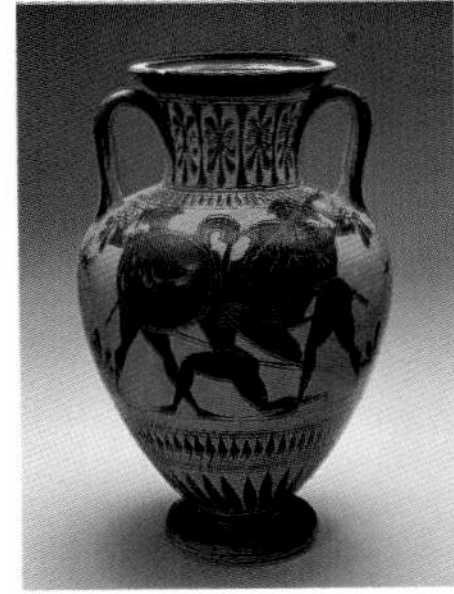

Approximate Roman Volume Equivalents

2 acetabulum = 1 quartarius
2 quartarius = 1 hemina
25 quartarius = 1 congius
4 congius = 1 urna
2 urna = 1 amphora

Complete.

1. 420 grains = ? qedet 3
2. 50 qedet = ? deben 5
3. 60 deben = ? sep 6
4. 3 sep = ? grains 42,000
5. 6 qedet = ? drachm 12
6. 2 sep = ? qedet 200
7. 630 grains = ? drachm 9
8. 60 drachm = ? deben 3
9. 28,000 grains = ? sep 2
10. 1 congius = ? acetabulum 50
11. 1 urna = ? quartarius 100
12. 1 amphora = ? congius 8
13. 75 quartarius = ? congius 3
14. 6 congius = ? hemina 75
15. 4 urna = ? acetabulum 800
16. Make up your own system of nonstandard weights or volume. Use at least three different units. Answers may vary.

Applying Mathematics to the Real World

This optional page shows the historical development of mathematics and also provides a change of pace.

Objective Familiarize students with the historical development of measures of weight and capacity.

Using Questioning

- Which is larger?
  a. 1 urna or 1 amphora? amphora
  b. 1 congius or 1 quartarius? congius
  c. 1 acetabulum or 1 hemina? hemina
- What is the largest unit in the Roman system of volume? amphora the smallest? acetabulum
- Which is larger?
  a. the qedet or grain? qedet
  b. the deben or sep? sep
  c. the qedet or deben? deben
- What is the smallest unit of measure in the qedet system? grain the largest? sep

Activity

Using Research Have students do a report or poster on units of weight or volume that were used in Ancient Egypt or Ancient Greece.

Using the Cumulative Review

This page provides an aid for maintaining skills and concepts presented thus far in the text.

A Cumulative Review is also provided in the Evaluation Masters Booklet as shown below.

Evaluation Masters Booklet, p. 30

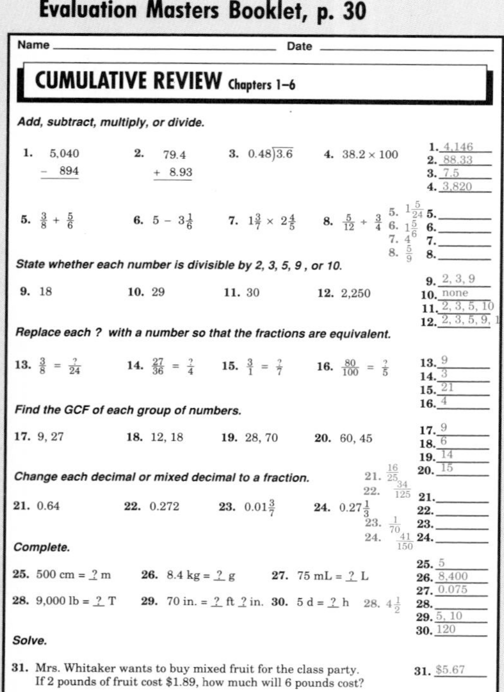

Name ______ Date ______

CUMULATIVE REVIEW Chapters 1–6

Add, subtract, multiply, or divide.

1. 5,040 − 894
2. 79.4 + 8.93
3. $0.48\overline{)3.6}$
4. 38.2×100
5. $\frac{3}{8} + \frac{5}{6}$
6. $5 - 3\frac{1}{6}$
7. $1\frac{3}{7} \times 2\frac{4}{5}$
8. $\frac{5}{12} + \frac{3}{4}$

State whether each number is divisible by 2, 3, 5, 9, or 10.

9. 18
10. 29
11. 30
12. 2,250

Replace each ? with a number so that the fractions are equivalent.

13. $\frac{3}{8} = \frac{?}{24}$
14. $\frac{27}{36} = \frac{?}{4}$
15. $\frac{3}{1} = \frac{?}{7}$
16. $\frac{80}{100} = \frac{?}{5}$

Find the GCF of each group of numbers.

17. 9, 27
18. 12, 18
19. 28, 70
20. 60, 45

Change each decimal or mixed decimal to a fraction.

21. 0.64
22. 0.272
23. $0.01\frac{3}{7}$
24. $0.27\frac{1}{3}$

Complete.

25. 500 cm = ? m
26. 8.4 kg = ? g
27. 75 mL = ? L
28. 9,000 lb = ? T
29. 70 in. = ? ft ? in.
30. 5 d = ? h

Solve.

31. Mrs. Whitaker wants to buy mixed fruit for the class party. If 2 pounds of fruit cost \$1.89, how much will 6 pounds cost?

Answers: 1. 4,146 2. 88.33 3. 7.5 4. 3,820 5. $1\frac{5}{24}$ 6. $1\frac{5}{6}$ 7. 4 8. $\frac{5}{9}$ 9. 2, 3, 9 10. none 11. 2, 3, 5, 10 12. 2, 3, 5, 9, 1 13. 9 14. 3 15. 21 16. 4 17. 9 18. 6 19. 14 20. 15 21. $\frac{16}{25}$ 22. $\frac{34}{125}$ 23. $\frac{1}{70}$ 24. $\frac{41}{150}$ 25. 5 26. 8,400 27. 0.075 28. $4\frac{1}{2}$ 29. 5, 10 30. 120 31. \$5.67

Cumulative REVIEW/TEST

Free Response

Lesson 1-6

Estimate.

1. 4,327 + 2,294 **6,000**　**2.** 7,468 − 4,923 **2,000**　**3.** 750,290 − 61,499 **690,000**

Lesson 2-4

4. Jovita adds 12.5 gallons of gasoline to fill her car. The tank holds 15 gallons. How much gasoline was in the car's gas tank before she added any? **2.5 gal gasoline**

Lesson 2-6

5. Write the next three numbers in the arithmetic sequence 8.7, 7.5, 6.3. **5.1, 3.9, 2.7**

Lesson 3-4

6. Dave wants to earn \$76 this week. He can make \$4 an hour at Colony's Mini-Golf. How many hours does Dave need to work? **19 h**

Lesson 4-1

State whether each number is divisible by 2, 3, 5, 9, or 10.

7. 36 **2, 3, 9**　**8.** 136 **2**　**9.** 180 **2, 3, 5, 9, 10**

Lesson 4-7

Replace each ● with <, >, or = to make a true sentence.

10. $\frac{4}{5}$ ● $\frac{3}{10}$ **>**　**11.** $\frac{6}{7}$ ● $\frac{3}{14}$ **>**　**12.** $\frac{1}{2}$ ● $\frac{1}{10}$ **>**　**13.** $\frac{2}{3}$ ● $\frac{13}{14}$ **<**

Lesson 4-9

Estimate.

14. $\frac{4}{4} + \frac{3}{10}$ **1**　**15.** $\frac{7}{15} + \frac{1}{10}$ **$\frac{1}{2}$**　**16.** $\frac{43}{45} - \frac{2}{5}$ **$\frac{1}{2}$**　**17.** $\frac{11}{12} - \frac{7}{8}$ **0**

Lesson 5-3

Multiply.

18. $3 \times 19\frac{1}{3}$ **58**　**19.** $2\frac{1}{2} \times \frac{1}{5}$ **$\frac{1}{2}$**　**20.** $15 \times 2\frac{4}{5}$ **42**　**21.** $3\frac{1}{2} \times 6\frac{3}{4}$ **$23\frac{5}{8}$**

22. Alberto has 230 albums. He sells $\frac{2}{5}$ of them at \$1.50 each. How much money does he make? **\$138.00**

Lessons 5-6, 5-7

Divide.

23. $\frac{3}{4} \div \frac{1}{3}$ **$2\frac{1}{4}$**　**24.** $\frac{5}{16} \div \frac{3}{8}$ **$\frac{5}{6}$**　**25.** $\frac{3}{10} \div 10$ **$\frac{3}{100}$**　**26.** $\frac{7}{8} \div 4$ **$\frac{7}{32}$**

27. $5\frac{1}{2} \div \frac{1}{2}$ **11**　**28.** $7\frac{3}{4} \div 1\frac{1}{2}$ **$5\frac{1}{6}$**　**29.** $\frac{1}{5} \div \frac{4}{5}$ **$\frac{1}{4}$**　**30.** $\frac{3}{4} \div \frac{9}{10}$ **$\frac{5}{6}$**

Lesson 6-3

Complete.

31. 5,000 g = ? kg **5**　**32.** 5,000 g = ? mg **5,000,000**　**33.** 75 L = ? mL **75,000**

34. 1,500 m = ? km **1.5**　**35.** 32 km = ? m **32,000**　**36.** 7 m = ? cm **700**

Lesson 6-10

37. Jack has 8 minutes of blank tape left. How much tape will be left blank after he records a song that lasts 3 minutes 17 seconds? **4 min 43 sec**

APPLYING THE LESSON

Independent Practice

Homework Assignment	
Minimum	Ex. 1-37 odd
Average	Ex. 1-37
Maximum	Ex. 1-37

Chapter 6

Multiple Choice

Choose the letter of the correct answer for each item.

1. What is 6,785,201 rounded to the nearest ten thousand?
▶ **a.** 6,790,000
b. 6,800,000
c. 6,890,000
d. 7,000,000

Lesson 1-4

2. Find the quotient when 79,272 is divided by 8.
e. 899
f. 999
▶ **g.** 9,909
h. 9,999

Lesson 3-4

3. Three fourths of a pizza added to seven twelfths of a pizza gives you how much pizza?
a. $\frac{10}{16}$
b. $\frac{5}{6}$
c. $1\frac{1}{4}$
▶ **d.** $1\frac{1}{3}$

Lesson 4-10

4. Lewis pole-vaulted $16\frac{3}{4}$ ft, $17\frac{3}{4}$ ft, and $17\frac{1}{4}$ ft. Find the average height of his vaults.
e. $16\frac{31}{36}$ ft
f. 17 ft
▶ **g.** $17\frac{1}{4}$ ft
h. $50\frac{7}{4}$ ft

Lessons 3-12, 4-11

5. In 1988, the USA's population was said to be increasing at a rate of 100,000 people a day. What is the rate of increase per minute?
a. 15
b. 12
▶ **c.** 69
d. 24

Lesson 6-10

6. What is the difference of 5,421 and 372?
e. 1,701
f. 4,049
▶ **g.** 5,049
h. 5,059

Lesson 2-3

7. What is the GCF of 12 and 8?
a. 2
▶ **b.** 4
c. 24
d. 96

Lesson 4-3

8. Find the result of dividing $7\frac{3}{4}$ by $\frac{1}{2}$.
e. $3\frac{7}{8}$
f. $7\frac{3}{8}$
g. $7\frac{3}{4}$
▶ **h.** $15\frac{1}{2}$

Lesson 5-7

9. A recipe calls for $\frac{3}{4}$ cup of sugar for one batch of cookies. How much sugar would half of a batch of cookies require?
▶ **a.** $\frac{3}{8}$ cup
b. $\frac{1}{2}$ cup
c. $\frac{6}{8}$ cup
d. $\frac{3}{2}$ cup

Lesson 5-3

10. Choose the most reasonable unit for measuring the mass of a science textbook.
e. centimeter
f. gram
▶ **g.** kilogram
h. meter

Lesson 6-4

Using the Cumulative Test

This test familiarizes students with a standardized format while testing skills and concepts presented up to this point.

APPLYING THE LESSON

Independent Practice

Homework Assignment	
Minimum	**Ex. 1-10**
Average	**Ex. 1-10**
Maximum	**Ex. 1-10**

Evaluation Masters Booklet, pp. 26-27

CHAPTER 6 TEST, FORM 1

Write the letter of the correct answer on the blank at the right of the page.

1. Name the metric unit for 0.01 meter.
A. centimeter **B.** kilometer **C.** dekameter **D.** millimeter — 1. A

2. Use a metric ruler to measure the stick below in centimeters.
A. 5 cm **B.** 6 cm **C.** 7 cm **D.** 8 cm — 2. C

3. Complete. 30 cm = ? m
A. 300 **B.** 3 **C.** 0.3 **D.** 0.03 — 3. C

4. Complete. 400 cm = ? mm
A. 0.4 **B.** 4 **C.** 40 **D.** 4,000 — 4. D

5. Complete. 6 mg = ? g
A. 0.06 **B.** 0.006 **C.** 6,000 **D.** 600 — 5. B

6. Complete. 8.7 g = ? kg
A. 8,700 **B.** 870 **C.** 0.0087 **D.** 0.087 — 6. C

7. Complete. 400 mL = ? L
A. 4,000 **B.** 0.4 **C.** 40,000 **D.** 0.004 — 7. B

8. Complete. 6.8 L = ? mL
A. 6,800 **B.** 680 **C.** 0.068 **D.** 0.0068 — 8. A

9. Find the order. M is before G. L is last. G is right after H.
A. M, G, H, L **B.** H, M, G, L **C.** M, H, G, L **D.** H, G, M, L — 9. C

10. Use a customary ruler to measure the nail to the nearest $\frac{1}{2}$ inch.
A. $1\frac{1}{2}$ in. **B.** 2 in. **C.** $2\frac{1}{2}$ in. **D.** 3 in. — 10. C

11. Complete. 30 in. = ? ft
A. $2\frac{1}{2}$ **B.** 10 **C.** $2\frac{1}{6}$ **D.** 360 — 11. A

Chapter

7 GEOMETRY

PREVIEWING the CHAPTER

This chapter presents the definitions and symbols of the most basic elements of geometry: point, line, line segment, ray, and plane. As an extension of the basic definitions, angles are defined and measured and parallel and perpendicular lines are introduced. Many different polygons are named with special emphasis given to classifying triangles and quadrilaterals. Finally, students are introduced to three-dimensional figures and their properties.

Problem-Solving Strategy Students learn to solve problems by following a four-step heuristic and using logical reasoning.

Lesson (Pages)	Lesson Objectives	State/Local Objectives
7-1 (212-213)	7-1: Identify, name, and draw points, lines, and planes.	
7-2 (214-215)	7-2: Name, draw, and measure angles.	
7-3 (216-217)	7-3: Classify angles.	
7-4 (218-219)	7-4: Make constructions using a compass and a straightedge.	
7-5 (220-221)	7-5: Identify and name parallel and perpendicular lines.	
7-6 (224-225)	7-6: Identify and name various types of polygons.	
7-7 (226-227)	7-7: Identify and name various types of triangles.	
7-8 (228-229)	7-8: Identify and name quadrilaterals.	
7-9 (230-231)	7-9: Identify and name three-dimensional figures.	
7-10 (232-233)	7-10: Solve problems using logical reasoning.	

ORGANIZING the CHAPTER

You may refer to the ***Course Planning Calendar*** on page T10.

Planning Guide

Lesson (Pages)	Extra Practice (Student Edition)	Blackline Masters Booklets: Reteaching	Practice	Enrichment	Activity	Multi-cultural Activity	Technology	Tech Prep	Evaluation	Other Resources
7-1 (212-213)	p. 466	p. 66	p. 72	p. 66			p. 21			Transparency 7-1
7-2 (214-215)	p. 466	p. 67	p. 73	p. 67						Transparency 7-2
7-3 (216-217)	p. 466	p. 68	p. 74	p. 68						Transparency 7-3
7-4 (218-219)		p. 69	p. 75	p. 69	p. 13					Transparency 7-4
7-5 (220-221)	p. 467	p. 70	p. 76	p. 70					p. 34	Transparency 7-5
7-6 (224-225)	p. 467	p. 71	p. 77	p. 71						Transparency 7-6
7-7 (226-227)	p. 467	p. 72	p. 78	p. 72	p. 35	p. 7	p. 7			Transparency 7-7
7-8 (228-229)	p. 468	p. 73	p. 79	p. 73				p. 13		Transparency 7-8
(7-9) (230-231)	p. 468	p. 74	p. 80	p. 74	p. 14			p. 14		Transparency 7-9
7-10 (232-233)		p. 81							p. 34	Transparency 7-10
Ch. Review (234-235)									pp. 31-33	Test and Review Generator
Ch. Test (236)										
Cumulative Review/Test (238-239)									p. 35	

OTHER CHAPTER RESOURCES

Student Edition
Chapter Opener, p. 210
Activity, p. 211
Computer Application, p. 222
Health Connections, p. 223
Journal Entry, p. 225
Portfolio Suggestion, p. 231
Math Connections, p. 237
Tech Prep Handbook, pp. 525-527

Teacher's Classroom Resources
Transparency 7-0
Lab Manual, pp. 35-36
Performance Assessment, pp. 13-14
Lesson Plans, pp. 72-81

Other Supplements
Overhead Manipulative Resources, Labs 20-22, pp. 15-16
Glencoe Mathematics Professional Series

ENHANCING the CHAPTER

Some of the blackline masters for enhancing this chapter are shown below.

COOPERATIVE LEARNING

This activity can be performed by groups of 2-4 students. Provide each group with a copy of a tangram puzzle, Lab Manual, p. 12. Have students cut apart the puzzle pieces. Have students classify each puzzle piece by the shape it is. Then have them arrange the pieces to form a parallelogram and a rectangle.

USING MODELS/MANIPULATIVES

Have each student make a dodecahedron and an octahedron from the patterns below. Have them identify the faces and tell the number of faces, vertices, and edges for each polyhedron.

Dodecahedron

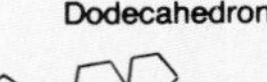

Octahedron

Cooperative Problem Solving, p. 35

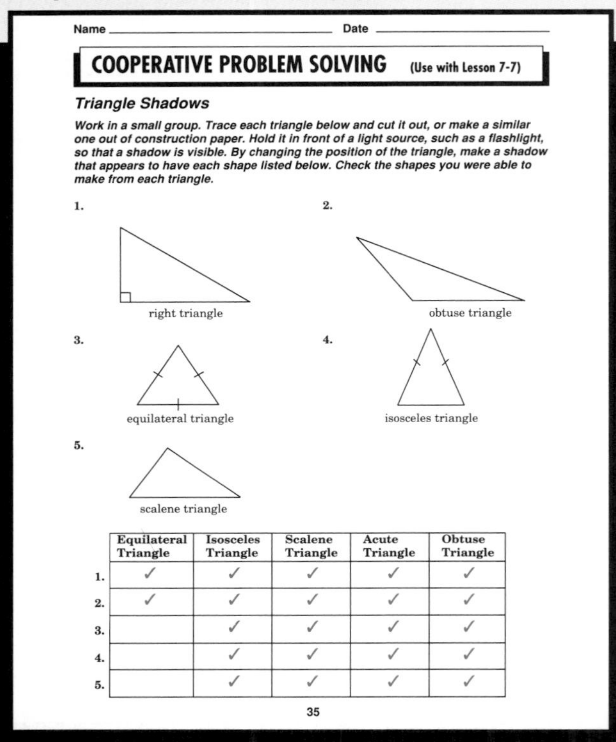

Name ______________________ Date ______________

COOPERATIVE PROBLEM SOLVING (Use with Lesson 7-7)

Triangle Shadows

Work in a small group. Trace each triangle below and cut it out, or make a similar one out of construction paper. Hold it in front of a light source, such as a flashlight, so that a shadow is visible. By changing the position of the triangle, make a shadow that appears to have each shape listed below. Check the shapes you were able to make from each triangle.

1. right triangle

2. obtuse triangle

3. equilateral triangle

4. isosceles triangle

5. scalene triangle

	Equilateral Triangle	Isosceles Triangle	Scalene Triangle	Acute Triangle	Obtuse Triangle
1.	✓	✓	✓	✓	✓
2.	✓	✓	✓	✓	✓
3.		✓	✓	✓	✓
4.		✓	✓	✓	✓
5.		✓	✓	✓	✓

35

Lab Activity, p. 36

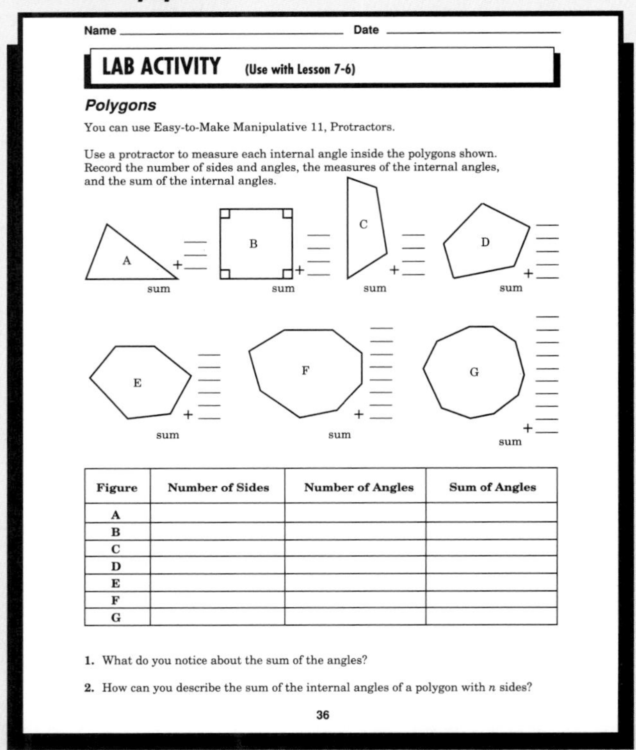

Name ______________________ Date ______________

LAB ACTIVITY (Use with Lesson 7-6)

Polygons

You can use Easy-to-Make Manipulative 11, Protractors.

Use a protractor to measure each internal angle inside the polygons shown. Record the number of sides and angles, the measures of the internal angles, and the sum of the internal angles.

Figure	Number of Sides	Number of Angles	Sum of Angles
A			
B			
C			
D			
E			
F			
G			

1. What do you notice about the sum of the angles?

2. How can you describe the sum of the internal angles of a polygon with n sides?

36

MEETING INDIVIDUAL NEEDS

Mainstreaming

Students with perceptual problems may have a difficult time "seeing" the hidden sides of figures often shown on diagrams with broken lines. Have such students construct models of three-dimensional figures or have such models on display. Allow students to examine the figures from all perspectives so that they can actually see and touch the faces and edges and gain a direct knowledge of the various relationships.

CRITICAL THINKING/PROBLEM SOLVING

Provide students with a copy of the figure below. Challenge them to divide the figure into four pieces that have the same size and shape.

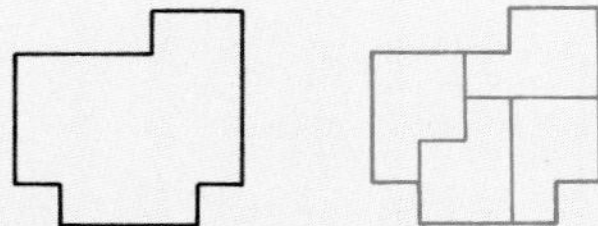

COMMUNICATION

Writing

Have students keep a written record of the types and relationships among triangles and among quadrilaterals. You may want to have students use these relationships to draw and label a family tree for triangles and a family tree for quadrilaterals. For example, the statement "all rectangles are not squares" is known to be true because squares are the "offspring" of rectangles.

Applications, p. 13

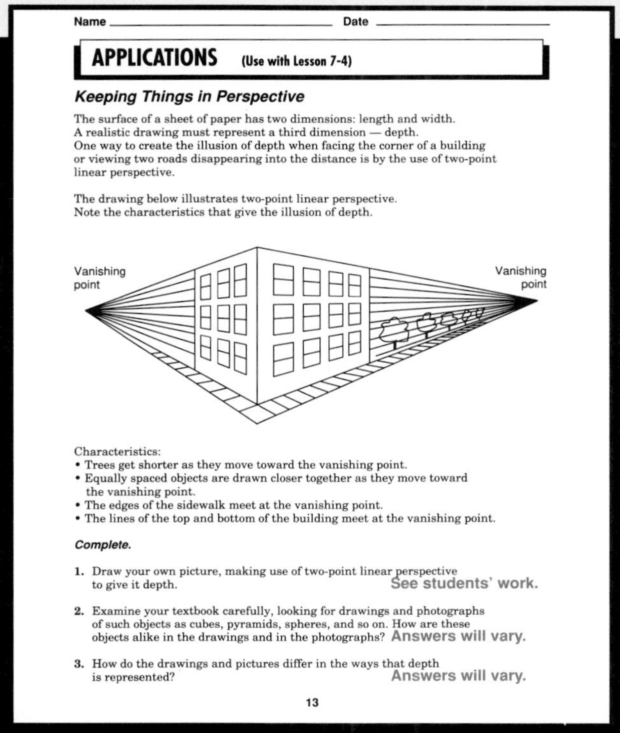

Name ______________ Date ______________

APPLICATIONS (Use with Lesson 7-4)

Keeping Things in Perspective

The surface of a sheet of paper has two dimensions: length and width. A realistic drawing must represent a third dimension — depth. One way to create the illusion of depth when facing the corner of a building or viewing two roads disappearing into the distance is by the use of two-point linear perspective.

The drawing below illustrates two-point linear perspective. Note the characteristics that give the illusion of depth.

Characteristics:
- Trees get shorter as they move toward the vanishing point.
- Equally spaced objects are drawn closer together as they move toward the vanishing point.
- The edges of the sidewalk meet at the vanishing point.
- The lines of the top and bottom of the building meet at the vanishing point.

Complete.

1. Draw your own picture, making use of two-point linear perspective to give it depth. **See students' work.**
2. Examine your textbook carefully, looking for drawings and photographs of such objects as cubes, pyramids, spheres, and so on. How are these objects alike in the drawings and in the photographs? **Answers will vary.**
3. How do the drawings and pictures differ in the ways that depth is represented? **Answers will vary.**

13

Calculator Activity, p. 7

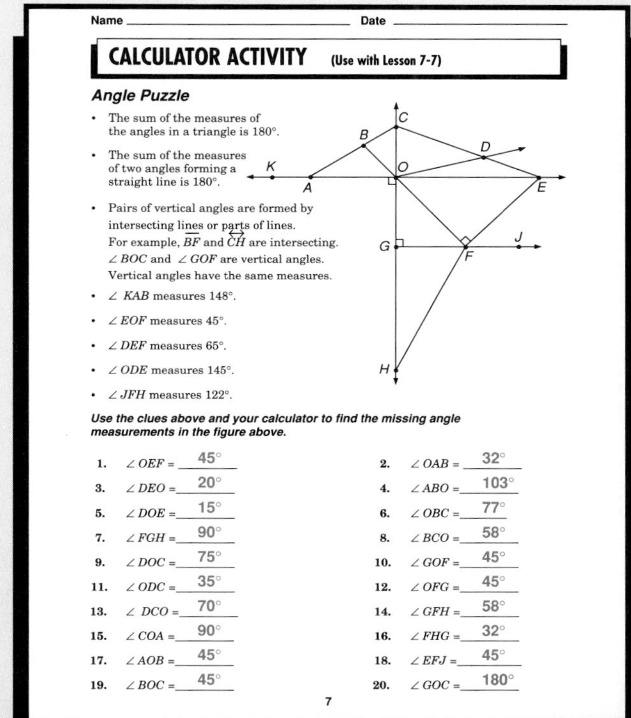

Name ______________ Date ______________

CALCULATOR ACTIVITY (Use with Lesson 7-7)

Angle Puzzle

- The sum of the measures of the angles in a triangle is 180°.
- The sum of the measures of two angles forming a straight line is 180°.
- Pairs of vertical angles are formed by intersecting lines or parts of lines. For example, $\overline{BF}$ and $\overleftrightarrow{CH}$ are intersecting. $\angle BOC$ and $\angle GOF$ are vertical angles. Vertical angles have the same measures.
- $\angle KAB$ measures 148°.
- $\angle EOF$ measures 45°.
- $\angle DEF$ measures 65°.
- $\angle ODE$ measures 145°.
- $\angle JFH$ measures 122°.

Use the clues above and your calculator to find the missing angle measurements in the figure above.

1. $\angle OEF$ = 45°
2. $\angle OAB$ = 32°
3. $\angle DEO$ = 20°
4. $\angle ABO$ = 103°
5. $\angle DOE$ = 15°
6. $\angle OBC$ = 77°
7. $\angle FGH$ = 90°
8. $\angle BCO$ = 58°
9. $\angle DOC$ = 75°
10. $\angle GOF$ = 45°
11. $\angle ODC$ = 35°
12. $\angle OFG$ = 45°
13. $\angle DCO$ = 70°
14. $\angle GFH$ = 58°
15. $\angle COA$ = 90°
16. $\angle FHG$ = 32°
17. $\angle AOB$ = 45°
18. $\angle EFJ$ = 45°
19. $\angle BOC$ = 45°
20. $\angle GOC$ = 180°

7

CHAPTER PROJECT

Have students research and draw optical illusions based on the geometric figures they are currently studying. Have students carefully explain each optical illusion. Optical illusions can be drawn based on line segments, lines, angles, and three-dimensional figures. Students may find the works of M.C. Escher helpful in drawing three-dimensional optical illusions. Two examples of optical illusions appear at the right.

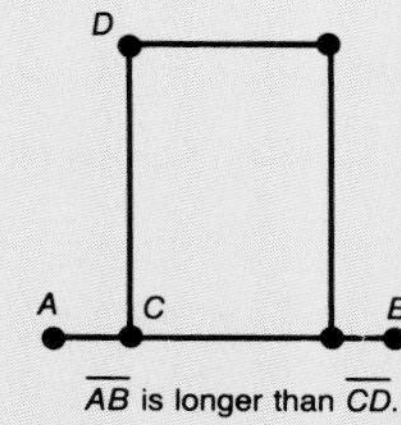

$\overline{AB}$ is longer than $\overline{CD}$.

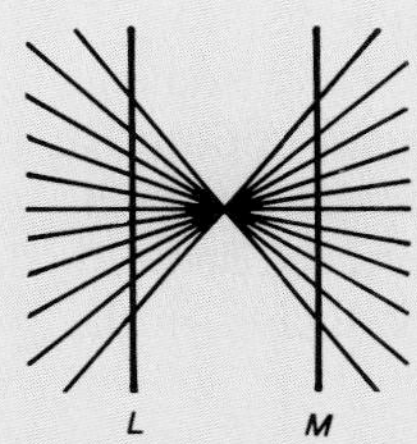

Lines L and M are parallel.

Using the Chapter Opener

Transparency 7-0 is available in the Transparency Package. It provides a full-color visual and motivational activity that you can use to engage students in the mathematical content of the chapter.

Using Discussion

- Have students read the opening paragraph. Ask students if they know what a planned city is.
- Ask students to find a map of their city and bring it to class or you may want to provide copies of their city map. Discuss the layout of the city. Are the streets organized using any certain pattern?
- Encourage students to research the planned city of Savannah. Who planned it and who planned Washington, D.C.?

William Bull and James Oglethorpe planned the city of Savannah. Washington D.C. was planned by a Frenchman, Pierre L'Enfant.

A map of Savannah is shown below. Most streets run north and south or east and west. Savannah is made up of wide tree lined streets that are crossed at intervals by small parks and squares.

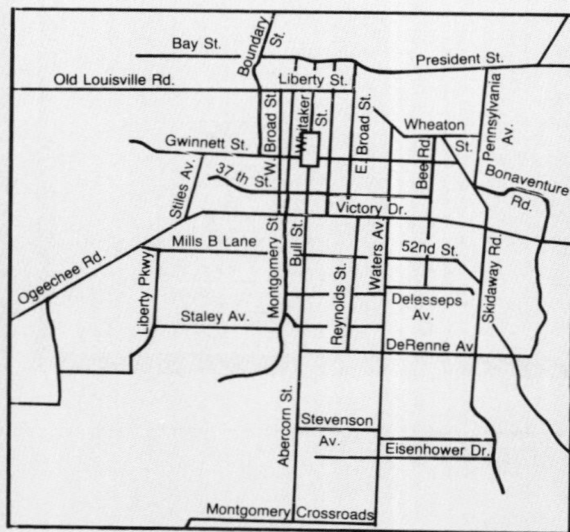

The three bulleted problems are motivational problems. Have students answer them, but it is not critical that they answer the second part of each problem. They will learn the names and definitions of geometric figures in the chapter.

Chapter

7 GEOMETRY

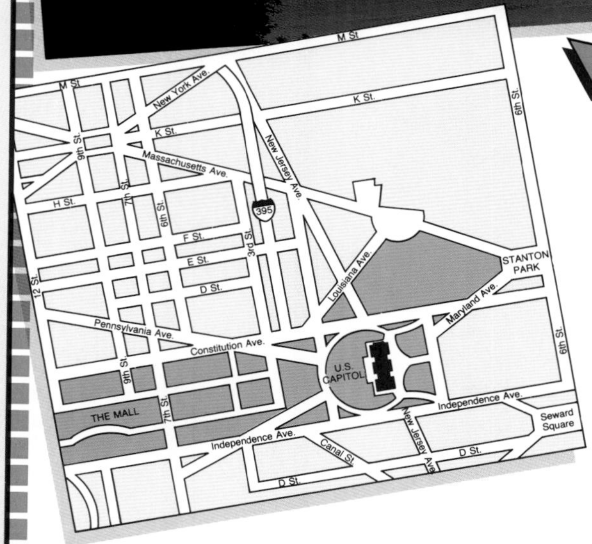

How do I get there from here?

How is your city laid out? Was it a planned city? Washington, D.C., and Savannah, Georgia, are both planned cities. A partial map of Washington, D.C., is shown at the right. Notice how most streets run east and west or north and south.

Can you find out how Savannah is laid out?

- Using the map, name two streets that will never cross. Do you know what these represent in geometry? **parallel lines**
- Name two streets that form square corners. Do you know what these represent in geometry? **right angles**
- Does Maryland Avenue represent a line or a line segment? Explain. **line segment, has a beginning and end**

ACTIVITY: Exploring Constructions

In geometry the compass and straightedge are two important tools. A **straightedge** is any object that can be used to draw a straight line. A **compass** is used to draw circles or parts of circles, called **arcs.**

Constructions are drawings made with only a compass and a straightedge.

Materials: compass, straightedge, and paper

Cooperative Groups

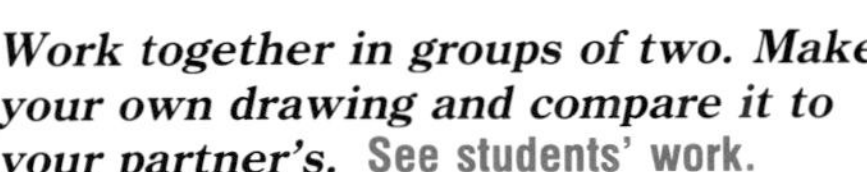

Work together in groups of two. Make your own drawing and compare it to your partner's. **See students' work.**

1. Follow these steps to draw a six-sided figure.

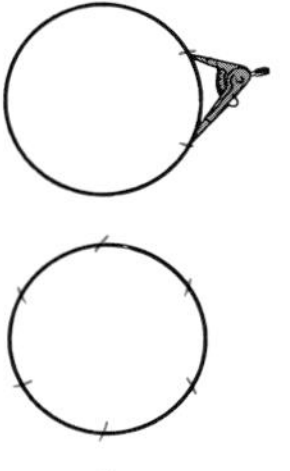

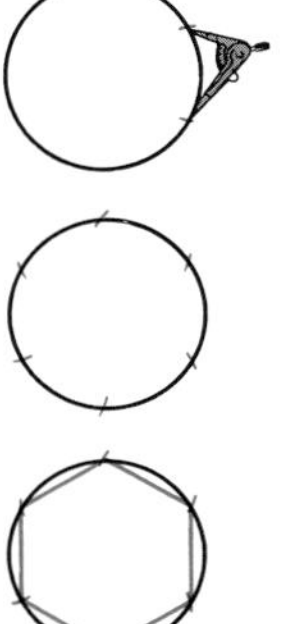

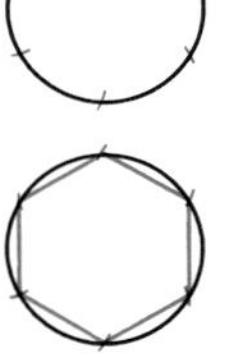

   a. Use the compass to draw a circle on your paper. Do not change the compass setting.

   b. Using the same compass setting, put the metal point on the circle and draw a small arc across the circle.

   c. Move the compass point to the arc and then draw another arc across the circle. Continue doing this until you have six arcs.

   d. Use a straightedge to connect the intersection points in order.

2. What is the name of this six-sided figure? **hexagon**

3. Explain how you would compare the distance between each arc using a compass. What do you notice about the distance between the arcs? **Use the same compass setting to measure the distance between each arc. Distance appears equal.**

4. Discuss with your partner how you could complete steps **a–d** above and then draw a triangle that has sides of equal measure. **Connect three points, skipping every other arc.**

5. Discuss with your partner how to draw the six-petaled flower shown at the right. Then draw the figure using your compass. **See students' work.**

Communicate Your Ideas

6. Why is a construction different from some other types of drawings? **No rulers are used to measure. Only a compass and straightedge are used.**

7. Make up your own design using a compass and straightedge. **See students' work.**

▶ INTRODUCING THE LESSON

Objective Draw geometric figures using a compass and straightedge.

▶ TEACHING THE LESSON

Using Cooperative Groups Remind students not to change the compass setting for a-d in Exercise 1.

Have students discuss Exercises 1-5 with their partners until they are satisfied that their constructions are correct.

▶ EVALUATING THE LESSON

Encourage students to make several attempts at drawing each of the constructions if they are having difficulty.

Communicate Your Ideas Once students have completed their designs, from Exercise 7, they may want to add color.

▶ EXTENDING THE LESSON

Closing the Lesson Encourage students to tell what they have learned about using a compass and straightedge.

Activity Worksheet, Lab Manual, p. 35

Name ______________ Date ______________

ACTIVITY WORKSHEET (Use with page 211)

Exploring Constructions

Be a star! How could you make this design with a compass and a straightedge?

Write a short paragraph to explain why swinging six arcs on a circle each congruent to the radius divides a circle into six congruent arcs.

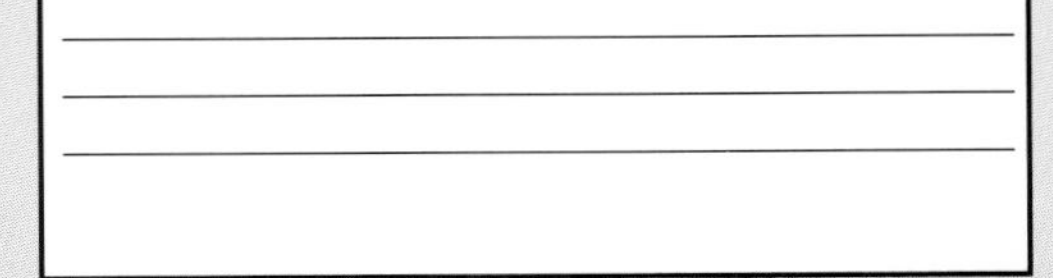

5-MINUTE CHECK-UP

(over Lesson 6-12)

Available on TRANSPARENCY 7-1.

1. Change 50° C to degrees Fahrenheit. 122° F
2. Change 86° F to degrees Celcius. 30° C
3. Change 5 hours to minutes. 300 min
4. Change 390 seconds to minutes. 6.5 min
5. Change 0.5 hours to minutes. 30 min

INTRODUCING THE LESSON

Have students look at the floor, the front wall, the edges of the chalkboard, and so on. Ask them to name the kinds of lines they see.

TEACHING THE LESSON

Using Discussion Discuss with students that points are the basic elements of geometry and that a point is an undefined term. We are only describing it here.

Have students explain the difference between a line, a line segment, and a ray.

Practice Masters Booklet, p. 72

PRACTICE WORKSHEET 7-1

Basic Terms of Geometry

Use symbols to name the line segment between each pair of cities in as many ways as possible.

1. Rochester and Tomah $\overline{RT}$, $\overline{TR}$
2. Chicago and Milwaukee $\overline{CM}$, $\overline{MC}$
3. Eau Claire and Green Bay $\overline{EG}$, $\overline{GE}$
4. St. Paul and Rochester $\overline{SR}$, $\overline{RS}$

Name two other real-life models for each figure. Answers may vary. Sample answers are given.

5. point: pencil tip
   pen point; nail point
6. ray: flashlight beam
   sunbeam; radar beam
7. line segment: pencil
   flag pole; crossbar on goalpost
8. part of a plane: desktop
   door; sidewalk

Use symbols to name each figure in the drawing at the right. Some rays may be named in different ways.

9. 12 rays
   $\overrightarrow{OA}$, $\overrightarrow{CA}$, $\overrightarrow{OB}$, $\overrightarrow{DB}$, $\overrightarrow{AO}$, $\overrightarrow{CO}$, $\overrightarrow{OC}$, $\overrightarrow{AC}$, $\overrightarrow{OD}$, $\overrightarrow{BD}$, $\overrightarrow{BO}$, $\overrightarrow{DO}$
10. 6 line segments
   $\overline{AO}$, $\overline{OC}$, $\overline{AC}$, $\overline{BO}$, $\overline{OD}$, $\overline{BD}$

Use words and symbols to name each figure in as many ways as possible.

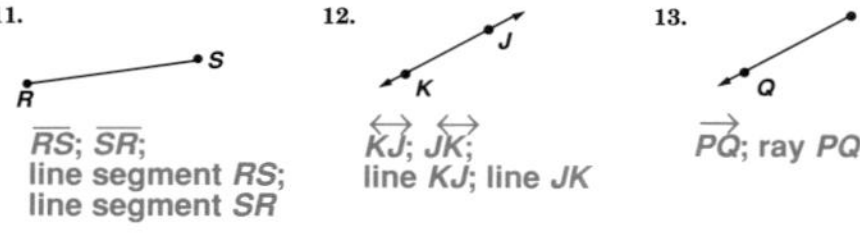

11. $\overline{RS}$; $\overline{SR}$; line segment RS; line segment SR
12. $\overleftrightarrow{KJ}$; $\overleftrightarrow{JK}$; line KJ; line JK
13. $\overrightarrow{PQ}$; ray PQ

7-1 BASIC TERMS OF GEOMETRY

Objective
Identify, name, and draw points, lines, and planes.

Erin is helping her father build a short brick wall around the end of their patio. Erin notices that parts of the wall illustrate a point, a line, a line segment, a ray, and a plane.

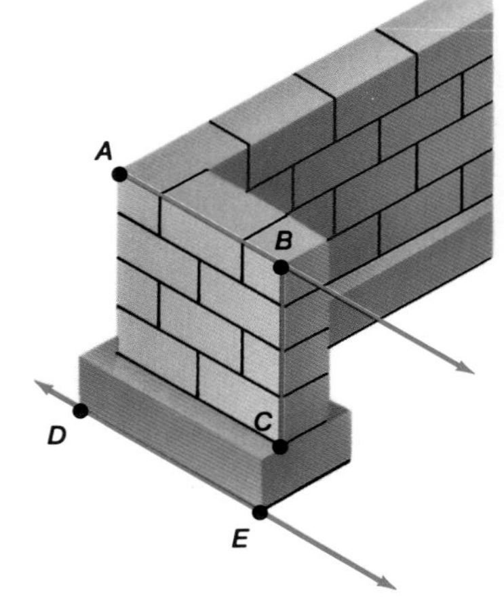

A **point** is an exact location in space. A point is named by a capital letter and is represented by a dot. Point A and point B are two points shown in the drawing at the right.

A **line** is formed by all the points on a never-ending straight path. A line is named by two points on the line. Line DE or line ED, shown in red, is written as $\overleftrightarrow{DE}$ or $\overleftrightarrow{ED}$.

A **line segment** is formed by two endpoints and the straight path between them. Line segment BC or line segment CB, shown in green, is written as $\overline{BC}$ or $\overline{CB}$.

A **ray** is formed by all the points in a never-ending straight path extending in only one direction. A ray is named by its endpoint (*always* given first) and a point on the ray. Ray AB, shown in blue, is written as $\overrightarrow{AB}$.

A **plane** is formed by all the points in a never-ending flat surface. A plane has no boundaries. A plane is named by three points not on a line. Part of plane ABC is shown in gray.

Name three points, two line segments, and two rays in the drawing of the wall that were not named above. **points C, D, E; $\overline{AB}$, $\overline{DE}$; $\overrightarrow{DE}$, $\overrightarrow{ED}$**

Answers may vary. Sample answers are given.
7. pin point; thumbtack
8. pencil; intersection of wall and floor
9. flashlight beam; laser beam
10. wall; floor

Guided Practice

Use symbols to name each figure.

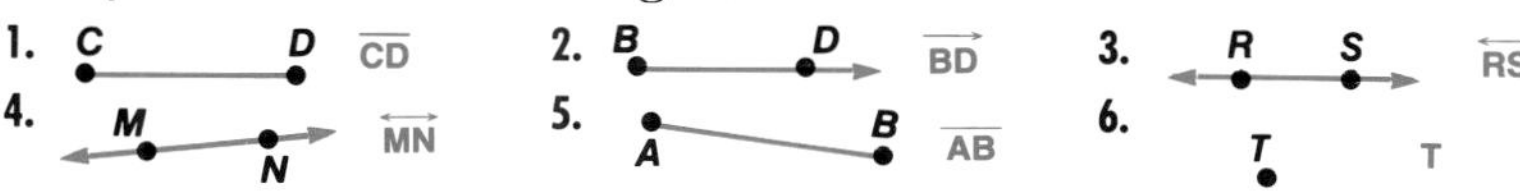

1. $\overline{CD}$
2. $\overrightarrow{BD}$
3. $\overleftrightarrow{RS}$
4. $\overrightarrow{MN}$
5. $\overline{AB}$
6. T

Name two other real-life models for each geometric term.

7. point: pencil tip
8. line segment: ruler
9. ray: radar beam
10. part of a plane: sheet of paper

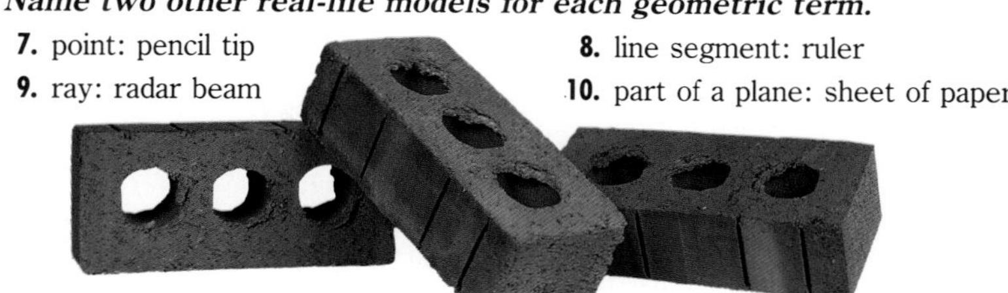

RETEACHING THE LESSON

In the drawing:

1. How many points are named? **3** Names? *A, B, C*
2. How many different lines can be named? **1** Names? $\overleftrightarrow{AB}$
3. How many different line segments can be named? **2** Names? $\overline{AB}$, $\overline{AC}$
4. How many different rays can be named? **3** Names? $\overrightarrow{AC}$, $\overrightarrow{AB}$, $\overrightarrow{BA}$

Reteaching Masters Booklet, p. 66

RETEACHING WORKSHEET 7-1

Basic Terms of Geometry

Figure	Description	Example	Word Name	Symbol
point	indicates exact location	P	point P	P
line	all points on a never-ending straight path	M N	line MN or line NM	$\overleftrightarrow{MN}$ or $\overleftrightarrow{NM}$
line segment	consists of two endpoints and the straight path between them	R S	line segment RS or line segment SR	$\overline{RS}$ or $\overline{SR}$
ray	consists of one endpoint and a never-ending straight path in one direction	Y X	ray YX	$\overrightarrow{YX}$
plane	never-ending flat surface	A C B	plane ABC	ABC

Exercises

Practice

Where and when did geometry originate?
See page 240.

Use words and symbols to name each figure in as many ways as possible. See margin.

15. Name three line segments.

$\overline{AB}$; $\overline{BC}$; $\overline{AC}$

16. Name three rays.

M N O P

$\overrightarrow{MP}$; $\overrightarrow{NP}$; $\overrightarrow{OP}$

Make a drawing of each figure. See answers on p. T522.

17. line AB **18.** point K **19.** line segment MN
20. ray DE **21.** $\overleftrightarrow{AD}$ **22.** $\overrightarrow{RS}$

Critical Thinking
23. 1 line segment
25. See answers on p. T522.

23. How many line segments can you draw through two points?
24. How many lines can you draw through one point? infinitely many
25. Draw models to illustrate your answers in Exercises 23–24.
26. How many rays can have the same endpoint? infinitely many
27. How many planes can contain the same line? infinitely many
28. How many points, not in a line, determine a plane? 3 points

Show Math

29. Make a drawing of a desk using a ruler and pencil. Label the corner points with letters. Name a point, a line segment, a ray, and a plane. See students' work.

Estimation
33. $\frac{14}{4}$ or $3\frac{1}{2}$ inches

Estimate the length of each line segment to the nearest $\frac{1}{4}$ inch.

30. $\overline{BC}$ $\frac{2}{4}$ or $\frac{1}{2}$ inch **31.** $\overline{CD}$ $\frac{3}{4}$ inch **32.** $\overline{BD}$ $\frac{5}{4}$ or $1\frac{1}{4}$ inches **33.** $\overline{AD}$

Interpret Data

34. Draw a straight line. Display the data from the table on the line using points, line segments, and rays. See answer on p. T522.

SIT-UPS (MEN)	
Number	Fitness Level
50+	very high
40–49	high
30–39	average
20–29	low
0–19	very low

Mixed Review

Lesson 2-3

Subtract.

35. $89 - 37$ = 52 **36.** $223 - 147$ = 76 **37.** $4{,}621 - 858$ = 3,763

Lesson 4-2

State whether each number is prime or composite.

38. 18 C **39.** 23 P **40.** 31 P **41.** 49 C

Lesson 4-14

42. Brandon has 15 gallons of gasoline in his car. A trip to his aunt's house uses ten gallons. Then Brandon puts three gallons more in the tank. A trip to a ballgame uses four gallons and then Brandon puts another eight gallons in the tank. How many gallons of gasoline are in the tank now? 12 gal

Chalkboard Examples

Use symbols to name each figure.

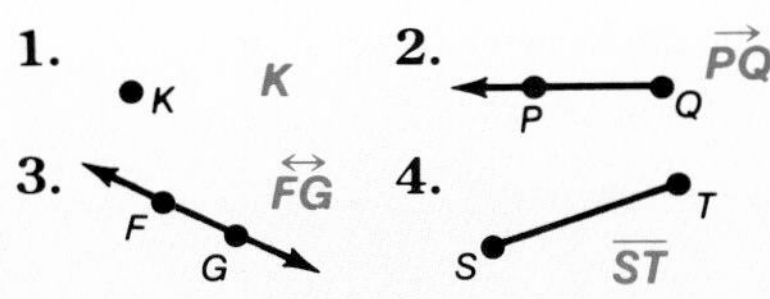

Additional examples are provided on TRANSPARENCY 7-1.

EVALUATING THE LESSON

Check for Understanding Use Guided Practice Exercises to check for student understanding.

Closing the Lesson Tell students that they will use the basic terms of geometry in working with angles, constructions, and polygons.

EXTENDING THE LESSON

Enrichment Give each student a copy of a ruler from **Lab Manual, p. 10** and a copy of a section of a map that includes the mileage scale. Have students estimate the mileage between two towns.

Enrichment Masters Booklet, p. 66

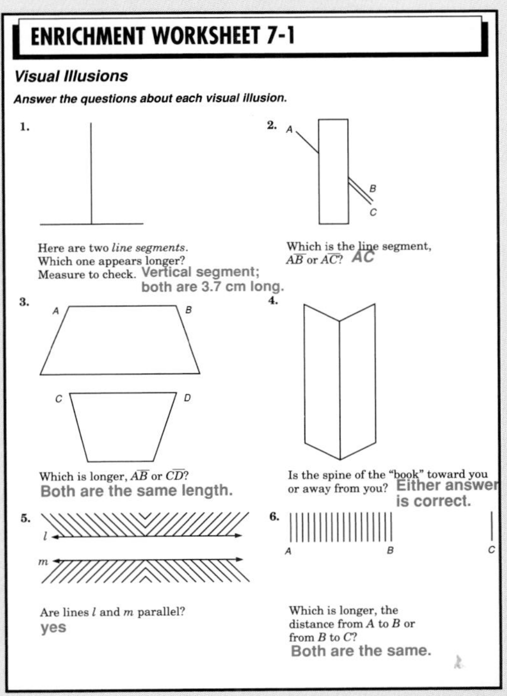
ENRICHMENT WORKSHEET 7-1

Visual Illusions

Answer the questions about each visual illusion.

1. Here are two *line segments*. Which one appears longer? Measure to check. Vertical segment; both are 3.7 cm long.

2. Which is the line segment, $\overline{AB}$ or $\overline{AC}$? $\overline{AC}$

3. Which is longer, $\overline{AB}$ or $\overline{CD}$? Both are the same length.

4. Is the spine of the "book" toward you or away from you? Either answer is correct.

5. Are lines l and m parallel? yes

6. Which is longer, the distance from A to B or from B to C? Both the same.

APPLYING THE LESSON

Independent Practice

Homework Assignment	
Minimum	Ex. 1-35 odd; 36-47
Average	Ex. 2-24 even; 25-42
Maximum	Ex. 1-42

Additional Answers

11. $\overline{GH}$; $\overline{HG}$; line segment GH; line segment HG
12. $\overleftrightarrow{TQ}$; $\overleftrightarrow{QT}$ line TQ; line QT
13. point M M
14. $\overrightarrow{HK}$ ray HK

(over Lesson 7-1)

Available on TRANSPARENCY 7-2.
Use symbols to name each figure.

1. B C $\overline{BC}$ 2. M M
3. P T $\overrightarrow{PT}$ 4. A D $\overleftrightarrow{AD}$

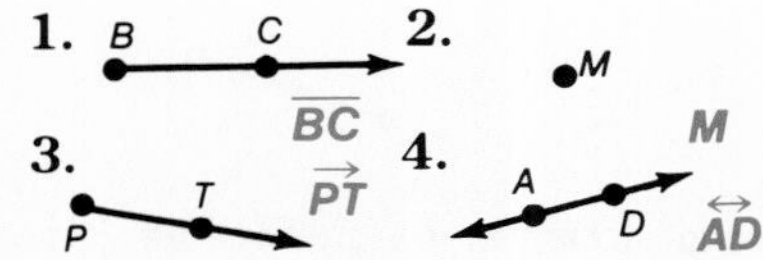

Extra Practice, Lesson 7-1, p. 466

▶ INTRODUCING THE LESSON

Have students look around the classroom and name items that contain an angle.

▶ TEACHING THE LESSON

Using Critical Thinking Ask students if the length of the ray affects the size of the angle. Have students explain their answer. **No The size of the opening between the rays determines the size of the angle.**

Ask students why a protractor has two sets of numbers. **So you don't have to turn the protractor if the angle faces the opposite direction.**

Practice Masters Booklet, p. 73

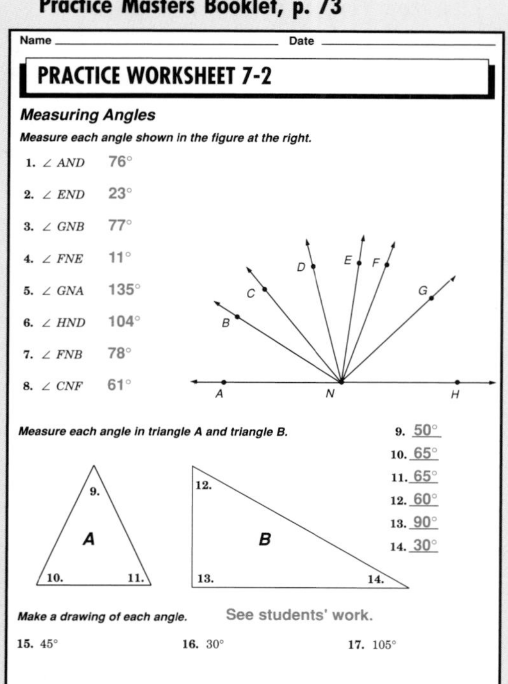

Name ______ Date ______

PRACTICE WORKSHEET 7-2

Measuring Angles

Measure each angle shown in the figure at the right.

1. ∠ AND 76°
2. ∠ END 23°
3. ∠ GNB 77°
4. ∠ FNE 11°
5. ∠ GNA 135°
6. ∠ HND 104°
7. ∠ FNB 78°
8. ∠ CNF 61°

Measure each angle in triangle A and triangle B.

9. 50°
10. 65°
11. 65°
12. 60°
13. 90°
14. 30°

Make a drawing of each angle. See students' work.

15. 45° 16. 30° 17. 105°

7-2 MEASURING ANGLES

Objective
Name, draw, and measure angles.

Tom Andres is a drafter. He is preparing a detailed drawing of a bridge from an engineer's rough sketch. One tool Tom uses is a protractor. Using a protractor helps Tom measure and draw angles to the correct size.

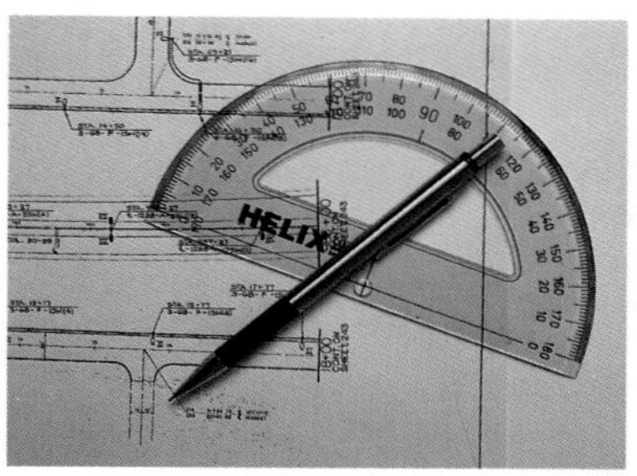

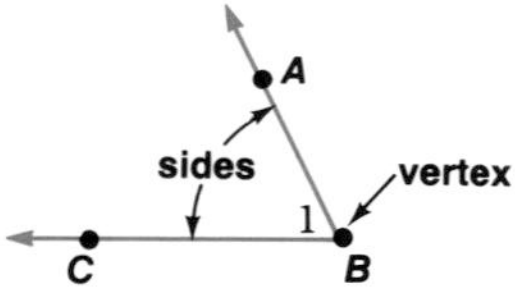

Two rays with a common endpoint form an **angle.** The common endpoint is called the **vertex.** The rays form the **sides** of the angle. In the angle shown at the left, Point B is the vertex. $\overrightarrow{BA}$ and $\overrightarrow{BC}$ are the sides of the angle.

Angle ABC is symbolized as $\angle ABC$. The letter naming the vertex must be in the middle. The angle can also be named as $\angle CBA$, $\angle B$, or $\angle 1$.

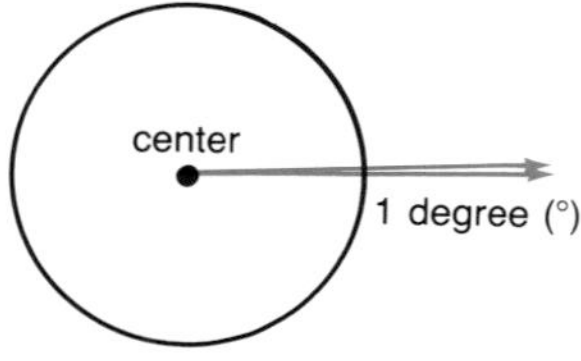

The most common unit used in measuring angles is the **degree.** Imagine a circle separated into 360 equal-sized angles with their vertex at the center. Each angle would measure one degree (1°) as shown at the left.

A **protractor** is used to measure angles.

Method

Use a protractor to measure angle **DEF.**

1. Place the center of the protractor on the vertex *(E)* and the 0° mark on one side of the angle *($\overrightarrow{ED}$)*.
2. Use the scale that begins with 0 at $\overrightarrow{ED}$. Read where the other ray *($\overrightarrow{EF}$)* crosses this scale.

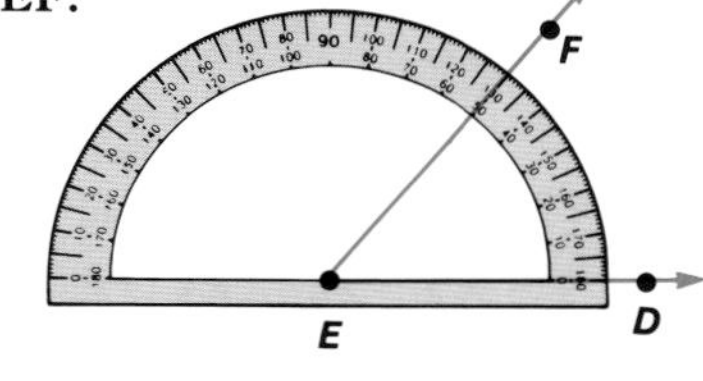

$\angle DEF$ measures 50°. This is symbolized $m\angle DEF = 50°$.

Guided Practice

Use symbols to name each angle in three ways.

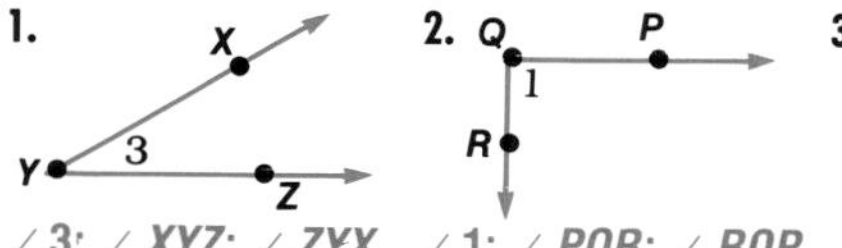

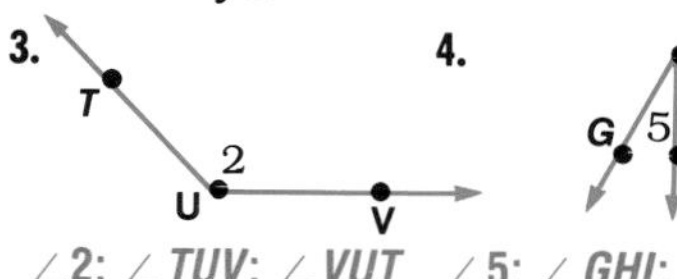

1. **∠3; ∠XYZ; ∠ZYX** 2. **∠1; ∠PQR; ∠RQP** 3. **∠2; ∠TUV; ∠VUT** 4. **∠5; ∠GHI; ∠IHG**

RETEACHING THE LESSON

Students work in small groups. *Make a drawing of each angle.* **See students' work.**

1. acute $\angle X$ 2. obtuse $\angle EFG$
3. right $\angle 3$

Measure each angle.

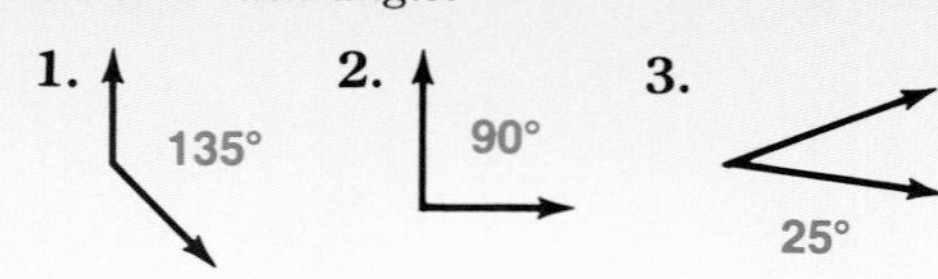

Reteaching Masters Booklet, p. 67

Name ______ Date ______

RETEACHING WORKSHEET 7-2

Measuring Angles

A protractor is used to measure angles. To measure ∠ ABC, place the center of the protractor on the vertex of the angle and the 0° mark on $\overrightarrow{BC}$. Read the scale that begins with 0. The degree measure of the angle is where $\overrightarrow{AB}$ crosses that scale.

The measure of ∠ ABC is 45°.

Measure each angle.

Exercises

Trace each angle and extend the sides. Then find the measure of each angle.

Practice

5. 74° 6. 123° 7. 90°

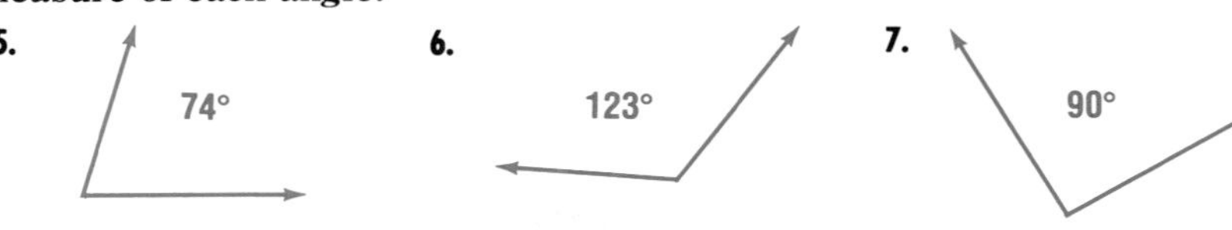

8. 24° 9. 136° 10. 80°

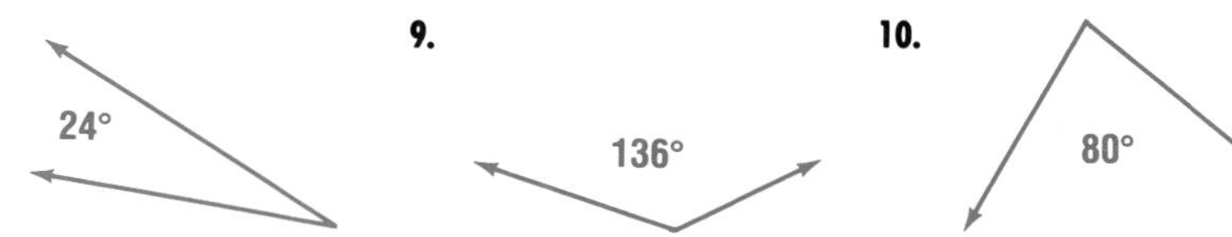

Talk Math

Use the figure at the right.

11. Name five angles. Explain why none of the angles can be named ∠I. ∠HIL; ∠HIK; ∠LIJ; ∠LIK; ∠KIJ

11. There is more than one angle with vertex I.

Estimation

12. Use the figure in Exercise 11. Estimate the degree measure of ∠HIL, ∠LIK, and ∠KIJ. Then use a protractor to measure ∠HIL, ∠LIK, and ∠KIJ. How close are your estimates to the actual measures?

12. Answers may vary. 55°, 90°, 35° Answers may vary.

Applications

13. Use the figure in Exercise 11. What is m∠HIL + m∠LIK + m∠KIJ? 180°

14. When you do sit-ups in an exercise program, describe the angle or angles your body makes. Answers may vary.

15. When you do push-ups in an exercise program describe the angle or angles your body makes with the floor. Answers may vary.

Show Math

16. Draw two lines as shown at the right. Measure each of the four angles. What do you notice about the sum of the four measures? The sum is 360°.

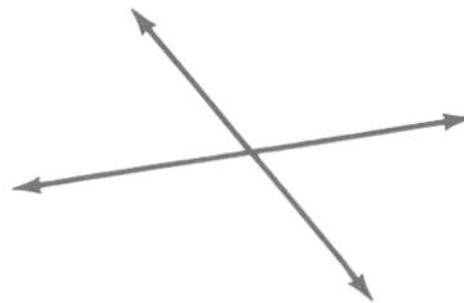

Mixed Review

Find the common ratio and write the next three terms in each sequence.

Lesson 3-9

17. 112, 56, 28, . . . 2; 14; 7; 3.5

18. 192, 48, 12, . . . 4; 3; $\frac{3}{4}$; $\frac{3}{16}$

Lesson 4-5

Replace each ■ with a number so that the fractions are equivalent.

19. $\frac{3}{5} = \frac{■}{15}$ 9

20. $\frac{8}{9} = \frac{■}{27}$ 24

21. $\frac{4}{7} = \frac{■}{35}$ 20

22. $\frac{1}{12} = \frac{■}{48}$ 4

Lesson 5-5

23. Six employees can make 24 pizzas in 30 minutes. How many pizzas can 10 employees make in $1\frac{1}{2}$ hours? 120 pizzas

APPLYING THE LESSON

Independent Practice

Homework Assignment	
Minimum	Ex. 1-23
Average	Ex. 1-23
Maximum	Ex. 1-23

Chalkboard Examples

Use symbols to name each angle in three ways.

1. ∠P, ∠APB, ∠3 2. ∠Q, ∠SQR, ∠2

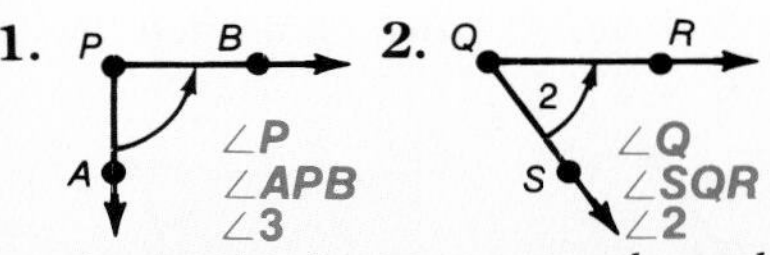

Use a protractor to measure each angle.

3. 90° 4. 48°

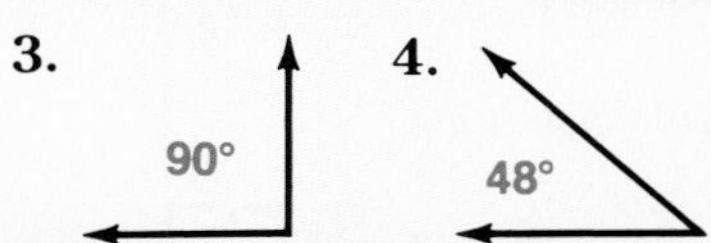

Additional examples are provided on TRANSPARENCY 7-2.

EVALUATING THE LESSON

Check for Understanding Use Guided Practice Exercises to check for student understanding.

Error Analysis Some students may read the wrong scale when measuring an angle.

Closing the Lesson Ask students to state in their own words how to measure an angle.

EXTENDING THE LESSON

Enrichment Masters Booklet, p. 67

ENRICHMENT WORKSHEET 7-2

Line Symmetry

Some geometric figures can be folded in half so that one part fits exactly on the other. This property is called **line symmetry**. Figures that have line symmetry are often used in designs ranging from building plans to corporate logos (those designs that companies use to keep their images in our minds).

When a figure is folded, the fold itself is part of a straight line. Thus, it is referred to as the line of symmetry. The figure is symmetrical with respect to that line.

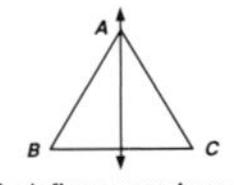

This isosceles triangle is symmetrical to a line that is perpendicular to its base. Notice that if we folded the triangle on this line, point *B* would coincide exactly with point *C*.

Each figure may have one, more than one, or no line of symmetry. For each figure, draw each line of symmetry. If a figure has no line of symmetry, write none.

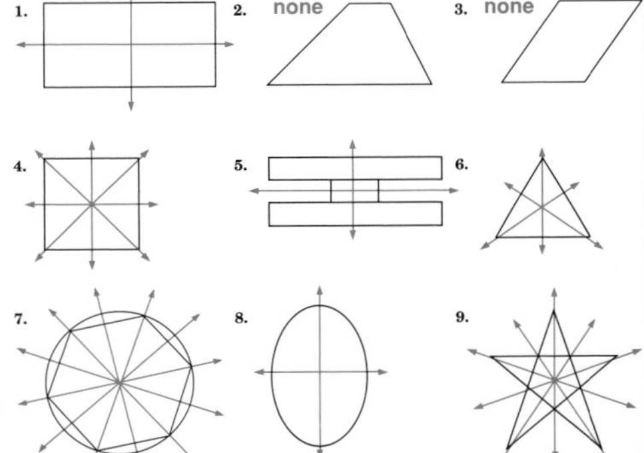

5-MINUTE CHECK-UP

(over Lesson 7-2)

Available on TRANSPARENCY 7-3.

Use symbols to name each angle in three ways.

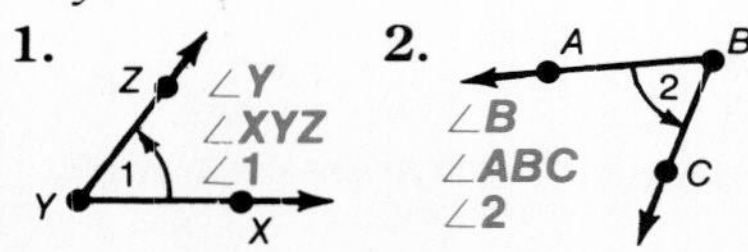

Measure each angle.

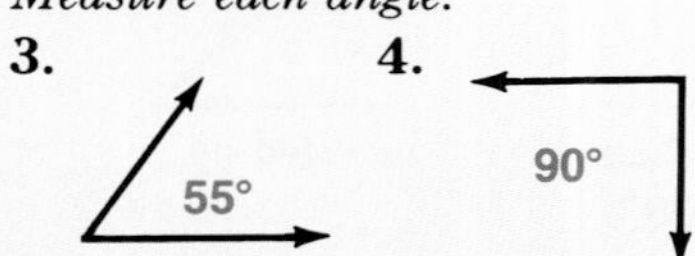

Extra Practice, Lesson 7-2, p. 466

INTRODUCING THE LESSON

Give students a copy of a city map with certain streets and intersections marked. Have students identify right, acute, and obtuse angles.

TEACHING THE LESSON

Using Manipulatives Have students construct a right angle on tracing paper. This can be used to classify angles rather than a protractor.

Practice Masters Booklet, p. 74

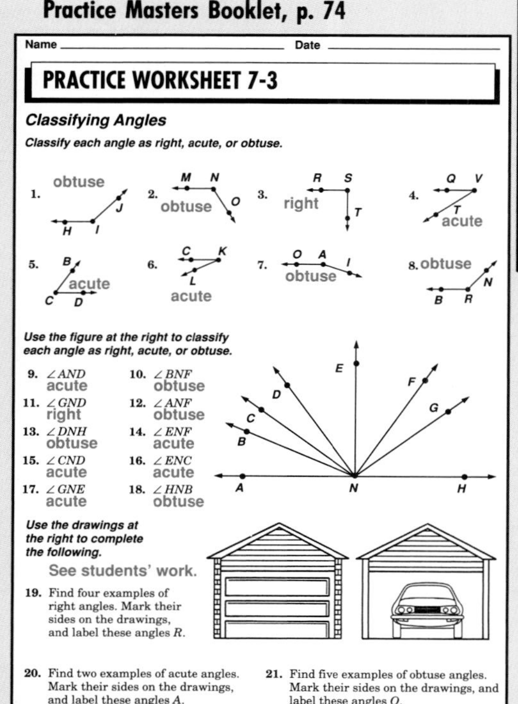
Name ____________ Date ____________

PRACTICE WORKSHEET 7-3

Classifying Angles

Classify each angle as right, acute, or obtuse.

1. obtuse 2. obtuse 3. right 4. acute

5. acute 6. acute 7. obtuse 8. obtuse

Use the figure at the right to classify each angle as right, acute, or obtuse.

9. ∠AND acute 10. ∠BNF obtuse
11. ∠GND right 12. ∠ANF obtuse
13. ∠DNH obtuse 14. ∠ENF acute
15. ∠CND acute 16. ∠ENC acute
17. ∠GNE acute 18. ∠HNB obtuse

Use the drawings at the right to complete the following.

See students' work.

19. Find four examples of right angles. Mark their sides on the drawings, and label these angles *R*.

20. Find two examples of acute angles. Mark their sides on the drawings, and label these angles *A*.

21. Find five examples of obtuse angles. Mark their sides on the drawings, and label these angles *O*.

7-3 CLASSIFYING ANGLES

Objective *Classify angles.*

Bridges are classified according to the way they are constructed. The Golden Gate Bridge in San Francisco is a suspension bridge. Likewise, an angle can be classified according to its degree measure.

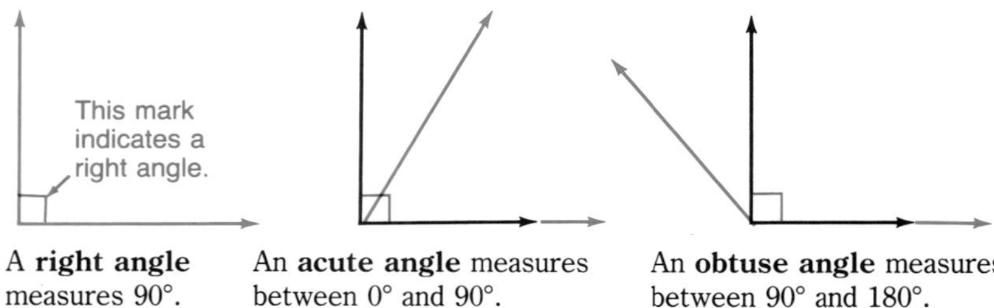

A **right angle** measures 90°. An **acute angle** measures between 0° and 90°. An **obtuse angle** measures between 90° and 180°.

Method

1. Measure the angle.
2. Determine if the measure is greater than 90°, less than 90°, or equal to 90°.
3. Classify the angle. If the measure is greater than 90° it is an obtuse angle. If less than 90°, and less than 180°, it is an acute angle. If equal to 90°, it is a right angle.

Example A ***Classify*** $\angle$**PQR.**

1. $m\angle PQR = 75°$
2. $75° < 90°$
3. Since 75° is less than 90°, $\angle PQR$ is an acute angle.

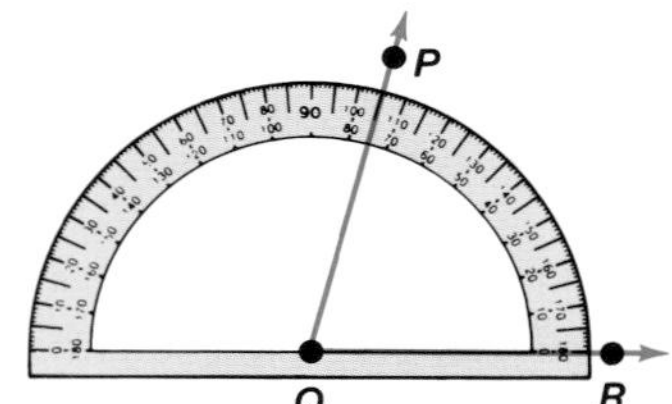

Guided Practice ***Classify each angle.***
Example A

1. 65°; acute
2. 90°; right
3. 120°; obtuse

RETEACHING THE LESSON

Have students show each time on a clock face and tell whether the angle formed by the hands of the clock is a *right angle, acute angle,* or *obtuse angle.*

1. 9:00 Right
2. 7:00 Obtuse
3. 12:10 Acute
4. 4:00 Obtuse

Reteaching Masters Booklet, p. 68

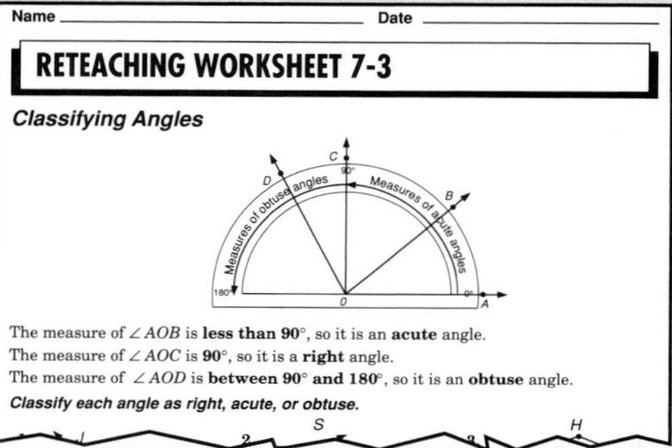
Name ____________ Date ____________

RETEACHING WORKSHEET 7-3

Classifying Angles

The measure of ∠*AOB* is **less than 90°**, so it is an **acute** angle.
The measure of ∠*AOC* is **90°**, so it is a **right** angle.
The measure of ∠*AOD* is **between 90° and 180°**, so it is an **obtuse** angle.

Classify each angle as right, acute, or obtuse.

Exercises

Practice

Classify each angle as right, acute, or obtuse.

4. 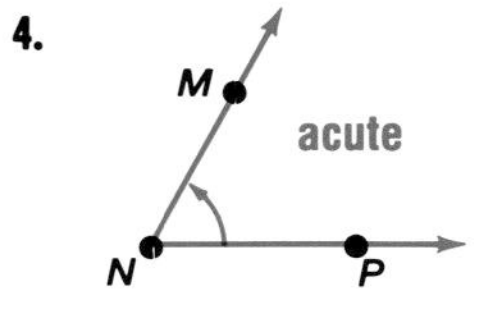

acute

5. 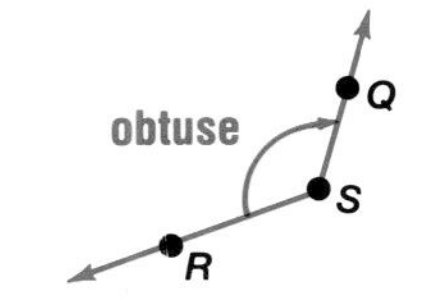

obtuse

6. 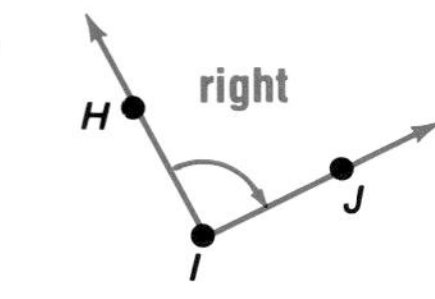

right

7. 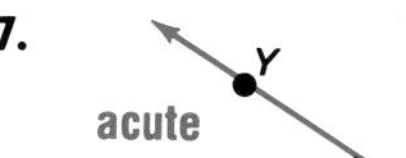

acute

8. 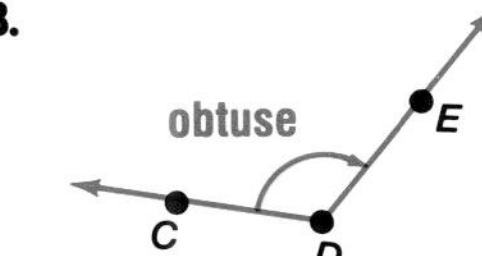

obtuse

9. 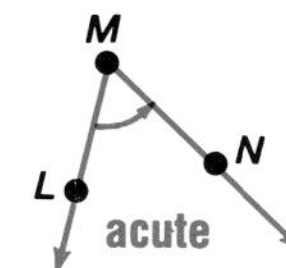

acute

Use the figure at the right to classify each angle.

10. $\angle BFA$ acute **11.** $\angle AFC$ right
12. $\angle BFD$ right **13.** $\angle DFA$ obtuse
14. $\angle DFE$ acute **15.** $\angle BFE$ obtuse

Talk Math

16. In the figure at the right, explain why none of the angles are named $\angle F$. All angles have F as the vertex.

Make Up a Problem

17. Make up a problem using the measures of two acute angles that, when added, equal 90°. See students' work.

Show Math

18. Make a drawing that shows that the measure of one acute angle plus the measure of one obtuse angle can equal exactly 180°. Answers may vary. See margin.

19. 30°; 60° You may want to include supplementary angles with Exercise 20.

Problem Solving

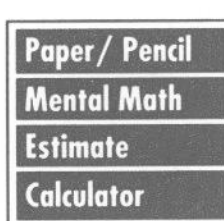

19. Two angles are **complementary** if the sum of their measures is 90°. If the measure of an angle is half of its complement, what is the measure of each angle?

20. From 12:00 noon to 3:00 P.M., how many times do the hands of a standard clock form a right angle?
6 times: 12:16, 12:49, 1:22, 1:54, 2:27, 3:00

Critical Thinking

21. Each figure below shows rays with a common endpoint. Count the number of angles in each figure.

FUN with MATH
Try a few magic squares.
See page 241.

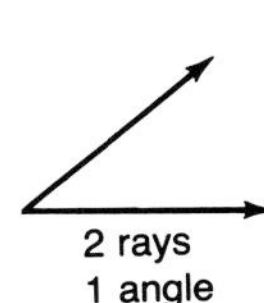
2 rays
1 angle

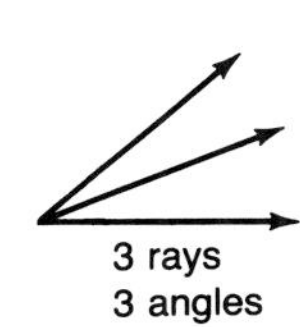
3 rays
3 angles

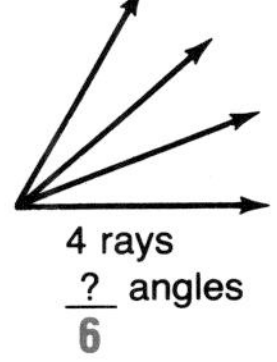
4 rays
? angles
6

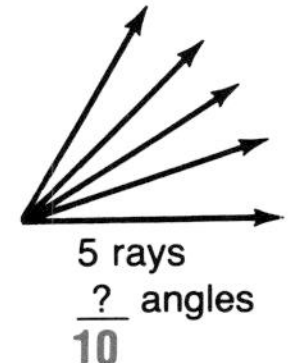
5 rays
? angles
10

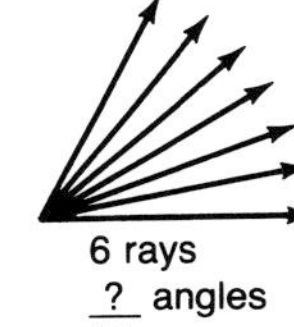
6 rays
? angles
15

Do you see a pattern? Predict the number of angles that are formed by 7 rays. 21 angles

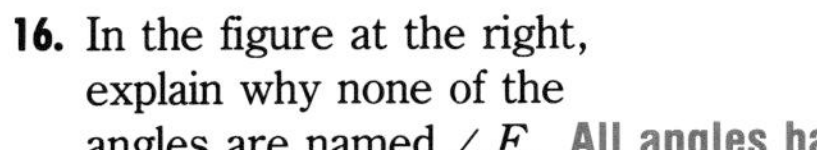

APPLYING THE LESSON

Independent Practice

Homework Assignment	
Minimum	Ex. 1-18; 20; 21
Average	Ex. 1-21
Maximum	Ex. 1-21

Additional Answers

18.

Chalkboard Examples

Example A *Classify each angle.*

1. obtuse **2.** right

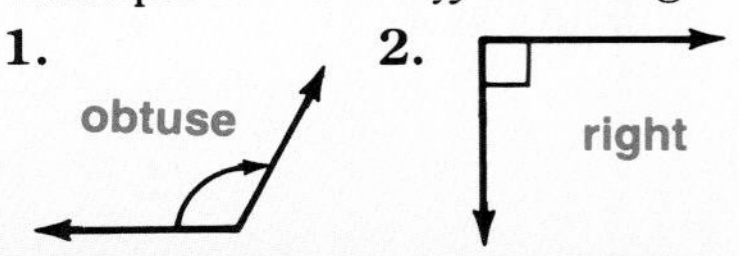

Additional examples are provided on TRANSPARENCY 7-3.

EVALUATING THE LESSON

Check for Understanding Use Guided Practice Exercises to check for student understanding.

Closing the Lesson Have students tell in their own words how to determine if an angle is right, acute, or obtuse.

EXTENDING THE LESSON

Enrichment When two half-planes share the same edge, they form a **dihedral angle.** A dihedral angle is named by its edge and a point in each half-plane. For example, the dihedral angle shown is $\angle D\text{-}\overline{XY}\text{-}E$.

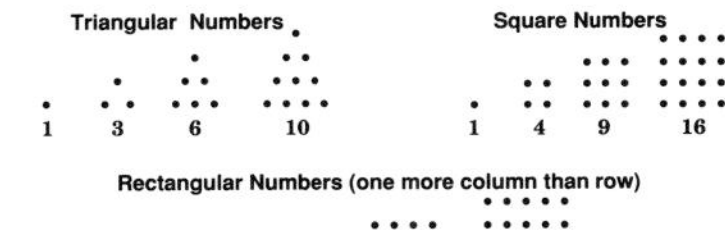

Have students draw and name their own dihedral angle.

Enrichment Masters Booklet, p. 68

ENRICHMENT WORKSHEET 7-3

Figurate Numbers

A **figurate number** is a number whose geometric representation looks like a triangle, a square, a rectangle, or another polygon.

Triangular Numbers: 1 3 6 10
Square Numbers: 1 4 9 16
Rectangular Numbers (one more column than row): 2 6 12 20

Study the patterns above to find the next ten numbers for each of the following kinds of numbers.

1. triangular 15, 21, 28, 36, 45, 55, 66, 78, 91, 105
2. square 25, 36, 49, 64, 81, 100, 121, 144, 169, 196
3. rectangular 30, 42, 56, 72, 90, 110, 132, 156, 182, 210

Add.

4. 1 + 3 + 5 9
5. 1 + 3 + 5 + 7 + 9 25
6. 1 + 3 + 5 + 7 + 9 + 11 + 13 49
7. 1 + 3 + 5 + 7 + 9 + 11 36
8. 1 + 3 + 5 + 7 + 9 + 11 + 13 + 15 + 17 81
9. The sum of consecutive odd numbers beginning with one is a square number.

Find the value of each of the following.

10. (8 × 1) + 1 9
11. (8 × 3) + 1 25
12. (8 × 6) + 1 49
13. (8 × 10) + 1 81
14. (8 × 15) + 1 121
15. (8 × 21) + 1 169
16. One greater than 8 times any triangular number is a square number.
17. Illustrate that any rectangular number is 2 times a triangular number.
2 × 1 = 2 2 × 3 = 6 2 × 6 = 12 2 × 10 = 20

(over Lesson 7-3)

Available on TRANSPARENCY 7-4.

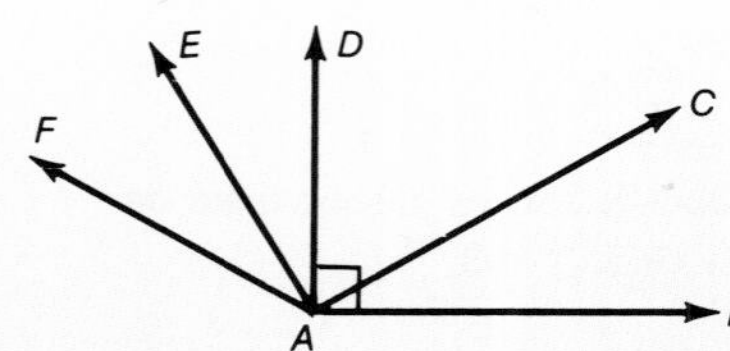

Use the figure above to classify each angle.

1. $\angle BAE$ obtuse **2.** $\angle BAD$ right
3. $\angle FAB$ obtuse **4.** $\angle EAD$ acute

Extra Practice, Lesson 7-3, p. 466

INTRODUCING THE LESSON

Have students name different types of tools they have used around the house. Ask students what they used them for and how they used them.

TEACHING THE LESSON

Using Questioning

- What does congruent mean? having the same measure
- What does bisect mean? to divide in half

Practice Masters Booklet, p. 75

Name ______ Date ______

PRACTICE WORKSHEET 7-4

Applications: Congruent Figures and Constructions

Construct a line segment congruent to each line segment. Then bisect each line segment.

See students' work.

1. A B
2. C D

Construct an angle congruent to each angle.

See students' work.

3. P Q R
4. M N O
5. D E F

Applications

7-4 CONGRUENT FIGURES AND CONSTRUCTIONS

Objective
Make constructions using a compass and a straightedge.

Mika Han is an architect. She is working on a design for a new building. Sometimes she uses a compass and straightedge to draw geometric constructions.

Congruent line segments and congruent angles can be constructed using a compass and a straightedge.

Congruent line segments have the same measure.
Congruent angles have the same measure.

Construction A ***Construct a line segment congruent to a given line segment.***

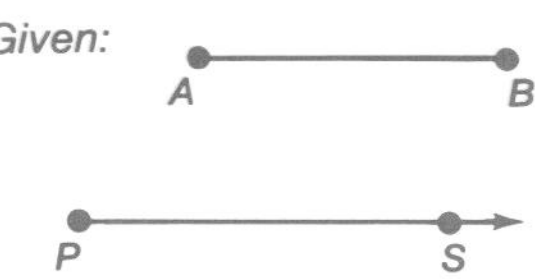

1. Use a straightedge to draw $\overrightarrow{PS}$ longer than $\overline{AB}$.

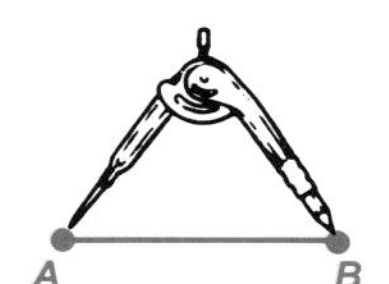

2. Open the compass to match $\overline{AB}$.

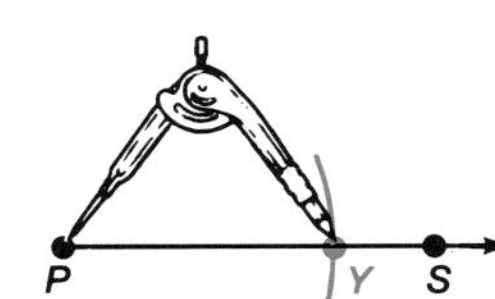

3. Use the same compass setting. With the compass point at *P*, draw an arc that intersects $\overrightarrow{PS}$ at *Y*, $\overline{PY} \cong \overline{AB}$.

Construction B ***Construct an angle congruent to a given angle.***

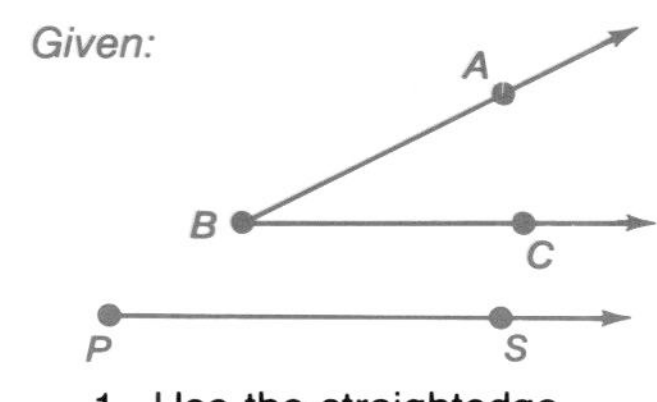

1. Use the straightedge to draw $\overrightarrow{PS}$.

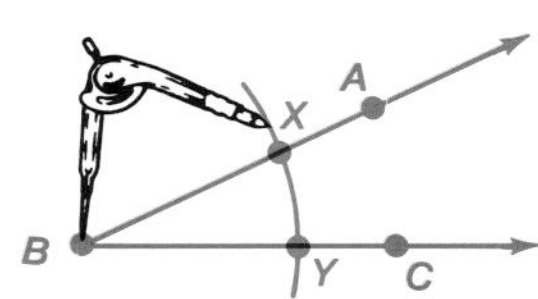

2. With the compass point at *B*, draw an arc that intersects $\angle ABC$ at *X* and *Y*.

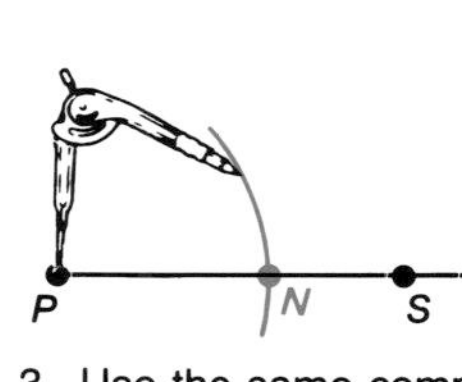

3. Use the same compass setting. With the compass point as P, draw an arc that intersects $\overrightarrow{PS}$ at *N*.

4. Open the compass to match *XY*. With the compass point at *N*, draw an arc. Label the intersection of the arcs *L*. Use a straightedge to draw $\overrightarrow{PL}$. $\angle LPS \cong \angle ABC$.

RETEACHING THE LESSON

For those students who need extra practice, have them make each construction.

1. Construct a line segment congruent to $\overline{AB}$.
2. Bisect $\overline{AB}$.
3. Construct an angle congruent to $\angle XYZ$.

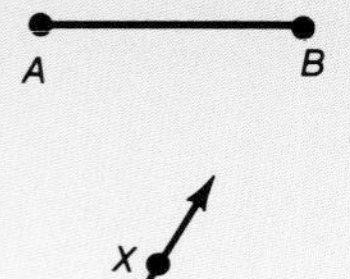

Reteaching Masters Booklet, p. 69

Name ______ Date ______

RETEACHING WORKSHEET 7-4

Applications: Congruent Figures and Constructions

Follow the steps for Construction A on page 218 of your student book to construct a line segment congruent to each segment.

1. H I
2. P Q

See students' work.

Follow the steps for Construction B on page 218 of your student book to construct an angle congruent to each angle.

3.
4.

You can also use a compass and a straightedge to bisect a line segment. **Bisect** means to separate into two congruent parts.

Construction C ***Bisect a given line segment.***

Given:

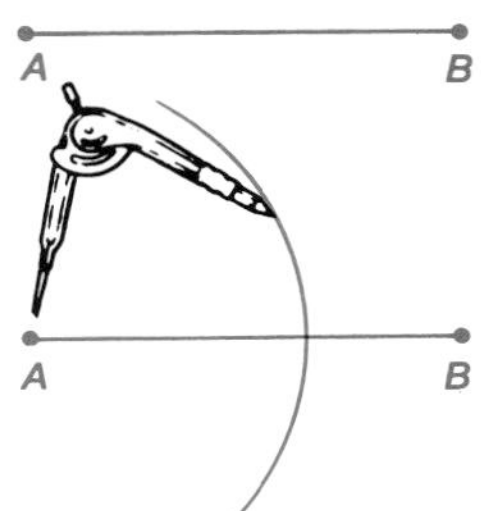

1. Open the compass to more than half the length of $\overline{AB}$. With the compass point at A, draw an arc.

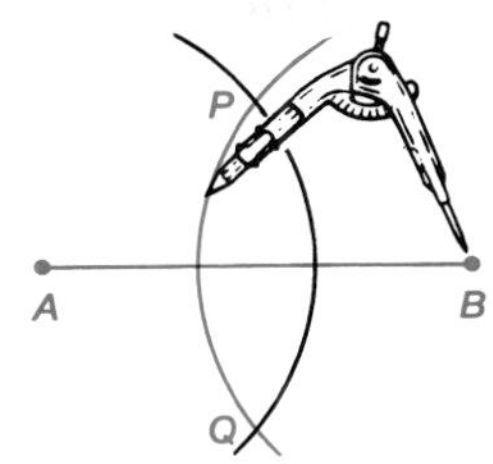

2. Use the same compass setting. With the compass point at B, draw another arc. Label the intersection points P and Q.

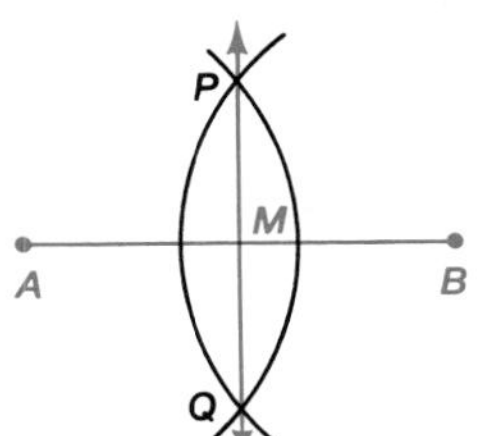

3. Draw $\overleftrightarrow{PQ}$. $\overleftrightarrow{PQ}$ bisects $\overline{AB}$. Point M is called the midpoint of $\overline{AB}$. $\overline{AM} \cong \overline{MB}$

Exercises

Trace each figure. Construct a figure congruent to each figure given. See students' work.

Practice

1. C ——— D

2. Z ——— W

3.

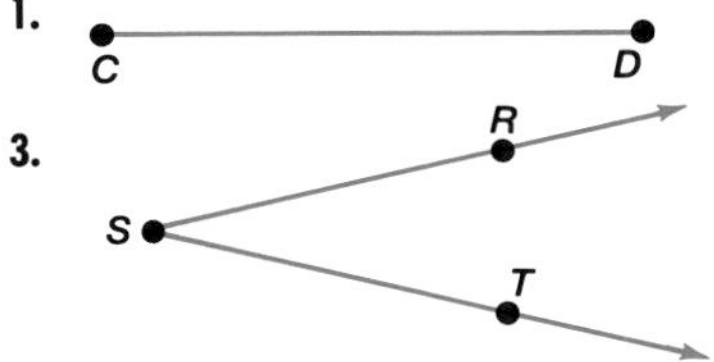

4.

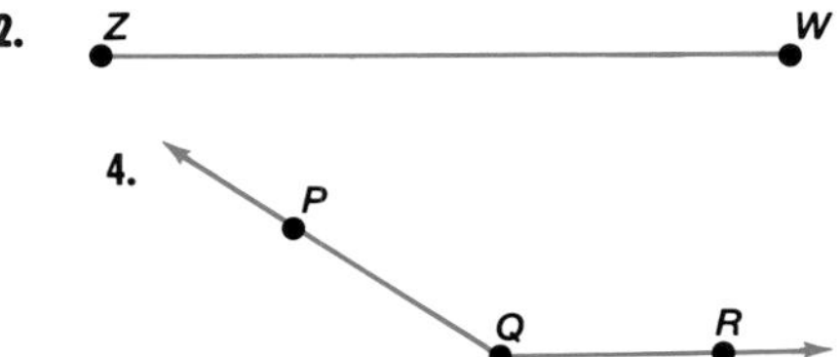

Trace each line segment. Bisect each line segment. See students' work.

5. P ——— Q

6. A ——— B

7. Trace $\overline{AB}$ from page 218 and perform Construction A. See students' work.

8. Trace $\overline{AB}$ from page 218 and perform Construction C. See students' work.

Problem Solving

9. Draw an obtuse angle. Separate it into four congruent angles by construction. See students' work.

10. Suppose you bisect an acute angle. What type of angle is each of the two angles formed? acute

11. If you bisect a right angle, what type of angle is each of the two angles formed? acute

12. Draw a line segment. Separate it into four congruent parts by construction. See students' work.

Talk Math

13. Explain how to construct a line segment that is twice as long as $\overline{CD}$ in exercise 1. Then construct the line segment. See margin.

Chalkboard Examples

Construction A 1. Trace $\overline{AB}$ from page 219 and perform construction A.

Construction B 2. Trace angle ABC from page 218 and perform Construction B.

Construction C 3. Trace $\overline{CD}$ from page 219 and perform Construction C.

Additional examples are provided on TRANSPARENCY 7-4.

EVALUATING THE LESSON

Check for Understanding Use Guided Practice Exercises to check for student understanding.

Closing the Lesson Ask students to explain in their own words how to bisect a line segment.

EXTENDING THE LESSON

Enrichment Complete using $\overline{a}$ and $\overline{b}$.

1. Construct a segment whose length is $2b$.
2. Construct a line segment whose length is $3a$.

Enrichment Masters Booklet, p. 69

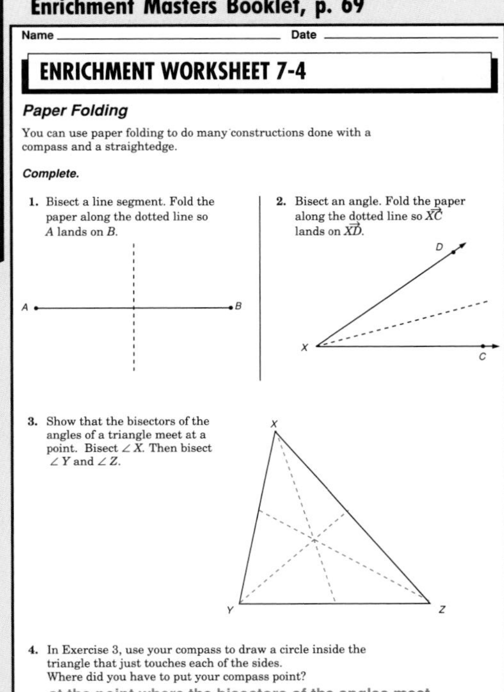
Name ______ Date ______

ENRICHMENT WORKSHEET 7-4

Paper Folding

You can use paper folding to do many constructions done with a compass and a straightedge.

Complete.

1. Bisect a line segment. Fold the paper along the dotted line so A lands on B.
2. Bisect an angle. Fold the paper along the dotted line so $\overrightarrow{XC}$ lands on $\overrightarrow{XD}$.
3. Show that the bisectors of the angles of a triangle meet at a point. Bisect $\angle X$. Then bisect $\angle Y$ and $\angle Z$.
4. In Exercise 3, use your compass to draw a circle inside the triangle that just touches each of the sides. Where did you have to put your compass point? at the point where the bisectors of the angles meet

APPLYING THE LESSON

Independent Practice

Homework Assignment	
Minimum	Ex. 1-13
Average	Ex. 1-13
Maximum	Ex. 1-13

Additional Answers

13. Draw a line. Construct a segment congruent to $\overline{CD}$. Construct a second segment congruent to $\overline{CD}$ on the line, extending from the end of the first segment.

5-MINUTE CHECK-UP

(over Lesson 7-4)

Available on TRANSPARENCY 7-5.

Construction A Draw any line segment. Then construct a line segment congruent to your line segment.

Construction B Draw any angle. Then construct an angle congruent to your angle.

Construction C Draw any line segment. Bisect your line segment.

INTRODUCING THE LESSON

Ask students to tell what they know about railroad tracks. **A good answer should refer to tracks remaining the same distance apart.**

TEACHING THE LESSON

Using Questioning

- Have students use a protractor to draw a 90° angle. Then ask what is true about the sides. **perpendicular**
- Have students use a ruler to draw two lines, one on each side of the ruler without moving the ruler. Then ask what is true about the two lines. **parallel**

Practice Masters Booklet, p. 76

PRACTICE WORKSHEET 7-5

Parallel and Perpendicular Lines

State whether each pair of lines is parallel, perpendicular, or skew. Use symbols to name all parallel and perpendicular lines.

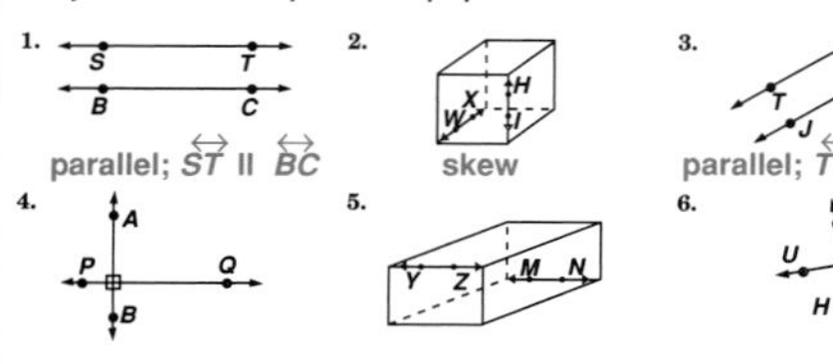

1. parallel; $\overleftrightarrow{ST} \parallel \overleftrightarrow{BC}$
2. skew
3. parallel; $\overleftrightarrow{TU} \parallel \overleftrightarrow{JK}$
4. perpendicular; $\overleftrightarrow{AB} \perp \overleftrightarrow{PQ}$
5. parallel; $\overleftrightarrow{MN} \parallel \overleftrightarrow{YZ}$
6. perpendicular; $\overleftrightarrow{GH} \perp \overleftrightarrow{UV}$

Use symbols to name the figures in each drawing.

7. all pairs of parallel lines $\overleftrightarrow{AE} \parallel \overleftrightarrow{BD}$
8. all pairs of perpendicular lines $\overleftrightarrow{AE} \perp \overleftrightarrow{FC}$; $\overleftrightarrow{BD} \perp \overleftrightarrow{FC}$
9. all line segments parallel to $\overline{LM}$ $\overline{ON}$, $\overline{QR}$, $\overline{PS}$
10. all line segments perpendicular to $\overline{OP}$ $\overline{QP}$, $\overline{PS}$, $\overline{LO}$, $\overline{NO}$
11. all line segments skew to $\overline{QP}$ $\overline{LM}$, $\overline{NO}$, $\overline{MR}$, $\overline{NS}$

7-5 PARALLEL AND PERPENDICULAR LINES

Objective
Identify and name parallel and perpendicular lines.

Katie was reading an office building floor plan for industrial technology class, when she noticed that many of the hallways were examples of parallel and perpendicular lines.

Lines in the same plane that *do not* intersect are called **parallel lines.**

Example A Line *MN* is parallel to line *OP*.

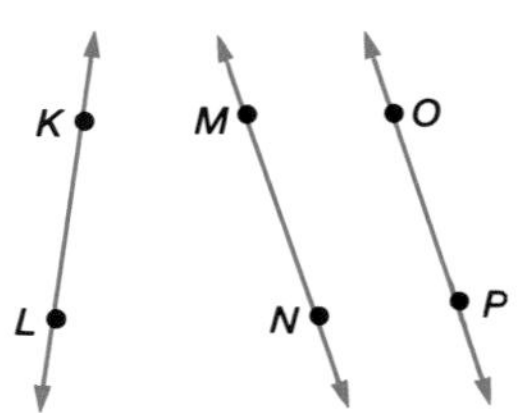

You can write this using symbols.

$\overleftrightarrow{MN} \parallel \overleftrightarrow{OP}$ Read as: Line *MN* is *parallel* to line *OP*.

Imagine that lines *KL* and *MN* are extended. They do intersect, so they are not parallel.

Two lines that intersect to form right angles are called **perpendicular lines.**

Example B Angles 1, 2, 3, and 4 are right angles. So, line *XZ* is perpendicular to line *YW*.

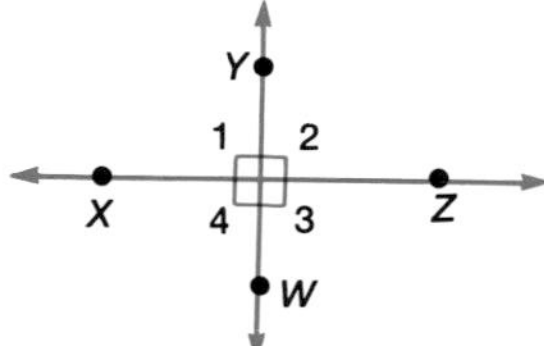

You can write this using symbols.

$\overleftrightarrow{XZ} \perp \overleftrightarrow{YW}$ Read as: Line *XZ* is *perpendicular* to line *YW*.

Lines that *do not* intersect and *are not* parallel are called **skew lines.**

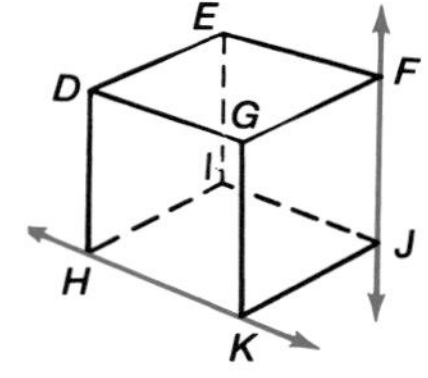

Example C In the figure at the right, line *FJ* and line *HK* are skew lines.

Guided Practice

State whether each pair of blue lines is* parallel, perpendicular, *or* skew. *Use symbols to name all parallel and perpendicular lines.

Examples A–C

1. 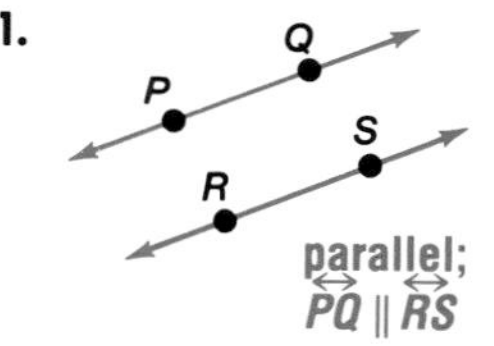

parallel; $\overleftrightarrow{PQ} \parallel \overleftrightarrow{RS}$

2. **perpendicular; $\overleftrightarrow{ZW} \perp \overleftrightarrow{XY}$**

3. 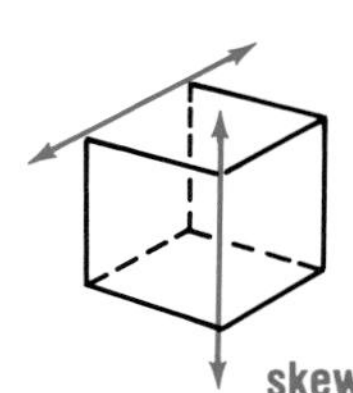
skew

RETEACHING THE LESSON

Have students draw representations of each of the following.

1. a pair of parallel lines
2. skew lines **See students' work.**
3. perpendicular lines
4. Have students name parallel lines, skew lines, and perpendicular lines on a box, a file drawer, and other objects in the classroom.

Reteaching Masters Booklet, p. 70

Name ____________ Date ____________

RETEACHING WORKSHEET 7-5

Parallel and Perpendicular Lines.

Lines in the same plane that *do not* intersect are called **parallel lines.**

Read $\overleftrightarrow{AB} \parallel \overleftrightarrow{CD}$ as line *AB* is parallel to line *CD*.

Two lines that intersect to form right angles are called **perpendicular lines.**

Angles 1, 2, 3, and 4 are right angles. So, line *IJ* is perpendicular to line *KL*. $\overleftrightarrow{IJ} \perp \overleftrightarrow{KL}$

Read $\overleftrightarrow{IJ} \perp \overleftrightarrow{KL}$ as line *IJ* is perpendicular to line *KL*.

State whether each of the following suggests parallel or perpendicular lines.

Exercises

State whether each pair of blue lines is parallel, perpendicular, or skew. Use symbols to name all parallel and perpendicular lines.

4. 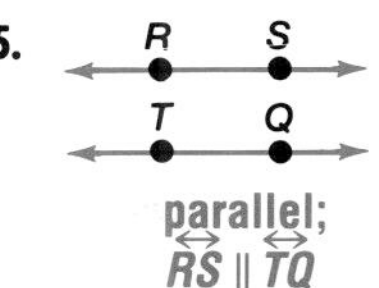

perpendicular; $\overleftrightarrow{OP} \perp \overleftrightarrow{MN}$

5. 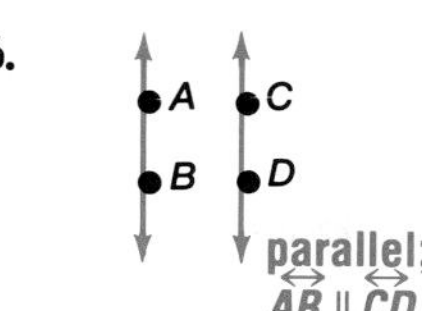

parallel; $\overleftrightarrow{RS} \parallel \overleftrightarrow{TQ}$

6. A C B D parallel; $\overleftrightarrow{AB} \parallel \overleftrightarrow{CD}$

7. 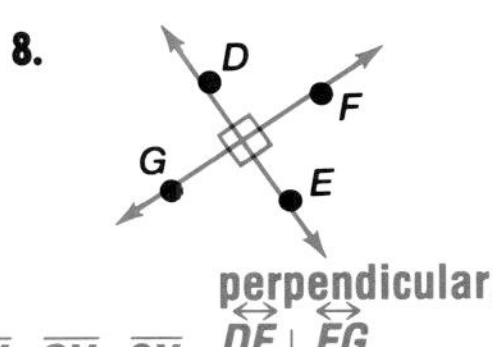

skew

8. 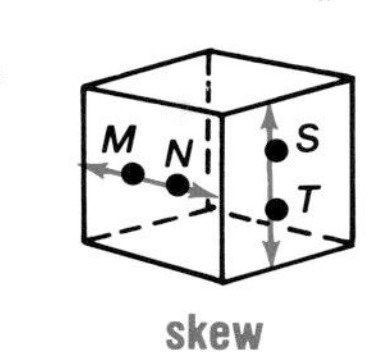

perpendicular; $\overleftrightarrow{DE} \perp \overleftrightarrow{FG}$

9. M N S T skew

10. $\overline{QR}$, $\overline{ST}$, $\overline{QV}$, $\overline{SX}$ 11. $\overline{TY}$, $\overline{QV}$, $\overline{SX}$

Complete. Use the cube at the right.

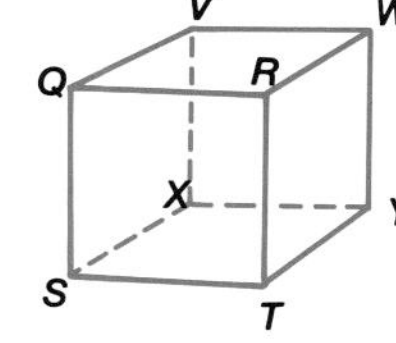

10. Name all the line segments that are perpendicular to line segment QS.

11. Name all the line segments that are parallel to line segment RW.

12. Name all the line segments that are skew to line segment QS. $\overline{VW}$, $\overline{XY}$, $\overline{RW}$; $\overline{TY}$

> **TECH PREP** On pages 525–527, you can learn how robotics technicians use mathematics in their jobs.

13. Use a compass and a straightedge to draw two line segments that are perpendicular. Hint: Think about what happens when you bisect a line. See students' work. See answers on pg. T522.

14. Find out how to use a compass and a protractor to draw two line segments that are parallel. See answer on p. T522.

15. Suppose you are in the starting position for doing push-ups. How is your body positioned relative to the floor? parallel

Critical Thinking

16. Explain how you would compare the lengths of $\overline{DE}$ and $\overline{EF}$ if you didn't have a ruler. Which looks longer $\overline{DE}$ or $\overline{EF}$?

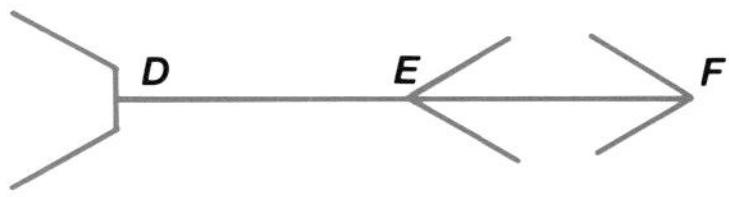

16. Answers may vary. Trace $\overline{DE}$ and put it over $\overline{EF}$ to compare their lengths. The lengths are the same.

Mixed Review

Lesson 1-4 ***Round each number to the nearest hundredth.***

17. 0.378 0.38 **18.** 0.844 0.84 **19.** 4.0949 4.09 **20.** 9.982 9.98 **21.** 13.004 13.00

Lesson 5-7 ***Divide.***

22. $4\frac{1}{4} \div 1\frac{1}{2}$ $2\frac{5}{6}$ **23.** $3\frac{4}{5} \div 5$ $\frac{19}{25}$ **24.** $2\frac{3}{8} \div 1\frac{2}{3}$ $1\frac{17}{40}$ **25.** $8 \div 1\frac{1}{2}$ $5\frac{1}{3}$

Lesson 6-9 **26.** The temperature at 5 A.M. was 2°F. At 3:00 P.M. the temperature reached a high of 44°F. What was the change in temperature? 42°F

APPLYING THE LESSON

Independent Practice

Homework Assignment	
Minimum	Ex. 1-26
Average	Ex. 1-26
Maximum	Ex. 1-26

Chapter 7, Quiz B (Lessons 7-6 through 7-10) is available in the Evaluation Masters Booklet, p. 34.

Chalkboard Examples

Examples A-C *State whether each pair of lines is parallel, perpendicular, or skew. Use symbols to name all parallel and perpendicular lines.*

1. **2.** **3.**

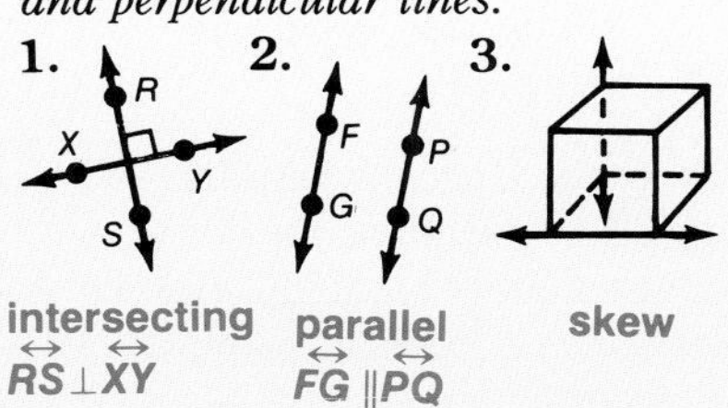

1. intersecting $\overleftrightarrow{RS} \perp \overleftrightarrow{XY}$ 2. parallel $\overleftrightarrow{FG} \parallel \overleftrightarrow{PQ}$ 3. skew

Additional examples are provided on TRANSPARENCY 7-5.

EVALUATING THE LESSON

Check for Understanding Use Guided Practice Exercises to check for student understanding.

Error Analysis Make sure students check the distance between parallel lines. It could be an illusion.

Closing the Lesson Ask students to summarize what they have learned about a pair of lines in a plane.

EXTENDING THE LESSON

Enrichment Masters Booklet, p. 70

ENRICHMENT WORKSHEET 7-5

Constructions: Perpendicular and Parallel Lines

You can construct perpendicular and parallel lines using only a compass and a straightedge.
Construct a line perpendicular to $\overleftrightarrow{AB}$ at P.

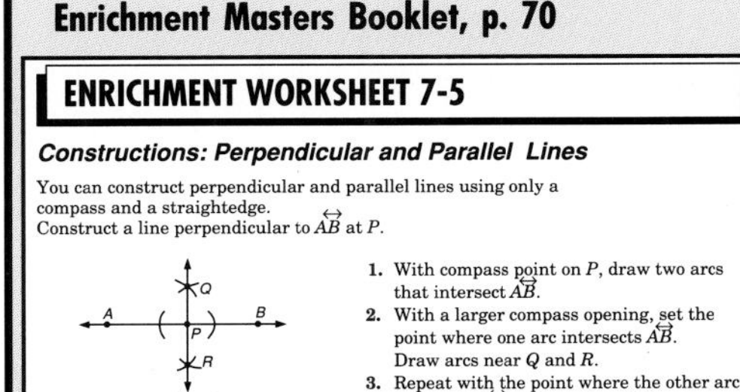

1. With compass point on P, draw two arcs that intersect $\overleftrightarrow{AB}$.
2. With a larger compass opening, set the point where one arc intersects $\overleftrightarrow{AB}$. Draw arcs near Q and R.
3. Repeat with the point where the other arc intersects $\overleftrightarrow{AB}$ to locate Q and R.
4. Draw $\overleftrightarrow{QR}$.

Construct parallel lines through J and K.

1. Following the steps above, construct a line perpendicular to $\overleftrightarrow{JK}$ at J.
2. Construct a line perpendicular to $\overleftrightarrow{JK}$ at K.

Construct a line perpendicular to each line at the point on the line.

1. **2.** See students' work.

A K

Construct parallel lines through the two points on each line.

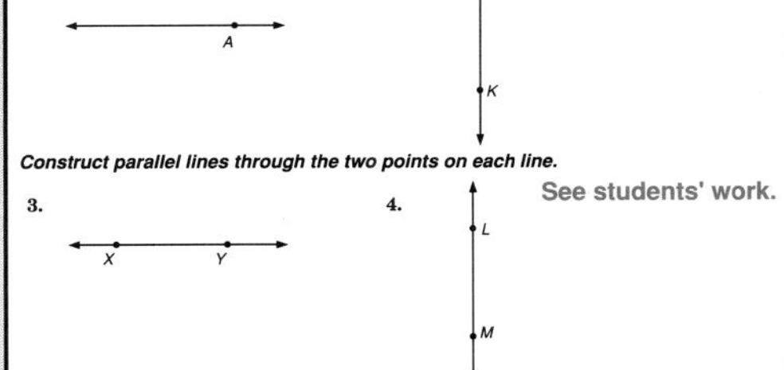

Applying Mathematics to Technology

This optional page provides students with an experience with computer literacy and also provides a change of pace.

Objective Identify and solve expressions in BASIC.

Using Discussion Discuss with the students BASIC symbols for addition (+), subtraction (−), multiplication (*), division (/), exponent or power (↑, followed by exponent or power).

All variables are represented by capitals in BASIC.

Review standard order of operations.

Using Vocabulary Discuss the definition of acronym. Point out that the word BASIC is an acronym.

Activity

Using Computers Have students write mathematical expressions in BASIC and solve the expression using a computer.

Computer Applications

INTRODUCTION TO BASIC

BASIC (Beginner's All-purpose Symbolic Instruction Code) is one of the most widely-used languages for microcomputers. This language uses the operational symbols shown at the right.

Operation	Mathematical Expression	BASIC Expression
addition	$a + b$	A+B
subtraction	$a - b$	A−B
multiplication	$a \times b$, $a \cdot b$, or ab	A*B
division	$a \div b$ or $\frac{a}{b}$	A/B
raising to a power	n^2	N↑2

All variables are represented by capital letters in BASIC.

BASIC expressions are evaluated using the standard order of operations.

Write each expression in BASIC.

Mathematical	BASIC
$\frac{1}{2}h \times (a + b)$	1/2*H*(A+B)
$4^3 - 2 \times 5 \div (2 + 3)$	4↑3−2*5/(2+3)

Find the value of 6*4+(2+3)↑2/5−7.

6*4+(2+3)↑2/5−7 = 6*4+5↑2/5−7

Do operations in parentheses.
Evaluate all powers. = 6*4+25/5−7

Do all multiplications and divisions from left to right. = 24+25/5−7
= 24+5−7

Do all additions and subtractions from left to right. = 29−7
= 22

Write each expression in BASIC.

1. $\frac{1}{4}y + 5$ **1/4*Y+5**

2. z^3 **Z↑3**

3. $(4 - p) \times 6$ **(4−P)*6**

Find the value of each expression.

4. 2*(4+8)/6 **4**

5. (2*3)↑2 **36**

6. 3+4+2↑2−7 **4**

7. Write an expression in BASIC that will give your age in five years. **Answers may vary. A sample answer is (1991−1977) + 5**

8. Beth found the value of the expression $59 - 4 \times 3^2$ as 495. Write the expression in BASIC. Then determine the answer given by a computer. Is Beth's answer correct? **59 − [4 × (3↑2)] = 23**

PHYSICAL FITNESS

Chris is a member of the high school soccer team. In order to stay in shape, she stays active and counts the calories in the food she eats. A calorie is a heat-producing or energy-producing unit.

To remain at her present weight, Chris must use the same number of calories that her meals contain.

The chart below gives the number of calories used per minute for each activity.

Activity	Calories Used per Minute
Bicycling	8.2
Running	13.0
Skiing	9.5
Swimming	8.7
Walking	5.2

Chris runs for 45 minutes each Saturday. How many calories does she use?

Multiply. 13 calories per minute
× 45 minutes
585 Chris uses 585 calories.

Determine the number of calories used in each activity.

1. 90 minutes of skiing **855**

2. 1 hour of swimming **522**

3. 2 hours of bicycling **984**

4. $1\frac{1}{2}$ hours of running **1,170**

5. Jennifer walked for 40 minutes and ran for 20 minutes. How many calories were used in these two activities? **468 calories**

6. How many minutes must Jill ski to use the calories in a 532-calorie hamburger? **56 minutes**

7. How many calories are used by Corey if he swims for 45 minutes, bicycles for 45 minutes, and walks for 45 minutes? **994.5 calories**

8. How long would Shawn have to run in order to use the same number of calories that are used in 40 minutes of walking? **11 minutes**

9. Michael had the following for lunch: a hot dog with cheese, 330 calories; a pear, 121 calories; a glass of milk, 164 calories. How many minutes must he bicycle to use these calories? **75 minutes**

10. If Jason swims $1\frac{1}{2}$ hours every day, how many extra calories a day must he consume in order to maintain his weight? **783 calories**

Applying Mathematics to the Real World

This optional page shows how mathematics is used in the real world and also provides a change of pace.

Objective Determine the number of calories used in an activity.

Using Critical Thinking Why do people gain weight? If your caloric intake remains the same, can you lose weight by exercising? Why? What do you have to do if you exercise everyday and want to maintain your weight?

Using Questioning If your body uses 6.8 cal/min playing tennis, how many calories would you use in 10 min? 40 min? 20 min? 30 min? **68 cal, 272 cal, 136 cal, 204 cal**

Activity

Using Data and Calculators Have students use a calculator and a calorie chart to complete the following. Keep a list of all food eaten in one day. Find the difference between the calories consumed and the calories needed to maintain their weight. **average calories needed to maintain weight: 2800-boys; 2200-girls**

Then calculate how long they would have to perform an activity of their choice from the chart on p. 223 to maintain their present weight.

5-MINUTE CHECK-UP

(over Lesson 7-5)

Available on TRANSPARENCY 7-6.
State whether each pair of lines is parallel, perpendicular, or skew. Use symbols to name all parallel and perpendicular lines.

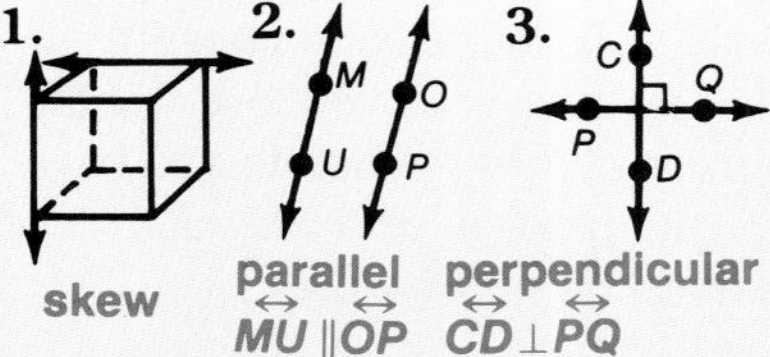

1. skew
2. parallel $\overleftrightarrow{MU} \parallel \overleftrightarrow{OP}$
3. perpendicular $\overleftrightarrow{CD} \perp \overleftrightarrow{PQ}$

Extra Practice, Lesson 7-5, p. 467

INTRODUCING THE LESSON

Have students draw plane figures on a sheet of paper. Tell students that line segments must intersect at endpoints. Discuss the names of these figures.

TEACHING THE LESSON

Using Vocabulary

- What nongeometric words begin with tri? quadri? penta? octa? deca? tricycle, quadrillion, pentathlon, octapus, decade

Practice Masters Booklet, p. 77

PRACTICE WORKSHEET 7-6

Polygons

Complete the chart.

	Name	Prefix	Number of Sides	Number of Angles
1.	triangle	tri-	3	3
2.	quadrilateral	quad-	4	4
3.	pentagon	penta-	5	5
4.	hexagon	hexa-	6	6
5.	octagon	octa-	8	8
6.	decagon	deca-	10	10

Draw an example of each polygon.

7. not regular decagon
8. not regular pentagon
9. regular quadrilateral
10. regular hexagon

Name each by the number of sides. Then state whether it is regular or not regular.

11. pentagon; not regular

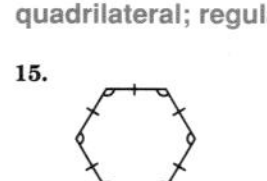

12. quadrilateral; regular

13. quadrilateral; not regular

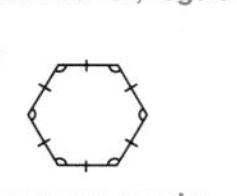

14. triangle; regular

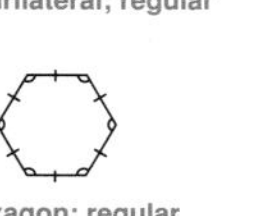

15. hexagon; regular
16. octagon; not regular

7-6 POLYGONS

Objective

Identify and name various types of polygons.

Street signs have many shapes. Simple, closed figures were used in the design of these signs.

Polygons are closed plane figures formed by line segments called **sides.** The line segments intersect at their endpoints, but do not cross. The intersection points are called **vertices** (plural of vertex).

Polygons are named by the number of sides they have. Some common polygons are listed in the chart at the right and examples are shown below.

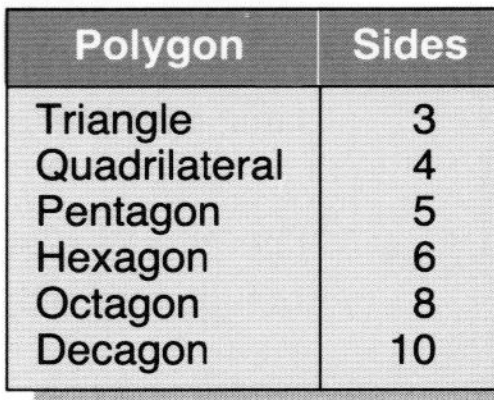

Polygon	Sides
Triangle	3
Quadrilateral	4
Pentagon	5
Hexagon	6
Octagon	8
Decagon	10

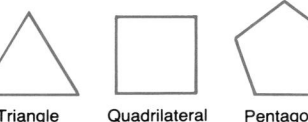

Triangle

Quadrilateral

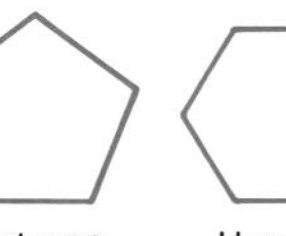
Pentagon

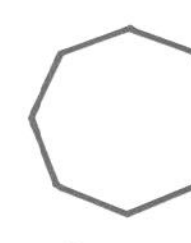
Hexagon

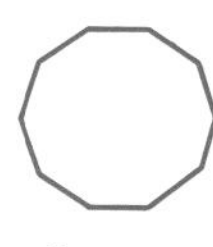
Octagon

Decagon

A polygon in which all sides are congruent and all angles are congruent is called a **regular polygon.** All angles referred to are interior angles.

3 congruent angles
3 congruent sides

4 congruent angles
4 congruent sides

5 congruent angles
5 congruent sides

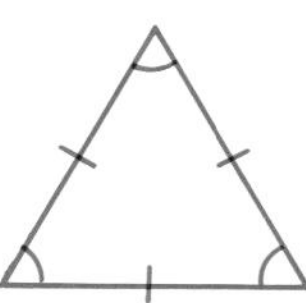
Regular Triangle or Equilateral Triangle

The matching red marks indicate congruent parts.

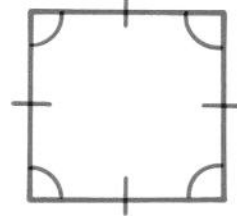
Regular Quadrilateral or Square

Regular Pentagon

Guided Practice

Name each polygon by the number of sides. Then state whether it is regular or not regular.

1. quadrilateral; not regular
2. pentagon; not regular

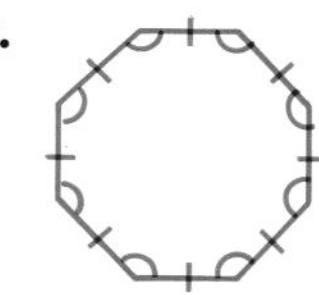

3. octagon; regular

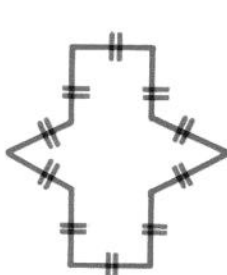

4. decagon; not regular

RETEACHING THE LESSON

Count the number of sides in each polygon. Name the polygon. Then state whether it is regular or not regular.

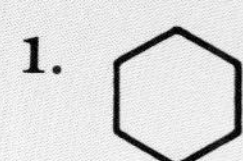

1. 6; hexagon regular
2. 4; quadrilateral not regular
3. 5; pentagon not regular
4. 3; triangle regular

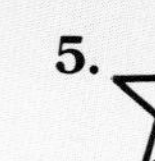

5. 10; decagon regular

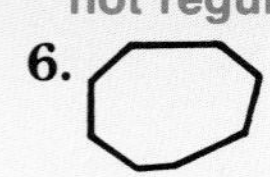

6. 8; octagon not regular

Reteaching Masters Booklet, p. 71

RETEACHING WORKSHEET 7-6

Polygons

A **polygon** is a closed figure in a plane formed by line segments called **sides**. Two sides of a polygon meet at a **vertex,** but do not cross.

Polygons are named by the number of sides they have. A **regular polygon** has all sides congruent and all angles congruent.

Polygon	Sides
Triangle	3
Quadrilateral	4
Pentagon	5
Hexagon	6
Octagon	8
Decagon	10

Tell whether each figure is a polygon. If it is, name it by the number of sides, then state whether it is regular or not regular. If it is not a polygon, explain your answer.

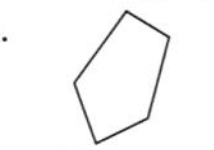

1. pentagon; not

2. not a polygon, since

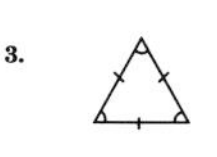

3. triangle; regular

Exercises

Practice

Name each polygon by the number of sides. Then state whether it is regular or not regular. 8. triangle; regular

5.  quadrilateral; not regular

6.   quadrilateral not regular

7.  triangle; not regular

8. 

Explain why each figure is not a polygon. 9. not closed; 10. line segments cross; 11. not made up of line segments

9. 10. 11.

Draw an example of each polygon. See answers on p. T522.

12. regular triangle **13.** regular quadrilateral **14.** regular hexagon

15. not regular triangle **16.** not regular pentagon **17.** not regular octagon

18. Name two conditions that are necessary for a hexagon to be called a regular hexagon. 6 congruent sides, 6 congruent angles

JOURNAL ENTRY

19. Explain why the words *closed* and *plane* are used to define a polygon. Closed means all line segments must intersect. Plane means the figures lie in one plane. They have only 2 dimensions, not 3.

Applications 

20. Cal Walker's patio is in the shape of a decagon. Each side is $2\frac{1}{2}$ feet long. What is the perimeter of his patio? 25 feet

21. Janet Guthrie drove a lap at the Indianapolis 500 in 47 seconds. One lap is 13,200 feet. Find the speed to the nearest tenth of a foot per second. 280.9 feet per second

24. 536 million barrels 25. Texas, Alaska

22. What were the top five oil producing states in the United States in 1991? See graph.

23. Which state produced the most oil? the least oil? Texas, Oklahoma

24. How much more oil did Texas produce than Louisiana?

25. Which two states produced *most closely* the same amount of oil?

26. *About* how much oil did the top oil producing states in the United States produce?

27. If the average price per barrel was $21, what was the value of the oil produced in Alaska?

26. about 3 billion barrels 27. $13,776,000,000

Top Oil States in 1991

In 1991, the United States produced 2.7 billion barrels of crude oil.

In millions of barrels

Texas	683
Alaska	656
California	320
Louisiana	147
Oklahoma	108

APPLYING THE LESSON

Independent Practice

Homework Assignment	
Minimum	Ex. 1-11; 12-26 even
Average	Ex. 1-27
Maximum	Ex. 1-27

Chalkboard Examples

Name each polygon by the number of sides. Then state whether it is regular or not regular.

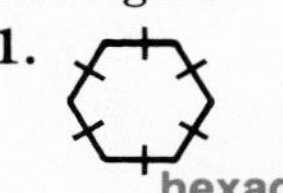

1. hexagon, regular

2. pentagon, not regular

Additional examples are provided on TRANSPARENCY 7-6.

EVALUATING THE LESSON

Check for Understanding Use Guided Practice Exercises to check for student understanding.

Closing the Lesson Ask students to state in their own words what they have learned about polygons.

EXTENDING THE LESSON

Enrichment Use the figure below to list all the geometric facts shown.

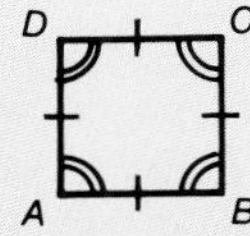

Sample answers are: infinitely many points; 4 angles are right angles; figure is a quadrilateral.

Enrichment Masters Booklet, p. 71

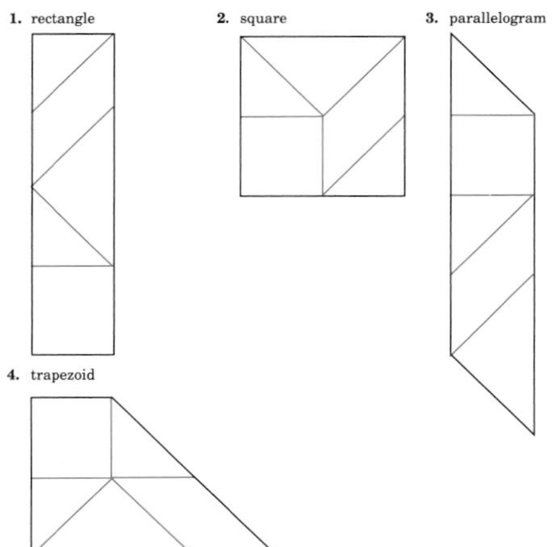

ENRICHMENT WORKSHEET 7-6

Tangrams

The five polygons below can be arranged to form other polygons.

Trace or cut out each polygon above. Use each piece once and only once to form each of the polygons below.

1. rectangle 2. square 3. parallelogram

4. trapezoid

5-MINUTE CHECK-UP

(over Lesson 7-6)

Available on TRANSPARENCY 7-7.

Name each polygon by the number of sides. Then state whether it is regular or not regular.

1.  pentagon, regular
2.  triangle, not regular
3.  octagon, regular
4.  decagon, not regular

Extra Practice, Lesson 7-6, p. 467

INTRODUCING THE LESSON

Ask students to classify and name different kinds of shoes. high tops, sports; high heels, dress; flats, casual; satin, formal; and so on

TEACHING THE LESSON

Using Manipulatives Have students make a *scalene* triangle on a geoboard. Change the vertices to make a triangle with two congruent sides and call it an *isosceles* triangle. Discuss the properties of each. You may want to use Geoboard, **Lab Manual, p. 13.**

Practice Masters Booklet, p. 78

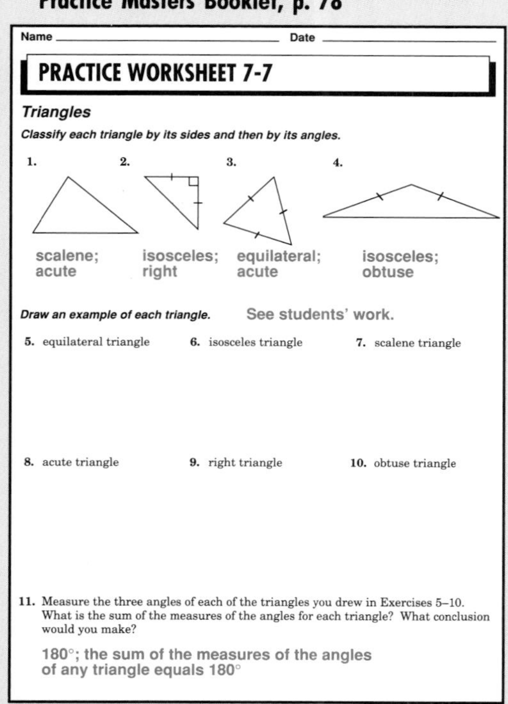

Name ______ Date ______

PRACTICE WORKSHEET 7-7

Triangles

Classify each triangle by its sides and then by its angles.

1. scalene; acute
2. isosceles; right
3. equilateral; acute
4. isosceles; obtuse

Draw an example of each triangle. See students' work.

5. equilateral triangle
6. isosceles triangle
7. scalene triangle
8. acute triangle
9. right triangle
10. obtuse triangle
11. Measure the three angles of each of the triangles you drew in Exercises 5–10. What is the sum of the measures of the angles for each triangle? What conclusion would you make?

180°; the sum of the measures of the angles of any triangle equals 180°

7-7 TRIANGLES

Objective

Identify and name various types of triangles.

The fastest roller coaster in the United States is the Steel Phantom in West Mifflin, Pennsylvania. Triangular braces are used in the construction of roller coasters. The triangular braces add rigidity to the structure.

Triangles can be classified by the number of congruent sides.

Example A

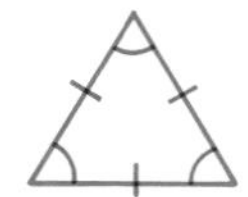

An **equilateral triangle** has 3 congruent sides.

An **isosceles triangle** has at least 2 congruent sides.

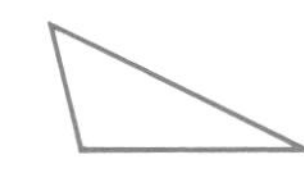

A **scalene triangle** has no congruent sides.

Triangles can also be classified by angles. All triangles have at least two acute angles. The third angle is used to classify the triangle.

Example B

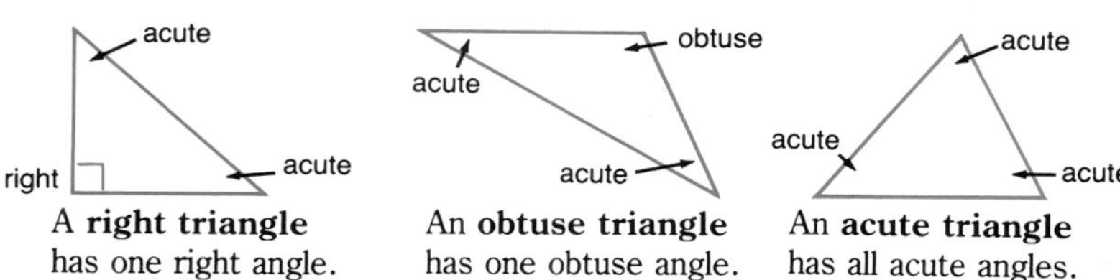

A **right triangle** has one right angle.

An **obtuse triangle** has one obtuse angle.

An **acute triangle** has all acute angles.

Guided Practice

Example A *Classify each triangle by its sides.*

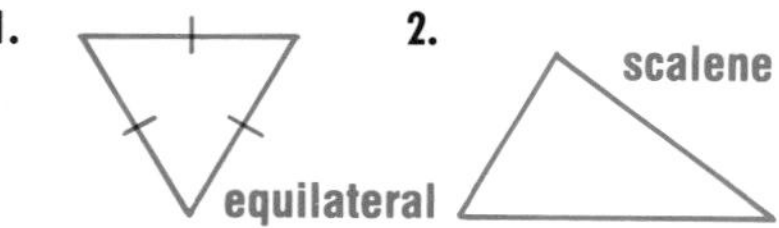

1. equilateral
2. scalene

3. isosceles

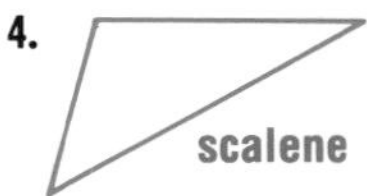

4. scalene

Example B *Classify each triangle by its angles.*

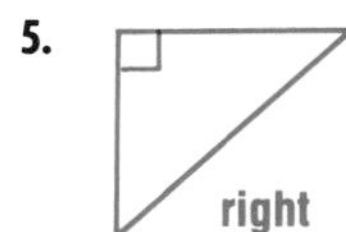

5. right
6. acute

7. acute

8. right

RETEACHING THE LESSON

One side of a triangle is given. Have students draw the other 2 sides to form the triangle named.

1. 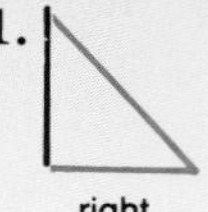 right
2. 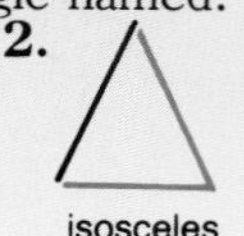 isosceles
3. 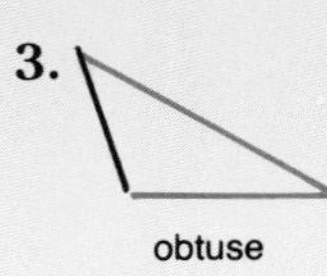 obtuse
4.  acute
5. 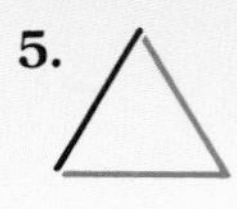 equilateral
6. 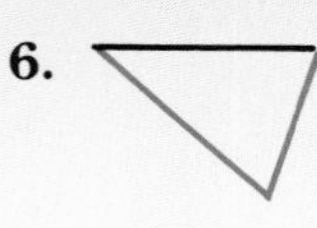 scalene

Reteaching Masters Booklet, p. 72

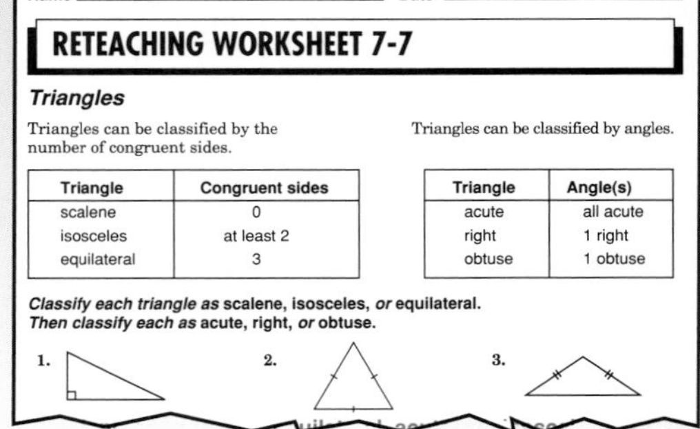

Name ______ Date ______

RETEACHING WORKSHEET 7-7

Triangles

Triangles can be classified by the number of congruent sides.

Triangle	Congruent sides
scalene	0
isosceles	at least 2
equilateral	3

Triangles can be classified by angles.

Triangle	Angle(s)
acute	all acute
right	1 right
obtuse	1 obtuse

***Classify each triangle as* scalene, isosceles, *or* equilateral. *Then classify each as* acute, right, *or* obtuse.**

1. 2. 3.

Exercises

Practice

Classify each triangle by its sides and then by its angles.

9. 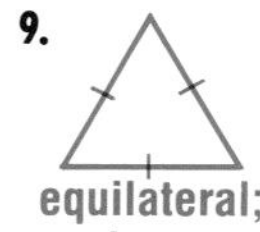 equilateral; acute

10.  scalene; right

11. 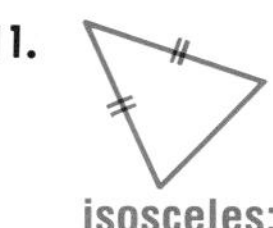 isosceles; acute

12. 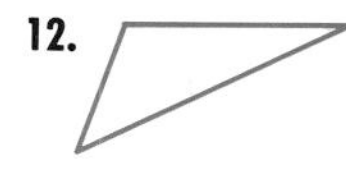 scalene; obtuse

13. 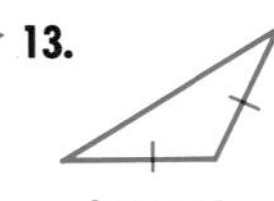 isosceles; obtuse

Draw an example of each triangle. See answers on p. T523.

14. scalene triangle	**15.** isosceles triangle	**16.** right triangle
17. equilateral triangle	**18.** acute triangle	**19.** obtuse triangle

See students' work.

Using Reasoning

State whether each statement is true *or* false. *Draw an example of each figure when the statement is true.* See students' work.

20. An equilateral triangle must be an acute triangle. true

21. An isosceles triangle can have a right angle. true

22. A scalene triangle may also be a right triangle. true

23. An obtuse triangle may also be an equilateral triangle. false

24. An equilateral triangle is a regular triangle. true

25. An acute triangle may also be an isosceles triangle. true

fUN with MATH

What did the acorn say when he finally grew up?

See page 240.

Use the figure at the right to name each of the following. Use the symbol △ for triangle.

26. five right triangles. See margin.

27. two acute triangles

28. three obtuse triangles

29. one isosceles triangle

30. seven scalene triangles

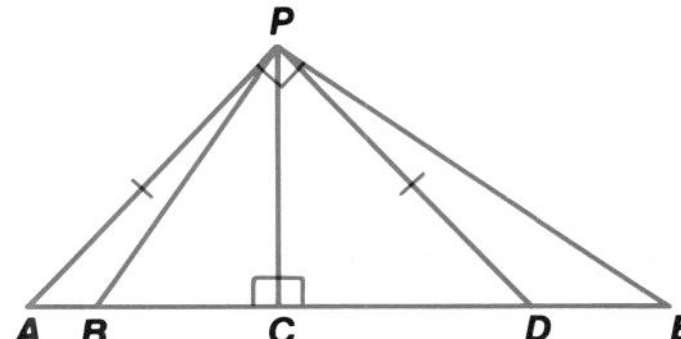

Collect Data

31. List the road signs you see on the way to school that are shaped like triangles. Answers may vary.

Cooperative Groups

32. The angles form a straight line. 180°

32. Draw and cut out any triangle. Label the angles 1, 2, and 3. Tear off each angle. Draw a line and place point P on the line. Place angles 1, 2, and 3 on the line so that their vertices are on Point P. Explain what you discover. What is the sum of the measures of the angle?

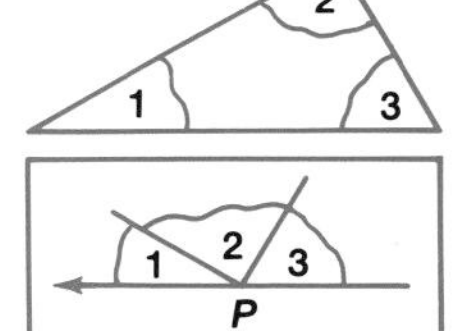

33. Suppose that a right triangle has a 42° angle. Find the measure of the other angles. 48°; 90°

34. What is the measure of each of the angles in an equilateral triangle? 60°

35. The sum of the measures of the two right angles is 180°. Since the sum of the measures of the angles in a triangle is 180°, perpendicular rays will not intersect to form the third vertex.

Write Math

35. Explain why you cannot draw a triangle with two right angles.

APPLYING THE LESSON

Independent Practice

Homework Assignment	
Minimum	Ex. 1-13; 15-35 odd
Average	Ex. 2-30 even; 31-35
Maximum	Ex. 1-35

Additional Answers

26. △*ACP*, △*ECP*, △*BPE*, △*BCP*, △*DCP*
27. △*BPD*, △*APD*
28. △*ABP*, △*EDP*, △*APE*
29. △*APD*
30. △*ABP*, △*ACP*, △*AEP*, △*BCP*, △*CDP* △*CEP*, △*DEP*

Chalkboard Examples

Example A *Classify each triangle by its sides.*

1.  equilateral

2. 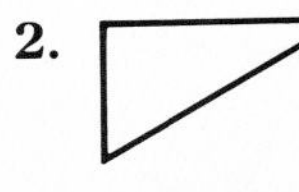 scalene

Example B *Classify each triangle by its angles.*

3. 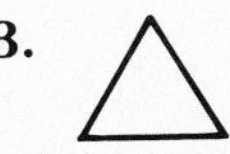 acute

4. 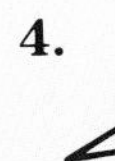 obtuse

Additional examples are provided on TRANSPARENCY 7-7.

EVALUATING THE LESSON

Check for Understanding Use Guided Practice Exercises to check for student understanding.

Error Analysis Emphasize that three types of triangles describe the sides and three types describe the angles.

Closing the Lesson Ask students to summarize what they have learned about triangles.

EXTENDING THE LESSON

Enrichment Masters Booklet, p. 72

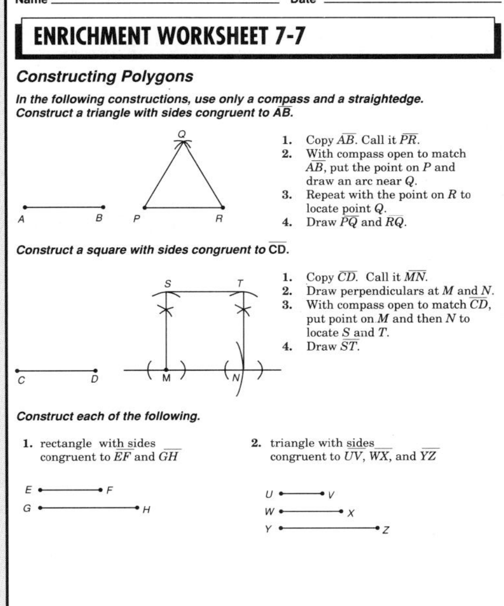

Name ______ Date ______

ENRICHMENT WORKSHEET 7-7

Constructing Polygons

In the following constructions, use only a compass and a straightedge.
Construct a triangle with sides congruent to $\overline{AB}$.

1. Copy $\overline{AB}$. Call it $\overline{PR}$.
2. With compass open to match $\overline{AB}$, put the point on P and draw an arc near Q.
3. Repeat with the point on R to locate point Q.
4. Draw $\overline{PQ}$ and $\overline{RQ}$.

Construct a square with sides congruent to $\overline{CD}$.

1. Copy $\overline{CD}$. Call it $\overline{MN}$.
2. Draw perpendiculars at M and N.
3. With compass open to match $\overline{CD}$, put point on M and then N to locate S and T.
4. Draw $\overline{ST}$.

Construct each of the following.

1. rectangle with sides congruent to $\overline{EF}$ and $\overline{GH}$
2. triangle with sides congruent to $\overline{UV}$, $\overline{WX}$, and $\overline{YZ}$

See students' work.

5-MINUTE CHECK-UP

(over Lesson 7-7)

Available on TRANSPARENCY 7-8.

Classify each triangle by its sides and then by its angles.

1. 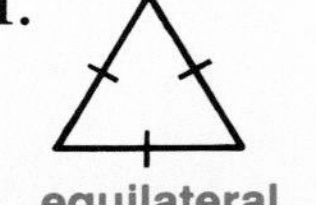 equilateral, acute

2. 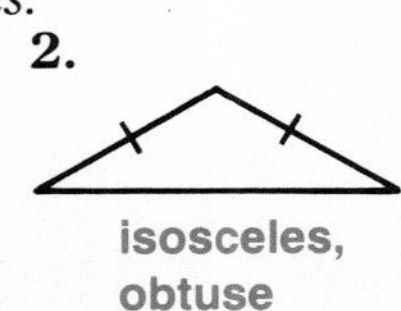 isosceles, obtuse

Extra Practice, Lesson 7-7, p. 467

INTRODUCING THE LESSON

Have students draw different kinds of four-sided plane figures. Have students suggest names for these four-sided figures.

TEACHING THE LESSON

Using Critical Thinking State whether each statement is true or false.

- All parallelograms are trapezoids. false
- Not every rectangle is a square. true
- Every rhombus is a parallelogram. true

Practice Masters Booklet, p. 79

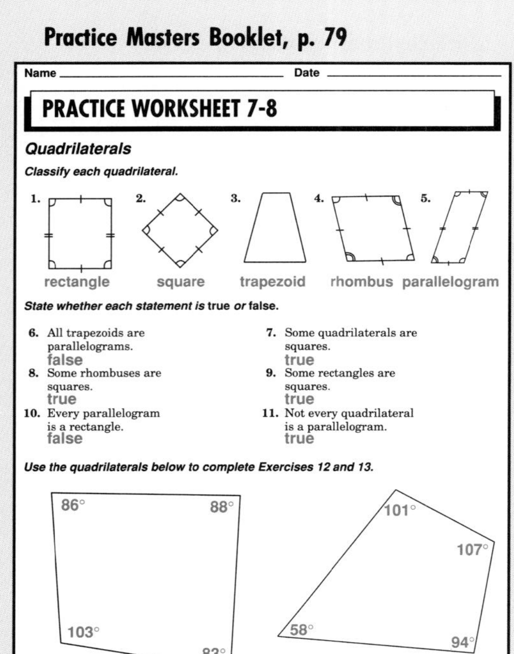

Name ______ Date ______

PRACTICE WORKSHEET 7-8

Quadrilaterals

Classify each quadrilateral.

1. rectangle 2. square 3. trapezoid 4. rhombus 5. parallelogram

State whether each statement is true or false.

6. All trapezoids are parallelograms. false
7. Some quadrilaterals are squares. true
8. Some rhombuses are squares. true
9. Some rectangles are squares. true
10. Every parallelogram is a rectangle. false
11. Not every quadrilateral is a parallelogram. true

Use the quadrilaterals below to complete Exercises 12 and 13.

12. Measure and record the measure of each angle of both quadrilaterals.
13. What is the sum of the measures of the angles of each quadrilateral? 360°

7-8 QUADRILATERALS

Objective

Identify and name quadrilaterals.

A tennis court is in the shape of a rectangle. A net and a white line divide the court in half. The two halves are also rectangles. Another white line is drawn lengthwise to allow for four players. This line divides the court into four congruent rectangles.

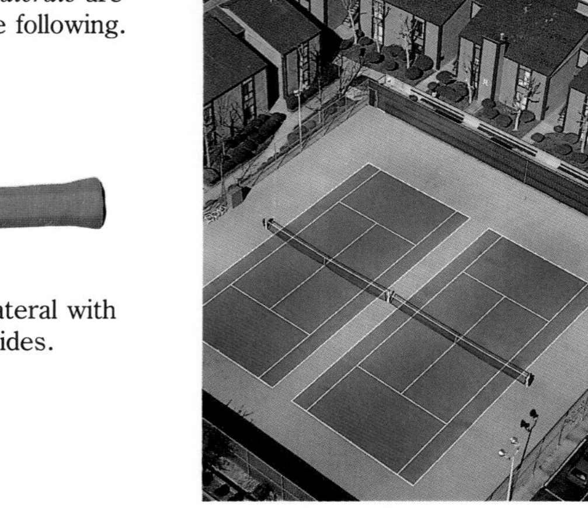

A rectangle is a quadrilateral because it has four sides. *Quadrilaterals* are classified according to the following.

- parallel sides
- congruent angles
- congruent sides

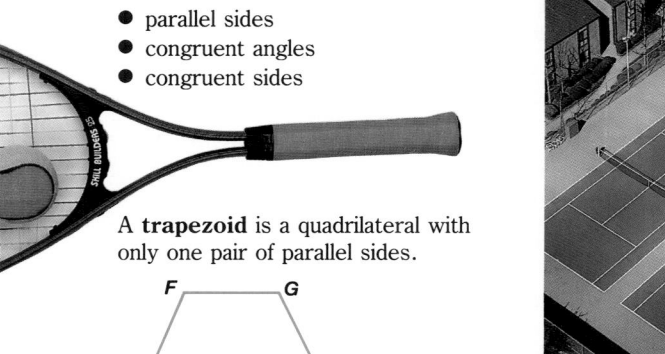

A **trapezoid** is a quadrilateral with only one pair of parallel sides.

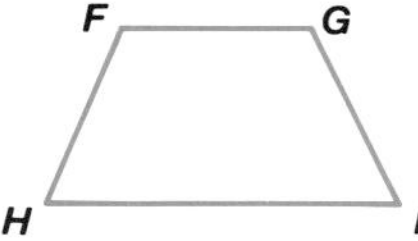

A **parallelogram** is a quadrilateral with two pairs of parallel sides.

Opposite sides are congruent.

Opposite angles are congruent.

A **rectangle** is a parallelogram with all angles congruent.

Opposite sides are congruent.

A **square** is a parallelogram with all sides and angles congruent.

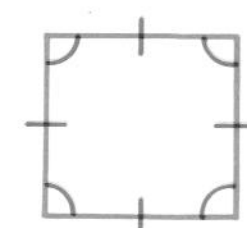

A square is a regular quadrilateral.

A **rhombus** is a parallelogram with all sides congruent.

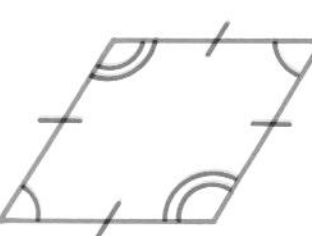

Opposite angles are congruent.

Guided Practice

The figures in Exercises 3 and 8 appear to be trapezoids.

Classify each quadrilateral.

1. rectangle **2.** square **3.** trapezoid **4.**  rhombus

RETEACHING THE LESSON

The figure below contains eight quadrilaterals. Classify each quadrilateral. 1 square, 1 trapezoid, 2 parallelograms, 2 rectangles, and 2 rhombuses

Reteaching Masters Booklet, p. 73

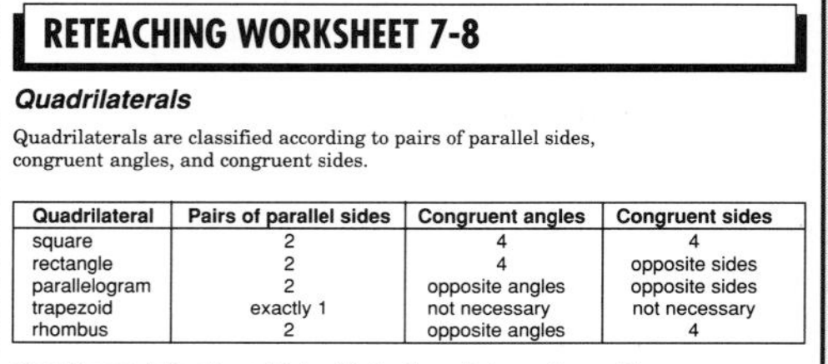

Name ______ Date ______

RETEACHING WORKSHEET 7-8

Quadrilaterals

Quadrilaterals are classified according to pairs of parallel sides, congruent angles, and congruent sides.

Quadrilateral	Pairs of parallel sides	Congruent angles	Congruent sides
square	2	4	4
rectangle	2	4	opposite sides
parallelogram	2	opposite angles	opposite sides
trapezoid	exactly 1	not necessary	not necessary
rhombus	2	opposite angles	4

Classify each lettered quadrilateral in the figure below as trapezoid, parallelogram, rectangle, rhombus, or square. Use the word that best describes what each quadrilateral appears to be.

1. *A* rectangle 2. *B* trapezoid

Exercises

Practice

Classify each quadrilateral.

5. 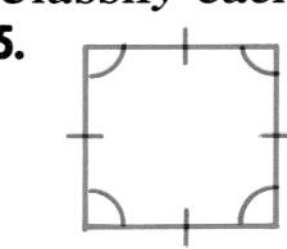 square **6.**  rectangle **7.** 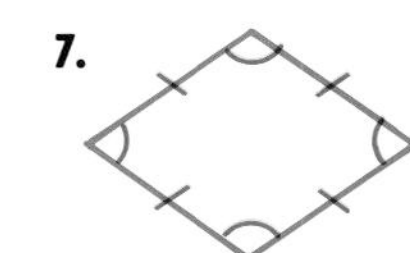 rhombus **8.** 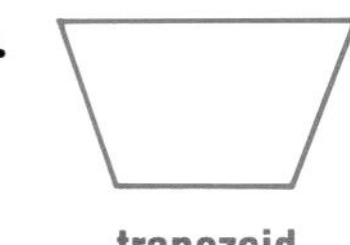 trapezoid

Show Math

Draw an example of each quadrilateral. See students' work.

9. square **10.** parallelogram **11.** rhombus **12.** trapezoid

Using Reasoning

State whether each statement is true *or* false. *Draw an example of each figure when the statement is true.* See students' work.

13. All rectangles are squares. false
14. All squares are rectangles. true
15. All parallelograms are squares. false
16. A rectangle is a quadrilateral. true
17. A rhombus is a rectangle. false
18. Not every parallelogram is a rectangle. true
19. A square is a rhombus. true

Problem Solving

20. Is this kite shape a parallelogram? Explain in your own words why or why not. Yes, 2 pairs of parallel sides.

21. The wood trim on a cabinet is in the shape of a rhombus. If the total length of the trim is 148 cm, what is the length of one side of the trim? 37 cm

Cooperative Groups

22. Draw and cut any rectangle. Label the angles 1, 2, 3, and 4. Tear off each angle. Place a point *D* on your paper. Place angles 1, 2, 3, and 4 so that their vertices are on Point *D*.

a. Explain what you discover. What is the sum of the measures of the angles? The angles form a circle. 360°

b. If a rhombus has an angle that measures 122°, what are the measures of the other angles? 122°; 58°; 58°

Critical Thinking

23. How many squares are in the figure at the right? Count only those squares that are outlined. 11 squares

Mixed Review

Change each fraction to a mixed number in simplest form.

Lesson 4-8

24. $\frac{6}{5}$ $1\frac{1}{5}$ **25.** $\frac{9}{5}$ $1\frac{4}{5}$ **26.** $\frac{18}{7}$ $2\frac{4}{7}$

Lesson 5-9

Change each decimal to a fraction.

27. 0.6 $\frac{3}{5}$ **28.** 0.42 $\frac{21}{50}$ **29.** 0.325 $\frac{13}{40}$ **30.** 0.638 $\frac{319}{500}$

Lesson 6-10

31. Alisa has to be at school at 8:00 A.M. It takes her 35 minutes to shower and get dressed. She needs 20 minutes to eat breakfast and 15 minutes to drive to school. What time should she get up in order to be at school on time? 6:50 A.M.

APPLYING THE LESSON

Independent Practice

Homework Assignment	
Minimum	Ex. 1-12; 14-22 even; 23-31
Average	Ex. 1-31
Maximum	Ex. 1-31

Chalkboard Examples

Classify each quadrilateral.

1. 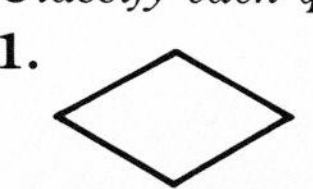 rhombus **2.** parallelogram

3. square **4.** trapezoid

Additional examples are provided on TRANSPARENCY 7-8.

EVALUATING THE LESSON

Check for Understanding Use Guided Practice Exercises to check for student understanding.

Closing the Lesson Ask students to summarize what they have learned about quadrilaterals.

EXTENDING THE LESSON

Enrichment Draw and classify each quadrilateral.

- one pair of parallel sides trapezoid
- two pairs of parallel sides parallelogram

Enrichment Masters Booklet, p. 73

Name ______ Date ______

ENRICHMENT WORKSHEET 7-8

Tessellations

A tessellation is a repetitive pattern using one or more geometric shapes. The shapes fit together without holes or gaps. The word *tessellate* comes from the Latin word for "tile." Tessellations are sometimes called tiling patterns.

1. Trace each regular polygon below. Cut each shape from cardboard. Use the cardboard pieces to try to draw a tessellation for each regular polygon. (The regular hexagon pattern is shown at the right.) Which regular polygons tessellate? triangle, square, hexagon

Can you make tessellations from the following combinations?

2. square and octagon yes
3. square, triangle, and hexagon yes
4. square, triangle, and dodecagon yes
5. another combination you choose Answers will vary.

equilateral triangle square regular hexagon

regular octagon regular dodecagon

5-MINUTE CHECK-UP

(over Lesson 7-8)

Available on TRANSPARENCY 7-9.

Classify each quadrilateral.

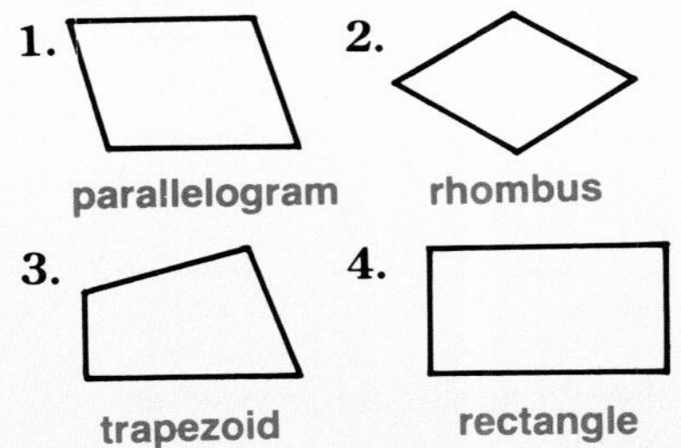

Extra Practice, Lesson 7-8, p. 468

▶ INTRODUCING THE LESSON

Show students a cereal box, an empty ice cream cone, a soft drink can, a ball, and so on. Ask students to suggest a mathematical name for each object.

▶ TEACHING THE LESSON

Using Models Have students make models from the rectangular prism, cube, cylinder, pyramid, and cone patterns, **Lab Manual, pp. 14, 15, 16, 17, and 18.** Have students examine their models.

Practice Masters Booklet, p. 80

PRACTICE WORKSHEET 7-9

Three-Dimensional Figures

Name each shape.

1. cylinder
2. pentagonal pyramid
3. triangular prism
4. cone

Make a drawing of each three-dimensional figure.

5. rectangular pyramid
6. cone
7. hexagonal prism

See students' work.

Copy and complete.

	Polyhedron	Number of Faces (*F*)	Number of Vertices (*V*)	Number of Edges (*E*)
8.	Rectangular Prism	6	8	12
9.	Triangular Pyramid	4	4	6
10.	Hexagonal Prism	8	12	18
11.	Rectangular Pyramid	5	5	8
12.	Hexagonal Pyramid	7	7	12
13.	Triangular Prism	5	6	9
14.	Octagonal Prism	10	16	24
15.	Octagonal Pyramid	9	9	16

7-9 THREE-DIMENSIONAL FIGURES

Objective

Identify and name three-dimensional figures.

Sally Cooper and her family camp in a tent that looks like a rectangular pyramid. The sides of her tent are triangular shaped and the floor of her tent has a rectangular shape.

Three-dimensional figures are solid figures that enclose part of space.

A solid figure with flat surfaces is a **polyhedron.** The flat surfaces are called **faces.** The faces intersect to form **edges.** The edges intersect to form **vertices.**

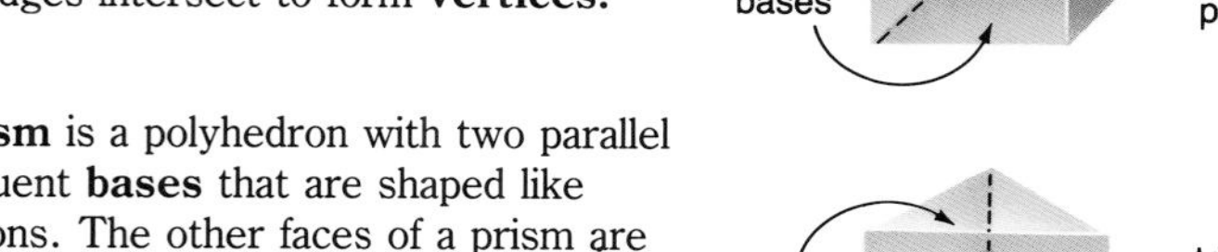

A **prism** is a polyhedron with two parallel congruent **bases** that are shaped like polygons. The other faces of a prism are shaped like parallelograms. A prism is named by the shape of its bases.

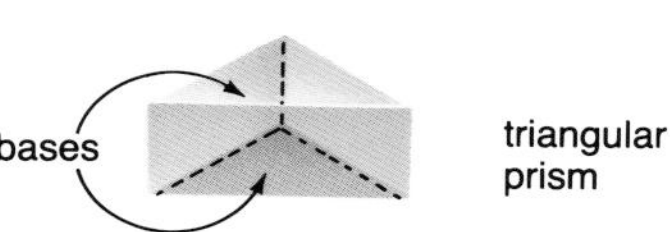

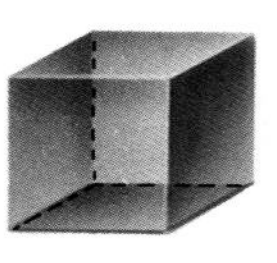

cube

A **cube** is a rectangular prism that has six square faces. All edges of a cube are the same length. How many edges are there? 12

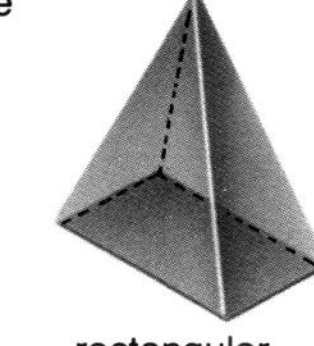

triangular pyramid

rectangular pyramid

A **pyramid** is a polyhedron with a single base shaped like a polygon. The faces of a pyramid are triangular and meet at a point. A pyramid is named by the shape of its base.

Some three-dimensional figures have curved surfaces.

A **cone** has a circular base and one vertex.

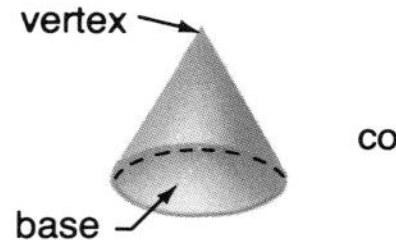

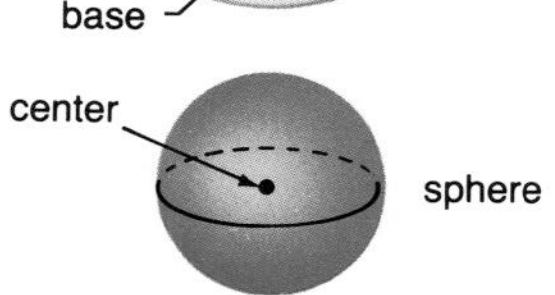

A **cylinder** has two parallel congruent circular bases.

center

sphere

A **sphere** is a solid with all points the same distance from a given point called the **center.**

RETEACHING THE LESSON

Match.

1. Cylinder b
2. prism c
3. sphere a

a.

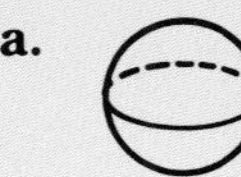

b.

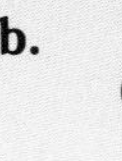

c.

Reteaching Masters Booklet, p. 74

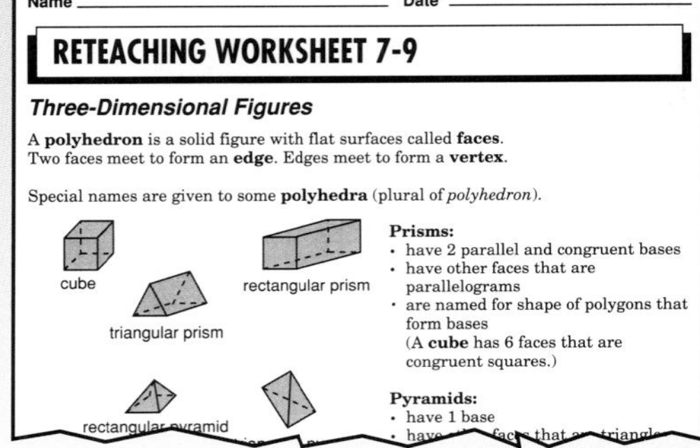

Name ______ Date ______

RETEACHING WORKSHEET 7-9

Three-Dimensional Figures

A **polyhedron** is a solid figure with flat surfaces called **faces**. Two faces meet to form an **edge**. Edges meet to form a **vertex**.

Special names are given to some **polyhedra** (plural of *polyhedron*).

cube

triangular prism

rectangular prism

rectangular pyramid

Prisms:
- have 2 parallel and congruent bases
- have other faces that are parallelograms
- are named for shape of polygons that form bases (A **cube** has 6 faces that are congruent squares.)

Pyramids:
- have 1 base

Guided Practice

Name each shape.

1. 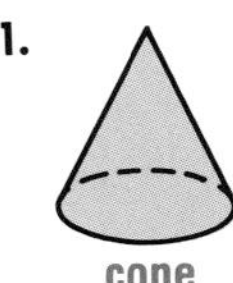 cone
2. 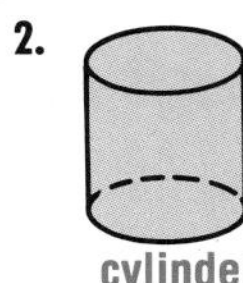 cylinder
3. 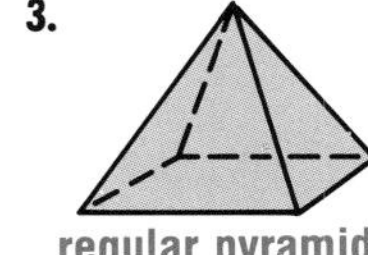 regular pyramid
4.  pentagonal prism

Exercises

Practice

5. Copy and complete.

Polyhedron	Number of Faces *(F)*	Number of Vertices *(V)*	Number of Edges *(E)*
Rectangular Prism	6	8	12
Triangular Pyramid	■ 4	■ 4	■ 6
Pentagonal Prism	■ 7	■ 10	■ 15
Triangular Prism	■ 5	■ 6	■ 9
Rectangular Pyramid	■ 5	■ 5	■ 8
Pentagonal Pyramid	■ 6	■ 6	■ 10

6. Select an assignment from this chapter that shows photos or sketches of physical models you made. **See students' work.**

Using Algebra

7. Write a formula called Euler's formula to show the relationship among the number of faces *(F)*, vertices *(V)*, and edges *(E)* of a polyhedron. $F + V = E + 2$

Problem Solving

Paper/ Pencil
Mental Math
Estimate
Calculator

8. Name a figure that has exactly one base and one vertex. **cone**
9. If a prism has 100 sides and two of them are bases, how many vertices would it have? **196 vertices**
10. The running track at the high school is $\frac{1}{4}$-mile long. *About* how many feet is this? ***about* 1,250 feet**

Cooperative Groups

11. See students' work. See margin.

13. 2; No, a cylinder does not have two line segments that intersect.

Materials: gumdrops, straws, paper, pencil

11. Make a model of a triangular prism, and a rectangular pyramid. Use the figures at the right as a guide. Then see if you can make the rest of the prisms and pyramids on page 230. Determine how each figure got its name.

Materials: paper towel tubes, paper, pencil, scissors, tape

12. Discuss with your group how to make a cylinder using a tube, paper, and tape. Then make the cylinder. **See students' work.**
13. How many bases does a cylinder have? Does the cylinder have vertices? Why or why not?

Critical Thinking

14. Copy and enlarge the figure shown at the right. Cut it out and fold it to make a cube. Three faces meet at each corner. What is the greatest sum of three numbers whose faces meet at a corner? **14 See margin.**

	1		
6	2	4	5
	3		

Chalkboard Examples

Name each shape.

1. 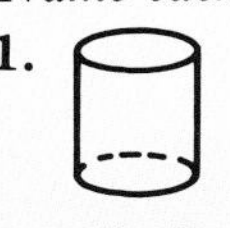 cylinder
2.  triangular prism
3. 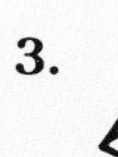 rectangular pyramid
4. 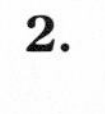 sphere

Additional examples are provided on TRANSPARENCY 7-9.

EVALUATING THE LESSON

Check for Understanding Use Guided Practice Exercises to check for student understanding.

Error Analysis Watch for students that call *faces* of polyhedra *sides.* Be sure students refer to the actual line segments as *edges.*

Closing the Lesson Have students explain in their own words why it is important to know about three-dimensional figures.

EXTENDING THE LESSON

Enrichment Masters Booklet, p. 74

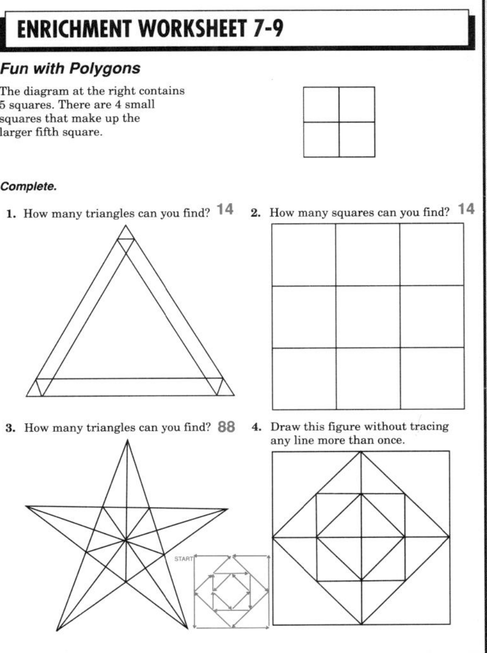
Name ______ Date ______

ENRICHMENT WORKSHEET 7-9

Fun with Polygons

The diagram at the right contains 5 squares. There are 4 small squares that make up the larger fifth square.

Complete.

1. How many triangles can you find? 14
2. How many squares can you find? 14
3. How many triangles can you find? 88
4. Draw this figure without tracing any line more than once.

APPLYING THE LESSON

Independent Practice

Homework Assignment	
Minimum	Ex. 1-7; 8-14 even
Average	Ex. 1-14
Maximum	Ex. 1-14

Additional Answers

11. Prisms are named by the shape of the figure. Pyramids are named by the shape of their base.
14. Each number shares a corner with every number on the cube except the one on the opposite face. Thus the combinations involving 1-3, 2-5, and 4-6 cannot occur. Therefore, the largest sum is $6 + 5 + 3 = 14$.

5-MINUTE CHECK-UP

(over Lesson 7-9)

Available on TRANSPARENCY 7-10.

Name each shape.

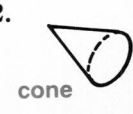

1.  triangular prism
2.  cone
3. cube
4.  rectangular pyramid

Extra Practice, Lesson 7-9, p. 468

INTRODUCING THE LESSON

Ask students to name times when classification skills are used in daily living. using yellow pages; finding items in a grocery store or department store

TEACHING THE LESSON

Using Critical Thinking Draw four objects on the board or use Guided Practice and ask students which object does not belong. Ask students to explain how they chose their answer.

Practice Masters Booklet, p. 81

Name __________ Date __________

PRACTICE WORKSHEET 7-10

Problem – Solving Strategy: Use Logical Reasoning

Which one does not belong? Explain your answer.

1. a. b. c. d.
b; has a curved surface

2. a. b. c. d.
c; does not have 2 pairs of parallel sides

3. a. vertex b. face c. ray d. edge
c; not part of a three-dimensional figure

4. a. rhombus b. rectangle c. scalene triangle d. obtuse triangle
a; always has congruent sides

Choose the letter of the best answer. Explain your answer.

5. line segment is to ruler as angle is to
a. triangle b. compass c. protractor d. ray
c; a protractor is used to measure an angle.

6. triangle is 180° as quadrilateral is to
a. 360° b. 90° c. 180° d. 400°
a; there are 360° in the measures of the angles of a quadrilateral.

7. prism is to parallellogram as pyramid is to
a. point b. triangle c. base d. vertex
b; the sides of a pyramid are triangles.

8. line is to ray as line segment is to
a. line b. ray c. point d. angle
c; a point is part of a line segment.

Problem-Solving Strategy

7-10 USE LOGICAL REASONING

Explore · Plan · Solve · Examine

Objective
Solve problems using logical reasoning.

Sean must work one more problem to complete his geometry assignment. He must decide which one of the following polygons does not belong in the group.

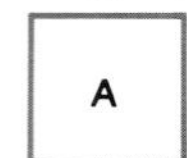
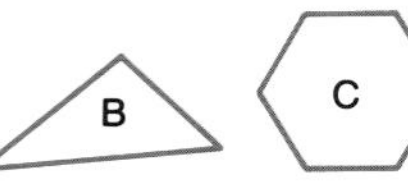

Explore

What is given?

- four different polygons

What is asked?

- Which polygon does not belong?

Plan

Look at the figures to decide what attributes they have in common, and what attributes distinguish one from the other three. Use logical reasoning.

Solve

- All the figures are about the same size.
- All the figures are the same color.
- All the figures are polygons.
- Figures A and C are regular polygons.
- Figure B is not a regular polygon.

So figure B does not belong in the group.

Examine

By definition, a regular polygon is one in which all sides are congruent and all angles are congruent. Figure B is the only polygon that does not meet these requirements. So the answer is reasonable.

Guided Practice

Which one does not belong? Explain your answer.

1.    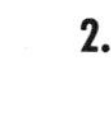
a. b. c. d.
b, not a ball

2. 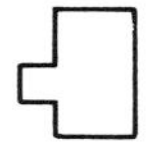 
a. b. c. d.
c, not a polygon

3.     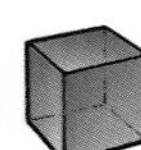 
a. b. c. d.
c, not three dimensional

4. 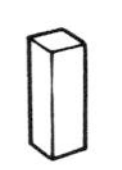 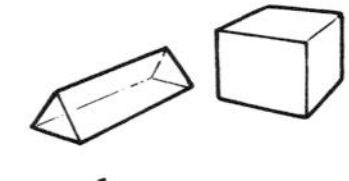
a. b. c. d.
d, not a prism

RETEACHING THE LESSON

Which does not belong? Explain your answer.

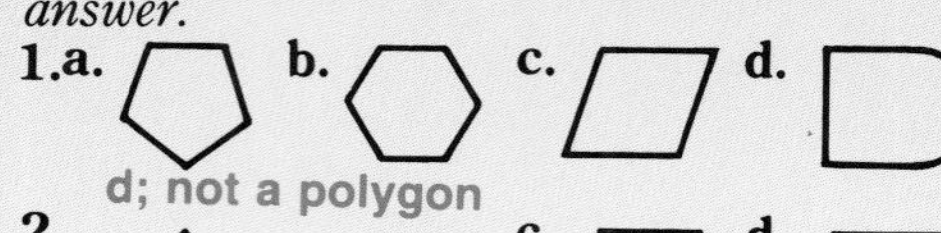

1. a. b. c. d.
d; not a polygon

2. a. b. c. d.
c; not a parallelogram

3. 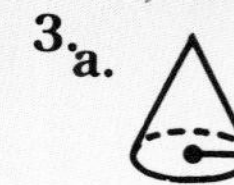 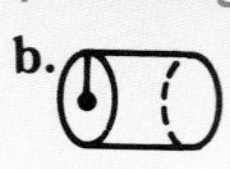 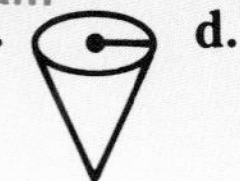 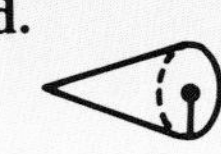
a. b. c. d.
b; not a cone

Problem Solving

Choose the letter of the best answer. Explain your answer. See margin.

5. shoe is to foot as right angle is to
   a. <u>square</u> b. circle c. pentagon d. triangular prism
6. hot is to cold as quadrilateral is to
   a. square b. decagon c. parallelogram d. <u>circle</u>
7. [figure] is to [figure] as [figure] is to
   a. [figure] <u>b.</u> [figure] c. [figure] d. [figure]

Solve. Use any strategy.

8. Two adjacent elevators are on the same floor. The first elevator goes up 4 floors and down 7. The second elevator goes down 11 floors and up 9. How many floors apart are the elevators when they last stopped? **1 floor**
9. Mr. Harding tells his students that he is thinking of a number. If you multiply the number by 5, subtract 15, and add 4, the result is 44. Find the number. **11**
10. Chad wants to take a vacation after he graduates. In order to visit three friends, he drives 300 miles east, 100 miles northwest, 300 miles west, and then returns home. Name the geometric figure that represents his path. **parallelogram**
11. What value of b makes $6^2 + b^2 = 10^2$ true? **8**
12. The number 16 is to 256 as 35 is to _?_. **1,225**
13. These are. . . . These are not. . . . Which of these are?

13. 49, 81 square numbers

FUN with MATH
When and where was the wheel first used?
See page 240.

Critical Thinking

14. **Symmetry** means that a line can be drawn on an object so that the shape on one side of the line matches the shape on the other or that the shapes are congruent. Print all the letters of the alphabet as capital letters. How many letters have a vertical line of symmetry? **11; A, H, I, M, O, T, U, V, W, X, Y**

Mixed Review

Add. (Lesson 4-10)

15. $\frac{3}{5} + \frac{4}{5}$ **$1\frac{2}{5}$** 16. $\frac{2}{3} + \frac{3}{4}$ **$1\frac{5}{12}$** 17. $\frac{3}{5} + \frac{4}{7}$ **$1\frac{6}{35}$** 18. $\frac{6}{8} + \frac{1}{4}$ **1**

Complete. (Lesson 6-4)

19. 42g = _?_ kg **0.042** 20. 9.5 g = _?_ mg **9,500** 21. 0.25 kg = _?_ g **250**

(Lesson 7-3)

22. Recall that two angles are complementary if the sum of their measures is 90°. Find the measure of an angle if it is 50° less than that of its complement. **20°**

Chalkboard Examples

Which one does not belong? Explain your answer.

1.
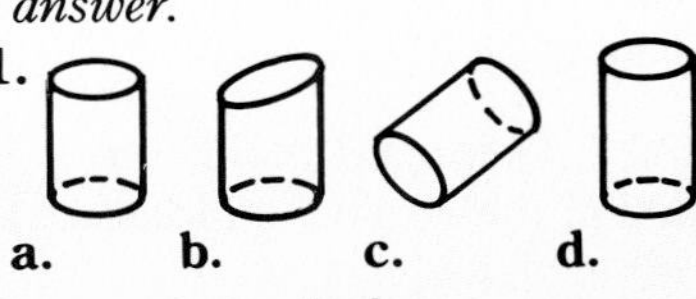
a. b. c. d.
b, not a cylinder

2.

a. b. c. d.
d, not a garden tool

Additional examples are provided on TRANSPARENCY 7-10.

EVALUATING THE LESSON

Check for Understanding Use Guided Practice Exercises to check for student understanding.

Error Analysis Watch for students who have trouble seeing similarities and differences in groups. Have them work with partners who understand the strategy.

Closing the Lesson Ask students to name the steps used in solving problems using logical reasoning.

APPLYING THE LESSON

Independent Practice

Homework Assignment	
Minimum	Ex. 1-22
Average	Ex. 1-22
Maximum	Ex. 1-22

Chapter 7, Quiz A (Lessons 7-1 through 7-5) is available in the Evaluation Masters Booklet, p. 34.

Additional Answers

5. looking for things that go together; A square requires a right angle. None of the other figures requires a right angle.
6. looking for opposites; A quadrilateral is made up of line segments. A circle is made up of a curved line.
7. looking for objects that are inside out; The square is inside the triangle. In b the triangle is inside the square.

Using the Chapter Review

The Chapter Review is a comprehensive review of the concepts presented in this chapter. This review may be used to prepare students for the Chapter Test.

Chapter 7, Quizzes A and B, are provided in the Evaluation Masters Booklet as shown below.

Quiz A should be given after students have completed Lessons 7-1 through 7-5. Quiz B should be given after students have completed Lessons 7-6 through 7-10.

These quizzes can be used to obtain a quiz score or as a check of the concepts students need to review.

Evaluation Masters Booklet, p. 34

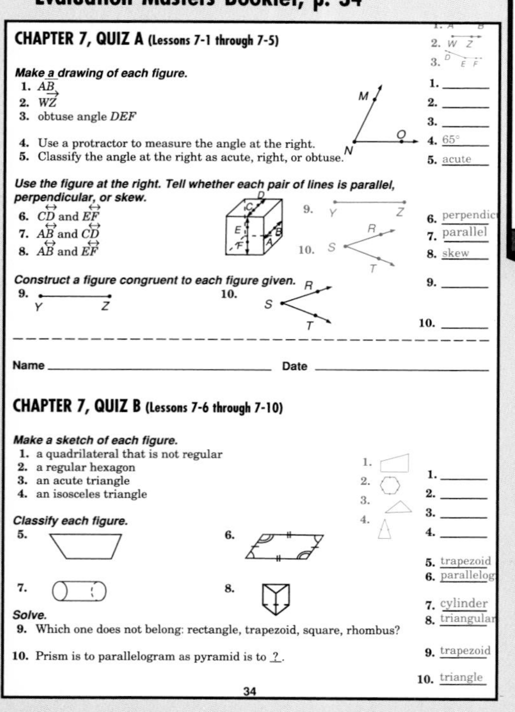

CHAPTER 7, QUIZ A (Lessons 7-1 through 7-5)

Make a drawing of each figure.
1. $\overline{AB}$
2. $\overrightarrow{WZ}$
3. obtuse angle DEF

4. Use a protractor to measure the angle at the right.
5. Classify the angle at the right as acute, right, or obtuse.

1. ____ 2. ____ 3. ____ 4. 65° 5. acute

Use the figure at the right. Tell whether each pair of lines is parallel, perpendicular, or skew.
6. $\overleftrightarrow{CD}$ and $\overleftrightarrow{EF}$
7. $\overleftrightarrow{AB}$ and $\overleftrightarrow{CD}$
8. $\overleftrightarrow{AB}$ and $\overleftrightarrow{EF}$

6. perpendic 7. parallel 8. skew

Construct a figure congruent to each figure given.
9. (segment YZ)
10. (angle RST)

9. ____ 10. ____

Name ____ Date ____

CHAPTER 7, QUIZ B (Lessons 7-6 through 7-10)

Make a sketch of each figure.
1. a quadrilateral that is not regular
2. a regular hexagon
3. an acute triangle
4. an isosceles triangle

1. ____ 2. ____ 3. ____ 4. ____

Classify each figure.
5. 6. 7. 8.

5. trapezoid 6. parallelog 7. cylinder 8. triangular

Solve.
9. Which one does not belong: rectangle, trapezoid, square, rhombus?
10. Prism is to parallelogram as pyramid is to ?.

9. trapezoid 10. triangle

34

Chapter 7 REVIEW

Vocabulary / Concepts

Choose the word or number from the list at the right to complete each sentence.

acute
angle
congruent
edges
isosceles
line segment
parallel
plane
point
scalene
square

1. A ? is formed by two endpoints and the straight path between them. **line segment**
2. Two rays with a common endpoint form an ?. **angle**
3. An ? angle measures between 0° and 90°. **acute**
4. Lines in the same plane that do not intersect are ?. **parallel**
5. An ? triangle has at least two congruent sides. **isosceles**
6. A ? is an exact location in space. **point**
7. An equilateral triangle has three ? sides. **congruent**
8. A ? is a quadrilateral with four congruent sides and four congruent angles. **square**
9. The faces of a polyhedron intersect to form ?. **edges**

Exercises / Applications

Lessons 7-1, 7-2, 7-3

Make a drawing of each figure. **See answers on p. T523.**

10. ray AB
11. line MN
12. line segment RS
13. $\overleftrightarrow{BC}$
14. two skew lines
15. $\angle DEF$
16. $\overline{HJ}$
17. $\overleftrightarrow{OP} \parallel \overleftrightarrow{XY}$
18. $\overleftrightarrow{RD} \perp \overleftrightarrow{MN}$

Lessons 7-2, 7-3

***Find the measure of each angle. Then classify each angle as* right, acute, *or* obtuse.**

19.

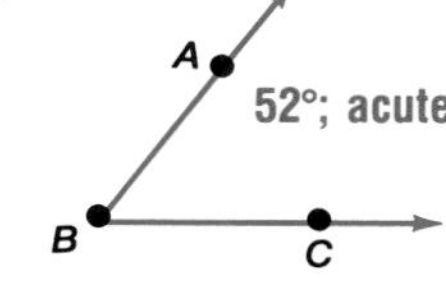

20.

21.

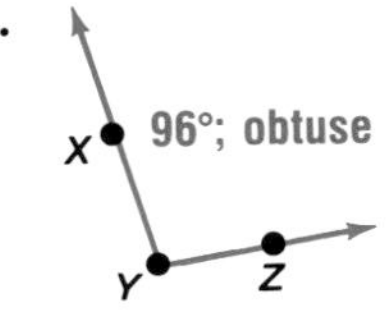

Lesson 7-4

Trace each figure. Construct a figure congruent to each figure given. **See students' work.**

22.

23.

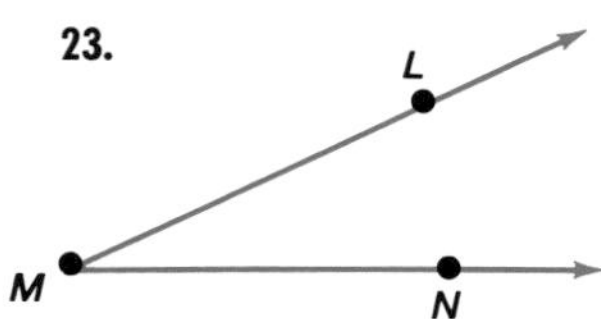

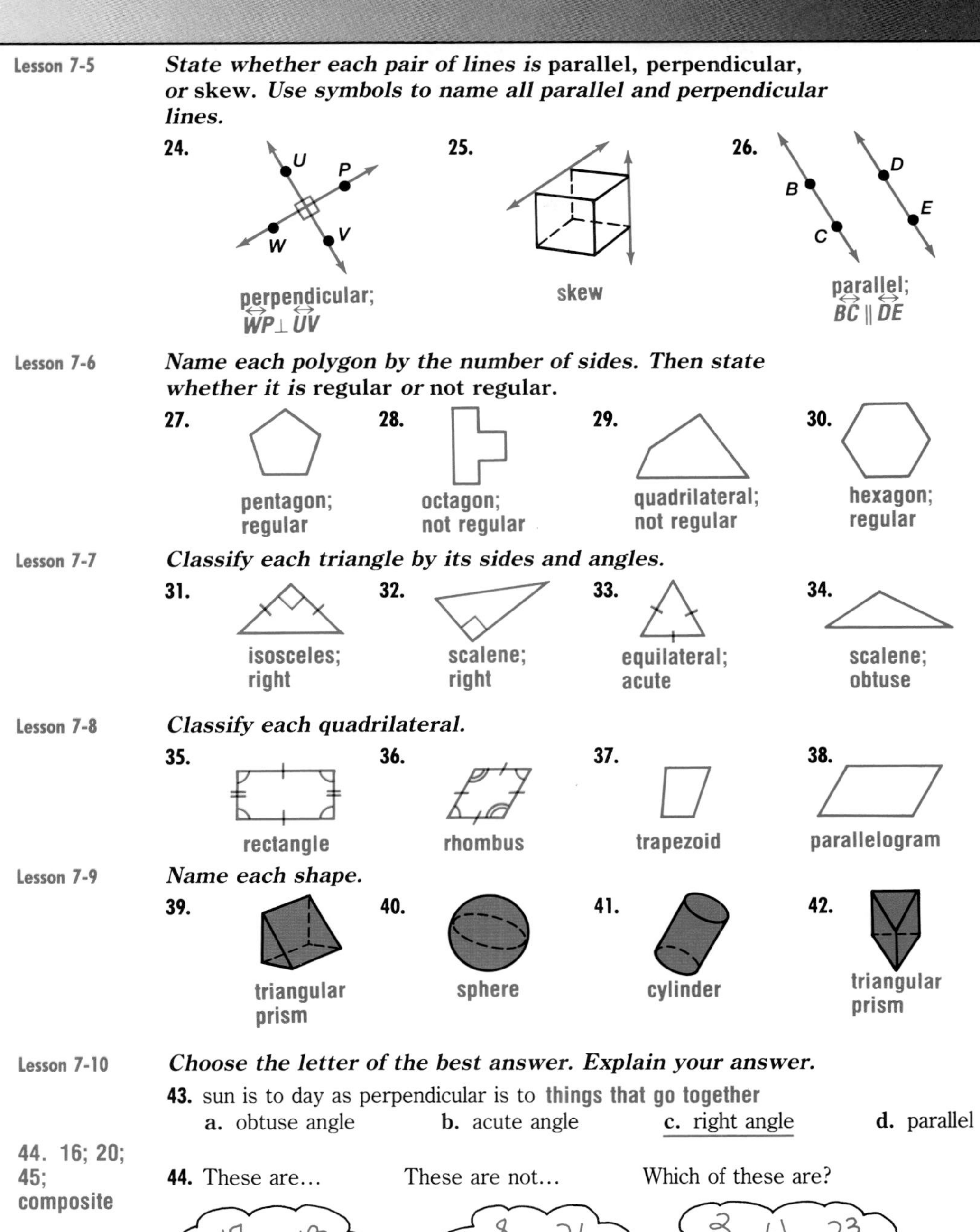

Lesson 7-5 **State whether each pair of lines is parallel, perpendicular, or skew. Use symbols to name all parallel and perpendicular lines.**

24. perpendicular; $\overleftrightarrow{WP} \perp \overleftrightarrow{UV}$

25. skew

26. parallel; $\overleftrightarrow{BC} \parallel \overleftrightarrow{DE}$

Lesson 7-6 **Name each polygon by the number of sides. Then state whether it is regular or not regular.**

27. pentagon; regular

28. octagon; not regular

29. quadrilateral; not regular

30. hexagon; regular

Lesson 7-7 **Classify each triangle by its sides and angles.**

31. isosceles; right

32. scalene; right

33. equilateral; acute

34. scalene; obtuse

Lesson 7-8 **Classify each quadrilateral.**

35. rectangle

36. rhombus

37. trapezoid

38. parallelogram

Lesson 7-9 **Name each shape.**

39. triangular prism

40. sphere

41. cylinder

42. triangular prism

Lesson 7-10 **Choose the letter of the best answer. Explain your answer.**

43. sun is to day as perpendicular is to things that go together
a. obtuse angle **b.** acute angle **c.** right angle **d.** parallel

44. 16; 20; 45; composite

44. These are... 17 53 31 7 These are not... 8 21 42 57 Which of these are? 2 61 23 45 16 20

APPLYING THE LESSON

Independent Practice

Homework Assignment	
Minimum	Ex. 1-9; 10-44 even
Average	Ex. 1-9; 11-43 odd; 44
Maximum	Ex. 1-44

Alternate Assessment Strategy

To provide a brief in-class review, you may wish to read the following questions to the class and require a verbal response.

1. How many endpoints are there in a line segment? **2**
2. What is the vertex of angle *XRT*? **R**
3. Classify an angle that has a measure of 120°. **obtuse**
4. What are two lines that intersect to form right angles? **perpendicular**
5. Name the polygon that has 8 congruent sides and 8 congruent angles. **regular octagon**
6. What is a triangle with no congruent sides? **scalene**
7. A triangle with all 3 angles less than 90° is what kind of triangle? **acute**
8. Is a trapezoid a quadrilateral? Explain. **yes**
9. How many faces does a rectangular prism have? **6**
10. What is the shape of the base of a cone? **circular**

Using the Chapter Test

This page may be used as a test or as an additional page of review if necessary. Two forms of a Chapter Test are provided in the Evaluation Masters Booklet. Form 2 (free response) is shown below, and Form 1 (multiple choice) is shown on page 239.

The **Tech Prep Applications Booklet** provides students with an opportunity to familiarize themselves with various types of technical vocations. The Tech Prep applications for this chapter can be found on pages 13–14.

Evaluation Masters Booklet, p. 33

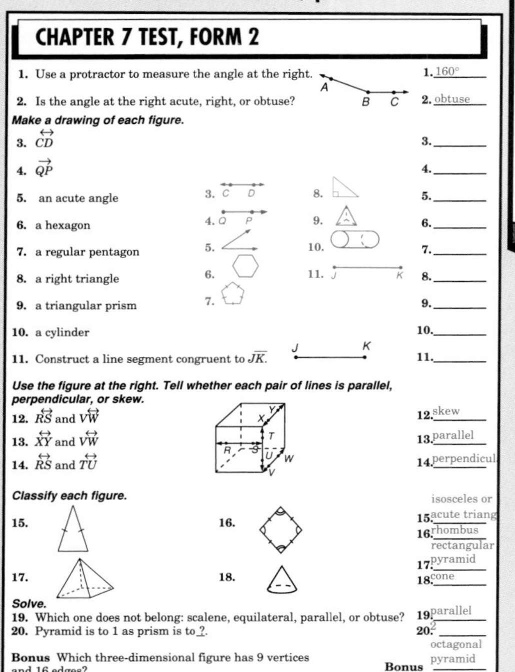

CHAPTER 7 TEST, FORM 2

1. Use a protractor to measure the angle at the right. 1. 160°
2. Is the angle at the right acute, right, or obtuse? 2. obtuse

Make a drawing of each figure.

3. $\overleftrightarrow{CD}$ 3. ____
4. $\overrightarrow{QP}$ 4. ____
5. an acute angle 5. ____
6. a hexagon 6. ____
7. a regular pentagon 7. ____
8. a right triangle 8. ____
9. a triangular prism 9. ____
10. a cylinder 10. ____
11. Construct a line segment congruent to $\overline{JK}$. 11. ____

Use the figure at the right. Tell whether each pair of lines is parallel, perpendicular, or skew.

12. $\overleftrightarrow{RS}$ and $\overleftrightarrow{VW}$ 12. skew
13. $\overleftrightarrow{XY}$ and $\overleftrightarrow{VW}$ 13. parallel
14. $\overleftrightarrow{RS}$ and $\overleftrightarrow{TU}$ 14. perpendicul

Classify each figure.

15. 15. isosceles or acute triang
16. 16. rhombus
17. 17. rectangular pyramid
18. 18. cone

Solve.

19. Which one does not belong: scalene, equilateral, parallel, or obtuse? 19. parallel
20. Pyramid is to 1 as prism is to __?__. 20. 2

Bonus Which three-dimensional figure has 9 vertices and 16 edges? **Bonus** octagonal pyramid

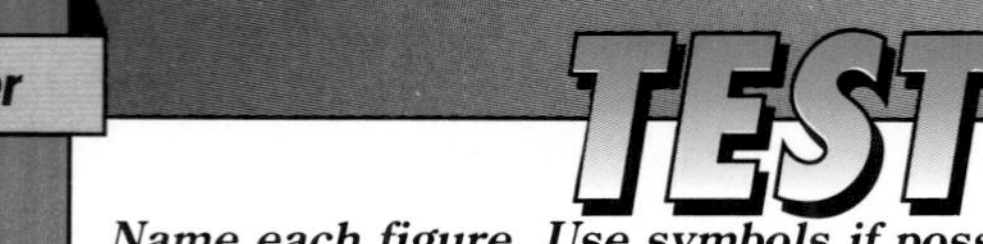

Chapter 7 TEST

Name each figure. Use symbols if possible. Be as specific as possible.

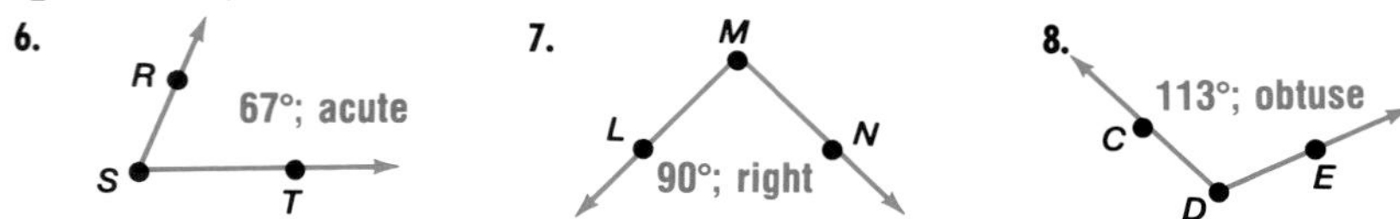

1. right triangle
2. ray RS or $\overrightarrow{RS}$
3. rectangular prism
4. quadrilateral; parallelogram
5. polygon; hexagon

Find the measure of each angle. Then classify each angle as right, acute, ***or*** obtuse.

6. 67°; acute
7. 90°; right
8. 113°; obtuse

Trace each figure. Construct a figure congruent to each figure given.

See students' work

9.

10.

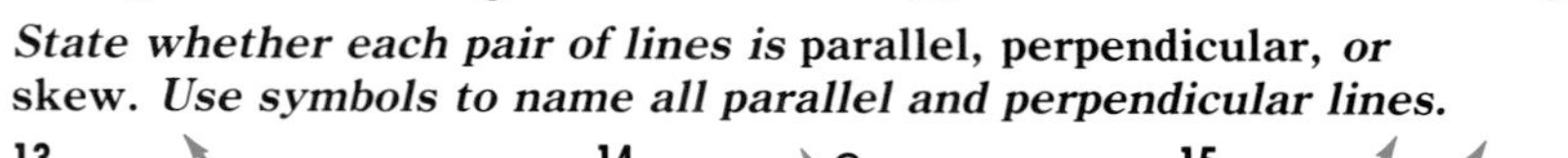

Trace each line segment. Bisect each line segment. See students' work.

11.

12.

State whether each pair of lines is parallel, perpendicular, ***or*** skew. ***Use symbols to name all parallel and perpendicular lines.***

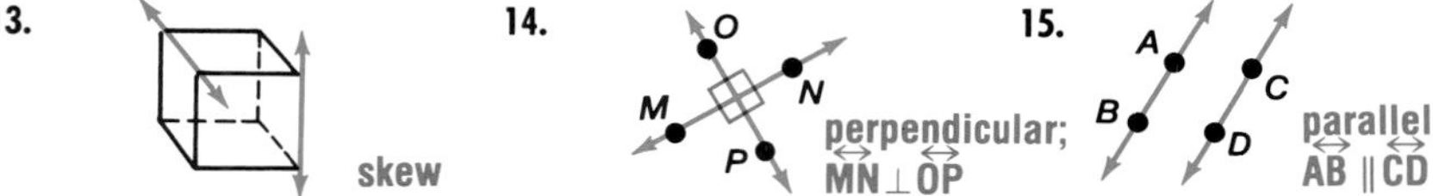

13. skew
14. perpendicular; $\overleftrightarrow{MN} \perp \overleftrightarrow{OP}$
15. parallel; $\overleftrightarrow{AB} \parallel \overleftrightarrow{CD}$

Classify each triangle by its sides and then by its angles.

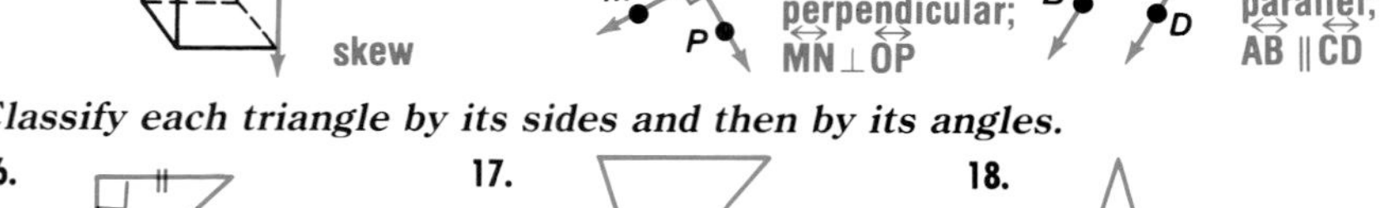

16. isosceles; right
17. scalene; acute
18. isosceles; acute

Solve. Use any strategy.

19. The number 3,000,000 is to 3×10^6 as 4×10^5 is to __?__ .

20. These are . . . These are not . . . Which of these are?

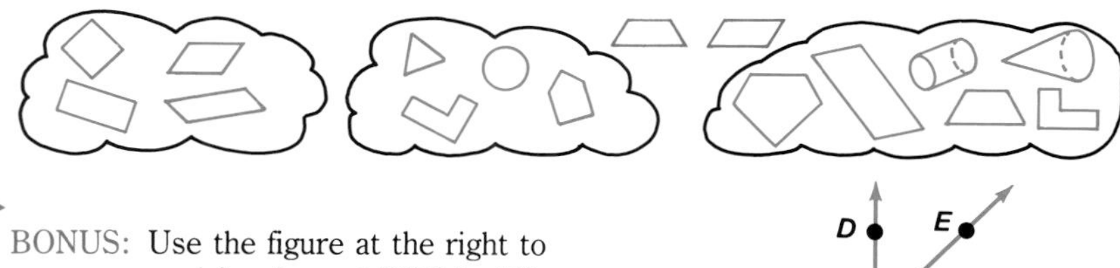

BONUS: Use the figure at the right to explain why m∠*DBE* is 45°.

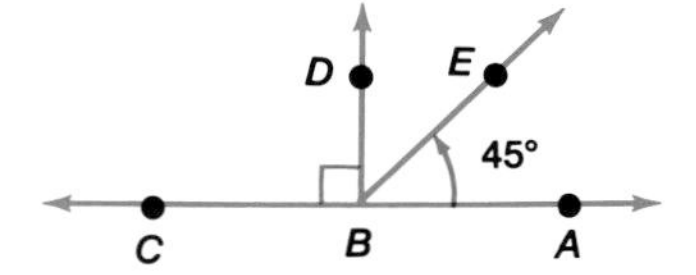

19. 400,000

BONUS: If m∠*DBC* is 90°, then m∠*DBA* is 90°. m∠*DBA* − m∠*EBA* = 90° − 45° or 45° m∠*DBE* is 45°

The **Performance Assessment Booklet** provides an alternative assessment for evaluating student progress. An assessment for this chapter can be found on pages 13–14.

A **Test and Review Generator** is provided in Apple, IBM, and Macintosh versions. You may use this software to create your own tests or worksheets, based on the needs of your students.

CONSTRUCTING TRIANGLES

You can use a compass and straightedge to construct triangles that are congruent to other triangles.

Construct a triangle congruent to a given triangle.

Given:

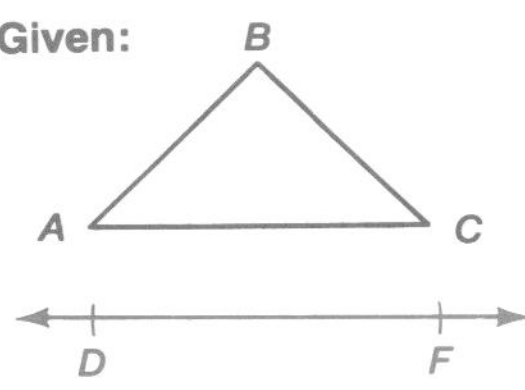

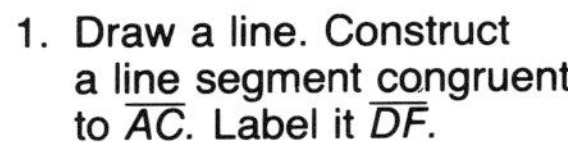

1. Draw a line. Construct a line segment congruent to $\overline{AC}$. Label it $\overline{DF}$.

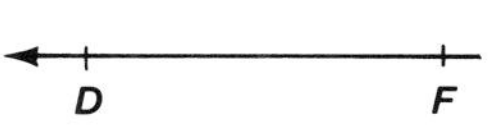

2. With compass open to the length of $\overline{AB}$, draw an arc from point D. With compass open to the length of $\overline{BC}$, draw an arc from point F.

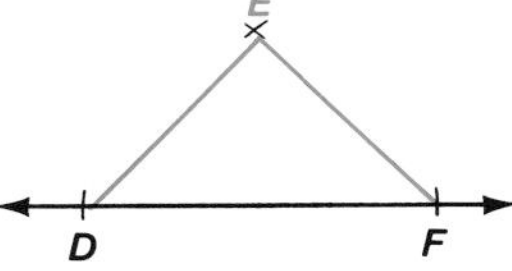

3. Label the intersection of the arcs E. Draw $\overline{DE}$ and $\overline{EF}$.

1. Trace $\triangle ABC$. Construct a triangle congruent to $\triangle ABC$.
See students' work.

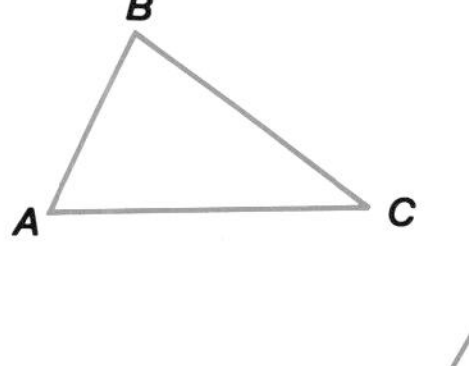

2. Trace $\triangle JKL$. Construct a triangle congruent to $\triangle JKL$.
See students' work.

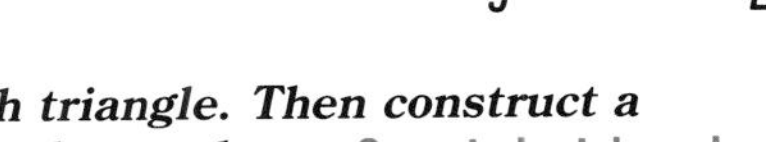

Use a straightedge to draw each triangle. Then construct a triangle congruent to each triangle you drew. **See students' work.**

3. scalene triangle

4. isosceles triangle

5. Trace each line segment at the right. Then construct a triangle having sides the length of line segments MN, PQ, and RS.
See students' work.

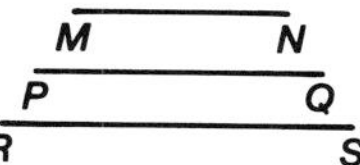

Applying Mathematics to the Real World

This optional page shows how mathematics is used in the real world and also provides a change of pace.

Objective Construct congruent triangles with a compass and straightedge.

Using Vocabulary Remind students that triangles can be named according to the lengths of their sides. Ask students to define scalene, isosceles, and equilateral triangles.

Using Critical Thinking Think about what you have studied in this chapter, try to find another way to construct a congruent angle. **Construct congruent side, angle, side to the given triangle or construct congruent angle, side, angle to the given triangle.**

Activity

Using Manipulatives Trace and cut out these figures. Try to arrange them to form an equilateral triangle without overlapping the figures.

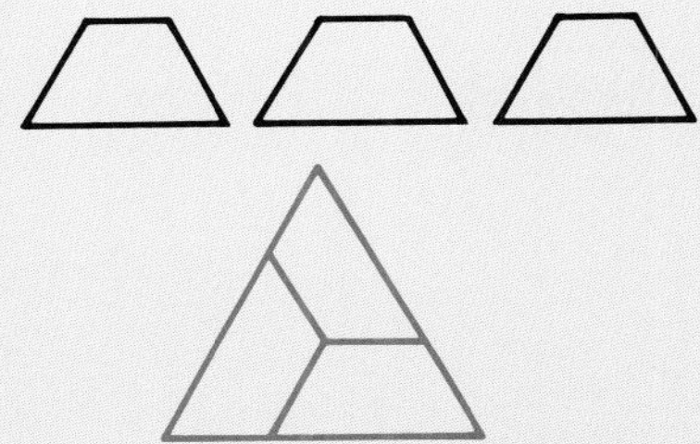

Using the Cumulative Review

This page provides an aid for maintaining skills and concepts presented thus far in the text.

A Cumulative Review is also provided in the Evaluation Masters Booklet as shown below.

Cumulative REVIEW/TEST

Free Response

Lesson 1-6

Estimate.

1. 3.568 + 9.237 13 **2.** 86.2 + 39.92 130 **3.** 0.42 + 1.7 2.0

4. 17.325 − 11.758 10 **5.** 541.2 − 38.7 500 **6.** 9.6 − 0.12 9.5

Lessons 2-1, 2-3

Add or subtract.

7. $\begin{array}{r} 563 \\ +103 \\ \hline 666 \end{array}$ **8.** $\begin{array}{r} 375 \\ -\ 25 \\ \hline 350 \end{array}$ **9.** $\begin{array}{r} 251 \\ -\ 34 \\ \hline 21 \end{array}$ **10.** $\begin{array}{r} 551 \\ +359 \\ \hline 910 \end{array}$

Lesson 3-1 $1,500,000

11. The cost of a 30-second commercial during Super Bowl XX was $500,000. What was the cost of three 30-second commercials?

Lesson 3-4

12. Mr. Benitez receives a bonus of $125 for each life insurance policy he sells. Last month he received a bonus of $1,000. How many policies did he sell? 8 policies

Lesson 4-6

Simplify each fraction.

13. $\frac{36}{40}$ $\frac{9}{10}$ **14.** $\frac{4}{28}$ $\frac{1}{7}$ **15.** $\frac{45}{105}$ $\frac{3}{7}$ **16.** $\frac{160}{224}$ $\frac{5}{7}$ **17.** $\frac{1,000}{40}$ 25 **18.** $\frac{95}{38}$ $\frac{5}{2}$

Lessons 5-3, 5-7

Multiply or divide.

19. $\frac{1}{2} \times \frac{1}{4}$ $\frac{1}{8}$ **20.** $\frac{7}{8} \times \frac{4}{3}$ $1\frac{1}{6}$ **21.** $\frac{36}{40} \div \frac{45}{100}$ 2 **22.** $\frac{15}{18} \div 2$ $\frac{5}{12}$

23. $20 \div \frac{1}{5}$ 100 **24.** $7\frac{1}{5} \times 8\frac{1}{3}$ 60 **25.** $12\frac{5}{6} \div 1\frac{3}{8}$ $9\frac{1}{3}$

Lesson 5-8

Change each fraction to a decimal.

26. $\frac{6}{8}$ 0.75 **27.** $\frac{17}{20}$ 0.85 **28.** $\frac{1}{5}$ 0.2 **29.** $\frac{3}{10}$ 0.30 **30.** $\frac{60}{10}$ 6.0 **31.** $\frac{18}{12}$ 1.5

Lesson 6-6

32. In two throws, a discus thrower hurls a discus 218 feet and 226 feet. What is the total length of the two throws in yards? 148 yards

Lesson 7-7

Classify each triangle by its sides and angles.

33.   equilateral; acute

34. 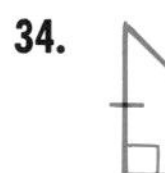 right; isosceles

35.  right; scalene

APPLYING THE LESSON

Independent Practice

Homework Assignment	
Minimum	Ex. 1-35 odd
Average	Ex. 1-35
Maximum	Ex. 1-35

Evaluation Masters Booklet, p. 35

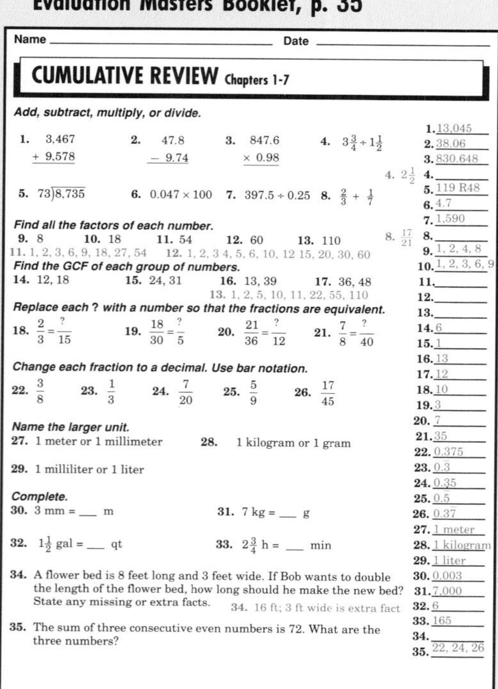

Name ______ Date ______

CUMULATIVE REVIEW Chapters 1-7

Add, subtract, multiply, or divide.

1. 3,467 + 9,578
2. 47.8 − 9.74
3. 847.6 × 0.98
4. $3\frac{3}{4} \div 1\frac{1}{2}$ (4. $2\frac{1}{2}$)
5. 73)8,735
6. 0.047 × 100
7. 397.5 ÷ 0.25
8. $\frac{2}{3} + \frac{1}{7}$ (8. $\frac{17}{21}$)

Find all the factors of each number.

9. 8 10. 18 11. 54 12. 60 13. 110

11. 1, 2, 3, 6, 9, 18, 27, 54 12. 1, 2, 3 4, 5, 6, 10, 12 15, 20, 30, 60 13. 1, 2, 5, 10, 11, 22, 55, 110

Find the GCF of each group of numbers.

14. 12, 18 15. 24, 31 16. 13, 39 17. 36, 48

Replace each ? with a number so that the fractions are equivalent.

18. $\frac{2}{3} = \frac{?}{15}$ 19. $\frac{18}{30} = \frac{?}{5}$ 20. $\frac{21}{36} = \frac{?}{12}$ 21. $\frac{7}{8} = \frac{?}{40}$

Change each fraction to a decimal. Use bar notation.

22. $\frac{3}{8}$ 23. $\frac{1}{3}$ 24. $\frac{7}{20}$ 25. $\frac{5}{9}$ 26. $\frac{17}{45}$

Name the larger unit.

27. 1 meter or 1 millimeter 28. 1 kilogram or 1 gram

29. 1 milliliter or 1 liter

Complete.

30. 3 mm = ___ m 31. 7 kg = ___ g

32. $1\frac{1}{2}$ gal = ___ qt 33. $2\frac{3}{4}$ h = ___ min

34. A flower bed is 8 feet long and 3 feet wide. If Bob wants to double the length of the flower bed, how long should he make the new bed? State any missing or extra facts. 34. 16 ft; 3 ft wide is extra fact

35. The sum of three consecutive even numbers is 72. What are the three numbers?

Answers: 1. 13,045 2. 38.06 3. 830.648 4. ___ 5. 119 R48 6. 4.7 7. 1,590 8. ___ 9. 1, 2, 4, 8 10. 1, 2, 3, 6, 9 11. ___ 12. ___ 13. ___ 14. 6 15. 1 16. 13 17. 12 18. 10 19. 3 20. 7 21. 35 22. 0.375 23. $0.\overline{3}$ 24. 0.35 25. $0.\overline{5}$ 26. $0.3\overline{7}$ 27. 1 meter 28. 1 kilogram 29. 1 liter 30. 0.003 31. 7,000 32. 6 33. 165 34. ___ 35. 22, 24, 26

Chapter 7

Multiple Choice

Choose the letter of the correct answer for each item.

1. Round 8,079 to the nearest hundred.
   a. 8,000
   b. 8,070
   c. 8,080
   ▶ d. 8,100

   Lesson 1-4

2. What is the product of 870 and 12?
   e. 2,610
   f. 8,874
   g. 9,440
   ▶ h. 10,440

   Lesson 3-1

3. Find all the factors of 20.
   a. 2, 5
   b. 1, 2, 5
   ▶ c. 1, 2, 4, 5, 10, 20
   d. 0, 20, 40, 60, 80, . . .

   Lesson 4-1

4. Akai ran a marathon in 3 h 48 min 10 s. Ross had a time of 4 h 27 min 35 s. How much faster than Ross was Akai?
   e. 21 min 25 s
   ▶ f. 39 min 25 s
   g. 1 h 21 min 25 s
   h. 1 h 39 min 25 s

   Lesson 6-10

5. An average envelope weighs about 4 grams and a sheet of typing paper weighs 3 grams. How many sheets of paper can you mail in an envelope and keep the total weight below 30 grams?
   a. 7
   b. 6
   ▶ c. 8
   d. 5

   Lesson 6-4

6. What is the result of dividing 22 by $\frac{9}{10}$?
   e. $\frac{9}{220}$
   f. $19\frac{4}{5}$
   g. $22\frac{4}{9}$
   ▶ h. $24\frac{4}{9}$

   Lesson 5-7

7. What is the prime factorization of 20?
   a. 1 × 20
   b. 4 × 5
   c. 2 × 5
   ▶ d. none of these

   Lesson 4-2

8. Estimate the sum of $\frac{7}{8}$ and $\frac{3}{4}$.
   e. 0
   f. 1
   ▶ g. 2
   h. 3

   Lesson 4-10

9. What word describes two lines that intersect to form right angles?
   a. congruent
   b. parallel
   ▶ c. perpendicular
   d. similar

   Lesson 7-5

10. Brian wanted to travel the shortest distance between each of his classes. Which of the following would not help him find the shortest distance?
    e. a map of his school
    f. the room numbers of his classes
    ▶ g. the school bell schedule
    h. the location of stairs

    Lesson 6-7

APPLYING THE LESSON

Independent Practice

Homework Assignment	
Minimum	Ex. 1-10
Average	Ex. 1-10
Maximum	Ex. 1-10

Using the Standardized Test

This test serves to familiarize students with standardized format while testing skills and concepts presented up to this point.

Evaluation Masters Booklet, pp. 31-32.

Name ____________ Date ____________

CHAPTER 7 TEST, FORM 1

Write the letter of the correct answer on the blank at the right of the page.

1. Find the symbol that names the figure at right.
   A. $\overrightarrow{BA}$ B. $\overleftrightarrow{AB}$ C. $\overline{BA}$ D. $\overrightarrow{AB}$ — 1. A
2. Find a representation of $\overline{CD}$.
   A. B. C. D. — 2. C
3. Use a protractor to measure the angle at the right.
   A. 112° B. 128° C. 64° D. 72° — 3. C
4. Classify the angle at right as right, acute, or obtuse.
   A. right B. acute C. obtuse — 4. B
5. Find an example of an obtuse angle.
   A. B. C. D. — 5. D
6. Which is always true about a bisector of $\overline{BC}$?
   A. It is congruent to $\overline{BC}$.
   B. It separates $\overline{BC}$ into two congruent angles.
   C. It separates $\overline{BC}$ into two congruent line segments.
   D. It forms an obtuse angle with $\overline{BC}$. — 6. C
7. Find the pair of lines that are parallel.
   A. B. C. D. — 7. D
8. Find the pair of lines that are perpendicular.
   A. B. C. D. — 8. A
9. Find the pair of lines that are skew.
   A. B. C. D. — 9. D
10. Find an example of a pentagon.
    A. B. C. D. — 10. A

Using Mathematical Connections

Objective To generate understanding and appreciation of connections between mathematics and real-life phenomena utilizing a number of other disciplines.

The following data is background information for each event on the time line.

Science / History The Wheel, Sumer & Syria (Babylon/Assyria), 3500 BC
The earliest record of a wheeled vehicle is from the cities of Sumer and Syria in 3500 BC in the country known today as Iraq. Sledges on rollers were used to transport heavy loads.

History Geometry, Babylon, 1800 BC
Clay tablets have been excavated near Baghdad in Iraq. One ancient tablet is inscribed with a square and diagonal lines that divide it into four right triangles. This tablet shows that the Babylonians, not the Greeks, were the first to use the formula we now call the Pythagorean Theorem. ($c^2 = a^2 + b^2$) The Babylonians had actually used this formula more than a thousand years before the Greek philosopher Pythagoras, for whom it was named.

Science / History First Compass, China, 300 BC
As early as 300 BC the Chinese discovered that the iron ore called lodestone attracts pieces of metal. They also found that a piece of lodestone carved into the shape of a spoon always pointed in the same direction. Chinese writings refer to a "south-pointing spoon." This was the world's first compass.

Science / History First Seismograph, China, AD 132
The Chinese astronomer and mathematician Chang Heng invented the first seismograph in AD 132 to detect earthquakes. His seismograph was eight feet high and had eight metal dragons arranged around a free-swinging pendulum in the center. Each dragon faced a different direction and held a coppper ball in its mouth. When an earthquake shook the seismograph, a ball dropped with a loud clang on the metal surface below. By drawing a line out from the dragon that had dropped the ball, an assistant could tell the direction of the quake.

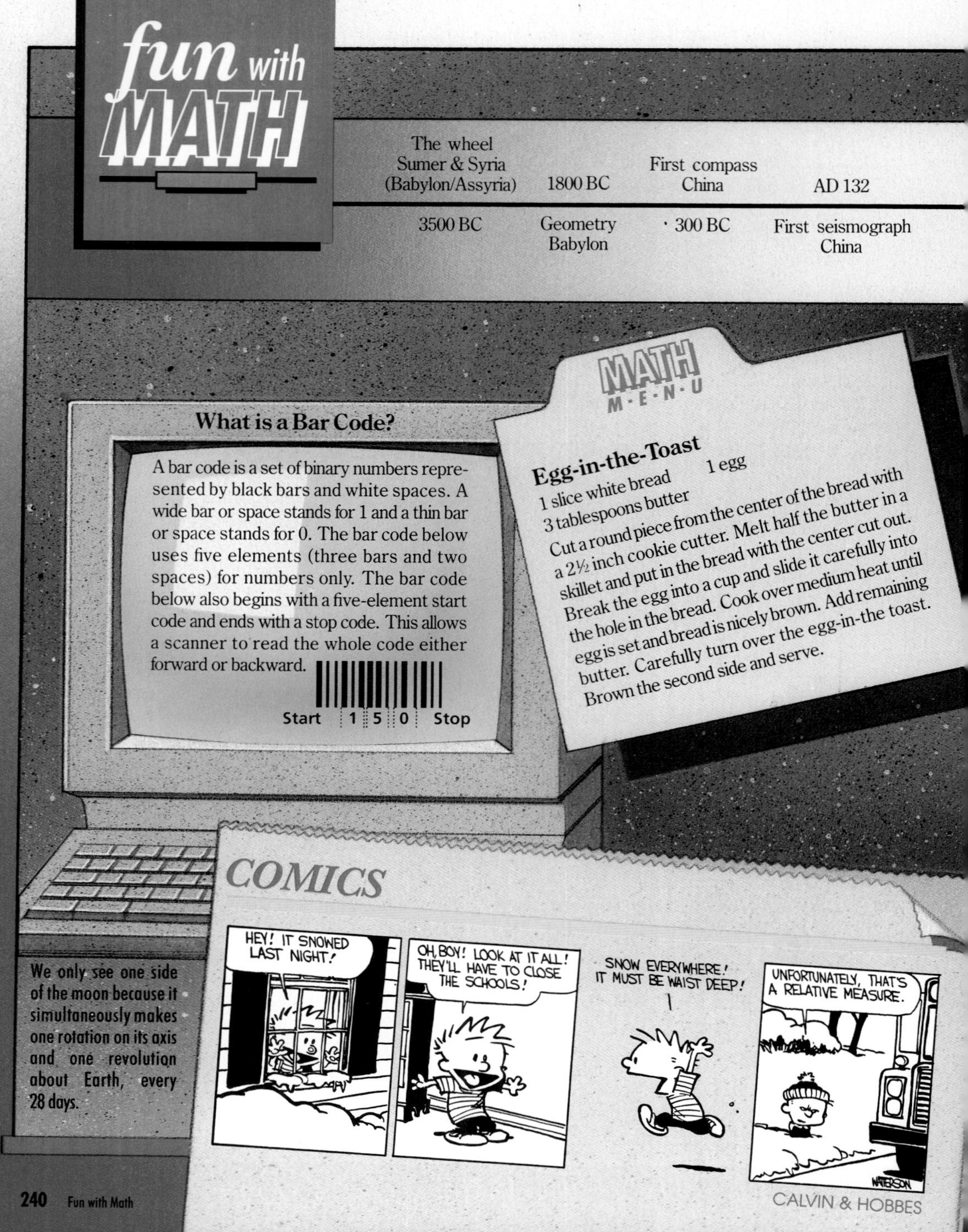

fun with MATH
The wheel
Sumer & Syria
(Babylon/Assyria)
3500 BC
1800 BC
Geometry
Babylon
First compass
China
· 300 BC
AD 132
First seismograph
China
MATH MENU
Egg-in-the-Toast
1 slice white bread 1 egg
3 tablespoons butter
Cut a round piece from the center of the bread with a 2½ inch cookie cutter. Melt half the butter in a skillet and put in the bread with the center cut out. Break the egg into a cup and slide it carefully into the hole in the bread. Cook over medium heat until egg is set and bread is nicely brown. Add remaining butter. Carefully turn over the egg-in-the toast. Brown the second side and serve.
What is a Bar Code?
A bar code is a set of binary numbers represented by black bars and white spaces. A wide bar or space stands for 1 and a thin bar or space stands for 0. The bar code below uses five elements (three bars and two spaces) for numbers only. The bar code below also begins with a five-element start code and ends with a stop code. This allows a scanner to read the whole code either forward or backward.
Start 1 5 0 Stop
We only see one side of the moon because it simultaneously makes one rotation on its axis and one revolution about Earth, every 28 days.
COMICS
HEY! IT SNOWED LAST NIGHT!
OH, BOY! LOOK AT IT ALL! THEY'LL HAVE TO CLOSE THE SCHOOLS!
SNOW EVERYWHERE! IT MUST BE WAIST DEEP!
UNFORTUNATELY, THAT'S A RELATIVE MEASURE.
WATERSON
CALVIN & HOBBES
240 Fun with Math

Quiz Answers

B.

6	1	8
7	5	3
2	9	4

C.

9	2	10
8	7	6
4	12	5

D.

14	3	10
5	9	13
8	15	4

The metric system France	AD 1887	The laser Miami, Florida	AD 1962
AD 1795	Induction telegraph Cincinnati, Ohio	AD 1960	*Silent Spring* Rachel Carson

JOKE!

Q: What did the acorn say when he finally grew up?

A: Geometry!

What is a carat anyway? The carat is a unit of measure. It is equivalent to 200 milligrams. It is based on the weight of the carob seed which was once a weighing standard used by jewelers in Africa and the Middle East. The word *carat* is believed to come from an Arabic word meaning *bean* or *seed.*

TEASER

Building roads is expensive, so civil engineers try to make them as short as possible while maintaining safety. A new interstate highway passes by the small towns of Apple Junction and Hermit Cove. One access ramp is proposed with two-lane roads connecting it to both towns. Where should the access ramp be placed so the total length of road A to R to H is as short as possible? (Hint: The shortest distance between two points is a straight line.)

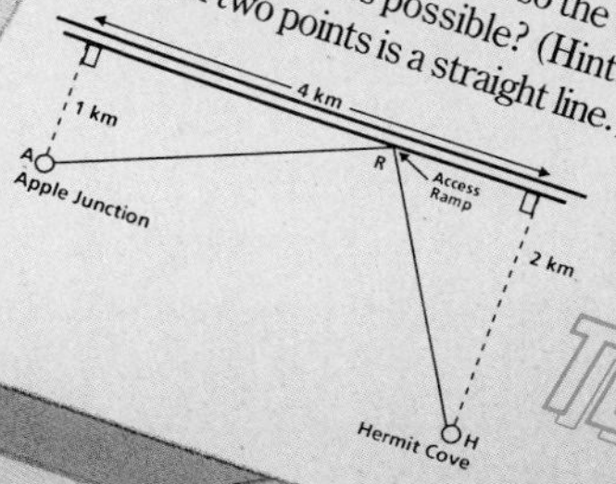

4 km; 1 km; 2 km; A Apple Junction; R Access Ramp; H Hermit Cove

(Flip Flop) Palindrome Riddles

1. Thigh ointment
   A: *LEG GEL*
2. A drink for tiny flies
   A: *GNAT TANG*
3. A hobo who tells untruths
   A: *RAIL LIAR*
4. Curves on a thread cylinder
   A: *SPOOL LOOPS*
5. Bedsheets
   A: *SLEEP PEELS*

QUIZ TIME

3 × 3 Magic Squares

A magic square is a square of numbers in which every row, column, and diagonal add up to the same total. In example A the magic number is 24. Copy and complete the remaining magic squares. Then try creating your own magic square and challenge a friend to complete it.

A

11	3	10
7	8	9
6	13	5

B

6	—	—
7	5	3
—	—	—

C

—	—	10
—	7	—
4	—	5

D

14	3	—
—	—	13
8	15	—

Teaser Answer

Have students imagine the interstate highway as the line of a plane mirror. Draw the image of Apple Junction behind the mirror, the same distance as it is in front of the mirror. Now, draw a line connecting Hermit Cove with the *image* of Apple Junction. The point where the line crosses the interstate highway is where the access ramp should be located. This gives the shortest route.

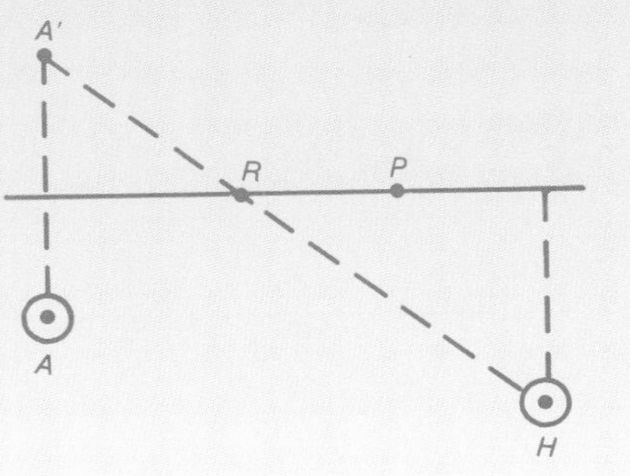

History The Metric System, France, AD 1795

Gabriel Mouton, vicar of St. Paul's Church in Lyon, France, founded the metric system in 1670. He used a base unit from the physical universe rather than the human body. After much investigation the French Academy of Sciences officially adopted the metric system for mandatory use in 1795. The recommended unit of length, the *metre*, was equal to one ten-millionth of the distance, on the surface of Earth, from the North pole to the equator.

Technology / History Induction Telegraph, Cincinnati, Ohio, AD 1887

Granville T. Woods was a black man born of free parents in Ohio in 1856. He has been referred to as the "Black Edison," and in 1887 the *American Catholic Tribune* called him "the greatest electrician in the world." Woods patented over 150 inventions, but one with a most significant impact was the Induction Telegraph.

Technology / Health The Laser, Miami, Florida, AD 1960

In 1960, Theodore Harold Maiman, an American physicist working for Hughes Research Laboratories designed the first ruby laser (*l*ight *a*mplification by *s*timulated *e*mission of *r*adiation.) Light from a laser can travel thousands of miles without spreading out. Today lasers are used to produce dazzling light shows, displays, and holograms. They are becoming commonly used for delicate surgery as well as for cutting steel.

Ecology *Silent Spring* by Rachel Carson, AD 1962

Rachel Carson's *Silent Spring* was published in 1962. The primary focus of her concern is the lethal effect of toxic chemicals in agriculture on mankind and environment. However, her work provides a thought-provoking focus for all our environmental efforts in attempting to find mutually beneficial ways of living in harmony with our planet.

Chapter

8 AREA AND VOLUME

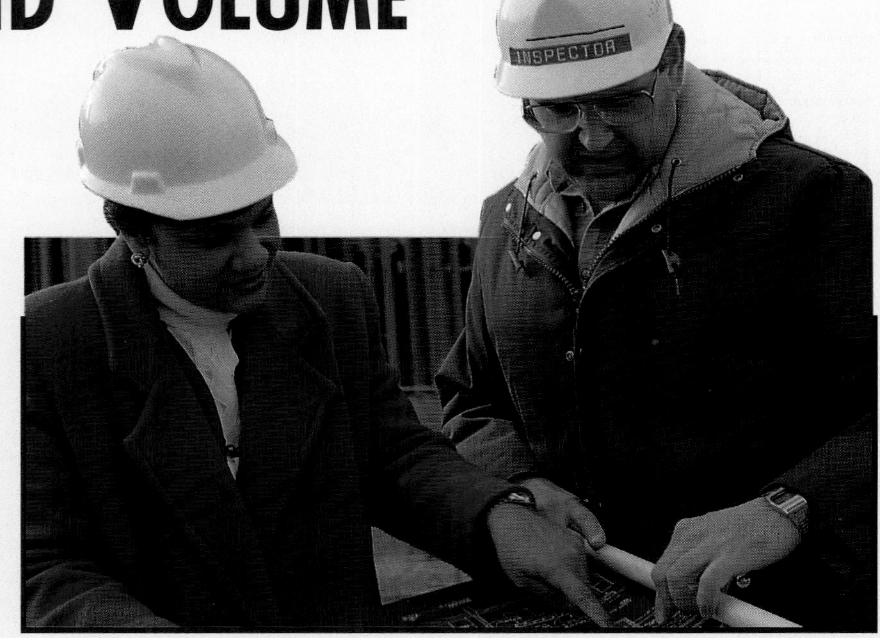

PREVIEWING the CHAPTER

This chapter builds upon the basic geometric concepts of the previous chapter by dealing with the perimeter, area, and volume of geometric figures. Students find the perimeters of polygons and circles. Formulas for the areas of rectangles, parallelograms, triangles, and circles are developed and applied. Students find the surface area of rectangular prisms and cylinders, and the volume of rectangular prisms, rectangular pyramids, cylinders, and cones. Verbal problems involving area and volume are then solved.

Problem-Solving Strategy Students learn to solve a problems by making a diagram.

Lesson (Pages)	Lesson Objectives	State/Local Objectives
8-1 (244-245)	8-1: Find the circumference of circles.	
8-2 (246-247)	8-2: Find the area of parallelograms.	
8-3 (248-249)	8-3: Find the area of triangles.	
8-4 (250-251)	8-4: Find the area of circles.	
8-5 (252-253)	8-5: Find the cost of carpeting a room.	
8-6 (254-255)	8-6: Solve problems by making a diagram.	
8-7 (258-259)	8-7: Find the surface area of rectangular prisms.	
8-8 (260-261)	8-8: Find the surface area of cylinders.	
8-9 (262-263)	8-9: Find the volume of rectangular prisms.	
8-10 (264-265)	8-10: Find the volume of pyramids.	
8-11 (266-267)	8-11: Find the volume of cylinders.	
8-12 (268-269)	8-12: Find the volume of cones.	
8-13 (270-271)	8-13: Solve problems using area and volume.	

ORGANIZING the CHAPTER

You may refer to the ***Course Planning Calendar*** on page T10.

Planning Guide

Lesson (Pages)	Extra Practice (Student Edition)	Blackline Masters Booklets								Other Resources
		Reteaching	Practice	Enrichment	Activity	Multi-cultural Activity	Technology	Tech Prep	Evaluation	
8-1 (244-245)	p. 468	p. 75	p. 82	p. 75			p. 22			Transparency 8-1
8-2 (246-247)	p. 469	p. 76	p. 83	p. 76			p. 8			Transparency 8-2
8-3 (248-249)	p. 469	p. 77	p. 84	p. 77						Transparency 8-3
8-4 (250-251)	p. 469	p. 78	p. 85	p. 78		p. 8		p. 15		Transparency 8-4
8-5 (252-253)		p. 79	p. 86	p. 79						Transparency 8-5
8-6 (254-255)			p. 87							Transparency 8-6
8-7 (258-259)	p. 470	p. 80	p. 88	p. 80	p. 15				p. 39	Transparency 8-7
8-8 (260-261)	p. 470	p. 81	p. 89	p. 81						Transparency 8-8
8-9 (262-263)	p. 470	p. 82	p. 90	p. 82	pp. 16, 36			p. 16		Transparency 8-9
8-10 (264-265)	p. 471	p. 83	p. 91	p. 83						Transparency 8-10
8-11 (266-267)	p. 471	p. 84	p. 92	p. 84						Transparency 8-11
8-12 (268-269)	p. 471	p. 85	p. 93	p. 85						Transparency 8-12
8-13 (270-271)		p. 86	p. 94	p. 86					p. 39	Transparency 8-13
Ch. Review (272-273)									pp. 36-38	Test and Review Generator
Ch. Test (274)										
Cumulative Review/Test (276-277)									p. 40	

OTHER CHAPTER RESOURCES

Student Edition
Chapter Opener, p. 242
Activity, p. 243
Computer Application, p. 256
Math Connections, p. 257
Journal Entry, p. 267
Portfolio Suggestion, p. 271
Consumer Connections, p. 275

Teacher's Classroom Resources
Transparency 8-0
Lab Manual, pp. 37-38
Performance Assesment, pp. 15-16
Lesson Plans, pp. 82-94

Other Supplements
Overhead Manipulative Resources, Labs 23-26, pp. 17-18
Glencoe Mathematics Professional Series

ENHANCING the CHAPTER

Some of the blackline masters for enhancing this chapter are shown below.

COOPERATIVE LEARNING

The following activity can be performed by groups of two students. Have students find the area of walking space in their classroom. Point out that desks, tables, and so on limit the available walking area. Students should find the area of the classroom and the area of each object that limits walking. Have students use graph paper to make a drawing of the classroom that corresponds to their measurements and calculations.

USING MODELS/MANIPULATIVES

Have small groups of students use ten unit cubes to build as many structures as possible such that at least one face of each cube joins one face of another cube. Have students find the surface area of each structure. Students should make a drawing of each structure and record its surface area. Then determine which structure of cubes has the greatest surface area and which has the least. Have students explain why the volume of the structures remains the same.

Cooperative Problem Solving, p. 36

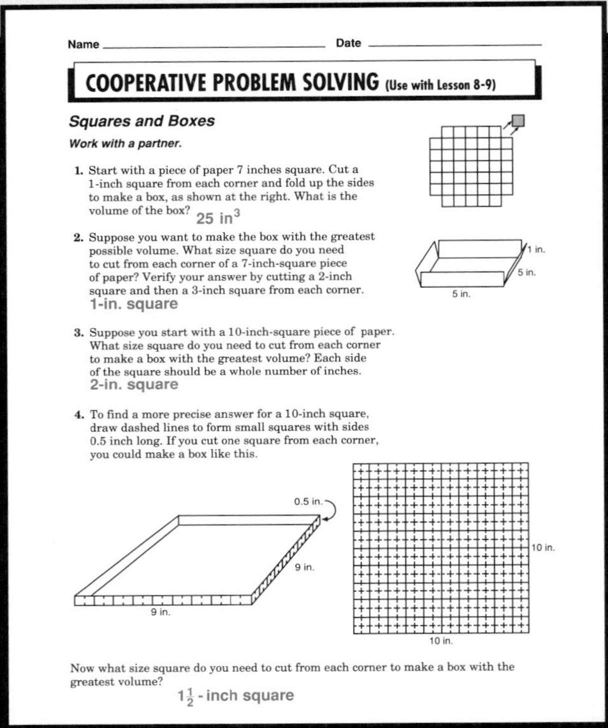

Name ______ Date ______

COOPERATIVE PROBLEM SOLVING (Use with Lesson 8-9)

Squares and Boxes

Work with a partner.

1. Start with a piece of paper 7 inches square. Cut a 1-inch square from each corner and fold up the sides to make a box, as shown at the right. What is the volume of the box? 25 in^3
2. Suppose you want to make the box with the greatest possible volume. What size square do you need to cut from each corner of a 7-inch-square piece of paper? Verify your answer by cutting a 2-inch square and then a 3-inch square from each corner. 1-in. square
3. Suppose you start with a 10-inch-square piece of paper. What size square do you need to cut from each corner to make a box with the greatest volume? Each side of the square should be a whole number of inches. 2-in. square
4. To find a more precise answer for a 10-inch square, draw dashed lines to form small squares with sides 0.5 inch long. If you cut one square from each corner, you could make a box like this.

Now what size square do you need to cut from each corner to make a box with the greatest volume? $1\frac{1}{2}$ - inch square

Multicultural Activities, p. 8

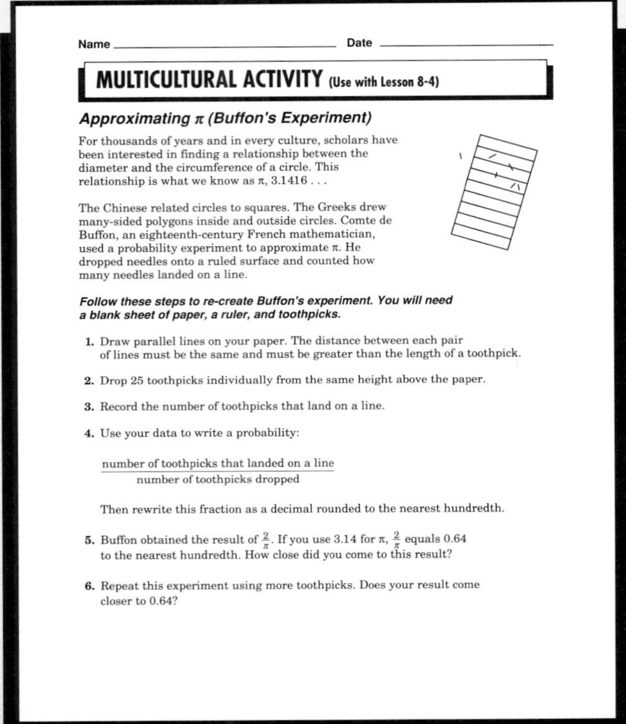

Name ______ Date ______

MULTICULTURAL ACTIVITY (Use with Lesson 8-4)

Approximating π (Buffon's Experiment)

For thousands of years and in every culture, scholars have been interested in finding a relationship between the diameter and the circumference of a circle. This relationship is what we know as π, 3.1416 . . .

The Chinese related circles to squares. The Greeks drew many-sided polygons inside and outside circles. Comte de Buffon, an eighteenth-century French mathematician, used a probability experiment to approximate π. He dropped needles onto a ruled surface and counted how many needles landed on a line.

Follow these steps to re-create Buffon's experiment. You will need a blank sheet of paper, a ruler, and toothpicks.

1. Draw parallel lines on your paper. The distance between each pair of lines must be the same and must be greater than the length of a toothpick.
2. Drop 25 toothpicks individually from the same height above the paper.
3. Record the number of toothpicks that land on a line.
4. Use your data to write a probability:

$$\frac{\text{number of toothpicks that landed on a line}}{\text{number of toothpicks dropped}}$$

Then rewrite this fraction as a decimal rounded to the nearest hundredth.

5. Buffon obtained the result of $\frac{2}{\pi}$. If you use 3.14 for π, $\frac{2}{\pi}$ equals 0.64 to the nearest hundredth. How close did you come to this result?
6. Repeat this experiment using more toothpicks. Does your result come closer to 0.64?

MEETING INDIVIDUAL NEEDS

Multicultural Activity

It has been remarked that "Western history is nothing more than footnotes to Plato." Have students research the Greek philosopher and mathematician, Plato. Have them describe any one important contribution to mathematics made by Plato. Also have them identify and draw the five possible Platonic solids: hexahedron, tetrahedron, octahedron, dodecahedron, and icosahedron.

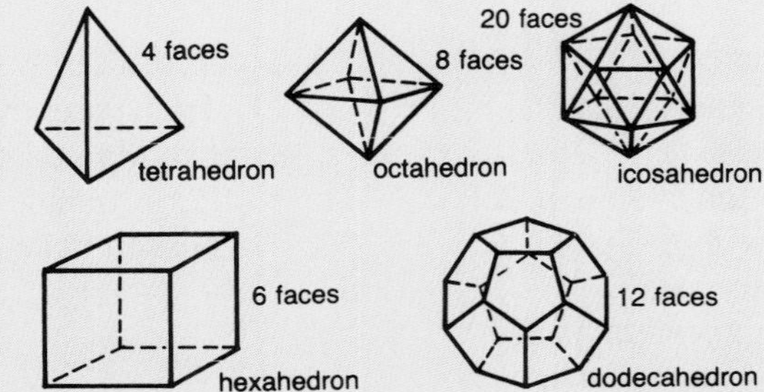

CRITICAL THINKING/PROBLEM SOLVING

Have small groups of students solve the following problem. Imagine a road running around the earth at the equator. Also imagine another road directly above it, elevated on pilings. If the length of the elevated road is 1 meter longer than the other road, how high are the pilings?

Have students use 3.14 for pi and let p represent the height of the pilings. ($C = 2\pi r$; $C_2 - C_1 = 2\pi p$; 1 m = $2\pi p$; $p \approx 0.159$ m) This problem can be approached in several ways; make a drawing, make tables, and solve a simpler problem.

COMMUNICATION

Speaking

Have students give oral presentations with drawings on finding the circumference of circles, the areas of parallelograms, triangles, and circles, the surface areas of rectangular prisms and cylinders, and the volumes of pyramids, cylinders, and cones. Have students describe the derivation of each formula and show how the perimeters, area, or volumes increase given incremental increases in a length, a radius, or a base.

Applications, p. 15

Name ______________________ Date ______________

APPLICATIONS (Use with Lesson 8-7)

Gift Wrapping

Two of the most common boxes wrapped by the gift wrap department of a store measure 10 in. by 15 in. by 2 in. and 11 in. by 17 in. by $2\frac{1}{2}$ in.

Solve each problem for boxes A and B.

	Box A 10 in. by 15 in. by 2 in.	Box B 11 in. by 17 in. by $2\frac{1}{2}$ in.
1. Each box is wrapped in a rectangular sheet of paper that has a $1\frac{1}{2}$-inch overlap on each side. ($1\frac{1}{2}$-in. overlap) What are the dimensions of the paper used to wrap one box?	27 in. by 22 in.	30 in. by 25 in.
2. How many boxes can be wrapped using a 20-yard roll of paper that is 50 inches wide?	52 boxes	48 boxes
3. A piece of ribbon is wrapped around the middle of the box in both directions with a 1-inch overlap for the knot. Another 48 inches of ribbon are used for a bow. How much ribbon is needed to wrap one box?	107 in.	115 in.
4. How many boxes can be wrapped using a 100-yard spool of ribbon?	33 boxes	31 boxes

Computer Activity, p. 22

Name ______________________ Date ______________

COMPUTER ACTIVITY (Use with Lesson 8-1)

Horizontal Tabs

Saul is working with a computer that displays text on a screen that is 40 characters wide. This means that 40 keystrokes (letters of the alphabet, punctuation marks, digits, or spaces) will fit on one line across the screen. A grid like the one below can be used to plan the layout of the text.

H	I	S	T	O	R	Y		F	I	N	A	L		E	X	A	M
M	A	Y		9	,		1	9	9	4							

Saul wants to center these lines. He uses the following steps for each line.

1. Count the number of keystrokes in line one. There are 18 keystrokes.
2. Subtract 18 from 40 to find the number of blank spaces before and after the text.
   $40 - 18 = 22$ blank spaces
3. Divide by 2 to distribute the blank spaces evenly before and after the text in the first line.
   $22 \div 2 = 11$
   There should be 11 spaces before and 11 spaces after the text.

`10 HTAB 12` (This tells the computer to begin printing at the 12th character position.)
`20 PRINT "HISTORY FINAL EXAM"`

Sometimes the number of keystrokes is an odd number. This is the case in line two above. There are 11 keystrokes in the date. Since $40 - 11 = 29$, Saul will need to put 14 extra spaces at one end of the text and 15 spaces at the other end. As a rule, the extra space is usually placed after the text.

`30 HTAB 15` (This tells the computer to begin printing at the 15th character position.)
`40 PRINT "MAY 9, 1994"`

Insert the correct character position to center each message.

1.
```
10 HTAB 12
20 PRINT "NORTH HIGH SCHOOL"
30 HTAB 17
40 PRINT "PRESENTS"
50 HTAB 16
60 PRINT "KING LEAR"
```
2.
```
10 HTAB 14
20 PRINT "ATTEND SCHOOL"
30 HTAB 14
40 PRINT "BE AN INVOLVED"
50 HTAB 17
60 PRINT "STUDENT"
```
3.
```
10 HTAB 13
20 PRINT "KENTUCKY DERBY:"
30 HTAB 10
40 PRINT "THE RUN FOR THE ROSES"
```
4.
```
10 HTAB 14
20 PRINT "STOREWIDE SALE"
30 HTAB 13
40 PRINT "BEGINS SATURDAY"
```

CHAPTER PROJECT

Give each small group of students the dimensions of a plot of land upon which they could construct a building. Tell them that a zoning board has ruled that no building may be more than 100 feet tall and no closer than 20 feet to the property lines. Have each group make a plan for a building that shows a side view and a top view. Have each student explain the reasoning behind the plan. Have them calculate the perimeter and the land area required by their building.

Using the Chapter Opener

Transparency 8-0 is available in the Transparency Package. It provides a full-color visual and motivational activity that you can use to engage students in the mathematical content of the chapter.

Using Discussion Have students read the opening paragraph. Allow students time to discuss various types of building they have observed.

Discuss with students some of the factors that go into determining the size of a building, such as cost of materials, cost of the property the building occupies, the type of building, requirements of the owner for space or special equipment, fire codes, building codes, architectural design, and so on.

You may wish to extend the discussion to careers that involve using area and volume other than the building trades. For example, interior designers need a good sense of space and proportion when filling a room with furniture.

The second paragraph is a motivational problem. Have students try to solve the problem and discuss how they arrived at the answer.

Chapter 8

AREA AND VOLUME

How big? How many?

Have you ever watched a house or an office building being built? Have you ever wondered who decides how large the building should be? Many factors determine the size of a building, such as the cost of construction or the size of the property the building occupies. Can you think of any other factors that affect the size of a building?

The Theatre Troupe at Central High is painting three walls for a backdrop for the fall play. The walls are 10 feet high. Two of the walls are 16 feet wide, and one wall is 22 feet wide. If one can of paint covers 250 square feet, how many cans of paint does the group need to paint the three walls? **3 cans**

ACTIVITY: Area of Irregular Figures

In this activity you will find the area of figures and shapes that do not have straight edges and square corners.

Materials: quarter-inch grid paper, pencil, straightedge, different colored pencils (optional)

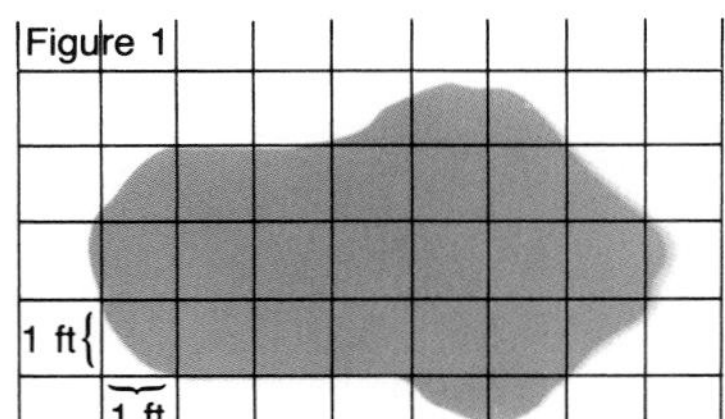

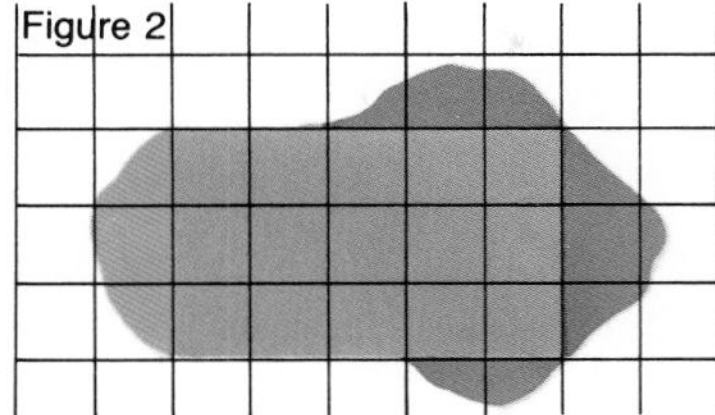

1. To find the area of an irregular shape, like the sand trap in Figure 1, separate it into regular and irregular parts.
2. In Figure 2, the brown rectangle encloses a regular part. What is the area of the rectangle in brown? **15 ft^2**
3. Estimate the area of the part of the sand trap shaded in orange by counting whole squares and partial squares. Estimate the area of the part of the sand trap shaded in blue. **2 ft^2; 4 ft^2**
4. Add the areas of the rectangle shaded in brown and the parts of the sand trap shaded in orange and blue. What is the approximate area of the whole sand trap? **21 ft^2**

Cooperative Groups

Work in groups of three or four.

5. Estimate the area of your hand in square inches. Then outline your hand on quarter-inch grid paper. Work in groups of three or four and find the area of each person's hand by separating the outline of the hand into regular and irregular shapes. Find the difference between your estimate and your result using the drawing. **Answers may vary.**

Variations You Can Try

6. Have a contest to see which group can best estimate areas of irregularly shaped objects that have been selected by the groups. Use objects already in the classroom, such as bookends, a calculator, a paperweight, and so on.

Answers may vary.

7. Try estimating the area of some circular objects, such as a clock or a plate. Outline the object and find an approximate area.

Communicate Your Ideas

8. Discuss your methods for separating the outline of a figure into regular and irregular parts with other groups.

INTRODUCING THE LESSON

Objective Find the area of irregularly shaped figures.

TEACHING THE LESSON

You may wish to discuss square units and the type of grid paper the students are using so that they will be aware of the correct label for square units.

Using Discussion Discuss with students the advantage of enclosing as many parts of an irregular figure as possible into regular figures to make the process of finding the area easier.

EVALUATING THE LESSON

Communicate Your Ideas If you wish to have groups compete against each other, have some irregularly shaped objects measured for area before the class begins.

EXTENDING THE LESSON

Closing the Lesson Allow students time to discuss their methods for finding area. Students will study more about area and volume in this chapter.

Activity Worksheet, Lab Manual, p. 37

ACTIVITY WORKSHEET (Use with page 243)

Finding the Area of Irregular Figures

Use copies of this grid for Exercises 5–7.

Object	Estimate (in^2)	Number of $\frac{1}{4}$-inch squares on the grid	Area (in^2)	Difference
5. Hand				
6a. Shoe Sole				
b. Teacher's Hand				
c.				
d.				

7. Explain how to estimate the area of irregular figures.

5-MINUTE CHECK-UP

(over Lesson 7-10)

Available on TRANSPARENCY 8-1.
Choose the letter of the best answer. Explain your answer.

1. 3 is to 27 as 4 is to
   a. 12   c. 16   b
   b. 64   d. 24

▶ INTRODUCING THE LESSON

Ask students how many times a day they use or determine the distance around something; for example, the coach says "Let's warm up, two laps around the field."

▶ TEACHING THE LESSON

Materials: stiff paper, ruler, scissors, compass

Using Manipulatives Have students use a compass to draw a circle, cut it out, and mark the edge of the circle at any point. Have students roll the circle on its edge along a straight line beginning with the mark on the circle, then measure the diameter and the length of the circumference that they rolled. Make three columns on the chalkboard—*d*, *C*, and *C* ÷ *d*. List students' findings, and find a class average for pi.

Practice Masters Booklet, p. 82

PRACTICE WORKSHEET 8-1

Circumference of Circles

Find the circumference of each circle described below. Use 3.14 for π. Round decimal answers to the nearest tenth.

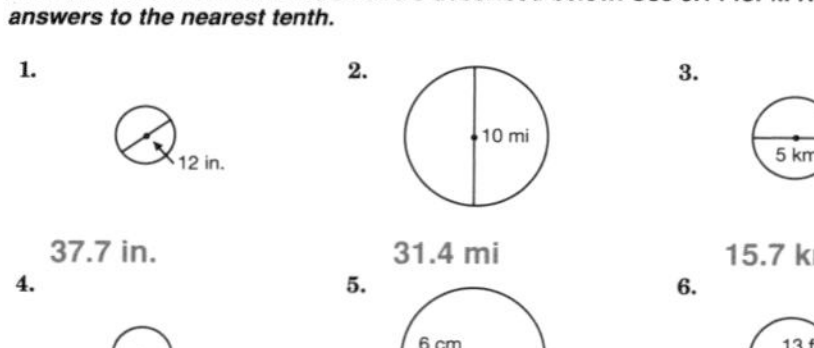

1. 37.7 in.   2. 31.4 mi   3. 15.7 km
4. 50.24 m   5. 37.7 cm   6. 81.6 ft

7. $d = 4.5$ cm 14.1 cm   8. $r = 15$ yd 94.2 yd   9. $d = 9$ ft 28.3 ft
10. $r = 12$ in. 75.4 in.   11. $d = 15$ in. 47.1 in.   12. $r = 50$ km 314 km
13. $d = 33$ m 103.6 m   14. $r = 5$ mm 31.4 mm   15. $d = 20$ mi 62.8 mi

Solve.

16. The West High School band is planning a formation for their halftime performance. They want to make a large circle with a diameter of 30 yards. If the 50 band members will be equally spaced, how many feet of circumference should there be between two consecutive positions? 1.9 ft
17. The garden at Central City Park has a large circular fountain in the center. The radius of the circle is 27 feet. The tiles outlining the fountain are 9 inches long. How many tiles are needed to complete the outline of the fountain? 226 tiles
18. The diameter of the earth is about 8,000 miles. What is the circumference of the earth? 25,120 mi
19. Suppose it were possible to string a wire around the earth's equator at a constant 100 feet above the earth. How much wire would be needed? 132,634,220 ft or 25,120.1 mi

8-1 CIRCUMFERENCE OF CIRCLES

Objective
Find the circumference of circles.

Did you ever wonder how far you travel on a Ferris wheel ride? Katie is riding on a Ferris wheel and is sitting 19 feet from the center. If the wheel goes around nine times for one ride, how far does Katie travel?

First we need to find the circumference of the Ferris wheel. The distance around a circle is called the **circumference.** To find circumference, the diameter or the radius of the circle must be known.

A line segment through the center of a circle with endpoints on the circle is called a *diameter.*

The ratio of the measure of the circumference *(C)* of a circle to the measure of its diameter *(d)* is the same for all circles. The Greek letter π (pi) stands for this ratio. An approximation for π is 3.14.

$$\frac{C}{d} = \pi \rightarrow C = \pi \times d \text{ or } C = \pi d$$

diameter (d)   radius (r)
d = 2r

The diameter *(d)* of a circle is twice the length of the radius *(r)*. Another formula for circumference can be found by substituting 2*r* for *d*.

$$C = \pi \times 2 \times r \text{ or } C = 2\pi r$$

A line segment from the center of a circle to any point on the circle is called a *radius.*

Method

1. Determine if the diameter or the radius is given.
2. If the diameter is given, use $C = \pi d$.
   If the radius is given, use $C = 2\pi r$.

Example A ***Find the distance that Katie travels in 9 rotations of the Ferris wheel. Use 3.14 for π. Round the answer to the nearest tenth.***

1. The radius is given.
2. Use $C = 2\pi r$.
   $C \approx 2 \times 3.14 \times 19$
   $C \approx 119.32$

≈ means is approximately equal to

19 feet

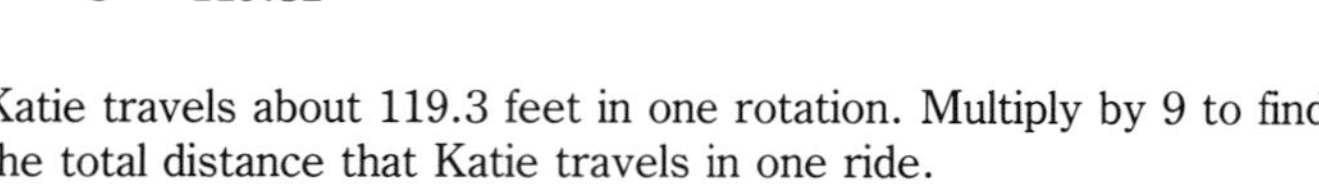

Katie travels about 119.3 feet in one rotation. Multiply by 9 to find the total distance that Katie travels in one ride.
119.3 × 9 = 1,073.88   Katie travels *about* 1,074 feet in one ride.

Example B ***Find the circumference of a circle whose diameter is 16 inches. Use 3.14 for π. Round the answer to the nearest tenth.***

1. The diameter is given.
2. Use $C = \pi d$.
   $C \approx 3.14 \times 16$
   $C \approx 50.24$

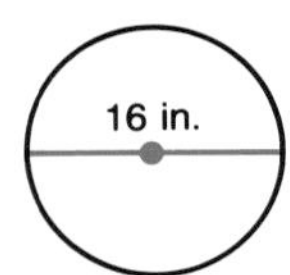

The circumference is about 50.2 inches.

RETEACHING THE LESSON

Students choose five circular objects and measure the diameter of each. Use the formula $C = \pi \times d$ to calculate the circumference. With a piece of string, find the circumference, then measure the string with a ruler. Check this measurement against the calculated circumference. The two values should be approximately equal.

Reteaching Masters Booklet, p. 75

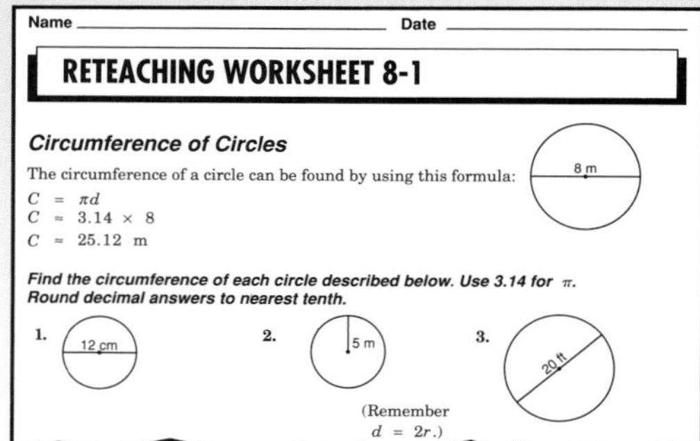

Name ____________ Date ____________

RETEACHING WORKSHEET 8-1

Circumference of Circles

The circumference of a circle can be found by using this formula:
$C = \pi d$
$C = 3.14 \times 8$
$C = 25.12$ m

Find the circumference of each circle described below. Use 3.14 for π. Round decimal answers to nearest tenth.

1. 12 cm   2. 5 m   3. 20 ft

(Remember $d = 2r$.)

Guided Practice

The radius of a circle is given. Find the circumference. Use 3.14 for π. Round decimal answers to the nearest tenth.

Example A

1. 4 cm 25.1 cm **2.** 3.5 m 22.0 m **3.** 13 in. 81.6 in. **4.** 240 mm 1,507.2 mm

The diameter of a circle is given. Find the circumference. Use 3.14 for π. Round decimal answers to the nearest tenth.

Example B

5. 1 m 3.1 m **6.** 6.7 ft 21.0 ft **7.** 51 cm 160.1 cm **8.** 23 in. 72.2 in.

Exercises

Find the circumference of each circle. Use 3.14 for π. Round decimal answers to the nearest tenth.

Practice

9.

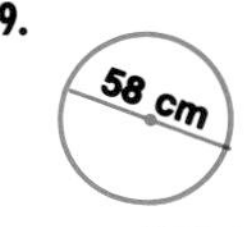

182.1 cm

10.

35.2 cm

11.

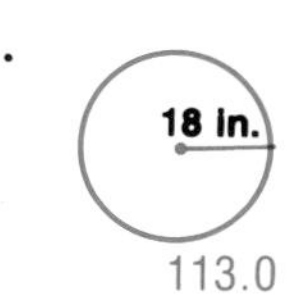

113.0 in.

12.

37.7 yd

Problem Solving

13. Shirley is making a tablecloth with lace trimming for a round table that has a diameter of 4 feet. She wants the tablecloth to extend 1 foot over the edge of the table all the way around. How much trimming does Shirley need to buy if she sews the trimming to the outside edge of the tablecloth? 18.8 feet or 6.3 yd

14. Stephanie is painting a 3-foot tall strip of blue around a silo for background for advertising. One gallon of paint will cover 250 square feet. If the diameter of the silo is 55 feet, how many gallons of paint does Stephanie need? 2.1 or 3 gal

3 feet

Mixed Review

Lesson 3-12

Find the average for each set of data. Round to the nearest tenth.

15. 16.2, 18.4, 19.1 17.9 **16.** 130.3, 132.7, 131.6, 128.5, 131.0 130.8

Lesson 7-8

Classify each quadrilateral.

17.

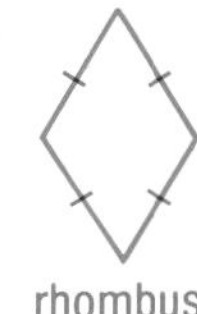

rhombus

18.

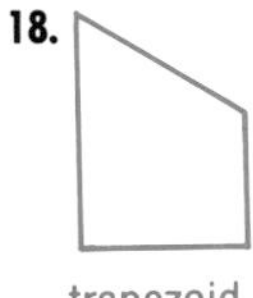

trapezoid

19.

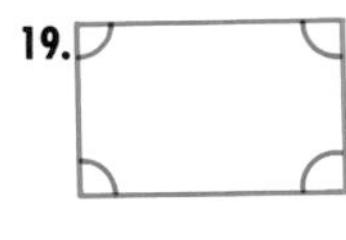

rectangle

20.

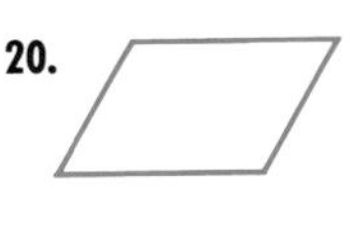

parallelogram

Lesson 5-3

21. On a trip to Erie, Marshall used $\frac{3}{4}$ of the gasoline in the tank. If he started with $14\frac{1}{2}$ gallons, how many gallons did he have left? 3.625 gallons

APPLYING THE LESSON

Independent Practice

Homework Assignment	
Minimum	Ex. 1-12; 13-21 odd
Average	Ex. 1-21
Maximum	Ex. 1-21

Chalkboard Examples

Examples A, B *Find the circumference. Use 3.14 for π. Round decimal answers to the nearest tenth.*

1. radius = 6 in. 37.7 in.
2. radius = 19 ft 119.3 ft
3. diameter = 3 ft 9.4 ft
4. diameter = 25 yd 78.5 yd

Additional examples are provided on TRANSPARENCY 8-1.

EVALUATING THE LESSON

Check for Understanding Use Guided Practice Exercises to check for student understanding.

Error Analysis Watch for students who do not differentiate between radius and diameter. Remind them that diameter is defined as twice the radius and they are not interchangeable.

Closing the Lesson Ask students to state in their own words how to calculate the circumference of a circle if the radius is known.

EXTENDING THE LESSON

Enrichment Masters Booklet, p. 75

ENRICHMENT WORKSHEET 8-1

Pi (π)

Prior to the seventeenth century, approximations for π were calculated using geometric techniques. These techniques involved the ratio of the circumference of a circle to its diameter.

However, in about 1658, the English mathematician Wallis proved that π equals the following product of an infinite number of fractions.

(A) $\pi = 4 \times \left(\frac{2}{3} \times \frac{4}{3} \times \frac{4}{5} \times \frac{6}{5} \times \frac{6}{7} \times \frac{8}{7} \times \frac{8}{9} \times \ldots\right)$

In 1674, the German mathematician Leibniz proved the following.

(B) $\pi = 4 \times \left(1 - \frac{1}{3} + \frac{1}{5} - \frac{1}{7} + \frac{1}{9} - \frac{1}{11} + \ldots\right)$

An approximation of π, to twenty-five decimal places, is given below.

3.1415926535897932384626433...

Use method (A) to compute approximations of π. Express your answer as a fraction in simplest form and as a decimal with five decimal places.

1. $4 \times \left(\frac{2}{3} \times \frac{4}{3}\right)$ $3\frac{5}{9}$; 3.55556

2. $4 \times \left(\frac{2}{3} \times \frac{4}{3} \times \frac{4}{5}\right)$ $2\frac{38}{45}$; 2.84444

3. $4 \times \left(\frac{2}{3} \times \frac{4}{3} \times \frac{4}{5} \times \frac{6}{5}\right)$ $3\frac{31}{75}$; 3.41333

4. $4 \times \left(\frac{2}{3} \times \frac{4}{3} \times \frac{4}{5} \times \frac{6}{5} \times \frac{6}{7}\right)$ $2\frac{162}{175}$; 2.92571

5. $4 \times \left(\frac{2}{3} \times \frac{4}{3} \times \frac{4}{5} \times \frac{6}{5} \times \frac{6}{7} \times \frac{8}{7} \times \frac{8}{9} \times \frac{10}{9} \times \frac{10}{11}\right)$ $3\frac{95}{43,659}$; 3.00218

Use method (B) to compute approximations of π. Express your answer as a fraction in simplest form and as a decimal with five decimal places.

6. $4 \times \left(1 - \frac{1}{3}\right)$ $2\frac{2}{3}$; 2.66667

7. $4 \times \left(1 - \frac{1}{3} + \frac{1}{5}\right)$ $3\frac{7}{15}$; 3.46667

8. $4 \times \left(1 - \frac{1}{3} + \frac{1}{5} - \frac{1}{7}\right)$ $2\frac{94}{105}$; 2.89524

9. $4 \times \left(1 - \frac{1}{3} + \frac{1}{5} - \frac{1}{7} + \frac{1}{9}\right)$ $3\frac{107}{315}$; 3.33968

10. $4 \times \left(1 - \frac{1}{3} + \frac{1}{5} - \frac{1}{7} + \frac{1}{9} - \frac{1}{11} + \frac{1}{13} - \frac{1}{15} + \frac{1}{17}\right)$ $3\frac{579,759}{2,297,295}$; 3.25237

11. Which expression gives the more accurate approximation of π, Problem 5 or Problem 10? Problem 10

5-MINUTE CHECK-UP

(over Lesson 8-1)

Available on TRANSPARENCY 8-2.

Find the circumference. Use 3.14 for π. Round decimal answers to the nearest tenth.

1. $d = 13$ in. 40.8 in.
2. $d = 8.2$ cm 25.7 cm
3. $r = 8$ km 50.2 km
4. $r = 16$ yd 100.5 yd

Extra Practice, Lesson 8-1, p. 468

▶ INTRODUCING THE LESSON

Ask students to list times when they have needed to find the area of a figure or room or place that did not have square corners or was not shaped like a square.

You may want to review Lesson 3-11 on pp. 92, 93 for area of rectangles before beginning this lesson.

▶ TEACHING THE LESSON

Materials: ruler, scissors, grid paper

Using Manipulatives Have students draw rectangles about 6 inches by 8 inches on grid paper, **Lab Manual, p. 8.** Have students change their rectangles to parallelograms by cutting off a triangular portion from one side and taping it to the other. Compare the areas of the rectangle and the square.

Practice Masters Booklet, p. 83

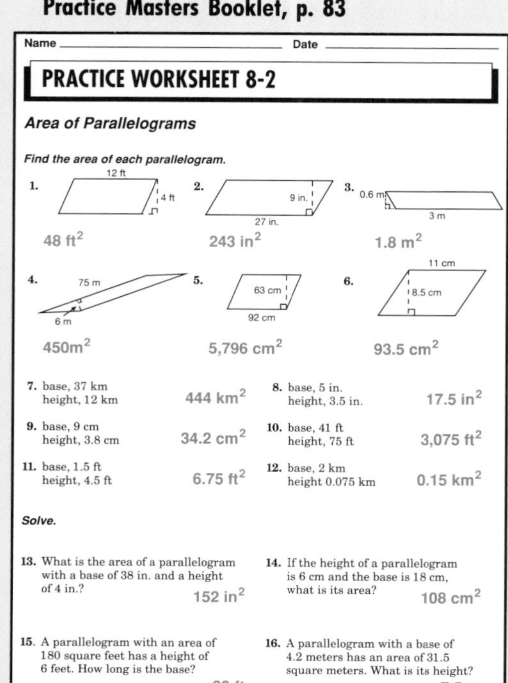

Name ______ Date ______

PRACTICE WORKSHEET 8-2

Area of Parallelograms

Find the area of each parallelogram.

1. 12 ft, 4 ft — 48 ft²
2. 9 in., 27 in. — 243 in²
3. 0.6 m, 3 m — 1.8 m²
4. 75 m, 6 m — 450m²
5. 63 cm, 92 cm — 5,796 cm²
6. 11 cm, 8.5 cm — 93.5 cm²
7. base, 37 km height, 12 km — 444 km²
8. base, 5 in. height, 3.5 in. — 17.5 in²
9. base, 9 cm height, 3.8 cm — 34.2 cm²
10. base, 41 ft height, 75 ft — 3,075 ft²
11. base, 1.5 ft height, 4.5 ft — 6.75 ft²
12. base, 2 km height 0.075 km — 0.15 km²

Solve.

13. What is the area of a parallelogram with a base of 38 in. and a height of 4 in.? 152 in²
14. If the height of a parallelogram is 6 cm and the base is 18 cm, what is its area? 108 cm²
15. A parallelogram with an area of 180 square feet has a height of 6 feet. How long is the base? 30 ft
16. A parallelogram with a base of 4.2 meters has an area of 31.5 square meters. What is its height? 7.5 m

8-2 AREA OF PARALLELOGRAMS

Objective
Find the area of parallelograms.

Four streets border an area of downtown Tulsa that is being renovated as a park. George Sands needs to find out the area of the new park to order sod. Main Street is 220 feet from Broad Street and Carver Street is 300 feet from Front Street. How many square feet of sod should Mr. Sands order?

Main St. is parallel to Broad St.
Carver is parallel to Front St.
220 ft
300 ft

You need to find the area of the park. You already know how to find the area of a rectangle. You can change a rectangle into a **parallelogram** that is not a rectangle by moving the triangle as shown below.

RECTANGLE
width
length

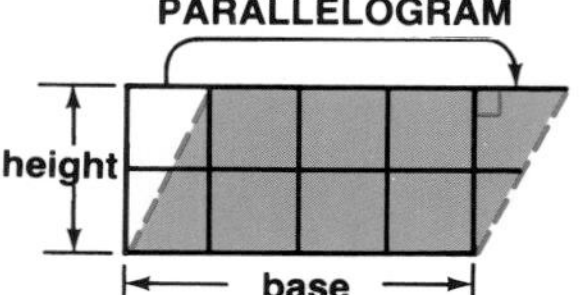

A *parallelogram* is a quadrilateral with two pairs of parallel sides. Explain how you know that the area of both figures is 8 square units.

Notice that the area of the rectangle is the same as the area of the parallelogram. So the area of any parallelogram can be found by multiplying the measures of the base *(b)* and the height *(h)*. The formula is $A = b \times h$ or $A = bh$.

Example A

Is the area of the park less than or greater than 1 acre? 1 acre is 43,560 ft²

Find the area of the new park.

$A = bh$ $\quad b = 300, h = 220$
$A = 300 \times 220$
$A = 66{,}000$

The area of the park is 66,000 square feet. Mr. Sands should order 66,000 square feet of sod.

Guided Practice
Example A

Find the area of each parallelogram.

1. 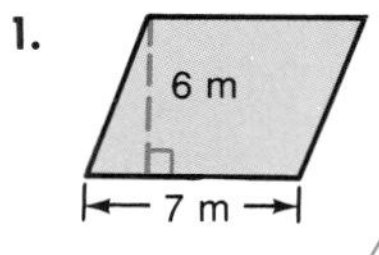

42 m²

2. 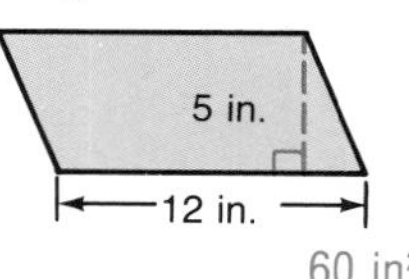

60 in²

3. 5 cm, 5.8 cm, 4.5 cm
22.5 cm²

4. base = 10 m
height = 4.5 m 45 m²
5. base = 8.5 cm
height = 35 cm 297.5 cm²
6. base = 4 ft
height = 12 ft 48 ft²

RETEACHING THE LESSON

On a sheet of grid paper have each student draw five parallelograms. Have students exchange papers and find the area of the parallelograms by counting the squares. Then have them find the area of the parallelograms using the formula $A = bh$. Compare answers.

Reteaching Masters Booklet, p. 76

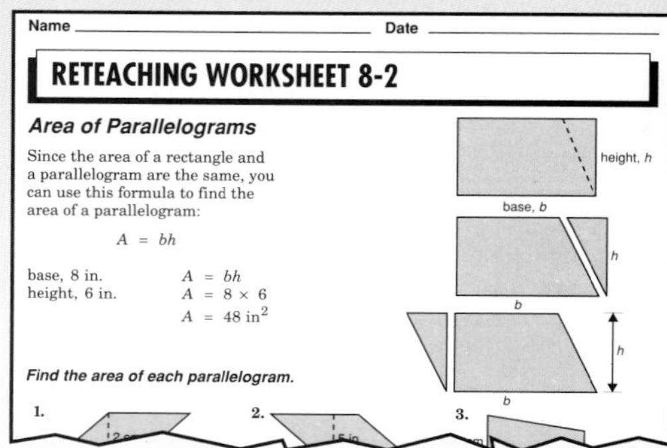

Name ______ Date ______

RETEACHING WORKSHEET 8-2

Area of Parallelograms

Since the area of a rectangle and a parallelogram are the same, you can use this formula to find the area of a parallelogram:

A = bh

base, 8 in. height, 6 in.
A = bh
A = 8 × 6
A = 48 in²

height, h
base, b

Find the area of each parallelogram.

1. 2. 3.

Exercises

Practice

Find the area of each parallelogram.

7\.

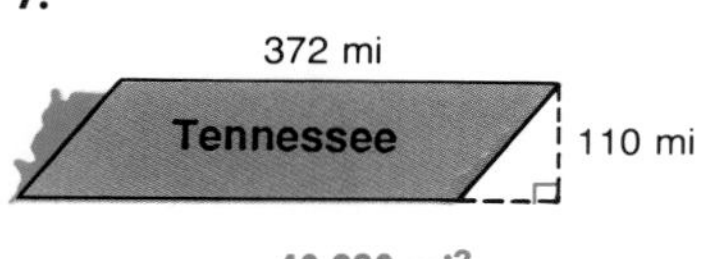

40,920 mi²

8\. **10 in²**

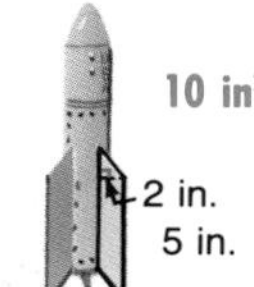

9\.

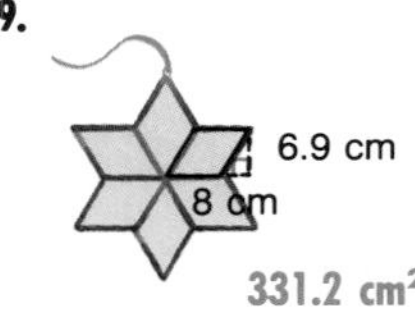

331.2 cm²

10\. base = 23 m
height = 14 m **322 m²**

11\. base = 1.5 yd
height = 3.7 yd **5.6 yd²**

12\. base = 10 m
height = 18.5 m **185 m²**

13\. What is the area of a parallelogram with a base of 4.5 m and a height of 6.7 m? **30.15 m²**

14\. If the height of a parallelogram is 43 inches and the base is 91 inches, what is its area? **3,913 in²**

Applications

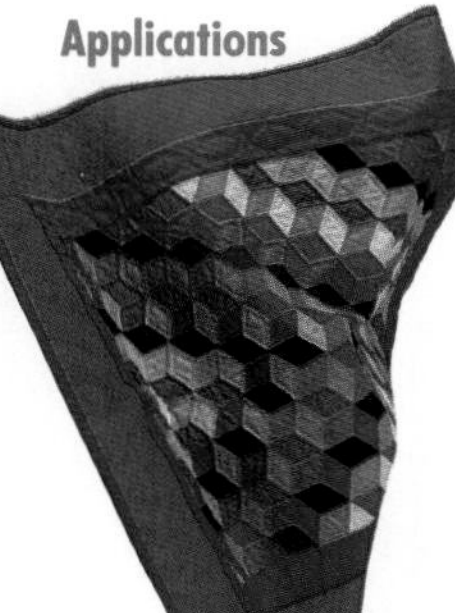

15\. The blue strip around the edge of Joan's quilt takes 48 parallelogram-shaped pieces. If the width of the strip is 8 in. and the length is 10 in., how many square inches of blue calico will Joan need? **3,840 in²** Ignore the extra material it will take to cut the pieces on the diagonal.

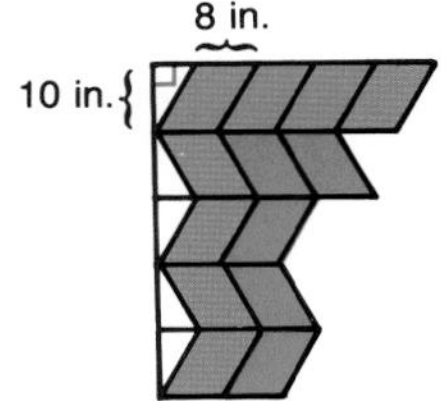

16\. Which parallelogram has the larger area, parallelogram A or parallelogram B? **B**

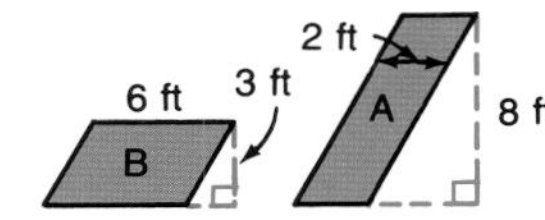

17\. The receivers of a communications satellite are 12 meters deep and 14.6 meters long. If there are 16 receivers on the satellite, how many square meters of receiver area does the satellite have? **2,803.2 m²**

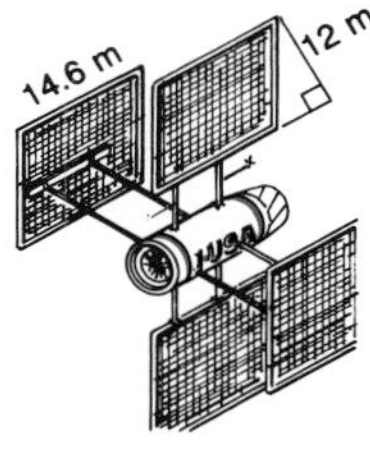

Show Math

18\. Cut 6 one-inch squares from a piece of paper. Use the squares to show two different rectangles with an area of 6 square inches. **See students' work.**

19\. Use your squares from Exercise 18 to show as many different designs as possible. The squares must be adjacent. **Answers may vary. There are thirty-five different arrangements.**

Using Formulas

Solve each problem using the formula for the area of a parallelogram.

20\. If the area of a parallelogram is 450 square centimeters and the base is 12.5 centimeters, what is the height of the parallelogram? **36 cm**

21\. The height of a parallelogram is 7.4 feet. If the area of the parallelogram is 59.2 square feet, find the length of the base. **8 ft**

APPLYING THE LESSON

Independent Practice

Homework Assignment	
Minimum	Ex. 1-12; 14-20 even
Average	Ex. 1-21
Maximum	Ex. 1-21

Chalkboard Examples

Find the area of each parallelogram.

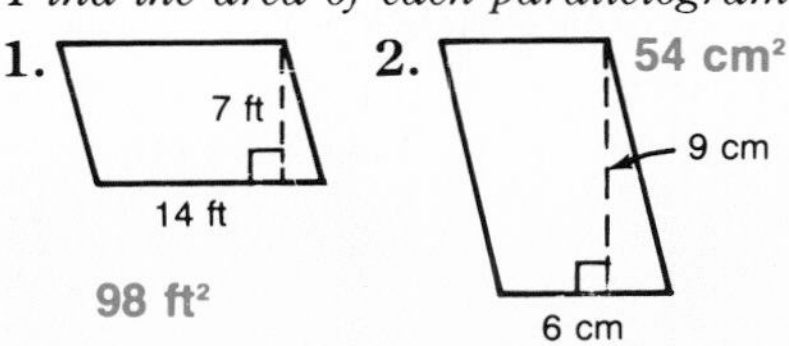

1\. **98 ft²** 2. **54 cm²**

Additional examples are provided on TRANSPARENCY 8-2.

EVALUATING THE LESSON

Check for Understanding Use Guided Practice Exercises to check for student understanding.

Closing the Lesson Have students describe what the height of a parallelogram is.

EXTENDING THE LESSON

Enrichment Have students draw a parallelogram on grid paper in a size of their own choosing but not larger than one-half of the sheet. Have students draw two more parallelograms, one with the same height and one-half the original base and one with one-half the height and one-half the original base. Have students compare the areas.

Enrichment Masters Booklet, p. 76

ENRICHMENT WORKSHEET 8-2

Pick's Law

There is a rule called **Pick's law** that can be used to find the area of irregular polygons. The polygon, or a scale drawing of it, must be drawn on dot paper.

Area = $\frac{1}{2}$(number of dots on the polygon) − 1 + (number of dots inside)

A = $\frac{1}{2}$(4) − 1 + (0)

A = 1

The area of the square is 1 square unit.

Find the area of each figure. Use Pick's law.

1\. 3 square units

2\. 4 square units

3\. 4 square units

4\. $7\frac{1}{2}$ square units

5\. 10 square units

6\. $7\frac{1}{2}$ square units

5-MINUTE CHECK-UP

(over Lesson 8-2)

Available on TRANSPARENCY 8-3.

Find the area of each parallelogram.

1. base, 20 m height, 65 m **1,300 m²**

2. base, 6.5 cm height, 40 cm **260 cm²**

Extra Practice, Lesson 8-2, p. 469

▶ INTRODUCING THE LESSON

Use a flat map of Earth and have students choose three cities. Have students connect the three cities with straight lines. Ask students if they know how to find the number of square miles enclosed by the triangle. (See Enrichment.)

▶ TEACHING THE LESSON

Materials: grid paper, pencil, scissors

Using Manipulatives Have students draw a parallelogram on grid paper, **Lab Manual, p. 8,** and cut it in half along one of its diagonals. Then have students write the formula for the area of a parallelogram. Ask them how they would adjust the formula to find the area of a triangle.

Practice Masters Booklet, p. 84

Name ____________ Date ____________

PRACTICE WORKSHEET 8-3

Area of Triangles

Find the area of each triangle.

1. 6 m, 52 m — 156 m²
2. 15 cm, 92 cm — 690 cm²
3. 3.8 cm, 8.5 cm — 16.15 cm²
4. 4 ft, 12 ft — 24 ft²
5. $2\frac{1}{2}$ ft, 3 ft — $3\frac{3}{4}$ ft²
6. 2 mi, 3 mi — 3 mi²
7. base, 6 km height, 5 km — 15 km²
8. base, 4.4 in. height, 3.5 in. — 7.7 in²
9. base, $\frac{2}{3}$ yd height, 1 yd — $\frac{1}{3}$ yd²
10. base, 24 mm height, 13 mm — 156 mm²
11. base, 5 mi height, 4 mi — 10 mi²
12. base, $3\frac{2}{3}$ yd height, $1\frac{2}{3}$ yd — $3\frac{1}{18}$ yd²

Solve.

13. A house has a triangular-shaped section of roof that measures 21 feet at the base and has a height of 14 feet. How many square feet of roofing will be needed for this section of roof? 147 ft²
14. Jack is making a tent from a pattern. Both ends are shaped like a triangle that has a base of 8 feet and a height of 6 feet, including excess for hems. How many square feet of material will Jack use for the ends? 48 ft²
15. A triangular-shaped section of lawn measures 15 meters at the base and has a height of 8 meters. How many square meters of grass is this? 60 m²
16. One section of town is bordered by three streets, forming a triangle that measures 6 kilometers at the base and has a height of 4 kilometers. How many square kilometers are contained in this section of town? 12 km²

8-3 AREA OF TRIANGLES

Objective
Find the area of triangles.

Chou Lin needs 250 square feet or less of mainsail to qualify for the class C Hampshire races. The base of the sail is 11 feet, and the height of the sail is 21 feet. Does Chou Lin qualify for the class C race?

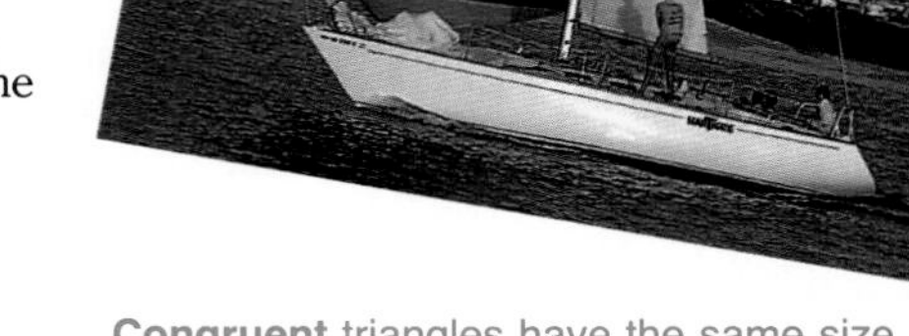

We can find the formula for the area of a triangle from the formula for the area of a parallelogram. Two **congruent** triangles form a parallelogram. The area of each triangle is one-half the area of the parallelogram.

The formula for the area of a parallelogram is $A = bh$. So the formula $A = \frac{1}{2} \times bh$ or $A = \frac{1}{2}bh$ can be used to find the area of a triangle.

Congruent triangles have the same size and shape. So they have the same area.

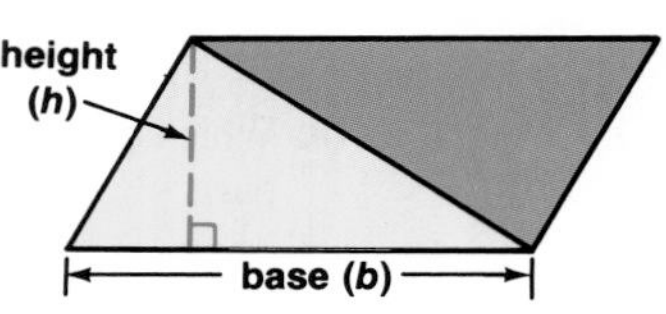

Example A *Find the area of Chou Lin's mainsail.*

$A = \frac{1}{2}bh \quad b = 11, h = 21$

$A = \frac{1}{2} \times 11 \times 21$

$A = 115.5$

To multiply by $\frac{1}{2}$ on a calculator, you can divide by 2.

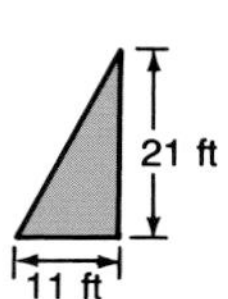

The area of the mainsail is 115.5 square feet. Chou Lin qualifies.

Guided Practice — Example A

Find the area of each triangle.

1. 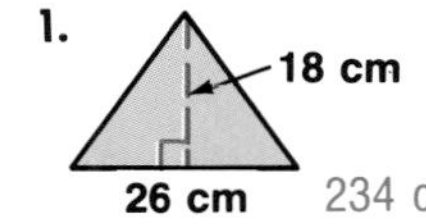

234 cm²

2. 9 ft, 8 ft — 36 ft²

3. 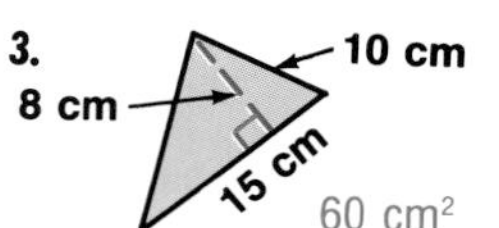

60 cm²

5. 11.2 in²
6. 1.575 m²

4. base = 7.5 ft height = 3 ft **11.25 ft²**
5. base = 6.4 in. height = 3.5 in.
6. base = 2.1 m height = 1.5 m
7. base = 22 yd height = 10 yd **110 yd²**
8. base = 14 mm height = 11 mm **77 mm²**
9. base = 3.9 cm height = 0.5 cm **0.975 cm²**

RETEACHING THE LESSON

Have each student draw five triangles on a sheet of grid paper. Have students exchange papers and find the area of the triangles by counting the squares. Then have them find the area of the triangles using the formula $A = \frac{1}{2}bh$. Compare answers.

Reteaching Masters Booklet, p. 77

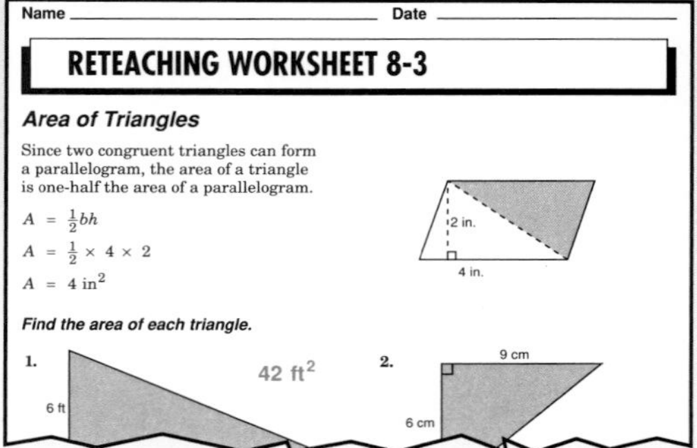
Name ____________ Date ____________

RETEACHING WORKSHEET 8-3

Area of Triangles

Since two congruent triangles can form a parallelogram, the area of a triangle is one-half the area of a parallelogram.

$A = \frac{1}{2}bh$

$A = \frac{1}{2} \times 4 \times 2$

$A = 4$ in²

Find the area of each triangle.

1. 6 ft — 42 ft²
2. 9 cm, 6 cm

Exercises

Practice

Find the area of each triangle. 15. 4.55 in² 18. $\frac{16}{9}$ or $1\frac{7}{9}$ yd²

10. 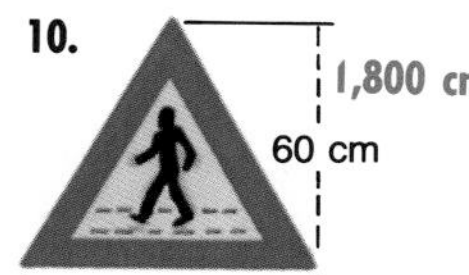

11. 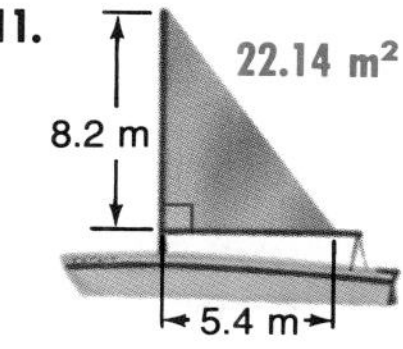

12. 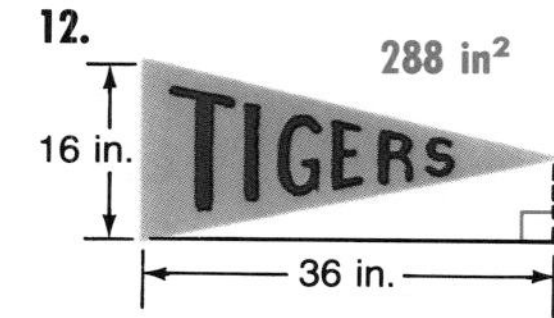

13. base = 17 cm, height = 6 cm **51 cm²**
14. base = 7.4 m, height = 12 m **44.4 m²**
15. base = 3.5 in., height = 2.6 in.
16. base = $2\frac{1}{2}$ ft, height = $1\frac{1}{4}$ ft **$1\frac{9}{16}$ ft²**
17. base = 21 mm, height = 32 mm **336 mm²**
18. base = $1\frac{1}{3}$ yd, height = $2\frac{2}{3}$ yd
19. Find the area of a triangle whose base is 4.5 meters and whose height is 7 meters. **15.75 m²**
20. If the base of a triangle is 52 feet and the height is 21 feet, what is its area? **546 ft²**

Applications

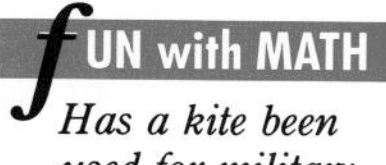

Has a kite been used for military reconnaissance? See page 308.

21. Jake is replacing the siding on a triangular portion of his house that was damaged by high winds. How many square feet does Jake need to replace? **104 ft²**

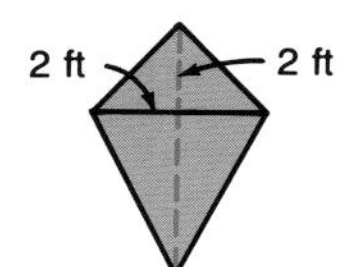

22. Tuygen is building a kite for Sunday's competition. The cross-shaped supports are 6 feet long and 4 feet long. What will be the total area of the finished kite? **12 ft²**

Show Math

23. Draw three triangles, one with an area of 8 square inches, one with an area of 16 square inches, and one with an area of 32 square inches. **See students' work.**
24. What is the greatest number of right triangles you can cut from a 4-foot by 8-foot sheet of plywood if the triangle has a base of 2 feet and a height of 2 feet? Use grid paper to show your answer. **16 triangles**

Mixed Review

Solve each equation. Check each solution.

Lesson 6-12

25. $x + 7 = 23$ **16**
26. $y - 14 = 89$ **103**
27. $t + 13.6 = 59.1$ **45.5**

Lesson 3-5

Multiply or divide.

28. 0.613×10^5 **61,300**
29. $426.5 \div 10^3$ **0.4265**
30. 0.21×10^4 **2,100**

Lesson 6-12

31. As a weather front passed Wheeling, West Virginia on Friday, the temperature fell 12°F, then rose 9°F. If the temperature before the front passed was 36°F, what was the temperature after the front passed? **33°F**

APPLYING THE LESSON

Independent Practice

Homework Assignment	
Minimum	Ex. 1-18; 19-31 odd
Average	Ex. 2-18 even; 19-31
Maximum	Ex. 1-31

Chalkboard Examples

Example A *Find the area of each triangle.*

1. (triangle: height 16 cm, base 16 cm) **128 cm²**
2. base, 24 ft; height, 12 ft **144 ft²**

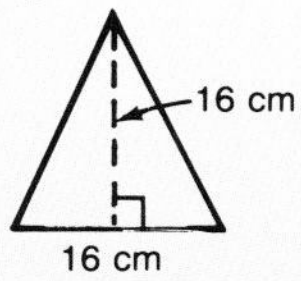

Additional examples are provided on TRANSPARENCY 8-3.

EVALUATING THE LESSON

Check for Understanding Use Guided Practice Exercises to check for student understanding.

Closing the Lesson Have students explain in their own words how to develop the formula for finding the area of a triangle.

EXTENDING THE LESSON

Enrichment Make copies of a map of the world. Have students choose three cities and connect the cities with straight lines. Using the scale of the map, have students find the number of square miles enclosed by their triangle.

Enrichment Masters Booklet, p. 77

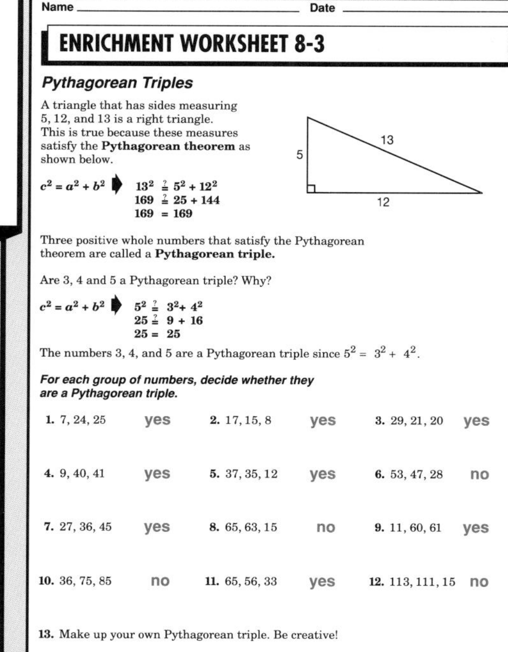

Name ______ Date ______

ENRICHMENT WORKSHEET 8-3

Pythagorean Triples

A triangle that has sides measuring 5, 12, and 13 is a right triangle. This is true because these measures satisfy the **Pythagorean theorem** as shown below.

$c^2 = a^2 + b^2$ ➧ $13^2 \stackrel{?}{=} 5^2 + 12^2$; $169 \stackrel{?}{=} 25 + 144$; $169 = 169$

Three positive whole numbers that satisfy the Pythagorean theorem are called a **Pythagorean triple.**

Are 3, 4 and 5 a Pythagorean triple? Why?

$c^2 = a^2 + b^2$ ➧ $5^2 \stackrel{?}{=} 3^2 + 4^2$; $25 \stackrel{?}{=} 9 + 16$; $25 = 25$

The numbers 3, 4, and 5 are a Pythagorean triple since $5^2 = 3^2 + 4^2$.

For each group of numbers, decide whether they are a Pythagorean triple.

1. 7, 24, 25 **yes**
2. 17, 15, 8 **yes**
3. 29, 21, 20 **yes**
4. 9, 40, 41 **yes**
5. 37, 35, 12 **yes**
6. 53, 47, 28 **no**
7. 27, 36, 45 **yes**
8. 65, 63, 15 **no**
9. 11, 60, 61 **yes**
10. 36, 75, 85 **no**
11. 65, 56, 33 **yes**
12. 113, 111, 15 **no**
13. Make up your own Pythagorean triple. Be creative!

5-MINUTE CHECK-UP

(over Lesson 8-3)

Available on TRANSPARENCY 8-4.

Find the area of each triangle.

1. 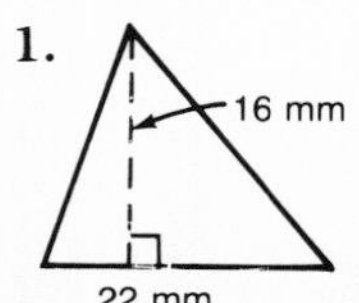

176 mm²

2.  

106.5 ft²

3. base, 8.4 ft height, 7 ft 29.4 ft²
4. base, 9.2 cm height, 10.3 cm 47.4 cm²

Extra Practice, Lesson 8-3, p. 469

INTRODUCING THE LESSON

Have students use a compass to draw a circle on a piece of grid paper. Ask them to design a video or music store or a movie theater using the circle in the floor plan. Ask students how they plan the size of a room or counter.

TEACHING THE LESSON

Using Calculators Have students use calculators for Exercises 1-17. When the diameter is given, have students enter the diameter, divide by 2, square the result, then multiply by π.

Practice Masters Booklet, p. 85

Name ______________ Date ______________

PRACTICE WORKSHEET 8-4

Area of Circles

Find the area of each circle whose radius is given. Use 3.14 for π. Round decimal answers to the nearest tenth.

1. 12 in 452.2 in²
2. 15 cm 706.5 cm²
3. 5 m 78.5 m²
4. 2.6 ft 21.2 ft²
5. 18 yd 1,017.4 yd²
6. 80 mm 20,096 mm²

Find the area of each circle whose diameter is given. Use 3.14 for π. Round decimal answers to the nearest tenth.

7. 56 m 2,461.8 m²
8. 4 in. 12.6 in²
9. 12 ft 113.0 ft²
10. 9 cm 63.6 cm²
11. 2.8 km 6.2 km²
12. 40 mi 1,256 mi²

Find the area of each circle described below. Use 3.14 for π. Round decimal answers to the nearest tenth.

13. 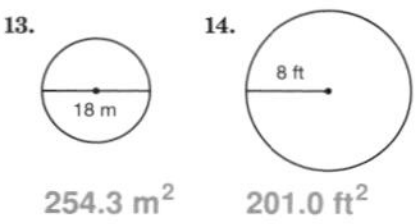

254.3 m²

14. 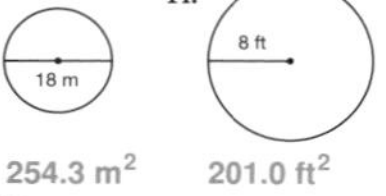

201.0 ft²

15. 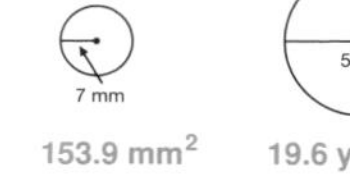

153.9 mm²

16.

19.6 yd²

Solve.

17. A circular fountain in Washington Park has a radius of 6 meters. How many square meters of tile are needed for the bottom of the fountain? 113.0 m²

18. A reading room at the library is circular in shape, with a diameter of 28 feet. How many square feet of carpet are needed for this room? 615.4 ft²

19. An old mansion has a large window in the shape of a semicircle. If the bottom edge is 8 feet long, what is the area of the window? 25.12 ft²

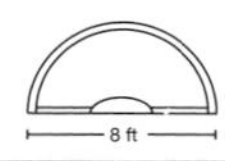

20. A sprinkler is set up at the corner of a building and is set to spray a radius of 14 meters. How many square meters of grass will the sprinkler cover? 461.6 m²

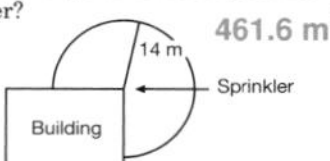

8-4 AREA OF CIRCLES

Objective
Find the area of circles.

As station manager at KCTY, Elena Simon tells potential advertisers that the listening area covers 10,560 square miles. If the tower can broadcast 58 miles in any direction, is Ms. Simon correct in her statement of the area covered by the broadcast? In order to answer this question, you need to find the area of a circle.

You can find a formula for the approximate area of a circle by starting with the formula for the area of a parallelogram. Suppose the circle is separated into parts. Then the parts are put together to form a figure that looks like a parallelogram. The circumference of the circle can be represented by C units, so the base of the parallelogram is $\frac{1}{2}C$ units. The height is r.

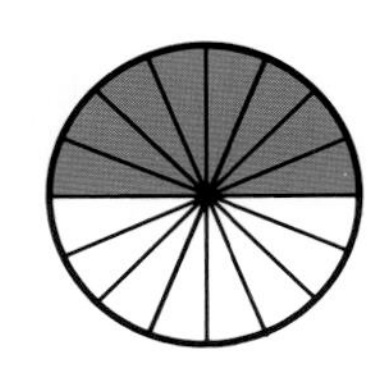

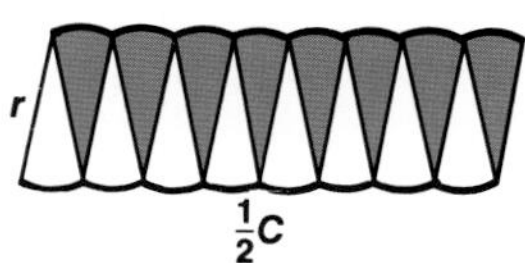

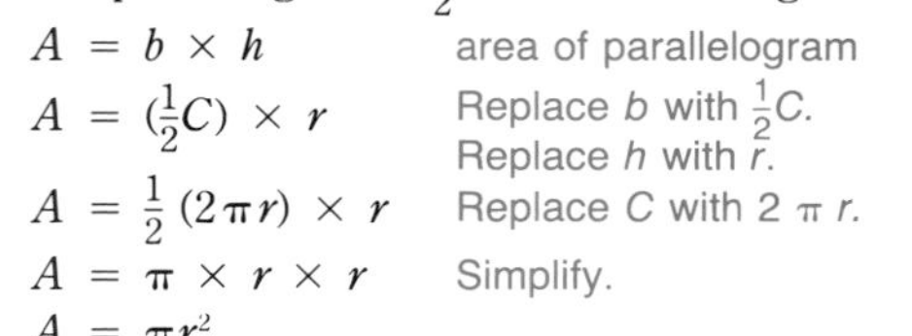

$A = b \times h$	area of parallelogram
$A = (\frac{1}{2}C) \times r$	Replace b with $\frac{1}{2}C$. Replace h with r.
$A = \frac{1}{2}(2\pi r) \times r$	Replace C with $2\pi r$.
$A = \pi \times r \times r$	Simplify.
$A = \pi r^2$	

Method

1. If the diameter is given, multiply it by $\frac{1}{2}$ to find the radius.
2. Use $A = \pi r^2$ to find the area.

Examples

A ***Find the area of a circle with a radius of 58 miles, KCTY radio's broadcast area.***

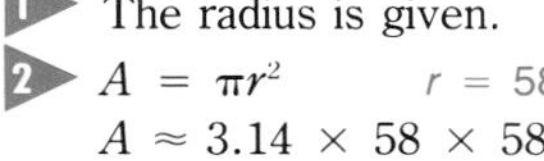

1. The radius is given.
2. $A = \pi r^2$ $\quad r = 58$
$A \approx 3.14 \times 58 \times 58$
$A \approx 10{,}562.96$

Ms. Simon is correct that the broadcast covers 10,560 square miles.

B ***Find the area of a circle with a diameter 32 inches long. Use 3.14 for π.***

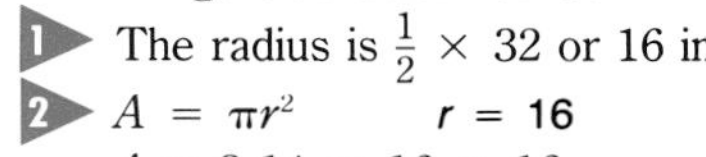

1. The radius is $\frac{1}{2} \times 32$ or 16 in.
2. $A = \pi r^2$ $\quad r = 16$
$A \approx 3.14 \times 16 \times 16$
$A \approx 803.84$

The area of the circle is *about* 804 in².

Guided Practice

Example A ***Find the area of each circle whose radius is given. Use 3.14 for π. Round decimal answers to the nearest tenth.***

1. 5 cm 78.5 cm² 2. 9 m 254.3 m² 3. 20 yd 1,256 yd² 4. 30 mm 2,826 mm²

Example B ***Find the area of each circle whose diameter is given. Use 3.14 for π. Round decimal answers to the nearest tenth.***

5. 8 cm 50.2 cm² 6. 10 in. 78.5 in² 7. 30 yd 706.5 yd² 8. 22 m 379.9 m²

RETEACHING THE LESSON

Have students draw circles on grid paper using the following measurements for the radii.
1. 1 cm 3.14 cm² 2. 2 cm 12.56 cm²
3. 3 cm 28.26 cm² 4. 4 cm 50.24 cm²

Have them estimate the area by counting the squares. Then have students use the area formula to compute the area, using 3.14 for π. Then compare.

Reteaching Masters Booklet, p. 78

Name ______________ Date ______________

RETEACHING WORKSHEET 8-4

Area of Circles

The area of a circle can be found by using this formula:

$A = \pi r^2$

The area of this circle is

$A = \pi \times r \times r$
$A \approx 3.14 \times 4 \times 4$ (Remember $\pi \approx 3.14$.)
$A \approx 50.24\ \text{cm}^2$

Find the area of each circle described below. Use 3.14 for π. Round decimal answers to the nearest tenth.

1. 2. 3.

Exercises

14. 28.3 cm²

Find the area of each circle described below. Use 3.14 for π. Round decimal answers to the nearest tenth.

9. 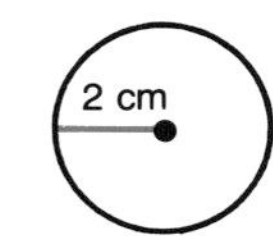

12.6 cm²

10. 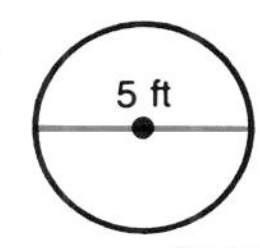

19.6 ft²

11. 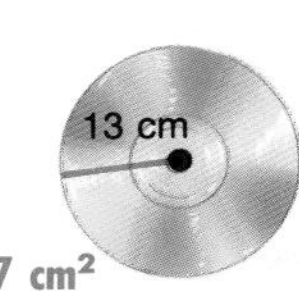

132.7 cm²

12. radius 3.2 in. 32.2 in² **13.** radius, 9 yd 254.3 yd² **14.** diameter, 6 cm

15. What is the area of a circle with a diameter of 19 cm? 283.4 cm²

16. *True* or *False* The area of a circle with a diameter of 6.4 inches is 32.2 in². true

Applications

17. Each lawn sprinkler at Markos, Inc. can reach 18 feet in any direction as it rotates. What is the area one sprinkler can water? 1,017.4 ft²

Problem Solving

18. Meg wanted to give her dog King as much space as possible in which to play. She has enough fencing to make a square pen that would have 6 feet on each side. She can also put King on a 6-foot chain. Will King have more room to play in the pen or on the chain? How much more? 77 square feet more

19. The circular flower garden in Ascott Gardens, Buckinghamshire, has a radius of 23 feet. If half of the area of the garden is planted with evergreens and the other half is planted with annuals, how many square feet has been planted with evergreens? 830.5 ft²

20. If the area of a compact disc is *about* 19.6 square inches, what is the diameter of the disc? *about* 5 in.

Critical Thinking

21. Draw as many four-sided figures as possible using any of the five points on the circle as vertices. 5 figures

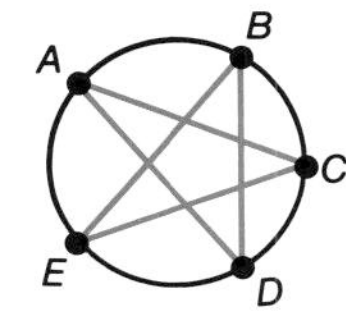

Calculator

Some calculators have a key labeled π. You can use this key to find the area of a circle.

Find the area of a circle with a radius of 3 m.

[π] [×] 3 [x²] 2 [=] 28.274334

The area is *about* 28.3 m².

25. 18,544.3 yd²

Use a calculator to find the area of each circle whose radius is given. Round decimal answers to the nearest tenth.

22. 6 mm 113.1 mm² **23.** 56 in. 9,852.0 in² **24.** 49.8 cm 7,791.3 cm² **25.** 76.83 yd

APPLYING THE LESSON

Independent Practice

Homework Assignment	
Minimum	**Ex. 1-11; 15-25 odd**
Average	**Ex. 2-14 even; 15-25**
Maximum	**Ex. 1-25**

Chalkboard Examples

Examples A, B *Find the area of each circle. Use 3.14 for π. Round decimal answers to the nearest tenth.*

1. r = 26 ft 2,122.6 ft²
2. r = 31 in. 3,017.5 in²
3. r = 15 mm 706.5 mm²
4. d = 34 yd 907.5 yd²
5. d = 25 km 490.6 km²
6. d = 17 cm 226.9 cm²

Additional examples are provided on TRANSPARENCY 8-4.

EVALUATING THE LESSON

Check for Understanding Use the Guided Practice Exercises to check for student understanding.

Error Analysis Watch for students who square the diameter instead of the radius when using the formula $A = \pi r^2$. Remind students to look at the formula and the data given to see if they are working with the radius (r) or the diameter (d).

Closing the Lesson Have students describe everyday situations that involve finding the area of a circle.

EXTENDING THE LESSON

Enrichment Masters Booklet, p. 78

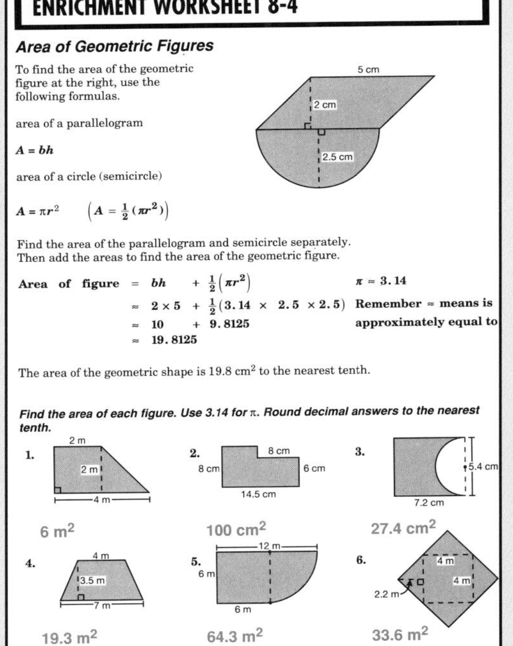

ENRICHMENT WORKSHEET 8-4

Area of Geometric Figures

To find the area of the geometric figure at the right, use the following formulas.

area of a parallelogram

$A = bh$

area of a circle (semicircle)

$A = \pi r^2$ $\left(A = \frac{1}{2}(\pi r^2)\right)$

Find the area of the parallelogram and semicircle separately. Then add the areas to find the area of the geometric figure.

Area of figure $= bh + \frac{1}{2}(\pi r^2)$ $\pi = 3.14$
$\approx 2 \times 5 + \frac{1}{2}(3.14 \times 2.5 \times 2.5)$ Remember $\approx$ means is approximately equal to
$\approx 10 + 9.8125$
≈ 19.8125

The area of the geometric shape is 19.8 cm² to the nearest tenth.

Find the area of each figure. Use 3.14 for π. Round decimal answers to the nearest tenth.

1. 6 m²
2. 100 cm²
3. 27.4 cm²
4. 19.3 m²
5. 64.3 m²
6. 33.6 m²

5-MINUTE CHECK-UP

(over Lesson 8-4)

Available on TRANSPARENCY 8-5.

Find the area of each circle described below. Use 3.14 for π. Round decimal answers to the nearest tenth.

1. radius, 24 in. **1,808.6 in.**
2. radius, 16 ft **803.8 ft**
3. diameter, 6 m **28.3 m**
4. diameter, 9 cm **63.6 cm**

Extra Practice, Lesson 8-4, p. 469

INTRODUCING THE LESSON

Have students estimate the number of square feet in the classroom. Then tell students carpet for the classroom costs $20 per square yard. Have students find the cost of carpeting the classroom.

TEACHING THE LESSON

Using Discussion In the Guided Practice, Exercises 2 and 4 on p. 252, you may wish to show students a three by three grid and have them use 1 yd² = 9 ft² to divide the answer by 9 when multiplying length and width measured in feet.

Practice Masters Booklet, p. 86

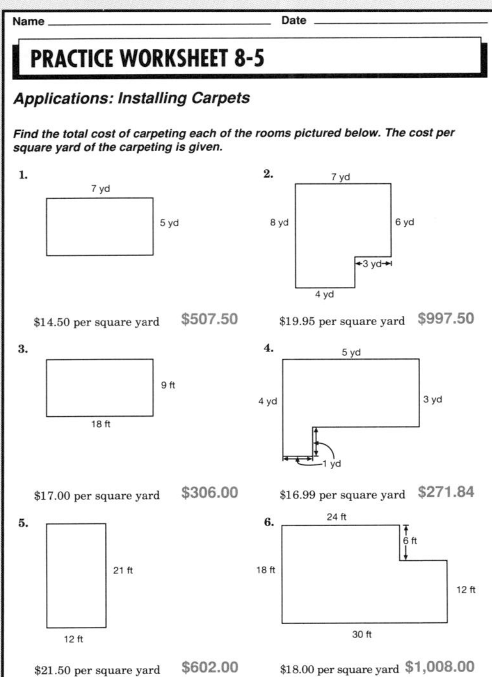

Name ______ Date ______

PRACTICE WORKSHEET 8-5

Applications: Installing Carpets

Find the total cost of carpeting each of the rooms pictured below. The cost per square yard of the carpeting is given.

1. 7 yd, 5 yd — $14.50 per square yard **$507.50**
2. 7 yd, 8 yd, 6 yd, 3 yd, 4 yd — $19.95 per square yard **$997.50**
3. 9 ft, 18 ft — $17.00 per square yard **$306.00**
4. 5 yd, 4 yd, 3 yd, 1 yd — $16.99 per square yard **$271.84**
5. 21 ft, 12 ft — $21.50 per square yard **$602.00**
6. 24 ft, 6 ft, 18 ft, 12 ft, 30 ft — $18.00 per square yard **$1,008.00**

8-5 INSTALLING CARPET

Objective
Find the cost of carpeting a room.

Diana Shultz is a designer and sales associate for The Carpetman. Mrs. Shultz arranges for installation, and computes the number of square yards and the cost of the order.

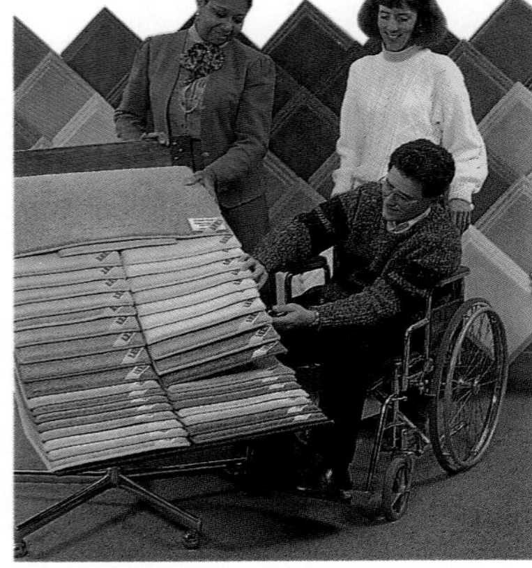

To compute the cost of carpet for a room that measures 4 yards by 6 yards, Mrs. Shultz finds the area of the floor in square yards and multiplies by the price per square yard.

4 yd, 6 yd

$A = \ell w$
$4 \times 6 = 24$
The area is 24 square yards.

Then multiply the area by the price per square yard.

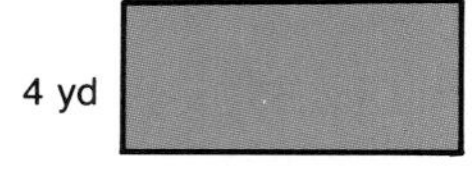

The cost of the carpet is $406.80.

Understanding the Data

- What are the dimensions of the room? **4 yd by 6 yd**
- What is the cost of the carpet per square yard? **$16.95**

Guided Practice

Find the total cost of carpeting each room pictured below. The cost per square yard of the carpeting is given.

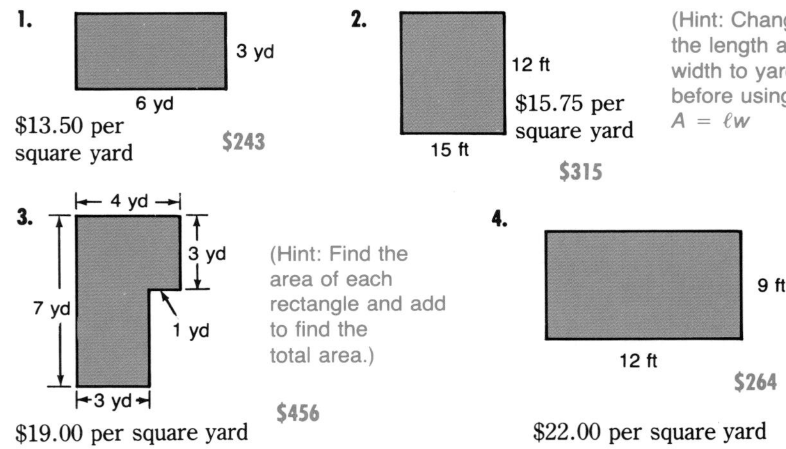

1. 6 yd, 3 yd — $13.50 per square yard **$243**

2. 12 ft, 15 ft — $15.75 per square yard **$315** (Hint: Change the length and width to yards before using $A = \ell w$)

3. 4 yd, 3 yd, 7 yd, 1 yd, 3 yd (Hint: Find the area of each rectangle and add to find the total area.) $19.00 per square yard **$456**

4. 9 ft, 12 ft — $22.00 per square yard **$264**

RETEACHING THE LESSON

Have students work in small groups of two or three to discuss the operations needed to solve the exercises on page 253. Then have students solve the problems independently.

Reteaching Masters Booklet, p. 79

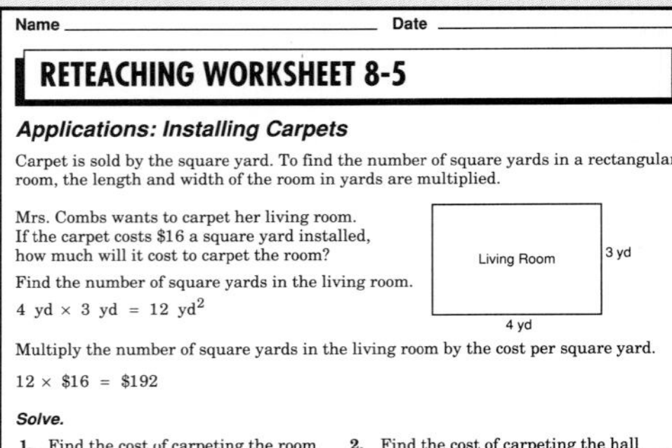

Name ______ Date ______

RETEACHING WORKSHEET 8-5

Applications: Installing Carpets

Carpet is sold by the square yard. To find the number of square yards in a rectangular room, the length and width of the room in yards are multiplied.

Mrs. Combs wants to carpet her living room. If the carpet costs $16 a square yard installed, how much will it cost to carpet the room?

Living Room — 3 yd, 4 yd

Find the number of square yards in the living room.

4 yd × 3 yd = 12 yd²

Multiply the number of square yards in the living room by the cost per square yard.

12 × $16 = $192

Solve.

1. Find the cost of carpeting the room
2. Find the cost of carpeting the hall

Problem Solving

Mrs. Shultz is ordering the carpet for the Hernandez's new house. The floor plan is shown at the right. ***Use the floor plan to answer Exercises 6–10.***

15′ 12′ 12′ 15′
Bedroom 1 | Bedroom 2 | Bedroom 3 | Kitchen
12′ 12′ 12′ 12′ 18′
6 ft Bath
18′ Living Room | Hall | Dining Room 18′
12′
27′ 15′

6. Mrs. Hernandez is carpeting the living room, dining room, and hall in the same color and style. The beige plush carpet costs $23.95 per square yard. What is the cost of the carpet? **127.3 yd²; $3,048.84**

7. Mrs. Hernandez has chosen the same sculptured style carpet for the three bedrooms, each in a different color, but at the same price. How much will it cost to carpet the three bedrooms if the carpet costs $21.95 per square yard? **52 yd²; $1,141.40**

8. The kitchen carpet is an indoor-outdoor carpet that costs $14.95 per square yard. How much will the kitchen carpet cost? **30 yd²; $448.50**

9. The bathroom carpet is a washable shag that comes on a roll 6 feet wide. The cost of the carpet is $13.95 for a length of carpet 1 yard long. How much will the carpet for the bathroom cost? **$3\frac{1}{3}$ yd; $46.50**

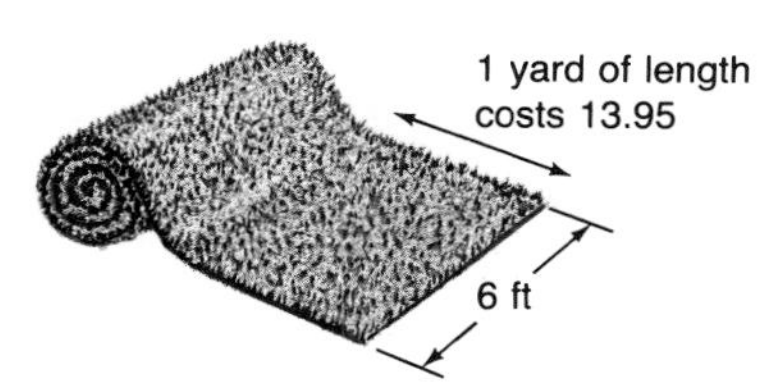

10. What is the total cost of carpeting the Hernandez's new house using the dollar amounts from Exercises 6–9? **$4,685.24**

Suppose

11. Suppose The Carpetman pays a wholesale price of $12.95 per square yard and charges $23.95 per square yard for plush carpet. How much profit does the store make when carpeting a room 4 yards by 6 yards? **$264**

Mixed Review

Lesson 1-7

Estimate. **Answers may vary. Typical answers are given.**

12. 45 × 9 **450**
13. 29.7 × 4.5 **150**
14. 33 × 23.6 **600**
15. 509 × 1.8 **1,000**

Lesson 7-5

16. Name all the line segments that are parallel to segment AB in the cube shown at the right. ***DC, EF, HG***

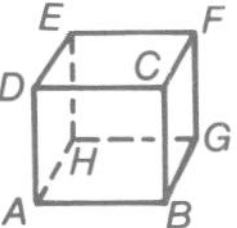

Lesson 6-6

17. Mr. Harold has a grandfather clock that chimes the number of the hour at the beginning of every hour and once every half hour. Ten minutes after Mr. Harold comes home, he hears one chime, a half hour later he hears one chime, and a half hour later he hears one chime. Between what two times did Mr. Harold come home? **12:00 and 12:30**

APPLYING THE LESSON

Independent Practice

Homework Assignment	
Minimum	**Ex. 2-10 even; 11-17**
Average	**Ex. 1-17**
Maximum	**Ex. 1-17**

Chalkboard Examples

Find the total cost of carpeting each room pictured below. The cost per square yard is given.

1.

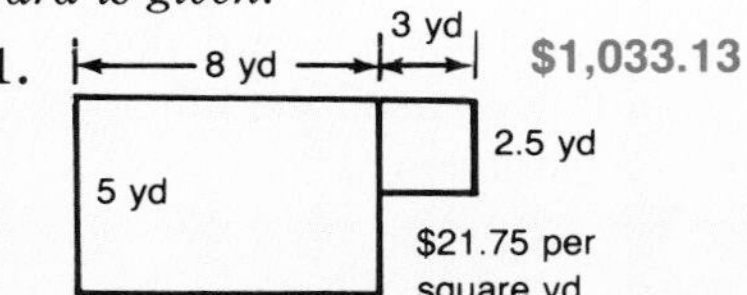

$1,033.13

Additional examples are provided on TRANSPARENCY 8-5.

EVALUATING THE LESSON

Check for Understanding Use Guided Practice Exercises to check for student understanding.

Closing the Lesson Have students explain in their own words how to find the cost of carpeting a room.

EXTENDING THE LESSON

Enrichment Have students assume that in Exercises 6-8 the cost of padding is not included. Give them a price of $4.95 a square yard for padding, except the bathroom and kitchen, and have them find the cost of carpeting the house. **179.3 yd², + $887.54 or $5,572.78**

Enrichment Masters Booklet, p. 79

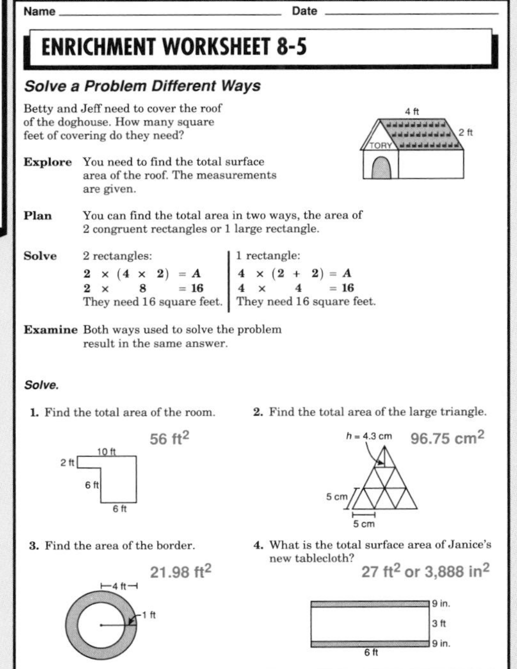

Name ______ Date ______

ENRICHMENT WORKSHEET 8-5

Solve a Problem Different Ways

Betty and Jeff need to cover the roof of the doghouse. How many square feet of covering do they need?

Explore You need to find the total surface area of the roof. The measurements are given.

Plan You can find the total area in two ways, the area of 2 congruent rectangles or 1 large rectangle.

Solve

2 rectangles:	1 rectangle:
2 × (4 × 2) = A	4 × (2 + 2) = A
2 × 8 = 16	4 × 4 = 16
They need 16 square feet.	They need 16 square feet.

Examine Both ways used to solve the problem result in the same answer.

Solve.

1. Find the total area of the room. **56 ft²**
2. Find the total area of the large triangle. **96.75 cm²**
3. Find the area of the border. **21.98 ft²**
4. What is the total surface area of Janice's new tablecloth? **27 ft² or 3,888 in²**

5-MINUTE CHECK-UP

(over Lesson 8-5)

Available on TRANSPARENCY 8-6.

1. What is the cost of carpeting two rooms if one room is 4 yards long and 3 yards wide, and the other room is 5 yards long and 4 yards wide? The carpeting costs $21.95 per square yard. **$702.40**

INTRODUCING THE LESSON

Ask students if they have traced and drawn a family tree. Have students discuss how to draw the two previous generations. Discuss the need for accuracy in applying charts and diagrams to real-life problems.

TEACHING THE LESSON

Using Critical Thinking Ask students why and when they should use diagrams to solve problems. **Answers may vary. A good answer should include problems that involve a chain of events or problems that present options.**

Practice Masters Booklet, p. 87

Name ____________________ Date ____________________

PRACTICE WORKSHEET 8-6

Problem-Solving Strategy: Making a Diagram

Solve by making a diagram.

1. Darryl, Marcia, and José each work out at Sam's Gym. Darryl works out once every 2 days, Marcia every 3 days, and José every 5 days. If they are all at the gym on March 4, what is the next date they will all be at the gym? **April 3**
2. It takes 5 minutes for the director to call 3 choir members and tell them the rehearsal is canceled. It takes 5 minutes for those choir members to call another 3 members each, and so on. How many choir members can be called in 20 minutes? **120 choir members**
3. Of the 90 members of the freshman class, $\frac{1}{2}$ participate in sports and $\frac{3}{5}$ participate in music activities. If $\frac{1}{3}$ do not participate in sports or music, how many participate in both sports and music? **39 students**
4. A gardener divides his seedlings equally among 5 gardens. In one garden he plants $\frac{3}{4}$ of the seedlings as a border and has 8 seedlings left over. How many seedlings did he start with? **160 seedlings**
5. A car is available in white, black, or gray; with an automatic or manual transmission; and with 2 doors or 4 doors. In how many different combinations of color, transmission, and doors is this car available? **12 combinations**
6. Hassan used part of his savings to buy a tape player for $199.95, a compact disc player for $254.99, and speakers for $175.50. He paid sales tax of $31.52 and still had $\frac{1}{3}$ of his savings left. How much were Hassan's savings before he made these purchases? **$992.94**

Problem-Solving Strategy

8-6 MAKING A DIAGRAM

▶ Explore ▶ Plan ▶ Solve ▶ Examine

Objective
Solve problems by making a diagram.

Three trains run on different schedules. The Express leaves the Fortieth Street station every 10 minutes, the Local leaves every 15 minutes, and the Commuter leaves every 20 minutes. If all three trains left the Fortieth Street station at noon, when will all three trains leave the station together again?

▶ Explore

What is given?

- One train runs every 10 minutes.
- One train runs every 15 minutes.
- One train runs every 20 minutes.
- Three trains left at noon from the same station.

What is asked?

- When will all three trains again leave Fortieth Street station at the same time?

▶ Plan

Make a diagram to represent the three trains and the times they leave the station. The solution will be the time when all three trains are represented at the same time.

▶ Solve

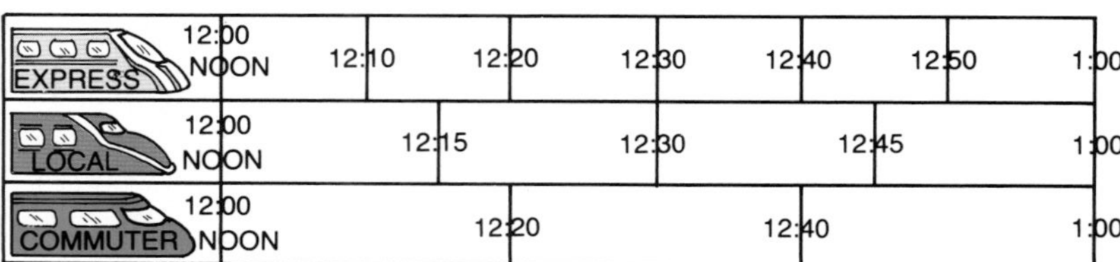

The trains will leave Fortieth Street station together at 1:00 P.M.

▶ Examine

The three trains leave together on the hour. The LCM of 10, 15, and 20 is sixty.

Guided Practice

Solve by making a diagram.

1. The advisor of the Freshman Class informs three students about a meeting. It takes 1 minute for her to tell three students and one minute for each of those three students to tell three other students, and so on. How many students will know about the meeting in three minutes? **39 students**

RETEACHING THE LESSON

Work with students to solve each problem.

1. One person enters an auditorium the first minute, 2 people enter the second minute, and so on. How many minutes will it take for 105 people to enter the auditorium? **14 min**
2. A person mails a recipe to 5 friends. Each of the five friends mails the recipe to five other friends and so on. How many recipes are in the fourth mailing? **625 recipes**

Problem Solving

Paper/ Pencil
Mental Math
Estimate
Calculator

2. Joan and Sally share a dorm room at college and both alarms are set for 6:30 A.M. Suppose Joan's snooze alarm is set to ring every 4 minutes while Sally's rings every 10 minutes. If they both forget to turn off their alarms and leave at 6:45 A.M. how many times will the alarms ring at the same time before they return at 9:00 A.M.? 6 times
3. Tim divides his entire stamp collection equally among Sarah, Scott, Dekker, and Akiko. Akiko gives her brother and sister each one third of her stamps. If they each received 5 stamps, how many stamps did Tim have in his collection? 60 stamps
4. Sixteen students in a math class are standing. If one fourth of these students sit down, one third of those still standing sit down, and 3 rise from their seats, how many are standing? 11 students
5. If three out of five doctors recommend walking instead of jogging, how many doctors out of 600 would probably recommend jogging? 240 doctors

FUN with MATH
Who was the first known woman doctor?
See page 308.

6. 8 outfits

6. Julie packs a red blouse, white blouse, black slacks, gray slacks, a blue blazer, and a black blazer for the weekend. Make a list to find how many different outfits she can make consisting of a blazer, slacks, and a blouse.
7. Kevin's beginning checking account balance is $197.53. If he writes checks for $21.35, $15.87, and $36.22, how much must he deposit in his account to bring the balance up to $200? Name the method of computation used. $75.91; calculator

Critical Thinking

8. Name a year in the twentieth century that reads the same backward and forward. Numbers such as this are called palindromes. 1991

Mixed Review

Subtract.

Lesson 2-4

9. $28 − $4.52 $23.48
10. 8.97 − 3.66 5.31
11. 43.087 − 21.98 21.107

Lesson 7-1

Use symbols to name each figure.

12.
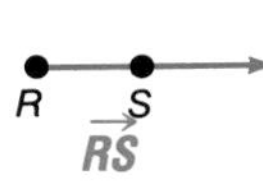

$\overrightarrow{RS}$

13.
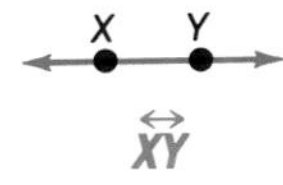

$\overleftrightarrow{XY}$

14.

$\overline{MN}$

15.
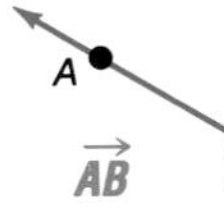

$\overrightarrow{AB}$

Lesson 8-1

16. Jackie rides her horse around a circular pen. The distance across the pen is 50 feet. How many feet will she ride her horse if she rides around the pen twenty times? 3,140 ft

Chalkboard Examples

1. Will's champion beagle had 5 puppies. If each of the five puppies has five puppies, and so on, which generation after Will's champion will have 3,125 puppies? 4th generation

Additional examples are provided on TRANSPARENCY 8-6.

EVALUATING THE LESSON

Check for Understanding Use the Guided Practice Exercises to check for student understanding.

Error Analysis Watch for students who do not complete a diagram. Remind students that the purpose of the diagram is to demonstrate the entire problem.

Closing the Lesson Have students suggest problems in real-life situations that can be solved using diagrams.

APPLYING THE LESSON

Independent Practice

Homework Assignment	
Minimum	Ex. 2-8 even; 9-16
Average	Ex. 1-16
Maximum	Ex. 1-16

Applying Mathematics to Technology

This optional page provides students with an experience with computer literacy and also provides a change of pace.

Objective Interpret simple BASIC statements.

Using Discussion Discuss the origin of the computer. The name of the first computer was ENIAC (Electrical Numerical Integrator and Computer). It was completed in 1945 at the University of Pennsylvania. It weighed about 30 tons and occupied the entire basement of a building.
Point out that the variables and data in the READ and DATA statements must be given in the same order.

Activity

Using Cooperative Groups and Computers In groups of two or three, have students write a simple BASIC program and run it on a computer. Examples: **1.** Given the number of hours worked and the rate per hour, compute and print a person's weekly pay. **2.** Given the number of wins and losses of a school team, compute and print the percentage of wins. Have students make up the data or use the record of the school's football team.

Computer Application

BASIC PROGRAMS

A **computer program** is a list of numbered statements that gives instructions to a computer. The number given to each statement is called a **line number.** Multiples of 10 are often used for line numbers so statements can be inserted later if necessary.

```
10 READ A,B
20 LET C=2*A+B↑3
30 PRINT C
40 DATA 5,2
50 END
RUN
18
```

Consider the program, RUN command, and output for the program shown at the right.

The **READ statement** (line 10) assigns values from the **DATA statement** (line 40) to the variables A and B. After line 10 is performed, the value of A is 5 and the value of B is 2. Every program that has a READ statement must have a DATA statement.

The **LET statement** (line 20) tells the computer to assign the value of the expression on the right to the variable on the left. For example, in the program above, C is assigned the value of 2*5+2↑3 or 18.

The **PRINT statement** (line 30) tells the computer to print the value assigned to C. The output for the program above is 18.

The **END statement** (line 50) tells the computer that the program is complete. The **RUN command** tells the computer to execute the program. It has no line number.

For each READ and DATA statement, state the value that the computer will assign to each variable.

1.
```
10 READ M
20 DATA 3
```
M = 3

2.
```
10 DATA 1.2,7
20 READ X, Y
```
X = 1.2; Y = 7

3.
```
30 READ A,B,C
95 DATA 3.2, 0.6, 5
```
3. A = 3.2; B = 0.6; C = 5

Determine the output for each PRINT statement.

4.
```
10 LET X=3.8
20 LET Y=6.81
30 PRINT X+Y
```
10.61

5.
```
10 READ N
20 LET S=5*N-6
30 PRINT S
40 DATA 27.3
```
130.5

6.
```
10 LET X=3+5
20 LET Y=2
30 LET Z=X↑Y
40 PRINT Z
```
64

PYTHAGOREAN THEOREM

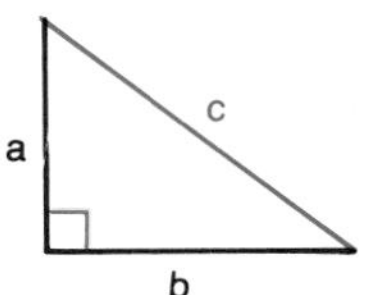

The longest side of a right triangle is called the **hypotenuse.** To find the length of the hypotenuse when you know the lengths of the other two sides, you can use the **Pythagorean Theorem.**

$c^2 = a^2 + b^2$, where c is the measure of the hypotenuse, and a and b are the measures of the other two sides

Find the length of the hypotenuse of the triangle at the right.

$c^2 = a^2 + b^2$ $\quad a = 9, b = 12$

Enter:  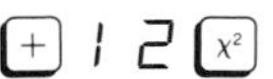  9 x^2 + 12 x^2 = 225

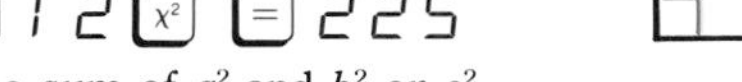

The display shows the sum of a^2 and b^2 or c^2.

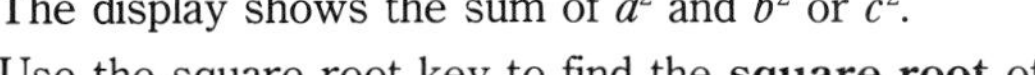

Use the square root key to find the **square root** of 225.

Enter:  $\sqrt{x}$ 15

The display shows the square root of 225.

The hypotenuse is 15 m long.

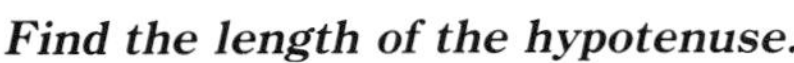

Find the length of the hypotenuse.

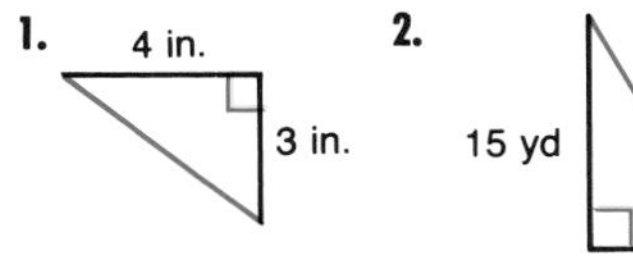

2. 15 yd, 8 yd

Solve.

3. If the foot of a ladder is 10 feet from the wall, how high up on the wall does a 50-foot ladder reach? **48.99 ft**

4. If a television screen measures 15 inches diagonally and 12 inches across the bottom, how high is the screen? **9 in.**

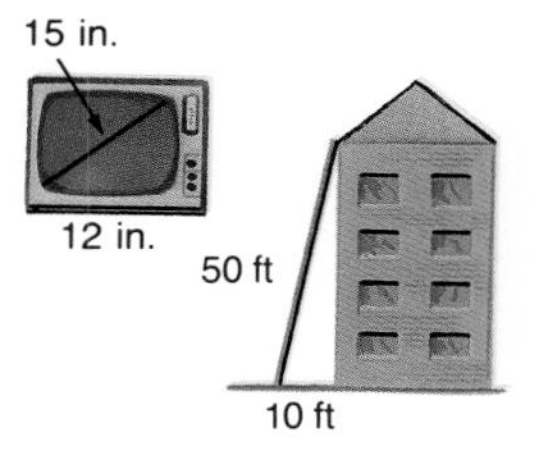

Applying Mathematics to the Real World

This optional page shows how mathematics is used in the real world and also provides a change of pace.

Objective Use the Pythagorean Theorem to find the length of a hypotenuse or side of a triangle.

Using Discussion As a historical note, discuss with students that Pythagoras, a Greek mathematician and philosopher, is believed to have proved this theorem about twenty-five hundred years ago.

Point out that the hypotenuse is always labeled 'c'. Either of the remaining sides (or legs) are labeled 'a' and 'b'.

Discuss the various occupations that use geometry, such as, surveyor, draftsman, engineer, carpenter, and so on.

Activity

Using Data Have students write a report about Pythagoras or do a poster explaining the Pythagorean Theorem.

5-MINUTE CHECK-UP

(over Lesson 8-6)

Available on TRANSPARENCY 8-7.

Solve by making a diagram.

1. The bicycle trail begins at the lighthouse and splits into two trails at the bend. Each of the two trails splits into two more trails, and so on. After four splits, the trails end at the ranger station. How many trails meet at the ranger station? **16 trails**

INTRODUCING THE LESSON

Have students give examples of presents they have wrapped or textbooks they have covered. Ask them how they determined the size of the wrapping.

TEACHING THE LESSON

Using Manipulatives In Exercise 13 have groups use construction paper to draw their pattern so that it will not collapse easily when they try to fold and tape it together. You may wish to provide more than one sheet for students who need a second try. You may wish to use the Rectangular Prism Pattern, **Lab Manual, p. 14.**

Practice Masters Booklet, p. 88

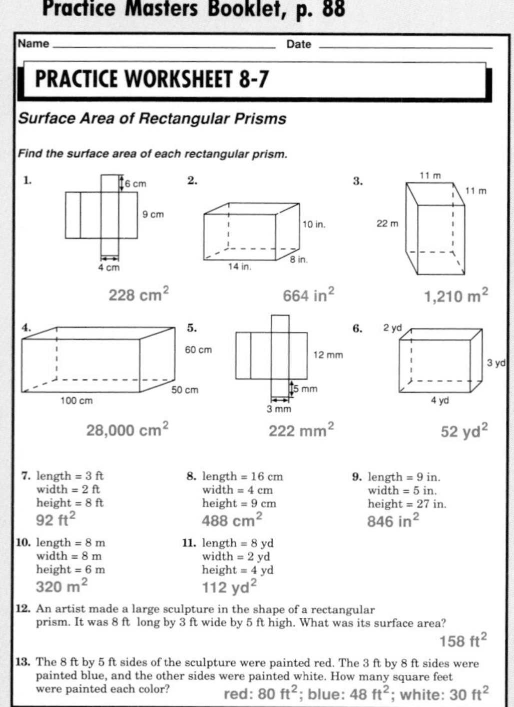

Name ______ Date ______

PRACTICE WORKSHEET 8-7

Surface Area of Rectangular Prisms

Find the surface area of each rectangular prism.

1. 6 cm, 9 cm, 4 cm — **228 cm²**
2. 10 in., 8 in., 14 in. — **664 in²**
3. 11 m, 11 m, 22 m — **1,210 m²**
4. 100 cm, 60 cm, 50 cm — **28,000 cm²**
5. 12 mm, 5 mm, 3 mm — **222 mm²**
6. 2 yd, 3 yd, 4 yd — **52 yd²**
7. length = 3 ft, width = 2 ft, height = 8 ft — **92 ft²**
8. length = 16 cm, width = 4 cm, height = 9 cm — **488 cm²**
9. length = 9 in., width = 5 in., height = 27 in. — **846 in²**
10. length = 8 m, width = 8 m, height = 6 m — **320 m²**
11. length = 8 yd, width = 2 yd, height = 4 yd — **112 yd²**
12. An artist made a large sculpture in the shape of a rectangular prism. It was 8 ft long by 3 ft wide by 5 ft high. What was its surface area? **158 ft²**
13. The 8 ft by 5 ft sides of the sculpture were painted red. The 3 ft by 8 ft sides were painted blue, and the other sides were painted white. How many square feet were painted each color? **red: 80 ft²; blue: 48 ft²; white: 30 ft²**

8-7 SURFACE AREA OF RECTANGULAR PRISMS

Objective
Find the surface area of rectangular prisms.

The Home Economics club is covering tissue boxes with quilted fabric to sell at their craft fair. How much fabric does Sheila need to cover a tissue box if the box is $9\frac{1}{2}$ inches long, $4\frac{1}{2}$ inches wide, and 4 inches high?

Sheila finds the **surface area** of the box by adding the areas of the faces.

Method

1. Use the formula $A = \ell w$ to find the area of each face.
2. Add the areas to find the surface area.

Example A ***Find the surface area of the tissue box.***

1 Front	$9\frac{1}{2} \times 4$	=	38
Back	$9\frac{1}{2} \times 4$	=	38
Top	$9\frac{1}{2} \times 4\frac{1}{2}$	=	$42\frac{3}{4}$
Bottom	$9\frac{1}{2} \times 4\frac{1}{2}$	=	$42\frac{3}{4}$
Right side	$4 \times 4\frac{1}{2}$	=	18
Left side	$4 \times 4\frac{1}{2}$	=	+ 18
2			$197\frac{1}{2}$

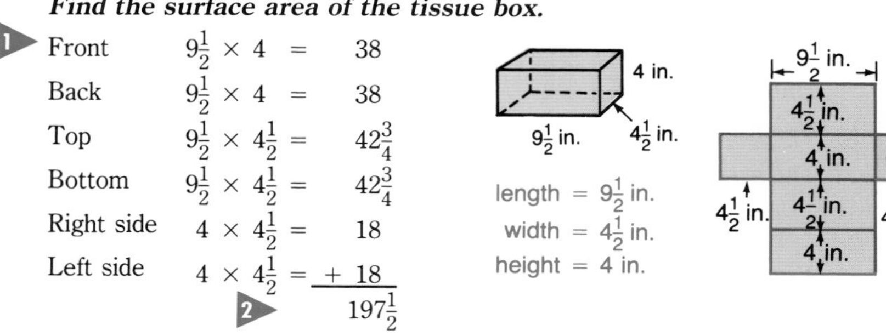

Sheila needs $197\frac{1}{2}$ square inches of fabric to cover a box.

Guided Practice ***Find the surface area of each figure by adding the areas.***

Example A

1. 2.

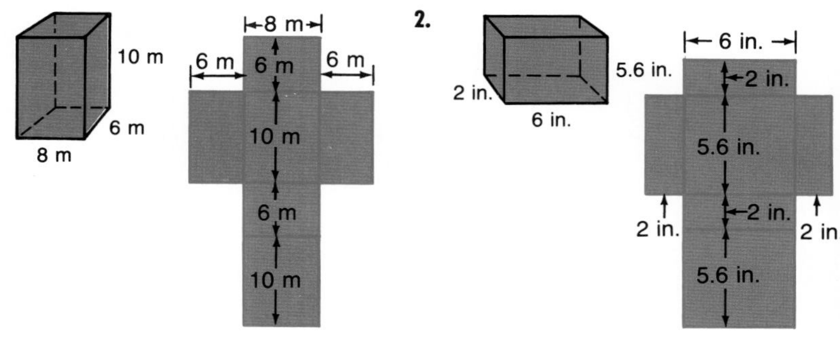

1. 80 m², 80 m², 48 m², 48 m², 60 m², 60 m², 376 m²
2. 33.6 in², 33.6 in², 12 in², 12 in², 11.2 in², 11.2 in², 113.6 in²

RETEACHING THE LESSON

Have students sketch flat layouts of each box and write the dimensions for the faces.

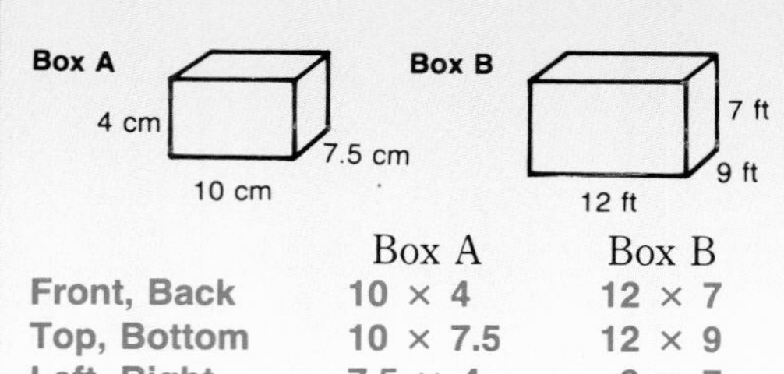

	Box A	Box B
Front, Back	**10 × 4**	**12 × 7**
Top, Bottom	**10 × 7.5**	**12 × 9**
Left, Right	**7.5 × 4**	**9 × 7**

Reteaching Masters Booklet, p. 80

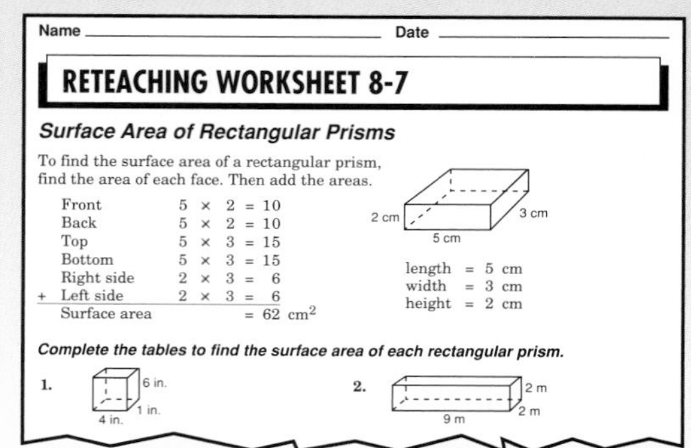

Name ______ Date ______

RETEACHING WORKSHEET 8-7

Surface Area of Rectangular Prisms

To find the surface area of a rectangular prism, find the area of each face. Then add the areas.

Front	5 × 2	= 10
Back	5 × 2	= 10
Top	5 × 3	= 15
Bottom	5 × 3	= 15
Right side	2 × 3	= 6
+ Left side	2 × 3	= 6
Surface area		= 62 cm²

length = 5 cm
width = 3 cm
height = 2 cm

Complete the tables to find the surface area of each rectangular prism.

1. 6 in., 1 in., 4 in.
2. 2 m, 2 m, 9 m

Exercises

Practice

Find the surface area of each rectangular prism.

3. 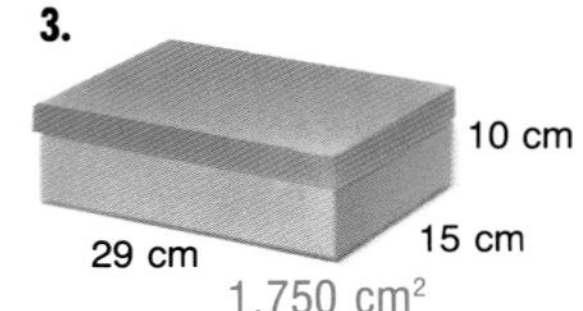

1,750 cm²

4. 

910 cm²

5. 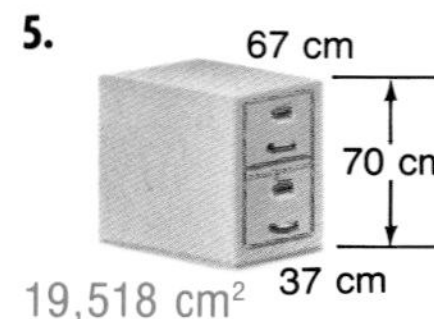

19,518 cm²

6. 136 cm² 7. 1,181 in² 8. 255.9 ft²

6. length = 8 cm
   width = 4 cm
   height = 3 cm

7. length = 20 in.
   width = 8.3 in.
   height = 15 in.

8. length = 2.7 ft
   width = 6.5 ft
   height = 12 ft

9. What is the surface area of a carton with a length of 45 cm, a width of 22 cm, and a height of 25 cm? 5,330 cm²

Applications

10. One box is 10 in. long, 4 in. wide, and 8 in. high. Another box is 10 in. long, 6 in. wide, and 6 in. high. Which box has the greater surface area?

11. Frank is building three wooden storage cubes for his stereo system. Each cube will be $2\frac{1}{2}$ feet long, $1\frac{1}{2}$ wide, and 1 foot high. How many square feet of plywood does Frank need for three cubes if the front of each cube is left open? 39 ft²

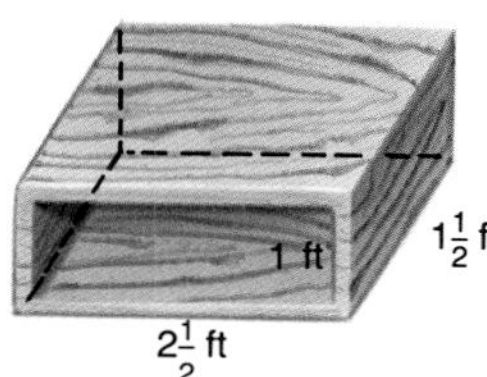

Write Math

12. In your own words, explain the change in the surface area of a rectangular prism if you double the length, double the width, and double the height. The surface area will be 4 times greater.

Cooperative Groups

13. In groups of three or four, use quarter-inch grid paper and draw a flat pattern that when cut out, folded, and taped will form a rectangular prism with a surface area of 88 square inches. Use 4 squares on the grid to represent 1 square inch. Answers may vary. A sample answer is a 2 × 4 × 6-inch prism.

Critical Thinking

14. A fruitbowl at the salad bar is full of apples and oranges. After 8 apples are eaten, 2 oranges remain for each apple. Then 12 oranges are eaten and 2 apples remain for each orange. How many apples were in the bowl before any were eaten? 16 apples

Using Variables

Find the value of each expression if $V = 24$, $\ell = 2$, $w = 3$, and $h = 4$.

15. $V \div (\ell \times w)$ 4

16. $(\ell \times w \times h) \div V$ 1

Chalkboard Examples

Example A *Find the surface area of the rectangular prism by adding the areas.*

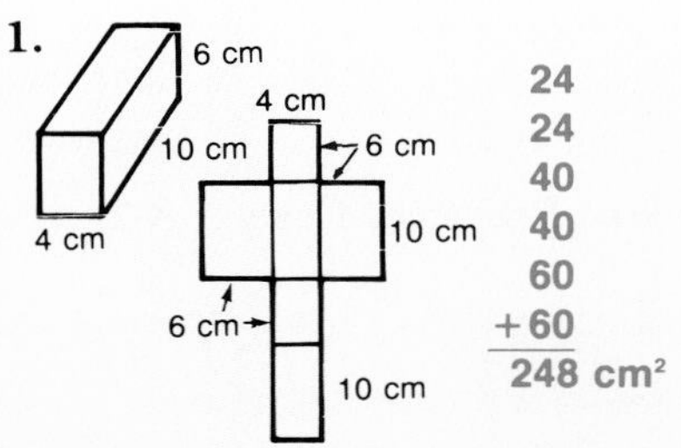

Additional examples are provided on TRANSPARENCY 8-7.

EVALUATING THE LESSON

Check for Understanding Use the Guided Practice Exercises to check for student understanding.

Error Analysis Watch for students who confuse measurements when finding areas of faces.

Closing the Lesson Have students state in their own words how to find the surface area of a rectangular prism.

EXTENDING THE LESSON

Enrichment Masters Booklet, p. 80

ENRICHMENT WORKSHEET 8-7

Surface Area of Pyramids

Find the surface area of each pyramid. The base of each pyramid is a regular polygon and the other faces are all the same size and shape.

1. 144 cm² 6 cm 9 cm

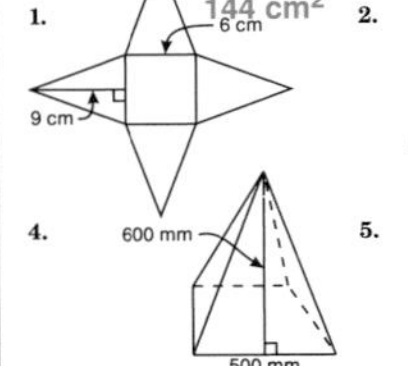

2. 476 in² 10 in. 14 in.

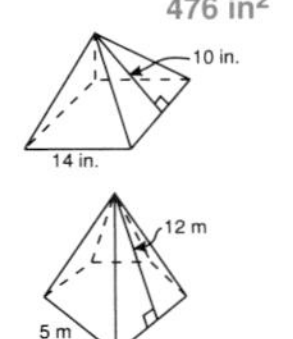

3. 572.6 m² 11 m 22 m $B = 209.6$ m²

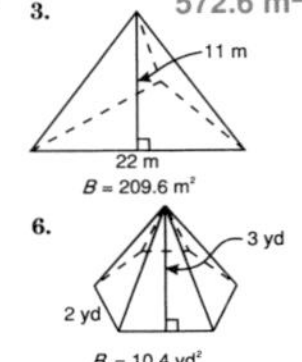

4. 600 mm 500 mm — 850,000 mm²

5. 12 m 5 m $B = 43.01$ m² — 193.01 m²

6. 3 yd 2 yd $B = 10.4$ yd² — 28.4 yd

Complete the chart to find the surface are of each square pyramid described below.

	length of side of square	height of triangular face	area of base	area of one triangular face	area of four triangular faces	surface area
7.	3 ft	8 ft	9 ft²	12 ft²	48 ft²	57 ft²
8.	16 cm	9 cm	256 cm²	72 cm²	288 cm²	544 cm²
9.	9 in.	27 in.	81 in²	121.5 in²	486 in²	567 in²
10.	8 m	6 m	64 m²	24 m²	96 m²	160 m²
11.	8 yd	4 yd	64 yd²	16 yd²	64 yd²	128 yd²

APPLYING THE LESSON

Independent Practice

Homework Assignment	
Minimum	Ex. 1-9 odd; 10-16
Average	Ex. 1-16
Maximum	Ex. 1-16

Chapter 8, Quiz A (Lessons 8-1 through 8-7) is available in the Evaluation Masters Booklet, p. 39.

5-MINUTE CHECK-UP

(over Lesson 8-7)

Available on TRANSPARENCY 8-8.

Find the surface area of each rectangular prism. 330 in²

1. length, 6 cm width, 6 cm 360 cm² height, 12 cm
2. length, 14 in. width, 5 in. height, 5 in.

Extra Practice, Lesson 8-7, p. 470

▶ INTRODUCING THE LESSON

Ask students if they have used cylindrical containers as a base for crafts, for example, covering coffee cans, puppets from cardboard tubes, and how they determined the amount of material needed to cover a container.

▶ TEACHING THE LESSON

Materials: paper, pencil, tape, scissors, ruler

Using Manipulatives Have students roll a piece of paper (notebook or construction) and tape it closed to form a cylinder and find the surface area assuming the ends are covered. You may wish to use the Cylinder Pattern, **Lab Manual, p. 16.**

Practice Masters Booklet, p. 89

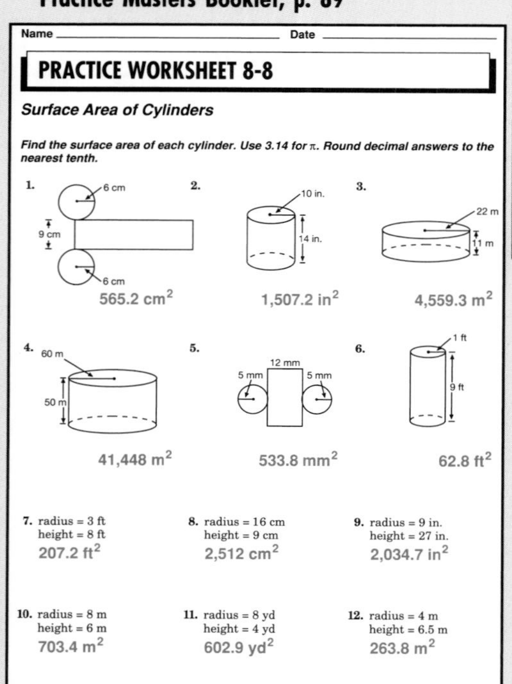

Name ______________ Date ______________

PRACTICE WORKSHEET 8-8

Surface Area of Cylinders

Find the surface area of each cylinder. Use 3.14 for π. Round decimal answers to the nearest tenth.

7. radius = 3 ft height = 8 ft 207.2 ft²
8. radius = 16 cm height = 9 cm 2,512 cm²
9. radius = 9 in. height = 27 in. 2,034.7 in²
10. radius = 8 m height = 6 m 703.4 m²
11. radius = 8 yd height = 4 yd 602.9 yd²
12. radius = 4 m height = 6.5 m 263.8 m²

8-8 SURFACE AREA OF CYLINDERS

Objective
Find the surface area of cylinders.

Mr. Brooke's fourth grade class is making banks to sell at the school fair. The students cover clean, empty coffee cans with fabric. Mr. Brooke cuts the slot in the top. How much fabric is needed to cover one can if the radius is 2 inches and the height is 5 inches? You need to find the surface area of the coffee can.

The surface area of a cylinder is the sum of the area of its two circular bases and the rectangle that forms its curved surface. The length of the rectangle is equal to the circumference of each circle.

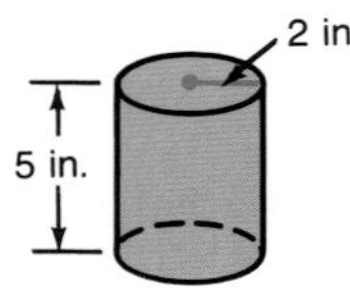

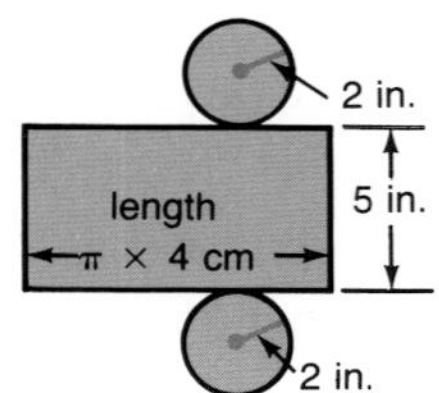

The radius of each circle is 2 in. So the diameter is 4 in. The circumference is πd or $\pi \times 4$.

The curved surface, when flat, forms a rectangle.

Method

1. Use the formula $A = \pi r^2$ to find the area of each circular base.
2. Use the formula $A = \ell w$ to find the area of the curved surface.
3. Add the areas to find the surface area.

Example A

Find the surface area of the coffee can. Use 3.14 for π.

1. $A = \pi r^2$
   $A \approx 3.14 \times 2^2$
   $A \approx 3.14 \times 4$
   $A \approx 12.56$

2. $A = \ell w$
   $A = \pi \times 4 \times 5$
   $A \approx 3.14 \times 20$
   $A \approx 62.8$

3. 12.56
   12.56
   + 62.80
   87.92 in²

The surface area of the coffee can is about 88 square inches.

Guided Practice

Find the surface area of each cylinder by adding the areas of the curved surface and each base. Use 3.14 for π. Round decimal answers to the nearest tenth.

Example A

1. **2.**

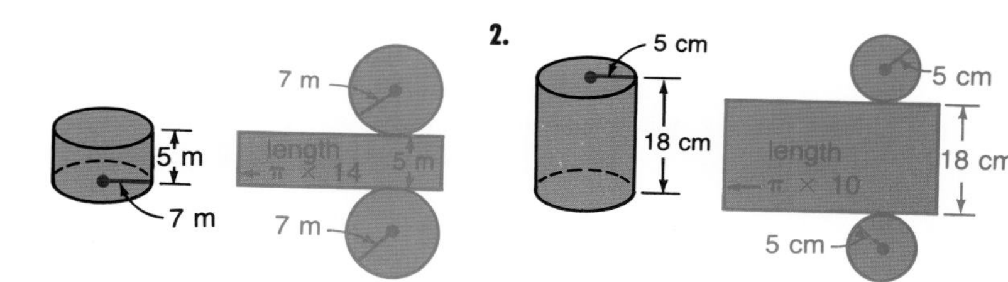

153.9 m², 153.9 m², 219.8 m², 527.6 m² 78.5 cm², 78.5 m², 565.2 cm², 722.2 cm²

RETEACHING THE LESSON

Have students sketch a flat layout of the cylinder shown below. Label the dimensions of each surface, find each area and the total surface area.

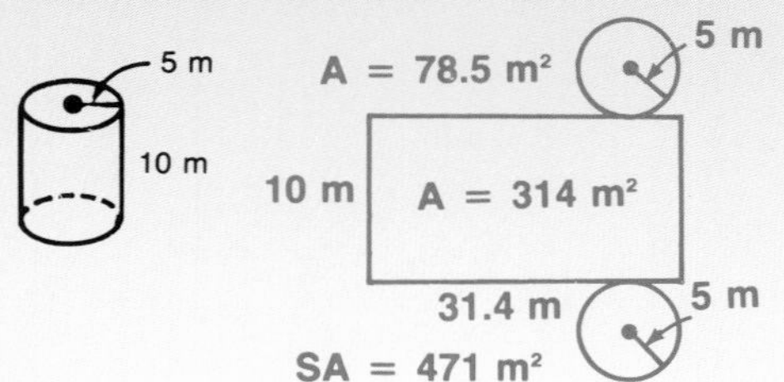

Reteaching Masters Booklet, p. 81

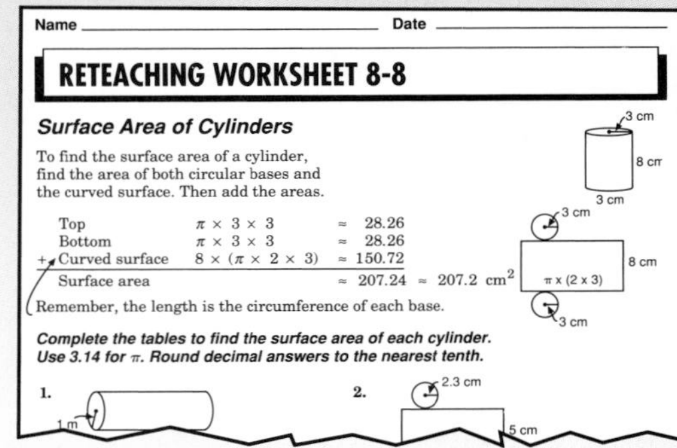

Name ______________ Date ______________

RETEACHING WORKSHEET 8-8

Surface Area of Cylinders

To find the surface area of a cylinder, find the area of both circular bases and the curved surface. Then add the areas.

Top	π × 3 × 3	≈ 28.26
Bottom	π × 3 × 3	≈ 28.26
+ Curved surface	8 × (π × 2 × 3)	≈ 150.72
Surface area		≈ 207.24 ≈ 207.2 cm²

Remember, the length is the circumference of each base.

Complete the tables to find the surface area of each cylinder. Use 3.14 for π. Round decimal answers to the nearest tenth.

1.
2.

Exercises

Find the surface area of each cylinder. Use 3.14 for π. Round decimal answers to the nearest tenth.

Practice

3. 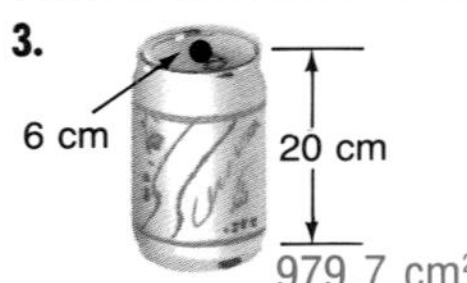

979.7 cm²

4. 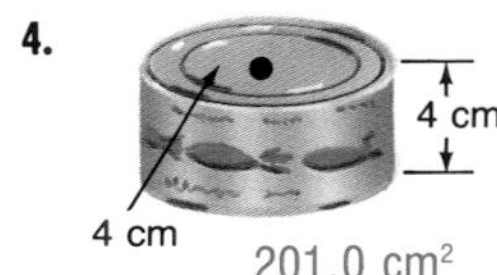

201.0 cm²

5. 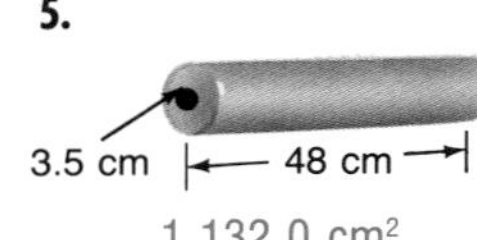

1,132.0 cm²

6. radius = 3.3 cm
   height = 10 cm
7. radius = 6 in.
   height = 16 in.
8. diameter = 19 m
   height = 10.5 m
9. One can has a radius of 5 cm and a height of 20 cm. Another can has a radius of 10 cm and height of 5 cm. Which can has more surface area? Use 3.14 for π.

Applications

10. The curved surface and top of each of Hawley's 3 cylindrical water towers are being painted. If the diameter of a tower is 30 feet and the height is 70 feet, how many gallons of paint are needed to paint 3 towers? One gallon covers 250 square feet. **96.1 gal**

11. Jenny orders the aluminum sheets from which Canco stamps the parts of aluminum cans for soft drinks. If the cans are 12 cm high and have a diameter of 6 cm, how many cans can be stamped from a sheet of aluminum that is 1,000 cm long and 200 cm wide? Use 3.14 for π. Subtract 1 can from your answer for scrap area around tops and bases of cans. **706 cans**

Critical Thinking

12. Mike's archery target has the scoring circles shown at the right. If Mike's score is 65 with four shots, write three different ways he could have scored 65. **See margin.**

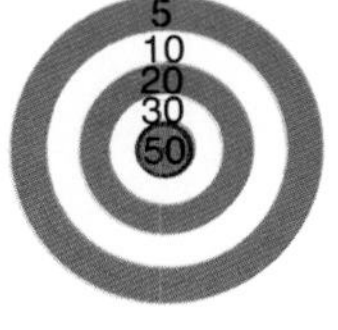

Show Math

13. Draw and label a diagram of a cylinder with a height of 11 feet that has a surface area of *about* 395 square feet. **Answers may vary. The diameter is about 8 ft.**

APPLYING THE LESSON

Independent Practice

Homework Assignment	
Minimum	Ex. 2-10 even; 11-13
Average	Ex. 1-13
Maximum	Ex. 1-13

Additional Answers

12. Answers may vary. Sample answers are 50, 10, 5, 0; 30, 20, 10, 5; 20, 20, 20, 5; 30, 30, 5, 0

Chalkboard Examples

Example A *Find the surface area of the cylinder. Use 3.14 for π. Round decimal answers to the nearest tenth.*

1. 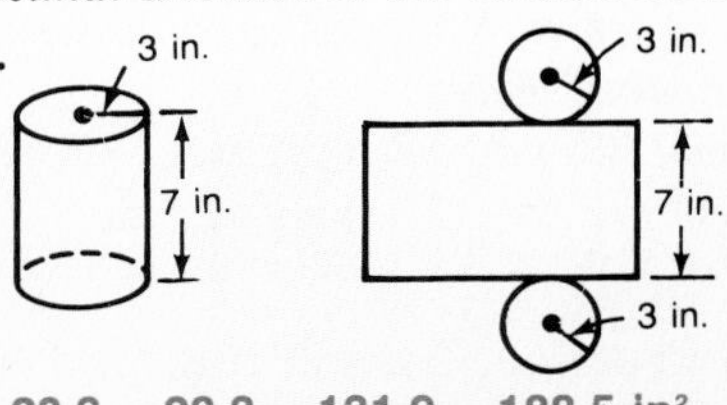

28.3, 28.3, 131.9, 188.5 in²

Additional examples are provided on TRANSPARENCY 8-8.

EVALUATING THE LESSON

Check for Understanding Use the Guided Practice Exercises to check for student understanding.

Error Analysis Watch for students who forget to use the area of the top twice in computing the surface area.

Closing the Lesson Have students explain in their own words how to find the circumference when computing the surface area of a cylinder.

EXTENDING THE LESSON

Enrichment Masters Booklet, p. 81

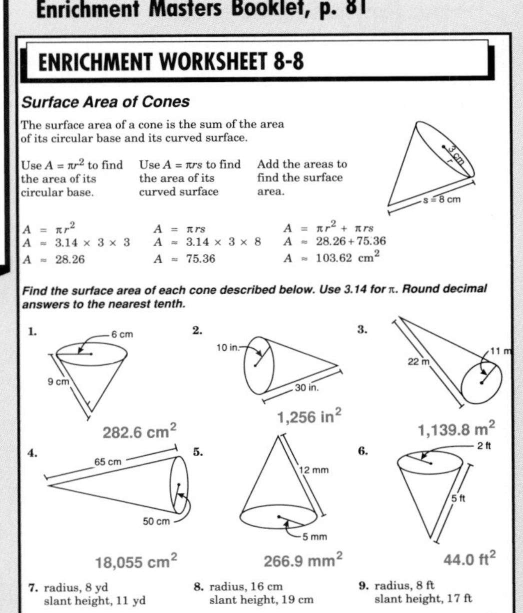

ENRICHMENT WORKSHEET 8-8

Surface Area of Cones

The surface area of a cone is the sum of the area of its circular base and its curved surface.

Use $A = \pi r^2$ to find the area of its circular base. | Use $A = \pi rs$ to find the area of its curved surface | Add the areas to find the surface area.

$A = \pi r^2$
$A = 3.14 \times 3 \times 3$
$A = 28.26$

$A = \pi rs$
$A = 3.14 \times 3 \times 8$
$A = 75.36$

$A = \pi r^2 + \pi rs$
$A = 28.26 + 75.36$
$A = 103.62$ cm²

Find the surface area of each cone described below. Use 3.14 for π. Round decimal answers to the nearest tenth.

1. 6 cm, 9 cm — 282.6 cm²
2. 10 in., 30 in. — 1,256 in²
3. 22 m, 11 m — 1,139.8 m²
4. 65 cm, 50 cm — 18,055 cm²
5. 12 mm, 5 mm — 266.9 mm²
6. 2 ft, 5 ft — 44.0 ft²
7. radius, 8 yd; slant height, 11 yd — 477.3 yd²
8. radius, 16 cm; slant height, 19 cm — 1,758.4 cm²
9. radius, 8 ft; slant height, 17 ft — 628.0 ft²
10. radius, 80 cm; slant height, 90 cm — 42,704 cm²
11. radius, 8 m; slant height, 9 m — 427.0 m²
12. radius, 9 in.; slant height, 8 in. — 480.4 in²

5-MINUTE CHECK-UP

(over Lesson 8-8)

Available on TRANSPARENCY 8-9.

Find the surface area of each cylinder. Use 3.14 for π. Round decimal answers to the nearest tenth.

1. radius, 3.5 cm height, 12 cm **340.7 cm²**
2. diameter, 2.5 in. height, 6 in. **56.9 in²**

Extra Practice, Lesson 8-8, p. 470

INTRODUCING THE LESSON

Ask students if they have packed a carton, or if they have fitted books on a shelf. Ask students to suggest other situations when they need to know the exact volume of a rectangular prism.

TEACHING THE LESSON

Using Critical Thinking Ask students how the volume of a rectangular prism changes if you double one of the dimensions. Ask students how the volume changes if you double two of the dimensions. How does the volume change if you double all three dimensions? **The volume is two times the original, four times the original, eight times the original.**

Practice Masters Booklet, p. 90

PRACTICE WORKSHEET 8-9

Volume of Rectangular Prisms

Find the volume of each rectangular prism described below.

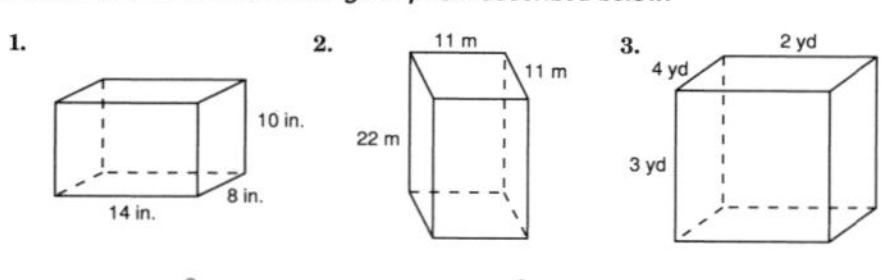

1. 1,120 in³
2. 2,662 m³
3. 24 yd³

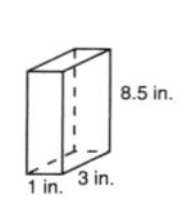

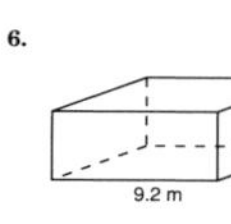

4. 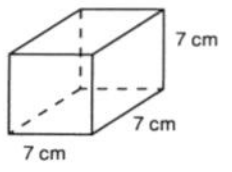 343 cm³
5. 25.5 in³
6. 92 m³

7. $l = 12$ ft, $w = 8$ ft, $h = 7$ ft — 672 ft³
8. $l = 5.2$ mm, $w = 12$ mm, $h = 4$ mm — 249.6 mm³
9. $l = 26$ cm, $w = 19$ cm, $h = 21$ cm — 10,374 cm³
10. $l = 4$ in., $w = 17$ in., $h = 3$ in. — 204 in³
11. $l = 1.3$ m, $w = 7$ m, $h = 2$ m — 18.2 m³
12. $l = 4$ yd, $w = 4$ yd, $h = 6$ yd — 96 yd³

8-9 VOLUME OF RECTANGULAR PRISMS

Objective
Find the volume of rectangular prisms.

The inner dimensions of Hank's freezer are 2 feet long, 2 feet wide, and 3 feet high. How many cubic feet of space does Hank have in the freezer?

Hank needs to find the volume inside the freezer. *Volume* is the amount of space that a solid contains. Volume is measured in cubic units.

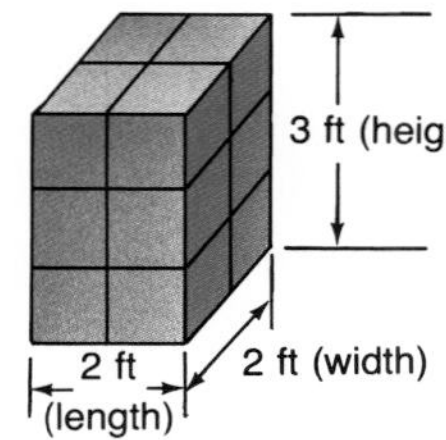

The volume *(V)* of a rectangular prism can be found by multiplying the measures of the length (ℓ), width *(w)*, and height *(h)*. The formula is $V = \ell \times w \times h$ or $V = \ell wh$.

Example A ***Find the volume of Hank's freezer.***

$V = \ell wh \quad \ell = 2, w = 2, h = 3$
$V = 2 \times 2 \times 3$
$V = 12$ or 12 ft^3

The volume of the freezer is 12 cubic feet.

Example B ***Find the volume of the rectangular prism.***

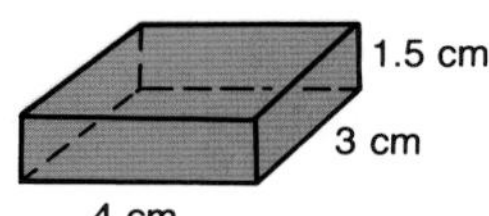

$V = \ell wh \quad \ell = 4, w = 3, h = 1.5$
$V = 4 \times 3 \times 1.5$
$V = 18$ or 18 cm^3

Guided Practice
Example A

Find the volume of each rectangular prism.

1.
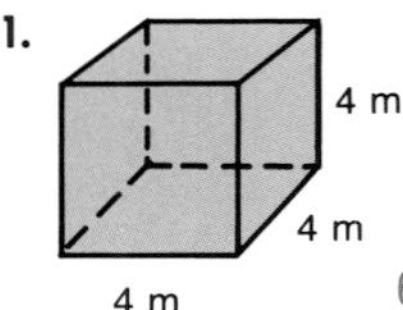
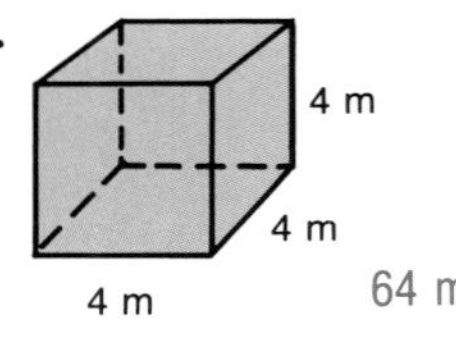

64 m³

2.
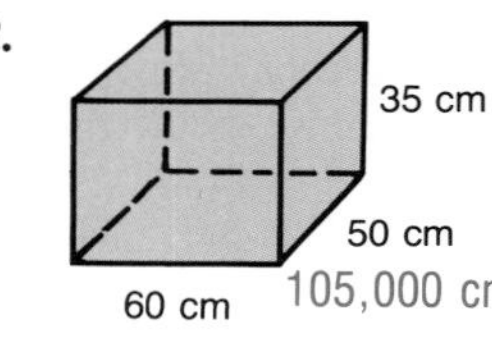

105,000 cm³

3.
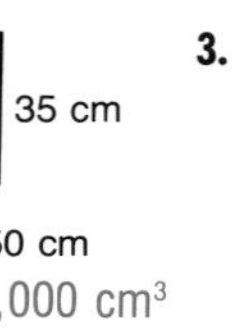
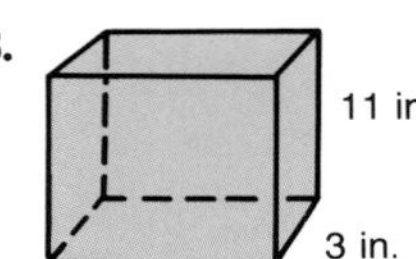

297 in³

4. $\ell = 7$ in., $w = 2.5$ in., $h = 11$ in. **192.5 in³**
5. $\ell = 14$ ft, $w = 15$ ft, $h = 0.5$ ft **105 ft³**
6. $\ell = 60$ in., $w = 5$ in., $h = 30$ in. **9,000 in³**

RETEACHING THE LESSON

Have students bring in various sized cardboard boxes. Have them find the length, width, height, and volume of each box. **Answers may vary.**

Reteaching Masters Booklet, p. 82

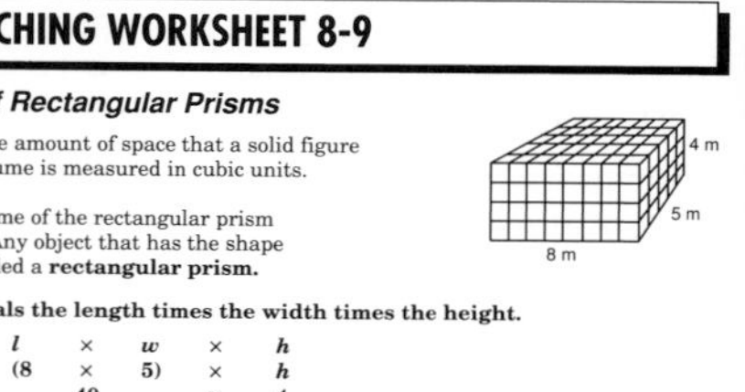

Name ______________ Date ______________

RETEACHING WORKSHEET 8-9

Volume of Rectangular Prisms

Volume is the amount of space that a solid figure contains. Volume is measured in cubic units.

Find the volume of the rectangular prism at the right. Any object that has the shape of a box is called a **rectangular prism.**

Volume equals the length times the width times the height.

V	=	l	×	w	×	h
V	=	(8	×	5)	×	h
V	=	40			×	4
V	=	160 cubic meters (m³)				

Find the volume of each rectangular prism. Some steps are given. Complete the remaining steps.

Exercises

Practice

Find the volume of each rectangular prism described below.

7. 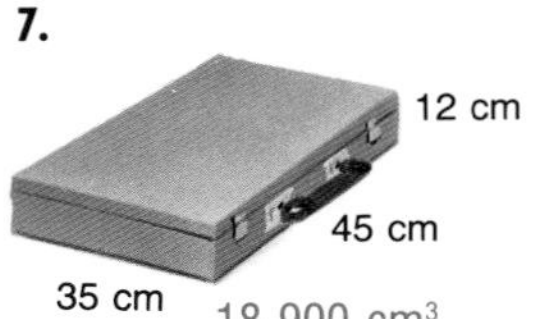

18,900 cm³

8. 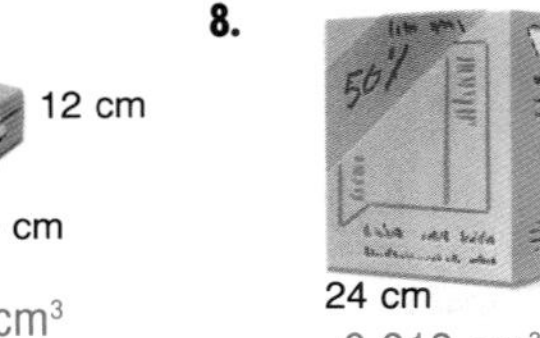 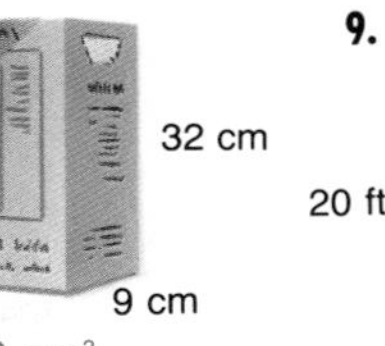

6,912 cm³

9. 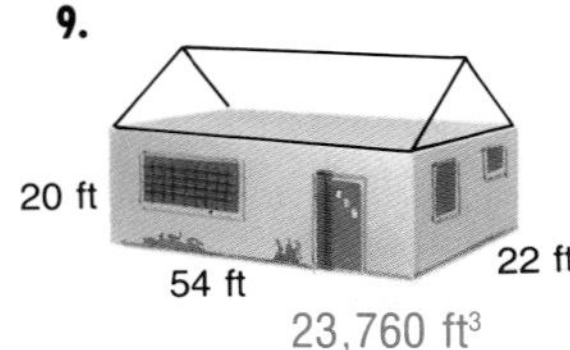

23,760 ft³

10. $\ell = 8$ cm, $w = 5$ cm, $h = 6$ cm **240 cm³**

11. $\ell = 8.5$ m, $w = 3$ m, $h = 9$ m **229.5 m³**

12. $\ell = 6$ in., $w = 2$ in., $h = 5.6$ in. **67.2 in³**

13. Each edge of a cube is 15 m long. What is the volume of the cube? **3,375 m³**

FUN with MATH

How well can you concentrate on a moving cube? See page 309.

Applications

14. The dimensions of an aquarium are 20 inches long, 14 inches wide, and 18 inches high. How many cubic inches of water does it hold? **5,040 in³**

15. A water bed measures $6\frac{1}{2}$ feet long, 4 feet wide, and $\frac{3}{4}$ foot deep. How many cubic feet of water does it take to fill the bed? **19.5 ft³**

16. If water weighs *about* 62 pounds per cubic foot, how much does the water for the bed in Exercise 15 weigh? **about 1,209 lb**

17. A rectangular greenhouse has a square base 64 feet on each side. The height of the ceiling is 12 feet. It takes a 6 horsepower pump to humidify 25,000 cubic feet and a 10 horsepower pump to humidify 50,000 cubic feet. Which pump is needed to humidify the greenhouse?

18. A rectangular carton holds 800 cubic inches of grass seed. It has a height of 10 inches and a width of 10 inches. How long is it? **8 in.**

Collect Data

19. In your classroom or home, find three different rectangular prisms and measure the length, width, and height. Then find their volume. **Answers may vary.**

Using Data

Use the chart for Exercises 20-23.

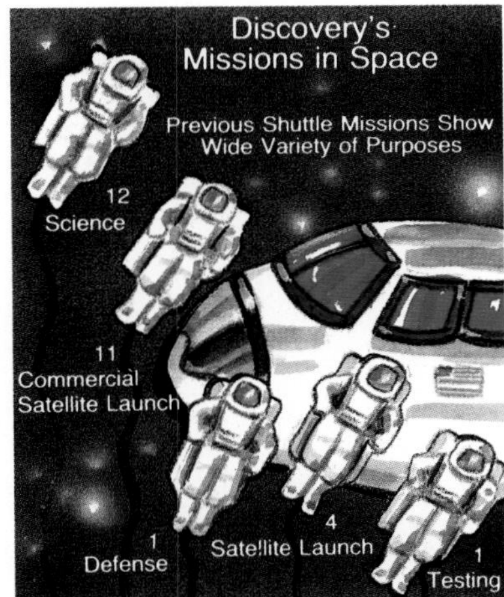

20. The space shuttle *Discovery* has been used for several different missions. How many missions has *Discovery* completed? **29 missions**

21. How many defense and commercial missions has *Discovery* had? **12 launches**

22. What is the most frequent purpose for *Discovery's* missions? **Science**

Research

23. In the library find some of the actual missions on which *Discovery* has been sent and the results of those missions. **See students' work.**

APPLYING THE LESSON

Independent Practice

Homework Assignment	
Minimum	Ex. 1-13 odd; 14-23
Average	Ex. 1-23
Maximum	Ex. 1-23

Chalkboard Examples

Example A *Find the volume of each rectangular prism.*

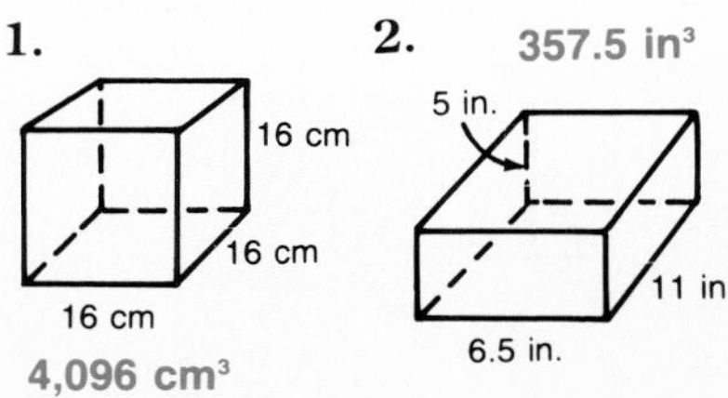

Additional examples are provided on TRANSPARENCY 8-9.

EVALUATING THE LESSON

Check for Understanding Use the Guided Practice Exercises to check for student understanding.

Error Analysis Watch for students who confuse the measures of surface area and volume. Remind students that surface area involves only two dimensions, volume involves three.

Closing the Lesson Have students suggest situations when they need to know the volume of a rectangular prism.

EXTENDING THE LESSON

Enrichment Masters Booklet, p. 82

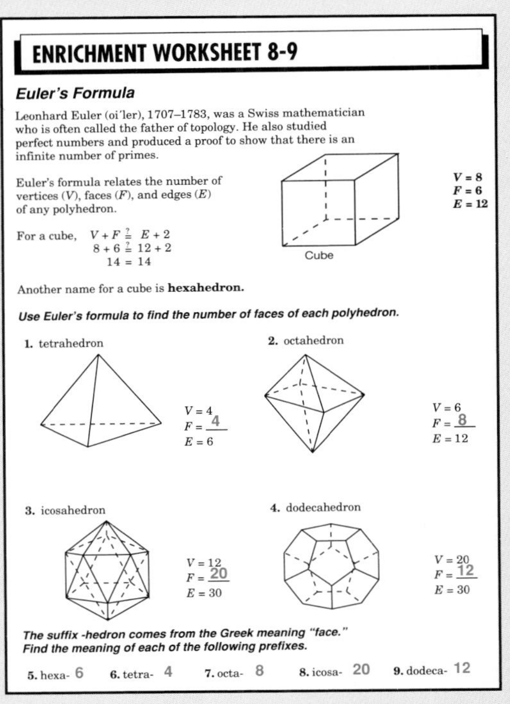

ENRICHMENT WORKSHEET 8-9

Euler's Formula

Leonhard Euler (oi'ler), 1707–1783, was a Swiss mathematician who is often called the father of topology. He also studied perfect numbers and produced a proof to show that there is an infinite number of primes.

Euler's formula relates the number of vertices (*V*), faces (*F*), and edges (*E*) of any polyhedron.

For a cube, $V + F \stackrel{?}{=} E + 2$
$8 + 6 \stackrel{?}{=} 12 + 2$
$14 = 14$

Cube: $V = 8$, $F = 6$, $E = 12$

Another name for a cube is **hexahedron.**

Use Euler's formula to find the number of faces of each polyhedron.

1. tetrahedron — $V = 4$, $F = $ 4, $E = 6$
2. octahedron — $V = 6$, $F = $ 8, $E = 12$
3. icosahedron — $V = 12$, $F = $ 20, $E = 30$
4. dodecahedron — $V = 20$, $F = $ 12, $E = 30$

The suffix -hedron comes from the Greek meaning "face." Find the meaning of each of the following prefixes.

5. hexa- 6 6. tetra- 4 7. octa- 8 8. icosa- 20 9. dodeca- 12

5-MINUTE CHECK-UP

(over Lesson 8-9)

Available on TRANSPARENCY 8-10.

Find the volume of the rectangular prism.

360 in³

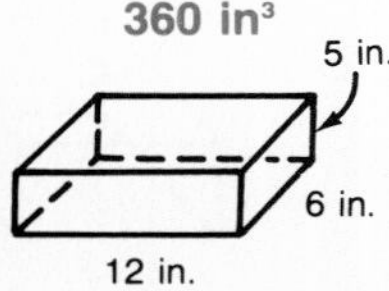

Extra Practice, Lesson 8-9, p. 470

INTRODUCING THE LESSON

Ask students if they have seen pictures of the pyramids in Egypt. Ask them to suggest ideas about the mechanics involved in building the pyramids.

TEACHING THE LESSON

Using Questioning

- How can you prove that the volume of a pyramid is one-third of the volume of a rectangular prism? **Build models of a pyramid and a rectangular prism with the same width, height, and length. Use sand or salt to show that the volume of 3 pyramids equals the volume of 1 rectangular prism.**

Practice Masters Booklet, p. 91

Name ______ Date ______

PRACTICE WORKSHEET 8-10

Volume of Pyramids

Find the volume of each pyramid.

1. 14 in., 8 in., 8 in. — 298.7 in³
2. 9 m, 6 m, 4 m — 72 m³
3. 8 cm, 12 cm, 12 cm — 384 cm³
4. 6 ft, 7 ft, 2 ft — 28 ft³
5. 12 mm, 3 mm, 3 mm — 36 mm³
6. 10 in., 4.6 in., 5.2 in. — 79.7 in³
7. $l = 12$ ft, $w = 12$ ft, $h = 7$ ft — 336 ft³
8. $l = 5.8$ m, $w = 4$ m, $h = 9$ m — 69.6 m³
9. $l = 9$ in., $w = 9$ in., $h = 16$ in. — 432 in³
10. $l = 8$ m, $w = 3$ m, $h = 11$ m — 88 m³
11. $l = 13$ cm, $w = 7.5$ cm, $h = 10$ cm — 325 cm³
12. $l = 15$ mm, $w = 15$ mm, $h = 17$ mm — 1,275 mm³

8-10 VOLUME OF PYRAMIDS

Objective
Find the volume of pyramids.

Bob's tent has a square base 7 feet on each side. If the central (and only) tent pole is 6 feet high, how many cubic feet of space are there in the tent?

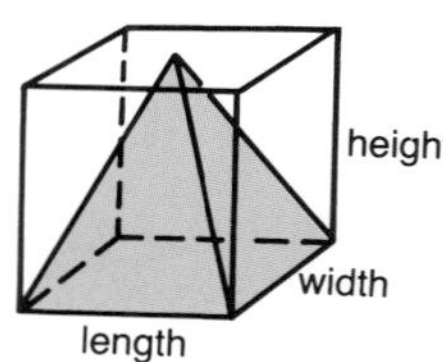

Bob's tent is shaped like a pyramid. The volume of a pyramid is one-third the volume of a prism with the same base and height as the pyramid. The formula is $V = \frac{1}{3}\ell wh$.

Example A ***Find the volume of Bob's tent.***

$V = \frac{1}{3}\ell wh \qquad \ell = 7, w = 7, h = 6$

$V = \frac{1}{3} \times 7 \times 7 \times 6$ — To multiply by $\frac{1}{3}$ on a calculator, you can divide by 3.

$V = 98$ or 98 ft³

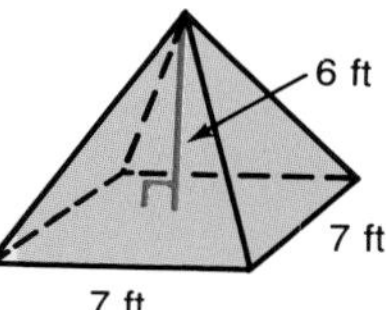

The volume of Bob's tent is about 131 cubic feet.

Example B ***Find the volume of a pyramid with length 200 meters, width 150 meters, and height 200 meters.***

$V = \frac{1}{3}\ell wh \qquad \ell = 200, w = 150, h = 200$

$V = \frac{1}{3} \times 200 \times 150 \times 200$

$V = 2{,}000{,}000$ or 2,000,000 m³

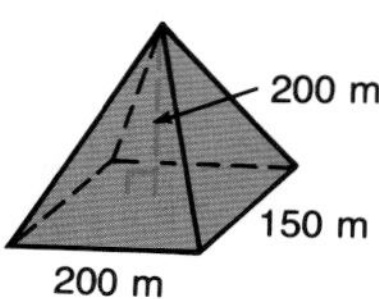

Guided Practice

Find the volume of each pyramid. Round decimal answers to the nearest tenth.

Example A

1. 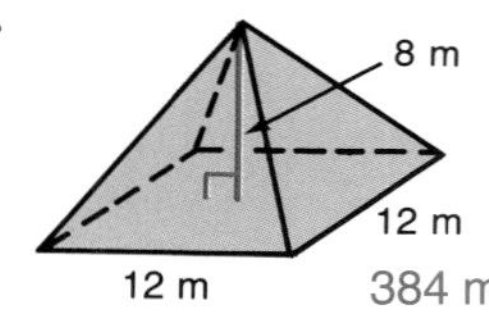

384 m³

2. 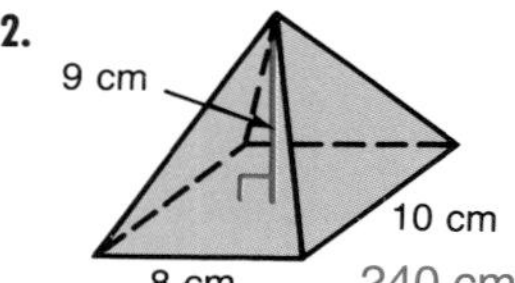

240 cm³

3. 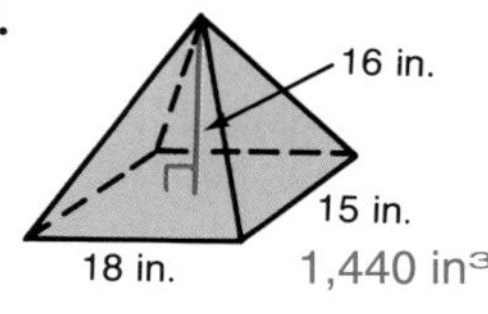

1,440 in³

Example B

4. $\ell = 11$ m, $w = 6$ m, $h = 13$ m **286 m³**
5. $\ell = 4$ cm, $w = 3$ cm, $h = 2.5$ cm **10 cm³**
6. $\ell = 11$ yd, $w = 10.5$ yd, $h = 9.5$ yd **365.8 yd³**

RETEACHING THE LESSON

Work with students to find the volume of each pyramid.

1. 6 yd, 4 yd, 5 yd — **40 yd³**
2. 10 cm, 12 cm, 15 cm — **600 cm³**
3. 20 in., 15 in., 12 in. — **1,800 in³**

Reteaching Masters Booklet, pg. 83

Name ______ Date ______

RETEACHING WORKSHEET 8-10

Volume of Pyramids

The volume of a *pyramid* is one-third the volume of a prism with the same length, width, and height.

$V = \frac{1}{3} \times (l \times w \times h)$

$V = \frac{1}{3} \times (10 \times 11 \times 18)$ — 18 cm, 11 cm, 10 cm

$V = \frac{1}{3} \times 1{,}980$

$V = 660$ cm³

Find the volume of each pyramid. Some steps are given. Complete the remaining steps.

1. $V = \frac{1}{3} \times (l \times w \times h)$
   $V = \frac{1}{3} \times (15 \times 12 \times 23)$
2. $V = \frac{1}{3} \times (l \times w \times h)$
   $V = \frac{1}{3} \times (29 \times 17 \times 42)$

Exercises

Practice

Find the volume of each pyramid.

7. **8.** **9.**

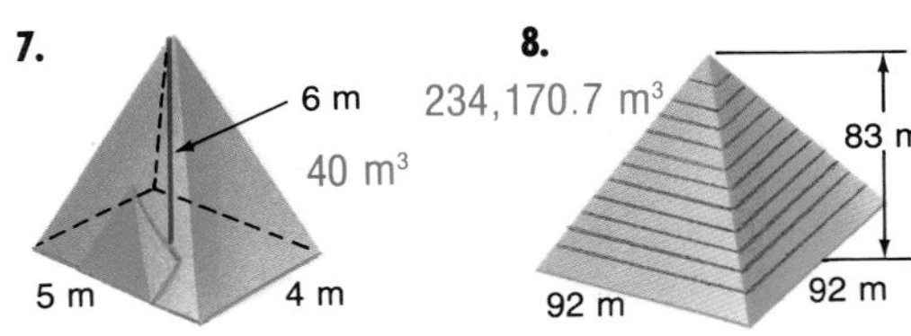

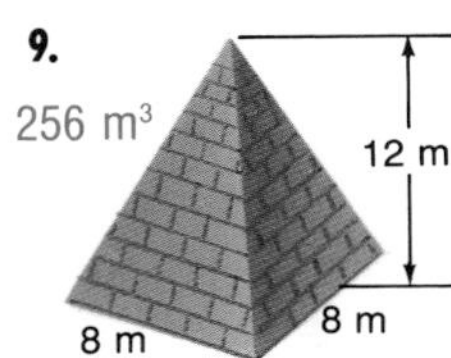

10. $\ell = 12$ m
$w = 6$ m
$h = 7$ m **168 m³**

11. $\ell = 4$ cm
$w = 5$ cm
$h = 2.4$ cm
16 cm³

12. $\ell = 9$ in.
$w = 10$ in.
$h = 12.3$ in.
369 in³

Applications

***f*UN with MATH**
Would you taste honey found in an Egyptian Pyramid?
See page 308.

13. The pyramid at the top of the Washington Monument has a square base 34 feet 5.5 inches on each side. The height of the pyramid is 55 feet. What is the volume of the pyramid to the nearest cubic foot?
21,769 ft³

14. Right Time Corporation produces a clock that is set in a clear acrylic pyramid. The clock has a square base 5 inches on each side and a height of 5 inches. It is shipped in a cube-shaped carton with inside dimensions of 5 inches by 5 inches by 5 inches. How many cubic inches of plastic filler are needed to fill the carton after the clock is placed inside? **125 in³ − 41.7 in³ or 83.3 in³**

Critical Thinking

15. If each dimension of a pyramid is doubled, how is the volume changed? **The volume is eight times greater.**

Using Equations

Solve for x, y, or z using the equation $x = yz$.

16. $x = 21$, $y = 7$, $z = \underline{?}$ **3**

17. $y = 5$, $z = 3$, $x = \underline{?}$ **15**

18. $z = 45$, $x = 180$, $y = \underline{?}$ **4**

19. $x = 1.2$, $z = 6$, $y = \underline{?}$ **0.2**

APPLYING THE LESSON

Independent Practice

Homework Assignment	
Minimum	**Ex. 1-19 odd**
Average	**Ex. 2-12 even; 13-19**
Maximum	**Ex. 1-19**

Chalkboard Examples

Example A *Find the volume of each pyramid.*

1.

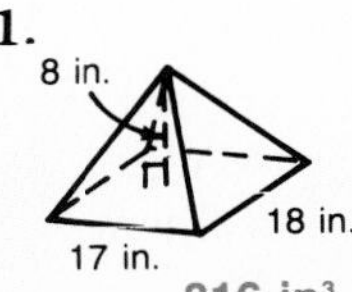

816 in³

2. $\ell = 14$ yd
$w = 12$ yd
$h = 10$ yd
560 yd³

Additional examples are provided on TRANSPARENCY 8-10.

EVALUATING THE LESSON

Check for Understanding Use the Guided Practice Exercises to check for student understanding.

Closing the Lesson Have students explain in their own words how to find the volume of a pyramid.

EXTENDING THE LESSON

Enrichment Have students research the formulas for area and volume that were known to the peoples of ancient Egypt.

Enrichment Masters Booklet, p. 83

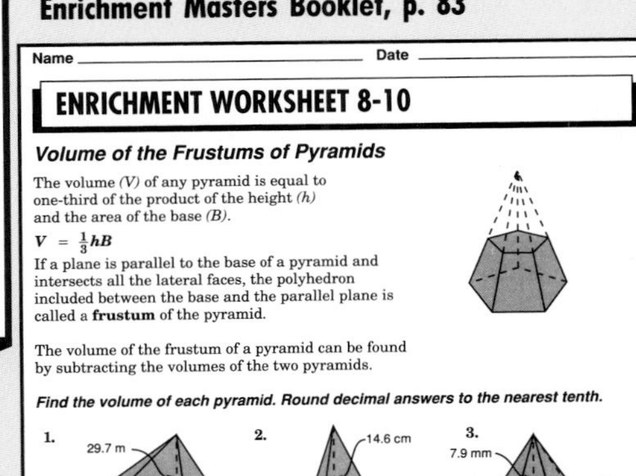

Name ______________________ Date ______________

ENRICHMENT WORKSHEET 8-10

Volume of the Frustums of Pyramids

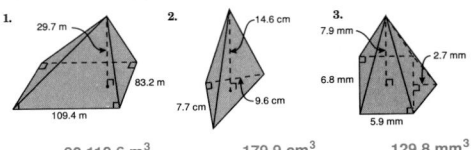

The volume *(V)* of any pyramid is equal to one-third of the product of the height *(h)* and the area of the base *(B)*.

$V = \frac{1}{3}hB$

If a plane is parallel to the base of a pyramid and intersects all the lateral faces, the polyhedron included between the base and the parallel plane is called a **frustum** of the pyramid.

The volume of the frustum of a pyramid can be found by subtracting the volumes of the two pyramids.

Find the volume of each pyramid. Round decimal answers to the nearest tenth.

1. 29.7 m, 83.2 m, 109.4 m — 90,110.6 m³

2. 14.6 cm, 7.7 cm, 9.6 cm — 179.9 cm³

3. 7.9 mm, 2.7 mm, 6.8 mm, 5.9 mm — 129.8 mm³

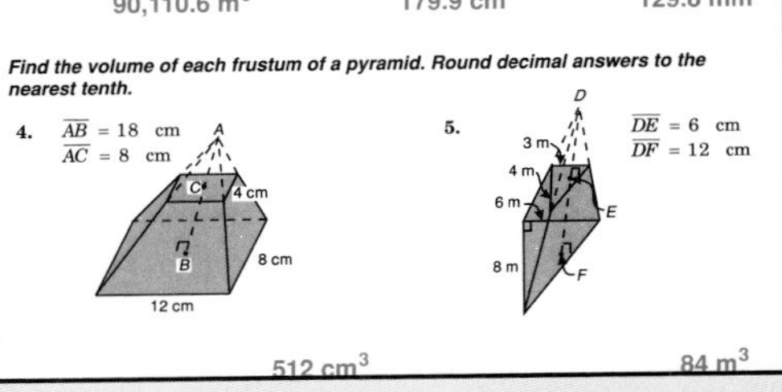

Find the volume of each frustum of a pyramid. Round decimal answers to the nearest tenth.

4. $\overline{AB} = 18$ cm, $\overline{AC} = 8$ cm; 4 cm, 8 cm, 12 cm — 512 cm³

5. $\overline{DE} = 6$ cm, $\overline{DF} = 12$ cm; 3 m, 4 m, 6 m, 8 m — 84 m³

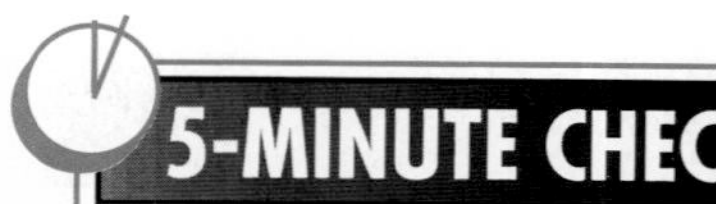

(over Lesson 8-10)

Available on TRANSPARENCY 8-11.

Find the volume of each pyramid.

1.

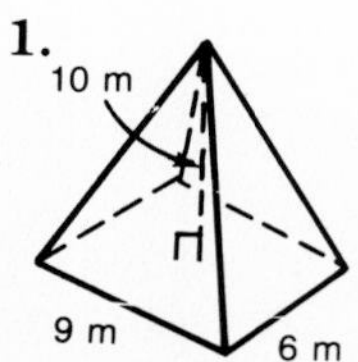

180 cm³

2. $\ell = 6$ mm
$w = 8$ mm
$h = 10$ mm
160 mm³

Extra Practice, Lesson 8-10, p. 471

▶ INTRODUCING THE LESSON

Ask students how many different sized cans they have seen in grocery stores. Ask students to describe how to find the volume of a cylinder without computing; for example, filling it with measures of a known volume.

▶ TEACHING THE LESSON

Using Discussion Discuss with students that the volume of a cylinder can be thought of as layers of the same circle stacked on top of each other.

Practice Masters Booklet, p. 92

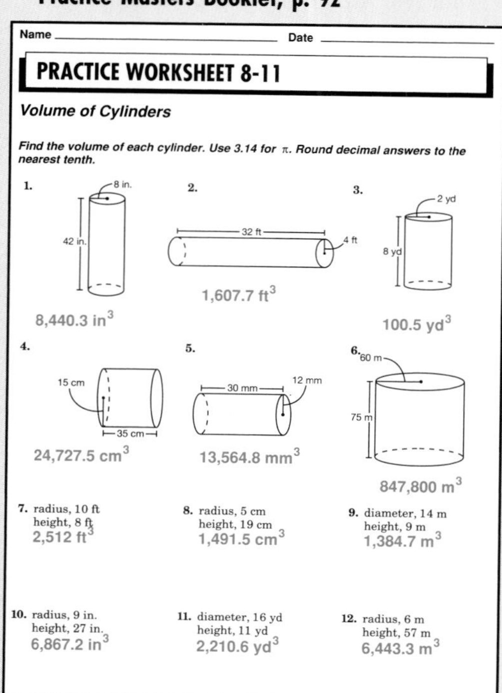
Name ____________ Date ____________

PRACTICE WORKSHEET 8-11

Volume of Cylinders

Find the volume of each cylinder. Use 3.14 for π. Round decimal answers to the nearest tenth.

1. 8 in.; 42 in. — 8,440.3 in³
2. 32 ft; 4 ft — 1,607.7 ft³
3. 2 yd; 8 yd — 100.5 yd³
4. 15 cm; 35 cm — 24,727.5 cm³
5. 30 mm; 12 mm — 13,564.8 mm³
6. 60 m; 75 m — 847,800 m³
7. radius, 10 ft height, 8 ft — 2,512 ft³
8. radius, 5 cm height, 19 cm — 1,491.5 cm³
9. diameter, 14 m height, 9 m — 1,384.7 m³
10. radius, 9 in. height, 27 in. — 6,867.2 in³
11. diameter, 16 yd height, 11 yd — 2,210.6 yd³
12. radius, 6 m height, 57 m — 6,443.3 m³

8-11 VOLUME OF CYLINDERS

Objective

Find the volume of cylinders.

Party Potato Chips packs their potato chips in cylindrical cartons. The new dip-style chip needs a container with *at least* 112 cubic inches of volume. Ginger Gomez's container design has a diameter of 4 inches and a height of 9 inches. Does the new container meet the requirement for volume?

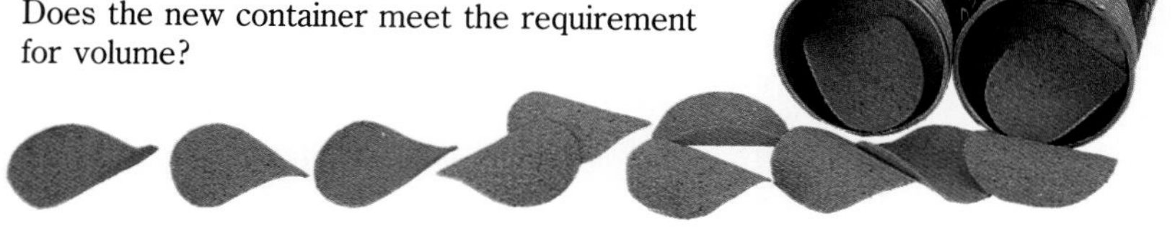

The volume of a cylinder is found by multiplying the measure of the area of the base by the height. Since the base is a circle, the area of the base is πr^2. So the formula is $V = \pi r^2 \times h$ or $V = \pi r^2 h$.

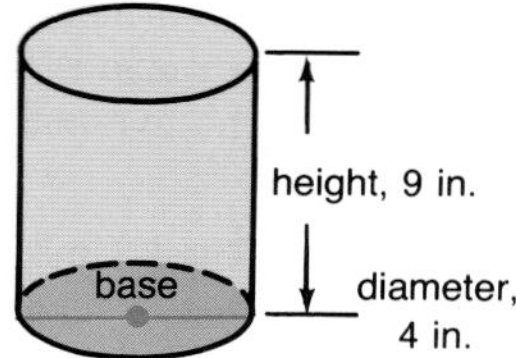

Example A ***Find the volume of the container Ms. Gomez designed. Use 3.14 for π.***

$V = \pi r^2 h$ $\qquad r = \frac{d}{2}$ or 2, $h = 9$

$V \approx 3.14 \times 2 \times 2 \times 9$

$V \approx 113.04$ or 113.04 in³

The volume of the cylinder is 113.0 cubic inches. Ms. Gomez's design meets the requirement for at least 112 cubic inches of volume.

Guided Practice

Practice ***Find the volume of each cylinder. Use 3.14 for π. Round decimal answers to the nearest tenth.***

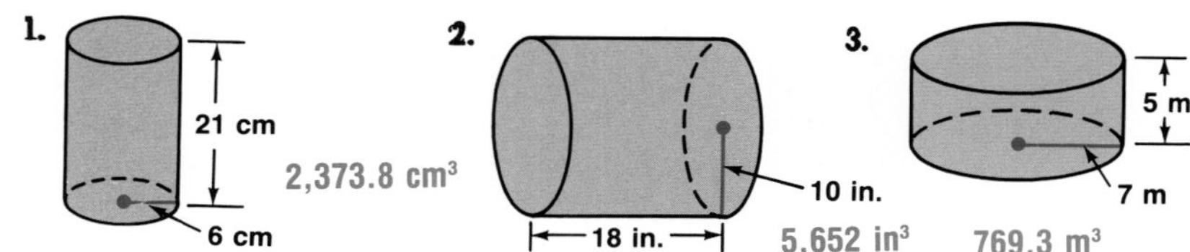

1. 2,373.8 cm³
2. 5,652 in³
3. 769.3 m³

4. radius = 2 in.
height = 5 in.
62.8 in³

5. diameter = 12 cm
height = 12 cm
1,356.5 cm³

6. radius = 9 ft
height = 5 ft
1,271.7 ft³

RETEACHING THE LESSON

Work with students to find the volume of each cylinder.

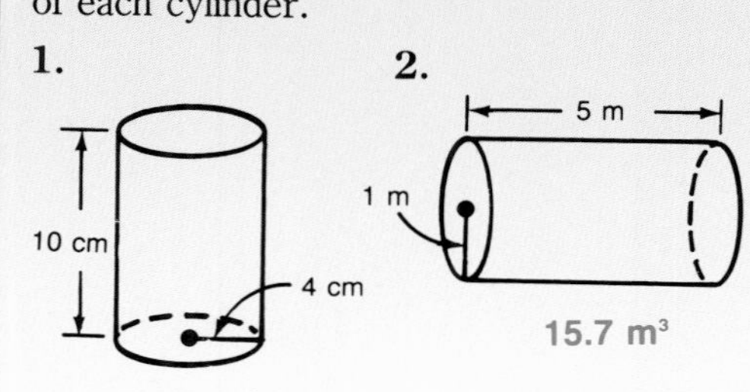

1. 502.4 cm³
2. 15.7 m³

Reteaching Masters Booklet, p. 84

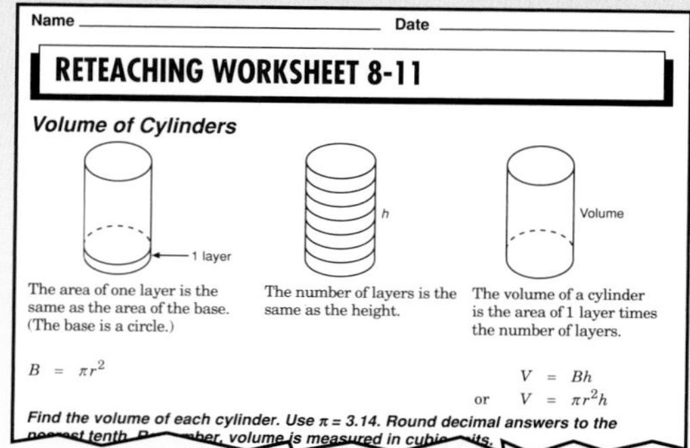
Name ____________ Date ____________

RETEACHING WORKSHEET 8-11

Volume of Cylinders

1 layer — The area of one layer is the same as the area of the base. (The base is a circle.) $B = \pi r^2$

h — The number of layers is the same as the height.

Volume — The volume of a cylinder is the area of 1 layer times the number of layers. $V = Bh$ or $V = \pi r^2 h$

Find the volume of each cylinder. Use π = 3.14. Round decimal answers to the nearest tenth. Remember, volume is measured in cubic units.

Exercises

Find the volume of each cylinder. Use 3.14 for π. Round decimal answers to the nearest tenth.

Practice

7. 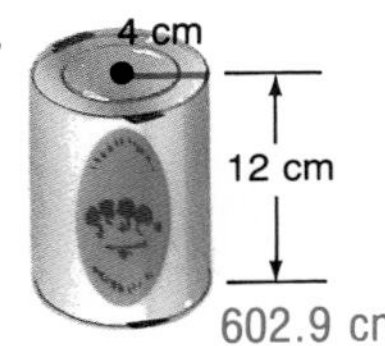

602.9 cm³

8. 

3,140 cm³

9. 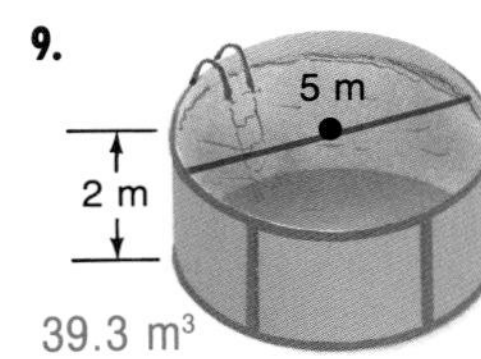

39.3 m³

10. radius = 20 mm
height = 20 mm
25,120 mm³

11. radius = 19 in.
height = 10 in.
11,335.4 in³

12. diameter = 14 cm
height = 17 cm
2,615.6 cm³

Applications

13. A storage silo on a farm has the shape of a cylinder with a diameter of 15 meters and a height of 20 meters. What is the volume of the silo? **≈3,532.5 m³**

14. Dottie's swimming pool has a diameter of 12 feet and a depth of 4 feet. If water costs $0.08 a cubic foot, by how much should Dottie expect her normal water bill to increase when she fills the pool?
≈452.2 ft³; $36.18

Critical Thinking

15. At Mystic Seaquarium, the starfish moves up its rock 2 inches each day and down 1 inch at night. At this rate how many days will it take the starfish to reach the top if it is 15 inches tall?
14 days

16. Draw a diagram of a cylinder with a height of 4 inches and a radius of 1 inch. Use the diagram to show the change in the volume if the height is cut in half. Describe the volume if the height is cut in half. **The volume is halved.**

17. Explain why it is always important to examine your answers when working with word problems. **See students' work.**

Mixed Review

Trace each angle and extend the sides. Then find the measure of each angle.

Lesson 7-2

18. 
65°

19. 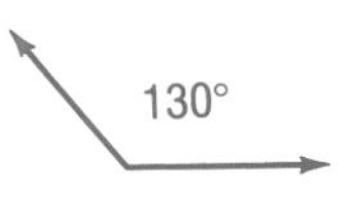
130°

20. 28°

Lesson 6-5

Complete.

21. 18 kL = __?__ L
18,000

22. 9.5 L = __?__ mL
9,500

23. 28 mL = __?__ kL
0.000028

Lesson 8-2

24. The walkway to Samuel's house is in the shape of a parallelogram. The width of the walkway near the house is 3 feet. The distance from the house to the sidewalk is 12 feet. What is the area of the walkway?
36 ft²

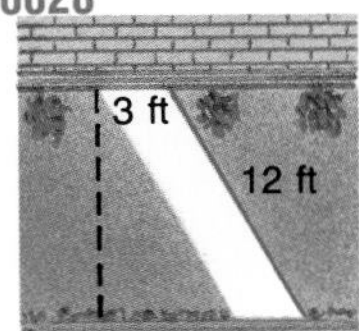

APPLYING THE LESSON

Independent Practice

Homework Assignment	
Minimum	Ex. 2-14 even; 15-24
Average	Ex. 1-11 odd; 13-24
Maximum	Ex. 1-24

Chalkboard Examples

Example A *Find the volume of each cylinder. Use 3.14 for π. Round decimal answers to the nearest tenth.*

1. 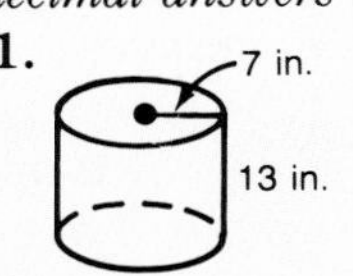

2,000.2 in³

2. diameter = 12 cm
height = 10 cm
1,130.4 cm³

Additional examples are provided on TRANSPARENCY 8-11.

▶ EVALUATING THE LESSON

Check for Understanding Use the Guided Practice Exercises to check for student understanding.

Closing the Lesson Have students explain in their own words why the formula for the volume of a cylinder is correct.

▶ EXTENDING THE LESSON

Enrichment Give students the following volumes and have them find dimensions for cylinders that might have the volumes: 285 ft³; 250 m³; 760 yd³. **Answers may vary. Sample answers are r = 3 ft, h = 10 ft; r = 2 m, h = 20 m; r = 11 yd, h = 2 yd**

Enrichment Masters Booklet, p. 84

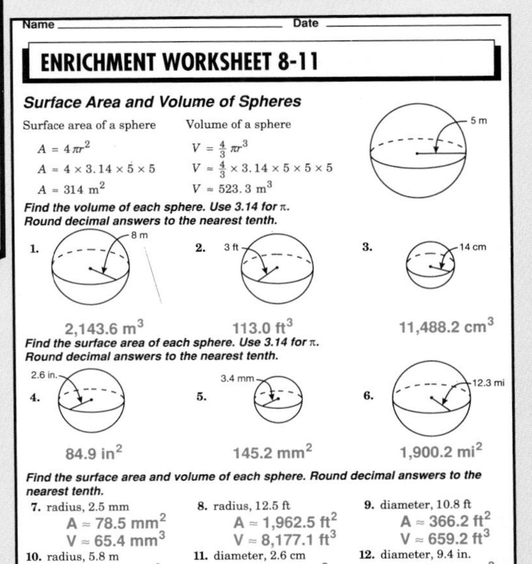

Name ________________ Date ________________

ENRICHMENT WORKSHEET 8-11

Surface Area and Volume of Spheres

Surface area of a sphere

$A = 4\pi r^2$

$A = 4 \times 3.14 \times 5 \times 5$

$A = 314 \text{ m}^2$

Volume of a sphere

$V = \frac{4}{3}\pi r^3$

$V \approx \frac{4}{3} \times 3.14 \times 5 \times 5 \times 5$

$V \approx 523.3 \text{ m}^3$

5 m

Find the volume of each sphere. Use 3.14 for π. Round decimal answers to the nearest tenth.

1. 8 m — **2,143.6 m³**
2. 3 ft — **113.0 ft³**
3. 14 cm — **11,488.2 cm³**

Find the surface area of each sphere. Use 3.14 for π. Round decimal answers to the nearest tenth.

4. 2.6 in. — **84.9 in²**
5. 3.4 mm — **145.2 mm²**
6. 12.3 mi — **1,900.2 mi²**

Find the surface area and volume of each sphere. Round decimal answers to the nearest tenth.

7. radius, 2.5 mm
A ≈ 78.5 mm²
V ≈ 65.4 mm³

8. radius, 12.5 ft
A ≈ 1,962.5 ft²
V ≈ 8,177.1 ft³

9. diameter, 10.8 ft
A ≈ 366.2 ft²
V ≈ 659.2 ft³

10. radius, 5.8 m
A ≈ 422.5 m²
V ≈ 816.9 m³

11. diameter, 2.6 cm
A ≈ 21.2 cm²
V ≈ 9.2 cm³

12. diameter, 9.4 in.
A ≈ 277.5 in²
V ≈ 434.7 in³

13. radius, 9.3 in.
A ≈ 1,086.3 in²
V ≈ 3,367.6 in³

14. diameter, 2.8 m
A ≈ 24.6 m²
V ≈ 11.5 m³

15. radius, 8.7 m
A ≈ 950.7 m²
V ≈ 2,756.9 m³

16. radius, 1.3 in.
A ≈ 21.2 in²
V ≈ 9.2 in³

17. diameter, 6.8 mi
A ≈ 145.2 mi²
V ≈ 164.6 mi³

18. diameter, 3.6 yd
A ≈ 40.7 yd²
V ≈ 24.4 yd³

5-MINUTE CHECK-UP

(over Lesson 8-11)

Available on TRANSPARENCY 8-12.

Find the volume of each cylinder. Use 3.14 for π. Round decimal answers to the nearest tenth.

1. 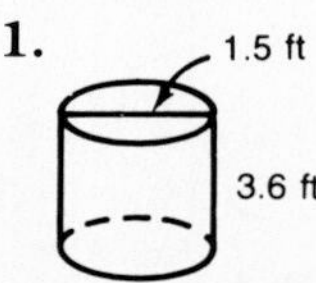

6.4 ft^3

2. radius = 5.2 m
height = 3 m
254.7 m^3

Extra Practice, Lesson 8-9, p. 470

INTRODUCING THE LESSON

Ask students how many different kinds of ice cream novelties they have had that are shaped like cones. Ask students to suggest other things that are shaped like cones.

TEACHING THE LESSON

Using Discussion Discuss with students the importance of using the radius in the computation and not the diameter. The formula uses the area of the circular base and the area of a circle is based on π and the radius.

Practice Masters Booklet, p. 93

Name ______________________ Date ______________

PRACTICE WORKSHEET 8-12

Volume of Cones

Find the volume of each cone described below. Round decimal answers to the nearest tenth.

1. 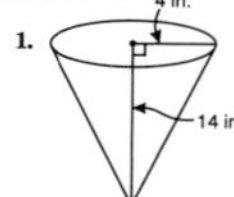

234.5 in^3

2. 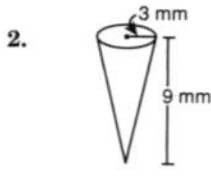

84.8 mm^3

3. 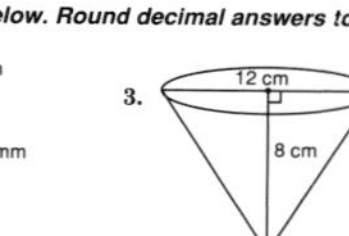

301.4 cm^3

4. 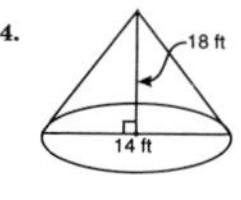

923.2 ft^3

5. 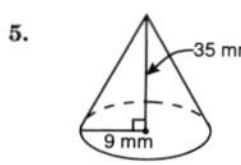

2,967.3 mm^3

6. 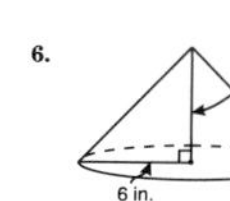

565.2 in^3

7. radius, 12 ft
height, 8 ft
1,205.8 ft^3

8. diameter, 28 m
height, 9 m
1,846.3 m^3

9. radius, 10 in.
height, 25 in.
2,616.7 in^3

10. diameter, 16 yd
height, 11 yd
736.9 yd^3

11. diameter, 10 cm
height, 5 cm
130.8 cm^3

12. radius, 27 cm
height, 19 cm
14,497.4 cm^3

8-12 VOLUME OF CONES

Objective

Find the volume of cones.

Steve works at the snow cone stand at Wyandot Park. The cone he fills with shaved ice is 6 inches tall and has a radius of 2 inches at the top. How many cubic inches of ice are needed to fill the cone?

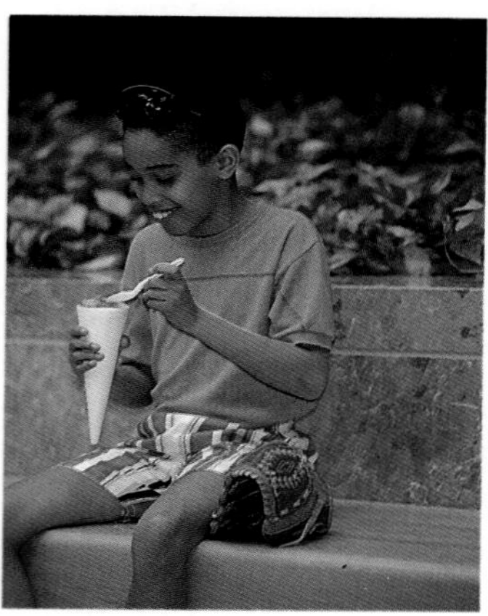

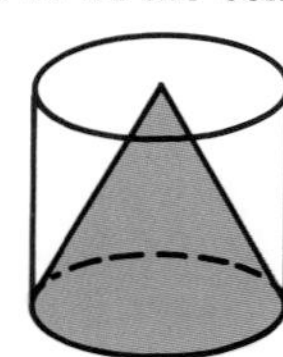

The volume of a cone is one-third the volume of a cylinder with the same radius and height as the cone. The formula is $V = \frac{1}{3}\pi r^2 h$.

Example A

Find the volume of the snow cone that Steve fills with shaved ice. Use 3.14 for π.

$V = \frac{1}{3}\pi r^2 h$ $r = 2, h = 6$

$V \approx \frac{1}{3} \times 3.14 \times 2 \times 2 \times 6$

$V \approx 25.12$ or 25.12 in^3

To multiply by $\frac{1}{3}$ on a calculator, you can divide by 3.

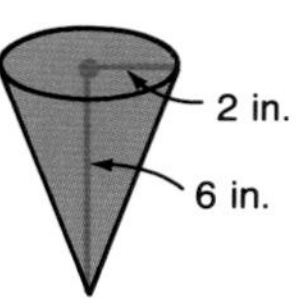

The volume of the cone is *about* 25 cubic inches.

Guided Practice

Example A

Find the volume of each cone. Use 3.14 for π. Round decimal answers to the nearest tenth.

1. 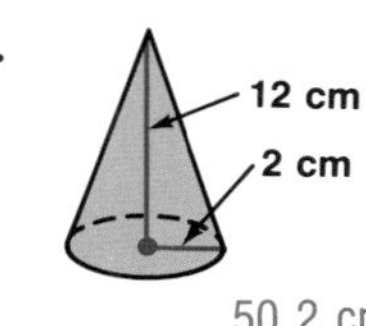

50.2 cm^3

2. 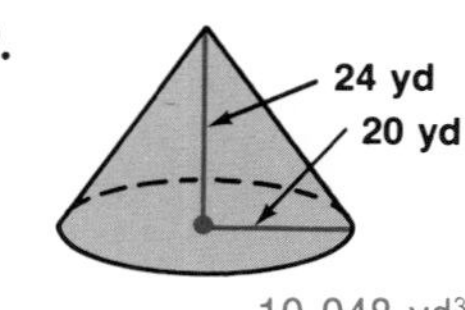

10,048 yd^3

3. 9 m, 4 m
150.7 in^3

4. radius = 8 m
height = 6 m
401.9 m^3

5. radius = 2 mm
height = 24 mm
100.5 mm^3

6. diameter = 12 in.
height = 10 in.
376.8 in^3

RETEACHING THE LESSON

Have students work together in small groups. Discuss how to use the formula $V = \frac{1}{3}\pi r^2 h$. Find the volume.

1. 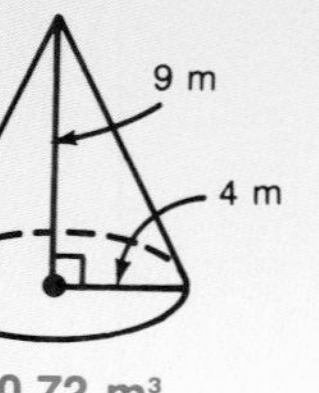

150.72 m^3

2. 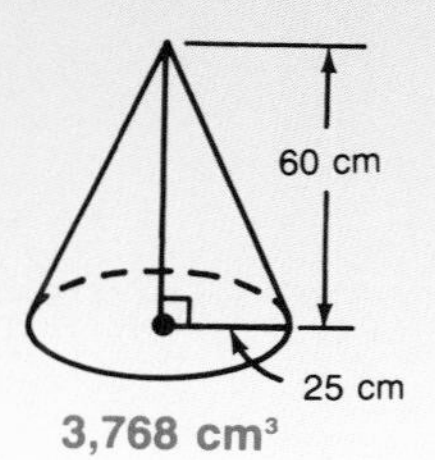

3,768 cm^3

Reteaching Masters Booklet, p. 85

Name ______________________ Date ______________

RETEACHING WORKSHEET 8-12

Volume of Cones

The volume of a cone is one-third the volume of a cylinder with the same radius and height as the cone.

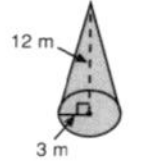

$V = \frac{1}{3}\pi r^2 h$

$V = \frac{1}{3} \times 3.14 \times 3 \times 3 \times 12$

$V = 113.04$ The volume is about 113.04 m^3.

Find the volume of each cone. Some steps are given. Complete the remaining steps. Use 3.14 for π. Round decimal answers to the nearest tenth.

1. $V = \frac{1}{3} \times \pi \times r \times r \times h$
$V = \frac{1}{3} \times 3.14 \times 3 \times 3 \times 7.1$
$V = 66.9$ m^3

2. $V = \frac{1}{3} \times \pi \times r \times r \times h$
$V = \frac{1}{3} \times$ 3.14 $\times$ 0.5 $\times$ 0.5 $\times$ 9
$V = 2.4$ m^3

Exercises

Find the volume of each cone. Use 3.14 for π. Round decimal answers to the nearest tenth.

Practice

7. 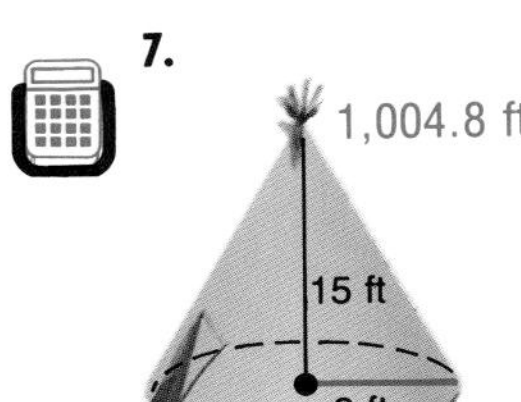

8. 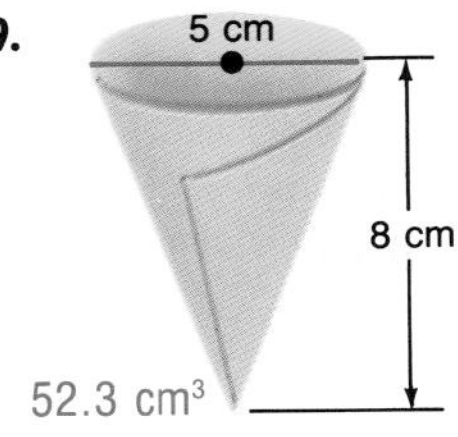

9. 5 cm, 8 cm, **52.3 cm³**

10. radius = 6 m
 height = 6 m
 226.1 m³

11. radius = 4 mm
 height = 24 mm
 401.9 mm³

12. diameter = 5 in.
 height = 10 in.
 65.4 in³

Applications

13. If the top of a silo is cone-shaped and has a radius of 12 feet and a height of 18 feet, what is the volume of the cone-shaped portion of the silo? **2,713 ft³**

14. Goody's Ice Cream Novelties makes an ice cream cone from a sugar cone filled with vanilla ice cream. Then the cone is dipped in chocolate. Each cone is 5 inches high and has a diameter of 2 inches across the top. Find the volume of the cone. **5.2 in³**

Problem Solving

15. A cone has a volume of 376.8 cubic centimeters. If the radius of the cone is 6 centimeters, what is the height of the cone? **10 cm**

Talk Math

16. How does the volume of a cone change if the height is cut in half? What happens to the volume of a cone if the height is doubled? **The volume is halved. The volume is doubled.**

Critical Thinking

17. If four days before tomorrow is Sunday, what is three days after yesterday? **Friday**

Cooperative Groups

18. In groups of three or four, use a clean, empty soup can or a cylindrical drinking glass, a piece of construction paper, and some sand or salt to prove that the volume of a cone is one-third the volume of a cylinder with the same height and base. Roll the construction paper into a cone shape with a sharp point. Place the point down into the can or glass. The base of the cone must have the same diameter as the top of the soup can. Mark the paper cone on the inside at the height of the soup can. Fill the cone with sand or salt and pour it into the can. How many conefuls does it take to fill the can? Explain your findings. **3 conefuls**

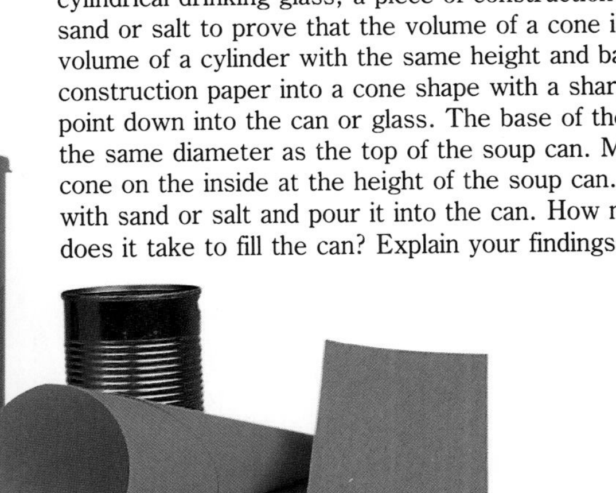

APPLYING THE LESSON

Independent Practice

Homework Assignment	
Minimum	Ex. 1-15 odd; 17-18
Average	Ex. 2-12 even; 13-18
Maximum	Ex. 1-18

Chalkboard Examples

Example A *Find the volume of each cone. Use 3.14 for π. Round decimal answers to the nearest tenth.*

1. 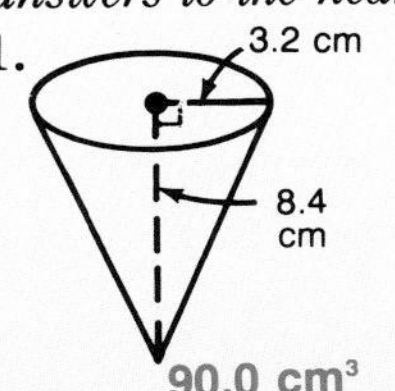

90.0 cm³

2. diameter = 2 mm
 height = 30 mm
 31.4 mm³

Additional examples are provided on TRANSPARENCY 8-12.

EVALUATING THE LESSON

Check for Understanding Use Guided Practice Exercises to check for student understanding.

Error Analysis Watch for students who fail to divide by three when computing the volume of a cone.

Closing the Lesson Ask students to state the relationship between the volume of a cylinder and the volume of a cone.

EXTENDING THE LESSON

Enrichment Masters Booklet, p. 85

Name ____________ Date ____________

ENRICHMENT WORKSHEET 8-12

Volume

The diameter of the sphere below is 14 cm. A regular hexagonal* pyramid is inscribed in the upper half of the sphere. The height and each side of the base of the pyramid have a measure equal to the length of the radius. A cone is inscribed in the lower half of the sphere.

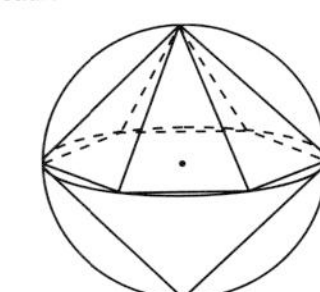

* Note: The area of a hexagon can be found by dividing it into 6 equilateral triangles. The height of one of the equilateral triangles is approximately equal to 0.87 times the radius.

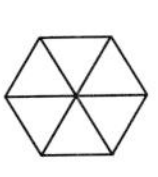

Find each of the following for the figure above. Use 3.14 for π. Round decimal answers to the nearest tenth.

1. the volume of the sphere **1,436.0 cm³**

2. the volume of the pyramid **298.4 cm³**

3. the volume of the cone **359.0 cm³**

4. the surface area of the sphere **615.4 cm²**

5. the volume of the smallest cylinder that can contain the sphere **2,154.0 cm³**

5-MINUTE CHECK-UP

(over Lesson 8-12)

Available on TRANSPARENCY 8-13.

Find the volume of each cone. Use 3.14 for π. Round decimal answers to the nearest tenth.

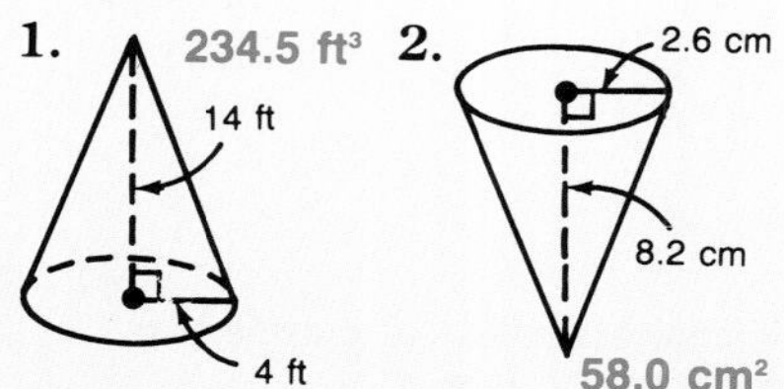

Extra Practice, Lesson 8-12, p. 471

INTRODUCING THE LESSON

Have students suggest real-life situations or careers that involve using area and volume.

TEACHING THE LESSON

Using Questioning

- When finding the area of a figure, how is the answer labeled? **square units**
- When finding the volume of a figure, how is the answer labeled? **cubic units**

Practice Masters Booklet, p. 94

Name ______________________ Date ______________

PRACTICE WORKSHEET 8-13

Problem Solving: Using Area and Volume

Solve. Round decimal answers to the nearest tenth.

1. Rita is planting English ivy. She needs 6 plants to cover 10 square feet of ground. How many plants does she need to cover a rectangular area 20 feet long and 15 feet wide? **180 plants**

2. Twelve boxes of detergent are to be placed in a carton. Each box is 8 in. by 3 in. by 11 in. How much space must the carton contain? Give possible dimensions of the carton. **3,168 in^3; possible dimensions: 16 in. by 9 in. by 22 in.**

3. A circular swimming pool is to be dug It is to have a diameter of 20 ft and a depth of 6 ft. How much dirt must be removed? **1,884 ft^3**

4. The weather service issued a severe storm warning for all counties within a 50-mile radius of Plainview. What is the area covered by the warning? **7,850 mi^2**

5. A cord of wood is equivalent to 128 cubic feet and is usually described as 4 ft by 4 ft by 8 ft. Herman helps his dad cut wood, which they sell. They have a stack 16 ft by 16 ft by 12 ft. How may cords of wood do they have ready for sale? **24 cords**

6. 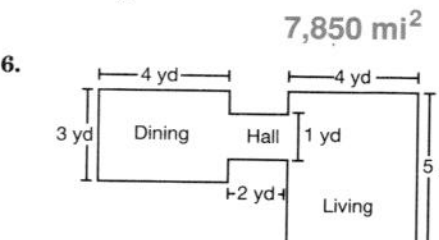

The dining, living, and hall areas are to be carpeted. How much will it cost if the carpet is priced at 412.89 per square yard? **$438.26**

Explore · Plan · Solve · Examine

8-13 USING AREA AND VOLUME

Objective

Solve problems using area and volume.

Alicia has 160 feet of fencing to place around the rectangular garden she is making. She wants the area of the garden to be as large as possible. What should the length and width of the garden be?

Explore

What is given?

- the perimeter of the garden, 160 feet

What is asked?

- With a perimeter of 160 feet, what width and length will make the largest area?

Plan

Draw various rectangles with a perimeter of 160 feet. Then find the area of each rectangle.

If the rectangles have different areas, look for a pattern to determine the greatest area.

Solve

Rectangles with a perimeter of 160 feet include:

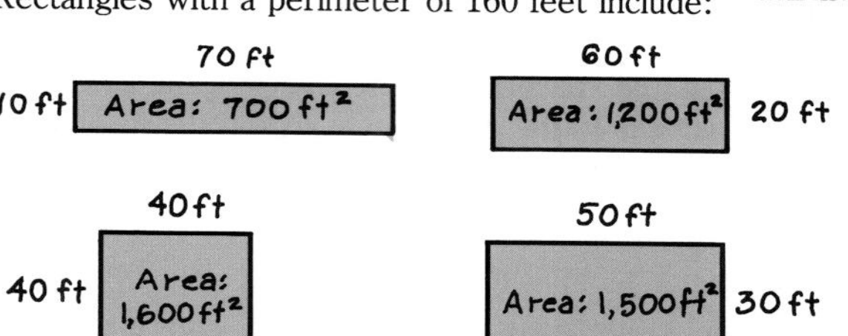

40 ft
40 ft
Area: 1,600 ft²

50 ft
Area: 1,500 ft²
30 ft

A 40-ft × 40-ft garden with 1,600 square feet of garden space seems to have the greatest area for 160 feet of fence.

Examine

Here's another way to solve the problem. Length plus width must always be 80 feet. But length times width varies.

80 = 1 + 79	1 × 79 = 79
80 = 2 + 78	2 × 78 = 156
80 = 3 + 77	3 × 77 = 231
⋮	⋮
80 = 39 + 41	39 × 41 = 1,599
80 = 40 + 40	40 × 40 = 1,600
80 = 41 + 39	41 × 39 = 1,599

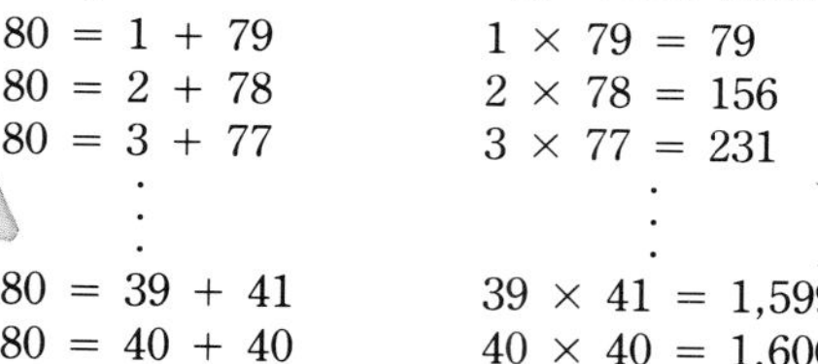

The greatest product is 1,600. So the answer is correct.

RETEACHING THE LESSON

Have students work in pairs. Discuss formulas necessary to solve the Exercises on page 271. Then, have students solve the problems independently.

Reteaching Masters Booklet, p. 86

Name ______________________ Date ______________

RETEACHING WORKSHEET 8-13

Problem Solving: Using Area and Volume

A hole with a 2-in. radius is drilled through a 5-in. cube of wood. What is the volume of the part of the cube that is left?

2 in.
5 in.
5 in.
5 in.

Explore What is given?
- radius of hole, 2 in.; height of hole, 5 in.
- length, width, and height of cube, 5 in.

What is asked?
- What is the volume of the part of the cube that is left?

Plan A hole in the shape of a cylinder is drilled through a cube.

Solve Subtract the volume of the cylinder from the volume of the cube.

$V = (l \times w \times h) - (\pi \times r \times r \times h)$

$V = (5 \times 5 \times 5) - (3.14 \times 2 \times 2 \times 5)$

Guided Practice

Solve. Round decimal answers to the nearest tenth.

1. Kayray's department store provides free delivery within a 60-mile radius of the store. What is the area covered by free delivery? **11,304 mi²**

Problem Solving

Paper/ Pencil
Mental Math
Estimate
Calculator

Solve. Round decimal answers to the nearest tenth.

2. Neil plants 12 tomato plants. Each plant will have 1 square foot of area. If the bed for the tomatoes is 2 feet wide, how long should the bed be? **6 ft**

1 in.
1 in.

3. Find the area of the figure shown on the geoboard. **2.5 in²**

4. If it takes 25 centimeters of speaker wire to go around a spool once, what is the diameter of the spool? ***about* 8 cm**

5. Kate is making a circular strawberry patch three tiers high. The radius of the center circle is 1.5 feet, the radius of the inner circle is 3 feet, and the radius of the outer circle is 6 feet. If each tier is 0.5 feet high, how many cubic feet of garden soil should Kate buy to fill the tiers? **74.2 ft³**

6. Draw a diagram of a 4-foot by 6-foot rectangle on a piece of paper. Choose 5 different vegetables or flowers to plant and separate the garden into 5 areas for each vegetable or flower. Find the area of each section of your garden and make sure that the five areas total the area of the whole garden. **See students' work.**

7. Select an assignment from this chapter that shows your favorite mathematical artwork. **See students' work.**

Mixed Review

Lesson 4-9

Estimate.

8. $\frac{4}{5} + \frac{7}{8}$ **2**
9. $\frac{1}{6} + \frac{3}{7}$ **$\frac{1}{2}$**
10. $\frac{7}{8} - \frac{1}{4}$ **$\frac{1}{2}$**
11. $\frac{8}{15} - \frac{2}{9}$ **$\frac{1}{4}$**

Lesson 8-3

12. Eli has a garden that is triangular in shape. Eli wants to fertilize the garden with fertilizer that comes in bags that cover 500 square feet. How many bags of fertilizer does he need to buy? **2 bags**

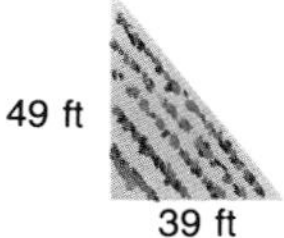

APPLYING THE LESSON

Independent Practice

Homework Assignment	
Minimum	**Ex. 2-6 even; 7-12**
Average	**Ex. 1-12**
Maximum	**Ex. 1-12**

Chapter 8, Quiz B (Lessons 8-8 through 8-13) is available in the Evaluation Masters Booklet, p. 39.

Chalkboard Examples

1. Carrie builds a rectangular bird feeder with a length of 12 inches and a width of 9 inches. If Carrie wants the feeder to hold 500 cubic inches of seed, how tall should the feeder be? **4.7 inches**

Additional examples are provided on TRANSPARENCY 8-13.

EVALUATING THE LESSON

Check for Understanding Use Guided Practice Exercises to check for student understanding.

Closing the Lesson Have students give examples of real-life problems they have solved using area and volume.

EXTENDING THE LESSON

Enrichment Have students make up two problems. The answers must be 100 square feet and 500 cubic feet.

Enrichment Masters Booklet, p. 86

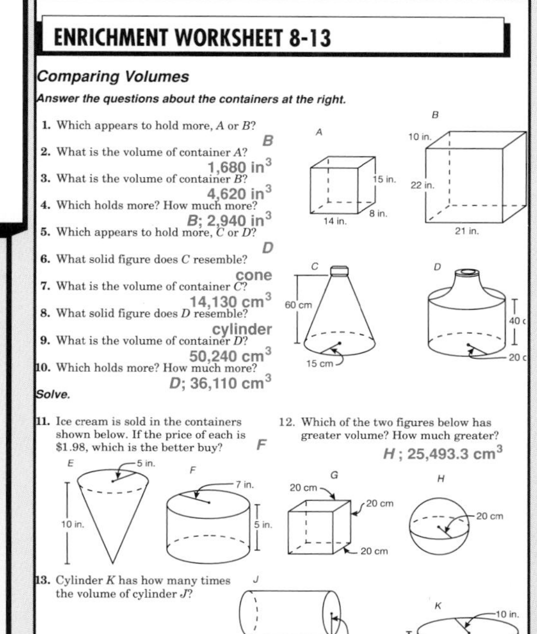

ENRICHMENT WORKSHEET 8-13

Comparing Volumes

Answer the questions about the containers at the right.

1. Which appears to hold more, *A* or *B*? ***B***
2. What is the volume of container *A*? **1,680 in³**
3. What is the volume of container *B*? **4,620 in³**
4. Which holds more? How much more? ***B*; 2,940 in³**
5. Which appears to hold more, *C* or *D*? ***D***
6. What solid figure does *C* resemble? **cone**
7. What is the volume of container *C*? **14,130 cm³**
8. What solid figure does *D* resemble? **cylinder**
9. What is the volume of container *D*? **50,240 cm³**
10. Which holds more? How much more? ***D*; 36,110 cm³**

Solve.

11. Ice cream is sold in the containers shown below. If the price of each is $1.98, which is the better buy? ***F***

12. Which of the two figures below has greater volume? How much greater? ***H*; 25,493.3 cm³**

13. Cylinder *K* has how many times the volume of cylinder *J*? **$3\frac{1}{3}$ times**

Using the Chapter Review

The Chapter Review is a comprehensive review of the concepts presented in this chapter. This review may be used to prepare students for the Chapter Test.

Quiz A should be given after students have completed Lessons 8-1 through 8-7. Quiz B should be given after students have completed Lessons 8-8 through 8-13.

Chapter 8, Quizzes A and B, are provided in the Evaluation Masters Booklet as shown below.

These quizzes can be used to obtain a quiz score or as a check of the concepts students need to review.

Evaluation Masters Booklet, p. 39

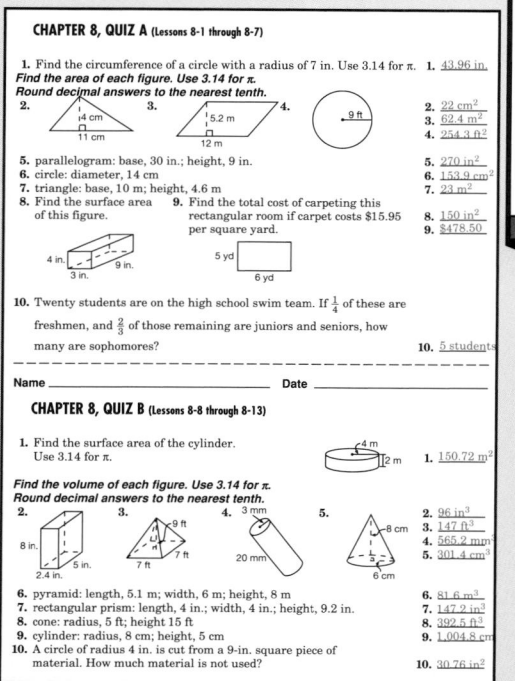

CHAPTER 8, QUIZ A (Lessons 8-1 through 8-7)

1. Find the circumference of a circle with a radius of 7 in. Use 3.14 for π. 1. 43.96 in.

Find the area of each figure. Use 3.14 for π.
Round decimal answers to the nearest tenth.

2. 3. 4.

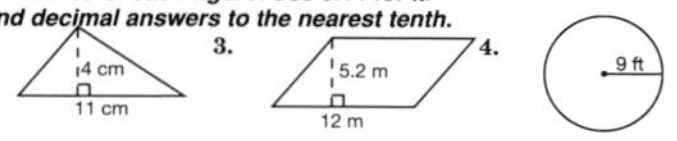

2. 22 cm^2
3. 62.4 m^2
4. 254.3 ft^2

5. parallelogram: base, 30 in.; height, 9 in. 5. 270 in^2
6. circle: diameter, 14 cm 6. 153.9 cm^2
7. triangle: base, 10 m; height, 4.6 m 7. 23 m^2
8. Find the surface area of this figure.
9. Find the total cost of carpeting this rectangular room if carpet costs \$15.95 per square yard.

8. 150 in^2
9. \$478.50

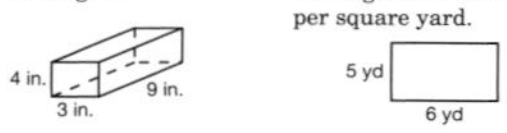

10. Twenty students are on the high school swim team. If $\frac{1}{4}$ of these are freshmen, and $\frac{2}{3}$ of those remaining are juniors and seniors, how many are sophomores? 10. 5 students

Name ________________ Date ________________

CHAPTER 8, QUIZ B (Lessons 8-8 through 8-13)

1. Find the surface area of the cylinder. Use 3.14 for π.

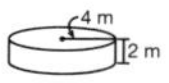

1. 150.72 m^2

Find the volume of each figure. Use 3.14 for π.
Round decimal answers to the nearest tenth.

2. 3. 4. 5.

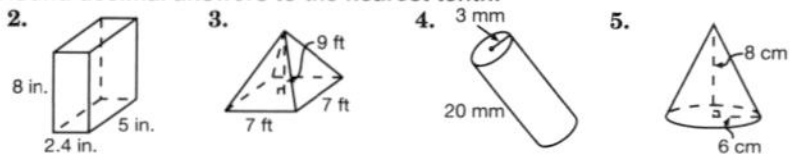

2. 96 in^3
3. 147 ft^3
4. 565.2 mm^3
5. 301.4 cm^3

6. pyramid: length, 5.1 m; width, 6 m; height, 8 m 6. 81.6 m^3
7. rectangular prism: length, 4 in.; width, 4 in.; height, 9.2 in. 7. 147.2 in^3
8. cone: radius, 5 ft; height 15 ft 8. 392.5 ft^3
9. cylinder: radius, 8 cm; height, 5 cm 9. 1,004.8 cm^3
10. A circle of radius 4 in. is cut from a 9-in. square piece of material. How much material is not used? 10. 30.76 in^2

Chapter 8 REVIEW

Vocabulary / Concepts

Write the letter of the term that best completes each sentence.

1. The distance around a circle is the _?_. b
2. Volume is expressed in _?_ units. d
3. The ratio of the measure of the circumference of a circle to the measure of its diameter is _?_. j
4. To find the area of a parallelogram you multiply the _?_ times the _?_. a, g
5. Two _?_ triangles form a parallelogram. c
6. To find the area of a circle you multiply _?_ times the _?_. j, m
7. The surface area of a rectangular prism is the sum of the area of its _?_. f
8. The volume of a rectangular prism is found by multiplying the measures of the _?_, _?_ and _?_. h, o, g
9. The volume of a _?_ is one-third the volume of a rectangular prism with the same base and height. k
10. The volume of a cylinder is found by _?_ the measure of the area of the base times the height. i

a. base
b. circumference
c. congruent
d. cubic
e. cylinder
f. faces
g. height
h. length
i. multiplying
j. π
k. pyramid
l. radius
m. radius squared
n. rectangular prism
o. width

Exercises / Applications

Lesson 8-1

Find the circumference of each circle. Use 3.14 for π. Round decimal answers to the nearest tenth.

11.

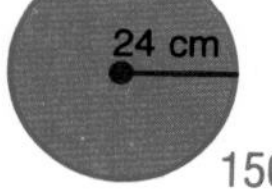

150.7 cm

12.

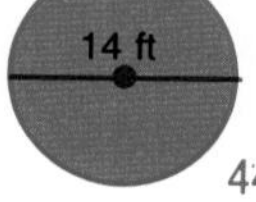

44 ft

13.

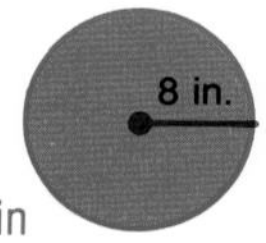

50.2 in

Lessons 8-2, 8-3, 8-4

Find the area of each figure. Use 3.14 for π.

14.

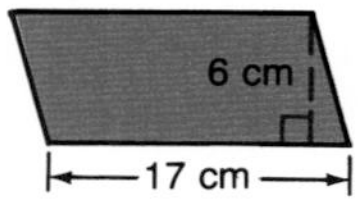

102 cm^2

15.

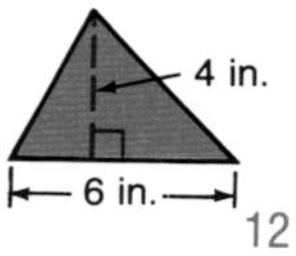

12 in^2

16.

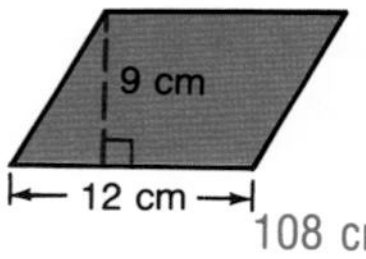

108 cm^2

17.

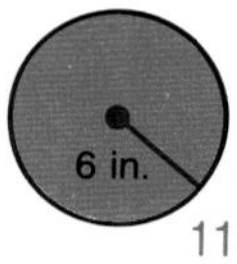

113.0 in^2

18.

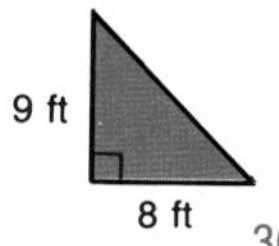

36 ft^2

19.

3215.4 cm^2

Lesson 8-5

20. Find the total cost of carpet for a room that is 15 feet long and 12 feet wide if the carpet includng padding costs \$22.95 a square yard? **\$459**

Lesson 8-6

21. Chang is reading a choose-your-own mystery book. At the end of the first chapter, there are two different chapter 2s from which to choose. At the end of each of the two different chapter 2s, there are two different chapter 3s from which to choose, and so on. How many different chapter 4s are there in the book? **8 chapter 4s**

Lessons 8-7, 8-8

Find the surface area of each rectangular prism or cylinder. Use 3.14 for π. Round decimal answers to the nearest tenth.

22. 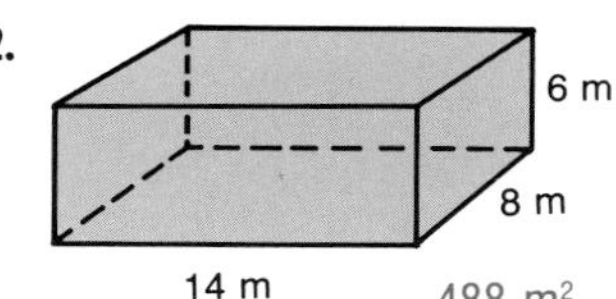

488 m²

23. 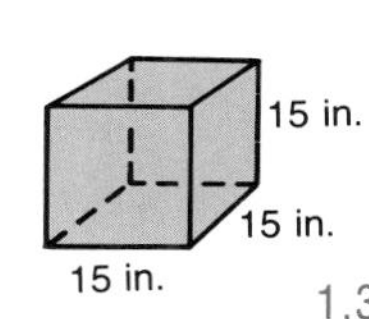

1,350 in²

24. 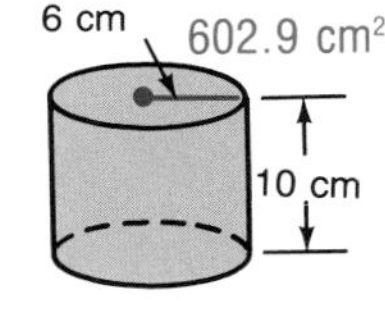

602.9 cm²

Lessons 8-9, 8-10, 8-11, 8-12

Find the volume of each rectangular prism, pyramid, cylinder, or cone. Use 3.14 for π. Round decimal answers to the nearest tenth.

25. rectangular prism
$\ell = 7$ cm, $w = 4$ cm, $h = 3$ cm **84 cm³**

26. pyramid
$\ell = 8$ in., $w = 10$ in., $h = 9$ in. **240 in³**

27. cylinder
$r = 3$ ft, $h = 7$ ft **197.8 ft³**

28. cone
$r = 11$ cm, $h = 12$ cm **1,519.8 cm³**

29. rectangular prism
$\ell = 10$ ft, $w = 8$ ft, $h = 5$ ft **400 ft³**

30. cylinder
$d = 4$ m, $h = 6$ m **75.4 m³**

Lesson 8-13

31. Applegate Farm ships milk to market daily in a tanker truck that is 23 feet long and has a diameter of 10 feet. How many gallons of milk do they ship in a full tanker if there are about 7.5 gallons in a cubic foot? ***about* 13,541 gal of milk**

32. Find the area of the piece of wood shown at the right. **2,280 cm²**

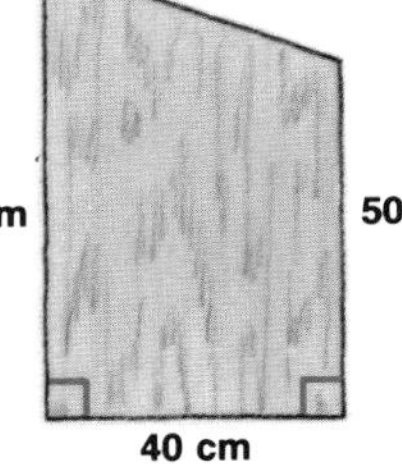

APPLYING THE LESSON

Independent Practice

Homework Assignment	
Minimum	Ex. 1-10; 12-32 even
Average	Ex. 1-32
Maximum	Ex. 1-32

Alternate Assessment Strategy

To provide a brief in-class review, you may wish to read the following questions to the class and require a verbal response.

1. What is the circumference of a circle that has a radius of 3 in.? **18.84 in.**
2. Find the area of a parallelogram with base of 5 mm and height of 4 mm. **20 mm²**
3. If the base of a triangle is 10 cm and the height is 8 cm, what is the area? **40 cm²**
4. What is the area of a circle with a radius of 10 mm? **314 mm²**
5. Find the surface area of a rectangular prism with length of 4 cm, width of 2 cm, and height of 10 cm. **136 cm²**
6. What is the surface area of a cylinder if the area of its bases are 48 cm² and the area of the curved surface equals 92 cm²? **140 cm²**
7. Find the volume of a rectangular prism that has a length of 4 mm, width of 12 mm and a base of 5 mm. **240 mm³**
8. Find the volume of a pyramid with length of 9 m, height of 6 m and width of 4 m. **72 m³**
9. What is the formula for finding the volume of a cylinder. $V = \pi r^2 h$
10. Find the volume of a cone with a radius of 6 in. and height of 5 in. **188.4 in³**

Using the Chapter Test

This page may be used as a test or as an additional page of review if necessary. Two forms of a Chapter Test are provided in the Evaluation Masters Booklet. Form 2 (free response) is shown below, and Form 1 (multiple choice) is shown on page 277.

The **Tech Prep Applications Booklet** provides students with an opportunity to familiarize themselves with various types of technical vocations. The Tech Prep applications for this chapter can be found on pages 15–16.

The **Performance Assessment Booklet** provides an alternative assessment for evaluating student progress. An assessment for this chapter can be found on pages 15–16.

Evaluation Masters Booklet, p. 38

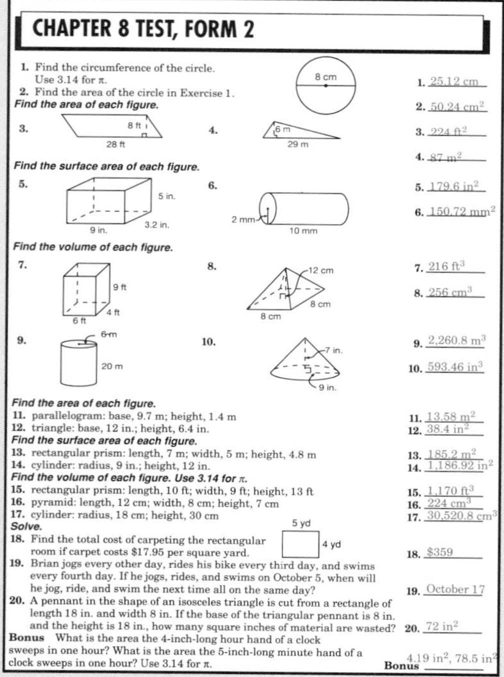

CHAPTER 8 TEST, FORM 2

1. Find the circumference of the circle. Use 3.14 for π.
2. Find the area of the circle in Exercise 1.

Find the area of each figure.

3. (parallelogram: 8 ft, 28 ft)
4. (triangle: 6 m, 29 m)

Find the surface area of each figure.

5. (rectangular prism: 5 in., 3.2 in., 9 in.)
6. (cylinder: 2 mm, 10 mm)

Find the volume of each figure.

7. (rectangular prism: 9 ft, 4 ft, 6 ft)
8. (pyramid: 12 cm, 8 cm, 8 cm)
9. (cylinder: 6 m, 20 m)
10. (cone: 7 in., 9 in.)

Find the area of each figure.

11. parallelogram: base, 9.7 m; height, 1.4 m
12. triangle: base, 12 in.; height, 6.4 in.

Find the surface area of each figure.

13. rectangular prism: length, 7 m; width, 5 m; height, 4.8 m
14. cylinder: radius, 9 in.; height, 12 in.

Find the volume of each figure. Use 3.14 for π.

15. rectangular prism: length, 10 ft; width, 9 ft; height, 13 ft
16. pyramid: length, 12 cm; width, 8 cm; height, 7 cm
17. cylinder: radius, 18 cm; height, 30 cm

Solve.

18. Find the total cost of carpeting the rectangular room if carpet costs \$17.95 per square yard. (5 yd, 4 yd)
19. Brian jogs every other day, rides his bike every third day, and swims every fourth day. If he jogs, rides, and swims on October 5, when will he jog, ride, and swim the next time all on the same day?
20. A pennant in the shape of an isosceles triangle is cut from a rectangle of length 18 in. and width 8 in. If the base of the triangular pennant is 8 in. and the height is 18 in., how many square inches of material are wasted?

Bonus What is the area the 4-inch-long hour hand of a clock sweeps in one hour? What is the area the 5-inch-long minute hand of a clock sweeps in one hour? Use 3.14 for π.

1. 25.12 cm
2. 50.24 cm^2
3. 224 ft^2
4. 87 m^2
5. 179.6 in^2
6. 150.72 mm^2
7. 216 ft^3
8. 256 cm^3
9. 2,260.8 m^3
10. 593.46 in^3
11. 13.58 m^2
12. 38.4 in^2
13. 185.2 m^2
14. 1,186.92 in^2
15. 1,170 ft^3
16. 224 cm^3
17. 30,520.8 cm^3
18. \$359
19. October 17
20. 72 in^2

Bonus 4.19 in^2, 78.5 in^2

Chapter 8 TEST

Find the circumference of each circle. Use 3.14 for π. Round decimal answers to the nearest tenth.

1. $r = 4$ cm **25.1 cm**
2. $r = 15$ ft **94.2 ft**
3. $d = 22$ in. **69.1 in.**
4. $d = 6$ m **18.8 m**
5. $r = 2.2$ m **13.8 m**

Find the area of each figure. Use 3.14 for π. Round decimal answers to the nearest tenth.

6. 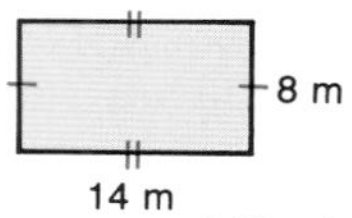 8 m, 14 m — 112 m^2
7. 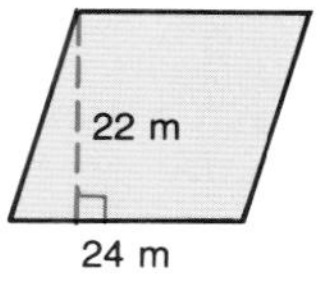 22 m, 24 m — 528 m^2
8. 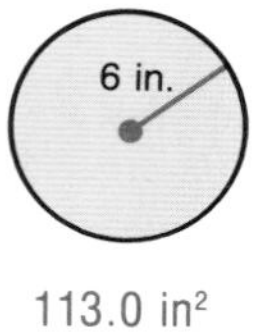 6 in. — 113.0 in^2
9. 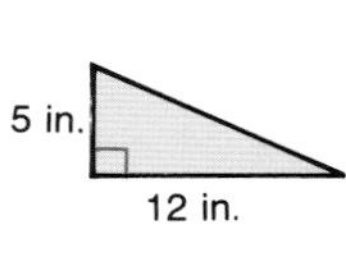

30 in^2

10. Jill is carpeting the kitchen with carpet that costs \$15.95 per square yard. The kitchen is 10 feet long and 9 feet wide. How much does Jill pay for the carpet? **\$159.50**

Find the surface area of each solid. Use 3.14 for π. Round decimal answers to the nearest tenth.

11. 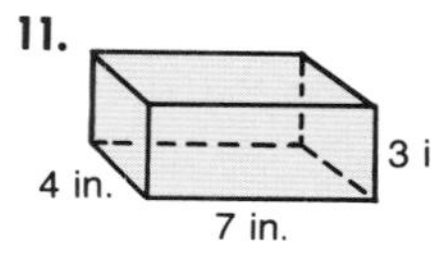

122 in^2
12. 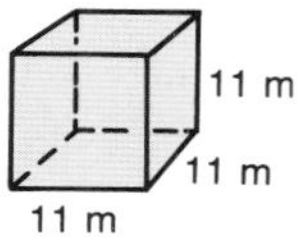 11 m, 11 m, 11 m — 726 m^2
13. 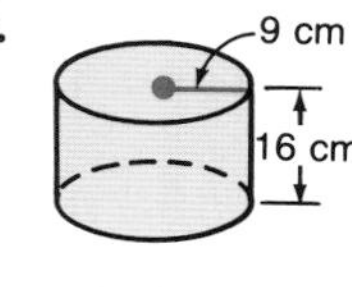

1,413 cm^2
14. 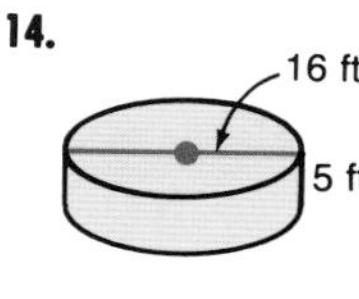

653.2 ft^2

Find the volume of each solid. Use 3.14 for π. Round decimal answers to the nearest tenth.

15. 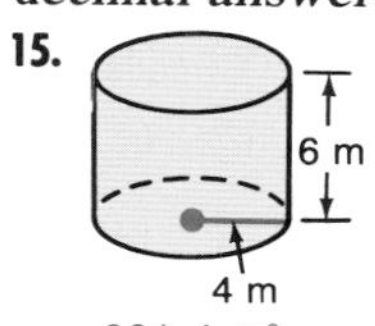

301.4 m^3
16. 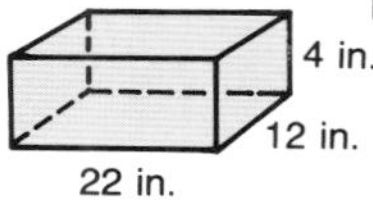 4 in., 12 in., 22 in. — 1,056 in^3
17. 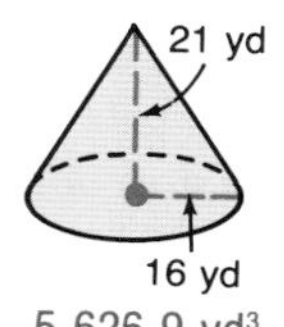

5,626.9 yd^3
18. 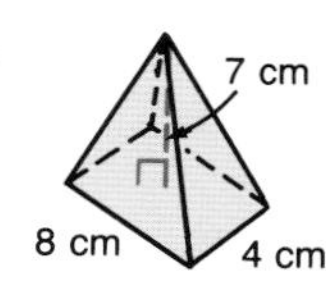

74.6 cm^3

19. If Jane sends a letter to 4 cousins and the cousins each send letters to 4 friends, how many letters will have been sent in all when the 4 friends each send letters to 4 more friends? **84 letters**
20. How many cubic feet of concrete are needed for a driveway that is 14 feet wide, 50 feet long, and 4 inches deep? **233.3 ft^3**

BONUS: If a square and a rectangle have the same perimeter, explain why the square has the larger area. **See margin.**

Additional Answers

Bonus: The greatest product of length and width occurs when length equals width. When length equals width, a rectangle is a square.

A **Test and Review Generator** is provided in Apple, IBM, and Macintosh versions. You may use this software to create your own tests or worksheets, based on the needs of your students.

TELEPHONE RATES

The phone company charges different rates for phone calls. The rates listed below are sample rates from a long distance company.

Dial-direct	Weekday full rate		Evening 35% discount		Night & weekend 60% discount		Operator-assisted	
	8 A.M. to 5 P.M.		to 11 P.M.		to 8 A.M.			
Sample Rates from City of **Columbus, Oh.** to:	First minute	Each additional minute	First minute	Each additional minute	First minute	Each additional minute	First three minutes	Each additional minute
Atlanta, Ga.	.53	.36	.34	.24	.21	.15	2.15	See dial-direct tables at left under time period that applies (weekday, evening or night & weekend).
Boston, Mass.	.53	.36	.34	.24	.21	.15	2.15	
Chicago, Ill.	.48	.34	.31	.23	.19	.14	2.05	
Denver, Colo.	.55	.38	.35	.25	.22	.16	2.25	
Houston, Tex.	.55	.38	.35	.25	.22	.16	2.25	
Philadelphia, Pa.	.50	.36	.32	.24	.20	.15	2.10	
St. Louis, Mo.	.50	.36	.32	.24	.20	.15	2.10	
Washington, D.C.	.50	.36	.32	.24	.20	.15	2.10	

Shannon lives in Columbus. On Thursday she calls her sister in Boston and talks from 8:00 P.M. to 8:15 P.M. What is she charged for this 15-minute call?

She finds the rate for an evening call to Boston.

Dial-direct	Evening 35% discount	
Chicago, Ill.	.31	.23
Denver, Colo.	.35	.25
Detroit, Mich.	.30	.21

Then she calculates the charge for the 15-minute call.

first minute — $0.34

next 14 minutes (15 − 1) — .24 [×] 14 [=] 3.36

total charge — [+] .34 [=] 3.70

Shannon is charged $3.70 for the call.

Use the rate table above to find the charge for each call described below. Assume each call is from Columbus.

1. direct-dial to Houston, 11:00 P.M. to 11:09 P.M. **$3.59**

2. operator-assisted to Atlanta, 9:00 A.M. to 9:05 A.M. **$3.21**

3. Arlanda calls her brother in Philadelphia on Monday and talks from 5:30 P.M. to 6:15 P.M. What is she charged for the call? **$10.88**

4. Darrell calls his parents in Denver on Sunday. They talk from 10:04 A.M. to 10:25 A.M. How much is Darrell charged for the call? **$3.42**

Applying Mathematics to the Real World

This optional page shows how mathematics is used in the real world and also provides a change of pace.

Objective Find the cost of telephone calls at different times of the day.

Using Discussion Discuss the various options long distance phone companies offer, such as call waiting, call forwarding, three-way calling. Discuss the effect these services have on the phone bill. Discuss the charges for long-distance service and the time of day that offers the more reasonable rate.

You may want to have students research competing long-distance companies for price and services offered.

Using Calculators Have students use calculators to solve problems.

Activity

Using Data Have students research the various charges that appear on their phone bill. These may be found in the customer guide in the front of the phone book or from the local telephone company.

Using the Cumulative Review

This page provides an aid for maintaining skills and concepts presented thus far in the text.

A Cumulative Review is also provided in the Evaluation Masters Booklet as shown below.

Evaluation Masters Booklet, p. 40

CUMULATIVE REVIEW Chapters 1-8

Estimate. (* Accept any reasonable estimate.)

1. 8,645 − 774 2. $7.39\overline{)3,628.1}$ 3. $7\frac{1}{8} \times 2\frac{3}{4}$ 4. \$3,295 + \$472 + \$6,431

Add, multiply, or divide.

5. $73\overline{)8,735}$ 6. 0.047 × 100 7. 397.5 ÷ 0.25 8. $\frac{2}{3} + \frac{1}{7}$

Find all the factors of each number.

9. 8 10. 18 11. 54 12. 60 13. 110

11. 1, 2, 3, 6, 9, 18, 27, 54 12. 1, 2, 3, 4, 5, 6, 10, 12, 15, 20, 30, 60
13. 1, 2, 5, 10, 11, 22, 55, 110

Fill each □ with a number so that the fractions are equivalent.

14. $\frac{2}{3} = \frac{\square}{15}$ 15. $\frac{18}{30} = \frac{\square}{5}$ 16. $\frac{21}{36} = \frac{\square}{12}$ 17. $\frac{7}{8} = \frac{\square}{40}$

Underline the larger unit.

18. 1 meter or 1 millimeter 19. 1 kilogram or 1 gram
20. 1 milliliter or 1 liter

Complete

21. 3 mm = ______ m 22. 7 kg = ______ g
23. $3\frac{1}{2}$ gal = ______ qt 24. $4\frac{1}{4}$ h = ______ min

Draw representations of each of the following.

25. $\overrightarrow{AB}$ 26. $\overleftrightarrow{CD} \perp \overleftrightarrow{EF}$ 27. octagon 28. obtuse triangle

Find the perimeter or circumference of each figure. Use 3.14 for π.

29. 30. 31.

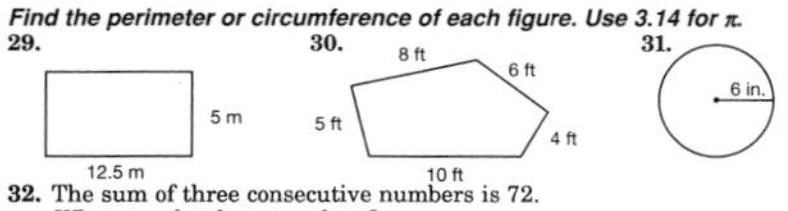

32. The sum of three consecutive numbers is 72. What are the three numbers?
33. A garden is 12 feet long and 6 feet wide. If Bob wants to double the length of the garden, how long should he make the new garden? State any missing or extra facts.

Answers: *1. 7,800 *2. 500 *3. 21 *4. \$10,000 5. 119 R48 6. 4.7 7. 1,590 8. $\frac{17}{21}$ 9. 1, 2, 4, 8 10. 1, 2, 3, 6, 9 11. ____ 12. ____ 13. ____ 14. 10 15 3 16. 7 17. 35 18. m 19. kg 20. L 21. 0.003 22 7,000 23. 6 24. 165 25. A B 26. E C F D 27. 28. 29. 35 m 30. 33 ft 31. 37.68 in. 32. 22, 24, 26 33. 16 ft; extra 3 ft wide

Cumulative REVIEW/TEST

Free Response

Lessons 1-11, 1-12

Solve each equation. Check your solution.

1. $8n = 48$ **6** 2. $26y = 13$ **0.5** 3. $17 + m = 103$ **86** 4. $n - 3.5 = 2$ **5.5**

Lesson 2-3

Subtract.

5. 5,476 − 1,362 **4,114** 6. 46 − 39 **7** 7. 3,762 − 553 **3,209**

Lesson 3-12

8. Bridget received the following test scores: 84, 93, 77, 89, 97, and 82. What is her average test score? **87**

Lesson 4-8

Change each fraction to a mixed number in simplest form.

9. $\frac{118}{12}$ $9\frac{5}{6}$ 10. $\frac{250}{12}$ $20\frac{5}{6}$ 11. $\frac{94}{24}$ $3\frac{11}{12}$

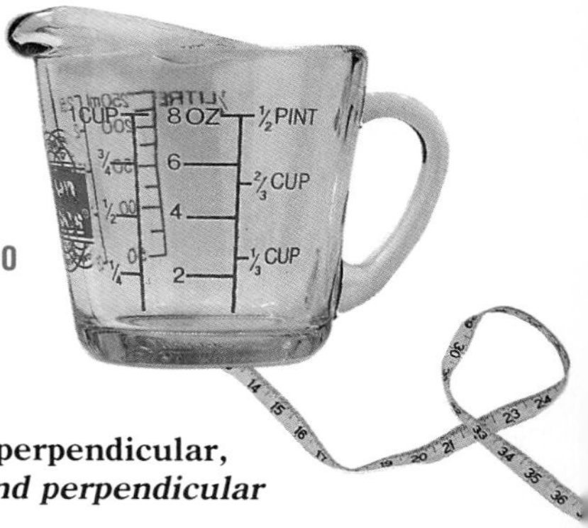

Lesson 5-9

Change each decimal to a fraction.

12. 0.07 $\frac{7}{100}$ 13. 0.7 $\frac{7}{10}$ 14. 0.55 $\frac{11}{20}$

Lesson 6-3

Complete.

15. 1,000 m = ? km **1** 16. 1 m = ? cm **100**

Lessons 6-7, 6-8

Complete.

17. 3 ft = ? in. **36 in.** 18. 8 ft = ? yd $2\frac{2}{3}$

Lesson 7-5

State whether each pair of lines is* parallel, perpendicular, *or* skew. *Use symbols to name all parallel and perpendicular lines.

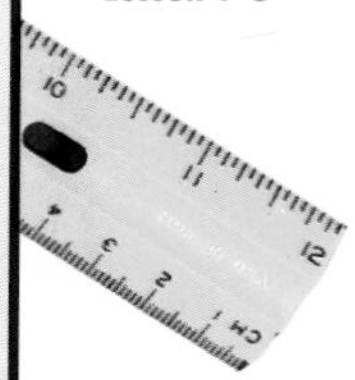

19.

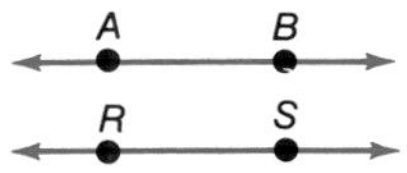

AB ∥ RS

20.

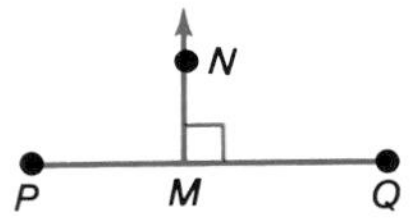

MN ⊥ PQ

21.

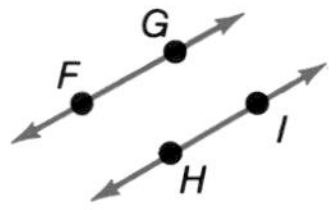

FG ∥ HI

Lesson 7-8

Classify each quadrilateral.

22.

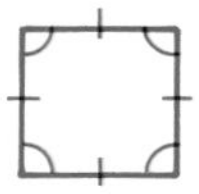

square

23.

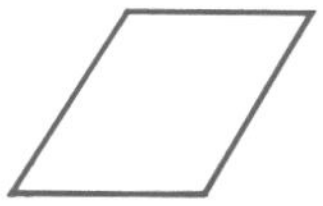

parallelogram

24.

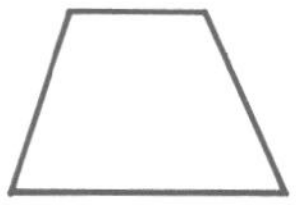

trapezoid

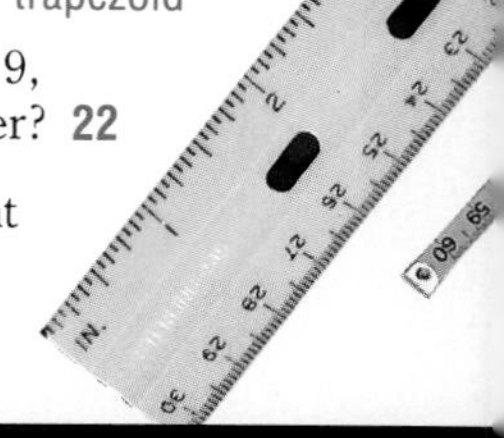

Lesson 7-10

25. A certain number is increased by 8, multiplied by 3, divided by 9, and decreased by 6. The result is 4. What is the original number? **22**

Lesson 8-3

26. Find the area of a triangle whose base is 8 cm and whose height is 4.5 cm. **18 cm**

APPLYING THE LESSON

Independent Practice

Homework Assignment	
Minimum	Ex. 1-25 odd; 26
Average	Ex. 1-26
Maximum	Ex. 1-26

Chapter 8

Multiple Choice

Choose the letter of the correct answer for each item.

1. Reba has 100 magazines in her collection. Each magazine is 0.343 cm thick. If Reba stacks the magazines in a pile, how thick will the pile be?
   a. 0.00343 cm
   b. 3.43 cm
   ▶ c. 34.3 cm
   d. *none of the above*

   Lesson 3-7

2. *About* how much is 8.982 divided by 2.3?
   e. 0.4
   ▶ f. 4
   g. 40
   h. 400

   Lesson 1-7

3. Change $2\frac{5}{6}$ to an improper fraction.
   a. $\frac{16}{6}$ c. $\frac{16}{5}$
   ▶ b. $\frac{17}{6}$ d. $\frac{17}{5}$

   Lesson 4-8

4. How many different outfits can be made from 2 pairs of pants, 3 shirts, and 5 ties?
   e. 5
   f. 10
   g. 25
   ▶ h. 30

   Lesson 5-5

5. Estimate the quotient when 5 is divided by $\frac{3}{4}$.
   a. 1 ▶ c. 6
   b. 2 d. 20

   Lesson 5-6

6. Cars were stalled bumper to bumper in two lanes for 4 miles as they approached the bridge. If cars occupy *about* 20 feet of length, *about* how many cars were stalled in both lanes?
   e. 2,640 g. 1,584
   f. 3,680 ▶ h. 2,112

   Lesson 6-7

7. Do not use a protractor. Estimate the measure of angle *ABC*.
   a. *about* 10°
   ▶ b. *about* 50°
   c. *about* 100°
   d. *about* 150°

   A B C

   Lesson 7-2

8. Complete. 7 qt = ? gal
   e. 1 g. 7
   ▶ f. $1\frac{3}{4}$ h. $2\frac{1}{2}$

   Lesson 6-8

9. The formula of the area of a circle is $A = \pi r^2$. Find the area of the circle. Use 3.14 for π.
   a. 9.42 cm²
   b. 18.84 cm²
   ▶ c. 28.26 cm²
   d. 113.04 cm²

   6 cm

   Lesson 8

10. Which of the following is equivalent to 1 kilogram?
    e. 100 mg g. 100 g
    f. 1,000 mg ▶ h. 1,000 g

    Lesson 6-1

Using the Standardized Test

This test familiarizes students with a standardized format while testing skills and concepts presented up to this point.

APPLYING THE LESSON

Independent Practice

Homework Assignment	
Minimum	Ex. 1-10
Average	Ex. 1-10
Maximum	Ex. 1-10

Evaluation Masters Booklet, pp. 36-37

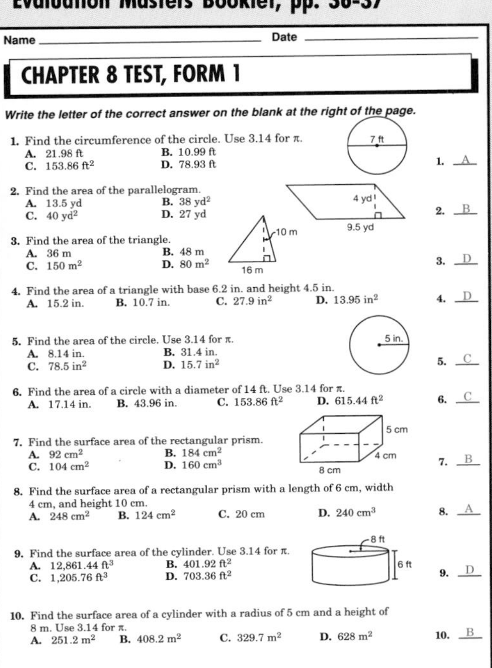

Name ______ Date ______

CHAPTER 8 TEST, FORM 1

Write the letter of the correct answer on the blank at the right of the page.

1. Find the circumference of the circle. Use 3.14 for π. (7 ft)
   A. 21.98 ft B. 10.99 ft
   C. 153.86 ft² D. 78.93 ft
   1. A
2. Find the area of the parallelogram. (4 yd, 9.5 yd)
   A. 13.5 yd B. 38 yd²
   C. 40 yd² D. 27 yd
   2. B
3. Find the area of the triangle. (10 m, 16 m)
   A. 36 m B. 48 m
   C. 150 m² D. 80 m²
   3. D
4. Find the area of a triangle with base 6.2 in. and height 4.5 in.
   A. 15.2 in. B. 10.7 in. C. 27.9 in² D. 13.95 in²
   4. D
5. Find the area of the circle. Use 3.14 for π. (5 in.)
   A. 8.14 in. B. 31.4 in.
   C. 78.5 in² D. 15.7 in²
   5. C
6. Find the area of a circle with a diameter of 14 ft. Use 3.14 for π.
   A. 17.14 in. B. 43.96 in. C. 153.86 ft² D. 615.44 ft²
   6. C
7. Find the surface area of the rectangular prism. (5 cm, 4 cm, 8 cm)
   A. 92 cm² B. 184 cm²
   C. 104 cm² D. 160 cm³
   7. B
8. Find the surface area of a rectangular prism with a length of 6 cm, width 4 cm, and height 10 cm.
   A. 248 cm² B. 124 cm² C. 20 cm D. 240 cm³
   8. A
9. Find the surface area of the cylinder. Use 3.14 for π. (8 ft, 6 ft)
   A. 12,861.44 ft³ B. 401.92 ft²
   C. 1,205.76 ft³ D. 703.36 ft²
   9. D
10. Find the surface area of a cylinder with a radius of 5 cm and a height of 8 m. Use 3.14 for π.
    A. 251.2 m² B. 408.2 m² C. 329.7 m² D. 628 m²
    10. B

Chapter

9 RATIO, PROPORTION, AND PERCENT

PREVIEWING the CHAPTER

In this chapter, ratios and proportions are discussed. Students determine whether two ratios are equal and solve proportions. Students are shown practical applications of ratios and proportions by using simple probability and determining actual distances from scale drawings. Ratios and proportions are also used to write percents as fractions and vice versa. This is followed by writing percents as decimals and vice versa.

Problem-Solving Strategy Students learn to solve a problem by using the guess-and-check strategy.

Lesson (Pages)	Lesson Objectives	State/Local Objectives
9-1 (280-281)	9-1: Write ratios.	
9-2 (282-283)	9-2: Find the probability of an event.	
9-3 (284-285)	9-3: Determine if two ratios form a proportion.	
9-4 (286-287)	9-4: Solve proportions by using cross products.	
9-5 (288-289)	9-5: Interpret scale drawings.	
9-6 (292-293)	9-6: Find the measure of a side of a similar figure.	
9-7 (294-295)	9-7: Locate distances and driving times on a map.	
9-8 (296-297)	9-8: Write percents as fractions and fractions as percents.	
9-9 (298-299)	9-9: Write percents as decimals and decimals as percents.	
9-10 (300-301)	9-10: Solve problems using guess-and-check.	

ORGANIZING the CHAPTER

You may refer to the ***Course Planning Calendar*** on page T10.

Planning Guide

Lesson (Pages)	Extra Practice (Student Edition)	Blackline Masters Booklets: Reteaching	Practice	Enrichment	Activity	Multicultural Activity	Technology	Tech Prep	Evaluation	Other Resources
9-1 (280-281)	p. 472	p. 87	p. 95	p. 87	pp. 17, 37					Transparency 9-1
9-2 (282-283)	p. 472	p. 88	p. 96	p. 88			p. 23			Transparency 9-2
9-3 (284-285)	p. 472	p. 89	p. 97	p. 89				p. 17		Transparency 9-3
9-4 (286-287)	p. 473	p. 90	p. 98	p. 90	pp. 18, 51			p. 18		Transparency 9-4
9-5 (288-289)	p. 473	p. 91	p. 99	p. 91			p. 9		p. 44	Transparency 9-5
9-6 (292-293)	p. 473	p. 92	p. 100	p. 92						Transparency 9-6
9-7 (294-295)		p. 93	p. 101	p. 93						Transparency 9-7
9-8 (296-297)	p. 474	p. 94	p. 102	p. 94						Transparency 9-8
9-9 (298-299)	p. 474	p. 95	p. 103	p. 95						Transparency 9-9
9-10 (300-301)			p. 104						p. 44	Transparency 9-10
Ch. Review (302-303)									pp. 41-43	Test and Review Generator
Ch. Test (304)										
Cumulative Review/Test (306-307)									p. 45	

OTHER CHAPTER RESOURCES

Student Edition
Chapter Opener, p. 278
Activity, p. 279
Consumer Connections, p. 290
On-the-Job Application, p. 297
Journal Entry, p. 297
Portfolio Suggestion, p. 289
Consumer Connections, p. 305
Tech Prep Handbook, pp. 528-530

Teacher's Classroom Resources
Transparency 9-0
Lab Manual, pp. 39-40
Performance Assessment, pp. 17-18
Lesson Plans, pp. 95-104

Other Supplements
Overhead Manipulative Resources, Labs 27-29, pp. 19-21
Glencoe Mathematics Professional Series

ENHANCING the CHAPTER

Some of the blackline masters for enhancing this chapter are shown below.

COOPERATIVE LEARNING

The following game can be played by groups of four students. Have students make a set of 36 cards, 12 with percents, 12 with equivalent decimals, and 12 with equivalent fractions. Shuffle the cards and place them face down in a 6 x 6 array. Each of three students choose a card. When the cards show equivalent percents, decimals, and fractions, they are removed from the array. When they are not equivalent, they are turned face down in their places. Students continue until all cards are matched. The fourth student counts the number of turns it takes to match the cards.

USING MODELS/MANIPULATIVES

Students can use meter sticks to show ratios. Have students measure their desk to the nearest centimeter to find the ratio of its length to width. To solve proportions such as $\frac{3}{5} = \frac{x}{20}$, have students count sets of 5 on the meter stick, beginning at 0, until they reach 20. (4 sets) Next have them mark off an equal number of sets of 3 on the meter stick to find the missing term. Have students show percents by using 100 cm as "per hundred." Thus, 35 cm is $\frac{35}{100}$ or 35%.

Cooperative Problem Solving, p. 37

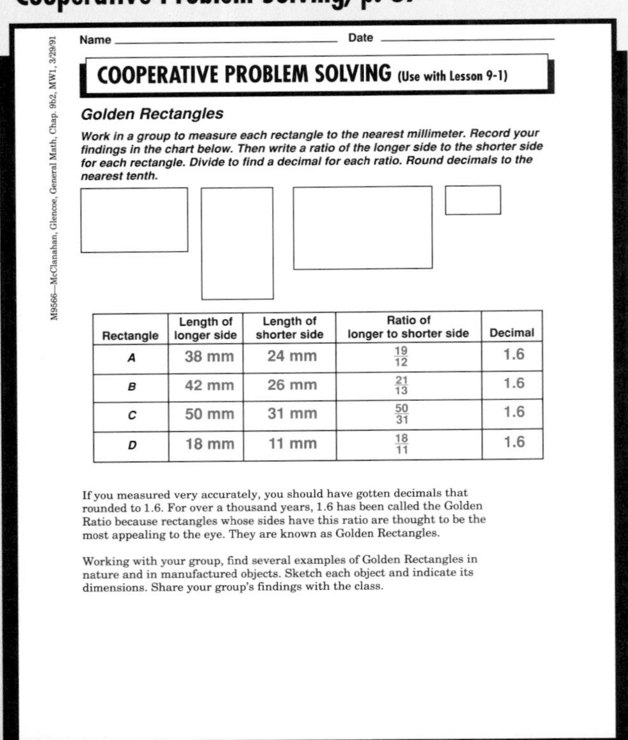

M9566—McClanahan, Glencoe, General Math, Chap. 9b2, MW1, 3/29/91

Name ______________ Date ______________

COOPERATIVE PROBLEM SOLVING (Use with Lesson 9-1)

Golden Rectangles

Work in a group to measure each rectangle to the nearest millimeter. Record your findings in the chart below. Then write a ratio of the longer side to the shorter side for each rectangle. Divide to find a decimal for each ratio. Round decimals to the nearest tenth.

Rectangle	Length of longer side	Length of shorter side	Ratio of longer to shorter side	Decimal
A	38 mm	24 mm	$\frac{19}{12}$	1.6
B	42 mm	26 mm	$\frac{21}{13}$	1.6
C	50 mm	31 mm	$\frac{50}{31}$	1.6
D	18 mm	11 mm	$\frac{18}{11}$	1.6

If you measured very accurately, you should have gotten decimals that rounded to 1.6. For over a thousand years, 1.6 has been called the Golden Ratio because rectangles whose sides have this ratio are thought to be the most appealing to the eye. They are known as Golden Rectangles.

Working with your group, find several examples of Golden Rectangles in nature and in manufactured objects. Sketch each object and indicate its dimensions. Share your group's findings with the class.

Lab Activity, p. 40

M9566—McClanahan, Glencoe, Chapter MSC, MW1, 4/1/91

Name ______________ Date ______________

LAB ACTIVITY (Use with Lesson 9-5)

Reductions and Enlargements

You can use Easy-to-Make Manipulative 8, Quarter-Inch Grid Paper.

Choose a fairly large object in your classroom whose dimensions are between 1 and 6 feet. Make a scale drawing of your object using the following ratio:

$\frac{1}{4}$ in. on grid = 1 ft actual

1. Is your drawing an enlargement or a reduction?

____

Choose a small object in your classroom whose dimensions are between 1 and 6 inches. Make a scale drawing using the following ratio:

4 in. on grid = $\frac{1}{16}$ in. actual

2. Is your drawing an enlargement or a reduction?

____

3. Make a scale drawing of other objects. When would you make an enlargement of an object?

____

4. When would you make a reduction of an object?

____

MEETING INDIVIDUAL NEEDS

Limited English Proficiency Students

Encourage LEP students to write as much as possible. This chapter contains several key concepts of mathematics that can be better understood by having students write their own verbal problems involving ratios, probability, proportion, and similarity. Have them exchange problems with a partner and discuss the meanings and solutions.

CRITICAL THINKING/PROBLEM SOLVING

Critical Thinking

Students should use any clues given and logical reasoning before guessing the answer to a problem. Present this problem. A three-digit number is between 200 and 400. The sum of the digits is 10. The sum of the hundred's digit and the one's digit equals the ten's digit. Find the number. **The hundred's digit must be either a 2 or a 3. List the 3-digit numbers beginning with a 2 or a 3 in which the sum of the digits is 10. The numbers are 253 and 352.**

COMMUNICATION

Writing

Have students write a short story based on the mathematical content of the chapter. The stories should include probability, a scale drawing, enlargements or reductions in size, and percentages. Students may want to incorporate the problem-solving strategy guess and check into their story as a method of solving a mystery. Encourage students to be creative.

Applications, p. 17

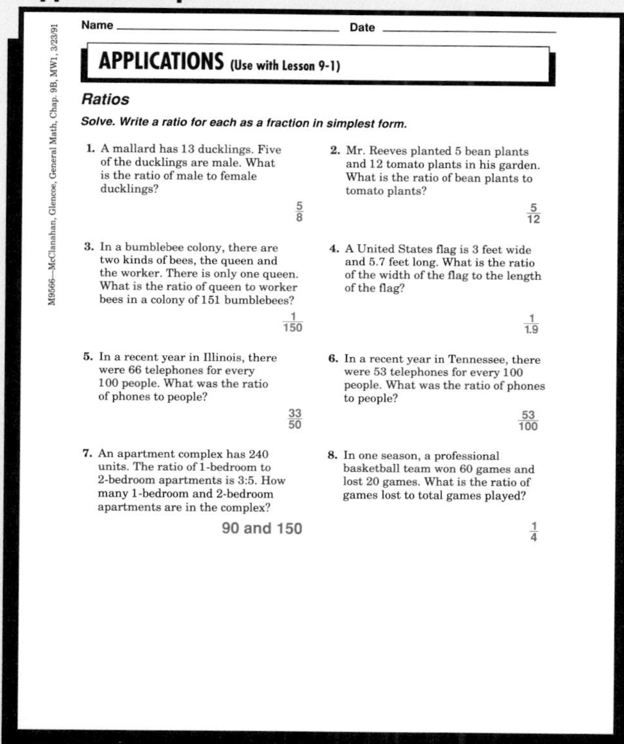

M9566—McClanahan, Glencoe, General Math, Chap. 9B, MW1, 3/23/91

Name ______________________ Date ______________

APPLICATIONS (Use with Lesson 9-1)

Ratios

Solve. Write a ratio for each as a fraction in simplest form.

1. A mallard has 13 ducklings. Five of the ducklings are male. What is the ratio of male to female ducklings? $\frac{5}{8}$

2. Mr. Reeves planted 5 bean plants and 12 tomato plants in his garden. What is the ratio of bean plants to tomato plants? $\frac{5}{12}$

3. In a bumblebee colony, there are two kinds of bees, the queen and the worker. There is only one queen. What is the ratio of queen to worker bees in a colony of 151 bumblebees? $\frac{1}{150}$

4. A United States flag is 3 feet wide and 5.7 feet long. What is the ratio of the width of the flag to the length of the flag? $\frac{1}{1.9}$

5. In a recent year in Illinois, there were 66 telephones for every 100 people. What was the ratio of phones to people? $\frac{33}{50}$

6. In a recent year in Tennessee, there were 53 telephones for every 100 people. What was the ratio of phones to people? $\frac{53}{100}$

7. An apartment complex has 240 units. The ratio of 1-bedroom to 2-bedroom apartments is 3:5. How many 1-bedroom and 2-bedroom apartments are in the complex? 90 and 150

8. In one season, a professional basketball team won 60 games and lost 20 games. What is the ratio of games lost to total games played? $\frac{1}{4}$

Calculator Activity, p. 9

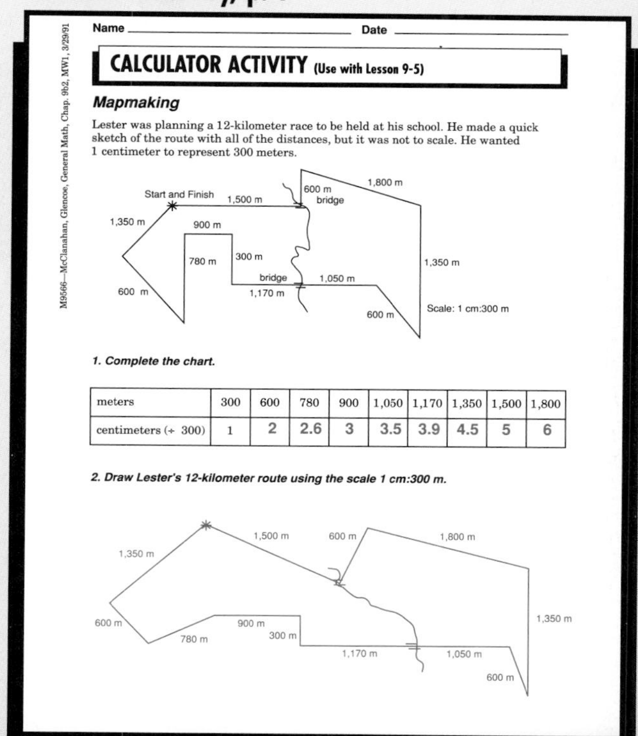

M9566—McClanahan, Glencoe, General Math, Chap. 9b2, MW1, 3/29/91

Name ______________________ Date ______________

CALCULATOR ACTIVITY (Use with Lesson 9-5)

Mapmaking

Lester was planning a 12-kilometer race to be held at his school. He made a quick sketch of the route with all of the distances, but it was not to scale. He wanted 1 centimeter to represent 300 meters.

1. Complete the chart.

meters	300	600	780	900	1,050	1,170	1,350	1,500	1,800
centimeters (÷ 300)	1	2	2.6	3	3.5	3.9	4.5	5	6

2. Draw Lester's 12-kilometer route using the scale 1 cm:300 m.

CHAPTER PROJECT

Have four groups of students conduct an orienteering rally by each creating a course on the school premises with designated checkpoints monitored by group members. Provide each group with a compass and any necessary instruction in its use. Have each group draw a scale map of the course showing the start, the checkpoints, and the finish. Give orientations as compass directions in degrees. Have groups exchange maps. Let a student from each group walk the course using the map and compass. Have each group present its results to the class.

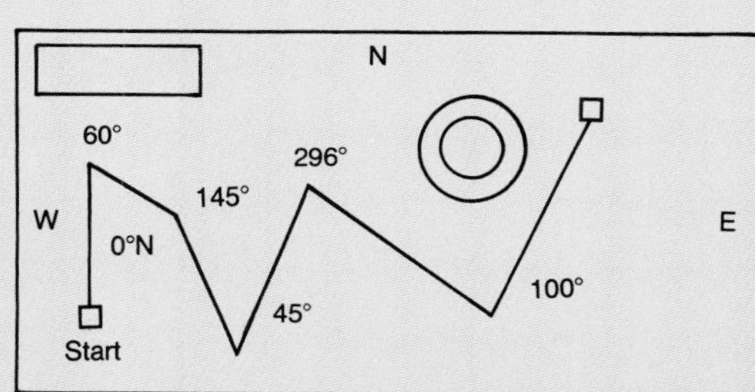

Transparency 9-0 is available in the Transparency Package. It provides a full-color visual and motivational activity that you can use to engage students in the mathematical content of the chapter.

Using Discussion

- Have students read the opening paragraph. Then discuss the first two questions with students. Encourage everyone to participate in the discussion.
- Allow students the opportunity to discuss their favorite teams and their win-loss records. If students don't know the win-loss record, encourage them to research this information.
- If students belong on a team, ask what their win-loss record is. Ask students why these records are important to keep.

The second paragraph is a motivational problem. Have students determine the answer and discuss ways in which they arrived at their answer.

Chapter 9

RATIO, PROPORTION, AND PERCENT

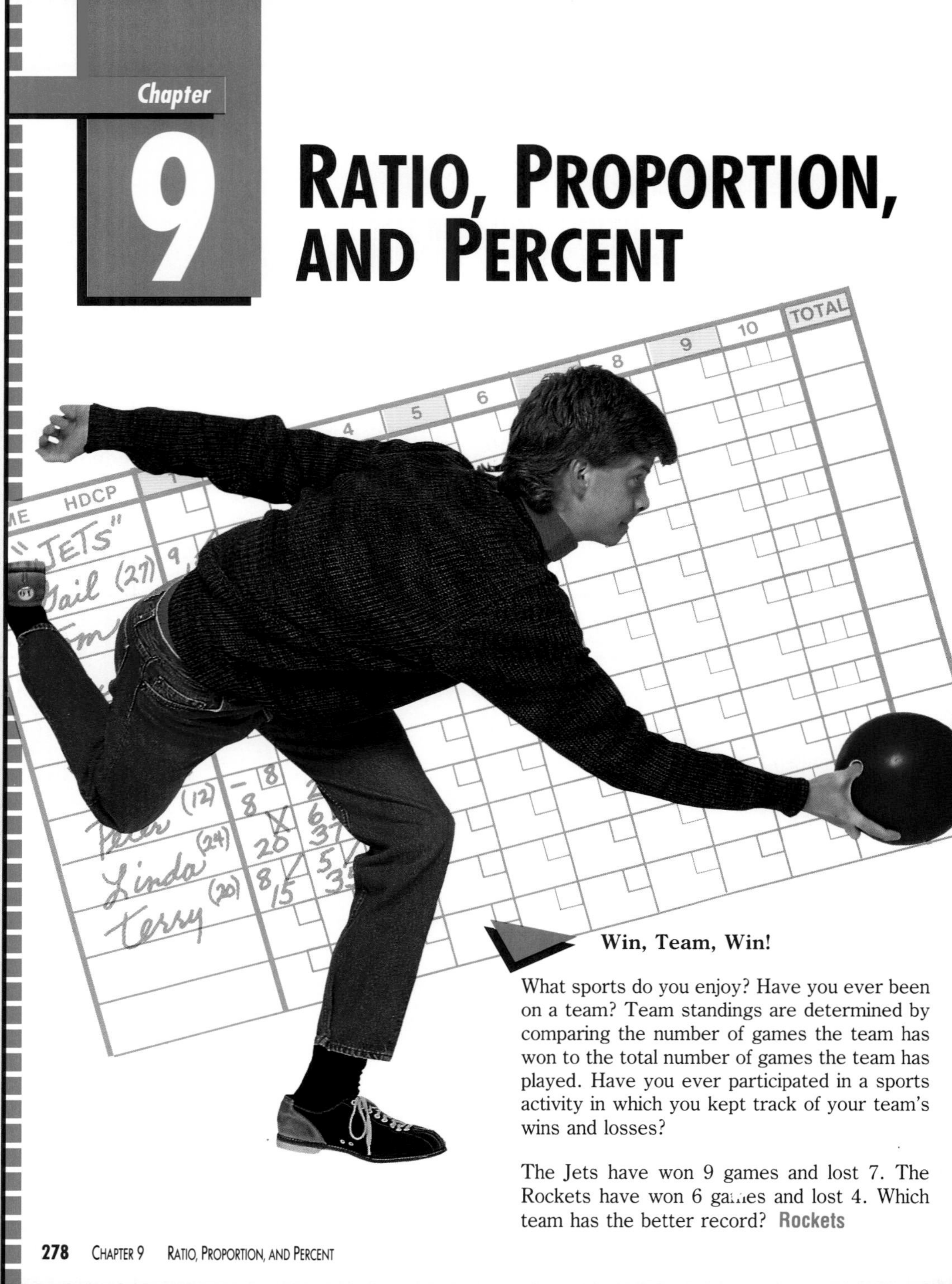

Win, Team, Win!

What sports do you enjoy? Have you ever been on a team? Team standings are determined by comparing the number of games the team has won to the total number of games the team has played. Have you ever participated in a sports activity in which you kept track of your team's wins and losses?

The Jets have won 9 games and lost 7. The Rockets have won 6 games and lost 4. Which team has the better record? **Rockets**

ACTIVITY: Similar Figures

Suppose you don't draw very well but you want to enlarge a picture of your favorite cartoon. There is a way you can draw it and make it look like the original.

Materials: pencil, half-inch grid paper, ruler

$\frac{1}{4}$-inch grid

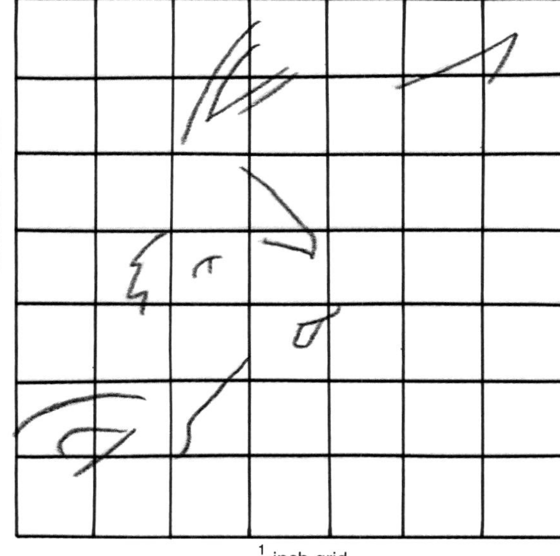

$\frac{1}{2}$-inch grid

1. 7 squares long and 7 squares wide

3. Transfer the lines that appear in each square on the cartoon to the corresponding square on the larger grid.

Procedures

1. A quarter-inch grid has been placed over the cartoon. How many squares long and how many squares wide is the grid?
2. Make sure the half-inch grid has the same number of squares.
3. Study the drawings above and explain how the grids are used to enlarge the cartoon.

On Your Own

4. Now trace the larger grid and all the lines within each square. Then complete the drawing. **See students' work.**

Cooperative Groups

Materials: a favorite cartoon, pencil, half-inch grid paper

5. Make an enlargement of your favorite cartoon. Decide what you need to do to get started. **See students' work.**
6. Does your cartoon look like the original? If so, congratulations!

Communicate Your Ideas

7. What is the same about your drawing and the original? How is your drawing different from the original? **same picture but larger**

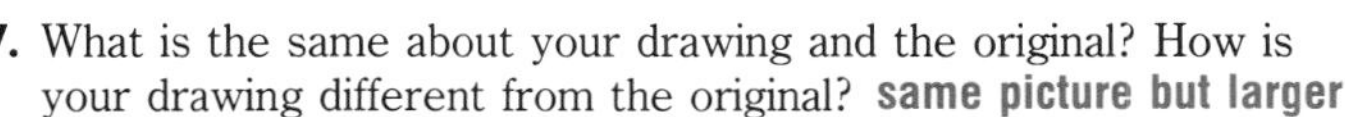

INTRODUCING THE LESSON

Objective Draw similar figures using a quarter-inch grid and a half-inch grid.

TEACHING THE LESSON

Using Cooperative Groups Divide the students into groups of two. Have students begin by discussing how they will start. Use half-inch grid, **Lab Manual, p. 9.**

Make sure students understand that it is important to have the same number of squares on both grids.

EVALUATING THE LESSON

Encourage students to make several attempts at making the grids over the original cartoon if they have difficulty.

Communicate Your Ideas Have students reduce their favorite cartoon. Have them determine what they need to get started.

EXTENDING THE LESSON

Closing the Lesson Encourage students to tell what they have learned about similar figures.

Activity Worksheet, Lab Manual, p. 39

Name ____________ Date ____________

ACTIVITY WORKSHEET Use with page 279

Similar Figures

Copy the figure shown on quarter-inch grid paper as accurately as you can on the half-inch grid paper below.

5-MINUTE CHECK-UP

(over Lesson 8-13)

Available on TRANSPARENCY 9-1.

1. A storage box is 4 ft long, 3 ft wide, and 2 ft 6 in. deep. What is the volume of the storage box? **30 ft³**
2. How would doubling the dimensions of the storage box affect the volume? **8 times as great**
3. What is the volume of a cone with a radius of 3 cm and a height of 20 cm? Use 3.14 for π. **188.4 cm³**

INTRODUCING THE LESSON

Determine the number of boys and number of girls in the class. Discuss ways in which the two numbers can be compared.

TEACHING THE LESSON

Using Models Using base-ten blocks, **Lab Manual, p. 1,** ask students to represent the ratio for 8 to 10. Have students reverse the models and state another ratio.

Practice Masters Booklet, p. 95

PRACTICE WORKSHEET 9-1

Ratio

Write each ratio as a fraction in simplest form.

1. 6 losses to 13 wins $\frac{6}{13}$
2. 4 inches of snow in 9 days $\frac{4}{9}$
3. 21 wins to 14 losses $\frac{3}{2}$
4. 15 children out of 60 passengers $\frac{1}{4}$
5. 12 losses in 52 games $\frac{3}{13}$
6. 24 passengers in 8 cars $\frac{3}{1}$
7. 18 tickets for $54 $\frac{1}{3}$
8. 27 wins to 18 losses $\frac{3}{2}$
9. 32 wins in a total of 80 games $\frac{2}{5}$
10. 144 bottles in 36 cartons $\frac{4}{1}$
11. 47 women out of 94 adults $\frac{1}{2}$
12. 360 miles in 6 hours $\frac{60}{1}$

Andy Peabody works for Widgets, Inc. Use his check stub to write a ratio that compares the following. (Do not write in simplest form.)

13. overtime hours to regular hours $\frac{2.5}{80}$
14. FICA tax to gross pay $\frac{22.38}{334.02}$
15. total deductions to gross pay $\frac{100.63}{334.02}$
16. United Fund to take-home pay $\frac{2.00}{233.39}$

Widgets, Inc.		Andy Peabody
Check Number	Tax Deductions	
12546	Federal Tax	State Tax
Pay Period Ending	41.40	6.38
2/14	FICA 22.38	City Tax 3.27
Regular Hours	Other Deductions	
80.0	Union Dues 10.00	Insurance 5.20
Overtime Hours 2.5	United Fund 2.00	Bonds 8.00
Gross Pay	Total Deductions	Take-Home Pay
334.02	100.63	233.39

Solve.

17. Bolton High School's basketball team has won 12 games and lost 6 games. What is their ratio of games won to games lost? $\frac{2}{1}$
18. Brandon got 12 problems correct and 3 problems wrong on today's quiz. What was his ratio of problems correct to the total number of problems? $\frac{4}{5}$

9-1 RATIO

Objective
Write ratios.

The Texas Rangers won 19 baseball games and lost 25 games.

American League Western Division		
Club	**W**	**L**
California	22	26
Seattle	20	25
Texas	19	25
Oakland	13	33

You can compare their wins and losses by writing a ratio. A **ratio** is a comparison of two numbers by division. Ratios can be written three ways.

19 to 25 19:25 $\frac{19}{25}$ Each ratio is read 19 to 25.

The Rangers' win-loss ratio is 19:25.

Method

1. Write the two numbers in the order you want to compare them.
2. Write ratios as fractions in simplest form.

Example A ***Write the win-loss ratio for California as a fraction in simplest form.***

1. 22 wins, 26 losses
   $\frac{22}{26}$ ← wins / ← losses
2. $\frac{22}{26} = \frac{11}{13}$ (÷ 2)

California has an $\frac{11}{13}$ win-loss ratio.

Example B ***Write the ratio for games won by Seattle to total games played.***

1. 20 wins, 20 + 25 or 45 total games
   $\frac{20}{45}$ ← wins / ← total games
2. $\frac{20}{45} = \frac{4}{9}$ (÷ 5) Seattle has a $\frac{4}{9}$ ratio for games won to total games.

Example C ***Phil drives about 300 miles with a full 15-gallon tank of gas. Write the ratio for miles Phil drives to gallons of gas used.***

1. 300 miles, 15 gallons 300 to 15 = $\frac{300}{15}$
2. $\frac{300}{15} = \frac{20}{1}$ (÷ 15) Phil has a miles-to-gallons ratio of about $\frac{20}{1}$.

Keep the 1 as the denominator.

Guided Practice

Write each ratio as a fraction in simplest form.

Example A
1. New York: wins to losses $\frac{30}{11}$
2. Philadelphia: losses to wins $\frac{28}{13}$

Example B
3. Boston: wins to total games played $\frac{24}{41}$
4. 6 absences in 180 school days $\frac{1}{30}$

Example C
5. 300 students to 20 teachers $\frac{15}{1}$
6. 76 heartbeats per 60 seconds $\frac{19}{15}$

NBA Eastern Conference Atlantic Division		
Club	**W**	**L**
New York	60	22
Boston	48	34
New Jersey	43	39
Orlando	41	41
Miami	36	46
Philadelphia	26	56
Washington	22	60

RETEACHING THE LESSON

Write each ratio as a fraction in simplest form.

1. 5 out of 35 $\frac{1}{7}$
2. 9 to 24 $\frac{3}{8}$
3. number of p's to number of i's in Mississippi $\frac{1}{2}$
4. number of O's to total number of letters in PROPORTION $\frac{3}{10}$
5. number of days in a week to number of months in a year $\frac{7}{12}$

Reteaching Masters Booklet, p. 87

Name ______ Date ______

RETEACHING WORKSHEET 9-1

Ratio

The number of wheels and the number of hubcaps can be compared in two ways.

By Subtraction	By Division as a Ratio
8 − 2 = 6	$\frac{8}{2}$
There are 6 more wheels than hubcaps.	The ratio of wheels to hubcaps is 8 to 2.

A ratio can be written three ways: $\frac{8}{2}$, 8 to 2, 8:2.

A ratio such as $\frac{8}{2}$ can be written in simplest form: $\frac{4}{1}$.

Exercises

Practice

9. $\frac{7}{11}$ 14. $\frac{\$5.12}{1}$ 15. $\frac{1}{7}$

Write each ratio as a fraction in simplest form.

7. 20 wins to 15 losses $\frac{4}{3}$
8. 24 losses to 32 wins $\frac{3}{4}$
9. 35 wins in a total of 55 games
10. 8 tickets for $48 $\frac{1}{\$6}$
11. 4 inches of rain in 30 days $\frac{2}{15}$
12. 2 pounds of bananas for 88¢ $\frac{1}{44¢}$
13. 156 kilometers in 3 hours $\frac{52}{1}$
14. $30.72 interest for 6 months
15. 4 students representing a class of 28 students
16. 102 passengers on 3 buses $\frac{34}{1}$
17. 27 losses to 36 wins $\frac{3}{4}$
18. 42 wins to 18 losses $\frac{7}{3}$

Interpreting Data

19. Which Central Division team has a win-loss ratio of $\frac{5}{11}$? **Tampa Bay**
20. Which Central Division team has a loss-win ratio of $\frac{1}{3}$? **Chicago**
21. Suppose Chicago had won two more of the 16 games they played. What would their win-loss ratio have been? $\frac{14}{2}$ or $\frac{7}{1}$

NFL National Conference Central Division

Club	W	L	T
Chicago	12	4	0
Minnesota	11	5	0
Tampa Bay	5	11	0
Detroit	4	12	0
Green Bay	4	12	0

Applications

22. 6 games

22. Carla's team has a 3:2 win-loss ratio. They lost four games. How many games did they win?
23. In 1988, Minnesota Viking's quarterback Wade Wilson completed 204 passes of 332 attempts. What was his ratio of completed passes to incomplete passes? $\frac{51}{32}$
24. Quan puts 2 quarts of antifreeze and 5 quarts of water into her radiator. What is the ratio of antifreeze to water that she added? **2 to 5**

Write Math

25. Explain in your own words what the ratio 10 fingers to 2 hands means. **Every 10 fingers has 2 hands. There are 5 times as many fingers as there are hands. Every hand has 5 fingers.**

Collect Data

26. Survey members of your class to see how many walk to school. Write a ratio to show walkers as a part of all students in your class. **Answers may vary.**

Mixed Review

Find the LCM of each group of numbers.

Lesson 4-3

27. 4, 6 **12**
28. 7, 13 **91**
29. 9, 12 **36**
30. 16, 20 **80**

Name the reciprocal of each number.

Lesson 5-6

31. $\frac{5}{9}$ $\frac{9}{5}$
32. $\frac{15}{31}$ $\frac{31}{15}$
33. $\frac{8}{13}$ $\frac{13}{8}$
34. $\frac{7}{3}$ $\frac{3}{7}$

Lesson 6-4

35. A $10 roll of quarters has a mass of 240 grams. What is the mass of one quarter in kilograms? **0.006 kg**

Chalkboard Examples

Write each ratio as a fraction in simplest form.

1. Example A Denver won 44 games in the NBA and lost 38 games. Write their win-loss ratio. $\frac{22}{19}$
2. Example B The New York Giants won 10 games and lost 6 games. Write their ratio of games won to total games played. $\frac{5}{8}$
3. Example C Jill works 8 hours each 24-hour period. Write the ratio of hours worked per 24 hours. $\frac{1}{3}$

Additional examples are provided on TRANSPARENCY 9-1.

EVALUATING THE LESSON

Check for Understanding Use Guided Practice Exercises to check for student understanding.

Closing the Lesson Have students tell about times when they use ratios.

EXTENDING THE LESSON

Enrichment Have students use a cube with A-F on the sides and roll it 30 times. Have students write ratios about the number of times one letter came up in relation to another.

Enrichment Masters Booklet, p. 87

Name ______ Date ______

ENRICHMENT WORKSHEET 9-1

Rates

A **rate** is a ratio of two measurements having different units:

rate: $\frac{196 \text{ miles}}{7 \text{ gallons}}$

A rate with a denominator of 1 is called a **unit rate.**

unit rate: $\frac{196 \text{ miles}}{7 \text{ gallons}} = \frac{28 \text{ miles}}{1 \text{ gallon}}$

The unit rate is 28 miles per gallon (mpg).

Write each of the following as a rate.

1. 90 sit-ups in 3 minutes $\frac{90 \text{ sit-ups}}{3 \text{ minutes}}$
2. 320 miles in 5 hours $\frac{320 \text{ miles}}{5 \text{ hours}}$
3. $42 earned in 8 hours $\frac{\$42}{8 \text{ hours}}$
4. 76 calories in 12 ounces $\frac{76 \text{ calories}}{12 \text{ ounces}}$

Write each of the following as a unit rate.

5. 21 meals in 7 days $\frac{3 \text{ meals}}{1 \text{ day}}$ or 3 meals per day
6. $12 in 3 hours $\frac{\$4}{1 \text{ hour}}$ or $4 per hour
7. 1,500 miles in 10 days $\frac{150 \text{ miles}}{1 \text{ day}}$ or 150 miles per day
8. $17 for 2 ties $\frac{\$8.50}{1 \text{ tie}}$ or $8.50 per tie
9. $6 bus fare in 5 days $\frac{\$1.20}{1 \text{ day}}$ or $1.20 per day
10. 48 hours in 4 days $\frac{12 \text{ hours}}{1 \text{ day}}$ or 12 hours per day

Solve.

11. Mr. Weber lost 12 pounds in 4 months. At that rate, how long will it take him to lose 21 pounds? 7 months
12. Jacqui saved $75 in 5 months. At that rate, how long will it take her to save $270? 18 months
13. An 18-ounce jar of peanut butter is $1.89. A 12-ounce jar of peanut butter is $1.39. Which is the better buy? 18-oz for $1.89
14. If apples are selling 8 for 99¢, how many can you buy for $1.86? 15 apples

APPLYING THE LESSON

Independent Practice

Homework Assignment	
Minimum	Ex. 2-24 even; 25-35
Average	Ex. 1-21 odd; 22-35
Maximum	Ex. 1-35

5-MINUTE CHECK-UP

(over Lesson 9-1)

Available on TRANSPARENCY 9-2.

Write each ratio as a fraction in simplest form.

1. 36 wins to 27 losses $\frac{4}{3}$
2. 18 wins in a total of 30 games $\frac{3}{5}$
3. 242 miles on 11 gallons of gas $\frac{22}{1}$
4. \$200 earned every 5 days $\frac{\$40}{1}$

Extra Practice, Lesson 9-1, p. 472

INTRODUCING THE LESSON

Show students examples of weather forecasts that include the probability of precipitation. Ask students what it means if the chances of rain are 60 out of 100.

TEACHING THE LESSON

Using Models Have students find the theoretical probability of each outcome on the spinner, **Lab Manual, p. 21.**

Practice Masters Booklet, p. 96

PRACTICE WORKSHEET 9-2

An Introduction to Probability

A date is chosen at random from the month of February. Find the probability of choosing each date.

1. The date is the fifteenth. $\frac{1}{28}$
2. The date is a Wednesday. $\frac{1}{7}$
3. It is after the twenty-fourth. $\frac{1}{7}$
4. It is before the sixth. $\frac{5}{28}$
5. It is an even-numbered date. $\frac{1}{2}$

February

S	M	T	W	T	F	S
	1	2	3	4	5	6
7	8	9	10	11	12	13
14	15	16	17	18	19	20
21	22	23	24	25	26	27
28						

A die is rolled once. Find the probability rolling each of the following.

6. a 5 $\frac{1}{6}$
7. a 2 $\frac{1}{6}$
8. an odd number $\frac{1}{2}$
9. a 5 or a 6 $\frac{1}{3}$
10. a number less than 3 $\frac{1}{3}$
11. *not* a 4 $\frac{5}{6}$
12. a number less than 1 0
13. *not* a 3 or a 4 $\frac{2}{3}$
14. a number less than 10 1
15. a 2, a 3, or a 4 $\frac{1}{2}$

Two dice are rolled once. The possible outcomes are listed at the right. Find each probability.

6,1	6,2	6,3	6,4	6,5	6,6
5,1	5,2	5,3	5,4	5,5	5,6
4,1	4,2	4,3	4,4	4,5	4,6
3,1	3,2	3,3	3,4	3,5	3,6
2,1	2,2	2,3	2,4	2,5	2,6
1,1	1,2	1,3	1,4	1,5	1,6

16. P(5,5) $\frac{1}{36}$
17. P(2,1) $\frac{1}{36}$
18. P(an odd sum) $\frac{1}{2}$
19. P(a sum of 7) $\frac{1}{6}$
20. P(a sum of 11) $\frac{1}{18}$
21. P(neither number 4) $\frac{25}{36}$
22. P(a product less than 8) $\frac{7}{18}$
23. P(neither number 3 or 4) $\frac{4}{9}$
24. P(a sum of 4 or 10) $\frac{1}{6}$
25. P(both numbers different) $\frac{5}{6}$

Solve.

26. A dish of nuts has 12 walnuts, 8 pecans, and 4 cashews. Bhatti takes one without looking. What is the probability it is a pecan? $\frac{1}{3}$
27. There are 15 girls and 11 boys in the class. One is chosen at random to attend a play. What is the probability that a girl is chosen? $\frac{15}{26}$

Probability Connection

9-2 AN INTRODUCTION TO PROBABILITY

Objective
Find the probability of an event.

Without looking, Jerry takes a marble out of a bag containing 3 red, 4 green, and 1 blue marble. What are the chances that the marble will be red? **The marbles are the same size and shape and are mixed up.**

The **probability** of an event is a ratio that describes how likely it is that the event will occur.

An **event** is a specific outcome or type of outcome.

Method

1. Determine the number of ways the desired event can occur.
2. Determine the number of ways all possible outcomes can occur.
3. Write the ratio the number of ways the desired event can occur to the number of ways all possible outcomes can occur.

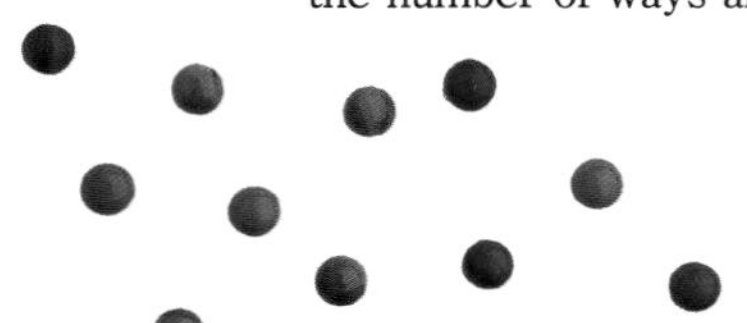

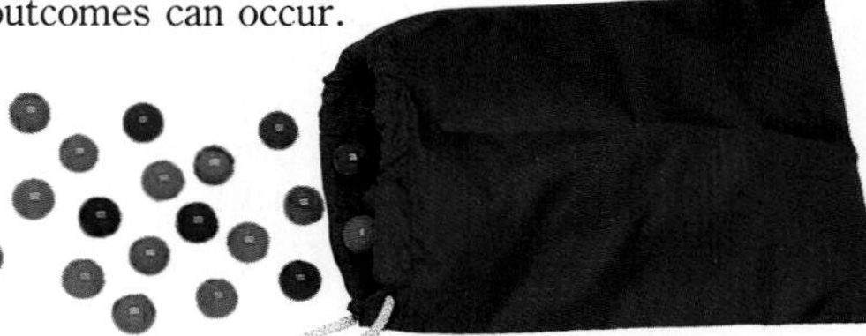

Example A ***What is the probability that Jerry will take out a red marble?***

1. The event is choosing a red marble. There are 3 red marbles. So, there are 3 ways to choose a red marble.
2. Since there are 8 marbles, there are 8 possible outcomes.
3. The probability of choosing a red marble is $\frac{3}{8}$. $P(\text{red}) = \frac{3}{8}$.

Example B ***What is the probability of taking out a green or blue marble?***

1. The event is choosing a green or a blue marble. There are 4 green marbles and 1 blue marble. So, there are 5 ways to choose either a green or a blue marble.
2. There are 8 possible outcomes.
3. $P(\text{green or blue}) = \frac{5}{8}$.

Example C ***What is the probability that Jerry will take out a white marble?***

1. The event is choosing a white marble. Since there are 0 white marbles, the event cannot occur.
2. There are 8 possible outcomes.
3. $P(\text{white}) = \frac{0}{8}$ or 0.

Guided Practice

Find the probability of choosing each marble from the bag described above. Write the probability in simplest form.

Examples A, B

1. green marble $\frac{1}{2}$
2. blue marble $\frac{1}{8}$
3. green or red marble $\frac{7}{8}$

Examples B, C

A drawer contains 3 black socks, 4 navy socks, and 5 white socks. One sock is chosen. Find the probability of each event.

4. white sock $\frac{5}{12}$
5. black sock $\frac{3}{12}$ or $\frac{1}{4}$
6. navy sock $\frac{4}{12}$ or $\frac{1}{3}$
7. white or navy sock $\frac{9}{12}$ or $\frac{3}{4}$
8. not a navy sock $\frac{8}{12}$ or $\frac{2}{3}$
9. brown sock 0

RETEACHING THE LESSON

State the number of possible outcomes.

1. rolling a die 6
2. spinning the spinner 3

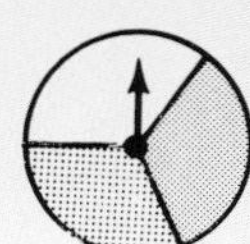

3. letters in the alphabet 26
4. hours in a day 24
5. digits in 9,999 4

Reteaching Masters Booklet, p. 88

Name ______________ Date ______________

RETEACHING WORKSHEET 9-2

An Introduction to Probability

In mathematics, the study of chance is called **probability.**

$$\text{probability of an event} = \frac{\text{number of ways the event can occur}}{\text{number of possible outcomes}}$$

When tossing a coin, there is one way that heads can occur. → 1
There are two possible outcomes—heads or tails. → 2

The probability of heads appearing is $\frac{1}{2}$.

A cooler contains 2 cans of grape juice, 3 cans of pineapple juice, and 7 cans of orange juice. A can of juice is chosen without looking. Find the probability of each event.

1. P(pineapple) $\frac{1}{4}$
2. P(orange) $\frac{7}{12}$
3. P(grape) $\frac{1}{6}$

A die is rolled once. Find the probability of each event.

Exercises

Practice

A die is rolled once. Find the probability of rolling each of the following.

10. a 4 $\frac{1}{6}$ **11.** a 2 or 4 $\frac{1}{3}$ **12.** a 2, 3, or 5 $\frac{1}{2}$ **13.** *not* a 4 $\frac{5}{6}$

14. a number less than 3 $\frac{1}{3}$ **15.** a number greater than 0 $\frac{6}{6}$

A box contains two red pencils, four yellow pencils, and three green pencils. One pencil is chosen. Find the probability of each event.

16. red pencil $\frac{2}{9}$ **17.** yellow pencil $\frac{4}{9}$ **18.** black pencil 0

19. not a red pencil $\frac{7}{9}$ **20.** red, yellow, or green pencil $\frac{12}{12}$

Interpreting Data

The possible outcomes for a roll of two dice are listed at the right. One die is red and the other is blue. Suppose the dice are rolled once. Find each probability.

6,1	6,2	6,3	6,4	6,5	6,6
5,1	5,2	5,3	5,4	5,5	5,6
4,1	4,2	4,3	4,4	4,5	4,6
3,1	3,2	3,3	3,4	3,5	3,6
2,1	2,2	2,3	2,4	2,5	2,6
1,1	1,2	1,3	1,4	1,5	1,6

21. $P(6,1)$ $\frac{1}{36}$ **22.** $P(2,3)$ $\frac{1}{36}$

23. P(even, odd) $\frac{1}{4}$ **24.** P(a sum of 3) $\frac{1}{18}$

25. P(both numbers even) $\frac{1}{4}$ **26.** P(*not* a sum of 7) $\frac{5}{6}$

27. P(a sum of 5 or 11) $\frac{1}{6}$ **28.** P(a sum of at least 10) $\frac{1}{6}$

Applications

29. A box contains 3 red pencils and 2 blue pencils. Brent chooses a red pencil and keeps it. Then Seth chooses a pencil. What is the probability that Seth's pencil is blue? $\frac{1}{2}$

FUN with MATH
Can you sail faster than the wind?
See page 309.

30. A test has three true-false questions. Becky guesses on each question. What is the probability that Becky answers all of the questions correctly? Hint: List all possible sets of three guesses.

31. A tour boat can hold 12 passengers. If 80 people are waiting for the tour, how many boats are needed? 7 boats

30. $\frac{1}{8}$; TTT, TTF, TFF, TFT, FTT, FFT, FFF, FTF

Critical Thinking

32. There are seven basketball teams in the Central Hoosier League. Every year each team plays every other team in the league. How many league games must be scheduled every year? 21 games

Suppose

33. Suppose you have three types of outcomes: *a, b* and *c*. If the probability of either *a* or *b* is $\frac{7}{12}$, how many outcomes of each type are there? Can you find more than one answer? Answers may vary. A sample answer is 3 *a*'s, 4 *b*'s, and 5 *c*'s.

Cooperative Groups

34. Predict how many times a coin will land with heads showing if it is tossed 4 times. Then toss a coin 4 times and record the results. Do the same for 8 tosses, 16 tosses, and 32 tosses. Is tossing a coin close to your prediction when you toss a lesser or greater number of times? Why? Answers may vary. When the number of tosses increases, the experimental probability approaches the theoretical probability.

APPLYING THE LESSON

Independent Practice

Homework Assignment	
Minimum	Ex. 1-31 odd; 32-34
Average	Ex. 2-28 even; 29-34
Maximum	Ex. 1-34

Chalkboard Examples

A bag contains seven pennies. They are dated 1925, 1932, 1936, 1940, 1951, 1954, and 1963. Find the probability of choosing each penny from the bag.

1. Example A an even-numbered year $\frac{4}{7}$

2. Example B before 1940 or after 1960 $\frac{4}{7}$

3. Example C a 1970 penny 0

Additional examples are provided on TRANSPARENCY 9-2.

EVALUATING THE LESSON

Check for Understanding Use Guided Practice Exercises to check for student understanding.

Error Analysis Be alert for students who write the number of remaining outcomes for the denominator, rather than the number of ways all possible outcomes can occur.

Closing the Lesson Have students tell about times when they could use knowledge of a probability.

EXTENDING THE LESSON

Enrichment Masters Booklet, p. 88

Name ______ Date ______

ENRICHMENT WORKSHEET 9-2

More on Probability

All possible outcomes for a roll of two dice are listed below.

1,1	1,2	1,3	1,4	1,5	1,6
2,1	2,2	2,3	2,4	2,5	2,6
3,1	3,2	3,3	3,4	3,5	3,6
4,1	4,2	4,3	4,4	4,5	4,6
5,1	5,2	5,3	5,4	5,5	5,6
6,1	6,2	6,3	6,4	6,5	6,6

Find each probability.

1. P(a sum of 4) $\frac{1}{12}$ **2.** P(a sum less than 7) $\frac{5}{12}$

3. P(a sum greater than 10) $\frac{1}{12}$ **4.** P(a sum less than 15) 1

5. P(a sum greater than 15) 0 **6.** P(a sum of 11) $\frac{1}{18}$

7. P(at least one 2) $\frac{11}{36}$ **8.** P(only one 6) $\frac{5}{18}$

9. P(both numbers the same) $\frac{1}{6}$ **10.** P(a sum of 8) $\frac{5}{36}$

11. P(*not* a sum of 8) $\frac{31}{36}$ **12.** P(a sum of 2 or 3) $\frac{1}{12}$

13. P(a sum less 5 or greater than 10) $\frac{1}{4}$ **14.** P(a sum *not* less than 5 or greater than 10) $\frac{3}{4}$

15. P(number on first die less than number on second die) $\frac{5}{12}$ **16.** P(number on first die greater than number on second die) $\frac{5}{12}$

17. P(both numbers even) $\frac{1}{4}$ **18.** P(first even, second odd) $\frac{1}{4}$

5-MINUTE CHECK-UP

(over Lesson 9-2)

Available on TRANSPARENCY 9-3.

A die is rolled once. Write the probability of rolling each of the following:

1. a 5 $\frac{1}{6}$ **2.** a 1 or 6 $\frac{1}{3}$

3. an even number $\frac{1}{2}$ **4.** a 0 or 7 0

5. a number greater than 4 $\frac{1}{3}$

Extra Practice, Lesson 9-2, p. 472

INTRODUCING THE LESSON

List these records: 6 won, 4 lost; 12 won, 5 lost; and 15 won, 8 lost. Then have students discuss which is the best won-lost record. Why is it best?

TEACHING THE LESSON

Using Questioning

- When determining if two ratios are equivalent, why are fractions changed to simplest form? easier to compare smaller numbers
- What other math topic have you studied that uses the same process as equivalent ratios? equivalent fractions

Practice Masters Booklet, p. 97

Name ______________________ Date ______________

PRACTICE WORKSHEET 9-3

Proportion

Determine if each pair of ratios forms a proportion.

1. 3 to 6, 4 to 5 N	2. 2 to 3, 1 to 2 N	3. 4 to 3, 3 to 4 N
4. 5:25, 2:10 Y	5. 19:20, 38:40 Y	6. 2:5, 11:25 N
7. 4 to 5, 24 to 25 N	8. 28:50, 43:53 N	9. 8 to 13, 32 to 52 Y
10. 30 to 24, 48 to 40 N	11. 35:21, 5:3 Y	12. 6:10, 20:12 N
13. 3 to 6, 30 to 60 Y	14. 5:4, 25:16 N	15. 36:18, 33:15 N
16. 9:2, 27:6 Y	17. 400 to 4, 50 to 5 N	18. 9:15, 36:60 Y
19. 100 to 20, 10 to 2 Y	20. 2 to 7, 6 to 28 N	21. 125:75, 30:18 Y

9-3 PROPORTION

Objective
Detemine if two ratios form a proportion.

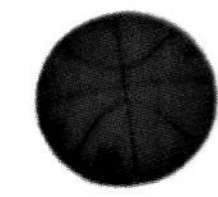

A Tampa High School basketball team won 9 of the first 12 games they played. During the entire season the team won 18 of 24 games. Do the mid-season and end-of-season records form a proportion?

If two ratios are equivalent, they form a **proportion**. The method below shows how to determine if two ratios form a proportion.

Method

1. Write each ratio as a fraction in simplest form.
2. Compare the fractions. Are they equivalent?

Example A

Determine if the ratios 9 to 12 and 18 to 24 form a proportion.

1. $\frac{9}{12} = \frac{3}{4}$ (÷ 3) $\frac{18}{24} = \frac{3}{4}$ (÷ 6)

2. $\frac{3}{4} = \frac{3}{4}$ or $\frac{9}{12} = \frac{18}{24}$

The ratios 9 to 12 and 18 to 24 are equivalent. So, the records form a proportion.

Example B

Determine if the ratios 20:8 and 16:4 form a proportion.

1. $\frac{20}{8} = \frac{5}{2}$ $\frac{16}{4} = \frac{4}{1}$

2. $\frac{5}{2} \neq \frac{4}{1}$ So, it follows that $\frac{20}{8} \neq \frac{16}{4}$

The ratios 20:8 and 16:4 are not equivalent. They do not form a proportion.

Guided Practice

Example A
Example B

Determine if each pair of ratios forms a proportion.

1. 3 to 6, 4 to 8 yes **2.** 12 to 18, 20 to 24 no
3. 9 to 3, 6 to 2 yes **4.** 15:45, 7:28 no
5. 20:16, 35:28 yes **6.** 16:12, 24:20 no

RETEACHING THE LESSON

Write each fraction in simplest form and then read as a ratio.

1. $\frac{8}{10}$ $\frac{4}{5}$ **2.** $\frac{15}{18}$ $\frac{5}{6}$ **3.** $\frac{22}{28}$ $\frac{11}{14}$

4. $\frac{10}{6}$ $\frac{5}{3}$ **5.** $\frac{18}{36}$ $\frac{1}{2}$ **6.** $\frac{19}{57}$ $\frac{1}{3}$

7. $\frac{24}{21}$ $\frac{8}{7}$ **8.** $\frac{15}{41}$ $\frac{15}{41}$ **9.** $\frac{17}{34}$ $\frac{1}{2}$

Reteaching Masters Booklet, p. 89

Name ______________________ Date ______________

RETEACHING WORKSHEET 9-3

Proportion

A sentence that states that two ratios are equivalent is called a **proportion**.

Method to determine if two ratios are equivalent:
① Write each ratio as a fraction in simplest form.
② Compare the fractions.

Examples: Determine if each pair of ratios forms a proportion.

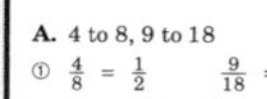

A. 4 to 8, 9 to 18
① $\frac{4}{8} = \frac{1}{2}$ $\frac{9}{18} = \frac{1}{2}$
② $\frac{1}{2} = \frac{1}{2}$
These ratios form a proportion.

B. 12:15, 15:20
① $\frac{12}{15} = \frac{4}{5}$ $\frac{15}{20} = \frac{3}{4}$
② $\frac{4}{5} \neq \frac{3}{4}$
These ratios do not form a proportion.

Write each ratio as a fraction in simplest form. Then determine if each pair of ratios form a proport...

Exercises

Practice

Determine if each pair of ratios forms a proportion.

7. 3 to 4, 9 to 12 **yes** **8.** 4 to 5, 5 to 4 **no** **9.** 8:10, 16:20 **yes**

10. 5 to 8, 2 to 3 **no** **11.** 3:7, 21:49 **yes** **12.** 5:24, 10:12 **no**

13. 14 to 8, 8 to 2 **no** **14.** 2:50, 4:100 **yes** **15.** 18:27, 18:36 **no**

16. 80 to 100, 8 to 10 **17.** 3 to $\frac{1}{3}$, 2 to $\frac{1}{2}$ **no** **18.** 1 to $\frac{2}{3}$, 9 to 6 **yes**

16. yes

19. Explain why the ratios 8:20 and 40:100 form a proportion.

19. They are equivalent. $\frac{8}{20} = \frac{2}{5}$ and $\frac{40}{100} = \frac{2}{5}$

20. Write *true* or *false*. 12:18 and 21:14 form a proportion. **false** Defend your answer. **They are not equivalent.** $\frac{12}{18} = \frac{2}{3}$ and $\frac{21}{14} = \frac{3}{2}$

Mental Math

21. Barb adds 2 quarts of antifreeze to the 5 quarts of water in her radiator. Which of these is an equivalent ratio? 5 to 2, 7 to 10, 8 to 20, 10 to 28.

22. Brian is 16 years old and his brother is 12. Steve is 20 years old and his brother is 15. Len is 24 years old and his brother is 19. Which ratio is not equivalent to the others?

23. Office workers often use 3 in. by 5 in. cards, 4 in. by 6 in. cards, and 5 in. by 8 in. cards. The dimensions of each card form a ratio. Are the ratios for any two cards equivalent? **no**

NFL American Conference Final 1988 Standings

Club	W	L	T
Eastern Division			
Buffalo	12	4	0
Indianapolis	9	7	0
New England	9	7	0
New York Jets	8	7	1
Miami	6	10	0
Central Division			
Cincinnati	12	4	0
Cleveland	10	6	0
Houston	10	6	0
Pittsburgh	5	11	0

Applications

24. Explain whether the 1988 win-loss ratios for the Buffalo Bills and the Cincinnati Bengals were equivalent.

24. yes; $\frac{12}{4} = \frac{12}{4}$

25. Compare the ratio of games lost to total games played for the Indianapolis Colts with the ratio for the New York Jets. Are they equivalent? **yes**

26. The Eastland High School football team won 3 of the 8 games they played. Which NFL team had an equivalent win-loss ratio for 1988? **Miami**

Science Connections

27. Light travels almost 1,900,000 miles in 10 seconds. Would it travel the 93,000,000 miles from the sun to Earth in 5 minutes? **no,** $\frac{190,000}{1} \neq \frac{310,000}{1}$

Using Equations

Solve each equation. Check your solution.

28. $12p = 120$ **10** **29.** $m - 56 = 32$ **88** **30.** $24 = 11 + x$ **13**

31. $68 \div t = 17$ **4** **32.** $r + 234 = 823$ **589** **33.** $8f = 2$ $\frac{1}{4}$

Critical Thinking

34. While cleaning the attic, Mrs. Stamos finds ten boxes. Five contain letters, four contain postcards, two contain both, and the others are empty. How many boxes are empty? **3 boxes**

APPLYING THE LESSON

Independent Practice

Homework Assignment	
Minimum	Ex. 1-27 odd; 28-34
Average	Ex. 2-18 even; 19-34
Maximum	Ex. 1-34

Chalkboard Examples

Example A *Determine if each pair of ratios forms a proportion.*

1. 18 to 24, 27 to 36 **Y**

2. 15 to 10, 10 to 5 **N**

Example B *Determine if each pair of ratios forms a proportion.*

3. 32:20, 20:6 **N**

4. 40:16, 30:12 **Y**

Additional examples are provided on TRANSPARENCY 9-3.

EVALUATING THE LESSON

Check for Understanding Use Guided Practice Exercises to check for student understanding.

Closing the Lesson Ask students to summarize what they have learned about comparing ratios.

EXTENDING THE LESSON

Enrichment Work in small groups. Have students determine the probability of rolling a 1 or an even number. Then have students roll a die 12 times and record the results. Ask students if the ratio obtained is equivalent to the probability.

Enrichment Masters Booklet, p. 89

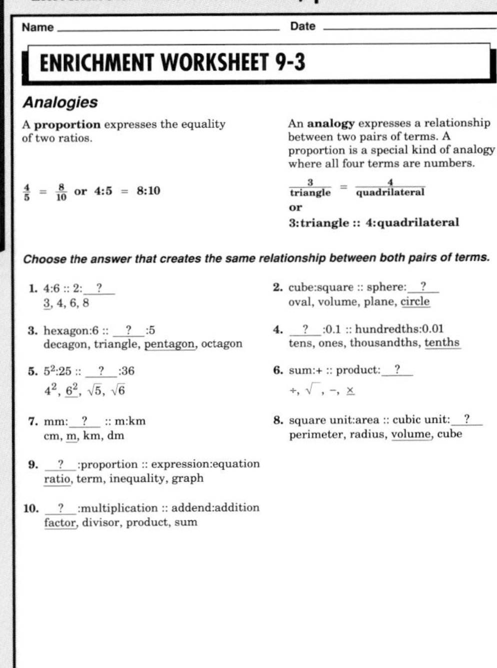

Name ____________ Date ____________

ENRICHMENT WORKSHEET 9-3

Analogies

A **proportion** expresses the equality of two ratios.

$\frac{4}{5} = \frac{8}{10}$ or 4:5 = 8:10

An **analogy** expresses a relationship between two pairs of terms. A proportion is a special kind of analogy where all four terms are numbers.

$\frac{3}{\text{triangle}} = \frac{4}{\text{quadrilateral}}$ or 3:triangle :: 4:quadrilateral

Choose the answer that creates the same relationship between both pairs of terms.

1. 4:6 :: 2:_?_
3, 4, 6, 8

2. cube:square :: sphere:_?_
oval, volume, plane, circle

3. hexagon:6 :: _?_:5
decagon, triangle, pentagon, octagon

4. _?_:0.1 :: hundredths:0.01
tens, ones, thousandths, tenths

5. 5^2:25 :: _?_:36
4^2, 6^2, $\sqrt{5}$, $\sqrt{6}$

6. sum:+ :: product:_?_
÷, $\sqrt{\ }$, −, ×

7. mm:_?_ :: m:km
cm, m, km, dm

8. square unit:area :: cubic unit:_?_
perimeter, radius, volume, cube

9. _?_:proportion :: expression:equation
ratio, term, inequality, graph

10. _?_:multiplication :: addend:addition
factor, divisor, product, sum

5-MINUTE CHECK-UP

(over Lesson 9-3)

Available on TRANSPARENCY 9-4.

Determine if each pair of ratios forms a proportion.

1. 4 to 5, 12 to 15 Y
2. 12 to 8, 4 to 3 N
3. 16:10, 10:4 N
4. 70:100, 7:10 Y

Extra Practice, Lesson 9-3, p. 472

▶ INTRODUCING THE LESSON

Show students a framed picture and ask them to find the dimensions of the frame. Have students describe how they would determine the dimensions for a frame with a larger picture but the same shape.

▶ TEACHING THE LESSON

Using Cooperative Groups Have students roll a pair of dice three times. Then have them create a proportion that includes the three numbers on the dice and solve. Repeat as many times as there are students in the group. Before students begin to solve the proportions, ask them to set a rule for numbers that do not divide evenly.

Practice Masters Booklet, p. 98

Name ______________ Date ______________

PRACTICE WORKSHEET 9-4

Solving Proportions

Solve each proportion.

1. $\frac{3}{4} = \frac{t}{12}$ 9
2. $\frac{5}{g} = \frac{50}{80}$ 8
3. $\frac{b}{8} = \frac{7}{10}$ 5.6
4. $\frac{5}{9} = \frac{70}{r}$ 126
5. $\frac{8}{11} = \frac{c}{44}$ 32
6. $\frac{15}{17} = \frac{3}{h}$ 3.4
7. $\frac{6}{18} = \frac{z}{9}$ 3
8. $\frac{x}{35} = \frac{4}{7}$ 20
9. $\frac{18}{p} = \frac{5}{11}$ 39.6
10. $\frac{21}{24} = \frac{14}{e}$ 16
11. $\frac{16}{d} = \frac{12}{9}$ 12
12. $\frac{s}{14} = \frac{6}{70}$ 1.2
13. $\frac{f}{18} = \frac{24}{9}$ 48
14. $\frac{32}{48} = \frac{4}{z}$ 6
15. $\frac{2}{21} = \frac{g}{84}$ 8
16. $\frac{36}{y} = \frac{8}{12}$ 54
17. $\frac{t}{4} = \frac{51}{18}$ $11.\overline{3}$ or $11\frac{1}{3}$
18. $\frac{6}{7} = \frac{24}{c}$ 28
19. $\frac{13}{52} = \frac{k}{8}$ 2
20. $\frac{35}{w} = \frac{25}{5}$ 7
21. $\frac{m}{90} = \frac{31}{9}$ 310

Algebra Connection

9-4 SOLVING PROPORTIONS

Objective
Solve proportions by using cross products.

Valerie is making pudding using a recipe that calls for 2 cups of milk for 5 servings. Will 6 cups of milk be enough for 15 servings? To find out, see if $\frac{2}{5}$ and $\frac{6}{15}$ form a proportion.

Besides using fractions in simplest form, another way to determine if two ratios form a proportion is to use **cross products.** If the cross products of two ratios are equal, then the ratios are equivalent and form a proportion.

Method

1. Find the cross products.
2. If the cross products are equal, the ratios form a proportion.

Example A ***Use cross products to determine if $\frac{2}{5}$ and $\frac{6}{15}$ form a proportion.***

1.
$$\frac{2}{5} \stackrel{?}{=} \frac{6}{15}$$
$$2 \times 15 \stackrel{?}{=} 5 \times 6$$ The cross products are 2×15 and 5×6.
$$30 = 30$$ Both equal 30.

2. The cross products are equal. $\frac{2}{5}$ and $\frac{6}{15}$ form a proportion. Note that $\frac{6}{15}$ in simplest form is $\frac{2}{5}$.

Six cups of milk will be enough for 15 servings.

Some proportions contain variables. To solve a proportion, use cross products. Then solve the resulting equation.

Method

1. Find the cross products.
2. Solve the equation.

Example B ***Solve the proportion $\frac{7}{8} = \frac{p}{32}$.***

1.
$$\frac{7}{8} = \frac{p}{32}$$
$$7 \times 32 = 8 \times p$$ The cross products are 7×32 and $8 \times p$.

2.
$$224 = 8 \times p$$ Multiply.
$$\frac{224}{8} = \frac{8 \times p}{8}$$ Divide each side by 8.
$$28 = p$$ **Check:** $7 \times 32 \stackrel{?}{=} 8 \times 28$, $224 = 224$ ✓ Replace p with 28 in the equation.

Guided Practice ***Use cross products to determine if each pair of ratios forms a proportion.***

Example A

1. $\frac{4}{6}, \frac{2}{3}$ yes
2. $\frac{4}{2}, \frac{3}{6}$ no
3. $\frac{16}{4}, \frac{20}{8}$ no
4. $\frac{15}{21}, \frac{5}{7}$ yes

RETEACHING THE LESSON

Use cross products to determine if each pair of ratios forms a proportion.

1. $\frac{7}{8}, \frac{6}{7}$ N
2. $\frac{4}{9}, \frac{12}{27}$ Y

Solve each proportion.

3. $\frac{1}{3} = \frac{5}{n}$ $n = 15$
4. $\frac{3}{8} = \frac{9}{x}$ $x = 24$

Reteaching Masters Booklet, p. 90

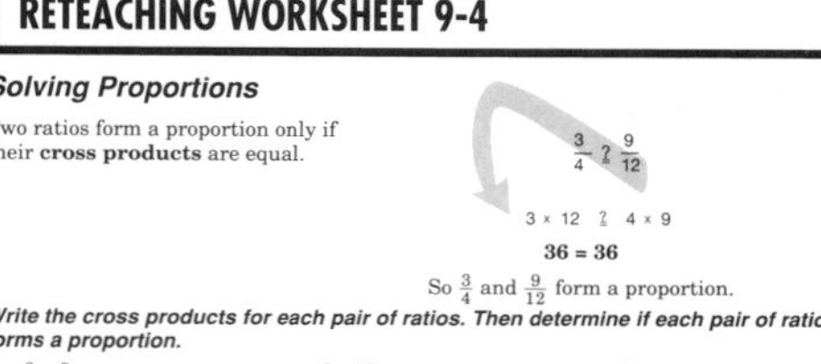

Name ______________ Date ______________

RETEACHING WORKSHEET 9-4

Solving Proportions

Two ratios form a proportion only if their **cross products** are equal.

$\frac{3}{4} \stackrel{?}{=} \frac{9}{12}$

$3 \times 12 \stackrel{?}{=} 4 \times 9$

36 = 36

So $\frac{3}{4}$ and $\frac{9}{12}$ form a proportion.

Write the cross products for each pair of ratios. Then determine if each pair of ratios forms a proportion.

1. $\frac{3}{5}, \frac{9}{15}$ $3 \times 15 \stackrel{?}{=} 5 \times 9$, $45 = 45$, yes
2. $\frac{9}{2}, \frac{18}{5}$ $9 \times 5 \stackrel{?}{=} 2 \times 18$, $45 \neq 36$, no
3. $\frac{20}{50}, \frac{4}{10}$ $20 \times 10 \stackrel{?}{=} 50 \times 4$, $200 = 200$, yes

Example B

Solve each proportion.

5. $\frac{6}{8} = \frac{n}{24}$ 18
6. $\frac{6}{7} = \frac{36}{r}$ 42
7. $\frac{t}{5} = \frac{9}{15}$ 3
8. $\frac{7}{a} = \frac{21}{28}$ 8. $9.\overline{3}$ or $9\frac{1}{3}$

Exercises

Practice

Use cross products to determine if each pair of ratios forms a proportion.

9. $\frac{8}{12}, \frac{9}{15}$ no
10. $\frac{4}{7}, \frac{36}{70}$ no
11. $\frac{21}{28}, \frac{3}{4}$ yes
12. $\frac{4}{15}, \frac{80}{300}$ yes

Solve each proportion.

13. $\frac{1}{2} = \frac{a}{14}$ 7
14. $\frac{3}{4} = \frac{c}{16}$ 12
15. $\frac{f}{15}, \frac{4}{5}$ 12
16. $\frac{g}{32} = \frac{3}{8}$ 12
17. $\frac{5}{9} = \frac{40}{a}$ 72
18. $\frac{14}{16} = \frac{7}{c}$ 8
19. $\frac{3}{5} = \frac{y}{42}$ 25.2
20. $\frac{1}{2} = \frac{x}{15}$ 7.5
21. $\frac{5}{4} = \frac{u}{100}$ 125
22. $\frac{6}{33} = \frac{2}{a}$ 11
23. $\frac{2}{3} = \frac{x}{100}$ 23. $66.\overline{6}$ or $66\frac{2}{3}$
24. $\frac{1}{8} = \frac{n}{100}$ 12.5

Determining Reasonable Answers

25. Tell if the solution is correct. If it is not correct, explain why not.

$$\frac{5}{8} = \frac{n}{24} \qquad \begin{aligned} 5 \times 24 &= 8 \times n \\ 120 &= 8 \times n \\ 120 \times 8 &= n \\ 960 &= n \end{aligned}$$

No, should have divided by 8, not multiplied

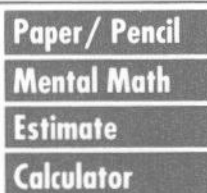
FUN with MATH
Try our recipe for Mexican cocoa. See page 308.

26. Harry uses a recipe that requires 32 ounces of fruit juice to make 2 gallons of punch. He wants to make 6 gallons, so he estimates that he'll need almost 100 ounces of fruit juice. Is this amount reasonable? Why? Yes $2 \times 3 = 6$ and 32×3 is about 100

Applications

Paper/ Pencil
Mental Math
Estimate
Calculator

27. A park ranger stocks a pond with 4 sunfish for every 3 perch. Suppose 296 sunfish are put in the pond. How many perch should be stocked? 222 perch

28. The ratio of land area to total area of Earth's surface is about 3 to 10. If the area of Earth's surface is about 197,000,000 square miles, *about* how many square miles of land are on Earth's surface? Answers may vary. A sample answer is about 60,000,000 sq mi

Cooperative Groups

29. For the three variables *a*, *b*, and *c*, $\frac{a}{b} = \frac{b}{c}$. Find a set of values for *a*, *b*, and *c*. Then find at least one other set of values. Answers may vary. Sample answers are 2, 4, 8, or 3, 6, 12.

Make Up a Problem

30. Make up your own problem that requires solving a proportion. Use three of these numbers: 3, 4, 9, and 15. Answers may vary.

Mixed Review

Subtract.

Lesson 4-13

31. $2\frac{3}{4} - 1\frac{1}{4}$ $1\frac{1}{2}$
32. $5\frac{8}{9} - 3\frac{2}{3}$ $2\frac{2}{9}$
33. $6\frac{5}{8} - 2\frac{1}{6}$ $4\frac{11}{24}$

Lesson 1-13

34. A furniture store bought 8 identical sofas for $4,000. How much did each sofa cost? $500

APPLYING THE LESSON

Independent Practice

Homework Assignment	
Minimum	Ex. 2-28 even; 29-34
Average	Ex. 1-23 odd; 14-34
Maximum	Ex. 1-34

Chalkboard Examples

Example A *Use cross products to determine if each pair of ratios forms a proportion.*

1. $\frac{6}{4}, \frac{3}{2}$ Y
2. $\frac{2}{5}, \frac{40}{100}$ Y

Example B *Solve each proportion.*

3. $\frac{4}{5} = \frac{36}{n}$ 45
4. $\frac{15}{25} = \frac{f}{10}$ 6

Additional examples are provided on TRANSPARENCY 9-4.

EVALUATING THE LESSON

Check for Understanding Use Guided Practice Exercises to check for student understanding.

Error Analysis Be alert for students who cross multiply correctly, but do not divide to solve the resulting equation. Review procedures for solving equations that use multiplication and division.

Closing the Lesson Have students tell about times they could use proportions to solve problems.

EXTENDING THE LESSON

Enrichment Masters Booklet, p. 90

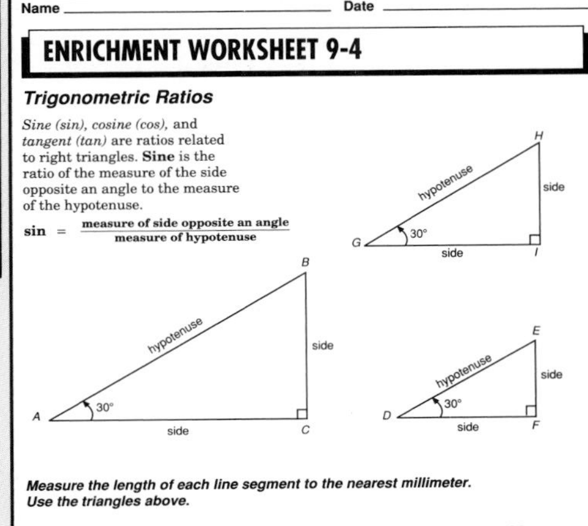

Name ______ Date ______

ENRICHMENT WORKSHEET 9-4

Trigonometric Ratios

Sine (sin), cosine (cos), and *tangent (tan)* are ratios related to right triangles. **Sine** is the ratio of the measure of the side opposite an angle to the measure of the hypotenuse.

$$\sin = \frac{\text{measure of side opposite an angle}}{\text{measure of hypotenuse}}$$

Measure the length of each line segment to the nearest millimeter. Use the triangles above.

1. $\overline{AB}$ 88 mm
2. $\overline{BC}$ 44 mm
3. $\overline{DE}$ 48 mm
4. $\overline{EF}$ 24 mm
5. $\overline{GH}$ 64 mm
6. $\overline{HI}$ 32 mm

Use the measurements in Exercises 1–6 to compute sin 30° for each triangle.

7. $\triangle ABC$: $\sin 30° = \frac{\text{measure of } \overline{BC}}{\text{measure of } \overline{AB}}$ 0.5
8. $\triangle GHI$: $\sin 30° = \frac{\text{measure of } \overline{HI}}{\text{measure of } \overline{GH}}$ 0.5
9. $\triangle DEF$: $\sin 30° = \frac{\text{measure of } \overline{EF}}{\text{measure of } \overline{DE}}$ 0.5
10. What do you notice about your answers to Problems 7, 8, and 9? Would you have expected this? They are the same.

(over Lesson 9-4)

Available on TRANSPARENCY 9-5.

Solve each proportion.

1. $\frac{2}{3} = \frac{n}{36}$ 24
2. $\frac{1}{24} = \frac{1.5}{x}$ 36
3. $\frac{b}{10} = \frac{18}{4}$ 45
4. $\frac{98}{p} = \frac{7}{3}$ 42

Extra Practice, Lesson 9-4, p. 473

▶ INTRODUCING THE LESSON

Have students name items they have seen that were made to scale. **model cars, model houses, maps**

▶ TEACHING THE LESSON

Using Critical Thinking Ask students why order is important when setting up a proportion.

Practice Masters Booklet, p. 99

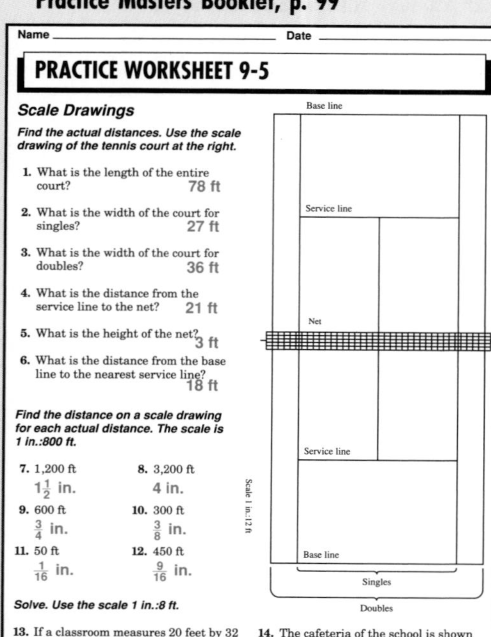
Name ______ Date ______

PRACTICE WORKSHEET 9-5

Scale Drawings

Find the actual distances. Use the scale drawing of the tennis court at the right.

1. What is the length of the entire court? 78 ft
2. What is the width of the court for singles? 27 ft
3. What is the width of the court for doubles? 36 ft
4. What is the distance from the service line to the net? 21 ft
5. What is the height of the net? 3 ft
6. What is the distance from the base line to the nearest service line? 18 ft

Find the distance on a scale drawing for each actual distance. The scale is 1 in.:800 ft.

7. 1,200 ft $1\frac{1}{2}$ in.
8. 3,200 ft 4 in.
9. 600 ft $\frac{3}{4}$ in.
10. 300 ft $\frac{3}{8}$ in.
11. 50 ft $\frac{1}{16}$ in.
12. 450 ft $\frac{9}{16}$ in.

Solve. Use the scale 1 in.:8 ft.

13. If a classroom measures 20 feet by 32 feet, what are the dimensions of the room on the scale drawing? $2\frac{1}{2}$ in. by 4 in.
14. The cafeteria of the school is shown on the drawing as being 10 inches by $12\frac{1}{2}$ inches. What are the actual dimensions of the cafeteria? 80 ft by 100 ft

9-5 SCALE DRAWINGS

Objective
Interpret scale drawings.

Mr. Upton uses a scale drawing as he lays out a baseball field for the new City Park. He determines the actual distances by measuring on the drawing and then solving a proportion.

The scale drawing shows the new baseball field. On the drawing 1 inch represents 144 feet.

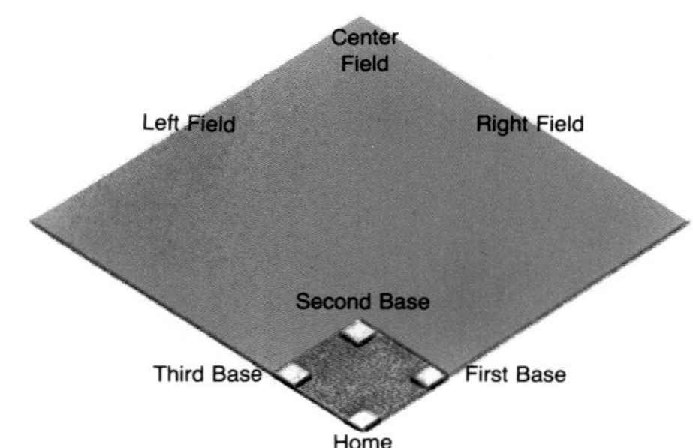

To determine the actual distance, measure the distance on the drawing. Then write and solve the proportion.

Method

1. Write the scale as a ratio.
2. Set up a proportion.
3. Solve the proportion.

Example A ***Find the actual length of the right field line.***

1. $\frac{1}{144}$ ← in. ← ft 1 in. represents 144 feet.
2. On the drawing, the right field line measures $2\frac{1}{2}$ or 2.5 in.

$$\frac{1}{144} = \frac{2.5}{x} \begin{matrix} \leftarrow \text{in.} \\ \leftarrow \text{ft} \end{matrix}$$

3. $1 \times x = 144 \times 2.5$ Use cross products to solve.

$x = 360$ The actual distance is 360 feet.

Example B Ricardo makes a scale drawing of his house. On his drawing 1.5 centimeters represents 5 meters. If the actual dimensions of the living room are 8 m by 6 m, what should be the dimensions on the drawing?

1. $\frac{1.5}{5}$ ← cm ← m
2. $\frac{1.5}{5} = \frac{x}{8}$ ← cm ← m ⟵ (Two proportions must be solved.) ⟶ $\frac{1.5}{5} = \frac{y}{6}$ ← cm ← m
3. $1.5 \times 8 = 5 \times x$ Use cross products. $1.5 \times 6 = 5 \times y$

$12 = 5 \times x$ $9 = 5 \times y$

$\frac{12}{5} = \frac{\overset{1}{\cancel{5}} \times x}{\underset{1}{\cancel{5}}}$ Divide each side by 5. $\frac{9}{5} = \frac{\overset{1}{\cancel{5}} \times y}{\underset{1}{\cancel{5}}}$

$2.4 = x$ $1.8 = y$

On the drawing, the dimensions should be 2.4 cm by 1.8 cm.

RETEACHING THE LESSON

Complete the following table.

	Scale	Actual Length	Scale Drawing Length
1.	5 ft = 1 in.	20 ft	4 in.
2.	7 m = 1 cm	42 m	6 cm
3.	8 yd = 1 in.	56 yd	7 in.
4.	12 km = 1 cm	60 km	5 cm
5.	10 cm = 1 mm	180 cm	18 mm

Reteaching Masters Booklet, p. 91

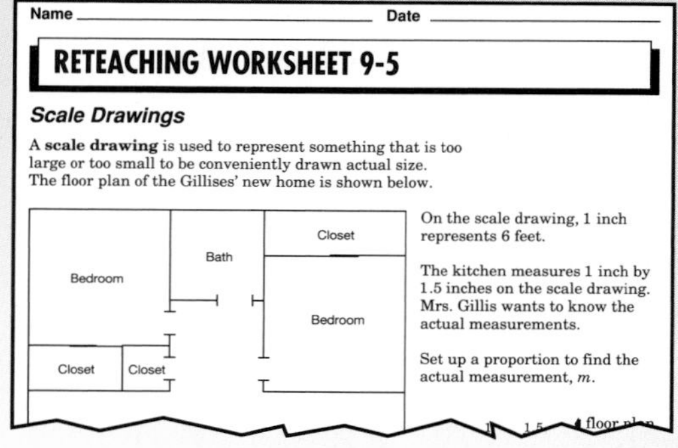
Name ______ Date ______

RETEACHING WORKSHEET 9-5

Scale Drawings

A **scale drawing** is used to represent something that is too large or too small to be conveniently drawn actual size. The floor plan of the Gillises' new home is shown below.

On the scale drawing, 1 inch represents 6 feet.

The kitchen measures 1 inch by 1.5 inches on the scale drawing. Mrs. Gillis wants to know the actual measurements.

Set up a proportion to find the actual measurement, *m*.

Guided Practice *Find the actual distance. Use the scale drawing on page 288.*

Example A

1. the distance from home plate to the center field fence **396 ft**
2. the distance between first base and second base **90 ft**
3. the distance from home plate to second base **117 ft**

Example B

Find the distance on a scale drawing for each actual distance. The scale is 1 in.:50 ft. (1 in.:50 ft means 1 in. represents 50 ft.)

4. 150 ft **3 in.** **5.** 125 ft **$2\frac{1}{2}$ in.** **6.** 75 ft **$1\frac{1}{2}$ in.** **7.** 20 ft **$\frac{2}{5}$ in.** **8.** $6\frac{1}{4}$ ft **$\frac{1}{8}$ in.**

Exercises

Find the actual distances. Use the scale drawing at the right.

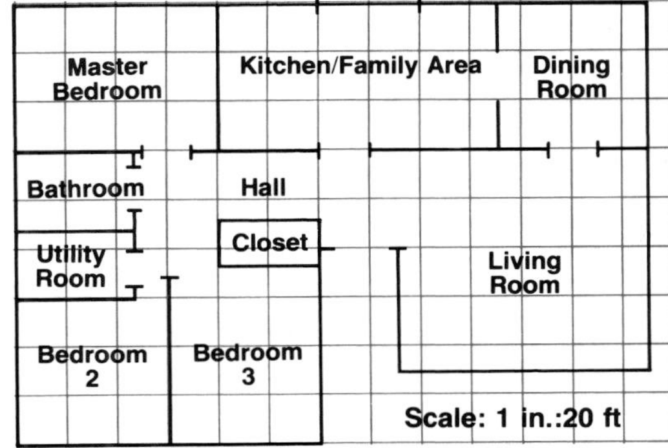

Practice

9. 20 ft × 15 ft
10. 15 ft × 15 ft

9. What are the dimensions of the master bedroom?
10. What are the dimensions of the dining room?
11. What are the dimensions of bedroom 2? **15 ft × 15 ft**
12. How many feet does $\frac{1}{4}$ inch represent on the scale drawing? **5 ft**

Find the distance on a scale drawing for each actual distance. The scale is 1 in.: 200 ft.

13. 300 feet **$1\frac{1}{2}$ in.** **14.** 1,000 feet **5 in.** **15.** 250 feet **$1\frac{1}{4}$ in.** **16.** 175 feet **$\frac{7}{8}$ in.**

17. 50 feet **$\frac{1}{4}$ in.** **18.** 1 foot **$\frac{1}{200}$ in.** **19.** 50 yards **$\frac{3}{4}$ in.** **20.** 200 yards **3 in.**

Applications

21. Use the scale drawing shown above. Tile is sold in pieces that measure 1 foot by 1 foot. How many pieces of tile are needed to cover the bathroom floor? **96 pieces**
22. A car travels 144 miles on 4 gallons of gasoline. At this rate, how many gallons are needed to drive 450 miles? **12.5 gal**

23. Select an exercise set from this chapter that you feel shows your best work. Explain why it is your best work. **See students' work.**
24. To make a scale drawing as large as possible of your home on an $8\frac{1}{2}$ in. by 11 in. sheet of paper, how would you decide what ratio to use for a scale?

24. Answers may vary. A good answer refers to the dimensions of the home and of the paper.

APPLYING THE LESSON

Independent Practice

Homework Assignment	
Minimum	**Ex. 1-23 odd; 24**
Average	**Ex. 2-20 even; 21**
Maximum	**Ex. 1-24**

Chapter 9, Quiz A (Lessons 9-1 through 9-5) is available in the Evaluation Masters Booklet, p. 44.

Chalkboard Examples

Example A *Find the actual distance. Use the scale 1 cm:20 km.*

1. 5 cm **100 km** **2.** 1.5 cm **30 km** **3.** 2.25 cm **45 km**

Example B *Find the distance on a scale drawing for each actual distance. The scale is 1 cm:20 km.*

4. 80 km **4 cm** **5.** 250 km **12.5 cm** **6.** 1 km **0.05 cm**

Additional examples are provided on TRANSPARENCY 9-5.

EVALUATING THE LESSON

Check for Understanding Use Guided Practice Exercises to check for student understanding.

Closing the Lesson Have students tell about times when they could use scale drawings.

EXTENDING THE LESSON

Enrichment Have students make scale drawings of the classroom, a gymnasium, or another room. Be sure the scale they select is appropriate for the sheet of paper they are using.

Enrichment Masters Booklet, p. 91

ENRICHMENT WORKSHEET 9-5

Scale Drawings

On the scale drawing below, 1 cm represents 10 m. Measure for each of the following questions and compute the actual measurement.

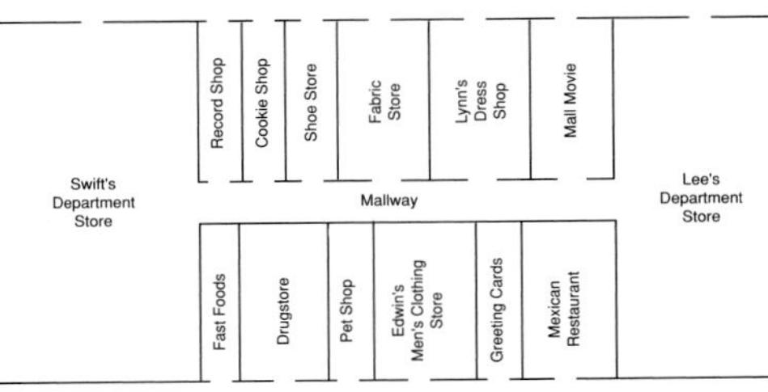

1. What are the dimensions of Swift's Department Store? length 81 m width 45 m
2. What are the dimensions of Lee's Department Store? length 81 m width 35 m
3. What are the dimensions of the mallway? length 92 m width 10 m
4. What is the length of each shop on the mallway? 36 m
5. What is the width of the pet shop? 10 m
6. What is the width of the movie? 18 m
7. What is the width of the drugstore? 19 m
8. What is the width of the Mexican Restaurant? 20 m
9. What is the width of Lynn's Dress Shop? 22 m

Applying Mathematics to the Real World

This optional page shows how mathematics is used in the real world and also provides a change of pace.

Objective Compute annual insurance premiums from a chart.

Using Vocabulary Discuss the terms cash value and beneficiary.

Discuss the following types of insurance: term, straight life, limited pay-life, and endowment.

Using Critical Thinking Why are insurance rates based on age? Why do insurance companies offer so many different types of insurance?

Activity

Using Questioning Have each student choose one of the limited pay-life or endowment policies that they would consider purchasing for themselves. Have them answer the following questions, first using 20 as the age at issue and second using 40 as the age at issue. Then compare answers.

1. What will be the total amount paid at the end of the 20 or 30 years?
2. Will they or their beneficiary receive more or less than the amount paid? How much?

Making Consumer Connections

Life Insurance

Life insurance protects against a financial loss due to death. Three of the four types of life insurance described build up a cash value. **Cash value** is money that the policy accumulates as the premiums are paid. This money may be returned to the policyholder upon retirement age or paid to the **beneficiary** if the policyholder dies before retirement age.

Insurance rates vary according to the age of the insured person and the type of policy. The younger the insured person is when the policy is purchased, the smaller the premium.

Term insurance gives temporary protection. The beneficiary, or beneficiaries, receive the money only if the insured person dies during a certain period of time.

Straight life insurance gives permanent lifetime protection. It builds up a cash value.

Limited pay-life insurance gives permanent lifetime protection. It builds up a cash value. The premiums are paid only for a specific number of years.

Endowment insurance builds up a cash value. The premiums are paid only for a specific number of years. After all the premiums are paid, the policyholder receives the money.

Denise is 20 years old. Use the table below to find her annual premium for a $15,000 straight life policy.

$17.60	premium per $1,000 of insurance
× 15	15 × $1,000 = $15,000 insurance
$264.00	Denise's annual premium is $264.

Annual Premiums Per $1,000 of Insurance

Age at Issue	Term 5-Year	Term 10-Year	Straight Life	Limited Payment 20-Year	Limited Payment 30-Year	Endowment 20-Year	Endowment 30-Year
15			$15.75	$28.10	$22.10	$49.50	$31.90
20	$ 7.75	$ 8.10	17.60	30.60	24.00	49.80	32.40
25	8.50	9.10	20.00	33.30	26.25	50.30	33.20
30	9.70	10.60	22.80	36.50	28.90	51.10	34.40
35	11.50	12.80	26.40	40.15	32.15	52.30	36.20
40	14.30	16.30	31.00	44.50	36.25	54.15	39.00

1. What is Denise's annual premium for a $15,000, 10-year term policy? **$121.50**
2. What is Denise's annual premium for a $20,000, 20-year endowment policy? **$996**
3. At the age of 25, Mr. Gerard bought a $25,000, 30-year limited payment policy. What is the total amount he will pay at the end of 30 years? Will the beneficiary receive more or less than Mr. Gerard paid? How much? **$19,687.50; more; $5,312.50**
4. At age 30, how much 10-year term insurance could you buy for approximately the same annual premium as the straight life insurance? **$2,000 term, $1,000 straight life**

AERIAL PHOTOGRAPHER

Tony Cecil is an aerial photographer. He uses special equipment to take photographs from the air. The photographs are a scale model of what is actually on land. They are used to make maps for travel, to study soil erosion and water pollution, or for military purposes.

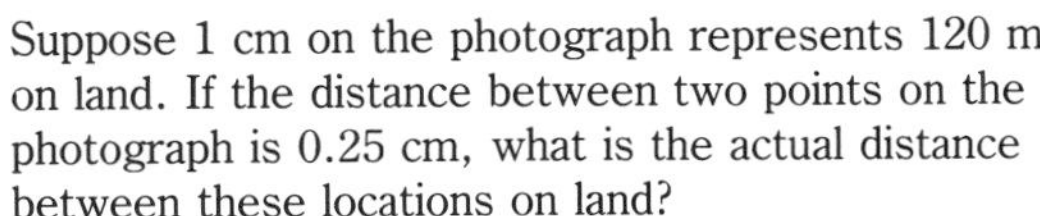

Suppose 1 cm on the photograph represents 120 m on land. If the distance between two points on the photograph is 0.25 cm, what is the actual distance between these locations on land?

Write the scale as a ratio. $\frac{1}{120}$ ← cm ← m

Set up a proportion. $\frac{1}{120} = \frac{0.25}{n}$ 0.25 cm represents n meters.

Solve the proportion. $1 \times n = 120 \times 0.25$

The actual distance is 30 m. $n = 30$

The distance on an aerial photograph is given. Find the actual distance if the scale is 1 cm:360 m.

1. 1.5 cm **540 m** **2.** 0.6 cm **216 m** **3.** 2.8 cm **1,008 m**

The actual distance is given. Find the distance on an aerial photograph if the scale is 1 cm:240 m.

4. 156 m **0.65 cm** **5.** 864 m **3.6 cm** **6.** 1,200 m **5 cm**

The distance on an aerial photograph and the actual distance are given. Find the scale.

7. 100 cm on photograph, 400 m actual **1 cm:4 m**

8. 12 cm on photograph, 360 m actual **1 cm:30 m**

9. On a radar screen the distance between two planes is $3\frac{1}{2}$ inches. If the scale is 1 inch on the screen to 2 miles in the air, what is the actual distance between the two planes? **7 miles**

10. Two planes in flight are 16 km apart. How far apart are they on a radar screen if the scale is 1 cm:1.6 km? **10 cm**

Applying Mathematics to the Work World

This optional page shows how mathematics is used in the work world and also provides a change of pace.

Objective Use a scale to find actual distance and distance on an aerial photograph.

Using Discussion Discuss who would employ an aerial photographer. Also discuss what type of education or training would be required.

Using Calculators Have students solve problems using calculators.

Activity

Using Cooperative Groups Using a state map, have students calculate the mileage from one point to another or have students plan a trip through the state and calculate the mileage.

5-MINUTE CHECK-UP

(over Lesson 9-5)

Available on TRANSPARENCY 9-6.

Find the actual distance. Use the scale 1 cm:20 km.

1. 1.5 cm **30 km** **2.** 10 cm **200 km**

Find the distance on a scale drawing for each actual distance. The scale is 1 cm: 20 km.

3. 100 km **5 cm** **4.** 5 km **0.25 cm**

Extra Practice, Lesson 9-5, p. 473

INTRODUCING THE LESSON

Ask students to think of distances they want to know that cannot be measured directly with a measuring tape; for example height of a building, distance across a river. Can we find such distances? How? **yes, similar figures**

TEACHING THE LESSON

Using Questioning

- How will you know that you have identified the ratios correctly? **height to height, shadow to shadow.**
- If you have difficulty understanding the problem, what could you do? **Make a drawing or diagram.**

Practice Masters Booklet, p. 92

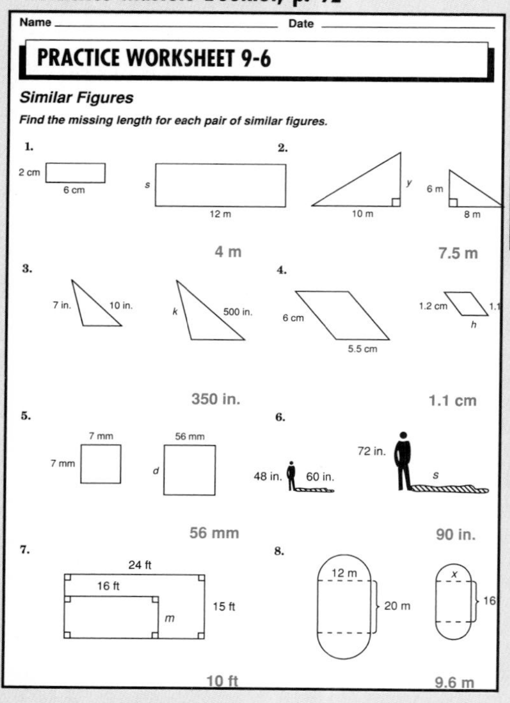

Name ______ Date ______

PRACTICE WORKSHEET 9-6

Similar Figures

Find the missing length for each pair of similar figures.

1. 2 cm, 6 cm; s, 12 m — **4 m**
2. 10 m, y; 6 m, 8 m — **7.5 m**
3. 7 in., 10 in.; k, 500 in. — **350 in.**
4. 6 cm, 5.5 cm; 1.2 cm, h — **1.1 cm**
5. 7 mm, 7 mm; 56 mm, d — **56 mm**
6. 48 in., 60 in.; 72 in., s — **90 in.**
7. 24 ft, 16 ft, 15 ft, m — **10 ft**
8. 12 m, 20 m; x, 16 — **9.6 m**

Geometry Connection

9-6 SIMILAR FIGURES

Objective

Find the measure of a side of a similar figure.

Alissa is a film processor for Speedy Photo Labs. She often enlarges photos during processing. If a figure is enlarged or reduced, the new figure is *similar* to the original.

The measures of corresponding sides of similar figures are proportional. That is, the ratios formed using pairs of corresponding sides are equivalent.

The triangles formed by the pole, tree, and shadows are similar.

Method

1. Identify two corresponding ratios.
2. Set up a proportion.
3. Solve the proportion.

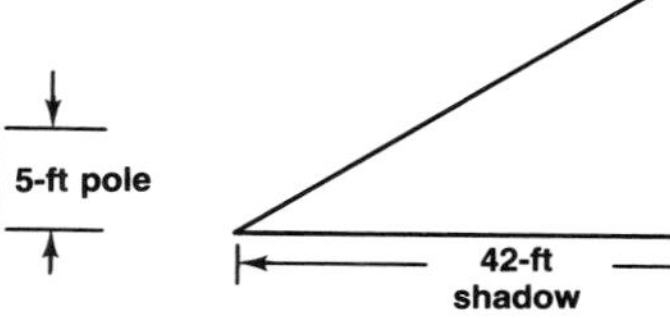

Example A ***Find the height (h) of the tree.***

1. ratio of shadows: 8 to 42 ratio of heights: 5 to h
2. shadow of pole → $\frac{8}{42} = \frac{5}{h}$ ← height of pole; shadow of tree → 42, h ← height of tree
3. $8 \times h = 42 \times 5$ Find cross products.

$8 \times h = 210$ Solve for h.

$\frac{\overset{1}{\cancel{8}} \times h}{\underset{1}{\cancel{8}}} = \frac{210}{8}$ Divide each side by 8.

$h = 26.25$ The height of the tree is 26.25 feet or 26 feet 3 inches.

Guided Practice

Example A

Find the missing length for each pair of similar figures.

1. 5 cm, 3 cm; n, 6 cm **10 cm**

2. 12 in., 9 in.; m, 12 in. **16 in.**

3. 16 cm, 8 cm; 25 cm, p **12.5 cm**

RETEACHING THE LESSON

Give students a copy of the figures shown. Have them find the similar pairs and their corresponding sides.

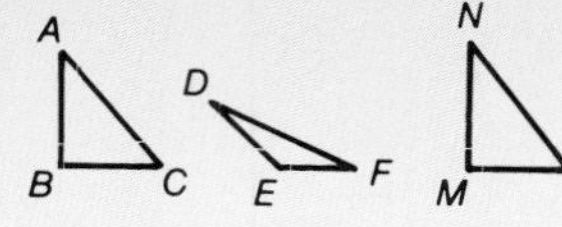

△ABC and △NMP
$\overline{AB}$ corresponds to $\overline{NM}$
$\overline{BC}$ corresponds to $\overline{MP}$
$\overline{AC}$ corresponds to $\overline{NP}$

△DEF and △RST
$\overline{DE}$ corresponds to $\overline{RS}$
$\overline{EF}$ corresponds to $\overline{ST}$
$\overline{DF}$ corresponds to $\overline{RT}$

Reteaching Masters Booklet, p. 100

Name ______ Date ______

RETEACHING WORKSHEET 9-6

Similar Figures

These triangles are similar. The measures of their corresponding sides are proportional, so you can solve a proportion to find the missing length.

6 in., 10 in., 8 in.; 3 in., x, 4 in.

$\frac{6}{3} = \frac{10}{x}$

$6x = 30$

$x = 5$

The missing length is 5 inches.

$\frac{6}{3} = \frac{10}{5} = \frac{8}{4}$

Write three ratios of measures of corresponding sides.

1. 2.

Exercises

Practice

Find the missing length for each pair of similar figures.

4.

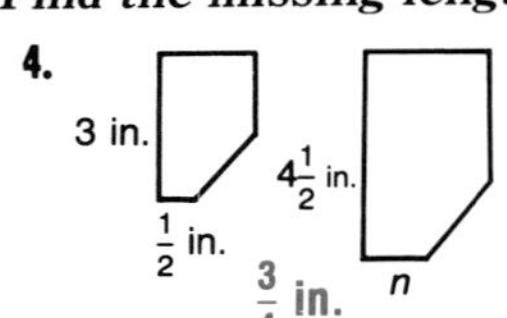

$\frac{3}{4}$ in.

5.

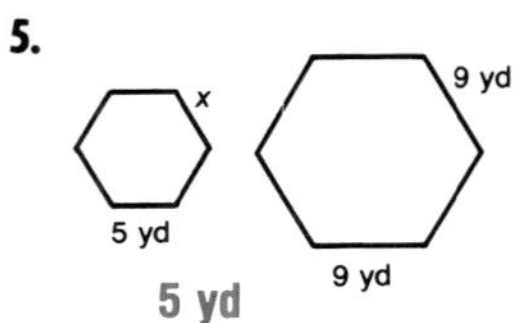

5 yd

6.

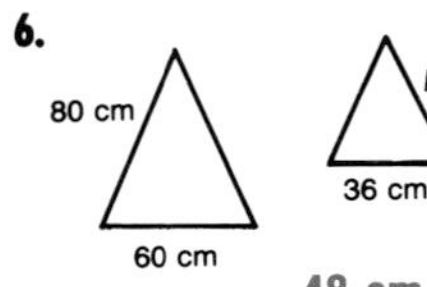

48 cm

7.

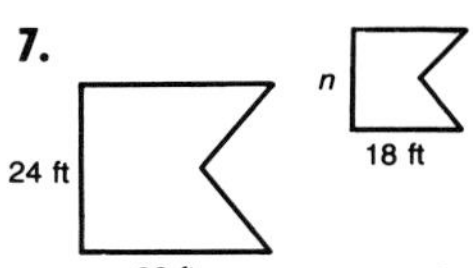

13.5 ft

8. 19 cm

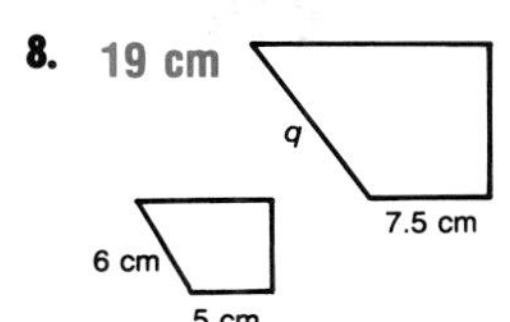

9. 15 m

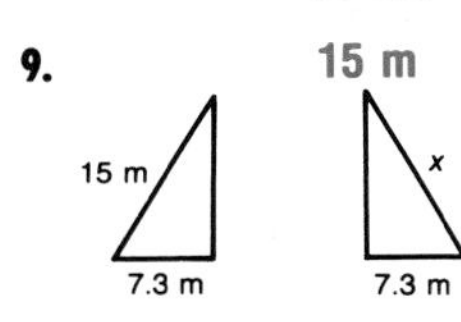

10. 9 ft

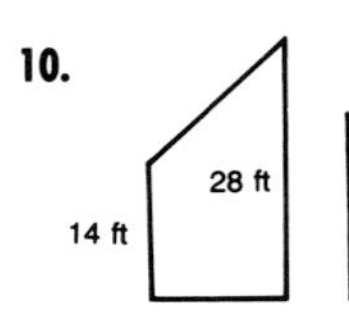

11. 14 cm

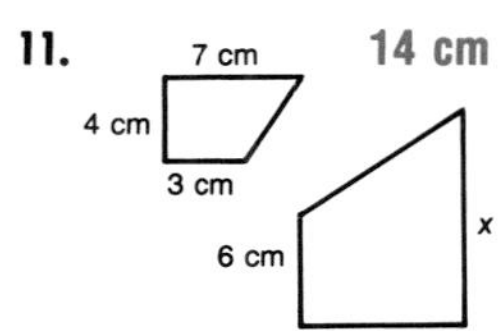

12. 21 ft

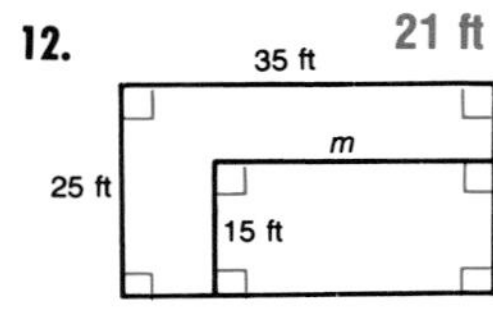

Applications
13. 15.75 m
14. 48 ft
15. 24 m
16. 48 ft

	Height of Pole	Shadow of Pole	Shadow of Tree
13.	3 m	8 m	42 m
14.	4 ft	5 ft	60 ft
15.	36 cm	60 cm	40 m
16.	32 in.	4 ft 2 in.	75 ft

Use the chart at the left to find the height of each tree given the height of the pole and the length of the shadows.

17. The shadow of a 4-foot pole is 6 feet long at the same time that the shadow of a tower is 52.5 feet long. How tall is the tower? 35 ft

18. Sara is 5 feet tall. Her shadow is 9 feet long and the shadow of a building is 36 feet long. How tall is the building? 20 ft

19. Brian is carving a model of a friend's car. The car is 172 in. long and 66 in. high. If the model is to be 6 in. long, how high should it be? 2.3 in.

Art Connection

20. Craig drew a design measuring 5 in. wide by 8 in. long. He wants to enlarge the design to make prints that will be $12\frac{1}{2}$ in. wide. Find the dimensions for the enlarged design. $12\frac{1}{2}$ in. × 20 in.

Critical Thinking

21. Suzanne ran for 20 minutes the first day of her fitness program. Each day she increased her time by a ratio of 2:3; that is, she ran half again as much. How many minutes did she run the fourth day? 67.5 min

Cooperative Groups

22. Find the height of the flagpole at your school by creating similar triangles. Measure the height of one student and the shadows of both the student and the flagpole. Answers may vary.

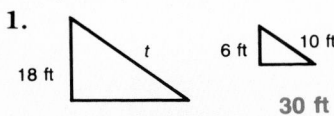

Chalkboard Examples

Example A *Find the missing length for each pair of similar figures.*

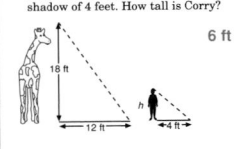

1. 18 ft, t; 6 ft, 10 ft — 30 ft

2. The shadow of the building is 15.5 m, and a 2 m pole has a shadow of 0.25 m. 124 m

Additional examples are provided on TRANSPARENCY 9-6.

EVALUATING THE LESSON

Check for Understanding Use Guided Practice Exercises to check for student understanding.

Error Analysis Watch for students who identify incorrect ratios. Emphasize the need to build each ratio with corresponding parts of the two figures.

Closing the Lesson Ask students to summarize what they have learned about using similar figures.

EXTENDING THE LESSON

Enrichment Masters Booklet, p. 92

Name ______ Date ______

ENRICHMENT WORKSHEET 9-6

Using Indirect Measurement

Solve. Use a proportion.

1. A giraffe is 18 feet tall and casts a shadow of 12 feet. Corry casts a shadow of 4 feet. How tall is Corry? 6 ft

18 ft, h, 12 ft, 4 ft

2. An oak tree casts a shadow of 10 meters, while a 45-meter pine tree casts a shadow of 15 meters. How tall is the oak tree? 30 m

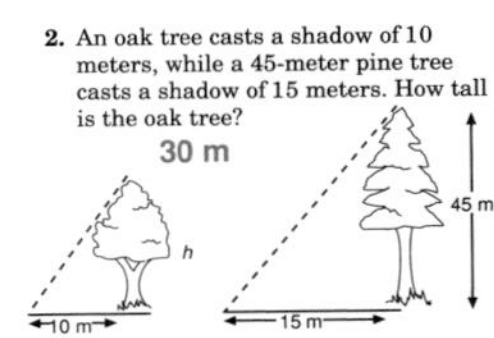

3. A fire hydrant 2.5 feet tall casts a shadow 1.5 feet long. A nearby street sign casts a shadow 6.9 feet long. How tall is the street sign? 11.5 ft

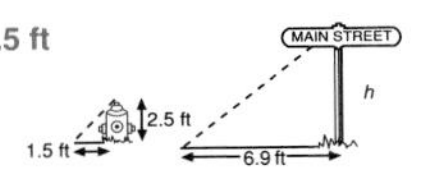

4. A fire truck with its extension ladder raised casts a shadow of 16 meters. A 10-meter house casts a shadow of 5 meters. How tall is the fire truck and ladder? 32 ft

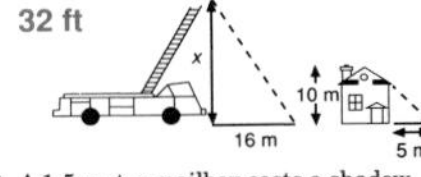

5. A 12-foot basketball pole casts a shadow 6 feet long. A nearby lawn chair is 3 feet high. How long is the chair's shadow? 1.5 ft

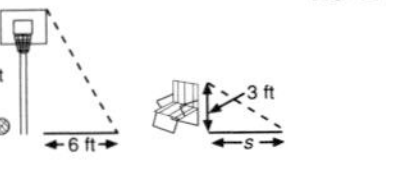

6. A 1.5-meter mailbox casts a shadow 1.2 meters long. A nearby office building casts a shadow 20 meters long. How tall is the building? 25 m

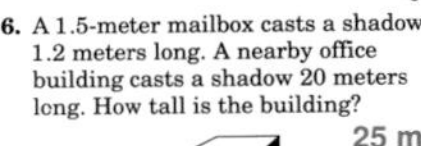

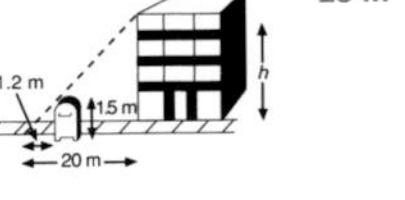

APPLYING THE LESSON

Independent Practice

Homework Assignment	
Minimum	Ex. 1-19 odd; 20-22
Average	Ex. 2-12 even; 13-22
Maximum	Ex. 1-22

5-MINUTE CHECK-UP

(over Lesson 9-6)

Available on TRANSPARENCY 9-7.

Find the missing length for each pair of similar figures.

1. 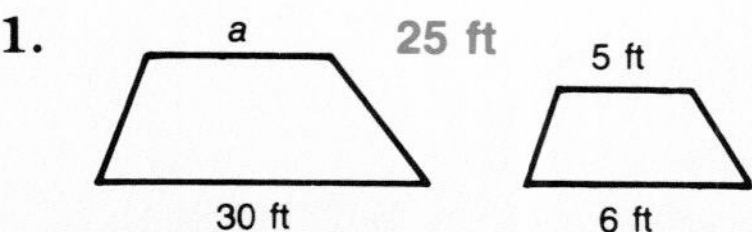

2. The shadow of the flagpole is 9 ft, and a 5 ft pole has a shadow measuring $1\frac{1}{2}$ ft. Find the height of the flagpole. **30 ft**

Extra Practice, Lesson 9-6, p. 473

INTRODUCING THE LESSON

Have students tell how to plan a short trip by car from school to a popular vacation spot or park in your area.

TEACHING THE LESSON

Using Models You may want to bring to class a road atlas. Have students plan a trip from their town to another town in your state. Ask students to name the other towns they would go through, the distance and driving time.

Practice Masters Booklet, p. 101

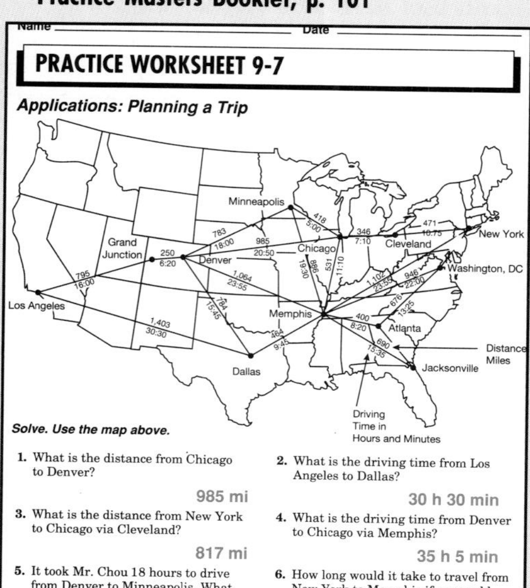

Name ______ Date ______

PRACTICE WORKSHEET 9-7

Applications: Planning a Trip

Solve. Use the map above.

1. What is the distance from Chicago to Denver? **985 mi**
2. What is the driving time from Los Angeles to Dallas? **30 h 30 min**
3. What is the distance from New York to Chicago via Cleveland? **817 mi**
4. What is the driving time from Denver to Chicago via Memphis? **35 h 5 min**
5. It took Mr. Chou 18 hours to drive from Denver to Minneapolis. What was his average speed? **43.5 mph**
6. How long would it take to travel from New York to Memphis if you could average 50 mph? **22.04 h or 22 h 2.4 min**
7. What is the shortest route shown from Minneapolis to Dallas? **Minneapolis to Memphis to Dallas**
8. It took the Sanchez family 34 hours to drive from Minneapolis to Dallas. What route did they most likely take? **Minneapolis to Denver to Dallas**

Applications

9-7 PLANNING A TRIP

Objective

Locate distances and driving times on a map.

Christy and Zac Taylor are planning to drive from Jacksonville, FL, to Denver, CO. They want to visit friends in Memphis and Dallas before going on to Denver. About how many miles will the Taylors drive from Jacksonville to Denver?

Christy and Zac use a map that shows approximate distances and driving times between cities.

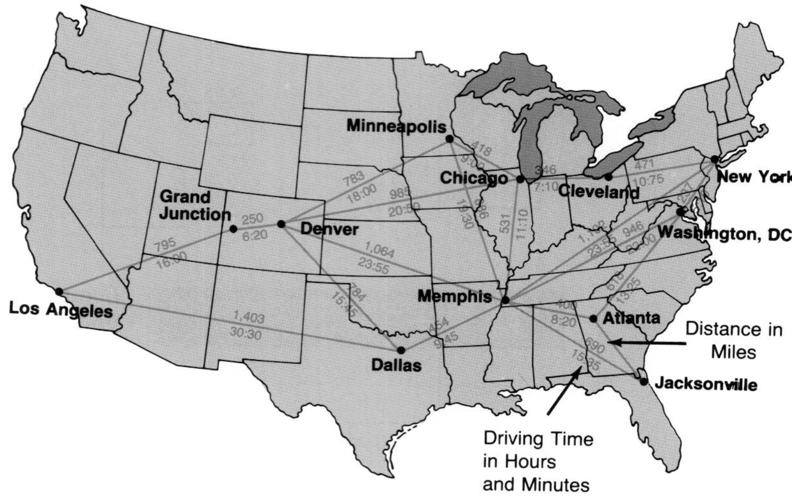

The map shows that the distance from Jacksonville to Memphis is 690 miles, from Memphis to Dallas is 454 miles, and from Dallas to Denver is 784 miles. Add to find the total mileage.

$$690 + 454 + 784 = 1{,}928$$

Christy and Zac will drive 1,928 miles from Jacksonville to Denver.

Understanding the Data

- On the map, what does the number above the line represent? **mileage**
- On the map, what does the number below the line represent? **driving time**
- What does the driving time 5:10 mean? **5 h 10 min**

Guided Practice

Solve. Use the map above.

1. What is the distance from Chicago to Memphis? **531 miles**
2. What is the driving time from Washington, D.C., to Atlanta? **13 h 25 min**
3. Between what two cities (no cities in between) is the driving time and distance the greatest? **Los Angeles and Dallas**

RETEACHING THE LESSON

Solve. Use the map on page 294.

1. What is the shortest route from Los Angeles to Minneapolis? **Los Angeles to Grand Junction to Denver to Minneapolis**
2. What is the distance from Los Angeles to Minneapolis along the route in Problem 1? **1,828 mi**
3. How long would it take you to drive from Los Angeles to Minneapolis using the route in Problem 1? **40 hr 20 min**

Reteaching Masters Booklet, p. 93

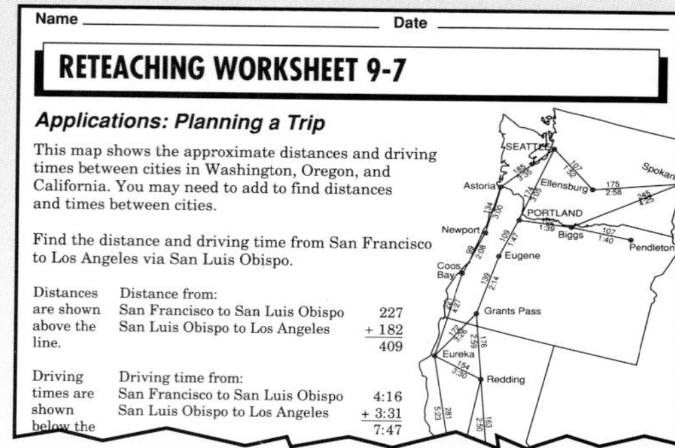

Name ______ Date ______

RETEACHING WORKSHEET 9-7

Applications: Planning a Trip

This map shows the approximate distances and driving times between cities in Washington, Oregon, and California. You may need to add to find distances and times between cities.

Find the distance and driving time from San Francisco to Los Angeles via San Luis Obispo.

Distances are shown above the line. Distance from:
San Francisco to San Luis Obispo 227
San Luis Obispo to Los Angeles + 182
409

Driving times are shown below the line. Driving time from:
San Francisco to San Luis Obispo 4:16
San Luis Obispo to Los Angeles + 3:31
7:47

Applications

Paper/ Pencil
Mental Math
Estimate
Calculator

Solve. Use the map on page 294.

4. How long will it take the Taylors to drive from Jacksonville to Denver via Memphis and Dallas? **41 h 5 min**
5. Name a different route that the Taylor's can take to return home to Jacksonville. **Denver to Memphis to Jacksonville**
6. Mrs. Salgado leaves Chicago at 7:20 A.M. and arrives in Minneapolis at 3:20 P.M. What is her average speed? **52.25 mph**
7. How long would it take you to travel from Los Angeles to Grand Junction at an average speed of 50 miles per hour? **15.9 h or 15 h 54 min**

Cooperative Groups

8. Use the map to plan a trip. Determine how many days the trip will take. If gasoline costs $1.39 a gallon, determine how much you will spend for gasoline. Remember to include your return trip. **See students' work.**

Using Formulas

Solve. Use the formula $d = rt$.

9. A race car driver covers 1,043 kilometers in 3.5 hours. What is the average speed of the driver? **298 kilometers per hour**
10. Min Toshio drove from Cleveland to New York in 10 hours 15 minutes at an average speed of 49 miles per hour. About how many miles did Min travel? **about 500 miles**

Calculator

On the map at the right, one inch represents 50 miles (1 in.:50 mi). Measure with a ruler the map distance between Glendale and Ashville. The map distance is 2.25 inches. What is the actual distance between the towns?

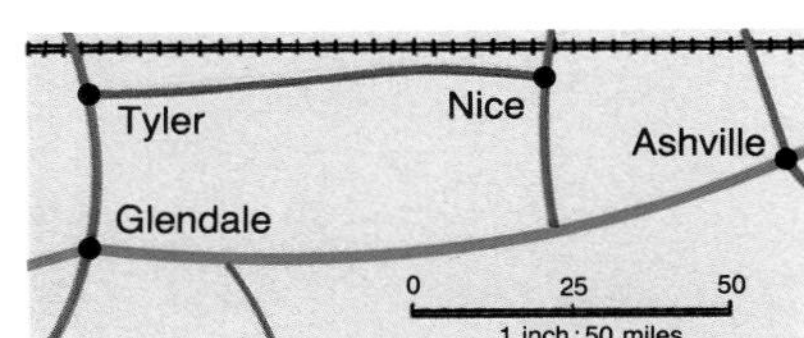

To find the actual distance, multiply 50 miles by the number of inches.

2.25 [×] 50 [=] 112.5 The actual distance is 112.5 miles.

Use the map, the scale, and a ruler to find the actual distance between each city.

11. Nice to Ashville **62.5 mi**
12. Tyler to Nice **75 mi**
13. Glendale to Tyler **25 mi**

Mixed Review

Lesson 2-3

Subtract.

14. 35 − 16 = **19**
15. 435 − 122 = **313**
16. 6,759 − 467 = **6,292**
17. 4,378 − 3,188 = **1,190**

Lesson 6-2

***Name the most reasonable unit for measuring each item. Use* kilometer, meter, centimeter, *or* millimeter.**

18. length of a car **m**
19. the width of a shoelace **mm**

Lesson 6-8

20. Sandra bought milk for $2.19 a gallon at Convenient Mart. She later found milk for $1.20 for 64 ounces at Quick Stop 'n Shop. Which store had the better buy on milk? **Convenient Mart**

APPLYING THE LESSON

Independent Practice

Homework Assignment	
Minimum	Ex. 2-20 even
Average	Ex. 1-20
Maximum	Ex. 1-20

Chalkboard Examples

Solve. Use the map on page 294.

1. The Hays family is planning a trip from New York to Chicago with a stop in Cleveland. What is the distance from New York to Chicago? What is the driving time? **817 miles, 18:25**

Additional examples are provided on TRANSPARENCY 9-7.

EVALUATING THE LESSON

Check for Understanding Use Guided Practice Exercises to check for student understanding.

Error Analysis Make sure students know that the number above the line is distance in miles and the number below the line is time in hours and minutes. Also, reinforce that a time of 15:35 means 15 hours and 35 minutes.

Closing the Lesson Ask students to summarize what they have learned about using a time and distance map.

EXTENDING THE LESSON

Enrichment Masters Booklet, p. 93

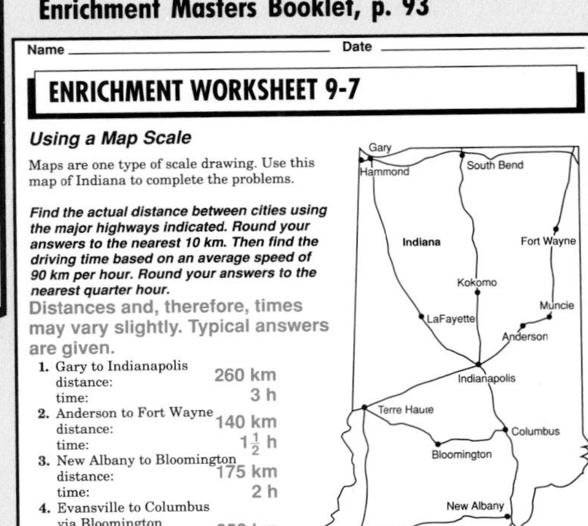

Name ______ Date ______

ENRICHMENT WORKSHEET 9-7

Using a Map Scale

Maps are one type of scale drawing. Use this map of Indiana to complete the problems.

Find the actual distance between cities using the major highways indicated. Round your answers to the nearest 10 km. Then find the driving time based on an average speed of 90 km per hour. Round your answers to the nearest quarter hour.

Distances and, therefore, times may vary slightly. Typical answers are given.

1. Gary to Indianapolis distance: **260 km** time: **3 h**
2. Anderson to Fort Wayne distance: **140 km** time: **$1\frac{1}{2}$ h**
3. New Albany to Bloomington distance: **175 km** time: **2 h**
4. Evansville to Columbus via Bloomington distance: **350 km** time: **4 h**
5. Shortest route from Fort Wayne to Hammond distance: **330km** time: **$3\frac{3}{4}$ h**

Solve.

6. Laura lives in Kokomo and Jenny lives in Fort Wayne. They plan to meet in Indianapolis at 2:00 P.M. If each drives an average speed of 90 km per hour, at what time should each leave her house? Round your answer to the nearest quarter hour. **Laura, 1:00 P.M.; Jenny, 11:45 A.M.**
7. Ms. Merman lives in Indianapolis and must visit each city shown south of Indianapolis once each week. What route should she follow to drive the fewest kilometers? To the nearest 10 kilometers, how far will she drive? **Indianapolis to Terre Haute to Evansville to New Albany to Columbus to Bloomington to Columbus to Indianapolis (or reverse order); 820 km**

5-MINUTE CHECK-UP

(over Lesson 9-7)

Available on TRANSPARENCY 9-8.

1. It took Suzanne 18 hours to drive 795 miles from Grand Junction, Colorado, to Los Angeles, California. What was her average speed? **44.2 mph**
2. If it takes you 13 hours and 25 minutes to drive 618 miles from Atlanta to Washington, D.C., about how fast would you be driving? **about 46 mph**

INTRODUCING THE LESSON

Determine what fractional part of the students in each of two classes rides the bus to school. Explain that we can compare the classes more easily if we use percents.

TEACHING THE LESSON

Using Manipulatives Use Fraction Models, **Lab Manual, pp. 3-4** to make a complete unit with no two pieces the same. Then find the percent for each fraction. Do the percents total 100?

Practice Masters Booklet, p. 102

Name ______________________ Date ______________

PRACTICE WORKSHEET 9-8

Ratios, Percents, and Fractions

Write each fraction as a percent.

1. $\frac{31}{100}$ 31%
2. $\frac{3}{100}$ 3%
3. $\frac{2}{5}$ 40%
4. $\frac{7}{4}$ 175%
5. $\frac{3}{10}$ 30%
6. $\frac{6}{25}$ 24%
7. $\frac{7}{8}$ $87\frac{1}{2}\%$
8. $\frac{9}{20}$ 45%
9. $\frac{4}{5}$ 80%
10. $\frac{17}{50}$ 34%
11. $\frac{3}{4}$ 75%
12. $\frac{19}{25}$ 76%
13. $1\frac{1}{4}$ 125%
14. $1\frac{1}{2}$ 150%
15. $\frac{5}{6}$ $83\frac{1}{3}\%$
16. $\frac{9}{50}$ 18%

Write each percent as a fraction in simplest form.

17. 13% $\frac{13}{100}$
18. 3% $\frac{3}{100}$
19. 65% $\frac{13}{20}$
20. 46% $\frac{23}{50}$
21. 300% 3
22. 55% $\frac{11}{20}$
23. 175% $1\frac{3}{4}$
24. 96% $\frac{24}{25}$
25. 60% $\frac{3}{5}$
26. 37.5% $\frac{3}{8}$
27. 250% $2\frac{1}{2}$
28. $83\frac{1}{3}\%$ $\frac{5}{6}$
29. $33\frac{1}{3}\%$ $\frac{1}{3}$
30. 100% 1
31. 12% $\frac{3}{25}$
32. 0.5% $\frac{1}{200}$

Solve.

33. If $\frac{1}{8} = 12\frac{1}{2}\%$, what percent is equivalent to $\frac{5}{8}$? 62.5%
34. On the average, 7 people out of 28 ride the bus to work. What percent do not ride the bus to work? 75%

9-8 RATIOS, PERCENTS, AND FRACTIONS

Objective
Write percents as fractions, and fractions as percents.

Blake hears the weather forecaster say there is an 80 percent chance of rain during the day. Should Blake take his umbrella with him? An 80 percent chance means that out of 100 days, it would probably rain during 80. Since the chance of rain is high, Blake should take his umbrella.

A ratio that compares a number to 100 may be written as a percent. **Percent** means *hundredths,* or *per 100.*

$$80 \text{ to } 100 = \frac{80}{100} = 80\%$$

The symbol % is the percent symbol. It means that a number is being compared to 100.

A percent can also be written as a fraction.

Method

1. Write the percent in the numerator and 100 in the denominator.
2. Write the fraction in simplest form.

Examples ***Write each percent as a fraction in simplest form.***

A 75%

1. $75\% = \frac{75}{100}$
2. $= \frac{\overset{3}{\cancel{75}}}{\underset{4}{\cancel{100}}}$
   $= \frac{3}{4}$

B 9.5%

1. $9.5\% = \frac{9.5}{100}$
2. $= \frac{9.5 \times 10}{100 \times 10}$
   $= \frac{95}{1{,}000}$
   $= \frac{19}{200}$

C $66\frac{2}{3}\%$

1. $66\frac{2}{3}\% = \frac{66\frac{2}{3}}{100}$
2. $= 66\frac{2}{3} \div 100$
   $= \frac{\overset{2}{\cancel{200}}}{3} \times \frac{1}{\underset{1}{\cancel{100}}}$
   $= \frac{2}{3}$

A fraction can be written as a percent.

Method

1. Set up a proportion with the fraction equal to a ratio with a denominator of 100, $\frac{n}{100}$.
2. Use cross products to solve the proportion.

Examples ***Write each fraction as a percent.***

D $\frac{4}{5}$

1. $\frac{4}{5} = \frac{n}{100}$
2. $4 \times 100 = 5 \times n$
   $400 = 5 \times n$
   $\frac{400}{5} = \frac{5 \times n}{5}$
   $80 = n \qquad \frac{4}{5} = 80\%$

$$\begin{array}{r} 80 \\ 5\overline{)400} \\ -40 \\ \hline 0 \end{array}$$

E $\frac{3}{8}$

1. $\frac{3}{8} = \frac{n}{100}$
2. $3 \times 100 = 8 \times n$
   $300 = 8 \times n$
   $\frac{300}{8} = \frac{8 \times n}{8}$
   $37\frac{1}{2} = n \qquad \frac{3}{8} = 37\frac{1}{2}\%$

$$\begin{array}{r} 37 \text{ R4} \\ 8\overline{)300} \\ -24 \\ \hline 60 \\ -56 \\ \hline 4 \end{array}$$

RETEACHING THE LESSON

Write each fraction as a percent.

1. $\frac{4}{5}$ **80%**
2. $1\frac{3}{4}$ **175%**
3. $\frac{7}{8}$ **87.5%**

Write each percent as a fraction in simplest form.

4. 40% $\frac{2}{5}$
5. $33\frac{1}{3}\%$ $\frac{1}{3}$
6. 112.5% $1\frac{1}{8}$ or $\frac{9}{8}$

Reteaching Masters Booklet, p. 94

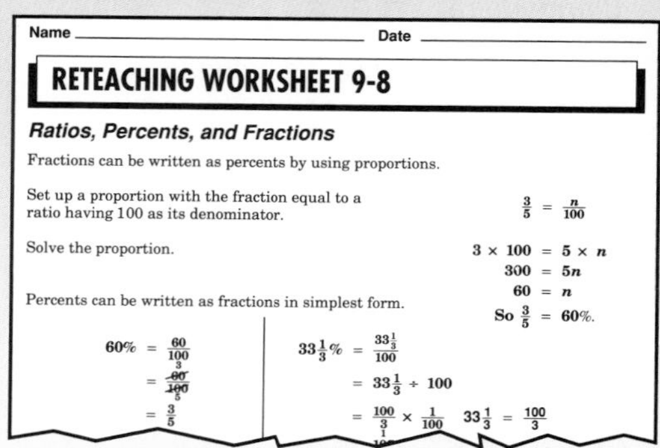

Name ______________________ Date ______________

RETEACHING WORKSHEET 9-8

Ratios, Percents, and Fractions

Fractions can be written as percents by using proportions.

Set up a proportion with the fraction equal to a ratio having 100 as its denominator. $\frac{3}{5} = \frac{n}{100}$

Solve the proportion.
$3 \times 100 = 5 \times n$
$300 = 5n$
$60 = n$
So $\frac{3}{5} = 60\%$.

Percents can be written as fractions in simplest form.

$60\% = \frac{60}{100} = \frac{\overset{3}{\cancel{60}}}{\underset{5}{\cancel{100}}} = \frac{3}{5}$

$33\frac{1}{3}\% = \frac{33\frac{1}{3}}{100} = 33\frac{1}{3} \div 100 = \frac{100}{3} \times \frac{1}{100} \qquad 33\frac{1}{3} = \frac{100}{3}$

Guided Practice

Examples A-C

Write each percent as a fraction in simplest form.

1. 50% $\frac{1}{2}$ **2.** 10% $\frac{1}{10}$ **3.** 200% 2 **4.** 150% $1\frac{1}{2}$ **5.** $12\frac{1}{2}$% $\frac{1}{8}$ **6.** 0.1% $\frac{1}{1,000}$

Examples D, E

Write each fraction as a percent.

7. $\frac{9}{10}$ 90% **8.** $\frac{1}{4}$ 25% **9.** $\frac{5}{8}$ 62.5% **10.** $\frac{17}{20}$ 85% **11.** $\frac{1}{3}$ $33\frac{1}{3}$% **12.** $\frac{1}{8}$ $12\frac{1}{2}$%

Exercises

13. $\frac{29}{100}$ 14. $\frac{1}{100}$ 15. $\frac{13}{20}$ 16. $\frac{7}{20}$ 17. $1\frac{1}{4}$ 18. 1 19. 2

Write each percent as a fraction in simplest form.

13. 29% **14.** 1% **15.** 65% **16.** 35% **17.** 125% **18.** 100%

19. 200% **20.** 3% $\frac{3}{100}$ **21.** 25% $\frac{1}{4}$ **22.** $16\frac{2}{3}$% $\frac{1}{6}$ **23.** 37.5% **24.** 0.3%

FUN with MATH

Try fractions if you come home late from a date.

See page 309.

Write each fraction as a percent. 23. $\frac{3}{8}$ 24. $\frac{3}{1,000}$

25. $\frac{21}{100}$ 21% **26.** $\frac{1}{5}$ 20% **27.** $\frac{1}{4}$ 25% **28.** $\frac{2}{25}$ 8% **29.** $\frac{41}{50}$ 82% **30.** $\frac{19}{20}$ 95%

31. $\frac{1}{8}$ $12\frac{1}{2}$% **32.** $\frac{3}{8}$ $37\frac{1}{2}$% **33.** $\frac{1}{6}$ $16\frac{2}{3}$% **34.** $\frac{5}{4}$ 125% **35.** $\frac{4}{1}$ 400% **36.** $1\frac{1}{2}$ 150%

Applications

Use the circle graph at the right.

37. What percent of 1989 car production in U.S. plants is represented by the graph? 100%

38. What fraction of new cars were produced by General Motors or Ford? $\frac{18}{25}$

39. What fraction of new cars were *not* produced by General Motors or Ford? $\frac{7}{25}$

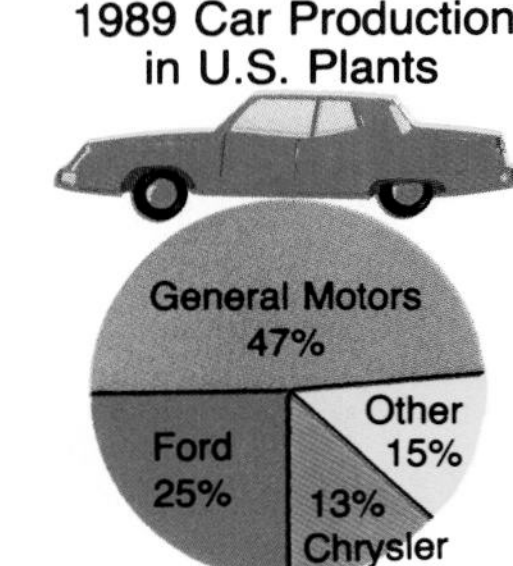

40. Suppose you earned $24 in one week by doing odd jobs. If during the second week you increased your earnings by 150%, how much would you earn the second week? $60

JOURNAL ENTRY

41. Describe one topic in this chapter that you would liked to have spent more time on and explain why. See students' work.

Mental Math

Some percents are used so often in computations that it will save time if you memorize their fractional equivalents. 45. a. $\frac{1}{3}$ b. $\frac{2}{3}$ c. $\frac{3}{3}$ or 1

Write each percent as a fraction. Study each column for a pattern.

42.	**43.**	**44.**	**45.**
a. 10%	**a.** 20%	**a.** 25%	**a.** $33\frac{1}{3}$
b. 30%	**b.** 40%	**b.** 50%	**b.** $66\frac{2}{3}$
c. 70%	**c.** 60%	**c.** 75%	**c.** 100%
d. 90%	**d.** 80%	**d.** 100%	
e. 100%	**e.** 100%		

42. a. $\frac{1}{10}$ b. $\frac{3}{10}$ c. $\frac{7}{10}$ d. $\frac{9}{10}$ e. $\frac{10}{10}$ or 1

43. a. $\frac{1}{5}$ b. $\frac{2}{5}$ c. $\frac{3}{5}$ d. $\frac{4}{5}$ e. $\frac{5}{5}$ or 1

44. a. $\frac{1}{4}$ b. $\frac{2}{4}$ or $\frac{1}{2}$ c. $\frac{3}{4}$ d. $\frac{4}{4}$ or 1

APPLYING THE LESSON

Independent Practice

Homework Assignment	
Minimum	Ex. 1-39 odd; 40-45
Average	Ex. 2-36 even; 37-45
Maximum	Ex. 1-45

Chalkboard Examples

Examples A-C *Write each percent as a fraction in simplest form.*

1. 93% $\frac{93}{100}$ **2.** 20% $\frac{1}{5}$ **3.** 250% $2\frac{1}{2}$

Examples D, E *Write each fraction as a percent.*

4. $\frac{4}{25}$ 16% **5.** $\frac{3}{4}$ 75% **6.** $\frac{5}{6}$ $83\frac{1}{3}$%

Additional examples are provided on TRANSPARENCY 9-8.

EVALUATING THE LESSON

Check for Understanding Use Guided Practice Exercises to check for student understanding.

Closing the Lesson Have students tell about times when they could use percents.

EXTENDING THE LESSON

Enrichment Have students collect data using percents. Do any of the lists of percents total more than 100% If so, why?

Enrichment Masters Booklet, p. 94

Name ______________ Date ______________

ENRICHMENT WORKSHEET 9-8

Permillage

A fraction with a denominator of 100, such as $\frac{25}{100}$, can be thought of as 25 hundredths (0.25) or 25 percent.

A fraction with a denominator of 1,000, such as $\frac{500}{1,000}$, can be thought of as 500 thousandths (0.500) or 500 **permill**. The permill has been used for a long time, particularly by German merchants.

The symbol for percent is %; the symbol for permill is ‰.

Write each as a decimal.

1. 325‰ 0.325 **2.** 405‰ 0.405 **3.** 770‰ 0.77

4. 83‰ 0.083 **5.** 50‰ 0.05 **6.** 9‰ 0.009

7. 1,200‰ 1.2 **8.** 1,050‰ 1.05 **9.** 1,000‰ 1

Write each permill as a fraction in simplest form.

10. 500‰ $\frac{1}{2}$ **11.** 800‰ $\frac{4}{5}$ **12.** 750‰ $\frac{3}{4}$

13. 50‰ $\frac{1}{20}$ **14.** 80‰ $\frac{2}{25}$ **15.** 75‰ $\frac{3}{40}$

16. 5‰ $\frac{1}{200}$ **17.** 8‰ $\frac{1}{125}$ **18.** 7.5‰ $\frac{3}{400}$

Solve. Use a proportion.

19. 500‰ of $60 $30 **20.** 10‰ of $1,000 $10

21. 750‰ of $200 $150 **22.** 250‰ of $7,000 $1,750

23. 5‰ of $50 $0.25 **24.** 1‰ of $400 $0.40

5-MINUTE CHECK-UP

(over Lesson 9-8)

Available on TRANSPARENCY 9-9.

Write each percent as a fraction in simplest form.

1. 60% $\frac{3}{5}$ **2.** 8.5% $\frac{17}{200}$ **3.** $33\frac{1}{3}\%$ $\frac{1}{3}$

Write each fraction as a percent.

4. $\frac{3}{4}$ 75% **5.** $\frac{7}{8}$ $87\frac{1}{2}\%$ **6.** $\frac{5}{6}$ $83\frac{1}{3}\%$

Extra Practice, Lesson 9-8, p. 474

INTRODUCING THE LESSON

Ask students how many cents they pay for sales tax when they buy something for a dollar. Then ask them how they would write that as a percent. For example, 6¢ per dollar is 6%.

TEACHING THE LESSON

Using Critical Thinking Ask students how it is possible to have more than 100%. A good answer will refer to the price of something doubling or going up in price 200%.

Practice Masters Booklet, p. 103

Name ______________________ Date ______________

PRACTICE WORKSHEET 9-9

Percents and Decimals

Write each decimal as a percent.

1. 0.34 34%	**2.** 0.715 71.5%	**3.** 0.04 4%	**4.** 1.52 152%
5. 3.1 310%	**6.** 0.605 60.5%	**7.** 1.2 120%	**8.** 0.6 60%
9. 0.79 79%	**10.** 0.004 0.4%	**11.** 7.7 770%	**12.** 0.09 9%
13. 0.349 34.9%	**14.** 0.1 10%	**15.** 0.076 7.6%	**16.** 5.525 552.5%

Write each percent as a decimal.

17. 65% 0.65	**18.** 26.3% 0.263	**19.** 2% 0.02	**20.** 1.52% 0.0152
21. $\frac{7}{10}\%$ 0.007	**22.** 450% 4.5	**23.** 0.9% 0.009	**24.** 90% 0.9
25. 300% 3	**26.** 0.2% 0.002	**27.** 79% 0.79	**28.** $34\frac{9}{10}\%$ 0.349
29. 7.7% 0.077	**30.** $7\frac{1}{2}\%$ 0.075	**31.** 175% 1.75	**32.** $4\frac{1}{4}\%$ 0.0425

Solve.

33. In a group of 100 people, 52 have brown hair, 29 have blond hair, and 19 people have red hair. What percent of the people have either blond or red hair? 48%

9-9 PERCENTS AND DECIMALS

Objective
Write percents as decimals and decimals as percents.

Mrs. Ohnishi uses a calculator when computing her taxes. When she finds the percent of an amount, she changes the percent to an equivalent decimal.

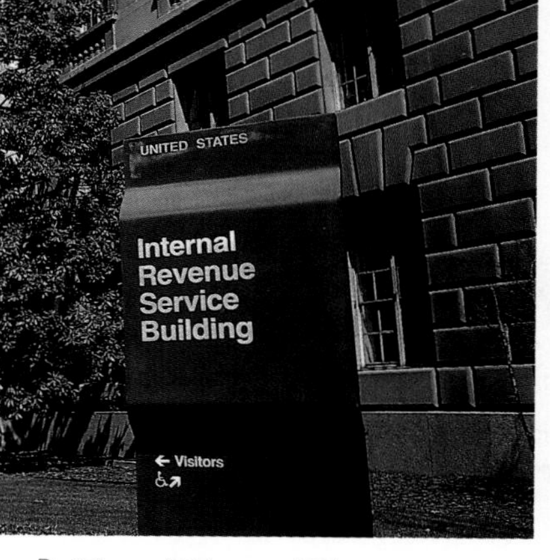

Since percent means hundredths, a percent may be written as a decimal by dividing by 100.

Method

1. Divide the percent by 100. (Move the decimal point two places to the left.)
2. Omit the % symbol.

Examples

Write each percent as a decimal.

A 12.5%
12.5% → 12.5% → 0.125
12.5% = 0.125

B 8%
8% → 08.% → 0.08
8% = 0.08

To write a decimal as a percent, multiply by 100.

Method

1. Multiply the decimal by 100. (Move the decimal point two places to the right.)
2. Write the % symbol.

Examples

Write each decimal as a percent.

C 0.06
0.06 → 0.06 → 6%
0.06 = 6%

D 3.5
3.5 → 3.50 → 350%
3.5 = 350%

Guided Practice

Write each percent as a decimal.

Example A: **1.** 2.5% 0.025 **2.** 73.2% 0.732 **3.** 0.6% 0.006 **4.** 9.9% 0.099 **5.** $\frac{1}{4}\%$ 0.0025

Example B: **6.** 75% 0.75 **7.** 7% 0.07 **8.** 100% 1 **9.** 9% 0.09 **10.** 50% 0.5

13. 1.5% 14. 10.5% 19. 615%

Write each decimal as a percent.

Example C: **11.** 0.12 12% **12.** 0.07 7% **13.** 0.015 **14.** 0.105 **15.** 0.9 90%

Example D: **16.** 7.5 750% **17.** 1.0 100% **18.** 2.0 200% **19.** 6.15 **20.** 8.1 810%

RETEACHING THE LESSON

Choose the percent that is equivalent to each decimal.

1. 0.04 400% 4%
2. 4.36 436% 43.6%
3. 0.007 7% 0.7%
4. 0.6 60% 6%

Choose the decimal that is equivalent to each percent.

5. 12% 0.12 1.2
6. 7% 7 0.07
7. 203% 203 2.03

Reteaching Masters Booklet, p. 95

Name ______________________ Date ______________

RETEACHING WORKSHEET 9-9

Percents and Decimals

Percent means hundredths.

To write a decimal as a percent, multiply by 100. Then write the percent sign (%).
0.75 ▸ 0.75 ▸ 75%

To write a percent as a decimal, divide by 100. Then omit the percent sign (%).
80% ▸ 0.80% ▸ 0.8

Write each decimal as a percent.

1. 0.71 71%	**2.** 0.05 5%	**3.** 0.97 97%	**4.** 0.36 36%
5. 0.333 33.3%	**6.** 0.175 17.5%	**7.** 2.10 210%	**8.** 1.5 150%
9. 0.8 80%	**10.** 0.075 7.5%	**11.** 2.03 203%	**12.** 1.0 100%

Exercises

Practice

Write each percent as a decimal.

21. 12% 0.12 **22.** 73% 0.73 **23.** 30% 0.3 **24.** 8% 0.08

25. 8.1% 0.081 **26.** 200% 2 **27.** 850% 8.5 **28.** 125% 1.25

29. 38.2% 0.382 **30.** $\frac{3}{10}$% 0.003 **31.** $1\frac{1}{2}$% 0.015 **32.** $2\frac{4}{10}$% 0.024

Write each decimal as a percent.

33. 0.68 68% **34.** 0.07 7% **35.** 0.7 70% **36.** 0.15 15%

37. 0.01 1% **38.** 0.009 0.9% **39.** 0.056 5.6% **40.** 0.775 77.5%

41. 1.38 138% **42.** 4.2 420% **43.** 3.0 300% **44.** 0.001 0.1%

Applications

45. The Warriors won 7 out of 15 games. What percent of their games did they win? $46\frac{2}{3}$% or $46.\overline{6}$%

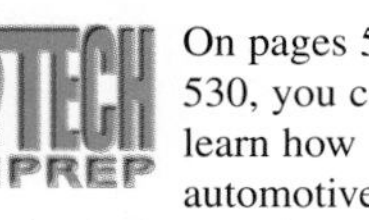

On pages 528–530, you can learn how automotive mechanics use mathematics in their jobs.

46. The Tigers won about 55% of their games. If they played 18 games, how many did they win? 10 games

47. Jackie made 89 of 112 free throws. How many free throws did she miss? 23 free throws

48. Explain how to change $\frac{4}{5}$ to a decimal and then to a percent.

Answers may vary. A good answer refers to division.

49. A hiker is at the campground. If there is an equal probability of choosing each path when a path divides, what percent of the time will the hiker end up in The Swamp?

$33\frac{1}{3}$%

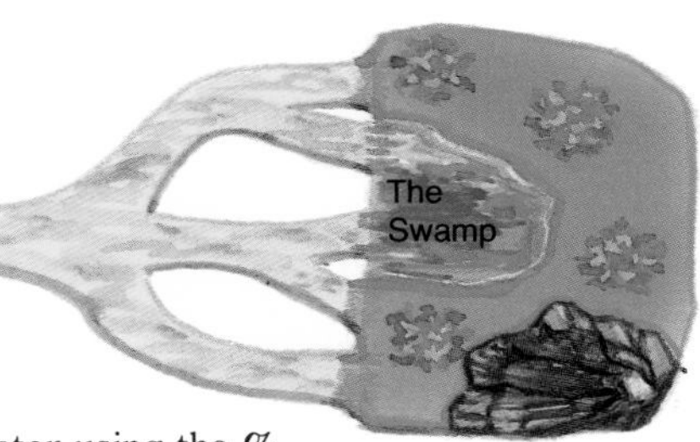

Calculator

You can enter a percent into a calculator using the % key to obtain its decimal equivalent.

Enter: 6 % The calculator divides the
Display: 6 0.06 percent by 100.

Use a calculator to write each percent as a decimal.

50. 25% 0.25 **51.** 5% 0.05 **52.** 30% 0.3 **53.** 100% 1 **54.** 350% 3.5 **55.** 0.5% 0.005

Mixed Review

Lesson 3-1

Multiply.

56. 23 × 12 = 276 **57.** 320 × 36 = 11,520 **58.** 435 × 20 = 8,700 **59.** 701 × 55 = 38,555 **60.** 1,002 × 102 = 102,204

Lesson 7-1

Make a drawing of each figure. See students' work.

61. ray XY **62.** line RS **63.** $\overline{DG}$

Lesson 8-3

64. Find the area of a triangle that has a base of 4.6 cm and has a height of 6.4 cm. 14.72 cm^2

APPLYING THE LESSON

Independent Practice

Homework Assignment	
Minimum	Ex. 2-50 even; 53-64
Average	Ex. 1-47 odd; 49-64
Maximum	Ex. 1-64

Chalkboard Examples

Examples A, B *Write each percent as a decimal.*

1. 37.5% 0.375 **2.** 4% 0.04

Examples C, D *Write each decimal as a percent.*

3. 0.95 95% **4.** 3.0 300%

Additional examples are provided on TRANSPARENCY 9-9.

EVALUATING THE LESSON

Check for Understanding Use Guided Practice Exercises to check for student understanding.

Error Analysis Be alert for students who do not move the decimal point when rewriting a decimal as a percent. Both show hundredths but in different ways.

Closing the Lesson Have students state in their own words how to write a percent as a decimal and how to write a decimal as a percent.

EXTENDING THE LESSON

Enrichment Masters Booklet, p. 95

Name ______ Date ______

ENRICHMENT WORKSHEET 9-9

Percents, Decimals, and Fractions

Solve. Use the chart.

Grade Book — Math Test Scores

Name	Number Right	Number Wrong
Alexander B.	15	5
Alicia F.	20	0
Christopher H.	19	1
Melissa K.	17	3

1. How many questions were on the math test? 20 questions
2. What percent of the questions did Melissa answer correctly? 85%
3. What fraction of the questions did Melissa answer correctly? $\frac{17}{20}$
4. What fraction of the questions did Alexander answer correctly? $\frac{1}{4}$
5. What percent of the questions did Christopher get right? 95%
6. What percent of the questions did Alicia get wrong? 0%
7. What percent of the total questions were answered correctly by the four students? 88.75%
8. What percent of the total questions were answered incorrectly by the four students? 11.25%

Solve. Round your answers to the nearest tenth of a percent.

9. Seven out of the 41 United States presidents were born in Ohio. What percent were born in Ohio? 17.1%
10. Eight of the 41 United States presidents were born in Virginia. What percent were not born in Virginia? 80.5%

5-MINUTE CHECK-UP

(over Lesson 9-9)

Available on TRANSPARENCY 9-10.

Write each percent as a decimal.

1. 65% 0.65 **2.** 8.9% 0.089

Write each decimal as a percent.

3. 4.6 460% **4.** 0.036 3.6%

Extra Practice, Lesson 9-9, p. 474

INTRODUCING THE LESSON

Show students a clear jar filled with beans and ask them to guess how many beans are in the jar. After everyone has guessed, count the beans to find who was closest to the actual answer.
Discuss ways that students made their guesses.

TEACHING THE LESSON

Using Cooperative Groups Divide students into small groups. Groups should use each of the digits 3, 4, 5, and 6 only once to write two 2-digit whole numbers that have the greatest possible product. 63 and 54

Practice Masters Booklet, p. 104

Name ____________ Date ____________

PRACTICE WORKSHEET 9-10

Problem-Solving Strategy: Guess and Check

Solve. Use the guess-and-check strategy.

1. Mrs. Alvirez is three times as old as her daughter. In 12 years, she will be twice as old as her daughter. What are their ages now? 36 and 12
2. Notebook paper can be purchased in packages of 75 or 100 sheets of paper. Niki buys 6 packages and gets 475 sheets. How many packages of 100 sheets of paper does she buy? 1 package
3. Bus fare is 75¢. Rob needed change to ride the bus home from the shopping center. He reached in his pocket and pulled out 5 coins, none larger than a quarter, that totaled 75¢. What were the coins? 2 quarters, 2 dimes, 1 nickel
4. The Murphys spent exactly $13 on movie tickets. Adult tickets cost $3 each. Children's tickets cost $1.75 each. They bought more children's tickets. How many of each did they buy? 2 adult and 4 children's tickets
5. The difference between two whole numbers is 10. Their product is 375. Find the two numbers. 15 and 25
6. Brian is four times as old as his sister Jan. In six years, Brian will be twice as old as Jan. How old is Brian now? 12
7. Jack has 6 coins in his pocket. The coins are nickels, dimes, and quarters. The total value of the coins is 60¢. How many of each coin does Jack have? 1 quarter, 2 dimes, 3 nickels
8. The sum of four consecutive whole numbers is 54. Find the four numbers. 12, 13, 14, 15

Problem-Solving Strategy

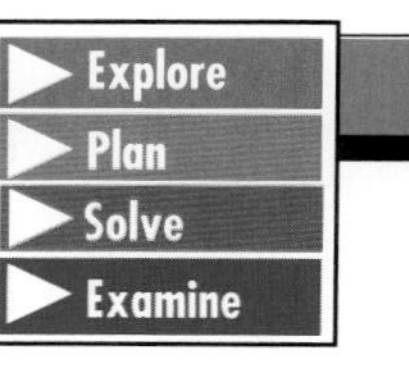

9-10 GUESS AND CHECK

Objective
Solve problems using guess and check.

Ms. Jenkins tells her math class that when a certain number is multiplied by itself, the product is 2,209. What is the number?

The guess-and-check strategy can be used to solve many problems. Here is an outline of how to use this strategy.

- Make a reasonable guess for a solution.
- Check the guess to see if it is correct.
- If it is not correct, decide how to improve the next guess.
- Make another guess and check it.
- Repeat these steps until the problem is solved.

Explore

What is given?

- The product of a certain number multiplied by itself is 2,209.

What is asked?

- What is the number?

Plan

Try multiples of ten to find a reasonable range for the number. Guess again using the ones digit as a clue.

Solve

Think: $40 \times 40 = 1{,}600$
$50 \times 50 = 2{,}500$

The number is between 40 and 50.

Try 45: 45 [×] 45 [=] 2025

The number is between 45 and 50.

Think: The product is 2,209, and we know $7 \times 7 = 49$.

Try a number that ends in a 7 like 47: 47 [×] 47 [=] 2209

The number is 47.

Examine

Since $47 \times 47 = 2{,}209$, the number is 47.

Guided Practice

Solve. Use the guess-and-check strategy.

1. Carla buys film in rolls of 24 and 36 exposures. She buys 5 rolls and gets 132 exposures. How many rolls of each film did she buy? 1-36 roll; 4-24 rolls
2. The sum of three consecutive whole numbers is 54. What are the three numbers? 17, 18, and 19

RETEACHING THE LESSON

Tell whether the next guess should be higher or lower. Then solve.

1. The sum of three consecutive numbers is 66. What are the numbers? First guess, 23, 24, 25. lower; 21; 22; 23
2. When a certain number is multiplied by itself, the product is 3,249. Find the number. First guess, 53. higher; 57

Problem Solving

Solve. Use any strategy.

3. Maggie has eight coins. They are quarters, dimes, and nickels. Their total value is $1.10. How many of each coin does Maggie have? **3 quarters, 2 dimes, and 3 nickels**

4. The sum of three consecutive even numbers is 138. What are the three numbers? **44, 46, and 48**

5. When a certain number is multiplied by itself, the product is 1,521. What is the number? **39**

6. Mike is four times as old as his brother Ben. In eight years, Mike will be twice as old as Ben. How old is Ben now? **4 years old**

7. The difference between two whole numbers is 15. Their product is 1,350. What are the numbers? **30 and 45**

8. On a map, 3 inches represents 450 miles. How many miles are represented by $2\frac{1}{2}$ inches? **375 miles**

9. A basket filled with special candies costs $24. The basket and the candies can be purchased separately. The basket costs $10 less than the candies. How much does the basket cost? **$7**

10. Concert tickets cost $7.50 for adults and $3.50 for children. Aaron's family spent $48.00 for tickets. More adult tickets were bought than children's tickets. How many of each did they buy? **5 adult tickets; 3 children's tickets**

Critical Thinking

11. On a desk calendar like the one at the right, the dates must range from 01-31. What numbers are on the two cubes? **left: 0, 1, 2, 6 (also 9), 7, 8; right: 0, 1, 2, 3, 4, 5**

Mixed Review

Lesson 6-7

Complete.

12. 4 ft = ___?___ in. **48**
13. 72 in. = ___?___ ft **6**
14. $3\frac{1}{3}$ yd = ___?___ in. **120**

Lesson 7-3

***Classify each angle as* right, acute, *or* obtuse.**

15. **obtuse**
16. **acute**
17. **right**

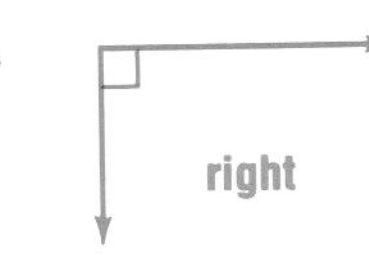

Lesson 8-4

18. The top of a circular plastic lid measures 8 cm from the center to the edge. What is the area of the lid? **201 cm²**

Chalkboard Examples

1. Lisa is three times as old as her niece. In five years, Lisa will be twice as old as her niece. How old are Lisa and her niece now? **15 yr; 5 yr**
2. Luis has nickels, dimes, and quarters in his pocket. If he has nine coins that total $1.05, how many of each coin does Luis have? **2 quarters, 4 dimes, and 3 nickels**

Additional examples are provided on TRANSPARENCY 9-10.

EVALUATING THE LESSON

Check for Understanding Use Guided Practice Exercises to check for student understanding.

Closing the Lesson Ask students to summarize what they have learned about the strategy of guess and check.

EXTENDING THE LESSON

Enrichment Use guess and check to replace each n to make the equation true.

$\frac{n}{81} = \frac{4}{n}$ **18**

APPLYING THE LESSON

Independent Practice

Homework Assignment	
Minimum	Ex. 1-11 odd; 12-18
Average	Ex. 1-18
Maximum	Ex. 1-18

Chapter 9, Quiz B (Lessons 9-6 through 9-10) is available in the Evaluation Masters Booklet, p. 44.

Using the Chapter Review

The Chapter Review is a comprehensive review of the concepts presented in this chapter. This review may be used to prepare students for the Chapter Test.

Chapter 9, Quizzes A and B, are provided in the Evaluation Masters Booklet as shown below.

Quiz A should be given after students have completed Lessons 9-1 through 9-5. Quiz B should be given after students have completed Lessons 9-6 through 9-10.

These quizzes can be used to obtain a quiz score or as a check of the concepts students need to review.

Evaluation Masters Booklet, p. 44

Name ______ Date ______

CHAPTER 9, QUIZ A (Lessons 9-1 through 9-5)

Members of the Ski Club include 14 girls and 12 boys. Write each ratio as a fraction in simplest form.

1. boys to girls 2. girls to total club members — 1. $\frac{6}{7}$ 2. $\frac{7}{13}$

Determine if each pair of ratios forms a proportion.

3. 3:5, 5:8 4. 8 to 6, 12 to 9 — 3. no 4. yes

Solve each proportion.

5. $\frac{x}{21} = \frac{2}{7}$ 6. $\frac{6}{3} = \frac{15}{a}$ — 5. 6 6. 7.5 or $7\frac{1}{2}$

The scale used for a drawing is 1 in.:75 ft.

7. The actual distance is 125 ft. What is the distance on the scale drawing? 7. $1\frac{2}{3}$ in.
8. The distance on the scale drawing is 3 in. What is the actual distance? 8. 225 ft

A box contains 2 blue, 2 yellow, and 4 green marbles. One marble is chosen. Find the probability of each.

9. yellow marble 10. blue or yellow marble — 9. $\frac{1}{4}$ 10. $\frac{1}{2}$

Name ______ Date ______

CHAPTER 9, QUIZ B (Lessons 9-6 through 9-10)

1. Write 65% as a fraction in simplest form. 1. $\frac{13}{20}$

Write each fraction as a percent.

2. $\frac{4}{5}$ 3. $\frac{3}{8}$ — 2. 80% 3. $37\frac{1}{2}$% or 37.5%

Write each percent as a decimal.

4. 7% 5. 73.5% — 4. 0.07 5. 0.735

Write each decimal as a percent.

6. 0.39 7. 0.586 — 6. 39% 7. 58.6%

8. Find the missing length for this pair of similar figures. 8. 3 in.

Solve.

9. It took Mrs. Sanchez 7 hours to drive a distance indicated on a map as 339 miles. What is her average speed to the nearest tenth mile? 9. 48.4 mph
10. Hamburgers cost $1.50 and drinks cost $0.49. If Larry bought only hamburgers and drinks and his bill is $8.94, how many of each did he buy? 10. 4 hamburgers, 6 drinks

---

Chapter 9 REVIEW

Vocabulary / Concepts

Write the letter of the word or phrase that best matches each description.

1. a symbol used to stand for a quantity **j**
2. a ratio that describes how likely it is that an event will occur **c**
3. a comparison of two numbers **f**
4. a ratio that compares a number to 100 **d**
5. a ratio that compares distances **g**
6. a sentence that states that two ratios are equivalent **e**
7. geometric figures with corresponding sides that are proportional **h**
8. used to determine if two ratios form a proportion **a**

a. cross products
b. equivalent
c. probability
d. percent
e. proportion
f. ratio
g. scale
h. similar
i. solution
j. variable

Exercises / Applications

Lesson 9-1

Write each ratio as a fraction in simplest form.

9. 24 wins to 18 losses $\frac{4}{3}$
10. 14 wins in 35 total games $\frac{2}{5}$
11. 336 miles with 16 gallons $\frac{21}{1}$
12. 16 passengers on 4 boats $\frac{4}{1}$

Lesson 9-2

A die is rolled once. Write the probability of rolling each of the following.

13. a 0 or 6 $\frac{1}{6}$
14. a 2 or 4 $\frac{1}{3}$
15. a number greater than 4 $\frac{1}{3}$

Lesson 9-3

Determine if each pair of ratios forms a proportion.

16. 2 to 3, 3 to 4 **N**
17. 16 to 12, 48 to 36 **Y**
18. 150:450, 13:39 **Y**

Lesson 9-4

Solve each proportion.

19. $\frac{3}{8} = \frac{m}{40}$ **15**
20. $\frac{2}{5} = \frac{10}{b}$ **25**
21. $\frac{t}{7} = \frac{54}{63}$ **6**
22. $\frac{81}{n} = \frac{150}{100}$ **54**
23. A 16-ounce box of cereal costs $1.80. At that rate, how much should a 20-ounce box cost? **$2.25**
24. Leslie's car can go 95 miles on 5 gallons of gasoline. How many gallons of gasoline does Leslie need to drive 304 miles? **16 gal**

Lesson 9-5

Find the distance on a scale drawing for each actual distance. The scale is 1 in.:12 mi.

25. 36 mi **3 in.**
26. 54 mi **$4\frac{1}{2}$ in.**
27. 96 mi **8 in.**
28. 3 mi **$\frac{1}{4}$ in.**

Lesson 9-6 ***Find the missing length for each pair of similar figures.***

29.

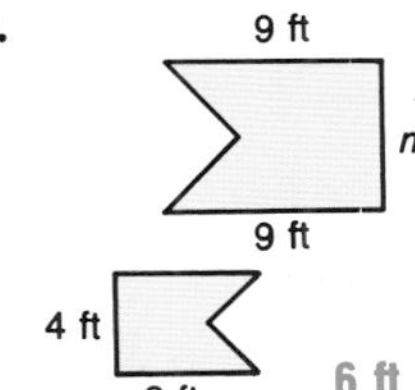

60 cm

30.

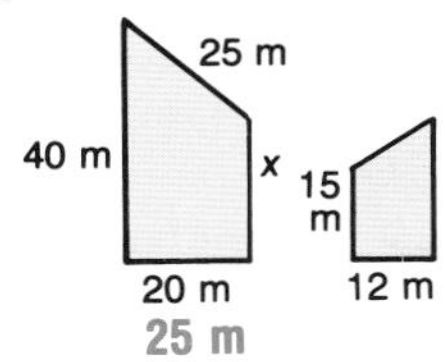

6 ft

31.

25 m
40 m
x
15 m
20 m
12 m

25 m

32. The shadow of a tree is 32 meters long at the same time that the shadow of a meterstick is 1.6 meters long. How tall is the tree? 20 m

Lesson 9-7 **33.** Anthony drives at an average speed of 52.5 miles per hour. After 4 hours 30 minutes, how many miles has he driven? 236.25 mi

34. A race car driver drives 426 miles at an average speed of 142 miles per hour. How long does it take her to drive the 426 miles? 3 hours

Lessons 9-8, 9-9 ***Copy and complete each table. Write fractions in simplest form.***

	Percent	Fraction
35.	16%	$\frac{4}{25}$
36.	92.3%	$\frac{923}{1,000}$
37.	9%	$\frac{9}{100}$
38.	35%	$\frac{7}{20}$

	Percent	Decimal
39.	16%	0.16
40.	3.1%	0.031
41.	89%	0.89
42.	235%	2.35

	Percent	Fraction
43.	50%	$\frac{1}{2}$
44.	$33\frac{1}{3}$%	$\frac{1}{3}$
45.	25%	$\frac{1}{4}$
46.	20%	$\frac{1}{5}$

Lesson 9-10 ***Solve. Use the guess-and-check strategy.***

47. The difference between two whole numbers is 7. Their sum is 99. What are the numbers? 53, 46

48. Mr. Ling is four times as old as his daughter. In 20 years he will be twice as old as his daughter. What are their ages now? 10 yr, 40 yr

APPLYING THE LESSON

Independent Practice

Homework Assignment	
Minimum	**Ex. 1-8; 10-48 even**
Average	**Ex. 1-47 odd**
Maximum	**Ex. 1-48**

Alternate Assessment Strategy

To provide a brief in-class review, you may wish to read the following questions to the class and require a verbal response.

1. What is the ratio of 30 wins to 10 losses, in lowest terms? $\frac{3}{1}$
2. What is the probability of rolling a 6 when a die is rolled? $\frac{1}{6}$
3. Does the pair of ratios, 6 to 9 and 2 to 3, form a proportion? yes
4. Solve the proportion where 1 to 2 equals x to 8. 4
5. If the scale on a map is 1 in. to 200 miles, how many miles are in 3.5 inches? 700 miles
6. Two rectangles are similar. One rectangle measures 12 in. by 6 in. What is the length of the shorter side of the small rectangle if the longer side is 10 in.? 5 in.
7. What is $\frac{4}{5}$ as a percent? 80%
8. What is 35% as a fraction in simplest form? $\frac{7}{20}$
9. What is 0.45 as a percent? 45%
10. What is 148% as a decimal? 1.48

Using the Chapter Test

This page may be used as a test or as an additional page of review if necessary. Two forms of a Chapter Test are provided in the Evaluation Masters Booklet. Form 2 (free response) is shown below, and Form 1 (multiple choice) is shown on page 307.

The **Tech Prep Applications Booklet** provides students with an opportunity to familiarize themselves with various types of technical vocations. The Tech Prep applications for this chapter can be found on pages 17–18.

Evaluation Masters Booklet, p. 43

CHAPTER 9 TEST, FORM 2

The Wildcats won 15 basketball games and lost 10 games. Write each ratio as a fraction in simplest form.

1. wins to losses **2.** wins to total games played

1. $\frac{3}{2}$ 2. $\frac{3}{5}$

Determine if each pair of ratios forms a proportion.

3. 3 to 5, 12 to 16 **4.** 32:8, 12:3

3. no 4. yes

Solve each proportion.

5. $\frac{5}{4} = \frac{10}{a}$ **6.** $\frac{3}{4} = \frac{x}{5}$

5. 8 6. 3.75 or $3\frac{3}{4}$

The scale used for a drawing is 1 in.:40 ft.

7. The actual distance is 120 ft. What is the distance on the scaled drawing? 7. 3 in.

8. The distance on the scale drawing is $4\frac{1}{4}$ in. What is the actual distance? 8. 170 ft.

Write each fraction as a percent.

9. $\frac{2}{5}$ **10.** $\frac{7}{8}$

9. 40% 10. $87\frac{1}{2}$ or 87.5%

11. Write 56% as a fraction in simplest form. 11. $\frac{14}{25}$

Write each percent as a decimal.

12. 52% **13.** 9%

12. 0.52 13. 0.09

14. Write 0.485 as a percent. 14. 48.5%

15. Find the missing length for this pair of similar figures. (n, 5 m; 3 m, 12 m) 15. 1.25 m or $1\frac{1}{4}$

A die is rolled once. Find the probability of rolling each of the following.

16. a 2 **17.** a 4 or 5

16. $\frac{1}{6}$ 17. $\frac{1}{3}$

Solve.

18. A map indicates that the driving time from Bluffton to Prairie City is 5:30. If this is based on driving an average speed of 50 mph, what is the approximate distance from Bluffton to Prairie City? 18. 275 mi

19. The sum of two numbers is 22. The product of the two numbers is 72. What are the two numbers? 19. 4, 18

20. David gave Mark dimes and quarters for 18 nickels. If David gave Mark 6 coins, how many of each coin did Mark receive? 20. 4 dimes, 2 quarters

Bonus The Cougars won 5 out of 13 games. Which is larger, their win-to-loss ratio or their losses-to-games played ratio? How much larger? **Bonus** Their win-to-loss ratio.

Chapter 9 TEST

Two pennies are tossed. Write the probability of each toss.

1. 2 heads $\frac{1}{4}$ **2.** 1 head and 1 tail $\frac{1}{2}$ **3.** no heads $\frac{1}{4}$

Solve each proportion.

4. $\frac{5}{8} = \frac{n}{44}$ 27.5 **5.** $\frac{3}{7} = \frac{48}{m}$ 112 **6.** $\frac{72}{b} = \frac{87}{29}$ 24

7. Greg's team ended the season with a 5:4 won-lost ratio. They lost 12 games. How many games did they win? 15

Find the distance on a scale drawing for each actual distance. The scale is 1 cm:20 km.

8. 160 km 8 cm **9.** 10 km 0.5 cm **10.** 65 km 3.25 cm

Find the missing length for each pair of similar figures.

11. (10 cm, 8 cm, 6 cm; 75 cm, n) 45 cm

12.

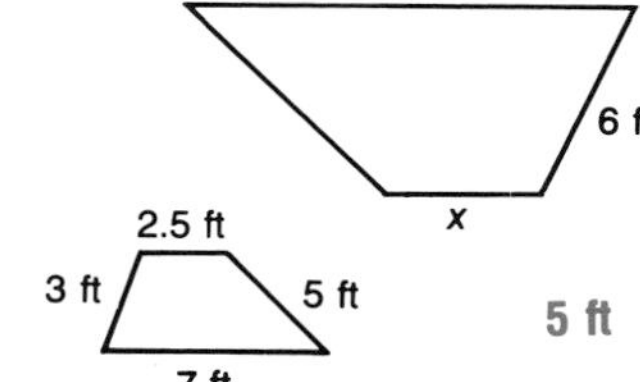

(3 ft, 2.5 ft, 5 ft, 7 ft; 6 ft, x) 5 ft

Solve.

13. The Dillers drove 1,102 miles from Memphis to New York. If it took them 24 hours, about what was their average speed? about 50 mph

Copy and complete each table. Write fractions in simplest form.

	Percent	Fraction
14.	32%	$\frac{8}{25}$
15.	68.7%	$\frac{687}{1,000}$
16.	20%	$\frac{1}{5}$

	Percent	Decimal
17.	32%	0.32
18.	2.8%	0.028
19.	280%	2.8

	Percent	Fraction
20.	$66\frac{2}{3}\%$	$\frac{2}{3}$
21.	75%	$\frac{3}{4}$
22.	$12\frac{1}{2}\%$	$\frac{1}{8}$

23. Out of 50 students in the band, 16 are also in the chorus. What percent are also in the chorus? 32%

Solve. Use the guess-and-check strategy. 24. 1 nickel, 2 dimes, 3 quarters

24. Manuel has only nickels, dimes, and quarters. The six coins have a total value of $1.00. How many of each coin does Manuel have?

25. A whole number is multiplied by itself. When the whole number is subtracted from the product, the result is 182. What is the number? 14

BONUS: Create a proportion in which *a*, *b*, and *c* are whole numbers, $a:b = b:c$, and $c = 16a$. $\frac{1}{4} = \frac{4}{16}$

The **Performance Assessment Booklet** provides an alternative assessment for evaluating student progress. An assessment for this chapter can be found on pages 17–18.

A **Test and Review Generator** is provided in Apple, IBM, and Macintosh versions. You may use this software to create your own tests or worksheets, based on the needs of your students.

GASOLINE MILEAGE

Gasoline mileage is the number of miles a car travels per gallon of gasoline.

Each week Tamara computes her gas mileage (mpg). She starts each week with a full tank of gasoline. She records the odometer reading each time she fills her gas tank.

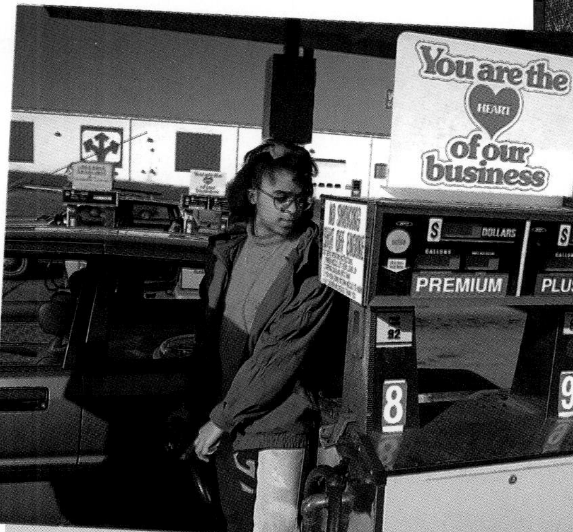

Start of week

End of week

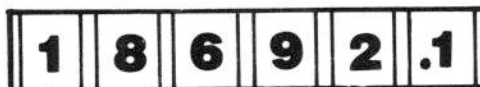

Tamara subtracts the odometer readings to find the miles driven.

18,692.1	end of week
− 18,398.1	start of week
294.0	miles driven

Tamara drove 294 miles.

The fill-up took 14.7 gallons. She uses this formula to find the gas mileage.

$$\text{mpg} = \frac{\text{miles driven}}{\text{number of gallons}}$$

$$= \frac{294}{14}$$ 294 [÷] 14.7 [=] 20

Tamara's car got 20 miles per gallon.

Find the miles per gallon for each car. Round decimal answers to the nearest tenth.

	Car	Miles Driven	Number of Gallons	
1.	Dave's Honda	524.4	15.2	34.5 mpg
2.	Kim's Ford	156.0	7.5	20.8 mpg

3. Julio drove a total of 320.6 miles and used 18.2 gallons of gasoline. What is his gas mileage? 17.6 mpg

4. How far can Jason travel on a tankful of gasoline if his car averages 18.5 mpg and the tank holds 12.5 gallons? 231.3 miles

5. If Jon drives from New York to Cincinnati, a distance of 660 miles, and his car averages 25 mpg, to the nearest gallon how much gasoline does he need? 27 gallons

Applying Mathematics to the Real World

This optional page shows how mathematics is used in the real world and also provides a change of pace.

Objective Compute miles per gallon.

Using Discussion Discuss variables that would affect the mileage a car would get.

Using Calculators Have students compute mpg using calculators.

Activity

Using Data Have students calculate the mileage for their own car or the family car. This could be done for any length of time—from a day to a week.

Using the Cumulative Review

This page provides an aid for maintaining skills and concepts presented thus far in the text.

A Cumulative Review is also provided in the Evaluation Masters Booklet as shown below.

Evaluation Masters Booklet, p. 45

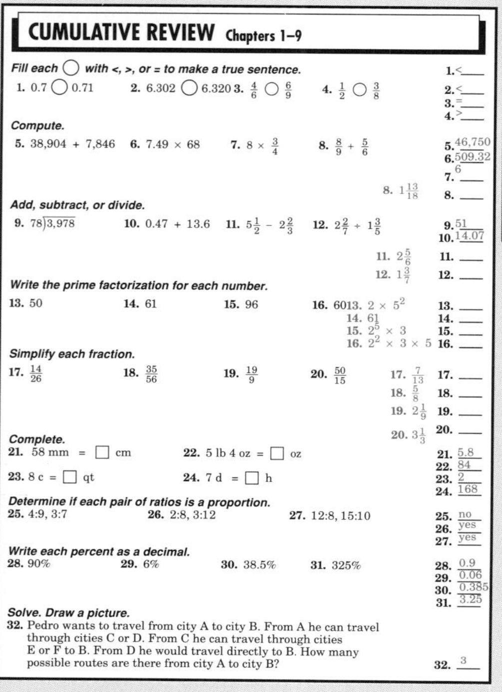

CUMULATIVE REVIEW Chapters 1–9

Fill each ◯ with <, >, or = to make a true sentence.

1. 0.7 ◯ 0.71 2. 6.302 ◯ 6.320 3. $\frac{4}{6}$ ◯ $\frac{6}{9}$ 4. $\frac{1}{2}$ ◯ $\frac{3}{8}$

1. < 2. < 3. = 4. >

Compute.

5. 38,904 + 7,846 6. 7.49 × 68 7. $8 \times \frac{3}{4}$ 8. $\frac{8}{9} + \frac{5}{6}$

5. 46,750 6. 509.32 7. 6 8. $1\frac{13}{18}$

Add, subtract, or divide.

9. 78)3,978 10. 0.47 + 13.6 11. $5\frac{1}{2} - 2\frac{2}{3}$ 12. $2\frac{2}{7} \div 1\frac{3}{5}$

9. 51 10. 14.07 11. $2\frac{5}{6}$ 12. $1\frac{3}{7}$

Write the prime factorization for each number.

13. 50 14. 61 15. 96 16. 60

13. 2×5^2 14. 61 15. $2^5 \times 3$ 16. $2^2 \times 3 \times 5$

Simplify each fraction.

17. $\frac{14}{26}$ 18. $\frac{35}{56}$ 19. $\frac{19}{9}$ 20. $\frac{50}{15}$

17. $\frac{7}{13}$ 18. $\frac{5}{8}$ 19. $2\frac{1}{9}$ 20. $3\frac{1}{3}$

Complete.

21. 58 mm = ☐ cm 22. 5 lb 4 oz = ☐ oz

23. 8 c = ☐ qt 24. 7 d = ☐ h

21. 5.8 22. 84 23. 2 24. 168

Determine if each pair of ratios is a proportion.

25. 4:9, 3:7 26. 2:8, 3:12 27. 12:8, 15:10

25. no 26. yes 27. yes

Write each percent as a decimal.

28. 90% 29. 6% 30. 38.5% 31. 325%

28. 0.9 29. 0.06 30. 0.385 31. 3.25

Solve. Draw a picture.

32. Pedro wants to travel from city A to city B. From A he can travel through cities C or D. From C he can travel through cities E or F to B. From D he would travel directly to B. How many possible routes are there from city A to city B?

32. 3

Cumulative REVIEW/TEST

Free Response

Lesson 1-9

Find the value of each expression.

1. 25 × 4 ÷ 2 **50**
2. 3 × (8 − 6) × 4 **24**
3. 3 × [6 × (5 + 3)] **144**

Lessons 2-1, 2-2, 2-3, 2-4

Add or subtract.

4. 4,065 + 178 **4,243**
5. 246 + 93 + 10.7 **349.7**
6. 14,078 − 5,928 **8,150**
7. 85.22 − 3.17 **82.05**
8. 7.17 + 68.9 **76.07**
9. 6.03 − 1.945 **4.085**

Lessons 3-1, 3-4, 3-7

Multiply or divide.

10. 1,581 ÷ 23 **68 R17**
11. 39,096 ÷ 362 **108**
12. 86 × 250 **21,500**
13. 783 × 207 **162,081**
14. 14.05 × 1,000 **14,050**
15. 0.7 ÷ 1,000 **0.0007**

Lesson 5-10

16. Carol Kerns orders a truckload of topsoil. She uses $\frac{1}{2}$ in the garden, $\frac{1}{3}$ in the flower beds, and $\frac{1}{6}$ in the yard. How much topsoil is left? **none**

Lesson 7-6

Name each polygon by the number of sides. Then state whether it is regular or not regular.

17. **hexagon regular**

18. 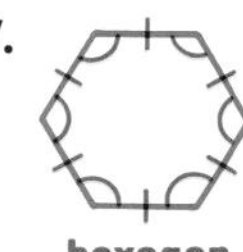 **triangle regular**

19. 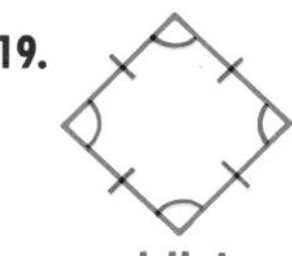 **quadrilateral regular**

20. **hexagon not regular**

Lesson 8-1

21. Steve uses 50 feet of fence to enclose a circular area for a dog pen. To the nearest foot, what is the radius of the dog pen? **8 ft**

Lesson 8-4

Find the area of each circle whose radius is given. Use 3.14 for π. Round decimal answers to the nearest tenth.

22. 2 in. **12.6 in²**
23. 10 in. **314 in²**

Lesson 8-11

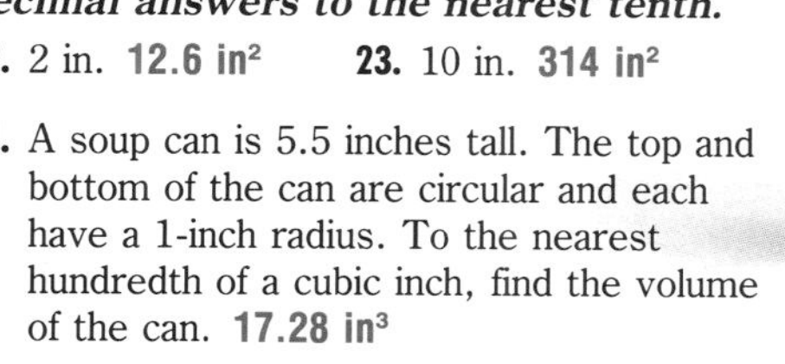

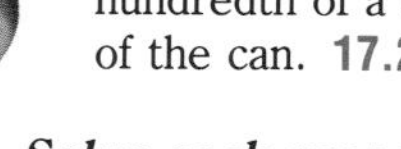

24. A soup can is 5.5 inches tall. The top and bottom of the can are circular and each have a 1-inch radius. To the nearest hundredth of a cubic inch, find the volume of the can. **17.28 in³**

Lesson 9-4

Solve each proportion.

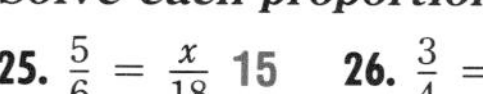

25. $\frac{5}{6} = \frac{x}{18}$ **15**
26. $\frac{3}{4} = \frac{x}{100}$ **75**
27. $\frac{x}{7} = \frac{9}{100}$ **0.63**
28. $\frac{x}{1,000} = \frac{1}{8}$ **125**
29. $\frac{25}{x} = \frac{20}{100}$ **125**

APPLYING THE LESSON

Independent Practice

Homework Assignment	
Minimum	**Ex. 2-20 even; 21-29**
Average	**Ex. 1-29**
Maximum	**Ex. 1-29**

Chapter 9

Multiple Choice

Choose the letter of the correct answer for each item.

1. Name the digit in the thousandths place-value position in 97,842.013.
a. 1 **c.** 7
▶ **b.** 3 **d.** 9

Lesson 1-1

2. Find the difference between 1,081 and 75.
e. 906
▶ **f.** 1,006
g. 1,014
h. 1,016

Lesson 2-3

3. What is the product of 6.5 and 7.02?
a. 42.471
▶ **b.** 45.63
c. 424.71
d. 456.3

Lesson 3-2

4. Replace the variable with a number to make a true statement. $\frac{3}{18} = \frac{n}{90}$
e. 5 **g.** 75
▶ **f.** 15 **h.** *none of these*

Lesson 4-5

5. Change 13.5% to a fraction.
a. 0.135
▶ **b.** $\frac{27}{200}$
c. $\frac{27}{20}$
d. $\frac{27}{2}$

Lesson 5-9

6. Which measurement is equivalent to 76 inches?
▶ **e.** 6 ft 4 in. **g.** 2 yd
f. 6 ft 6 in. **h.** 2 yd 1 ft

Lesson 6-7

7. Classify the triangle shown below.
▶ **a.** isosceles, acute
b. isosceles, obtuse
c. equilateral, acute
d. scalene, acute

Lesson 7-7

8. The formula for the volume of a pyramid is $V = \frac{1}{3}\ell wh$.
Find the volume.
e. 18 cm^3
▶ **f.** 600 cm^3
g. 720 cm^3
h. 1,800 cm^3

10 cm
12 cm
15 cm

Lesson 8-10

9. A bag contains 4 green marbles, 2 white marbles, and 3 black marbles. One marble is drawn. Find the probability that it is black or white.
a. $\frac{6}{81}$ **c.** $\frac{3}{7}$
b. $\frac{2}{7}$ ▶ **d.** $\frac{5}{9}$

Lesson 9-2

10. Albert is considering purchasing carpeting for his home. Which type of measurement will he need to be most concerned about?
e. length of hallway
f. total cubic feet
▶ **g.** total square feet
h. width of living room

Lesson 8-5

APPLYING THE LESSON

Independent Practice

Homework Assignment	
Minimum	**Ex. 1-10**
Average	**Ex. 1-10**
Maximum	**Ex. 1-10**

Using the Cumulative Test

This test serves to familiarize students with standardized format while testing skills and concepts presented up to this point.

Evaluation Masters Booklet, pp. 41-42

Name ______ Date ______

CHAPTER 9 TEST, FORM 1

Write the letter of the correct answer on the blank at the right of the page.

1. Every 6 minutes the water drips 27 times. Find the ratio of drips to minutes in simplest form.
A. $\frac{3}{7}$ **B.** $\frac{9}{2}$ **C.** $\frac{7}{3}$ **D.** $\frac{2}{9}$ 1. B

2. Determine which pair of ratios forms a proportion.
A. 3:2, 4:1 **B.** 8:2, 16:4 **C.** 1:5, 4:15 **D.** 1:5, 15:3 2. B

3. Solve the proportion $\frac{x}{4} = \frac{8}{6}$.
A. $5\frac{1}{3}$ **B.** 12 **C.** 3 **D.** $7\frac{1}{2}$ 3. A

4. Solve the proportion $\frac{4}{5} = \frac{r}{15}$.
A. 5 **B.** $18\frac{3}{4}$ **C.** 9 **D.** 12 4. D

5. Solve the proportion $\frac{2}{5} = \frac{7}{c}$.
A. $17\frac{1}{2}$ **B.** $2\frac{4}{5}$ **C.** $1\frac{3}{7}$ **D.** 3 5. A

6. The actual distance is 75 ft. Find the distance on a scale drawing if the scale is 1 in.:25 ft.
A. $\frac{1}{3}$ in. **B.** 3 in. **C.** $8\frac{1}{3}$ in. **D.** $1\frac{1}{2}$ in. 6. B

7. The distance on a scale drawing is 2 in. Find the actual distance if the scale is 1 in.:300 ft.
A. 150 ft **B.** 300 ft **C.** 600 ft **D.** 900 ft 7. C

8. Find 35% written as a fraction in simplest form.
A. $\frac{1}{4}$ **B.** $\frac{1}{3}$ **C.** $\frac{5}{7}$ **D.** $\frac{7}{20}$ 8. D

9. Find $\frac{4}{5}$ written as a percent.
A. 8% **B.** 4% **C.** 80% **D.** 40% 9. C

10. Find $\frac{5}{8}$ written as a percent.
A. 5% **B.** 50% **C.** 625% **D.** 62.5% 10. D

fun with MATH

First schools Sumer (Babylon)	1800 BC	Reconnaissance kite China	AD 399
2500 BC	Stonehenge England	206 BC	Fabiola Roman physician

Why all the fuss about acid rain?

The burning of fossil fuels by industrial plants releases pollutants into the air. These pollutants often form acid compounds and travel vast distances before falling back to Earth as acid rain. Making a fuss may be the only way to get the polluters to listen and to get effective national and international laws passed to stop the production of the deadly pollutants.

MATH M·E·N·U

Mexican Cocoa

4 cups of milk
1 4-ounce milk chocolate candy bar
1 teaspoon ground cinnamon
1 teaspoon vanilla

Pour milk into a saucepan and heat at medium temperature. Add cut up chocolate and stir until it is melted. Remove from heat and add cinnamon and vanilla. Set aside. At serving time, return to heat and beat mixture with a whisk until frothy and hot. Pour into cups and serve.

COMICS

SNIFF! SNIFF!
SIGH
EVERY NOW AND THEN YOU RUN INTO A REALLY TOUCHING STORY PROBLEM!
BATIUK
4-23
FUNKY WINKERBEAN

Honey is the only food that does not spoil. Honey, found in the tombs of Egyptian pharaohs, that was thousands of years old, was found safe to eat.

308 Fun with Math

Using Mathematical Connections

Objective To generate understanding and appreciation of connections between mathematics and real-life phenomena utilizing a number of other disciplines.

Social Studies First Schools, Sumer (Babylon), 2500 BC
The Sumerian school was a direct outgrowth of the development of Sumer's cunieform system of writing. Cunieform writing consists of wedge-shaped symbols that represent sounds as in an alphabet. The Sumerian school was established for the purpose of training scribes, or professionals. Only the sons of wealthy families could afford the cost and time of school. The school's headmaster was called the *school father* and a student was referred to as a *school son.*

Social Studies Stonehenge, England, 1800 BC
On the flat Salisbury Plain in the south of England stands a circle of huge stone columns called megaliths. Some of the stones are still joined together by huge horizontal slabs that lie across their tops. This strange circle of stones is called Stonehenge. *Henge* is an old English word meaning *to hang.*
Stonehenge is believed to be a very accurate solar calendar. It is believed that the megalith builders developed a 16-month calendar, used a year of 365 days, and had a leap year every fourth year.
Extensive study of Stonehenge has led present-day mathematicians to believe that the megalith builders were skilled in the use of the Pythagorean Theorem at least 1,500 years before Pythagoras.

Art / Social Studies / Technology
Reconnaissance Kite, China, 206 BC
Kites are the oldest form of aircraft. It is estimated that they probably originated in China about 1000 BC.
A general named Han Hsin made the first known military use of a kite about 206 BC. Han Hsin's troops had been unable to conquer an army that had fortified itself behind the walls of a city. He decided to have his troops dig a tunnel under the walls. He flew a kite over the walled city and measured the amount of string that he let out. From this measurement he was able to calculate the distance that his men would have to dig in order to come up inside the city.

Home Economics / History First Chocolate as Food, England, AD 1847
The Mayans and the Aztecs were the first to use cocoa beans to make a cold, bitter drink called *chocolatl.*
Christopher Columbus became the first European to see and taste chocolate in 1502.
In 1828, a Dutch chemist filed a patent for chocolate powder. In 1847, an English chocolate company, Fry and Sons, introduced "eating chocolate." In 1876, Daniel Peter and Henri Nestle in Vevey, Switzerland, created milk chocolate.
In 1880, a Swiss manufacturer came up with the precise combination of ingredients for a chocolate that would melt in your mouth. Later developments included the first candy bars produced by Hershey in 1911 and filled chocolates produced in 1913 by a Swiss chocolatier.

Multiplication sign (×)
England
AD 1631
AD 1847
First chocolate as food
England
Photocopier
USA
AD 1938
AD 1940
Blood plasma program
USA
JOKE!
Q: Did you hear what happened to the plants in the math room?
A: They grew square roots!
RIDDLE
Q: A father told his daughter to be home from a date at a quarter of 12. She came home at 3. How did she explain the late hour of her arrival to her dad?
A: She reminded him that in math class she had learned that ¼ of 12 is 3.
How is it possible to sail faster than the wind?
When wind fills the sail of a sailboat, the sail takes on a convex shape. As a sailboat sails at an angle into the wind, the sail splits the air flowing past. The air is squeezed and must speed up to get around the front of the sail. This reduces the air pressure outside and in front of the sail creating a suction force which pulls the boat along, in addition to its being pushed. It is because of the extra pulling force that a sailboat can actually sail faster than the wind.
QUESTION and ANSWER
How does a photocopier make copies?
A light, lens, and mirrors operate to project an image of the item to be copied onto a metal drum. The drum receives a negative electric charge that disappears wherever light-colored areas of the image strike the metal surface. The dark areas remain negatively charged. Positively-charged particles of dark toner powder are attracted to the negatively-charged dark areas. The image made up of dark toner is transferred to paper and sealed by a heater. A warm copy of your original item emerges.
QUIZ TIME
The following are three views of the same cube. Name the side opposite A by carefully studying the three views. Then, name the side opposite B and the side opposite C.
A
B
C
D
E
A
F
E
D
Fun with Math 309

Health Blood Plasma Program, USA, AD 1940
Before Dr. Charles Richard Drew (1904-1950) was forty years old his contribution to medicine had saved hundreds of thousands of lives during World War II. Dr. Drew, a black physician, was a pioneer in blood plasma preservation.
Dr. Drew discovered ways and means of preserving blood plasma in what are commonly known as blood banks. At the time of his death, in 1950, he was chief surgeon and chief of staff at Freedman's Hospital in Washington, D.C.

Quiz Answers

F is opposite side A
D is opposite side B
E is opposite side C

Health Fabiola, Roman Physician, AD 399
Fabiola was one of fifteen female followers of St. Jerome who practiced medicine and offered their services free to the poor. Fabiola established a hospital and treated those rejects from society who suffered from *loathsome diseases.* Although Fabiola's approach to medicine was pragmatic rather than theoretical, her work represents the involvement of early women in medicine.

History Multiplication Sign (X), England, AD 1631
William Oughtred (aw' tred) was a minister and not a professional mathematician. However, he spent almost all the time he could spare on mathematics. Oughtred published a textbook on mathematics in 1631 in which he introduced the multiplication sign (X). His greatest innovation, however, came in 1622. We know it today as a slide rule, and, although it is obsolete now, for centuries, engineers carried slide rules.

Technology / History Electrostatic photocopying, USA, AD 1938
Electrostatic photocopying was invented in 1938 by Chester F. Carlson, an American physicist. Unlike earlier methods, which require liquid developers, Carlson's process is completely dry. It is known as *xerography* which comes from two Greek words—*xeros* meaning *dry* and *graphia* meaning *writing.* Carlson's process was perfected in 1944 by Roland M. Schaffert, research physicist for the Battelle Memorial Institute, Columbus, Ohio, and the Haloid Company (now Xerox Corporation). Electrostatic photocopying is one of the three chief photocopying methods. The others are projection and contact photocopying.

Chapter

10 APPLYING PERCENTS

PREVIEWING the CHAPTER

The major focus of this chapter is on solving percent problems using equations. The general formula, rate × base = percentage, is used in finding the percent of a number, finding what percent one number is of another, and finding a number when a percent of it is known. Students also learn to estimate the percent of a number. Consumer applications include percent of change, discount, interest, and compound interest. The chapter includes problems involving percent.

Problem-Solving Strategy Students solve problems using Venn diagrams.

Lesson (Pages)	Lesson Objectives	State/Local Objectives
10-1 (312-313)	10-1: Find the percent of a number.	
10-2 (314-315)	10-2: Find the percent one number is of another.	
10-3 (316-317)	10-3: Find a number when a percent of it is known.	
10-4 (318-319)	10-4: Solve problems using Venn diagrams.	
10-5 (320-321)	10-5: Estimate the percent of a number.	
10-6 (324-325)	10-6: Find the percent of increase or decrease in a quantity or price.	
10-7 (326-327)	10-7: Find the discount, the rate of discount, and the sale price.	
10-8 (328-329)	10-8: Find simple interest.	
10-9 (330-331)	10-9: Find compound interest.	
10-10 (332-333)	10-10: Solve problems using percents.	

ORGANIZING the CHAPTER

You may refer to the ***Course Planning Calendar*** on page T10.

Planning Guide

Lesson (Pages)	Extra Practice (Student Edition)	Blackline Masters Booklets: Reteaching	Practice	Enrichment	Activity	Multi-cultural Activity	Technology	Tech Prep	Evaluation	Other Resources
10-1 (312-313)	p. 474	p. 96	p. 105	p. 96	p. 19		p. 24			Transparency 10-1
10-2 (314-315)	p. 475	p. 97	p. 106	p. 97		p. 10		p. 19		Transparency 10-2
10-3 (316-317)	p. 475	p. 98	p. 107	p. 98						Transparency 10-3
10-4 (318-319)			p. 108							Transparency 10-4
10-5 (320-321)	p. 475	p. 99	p. 109	p. 99					p. 49	Transparency 10-5
10-6 (324-325)	p. 476	p. 100	p. 110	p. 100				p. 20		Transparency 10-6
10-7 (326-327)	p. 476	p. 101	p. 111	p. 101	p. 20		p. 10			Transparency 10-7
10-8 (328-329)	p. 476	p. 102	p. 112	p. 102						Transparency 10-8
10-9 (330-331)		p. 103	p. 113	p. 103	p. 38					Transparency 10-9
10-10 (332-333)		p. 104	p. 114	p. 104					p. 49	Transparency 10-10
Ch. Review (334-335)									pp. 46-48	Test and Review Generator
Ch. Test (336)										
Cumulative Review/Test (338-339)									p. 50	

OTHER CHAPTER RESOURCES

Student Edition
Chapter Opener, p. 310
Activity, p. 311
Consumer Connections, p. 322
On-the-Job Application, p. 323
Journal Entry, p. 319
Portfolio Suggestion, p. 329
Consumer Connections, p. 337

Teacher's Classroom Resources
Transparency 10-0
Lab Manual, pp. 41-42
Performance Assessment, pp. 19-20
Lesson Plans, pp. 105-114

Other Supplements
Overhead Manipulative Resources, Labs 30-31, p. 22
Glencoe Mathematics Professional Series

ENHANCING the CHAPTER

Some of the blackline masters for enhancing this chapter are shown below.

COOPERATIVE LEARNING

Have students work in groups of two to solve the following problem.
An employee of the Acme Company is offered two salary plans.

Plan A: Receive $500 weekly.
Get an increase of 40% every 2 years.

Plan B: Receive $200 weekly.
Get an increase of 50% every year.

With which plan would you earn more money after 3 years? **A** After 5 years? **B**

USING MODELS/MANIPULATIVES

Have students use counters to find the percent of a number. For example, to find 25% of 30, draw and divide a large rectangle into 4 equal sections. Each section is $\frac{1}{3}$ or 25% of the whole. Take 30 counters and place one counter in each section in turn. Each section contains 7 counters with 2 counters left over. Explain that 25% of 30 is 7.5 since 2 out of 4 sections would contain an extra counter. Expand this activity to find other percents based on thirds, fourths, fifths, eighths, and tenths.

Cooperative Problem Solving, p. 38

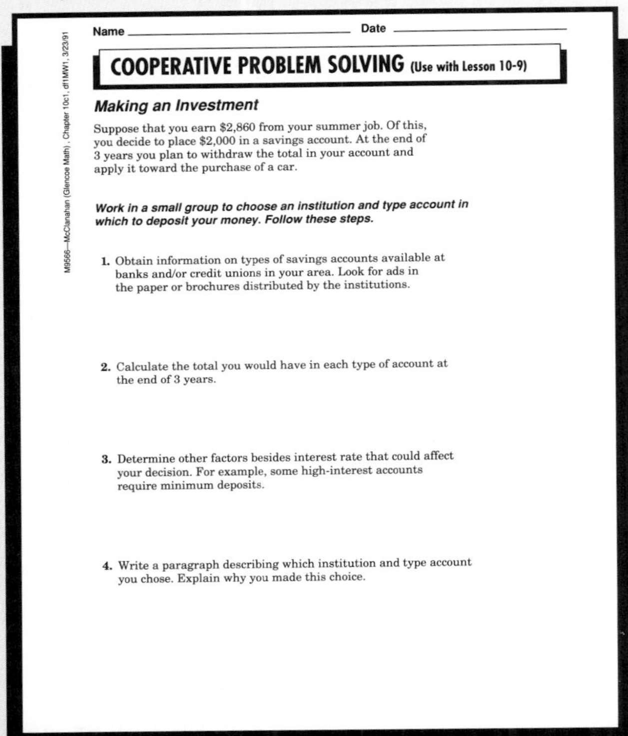

M9566—McClanahan (Glencoe Math), Chapter 10c1, d1MW1, 3/23/91

Name ____________ Date ____________

COOPERATIVE PROBLEM SOLVING (Use with Lesson 10-9)

Making an Investment

Suppose that you earn $2,860 from your summer job. Of this, you decide to place $2,000 in a savings account. At the end of 3 years you plan to withdraw the total in your account and apply it toward the purchase of a car.

Work in a small group to choose an institution and type account in which to deposit your money. Follow these steps.

1. Obtain information on types of savings accounts available at banks and/or credit unions in your area. Look for ads in the paper or brochures distributed by the institutions.
2. Calculate the total you would have in each type of account at the end of 3 years.
3. Determine other factors besides interest rate that could affect your decision. For example, some high-interest accounts require minimum deposits.
4. Write a paragraph describing which institution and type account you chose. Explain why you made this choice.

Lab Activity, p. 42

M9566—McClanahan, Glencoe, Chapter MSC, MW1, 4/1/91

Name ____________ Date ____________

LAB ACTIVITY (Use with Lesson 10-3)

Percents of All Types

You can use a set of five index cards to help you solve each of three types of percent problems.

I To find the part (p) of a number, multiply the rate (r) by the base (b). | p | = | r | × | b |

II To find the rate (r) or percent, divide the part (p) by the base (b). | r | = | p | ÷ | b |

III To find a number (b) when a part (p) of it is known, divide p by the rate (r). | b | = | p | ÷ | r |

To use the cards, turn the letter cards over, write your information from the problem on the blank side of the appropriate card, and then solve the problem.

Examples:

Example A p. 312
40% of $90 $p = 36$ | p | = | 0.40 | × | 90 |

Example A p. 314
What percent of 120 is 90? $r = 75\%$ | r | = | 90 | ÷ | 120 |

Example A p. 316
40% of what number is 48? $b = 120$ | b | = | 48 | ÷ | 0.40 |

Use your cards to answer each question.

1. What is 23% of 48?
2. 13 is what percent of 52?
3. 36% of what number is 27?
4. What is 105% of 37?
5. 85 is what percent of 17?
6. 320% of what number is 80?

MEETING INDIVIDUAL NEEDS

Students at Risk

Allow students to discover patterns involving percents that will help them become more comfortable with percents. Distribute calculators. List amounts of money on the chalkboard and make a column for 10% of the amounts. Have students find the pattern between the original amount and 10% of that amount. Expand this activity by developing students facility with 15%, 25%, $33\frac{1}{3}$% and 50%. Students should develop the ability to perform mental computations with some percents.

CRITICAL THINKING/PROBLEM SOLVING

Have students pretend that they are clerks at a store. As employees, they receive a 20% discount on the selling price of all merchandise. Suppose all merchandise is marked down by 10%. If the sale price is determined after each discount rate is applied, is that the same as adding the two discount rates together and then finding the sale price?
Have students answer without performing any computations and explain their reasoning. Students should note that by first adding the percents they are finding the percents of greater amounts.

COMMUNICATION

Speaking

Have students give an oral presentation of applications of percents in everyday life to the class. Have them use any visual aids necessary to support their examples. Students should show sample calculations with percents. Applications may include grade-point averages, income tax forms, batting averages, survey statistics, savings accounts, tips, discount, making circle graphs, bank loans, sales tax, and so on.

Applications, p. 19

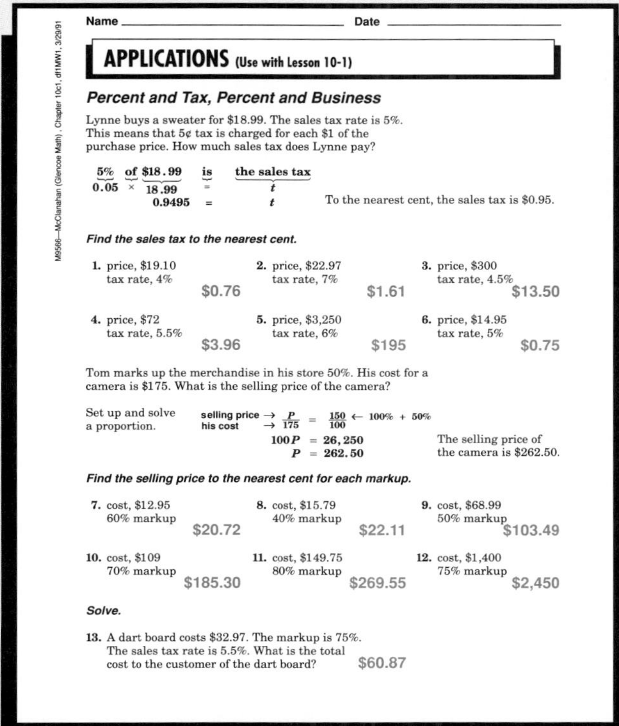

M9566—McClanahan (Glencoe Math), Chapter 10c1, df1MW1, 3/20/91

Name ______________________ Date ______________

APPLICATIONS (Use with Lesson 10-1)

Percent and Tax, Percent and Business

Lynne buys a sweater for $18.99. The sales tax rate is 5%. This means that 5¢ tax is charged for each $1 of the purchase price. How much sales tax does Lynne pay?

$\underbrace{5\%}_{0.05}$ $\underbrace{\text{of}}_{\times}$ $\underbrace{\$18.99}_{18.99}$ $\underbrace{\text{is}}_{=}$ $\underbrace{\text{the sales tax}}_{t}$

$0.9495 = t$ To the nearest cent, the sales tax is $0.95.

Find the sales tax to the nearest cent.

1. price, $19.10, tax rate, 4% — $0.76
2. price, $22.97, tax rate, 7% — $1.61
3. price, $300, tax rate, 4.5% — $13.50
4. price, $72, tax rate, 5.5% — $3.96
5. price, $3,250, tax rate, 6% — $195
6. price, $14.95, tax rate, 5% — $0.75

Tom marks up the merchandise in his store 50%. His cost for a camera is $175. What is the selling price of the camera?

Set up and solve a proportion.

selling price → / his cost → $\frac{P}{175} = \frac{150}{100}$ ← 100% + 50%

$100P = 26,250$

$P = 262.50$

The selling price of the camera is $262.50.

Find the selling price to the nearest cent for each markup.

7. cost, $12.95, 60% markup — $20.72
8. cost, $15.79, 40% markup — $22.11
9. cost, $68.99, 50% markup — $103.49
10. cost, $109, 70% markup — $185.30
11. cost, $149.75, 80% markup — $269.55
12. cost, $1,400, 75% markup — $2,450

Solve.

13. A dart board costs $32.97. The markup is 75%. The sales tax rate is 5.5%. What is the total cost to the customer of the dart board? — $60.87

Calculator Activity, p. 10

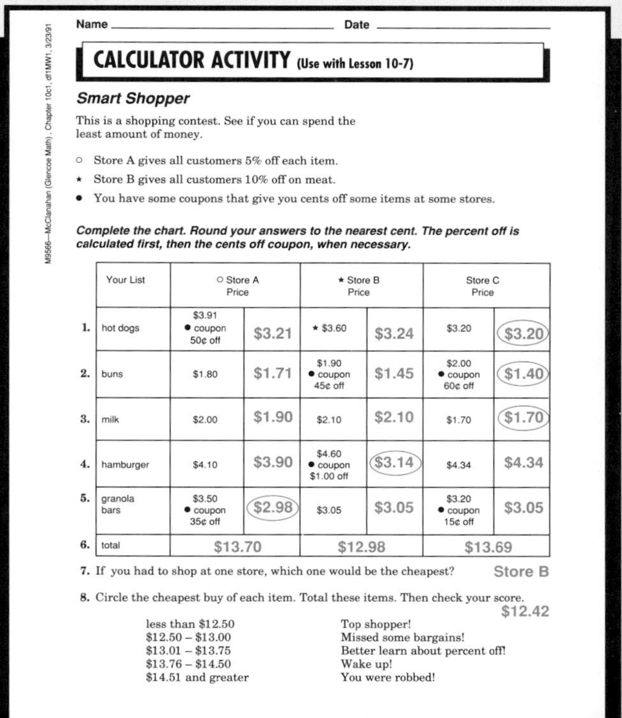

M9566—McClanahan (Glencoe Math), Chapter 10c1, df1MW1, 3/23/91

Name ______________________ Date ______________

CALCULATOR ACTIVITY (Use with Lesson 10-7)

Smart Shopper

This is a shopping contest. See if you can spend the least amount of money.

- ○ Store A gives all customers 5% off each item.
- ★ Store B gives all customers 10% off on meat.
- ● You have some coupons that give you cents off some items at some stores.

Complete the chart. Round your answers to the nearest cent. The percent off is calculated first, then the cents off coupon, when necessary.

	Your List	○ Store A Price		★ Store B Price		Store C Price	
1.	hot dogs	$3.91 ● coupon 50¢ off	$3.21	★ $3.60	$3.24	$3.20	($3.20)
2.	buns	$1.80	$1.71	$1.90 ● coupon 45¢ off	$1.45	$2.00 ● coupon 60¢ off	($1.40)
3.	milk	$2.00	$1.90	$2.10	$2.10	$1.70	($1.70)
4.	hamburger	$4.10	$3.90	$4.60 ● coupon $1.00 off	($3.14)	$4.34	$4.34
5.	granola bars	$3.50 ● coupon 35¢ off	($2.98)	$3.05	$3.05	$3.20 ● coupon 15¢ off	$3.05
6.	total	$13.70		$12.98		$13.69	

7. If you had to shop at one store, which one would be the cheapest? Store B
8. Circle the cheapest buy of each item. Total these items. Then check your score. $12.42

less than $12.50	Top shopper!
$12.50 – $13.00	Missed some bargains!
$13.01 – $13.75	Better learn about percent off!
$13.76 – $14.50	Wake up!
$14.51 and greater	You were robbed!

CHAPTER PROJECT

Tell students that they have $500 to spend. Have them cut out ads from a newspaper or magazine for items on sale. The ads should show two or more of the rate of discount, the regular price, and the sale price. Compute and show any missing information. Have students make a display showing as many different ads as possible for which the total sale price of the items is slightly less than or equal to $500.

Using the Chapter Opener

Transparency 10-0 is available in the Transparency Package. It provides a full-color visual and motivational activity that you can use to engage students in the mathematical content of the chapter.

Using Discussion

- Have students read the opening paragraph. Allow students time to discuss the various types of sales they are familiar with or special bargains they have found.
- Discuss with students any other areas in business that use percents with which they are familiar. Ask them to speculate about how businesses record and report total sales and expenses.
- Ask students how some sports statistics use percents. You may wish to have them explain how batting averages are reported. Ask them if they know any other areas in everyday life that use percents, such as grading systems.

The last paragraph is a motivational problem. Have students try to solve the problem and discuss how they arrived at the answer.

Chapter 10

APPLYING PERCENTS

SALE!!! 30% off everything

Do you check for advertised sales on items you buy? Do you watch for special savings on expensive items you would like to buy? Many businesses sell items at a certain percent off during seasonal sales and clearance sales. Businesses also use percents to express profits, losses, discounts, commissions, and interest.

Shirley Brobeck sells sweatshirts for Harper's Sweatshirts. In addition to her salary, she earns 2% of the retail price of any goods she sells. These earnings, based on sales, are called a commission. How much commission does Mrs. Brobeck earn for selling $189 worth of sweatshirts? **$3.78**

ACTIVITY: Percent

In this activity you will find the percent of a number. Suppose you surveyed your class to find the number of students who rented a video tape last week. Of the 100 students surveyed, 33 students said they rented a video tape last week. What percent of the students rented a video tape?

Materials: 10 × 10 grids, colored pencil (optional)

Cooperative Groups

Work together in groups of three or four.

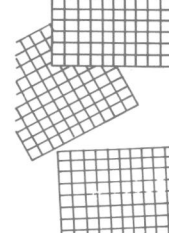

1. Shade 33 squares on a 10 × 10 grid. Write a fraction with a denominator of 100 that represents the shaded part of the grid. $\frac{33}{100}$
2. Percents are ratios that compare numbers to 100. The word *percent* means per 100 or hundredths. The percent symbol (%) means that a number is being compared to 100. If 33 out of 100 students had rented a video last week, write the percent of students who had rented a video using the percent symbol (%). **33%**
3. Percents can be expressed as fractions without the percent symbol. Look at the fraction you wrote in Exercise 1. What is the fraction in simplest form? How would you write this fraction as a decimal? $\frac{33}{100}$, **0.33**
4. Numbers can be expressed as a percent, as a fraction, as a decimal, and as a ratio. Write the number of students who rented a video tape in the last week in four ways. Each of the four ways represents the same percent. **33%,** $\frac{33}{100}$, **0.33, 33 out of 100**
5. Shade 46 squares on a 10 × 10 grid. Write the number of shaded squares as a fraction, a decimal, a ratio, and as a percent.

5. $\frac{46}{100}$, **0.46, 46 out of 100, 46%** **6.** $\frac{1}{100}$, **0.01, 1%;** $\frac{5}{1{,}000}$, **0.005, 0.5%**

Extend the Concept

6. Decide with your group how to express 1 out of 100 and 0.5 out of 100 as a fraction, a decimal, and a percent.
7. Decide with your group how to express 2,000 out of 100 as a fraction, a decimal, and a percent. $\frac{20}{1}$, **20.0, 2,000%**
8. Suppose you read an article that says the cost of oranges will be 240% of the usual price because of frost in Florida. With your group, write a sentence describing the increase in the cost of oranges. **See margin.**

You will study more about percent in this chapter.

▶ INTRODUCING THE LESSON

Objective Write percents in equivalent forms.

You may want to review proportion with students before beginning the activity.

▶ TEACHING THE LESSON

Using Discussion Discuss with students that a percent written with the % symbol is usually found in verbal sentences. To compute with percents, the symbol is dropped and another form is used.

▶ EVALUATING THE LESSON

Communicate Your Ideas Choosing a convenient equivalent often aids in computation with percents. Decimals are more convenient to use with calculators. Fractions are often used for mental computation or when the base is divisible by the denominator of the fraction.

▶ EXTENDING THE LESSON

Closing the Lesson Allow groups time to share their ideas about percents with other groups.

Additional Answers

8. Answers may vary. A good answer should refer to the increased price as *about* $2\frac{1}{2}$ times the original price.

Activity Worksheet, Lab Manual, p. 41

Name ______ Date ______

ACTIVITY WORKSHEET (Use with page 311)

Percent

Use for Exercises 1–4. Use for Exercise 5.

Fraction ___ Percent ___ Fraction ___ Percent ___

Decimal ___ Out of 100 ___ Decimal ___ Out of 100 ___

6. Fraction ___ Decimal ___ Percent ___
7. Fraction ___ Decimal ___ Percent ___
8. ______

5-MINUTE CHECK-UP

(over Lesson 9-10)

Available on TRANSPARENCY 10-1.

Solve.

1. If a certain number is added to 17, the result is 4 times 10. What is the number? **23**

2. Jim has 9 coins with a value of $1.10. Jim has nickels, dimes, and quarters. If he has more nickels than dimes, how many of each coin does he have? **3 quarters, 1 dime, 5 nickels**

▶ INTRODUCING THE LESSON

Ask students to name any areas that use percents: sports statistics, sale advertisements, commissions, and so on

▶ TEACHING THE LESSON

Using Vocabulary Have students explain the terms rate, base, and percentage in their own words. Emphasize that the rate must be changed to a decimal or a fraction before computing.

Practice Masters Booklet, p. 105

Name ______________________ Date ______________

PRACTICE WORKSHEET 10-1

Finding the Percent of a Number

Find each percentage. Use a decimal for the percent.

1. 31% of 600	186	**2.** 9% of 70	6.3
3. 12% of 1,875	225	**4.** 38% of 4,250	1,615
5. 1% of 400	4	**6.** 15% of 72	10.8
7. 43% of 9,200	3,956	**8.** 3% of 150	4.5
9. 52% of 400	208	**10.** 4% of 20	0.8

Find each percentage. Use a fraction for the percent.

11. 50% of 30	15	**12.** 40% of 65	26
13. 20% of 70	14	**14.** 25% of 160	40
15. $66\frac{2}{3}$% of 360	240	**16.** 60% of 45	27
17. 75% of 64	48	**18.** 10% of 210	21
19. $33\frac{1}{3}$% of 99	33	**20.** 80% of 20	16

The Correa family's monthly take-home pay is $2,125. This table shows their monthly budget. Find how much the Correas spend on each expense.

Expense	Percent of take-home pay
Housing	31%
Food	37%
Clothing	15%
Entertainment	6%
Transportation	9%
Savings	2%

21. housing $658.75 **22.** food $786.25
23. clothing $318.75 **24.** entertainment $127.50
25. transportation $191.25 **26.** savings $42.50

10-1 FINDING THE PERCENT OF A NUMBER

Objective

Find the percent of a number.

You may wish to review Lessons 9-8 and 9-9 before beginning this lesson.

Amanda has been shopping for a combination radio/tape player. She found the one she likes on sale at 40% off the regular price. If the radio usually sells for $90, how much would Amanda save if she buys the radio on sale? To answer the question, find 40% of $90.

To find the percent of a number, multiply the number by the percent. The equation below shows the relationship between the percent or rate r, the number or base b, and the percentage p.

$$\text{rate} \times \text{base} = \text{percentage}$$
$$r \times b = p$$

The percent of a number is called a percentage.

Method

1. Write the percent as a fraction or decimal.
2. Multiply the base by the decimal or fraction.

Examples

A ***How much will Amanda save? Find 40% of $90.***

1. 40% → 0.40
2. 0.40 × 90 = 36

40% of 90 = 36

Amanda will save $36.

B ***Find 105% of 72.***

1. 105% → 1.05
2. 1.05 × 72 = 75.60

105% of 72 is 75.6.

It is easier to change some percents to fractions rather than decimals. Some convenient facts are given in the table at the right.

$25\% = \frac{1}{4}$	$33\frac{1}{3}\% = \frac{1}{3}$	$20\% = \frac{1}{5}$	$12\frac{1}{2}\% = \frac{1}{8}$
$50\% = \frac{1}{2}$	$66\frac{2}{3}\% = \frac{2}{3}$	$40\% = \frac{2}{5}$	$37\frac{1}{2}\% = \frac{3}{8}$
$75\% = \frac{3}{4}$		$60\% = \frac{3}{5}$	$62\frac{1}{2}\% = \frac{5}{8}$
		$80\% = \frac{4}{5}$	$87\frac{1}{2}\% = \frac{7}{8}$

Example C ***Find $33\frac{1}{3}$% of 45.***

1. $33\frac{1}{3}\% = \frac{1}{3}$
2. $\frac{1}{3} \times 45 = \frac{1}{\cancel{3}_1} \times \cancel{45}^{15}$
   $= 15$

Guided Practice

Find each percentage. Use a decimal for the percent.

Examples A, B

1. 12% of 50 **6** **2.** 18% of 70 **12.6** **3.** 113% of 70 **79.1** **4.** 3% of 45 **1.35**

Find each percentage. Use a fraction for the percent.

Example C

5. 10% of 50 **5** **6.** $12\frac{1}{2}$% of 16 **2** **7.** $66\frac{2}{3}$% of 27 **18** **8.** 40% of 16 **6.4**

RETEACHING THE LESSON

Have students work in pairs and write a step-by-step analysis to each problem in the Guided Practice. Then solve. (Example: 40% of 60 is what number?)

$$r = 40\% = 0.40$$
$$b = 60$$
$$p = ?$$
$$r \times b = p$$
$$0.40 \times 60 = p$$
$$24.00 = p$$

Reteaching Masters Booklet, p. 96

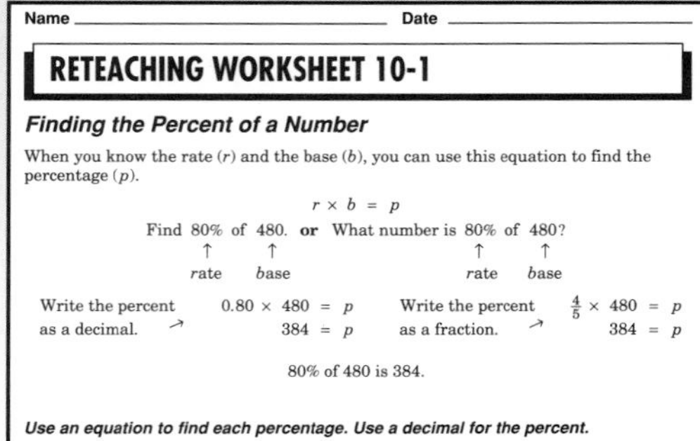

Name ______________________ Date ______________

RETEACHING WORKSHEET 10-1

Finding the Percent of a Number

When you know the rate (r) and the base (b), you can use this equation to find the percentage (p).

$$r \times b = p$$

Find 80% of 480. **or** What number is 80% of 480?
(80% = rate, 480 = base)

Write the percent as a decimal. $0.80 \times 480 = p$; $384 = p$

Write the percent as a fraction. $\frac{4}{5} \times 480 = p$; $384 = p$

80% of 480 is 384.

Use an equation to find each percentage. Use a decimal for the percent.

Exercises

Practice

Find each percentage. Use a decimal for the percent.

9. 15% of 20 **3** **10.** 12% of 50 **6** **11.** 22% of 500 **110** **12.** 36% of 300 **108**

13. 1% of 200 **2** **14.** 5% of 40 **2** **15.** 6% of 9 **0.54** **16.** 8% of 7,250 **580**

Find each percentage. Use a fraction for the percent.

17. 25% of 40 **10** **18.** 80% of 200 **160** **19.** 50% of 86 **43** **20.** 60% of 95 **57**

21. 80% of 915 **732** **22.** $37\frac{1}{2}$% of 320 **120** **23.** $87\frac{1}{2}$% of 8 **7** **24.** 10% of 7.2 **0.72**

Find each percentage.

25. 75% of 16 **12** **26.** 7% of 600 **42** **27.** 240% of 25 **60** **28.** 23% of 90 **20.7**

29. $37\frac{1}{2}$% of 120 **45** **30.** 4% of 75 **3** **31.** 15% of 200 **30** **32.** $12\frac{1}{2}$% of 300 **37.5**

33. *True* or *False*: $37\frac{1}{2}$% of 24 is 8. **false**

Talk Math

34. Describe the advantages of using a fraction instead of a decimal to find $33\frac{1}{3}$% of 45? **Answers may vary. See margin.**

35. You know 50% $= \frac{1}{2}$ and $12\frac{1}{2}$% $= \frac{1}{8}$. How would you find the percents that are equivalent to $\frac{5}{8}$ and $\frac{3}{8}$? **See margin.**

Applications

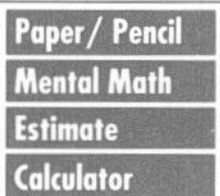

36. A mountain pack usually sells for $48. The sale price is 80% of the usual price. What is the sale price? **$38.40**

37. Of the light bulbs produced at Manual Lighting, 2% are defective. If 2,500 light bulbs are produced each hour, how many defective bulbs are produced in an 8-hour shift? **400 light bulbs**

38. Mrs. Roberts sells stock for her client. The total sale price of the stock is $25,000. If her commission is 3% of the sale price, how much commission does she receive? **$750**

Critical Thinking

39. Make 1,001 using only 7s and any combination of operations: +, −, ×, ÷. **Answers may vary. A sample answer is 7 × 77 + 7 × 77 − 77.**

Mixed Review

Find the perimeter.

Lesson 2-8

40. square; side, 13.3 cm **53.2 cm** **41.** rectangle; length, 4 in. width, 3.2 in. **14.4 in.**

Lesson 5-7

Divide.

42. $3\frac{1}{4} \div 2\frac{1}{2}$ **$1\frac{3}{10}$** **43.** $8\frac{2}{3} \div 2\frac{3}{4}$ **$3\frac{5}{33}$** **44.** $10\frac{3}{5} \div 3\frac{1}{3}$ **$3\frac{9}{50}$**

Lesson 5-5

45. Andrew and Lamont can load 28 cartons a minute. If three friends help and all three work at the same rate, how long will it take all five friends to load 1,260 cartons? **18 minutes**

Chalkboard Examples

Examples A, B *Find each percentage. Use a decimal for the percent.*

1. 40% of 96 **38.4** **2.** 15% of 60 **9**

3. 110% of 12 **13.2** **4.** 250% of 6 **15**

Example C *Find each percentage. Use a fraction for the percent.*

5. $12\frac{1}{2}$% of 24 **3** **6.** 75% of 32 **24**

Additional examples are provided on TRANSPARENCY 10-1.

EVALUATING THE LESSON

Check for Understanding Use Guided Practice Exercises to check for student understanding.

Error Analysis Watch for students who change 8% to 0.8. Remind students that percents are based on hundredths. The decimal point needs to be moved two decimal places, and this may require that zeros be inserted.

Closing the Lesson Tell the students that they will use percents when finding discounts, interest, sales taxes, and commissions.

EXTENDING THE LESSON

Enrichment Masters Booklet, p. 96

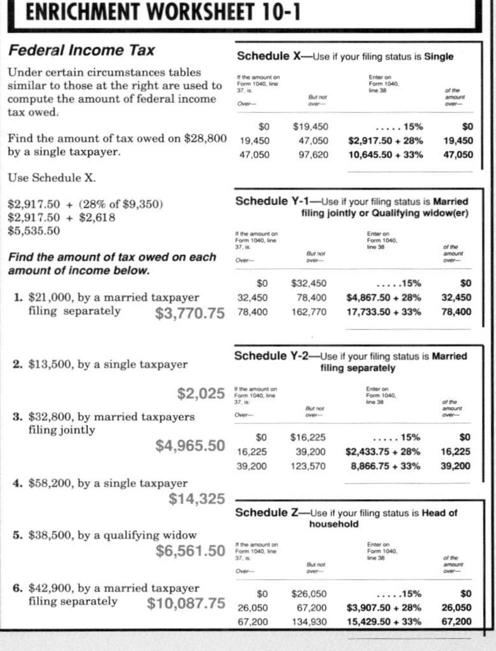

Name ______________ Date ______________

ENRICHMENT WORKSHEET 10-1

Federal Income Tax

Under certain circumstances tables similar to those at the right are used to compute the amount of federal income tax owed.

Find the amount of tax owed on $28,800 by a single taxpayer.

Use Schedule X.

$2,917.50 + (28% of $9,350)
$2,917.50 + $2,618
$5,535.50

Find the amount of tax owed on each amount of income below.

1. $21,000, by a married taxpayer filing separately **$3,770.75**
2. $13,500, by a single taxpayer **$2,025**
3. $32,800, by married taxpayers filing jointly **$4,965.50**
4. $58,200, by a single taxpayer **$14,325**
5. $38,500, by a qualifying widow **$6,561.50**
6. $42,900, by a married taxpayer filing separately **$10,087.75**

Schedule X—Use if your filing status is **Single**

If the amount on Form 1040, line 37, is: Over—	But not over—	Enter on Form 1040, line 38	of the amount over—
$0	$19,450	.....15%	$0
19,450	47,050	$2,917.50 + 28%	19,450
47,050	97,620	10,645.50 + 33%	47,050

Schedule Y-1—Use if your filing status is **Married filing jointly or Qualifying widow(er)**

If the amount on Form 1040, line 37, is: Over—	But not over—	Enter on Form 1040, line 38	of the amount over—
$0	$32,450	.....15%	$0
32,450	78,400	$4,867.50 + 28%	32,450
78,400	162,770	17,733.50 + 33%	78,400

Schedule Y-2—Use if your filing status is **Married filing separately**

If the amount on Form 1040, line 37, is: Over—	But not over—	Enter on Form 1040, line 38	of the amount over—
$0	$16,225	.....15%	$0
16,225	39,200	$2,433.75 + 28%	16,225
39,200	123,570	8,866.75 + 33%	39,200

Schedule Z—Use if your filing status is **Head of household**

If the amount on Form 1040, line 37, is: Over—	But not over—	Enter on Form 1040, line 38	of the amount over—
$0	$26,050	.....15%	$0
26,050	67,200	$3,907.50 + 28%	26,050
67,200	134,930	15,429.50 + 33%	67,200

APPLYING THE LESSON

Independent Practice

Homework Assignment	
Minimum	Ex. 1-45 odd
Average	Ex. 2-32 even; 34-45
Maximum	Ex. 1-45

Additional Answers

34. Using $\frac{1}{3}$ as a multiplier is like dividing by 3. Since 45 is divisible by 3, $\frac{1}{3}$ of 45 is 15.

35. $\frac{1}{2} + \frac{1}{8} = \frac{5}{8}$, so 50% + $12\frac{1}{2}$% = $62\frac{1}{2}$%;
$\frac{1}{2} - \frac{1}{8} = \frac{3}{8}$, so 50% − $12\frac{1}{2}$% = $37\frac{1}{2}$%.

5-MINUTE CHECK-UP

(over Lesson 10-1)

Available on TRANSPARENCY 10-2.

Find each percentage.

1. 80% of 200 160 **2.** 150% of 60 90
3. 37.5% of 240 90
4. 20% of 14.4 2.88
5. 62% of 50 31
6. $33\frac{1}{3}$% of 123 41

Extra Practice, Lesson 10-1, p. 474

INTRODUCING THE LESSON

Ask students if they have an allowance and if they budget their money by the week or month. Then ask students if they have already spent part of their money. Ask students what part or percent of their allowance they have spent.

TEACHING THE LESSON

Using Manipulatives Use quarter-inch grid paper, **Lab Manual, p. 8,** and shade portions of a 10 by 10 area to help students visualize a part of 100. Have them shade 25, 50, 60, and 75 squares and ask students what percent of 100 squares these are.

Practice Masters Booklet, p. 106

Name ____________ Date ____________

PRACTICE WORKSHEET 10-2

Finding What Percent One Number Is of Another

Find each percent. Use an equation.

1. What percent of 50 is 7? 14%
2. 16 is what percent of 30? 53.3% or $53\frac{1}{3}$%
3. What percent of 80 is 120? 150%
4. What percent of 40 is 90? 225%
5. 6 is what percent of 24? 25%
6. What percent of 15 is 5? 33.3% or $33\frac{1}{3}$%
7. What percent of 20 is 29? 145%
8. 15 is what percent of 60? 25%
9. 63 is what percent of 42? 150%
10. What percent of 72 is 9? 12.5% or $12\frac{1}{2}$%
11. What percent of 80 is 4? 5%
12. 6 is what percent of 1,200? 0.5% or $\frac{1}{2}$%
13. What percent of 60 is 12? 20%
14. 8 is what percent of 32? 25%
15. What percent of 20 is 25? 125%
16. 2 is what percent of 1,000? 0.2% or $\frac{1}{5}$%

Solve.

17. A model rocket usually sells for $4.00. Jamie got an $0.80 discount. What percent of the original price is the discount? 20%
18. Dinah bought a $50 coat for $27. What percent of the original price is the sale price? 54%
19. Josephine put $75 down on the purchase of a $300 stereo system. She will pay the rest when it is delivered. What percent of the total price is her down payment? 25%
20. David bought a backpack for $14. The sales tax on his purchase was $0.70. What percent of the purchase price is the sales tax? 5%

10-2 FINDING WHAT PERCENT ONE NUMBER IS OF ANOTHER

Objective
Find the percent one number is of another.

Josh is buying a skateboard. The one he likes was originally priced for $120. The sale price is $90. $90 is what percent of the original amount?

In this problem you need to find the rate. To find the rate (the percent) when the base $120 and percentage $90 are known, use the equation $r \times b = p$ and solve for r.

Method

1. Write the equation $r \times b = p$.
2. Solve for r.

Example A

What percent of $120 is $90?

1. $r \times 120 = 90$
2. $\frac{r \times 120}{120} = \frac{90}{120}$ $b = 120$, $p = 90$

$r =$ 90 ÷ 120 = 0.75 or 75%

$90 is 75% of $120.

Example B

80 is what percent of 64?

1. $r \times 64 = 80$
2. $\frac{r \times 64}{64} = \frac{80}{64}$ $b = 64, p = 80$

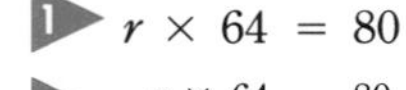

$r =$ 80 ÷ 64 = 1.25 or 125%

80 is 125% of 64.

Guided Practice

Find each percent. Use an equation.

Example A

1. What percent of 90 is 72? 80%
2. What percent of 80 is 72? 90%
3. What percent of 200 is 30? 15%
4. What percent of 16 is 24? 150%

Example B

5. 15 is what percent of 40? 37.5% or $37\frac{1}{2}$%
6. 14 is what percent of 28? 50%
7. 40 is what percent of 25? 160%
8. 4 is what percent of 5? 80%

RETEACHING THE LESSON

Have students work in pairs and write a step-by-step analysis to each problem in Guided Practice. Then solve.

Example: 16 is what percent of 20?

$r = ?;\ b = 20;\ p = 16$

$r \times b = p$

$r \times 20 = 16$

$\frac{r \times 20}{20} = \frac{16}{20}$

$r = \frac{4}{5} = 80\%$

Reteaching Masters Booklet, p. 97

Name ____________ Date ____________

RETEACHING WORKSHEET 10-2

Finding What Percent One Number Is of Another

When you know the base (b) and the percentage (p), you can use this equation to find the rate (r).

$r \times b = p$

What percent of 160 is 40? **or** 40 is what percent of 160?

(160 = base, 40 = percentage; 40 = percentage, 160 = base)

$r \times 160 = 40$

$\frac{r \times 160}{160} = \frac{40}{160}$

$r = \frac{40}{160} = \frac{1}{4}$ or 25%

9. 66.6% or $66\frac{2}{3}\%$ 12. 833.3% or $833\frac{1}{3}\%$ 13. 2.5% or $2\frac{1}{2}\%$

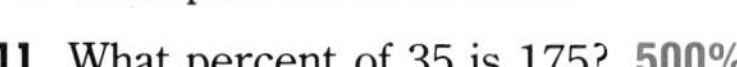

Exercises

Practice

Find each percent. Use an equation.

9. What percent of 48 is 32?

10. What percent of 200 is 60? 30%

11. What percent of 35 is 175? 500%

12. What percent of 36 is 300?

13. What percent of 80 is 2?

14. What percent of 75 is 3? 4%

15. 180 is what percent of 120? 150%

16. 16 is what percent of 12?

17. 150 is what percent of 200? 75%

18. 32 is what percent of 128? 25%

19. 12 is what percent of 16? 75%

20. 15 is what percent of 75? 20%

16. 133.3% or $133\frac{1}{3}\%$

Applications

21. The mountain bike that Sue wants usually sells for \$300. The sale price is \$240. What percent of the original price is the sale price? 80%

22. Troy missed 20 of the 125 points on a science test. What percent of the total points did Troy miss? 16%

23. Hazel made 18 of her 24 free throw attempts in her last 4 games. What percent of her free throws did Hazel make? 75%

24. Sam Strothers is a sales representative for Beacon Fashions, Inc. If he sells 115% of his first six months' quota (sales goal), Mr. Strothers will earn a three-day trip to Miami. If his quota is \$18,500 and Mr. Strothers sold \$22,200 worth of goods, what percent of his quota did he sell? 120%

Suppose

25. Suppose 80 out of 200 people between the ages of 18 and 34 would find it exciting to visit another solar system. What percent of people between the ages of 18 and 34 would find it exciting to visit another solar system? 40%

Collect Data

26. Survey your class. Find what percent would like to visit another solar system. Find what percent would like to visit a space station. Answers may vary.

Estimation

At a restaurant, it is customary to tip the waiter or waitress. Many people leave 15% of the total bill as a tip. It is easy to calculate 15% of a number using the fact that 15% = 10% + 5%.

To estimate 15% of \$24.80, take 10% of \$25 and then add 5% of \$25.

10% equals $\frac{1}{10}$, so $\frac{1}{10}$ of \$25 equals \$2.50.

5% is half of 10%, so $\frac{1}{2}$ of \$2.50 equals \$1.25.

15% of \$24.80 is about \$2.50 + \$1.25 or about \$3.75.

Answers may vary. Typical answers are given.

Estimate a 15% tip for each amount.

27. \$12 \$1.80 **28.** \$28 \$4.20 **29.** \$13.60 \$2.10 **30.** \$9.20 \$1.35

APPLYING THE LESSON

Independent Practice

Homework Assignment	
Minimum	**Ex. 2-24 even; 25-30**
Average	**Ex. 1-19 odd; 21-30**
Maximum	**Ex. 1-30**

Chalkboard Examples

Example A *Find each percent. Use an equation.*

1. What percent of 150 is 75? 50%

2. What percent of 25 is 5? 20%

Example B *Find each percent. Use an equation.*

3. 60 is what percent of 40? 150%

4. 150 is what percent of 200? 75%

Additional examples are provided on TRANSPARENCY 10-2.

EVALUATING THE LESSON

Check for Understanding Use Guided Practice Exercises to check for student understanding.

Error Analysis Watch for students who think that the percentage is always less than the base. Remind students that percents greater than 100 yield percentages greater than the base.

Closing the Lesson Have students explain in their own words how to find what percent one number is of another.

EXTENDING THE LESSON

Enrichment Masters Booklet, p. 97

Name ______ Date ______

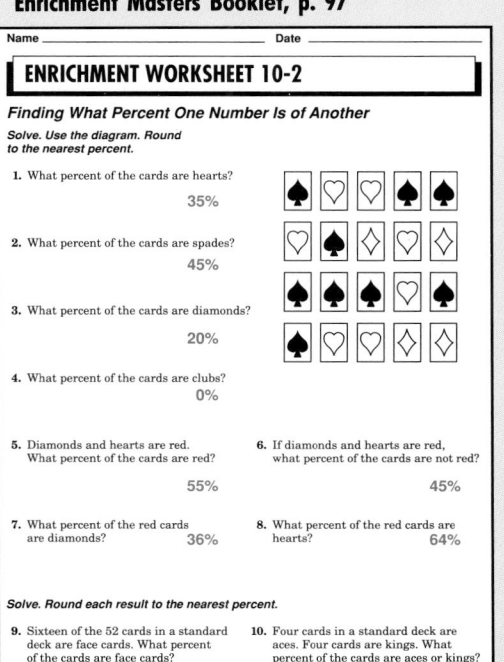

ENRICHMENT WORKSHEET 10-2

Finding What Percent One Number Is of Another

Solve. Use the diagram. Round to the nearest percent.

1. What percent of the cards are hearts? 35%
2. What percent of the cards are spades? 45%
3. What percent of the cards are diamonds? 20%
4. What percent of the cards are clubs? 0%
5. Diamonds and hearts are red. What percent of the cards are red? 55%
6. If diamonds and hearts are red, what percent of the cards are not red? 45%
7. What percent of the red cards are diamonds? 36%
8. What percent of the red cards are hearts? 64%

Solve. Round each result to the nearest percent.

9. Sixteen of the 52 cards in a standard deck are face cards. What percent of the cards are face cards? 31%
10. Four cards in a standard deck are aces. Four cards are kings. What percent of the cards are aces or kings? 15%

5-MINUTE CHECK-UP

(over Lesson 10-2)

Available on TRANSPARENCY 10-3.

Find each percent. Use an equation.

1. What percent of 80 is 24? 30%
2. 75 is what percent of 50? 150%
3. What percent of 350 is 70? 20%
4. 1 is what percent of 100? 1%
5. What percent of 7 is 21? 300%

Extra Practice, Lesson 10-2, p. 475

▶ INTRODUCING THE LESSON

Have students pretend they are selling greeting cards to earn money for a class trip to Washington, D.C. Tell students the class adviser has announced they have made $4,200 or 60% of their goal. Ask students to find their goal. Then have students explain how they found the answer.
goal: $7,000

▶ TEACHING THE LESSON

Using Discussion Discuss with students when it is appropriate to use a fraction instead of a percent.

Practice Masters Booklet, p. 107

Name ______________________ Date ______________

PRACTICE WORKSHEET 10-3

Finding a Number When a Percent of It Is Known

Find each number. Use an equation.

1. 30% of what number is 120?	400	**2.** 25% of what number is 60?	240
3. 8 is 5% of what number?	160	**4.** 100 is 100% of what number?	100
5. 80% of what number is 40?	50	**6.** 12% of what number is 42?	350
7. 420 is 60% of what number?	700	**8.** 25% of what number is 62.5?	250
9. 25 is $62\frac{1}{2}$% of what number?	40	**10.** 40% of what number is 80?	200
11. 50% of what number is 350?	700	**12.** 200% of what number is 800?	400
13. 54 is 75% of what number?	72	**14.** 40% of what number is 400?	1,000
15. $33\frac{1}{3}$% of what number is 80?	240	**16.** 18 is $37\frac{1}{2}$% of what number?	48
17. 28% of what number is 7?	25	**18.** 90 is 150% of what number?	60
19. $87\frac{1}{2}$% of what number is 70?	80	**20.** 28 is 80% of what number?	35

Solve.

21. In a school survey, 60% of the students said they own calculators. 210 sudents said they own calculators. How many students were surveyed? 350 students

22. The Bulldogs have won 80% of their basketball games. If they lost 4 games, how many games have they played? 20 games

10-3 FINDING A NUMBER WHEN A PERCENT OF IT IS KNOWN

Objective
Find a number when a percent of it is known.

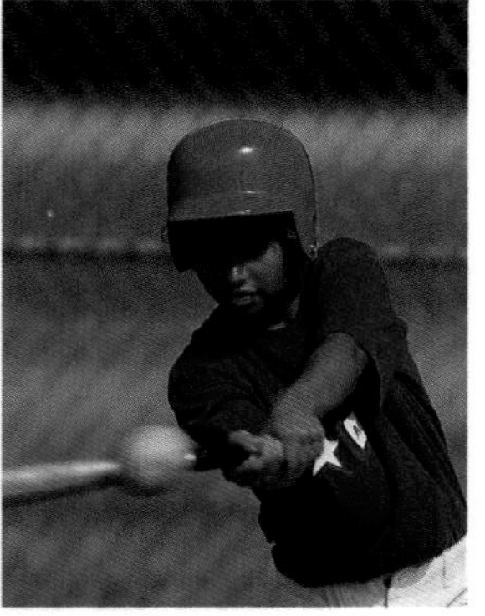

After softball season, Stacy's coach told her that she got a hit 40% of the times she was at bat. If she had 48 hits, how many times was Stacy at bat? You need to find the number of which 48 is 40%.

To find a number when a percent of it is known, use $r \times b = p$ and solve for b.

Method

1. Write the percent as a decimal or fraction.
2. Write the equation using the variable b for the base.
3. Solve for b.

Example A

How many times was Stacy at bat? 40% of a number is 48. What is the number?

1. 40% → 0.40 $r = 40\%,\ p = 48$
2. $0.40 \times b = 48$
3. $\frac{0.40 \times b}{0.40} = \frac{48}{0.40}$

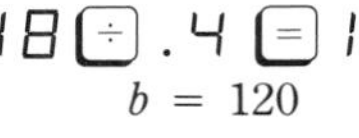

48 ÷ .4 = 120

$b = 120$

40% of 120 is 48.
Stacy was at bat 120 times.

Examples

B ***$66\frac{2}{3}$% of what number is 8?***

1. $66\frac{2}{3}\% \to \frac{2}{3}$
2. $\frac{2}{3} \times b = 8$
3. $\frac{\overset{1}{\cancel{3}}}{\underset{1}{\cancel{2}}} \times \frac{\overset{1}{\cancel{2}}}{\underset{1}{\cancel{3}}} \times b = \frac{3}{\underset{1}{\cancel{2}}} \times \overset{4}{\cancel{8}}$

$b = 12$

$66\frac{2}{3}$% of 12 is 8.

Point out that multiplying each side by $\frac{3}{2}$ isolates the variable.

C ***11.7 is 200% of what number?***

1. 200% → 2.00 or 2
2. $2 \times b = 11.7$
3. $\frac{2 \times b}{2} = \frac{11.7}{2}$

$b = 5.85$

11.7 is 200% of 5.85.

Guided Practice

Example A

Find each number. Use an equation.

1. 15% of the number is 3. 20
2. 125% of the number is 50. 40

RETEACHING THE LESSON

Have students write a step-by-step analysis to each problem in Guided Practice. Then solve.

Example: 60% of the number is 540

$r = 60\% = 0.60;\ b = ?;\ p = 540$

$r \times b = p$

$0.60 \times b = 540$

$\frac{0.60 \times b}{0.60} = \frac{540}{0.60}$

$b = 900$

Reteaching Masters Booklet, p. 98

Name ______________________ Date ______________

RETEACHING WORKSHEET 10-3

Finding a Number When a Percent of It Is Known

When you know the rate (r) and the percentage (p), you can use this equation to find the base (b).

$r \times b = p$

315 is 42% of what number? **or** 42% of what number is 315?

(315 = *percentage*, 42% = *rate*; 42% = *rate*, 315 = *percentage*)

Write the percent as a decimal. → $0.42 \times b = 315$

$\frac{0.42 \times b}{0.42} = \frac{315}{0.42}$

$b = 750$

So 315 is 42% of 750.

Example B 3. 50% of what number is 12? 24 4. $33\frac{1}{3}\%$ of what number is 18? 54

Example C 5. 45 is 300% of what number? 15 6. 69.3 is 110% of what number? 63

Exercises

Practice

Find each number. Use an equation.

7. 15% of the number is 6. 40
8. 27% of the number is 54. 200
9. 28% of the number is 14. 50
10. 7% of the number is 6.3 90
11. 25% of the number is 17. 68
12. 50% of the number is 29. 58
13. 30% of what number is 30? 100
14. 45% of what number is 18? 40
15. $33\frac{1}{3}\%$ of what number is 20? 60
16. 125% of what number is 15? 12
17. 200% of what number is 35? 17.5
18. 300% of what number is 9? 3
19. 10 is 20% of what number? 50
20. 6.2 is 40% of what number? 15.5
21. 4 is $12\frac{1}{2}\%$ of what number? 32
22. 18 is $66\frac{2}{3}\%$ of what number? 27
23. 150 is 200% of what number? 75
24. 8 is 100% of what number? 8

Application

25. In the election for freshman class officers, Joann received 52% of the votes. If Joann received 104 votes, how many votes were cast? 200 votes

Using Data

Use the chart for Exercises 26 through 28.

Business and Women

Number of firms owned by women: 1982 2.6 million; 1987 4.1 million

Revenue from firms owned by women: 1982 $98.3 billion; 1987 billion

26. Businesses owned and operated by women in 1987 accounted for 13.9% of business revenue in the United States. What was the total business revenue in the United States for firms owned by both men and women? $2,000.7 billion

fUN with MATH

Where and when was statistics established?
See page 373.

27. If women owned 30% of the number of United States businesses in 1987, how many businesses were owned by either men or both men and women in 1987? 13.67 million

28. How many more businesses were owned by women in 1987 than in 1982? This increase is what percent of the number of businesses owned by women in 1982? 1.5 million; 57.7%

Critical Thinking

29. Swapping two adjacent (side-by-side) markers constitutes one move in a game. What is the least number of moves required to change 6 moves

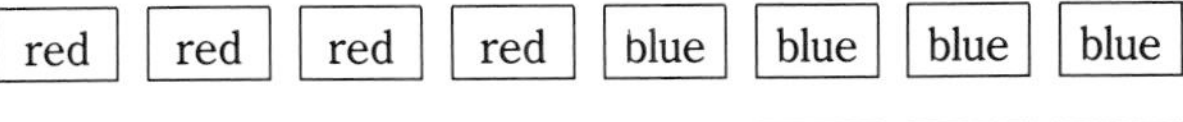

to red blue red blue red blue red blue ?

APPLYING THE LESSON

Independent Practice

Homework Assignment	
Minimum	Ex. 1-29 odd
Average	Ex. 2-24 even; 25-29
Maximum	Ex. 1-29

Chalkboard Examples

Examples A, B, C *Find each number. Use an equation.*

1. 40% of a number is 20. 50
2. 5% of a number is 7.5. 150
3. $12\frac{1}{2}\%$ of what number is 8? 64
4. $33\frac{1}{3}\%$ of what number is 9? 27
5. 3.4 is 200% of what number? 1.7
6. 57.2 is 110% of what number? 52

Additional examples are provided on TRANSPARENCY 10-3.

EVALUATING THE LESSON

Check for Understanding Use Guided Practice Exercises to check for student understanding.

Closing the Lesson Ask the students to tell in their own words how they would find a number when a percent of it is known.

EXTENDING THE LESSON

Enrichment Using a newspaper, have students make up a problem about finding a number when a percent of it is known. Have students trade problems and solve.

Enrichment Masters Booklet, p. 98

Name ______ Date ______

ENRICHMENT WORKSHEET 10-3

Sales Tax Table

Many stores use a table like the one at the right to help cashiers figure the amount of sales tax on purchases.

BRACKETED TAX COLLECTION SCHEDULE FOR STATE SALES AND USE TAX
TOTAL 5% TAX LEVY
SALES 15¢ AND UNDER - NO TAX

Each Sale	Tax	Each Sale	Tax
.16 to .20	.01	9.01 to 9.20	.46
.21 to .40	.02	9.21 to 9.40	.47
.41 to .60	.03	9.41 to 9.60	.48
.61 to .80	.04	9.61 to 9.80	.49
.81 to 1.00	.05	9.81 to 10.00	.50
1.01 to 1.20	.06	10.01 to 10.20	.51
1.21 to 1.40	.07	10.21 to 10.40	.52
1.41 to 1.60	.08	10.41 to 10.60	.53
1.61 to 1.80	.09	10.61 to 10.80	.54
1.81 to 2.00	.10	10.81 to 11.00	.55
2.01 to 2.20	.11	11.01 to 11.20	.56
2.21 to 2.40	.12	11.21 to 11.40	.57
2.41 to 2.60	.13	11.41 to 11.60	.58
2.61 to 2.80	.14	11.61 to 11.80	.59
2.81 to 3.00	.15	11.81 to 12.00	.60
3.01 to 3.20	.16	12.01 to 12.20	.61
3.21 to 3.40	.17	12.21 to 12.40	.62
3.41 to 3.60	.18	12.41 to 12.60	.63
3.61 to 3.80	.19	12.61 to 12.80	.64
3.81 to 4.00	.20	12.81 to 13.00	.65
4.01 to 4.20	.21	13.01 to 13.20	.66
4.21 to 4.40	.22	13.21 to 13.40	.67
4.41 to 4.60	.23	13.41 to 13.60	.68
4.61 to 4.80	.24	13.61 to 13.80	.69
4.81 to 5.00	.25	13.81 to 14.00	.70
5.01 to 5.20	.26	14.01 to 14.20	.71
5.21 to 5.40	.27	14.21 to 14.40	.72
5.41 to 5.60	.28	14.41 to 14.60	.73
5.61 to 5.80	.29	14.61 to 14.80	.74
5.81 to 6.00	.30	14.81 to 15.00	.75

Use the tax table to find the sales tax on each purchase.

1. $0.53 $0.03
2. $0.92 $0.05
3. $0.79 $0.04
4. $0.16 $0.01
5. $0.35 $0.02
6. $2.43 $0.13
7. $0.29 $0.02
8. $1.05 $0.06
9. $0.05 $0.00
10. $5.95 $0.30
11. $1.29 $0.07
12. $10.42 $0.53

Solve. Use the tax table.

13. What is the rate of sales tax for the table? 5%
14. When is there no tax charged on a purchase? when $0.15 or less
15. Sarah wants to buy a tape for $5.95. She has $6.50. Does she have enough money to buy the tape? yes
16. Bill wants to buy a backpack for $12.95. He has $13.00. Does he have enough money to buy the backpack? no
17. Paul bought lunch for $1.33 plus 5% sales tax. How much tax did he pay? $0.07
18. Mikki bought a card for 75¢ plus tax. How much was the tax? $0.04
19. Lia wants to buy a poster for $4.99. She has $5.20. Does she have enough money to buy the poster? no
20. Raul wants to buy a paperback book for $4.49. He has $4.70. Does he have enough money to buy the book? no
21. Kalli bought a roll of film for $2.03 plus 5% sales tax. How much tax did Kalli pay? $0.11
22. Andy bought his mother a plant for her office. It cost $9.79 plus tax. How much was the tax? $0.49

5-MINUTE CHECK-UP

(over Lesson 10-3)

Available on TRANSPARENCY 10-4.

Find each number. Use an equation.

1. 40% of the number is 3.2 8
2. 100 is 125% of what number? 80
3. Jason sold 27 boxes of popcorn at lunchtime. If Jason sold $37\frac{1}{2}\%$ of his supply, how many boxes did Jason have to sell? 72 boxes

Extra Practice, Lesson 10-3, p. 475

INTRODUCING THE LESSON

Ask students to give examples of problems that are best solved using a diagram.

TEACHING THE LESSON

Using Discussion Discuss with students each part of a Venn diagram and the information each part represents. Remind students that the number outside the rectangle is the total of all the parts within the rectangle.

Practice Masters Booklet, p. 108

PRACTICE WORKSHEET 10-4

Problem-Solving Strategy: Using Venn Diagrams

Solve. Use a Venn diagram.

1. How many students take a non-English language class? 87 students
2. How many students take both Spanish and French? 8 students
3. How many students take both Spanish and German? 3 students
4. How many students take all three languages? 2 students
5. Suppose the school has 300 students. How many students do not take a language other than English? 213 students
6. Suppose the school has 300 students. What percent of students take a non-English language? 29%

French 18 | 6 | Spanish 45 | 2 | 5 | 1 | 10 German

At Highwood High School, 36 freshmen are taking woodworking, 25 freshmen are taking cooking, and 9 freshmen are taking both courses. There are 100 students in the freshmen class.

Freshman Class: Woodworking 27 | 9 | Cooking 16

Solve.

7. Draw the Venn diagram.
8. How many freshmen are taking only cooking? 16 freshmen
9. How many freshmen are taking only woodworking? 27 freshmen
10. How many freshmen are taking neither woodworking nor cooking? 48 freshmen

Solve. Use any strategy.

11. A pizza shop offers 6 toppings on its pizzas. How many combinations of 2 toppings are possible? 15
12. One number is $\frac{2}{3}$ of another. Their sum is 60. What are the two numbers? 24 and 36

Problem-Solving Strategy

Explore | Plan | Solve | Examine

10-4 USING VENN DIAGRAMS

Objective
Solve problems using Venn diagrams.

The Blake High School *Gazette* reported the results of a survey of 46 students. Students were asked how many like to drive a car or a van. Jan wants to know how many like to drive neither type of vehicle.

Results of Car Survey	
Like cars	22
Like vans	28
Like both	7

Explore

What is given?

- Seven students like both cars and vans.
- Twenty-two students like cars.
- Twenty-eight students like vans.
- The total number of students surveyed is 46.

What is asked?

- How many students like neither cars nor vans?

Plan Use a Venn diagram.

Solve The rectangle represents the total number of students surveyed, 46. Each circle represents a type of vehicle. The overlapping area of the two circles represents the number of students who like both. Write 7 in this area.

The number of students who like *only* vans is 28 − 7 or 21. Write 21 as shown.

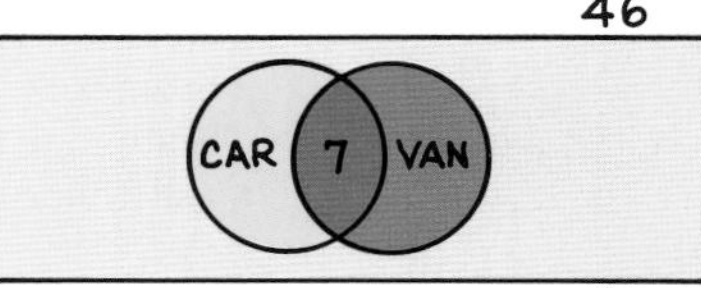

The number of students who like *only* cars is 22 − 7 or 15. Write 15 as shown.

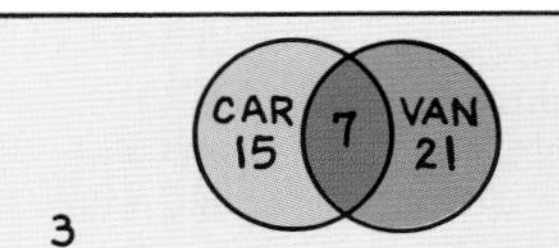

The number of students who like one type or both types of vehicle is found by adding 21 + 15 + 7 or 43.

The number of students who like neither type of vehicle is 46 − 43 or 3.

Examine Check your solution by adding all the possibilities from each part of the diagram. 15 + 21 + 7 + 3 = 46 The solution is correct.

Guided Practice

See answer on p. T523.

Solve. Use a Venn diagram.

1. Information Now, Inc. mailed 125 families a survey asking which types of movies they attended in the last six months. How many families did not return the survey? 8 families

Type of Movie	Attended
Comedy	79
Adventure	61
Cartoon	22
Comedy and Cartoon	6
Cartoon and Adventure	5
Comedy and Adventure	26
Comedy, Cartoon, and Adventure	4

RETEACHING THE LESSON

Have students draw Venn diagrams of each of the following. Have them place numbers and categories in the correct sections of each diagram. See students' work.

1. There are 24 cookies. Seven cookies have nuts, six cookies have raisins, and the rest have both.
2. A dog has six puppies. Three are brown, two are black, and the other is both brown and black.

Problem Solving

Paper/ Pencil
Mental Math
Estimate
Calculator

Solve. Use any strategy. See answer on p. T523.

2. Renley High School intramural participants were surveyed about the sports in which they participate. Thirty-five play baseball, 31 play basketball, 28 play soccer, 7 play baseball and basketball, 9 play basketball and soccer, 6 play baseball and soccer, and 5 play all three sports. How many students were surveyed? 62 students
3. Find the least common multiple of 2, 3, and 9. 18
4. Julie offers free samples of apple juice in Star's grocery store to encourage customers to buy the juice. She has 10 cases of juice to sell. If each case contains 12 bottles of juice and Julie sells 72 bottles, what percent of her stock does she sell? 60%
5. Nick is a waiter at Cook's Family Restaurant. In 1 hour on Thursday night, he waited on 4 tables. The bills for the 4 tables came to \$24.30, \$26.80, \$25.75, and \$29.90. If Nick's customers left the customary 15% tip, *about* how much should Nick expect in tips for the 4 tables? \$15
6. Jonelle delivers newspapers to 53 families 6 days a week. The Globe offers a flat rate of \$21.50 per week or a rate of \$0.07 a paper to paper carriers. Should Jonelle accept the flat rate or the per paper rate? \$0.07 per paper

7. Describe the strategies you used to solve problems 2 through 6 above. See students' work.

Critical Thinking

8. Count all the rectangles of any size in the figure. 15 rectangles

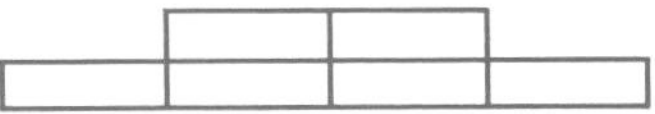

Mixed Review

Replace each ● with <, >, or = to make a true statement.

Lesson 4-7

9. $\frac{4}{5} \bullet \frac{8}{10}$ =
10. $\frac{2}{3} \bullet \frac{3}{4}$ <
11. $\frac{5}{6} \bullet \frac{4}{5}$ >
12. $\frac{8}{9} \bullet \frac{10}{12}$ >

Lesson 6-1

Complete.

13. 400 mm = __?__ cm 40
14. 0.5 m = __?__ cm 50
15. 3.9 cm = __?__ mm 39

Lesson 7-5

16. If you were to choose an edge in a cube, at random, how many other edges would be parallel to the edge? How many edges in the same plane would be perpendicular to the edge? 3, 2

APPLYING THE LESSON

Independent Practice

Homework Assignment	
Minimum	Ex. 1-5 odd; 7-16
Average	Ex. 1-16
Maximum	Ex. 1-16

Chalkboard Examples

1. From a television survey of 1,000 families, 630 said they watch television on Sunday evening, 650 said they watch television on Wednesday, and 300 said they watch television on both Sunday and Wednesday. How many families watch television on neither Sunday nor Wednesday? 20 families

Additional examples are provided on TRANSPARENCY 10-4.

EVALUATING THE LESSON

Check for Understanding Use Guided Practice Exercises to check for student understanding.

Error Analysis Watch for students who use the total number for a category and then enter the *both* number in the intersection of the two circles. Remind students that unless the number that belongs in both sets is unknown, the *both* number should be entered first.

Closing the Lesson Have students explain in their own words the parts of a Venn diagram and the part of the data that each represents.

5-MINUTE CHECK-UP

(over Lesson 10-4)

Available on TRANSPARENCY 10-5.

Solve. Use a Venn diagram.

1. Of 28 students, 10 take science, 12 take industrial tech., and 7 take both science and industrial tech. How many take neither science nor industrial tech.? 13

▶ INTRODUCING THE LESSON

Ask students how many times they have guessed at a sale price when shopping and have found merchandise offered at a percent off. Tell students this lesson will help them estimate percents.

▶ TEACHING THE LESSON

Using Questioning

- How does knowing the fractional equivalents help in estimating percentages? Fractional parts are quicker to compute mentally when fractional equivalents do not have to be computed.

Practice Masters Booklet, p. 109

Name ______________ Date ______________

PRACTICE WORKSHEET 10-5

Estimating the Percent of a Number

Estimate.

Answers may vary. Typical answers are given.

1. 21% of 38	8	2. 38% of 120	50	3. 27% of 82	20
4. 32% of 110	40	5. 146% of 52	75	6. 42% of 91	36
7. 79% of 203	160	8. 53% of 300	150	9. 91% of 500	450
10. 64% of 178	120	11. 0.5% of 600	3	12. 18% of 44	9
13. 41% of 60	25	14. 34% of 16	5	15. 24% of 64	16
16. 17% of 36	7	17. 86% of 25	21	18. 11% of 207	20
19. 27% of 46	12	20. 33% of 125	40	21. 48% of 76	40
22. 68% of 66	44	23. 39% of 52	20	24. 89% of 298	270
25. 63% of 40	24	26. 173% of 84	140	27. 85% of 72	63

Solve.

28. 24% of the cars in the parking lot were blue. If there were 87 cars in the lot, about how many of them were blue? about 20 cars

29. Sharon got 85% of the test questions done correctly. If there were 40 test questions, how many did Sharon get right? 34 questions

10-5 ESTIMATING THE PERCENT OF A NUMBER

Objective
Estimate the percent of a number.

Mary McCoy bought a VCR so she could tape her favorite television show on Thursday night while she is at work. The sales tax on the VCR is 8%. If the VCR costs $299, about how much tax did Mrs. McCoy pay?

Percentages can be estimated when the exact amount is not necessary or when you wish to check the reasonableness of a calculated result.

Method

1. Round the percent to a familiar fractional equivalent.
2. Round any other factors to numbers that can be multiplied mentally.
3. Multiply.

Examples

Explain why Mrs. McCoy will actually pay less than $30 sales tax.

A ***About how much tax did Mrs. McCoy pay? Estimate 8% of $299.***

1. 8% is a little less than 10%, which is $\frac{1}{10}$.
2. 299 rounds to 300, which is a multiple of 10.
3. $\frac{1}{10} \times 300 = 30$

Mrs. McCoy paid *about* $30 in sales tax.

B ***Estimate 40% of 86.***

1. $40\% \rightarrow \frac{2}{5}$
2. Round 86 to 85 because 85 is a multiple of 5.
3. $\frac{2}{\not 5_1} \times \not{85}^{17} = 34$

40% of 86 is *about* 34.

Example C

Estimate 22% of $18.27.

1. 22% is a little more than 20%, which is $\frac{1}{5}$.
2. Round 18.27 to 20 because 20 is a multiple of 5.
3. $\frac{1}{5} \times 20 = 4$

22% of $18.27 is *about* $4.00.

Guided Practice

Answers may vary.
Estimate. Typical answers are given.

Example A	1. 9% of 50 5	2. 18% of 150 30
Example B	3. 10% of 19 2	4. 25% of 43 10
Example C	5. 77% of 315 240	6. 35% of $8.89 $3

RETEACHING THE LESSON

Name a fraction that can be used as an estimate for each percent.

1. 72% $\frac{3}{4}$
2. 12% $\frac{1}{10}$
3. 0.9% $\frac{1}{100}$

Estimate each answer. Answers may vary.

4. 72% of 810 600
5. 12% of 1,650 200
6. 0.9% of 500 5
7. 33% of 72 25

Reteaching Masters Booklet, p. 99

Name ______________ Date ______________

RETEACHING WORKSHEET 10-5

Estimating the Percent of a Number

You can estimate the percent of a number by substituting a rounded percent and a rounded number in the equation $r \times b = p$.

Estimate 24% of 39.
rate: about 25%, or $\frac{1}{4}$
base: about 40, a multiple of 4

$\frac{1}{4} \times 40 = p$
$10 = p$

So 24% of 39 is about 10.

Estimate 68% of 37.
rate: about $66\frac{2}{3}\%$, or $\frac{2}{3}$
base: about 36, a multiple of 3

$\frac{2}{3} \times 36 = p$
$24 = p$

So 68% of 37 is about 24.

Exercises

Practice

Estimate. **Answers may vary. Typical answers are given.**

7. 11% of 108.9 **11**   **8.** 9% of 315 **31**   **9.** 63% of 120 **72**

10. 70% of 82 **56**   **11.** 15% of 58 **9**   **12.** 6% of 2,459 **150**

13. 82% of 27 **24**   **14.** 73% of 27 **21**   **15.** 22% of 27 **5**

16. 110% of 89 **100**   **17.** 121% of 70 **84**   **18.** $\frac{7}{8}$% of 200 **2**

19. $1\frac{1}{4}$% of 100 **1**   **20.** 0.5% of 800 **4**   **21.** 0.8% of 500 **5**

22. 0.1% of 90 **0.1**   **23.** 65% of 27 **18**   **24.** 1.2% of $340 **$4.08**

25. Estimate 12% of $18 billion. **2 billion**

26. *True* or *false*: 19% of 26,000 is about 5,000. **true**

Applications

27. Bill and Marie, and John and Sherri went to the Tiki Polynesian restaurant after the prom. The bill for dinner came to $49.56. *About* how much should they have left as a customary 15% tip? ***about*** **$7.50**

28. Bev has started a savings account for college. She has $250 in the account. If Bev's money earns 6% annual interest, *about* how much will she receive in interest if she leaves the money in her account for 1 year? ***about*** **$12.50**

29. Mack put $900 in a 12-month certificate of deposit that earns 9.5%. *About* how much interest will Mack receive after 1 year? ***about*** **$90**

30. Lucy found a blouse on sale for $18.99 and slacks on sale for $31.90. The sales tax is 5%. Lucy has $60 in her purse. *About* how much sales tax would Lucy pay? Will Lucy be able to buy the blouse and slacks? **$2.50; yes**

Make Up a Problem

31. Make up a percent problem that has an estimated answer of *about* $25. **Answers may vary.**

32. Make up a percent problem that has an estimated answer of *about* 15%. **Answers may vary.**

Show Math

33. Draw a rectangle about 2 inches long and 1 inch wide. Separate the area of the rectangle into 4 parts as follows. One part should be about 49% of the area. Each of the other 3 parts should be about 17% of the area. **See margin.**

Estimation

34. $1.60; $20.60
35. $0.80; $8.80
36. $4.00; $56.00
37. $6.40; $89.40
38. $16; $254

One way to mentally estimate sales tax is as follows. Round the price to the nearest ten dollars. Then multiply the tax by the rounded price.

The sales tax is 6%. Estimate the total price, tax included, of a dress that costs $57.99.

A tax rate of 6% means 6¢ tax on every dollar.

$57.99 rounds to $60.   60 × 6¢ = 360¢ or $3.60

The total cost is *about* $58 + $3.60 or $61.60.

Estimate the sales tax and total cost for each price. The tax rate is 8%. **Answers may vary. Typical answers are given.**

34. $18.99   **35.** $7.89   **36.** $51.99   **37.** $83.25   **38.** $238

APPLYING THE LESSON

Independent Practice

Homework Assignment	
Minimum	Ex. 2-32 even; 33-38
Average	Ex. 1-26 odd; 27-38
Maximum	Ex. 1-38

Chapter 10, Quiz A (Lessons 10-1 through 10-5) is available in the Evaluation Masters Booklet, p. 49.

Additional Answers

33. Answers may vary. A good answer will show a rectangle divided in half with one of the halves divided into thirds.

Chalkboard Examples

Example A *Estimate.*
1. 52% of 80 **40**
2. 14% of 200 **25**

Example B *Estimate.*
3. 33% of 75 **25**
4. 40% of 35 **14**

Example C *Estimate.*
5. 23% of 39 **10**
6. 53% of 128 **64**

Additional examples are provided on TRANSPARENCY 10-5.

EVALUATING THE LESSON

Check for Understanding Use Guided Practice Exercises to check for student understanding.

Closing the Lesson Have students summarize the steps to use to estimate the percent of a number.

EXTENDING THE LESSON

Enrichment Have students use the newspaper and estimate the sales tax or sale price of items advertised at a percent-off sale.

Enrichment Masters Booklet, p. 99

Name ______________ Date ______________

ENRICHMENT WORKSHEET 10-5

Estimating with Percent

Find the most reasonable answer.

1. 40% of 75	25	50	**30**
2. 80% of 762	500	**610**	675
3. 35% of 93	**30**	50	3
4. 12% of 154	40	**20**	30
5. 18% of 630	**115**	150	12
6. 54% of 429	175	150	**230**
7. 89% of 350	**300**	200	30
8. 75% of 450	300	400	**340**
9. 98% of 511	550	**500**	400
10. 200% of 134	**260**	175	200
11. $5\frac{1}{4}$% of 280	150	100	**15**
12. $1\frac{1}{10}$% of 472	**5**	25	50
13. 2% of 513	50	100	**10**
14. 120% of 375	500	**450**	600
15. $1\frac{9}{10}$% of 498	**10**	50	100
16. $8\frac{1}{10}$% of 65	15	25	**5**
17. 125% of 675	300	1,000	**800**
18. $133\frac{1}{3}$% of 270	450	**360**	400

Applying Mathematics to the Real World

This optional page shows how mathematics is used in the real world and also provides a change of pace.

Objective Find sales tax using a tax schedule or multiplication.

Using Discussion Discuss how the sales tax varies in different states. In some states the sales tax varies in different counties. Also, point out that the sales tax base differs among states. Some states tax clothing or food, and others do not.

Activity

Using Data Have students research sales tax in your state. Determine the rate of tax and what is taxable and is not taxable. Have students make a chart of their findings.

Sales Tax

Jacob works part time for Henley's Pharmacy. For amounts under $11.00, Jacob uses a tax table to find the sales tax rather than computing the tax. If Joan's purchases totaled $9.68, what is the amount of the sales tax from the tax table shown below?

In most states, a sales tax is charged on purchases. The sales tax provides a way to raise money for states and cities. It is a percent of each sale.

SALES & USE TAX FOR STATE, COUNTY AND/OR TRANSIT TAX—TOTAL 5 1/2% TAX LEVY
SALES 15¢ AND UNDER—NO TAX

Each Sale	Tax	Each Sale	Tax
.16 to .18	.01	5.47 to 5.64	.31
.19 to .36	.02	5.65 to 5.82	.32
.37 to .54	.03	5.83 to 6.00	.33
.55 to .72	.04	6.01 to 6.18	.34
.73 to .90	.05	6.19 to 6.36	.35
.91 to 1.09	.06	6.37 to 6.54	.36
1.10 to 1.27	.07	6.55 to 6.72	.37
1.28 to 1.46	.08	6.73 to 6.90	.38
1.47 to 1.64	.09	6.91 to 7.09	.39
1.65 to 1.82	.10	7.10 to 7.27	.40
1.83 to 2.00	.11	7.28 to 7.46	.41
2.01 to 2.18	.12	7.47 to 7.64	.42
2.19 to 2.36	.13	7.65 to 7.82	.43
2.37 to 2.54	.14	7.83 to 8.00	.44
2.55 to 2.72	.15	8.01 to 8.18	.45
2.73 to 2.90	.16	8.19 to 8.36	.46
2.91 to 3.09	.17	8.37 to 8.54	.47
3.10 to 3.27	.18	8.55 to 8.72	.48
3.28 to 3.46	.19	8.73 to 8.90	.49
3.47 to 3.64	.20	8.91 to 9.09	.50
3.65 to 3.82	.21	9.10 to 9.27	.51
3.83 to 4.00	.22	9.28 to 9.46	.52
4.01 to 4.18	.23	9.47 to 9.64	.53
4.19 to 4.36	.24	9.65 to 9.82	.54
4.37 to 4.54	.25	9.83 to 10.00	.55
4.55 to 4.72	.26	10.01 to 10.18	.56
4.73 to 4.90	.27	10.19 to 10.36	.57
4.91 to 5.09	.28	10.37 to 10.54	.58
5.10 to 5.27	.29	10.55 to 10.72	.59
5.28 to 5.46	.30	10.73 to 10.90	.60

Find the sales tax on a $9.68 purchase.

$9.68 is more than $9.55 but less than $9.72. The tax on $9.68 is $0.54.

For larger purchases, multiplication can be used to find the amount of tax. **Remind students that any part of a cent is treated as a whole cent when computing the amount of tax.**

Find the tax on a $180 purchase with a $5\frac{1}{2}$% sales tax.

180 [×] .055 [=] 9.9 — The tax is $9.90.

9.9 [+] 180 [=] 189.9 — The total cost is $189.90

1. $.55; $10.50 2. $1.38; $26.37 3. $6.33; $121.28

Find the sales tax and total cost for each price. The tax rate is $5\frac{1}{2}$%. Use the tax schedule or multiplication. **4. $8.67; $166.17**

1. $9.95 **2.** $24.99 **3.** $114.95 **4.** $157.50

5. Jackie bought a blouse for $20, jeans for $32, socks for $4.95 and a sweater for $38. Find the total sales tax and the total cost.

5. $5.22; $100.17

SALES REPRESENTATIVE

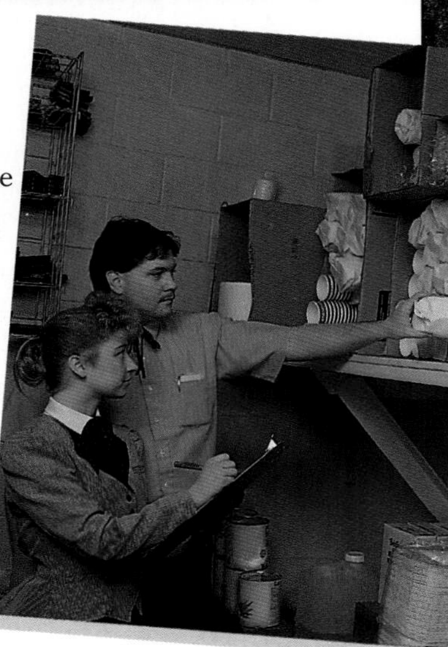

Kelly Chapin sells paper goods to restaurants. She earns 3.5% **commission** on sales. This means she earns \$0.035 on each \$1.00 in sales. The sales for one week total \$4,575. Estimate her commission.

To estimate a product, round each factor to its greatest place-value position. Then multiply.

$$\begin{array}{r} \$4{,}575 \\ \times\ 0.035 \\ \hline \end{array} \rightarrow \begin{array}{r} \$5{,}000 \\ \times\ \ \ 0.04 \\ \hline \$200.00 \end{array}$$

Kelly estimates that her commission earnings are *about* \$200.

Compute the exact amount she earned.

4575 [×] .035 [=] 160.125

Kelly earned \$160.13 in commission.

Answers may vary. Typical answers are given.

Estimate. Then compute the exact amount.

1. \$867.50 × 0.09 = \$78.08 (\$81)

2. 5,472 × 0.055 = \$300.96 (\$300)

3. \$34,920 × 0.085 = \$2,968.20 (\$2,700)

4. Jan Hille sells used cars. She earns 4.5% commission on sales. Her sales for last week were \$32,854. Estimate her commission. Then compute the exact amount. \$1,500; \$1,478.43

5. Jill Kunkle sells jewelry for Kingman's Diamond Emporium. Jill earns 8% commission and her monthly sales total is \$29,450. Estimate her commission. Then compute the exact amount. \$2,400; \$2,356

6. Refer to Exercise 5. Estimate Ms. Kunkle's annual earnings assuming sales continue at the same rate. \$24,000 to \$30,000

7. If Alex Higgins' commission is 8%, how much will he receive for selling 25 boxes of computer disks at \$32.25 per box? \$64.50

8. Jose Clemente received a commission of \$9,750 for selling a house for \$150,000. What is his rate of commission? 6.5%

9. Jason Baxter earns a salary of \$400 plus 7% commission on sales *over* \$4,500. If his sales are \$6,218, what are his total earnings? \$520.26

Applying Mathematics to the Work World

This optional page shows how mathematics is used in the work world and also provides a change of pace.

Objective Estimate and compute commissions.

Using Discussion Discuss the term commission and the different combinations of salary and commission. Ask students if they are familiar with variable commission schedules in which the percent of commission changes as the total sales change.

Activity

Using Research Have students research the help wanted section of the newspaper to find jobs that offer commissions as reimbursement. Have students list their findings.

5-MINUTE CHECK-UP

(over Lesson 10-5)

Available on TRANSPARENCY 10-6.

Estimate.

1. 35% of 120 40 **2.** 60% of 62 36

3. 42% of 160 64 **4.** 28% of 21.5 5

Extra Practice, Lesson 10-5, p. 475

INTRODUCING THE LESSON

Ask students to assume the amount of time they spend exercising each week has increased from 3 hours to 5 hours. Ask students to find the percent of increase and explain how they found the answer.

TEACHING THE LESSON

Using Discussion Discuss with students that the change is always compared to the original number. If the new number is greater than the original, the change is an increase. If the new number is less than the original, the change is a decrease.

Practice Masters Booklet, p. 110

Name ______________ Date ______________

PRACTICE WORKSHEET 10-6

Percent of Change

Find the percent of increase. Round to the nearest percent.

1. original weight, 120 lb; new weight, 125 lb 4%
2. original volume, 3 L; new volume, 3.09 L 3%
3. original price, $12; new price, $13.50 13%
4. original number, 520; new number, 640 23%
5. original weight, 3.2 oz; new weight, 4.2 oz 31%
6. original price, $2.99; new price, $3.29 10%

Find the percent of decrease. Round to the nearest percent.

7. original price, $15; new price, $11 27%
8. original price, $500; new price, $450 10%
9. original weight, 85 lb; new weight, 78 lb 8%
10. original number, 650; new number, 575 12%
11. original price, $9.95; new price, $6.75 32%
12. original weight, 8 kg; new weight, 7.5 kg 6%

Solve. Round to the nearest percent.

13. A vitamin used to be packaged in bottles of 60. Now there are 75 vitamins per bottle. What is the percent of increase? 25%
14. Larry and Sue used to have 65 customers on their paper route. Now they have 59. What is the percent of decrease? 9%
15. A pair of shoes was $50 one year and $52.50 the next year. What is the percent of increase?
16. Groceries cost $60 but, after redeeming coupons, the bill was $55.70. What is the percent of decrease?

10-6 PERCENT OF CHANGE

Objective
Find the percent of increase or decrease in a quantity or price.

Steve works for the Green Thumb garden shop after school and on Saturdays. Last week his hourly wage was raised from $5.00 to $5.45. What is the percent of increase in Steve's hourly wage?

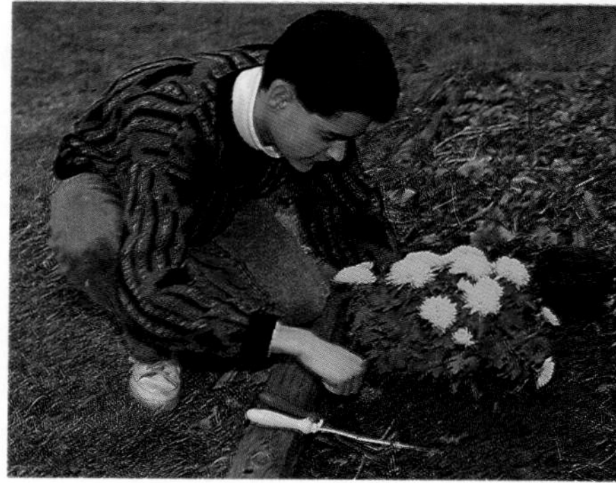

When the quantity or price of an item changes, the amount of increase or decrease from the original amount can be written as a percent. This percent is called the percent of change.

Method

1. Find the amount of increase or decrease.
2. Find what percent the amount of increase or decrease is of the original amount.

Example A ***Find the percent of increase in Steve's hourly wage.***

1 5.45 [−] 5.00 [=] 0.45 [÷] 5.00 [=] 0.09

amount of increase ↑ 2

Write the result as a percent. The percent of increase in Steve's hourly wage is 9%.

Example B For the first month at college, Ann's phone bill was $48.50. Ann's phone bill for the second month was $38.80. Find the percent of decrease in Ann's bill.

Estimate: $38.80 is about $40 and $48.50 is about $50.
$50 − $40 = $10. $10 is $\frac{1}{5}$ of $50 or 20%.

1 48.50 [−] 38.80 [=] 9.70 [÷] 48.50 [=] 0.2

amount of decrease ↑ 2

Write the result as a percent. The percent of decrease in Ann's phone bill is 20%.

Guided Practice ***Find the percent of increase or decrease. Round to the nearest percent.***

Examples A, B

		Item	Original Price	New Price
13%	**1.**	Shirt	$18	$20.25
25%	**2.**	Tie	$12	$15
6%	**3.**	Shoes	$36	$38.16
11%	**4.**	Jeans	$39.50	$43.90

	City	Rainfall (in.) 1985	Rainfall (in.) 1987	
5.	Buffalo	46.00	38.61	16%
6.	Ashville	35.94	26.50	26%
7.	Anchorage	15.51	14.32	8%
8.	Norfolk	44.81	38.68	14%

RETEACHING THE LESSON

Find the difference in price and the percent of increase or decrease given the original and new price.

1. $8; $10 $2; 25%
2. $10; $8 $2; 20%
3. $40; $45 $5; 12.5%
4. $170; $136 $34; 20%

Reteaching Masters Booklet, p. 100

Name ______________ Date ______________

RETEACHING WORKSHEET 10-6

Percent of Change

To find a percent of change, compare the amount of increase or decrease to the original amount.

Find the percent of increase if the original price was $50 and the new price is $52.50.

increase → / original → $\frac{52.50 - 50}{50} = \frac{2.50}{50}$ = 0.05 or 5%

The percent of increase is 5%.

Find the percent of decrease if the original price was $10 and the new price is $9.30.

decrease → / original → $\frac{10 - 9.30}{10} = \frac{0.70}{10}$ = 0.07 or 7%

The percent of decrease is 7%.

Find the percent of increase or decrease. Round to the nearest percent.

1. original weight, 120 lb; new weight, 125 lb 4%
2. original price, $12; new price, $13 8%

Exercises

See margin.

Practice

Find the percent of increase. Round to the nearest percent.

	Item	Original Price	New Price
9.	Milk	\$1.50/gal	\$1.79/gal
10.	Bread	\$0.94/loaf	\$1.09/loaf
11.	Yogurt	48¢/cup	59¢/cup

	Item	Original Price	New Price
12.	Grapes	80¢/pound	99¢/lb
13.	Candy	35¢/bar	45¢/bar
14.	Juice	69¢/bottle	79¢/bottle

Find the percent of decrease. Round to the nearest percent. See margin.

	Item	Original Price	New Price
15.	Calculator	\$50	\$15
16.	Radio	\$65	\$39
17.	Computer	\$10,300	\$2,884

	City	January Rainfall	April Rainfall
18.	New York	6.4 in.	5.44 in.
19.	Phoenix	0.44 in.	0.33 in.
20.	Chicago	3 in.	2.88 in.

Applications

21. Mrs. Hsu manages the coffee shop-deli at the Corporate Corners office building. In the last six months, sales of regular coffee have decreased from 160 cups to 85 cups per morning. Sales of decaffeinated coffee have increased from 48 cups to 72 cups per morning. What is the percent of decrease in sales of regular coffee? **47%**

22. A catalog ad from a 1909 issue advertised a disc talking machine (phonograph) for \$14.90. A catalog ad from a 1990 issue advertised a compact disc player for \$179.90. What is the percent of increase in price? **1,107%**

Research

23. In the library, from old newpapers or magazines at least thirty years old, research the price of a shirt, a pair of shoes, or a band instrument, such as a trumpet. Find the percent of increase in price. **Answers may vary.**

Suppose

24. Suppose during the oil crisis in 1990, the price of a gallon of gasoline at your favorite gas station went up 34%. Then it dropped 10% after three months. If the price of a gallon of gasoline before the crisis was \$1.19, what was the price of a gallon three months after the crisis began? **\$1.44 per gallon**

Mixed Review

Name the larger unit.

Lesson 6-1

25. 1 mm or 1 m **1 m**

26. 1 liter or 1 kiloliter **1 kL**

Lesson 6-10

Find the elapsed time. **27. 4 h 10 m 28. 11 h 55 m**

27. from 2:40 AM to 6:50 AM

28. from 10:45 AM to 10:40 PM

Lesson 8-1

29. Joni wants to decorate an empty cookie tin to hold odds and ends on her desk. If the can is 8 inches across, how long does the piece of fabric have to be to fit around the can? **at least 25.12 inches**

Chalkboard Examples

Example A *Find the percent of increase per pound. Round to the nearest percent.*

1. Raisins \$1.69; \$1.99 **18%**
2. Peanut Butter \$2.29; \$2.79 **22%**

Example B *Find the percent of decrease. Round to the nearest percent.*

3. 14,049; 12,644 **10%**
4. 5,681; 5,397 **5%**

Additional examples are provided on TRANSPARENCY 10-6.

EVALUATING THE LESSON

Check for Understanding Use Guided Practice Exercises to check for student understanding.

Closing the Lesson Have the students tell about the times when they would use a percent of change.

EXTENDING THE LESSON

Enrichment Discuss this method for finding the percent of increase or decrease.

Use Example A on p. 324. Have students divide 5.45 by 5.00. The result is 1.09 or 109%. Subtract 100% from the result. The percent of increase is 9%.

Enrichment Masters Booklet, p. 100

Name ______ Date ______

ENRICHMENT WORKSHEET 10-6

Percent of Change

Solve.

1. Janice bought a bathing suit for \$38.80 that originally sold for \$48.50. What was the percent of decrease? **20%**
2. The population of a city grew from 75,000 to 100,000 in a ten-year period. Find the percent of increase. **33.$\overline{3}$% or $33\frac{1}{3}$%**
3. The employment of an aircraft factory dropped from 3,200 to 2,400 in one year. Find the percent of decrease. **25%**
4. The Sunny Side Day Camp enrollment increased from 120 campers to 130 campers. Find the percent of increase. **8.$\overline{3}$% or $8\frac{1}{3}$%**
5. This week's absences at Wilson High School decreased by 24 over the previous week's figure of 72. Find the percent of decrease. **33.3% or $33\frac{1}{3}$%**
6. A lot was sold for \$45,000 that was valued at \$15,000 ten years ago. What was the percent of increase in the ten-year period? **200%**
7. Last year Mr. Black purchased a new automobile for \$12,900. If the rate of depreciation is 20% for the first year, what is the current book value of his car? **\$10,320**
8. During one winter, 24 inches of snow fell in Fairfield. This was an increase of 20% over the previous year. Find the amount of snowfall for the previous year. **20 in.**

APPLYING THE LESSON

Independent Practice

Homework Assignment	
Minimum	Ex. 2-22 even; 23-29
Average	Ex. 1-19 odd; 21-29
Maximum	Ex. 1-29

Additional Answers

9. 19%; 10. 16%; 11. 23%; 12. 24%; 13. 29%; 14. 14%; 15. 70%; 16. 40%; 17. 72%; 18. 15%; 19. 25%; 20. 4%;
23. Answers may vary. A good answer should refer to the new price as about a little more than twice the old price.

5-MINUTE CHECK-UP

(over Lesson 10-6)

Available on TRANSPARENCY 10-7.

Find the percent of increase or decrease. Round to the nearest percent.

1. The original price of a suit was $200. The price was increased $50. What was the percent of increase? **25%**
2. A stereo was priced at $200. The sale price was $130. What was the percent of decrease? **35%**

Extra Practice, Lesson 10-6, p. 476

INTRODUCING THE LESSON

Stores have sales during which time the prices are discounted. Why do you think stores give discounts? From ads that you have seen, name several words or phrases that indicate that a discount is being offered.

TEACHING THE LESSON

Using Questioning

- How is the discount rate like a percent of decrease? **A discount rate is another name for a percent of decrease. It is applied to prices that are being reduced.**

Practice Masters Booklet, p. 111

Name ________ Date ________

PRACTICE WORKSHEET 10-7

Discount

Find the discount and the sale price.

1. television, $500 discount rate, 15% — $75, $425
2. bicycle, $149.50 discount rate, 20% — $29.90, $119.60
3. typewriter, $240 discount rate, 10% — $24, $216
4. watch, $35.75 discount rate, 15% — $5.36, $30.39

Find the discount and the discount rate. Round the discount rate to the nearest percent.

5. violin, $160 sale price, $140 — $20, 13%
6. golf clubs, $125 sale price, $99.95 — $25.05, 20%
7. fishing rod, $35 sale price, $28.50 — $6.50, 19%
8. computer, $795 sale price, $725 — $70, 9%
9. roller skates, $179 sale price, $150 — $29, 16%
10. video cassette, $20 sale price, $13.50 — $6.50, 33%
11. encyclopedia, $699 sale price, $599 — $100, 14%
12. cordless phone, $125 sale price, $89.99 — $35.01, 28%

10-7 DISCOUNT

Objective
Find the discount, the rate of discount, and the sale price.

Barb Bennett is moving into her own apartment. She wants to buy a microwave oven, a toaster, and a television. Kitchen Korner offers a microwave at a discount of 20%. If the regular price of the microwave is $349, find the discount and the sale price.

A **discount** is an amount subtracted from the regular price of an item. A **discount rate** is a percent of decrease in the regular price.

Method

1. Multiply the regular price by the rate of discount to find the discount.
2. Subtract the discount from the regular price to find the sale price.

Example A

What is the discount and the sale price of the microwave oven selling at a discount rate of 20% if the regular price is $349?

Estimate: 20% or $\frac{1}{5}$ of 350 is 70. The discount is about $70.

1. 349 [−] 20 [%] 69.8 [=] 279.2 or $279.20
   (69.8 = discount; 2)

The discount is $69.80 and the sale price is $279.20.

If you know the regular price and the sale price, you can find the discount and the discount rate.

Method

1. Subtract the sale price from the regular price to find the discount.
2. Divide the discount by the regular price to find the discount rate.

Example B

A toaster is on sale for $23.95. Find the discount and the discount rate if the regular price is $32.00. Round the discount rate to the nearest percent.

1. 32. [−] 23.95 [=] 8.05 [÷] 32. [=] 0.2515625
   (8.05 = discount; 2)

The discount is $8.05 and the discount rate is 25%.

Guided Practice

Example A

Find the discount and the sale price.

1. bus fare, $8.50, discount rate, 10% **$0.85, $7.65**
2. picture, $55.60, discount rate, 25% **$13.90, $41.70**
3. stereo, $900 discount rate, 35% **$315, $585**
4. cassette tape, $6.50 discount rate, 50% **$3.25, $3.25**

Find the discount and the discount rate. Round the discount rate to the nearest percent.

Example B

5. jacket, $56, sale price, $44.80 **$11.20, 20%**
6. car, $12,650, sale price, $10,890 **$1,760, 14%**

RETEACHING THE LESSON

Have students bring in sale ads and discount coupons. Distinguish between those ads that give a rate of discount and those that give a discount. Emphasize that the discount is an amount and the rate of discount is a percent. Demonstrate finding the sale price from the regular price and the discount.

Reteaching Masters Booklet, p. 101

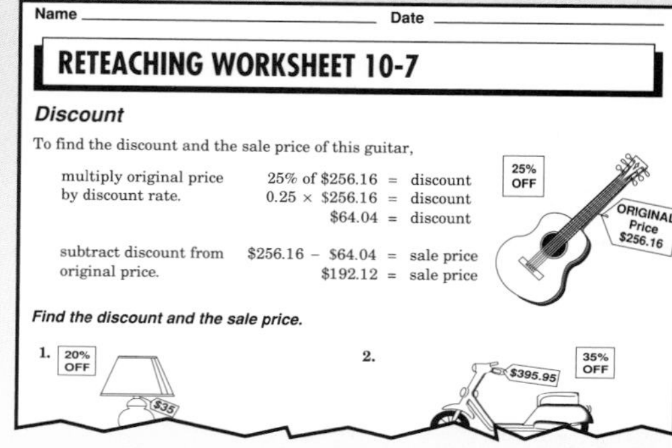

Name ________ Date ________

RETEACHING WORKSHEET 10-7

Discount

To find the discount and the sale price of this guitar,

multiply original price by discount rate.	25% of $256.16 = discount 0.25 × $256.16 = discount $64.04 = discount	
subtract discount from original price.	$256.16 − $64.04 = sale price $192.12 = sale price	

Find the discount and the sale price.

1.

2.

Exercises

Practice

Find the discount and the sale price.

7. album, $9.80 **$0.98, $8.82**
discount rate, 10%

8. blouse, $22 **$4.40, $17.60**
discount rate, 20%

9. textbook, $28.75 **$1.44, $27.31**
discount rate, 5%

10. car, $9,200 **$690, $8,510**
discount rate, 7.5%

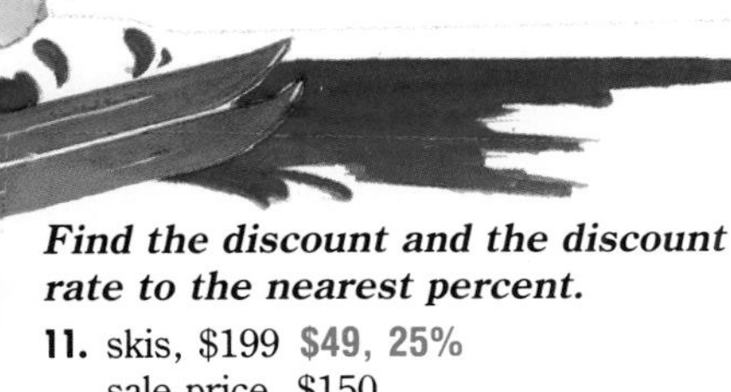

Find the discount and the discount rate. Round the discount rate to the nearest percent.

11. skis, $199 **$49, 25%**
sale price, $150

12. boat, $2,780 **$347.50, 13%**
sale price, $2,432.50

13. backpack, $20 **$4.34, 21.7%**
sale price, $15.66

14. video recorder, $1,100 **$308, 28%**
sale price, $792

Applications

15. Laluca's is having a 30%-off sale on all decorative living room accessories. What is the sale price of a vase that sells for $45? **$31.50**

16. Craig wants to buy a keyboard tray for his computer keyboard. He found one at Discount Office Warehouse for $28.49. Discount Office Warehouse advertises all merchandise always 30% off regular price. What is the regular price of the keyboard tray that Craig wants? **$40.70**

Suppose

17. Suppose Watkin's Department Store pays $48 for a denim jacket from the manufacturer. Watkin's added 25% to the cost, pricing the jacket at $60. Later Watkin's offered the jacket on sale at 25% off. Why is the sale price not equal to $48? What is the sale price? **25% of 60 is larger than 25% of 48, $45**

Talk Math

18. If the discount on an item is 20%, what percent of the original price is the sale price? **80%**

Cooperative Groups

19. ***In groups of three or four, copy and complete the table.***

Original Price	Discount Rate	Sale Price	
$100	25%	■	**$75**
$100	50%	■	**$50**
$100	20%	■	**$80**
$100	30%	■	**$70**
$100	40%	■	**$60**

Discuss with your group the relationship between the discount rate and the sale price. Describe the relationship. Think of another way to find the sale price when you know the discount rate without finding the discount first. Explain your method. **See margin.**

Critical Thinking

20. Suppose February 10, 1965 was Wednesday. What day of the week was March 10, 1965? **Wednesday**

APPLYING THE LESSON

Independent Practice

Homework Assignment	
Minimum	Ex. 1-17 odd; 18-20
Average	Ex. 1-20
Maximum	Ex. 1-20

Additional Answers

19. The sum of the discount and the sale price is the regular price. To find the sale price without finding the discount, subtract the discount rate from 100% and multiply by the regular price.

Chalkboard Examples

Examples A and B *Find the discount, the sale price, or the discount rate. Round the discount rate to the nearest percent.*

1. concert ticket, $35
   discount rate, 13% **$4.55, $30.45**
2. CD, $16.99
   discount rate, 20% **$3.40, $13.59**
3. sweatsuit, $50
   sale price, $38 **$12, 24%**
4. running shoes, $65
   sale price, $42.50 **$22.50, 35%**

Additional examples are provided on TRANSPARENCY 10-7.

▶ EVALUATING THE LESSON

Check for Understanding Use Guided Practice Exercises to check for student understanding.

Error Analysis Watch for students who divide the sale price by the original price and use the result for the discount rate.

Closing the Lesson Have students explain in their own words how to find a discount.

▶ EXTENDING THE LESSON

Enrichment Masters Booklet, p. 101

Name ______ Date ______

ENRICHMENT WORKSHEET 10-7

Discount

Solve.

1. Jennifer bought a television set listed at $752 with a $12\frac{1}{2}$% discount. How much did she pay for the television set? **$658**
2. A football with a catalog price of $24 sold for $21. Find the percent of discount. **$12\frac{1}{2}$%**
3. A refrigerator was bought for $630 after a 10% discount. Find the regular price. **$700**
4. Mr. McAuliffe bought a round-trip ticket to Florida at a special super saver rate for $135. If the rate of discount was 25%, what was the original price of the ticket? **$180**
5. Mr. Lomax bought a watch for $507.60 after a 6% disount. What was the original price of the watch? **$540**
6. The Resortique Gift Shop ordered merchandise listed at $450. A discount of $67.50 was allowed. Find the rate of discount. **15%**
7. Molly bought a sweater for $19. She found it on a sale rack that was marked $33\frac{1}{3}$% off. What was the original price of the sweater? **$28.50**
8. The regular price of an automobile was $10,550. What would the price be if there were a 10% off sale? **$9,495**
9. A lawn mower listed for $180 was sold with successive discounts of 10% and 8%. What was the sales price? **$149.04**
10. The clerk marked the price of a chair $382.50. This was after he had deducted successive discounts of 15% and 10%. find the original price of the chair. **$500**

5-MINUTE CHECK-UP

(over Lesson 10-7)

Available on TRANSPARENCY 10-8.

Find the discount and sale price.

1. skates, \$92 — \$18.40; \$73.60
   discount rate, 20%

Find the discount and the discount rate.

2. sailboat, \$6,000 — \$2,100; 35%
   sale price, \$3,900

Extra Practice, Lesson 10-7, p. 476

▶ INTRODUCING THE LESSON

Ask students to explain what they think a bank does with the money people deposit.

▶ TEACHING THE LESSON

Using Discussion Discuss with students the need to express time other than one year as a part of a year. Remind students that the relationship expressed by the formula is based on one year.

Practice Masters Booklet, p. 112

Name ______________________ Date ______________

PRACTICE WORKSHEET 10-8

Interest

Find the interest owed on each loan. Then find the total amount to be repaid.

1. principal: \$300
   annual rate: 12%
   time: 2 years
   \$72; \$372
2. principal: \$750
   annual rate: 8%
   time: 6 months
   \$30; \$780
3. principal: \$4,000
   annual rate: $10\frac{1}{2}$%
   time: 4 years
   \$1,680; \$5,680
4. principal: \$980
   annual rate: 11%
   time: 24 months
   \$215.60; \$1,195.60
5. principal: \$30,000
   annual rate: $12\frac{1}{2}$%
   time: 25 years
   \$93,750; \$123,750
6. principal: \$1,250
   rate: $1\frac{1}{2}$% per month
   time: 2 months
   \$37.50; \$1,287.50

Find the interest earned on each deposit.

7. principal: \$3,340
   annual rate: $5\frac{1}{2}$%
   time: 8 months
   \$122.47
8. principal: \$875
   annual rate: 6%
   time: 4 years
   \$210
9. principal: \$1,350
   annual rate: $6\frac{1}{4}$%
   time: 6 months
   \$42.19
10. principal: \$7,200
    annual rate: $9\frac{1}{2}$%
    time: $1\frac{1}{4}$ years
    \$855
11. principal: \$5,000
    annual rate: $7\frac{1}{4}$%
    time: 3 years
    \$1,087.50
12. principal: \$8,900
    annual rate: $10\frac{1}{2}$%
    time: 18 months
    \$1,401.75

Solve.

13. Chin borrowed \$1,500 from the bank at $9\frac{1}{2}$% interest per year for 6 months. How much did he have to pay back at the end of the 6-month period? \$1,571.25
14. Sarah lent Nathan \$450 at 6% annual interest to be paid back in 4 months. How much should Sarah receive when Nathan repays the loan? \$459

10-8 INTEREST

Objective
Find simple interest.

Note that this lesson covers only simple interest. Compound interest is presented on page 330.

Pat borrowed \$1,200 from his mother to pay part of the tuition for his last year at Carver State. He will repay his mother in one year at an interest rate of $9\frac{1}{4}$%. How much interest will Pat pay his mother?

If you borrow money from a bank or a person, you must pay **interest** on the amount of the loan. The amount borrowed is called the **principal**. Interest is a percent of the principal. Principal plus interest is the amount that must be repaid.

The interest you must pay depends on how much money you borrow, the yearly interest rate, and how long you borrow the money. The formula shows the relationship of interest to principal, rate, and time.

Interest = principal × rate × time

$$I = p \times r \times t$$

Method

1. Write interest rate (r) as a decimal.
2. Write the time (t) in years.
3. Use the formula $I = p \times r \times t$.

The interest rate is a yearly rate. So time must be given in years.

Examples

A ***Find the interest charged on Pat's loan of \$1,200 for 1 year at $9\frac{1}{4}$%.***

1. $9\frac{1}{4}\% = 9.25\% \rightarrow 0.0925$
2. 1 year
3. $I = p \times r \times t$
   $= 1{,}200 \times 0.0925 \times 1$
   $= 111$

Pat will pay his mother \$111 in interest.

B ***Find the interest earned on a deposit of \$200 for 20 months at $7\frac{3}{4}$%. Round to the nearest cent.***

1. $7\frac{3}{4}\% = 7.75\% \rightarrow 0.0775$
2. 20 months $\rightarrow \frac{20}{12}$ or $\frac{5}{3}$ years
3. $I = p \times r \times t$
   $= 200 \times 0.0775 \times 5 \div 3$
   $= 25.833333$

The interest earned is \$25.83.

Guided Practice

Find the interest owed on each loan. Then find the total amount to be repaid. Round to the nearest cent.

Example A

1. principal: \$500
   annual rate: 12%
   time: 1 year
   \$60, \$560
2. principal: \$1,200
   annual rate: 18%
   time: 3 years
   \$648, \$1,848
3. principal: \$620
   annual rate: $9\frac{3}{4}$%
   time: 6 months
   \$30.23, \$650.23

RETEACHING THE LESSON

Find the interest. Use the rule interest = principal × rate × time.

1. principal, \$100
   rate, 10%
   time, 1 year \$10
2. principal, \$500
   rate, 10%
   time, 1 year \$50
3. principal, \$1,000
   rate, 20%
   time, 1 year
   \$200
4. principal, \$100
   rate, 10%
   time, $\frac{1}{2}$ year \$5

Reteaching Masters Booklet, p. 102

Name ______________________ Date ______________

RETEACHING WORKSHEET 10-8

Interest

Money deposited in a savings account earns interest. If you borrow money from a bank, you must pay interest.
The amount of **interest (*I*)** depends on the following.

- the **principal (*p*)**—the money deposited or borrowed
- the **rate (*r*)**—a percent per year
- the **time (*t*)**—given in years

$$\underbrace{\text{interest}}_{I} = \underbrace{\text{principal}}_{p} \times \underbrace{\text{rate}}_{r} \times \underbrace{\text{time}}_{t}$$

Chuck deposits \$200 in a bank that pays 7% interest per year. In $2\frac{1}{2}$ years, how much interest does Chuck's money earn?

$I = p \times r \times t$
$I = \$200 \times 0.07 \times 2.5$
$I = \$35$
Chuck's money earns \$35 interest.

Find the interest earned on each deposit.

Example B

Find the interest earned on each deposit. $32.81

4. principal: \$230
annual rate: 6%
time: 1 year $13.80

5. principal: \$80
annual rate: 8%
time: 18 months $9.60

6. principal: \$500
annual rate: $8\frac{3}{4}$%
time: 9 months

7. \$21.60, \$111.60 8. \$1,050, \$3,550 9. \$1,200, \$3,700

Exercises

Practice

Find the interest owed on each loan. Then find the total amount to be repaid. 10. \$712.50, \$3,212.50 11. \$18.45, \$223.45

7. principal: \$90
annual rate: 12%
time: 2 years

8. principal: \$2,500
annual rate: 14%
time: 36 months

9. principal: \$2,500
annual rate: 12%
time: 48 months

10. principal: \$2,500
annual rate: $14\frac{1}{4}$%
time: 2 years

11. principal: \$205
annual rate: 18%
time: 6 months

12. principal: \$205
rate: $1\frac{1}{2}$% per month
time: 6 months

Find the interest earned on each deposit. 12. \$1.54, \$206.54 15. \$544.28

13. principal: \$120
annual rate: $9\frac{1}{2}$%
time: 3 years $34.20

14. principal: \$10,500
annual rate: 10.25%
time: $1\frac{1}{2}$ years $1,614.38

15. principal: \$6,700
annual rate: $9\frac{3}{4}$%
time: 10 months

16. principal: \$1,000
annual rate: $9\frac{1}{4}$%
time: 10 years $925

17. principal: \$150
annual rate: 8%
time: 18 months $18

18. principal: \$750
annual rate: $9\frac{1}{2}$%
time: 10 months $59.38

Applications

19. Adel is saving for a car. Last month she deposited \$350 in her savings account. If the account pays $5\frac{1}{2}$% interest per year, how much interest will Adel earn on \$350 in 2 years? $38.50

20. Luther put \$1,000 in a certificate of deposit for 24 months. The certificate pays 10.73% interest per year. How much interest will Luther earn in 24 months? What is the total amount Luther will have at the end of 24 months? $214.60; $1,214.60

PORTFOLIO

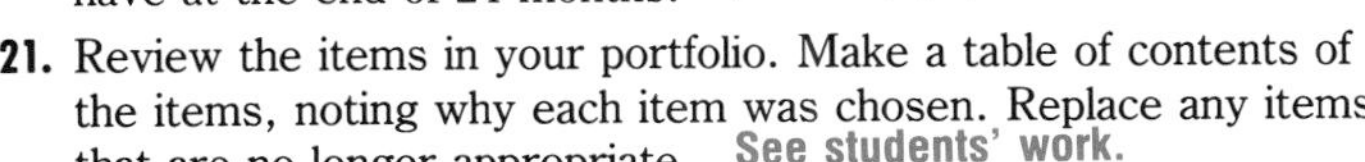

21. Review the items in your portfolio. Make a table of contents of the items, noting why each item was chosen. Replace any items that are no longer appropriate. See students' work.

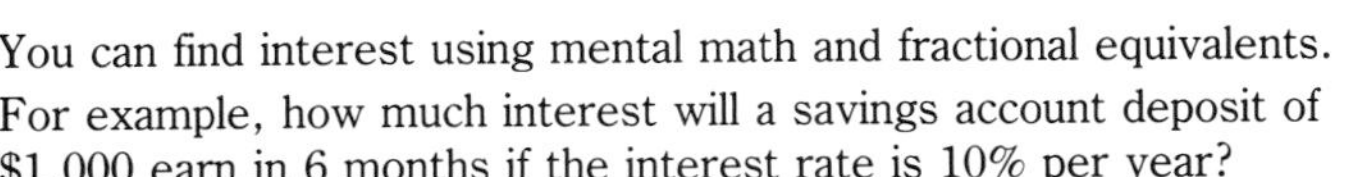

You can find interest using mental math and fractional equivalents. For example, how much interest will a savings account deposit of \$1,000 earn in 6 months if the interest rate is 10% per year?

Think: 10% is the same as $\frac{1}{10}$. 6 months is $\frac{1}{2}$ of a year.

$I = 1{,}000 \times \frac{1}{10} \times \frac{1}{2} \rightarrow I = 100 \times \frac{1}{2}$ or 50

The interest for 6 months on \$1,000 at 10% per year is \$50.

Find the interest earned using mental math.

22. p, \$750
r, 10%
t, 8 months $50

23. p, \$2,000
r, 5%
t, 2 years $200

24. p, \$4,800
r, $12\frac{1}{2}$%
t, 6 months $300

APPLYING THE LESSON

Independent Practice

Homework Assignment	
Minimum	**Ex. 1-23 odd**
Average	**Ex. 2-18 even; 19-24**
Maximum	**Ex. 1-24**

Chalkboard Examples

Example A *Find the interest owed on the loan. Then find the total amount to be repaid.*

1. principal: \$200
annual rate: 15%
time: 6 months $15, $215

Example B *Find the interest earned on each deposit.*

2. principal: \$1,000
annual rate: 7%
time: 2 years $140

Additional examples are provided on TRANSPARENCY 10-8.

EVALUATING THE LESSON

Check for Understanding Use Guided Practice Exercises to check for student understanding.

Closing the Lesson Have students tell when they could use this lesson.

EXTENDING THE LESSON

Enrichment Refer to Exercise 21. Have students make up three problems using the information they have collected from banks or newspapers. Have students exchange their problems and solve.

Enrichment Masters Booklet, p. 102

Name ________ Date ________

ENRICHMENT WORKSHEET 10-8

Interest

Solve. Round your answers to the nearest cent.

1. Seth Tyler opens a savings account with \$950. The bank pays $6\frac{1}{4}$% annual interest. How much interest will the account earn in 1 year? $59.38

Name Seth Tyler Account No. E210598

	Date	Interest	Withdrawal	Deposit	Balance
1	Apr. 30			\$950	\$950
2					
3					
4					

2. Dawn Larson deposits \$1,530 in an account that pays $5\frac{3}{4}$% interest annually. How much money is in Dawn's account after 1 year? $1,617.98

3. Tana's savings account earns \$50.63 interest in 9 months. The annual interest rate is $7\frac{1}{2}$%. How much did Tana have in her account before interest? (Round answer to the nearest dollar.) $900

4. Find the principal plus interest on \$200 left on deposit for 18 months at an annual interest rate of $5\frac{1}{2}$%. $216.50

5. Find the principal plus interest on \$5,000 left on deposit for 30 months at an annual interest rate of $6\frac{1}{2}$%. $5,812.50

6. The monthly interest rate for the use of a credit card is 1.8% on the first \$200 and 1.2% on the amount over \$200. What is the interest charged on an unpaid balance of \$318.50? $5.02

7. Troy deposits some money in a savings account that pays 7% interest annually. At the end of 1 year he has \$428 in his account. How much did Troy have in his account before interest? $400

8. The monthly interest for the use of a credit card is $1\frac{1}{2}$% on the first \$200 and $1\frac{1}{4}$% on the amount over \$200. What is the interest charged on an unpaid balance of \$425.88? $5.82

9. Larry deposits some money in his savings account. The interest rate is $6\frac{1}{4}$% annually. After 3 months his account earns \$11.78. How much did Larry have in his account before interest? (Round answer to the nearest dollar.) $754

5-MINUTE CHECK-UP

(over Lesson 10-8)

Available on TRANSPARENCY 10-9.

Find the interest owed on each loan. Then find the total amount to be repaid.

1. principal: \$5,000
annual rate: 13%
time: 2 years **\$1,300; \$6,300**

Find the interest earned on each deposit.

2. principal: \$3,500
annual rate: 6.5%
time: 6 months **\$113.75**

Extra Practice, Lesson 10-8, p. 476

▶ INTRODUCING THE LESSON

Ask students if they have a savings account and if they know the rate of interest they are earning. Ask students if they know how compound interest is computed.

▶ TEACHING THE LESSON

Using Cooperative Groups In groups of two or three, have students find the interest earned in one year on a deposit of \$1,000 at a 6% annual interest rate and at a 6% annual rate compounded quarterly. Have each group explain which rate produces more interest.

Practice Masters Booklet, p. 113

Name ______________________ Date ______________________

PRACTICE WORKSHEET 10-9

Applications: Compound Interest

Find the savings total for each account.

1. principal: \$400
annual rate: 6%
compounded quarterly
time: 1 year
\$424.54

2. principal: \$600
annual rate: 5%
compounded quarterly
time: 6 months
\$615.09

3. principal: \$750
annual rate: 7%
compounded semiannually
time: 1 year
\$803.42

4. principal: \$2,000
annual rate: 8%
compounded semiannually
time: 2 years
\$2,339.72

Solve.

5. Daniel is planning to deposit \$500 in one of two savings accounts. One account pays 6% simple interest. The other pays $5\frac{1}{2}$% interest compounded quarterly. In which account would he have the greater total at the end of 1 year? How much greater?
6% simple interest; \$1.92

6. Rita has \$500 to deposit. The Sun Bank pays 6% interest compounded quarterly. The Star Bank pays 6% interest compounded semiannually. The Moon Bank pays 6% compounded annually. Which bank pays the most interest?
Sun Bank

7. How much interest does each bank pay? What is the difference between the interest paid by the highest and lowest paying banks?
Sun: \$30.68; Star: \$30.45; Moon: \$30; \$0.68

Applications

10-9 COMPOUND INTEREST

Objective
Find compound interest.

Mark's savings account pays 6% annual interest compounded quarterly. At the end of 1 year, how much interest will Mark earn on \$1,000?

Most savings accounts pay **compound interest.** Compound interest is computed at stated intervals. At the end of the first interval, the interest earned is added to the account. At the end of the next interval, interest is earned on the balance that includes the previous interest. Interest may be added annually, semiannually (2 times a year), quarterly (4 times a year), monthly, daily, or continuously.

Example

Find the savings total at the end of one year if \$1,000 is deposited in a savings account at 6% annual interest compounded quarterly.

first quarter: $I = p \times r \times t$
$I =$ 1000 [×] 0.06 [÷] 4 [=] 15
$I =$ \$15.00 — Each quarter is $\frac{1}{4}$ of a year. To multiply by $\frac{1}{4}$ on the calculator, divide by 4.

The new balance or principal is 1000 [+] 15 [=] 1015 [STO] or \$1,015 after one quarter. Use the [STO] key to save the new balance after each computation.

After each of the next three quarters, interest is computed and added to the account.

second quarter: [×] 0.06 [÷] 4 [+] [RCL] [=] 1030.225 [STO]
new balance: \$1,030.225 — ↑ This recalls the last value put in memory; in this case, 1015.

third quarter [×] 0.06 [÷] 4 [+] [RCL] [=] 1045.6784 [STO]
new balance: \$1,045.68

fourth quarter: [×] 0.06 [÷] 4 [+] [RCL] [=] 1061.3636
new balance: \$1,061.3636

The savings total at the end of one year is \$1,061.36

Understanding the Data

- How much interest is earned at the end of the second quarter? **\$15.23**
- What is the new balance at the end of the second quarter? **\$1,030.23**
- How much interest is earned at the end of the fourth quarter? **\$15.69**

RETEACHING THE LESSON

Work with students to solve each problem.

1. Troy deposits \$200 in a savings account that pays 7% interest compounded quarterly. What is Troy's savings total at the end of one year? **\$214.37**
2. Martha opens a savings account with \$1,000. It earns 6.5% interest compounded semi-annually. What is Martha's savings total at the end of one year? **\$1066.06**

Reteaching Masters Booklet, p. 103

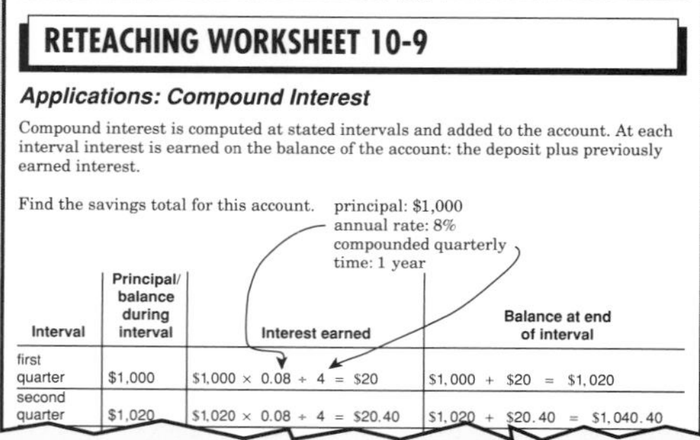

Name ______________________ Date ______________________

RETEACHING WORKSHEET 10-9

Applications: Compound Interest

Compound interest is computed at stated intervals and added to the account. At each interval interest is earned on the balance of the account: the deposit plus previously earned interest.

Find the savings total for this account. principal: \$1,000
annual rate: 8%
compounded quarterly
time: 1 year

Interval	Principal/ balance during interval	Interest earned	Balance at end of interval
first quarter	\$1,000	\$1,000 × 0.08 ÷ 4 = \$20	\$1,000 + \$20 = \$1,020
second quarter	\$1,020	\$1,020 × 0.08 ÷ 4 = \$20.40	\$1,020 + \$20.40 = \$1,040.40

Guided Practice

Find the savings total for each account.

1. principal: \$250
   annual rate: 6%
   compounded quarterly
   time: 1 year **\$265.34**
2. principal: \$500
   annual rate: 9%
   compounded annually
   time: 1 year **\$545**

You can use the formula $S = P\left(1 + \frac{r}{n}\right)^{n \times t}$ and the y^x key on a calculator to compute powers. Find the savings total after 1 year for a deposit of \$500 at an annual rate of 9% compounded quarterly.

$S = 500 \times \left(1 + \frac{0.09}{4}\right)^{4 \times 1}$

S = savings total
P = beginning principal
r = annual rate
n = number of times compounded annually
t = number of years

Enter: 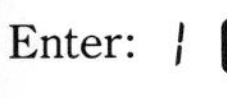 1 + .09 ÷ 4 = The result of the operations inside the parentheses is 1.0225.

Enter: y^x 4 = The value of $(1.0225)^4$ is 1.0930833.

Enter: × 500 = The product is 546.54166.

The savings total is \$546.54.

Use the formula to find the savings total.

3. principal: \$250
   annual rate: 6%
   compounded semiannually
   time: 1 year **\$265.23**
4. principal: \$500
   annual rate: 9%
   compounded quarterly
   time: 1 year **\$546.54**

Applications

5. Tammara deposits \$250 in her savings account that pays 8% interest compounded quarterly. After 6 months, how much will Tammara have in the account? **\$260.10**
6. Khalid has \$200 to start a savings account for a car. One account pays 6% annual interest compounded annually while another pays 5% interest compounded quarterly. What is the savings total Khalid can expect from each account after 1 year? Which account pays more interest? **\$212; \$210.19; 6% annually pays more**
7. Mrs. Schoby put \$1,200 in a savings account for her daughter's college expenses. What savings total will she have in three years if the account earns 6.5% interest compounded semiannually? **\$1,453.86**

FUN with MATH
How does an electronic calculator add?
See page 373.

Mixed Review

Lesson 8-3 — ***Find the area of each triangle described below.*** **9. 12.96 mm²**

8. base, 12 cm; height, 10 cm **60 cm²**
9. base, 3.2 mm; height, 8.1 mm

Lesson 9-3 — ***Determine if each pair of ratios forms a proportion.***

10. 5 to 9, 6 to 10 **no**
11. 1 to 3, 4 to 12 **yes**
12. 6 to 10, 12 to 20 **yes**

Lesson 9-2

13. A bag contains 2 red marbles, 3 green marbles, and 4 yellow marbles. What is the probability of choosing a red marble? $\frac{2}{9}$

Chalkboard Examples

Find the savings total.

1. principal: \$500
   annual rate: 8%
   compounded semiannually
   time: 18 months
   \$562.43

Additional examples are provided on TRANSPARENCY 10-9.

EVALUATING THE LESSON

Check for Understanding Use Guided Practice Exercises to check for student understanding.

Error Analysis Watch for students who fail to add the interest to the principal after each compounding period. Remind students that the principal should be greater than in the previous computation each time they compound the interest.

Closing the Lesson Have students explain in their own words how to find interest compounded semiannually.

EXTENDING THE LESSON

Enrichment Masters Booklet, p. 103

ENRICHMENT WORKSHEET 10-9

Intersection and Union of Sets

The set of all factors of 8 and 12 can be listed in set notation as follows:

factors of 8 A = {1, 2, 4, 8}
factors of 12 B = {1, 2, 3, 4, 6, 12}

The **intersection** of sets A and B is the set of elements that are in both A and B. The diagram below shows the intersection of A and B

A intersection B = the set containing 1,2 and 4
A ∩ B = {1,2,4}

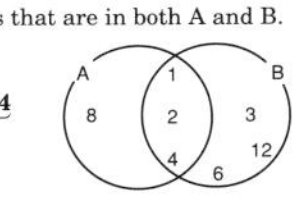

The **union** of sets A and B is the set of elements that are in A or in B or in both.

A union B = the set containing 1, 2, 3, 4, 6, 8, and 12
A ∪ B = {1, 2, 3, 4, 6, 8, 12}

The empty set has no elements. The empty set is written ∅.

Give the intersection and union for each of the following.

1. X ∩ Y {6, 8}
   X ∪ Y {0, 1, 2, 3, 4, 5, 6, 7, 8, 9, 10, 12}
2. M ∩ N {1, 2, 6}
   M ∪ N {1, 2, 4, 6, 8, 16, 24}

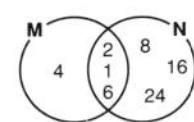

3. C = {2, 4, 6, 8, 10, 12}
   D = {0, 3, 6, 9, 12, 15, 18}
   {6, 12}
   {0,2,3,4,6,8,9,10,12,15,18}
4. E = {a, e, i, o, u}
   F = {c, a, k, e}
   {a, e}
   {a,c,e,i,k,o,u}
5. G = {1, 2, 3, 4, 5}
   H = {2, 3, 5, 7, 11}
   I = {1, 3, 5, 7, 9} {3, 5}
   {1,2,3,4,5,7,9,11}
6. Let J = {3, 12, 21, 30} and R = {2, 11, 20}. What is J ∩ R? ∅

Let K be the set of all counting numbers less than 60. Let L be the set of all counting numbers greater than 50.

7. Find K ∩ L. {51,52,53,54,55,56,57,58,59}
8. Find K ∪ L. {1,2,3,...}
9. Describe K ∩ L using the words greater than and less than.
   all counting numbers greater than 50 and less than 60

APPLYING THE LESSON

Independent Practice

Homework Assignment	
Minimum	Ex. 1-13
Average	Ex. 1-13
Maximum	Ex. 1-13

5-MINUTE CHECK-UP

(over Lesson 10-9)

Available on TRANSPARENCY 10-10.

Find the savings total.

1. principal: $750
annual rate: 8.5%
compounded quarterly
time: 1 year
$815.81

▶ INTRODUCING THE LESSON

Ask students how often they use percents in everyday situations. Ask students to name situations in business that they expect to use percents.

▶ TEACHING THE LESSON

Using Discussion Discuss with students the need to accurately identify the original amount and whether the amounts are increasing or decreasing. Remind students that careful reading will help prevent errors.

Practice Masters Booklet, p. 114

Name ______________________ Date ______________

PRACTICE WORKSHEET 10-10

Problem Solving: Using Percent

Solve.

1. Martin saves $63 each week. If he saves 15% of his wages, how much does he earn per week? **$420**

2. Stuart wants to gain 15 pounds. This is 12% of his present weight. How much does Stuart weigh now? **125 lb**

3. During a sale, Ms. Trevino puchased a blender for 75% of the regular price. She paid $27 for the blender. What was the regular price? **$36**

4. When a truck is loaded to 45% of its capacity, there are 108 cases on the truck. How many cases will be on the truck when it is loaded to capacity? **240 cases**

5. Betty bought a camera at a 20% reduction sale. If she paid $18 for it, what was the regular price? **$22.50**

6. If 48% of the students in a school are boys, and the girls number 468, how many students are enrolled? **900 students**

7. If an ore contains 16% copper, how many tons of ore are needed to get 20 tons of copper? **125 tons**

8. Patricia Walsh receives a base salary of $125 plus a commission of 6% of sales. If her pay for one week was $441.80, what were her total sales? **$5,280**

9. Christopher Cernami answered 4 test questions incorrectly and scored 90%. How many questions did he answer correctly? **36 questions**

10. The Sluggerville baseball team won 32 games. Their percent of games lost was 20%. Find the number of games the team played. **40 games**

Problem Solving

10-10 USING PERCENT

Objective
Solve problems using percents.

As marketing manager for All-Sports Equipment, Bob Henderson keeps track of sales for each product. Sales are recorded for each month, quarter, half, and entire year. For the annual meeting, Bob needs to find the percent of increase in annual sales of headbands from 1989 to 1994.

Headband Sales per Year	
1989	$44,235
1990	46,889
1991	51,109
1992	53,153
1993	57,937
1994	63,831

▶ Explore

What is given?
- annual headband sales for 1989 through 1994

What is asked?
- What is the percent of increase in annual sales of headbands?

▶ Plan

Subtract the 1989 sales from the 1994 sales and find the percent of increase in sales.

▶ Solve

63831 [−] 44235 [=]
19596 [÷] 44235 [=] 0.4429976
↑ amount of increase

The percent of increase in headband sales rounded to the nearest percent is 44%.

▶ Examine

The amount of increase in sales is about 20,000. 20,000 is a little less than half of 45,000 and 44% is a little less than half. The answer is reasonable.

Guided Practice

Solve. Use the chart. Round to the nearest percent.

1. The *amount* of increase from 1993 to 1994 is greater than the *amount* of increase from 1990 to 1991. Is the percent of increase from 1993 to 1994 *less than, greater than,* or *the same as* the percent of increase from 1990 to 1991? **greater**
2. If the estimated increase in sales for 1995 and 1996 is 8% each year, what is the amount of estimated sales for 1996 to the nearest dollar? **$74,452**
3. The research consultant Sally Martin estimated that sales will increase 12% from 1994 to 1995 if All-Sports introduces a new line of tri-colored headbands. What is the amount of estimated sales for 1995 if the estimated percent of increase is 12%? **$71,491**

RETEACHING THE LESSON

Have students work together in small groups to discuss how to solve the exercises on page 333.

Reteaching Masters Booklet, p. 104

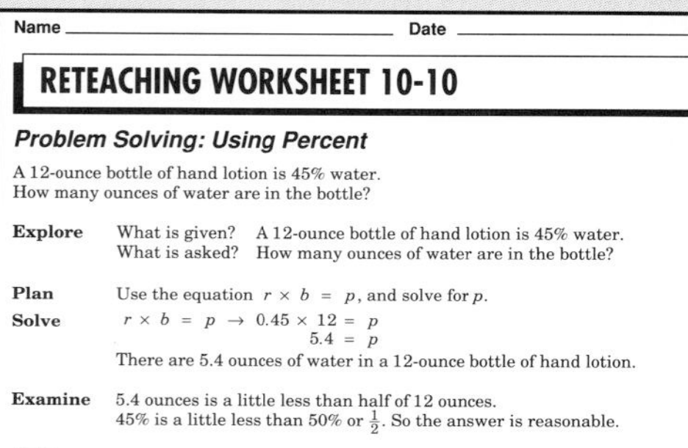

Name ______________________ Date ______________

RETEACHING WORKSHEET 10-10

Problem Solving: Using Percent

A 12-ounce bottle of hand lotion is 45% water. How many ounces of water are in the bottle?

Explore What is given? A 12-ounce bottle of hand lotion is 45% water.
What is asked? How many ounces of water are in the bottle?

Plan Use the equation $r \times b = p$, and solve for p.

Solve $r \times b = p \rightarrow 0.45 \times 12 = p$
$5.4 = p$
There are 5.4 ounces of water in a 12-ounce bottle of hand lotion.

Examine 5.4 ounces is a little less than half of 12 ounces.
45% is a little less than 50% or $\frac{1}{2}$. So the answer is reasonable.

Solve.

Problem Solving

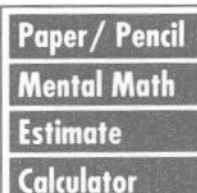

Paper/ Pencil
Mental Math
Estimate
Calculator

4. A newspaper survey at Jason City High showed that 5 out of 6 students ride the bus to school. What percent of students ride the bus? **83%**

5. Don played in 60% of the football games this season and 50% of the games last season. How many games were there during this season if Don played in 6 games? **10 games**

6. Dionne has $1,200 in her savings account for a car. If the account earns 6.5% compounded semiannually, how much will Dionne have in 3 years? **$1,453.86**

Using Formulas

7. The formula for the volume of a cylinder is $V = \pi r^2 h$. Mr. McCullogh is expanding his silo that measures 20 meters high and has a diameter of 15 meters. If he adds 4 meters of height to his silo, what is the percent of increase in the volume the silo can hold? Use 3.14 for π. **20%**

Using Data

Use the chart for Exercises 8-9.

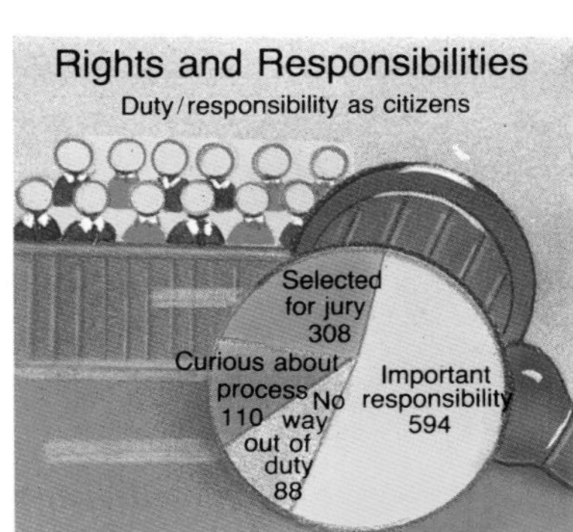

8. There were 1,100 people surveyed. What percent of the people surveyed said they served on a jury because they feel it is a duty or a responsibility? **54%**

9. Of the 1,100 people surveyed, what percent served on a jury because they couldn't get out of it or because they wanted to serve? **18%**

FUN with MATH

When and where was democracy born?
See page 372.

Collect Data

10. Survey the members of your class and find the percent who would be willing to serve on a jury and the percent who would *not* be willing to serve on a jury. **Answers may vary.**

Critical Thinking

11. What symbol comes next?

Mixed Review

Lesson 7-9

Classify each quadrilateral.

12.  **rhombus**

13. 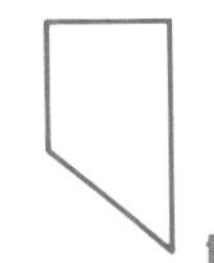 **trapezoid**

14. 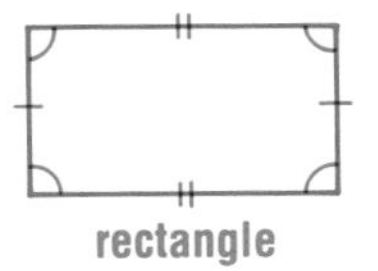 **rectangle**

Lesson 8-9

Find the volume of each rectangular prism.

15. length = 4 cm, width = 3 cm, height = 2 cm **24 cm³**

16. length = 10 in., width = 7 in., height = 5 in. **350 in³**

17. length = 4.5 m, width = 3.2 m, height = 1.7 m **24.48 m³**

Lesson 9-5

18. On a map the scale is 1 inch:200 miles. How far is it between Dallas and Detroit if the distance on the scale drawing is 5.8 inches? **1,160 miles**

11. The symbols are the counting numbers placed back-to-back.

APPLYING THE LESSON

Independent Practice

Homework Assignment	
Minimum	Ex. 1-11 odd; 12-18
Average	Ex. 1-18
Maximum	Ex. 1-18

Chapter 10, Quiz B (Lessons 10-6 through 10-10) is available in the Evaluation Masters Booklet, p. 49.

Chalkboard Examples

1. Find the percent of increase in headband sales from 1987 to 1988. Is the percent of increase greater or less than average percent of increase per year for the six years sales shown? **4%; less than the average**

Additional examples are provided on TRANSPARENCY 10-10.

EVALUATING THE LESSON

Check for Understanding Use Guided Practice Exercises to check for student understanding.

Closing the Lesson Have students tell about times when they used percents to solve problems.

EXTENDING THE LESSON

Enrichment Explain why a 10% increase followed by a 10% decrease is always greater than the original amount. Explain why a 10% decrease followed by a 10% increase is always less than the original amount.

Enrichment Masters Booklet, p. 104

Name ______ Date ______

ENRICHMENT WORKSHEET 10-10

Using Percent

Solve.

1. In a survey of 3,000 people, 41% listed speaking before a group as one of their biggest fears. How many people gave that response? **1,230 people**

2. A standard tip in a restaurant is sometimes considered 15% of the bill. How much money should you leave as a tip if the bill is $12.80? **$1.92**

3. A pair of jeans sells for $24. There is an 8% sales tax on clothing. How much will the jeans cost with the tax? **$25.92**

4. The Acme Service Center gives a 10% discount for cash payments. If Mike pays $16.65 in cash for his repairs, how much is he saving? **$1.85**

5. A gasoline service station gives a 3% discount for cash payments. How much will John save on a fill-up of $15.00 if he pays cash? **$0.45**

6. A mixed nut assortment contains 50% peanuts, 25% almonds, 15% cashews, and 10% Brazil nuts. If there are 8 oz. of peanuts in the can, how many ounces of the other kinds are there? **4 oz almonds, 2.4 oz cashews, 1.6 oz Brazil nuts**

7. As an employee of the Better Dress shop, Marlo receives a 20% discount on all her purchases. How much will she pay for a dress that was regularly priced at $72 but is on sale for 30% off? **$40.32**

Using the Chapter Review

The Chapter Review is a comprehensive review of the concepts presented in this chapter. This review may be used to prepare students for the Chapter Test.

Chapter 10, Quizzes A and B, are provided in the Evaluation Masters Booklet as shown below.

Quiz A should be given after students have completed Lessons 10-1 through 10-5. Quiz B should be given after students have completed Lessons 10-6 through 10-10.

These quizzes can be used to obtain a quiz score or as a check of the concepts students need to review.

Evaluation Masters Booklet, p. 49

Name ______________ Date ______________

CHAPTER 10, QUIZ A (Lessons 10-1 through 10-5)

Find each percentage.

1. 35% of 120 2. $66\frac{2}{3}$% of 54

Find each percent.

3. What percent of 105 is 84? 4. 35 is what percent of 280?

Find each number.

5. 35 is 500% of what number? 6. $33\frac{1}{3}$% of what number is 42?

Estimate. *Accept any reasonable estimate.

7. 13% of 95 8. 31% of 74

Solve. Use the Venn diagram showing results of a survey on exercise preference.

Jog 22, Cycle 16, 7, 3, 10, 8, 12 Swim

9. How many prefer jogging and cycling?
10. How many were surveyed?

Answers: 1. 42 2. 36 3. 80% 4. 12.5% or 5. 7 6. 126 *7. 12 *8. 25 9. 10 10. 78

Name ______________ Date ______________

CHAPTER 10, QUIZ B (Lessons 10-6 through 10-10)

Find the percent of increase.

1. new price: \$81; original price: \$72

Find the percent of decrease.

2. original weight: 8 kg; new weight: 6.8 kg

Find the sale price.

3. blouse: \$18.96, discount rate: 20% 4. truck: \$8,900, discount rate: 8.5%

Find the discount rate.

5. coat: \$29.50, sale price: \$26.55 6. tray: \$48, sale price: \$44.40

Find the total amount to be repaid on this loan.

7. principal: \$2,500; annual rate: 9%; time: 18 months

Find the interest earned on this deposit.

8. principal: \$5,600; annual rate: 12%; time: 5 years

Find the savings total for this account.

9. principal: \$400; annual rate: $5\frac{1}{2}$% compounded annually; time: 1 year

Solve.

10. Larry made 85% on his exam. If there were 140 questions, how many did he answer correctly?

Answers: 1. 12.5% or 2. 15% 3. \$15.17 4. \$8,143.50 5. 10% 6. 7.5% or 7 7. \$2,837.50 8. \$3,360 9. \$422 10. 119

Chapter 10 REVIEW

2. commission 5. discount rate 6. interest 7. principal

Vocabulary / Concepts

Use a word or number from the list at the right to complete each sentence.

5%, 20%, commission, compound, discount, discount rate, four times, interest, multiply, once, principal, sales tax, simple, twice

1. To find the percent of a number, ? the number by the percent. **multiply**
2. Salespersons may earn ? on each sale.
3. Written as a percent, $\frac{1}{5}$ is ? . **20%**
4. The ? is a percent of each purchase collected by the state or city. **sales tax**
5. The ? is a percent of decrease in price.
6. The cost of borrowing money is the ? .
7. ? is the amount of money borrowed.
8. Quarterly means ? a year. **four times**
9. ? interest is computed on the principal plus previous interest. **compound**

Exercises / Applications

Lesson 10-1 ***Find each percentage. Use r × b = p.***

10. 20% of 15 **3**
11. 16% of 80 **12.8**
12. 125% of 56 **70**
13. $33\frac{1}{3}$% of 1,890 **630**
14. $87\frac{1}{2}$% of 640 **560**
15. 6.2% of 115 **7.13**

Lesson 10-2 ***Find each percent. Use an equation.***

16. What percent of 56 is 28? **50%**
17. What percent of 24 is 31.2? **130%**
18. What percent of 15 is 10? **$66\frac{2}{3}$%**
19. 25 is what percent of 80? **31.25%**
20. 100 is what percent of 40? **250%**
21. 6 is what percent of 27? **22.2%**

Lesson 10-3 ***Find each number. Use an equation.***

22. 29% of what number is 58? **200**
23. 40% of what number is 8? **20**
24. 150% of what number is 72? **48**
25. 18 is 60% of what number? **30**

Lesson 10-4 ***Solve. Use a Venn diagram.***

26. At their 5-year class reunion, the class of '87 held a buffet dinner. If 43 people chose roast beef, 27 people chose chicken, 13 people chose both beef and chicken, and 6 chose neither, how many people attended the reunion? **63 people**

27. 70 29. \$0.60 30. 13

Lesson 10-5 ***Estimate.*** **Answers may vary. Typical answers are given.**

27. 34% of 210
28. 63% of 95 **60**
29. 15% of \$3.50
30. $9\frac{3}{4}$% of 132

Lesson 10-6 ***Find the percent of increase or decrease. Round decimal answers to the nearest percent.***

	Item	1987	1989
31.	plane ticket	$320	$360
32.	house	$85,600	$93,200
33.	computer	$3,000	$2,848

	Name	1988 Salary	1990 Salary
34.	Snyder	$21,500	$20,000
35.	Jensen	$29,200	$31,500
36.	Holt	$42,000	$47,900

31. 13% incr. 32. 9% incr. 33. 5% decr. 34. 7% decr. 35. 8% incr. 36. 14% incr.

Lesson 10-7 ***Find the discount, the sale price, or the discount rate.***

37. coat, $120
discount rate, 22%
sale price, ? **$93.60**

38. theater ticket, $24
discount, $2.40
discount rate, ? **10%**

39. car, $9,400
discount rate, 7%
discount, ? **$658**

Lesson 10-8 ***Find the interest owed on each loan or the interest earned on each deposit.***

40. deposit, $340
annual rate, 9%
time, 15 months **$38.25**

41. deposit, $1,200
annual rate, 9.75%
time, 2 years **$234**

42. loan, $500
annual rate, 12%
time, 3 months **$15**

43. loan, $200
annual rate, 10%
time, 8 months **$13.33**

Lesson 10-9 **44.** Find the savings total for a deposit of $500 in an account that pays 6.5% compounded quarterly for two years. **$568.82**

Lesson 10-10 **45.** The average American consumed 95.3 pounds of fresh vegetables in 1986 and 100.3 pounds in 1988. What was the percent of increase in consumption of fresh vegetables? **5%**

46. Mary Ann deposits $220 in a savings account that pays $9\frac{3}{4}\%$ annual interest. What will her savings total be at the end of 3 years? **$284.35**

47. $3,594.71 **47.** Rita and Dan McDonald borrowed $3,000 at an interest rate of $9\frac{1}{4}\%$ compounded semiannually for 2 years. Find the total amount to be repaid.

48. In November, 1988, *TV Review* magazine reported that about 18% of U.S. households with televisions watched "Head of the Class" regularly. If 15,948,000 households watched "Head of the Class," *about* how many households with television are there in the U.S.? **88,600,000 households**

APPLYING THE LESSON

Independent Practice

Homework Assignment	
Minimum	**Ex. 1-9; 10-48 even**
Average	**Ex. 1-43 odd; 44-48**
Maximum	**Ex. 1-48**

Alternate Assessment Strategy

To provide a brief in-class review, you may wish to read the following questions to the class and require a verbal response.

1. What is 20% of 50? **10**
2. 24 is what percent of 30? **80%**
3. 50% of what number is 18? **36**
4. Estimate 48% of 81. **40**
5. If gas sold for $1.27 a gallon and now sells for $1.18 a gallon, what is the percent of decrease? **7%**
6. A ski jacket regularly sells for $85. What is the sale price if the discount rate is 30%? **$59.50**
7. Find the interest earned on $3,000 for 5 years at 10% annual interest. **$1,500**
8. A $25 pair of jeans is on sale for $17.50. What is the rate of discount? **30%**
9. 22 is what percent of 55? **40%**
10. 80% of what number is 48? **60**

Using the Chapter Test

This page may be used as a test or as an additional review if necessary. Two forms of the Chapter Test are provided in the Evaluation Masters Booklet. Form 2 (free response) is shown below, and Form 1 (multiple choice) is shown on page 339.

The **Tech Prep Applications Booklet** provides students with an opportunity to familiarize themselves with various types of technical vocations. The Tech Prep applications for this chapter can be found on pages 19–20.

The **Performance Assessment Booklet** provides an alternative assessment for evaluating student progress. An assessment for this chapter can be found on pages 19–20.

Evaluation Masters Booklet, p. 48

Name ______________ Date ______________

CHAPTER 10 TEST, FORM 2

Find each percentage.

1. 7% of 60 — 2. $62\frac{1}{2}$% of 480

Find each percent.

3. What percent of 84 is 63? — 4. What percent of 50 is 125?
5. 36 is what percent of 60?

Find each number.

6. 42 is 35% of what number? — 7. 400% of what number is 28?
8. $87\frac{1}{2}$% of the number is 140. Find the number.

Estimate. *Accept any reasonable estimate.

9. 24% of 79 — 10. 1.2% of 305

Find the percent of increase.

11. item: tablet; original price: $0.85; new price: $1.02

Find the percent of decrease.

12. item: book; original price: $24; new price: $15

Find the sale price.

13. item: table; regular price: $199; discount rate: 20%

Find the discount rate.

14. item: boat; regular price: $4,200; sale price: $3,570

Find the interest owed on the following loan.

15. principal: $4,500; annual rate; 12%; time: 3 years

Find the interest earned on the following deposit.

16. principal: $3,700; annual rate: 9%; time: 6 months

Find the savings total for the following account.

17. principal: $2,000; annual rate: 8% compounded semiannually; time: 1 year

Solve.

18. Mr. Rodriguez earned $18,500 this year. He saved $370. What percent of his earnings did he save?
19. What is the sales tax on a car which sells for $11,900 if the tax rate is 5%?
20. In a survey, teens were asked what type of movies they saw last month. Twenty-four saw action, 22 saw thrillers, 26 saw comedies, 9 saw action and comedies, 8 saw action and thrillers, 5 saw comedies and thrillers, and 3 saw all three types. How many teens were surveyed?

Bonus E.Z. Credit charges 18% simple interest annually on monthly balances and a $25 annual fee. If Ms. Choi has an average monthly balance of $1,000, what is her annual cost for E.Z. Credit?

Answers: 1. 4.2 2. 300 3. 75% 4. 250% 5. 60% 6. 120 7. 7 8. 160 *9. 20 *10. 3 11. 20% 12. $37\frac{1}{2}$% 13. $159.20 14. 15% 15. $1,620 16. $166.50 17. $2,163.20 18. 2% 19. $595 20. 53 teens Bonus $205

Chapter 10 TEST

Find each missing value. Use an equation.

1. 18% of 320 is what number? **57.6**
2. 13 is $12\frac{1}{2}$% of what number? **104**
3. Ninety-six students responded to a survey. Forty students said they like both rock and country, 55 students like rock, and 47 students like country. How many students like neither rock nor country? Use a Venn diagram. **34 students**

Estimate. **Answers may vary. Typical answers are given.**

4. 9.8% of $127
5. 22% of $80
6. 52% of $1,000
7. $5\frac{3}{4}$% of $200

4. $13 5. $16 6. $500 7. $10 8. $0.77, $8.86 9. $43.75, $306.25

Find the discount and the sale price.

8. album, $9.63; discount rate, 8%
9. ring, $350; discount rate, $12\frac{1}{2}$%

Find the interest earned on each deposit.

10. principal: $90
annual rate: 8%
time: 8 months **$4.80**
11. principal: $375
annual rate: $7\frac{3}{4}$%
time: 2 years **$58.13**
12. principal: $1,888
annual rate: $7\frac{1}{4}$%
time: 15 months **12. $171.10**

Find the savings total for each account.

13. principal: $500
6%, compounded semiannually
time: 2 years
14. principal: $3,000
5%, compounded quarterly
time: 1 year
15. principal: $1,000
$8\frac{1}{2}$%, compounded semiannually
time: 2 years

13. $562.75 14. $3,152.84 15. $1,181.15

Solve.

16. Sung Lee bought a beach volleyball set for $45.00. If the original price was $62.50, what was the discount rate? **28%**
17. On a test with 200 questions, Sharon missed 32 questions. What percent of the questions did she answer correctly? **84%**
18. What is the sales tax and the total cost of a stereo that sells for $890? The tax rate is 5.5%. **$48.95, $938.95**
19. As a handling fee, Phillips Corporation charges 2% of the total sale price for all stock transactions. What is the handling fee for a sale of $18,650? **$373**
20. Larry Bird made 3,356 free throws during his basketball career through the 1988-89 season. This is 89.6% of his free throw attempts. How many attempts did Larry Bird make? **3,745 attempts**

BONUS: In your own words, explain why an account that earns 6% compounded quarterly will earn more interest in one year than an account that earns 6% interest per year. **See margin.**

Additional Answers

BONUS: The account that earns 6% compounded quarterly will earn more interest because the interest for each quarter earns interest.

A **Test and Review Generator** is provided in Apple, IBM, and Macintosh versions. You may use this software to create your own tests or worksheets, based on the needs of your students.

MAIL ORDERING

Glenda Steele ordered three stove-top popcorn poppers weighing 2 pounds 8 ounces each as gifts for her nephews and niece. She finds the weight of her total order to compute shipping and handling charges.

The customer sometimes pays shipping and handling charges. The rates for one company are shown below.

Delivery Weight	Local	Delivery Weight	Local	Delivery Weight	Local	Delivery Weight	Local
0– 0.5 lb	1.25	13.1–14 lb	2.51	27.1–28 lb	3.25	41.1–42 lb	4.00
0.6– 1 lb	1.50	14.1–15 lb	2.57	28.1–29 lb	3.31	42.1–43 lb	4.05
1.1– 2 lb	1.86	15.1–16 lb	2.62	29.1–30 lb	3.36	43.1–44 lb	4.10
2.1– 3 lb	1.98	16.1–17 lb	2.67	30.1–31 lb	3.41	44.1–45 lb	4.16
3.1– 4 lb	2.09	17.1–18 lb	2.72	31.1–32 lb	3.47	45.1–46 lb	4.21
4.1– 5 lb	2.19	18.1–19 lb	2.78	32.1–33 lb	3.52	46.1–47 lb	4.26
5.1– 6 lb	2.23	19.1–20 lb	2.83	33.1–34 lb	3.57	47.1–48 lb	4.31
6.1– 7 lb	2.26	20.1–21 lb	2.88	34.1–35 lb	3.63	48.1–49 lb	4.37
7.1– 8 lb	2.29	21.1–22 lb	2.94	35.1–36 lb	3.68	49.1–50 lb	4.42
8.1– 9 lb	2.32	22.1–23 lb	2.99	36.1–37 lb	3.73	Each additional lb	0.07
9.1–10 lb	2.35	23.1–24 lb	3.04	37.1–38 lb	3.78		
10.1–11 lb	2.39	24.1–25 lb	3.10	38.1–39 lb	3.84		
11.1–12 lb	2.42	25.1–26 lb	3.15	39.1–40 lb	3.89	*0.1 → means one tenth of a lb	
12.1–13 lb	2.46	26.1–27 lb	3.20	40.1–41 lb	3.94		

To find the total weight of her order, Mrs. Steele multiplies and adds the weights.

```
   2 lb  8 oz      3 popcorn poppers at
×        3         2 lb 8 oz each
   6 lb 24 oz      Rename 24 oz
+1 lb  8 oz        as 1 lb 8 oz.
   7 lb  8 oz
```

The delivery weight for 3 popcorn poppers is 7 pounds 8 ounces. This is between 7.1 and 8 pounds. So the shipping and handling charge is $2.29.

Find the total weight for each order. Then find the shipping and handling charges.

1. 1 suitcase at 9 lb 8 oz and 2 flight bags at 4 lb 8 oz each **18 lb 8 oz; $2.78**

2. 3 necklaces at 4 oz each and 6 bracelets at 3.5 oz each **2 lb 1 oz; $1.98**

3. Jake placed an order for toys that weighed 6 lb 14 oz. The next week he ordered a set of dishes that weighed 12 lb 12 oz. How much would Jake save in shipping and handling charges if he ordered the toys and dishes at the same time? **$1.89**

Applying Mathematics to the Real World

This optional page shows how mathematics is used in the real world and also provides a change of pace.

Objective Find shipping weights and charges for ordering goods by mail.

Using Discussion Discuss the pros and cons of catalog shopping. Ask students if shopping by mail is always more economical than local shopping.

Using Data Have students bring catalogs to class. Make a list of items you want them to purchase through the catalog. Have students calculate the total cost of the order.

Using the Cumulative Review

This page provides an aid for maintaining skills and concepts presented thus far in the text.

A Cumulative Review is also provided in the Evaluation Masters Booklet as shown below.

Evaluation Masters Booklet, p. 50

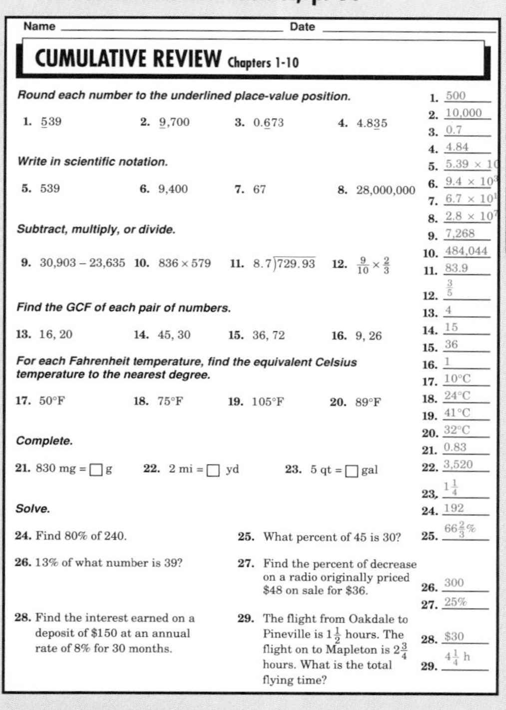

Name ______ Date ______

CUMULATIVE REVIEW Chapters 1-10

Round each number to the underlined place-value position.

1. 539 2. 9,700 3. 0.673 4. 4.835

Write in scientific notation.

5. 539 6. 9,400 7. 67 8. 28,000,000

Subtract, multiply, or divide.

9. 30,903 – 23,635 10. 836 × 579 11. $8.7\overline{)729.93}$ 12. $\frac{9}{10} \times \frac{2}{3}$

Find the GCF of each pair of numbers.

13. 16, 20 14. 45, 30 15. 36, 72 16. 9, 26

For each Fahrenheit temperature, find the equivalent Celsius temperature to the nearest degree.

17. 50°F 18. 75°F 19. 105°F 20. 89°F

Complete.

21. 830 mg = □ g 22. 2 mi = □ yd 23. 5 qt = □ gal

Solve.

24. Find 80% of 240.
25. What percent of 45 is 30?
26. 13% of what number is 39?
27. Find the percent of decrease on a radio originally priced $48 on sale for $36.
28. Find the interest earned on a deposit of $150 at an annual rate of 8% for 30 months.
29. The flight from Oakdale to Pineville is $1\frac{1}{2}$ hours. The flight on to Mapleton is $2\frac{3}{4}$ hours. What is the total flying time?

1. 500
2. 10,000
3. 0.7
4. 4.84
5. 5.39 × 10
6. 9.4 × 10
7. 6.7 × 10
8. 2.8 × 10
9. 7,268
10. 484,044
11. 83.9
12. $\frac{3}{5}$
13. 4
14. 15
15. 36
16. 1
17. 10°C
18. 24°C
19. 41°C
20. 32°C
21. 0.83
22. 3,520
23. $1\frac{1}{4}$
24. 192
25. $66\frac{2}{3}\%$
26. 300
27. 25%
28. $30
29. $4\frac{1}{4}$ h

Cumulative REVIEW/TEST

Free Response

3. 35 4. 18 5. 20

Lesson 1-1 ***Replace each ▩ with a number to make a true sentence.***

1. 700 = ▩ tens **70** **2.** 200 = ▩ ones **200** **3.** 3,500 = ▩ hundreds
4. 0.018 = ▩ thousandths **5.** 0.2 = ▩ hundredths **6.** 0.7 = ▩ tenths **7**

Lesson 3-1 ***Multiply.***

7. 6,413 × 65 = **416,845** **8.** 2,117 × 39 = **82,563** **9.** 1,318 × 63 = **83,034** **10.** 8,185 × 225 = **1,841.625** **11.** 4,347 × 381 = **1,656,207**

12. 121 × 7 **847** **13.** 8 × 6,469 **51,752** **14.** 444 × 16 **7,104**
15. 639 × 132 **84,348** **16.** 671 × 669 **448,899** **17.** 263 × 9,188 **2,416,444**

Lesson 4-9 Typical answers are given.
Estimate. Answers may vary. 20. 8

18. $\frac{17}{20} + \frac{9}{12}$ 2 **19.** $\frac{4}{5} - \frac{1}{10}$ 1 **20.** $10\frac{8}{10} - 3\frac{1}{11}$

Lesson 5-1 Typical answers are given.
Estimate. Answers may vary. 22. 2 23. 3

21. $\frac{3}{7} \times 9\frac{1}{10}$ 5 **22.** $4\frac{4}{5} \div 2\frac{5}{11}$ **23.** $10\frac{3}{5} \div 3\frac{1}{2}$

Lessons 6-2, 6-4, 6-5 ***Choose the most reasonable unit of measure.***

24. length of a safety pin a. <u>millimeter</u> b. meter c. kilometer
25. mass of a chair a. milligram b. gram c. <u>kilogram</u>
26. capacity of a teaspoon a. <u>milliliter</u> b. liter c. kiloliter

Lesson 7-8 **27.** A rectangle has a perimeter of 48 inches. The length of one side is 15 inches. What are the measures of the other three sides? **15 in.; 9 in.; 9 in.**

Lesson 8-4 **28.** The diameter of a circular go-cart racetrack is 250 meters. To the nearest meter, what is the area of the racetrack and the space in the center? **49,063 m²**

Lesson 9-4 ***Solve each proportion.***

29. $\frac{2}{3} = \frac{x}{9}$ **6** **30.** $\frac{5}{7} = \frac{y}{35}$ **25** **31.** $\frac{1}{9} = \frac{n}{36}$ **4**

Lesson 9-8 ***Write each percent as a fraction in simpiest form.***

32. 80% $\frac{4}{5}$ **33.** 70% $\frac{7}{10}$ **34.** 58% $\frac{29}{50}$

Lesson 10-7 ***Find the discount and the sale price.***

35. radio, $55
discount rate, 20% **$11, $44**

36. compact disc, $14.95
discount rate, 15% **$2.24, $12.71**

APPLYING THE LESSON

Independent Practice

Homework Assignment	
Minimum	Ex. 2-36 even
Average	Ex. 1-36
Maximum	Ex. 1-36

Chapter 10

Multiple Choice

Choose the letter of the correct answer for each item.

1. What is the sum of one-fourth and one-eighth?

a. $\frac{1}{12}$ ▶ **c.** $\frac{3}{8}$
b. $\frac{1}{6}$ **d.** $\frac{3}{4}$

Lesson 4-10

2. Estimate the product of 0.93 and 17.2.

e. 7
▶ **f.** 17
g. 170
h. 1,700

Lesson 3-2

3. Residents must pay 30% of the cost of putting curbs on their streets. How much will 50 feet of curbing at $3.25 per foot cost for a resident?

a. $52.50 **c.** $49.25
▶ **b.** $48.75 **d.** $162.50

Lesson 10-10

4. What is the product of $7\frac{3}{14}$ and $10\frac{8}{9}$?

e. $70\frac{4}{21}$
f. $76\frac{23}{30}$
▶ **g.** $78\frac{5}{9}$
h. $157\frac{1}{9}$

Lesson 5-3

5. Jamall earns $45 for every $1,000 worth of tires he sells. Which proportion can be used to find earnings on sales of $5,580?

a. $\frac{45}{1,000} = \frac{5,580}{x}$
b. $\frac{45}{5,580} = \frac{1,000}{x}$
c. $\frac{1,000}{5,580} = \frac{x}{45}$
▶ **d.** $\frac{45}{1,000} = \frac{x}{5,580}$

Lesson 9-3

6. What is the area of a square with sides 3 inches long?

e. 6 in^2 **g.** 12 in^2
▶ **f.** 9 in^2 **h.** 27 in^2

Lesson 8-2

7. Solve the equation $\frac{x}{36} = 4$.

a. 8
b. 9
c. 124
▶ **d.** 144

Lesson 1-12

8. If there are 3 girls for every 8 boys, how many girls will there be if there are 1,000 boys?

e. 37.5 **g.** $2,666.\overline{6}$
▶ **f.** 375 **h.** 3,000

Lesson 9-3

9. If Steven missed 14 out of 50 questions on an exam, what percent of the questions did he answer correctly?

a. 14%
b. 28%
▶ **c.** 72%
d. 93%

Lesson 10-2

10. Debbie weighed 7 pounds at birth. She lost 4 ounces during the first week. How much did she weigh then?

e. 3 pounds
f. 6 pounds 4 ounces
g. 6 pounds 8 ounces
▶ **h.** 6 pounds 12 ounces

Lesson 6-8

Using the Cumulative Test

This test serves to familiarize students with standardized format while testing skills and concepts presented up to this point.

APPLYING THE LESSON

Independent Practice

Homework Assignment	
Minimum	**Ex. 1-10**
Average	**Ex. 1-10**
Maximum	**Ex. 1-10**

Evaluation Masters Booklet, pp. 46-47

CHAPTER 10 TEST, FORM 1

Write the letter of the correct answer on the blank at the right of the page.

1. Find the percentage. 35% of 40 — 1. B
A. 114 **B.** 14 **C.** 11.4 **D.** 1.4

2. Find the percentage. 130% of 70 — 2. D
A. 2.1 **B.** 21 **C.** 9.1 **D.** 91

3. Find the percentage. $62\frac{1}{2}$% of 96 — 3. C
A. 12 **B.** 36 **C.** 60 **D.** 84

4. What percent of 200 is 32? — 4. A
A. 16% **B.** 8% **C.** 64% **D.** 6.4%

5. What percent of 48 is 36? — 5. A
A. 75% **B.** $133\frac{1}{2}$% **C.** 50% **D.** $66\frac{2}{3}$%

6. 24 is what percent of 64? — 6. D
A. $33\frac{1}{3}$% **B.** 50% **C.** 25% **D.** 37.5%

7. 20% of the number is 3. Find the number. — 7. C
A. 0.3 **B.** 0.15 **C.** 15 **D.** 60

8. $66\frac{2}{3}$% of what number is 18? — 8. B
A. 12 **B.** 27 **C.** 24 **D.** 15

9. 175% of what number is 21? — 9. B
A. 9 **B.** 12 **C.** 15 **D.** 28

10. 28 is $87\frac{1}{2}$% of what number? — 10. A
A. 32 **B.** 24 **C.** 27 **D.** 42

11. Choose the best estimate. 26% of 81 — 11. D
A. 16 **B.** 400 **C.** 320 **D.** 20

12. Choose the best estimate. 37% of 55 — 12. C
A. 88 **B.** 75 **C.** 21 **D.** 24

13. Find the percent of increase. Original price: $4, new price: $4.50 — 13. A
A. $12\frac{1}{2}$% **B.** 25% **C.** 15% **D.** $37\frac{1}{2}$%

Chapter

STATISTICS AND GRAPHS

PREVIEWING the CHAPTER

This chapter discusses different ways of describing, organizing, and presenting data. Finding the mean, median, mode, and range of a set of data provides important information about central tendencies. Frequency tables help organize data. Bar graphs, line graphs, pictographs, and circle graphs are used to emphasize different characteristics of a set of data. In addition, stem-and-leaf plots, measures of variation, and box-and-whisker plots are presented.

Problem-Solving Strategy Students solve verbal problems by looking for a pattern.

Lesson (Pages)	Lesson Objectives	State/Local Objectives
11-1 (342-343)	11-1: Find the median, mode, and range of a set of data.	
11-2 (344-345)	11-2: Construct and interpret frequency tables.	
11-3 (346-347)	11-3: Interpret data.	
11-4 (348-349)	11-4: Construct bar graphs.	
11-5 (350-351)	11-5: Construct line graphs.	
11-6 (354-355)	11-6: Construct pictographs.	
11-7 356-357)	11-7: Construct circle graphs.	
11-8 (358-359)	11-8: Interpret and construct a stem-and-leaf plot.	
11-9 (360-361)	11-9: Find quartiles and interquartile range.	
11-10 (362-363)	11-10: Construct and interpret box-and-whisker plots.	
11-11 (364-365)	11-11: Solve verbal problems by looking for a pattern.	

ORGANIZING the CHAPTER

You may refer to the ***Course Planning Calendar*** on page T10.

Planning Guide

Lesson (Pages)	Extra Practice (Student Edition)	Blackline Masters Booklets: Reteaching	Practice	Enrichment	Activity	Multi-cultural Activity	Technology	Tech Prep	Evaluation	Other Resources
11-1 (342-343)	p. 477	p. 105	p. 115	p. 105	p. 21		p. 11			Transparency 11-1
11-2 (344-345)		p. 106	p. 116	p. 106	p. 39					Transparency 11-2
11-3 (346-347)		p. 107	p. 117	p. 107						Transparency 11-3
11-4 (348-349)	p. 477	p. 108	p. 118	p. 108		p. 11				Transparency 11-4
11-5 (350-351)	p. 478	p. 109	p. 119	p. 109				p. 21		Transparency 11-5
11-6 (354-355)	p. 478	p. 110	p. 120	p. 110			p. 25		p. 54	Transparency 11-6
11-7 (356-357)	p. 478	p. 111	p. 121	p. 111	p. 22			p. 22		Transparency 11-7
11-8 (358-359)	p. 479	p. 112	p. 122	p. 112						Transparency 11-8
11-9 (360-361)	p. 479	p. 113	p. 123	p. 113						Transparency 11-9
11-10 (362-363)	p. 479	p. 114	p. 124	p. 114						Transparency 11-10
11-11 (364-365)			p. 125						p. 54	Transparency 11-11
Ch. Review (366-367)									pp. 51-53	Test and Review Generator
Ch. Test (368)										
Cumulative Review/Test (370-371)									p. 55	

OTHER CHAPTER RESOURCES

Student Edition
Chapter Opener, p. 340
Activity, p. 341
Computer Application, p. 352
On-the-Job Application, p. 353
Journal Entry, p. 357
Portfolio Suggestion, p. 365
Consumer Connections, p. 369
Tech Prep Handbook, pp. 531-533

Teacher's Classroom Resources
Transparency 11-0
Lab Manual, pp. 43-44
Performance Assessment, pp. 21-22
Lesson Plans, pp. 115-125

Other Supplements
Overhead Manipulative Resources, Labs 32-33, p. 23
Glencoe Mathematics Professional Series

ENHANCING the CHAPTER

Some of the blackline masters for enhancing this chapter are shown below.

COOPERATIVE LEARNING

The following data are scores for three math tests.
Test 1: 86, 77, 90, 88, 93, 84, 85, 93, 69, 81, 72, 90
Test 2: 80, 82, 62, 68, 64, 70, 84, 100, 91, 100, 75, 90
Test 3: 100, 60, 82, 91, 78, 75, 100, 64, 58, 85, 90, 87
For each set of scores, have students find the mean, median, mode, and range; show the data in a bar graph; show the number of A's, B's, C's, D's, and E's by your grade scale as a circle graph; show the change in class averages as a line graph; and compare the tests scores with box-and-whisker plots.

USING MODELS/MANIPULATIVES

Have students draw the axes for a bar graph. Label the vertical axis from 0-10 in 1-inch intervals. Give students 8 bars of lengths 2 in., 3 in., 4 in., 5 in., 6 in., 8 in., 5 in., and 7 in. Have them place the bars on the graph in any order. Then have them find the average number of inches by cutting lengths from some bars and adding the lengths to other bars so that each bar has the same height. **5 in.**

Cooperative Problem Solving, p. 39

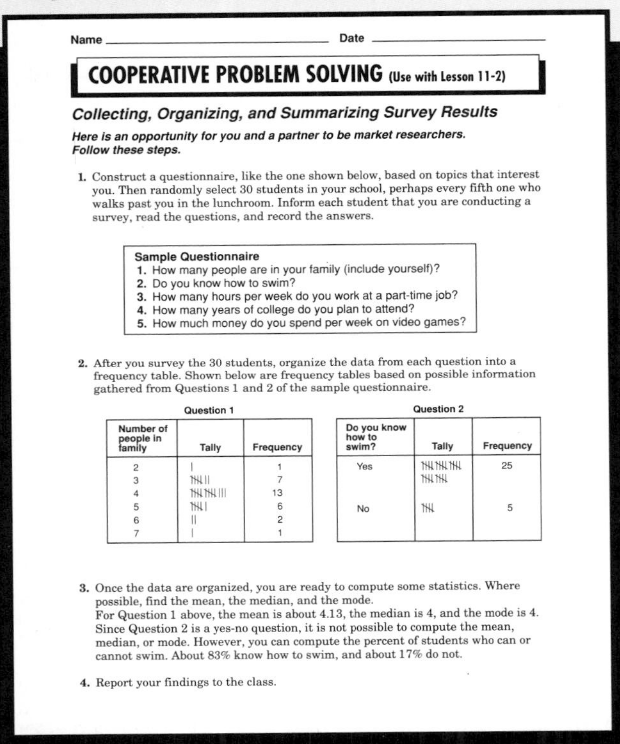

Name ______________ Date ______________

COOPERATIVE PROBLEM SOLVING (Use with Lesson 11-2)

Collecting, Organizing, and Summarizing Survey Results

Here is an opportunity for you and a partner to be market researchers. Follow these steps.

1. Construct a questionnaire, like the one shown below, based on topics that interest you. Then randomly select 30 students in your school, perhaps every fifth one who walks past you in the lunchroom. Inform each student that you are conducting a survey, read the questions, and record the answers.

Sample Questionnaire
1. How many people are in your family (include yourself)?
2. Do you know how to swim?
3. How many hours per week do you work at a part-time job?
4. How many years of college do you plan to attend?
5. How much money do you spend per week on video games?

2. After you survey the 30 students, organize the data from each question into a frequency table. Shown below are frequency tables based on possible information gathered from Questions 1 and 2 of the sample questionnaire.

Question 1

Number of people in family	Tally	Frequency
2	|	1
3	𝍸 ||	7
4	𝍸 𝍸 |||	13
5	𝍸 |	6
6	||	2
7	|	1

Question 2

Do you know how to swim?	Tally	Frequency
Yes	𝍸 𝍸 𝍸 𝍸 𝍸	25
No	𝍸	5

3. Once the data are organized, you are ready to compute some statistics. Where possible, find the mean, the median, and the mode.
For Question 1 above, the mean is about 4.13, the median is 4, and the mode is 4. Since Question 2 is a yes-no question, it is not possible to compute the mean, median, or mode. However, you can compute the percent of students who can or cannot swim. About 83% know how to swim, and about 17% do not.

4. Report your findings to the class.

Lab Activity, p. 44

Name ______________ Date ______________

LAB ACTIVITY (Use with Lesson 11-8)

Display Your Data

You can use Easy-to-Make Manipulative 8, Quarter-Inch Grid Paper.

Use the following information to make a bar graph, stem-and-leaf plot, frequency table, and line graph. Give one advantage and one disadvantage for using each to display these data.

During one month, the temperatures in degrees Fahrenheit in Ideal City, Vacationland, were as shown on this calendar.

S	M	T	W	TH	F	S
				1 / 70	2 / 72	3 / 73
4 / 76	5 / 73	6 / 72	7 / 75	8 / 75	9 / 75	10 / 76
11 / 71	12 / 70	13 / 73	14 / 73	15 / 72	16 / 77	17 / 77
18 / 75	19 / 77	20 / 79	21 / 73	22 / 72	23 / 75	24 / 75
25 / 75	26 / 73	27 / 73	28 / 79	29 / 70	30 / 75	

Bar Graph, A: ______________
D: ______________
Stem-and-Leaf Plot, A: ______________
D: ______________
Frequency Table, A: ______________
D: ______________
Line Graph, A: ______________
D: ______________

MEETING INDIVIDUAL NEEDS

Limited English Proficiency Students

Give LEP students simple bar graphs, line graphs, pictographs, and circle graphs. Ask them to write short articles that might accompany the graph in a newspaper or magazine. Have students share their articles with others in the class, and have the others check the authors' interpretations of the graphs.

CRITICAL THINKING/PROBLEM SOLVING

The following problem applies logical reasoning. The mean of five one-digit numbers is 5.2. the mode of the five numbers is 2, and the median is 5. What are the numbers?

Assume that the numbers are arranged in order from least to greatest. Since the median is 5, the third number is 5. Since the mode is 2, the first two numbers are 2. By multiplying the number of one-digit numbers by the mean, you know that the sum of the numbers is 26. The remaining numbers are 8, 9.

COMMUNICATION

Writing

Have students keep a journal throughout this chapter and write in their own words the definitions of mean, median, mode, and range, and explain under what conditions each measure of central tendency would give the "better" average. Have students describe the advantages and disadvantages of using bar graphs, line graphs, pictographs, circle graphs, stem-and-leaf plots, and box-and-whisker plots to describe sets of data.

Applications, p. 21

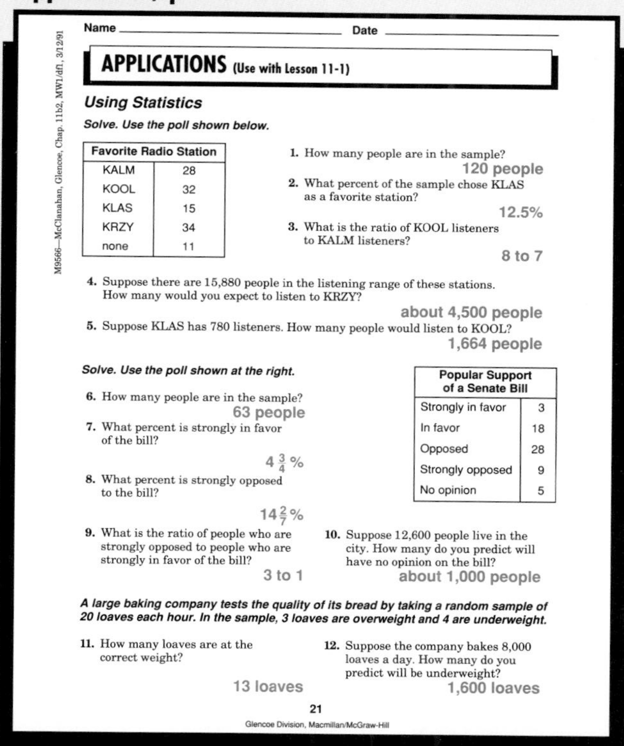

M9566—McClanahan, Glencoe, Chap. 11b2, MW1/dfl, 3/12/91

Name ______________________ Date ______________

APPLICATIONS (Use with Lesson 11-1)

Using Statistics

Solve. Use the poll shown below.

Favorite Radio Station	
KALM	28
KOOL	32
KLAS	15
KRZY	34
none	11

1. How many people are in the sample? 120 people
2. What percent of the sample chose KLAS as a favorite station? 12.5%
3. What is the ratio of KOOL listeners to KALM listeners? 8 to 7
4. Suppose there are 15,880 people in the listening range of these stations. How many would you expect to listen to KRZY? about 4,500 people
5. Suppose KLAS has 780 listeners. How many people would listen to KOOL? 1,664 people

Solve. Use the poll shown at the right.

Popular Support of a Senate Bill	
Strongly in favor	3
In favor	18
Opposed	28
Strongly opposed	9
No opinion	5

6. How many people are in the sample? 63 people
7. What percent is strongly in favor of the bill? $4\frac{3}{4}$%
8. What percent is strongly opposed to the bill? $14\frac{2}{7}$%
9. What is the ratio of people who are strongly opposed to people who are strongly in favor of the bill? 3 to 1
10. Suppose 12,600 people live in the city. How many do you predict will have no opinion on the bill? about 1,000 people

A large baking company tests the quality of its bread by taking a random sample of 20 loaves each hour. In the sample, 3 loaves are overweight and 4 are underweight.

11. How many loaves are at the correct weight? 13 loaves
12. Suppose the company bakes 8,000 loaves a day. How many do you predict will be underweight? 1,600 loaves

21

Glencoe Division, Macmillan/McGraw-Hill

Calculator Activity, p. 11

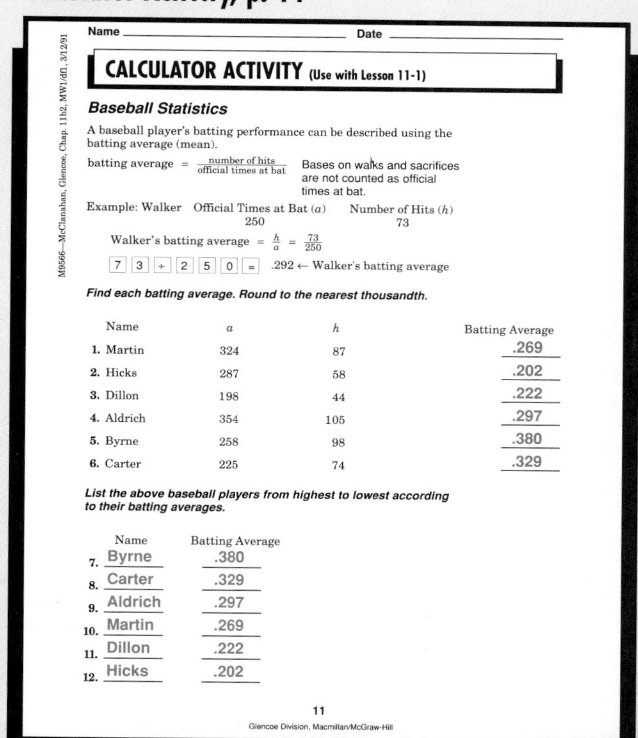

M9566—McClanahan, Glencoe, Chap. 11b2, MW1/dfl, 3/12/91

Name ______________________ Date ______________

CALCULATOR ACTIVITY (Use with Lesson 11-1)

Baseball Statistics

A baseball player's batting performance can be described using the batting average (mean).

batting average = $\frac{\text{number of hits}}{\text{official times at bat}}$ Bases on walks and sacrifices are not counted as official times at bat.

Example: Walker Official Times at Bat (a) 250 Number of Hits (h) 73

Walker's batting average = $\frac{h}{a} = \frac{73}{250}$

7 3 ÷ 2 5 0 = .292 ← Walker's batting average

Find each batting average. Round to the nearest thousandth.

	Name	a	h	Batting Average
1.	Martin	324	87	.269
2.	Hicks	287	58	.202
3.	Dillon	198	44	.222
4.	Aldrich	354	105	.297
5.	Byrne	258	98	.380
6.	Carter	225	74	.329

List the above baseball players from highest to lowest according to their batting averages.

	Name	Batting Average
7.	Byrne	.380
8.	Carter	.329
9.	Aldrich	.297
10.	Martin	.269
11.	Dillon	.222
12.	Hicks	.202

11

Glencoe Division, Macmillan/McGraw-Hill

CHAPTER PROJECT

You may want to have students form cooperative learning groups. Have students gather the following weather data from a local newspaper each day for a 2-week period: daily high and low temperatures, rainfall, pollution index, wind direction, highest wind speed, and overall conditions. Have each group present the data to the class in the form of tables, graphs, and statistics. Each group should discuss how and why the data is displayed as it is.

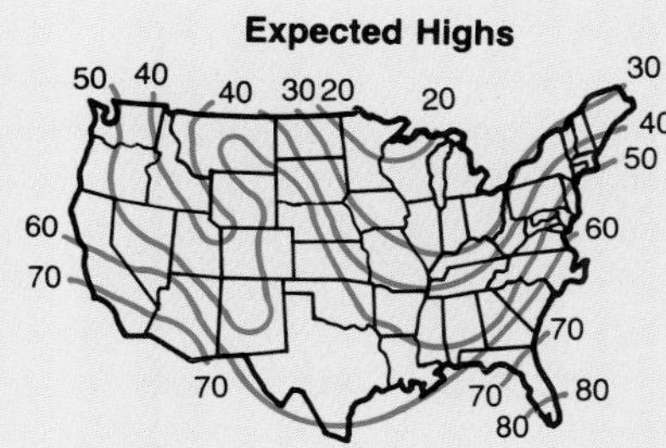

Using the Chapter Opener

Transparency 11-0 is available in the Transparency Package. It provides a full-color visual and motivational activity that you can use to engage students in the mathematical content of the chapter.

Using Discussion

- Have students read the opening paragraph. Then ask, "By show of hands, how many students have moved in their lifetime?"
- Allow students the opportunity to speculate on why people move so often.
- If students are unable to answer the last question in the opening paragraph, return to this question after students have completed the problem in the second paragraph.

The second paragraph is a motivational problem. Have students solve the problem and discuss the ways in which they arrived at their answer.

- Has Melissa's family moved more or less than the average? **more**

You may want to point out that finding an average falls under the category of statistics.

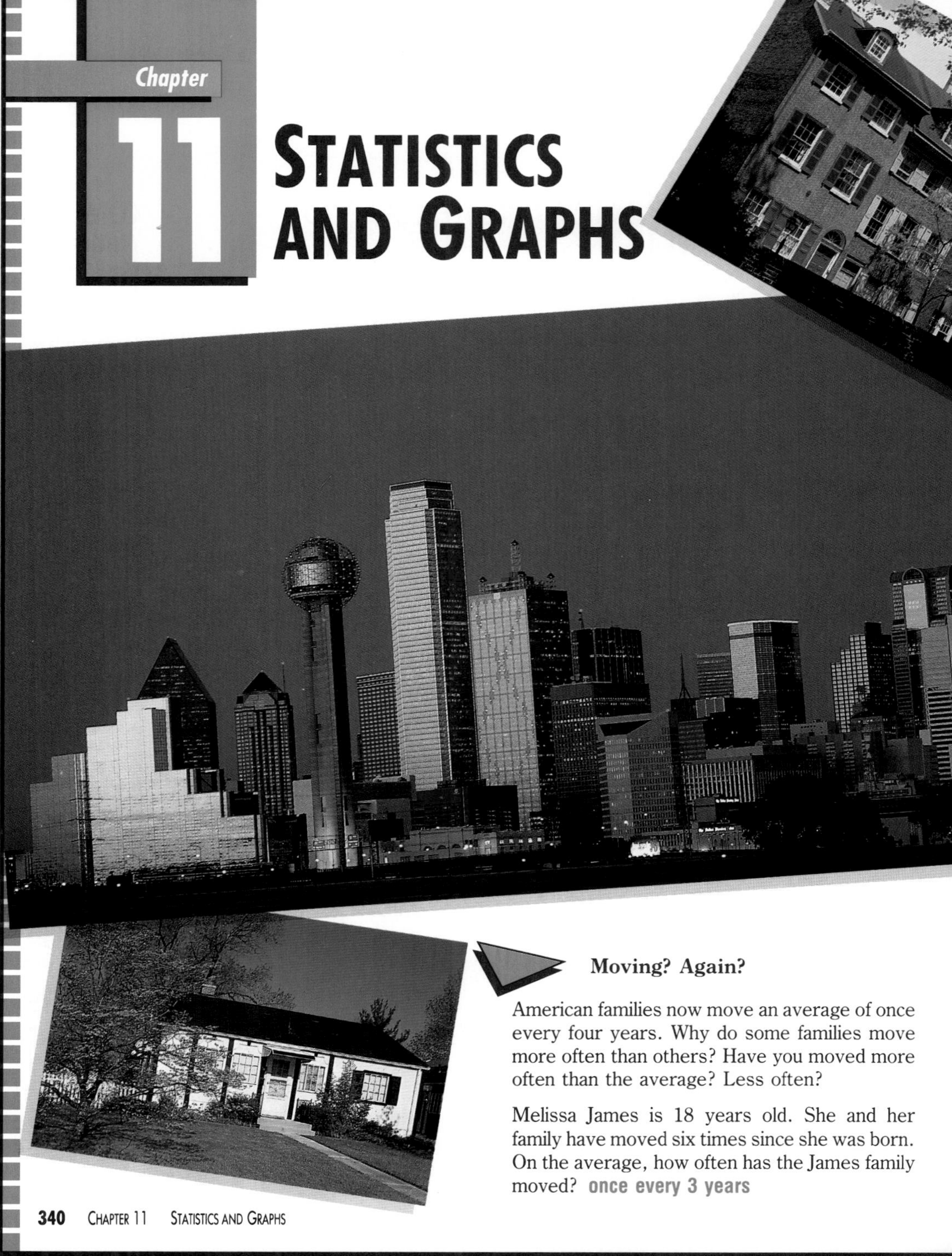

Chapter 11

STATISTICS AND GRAPHS

Moving? Again?

American families now move an average of once every four years. Why do some families move more often than others? Have you moved more often than the average? Less often?

Melissa James is 18 years old. She and her family have moved six times since she was born. On the average, how often has the James family moved? **once every 3 years**

340 CHAPTER 11 STATISTICS AND GRAPHS

ACTIVITY: Averages

There are three types of averages. Do you know what they are called? In Chapter 3, you studied the most common average, the mean. You can find an average without having to compute. Use the following activity to find the average score on a test. **mean, median, mode**

Materials: 9 index cards per student (or paper cut into 9 equal pieces) pencil

1. On each index card, write a test score. You can write any scores you want. The scores must range from 0 to 100. Each person should have a set of nine cards.
2. Shuffle the cards.

Cooperative Groups

Work together in small groups.

3. Exchange your nine test-score cards with those of another member of your group.
4. Discuss with your group ways in which you could find an average test score without having to compute. Write a short paragraph about your ideas. **Answers may vary.**
5. Use your index cards to arrange the scores in order from least to greatest. What is the middle score? The middle score is one kind of average for all the test scores. **Answers may vary.**
6. Can you think of another way to find an average without having to compute? Discuss this with your group. (Hint: Do any of the scores in your set of cards appear more than once? If so, you have found another kind of average.) **Answers may vary.**

Communicate Your Ideas

7. Discuss with your class the ideas from Problem 4.
8. Do you know what name is given to the average you found in Problem 5? in Problem 6? **median, mode**

INTRODUCING THE LESSON

Objective To discover the mean, median, and mode of data.

TEACHING THE LESSON

It is not necessary, at this time, for students to be able to name the three averages or to know how to find an average.

Using Cooperative Groups Have students work in groups of four. Ask students to work with their group on problems 3-6. As they solve problems 5-6, they can check each others' work to make sure they understand the problem.

EVALUATING THE LESSON

Using Manipulatives Encourage students to arrange the score cards various ways to see if they can come up with ideas for problems 4-6.

EXTENDING THE LESSON

Closing the Lesson Allow time for students to discuss with the class their findings from problem 2.

Activity Worksheet, Lab Manual, p. 43

Name ______ Date ______

ACTIVITY WORKSHEET (Use with page 341)

Average

Cut out the squares below to make nine test-score cards.

Exercise 4. ______

Exercise 5. ______

Exercise 6. ______

Exercise 8. ______

5-MINUTE CHECK-UP

(over Lesson 10-10)

Available on TRANSPARENCY 11-1.

1. Jan scored 90% on a math quiz. If there were 10 problems, how many did she get wrong? 1
2. If Bob missed 3 problems on a 15-point quiz, what was his score as a percent? 80%
3. Barb borrows $3,500 from the bank at 13% interest for 2 years. How much interest does she pay? $910

INTRODUCING THE LESSON

Have students write their telephone number on a piece of paper. Ask students how many have a telephone number with a digit that appears more than once, twice, three times, and so on. Ask students if they know what the number that appears most often in a set of numbers is called.

TEACHING THE LESSON

Using Critical Thinking Means, medians, and modes are "averages" for a set of data. Have students describe situations in which one is more appropriate than the other.

Practice Masters Booklet, p. 115

Name ________ Date ________

PRACTICE WORKSHEET 11-1

Median, Mode, and Range

Find the median, mode, and range for each set of data.

1. 51, 47, 48, 51, 49, 51, 52 — 51, 51, 5
2. 114, 117, 114, 119, 115, 116, 112 — 115, 114, 7
3. 10, 15, 10, 15, 4, 1, 4, 15, 10, 15, 11, 11, 7 — 10, 15, 14
4. 49, 52, 54, 51, 60, 57, 73, 51, 55, 57, 53, 57 — 54.5, 57, 24
5. 172, 176, 172, 177, 177, 175, 175, 178, 172, 174, 177 — 175, 172 and 177, 6
6. 9.2, 8.6, 9.3, 9.7, 7.2, 9.6, 9.9, 9.5, 9.4 — 9.4, no mode, 2.7
7. 0.64, 1.62, 1.66, 0.62, 1.66, 0.64, 0.62, 1.62 — 1.13, no mode, 1.04
8. 160, 156, 160, 315, 159, 160, 153, 251, 158, 150 — 159, 160, 165

Solve.

9. Allan scored 87, 92, 85, 88, and 84 on math tests this semester. What is the median of these scores? The range? 87; 8
10. On his next test, Allan scored 98. How does this affect the median? How does it affect the range? increased to 87.5; increased to 14

11-1 MEDIAN, MODE, AND RANGE

Objective
Find the median, mode, and range of a set of data.

For statistics class, Koji Ott needs to find the median, mode, and range for a set of data of his own choosing. To gather his data, he asked 11 classmates to estimate how many hours they talked on the telephone last week.

Number of Hours on the Telephone			
Sue	3	Lyn	7
Bob	6	Katy	1
Tim	2	John	4
Joe	6	Carole	8
Pete	0	Ida	6
Jill	3		

You may wish to review *mean* on page 94 with students.

The **median** of a set of data is the middle number when the data are listed in order. The **mode** is the number that appears most often. The **range** is the difference between the greatest and the least number.

Method

1. List the numbers in order from least to greatest.
2. To find the median, locate the middle number or numbers. If there are two middle numbers, find their mean.
3. To find the mode, determine the number that appears most often. There can be more than one mode or no mode.
4. To find the range, subtract the least number from the greatest number.

Example A ***Find the median, mode, and range for Koji's data.***

1. List the numbers in order. 0,1,2,3,3,4,6,6,6,7,8 (5 numbers ↑ 5 numbers)
2. The middle number is 4. So the median is 4.
3. The number 6 appears most often. So the mode is 6.
4. $8 - 0 = 8$ The range is 8.

The median number of hours that Koji's classmates talked on the telephone in one week is 4. The mode is 6 hours, and the range is 8 hours.

Example B ***Find the median, mode, and range for the set of data.***

343, 437, 265, 419, 356, 302

1. 265, 302, 343, 356, 419, 437
2. There are two middle numbers, 343 and 356.
   343 [+] 356 [÷] 2 [=] 349.5 The median is 349.5.
3. Each number appears only once. So there is no mode.
4. 437 [−] 265 [=] 172 The range is 172.

Guided Practice ***Find the median, mode, and range for each set of data.***
Example A

1. 6, 4, 5, 8, 2, 6, 9 — 6, 6, 7
2. 11, 17, 14, 15, 11 — 14, 11, 6
3. 75, 72, 71, 78, 75 — 75, 75, 7

RETEACHING THE LESSON

Find the range, median, and mode for each set of data.

1. 2, 3, 2, 1, 8, 2 7; 2; 2
2. 3, 5, 1, 5, 4, 1, 2 4; 3; 1 and 5
3. 9, 6, 9, 7, 9 3; 9; 9
4. 10, 200, 10, 10, 100 190; 10; 10

Reteaching Masters Booklet, p. 105

Name ________ Date ________

RETEACHING WORKSHEET 11-1

Median, Mode, and Range

Member	Height in Inches
JC	57
KL	60
MW	61
NL	62
PK	62
LV	70

The heights of the ensemble members are listed at the left.

The **mean** height is the sum of all the heights divided by the number of addends. The sum is 372 inches. There are 6 addends. The mean is 62 inches.

To find the median, mode, and range of the heights, list them in order.

57, 60, 61, 62, 62, 70

range: the difference between the tallest and shortest heights
median: middle height or mean of two middle heights
mode: the height that appears most often, 62 in.

4. 35.5, 43, 20 5. 53, no mode, 72 6. 278.5, no mode, 93

Example B **4.** 43, 28, 43, 23 **5.** 96, 51, 24, 55 **6.** 322, 325, 235, 232

7. 2, 2, 8 8. 4, 6, 8 9. 29, 12, 86 10. 54, 54, 54 11. 68.5, 91, 66

Exercises

Practice

Find the median, mode, and range for each set of data.

7. 1, 5, 9, 1, 2, 5, 8, 2, 7, 2, 2 **8.** 3, 2, 4, 3, 6, 6, 6, 9, 1, 8, 4

9. 12, 98, 29, 12, 71, 37, 15 **10.** 53, 31, 54, 63, 84, 54, 30

11. 64, 25, 46, 50, 91, 86, 73, 91 **12.** 47, 49, 90, 6, 69, 89, 47, 66

13. 40, 42, 41, 43, 41, 40, 42, 43 **14.** 314, 179, 275, 341, 725

12. 57.5, 47, 83

15. 3.4, 1.8, 2.6, 1.8, 2.3, 3.1 **16.** 0.6, 0.7, 0.7, 1.0, 0.9, 1.0, 1.2

13. 41.5, no mode, 3

17. 7.5, 8.3, 8.8, 8.3, 7.4, 8.35, 8.1, 7.6, 8.3, 7.59 8.2, 8.3, 1.4

18. 5, 7.4, 6.5, 5.25, 6, 4.5, 6, 6.25, 6.5, 4.6, 6.25, 8, 4.5, 6 6, 6, 3.5

14. 314, no mode, 546 15. 2.45, 1.8, 1.6 16. 0.9; 0.7 and 0.1; 0.6

Applications

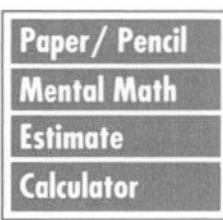

Solve. Use the chart for Exercises 19–21.

19. Find the range for the high temperatures in each city for the year. 34.5, 48.5, 8.2

20. Find the median for high temperatures in each city for the year. 46.85, 68.4, 83.4

21. no mode, no mode, 87.4

21. Find the mode for the high temperatures in each city for the year.

Average Daily High Temperatures (°F)			
	Juneau	**St. Louis**	**Honolulu**
Jan.	29.1	39.9	79.3
Feb.	33.9	44.2	79.2
Mar.	38.2	53.0	79.7
Apr.	46.5	67.0	81.4
May	55.4	76.0	83.6
June	62.0	84.9	85.6
July	63.6	88.4	86.8
Aug.	62.3	87.2	87.4
Sept.	56.1	80.1	87.4
Oct.	47.2	69.8	85.8
Nov.	37.3	54.1	83.2
Dec.	32.0	42.7	80.3

Suppose

22. Suppose Mark scored 81 and 78 on his first two science quizzes. What must his score be on the third quiz so that his mean score is 85? 96

Using Equations
23. $g - 256 = 102$, $g = 358$

23. If the range of a set of data is 102 and the least number is 256, what is the greatest number in the set? Write an equation and solve.

Make Up a Problem
24. See students' work.

24. Make up a set of data in which the mean does not equal either the mode or the median.

Estimation

The *clustering strategy* can be used to estimate the sum of a group of numbers. For example, the numbers in the chart *cluster* around 10,000. An estimate of the total number of births each week in the U.S. is 7 × 10,000 or 70,000.

Average U.S. Births per Day	
Sun.	8,532
Mon.	10,243
Tues.	10,730
Wed.	10,515
Thurs.	10,476
Fri.	10,514
Sat.	8,799

Use the clustering strategy to estimate each sum.

25. 82 + 77 + 82 + 76 + 79 400 **26.** 396 + 391 + 411 + 407 + 389 2,000

27. 6,637 + 5,952 + 5,848 18,000 **28.** 8.46 + 4.69 + 7.38 + 4.0 6

APPLYING THE LESSON

Independent Practice

Homework Assignment	
Minimum	**Ex. 1-21 odd; 22-28**
Average	**Ex. 2-18 even; 19-28**
Maximum	**Ex. 1-28**

Chalkboard Examples

Examples A, B *Find the median, mode, and range for each set of data.*

1. 2, 5, 7, 3, 4, 4, 8, 7, 7 5, 7, 6

2. 27, 28, 25, 29, 28, 27, 28 28, 28, 4

3. 12, 8, 13, 11, 15, 7, 13, 5 11.5, 13, 10

4. 2.9, 3.3, 3.5, 2.7, 3.3, 3.1 3.2, 3.3, 0.8

Additional examples are provided on TRANSPARENCY 11-1.

EVALUATING THE LESSON

Check for Understanding Use Guided Practice Exercises to check for student understanding.

Error Analysis When finding the median of an even-numbered set of data, watch for students who use the mean of the data instead of the mean of the two middle numbers. Remind students that the number of addends in the computation is two.

Closing the Lesson Have students tell when they would use the median, mode, and range.

EXTENDING THE LESSON

Enrichment Masters Booklet, p. 105

Name ______________________ Date ______________

ENRICHMENT WORKSHEET 11-1

Choosing the Best Average

Solve. Name the average you used: mean, median, or mode.

Name	Age	Weight (lb)	Hourly Salary	Hair Color	Favorite Sport
Jim	23	175	$4.52	blond	tennis
Fay	19	112	$3.75	black	golf
Kim	20	126	$4.12	blond	tennis
Tom	22	168	$4.50	red	baseball
Bill	22	175	$4.71	black	tennis
Bev	22	121	$4.45	black	golf
Ron	22	159	$4.55	brown	skiing
Amy	13	102	$2.50	black	tennis
Mack	21	158	$4.53	black	tennis

1. Find the average age. 22; mode
2. Find the average weight. 144; mean
3. Find the average salary. $4.50; median
4. Find the average hair color. black; mode
5. Find the average favorite sport. tennis; mode
6. Describe the average person in the group. 22 years old; weighs 144 lb; earns $4.50 per hour; has black hair; favorite sport is tennis

Solve. Use the quiz scores at the right.

Quiz Scores							
3,	2,	7,	2,	2,	7,	4,	6
8,	7,	7,	3,	6,	3,	8,	7
2,	1,	8,	6,	2,	8,	6,	8
5,	8,	6,	8,	8,	8,	7,	6
7,	4,	5,	1,	1,	7,	6,	6

7. Complete the frequency table for the data.
8. Find the mean. 5.4
9. Find the mode. 8
10. Find the median. 6
11. Which average best describes the data? median

Score	Tally	Frequency
1	|||	3
2	~~||||~~	5
3	|||	3
4	||	2
5	||	2
6	~~||||~~ |||	8
7	~~||||~~ |||	8
8	~~||||~~ ||||	9

5-MINUTE CHECK-UP

(over Lesson 11-1)

Available on TRANSPARENCY 11-2.

Find the numbers indicated for the set of data. 21, 19, 19, 24, 21, 13, 22, 21, 10

1. median 21 **2.** mode 21 **3.** range 14

Extra Practice, Lesson 11-1, p. 477

▶ INTRODUCING THE LESSON

Show students examples of tables or charts from newspapers or magazines. Ask them to hypothesize about ways the data was collected and organized.

▶ TEACHING THE LESSON

Using Questioning

- How can you check that you have tallied all the data in a set? The sum of the frequency column equals the number of data in a set.
- In Example A, why is each score multiplied by the frequency when computing the mean? It is a quicker method than adding each score to find the sum of the points scored.

11-2 FREQUENCY TABLES

Objective
Make and interpret frequency tables.

Mrs. Simms lists the scores of her gymnastics team. She wants to organize the data in a way that makes it easy to study the scores.

A **frequency table** is a table for organizing a set of data.

Method

1. Make a table with three columns.
2. In the first column, list the items or numbers in the set of data.
3. In the second column, make a tally mark each time the item or number appears in the set of data.
4. In the third column, write the frequency, the number of times the item or number appears.

Example A

Make a frequency table for the scores of the gymnastics team given in Mrs. Simms' list. Then find the mean, median, and mode of the scores.

Score	Tally	Frequency
2	|||	3
3	||	2
4	卌 ||	7
5	卌 |||	8

To find the mean, compute the total number of points scored and divide by the total number of scores.

The total number of points scored equals $(2 \times 3) + (3 \times 2) + (4 \times 7) + (5 \times 8)$ or 80.

$$\text{mean} = \frac{80}{20} \text{ or } 4$$

← The total number of points scored is 80.
← The number of scores is 20.

Since there are 20 scores, the median is the mean of the tenth and eleventh scores when listed in order.

$$\text{median} = \frac{4 + 4}{2} \text{ or } 4$$

The mode is 5. Why? occurs 8 times

Guided Practice

Example A

Copy the table below and complete the frequency column.

1.

Height	Tally	Frequency
59 in.	卌 卌	10
62 in.	卌 卌 |||	13
65 in.	卌 卌 卌 ||	17
68 in.	卌 卌	10
70 in.	卌 ||||	9
72 in.	卌 卌 卌 |	16

Practice Masters Booklet, p. 116

Name ____________ Date ____________

PRACTICE WORKSHEET 11-2

Frequency Tables

1. Complete the frequency column of this table.

Score	Tally	Frequency
1	卌 ||||	9
2	卌 卌 ||||	14
3	卌 |||	8
4	卌 卌 卌 卌 ||	22
5	卌 |	6
6	卌 卌 卌 |||	18

Find each of the following. Round to the nearest tenth.

2. mean 3.7
3. median 4
4. mode 4
5. range 5

Solve. Use the data at the right. Round to the nearest tenth.

6. Make a frequency table for this set of data.
7. Find the mean. 2.6
8. Find the median. 2
9. Find the mode. 2
10. Find the range. 4

Number of Children in each family				
2	2	5	2	2
5	3	1	3	1
4	2	1	5	2
2	1	3	4	2

Number	Tally	Frequency
1	||||	4
2	卌 |||	8
3	|||	3
4	||	2
5	|||	3

11. Make a frequency table for this set of data.
12. Find the mean. 5.5 h
13. Find the median. 5 h
14. Find the mode. 4 h
15. Find the range. 6 h

Hours Practiced Each Week by Band Members				
4	3	7	7	4
4	7	4	9	3
3	5	9	9	9
4	4	5	3	7
7	7	5	4	4

Number	Tally	Frequency
3	||||	4
4	卌 |||	8
5	|||	3
7	卌 |	6
9	||||	4

RETEACHING THE LESSON

Mr. Hawkins recorded the number of days each homeroom student was absent during one grading period as follows: 2, 1, 3, 2, 0, 3, 2, 3, 1, 2, 0, 4, 1, 3, 2, 1, 2, 2, 4, 0, 3, 1, 2, 1, 4, 3, 0, 2, 1, 2

See students' work.

1. Make a frequency table for this data.
2. Find the mean. 1.9
3. Find the median. 2
4. Find the mode. 2

Reteaching Masters Booklet, p. 106

Name ____________ Date ____________

RETEACHING WORKSHEET 11-2

Frequency Tables

The **frequency table** at the right shows the number of hours 29 students spend doing homework each school night.

Hours	Tally	Frequency
0	||	2
$\frac{1}{2}$	||||	4
1	卌 |||	8
$1\frac{1}{2}$	卌 |	6
2	卌	5
$2\frac{1}{2}$	|||	3
3	|	1

The range is the difference between the greatest number of hours, 3, and the least number of hours, 0. The range is 3 hours.

The median is the middle or 15th amount of time. The median is $1\frac{1}{2}$ hours.

The mode is the amount of time with the highest frequency. The mode is 1 hour.

The mean is found by multiplying each amount of time by its frequency and finding the total. Then the total is divided by the sum of the frequencies.

$(0 \times 2) + (\frac{1}{2} \times 4) + (1 \times 8) + (1\frac{1}{2} \times 6) + (2 \times 5) + (2\frac{1}{2} \times 3) + (3 \times 1) = 80$

Exercises

2. 66.2 in. 3. 65 in.

Use the completed frequency table in Exercise 1 to find each of the following. Round to the nearest tenth.

Practice

2. mean **3.** median **4.** mode 65 in. **5.** range 13 in.

Solve. Use the data at the right.

6. Make a frequency table for the set of data. See students' work.
7. Find the mean. 3.75 mi
8. Find the median. 3.5 mi
9. Find the mode. 3 mi
10. Find the range. 4 mi

Number of miles run each day by members of the track team				
3	5	2	6	3
2	4	3	5	5
6	2	3	3	4
3	5	5	4	2

Collect Data

11. Collect data from your classmates about their favorite season or the kind of pet they own. Then organize the data into a frequency table. See students' work.

Decision Making

12. The number of each size of track shoe sold last month by the Athlete's Locker is given in the table at the right. Which size should the store stock the most? the least? size 10, size $10\frac{1}{2}$

Size	Tally	Frequency
9	卌卌 II	12
$9\frac{1}{2}$	卌卌卌 I	16
10	卌卌卌卌 III	23
$10\frac{1}{2}$	卌卌 I	11
11	卌卌卌	15
$11\frac{1}{2}$	卌卌卌 IIII	19

Applications

13. Make a frequency table for the data: Friday Video rentals—comedy, drama, comedy, drama, adventure, comedy, comedy, adventure, drama, comedy, adventure, comedy, drama, drama, comedy, comedy, drama, comedy, comedy.

If the rental price of a comedy video is $2.00 and the rental price for all other videos is $1.50, what is the mean rental price for Friday's rentals? $1.76

Critical Thinking

14. Suppose you are responsible for the security of an ancient artifact being displayed in a rectangular room at the local museum. How would you position 10 guards so that there are the same number of guards along each wall?

Calculator

To find the total of the items in a frequency table, multiply before adding. When finding the mean, parentheses show that the sum is found before dividing by the total number of data values.

Score	Frequency
9.5	5
9	7
8.5	11
8	17

(5 × 9.5 + 7 × 9 + 11 × 8.5 + 17 × 8) ÷ (5 + 7 + 11 + 17) = 8.5

the sum of the scores ÷ the sum of the frequencies

The mean is 8.5.

15. size $10\frac{1}{2}$

15. From Exercise 12, use a calculator to find the mean shoe size (to the nearest half-size) sold last month.

Chalkboard Examples

Example A *Make a frequency table for the set of scores. Then find the mean, median, and mode.*
5, 8, 3, 6, 4, 7, 6, 6, 4, 7, 5, 6, 8, 7, 8, 5, 6, 3, 7, 4, 6, 7, 6, 4, 7, 5, 6, 4, 5 5.7, 6, 6

Additional examples are provided on TRANSPARENCY 11-2.

▶ EVALUATING THE LESSON

Check for Understanding Use Guided Practice Exercises to check for student understanding.

Closing the Lesson Have students tell why organizing data in a frequency table is helpful.

▶ EXTENDING THE LESSON

Enrichment Work in small groups. Have students roll a number cube 50 times, tally the results, and make a frequency table for the data. Have groups compare their results and display their tables.

Enrichment Masters Booklet, p. 106

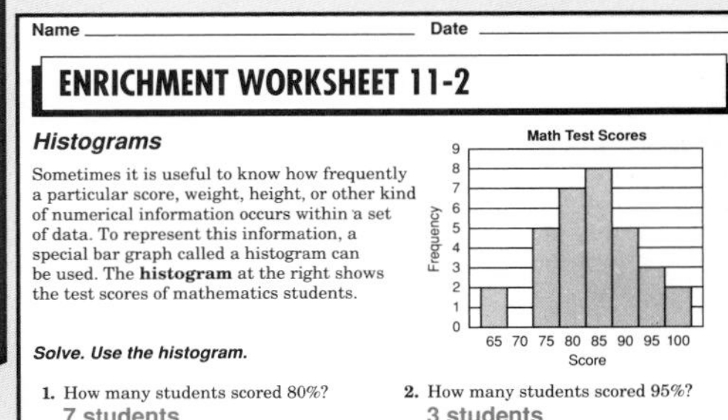

Name ________ Date ________

ENRICHMENT WORKSHEET 11-2

Histograms

Sometimes it is useful to know how frequently a particular score, weight, height, or other kind of numerical information occurs within a set of data. To represent this information, a special bar graph called a histogram can be used. The **histogram** at the right shows the test scores of mathematics students.

Solve. Use the histogram.

1. How many students scored 80%? 7 students
2. How many students scored 95%? 3 students
3. What is the range of scores? 35
4. How many students took the test? 32 students
5. What is the mode? 85
6. What is the median score? 85
7. What is the average score? 83.75
8. What does a "frequency of zero" mean? How is this shown on the histogram? A particular piece of data does not occur; space above that data is left blank. There are 0 scores of 70.

Mr. Hawkins recorded the number of days that each homeroom student was absent during one grading period as follows: 2, 1, 3, 2, 0, 3, 2, 3, 1, 2, 0, 3, 1, 3, 2, 1, 2, 2, 4, 0, 3, 1, 2, 1, 4, 3, 0, 2, 0, 2. Complete.

9. Make a frequency table for this data.
10. Make a histogram for this data.

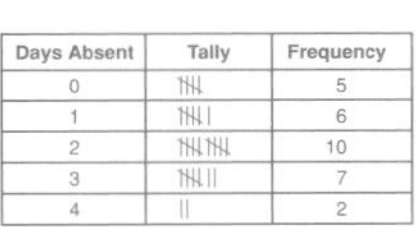

Days Absent	Tally	Frequency
0	卌	5
1	卌 I	6
2	卌卌	10
3	卌 II	7
4	II	2

Student Absences
Frequency
0 2 4 6 8 10 12
0 1 2 3 4
Days

APPLYING THE LESSON

Independent Practice

Homework Assignment	
Minimum	Ex. 2-10 even; 11-15
Average	Ex. 1-15
Maximum	Ex. 1-15

5-MINUTE CHECK-UP

(over Lesson 11-2)

Available on TRANSPARENCY 11-3.
The scores on a 10-point quiz for Mrs. Leonard's math class are 8, 6, 9, 10, 8, 7, 9, 9, 8, 10, 7, 5, 3, 8, 9, 6, 10, 7, 10, 9, 8, 10, 9, 7, and 8.

1. Make a frequency table for the set of data. **See students' work.**
2. Find the mean. **8**
3. Find the median. **8**
4. Find the mode. **8 and 9**

Extra Practice, Lesson 11-2, p. 477

INTRODUCING THE LESSON

Ask students to name situations in which misusing statistics can be to someone's advantage.

TEACHING THE LESSON

Using Discussion Point out that there are different "averages" to describe what is typical. Discuss why we often need to know whether a given "average" is a mean, a median, or a mode in order to interpret the situation.

Practice Masters Booklet, p. 117

Name ______________ Date ______________

PRACTICE WORKSHEET 11-3

Applications: Misusing Statistics

The commissions earned last month by salespeople at Electronics City are shown at the right.

Commission	Number of Salespeople
$ 500	7
$ 800	5
$1,000	3
$2,200	1
$2,800	2
$5,400	2

1. Find the mean, median, and mode of the commission amounts earned.
$1,455; $800; $500
2. Which "average" would you use in an advertisement to hire new salespeople? Why?
The mean; it is the highest.
3. Which "average" best describes the commission earned by all salespeople? Why?
The median; only $\frac{1}{4}$ of the salespeople earn more than the mean and more than $\frac{1}{2}$ earn more than the mode.

The number of yards a football player gained by rushing during each game of a season are shown at the right.

38	42	54	47
46	62	64	58

4. Find the mean and median of the yards rushed per game to the nearest tenth.
51.4; 50.5
5. Which "average" best describes the yards rushed per game? Why?
Accept either mean or median.
6. Instead of rushing for 38 yards in one game, suppose the player rushed for 102 yards. How would the mean and median be affected?
higher mean—59.4; higher median—56
7. Now which "average" best describes the yards rushed per game? Why?
The median; in only $\frac{1}{4}$ of the games did the player rush for more than the mean.

The prices of different camcorders carried by a store are listed at the right.

$ 899
$ 999
$ 950
$ 1,950
$ 1,099

8. Find the mean and median prices.
$1,179.40; $999
9. Which "average" best describes the price of a camcorder? Why?
The median; only one is priced higher than the mean.
10. Suppose that the store replaces the most expensive camcorder with a model priced at $1,100. How would the mean and median be affected?
lower mean—$1,009.40; median not affected

Applications

11-3 MISUSING STATISTICS

Objective
Interpret data.

The "average" salary of workers is of interest to both employers and employees. The chart at the right lists the income of each job classification at the JCH Compact Disc Company.

Job	Number of Employees	Salary
plant workers	20	$15,600
skilled workers	9	$20,600
supervisors	6	$25,000
managers	3	$32,500
vice-presidents	2	$56,500
president	1	$81,000

Understanding the Data

- What is the salary earned by a vice-president? **$56,500**
- Which job classification pays $25,000? **supervisors**
- How many employees are there at the JCH CD Company? **41 employees**

The "average" salary can be reported using the mean, median, or mode. The average chosen depends on the point of view a person takes.

Mean:

$$\frac{20(15{,}600) + 9(20{,}600) + 6(25{,}000) + 3(32{,}500) + 2(56{,}500) + 81{,}000}{41} = \$22{,}900$$

Median: The middle salary is $20,600 earned by a skilled worker.

Mode: The salary that appears most often is $15,600.

5. The president, vice-president, managers, and supervisors would probably be excluded. They are probably not union members.

Applications

1. The company needs to hire more employees. Which "average" is likely to be used in advertisements? Why? **the mean; It is the greatest.**
2. Which "average" *best* describes the salary for plant and skilled workers at the company? Why? **the mode; It is the most common.**
3. Which "average" *best* describes the salary for all employees at the company? Why? **the median; Only three earn large salaries.**
4. Which "average" *best* describes the salary for supervisors and managers? **the mean**
5. If a union representative negotiates for higher salaries, which job classifications are likely to be used to calculate the "averages"? Why?

RETEACHING THE LESSON

The following scores were received on a math test: 60, 35, 96, 85, 80, 70, 85, 65, 85, and 63.

1. Find the mean. **72.4**
2. Find the median. **75**
3. Find the mode. **85**
4. Which average makes the scores seem the highest? **mode**
5. Which average makes the scores seem the lowest? **mean**
6. Which do you think is the best average? **median**

Reteaching Masters Booklet, p. 107

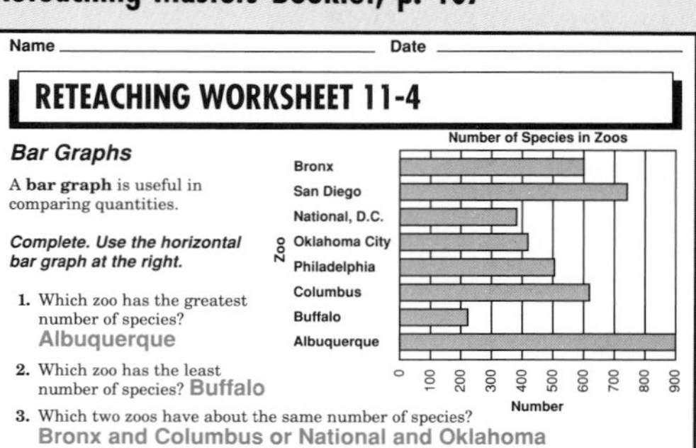

Name ______________ Date ______________

RETEACHING WORKSHEET 11-4

Bar Graphs

A **bar graph** is useful in comparing quantities.

Complete. Use the horizontal bar graph at the right.

1. Which zoo has the greatest number of species? **Albuquerque**
2. Which zoo has the least number of species? **Buffalo**
3. Which two zoos have about the same number of species? **Bronx and Columbus or National and Oklahoma**
4. About how many species does the Bronx Zoo have? **about 600 species**

Problem Solving

6. Suppose the JCH CD Company adds a second shift of seven plant workers, three skilled workers, one supervisor, and one manager. Would you expect the mean, median, and mode, including the new employees, to increase or decrease? Why? **The mean and median will be lower. The mode is the same. See margin.**

Talk Math

7. When a company president quotes one "average" and a union negotiator quotes another "average," they are using certain data for a specific reason. Describe a situation in which choosing the median would be better than choosing the mode. **Answers may vary. A good answer would refer to example and exercises on page 346.**

Interpreting Data

The ages of the guests at a birthday party are given at the right.

19 25 88 23 24
26 20 23 22

8. Find the mean age and the median age of the guests. **30, 23**

9. If the youngest guest had been 13 and the oldest 45, how would the mean and median be affected? **lower mean–25, median–not affected**

The prices of several homes listed by a real estate agent are given at the right.

$87,500	$49,000
$200,000	$78,000
$62,500	$69,800

10. Find the mean price and the median price. **$91,133; $73,900**

11. If the most expensive home costs $100,000 instead of $200,000, how would the median be affected? the mean? **median–not affected; lower mean–$74,467**

Critical Thinking

12. Can you figure out what year the first professional football game was played if the year rounded to the nearest 10 is 1900 and the sum of its digits is 23? **1895**

13. 22 14. 15 15. 3.98 16. 17

Mixed Review

Solve each equation. Check your solution.

Lesson 1-11

13. $x - 4 = 18$
14. $19 = 4 + y$
15. $0.23 + z = 4.21$
16. $10 = b - 7$

17. 0.0048 18. 52.3 19. 0.63 20. 3,000

Lesson 6-3

Complete.

17. 4.8 m = _?_ km
18. 523 mm = _?_ cm
19. 63 cm = _?_ m
20. 3 km = _?_ m

Lesson 7-7

21. Albert is cutting a piece of wood to make a pedestal for a lamp. The pedestal is to be in the shape of an equilateral triangle. One side is to have a length of 15 inches. What are the lengths of the other two sides? **15 in. for both**

Chalkboard Examples

Find the mean, median and mode for each set of data. Round to the nearest tenth.

1. 75, 85, 60, 95, 75, 80, 65, 75
   mean = 76.3 median = 75
   mode = 75

Additional examples are provided on TRANSPARENCY 11-3.

EVALUATING THE LESSON

Check for Understanding Use Guided Practice Exercises to check for student understanding.

Closing the Lesson Ask students to summarize what they have learned about misusing statistics.

EXTENDING THE LESSON

Enrichment Divide students into groups. Have each group create sets of 20 numbers which meet the following qualifications.

Set A: mean > median > mode
Set B: mode > median > mean
Set C: the mean is greater than 95% of the numbers

Enrichment Masters Booklet, p. 107

ENRICHMENT WORKSHEET 11-3

Using and Misusing Graphs

Complete the following to discover whether graph A or B is misleading.

1. Do graphs A and B give the same information on sales? **yes**
2. Find the ratio of Hilly's sales to Valley's sales. $\frac{3}{1}$
3. In graph A, the Hilly van is about 2.5 cm high by 6 cm long. What is its area? **about 15 cm²**
4. The Valley van is about 0.75 cm high by 2 cm long. What is its area? **about 1.5 cm²**
5. In graph B, both vans are about 1.5 cm high. The Hilly van is about 6 cm long. What is its area? **9 cm²**
6. The Valley van is about 2 cm long. What is its area? **3 cm²**
7. Compute the following ratios.
   graph A: $\frac{\text{Area of Hilly}}{\text{Area of Valley}}$ $\frac{10}{1}$
   graph B: $\frac{\text{Area of Hilly}}{\text{Area of Valley}}$ $\frac{3}{1}$
8. Compare the results of Exercises 2 and 7. Which graph is misleading? Explain your answer. **A; the ratio of actual sales is $\frac{3}{1}$, the ratio of representations is $\frac{10}{1}$.**

Compare graphs C and D to answer each question.

9. Which graph is easier to read? **C**
10. Compare the vertical scales. How do they differ? **C: intervals of 1 D: intervals of 10**
11. Which graph gives a more accurate impression of the trend in sales? Explain. **D; increase is steady and very gradual.**

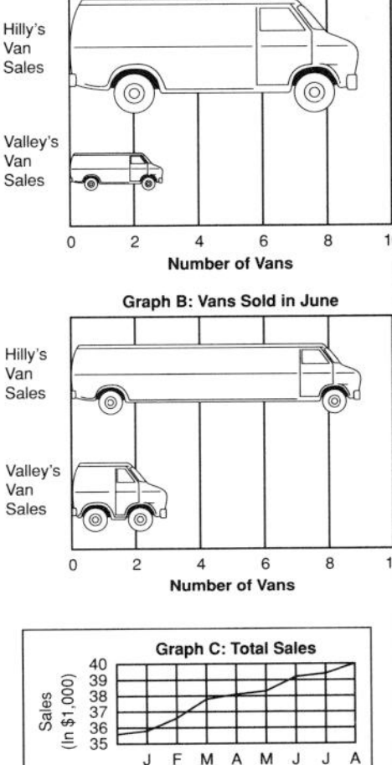

APPLYING THE LESSON

Independent Practice

Homework Assignment	
Minimum	Ex. 1-21
Average	Ex. 1-21
Maximum	Ex. 1-21

Additional Answers

6. More employees with lower salaries were hired than employees with greater salaries. When this happens the mean and median become lower.

5-MINUTE CHECK-UP

(over Lesson 11-3)

Available on TRANSPARENCY 11-4.
The number of cars sold by a dealership in one month: 45 subcompact cars sold at $9,000 each; 30 compact cars sold at $12,000 each; 25 mid-sized cars sold at $20,000 each; and 10 full-sized cars sold at $25,000 each.

1. Find the mean, median, and mode. **$13,773; $12,000, $9,000**
2. Which "average" best describes the price of their cars? **mean**
3. Which "average" is likely to be used to advertise the lowest prices in town? **mode**

INTRODUCING THE LESSON

Ask students which subject they like best. Make a frequency table to show their responses. Then find a scale with equal quantities that would best represent their frequencies.

TEACHING THE LESSON

Using Cooperative Groups Use the chart in Exercise 3 on page 349. Have groups determine the scale, the size of intervals for the scale, and whether to make the graph vertical or horizontal.

Practice Masters Booklet, p. 118

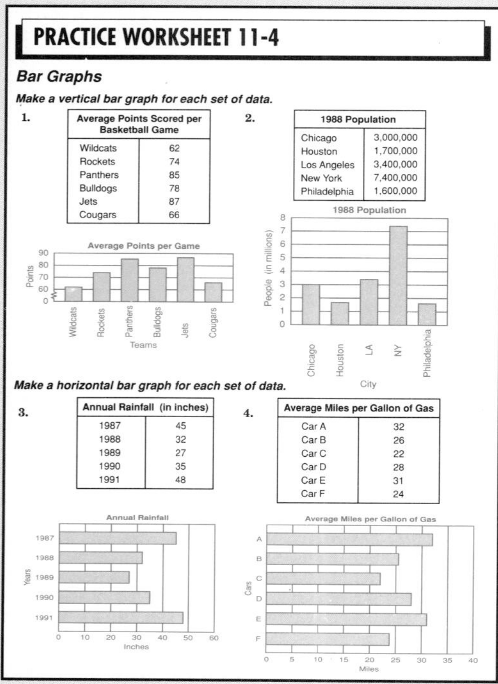

PRACTICE WORKSHEET 11-4

Bar Graphs

Make a vertical bar graph for each set of data.

1.

Average Points Scored per Basketball Game	
Wildcats	62
Rockets	74
Panthers	85
Bulldogs	78
Jets	87
Cougars	66

2.

1988 Population	
Chicago	3,000,000
Houston	1,700,000
Los Angeles	3,400,000
New York	7,400,000
Philadelphia	1,600,000

Make a horizontal bar graph for each set of data.

3.

Annual Rainfall (in inches)	
1987	45
1988	32
1989	27
1990	35
1991	48

4.

Average Miles per Gallon of Gas	
Car A	32
Car B	26
Car C	22
Car D	28
Car E	31
Car F	24

11-4 BAR GRAPHS

Objective
Make bar graphs.

Mr. Delano is the owner of the Action Car Dealership. To increase sales, he uses a *bar graph* to show his sales representatives the number of cars a competitor sold that month.

A **bar graph** is used to compare quantities. The length of each bar represents a number. Use the steps below to make a bar graph.

Method

1. Label the graph with a title.
2. Draw and label the vertical axis and the horizontal axis.
3. Mark off equal spaces on one of the axes and label it with the scale that best represents the data.
4. Draw bars to show the quantities. Label each bar.

Axes is the plural form of axis.

Example A ***Make a horizontal bar graph for the data in the table at the right.***

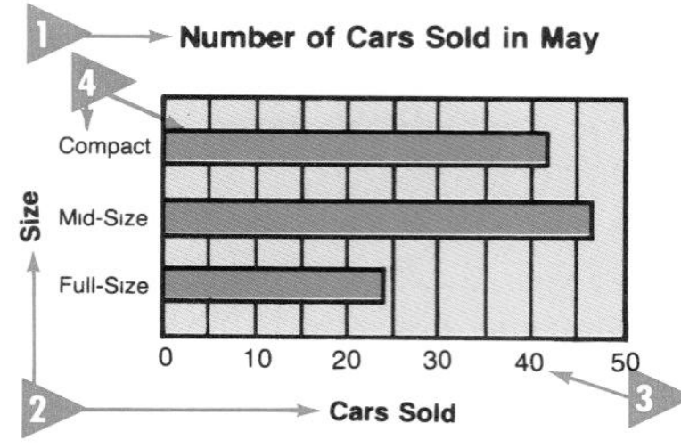

Number of Cars Sold in May	
Compact	42
Mid-size	47
Full-size	24

Example B ***Make a vertical bar graph for the data in the table at the right.***

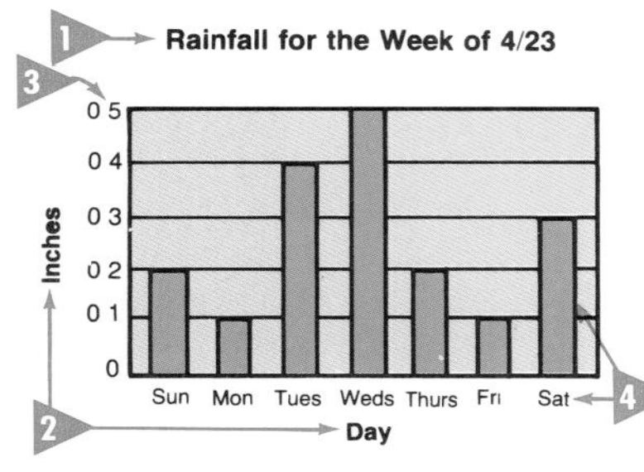

Rainfall for the Week of 4/23	
Sunday	0.2 inches
Monday	0.1 inches
Tuesday	0.4 inches
Wednesday	0.5 inches
Thursday	0.2 inches
Friday	0.1 inches
Saturday	0.3 inches

Guided Practice

Make the following bar graphs. **See answers on p. T524.**

Example A — 1. a vertical bar graph for the data in Example A

Example B — 2. a horizontal bar graph for the data in Example B

RETEACHING THE LESSON

Information: Pets Owned
Fish, 4; Cats, 8; Birds, 2; Dogs, 12
Have students answer each question. Then make a horizontal bar graph.

1. What is the title of the graph? **Pets Owned**
2. How would you label the vertical axis? **Type of Pet**
   The horizontal axis? **Number of Pets**
3. Mark off equal spaces on the horizontal axis. Each mark will represent how many pets? **(1 or 2)**

Reteaching Masters Booklet, p. 108

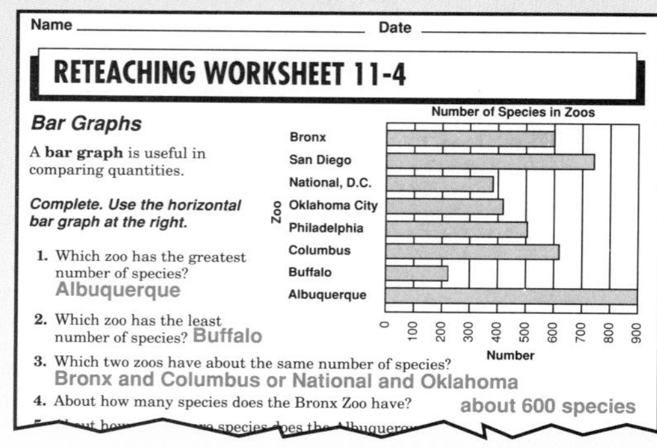

Name ______________ Date ______________

RETEACHING WORKSHEET 11-4

Bar Graphs

A **bar graph** is useful in comparing quantities.

Complete. Use the horizontal bar graph at the right.

1. Which zoo has the greatest number of species? **Albuquerque**
2. Which zoo has the least number of species? **Buffalo**
3. Which two zoos have about the same number of species? **Bronx and Columbus or National and Oklahoma**
4. About how many species does the Bronx Zoo have? **about 600 species**

Exercises

Practice

FUN with MATH *Where and when did the zero originate?* See page 372.

Make a vertical bar graph for each set of data. See answers on p. T524.

3.

Highest Mountain on Each Continent (in feet above sea level)	
Asia	29,000
Europe	19,000
Africa	20,000
Australia	8,000
South America	23,000
Antarctica	17,000
North America	20,000

4.

Average High Temperature (°F)	
March	76°
April	81°
May	86°
June	90°
July	92°
August	93°
September	85°

See answers on p. T524.

Make a horizontal bar graph for each set of data.

5.

Runs Scored in 1 Week	
Reds	23
Braves	15
Padres	12
Dodgers	19
Astros	8
Giants	11

6.

Dollars Saved Each Week	
Week 1	$10
Week 2	$15
Week 3	$16
Week 4	$ 8
Week 5	$13

7. New York and Tulsa; Chicago and Miami

Applications

Solve. Use the bar graph.

Number of Television Stations in Selected Cities

Number: 0, 2, 4, 6, 8, 10, 12, 14, 16, 18

City: New York, Chicago, Los Angeles, Tulsa, Miami, Detroit, Dallas, Charlotte

7. Which cities have the same number of television stations?
8. Which city has the greatest number of television stations? Los Angeles
9. Which city has the least number of television stations? Charlotte
10. How many television stations are in Dallas? 11 stations
11. How many more stations are there in Chicago than in New York? 3 stations
12. Amy bought a blouse for $30, a skirt for $36, a vest for $29, and a pair of shoes for $29. How much did Amy spend in all? $124
13. Dan ran five miles each day for five days and two miles each day for two days. How many miles did he run in all? 29 mi

Collect Data

14. Survey your classmates to find their favorite rock star. Make a horizontal bar graph to illustrate the data. See students' work.
15. Survey your classmates to find their favorite radio station. Make a vertical bar graph to illustrate the data. See students' work.

Chalkboard Examples

Examples A-B *Make a horizontal and/or vertical bar graph for the data.* The average low temperature, in degreees Fahrenheit, is 43° for November, 38° for December, 29° for January, 32° for February, and 39° for March. See graphs below.

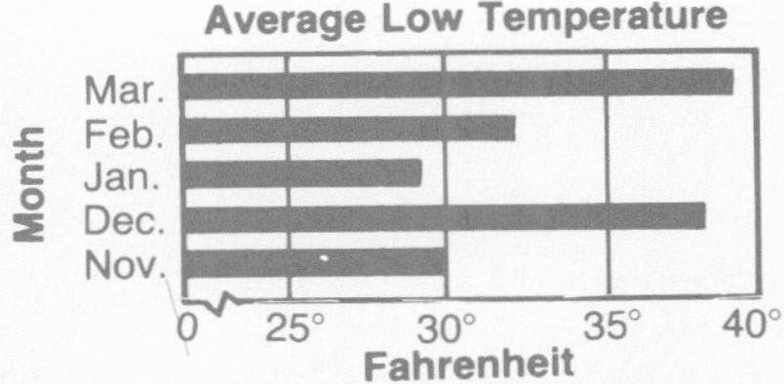

Additional examples are provided on TRANSPARENCY 11-4.

EVALUATING THE LESSON

Check for Understanding Use Guided Practice Exercises to check for student understanding.

Closing the Lesson Have students tell about times when they could make a bar graph to show their data.

EXTENDING THE LESSON

Enrichment Have students bring in results of a survey from a newspaper or magazine. Then make a bar graph for these results.

Enrichment Masters Booklet, p. 108

Name ________________ Date ________________

ENRICHMENT WORKSHEET 11-4

Mean Variation

The chart lists the scores of 7 students on a math test. The mean (or average) score on the test was 85.

The last column shows the difference from the mean of each test score.

The **mean variation** is the average of the differences in the last column. To find the mean variation, add the differences from the mean. Then, divide the sum by the number of addends.

Score	Difference from Mean
78	85 − 78 = 7
86	86 − 85 = 1
95	95 − 85 = 10
82	85 − 82 = 3
88	88 − 85 = 3
76	85 − 76 = 9
90	90 − 85 = 5

Add $7 + 1 + 10 + 3 + 3 + 9 + 5 = 38$ Divide $7\overline{)38.00}$ = 5.42

To the nearest tenth, the mean variation is 5.4.

Find the mean for each set of data. Round to the nearest tenth or cent. Then find the mean variation to the nearest tenth or cent.

1. 47, 53, 58, 39, 40, 55 48.7; 6.7
2. 8.6, 9.3, 10.5, 6.5, 11.8, 14.7 10.2; 2.1
3. 75, 82, 93, 64, 59, 77, 92, 54, 80 75.1; 10.8
4. $159, $108, $144, $109, $129, $142 $131.83; $16.50

Solve.

5. Karen's bowling average for each of five weeks was 149, 155, 132, 145, and 140. Find the mean variation to the nearest whole number. 7
6. The 8:00 A.M. temperature readings for Cleveland for one week were 19°, 18°, 21°, 20°, 18°, 18°, and 21°. Find the mean variation to the nearest tenth. 1.2°
7. Which set of data has the greater mean variation, a set with numbers grouped closely or a set with numbers spaced far apart? set with numbers spaced far apart

APPLYING THE LESSON

Independent Practice

Homework Assignment	
Minimum	Ex. 1-15 odd
Average	Ex. 2-6 even; 7-15
Maximum	Ex. 1-15

Example B Answer

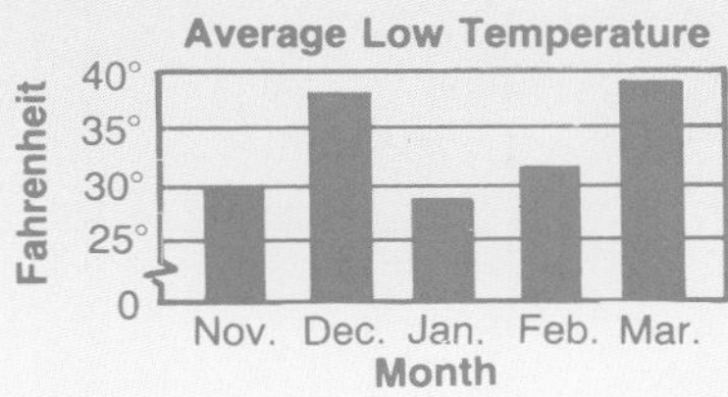

5-MINUTE CHECK-UP

(over Lesson 11-4)

Available on TRANSPARENCY 11-5.

Use Example B on page 348 to answer these questions.

1. Which day had twice as much rain as Sunday? **Tuesday**
2. How much more rainfall was there on Wednesday than on Thursday? **0.3 inches**
3. On which days was the amount of rainfall the same? **Mon., Fri.; Sun., Thurs.**

Extra Practice, Lesson 11-4, p. 477

INTRODUCING THE LESSON

Have students name situations, other than math class, in which they have used bar graphs. What advantages did they find in using a graph?

TEACHING THE LESSON

Using Calculators Use calculators to determine the percent of change in population for each 40 year period reported in Exercise 3. What will the population be in the year 2020 if there is the same percent of change from 1980 to 2020 as from 1940 to 1980? **390 million**

Practice Masters Booklet, p. 119

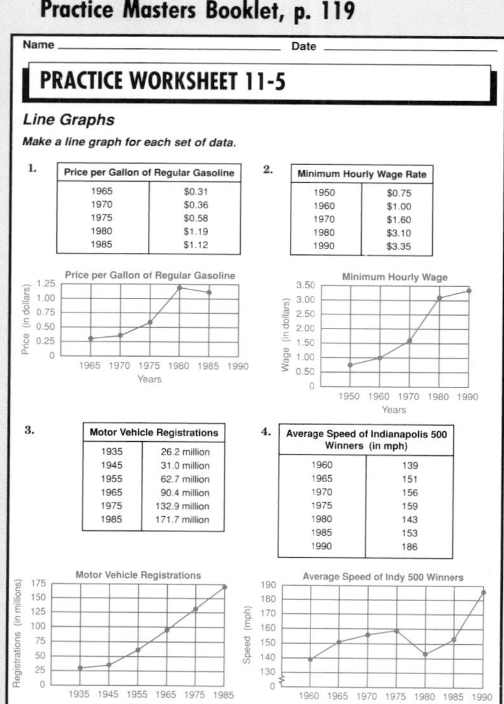

Name ______ Date ______

PRACTICE WORKSHEET 11-5

Line Graphs

Make a line graph for each set of data.

1.

Price per Gallon of Regular Gasoline	
1965	$0.31
1970	$0.36
1975	$0.58
1980	$1.19
1985	$1.12

2.

Minimum Hourly Wage Rate	
1950	$0.75
1960	$1.00
1970	$1.60
1980	$3.10
1990	$3.35

3.

Motor Vehicle Registrations	
1935	26.2 million
1945	31.0 million
1955	62.7 million
1965	90.4 million
1975	132.9 million
1985	171.7 million

4.

Average Speed of Indianapolis 500 Winners (in mph)	
1960	139
1965	151
1970	156
1975	159
1980	143
1985	153
1990	186

11-5 LINE GRAPHS

Objective
Make line graphs.

Ms. Logan is a manager of the coat department for Mear's Department Store. Once a month she prepares a sales report for her supervisor. Ms. Logan uses a *line graph* as part of her report.

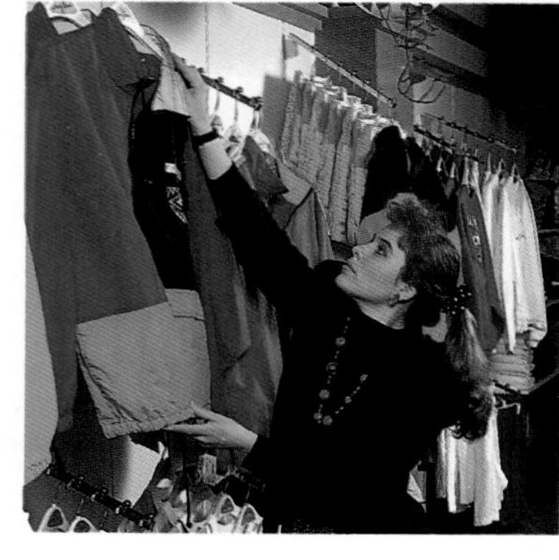

A **line graph** is used to show change and direction of change over a period of time. Use the steps listed below to make a line graph.

Method

1. Label the graph with a title.
2. Draw and label the vertical axis and the horizontal axis.
3. Mark off equal spaces on the vertical axis and label it with the scale that best represents the data.
4. Mark off equal spaces on the horizontal scale and label it with the appropriate time period.
5. Draw a dot to show each data point. Draw line segments to connect the dots.

Explain the meanings of *vertical* and *horizontal.*

Example A ***Make a line graph for the data in the table below.***

Monthly Coat Sales	
Jan.	$2,000
Feb.	$1,500
Mar.	$2,500
Apr.	$3,000
May	$3,500
June	$3,000

Guided Practice
Example A

Make a line graph for each set of data. **See answers on p. T524.**

1.

Absences in Math Class	
Monday	3
Tuesday	1
Wednesday	4
Thursday	5
Friday	1

2.

Temperatures on Dec. 7	
10 A.M.	1°C
12 P.M.	3°C
2 P.M.	5°C
4 P.M.	6°C
6 P.M.	4°C

RETEACHING THE LESSON

Have students answer each question. Then make a line graph.

Gail's Growth in Inches

age (years)	0	1	2	3	4
height (in.)	20	29	33	37	39

1. What is the title of the line graph? **Gail's Growth in Inches**
2. How would you label the vertical axis? the horizontal axis? **Answers may vary.**

Reteaching Masters Booklet, p. 109

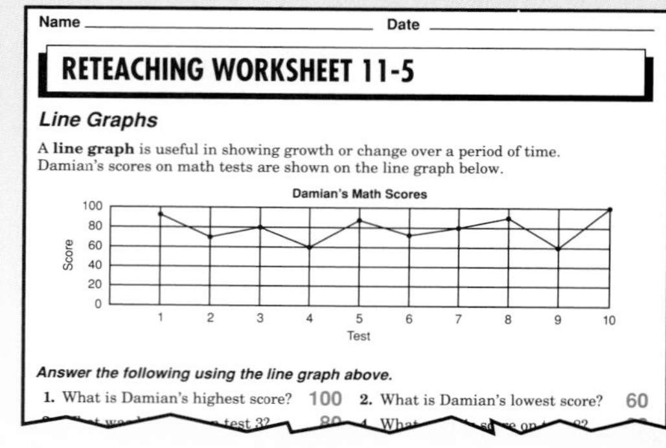

Name ______ Date ______

RETEACHING WORKSHEET 11-5

Line Graphs

A **line graph** is useful in showing growth or change over a period of time. Damian's scores on math tests are shown on the line graph below.

Answer the following using the line graph above.

1. What is Damian's highest score? **100** 2. What is Damian's lowest score? **60**

Exercises

Practice

Paper/ Pencil
Mental Math
Estimate
Calculator

Make a line graph for each set of data. See answers on p. T525.

3.

Estimated Population of the United States	
1820	10 million
1860	31 million
1900	76 million
1940	132 million
1980	227 million

4.

6 A.M. Barometer Readings	
Monday	29.5 inches
Tuesday	29.7 inches
Wednesday	30.1 inches
Thursday	30.2 inches
Friday	29.8 inches

5.

Growth of Saguaro Cactus (in feet)	
25 years	2
50 years	6
75 years	20
100 years	30
125 years	35

6.

Price for a Share of Ajax Stock	
Jan. 1	$23.00
Feb. 1	$21.00
Mar. 1	$22.50
Apr. 1	$24.25
May 1	$27.00

Applications

Solve. Use the line graph.

7. What was the height of the corn plant after 30 days? 20 cm

8. about 45 days

8. About how many days after planting was the height 60 cm?

9. During which 10 days did the height increase the most? 40-50 days

10. During which 10 days did the height increase the least? 10-20 days

11. How much higher was the corn plant after 90 days than after 60 days? 42 cm

Growth of a Corn Plant
Height in Centimeters: 0, 20, 40, 60, 80, 100, 120, 140, 160
Days: 10, 20, 30, 40, 50, 60, 70, 80, 90

Critical Thinking

12. Choose a number. Multiply that number by 2. Then add 5 and multiply by 5. Subtract 25 from the result, and divide by 10. Try this with three different numbers. What can you say about the result each time? Explain. See margin.

13. 4.3×10^6 14. 4.21×10^5 15. 8.21×10^4 16. 9.42×10^2

Mixed Review

Lesson 3-8

Write in scientific notation.

13. 4,300,000 **14.** 421,000 **15.** 82,100 **16.** 942

Lesson 7-9

Name each shape.

17.

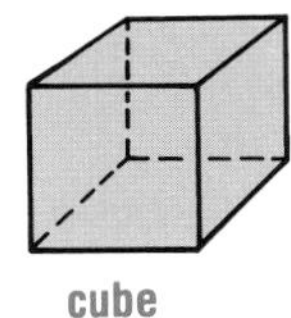

cube

18.

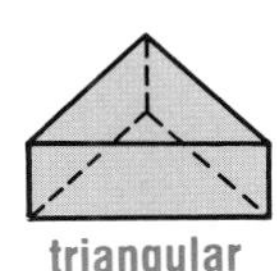

triangular prism

19.

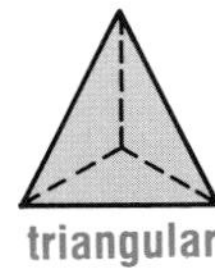

triangular pyramid

Lesson 8-6

20. Monica rides her bike 2 miles due east from her home. She then rides 1 mile south, then turns and rides 5 miles west, and turns again and rides 1 mile north. How far from home is she? 3 miles

APPLYING THE LESSON

Independent Practice

Homework Assignment	
Minimum	Ex. 2-20 even
Average	Ex. 1-5 odd; 7-20
Maximum	Ex. 1-20

Additional Answers

12. It is the same as the original number. Since $\times 2$, $+5$, $\times 5$ is the same as $\times 10$ and $+25$, you are undoing what you did to the number ($\times 10$, $+25$) when you -25 and -10.

Chalkboard Examples

Example A *Make a line graph for the data below.*

The price for a share of UPC stock from July to November is: July, $6.50; August, $7.00; September, $7.25; October, $6.75; and November, $6.25.

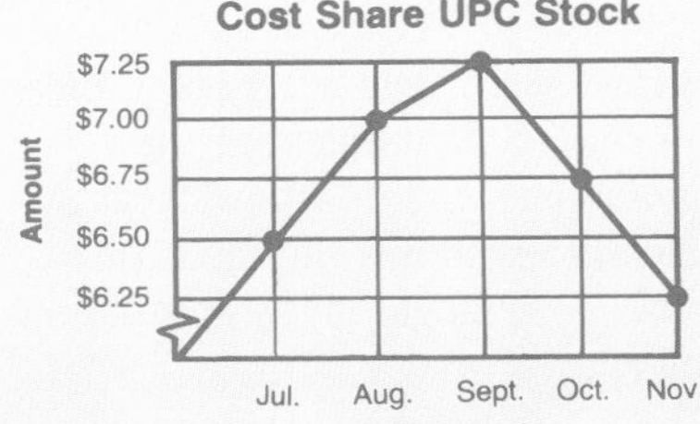

Additional examples are provided on TRANSPARENCY 11-5.

EVALUATING THE LESSON

Closing the Lesson Ask students to tell about bar and line graphs.

EXTENDING THE LESSON

Enrichment Have students make a double line graph that shows the boys' attendance and the girls' attendance in your class for one week.

Enrichment Masters Booklet, p. 109

ENRICHMENT WORKSHEET 11-5

Using Graphs

A-1 Repair Shop charges by the half hour for repair time. The first half-hour charge is $15. The charge for each additional half hour or part of a half hour is $8. A graph of the charge for each half hour is shown at the right.

A-1 Repair Charges
Charges: $10, $20, $30, $40, $50
Hours: 0, 1, 2, 3, 4, 5

Suppose a customer has a washing machine repaired at a total cost of $56.55. The customer was charged for two hours' repair time plus parts. How much did the parts cost?

Explore What is given? total cost, $56.55; repair time, 2 hours
What is asked? the cost for parts

Plan First use the graph to find the charge for 2 hours of repair time. Then subtract this amount from the total cost.

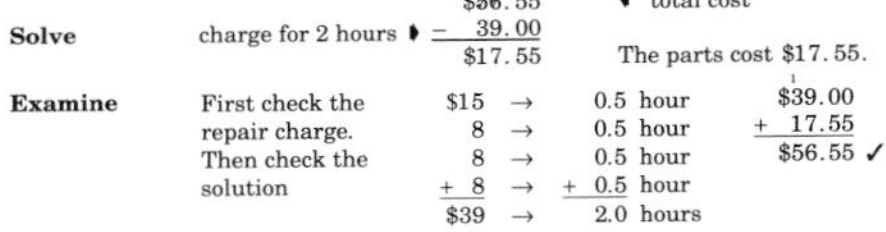

Solve $56.55 ← total cost
charge for 2 hours → − 39.00
$17.55
The parts cost $17.55.

Examine First check the repair charge. Then check the solution

$15 → 0.5 hour
8 → 0.5 hour
8 → 0.5 hour
+ 8 → + 0.5 hour
$39 → 2.0 hours

$39.00
+ 17.55
$56.55 ✓

Use the graph shown above.

1. What is the charge for 3 hours of repair time? $55

2. The repair time for a toaster was 1 hour 15 minutes. What was the repair charge? $31

Solve. Use the graph at the right.

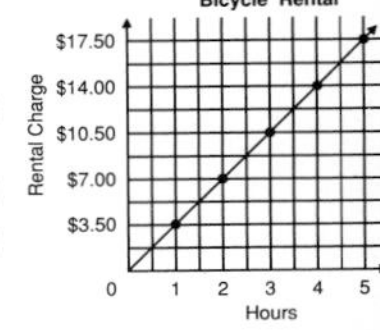

3. How much would it cost to rent a bicycle for 3 hours? $10.50

4. Suppose you had to leave a $10 deposit on the bicycle. How much money would you need to rent a bicycle for 2 hours? $17.00

5. How much would it cost to rent 2 bicycles for 5 hours each? $35.00

Applying Mathematics to Tehnology

This optional page provides students with an experience with computer literacy and also provides a change of pace.

Objective Make a flowchart for a set of data.

Using Discussion Discuss with students the three basic shapes. An oval means stop or start and is used at the beginning and end of a flowchart. A parallelogram shows the input or output. A rectangle gives instructions to follow.

Discuss with students the importance of setting rules before beginning and following rules during a project.

Using Questioning Which should be done first, the program or the flowchart? Why? **The flowchart should be done first. The purpose of a flowchart is to organize the rules.**

Activity

Using Cooperative Groups Have students draw a flowchart for the steps they would follow to play their stereo or draw a flowchart for a process of their own choosing. These could be placed on the bulletin board when complete.

Computer Application

FLOWCHARTS

A **flowchart** can be used to plan a computer program. Each shape used in a flowchart has a special meaning. A *parallelogram* shows input or output. A *rectangle* shows assignment of variables. An *oval* shows the end of a program.

Make a flowchart for the following program.

```
10 READ M,T,W,R,F
20 DATA 10,6,8,9,7
30 LET SUM=M+T+W+R+F
40 LET MEAN=SUM/5
50 PRINT SUM
60 PRINT MEAN
70 END
```

DATA statements are usually not included in a flowchart.

A word can be used as a variable.

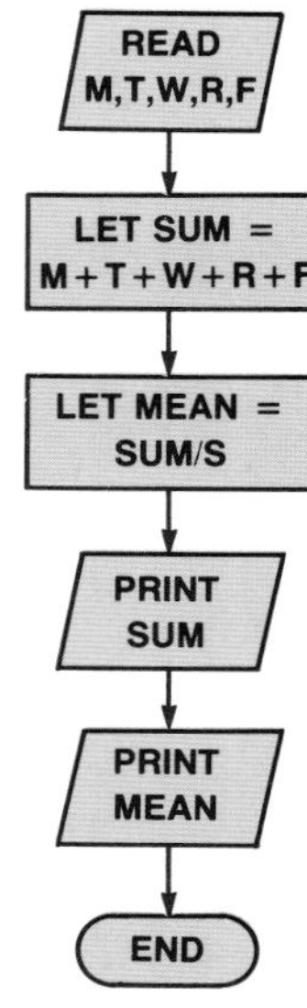
READ M,T,W,R,F
LET SUM = M+T+W+R+F
LET MEAN = SUM/S
PRINT SUM
PRINT MEAN
END

Make a flowchart for each program. See answers on p. T525.

1.
```
10 LET Y=1
20 PRINT Y
30 PRINT Y↑3
40 LET Y=Y+1
50 PRINT Y
60 PRINT Y↑3
70 END
```

2.
```
10 READ D,E,F
20 PRINT D
30 PRINT E
40 PRINT F
50 PRINT D↑2+4*E+F
60 DATA 2,7,8
70 END
```

3. Draw a flowchart for a program to change any given number of feet to inches. Then write a program from the flowchart.
4. Draw a flowchart for a program to find the surface area of a rectangular prism. Then write a program from the flowchart.

INVENTORY SPECIALIST

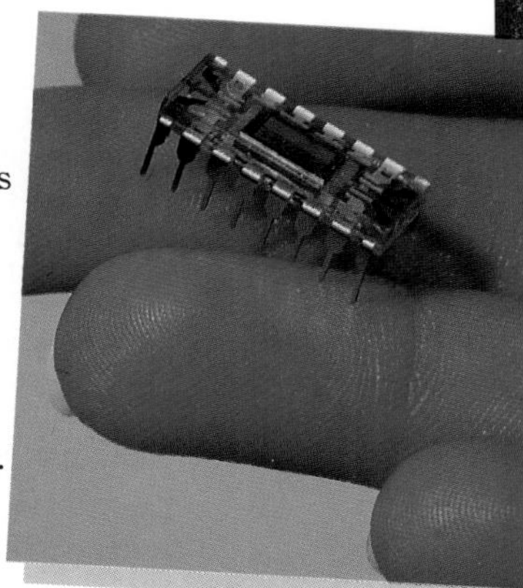

Kelly Woode is an inventory specialist for a computer firm. Each week she counts the items on hand and writes a report showing the number of items in stock and the number moving out of stock. She also makes sure that enough parts are in stock.

Kelly's company uses a certain kind of microchip. In the last six weeks, 35, 62, 44, 73, 36, and 68 microchips were used. Currently, there are 238 microchips in stock. Kelly approximates how long the supply will remain in stock.

1. First, she finds the mean number of microchips used per week. P-11-353

   (35 + 62 + 44 + 73 + 36 + 68) ÷ 6 = 53

2. Then she divides the number in stock by the mean, 53. 238 ÷ 53 = 4.490566

Thus, the supply should remain in stock about $4\frac{1}{2}$ weeks.

Approximate the number of weeks (to the nearest $\frac{1}{2}$ week) each supply of parts will remain in stock.

Part	Number in Stock		Usage—Each of Last 6 Weeks					
1. circuit board	82	8 weeks	8	7	13	6	14	13
2. terminal	26	8 weeks	3	3	1	6	3	4
3. connector	112	18 weeks	10	2	6	5	0	14
4. port	19	8 weeks	3	1	4	0	5	2
5. switch	814	21 weeks	15	46	10	38	74	51

6. Since the shipping time for circuit boards is six weeks, how many weeks after making this report should Kelly order circuit boards so that the supply will always be in stock? 2 weeks
7. If the firm doubles their production for the last week of the month, how many connectors would Kelly expect to be used during that week? 12 connectors

Applying Mathematics to the Work World

This optional page shows how mathematics is used in the work world and also provides a change of pace.

Objective Predict how long supplies will remain in stock.

Using Discussion Discuss the disadvantages of having too little stock on hand and the disadvantages of having too much stock on hand.

Also discuss why the mean is used for the typical number of items moving out of stock rather than the mode or median.

Using a Calculator After computing the number of weeks each part will remain in stock, determine how many additional microchips need to be ordered so that there are enough microchips in stock for a total of six months. Suppose the microchips are packed and ordered in quantities of one dozen. 6 dozen

Activity

Using Cooperative Groups Have students determine the average number of pencils they use in a month at school. Then determine how many pencils they would need to keep in stock to supply their math class with enough pencils to last the entire school year.

5-MINUTE CHECK-UP

(over Lesson 11-5)

Available on TRANSPARENCY 11-6.

Use Example A on page 350 to answer each question.

1. Which months had twice as many sales as February? April, June
2. April sales were what percent of the sales for February? about 200%

Extra Practice, Lesson 11-5, p. 478

▶ INTRODUCING THE LESSON

Look at the graph in Example A. How did the pictograph get its name? uses pictures Why is there no scale on a pictograph? pictures replace it How do you know what each picture represents? symbol and definition at bottom of graph

▶ TEACHING THE LESSON

Using Discussion Can a pictograph and bar graph show the same data? Explain. yes, both are used to compare quantities

Discuss with students the partial symbols in Example A and which data in the Guided Practice will require part of a symbol.

Practice Masters Booklet, p. 120

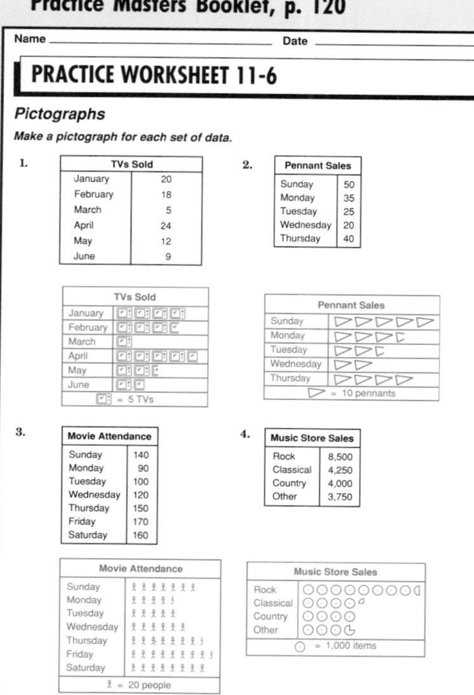

Name ______________________ Date ______________

PRACTICE WORKSHEET 11-6

Pictographs

Make a pictograph for each set of data.

1.

TVs Sold	
January	20
February	18
March	5
April	24
May	12
June	9

2.

Pennant Sales	
Sunday	50
Monday	35
Tuesday	25
Wednesday	20
Thursday	40

3.

Movie Attendance	
Sunday	140
Monday	90
Tuesday	100
Wednesday	120
Thursday	150
Friday	170
Saturday	160

4.

Music Store Sales	
Rock	8,500
Classical	4,250
Country	4,000
Other	3,750

11-6 PICTOGRAPHS

Objective
Construct pictographs.

Mr. Bailey, a county extension agent, volunteered to give a speech on the decrease in the number of farms over the past few years. He used a *pictograph* to present the data visually and make it easier to remember.

A **pictograph** is used to compare data in a visually appealing way. Use the steps listed below to make a pictograph.

Method

1. Label the graph with a title. Write the scale on the vertical axis of the graph.
2. Choose a symbol and what it will represent. Write the definition of the symbol at the bottom of the graph.
3. Determine how many symbols will be used for each item by dividing.
4. Draw the symbols.

Example A

Make a pictograph for the data in the table at the right.

Number of Farms in the U.S.	
1965	3,300,000
1970	3,000,000
1975	2,500,000
1980	2,500,000
1985	2,300,000
1990	2,250,000

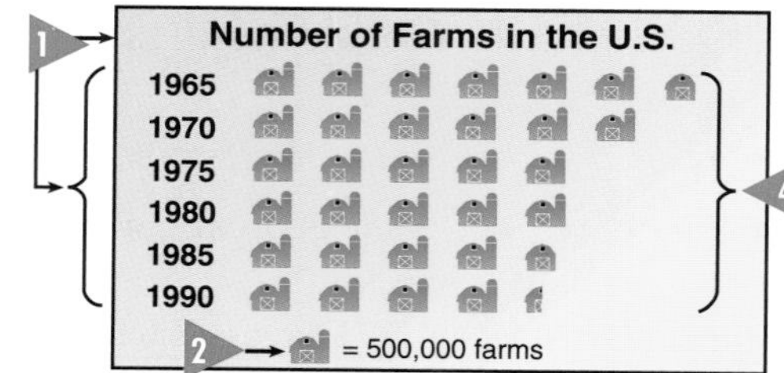

3 Divide 3,000,000 by 500,000. The result is 6. Draw 6 symbols next to 1970.

Guided Practice
Example A

Use the pictograph in Example A to answer each question.

1. How many farms does each symbol represent? 500,000 farms
2. Which 5-year period had the greatest decrease in the number of farms? 1970-1975
3. Which 5-year period had no change in the number of farms? 1975-1980
4. How many more farms were there in 1965 than in 1990? 1,050,000 farms

Make a pictograph for each set of data. See answers on p. T526.

5.

Apple Production in 4 Orchards	
Adams	100,000
Lynd	275,000
Cooley	250,000
Smith	150,000

6.

Boxes of Cereal Sold	
Corn Flakes	550
Puffed Rice	250
Bran Flakes	300
Raisin Bran	400

RETEACHING THE LESSON

Make a pictograph.

Students' Favorite Sport

Football	540	Swimming	216
Basketball	360	Tennis	72
Baseball	432	Other	180

1. How many students will one symbol represent? Answers may vary.
2. How many symbols will you need to represent the number of students who like football? basketball? baseball? swimming? tennis? other? Answers may vary.

Reteaching Masters Booklet, p. 110

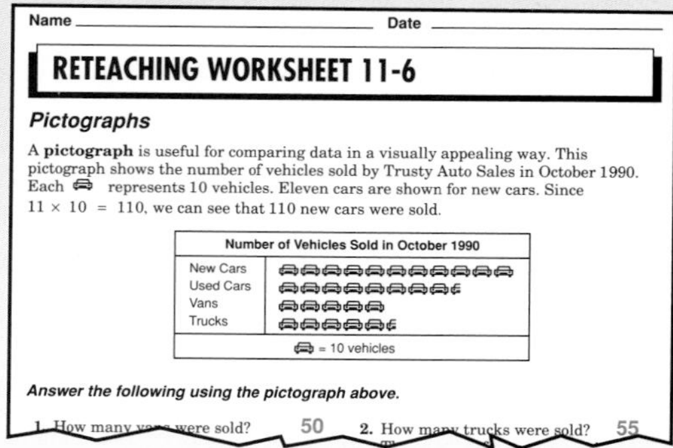

Name ______________________ Date ______________

RETEACHING WORKSHEET 11-6

Pictographs

A **pictograph** is useful for comparing data in a visually appealing way. This pictograph shows the number of vehicles sold by Trusty Auto Sales in October 1990. Each [car symbol] represents 10 vehicles. Eleven cars are shown for new cars. Since 11 × 10 = 110, we can see that 110 new cars were sold.

Answer the following using the pictograph above.

1. How many vans were sold? 50
2. How many trucks were sold? 55

Exercises

Practice

Make a pictograph for each set of data. See answers on p. T526.

7.

Votes Received	
Johnson	800
Kuhn	850
Pitkin	1200
Van Dyke	600

8.

Number of Library Books Loaned	
Monday	60
Tuesday	40
Wednesday	45
Thursday	25
Friday	80

Mental Math

Look for a pattern. Divide mentally. Write only your answers.

9. 50 ÷ 5 10 **10.** 280 ÷ 40 7 **11.** 4,800 ÷ 60 80 **12.** 2,700 ÷ 9 300

15. 9,000

13. 63,000 ÷ 700 90 **14.** 250,000 ÷ 5,000 50 **15.** 7,200,000 ÷ 800

16. 3,500,000 ÷ 500,000 7 **17.** 24,000,000 ÷ 30,000 800

Applications

18. *about* 5,000,000 people

20. *about* 1,000,000 people

21. *about* 500,000 people

Solve. Use the pictograph.

18. *About* how many people lived in North Carolina in 1970?

19. Did the population of North Carolina increase or decrease from 1960 to 1970?

20. *About* how many more people lived in North Carolina in 1980 than in 1970?

21. *About* how many more people lived in North Carolina in 1960 than in 1950?

Population of North Carolina

1950
1960
1970
1980
1990

= 1,000,000 people

Collect Data

22. Survey your classmates to find their favorite sport. Make a pictograph to illustrate the data. See students' work.

23. Survey the families in your neighborhood to find how many pets they have. Make a pictograph to illustrate the data. See students' work.

Mixed Review

Divide.

Lesson 5-6

24. $\frac{8}{9} \div \frac{2}{3}$ $\frac{4}{3}$ **25.** $\frac{3}{5} \div \frac{1}{10}$ 6 **26.** $\frac{6}{13} \div \frac{4}{7}$ $\frac{21}{26}$ **27.** $4 \div \frac{3}{5}$ $6\frac{2}{3}$

Lesson 8-7

Find the surface area of each rectangular prism.

28.

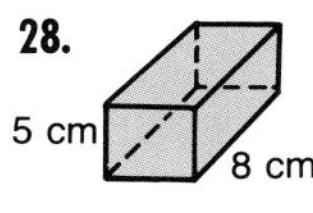

210 cm

29. length = 14 in.
width = 8 in.
height = 2 in.
480 in²

30. length = 9 mm
width = 4 mm
height = 3.5 mm
172 mm²

Lesson 7-10

31. At a busy intersection a stop light has a green arrow that allows 15 cars to turn left before changing. However, a semi-truck takes as much time as 3 cars in turning. How many cars can turn when there are 3 semi-trucks in line to turn? 6 cars

Chalkboard Examples

Chalkboard Examples

Example A *Make a pictograph for the data in the table.* See bottom margin.

Number of Cars Produced	
1981	4,500,000
1982	5,000,000
1983	7,000,000
1984	7,500,000
1985	7,000,000

Additional examples are provided on TRANSPARENCY 11-6.

EVALUATING THE LESSON

Check for Understanding Use Guided Practice Exercises to check for student understanding.

Error Analysis Watch for students who associate one symbol with one hundred, one thousand, and so on.

Closing the Lesson Ask students to state in their own words how to make a pictograph.

EXTENDING THE LESSON

Enrichment Masters Booklet, p. 110

ENRICHMENT WORKSHEET 11-6

Relative Frequency

In an English language text, on the average, 1,000 letters include the following number of particular letters.

E: 131 **T:** 105 **O:** 80
R: 68 **I:** 63 **S:** 61

The estimate of the probability of an event occurring is called **relative frequency.** It is equal to the number of successes divided by the number of trials. In the case of occurrence of letters, it is the number of times the letter occurs (a success) divided by the total number of letters (trials).

From the information above, write the relative frequency of each of the following letters as a decimal.

1. I 0.063 **2.** R 0.068 **3.** T 0.105

4. O 0.08 **5.** E 0.131 **6.** S 0.061

For the newspaper article at the right, find the number of times each letter appears.

7. I 76 **8.** S 68
9. T 99 **10.** R 38
11. O 71 **12.** E 107

Find, to three decimal places, the relative frequency of each letter in the newspaper article. Assume there are a total of 908 letters. Round answers to the nearest thousandth.

13. I 0.084 **14.** S 0.075
15. T 0.109 **16.** R 0.042
17. O 0.078 **18.** E 0.118

19. How are the answers for Exercises 1–6 and 13–18 likely to compare with the combined results from ten similar-sized articles?
They are likely to be close.

Music is Everybody's Language was the theme of the special week set aside this month to recognize Music in Our Schools.

Long interested in the attention given to learning about the importance of music in daily life, the sponsoring organization, Music Educators National Conference, consists of a membership of 64,000 music teachers. The influence of this organization has tremendously upgraded the acceptance of music by both students and parents for many years.

All this brings to mind my own experience in Evening High School with the subject of music and its appreciation. As a student I had unfortunately selected two courses with a gap of one period between them. This was a catastrophe in night school because the valuable time was yours to use or waste, and those students attending this kind of schooling were definitely not attracted by waste. Therefore, after a bit of self-condemnation, I decided to take this elective course titled *Music Appreciation.* With my scant knowledge of music and an attentive ear, my curiosity was aroused to the point that I thought it might at least be fun—a sort of vacation.

APPLYING THE LESSON

Independent Practice

Homework Assignment	
Minimum	Ex. 1-31 odd
Average	Ex. 2-8 even; 9-31
Maximum	Ex. 1-31

Chapter 11, Quiz A (Lessons 11-1 through 11-6) is available in the Evaluation Masters Booklet, p. 54.

Example A Answer

Cars Produced

1981
1982
1983
1984
1985

= 1,000,000 cars

5-MINUTE CHECK-UP

(over Lesson 11-6)

Available on TRANSPARENCY 11-7.

Use Example A on page 354 to find the percent of change to the nearest tenth for each period.

1. from 1965 to 1970 9.1%
2. from 1970 to 1975 16.7%
3. from 1975 to 1980 0%
4. from 1965 to 1985 30.3%

Extra Practice, Lesson 11-6, p. 478

INTRODUCING THE LESSON

Show students a bar graph, line graph, pictograph, and circle graph from newspapers or magazines. Compare the various types of graphs.

TEACHING THE LESSON

Using Questioning

- If two radii are drawn to form an angle, how many degrees are in that angle for $\frac{1}{2}$ of a circle? 180° $\frac{1}{4}$ of a circle? 90° $\frac{1}{6}$ of a circle? 60°
- What operation did you use to find the degrees? multiplication

Practice Masters Booklet, p. 121

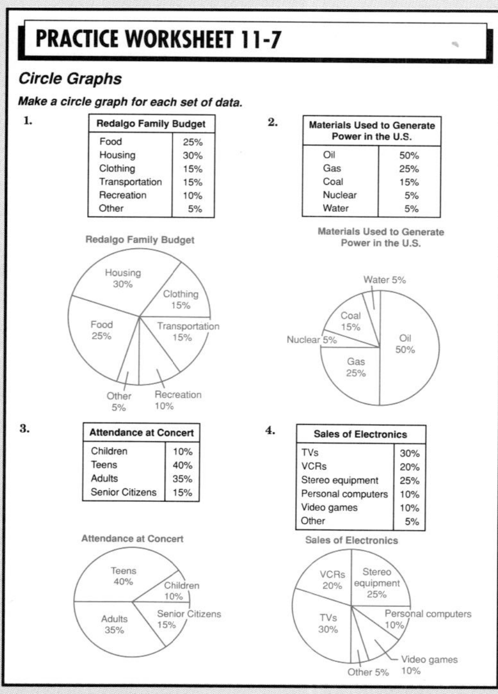

PRACTICE WORKSHEET 11-7

Circle Graphs

Make a circle graph for each set of data.

1.

Redalgo Family Budget	
Food	25%
Housing	30%
Clothing	15%
Transportation	15%
Recreation	10%
Other	5%

Redalgo Family Budget: Housing 30%; Clothing 15%; Transportation 15%; Recreation 10%; Other 5%; Food 25%

2.

Materials Used to Generate Power in the U.S.	
Oil	50%
Gas	25%
Coal	15%
Nuclear	5%
Water	5%

Materials Used to Generate Power in the U.S.: Water 5%; Coal 15%; Nuclear 5%; Oil 50%; Gas 25%

3.

Attendance at Concert	
Children	10%
Teens	40%
Adults	35%
Senior Citizens	15%

Attendance at Concert: Teens 40%; Children 10%; Adults 35%; Senior Citizens 15%

4.

Sales of Electronics	
TVs	30%
VCRs	20%
Stereo equipment	25%
Personal computers	10%
Video games	10%
Other	5%

Sales of Electronics: VCRs 20%; Stereo equipment 25%; TVs 30%; Personal computers 10%; Video games 10%; Other 5%

11-7 CIRCLE GRAPHS

Objective
Construct circle graphs.

For social studies class, Mike made a *circle graph* to compare the number of persons in U.S. households for 1988.

A **circle graph** is used to compare parts of a whole. The whole amount is shown as a circle. Each part is shown as a percent of the whole. The percents should total 100%. Mike used the steps listed below to make a circle graph.

Method

1. Label the graph with a title.
2. Use a compass to draw a circle.
3. Multiply 360° by each percent to find the angle measure for each part. Round to the nearest degree. There are 360° in a circle.
4. Use a protractor to draw the angles by placing the center of the protractor at the center of the circle.
5. Label each part of the circle.

Example A

Make a circle graph for the data in the table at the right.

Persons in U.S. Households in 1988	
1 person	28%
2 persons	30%
3 persons	17%
4 persons	15%
5 or more persons	10%

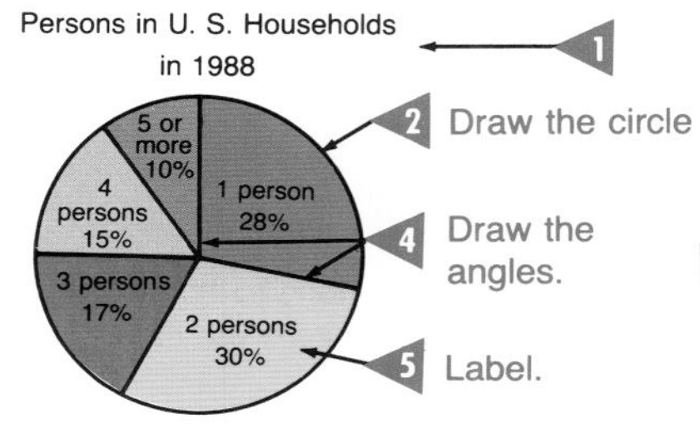

3
28% of 360 = 0.28 × 360 ≈ 101°
30% of 360 = 0.30 × 360 ≈ 108°
17% of 360 = 0.17 × 360 ≈ 61°
15% of 360 = 0.15 × 360 ≈ 54°
10% of 360 = 0.10 × 360 ≈ 36°

Guided Practice

Example A

Use the circle graph in Example A to answer each question.

1. What percent of households have 4 persons? 15%
2. The number of 4-person households is half of what size households? 2. 2-person households
3. What percent of households have 4 or more persons? 25%
4. What percent of households have less than 3 persons? 58%

Make a circle graph for the set of data.

5.

Earth's Surface	
Water	70%
Land	30%

5. See answer on p. T527.

RETEACHING THE LESSON

Make a circle graph for the following data.
Favorite TV shows: Movies, 12%; Sports, 20%; News, 4%; Drama, 16%; Comedy, 20%; Music, 28%.
How many degrees of the circle would represent movies? sports? news? drama? comedy? music? 43°; 72°; 14°; 58°; 72°; 101°

Reteaching Masters Booklet, p. 111

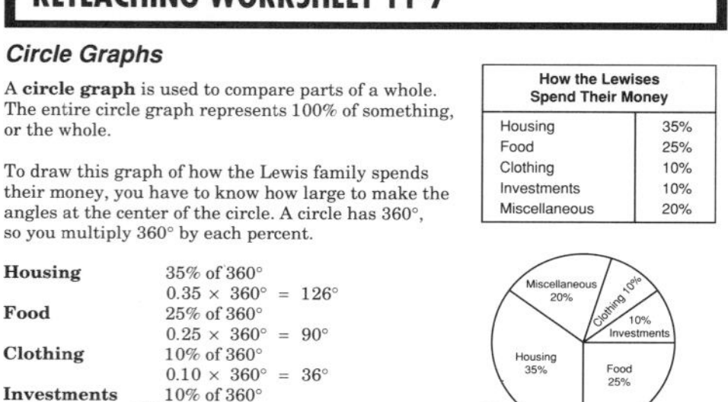

Name ____________ Date ____________

RETEACHING WORKSHEET 11-7

Circle Graphs

A **circle graph** is used to compare parts of a whole. The entire circle graph represents 100% of something, or the whole.

To draw this graph of how the Lewis family spends their money, you have to know how large to make the angles at the center of the circle. A circle has 360°, so you multiply 360° by each percent.

How the Lewises Spend Their Money	
Housing	35%
Food	25%
Clothing	10%
Investments	10%
Miscellaneous	20%

Housing 35% of 360°
0.35 × 360° = 126°
Food 25% of 360°
0.25 × 360° = 90°
Clothing 10% of 360°
0.10 × 360° = 36°
Investments 10% of 360°
× 360° = 36°

Miscellaneous 20%; Clothing 10%; 10% Investments; Housing 35%; Food 25%

Exercises

Practice

Make a circle graph for each set of data. **See answers on p. T527.**

6.

Water Use in the U.S.	
Agriculture	35%
Public Water	10%
Utilities	7%
Industry	48%

7.

Elements of Earth's Crust	
Oxygen	47%
Silicon	28%
Aluminum	8%
Iron	5%
Other	12%

8.

Chemical Composition of the Human Body	
Oxygen	65%
Carbon	18%
Hydrogen	10%
Nitrogen	3%
Other	4%

9.

Earth's Water	
Pacific Ocean	46%
Atlantic Ocean	23%
Indian Ocean	20%
Arctic Ocean	4%
Other	7%

10. Money for building homes comes from the following sources: savings and loans, 38.8%; commercial banks, 23.5%; life insurance, 18.8%; mutual savings banks, 16.3%; pension funds, 2.6%. Make a circle graph to illustrate these data.

Applications

Solve. Use the circle graph.

Sales of Men's Clothing

Other	12%
Suits	32%
Sport Coats	21%
Pants	15%
Shirts	12%
Ties	8%

11. What percent of the sales is shirts? **12%**

12. Which item provided almost one-third of the sales? **suits**

13. The sales for suits is how many times greater than the sales for ties? **4 times**

14. How many degrees is the angle labeled *sport coats?* **75.6°**

15. Which part has an angle of 115°? **suits**

16. Suppose the total sales for men's clothing is $2,500. How much was sales of pants? **$375**

17. Suppose Cindy Gannett budgets $240 per month, or 12% of her monthly salary, for food. What is her monthly salary? **$2,000**

> **TECH PREP** On pages 531–533, you can learn how biomedical ipment specialists use hematics in their jobs.

Cooperative Groups

18. Make a frequency table of the hair color of your classmates. Then make a circle graph to illustrate the data. What conclusions can you make? **See students' work.**

19. Make a circle graph to illustrate how you spend an average 24-hour day. Include the following categories: school, sleep, job, eating, leisure, and miscellaneous. **See students' work.**

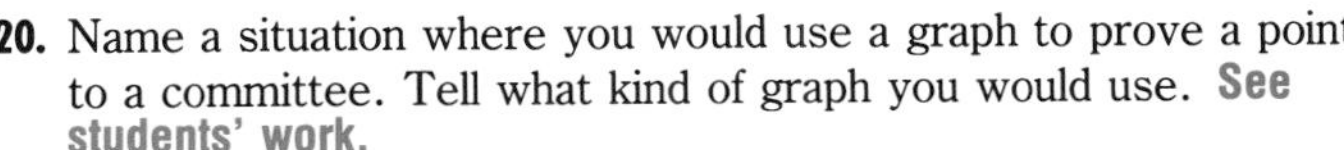

20. Name a situation where you would use a graph to prove a point to a committee. Tell what kind of graph you would use. **See students' work.**

APPLYING THE LESSON

Independent Practice

Homework Assignment	
Minimum	Ex. 1-17 odd; 18-20
Average	Ex. 2-10 even; 11-20
Maximum	Ex. 1-20

Example A Answer

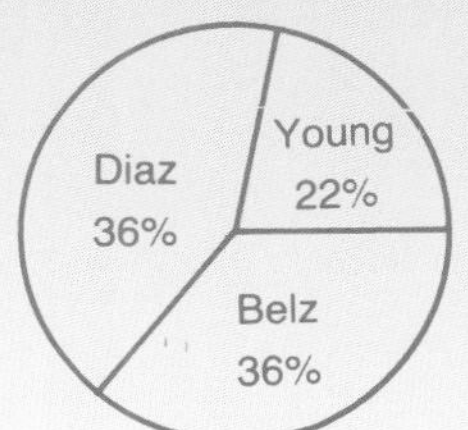

Chalkboard Examples

Example A *Make a circle graph for the data below.*

The percent of votes for class president are: Belz, 36%; Diaz, 42%; and Young, 22%. **See bottom margin.**

$0.36 \times 360 \approx 130°$
$0.42 \times 360 \approx 151°$
$0.22 \times 360 \approx 79°$

Additional examples are provided on TRANSPARENCY 11-7.

EVALUATING THE LESSON

Check for Understanding Use Guided Practice Exercises to check for student understanding.

Closing the Lesson Ask students to tell about times when they could use a circle graph.

EXTENDING THE LESSON

Enrichment Have students make up a problem involving a circle graph and solve it. Then give their problem to a classmate to solve. Students may want to use a circle graph with the percents missing as their problem.

Enrichment Masters Booklet, p. 111

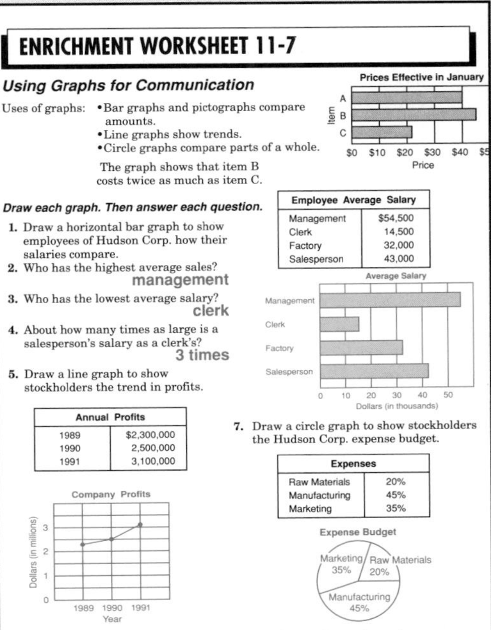

ENRICHMENT WORKSHEET 11-7

Using Graphs for Communication

Uses of graphs:
- Bar graphs and pictographs compare amounts.
- Line graphs show trends.
- Circle graphs compare parts of a whole.

The graph shows that item B costs twice as much as item C.

Draw each graph. Then answer each question.

1. Draw a horizontal bar graph to show employees of Hudson Corp. how their salaries compare.
2. Who has the highest average sales? **management**
3. Who has the lowest average salary? **clerk**
4. About how many times as large is a salesperson's salary as a clerk's? **3 times**
5. Draw a line graph to show stockholders the trend in profits.

Employee Average Salary	
Management	$54,500
Clerk	14,500
Factory	32,000
Salesperson	43,000

Annual Profits	
1989	$2,300,000
1990	2,500,000
1991	3,100,000

6. Does the graph show an increasing or decreasing trend in profits?

7. Draw a circle graph to show stockholders the Hudson Corp. expense budget.

Expenses	
Raw Materials	20%
Manufacturing	45%
Marketing	35%

8. Use the graph to estimate how many times as large as the raw materia budget the manufacturing budget is.

5-MINUTE CHECK-UP

(over Lesson 11-7)

Available on TRANSPARENCY 11-8.

Use each percent to find the number of degrees for the angles in a circle graph. Round to the nearest whole number.

1. 10% **36** **2.** 20% **72** **3.** 15% **54**
4. 29% **104** **5.** 26% **94**

Extra Practice, Lesson 11-7, p. 478

INTRODUCING THE LESSON

Order from least to greatest. Then find the median and mode.

1. 98, 89, 95, 76, 32, 89, 94, 89
32, 76, 89, 89, 89, 94, 95, 98; 89, 89

TEACHING THE LESSON

Using Manipulatives and Cooperative Groups Have each group write 9 two-digit numbers on 3″ × 5″ cards and give them to another group. Have each group make a large stem-and-leaf plot. Then cut the cards between the two digits and place them on the stem-and-leaf plot. They can arrange the order of each row once the cards are placed. Ask students to find the median and mode of each set of data.

Practice Masters Booklet, p. 122

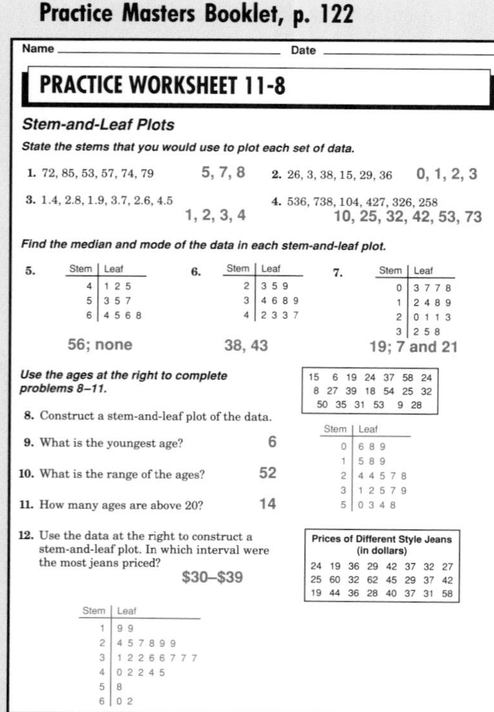

Name ______________________ Date ______________

PRACTICE WORKSHEET 11-8

Stem-and-Leaf Plots

State the stems that you would use to plot each set of data.

1. 72, 85, 53, 57, 74, 79 **5, 7, 8**
2. 26, 3, 38, 15, 29, 36 **0, 1, 2, 3**
3. 1.4, 2.8, 1.9, 3.7, 2.6, 4.5 **1, 2, 3, 4**
4. 536, 738, 104, 427, 326, 258 **10, 25, 32, 42, 53, 73**

Find the median and mode of the data in each stem-and-leaf plot.

5.

Stem	Leaf
4	1 2 5
5	3 5 7
6	4 5 6 8

56; none

6.

Stem	Leaf
2	3 5 9
3	4 6 8 9
4	2 3 3 7

38, 43

7.

Stem	Leaf
0	3 7 7 8
1	2 4 8 9
2	0 1 1 3
3	2 5 8

19; 7 and 21

Use the ages at the right to complete problems 8–11.

15	6	19	24	37	58	24
8	27	39	18	54	25	32
50	35	31	53	9	28	

8. Construct a stem-and-leaf plot of the data.

Stem	Leaf
0	6 8 9
1	5 8 9
2	4 4 5 7 8
3	1 2 5 7 9
5	0 3 4 8

9. What is the youngest age? **6**
10. What is the range of the ages? **52**
11. How many ages are above 20? **14**
12. Use the data at the right to construct a stem-and-leaf plot. In which interval were the most jeans priced? **$30–$39**

Prices of Different Style Jeans (in dollars)							
24	19	36	29	42	37	32	27
25	60	32	62	45	29	37	42
19	44	36	28	40	37	31	58

Stem	Leaf
1	9 9
2	4 5 7 8 9 9
3	1 2 2 6 6 7 7 7
4	0 2 2 4 5
5	8
6	0 2

11-8 STEM-AND-LEAF PLOTS

Objective
Interpret and construct a stem-and-leaf plot.

Candice Brown checks the inventory at a video store. Each day she marks the number of tapes sold. The amounts sold each day during a three-week period are given at the right.

19	30	27	34	23	
31	46	43	36	27	
34	39	48	15	47	
35	33	21	37	27	42

Numerical data can be organized in a **stem-and-leaf plot.** The greatest place value of the data is used for the stem. The next greatest place value is used for the leaves.

Method

1. Draw and label a plot with two columns.
2. In the first column, list the greatest place values of the data. Use each digit only once.
3. In the second column, record the next greatest place value of each number next to the correct stem.
4. Make a second stem-and-leaf plot to arrange the leaves in order from least to greatest.

Example A ***Make a stem-and-leaf plot for the data above.***

1

Stem	Leaf
1	9 5
2	7 3 7 1 7
3	0 5 1 6 3 9 4 4 7
4	6 3 8 7 2

2 The data range from 15 to 48. So the stems range from 1 to 4.

3 (leaves recorded next to the stems)

4|6 represents the number 46.

4

Stem	Leaf
1	5 9
2	1 3 7 7 7
3	0 1 3 4 4 5 6 7 9
4	2 3 6 7 8

Guided Practice

Example A ***Use the stem-and-leaf plot in Example A to answer each question.***

1. What is the least number of videos rented in one day? **15 videos**
2. What is the greatest number of videos rented in one day? **48 videos**
3. What does 2|1 represent? **21 videos**
4. How many days had 27 rentals? **3 days**
5. What is the mode of the data? **27**
6. What is the median of the data? **34**
7. Each row of numbers represents an interval of 10. If the first line 1|5 9 represents the interval 10–19, in which interval do the days with the most rentals fall? **30–39**

RETEACHING THE LESSON

1. State the stems that you would use to plot the data. **1; 2; 3; 4; 5**
29, 36, 43, 22, 51, 19, 23, 34
50, 41, 22, 18, 12, 35, 36, 38

Reteaching Masters Booklet, p. 112

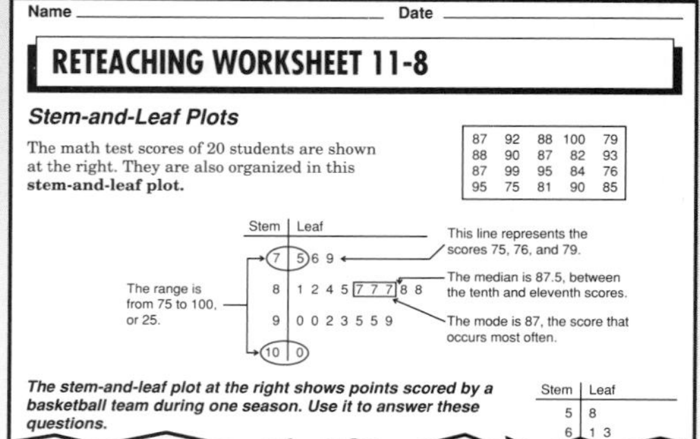

Name ______________________ Date ______________

RETEACHING WORKSHEET 11-8

Stem-and-Leaf Plots

The math test scores of 20 students are shown at the right. They are also organized in this **stem-and-leaf plot.**

87	92	88	100	79
88	90	87	82	93
87	99	95	84	76
95	75	81	90	85

Stem	Leaf
7	5 6 9
8	1 2 4 5 7 7 7 8 8
9	0 0 2 3 5 5 9
10	0

The range is from 75 to 100, or 25.

This line represents the scores 75, 76, and 79.

The median is 87.5, between the tenth and eleventh scores.

The mode is 87, the score that occurs most often.

The stem-and-leaf plot at the right shows points scored by a basketball team during one season. Use it to answer these questions.

Stem	Leaf
5	8
6	1 3

Exercises

Practice

State the stems that you would use to plot each set of data.

8. 18, 36, 43, 25, 32, 4, 27 **9.** 12, 36, 24, 57, 10, 28, 39, 52

10. 158, 581, 182, 368, 404, 545 **11.** 6.4, 5.5, 7.6, 8.4, 5.1, 8.9, 7.2

8. 0, 1, 2, 3, 4 9. 1, 2, 3, 5 10. 15, 18, 36, 40, 54, 58 11. 5, 6, 7, 8

Find the median and mode of the data in each stem-and-leaf-plot.

12.

Stem	Leaf
2	4 5
3	2 3 6
4	0 1 2 3 8 9

40, none

13.

Stem	Leaf
8	0 0 2 1
9	4 8 9 9
10	0 6 7 9

98.5; 80 and 99

14.

Stem	Leaf
4	2 4 6
5	0 1 3 8 9
6	3 4 4 5 8
7	1 7

59, 64

16. 24

Applications

Use the science test scores at the right to complete Exercises 15–20.

100	62	85	72	99	87
87	93	77	86	96	79
100	86	94	68	75	90
88	99	87	73	66	89

15. Construct a stem-and-leaf plot of the data. See margin.

16. How many students took the test?

17. What is the lowest test score? 62

18. Find the range of the scores. 38

19. How many students had scores above 89? 8

20. In which interval did most students score? 80–89

21. Use the data at the right to construct a stem-and-leaf plot. With what age group was the concert the most popular? 10–19 year olds

Ages of People Attending a Concert

16	23	13	22	9	11	26	16	35	42
29	24	38	14	6	17	18	12	7	19
26	17	25	15	7	24	15	18	33	25

Critical Thinking

22. Carla ate 100 peanuts in five days. Each day she ate five more peanuts than she did the day before. How many peanuts did she eat the first day? 10 peanuts

Mixed Review

Find the surface area of each cylinder described below. Use 3.14 for π. Round decimal answers to the nearest tenth.

Lesson 8-3

23. 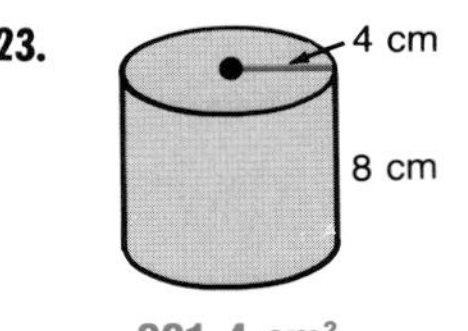

301.4 cm^2

24. 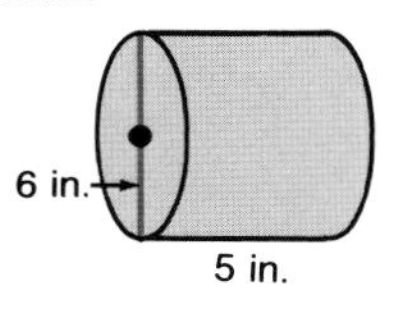

395.6 mm^2

25. 6 in. 5 in.

150.7 in^2

Lesson 9-1

Write each ratio as a fraction in lowest terms.

26. 6 teams for 7 coaches $\frac{6}{7}$ **27.** 40 rooms for 1,000 students $\frac{1}{25}$

Lesson 9-9

28. During the softball season, Carrie got on base 51 out of 85 times at bat. What percent of the times at bat did she get on base? 60%

Chalkboard Examples

Example A *Make a stem-and-leaf plot for each set of data.*

1. 45, 32, 46, 29, 37, 20, 34, 20, 42, 25

2. 67, 66, 78, 89, 74, 92, 96, 89, 75, 84, 90

1.

Stem	Leaf
2	0 0 5 9
3	2 4 7
4	2 5 6

2.

Stem	Leaf
6	6 7
7	4 5 8
8	4 9 9
9	0 2 6

Additional examples are provided on TRANSPARENCY 11-8.

▶ EVALUATING THE LESSON

Check for Understanding Use Guided Practice Exercises to check for student understanding.

Error Analysis Watch for students who do not put the numbers in order from least to greatest on the plot.

Closing the Lesson Ask students to tell when they will use a stem-and-leaf plot.

▶ EXTENDING THE LESSON

Enrichment Masters Booklet, p. 112

ENRICHMENT WORKSHEET 11-8

Back-to-Back Stem-and-Leaf Plots

The table at the right shows the highest and the lowest math test scores received by several students during one semester. These data can also be organized in a **back-to-back stem-and-leaf plot.**

Student	Highest Score	Lowest Score
Jana	82	71
Miguel	81	64
Hai	87	73
Tawanna	94	86
Erica	89	72
Marlon	84	68

Leaves for lowest scores	Stem	Leaves for highest scores
8 4	6	
3 2 1	7	
6	8	1 2 4 7 9
	9	4

Leaves on the left are arranged in order from least to greatest going from **right to left.**

Leaves on the right are arranged in order from least to greatest going from **left to right.**

Construct a back-to-back stem-and-leaf plot for each set of data.

1. **July Temperatures (°F)**

City	High	Low
Athens, Greece	90	72
Berlin, Germany	74	55
Cairo, Egypt	96	70
Calcutta, India	90	79
Dublin, Ireland	67	51
Lima, Peru	67	57
London, UK	73	55
Madrid, Spain	87	62
Manila, Philippines	88	73
Mexico City, Mexico	74	54
Montreal, Canada	78	61
Moscow, USSR	76	55
Nairobi, Kenya	69	51
Oslo, Norway	73	56
Reykjavik, Iceland	58	48
Riyadh, Saudi Arabia	107	78
Santiago, Chile	59	37
Sydney, Australia	50	46
Tokyo, Japan	83	70
Warwaw, Poland	75	56

Leaves for low temperatures	Stem	Leaves for high temperatures
7	3	
8 6	4	
7 7 6 6 5 5 5 4 1	5	0 8 9
2 1	6	7 7 9
9 8 3 2 0 0	7	3 3 4 4 5 6 8
	8	3 7 8
	9	0 0 6
	10	7

2. **Super Bowl Scores**

Game	Winning Team	Losing Team
I	35	10
II	22	14
III	16	7
IV	23	7
V	16	13
VI	24	3
VII	14	7
VIII	24	7
IX	16	6
X	21	17
XI	32	14
XII	27	10
XIII	35	31

Game	Winning Team	Losing Team
XIV	31	19
XV	27	10
XVI	26	21
XVII	27	17
XVIII	38	9
XIX	28	16
XX	46	10
XXI	39	20
XXII	42	10
XXIII	20	16
XXIV	55	10
XXV	20	19

Leaves for losing scores	Stem	Leaves for winning scores
9 7 7 7 7 6 3	0	
9 9 7 7 6 6 4 4 3 0 0 0 0 0 0	1	4 6 6 6
1 0	2	0 0 1 2 3 4 4 6 7 7 7 8
1	3	1 2 5 5 8 9
	4	2 6
	5	5

APPLYING THE LESSON

Independent Practice

Homework Assignment	
Minimum	Ex. 2-14 even; 15-28
Average	Ex. 1-13 odd; 15-28
Maximum	Ex. 1-28

Additional Answers

15.

Stem	Leaf
6	2 6 8
7	2 3 5 7 9
8	5 6 6 7 7 7 8 9
9	0 3 4 6 9 9
10	0 0

5-MINUTE CHECK-UP

(over Lesson 11-8)

Available on TRANSPARENCY 11-9.

Make a stem-and-leaf plot for the data.
34, 37, 43, 46, 22, 48, 52, 48, 37, 39, 26

1. What is the median of the data? 39
2. What is the range of the data? 30
3. How many values are in the set? 11

Extra Practice, Lesson 11-8, p. 479

INTRODUCING THE LESSON

Ask students how many hours (to the nearest half hour) of television they watch in one week. List responses on the chalkboard and have students find the range. Ask how the range is affected by the values.

TEACHING THE LESSON

Using Manipulatives Give each student a 1″ × 8″ strip of paper. Have students write the first twelve odd numbers evenly spaced across one side of the strip. Ask them to tear the strip at the median and at the median of each half. Then have students tape the two middle quarters together. Ask them what part of the data this represents.

Practice Masters Booklet, p. 123

Name ______________________ Date ______________

PRACTICE WORKSHEET 11-9

Measures of Variation

Find the upper quartile, lower quartile, and interquartile range for each set of data.

1. 2, 5, 7, 9, 10, 12, 14, 18 **13; 6; 7**
2. 17, 18, 54, 57, 60, 62, 74, 76, 80, 81 **76; 54; 22**
3. 39, 48, 57, 24, 35, 32, 42, 56 **52; 33.5; 18.5**
4. 64, 27, 39, 56, 57, 16, 60, 72, 38, 41 **60; 38; 22**

5.

Stem	Leaf
2	4 5
3	0 1 2
4	5 8
5	2 4 9

2 | 4 represents 24.

52; 30; 22

6.

Stem	Leaf
0	4
1	1 1 5 6 8
2	3 4
3	2 7

3 | 2 represents 32.

24; 11; 13

7.

Stem	Leaf
5	0 0 1 3
6	5 8 8
7	1 2 4
8	7 9

5 | 0 represents 50.

73; 52; 21

8.

Stem	Leaf
8	2 4
9	3 5 6 8 9
10	2 4 7 8
11	2

11 | 2 represents 112.

105.5; 94; 11.5

Solve. Use the data at the right.

Customers per Hour		
Time	Rags & Riches	Jean Joint
10:00–10:59	24	15
11:00–11:59	36	32
12:00–12:59	54	57
1:00– 1:59	42	38
2:00– 2:59	38	29
3:00– 3:59	14	28
4:00– 4:59	52	35
5:00– 5:59	49	39
6:00– 6:59	22	40
7:00– 7:59	48	32
8:00– 8:59	25	37
9:00– 9:59	16	24

9. Make a stem-and-leaf plot for each store's data.

Rags & Riches

Stem	Leaf
1	4 6
2	2 4 5
3	6 8
4	2 8 9
5	2 4

Jean Joint

Stem	Leaf
1	5
2	4 8 9
3	2 2 5 7 8 9
4	0
5	7

10. Find the interquartile range of each store's data.
Rags & Riches—25.5; Jean Joint—10
11. From your findings in Problem 10, which store has the more consistent number of customers per hour?
Jean Joint

11-9 MEASURES OF VARIATION

Objective
Find quartiles and interquartile range.

Jamie Brooks is moving from San Francisco to Chicago. He compares the temperatures in both cities.

Mean Daily Temperature (°F)		
	San Francisco	Chicago
January	49	21
February	52	26
March	53	36
April	55	49
May	58	59
June	61	69
July	62	73
August	63	72
September	64	65
October	61	54
November	55	40
December	49	28

In some sets of data, the values are close together. In other sets, the values are far apart. The spread of values in a set of data is called the **variation.** The *range* is one such measure of variation. What are the ranges of the mean daily temperatures for San Francisco and Chicago?
15, 52

Another measure of variation is the **interquartile range. Quartiles** are values that divide the ordered data set into four equal parts. The interquartile range is the difference between the upper and lower quartiles of the data.

Method

1. Find the median of the data.
2. Find the median of the upper half (upper quartile) and the median of the lower half (lower quartile) of the data.
3. To find the interquartile range, subtract the lower quartile value from the upper quartile value.

Example A ***Find the interquartile range of the mean daily temperatures in San Francisco.***

1. 49 49 52 53 55 55 ↑ 58 61 61 62 63 64

The median of the data is $\frac{55 + 58}{2}$ or 56.5.

2. 49 49 52 ↑ 53 55 55 | 58 61 61 ↑ 62 63 64
lower quartile — upper quartile

The upper quartile is $\frac{61 + 62}{2}$ or 61.5.

The lower quartile is $\frac{52 + 53}{2}$ or 52.5.

3. The interquartile range is 61.5 − 52.5 or 9. The middle half of the mean daily temperatures varies 9°F.

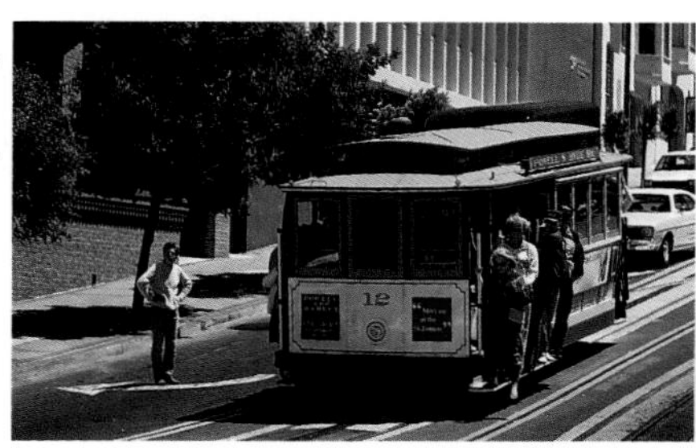

RETEACHING THE LESSON

Use the data below to complete each sentence.

75 78 ↓ 80 81 ↓ 83 84 ↓ 85 86

A. The median is $\frac{81 + 83}{2}$ or 82.

B. The lower quartile is $\frac{78 + 80}{2}$ or 79.

C. The upper quartile is $\frac{84 + 85}{2}$ or 84.5.

D. The interquartile range equals the upper quartile minus the lower quartile. Find the interquartile range. 5.5

Reteaching Masters Booklet, p. 113

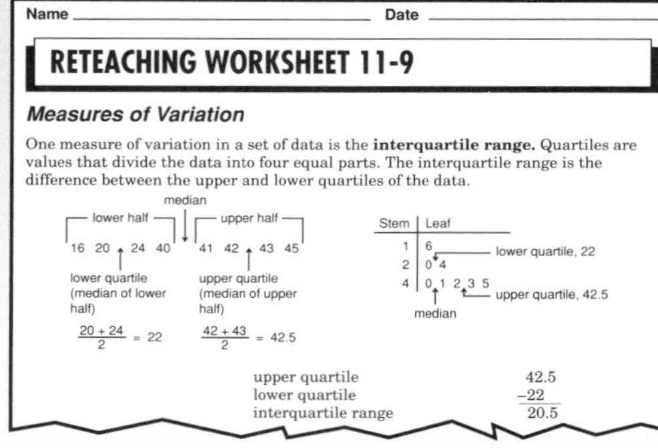

Name ______________________ Date ______________

RETEACHING WORKSHEET 11-9

Measures of Variation

One measure of variation in a set of data is the **interquartile range.** Quartiles are values that divide the data into four equal parts. The interquartile range is the difference between the upper and lower quartiles of the data.

upper quartile	42.5
lower quartile	−22
interquartile range	20.5

Guided Practice

Example A

1. Find the median, upper quartile, lower quartile, and the interquartile range for the temperatures in Chicago. **51.5, 67, 32, 35**
2. Compare the interquartile ranges for San Francisco and Chicago. Does the temperature in Chicago vary more or less than the temperature in San Francisco? **more**

Exercises

3. 19, 9, 10 4. 10, 2, 8

Practice

Find the upper quartile, lower quartile, and interquartile range for each set of data.

3. 2, 6, 12, 16, 16, 18, 20, 22
4. 13, 2, 13, 2, 5, 4, 3, 10, 1, 2

5.

Stem	Leaf
1	2
2	1 (3) 5
3	3 7 8 (9)
4	2 2

4|2 represents 42 **16**

6.

Stem	Leaf
8	2 5 (6) 8
9	0 2 4
10	(3) 7
11	9

11|9 represents 119 **17**

7. What do the range and the interquartile range show about the San Francisco and Chicago temperatures that the mean does not? **the spread of the values**

Applications

8. See margin.

8. The girls' basketball team has 15 wins. The points scored in each win are shown at the right. Make a stem-and-leaf plot for the data and find the interquartile range.

52	64	55
74	62	50
58	65	60
57	59	64
52	62	65

9. From the data in Exercise 8, would you conclude that the girls' basketball team is consistent or inconsistent in the number of points they score during a game? **consistent**

The table below gives the milligrams of sodium in a slice for several kinds of pizza.

10. Find the range and the interquartile range of the milligrams of sodium in a slice for the brands of pizza in the table. **1,072; 448**
11. If the recommended daily allowance of sodium is 2,400 mg, how many slices of Candu's pizza could you eat without going over the recommended daily allowance? **3 slices**

Product	Sodium (mg)
Celentano Cheese Pizza	364
Chef Boyardee pizza mix	1,145
Candu's Cheese Pizza	697
Stouffer's French Bread Pizza	1,064
Totino's Microwave Pizza	1,436
Celeste Pizza	1,112
Croissant Pastry Pizza	953

12. Which brand could you eat the most slices of without going over the recommended daily allowance? **Celentano Cheese**

Using Algebra

Evaluate each expression if $n = 4$.

13. $4n - 2$ **14**
14. $3(n + 6)$ **30**
15. $(n - 3) + 12$ **13**

Chalkboard Examples

Example A *Find the interquartile range.*

1. 2, 3, 5, 6, 8, 8, 10, 12 **5**
2. 3, 5, 5, 6, 7, 9, 13, 15, 17 **9**
3. 36, 37, 39, 41, 43, 44, 45, 47, 48 **8**

Additional examples are provided on TRANSPARENCY 11-9.

EVALUATING THE LESSON

Check for Understanding Use Guided Practice Exercises to check for student understanding.

Error Analysis Watch for students who do not list numbers in order before finding quartiles and interquartile ranges. Prevent by making a stem-and-leaf plot.

Closing the Lesson Have students state in their own words how to find the interquartile range.

EXTENDING THE LESSON

Enrichment Masters Booklet, p. 113

ENRICHMENT WORKSHEET 11-9

Scattergrams

Scattergrams are useful in determining whether there is a relationship, or correlation, between variables.

The data from the table at the right are graphed in the scattergram by making a dot for each datum. Then a **line of best fit** is drawn as close to the dots as possible. On this scattergram, the line of best fit slants upward, indicating a positive correlation between the year and the height jumped.

Olympic Records for Men's High Jump	
Year	Height
1968	7' 4 1/4"
1972	7' 3 3/4"
1976	7' 4 1/2"
1980	7' 8 3/4"
1984	7' 8 1/2"
1988	7' 9 1/2"

Olympic Records for Mens High Jump

line of best fit

A negative correlation is indicated by a line of best fit that slants downward. Sometimes there is no correlation. In these cases, the dots are so scattered that a line of best fit cannot be drawn.

Make a scattergram for each set of data. Then state whether there is a positive, negative, or no correlation among the variables.

1.

Electoral Votes Received by President	
Year	Votes
1952	442
1956	457
1960	303
1964	486
1968	301
1972	520
1976	297
1980	489
1984	525
1988	426

2.

November High Temperatures	
Date	Temperature (°F)
3	50
6	47
9	51
12	46
15	44
18	47
21	46
24	45
27	42
30	44

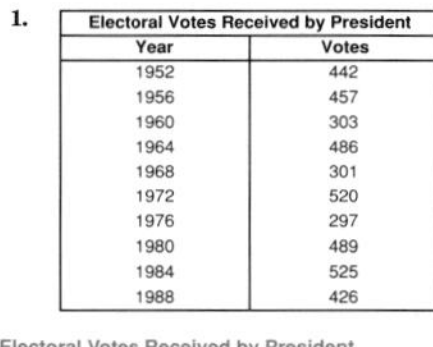

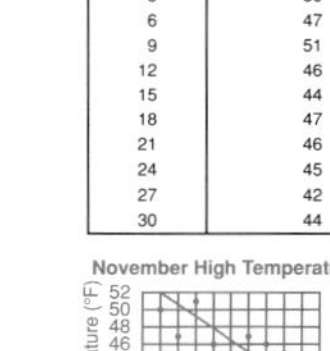

negative correlation

APPLYING THE LESSON

Independent Practice

Homework Assignment	
Minimum	Ex. 1-15
Average	Ex. 1-15
Maximum	Ex. 1-15

Additional Answers

8. 9,

Stem	Leaf
5	0 2 2 5 7 8 9
6	0 2 2 4 4 5 5
7	4

5-MINUTE CHECK-UP

(over Lesson 11-9)

Available on TRANSPARENCY 11-10.

Given the set of data, answer each question.

30, 28, 12, 14, 27, 18, 27, 26, 23, 15, 18, 16, 15, 22

1. Find the upper and lower quartiles. **27, 15**
2. Find the range and interquartile range. **18, 12**
3. Three-fourths of the data falls above which value? **lower quartile, 15**

Extra Practice, Lesson 11-9, p. 479

▶ INTRODUCING THE LESSON

Discuss with students the need to display data in newspapers, magazines, and so on. Have them list types of data for which the spread of values is an important factor. **durability, price, temperature, growth rate**

▶ TEACHING THE LESSON

Using Critical Thinking Ask students why the length of the box and whiskers on the graph in Example A is not equally divided into fourths. **The values are being plotted, not the number of values.**

Practice Masters Booklet, p. 124

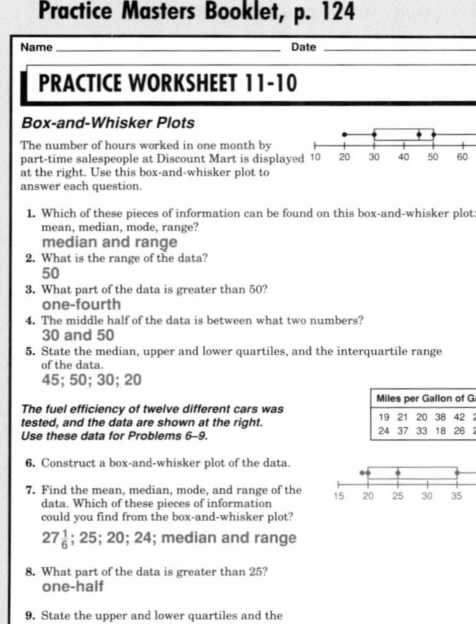

Name ______________________ Date ______________

PRACTICE WORKSHEET 11-10

Box-and-Whisker Plots

The number of hours worked in one month by part-time salespeople at Discount Mart is displayed at the right. Use this box-and-whisker plot to answer each question.

1. Which of these pieces of information can be found on this box-and-whisker plot: mean, median, mode, range?
   median and range
2. What is the range of the data?
   50
3. What part of the data is greater than 50?
   one-fourth
4. The middle half of the data is between what two numbers?
   30 and 50
5. State the median, upper and lower quartiles, and the interquartile range of the data.
   45; 50; 30; 20

The fuel efficiency of twelve different cars was tested, and the data are shown at the right. Use these data for Problems 6–9.

Miles per Gallon of Gas					
19	21	20	38	42	28
24	37	33	18	26	20

6. Construct a box-and-whisker plot of the data.

7. Find the mean, median, mode, and range of the data. Which of these pieces of information could you find from the box-and-whisker plot?
   $27\frac{1}{6}$; 25; 20; 24; median and range
8. What part of the data is greater than 25?
   one-half
9. State the upper and lower quartiles and the interquartile range of the data.
   35; 20; 15

11-10 BOX-AND-WHISKER PLOTS

Objective

Construct and interpret box-and-whisker plots.

Number of Hours of Satisfactory Performance	
38	44
22	46
57	60
58	53
44	33
42	48

Andi Greer tested twelve brands of batteries for the number of hours of satisfactory performance. She will display the data in a box-and-whisker plot for her article in *Wise Consumer.*

A **box-and-whisker plot** is a graph that shows the median, quartiles, and extremes (the least and greatest values) of a set of data.

Method

1. Draw a horizontal line and mark it with a number scale for the set of data.
2. Plot the median, the quartiles, and the extremes on the line.
3. Draw a *box* around the middle half of the data from the lower quartile to the upper quartile.
4. Indicate the median by drawing a vertical line through its point.
5. Draw the whiskers by connecting the lower extreme to the lower quartile and the upper quartile to the upper extreme with a line.

Example A

Draw a box-and-whisker plot to display the data from Andi's battery test results.

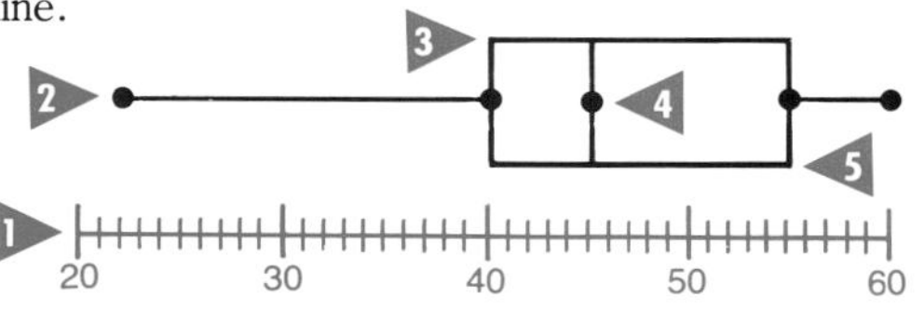

Guided Practice

Use this box-and-whisker plot to answer each question.

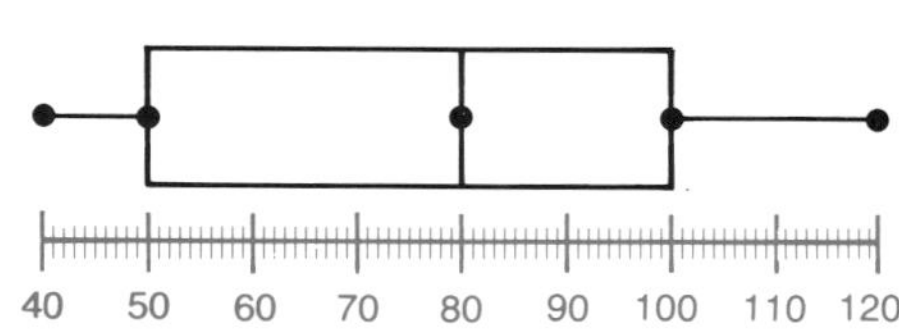

Example A

1. What is the range and interquartile range of the data? **80, 50**
2. The middle half of the data is between which two values? **50, 100**
3. What part of the data is greater than 100? **one-fourth of the data**
4. What part of the data is greater than 80? **one-half of the data**
5. What part of the data is less than 80? **one-half of the data**

RETEACHING THE LESSON

Complete. Use the box-and-whisker plot.

15, 12, 10, 18, 12, 12, 15, 14

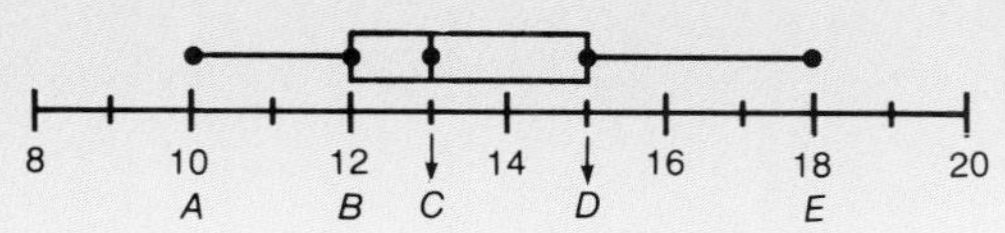

1. The median is located at point **C**.
2. The **upper quartile** is located at point D.
3. Points A and E are the **extremes**.
4. The **lower quartile** is located at point B.

Reteaching Masters Booklet, p. 114

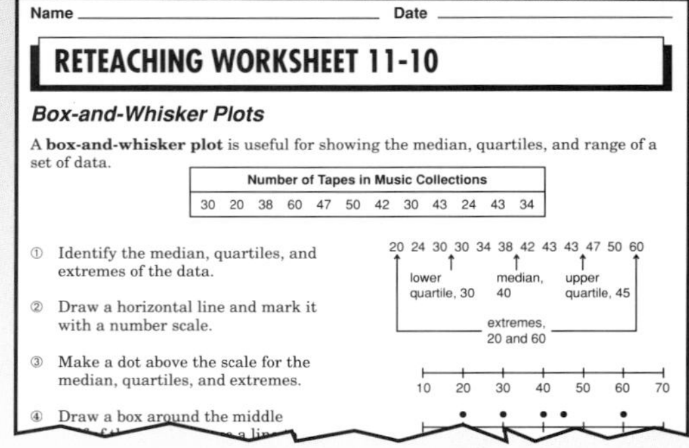

Name ______________________ Date ______________

RETEACHING WORKSHEET 11-10

Box-and-Whisker Plots

A **box-and-whisker plot** is useful for showing the median, quartiles, and range of a set of data.

Number of Tapes in Music Collections											
30	20	38	60	47	50	42	30	43	24	43	34

① Identify the median, quartiles, and extremes of the data.

② Draw a horizontal line and mark it with a number scale.

③ Make a dot above the scale for the median, quartiles, and extremes.

④ Draw a box around the middle

Exercises

6. the median, extremes, upper and lower quartiles, measures of variation

Use the box-and-whisker plot in Example A to answer each question. 7. the mode, the exact values of some of the data, the mean

Practice 9. one-fourth of the number of the values, yes, Quartiles divide the data into four equal parts.

6. What information is easier to find on a box-and-whisker plot than on a stem-and-leaf plot?

7. What information is harder to find on a box-and-whisker plot than on a stem-and-leaf plot?

8. State the median, upper and lower quartiles, the range, and interquartile range of the data in Example A. **45, 55, 40, 15, 38**

9. What fractional part of the data is included in each whisker? Is this true for every box-and-whisker plot? Why?

10. range, upper quartile, median

Applications *Compare the two box-and-whisker plots at the right. Then solve Exercises 10–12.*

Plot A

Plot B

5 10 15 20 25 30

10. What is similar about the data in the two plots?

11. What is different about the data in the two plots?

11. the lower quartile, interquartile range

Interpreting Data **12.** If Plot A represents the number of hours of light from a test of ten Brand X light bulbs and Plot B represents the number of hours of light from a test of ten Brand Y light bulbs, which brand is the better buy? Why? **Brand Y, three-fourths of the test values were above 13 hours.**

Collect Data **13.** In a grocery store or consumer magazine, find the number of calories in a serving of potato chips for six different types or brands.

a. Construct a box-and-whisker plot of the data. **See students' work.**

b. What is the range of the number of calories in a serving?

c. Does the data indicate that different brands of potato chips are similar in the number of calories they have?

Critical Thinking **14.** Rachel has 18 times more nickels than quarters. If Rachel's coins have a total value of $6.90, how many nickels does Rachel have? **108 nickels**

Mixed Review *Find the equivalent Celsius temperature to the nearest degree.*

Lesson 6-9

15. 82°F **28°C** **16.** 60°F **16°C** **17.** 43°F **6°C** **18.** 110°F **43°C**

Lesson 8-10 *Find the volume of each pyramid.*

19.

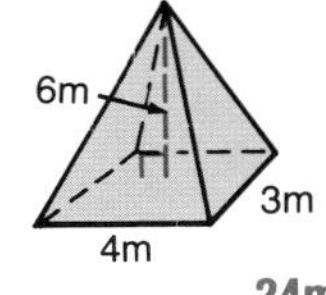

24m³

20.

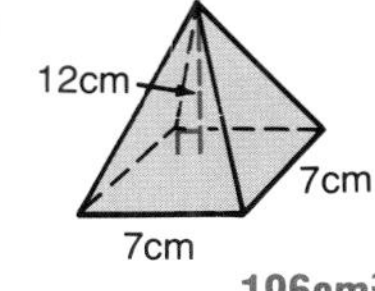

196cm³

21. length = 10 in.
width = 12 in.
height = 11.4 in.
456 in³

Lesson 9-4 **22.** Denise uses 3 gallons of gasoline for every 100 miles she drives her car. How much gasoline does she use if she drives 175 miles? **5.25 gallons**

APPLYING THE LESSON

Independent Practice

Homework Assignment	
Minimum	Ex. 2-12 even; 13-22
Average	Ex. 1-22
Maximum	Ex. 1-22

Example A Answers

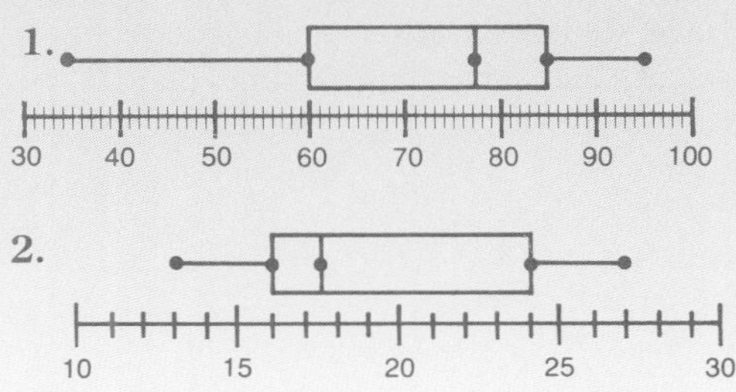

Chalkboard Examples

Example A *Construct a box-and-whisker plot for each set of data.* **See bottom margin.**

1. 55, 85, 95, 40, 65, 70, 35, 75, 90, 80, 85, 80, 75, 85

2. 16, 17, 15, 16, 18, 16, 13, 25, 27, 26, 21, 16, 18, 23

Additional examples are provided on TRANSPARENCY 11-10.

EVALUATING THE LESSON

Check for Understanding Use Guided Practice Exercises to check for student understanding.

Closing the Lesson Have students tell in their own words which parts of the data are contained in each section of a box-and-whisker plot.

EXTENDING THE LESSON

Enrichment Divide the class into groups of three or four. Research the mean daily temperature for two cities that begin with the same letter and construct two box-and-whisker plots on the same scale.

Enrichment Masters Booklet, p. 114

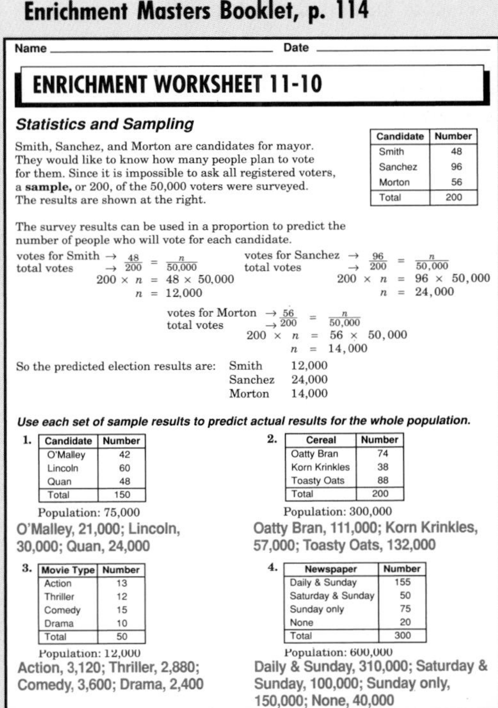

Name ______ Date ______

ENRICHMENT WORKSHEET 11-10

Statistics and Sampling

Smith, Sanchez, and Morton are candidates for mayor. They would like to know how many people plan to vote for them. Since it is impossible to ask all registered voters, a **sample,** or 200, of the 50,000 voters were surveyed. The results are shown at the right.

Candidate	Number
Smith	48
Sanchez	96
Morton	56
Total	200

The survey results can be used in a proportion to predict the number of people who will vote for each candidate.

votes for Smith → $\frac{48}{200} = \frac{n}{50{,}000}$ ← total votes
$200 \times n = 48 \times 50{,}000$
$n = 12{,}000$

votes for Sanchez → $\frac{96}{200} = \frac{n}{50{,}000}$ ← total votes
$200 \times n = 96 \times 50{,}000$
$n = 24{,}000$

votes for Morton → $\frac{56}{200} = \frac{n}{50{,}000}$ ← total votes
$200 \times n = 56 \times 50{,}000$
$n = 14{,}000$

So the predicted election results are: Smith 12,000; Sanchez 24,000; Morton 14,000

Use each set of sample results to predict actual results for the whole population.

1.

Candidate	Number
O'Malley	42
Lincoln	60
Quan	48
Total	150

Population: 75,000
O'Malley, 21,000; Lincoln, 30,000; Quan, 24,000

2.

Cereal	Number
Oatty Bran	74
Korn Krinkles	38
Toasty Oats	88
Total	200

Population: 300,000
Oatty Bran, 111,000; Korn Krinkles, 57,000; Toasty Oats, 132,000

3.

Movie Type	Number
Action	13
Thriller	12
Comedy	15
Drama	10
Total	50

Population: 12,000
Action, 3,120; Thriller, 2,880; Comedy, 3,600; Drama, 2,400

4.

Newspaper	Number
Daily & Sunday	155
Saturday & Sunday	50
Sunday only	75
None	20
Total	300

Population: 600,000
Daily & Sunday, 310,000; Saturday & Sunday, 100,000; Sunday only, 150,000; None, 40,000

5-MINUTE CHECK-UP

(over Lesson 11-10)

Available on TRANSPARENCY 11-11.

Determine the size of the angle for each percent in a circle graph.

1. 10% 36° **2.** 15% 54° **3.** 20% 72°
4. 30% 108° **5.** 25% 90°

Extra Practice, Lesson 11-10, p. 479

INTRODUCING THE LESSON

Identify and continue each pattern.

1. 190, 224, 258, 292, 326, 360 +34
2. 4, 16, 64, 256, 1,024, 4,096 ×4
3. 3, 18, 6, 36, 12, 72, 24, 144 ×6, ÷3

TEACHING THE LESSON

Using Models Have students draw the staircase shown below. By drawing (adding) more blocks to the figure, ask students how many blocks would be needed for a staircase with eleven steps? 66 blocks

Practice Masters Booklet, p. 125

Name ______________ Date ______________

PRACTICE WORKSHEET 11-11

Problem-Solving Strategy: Look for a Pattern

Solve.

1. Rhonda swam 4 laps the first day, 6 laps the second day, 8 laps the third day, and so on until she could swim 20 laps. How many days did it take her to reach 20 laps?
9 days

2. A dropped ball from a high place falls 32 ft the first second, 32 ft × 4 the second second, and 32 ft × 9 the third second. How far does it fall the fifth second?
32 ft × 25, or 800 ft

3. From one corner of a four-sided figure, one diagonal can be drawn. From one corner of a five-sided figure, two diagonals can be drawn. How many diagonals can be drawn from one corner of a nine-sided figure?

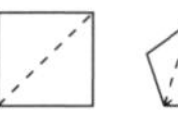 

6 diagonals

4. Tina started working at an hourly wage of $4.40. She will get a 20¢ raise every 6 months. What will be her hourly rate after working 3 years?
$5.60

5. Each of the 8 people at the party exchanges greetings with each other person at the party. How many exchanges of greeting are there?
28 exchanges

6. A new music store had 20 customers the first day it was open. The number of customers doubled each day for the next four days. How many customers did the store have on the fifth day?
320 customers

7. Sixteen teams are involved in a single-elimination tournament. When a team loses, it is eliminated. How many games are needed to determine this tournament's winner?
15 games

8. Danny plans to do 5 more push-ups each day until he can do 75. He begins with 15. How many days will it take him?
13 days

Problem-Solving Strategy

Explore / Plan / Solve / Examine

11-11 LOOK FOR A PATTERN

Objective
Solve verbal problems by looking for a pattern.

Debbie begins a physical fitness program. Debbie's goal is to do 100 sit-ups a day. On the first day of the program, she does 20 sit-ups. Every fifth day of the program, she increases the number of sit-ups by 10. After how many days will she reach her goal?

You can use a list, a table, or a drawing to find a pattern.

Explore

What is given?
- Debbie begins with 20 sit-ups.
- Every fifth day the number increases by 10.

What is asked?
- On what day does Debbie do 100 sit-ups?

Plan

First make a list showing the information you are given. Then look for a pattern and apply it to 100 sit-ups.

Solve

Day	Increase	Number of Sit-Ups
1	0	20
5	0 + 10 = 10	20 + 10 = 30
10	10 + 10 = 20	20 + 20 = 40
15	20 + 10 = 30	20 + 30 = 50

Notice that the increase in sit-ups is twice the day number. Subtract 20 from 100 to find the increase. Then divide by 2 to find the day. Debbie will do 100 sit-ups on day (100 − 20) ÷ 2 or day 40.

Examine

You can check the solution by extending the list to day 40. The solution is correct.

Guided Practice

1. Gena plants five strawberry plants in her garden. The number of plants triples every year. How many plants will Gena have in her garden in six years? 1,215 plants

2. A volleyball team has 6 players. Suppose each player does a high five with every other player. How many high fives take place? 15 high fives

RETEACHING THE LESSON

Have students work in small groups to discuss and solve the problems on page 365.

Problem Solving

FUN with MATH

How about a recipe for Campfire Doughboys?
See page 373.

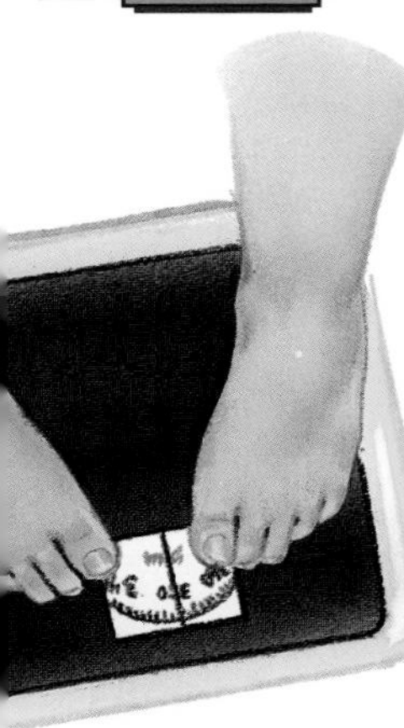

6. 8 weeks

3. The pages in a book are numbered starting with 1. To number all the pages, the printer uses a total of 381 digits. How many pages are in the book? **163 pages**

4. Pat mails a recipe to five friends. Each of the five friends mails the recipe to five more friends, and so on. What is the total number of recipes in the sixth mailing? **15,625 recipes**

5. Mrs. Burns buys numerals to put on the door of each apartment in a 99-unit apartment building. The apartments are numbered 1 through 99. How many of each digit (0, 1, 2, 3, 4, 5, 6, 7, 8, and 9) should Mrs. Burns buy? **9 zeros and 20 of every other digit**

6. Jessie wants to save $25 for a gift. She begins by saving 10¢ the first week. Each week she saves twice as much as the week before. In how many weeks will she have saved at least $25?

7. Mr. Mason weighed 170 pounds on his 40th birthday. Then he began gaining about 2 pounds every year. At this rate, how much will he weigh when he is 50 years old? **190 pounds**

8. In a single-elimination softball tournament, teams are eliminated when they lose. If 8 teams are involved, how many games must be played to determine a tournament winner? **7 games**

9.

Date	Oct.–May	June–Sept.
3 years ago	$35	$40
2 years ago	$38	$44
1 year ago	$41	$48
this year	$44	?

Motel rates change with the time of year. What do you expect the June–Sept. rates to be for this year if the same pattern continues? **$52**

Critical Thinking

10. The map shows the streets between the Tabor house and the Lee house. They are all one-way streets. How many different ways are there to go from the Tabor house to the Lee house? **21 ways**

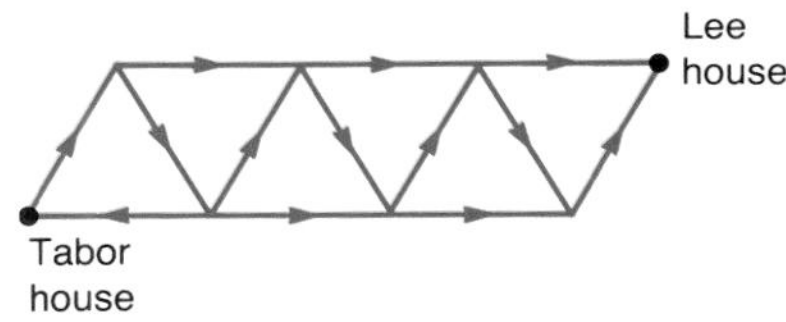

Mixed Review

Write each fraction as a percent.

Lesson 9-8

11. $\frac{15}{30}$ **50%** **12.** $\frac{9}{36}$ **25%** **13.** $\frac{17}{20}$ **85%** **14.** $\frac{4}{20}$ **20%** **15.** $\frac{3}{9}$ **$33\frac{1}{3}$%**

Lesson 10-1

Find the percent of each number.

16. 5% of 50 **2.5** **17.** 20% of 340 **68** **18.** 45% of 540 **243**

19. Select some of your work from this chapter that shows how you used a calculator or computer. **See students' work.**

APPLYING THE LESSON

Independent Practice

Homework Assignment	
Minimum	Ex. 2-10 even; 11-19
Average	Ex. 1-19
Maximum	Ex. 1-19

Chapter 11, Quiz B (Lessons 11-7 through 11-11) is available in the Evaluation Masters Booklet, p. 54.

Chalkboard Examples

Solve. Find a pattern.

1. Greg spends half of his 512 dimes the first day and half the remaining dimes each day after that. How many dimes does he spend on day 8? **2 dimes**

2. Tear a sheet of paper in half. Place the two pieces of paper on top of each other, and tear them in half. This gives four pieces of paper. How many pieces of paper would you have after 6 such tears? **64 pieces**

Additional examples are provided on TRANSPARENCY 11-11.

EVALUATING THE LESSON

Check for Understanding Use Guided Practice Exercises to check for student understanding.

Error Analysis Watch for students who generalize a pattern from too little data. Have students observe at least 3 instances before naming a pattern.

Closing the Lesson Have students solve. The first day Greg gives a message to two friends. The next day each friend gives the message to two of their friends. If the pattern continues, how many hear the message on the fifth day? **32 people**

Using the Chapter Review

The Chapter Review is a comprehensive review of the concepts presented in this chapter. This review may be used to prepare students for the Chapter Test.

Chapter 11, Quizzes A and B, are provided in the Evaluation Masters Booklet as shown below.

Quiz A should be given after students have completed Lessons 11-1 through 11-6. Quiz B should be given after students have completed Lessons 11-7 through 11-11.

These quizzes can be used to obtain a quiz score or as a check of the concepts students need to review.

Evaluation Masters Booklet, p. 54

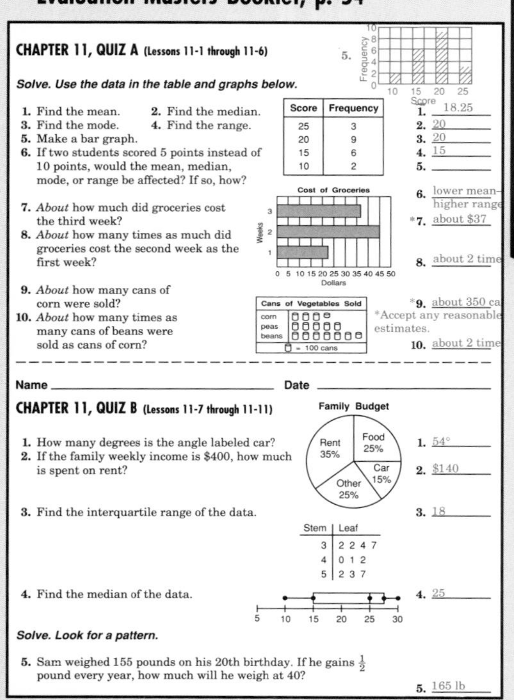

CHAPTER 11, QUIZ A (Lessons 11-1 through 11-6)

Solve. Use the data in the table and graphs below.

1. Find the mean.
2. Find the median.
3. Find the mode.
4. Find the range.
5. Make a bar graph.
6. If two students scored 5 points instead of 10 points, would the mean, median, mode, or range be affected? If so, how?

Score	Frequency
25	3
20	9
15	6
10	2

7. *About* how much did groceries cost the third week?
8. *About* how many times as much did groceries cost the second week as the first week?

9. *About* how many cans of corn were sold?
10. *About* how many times as many cans of beans were sold as cans of corn?

1. 18.25
2. 20
3. 20
4. 15
5.
6. lower mean–higher range
*7. about $37
8. about 2 time
*9. about 350 ca
*Accept any reasonable estimates.
10. about 2 time

Name ______________ Date ______________

CHAPTER 11, QUIZ B (Lessons 11-7 through 11-11)

1. How many degrees is the angle labeled car?
2. If the family weekly income is $400, how much is spent on rent?
3. Find the interquartile range of the data.

Stem	Leaf
3	2 2 4 7
4	0 1 2
5	2 3 7

4. Find the median of the data.

Solve. Look for a pattern.

5. Sam weighed 155 pounds on his 20th birthday. If he gains $\frac{1}{2}$ pound every year, how much will he weigh at 40?

1. 54°
2. $140
3. 18
4. 25
5. 165 lb

Chapter 11 REVIEW

Vocabulary / Concepts

Write the letter of the term that best matches each description.

1. the difference between the greatest and the least number k
2. used to show change and direction of change over a period of time f
3. used to compare parts of a whole c
4. the average of a set of data g
5. the middle number of a set of data when the data is listed in order h
6. a way of organizing numbers or items in a set of data d
7. used to compare data in a visually appealing way j
8. the number that appears most often in a set of data i
9. used to show the median, quartiles, and extremes of a set of data b
10. the median of the upper half of a set of data m
11. the difference between the upper quartile and the lower quartile in a set of data e

a. bar graph
b. box-and-whisker plot
c. circle graph
d. frequency table
e. interquartile range
f. line graph
g. mean
h. median
i. mode
j. pictograph
k. range
l. stem-and-leaf plot
m. upper quartile

Exercises / Applications

Find the median, mode, and range for each set of data.

Lesson 11-1

12. 8, 4, 8, 9, 6, 7, 4, 2, 5, 4, 8, 9, 8 **median: 7, mode: 8, range: 7**
13. 14, 12, 19, 14, 20, 14, 13, 15, 16, 16, 14, 12 **median: 14, mode: 14, range: 8**

Lesson 11-2

Solve. Use the data at the right.

14. Make a frequency table for the set of data. **See margin.**
15. Find the range. **40**
16. Find the median. **82.5**
17. Find the mode. **75**

Scores on a History Quiz				
100	95	75	85	90
75	75	65	90	100
85	75	60	90	80
70	95	90	70	75

Lesson 11-3

The ages of the workers at a fast food restaurant are given at the right.

16 16 17 15 40
18 19 39 63

18. Find the mean age and the median age of the workers. **27, 18**
19. Which average age should the restaurant use if it wants to attract college age students? **median age**

Additional Answers

14.

Score	Tally	Frequency
100	||	2
95	||	2
90	||||	4
85	||	2
80	|	1
75	~~||||~~	5
70	||	2
65	|	1
60	|	1

See answers for exercises 20–24 on p. T527-T528.

Lessons 11-4, 11-5, 11-6, 11-8, 11-10

Make the type of graph indicated for each set of data.

20. bar graph

High Temperatures in Tulsa, Oklahoma	
Sun.	94°F
Mon.	91°F
Tues.	87°F
Wed.	89°F
Thurs.	91°F
Fri.	95°F
Sat.	97°F

21. line graph

Average Monthly Rainfall (in inches)	
Mar.	1
Apr.	5
May	3
June	2
July	3
Aug.	1
Sept.	4
Oct.	3

22. pictograph

Favorite Radio Station	
WXNY	5
WNCI	12
WQFM	8
WRFD	9
WBBY	2

23. stem-and-leaf plot

The Ages of Each U.S. President on His First Inauguration						
57	61	57	57	58	57	61
65	52	56	46	54	49	50
55	51	54	51	60	62	43
54	68	51	49	64	50	48
47	55	55	54	42	51	56
55	56	61	52	69	64	

24. Construct a box-and-whisker plot for the data in Exercise 23.

Lesson 11-7

Members participating in Sports Club Activities

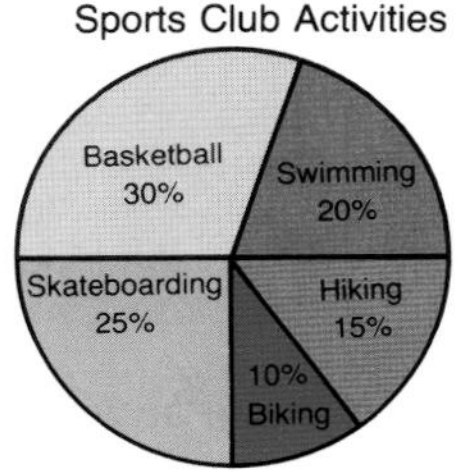

25. What percent of the members bike? **10%**

26. In which activity did over one-fourth of the members participate? **basketball**

27. Suppose there are 200 members. How many participants are there in hiking? **30 participants**

28. How many degrees is the angle labeled swimming? **72°**

Lesson 11-9

Find the upper quartile, lower quartile, and interquartile range for each set of data.

29. 4, 7, 8, 8, 10, 10, 12 **10; 7; 3**

30. 16, 22, 15, 18, 15, 22, 20, 16 **21; 15.5; 5.5**

Lesson 11-11

Solve. Use look for a pattern.

31. Toby mails a math puzzle to three friends. Each of the three friends mails the puzzle to three more friends, and so on. What is the total number of puzzles in the fifth mailing? **243 puzzles**

APPLYING THE LESSON

Independent Practice

Homework Assignment	
Minimum	Ex. 1-11; 2-30 even; 31
Average	Ex. 1-31
Maximum	Ex. 1-31

Alternate Assessment Strategy

To provide a brief in-class review, you may wish to read the following questions to the class and require a verbal response.

1. Find the median and mode for the set of data, 10, 12, 16, 16, 20. **16; 16**
2. Describe the use of a frequency table in collecting data. **organizes data**
3. True or false. A bar graph is used to compare qualities. **True**
4. How is a line graph useful? **Show change over a period of time.**
5. What is the most visually appealing type graph? **pictorgraph.**
6. What type of graph would you use to show the percentage of different elements in the air? **circle**
7. What are the stems you would use to plot this data, 16, 25, 36, 41, 42? **1, 2, 3, 4**
8. In a stem and leaf plot what number does 3|7 represent? **37**
9. In a box and whisker plot, what do the whiskers represent? **extremes**
10. Find the range of the data, 13, 47, 18, 26. **34**

Using the Chapter Test

This page may be used as a test or as an additional page of review if necessary. Two forms of a Chapter Test are provided in the Evaluation Masters Booklet. Form 2 (free response) is shown below, and Form 1 (multiple choice) is shown on page 371.

The **Tech Prep Applications Booklet** provides students with an opportunity to familiarize themselves with various types of technical vocations. The Tech Prep applications for this chapter can be found on pages 21–22.

The **Performance Assessment Booklet** provides an alternative assessment for evaluating student progress. An assessment for this chapter can be found on pages 21–22.

Evaluation Masters Booklet, p. 53

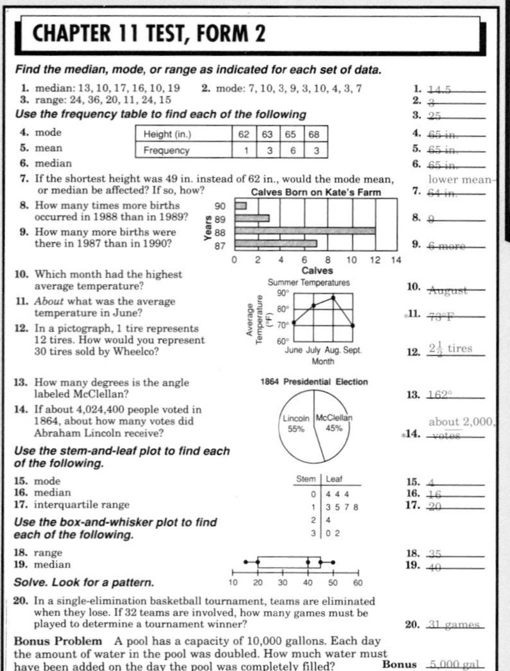

CHAPTER 11 TEST, FORM 2

Find the median, mode, or range as indicated for each set of data.

1. median: 13, 10, 17, 16, 10, 19
2. mode: 7, 10, 3, 9, 3, 10, 4, 3, 7
3. range: 24, 36, 20, 11, 24, 15

Use the frequency table to find each of the following

4. mode
5. mean
6. median

Height (in.)	62	63	65	68
Frequency	1	3	6	3

7. If the shortest height was 49 in. instead of 62 in., would the mode mean, or median be affected? If so, how?
8. How many times more births occurred in 1988 than in 1989?
9. How many more births were there in 1987 than in 1990?

10. Which month had the highest average temperature?
11. *About* what was the average temperature in June?
12. In a pictograph, 1 tire represents 12 tires. How would you represent 30 tires sold by Wheelco?

13. How many degrees is the angle labeled McClellan?
14. If about 4,024,400 people voted in 1864, about how many votes did Abraham Lincoln receive?

Use the stem-and-leaf plot to find each of the following.

15. mode
16. median
17. interquartile range

Stem	Leaf
0	4 4 4
1	3 5 7 8
2	4
3	0 2

Use the box-and-whisker plot to find each of the following.

18. range
19. median

Solve. Look for a pattern.

20. In a single-elimination basketball tournament, teams are eliminated when they lose. If 32 teams are involved, how many games must be played to determine a tournament winner?

Bonus Problem A pool has a capacity of 10,000 gallons. Each day the amount of water in the pool was doubled. How much water must have been added on the day the pool was completely filled?

Answers: 1. 14.5 2. 3 3. 25 4. 65 in. 5. 65 in. 6. 65 in. 7. lower mean–64 in. 8. 9 9. 6 more 10. August 11. 73°F 12. $2\frac{1}{2}$ tires 13. 162° 14. about 2,000,[illegible] votes 15. 4 16. 16 17. 20 18. 35 19. 40 20. 31 games Bonus 5,000 gal

Chapter 11 TEST

Solve. Use the data at the right.

Wages of Workers at Fouse Co.			
$14,500	$19,400	$13,000	$14,500
$11,300	$28,600	$14,500	$11,300
$13,000	$14,500	$16,000	$19,400
$13,000	$11,300	$13,000	$14,500
$16,000	$14,500	$16,000	$13,000

1. Make a frequency table for the set of data. **See margin.**
2. Find the range. **$17,300**
3. Find the median. **$14,500**
4. Find the mode. **$14,500**

The prices of several used cars are listed at the right.

$2,500 $3,500 $4,000
$3,500 $5,000 $10,000

5. Find the mean price and the median price. **$4,750; $3,750**
6. If the most expensive used car costs $20,000 instead of $10,000, how would the median be affected? the mean? **median not affected; greater mean**

Make the type of graph indicated for each set of data. **See answers on p. T528.**

7. bar graph

Height of Skyscrapers (in feet)	
Chrysler Building	1,046
John Hancock Center	1,127
Empire State Building	1,250
World Trade Center	1,350
Sears Tower	1,454

8. line graph

Average Monthly Temperatures in Oklahoma City, Oklahoma (°F)			
Jan.	35.9	July	82.1
Feb.	40.8	Aug.	81.1
March	49.1	Sept.	73.3
April	60.2	Oct.	62.3
May	68.4	Nov.	48.8
June	77.0	Dec.	39.9

9. stem-and-leaf plot

Height in Inches of Track Team Members					
61	60	67	62	59	72
68	63	72	78	60	65
72	55	66	70	59	77

10. circle graph

Monthly Budget	
Food	20%
Clothes	10%
Rent	40%
Other	15%
Transportation	15%

Use the box-and-whisker plot to answer each question.

30 40 50 60 70 80 90 100

11. What is the range? **65**
12. What is the interquartile range? **30**
13. What is the median? **60**
14. A business club has 8 members. Suppose each member shakes hands with every other member. How many handshakes take place? **28 handshakes**

BONUS: Write two questions that can only be answered by using the box-and-whisker plot used in Exercises 11–13. **Answers may vary.**

Additional Answers

1.

Wage	Tally	Frequency
$11,300	|||	3
$13,000	𝍸	5
$14,500	𝍸		6
$16,000	|||	3
$19,400	||	2
$28,600			1

A **Test and Review Generator** is provided in Apple, IBM, and Macintosh versions. You may use this software to create your own tests or worksheets, based on the needs of your students.

MISLEADING GRAPHS

Both graphs below show monthly sales for a small business. Notice that both graphs show the same data. However, the graphs look different because of the scales used along the vertical axes.

Graph A: Monthly Sales

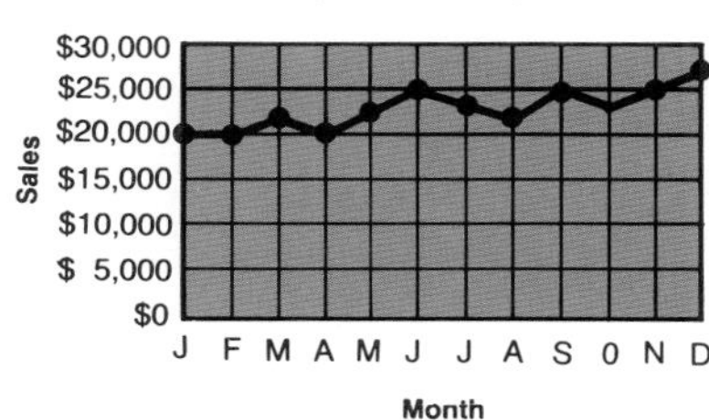

Graph B: Monthly Sales

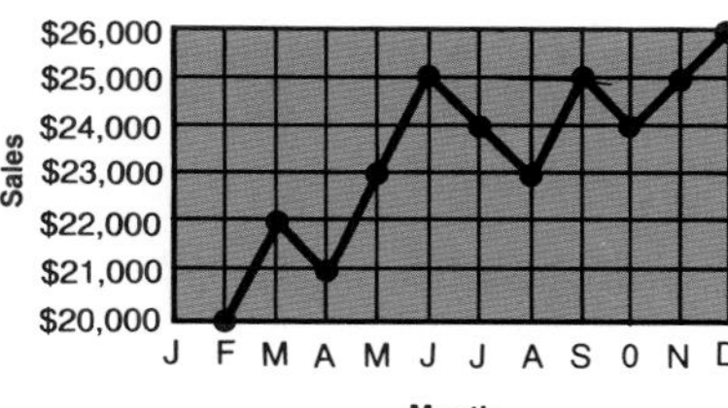

Graph B is *misleading.* It seems to show a very large increase in sales because the scale does not begin at zero.

The graphs below show the results of a survey on favorite restaurants.

Graph D: Favorite Restaurant

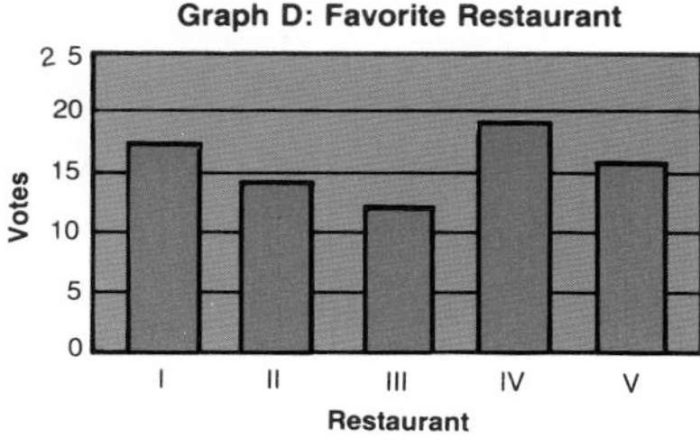

Graph D: Favorite Restaurant

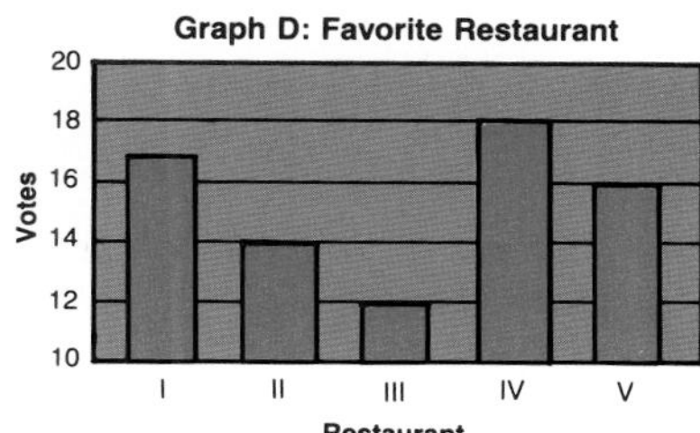

1. Do graphs C and D display the same data? **See margin.**
2. Find the number of votes for each restaurant.
3. In graph D, the bar for restaurant II is twice as long as the bar for restaurant III. Does this mean there were twice as many votes?
4. What causes the difference in voting to appear greater in graph D?
5. Why would restaurant III prefer graph C?
6. Why would restaurant IV prefer graph D?
7. Which graph best represents the result? Explain.

Applying Mathematics to the Real World

This optional page shows how mathematics is used in the real world and also provides a change of pace.

Objective Identify a misleading graph.

Using Discussion Discuss differences implied by each graph in each pair of graphs.

Discuss why different people who are making the graphs might want to create different impressions.

Stress the need for the number scale to start at zero if the graph is to show data accurately. When it is not practical to start at zero, a break in the line just above zero should be used.

Activity

Using Data Have students collect misleading graphs from newspapers and magazines. Then ask them to use the data from that graph to make a graph that is not misleading. How do the two graphs compare?

Additional Answers

1. yes
2. See graph.
3. no
4. The vertical axis does not start at 0.
5. It has the least votes, and the difference looks smaller in graph C.
6. It has the most votes, and the difference looks greater in graph D.
7. Graph C, since there was not a wide range in number of votes.

Using the Cumulative Review

This page provides an aid for maintaining skills and concepts presented thus far in the text. A Cumulative Review is also provided in Evaluation Masters Booklet as shown below.

Evaluation Masters Booklet, p. 55

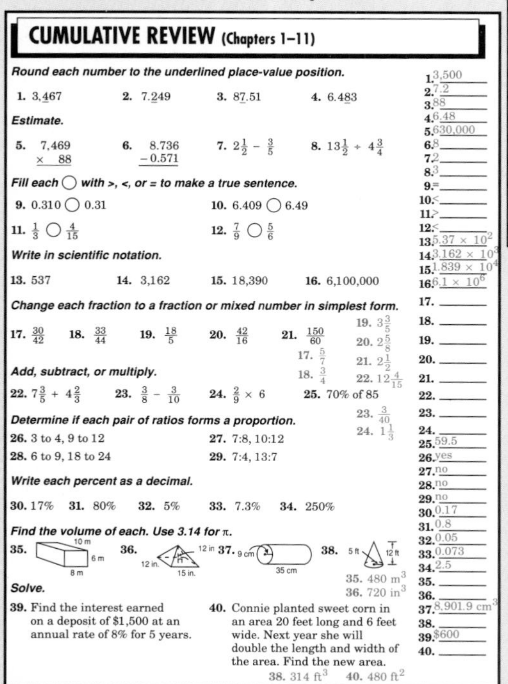

CUMULATIVE REVIEW (Chapters 1–11)

Round each number to the underlined place-value position.

1. 3,467 2. 7.249 3. 87.51 4. 6.483

Estimate.

5. 7,469 × 88 6. 8.736 − 0.571 7. $2\frac{1}{2} - \frac{3}{5}$ 8. $13\frac{1}{2} + 4\frac{3}{4}$

Fill each ◯ with >, <, or = to make a true sentence.

9. 0.310 ◯ 0.31 10. 6.409 ◯ 6.49

11. $\frac{1}{3}$ ◯ $\frac{4}{15}$ 12. $\frac{7}{9}$ ◯ $\frac{5}{6}$

Write in scientific notation.

13. 537 14. 3,162 15. 18,390 16. 6,100,000

Change each fraction to a fraction or mixed number in simplest form.

17. $\frac{30}{42}$ 18. $\frac{33}{44}$ 19. $\frac{18}{5}$ 20. $\frac{42}{16}$ 21. $\frac{150}{60}$

Add, subtract, or multiply.

22. $7\frac{3}{5} + 4\frac{2}{3}$ 23. $\frac{3}{8} - \frac{3}{10}$ 24. $\frac{2}{9} \times 6$ 25. 70% of 85

Determine if each pair of ratios forms a proportion.

26. 3 to 4, 9 to 12 27. 7:8, 10:12

28. 6 to 9, 18 to 24 29. 7:4, 13:7

Write each percent as a decimal.

30. 17% 31. 80% 32. 5% 33. 7.3% 34. 250%

Find the volume of each. Use 3.14 for π.

35. (10 m, 6 m, 8 m) 36. (12 in, 12 in., 15 in.) 37. (9 cm, 35 cm) 38. (5 ft, 12 ft)

Solve.

39. Find the interest earned on a deposit of $1,500 at an annual rate of 8% for 5 years.

40. Connie planted sweet corn in an area 20 feet long and 6 feet wide. Next year she will double the length and width of the area. Find the new area.

17. $\frac{5}{7}$ 18. $\frac{3}{4}$ 19. $3\frac{3}{5}$ 20. $2\frac{5}{8}$ 21. $2\frac{1}{2}$ 22. $12\frac{4}{15}$ 23. $\frac{3}{40}$ 24. $1\frac{1}{3}$ 35. 480 m^3 36. 720 in^3 38. 314 ft^3 40. 480 ft^2

1. 3,500 2. 7.2 3. 88 4. 6.48 5. 630,000 6. 8 7. 2 8. 3 9. = 10. < 11. > 12. < 13. 5.37×10^2 14. 3.162×10^3 15. 1.839×10^4 16. 6.1×10^6 17. 18. 19. 20. 21. 22. 23. 24. 25. 59.5 26. yes 27. no 28. no 29. no 30. 0.17 31. 0.8 32. 0.05 33. 0.073 34. 2.5 35. 36. 37. 8,901.9 cm^3 38. 39. $600 40.

Cumulative REVIEW/TEST

Free Response

Lessons 1-11, 1-12

Solve each equation.

1. $23 + y = 130$ 107 **2.** $x - 28 = 10$ 38 **3.** $4 = \frac{x}{20}$ 80

4. $18 = 3y$ 6 **5.** $100 - 52 = y$ 48 **6.** $8x = 7$ $\frac{7}{8}$

Lessons 2-1, 2-3

Add or subtract. 10. 52,137 12. 532.587

7. 6,354 − 617 5,737 **8.** 7,292 + 82 7,374 **9.** 1,979 + 876 2,855

10. 54,685 − 2,548 **11.** 1,983 − 1,765 218 **12.** 62 + 3.587 + 467

Lesson 4-7

Replace each ● with <, >, or = to make a true sentence.

13. $\frac{3}{7}$ ● $\frac{4}{7}$ < **14.** $\frac{1}{2}$ ● $\frac{4}{8}$ = **15.** $\frac{2}{3}$ ● $\frac{7}{9}$ < **16.** $\frac{7}{15}$ ● $\frac{5}{12}$ > **17.** $\frac{7}{8}$ ● $\frac{8}{9}$ <

Lesson 5-7

18. $\frac{3}{8}$ or 3 pieces

18. Four pizzas cut into eight pieces each were eaten by nine friends. If everybody had equal amounts, how much did each person eat?

Lessons 6-7, 6-8

Complete. 23. 3, 4

19. 1 ft = ■ in. 12 **20.** 48 in. = ■ ft 4

21. 1 mi = ■ ft 5,280 **22.** 1 gal = ■ qt 4

23. 100 oz = ■ qt ■ oz **24.** 10,000 lb = ■ T 5

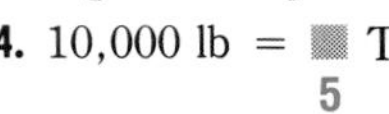

Lesson 7-1

Use words and symbols to name each figure in as many ways possible.

25.

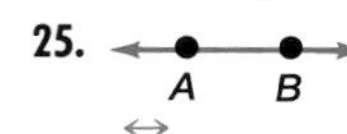

$\overleftrightarrow{AB}$; Line AB

26.

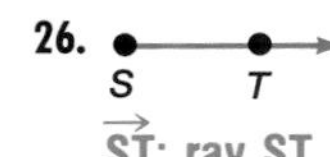

$\overrightarrow{ST}$; ray ST

27. M N

line segment MN, $\overline{MN}$

28. Q

Point Q

Lesson 8-12

Find the volume of each cone. Use 3.14 for π. Round decimal answers to the nearest tenth.

29.

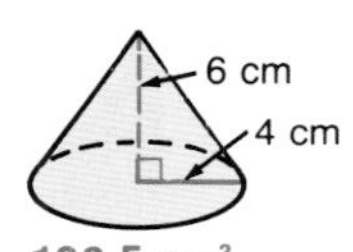

100.5 cm^3

30.

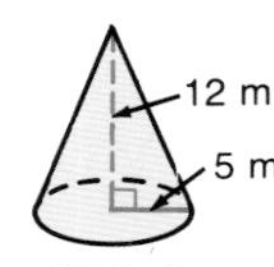

314m^3

31.

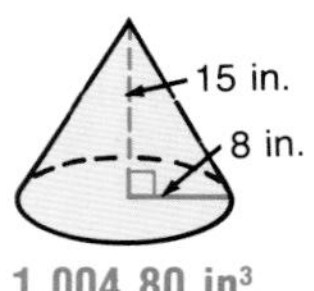

1,004.80 in^3

Lesson 9-8

32. Ben's Bakery gives an extra donut with each dozen you buy. If you buy 80 donuts to take to a breakfast meeting, how many donuts will you get free? 6 donuts

Lesson 11-1

33. Jacob's quiz scores are 18, 10, 25, 13, and 19. Find the mean, median, and range of his quiz scores. 17; 18; 15

APPLYING THE LESSON

Independent Practice

Homework Assignment	
Minimum	Ex. 2-32 even; 33
Average	Ex. 1-33
Maximum	Ex. 1-33

Multiple Choice

Choose the letter of the correct answer for each item.

1. Find the value of $4 \times 5^2 - 3 \times 2$.
- **a.** 91
- ▶ **b.** 94
- **c.** 176
- **d.** 394

Lesson 1-9

2. What is the sum of 40,592 and 3,287?
- **e.** 37,305
- ▶ **f.** 43,879
- **g.** 43,889
- **h.** 73,462

Lesson 2-1

3. Find the product when 6 is multiplied by 2,220.
- **a.** 1,320
- **b.** 1,332
- **c.** 12,320
- ▶ **d.** 13,320

Lesson 3-1

4. Find the lowest common multiple (LCM) of 24, 42, and 48.
- **e.** 6
- **f.** 96
- ▶ **g.** 336
- **h.** 48,384

Lesson 4-3

5. Janet needs $2\frac{7}{8}$ yards of fabric. She buys $3\frac{1}{4}$ yards. How much extra fabric did she buy?
- **a.** $\frac{1}{4}$ yard
- **c.** $1\frac{3}{8}$ yards
- **b.** $\frac{5}{8}$ yard
- ▶ **d.** *none of these*

Lesson 6-7

6. Change $0.12\frac{1}{2}$ to a fraction.
- **e.** $\frac{1}{80}$
- ▶ **f.** $\frac{1}{8}$
- **g.** $\frac{1}{4}$
- **h.** $\frac{25}{2}$

Lesson 5-9

7. A salt solution has 3 cups salt for every 8 cups water. What is the ratio of salt to water?
- ▶ **a.** 3:8
- **b.** 8:3
- **c.** 3:11
- **d.** 8:11

Lesson 9-1

8. In a class election, Brady received 60% of the 30 votes cast. How many more votes did Brady receive than the other candidate?
- **e.** 8
- **g.** 10
- ▶ **f.** 6
- **h.** 9

Lesson 10-1

9. 88 is what percent of 110?
- **a.** 8%
- ▶ **b.** 80%
- **c.** 88%
- **d.** 125%

Lesson 10-2

10. Ed deposits $150 in an account that pays 7% simple annual interest. How much interest will he earn in 2 years?
- **e.** $7
- ▶ **g.** $21
- **f.** $14
- **h.** $210

Lesson 10-8

APPLYING THE LESSON

Independent Practice

Homework Assignment	
Minimum	**Ex. 1-10**
Average	**Ex. 1-10**
Maximum	**Ex. 1-10**

Using the Cumulative Test

This test serves to familiarize students with standardized format while testing skills and concepts presented up to this point.

Evaluation Masters Booklet, pp. 51-52

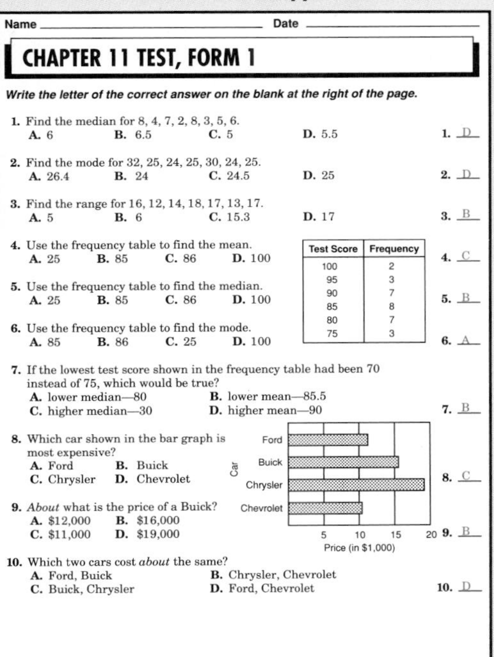

Name ________ Date ________

CHAPTER 11 TEST, FORM 1

Write the letter of the correct answer on the blank at the right of the page.

1. Find the median for 8, 4, 7, 2, 8, 3, 5, 6.
A. 6 **B.** 6.5 **C.** 5 **D.** 5.5 — 1. D

2. Find the mode for 32, 25, 24, 25, 30, 24, 25.
A. 26.4 **B.** 24 **C.** 24.5 **D.** 25 — 2. D

3. Find the range for 16, 12, 14, 18, 17, 13, 17.
A. 5 **B.** 6 **C.** 15.3 **D.** 17 — 3. B

4. Use the frequency table to find the mean.
A. 25 **B.** 85 **C.** 86 **D.** 100 — 4. C

5. Use the frequency table to find the median.
A. 25 **B.** 85 **C.** 86 **D.** 100 — 5. B

6. Use the frequency table to find the mode.
A. 85 **B.** 86 **C.** 25 **D.** 100 — 6. A

Test Score	Frequency
100	2
95	3
90	7
85	8
80	7
75	3

7. If the lowest test score shown in the frequency table had been 70 instead of 75, which would be true?
A. lower median—80 **B.** lower mean—85.5
C. higher median—30 **D.** higher mean—90 — 7. B

8. Which car shown in the bar graph is most expensive?
A. Ford **B.** Buick
C. Chrysler **D.** Chevrolet — 8. C

9. *About* what is the price of a Buick?
A. $12,000 **B.** $16,000
C. $11,000 **D.** $19,000 — 9. B

10. Which two cars cost *about* the same?
A. Ford, Buick **B.** Chrysler, Chevrolet
C. Buick, Chrysler **D.** Ford, Chevrolet — 10. D

Using Mathematical Connections

Objective Understand, appreciate, and enjoy the connections between mathematics and real-life phenomena and other disciplines.
The following data is background information for each event on the time line.

Social Studies Noah's Ark, Turkey, 2368? BC
The story of Noah's ark tells about a flood that covered Earth. Noah built an Ark to save himself, his family, and selected species of animals. The story of such a flood is universal. The story exists throughout the Middle East, Europe, Asia, Africa and the Americas. Records of Hopi Indians of our own country say he escaped to a high mountain. Noah's Ark is claimed to exist on Mount Ararat in eastern Turkey. However, the question remains, "Is it Noah's Ark, or isn't it?"

Social Studies The Birth of Democracy, Athens, Greece, 510 BC
Democracy began in ancient Greece as early as the 500s BC. In 594 BC, a man named Solon became chief magistrate of Athens and established the early beginnings of democracy. He did away with the existing system of sharecropping, forgave debts, and provided a citizen assembly with more authority. In 546 BC, Pisistratus gained power from Solon and re-established tyranny for himself and former aristocrats. In 510 BC Cleisthenes solidified democracy as the new political system for Athens and Attica. Athenian democracy differed somewhat from democracy today. It was a direct democracy rather than our representative one.

Technology The Rotating Fan, China, 200 BC
The first rotating fan was built in China during the second century BC. It was built for separating the husks and stalks (chaff) from the grain. The earlier method of separating the grain was to throw it into the air in a strong wind. The Chinese were not satisfied with waiting for a strong wind, so they came up with a brilliant invention. Grain was put into a hopper and a hand-operated fan blew continuous streams of air over it. The air from the fan blew the chaff away. Grain then fell down into a chaff-free pile.

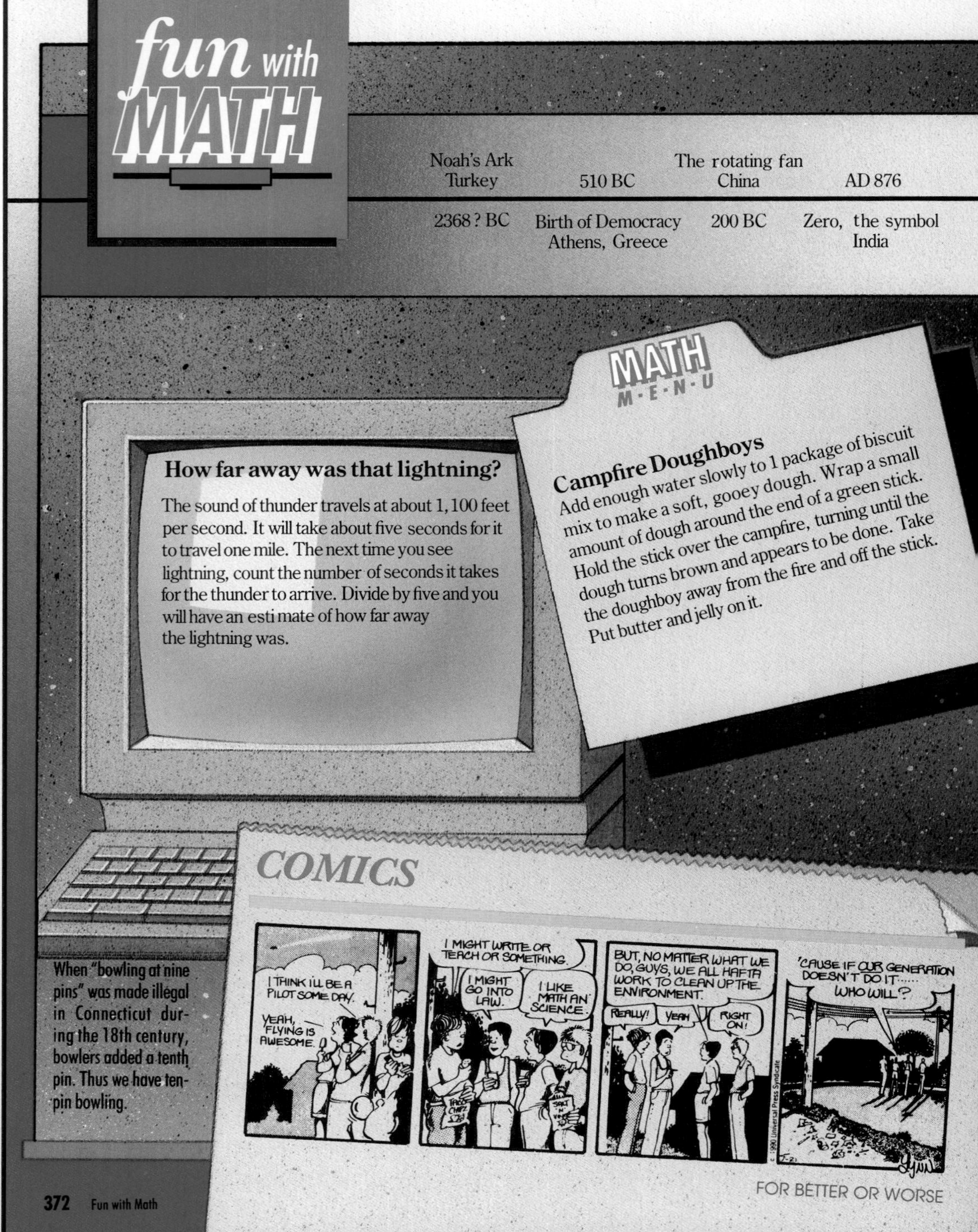

fun with MATH
Noah's Ark Turkey
2368 ? BC
510 BC
Birth of Democracy Athens, Greece
The rotating fan China
200 BC
AD 876
Zero, the symbol India
MATH M-E-N-U
Campfire Doughboys
Add enough water slowly to 1 package of biscuit mix to make a soft, gooey dough. Wrap a small amount of dough around the end of a green stick. Hold the stick over the campfire, turning until the dough turns brown and appears to be done. Take the doughboy away from the fire and off the stick. Put butter and jelly on it.
How far away was that lightning?
The sound of thunder travels at about 1,100 feet per second. It will take about five seconds for it to travel one mile. The next time you see lightning, count the number of seconds it takes for the thunder to arrive. Divide by five and you will have an esti mate of how far away the lightning was.
When "bowling at nine pins" was made illegal in Connecticut during the 18th century, bowlers added a tenth pin. Thus we have ten-pin bowling.
COMICS
I THINK I'LL BE A PILOT SOME DAY.
YEAH, FLYING IS AWESOME.
I MIGHT WRITE OR TEACH OR SOMETHING.
I MIGHT GO INTO LAW.
I LIKE MATH AN' SCIENCE.
BUT, NO MATTER WHAT WE DO, GUYS, WE ALL HAFTA WORK TO CLEAN UP THE ENVIRONMENT.
REALLY!
YEAH
RIGHT ON!
'CAUSE IF OUR GENERATION DOESN'T DO IT...... WHO WILL?
© 1990 Universal Press Syndicate
FOR BETTER OR WORSE
372 Fun with Math

History Zero, the Symbol, India, AD 876
The earliest occurrence of zero, as we know it was found on an inscription at Gwalior, India, dated AD 876. However, a booklet on Indian arithmetic written by Arabian mathematician Al-Khowarizmi describes the symbol for zero and may have appeared as early as AD 820. In his booklet, he says, "When after subtracting, nothing is left, write a small circle, lest the place remains empty. The small circle must occupy the place, lest there will be fewer places and, for instance, the second is taken for the first."

Statistics England AD 1662
First electronic computer Pennsylvania, USA AD 1946
Symmetry in the universe? New York, USA AD 1956
Genetic engineering Pennsylvania, USA AD 1983
JOKE!
Q: What do you call a person who counts on her fingers?
A: A digital computer
RIDDLE
Q: Use one word to describe what is shown here.
A: Hy pot en use
How does an electronic calculator add 4 plus 5?
Pressing the three keys, 4, +, and 5, on your calculator closes a set of contacts for each key on the circuit board beneath the keyboard. This sends signals to the calculator's processing unit to store the codes for each key in memory. Pressing the equals key tells the processing unit to take the three codes from memory and to perform the addition operation. The answer is decoded and sent to the display as 9.
TEASER
Place four coins on your desk, all with tails up. A move consists of turning three coins over at a time. How many such moves do you need to arrange for all four heads to be up?
T T T T
Once you have solved this teaser you may like to investigate another teaser with five coins all tails up. In this teaser, a move consists of turning over any four coins.
QUIZ TIME
Try these number patterns to start you thinking.
1. Choose any single-digit number and multiply it by 9. The multiply the product by 12345679 on your calculator. Are you surprised at the answer? Try more numbers and see what happens. Then explain your answers.
2. What number would you substitute for 12345679 if you first multiplied your single-digit number by 7 instead of 9?
Fun with Math 373

Quiz Answers

1. If the single-digit number is *a*, the answer is *aaa aaa aaa*. This is because 123455679 = 111 111 111 ÷ 9.
2. The answer is 15873 because 111 111 ÷ 7 = 15873. Then, if the single-digit number is *a*, 15873 × (7 × *a*) = *aaa* aaa.

Teaser Answers

This activity is possible in four moves. One solution is as follows:

T	T	T	T	Start
T*	H	H	H	First move
H	H*	T	T	Second move
T	T	T*	H	Third move
H	H	H	H	Fourth move

*The asterisk indicates the coin which was not turned over in the move.

History Statistics, England, AD 1662
Statistics is the science of collecting and analyzing data. The founding of the science of statistics and demography is credited to John Graunt. In 1662 he published his *Natural and Political Observations* . . . in collaboration with economist Sir William Petty. It wasn't until 1746 that a professor from Gottingen, Germany, Gottfried Achenwall, created the word *statistics.*

High-Tech First Electronic Computer, Pennsylvania, USA, AD 1946
In 1946 Dr. John W. Mauchly and J. Presper Eckert of the University of Pennsylvania designed and built the first all electronic computer. It was called ENIAC *E*lectronic *N*umerical *I*ntegrator *a*nd *C*alculator. ENIAC was as large as a house and contained no less than 18,000 vacuum tubes. It was not a real computer but a calculator since it did not have a memory.

Science Symmetry in the Universe?, New York, USA, AD 1956
The theory, "the conservation of parity," states that the universe makes no distinction between left and right. If you were to step into a looking-glass house in which left and right were reversed, you would not know the difference. The laws of nature would remain unchanged. In 1956 Dr. Chien-Shiung Wu, a woman physicist at Columbia University, proved that there is "right-handedness" and "left handedness" in the universe, and they each obey distinct laws of nature. That is, you would in fact be able to tell reality from a looking glass.

Health Genetic Engineering, Pennsylvania, USA, AD 1983
Genetics is the science of heredity founded by Gregory Mendel in 1865. Each hereditary characteristic, such as eye color and hair color is transferred from parents to offspring by a specific gene in each cell.
In 1983, Professor R. L. Brinster and his colleagues at the University of Pennsylvania began a field of genetic engineering. They succeeded in splicing, or removing, genes that control growth in human cells and grafted them into mouse embryos. The mice developed into giants.

12 INTEGERS

PREVIEWING the CHAPTER

In this chapter, the number line is used to introduce integers, opposites, and absolute value. Students use the definition of absolute value to apply the rules for adding integers with the same or with different signs. Subtraction is presented as a variation of addition. Number patterns are used to introduce the rules for multiplying and dividing integers.

Problem-Solving Strategy Students learn to solve problems by working backwards.

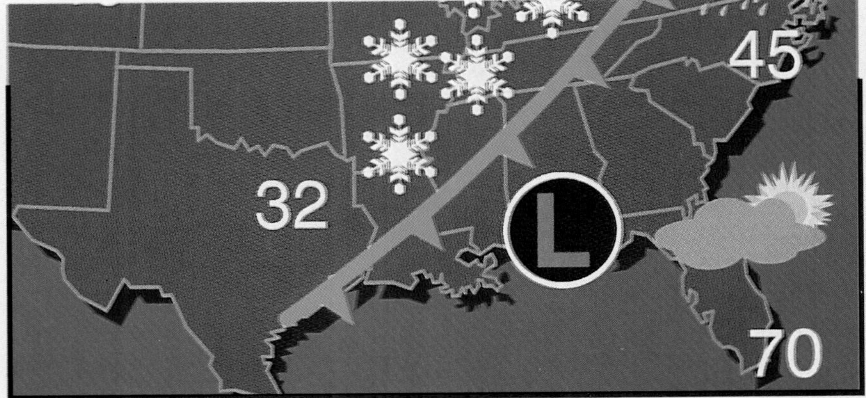

Lesson (Pages)	Lesson Objectives	State/Local Objectives
12-1 (376-377)	12-1: Identify, order, and compare integers.	
12-2 (378-379)	12-2: Add integers.	
12-3 (380-381)	12-3: Subtract integers.	
12-4 (382-383)	12-4: Use a windchill chart to determine equivalent temperatures.	
12-5 (386-387)	12-5: Multiply integers.	
12-6 (388-389)	12-6: Divide integers	
12-7 (390-391)	12-7: Solve verbal problems by working backwards.	

ORGANIZING the CHAPTER

You may refer to the ***Course Planning Calendar*** on page T10.

Planning Guide

Lesson (Pages)	Extra Practice (Student Edition)	Blackline Masters Booklets: Reteaching	Practice	Enrichment	Activity	Multicultural Activity	Technology	Tech Prep	Evaluation	Other Resources
12-1 (376-377)	p. 480	p. 115	p. 126	p. 115			p. 26			Transparency 12-1
12-2 (378-379)	p. 480	p. 116	p. 127	p. 116	p. 23					Transparency 12-2
12-3 (380-381)	p. 480	p. 117	p. 128	p. 117		p. 54		p. 23		Transparency 12-3
12-4 (382-383)		p. 118	p. 129	p. 118			p. 12		p. 59	Transparency 12-4
12-5 (386-387)	p. 481	p. 119	p. 130	p. 119	p. 40			p. 24		Transparency 12-5
12-6 (388-389)	p. 481	p. 120	p. 131	p. 120	p. 24					Transparency 12-6
12-7 (390-391)			p. 132						p. 59	Transparency 12-7
Ch. Review (392-393)									pp. 56-58	Test and Review Generator
Ch. Test (394)										
Cumulative Review/Test (396-397)									p. 60	

OTHER CHAPTER RESOURCES

Student Edition
Chapter Opener, p. 374
Activity, p. 375
On-the-Job Application, p. 384
Consumer Connections, p. 385
Journal Entry, p. 381
Portfolio Suggestion, p. 391
Consumer Connections, p. 395

Teacher's Classroom Resources
Transparency 12-0
Lab Manual, pp. 45-46
Performance Assessment, pp. 23-24
Lesson Plans, pp. 126-132

Other Supplements
Overhead Manipulative Resources, Labs 34-36, pp. 24-26
Glencoe Mathematics Professional Series

ENHANCING the CHAPTER

Some of the blackline masters for enhancing this chapter are shown below.

COOPERATIVE LEARNING

The following game can be played by pairs of students. Have each group draw a number line from –30 to 30. Use two different colored dice. Designate one die as positive and the other as negative. Each student places a marker on 0 and takes turns rolling the dice. Students may add, subtract, multiply, or divide the integers. If the result is positive, move to the right along the number line. If the result is negative, move to the left. The goal is for each player to land on 30 by exact count.

USING MODELS/MANIPULATIVES

Have students use counters to represent positive and negative integers. Suppose a white counter represents 1 and a blue counter represents –1. Have students demonstrate addition of integers by joining two groups of counters. Note that pairing a 1 and –1 gives 0. Have students demonstrate subtraction of integers by first writing a related addition expression. Have them demonstrate multiplication of integers by forming groups of counters. Multiplication by a negative integer can be explained as changing the sign (counter color) of the groups.

Cooperative Problem Solving, p. 40

M9566—McClanahan, Glencoe, Chap. 12b2, MW1, 3/25/91

Name ____________________ Date ____________________

COOPERATIVE PROBLEM SOLVING (Use with Lesson 12-5)

Investing in the Stock Market

The business section of the daily newspaper usually contains stock market reports of the previous day's activities with listings such as this.

Company ⟶ ABCorp .52 250 $15\frac{1}{8}$ $-\frac{1}{4}$
Dividend (.52)
Shares sold (250)
Closing price (in dollars) ($15\frac{1}{8}$)
Change in value over previous day (in dollars) ($-\frac{1}{4}$)

Imagine that you and a partner invest up to $500 in the stock market. Complete the following steps to calculate how much you gain or lose over a 5-day period.

1. Select the stocks to buy using the closing prices shown in today's stock market reports. You may buy more than one share of each stock. Record your purchases in the first four columns of a table like the one below.

Company	Purchase Price per Share	Shares Purchased	Total Purchase Price	Net Change per Share	Net Change for Total Shares Purchased
Totals					

2. For each of the next 5 business days, record the change per share of your stocks in a table like this.

Company	Change per Share					Net Change per Share
	Day 1	Day 2	Day 3	Day 4	Day 5	

3. After following your stocks' activities, calculate the net change per share and record this in the fifth column of the table you started in Step 1.
4. Complete the last column of your table. Subtract the amount of your original investment from the total of this column to find your net gain or loss.

40

Lab Activity, p. 46

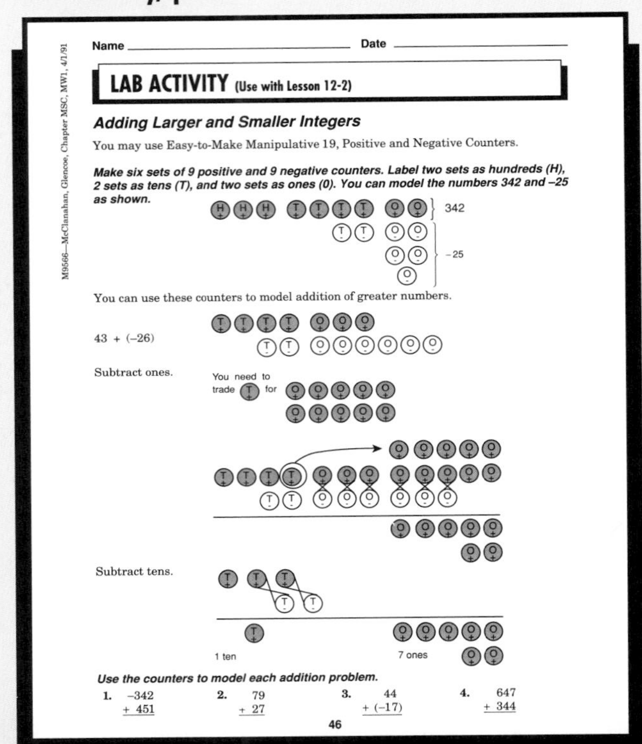
M9566—McClanahan, Glencoe, Chapter MSC, MW1, 4/1/91

Name ____________________ Date ____________________

LAB ACTIVITY (Use with Lesson 12-2)

Adding Larger and Smaller Integers

You may use Easy-to-Make Manipulative 19, Positive and Negative Counters.

Make six sets of 9 positive and 9 negative counters. Label two sets as hundreds (H), 2 sets as tens (T), and two sets as ones (0). You can model the numbers 342 and –25 as shown.

You can use these counters to model addition of greater numbers.

43 + (–26)

Subtract ones.

Subtract tens.

Use the counters to model each addition problem.

1. –342 + 451
2. 79 + 27
3. 44 + (–17)
4. 647 + 344

46

MEETING INDIVIDUAL NEEDS

Mainstreaming

Use descriptive words and visual images to continually reinforce the idea that integers are numbers that have both magnitude and direction. Such words are decrease/increase, up/down, below/above, find/lose, before/after, and so on. Allow students to use number lines when performing operations with integers. Students will benefit from doing simple exercises such as 4(–3); moving –3 units 4 times for a product of –12. Remind students that division is the inverse of multiplication so change division to multiplication and use a number line.

CRITICAL THINKING/PROBLEM SOLVING

Have students use a flowchart and a reverse flowchart to model steps in a problem that is solved by working backwards. The flowchart shows the initial unknown and each step used to reach the given result. The reverse flowchart should show the given result and the inverse of each step leading to the initial unknown. Use flowcharts to solve the following problem. When Stacy received her paycheck, she spent $6.75 of it on a book and twice that amount on a radio. She had $9.75 left. How much was her paycheck? $30.00

COMMUNICATION

Writing

Have students write problems or descriptions of situations that illustrate addition, subtraction, multiplication, and division with integers. A different problem or situation should be written for operations with the same sign and with different signs. Situations may involve money, distance and direction, stock prices,temperatures,elevations, and so on.

Applications, p. 23

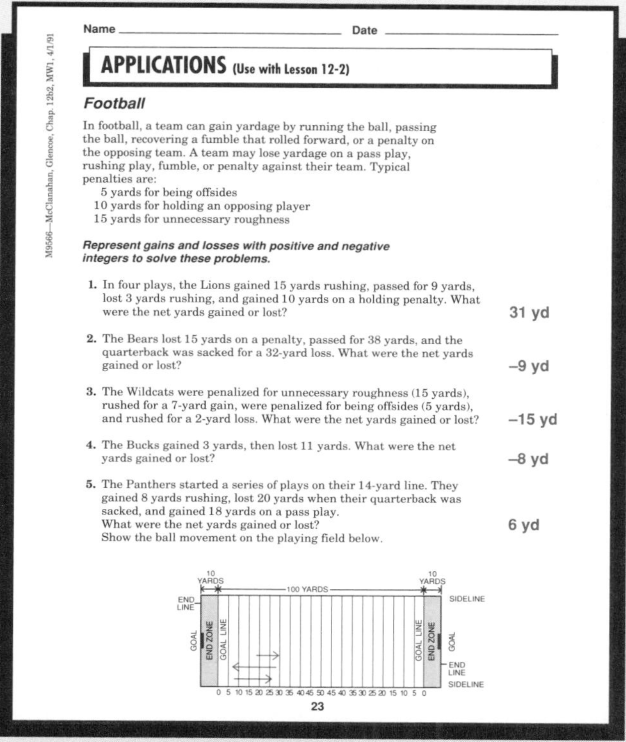

M9566—McClanahan, Glencoe, Chap. 12b2, MW1, 4/1/91

Name ______________ Date ______________

APPLICATIONS (Use with Lesson 12-2)

Football

In football, a team can gain yardage by running the ball, passing the ball, recovering a fumble that rolled forward, or a penalty on the opposing team. A team may lose yardage on a pass play, rushing play, fumble, or penalty against their team. Typical penalties are:

5 yards for being offsides
10 yards for holding an opposing player
15 yards for unnecessary roughness

Represent gains and losses with positive and negative integers to solve these problems.

1. In four plays, the Lions gained 15 yards rushing, passed for 9 yards, lost 3 yards rushing, and gained 10 yards on a holding penalty. What were the net yards gained or lost? 31 yd
2. The Bears lost 15 yards on a penalty, passed for 38 yards, and the quarterback was sacked for a 32-yard loss. What were the net yards gained or lost? –9 yd
3. The Wildcats were penalized for unnecessary roughness (15 yards), rushed for a 7-yard gain, were penalized for being offsides (5 yards), and rushed for a 2-yard loss. What were the net yards gained or lost? –15 yd
4. The Bucks gained 3 yards, then lost 11 yards. What were the net yards gained or lost? –8 yd
5. The Panthers started a series of plays on their 14-yard line. They gained 8 yards rushing, lost 20 yards when their quarterback was sacked, and gained 18 yards on a pass play. What were the net yards gained or lost? 6 yd
Show the ball movement on the playing field below.

23

Calculator Activity, p. 12

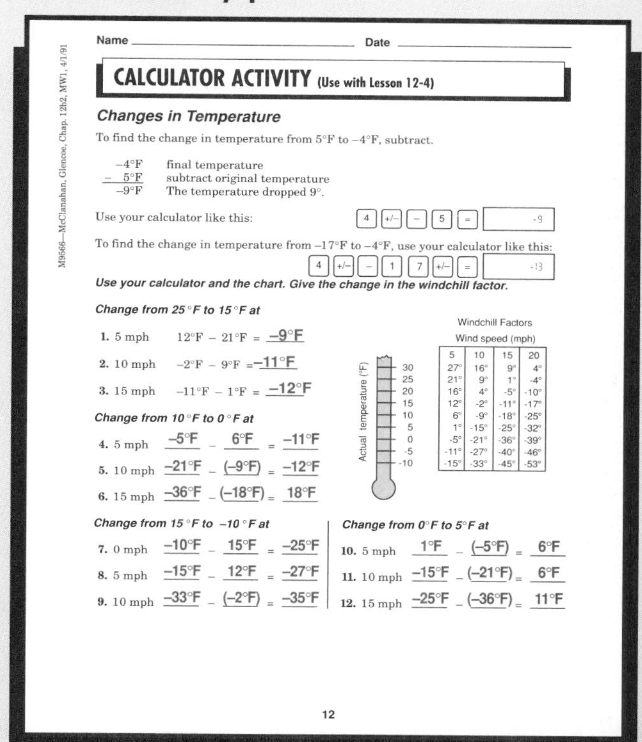

M9566—McClanahan, Glencoe, Chap. 12b2, MW1, 4/1/91

Name ______________ Date ______________

CALCULATOR ACTIVITY (Use with Lesson 12-4)

Changes in Temperature

To find the change in temperature from 5°F to –4°F, subtract.

–4°F	final temperature
– 5°F	subtract original temperature
–9°F	The temperature dropped 9°.

Use your calculator like this: [4] [+/–] [–] [5] [=] -9

To find the change in temperature from –17°F to –4°F, use your calculator like this: [4] [+/–] [–] [1] [7] [+/–] [=] -13

Use your calculator and the chart. Give the change in the windchill factor.

Change from 25 °F to 15 °F at

1. 5 mph 12°F – 21°F = –9°F
2. 10 mph –2°F – 9°F = –11°F
3. 15 mph –11°F – 1°F = –12°F

Change from 10 °F to 0 °F at

4. 5 mph –5°F – 6°F = –11°F
5. 10 mph –21°F – (–9°F) = –12°F
6. 15 mph –36°F – (–18°F) = 18°F

Change from 15 °F to –10 °F at

7. 0 mph –10°F – 15°F = –25°F
8. 5 mph –15°F – 12°F = –27°F
9. 10 mph –33°F – (–2°F) = –35°F

Change from 0°F to 5°F at

10. 5 mph 1°F – (–5°F) = 6°F
11. 10 mph –15°F – (–21°F) = 6°F
12. 15 mph –25°F – (–36°F) = 11°F

Windchill Factors

Actual temperature (°F)	Wind speed (mph) 5	10	15	20
30	27°	16°	9°	4°
25	21°	9°	1°	-4°
20	16°	4°	-5°	-10°
15	12°	-2°	-11°	-17°
10	6°	-9°	-18°	-25°
5	1°	-15°	-25°	-32°
0	-5°	-21°	-36°	-39°
-5	-11°	-27°	-40°	-46°
-10	-15°	-33°	-45°	-53°

12

CHAPTER PROJECT

Have students make family time lines. Let the year of their birth represent the origin, 0. Events occurring prior to their birth are placed to the left of 0; events occurring after birth are placed to the right of 0. Have students write addition, subtraction, and multiplication problems based on events on their time line.

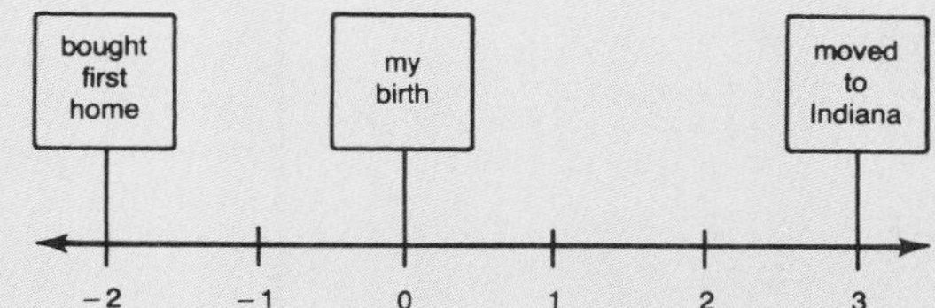

Using the Chapter Opener

Transparency 12-0 is available in the Transparency Package. It provides a full-color visual and motivational activity that you can use to engage students in the mathematical content of the chapter.

Using Discussion

- Have students read the opening paragraph. See if students know about how many days a year the temperature drops below zero in their community. If temperatures in your area do not drop below zero, have students research areas like Alaska or Antarctica.
- Ask students to answer the second and third questions. Ask if anyone can read the numbers correctly. **negative 2; negative 10**
- Allow students the opportunity to express reasons for needing to know the temperature.

The second paragraph is a motivational problem. Have students solve the problem and discuss the ways in which they arrived at their answers.

Chapter 12

INTEGERS

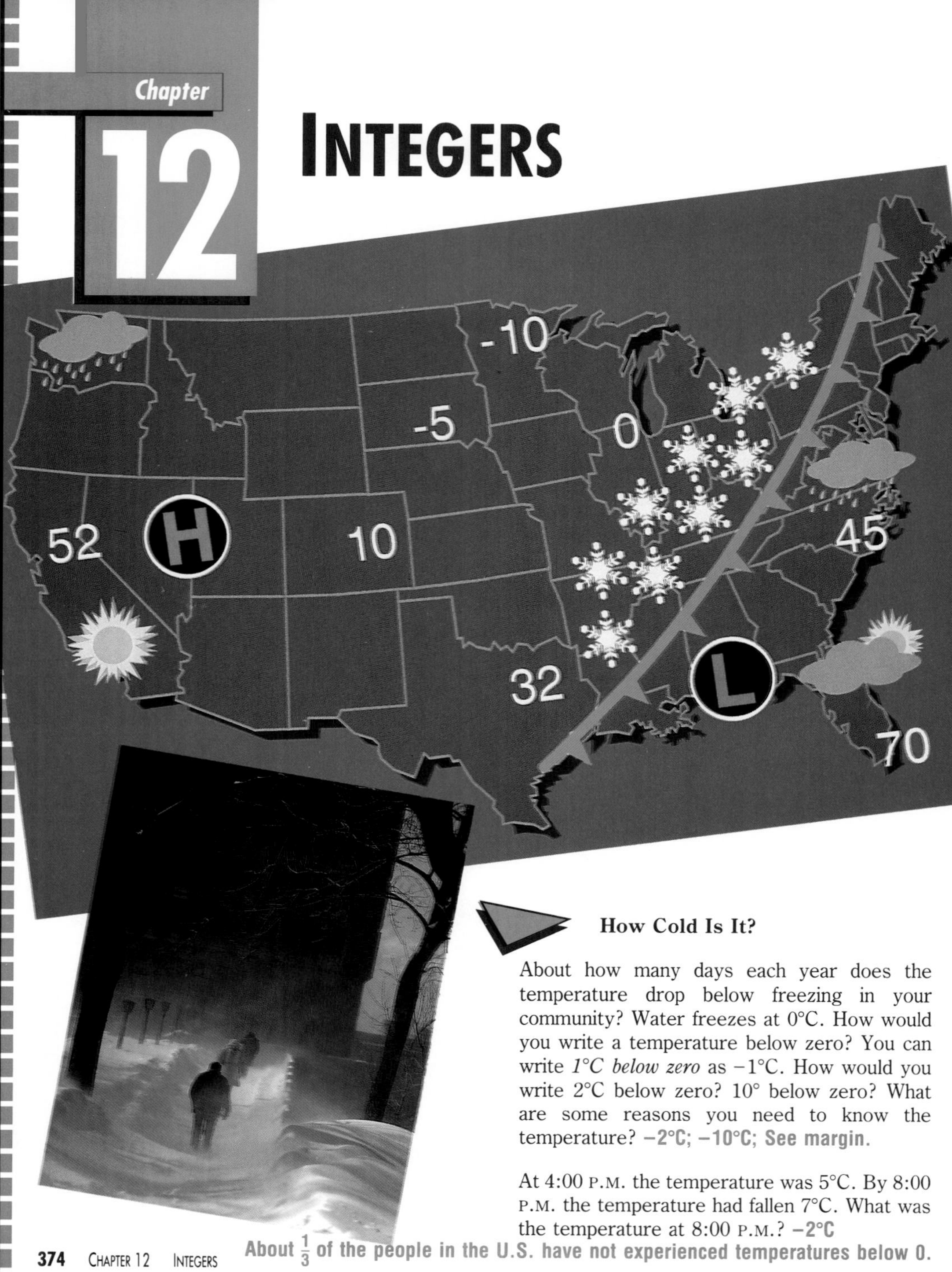

How Cold Is It?

About how many days each year does the temperature drop below freezing in your community? Water freezes at 0°C. How would you write a temperature below zero? You can write *1°C below zero* as −1°C. How would you write 2°C below zero? 10° below zero? What are some reasons you need to know the temperature? **−2°C; −10°C; See margin.**

At 4:00 P.M. the temperature was 5°C. By 8:00 P.M. the temperature had fallen 7°C. What was the temperature at 8:00 P.M.? **−2°C**

About $\frac{1}{3}$ of the people in the U.S. have not experienced temperatures below 0.

Additional Answers

Some reasons are how to dress; if cold enough to ice skate safely; warm enough to walk to a destination; cool enough to do outside strenuous activity

ACTIVITY: Adding Integers

The temperature *6° below zero* can be written as −6° and the temperature *6° above zero* can be written as +6°. −6 is read as *negative 6* and +6 is read as *positive 6*.

Usually, the plus signs on numbers like +1 and +2 are omitted.

Numbers such as −6 and +6 are called integers. The set of integers is listed below.

. . . , −6, −5, −4, −3, −2, −1, 0, +1, +2, +3, +4, +5, +6, . . .

In the activities that follow, positive and negative counters will represent positive and negative integers. When the counters are combined, every positive counter cancels exactly one negative counter; that is, a negative paired with a positive results in zero. The models below will help you understand how to add integers.

Materials: counters

Cooperative Groups

Work with a partner. Study these models for each addition problem.

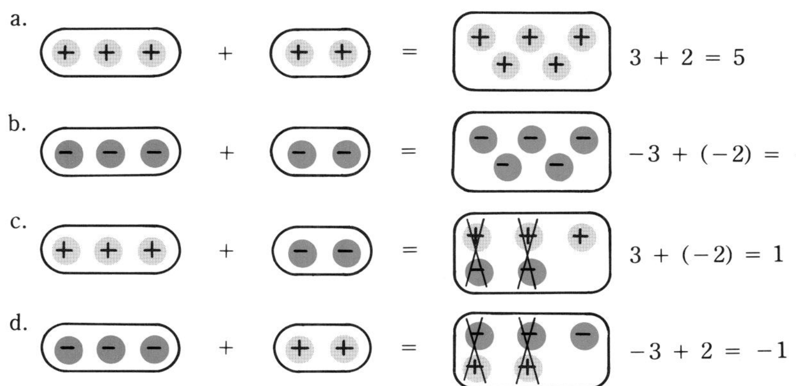

a. 3 + 2 = 5

b. −3 + (−2) = −5

c. 3 + (−2) = 1

d. −3 + 2 = −1

Communicate Your Ideas See students' work.

Use counters to model each problem. Then write the sum.

1. 4 + 3 7　**2.** −5 + (−4) −9　**3.** −6 + 2 −4　**4.** 3 + (−8) −5

5. −2 + (−2) −4　**6.** −8 + 4 −4　**7.** −7 + 7 0　**8.** 1 + (−4) −3

Playing a Game

9. The "Game of 6" can be played with 2 to 5 players. Make a deck of 52 cards by writing each integer from −6 through 6 on four differently colored sets of thirteen cards. The dealer deals two cards to each player. Each player mentally finds the sum of the numbers on the cards. If the sum is close to 6, the player may pass. If the sum is not close to 6, the player may receive up to three more cards, one at a time, to try to reach that goal. The winner is the player who has a sum of 6 or closest to 6.

INTRODUCING THE LESSON

Objective Add integers using counters.

TEACHING THE LESSON

Ask students to name times other than reporting temperature when an integer is needed. **elevation, time, money, yards lost or gained in football, stock market**

Using Cooperative Groups Each student should have their own counters to work Exercises 1-8. Make sure students discuss their findings with their partners for Exercises 1-8.

EVALUATING THE LESSON

Playing the Game Encourage students to play the game. If they have trouble finding the sums mentally, allow them to use the counters.

EXTENDING THE LESSON

Closing the Lesson Allow time for students to discuss their findings from playing the game. Students will learn if their findings are correct in the following lessons.

Activity Worksheet, Lab Manual, p. 45

Name ____________ Date ____________

ACTIVITY WORKSHEET (Use with page 375)

Adding Integers

Use for Exercise 9.

13 ♡	1	2	3	4	5	6	0
	−1	−2	−3	−4	−5	−6	
13 ♤	1	2	3	4	5	6	0
	−1	−2	−3	−4	−5	−6	
13 ♧	1	2	3	4	5	6	0
	−1	−2	−3	−4	−5	−6	
13 ◇	1	2	3	4	5	6	0
	−1	−2	−3	−4	−5	−6	

Explain what you have learned about adding integers.

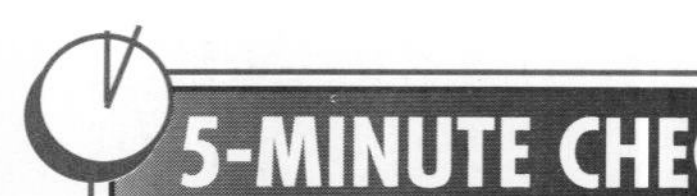

5-MINUTE CHECK-UP

(over Lesson 11-11)

Available on TRANSPARENCY 12-1.

1. Nancy mails a recipe to 4 friends. Each of them mails the recipe to 4 of their friends. What is the total number of recipes in the fourth mailing? **256 recipes**
2. Elena weighed 105 pounds on her 30th birthday. She began gaining $1\frac{1}{2}$ pounds every year. At this rate, how much will she weigh when she is 50 years old? **135 pounds**

INTRODUCING THE LESSON

Have students say a word and then have another student name the opposite of that word. For example, hot–cold, wet–dry, or sunny–cloudy.

TEACHING THE LESSON

Using Models Display a number line either on the chalkboard or on the overhead projector. Have students show the location of an integer, its opposite, and its absolute value.

Practice Masters Booklet, p. 126

Name ______________________ Date ______________

PRACTICE WORKSHEET 12-1

Integers

Replace ○ with <, >, or = to make a true sentence.

1. −12 ○ 15 **<**	2. 6 ○ 18 **<**	3. 13 ○ −21 **>**
4. 9 ○ 6 **>**	5. −2.3 ○ −2.2 **<**	6. −4 ○ −12 **>**
7. −7 ○ 7 **<**	8. −7 ○ 5 **<**	9. 9 ○ −2 **>**
10. |18| ○ |−14| **>**	11. |−13| ○ |−15| **<**	12. |−12| ○ |12| **=**
13. |−3| ○ |7| **<**	14. |0| ○ |−5| **<**	15. |−10| ○ |4| **>**
16. |8| ○ |−8| **=**	17. |−13| ○ |−6| **>**	18. |4| ○ |−11| **<**

Order from least to greatest.

19. 5, 8, −2, 0, −4 **−4, −2, 0, 5, 8**
20. −1, 6, 2, 5, −3 **−3, −1, 2, 5, 6**
21. 1, 0, −1, −8, −7, −3 **−8, −7, −3, −1, 0, 1**
22. 1, −6, 4, −4, 7, −1 **−6, −4, −1, 1, 4, 7**
23. 0, −7, 2, −9, 5, 3 **−9, −7, 0, 2, 3, 5**
24. 1, 4, −2, 8, −5, −9 **−9, −5, −2, 1, 4, 8**
25. 4, −3, 7, 1, 0, −2 **−3, −2, 0, 1, 4, 7**
26. 5, −2, −5, 0, 2, 1 **−5, −2, 0, 1, 2, 5**
27. −7, 2, −1, −9, 3 **−9, −7, −1, 2, 3**
28. 1, −3, −5, 4, −2, 6 **−5, −3, −2, 1, 4, 6**

Complete.

29. If −6 indicates a drop of 6°C in temperature, what does 6 indicate? **a rise of 6°C**
30. If 31 indicates 31 seconds after a rocket launch, what does −31 indicate? **31 s before launch**

12-1 INTEGERS AND THE NUMBER LINE

Objective
Identify, order, and compare integers.

Fran Cantu was a contestant on a TV quiz show. Her final dollar score was \$1,050. Her opponent's final dollar score was −\$1,050. Fran was declared the winner. In algebra, pairs of numbers like 1,050 and −1,050 are called **opposites.**

Every number has an opposite. On the number line a number and its opposite are the same distance from zero. −3 and 3 are opposites.
The opposite of 0 is 0.

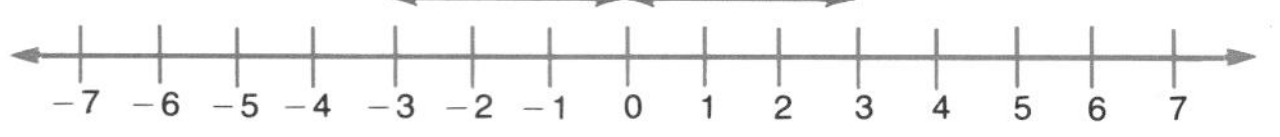

Zero is neither positive nor negative.

The whole numbers and their opposites are called **integers.**

The number of units a number is from zero on the number line is called its **absolute value.** The absolute value of −3 is 3. It is written $|-3| = 3$. All absolute values are positive or zero.

You can compare two integers using the number line.

Method

1. Draw or imagine a number line.
2. The greater number is farther to the right on a horizontal number line. On a vertical number line, the greater number is above the lesser number.

Examples

A ***Compare −4 and 3.***

1.
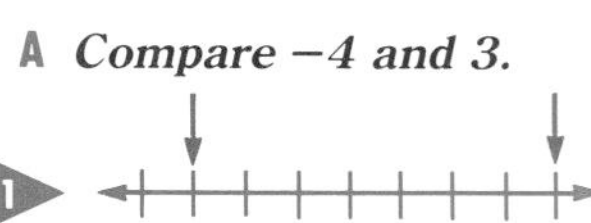

2. −4 is to the left of 3. So $-4 < 3$.

B ***Compare −2 and −7.***

1.

2. $-2 > -7$

C ***Compare 0 and −3.***

1.
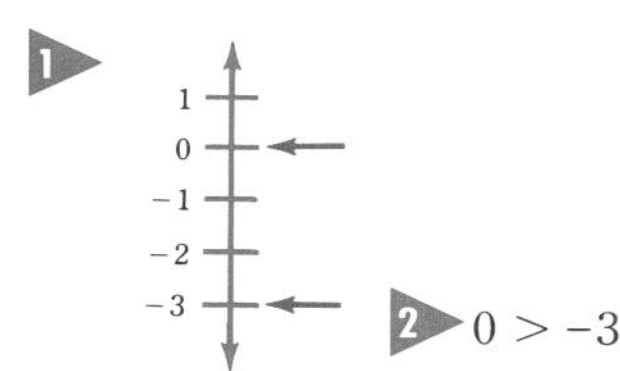

2. $0 > -3$

D ***Order −6, 6, −4, 4, 0 from least to greatest.***

1.
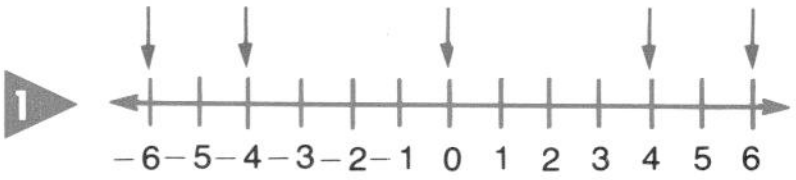

2. The order from least to greatest is −6, −4, 0, 4, 6.

RETEACHING THE LESSON

Use a number line to order the numbers from least to greatest.

1. −4, 6, −2 **−4, −2, 6**
2. 14, −10, −8, 6 **−10, −8, 6, 14**
3. −6, 1, −4, 2 **−6, −4, 1, 2**
4. −5, −7, 0, −1 **−7, −5, −1, 0**
5. −4.621, −4.126, −4.612 **−4.621, −4.612, −4.126**

Reteaching Masters Booklet, p. 115

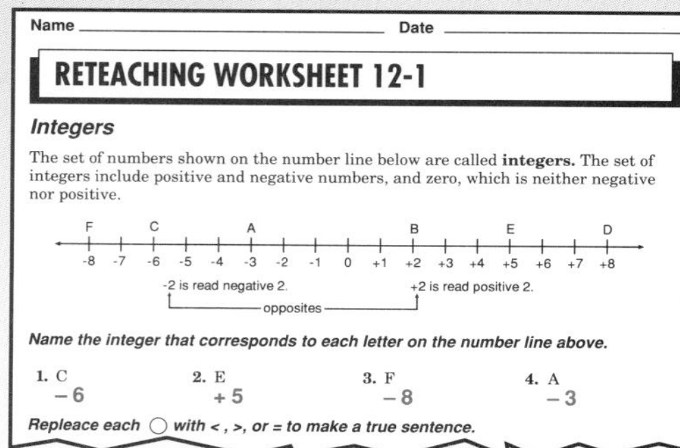

Name ______________________ Date ______________

RETEACHING WORKSHEET 12-1

Integers

The set of numbers shown on the number line below are called **integers.** The set of integers include positive and negative numbers, and zero, which is neither negative nor positive.

Name the integer that corresponds to each letter on the number line above.

1. C **−6**
2. E **+5**
3. F **−8**
4. A **−3**

Repleace each ○ with < , >, or = to make a true sentence.

Guided Practice

Examples A, B, C

Replace each ● with <, >, or = to make a true sentence.

1. 3 ● 7 <   **2.** 5 ● 0 >   **3.** −2 ● 2 <   **4.** −5 ● 0 <
5. −4 ● −6 >   **6.** −3 ● 2 <   **7.** −1 ● −3 >   **8.** −14 ● 2 <

Example D

Order from least to greatest.

9. 4, −3, 0, −9, 5 **−9, −3, 0, 4, 5**   **10.** 6, −6, 2, −8, 0 **−8, −6, 0, 2, 6**
11. −1, 0, 11, −7, 6 **−7, −1, 0, 6, 11**   **12.** −5, 0, 5, 3, −3 **−5, −3, 0, 3, 5**

Exercises

Practice

Replace each ● with <, >, or = to make a true sentence.

13. −1 ● 0 <   **14.** 0 ● −4 >   **15.** −3 ● −4 >   **16.** −7 ● −1 <
17. 5 ● −2 >   **18.** 3 ● 8 <   **19.** −2 ● 3 <   **20.** 3 ● 4 <
21. −10 ● −15 >   **22.** −1,000 ● −1   **23.** $-\frac{1}{2}$ ● $-\frac{1}{4}$ <   **24.** −2 ● −2.1 >

22. <

25. |−4| ● |12| <   **26.** |−7| ● |15| <   **27.** |24| ● |−17| >   **28.** |2| ● |−8| <
29. |0| ● |−9| <   **30.** |10| ● |−10| =   **31.** |15| ● |−4| >   **32.** |−3| ● |5| <

Order from least to greatest.

33. 8, 3, −4, 9, 0 **−4, 0, 3, 8, 9**   **34.** −7, 5, 0, −2, 2 **−7, −2, 0, 2, 5**
35. −2, 0, −4, −1, 1 **−4, −2, −1, 0, 1**   **36.** 0, −11, −20, 10, −15
37. 3.5, −3, 0, −3.5, 3   **38.** $-\frac{1}{2}, -\frac{1}{4}, \frac{1}{4}, \frac{1}{2}, 0$ $-\frac{1}{2}, -\frac{1}{4}, 0, \frac{1}{4}, \frac{1}{2}$

36. **−20, −15, −11, 0, 10**

37. **−3.5, −3, 0, 3, 3.5**

Number Sense

39. If +10 indicates 10 seconds after takeoff, what integer indicates 10 seconds before takeoff? What integer represents the takeoff time? **−10; 0**

Applications

40. **−63°C**

Solve. Use the table at the right for Problems 40–42.

40. What is the lowest recorded temperature in North America?

41. Where is the world's lowest recorded temperature? **Antarctica**

42. Order the lowest recorded temperatures of the countries and continents from 1 to 8, with 1 being the lowest. **See chart.**

43. The average daily low temperature for January in Great Falls, Montana, is about −5°F. In Chicago, it is about 2°F. Which city has the colder average?

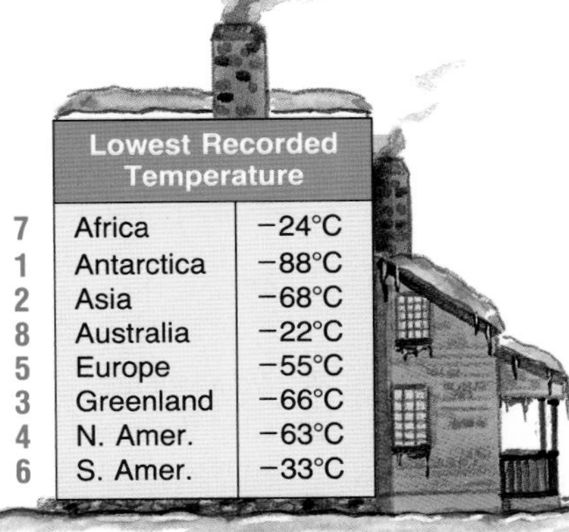

	Lowest Recorded Temperature	
7	Africa	−24°C
1	Antarctica	−88°C
2	Asia	−68°C
8	Australia	−22°C
5	Europe	−55°C
3	Greenland	−66°C
4	N. Amer.	−63°C
6	S. Amer.	−33°C

Critical Thinking

44. At a party, everyone shook hands with everyone else exactly once. There were 36 handshakes. Find the number of people at the party. **9 people**

Research

45. Look through magazines, newspapers, and journals to make a list of the different ways integers are used in our lives. **Answers may vary.**

APPLYING THE LESSON

Independent Practice

Homework Assignment	
Minimum	Ex. 1-43 odd; 44-45
Average	Ex. 2-38 even; 39-45
Maximum	Ex. 1-45

Chalkboard Examples

Example A, B, C *Replace each ● with <, >, or = to make a true sentence.*

1. −5 ● 3 <   **2.** −3 ● −8 >
3. 0 ● 4 <

Example D *Order from least to greatest.*

4. −8, 6, 3, 0, −5 **−8, −5, 0, 3, 6**   **5.** 0, 4, 7, −2, −5 **−5, −2, 0, 4, 7**

Additional examples are provided on TRANSPARENCY 12-1.

EVALUATING THE LESSON

Check for Understanding Use Guided Practice Exercises to check for student understanding.

Closing the Lesson Have students tell about times when they could use integers and their absolute values.

EXTENDING THE LESSON

Enrichment Have students complete the following statements.

1. If $n > 0$, then $|n| = \underline{?}$. **n**
2. If $n > 0$, then $|-n| = \underline{?}$. **n**
3. If $n < 0$, then $|n| = \underline{?}$. **−n**
4. If $n < 0$, then $|-n| = \underline{?}$. **−n**

Enrichment Masters Booklet, p. 115

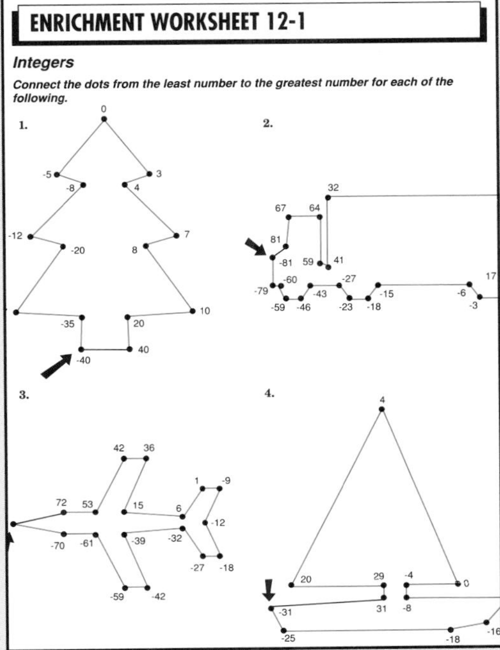

Name ______ Date ______

ENRICHMENT WORKSHEET 12-1

Integers

Connect the dots from the least number to the greatest number for each of the following.

5-MINUTE CHECK-UP

(over Lesson 12-1)

Available on TRANSPARENCY 12-2.

Replace each ● with <, >, or = to make a true sentence.

1. 5 ● 9 <   **2.** −2 ● −5 >

3. 0 ● −6 >

Order from least to greatest.

4. −4, −7, 0, 3, 6, 5 **−7, −4, 0, 3, 5, 6**

5. −2, 0, −5, 1, 4 **−5, −2, 0, 1, 4**

Extra Practice, Lesson 12-1, p. 480

INTRODUCING THE LESSON

Have students demonstrate how to add integers using positive and negative counters on the overhead or chalkboard.

1. −6 + (−3) **−9**   **2.** 7 + (−9) **−2**

TEACHING THE LESSON

Using Cooperative Groups Divide students into groups of two. Have them work together with a number line. One student gives two addends similar to those in exercises 9-28. The other student uses the number line to find the sum.

Practice Masters Booklet, p. 127

Name ______________________ Date ______________

PRACTICE WORKSHEET 12-2

Adding Integers

Write an equation for each number line. Then solve.

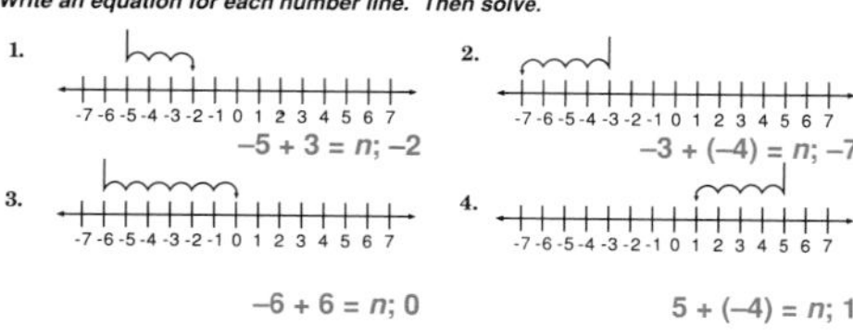

1. −5 + 3 = *n*; −2
2. −3 + (−4) = *n*; −7
3. −6 + 6 = *n*; 0
4. 5 + (−4) = *n*; 1

Add. Use a number line if necessary.

5. 21 + (−8)	13	**6.** −6 + 4	−2	**7.** 13 + (−19)	−6
8. −2 + 6	4	**9.** 41 + 3	44	**10.** −7 + (−5)	−12
11. −12 + (−15)	−27	**12.** 3 + (−3)	0	**13.** −1.1 + (−2)	−3.1
14. 4.3 + 1.3	5.6	**15.** −5.2 + 3.4	−1.8	**16.** 9 + (−7)	2
17. −7 + 11	4	**18.** −4.2 + (−8.5)	−12.7	**19.** 3 + 5	8
20. −3 + 15	12	**21.** $\frac{4}{5} + \left(-\frac{3}{10}\right)$	$\frac{1}{2}$	**22.** 1 + (−7)	−6
23. $-\frac{1}{2} + \left(-\frac{3}{4}\right)$	$-1\frac{1}{4}$	**24.** $-\frac{2}{3} + \frac{5}{12}$	$-\frac{1}{4}$	**25.** 6 + (−8)	−2

Solve.

26. The balance in Sally's checking account was $273.25. She wrote a check for $72.15 and made a deposit of $55.20. What is her new balance? **$256.30**

27. The plane was flying at 7,800 feet when it had to climb 2,500 feet to clear a mountain range. What was its new altitude? **10,300 ft**

28. Mark had business on the twenty-seventh floor. Later he went up 6 floors to the rooftop observatory. On what floor is the observatory? **thirty-third floor**

29. Starting on their own 25-yard line, the Packers gain 8 yards, lose 5 yards, and gain 7 yards. What yard line are they on now? **35-yard line**

12-2 ADDING INTEGERS

Objective
Add integers.

In the spring after a period of heavy rain, the water level in Smyth Reservoir was 3 feet above its normal level of 0. In the summer, after a drought, the water level was 7 feet below the level of the heavy rains. What was the water level after the drought?

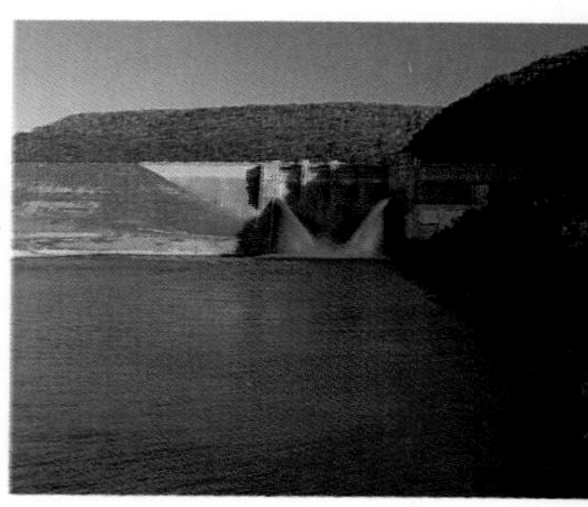

The number line can be used to show addition of integers. Move up when adding a positive number. Move down when adding a negative number.

Example A ***Find 3 + (− 7) using the number line.***

Start at 3. Then move down 7 units. The sum is −4.

3 + (−7) = −4

The water level in Smyth Reservoir is −4 or 4 feet below normal after the drought.

5, 4, 3, 2, 1, 0, −1, −2, −3, −4, −5 ← normal water level

Usually it is not convenient to use a number line. Use the following rules for adding integers.

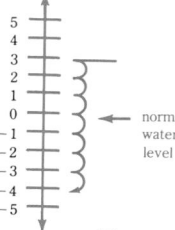

These rules also apply to positive and negative decimals and fractions.

Method

1. To add integers with the *same sign,* add their absolute values.
2. Give the result the same sign as the integers.

Examples

B ***5 + 23***

Both integers are positive.

1. |5| + |23| = 5 + 23
   = 28
2. Give the result a positive sign.
   5 + 23 = 28

C ***−5 + (−23)***

Both integers are negative.

1. |−5| + |−23| = 5 + 23
   = 28
2. Give the result a negative sign.
   −5 + (−23) = −28

Method

1. To add integers with *different signs,* subtract their absolute values.
2. Give the result the same sign as the integer with the greater absolute value.

Point out that students must determine the greater absolute value before subtracting.

Examples

D ***2 + (− 7)***

One integer is positive and one is negative.

1. |−7| − |2| = 7 − 2
   = 5
2. Since |−7| > |2|, the result is negative.
   2 + (−7) = −5

E ***13 + (− 6)***

One integer is positive and one is negative.

1. |13| − |−6| = 13 − 6
   = 7
2. Since |13| > |−6|, the result is positive.
   13 + (−6) = 7

RETEACHING THE LESSON

Add. Use a number line.

1. 5 + 7 = **12**   **2.** −5 + 7 = **2**

3. 5 + −7 = **−2**   **4.** −7 + −5 = **−12**

5. −2 + 2 = **0**   **6.** 0 + −6 = **−6**

7. −8 + −8 = **−16**

8. −9 + 6 = **−3**

Reteaching Masters Booklet, p. 116

Name ______________________ Date ______________

RETEACHING WORKSHEET 12-2

Adding Integers

A number line can be used to add integers. Move to the **right** when the sign is **positive.** Move to the **left** when the sign is **negative.**

Find −2 + (−5) using the number line.   Find 6 + (−8) using the number line.

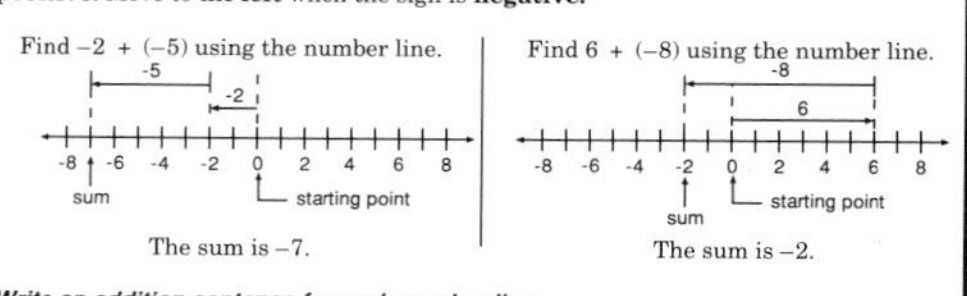

The sum is −7.   The sum is −2.

Write an addition sentence for each number line.

1.   2.

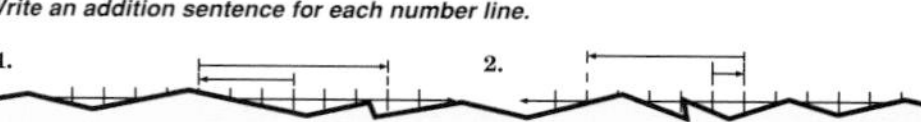

Guided Practice

Example A

Write an addition sentence for each number line.

1.

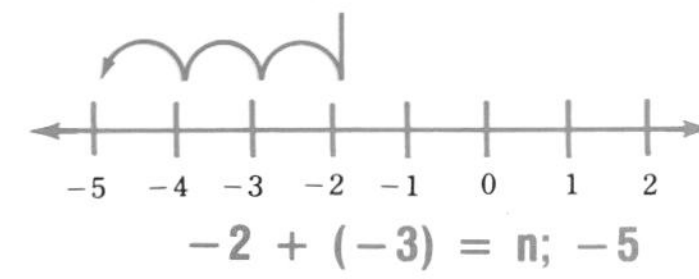

$-2 + (-3) = n; -5$

2.

−3 −2 −1 0 1 2 3 4

$-3 + 6 = n; -3$

Add. Show your answer on a number line. See answers on p. T528.

Examples B, C

3. $5 + 3$ 8 **4.** $-4 + (-3)$ −7 **5.** $-2 + (-4)$ −6

Examples D, E

6. $2 + (-4)$ −2 **7.** $8 + (-6)$ 2 **8.** $9 + (-10)$ −1

Exercises

11. −15 12. −13 15. −4 16. −16 19. −33 20. −20

Add. Use a number line if necessary. 23. −38 24. −53 25. −0.8

Practice

9. $-8 + 7$ −1 **10.** $12 + 13$ 25 **11.** $-8 + (-7)$ **12.** $-7 + (-6)$

13. $8 + (-8)$ 0 **14.** $-2 + 2$ 0 **15.** $-2 + (-2)$ **16.** $-8 + (-8)$

17. $24 + (-13)$ 11 **18.** $-17 + 12$ −5 **19.** $-14 + (-19)$ **20.** $-9 + (-11)$

21. $-3 + 16$ 13 **22.** $-28 + 50$ 22 **23.** $13 + (-51)$ **24.** $-74 + 21$

Number Sense

25. $-4.5 + 3.7$ **26.** $-5.2 + (-6)$ −11.2 **27.** $-17 + (-31)$ −48 **28.** $-\frac{4}{3} + \frac{1}{2}$ $-\frac{5}{6}$

Write Math

29. Write a rule for adding two integers with different signs. See margin.

Applications

30. In a football game, the Denver Broncos lost 6 yards on one play and then lost 3 yards on the next. What was the net gain? −9 yards

31. The temperature at dawn was −12°C. By noon the temperature was 18° higher. What was the temperature at noon? 6°C

32. A submarine at 1,150 feet below sea level descends an additional 1,250 feet. How far below sea level is the submarine now? −2,400 feet

FUN with MATH

What is unusual about the day and the year on Venus? See page 422.

33. At night the average surface temperature on the planet Saturn is −150°C. During the day the temperature rises 27°C. What is the temperature on the planet's surface during the day? −123°C

34. A weather balloon rises 300 feet from the ground, drops 125 feet, and then rises 450 feet. Write an equation. Then find the height of the balloon. $300 + (-125) + 450 = n$; 625 feet

Critical Thinking

35. Find the value of each letter.

$$\begin{array}{r} \text{ODD} \\ +\text{ODD} \\ \hline \text{EVEN} \end{array}$$

Two possible solutions: O = 6, D = 5, E = 1, V = 3, N = 0; O = 8, D = 5, E = 1, V = 7, N = 0

APPLYING THE LESSON

Independent Practice

Homework Assignment	
Minimum	Ex. 2-28 even; 29-35 odd
Average	Ex. 1-29 odd; 30-35
Maximum	Ex. 1-35

Additional Answers

29. To add integers with different signs, find the difference of their absolute values. Give the result the same sign as the integer with the greater absolute value.

Chalkboard Examples

Example A *Write an addition sentence for the number line. Then solve.*

1.

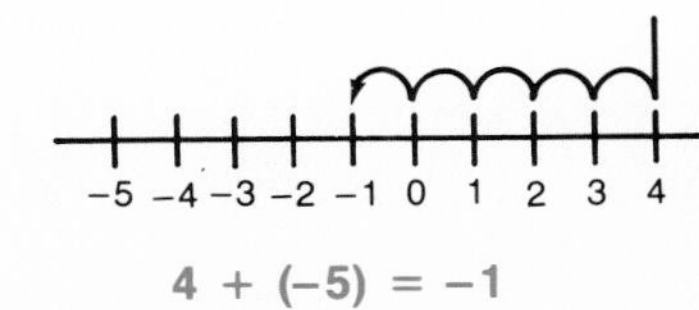

$4 + (-5) = -1$

Examples B, C *Add. Show your answer on a number line.*

2. $4 + 5$ 9 **3.** $-6 + (-4)$ −10

Examples D, E *Add. Show your answer on a number line.*

4. $3 + (-5)$ −2 **5.** $9 + (-3)$ 6

Additional examples are provided on TRANSPARENCY 12-2.

EVALUATING THE LESSON

Check for Understanding Use Guided Practice Exercises to check for student understanding.

Closing the Lesson Ask students to state in their own words how to add integers.

EXTENDING THE LESSON

Enrichment Masters Booklet, p. 116

ENRICHMENT WORKSHEET 12-2

Addition of Matrices

A rectangular array of numbers (elements) arranged in rows and columns is called a **matrix.** A matrix is usually enclosed by brackets.

columns

rows $\begin{bmatrix} 2 & 4 \\ -5 & 3 \end{bmatrix}$ 2 × 2 matrix

$\begin{bmatrix} 3 & -1 & 4 & 2 \\ 1 & 0 & 5 & -6 \end{bmatrix}$ 2 × 4 matrix

$\begin{bmatrix} 3\frac{1}{2} & -2 \\ -1 & 0 \\ 2 & \frac{1}{2} \end{bmatrix}$ 3 × 2 matrix

The *sum* of two $m \times n$ matrices is an $m \times n$ matrix in which each element is the sum of the corresponding elements of the given matrices.

$$\begin{bmatrix} 1 & 3 & -2 \\ -4 & 0 & 5 \end{bmatrix} + \begin{bmatrix} 2 & -1 & -2 \\ 3 & 4 & -3 \end{bmatrix} = \begin{bmatrix} 1+2 & 3+(-1) & -2+(-2) \\ -4+3 & 0+4 & 5+(-3) \end{bmatrix} = \begin{bmatrix} 3 & 2 & -4 \\ -1 & 4 & 2 \end{bmatrix}$$

Write the sum of the matrices.

1. $\begin{bmatrix} 4 & 5 \\ 3 & 8 \end{bmatrix} + \begin{bmatrix} 2 & 0 \\ 1 & 6 \end{bmatrix}$ $\begin{bmatrix} 6 & 5 \\ 4 & 14 \end{bmatrix}$

2. $\begin{bmatrix} -1 & 3 \\ 5 & -4 \end{bmatrix} + \begin{bmatrix} -2 & 2 \\ -3 & 4 \end{bmatrix}$ $\begin{bmatrix} -3 & 5 \\ 2 & 0 \end{bmatrix}$

3. $\begin{bmatrix} 6 & 3 & -7 \\ 4 & 0 & -2 \end{bmatrix} + \begin{bmatrix} -6 & 2 & -1 \\ -2 & 3 & 3 \end{bmatrix}$ $\begin{bmatrix} 0 & 5 & -8 \\ 2 & 3 & 1 \end{bmatrix}$

4. $\begin{bmatrix} 6 & 5 \\ -7 & 4 \\ 8 & -3 \end{bmatrix} + \begin{bmatrix} -6 & -5 \\ 7 & -4 \\ -8 & 3 \end{bmatrix}$ $\begin{bmatrix} 0 & 0 \\ 0 & 0 \\ 0 & 0 \end{bmatrix}$

5. $\begin{bmatrix} -1 & 4 \\ 0 & 7 \end{bmatrix} + \begin{bmatrix} 8 & -1 \\ -2 & -5 \end{bmatrix} + \begin{bmatrix} -6 & 3 \\ 9 & -2 \end{bmatrix}$ $\begin{bmatrix} 1 & 6 \\ 7 & 0 \end{bmatrix}$

Write the missing elements.

6. $\begin{bmatrix} 6 & 3 \\ 0 & -4 \end{bmatrix} + \begin{bmatrix} 2 & -2 \\ -3 & -1 \end{bmatrix} = \begin{bmatrix} 8 & 1 \\ -3 & -5 \end{bmatrix}$

7. $\begin{bmatrix} 13 & 6 & -10 \\ 2 & 15 & -8 \end{bmatrix} + \begin{bmatrix} -4 & 0 & 8 \\ 1 & -7 & 3 \end{bmatrix} = \begin{bmatrix} 9 & 6 & -2 \\ 3 & 8 & -5 \end{bmatrix}$

5-MINUTE CHECK-UP

(over Lesson 12-2)

Available on TRANSPARENCY 12-3.

Add.

1. 8 + 9 **17** **2.** −7 + (−6) **−13**

3. −4 + 8 **4** **4.** 6 + (−9) **−3**

Extra Practice, Lesson 12-2, p. 480

INTRODUCING THE LESSON

Have students write each subtraction sentence as an addition sentence.

1. 5 − (−2) = 7
5 + 2 = 7

2. −8 − 4 = −12
−8 + (−4) = −12

3. 4 − 6 = −2
4 + (−6) = −2

4. −3 − (−4) = 1
−3 + 4 = 1

TEACHING THE LESSON

Using Models Have students draw a number line that ranges from −10 to 10. Then use the number line to find the distance in units between each pair of numbers.

1. 3, −5 **8** **2.** −3, −5 **2** **3.** 0, −3 **3**

4. −3, 5 **8** **5.** 3, 5 **2** **6.** 0, 5 **5**

Practice Masters Booklet, p. 128

Name ______________________ Date ______________

PRACTICE WORKSHEET 12-3

Subtracting Integers

Subtract.

1.	9 − (−2)	11	**2.**	−12 − (−5)	−7	**3.**	−3 − 16	−19
4.	2 − (−8)	10	**5.**	15 − 6	9	**6.**	−14 − 3	−17
7.	−11 − (−7)	−4	**8.**	1 − 13	−12	**9.**	9 − 2	7
10.	−25 − (−16)	−9	**11.**	32 − (−7)	39	**12.**	−3 − 21	−24
13.	72 − (−8)	80	**14.**	61 − (−55)	116	**15.**	1.9 − (−2.1)	4
16.	−3 − (−5)	2	**17.**	−1 − 4	−5	**18.**	−1.6 − 1.6	−3.2
19.	2 − 8	−6	**20.**	0.3 − (−1.9)	2.2	**21.**	4 − (−1)	5
22.	5 − (−5)	10	**23.**	$\frac{1}{6} - \frac{3}{4}$	$-\frac{7}{12}$	**24.**	$-\frac{5}{6} - \frac{1}{3}$	$-1\frac{1}{6}$
25.	$\frac{7}{8} - \left(-\frac{1}{4}\right)$	$1\frac{1}{8}$	**26.**	−10 − 5	−15			

Solve.

27. Rocky's Flower Shop deposited checks and cash in the amount of $2,452.60 on Friday. On Monday the bank called to say that one of the checks had been returned. If that check was for $45.87, what was the actual deposit on Friday? **$2,406.73**

28. When Lucy received her checking account statement, it showed a balance of $515.63. Lucy's check register showed a deposit of $253.62 and two checks for $21.75 and $37.89 which were not included in the statement. What is Lucy's actual account balance? **$709.61**

29. Bob wrote a check for $15.75 to pay for a ticket to the symphony. The check was returned because the tickets had all been sold. Bob's check register showed a balance of $372.15. What is his new balance? **$387.90**

30. Jenny's balance was $349.95. She deposited three checks, $45.00, $125.63, and $38.25. She also withdrew $50.00 in cash. What is her new balance after these transactions? **$508.83**

12-3 SUBTRACTING INTEGERS

Objective
Subtract integers.

Miami is 12 feet above sea level. New Orleans is 5 feet below sea level. To find the difference in their elevations subtract 12 − (−5).

Remember that every subtraction expression can be written as an addition expression. So, 12 − (−5) can be written as 12 + 5.

12 + 5 = 17 12 − (−5) = |12| + |−5|
= 12 + 5
= 17

Adding an integer gives the same result as subtracting its opposite.

The integers have different signs. Add their absolute values. Since |12| > |−5|, the sum is positive.

What number represents sea level? **0**

The difference in elevation is 17 feet.

Method

1. Write the subtraction expression as an addition expression.
2. Add.

These rules also apply to positive and negative decimals and fractions.

Examples

A ***12 − 8***

To subtract 8, add −8

1 12 − 8 = 12 + (−8)
2 = 4

12 − 8 = 4

B ***−13 − (−4)***

To subtract −4, add 4.

1 −13 − (−4) = −13 + 4
2 = −9

−13 − (−4) = −9

C ***−18 − 5***

1 −18 − 5 = −18 + (−5)
2 = −23

The difference is −23.

Check. $-23 + 5 \stackrel{?}{=} -18$
$-18 = -18$ ✓

D ***9 − 14***

1 9 − 14 = 9 + (−14)
2 = −5

So, 9 − 14 = −5.

Check. $-5 + 14 \stackrel{?}{=} 9$
$9 = 9$ ✓

Guided Practice

Subtract.

Example A	**1.** 12 − (−5) **17**	**2.** 2 − (−2) **4**	**3.** 14 − (−6) **20**	**4.** −3 − (−7) **4**
Example B	**5.** −9 − (−4) **−5**	**6.** −6 − (−5) **−1**	**7.** −12 − (−7) **−5**	**8.** −5 − (−6) **1**
Example C	**9.** −12 − 3 **−15**	**10.** −8 − 8 **−16**	**11.** −9 − 3 **−12**	**12.** −5 − 3 **−8**
Example D	**13.** 15 − 15 **0**	**14.** 18 − 28 **−10**	**15.** 19 − 11 **8**	**16.** 9 − 10 **−1**

RETEACHING THE LESSON

Write an addition expression for each subtraction problem.

1. 4 − 11
4 + (−11)

2. 8 − (−5)
8 + 5

3. −7 − (−7)
−7 + 7

4. −8 − 2
−8 + (−2)

5. 24 − (−24)
24 + 24

6. 32 − 23
32 + (−23)

7. −3.7 − (−6.2)
−3.7 + 6.2

8. −5.38 − 6.92
−5.38 + (−6.92)

Reteaching Masters Booklet, p. 117

Name ______________________ Date ______________

RETEACHING WORKSHEET 12-3

Subtracting Integers

To subtract an integer, add its opposite.

Subtract. 8 − 11 8 − 11 = 8 + (−11) = −3	Subtract. 6 − 2 6 − 2 = 6 + (−2) = 4	Subtract. −11 − (−4) −11 − (−4) = −11 + 4 = −7

Name the opposite of each integer.

1. −14 **14** **2.** 33 **−33** **3.** −1 **1** **4.** 0 **0** **5.** 1,187 **−1,187**

Write a related addition expression for each subtraction expression.

6. 6 − 1 **6 + (−1)** **7.** −6 − 1 **−6 + (−1)** **8.** −6 − (−1) **−6 + 1** **9.** 6 − (−1) **6 + 1**

Exercises

Practice

Subtract.

17. $-5 - 4$ **−9**　**18.** $-6 - (-8)$ **2**　**19.** $7 - 13$ **−6**　**20.** $-4 - 9$ **−13**

21. $8 - 14$ **−6**　**22.** $9 - (-2)$ **11**　**23.** $-8 - 5$ **−13**　**24.** $11 - 7$ **4**

25. $15 - 19$ **−4**　**26.** $-1 - 15$ **−16**　**27.** $-3 - (-7)$ **4**　**28.** $2 - 9$ **−7**

29. $-17 - (-17)$ **0**　**30.** $-23 - (-23)$ **0**　**31.** $-6 - 6$ **−12**　**32.** $0 - 11$ **−11**

Number Sense

33. $-57.3 - 27$　**34.** $-5.3 - 8$　**35.** $4.2 - 6.7$　**36.** $-\frac{1}{2} - \frac{3}{4}$

37. Find the sum of 8 and −14. **−6**

38. Find the difference of 59 − 74. **−15**

33. −84.3　34. −13.3　35. −2.5　36. $-1\frac{1}{4}$

Applications

Paper/ Pencil
Mental Math
Estimate
Calculator

39. At 8:30 P.M. the temperature was 5°C. What was the temperature at midnight if the temperature had dropped 12°? **−7°C**

40. On Monday the price of a share of stock is quoted as $28\frac{3}{8}$ points. It falls $\frac{1}{2}$ point on Tuesday and $\frac{1}{8}$ point on Wednesday. What is the new price? **$27\frac{3}{4}$ points**

41. 14,776 feet

41. In California, Mt. Whitney is the highest point and Death Valley is the lowest point. Their elevations are 14,494 feet and −282 feet respectively. Find the difference in their elevations.

42. Fred and Bob are both running backs for their high school football team. Fred ran for a total gain of 67 yards while Bob ran for a total of −14 yards. What is the difference in their yards run? **81 yards**

JOURNAL ENTRY

43. Tell how you would explain adding and subtracting integers to a friend. **See students' work.**

Using Expressions

Find the value of each expression if $n = -4$ and $p = -2$.

44. $n - 2$ **−6**　**45.** $n - (-2)$ **−2**　**46.** $p - 0$ **−2**　**47.** $0 - p$ **2**

Calculator

Use the steps below to enter −14 on a calculator. Note that the change-sign key is pressed *after* the absolute value of the integer is entered.

Enter 14.　Then, press the change-sign key, [+/−]. −14

To add −14 + 23, enter:　14 [+/−] [+] 23 [=] 9

To subtract −14 − 23, enter:　14 [+/−] [−] 23 [=] −37

Use a calculator to find each sum or difference. **50. −37　51. −7**

48. $4 + (-17)$ **−13**　**49.** $-8 + 2$ **−6**　**50.** $-21 + (-16)$　**51.** $-44 + 37$

52. $3 - 10$ **−7**　**53.** $6 - (-22)$ **28**　**54.** $-57 - 60$　**55.** $-67 - (-96)$

55. 29　**−117**

APPLYING THE LESSON

Independent Practice

Homework Assignment	
Minimum	Ex. 1-55 odd
Average	Ex. 2-36 even; 37-55
Maximum	Ex. 1-55

Chalkboard Examples

Example A *Subtract.*

1. $8 - (-4)$ **12**　**2.** $3 - (-7)$ **10**

Example B *Subtract.*

3. $-6 - (-3)$ **−3**　**4.** $-2 - (-5)$ **3**

Example C *Subtract.*

5. $-9 - 4$ **−13**　**6.** $-1 - 5$ **−6**

Example D *Subtract.*

7. $7 - 4$ **3**　**8.** $5 - 8$ **−3**

Additional examples are provided on TRANSPARENCY 12-2.

EVALUATING THE LESSON

Check for Understanding Use Guided Practice Exercises to check for student understanding.

Closing the Lesson Have students state in their own words how to subtract integers.

EXTENDING THE LESSON

Enrichment Have each student write four exercises like exercises 44-47 and solve them. Students exchange exercises and evaluate each other's expressions and answers.

Enrichment Masters Booklet, p. 117

Name ______________________ Date ______________

ENRICHMENT WORKSHEET 12-3

Subtraction of Matrices

The *difference* of two $m \times n$ matrices is an $m \times n$ matrix in which each element is the difference of the corresponding elements of the given matrices.

$$\begin{bmatrix}2 & 4\\1 & -3\end{bmatrix} - \begin{bmatrix}5 & 0\\-4 & 7\end{bmatrix} = \begin{bmatrix}2-5 & 4-0\\1-(-4) & -3-7\end{bmatrix} = \begin{bmatrix}-3 & 4\\5 & -10\end{bmatrix}$$

Write the difference of the matrices.

1. $\begin{bmatrix}4 & -7\\-9 & 3\end{bmatrix} - \begin{bmatrix}3 & 8\\-2 & 4\end{bmatrix}$　$\begin{bmatrix}1 & -15\\-7 & -1\end{bmatrix}$

2. $\begin{bmatrix}1 & 5 & 7\\5 & 2 & -6\\3 & 0 & -2\end{bmatrix} - \begin{bmatrix}-3 & 6 & -9\\4 & -3 & 0\\8 & -2 & 3\end{bmatrix}$　$\begin{bmatrix}4 & -1 & 16\\1 & 5 & -6\\-5 & 2 & -5\end{bmatrix}$

3. $\begin{bmatrix}3 & 8\\-2 & 4\end{bmatrix} - \begin{bmatrix}-1 & -5\\2 & -8\end{bmatrix}$　$\begin{bmatrix}4 & 13\\-4 & 12\end{bmatrix}$　**4.** $\begin{bmatrix}-2 & 1\\0 & -3\end{bmatrix} - \begin{bmatrix}8 & -3\\4 & -5\end{bmatrix}$　$\begin{bmatrix}-10 & 4\\-4 & 2\end{bmatrix}$

5. $\begin{bmatrix}6 & 5\\8 & 4\end{bmatrix} - \begin{bmatrix}-3 & -2\\-5 & -9\end{bmatrix}$　$\begin{bmatrix}9 & 7\\13 & 13\end{bmatrix}$

6. $\begin{bmatrix}3 & 4 & -7\\-2 & 0 & 4\end{bmatrix} - \begin{bmatrix}-2 & 7 & 3\\5 & -9 & 1\end{bmatrix}$　$\begin{bmatrix}5 & -3 & -10\\-7 & 9 & 3\end{bmatrix}$

Write the missing elements.

7. $\begin{bmatrix}6 & 9 & -4\\-11 & 13 & -8\\20 & 4 & -2\end{bmatrix} - \begin{bmatrix}4 & 6 & -8\\-14 & 14 & -12\\14 & -3 & -2\end{bmatrix} = \begin{bmatrix}2 & 3 & 4\\3 & -1 & 4\\6 & 7 & 0\end{bmatrix}$

8. $\begin{bmatrix}4 & -5 & 6\\-7 & 0 & 0\\8 & -3 & 1\end{bmatrix} - \begin{bmatrix}0 & -2 & 7\\-10 & 1 & 3\\3 & 0 & -6\end{bmatrix} = \begin{bmatrix}4 & -3 & -1\\3 & -1 & -3\\5 & -3 & 7\end{bmatrix}$

5-MINUTE CHECK-UP

(over Lesson 12-3)

Available on TRANSPARENCY 12-4.

Subtract.

1. $3 - 6$ **−3**
2. $-5 - (-8)$ **3**
3. $16 - (-15)$ **31**
4. $20 - 13$ **7**

Extra Practice, Lesson 12-3, p. 480

INTRODUCING THE LESSON

Have students describe the coldest day they can remember. Ask them what the temperature was that day and what temperature it felt like.

TEACHING THE LESSON

Using Discussion Have students tell what a thermometer measures. Also, have them tell what two measurements are needed to find the windchill factor. Point out that the windchill factor is not measured by an instrument, but is a mathematical calculation involving the actual temperature and the speed of the wind. A chart is developed to find the windchill factor.

Practice Masters Booklet, p. 129

Name ______________________ Date ______________

PRACTICE WORKSHEET 12-4

Applications: Windchill Factor

Windchill Chart

Wind speed in mph	Actual temperature (°Fahrenheit) 50	40	30	20	10	0	-10	-20	-30
	Equivalent temperature (°Fahrenheit)								
0	50	40	30	20	10	0	-10	-20	-30
5	48	37	27	16	6	-5	-15	-26	-36
10	40	28	16	4	-9	-21	-33	-46	-58
15	36	22	9	-5	-18	-36	-45	-58	-72
20	32	18	4	-10	-25	-39	-53	-67	-82
25	30	16	0	-15	-29	-44	-59	-74	-88
30	28	13	-2	-18	-33	-48	-63	-79	-94

Use the chart above to find each equivalent temperature.

1. 40°F, 5 mph **37°F**
2. –10°F, 20 mph **–53°F**
3. 0°F, 15 mph **–36°F**
4. 50°F, 10 mph **40°F**
5. –20°F, 30 mph **–79°F**
6. 20°F, 10 mph **4°F**
7. 40°F, 25 mph **16°F**
8. 30°F, 0 mph **30°F**
9. 10°F, 5 mph **6°F**
10. –20°F, 15 mph **–58°F**
11. 0°F, 30 mph **–48°F**
12. –30°F, 0 mph **–30°F**

What is the difference between each actual temperature and the equivalent temperature when the wind speed is 10 mph?

13. 40°F **12°F**
14. –20°F **26°F**
15. 0°F **21°F**
16. –30°F **28°F**
17. 10°F **19°F**
18. 20°F **16°F**

Solve.

19. The actual temperature is –20°F. The equivalent temperature with the windchill factor is –74°F. What is the wind speed? **25 mph**

20. The wind speed is 15 mph. The equivalent temperature with the windchill factor is 9°F. What is the actual temperature? **30°F**

Applications

12-4 WINDCHILL FACTOR

Objective

Use a windchill chart to determine equivalent temperatures.

Howie heard the weather forecaster on the radio say the low temperature in Chicago on January 11 was −10°F, but the wind made it feel like it was −45°F.

A thermometer measures the temperature of the air. However, if the wind is blowing, the temperature may feel much colder than the thermometer reading. The windchill factor depends on the actual temperature and the speed of the wind.

The chart below can be used to predict the equivalent temperature taking the windchill factor into account.

In Chicago on January 11, the low temperature was −10°F. The wind speed was 15 mph. The windchill factor made the temperature feel equivalent to −45°F with no wind.

Windchill Chart

Wind speed in mph	Actual temperature (°Fahrenheit) 50	40	30	20	10	0	−10	−20	−30
	Equivalent temperature (°Fahrenheit)								
0	50	40	30	20	10	0	−10	−20	−30
5	48	37	27	16	6	−5	−15	−26	−36
10	40	28	16	4	−9	−21	−33	−46	−58
15	36	22	9	−5	−18	−36	−45	−58	−72
20	32	18	4	−10	−25	−39	−53	−67	−82
25	30	16	0	−15	−29	−44	−59	−74	−88
30	28	13	−2	−18	−33	−48	−63	−79	−94

Understanding the Data

- Explain in your own words what windchill factor is.
- Explain what a temperature of −10°F and an equivalent temperature of −45°F with the windchill factor means. **The temperature is the same as −45°F with no wind.**
- If the actual temperature is −10°F and the equivalent temperature is −45°F with the windchill factor, what is the wind speed? **15 mph**

Guided Practice

Solve. Use the windchill chart above.

1. On February 10, the actual temperature in New York was 0°F. The wind speed was 20 mph. What was the equivalent temperature with the windchill factor? **−39°F**

2. In Pittsburgh on December 20, the wind speed was 30 mph. The actual temperature was 10°F. What was the equivalent temperature with the windchill factor? **−33°F**

 ● **The actual temperature and the speed of the wind may make the temperature feel much colder than it is.**

RETEACHING THE LESSON

Solve. Use the windchill chart on page 382.

1. Actual temperature is 30°F. Wind speed is 20 mph. Find the windchill factor. **4°F**

2. Actual temperature is −10°F. Wind speed is 10 mph. Find the windchill factor. **−33°F**

3. In problem 1, what is the difference between the actual temperature and the windchill factor? **26°**

Reteaching Masters Booklet, p. 118

Name ______________________ Date ______________

RETEACHING WORKSHEET 12-4

Applications: Windchill Factor

When the wind blows, the temperature may feel colder than the actual temperature indicated on a thermometer. This is known as the **windchill factor.**

Actual temperature: 20°F

Wind speed: 15 mph

Equivalent temperature is –5°F.

Windchill Chart

Wind speed in mph	Actual temperature (°Fahrenheit) 50	40	30	20	10	0	-10	-20	-30
	Equivalent temperature (°Fahrenheit)								
0	50	40	30	20	10	0	-10	-20	-30
5	48	37	27	16	6	-5	-15	-26	-36
10	40	28	16	4	-9	-21	-33	-46	-58
15	36	22	9	-5	-18	-36	-45	-58	-72
20	32	18	4	-10	-25	-39	-53	-67	-82
25	30	16	0	-15	-29	-44	-59	-74	-88
30	28	13	-2	-18	-33	-48	-63	-79	-94

Use the chart above to find each equivalent temperature.

Exercises

Use the windchill chart on page 382 to find each equivalent temperature.

Practice

3. 40°F, 5 mph **37°F** **4.** 10°F, 15 mph **−18°F** **5.** 0°F, 30 mph **−48°F**
6. −10°F, 10 mph **−33°F** **7.** −20°F, 25 mph **−74°F** **8.** 20°F, 20 mph **−10°F**
9. 20°F, 0 mph **20°F** **10.** −30°F, 30 mph **−94°F** **11.** −20°F, 30 mph **−79°F**

What is the difference between each actual temperature and the equivalent temperature when the wind speed is 20 mph?

12. 50°F **18°F** **13.** 30°F **26°F** **14.** 10°F **35°F** **15.** 0°F **39°F** **16.** −30°F **52°F**

Applications

17. In Detroit on December 15, the actual temperature was 20°F. The equivalent temperature with the windchill factor was −15°F. What is the wind speed? **25 mph**

18. In Buffalo on February 3, the wind speed was 25 mph. The temperature with the windchill factor was −44°F. What was the actual temperature? **0°F**

19. Suppose the actual temperature is −30°F and the wind speed is 20 mph. What is the difference between the actual temperature and the equivalent temperature? **52°F**

Interpreting Data
20. yes, 1960
22. $192.7 billion

Solve. Use the graph.

U.S.A. Government Budget Surplus/Deficit
(in billions)
(in 1990 dollars)
Fiscal Year
'40 '50 '60 '70 '80 '90
−$27.3 −$16.9 +$1.3 −$9.5 −$117.9 −$220

20. Did the government have a surplus in the budget? If so, in which year?

21. What is the difference in the budget surplus or deficit between 1960 and 1990? **$221.3 billion**

22. How much greater is the budget surplus or deficit in 1990 than in 1940?

23. Does the budget tend to have a surplus or a deficit? **deficit**

24. In which year was the deficit the least? the most? **1950; 1990**

Mixed Review

Lesson 2-2

Add.

25. 52.3 + 14.1 = **66.4** **26.** 91.47 + 23.59 = **115.06** **27.** 143.05 + 93.61 = **236.66** **28.** 45.003 + 184.297 = **229.3**

Lesson 7-6

Draw an example of each polygon. **See students' work.**

29. regular octagon **30.** not regular triangle

APPLYING THE LESSON

Independent Practice

Homework Assignment	
Minimum	Ex. 1-29 odd
Average	Ex. 2-16 even; 17-30
Maximum	Ex. 1-30

Chapter 12, Quiz A (Lessons 12-1 through 12-4) is available in the Evaluation Masters Booklet, p. 59.

Chalkboard Examples

Use the windchill chart on page 382 to find each equivalent temperature.

1. 30°F, 10 mph **16°F**
2. −20°F, 20 mph **−67°F**
3. If the temperature with the windchill factor is −10°F and the wind speed is 20 mph, what is the actual temperature? **20°F**

Additional examples are provided on TRANSPARENCY 12-4.

EVALUATING THE LESSON

Check for Understanding Use Guided Practice Exercises to check for student understanding.

Error Analysis Make sure students read the Windchill Chart correctly. Emphasize that wind speed is given vertically on the left side of the chart and actual temperature is given across the top.

Closing the Lesson Ask students to summarize what they have learned about the windchill factor.

EXTENDING THE LESSON

Enrichment Masters Booklet, p. 118

Name ______ Date ______

ENRICHMENT WORKSHEET 12-4

Rational Numbers and Absolute Value

Rational numbers are any numbers that can be written as a quotient of two integers, when the divisor is not 0.

0.25 $\frac{-3}{2}$ −7 $5\frac{1}{3}$ $0.16161\overline{6}$
$\frac{1}{4}$ $\frac{-7}{1}$ $\frac{16}{3}$ $\frac{16}{99}$

Opposites	
−4	4
$\frac{1}{2}$	$-\frac{1}{2}$
0	0

Absolute Value
$|3| = 3$
$|-5| = 5$
$|0| = 0$

Complete.

1. A rational number and its **opposite** are the same distance from 0 on the number line.
2. Every rational number has an **opposite**.
3. The number of units a number is from 0 is called its **absolute value**.
4. The symbol for absolute value is **| |**.

Write each number in the form of a quotient of two integers.

5. 5 $\frac{5}{1}$ 6. $2\frac{1}{4}$ $\frac{9}{4}$ 7. $0.33\overline{3}$ $\frac{1}{3}$ 8. −9 $-\frac{9}{1}$

Write the opposite of each number.

9. $\frac{1}{2}$ $-\frac{1}{2}$ 10. −6 **6** 11. $-\frac{7}{2}$ $\frac{7}{2}$ 12. 0.75 **−0.75**

Write the absolute value of each number.

13. $|6|$ **6** 14. $|-1|$ **1** 15. $|-10|$ **10** 16. $|\frac{1}{4}|$ $\frac{1}{4}$
17. $|0.7|$ **0.7** 18. $|\frac{10}{3}|$ $\frac{10}{3}$ 19. $|-0.35|$ **0.35** 20. $|\frac{-2}{5}|$ $\frac{2}{5}$

Applying Mathematics to the Real World

This optional page shows how mathematics is used in the real world and also provides a change of pace.

Objective Determine the cost of catered snacks.

Using Discussion Discuss other situations where a catering manager would use math, such as computing the cost of the menu offerings, salaries, and so on.

Activity

Using Data Have students collect menus from different restaurants. Divide students into small groups of three or four. Have each student order from a menu. Exchange orders and calculate the cost of the order.

On-the-Job Application

CATERING MANAGER

Matthew Banks is the catering manager at a hotel. He organizes banquets for wedding receptions, parties, conventions, and business meetings.

Part of the menu available for business meetings is shown below.

Afternoon Break
Freshly Brewed Coffee, Regular or Decaffeinated, Tea, Assorted Chilled Soft Drinks, Mineral Water
$2.95 per person

Health Break
Assorted Fruit, Muffins, Chilled Assorted Fruit Juices, and Mineral Water
$4.95 per person

Harvest Break
Assorted Donuts and Hot Spiced Cider
$3.50 per person

To add a little variety to your mid-meeting breaks, may we suggest one or more of the following.

Assorted Fruit Yogurt	$1.95 per person
Whole Fresh Fruit	$2.25 per piece
Assorted Cookies	$3.95 per dozen
Ice Cream	$1.50 per serving

The Jackson Corporation is holding a business meeting for thirty people.

How much will the Afternoon Break and five dozen Assorted Cookies cost?

Afternoon Break: $2.95 per person × 30 people = $88.50

Assorted Cookies: $3.95 per dozen × 5 dozen = $19.75

Total: $88.50 + 19.75 = $108.25

The total cost is $108.25.

Find the cost of each order. Use the menu above.

1. Harvest Break and Assorted Fruit Yogurt, 40 people **$218**
2. Afternoon Break, 45 people, and 30 pieces Whole Fresh Fruit **$200.25**
3. Health Break, 60 people, and 30 servings Ice Cream **$342**
4. Mrs. Valdez is having an afternoon tea for 25 people. She orders the Health Break, Assorted Fruit Yogurt, and four dozen Assorted Cookies. Find the cost of Mrs. Valdez's order. **$188.30**
5. You are in charge of refreshments for the Class Party. $700 is budgeted for refreshments for 115 people. Plan a menu and calculate the cost. **Answers may vary.**

INSULATION COSTS

Jeff Bayley wants to put insulation in his home. Insulation helps prevent heating and cooling loss. Thus, less energy is used and more money is saved. On the photo red indicates heat loss.

Suppose insulation costs $990. The average monthly fuel savings are $33. Mr. Bayley wants to find the number of years it will take for the fuel costs saved to pay for the insulation.

$$\begin{array}{rl} \$\ 33 & \text{monthly savings} \\ \times\ 12 & \text{months in a year} \\ \hline \$396 & \text{yearly savings} \end{array}$$

$$\text{yearly savings}\quad 396\overline{)\$990.0}\ \text{cost of insulation} \qquad 2.5\ \text{years}$$

It will take 2.5 years for the fuel savings to pay for the insulation.

Find the number of years it will take for the fuel cost savings to pay for the insulation. The charge for insulation and the average fuel savings per month are given. Express answers to the nearest tenth of a year.

1. $787.50; $18.75 3.5 **2.** $2,112; $42 4.2 **3.** $1,987; $32.50 5.1

4. $918.55; $21.20 3.6 **5.** $648.85; $19.20 2.8 **6.** $2,018; $47.75 3.5

7. The Alfred's monthly fuel bill is $175. They estimate that insulation will cost $885. What must the average monthly fuel cost savings be to pay for the insulation in 2 years? **$36.88**

8. If insulation costs $1,500 and the fuel costs saved will pay for the insulation in 4 years, what is the average monthly fuel savings? **$31.25**

9. An insulation company advertises that their insulation will result in average monthly fuel savings of 25%. The Carter's average monthly fuel bill is $185. The insulation costs $1,110. How many years will it take for the fuel cost savings to pay for the insulation? **2 years**

10. An insulation company advertises that their insulation will result in average monthly fuel savings of $42.50. The company also claims that fuel costs saved will pay for the insulation in 2.5 years. How much does their insulation cost? **$1,275**

Applying Mathematics to the Real World

This optional page shows how mathematics is used in the real world and also provides a change of pace.

Objective Determine insulation costs and fuel savings.

Using Discussion Attic insulation is what comes to mind when we think of insulation. Discuss other ways we can insulate to save money. **blanket insulation for a hot water tank, weather stripping, storm doors, windows, and so on**

Using Calculators Have students use calculators to solve problems.

Activity

Using Data Have students make a poster or a collage illustrating ways to save energy.

5-MINUTE CHECK-UP

(over Lesson 12-4)

Available on TRANSPARENCY 12-5.
Use the windchill chart on page 382 to find each equivalent temperature.

1. 30°F, 20 mph 4°F
2. −30°F, 10 mph −58°F

INTRODUCING THE LESSON

Have students name ways in which they have seen integers used. How is multiplication of integers helpful? temperature, above and below sea level; multiplication is a shortcut to addition and the rules for multiplication of integers tells which sign to use quickly

TEACHING THE LESSON

Using Questioning

- What is the sign of the product if both factors are positive? positive
- What is the sign of the product if one factor is positive and one factor is negative? negative
- What is the sign of the product if one factor is zero? neither positive nor negative

Practice Masters Booklet, p. 130

Name ____________ Date ____________

PRACTICE WORKSHEET 12-5

Multiplying Integers

Multiply.

1.	9 × (−2)	−18	**2.**	−12 × (−5)	60
3.	−3 × 16	−48	**4.**	2 × (−8)	−16
5.	15 × 6	90	**6.**	−14 × 3	−42
7.	−11 × (−7)	77	**8.**	1 × 13	13
9.	9 × 15	135	**10.**	−25 × (−16)	400
11.	30 × (−7)	−210	**12.**	−3 × 21	−63
13.	70 × (−80)	−5,600	**14.**	61 × (−59)	−3,599
15.	9.1 × (−1.9)	−17.29	**16.**	−3.8 × (−5.2)	19.76
17.	−6 × 0	0	**18.**	−1.5 × 1.5	−2.25
19.	−7 × 8	−56	**20.**	7 × (−21)	−147
21.	0 × (−2)	0	**22.**	$-\frac{2}{3} \times \left(-\frac{3}{4}\right)$	$\frac{1}{2}$
23.	$-\frac{3}{4} \times \frac{1}{6}$	$-\frac{1}{8}$	**24.**	6 × (−10)	−60
25.	$\frac{8}{9} \times \left(-\frac{1}{4}\right)$	$-\frac{2}{9}$	**26.**	−4 × (−25)	100

Solve.

27. A deep-sea exploring ship is pulling up a diver at the rate of 25 feet per minute. The diver is 200 feet below sea level. How deep was the diver 10 minutes ago?
450 feet below sea level

28. Joe is playing a game with a regular die. If the number that turns up is even, he will gain 5 times the number that comes up. If it is odd, he will lose 10 times the number that comes up. He tosses a 3. Express the result as an integer. −30

29. Barb's Swimsuit Outlet sold 312 swimsuits during a special sale where every suit was marked down to $9.99. Twenty-nine suits were returned for refunds. What was the actual income from the sale? $2,827.17

30. After a concert was canceled, ticket holders could exchange tickets for a future date or get a refund of the $20 price. Seven hundred tickets were returned for refunds. Express this income as an integer. −14,000

12-5 MULTIPLYING INTEGERS

Objective
Multiply integers.

Thick masses of ice that move slowly on land are called glaciers. Glaciers will melt and appear to move backwards when the temperature rises. The movement of the glacier can range from a few centimeters to a meter a day. Suppose a glacier melts 6 centimeters a day. What was its position 3 days ago? (The solution is in Example A.)

Look at some multiples of 3.

The product of two positive integers is positive. → 2 × 3 = 6, 1 × 3 = 3

0 × 3 = 0

One factor is positive and one is negative. The product is negative. → −1 × 3 = −3, −2 × 3 = −6

Describe the pattern you see. The products decrease by 3. 0 − 3 = −3

Look at some multiples of −3.

2 × −3 = −6
1 × −3 = −3
0 × −3 = 0

Both factors are negative. The product is positive. → −1 × −3 = 3, −2 × −3 = 6

Describe the pattern you see. The products increase by 3. 0 + 3 = 3

These rules also apply to positive and negative decimals and fractions.

Method

1. Find the product of the absolute values.
2. The product of two integers with the same sign is positive. The product of two integers with different signs is negative.

Examples

A ***−3 × (−6)***

1. |−3| × |−6| = 18
2. The factors have the same sign. The product is positive.

−3 × (−6) = 18

Three days ago the glacier was 18 centimeters forward of where it is now.

B ***5 × 14***

1. |5| × |14| = 70
2. The factors have the same sign. The product is positive.

5 × 14 = 70

C ***−6 × 3***

1. |−6| × |3| = 18
2. The factors have different signs. The product is negative.

−6 × 3 = −18

D ***0 × (−38)***

|0| × |−38| = 0

Any number multiplied by zero is zero.

RETEACHING THE LESSON

Multiply.

1. −18 × 5 −90
2. −5 × (−3) 15
3. 7 × (−3) −21
4. 80 × 5 400
5. 0 × (−12) 0
6. −3 × (−200) 600
7. −12 × 11 −132
8. −100 × 5 −500

Reteaching Masters Booklet, p. 119

Name ____________ Date ____________

RETEACHING WORKSHEET 12-5

Multiplying Integers

Method: ① Find the product of the absolute values.
② The product of two integers with the same sign is positive. The product of two integers with different signs is negative. (Zero times any number is zero.)

Examples: Multiply.

A. 5 × 7
① |5| × |7|
② The factors have the same sign. The product is positive.
5 × 7 = 35

B. −3 × (−9)
① |−3| × |−9| = 27
② The factors have the same sign. The product is positive.
−3 × (−9) = 27

C. 4 × (−6)
① |4| × |−6| = 24
② The factors have different signs. The product is negative.
4 × (−6) = −24

Multiply.

Guided Practice

Multiply.

Example A

1. $-5 \times (-3)$ **2.** $-3 \times (-12)$ **3.** $-1 \times (-9)$ **4.** $-45 \times (-2)$

Example B

5. 7×5 **6.** 21×4 **7.** 8×9 **8.** 6×5

Examples C, D

9. $4 \times (-3)$ **10.** -5×6 **11.** -1×32 **12.** -98×1

13. $7 \times (-11)$ **14.** $54 \times (-10)$ **15.** 0×15 **16.** -8×0

Exercises

Practice

Multiply.

17. 5×8 **18.** $9 \times (-2)$ **19.** -3×2 **20.** $-5 \times (-4)$

21. $6 \times (-7)$ **22.** $0 \times (-5)$ **23.** -10×7 **24.** 9×21

25. $9 \times (-6)$ **26.** 5×15 **27.** $-8 \times (-7)$ **28.** $-1 \times (-34)$

29. -3×123 **30.** $-14 \times (-4)$ **31.** $35 \times (-6)$ **32.** -100×0

33. $6 \times (-2.1)$ **34.** $-0.3 \times (-0.3)$ **35.** $\frac{2}{3} \times (-\frac{6}{7})$ **36.** $-1\frac{2}{5} \times (-2\frac{2}{9})$

37. What is the product of -18 and 6?

Number Sense

38. If an odd number of negative factors are multiplied, is the product positive or negative?

Applications

39. On Cape Cod several cliffs recede 3.3 feet each year. How much do the cliffs recede in 5 years?

40. At 5:00 P.M. the temperature was 0°C. For the next 6 hours it dropped 2°C each hour. What was the temperature at 11:00 P.M.?

41. The record low temperature in Ohio is -39°C. The record high temperature is 45°C. What is the difference in the temperatures?

Using Expressions

Find the value of each expression if $x = -8$.

42. $4x$ **43.** $0x$ **44.** $-3x$ **45.** $-6x$

Critical Thinking

46. Suppose one person starts a story and tells it to four other people within 20 minutes. Each person then tells four other people within 20 minutes and so on. How long will it take for one million people to hear the story?

Mixed Review

Lesson 4-6

Simplify each fraction.

47. $\frac{12}{18}$ **48.** $\frac{9}{21}$ **49.** $\frac{6}{24}$

Lesson 9-6

Find the missing length for each pair of similar figures.

50.

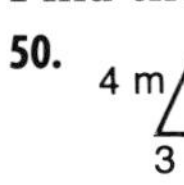

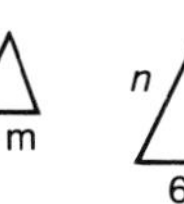

51.

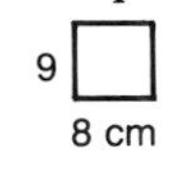

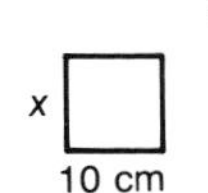

52.

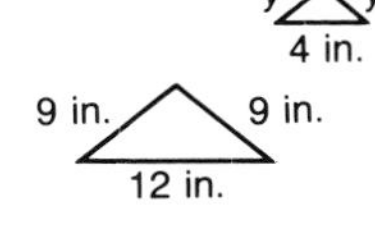

Lesson 10-2

53. A store donated money to a charity based upon a percentage of total sales for a given day. If the total sales were above \$5,000, the store would donate 2%. If the sales were above \$7,500, the store would donate 3% of total sales. How much money was donated if the total sales for the day were \$8,258?

APPLYING THE LESSON

Independent Practice

Homework Assignment	
Minimum	Ex. 2-52 even; 53
Average	Ex. 1-37 odd; 39-53
Maximum	Ex. 1-53

Chalkboard Examples

Example A *Multiply.*

1. $-2 \times (-4)$ 8 **2.** $-4 \times (-6)$ 24

Example B *Multiply.*

3. 6×5 30 **4.** 8×6 48

Example C *Multiply.*

5. -8×9 -72 **6.** -7×6 -42

Example D *Multiply.*

7. -6×0 0 **8.** $0 \times (-3)$ 0

Additional examples are provided on TRANSPARENCY 12-5.

EVALUATING THE LESSON

Check for Understanding Use Guided Practice Exercises to check for student understanding.

Closing the Lesson Ask students to summarize what they have learned about multiplying integers.

EXTENDING THE LESSON

Enrichment Have students make up a game in which expressions like exercises 42-45 are used.

Enrichment Masters Booklet, p. 119

Name ______________________ Date ______________

ENRICHMENT WORKSHEET 12-5

Multiplying Integers

Compete each square. The product of each row and column is given.

1.

2	6	12
4	3	12
8	18	144

or

−2	−6
−4	−3

2.

5	1	5
−3	−2	6
−15	−2	30

or

−5	−1
3	2

3.

−4	7	−28
−3	2	−6
12	14	168

or

4	−7
3	−2

4.

0	3	0
−4	2	−8
0	6	0

or

0	−3
4	−2

5.

5	5	25
5	5	25
25	25	625

or

−5	−5
−5	−5

6.

1	−2	−2
−1	2	−2
−1	−4	4

or

−1	2
1	−2

7.

3	4	12
−2	5	−10
−6	20	−120

or

−3	−4
2	−5

8.

8	−3	−24
−2	1	−2
−16	−3	48

or

−8	3
2	−1

9.

1	1	1
−1	1	−1
−1	1	−1

or

−1	−1
1	−1

5-MINUTE CHECK-UP

(over Lesson 12-5)

Available on TRANSPARENCY 12-6.

Multiply.

1. 6×12 72 **2.** $-4 \times (-8)$ 32
3. -7×4 −28 **4.** $0 \times (-2)$ 0

Extra Practice, Lesson 12-5, p. 481

▶ INTRODUCING THE LESSON

Have students give the two related division sentences for each multiplication sentence.

1. $4 \times (-3) = -12$ $-12 \div (-3) = 4$ $-12 \div 4 = -3$

2. $-8 \times (-2) = 16$ $16 \div (-2) = -8$ $16 \div (-8) = -2$

▶ TEACHING THE LESSON

Using Discussion Discuss with students that dividing by a number is the same as multiplying by its reciprocal or multiplicative inverse. Notice that the rules for multiplication and division are the same.

12-6 DIVIDING INTEGERS

Objective *Divide integers.*

Joe Johnson is a halfback for the community college he attends. In the latest game, he carried the ball the last two plays and his net gain was −12 yards. What was Joe's average net gain? (The problem is solved in Example D.)

Every multiplication sentence has two related division sentences.	$3 \times 4 = 12$ has the related sentences: $12 \div 4 = 3$ and $12 \div 3 = 4$	$5 \times (-2) = -10$ has the related sentences: $-10 \div (-2) = 5$ and $-10 \div 5 = -2$

You can use your knowledge about multiplying integers when you divide integers.

These rules also apply to positive and negative decimals and fractions.

Method

1. Divide the absolute values.
2. The quotient of two integers with the same sign is positive. The quotient of two integers with different signs is negative.

Examples

A $45 \div 15$

1. $|45| \div |15| = 3$
2. The divisor and dividend have the same sign. So the quotient is positive.
   $45 \div 15 = 3$
   Check: $3 \times 15 \stackrel{?}{=} 45$
   $45 = 45$ ✓

B $-16 \div (-2)$

1. $|-16| \div |-2| = 8$
2. The divisor and dividend have the same sign. So the quotient is positive.
   $-16 \div (-2) = 8$
   Check: $8 \times (-2) \stackrel{?}{=} -16$
   $-16 = -16$ ✓

C $45 \div (-15)$

1. $|45| \div |-15| = 3$
2. The divisor and dividend have different signs. The quotient is negative.
   $45 \div (-15) = -3$
   Check: $-3 \times (-15) \stackrel{?}{=} 45$
   $45 = 45$ ✓

D $-12 \div 2$

1. $|-12| \div |2| = 6$
2. The divisor and dividend have different signs. The quotient is negative.
   $-12 \div 2 = -6$
   Joe's average net gain was −6 yards.

Guided Practice

Divide.

Example A	**1.** $14 \div 2$ 7	**2.** $25 \div 5$ 5	**3.** $15 \div 60$ 0.25	**4.** $5 \div 25$ 0.2
Example B	**5.** $-18 \div (-3)$ 6	**6.** $-24 \div (-4)$ 6	**7.** $-2 \div (-4)$ 0.5	**8.** $-30 \div (-4)$ 7.5
Example C	**9.** $36 \div (-3)$ −12	**10.** $45 \div (-5)$ −9	**11.** $3 \div (-6)$ −0.5	**12.** $0 \div (-7)$ 0
Example D	**13.** $-56 \div 4$ −14	**14.** $-81 \div 9$ −9	**15.** $-1 \div 10$ −0.1	**16.** $0 \div 5$ 0

Practice Masters Booklet, p. 131

Name ____________ Date ____________

PRACTICE WORKSHEET 12-6

Dividing Integers

Divide.

1. $18 \div (-3)$	−6	**2.** $-48 \div (-12)$	4
3. $-45 \div 15$	−3	**4.** $40 \div (-8)$	−5
5. $0 \div 0$	0	**6.** $-63 \div 3$	−21
7. $-140 \div (-7)$	20	**8.** $52 \div 13$	4
9. $-40 \div (-5)$	8	**10.** $-120 \div (-6)$	20
11. $168 \div (-7)$	−24	**12.** $-210 \div 21$	−10
13. $5{,}680 \div (-80)$	−71	**14.** $27 \div (-9)$	−3
15. $0 \div (-1)$	0	**16.** $-7.2 \div (-0.9)$	8
17. $-14.64 \div 4$	−3.66	**18.** $-15 \div 1.5$	−10
19. $14 \div (-10)$	−1.4	**20.** $-25 \div (5)$	−5
21. $4.4 \div (-2.2)$	−2	**22.** $-\frac{2}{3} \div \left(-\frac{4}{9}\right)$	$1\frac{1}{2}$
23. $-\frac{3}{4} \div \frac{1}{8}$	−6	**24.** $-80 \div (-4)$	20
25. $\frac{8}{9} \div \left(-\frac{4}{15}\right)$	$-3\frac{1}{3}$	**26.** $-4 \div 8$	−0.5
27. $-18 \div 6$	−3	**28.** $51 \div (-3)$	−17
29. $-91 \div (-13)$	7	**30.** $-42 \div 14$	−3
31. $-96 \div (-24)$	4	**32.** $90 \div (-15)$	−6

Solve.

33. Stan weighed 225 pounds on March 1. Four months later he weighed 197 pounds. What was his average change in weight per month?
loss of 7 lb or −7 lb

34. In 1980, the population of Clayborn was 15,792. In 1986 the population was 19,296. What was the average change in population per year?
increase of 584 or 584

RETEACHING THE LESSON

Divide.

1. $-56 \div (-7) =$ 8 **2.** $76 \div 4 =$ 19
3. $-305 \div 5 =$ −61 **4.** $-502 \div 2 =$ −251
5. $128 \div (-8) =$ −16 **6.** $-189 \div (-9) =$ 21

Reteaching Masters Booklet, p. 120

Name ____________ Date ____________

RETEACHING WORKSHEET 12-6

Dividing Integers

Method: ① Divide the absolute values.
② The quotient of two integers with the same sign is positive. The quotient of two integers with different signs is negative.

Examples: Divide.

A. $32 \div 8$
① $|32| \div |8| = 4$
② The divisor and dividend have the same sign. The quotient is positive.
$32 \div 8 = 4$
Check. $4 \times 8 = 32$

B. $-18 \div (-6)$
① $|-18| \div |-6| = 3$
② The divisor and dividend have the same sign. The quotient is positive.
$-18 \div -6 = 3$
Check. $3 \times (-6) = -18$

C. $27 \div (-3)$
① $|27| \div |-3| = 9$
② The divisor and dividend have different

D. $-20 \div 4$
① $|-20| \div |4| = 5$
② The divisor and dividend have

Practice

Divide.

17. $-12 \div (-3)$ **18.** $24 \div 4$ **19.** $18 \div (-2)$ **20.** $-27 \div (-3)$

21. $15 \div (-3)$ **22.** $-48 \div 6$ **23.** $-12 \div 6$ **24.** $0 \div 2$

25. $0 \div (-4)$ **26.** $36 \div 4$ **27.** $-42 \div 7$ **28.** $-133 \div 7$

29. $-56 \div (-4)$ **30.** $-200 \div 40$ **31.** $8 \div (-1)$ **32.** $45 \div (-9)$

33. $37 \div (-10)$ **34.** $-9 \div (-72)$ **35.** $4 \div (-4)$ **36.** $-4 \div 8$

37. $-75 \div (-7)$ **38.** $12 \div (-0.5)$ **39.** $-4.9 \div 1.4$ **40.** $\frac{1}{2} \div (-\frac{3}{4})$

41. Find the quotient of 165 and -3.

Number Sense

42. *True* or *false*. It is possible to have a positive integer as the quotient when dividing a negative integer by a positive integer.

Replace each ● with >, <, or = to make a true sentence.

43. $4 - 28 \div (-7)$ ● $4 - 4$ **44.** $-8 + 3 \times (-2)$ ● -8×2

45. $36 - 4 + 8 \div 4$ ● $-6 \times (-7)$ **46.** $-31 + 6 \times -3$ ● -7×7

Using Equations

Solve each equation.

47. $a + 7 = 13$ **48.** $n \div (-3) = 4$ **49.** $6 \times b = -18$

50. $-24 \div g = -3$ **51.** $12 + t = -11$ **52.** $h - (-4) = 7$

Applications

53. During a 5-day period, a stock price had a change of -2. What was the average change per day?

FUN with MATH

Try our recipe for Pineapple Chicken. See page 422.

54. From noon until 6:00 P.M., the temperature change was -12°C. Find the average change per hour.

55. The first recorded Olympic Games were held in 776 B.C. According to legend, Romulus founded Rome in 753 B.C. Which event occurred first?

Make Up a Problem

56. Make up a problem that matches the number sentence in Exercise 27.

Calculator

What is the quotient, *q*, when 6 is divided by 0?

Every division sentence has a related multiplication sentence. ($14 \div 2 = 7$; $7 \times 2 = 14$) So $6 \div 0 = q$ has the related sentence $q \times 0 = 6$. Since there is no number that can replace *q* to make a true statement, we say the quotient $6 \div 0$ is *undefined.* Your calculator will give an ERROR (E) message if you try to divide by 0.

Use a calculator to find each quotient.

57. $-8 \div 0$ **58.** $\frac{0}{3}$ **59.** $\frac{3}{0}$ **60.** $\frac{18}{-6 + 6}$ **61.** $0\overline{)12}$

APPLYING THE LESSON

Independent Practice

Homework Assignment	
Minimum	Ex. 2-60 even
Average	Ex. 1-41 odd; 43-61
Maximum	Ex. 1-61

Chalkboard Examples

Examples A, B *Divide.*

1. $54 \div 6$ **9** **2.** $-24 \div (-2)$ **12**

Examples C, D *Divide.*

3. $36 \div (-9)$ **−4** **4.** $-14 \div 2$ **−7**

Additional examples are provided on TRANSPARENCY 12-6.

▶ EVALUATING THE LESSON

Check for Understanding Use Guided Practice Exercises to check for student understanding.

Error Analysis With this lesson students will have had experiences with all four operations on integers, and some students may confuse the rules for signs. Emphasize the fact that the rules for multiplication and division are the same; contrast these rules with those for addition and subtraction.

Closing the Lesson Ask students to state in their own words how to divide integers.

▶ EXTENDING THE LESSON

Enrichment Masters Booklet, p. 120

Name ____________ Date ____________

ENRICHMENT WORKSHEET 12-6

Commutative Property

If a pair of numbers is commutative under a given operation, then the order in which the two numbers are operated on does not matter. The result will be the same in either case.

Which of the statements at the right are true?

a $2 + (-5) = -5 + 2$
b $5 - (-2) = -2 - 5$
c $-2 \times 5 = 5 \times (-2)$
d $10 \div (-2) = -2 \div 10$

Statements a and c are true.

Add, subtract, multiply, or divide. Write each answer in simplest form.

1. a $0.5 + (-0.6)$ −0.1 **b** $-0.6 + 0.5$ −0.1

2. a $-4 + (-2)$ −6 **b** $-2 + (-4)$ −6

3. a $0.5 - 0.6$ −0.1 **b** $0.6 - 0.5$ 0.1

4. a $-0.8 \div 0.2$ −4 **b** $0.2 \div (-0.8)$ $-\frac{1}{4}$

5. a -3×2 −6 **b** 2×-3 −6

6. a -5×-7 35 **b** -7×-5 35

7. Which of Exercises 1–6 have the same answer for parts **a** and **b**? 1, 2, 5, 6
Which exercises have a different answer? 3, 4
What conclusions can you draw about the commutative property for integers from these exercises?
Addition and multiplication are commutative; subtraction and division are not.

5-MINUTE CHECK-UP

(over Lesson 12-6)

Available on TRANSPARENCY 12-7.

Divide.

1. $36 \div 9$ 4 **2.** $-24 \div (-4)$ 6
3. $64 \div (-8)$ −8 **4.** $-24 \div 24$ −1

Extra Practice, Lesson 12-6, p. 481

INTRODUCING THE LESSON

Ask students to think about what they normally do from the time they get up in the morning until they leave for school. Ask them to describe their routine in reverse order.

TEACHING THE LESSON

Using Questioning

- Start with a number. Multiply it by 3, add 8, and divide by 5. If the answer is 7, what was the original number? 9
- What operations did you use to solve problem 1? × 5; − 8; ÷ 3

Practice Masters Booklet, p. 132

Name ______ Date ______

PRACTICE WORKSHEET 12-7

Problem-Solving Strategy: Work Backwards

Solve. Work backwards.

1. Eight times a number plus 7 is equal to 79. Find the number. 9
2. Twice a number decreased by 12 is equal to 10. Find the number. 11
3. Seven more than 3 times a number is equal to 49. Find the number. 14
4. If 8 is added to a number and then the number is divided by 3, the result is 12. Find the number. 28
5. Nine less than 6 times a number is equal to 93. Find the number. 17
6. If the sum of a number and 10 is divided by 3, the result is 13. Find the number. 29
7. If a number is multiplied by 4, then 6 is added to it, and the result is divided by 7, the answer is 6. Find the number. 9
8. If a number is divided by 8 and then 12 is added to the quotient, the result is 17. Find the number. 40

Problem-Solving Strategy

Explore / Plan / Solve / Examine

12-7 WORK BACKWARDS

Objective
Solve verbal problems by working backwards.

A store is having a clearance sale to make room for new inventory. The store owner reduces the price of a stereo system by 50%. When the stereo still does not sell, that price is reduced another 25%. The stereo finally sells for half of the last price, or \$150. What was the original price of the stereo system?

Explore

What is given?

- A stereo sells for \$150 after the original price is reduced by 50%, by 25%, and finally for half of the last price.

What is asked?

- What was the original price of the stereo?

Plan

When you know the final price you can often work backwards step by step to find the original price.

Solve

The store sells the stereo system for \$150, one-half of the previous price. So its previous price was \$150 × 2 or \$300. This price had been reduced 25%, \$300 is therefore 75% of the previous price. Before this it was reduced 50% from the original price. Use proportions to find the prices.

end result
$\frac{1}{2}$ \$150 $\frac{1}{2}$ \$150
\$300
a reduction of 25% from
\$400
a reduction of 50% from
\$800
original price

$$\frac{75}{100} = \frac{300}{s}$$
$$75 \times s = 100 \times 300$$
$$75s = 30{,}000$$
$$s = 400$$

$$\frac{50}{100} = \frac{400}{p}$$
$$50 \times p = 100 \times 400$$
$$50p = 40{,}000$$
$$p = 800$$

The original price was \$800.

Examine

Start with \$800. Fifty percent of \$800 is \$400. Since twenty-five percent of \$400 is \$100, the next price is \$300. One-half of \$300 is \$150, so the solution is correct.

Guided Practice

Solve. Work backwards.

1. Marla picks some apples. She gives half to her mother and half of the remaining apples to her friend Andy. Andy receives 7 apples. How many apples did Marla pick? **28 apples**

RETEACHING THE LESSON

Emphasize that in working backwards through a problem, inverse operations are used.

$x + 20 - 15 = 37$ Find x.

$$37 + 15 = 52 \qquad 52 - 20 = 32 \rightarrow x = 32$$

Have students solve each equation.

1. $a + 27 - 82 = 104$ $a = 159$
2. $t \div 5 \times 9 = 90$ $t = 50$
3. $r \times 16 - 43 = 21$ $r = 4$

Problem Solving

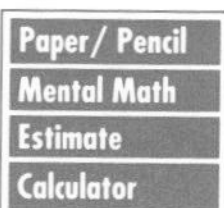

Solve. Use any strategy.

2. Rico tries to sell his bicycle at a garage sale. He reduces the price by half. It still does not sell, so he reduces that price by 20%. He has to reduce the last price by 25% before it sells for $18. What was the original price? **$60**
3. A certain kind of microbe doubles its population every 12 hours. After 3 days there are 640 microbes. How many microbes were there at the beginning of the first day? **10 microbes**
4. Forty percent of the cars Mr. Tibbs sells are compacts. Twenty percent of the compact cars he sells are hatchbacks. If Mr. Tibbs sells 34 hatchbacks, how many cars has he sold in all? **425 cars**
5. Sandy mailed letters and postcards for a total of $1.82. If it costs 29¢ to mail each letter and 19¢ to mail each postcard, how many of each did she mail? **3 letters; 5 postcards**
6. Local library fines for overdue books are 12¢ for the first day, 6¢ for each of the next two days, and 4¢ each day thereafter. If Gail paid a fine of 80¢, how many days was her book overdue? **17 days**
7. Four painters can paint 4 rooms in 4 hours. How many rooms of the same size can 8 painters paint in 8 hours? **16 rooms**
8. Half of the students in a class are girls. One-third of the girls have blond hair. If five girls have blond hair, how many students are in the class? **30 students**
9. Ms. James is making eggrolls for a big party. She needs a pound of cornstarch to seal the wrappers. It comes in 3-oz boxes for 36¢ and 6-oz boxes for 67¢. How many of each size should she purchase to get the best buy? **3 of the 6-oz boxes**

Critical Thinking

10. A magic square is a number square in which the numbers in all rows, columns, and diagonals have the same sum. This sum is called the magic sum. Find the missing numbers if the magic sum is -34.

-16	-2	-3		**−13**
-5	-16			**−10; −8**
-9		-6		**−7; −12**
	-14		-1	**−4; −15**

11. Name your favorite lesson from this chapter. Then write a paragraph explaining why it was your favorite. **See students' work.**

Mixed Review

Complete.

Lesson 6-5

12. 7 L = ? mL **7,000**
13. 148.1 mL = ? L **0.1481**
14. 91.2 L = ? kL **0.0912**

Lesson 11-11

15. Monica's soccer team enters a double-elimination tournament with three other teams. That is, a team is eliminated from winning the tournament only after two losses. Monica's team loses their first game, yet they win the tournament. How many games did they have to play? **4 games**

Chalkboard Examples

1. Sasha gets up in the morning and spends 25 minutes in the bathroom. He eats breakfast for 20 minutes and then spends 25 more minutes in his room. He takes 10 minutes to collect his books and coat. He walks to the bus stop in 15 minutes to catch the bus at 7:05 A.M. What time did he get up? **5:30 A.M.**

Additional examples are provided on TRANSPARENCY 12-7.

EVALUATING THE LESSON

Check for Understanding Use Guided Practice Exercises to check for student understanding.

Error Analysis Watch for students who do not correctly "undo" operations when working backwards to solve a problem. Stress that they should use the opposite operations as given in the problem.

Closing the Lesson Ask students to summarize what they have learned about solving problems when they must work backwards to reach a conclusion.

APPLYING THE LESSON

Independent Practice

Homework Assignment	
Minimum	Ex. 1-15
Average	Ex. 1-15
Maximum	Ex. 1-15

Chapter 12, Quiz B (Lessons 12-5 through 12-7) is available in the Evaluation Masters Booklet, p.59.

Using the Chapter Review

The Chapter Review is a comprehensive review of the concepts presented in this chapter. This review may be used to prepare students for the Chapter Test.

Chapter 12, Quizzes A and B, are provided in the Evaluation Masters Booklet as shown below.

Quiz A should be given after students have completed Lessons 12-1 through 12-4. Quiz B should be given after students have completed Lessons 12-5 through 12-7.

These quizzes can be used to obtain a quiz score or as a check of the concepts students need to review.

Evaluation Masters Booklet, p. 59

CHAPTER 12, QUIZ A (Lessons 12-1 through 12-4)

Replace each ○ with $<$, $>$, or $=$ to make a true sentence.

1. $7 \bigcirc -11$
2. $|-3| \bigcirc |3|$

Order from least to greatest.

3. −6, 1, 0, −8, 3
4. 3, −1, −3, 2, −4

Add.

5. −7 + (−4)
6. 6 + (−10)

Subtract.

7. −9 − 5
8. −1 − (−8)

Use the chart to solve.

Windchill Chart

Wind speed in mph	Actual temperature (°Fahrenheit)								
	50	40	30	20	10	0	-10	-20	-30
	Equivalent temperature (°Fahrenheit)								
10	40	28	16	4	-9	-21	-33	-46	-58
15	36	22	9	-5	-18	-36	-45	-58	-72
20	32	18	4	-10	-25	-39	-53	-67	-82

9. Find the equivalent temperature when the actual temperature is 10°F and the wind speed is 20 mph.
10. What is the difference between an actual temperature of 10°F and the equivalent temperature when the wind speed is 15 mph?

1. $>$
2. $=$
3. −8, −6, 0, 1,
4. −4, −3, −1, 2
5. −11
6. −4
7. −14
8. 7
9. −25°F
10. 28°F

Name ________________ Date ________________

CHAPTER 12, QUIZ B (Lessons 12-5 through 12-7)

Multiply.

1. −6 × (−9)
2. 5 × (−3)
3. 0 × (−4)
4. −8 × 7

Divide.

5. −18 ÷ (−6)
6. $7\overline{)-42}$
7. 0 ÷ (−3)
8. $-12\overline{)60}$
9. Joel put half the money he earned one summer in savings. He also spent $325 for a stereo. If he still had $440 left that was not in savings, how much did Joel earn that summer?
10. Joyce Terrio sold her car for $50 more than half what she paid for it. If she sold the car for $1,300, how much did she pay for it?

1. 54
2. −15
3. 0
4. −56
5. 3
6. −6
7. 0
8. −5
9. $1,530
10. $2,500

Chapter 12 REVIEW

1. opposite 2. absolute value

Choose the word or phrase from the list at the right that best completes each sentence. You may use a word more than once.

Vocabulary / Concepts

1. To subtract an integer, add its _?_ .
2. On a number line, the distance that a number is from zero is its _?_ .
3. The _?_ and their opposites make up the set of _?_ . **whole numbers, integers**
4. The product of two negative numbers is _?_ . **positive**
5. The quotient of a negative number and a positive number is _?_ . **negative**
6. To add integers with different signs, _?_ their absolute values. Then give the same sign as the integer with the _?_ absolute value. **subtract, greater**
7. Subtracting an integer gives the same result as _?_ its opposite. **adding**
8. The product of a negative integer and a positive integer is _?_ . **negative**

absolute value
adding
divide
greater
integers
lesser
multiply
negative
opposite
positive
subtract
whole numbers

Exercises / Applications

Lesson 12-1

Replace each ● with $<$, $>$, or $=$ to make a true sentence.

9. $0 \bullet -2$ $>$
10. $-3 \bullet -5$ $>$
11. $-7 \bullet 7$ $<$
12. $2 \bullet -4$ $>$
13. $|-5| \bullet 5$ $=$
14. $|-3| \bullet 3$ $=$
15. $|-11| \bullet |7|$ $>$
16. $|-9| \bullet |9|$ $=$

Order from least to greatest.

17. −7, 7, 9, −9, 0 **−9, −7, 0, 7, 9**
18. 0, 100, 101, −100, −101 **−101, −100, 0, 100, 101**

Lesson 12-2

Add. **20. −20 22. −35 25. −22**

19. 5 + (−8) **−3**
20. −3 + (−17)
21. −5 + 5 **0**
22. −68 + 33
23. −8 + 12 **4**
24. 0 + (−7) **−7**
25. −9 + (−13)
26. −14 + 0 **−14**

Lesson 12-3

Subtract.

27. 9 − 16 **−7**
28. 12 − 4 **8**
29. 14 − (−3) **17**
30. 0 − 8 **−8**
31. −9 − 0 **−9**
32. 16 − (−16) **32**
33. −5 − (+8) **−13**
34. −14 − (−8) **−6**
35. Water in a reservoir was 6 feet below normal. Water was pumped into the reservoir until the water level was 5 feet above normal. How many feet, in all, did the water rise or fall? **11 ft rise**

Lesson 12-4

Solve. Use the windchill chart on page 382.

36. In Baltimore on January 24, the wind speed was 10 mph. The actual temperature was −10°F. What was the equivalent temperature with the windchill factor? **−33°F**

37. Suppose the actual temperature is 0°F and the wind speed is 25 mph. What is the difference between the actual temperature and the equivalent temperature? **44°F**

Lesson 12-5

Multiply.

38. −9 × 5 **−45** **39.** −3 × (−18) **54** **40.** 7 × 8 **56** **41.** −11 × (−9) **99**

42. 14 × (−7) **−98** **43.** +7 × (−12) **−84** **44.** −16 × (8) **−128** **45.** −7 × (−12) **84**

Lesson 12-6

Divide.

46. 63 ÷ 7 **9** **47.** −36 ÷ 4 **−9** **48.** 39 ÷ (−13) **−3** **49.** −15 ÷ 12 $-1\frac{1}{4}$

50. −5 ÷ 10 $-\frac{1}{2}$ **51.** −48 ÷ (−6) **8** **52.** −63 ÷ (−9) **7** **53.** 0 ÷ (−4) **0**

Lesson 12-7

Solve. Use any strategy.

54. Joan spends 10 minutes eating a snack when she gets home from school. Then she studies math, science, and English for 25 minutes each. When she finishes, it is 5:30 P.M. At what time did she get home from school? **4:05 P.M.**

55. Pick a number, triple it, add 8, and divide by 5. If the result is 7, what was the original number? **9**

56. The water level of a tank fell at the rate of $1\frac{1}{2}$ feet per hour for 6 hours. What is the water level now if the water level had been 38 feet? **29 feet**

57. The temperature rose for 8 hours at a rate of 2° per hour. What was the original temperature if the final temperature was −11°F? **−27°F**

Alternate Assessment Strategy

To provide a brief in-class review, you may wish to read the following questions to the class and require a verbal response.

1. Which number is greater, −5 or −8? **−5**
2. Order these numbers from least to greatest: −8, 7, 2, −1, −3 **−8, −3, −1, 2, 7**
3. Add negative 14 and negative 8. **−22**
4. Find the sum of 16 and negative 9. **7**
5. Subtract negative 24 from 6. **30**
6. What is the difference when negative 4 is subtracted from negative 8? **−4**
7. Multiply negative 5 by negative 7. **35**
8. What is the product when negative 9 is multiplied by 6? **−54**
9. What is the result when negative 25 is divided by 5? **−5**
10. Divide negative 80 by negative 8. What is the quotient? **10**

APPLYING THE LESSON

Independent Practice

Homework Assignment	
Minimum	Ex. 1-8; 9-57 odd
Average	Ex. 1-57 odd
Maximum	Ex. 1-57

Using the Chapter Test

This page may be used as a test or as an additional page of review if necessary. Two forms of a Chapter Test are provided in the Evaluation Masters Booklet. Form 2 (free response) is shown below, and Form 1 (multiple choice) is shown on page 397.

The **Tech Prep Applications Booklet** provides students with an opportunity to familiarize themselves with various types of technical vocations. The Tech Prep applications for this chapter can be found on pages 23–24.

Evaluation Masters Booklet, p. 58

Name ______________ Date ______________

CHAPTER 12 TEST, FORM 2

Replace each ○ with <, >, or = to make a true sentence.

1. $-4 \bigcirc 2$ 2. $-3 \bigcirc -7$ 3. $|5| \bigcirc |-5|$

1. <
2. >
3. =

Order from least to greatest.

4. 5, 0, −7, 2, −4 5. 3, 1, −3, 4, −5

4. −7, −4, 0, 2, 5
5. −5, −3, 1, 3, 4

Add.

6. −8 + 5 7. 8 + (−20) 8. −7 + (−4)

6. −3
7. −12
8. −11

Subtract.

9. −4 − (−16) 10. −8 − 7 11. 12 − (−5)

9. 12
10. −15
11. 17

Multiply.

12. −3 × (−8) 13. −2 × 4 14. 6 × (−7)

12. 24
13. −8
14. −42

Divide.

15. −35 ÷ 7 16. $-8\overline{)-256}$ 17. 72 ÷ (−9)

15. −5
16. 32
17. −8

Use the chart to solve.

Windchill Chart

Wind speed in mph	Actual temperature (°Fahrenheit) 50	40	30	20	10	0	-10	-20	-30
	Equivalent temperature (°Fahrenheit)								
20	32	18	4	-10	-25	-39	-53	-67	-82
25	30	16	0	-15	-29	-44	-59	-74	-88
30	28	13	-2	-18	-33	-48	-63	-79	-94

18. Find the equivalent temperature when the actual temperature is 30°F and the wind speed is 20 mph.
19. What is the difference between an actual temperature of 20°F and the equivalent temperature when the wind speed is 30 mph?
20. During one month Toya wrote checks for one-third of the amount she had in her checking account. She made no deposits. After paying a service charge of $8, her balance was $82. How much did she have in her account at the beginning of the month?

18. 4°F
19. 38°F
20. $135

Bonus Compute. $\frac{4 \times (-7 - 2)}{-3}$ **Bonus** 12

Chapter 12 TEST

Replace each ● with <, >, or = to make a true sentence.

1. 2 ● 8 <
2. 3 ● −5 >
3. −3 ● −11 >
4. −9 ● −5 <
5. −8 ● 9 <
6. |−4| ● |2| >

Add or subtract.

7. −3 + (−7) **−10**
8. −8 + (−1) **−9**
9. 14 + (−16) **−2**
10. 19 + (−13) **6**
11. 20 − (−5) **25**
12. −15 − 15 **−30**
13. −6 − 4 **−10**
14. 0 − (−3) **3**
15. −8 − (−8) **0**
16. 13 + (−14) **−1**
17. 2 − 3 **−1**
18. −3 + 8 **5**

Solve. Use the chart.

19. In Terre Haute on December 28, the low temperature was 20°F. The wind speed was 10 mph. With the windchill factor, what was the equivalent temperature? **4°F**

Windchill Chart

Wind speed in mph	Actual temperature (°Fahrenheit) 50	40	30	20	10	0	−10
	Equivalent temperature (°Fahrenheit)						
0	50	40	30	20	10	0	−10
5	48	37	27	16	6	−5	−15
10	40	28	16	4	−9	−21	−33

Multiply or divide.

20. 72 ÷ (−8) **−9**
21. −15 ÷ 3 **−5**
22. 42 ÷ (−3) **−14**
23. −88 ÷ (−4) **22**
24. −18 × 3 **−54**
25. −5 × (−3) **15**
26. 80 × 5 **400**
27. 0 × (−12) **0**
28. −14 ÷ 2 **−7**
29. 20 ÷ (−5) **−4**
30. −2 × 5 **−10**
31. 5 × (−7) **−35**

Solve. Use working backwards.

32. Kenny entered a triathlon in which he biked twice as far as he ran. He ran five times as far as he swam. He biked 12 miles. How far did he swim? **1.2 miles**
33. Jolie picked some apples. She gave half to her mother and half of the remaining apples to her friend Tracey. Tracey received 9 apples. How many apples did Jolie pick? **36 apples**

BONUS: If $n < 0$, is $n^2 < 0$? Explain. **BONUS: no, A number less than zero is a negative number. When two negative numbers are multiplied, the result is a positive number.**

The **Performance Assessment Booklet** provides an alternative assessment for evaluating student progress. An assessment for this chapter can be found on pages 23–24.

A **Test and Review Generator** is provided in Apple, IBM, and Macintosh versions. You may use this software to create your own tests or worksheets, based on the needs of your students.

City and State Income Taxes

Many city and state governments levy their own income tax. Ned Davis lives in Westerville which levies both a city and state income tax. His employer is required to deduct these taxes from his wages.

The city rate is 1%.
Find 1% of $299.55.

$299.55	weekly wage
× 0.01	rate (1% = 0.01)
$2.9955	city tax

To the nearest cent, the weekly city tax deduction is $3.00.

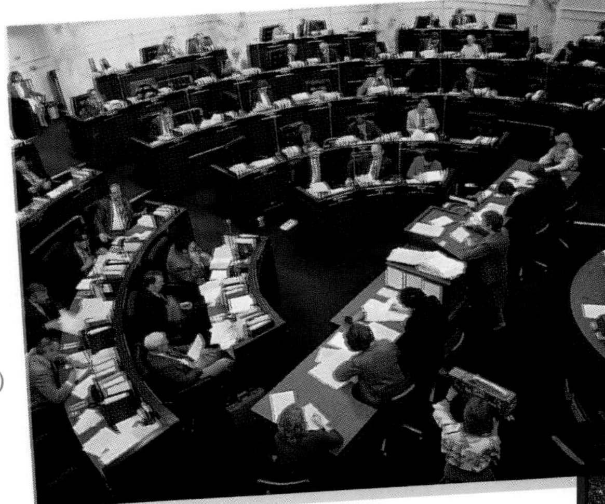

The state rate is 2.5%.
Find 2.5% of $299.55.

$299.55	weekly wage
× 0.025	rate (2.5% = 0.025)
$7.48875	state tax

To the nearest cent, the weekly state tax deduction is $7.49.

Find the city tax deduction for each wage to the nearest cent. Use a 1% rate.

1. $237.35 **$2.37** **2.** $179.29 **$1.79** **3.** $322.56 **$3.23** **4.** $262.50 **$2.63**

Find the state tax deduction for each wage to the nearest cent. Use a 2.5% rate.

5. $179.29 **$4.48** **6.** $256 **$6.40** **7.** $314.56 **$7.86** **8.** $278.50 **$6.96**

9. Ms. Kosar pays 1.5% city tax and 2.75% state tax. Her annual salary is $36,000. How much is deducted from her monthly salary for city and state taxes? **$127.50**

10. Mr. Walters makes $650 a week. His employer deducts 1.25% for city tax and 2% for state tax. How much more state tax does he pay than city tax? **$4.87**

Applying Mathematics to the Real World

This optional page shows how mathematics is used in the real world and also provides a change of pace.

Objective Find the amount for city and state income tax.

Using Discussion Discuss the various services provided by the city and state that are paid for with the income tax revenues. **police and fire protection, and so on**

Using Calculators Have students calculate taxes using a calculator.

Activity

Using Questioning Have students find the income tax rate levied by your town or city and state. Also, have them compare this rate with neighboring cities and/or states. Decide whether your tax rate is high, low, or average.

Using the Cumulative Review

This page provides an aid for maintaining skills and concepts presented thus far in the text.

A Cumulative Review is also provided in Evaluation Masters Booklet as shown below.

Evaluation Masters Booklet, p. 60

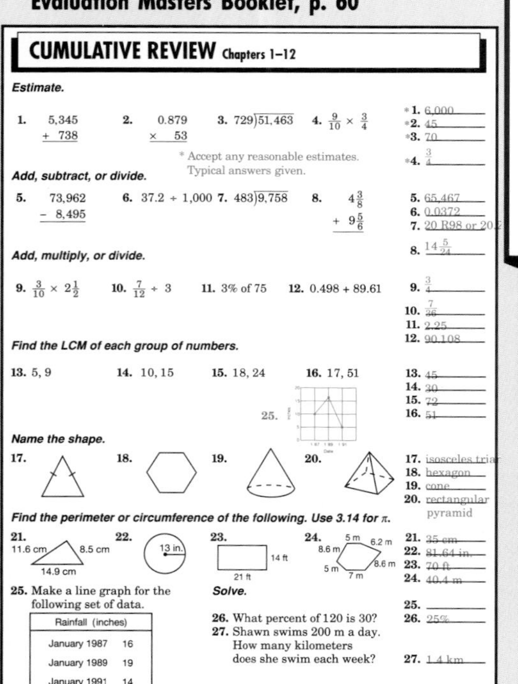

CUMULATIVE REVIEW Chapters 1–12

Estimate.

1. 5,345 + 738
2. 0.879 × 53
3. 729)51,463
4. $\frac{9}{10} \times \frac{3}{4}$

* Accept any reasonable estimates. Typical answers given.

Add, subtract, or divide.

5. 73,962 − 8,495
6. 37.2 ÷ 1,000
7. 483)9,758
8. $4\frac{3}{8} + 9\frac{5}{6}$

Add, multiply, or divide.

9. $\frac{3}{10} \times 2\frac{1}{2}$
10. $\frac{7}{12} \div 3$
11. 3% of 75
12. 0.498 + 89.61

Find the LCM of each group of numbers.

13. 5, 9
14. 10, 15
15. 18, 24
16. 17, 51

Name the shape.

17. 18. 19. 20.

Find the perimeter or circumference of the following. Use 3.14 for π.

21. 11.6 cm, 8.5 cm, 14.9 cm
22. 13 in.
23. 14 ft, 21 ft
24. 5 m, 6.2 m, 8.6 m, 8.6 m, 5 m, 7 m

25. Make a line graph for the following set of data.

Rainfall (inches)	
January 1987	16
January 1989	19
January 1991	14

Solve.

26. What percent of 120 is 30?
27. Shawn swims 200 m a day. How many kilometers does she swim each week?

Answers: *1. 6,000 *2. 45 *3. 70 *4. $\frac{3}{4}$ 5. 65,467 6. 0.0372 7. 20 R98 or 20.2 8. $14\frac{5}{24}$ 9. $\frac{3}{4}$ 10. $\frac{7}{36}$ 11. 2.25 12. 90.108 13. 45 14. 30 15. 72 16. 51 17. isosceles triangle 18. hexagon 19. cone 20. rectangular pyramid 21. 35 cm 22. 81.64 in. 23. 70 ft 24. 40.4 m 25. (graph) 26. 25% 27. 1.4 km

Cumulative REVIEW/TEST

Free Response

Lesson 1-12 *Solve each equation. Check your solution.*

1. $5n = 25$ 5 **2.** $\frac{t}{4} = 9$ 36 **3.** $8y = 72$ 9

Lesson 2-3 *Add.*

4. 93 + 89 182 **5.** 232 + 295 527 **6.** 1,751 + 156 1,907

7. 1,638 + 4,534 6,172 **8.** 83,346 + 1,714 85,060 **9.** 54,648 + 77,059 131,707

Lesson 4-9 *Estimate.*

10. $2\frac{1}{2} + 19\frac{1}{8}$ $21\frac{1}{2}$ **11.** $27\frac{3}{4} + 4\frac{1}{8}$ 32 **12.** $30 - 7\frac{9}{11}$ 22 **13.** $10\frac{1}{8} - 6\frac{2}{3}$ 3

Lesson 5-2 *Multiply.*

14. $\frac{1}{4} \times \frac{1}{4}$ $\frac{1}{16}$ **15.** $\frac{1}{2} \times \frac{1}{4}$ $\frac{1}{8}$ **16.** $\frac{3}{8} \times \frac{5}{6}$ $\frac{5}{16}$ **17.** $\frac{1}{6} \times \frac{1}{2}$ $\frac{1}{12}$

Lesson 8-2

18. Rudy's yard is in the shape of a parallelogram. How much area does his yard cover if the base length is 45 feet and the height is 120 feet? 5,400 ft²

Lesson 9-5 *Find the distance on a scale drawing for each actual distance. The scale is 1 in.: 75 ft.*

19. 300 ft 4 in. **20.** 50 ft $\frac{2}{3}$ in.

Lesson 10-7

21. Andy buys hockey equipment on sale for 30% off the original price. He buys skates that regularly sell for $85 and he buys a stick that had an original price of $58. How much did he pay for the two items? $100.10

Lesson 10-8 *Find the simple interest earned on each deposit.*

22. principal: $220
annual rate: 8%
time: 1 year $17.60

23. principal: $220
annual rate: 8%
time: 4 years $70.40

Lesson 11-1 *Find the mean, median, and mode for each set of data.*

24. 75, 90, 82, 79, 81, 82 81.5; 81.5; 82

25. 2, 6, 7, 9, 5, 6, 7, 9, 9, 4 6.4; 6.5; 9

26. −4, 7, 4, −7, 2, 8, 4 2; 4; 4

Lesson 12-2

27. At 6 A.M. the temperature was −4°F. At noon the temperature was 28°F. What was the change in temperature? 32°

APPLYING THE LESSON

Independent Practice

Homework Assignment	
Minimum	Ex. 1-17 odd; 18-27
Average	Ex. 1-27
Maximum	Ex. 1-27

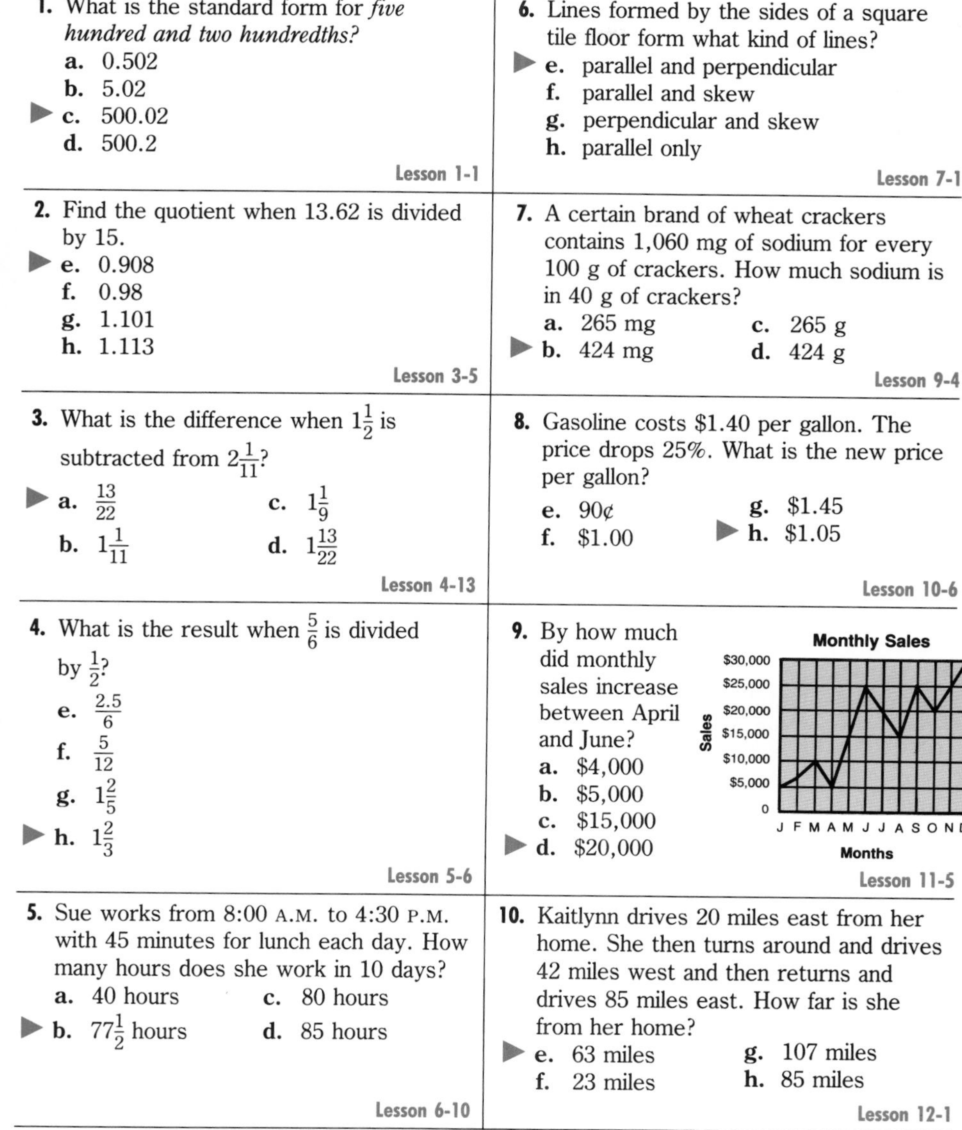

Chapter 12

Multiple Choice

Choose the letter of the correct answer for each item.

1. What is the standard form for *five hundred and two hundredths?*
   a. 0.502
   b. 5.02
   ▶ c. 500.02
   d. 500.2

   Lesson 1-1

2. Find the quotient when 13.62 is divided by 15.
   ▶ e. 0.908
   f. 0.98
   g. 1.101
   h. 1.113

   Lesson 3-5

3. What is the difference when $1\frac{1}{2}$ is subtracted from $2\frac{1}{11}$?
   ▶ a. $\frac{13}{22}$ c. $1\frac{1}{9}$
   b. $1\frac{1}{11}$ d. $1\frac{13}{22}$

   Lesson 4-13

4. What is the result when $\frac{5}{6}$ is divided by $\frac{1}{2}$?
   e. $\frac{2.5}{6}$
   f. $\frac{5}{12}$
   g. $1\frac{2}{5}$
   ▶ h. $1\frac{2}{3}$

   Lesson 5-6

5. Sue works from 8:00 A.M. to 4:30 P.M. with 45 minutes for lunch each day. How many hours does she work in 10 days?
   a. 40 hours c. 80 hours
   ▶ b. $77\frac{1}{2}$ hours d. 85 hours

   Lesson 6-10

6. Lines formed by the sides of a square tile floor form what kind of lines?
   ▶ e. parallel and perpendicular
   f. parallel and skew
   g. perpendicular and skew
   h. parallel only

   Lesson 7-1

7. A certain brand of wheat crackers contains 1,060 mg of sodium for every 100 g of crackers. How much sodium is in 40 g of crackers?
   a. 265 mg c. 265 g
   ▶ b. 424 mg d. 424 g

   Lesson 9-4

8. Gasoline costs $1.40 per gallon. The price drops 25%. What is the new price per gallon?
   e. 90¢ g. $1.45
   f. $1.00 ▶ h. $1.05

   Lesson 10-6

9. By how much did monthly sales increase between April and June?
   a. $4,000
   b. $5,000
   c. $15,000
   ▶ d. $20,000

   Lesson 11-5

10. Kaitlynn drives 20 miles east from her home. She then turns around and drives 42 miles west and then returns and drives 85 miles east. How far is she from her home?
    ▶ e. 63 miles g. 107 miles
    f. 23 miles h. 85 miles

    Lesson 12-1

CUMULATIVE REVIEW/TEST 397

APPLYING THE LESSON

Independent Practice

Homework Assignment	
Minimum	Ex. 1-10
Average	Ex. 1-10
Maximum	Ex. 1-10

Using the Cumulative Test

This test serves to familiarize students with standardized format while testing skills and concepts presented up to this point.

Evaluation Masters Booklet, pp. 56-57

Name ____________ Date ____________

CHAPTER 12 TEST, FORM 1

Write the letter of the correct answer on the blank at the right of the page.

1. Replace ○ with >, <, or = to make a true sentence. –7 ○ 5
   A. > B. < C. = 1. B
2. Replace ○ with >, <, or = to make a true sentence. –6 ○ –8
   A. > B. < C. = 2. A
3. Replace ○ with >, <, or = to make a true sentence. |–3| ○ |2|
   A. > B. < C. = 3. A
4. Order 2, –6, 7, 0, – 4 from least to greatest.
   A. –4, –6, 0, 2, 7 B. –6, –4, 0, 2, 7 C. 7, 2, 0, –4, –6 D. 7, 2, 0, –6, –4 4. B
5. Order 1, –1, 2, 0, –3 from least to greatest.
   A. 0, –1, 1, 2, –3 B. 2, 1, 0, –1, –3
   C. –1, 0, 1, 2, –3 D. –3, –1, 0, 1, 2 5. D
6. Add. –9 + 4
   A. –5 B. –13 C. 5 D. 13 6. A
7. Add. –7 + (–6)
   A. –1 B. 1 C. –13 D. 13 7. C
8. Subtract. –3 – (–9)
   A. 6 B. –11 C. –6 D. 11 8. A
9. Subtract. 13 – 18
   A. –31 B. –5 C. 31 D. 5 9. B
10. Subtract. 8 – (–7)
    A. 1 B. –1 C. 15 D. –15 10. C
11. Multiply. 7 × (–5)
    A. 35 B. 2 C. –2 D. –35 11. D
12. Multiply. –4 × (–6)
    A. –10 B. –24 C. 2 D. 24 12. D
13. Multiply. –5 × 3
    A. –15 B. 2 C. –2 D. 15 13. A
14. Divide. –20 ÷ (–5)
    A. –4 B. 25 C. 4 D. –25 14. C

Chapter

13 EXTENDING ALGEBRA

PREVIEWING the CHAPTER

In this chapter, students solve one-step equations containing integers using addition, subtraction, multiplication, or division. The standard order of operations is applied in reverse to solve two-step equations. The chapter continues with a discussion of graphing ordered pairs of numbers and equations on the coordinate plane. Reading a grid map provides students with a real-world application of the coordinate plane.

Problem-Solving Strategy Students learn to solve verbal problems by writing equations.

Lesson (Pages)	Lesson Objectives	State/Local Objectives
13-1 (400-401)	13-1: Solve equations involving integers and addition or subtraction.	
13-2 (402-403)	13-2: Solve equations involving integers and multiplication or division.	
13-3 (404-405)	13-3: Solve equations involving integers and two operations.	
13-4 (408-409)	13-4: Name and graph points on the coordinate plane by using ordered pairs.	
13-5 (410-411)	13-5: Graph linear equations.	
13-6 (412-413)	13-6: Locate places on a grid map.	
13-7 (414-415)	13-7: Solve verbal problems by writing an equation.	

ORGANIZING the CHAPTER

You may refer to the ***Course Planning Calendar*** on page T10.

Planning Guide

Lesson (Pages)	Extra Practice (Student Edition)	Blackline Masters Booklets: Reteaching	Practice	Enrichment	Activity	Multi-cultural Activity	Technology	Tech Prep	Evaluation	Other Resources
13-1 (400-401)	p. 481	p. 121	p. 133	p. 121	p. 25					Transparency 13-1
13-2 (402-403)	p. 482	p. 122	p. 134	p. 122			p. 13	p. 25		Transparency 13-2
13-3 (404-405)	p. 482	p. 123	p. 135	p. 123						Transparency 13-3
13-4 (408-409)	p. 482	p. 124	p. 136	p. 124	p. 26		p. 27		p. 64	Transparency 13-4
13-5 (410-411)	p. 483	p. 125	p. 137	p. 125				p. 26		Transparency 13-5
13-6 (412-413)		p. 126	p. 138	p. 126						Transparency 13-6
13-7 (414-415)			p. 139		p. 41	p. 13			p. 64	Transparency 13-7
Ch. Review (416-417)									pp. 61-63	Test and Review Generator
Ch. Test (418)										
Cumulative Review/Test (420-421)									p. 65	

OTHER CHAPTER RESOURCES

Student Edition

Chapter Opener, p. 398
Activity, p. 399
Consumer Connections, p. 406
On-the-Job Application, p. 407
Journal Entry, p. 401
Portfolio Suggestion, p. 415
Social Studies Connections, p. 419
Tech Prep Handbook, pp. 534-536

Teacher's Classroom Resources

Transparency 13-0
Lab Manual, pp. 47-48
Performance Assessment, pp. 25-26
Lesson Plans, pp. 133-139

Other Supplements

Overhead Manipulative Resources, Labs 37-38, pp. 27-28
Glencoe Mathematics Professional Series

ENHANCING the CHAPTER

Some of the blackline masters for enhancing this chapter are shown below.

COOPERATIVE LEARNING

This game can be played by groups of two students. Have students draw a 10 x 10 coordinate plane. Each player in turn names an ordered pair and marks that point with a chosen symbol (X or O). The player who gets 5 points in a row, column, or diagonal is the winner. Have that student draw a straight line through the points. Both students then make a table of the five ordered pairs that lie on the line and use those values to find the equation that represents the line.

USING MODELS/MANIPULATIVES

Have students use algebra tiles to model the steps in the solution of one-step and of two-step equations. Students should demonstrate the use of inverse operations to get the variable alone on one side of the equation. Remind students that a 1-piece and a –1-piece added together equal 0. When modeling two-step equations, have students use several variable pieces to represent the number of times the variable is used as a factor. Students may work in small groups and explain their models to each other.

Cooperative Problem Solving, p. 41

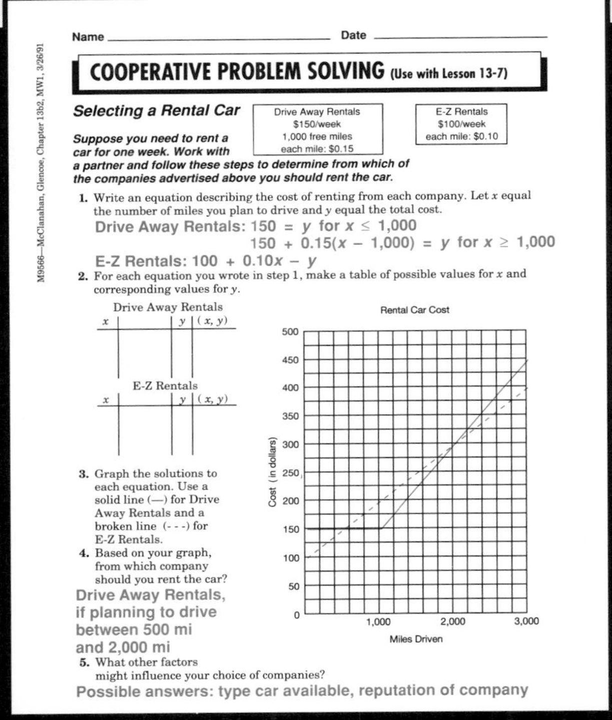

M9566—McClanahan, Glencoe, Chapter 13b2, MW1, 3/26/91

Name ______ Date ______

COOPERATIVE PROBLEM SOLVING (Use with Lesson 13-7)

Selecting a Rental Car

Drive Away Rentals
$150/week
1,000 free miles
each mile: $0.15

E-Z Rentals
$100/week
each mile: $0.10

Suppose you need to rent a car for one week. Work with a partner and follow these steps to determine from which of the companies advertised above you should rent the car.

1. Write an equation describing the cost of renting from each company. Let x equal the number of miles you plan to drive and y equal the total cost.
   Drive Away Rentals: $150 = y$ for $x \leq 1{,}000$
   $150 + 0.15(x - 1{,}000) = y$ for $x \geq 1{,}000$
   E-Z Rentals: $100 + 0.10x - y$
2. For each equation you wrote in step 1, make a table of possible values for x and corresponding values for y.

Drive Away Rentals

x	y	(x, y)

E-Z Rentals

x	y	(x, y)

3. Graph the solutions to each equation. Use a solid line (—) for Drive Away Rentals and a broken line (- - -) for E-Z Rentals.
4. Based on your graph, from which company should you rent the car?
   Drive Away Rentals, if planning to drive between 500 mi and 2,000 mi
5. What other factors might influence your choice of companies?
   Possible answers: type car available, reputation of company

Lab Activity, p. 48

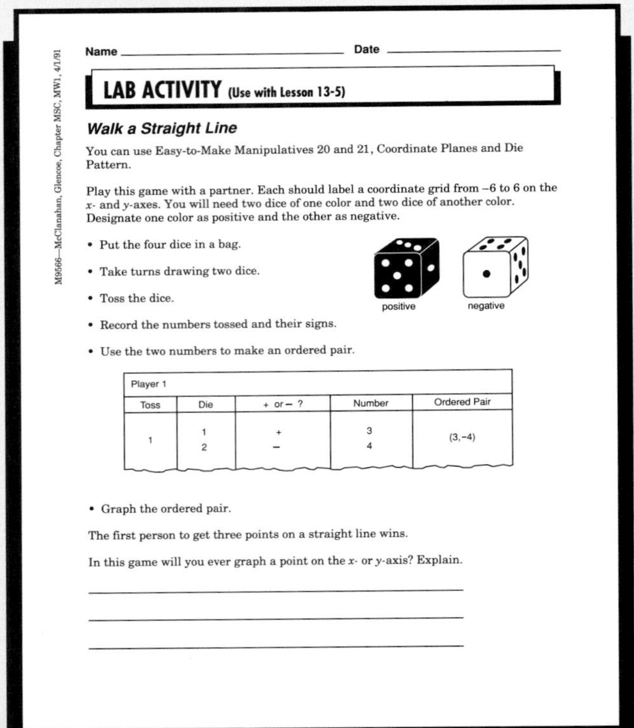

M9566—McClanahan, Glencoe, Chapter MSC, MW1, 4/1/91

Name ______ Date ______

LAB ACTIVITY (Use with Lesson 13-5)

Walk a Straight Line

You can use Easy-to-Make Manipulatives 20 and 21, Coordinate Planes and Die Pattern.

Play this game with a partner. Each should label a coordinate grid from –6 to 6 on the x- and y-axes. You will need two dice of one color and two dice of another color. Designate one color as positive and the other as negative.

- Put the four dice in a bag.
- Take turns drawing two dice.
- Toss the dice.
- Record the numbers tossed and their signs.
- Use the two numbers to make an ordered pair.

Player 1				
Toss	Die	+ or – ?	Number	Ordered Pair
1	1 2	+ –	3 4	(3, –4)

- Graph the ordered pair.

The first person to get three points on a straight line wins.

In this game will you ever graph a point on the x- or y-axis? Explain.

______

______

______

MEETING INDIVIDUAL NEEDS

Limited English Proficiency Students

When presenting verbal problems that can be solved with equations, give variable, number, and operation sign cards to limited English proficiency students. As you work on the problems, have students hold up the appropriate cards to make algebraic sentences. When assigning verbal problem exercises, have limited English proficiency students work in pairs with non-limited English proficiency students translating English sentences into algebraic sentences. Encourage them to use synonyms such as minus and less. Students may need help noting the difference between "less" and "less than."

CRITICAL THINKING/PROBLEM SOLVING

Point to students that in order to solve a verbal problem by writing and solving an equation they must be able to translate the words into corresponding numbers and symbols. One method for doing so is to underline the unknown and the mathematical relationship in the problem. To write an equation, students should pay attention to key words that describe the operations of addition, subtraction, multiplication, and division. Then they should translate from the problem words to algebraic symbols.

COMMUNICATION

Speaking

Have students solve one-step and two-step equations at the chalkboard and describe aloud to the class the procedure used and the reasons for each step. If you choose to conduct the Problem-Solving Strategy lesson as an in-class activity, have students read each problem aloud and discuss possible methods of solution.

Applications, p. 25

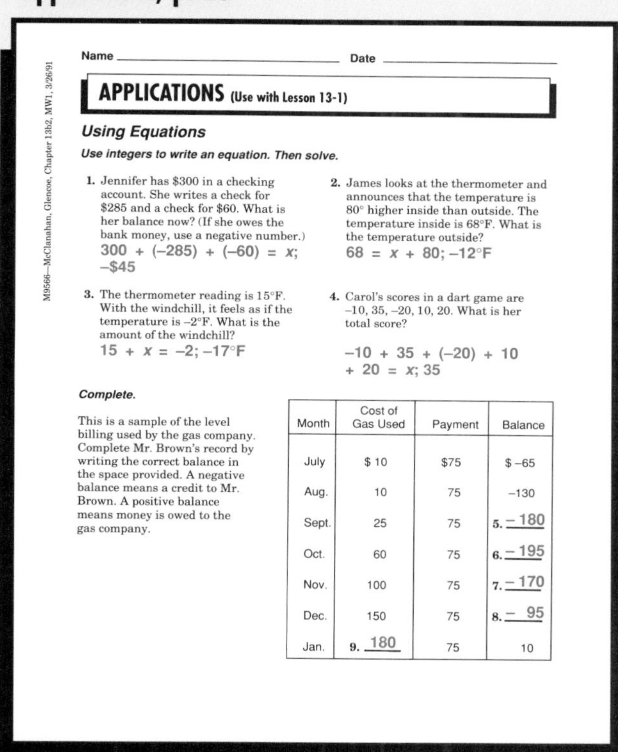

M9566—McClanahan, Glencoe, Chapter 13b2, MW1, 3/26/91

Name ______________ Date ______________

APPLICATIONS (Use with Lesson 13-1)

Using Equations

Use integers to write an equation. Then solve.

1. Jennifer has $300 in a checking account. She writes a check for $285 and a check for $60. What is her balance now? (If she owes the bank money, use a negative number.)
   $300 + (-285) + (-60) = x$; -$45

2. James looks at the thermometer and announces that the temperature is 80° higher inside than outside. The temperature inside is 68°F. What is the temperature outside?
   $68 = x + 80$; -12°F

3. The thermometer reading is 15°F. With the windchill, it feels as if the temperature is -2°F. What is the amount of the windchill?
   $15 + x = -2$; -17°F

4. Carol's scores in a dart game are -10, 35, -20, 10, 20. What is her total score?
   $-10 + 35 + (-20) + 10 + 20 = x$; 35

Complete.

This is a sample of the level billing used by the gas company. Complete Mr. Brown's record by writing the correct balance in the space provided. A negative balance means a credit to Mr. Brown. A positive balance means money is owed to the gas company.

Month	Cost of Gas Used	Payment	Balance
July	$ 10	$75	$ -65
Aug.	10	75	-130
Sept.	25	75	5. -180
Oct.	60	75	6. -195
Nov.	100	75	7. -170
Dec.	150	75	8. - 95
Jan.	9. 180	75	10

Calculator Activity, p. 13

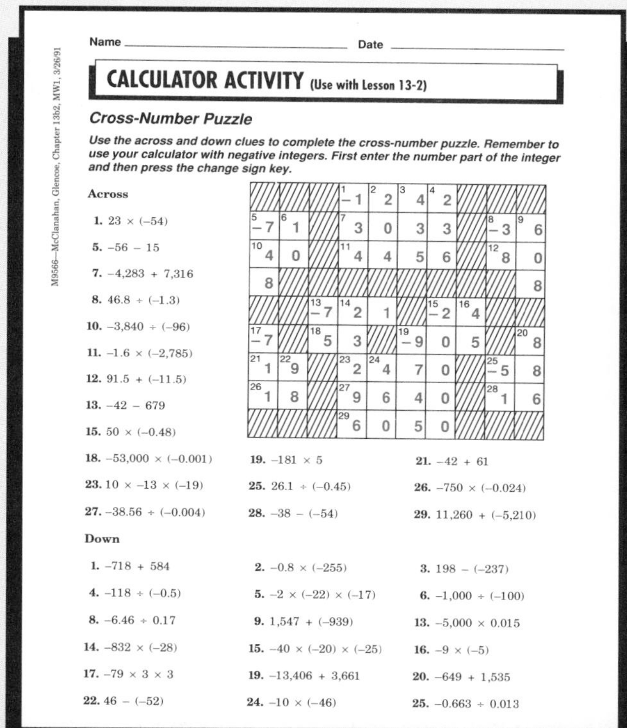

M9566—McClanahan, Glencoe, Chapter 13b2, MW1, 3/26/91

Name ______________ Date ______________

CALCULATOR ACTIVITY (Use with Lesson 13-2)

Cross-Number Puzzle

Use the across and down clues to complete the cross-number puzzle. Remember to use your calculator with negative integers. First enter the number part of the integer and then press the change sign key.

			[1] −1	[2] 2	[3] 4	[4] 2			
[5] −7	[6] 1		[7] 3	0	3	3		[8] −3	[9] 6
[10] 4	0		[11] 4	4	5	6		[12] 8	0
8									8
		[13] −7	[14] 2	1		[15] −2	[16] 4		
[17] −7		[18] 5	3		[19] −9	0	5		[20] 8
[21] 1	[22] 9		[23] 2	[24] 4	7	0		[25] −5	8
[26] 1	8		[27] 9	6	4	0		[28] 1	6
			[29] 6	0	5	0			

Across

1. $23 \times (-54)$
5. $-56 - 15$
7. $-4{,}283 + 7{,}316$
8. $46.8 \div (-1.3)$
10. $-3{,}840 \div (-96)$
11. $-1.6 \times (-2{,}785)$
12. $91.5 + (-11.5)$
13. $-42 - 679$
15. $50 \times (-0.48)$
18. $-53{,}000 \times (-0.001)$
19. -181×5
21. $-42 + 61$
23. $10 \times -13 \times (-19)$
25. $26.1 \div (-0.45)$
26. $-750 \times (-0.024)$
27. $-38.56 \div (-0.004)$
28. $-38 - (-54)$
29. $11{,}260 + (-5{,}210)$

Down

1. $-718 + 584$
2. $-0.8 \times (-255)$
3. $198 - (-237)$
4. $-118 \div (-0.5)$
5. $-2 \times (-22) \times (-17)$
6. $-1{,}000 \div (-100)$
8. $-6.46 \div 0.17$
9. $1{,}547 + (-939)$
13. $-5{,}000 \times 0.015$
14. $-832 \times (-28)$
15. $-40 \times (-20) \times (-25)$
16. $-9 \times (-5)$
17. $-79 \times 3 \times 3$
19. $-13{,}406 + 3{,}661$
20. $-649 + 1{,}535$
22. $46 - (-52)$
24. $-10 \times (-46)$
25. $-0.663 \div 0.013$

CHAPTER PROJECT

Have students collect formulas in two variables that correspond to one-step and two-step equations. Have them make a table of values for each formula and graph the values on a coordinate plane. Then students should draw the corresponding graph. Possible topics that can be modeled by formulas are interest earned on principal, distances sound travels in water, postage costs, braking distance and so on.

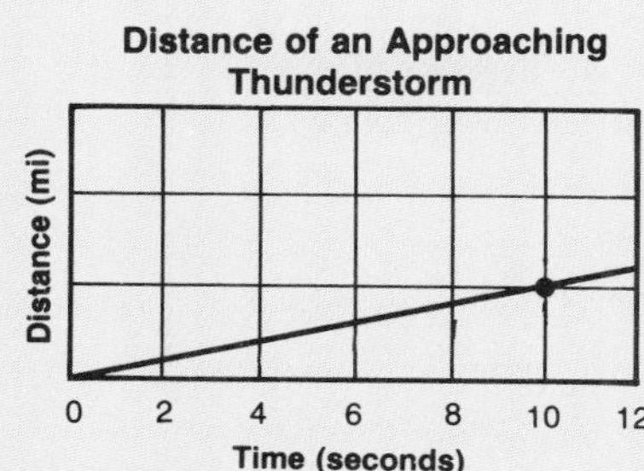

Using the Chapter Opener

Transparency 13-0 is available in the Transparency Package. It provides a full-color visual and motivational activity that you can use to engage students in the mathematical content of the chapter.

Using Discussion Have students read the opening paragraph.

- Discuss with students how and when they use simple equations in everyday life. Their own activities are a good source for this discussion. For example, the photography club has some money and wants to sponsor an event. They need to know how much they have, how much they need, and how much more they have to earn to sponsor the event.

- Point out to students that they use equations often without being aware of it. Some equation-solving operations have become so automatic that they are used without conscious effort.

The second paragraph is a motivational problem. Have students solve the problem by writing an equation first and then solving. Have students discuss the equation they used.

Chapter 13

EXTENDING ALGEBRA

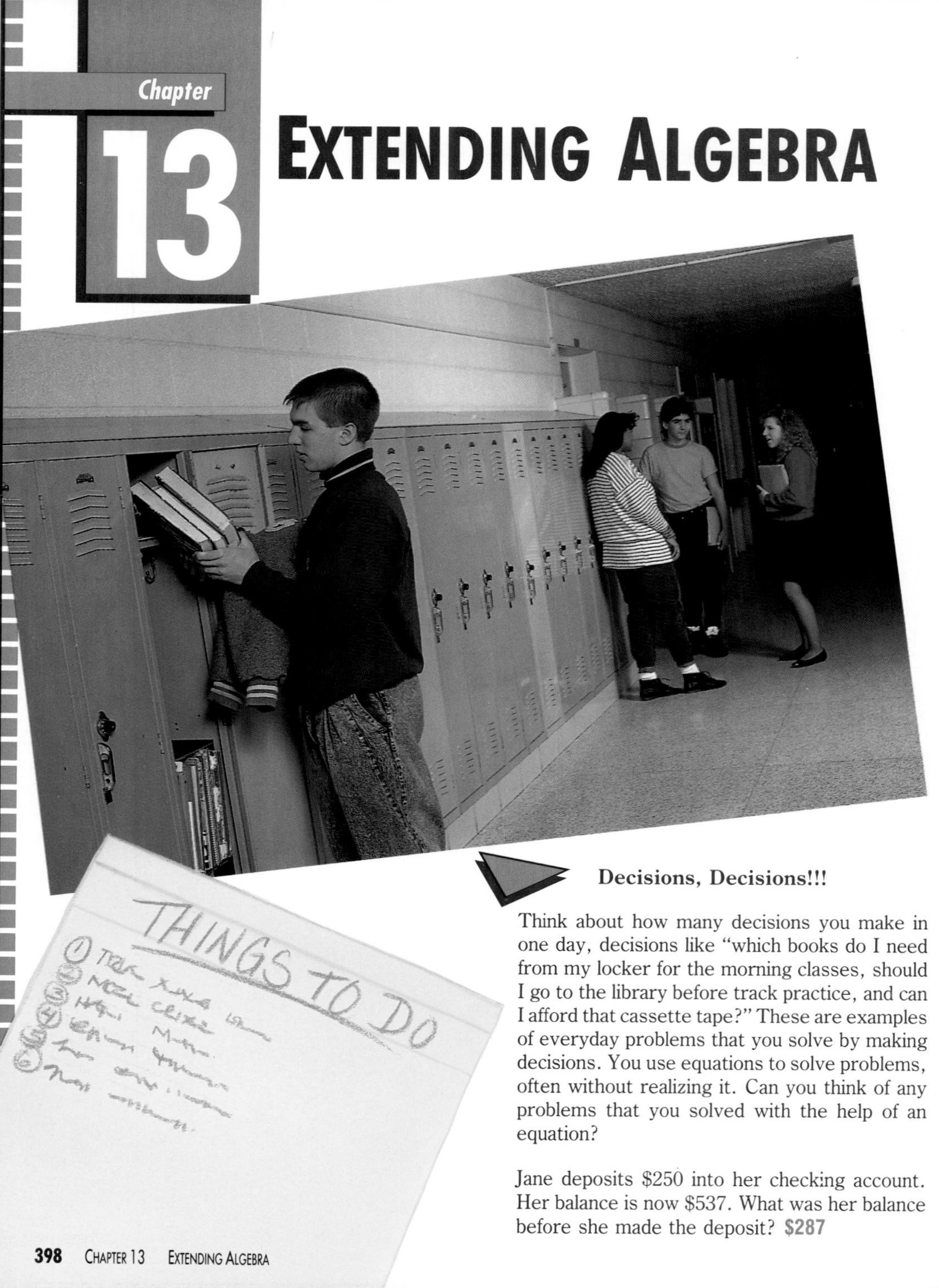

Decisions, Decisions!!!

Think about how many decisions you make in one day, decisions like "which books do I need from my locker for the morning classes, should I go to the library before track practice, and can I afford that cassette tape?" These are examples of everyday problems that you solve by making decisions. You use equations to solve problems, often without realizing it. Can you think of any problems that you solved with the help of an equation?

Jane deposits \$250 into her checking account. Her balance is now \$537. What was her balance before she made the deposit? **\$287**

ACTIVITY: Equations

In this activity you will translate verbal sentences to mathematical sentences. You will also discover an "equational" magic trick to amaze your family and friends.

Cooperative Groups

Work in groups of three or four.

1. With your group, discuss how these two sentences are alike and how they are different.

Art threw the football.
The football was thrown by Art.

2. Just as different sentences may have the same meaning, different equations may have the same meaning or *solution*. Look at the equation $2x + 6 = 10$. Another equation that has the same solution is $2x = 4$.

a. What operation, addition, subtraction, multiplication, or division, is used to get $2x = 4$ from the first equation? **subtraction of 6**

b. What operation is used to get $x = 2$? **division by 2**

Answers may vary. Typical answers are given.

3. With your group, write another equation that has the same solution as each given equation. **a. $2t = 9$ b. $6y = 30$ c. $d = 10$**

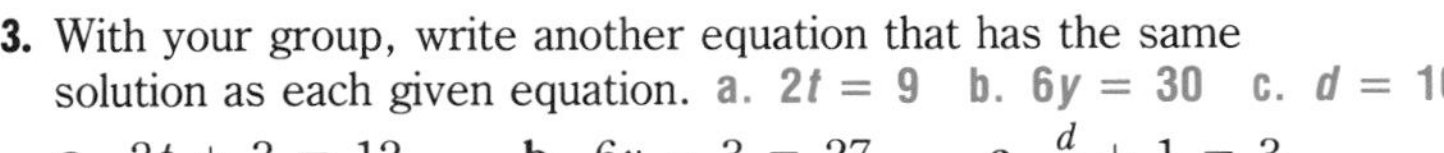

a. $2t + 3 = 12$ **b.** $6y - 3 = 27$ **c.** $\frac{d}{5} + 1 = 3$

When did Houdini live?
See page 423.

4. Designate one member of your group as "Houdini" (the master magician). Houdini reads these instructions. "Choose any number. Add 3 to it. Multiply the sum by 100. Subtract 300 from the product. Divide the difference by 10. Write the result. I can tell you the original number." Houdini mentally divides the result by 10 and tells the person his or her original number.

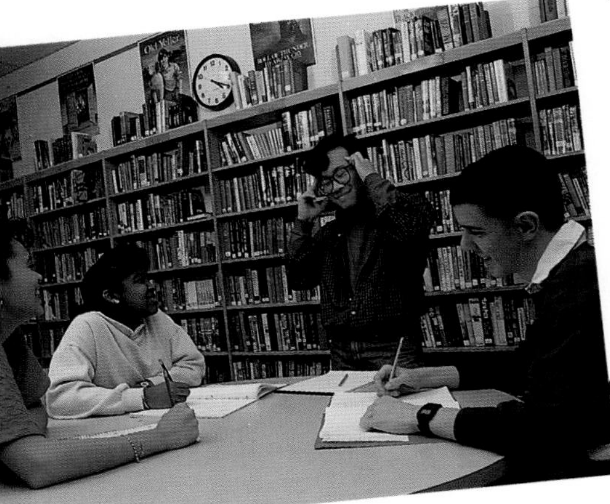

5. The "secret" is that Houdini used an equation that has the same solution to create his magic. Look at the instructions.

Choose a number. $\rightarrow n$
Add 3 to it. $\rightarrow n + 3$
Multiply by 100. $\rightarrow 100(n + 3)$ or $100n + 300$
Subtract 300. $\rightarrow 100n$
Divide by 10. $\rightarrow 10n$

Note that under these specific conditions, the result will always be 10 times the original number. It's easy to divide mentally by 10 and give the original number as if by magic.

6. With your group, create your own magic instructions. Test it on your group. See if other groups can figure out your equivalent equations.

INTRODUCING THE LESSON

Objective Write equivalent equations.

TEACHING THE LESSON

Using Cooperative Groups After completing Exercise 3, have each student in a group make up three equations. Then have other students in the group write equivalent equations.

EVALUATING THE LESSON

Using Discussion Discuss with students that each form of an equation they write during the solving process is an equivalent form of the equation. It is often convenient to use an equation in a form other than the one that is given. In the Teacher margin for Lesson 13-5 on p. 411, the Enrichment Activity suggests finding the slope of a line. This activity can use equivalent equations, that is, changing $Ax + By = C$ to $y = mx + b$.

EXTENDING THE LESSON

Closing the Lesson Allow time for students to discuss their magic equations and how they created them.

Activity Worksheet, Lab Manual, p. 48

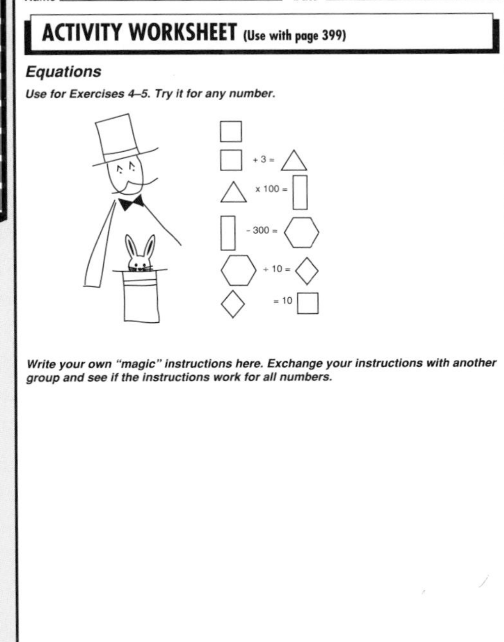
Name ____________ Date ____________

ACTIVITY WORKSHEET (Use with page 399)

Equations

Use for Exercises 4–5. Try it for any number.

Write your own "magic" instructions here. Exchange your instructions with another group and see if the instructions work for all numbers.

5-MINUTE CHECK-UP

(over Lesson 12-7)

Available on TRANSPARENCY 13-1.

1. David thinks of a number. If he multiplies it by 4, then subtracts 20, and finally adds 5, he gets 9. What is the number? **6**
2. Kristie has a paper route. Each week she earns $8.00 plus 5 cents for each paper she delivers. Last week she earned $13.75. How many papers did she deliver? **115 papers**

INTRODUCING THE LESSON

Show students a checkbook register with a negative balance. Is it possible to add money to the account and have a negative balance? **yes** What happens when a checking account has a negative balance? **Checks bounce.**

TEACHING THE LESSON

Using Manipulatives Give each student ten positive disks and ten negative disks, or use the Integer Counters from the **Lab Manual, p. 19.** Have students use the counters to check the results in Exercises 1-3 and 7-9.

Practice Masters Booklet, p. 133

Name ______________________ Date ______________

PRACTICE WORKSHEET 13-1

Solving Equations Using Addition or Subtraction

Solve each equation. Check your solution.

1. $13 = 9 + a$ **4**	**2.** $15 + b = -38$ **−53**	**3.** $72 = a - 7$ **79**	**4.** $b - 15 = -38$ **−23**
5. $t + (-27) = -10$ **17**	**6.** $4 + c = 9$ **5**	**7.** $c - (-8) = -17$ **−25**	**8.** $-26 + r = 215$ **241**
9. $d - 4 = 9$ **13**	**10.** $e - 117 = -215$ **−98**	**11.** $28 + x = 9$ **−19**	**12.** $f - 57 = -121$ **−64**
13. $8 = s + 1\frac{3}{5}$ **$6\frac{2}{5}$**	**14.** $k + (-0.7) = -1.1$ **−0.4**	**15.** $4 = g - 3\frac{3}{5}$ **$7\frac{3}{5}$**	**16.** $h - 35 = 27$ **62**
17. $-13.2 + y = 10$ **23.2**	**18.** $n - 2.7 = 2.7$ **5.4**	**19.** $p - (-0.3) = -0.8$ **−1.1**	**20.** $g + 0.3 = 0.8$ **0.5**
21. $q - 6.6 = -1.4$ **5.2**	**22.** $t + 86 = -14$ **−100**	**23.** $v + (-1.7) = 3.2$ **4.9**	**24.** $r - 7.1 = 3.2$ **10.3**

Solve.

25. The high temperature today after a morning low of −10°F, is 25°F. How many degrees did the temperature rise? **35°F**

26. There were 27 students in the library. Four left and then another group came in, bringing the total to 45. How many were in that group? **22 students**

27. Bob is on a weight-loss program. He has lost 15.2 pounds so far. If he now weighs 197 pounds, how much did he weigh when he began? **212.2 lb**

28. Jane has $382.57 in her checking account. After writing a check for $451.25, what is her balance? **−$68.68**

13-1 SOLVING EQUATIONS USING ADDITION OR SUBTRACTION

Objective *Solve equations involving integers and addition or subtraction.*

Dawn was playing a game. In one turn she lost 17 points. Her accumulated score after losing 17 points was −9. What was her score before losing 17 points?

Solving an equation means finding the replacement for the variable that results in a true sentence. To solve an equation, get the variable by itself on one side of the equals sign. If a number has been added to the variable, subtract. If a number has been subtracted from the variable, add.

Method

1. Add or subtract the same number on each side of the equation to get the variable by itself.
2. Check the solution.

Example A ***Find Dawn's score before losing 17 points. Let x = Dawn's score before losing 17 points.***

$x - 17 = -9$ — 17 is subtracted from x.

1. $x - 17 + 17 = -9 + 17$ — Add 17 to each side.

   $x = 8$

2. **Check:** $x - 17 = -9$ — In the original equation,

   $8 - 17 \stackrel{?}{=} -9$ — replace x with 8.

   $-9 = -9$ ✓ — The solution is 8.

Dawn's score before losing 17 points was 8.

Example B ***Solve −210 + x = 100.***

$-210 + x = 100$ — −210 is added to x.

1. $-210 - (-210) + x = 100 - (-210)$ — Subtract −210 from each side.

   $x = 310$

2. **Check:** $-210 + x \stackrel{?}{=} 100$

   $-210 + 310 \stackrel{?}{=} 100$ — Replace x with 310.

   $100 = 100$ ✓ — The solution is 310.

Guided Practice

Solve each equation. Check your solution.

Example A

1. $x - 7 = 2$ **9** **2.** $-1 = x - 7$ **6** **3.** $y - 3 = -8$ **−5**

4. $t - 25 = -5$ **20** **5.** $-8 = p - 8$ **0** **6.** $40 = y - 117$ **157**

Example B

7. $x + 7 = 4$ **−3** **8.** $7 + t = -2$ **−9** **9.** $4 = p + 8$ **−4**

10. $1 + y = -7$ **−8** **11.** $x + 0 = -35$ **−35** **12.** $0 = y + (-29)$ **29**

RETEACHING THE LESSON

Complete.

1. $a - 9 = 3$

$a - 9 + \underline{9} = 3 + \underline{9}$

$a = \underline{12}$

2. $x + 7 = 5$

$x + 7 - \underline{7} = 5 - \underline{7}$

$x = \underline{-2}$

Solve each equation. Check your solution.

3. $d - 5 = -14$ **−9**

4. $-3.6 + s = -8.5$ **−4.9**

Reteaching Masters Booklet, p. 121

Name ______________________ Date ______________

RETEACHING WORKSHEET 13-1

Solving Equations Using Addition or Subtraction

Method:
① Identify the variable.
② To get the variable by itself, add the same number to, or subtract the same number from, each side of the equation.
③ Check the solution.

Examples: Solve each equation.

A. ① $x - 9 = -13$ — 9 is subtracted from x, so
② $x - 9 + 9 = -13 + 9$ — add 9 to each side.
$x = -4$ — The solution is −4.
③ **Check.** $x - 9 = -4 - 9 = -13$ — Replace x with −4.

B. ① $-75 + x = 50$ — −75 is added to x, so
② $-75 - (-75) + x = 50 - (-75)$ — subtract −75 from each side.
$x = 125$ — The solution is 125.
③ **Check:** $-75 + x = -75 + 125 = 50$ — Replace x with 125.

Exercises

Practice

25. 383 27. 0
29. −0.8

Solve each equation. Check your solution.

13. $y + 7 = 5$ **−2**
14. $-8 + p = -12$ **−4**
15. $-15 + a = 19$ **34**
16. $y - 6 = -15$ **−9**
17. $-17 = 14 + x$ **−31**
18. $-8 = y + 6$ **−14**
19. $y + (-2) = 5$ **7**
20. $b - (-7) = -3$ **−10**
21. $x + (-4) = -7$ **−3**
22. $-1 + y = -1$ **0**
23. $-2 = a - 17$ **15**
24. $x + (-20) = 6$ **26**
25. $283 = x + (-100)$
26. $-157 + x = 1$ **158**
27. $a = -32 - (-32)$
28. $x + 3.2 = 1.5$ **−1.7**
29. $-1.5 + y = -2.3$
30. $p - 1 = -\frac{1}{3}$ **$\frac{2}{3}$**

Applications

31. Alma made a \$300 deposit to her checking account. After making the deposit, her balance was \$287. What was her balance before she made the deposit? **−\$13**

When was the first love song written? See page 422.

32. The temperature in October at the base of Mt. Washington was 31°F. The temperature at the weather station on the top of the mountain was −7°F. What is the difference in temperature from the base to the top of the mountain? **38°F**

33. Explain how equations are solved using addition or subtraction. Give an example of each. **See students' work.**

34. +0.02, +0.04, −0.10

34. Write the change in the average mortgage interest rate from June 28 to July 5, from July 5 to July 12, from July 12 to July 19.

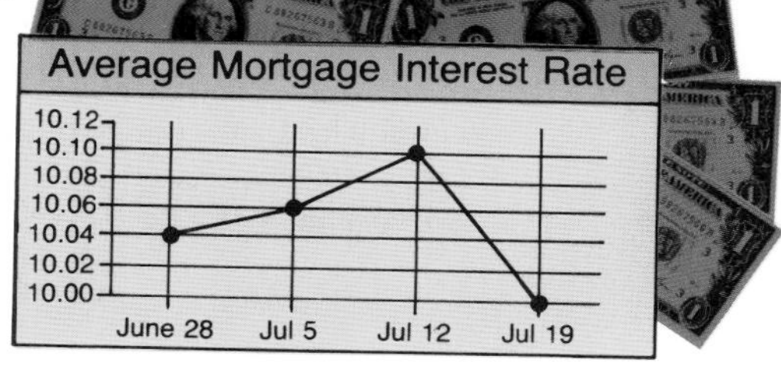

35. Find the change from June 28 to July 19. Compare this change to the sum of the changes from Exercise 34.
−0.04; They are the same.

Mixed Review

Multiply.

Lesson 3-2

36. 4.5×1.8 = **8.1**
37. 16.3×4.7 = **76.61**
38. 19.5×21.6 = **421.2**
39. 0.082×4.1 = **0.3362**

Lesson 8-2

Find the area of each parallelogram described below.

40. base, 12 m
height, 2.1 m
25.2 m²

41. base, 9.3 cm
height, 6.5 cm
60.45 cm²

42. base, 18.1 mm
height, 23.5 mm
425.35 mm²

Lesson 10-3

43. In last year's basketball season, Trevor made 85% of his free throws. How many total free throws did he shoot if he made 34? **40 free throws**

APPLYING THE LESSON

Independent Practice

Homework Assignment	
Minimum	Ex. 1-43 odd
Average	Ex. 2-30 even; 31-43
Maximum	Ex. 1-43

Chalkboard Examples

Example A *Solve each equation. Check your solution.*

1. $n - 8 = -3$ **5**
2. $-4 = -1(-x)$ **−4**

Example B *Solve each equation. Check your solution.*

3. $y + 6 = -2$ **−8**
4. $-3 = 12 + m$ **−15**

Additional examples are provided on TRANSPARENCY 13-1.

EVALUATING THE LESSON

Check for Understanding Use Guided Practice Exercises to check for student understanding.

Error Analysis Watch for students who use the "sign" of a number as an operation. Remind students that the parentheses in $x(-3)$ means multiply a negative 3 and x, not the difference of x and 3.

Closing the Lesson Have students tell about times when they could use equations to solve problems.

EXTENDING THE LESSON

Enrichment Masters Booklet, p. 121

Name ______________ Date ______________

ENRICHMENT WORKSHEET 13-1

Adding Binomials and Trinomials

A **mononomial** is a number, a variable, or a product of numbers and variables. An expression that can be written as a sum of monomials is called a **polynomial.** A **binomial** is a polynomial with two terms, and a **trinomial** is a polynomial with three terms.

Polynomials

Monomial	Binomial	Trinomial
$4x$	$3x + 2$	$5x - 2y + 7$

To add binomials and trinomials, add their like terms.

Examples:

$(5x + 1) + (4x + 3) = 9x + 4$
$(x + 2y + 6) + (3x + y - 2) = 4x + 3y + 4$
$(3x + 2y + 1) + (5x + 2) = 8x + 2y + 3$

Add.

1. $(2x + 3) + (3x + 9)$ **$5x + 12$**
2. $(6x - 1) + (x + 8)$ **$7x + 7$**
3. $(4y + 2) + (6y - 1)$ **$10y + 1$**
4. $(2y - 7) + (7y - 2)$ **$9y - 9$**
5. $(x + y + 8) + (5x + 2y + 3)$ **$6x + 3y + 11$**
6. $(8x + 9) + (x + 4y + 11)$ **$9x + 4y + 20$**
7. $(x + 2y + 4) + (x + 6y - 3)$ **$2x + 8y + 1$**
8. $(3x - y + 7) + (2x + 3y + 3)$ **$5x + 2y + 10$**
9. $(5x + 3y + 2) + (2x + 9)$ **$7x + 3y + 11$**
10. $(4x - y + 6) + (3y - 4)$ **$4x + 2y + 2$**
11. $(6x + 12) + (8y - 2)$ **$6x + 8y + 10$**
12. $(8y + 7) + (x - 13)$ **$x + 8y - 6$**

5-MINUTE CHECK-UP

(over Lesson 13-1)

Available on TRANSPARENCY 13-2.

Solve each equation. Check your solution.

1. $n - 7 = -14$ -7
2. $-1.2 = u - 2.5$ 1.3
3. $p - (-3) = -7$ -10
4. $q + 13 = 6$ -7
5. $9 + x = -5$ -14
6. $-1\frac{2}{3} = y + \frac{2}{3}$ $-2\frac{1}{3}$

Extra Practice, Lesson 13-1, p. 481

▶ INTRODUCING THE LESSON

Show a nickel. Ask how many times you would need to lose a nickel to lose 50 cents. If we think of a lost nickel as -5, and n as the number of times, then $n \times -5 = -50$. **10**

▶ TEACHING THE LESSON

Using Discussion Ask students how their understanding of inverse operations helps them in this lesson. In effect, multiplication and division "undo" each other. Discuss with students that to solve equations they need to "undo" the operations in the equation.

Practice Masters Booklet, p. 134

Name ____________ Date ____________

PRACTICE WORKSHEET 13-2

Solving Equations Using Multiplication or Division

Solve each equation. Check your solution.

1. $9 = \frac{a}{3}$ 27	**2.** $84 = 7a$ 12	**3.** $15b = -60$ −4	**4.** $25 = \frac{b}{10}$ 250
5. $c + (-5) = 9$ −45	**6.** $-4d = 0$ 0	**7.** $\frac{d}{6} = -15$ −90	**8.** $-4 = \frac{e}{-12}$ 48
9. $-11e = -77$ 7	**10.** $10 = g + 7$ 70	**11.** $6 = \frac{3}{4}g$ 8	**12.** $-\frac{2}{3}j = 22$ −33
13. $h + 8 = 25$ 200	**14.** $39 = -13k$ −3	**15.** $75 = 25l$ 3	**16.** $-7 = \frac{j}{2}$ −14
17. $-56 = \frac{k}{-10}$ 560	**18.** $-24 = 2m$ −12	**19.** $\frac{m}{-5} = 2.25$ −11.25	**20.** $27n = 27$ 1
21. $3p = -45$ −15	**22.** $12 = \frac{p}{-6}$ −72	**23.** $11 = \frac{q}{9}$ 99	**24.** $-6q = 3.6$ −0.6

Solve.

25. The product of two numbers is 12. Their sum is −7. What are the numbers?
−3 and −4

26. The quotient of two numbers is −1. Their difference is 8. What are the numbers?
−4 and 4

27. When the cold front moved through, the temperature fell 24°F in 4 hours. What was the average temperature rise or fall per hour?
fall of 6°F

28. Brandon, Maria, Tonya, and Conrad shared the cost of a $12.76 pizza equally. What was each person's share?
$3.19

13-2 SOLVING EQUATIONS USING MULTIPLICATION OR DIVISION

Objective
Solve equations involving integers and multiplication or division.

Dan hopes to lose 12 pounds. If he loses 2 pounds a week, how many weeks will it take for him to lose 12 pounds? This problem can be expressed by the equation $-2x = -12$.

To solve an equation in which the variable is multiplied by a number, divide each side by that number. If the variable is divided by a number, multiply each side by that number.

Method

1. Multiply or divide each side of the equation by the same nonzero number to get the variable by itself.
2. Check the solution.

Example A

Point out that x = the number of weeks.

Find how many weeks it will take Dan to lose 12 pounds.

$-2x = -12$ — x is multiplied by -2.

1. $\frac{-2x}{-2} = \frac{-12}{-2}$ — Divide each side by -2.
$x = 6$

2. **Check:** $-2x = -12$ — In the original equation, replace x with 6.
$-2 \times 6 \stackrel{?}{=} -12$
$-12 = -12$ ✓ The solution is 6.

It will take Dan 6 weeks to lose 12 pounds.

Example B

Solve $\frac{y}{4} = -48$.

$\frac{y}{4} = -48$ — y is divided by 4.

1. $\frac{y}{4} \times 4 = -48 \times 4$ — Multiply each side by 4.
$y = -192$

2. **Check:** $\frac{y}{4} = -48$ — Replace the variable with -192.
$\frac{-192}{4} \stackrel{?}{=} -48$
$-48 = -48$ ✓ The solution is -192.

Guided Practice

Solve each equation. Check your solution.

Example A
1. $2x = 5$ 2.5 **2.** $-60 = 12y$ -5 **3.** $6 = 0.4x$ 15 **4.** $\frac{3y}{5} = -45$ -75

Example B
5. $\frac{y}{10} = -5$ -50 **6.** $\frac{p}{4} = 10$ 40 **7.** $y \div 8 = -24$ -192 **8.** $x \div 5 = 40$ 200

RETEACHING THE LESSON

Complete.

1. $3b = 12$
$3b \div \underline{3} = 12 \div \underline{3}$
$b = \underline{4}$

2. $n \div 3 = 2$
$\underline{3} \times n \div 3 = \underline{3} \times 2$
$n = \underline{6}$

Solve each equation. Check your solution.

3. $0.9m = 27$ 30
4. $x \div (-22) = 7$ -154

Reteaching Masters Booklet, p. 122

RETEACHING WORKSHEET 13-2

Solving Equations Using Multiplication or Division

Method:
① Identify the variable.
② Multiply or divide each side of the equation by the same nonzero number to get the variable by itself.
③ Check the solution.

Examples: Solve each equation.

A.
① $-7x = 42$ — x is multiplied by −7, so
② $\frac{-7x}{-7} = \frac{42}{-7}$ — divide each side by −7.
$x = -6$ — The solution is −6.
③ **Check.** $-7x = -7 \times (-6) = 42$ — Replace the variable with −6.

B.
① $\frac{y}{-3} = -29$ — y is divided by −3, so
② $\frac{y}{-3} \times -3 = -29 \times -3$ — multiply each side by −3.
$y = 87$ — The solution is 87.
③ **Check.** $\frac{y}{-3} = \frac{87}{-3} = -29$ — Replace the variable with 87.

16. $-22\frac{1}{2}$ 17. 8.3 22. −27 24. $\frac{1}{2}$ 28. −0.2

Exercises

Solve each equation. Check your solution.

Practice

9. $-7q = 5$ $-\frac{5}{7}$
10. $\frac{x}{-2} = 13$ −26
11. $4x = -16$ −4
12. $10 = -7x$ $-1\frac{3}{7}$
13. $\frac{1}{2}a = -2$ −4
14. $5 = -35x$ $-\frac{1}{7}$
15. $-36 = \frac{1}{4}x$ −144
16. $15 = -\frac{2}{3}p$
17. $0.1x = 0.83$
18. $3t = \frac{1}{2}$ $\frac{1}{6}$
19. $\frac{1}{3}y = -1$ −3
20. $13y = 0$ 0
21. $5y = -4$ $-\frac{4}{5}$
22. $y \div 9 = -3$
23. $-4 = -16x$ $\frac{1}{4}$
24. $-18p = -9$
25. $\frac{x}{6} = -7$ −42
26. $-4 = \frac{t}{2}$ −8
27. $\frac{2}{3}y = \frac{4}{9}$ $\frac{2}{3}$
28. $-4x = 0.8$
29. $-\frac{3}{2}x = -1$ $\frac{2}{3}$
30. $-0.1p = 10$
31. $-2 = -10y$ $\frac{1}{5}$
32. $-\frac{4}{5}x = -3\frac{1}{5}$

30. −100 32. 4

Applications

33. During the state playoffs, the Wildcats lost 18 yards in 3 plays. What was their average loss per play? loss of 6 yards

34. Troy, Deanna, and Jenny stopped selling T-shirts because their business was losing money. They shared the $141 loss equally. How much did each lose? $47

Number Sense

35. The product of two numbers is −20. Their sum is 1. What are the numbers? 5, −4

Research

36. Find out the average low temperature for each month of the year where you live and make a table to show the temperatures. Determine the amount of change from month to month. See students' work.

Mental Math

Study these examples.

- The square of a negative number is positive. $(-3)^2 = -3 \times (-3) = 9$
- The cube of a negative number is negative. $(-3)^3 = \underbrace{-3 \times (-3)}_{9} \times (-3) = -27$
- The fourth power of a negative number is positive. $(-3)^4 = \underbrace{-3 \times (-3) \times (-3)}_{-27} \times (-3) = 81$

From these examples, we can conclude that:

a. An even power of a negative number is positive.

b. An odd power of a negative number is negative.

Find each product.

37. $-2 \times (-2) \times (-2)$ −8
38. $-4 \times (-4)$ 16
39. $-1 \times (-1) \times (-1) \times (-1) \times (-1)$
40. $-2 \times (-3) \times (-1)$ −6
41. $(-3)^5$ −243
42. $(-7)^2$ 49
43. $(-2)^4$ 16

39. −1

APPLYING THE LESSON

Independent Practice

Homework Assignment	
Minimum	Ex. 1-35 odd; 36-43
Average	Ex. 2-32 even; 33-43
Maximum	Ex. 1-43

Chalkboard Examples

Example A *Solve each equation. Check your solution.*

1. $n(-3) = 15$ −5
2. $-\frac{3}{4} = p\left(\frac{1}{4}\right)$ −3

Example B *Solve each equation. Check your solution.*

3. $\frac{y}{3} = -27$ −81
4. $3.4 = \frac{m}{-2.5}$ −8.5

Additional examples are provided on TRANSPARENCY 13-2.

▶ EVALUATING THE LESSON

Check for Understanding Use Guided Practice Exercises to check for student understanding.

Error Analysis Watch for students who do the operation indicated with the numbers available, instead of applying the inverse operation to each side of the equation.

Closing the Lesson Ask students to state in their own words how to solve an equation that uses multiplication or division.

▶ EXTENDING THE LESSON

Enrichment Masters Booklet, p. 122

Name ______________________ Date ______________________

ENRICHMENT WORKSHEET 13-2

Subtracting Binomials and Trinomials

A rational number can be subtracted by adding its **additive inverse** or opposite. You can find the additive inverse of a number by multiplying the number by –1. For example, the additive inverse of 3 is (–1) × 3, or –3.

To find the additive inverse of a binomial or trinomial, replace each term by its additive inverse.

binomial	additive inverse	trinomial	additive inverse
$x + 2y$	$-x - 2y$	$2x - 3y + 5$	$-2x + 3y - 5$
$2x - 4y$	$-2x + 4y$	$-8x + 5y - 7z$	$8x - 5y + 7z$

You can subtract a binomial or trinomial by adding its additive inverse.

Examples.

A. $4x + 5y - (6x + 8y)$; additive inverse $-6x - 8y$; $4x + 5y + (-6x - 8y) = -2x - 3y$

B. $12a - 3b + 3 - (6a - 12b - 4)$; additive inverse $-6a + 12b + 4$; $12a - 3b + 3 + (-6a + 12b + 4) = 6a + 9b + 7$

Subtract.

1. $2x + 6 - (x + 4)$ $x + 2$
2. $6x - 1 - (2x + 5)$ $4x - 6$
3. $14y + 2 - (6y - 1)$ $8y + 3$
4. $12y - 7 - (-7y - 2)$ $19y - 5$
5. $9x + y + 8 - (5x + 3y - 3)$ $4x - 2y + 11$
6. $6x + 7 - (-x - 4y - 12)$ $7x + 4y + 19$
7. $x + 2y + 4 - (x - 6y + 3)$ $8y + 1$
8. $5x - y + 8 - (2x + 3y - 4)$ $3x - 4y + 12$
9. $5x + 3y + 2 - (2x - 9)$ $3x + 3y + 11$
10. $4x - y + 6 - (-3y - 5)$ $4x + 2y + 11$
11. $9x + 12 - (5y + 3)$ $9x - 5y + 9$
12. $18y - 7 - (-2x + 13)$ $2x + 18y - 20$

5-MINUTE CHECK-UP

(over Lesson 13-2)

Available on TRANSPARENCY 13-3.

Solve each equation. Check your solution.

1. $0.6x = 10$ $16\frac{2}{3}$
2. $-0.92 = 0.04y$ -23
3. $\frac{x}{4} = -1$ -4
4. $y \div 9 = -3$ -27

Extra Practice, Lesson 13-2, p. 482

INTRODUCING THE LESSON

Ask students to list the steps necessary to make a peanut butter sandwich. Then ask how many listed more than one step. Have students give examples of real-life problems that need more than one step to solve.

TEACHING THE LESSON

Using Discussion Discuss the difference in the order of operations for solving equations from the order for evaluating expressions. Emphasize that it is usually best to "undo" addition and subtraction first. Then undo multiplication and division.

Practice Masters Booklet, p. 135

Name ____________ Date ____________

PRACTICE WORKSHEET 13-3

Solving Two-Step Equations

Solve each equation. Check your solution.

1. $\frac{d}{6} + 7 = 18$ 66
2. $2a + 7 = -3$ −5
3. $3(p - 5) = 75$ 30
4. $5t + (-32) = 43$ 15
5. $2 + 5c = 17$ 3
6. $\frac{j}{3} + 4 = -2$ −18
7. $\frac{b}{-5} - 9 = 16$ −125
8. $-2(2 + y) = 6$ −5
9. $-42 = \frac{a}{3} - 9$ −99
10. $25 = -3b - 2$ −9
11. $\frac{c - 20}{-5} = -2$ 30
12. $-18 = 5g + (-3)$ −3
13. $-81 = 5e - 6$ −15
14. $5 = \frac{f}{-4} - 3$ −32
15. $3(d - 3) = -15$ −2
16. $42 = -5k + 2$ −8
17. $-11 = \frac{j}{3} - 2$ −27
18. $\frac{2}{7}(h + 5) = 4$ 9
19. $25 = \frac{l}{3} + 26$ −3
20. $-2 + 6m = 22$ 4

Solve.

21. Four times a number is added to 12. The result is −28. What is the number? −10
22. Half of a number is subtracted from 19. The result is 13. What is the number? 12
23. If Marilyn buys 3 more compact discs, she will have twice as many discs as Sharon. Sharon has 12 discs. How many discs does Marilyn have? 21 discs
24. Milton is $30 short of having half the money he needs to buy a $400 bicycle. How much money does Milton have? $170

13-3 SOLVING TWO-STEP EQUATIONS

Objective

Solve equations involving integers and two operations.

Mrs. O'Hara ordered four pizzas for Jenny's after-prom party. She used a coupon for $3 off one pizza. The total bill after subtracting the value of the coupon was $39. What was the regular price of one pizza? This problem can be solved using the equation $4n - 3 = 39$.

There is often more than one operation in an equation. To solve a two-step equation, you must apply the *standard order of operations* in reverse.

Method

1. Add or subtract the appropriate number to each side of the equation to get all terms involving the variable on one side of the equation. Next, multiply or divide each side of the equation by the appropriate nonzero number.
2. Check the solution.

Example A ***Find the regular price of one pizza.***

Point out that n = the regular price of one pizza.

$4n - 3 = 39$ First, n is multiplied by 4. Then 3 is subtracted from the product.

1. $4n - 3 + 3 = 39 + 3$ Add 3 to each side.

$4n = 42$

$\frac{4n}{4} = \frac{42}{4}$ Divide each side by 4.

$n = 10.5$ The regular price of one pizza is $10.50.

2. **Check:** $4n - 3 = 39$

$4(10.5) - 3 \stackrel{?}{=} 39$ In the original equation, replace n with 10.5.

$42 - 3 \stackrel{?}{=} 39$

$39 = 39$ ✓

Example B ***Solve $-2(n - 3) = 8$.***

$-2(n - 3) = 8$ First, 3 is subtracted from n. Then the difference is multiplied by −2.

1. $\frac{-2(n - 3)}{-2} = \frac{8}{-2}$ Divide each side by −2.

$n - 3 = -4$

$n - 3 + 3 = -4 + 3$ Add 3 to each side.

$n = -1$

2. **Check:** $-2(n - 3) = 8$ In the original equation, replace n with −1.

$-2(-1 - 3) \stackrel{?}{=} 8$

$-2(-4) \stackrel{?}{=} 8$

$8 = 8$ ✓ The solution is correct.

Guided Practice ***Solve each equation. Check your solution.***

Example A

1. $2x + 3 = -5$ −4
2. $3y - 4 = 14$ 6
3. $7y + (-2) = -51$ −7

RETEACHING THE LESSON

Complete.

1. $-4x + 3 = 19$

$-4x + 3 - \underline{3} = 19 - \underline{3}$

$-4x = \underline{16}$

$-4x \div \underline{-4} = 16 \div \underline{-4}$

$x = \underline{-4}$

Solve each equation. Check your answer.

2. $2 + 5m = -18$ −4
3. $-8k - 21 = 75$ −12

Reteaching Masters Booklet, p. 123

Name ____________ Date ____________

RETEACHING WORKSHEET 13-3

Two-Step Equations

Method:
① Identify the variable.
② Add or subtract the same number on each side of the equation. Multiply or divide each side of the equation by the same number.
③ Check the solution.

Examples: Solve each equation.

A. ① $3x - 4 = 11$ x is multiplied by 3, and 4 is subtracted from the product.

② $3x - 4 + 4 = 11 + 4$ Add 4 to each side.

$3x = 15$

$\frac{3x}{3} = \frac{15}{3}$ Divide each side by 3.

$x = 5$ The solution is 5.

③ **Check.** $3x - 4 = 3(5) - 4 = 15 - 4 = 11$ Replace the variable with 5.

Example B

4. $4(n + 2) = -20$ **−7** 5. $-5(x - 3) = 40$ **−5** 6. $-15 = 2(x + 3)$ **−10.5**

Solve each equation. Check your solution.

7. $49 = 7t + 14$ **5** 8. $2 - 3a = 8$ **−2** 9. $8x + 5 = -45$ **$-6\frac{1}{4}$**

10. $\frac{x}{5} + 1 = -1$ **−10** 11. $5(x + 1) = -15$ **−4** 12. $6(x - 0.2) = 3$ **0.7**

13. $-2y - 1 = 9$ **−5** 14. $-11 = 3d - 2$ **−3** 15. $\frac{5}{9}(F - 32) = 100$

16. $0 = -3x + (-15)$ **−5** 17. $\frac{2}{3}(y - 4) = 12$ **22** 18. $500 = 10(x + 30)$ **20**

19. $284 + -35y = 4$ **8** 20. $2y + 7 = -42$ **$-24\frac{1}{2}$** 21. $-6 = \frac{4 - y}{2}$ **16**

15. 212

Applications

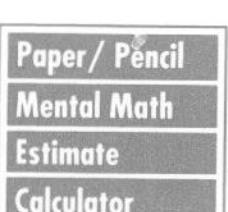

22. Henry answered all six questions correctly on the history quiz. But he lost two points on each question because he did not answer using complete sentences. Henry's grade on the quiz is 48. If each question is worth the same number of points, how much is each question worth? **10 points**

23. Leaman's is having a 30%-off everything sale. Chang buys a shirt at a sale price of $21. What is the original price of the shirt? **$30**

24. Mr. Thomas divides a case of floppy discs equally among the six members of the Computer Club. Kathy already has three discs of her own. With the new discs from Mr. Thomas, Kathy has seven discs. How many discs were there in the case? **24 discs**

Make Up a Problem

25. Make up an equation that will require two steps to solve. Exchange with a classmate. Solve your classmate's equation. Return the equation and check your classmate's solution. **Answers may vary.**

Talk Math

26. In your own words, describe how you would solve $\frac{2}{3}(x + 1) = 4$ using the least number of steps. **See margin.**

Mixed Review

Divide.

Lesson 5-6

27. $\frac{4}{5} \div \frac{1}{4}$ **$3\frac{1}{5}$** 28. $\frac{3}{8} \div \frac{4}{7}$ **$\frac{21}{32}$** 29. $\frac{1}{5} \div \frac{2}{15}$ **$1\frac{1}{2}$** 30. $6 \div \frac{3}{8}$ **16**

Lesson 11-8

List the stems that you would use to plot the data on a stem-and-leaf plot.

31. 23, 25, 29, 31, 26, 43, 15 **1, 2, 3, 4** 32. 6, 25, 19, 16, 46, 9, 33, 18 **0, 1, 2, 3, 4**

Lesson 9-10

33. The sum of two whole numbers is 54. Their product is 680. What are the numbers? **20, 34**

APPLYING THE LESSON

Independent Practice

Homework Assignment	
Minimum	Ex. 1-25 odd; 26-33
Average	Ex. 2-20 even; 22-33
Maximum	Ex. 1-33

Additional Answers

26. First multiply each side by $\frac{3}{2}$. Then subtract 1 from each side.

Chalkboard Examples

Example A *Solve each equation. Check your solution.*

1. $-9 = 2x + 3$ **−6**
2. $4 - 3x = 6$ **$-\frac{2}{3}$**

Example B *Solve each equation. Check your solution.*

3. $-3(p - 4) = -18$ **10**
4. $\frac{2}{3}(y - 3) = 6$ **12**

Additional examples are provided on TRANSPARENCY 13-3.

EVALUATING THE LESSON

Check for Understanding Use Guided Practice Exercises to check for student understanding.

Closing the Lesson Ask students to state in their own words how to solve an equation involving two operations.

EXTENDING THE LESSON

Enrichment Have students make up a problem involving three different operations. Then give their problem to a classmate to solve. Have the student who made up the problem verify the solution.

Enrichment Masters Booklet, p. 123

Name ____________ Date ____________

ENRICHMENT WORKSHEET 13-3

Multiplying Binomials

Two binomials can be multiplied by using a memory device called the **FOIL method** as shown below.

To multiply two binomials, find the sum of the products of
F the first terms,
O the outer terms,
I the inner terms, and
L the last terms.

Example: $(2x + 3)(5x + 8)$

Multiply the First terms.	$2x \cdot 5x$: $(2x + 3)(5x + 8)$	$10x^2$ +
Multiply the Outer terms.	$2x \cdot 8$: $(2x + 3)(5x + 8)$	$16x$ +
Multiply the Inner terms.	$3 \cdot 5x$: $(2x + 3)(5x + 8)$	$15x$ +
Multiply the Last terms.	$3 \cdot 8$: $(2x + 3)(5x + 8)$	24

$(2x + 3)(5x + 8) = 10x^2 + 16x + 15x + 24$
$= 10x^2 + 31x + 24$ Combine the like terms $16x$ and $15x$.

Multiply.

1. $(x + 3)(x + 5)$ $x^2 + 8x + 15$
2. $(2x + 4)(x + 7)$ $2x^2 + 18x + 28$
3. $(3x + 1)(3x + 1)$ $9x^2 + 6x + 1$
4. $(x + 4)(x + 4)$ $x^2 + 8x + 16$
5. $(2x + 1)(2x + 3)$ $4x^2 + 8x + 3$
6. $(x - 3)(x + 2)$ $x^2 - x - 6$
7. $(x + 3)(x - 2)$ $x^2 + x - 6$
8. $(x - 4)(x - 8)$ $x^2 - 12x + 32$
9. $(2x + 1)(x + 8)$ $2x^2 + 17x + 8$

Applying Mathematics to the Real World

This optional page shows how mathematics is used in the real world and also provides a change of pace.

Objective Compute the finance charge on credit card purchases.

Using Discussion Point out that using a bank card is a form of a loan and finance charges are interest.

Also, point out that some credit card companies give you a certain length of time to pay the full amount (usually 30 days) without paying finance charges, while some start assessing finance charges as soon as you charge an item. Some credit card companies compute the finance charge based on an average daily balance. Emphasize to students the importance of knowing the terms of credit when choosing a credit card.

Activity

Using Data Have students find the rate of interest charged by various types of credit cards. The lobby of local banks usually offers free pamphlets describing the banks' services. Have the students make a poster that shows this information.

Making Consumer Connections

FINANCE CHARGES

Many people often buy items on credit by borrowing money from a bank or by using a credit card. Most credit card companies charge a fee on the monthly balance of an account. This fee is a **finance charge.**

In April, Meagan Williams bought a television set for \$525. She charged the bill to her credit card account. Ms. Williams' credit card charges a monthly finance charge of 1.54%. The first payment was due in May. Ms. Williams paid the minimum amount due, \$20. What will the balance, including interest, be on her next statement if she does not add any more charges?

525 [−] 20 [=] 505 [×] .0154 [=] 7.777 [+] 505 [=] 512.777

The balance on the next statement will be \$512.78.

If you do not need to know the amount of the finance charge, you can multiply by 1.0154.

525 [−] 20 [=] 505 [×] 1.0154 [=] 512.777

Multiplying 505 by 1.0154 is the same as multiplying by 0.0154 and adding 505.

Copy and complete the table. The finance charge is 1.5% a month.

	Previous Balance	Monthly Payment	Finance Charge	New Balance
1.	\$125	\$20	■	■
2.	\$505	\$50	■	■
3.	\$752	\$152	■	■
4.	\$421	\$40	■	■

\$1.58 \$106.58
\$6.83 \$461.83
\$9.00 \$609.00
\$5.72 \$386.72

5. Mrs. West has a balance of \$680 on her credit card account. The monthly finance charge is 1.12%. One month she makes a payment of \$75. The next month she makes a payment of \$150. What is her new balance? **\$466.95**

6. Refer to Exercise 5. Suppose Mrs. West pays \$100 each month. How long will it take her to pay off the bill? How much will she pay in finance charges? **7 months, \$23.47**

On-the-Job Application

REGISTERED NURSE

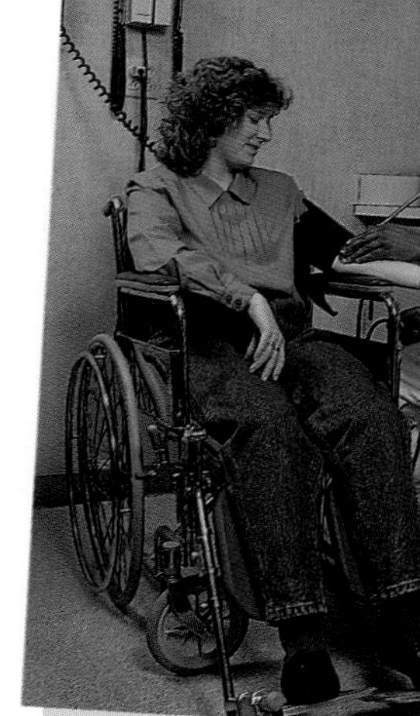

Jon Farrell is a registered nurse (RN). He takes temperatures, blood pressures, gives injections, and administers medication. Mr. Farrell supervises licensed practical nurses (LPNs) and is under the direction of doctors. Sometimes he does special duty nursing in a private home, a burn unit, or a nursing home.

A formula for determining normal blood pressure *(B.P.)* is $110 + (\frac{1}{2} \times A)$ where *A* represents age.

Compute normal blood pressure for an 18-year-old.

$B.P. = 110 + (\frac{1}{2} \times A)$

$= 110 + (\frac{1}{2} \times 18)$ — Substitute 18 for *A*.

$= 110 + 9$ — Multiply $\frac{1}{2}$ times 18.

$= 119$ — Add.

Normal blood pressure for an 18-year-old is 119.

1. Answers may vary.

Calculate a normal blood pressure reading for the following ages.

1. your age **2.** 57 139 **3.** 26 123 **4.** 65 143

5. Mr. Smith has a normal blood pressure for his age. How old is he if his blood pressure is 135? 50 years old

Type	Percent
A positive	38%
O positive	36%
B positive	8%
O negative	6%
A negative	6%
AB positive	3.5%
B negative	2%
AB negative	0.5%

The chart on the left shows different blood types and the percent of the population that has the blood type.

Find the approximate number of people out of 3,000 who have a blood type of O positive.

3000 [×] .36 [=] 1080

Out of 3,000 people, about 1,080 have a blood type that is O positive.

Use the chart and find the approximate number of people out of a group of 1,500 who have each blood type. 6. 8 people 7. 570 people

6. AB negative **7.** A positive **8.** AB positive **9.** O negative

10. Out of a group of 10,000 people, about how many have a blood type that is A positive or A negative? 4,400 people

8. 53 people
9. 90 people

Applying Mathematics to the Work World

This optional page shows how mathematics is used in the work world and also provides a change of pace.

Objective Compute normal blood pressure.

Using Discussion Discuss the difference between a registered nurse (RN) and a licensed practical nurse (LPN). Discuss differences in education and training as well as on-the-job duties.

In Exercise 10 have students solve the problem using two different methods. Have students find percentage of people that have type A+ and type A–, then add to find the total. Also, have students add the percent of people (rate) that have type A+ and A–, then take that percentage of the 10,000 to find the number that have A+ or A–. Have students decide which method is easier for them and use that method to solve Exercise 11.

Using a Calculator You may wish to make this a calculator activity to facilitate the computation. Discuss with students that nurses do not carry calculators. They memorize appropriate blood pressures so that an abnormal reading can be related to the doctor.

Activity

Using Data Have students find their blood type. Make a chart showing what percent of the class has each blood type. Compare with the percents in the chart.

5-MINUTE CHECK-UP

(over Lesson 13-3)

Available on TRANSPARENCY 13-4.
Solve each equation. Check your solution.

1. $5y - 6 = -4$ $\frac{2}{5}$

2. $-4(m - 3) = 6$ $1\frac{1}{2}$

3. $-3 = \frac{n + 3}{5}$ -18

4. $(5 + w) \div 7 = -4$ -33

Extra Practice, Lesson 13-3, p. 482

INTRODUCING THE LESSON

Ask students to describe the exact location of your desk and their desks in the classroom. Tell students that in mathematics we can describe a location more exactly and more briefly by using the coordinate plane.

TEACHING THE LESSON

Using Discussion Emphasize that the origin does not count as *one* itself; it is one away from the origin. Emphasize that order is important when reading an ordered pair. Have students locate (2, 4) and (4, 2).

Practice Masters Booklet, p. 136

Name ______________________ Date ______________

PRACTICE WORKSHEET 13-4

The Coordinate Plane

Find the ordered pair for each point labeled on the coordinate plane.

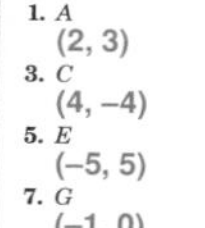
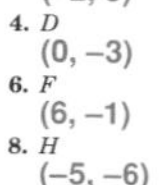

1. A (2, 3)
2. B (−2, 5)
3. C (4, −4)
4. D (0, −3)
5. E (−5, 5)
6. F (6, −1)
7. G (−1, 0)
8. H (−5, −6)

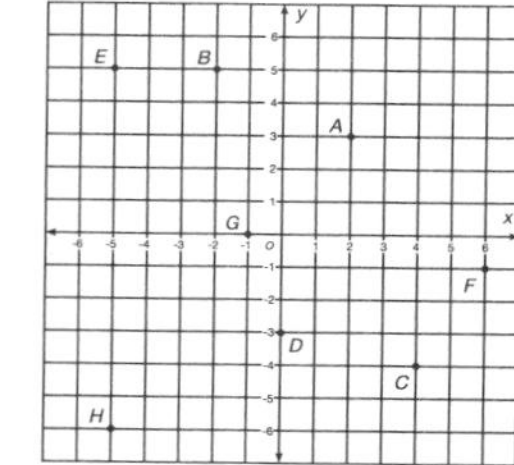

Draw coordinate axes on the grid at the right. Graph each ordered pair. Label each point with the given letter.

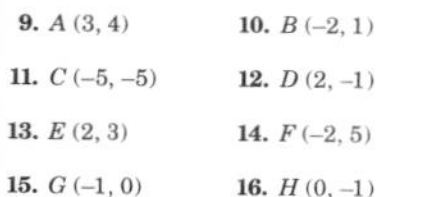
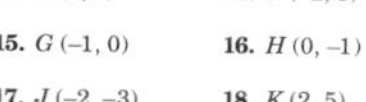

9. A (3, 4)
10. B (−2, 1)
11. C (−5, −5)
12. D (2, −1)
13. E (2, 3)
14. F (−2, 5)
15. G (−1, 0)
16. H (0, −1)
17. J (−2, −3)
18. K (2, 5)
19. L (−5, 2)
20. M (5, 5)
21. N (1, 2)
22. P (−1, −1)
23. Q (4, 5)
24. R (5, −4)

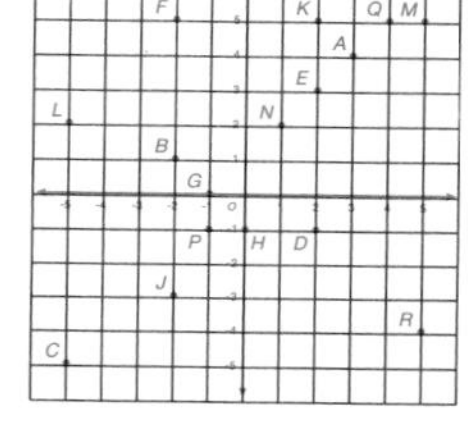

13-4 THE COORDINATE PLANE

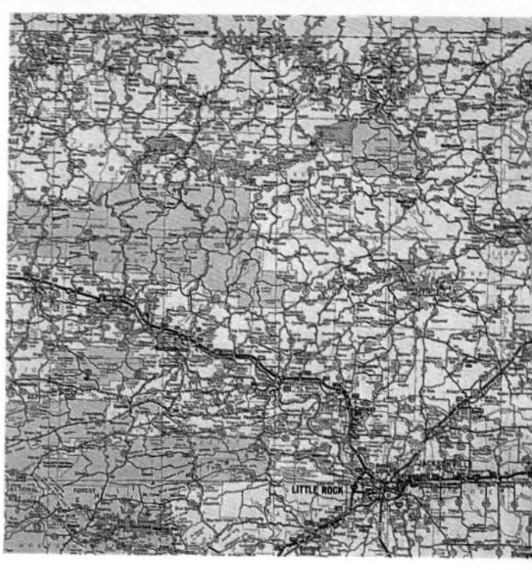

Objective
Name and graph points on the coordinate plane by using ordered pairs.

Axes is the plural form of axis.

Numbers are graphed on the number line. Ordered pairs of numbers, such as (2, 4), (−1, 5), (−3, −8), are graphed on the **coordinate plane.** The coordinate plane has two perpendicular number lines. Each number line is called an **axis.** The axes intersect at the **origin** labeled *O*. Every point on the coordinate plane has two coordinates. One corresponds to the horizontal axis and the other to the vertical axis. The coordinates are written as an **ordered pair.**

Method

1. The first coordinate in the ordered pair corresponds to a number on the horizontal axis. Numbers to the right of the origin are positive and to the left of the origin are negative.
2. The second coordinate in the ordered pair corresponds to a number on the vertical axis. Numbers above the origin are positive, numbers below the origin are negative.
3. Write the numbers that correspond to the location of the point.

Example A ***Find the ordered pair for each point labeled on the coordinate plane.***

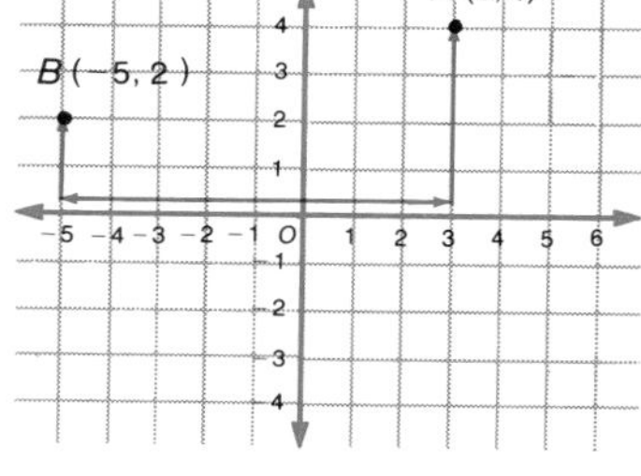

1. Point *A* is 3 units to the right of the origin on the horizontal axis.
2. Point *A* is 4 units above the origin on the vertical axis.
3. The ordered pair is (3, 4).

1. Point *B* is 5 units to the left of the origin on the horizontal axis.
2. Point *B* is 2 units above the origin on the vertical axis.
3. The ordered pair is (−5, 2).

Example B ***Graph the ordered pairs.***

Point *D* (5, −3) Start at the origin.

(5, −3)

1. Locate 5 on the horizontal axis.
2. Then move down 3 units and draw a dot.

Point *E* (−2, 0) Start at the origin.

(−2, 0)

1. Locate −2 on the horizontal axis.
2. (Move 0 units.) Draw a dot.

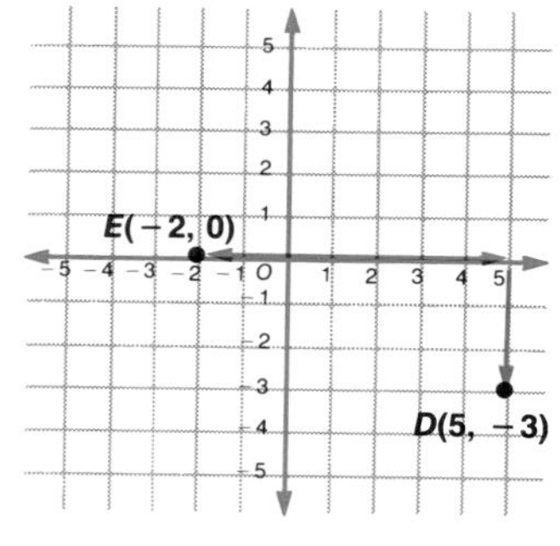

RETEACHING THE LESSON

Complete.

1. (4, 4) Move right 4 units. Then move up 4 units.
2. (−5, −3) Move left 5 units. Then move down 3 units.
3. (−3, 0) Move left 3 units. Then move up 0 units.
4. (2, −1) Move right 2 units. Then move down 1 unit.

Reteaching Masters Booklet, p. 124

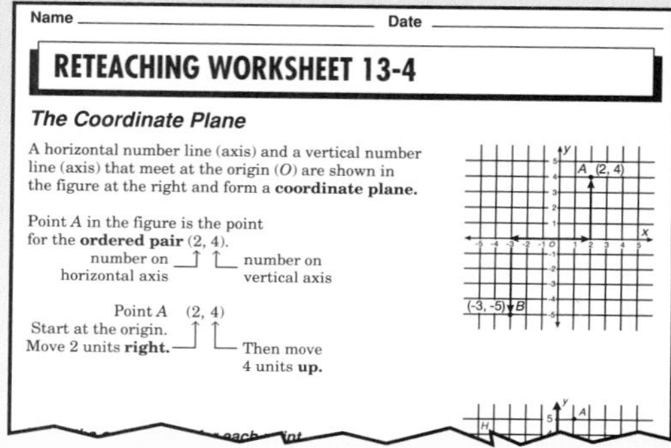

Name ______________________ Date ______________

RETEACHING WORKSHEET 13-4

The Coordinate Plane

A horizontal number line (axis) and a vertical number line (axis) that meet at the origin (*O*) are shown in the figure at the right and form a **coordinate plane.**

Point *A* in the figure is the point for the **ordered pair** (2, 4).
number on horizontal axis — number on vertical axis

Point *A* (2, 4)
Start at the origin. Move 2 units **right.** — Then move 4 units **up.**

Guided Practice

Example A

Find the ordered pair for each point labeled on the coordinate plane.

1. G (2, 1) **2.** H (−3, 2) **3.** I (−3, −2)
4. J (3, −3) **5.** K (1, 3) **6.** L (1, 0)

See answers on p. T529.

Example B

Draw a coordinate plane. Graph each ordered pair. Label each point with the given letter.

7. M (2, 3) **8.** N (−4, 5) **9.** O (0, 0)

10. P (4, −3) **11.** Q (−2, −5) **12.** R (4, 0)

13. S (0, 4) **14.** T (0, −4) **15.** U (−4, 0)

Exercises

Find the ordered pair for each point labeled on the coordinate plane.

16. U (0, 3) **17.** V (−2, 3) **18.** A (−3, 0)

19. B (−3, −2) **20.** E (0, −3) **21.** D (4, −4)

Draw a coordinate plane. Graph each ordered pair. Label each point with the given letter. See answers on p. T529.

22. H (0, −1) **23.** I (2, 5) **24.** J (7, 0) **25.** K (−3, 6)

26. L (5, −3) **27.** M (−7, 3) **28.** N (3, −4) **29.** P (−2, −4)

30. Q (5, −1) **31.** R (−3, −3) **32.** S (4.5, 2.5) **33.** T (−4.5, 5)

Show Math

34. Graph the following ordered pairs and connect the points in order: (2,1), (4,3), (6, 1), (6, −2), and (2, −2). What geometric figure is formed? pentagon

FUN with MATH
Where was algebra born?
See page 422.

35. The ordered pairs for three vertices of a square are (2, 1), (−3, 1), and (−3, −4). What is the ordered pair for the other vertex? (2, −4)

36. The ordered pairs for three vertices of a parallelogram are (4, 2), (−2, 2), and (−4, −2). What is an ordered pair for another vertex? Answers may vary. Typical answers are (+2, −2), (6, 6).

Talk Math

37. Examine a map in which locations are identified by a letter and a number. Describe how locating a point on the map is similar to locating a point on the coordinate plane. It is similar in moving horizontally for one number and vertically for the other.

Critical Thinking

38. A line in the coordinate plane passes from the upper left, through the origin, and on to the lower right. What do the ordered pairs for all points except (0,0) along this line have in common? See margin.

Chalkboard Examples

Example A *Find the ordered pair for each point labeled on the coordinate plane.*

A (2, 3)
B (4, −3)
C (−3, −3)
D (−2, 0)

Example B *Draw a coordinate plane. Graph each ordered pair. Label each point with the given letter.* See answers on graph above.

E. (3, 4) **F.** (−3, −4) **G.** (2, 0)

Additional examples are provided on TRANSPARENCY 13-4.

EVALUATING THE LESSON

Check for Understanding Use Guided Practice Exercises to check for student understanding.

Closing the Lesson Have students describe an ordered pair.

EXTENDING THE LESSON

Enrichment Masters Booklet, p. 124

Name ______ Date ______

ENRICHMENT WORKSHEET 13-4

Using Coordinates

Use the points (1,1), (4, 4), and (2, 4) for Problems 1–3.

1. Graph and connect the points. This polygon is a triangle.

2. You can draw a reflection of your original figure. Multiply each of the first coordinates by −1. Graph and connect the points. This triangle is a reflection of the one in Problem 1.

3. You can draw another reflection of your original figure. Multiply each of the second coordinates by −1. Graph and connect the points.

Use the points (0, −4), (2, 2), (6, 2), and (4, −4) for Problems 4–6.

4. Graph and connect the points. This polygon is a parallelogram.

5. Add 2 to both coordinates. Graph and connect the new points. How does this figure compare to the polygon you drew in Problem 4? same size and shape

6. Multiply each coordinate in Problem 4 by 2. Graph and connect the new points. How does this figure compare to the polygon you drew in Problem 4? same shape but 4 times as large

7. Multiply each coordinate in Problem 4 by $\frac{1}{2}$. Graph and connect the new points. How does this figure compare to the polygon you drew in Problem 4? same shape but $\frac{1}{4}$ as large

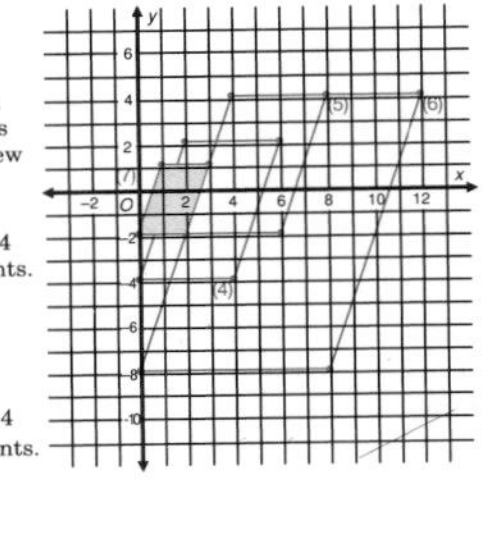

APPLYING THE LESSON

Independent Practice

Homework Assignment	
Minimum	Ex. 1-38
Average	Ex. 1-38
Maximum	Ex. 1-38

Chapter 13, Quiz A (Lessons 13-1 through 13-4) is available in the Evaluation Masters Booklet, p. 64.

Additional Answers

38. Each ordered pair has one negative and one positive coordinate.

5-MINUTE CHECK-UP

(over Lesson 13-4)

Available on TRANSPARENCY 13-5.
Draw a coordinate plane. Graph each ordered pair. Label each point with the given letter.

A (1, 3)
B (−4, 2)
C (0, −2)
D (2, −2)

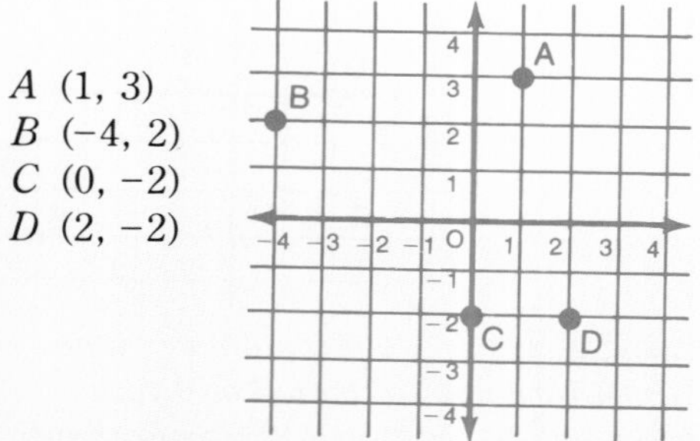

Extra Practice, Lesson 13-4, p. 482

INTRODUCING THE LESSON

Ask students how many different ways there are to make change for a nickel using only standard US currency.

TEACHING THE LESSON

Using Questioning

- Describe how to graph (2.5, 2.5) on the coordinate plane. From the origin, move horizontally 2 plus $\frac{1}{2}$ units to the right. Then vertically 2 plus $\frac{1}{2}$ units up.

Practice Masters Booklet, p. 137

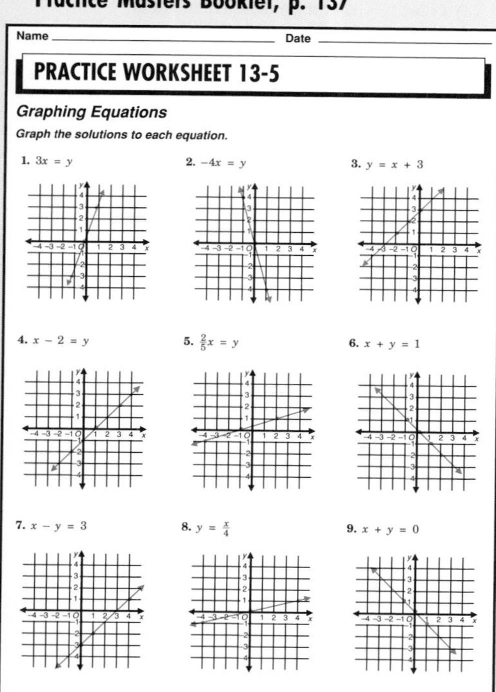

Name ________ Date ________

PRACTICE WORKSHEET 13-5

Graphing Equations

Graph the solutions to each equation.

1. $3x = y$
2. $-4x = y$
3. $y = x + 3$
4. $x - 2 = y$
5. $\frac{2}{5}x = y$
6. $x + y = 1$
7. $x - y = 3$
8. $y = \frac{x}{4}$
9. $x + y = 0$

13-5 GRAPHING EQUATIONS

Objective
Graph linear equations.

Jay wants to know what pairs of numbers have a sum of 4. The equation $x + y = 4$ describes this situation. There are two variables, x and y.

You have worked with equations that have one variable and one solution. An equation with two or more variables has more than one solution. You can use the coordinate plane to graph the solutions for an equation with two variables. Each solution can be written as an ordered pair.

Method

1. Find several ordered pairs that satisfy the equation.
2. Graph the ordered pairs.
3. Draw a line through the points. All of the ordered pairs for points on the line are solutions to the equation.

Example A

Graph the solution to $x + y = 4$. Rewrite the equation as $y = 4 - x$ so it is easier to find values of y.

1. Make a table of possible values for x. Find the corresponding values for y.

x	$4 - x$	y	(x, y)
−2	4 − (−2)	6	(−2, 6)
−1	4 − (−1)	5	(−1, 5)
0	4 − 0	4	(0, 4)
1	4 − 1	3	(1, 3)

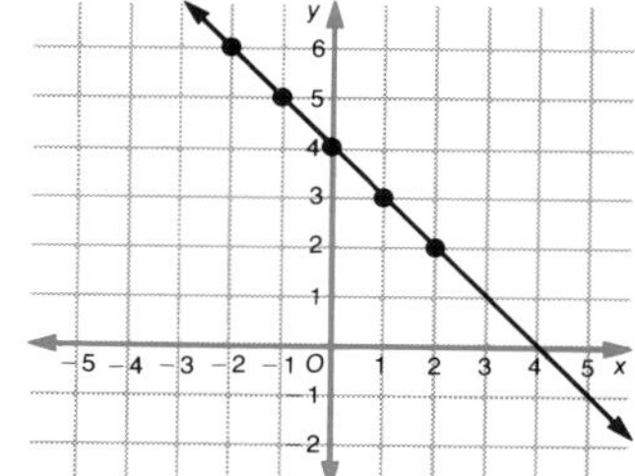

2. Graph the ordered pairs.
3. Use a ruler to draw a line through the points.

Solutions include (3, 1), (4, 0) and (5, −1).

The coordinates of any ordered pair for points on the line $y = 4 - x$ have a sum of 4. All of these ordered pairs are solutions.

Example B

Graph the solution to $y = \frac{1}{2}x + 1$.

1.

x	$\frac{1}{2}x + 1$	y	(x, y)
−4	$\frac{1}{2}(-4) + 1$	−1	(−4, −1)
−2	$\frac{1}{2}(-2) + 1$	0	(−2, 0)
0	$\frac{1}{2}(0) + 1$	1	(0, 1)
2	$\frac{1}{2}(2) + 1$	2	(2, 2)

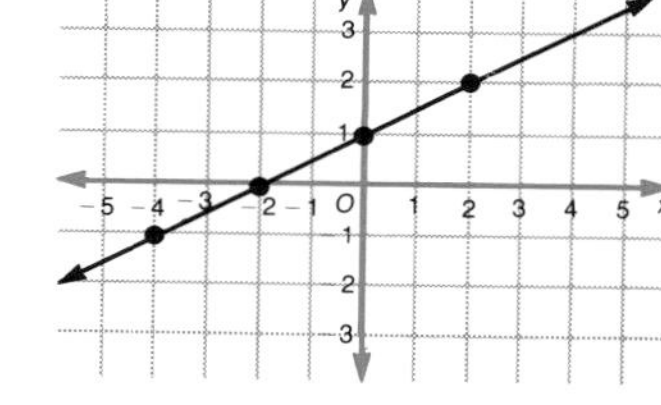

2. Graph the ordered pairs.
3. All of the points on the line are solutions to the equation $y = \frac{1}{2}x + 1$.

RETEACHING THE LESSON

Complete each table.

1. $y = x + 5$

x	$x + 5$	y	(x, y)
1	1 + 5	6	(1, 6)
9	9 + 5	14	(9, 14)

2. $y = 11 - x$

x	$11 - x$	y	(x, y)
9	11 − 9	2	(9, 2)
7	11 − 7	4	(7, 4)
5	11 − 5	6	(5, 6)

Reteaching Masters Booklet, p. 125

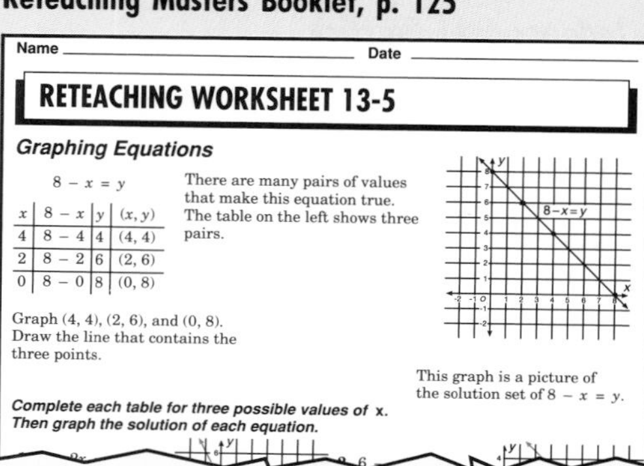

Name ________ Date ________

RETEACHING WORKSHEET 13-5

Graphing Equations

$8 - x = y$

x	$8 - x$	y	(x, y)
4	8 − 4	4	(4, 4)
2	8 − 2	6	(2, 6)
0	8 − 0	8	(0, 8)

There are many pairs of values that make this equation true. The table on the left shows three pairs.

Graph (4, 4), (2, 6), and (0, 8). Draw the line that contains the three points.

This graph is a picture of the solution set of $8 - x = y$.

Complete each table for three possible values of x. Then graph the solution of each equation.

Guided Practice

Example A, B

Graph the solutions to each equation. See answers on p. T529.

1. $x - 5 = y$ 2. $x + 3 = y$ 3. $\frac{x}{2} = y$ 4. $\frac{1}{3}x = y$

Exercises

Practice

Graph the solutions to each equation. See answers on pp. T529-T530.

5. $x - 3 = y$
6. $2x = y$
7. $\frac{2}{3}x = y$
8. $x + 1 = y$
9. $1.5x = y$
10. $x + 2.5 = y$
11. $x - 1.5 = y$
12. $x = y$
13. $x + y = 5$
14. $x + y = -3$
15. $x - y = 4$
16. $x - y = 0$
17. Draw a graph to show all possible pairs of numbers whose sum is 13. See answer on p. T530.

Applications

18. An oscilloscope is an electronic instrument that is used to show any kind of vibration, such as sound waves. The screen of an oscilloscope is marked with grids like the coordinate plane. The shape of a vibration is shown on the screen and can be measured using the grid. The amplitude (loudness) of a sound wave is the vertical distance from the rest position (the horizontal axis) to the highest point. The wave shown in blue has an amplitude of 20. What is the amplitude of the wave shown in red? 30

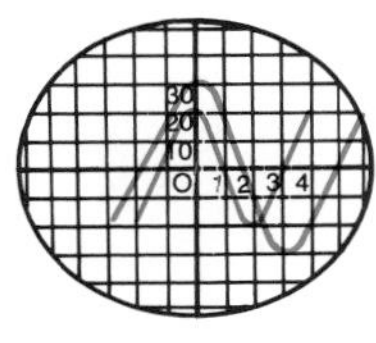

19. In Tuesday's *Daily News*, the financial page used a graph to show the increase in price per share of Americomp stock. If the price per share continues to increase at the rate shown on the chart, in which month will the price per share be $40? October

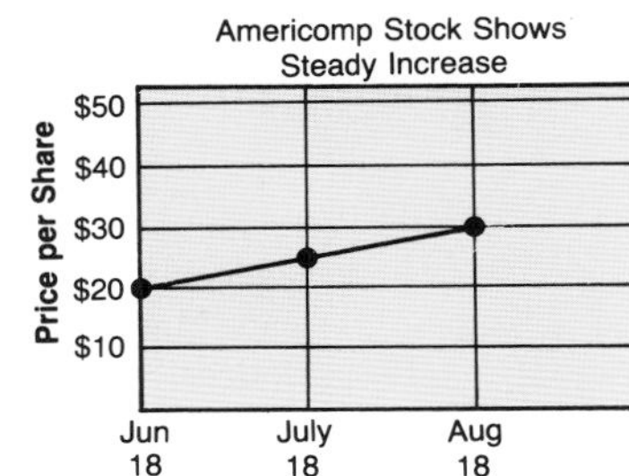

Problem Solving

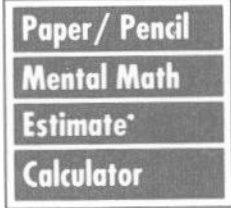

20. Beth wants to make a rectangular garden with an area of 900 square feet. What dimensions should she use in order to use as little fence as possible around the garden? 30 ft by 30 ft
21. A 50-pound bag of plant food contains 16 pounds of phosphoric acid. How many pounds of phosphoric acid are in a 25-pound bag of the same plant food? 8 pounds
22. Simon's age is 3 more than twice his brother's age. If Simon is 11, how old is his brother? 4 years old

Critical Thinking

23. Graph the two ordered pairs (5, 2) and (10, 7) on a coordinate plane. Draw a straight line through the two points. Make a table of 5 ordered pairs that lie on the line. Use the values in the table to find the equation that represents the line. $x - 3 = y$

Chalkboard Examples

Example A *Graph the solutions to each equation.* See answers in margin.

1. $2 - x = y$ 2. $x + 5 = y$

Example B *Graph the solutions to each equation.* See answers in margin.

3. $3x = y$ 4. $x + y = 4$

Additional examples are provided on TRANSPARENCY 13-5.

EVALUATING THE LESSON

Check for Understanding Use Guided Practice Exercises to check for student understanding.

Closing the Lesson Have students state the steps needed to graph an equation.

EXTENDING THE LESSON

Enrichment Define the slope of a line for students. Use "rise over run." Show students how to find the slope of the line in Example A (−1) and Example B $(+\frac{1}{2})$ on p. 410. Point out the difference in direction of the lines in Examples A and B. Have students find the slope of the lines in the Guided Practice Exercises 1-3 on p. 411.

Enrichment Masters Booklet, p. 125

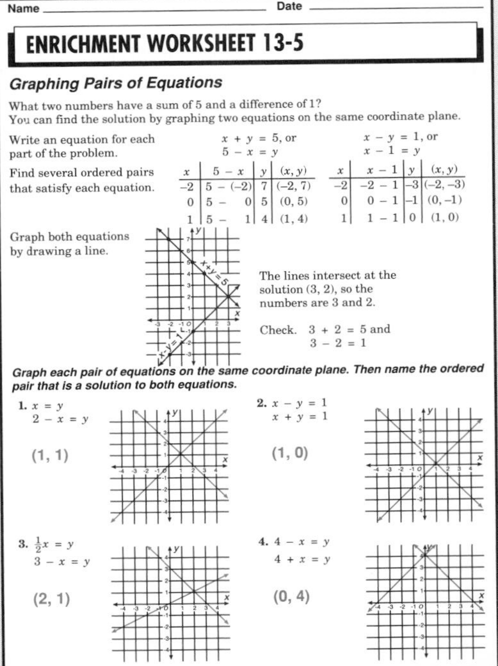

Name ______ Date ______

ENRICHMENT WORKSHEET 13-5

Graphing Pairs of Equations

What two numbers have a sum of 5 and a difference of 1?
You can find the solution by graphing two equations on the same coordinate plane.

Write an equation for each part of the problem. $x + y = 5$, or $5 - x = y$ $x - y = 1$, or $x - 1 = y$

Find several ordered pairs that satisfy each equation.

x	$5 - x$	y	(x, y)
−2	5 − (−2)	7	(−2, 7)
0	5 − 0	5	(0, 5)
1	5 − 1	4	(1, 4)

x	$x - 1$	y	(x, y)
−2	−2 − 1	−3	(−2, −3)
0	0 − 1	−1	(0, −1)
1	1 − 1	0	(1, 0)

Graph both equations by drawing a line.

The lines intersect at the solution (3, 2), so the numbers are 3 and 2.

Check. $3 + 2 = 5$ and $3 - 2 = 1$

Graph each pair of equations on the same coordinate plane. Then name the ordered pair that is a solution to both equations.

1. $x = y$, $2 - x = y$ (1, 1)
2. $x - y = 1$, $x + y = 1$ (1, 0)
3. $\frac{1}{2}x = y$, $3 - x = y$ (2, 1)
4. $4 - x = y$, $4 + x = y$ (0, 4)

APPLYING THE LESSON

Independent Practice

Homework Assignment	
Minimum	Ex. 1-23 odd
Average	Ex. 1-23
Maximum	Ex. 1-23

Examples Answers 1-4

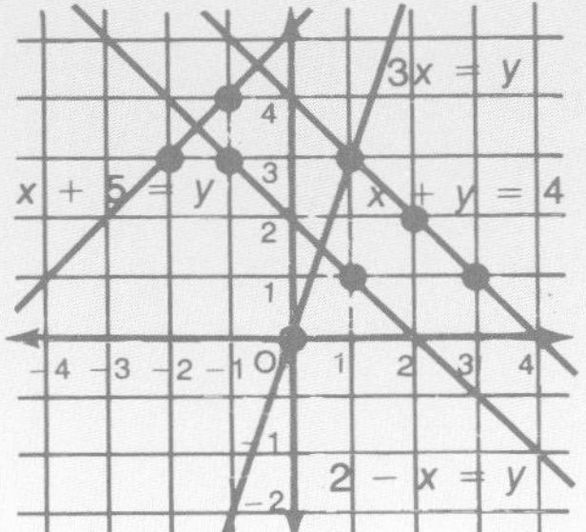

5-MINUTE CHECK-UP

(over Lesson 13-5)

Available on TRANSPARENCY 13-6.

Graph the solution to the equation.

1. $x + 3 = -y$ **2.** $3x = y$

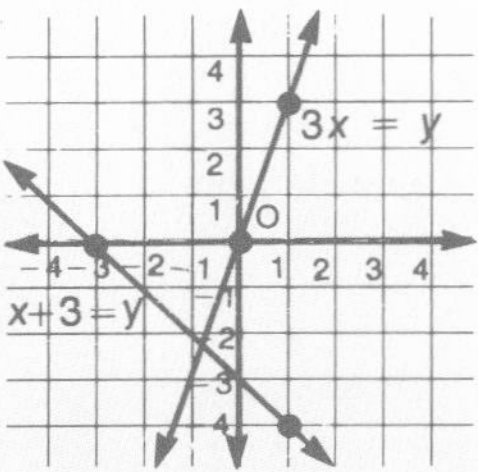

Extra Practice, Lesson 13-5, p. 483.

INTRODUCING THE LESSON

Have students explain how they would find the location of a street with which they are unfamiliar.

TEACHING THE LESSON

Using Models Have students look at the grid map on page 412. Have them locate Dodger Stadium by following the instructions in the paragraph following the grid map.

Practice Masters Booklet, p. 138

PRACTICE WORKSHEET 13-6

Applications: Reading a Grid Map

SAN ANTONIO
Univ. of Texas at San Antonio
South Texas Medical Center
Castle Hills
San Antonio Int'l. Airport
Alamo Heights
Fort Sam Houston
San Antonio Mus. of Art
City Hall
The Alamo
Joe Freeman Coliseum
Gonzales Hwy.
Lackland A.F.B.
San Jose Mission Nat'l. Historic Site
Stinson Field
Brooks Aero Med. Center

Use the map above.

1. What hospital is located in B-5? **South Texas Medical Center**
2. What airport is located in D-1? **Stinson Field**
3. What highway is the University of Texas at San Antonio near? **10/87**
4. What highway passes through Lackland Air Force Base? **410**
5. Which highways would you travel to go from City Hall to San Antonio International Airport? **81/87 and 281**

Give the location of each as a letter-number pair.

6. Joe Freeman Coliseum **E-3**
7. Castle Hills **C-5**
8. San Antonio Museum of Art **D-3**
9. Alamo Heights **D-4**
10. San Jose Mission National Historic Site **D-2**
11. Gonzales Highway **E-2, F-2**

Applications

13-6 READING A GRID MAP

Objective
Locate places on a grid map.

Marie is visiting her cousin Sharyn in California. They are going to a baseball game at Dodger Stadium. Marie and Sharyn locate Dodger Stadium on a grid map to check their directions before driving to the stadium.

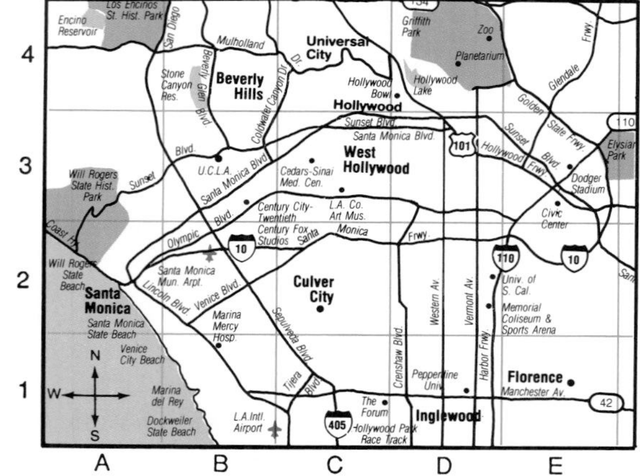

Index
Cedars-Sinai Medical Center, C-3
Dodger Stadium, E-3
The Forum, C-1
Griffith Park, D-4
Griffith Planetarium, D-4
Hollywood Park Race Track, D-1
LA International Airport, B-1
Los Angeles County Art Museum, C-3
Marina del Rey, B-1
Marina Mercy Hospital, B-1
Pepperdine University, D-1
University of California LA (UCLA), B-3
University of Southern Calif. (USC), D-2

Each square on a grid map can be identified by a letter-number pair. To find Dodger Stadium, check the index for the location. Dodger Stadium is in square E-3. Find the letter E along the bottom of the map. Move up to the square labeled 3 at the left of the map. Locate Dodger Stadium within square E-3.

Understanding the Data

- Where do you find the letters on the grid map? **along the bottom edge**
- Where do you find the numbers on the grid map? **along the left edge**
- What do you find in the index? **the names of locations, the letter and number of locations**
- What does a letter-number pair indicate? **the location of the square in which a place can be found**

Check for Understanding

Use the map and the index above.

1. What park is located in A-4? **Los Encinos St. Hist. Park**
2. What highway is near University of Southern California? **Hwy 110**
3. What airport is located in B-2? **Santa Monica Mun. Arpt.**
4. Within what square do you find Marina del Rey? **B-1**

Give the location of each as a letter-number pair.

5. Stone Canyon Reservoir **B-4**
6. Hollywood Bowl **D-4**
7. Twentieth Century Fox Studios **B-3**
8. Florence **E-1**

RETEACHING THE LESSON

Copy the map on page 412 and give each student a map. Have students use colored pencils to shade square A-1. Point out that any place labeled A-1 will be found in this square. Using different colors, do the same for all the other squares B-1 thru E-4.

Reteaching Masters Booklet, p. 126

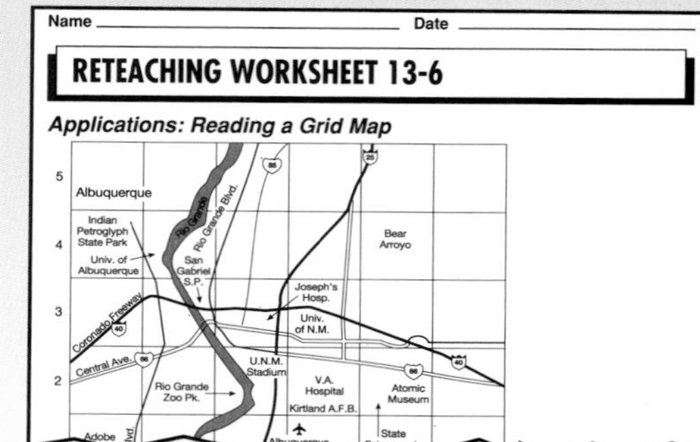
Name ______ Date ______

RETEACHING WORKSHEET 13-6

Applications: Reading a Grid Map

Exercises

9. Griffith Park Zoo 11. LA International

Practice

Use the map and index on page 412.

9. What zoo is located in D-4?
10. What highway is near Pepperdine University? Hwy 42
11. What airport is located in B-1?
12. What medical center is located in C-3? Cedars-Sinai

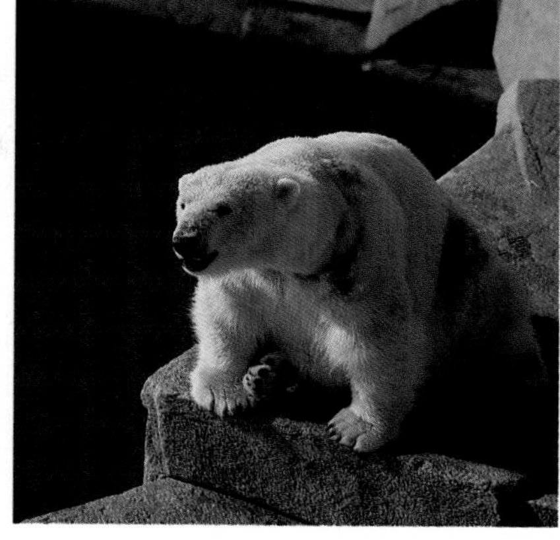

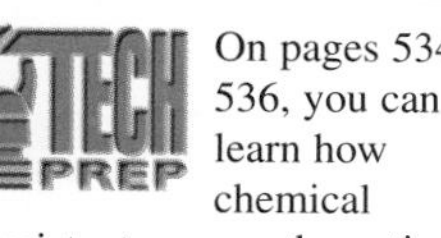

On pages 534–536, you can learn how chemical assistants use mathematics in their jobs.

Give the location of each with a letter-number pair.

13. Santa Monica State Beach A-2
14. LA County Art Museum C-3
15. Univ. of Southern California D-2
16. Griffith Planetarium D-4

Applications

17. Give directions from LA International Airport to The Forum. See margin.
18. Give directions from UCLA to LA International Airport. See margin.
19. Scott drove from the University of Southern California toward Dodger Stadium on Highway 110. After driving past what road did Scott begin to watch for Dodger Stadium? Sunset Blvd.
20. If you are traveling from E-1 to A-4, are you traveling northeast or northwest? northwest

FUN with MATH

Are there women heart surgeons?
See page 423.

21. Find a grid map of your area in the library or city hall. Make a list of four places and their letter-number location. Answers may vary.

Mixed Review

Lesson 7-2

Trace each angle and extend the sides. Then find the measure of each angle.

22. 

23. 

24. 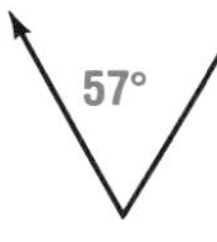

Lesson 12-2

Add. Use a number line if necessary.

25. $-6 + 4$ -2
26. $-10 + (-8)$ -18
27. $18 + (-24)$ -6

Lesson 10-4

28. At Ricardo's Pizza Shop one night, they sold 32 pizzas with pepperoni, 24 with mushrooms, and 21 with sausage. Of these, 7 had pepperoni and mushroom on one pizza, 3 had sausage and mushrooms, 6 had pepperoni and sausage and 4 had all three toppings. How many pizzas were sold? 53 pizzas

Chalkboard Examples

Have students explain how to locate each place using the grid map and index on p. 412

1. Cedars-Sinai Medical Center
2. University of Southern Calif.
3. University of California LA
4. The Forum

Additional examples are provided on TRANSPARENCY 13-6.

EVALUATING THE LESSON

Check for Understanding Use Guided Practice Exercises to check for student understanding.

Closing the Lesson Ask students to explain how to use a grid map.

EXTENDING THE LESSON

Enrichment Find a grid map of your state in the library. Have students list the letter-number location of six points of interest. For example, the state capital, colleges or universities, tourist attractions, and so on.

Enrichment Masters Booklet, p. 126

ENRICHMENT WORKSHEET 13-6

Polar Coordinates

In a rectangular coordinate system, the ordered pair (x, y) describes the location of a point x units from the origin along the x-axis and y units from the origin along the y-axis.

In a polar coordinate system, the ordered pair (r, θ) describes the location of a point r units from the pole on the ray (vector) whose endpoint is the pole and which forms an angle of θ with the polar axis.

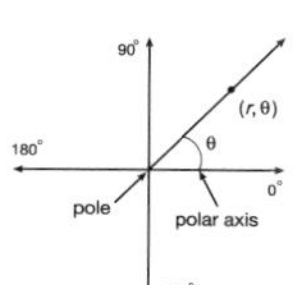

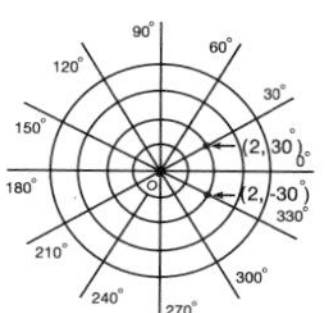

Locate each point on the polar grid below.

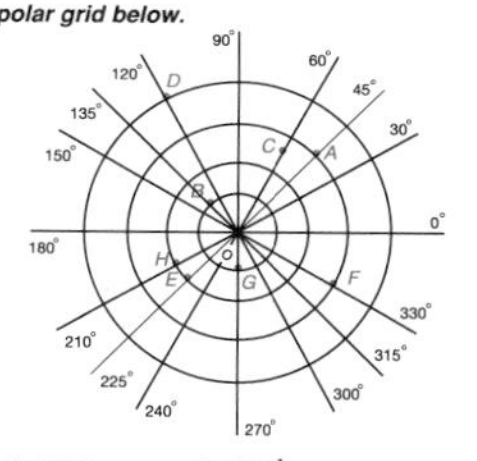

1. $A\ (3, 45°)$ 2. $B\ (1, 135°)$ 3. $C\ (2\frac{1}{2}, 60°)$ 4. $D\ (4, 120°)$
5. $E\ (2, 225°)$ 6. $F\ (3, -30°)$ 7. $G\ (1, -90°)$ 8. $H\ (-2, 30°)$

APPLYING THE LESSON

Independent Practice

Homework Assignment	
Minimum	Ex. 1-28
Average	Ex. 1-28
Maximum	Ex. 1-28

Additional Answers

17. N. on Lincoln Blvd. to Tijera Blvd., N. on Tijera Blvd. to Manchester Ave., N. on Manchester Ave to the Forum.
18. W. on Sunset Blvd. to Sepulveda Blvd., S. on Sepulveda Blvd. to Tijera Blvd., to LA International Airport

5-MINUTE CHECK-UP

(over Lesson 13-6)

Available on TRANSPARENCY 13-7.

1. What does a letter-number pair in a grid map index indicate? **the square in which a place is located**

▶ INTRODUCING THE LESSON

Have students imagine they are part of a tour group in a nonEnglish-speaking country. Ask them to describe what would happen if their guide translated directions incorrectly and they found themselves at the railroad station instead of the airport. Emphasize the need for accurate translation when writing equations.

▶ TEACHING THE LESSON

Using Discussion Discuss with students the importance of careful reading before writing an equation. It is helpful to break sentences into phrases as shown in the Solve step of the example.

Practice Masters Booklet, p. 139

Name ______________________ Date ______________

PRACTICE WORKSHEET 13-7

Problem-Solving Strategy: Writing an Equation

Solve. Write an equation.

1. Sheila has 5 fewer tapes than Cheryl. Together they have 33 tapes. How many tapes does each girl have?
   $(x - 5) + x = 33$; Sheila—14 tapes; Cheryl—19 tapes
2. Ted's highest bowling scores is 236. This is 10 less than three times his lowest score. What is Ted's lowest score?
   $236 = 3x - 10$; 82
3. Movie attendance at the 7:00 show was 342. This is 25 more than half the number of people who attended the 9:00 show. How many attended the 9:00 show?
   $342 = \frac{1}{2}x + 25$; 634 people
4. Jody jogged and cycled a total of 130 miles last week. The number of miles she jogged is 2 more than one-third the number of miles she cycled. How many miles did she jog? Cycle?
   $130 = x + \frac{1}{3}x + 2$; jog—34 mi; cycle—96 mi
5. A bookstore displayed 110 copies of the #1 bestseller in 12 rows. Five rows contained 2 fewer books than the other rows. How many books were in each row?
   $110 = 7x + 5(x - 2)$; 5 rows had 8 books; 7 rows had 10 books
6. Marla earned $39 baby-sitting this week. On Saturday she baby-sat 1 hour more than twice the number of hours she baby-sat on Friday. If she earns $3 per hour, how many hours did she baby-sit each day?
   $39 = 3(x + 2x + 1)$; Friday—4 h; Saturday—9 h

Problem-Solving Strategy

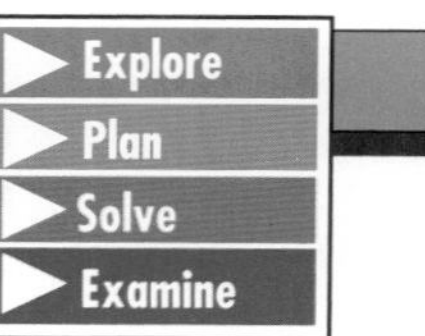

13-7 WRITING AN EQUATION

Objective
Solve problems using equations.

Adam and Belinda compare projects for the science fair. Adam raised six hamsters on a special diet he created. The number of fish Belinda raised divided by six is two less than the number of hamsters Adam raised. How many fish did Belinda raise?

▶ Explore

What is given?

- Adam has six hamsters.
- The number of fish Belinda raised divided by six is two less than the number of hamsters.

What is asked?

- How many fish did Belinda raise?

▶ Plan

Let f equal the number of fish. Translate the problem into an equation using the variable f.

▶ Solve

The number of fish divided by six	is	two less than	the number of hamsters
$f \div 6$	$=$	-2	$+6$

$\frac{f}{6} = -2 + 6$ — $f \div 6$ can be expressed as $\frac{f}{6}$.

$\frac{f}{6} = 4$ — $-2 + 6 = 4$

$6 \times \frac{f}{6} = 6 \times 4$ — Multiply each side by 6.

$f = 24$

Belinda raised 24 fish.

▶ Examine

Read the problem again to see if the answer makes sense. Is 24 divided by 6 equal to 2 less than 6? Yes, the solution is correct.

Guided Practice

1. $2x + x = 81$; 54 tapes, 27 compact discs

Solve. Write an equation.

1. Susan has a total of 81 compact discs and tapes. If she has twice as many tapes as compact discs, how many of each does she have? Hint: Let x equal compact discs.
2. Glenn sells his stereo for $212. The sale price is $10.50 more than half the amount he originally paid for it. How much did he pay for his stereo?

2. $212 = 10.50 + \frac{1}{2}x$; $403

RETEACHING THE LESSON

Have students translate each sentence into an equation.

1. Some number x increased by ten is equal to thirty-eight. $x + 10 = 38$
2. A number n decreased by four is equal to eighteen. $n - 4 = 18$
3. Nine times a number y plus two is equal to fifty. $9y + 2 = 50$
4. A number b increased by five is equal to nine. $b + 5 = 9$
5. A number m divided by seven is equal to fourteen. $\frac{m}{7} = 14$

Problem Solving

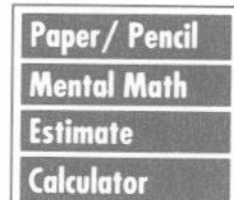

Solve. Use any strategy.

3. The student council is selling school pins and mums for homecoming. They sell three times as many pins as mums. If they sell 117 pins, how many mums did they sell? **39 mums**

4. In triangle ABC, $\angle C$ measures 63°. Find the measures of the other two angles if the measure of $\angle B$ is twice the measure of $\angle A$. Remember, the sum of the measures of the angles in a triangle is 180°. $\angle B$, **78°**; $\angle A$, **39°**

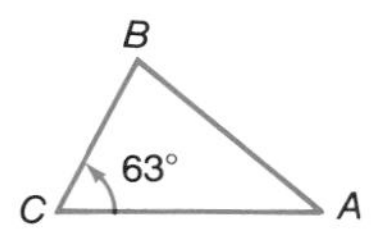

5. Pet Party pet shop has only parakeets and kittens located in the front corner of the store. If the animals in this corner have 18 heads and 52 feet, how many parakeets and kittens are there? **10 parakeets, 8 kittens**

6. Write a mathematical autobiography that describes your mathematical growth over the past year. Use correct punctuation, spelling, and good grammar. **See students' work.**

Critical Thinking

7. Nicole read about Hartley's 25%-off sale. Hartley's has a blouse and a pair of slacks that Nicole wants. If the original price of the blouse is $28 and the original price of the slacks is $30, *about* how much money should Nicole take to the store? ***about* $45**

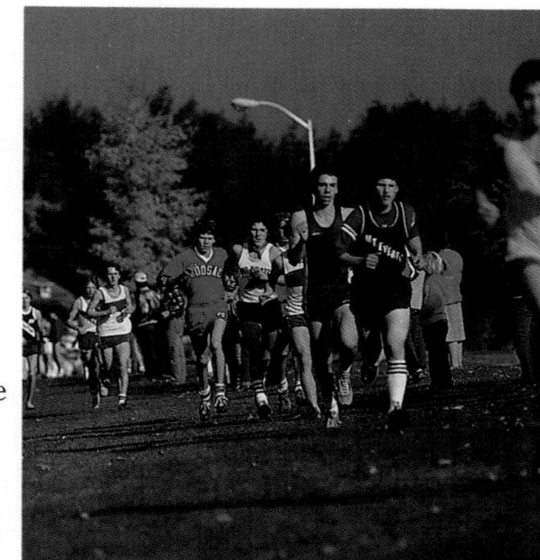

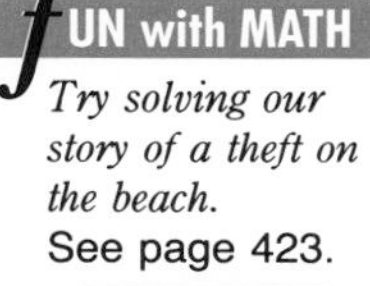

Try solving our story of a theft on the beach.
See page 423.

8. Four members of the track team finished the cross-country race at state competition. Don placed third. Lou finished six places behind Don. Keith finished three places in front of Lou, and George finished two places behind Keith. In which places did Lou, Keith, and George finish? **9th, 6th, 8th**

Mixed Review

Use the box-and-whisker plot to answer each question.

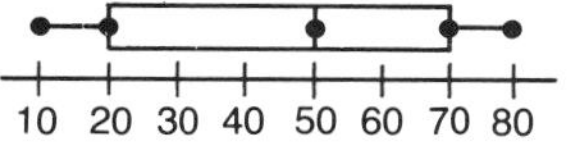

Lesson 11-10

9. The middle half of the data is between which two values? **20, 70**

10. What part of the data is greater than 50?

10. 1/2 of data

Lesson 12-3

Subtract.

11. 15 − (−6) **21**

12. −21 − (−16) **−5**

13. −23 − 7 **−30**

Lesson 2-5

14. A mountain climber can descend 50 feet every minute by rappelling. How far can the climber descend in 8.5 minutes? **425 feet**

Chalkboard Examples

Solve. Write an equation.

1. James bought a snowboard from the Ski Smith at 30% off. He paid $20 less than Geoff who paid $216 at the Alpine Haus. What was the original price of James's board?
0.7x = 216 − 20; $280

Additional examples are provided on TRANSPARENCY 13-7.

EVALUATING THE LESSON

Check for Understanding Use Guided Practice Exercises to check for student understanding.

Error Analysis Watch for students who forget or skip phrases such as *less than* and *more than.* Encourage students to examine the problem for clues such as which quantity should be greater, which should be lesser, which quantity is being increased, or which is being decreased.

Closing the Lesson Have students explain in their own words why careful reading is important for writing equations.

APPLYING THE LESSON

Independent Practice

Homework Assignment	
Minimum	Ex. 2-14 even
Average	Ex. 1-14
Maximum	Ex. 1-14

Chapter 13, Quiz B (Lessons 13-5 through 13-7) is available in the Evaluation Masters Booklet, p. 64.

Using the Chapter Review

The Chapter Review is a comprehensive review of the concepts presented in this chapter. This review may be used to prepare students for the Chapter Test.

Chapter 13, Quizzes A and B, are provided in the Evaluation Masters Booklet as shown below.

Quiz A should be given after students have completed Lessons 13-1 through 13-4. Quiz B should be given after students have completed Lessons 13-5 through 13-7.

These quizzes can be used to obtain a quiz score or as a check of the concepts students need to review.

Evaluation Masters Booklet, p. 64

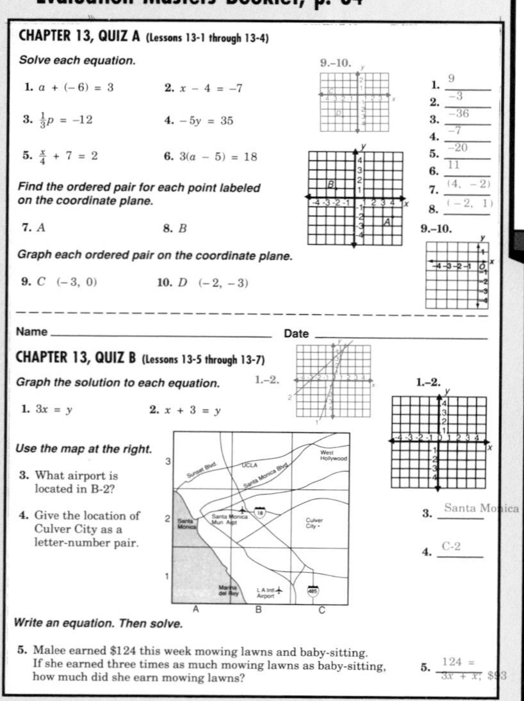

CHAPTER 13, QUIZ A (Lessons 13-1 through 13-4)

Solve each equation.

1. $a + (-6) = 3$ 2. $x - 4 = -7$
3. $\frac{1}{3}p = -12$ 4. $-5y = 35$
5. $\frac{x}{4} + 7 = 2$ 6. $3(a - 5) = 18$

Find the ordered pair for each point labeled on the coordinate plane.

7. A 8. B

Graph each ordered pair on the coordinate plane.

9. C $(-3, 0)$ 10. D $(-2, -3)$

1. 9
2. −3
3. −36
4. −7
5. −20
6. 11
7. (4, −2)
8. (−2, 1)
9.–10.

Name ______________ Date ______________

CHAPTER 13, QUIZ B (Lessons 13-5 through 13-7)

Graph the solution to each equation.

1. $3x = y$ 2. $x + 3 = y$

Use the map at the right.

3. What airport is located in B-2?
4. Give the location of Culver City as a letter-number pair.

Write an equation. Then solve.

5. Malee earned $124 this week mowing lawns and baby-sitting. If she earned three times as much mowing lawns as baby-sitting, how much did she earn mowing lawns?

1.–2.
3. Santa Monica
4. C-2
5. $124 = 3x + x$; $93

Chapter 13 REVIEW

Vocabulary / Concepts

Choose the word or phrase from the list at the right that best completes each sentence.

1. To solve an equation with one variable, add, subtract, multiply, or divide the __?__ on each side of the equation. **same number**
2. To check a solution, replace the __?__ with the possible solution. **variable**
3. A solution to an equation with two variables can be written as a(n) __?__. **3. ordered pair**
4. A solution to an equation with two variables can be graphed on a(n) __?__. **4. coordinate plane**
5. The first coordinate in an ordered pair represents a value along the __?__ axis. **5. horizontal**
6. An equation in two variables has __?__ solution. **more than one**
7. On a coordinate plane, the point that has coordinates (0,0) is called the __?__. **origin**
8. The graph of the ordered pair (0,5) lies on the __?__ axis. **vertical**

coordinate plane
horizontal
more than one
negative
one
opposite
ordered pair
origin
positive
same number
variable
vertical

Exercises / Applications

Lesson 13-1

Solve each equation. Check your solution.

9. $y + (-7) = 8$ **15** **10.** $-11 = x + 9$ **−20** **11.** $m - 9 = 10$ **19**

12. $-18 = y - 2$ **−16** **13.** $x - 30 = -5$ **25** **14.** $y + 12 = -12$ **−24**

Lesson 13-2

Solve each equation. Check your solution.

15. $4p = -8$ **−2** **16.** $-3t = 18$ **−6** **17.** $-104 = 8r$ **−13**

18. $42 = \frac{y}{-3}$ **−126** **19.** $\frac{x}{8} = -24$ **−192** **20.** $\frac{p}{-2} = 14$ **−28**

Lesson 13-3

Solve each equation. Check your solution.

21. $5m + 2 = -18$ **−4** **22.** $2y - 7 = 8$ **7.5**

23. $-23 = 6x + (-5)$ **−3** **24.** $3(n + 4) = -12$ **−8**

25. $5(x - 2) = -18$ **$-1\frac{3}{5}$** **26.** $8.75 = -2.5(7 + y)$ **−10.5**

Lesson 13-4

Find the ordered pair for each point labeled on the coordinate plane.

27. A (−4, 3) **28.** B (3, −2) **29.** C (−2, −3)

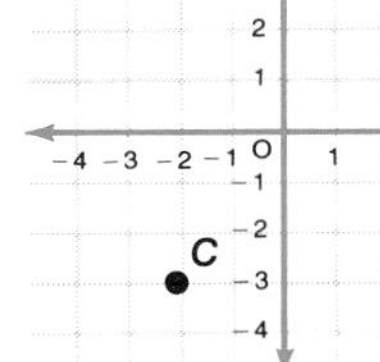

Draw a coordinate plane. Then graph each ordered pair. Label each point with the given letter. See answers on p. T530.

30. I (1, 4) **31.** J (1, −4) **32.** K (−3, 5)

33. M (2, −3) **34.** P (4, 1) **35.** Q (0, 6)

Lesson 13-5

Graph the solutions to each equation. See answers on p. T530.

36. $12 - x = y$ **37.** $x + 2 = y$ **38.** $x - 7 = y$

39. $3x = y$ **40.** $\frac{1}{2}x = y$ **41.** $x + y = 5$

42. American University 43. River Road

Lesson 13-6

Use the map to answer these questions.

42. What university is located in E-4?

43. What road runs from C-6 to E-5?

44. F2 46. B3

Give the location of each as a letter-number pair.

44. Georgetown U. **45.** McLean A4

46. Chesterbrook **47.** Cabin John B6

Lesson 13-7

Solve. Write an equation.

48. Mr. York bought 200 shares of stock at \$31 per share. Then the price of each share went up \$2.50. What was the total value of the 200 shares after the price rise? \$6,700

49. From the surface, a diver descended to 30 meters below sea level where she obtained a water sample. She then rose 12 meters and collected another water sample. How far below the surface was she when she collected the second sample? 18 m

50. Frank and David buy plain sweatshirts from a wholesale dealer. They print their school's name and team's name on the sweatshirts and sell them for \$7 more than the wholesale price. They sold 40 sweatshirts and collected \$640. What is the wholesale price for each? \$9

51. Mrs. Eearitt has two jobs. The first pays \$50 more per week than the second. Her total earnings are \$320 per week. How much does she earn from each job? \$135, \$185

APPLYING THE LESSON

Independent Practice

Homework Assignment	
Minimum	Ex. 1-8; 9-25 odd; 27-51
Average	Ex. 1-51
Maximum	Ex. 1-51

Alternate Assessment Strategy

To provide a brief in-class review, you may wish to read the following questions to the class and require a verbal response.

1. Solve the equation y minus 5 equals 10. $y = 15$
2. If negative 8 is added to a number, the result is 12. What is the number? 20
3. Solve the equation 4 times x equals negative 32. $x = -8$
4. What number multiplied by negative 2 equals 14? −7
5. What number when multiplied by 2, then increased by 3 equals 11? 4
6. In the ordered pair, (−5,6), which number is on the horizontal axis? −5
7. Is the point (2,3) on the graph of the line $8 - x = y$? no
8. Does the line $3x = y$ pass through the origin, (0,0)? yes
9. Find the number that if multiplied by 6 is equal to 4 plus 20. 4
10. Where do the horizontal and vertical axes intersect in the coordinate plane? origin

Using the Chapter Test

This page may be used as a test or as an additional review if necessary. Two forms of the Chapter Test are provided in the Evaluation Masters Booklet. Form 2 (free response) is shown below, and Form 1 (multiple choice) is shown on page 421.

The **Tech Prep Applications Booklet** provides students with an opportunity to familiarize themselves with various types of technical vocations. The Tech Prep applications for this chapter can be found on pages 25–26.

Evaluation Masters Booklet, p. 63

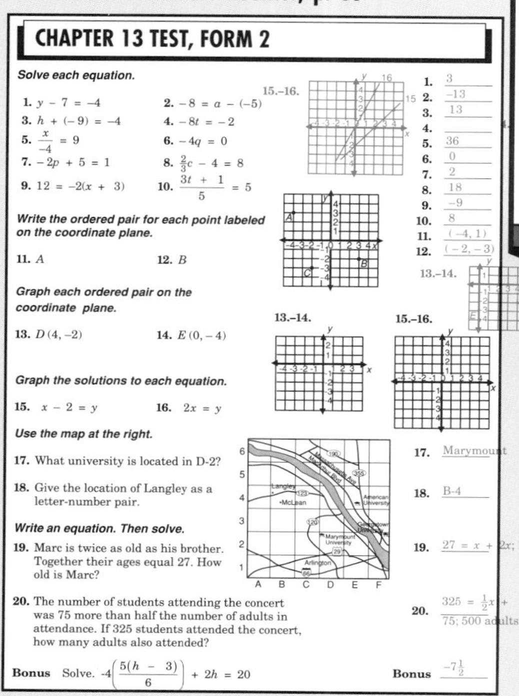

CHAPTER 13 TEST, FORM 2

Solve each equation.

1. $y - 7 = -4$
2. $-8 = a - (-5)$
3. $h + (-9) = -4$
4. $-8t = -2$
5. $\frac{x}{-4} = 9$
6. $-4q = 0$
7. $-2p + 5 = 1$
8. $\frac{2}{3}c - 4 = 8$
9. $12 = -2(x + 3)$
10. $\frac{3t + 1}{5} = 5$

Write the ordered pair for each point labeled on the coordinate plane.

11. *A*
12. *B*

Graph each ordered pair on the coordinate plane.

13. *D* (4, −2)
14. *E* (0, −4)

Graph the solutions to each equation.

15. $x - 2 = y$
16. $2x = y$

Use the map at the right.

17. What university is located in D-2?
18. Give the location of Langley as a letter-number pair.

Write an equation. Then solve.

19. Marc is twice as old as his brother. Together their ages equal 27. How old is Marc?
20. The number of students attending the concert was 75 more than half the number of adults in attendance. If 325 students attended the concert, how many adults also attended?

Bonus Solve. $-4\left(\frac{5(h - 3)}{6}\right) + 2h = 20$

Answers: 1. 3 2. −13 3. 13 4. $\frac{1}{4}$ 5. 36 6. 0 7. 2 8. 18 9. −9 10. 8 11. (−4, 1) 12. (−2, −3) 13.–14. 15.–16. 17. Marymount 18. B-4 19. $27 = x + 2x$; 18 20. $325 = \frac{1}{2}x + 75$; 500 adults Bonus $-7\frac{1}{2}$

Chapter 13 TEST

Solve each equation. Check your solution.

1. $m - 6 = 4$ **10**
2. $-8 + x = 2$ **10**
3. $8x = -32$ **−4**
4. $-36 = y - 4$ **32**

Find the ordered pair for each point labeled on the coordinate plane.

5. *P* **(−2, 3)**
6. *Q* **(0, 3)**
7. *R* **(4, 1)**
8. *S* **(4, −1)**
9. *T* **(1, −3)**
10. *U* **(0, −2)**
11. *V* **(−2, −3)**
12. *W* **(−3, −2)**
13. *Z* **(−3, 0)**

Draw a coordinate plane. Then graph each ordered pair. Label each point with the given letter. **See answers on p. T530.**

14. *A* (5,3)
15. *B* (−2, −4)
16. *C* (1, −6)
17. *D* (−2, 3)

Graph the solutions to each equation. **See answers on p. T530.**

18. $3 + x = y$
19. $\frac{1}{2}x + 4 = y$
20. $3x = y$
21. $\frac{1}{3}x = y$
22. Assume the letters of your grid map are on the horizontal edge and the numbers are on the vertical edge. Explain how to locate area C-5. **Move horizontally to the square labeled C. Move vertically to the square across from 5.**

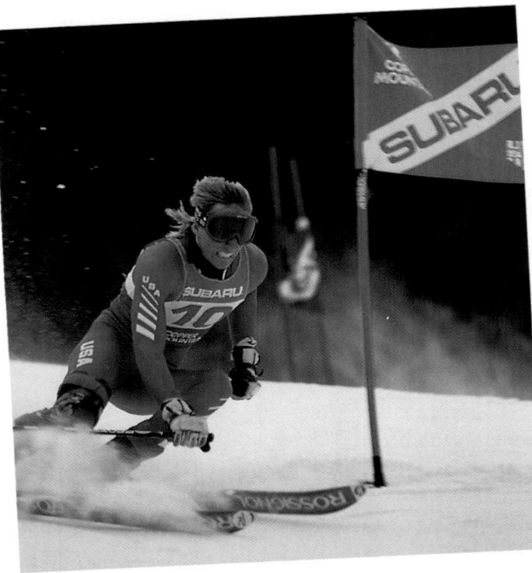

Solve. Write an equation.

23. Bill Raymond skied the slalom course in 138 seconds. This was 15 seconds faster than his father. What was his father's time? **23. 153 s, $138 + 15 = t$**
24. Janet gave each of her three sisters an equal number of pictures from her collection of popular actors and actresses. Karrie already had 2 pictures. She has 8 pictures after getting some from Janet. How many pictures did Janet give away in all? **24. $\frac{x}{3} + 2 = 8$, 18 pictures**
25. Doug weighs 180 pounds. He loses 2 pounds each week for 5 weeks, and then gains 1 pound each week for 3 weeks. How much does he weigh after the losses and gains? **25. 173 lb; $180 + (-2)5 + (1)3 = w$**

BONUS: Find two integers that have a sum of 6 and a difference of −20. **−7, 13**

The **Performance Assessment Booklet** provides an alternative assessment for evaluating student progress. An assessment for this chapter can be found on pages 25–26.

A **Test and Review Generator** is provided in Apple, IBM, and Macintosh versions. You may use this software to create your own tests or worksheets, based on the needs of your students.

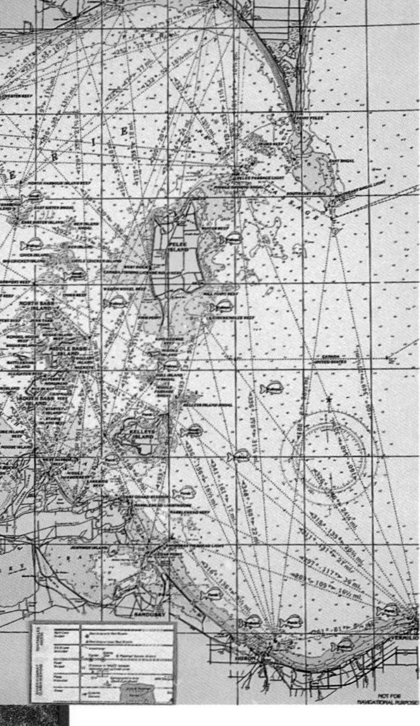

LATITUDE AND LONGITUDE

Most globes and maps of Earth show grid lines, called latitude lines and longitude lines. Because Earth is shaped like a sphere, these grid lines are actually circles or parts of circles. Latitude and longitude are measured in degrees because circles may be divided into degrees.

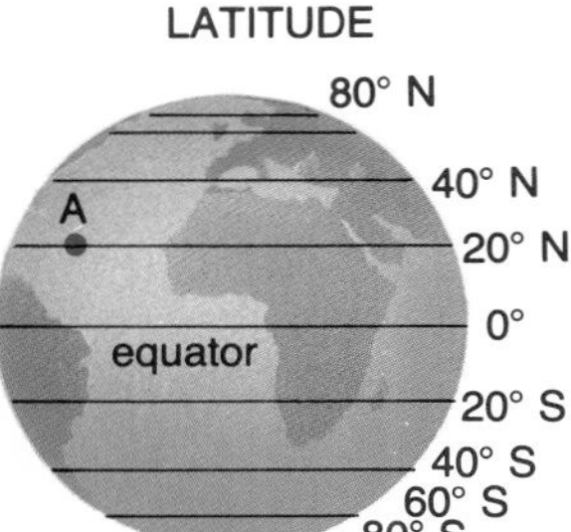

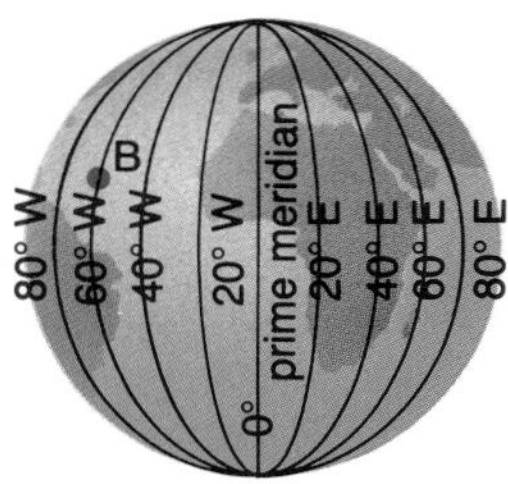

Any point on Earth is a certain number of degrees north or south of the equator. For example, point A is located at 20°N latitude.

Any point on Earth is within 180° east or west of the prime meridian. For example, point B is located at 40°W longitude.

Use the map at the right to find the latitude and longitude of each location to the nearest degree.

1. Providence, Rhode Island **41°N, 71°W**
2. Portsmouth, New Hampshire **43°N, 71°W**
3. Portland, Maine **44°N, 70°W**
4. Which city is farthest west? **New York City** What is its latitude and longitude? **41°N, 74°W**
5. Which ship is farthest north? **X** Which ship is farthest south? **Z** What is the approximate difference between their latitudes? **3°**

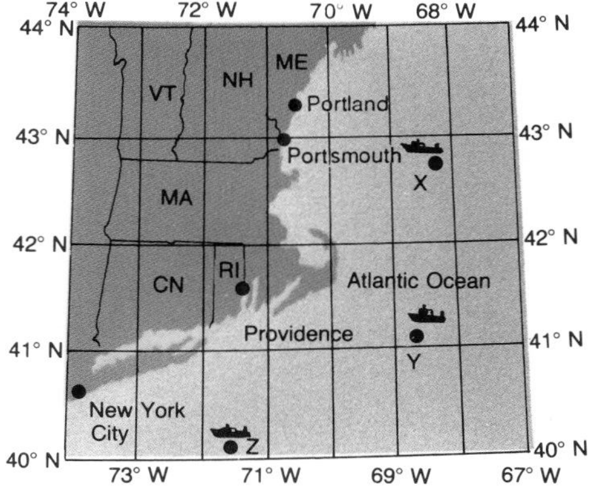

Applying Mathematics to the Real World

This optional page shows how mathematics is used in the real world and also provides a change of pace.

Objective Locate places on Earth using latitude and longitude.

Using Discussion Point out that the Prime Meridian runs through the original site of the Royal Observatory at Greenwich, England.

Discuss with students that the equator divides Earth into the northern hemisphere and the southern hemisphere. Discuss the similarity between the latitude and longitude system and the coordinate plane.

Activity

Using Questioning Have students research and write a report on the relationship between longitude and time zones.

Using the Cumulative Review

This page provides an aid for maintaining skills and concepts presented thus far in the text.

A Cumulative Review is also provided in the Evaluation Masters Booklet as shown below.

Cumulative REVIEW/TEST

Free Response

Lesson 1-7 — 1. 2,100 2. 4,000 4. 70

Estimate.

1. 318 × 7 **2.** 43 × 112 **3.** 285 ÷ 72 4 **4.** 4,128 ÷ 59

Lesson 2-1

5. A car dealer sold 1,774 cars one year and 1,967 cars the next year. How many were sold in the two years? **3,741 cars**

Lesson 4-10

Add.

6. $\frac{1}{5} + \frac{3}{5}$ $\frac{4}{5}$ **7.** $\frac{3}{7} + \frac{5}{7}$ $1\frac{1}{7}$ **8.** $\frac{1}{6} + \frac{2}{3}$ $\frac{5}{6}$ **9.** $\frac{3}{4} + \frac{7}{8}$ $1\frac{5}{8}$

Lesson 6-8

Complete. 10. 16 11. 64 12. 5

10. 4 gal = _?_ qt **11.** 2 qt = _?_ oz **12.** 80 oz = _?_ lb

Lessons 8-9, 8-10

Find the volume of each solid.

13. 1.8 ft, 2 ft, 5 ft — **18 ft³**

14. 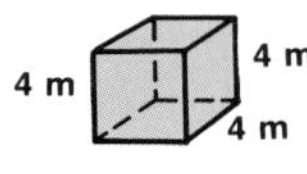

64 m³

15. 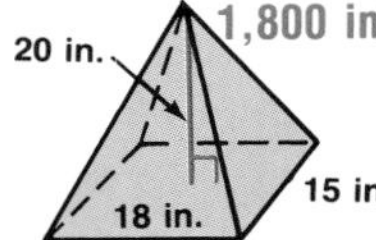

1,800 in³

Lesson 10-7

16. Janis buys 3 sweatshirts that normally sell for $24 each. Two of the shirts are on sale for 25% off and one shirt is discounted for 35% off regular price. How much does she pay for the 3 sweatshirts? **$51.60**

Lessons 11-2, 11-4

See answers on p. T531.

Use the data at the right.

17. Make a frequency-table for the data.

18. Find the mean, median, mode, and range.

19. Make a bar graph for the data.

Ring Sizes of Senior Girls					
5	6	6	7	7	$6\frac{1}{2}$
7	5	$7\frac{1}{2}$	$6\frac{1}{2}$	$7\frac{1}{2}$	6
$7\frac{1}{2}$	$6\frac{1}{2}$	6	7	6	7

Lessons 12-2, 12-3

Add or subtract.

20. 30 − 45 **−15** **21.** 10 + (−17) **−7** **22.** 28 + (−57) **−29** **23.** −71 + 100 **29**

Lessons 12-5, 12-6

Multiply or divide.

24. 3 × (−8) **−24** **25.** 2 × 14 **28** **26.** 20 ÷ (−2) **−10** **27.** −45 ÷ 3 **−15**

Lesson 13-5

Graph the solutions to each equation.

See answers on p. T531.

28. $x + y = 0$ **29.** $3x + 2 = y$

APPLYING THE LESSON

Independent Practice

Homework Assignment	
Minimum	Ex. 1-29 odd
Average	Ex. 1-29 odd
Maximum	Ex. 1-29

Evaluation Masters Booklet, p. 65

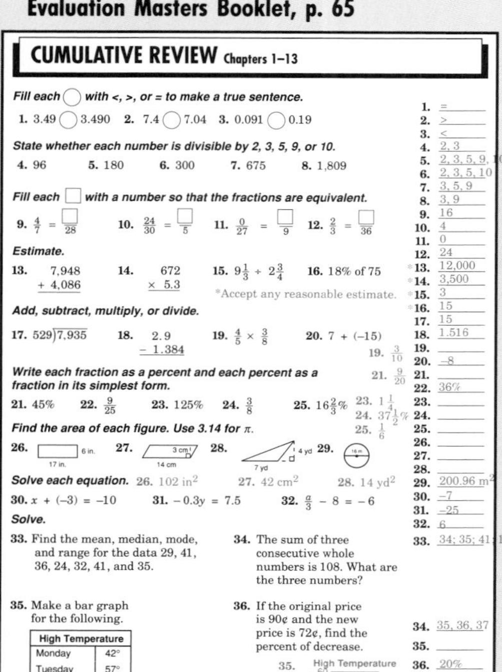

CUMULATIVE REVIEW Chapters 1–13

Fill each ◯ with <, >, or = to make a true sentence.

1. 3.49 ◯ 3.490 **2.** 7.4 ◯ 7.04 **3.** 0.091 ◯ 0.19

State whether each number is divisible by 2, 3, 5, 9, or 10.

4. 96 **5.** 180 **6.** 300 **7.** 675 **8.** 1,809

Fill each ☐ with a number so that the fractions are equivalent.

9. $\frac{4}{7} = \frac{☐}{28}$ **10.** $\frac{24}{30} = \frac{☐}{5}$ **11.** $\frac{0}{27} = \frac{☐}{9}$ **12.** $\frac{2}{3} = \frac{☐}{36}$

Estimate.

13. 7,948 + 4,086 **14.** 672 × 5.3 **15.** $9\frac{1}{3} \div 2\frac{3}{4}$ **16.** 18% of 75

*Accept any reasonable estimate.

Add, subtract, multiply, or divide.

17. 529)7,935 **18.** 2.9 − 1.384 **19.** $\frac{4}{5} \times \frac{3}{8}$ **20.** 7 + (−15)

Write each fraction as a percent and each percent as a fraction in its simplest form.

21. 45% **22.** $\frac{9}{25}$ **23.** 125% **24.** $\frac{3}{8}$ **25.** $16\frac{2}{3}\%$

Find the area of each figure. Use 3.14 for π.

26. 6 in., 17 in. **27.** 3 cm, 14 cm **28.** 4 yd, 7 yd **29.** 16 m

26. 102 in² 27. 42 cm² 28. 14 yd²

Solve each equation.

30. $x + (-3) = -10$ **31.** $-0.3y = 7.5$ **32.** $\frac{a}{3} - 8 = -6$

Solve.

33. Find the mean, median, mode, and range for the data 29, 41, 36, 24, 32, 41, and 35.

34. The sum of three consecutive whole numbers is 108. What are the three numbers?

35. Make a bar graph for the following.

High Temperature	
Monday	42°
Tuesday	57°
Wednesday	51°

36. If the original price is 90¢ and the new price is 72¢, find the percent of decrease.

35. High Temperature (Temp. 0–60; Days M T W)

Answers: 1. = 2. > 3. < 4. 2, 3 5. 2, 3, 5, 9, 10 6. 2, 3, 5, 10 7. 3, 5, 9 8. 3, 9 9. 16 10. 4 11. 0 12. 24 *13. 12,000 *14. 3,500 *15. 3 *16. 15 17. 15 18. 1.516 19. $\frac{3}{10}$ 20. −8 21. $\frac{9}{20}$ 22. 36% 23. $1\frac{1}{4}$ 24. $37\frac{1}{2}\%$ 25. $\frac{1}{6}$ 26. 102 in² 27. 42 cm² 28. 14 yd² 29. 200.96 m² 30. −7 31. −25 32. 6 33. 34; 35; 41; 17 34. 35, 36, 37 35. (graph) 36. 20%

Chapter 13

Multiple Choice

Choose the letter of the correct answer for each item.

1. Write 6^3 in standard form.
   a. 18
   b. 36
   ▶ c. 216
   d. 729

   Lesson 1-2

2. What is the quotient when 45 is divided by 100?
   e. 0.0045
   ▶ f. 0.45
   g. 0.045
   h. 4,500

   Lesson 3-7

3. Divide $6\frac{1}{4}$ by $2\frac{1}{8}$. What is the result?
   a. $\frac{17}{50}$
   ▶ b. $2\frac{16}{17}$
   c. $3\frac{1}{2}$
   d. $13\frac{9}{32}$

   Lesson 5-7

4. The formula for changing Fahrenheit degrees to Celsius degrees is $C = \frac{5}{9}(F - 32)$. Find the Celsius reading for 68°F.
   e. 30°C
   f. 5.7°C
   ▶ g. 20°C
   h. *none of these*

   Lesson 6-9

5. The surface area of a solid is the sum of the areas of each surface. What is the surface area of the solid?
   a. 6 ft^2
   b. $6\frac{1}{2}$ ft^2
   c. $11\frac{1}{2}$ ft^2
   ▶ d. 23 ft^2

   1 ft
   $1\frac{1}{2}$ ft
   4 ft

   Lesson 8-7

6. 32 is 50% of what number?
   e. 16
   ▶ f. 64
   g. 160
   h. 1,600

   Lesson 10-2

7. The frequency table lists the ages of sophomores at East High. How many sophomores are at East High?
   a. 48
   b. 162
   ▶ c. 489
   d. 537

Age	Frequency
15	112
16	324
17	53

   Lesson 11-2

8. Which equation is graphed at the right?
   e. $x = y$
   f. $x = 2y$
   ▶ g. $x + 2 = y$
   h. $x - 2 = y$

   Lesson 13-5

9. Rachael spent half of her money for dinner and half of what remained for earrings. She had $4 left. How much did she spend for dinner?
   a. $20
   b. $16
   ▶ c. $8
   d. $4

   Lesson 12-7

10. Suppose you were to read the numbers −27, 17, −17, 27, and −16 on a number line. In which order would they appear from left to right?
    e. −16, −17, −27, 17, 27
    ▶ f. −27, −17, −16, 17, 27
    g. −27, −16, −17, 17, 27
    h. *none of the above*

    Lesson 12-1

Using the Cumulative Test

This test serves to familiarize students with standardized format while testing skills and concepts presented up to this point.

APPLYING THE LESSON

Independent Practice

Homework Assignment	
Minimum	Ex. 1-10
Average	Ex. 1-10
Maximum	Ex. 1-10

Evaluation Masters Booklet, pp. 61-62

Name ____________ Date ____________

CHAPTER 13 TEST, FORM 1

Write the letter of the correct answer on the blank at the right of the page.

1. Solve. $y + (-3) = -7$ A. 4 B. 10 C. −4 D. −10 — 1. C
2. Solve. $-3 = p - 8$ A. 5 B. −11 C. −5 D. 11 — 2. A
3. Solve. $-7 - b = -7$ A. 7 B. −14 C. 14 D. 0 — 3. D
4. Solve. $-6x = 24$ A. −18 B. 18 C. 4 D. −4 — 4. D
5. Solve. $\frac{y}{-2} = -5$ A. 2.5 B. 10 C. −2.5 D. −10 — 5. B
6. Solve. $\frac{3}{4}h = -12$ A. 16 B. −9 C. −16 D. 9 — 6. C
7. Solve. $17 = 3t - 4$ A. 7 B. −7 C. $4\frac{1}{3}$ D. $-4\frac{1}{3}$ — 7. A
8. Solve. $\frac{m}{5} + 8 = -7$ A. −3 B. $\frac{-1}{5}$ C. $\frac{1}{5}$ D. −75 — 8. D
9. Solve. $6 = -3(x - 4)$ A. −6 B. 2 C. −14 D. $-3\frac{1}{3}$ — 9. B
10. Find the ordered pair for point A on the coordinate plane at the right. A. (4, −1) B. (1, 4) C. (−1, −4) D. (−4, −1) — 10. C
11. Find the ordered pair for point B on the coordinate plane at the right. A. (−3, 4) B. (4, −3) C. (−4, −3) D. (3, 4) — 11. B
12. Find the letter of the graph of the ordered pair (−2, 3) on the coordinate plane at the right. A. E B. F C. G D. H — 12. A

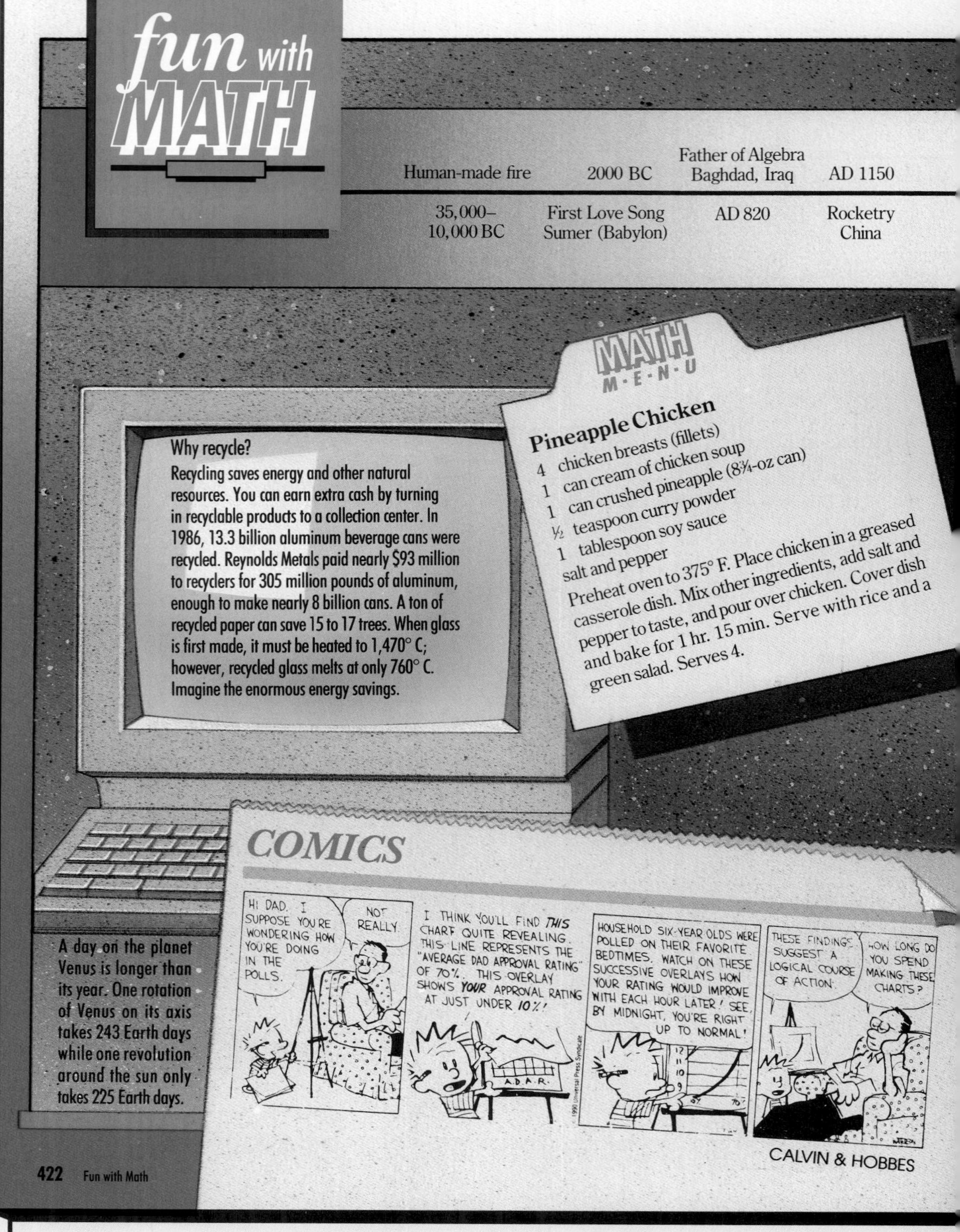

fun with MATH
Human-made fire
35,000–10,000 BC
2000 BC
First Love Song Sumer (Babylon)
Father of Algebra Baghdad, Iraq
AD 820
AD 1150
Rocketry China
Why recycle?
Recycling saves energy and other natural resources. You can earn extra cash by turning in recyclable products to a collection center. In 1986, 13.3 billion aluminum beverage cans were recycled. Reynolds Metals paid nearly $93 million to recyclers for 305 million pounds of aluminum, enough to make nearly 8 billion cans. A ton of recycled paper can save 15 to 17 trees. When glass is first made, it must be heated to 1,470° C; however, recycled glass melts at only 760° C. Imagine the enormous energy savings.
MATH M·E·N·U
Pineapple Chicken
4 chicken breasts (fillets)
1 can cream of chicken soup
1 can crushed pineapple (8¾-oz can)
½ teaspoon curry powder
1 tablespoon soy sauce
salt and pepper
Preheat oven to 375° F. Place chicken in a greased casserole dish. Mix other ingredients, add salt and pepper to taste, and pour over chicken. Cover dish and bake for 1 hr. 15 min. Serve with rice and a green salad. Serves 4.
A day on the planet Venus is longer than its year. One rotation of Venus on its axis takes 243 Earth days while one revolution around the sun only takes 225 Earth days.
COMICS
HI DAD. I SUPPOSE YOU'RE WONDERING HOW YOU'RE DOING IN THE POLLS.
NOT REALLY.
I THINK YOU'LL FIND THIS CHART QUITE REVEALING. THIS LINE REPRESENTS THE "AVERAGE DAD APPROVAL RATING" OF 70%. THIS OVERLAY SHOWS YOUR APPROVAL RATING AT JUST UNDER 10%!
A.D.A.R.
HOUSEHOLD SIX-YEAR OLDS WERE POLLED ON THEIR FAVORITE BEDTIMES. WATCH ON THESE SUCCESSIVE OVERLAYS HOW YOUR RATING WOULD IMPROVE WITH EACH HOUR LATER! SEE, BY MIDNIGHT, YOU'RE RIGHT UP TO NORMAL!
THESE FINDINGS SUGGEST A LOGICAL COURSE OF ACTION.
HOW LONG DO YOU SPEND MAKING THESE CHARTS?
CALVIN & HOBBES
422 Fun with Math

Using Mathematical Connections

Objective Understand, appreciate, and enjoy the connections between mathematics and real-life phenomena and other disciplines.
The following data is background information for each event on the time line.

History/Technology Human-made Fire, Africa, Asia, Europe; 35,000–10,000 B.C.
The first evidence connecting man with fire comes from Lake Turkana (Lake Rudolf) in Kenya, Africa. Evidence of cooking fires dating back about 500,000 years has been found near Peking, China and near Budapest, Hungary. Much more evidence attests to the use of fires by the *Neanderthals* of 75,000 years ago. It is difficult to know with certainty whether these humans made fire or simply kept fire alive that had started by natural phenomena, such as lightning and volcanoes. Charred animal bones found in camps, plus evidence of pyrites and flint, lead experts to believe that humans could make fire around 30,000 to 10,000 B.C.

Art/Music/Poetry/History First Love Song, Sumer (Babylon), 2000 B.C.
The earliest known love song has been found on a clay tablet in Turkey. It has been translated as a love song to a Sumerian king, Shu-Sin, from his selected bride. The following is the translation of the first verse.

Bridegroom, dear to my heart,
Goodly is your beauty, honeysweet,
Lion, dear to my heart,

Goodly is your beauty, honeysweet.

History Poseidonius Measures Earth—Incorrectly! Rhodes, Greece, 100 B.C.
Poseidonius was a Greek philosopher who is known because of a mistake. He attempted to duplicate the work of an earlier Greek astronomer Eratosthenes, who calculated the circumference of Earth using the shift of the sun.
Poseidonius decided to use the position of the star Canopus instead of the sun. He incorrectly calculated the circumference of Earth as only 180,000 stadia (18,000 miles). Two hundred years later, the famous Greek astronomer Ptolemy accepted Poseidonius' measurement over that of Eratosthenes.
Columbus used Poseidonius', and Ptolemy's measurement and estimated that Asia lay only three or four thousand miles westward. Had he known that Asia actually lay 12,000 miles away, he may never have dreamed of sailing.

History Father of Algebra, Baghdad, Iraq, AD 820
Around 250 B.C., the Greek mathematician Diophantus published his *Arithmetica* with 150 algebra problems. He advanced algebra beyond the use of entire words in equations to using abbreviations. However, the true father of algebra is considered to be an Arabian mathematician, Al-Khowarizani. In fact, the name *algebra* has been taken from the title of his *Al-jabr w'almuqabala,* published in AD 820.

Harry Houdini — King of Handcuffs — AD 1874

First woman heart surgeon — USA — AD 1908–1977

Computer on a chip — USA — AD 1969

Hero 1—Robot — USA — AD 1982

ow does a robot work?

robot is a machine that can be ogrammed to do different tasks. To ogram a robot to pick something up d put it down somewhere else,

e programmer moves each of the robot's ints by pressing buttons on a *teach unit* tached to the robot's arm. At the same ne, the programmer engages a record itton. This enters the program into e robot's computer memory. he robot can carry out the entire sk on its own from memory.

RIDDLE

Q: Why is simplifying a fraction like powdering your nose?

A: It improves the appearance without changing the value.

Find the indicated average for each letter.

ACROSS

A. Mean: 5, 6, 9, 11, 2, 11, 40
B. Median: 36, 37, 41, 43, 43
D. Mode: 84, 10, 71, 10, 10, 69
E. Mean: 421; 98; 602; 935; 1,084
G. Mode: 256, 286, 95, 12, 286, 76

DOWN

A. Mode: 105, 116, 78, 87, 116, 125
B. Mean: 36, 37, 41, 43, 43
C. Median: 4; 2,067; 583; 5,300; 2,419
D. Median: 45, 18, 67, 9, 3, 18
F. Mode: 22, 23, 32, 28, 23, 33
G. Mean: 28, 61, 0, 35, 19, 14, 28, 7

TEASER

QUIZ TIME

It's Saturday evening. There's been a theft. On Monday, the victim, Leonardo Fuentes, left his wallet in a towel on the beach and went in swimming. "That was at 3 P.M.," he said. "I came out at 3:30 and everything was gone." A local surfer called Wave is suspected, so I talk with her. "I hang out on the beach," she said. "But I didn't take that guy's stuff. The high tide must have washed it away. At 3:30 the tide was far up the beach." The lifeguard tells me that he couldn't remember what time high tide occurred on Monday. "Today, it was at 4:30," he said. "And it's gotten about 25 minutes later each day during this week."

Was Wave telling the truth? Why, or why not?

JOKE!

Q: What will happen to the inch worm when we go metric?

A: He'll become a centipede.

Fun with Math 423

Teaser Answer

A 1	2		B 4	1		C 2
1		D 1	0			0
E 6	F 2	8		G 2	8	6
	3			4		7

Quiz Answer

Wave could not have been telling the truth. The high tide on Monday occurred at about 2:25 PM. Between 3:00 and 3:30 PM the tide would have been going out, not coming in.

Science / Technology Rocketry, China, A.D. 1150
Inspiration for rocketry is believed to have come from a type of Chinese firework known as the "ground rat." Around A.D. 1150 a "ground rat" attached to a feathered stick constituted a rocket-arrow. By A.D. 1300, the Chinese refined the rocket tube by constricting the rear flow of gases. As such, they invented the nozzle, one of the most important principles of aerodynamics.

History Harry Houdini (1874-1926)
Young Harry Houdini became fascinated with magic. Harry became more popular as an escapist than as a magician and became known as the "King of Handcuffs." Harry Houdini died after surgery on October 31, 1926—Halloween.

Health First Woman Heart Surgeon, USA, 1908-1977
Dr. Myra A. Logan was the first woman surgeon to operate on the heart in the world's ninth such operation. She was also the first black woman to be elected a Fellow of the American College of Surgeons. Dr. Logan served the Harlem and wider New York City community for more than thirty-five years. She was a pioneer in the practice of group medicine.

Technology Computer on a chip, USA, 1969
In 1969, Marcian E. "Ted" Hoff of Intel Corporation invented the microprocessor, or the "computer on a chip." In doing so, he revolutionized the entire field of microelectronics. The first microprocessor was Intel's 4004 and was not much larger than a pencil point.

Industrial Technology / Home Economics
Hero 1—Robot, USA, 1982
The first industrial robots, called Unimates, appeared in 1962 in Danbury, Connecticut. The first Nursing Robot was invented in 1983 by Professor Hiroyasu Funakubo of Japan. The most humanoid robot is the Marilyn Monroe robot created in 1983 by Shumeihi Mizuno of Japan. The first guide-dog robot was invented in 1983 by Susumu Tachi of Japan. Hero 1 was the first household robot. It was invented in 1982 by the American company Heath Kit.

Chapter

14 PROBABILITY

PREVIEWING the CHAPTER

The chapter begins by clarifying the concept of equally likely outcomes. Students find the number of outcomes using tree diagrams and multiplication. Probability is introduced as a ratio that describes how likely it is that an event will occur. Students add probabilities to find the probability of two mutually exclusive events. Dependent and independent events are discussed. The chapter concludes with samples to predict outcomes and finding odds.

Problem-Solving Strategy Students learn to use samples to predict.

Lesson (Pages)	Lesson Objectives	State/Local Objectives
14-1 (426-427)	14-1: Find the number of possible outcomes using tree diagrams or multiplication.	
14-2 (428-429)	14-2: Express the probability of an event as a percent.	
14-3 (430-431)	14-3: Find the probability of independent and dependent events	
14-4 (432-433)	14-4: Find the probability of one event or another.	
14-5 (436-437)	14-5: Find probabilities from market samples.	
14-6 (438-439)	14-6: Find the odds for an event.	
14-7 (440-441)	14-7: Solve verbal problems using samples to make predictions.	

ORGANIZING the CHAPTER

You may refer to the ***Course Planning Calendar*** on page T10.

Planning Guide

Lesson (Pages)	Extra Practice (Student Edition)	Blackline Masters Booklets: Reteaching	Practice	Enrichment	Activity	Multi-cultural Activity	Technology	Tech Prep	Evaluation	Other Resources
14-1 (426-427)	p. 483	p. 127	p. 140	p. 127	p. 27	p. 56	p. 14			Transparency 14-1
14-2 (428-429)	p. 483	p. 128	p. 141	p. 128	p. 42			p. 27		Transparency 14-2
14-3 (430-431)	p. 484	p. 129	p. 142	p. 129						Transparency 14-3
14-4 (432-433)	p. 484	p. 130	p. 143	p. 130	p. 28				p. 69	Transparency 14-4
14-5 (436-437)		p. 131	p. 144	p. 131			p. 28			Transparency 14-5
14-6 (438-439)	p. 484	p. 132	p. 145	p. 132						Transparency 14-6
14-7 (440-441)			p. 146					p. 28	p. 69	Transparency 14-7
Ch. Review (442-443)									pp. 66-68	Test and Review Generator
Ch.Test (444)										
Cumulative Review/Test (446-447)									pp. 70, 77-83	

OTHER CHAPTER RESOURCES

Student Edition
Chapter Opener, p. 424
Activity, p. 425
Computer Applications, p. 434
On-the-Job Application, p. 435
Journal Entry, p. 437
Portfolio Suggestion, p. 441
Consumer Connections, p. 445

Teacher's Classroom Resources
Transparency 14-0
Lab Manual, pp. 49-50
Performance Assessment, pp. 27-28
Lesson Plans, pp. 140-146

Other Supplements
Overhead Manipulative Resources, Labs 39-40, pp. 29-30
Glencoe Mathematics Professional Series

ENHANCING the CHAPTER

Some of the blackline masters for enhancing this chapter are shown below.

COOPERATIVE LEARNING

The following activity can be performed by groups of two. Explain that random numbers have no predictable order and any number is equally likely to occur. Have student pairs use a phone book to generate 100 random numbers by opening the book and selecting any name on the page. Write the last four digits of the person's phone number. Then write the last four digits of the next 24 persons' numbers listed. Record the number of times each digit appears. If the 100 digits are truly random, each digit should appear about 10 times. Repeat.

USING MODELS/MANIPULATIVES

Give a set of 25 cards marked with 5 As, 5 Bs, 5 Cs, 5 Ds, and 5 Es to each group of 2 students. The experimenter shuffles the cards and places them face down in a pile. The subject guesses the letter on the top card. Without the subject seeing, the experimenter records the trial number, the letter, and the subject's guess. Perform 10 trials. Find the mathematical probability of correctly guessing a letter. Predict the number of correct guesses. Compare the prediction with the actual result. Repeat the experiment for 20 trials, 50 trials, and 100 trials. Compare the results.

Cooperative Problem Solving, p. 42

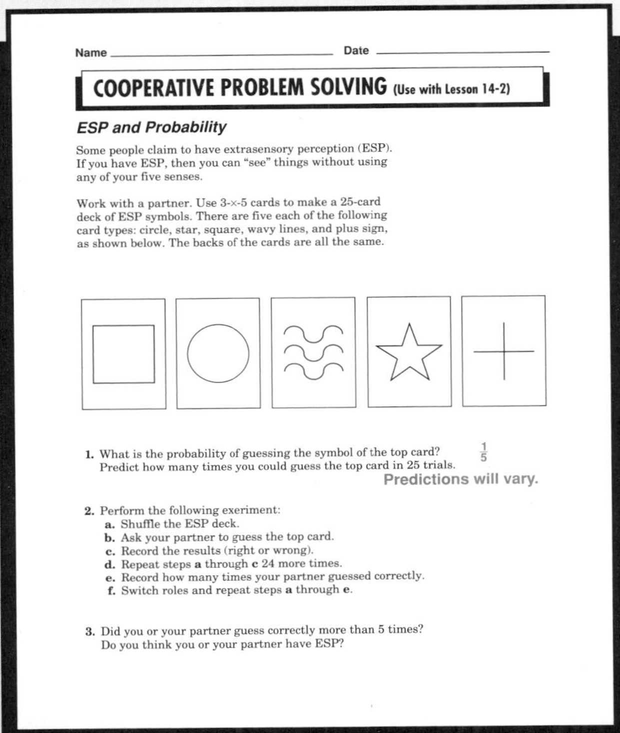

Name ______________ Date ______________

COOPERATIVE PROBLEM SOLVING (Use with Lesson 14-2)

ESP and Probability

Some people claim to have extrasensory perception (ESP). If you have ESP, then you can "see" things without using any of your five senses.

Work with a partner. Use 3-×-5 cards to make a 25-card deck of ESP symbols. There are five each of the following card types: circle, star, square, wavy lines, and plus sign, as shown below. The backs of the cards are all the same.

1. What is the probability of guessing the symbol of the top card? $\frac{1}{5}$
   Predict how many times you could guess the top card in 25 trials. Predictions will vary.

2. Perform the following exeriment:
   a. Shuffle the ESP deck.
   b. Ask your partner to guess the top card.
   c. Record the results (right or wrong).
   d. Repeat steps **a** through **c** 24 more times.
   e. Record how many times your partner guessed correctly.
   f. Switch roles and repeat steps **a** through **e**.

3. Did you or your partner guess correctly more than 5 times? Do you think you or your partner have ESP?

Multicultural Activities, p. 14

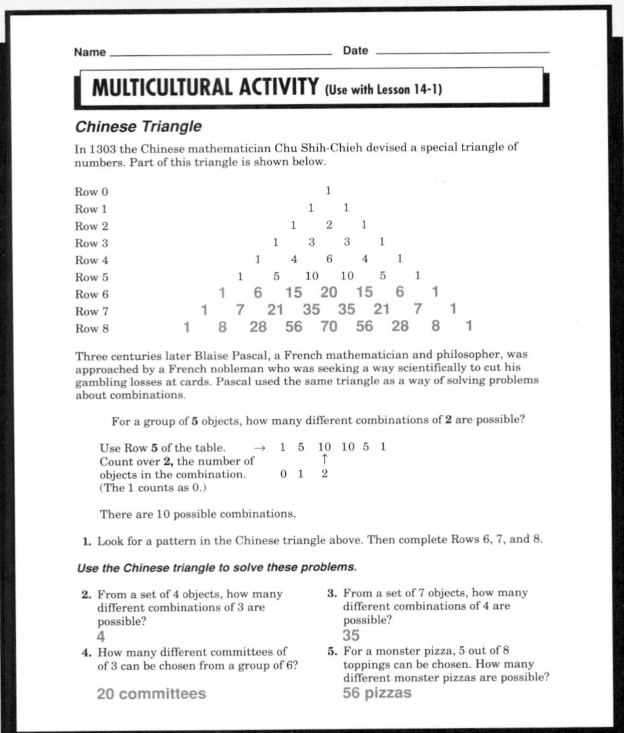

Name ______________ Date ______________

MULTICULTURAL ACTIVITY (Use with Lesson 14-1)

Chinese Triangle

In 1303 the Chinese mathematician Chu Shih-Chieh devised a special triangle of numbers. Part of this triangle is shown below.

Row 0	1
Row 1	1 1
Row 2	1 2 1
Row 3	1 3 3 1
Row 4	1 4 6 4 1
Row 5	1 5 10 10 5 1
Row 6	1 6 15 20 15 6 1
Row 7	1 7 21 35 35 21 7 1
Row 8	1 8 28 56 70 56 28 8 1

Three centuries later Blaise Pascal, a French mathematician and philosopher, was approached by a French nobleman who was seeking a way scientifically to cut his gambling losses at cards. Pascal used the same triangle as a way of solving problems about combinations.

For a group of **5** objects, how many different combinations of **2** are possible?

Use Row **5** of the table. → 1 5 10 10 5 1
Count over **2**, the number of objects in the combination. ↑
(The 1 counts as 0.) 0 1 2

There are 10 possible combinations.

1. Look for a pattern in the Chinese triangle above. Then complete Rows 6, 7, and 8.

Use the Chinese triangle to solve these problems.

2. From a set of 4 objects, how many different combinations of 3 are possible? 4
3. From a set of 7 objects, how many different combinations of 4 are possible? 35
4. How many different committees of of 3 can be chosen from a group of 6? 20 committees
5. For a monster pizza, 5 out of 8 toppings can be chosen. How many different monster pizzas are possible? 56 pizzas

MEETING INDIVIDUAL NEEDS

Students at Risk

The study of probability includes some very subtle concepts about the nature of events and a special vocabulary to describe those concepts. Promote class discussions in which everyday expressions for the probability of an event (including slang) are used. Examples may include "hardly a chance" for low probabilities to "perhaps" for moderate probabilities to "a sure thing" for high probabilities. Construct a large number line labeled from 0 to 1. Relate numerical values for probabilities on the number line to the slang expressions.

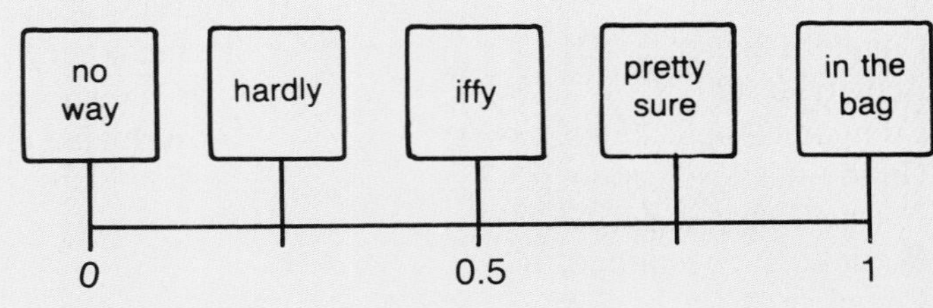

CRITICAL THINKING/PROBLEM SOLVING

Critical Thinking

Have students solve the following problem.

A building has 6 doors. Find the number of ways a person can enter the building by one door and leave by another. **For each door entered, there are 5 by which to leave. Thus, for 6 doors, there are 6 x 5 = 30 ways to enter and leave.**

COMMUNICATION

Writing

Have students write research papers on any topic concerning probability. Possible topics are: probability and genetics, random numbers, binomial probability, sampling techniques, games of chance, weather prediction, quality control, quantum theory and probability, empirical probability, and so on.

Applications, p.27

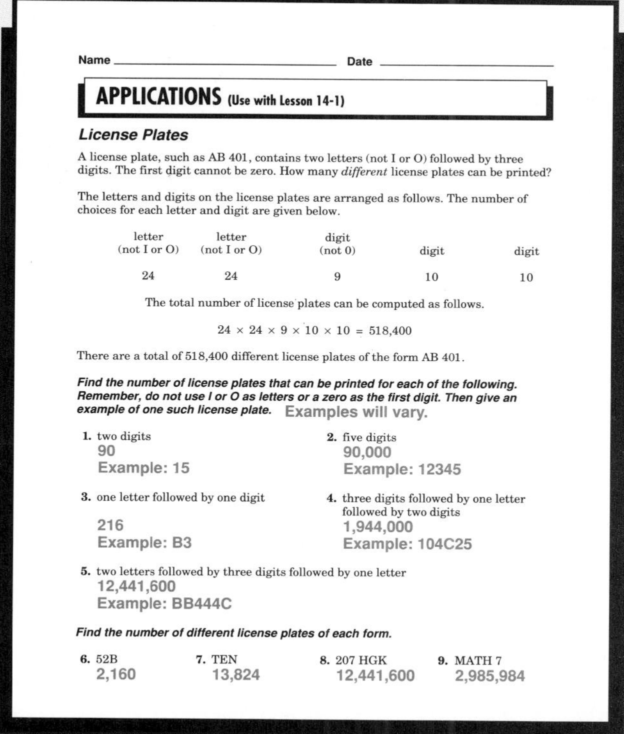

Name ____________ Date ____________

APPLICATIONS (Use with Lesson 14-1)

License Plates

A license plate, such as AB 401, contains two letters (not I or O) followed by three digits. The first digit cannot be zero. How many *different* license plates can be printed?

The letters and digits on the license plates are arranged as follows. The number of choices for each letter and digit are given below.

letter (not I or O)	letter (not I or O)	digit (not 0)	digit	digit
24	24	9	10	10

The total number of license plates can be computed as follows.

$24 \times 24 \times 9 \times 10 \times 10 = 518{,}400$

There are a total of 518,400 different license plates of the form AB 401.

Find the number of license plates that can be printed for each of the following. Remember, do not use I or O as letters or a zero as the first digit. Then give an example of one such license plate. Examples will vary.

1. two digits
   90
   Example: 15
2. five digits
   90,000
   Example: 12345
3. one letter followed by one digit
   216
   Example: B3
4. three digits followed by one letter followed by two digits
   1,944,000
   Example: 104C25
5. two letters followed by three digits followed by one letter
   12,441,600
   Example: BB444C

Find the number of different license plates of each form.

6. 52B
   2,160
7. TEN
   13,824
8. 207 HGK
   12,441,600
9. MATH 7
   2,985,984

Computer Activity, p.28

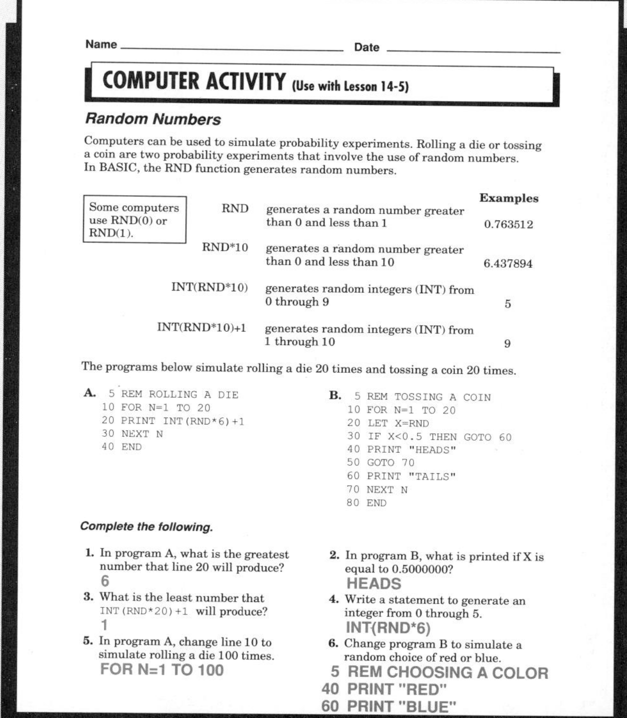

Name ____________ Date ____________

COMPUTER ACTIVITY (Use with Lesson 14-5)

Random Numbers

Computers can be used to simulate probability experiments. Rolling a die or tossing a coin are two probability experiments that involve the use of random numbers. In BASIC, the RND function generates random numbers.

Some computers use RND(0) or RND(1).

		Examples
RND	generates a random number greater than 0 and less than 1	0.763512
RND*10	generates a random number greater than 0 and less than 10	6.437894
INT(RND*10)	generates random integers (INT) from 0 through 9	5
INT(RND*10)+1	generates random integers (INT) from 1 through 10	9

The programs below simulate rolling a die 20 times and tossing a coin 20 times.

A.
```
5 REM ROLLING A DIE
10 FOR N=1 TO 20
20 PRINT INT(RND*6)+1
30 NEXT N
40 END
```

B.
```
5 REM TOSSING A COIN
10 FOR N=1 TO 20
20 LET X=RND
30 IF X<0.5 THEN GOTO 60
40 PRINT "HEADS"
50 GOTO 70
60 PRINT "TAILS"
70 NEXT N
80 END
```

Complete the following.

1. In program A, what is the greatest number that line 20 will produce?
   6
2. In program B, what is printed if X is equal to 0.5000000?
   HEADS
3. What is the least number that INT(RND*20)+1 will produce?
   1
4. Write a statement to generate an integer from 0 through 5.
   INT(RND*6)
5. In program A, change line 10 to simulate rolling a die 100 times.
   FOR N=1 TO 100
6. Change program B to simulate a random choice of red or blue.
   5 REM CHOOSING A COLOR
   40 PRINT "RED"
   60 PRINT "BLUE"

CHAPTER PROJECT

Have each student devise and conduct a project on some aspect of probability. Each project should involve an empirical investigation and a determination of probabilities involved. Examples of possible projects are: conducting a survey to make an empirical prediction, determining the empirical probability and expected outcomes of an event of their choice, demonstrate the relationship between predicted outcome based on mathematical probability and the number of trials, and finding the probability of rain and the actual amount of rain each day for a week.

Using the Chapter Opener

Transparency 14-0 is available in the Transparency Package. It provides a full-color visual and motivational activity that you can use to engage students in the mathematical content of the chapter.

Using Questioning Have students read the opening paragraph.

- Do all raffles sell a limited number of tickets? **no**
- What happens to the probability of winning if an unlimited number of raffle tickets are sold? **probability of winning becomes less**
- What other types of games or contests depend on chance? **Answers may vary.**

The second paragraph is a motivational problem. Have students try to solve the problem and discuss how they arrived at the answer.

You may want to discuss the probability of winning a game or contest sponsored by a national fast food restaurant or a national soft drink company or the probability of winning a state lottery.

Chapter

14 PROBABILITY

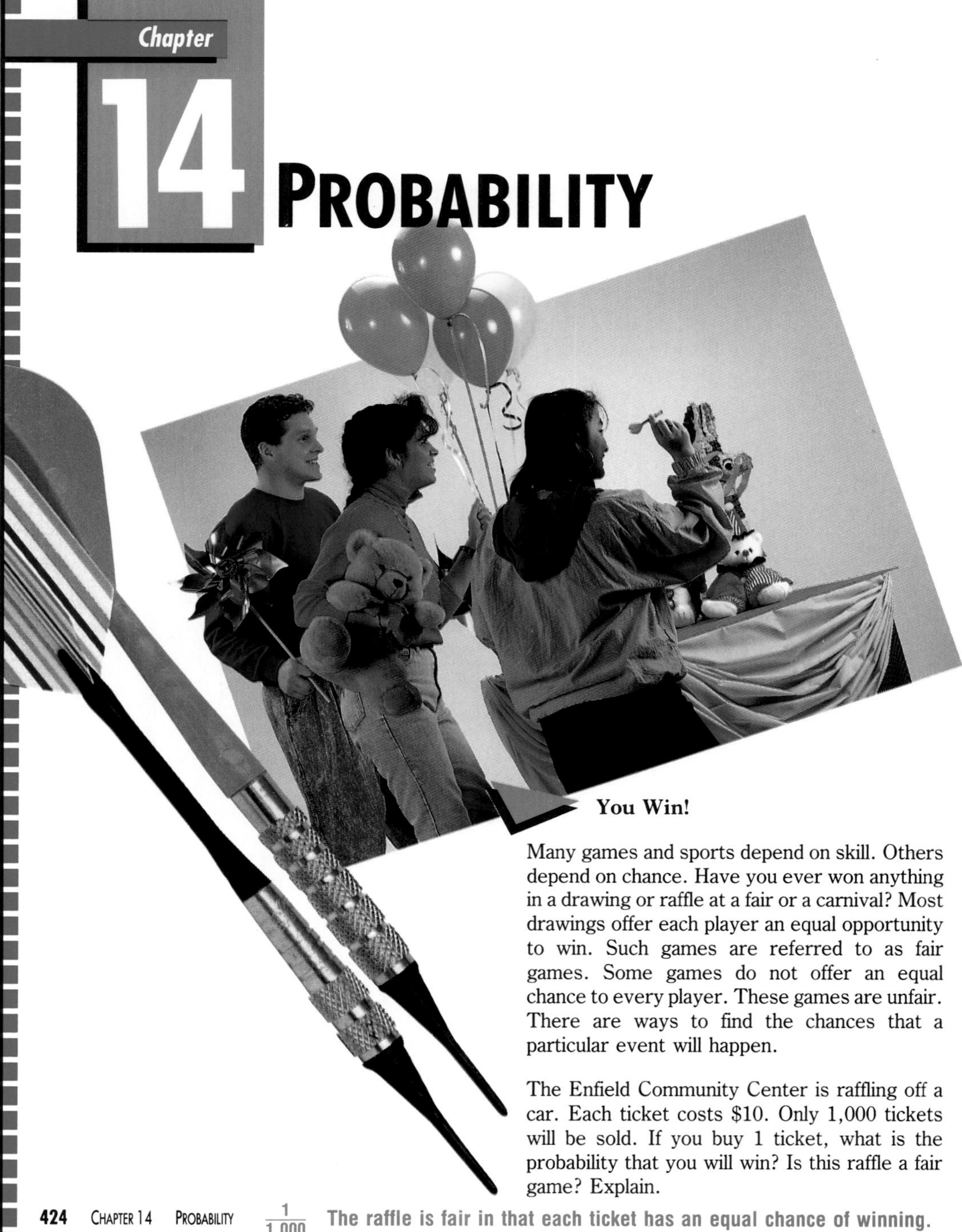

You Win!

Many games and sports depend on skill. Others depend on chance. Have you ever won anything in a drawing or raffle at a fair or a carnival? Most drawings offer each player an equal opportunity to win. Such games are referred to as fair games. Some games do not offer an equal chance to every player. These games are unfair. There are ways to find the chances that a particular event will happen.

The Enfield Community Center is raffling off a car. Each ticket costs $10. Only 1,000 tickets will be sold. If you buy 1 ticket, what is the probability that you will win? Is this raffle a fair game? Explain.

$\frac{1}{1,000}$ **The raffle is fair in that each ticket has an equal chance of winning. Holders of more than one ticket have a greater chance of winning than a holder of one ticket.**

ACTIVITY: Probability and Games

In this activity you will explore probability and its uses in games involving draws and chance. You will also discover what makes a game *fair* or *unfair*. Many people enjoy playing games that depend on chance. Knowing whether a game is fair or unfair will help you decide if you wish to take a chance.

Materials: coins, dice, spinners, paper and pencil

1. 12 outcomes, $\frac{1}{12}$

Cooperative Groups

Work in groups of three or four.

1. What outcomes are possible if you toss a coin and roll a die. Draw a tree diagram to show the possible outcomes. How many different outcomes are there? What is the probability of tossing a head and rolling a 5?

2. With your group play this game 50 times. Roll two dice. Add the numbers that face up on the dice. Player 1 wins if the sum of the numbers is odd. Player 2 wins if the sum of the numbers is even. One member of the group records the outcomes. How many wins did each player have out of 50? **Answers may vary.**

3. 36 outcomes, 18 even, 18 odd. The game is fair.

3. Make a table with a list of the possible outcomes for the game in Exercise 2. How many different possible outcomes are there? How many sums will be even? How many will be odd? Decide with your group if the game is fair or unfair.

4. In a sentence, describe what makes a game fair or unfair. **The probability for each player of a winning outcome must be the same.**

Varying the Game

5. Make a list of the outcomes for this game—Roll two dice. Multiply the two numbers on the dice. Player 1 wins if the product is odd. Player 2 wins if the product is even. Decide with your group if the game is fair. Explain your reasons. **See margin.**

Extending the Idea

6. With your group create a game for two players using a spinner and a coin. Make the game fair. Then play the game and record the results. Were there an even number of wins by each player? **Answers may vary.**

Communicate Your Ideas

7. Discuss with your group why a fair game does not always assure that each player will have the same number of wins in a specific number of games. **Answers may vary. See margin.**

Additional Answers

5. There are 9 odd products and 27 even products. For each desired outcome, in this case an odd or an even product, to be equally likely, the probability of an odd product must equal the probability of an even product. The game is unfair.
7. A good answer will refer to the fact that the theoretical probability does not always occur in the practical application of the activity.

▶ INTRODUCING THE LESSON

Objective Determine fair and unfair games.

▶ TEACHING THE LESSON

You may want to review Lesson 9-2 on pp. 282, 283 before beginning this activity.

Using Discussion Discuss with students what it means for each player to have an equal chance of winning a game. If the probability of a winning outcome is the same for each player, a game is fair.

▶ EVALUATING THE LESSON

Communicate Your Ideas Allow groups time to discuss the fair and unfair games they have created with other groups. Groups may want to try playing the games that other groups have created.

▶ EXTENDING THE LESSON

Closing the Lesson Have students explain in their own words what makes a game fair.

Activity Worksheet, Lab Manual, p. 49

Name ______________ Date ______________

ACTIVITY WORKSHEET (Use with page 425)

Probability and Games

Use with Exercise 2.

Outcome that occurred most often ______

Outcome that occurred least often ______

Use with Exercise 3.

Player 1

	tally	total
odd		
even		

Player 2

	tally	total
odd		
even		

Use with Exercises 4 and 6.

Die 1	Die 2	Sum	Product

Die 1	Die 2	Sum	Product

5-MINUTE CHECK-UP

(over Lesson 13-7)

Available on TRANSPARENCY 14-1.

Solve. Write an equation.

1. At Quik Parcel delivery, Mary makes $10 more than twice Jody's earnings if she works on Saturday. Last week Mary worked on Saturday. Jody earned $82 last week. How much did Mary earn? **$M = 2(82) + 10$; $174**

INTRODUCING THE LESSON

Have students give examples of probabilities that occur in everyday living. Ask them what a 50% chance of rain means. Tell students they will study about probability, chance, and odds in this chapter.

TEACHING THE LESSON

Using Discussion Discuss with students the usefulness of using a tree diagram to find the number of outcomes. The tree diagram also shows each outcome so that a specific event can be counted. This is particularly helpful when finding probabilities.

Practice Masters Booklet, p. 140

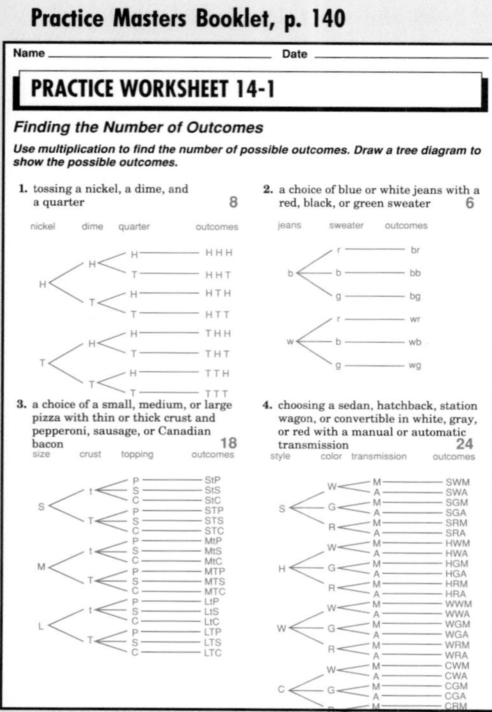

Name ______________ Date ______________

PRACTICE WORKSHEET 14-1

Finding the Number of Outcomes

Use multiplication to find the number of possible outcomes. Draw a tree diagram to show the possible outcomes.

1. tossing a nickel, a dime, and a quarter **8**

nickel dime quarter outcomes

HHH, HHT, HTH, HTT, THH, THT, TTH, TTT

2. a choice of blue or white jeans with a red, black, or green sweater **6**

jeans sweater outcomes

br, bb, bg, wr, wb, wg

3. a choice of a small, medium, or large pizza with thin or thick crust and pepperoni, sausage, or Canadian bacon **18**

size crust topping outcomes

StP, StS, StC, STP, STS, STC, MtP, MtS, MtC, MTP, MTS, MTC, LtP, LtS, LtC, LTP, LTS, LTC

4. choosing a sedan, hatchback, station wagon, or convertible in white, gray, or red with a manual or automatic transmission **24**

style color transmission outcomes

SWM, SWA, SGM, SGA, SRM, SRA, HWM, HWA, HGM, HGA, HRM, HRA, WWM, WWA, WGM, WGA, WRM, WRA, CWM, CWA, CGM, CGA, CRM, CRA

14-1 FINDING THE NUMBER OF OUTCOMES

Objective

Find the number of possible outcomes using tree diagrams or multiplication.

Melissa King is a finalist on the Perils game show. The show host will toss a nickel and a dime for the bonus. If Melissa calls the outcome correctly, she'll win a $500 bonus. Melissa wants to know the number of possible outcomes.

When more than one coin is tossed, you can use a tree diagram to determine the number of ways each outcome can occur.

A list of all the possible outcomes is a **sample space.** Any specific type of outcome is an **event.**

Example A

You can draw a tree diagram to show the possible outcomes when a nickel and a dime are tossed at the same time.

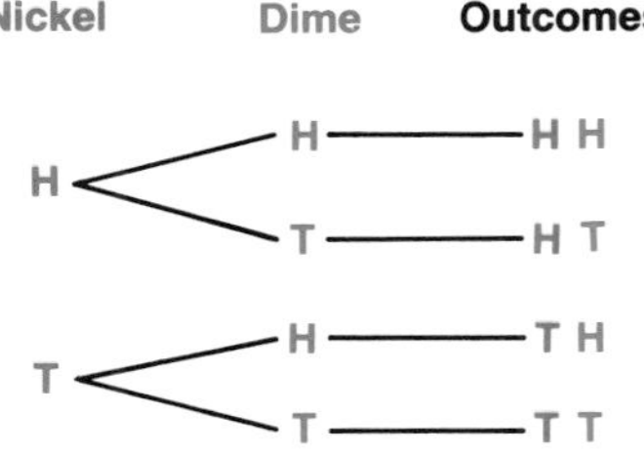

In this tree diagram, the top branch shows the outcome of a head and a head. This is symbolized HH.

HT means that the nickel came up heads and the dime came up tails. TH means that the nickel came up tails and the dime came up heads.

The tree diagram shows four possible outcomes.

Example B

You can also use multiplication to find the number of possible outcomes when a nickel and a dime are tossed at the same time.

nickel		dime		nickel and dime
2	×	2	=	4
possible outcomes		possible outcomes		possible outcomes

Guided Practice

Example A

See answers on p. T531.

Draw a tree diagram to show the possible outcomes.

1. tossing the coin and spinning the spinner
2. spinning each spinner once

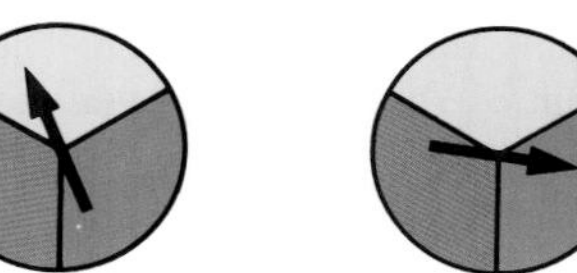

Example B

Use multiplication to find the number of possible outcomes.

3. rolling a red die and a blue die **$6 \times 6 = 36$**
4. tossing a coin and rolling a die **$2 \times 6 = 12$**

RETEACHING THE LESSON

Draw a tree diagram to show the possible outcomes. **6 outcomes**

1. a choice of beef, pork or fish with a choice of baked potatoes or fries

B — BP — BBP
B — FF — BFF
P — BP — PBP
P — FF — PFF
F — BP — FBP
F — FF — FFF

Reteaching Masters Booklet, p. 127

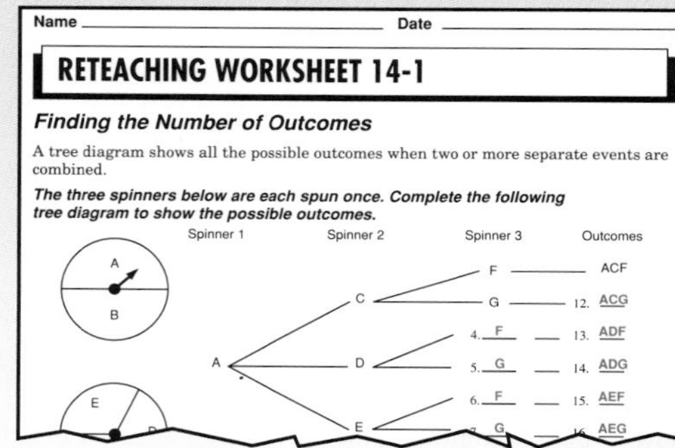

Name ______________ Date ______________

RETEACHING WORKSHEET 14-1

Finding the Number of Outcomes

A tree diagram shows all the possible outcomes when two or more separate events are combined.

The three spinners below are each spun once. Complete the following tree diagram to show the possible outcomes.

Spinner 1 Spinner 2 Spinner 3 Outcomes

A — C — F — ACF
A — C — G — 12. ACG
A — D — 4. F — 13. ADF
A — D — 5. G — 14. ADG
A — E — 6. F — 15. AEF
A — E — G — AEG

Exercises

See answers on pp. T531-T532.

Use multiplication to find the number of possible outcomes. Draw a tree diagram to show the possible outcomes.

Practice

5. tossing a penny, a nickel, and a dime

6. spinning each spinner once

7. tossing a quarter, rolling a die, and tossing a penny

8. a choice of a desk or a wall phone in white, black, green, or tan

Applications

9. Mrs. Jergen's history quiz has four multiple choice questions. Each question has four answer choices. How many possible sets of answers are there for the quiz? **$4 \times 4 \times 4 \times 4 = 256$ choices**

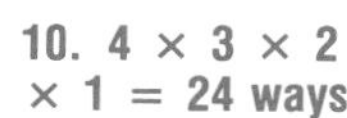

10. $4 \times 3 \times 2 \times 1 = 24$ ways

10. Four people are to sit in four chairs in a row. How many ways can the four people be seated? (Hint: once a person is seated, he or she is no longer a possible choice for the other chairs.)

11. $\frac{1}{100}$

11. A jar of marbles contains 490 blue marbles, 500 red marbles, and 10 green marbles. If Karen draws a green marble while she is blindfolded, she wins 100 free gallons of gasoline from the Corner-Stop station. What is the probability that Karen will draw a green marble?

Cooperative Groups

12. In groups of three or four, make two spinners with regions of different sizes. One must be one-half red, one-fourth green, and one-fourth yellow. The other must be two-thirds blue and one-third orange. Make a list of the possible outcomes for each spinner. With your group, explain which outcomes on each spinner should occur most often. **See margin.**

Mixed Review

Lesson 2-4

Subtract.

13. $18.4 - 12.6$ **5.8** **14.** $25.7 - 14.6$ **11.1** **15.** $38.2 - 29.7$ **8.5** **16.** $183.01 - 45.98$ **137.03** **17.** $44.878 - 32.909$ **11.969**

Lesson 9-2

A die is rolled once. Write the probability of each roll.

18. a 2 or a 5 $\frac{1}{3}$ **19.** an even number $\frac{1}{2}$ **20.** a 1 or a 6 $\frac{1}{3}$

Lesson 12-7

21. Joel currently has 2,880 classic rock singles. Joel began his collection 7 years ago and has doubled the number in his collection each year. How many singles did Joel collect during the first year? **45 singles**

Chalkboard Examples

Example A *Draw a tree diagram to show the possible outcomes.*

1. apple, cherry, or peach pie with a choice of coffee or tea

A — C — AC
A — T — AT
C — C — CC
C — T — CT
P — C — PC
P — T — PT

Example B *Use multiplication to find the number of possible outcomes.*

2. tossing three dice **216**

Additional examples are provided on TRANSPARENCY 14-1.

EVALUATING THE LESSON

Check for Understanding Use Guided Practice Exercises to check for student understanding.

Closing the Lesson Have students explain how to draw a tree diagram.

EXTENDING THE LESSON

Enrichment Have students make a list of 2 choices each of shirts, slacks, and shoes from their dream wardrobe and make a tree diagram.

Enrichment Masters Booklet, p. 127

Name ________ Date ________

ENRICHMENT WORKSHEET 14-1

Many Outcomes

Solve.

1. Four dice are rolled. How many outcomes are possible? **1,296**

2. Six coins are tossed. How many outcomes are possible? **64**

3. How many outcomes are possible when you toss a coin and then roll two dice? **72**

4. How many outfits are possible with 4 shirts, 2 pairs of slacks, 4 ties, and 4 sweaters? **128 outfits**

5. If your teacher gives you a 5-question multiple-choice quiz, where each question has 3 possible answers, how many sets of answers are possible? **243 sets**

6. How many different lunches can be selected by taking one choice from each group?
Group 1: fish, chicken, ham
Group 2: soup, salad, fruit
Group 3: milk, tea
18 lunches

7. Each spinner is spun once. How many outcomes are possible? **24**

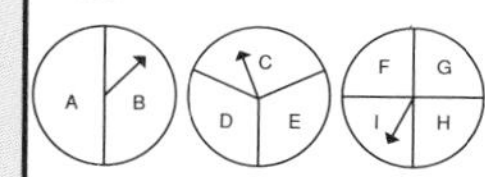

8. Each spinner is spun once. How many outcomes are possible? **160**

9. The Soup 'n' Salad Special sells 15 different salads and 12 different soups. How many different combinations of one salad and one soup can be chosen? **180 combinations**

10. A frame shop has 30 types of frames, 18 mat colors, and both regular and nonglare glass. How many combinations of frame, mat, and glass are available? **1,080 combinations**

APPLYING THE LESSON

Independent Practice

Homework Assignment	
Minimum	Ex. 2-12 even; 13-21
Average	Ex. 1-21
Maximum	Ex. 1-21

Additional Answers

12. First spinner, red, green, yellow; Second spinner, blue, orange Red should occur most often on the first spinner since it has the largest area, $\frac{1}{2}$. Blue should occur most often on the second spinner since it has the larger area, $\frac{2}{3}$.

5-MINUTE CHECK-UP

(over Lesson 14-1)

Available on TRANSPARENCY 14-2.

Draw a tree diagram to show the possible outcomes.

1. a choice of white or green skirt with a pink, white or blue sweater

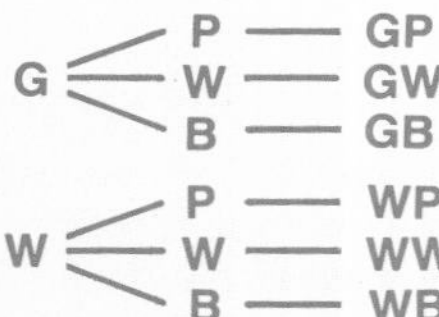

Extra Practice, Lesson 14-1, p. 483

INTRODUCING THE LESSON

You may want to review Lesson 9-2 pp. 282, 283, before this lesson.

Ask students what the probability is of: rain today, snow today, a visit to the school cafeteria today, and a walk home after school today.

TEACHING THE LESSON

Using Discussion Discuss the definition of probability of an event and the notation for writing probability. Remind students probability ranges from 0-1, so the percent will range from 0% to 100%.

Practice Masters Booklet, p. 141

Name ______________ Date ______________

PRACTICE WORKSHEET 14-2

Probability and Percents

There are two red marbles, one green marble, and one yellow marble in a bag. One marble is chosen. Find the probability that each event will occur.

1. a red marble 50%
2. a yellow marble 25%
3. a red or a green marble 75%
4. a yellow or a green marble 50%
5. *not* a green marble 75%
6. a blue marble 0%

There are two black, four red, three purple, and one orange marker in a box. One marker is chosen. Find the probability the event will occur.

7. a red marker 40%
8. a black or a purple marker 50%
9. *not* an orange marker 90%
10. a green marker 0%
11. a red, a purple, or an orange marker 80%
12. *not* a red or a purple marker 30%

Each spinner at the right is spun once. Find the probability of spinning each of the following. Round to the nearest whole percent. (Hint: First make a list of the possible combinations of spins.)

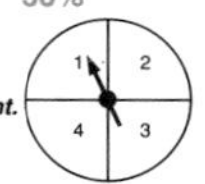

13. a 3 on both spinners 6%
14. a 2 on the first spinner and a 4 on the second spinner 6%
15. a 1 on one spinner and a 3 on the other 13%
16. both numbers the same 25%
17. both numbers less than 4 56%
18. a 3 on exactly one spinner 38%
19. a 3 on at least one spinner 44%
20. one number less than the other 75%

14-2 PROBABILITY AND PERCENTS

Objective
Express the probability of an event as a percent.

At the summer carnival in Shillington, Dr. Gomez has a chance at the $1,000 prize by spinning a spinner. The spinner is divided into ten equal-sized parts. Three parts are blue, three are red, two are yellow, and two are black. What is the probability Dr. Gomez will win if he chooses blue?

You may want to review Lesson 9-2, pp. 282, 283, before beginning this lesson.

Remember that the probability of an event is a ratio.

$$\text{probability} = \frac{\text{number of ways event can occur}}{\text{number of ways all possible outcomes can occur}}$$

Probabilities are often written as percents.

Method

1. Find the probability as a ratio.
2. Express the probability as a percent.

Example A

Find the probability that Dr. Gomez will win if he chooses blue. Express the probability as a percent.

1. The probability of the spinner landing on blue is $\frac{3}{10}$.

$\frac{3}{10} = \frac{n}{100}$	Remember that percent means *per hundred.*
$3 \times 100 = 10 \times n$	Find the cross products.
$\frac{300}{10} = \frac{10 \times n}{10}$	Solve for *n*.
$30 = n$	

The probability that Dr. Gomez will win if he chooses blue is 30%.

Example B

There are three blue pens, five red pens, and two black pens in a drawer. One pen is chosen without looking. Find the probability that a green pen will be chosen.

1. The probability of choosing a green pen is $\frac{0}{10}$ or 0.
2. A probability of 0 can be written as 0%.

There is a 0% probability that a green pen will be chosen.

Guided Practice

Examples A, B

Use the spinner above. Find the probability that each event will occur. Express as a percent.

1. landing on red 30%
2. landing on yellow 20%
3. landing on black 20%

RETEACHING THE LESSON

Give the meaning of each expression.

1. 10% chance of rain
   The probability of rain is $\frac{1}{10}$.
2. 50% chance of snow
   The probability of snow is $\frac{1}{2}$.
3. 100% chance of sunshine
   The probability of sunshine is 1.
4. 0% chance of rain
   The probability of rain is 0.

Reteaching Masters Booklet, p. 128

Name ______________ Date ______________

RETEACHING WORKSHEET 14-2

Probability and Percents

There are 4 red, 1 green, 7 blue, and 8 yellow marbles in a jar. The probability of choosing a blue one without looking is $\frac{7}{20}$. This probability can also be written as a percent.

Remember how to write a fraction as a percent.

probability → $\frac{7}{20} = \frac{r}{100}$ ←	ratio with a denominator of 100
$7 \times 100 = 20 \times r$	Use cross products.
$700 = 20 \times r$	
$\frac{700}{20} = \frac{20 \times r}{20}$	
$35 = r$	So $\frac{7}{20} = \frac{35}{100}$, or 35%.

The probability of choosing a blue marble is 35%.

Write each probability as a percent.

1. $\frac{3}{4}$
2. $\frac{1}{5}$
3. $\frac{3}{20}$
4. $\frac{2}{3}$

Examples A, B
6. 80% 8. 20%
9. 60%

There are two quarters, two dimes, and one penny in a jar. One coin is chosen. Find the probability of choosing each event. Express as a percent.

4. a quarter **40%** **5.** a penny **20%** **6.** a dime or a quarter
7. a half-dollar **0%** **8.** a penny or a nickel **9.** a penny or a dime

Exercises

Practice
15. 100%

There are three red balls and two blue balls in a box. One ball is chosen. Find the probability that each event will occur. Express as a percent.

10. a red ball **60%** **11.** a blue ball **40%** **12.** a green ball **0%**
13. *not* a blue ball **60%** **14.** *not* a red ball **40%** **15.** *not* a green ball
16. a red or a blue ball **100%** **17.** a yellow or a green ball **0%**

A red die and a blue die are rolled together. See the list of possible outcomes on page 283. Find the probability of rolling each event. Round to the nearest whole percent.

18. a red 6 and a blue 1 **3%** **19.** a blue 4 and a red 3 **3%**
20. a red 4 and a blue 3 **3%** **21.** a 3 and a 2 **6%**
22. a 2 on both dice **3%** **23.** both numbers the same **17%**
24. both numbers odd **25%** **25.** both numbers less than 3 **11%**

Applications
26. $\frac{1}{200}$ or 0.5%
27. $\frac{1}{25}$ or 4%
28. $\frac{3}{5}$ or 60%

26. The Long High School marching band sold 2,000 raffle tickets. Gerri bought 10 tickets. What is the probability she will win the raffle?

27. Special stickers have been attached to 1,000 cassette tapes in Mister Music's stock. Ten out of every 250 stickers win a free cassette. If all 1,000 tapes are sold and you buy one of them, what is the probability that you will win a free cassette?

28. The weather report states that there is a 40% chance of rain today. What is the probability that it will not rain today?

Write Math

29. Write a word problem that uses this tree diagram to solve the problem. Then solve the problem. **Answers may vary.**

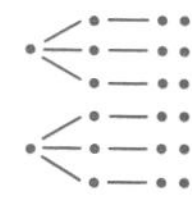

Critical Thinking

30. Suppose @ represents a certain phrase. If 12 @ 3 and 4, 8 @ 4 and 8, 6 @ 2 and 3, and 15 @ 3 and 5, what does @ represent? **is the least common multiple of**

APPLYING THE LESSON

Independent Practice

Homework Assignment	
Minimum	Ex. 1-27 odd; 29-30
Average	Ex. 2-24 even; 26-30
Maximum	Ex. 1-30

Chalkboard Examples

Jason is choosing school supplies from two large bins at Value Mart. He can get a red, blue, yellow, or green notebook and a black, blue, or red pen. Find the probability that each event will occur.

1. a yellow notebook and red pen $8\frac{1}{3}\%$
2. a pen and notebook of the same color $16\frac{2}{3}\%$
3. a black pen and a green notebook $8\frac{1}{3}\%$

Additional examples are provided on TRANSPARENCY 14-2.

EVALUATING THE LESSON

Check for Understanding Use Guided Practice Exercises to check for student understanding.

Closing the Lesson Ask students to state in their own words how to state probability as a percent.

EXTENDING THE LESSON

Enrichment Have students find the probability of rain in the newspaper for each of the next two days in three cities around the world.

Enrichment Masters Booklet, p. 128

ENRICHMENT WORKSHEET 14-2

Using a Random Number Table

When actual experiments cannot be performed or time is not available to perform them, a random number table like the one below is useful for simulating the experiments.

3 2 9 4 2 9 5 4 1 6 4 2 3 3 9 5 9 0 4 5 2 6 6 9 3 4 9 0 5 7
0 7 4 1 0 9 9 8 5 9 8 3 8 2 8 2 1 4 0 9 2 9 0 9 4 6 5 1 1 4
5 9 9 8 1 6 8 1 5 5 4 5 6 7 3 7 6 2 1 0 5 8 2 1 9 4 5 7 3 8
4 6 2 5 1 2 5 4 3 7 6 9 6 5 4 9 9 7 1 6 1 1 5 6 3 0 8 8 0 3
6 5 5 5 8 5 1 9 0 4 9 3 1 2 3 2 7 8 8 7 5 3 1 3 8 2 1 4 8 8

Here's how to use the random number table to predict how many times this spinner will land on C in 100 spins.

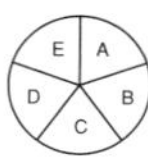

- There are 10 different digits in the random number table, 0 to 9. On the spinner, $\frac{1}{5}$ of the spaces have a C. So select $\frac{1}{5}$ of the 10 different digits in the table to represent landing on C. Let's use 1 and 2.
- Count the number of times 1 or 2 occurs in the first 100 numbers of the table. They occur 21 times.

The probability of landing on C is $\frac{1}{5}$, and $\frac{1}{5}$ of 100 is 20. So the prediction of landing on C 21 times out of 100 is reasonable.

Make the following predictions using the random number table above.
Answers may vary due to numbers selected. Approximations are given.

1. The probability that a customer at Peg's Gift Shop will make a purchase is $\frac{7}{10}$. How many customers out of 80 are likely to make a purchase? **56 customers**
2. The probability that a student is wearing jeans is $\frac{1}{2}$. How many students out of 120 are likely to be wearing jeans? **60 students**
3. It is equally likely that each person to enter a certain movie theater is male or female. Out of 150 people to enter this theater, how many will be males? **75 people**
4. During winter months, the probability of snow on any day is $\frac{3}{5}$. How many days is it likely to snow out of 90 winter days? **54 days**

5-MINUTE CHECK-UP

(over Lesson 14-2)

Available on TRANSPARENCY 14-3.

Bill has 4 cans of meat soups (beef vegetable, chicken noodle, chili beef, and ham with beans) and 3 cans of cream soups (celery, mushroom, and broccoli). Find the probability of choosing, without looking, each soup. Express as a whole percent.

1. a soup with chicken 14%
2. a soup with beef 29%
3. a soup that is not creamed 57%

Extra Practice, Lesson 14-2, p. 483

INTRODUCING THE LESSON

Ask students if, at the same time, they can be 16 years old and on the football team. Ask students to name other situations that involve *not at the same time* and *at the same time.*

TEACHING THE LESSON

Using Questioning

- When two events are dependent, how can you find the probability of the second event? The number of outcomes used in the ratio for the probability of the second event is reduced by subtracting the first event from the sample space.

Practice Masters Booklet, p. 142

PRACTICE WORKSHEET 14-3

Multiplying Probabilities

A bag contains four red, five white, and three blue marbles. Suppose you choose a marble from the bag and then choose another marble without replacing the first one. Find the probability of choosing each event.

1. white both times $\frac{5}{33}$
2. blue both times $\frac{1}{22}$
3. red both times $\frac{1}{11}$
4. same color both times $\frac{19}{66}$
5. red, then white $\frac{5}{33}$
6. white, then red $\frac{5}{33}$
7. white, then blue $\frac{5}{44}$
8. red, then green 0

Suppose you choose a marble from the bag described above, replace it, and then choose another. Find the probability of choosing each event.

9. white both times $\frac{25}{144}$
10. blue both times $\frac{1}{16}$
11. red both times $\frac{1}{9}$
12. same color both times $\frac{25}{72}$
13. red, then white $\frac{5}{36}$
14. white, then red $\frac{5}{36}$
15. white, then blue $\frac{5}{48}$
16. red, then green 0

Solve.

17. A card is drawn from the deck shown below. Then another card is drawn without replacing the first. What is the probability that the cards drawn will be 2, 3 or 3, 2? $\frac{4}{15}$
18. A card is selected at random from the group shown below. A second card is drawn without replacing the first. What is the probability that both cards will have odd numbers on them? $\frac{2}{5}$

14-3 MULTIPLYING PROBABILITIES

Objective

Find the probability of independent and dependent events.

Gary won third prize in New Mall's Grand Opening. He will draw a dime, not replace it, and draw another dime from the pot o' gold containing six very valuable dimes. One was minted in 1920, two in 1924, and three in 1925. What is the probability Gary will draw a 1925 dime both times?

You can multiply two probabilities to find the probability that one event *and* another event will occur. The key word is "and."

If one event is affected by the occurrence of another event, the events are **dependent.** Two events are **independent** if the occurrence of one event does not affect whether the other event occurs.

Method

1. Find the probability of the first event.
2. Find the probability of the second event.
3. Multiply the probabilities.

Example A

Suppose Gary chooses a dime, does not replace it, and draws another dime. What is the probability that he chooses a 1925 dime both times?

1. $P(1925) = \frac{3}{6}$ or $\frac{1}{2}$
2. $P(1925) = \frac{2}{5}$

These events are dependent. Since the first dime is not replaced, there are only five dimes left on the second draw. Also, we are assuming that a 1925 dime is chosen the first time. So there are only two 1925 dimes left.

3. $P(1925 \text{ both times}) = \frac{1}{2} \times \frac{2}{5} = \frac{1}{5}$

The probability that Gary will draw a 1925 dime both times is $\frac{1}{5}$.

Example B

Suppose Gary draws a dime, replaces it and draws another dime. What is the probability that he will draw a 1924 dime both times?

1. $P(1924) = \frac{2}{6}$ or $\frac{1}{3}$
2. $P(1924) = \frac{2}{6}$ or $\frac{1}{3}$

These events are independent since the first dime is replaced.

3. $P(1924 \text{ both times}) = \frac{1}{3} \times \frac{1}{3} = \frac{1}{9}$

Guided Practice

Find the probability of each event. Use the pot o' gold full of dimes. Each dime drawn is not replaced.

Example A

2. $\frac{1}{10}$ 4. $\frac{1}{5}$

1. P(a 1924 dime both times) $\frac{1}{15}$
2. P(a 1920 dime, then a 1925 dime)
3. P(a 1920 dime both times) 0
4. P(a 1925 dime, then a 1924 dime)

RETEACHING THE LESSON

There are 4 red, 1 green, 7 blue and 8 yellow marbles in a jar. Find each probability.

1. choosing a green marble $\frac{1}{20}$
2. choosing a red or blue marble $\frac{11}{20}$
3. choosing a green or a yellow marble $\frac{9}{20}$
4. choosing a marble that is not red $\frac{4}{5}$

Reteaching Masters Booklet, p. 129

Name ______ Date ______

RETEACHING WORKSHEET 14-3

Multiplying Probabilities

Suppose 10 buttons are put in a bag. Then you draw one without looking. What is the probability that you selected a white button?

P(white) = $\frac{3}{10}$ ← number of white buttons / ← total number of buttons

Now suppose you replace the button and draw another one. What is the probability that you draw a white button twice?

P(white) 1st draw		P(white) 2nd draw		P(two white buttons)
$\frac{3}{10}$	×	$\frac{3}{10}$	=	$\frac{9}{100}$

Now suppose you do not replace the first button. You draw another button. What is the probability that you draw two white buttons?

P(white) 1st draw		P(white) 2nd draw		P(two white buttons)
$\frac{3}{10}$	×	$\frac{2}{9}$	=	$\frac{6}{90}$ or $\frac{1}{15}$

Since you did not replace the first

Example B *Find the probability of each event. Use the pot o' gold full of dimes on page 430. Each dime drawn is replaced.*

5. P(a 1925 dime both times) $\frac{1}{4}$ **6.** P(a 1920 dime both times) $\frac{1}{36}$

7. P(a 1925, then a 1924 dime) $\frac{1}{6}$ **8.** P(a 1920, then a 1924 dime) $\frac{1}{18}$

Exercises

Practice

A bag contains two white, three green, and five yellow marbles. Suppose you choose a marble and then without replacing it, choose another marble. Find the probability of choosing each event.

9. P(white, then green) $\frac{1}{15}$ **10.** P(yellow, then white) $\frac{1}{9}$

11. P(green, then blue) 0 **12.** P(white, then yellow) $\frac{1}{9}$

13. P(white both times) $\frac{1}{45}$ **14.** P(green both times) $\frac{1}{15}$

15\. $\frac{2}{9}$

15. P(yellow both times) **16.** P(same color both times) $\frac{14}{45}$

Suppose you choose a marble from the bag for Exercises 9–16, replace it, and then choose another marble. Find the probability of choosing each event.

17. P(white, then green) $\frac{3}{50}$ **18.** P(yellow, then white) $\frac{1}{10}$

19. P(green, then blue) 0 **20.** P(white, then yellow) $\frac{1}{10}$

21. P(white both times) $\frac{1}{25}$ **22.** P(green both times) $\frac{9}{100}$

23\. $\frac{1}{4}$

23. P(yellow both times) **24.** P(same color both times) $\frac{19}{50}$

Applications

25\. $\frac{1}{625}$

25. In Mr. Morris's department, a jar contains twenty-five index cards with one name on each card. One card is drawn each month and returned to the jar. What is the probability that Mr. Morris's name will be drawn two months in a row?

26\. $\frac{1}{30}$

26. Six members of the swim team qualified for the drawing for free swimsuits at Sport Haus. The coach put their names on a slip of paper and dropped them in a box. Two team members will receive a prize. The students' names are Linda, Lisa, David, Marianne, Shirley, and Keith. What is the probability that Lisa and David will win? Once a name is drawn, it is not replaced.

27\. $\frac{1}{2}$, $\frac{1}{2}$

27. Use two nickels. Mark an X on each side of one. On the other mark an X on one side and a Y on the other. What is the probability of getting an XX on a toss? an XY?

28\. $\frac{1}{2}$, $\frac{1}{2}$ Yes, the chances for either partner are the same.

28. Refer to Exercise 27. If you were playing with a partner and you get a point for every XX and your partner gets a point for every XY, would it be a fair game? Explain your answer.

Make Up a Problem

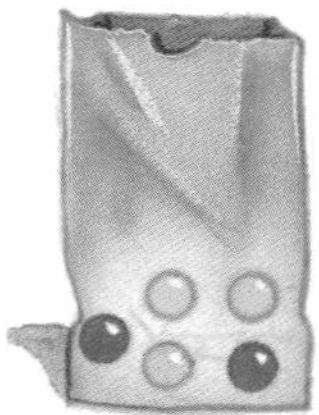
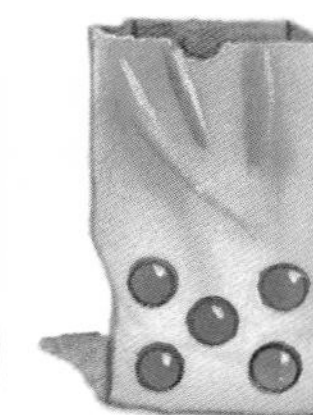
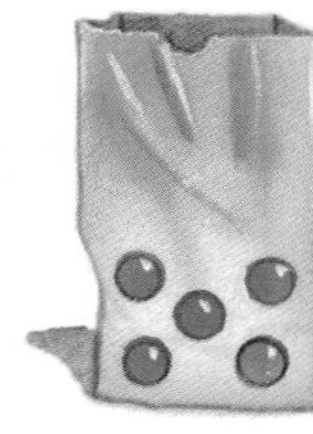

29. Look at the picture. Make up a problem using the information in the picture. Then solve the problem. **Answers may vary.**

Chalkboard Examples

Example A *Find the probability of each event. Use a standard deck of 52 cards. Each card drawn is not replaced.*

1. an ace and the Queen of Hearts $\frac{1}{663}$

2. a red card and a black four $\frac{1}{52}$

Example B *Find the probability of each event. Use a standard deck of 52 cards. Each card drawn is replaced.*

3. a black card and the Jack of Diamonds $\frac{1}{104}$

4. a ten and a Jack $\frac{1}{169}$

Additional examples are provided on TRANSPARENCY 14-3.

EVALUATING THE LESSON

Check for Understanding Use Guided Practice Exercises to check for student understanding.

Error Analysis Watch for students who fail to subtract the outcomes from the sample space before finding the probability of the second event for dependent events.

Closing the Lesson Have students tell when to multiply probabilities.

EXTENDING THE LESSON

Enrichment Masters Booklet, p. 129

Name ____________ Date ____________

ENRICHMENT WORKSHEET 14-3

Multiplying Probabilities

Jacks, or jackstones, is a popular children's game played with a ball and small metal objects called jacks.

To play: 1. Throw the ball up in the air.
2. Without looking, pick up a jack.
3. Catch the ball after one bounce.

blue blue blue red red gold blue blue red gold

Suppose Jessica plays with the jacks shown at the right. Each time she throws the ball in the air, she picks up a jack. She does not replace it. It is equally likely that she will pick up any jack during each throw.

Find the probability of each event.

1\. Jessica throws the ball three times. She picks up all red jacks. $\frac{1}{120}$

2\. Jessica throws the ball three times. She picks up all blue jacks. $\frac{1}{12}$

3\. Jessica throws the ball three times. She picks up a blue, then a red, and then a gold jack. $\frac{1}{24}$

4\. Jessica throws the ball three times. None of the jacks she picks up is gold. $\frac{7}{15}$

5\. Jessica throws the ball five times. She picks up all blue jacks. $\frac{1}{252}$

6\. Jessica throws the ball five times. She picks up two red jacks and then three blue jacks. $\frac{1}{84}$

7\. Jessica throws the ball eight times. She picks up five blue and then three red jacks. $\frac{1}{2,520}$

8\. Jessica throws the ball ten times. She picks up two gold, then three red, and then five blue jacks. $\frac{1}{2,520}$

APPLYING THE LESSON

Independent Practice

Homework Assignment	
Minimum	Ex. 1-29 odd
Average	Ex. 1-23 odd; 25-29
Maximum	Ex. 1-29

5-MINUTE CHECK-UP

(over Lesson 14-3)

Available on TRANSPARENCY 14-4.

Chris has three dimes in her purse. One was minted in 1930, and the other two were minted in 1973.

1. If she draws two dimes out of the bag without replacing the first, what is the probability of drawing the two minted in 1973? $\frac{1}{3}$ **2.** $\frac{2}{9}$
2. What is the probability of drawing a coin minted in 1930, replacing it, and then drawing a dime minted in 1973?

Extra Practice, Lesson 14-3, p. 484

INTRODUCING THE LESSON

Have students give examples of probabilities in real life that involve more than one event, such as rain and temperature higher than 50°F. Ask students how often things they need to know involve only one event.

TEACHING THE LESSON

Using Discussion Emphasize for students that subtraction is only used when both conditions can be satisfied simultaneously.

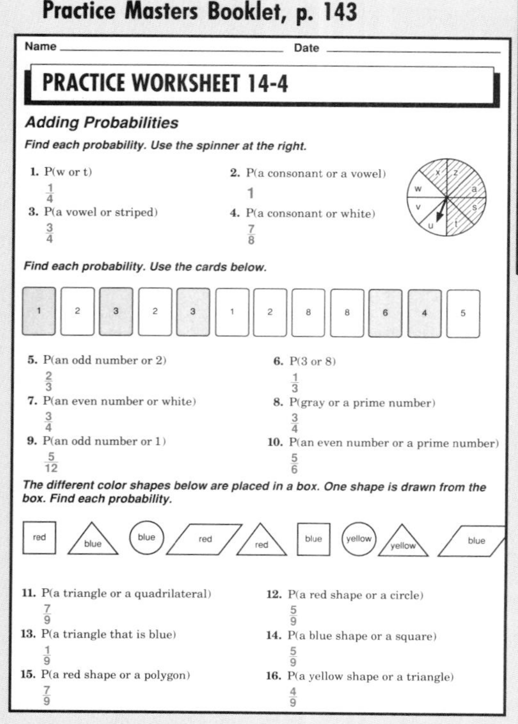

Practice Masters Booklet, p. 143

Name ______________ Date ______________

PRACTICE WORKSHEET 14-4

Adding Probabilities

Find each probability. Use the spinner at the right.

1. P(w or t) $\frac{1}{4}$
2. P(a consonant or a vowel) 1
3. P(a vowel or striped) $\frac{3}{4}$
4. P(a consonant or white) $\frac{7}{8}$

Find each probability. Use the cards below.

1 2 3 2 3 1 2 8 8 6 4 5

5. P(an odd number or 2) $\frac{2}{3}$
6. P(3 or 8) $\frac{1}{3}$
7. P(an even number or white) $\frac{3}{4}$
8. P(gray or a prime number) $\frac{3}{4}$
9. P(an odd number or 1) $\frac{5}{12}$
10. P(an even number or a prime number) $\frac{5}{6}$

The different color shapes below are placed in a box. One shape is drawn from the box. Find each probability.

red blue blue red red blue yellow yellow blue

11. P(a triangle or a quadrilateral) $\frac{7}{9}$
12. P(a red shape or a circle) $\frac{5}{9}$
13. P(a triangle that is blue) $\frac{1}{9}$
14. P(a blue shape or a square) $\frac{5}{9}$
15. P(a red shape or a polygon) $\frac{7}{9}$
16. P(a yellow shape or a triangle) $\frac{4}{9}$

14-4 ADDING PROBABILITIES

Objective
Find the probability of one event or another.

The spinner shown below is divided into six equal regions. Every region is shaded alternately blue or red. You want to know the probability that you will spin an even number or a 5.

You can add two probabilities to find the probability that one event *or* another event will occur. The key word is "or."

Method

1. Find the probability of the first event. **Point out that this is not the same as the probability of *both* events occurring.**
2. Find the probability of the second event.
3. Add the probabilities. If the two events can occur at the same time, subtract the probability of both events occurring.

Example A ***Find the probability of spinning an even number or a 5.***

1. $P(\text{even number}) = P(2, 4, \text{ or } 6) \text{ or } \frac{3}{6}$
2. $P(5) = \frac{1}{6}$
3. $P(\text{even number or } 5) = \frac{3}{6} + \frac{1}{6}$
$= \frac{4}{6} \text{ or } \frac{2}{3}$

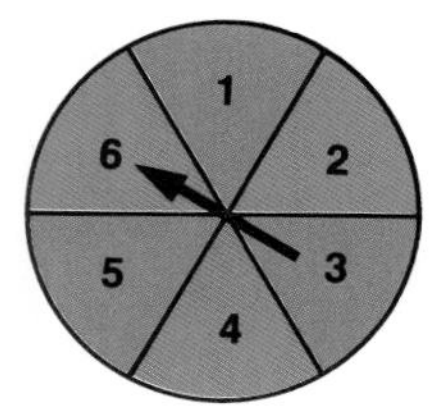

The probability of spinning an even number or 5 is $\frac{2}{3}$.

Example B ***Find the probability of spinning blue or a number greater than 2.***

1. $P(\text{blue}) = P(1, 3, \text{ or } 5) \text{ or } \frac{3}{6}$
2. $P(\text{number greater than } 2) = P(3, 4, 5, \text{ or } 6) \text{ or } \frac{4}{6}$
3. Add. $\frac{3}{6} + \frac{4}{6} = \frac{7}{6}$ Remember that the probability of an event will be from 0 to 1, inclusive.

Since some of the blue numbers are greater than 2, some outcomes, namely spinning a 3 or a 5, have been counted twice. Subtract the probability of these events (blue 3 and blue 5) occurring.

$$\frac{7}{6} - \frac{2}{6} = \frac{5}{6}$$

The probability of blue or a number greater than 2 is $\frac{5}{6}$.

Guided Practice

Example A ***Find each probability. Use the spinner shown above.***

1. $P(2 \text{ or } 3)$ $\frac{1}{3}$
2. $P(1, 2, \text{ or } 3)$ $\frac{1}{2}$
3. $P(\text{an odd number or } 2)$ $\frac{2}{3}$
4. $P(3 \text{ or } 5)$ $\frac{1}{3}$
5. $P(\text{red or } 1)$ $\frac{2}{3}$
6. $P(\text{an even number or blue})$ 1

Example B ***Find each probability. Use the spinner shown above.***

7. $P(\text{blue or a number less than } 4)$ $\frac{2}{3}$
8. $P(\text{red or a number less than } 3)$ $\frac{2}{3}$

RETEACHING THE LESSON

A bag contains 2 red marbles and 1 blue marble. Find the probability for each draw. Each marble drawn is replaced.

1. a red marble, then a blue marble $\frac{2}{9}$
2. a red marble two times $\frac{4}{9}$

Find the probability for each draw. Each marble drawn is not replaced.

3. a red marble, then a blue marble $\frac{1}{3}$
4. a red marble two times $\frac{1}{3}$

Reteaching Masters Booklet, p. 130

Name ______________ Date ______________

RETEACHING WORKSHEET 14-4

Adding Probabilities

A card is selected from a deck of 52 playing cards. What is the probability that it is either a red card *or* a club?

P(red card) + P(club) = P(red card or a club)

$\frac{26}{52} + \frac{13}{52} = \frac{39}{52} \text{ or } \frac{3}{4}$

The probability of selecting either a red card or a club is $\frac{3}{4}$.

The first card is replaced in the deck, and another card is selected. What is the probability that it is either a red card or a queen?

Notice that the queen of hearts and the queen of diamonds are both red cards and queens. The probability of outcomes that are counted twice must be subtracted.

P(red card) P(queen) P(red card *and* a queen) P(red card *or* a queen)

Exercises

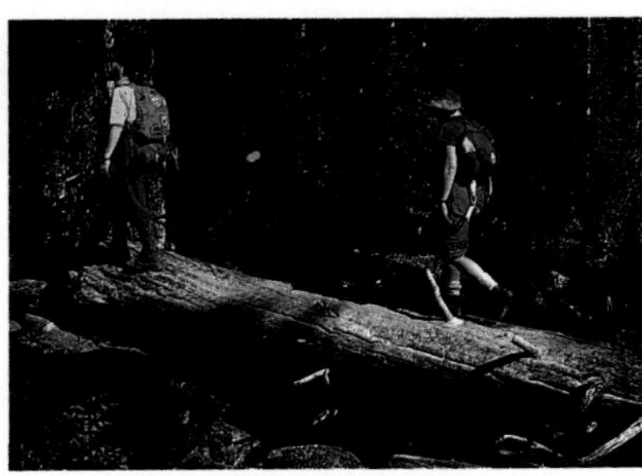

The names of the students in Ms. Sparver's class who have qualified for the survival course are given below. Each name is written on a slip of paper and placed in a box. A slip of paper is drawn from the box. Find each probability.

CHARLES DOTTIE DREW SUKI RACHEL CATHY CORY DON

9. $\frac{3}{4}$ 10. $\frac{3}{8}$ 11. $\frac{1}{8}$ 12. $\frac{1}{4}$ 13. $\frac{3}{8}$ 14. $\frac{1}{4}$

9. P(a name beginning with C or D)

10. P(a name ending with S or Y)

11. P(a 4-letter name beginning with C)

12. P(a 4-letter name beginning with S or D)

13. P(a name containing 4 letters or beginning with S)

14. P(a name containing 6 letters or beginning with R)

Find each probability. Use the spinner at the right.

15. P(green or blue) $\frac{1}{2}$

16. P(green or red) $\frac{1}{2}$

17. P(blue or a number greater than 5) $\frac{3}{4}$

18. P(red or an even number) $\frac{5}{8}$

19. P(a prime number or green) $\frac{5}{8}$

20. P(a prime number or an odd number) $\frac{9}{16}$

Applications

21. $\frac{1}{3}$

21. Sharon and Jeff are playing Travelog. Sharon rolls a die. She needs to roll a 3 or a 5 to win. What is the probability Sharon will roll a 3 or a 5?

Suppose

22. Suppose you make a spinner like the one used in Exercises 15–20, except that you place the four wedges of each color next to each other. Does this arrangement change the probability of spinning any one of the colors from the original probability? Explain. **No, there are still 4 out of 16 chances to spin each color.**

23. Refer to Exercise 22. Does the new arrangement of colors affect the probability of spinning a blue or a number greater than 5? Explain. **See margin.**

Cooperative Groups

24. In small groups, copy the chart from Exercises 21–28 on page 283. Then find each probability.

a. P(doubles or a sum of 7) $\frac{1}{3}$

b. P(a sum of 8 or sum of 9) $\frac{1}{4}$

c. Now set up a table like the one shown at the right. Roll two dice 36 times and record your results. Find the probability of 24a and 24b using the results from your table. Compare your experimental probability with the probability you found in 24a and 24b. **Answers may vary.**

Roll #	Result
1	?
2	?
⋮	⋮

Chalkboard Examples

Example A *Find each probability. Use one roll of one die.*

1. rolling a 5 or 6 $\frac{1}{3}$

2. rolling a 5 and a 6 0

3. rolling a 3 or an even number $\frac{4}{6}$ or $\frac{2}{3}$

Example B *Find each probability. Use one roll of one die.*

4. rolling a 2 or a prime number $\frac{1}{2}$

5. rolling a 2 or a composite number $\frac{1}{2}$

6. rolling a 2 and a prime number $\frac{1}{6}$

Additional examples are provided on TRANSPARENCY 14-4.

EVALUATING THE LESSON

Check for Understanding Use Guided Practice Exercises to check for student understanding.

Error Analysis Watch for students who do not reduce to simplest form after computing probabilities.

Closing the Lesson Ask students to summarize what they have learned about adding probabilities.

EXTENDING THE LESSON

Enrichment Masters Booklet, p. 130

ENRICHMENT WORKSHEET 14-4

Adding Probabilities

Solve.

1. Ben's wallet contains 1 twenty-dollar bill, 1 ten-dollar bill, 2 five-dollar bills, and 5 one-dollar bills. He chooses one bill at random. What is the probability that he chooses a one-dollar bill or a ten-dollar bill? $\frac{2}{3}$

2. Ben returns the bill to his wallet and then chooses another one at random. What is the probability that he chooses a ten-dollar bill or a bill larger than a five-dollar bill? $\frac{2}{9}$

3. Bonnie's key chain has 1 bicycle key, 2 house keys, and 2 car keys on it. Suppose the chain breaks and all the keys fall to the ground. She picks one up at random. What is the probability that she picks up a bicycle key or a house key? $\frac{3}{5}$

bicycle key

car keys

house keys

4. What is the probability that Bonnie picks up a house key or not a bicycle key? $\frac{4}{5}$

Contestants spin for prizes on the game show Spin to Win. Find the probability of spinning each of the following.

5. a car or $250 $\frac{7}{16}$

6. a vacation or $1,000 $\frac{5}{16}$

7. $500 or a cash prize greater than $250 $\frac{7}{16}$

8. a vacation or prize with a value greater than $1,000 $\frac{5}{16}$

Spin to Win Spinner

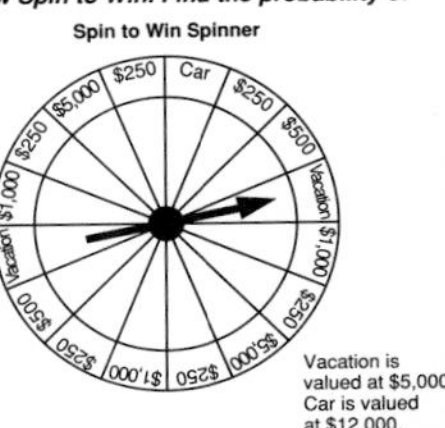

Vacation is valued at $5,000. Car is valued at $12,000.

APPLYING THE LESSON

Independent Practice

Homework Assignment	
Minimum	Ex. 2-24 even
Average	Ex. 2-20 even; 21-24
Maximum	Ex. 1-24

Chapter 14, Quiz A (Lessons 14-1 through 14-4) is available in the Evaluation Masters Booklet, p. 69.

Additional Answers

23. The probability of spinning blue or a number greater than 5 will remain the same if the four adjacent blue wedges are numbered 14, 15, 16 and 1. Any other placement of four adjacent wedges will result in a different probability.

Applying Mathematics to Technology

This optional page provides students with an experience with computer literacy and also provides a change of pace.

Objective Work with computer simulations.

This program is designed for Apple users. Radio Shack users replace RND(1) with RND(0). Other computers may use RND. Check the computer manual for the use of the random command for your computer.

Using Discussion Discuss with students the types of games in a video arcade that are simulations.

Review the terms average (mean) and range.

Activity

Using Computers Have students change line 200 to “IF RND(1) < 0.8 THEN GOTO 500”. Run the program and find how the output changes (more heads are tossed).

Let students change some of the lines in the program and tell how it affects the output.

Computer Application

SIMULATIONS

Computers are often used to simulate real-life situations. Sometimes pilots use computer simulations to practice maneuvers without endangering life and equipment. Scientists and engineers use simulations to test materials under extreme temperature and weight conditions.

The BASIC program below simulates an experiment in probability. Type it into a computer. To stop the output, push CTRL-C or BREAK.

```
100 FOR N=1 TO 10
200 IF RND (1) < 0.5 THEN GO TO 500
300 PRINT "-";
400 GOTO 600
500 PRINT "H";
600 NEXT N
700 PRINT
800 GOTO 100
RUN
H-HHHHHH-H
----H---HH
-H--H-----
-H--HHH-HH
H-HH--HHHH
HH--H-----
H-H-H--H-H
```

The command RND(1) tells the computer to choose a number between 0 and 1 at random.

Your output may look similar to this output.

Each row shows ten flips of a coin. The computer prints “H” for heads and “-” for tails.

See margin.

Use the program and output shown above. Solve.

1. If you flip a coin ten times, is it possible to get nine heads in a row? How often would it happen?

2. If you flip several tails in a row, will the next few tosses be heads to make up for many tails?

3. Each row in the output shown above shows the outcomes for ten flips of a coin. In the first row, the experimental probability of tossing a tail is $\frac{1}{5}$ and a head is $\frac{4}{5}$. Will the experimental probability ever match the theoretical probability?

Additional Answers

1. Yes. The probability of nine heads in a row is $\frac{1}{512}$.
2. No. The probability of a head is always $\frac{1}{2}$. As long as the random command is left as it is in the program, these answers are correct. Students have not previously seen the random command and may not know that the probability has been set at one-half by the command.
3. In the output shown in the example, the theoretical probability is matched after nine rows, then changes as more rows are produced. The experimental probability may never match the theoretical probability.

Car Rental Agency Manager

Ramon Perez manages a car rental agency. One of his job responsibilities is to make sure that each car is in proper running condition. Another is to compute the rental charges for customers.

The charge for a mid-size car is $29 a day and 28¢ per mile. Tax is an additional 5.5%. If a customer keeps a car for three days and drives it 286 miles, what is the rental charge?

29 ⊠ 3 ⊞ 286 ⊠ .28 ⊟ 167.08

The rental charge for the mid-size car is $167.08.

To determine the total including tax, multiply $167.08 by 1.055. (100% + 5.5% = 105.5% or 1.055)

167.08 is already showing in the calculator.

⊠ 1.055 ⊟ 176.2694

The total cost including tax is $176.27 **Remind students that a part of a cent is usually charged as a whole cent at most stores, restaurants, and businesses.**

In Exercises 1–3, the number of miles and days are given. Use $29 a day and 22¢ a mile to find the rental charge. Assume there is no tax. **1. $82.86 2. $401.34 3. $275.10**

1. 113 miles, 2 days **2.** 1,297 miles, 4 days **3.** 853 miles, 3 days

4. Mrs. Gandi rented a compact car for 5 days at $19.95 a day and 19¢ a mile. When she rented the car, the mileage was 44,264.1. When she returned the car, the mileage was 45,179.5. What is the rental charge? Assume there is no tax. **$273.79**

5. Tom Bellisimo rented a full-size car for 2 weeks at $131.25 a week and 23¢ for every mile over 250. If Tom drove 1,987.3 miles and the tax is 5.5%, what was the total cost of renting the car? What was the rental charge? **$698.67; 662.24**

6. Jennifer Norris needs to rent a car for a 5-day business trip. The trip is 1,248.5 miles. Action Rental charges $31 a day and 17.5¢ a mile. Fulton Rental charges $29.95 a day and 20¢ a mile. Which rental agency offers the better buy for this trip? **Action Rental**

Applying Mathematics to the Work World

This optional page shows how mathematics is used in the work world and also provides a change of pace.

Objective Find the cost of renting a car.

Using Discussion Discuss rental car charges. Point out that you must take into account the daily charge plus the mileage charge. The company with the lowest daily charge is not always the better deal. If a company has a low daily charge, they generally have a high mileage charge.

Using Calculators You may wish to have students use a calculator for ease of computation.

Activity

Using Cooperative Groups Have students collect ads for three or four different rental agencies. Divide students into small groups. Have them plan a trip including a certain number of days and miles. Compute the rental charges for each of the rental agencies. Determine which company offers the best price.

5-MINUTE CHECK-UP

(over Lesson 14-4)

Available on TRANSPARENCY 14-5.

Use a deck of 26 alphabet cards. Each card has a different letter of the alphabet on it. Find each probability.

1. P (A or Z) $\frac{1}{13}$
2. P (a vowel or consonant) 1
3. P (M or a vowel) $\frac{3}{13}$
4. P (H or any letter from S-Z) $\frac{9}{26}$

Extra Practice, Lesson 14-4, p. 484

INTRODUCING THE LESSON

Ask students to list their favorite health care products, such as shampoo or toothpaste. Ask them if packaging, advertising, performance, or price influences their shopping choices.

TEACHING THE LESSON

Using Discussion Ask students which would be the best way to find how many students in school own a bicycle. Discuss with students the importance of taking a *representative* sample. Ask students to describe a situation that is not *representative*.

Practice Masters Booklet, p. 144

Name ______________________ Date ______________

PRACTICE WORKSHEET 14-5

Applications: Marketing Research

A company surveyed 120 people about their favorite radio station and the amount of time they listen to the radio each day. The results are shown in the tables below.

Favorite Radio Station	Number of People
KALM	28
KOOL	32
KLAS	15
KRZY	34
none	11

Time Spent Listening to Radio Daily	Number of People
more than 2 h	28
61 min–2 h	30
1 min–1 h	50
0 min	12

Find the probability of a person liking each station best.

1. KOOL $\frac{4}{15}$
2. KOOL or KRZY $\frac{11}{20}$
3. not KALM $\frac{23}{30}$
4. neither KALM nor KOOL $\frac{1}{2}$

Find the probability of a person most likely listening to the radio for each time.

5. 1 min–1 h $\frac{5}{12}$
6. more than 1 h $\frac{29}{60}$
7. 2 h or less $\frac{23}{30}$
8. not at all $\frac{1}{10}$

Assuming that the choice of station and the amount of time spent listening are independent, find the probability of each combination of favorite station and time spent listening to the radio each day.

9. KRZY and more than 2 h $\frac{119}{1{,}800}$
10. KALM and 61 min–2 h $\frac{7}{120}$
11. not KOOL and 1 min–1 h $\frac{11}{36}$
12. no favorite station and not listening at all $\frac{11}{1{,}200}$

Applications

14-5 MARKETING RESEARCH

Objective
Find probabilities from market samples.

Joanie Mock is a marketing research assistant. She collects data on the interests and tastes of people who use products that her company makes and sells. From the data she collects, her company decides which products to make, how many to make, how to package the products, and how to advertise the products.

Ms. Mock's company, National Interior Design, wants to know which colors and patterns of wallpaper are liked best. Of course, every person cannot be questioned. A smaller, representative group called a **sample** is surveyed. A sample is representative when it has the same characteristics as the larger group. The results of Ms. Mock's sample are shown below.

Favorite Color	Number of People
Black	11
Blue	41
Green	25
Red	34
Yellow	39

Favorite Pattern	Number of People
Florals	46
Small prints	24
Solid	18
Stripes	62

Understanding the Data

- What is the total number of people surveyed? **150 people**
- What color is liked by the most people? **blue**
- What pattern is liked by the most people? **stripes**
- What color and pattern is liked by the least number of people? **black solid**

Check for Understanding

Use the data from the tables above.

1. How many people don't prefer green? **125**
2. How many people like small prints or stripes? **86**
3. How many people don't prefer stripes? **88**
4. What is the probability of a person liking green? $\frac{1}{6}$
5. Last year, Ms. Mock's sample showed the following: florals, 28; small prints, 44; solids, 23; and stripes, 55. Based on the change in the sample, should Ms. Mock recommend adding a new stripe or a new floral design? **a floral design**

RETEACHING THE LESSON

A box contains four red chips, six blue chips, five white chips, and three yellow chips.
Suppose you choose a chip (without looking), replace it, then choose another chip. Find the probability of choosing each of the following.

1. a blue chip, then a red chip $\frac{2}{27}$
2. a yellow chip both times $\frac{1}{36}$

Reteaching Masters Booklet, p. 131

Name ______________________ Date ______________

RETEACHING WORKSHEET 14-5

Applications: Marketing Research

A company surveyed 150 people about their favorite spectator sport and what they prefer to eat while watching. The results are shown in the tables below.

Favorite Spectator Sport	Number of People
Baseball	55
Football	30
Basketball	45
Tennis	20

Favorite Snack While Watching a Game	Number of People
Hot dogs	50
Popcorn	60
Pretzels	15
Peanuts	25

Probability of a person liking football

$\frac{\text{people who like football}}{\text{people surveyed}} \rightarrow \frac{30}{150} = \frac{1}{5}$

Probability of a person liking pretzels

$\frac{\text{people who like pretzels}}{\text{people surveyed}} \rightarrow \frac{15}{150} = \frac{1}{10}$

Exercises

Practice

Find the probability of a person liking each of the following colors best. Use the survey on page 436.

6. blue $\frac{41}{150}$ 7. not yellow $\frac{111}{150}$ 8. neither red nor blue $\frac{1}{2}$

Find the probability of a person liking each pattern best.

9. florals $\frac{23}{75}$ 10. solid or florals $\frac{32}{75}$ 11. not stripes $\frac{44}{75}$

Assuming that the choices of color and pattern are independent, find the probability of a person liking each of the following color and pattern combinations.

12. black and florals $\frac{253}{11,250}$ 13. red and stripes $\frac{527}{5,625}$ 14. blue and solid $\frac{369}{11,250}$

15. small prints and *not* yellow $\frac{1,332}{11,250}$ 16. blue and *not* stripes $\frac{1,804}{11,250}$

Applications

17. Brad surveyed 50 grocery store customers. Tastee fishsticks were favored by 26 of the 50 customers. What is the probability that a customer favors Tastee? $\frac{13}{25}$

18. The table below shows the results of surveying 40 people about their favorite brand of shampoo and conditioner.

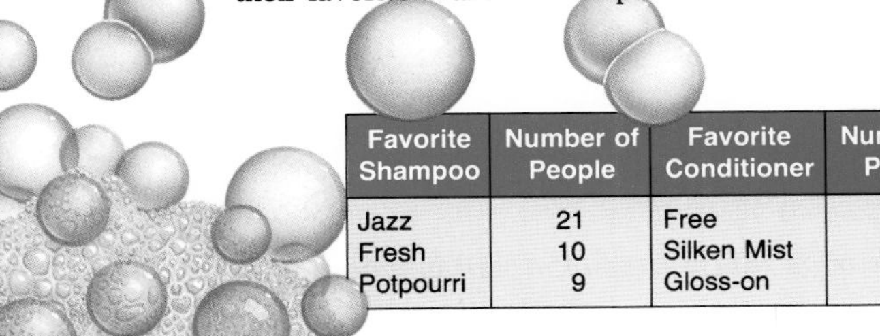

Favorite Shampoo	Number of People	Favorite Conditioner	Number of People
Jazz	21	Free	18
Fresh	10	Silken Mist	15
Potpourri	9	Gloss-on	7

a. What is the probability that a person likes Fresh and Free? Assume that the choices are independent. $\frac{9}{80}$

b. Would the results of this survey affect the brand of shampoo the store should sell? See margin.

19. Explain how probability can be used in marketing research. See students' work.

Critical Thinking

20. What is one half of 2^{60}? 2^{59}

Mixed Review

Estimate.

Lesson 10-5

21. 16% of 80 12 22. 25% of 121 30 23. 62% of 364 240

Lesson 13-3

24. Claudia bought 4 turtlenecks. She paid $12 cash and wrote a $20 check for the remaining amount. How much was each turtleneck? $8

Chalkboard Examples

1. Three out of five seniors at Central High plan to go to college. If there are 250 in the senior class, how many plan to go to college? 150
2. Three out of ten voters registered in a precinct cast their ballots on election day. What is the percent of people that voted? 30%

Additional examples are provided on TRANSPARENCY 14-5.

EVALUATING THE LESSON

Check for Understanding Use Guided Practice Exercises to check for student understanding.

Closing the Lesson Have students summarize how to find the probability of an event based on a sample of events.

EXTENDING THE LESSON

Enrichment Choose 10 students and ask how many like rap music. From their responses, find the probability that a person in your class likes rap music. Predict how many in your class and school like rap music.

Enrichment Masters Booklet, p. 131

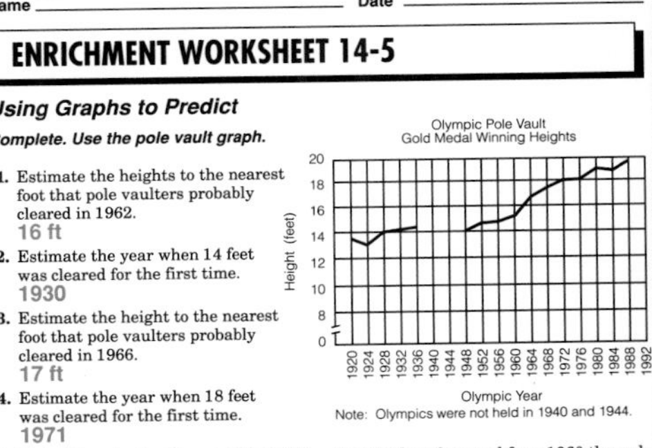

Name ____________ Date ____________

ENRICHMENT WORKSHEET 14-5

Using Graphs to Predict

Complete. Use the pole vault graph.

1. Estimate the heights to the nearest foot that pole vaulters probably cleared in 1962. 16 ft
2. Estimate the year when 14 feet was cleared for the first time. 1930
3. Estimate the height to the nearest foot that pole vaulters probably cleared in 1966. 17 ft
4. Estimate the year when 18 feet was cleared for the first time. 1971
5. If the Olympics had been held in 1940, predict what the winning height would have been (to the nearest foot). 14 ft
6. Based on the trend from 1960 through 1988, would you predict the winning height in 1992 to be over or under 20 feet? over

Complete. Use the savings graph.

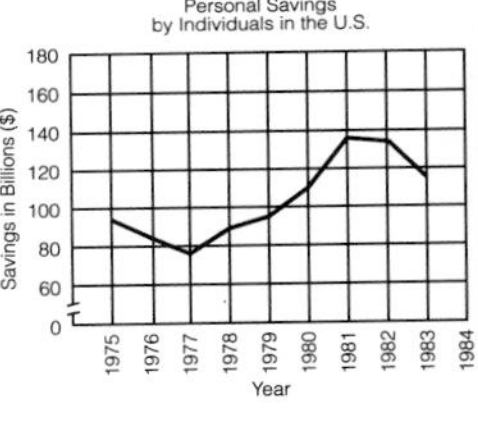

7. Based on the trend from 1977 through 1981, what level of savings would you predict for 1982? about $150 billion
8. How does your prediction for 1982 compare with the actual level of savings for 1982? prediction was about $15 billion higher
9. Based on the trend from 1981 through 1983, what level of savings would you predict for 1984? about $105 billion
10. The actual level of savings in 1984 was $156.1 billion. How does your prediction for 1984 compare with this actual level? prediction was

APPLYING THE LESSON

Independent Practice

Homework Assignment	
Minimum	Ex. 2-24 even
Average	Ex. 2-16 even; 17-24
Maximum	Ex. 1-24

Additional Answers

18b. not enough data, You need to know the details of the people sampled: age, sex, income level, and so on.

5-MINUTE CHECK-UP

(over Lesson 14-5)

Available on TRANSPARENCY 14-6.

1. In Mr. Herr's freshman homeroom, 6 students walk to school, 12 ride the bus, and 7 ride with older students. If there are 125 students in the freshman class, predict how many walk and how many ride with friends. **30 walk, 35 ride with friends**

INTRODUCING THE LESSON

Ask students to give examples of times they have heard the odds of an event given. **Although odds are common with sporting events, they are often found in other areas, such as state lotteries and in election news.**

TEACHING THE LESSON

Using Discussion Have students discuss the difference between odds and probability. Emphasize that the sum of the numerator and denominator of the ratio that expresses the odds is the total of all possible outcomes. Although probability ranges from 0 to 1, the fraction that expresses odds may be greater than 1.

Practice Masters Booklet, p. 145

Name ______________________ Date ______________

PRACTICE WORKSHEET 14-6

Odds

A die is rolled. Find the odds for each roll.

1. a 6 **1 to 5**
2. not an odd number **3 to 3**
3. a number less than 4 **3 to 3**
4. a 5 or a 6 **2 to 4**
5. a 9 **0 to 6**
6. a 6 or a 1 **2 to 4**
7. a 4, 3, or 1 **3 to 3**
8. not a 2 **5 to 1**
9. a 2 or an odd number **4 to 2**
10. a 3 or an 8 **1 to 5**

Two dice are rolled. Find the odds for each roll.

11. both dice different **30 to 6**
12. a 3 and a 5 **2 to 34**
13. a sum of 12 **1 to 35**
14. a sum of 7 **6 to 30**
15. a sum of 5 or 10 **7 to 29**
16. a sum less than 5 **6 to 30**
17. a sum of 1 **0 to 36**
18. *not* a sum of 8 **31 to 5**
19. a sum of 4 and a 2 on at least one die **1 to 35**
20. a sum of 5 and a 3 on one die **2 to 34**

Solve.

21. On a TV game show, the contestant is supposed to match the correct price with a prize. If four prices are shown, what are the odds of selecting the correct one at random? **1 to 3**
22. If two coins are tossed, what are the odd for both coins landing with the same side up? What are the odds for both coins landing heads? **2 to 2; 1 to 3**

14-6 ODDS

Objective
Find the odds for an event.

The student committee for the Read-a-thon fund raiser consists of three freshmen, four sophomores, a junior, and four seniors. Each student's name is placed in a hat and one name is chosen to be the chairperson of the committee. What are the odds that a freshman will be chosen?

The ratio of the number of ways an event can occur to the number of ways the event cannot occur is called the **odds** for the event.

$$\text{odds} = \frac{\text{number of ways an event can occur}}{\text{number of ways the event cannot occur}}$$

Method

1. Find the number of ways the event can occur.
2. Find the number of ways the event cannot occur.
3. Write the ratio of 1 to 2.

Example A ***Find the odds that a freshman will be chosen.***

1. There are 3 ways a freshman could be chosen.
2. There are 9 ways a freshman could not be chosen.
3. The odds are $\frac{3}{9}$ or 3 to 9.

The odds are 3 to 9 that a freshman will be chosen.

Example B ***Find the odds that a freshman or a sophomore will be chosen.***

1. There are 7 students who are freshmen or sophomores.
2. There are 5 students who are not freshmen or sophomores.
3. The odds are $\frac{7}{5}$ or 7 to 5 that a freshman or a sophomore will be chosen.

Guided Practice

A die is rolled. Find the odds for each roll.

1. a 1 **1 to 5**
2. a 4 **1 to 5**
3. a 3 **1 to 5**
4. a 7 **0 to 6**
5. an odd number **3 to 3**
6. an even number **3 to 3**
7. a prime number **3 to 3**
8. a multiple of 3 **2 to 4**
9. a number less than 3 **2 to 4**
10. a multiple of 2 **3 to 3**

RETEACHING THE LESSON

The numbers 1 through 10 are written on small cards and placed in a jar. One card is chosen at random. Find the odds for each of the following.

1. a 5 $\frac{1}{9}$
2. an even number $\frac{5}{5}$
3. a number greater than 8 $\frac{2}{8}$

The odds for an event are given. Find the probability of the event.

4. $\frac{2}{3}$ $\frac{2}{5}$
5. $\frac{7}{3}$ $\frac{7}{10}$
6. 5 to 1 $\frac{5}{6}$

Reteaching Masters Booklet, p. 132

Name ______________________ Date ______________

RETEACHING WORKSHEET 14-6

Odds

odds for an event = number of ways that event can occur to number of ways that event cannot occur

If a die is rolled once, find the odds for rolling a 4.

A 4 can only occur in one way. The other events are 1, 2, 3, 5, and 6.
1 to 5

The odds of rolling a 4 are 1 to 5.

Find the odds for each event. Use the spinner at the right.

1. a 6 **1 to 15**
2. *not* a 13 **15 to 1**

13. 4 to 2 17. 3 to 3 18. 4 to 2

Example B

11. a 2 or a 6 2 to 4 12. a 3 or a 5 2 to 4 13. a 1, 2, 4, or 6
14. a 3 or an even number 4 to 2 15. a 1 or a 7 1 to 5
16. a 6 or an odd number 4 to 2 17. a 1 or a number greater than 4
18. a 6 or a number less than 4 19. a 1 or a prime number 4 to 2

21. 3 to 33 23. 31 to 5 24. 34 to 2 25. 33 to 3

Exercises

Practice

Two dice are rolled. Find the odds for each roll.

20. a sum of 7 6 to 30 21. a sum of 10 22. a sum of 2 1 to 35
23. *not* a sum of 6 24. *not* a sum of 3 25. *not* a sum of 4
26. a sum of 8 with a 6 on one die 2 to 34
27. an odd sum or a sum less than 7 27 to 9

Suppose you choose a card from those shown at the left. Find the odds for each event. Then find the probability.

28. a blue card 5 to 15; $\frac{1}{4}$
29. a purple card 6 to 14; $\frac{3}{10}$
30. an orange card 9 to 11; $\frac{9}{20}$
31. *not* a purple card 14 to 6; $\frac{7}{10}$
32. *not* a blue card 15 to 5; $\frac{3}{4}$
33. a purple card or a blue card
34. *neither* a purple *nor* a blue card

33. 11 to 9; $\frac{11}{20}$ 34. 9 to 11; $\frac{9}{20}$

Applications

35. Two coins are tossed. Find the odds for tossing at least one tail. 3 to 1
36. Jason's coach told him the odds that he will qualify for the state finals are 4 to 1. What is the probability that Jason will qualify? $\frac{4}{5}$
37. For the gumball machine in Parker's Drugstore, the odds of getting a prize with the gum are 5 to 31. What is the probability of getting a prize? $\frac{5}{36}$

Collect Data

38. Read an article about some sporting events that gives the odds of winning. Why is sports a common application of odds? People speculate about the chances of winning in various sporting events, tournaments, and so on.

Mental Math

Subtracting from 100 can be done mentally as shown below.

Find 100 − 24.

$100 - 24 = \underbrace{100 - 20} - 4$
$= 80 - 4$ Think.
$= 76$

Find 100 − 53.

$100 - 53 = \underbrace{100 - 50} - 3$
$= 50 - 3$ Think.
$= 47$

Find each difference.

39. 100 − 42 58 40. 100 − 65 35 41. 100 − 33 67
42. 100 − 79 21 43. 100 − 54 46 44. 100 − 22 78

Chalkboard Examples

The swim team consists of 3 seniors, 4 juniors, and 5 sophomores. To select the captain, one name is drawn.

Example A

1. Find the odds that a sophomore will be selected. 5 to 7
2. Find the odds that a junior will be selected. 4 to 8

Example B

3. Find the odds that a junior or senior will be selected. 7 to 5

Additional examples are provided on TRANSPARENCY 14-6.

EVALUATING THE LESSON

Check for Understanding Use Guided Practice Exercises to check for student understanding.

Error Analysis Watch for students who use the total number of outcomes as the denominator in computing the odds. Remind students to check by adding the numerator and denominator.

Closing the Lesson Have students explain in their own words how to find the odds of an event.

EXTENDING THE LESSON

Enrichment Masters Booklet, p. 132

Name ________ Date ________

ENRICHMENT WORKSHEET 14-6

Permutations and Combinations

An arrangement of a set of objects in a particular order is called a **permutation** of those objects.

In how many ways can 4 pictures be hung in a row?

choices for 1st picture		choices for 2nd picture		choices for 3rd picture		choices for 4th picture		
4	×	3	×	2	×	1	=	24

Instead of writing 4 × 3 × 2 × 1, you can write 4! which is read *4 factorial*. The number of permutations of 4 pictures is 4! or 24.

An arrange of a set of objects in no particular order is called a **combination** of the objects.

From 5 pictures, in how many ways can 3 be chosen?

$\text{number of combinations} = \frac{\text{number of permutations of 5 pictures chosen 3 at a time}}{\text{number of permutations of 3 pictures chosen 3 at a time}} = \frac{5 \times 4 \times 3}{3!} = 10$

The number of combinations of 5 pictures chosen 3 at a time is 10.

Tell whether each arrangement is a permutation or combination. Then solve.

1. In how many ways can 9 record albums be placed in a row on display? permutation; 362,880 ways
2. A certain type bicycle lock has four numbers. There is a choice of 10 digits for each number. How many different locks are possible? combination; 10,000 locks
3. From a list of 10 books, how many lists of 6 books can be selected? combination; 210 lists
4. How many ways can a club of 15 people choose 5 different officers? permutation; 360,360
5. How many baseball teams of 9 members can be formed from 14 players? combination; 2,002 teams
6. There are 6 different toppings for pizza. How many different pizzas with 3 topping are there? combination; 20 pizzas

APPLYING THE LESSON

Independent Practice

Homework Assignment	
Minimum	Ex. 2-44 even
Average	Ex. 2-34 even; 35-44
Maximum	Ex. 1-44

5-MINUTE CHECK-UP

(over Lesson 14-6)

Available on TRANSPARENCY 14-7.

A die is rolled and the spinner is spun. Find the odds for each event.

1. rolling a 2 and spinning a 4 2 to 8

2. rolling an even number and spinning an odd number 5 to 5

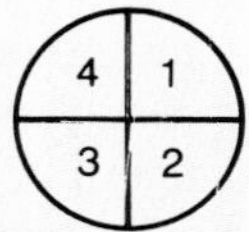

3. rolling a 5 and spinning an even number 3 to 7

Extra Practice, Lesson 14-6, p. 484

INTRODUCING THE LESSON

Ask students what kind of research they think advertising agencies do. Ask them if they have ever answered survey questionnaires and for what types of products.

TEACHING THE LESSON

Using Questioning

- How do you find the probability of an event from the data in a survey? The number who chose a certain event is the numerator. The number surveyed is the denominator.

Practice Masters Booklet, p. 146

Name ______________ Date ______________

PRACTICE WORKSHEET 14-7

Problem-Solving Strategy: Using Samples to Predict

Solve. Use the results of the survey below for Exercises 1–9. The sample is taken from 1,500 students.

What do you collect?	
Stamps	35
Coins	20
Baseball cards	80
Rocks	15

1. How many students are in the sample? 150 students
2. What is the probability that a student collects baseball cards? $\frac{8}{15}$
3. Predict how many of the 1,500 students collect baseball cards. 800 students
4. What is the probability that a student collects rocks? $\frac{1}{10}$
5. Predict how many of the 1,500 students collect rocks. 150 students
6. What is the probability that a student collects coins? $\frac{2}{15}$
7. Predict how many of the 1,500 students collect coins. 200 students
8. If there are 30 students in one class, predict how many collect coins. 4 students
9. Is it possible that a student will not collect stamps, coins, baseball cards, or rocks? yes

Solve.

10. In a survey of 200 people in Westville, 150 plan to vote in the town election. What is the probability that a person in Westville will vote? $\frac{3}{4}$
11. Refer to Exercise 10. If the population of Westville is 14,000, predict how many people will vote. 10,500 people
12. If the television ad that states five out of six people will buy a new toothpaste is correct, what is the probability that your neighbor will buy it?
13. Refer to Exercise 12. If 3,000 supermarket customers are planning to buy toothpaste, predict how many will buy the new toothpaste.

Problem-Solving Strategy

Explore / Plan / Solve / Examine

14-7 USING SAMPLES TO PREDICT

Objective

Solve verbal problems using samples to make predictions.

Johnson's Department Store conducts a survey of 75 shoppers in the mall and asks if they have a Johnson's credit card. Thirty of the 75 shoppers say that they have one. If about 3,200 people shop in the mall each day, predict how many of them probably have a Johnson's credit card.

Predictions about a large group of people based on the choices of a sample population are *estimates.* The best estimates are made when the sample population is representative of the larger group. For example, if you stand at the door of Johnson's Department Store to conduct the survey described above, the sample results would probably not be representative.

Explore

What is given?

- There are about 3,200 shoppers each day.
- In the sample of 75 shoppers, 30 have a credit card.

What is asked?

- How many shoppers in the mall probably have a Johnson's credit card?

Plan

Use the sample of 75 shoppers. Let n represent the total number of shoppers who probably have a Johnson's credit card. The ratio of n to 3,200 should be equivalent to the ratio of 30 to 75. Set up a proportion and solve for n.

Solve

$$\frac{n}{3{,}200} = \frac{30}{75}$$

3200 ⊗ 30 ÷ 75 = 1280

The number of shoppers that probably have a credit card is *about* 1,280.

Examine

30 out of 75 or less than half of the shoppers in the sample say they have a Johnson's credit card. Since half of 3,200 is 1,600, it seems reasonable to predict that about 1,280 shoppers have a Johnson's credit card.

Guided Practice

1. A sample of 100 seniors at Madison High School showed that 72 expect to continue their education after graduation. If there are 700 students in the class, predict the number of students who will continue their education. 504 students

RETEACHING THE LESSON

Have students work in pairs and solve the problems on page 441.

2. A cashier in a supermarket noticed that 3 out of her first 20 customers had at least three coupons. Based on this sample, how many customers out of 100 had *less* than three coupons? **85 customers**

Problem Solving

Paper/ Pencil
Mental Math
Estimate
Calculator

Sharp Cable Company has 4,800 customers. Use the results of a customer preference survey shown below for Exercises 3–6.

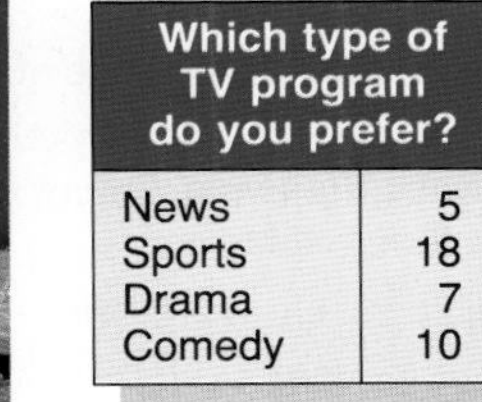

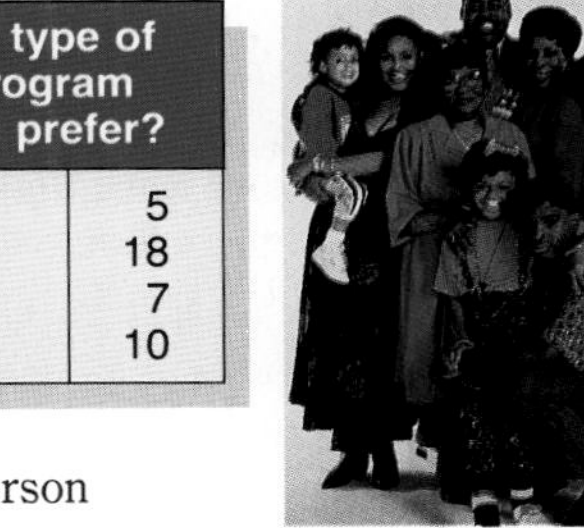

Which type of TV program do you prefer?	
News	5
Sports	18
Drama	7
Comedy	10

3. What is the probability that a person prefers comedy programs? $\frac{1}{4}$
4. Predict how many of the 4,800 cable customers prefer comedy programs. **1,200 customers**
5. What is the probability that a person prefers news programs? $\frac{1}{8}$
6. Predict how many of the 4,800 customers prefer news programs. **600 customers**
7. In a survey of 200 residents of Westville, 80 use Shine-on toothpaste. What is the probability that a Westville resident uses Shine-on toothpaste? $\frac{2}{5}$

8. Review the items in your portfolio. Make a table of contents of the items, noting why each item was chosen. Your portfolio should include a variety of assignments and reflect your progress over the year. Replace any items that are no longer appropriate. **See students' work.**

9. The choices of a large group may not agree with a prediction. Write two reasons why the actual choices may not agree with a prediction. **See margin.**

Critical Thinking

10. Arrange eight congruent equilateral triangles to form a parallelogram. **Answers may vary. A typical answer is given.**

Mixed Review

Find the area of each circle whose radius is given. Use 3.14 for π. Round decimal answers to the nearest tenth.

Lesson 8-4

11. 7 cm **153.9 cm²**
12. 8 in. **201.0 in²**
13. 24 mm **1808.6 mm²**
14. 3.5 cm **38.5 cm²**

Lesson 11-2

Solve. Use data at the right.

15. Find the mean. **41.2**
16. Find the median. **38**
17. Find the mode. **38**

Golf Team Scores				
49	42	38	38	43
38	36	50	37	

Chalkboard Examples

1. Gucci Poochi Pet Shops surveyed 70 dog owners about their preference for dog collars. There were 35 owners who preferred leather, 20 who preferred nylon, and 15 who preferred chain. Of the 580 dog owners on the shops mailing list, predict how many will prefer leather collars and how many will prefer nylon collars. **290, leather; 166, nylon**

Additional examples are provided on TRANSPARENCY 14-7.

EVALUATING THE LESSON

Check for Understanding Use Guided Practice Exercises to check for student understanding.

Closing the Lesson Have students summarize how to find the probability and the odds of an event from the data in a sample.

EXTENDING THE LESSON

Enrichment Have students design a questionnaire using three to five questions about student preferences for music or food. Have students survey a sample of ten students, then predict the preferences of the student body.

APPLYING THE LESSON

Independent Practice

Homework Assignment	
Minimum	Ex. 1-7 odd; 8-17
Average	Ex. 1-17
Maximum	Ex. 1-17

Chapter 14, Quiz B (Lessons 14-5 through 14-7) is available in the Evaluation Masters Booklet, p. 69.

Additional Answers

9. The sample was not representative. The larger population did not have the same preferences as the sample. Too much time elapsed between taking sample and choice of population.

Using the Chapter Review

The Chapter Review is a comprehensive review of the concepts presented in this chapter. This review may be used to prepare students for the Chapter Test.

Chapter 14, Quizzes A and B, are provided in the Evaluation Masters Booklet as shown below.

Quiz A should be given after students have completed Lessons 14-1 through 14-4. Quiz B should be given after students have completed Lessons 14-5 through 14-7.

These quizzes can be used to obtain a quiz score or as a check of the concepts students need to review.

Evaluation Masters Booklet, p. 69

CHAPTER 14, QUIZ A (Lessons 14-1 through 14-4)

Use multiplication to find the number of possible outcomes. 1. 24

1. tossing a dime, tossing a quarter, and rolling a die
2. a choice of cherry or apple pie with a choice of milk, tea, or coffee 2. 6

Two dice are rolled. Find the probability of rolling each of the following. Round to the nearest percent. 3. 3%

3. a 1 on the first die and a 3 on the second die
4. both odd numbers 4. 25%

Find the probability of spinning each of the following. 5. $\frac{1}{2}$

5. P(white or 5)
6. P(gray or an even number) 6. $\frac{5}{6}$

A bag contains 3 blue marbles, 2 green marbles, and 5 white marbles. Suppose you choose a marble, replace it, and then choose another marble. Find the probability of choosing each event. 7. $\frac{3}{50}$

7. blue, then green
8. white both times 8. $\frac{1}{4}$

Suppose you choose a marble from the bag described above and then choose another one without replacing the first one. Find the probability of choosing each event. 9. $\frac{1}{6}$

9. blue, then white
10. green both times 10. $\frac{1}{45}$

Name ________________ Date ________________

CHAPTER 12, QUIZ B (Lessons 14-5 through 14-7)

Two dice are rolled. Find the odds for rolling each of the following. 1. 5 to 31

1. a sum of 6
2. a sum of 7 and a 6 on one die 2. 2 to 34

Solve. Use the survey results shown. The sample is part of a group of 320 people. 3. 80 peopl

3. How many people are in the sample?
4. What is the probability that a person uses Brite toothpaste? 4. $\frac{3}{8}$
5. Predict how many of the 320 people use Clean toothpaste. 5. 60 peopl

Which kind of toothpaste do you use?	
Brite	30
Smile	35
Clean	15

Chapter 14 REVIEW

2. independent 6. event 7. dependent

Vocabulary / Concepts

Choose a word from the list at the right that best completes each sentence.

dependent, equally likely, event, independent, odds, outcome, random, ratio, replaced, tree

1. The probability of an event is a _?_ that describes how likely it is that the event will occur. **ratio**
2. Two events are _?_ if the occurrence of one event does not affect the occurrence of the other event.
3. The ratio of the number of ways an event can occur to the number of ways the event cannot occur is called the _?_ for the event. **odds**
4. An _?_ is a possible result. **outcome**
5. A _?_ diagram can be used to determine the possible outcomes. **tree**
6. An _?_ is a specific outcome or type of outcome.
7. Two events are _?_ if the occurrence of one of the events affects the occurrence of the other event.

Exercises / Applications

Lesson 14-1

See answers on p. T532.

Use multiplication to find the number of possible outcomes. Then draw a tree diagram to show the possible outcomes.

8. a choice of a Probe or a Beretta in blue, black, or white
9. a choice of a green, brown, or black skirt with a choice of a white, tan, or yellow blouse

Lesson 14-2

A box contains three red chips, five black chips, and two blue chips. One chip is chosen. Find the probability that each event will occur. Express as a percent.

10. a red chip **30%** **11.** a black chip **50%** **12.** a blue chip **20%**

13. *not* a red chip **70%** **14.** *not* a black chip **50%** **15.** *not* a blue chip **80%**

16. a white chip **0%** **17.** *not* a white chip **100%** **18.** a red or a blue chip **50%**

Lesson 14-3

A box contains six red, eight green, and two blue marbles. Find the probability of each event.

19. draw a red marble, replace it, draw a blue marble $\frac{3}{64}$

20. draw a green marble, replace it, draw a red marble $\frac{3}{16}$

21. draw a green marble, do not replace it, draw a blue marble $\frac{1}{15}$

22. draw a blue marble, do not replace it, draw a red marble $\frac{1}{20}$

Lesson 14-4

The following words are written on slips of paper and placed in a box.

PEPPER PARSNIP CARROT CABBAGE CUCUMBER

A slip of paper is chosen from the box. Find each probability.

23. P(a word beginning with P or C) 1

24. P(a word ending in R or P) $\frac{3}{5}$

25. P(a 6-letter or an 8-letter word) $\frac{3}{5}$

26. P(a word beginning with B or D) 0

Lesson 14-5

Know-it-All Marketing surveyed 150 people for Better Bakers about the flavors they like best. Find the probability of a person liking each flavor or combination of flavors.

Favorite Flavor	Number of People
Vanilla	45
Chocolate	46
Cherry	32
Peach	20
Raspberry	7

27. vanilla $\frac{9}{30}$

28. chocolate $\frac{23}{75}$

29. cherry or peach $\frac{26}{75}$

30. chocolate or raspberry $\frac{53}{150}$

Lesson 14-6

A box contains six red, eight green, and two blue marbles. Find the odds for choosing each event.

31. *not* a green marble 8 to 8

32. a red or green marble 14 to 2

Lesson 14-7

33. In a sample of 50 car owners, 35 said they prefer to have regular maintenance performed by the dealer from whom they bought the car. Predict how many car owners out of 1,000 will prefer to have the dealer perform regular maintenance. 700 owners

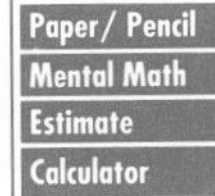

34. On the average Miss Chung, a sales representative for a publisher, sells a set of encyclopedias to 3 out of every 25 households she visits. If she visits 200 households, how many sets should she expect to sell? 24 sets

35. Mr. Crady said the odds that the University of Kansas would win the NCAA basketball tournament are 1 to 2. What is the probability that the University of Kansas will win? $\frac{1}{3}$

36. Andy Hayes and Becky Girard went to Paradise Cafe for lunch. The luncheon special has an entrée and 3 choices for a side dish. There are also some dessert choices. If there is a total of 18 different meals, how many dessert choices are there? 6 dessert choices

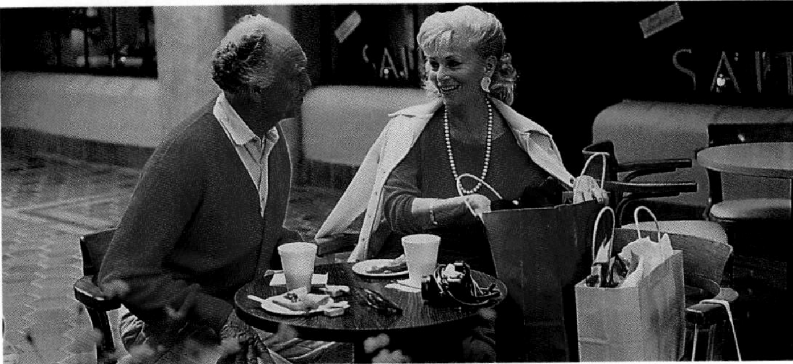

APPLYING THE LESSON

Independent Practice

Homework Assignment	
Minimum	Ex. 1-7; 8-26 even; 27-36
Average	Ex. 1-36
Maximum	Ex. 1-36

Alternate Assessment Strategy

To provide a brief in-class review, you may wish to read the following questions to the class and require a verbal response.

1. What are the number of outcomes possible when 2 coins are tossed? 4
2. How many outcomes are possible with a choice of shag or textured carpet in brown, grey, or blue? 6

There are two blue, three red, and five black notebooks. Choose one without looking.

3. What is the probability that a red notebook will be chosen? 30%
4. What is the probability that a black notebook will be chosen? 50%
5. Using a 6 spot spinner, what is the probability of spinning a 2 or a 3? $\frac{1}{3}$
6. Using a 6 spot spinner, what is the probability of spinning an even number? $\frac{1}{2}$
7. Albert has four blue and six yellow highlighters in his desk drawer. If he chooses two without looking, what is the probability that he chooses one blue and one yellow highlighter? $\frac{4}{15}$
8. A die is rolled. Find the odds that a 2 or a 6 is rolled. 2 to 4
9. A die is rolled. Find the odds that a 5 or an even number is rolled. 4 to 2
10. Two dice are rolled. What are the odds of rolling a sum of three? 2 to 34

Using the Chapter Test

This page may be used as a test or as an additional review if necessary. Two forms of the Chapter Test are provided in the Evaluation Masters Booklet. Form 2 (free response) is shown below, and Form 1 (multiple choice) is shown on page 447.

The **Tech Prep Applications Booklet** provides students with an opportunity to familiarize themselves with various types of technical vocations. The Tech Prep applications for this chapter can be found on pages 27–28.

Evaluation Masters Booklet, p. 68

CHAPTER 14 TEST, FORM 2

Use multiplication to find the number of possible outcomes.

1. rolling a die and tossing a penny
2. a choice of a red, blue, or yellow shirts with a choice of gray or black pants
3. spinning each spinner once

1, 2, 3, 4 | red, gray | blue, pink

1. 12
2. 6
3. 16

There are 4 white balls and 6 black balls in a box. One ball is chosen. Find the probability of choosing each of the following. Use percents.

4. a black ball
5. *not* a black ball
6. a red ball or a white ball

4. 60%
5. 40%
6. 40%

The radio call letters WPAL, WAKS, WHTE, WEDL, and WCAT are written on slips of paper and placed in a box. A slip of paper is drawn from the box. Find the probability of choosing each of the following.

7. call letters with T or S
8. call letters with A or E
9. call letters with A or T

7. $\frac{3}{5}$
8. 1
9. $\frac{4}{5}$

A bag contains 2 red, 1 blue, and 5 black marbles. Suppose you choose a marble, replace it, then choose another marble. Find the probability of choosing each of the following.

10. red, then blue
11. black, then red

10. $\frac{1}{32}$
11. $\frac{5}{32}$

Suppose you choose a marble from the bag described above and then choose another marble without replacing the first one. Find the probability of choosing each of the following.

12. blue, then black
13. red, then blue

12. $\frac{5}{56}$
13. $\frac{1}{28}$

Two dice are rolled. Find the odds for rolling each of the following.

14. a sum of 4
15. *not* a sum of 7
16. a sum of 5 and a 2 on one die
17. an even sum or a sum < 5

14. 3 to 33
15. 30 to 6
16. 2 to 34
17. 20 to 16

Solve. Use the survey results shown. The sample is part of a group of 160.

yogurt flavor preferred	
vanilla	12
banana	8

18. How many people are in the sample?
19. What is the probability that a person prefers vanilla?
20. Predict how many people of the 160 people prefer vanilla.

Bonus A spinner has 16 equal sections. It is spun once. The probability of landing on orange is $\frac{1}{4}$. The probability of landing on orange or green is $\frac{5}{8}$. How many sections are green?

18. 20 people
19. $\frac{3}{5}$
20. 96 people

Bonus 6 sections

A die is rolled once and the spinner shown at the right is spun once. 30 outcomes

1. Use multiplication to find the number of possible outcomes.
2. Draw a tree diagram to show the possible outcomes.
3. If you roll three dice instead of one and spin the spinner once, how many different outcomes are possible?

2. See answer on p. T532.
3. 1,080 outcomes

The letters in the word PAPER are written on slips of paper and placed in a box. Find the probability of each draw.

4. draw an A, replace it, draw an R $\frac{1}{25}$
5. draw an E, replace it, draw a P $\frac{2}{25}$
6. draw a P, do not replace it, draw an A $\frac{1}{10}$
7. draw a vowel both times without replacing the first one $\frac{1}{10}$

Use the letters placed in the box from Exercises 4–7 above. One slip is drawn from the box. Find the probability that each event will occur. Express as a percent.

8. a P 40%
9. not an R 80%
10. a P or a vowel 80%
11. an S 0%
12. a P or an E 60%
13. an R or a consonant

13. 80%

14. A survey of 200 registered voters in Fairfield showed 110 voters favor Jenkins for mayor. Predict how many will vote for Jenkins if 26,780 registered voters cast their votes. 14,729 voters

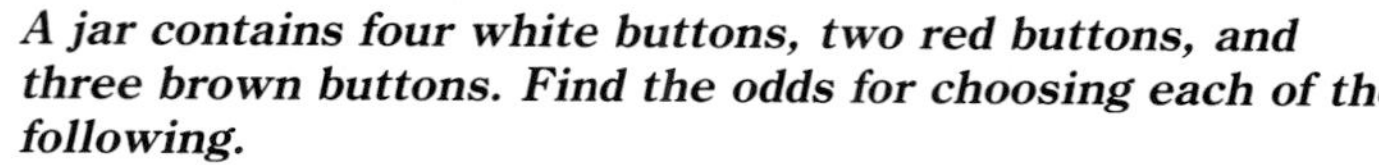

A jar contains four white buttons, two red buttons, and three brown buttons. Find the odds for choosing each of the following.

15. a brown button 3 to 6
16. a white button 4 to 5
17. a red button 2 to 7
18. *not* a white button 5 to 4
19. a white or a brown button 7 to 2
20. *neither* a white *nor* a brown button

20. 2 to 7

21. The odds that the Vikings will win are 2 to 3. Find the probability of the Vikings' winning. $\frac{2}{5}$
22. A nickel, a dime, and a quarter are tossed. Find the probability of obtaining all tails. $\frac{1}{8}$
23. In a survey of 300 voters in Huntsville, 120 favor passing the school levy. What is the probability that a voter favors passing the school levy? $\frac{2}{5}$
24. A poll of 500 households in Eastown shows that 245 households watch the local news at 6:00 P.M. Predict how many of the 12,700 households in Eastown watch the local news at 6:00 P.M.
25. A box contains four balls numbered 1 through 4. The balls are drawn from the box one at a time and not replaced. In how many possible orders can they be drawn? 24 orders

24. 6,223 households

BONUS: Write a situation represented by the tree diagram.

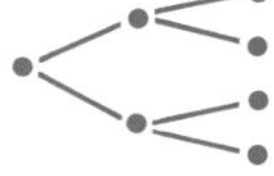

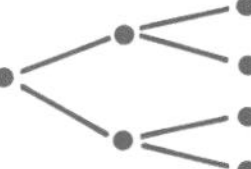

The **Performance Assessment Booklet** provides an alternative assessment for evaluating student progress. An assessment for this chapter can be found on pages 27–28.

A **Test and Review Generator** is provided in Apple, IBM, and Macintosh versions. You may use this software to create your own tests or worksheets, based on the needs of your students.

MARKUP

Tyson Bodey owns a sporting goods store. He sells most of the items in his store with at least a 50% markup. The amount Mr. Bodey pays for an item in his stock is the cost of an item. The **markup** is the amount added to the cost to obtain the selling price. The markup is usually a percent of the cost.

When Mr. Bodey sells an item, the markup is income for the store. This income is used to pay the expenses of operating the store, such as rent, salaries, insurance, and so on. A markup is necessary for a business to succeed.

Mr. Bodey pays $16.50 for a volleyball and the markup is 50% of the cost. Find the selling price.

The selling price is found by adding the amount of the markup to the cost.

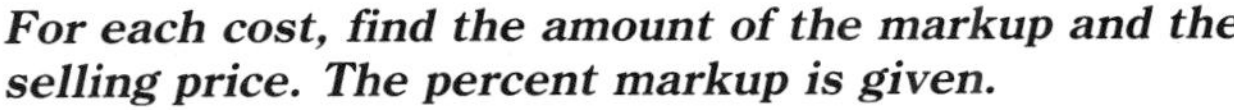

16.5 × 50 % + 16.5 = 24.75

The selling price of the volleyball is $24.75.

For each cost, find the amount of the markup and the selling price. The percent markup is given.

	Cost	% of Markup	Markup	Selling Price
1.	$25	25%	▩ **$6.25**	▩ **$31.25**
2.	$9.30	40%	▩ **$3.72**	▩ **$13.02**
3.	$3.95	100%	▩ **$3.95**	▩ **$7.90**

4. The markup on a television is $\frac{1}{3}$ of the cost. If the cost of the television is $345, find the markup and the selling price. **$115; $460**

5. The selling price of a sweatshirt is $21.75. If the cost is $15, what is the markup? What is the percent of markup? **$6.75; 45%**

6. If the selling price of a football is $27.70 and the markup is 55% of the cost, what is the cost of the football? Round to the nearest cent. **$17.87**

Applying Mathematics to the Real World

This optional page shows how mathematics is used in the real world and also provides a change of pace.

Objective Find the amount of markup, selling price, or cost of an item.

Using Questioning

- What is the cost? **the amount the retailer pays for an item**
- What is the markup? **the amount added to the cost to determine selling price**
- What is the selling price? **the sum of the cost and the markup, the price at which an item sells**

Activity

Collecting Data Have students survey local businesses to find the markup rate of their goods. Have students compare and discuss their findings.

Using the Cumulative Review

This page provides an aid for maintaining skills and concepts presented thus far in the text.

A Cumulative Review is also provided in the Evaluation Masters Booklet as shown below.

Evaluation Masters Booklet, p. 70

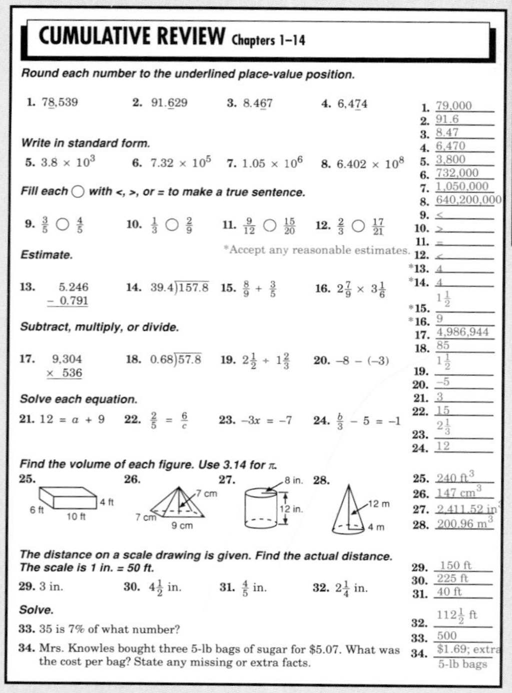

CUMULATIVE REVIEW Chapters 1–14

Round each number to the underlined place-value position.

1. 78,539 2. 91.629 3. 8.467 4. 6,474

Write in standard form.

5. 3.8×10^3 6. 7.32×10^5 7. 1.05×10^6 8. 6.402×10^8

Fill each ○ with <, >, or = to make a true sentence.

9. $\frac{3}{5}$ ○ $\frac{4}{5}$ 10. $\frac{1}{3}$ ○ $\frac{2}{9}$ 11. $\frac{9}{12}$ ○ $\frac{15}{20}$ 12. $\frac{2}{3}$ ○ $\frac{17}{21}$

Estimate. *Accept any reasonable estimates.

13. $5.246 - 0.791$ 14. $39.4\overline{)157.8}$ 15. $\frac{8}{9} + \frac{3}{5}$ 16. $2\frac{7}{9} \times 3\frac{1}{6}$

Subtract, multiply, or divide.

17. $9{,}304 \times 536$ 18. $0.68\overline{)57.8}$ 19. $2\frac{1}{2} + 1\frac{2}{3}$ 20. $-8 - (-3)$

Solve each equation.

21. $12 = a + 9$ 22. $\frac{2}{5} = \frac{6}{c}$ 23. $-3x = -7$ 24. $\frac{b}{3} - 5 = -1$

Find the volume of each figure. Use 3.14 for π.

25. (rectangular prism: 6 ft, 10 ft, 4 ft) 26. (pyramid: 7 cm, 9 cm, 7 cm) 27. (cylinder: 8 in., 12 in.) 28. (cone: 12 m, 4 m)

The distance on a scale drawing is given. Find the actual distance. The scale is 1 in. = 50 ft.

29. 3 in. 30. $4\frac{1}{2}$ in. 31. $\frac{4}{5}$ in. 32. $2\frac{1}{4}$ in.

Solve.

33. 35 is 7% of what number?

34. Mrs. Knowles bought three 5-lb bags of sugar for $5.07. What was the cost per bag? State any missing or extra facts.

Answers:
1. 79,000
2. 91.6
3. 8.47
4. 6,470
5. 3,800
6. 732,000
7. 1,050,000
8. 640,200,000
9. <
10. >
11. =
12. <
*13. 4
*14. 4
*15. $1\frac{1}{2}$
*16. 9
17. 4,986,944
18. 85
19. $1\frac{1}{2}$
20. −5
21. 3
22. 15
23. $2\frac{1}{3}$
24. 12
25. 240 ft^3
26. 147 cm^3
27. 2,411.52 in^3
28. 200.96 m^3
29. 150 ft
30. 225 ft
31. 40 ft
32. $112\frac{1}{2}$ ft
33. 500
34. $1.69; extra 5-lb bags

Cumulative REVIEW/TEST

Free Response

Lesson 1-4 — ***Round each number to the underlined place-value position.***

1. 5,009 **5,010** 2. 54,621 **55,000** 3. 30,082 **30,100** 4. 9,909 **9,900**

Lesson 2-8 — ***Find the perimeter of each figure.***

5. 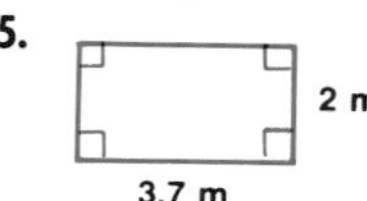

6. 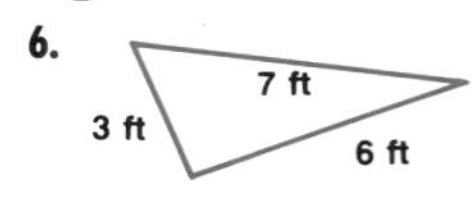

7. 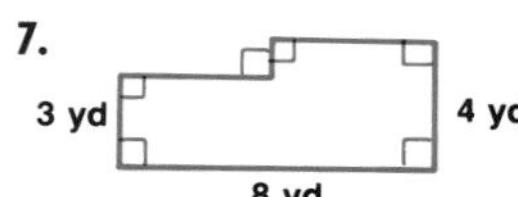

Lesson 3-8 — ***Write in standard form.***

8. 4.6×10^5 **460,000** 9. 7.01×10^3 **7,010** 10. 9×10^4 **90,000** 11. 2.83×10^2 **283**

Lesson 5-8 — ***Change each fraction to a decimal. Use bar notation to show a repeating decimal.***

12. $\frac{1}{5}$ **0.2** 13. $\frac{2}{3}$ **$0.\overline{6}$** 14. $\frac{5}{8}$ **0.625** 15. $\frac{1}{6}$ **$0.1\overline{6}$** 16. $\frac{8}{9}$ **$0.\overline{8}$**

Lesson 7-10 — 17. The number 13 is to 169 as 18 is to what number? **234**

Lesson 9-5 — 18. Darnell is 6 ft tall and his shadow is 15 ft long. Mark is standing next to him, and his shadow is 13 ft long. How tall is Mark? Round to the nearest inch. **5 ft 2 in.**

Lesson 11-8 — ***Find the median and mode of the data in each stem-and-leaf-plot.***

19. 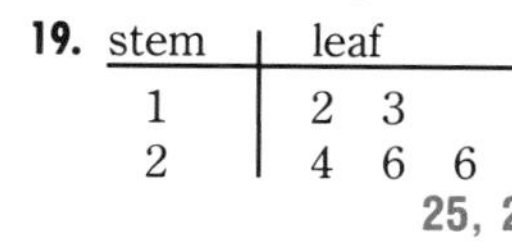

stem	leaf
1	2 3
2	4 6 6 7

25, 26

20. 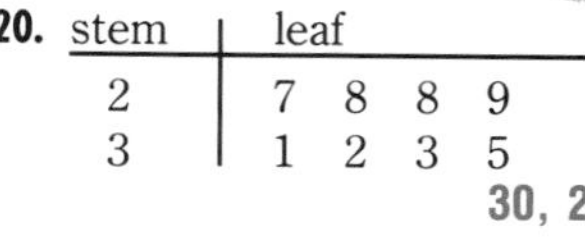

stem	leaf
2	7 8 8 9
3	1 2 3 5

30, 28

Lessons 13-1, 13-2 — ***Solve each equation.***

21. $x + 3 = -6$ **−9** 22. $4x = -20$ **−5**

23. $-5p = 10$ **−2** 24. $-3 = y + 1$ **−4**

25. 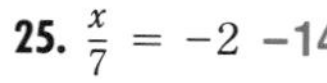 $\frac{x}{7} = -2$ **−14** 26. $\frac{y}{20} = -\frac{3}{4}$ **−15**

Lesson 14-3 — ***A drum contains forty balls, numbered 1 through 40. All of the even-numbered balls are red. The odd-numbered balls are blue. One ball is chosen. Find each probability.***

27. 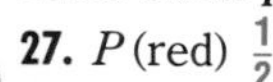 $P(\text{red})$ **$\frac{1}{2}$** 28. $P(\text{a multiple of 7})$ **$\frac{1}{8}$**

29. 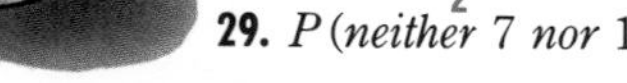 $P(\textit{neither } 7 \textit{ nor } 11)$ **$\frac{19}{20}$** 30. $P(\textit{not} \text{ a } 13)$ **$\frac{39}{40}$**

APPLYING THE LESSON

Independent Practice

Homework Assignment	
Minimum	Ex. 2-30 even
Average	Ex. 1-30
Maximum	Ex. 1-30

Chapter 14

Multiple Choice

Choose the letter of the correct answer for each item.

1. Find the value of $3 \times 4^2 - 2$.
a. 10 **b.** 42 ▶ **c.** 46 **d.** 142

Lesson 1-9

2. Mr. Kasten bought a 13.5-pound turkey for $7.83. What was the price per pound?
e. $0.17
▶ **f.** $0.58
g. $0.60
h. $1.73

Lesson 3-5

3. Twins are born weighing $5\frac{1}{2}$ pounds and $6\frac{3}{8}$ pounds. What is their total weight?
a. $11\frac{2}{5}$ pounds
b. $11\frac{7}{16}$ pounds
c. $11\frac{1}{2}$ pounds
▶ **d.** $11\frac{7}{8}$ pounds

Lesson 4-11

4. Which measurement is equivalent to 750 mL?
▶ **e.** 0.75 L **f.** 0.75 kL **g.** 7.5 L **h.** 7.5 kL

Lesson 6-5

5. How many grams of vitamin granules should Jerry mix with his horse's food if the correct amount is 250 mg for every 50 pounds of the horse's weight? The horse weighs 1,500 pounds.
▶ **a.** 7.5 g **b.** 6 g **c.** 7,500 g **d.** 6,000 g

Lesson 9-3

6. Solve the equation $x + 12 = -4$.
e. 8 ▶ **f.** −16 **g.** −8 **h.** 20

Lesson 13-1

7. Use the bar graph. How many runs were scored by the Braves?
a. 3
b. 12
▶ **c.** 15
d. 21

Runs Scored in 1 Week

Number: 0, 5, 10, 15, 20, 25

Team: Reds, Braves, Padres, Dodgers, Astros, Giants

Lesson 11-4

8. A $10.50 hat is on sale for 20% off. What is the sale price?
▶ **e.** $8.40
f. $10.30
g. $10.80
h. $12.60

Lesson 10-7

9. Jim makes 5 out of every 8 free throws. What is the probability that he will make a free throw?
a. 60% **b.** 37.5% **c.** 56% ▶ **d.** 62.5%

Lesson 14-2

10. If you buy 3 gallons of paint for $12.95 a gallon and then get a fourth gallon free, what is the percent of discount on 4 gallons of paint?
e. 18% ▶ **f.** 25% **g.** 35% **h.** 50%

Lesson 10-7

APPLYING THE LESSON

Independent Practice

Homework Assignment	
Minimum	Ex. 1-10
Average	Ex. 1-10
Maximum	Ex. 1-10

Using the Cumulative Test

This test serves to familiarize students with standardized format while testing skills and concepts presented up to this point.

Evaluation Masters Booklet, pp. 66-67

Name ______________ Date ______________

CHAPTER 14 TEST, FORM 1

Write the letter of the correct answer on the blank at the right of the page.

1. Use multiplication to find the number of possible outcomes for tossing a dime and rolling a die.
A. 2 **B.** 4 **C.** 8 **D.** 12 — 1. D

2. Use multiplication to find the number of possible outcomes for a choice of black, brown, blue, or white pants with a choice of a white, blue, or red shirt.
A. 12 **B.** 10 **C.** 7 **D.** 6 — 2. A

3. Use multiplication to find the number of possible outcomes for a choice of three gifts, a choice of two boxes, and a choice of three bows.
A. 8 **B.** 9 **C.** 18 **D.** 24 — 3. C

4. Two dice are rolled, one at a time. Find the probability of rolling a 5 on the first die and a 2 on the second. Round to the nearest whole percent.
A. 3% **B.** 17% **C.** 33% **D.** 100% — 4. A

5. Refer to Problem 4. Find the probability of rolling both numbers less than 4. Round to the nearest whole percent.
A. 13% **B.** 25% **C.** 50% **D.** 75% — 5. B

6. Four red balls numbered 1 through 4 and three blue balls numbered 3 through 5 are placed in a bag. A ball is drawn from the bag. Find the probability of choosing a red or 5 ball.
A. $\frac{2}{7}$ **B.** $\frac{5}{9}$ **C.** $\frac{4}{7}$ **D.** $\frac{5}{7}$ — 6. D

7. Refer to Problem 6. Find the probability of choosing a blue or 4 ball.
A. $\frac{1}{3}$ **B.** $\frac{5}{9}$ **C.** $\frac{4}{7}$ **D.** $\frac{5}{7}$ — 7. C

8. Find the probability of spinning white or 8.
A. $\frac{1}{8}$ **B.** $\frac{3}{8}$ **C.** $\frac{1}{2}$ **D.** $\frac{5}{8}$ — 8. C

9. Find the probability of spinning black or an even number.
A. $\frac{1}{4}$ **B.** $\frac{1}{2}$ **C.** $\frac{3}{4}$ **D.** 1 — 9. C

8 1 2 3 4 5 6 7

EXTRA PRACTICE

Lesson 1-1 *Replace each ■ with a number or a word to make a true sentence.*

1. 40 = ■ tens **4**
2. 600 = ■ hundreds **6**
3. 7,000 = ■ ones **7,000**
4. 9,000 = ■ hundreds **90**
5. 800 = ■ tens **80**
6. 80 = ■ tens **8**
7. 40,000 = ■ thousands **40**
8. 2,000 = ■ ones **2,000**
9. 500 = ■ hundreds **5**
10. 700,000 = ■ ten thousands **70**
11. 60,000 = ■ ten thousands **6**
12. 0.15 = ■ hundredths **15**
13. 0.2705 = ■ ten-thousandths **2,705**
14. 0.064 = ■ thousandths **64**
15. 3.47 = ■ and ■ hundredths **3; 47**
16. 8.0234 = ■ and ■ ten-thousandths **8; 234**
17. 56.071 = ■ and ■ thousandths **56; 71**
18. 0.13 = 13 ■ **hundredths**
19. 0.045 = 45 ■ **thousandths**
20. 2.053 = 2 ■ 53 ■ **and thousandths**
21. 9.0042 = 9 ■ 42 ■ **and ten-thousandths**
22. 57.003 = 57 ■ 3 ■ **and thousandths**

Lesson 1-2 *Find the number named.*

1. 5^2 **25**
2. 7^2 **49**
3. 2^4 **16**
4. 10^3 **1,000**
5. 12^2 **144**
6. 2^5 **32**
7. 9^2 **81**
8. 3^3 **27**
9. 15^2 **225**
10. 10^5 **100,000**
11. 5^3 **125**
12. 13^1 **13**

Write using exponents.

13. 8 × 8 × 8 **8^3**
14. 3 × 3 **3^2**
15. 2 × 2 × 2 × 2 × 2 **2^5**
16. 20 **20^1**
17. 9 × 9 **9^2**
18. 6 × 6 × 6 **6^3**
19. 11 squared **11^2**
20. 2 cubed **2^3**

Lesson 1-3 *Replace each ● with $<$, $>$, or $=$ to make a true sentence.*

1. 0.36 ● 0.63 $<$
2. 1.74 ● 1.7 $>$
3. 4.03 ● 4.003 $>$
4. 0.06 ● 0.066 $<$
5. 10.5 ● 10.05 $>$
6. 3.0 ● 3 $=$
7. 5.632 ● 5.623 $>$
8. 0.423 ● 0.5 $<$
9. 2.020 ● 2.202 $<$
10. 0.93 ● 0.9 $>$
11. 0.205 ● 0.025 $>$
12. 0.46 ● 0.49 $<$
13. 13.100 ● 13.1 $=$
14. 0.062 ● 0.62 $<$
15. 9.99 ● 9.099 $>$
16. 0.030 ● 0.03 $=$
17. 20 ● 19 $>$
18. 79 ● 97 $<$
19. 101 ● 98 $>$
20. 3,654 ● 3,645 $>$
21. 5,274 ● 5,638 $<$
22. 4,951 ● 5,307 $<$
23. 8,674 ● 8,659 $>$
24. 65,870 ● 65,834 $>$
25. 36,924 ● 36,851 $>$
26. 59,430 ● 59,438 $<$
27. 72,619 ● 73,835 $<$
28. 94,367 ● 92,538 $>$
29. 274,910 ● 274,605 $>$

31. four hundred eighty-five 32. ten and one hundred twenty-three thousandths

Lesson 1-4

Round each number to the underlined place-value position.

1. 18 20 **2.** 23 20 **3.** 33 30 **4.** 56 60 **5.** 98 100 **6.** 135 140

7. 834 830 **8.** 923 900 **9.** 1,201 1,200 **10.** 3,125 3,130 **11.** 6,346 6,350 **12.** 9,982 9,980

13. 110,723 110,700 **14.** 14,679 14,680 **15.** 25,620 25,600 **16.** 33,440 33,400 **17.** 38,471 38,500 **18.** 92,187 90,000

19. 0.728 0.7 **20.** 1.475 1.48 **21.** 3.281 3.3 **22.** 5.812 5.812 **23.** 6.777 7 **24.** 0.995 1.00

25. 0.079 0.08 **26.** 0.821 1 **27.** 16.581 16.6 **28.** 21.631 21.6 **29.** 19.741 20 **30.** 9.610 9.6

Write the name for each number in words. 33. one and fifty-three hundredths 34. six hundred seventy-three and two tenths

31. 485 **32.** 10.123 **33.** 1.53 **34.** 673.2 **35.** 4.03

36. 19,002 **37.** 9.0427 **38.** 58.71 **39.** 62.501 **40.** 1.001

35. four and three hundredths 36. nineteen thousand two 37. nine and four hundred twenty-seven ten thousandths 38. fifty-eight and seventy-one hundredths 39. sixty-two and five hundred one thousandths 40. one and one thousandth

Lesson 1-6

Estimate. Answers may vary. Typical answers are given.

1. 36 + 42 = 80 **2.** 56 + 89 = 150 **3.** 62 − 24 = 40 **4.** 91 − 75 = 10 **5.** 300 + 238 = 500 **6.** 539 + 251 = 800

7. 14.5 + 4.2 + 5.3 = 24 **8.** 5.63 + 8.5 + 2.149 = 17 **9.** 1.63 + 7.4 + 0.219 = 9.2 **10.** 0.92 + 0.6 + 0.475 = 2 **11.** \$5.26 − 3.95 = \$1 **12.** 17.3 − 5.86 = 11

13. 5,406 + 788 6,200 **14.** 4,511 − 653 3,800 **15.** 8,783 + 9,963 19,000

16. 19,023 + 4,449 23,000 **17.** 23,861 − 7,139 17,000 **18.** 4,232 − 1,776 2,000

19. 63,792 − 43,065 20,000 **20.** 25,565 + 19,536 50,000 **21.** 9,639 − 183 9,400

22. 14.9 − 9.4 6 **23.** 63.3 − 18.46 45 **24.** 0.473 − 0.29 0.2

25. \$4.19 + \$3.90 + \$1.75 \$10 **26.** 93.9 − 62 32 **27.** \$40 − \$6.80 \$33

Lesson 1-7

Estimate. Answers may vary. Typical answers are given.

1. 25 × 18 = 600 **2.** 92 × 125 = 9,000 **3.** 86 × 325 = 27,000 **4.** 123 × 84 = 8,000 **5.** 631 × 29 = 18,000 **6.** 856 × 51 = 45,000

7. 383 × 121 = 40,000 **8.** 456 × 702 = 350,000 **9.** 1,321 × 412 = 400,000 **10.** 3,502 × 1,619 = 8,000,000 **11.** 8,829 × 7,312 = 63,000,000 **12.** 9,900 × 331 = 3,000,000

13. 63 × 5 300 **14.** 93 × 46 4,500 **15.** 68 × 582 42,000 **16.** 986 × 31 30,000 **17.** 5,463 × 56 300,000

18. 74 × 8,561 630,000 **19.** 463 × 538 250,000 **20.** 965 × 349 300,000 **21.** 3,496 × 580 1,800,000 **22.** 249 × 5,892 1,200,000

23. 636 ÷ 19 30 **24.** 456 ÷ 26 15 **25.** 682 ÷ 65 10 **26.** 935 ÷ 42 23 **27.** 858 ÷ 88 9

28. 2,349 ÷ 23 100 **29.** 5,465 ÷ 38 140 **30.** 1,980 ÷ 82 20 **31.** 7,562 ÷ 64 120 **32.** 9,642 ÷ 55 160

33. 833 ÷ 15 40 **34.** 9,621 ÷ 120 80 **35.** 8,121 ÷ 61 130 **36.** 6,423 ÷ 56 100 **37.** 8,327 ÷ 72 120

38. 439 ÷ 63 7 **39.** 8,444 ÷ 131 80 **40.** 9,731 ÷ 82 125 **41.** 7,176 ÷ 43 180 **42.** 8,133 ÷ 99 80

Lesson 1-9 *Find the value of each expression.*

1. $12 + 3 - 6$ 9
2. $25 + 2 - 3$ 24
3. $19 - 6 + 7$ 20
4. $24 - 19 + 8$ 13
5. $4 \times 3 \div 2$ 6
6. $18 + 4 \div 2$ 20
7. $23 - 7 \times 3$ 2
8. $3 \times 12 + 4$ 40
9. $4 \div 1 + 10$ 14
10. $17 + 7 \times 4$ 45
11. $19 \times 2 + 7$ 45
12. $16 \times 4 \div 8$ 8
13. $16 + 2 - 4$ 14
14. $5 \times 2 + 3$ 13
15. $15 \div 3 - 2$ 3
16. $3 \times 2 \div 3$ 2
17. $5 \times 6 \div 10 + 1$ 4
18. $12 + 6 \div 3 - 5$ 9
19. $7 - 2 \times 8 \div 4$ 3
20. $27 \div 9 \times 2 + 6$ 12
21. $36 \div 9 \times 2 \div 4$ 2
22. $6 \times 7 \div 3 \div 2$ 7
23. $5^2 - 6 \times 2$ 13
24. $17 - 2^3 + 5$ 14
25. $27 \div 3^2 \times 2$ 6
26. $4 + 4^2 \times 2 - 8$ 28
27. $12 \div 3 - 2^2 + 6$ 6
28. $3^3 \times 2 - 5 \times 3$ 39
29. $(10 \times 2) + (7 \times 3)$ 41
30. $(14 \div 7) - (12 \div 6)$ 0
31. $3 \times [4 \times (7 + 1)]$ 96
32. $8 \times [7 + (5 - 1) \div 2]$ 72
33. $10 \times [(4 + 1) \times 6 \div 2]$ 150
34. $6 \times [7 \times (4 \times 4) + 2]$ 684

Lesson 1-10 *Write an expression for each phrase.*

1. 12 more than a number $n + 12$
2. 3 less than a number $n - 3$
3. a number divided by 4 $n \div 4$ or $\frac{n}{4}$
4. a number increased by 7 $n + 7$
5. a number decreased by 12 $n - 12$
6. 8 times a number $8n$
7. the product of a number and 2 $2n$
8. a number plus 13 $n + 13$
9. a number subtracted from 21 $21 - n$
10. 15 divided by a number $15 \div n$ or $\frac{15}{n}$
11. 28 multiplied by m $28m$
12. the sum of a number and 33 $n + 33$
13. 54 divided by n $54 \div n$ or $\frac{54}{n}$
14. 18 increased by y $18 + y$
15. q decreased by 20 $q - 20$
16. n times 41 $41n$
17. the product of 17 and b $17b$
18. c subtracted from 36 $36 - c$

Lesson 1-11 *Solve each equation.*

1. $q - 7 = 7$ 14
2. $g - 3 = 10$ 13
3. $b + 7 = 12$ 5
4. $a + 3 = 15$ 12
5. $r - 3 = 4$ 7
6. $t + 3 = 21$ 18
7. $s + 10 = 23$ 13
8. $7 + a = 10$ 3
9. $7 + a = 10$ 3
10. $14 + m = 24$ 10
11. $9 + n = 13$ 4
12. $13 + v = 31$ 18
13. $s - 0.4 = 6$ 6.4
14. $x - 1.3 = 12$ 13.3
15. $y + 3.4 = 18$ 14.6
16. $z + 0.34 = 3.1$ 2.76
17. $0.013 + h = 4.0$ 3.987
18. $63 + f = 71$ 8
19. $7 + g = 91$ 84
20. $19 + j = 29$ 10
21. $z - 12.1 = 14$ 26.1
22. $w - 0.1 = 0.32$ 0.42
23. $v - 18 = 13.7$ 31.7
24. $r - 12.2 = 1.3$ 13.5
25. $s + 1.3 = 18$ 16.7
26. $t + 3.43 = 7.4$ 3.97
27. $x + 7.4 = 23.5$ 16.1
28. $y - 7.1 = 6.2$ 13.3
29. $p + 3.1 = 18$ 14.9
30. $q - 2.17 = 21$ 23.17
31. $g - 0.12 = 7.1$ 7.22
32. $h + 15 = 18.4$ 3.4
33. $j - 3 = 7.4$ 10.4
34. $k - 6.23 = 8$ 14.23
35. $m + 6 = 10.01$ 4.01
36. $n - 0.05 = 23$ 23.05

Lesson 1-12 *Solve each equation.*

1. $4x = 36$ **9**
2. $3y = 39$ **13**
3. $4z = 16$ **4**
4. $9w = 54$ **6**
5. $2m = 18$ **9**
6. $42 = 6n$ **7**
7. $72 = 8k$ **9**
8. $20r = 20$ **1**
9. $420 = 5s$ **84**
10. $325 = 25t$ **13**
11. $14 = 2p$ **7**
12. $18q = 36$ **2**
13. $40 = 10a$ **4**
14. $100 = 20b$ **5**
15. $416 = 4c$ **104**
16. $45 = 9d$ **5**
17. $2g = 0.6$ **0.3**
18. $3h = 0.12$ **0.04**
19. $5k = 0.35$ **0.07**
20. $12x = 144$ **12**
21. $\frac{x}{6} = 6$ **36**
22. $\frac{z}{7} = 8$ **56**
23. $\frac{c}{10} = 8$ **80**
24. $\frac{x}{2} = 4$ **8**
25. $\frac{m}{7} = 5$ **35**
26. $\frac{n}{3} = 6$ **18**
27. $\frac{p}{4} = 4$ **16**
28. $\frac{r}{5} = 5$ **25**
29. $\frac{s}{9} = 8$ **72**
30. $\frac{t}{5} = 6$ **30**
31. $\frac{w}{7} = 8$ **56**
32. $\frac{c}{8} = 2$ **16**

Lesson 2-1 *Add.*

1. 38 + 21 = **59**
2. 836 + 74 = **910**
3. 64 + 718 = **782**
4. 439 + 199 = **638**
5. 834 + 689 = **1,523**
6. 593 + 152 = **745**
7. 4,762 + 5,851 = **10,613**
8. 5,967 + 533 = **6,500**
9. 72,752 + 15,601 = **88,353**
10. 6,120 + 19,932 = **26,052**
11. 83,496 + 815 = **84,311**
12. 26 + 85 **111**
13. 65 + 305 **370**
14. 438 + 226 **664**
15. 336 + 986 **1,322**
16. 2,382 + 4,609 **6,991**
17. 6,549 + 385 **6,934**
18. 83,402 + 74,379 **157,781**
19. 2,367 + 40,493 **42,860**
20. 30,260 + 979 **31,239**
21. 4,301 + 7,812 **12,113**
22. 3,111 + 4,222 **7,333**
23. 18,123 + 850 **18,973**
24. 42,017 + 287 **42,304**
25. 51,007 + 987 **51,994**
26. 25,435 + 720 **26,155**

Lesson 2-2 *Add.*

1. 0.46 + 0.72 + 0.81 = **1.99**
2. 13.7 + 2.6 + 4.9 = **21.2**
3. 48.3 + 0.91 + 9.85 = **59.06**
4. \$4.68 + 23.99 + 17.10 = **\$45.77**
5. 15 + 6.02 + 3.8 = **24.82**
6. 8.9 + 12 + 0.38 = **21.28**
7. 5.61 + 0.09 **5.70**
8. 0.38 + 2.46 + 0.19 **3.03**
9. 5.9 + 0.45 + 1.13 **7.48**
10. \$14.70 + \$3.65 + \$2.40 **\$20.75**
11. 17.9 + 18 **35.9**
12. 4.075 + 3.6 + 0.08 **7.755**
13. 3.06 + 0.17 + 0.097 **3.327**
14. 38.786 + 14.5 **53.286**
15. 7.92 + 0.792 + 9.2 **17.912**
16. 0.46 + 0.5 + 5 **5.96**
17. 14 + 7.41 + 0.6 **22.01**
18. 0.446 + 44 + 6.44 **50.886**
19. 0.51 + 1.3 + 6 **7.81**
20. 18 + 9.32 + 2.1 **29.42**
21. 6.7 + 3.2 + 5.05 **14.95**
22. 8.5 + 3.1 + 0.01 **11.61**
23. 19.2 + 7.36 + 4.2 **30.76**
24. 18 + 0.7 + 16.5 **35.2**

Lesson 2-3 *Subtract.*

1. 38 − 25 **13**
2. 436 − 178 **258**
3. 600 − 274 **326**
4. 381 − 56 **325**
5. 515 − 288 **227**
6. 642 − 375 **267**
7. 4,963 − 575 **4,388**
8. 8,902 − 4,266 **4,636**
9. 60,202 − 9,786 **50,416**
10. 36,436 − 8,718 **27,718**
11. 76,814 − 27,925 **48,889**
12. 57 − 38 **19**
13. 706 − 88 **618**
14. 531 − 64 **467**
15. 413 − 326 **87**
16. 633 − 208 **425**
17. 8,920 − 465 **8,455**
18. 3,636 − 1,597 **2,039**
19. 7,437 − 5,548 **1,889**
20. 46,462 − 4,483 **41,979**
21. 5,877 − 4,186 **1,691**
22. 12,145 − 387 **11,758**
23. 19,123 − 6,421 **12,702**
24. 23,411 − 3,418 **19,993**
25. 43,761 − 21,421 **22,340**
26. 56,776 − 32,411 **24,365**
27. 83,976 − 82,135 **1,841**
28. 95,163 − 4,832 **90,331**
29. 123,411 − 95,109 **28,302**

Lesson 2-4 *Subtract.*

1. 0.89 − 0.65 **0.24**
2. 7.3 − 4.9 **2.4**
3. 0.836 − 0.75 **0.086**
4. 6.3 − 4.28 **2.02**
5. 7 − 2.36 **4.64**
6. 4.91 − 2.32 **2.59**
7. 17.83 − 0.24 **17.59**
8. 0.763 − 0.49 **0.273**
9. 2.63 − 1.8 **0.83**
10. 0.563 − 0.08 **0.483**
11. 2.87 − 1.965 **0.905**
12. 5.19 − 0.238 **4.952**
13. 0.8 − 0.526 **0.274**
14. \$30 − \$14.16 **\$15.84**
15. 73 − 0.45 **72.55**
16. 63.6 − 48.48 **15.12**
17. 155 − 6.78 **148.22**
18. 24.96 − 8.088 **16.872**
19. \$12 − \$2.39 **\$9.61**
20. 15.9 − 0.999 **14.901**
21. \$138 − \$19.88 **\$118.12**
22. 19 − 3.076 **15.924**
23. 42.6 − 18.7 **23.9**
24. 23.61 − 14.77 **8.84**
25. 82.063 − 52.13 **29.933**
26. 92.6 − 26.071 **66.529**
27. 52.67 − 18.66 **34.01**
28. 26.007 − 18.3 **7.707**
29. 135.68 − 23.77 **111.91**
30. 25.68 − 2.007 **23.673**
31. 77.077 − 25.1 **51.977**
32. 147.002 − 96.01 **50.992**
33. 45.79 − 2.3 **43.49**

Lesson 2-6 *Write the next three numbers in each sequence.*

1. 14, 21, 28, ?, ?, ? **35; 42; 49**
2. 36, 42, 48, ?, ?, ? **54; 60; 66**
3. 2, 6, 10, ?, ?, ? **14; 18; 22**
4. 18, 16, 14, ?, ?, ? **12; 10; 8**
5. 3, 7, 11, ?, ?, ? **15; 19; 23**
6. 25, 20, 15, ?, ?, ? **10; 5; 0**
7. 80, 70, 60, ?, ?, ? **50; 40; 30**
8. 100, 85, 70, ?, ?, ? **55; 40; 25**
9. 25, 45, 65, ?, ?, ? **85; 105; 125**
10. 6, 12.5, 19, ?, ?, ? **25.5; 32; 38.5**
11. 36, 34, 32, ?, ?, ? **30; 28; 26**
12. 3, 7.2, 11.4, ?, ?, ? **15.6; 19.8; 24**
13. 7.2, 8.3, 9.4, ?, ?, ? **10.5; 11.6; 12.7**
14. 18.6, 18.5, 18.4, ?, ?, ? **18.3; 18.2; 18.1**
15. 93, 193, 293, ?, ?, ? **393; 493; 593**
16. 8, 128, 248, ?, ?, ? **368; 488; 608**

Lesson 2-8 *Find the perimeter of each figure.*

1. 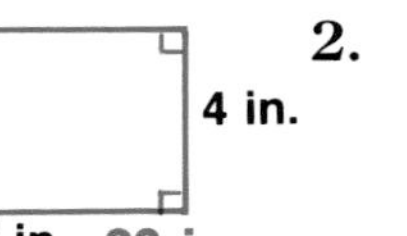

22 in.

2. 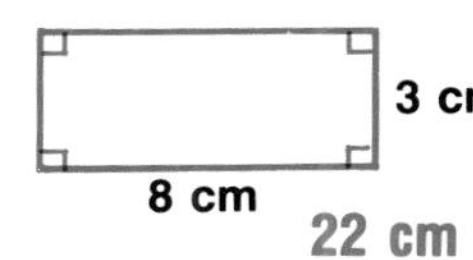

22 cm

3. 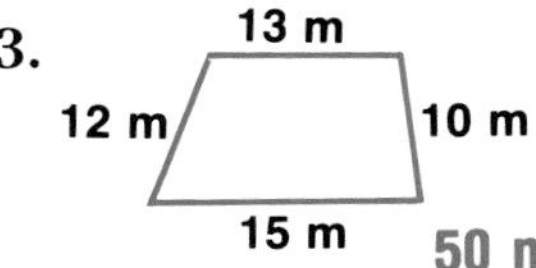

50 m

4. 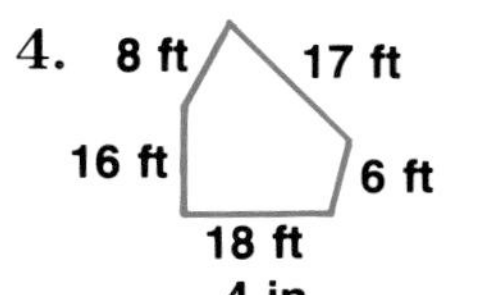

65 ft

5. 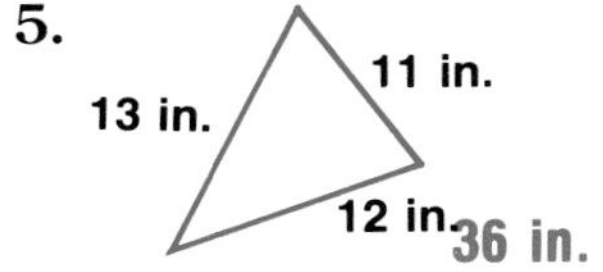

36 in.

6. 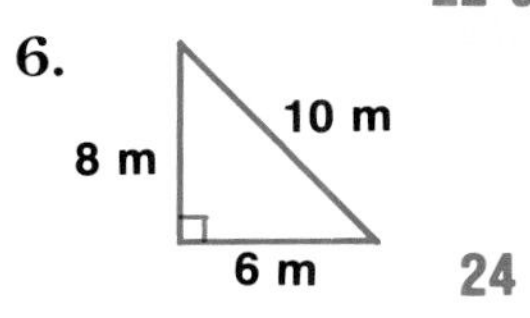

24 m

7. 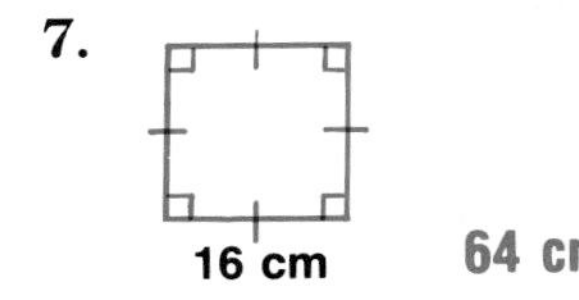

64 cm

8. 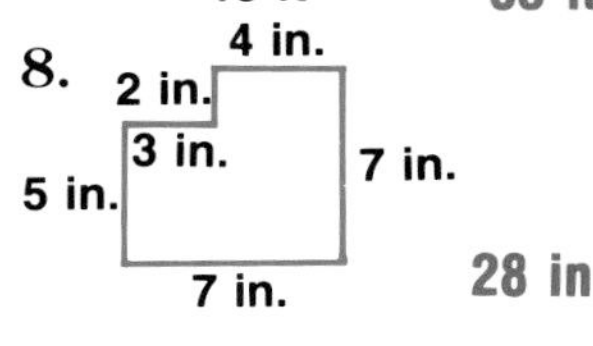

28 in.

9. 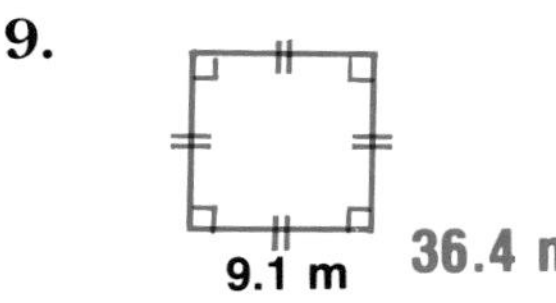

36.4 m

10. 10 yd, 3 yd, 6 yd, 7 yd, 4 yd, 4 yd
34 yd

11. 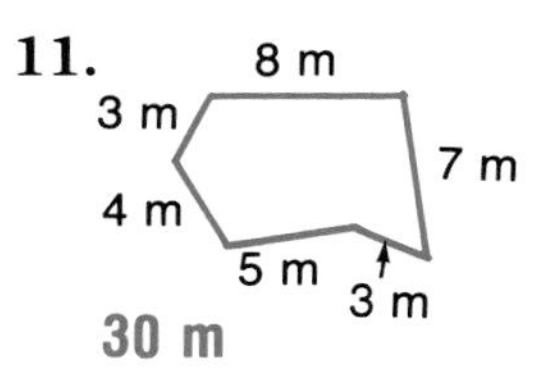

30 m

12. 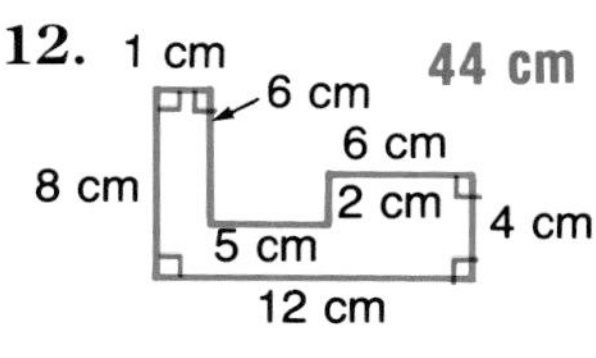

44 cm

Lesson 3-1 *Multiply.*

1. 5,338 × 9 **48,042**
2. 82 × 14 **1,148**
3. 496 × 60 **29,760**
4. 735 × 37 **27,195**
5. 5,302 × 53 **281,006**
6. 3,791 × 68 **257,788**
7. 7,482 × 78 **583,596**
8. 643 × 205 **131,815**
9. 491 × 372 **182,652**
10. 1,445 × 470 **679,150**
11. 2,339 × 608 **1,422,112**
12. 8,516 × 926 **7,885,816**
13. 6 × 36 **216**
14. 834 × 9 **7,506**
15. 7 × 4,566 **31,962**
16. 23 × 82 **1,886**
17. 28 × 416 **11,648**
18. 556 × 28 **15,568**
19. 65 × 790 **51,350**
20. 4,562 × 13 **59,306**
21. 87 × 1,762 **153,294**
22. 39 × 43,625 **1,701,375**
23. 409 × 600 **245,400**
24. 314 × 233 **73,162**
25. 4,820 × 701 **3,378,820**
26. 1,526 × 370 **564,620**
27. 8,614 × 129 **1,111,206**
28. 5,621 × 683 **3,839,143**
29. 3,183 × 425 **1,352,775**
30. 4,606 × 523 **2,408,938**
31. 8,660 × 4,320 **37,411,200**
32. 9,002 × 6,230 **56,082,460**
33. 8,463 × 987 **8,352,981**
34. 7,630 × 5,005 **38,188,150**

Lesson 3-2 *Multiply.*

1. 9.4 × 8 **75.2**
2. 14 × 0.3 **4.2**
3. 4.56 × 7 **31.92**
4. 7.3 × 11 **80.3**
5. $3.57 × 23 **$82.11**
6. 0.236 × 8 **1.888**
7. 0.5 × 0.8 **0.4**
8. 8.76 × 3.9 **34.164**
9. 43.68 × 2.07 **90.4176**
10. 0.004 × 5.6 **0.0224**
11. 4.8 × 32 **153.6**
12. 8.31 × 58 **481.98**
13. 46 × 2.4 **110.4**
14. 5 × 6.125 **30.625**
15. 4.304 × 16 **68.864**
16. $9.14 × 48 **438.72**
17. 346 × 0.67 **231.82**
18. 28 × $0.86 **$24.08**
19. 0.4 × 0.23 **0.092**
20. 4.59 × 0.3 **1.377**
21. 65.1 × 0.38 **24.738**
22. 4.13 × 2.09 **8.6317**
23. 214.2 × 0.8 **171.36**
24. 9.76 × 1.12 **10.9312**
25. 0.07 × 3.28 **0.2296**
26. 14.25 × 0.06 **0.855**
27. 83.3 × 4.5 **374.85**
28. 633.45 × 16.2 **10,261.89**
29. 18.073 × 14.2 **256.6366**
30. 73.45 × 18.73 **1,375.7185**
31. 23.63 × 43.85 **1,036.1755**
32. 96.081 × 19 **1,825.539**

Lesson 3-4 *Divide.*

1. 5)345 69
2. 8)416 52
3. 9)489 54R3
4. 4)2,354 588R2
5. 8)10,036 1,254R4
6. 14)126 9
7. 35)276 7R31
8. 42)2,730 65
9. 58)4,284 73R50
10. 92)8,470 92R6
11. 180)4,500 25
12. 287)3,784 13R53
13. 370 ÷ 5 74
14. 215 ÷ 6 35R5
15. 5,285 ÷ 9 587R2
16. 2,040 ÷ 8 255
17. 780 ÷ 60 13
18. 870 ÷ 36 24R6
19. 567 ÷ 96 5R87
20. 4,089 ÷ 47 87
21. 2,075 ÷ 28 74R3
22. 5,846 ÷ 16 365R6
23. 18,900 ÷ 54 350
24. 12,834 ÷ 22 583R8
25. 3,200 ÷ 200 16
26. 7,626 ÷ 305 25R1
27. 8,502 ÷ 654 13
28. 31,611 ÷ 123 257
29. 4,620 ÷ 400 11R220
30. 8,320 ÷ 35 237R25
31. 8,615 ÷ 56 153R47
32. 4,323 ÷ 173 24R171
33. 7,315 ÷ 26 281R9
34. 6,426 ÷ 12 535R6
35. 4,777 ÷ 77 62R3
36. 5,325 ÷ 70 76R5
37. 1,008 ÷ 58 17R22
38. 7,998 ÷ 2,600 3R198
39. 14,691 ÷ 36 408R3
40. 8,430 ÷ 121 69R81

Lesson 3-5 *Divide.*

1. 6)1.26 0.21
2. 8)23.2 2.9
3. 6)89.22 14.87
4. 15)54.75 3.65
5. 13)128.31 9.87
6. 9)2.583 0.287
7. 47)11.28 0.24
8. 26)32.5 1.25
9. 0.5)18.45 36.9
10. 0.08)5.2 65
11. 2.6)0.65 0.25
12. 1.3)12.831 9.87
13. 0.87)5.133 5.9
14. 2.54)24.13 9.5
15. 3.7)35.89 9.7
16. 14.5)142.1 9.8
17. 7.2 ÷ 8 0.9
18. 0.036 ÷ 9 0.004
19. 1.75 ÷ 7 0.25
20. 167.5 ÷ 25 6.7
21. 37.1 ÷ 14 2.65
22. 5.88 ÷ 0.4 14.7
23. 3.7 ÷ 0.5 7.4
24. 6.72 ÷ 2.4 2.8
25. 41.4 ÷ 18 2.3
26. 9.87 ÷ 0.3 32.9
27. 8.45 ÷ 2.5 3.38
28. 7.8 ÷ 2 3.9
29. 90.88 ÷ 14.2 6.4
30. 33.6 ÷ 8.4 4
31. 25.389 ÷ 4.03 6.3
32. 85.92 ÷ 4.8 17.9
33. 63.18 ÷ 16.2 3.9
34. 18.49 ÷ 4.3 4.3
35. 9.06 ÷ 0.003 3,020
36. 1.02 ÷ 0.3 3.4

Lesson 3-7 *Multiply or divide.*

1. 14.7 × 10 147
2. 3.36 × 10 33.6
3. 8.404 × 100 840.4
4. 7.9 × 100 790
5. 0.49 × 100 49
6. 0.643 × 100 64.3
7. 7.5 × 1,000 7,500
8. 260 × 100 26,000
9. 46.9 × 10^2 4,690
10. 18 × 10^3 18,000
11. 5.95 × 1,000 5,950
12. 0.37 × 10 3.7
13. 0.63 ÷ 10 0.063
14. 14.6 ÷ 100 0.146
15. 3.68 ÷ 10 0.368
16. 17.9 ÷ 100 0.179
17. 23.1 ÷ 1,000 0.0231
18. 436.6 ÷ 100 4.366
19. 42.9 ÷ 1,000 0.0429
20. 0.07 ÷ 10 0.007
21. 4.04 ÷ 10^2 0.0404
22. 80.9 ÷ 10^1 8.09
23. 7.73 ÷ 10^3 0.00773
24. 56,707 ÷ 100 567.07
25. 18.6 × 10 186
26. 25.6 × 100 2,560
27. 0.06 × 1,000 60
28. 83.4 × 10 834
29. 56.8 × 10^3 56,800
30. 6.07 × 10^4 60,700
31. 47.6 ÷ 100 0.476
32. 963 ÷ 1,000 0.963
33. 7.6 ÷ 10^2 0.076
34. 9.6 ÷ 10^3 0.0096
35. 4.63 ÷ 10^3 0.00463
36. 73.3 ÷ 10^2 0.733

Lesson 3-8

Write each in scientific notation.

1. 350 **3.5×10^2**
2. 628 **6.28×10^2**
3. 1,423 **1.423×10^3**
4. 800 **8.0×10^2**
5. 1,600 **1.6×10^3**
6. 3,450 **3.45×10^3**
7. 9,220 **9.22×10^3**
8. 7,100 **7.1×10^3**
9. 19,000 **1.9×10^4**
10. 25,500 **2.55×10^4**
11. 63,350 **6.335×10^4**
12. 123,300 **1.233×10^5**
13. 401,000 **4.01×10^5**
14. 898,000 **8.98×10^5**
15. 923,000 **9.23×10^5**
16. 4,500,000 **4.5×10^6**
17. 18,000,000 **1.8×10^7**
18. 967,000 **9.67×10^5**
19. 42,010,000 **4.201×10^7**
20. 8,000,000 **8.0×10^6**

Write in standard form.

21. 5.23×10^2 **523**
22. 6.8×10^3 **6,800**
23. 8.4×10^6 **8,400,000**
24. 9.1×10^5 **910,000**
25. 7.04×10^3 **7,040**
26. 8.001×10^7 **80,010,000**
27. 4.35×10^4 **43,500**
28. 8.2×10^3 **8,200**
29. 4×10^6 **4,000,000**
30. 6.07×10^2 **607**
31. 8.9×10^8 **890,000,000**
32. 6.7×10^4 **67,000**
33. 3.01×10^5 **301,000**
34. 8.102×10^4 **81,020**
35. 7.65×10^6 **7,650,000**
36. 8.3×10^6 **8,300,000**

Lesson 3-9

Find the common ratio and write the next three terms in each sequence.

1. 2, 6, 18, . . . **54; 162; 486**
2. 3, 9, 27, . . . **81; 243; 729**
3. 100, 50, 25, . . . **12.5; 6.25; 3.125**
4. 180, 144, 115.2, . . . **92.16; 73.728; 58.9824**
5. 5, 25, 125, . . . **625; 3,125; 15,625**
6. 1, 6, 36, . . . **216; 1,296; 7,776**
7. 900, 270, 81, . . . **24.3; 7.29; 2.187**
8. 1,600, 800, 400, . . . **200; 100; 50**
9. 90, 63, 44.1, . . . **30.87; 21.609; 15.1263**
10. 14, 28, 56, . . . **112; 224; 448**
11. 5, 15, 45, . . . **135; 405; 1,215**
12. 625, 125, 25, . . . **5; 1; 0.2**
13. 96, 48, 24, . . . **12; 6; 3**
14. 500, 375, 281.25, . . . **210.9375, 158.20312, 118.65234**
15. 1,530, 459, 137.7, . . . **41.31; 12.393; 3.7179**
16. 1,100, 990, 891, . . . **801.9, 721.71, 649.539**
17. 2, 8, 32, . . . **128; 512; 2,048**
18. 5, 10, 20, . . . **40; 80; 160**

Lesson 3-10

Solve for d, r, or t using the formula $d = rt$.

1. $d = 250$, $r = 50$, $t =$? **5**
2. $d = 300$, $r = 60$, $t =$? **5**
3. $d = 1{,}500$, $r = 60$, $t =$? **25**
4. $d = 2{,}000$, $r = 50$, $t =$? **40**
5. $d = 35$, $r = 35$, $t =$? **1**
6. $d = 1{,}800$, $r = 90$, $t =$? **20**
7. $d = 300$, $r = 2.5$, $t =$? **120**
8. $d = 500$, $t = 20$, $r =$? **25**
9. $d = 3{,}000$, $r = 40$, $t =$? **75**
10. $d = 600$, $t = 15$, $r =$? **40**
11. $r = 55$, $t = 3$, $d =$? **165**
12. $r = 65$, $t = 4$, $d =$? **260**

Solve for r, y, or n using the formula $r = y \div n$.

13. $y = 112$, $n = 16$, $r =$? **7**
14. $y = 231$, $n = 21$, $r =$? **11**
15. $y = 150$, $n = 20$, $r =$? **7.5**
16. $y = 315$, $n = 30$, $r =$? **10.5**

Lesson 3-11 *Find the area of each rectangle.*

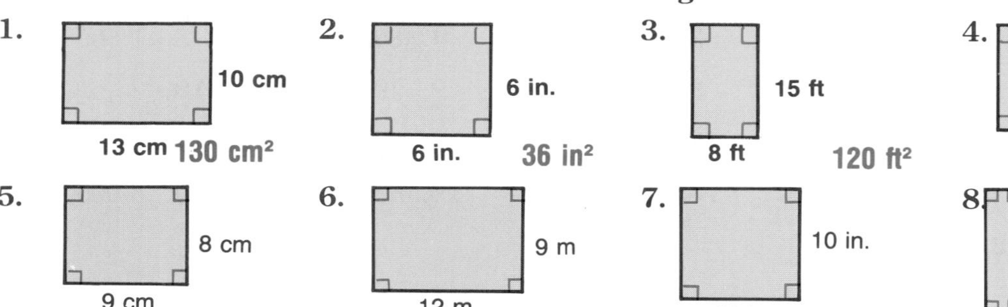

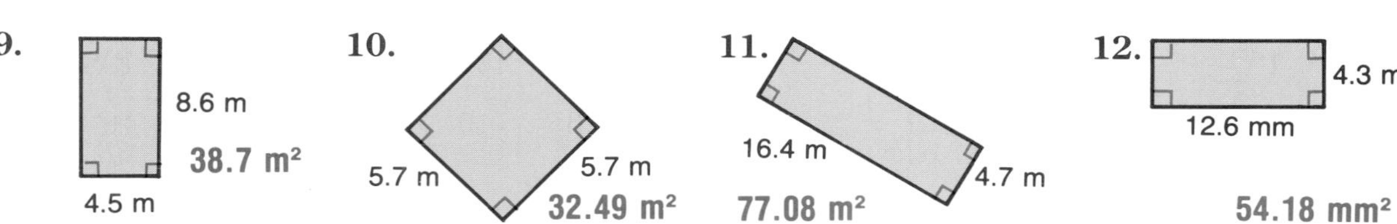

Lesson 3-12 *Find the mean for each set of data. Round to the nearest tenth.*

1. 6, 3, 5, 6, 7, 9 **6**
2. 8, 10, 7, 3, 2, 5, 7 **6**
3. 43, 36, 72, 15, 44 **42**
4. 63, 65, 66, 60, 71 **65**
5. 12, 15, 38, 11, 43, 46 **27.5**
6. 55, 56, 36, 59, 64 **54**
7. 24, 32, 42, 14, 12, 13, 10 **21**
8. 25, 52, 45, 54, 39, 93 **51.3**
9. 204, 430, 680, 190 **376**
10. 563, 560, 565, 600 **572**
11. 2.6, 3.5, 8.2, 5.3 **4.9**
12. 5.7, 7.8, 8.6, 6.5 **7.2**
13. 20.6, 17.9, 18.5 **19**
14. \$6.35, \$7.86, \$5.92, \$4.46 **\$6.10**
15. 103.6, 101.4, 190.8 **131.9**
16. 10.48, 12.62, 17.17 **13.4**
17. 12, 13, 9, 6, 7, 15, 13, 12, 10, 8 **10.5**
18. 23, 36, 65, 58, 70, 60, 30 **48.9**
19. 4, 8, 46, 39, 78, 6, 16, 17, 9, 11 **23.4**
20. 14, 96, 102, 17, 28, 84, 58 **57**
21. 556, 672, 783, 491, 186, 781 **578.2**
22. 93, 781, 430, 80, 760, 490 **439**
23. \$7.50, \$6.90, \$8.30, \$7.70, \$4.50 **\$7.00**
24. 5.8, 7.9, 6.3, 5.7, 7.6, 4.9 **6.4**

5. 1; 2; 3; 5; 6; 10; 15; 30 **8. 1; 2; 3; 4; 5; 6; 8; 10; 12; 15; 20; 24; 30; 40; 60; 120**
9. 1; 2; 3; 4; 5; 6; 10; 12; 15; 20; 30; 60 **10. 1; 2; 3; 4; 6; 8; 12; 16; 24; 32; 48; 96**

Lesson 4-1 *Find all the factors of each number.*

1. 4 **1; 2; 4**
2. 10 **1; 2; 5; 10**
3. 15 **1; 3; 5; 15**
4. 25 **1; 5; 25**
5. 30
6. 55 **1, 5, 11, 55**
7. 42 **1; 2; 3; 6; 7; 14; 21; 42**
8. 120
9. 60
10. 96

State whether each number is divisible by 2, 3, 5, 9, or 10.

11. 42 **2; 3**
12. 55 **5**
13. 63 **3; 9**
14. 72 **2; 3; 9**
15. 99 **3; 9**

Lesson 4-2 *State whether each number is prime or composite.*

1. 7 **P** **2.** 9 **C** **3.** 15 **C** **4.** 17 **P** **5.** 23 **P** **6.** 27 **C**

7. 33 **C** **8.** 39 **C** **9.** 41 **P** **10.** 55 **C** **11.** 27 **C** **12.** 61 **P**

13. 62 **C** **14.** 75 **C** **15.** 77 **C** **16.** 79 **P** **17.** 84 **C** **18.** 89 **P**

Write the prime factorization of each number.

19. 9 **20.** 12 **21.** 16 **22.** 24 **23.** 28 **24.** 31

19. 3×3 **20.** $2 \times 2 \times 3$ **21.** $2 \times 2 \times 2 \times 2$ **22.** $2 \times 2 \times 2 \times 3$ **23.** $2 \times 2 \times 7$ **24.** 1×31

25. 36 **26.** 38 **27.** 45 **28.** 48 **29.** 52 **30.** 56

25. $2 \times 2 \times 3 \times 3$ **26.** 2×19 **27.** $3 \times 3 \times 5$ **28.** $2 \times 2 \times 2 \times 2 \times 3$

31. 120 **32.** 135 **33.** 140 **34.** 144 **35.** 201 **36.** 203

37. 319 **38.** 420 **39.** 444 **40.** 600 **41.** 635 **42.** 725

29. $2 \times 2 \times 13$ **30.** $2 \times 2 \times 2 \times 7$ **31.** $2 \times 2 \times 2 \times 3 \times 5$ **32.** $3 \times 3 \times 3 \times 5$
33. $2 \times 2 \times 5 \times 7$ **34.** $2 \times 2 \times 2 \times 2 \times 3 \times 3$ **35.** 3×67 **36.** 7×29
37. 11×29 **38.** $2 \times 2 \times 3 \times 5 \times 7$ **39.** $2 \times 2 \times 3 \times 37$
40. $2 \times 2 \times 2 \times 3 \times 5 \times 5$ **41.** 5×127 **42.** $5 \times 5 \times 29$

Lesson 4-3 *Find the GCF and the LCM of each group of numbers.*

1. 6, 7 **1; 42** **2.** 3, 12 **3; 12** **3.** 5, 15 **5; 15** **4.** 10, 15 **5; 30** **5.** 8, 16 **8; 16**

6. 9, 12 **3; 36** **7.** 15, 20 **5; 60** **8.** 20, 25 **5; 100** **9.** 18, 36 **18; 36** **10.** 14, 21 **7; 42**

11. 16, 24 **8; 48** **12.** 30, 40 **10; 120** **13.** 35, 49 **7; 245** **14.** 27, 36 **9; 108** **15.** 19, 57 **19; 57**

16. 24, 46 **2; 552** **17.** 56, 16 **8; 112** **18.** 36, 42 **6; 252** **19.** 28, 20 **4; 140** **20.** 21, 35 **7; 105**

21. 3, 6, 12 **3; 12** **22.** 6, 8, 10 **2; 120** **23.** 10, 20, 30 **10; 60** **24.** 12, 16, 20 **4; 240**

25. 4, 8, 20 **4; 40** **26.** 9, 12, 15 **3; 180** **27.** 2, 9, 15 **1; 90** **28.** 2, 8, 13 **1; 104**

29. 4, 7, 9 **1; 252** **30.** 5, 10, 16 **1; 80** **31.** 8, 9, 11 **1; 792** **32.** 7, 9, 11 **1; 693**

33. 8, 12, 20 **4; 120** **34.** 3, 5, 7 **1; 105** **35.** 6, 7, 10 **1; 210** **36.** 10, 12, 15 **1; 60**

Lesson 4-5 *Replace each ■ with a number so that the fractions are equivalent.*

1. $\frac{1}{2} = \frac{\blacksquare}{12}$ **6** **2.** $\frac{3}{4} = \frac{\blacksquare}{16}$ **12** **3.** $\frac{2}{3} = \frac{\blacksquare}{9}$ **6** **4.** $\frac{4}{5} = \frac{\blacksquare}{20}$ **16**

5. $\frac{9}{10} = \frac{\blacksquare}{100}$ **90** **6.** $\frac{8}{12} = \frac{\blacksquare}{3}$ **2** **7.** $\frac{12}{16} = \frac{\blacksquare}{4}$ **3** **8.** $\frac{8}{9} = \frac{\blacksquare}{72}$ **64**

9. $\frac{25}{30} = \frac{\blacksquare}{6}$ **5** **10.** $\frac{15}{20} = \frac{\blacksquare}{4}$ **3** **11.** $\frac{25}{75} = \frac{\blacksquare}{3}$ **1** **12.** $\frac{19}{38} = \frac{\blacksquare}{2}$ **1**

13. $\frac{6}{7} = \frac{\blacksquare}{42}$ **36** **14.** $\frac{36}{40} = \frac{\blacksquare}{10}$ **9** **15.** $\frac{75}{100} = \frac{\blacksquare}{4}$ **3** **16.** $\frac{8}{3} = \frac{\blacksquare}{9}$ **24**

17. $\frac{4}{3} = \frac{\blacksquare}{12}$ **16** **18.** $\frac{50}{25} = \frac{\blacksquare}{1}$ **2** **19.** $\frac{18}{6} = \frac{\blacksquare}{1}$ **3** **20.** $\frac{15}{3} = \frac{\blacksquare}{1}$ **5**

21. $\frac{7}{3} = \frac{\blacksquare}{21}$ **49** **22.** $\frac{75}{70} = \frac{\blacksquare}{14}$ **15** **23.** $\frac{9}{3} = \frac{\blacksquare}{9}$ **27** **24.** $\frac{24}{8} = \frac{\blacksquare}{1}$ **3**

25. $\frac{12}{16} = \frac{\blacksquare}{4}$ **3** **26.** $\frac{4}{1} = \frac{\blacksquare}{4}$ **16** **27.** $\frac{10}{4} = \frac{\blacksquare}{12}$ **30** **28.** $\frac{17}{3} = \frac{\blacksquare}{9}$ **51**

Lesson 4-6

Simplify each fraction.

1. $\frac{4}{6}$ $\frac{2}{3}$ 2. $\frac{5}{10}$ $\frac{1}{2}$ 3. $\frac{10}{15}$ $\frac{2}{3}$ 4. $\frac{16}{24}$ $\frac{2}{3}$ 5. $\frac{7}{14}$ $\frac{1}{2}$ 6. $\frac{7}{21}$ $\frac{1}{3}$

7. $\frac{9}{16}$ $\frac{9}{16}$ 8. $\frac{12}{20}$ $\frac{3}{5}$ 9. $\frac{19}{38}$ $\frac{1}{2}$ 10. $\frac{15}{25}$ $\frac{3}{5}$ 11. $\frac{4}{10}$ $\frac{2}{5}$ 12. $\frac{8}{12}$ $\frac{2}{3}$

13. $\frac{15}{30}$ $\frac{1}{2}$ 14. $\frac{40}{60}$ $\frac{2}{3}$ 15. $\frac{42}{63}$ $\frac{2}{3}$ 16. $\frac{27}{36}$ $\frac{3}{4}$ 17. $\frac{36}{42}$ $\frac{6}{7}$ 18. $\frac{50}{75}$ $\frac{2}{3}$

19. $\frac{4}{8}$ $\frac{1}{2}$ 20. $\frac{3}{15}$ $\frac{1}{5}$ 21. $\frac{4}{20}$ $\frac{1}{5}$ 22. $\frac{5}{30}$ $\frac{1}{6}$ 23. $\frac{2}{20}$ $\frac{1}{10}$ 24. $\frac{11}{44}$ $\frac{1}{4}$

25. $\frac{7}{28}$ $\frac{1}{4}$ 26. $\frac{8}{64}$ $\frac{1}{8}$ 27. $\frac{9}{99}$ $\frac{1}{11}$ 28. $\frac{10}{10}$ 1 29. $\frac{28}{56}$ $\frac{1}{2}$ 30. $\frac{9}{21}$ $\frac{3}{7}$

31. $\frac{18}{24}$ $\frac{3}{4}$ 32. $\frac{18}{54}$ $\frac{1}{3}$ 33. $\frac{9}{63}$ $\frac{1}{7}$ 34. $\frac{10}{80}$ $\frac{1}{8}$ 35. $\frac{40}{90}$ $\frac{4}{9}$ 36. $\frac{36}{72}$ $\frac{1}{2}$

37. $\frac{12}{48}$ $\frac{1}{4}$ 38. $\frac{21}{28}$ $\frac{3}{4}$ 39. $\frac{33}{33}$ 1 40. $\frac{42}{77}$ $\frac{6}{11}$ 41. $\frac{64}{96}$ $\frac{2}{3}$ 42. $\frac{16}{60}$ $\frac{4}{15}$

43. $\frac{14}{30}$ $\frac{7}{15}$ 44. $\frac{12}{64}$ $\frac{3}{16}$ 45. $\frac{15}{63}$ $\frac{5}{21}$ 46. $\frac{17}{34}$ $\frac{1}{2}$ 47. $\frac{18}{22}$ $\frac{9}{11}$ 48. $\frac{22}{55}$ $\frac{2}{5}$

49. $\frac{43}{51}$ $\frac{43}{51}$ 50. $\frac{30}{100}$ $\frac{3}{10}$ 51. $\frac{13}{39}$ $\frac{1}{3}$ 52. $\frac{17}{27}$ $\frac{17}{27}$ 53. $\frac{16}{24}$ $\frac{2}{3}$ 54. $\frac{8}{28}$ $\frac{2}{7}$

Lesson 4-7

Replace each ● with $<$, $>$, or $=$ to make a true sentence.

1. $\frac{2}{3}$ ● $\frac{1}{3}$ $>$ 2. $\frac{3}{4}$ ● $\frac{1}{4}$ $>$ 3. $\frac{5}{11}$ ● $\frac{9}{11}$ $<$ 4. $\frac{1}{6}$ ● $\frac{5}{6}$ $<$ 5. $\frac{5}{8}$ ● $\frac{3}{8}$ $>$

6. $\frac{1}{4}$ ● $\frac{3}{12}$ $=$ 7. $\frac{3}{7}$ ● $\frac{6}{14}$ $=$ 8. $\frac{1}{2}$ ● $\frac{3}{4}$ $<$ 9. $\frac{3}{5}$ ● $\frac{7}{10}$ $<$ 10. $\frac{2}{3}$ ● $\frac{6}{7}$ $<$

11. $\frac{5}{8}$ ● $\frac{4}{5}$ $<$ 12. $\frac{2}{9}$ ● $\frac{1}{3}$ $<$ 13. $\frac{4}{7}$ ● $\frac{4}{11}$ $>$ 14. $\frac{3}{8}$ ● $\frac{3}{9}$ $>$ 15. $\frac{5}{12}$ ● $\frac{7}{16}$ $<$

16. $\frac{3}{3}$ ● $\frac{10}{10}$ $=$ 17. $\frac{1}{6}$ ● $\frac{1}{9}$ $>$ 18. $\frac{5}{24}$ ● $\frac{8}{18}$ $<$ 19. $\frac{3}{4}$ ● $\frac{5}{6}$ $<$ 20. $\frac{4}{9}$ ● $\frac{12}{16}$ $<$

21. $\frac{8}{10}$ ● $\frac{9}{12}$ $>$ 22. $\frac{3}{4}$ ● $\frac{4}{8}$ $>$ 23. $\frac{5}{15}$ ● $\frac{3}{9}$ $=$ 24. $\frac{7}{18}$ ● $\frac{6}{27}$ $>$ 25. $\frac{10}{25}$ ● $\frac{3}{15}$ $>$

27. $\frac{7}{10}$ ● $\frac{6}{5}$ $<$ 28. $\frac{8}{12}$ ● $\frac{6}{10}$ $>$ 29. $\frac{4}{6}$ ● $\frac{3}{8}$ $>$ 30. $\frac{9}{10}$ ● $\frac{11}{12}$ $<$ 31. $\frac{4}{7}$ ● $\frac{8}{11}$ $<$

32. $\frac{12}{15}$ ● $\frac{4}{5}$ $=$ 33. $\frac{9}{13}$ ● $\frac{8}{10}$ $<$ 34. $\frac{7}{22}$ ● $\frac{8}{23}$ $<$ 35. $\frac{5}{17}$ ● $\frac{6}{19}$ $<$ 36. $\frac{10}{24}$ ● $\frac{8}{20}$ $>$

37. $\frac{6}{24}$ ● $\frac{8}{32}$ $=$ 38. $\frac{7}{27}$ ● $\frac{8}{30}$ $<$ 39. $\frac{11}{13}$ ● $\frac{12}{17}$ $>$ 40. $\frac{9}{31}$ ● $\frac{6}{23}$ $>$ 41. $\frac{7}{29}$ ● $\frac{9}{31}$ $<$

Lesson 4-8

Change each number to a mixed number in simplest form.

1. $\frac{7}{3}$ $2\frac{1}{3}$ 2. $\frac{5}{2}$ $2\frac{1}{2}$ 3. $\frac{8}{3}$ $2\frac{2}{3}$ 4. $\frac{9}{4}$ $2\frac{1}{4}$ 5. $\frac{10}{5}$ 2 6. $\frac{8}{4}$ 2

7. $\frac{6}{2}$ 3 8. $\frac{7}{2}$ $3\frac{1}{2}$ 9. $\frac{18}{3}$ 6 10. $\frac{5}{4}$ $1\frac{1}{4}$ 11. $\frac{6}{5}$ $1\frac{1}{5}$ 12. $\frac{21}{4}$ $5\frac{1}{4}$

13. $\frac{23}{5}$ $4\frac{3}{5}$ 14. $\frac{18}{4}$ $4\frac{1}{2}$ 15. $\frac{35}{6}$ $5\frac{5}{6}$ 16. $\frac{25}{3}$ $8\frac{1}{3}$ 17. $\frac{26}{4}$ $6\frac{1}{2}$ 18. $\frac{19}{3}$ $6\frac{1}{3}$

19. $\frac{20}{6}$ $3\frac{1}{3}$ 20. $\frac{25}{7}$ $3\frac{4}{7}$ 21. $\frac{28}{9}$ $3\frac{1}{9}$ 22. $\frac{29}{7}$ $4\frac{1}{7}$ 23. $\frac{18}{7}$ $2\frac{4}{7}$ 24. $\frac{33}{4}$ $8\frac{1}{4}$

25. $\frac{85}{6}$ $14\frac{1}{6}$ 26. $\frac{81}{7}$ $11\frac{4}{7}$ 27. $\frac{65}{6}$ $10\frac{5}{6}$ 28. $\frac{62}{9}$ $6\frac{8}{9}$ 29. $\frac{45}{3}$ 15 30. $\frac{42}{6}$ 7

31. $\frac{95}{3}$ $31\frac{2}{3}$ 32. $\frac{72}{6}$ 12 33. $\frac{65}{4}$ $16\frac{1}{4}$ 34. $\frac{55}{10}$ $5\frac{1}{2}$ 35. $\frac{19}{9}$ $2\frac{1}{9}$ 36. $\frac{46}{7}$ $6\frac{4}{7}$

37. $\frac{53}{11}$ $4\frac{9}{11}$ 38. $\frac{58}{13}$ $4\frac{6}{13}$ 39. $\frac{67}{15}$ $4\frac{7}{15}$ 40. $\frac{72}{19}$ $3\frac{15}{19}$ 41. $\frac{96}{14}$ $6\frac{6}{7}$ 42. $\frac{25}{14}$ $1\frac{11}{14}$

43. $\frac{83}{21}$ $3\frac{20}{21}$ 44. $\frac{76}{31}$ $2\frac{14}{31}$ 45. $\frac{96}{25}$ $3\frac{21}{25}$ 46. $\frac{25}{9}$ $2\frac{7}{9}$ 47. $\frac{76}{22}$ $3\frac{5}{11}$ 48. $\frac{81}{33}$ $2\frac{15}{33}$

Lesson 4-9

Estimate. Answers may vary. Typical answers are given.

1. $\frac{1}{2} + \frac{1}{3}$ 1   2. $\frac{2}{5} + \frac{2}{3}$ $1\frac{1}{2}$ 3. $\frac{4}{5} + \frac{9}{13}$ $1\frac{1}{2}$ 4. $\frac{7}{10} + \frac{7}{8}$ 2

5. $\frac{2}{3} + \frac{10}{11}$ 2   6. $\frac{14}{15} + \frac{19}{20}$ 2   7. $\frac{2}{5} + \frac{1}{8}$ $\frac{1}{2}$ 8. $\frac{9}{10} + \frac{1}{6}$ 1

9. $\frac{8}{9} - \frac{1}{2}$ $\frac{1}{2}$ 10. $\frac{4}{5} - \frac{2}{3}$ $\frac{1}{2}$ 11. $\frac{16}{17} - \frac{4}{7}$ $\frac{1}{2}$ 12. $\frac{7}{16} - \frac{1}{12}$ $\frac{1}{2}$

13. $\frac{5}{8} + \frac{3}{4}$ $1\frac{1}{2}$ 14. $\frac{1}{6} + \frac{3}{8}$ $\frac{1}{2}$ 15. $\frac{7}{9} - \frac{5}{12}$ $\frac{1}{2}$ 16. $\frac{1}{4} + \frac{5}{8}$ $1\frac{1}{2}$

17. $5\frac{1}{4} + 6\frac{2}{3}$ 12   18. $3\frac{1}{4} + 7\frac{3}{7}$ $10\frac{1}{2}$ 19. $10\frac{1}{5} + 11\frac{1}{8}$ 21   20. $14\frac{3}{5} + 7\frac{1}{3}$ $21\frac{1}{2}$

21. $2\frac{7}{8} + 3\frac{9}{10}$ 7   22. $8\frac{1}{5} + 4\frac{1}{9}$ 12   23. $18\frac{2}{7} + 2\frac{4}{9}$ $20\frac{1}{2}$ 24. $12\frac{2}{3} + 14\frac{1}{8}$ 27

25. $7\frac{1}{7} - 4\frac{2}{3}$ 2   26. $16\frac{3}{5} - 14\frac{1}{4}$ $2\frac{1}{2}$ 27. $19\frac{2}{3} - 4\frac{1}{4}$ 16   28. $22\frac{1}{8} - 15\frac{7}{8}$ 6

29. $17 - \frac{5}{6}$ 16   30. $9\frac{5}{7} - 2\frac{1}{3}$ 8   31. $8\frac{9}{11} - 4\frac{2}{5}$ $4\frac{1}{2}$ 32. $6\frac{1}{5} - 2\frac{1}{3}$ $3\frac{1}{2}$

Lesson 4-10

Add.

1. $\frac{5}{11} + \frac{9}{11}$ $1\frac{3}{11}$ 2. $\frac{1}{8} + \frac{5}{8}$ $\frac{3}{4}$ 3. $\frac{7}{10} + \frac{7}{10}$ $1\frac{2}{5}$ 4. $\frac{5}{12} + \frac{9}{12}$ $1\frac{1}{6}$ 5. $\frac{1}{3} + \frac{1}{2}$ $\frac{5}{6}$

6. $\frac{2}{9} + \frac{1}{3}$ $\frac{5}{9}$ 7. $\frac{1}{2} + \frac{3}{4}$ $1\frac{1}{4}$ 8. $\frac{1}{4} + \frac{3}{12}$ $\frac{1}{2}$ 9. $\frac{3}{7} + \frac{6}{14}$ $\frac{6}{7}$ 10. $\frac{2}{5} + \frac{2}{3}$ $1\frac{1}{15}$

11. $\frac{1}{4} + \frac{3}{5}$ $\frac{17}{20}$ 12. $\frac{4}{9} + \frac{1}{2}$ $\frac{17}{18}$ 13. $\frac{5}{7} + \frac{4}{6}$ $1\frac{8}{21}$ 14. $\frac{3}{4} + \frac{1}{6}$ $\frac{11}{12}$ 15. $\frac{5}{12} + \frac{5}{16}$ $\frac{35}{48}$

16. $\frac{3}{5} + \frac{3}{4}$ $1\frac{7}{20}$ 17. $\frac{2}{3} + \frac{1}{8}$ $\frac{19}{24}$ 18. $\frac{9}{10} + \frac{1}{3}$ $1\frac{7}{30}$ 19. $\frac{8}{15} + \frac{2}{9}$ $\frac{34}{45}$ 20. $\frac{5}{6} + \frac{7}{8}$ $1\frac{17}{24}$

21. $\frac{6}{7} + \frac{6}{9}$ $1\frac{11}{21}$ 22. $\frac{3}{7} + \frac{3}{4}$ $1\frac{5}{28}$ 23. $\frac{5}{7} + \frac{5}{9}$ $1\frac{17}{63}$ 24. $\frac{7}{8} + \frac{5}{6}$ $1\frac{17}{24}$ 25. $\frac{3}{4} + \frac{4}{9}$ $1\frac{7}{36}$

26. $\frac{3}{10} + \frac{24}{25}$ $1\frac{13}{50}$ 27. $\frac{6}{7} + \frac{1}{3}$ $1\frac{4}{21}$ 28. $\frac{7}{9} + \frac{4}{5}$ $1\frac{26}{45}$ 29. $\frac{2}{3} + \frac{7}{10}$ $1\frac{11}{30}$ 30. $\frac{1}{4} + \frac{5}{6}$ $1\frac{1}{12}$

31. $\frac{7}{12} + \frac{11}{18}$ $1\frac{7}{36}$ 32. $\frac{9}{16} + \frac{13}{24}$ $1\frac{5}{48}$ 33. $\frac{8}{15} + \frac{2}{3}$ $1\frac{1}{5}$ 34. $\frac{5}{6} + \frac{13}{24}$ $1\frac{3}{8}$ 35. $\frac{5}{14} + \frac{11}{28}$ $\frac{3}{4}$

36. $\frac{11}{12} + \frac{7}{8}$ $1\frac{19}{24}$ 37. $\frac{3}{8} + \frac{1}{6}$ $\frac{13}{24}$ 38. $\frac{1}{2} + \frac{5}{7}$ $1\frac{3}{14}$ 39. $\frac{4}{9} + \frac{1}{6}$ $\frac{11}{18}$ 40. $\frac{5}{6} + \frac{7}{15}$ $1\frac{3}{10}$

41. $\frac{4}{9} + \frac{1}{6}$ $\frac{11}{18}$ 42. $\frac{5}{12} + \frac{7}{8}$ $1\frac{7}{24}$ 43. $\frac{4}{9} + \frac{5}{18}$ $\frac{13}{18}$ 44. $\frac{1}{2} + \frac{7}{18}$ $\frac{8}{9}$ 45. $\frac{5}{12} + \frac{3}{8}$ $\frac{19}{24}$

46. $\frac{9}{20} + \frac{2}{15}$ $\frac{7}{12}$ 47. $\frac{5}{6} + \frac{4}{5}$ $1\frac{19}{30}$ 48. $\frac{2}{3} + \frac{2}{7}$ $\frac{20}{21}$ 49. $\frac{7}{20} + \frac{4}{5}$ $1\frac{3}{20}$ 50. $\frac{4}{5} + \frac{17}{25}$ $1\frac{12}{25}$

10. $10\frac{4}{5}$ 11. $16\frac{11}{21}$ 12. $29\frac{5}{9}$ 13. $19\frac{1}{24}$ 14. $23\frac{1}{12}$ 15. $17\frac{31}{45}$ 16. $10\frac{19}{24}$ 18. $20\frac{83}{132}$

Lesson 4-11

Add. 19. $15\frac{11}{12}$ 20. $13\frac{17}{20}$ 21. $26\frac{72}{143}$ 22. $36\frac{17}{24}$ 23. $25\frac{11}{28}$

1. $2\frac{1}{3} + 1\frac{1}{3}$ $3\frac{2}{3}$ 2. $5\frac{2}{7} + 2\frac{3}{7}$ $7\frac{5}{7}$ 3. $6\frac{3}{8} + 7\frac{1}{8}$ $13\frac{1}{2}$ 4. $1\frac{3}{4} + 2\frac{1}{4}$ 4   5. $4\frac{6}{7} + 9\frac{6}{7}$ $14\frac{5}{7}$

6. $5\frac{1}{2} + 3\frac{1}{4}$ $8\frac{3}{4}$ 7. $2\frac{2}{3} + 4\frac{1}{9}$ $6\frac{7}{9}$ 8. $7\frac{4}{5} + 9\frac{3}{10}$ $17\frac{1}{10}$ 9. $3\frac{3}{4} + 5\frac{5}{8}$ $9\frac{3}{8}$ 10. $3\frac{2}{5} + 7\frac{6}{15}$

11. $10\frac{2}{3} + 5\frac{6}{7}$ 12. $17\frac{2}{9} + 12\frac{1}{3}$ 13. $6\frac{5}{12} + 12\frac{5}{8}$ 14. $7\frac{1}{4} + 15\frac{5}{6}$ 15. $8\frac{2}{15} + 9\frac{5}{9}$

16. $6\frac{1}{8} + 4\frac{2}{3}$ 17. $7 + 6\frac{4}{9}$ $13\frac{4}{9}$ 18. $8\frac{1}{12} + 12\frac{6}{11}$ 19. $7\frac{2}{3} + 8\frac{1}{4}$ 20. $9\frac{3}{5} + 4\frac{1}{4}$

21. $12\frac{3}{11} + 14\frac{3}{13}$ 22. $21\frac{1}{3} + 15\frac{3}{8}$ 23. $19\frac{1}{7} + 6\frac{1}{4}$ 24. $9\frac{2}{5} + 8\frac{1}{3}$ 25. $12\frac{1}{3} + 6\frac{1}{4}$

26. $21\frac{3}{8} + 17\frac{1}{5}$ 27. $6\frac{2}{5} + 8\frac{1}{9}$ 28. $13\frac{1}{2} + 14\frac{3}{8}$ 29. $23\frac{5}{6} + 2\frac{1}{5}$ 30. $16\frac{4}{7} + 12\frac{1}{8}$

24. $17\frac{11}{15}$ 25. $18\frac{7}{12}$ 26. $38\frac{23}{40}$ 27. $14\frac{23}{45}$ 28. $27\frac{7}{8}$ 29. $26\frac{1}{30}$ 30. $28\frac{39}{56}$

Lesson 4-12

Subtract.

1. $\frac{12}{13} - \frac{7}{13}$ **$\frac{5}{13}$** 2. $\frac{15}{18} - \frac{12}{18}$ **$\frac{1}{6}$** 3. $\frac{10}{14} - \frac{3}{14}$ **$\frac{1}{2}$** 4. $\frac{7}{20} - \frac{5}{20}$ **$\frac{1}{10}$** 5. $\frac{2}{3} - \frac{1}{2}$ **$\frac{1}{6}$**

6. $\frac{5}{9} - \frac{1}{3}$ **$\frac{2}{9}$** 7. $\frac{5}{8} - \frac{2}{5}$ **$\frac{9}{40}$** 8. $\frac{3}{4} - \frac{1}{2}$ **$\frac{1}{4}$** 9. $\frac{7}{8} - \frac{3}{16}$ **$\frac{11}{16}$** 10. $\frac{8}{9} - \frac{2}{6}$ **$\frac{5}{9}$**

11. $\frac{11}{12} - \frac{5}{18}$ **$\frac{23}{36}$** 12. $\frac{5}{6} - \frac{3}{14}$ **$\frac{13}{21}$** 13. $\frac{11}{15} - \frac{7}{25}$ **$\frac{34}{75}$** 14. $\frac{9}{12} - \frac{3}{18}$ **$\frac{7}{12}$** 15. $\frac{7}{9} - \frac{2}{15}$ **$\frac{29}{45}$**

16. $\frac{5}{8} - \frac{3}{5}$ **$\frac{1}{40}$** 17. $\frac{7}{9} - \frac{2}{3}$ **$\frac{1}{9}$** 18. $\frac{13}{16} - \frac{5}{8}$ **$\frac{3}{16}$** 19. $\frac{3}{4} - \frac{7}{12}$ **$\frac{1}{6}$** 20. $\frac{4}{5} - \frac{2}{7}$ **$\frac{18}{35}$**

21. $\frac{7}{8} - \frac{5}{6}$ **$\frac{1}{24}$** 22. $\frac{5}{7} - \frac{1}{4}$ **$\frac{13}{28}$** 23. $\frac{9}{10} - \frac{1}{2}$ **$\frac{2}{5}$** 24. $\frac{8}{9} - \frac{2}{3}$ **$\frac{2}{9}$** 25. $\frac{5}{6} - \frac{1}{3}$ **$\frac{1}{2}$**

26. $\frac{2}{3} - \frac{1}{6}$ **$\frac{1}{2}$** 27. $\frac{9}{16} - \frac{1}{2}$ **$\frac{1}{16}$** 28. $\frac{5}{8} - \frac{11}{20}$ **$\frac{3}{40}$** 29. $\frac{14}{15} - \frac{2}{9}$ **$\frac{32}{45}$** 30. $\frac{1}{4} - \frac{1}{6}$ **$\frac{1}{12}$**

31. $\frac{11}{12} - \frac{5}{6}$ **$\frac{1}{12}$** 32. $\frac{14}{15} - \frac{2}{3}$ **$\frac{4}{15}$** 33. $\frac{13}{16} - \frac{5}{8}$ **$\frac{3}{16}$** 34. $\frac{19}{20} - \frac{2}{5}$ **$\frac{11}{20}$** 35. $\frac{49}{100} - \frac{3}{25}$ **$\frac{37}{100}$**

36. $\frac{4}{5} - \frac{1}{6}$ **$\frac{19}{30}$** 37. $\frac{23}{25} - \frac{27}{50}$ **$\frac{19}{50}$** 38. $\frac{19}{25} - \frac{1}{2}$ **$\frac{13}{50}$** 39. $\frac{5}{6} - \frac{13}{16}$ **$\frac{1}{48}$** 40. $\frac{15}{64} - \frac{7}{32}$ **$\frac{1}{64}$**

1. $1\frac{3}{10}$ 3. $5\frac{1}{6}$ 4. $1\frac{1}{2}$ 6. $6\frac{3}{14}$ 7. $2\frac{3}{4}$ 8. $2\frac{1}{4}$ 11. $5\frac{32}{45}$ 12. $3\frac{11}{20}$ 13. $3\frac{1}{5}$ 14. $23\frac{5}{8}$

Lesson 4-13

Subtract. **16. $4\frac{1}{20}$ 17. $9\frac{3}{20}$ 18. $21\frac{1}{2}$ 19. $11\frac{19}{20}$ 22. $9\frac{1}{20}$ 23. $37\frac{4}{5}$**

1. $2\frac{7}{10} - 1\frac{4}{10}$ 2. $8\frac{6}{7} - 2\frac{5}{7}$ **$6\frac{1}{7}$** 3. $7\frac{5}{12} - 2\frac{3}{12}$ 4. $6\frac{13}{14} - 5\frac{6}{14}$ 5. $13\frac{7}{12} - 9\frac{1}{4}$ **$4\frac{1}{3}$**

6. $9\frac{4}{7} - 3\frac{5}{14}$ 7. $11\frac{2}{3} - 8\frac{11}{12}$ 8. $15\frac{6}{9} - 13\frac{5}{12}$ 9. $3\frac{4}{7} - 1\frac{2}{3}$ **$1\frac{19}{21}$** 10. $7\frac{1}{8} - 4\frac{1}{3}$ **$2\frac{19}{24}$**

11. $18\frac{1}{9} - 12\frac{2}{5}$ 12. $12\frac{3}{10} - 8\frac{3}{4}$ 13. $13\frac{1}{5} - 10$ 14. $29\frac{5}{8} - 6$ 15. $4 - 1\frac{2}{3}$ **$2\frac{1}{3}$**

16. $16\frac{1}{4} - 12\frac{1}{5}$ 17. $15\frac{2}{5} - 6\frac{1}{4}$ 18. $23\frac{1}{2} - 2$ 19. $18\frac{1}{5} - 6\frac{1}{4}$ 20. $23\frac{2}{3} - 4\frac{1}{2}$ **$19\frac{1}{6}$**

21. $5\frac{2}{3} - 3\frac{1}{2}$ **$2\frac{1}{6}$** 22. $16\frac{1}{4} - 7\frac{1}{5}$ 23. $43 - 5\frac{1}{5}$ 24. $16\frac{3}{5} - 7\frac{1}{7}$ 25. $6\frac{1}{2} - 5\frac{1}{4}$ **$1\frac{1}{4}$**

26. $8\frac{3}{5} - 2\frac{1}{5}$ **$6\frac{2}{5}$** 27. $21\frac{5}{8} - 3\frac{1}{4}$ 28. $26\frac{2}{3} - 6\frac{1}{5}$ 29. $8\frac{1}{5} - 4\frac{1}{4}$ **$3\frac{19}{20}$** 30. $6\frac{3}{7} - 2\frac{2}{9}$ **$4\frac{13}{63}$**

31. $14\frac{1}{6} - 3\frac{2}{3}$ 32. $25\frac{4}{7} - 21$ 33. $26 - 4\frac{1}{9}$ 34. $17\frac{3}{9} - 4\frac{3}{5}$ 35. $18\frac{3}{10} - 14\frac{1}{8}$ **$4\frac{7}{40}$**

36. $26\frac{1}{4} - 3$ 37. $19\frac{2}{3} - 3\frac{1}{4}$ 38. $18\frac{1}{9} - 1\frac{3}{7}$ 39. $6 - 4\frac{3}{5}$ **$1\frac{2}{5}$** 40. $12\frac{2}{3} - 10$ **$2\frac{2}{3}$**

24. $9\frac{16}{35}$ 27. $18\frac{3}{8}$ 28. $20\frac{7}{15}$ 31. $10\frac{1}{2}$ 32. $4\frac{4}{7}$ 33. $21\frac{8}{9}$ 34. $12\frac{11}{15}$ 36. $23\frac{1}{4}$ 37. $16\frac{5}{12}$ 38. $16\frac{43}{63}$

Answers may vary. Typical answers are given.

Lesson 5-1

Estimate each product or quotient.

1. $3\frac{1}{8} \times 4\frac{1}{4}$ **12** 2. $7\frac{1}{3} \times 3\frac{3}{4}$ **28** 3. $7\frac{1}{2} \times 8\frac{3}{5}$ **72** 4. $6\frac{1}{9} \times 4\frac{1}{2}$ **30** 5. $7\frac{2}{3} \times 1\frac{3}{4}$ **16**

6. $8\frac{3}{6} \times 5\frac{2}{3}$ **54** 7. $2\frac{3}{8} \times 4\frac{7}{8}$ **10** 8. $6\frac{5}{6} \times 5\frac{1}{4}$ **35** 9. $12 \times 5\frac{6}{7}$ **72** 10. $9\frac{2}{5} \times 2\frac{3}{7}$ **18**

11. $1\frac{3}{10} \times 4\frac{7}{9}$ **5** 12. $7\frac{6}{7} \times 8\frac{1}{5}$ **64** 13. $10\frac{1}{8} \times 5\frac{6}{7}$ 14. $4\frac{1}{10} \times 8\frac{1}{9}$ **32** 15. $1\frac{3}{5} \times 7\frac{6}{9}$ **16**

16. $2\frac{7}{11} \times 3\frac{10}{11}$ 17. $8\frac{2}{3} \times 8\frac{1}{5}$ **72** 18. $2\frac{6}{13} \times 1\frac{7}{10}$ **4** 19. $3\frac{1}{3} \times 3\frac{5}{6}$ **12** 20. $4\frac{2}{7} \times 4\frac{2}{5}$ **16**

21. $3\frac{1}{5} \div 3\frac{1}{9}$ **1** 22. $21\frac{2}{3} \div 2\frac{4}{9}$ **7** 23. $16\frac{1}{4} \div 4\frac{1}{5}$ **4** 24. $25\frac{1}{4} \div 4\frac{2}{3}$ **5** 25. $16\frac{1}{8} \div 3\frac{3}{4}$ **4**

26. $15\frac{1}{4} \div 4\frac{4}{5}$ **3** 27. $12\frac{1}{8} \div 6\frac{1}{9}$ **2** 28. $16\frac{2}{5} \div 8\frac{1}{4}$ **2** 29. $23\frac{2}{3} \div 3\frac{1}{5}$ **8** 30. $20\frac{2}{3} \div 7\frac{1}{4}$ **3**

31. $17\frac{1}{8} \div 5\frac{5}{6}$ **3** 32. $23\frac{5}{8} \div 11\frac{7}{9}$ 33. $19\frac{5}{8} \div 3\frac{2}{3}$ **5** 34. $12\frac{1}{4} \div 11\frac{5}{6}$ 35. $27\frac{5}{6} \div 6\frac{3}{4}$ **4**

13. 60 16. 12 32. 2 34. 1

Lesson 5-2

Multiply.

1. $\frac{2}{3} \times \frac{4}{5}$ **$\frac{8}{15}$** 2. $\frac{1}{6} \times \frac{2}{5}$ **$\frac{1}{15}$** 3. $\frac{4}{9} \times \frac{3}{7}$ **$\frac{4}{21}$** 4. $\frac{5}{12} \times \frac{6}{11}$ **$\frac{5}{22}$** 5. $\frac{7}{10} \times \frac{5}{14}$ **$\frac{1}{4}$**

6. $\frac{3}{8} \times \frac{8}{9}$ **$\frac{1}{3}$** 7. $\frac{3}{5} \times \frac{1}{12}$ **$\frac{1}{20}$** 8. $\frac{2}{5} \times \frac{5}{8}$ **$\frac{1}{4}$** 9. $\frac{7}{15} \times \frac{3}{21}$ **$\frac{1}{15}$** 10. $\frac{6}{10} \times \frac{2}{3}$ **$\frac{2}{5}$**

11. $\frac{5}{6} \times \frac{15}{16}$ **$\frac{25}{32}$** 12. $\frac{6}{14} \times \frac{12}{18}$ **$\frac{2}{7}$** 13. $\frac{2}{3} \times \frac{3}{13}$ **$\frac{2}{13}$** 14. $\frac{4}{9} \times \frac{1}{6}$ **$\frac{2}{27}$** 15. $\frac{1}{5} \times 4$ **$\frac{4}{5}$**

16. $\frac{3}{4} \times \frac{5}{6}$ **$\frac{5}{8}$** 17. $\frac{9}{10} \times \frac{3}{4}$ **$\frac{27}{40}$** 18. $\frac{8}{9} \times \frac{2}{3}$ **$\frac{16}{27}$** 19. $\frac{6}{7} \times \frac{4}{5}$ **$\frac{24}{35}$** 20. $\frac{3}{8} \times \frac{4}{5}$ **$\frac{3}{10}$**

21. $\frac{8}{11} \times \frac{11}{12}$ **$\frac{2}{3}$** 22. $\frac{5}{6} \times \frac{3}{5}$ **$\frac{1}{2}$** 23. $\frac{6}{7} \times \frac{7}{21}$ **$\frac{2}{7}$** 24. $\frac{8}{9} \times \frac{9}{10}$ **$\frac{4}{5}$** 25. $\frac{2}{3} \times \frac{5}{8}$ **$\frac{5}{12}$**

26. $\frac{2}{3} \times \frac{5}{7}$ **$\frac{10}{21}$** 27. $\frac{3}{4} \times \frac{5}{6}$ **$\frac{5}{8}$** 28. $\frac{1}{5} \times \frac{12}{13}$ **$\frac{12}{65}$** 29. $\frac{9}{10} \times \frac{1}{4}$ **$\frac{9}{40}$** 30. $\frac{1}{2} \times \frac{1}{2}$ **$\frac{1}{4}$**

31. $\frac{7}{11} \times \frac{12}{15}$ **$\frac{28}{55}$** 32. $\frac{7}{9} \times \frac{5}{7}$ **$\frac{5}{9}$** 33. $\frac{8}{13} \times \frac{2}{11}$ **$\frac{16}{143}$** 34. $\frac{4}{7} \times \frac{2}{9}$ **$\frac{8}{63}$** 35. $\frac{3}{11} \times \frac{7}{15}$ **$\frac{7}{55}$**

36. $\frac{4}{9} \times \frac{24}{25}$ **$\frac{32}{75}$** 37. $\frac{1}{9} \times \frac{6}{13}$ **$\frac{2}{39}$** 38. $\frac{4}{7} \times 6$ **$3\frac{3}{7}$** 39. $\frac{7}{10} \times 5$ **$3\frac{1}{2}$** 40. $\frac{4}{9} \times 6$ **$2\frac{2}{3}$**

Lesson 5-3

Multiply.

1. $3 \times \frac{1}{9}$ **$\frac{1}{3}$** 2. $5 \times \frac{6}{7}$ **$4\frac{2}{7}$** 3. $\frac{3}{5} \times 15$ **9** 4. $3\frac{1}{2} \times 4\frac{1}{3}$ **$15\frac{1}{6}$** 5. $2\frac{2}{5} \times 1\frac{1}{5}$ **$2\frac{22}{25}$**

6. $3\frac{5}{8} \times 4\frac{1}{2}$ 7. $\frac{4}{5} \times 2\frac{3}{4}$ **$2\frac{1}{5}$** 8. $6\frac{1}{8} \times 5\frac{1}{7}$ **$31\frac{1}{2}$** 9. $2\frac{2}{3} \times 2\frac{1}{4}$ **6** 10. $1\frac{4}{5} \times \frac{3}{5}$ **$1\frac{2}{25}$**

11. $6\frac{2}{3} \times 7\frac{3}{5}$ **$50\frac{2}{3}$** 12. $3\frac{1}{2} \times 2\frac{4}{7}$ **9** 13. $10 \times 2\frac{2}{3}$ 14. $8 \times 7\frac{1}{8}$ **57** 15. $3\frac{5}{6} \times 12$ **46**

16. $5\frac{1}{4} \times 6\frac{2}{3}$ **35** 17. $7\frac{1}{5} \times 3\frac{1}{4}$ **$23\frac{2}{5}$** 18. $8\frac{3}{4} \times 2\frac{2}{5}$ **21** 19. $4\frac{1}{3} \times 2\frac{1}{7}$ **$9\frac{2}{7}$** 20. $8\frac{1}{9} \times 2\frac{1}{4}$ **$18\frac{1}{4}$**

21. $1\frac{3}{5} \times 6\frac{2}{5}$ 22. $8\frac{2}{5} \times 3\frac{4}{7}$ **30** 23. $9\frac{1}{4} \times 3\frac{1}{3}$ **$30\frac{5}{6}$** 24. $4\frac{3}{4} \times 2\frac{2}{3}$ **$12\frac{2}{3}$** 25. $8\frac{3}{4} \times 3\frac{2}{7}$ **$28\frac{3}{4}$**

26. $12\frac{1}{4} \times 1\frac{1}{7}$ 27. $16\frac{1}{5} \times 2\frac{1}{3}$ 28. $3\frac{1}{4} \times 4\frac{1}{6}$ 29. $1\frac{9}{16} \times 4\frac{4}{5}$ **$7\frac{1}{2}$** 30. $9 \times 7\frac{1}{4}$ **$65\frac{1}{4}$**

31. $1\frac{2}{5} \times 6\frac{3}{4}$ **$9\frac{9}{20}$** 32. $5\frac{3}{7} \times 14$ **76** 33. $4\frac{1}{8} \times 2\frac{2}{3}$ **11** 34. $9\frac{1}{4} \times 9$ **$83\frac{1}{4}$** 35. $1\frac{1}{2} \times 1\frac{11}{14}$ **$2\frac{19}{28}$**

36. $11\frac{1}{9} \times 2\frac{1}{8}$ 37. $6\frac{1}{4} \times 2\frac{2}{3}$ **$16\frac{2}{3}$** 38. $7\frac{2}{7} \times 8$ **$58\frac{2}{7}$** 39. $3\frac{1}{5} \times 10$ **32** 40. $6\frac{3}{7} \times 8\frac{2}{5}$ **54**

6. $16\frac{5}{16}$ 13. $26\frac{2}{3}$ 21. $10\frac{6}{25}$ 26. 14 27. $37\frac{4}{5}$ 28. $13\frac{13}{24}$ 36. $23\frac{11}{18}$

Lesson 5-6

Divide.

1. $\frac{2}{3} \div \frac{1}{2}$ **$1\frac{1}{3}$** 2. $\frac{3}{5} \div \frac{2}{5}$ **$1\frac{1}{2}$** 3. $\frac{7}{10} \div \frac{3}{8}$ **$1\frac{13}{15}$** 4. $\frac{5}{9} \div \frac{2}{3}$ **$\frac{5}{6}$** 5. $\frac{7}{12} \div \frac{7}{9}$ **$\frac{3}{4}$**

6. $4 \div \frac{2}{3}$ **6** 7. $8 \div \frac{4}{5}$ **10** 8. $9 \div \frac{5}{9}$ **$16\frac{1}{5}$** 9. $\frac{2}{7} \div 2$ **$\frac{1}{7}$** 10. $\frac{4}{11} \div 4$ **$\frac{1}{11}$**

11. $\frac{1}{14} \div 7$ **$\frac{1}{98}$** 12. $\frac{2}{13} \div \frac{5}{26}$ **$\frac{4}{5}$** 13. $\frac{4}{7} \div \frac{6}{7}$ **$\frac{2}{3}$** 14. $\frac{7}{8} \div \frac{1}{3}$ **$2\frac{5}{8}$** 15. $\frac{10}{11} \div \frac{4}{5}$ **$1\frac{3}{22}$**

16. $15 \div \frac{3}{5}$ **25** 17. $\frac{9}{14} \div \frac{3}{4}$ **$\frac{6}{7}$** 18. $\frac{8}{9} \div \frac{5}{6}$ **$1\frac{1}{15}$** 19. $\frac{4}{9} \div 36$ **$\frac{1}{81}$** 20. $49 \div \frac{13}{14}$

21. $\frac{3}{5} \div \frac{2}{3}$ **$\frac{9}{10}$** 22. $\frac{8}{9} \div \frac{4}{5}$ **$1\frac{1}{9}$** 23. $\frac{3}{4} \div \frac{15}{16}$ **$\frac{4}{5}$** 24. $6 \div \frac{1}{5}$ **30** 25. $\frac{4}{3} \div \frac{1}{5}$ **$6\frac{2}{3}$**

26. $\frac{11}{12} \div \frac{5}{6}$ **$1\frac{1}{10}$** 27. $\frac{9}{10} \div \frac{5}{6}$ **$1\frac{2}{25}$** 28. $\frac{4}{5} \div 8$ **$\frac{1}{10}$** 29. $\frac{9}{11} \div \frac{3}{11}$ **3** 30. $\frac{4}{7} \div \frac{9}{14}$ **$\frac{8}{9}$**

31. $\frac{15}{16} \div 10$ **$\frac{3}{32}$** 32. $\frac{4}{5} \div \frac{7}{10}$ **$1\frac{1}{7}$** 33. $4 \div \frac{1}{3}$ **12** 34. $15 \div \frac{5}{7}$ **21** 35. $\frac{12}{13} \div \frac{11}{13}$ **$1\frac{1}{11}$**

36. $\frac{7}{15} \div \frac{7}{9}$ **$\frac{3}{5}$** 37. $\frac{4}{9} \div \frac{8}{21}$ **$1\frac{1}{6}$** 38. $\frac{5}{12} \div \frac{25}{36}$ **$\frac{3}{5}$** 39. $\frac{11}{12} \div 33$ **$\frac{1}{36}$** 40. $\frac{13}{20} \div \frac{39}{40}$ **$\frac{2}{3}$**

20. $52\frac{10}{13}$

Lesson 5-7

Divide. 31. $1\frac{83}{100}$

1. $\frac{3}{5} \div 1\frac{2}{3}$ **$\frac{9}{25}$**
2. $2\frac{1}{2} \div 1\frac{1}{4}$ **2**
3. $4\frac{3}{5} \div 4\frac{1}{5}$ **$1\frac{2}{21}$**
4. $3\frac{2}{9} \div \frac{3}{4}$ **$4\frac{8}{27}$**
5. $12 \div 2\frac{2}{5}$ **5**
6. $7 \div 4\frac{9}{10}$ **$1\frac{3}{7}$**
7. $5\frac{1}{9} \div 5$ **$1\frac{1}{45}$**
8. $1\frac{3}{7} \div 10$ **$\frac{1}{7}$**
9. $1\frac{3}{4} \div 2\frac{3}{8}$ **$\frac{14}{19}$**
10. $9\frac{1}{3} \div 5\frac{2}{5}$ **$1\frac{59}{81}$**
11. $3\frac{3}{5} \div \frac{4}{5}$ **$4\frac{1}{2}$**
12. $8\frac{2}{5} \div 4\frac{1}{2}$ **$1\frac{13}{15}$**
13. $6\frac{1}{3} \div 2\frac{1}{2}$ **$2\frac{8}{15}$**
14. $5\frac{1}{4} \div 2\frac{1}{3}$ **$2\frac{1}{4}$**
15. $6\frac{1}{5} \div 7\frac{1}{2}$ **$\frac{62}{75}$**
16. $4\frac{1}{8} \div 3\frac{2}{3}$ **$1\frac{1}{8}$**
17. $6\frac{1}{4} \div 2\frac{1}{5}$ **$2\frac{37}{44}$**
18. $2\frac{5}{8} \div \frac{1}{2}$ **$5\frac{1}{4}$**
19. $4\frac{2}{5} \div 1\frac{1}{9}$ **$3\frac{24}{25}$**
20. $6\frac{3}{7} \div 2\frac{1}{2}$ **$2\frac{4}{7}$**
21. $5\frac{2}{3} \div 4$ **$1\frac{5}{12}$**
22. $4\frac{1}{8} \div 1\frac{2}{3}$ **$2\frac{19}{40}$**
23. $12\frac{1}{4} \div 3\frac{1}{2}$ **$3\frac{1}{2}$**
24. $5\frac{1}{2} \div 3\frac{1}{4}$ **$1\frac{9}{13}$**
25. $7 \div 1\frac{1}{4}$ **$5\frac{3}{5}$**
26. $5\frac{1}{2} \div 3\frac{2}{3}$ **$1\frac{1}{2}$**
27. $7\frac{1}{5} \div 2$ **$3\frac{3}{5}$**
28. $9\frac{3}{7} \div 2\frac{1}{5}$ **$4\frac{2}{7}$**
29. $4 \div 3\frac{1}{3}$ **$1\frac{1}{5}$**
30. $6\frac{2}{3} \div 10$ **$\frac{2}{3}$**
31. $12\frac{1}{5} \div 6\frac{2}{3}$
32. $6\frac{2}{3} \div 5\frac{5}{6}$ **$1\frac{1}{7}$**
33. $12\frac{1}{4} \div 8$ **$1\frac{17}{32}$**
34. $7\frac{9}{16} \div 2\frac{3}{4}$ **$2\frac{3}{4}$**
35. $4\frac{3}{8} \div 5$ **$\frac{7}{8}$**
36. $1\frac{5}{6} \div 3\frac{2}{3}$ **$\frac{1}{2}$**
37. $21 \div 5\frac{1}{4}$ **4**
38. $18 \div 2\frac{1}{4}$ **8**
39. $12 \div 3\frac{3}{5}$ **$3\frac{1}{3}$**
40. $16\frac{3}{5} \div 4$ **$4\frac{3}{20}$**

Lesson 5-8

Change each fraction to a decimal. Use bar notation to show a repeating decimal.

1. $\frac{3}{4}$ **0.75**
2. $\frac{5}{8}$ **0.625**
3. $\frac{3}{25}$ **0.12**
4. $\frac{1}{20}$ **0.05**
5. $\frac{5}{11}$ **$0.\overline{45}$**
6. $\frac{11}{18}$ **$0.6\overline{1}$**
7. $\frac{1}{3}$ **$0.\overline{3}$**
8. $\frac{4}{9}$ **$0.\overline{4}$**
9. $\frac{37}{40}$ **0.925**
10. $\frac{1}{15}$ **$0.0\overline{6}$**
11. $\frac{7}{12}$ **$0.58\overline{3}$**
12. $\frac{3}{16}$ **0.1875**
13. $\frac{3}{50}$ **0.06**
14. $\frac{14}{45}$ **$0.3\overline{1}$**
15. $\frac{5}{12}$ **$0.41\overline{6}$**
16. $\frac{1}{16}$ **0.0625**
17. $\frac{10}{33}$ **$0.\overline{30}$**
18. $\frac{4}{15}$ **$0.2\overline{6}$**
19. $\frac{1}{5}$ **0.2**
20. $\frac{6}{11}$ **$0.\overline{54}$**
21. $\frac{4}{9}$ **$0.\overline{4}$**
22. $\frac{7}{10}$ **0.7**
23. $\frac{5}{8}$ **0.625**
24. $\frac{4}{25}$ **0.16**
25. $\frac{9}{16}$ **0.5625**
26. $\frac{17}{20}$ **0.85**
27. $\frac{1}{8}$ **0.125**
28. $\frac{13}{16}$ **0.8125**
29. $\frac{9}{20}$ **0.45**
30. $\frac{8}{30}$ **$0.2\overline{6}$**
31. $\frac{5}{18}$ **$0.2\overline{7}$**
32. $\frac{1}{6}$ **$0.1\overline{6}$**
33. $\frac{2}{3}$ **$0.\overline{6}$**
34. $\frac{7}{8}$ **0.875**
35. $\frac{9}{25}$ **0.36**
36. $\frac{5}{9}$ **$0.\overline{5}$**
37. $\frac{22}{45}$ **$0.4\overline{8}$**
38. $\frac{12}{25}$ **0.48**
39. $\frac{33}{50}$ **0.66**
40. $\frac{16}{25}$ **0.64**
41. $\frac{9}{11}$ **$0.\overline{81}$**
42. $\frac{2}{33}$ **$0.\overline{06}$**
43. $\frac{11}{25}$ **0.44**
44. $\frac{39}{40}$ **0.975**
45. $\frac{15}{16}$ **0.9375**
46. $\frac{5}{12}$ **$0.41\overline{6}$**
47. $\frac{17}{18}$ **$0.9\overline{4}$**
48. $\frac{41}{45}$ **$0.9\overline{1}$**

Lesson 5-9

Change each decimal to a fraction.

1. 0.6 **$\frac{3}{5}$**
2. 0.9 **$\frac{9}{10}$**
3. 0.45 **$\frac{9}{20}$**
4. 0.08 **$\frac{2}{25}$**
5. 0.96 **$\frac{24}{25}$**
6. 0.39 **$\frac{39}{100}$**
7. 0.55 **$\frac{11}{20}$**
8. 0.36 **$\frac{9}{25}$**
9. 0.79 **$\frac{79}{100}$**
10. 0.404 **$\frac{101}{250}$**
11. 0.565 **$\frac{113}{200}$**
12. 0.083 **$\frac{83}{1,000}$**
13. 0.208 **$\frac{26}{125}$**
14. 0.566 **$\frac{283}{500}$**
15. 0.734 **$\frac{367}{500}$**
16. 0.005 **$\frac{1}{200}$**
17. 0.004 **$\frac{1}{250}$**
18. 0.061 **$\frac{61}{1,000}$**
19. 0.072 **$\frac{9}{125}$**
20. 0.009 **$\frac{9}{1,000}$**
21. 0.085 **$\frac{17}{200}$**
22. 0.601 **$\frac{601}{1,000}$**
23. 0.432 **$\frac{54}{125}$**
24. 0.088 **$\frac{11}{125}$**
25. 0.074 **$\frac{37}{500}$**
26. $0.85\frac{5}{7}$ **$\frac{6}{7}$**
27. $0.83\frac{1}{3}$ **$\frac{5}{6}$**
28. $0.66\frac{2}{3}$ **$\frac{2}{3}$**
29. $0.55\frac{5}{9}$ **$\frac{5}{9}$**
30. $0.77\frac{7}{9}$ **$\frac{7}{9}$**
31. $0.28\frac{4}{7}$ **$\frac{2}{7}$**
32. $0.37\frac{1}{2}$ **$\frac{3}{8}$**
33. $0.31\frac{1}{4}$ **$\frac{5}{16}$**
34. $0.22\frac{2}{9}$ **$\frac{2}{9}$**
35. $0.16\frac{2}{3}$ **$\frac{1}{6}$**
36. $0.14\frac{2}{7}$ **$\frac{1}{7}$**
37. $0.85\frac{5}{7}$ **$\frac{6}{7}$**
38. $0.33\frac{1}{3}$ **$\frac{1}{3}$**
39. $0.09\frac{1}{11}$ **$\frac{1}{11}$**
40. $0.11\frac{1}{9}$ **$\frac{1}{9}$**
41. 0.56 **$\frac{14}{25}$**
42. 0.63 **$\frac{63}{100}$**
43. 0.09 **$\frac{9}{100}$**
44. 0.02 **$\frac{1}{50}$**
45. 0.72 **$\frac{18}{25}$**

Lesson 6-1 *Complete.*

1. 1 liter = ? deciliters 10
2. 1 centiliter = ? milliliters 10
3. 1 gram = ? milligrams 1,000
4. 1 dekagram = ? grams 10
5. 1 meter = ? millimeters 1,000
6. 1 hectometer = ? meters 100
7. 1 kilogram = ? grams 1,000
8. 1 millimeter = ? decimeters 0.01

Name the larger unit.

9. 1 meter or 1 decimeter
10. 1 kilogram or 1 gram
11. 1 milligram or 1 gram
12. 1 dekaliter or 1 hectoliter
13. 1 kilometer or 1 decimeter
14. 1 gram or centigram
15. 1 centiliter or 1 deciliter
16. 1 decimeter or 1 dekameter

Lesson 6-2 *Choose the most reasonable measurement.*

1. length of scissors	20 cm	20 m	20 km
2. length of house	15 m	15 cm	15 km
3. height of street sign	3 km	3 m	3 cm
4. width of door	120 mm	120 km	120 cm
5. length of child's shoe	100 mm	100 m	100 cm
6. width of shoelace	5 mm	5 cm	5 m
7. height of a tree	18 m	18 cm	18 km

Use a metric ruler to measure each of the following segments. Give the measurement in centimeters and millimeters.

1. 2.8 cm, 28 mm
2. 4.9 cm, 49 mm
3. 1.7 cm, 17 mm
4. 5.1 cm, 51mm
5. 3.5 cm, 35 mm
6. 4 cm, 40 mm
7. 6.7 cm, 67 mm
8. 5.4 cm, 54 mm

Lesson 6-3 *Complete.*

1. 1 kg = ▒ g 1,000
2. 1 mm = ▒ m 0.001
3. 1 cm = ▒ m 0.01
4. 400 mm = ▒ cm 40
5. 4 km = ▒ m 4,000
6. 660 cm = ▒ m 6.6
7. 0.3 km = ▒ m 300
8. 30 mm = ▒ cm 3
9. 84.5 m = ▒ km 0.0845
10. 4.8 cm = ▒ mm 48
11. 14 m = ▒ mm 14,000
12. 31.8 mm = ▒ m 0.0318
13. 36 km = ▒ m 36,000
14. 1,838 m = ▒ km 1.838
15. 415 m = ▒ cm 41,500
16. 43 mm = ▒ cm 4.3
17. 93 m = ▒ cm 9,300
18. 1.7 m = ▒ cm 170
19. 54 cm = ▒ m 0.54
20. 18 km = ▒ cm 1,800,000
21. 45 cm = ▒ mm 450

Lesson 6-4 *Complete.*

1. 4 kg = ■ g **4,000**
2. 632 mg = ■ g **0.632**
3. 4,497 g = ■ kg **4.497**
4. 15 g = ■ mg **15,000**
5. 30 g = ■ kg **0.030**
6. 3 mg = ■ g **0.003**
7. 7.82 g = ■ mg **7,820**
8. 38.6 kg = ■ g **38,600**
9. 8.9 g = ■ kg **0.0089**
10. 6 kg = ■ g **6,000**
11. 9.5 mg = ■ g **0.0095**
12. 63.4 kg = ■ g **63,400**
13. 12.6 g = ■ mg **12,600**
14. 5.6 g = ■ kg **0.0056**
15. 0.5 kg = ■ g **500**
16. 21 g = ■ mg **21,000**
17. 61.2 mg = ■ g **0.0612**
18. 61 g = ■ mg **61,000**
19. 1.02 kg = ■ g **1,020**
20. 10.3 mg = ■ g **0.0103**
21. 23.5 g = ■ mg **23,500**
22. 53 mg = ■ kg **0.000053**
23. 95 mg = ■ kg **0.000095**
24. 65 mg = ■ g **0.065**
25. 0.51 kg = ■ mg **510,000**
26. 0.63 kg = ■ g **630**
27. 563 g = ■ kg **0.563**
28. 96.4 mg = ■ g **0.0964**
29. 5,347 mg = ■ g **5.347**
30. 732 mg = ■ g **0.732**

Lesson 6-5 *Complete.*

1. 4 L = ■ kL **0.004**
2. 8 kL = ■ L **8,000**
3. 2.9 kL = ■ L **2,900**
4. 12 L = ■ kL **0.012**
5. 5 mL = ■ L **0.005**
6. 13.5 L = ■ kL **0.0135**
7. 1.3 L = ■ kL **0.0013**
8. 6.1 mL = ■ L **0.0061**
9. 3.5 kL = ■ L **3,500**
10. 3,330 L = ■ kL **3.33**
11. 4,301 mL = ■ L **4.301**
12. 16 L = ■ mL **16,000**
13. 0.351 kL = ■ L **351**
14. 16.35 kL = ■ L **16,350**
15. 18.5 mL = ■ L **0.0185**
16. 5.6 mL = ■ L **0.0056**
17. 63 L = ■ mL **63,000**
18. 853 kL = ■ L **853,000**
19. 321 L = ■ kL **0.321**
20. 6.53 kL = ■ L **6,530**
21. 485 mL = ■ L **0.485**
22. 0.538 kL = ■ mL **538,000**
23. 0.721 L = ■ mL **721**
24. 1,471 mL = ■ L **1.471**
25. 3.5 L = ■ kL **0.0035**
26. 402 L = ■ kL **0.402**
27. 6.5 kL = ■ mL **6,500,000**

Lesson 6-7 *Complete.*

1. 7 ft = ■ in. **84**
2. 3 yd = ■ ft **9**
3. 5 yd = ■ in. **180**
4. 5 mi = ■ yd **8,800**
5. 48 in. = ■ ft **4**
6. 31,680 ft = ■ mi **6**
7. $\frac{1}{4}$ mi = ■ ft **1,320**
8. 30 in. = ■ ft **$2\frac{1}{2}$**
9. $3\frac{1}{2}$ yd = ■ in. **126**
10. 69 in. = ■ ft ■ in. **5; 9**
11. 9 ft = ■ in. **108**
12. 24 in. = ■ ft **2**
13. 12 ft = ■ yd **4**
14. 36 ft = ■ yd **12**
15. 36 in. = ■ yd **1**
16. 5 yd = ■ ft **15**
17. 6 yd = ■ in. **216**
18. 2 yd = ■ in. **72**
19. 2 mi = ■ ft **10,560**
20. 1 mi = ■ in. **63,360**
21. 6 ft = ■ in. **72**
22. 72 in. = ■ ft **6**
23. 27 ft = ■ yd **9**
24. 15 ft = ■ yd **5**

Lesson 6-8 *Complete.*

1. 4,000 lb = ▒ T **2**
2. 5 T = ▒ lb **10,000**
3. 2 lb = ▒ oz **32**
4. 12,000 lb = ▒ T **6**
5. $\frac{1}{4}$ lb = ▒ oz **4**
6. 6 lb 2 oz = ▒ oz **98**
7. 122 oz = ▒ lb ▒ oz **7; 10**
8. 24 fl oz = ▒ c **3**
9. 8 pt = ▒ c **16**
10. 10 pt = ▒ qt **5**
11. $2\frac{1}{4}$ c = ▒ fl oz **18**
12. $1\frac{1}{2}$ pt = ▒ c **3**
13. 4 gal = ▒ qt **16**
14. 4 qt = ▒ fl oz **128**
15. 12 pt = ▒ c **24**
16. 9 lb = ▒ oz **144**
17. 15 qt = ▒ gal **$3\frac{3}{4}$**
18. 4 pt = ▒ c **8**
19. 2,000 lb = ▒ T **1**
20. 3 T = ▒ lb **6,000**
21. 6 lb = ▒ oz **96**
22. 2 gal = ▒ fl oz **256**
23. 20 pt = ▒ qt **10**
24. 18 qt = ▒ pt **36**
25. 3 gal = ▒ pt **24**
26. 24 pt = ▒ gal **3**
27. 20 lb = ▒ oz **320**

Lesson 6-9 *Find the equivalent Fahrenheit temperature to the nearest degree.*

1. 20°C **68°F**
2. 15°C **59°F**
3. 85°C **185°F**
4. 90°C **194°F**
5. 100°C **212°F**
6. 45°C **113°F**
7. 62°C **144°F**
8. 19°C **66°F**
9. 2°C **36°F**
10. 5°C **41°F**
11. 18°C **64°F**
12. 22°C **72°F**
13. 26°C **79°F**
14. 30°C **86°F**
15. 35°C **95°F**
16. 27°C **81°F**
17. 36°C **97°F**
18. 46°C **115°F**
19. 48°C **118°F**
20. 52°C **126°F**

Find the equivalent Celsius temperature to the nearest degree.

21. 4°F **−16°C**
22. 8°F **−13°C**
23. 20°F **−7°C**
24. 35°F **2°C**
25. 38°F **3°C**
26. 100°F **38°C**
27. 98°F **37°C**
28. 90°F **32°C**
29. 15°F **−9°C**
30. 150°F **66°C**
31. 180°F **82°C**
32. 21°F **−6°C**
33. 32°F **0°C**
34. 45°F **7°C**
35. 63°F **17°C**
36. 72°F **22°C**
37. 78°F **26°C**
38. 95°F **35°C**
39. 200°F **93°C**
40. 210°F **99°C**

Lesson 6-10 *Complete.*

1. 8 min = ▒ s **480**
2. 120 h = ▒ d **5**
3. 4 h = ▒ min **240**
4. 480 min = ▒ h **8**
5. 600 s = ▒ min **10**
6. 30 min = ▒ h **$\frac{1}{2}$**
7. $\frac{3}{4}$ h = ▒ min **45**
8. 4 h 40 min = ▒ min **280**
9. 9 min 20 s = ▒ s **560**
10. 5 d = ▒ h **120**
11. 12 h = ▒ min **720**
12. 16 h = ▒ min **960**
13. 12 min = ▒ s **720**
14. 360 s = ▒ h **0.1**
15. 4 h = ▒ s **14,400**
16. 12 d = ▒ h **288**
17. 72 h = ▒ d **3**
18. 180 min = ▒ h **3**
19. 12 h = ▒ d **$\frac{1}{2}$**
20. 14 h = ▒ min **840**
21. $5\frac{1}{4}$ h = ▒ min **315**
22. 30 d = ▒ h **720**
23. 20 h = ▒ min **1,200**
24. 360 s = ▒ min **6**
25. 3,600 s = ▒ h **1**
26. 18 min = ▒ s **1080**
27. 5.5 h = ▒ min **330**

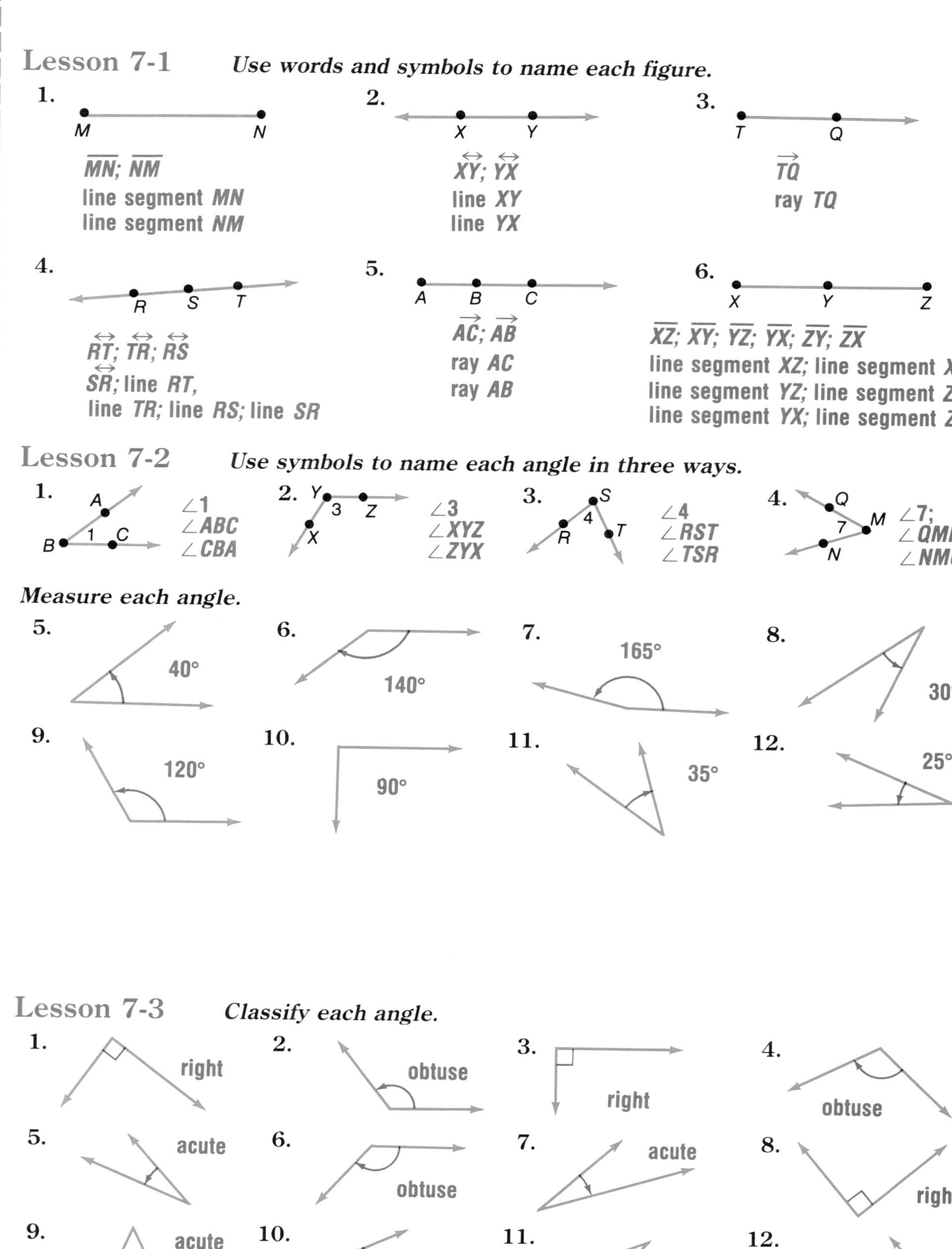

Lesson 7-1 *Use words and symbols to name each figure.*

1. $\overline{MN}$; $\overline{NM}$
line segment *MN*
line segment *NM*

2. $\overleftrightarrow{XY}$; $\overleftrightarrow{YX}$
line *XY*
line *YX*

3. $\overrightarrow{TQ}$
ray *TQ*

4. $\overleftrightarrow{RT}$; $\overleftrightarrow{TR}$; $\overleftrightarrow{RS}$
$\overleftrightarrow{SR}$; line *RT*,
line *TR*; line *RS*; line *SR*

5. $\overrightarrow{AC}$; $\overrightarrow{AB}$
ray *AC*
ray *AB*

6. $\overline{XZ}$; $\overline{XY}$; $\overline{YZ}$; $\overline{YX}$; $\overline{ZY}$; $\overline{ZX}$
line segment *XZ*; line segment *XY*;
line segment *YZ*; line segment *ZX*;
line segment *YX*; line segment *ZY*

Lesson 7-2 *Use symbols to name each angle in three ways.*

1. $\angle 1$
$\angle ABC$
$\angle CBA$

2. $\angle 3$
$\angle XYZ$
$\angle ZYX$

3. $\angle 4$
$\angle RST$
$\angle TSR$

4. $\angle 7$;
$\angle QMN$;
$\angle NMQ$

Measure each angle.

5. 40°
6. 140°
7. 165°
8. 30°
9. 120°
10. 90°
11. 35°
12. 25°

Lesson 7-3 *Classify each angle.*

1. right
2. obtuse
3. right
4. obtuse
5. acute
6. obtuse
7. acute
8. right
9. acute
10. obtuse
11. acute
12. obtuse

Lesson 7-5

State whether each pair of lines is parallel, perpendicular, *or* skew. *Use symbols to name all parallel and perpendicular lines.*

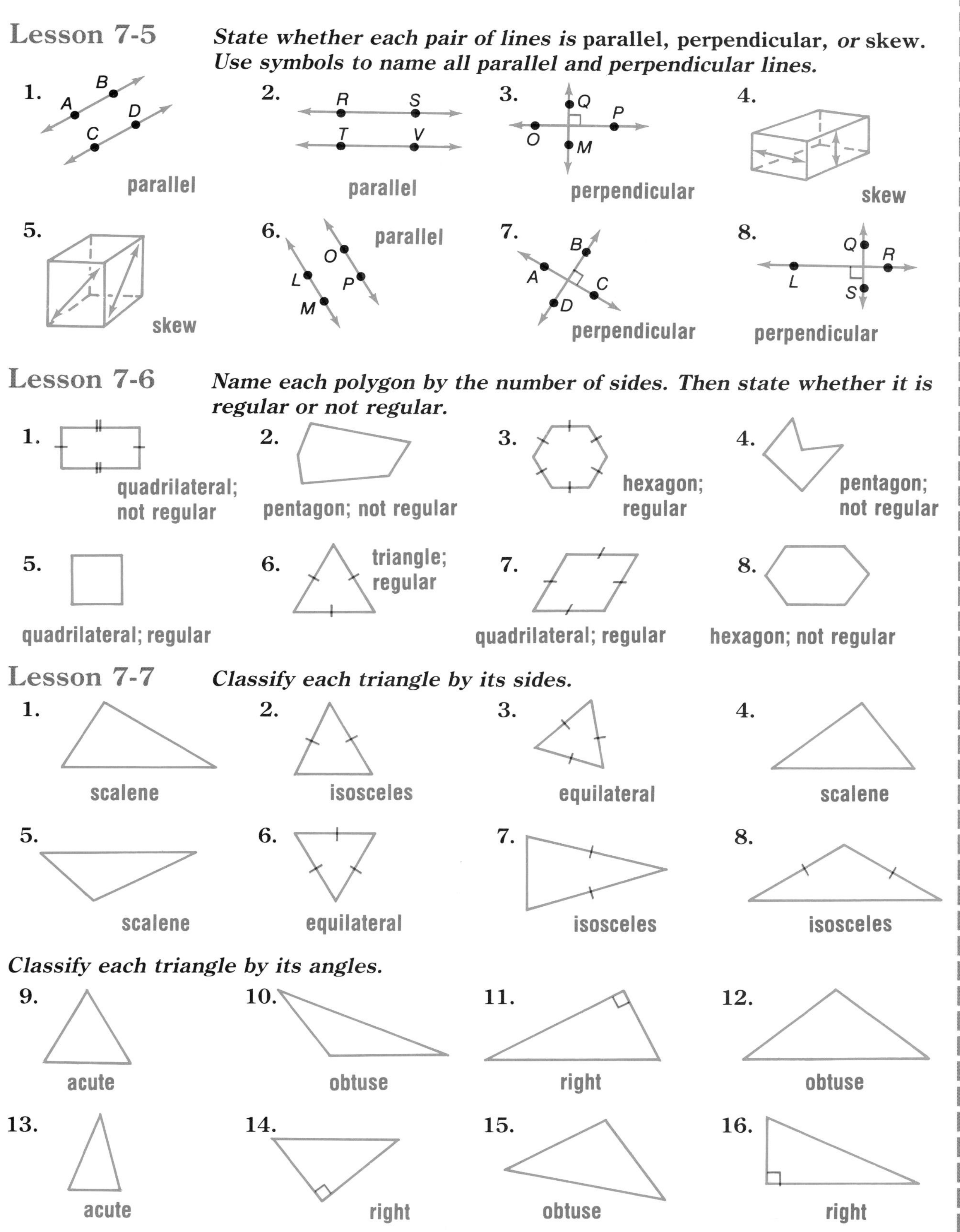

Lesson 7-6

Name each polygon by the number of sides. Then state whether it is regular or not regular.

Lesson 7-7

Classify each triangle by its sides.

Classify each triangle by its angles.

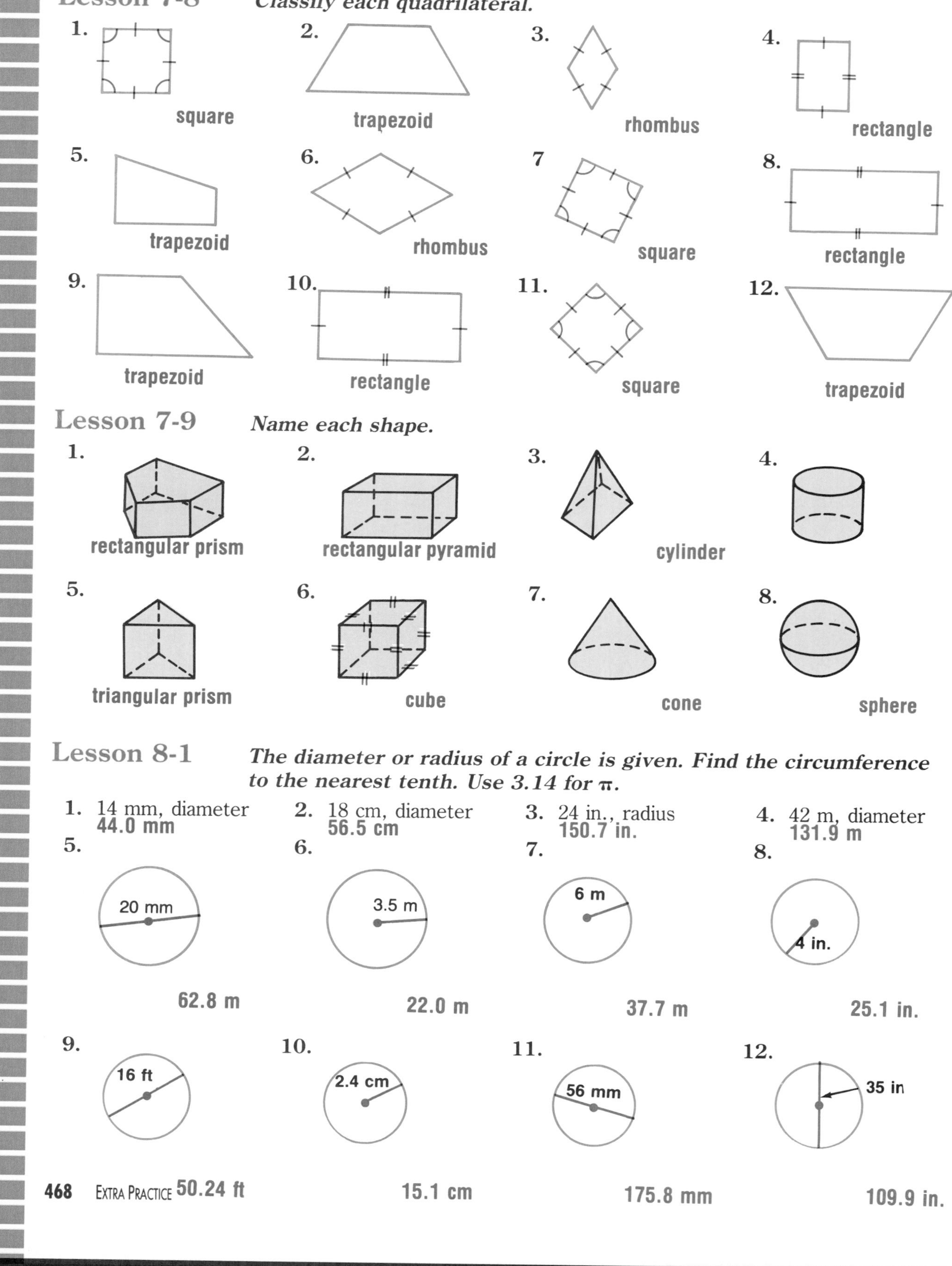

Lesson 7-8 *Classify each quadrilateral.*

1. square
2. trapezoid
3. rhombus
4. rectangle
5. trapezoid
6. rhombus
7. square
8. rectangle
9. trapezoid
10. rectangle
11. square
12. trapezoid

Lesson 7-9 *Name each shape.*

1. rectangular prism
2. rectangular pyramid
3. cylinder
4.
5. triangular prism
6. cube
7. cone
8. sphere

Lesson 8-1 *The diameter or radius of a circle is given. Find the circumference to the nearest tenth. Use 3.14 for π.*

1. 14 mm, diameter **44.0 mm**
2. 18 cm, diameter **56.5 cm**
3. 24 in., radius **150.7 in.**
4. 42 m, diameter **131.9 m**
5. 20 mm — **62.8 m**
6. 3.5 m — **22.0 m**
7. 6 m — **37.7 m**
8. 4 in. — **25.1 in.**
9. 16 ft — **50.24 ft**
10. 2.4 cm — **15.1 cm**
11. 56 mm — **175.8 mm**
12. 35 in — **109.9 in.**

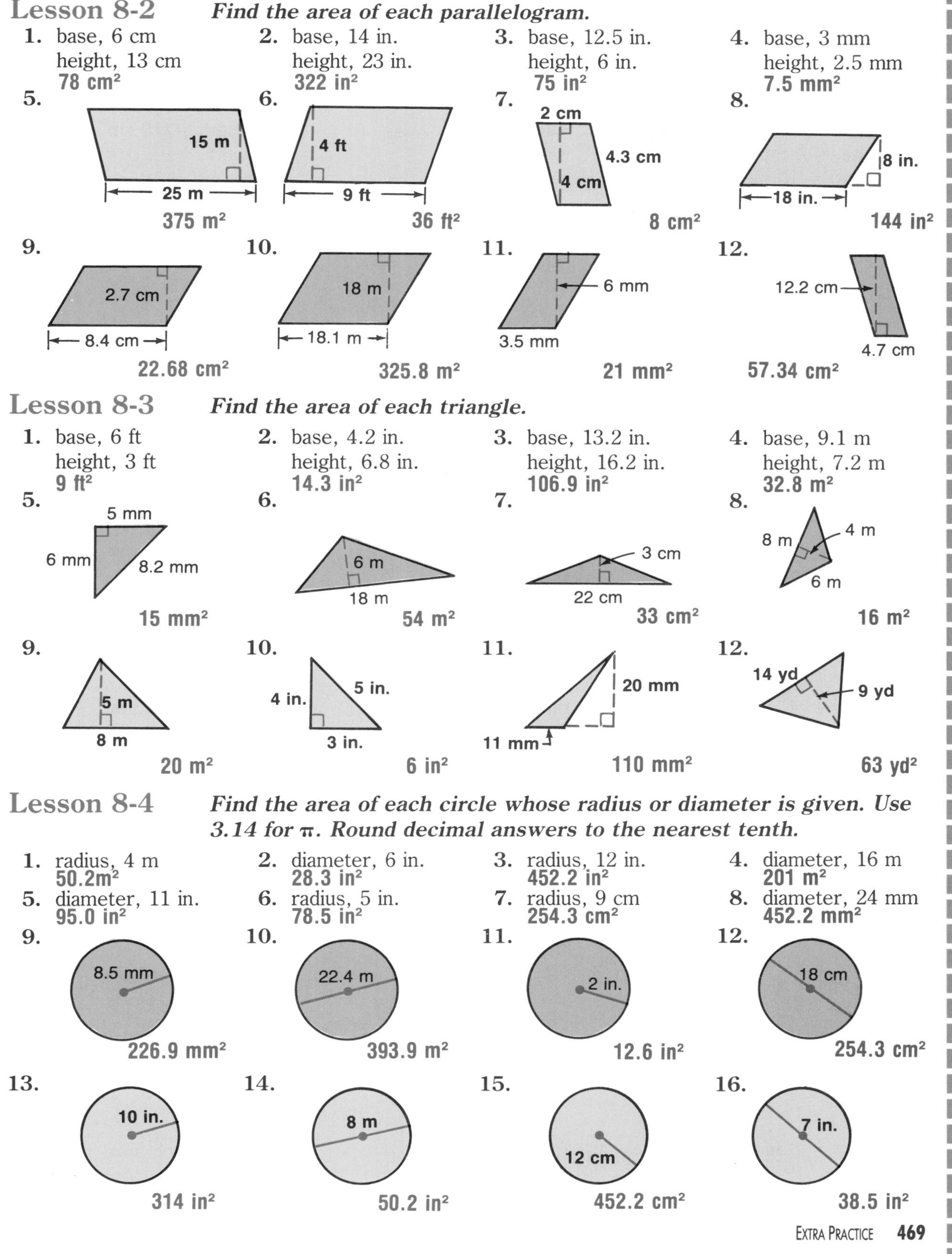

Lesson 8-2 *Find the area of each parallelogram.*

1. base, 6 cm
   height, 13 cm
   78 cm²
2. base, 14 in.
   height, 23 in.
   322 in²
3. base, 12.5 in.
   height, 6 in.
   75 in²
4. base, 3 mm
   height, 2.5 mm
   7.5 mm²
5. 15 m; 25 m
   375 m²
6. 4 ft; 9 ft
   36 ft²
7. 2 cm; 4.3 cm; 4 cm
   8 cm²
8. 8 in.; 18 in.
   144 in²
9. 2.7 cm; 8.4 cm
   22.68 cm²
10. 18 m; 18.1 m
    325.8 m²
11. 6 mm; 3.5 mm
    21 mm²
12. 12.2 cm; 4.7 cm
    57.34 cm²

Lesson 8-3 *Find the area of each triangle.*

1. base, 6 ft
   height, 3 ft
   9 ft²
2. base, 4.2 in.
   height, 6.8 in.
   14.3 in²
3. base, 13.2 in.
   height, 16.2 in.
   106.9 in²
4. base, 9.1 m
   height, 7.2 m
   32.8 m²
5. 5 mm; 6 mm; 8.2 mm
   15 mm²
6. 6 m; 18 m
   54 m²
7. 3 cm; 22 cm
   33 cm²
8. 8 m; 4 m; 6 m
   16 m²
9. 5 m; 8 m
   20 m²
10. 4 in.; 5 in.; 3 in.
    6 in²
11. 20 mm; 11 mm
    110 mm²
12. 14 yd; 9 yd
    63 yd²

Lesson 8-4 *Find the area of each circle whose radius or diameter is given. Use 3.14 for π. Round decimal answers to the nearest tenth.*

1. radius, 4 m
   50.2m²
2. diameter, 6 in.
   28.3 in²
3. radius, 12 in.
   452.2 in²
4. diameter, 16 m
   201 m²
5. diameter, 11 in.
   95.0 in²
6. radius, 5 in.
   78.5 in²
7. radius, 9 cm
   254.3 cm²
8. diameter, 24 mm
   452.2 mm²
9. 8.5 mm
   226.9 mm²
10. 22.4 m
    393.9 m²
11. 2 in.
    12.6 in²
12. 18 cm
    254.3 cm²
13. 10 in.
    314 in²
14. 8 m
    50.2 in²
15. 12 cm
    452.2 cm²
16. 7 in.
    38.5 in²

Lesson 8-7 *Find the surface area for each rectangular prism.*

1. length = 2 in.
   width = 1 in.
   height = 10 in. **64 in²**
2. length = 18 m
   width = 7 m
   height = 14 m **952 m²**
3. length = 2.5 cm
   width = 1 cm
   height = 4.5 cm **36.5 cm²**
4. length = 6 mm
   width = 4 mm
   height = 10 mm **248 mm²**
5. length = 14 in.
   width = 7 in.
   height = 14 in. **784 in²**
6. length = 10 cm
   width = 10 cm
   height = 10 cm **600 cm²**
7. length = 4.5 mm
   width = 3.6 mm
   height = 10.6 mm
   204.12 mm²
8. length = 18 cm
   width = 12 cm
   height = 11 cm
   1,092 cm²
9. length = 12.6 m
   width = 6.8 m
   height = 10.4 m
   574.88 m²

Lesson 8-8 *Find the surface area of each cylinder. Use 3.14 for π. Round decimal answers to nearest tenth.*

1. 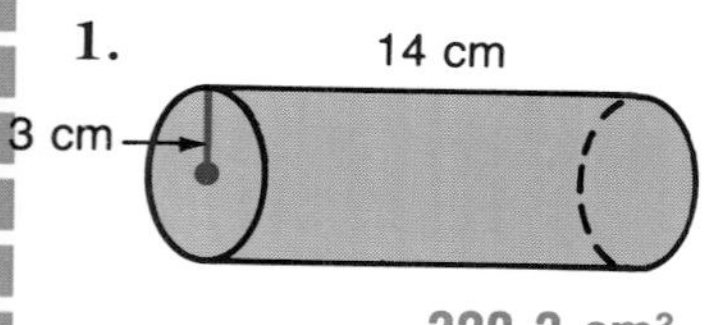

320.3 cm²

2. 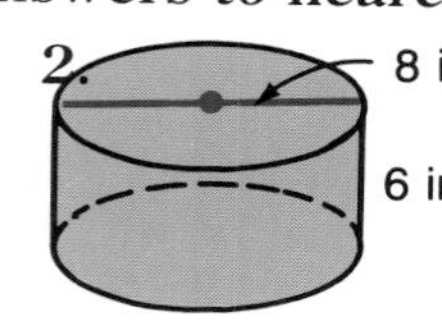

703.4 in²

3. 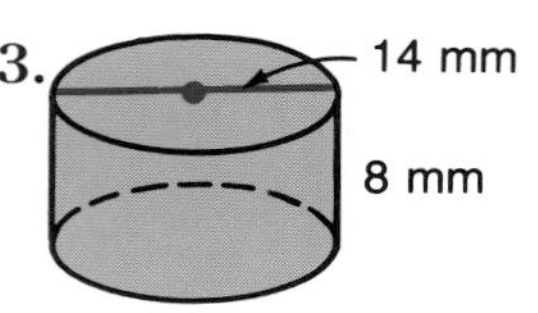

1,934.2 mm²

4. 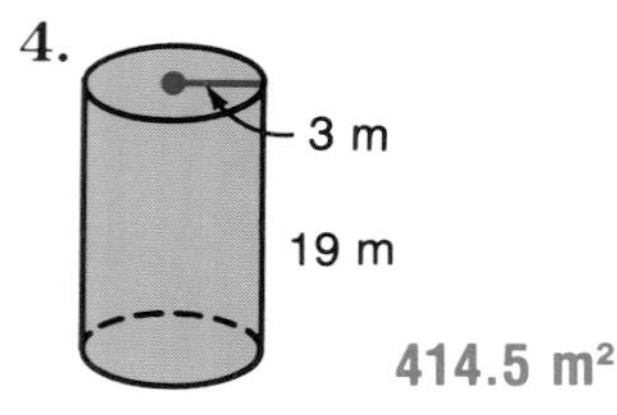

414.5 m²

5. 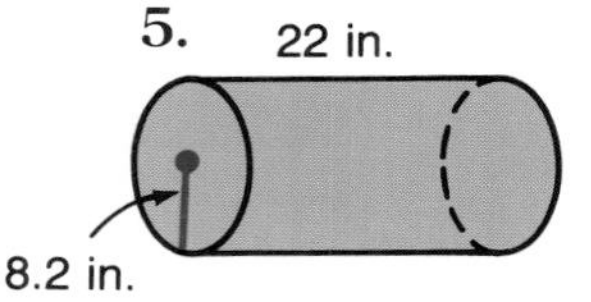

1,555.2 in²

6. 14.2 m
3.6 m
402.4 m²

7. radius = 4.2 cm
   height = 12.4 cm
   437.8 cm²
8. radius = 5 in.
   height = 10 in.
   471 in²
9. radius = 6.3 in.
   height = 4.6 in.
   431.2 in²

Lesson 8-9 *Find the volume of each rectangular prism.*

1. 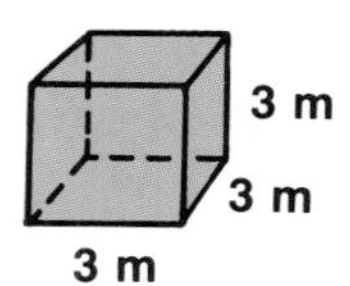

27 m³

2. 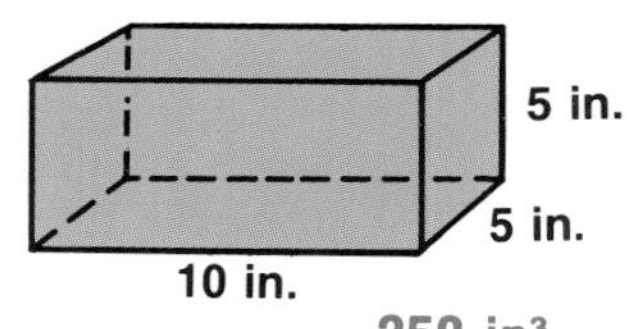

250 in³

3. 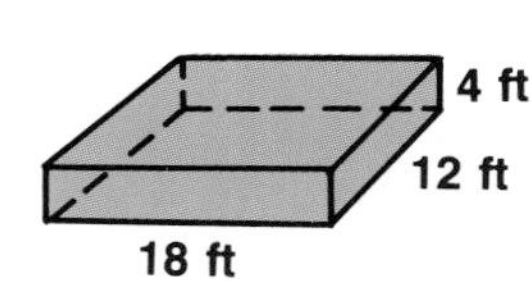

864 ft³

4. 7 cm
9 cm
8 cm
504 cm³

5. 2 in.
14 in.
12 in.
336 in³

6. 4 m
20 m
4 m
320 m³

7. length = 4 mm
   width = 12 mm
   height = 1.5 mm
   72 mm³
8. length = 16 cm
   width = 20 cm
   height = 20.4 cm
   6,528 cm³
9. length = 8.5 m
   width = 2.1 m
   height = 7.6 m
   135.66 m³

Lesson 8-10 *Find the volume of each pyramid.*

1. $\ell = 4$ cm
   $w = 12$ cm
   $h = 6.3$ cm **100.8 cm³**

2. $\ell = 8$ m
   $w = 10$ m
   $h = 2.3$ m **61.3 m³**

3. $\ell = 5$ in.
   $w = 8$ in.
   $h = 12$ in. **160 in³**

4. 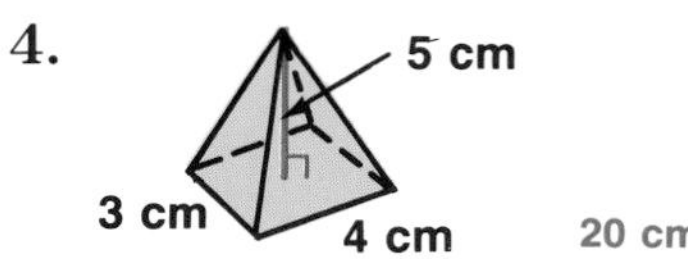

20 cm³

5. 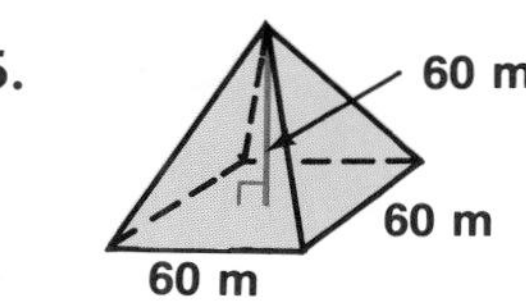

72,000 m³

6. 24 ft
12 ft
10 ft
960 ft³

7. 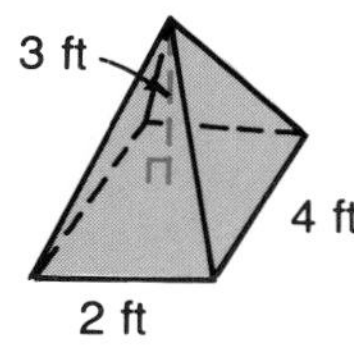

8 ft³

8. 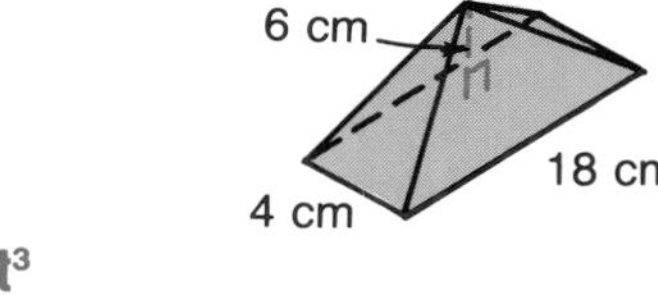

144 cm³

9. 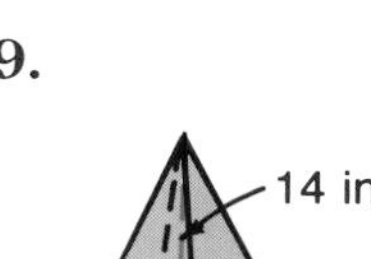

$228\frac{2}{3}$ in³

Lesson 8-11 *Find the volume of each cylinder. Use 3.14 for π. Round to the nearest tenth.*

1. $r = 6$ cm
   $h = 12$ cm **1,356.5 cm³**

2. $r = 4$ m
   $h = 12$ m **602.9 m³**

3. $r = 3.5$ mm
   $h = 4.2$ mm **161.6 mm³**

4. $r = 9$ in.
   $h = 13$ in. **3,306.4 in³**

5. $r = 15$ mm
   $h = 20$ mm **14,130 mm³**

6. $r = 8$ in.
   $h = 11$ in. **2,210.6 in³**

7. 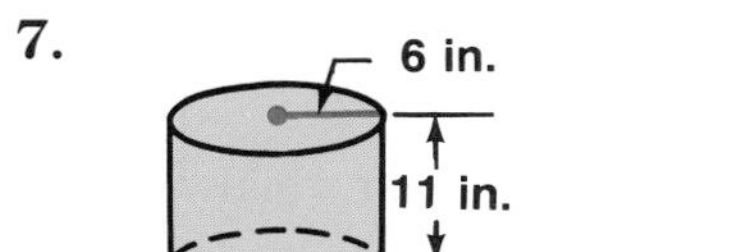

1,243.44 in³

8. 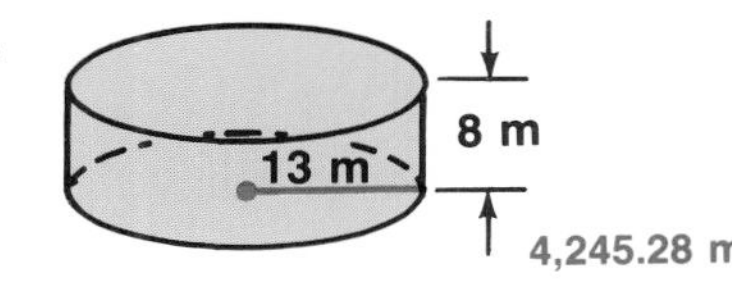

4,245.28 m³

9. 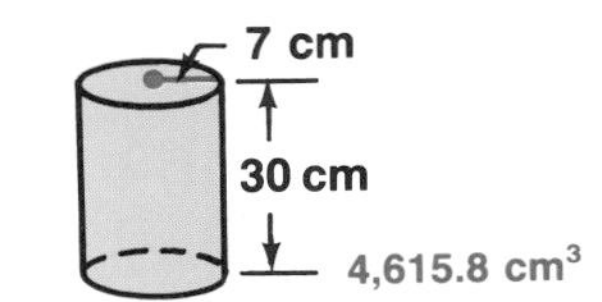

4,615.8 cm³

Lesson 8-12 *Find the volume of each cone. Use 3.14 for π. Round decimal answers to nearest tenth.*

1. $r = 6$ m
   $h = 10$ m **376.8 m³**

2. $r = 5$ in.
   $h = 13$ in. **340.2 in³**

3. $r = 1$ cm
   $h = 8.2$ cm **8.6 cm³**

4. $r = 8$ mm
   $h = 4.5$ mm **301.4 mm³**

5. $r = 4$ cm
   $h = 7.2$ cm **120.6 cm³**

6. $r = 9$ in.
   $h = 14.2$ in. **1,203.9 in³**

7. $r = 3$ cm
   $h = 7$ cm **65.9 cm³**

8. $r = 7$ m
   $h = 12.7$ m **651.3 m³**

9. $r = 2$ mm
   $h = 6.4$ mm **26.8 mm³**

10. 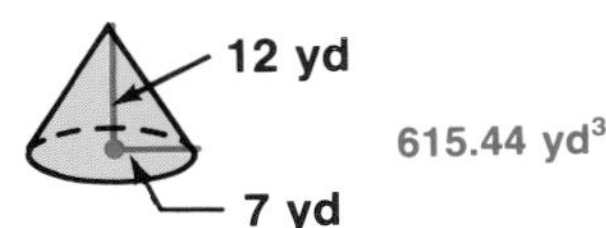

615.44 yd³

11. 15 in.
11 in.
1,899.7 in³

12. 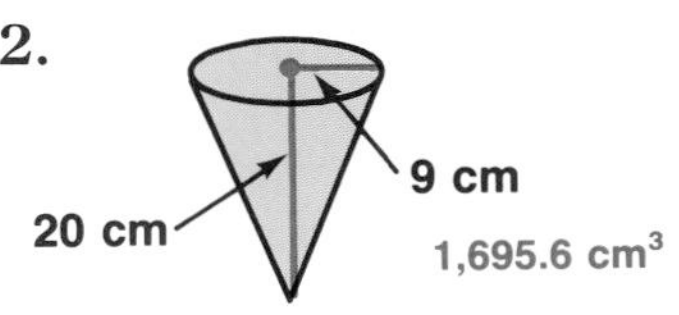

1,695.6 cm³

Lesson 9-1 *Write each ratio as a fraction in lowest terms.*

1. 14 wins to 10 losses $\frac{7}{5}$
2. 12 boys to 20 girls $\frac{3}{5}$
3. 30 pennies to 25 dimes $\frac{6}{5}$
4. 20 tickets for \$300 $\frac{1}{15}$
5. 5 in. of snow in 10 hours $\frac{1}{2}$
6. 18 cars on 3 trucks $\frac{6}{1}$
7. 285 students for 16 classrooms $\frac{285}{16}$
8. 300 miles in 6 hours $\frac{50}{1}$
9. 211 seniors in 2 schools $\frac{211}{2}$
10. 58 stores in 3 shopping malls $\frac{58}{3}$
11. 48 wins to 12 losses $\frac{4}{1}$
12. 24 pairs of socks to 2 feet $\frac{12}{1}$
13. 1,800 logs for 10 fireplaces $\frac{180}{1}$
14. 12 teachers for 300 students $\frac{2}{50}$
15. 2 tickets for \$35 $\frac{2}{35}$
16. 8 lb of fruit for \$3.20 $\frac{1}{0.40}$

Lesson 9-2 *A die is rolled once. Write the probability of rolling each.*

1. a 1 or 4 $\frac{1}{3}$
2. a 2 or 5 $\frac{1}{3}$
3. a 0 **0**
4. a 7 or 1 $\frac{1}{6}$
5. a 3 or 6 $\frac{1}{3}$
6. a 1 or 5 $\frac{1}{3}$
7. an even number $\frac{1}{2}$
8. a number larger than 3 $\frac{1}{2}$
9. a number less than 2 $\frac{1}{6}$
10. *not* a 3 $\frac{5}{6}$
11. *not* a 2 or 4 $\frac{2}{3}$
12. an odd number $\frac{1}{2}$

A box contains two yellow pencils, four red pencils, and three blue pencils. One pencil is chosen. Find the probability of each event.

13. P (red) $\frac{4}{9}$
14. P (blue) $\frac{1}{3}$
15. P (green) **0**
16. P (yellow) $\frac{2}{9}$
17. P (red or yellow) $\frac{2}{3}$
18. P (blue or yellow) $\frac{5}{9}$
19. P (red or blue) $\frac{7}{9}$
20. P (*not* blue) $\frac{2}{3}$
21. P (*not* red) $\frac{5}{9}$

In a bag there are three quarters, two nickels, and four dimes. One coin is chosen. What is the probability of each?

22. P (quarter) $\frac{1}{3}$
23. P (nickel) $\frac{2}{9}$
24. P (dime) $\frac{4}{9}$
25. P (penny) **0**
26. P (quarter or dime) $\frac{7}{9}$
27. P (dime or nickel) $\frac{2}{3}$
28. P (*not* quarter) $\frac{2}{3}$
29. P (*not* dime) $\frac{5}{9}$
30. P (quarter, nickel, or dime) **1**

Lesson 9-3 *Determine if each pair of ratios forms a proportion.*

1. 3 to 5, 5 to 10 **No**
2. 8 to 4, 6 to 3 **Yes**
3. 10 to 15, 5 to 3 **No**
4. 2 to 8, 1 to 4 **Yes**
5. 6 to 18, 3 to 9 **Yes**
6. 14 to 21, 12 to 18 **Yes**
7. 4 to 20, 5 to 25 **Yes**
8. 9 to 27, 1 to 3 **Yes**
9. 4 to 9, 5 to 10 **No**
10. 18 to 20, 27 to 30 **Yes**
11. 15 to 18, 5 to 6 **Yes**
12. 25 to 9, 42 to 8 **No**
13. 9 to 4, 9 to 5 **No**
14. 18 to 9, 20 to 10 **Yes**
15. 42 to 3, 28 to 2 **Yes**
16. 8 to 9, 16 to 18 **Yes**
17. 5 to 1, 7 to 2 **No**
18. 26 to 21, 21 to 26 **No**

Lesson 9-4

Solve each proportion.

1. $\frac{2}{3} = \frac{a}{12}$ **8**
2. $\frac{7}{8} = \frac{c}{16}$ **14**
3. $\frac{3}{7} = \frac{21}{d}$ **49**
4. $\frac{2}{5} = \frac{18}{x}$ **45**
5. $\frac{9}{10} = \frac{27}{m}$ **30**
6. $\frac{3}{5} = \frac{n}{21}$ **$12\frac{3}{5}$**
7. $\frac{5}{12} = \frac{b}{5}$ **$2\frac{1}{12}$**
8. $\frac{4}{36} = \frac{2}{y}$ **18**
9. $\frac{3}{10} = \frac{z}{36}$ **$10\frac{4}{5}$**
10. $\frac{4}{5} = \frac{r}{100}$ **80**
11. $\frac{4}{5} = \frac{8}{b}$ **10**
12. $\frac{12}{3} = \frac{x}{1}$ **4**
13. $\frac{9}{2} = \frac{y}{6}$ **27**
14. $\frac{14}{7} = \frac{7}{m}$ **$3\frac{1}{2}$**
15. $\frac{21}{3} = \frac{r}{15}$ **105**
16. $\frac{5}{4} = \frac{n}{100}$ **125**
17. $\frac{3}{2} = \frac{y}{18}$ **27**
18. $\frac{4}{7} = \frac{16}{r}$ **28**
19. $\frac{8}{11} = \frac{y}{33}$ **24**
20. $\frac{15}{17} = \frac{30}{w}$ **34**
21. $\frac{9}{2} = \frac{9}{n}$ **2**
22. $\frac{3}{8} = \frac{m}{40}$ **15**
23. $\frac{7}{11} = \frac{21}{y}$ **33**
24. $\frac{15}{21} = \frac{5}{z}$ **7**
25. $\frac{22}{25} = \frac{k}{10}$ **$8\frac{4}{5}$**
26. $\frac{24}{48} = \frac{j}{50}$ **25**
27. $\frac{9}{27} = \frac{t}{42}$ **14**
28. $\frac{3}{19} = \frac{c}{38}$ **6**
29. $\frac{40}{100} = \frac{2}{g}$ **5**
30. $\frac{18}{45} = \frac{4}{f}$ **10**
31. $\frac{5}{9} = \frac{p}{10}$ **$5\frac{5}{9}$**
32. $\frac{7}{9} = \frac{q}{24}$ **$18\frac{2}{3}$**
33. $\frac{18}{7} = \frac{z}{2}$ **$5\frac{1}{7}$**
34. $\frac{14}{1} = \frac{7}{y}$ **$\frac{1}{2}$**
35. $\frac{18}{19} = \frac{s}{38}$ **36**
36. $\frac{2}{3} = \frac{t}{4}$ **$2\frac{2}{3}$**
37. $\frac{9}{10} = \frac{r}{25}$ **$22\frac{1}{2}$**
38. $\frac{16}{19} = \frac{v}{48}$ **$40\frac{8}{19}$**
39. $\frac{7}{8} = \frac{a}{12}$ **$10\frac{1}{2}$**
40. $\frac{14}{3} = \frac{12}{m}$ **$2\frac{4}{7}$**

Lesson 9-5

Find the distance on a scale drawing for each actual distance. The scale is 1 in.: 20 ft.

1. 40 ft **2 in.**
2. 80 ft **4 in.**
3. 120 ft **6 in.**
4. 90 ft **4.5 in.**
5. 60 ft **3 in.**
6. 130 ft **$6\frac{1}{2}$ in.**
7. 10 ft **$\frac{1}{2}$ in.**
8. 5 ft **$\frac{1}{4}$ in.**
9. 15 ft **$\frac{3}{4}$ in.**
10. 220 ft **11 in.**
11. 350 ft **$17\frac{1}{2}$ in.**
12. 400 ft **20 in.**
13. 420 ft **21 in.**
14. 170 ft **8.5 in.**
15. 200 ft **10 in.**
16. 85 ft **$4\frac{1}{4}$ in.**
17. 75 ft **$3\frac{3}{4}$ in.**
18. 55 ft **$2\frac{3}{4}$ in.**
19. 125 ft **$6\frac{1}{4}$ in.**
20. 95 ft **$4\frac{3}{4}$ in.**
21. 2 ft **$\frac{1}{10}$ in.**
22. 18 ft **$\frac{9}{10}$ in.**
23. 24 ft **$1\frac{1}{5}$ in.**
24. 16 ft **$\frac{4}{5}$ in.**
25. 12 ft **$\frac{3}{5}$ in.**
26. 82 ft **$4\frac{1}{10}$ in.**
27. 66 ft **$3\frac{3}{10}$ in.**
28. 112 ft **$5\frac{3}{5}$ in.**
29. 225 ft **$11\frac{1}{4}$ in.**
30. 164 ft **$8\frac{1}{5}$ in.**

Lesson 9-6

Find the missing length for each pair of similar figures.

1.
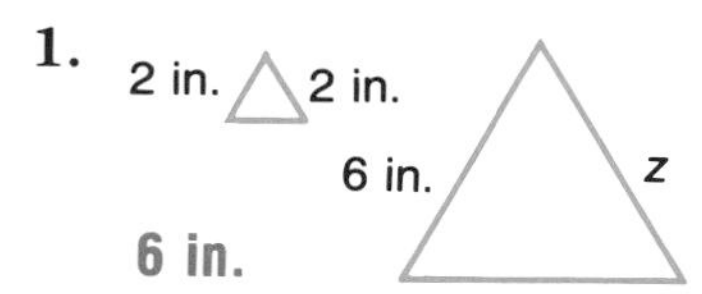

6 in.

2.

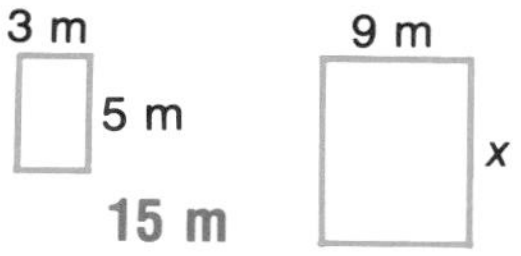

15 m

3.
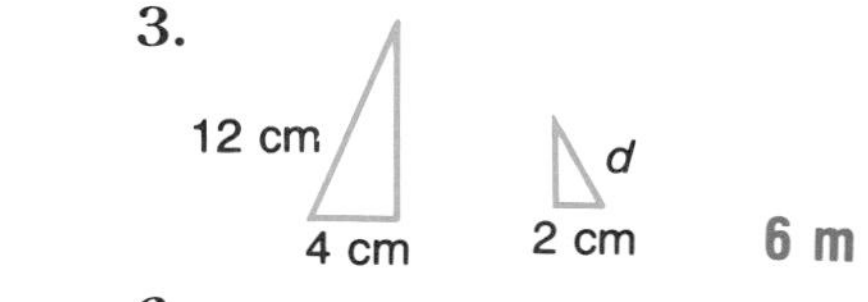

6 m

4.
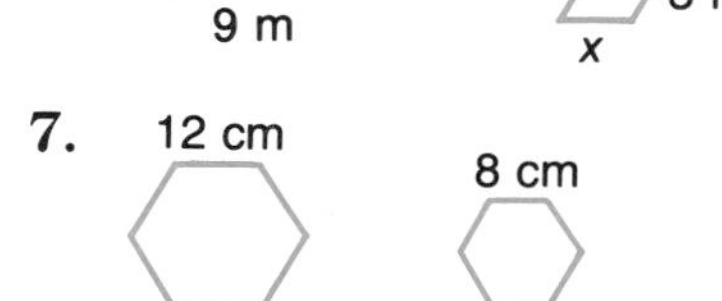

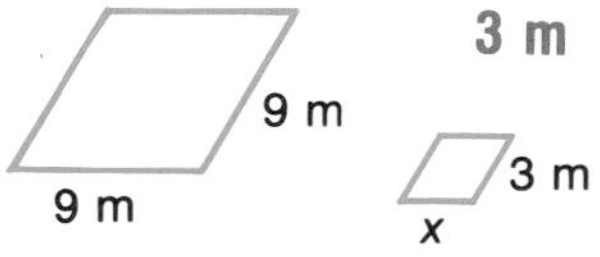

3 m

5.
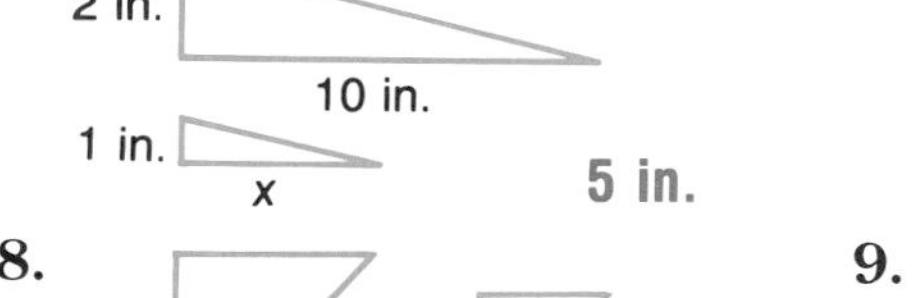

5 in.

6.
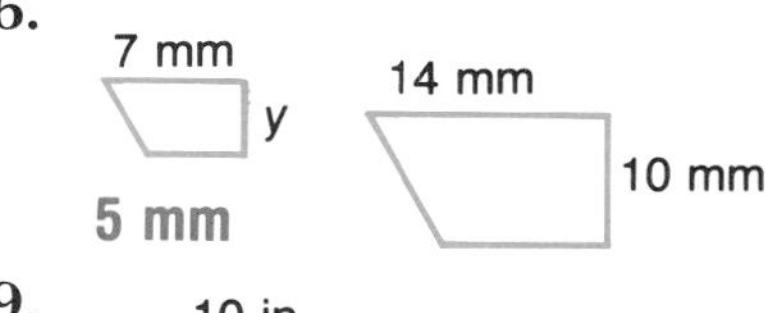

5 mm

7.
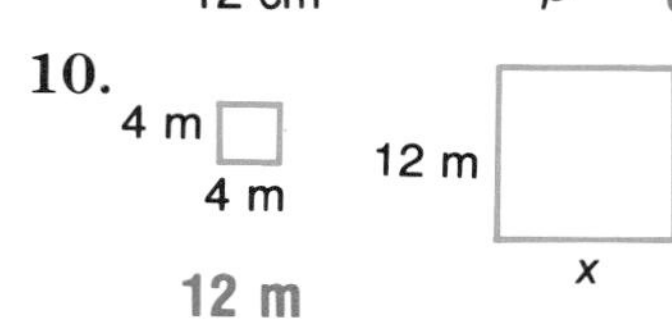

8 cm

8.
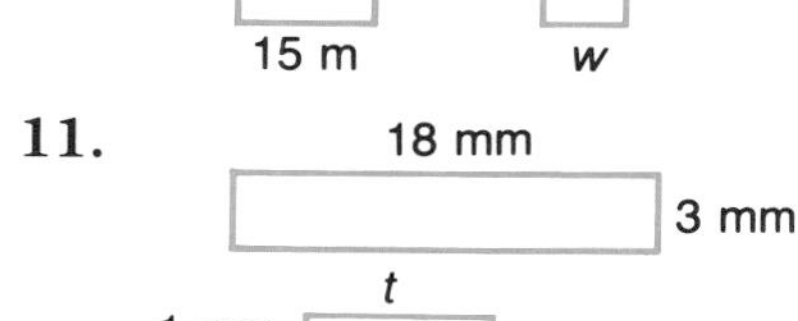

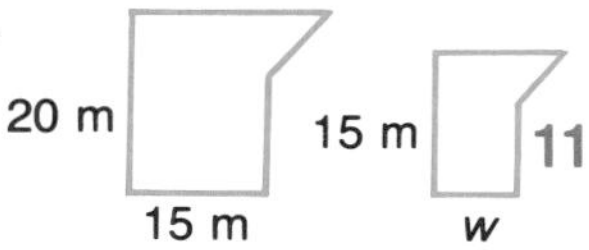

11.25 m

9.
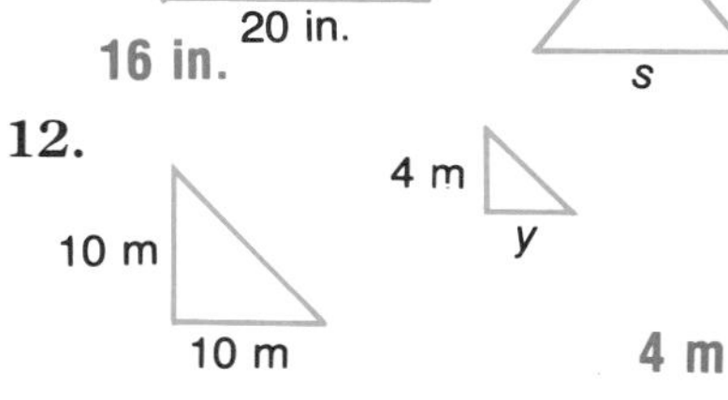

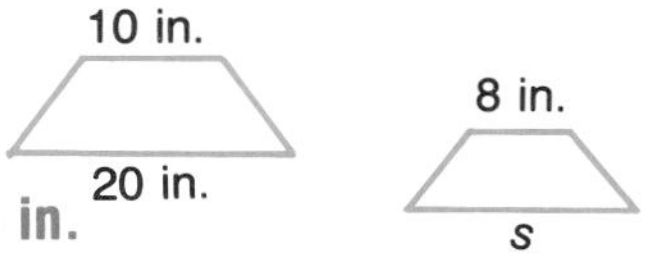

16 in.

10. 4 m, 4 m; 12 m, x
12 m

11. 18 mm, 3 mm; t, 1 mm
6 mm

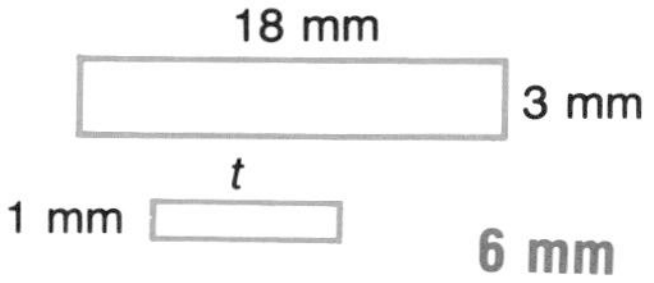

12. 10 m, 10 m; 4 m, y
4 m

Lesson 9-8

Write each percent as a fraction in simplest form.

1. 40% $\frac{2}{5}$
2. 10% $\frac{1}{10}$
3. 15% $\frac{3}{20}$
4. 23% $\frac{23}{100}$
5. 2% $\frac{1}{50}$
6. 8% $\frac{2}{25}$
7. 4% $\frac{1}{25}$
8. 80% $\frac{4}{5}$
9. 400% 4
10. 49% $\frac{49}{100}$
11. 310% $3\frac{1}{10}$
12. 550% $5\frac{1}{2}$
13. 24.5% $\frac{49}{200}$
14. 46.25% $\frac{37}{80}$
15. 13.75% $\frac{11}{80}$
16. 42% $\frac{21}{50}$
17. 4.5% $\frac{9}{200}$
18. 92.75% $\frac{371}{400}$
19. 16.5% $\frac{33}{200}$
20. 735% $\frac{147}{20}$ or $7\frac{7}{20}$

Write each fraction as a percent.

21. $\frac{2}{100}$ 2%
22. $\frac{26}{100}$ 26%
23. $\frac{2}{5}$ 40%
24. $\frac{1}{2}$ 50%
25. $\frac{3}{25}$ 12%
26. $\frac{3}{10}$ 30%
27. $\frac{2}{50}$ 4%
28. $\frac{3}{20}$ 15%
29. $\frac{20}{25}$ 80%
30. $\frac{21}{50}$ 42%
31. $\frac{8}{10}$ 80%
32. $\frac{10}{25}$ 40%
33. $\frac{10}{16}$ 62.5%
34. $\frac{7}{8}$ 87.5%
35. $\frac{5}{12}$ $41\frac{2}{3}$%
36. $\frac{3}{8}$ 37.5%
37. $\frac{4}{6}$ $66\frac{2}{3}$%
38. $\frac{1}{3}$ $33\frac{1}{3}$%
39. $\frac{7}{10}$ 70%
40. $\frac{4}{20}$ 20%
41. $\frac{1}{4}$ 25%
42. $\frac{2}{3}$ $66\frac{2}{3}$%
43. $\frac{9}{20}$ 45%
44. $\frac{19}{25}$ 76%

Lesson 9-9

Write each percent as a decimal.

1. 0.2% 0.002
2. 4.6% 0.046
3. 7.8% 0.078
4. 25.4% 0.254
5. 16.8% 0.168
6. 19% 0.19
7. 25% 0.25
8. 14% 0.14
9. 98% 0.98
10. 72% 0.72
11. 145% 1.45
12. 223% 2.23
13. 104% 1.04
14. 23.7% 0.237
15. 0.08% 0.0008
16. 0.45% 0.0045
17. 0.621% 0.00621
18. 2.56% 0.0256
19. 22.71% 0.2271
20. 14.06% 0.1406

Write each decimal as a percent.

21. 0.35 35%
22. 0.23 23%
23. 0.06 6%
24. 0.08 8%
25. 0.9 90%
26. 0.006 0.6%
27. 0.066 6.6%
28. 0.036 3.6%
29. 0.132 13.2%
30. 0.778 77.8%
31. 0.48 48%
32. 0.39 39%
33. 4.83 483%
34. 5.56 556%
35. 2.34 234%
36. 1.8 180%
37. 2.6 260%
38. 5.35 535%
39. 7.65 765%
40. 0.79 79%
41. 14.23 1,423%
42. 12.17 1,217%
43. 6.21 621%
44. 9.65 965%

Lesson 10-1

Find each percentage.

1. 25% of 20 5
2. 10% of 90 9
3. 16% of 30 4.8
4. 39% of 40 15.6
5. 250% of 100 250
6. 6% of 86 5.16
7. 78% of 50 39
8. 3% of 46 1.38
9. $66\frac{2}{3}$% of 60 40
10. $12\frac{1}{2}$% of 160 20
11. 9% of 29 2.61
12. 18% of 350 63
13. 74% of 600 444
14. 89% of 47 41.83
15. 435% of 30 130.5
16. 156% of 78 121.68
17. 19% of 200 38
18. 48% of 15 7.2
19. 28% of 4 1.12
20. 77% of 100 77
21. 34% of 38 12.92
22. 5% of 420 21
23. 55% of 134 73.7
24. 68% of 68 46.24
25. 25% of 48 12
26. 39% of 126 49.14
27. 14% of 40 5.6
28. 93% of 63 58.59
29. 40% of 45 18
30. 18% of 90 16.2
31. 31% of 13 4.03
32. 206% of 65 133.9
33. 22% of 300 66
34. 42% of 150 63
35. 24% of 340 81.6
36. 90% of 140 126

Lesson 10-2 *Find each percent. Use an equation.*

1. What percent of 10 is 5? **50%**
2. What percent of 16 is 4? **25%**
3. What percent of 25 is 5? **20%**
4. What percent of 12 is 3? **25%**
5. What percent of 56 is 14? **25%**
6. What percent of 63 is 56.7? **90%**
7. What percent of 80 is 20.8? **26%**
8. What percent of 400 is 164? **41%**
9. What percent of 550 is 61.6? **11.2%**
10. What percent of 42 is 14? **$33\frac{1}{3}\%$**
11. What percent of 4 is 20? **500%**
12. What percent of 5 is 45? **900%**
13. 15 is what percent of 10? **150%**
14. 28 is what percent of 7? **400%**
15. 27 is what percent of 54? **50%**
16. 21 is what percent of 84? **25%**
17. 23.4 is what percent of 65? **36%**
18. 111 is what percent of 148? **75%**
19. 24 is what percent of 72? **$33\frac{1}{3}\%$**
20. 61.5 is what percent of 600? **$10\frac{1}{4}\%$**

Lesson 10-3 *Find each number. Use an equation.*

1. 14% of the number is 63. Find the number. **450**
2. 75% of the number is 27. Find the number. **36**
3. 63% of what number is 63? **100**
4. 55% of what number is 33? **60**
5. 20% of what number is 5? **25**
6. 30% of what number is 27? **90**
7. $66\frac{2}{3}\%$ of what number is 40? **60**
8. $33\frac{1}{3}\%$ of what number is 15? **45**
9. 500% of what number is 45? **9**
10. 150% of what number is 54? **36**
11. 39 is 5% of what number? **780**
12. 30.8 is 35% of what number? **88**
13. 29.7 is 55% of what number? **54**
14. 72 is 24% of what number? **300**
15. 108 is 18% of what number? **600**
16. 3 is $37\frac{1}{2}\%$ of what number? **8**
17. 9 is $33\frac{1}{3}\%$ of what number? **27**
18. 57 is 300% of what number? **19**
19. 300 is 150% of what number? **200**
20. 125 is 500% of what number? **25**

Lesson 10-5 *Estimate.* **Typical answers are given. Answers may vary.**

1. 17% of 36 **6**
2. 86% of 24 **21**
3. 11% of 20 **2**
4. 27% of 48 **12**
5. 33% of 12 **4**
6. 48% of 76 **38**
7. 68% of 66 **40**
8. 63% of 40 **25**
9. 39% of 50 **20**
10. 89% of 200 **180**
11. 73% of 84 **63**
12. 85% of 72 **63**
13. 9% of 32 **3**
14. 24% of 84 **21**
15. 78% of 20 **16**
16. 65% of 85 **60**
17. 48% of \$23.95 **\$12**
18. 98% of \$5.50 **\$5.40**
19. 1.5% of 135 **2**
20. 125% of 100 **125**
21. 0.6% of 205 **1**

Lesson 10-6 *Find the percent of increase. Round to the nearest percent.*

	Item	Original Price	New Price		Item	Original Price	New Price
1.	soup	43¢/can	52¢/can	2.	apples	99¢/pound	$1.05/lb
3.	butter	88¢/lb	$1.09/lb	4.	gum	35¢/pack	48¢/pack
5.	cookies	$2.39/lb	$2.59/lb	6.	soda	99¢/liter	$1.19/liter

Find the percent of decrease. Round to the nearest percent.

	Item	Original Price	New Price		Item	Original Price	New Price
7.	phone	$35	$29	8.	toaster	$55	$46
9.	radio	$28	$19	10.	tool box	$88	$72
11.	TV	$550	$425	12.	shoes	$78	$44

1. 21% **2. 24%** **3. 8%** **4. 6%** **5. 37%** **6. 20%**
7. 17% **8. 16%** **9. 32%** **10. 18%** **11. 23%** **12. 44%**

Lesson 10-7 *Find the discount and sale price.*

1. piano, $4,220
   discount rate, 35% **$1,477; $2,743**
2. sweater, $38
   discount rate, 25% **$9.50; $28.50**
3. scissors, $14
   discount rate, 10% **$1.40; $12.60**
4. compact disc, $15.95
   discount rate, 20% **$3.19; $12.76**
5. book, $29
   discount rate, 40% **$11.60; $17.40**
6. answering machine, $69
   discount rate, 15% **$10.35; $58.65**
7. motorcycle, $3,540
   discount rate, 30% **$1,062; $2,478**
8. pants, $45
   discount rate, 50% **$22.50; $22.50**
9. tire, $65
   discount rate, 33% **$21.45; $43.55**
10. VCR, $280
    discount rate, 25% **$70; $210**

Lesson 10-8 *Find the interest owed on each loan. Then find the total amount to be repaid.*

1. principal, $300
   annual rate, 10% **$90; $390**
   time, 3 years
2. principal, $4,000
   annual rate, 12.5% **$2,000; $6,000**
   time, 4 years
3. principal, $3,200
   annual rate, 8% **$1,280; $4,480**
   time, 5 years
4. principal, $10,200
   annual rate, 9.5% **$5,814; $16,014**
   time, 6 years
5. principal, $20,000
   annual rate, 14% **$56,000; $76,000**
   time, 20 years
6. principal, $6,300
   annual rate, 6.5% **$819; $7,119**
   time, 24 months

Find the interest earned on each deposit.

7. principal, $500
   annual rate, 7% **$70**
   time, 2 years
8. principal, $2,500
   annual rate, 6.5% **$487.50**
   time, 3 years

Lesson 11-1 *Find the median, mode, and range for each set of data.*

1. 1, 2, 4, 6, 1, 2, 3, 1 **2; 1; 5**
2. 5, 4, 3, 8, 9, 7, 8, 8, 9 **8; 8; 6**
3. 38, 92, 92, 38, 46 **46; 38 and 92; 54**
4. 19, 17, 83, 82, 81, 80 **80.5; none; 66**
5. 236, 49, 55, 237, 49 **55; 49; 188**
6. 45, 46, 45, 45, 46, 47 **45.5; 45; 2**
7. 296, 926, 692, 296 **494; 296; 630**
8. 173, 171, 172, 171, 173 **172; 171 and 173; 2**
9. 6.6, 7.9, 8.3, 4.5, 8.3 **7.9; 8.3; 3.8**
10. 9.4, 3.8, 9.4, 4.9, 3.8 **4.9; 3.8 and 9.4; 5.6**
11. 12.1, 12.1, 13.2, 10.8 **12.1; 12.1; 2.4**
12. 14.4, 14.8, 14.3, 14.8, 14.9 **14.8; 14.8; 0.6**
13. 20.46, 20.64, 20.66 **20.64; none; 0.2**
14. 17.94; 18.86; 17.94 **17.94; 17.94; 0.92**
15. 15, 19, 91, 51, 51, 55, 55, 56, 55 **55; 55; 76**
16. 43, 34, 42, 45, 43, 43, 45, 42 **43; 43; 11**
17. 136, 163, 163, 136, 636, 136 **149.5; 136; 500**
18. 719, 983, 919, 917, 919, 719, 917 **917; none; 264**
19. 5.8, 8.9, 8.8, 8.6, 8.8, 8.8, 8.9 **8.8; 8.8; 3.1**
20. 0.4, 0.8, 1.3, 0.9, 2.0, 0.6, 0.8, 4.0 **0.85; 0.8; 3.6**
21. 1.9, 8.0, 6.3, 3.6, 1.0, 5.7, 5.6 **5.6; none; 7**
22. 4.95, 5.95, 1.39, 4.59, 4.59, 5.59 **4.77; 4.59; 4.56**

Lesson 11-2 *Solve. Use the data at the right.*

1. Make a frequency table for the set of data. **See students' work.**
2. Find the mean. $10\frac{3}{8}$
3. Find the median. $10\frac{1}{2}$
4. Find the mode. **11**
5. Find the range. $6\frac{1}{2}$

Shoe Sizes Sold at Athletic Shoe Store				
10	$10\frac{1}{2}$	13	$10\frac{1}{2}$	13
9	$10\frac{1}{2}$	$10\frac{1}{2}$	11	8
$6\frac{1}{2}$	11	11	11	$8\frac{1}{2}$
8	12	$9\frac{1}{2}$	12	11

Lesson 11-4 *Make a vertical bar graph for each set of data.* **See answers on p. T533.**

1.

Average Snowfall	
Nov.	1.2 inches
Dec.	2.6 inches
Jan.	4.8 inches
Feb.	5.3 inches
Mar.	4.2 inches
April	1.6 inches

2.

Average Points Scored per Game	
Lakers	123
Pistons	127
76ers	119
Bulls	120
Cavs	114
Pacers	118

Make a horizontal bar graph for each set of data. **See answers on p. T533.**

3.

Cars Sold per Month	
Jan.	52
Mar.	56
May	76
July	62
Sept.	95
Nov.	100

4.

Average Hours Worked per Day	
Monday	8
Tuesday	6
Wednesday	6
Thursday	7
Friday	7
Saturday	4

Lesson 11-5 *Make a line graph for each set of data.* For answers, see p. T533.

1.

Temperature on July 10	
8 AM	70° F
10 AM	75° F
12 noon	82° F
2 PM	86° F
4 PM	87° F
6 PM	87° F
8 PM	85° F

2.

Height of a Child	
2 yr	2′8″
4 yr	3′6″
6 yr	4′2″
8 yr	4′5″
10 yr	4′11″
12 yr	5′4″
14 yr	5′10″

3.

Number of Students at Central	
1950	850
1960	911
1970	948
1980	1,120
1990	1,450

4.

Inches of Rainfall	
1980	23 inches
1982	40 inches
1984	38 inches
1986	42 inches
1988	36 inches

Lesson 11-6 *Make a pictograph for each set of data.*

1.

Cans of Soda Sold	
Smith's	820
Carl's	510
Amy's	220
Green's	450

2.

Total Number of Coaching Wins	
Smith	421
Allen	310
Craig	120
Lewis	266

3.

Number of Refrigerators Sold	
Salesperson 1	26
Salesperson 2	32
Salesperson 3	18
Salesperson 4	21

4.

Number of Restaurants per City	
Calico	42
Benz	31
Smithfield	26
Longsville	27

Lesson 11-7 *Make a circle graph for each set of data.*

1.

Household Income	
primary job	82%
secondary job	9%
investments	5%
gifts	3%
other	1%

2.

Sporting Goods Sales	
shoes	44%
apparel	28%
equipment	22%
magazines	6%

3.

Energy Use in Home	
Heating/cooling	51%
Appliances	28%
Lights	15%
Other	6%

4.

Students in North High School	
White	30%
Black	28%
Hispanic	24%
Asian	18%

Lesson 11-8 *Find the median and mode of the data in each stem-and-leaf plot.*

1.

Stem	Leaf
0	2 4 6
1	6 6 6 7 9
2	6 6 6 5 5
3	2 2 3 3
4	1 2 2

26; 16 and 26

2.

Stem	Leaf
4	5 5 6
5	1 2 3 5
6	8 8 8 9 9
7	2 2 2 3 5
8	1 1 1 2

69; 68, 72, 81

3.

Stem	Leaf
0	3 6 7 8
1	2 4 6 7
2	1 1 1 3
3	7 7 8 9

19; 21

Use the weights of 24 elementary school students to complete.

86	53	80	71
76	58	44	74
78	61	58	76
55	65	68	78
62	72	82	56
92	73	90	95

4. Construct a stem-and-leaf plot of the data.

Stem	Leaf
4	4
5	3 5 6 8 8
6	1 2 5 8
7	1 2 3 4 6 6 8 8
8	0 2 6
9	0 2 5

5. What is the smallest weight for a student? 44
6. Find the range of the weights. 51
7. How many students weighed more than 75? 10
8. What is the median weight? 72.5

Lesson 11-9 *Find the upper quartile, lower quartile, and interquartile range for each set of data.*

1. 4, 6, 10, 11, 21, 25, 29, 37 27; 8; 19
2. 2, 9, 13, 17, 20, 36, 38, 51 37; 11; 26
3. 17, 27, 31, 38, 53, 42, 47 42; 27; 15
4. 38, 44, 59, 73, 84, 92, 95 92; 44; 48

5.

Stem	Leaf
1	0 1
2	0 3 4
3	1 1 2
4	2 5 6
5	8 9 9
6	0 1

58.5; 23.5; 35

6.

Stem	Leaf
3	5 7 9
4	5 5 5
5	2 4 4
6	8
7	2 3
8	1

70; 42; 28

7.

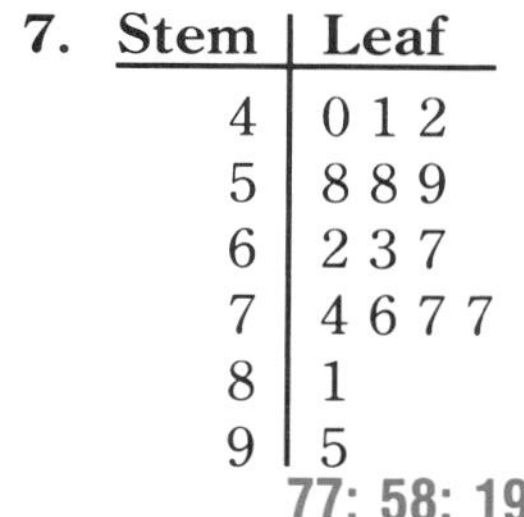

Stem	Leaf
4	0 1 2
5	8 8 9
6	2 3 7
7	4 6 7 7
8	1
9	5

77; 58; 19

Lesson 11-10 *Use the box and whisker plot to answer each question.*

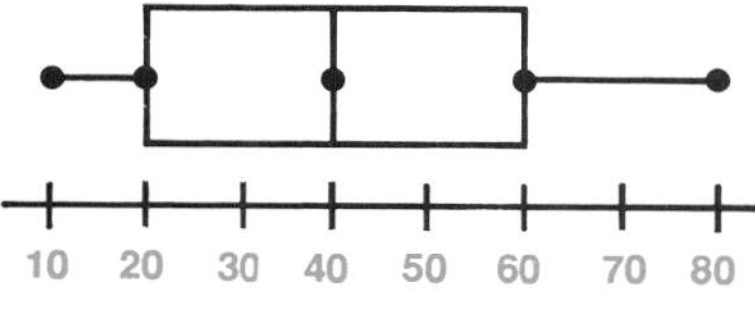

1. What is the range and interquartile range of the data? 70; 40
2. The middle half of the data is between which two values? 20 and 60
3. What part of the data is greater than 60? $\frac{1}{4}$
4. What part of the data is greater than 40? $\frac{1}{2}$
5. What part of the data is less than 40? $\frac{1}{2}$
6. What numbers do 10 and 80 represent? the extremes
7. What is the median? 40
8. What are the upper and lower quartiles? 60; 20

Lesson 12-1

Replace each ● with <, >, or = to make a true sentence.

1. −3 ● 0 <
2. −1 ● −2 >
3. −5 ● −4 <
4. 6 ● −7 >
5. 8 ● 10 <
6. −6 ● 6 <
7. −11 ● −20 >
8. −8 ● 2 <
9. −13 ● −12 <
10. 5 ● 2 >
11. 9 ● −8 >
12. 19 ● −19 >
13. |−2| ● |5| <
14. |13| ● |−19| <
15. |−6| ● |2| >
16. |14| ● |−14| =
17. |0| ● |−4| <
18. |23| ● |−20| >
19. |−75| ● |75| =
20. |−32| ● |30| >
21. −10 ● 18 <
22. −20 ● −38 >
23. −71 ● 72 <
24. −15 ● −35 >

Order from least to greatest.

25. −2, −8, 4, 10, −6, −12, **−12, −8, −6, −2, 4, 10**
26. 19, −19, −21, 32, −14, 18 **−21, −19, −14, 18, 19, 32**
27. −5, −3, −4.5, −6.6, 1.8 **−6.6, −5, −4.5, −3, 1.8**
28. 4.2, 5.6, −6.5, −6.6, −4.2 **−6.6, −6.5, −4.2, 4.2, 5.6**
29. 18, 23, 95, −95, −18, −23, 2 **−95, −23, −18, 2, 18, 23, 95**
30. 46, −48, −47, −52, −18, 12 **−52, −48, −47, −18, 12, 46**

Lesson 12-2

Add.

1. −4 + 8 **4**
2. 14 + 16 **30**
3. −7 + (−7) **−14**
4. −9 + (−6) **−15**
5. 5 + (−5) **0**
6. −3 + (−3) **−6**
7. −18 + 11 **−7**
8. −4 + 17 **13**
9. −13 + (−11) **−24**
10. −36 + 40 **4**
11. −23 + (−36) **−59**
12. −42 + 29 **−13**
13. 18 + (−32) **−14**
14. −26 + 74 **48**
15. 42 + (−18) **24**
16. −25 + 12 **−13**
17. −33 + (−12) **−45**
18. 18 + (−63) **−45**
19. −38 + (−39) **−77**
20. −15 + (−10) **−25**
21. 25 + (−32) **−7**
22. 62 + (−95) **−33**
23. 82 + (−63) **19**
24. 47 + 12 **59**
25. −96 + (−18) **−114**
26. −67 + (−14) **−81**
27. −91 + (−11) **−102**
28. 60 + (−42) **18**
29. −81 + (−17) **−98**
30. −69 + (−32) **−101**
31. −100 + 98 **−2**
32. 95 + (−5) **90**
33. −120 + (−2) **−122**
34. 120 + (−2) **118**
35. −120 + 2 **−118**
36. −46 + (−3) **−49**

Lesson 12-3

Subtract.

1. 3 − 7 **−4**
2. −5 − 4 **−9**
3. −6 − 2 **−8**
4. −4 − 15 **−19**
5. 8 − 13 **−5**
6. 6 − (−4) **10**
7. 12 − 9 **3**
8. −2 − 23 **−25**
9. 63 − 78 **−15**
10. 0 − (−45) **45**
11. −20 − 0 **−20**
12. −5 − (−9) **4**
13. −19 − (−19) **0**
14. −8 − (−12) **4**
15. −18 − (−26) **8**
16. 26 − (−14) **40**
17. 43 − (−18) **61**
18. −18 − 23 **−41**
19. −26 − 42 **−68**
20. −61 − 21 **−82**
21. −23 − (−42) **19**
22. −60 − (−36) **−24**
23. −18 − (−6) **−12**
24. −43 − (−41) **−2**
25. 55 − 33 **22**
26. 58 − (−10) **68**
27. 72 − (−19) **91**
28. 84 − (−61) **145**
29. −41 − 15 **−56**
30. −81 − 21 **−102**
31. −67 − 28 **−95**
32. −51 − 47 **−98**
33. −86 − (−61) **−25**
34. −4.5 − (−8.5) **4**
35. −16.4 − (−6.5) **−9.9**
36. −24.6 − 8.5 **−33.1**

Lesson 12-5 *Multiply.*

1. $5 \times (-2)$ **−10**	2. $6 \times (-4)$ **−24**	3. 4×21 **84**	4. -13×4 **−52**
5. $-11 \times (-5)$ **55**	6. -6×45 **−270**	7. $-9 \times (-38)$ **342**	8. -50×0 **0**
9. $64 \times (-10)$ **−640**	10. -8×114 **−912**	11. -3×14 **−42**	12. -18×6 **−108**
13. $4 \times (-20)$ **−80**	14. $-4 \times (-16)$ **64**	15. $-12 \times (-12)$ **144**	16. -15×12 **−180**
17. $16 \times (-5)$ **−80**	18. -3×16 **−48**	19. $18 \times (-10)$ **−180**	20. $-5 \times (-32)$ **160**
21. $-16 \times (-12)$ **192**	22. $-80 \times (-5)$ **400**	23. 5×-12 **−60**	24. $-13 \times (-3)$ **39**
25. 14×7 **98**	26. -29×10 **−290**	27. $-11 \times (-11)$ **121**	28. $15 \times (-8)$ **−120**
29. -9×15 **−135**	30. $-7 \times (-21)$ **147**	31. $9 \times (-12)$ **−108**	32. $-12 \times (-11)$ **132**
33. $(-8)(-8)(2)$ **128**	34. $5(-7)(-4)$ **140**	35. $6(3)(-2)$ **−36**	36. $(-4)(-2)(-7)$ **−56**

Lesson 12-6 *Divide.*

1. $4 \div (-2)$ **−2**	2. $16 \div (-8)$ **−2**	3. $-14 \div (-2)$ **7**	4. $32 \div 8$ **4**
5. $18 \div (-3)$ **−6**	6. $0 \div (-1)$ **0**	7. $-42 \div 6$ **−7**	8. $-63 \div (-9)$ **7**
9. $8 \div (-8)$ **−1**	10. $-100 \div (-20)$ **5**	11. $15 \div (-1)$ **−15**	12. $-45 \div (-10)$ **4.5**
13. $14 \div (-7)$ **−2**	14. $-21 \div 3$ **−7**	15. $25 \div (-5)$ **−5**	16. $-50 \div 10$ **−5**
17. $-42 \div (-7)$ **6**	18. $64 \div (-4)$ **−16**	19. $-32 \div (-8)$ **4**	20. $63 \div (-7)$ **−9**
21. $-81 \div (-9)$ **9**	22. $-100 \div (-20)$ **5**	23. $48 \div (-6)$ **−8**	24. $-72 \div 3$ **−24**
25. $-55 \div 11$ **−5**	26. $28 \div (-7)$ **−4**	27. $-40 \div 8$ **−5**	28. $-144 \div (-4)$ **36**
29. $-36 \div (-6)$ **6**	30. $18 \div (-2)$ **−9**	31. $99 \div (-9)$ **−11**	32. $-56 \div (-8)$ **7**
33. $-121 \div (-11)$ **11**	34. $-81 \div 9$ **−9**	35. $68 \div (-2)$ **−34**	36. $169 \div (-13)$ **−13**

Lesson 13-1 *Solve each equation. Check your solution.*

1. $x + 4 = 9$ **5**	2. $y - 3 = 15$ **18**	3. $-4 + b = 12$ **16**
4. $z - 10 = -8$ **2**	5. $-7 = x + 12$ **−19**	6. $m + (-2) = 6$ **8**
7. $r - (-8) = 14$ **6**	8. $t - 13 = -3$ **10**	9. $-2 + n = -2$ **0**
10. $-19 = y - 19$ **0**	11. $a + 6 = -9$ **−15**	12. $-14 + c = -12$ **2**

13. $m - 5 = 3$ **8**	14. $x - 2 = 14$ **16**	15. $r - 6 = 12$ **18**	16. $y - 9 = 15$ **24**
17. $s - 1.3 = 2.1$ **3.4**	18. $3.8 = t - 4.6$ **8.4**	19. $w - \frac{1}{4} = \frac{3}{4}$ **1**	20. $\frac{1}{3} = z - 2\frac{5}{6}$ **$3\frac{1}{6}$**
21. $a + 7 = 10$ **3**	22. $k + 5 = 6$ **1**	23. $q + 7 = 20$ **13**	24. $h + 12 = 14$ **2**
25. $m + 2.6 = 9.8$ **7.2**	26. $4.3 + b = 5.8$ **1.5**	27. $3.1 = 2.5 + n$ **0.6**	28. $a + \frac{1}{3} = \frac{1}{2}$ **$\frac{1}{6}$**
29. $\frac{2}{5} + c = 1\frac{1}{10}$ **$\frac{7}{10}$**	30. $2 = q + \frac{7}{8}$ **$1\frac{1}{8}$**	31. $1\frac{3}{4} = \frac{1}{4} + x$ **$1\frac{1}{2}$**	32. $2\frac{5}{8} = 1\frac{3}{5} + y$ **$1\frac{1}{40}$**
33. $\frac{3}{4} + m = -\frac{5}{8}$ **$-1\frac{3}{8}$**	34. $3 = t - 6\frac{1}{3}$ **$9\frac{1}{3}$**	35. $\frac{4}{7} = -\frac{8}{9} + s$ **$1\frac{29}{63}$**	36. $3\frac{1}{3} = -6\frac{1}{8} + r$ **$9\frac{11}{24}$**

Lesson 13-2 *Solve each equation.*

1. $3m = -15$ **−5**
2. $4x = 12$ **3**
3. $-324 = 30y$ **−10.8**
4. $\frac{c}{-4} = 10$ **−40**
5. $\frac{1}{2}w = -7$ **−14**
6. $8 = \frac{y}{-4}$ **−32**
7. $5q = -40$ **−8**
8. $-6h = -5$ **$\frac{5}{6}$**
9. $r \div 7 = -8$ **−56**
10. $\frac{2}{3}x = \frac{8}{15}$ **$\frac{4}{5}$**
11. $-12 = \frac{1}{5}y$ **−60**
12. $-1 = -\frac{4}{5}m$ **$\frac{5}{4}$**
13. $\frac{1}{3}d = 15$ **45**
14. $\frac{1}{5}t = 25$ **125**
15. $\frac{-3}{10}f = 12$ **−40**
16. $\frac{-3}{4}a = 6$ **−8**
17. $0 = 6r$ **0**
18. $\frac{y}{12} = -6$ **−72**
19. $\frac{-3}{8} = \frac{3}{8}k$ **−1**
20. $\frac{3}{8} = \frac{1}{2}x$ **$\frac{3}{4}$**
21. $-1 = \frac{a}{21}$ **−21**

Lesson 13-3 *Solve each equation. Check your solution.*

1. $2x + 4 = 14$ **5**
2. $5p - 10 = 0$ **2**
3. $5 - 6a = 41$ **−6**
4. $\frac{x}{3} - 7 = 2$ **27**
5. $\frac{2}{3}y + 10 = 22$ **18**
6. $15 = \frac{1}{4}m - 6$ **84**
7. $3(r - 1) = 9$ **4**
8. $-18 = -6(q - 4)$ **7**
9. $0 = -4x + (-28)$ **−7**
10. $3x - 1 = -5$ **$-\frac{4}{3}$**
11. $-2x + 3 = 1$ **1**
12. $\frac{1}{2}a - 3 = -14$ **−22**
13. $5 - \frac{3}{4}y = -7$ **16**
14. $\frac{1}{3}z + 5 = -3$ **−24**
15. $-3(x - 5) = 12$ **1**
16. $\frac{1}{2}(q - 1) = 10$ **21**
17. $\frac{1}{2}x - 3 = 7$ **20**
18. $\frac{3}{4}t + 6 = 9$ **4**
19. $9a - 8 = 73$ **9**
20. $5 - 2y = 15$ **−5**
21. $3(c - 4) = 3$ **5**

Lesson 13-4 *Find the ordered pair for each point labeled on the coordinate plane.*

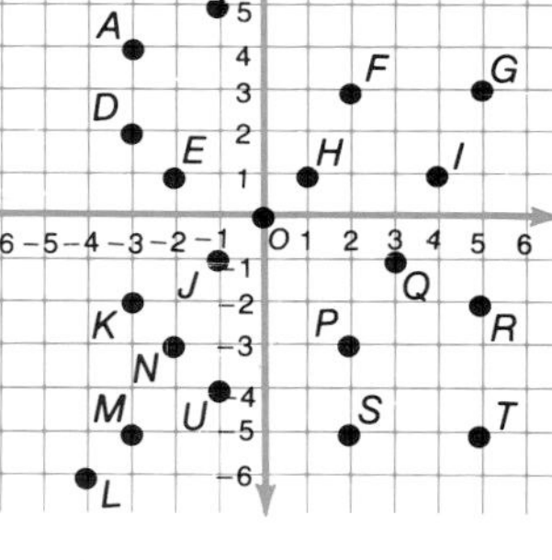

1. *A* **(−3, 4)**
2. *B* **(−1, 5)**
3. *C* **(2, 6)**
4. *D* **(−3, 2)**
5. *E* **(−2, 1)**
6. *F* **(2, 3)**
7. *G* **(5, 3)**
8. *H* **(1, 1)**
9. *I* **(4, 1)**
10. *J* **(−1, −1)**
11. *K* **(−3, −2)**
12. *L* **(−4, −6)**
13. *M* **(−3, −5)**
14. *N* **(−2, −3)**
15. *O* **(−1, −4)**
16. *P* **(2, −3)**
17. *Q* **(3, −1)**
18. *R* **(5, −2)**
19. *S* **(2, −5)**
20. *T* **(5, −5)**

Draw a coordinate plane. Graph each ordered pair. Label each point with the letter given. **See answers on p. T534.**

1. *A* (4, 2)
2. *B* (2, 0)
3. *C* (−3, 0)
4. *D* (0, −3)
5. *E* (4, 1)
6. *F* (−4, 0)
7. *G* (−2, 1)
8. *H* (−2, −1)
9. *I* (−4, 3)
10. *J* (0, 5)
11. *K* (0, −2)
12. *L* (6, 5)
13. *M* (−6, 5)
14. *N* (6, −5)
15. *O* (−6, −3)
16. *P* (1, 3)
17. *Q* (−1, −3)
18. *R* (3, 4)
19. *S* (4, 5)
20. *T* (2, −2)
21. *U* (−3, −6)
22. *V* (−2, −4)
23. *W* (3, −1)
24. *X* (1, −4)
25. *Y* (0, 0)

Lesson 13-5 *Graph the solutions to each equation.* See answers on p. T534.

1. $4x = y$
2. $x - 7 = y$
3. $5x - 2 = y$
4. $\frac{1}{2}x + 1 = y$
5. $3x + 2 = y$
6. $2x = y$
7. $4x + 2 = y$
8. $3x - 1 = y$
9. $\frac{1}{2}x = y$
10. $x + y = 6$
11. $x = y$
12. $x + 1 = y$
13. $\frac{1}{3}x = y$
14. $2 + x = y$
15. $3x = y$
16. $x - 2 = y$
17. $x - 5 = y$
18. $x - y = 0$

Lesson 14-1 *Draw a tree diagram to show the possible outcome.* See answers on p. T534.

1. tossing the two coins.

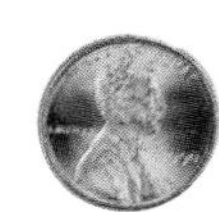

2. tossing the three coins

Use multiplication to find the number of possible outcomes.

3. tossing a coin and rolling a die 12
4. tossing a quarter and dime and rolling a die 24
5. a choice of strawberry, vanilla or chocolate for a three scoop ice cream cone 9
6. a choice of a queen or king size bed with a firm, or super firm mattress 4
7. a choice of blue or gray pants with a choice of white, yellow, or blue stripe oxford shirt 6
8. a choice of roses or carnations in red, yellow, pink, or white 8

Lesson 14-2 *A drawer contains two red socks, four blue socks, and six black socks. One sock is chosen from the drawer. Find the chance of choosing each of the following.*

1. a black sock 50%
2. a red sock $16\frac{2}{3}\%$
3. a blue sock $33\frac{1}{3}\%$
4. a brown sock 0%
5. *not* a red sock $83\frac{1}{3}\%$
6. *not* a blue sock $66\frac{2}{3}\%$
7. *not* a black sock 50%
8. a red or blue sock 50%
9. a black or red sock $66\frac{2}{3}\%$
10. a green or blue sock $33\frac{1}{3}\%$
11. a blue or black sock $83\frac{1}{3}\%$
12. a blue or a white sock $33\frac{1}{3}\%$

There are three quarters, five dimes, and twelve pennies in a bag. One coin is chosen. Find the chance of choosing each of the following.

13. a quarter 15%
14. a penny 60%
15. a dime 25%
16. a nickel 0%
17. a quarter or a dime 40%
18. a dime or a penny 85%
19. a nickel or a dime 25%
20. not a dime 75%
21. not a penny 40%
22. a quarter or a penny 75%
23. not a quarter 85%
24. not a nickel 100%

Lesson 14-3

A bag contains four blue chips, five red chips, and three green chips. A chip is chosen from the bag. Find the probability of each event.

1. P(a blue chip) $\frac{1}{3}$
2. P(a red chip) $\frac{5}{12}$
3. P(a green chip) $\frac{1}{4}$
4. P(a white chip) 0
5. P(a red or blue chip) $\frac{3}{4}$
6. P(a red or green chip) $\frac{2}{3}$
7. P(*not* a green chip) $\frac{3}{4}$
8. P(*not* a blue chip) $\frac{2}{3}$
9. P(*not* a red chip) $\frac{7}{12}$
10. P(*not* a white chip) 1
11. P(*not* a blue or red chip) $\frac{1}{4}$
12. P(*not* a green or blue chip) $\frac{5}{12}$

Lesson 14-4

Use the same situation in Lesson 14-3 to find the probability of each event.

1. draw a red chip, replace it, draw a green chip $\frac{5}{48}$
2. draw a blue chip, replace it, draw a blue chip $\frac{1}{9}$
3. draw a green chip, replace it, draw a red chip $\frac{5}{48}$
4. draw a green chip, replace it, draw a blue chip $\frac{1}{12}$
5. draw a blue chip, do not replace it, draw a green chip $\frac{1}{11}$
6. draw a red chip, do not replace it, draw a green chip $\frac{5}{44}$

A box contains three red, five blue, and two purple marbles. Supoose you choose a marble, replace it, and choose another marble. Find the probability of each event.

7. P (red, then blue) $\frac{3}{20}$
8. P (blue, then purple) $\frac{1}{10}$
9. P (purple, then red) $\frac{3}{50}$
10. P (blue, then red) $\frac{3}{20}$
11. P (red both times) $\frac{9}{100}$
12. P (blue both times) $\frac{1}{4}$
13. P (purple both times) $\frac{1}{25}$
14. P (same color both times) $\frac{19}{50}$

Lesson 14-6

A box contains ten blue blocks, three yellow blocks, and five purple blocks. One block is chosen from the box. Find the odds for each event.

1. P (a blue block) 10 to 8
2. P (a yellow block) 3 to 15
3. P (a purple block) 5 to 13
4. P (a white block) 0 to 18
5. P (a green block) 0 to 18
6. P (*not* a blue block) 8 to 10
7. P (*not* a yellow block) 15 to 3
8. P (*not* a purple block) 13 to 5
9. P (a blue or purple block) 15 to 3
10. P (a yellow or blue block) 13 to 5
11. P (a green or blue block) 10 to 8
12. P (a green or white block) 0 to 18
13. P (a purple or yellow block) 8 to 10
14. P (a purple or blue block) 15 to 3
15. P (a purple or white block) 5 to 13
16. P (*not* a purple or blue block) 3 to 15
17. P (*not* a blue or yellow block) 5 to 13
18. P (*not* a green or blue block) 8 to 10
19. P (*not* a yellow or purple block) 10 to 8
20. P (*not* a blue or purple block) 3 to 15
21. P (*not* a green or yellow block) 15 to 3
22. P (*not* a red, yellow, or blue block) 5 to 13
23. P (*not* a yellow, blue, or black block) 5 to 13

GLOSSARY

absolute value (376) The number of units a number is from zero on the number line.

acute angle (216) An angle with measure between 0° and 90°.

acute triangle (226) A triangle with all acute angles.

angle (214) Two rays with a common endpoint.

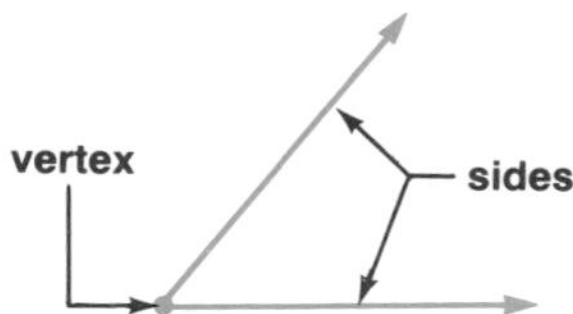

arc (211) A part of a circle.

area (94) The number of square units that cover a surface.

arithmetic sequence (54) A sequence where the difference between consecutive numbers is the same.

axis (408) Number line that is in the coordinate plane.

bar graph (348) A graph which is used to compare quantities. The length of each bar represents a number.

bases (230) The two polygon shaped parallel sides of a polyhedron.

BASIC (222) **B**eginner's **A**ll-**P**urpose **S**ymbolic **I**nstruction **C**ode. Language used for micro computers.

basic units (178) Units used in the metric system to show length, mass, and capacity (gram, liter, meter).

bisect (219) Separate into two congruent parts.

box-and-whisker plot (362) A graph that shows the median, quartiles, and extremes of a set of data.

center (230) A given point of a sphere from which all points on the sphere are an equal distance.

chance (428) The probability of an event expressed as a percent.

circle (244) A closed path of points in a plane, all the same distance from a fixed point called the center.

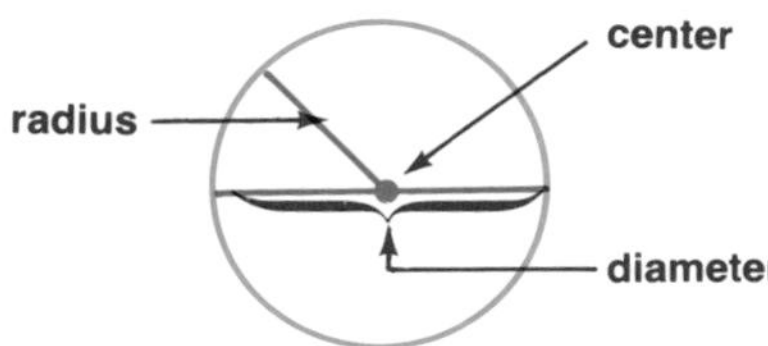

circle graph (356) A graph used to compare parts of a whole. The whole amount is shown as a circle and each part is shown as a percent of the whole.

circumference (244) The distance around a circle found by using formula $C = 2\pi r$.

commission (310) Earnings based on a percent of sales.

common ratio (88) The same number used in a geometric sequence to multiply or divide by to obtain consecutive numbers.

compass (211) A tool used to draw circles and parts of circles.

complementary (217) Two angles whose sum is 90°.

composite number (110) A number that has more than two factors.

compound interest (330) Interest computed at stated intervals and added to the principal at the end of the interval.

cone (230) A three dimensional figure with a circular base and one vertex.

congruent triangles (248) Triangles that have the same size and shape.

constructions (211) Drawings that are made with only a compass and a straight edge.

coordinate plane (408) A plane on which ordered pairs are graphed consisting of two perpendicular number lines that are the axes.

cross products (286) A way to determine if two ratios form a proportion. If the cross products are equal then the ratio forms a proportion. In the proportion $\frac{2}{3} = \frac{8}{12}$, the cross products are 2×12 and 3×8.

cube (230) A rectangular prism that has six square faces.

cup (194) A customary unit of capacity equal to 8 fluid ounces.

cylinder (230) A three dimensional figure with two parallel congruent circular bases.

data (96) Information in the form of numbers.

decagon (224) A polygon with 10 sides.

degree (214) Common unit used in measuring angles.

denominator (116) Tells the number of objects or equal-sized parts. The bottom number in a fraction.

dependent event (430) An event that is affected by the occurrence of another event.

diameter (244) A line segment through the center of a circle with endpoints on the circle. See *circle.*

difference (14) The result of subtracting one number from another.

$$9 - 4 = 5$$

↑ difference

discount (326) An amount subtracted from the regular price of an item.

discount rate (326) The percent of decrease in the regular price of an item.

dividend (76) The original number into which the divisor is divided.

$$40 \div 5 = 8 \qquad 5\overline{)40}^{\,8}$$

↑—dividend—↑

divisor (76) The number divided into the dividend.

$$40 \div 5 = 8$$

↑ divisor

down payment (74) A small portion of the total price to be paid at the time of purchase.

edges (230) The intersection of faces of a polyhedron.

equation (26) A mathematical sentence with an equals sign.

equilateral triangle (226) A triangle with 3 congruent sides.

equivalent fractions (116) Fractions that name the same number. $\frac{3}{4}$ and $\frac{6}{8}$ are equivalent fractions.

event (282) A specific outcome or type of outcome.

expanded form (6) A method of writing numbers using place value and addition.

$$739 = 700 + 30 + 9 \text{ or}$$
$$(7 \times 100) + (3 \times 10) + (9 \times 1) \text{ or}$$
$$(7 \times 10^2) + (3 \times 10^1) + (9 \times 10^0)$$

exponent (6) The number of times the base is used as a factor. In 10^3, the exponent is 3.

faces (230) The flat surfaces of a polyhedron.

factors (108) Numbers that divide into another number so that the remainder is zero.

finance charge (74) The difference between the credit price and the cash price of an item.

flowchart (352) A diagram used to help plan steps in a computer program.

fluid ounce (194) A customary unit of capacity.

foot (190) A customary unit of length equal to 12 inches.

formula (90) An equation that shows how certain quantities are related.

fraction (116) A number used to name part of a whole or group.

frequency table (344) A table for organizing a set of data that shows the number of times each item or number appears.

gallon (194) A customary unit of capacity equal to 4 quarts.

geometric sequence (88) A sequence of numbers in which succeeding terms are found by multiplying the preceeding term by the same number (common ratio.)

gram (178) A basic unit of mass in the metric system.

greatest common factor (GCF) (112) The greatest number that is a factor of two or more numbers. The greatest common factor of 24 and 30 is 6.

gross pay (114) Total wage before deductions are made.

hexagon (224) A polygon with six sides.

hypoteneuse (257) The longest side of a right triangle, the side opposite the right angle. See *right triangle.*

improper fraction (124) A fraction that has a numerator greater than or equal to the denominator.

inch (190) A customary unit of length.

independent event (430) An event that is *not* affected by the occurrence of another event.

integers (376) The whole numbers and their opposites.

interest (328) A percent of the principal that is an amount earned on a deposit or an amount owed on a loan.

interquartile range (360) A measure of variation, it is the difference between the upper and lower quartiles.

inverse operations (26) Pairs of operations that undo each other. Addition and subtraction are inverse operations. Multiplication and division are inverse operations.

isosceles triangle (226) A triangle that has at least 2 congruent sides.

kilowatt hour (kWh) (51) The unit used to measure electrical energy equal to 1 kilowatt of electricity used for 1 hour.

least common denominator (122) The least common multiple of the denominators of fractions. The least common denominator of $\frac{4}{5}$ and $\frac{5}{6}$ is 30.

least common multiple (LCM) (112) The least nonzero number that is a multiple of two or more numbers. The least common multiple of 6 and 8 is 24.

line (212) All the points on a never-ending straight path. A representation of line *AB* ($\overleftrightarrow{AB}$) is shown below.

line graph (350) A graph used to show change and direction of change over a period of time.

line segment (212) Two endpoints and the straight path between them. A representation of line segment *CD* ($\overline{CD}$) is shown below.

liter (178) A basic unit of capacity in the metric system.

mathematical expression (24) Variables and/or numbers combined by symbols of operations.

$x + 3 \quad a \times b \quad 4 - 2$

mean (94) An average of a set of data.

measure of variation (360) The spread of values in a set of data.

median (342) Middle number of the set of data when listed in order.

meter (178) A basic unit of length in the metric system.

mile (190) A customary unit of length equal to 5,280 feet or 1,760 yards.

mixed decimals (162) Fractions that are equivalent to repeating decimals can be written in this form.

$\frac{1}{3} = 0.33\frac{1}{3} \quad \frac{1}{6} = 0.16\frac{2}{3} \quad \frac{6}{7} = 0.85\frac{5}{7}$

mixed number (124) A number that indicates the sum of a whole number and a fraction.

mode (342) The number that appears most often in a set of data.

multiple (112) A number obtained by multiplying a given number by any whole number. A multiple of 12 is 24.

net pay (114) Take home pay found after subtracting total and personal deductions from gross pay.

numerator (116) Tells the number of objects or parts being considered. The top number in a fraction.

obtuse angle (216) An angle with measure between 90° and 180°.

obtuse triangle (226) A triangle with an obtuse angle.

octagon (224) A polygon with eight sides.

odds (438) A ratio of the number of ways an event can occur to the number of ways the event cannot occur.

opposites (376) Two integers are opposites if their sum is 0. The integers 2 and -2 are opposites because $2 + (-2) = 0$.

ordered pair (408) A pair of numbers where order is important. Ordered pairs may be graphed on a coordinate plane.

origin (408) The point at which the axes intersect in the coordinate plane.

ounce (194) A customary unit of weight.

outcome (282) A possible result of a probability experiment. When a coin is tossed, one outcome is tails.

output (50) The results of computer processing. Two output devices and the cathode-ray tube (CRT) and the printer.

parallel lines (220) Lines that are in the same plane that do not intersect.

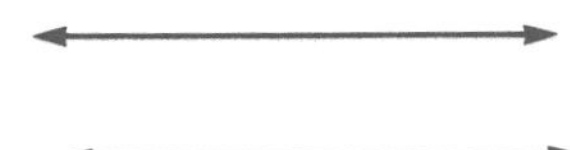

parallelogram (228) A quadrilateral with two pairs of parallel sides.

pentagon (224) A polygon with five sides.

percent (296) A way of expressing hundredths using the percent symbol (%). Thus, $7\% = \frac{7}{100}$ or 0.07.

perimeter ***(P)*** (58) The distance around a polygon.

perpendicular lines (220) Two lines that intersect to form right angles.

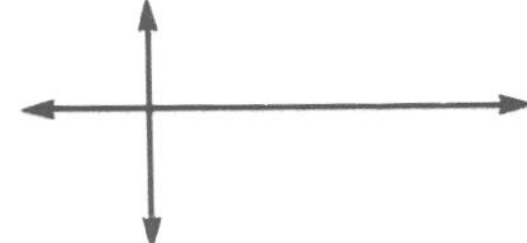

pi (π) (244) The ratio of the circumference of a circle to the diameter of a circle. π is approximately equal to 3.14.

pictograph (354) A graph used to compare data in a visually appealing way.

pint (194) A customary unit of capacity equal to 2 cups.

place value (4) A system for writing numbers. In this system the position of a digit determines its value.

plane (212) All the points in a never-ending flat surface.

point (212) An exact location in space.

polygons (224) Closed plane figures formed by line segments.

polyhedron (230) A solid figure with flat surfaces.

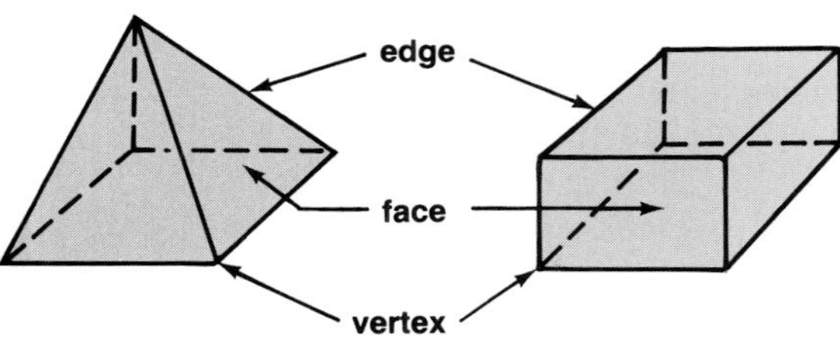

pound (194) A customary unit of weight equal to 16 ounces.

powers (6) Expressions written with exponents.

10^3 3^4 8^3

prime factorization (110) A composite number that is expressed as the product of prime numbers. The prime factorization of 12 is $2 \times 2 \times 3$.

prime number (110) A number that has exactly two factors, 1 and the number itself.

principal (328) The amount of money borrowed or invested on which interest is based.

prism (230) A polyhedron with two parallel congruent bases that are shaped like polygons.

probability (282) A ratio that describes how likely it is that an event will occur.

product (16) The result of multiplying two or more numbers.

$3 \times 5 = 15$

proper fraction (124) A fraction that has a numerator less than the denominator.

proportion (284) A mathematical sentence that states two ratios are equivalent.

protractor (214) A device used in measuring angles.

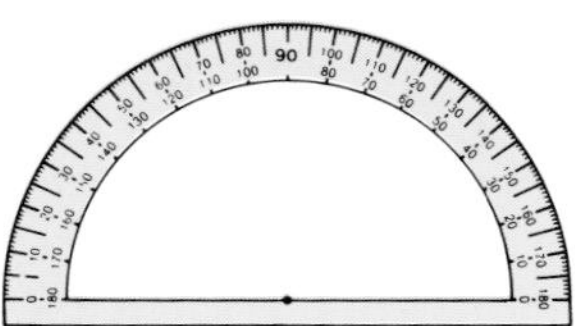

pyramid (230) A polyhedron with a single base shaped like a polygon.

quadrilateral (224) A polygon with four sides.

quart (194) A customary unit of capacity equal to 2 pints.

quartiles (360) Values that divide data into four equal parts.

quotient (76) The result of dividing one number by another.

radius (244) A line segment from the center of a circle to any point on the circle. See *circle.*

range (342) The difference between the greatest and least number in a set of data.

ratio (280) A comparison of two numbers by division. A comparison of 9 and 12 can be written as 9 to 12, 9:12, and $\frac{9}{12}$.

ray (212) All the points in a never-ending straight path extending in only one direction. A representation of ray *DE* ($\overrightarrow{DE}$) is shown below.

reciprocals (156) Two numbers whose product is 1. Since $\frac{5}{12} \times \frac{12}{5} = 1$, the reciprocal of $\frac{5}{12}$ is $\frac{12}{5}$.

rectangle (228) A parallelogram with four right angles.

regular polygon (224) A polygon in which all sides are congruent and all angles are congruent.

remainder (76) The whole number left after one number is divided into another number.

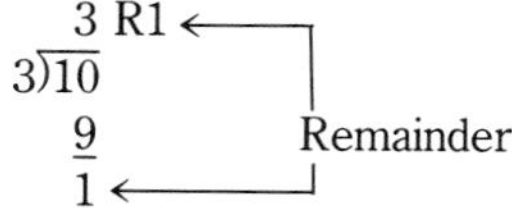

repeating decimal (162) A decimal whose digits repeat in groups of one or more.

$0.666\ldots$ or $0.\overline{6}$

rhombus (228) A parallelogram with all sides congruent.

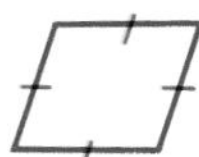

right angle (216) An angle that measures 90°.

right triangle (226) A triangle with one right angle.

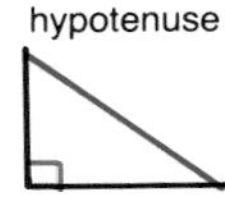

sample (436) A smaller, representative group of a larger group surveyed when it is unreasonable to survey the large group.

sample space (426) A list of all possible outcomes from an action, such as tossing a coin or rolling a die.

scale drawing (288) A geometrically similar representation of something too large or too small to be conveniently drawn to actual size.

scalene triangle (226) A triangle with no congruent sides.

scientific notation (86) A way of expressing numbers as the product of a number that is at least 1, but less than 10 and a power of 10. In scientific notation 4,100 is written 4.1×10^3.

sequence (54) A list of numbers that follows a certain pattern.

sides (214) The two rays that form an angle.

similar figures (292) Two geometric figures that have the same shape but may differ in size. Corresponding sides of similar figures are proportional.

simplest form (118) The form of a fraction when the GCF of the numerator and denominator is 1.

skew lines (220) Lines that do not intersect and are not parallel.

sphere (232) A solid with all points the same distance from a given point.

square (228) A parallelogram with all sides and angles congruent.

standard form (4) A number written using only the digits and place value. The standard form of seven hundred thirty nine is 739.

stem-and-leaf plot (358) A graph for organizing data in which each number is separated into two parts. One part forms the stem and one part forms the leaves. The data is displayed in two columns, the stem on the left and the leaves on the right.

straightedge (211) Any object that can be used to draw a straight line. Important tool in geometry.

sum (14) The result of adding two or more numbers.

$$5 + 3 = 8$$

surface area (258) The sum of the areas of each surface of a three-dimensional figure.

symmetry (233) A shape such that a line may be placed on a figure so that the shape on one side of the line matches the shape on the other side or that the parts are congruent.

terminating decimal (162) A quotient in which the division ends or terminates with a remainder of zero.

ton (194) A customary unit of weight equal to 2,000 pounds.

trapezoid (228) A quadrilateral with only one pair of parallel sides.

triangle (224) A polygon with three sides.

undefined quotient (389) Since multiplication and division are inverse operations, division by zero results in a contradiction. If $5 \div 0 = n$, then there is *no* related statement $0 \times n = 5$ because there is no number that can replace n and result in a true statement. Therefore, we say that division by zero is undefined.

unit fraction (151) A fraction that has a numerator of 1.

variable (24) A symbol, usually a letter, used to represent a number.

vertex (214) The common endpoint of two rays that form an angle.

volume (262) The number of cubic units required to fill a space.

yard (190) A customary unit of length equal to 3 feet or 36 inches.

INDEX

D

E

H

I

J

K

L

M

N

O

P

Q

R

S

T

U

V

W

Y

Z

SELECTED ANSWERS

CHAPTER 1 APPLYING NUMBERS AND VARIABLES

Pages 4-5 Lesson 1-1

1. tens **2.** thousands **3.** ten thousands **4.** tenths **5.** hundredths **6.** thousandths **7.** hundreds **8.** ones **9.** six hundred three **11.** two thousand, seven and eight tenths **13.** 0.420 **15.** 0.040 **17.** 50 **19.** 8 **21.** 0 **23.** 2 **25.** 7 **27.** 1 **29.** one hundred and five hundredths **31.** three hundred three and three hundredths **33.** 3,000 **35.** 0.48 **37.** 0.025 **39.** 4, 1, 3, 5, 2 **41.** seventy-three billion, seventy three thousand million **43.** 0.030

Pages 6-7 Lesson 1-2

1. 10 × 10 × 10; 1,000 **2.** 5 × 5; 25 **3.** 2 × 2 × 2 × 2; 16 **4.** 12 × 12; 144 **5.** 20 × 20 × 20; 8,000 **6.** 7^3 **7.** 25^2 **8.** 5^4 **9.** $(1 \times 10^2) + (5 \times 10^1) + (6 \times 10^0)$ **10.** $(2 \times 10^3) + (9 \times 10^2) + (3 \times 10^1) + (3 \times 10^0)$ **11.** $(7 \times 10^2) + (8 \times 10^0)$ **12.** $(4 \times 10^3) + (7 \times 10^1) + (6 \times 10^0)$ **13.** $(9 \times 10^4) + (6 \times 10^3)$ **15.** 3 × 3 × 3; 27 **17.** 8^2 **19.** 3^4 **21.** 11^2 **23.** 475 **25.** $(3 \times 10^2) + (4 \times 10^1) + (5 \times 10^0)$ **27.** $(3 \times 10^3) + (7 \times 10^2) + (5 \times 10^0)$ **29.** 3,136 **31.** 662,596 **33.** 181,476 **35.** 56,715,961 **37.** 25 tiles **39.** 10^{-1} **40.** 3 **41.** 1 **42.** 9 **43.** 8 **44.** 500 **45.** 1,600 **46.** 50 **47.** $28,000

Page 9 Lesson 1-3

1. < **2.** = **3.** > **4.** < **5.** < **6.** > **7.** = **8.** 5, 5.05, 5.105, 5.15, 5.51 **9.** 237, 273, 327, 372, 723, 732 **11.** = **13.** > **15.** < **17.** 0.128, 0.821, 1.28, 1.82 **19.** 1,002, 1,008, 1,035, 1,156 **21.** Ehrhardt, Komisova, Donkova **26.** tens **27.** hundredths **28.** thousands **29.** hundreds **30.** 4^4 **31.** 15^3 **32.** 5^3 **33.** $9

Page 11 Lesson 1-4

1. 4,000 **2.** 1,000 **3.** 1,000 **4.** 10,000 **5.** 45,000 **6.** 1 **7.** 101 **8.** 50 **9.** 0 **10.** 10 **11.** 0.78 **12.** 12.86 **13.** 0.10 **14.** $8.55 **15.** $0.40 **17.** 60 **19.** 410 **21.** 1,000 **23.** 6,730 **25.** 80,000 **27.** 93,000,000 **29.** $40.00 **31.** 0.0 **33.** 0.2700 **35.** 1.000 **37.** six thousand, thirty-five **39.** twenty-five thousandths **41.** $35 **43.** 7 shares **45.** 910 miles **47.** 240 miles **49.** exact **51.** exact

Pages 12-13 Lesson 1-5

1. $3,056 **2.** $2,930 **3.** $2,944 **4.** $3,126 **5.** $2,972 **6.** $3,084 **7.** $3,182 **8.** $3,070 **9.** $3,427 **11.** $2,989 **13.** $19,549 **17.** $2,951 **19.** They pay the same amount. **21.** > **22.** < **23.** > **24.** 20 **25.** 300 **26.** 50,000 **27.** 0.880 **28.** 115,124

Pages 14-15 Lesson 1-6

1. 500 **2.** 10,000 **3.** 4,600 **4.** 0.7 **5.** 11.4 **6.** $23 **7.** 148 **8.** 300 **9.** 6,200 **10.** 5.000 **11.** 0.1 **12.** $8.00 **13.** 20,000 **14.** $80 **15.** 900 **17.** 2,000 **19.** 100.00 **21.** 28.000 **23.** 2.00 **25.** 4,100 **27.** $3.70 **29.** 31 **31.** $2 **33.** no **35.** 7 minutes **37.** no The estimate is about $50. Even with tax $67.56 is too much.

Pages 16-17 Lesson 1-7

1. 1,800 **2.** 180,000 **3.** 0 **4.** 600,000 **5.** $77 **6.** 16 **7.** 2 **8.** 8 **9.** 4 **10.** 60 **11.** 200 **13.** 20 **15.** 120 **17.** 21,000 **19.** 68 **21.** 7 **23.** 55 **25.** $5 **27.** 20,000 **29.** $600 **31.** 30 **33.** 33 **35.** Did not multiply 4,000 × 30 correctly. Four zeros are required. **37.** 23 mpg **39.** **a.** 60 **41.** **b.** 225 million

Pages 20-21 Lesson 1-8

1. John-1; Aaron-2; Lyn-3; Al-5; Diane-6 **2.** Dan-teacher; William-lawyer; Laurie-pediatrician **3.** Paul-pepperoni and anchovies; David-sausage; Ben-pepperoni and sausage **5.** 28 pieces **7.** Jack-rock; Kevin-heavy metal; Jason-jazz **9.** 21st floor **11.** 3 **12.** 46 **13.** 9 **14.** 64 **15.** 1,200 **16.** 6 **17.** 5,200 **18.** 570 **19.** 200 miles

Pages 22-23 Lesson 1-9

1. 50 **2.** 18 **3.** 48.3 **4.** 35 **5.** 2.4 **6.** 0 **7.** 2 **8.** 8 **9.** 10 **10.** 54 **11.** 64 **12.** 16 **13.** 14 **15.** 12 **17.** 120 **19.** 20 **21.** 9 **23.** 80 **25.** 10 **27.** 60 **29.** $2 **31.** The tapes alone would weigh 360 pounds. **35.** 16 **37.** 20 **38.** 1.6 **39.** 110 **40.** $14

41. 48 **42.** 50 **43.** 6,400 **44.** Jenny-Ft. Myers, Betty-Daytona, Ashley-Miami

Page 25 Lesson 1-10

1. $7 + n$ **2.** $n - 5$ **3.** $16 \div n, \frac{16}{n}$ **4.** $8 \times n, 8n$ **5.** 39 **6.** 28 **7.** 30 **8.** 1 **9.** $n + 8$ **11.** $3 - n$ **13.** $7n$ **15.** $n - 3$ **17.** 17 **19.** 88 **21.** 6 **23.** 1 **25.** 12 **27.** seven more than a number **29.** a number divided by 9 **31.** a number decreased by 26 **33.** 4; 11 **35.** 4; 17 **37.** 5.61 **39.** 22.4196

Page 27 Lesson 1-11

1. 5; 14 **2.** 20; 50 **3.** 0.6; 0.8 **4.** 9; 18 **5.** 3; 2 **6.** 4.7; 3.2 **7.** 40; 50 **8.** 4; 5 **9.** 16 **11.** 14 **13.** 2 **15.** 11 **17.** 18 **19.** 1 **21.** 0.3 **23.** 4 **25.** 16 **27.** 50 tapes **29.** 3 empty backpacks

Page 29 Lesson 1-12

1. 3; 4 **2.** 7; 3 **3.** 20; 4 **4.** 3; 0.2 **5.** 5; 35 **6.** 10; 300 **7.** 2; 0.6 **8.** 25; 250 **9.** 8 **11.** 0 **13.** 0.04 **15.** 48 **17.** 42 **19.** 0.8 **21.** 400 **23.** 400 **25.** 12 **27.** 400 **29.** $20 **31.** $400,900,000,000

Pages 30-31 Lesson 1-13

1. a. $632 + 346 = y$; 978 people **2. a.** $20 \times 5 = z$; 100 audio tapes **3.** $590 **5.** 700 calories **7.** 2,100 seats **9.** 26 weeks **11.** $11,125 **14.** 216 **15.** 64 **16.** 144 **17.** 6,561 **18.** $6 + n$ **19.** $11 - n$ **20.** $12n$ **21.** $23 + n$ **22.** n = Frank's age now; $n + 14 = 30$; $n = 16$

Pages 32-33 Chapter 1 Review

1. $(6 \times 10^2) + (4 \times 10^0)$ **3.** estimating **5.** 8.030 **7.** equation **9.** variable **11.** 360 **13.** 50 **15.** $(2 \times 10^1) + (4 \times 10^0)$ **17.** $(2 \times 10^3) + (6 \times 10^2) + (5 \times 10^1)$ **19.** $(7 \times 10^3) + (5 \times 10^0)$ **21.** 10.708, 10.78, 10.87, 10.871 **23.** 10,000 **25.** 10.7 **27.** $3,056 **29.** $2,876 **31.** 3,200 **33.** $6.00 **35.** 24,000 **37.** 100 **39.** 90 **41.** 24.6 **43.** 7 **45.** 11 **47.** 2 **49.** 81 **51.** $y = 5$ **53.** $k = 7$ **55.** $p = 70$ **57.** $m = 7$ **59.** $y = 8$ **61.** $65 **63.** yes **65.** $220

Pages 36-37 Cumulative Review/Test

Free Response: **1.** 40 **3.** 200 **5.** 7 **7.** 2 **9.** 6 **11.** 200 **13.** 81 **15.** 343 **17.** 1,000,000 **19.** $(4 \times 10^3) + (3 \times 10^2) + (8 \times 10^1) + (1 \times 10^0)$ **21.** $(3 \times 10^4) + (3 \times 10^3) + (3 \times 10^2) + (3 \times 10^1) + (2 \times 10^0)$ **23.** 43, 44, 45 **25.** Bill Huck **27.** 10.500 **29.** 25.12 **31.** 0.1 **33.** 7.70 **35.** 6,700 **37.** 1 **39.** 0.7 **41.** 500 **43.** 7
Multiple Choice: **1.** d **3.** b **5.** c **7.** a **9.** b

CHAPTER 2 PATTERNS: ADDING AND SUBTRACTING

Pages 42-43 LESSON 2-1

1. 79 **2.** 38 **3.** 98 **4.** 128 **5.** 137 **6.** 919 **7.** 880 **8.** 673 **9.** 602 **10.** 33,264 **11.** 46,360 **12.** 76,886 **13.** 96 **15.** 83 **17.** 1,837 **19.** 5,100 **21.** 69,009 **23.** 1,102 **25.** 163,703 **27.** 19,084 **29.** 290 km **31.** $455 **34.** 5 **35.** 28 **36.** 20 **37.** 8 **38.** 34 **39.** 9 **40.** 82 **41.** $2.30

Pages 44-45 Lesson 2-2

1. 0.97 **2.** 7.978 **3.** 30.1 **4.** 30.79 **5.** 57.91 **6.** 92.378 **7.** 22.03 **8.** 197.28 **9.** 12.49 **10.** 21.38 **11.** 947.79 **12.** 21,259.2 **13.** 0.41 **15.** 2.71 **17.** $2.00 **19.** $51.59 **21.** 8.565 **23.** $942.46 **25.** $18.13 **27.** 1.886 **29.** 43.79 seconds **31.** 5.4% **33.** $t + 0.12$ **37.** less

Pages 46-47 Lesson 2-3

1. 25 **2.** 23 **3.** 39 **4.** 114 **5.** 108 **6.** 3,461 **7.** 4,705 **8.** 4,065 **9.** 3,409 **10.** 1,342 **11.** 32 **13.** 23 **15.** 418 **17.** 792 **19.** 10,102 **21.** 1,078 **23.** 964 **25.** 40,711 **27.** 1,182 feet **29.** $5.50 **32.** 3 **33.** 12 **34.** 108 **35.** 87 **36.** 184 **37.** 1,422 **38.** 33,608 **39.** 1,001

Pages 48-49 Lesson 2-4

1. 1.8 **2.** 8.18 **3.** 77.42 **4.** 0.337 **5.** 1.16 **6.** 2.95 **7.** 5.37 **8.** 213.75 **9.** 28.88 **10.** 2,210.5 **11.** 2.4 **13.** 0.722 **15.** 0.82 **17.** 2.668 **19.** 1.308 **21.** 0.11 **23.** 0.146 **25.** 534.327 **27.** 0.045 **29.** $341 million

Pages 52-53 Lesson 2-5

1. Numbers cannot be altered by someone else.
2. Words cannot be altered by someone else. **3.** and **5.** Twenty-nine and $\frac{00}{100}$ **7.** One hundred eight and $\frac{97}{100}$ **9.** $1,209.25 **11.** $451.26 **13.** $0.10 \times n = s$ **15.** $0.25 **16.** $13z$ **17.** $x - 5$ **18.** $18 \div n, \frac{18}{n}$ **19.** $n + 14$ **20.** 10.98 **21.** 28.278 **22.** 108.975 **23.** 19.062 **24.** 1,234 miles

Pages 54-55 Lesson 2-6

1. 2; 14, 16, 18 **2.** 3.0; 12.6, 15.6, 18.6 **3.** 1; 5, 4, 3 **4.** 3; 40, 37, 34 **5.** 68, 60, 52 **7.** 44, 54, 64 **9.** 175, 150, 125 **11.** 19.6, 21.3, 23 **13.** no **15.** yes; 21.7, 26.9, 32.1 **17.** 27, 45, 54 **19.** 42, 57.8, 89.4 **21.** 7.0, 5.6, 4.2 **23.** 3 weeks **25.** 36, 42; 3, 4, 5 **27.** $12 + (n - 1) \times 6$ **29.** 250 **31.** 1,000

Pages 56-57 Lesson 2-7

1. $189.75; calculator **2.** $9.00; estimate **3.** $603; mental computation **5.** square, blue; circle, red; triangle, green **7.** $13; estimate **9.** 36 chickens, 18 pigs **10.** 109 **11.** 1,576 **12.** 12,753 **13.** 71,793 **14.** 9 **15.** 455 **16.** 881 **17.** 3,329 **18.** $319.32

Pages 58-59 Lessson 2-8

1. 95 cm **2.** 27.1 m **3.** 22 m **4.** 140 mm **5.** 48 cm **6.** 88 ft **7.** 56.4 cm **9.** 24 ft **11.** 56 ft **13.** 218 inches **15.** 114 plants **17.** 350 ft **19.** 312 m **21.** Multiply the measure of one side by 6.

Pages 60-61 Lesson 2-9

1. $3.38; The cost of the gasoline is not needed. **2.** missing the cost of the darkroom supplies **3.** missing the original amount of flour **5.** $455 **7.** Pour 5 L into the 5 L bucket. There are 2 L left. Fill the 2-L bucket from the 5-L bucket. Add the 2 L to the 2 L in the largest bucket. **9.** 0.002 **10.** 200 **11.** 200 **12.** 1.8 **13.** $78.00, calculator or paper/pencil

Pages 62-63 Chapter 2 Review

1. e **3.** g **5.** f **7.** 84 **9.** 1,436 **11.** 0.510 **13.** 18.4 **15.** 4,741 **17.** 4.842 **19.** 774 **21.** 1.14 **23.** 2.286 **25.** 42,479 **27.** 37,537 **29.** $363.51 **31.** $794.37 **33.** yes; 17.9, 20.1, 22.3 **35.** yes; 55.1, 45, 34.9 **37.** 360 miles, estimate **39.** 34 inches **41.** 28 m **43.** 88 ft **45.** missing initial weight **47.** 33; do not need number of pompoms

Pages 66-67 Cumulative Review/Test

Free Response: 1. 750 **3.** 600 **5.** 4,800 **7.** 390,000 **9.** 780 **11.** 5,500 **13.** 100,000 **15.** 14 **17.** 62 **19.** 252 **21.** 79 **23.** 4,523 **25.** 62,608 **27.** 9,556 **29.** 161 **31.** 903 **33.** 198 **35.** 131 **37.** 1,661 **39.** 25 CDs

Multiple Choice: 1. c **3.** b **5.** b **7.** b **9.** c

CHAPTER 3 PATTERNS: MULTIPLYING AND DIVIDING

Pages 70-71 Lesson 3-1

1. 1,260 **2.** 930 **3.** 1,425 **4.** 2,336 **5.** 1,508 **6.** 7,810 **7.** 5,978 **8.** 33,640 **9.** 2,320 **10.** 22,260 **11.** 3,268 **13.** 960 **15.** 2,460 **17.** 31,122 **19.** 6,634 **21.** 197,756 **23.** 145,830 **25.** 392,424 **27.** 1,252,368 **29.** 448 pages **31.** 15 **33.** 7 **35.** 132 **39.** 126 **41.** 180

Pages 72-73 Lesson 3-2

1. 55.2 **2.** 58.8 **3.** 60.8 **4.** 4.75 **5.** 27.72 **6.** 0.06 **7.** 0.094 **8.** 0.045 **9.** 0.0014 **10.** 0.0012 **11.** 3.5 **13.** 0.40 **15.** 1.173 **17.** 0.021 **19.** 29.88 **21.** 4.045 **23.** 0.0453 **25.** 0.3410 **27.** 0.4710 **29.** 15.6 **31.** $97.75 **33.** 15,000 seats; yes; $1\frac{1}{3}$ of 44,702 is 59,454, so the Kingdome has more than $1\frac{1}{3}$ the seating **35.** 1.4 **37.** correct

Pages 74-75 Lesson 3-3

1. $442; $58 **2.** $2,301.64; 401.64 **3.** $7,112; $1,116 **4.** $3,738; $688 **5.** $1,594; $394 **7.** $24 **9.** $1,800 **13.** $9,455 **14.** 25 **15.** 28 **16.** 54 **17.** 12 cm **18.** 32 in. **19.** 24 mm **20.** 780 miles

Page 77 Lesson 3-4

1. 27 **2.** 96 **3.** 163 R2 **4.** 228 R1 **5.** 925 **6.** 3 R2, 1 digit **7.** 3 R3, 1 digit **8.** 6 R28, 1 digit **9.** 12 R29, 2 digits **10.** 7 R4, 1 digit **11.** 1 digit **13.** 2 digits **15.** 54 **17.** 64 **19.** 4 R20 **21.** 6 R1 **23.** 4 R18 **25.** 8 R150 **27.** 39 R48 **29.** 11 R51 **31.** approximately 52 weeks **33.** 19%, 190; 12%, 120; 8%, 80; 23%, 230

Pages 78-79 Lesson 3-5

1. 65.4 **2.** 8,204 **3.** 4,500 **4.** 340,000 **5.** 49 **6.** 512 **7.** 62,900 **8.** 370 **9.** 0.00448 **10.** 9.41 **11.** 0.145 **12.** 3.412 **13.** 0.637 **14.** 0.0239 **15.** 3.9 **16.** 0.048 **17.** 1,750 **19.** 372 **21.** 178 **23.** 14,700 **25.** 230.4 **27.** 28,400 **29.** 0.093 **31.** 0.529 **33.** 0.2817 **35.** 0.0118 **37.** 0.00605 **39.** 5.07 **43.** 1,000 $10 bills **45.** 0.345 **47.** 0.001 **49.** 4.71 **50.** 15.17 **51.** 80.996 **52.** 6.132 **53.** 35 books

Pages 80-81 Lesson 3-6

1. 55 blocks **2.** 135 **3.** 64 cans **5.** $46.50 **7.** 337 miles **9.** 7 min 49.7 sec **10.** # = 1, * = 0; 10.1 × 10 = 101.0 **11.** 0.84 **12.** 11.98 **13.** 0.49 **14.** 1.01 **15.** 70.24 **16.** 16.609 **17.** 4.581 **18.** 237.293 **19.** 10

Pages 82-83 Lesson 3-7

1. 5.1 **2.** 4.3 **3.** 18.9 **4.** 3.03 **5.** 0.013 **6.** 19 **7.** 2.8 **8.** 0.25 **9.** 24.3 **10.** 60 **11.** 1.12 **13.** 0.26 **15.** 20 **17.** 680 **19.** 2.6 **21.** 20.4 **23.** 40 **25.** 0.8 **27.** 30 **29.** 71 **31.** 0.7 **33.** 1.12 **35.** 2.0 **37.** 5.22 **39.** 0.4 **41.** 1.44 yards **43.** 2.56 **45.** 1.75 **47.** 4.3

Pages 86-87 Lesson 3-8
1. 4.68×10^2 **2.** 1.4×10^1 **3.** 1.543×10^4 **4.** 2.66×10^5 **5.** 5×10^4 **6.** 1.4×10^2 **7.** 2.1×10^7 **8.** 3.5×10^6 **9.** 2.07×10^3 **10.** 5.4006×10^5 **11.** 575 **12.** 3,100 **13.** 271,000 **14.** 76,800 **15.** 38,000,000 **16.** 1,090,000 **17.** 4,002 **18.** 100,000 **19.** 3.5×10^5 **21.** 9.2372×10^4 **23.** 6.7×10^6 **25.** 1.5×10^1 **27.** 6×10^6 **29.** 4.17×10^{11} **31.** 431 **33.** 53,000 **35.** 217,000 **37.** 750,000 **39.** 6,380,000,000 **41.** 5,350,000 **43.** 25,730,000 miles; 2.573×10^7 miles **47.** 1.42×10^8; 142,000,000 **49.** 1.6×10^8; 160,000,000 **51.** 1×10^{11}; 100,000,000,000

Page 89 Lesson 3-9
1. 2; 34.4, 68.8, 137.6 **2.** 10; 56,000, 560,000, 5,600,000 **3.** 5; 937.5 4,687.5 23,437.5 **4.** 0.1; 0.091, 0.0091, 0.00091 **5.** 0.4; 12.8, 5.12, 2.048 **6.** 0.5; 2.5, 1.25, 0.625 **7.** 4; 1,472, 5,888, 23,552 **9.** 5, 250, 1,250, 6,250 **13.** 1.2; 2.0736, 2.48832, 2.985984 **15.** 3; 54, 162, 486 **17.** 0.5; 8, 4, 2 **19.** 2,048 vibrations **21.** sixth birthday **23.** *about* 1 **25.** *about* 0.1 **27.** *about* 10 **29.** 77.8 **31.** 7,780 **33.** 12,309 **35.** 4.253 **37.** 0.04253

Pages 90-91 Lesson 3-10
1. 4.8 hr **2.** 5 miles **3.** 2.5 mph **4.** 1,430 miles **5.** 11 yards per carry **6.** 294 yards **7.** 9.5 yards per carry **8.** 25 carries **9.** 4 mph **11.** 1.5 hours **13.** 30 carries **15.** 6.6 yards per carry **17.** 22 mpg **19.** 24.1 mpg **21.** 44 mpg **23.** $s = 216 \div 8$, 27 mpg **25.** 30 ohms **27.** 12 **28.** 71.2 **29.** 5.47 **30.** 20 **31.** 70 **32.** 9 **33.** 10 **34.** $1.44

Pages 92-93 Lesson 3-11
1. 10 in^2 **2.** 32 in^2 **3.** 36 cm^2 **4.** 154 cm^2 **5.** 256 m^2 **6.** 90 in^2 **7.** 240 yd^2 **8.** 1,600 cm^2 **9.** 20 cm^2 **11.** 360 ft^2 **13.** 8,100 ft^2 **15.** $126 **17.** 8 feet **21.** 5.98×10^{15}; 5,980,000,000,000,000

Pages 94-95 Lesson 3-12
1. 6 **2.** 4 **3.** 6 **4.** 17 **5.** 38 **6.** 24 **7.** 59.7 **8.** 10.4 **9.** 119.7 **10.** 0.6 **11.** 6.1 **12.** 4.1 **13.** 115 **15.** 175 **17.** 1.2 **19.** 6 **21.** 646.5 **25.** $50,480 **29.** $22.99 **31.** 80 **33.** 6,000

Pages 96-97 Lesson 3-13
1. 1.1 m **2.** Larry **3.** $53.30 **5.** $1.00 **11.** $x \div 4$ **13.** $(t \div 3) + 10$ **14.** 128 **15.** 5,772 **16.** 14,513 **17.** 140,418 **18.** 15 R4 **19.** 9 **20.** 102 R41 **21.** 13 R330 **22.** $286

Pages 98-99 Chapter 3 Review
1. two, left **3.** dividend **5.** before **7.** 9.52×10^9 **9.** mean **11.** 522 **13.** 2,646 **15.** 22.47 **17.** 689.91 **19.** 1.392 **21.** 70 R4 **23.** 1.457 **25.** 6 **27.** 6,570 **29.** 0.0148 **31.** 0.625 **33.** 3.5 **35.** 6.8 **37.** 0.475 **39.** 0.5625 **41.** 7.99×10^4 **43.** 8.6×10^7 **45.** 490 **47.** 9,700 **49.** 21.6, 6.48, 1.944 **51.** 87.75, 131.625, 197.4375 **53.** $t = 3$ h **55.** 160 m^2 **57.** 5.5 **59.** $6.90 **61.** $1.83 **63.** 840,000 people

Pages 102-103 Cumulative Review/Test
Free Response: 1. $n + 7$ **3.** $6m$ **5.** $g + 32$ **7.** 21 **9.** 15 **11.** $\frac{1}{2}$ **13.** 1,200 miles **15.** 3.05 **17.** 26.13 **19.** 6.09 **21.** $11.95 **23.** 252 **25.** 924 **27.** 19,296 **29.** $11 **31.** 19.482 **33.** 0.36 **35.** 50.076 **37.** 0.75
Multiple Choice: 1. d **3.** d **5.** b **7.** d **9.** b

CHAPTER 4 FRACTIONS: ADDING AND SUBTRACTING

Page 109 Lesson 4-1
1. 1, 2, 3, 6 **2.** 1, 2, 5, 10 **3.** 1, 2, 3, 5, 6, 10, 15, 30 **4.** 1, 5, 11, 55 **5.** 1, 2, 3, 4, 5, 6, 8, 10, 12, 15, 20, 24, 30, 40, 60, 120 **6.** 2, 3, 9 **7.** 2 **8.** 2, 3 **9.** 3 **10.** 5 **11.** 1, 3, 9 **13.** 1, 2, 3, 6, 9, 18 **15.** 1, 3, 7, 21 **17.** 1, 2, 3, 4, 6, 9, 12, 18, 36 **19.** 1, 2, 3, 6, 7, 14, 21, 42 **21.** none **23.** 5 **25.** 2, 3, 9 **27.** 3, 5, 9 **29.** 2, 3 **31.** 1,050 and 1,260 **33.** 2003 **35.** 19 fish **37.** in packages of 1, 3, 9 or 81 **39.** 3

Pages 110-111 Lesson 4-2
1. P **2.** P **3.** C **4.** C **5.** P **6.** C **7.** $2^3 \times 3$ **8.** $2 \times 5 \times 7$ **9.** $2 \times 3 \times 17$ **10.** 11^2 **11.** $2^2 \times 41$ **12.** $5^2 \times 3^2$ **13.** P **15.** P **17.** P **19.** P **21.** P **23.** C **25.** $2^2 \times 5$ **27.** $2^2 \times 13$ **29.** $2^2 \times 7$ **31.** 5×31 **33.** $2^5 \times 3$ **35.** 2×5^4 **37.** 2, 3, 5, 7, 11, 13, 17, 19, 23, 29 **39.** 53 **41.** All even numbers are divisible by 2. **43.** 234; 156; 36; yes **45.** The number is divisible by the product of any combination of the relatively prime numbers. **46.** 49 **47.** 751 **48.** 289 **49.** 3,687 **50.** 21,429 **51.** 40 R3 **52.** 100 R10 **53.** 140 R4 **54.** 15 R399 **55.** A-$0.22, B-$0.21, Store B

Page 113 Lesson 4-3

1. 3 **2.** 10 **3.** 1 **4.** 27 **5.** 6 **6.** 12 **7.** 84 **8.** 50 **9.** 420 **10.** 24 **11.** 2 **13.** 4 **15.** 8 **17.** 5 **19.** 32 **21.** 30 **23.** 60 **25.** 15 **27.** 48 **29.** 42 **31.** $76.88 **33.** $44.97 **35.** $2 \times 5 \times 5$ **37.** June **39.** He suggested that every even whole number greater than or equal to 4 is the sum of two primes.

Pages 114-115 Lesson 4-4

1. $225.03, $344.16 **3.** $141.83 **5.** $201.84 **7.** $2,224.04 **9.** $105.66 **11.** gross pay—total earnings; net pay—take-home pay **13.** compare it to previous statements, find the error, contact your employer or payroll department immediately **15.** 410 **16.** 690 **17.** 130,312 **18.** 693,684 **19.** 40 mph **20.** 478.5 miles **21.** 3 hours **22.** 12 yd^2

Pages 116-117 Lesson 4-5

1. 24 **2.** 7 **3.** 4 **4.** 6 **5.** 14 **6.** 1 **7.** 2 **8.** 1 **9.** 2 **10.** 1 **11.** 15 **13.** 24 **15.** 9 **17.** 1 **19.** 5 **21.** 9 **23.** 1 **25.** 18 **27.** 0 **29.** 18 **31.** $\frac{4}{6}$ **33.** $\frac{2}{8}$ **35.** $\frac{1}{2}$ **37.** two fractions equivalent to $\frac{6}{20}$ **39.** two fractions equivalent to $\frac{9}{20}$ **41.** 15 million barrels **43.** decreases

Pages 118-119 Lesson 4-6

1. $\frac{3}{4}$ **2.** $\frac{7}{8}$ **3.** $\frac{3}{4}$ **4.** $\frac{1}{2}$ **5.** $\frac{1}{3}$ **6.** $\frac{3}{4}$ **7.** $\frac{1}{4}$ **8.** $\frac{1}{2}$ **9.** $\frac{1}{2}$ **10.** $\frac{3}{5}$ **11.** $\frac{2}{5}$ **12.** $\frac{1}{3}$ **13.** $\frac{4}{5}$ **14.** $\frac{2}{5}$ **15.** $\frac{3}{5}$ **16.** $\frac{14}{19}$ **17.** $\frac{3}{4}$ **18.** $\frac{3}{4}$ **19.** $\frac{5}{6}$ **20.** $\frac{9}{20}$ **21.** $\frac{8}{21}$ **22.** $\frac{2}{3}$ **23.** $\frac{3}{8}$ **24.** $\frac{6}{31}$ **25.** $\frac{9}{16}$ **27.** $\frac{3}{8}$ **29.** $\frac{3}{4}$ **31.** $\frac{5}{9}$ **33.** $\frac{3}{8}$ **35.** $\frac{1}{3}$ **37.** $\frac{4}{5}$ **39.** $\frac{8}{9}$ **41.** $\frac{3}{5}$ **43.** $\frac{2}{3}$ **45.** $\frac{22}{25}$ **47.** $(5 + 5) \times 5$ **49.** $5 \div (5 + 5)$ **53.** 150 **54.** 600 **55.** 7,100 **56.** 4,000 **57.** 0.128 **58.** 13.1066 **59.** 602.58 **60.** 38.28319 **61.** 32 pencils

Pages 122-123 Lesson 4-7

1. $>$ **2.** $>$ **3.** $<$ **4.** $<$ **5.** $>$ **6.** $>$ **7.** $>$ **8.** $<$ **9.** $=$ **10.** $<$ **11.** $<$ **13.** $>$ **15.** $>$ **17.** $<$ **19.** $=$ **21.** $=$ **23.** $>$ **25.** $>$ **27.** $<$ **29.** $<$ **31.** $\frac{1}{5}, \frac{1}{4}, \frac{1}{3}, \frac{1}{2}$ **33.** $\frac{1}{3}, \frac{3}{8}, \frac{5}{6}, \frac{7}{8}$ **35.** $\frac{1}{4}, \frac{5}{16}, \frac{1}{2}, \frac{11}{16}$ **37.** Africa **39.** about 5 pounds **41.** no **43.** no **45.** yes **47.** yes **49.** no

Page 125 Lesson 4-8

1. $1\frac{2}{3}$ **2.** $1\frac{1}{6}$ **3.** $1\frac{1}{2}$ **4.** $5\frac{1}{4}$ **5.** $2\frac{2}{3}$ **6.** $2\frac{4}{5}$ **7.** 1 **8.** 3 **9.** 2 **10.** 4 **11.** 6 **12.** 9 **13.** $\frac{11}{8}$ **14.** $\frac{19}{12}$ **15.** $\frac{11}{3}$ **16.** $\frac{22}{5}$ **17.** $\frac{26}{7}$ **18.** $\frac{35}{3}$ **19.** $1\frac{1}{7}$ **21.** 3 **23.** $1\frac{2}{7}$ **25.** $2\frac{1}{5}$ **27.** $2\frac{2}{3}$ **29.** $8\frac{4}{9}$ **31.** $\frac{7}{5}$ **33.** $\frac{15}{8}$ **35.** $\frac{29}{10}$ **37.** one-half **39.** two and seven-eighths **41.** $\frac{3n + 1}{3}$ **43.** $1\frac{7}{10}$ acres **45.** 11 weeks **46.** 24 **47.** 23 **48.** 44.8, 89.6, 179.2 **49.** 27, 9, 3 **50.** $\frac{1}{3}$ year

Page 127 Lesson 4-9

1. 1 **2.** $\frac{1}{2}$ **3.** $\frac{1}{2}$ **4.** 0 **5.** 0 **6.** 1 **7.** 7 **8.** 18 **9.** 28 **10.** 32 **11.** 13 **12.** 20 **13.** 1 **14.** $1\frac{1}{2}$ **15.** 4 **16.** 12 **17.** $\frac{1}{2}$ **18.** 1 **19.** $2\frac{1}{2}$ **20.** 1 **21.** $\frac{1}{2}$ **23.** $\frac{1}{2}$ **25.** $\frac{1}{2}$ **27.** $12\frac{1}{2}$ **29.** 7 **31.** 13 **35.** $1\frac{7}{16}$-mile race **37.** $4 \times 4 \times 4 = 64$ **38.** $2 \times 2 \times 2 \times 2 \times 2 \times 2 = 64$ **39.** $5 \times 5 = 25$ **40.** $8 \times 8 = 64$ **41.** 200 **42.** 12 **43.** 150,000 **44.** 22 **45.** 2 times

Page 129 Lesson 4-10

1. $\frac{3}{4}$ **2.** $\frac{4}{5}$ **3.** $\frac{7}{8}$ **4.** $\frac{2}{3}$ **5.** $1\frac{1}{4}$ **6.** $1\frac{1}{4}$ **7.** $1\frac{1}{2}$ **8.** $1\frac{3}{5}$ **9.** $\frac{7}{8}$ **10.** $\frac{7}{24}$ **11.** $\frac{31}{35}$ **12.** $1\frac{4}{21}$ **13.** $\frac{4}{5}$ **15.** $\frac{2}{3}$ **17.** $1\frac{1}{4}$ **19.** $1\frac{1}{12}$ **21.** $1\frac{1}{5}$ **23.** $1\frac{17}{24}$ **25.** $1\frac{5}{6}$ **27.** $1\frac{4}{9}$ **29.** $1\frac{13}{36}$ **31.** $\frac{7}{12}$ can **33.** $1\frac{1}{8}$ **35.** $1\frac{5}{8}$ **37.** $\frac{1}{2}$ inch

Pages 130-131 Lesson 4-11

1. $11\frac{3}{5}$ **2.** $13\frac{8}{11}$ **3.** $15\frac{1}{2}$ **4.** $9\frac{3}{4}$ **5.** $15\frac{1}{2}$ **6.** $7\frac{7}{8}$ **7.** $8\frac{5}{6}$ **8.** $12\frac{2}{3}$ **9.** $9\frac{5}{7}$ **11.** $8\frac{1}{4}$ **13.** $16\frac{19}{24}$ **15.** $11\frac{3}{4}$ **17.** $10\frac{1}{4}$ **19.** $11\frac{1}{2}$ **21.** $14\frac{5}{6}$ **23.** $24\frac{1}{2}$ **25.** $8\frac{3}{4}$ **27.** $5\frac{1}{4}$ **29.** $25\frac{7}{8}$ **31.** $8\frac{3}{8}$ **33.** $13\frac{7}{8}$ **35.** $8\frac{3}{8}$ yards **37.** Since all operations are addition, the sum is the same with or without parentheses. **39.** 161.8 **40.** $99.02 **41.** 73.63 **42.** 76.812 **43.** 21, 27, 30 **44.** 1.3, 11.2, 14.5 **45.** $17; do not need bold type cost

Pages 132-133 Lesson 4-12

1. $\frac{1}{3}$ **2.** $\frac{1}{9}$ **3.** $\frac{4}{7}$ **4.** $\frac{4}{11}$ **5.** $\frac{1}{5}$ **6.** $\frac{1}{6}$ **7.** $\frac{5}{12}$ **8.** $\frac{23}{90}$

9. $\frac{1}{12}$ **10.** $\frac{1}{6}$ **11.** $\frac{1}{2}$ **13.** $\frac{7}{20}$ **15.** $\frac{2}{3}$ **17.** $\frac{1}{20}$ **19.** 0
21. $\frac{1}{24}$ **23.** $\frac{1}{2}$ **25.** $\frac{7}{18}$ **27.** $\frac{11}{24}$ **29.** $\frac{2}{3}$ **31.** $\frac{1}{2}$ cup
33. 14 students liked both. **35.** 19

Page 135 Lesson 4-13

1. $2\frac{2}{3}$ **2.** $1\frac{1}{2}$ **3.** $5\frac{1}{4}$ **4.** $3\frac{3}{5}$ **5.** $3\frac{1}{20}$ **6.** $4\frac{2}{15}$ **7.** $11\frac{1}{8}$
8. $6\frac{4}{9}$ **9.** $3\frac{5}{6}$ **10.** $6\frac{17}{20}$ **11.** $4\frac{7}{24}$ **12.** $4\frac{23}{24}$ **13.** $3\frac{1}{3}$
14. $4\frac{3}{4}$ **15.** $9\frac{3}{8}$ **16.** $13\frac{3}{7}$ **17.** $10\frac{3}{8}$ **19.** $3\frac{2}{3}$ **21.** $8\frac{1}{3}$
23. $1\frac{33}{40}$ **25.** $4\frac{5}{6}$ **27.** $2\frac{1}{2}$ **29.** $13\frac{5}{7}$ **31.** $1\frac{3}{8}$ **33.** $5\frac{1}{2}$
35. $10\frac{5}{8}$ **37.** $1\frac{19}{24}$ **39.** $\frac{22}{75}$ **41.** $\frac{5}{12}$ **43.** 1, 5, 3, 2, 4

Pages 136-137 Lesson 4-14

1. 16 sections **2.** 4 pictures **3.** 51 inches **5.** 9 days
7. 20 persons **11.** 19 ways **12.** 31.2
13. 41.5 **14.** 765.5 **15.** 3.573 **16.** 1,840 **17.** 96
18. 3,720 **19.** 628 **20.** 0.084 **21.** 0.624
22. 0.0159
23. 4.6 **24.** no, $\frac{1}{8}$

Pages 138-139 Chapter 4 Review

1. 30 **3.** $\frac{2}{3}$ **5.** GCF **7.** mixed number **9.** 1, 2, 3, 4, 6, 12 **11.** 1, 29 **13.** 1, 2, 3, 6, 7, 14, 21, 42 **15.** 3
17. 2, 3, 9 **19.** $2 \times 7 \times 7$ **21.** $2 \times 3 \times 3 \times 5$
23. $2 \times 2 \times 3 \times 13$ **25.** 16 **27.** 2 **29.** 28
31. 24 **33.** 30 **35.** $5,701.80 **37.** 12 **39.** 2 **41.** 6
43. 21 **45.** 4 **47.** $\frac{3}{4}$ **49.** $\frac{1}{7}$ **51.** $\frac{19}{36}$ **53.** $>$ **55.** $<$
57. $7\frac{1}{3}$ **59.** $6\frac{3}{4}$ **61.** $7\frac{4}{5}$ **63.** $\frac{37}{8}$ **65.** $\frac{20}{3}$ **67.** $\frac{71}{9}$
69. 1 **71.** $8\frac{1}{2}$ **73.** 6 **75.** $1\frac{1}{4}$ **77.** $1\frac{1}{20}$ **79.** $6\frac{2}{9}$
81. $9\frac{9}{11}$ **83.** $12\frac{11}{12}$ **85.** $\frac{3}{5}$ **87.** $\frac{11}{18}$ **89.** $2\frac{2}{5}$ **91.** $3\frac{5}{12}$
93. $17\frac{1}{2}$ **95.** Cut the cake into thirds horizontally and then cut the cake one fourth of the way vertically.

Pages 142-143 Cumulative Review/Test

Free Response: **1.** 600 **3.** 4,000 **5.** 850,600
7. 5,400 **9.** 7 **11.** 8,700 **13.** 180,000 **15.** 300,000
17. $483 + n = 625$; 142 **19.** 929 **21.** 468 **23.** 248
25. 171 **27.** 4,758 **29.** 3,886 **31.** 146 **33.** 9,496
35. 9,940 **37.** 12 weeks **39.** $3 \times 2 \times 2 \times 17$
41. 2×37 **43.** 70 **45.** 4 **47.** 6
Multiple Choice: **1.** c **3.** b **5.** c **7.** b **9.** c

CHAPTER 5 FRACTIONS: MULTIPLYING AND DIVIDING

Pages 146-147 Lesson 5-1

1. 21 **2.** 2 **3.** 12 **4.** 9 **5.** 2 **6.** 5 **7.** 2 **8.** 4
9. less than **10.** greater than **11.** 15 **13.** 2 **15.** 7
17. 6 **19.** less than **21.** less than **23.** about $9
25. about 4 weeks **27.** 5 **29.** 36 **31.** 6.94
32. 5.42 **33.** 21.62 **34.** 48,000 **35.** 6,230
36. 901,000 **37.** $1\frac{5}{12}$ cups

Pages 148-149 Lesson 5-2

1. $\frac{3}{16}$ **2.** $\frac{5}{9}$ **3.** $\frac{8}{15}$ **4.** $\frac{3}{10}$ **5.** $\frac{1}{3}$ **6.** $\frac{3}{7}$ **7.** $1\frac{5}{16}$ **8.** 15
9. $4\frac{3}{8}$ **10.** 6 **11.** $\frac{4}{63}$ **13.** $\frac{40}{63}$ **15.** $\frac{4}{9}$ **17.** $\frac{7}{9}$ **19.** $3\frac{3}{4}$
21. 10 **23.** $\frac{3}{8}$ **25.** $\frac{9}{40}$ **27.** true **29.** $48.75
31. $1\frac{5}{12}$ **33.** $2\frac{2}{3}$ **35.** 65.2 **36.** 17.03 **37.** 4.05
38. 182 **39.** 360 **40.** 3,213 **41.** $33

Pages 150-151 Lesson 5-3

1. $4\frac{1}{8}$ **2.** $8\frac{2}{3}$ **3.** $7\frac{7}{8}$ **4.** $18\frac{2}{3}$ **5.** 39 **6.** 130 **7.** $1\frac{19}{30}$
8. $3\frac{31}{63}$ **9.** $1\frac{7}{9}$ **10.** 30 **11.** $12\frac{2}{3}$ **12.** 8 **13.** $4\frac{23}{28}$
15. $3\frac{3}{5}$ **17.** 25 **19.** $4\frac{7}{12}$ **21.** 1 **23.** $51\frac{1}{4}$ **25.** 45
27. $15\frac{1}{5}$ **29.** $42\frac{1}{4}$ **31.** 5 hours **35.** yes **37.** 35
39. 40

Pages 152-153 Lesson 5-4

1. $\frac{11}{20}$ ton **2.** $\frac{1}{5}$ ton **3.** $\frac{1}{8}$ ton **4.** 1 ton **5.** $\frac{1}{4}$ ton
6. $\frac{1}{4}$ ton **7.** $\frac{1}{2}$ ton **8.** $\frac{1}{6}$ ton **9.** $\frac{1}{40}$ ton
11. $\frac{13}{20}$ of the total coal deposits **13.** Appalachian Plateau **15.** $\frac{9}{50}$ of the total coal deposits
17. *about* $65 **19.** *about* 30 million tons **20.** 4^3, 8^2
21. 3^4, 9^2 **22.** 1^2, 1^3 **23.** 10^4, 100^2 **24.** 1, 2, 4, 8, 16, 32 **25.** 1, 3, 5, 9, 15, 45 **26.** 1, 2, 3, 4, 6, 8, 12, 24 **27.** 1, 2, 3, 6, 9, 18, 27, 54
28. 700 students

Pages 154-155 Lesson 5-5

1. 48 lawns **2.** Since the last cut yields two pieces, only 15 cuts are necessary. **3.** $(337 + 4) \times 100 = 34,100$ feet. **5.** 2,678,400 seconds, calculator **7.** 6 socks
11. 22.3 **12.** 5.5 **13.** 0.8 **14.** 2×3^2 **15.** $2^2 \times 5$
16. $2^2 \times 7$ **17.** $2 \times 3 \times 7$ **18.** $2 \times 3 \times 17$
19. $7\frac{5}{6}$ hours

Page 159 Lesson 5-6

1. $\frac{9}{8}$ **2.** $\frac{3}{2}$ **3.** $\frac{20}{3}$ **4.** $\frac{7}{5}$ **5.** $\frac{1}{5}$ **6.** $\frac{1}{12}$ **7.** $5\frac{1}{3}$ **8.** 16 **9.** 18 **10.** 14 **11.** $2\frac{2}{3}$ **12.** $\frac{14}{25}$ **13.** $\frac{3}{40}$ **14.** $\frac{1}{12}$ **15.** $\frac{20}{19}$ **17.** $\frac{16}{11}$ **19.** $1\frac{7}{8}$ **21.** $\frac{5}{8}$ **23.** 16 **25.** 4 **27.** $\frac{1}{36}$ **29.** $\frac{1}{20}$ **31.** $1\frac{3}{4}$ **33.** no; $\frac{7}{8} \div \frac{1}{6} = 5\frac{1}{4}$
35. Six rectangles, each divided into thirds shows a quotient of 18 smaller rectangles.
37. $\frac{2}{3}$ **39.** $\frac{1}{4}$

Pages 160-161 Lesson 5-7

1. $1\frac{7}{18}$ **2.** $1\frac{37}{62}$ **3.** 6 **4.** $\frac{2}{11}$ **5.** $\frac{2}{29}$ **6.** $1\frac{47}{88}$ **7.** $\frac{9}{14}$ **8.** $1\frac{1}{4}$ **9.** $\frac{5}{6}$ **10.** $\frac{7}{8}$ **11.** $\frac{1}{12}$ **12.** 5 **13.** $1\frac{7}{8}$ **14.** $7\frac{1}{2}$ **15.** $\frac{8}{11}$ **16.** $1\frac{1}{6}$ **17.** $\frac{3}{8}$ **19.** $2\frac{1}{2}$ **21.** $6\frac{2}{3}$ **23.** $\frac{3}{10}$ **25.** $\frac{2}{5}$ **27.** $1\frac{1}{10}$ **29.** 2 **31.** $2\frac{8}{13}$ **33.** 5
35. 30 pieces **37.** 20 boards **41.** \$18 **43.** \$29 **45.** \$34

Pages 162-163 Lesson 5-8

1. 0.4 **2.** 0.375 **3.** 0.75 **4.** 0.3125 **5.** $0.\overline{1}$ **6.** $0.\overline{15}$ **7.** $0.8\overline{3}$ **8.** $0.58\overline{3}$ **9.** $0.66\frac{2}{3}$ **10.** $0.53\frac{11}{13}$ **11.** $0.88\frac{8}{9}$ **12.** $0.54\frac{1}{6}$ **13.** 0.7 **15.** 0.55 **17.** $0.\overline{27}$ **19.** $0.\overline{6}$ **21.** $0.\overline{7}$ **23.** $0.6\overline{2}$ **25.** 0.1875 **27.** $0.91\overline{6}$ **29.** $0.63\frac{1}{3}$ **31.** $0.33\frac{1}{3}$ **33.** 0.008 **35.** 0.875 inch
37. Answers may vary. A good answer refers to prime factors other than 2 or 5. That is, denominators without prime factors 2 or 5 result in terminating decimals.
39. 89% = 0.89; 1,051 students

Pages 164-165 Lesson 5-9

1. $\frac{3}{500}$ **2.** $\frac{21}{250}$ **3.** $\frac{1}{8}$ **4.** $\frac{13}{20}$ **5.** $\frac{51}{125}$ **6.** $\frac{49}{50}$ **7.** $\frac{53}{100}$ **8.** $\frac{31}{100}$ **9.** $\frac{16}{25}$ **10.** $\frac{1}{20}$ **11.** $\frac{1}{7}$ **12.** $\frac{2}{3}$ **13.** $\frac{2}{9}$ **14.** $\frac{6}{7}$ **15.** $\frac{5}{7}$ **17.** $\frac{4}{5}$ **19.** $\frac{3}{4}$ **21.** $\frac{19}{50}$ **23.** $\frac{101}{1,000}$ **25.** $\frac{243}{500}$ **27.** $\frac{56}{125}$ **29.** $\frac{1}{125}$ **31.** $10\frac{9}{50}$ **33.** $\frac{1}{3}$ **35.** $\frac{1}{11}$ **37.** = **39.** < **41.** 15 pieces **43.** $\frac{15}{18}$ **45.** 0.675 **47.** 0.125 or $\frac{1}{8}$ **49.** 0.515625 or $\frac{33}{64}$

Pages 166-167 Lesson 5-10

1. $15\frac{5}{8}$ yd **2.** $10\frac{1}{2}$ feet **3.** 264 students **5.** about 14 minutes **7.** about 1,000 miles long **9.** $46\frac{7}{8}$ miles **11.** $2\frac{1}{8}$ bushels **13.** 5 monkeys **15.** 66 **16.** 22 **17.** 378 **18.** 8 **19.** 7 **20.** 15 **21.** 9 **22.** 17 **23.** Albert

Pages 168-169 Chapter 5 Review

1. i **3.** e **5.** d **7.** 45 **9.** 2 **11.** less than **13.** $\frac{1}{9}$ **15.** $6\frac{2}{3}$ **17.** $\frac{9}{25}$ **19.** $\frac{4}{9}$ **21.** $\frac{2}{3}$ **23.** $1\frac{1}{20}$ **25.** $27\frac{3}{5}$ **27.** $16\frac{2}{3}$ **29.** 1 **31.** $\frac{1}{10}$ ton **33.** $\frac{3}{5}$ ton **35.** 36 centerpieces **37.** $\frac{4}{3}$ **39.** $\frac{1}{6}$ **41.** $\frac{2}{9}$ **43.** $1\frac{3}{4}$ **45.** $\frac{3}{4}$ **47.** $\frac{3}{20}$ **49.** $1\frac{2}{9}$ **51.** $\frac{7}{12}$ **53.** $\frac{21}{40}$ **55.** 8 **57.** $\frac{10}{27}$ **59.** 2 **61.** 0.1 **63.** 0.52 **65.** $0.\overline{01}$ **67.** $\frac{17}{500}$ **69.** $\frac{1}{4}$ **71.** 3 scoops **73.** $\frac{19}{30}$ of her salary **75.** $6\frac{2}{3}$

Pages 172-173 Cumulative Review/Test

Free Response: 1. $3 \times 3 \times 3 \times 3 = 81$ **3.** $10 \times 10 \times 10 \times 10 \times 10 = 100,000$ **5.** $1 \times 1 \times 1 \times 1 \times 1 \times 1 \times 1 \times 1 \times 1 \times 1 = 1$ **7.** $8 \times 8 = 64$ **9.** 12 **11.** 10 **13.** 156 **15.** 7.8 **17.** 162.7 **19.** 3,888 **21.** 7,680 **23.** 63,000 **25.** 4 **27.** 4 **29.** 400 **31.** \$124.96 **33.** 1, 89; prime **35.** 1, 3, 13, 39 **37.** 4; 360 **39.** 1; 225 **41.** $1\frac{3}{4}$ cups **43.** $\frac{1}{21}$ **45.** $15\frac{5}{9}$ **47.** $1\frac{17}{27}$ **49.** $36\frac{2}{5}$
Multiple Choice: 1. c **3.** b **5.** b **7.** a **9.** b

CHAPTER 6 MEASUREMENT

Pages 178-179 Lesson 6-1

1. thousandths **2.** thousands **3.** hundredths **4.** tens **5.** tenths **6.** hundreds **7.** 10 **8.** 10 **9.** 1,000 **10.** 1,000 **11.** dekagram **12.** centigram **13.** millimeter **14.** decimeter **15.** 1,000 **17.** 0.001 **19.** 0.01 **21.** millimeter **23.** hectogram **25.** dekaliter **27.** kilometer **29.** liter **31.** 1 dekameter **33.** 7 spools **35.** Changes for road signs, record keeping for sports events, pricing per metric unit, packaging, scales, and so on **37.** The current metric system was developed in 1795 by the French Academy of Sciences and expanded in 1960. In the United States, the Metric Conversion Act of 1975 specified a national policy of *voluntary* use of the metric system and established the U.S. Metric Board to help ease the change. **39.** 463 **40.** 58.2 **41.** 16,400 **42.** 697 **43.** $5\frac{2}{3}$ **44.** $8\frac{1}{7}$ **45.** $3\frac{5}{18}$ **46.** $3\frac{1}{10}$ **47.** no; too short

Pages 180-181 Lesson 6-2

1. meter **2.** centimeter **3.** millimeter **4.** meter **5.** kilometer **6.** millimeter **7.** 66 cm **9.** 22 cm **11.** 120 cm **13.** 1.2 cm, 12 mm **15.** 25 bows **17.** 1,149 km **21.** *about* 5 m

Pages 182-183 Lesson 6-3

1. 300 **2.** 8,000 **3.** 20 **4.** 120 **5.** 4 **6.** 960 **7.** 2 **8.** 4 **9.** 0.0046 **10.** 0.275 **11.** 0.00978 **12.** 567 **13.** 70 **15.** 3.5 **17.** 400 **19.** 5 **21.** 0.3 **23.** 420 **25.** 0.0087 **27.** 36 **29.** 1,092 mps **31.** 49.4 km **35.** 7.01×10^6 **36.** 8.3×10^5 **37.** 4.8×10^7 **38.** 5.37×10^4 **39.** 2 **40.** 2, 3 **41.** 3 **42.** 3, 5, 9 **43.** 2, 3, 5, 9, 10 **44.** pizza

Pages 184-185 Lesson 6-4

1. 1 g **2.** 17 mg **3.** 1,000 kg **4.** 10 g **5.** 80 g **6.** 1.4 kg **7.** 0.0049 **8.** 5.862 **9.** 8.754 **10.** 67,000 **11.** 73,000 **12.** 9,700 **13.** 8,000 **15.** 0.752 **17.** 9,400 **19.** 0.912 **21.** 24,000 **23.** 0.005 **25.** 25,800 grams **27.** 2 kg **29.** 12 cans **31.** 129 **32.** 1,456 **33.** 7,623 **34.** 13,288 **35.** 78,800 **36.** 30 R3 **37.** 17 R9 **38.** 52 R62 **39.** 51 R51 **40.** 36 days

Pages 186-187 Lesson 6-5

1. liter **2.** liter **3.** liter **4.** milliliter **5.** 42,000 **6.** 900 **7.** 36,000 **8.** 0.791 **9.** 0.2356 **10.** 0.004 **11.** 6,000 **13.** 7,500 **15.** 0.234 **17.** 0.127 **19.** 28,000 **21.** 0.262 **23.** 17,000 **25.** 0.0968 **27.** 780,000 mL **29.** 30 mL **31.** 250 cups **33.** 12.3 **34.** 4.36 **35.** 18.67 **36.** 35 **37.** 36 **38.** 51 **39.** $37

Pages 188-189 Lesson 6-6

1. 13 pretzels **2.** $5.50 gain **3.** 240 flowers **5.** 4 blocks west **7.** 60 **9.** 2 hours longer **10.** 24 **11.** 63 **12.** 5 **13.** 4 **14.** $2\frac{2}{7}$ **15.** $\frac{1}{2}$ **16.** $\frac{2}{11}$ **17.** $\frac{7}{8}$ **18.** 20 rungs

Pages 190-191 Lesson 6-7

1. 36 **2.** 6 **3.** 3,520 **4.** 108 **5.** 10,560 **6.** 45 **7.** 2 **8.** 2 **9.** 5 **10.** 2, 6 **11.** 7, 2 **12.** 6, 4 **13.** 60 **15.** 72 **17.** 7, 2 **19.** 9 **21.** 5 **23.** 18 **25.** 2, 2 **27.** $1\frac{1}{2}$ **29.** 6 ft 4 in. **31.** Answers may vary. You may respond to the ease of computing in the metric system or the familiarity of using the customary system. **33.** Size 9

Pages 194-195 Lesson 6-8

1. 16 **2.** 8 **3.** 5 **4.** 3 **5.** 5 **6.** $1\frac{1}{2}$ **7.** 4,000 **8.** 3 **9.** 6 **10.** 80 **11.** 48 **12.** 8 **13.** 32 **15.** 2 **17.** 10 **19.** 11 **21.** 12 **23.** 8 **25.** 16 **27.** 16 **29.** 96 **31.** 7 pounds **33.** gallon **35.** cup, ounce **37.** 20 weeks **39.** about 59 pounds **41.** 32 **43.** 4 **45.** Find the number of ounces in one gallon, 128, and multiply by 8.

Pages 196-197 Lesson 6-9

1. 20°F **2.** 30°C **3.** 23°F **4.** 70°F **5.** 140°F **6.** 59°F **7.** 122°F **8.** 185°F **9.** 86°F **10.** 5°C **11.** 15°C **12.** 65°C **13.** 115°C **14.** 90°C **15.** 212°F **17.** 127°F **19.** 252°F **21.** 55°F **23.** 145°F **25.** 35°C **27.** 38°C **39.** 100°C **31.** 32°C **33.** 1°C **35.** 35.6°F **37.** 20.5° to 23.3°C **39.** 10 hours **41.** 158° **43.** 100°

Pages 198-199 Lesson 6-10

1. 1 h 5 min **2.** 3 days **3.** 6 h **4.** 500 s **5.** 2 days **6.** 300 min. **7.** 300 **8.** 480 **9.** 96 **10.** 2 **11.** 6 **12.** 1 **13.** 3 h 25 min **14.** 2 h 23 min **15.** 240 **17.** 3 **19.** $\frac{3}{4}$ **21.** 1,440 **23.** 270 **25.** 13 h 30 min **27.** 243 sec **31.** 300 times

Pages 200-201 Lesson 6-11

1. 13:35 **2.** 19:20 **3.** 02:10 **4.** 06:40 **5.** 16:50 **6.** 00:10 **7.** 8 h **8.** 8 h **9.** 8 h **11.** 6 h 30 min **13.** 8 h 10 min **15.** 7 h 10 min **17.** no; 20 minutes late **19.** about 278 households **22.** $10\frac{3}{8}$ **23.** $21\frac{11}{12}$ **24.** $9\frac{13}{45}$ **25.** $17\frac{20}{21}$ **26.** $15\frac{6}{15}$ **27.** 32 **28.** $7\frac{1}{2}$ **29.** $4\frac{1}{8}$ **30.** 0.36 km

Pages 202-203 Lesson 6-12

1. 7 ft 9 in. **2.** 24 glasses **3.** $3\frac{1}{3}$ h **4.** 8 yd **5.** 1 mm too large **7.** 113.5 grams **10.** 900 **11.** 4,900 **12.** 17.00 **13.** 1,000 **14.** 186 **15.** 1,716 **16.** 12,600 **17.** 150,381 **18.** 840 yards

Pages 204-205 Chapter 6 Review

1. 100 **3.** liter **5.** Celsius **7.** time **9.** customary **11.** foot **13.** centimeter **15.** milliliter **17.** milligram

19. 4.2 cm; 42 mm **21.** 5.1 cm; 51 mm **23.** 50 **25.** 1 **27.** 0.4 **29.** 34,000 **31.** 4.536 **33.** 8,500 **35.** 36 **37.** 10 **39.** 12 **41.** 30 **43.** 4,000 **45.** 40 **47.** 86°F **49.** 38°C **51.** 42°C **53.** 2°C **55.** 180 **57.** 168 **59.** 302 **61.** 02:33 **63.** 23:29 **65.** 6 h 59 min **67.** 9,600 kg **69.** 256 min

Pages 208-209 Cumulative Review/Test

Free Response: 1. 6,000 **3.** 690,000 **5.** 5.1, 3.9, 2.7 **7.** 2, 3, 9 **9.** 2, 3, 5, 9, 10 **11.** > **13.** < **15.** $\frac{1}{2}$ **17.** 0 **19.** $\frac{1}{2}$ **21.** $23\frac{5}{8}$ **23.** $2\frac{1}{4}$ **25.** $\frac{3}{200}$ **27.** 11 **29.** $\frac{1}{4}$ **31.** 5 **33.** 75,000 **35.** 32,000 **37.** 4 min 43 sec

Multiple Choice: 1. a **3.** d **5.** c **7.** b **9.** a

CHAPTER 7 GEOMETRY

Pages 212-213 Lesson 7-1

1. $\overline{CD}$ **2.** $\overrightarrow{BD}$ **3.** $\overleftrightarrow{RS}$ **4.** $\overleftrightarrow{MN}$ **5.** $\overline{AB}$ **6.** T **7.** pin point; thumbtack **8.** pencil; intersection of the wall and the floor **9.** flashlight beam; laser beam **10.** wall; floor **11.** $\overline{GH}$; $\overline{HG}$; line segment GH; line segment HG **13.** point M; M **23.** 1 line segment **27.** infinitely many **31.** $\frac{3}{4}$ inch **33.** $\frac{14}{4}$ or $3\frac{1}{2}$ inches **35.** 52 **36.** 76 **37.** 3,763 **38.** C **39.** P **40.** P **41.** C **42.** 12 gallons

Pages 214-215 Lesson 7-2

1. $\angle 3$; $\angle XYZ$; $\angle ZYX$ **2.** $\angle 1$; $\angle PQR$; $\angle RQP$ **3.** $\angle 2$; $\angle TUV$; $\angle VUT$ **4.** $\angle 5$; $\angle GHI$; $\angle IHG$ **5.** 74° **7.** 90° **9.** 136° **11.** $\angle HIL$; $\angle HIK$; $\angle LIJ$; $\angle LIK$; $\angle KIJ$; There is more than one angle with vertex I. **13.** 180° **17.** 2; 14; 7; 3.5 **18.** 4; 3; $\frac{3}{4}$; $\frac{3}{8}$ **19.** 9 **20.** 24 **21.** 20 **22.** 4 **23.** 120 pizzas

Pages 216-217 Lesson 7-3

1. 65°; acute **2.** 90°; right **3.** 120°; obtuse **5.** obtuse **7.** acute **9.** acute **11.** right **13.** obtuse **15.** obtuse **19.** 30°, 60° **21.** 6; 10; 15; 21 angles

Page 219 Lesson 7-4

11. acute **13.** Draw a line. Construct a segment congruent to $\overline{CD}$. Construct a second segment congruent to $\overline{CD}$ on the line, extending from the end of the first segment.

Pages 220-221 Lesson 7-5

1. parallel; $\overleftrightarrow{PQ} \parallel \overleftrightarrow{RS}$ **2.** perpendicular $\overleftrightarrow{ZW} \perp \overleftrightarrow{XY}$ **3.** skew **5.** parallel; $\overleftrightarrow{RS} \parallel \overleftrightarrow{TQ}$ **7.** skew **9.** skew **11.** $\overline{TY}$, $\overline{QV}$, $\overline{SX}$ **15.** parallel **17.** 0.38 **18.** 0.84 **19.** 4.09 **20.** 9.98 **21.** 13.00 **22.** $2\frac{5}{6}$ **23.** $\frac{19}{25}$ **24.** $1\frac{17}{40}$ **25.** $5\frac{1}{3}$ **26.** 42°F

Pages 224-225 Lesson 7-6

1. quadrilateral; not regular **2.** pentagon; not regular **3.** octagon; regular **4.** decagon; not regular **5.** quadrilateral; not regular **7.** triangle; not regular **9.** not closed **11.** not made up of line segments **19.** Closed means all line segments must intersect. Plane means the figures lie in one plane. They have only 2 dimensions, not 3. **21.** 280.9 feet per second **23.** Texas, Oklahoma **25.** Texas, Alaska **27.** $13,776,000,000

Pages 226-227 Lesson 7-7

1. equilateral **2.** scalene **3.** isosceles **4.** scalene **5.** right **6.** acute **7.** acute **8.** right **9.** equilateral; acute **11.** isosceles; acute **13.** isosceles; obtuse **21.** true **23.** false **25.** true **27.** $\triangle PBD$; $\triangle APD$ **29.** $\triangle APD$ **33.** 48°, 90° **35.** The sum of the measures of the two right angles is 180°. Since the sum of the measures of the angles in a triangle is 180°, perpendicular rays will not intersect to form the third vertex.

Pages 228-229 Lesson 7-8

1. rectangle **2.** square **3.** trapezoid **4.** rhombus **5.** square **7.** rhombus **13.** false **15.** false **17.** false **19.** true **21.** 37 cm **23.** 11 squares **24.** $1\frac{1}{5}$ **25.** $1\frac{4}{5}$ **26.** $2\frac{4}{7}$ **27.** $\frac{3}{5}$ **28.** $\frac{21}{50}$ **29.** $\frac{13}{40}$ **30.** $\frac{319}{500}$ **31.** 6:50 A.M.

Page 231 Lesson 7-9

1. cone **2.** cylinder **3.** rectangular pyramid **4.** pentagonal prism **5.** 4; 4; 6; 7; 10; 15; 5; 6; 9; 5; 5; 8; 6; 6; 10 **7.** $F + V = E + 2$ **9.** 200 vertices **11.** Prisms are named by the shape of the figure.

Pyramids are named by the shape of their base.
13. 2; No, a cylinder does not have two line segments that intersect.

Pages 232-233 Lesson 7-10

1. b, not a ball **2.** c, not a polygon **3.** c, not three dimensional **4.** d, not a prism **5.** a **7.** b **9.** 11 **11.** 8 **13.** 49, 81; square numbers **15.** $1\frac{2}{5}$ **16.** $1\frac{5}{12}$ **17.** $1\frac{6}{35}$ **18.** 1 **19.** 0.042 **20.** 9,500 **21.** 250 **22.** 20°

Pages 234-235 Chapter 7 Review

1. line segment **3.** acute **5.** isosceles **7.** congruent **9.** edges **19.** 52°; acute **21.** 96°; obtuse **25.** skew **27.** pentagon; regular **29.** quadrilateral; not regular **31.** isosceles; right **33.** equilateral; acute **35.** rectangle **37.** trapezoid **39.** triangular prism **41.** cylinder **43.** c

Pages 238-239 Cumulative Review/Test

Free Response: **1.** 13 **3.** 2.1 **5.** 500 **7.** 666 **9.** 217 **11.** $1,500,000 **13.** $\frac{9}{10}$ **15.** $\frac{3}{7}$ **17.** 25 **19.** $\frac{1}{8}$ **21.** 2 **23.** 100 **25.** $9\frac{1}{3}$ **27.** 0.85 **29.** 0.30 **31.** 1.5 **33.** equilateral, acute **35.** right; scalene
Multiple Choice; **1.** d **3.** c **5.** c **7.** d **9.** c

CHAPTER 8 AREA AND VOLUME

Page 245 Lesson 8-1

1. 25.1 cm **2.** 22.0 m **3.** 81.6 in. **4.** 1,507.2 mm **5.** 3.1 m **6.** 21.0 ft **7.** 160.1 cm **8.** 72.2 in. **9.** 182.1 cm **11.** 113.0 in. **13.** 18.8 ft or 6.3 yd **15.** 17.9 **16.** 130.8 **17.** rhombus **18.** trapezoid **19.** rectangle **20.** parallelogram **21.** 3.625 gallons

Pages 246-247 Lesson 8-2

1. 42 m^2 **2.** 60 in^2 **3.** 22.5 cm^2 **4.** 45 m^2 **5.** 297.5 cm^2 **6.** 48 ft^2 **7.** 40,920 miles^2 **9.** 55.2 cm^2 **11.** 5.6 yd^2 **13.** 30.15 m^2 **15.** 3,840 in^2 **17.** 2,803.2 m^2 **19.** There are thirty-five different arrangements. **21.** 8 ft

Pages 248-249 Lesson 8-3

1. 234 cm^2 **2.** 36 ft^2 **3.** 60 cm^2 **4.** 11.25 ft^2 **5.** 11.2 in^2 **6.** 1.575 m^2 **7.** 110 yd^2 **8.** 77 mm^2 **9.** 0.975 cm^2 **11.** 22.14 m^2 **13.** 51 cm^2 **15.** 4.55 in^2 **17.** 336 mm^2 **19.** 15.75 m^2 **21.** 104 ft^2 **25.** 16 **26.** 103 **27.** 45.5 **28.** 61,300 **29.** 0.4265 **30.** 2,100 **31.** 33°F

Pages 250-251 Lesson 8-4

1. 78.5 cm^2 **2.** 254.3 m^2 **3.** 1,256 yd^2 **4.** 2,826 mm^2 **5.** 50.2 cm^2 **6.** 78.5 in^2 **7.** 706.5 yd^2 **8.** 379.9 m^2 **9.** 12.6 cm^2 **11.** 132.7 cm^2 **13.** 254.3 yd^2 **15.** 283.4 cm^2 **17.** 1,017.4 ft^2 **19.** 830.5 ft^2 **21.** 5 figures **23.** 9,852.0 in^2

Pages 252-253 Lesson 8-5

1. $243 **2.** $315 **3.** $456 **4.** $264 **5.** 127.3 yd^2; $3,048.84 **7.** 30 yd^2; $448.50 **9.** $4,685.24 **11.** 450 **12.** 150 **13.** 600 **14.** 1,000 **15.** $\overline{DC}$, $\overline{EF}$, $\overline{HG}$ **16.** 12:00 and 12:30

Pages 254-255 Lesson 8-6

1. 39 students **3.** 60 stamps **5.** 240 doctors **7.** $75.91; calculator **9.** $23.48 **10.** 5.31 **11.** 21.107 **12.** $\overrightarrow{RS}$ **13.** $\overleftrightarrow{XY}$ **14.** $\overline{MN}$ **15.** $\overrightarrow{AB}$ **16.** 3,140 ft

Pages 258-259 Lesson 8-7

1. 80 m^2, 80 m^2, 48 m^2, 48 m^2, 60 m^2, 60 m^2, 376 m^2 **2.** 33.6 in^2, 33.6 in^2, 12 in^2, 12 in^2, 11.2 in^2, 11.2 in^2, 113.6 in^2 **3.** 1,750 cm^2 **5.** 19,518 cm^2 **7.** 1,181 in^2 **9.** 5,330 cm^2 **11.** 39 ft^2 **13.** 2 × 4 × 6 inch-prism **15.** 4

Pages 260-261 Lesson 8-8

1. 153.9 m^2, 153.9 m^2; 219.8 m^2, 527.6 m^2 **2.** 78.5 cm^2, 78.5 m^2, 565.2 cm^2, 722.2 cm^2 **3.** 979.7 cm^2 **5.** 1,132.0 cm^2 **7.** 829.0 in^2 **9.** 10 cm radius, 5 cm height **11.** 706 cans **13.** *about* 8 ft diameter

Pages 262-263 Lesson 8-9

1. 64 m^3 **2.** 105,000 cm^3 **3.** 297 in^3 **4.** 192.5 in^3 **5.** 105 ft^3 **6.** 9,000 in^3 **7.** 18,900 cm^3 **9.** 23,760 ft^3 **11.** 229.5 m^3 **13.** 3,375 m^3 **15.** 19.5 ft^3 **17.** 10 horsepower **21.** 12 launches

Pages 264-265 Lesson 8-10

1. 384 m^3 **2.** 300 cm^3 **3.** 1,200 in^3 **4.** 286 m^3 **5.** 10 cm^3 **6.** 367.8 yd^3 **7.** 40 m^3 **9.** 256 m^3 **11.** 16 cm^3 **13.** 21,769 ft^3 **15.** The volume is eight times greater. **17.** 15 **19.** 0.2

Pages 266-267 Lesson 8-11
1. 2,373.8 cm^3 **2.** 5,652 in^3 **3.** 769.3 m^3 **4.** 62.8 in^3 **5.** 1,356.5 cm^3 **6.** 1,271.7 ft^3 **7.** 602.9 cm^3 **9.** 39.3 m^3 **11.** 11,335.4 in^3 **13.** $\approx$ 3,532.5 m^3 **15.** 14 days **18.** 65° **19.** 130° **20.** 28° **21.** 18,000 **22.** 9,500 **23.** 0.000028 **24.** 36 ft^2

Pages 268-269 Lesson 8-12
1. 50.2 cm^3 **2.** 10,048 yd^3 **3.** 150.7 m^3 **4.** 401.9 m^3 **5.** 100.5 mm^3 **6.** 376.8 in^3 **7.** 1,004.8 ft^3 **9.** $52.\overline{3}$ cm^3 **11.** 401.9 mm^3 **13.** 2,713 ft^3 **15.** 10 cm **17.** Friday

Pages 270-271 Lesson 8-13
1. 11,304 mi^2 **2.** 6 ft **3.** 2.5 in^2 **5.** 74.2 ft^3 **8.** 2 **9.** $\frac{1}{2}$ **10.** $\frac{1}{2}$ **11.** $\frac{1}{4}$ **12.** 2 bags

Pages 272-273 Chapter 8 Review
1. b **3.** j **5.** c **7.** f **9.** k **11.** 150.7 cm **13.** 50.2 in. **15.** 12 in^2 **17.** 113.04 in^2 **19.** 3,215.36 cm^2 **21.** 8 chapter 4s **23.** 1,350 in^2 **25.** 84 cm^3 **27.** 197.8 ft^3 **29.** 400 ft^3 **31.** *about* 13,541 gallons of milk

Pages 276-277 Chapter 8 Cumulative Review/Test
Free Response: 1. 6 **3.** 86 **5.** 4,114 **7.** 3,209 **9.** $9\frac{5}{6}$ **11.** $3\frac{11}{12}$ **13.** $\frac{7}{10}$ **15.** 1 **17.** 36 in. **19.** $\overleftrightarrow{AB} \parallel \overleftrightarrow{RS}$ **21.** $\overleftrightarrow{FG} \parallel \overleftrightarrow{HI}$ **23.** parallelogram **25.** 22
Multiple Choice: 1. c **3.** b **5.** c **7.** b **9.** c

CHAPTER 9 RATIO, PROPORTION, & PERCENT

Pages 280-281 Lesson 9-1
1. $\frac{30}{11}$ **2.** $\frac{28}{13}$ **3.** $\frac{24}{41}$ **4.** $\frac{1}{30}$ **5.** $\frac{15}{1}$ **6.** $\frac{19}{15}$ **7.** $\frac{4}{3}$ **9.** $\frac{7}{11}$ **11.** $\frac{2}{15}$ **13.** $\frac{52}{1}$ **15.** $\frac{1}{7}$ **17.** $\frac{3}{4}$ **19.** Tampa Bay **21.** $\frac{14}{2}$ or $\frac{7}{1}$ **23.** $\frac{51}{32}$ **25.** Every 10 fingers has two hands. There are five times as many fingers as there are hands. Every hand has 5 fingers. **27.** 12 **28.** 91 **29.** 36 **30.** 80 **31.** $\frac{9}{5}$ **32.** $\frac{31}{15}$ **33.** $\frac{13}{8}$ **34.** $\frac{3}{7}$ **35.** 0.006 kg

Pages 282-283 Lesson 9-2
1. $\frac{1}{2}$ **2.** $\frac{1}{8}$ **3.** $\frac{7}{8}$ **4.** $\frac{5}{12}$ **5.** $\frac{3}{12}$ or $\frac{1}{4}$ **6.** $\frac{4}{12}$ or $\frac{1}{3}$ **7.** $\frac{9}{12}$ or $\frac{3}{4}$ **8.** $\frac{8}{12}$ or $\frac{2}{3}$ **9.** 0 **11.** $\frac{1}{3}$ **13.** $\frac{5}{6}$ **15.** $\frac{6}{6}$ **17.** $\frac{4}{9}$ **19.** $\frac{7}{9}$ **21.** $\frac{1}{36}$ **23.** $\frac{1}{4}$ **25.** $\frac{1}{4}$ **27.** $\frac{1}{6}$ **29.** $\frac{1}{2}$ **31.** 7 boats **33.** A sample answer is 3 *a*'s, 4 *b*'s, and 5 *c*'s.

Pages 284-285 Lesson 9-3
1. yes **2.** no **3.** yes **4.** no **5.** yes **6.** no **7.** yes **9.** yes **11.** yes **13.** no **15.** no **17.** no **19.** They are equivalent. $\frac{8}{20} = \frac{2}{5}$ and $\frac{40}{100} = \frac{2}{5}$ **21.** 8 to 20 **23.** no **25.** yes **27.** no, $\frac{190,000}{1} \neq \frac{316,667}{1}$ **29.** 88 **31.** 4 **33.** $\frac{1}{4}$

Pages 286-287 Lesson 9-4
1. yes **2.** no **3.** no **4.** yes **5.** 18 **6.** 42 **7.** 3 **8.** $9.\overline{3}$ or $9\frac{1}{3}$ **9.** no **11.** yes **13.** 7 **15.** 12 **17.** 72 **19.** 25.2 **21.** 125 **23.** $66.\overline{6}$ or $66\frac{2}{3}$ **25.** No—should divide by 8, not multiply **27.** 222 perch **29.** Sample answers are 2, 4, 8 or 3, 6, 12 **31.** $1\frac{1}{2}$ **32.** $2\frac{2}{9}$ **33.** $4\frac{11}{24}$ **34.** \$500

Page 289 Lesson 9-5
1. 396 ft **2.** 90 ft **3.** 117 ft **4.** 3 in. **5.** $2\frac{1}{2}$ in. **6.** $1\frac{1}{2}$ in. **7.** $\frac{2}{5}$ in. **8.** $\frac{1}{8}$ in. **9.** 20 ft × 15 ft **11.** 15 ft × 15 ft **13.** $1\frac{1}{2}$ in. **15.** $1\frac{1}{4}$ in. **17.** $\frac{1}{4}$ in. **19.** $\frac{3}{4}$ in. **21.** 96 pieces

Pages 292-293 Lesson 9-6
1. 10 cm **2.** 16 in. **3.** 12.5 cm **5.** 5 yd **7.** 13.5 ft **9.** 15 m **11.** 14 cm **13.** 15.75 m **15.** 24 m **17.** 35 ft **19.** 2.3 in. **21.** 67.5 min

Pages 294-295 Lesson 9-7
1. 531 miles **2.** 13 h 25 min **3.** Los Angeles & Dallas **5.** Denver to Memphis to Jacksonville **7.** 15.9 h or 15 h 54 min **9.** 298 kilometers per hour **11.** 62.5 mi **13.** 25 mi **14.** 19 **15.** 313 **16.** 6,292 **17.** 1,190 **18.** m **19.** mm **20.** Convenient Mart

Page 297 Lesson 9-8

1. $\frac{1}{2}$ **2.** $\frac{1}{10}$ **3.** 2 **4.** $1\frac{1}{2}$ **5.** $\frac{1}{8}$ **6.** $\frac{1}{1,000}$ **7.** 90% **8.** 25% **9.** 62.5% **10.** 85% **11.** $33\frac{1}{3}$% **12.** $12\frac{1}{2}$% **13.** $\frac{29}{100}$ **15.** $\frac{13}{20}$ **17.** $1\frac{1}{4}$ **19.** 2 **21.** $\frac{1}{4}$ **23.** $\frac{3}{8}$ **25.** 21% **27.** 25% **29.** 82% **31.** $12\frac{1}{2}$% **33.** $16\frac{2}{3}$% **35.** 400% **37.** 100% **39.** $\frac{7}{25}$

Pages 298-299 Lesson 9-9

1. 0.025 **2.** 0.732 **3.** 0.006 **4.** 0.099 **5.** 0.0025 **6.** 0.75 **7.** 0.07 **8.** 1 **9.** 0.09 **10.** 0.5 **11.** 12% **12.** 7% **13.** 1.5% **14.** 10.5% **15.** 90% **16.** 750% **17.** 100% **18.** 200% **19.** 615% **20.** 810% **21.** 0.12 **23.** 0.3 **25.** 0.081 **27.** 8.5 **29.** 0.382 **31.** 0.015 **33.** 68% **35.** 70% **37.** 1% **39.** 5.6% **41.** 138% **43.** 300% **45.** $46\frac{2}{3}$% or $46.\overline{6}$% **47.** 23 free throws **49.** $33\frac{1}{3}$% **51.** 0.05 **53.** 1 **55.** 0.005 **56.** 276 **57.** 11,520 **58.** 8,700 **59.** 38,555 **60.** 102,204 **64.** 14.72 cm^2

Pages 300-301 Lesson 9-10

1. 1-36 roll; 4-24 rolls **2.** 17, 18, and 19 **3.** 3 quarters, 2 dimes, and 3 nickels **5.** 39 **7.** 30 and 45 **9.** $7 **11.** left: 0, 1, 2, 6 (also 9), 7, 8; right: 0, 1, 2, 3, 4, 5 **12.** 48 **13.** 6 **14.** 120 **15.** obtuse **16.** acute **17.** right **18.** 201 cm^2

Pages 302-303 Chapter 9 Review

1. j **3.** f **5.** g **7.** h **9.** $\frac{4}{3}$ **11.** $\frac{21}{1}$ **13.** $\frac{1}{6}$ **15.** $\frac{1}{3}$ **17.** Y **19.** 15 **21.** 6 **23.** $2.25 **25.** 3 in. **27.** 8 in. **29.** 60 cm **31.** 25 m **33.** 236.25 mi **35.** $\frac{4}{25}$ **37.** 9% **39.** 0.16 **41.** 89% **43.** $\frac{1}{2}$ **45.** 25% **47.** 53, 46

Pages 306-307 Cumulative Review/Test

Free Response: **1.** 50 **3.** 144 **5.** 349.7 **7.** 82.05 **9.** 4.085 **11.** 108 **13.** 162,081 **15.** 0.0007 **17.** hexagon, regular **19.** quadrilateral, regular **21.** 8 ft **23.** 314 in^2 **25.** 15 **27.** 0.63 **29.** 125 **31.** 75 **33.** 125

Multiple Choice: **1.** b **3.** b **5.** b **7.** a **9.** d

CHAPTER 10 APPLYING PERCENTS

Pages 312-313 Lesson 10-1

1. 6 **2.** 12.6 **3.** 79.1 **4.** 1.35 **5.** 5 **6.** 2 **7.** 18 **8.** 6.4 **9.** 3 **11.** 110 **13.** 2 **15.** 0.54 **17.** 10 **19.** 43 **21.** 732 **23.** 7 **25.** 12 **27.** 60 **29.** 45 **31.** 30 **33.** false **35.** Using $\frac{1}{3}$ as a multiplier is like dividing by 3. Since 45 is divisible by 3, $\frac{1}{3}$ of 45 is 15. **37.** 400 light bulbs **39.** $7 \times 77 + 7 \times 77 - 77$ **40.** 53.2 cm **41.** 14.4 in. **42.** $1\frac{3}{10}$ **43.** $3\frac{5}{33}$ **44.** $3\frac{9}{50}$ **45.** 18 minutes

Pages 314-315 Lesson 10-2

1. 80% **2.** 90% **3.** 15% **4.** 150% **5.** 37.5% or $37\frac{1}{2}$% **6.** 50% **7.** 160% **8.** 80% **9.** $66.\overline{6}$% or $66\frac{2}{3}$% **11.** 500% **13.** 2.5% or $2\frac{1}{2}$% **15.** 150% **17.** 75% **19.** 75% **21.** 80% **23.** 75% **25.** 40% **27.** $1.80 **29.** $2.10

Pages 316-317 Lesson 10-3

1. 20 **2.** 40 **3.** 24 **4.** 54 **5.** 15 **6.** 63 **7.** 40 **9.** 50 **11.** 68 **13.** 100 **15.** 60 **17.** 17.5 **19.** 50 **21.** 32 **23.** 75 **25.** 200 votes **27.** 13.67 million **29.** 32 **31.** 15 **33.** 6 moves

Pages 318-319 Lesson 10-4

1. 8 families **3.** 18 **5.** $15 **9.** = **10.** < **11.** > **12.** > **13.** 40 **14.** 50 **15.** 39 **16.** 3, 2

Pages 320-321 Lesson 10-5

1. 5 **2.** 30 **3.** 2 **4.** 10 **5.** 240 **6.** $3 **7.** 11 **9.** 72 **11.** 9 **13.** 24 **15.** 5 **17.** 84 **19.** 1 **21.** 5 **23.** 18 **25.** 2 billion **27.** *about* $7.50 **29.** *about* $90 **35.** $0.80; $8.80 **37.** $6.40; $89.40

Pages 324-325 Lesson 10-6

1. 13% **2.** 25% **3.** 6% **4.** 11% **5.** 16% **6.** 26% **7.** 8% **8.** 14% **9.** 19% **11.** 23% **13.** 29% **15.** 70% **17.** 72% **19.** 25% **21.** 47% **25.** 1 m **26.** 1 kL **27.** 4 h 10 m **28.** 11 h 55 m **29.** at least 25.12 inches

Pages 326-327 Lesson 10-7

1. $0.85, $7.65 **2.** $13.90, $41.70 **3.** $315, $585 **4.** 3.25, $3.25 **5.** $11.20, 20% **6.** $1,760, 14% **7.** $0.98, $8.82 **9.** $1.44, $27.31 **11.** $49, 25%

13. \$4.34, 21.7% **15.** \$31.50 **17.** 25% of 60 is larger than 25% of 48, \$45 **19.** \$75, \$50, \$80, \$70, \$60

Pages 328-329 Lesson 10-8

1. \$60, \$560 **2.** \$648, \$1,848 **3.** \$30.23, \$650.23 **4.** \$13.80 **5.** \$9.60, \$32.81 **7.** \$21.60, \$111.60 **9.** \$1,200, \$3,700 **11.** \$18.45, \$223.45 **13.** \$34.20 **15.** \$544.38 **17.** \$18 **19.** \$38.50 **23.** \$200

Page 331 Lesson 10-9

1. \$265.34 **2.** \$545 **3.** \$265.23 **4.** \$546.54 **5.** \$260.10 **7.** \$1,453.86 **8.** 60 cm^2 **9.** 12.96 mm^2 **10.** no **11.** yes **12.** yes **13.** $\frac{2}{9}$

Page 332 Lesson 10-10

1. greater **2.** \$74,452 **3.** \$71,491 **5.** 10 games **7.** 20% **9.** 18% **12.** rhombus **13.** trapezoid **14.** rectangle **15.** 24 cm^3 **16.** 350 in^3 **17.** 24.48 m^3 **18.** 1,160 miles

Pages 334-335 Chapter 10 Review

1. multiply **3.** 20% **5.** discount rate **7.** principal **9.** compound **11.** 12.8 **13.** 630 **15.** 7.13 **17.** 130% **19.** 31.25% **21.** 22.2% **23.** 20 **25.** 30 **27.** 70 **29.** \$0.60 **31.** 13% increase **33.** 5% decrease **35.** 8% increase **37.** \$93.60 **39.** \$658 **41.** \$234 **43.** \$13.33 **45.** 5% **47.** \$3,594.71

Pages 338-339 Cumulative Review/Test

Free Response: 1. 70 **3.** 35 **5.** 20 **7.** 416,845 **9.** 83,034 **11.** 1,656,207 **13.** 51,752 **15.** 84,348 **17.** 2,416,444 **19.** 2 **21.** 8 **23.** 5 **25.** 3 **27.** kilogram **29.** 15 in., 9 in., 9 in. **31.** 6 **33.** 4 **35.** $\frac{7}{10}$ **37.** \$11, \$44

Multiple Choice: 1. c **3.** b **5.** d **7.** d **9.** c

CHAPTER 11 STATISTICS AND GRAPHS

Pages 342-343 Lesson 11-1

1. 6, 6, 7 **2.** 14, 11, 6 **3.** 75, 75, 7 **4.** 35.5, 43, 20 **5.** 53, no mode, 72 **6.** 278.5, no mode, 93 **7.** 2, 2, 8 **9.** 29, 12, 86 **11.** 68.5, 91, 66 **13.** 41.5, no mode, 3 **15.** 2.45, 1.8, 1.6 **17.** 8.2, 8.3, 1.4 **19.** 34.5, 48.5, 8.2 **21.** no mode, no mode, 87.4 **23.** $g - 256 = 102$, $g = 358$ **25.** 400 **27.** 18,000

Pages 344-345 Lesson 11-2

1. 10; 13; 17; 10; 9; 16 **3.** 65 in. **5.** 13 in. **7.** 3.75 mi **9.** 3 mi **13.** \$1.76 **15.** size $10\frac{1}{2}$

Pages 346-347 Lesson 11-3

1. mean, It is the greatest. **2.** mode, It is the most common. **3.** median, Only three earn large salaries. **4.** mean **5.** The president, vice-president, managers, and supervisors would probably be excluded. They are probably not union members. **7.** A good answer would refer to example and exercises on page 346. **9.** lower mean—25, median—not affected **11.** median—not affected; lower mean—\$74,467 **13.** 22 **14.** 15 **15.** 3.98 **16.** 17 **17.** 0.0048 **18.** 52.3 **19.** 0.63 **20.** 3,000 **21.** 15 in. for both

Pages 348-349 Lesson 11-4

7. New York and Tulsa; Chicago and Miami **9.** Charlotte **11.** 3 stations **13.** 29 mi

7. 20 cm **9.** 40-50 days **11.** 42 cm **13.** 4.3×10^6 **14.** 4.21×10^5 **15.** 8.21×10^4 **16.** 9.42×10^2 **17.** cube **18.** triangular prism **19.** triangular pyramid **20.** 3 miles

Pages 354-355 Lesson 11-6

1. 500,000 farms **2.** 1970-1975 **3.** 1975-1980 **4.** 1,050,000 farms **9.** 10 **11.** 80 **13.** 90 **15.** 9,000 **17.** 800 **19.** increase **21.** about 500,000 people **24.** $\frac{4}{3}$ **25.** 6 **26.** $\frac{21}{26}$ **27.** $6\frac{2}{3}$ **28.** 210 cm^2 **29.** 480 in^2 **30.** 172 mm^2 **31.** 6 cars

Pages 356-357 Lesson 11-7

1. 15% **2.** 2-person households **3.** 25% **4.** 58% **11.** 12% **13.** 4 times **15.** suits **17.** \$2,000

Pages 358-359 Lesson 11-8

1. 15 videos **2.** 48 videos **3.** 21 videos **4.** 3 days **5.** 27 **6.** 34 **7.** 30–39 **9.** 1, 2, 3, 5 **11.** 5, 6, 7, 8 **13.** 98.5; 80 and 99 **17.** 62 **19.** 8 **21.** 10-19 year olds **23.** 301.4 cm^2 **24.** 395.6 mm^2 **25.** 150.7 in^2 **26.** $\frac{6}{7}$ **27.** $\frac{1}{25}$ **28.** 60%

Page 361 Lesson 11-9

1. 51.5, 67, 32, 35 **2.** more **3.** 19, 9, 10 **5.** 3, 9, 16

7. the spread of the values 9. consistent 11. 3 slices 13. 14 15. 13

Pages 362-363 Lesson 11-10
1. 80, 50 **2.** 50, 100 **3.** one fourth of the data **4.** one half of the data **5.** one half of the data **7.** the mode, the exact values of some of the data, the mean **9.** one-fourth of the number of the values, yes, Quartiles divide the data into four equal parts. **11.** the lower quartile, interquartile range **15.** 28°C **16.** 16°C **17.** 6°C **18.** 43°C **19.** 24 m^3 **20.** 196 cm^3 **21.** 456 in^3 **22.** 5.25 gallons

Pages 364-365 Lesson 11-11
1. 1,215 plants **2.** 15 high fives **3.** 163 pages **5.** 9 zeros and 20 of every other digit **7.** 190 pounds **9.** $52 **11.** 50% **12.** 25% **13.** 85% **14.** 20% **15.** $33\frac{1}{3}$% **16.** 2.5 **17.** 68 **18.** 243

Pages 366-367 Chapter 11 Review
1. k **3.** c **5.** h **7.** j **9.** b **11.** e **13.** median: 14, mode: 14, range: 8 **15.** 40 **17.** 75 **19.** median age **25.** 10% **27.** 30 participants **29.** 10; 7; 3 **31.** 243 puzzles

Pages 370-371 Cumulative Review/Test
Free Response: 1. 107 **3.** 80 **5.** 48 **7.** 5,737 **9.** 2,855 **11.** 218 **13.** < **15.** < **17.** < **19.** 12 **21.** 5,280 **23.** 5 **25.** $\overleftrightarrow{AB}$; line *AB* **27.** line segment *MN*; $\overline{MN}$ **29.** 100.5 cm^3 **31.** 1,004.80 in^3 **33.** 17; 18; 15
Multiple Choice: 1. b **3.** d **5.** d **7.** a **9.** b

CHAPTER 12 INTEGERS

Page 377 Lesson 12-1
1. < **2.** > **3.** < **4.** < **5.** > **6.** < **7.** > **8.** < **9.** −9, −3, 0, 4, 5 **10.** −8, −6, 0, 2, 6 **11.** −7, −1, 0, 6, 11 **12.** −5, −3, 0, 3, 5 **13.** < **15.** > **17.** > **19.** < **21.** > **23.** < **25.** < **27.** > **29.** < **31.** > **33.** −4, 0, 3, 8, 9 **35.** −4, −2, −1, 0, 1 **37.** −3.5, −3, 0, 3, 3.5 **39.** −10; 0 **41.** Antarctica **43.** Great Falls, Montana

Page 379 Lesson 12-2
1. −2 + (−3) = −5 **2.** −3 + 6 = 3 **3.** 8 **4.** −7 **5.** −6 **6.** −2 **7.** 2 **8.** −1 **9.** −1 **11.** −15 **13.** 0 **15.** −4 **17.** 11 **19.** −33 **21.** 13 **23.** −38 **25.** −0.8 **27.** −48 **29.** To add integers with different signs, find the difference of their absolute values. Give the result the same sign as the integer with the greater absolute value. **31.** 6°C **33.** −123°C **35.** O = 6; D = 5; E = 1; V = 3; N = 0

Pages 380-381 Lesson 12-3
1. 17 **2.** 4 **3.** 20 **4.** 4 **5.** −5 **6.** −1 **7.** −5 **8.** 1 **9.** −15 **10.** −16 **11.** −12 **12.** −8 **13.** 0 **14.** −10 **15.** 8 **16.** −1 **17.** −9 **19.** −6 **21.** −6 **23.** −13 **25.** −4 **27.** 4 **29.** 0 **31.** −12 **33.** −84.3 **35.** −2.5 **37.** −6 **39.** −7°C **41.** 14,776 feet **45.** −2 **47.** 2 **49.** −6 **51.** −7 **53.** 28 **55.** 29

Pages 382-383 Lesson 12-4
1. −39°F **2.** −33°F **3.** 37°F **5.** −48°F **7.** −74°F **9.** 20°F **11.** −79°F **13.** 26°F **15.** 39°F **17.** 25 mph **19.** 52°F **21.** $221.3 billion **23.** deficit **25.** 66.4 **26.** 115.06 **27.** 236.66 **28.** 229.3

Page 387 Lesson 12-5
1. 15 **2.** 36 **3.** 9 **4.** 90 **5.** 35 **6.** 84 **7.** 72 **8.** 30 **9.** −12 **10.** −30 **11.** −32 **12.** −98 **13.** −77 **14.** −540 **15.** 0 **16.** 0 **17.** 40 **19.** −6 **21.** −42 **23.** −70 **25.** −54 **27.** 56 **29.** −369 **31.** −210 **33.** −12.6 **35.** $-\frac{4}{7}$ **37.** −108 **39.** 16.5 feet **41.** 84°C **43.** 0 **45.** 48 **47.** $\frac{2}{3}$ **48.** $\frac{3}{7}$ **49.** $\frac{1}{4}$ **50.** $n = 8$m **51.** $x = 11\frac{1}{4}$ cm **52.** $y = 3$ in. **53.** $247.74

Pages 388-389 Lesson 12-6
1. 7 **2.** 5 **3.** 0.25 **4.** 0.2 **5.** 6 **6.** 6 **7.** 0.5 **8.** 7.5 **9.** −12 **10.** −9 **11.** −0.5 **12.** 0 **13.** −14 **14.** −9 **15.** −0.1 **16.** 0 **17.** 4 **19.** −9 **21.** −5 **23.** −2 **25.** 0 **27.** −6 **29.** 14 **31.** −8 **33.** −3.7 **35.** −1 **37.** $10\frac{5}{7}$ **39.** −3.5 **41.** −55 **43.** > **45.** < **47.** 6 **49.** −3 **51.** −23 **53.** −0.4 or $-\frac{2}{5}$ **55.** Olympic Games **57.** E **59.** E **61.** E

Pages 390-391 Lesson 12-7

1. 28 apples **3.** 10 microbes **5.** 3 letters; 5 postcards **7.** 16 rooms **9.** 3 of the 6-oz boxes **12.** 7,000 **13.** 0.1481 **14.** 0.0912 **15.** 4 games

Pages 392-393 Chapter 12 Review

1. opposite **3.** whole numbers, integers **5.** negative **7.** adding **9.** > **11.** < **13.** = **15.** > **17.** −9, −7, 0, 7, 9 **19.** −3 **21.** 0 **23.** 4 **25.** −22 **27.** −7 **29.** 17 **31.** −9 **33.** −13 **35.** 11 ft rise **37.** 44°F **39.** 54 **41.** 99 **43.** −84 **45.** 84 **47.** −9 **49.** $-1\frac{1}{4}$ **51.** 8 **53.** 0 **55.** 9 **57.** −27°F

Pages 396-397 Cumulative Review/Test

Free Response: 1. 5 **3.** 9 **5.** 527 **7.** 6,172 **9.** 131,707 **11.** 32 **13.** 3 **15.** $\frac{1}{8}$ **17.** $\frac{1}{12}$ **19.** 4 in. **21.** $100.10 **23.** $70.40 **25.** 6.4; 6.5; 9 **27.** 32°
Multiple Choice: 1. c **3.** a **5.** b **7.** b **9.** d

CHAPTER 13 EXTENDING ALGEBRA

Pages 400-401 Lesson 13-1

1. 9 **2.** 6 **3.** −5 **4.** 20 **5.** 0 **6.** 157 **7.** −3 **8.** −9 **9.** −4 **10.** −8 **11.** −35 **12.** 29 **13.** −2 **15.** 34 **17.** −31 **19.** 7 **21.** −3 **23.** 15 **25.** 383 **27.** 0 **29.** −0.8 **31.** −$13 **35.** −0.04; They are the same. **36.** 8.1 **37.** 76.61 **38.** 421.2 **39.** 0.3362 **40.** 25.2 m^2 **41.** 60.45 cm^2 **42.** 425.35 mm^2 **43.** 40 free throws

Pages 402-403 Lesson 13-2

1. 2.5 **2.** −5 **3.** 15 **4.** −75 **5.** −50 **6.** 40 **7.** −192 **8.** 200 **9.** $-\frac{5}{7}$ **11.** −4 **13.** −4 **15.** −144 **17.** 8.3 **19.** −3 **21.** $-\frac{4}{5}$ **23.** $\frac{1}{4}$ **25.** −42 **27.** $\frac{2}{3}$ **29.** $\frac{2}{3}$ **31.** $\frac{1}{5}$ **33.** loss of 6 yards **35.** 5, −4 **37.** −8 **39.** −1 **41.** −243 **43.** 16

Pages 404-405 Lesson 13-3

1. −4 **2.** 6 **3.** −7 **4.** −20 **5.** −5 **6.** −10.5 **7.** 5 **9.** $-6\frac{1}{4}$ **11.** −4 **13.** −5 **15.** 212 **17.** 22 **19.** 8 **21.** 16 **23.** $30 **27.** $3\frac{1}{5}$ **28.** $\frac{21}{32}$ **29.** $1\frac{1}{2}$ **30.** 16 **31.** 1, 2, 3, 4 **32.** 0, 1, 2, 3, 4 **33.** 20, 34

Page 409 Lesson 13–4

1. (2,1) **2.** (−3,2) **3.** (−3,−2) **4.** (3,−3) **5.** (1,3) **6.** (1,0) **17.** (−2,3) **19.** (−3,−2) **21.** (4,−4) **35.** (2,−4) **37.** It is similar in moving horizontally for one number and vertically for the other.

Pages 410-411 Lesson 13-5

19. October **21.** 8 pounds **23.** $x - 3 = y$

Pages 412-413 Lesson 13-6

1. Los Encinos St. Historical Park **2.** Highway 110 **3.** Santa Monica Mun. Airport **4.** B-1 **5.** B-4 **6.** D-4 **7.** B-3 **8.** E-1 **9.** Griffith Park Zoo **11.** LA International **13.** A-2 **15.** D-2 **17.** N. on Lincoln Blvd. to Tijera Blvd.; N. on Tijera Blvd. to Manchester Ave.; E. on Manchester Ave. to the Forum **19.** Sunset Blvd. **22.** 75° **23.** 105° **24.** 57° **25.** −2 **26.** −18 **27.** −6 **28.** 53 pizzas

Pages 414-415 Lesson 13-7

1. $2x + x = 81$; 54 tapes, 27 compact discs **2.** $212 = 10.50 + \frac{1}{2}x$; $403 **3.** 39 mums **5.** 10 parakeets, 8 kittens **7.** *about* $45 **9.** 20, 70 **10.** $\frac{1}{2}$ of the data **11.** 21 **12.** −5 **13.** −30 **14.** 425 feet

Pages 416-417 Chapter 13 Review

1. same number **3.** ordered pair **5.** horizontal **7.** origin **9.** 15 **11.** 19 **13.** 25 **15.** −2 **17.** −13 **19.** −192 **21.** −4 **23.** −3 **25.** $-1\frac{3}{5}$ **27.** (−4,3) **29.** (−2,−3) **43.** River Road **45.** A4 **47.** B6 **49.** 18 m **51.** $135, $185

Pages 420-421 Cumulative Review/Test

Free Response: 1. 2,100 **3.** 4 **5.** 3,741 cars **7.** $1\frac{1}{7}$ **9.** $1\frac{5}{8}$ **11.** 64 **13.** 18 ft^3 **15.** 1,800 in^3 **21.** −7 **23.** 29 **25.** 28 **27.** −15 **31.** 11 miles, work; 15 miles, school
Multiple Choice: 1. c **3.** b **5.** d **7.** c **9.** c

CHAPTER 14 PROBABILITY

Pages 426-427 Lesson 14-1

3. $6 \times 6 = 36$ **4.** $2 \times 6 = 12$ **5.** 8 outcomes **7.** 24 outcomes **9.** $4 \times 4 \times 4 \times 4 = 256$ choices **11.** $\frac{1}{100}$ **13.** 5.8 **14.** 11.1 **15.** 8.5 **16.** 137.03

17. 11.969 **18.** $\frac{1}{3}$ **19.** $\frac{1}{2}$ **20.** $\frac{1}{3}$ **21.** 45 singles

Pages 428-429 Lesson 14-2

1. 30% **2.** 20% **3.** 20% **4.** 40% **5.** 20% **6.** 80% **7.** 0% **8.** 20% **9.** 60% **11.** 40% **13.** 60% **15.** 100% **17.** 0% **19.** 3% **21.** 6% **23.** 17% **25.** 11% **27.** $\frac{1}{25}$ or 4%

Pages 430-431 Lesson 14-3

1. $\frac{1}{15}$ **2.** $\frac{1}{10}$ **3.** 0 **4.** $\frac{1}{5}$ **5.** $\frac{1}{4}$ **6.** $\frac{1}{36}$ **7.** $\frac{1}{6}$ **8.** $\frac{1}{18}$ **9.** $\frac{1}{15}$ **11.** 0 **13.** $\frac{1}{45}$ **15.** $\frac{2}{9}$ **17.** $\frac{3}{50}$ **19.** 0 **21.** $\frac{1}{25}$ **23.** $\frac{1}{4}$ **25.** $\frac{1}{625}$ **27.** $\frac{1}{2}$, $\frac{1}{2}$

Pages 432-433 Lesson 14-4

1. $\frac{1}{3}$ **2.** $\frac{1}{2}$ **3.** $\frac{2}{3}$ **4.** $\frac{1}{3}$ **5.** $\frac{2}{3}$ **6.** 1 **7.** $\frac{2}{3}$ **8.** $\frac{2}{3}$ **9.** $\frac{3}{4}$ **11.** $\frac{1}{8}$ **13.** $\frac{3}{8}$ **15.** $\frac{1}{2}$ **17.** $\frac{3}{4}$ **19.** $\frac{5}{8}$ **21.** $\frac{1}{3}$

23. The probability of spinning blue or a number greater than 5 will remain the same if the four adjacent blue wedges are numbered 14, 15, 16 and 1. Any other placement of four adjacent wedges will result in a different probability.

Pages 436-437 Lesson 14-5

1. 125 **2.** 86 **3.** 88 **4.** $\frac{1}{6}$ **5.** a floral design **7.** $\frac{111}{150}$ **9.** $\frac{23}{75}$ **11.** $\frac{44}{75}$ **13.** $\frac{527}{5,625}$ **15.** $\frac{1,332}{11,250}$ **17.** $\frac{13}{25}$ **21.** 12 **23.** 240

Pages 438-439 Lesson 14-6

1. 1 to 5 **2.** 1 to 5 **3.** 1 to 5 **4.** 0 to 6 **5.** 3 to 3 **6.** 3 to 3 **7.** 3 to 3 **8.** 2 to 4 **9.** 2 to 4 **10.** 3 to 3 **11.** 2 to 4 **12.** 2 to 4 **13.** 4 to 2 **14.** 4 to 2 **15.** 1 to 5 **16.** 4 to 2 **17.** 3 to 3 **18.** 4 to 2 **19.** 4 to 2 **21.** 3 to 33 **23.** 31 to 5 **25.** 33 to 3 **27.** 27 to 9 **29.** 6 to 14; $\frac{3}{10}$ **31.** 14 to 6; $\frac{7}{10}$ **33.** 11 to 9; $\frac{11}{20}$ **35.** 3 to 1 **37.** $\frac{5}{36}$ **39.** 58 **41.** 67 **43.** 46

Pages 440-441 Lesson 14-7

1. 504 students **2.** 85 customers **3.** $\frac{1}{4}$ **5.** $\frac{1}{8}$ **7.** $\frac{2}{5}$

9. The sample was not representative. The larger population did not have the same preferences as the sample. Too much time elapsed between taking sample and choices of population. **11.** 153.9 cm^2 **12.** 201.0 in^2 **13.** 1,808.6 mm^2 **14.** 38.5 cm^2 **15.** 41.2 **16.** 38 **17.** 38

Pages 442-443 Chapter 14 Review

1. ratio **3.** odds **5.** tree **7.** dependent **9.** 9 outcomes **11.** 50% **13.** 70% **15.** 80% **17.** 100% **19.** $\frac{3}{64}$ **21.** $\frac{1}{15}$ **23.** 1 **25.** $\frac{3}{5}$ **27.** $\frac{9}{30}$ **29.** $\frac{26}{75}$ **31.** 8 to 8 **33.** 700 owners **35.** $\frac{1}{3}$

Pages 446-447 Cumulative Review/Test

Free Response: 1. 5,010 **3.** 30,100 **5.** 11.4 m **7.** 24 yd **9.** 7,010 **11.** 283 **13.** $0.\overline{6}$ **15.** $0.1\overline{6}$ **17.** 234 **19.** 25, 26 **21.** −9 **23.** −2 **25.** −14 **27.** $\frac{1}{2}$ **29.** $-\frac{19}{20}$

Multiple Choice: 1. c **3.** d **5.** a **7.** c **9.** d

Photo Credits

Cover, Aaron Haupt, Image Development/Design, Terry Anderson; **v-xi,** Ken Frick; **A1,** (t) Ken Frick, (b) Aaron Haupt; **A2-A11,** Ken Frick; **2,** (t) Randy Schieber, (o) Ken Frick; **4,** ©Farrell Grehan, FPG; **5,** Paolo Koch, Photo Researchers; **6,** Hickson & Associates Photography; **8,** Ken Frick; **10,** Steven E. Sutton/Duomo; **12,** Ken Frick; **14,** (t) Ken Frick, (b) Janet Adams; **15,** Ken Frick; **16,** Superstock, Inc; **18,** Doug Martin; **19,** Bob Hipp; **20,** Ken Frick; **21,** Doug Martin; **23-24,** Ken Frick; **27,** Studiohio; **28,** Randy Schieber; **30,** (t) Randy Schieber, (b) Ken Frick; **31,** Tim Courlas; **33,** (l) Steve Lissau, (r) Alan Carey; **35,** Bud Fowle; **36,** Studiohio; **38,** CALVIN & HOBBES Copyright 1988 Universal Press Syndicate. Reprinted with permission, all rights reserved; **40,** Kenji Kerins; **42,** Doug Martin; **44,** © Eugene Gebhardt, FPG; **45,** Mary Lou Uttermohlen; **46,** Bud Fowle; **48,** Studiohio; **50,** Doug Martin; **51,** Randy Schieber; **53,** Ken Frick; **54,** Doug Martin; **57,** ©Barry Rosenthal, FPG; **59,** John Griffin/The Image Works; **60,** Doug Martin; **61,** ©Michael Nelson, FPG; **63,** Doug Martin; **64,** ©Dennie Cody, FPG; **65,** Doug Martin; **66,** Ken Frick; **68,** Studiohio; **70,** Doug Martin; **71-72,** Ken Frick; **76,** Bud Fowle; **78, 81,** Ken Frick; **82,** Doug Martin; **83,** ©TRAVELPIX, FPG; **84,** Doug Martin; **85,** Randy Schieber; **89,** Jim Franck; **90,** ©Allsports USA/Brian Drake; **91,** Ken Frick; **94,** Bud Fowle; **95,** Janet Adams; **96,** (t,b) Ken Frick, (c) Jim Franck; **99,** ©Bob Taylor, FPG; **100,** Doug Martin; **101,** ©Lee Balterman, FPG; **104,** FRANK & ERNEST Reprinted by permission of NEA, INC.; **106,** (t) Studiohio, (b) Doug Martin; **109,** Ken Frick; **110,** ©John Terrence Turner, FPG; **111,** Ken Frick; **113,** Bud Fowle; **115,** Ken Frick; **116,** Tim Courlas; **118,** Michael Sargent/The White House; **119,** Ken Frick; **120,** Hank Morgan, Photo Researchers; **122,** Terje Kveen/The Image Bank; **123,** Nancy Durrell McKenna, Photo Researchers; **124,** ©Allsports USA/Rick Stewart; **127,** Ken Frick; **128, 130,** Bud Fowle; **132,** ©Allen B. Seiden, FPG; **133,** Bud Fowle; **134,** (l) Hickson & Associates Photography, (r) Doug Martin; **137,** (l) Steven Lissau, (r) Ken Frick; **139,** ©M. Kozlowski, FPG; **142,** Ken Frick; **144,** (l) Studiohio, (r) Ken Frick; **146,** Michel Tcherevkoff/The Image Bank; **147,** Brian Parker/Tom Stack & Associates; **150,** Doug Martin; **152,** G. Glod/Superstock, Inc.; **154,** Hickson & Associates Photography; **156,** Doug Martin; **157,** ©T.J. Florian/Rainbow; **158,** Ken Frick; **160,** Bud Fowle; **162,** Doug Martin; **165,** ©Allsports USA/Otto Greule; **166,** Tim Courlas; **169,** Lloyd Lemmermann; **170,** Ken Frick; **171,** Doug Martin; **172,** Frank Cezus; **174,** BROOM-HILDA ©Reprinted by permission: Tribune Media Services; **176,** Studiohio; **179,** Doug Martin; **182,** Lloyd Lemmermann; **183,** Ken Frick; **185,** (l) Ken Frick, (r) Superstock, Inc; **187,** Doug Martin; **188-189,** Ken Frick; **191,** Brian Parker/Tom Stack & Associates; **192,** Doug Martin; **193,** Superstock Inc; **194,** Hickson & Associates Photography; **195,** Frank Balthis; **197,** ©Allsports USA/Mike Powell; **199,** Larry Kolvoord/The Image Works; **200,** Ken Frick; **202,** ©Bob Daemmrich; **203,** (t) Ken Frick, (b) Bud Fowle; **205,** ©Allsports USA/Tony Duffy; **206,** Aaron Haupt; **207,** Smithsonian Institution; **208,** Ken Frick; **210,** ©Peter Grigley, FPG; **212, 214,** Ken Frick; **216,** Alan Pitcairn/Grant Heilman Photography; **217,** Ken Frick; **218,** (t) Hickson & Associates Photography, (b) Ken Frick; **222,** Doug Martin; **223,** ©Bob Daemmrich; **224,** ©Michael Tamborrino, FPG; **225,** Cletis Reaves, FPG; **226,** Cincinnati Convention and Visitor's Bureau; **228,** (l) Ken Frick, (r) ©Tom Carroll, FPG; **232,** Bud Fowle; **237,** Doug Martin; **238,** Ken Frick; **240,** CALVIN & HOBBES Copyright 1988 Universal Press Syndicate. Reprinted with permission, all rights reserved; **242,** Lloyd Lemmermann; **242,** Böhm-NBBJ Architects; **244,** Picture Unlimited; **245,** Ken Frick; **246,** Doug Martin; **247,** Janet Adams; **248,** Steve Lissau; **250,** Ted Rice; **251,** Ken Frick; **252,** Doug Martin; **253,** Ken Frick; **254,** Aaron Haupt; **255,** Studiohio; **256,** Hickson & Associates Photography; **257,** Larry Hamill; **258,** Ken Frick; **261,** Lloyd Lemmermann; **262,** Bud Fowle; **264,** ©Paul Degreve, FPG; **265,** ©David Sucsy, FPG; **266,** Ken Frick; **268,** Randy Schieber; **269,** Ken Frick; **270,** (t) Lloyd Lemmermann, (b) Ken Frick; **273,** John Colwell/Grant Heilman Photography; **274,** Studiohio; **276,** Ken Frick; **277,** file photo; **278,** Bud Fowle; **280,** Ken Frick; **281-284,** Ken Frick; **284, 286,** Doug Martin; **287,** David R. Frazier; **291,** Joe Munroe, Photo Researchers; **292,** Pictures Unlimited; **296,** Ted Rice; **298,** Pictures Unlimited; **299,** (t,b) Ken Frick, (c) Roger K. Burnard; **300,** Ken Frick; **303,** Superstock, Inc; **305,** Lloyd Lemmermann; **306,** (l) Bud Fowle, (r) Ken Frick; **308,** FUNKY WINKERBEAN Copyright 1990 North America Syndicate. Reprinted with permission, all rights reserved; **310,** (t) Doug Martin, (b) Jim Franck; **311,** Daniel Erickson; **313,** Grant Heilman/Grant Heilman Photography; **314,** Herbet Eisenberg/Superstock Inc; **315,** (t) Ken Frick, (b) Jim Franck; **316,** (t) Bob Dammerich/The Image Works, (b) Ken Frick; **318,** Superstock Inc; **319,** Hickson & Associates Photography; **320,** Doug Martin; **321,** Hickson & Associates Photography; **322-323,** Doug Martin; **324,** Lloyd Lemmermann; **324, 326,** Ken Frick; **327,** ©Dennis Hallinan, FPG; **329-330,** Doug Martin; **331,** Gerard Photography; **332,** Ken Frick; **335,** Janet Adams; **337,** Ken Frick; **338,** Doug Martin; **340,** (t) Superstock Inc, (c) ©TRAVELPIX, FPG; (b) Aaron Haupt; **342,** Ken Frick; **346,** W. Strode/Superstock Inc; **347,** file photo; **349,** Ken Frick; **350,** Doug Martin; **352,** Studiohio; **353,** David Frazier; **354,** Joseph A. Dichello; **356,** Superstock Inc; **358,** Doug Martin; **360,** Aaron Haupt; **362,** Ken Frick; **364, 366,** Doug Martin; **369,** Mary Lou Uttermohlen; **370,** Ken Frick; **372,** FOR BETTER OR WORSE Copyright 1990 Universal Press Syndicate. Reprinted with permission, all rights reserved; **374,** (t) file photo, (b) Superstock Inc; **376,** Doug Martin; **378,** Runk/Schoenberge/Grant Heilman Photography; **379,** Michael Melford/The Image Bank; **380,** Walter R. Marks/Metro Dade Tourism; **381,** Ken W. Davis/Tom Stack & Associates; **382,** The Image Works; **384,** Bud Fowle; **385,** Owens-Corning Fiberglass Corporation; **386,** ©Dave Bartuff, FPG; **388,** Ken Frick; **389,** ©T. Tracy, FPG; **390,** Studiohio; **392,** file photo; **393,** Superstock Inc; **394,** Pictures Unlimited; **395,** David Frazier; **396,** (l) Ken Frick, (r) Studiohio; **398,** Doug Martin; **399,** Randy Schieber; **400,** Doug Martin; **403,** First Image; **404,** Ken Frick; **405,** Doug Martin; **406,** Lloyd Lemmermann; **407,** Doug Martin; **408, 412,** Ken Frick; **413,** George H. Harrison/Grant Heilman Photography; **415,** ©Dan McCoy/Rainbow; **418,** Brian Parker/Tom Stack & Associates; **419,** Ken Frick; **420,** (l) Ken Frick, (r) Studiohio; **422,** CALVIN & HOBBES Copyright 1990 Universal Press Syndicate. Reprinted with permission, all rights reserved; **424,** (t) Studiohio, (b) Ken Frick; **427,** Cobalt Photography; **428,** Ken Frick; **429,** Brian Parker/Tom Stack & Associates; **430,** Bob Hipp; **431,** Ken Frick; **433,** Kent Dannen, Photo Researchers; **434,** Superstock Inc; **435,** Bud Fowle; **436, 438,** Doug Martin; **439,** Ken Frick; **440,** Ted Rice; **441,** (l) David R. Frazier, Photo Researchers, (cl) First Image, (cr,r) ©Henry Gris, FPG; **442,** Superstock Inc; **443,** ©T.J. Florian/Rainbow; **444,** Ken Frick; **445,** Lloyd Lemmermann; **483,** Ken Frick.

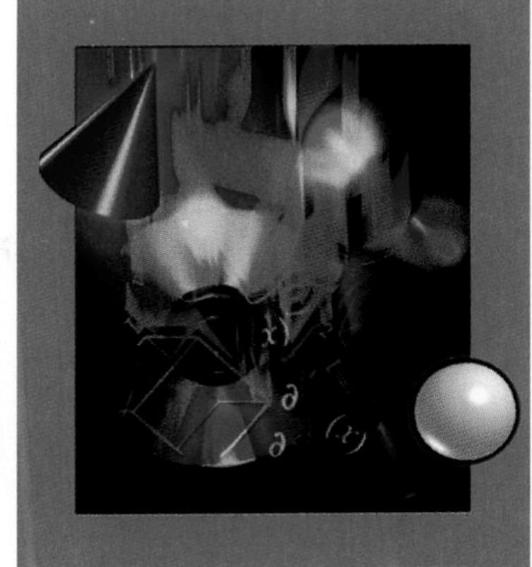

H A N D B O O K

What does a robotics technician do? To find out, you can read pages 525-527. The Tech Prep Handbook provides a career description, a description of technical equipment used in the occupation, information about the type of education required, and a list of resources for each occupation listed. This information can help you determine whether you would be interested in pursuing this occupation.

Applying Tech Prep to a Career

Objectives

- Familiarize students with the career description and education required for a career as a laser technician.
- Write numbers in expanded form using exponents.
- Read a circle graph.
- Determine interests and skills described in a career by expressing thoughts and feelings.
- Code and decode the Universal Product Code.

Mathematical Content Students should have completed Lesson 1-2 or know how to write numbers in expanded form using exponents before attempting the Exercises.

HANDBOOK

LASER TECHNICIAN

The narrow beams of light that zip across the stage at a rock concert are produced by lasers. Where do you think the word laser comes from? Each letter in the word laser stands for a word.

Light
Amplification by
Stimulated
Emission of
Radiation

Beams of laser light do not spread out because all their waves travel in the same direction.

CAREER DESCRIPTION

Laser technicians operate and adjust all the elements of laser devices to make them work properly in industry, supermarkets, offices, and homes. A laser technician troubleshoots and repairs systems that use lasers, performs tests and measurements using electronic testing equipment, operates laser systems and related equipment, assembles laser devices and systems, and researches and develops laser devices.

TOOLS OF THE TRADE

An oscilloscope, shown below, may be used to troubleshoot a malfunction in a laser system. It does not generate the usual sine wave (wavy line) that you may be familiar with. Instead it measures digital signals like those produced by electronic equipment with a series of ones and zeros.

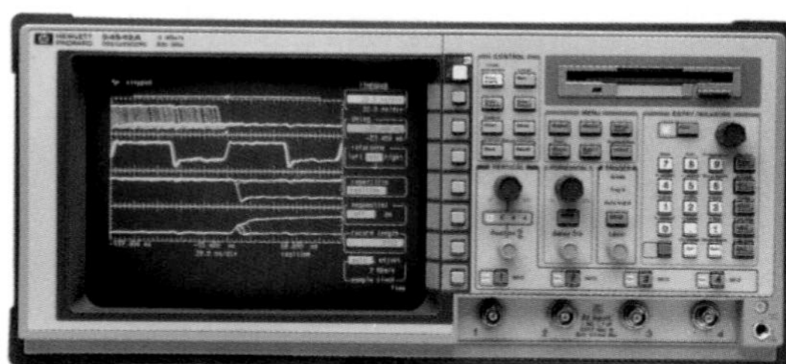

EDUCATION

A good background in mathematics and physics is required. Plan to attend a 2-year laser technician program. Such programs are offered at technical institutes, community colleges, and technical colleges. Also, laser manufacturers may hire graduates with a 2-year associate degree in physics or electronics.

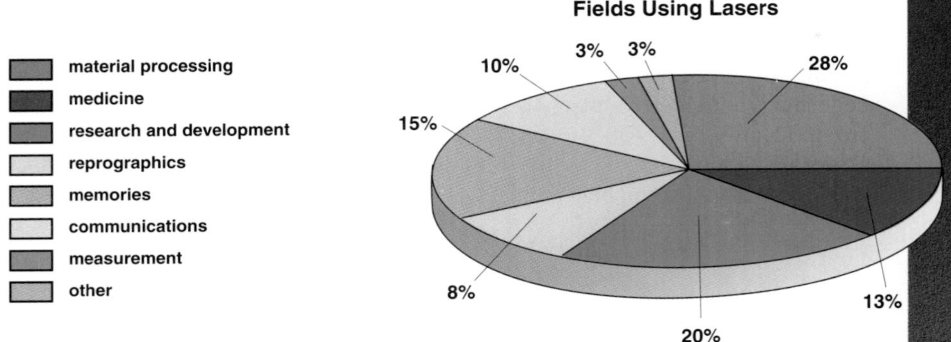

EXERCISES

Write in expanded form using 10^0, 10^1, 10^2, and so on.

The following numbers represent the speed at which light travels.

1. 300,000,000 meters per second
2. 186,000 miles per second

1. 3×10^8 2. $(1 \times 10^5) + (8 \times 10^4) + (6 \times 10^3)$

Use the circle graph above to answer each question.

3. What field is the largest user of laser applications? material processing
4. What percent of laser applications do medicine and communications make up? 23%

5. operating laser devices, reading an oscilloscope, mathematics, and physics

Express in your own words.

5. What kind of skills are required of a laser technician?
6. Do you think you would be interested in being a laser technician? Why or why not? Answers may vary.
7. What do you like or not like about the career of laser technician? Answers may vary.

Time Students will need 1-2 days to complete pages 516-517. Students will need about one week to complete the project on page 518.

Teaching Notes Have students read pages 516-517 and complete the Exercises.

Encourage students to examine their thoughts and feelings in Exercises 5-7 on page 517. These exercises provide students with an opportunity to think about their interests, skills, likes, and dislikes for a career that may last a lifetime. You may want to have class discussions so that students can share their thoughts about how this career relates to them. This may also help them to realize that people have different interests from their own. You may want to have students keep a list of the things they like and do not like about each career. Thus, when students complete the handbook, they will have a list of their interests and skills.

Encourage students to use the resources listed at the end of the project. They may want to do their own research at the library. Have students share their findings. Students could also interview a laser technician or you may want to have a laser technician make a presentation to the class.

Project Description Students use the Universal Product Code to make shopping lists. Then they use the bar code and an inventory list to decipher what they are buying and the cost. Essentially they are doing the same career that the scanner and the computer are doing at the grocery store.

Project Notes Have students work in pairs or in groups of three.

Groups should work on their projects at a regular pace rather than trying to finish at the end of the week. You may want to schedule class time for groups to work on their projects.

You can extend the project by having students decode UPCs from items not used in the shopping list using the UPC cipher set shown below. First remove the numbers from below the bars. Then list the numbers from below the bars on the chalkboard and their products. Have students match those numbers and name the products from the bar code numbers.

UPC Cipher Set

first character set	left cipher set	right cipher set
0	0001101	1110010
1	0011001	1100110
2	0010011	1101100
3	0111101	1000010
4	0100011	1011100
5	0110001	1001110
6	0101111	1010000
7	0111011	1000100
8	0110111	1001000
9	0001011	1110100

START and STOP codes: 101
CENTER code: 01010

Translate the bar code above.

1. Translate each single black bar as a 1, each double black bar as 11, and so on.
2. Translate each single white space as 0, each double white space as 00, and so on.
3. Following the START code 101, translate the cipher that designates the character set.
4. Translate the next five numbers in the left cipher set according to the left cipher column.
5. Following the six numbers of the left cipher set, translate the CENTER code as 01010.
6. Following the CENTER code 01010, translate the five numbers of the right cipher set according to the right cipher column.
7. Next, translate the modulo check cipher from the right cipher set.
8. Complete your translation with the STOP code 101.

PROJECT

UPC Coding and Decoding

Materials: 10 grocery store items each with a UPC bar code and its price

Overview Most grocery products are marked with a bar code system known as a Universal Product Code, or UPC, as shown at the right. Each UPC contains 12 numbers. The first six numbers include one to designate the character set used in the code and five to identify the maker of the product. The second six numbers include five to identify the product followed by one number called the modulo check cipher. For example, omitting the first and last number, the code 51000-00011 identifies a $10\frac{3}{4}$-ounce can of Campbell's tomato soup.

Procedure Make an inventory list like the one below. Remove each UPC code from the grocery items and place it beneath the UPC number on the inventory list.

UPC Number	Maker of Product	Product	Price
51000-00011	Campbell's	$10\frac{3}{4}$-ounce can tomato soup	\$0.89

Then make up a shopping list using only the UPC numbers. Use your inventory list to identify, on a separate sheet of paper, the maker of the product, identify the product, and name its price. Exchange shopping lists with other groups. Compare their answers to your shopping list with your answers. **See students' work.**

RESOURCES

For further information on a career as a laser technician, write or call the following association.

- Laser Institute of America
  12424 Research Parkway, Suite 130
  Orlando, FL 32826 (407) 380-1553

Check your local library for the following books or magazines.

- Hech, Jeff, The Laser Guidebook
- Laser Focus World
  PennWell Publishing Company
  1 Technology Park Drive
  Westford, MA 01886 (508) 692-0700

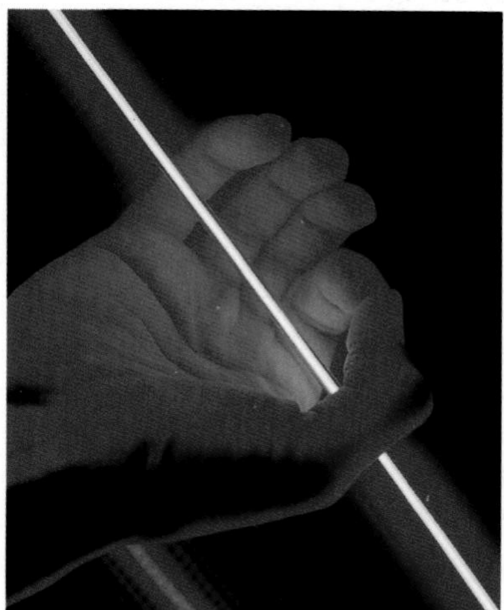

CAREER DESCRIPTION

An agribusiness specialist may work in business management. She or he may be employed by a large farm or dairy as a business manager, a credit institution as a lender, or a business as a buyer for farm products.

An agribusiness specialist may work in sales and service. This specialist might work for an aerial crop spraying company, a distribution company for farm products, an insurance company, or any company that offers services to farmers.

A third area is record keeping. Agribusiness specialists may set up and analyze record systems for farmers. He or she can be valuable in helping farmers get the most benefit from the records.

TOOLS OF THE TRADE

The major tool of the trade for an agribusiness specialist is the computer. The specialist needs programs for record keeping, inventory lists, data bases, spreadsheets, scheduling, payroll, bookkeeping,taxes, graphics, and so on.

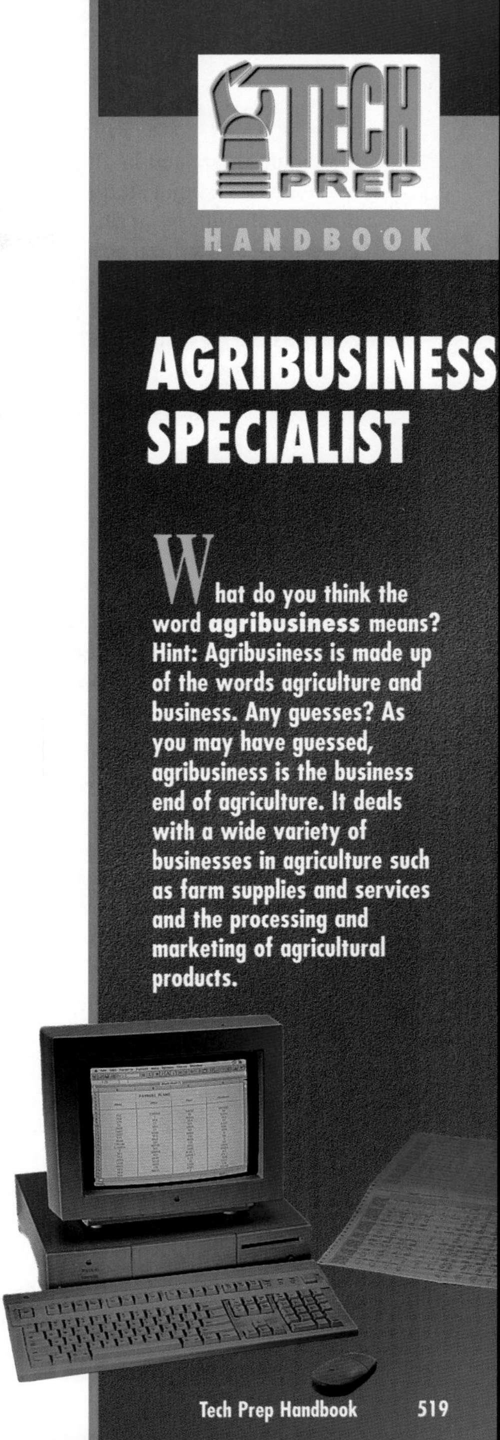

AGRIBUSINESS SPECIALIST

What do you think the word agribusiness means? Hint: Agribusiness is made up of the words agriculture and business. Any guesses? As you may have guessed, agribusiness is the business end of agriculture. It deals with a wide variety of businesses in agriculture such as farm supplies and services and the processing and marketing of agricultural products.

Applying Tech Prep to a Career

Objectives

- Familiarize students with the career description and education required for a career as an agribusiness specialist.
- Solve problems by making a list and using the pattern.
- Solve problems by using a graph.
- Determine interest and skills described in a career by expressing thoughts and feelings.
- Determine the handlers for products from farm to grocery store.

Mathematical Content Students should have completed Lesson 3-6 or know how to look for a pattern by making a list before attempting the Exercises. Students should have previous experience reading a graph.

Time Students will need 1-2 days to complete pages 519-520. Students will need about one week to complete the project on page 520.

Teaching Notes Have students read pages 519-520 and complete the Exercises.

If any student's family owns a farm or is familiar with the business of a friend's farm, they may want to describe the parts of agribusiness that are used on these farms.

You may want to assign a student to contact the county extension office in order to gather information (booklets, pamphlets, or videos) about careers in agribusiness or that describes the operation of a farm. You may want to invite the county extension agent to class to make a presentation.

Have students discuss their responses to Exercises 8-10 on page 520.

Encourage students to use the resources listed at the end of the project. They may want to do their own research at the library. Have students share their findings. Students could also interview an agribusiness specialist or you may want to have an agribusiness specialist make a presentation to the class.

EDUCATION

If you are interested in a career as an agribusiness specialist, you should take courses in mathematics, social studies, agriculture, business, and a laboratory science such as biology, chemistry, or physics. English literature and composition will be helpful since good oral and written communication skills are essential.

After high school, plan to attend a 2-year agricultural or technical college program. The college courses will include business, economics, science, agriculture, and supervised career experience.

EXERCISES

Corn yields in Michigan can be expected to decrease about one bushel per acre for each day corn is planted after May 10 and two bushels per acre for each day corn is planted after May 25. **3. Danson family: 7,000 bushels less; Sims: 13,500 bushels less**

1. If the Danson family plants corn on May 20, how many bushels per acre less can they expect than if they had planted by May 10? **10 bushels per acre less**
2. If the Sims family plants corn on May 31, how many bushels per acre less can the Sims expect than if they had planted on May 10? **27 bushels per acre less**
3. The Danson family plants 700 acres of corn and the Sims family plants 500 acres. How many total bushels of corn less can each family expect to harvest as a result of their late planting? See Exercises 1 and 2 for more information.
4. If the corn price is $3.15 per bushel, how much did the Danson and Sims families lose as a result of their late planting? See Exercise 3 for more information. **Danson: $22,050; Sims: $42,525**

Solve. Use the graph at the right.

5. How many sports items can be made from the hide of one cow? **224 items**
6. What fraction of the cow's hide is used to make baseballs? **$\frac{9}{14}$**
7. Name one sports item that makes up $\frac{3}{56}$ of the sports items from the cow's hide. **baseball gloves or basketballs**

Express in your own words.

8. Name the skills you have that prepare you to be an agribusiness specialist. **Answers may vary.**
9. Name one area that you would be interested in doing as an agribusiness specialist and tell why. **Answers may vary.**
10. Do you think you would pursue a career as an agribusiness specialist? Why or why not? **Answers may vary.**

PROJECT

Tracking Product Handlers from the Farm to the Grocery Store

Materials: rulers, paper

Overview You will be making a **flowchart** to show the different routes a farm product travels in making its way from the farm to the shelves of a grocery store. A flowchart is a diagram used to show the steps in a procedure.

Procedure Interview a farmer or an agribusiness specialist (at a grain elevator, a feed store, or so on), do research, or write to a food company to determine the handling of a farmer's product from the time the farmer begins working with the product until you take it home from the grocery store.

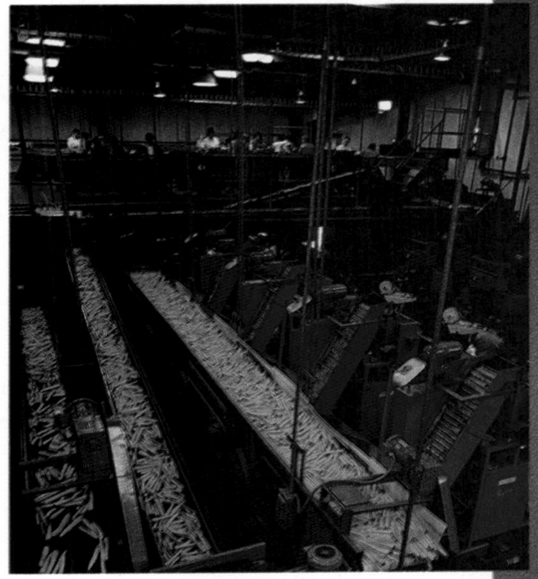

You may choose various types of products such as grains, animals, nuts, fruits, vegetables, lumber, flowers, trees, and so on.

You may even start backwards by starting with an item from the store and contacting the company on the label.

Once you have collected your data, make a flowchart. Compare your flowcharts with other groups in your class. See margin.

RESOURCES

For more information on a career as an agribusiness specialist, write or call the following associations.

- U.S. Department of Agriculture
  Department of Public Information
  Washington, D.C. 20250 (202) 447-3631
- Council for Agricultural Science and Technology
  137 Lynn Avenue
  Ames, IA 50010-7197 (515) 292-2125
- Agribusiness Council
  2550 M Street, NW, Suite 275
  Washington, D.C. 20037 (202) 296-4563
- Future Farmers of America
  National FFA Venter
  P.O. Box 15160
  5632 Mt. Vernon Memorial Hwy
  Alexandria, VA 22309 (703) 360-3600

Project Description Students track the stages of a product from the farm to the grocery store. Students will need to interview several people. Once they have collected their data, students are asked to make a flowchart.

Project Notes Have students work in pairs or in groups of three.

Groups should work on their projects at a regular pace rather than trying to finish at the end of the week. You may want to schedule class time for groups to work on their projects.

If students want more of a challenge, they could track the handlers by name. They might start by talking to the grocery store manager to find the names and locations of the handlers for one brand name of green beans. Students would then contact the wholesalers, distributors, and processors. The processor might be able to name a specific farming region from which the raw product was purchased.

Answers

Project Answers may vary. A sample answer for green beans: farmer purchases supplies → farmer grows and harvests beans → farmer sells to processing and cannery factory → factory sells to a distributor → distributor sells to wholesaler → wholesaler sells to retailer → retailer sells to consumer

Applying Tech Prep to a Career

Objectives

- Familiarize students with the career description and education required for a career as a graphic arts professional.
- Convert measures such as picas to points and picas to inches.
- Write a measure as a fraction or as a decimal.
- Determine interests and skills described in a career by expressing thoughts and feelings.
- Write and produce a children's book.

Mathematical Content Students should have completed Lessons 4-6, 5-8, and 6-3 or know how to write and simplify fractions, change a fraction to a decimal, and convert from one measure to another before attempting the Exercises.

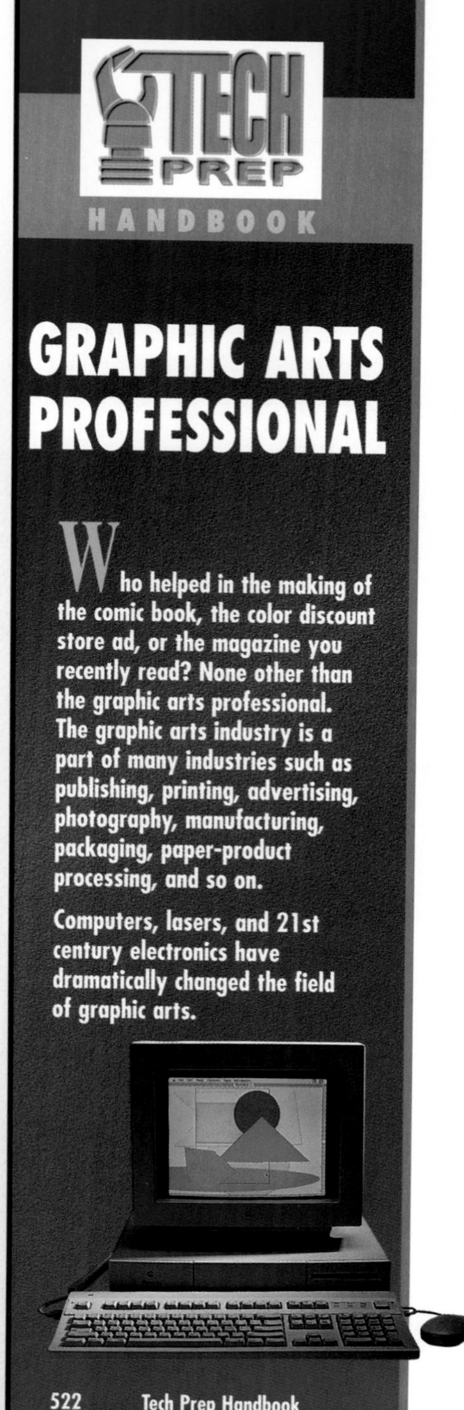

TECH PREP
HANDBOOK
GRAPHIC ARTS PROFESSIONAL
Who helped in the making of the comic book, the color discount store ad, or the magazine you recently read? None other than the graphic arts professional. The graphic arts industry is a part of many industries such as publishing, printing, advertising, photography, manufacturing, packaging, paper-product processing, and so on.
Computers, lasers, and 21st century electronics have dramatically changed the field of graphic arts.
522 Tech Prep Handbook

CAREER DESCRIPTION

Since graphic arts is normally done in a small business setting, the graphic arts professional must learn to serve in several different roles. The roles include estimator, process and quality controller, color laboratory technician, darkroom technician, technical illustrator, and so on. The estimator estimates the cost of a project for a customer. The process and quality controller makes sure that each step of the process is done correctly and that the quality of the end product is good. The color laboratory technician works with the color film, separating colors for art and photos, and checks the quality of the color. The darkroom technician works with various kinds of film. The technical illustrator uses the computer to draw technical art such as machinery parts.

TOOLS OF THE TRADE

Two of the many pieces of equipment used in the graphic arts industry are described below.

A technical illustrator uses a computer to draw technical art as seen on the computer screen on the left. The printer uses a color press as shown on the right.

EDUCATION

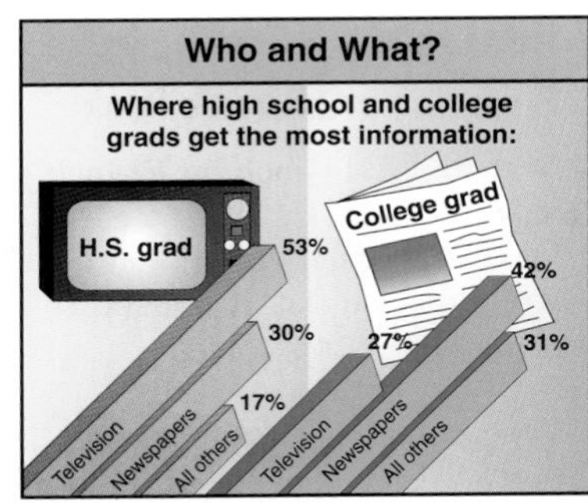

While in high school, 4 years of English, 2 years of mathematics, 1 year of chemistry or physics, mechanical drawing, basic electronics, and computer classes will prepare you to be a graphic arts professional. Upon completion of high school, enroll in a 2-year graphic arts program at a technical college. Students in technical colleges' graphic arts programs often find employment, or are recruited, before they graduate.

Study the graph on the right. Who gets more information from the newspaper? Who watches more TV? Why do you think this is so? What resources do you think might be in the All Others category? **college grads, H.S. grad, Answers may vary.**

EXERCISES

Measurement is one area that is common to most of the graphic arts industry. The measurement system for graphic arts uses points and picas. Points are used to measure small measures such as type (8-point type, 10-point type, and so on). Picas measure larger measures such as the dimensions of the page or a piece of art. There are 6 picas in 1 inch and 12 points in 1 pica.

Name the number of points in each pica.

1. 2 picas **24**
2. 3 picas **36**
3. 1½ picas **18**

Name the fraction of an inch that represents each pica.

4. 1 pica $\frac{1}{6}$
5. 3 picas $\frac{1}{2}$
6. 10 picas $1\frac{2}{3}$

Name the decimal equivalent in inches for each pica. Round to the nearest hundredth.

7. 1 pica **0.17**
8. 3 picas **0.50**
9. 10 picas **1.67**

Express in your own words.

10. Do you think you would be interested in being a graphic arts professional? Why or why not? **Answers may vary.**
11. If you are interested in another part of the graphic arts industry mentioned but not discussed in this section, what could you do? **Answers may vary.**
12. Have you ever discussed your work future with a guidance or vocational counselor at your school? Why or why not? **Answers may vary.**

Time Students will need 1-2 days to complete pages 522-523. Students will need about one week to complete the project on page 524.

Teaching Notes Have students read pages 522-523 and complete the Exercises.

Students may want to interview parents, friends, and relatives to see why they chose the careers they have. You may want students to make a visual presentation (graph, chart, or pictorial) for the bulletin board.

Have students discuss their responses to Exercises 10-12 on page 523 and continue to have students add to their list of interests, skills, likes, and dislikes.

You may want to use the following exercise as an introduction to the project.

Have students draw a diagram of a cover for a book. The book is $8\frac{1}{2}$ inches long by $10\frac{3}{4}$ inches wide, with a spine $\frac{3}{8}$ inches wide, and a surrounding binding edge that is $\frac{5}{8}$ inches wide. What is the total length and width of the cover before it goes on the book? **See diagram below. $9\frac{3}{4}$ inches by $23\frac{1}{8}$ inches**

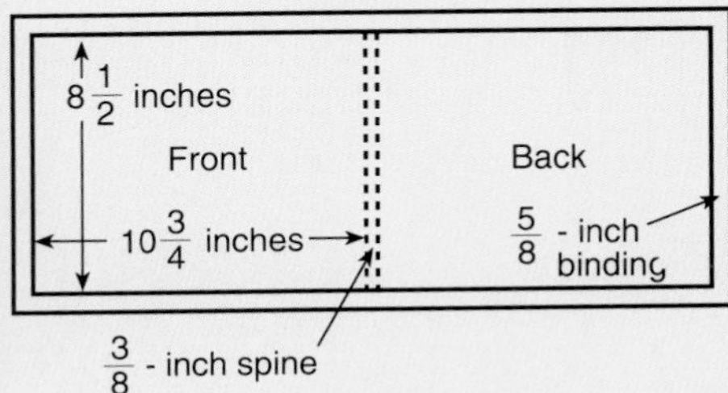

Encourage students to use the resources listed at the end of the project. Students could also interview a graphic arts professional or you may want to have a graphic arts professional make a presentation to the class.

Project Description Students are to write and produce a children's book. They will write the story, draw pictures, and make the book. Students will lay out the book on a single sheet of paper. This will give students an idea of how books are laid out by printers.

Project Notes Have students work in groups of four.

Groups should work on their projects at a regular pace rather than trying to finish at the end of the week. You may want to schedule class time for groups to work on their projects.

Each student should be assigned a task. They will need to prepare the front and back covers, write 14 pages of the children's story, type or write neatly the story after it is written, and include artwork, a photo, or a graph.

Now each group needs to bring the layout and text together. They must fit the text and illustrations on the layout, put the pages in the correct sequence, and tape the pages to the layout.

You may want to staple the book together. It will be easier to use the book if the staples are placed inside and at the middle of the book.

PROJECT

Producing a Children's Book

Materials: 11-inch by 17-inch white paper, scissors, tape, blue pencils, rulers

Overview You will be writing and producing a 16-page children's book. To begin, you will a make a flat. A flat shows the order of the pages as laid out on a single flat sheet of paper before it is folded.

Procedure Work with a group. Begin by folding the 11 × 17-inch sheet of paper into two $8\frac{1}{2} \times 11$-inch sheets. Fold the paper again into four $5\frac{1}{2} \times 8\frac{1}{2}$ -inch sheets. Then fold the paper again into eight $4\frac{1}{4} \times 5\frac{1}{2}$ -inch sheets. After the pages have been folded, label the front page as Front Cover, label the back page as Back Cover, and number the other pages 1 through 14. Remember to number both sides of the pages. Unfold the paper to see the page sequence. This is your flat. Leave the paper unfolded. Use the blue pencil to mark the same-size top, bottom, left, and right margins on every page. Now begin writing the 14 pages for your children's story and the front and back covers. Draw artwork for each page. When the story is written, either type, word process, or print the story onto a piece of paper the size of the pages on your flat. The story and art must fit inside the blue margin lines. Then tape the story and art to the correct pages of the flat.

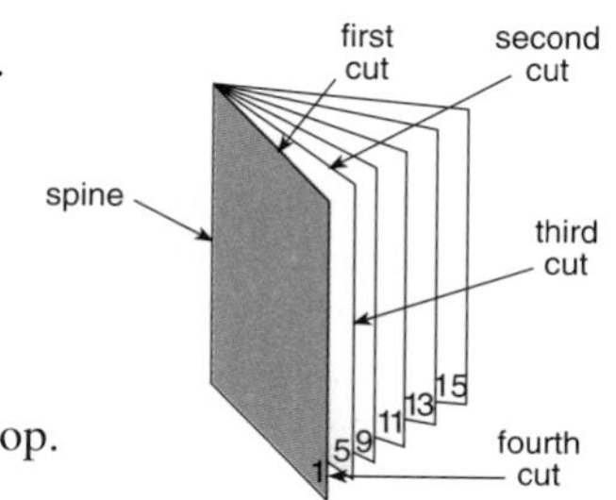

Now fold the paper as you did earlier. Holding the book, vertically cut the paper on the folds as shown in the figure on the right. *Do not cut the fold on the spine.*

The first and second cuts are parallel to the bottom or the top.

The third and fourth cuts are vertical to the spine. **See students' work.**

RESOURCES

For further information on a career as a graphic arts professional, write or call the following association.

- Graphic Arts Technical Foundation
  4615 Forbes Avenue
  Pittsburgh, PA 15213 (412) 621-6941

Check your local library for the following magazine.

- Graphic Arts Monthly
  249 W. 17th Street
  New York, NY 10011 (212) 463-6834

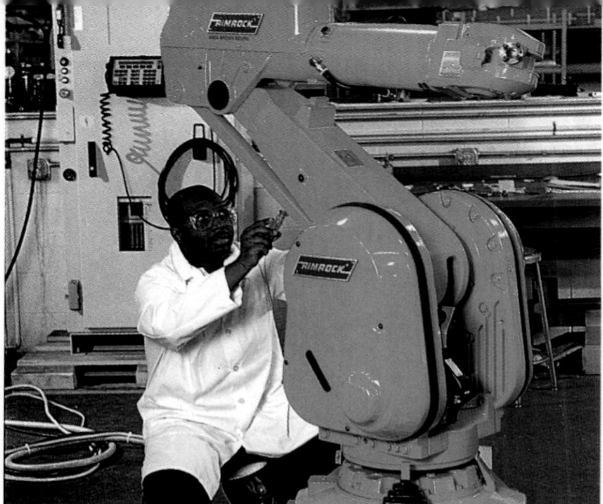

CAREER DESCRIPTION

Robotics technicians assist engineers in the design, development, production, testing, and operation of robots. Technicians who are also trained in computer programming sometimes perform programming and reprogramming of robots.

Two computer systems are important for robotics technicians to know. Computer Aided Manufacturing (CAM) is a system that is used to run factories and plants. Computer Aided Design (CAD) is a system used to design parts and products. CAM and CAD are often linked together so engineers and technicians can preview a simulated part or product before it is actually produced.

Robotics technicians often serve as a link between robotics engineers and customers. They may also be responsible for installing robots at manufacturing plants or other sites.

The demand for robotics technicians is high and is expected to remain high.

TOOLS OF THE TRADE

Besides the computer, the robot is also a tool of the trade.

HANDBOOK

ROBOTICS TECHNICIAN

Would it seem strange for a maid to be a robot? How would you feel if you walked into a warehouse and saw carts going up and down the aisles by themselves? Have you seen robots working on the assembly line in a car factory?

Robotics is a fast-growing technology. It includes the design, maintenance, and use of robots. Robots on an assembly line are machines that work in place of a human hand and arm. Some robots also work as walking machines or teleoperators. Robots usually contain and are operated by tiny computers called microprocessors.

Applying Tech Prep to a Career

Objectives

- Familiarize students with the career description and education required for a career as a robotics technician.
- Solve problems by using formulas.
- Solve problems by using a double-line graph.
- Determine interests and skills described in a career by expressing thoughts and feelings.
- Create artificial intelligence for a robot.

Mathematical Content Students should have completed Lessons 7-2 and 8-4 or know how to measure angles and how to use a formula to solve a problem before attempting the Exercises.

Time Students will need 1 day to complete pages 525-526. Students will need about one week to complete the project on page 527.

Teaching Notes Have students read pages 525-526 and complete the Exercises.

Check with factories or plants in your area to find out whether they use robots in their daily operations. If so, you may want to plan a tour so students can see what tasks they do and their physical form. If this is not possible, you may be able to find a video that shows robots in an automobile factory. You may want to contact the library or one of the resources in the student edition about videos.

Have students discuss their responses to Exercises 6-7 on page 526 and continue to have students add to their list of interests, skills, likes, and dislikes.

Encourage students to use the resources listed at the end of the project. Students could also interview a robotics technician or you may want to have a robotics technician make a presentation to the class.

EDUCATION

Most robotics technicians earn a 2-year associate degree in robot technology. Studies usually include hydraulics, pneumatics, electronics, CAD/CAM systems, and microprocessors. Robotics manufacturers generally provide additional on-the-career training.

EXERCISES

Solve. Use the diagram at the right. Use $\frac{22}{7}$ for π.

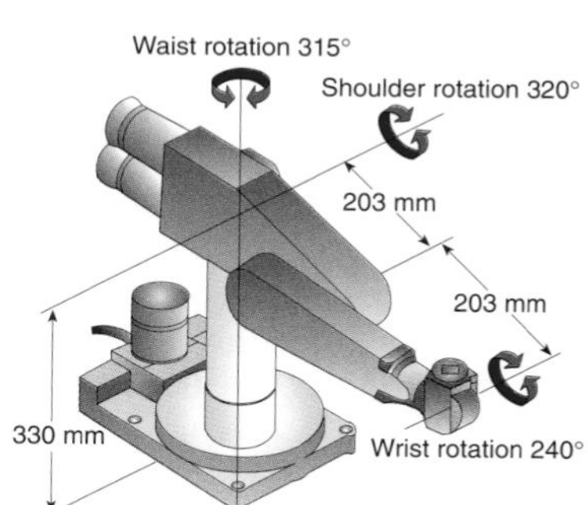

1. The robot's waist can rotate 315°. The total length of the robot's arm is 406 mm. Find the area the robot's arm will cover as its waist rotates 315°. Use the formula $A = \frac{7}{8} \cdot \pi \cdot r^2$. **453,299 mm²**
2. The shoulder of the robot will rotate vertically 320°. The total length of the robot's arm is 406 mm. Find the area the robot's arm will cover as its shoulder rotates 320°. Use the formula $A = \frac{8}{9} \cdot \pi \cdot r^2$. Round to the nearest whole number. **460,494 mm²**

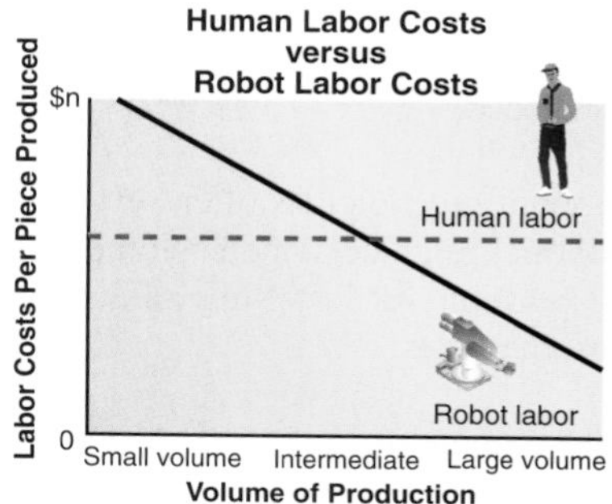

Solve. Use the graph at the right.

3. Is human labor or robot labor more cost effective when a manufacturing plant produces a small volume of products? **human labor**
4. Is human labor or robot labor more cost effective when a manufacturing plant produces a large volume of products? **robot labor**
5. As a manufacturer, what would you do if your plant produced an intermediate volume of products? **Answers may vary. A sample answer is that the costs are about the same so it would be better for the economy to hire human labor.**

Express in your own words. **7. Answers may vary.**

6. What do you like about this career? **Answers may vary.**
7. Do you think you would want to be a robotics technician? Why or why not?

PROJECT Artificial Intelligence

Materials: three-square by three-square board, one-half inch grid paper, 3 red checkers, 3 black checkers, red pen, black pen, 4 beads each a different color, 4 crayons or 4 colored pencils the same colors as the beads, 1 ketchup cup

Overview Few of today's robots are able to adjust to changing conditions. Providing robots with artificial intelligence may provide robots this ability. Teaching a robot to learn from its mistakes is shown by the following game of Six Piece Checkers.

Playing the Game Work in groups. Choose one member of your group to be the robot. Then choose a member of your group to be the human player.

Line up the checkers on opposite ends of the board as shown. You move first using the red checker. A checker piece can move forward one square or take an opponent's piece on the diagonal. A player wins by getting a checker to the opposite side of the board, taking all the opponent's pieces, or blocking so the opponent cannot move.

In order to create artificial intelligence for this robot, you need a record of the possible moves the robot can make. To do this, draw three-square by three-square boards on grid paper. Draw a diagram for each checker position after each odd-numbered move (your move) that can be made. Draw an arrow on the diagram to show the robot's legal move. Each arrow on the diagram is the same color as one of the beads. Keep all the (robot's) second-, fourth-, and sixth-move boards together. Do this for several games. **See students' work.**

When it's the robot's turn to move, find the board on your paper that matches the gameboard. Put the beads in the ketchup cup and have the robot select a bead without looking. Make the move shown by the arrow that is the same color as the bead. Since each turn can have 1-4 legal moves, place the same number of matching colored beads in the cup as the number of legal moves shown on the paper.

When the game is over, determine who won. If the robot lost, remove the bead from the cup of the last move the robot made. This is how the robot learns from its mistakes. Keep track of this next to the appropriate board on your paper each time the robot loses. The robot soon becomes the perfect player.

RESOURCES

For further information on a career as a robotics technician, write or call the following association.

- Robotics International
  One SME Drive
  P.O. Box 930
  Dearborn, MI 48121 (313) 271-1500

Project Description Students make gameboards to determine as many possible moves as they can on a 3-by-3 square. The purpose is to eliminate poorly chosen moves so a robot can become the perfect player. This creates artificial intelligence.

Project Notes Have students work in groups of four or five.

Groups should work on their projects at a regular pace rather than trying to finish at the end of the week. You may want to schedule class time for groups to work on their projects.

Remind students that a player can only move forward one square or take an opponent's place on the diagonal. Some positions allow the robot to make 1-4 legal moves.

Once students have finished playing several games, you may want them to switch games with other groups to test their "robot's" intelligence and also to see how their moves differed from the other groups.

Applying Tech Prep to a Career

Objectives

- Familiarize students with the career description and education required for a career as an automotive mechanic.
- Read a multiple line graph.
- Write fractions as percents.
- Determine interests and skills described in a career by expressing thoughts and feelings.
- Prepare a presentation and make a graph that supports the presentation.

Mathematical Content Students should have completed Lesson 9-8 or know how to write fractions as percents before attempting the Exercises.

HANDBOOK

AUTOMOTIVE MECHANIC

Almost every teenager in America dreams of having her or his own car. Did you know that the Arab Oil Embargo of 1973 raised oil prices and forced the U.S. auto industry to redesign cars for greater fuel efficiency? Also, Japan and West Germany became more efficient at manufacturing cars than the U.S. manufacturers. This forced the U.S. into using robots and computer-aided processes to improve their manufacturing and assembly-line operations.

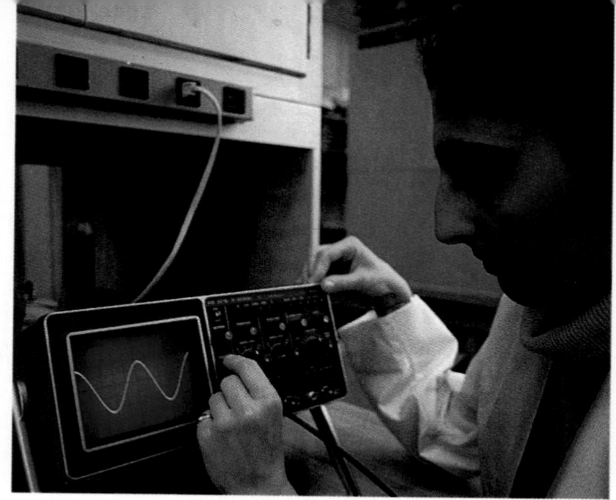

CAREER DESCRIPTION

Automotive mechanics help engineers design, develop, maintain, and repair automotive equipment.

There are five categories of automotive technicians.

1. *Research and Development Mechanics* prepare engines or related equipment for testing.
2. *Service and Sales Representatives* make sure that customers get the maximum performance from engines and advise customers in buying the best engines for their needs.
3. *Mechanics in Related Fields* work for transportation and other companies, oil companies, and insurance companies.
4. *High-Performance Engine Mechanics* are mostly in demand in the car dealerships.
5. *Manufacturing Mechanics* work in manufacturing plants at many different careers.

TOOLS OF THE TRADE

Automotive mechanics use oscilloscopes, as shown on the left, to diagnose engine problems. Oscilloscopes produce various types of sine waves, as shown on the screen. The sine waves represent levels of electrical signals that are received over a certain amount of time.

EDUCATION

To be an automotive mechanic, you must attend a 2-year program at a technical college. There are also several 4-year programs available.

Mathematics and science are important courses for you to take in high school to prepare for the technical college. One year of algebra and one year of geometry are recommended. Also one year of laboratory science such as physics or chemistry should be taken.

EXERCISES

Solve. Use the line graph on the right.

1. What percent of cars that are between two and three years old have a front-end alignment? **about 21%**

2. What percent of cars that are between four and five years old have their shock absorbers replaced? **about 17%**

3. What percent of cars that are between seven and eight years old have a muffler and tailpipe replaced? **about 34%**

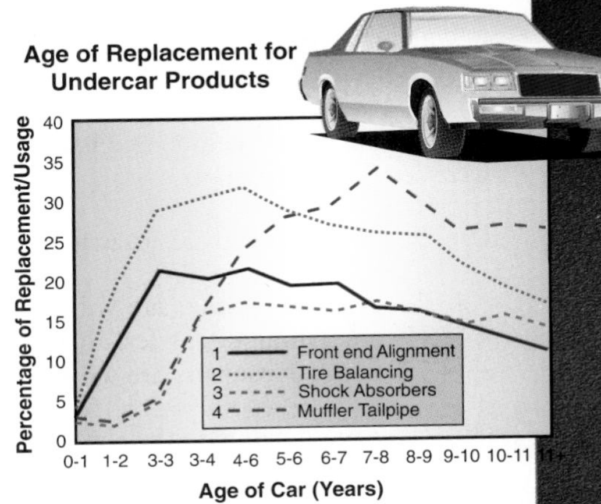

Solve.

4. Suppose a typical sedan weighs 4,000 pounds and 2,240 pounds of that weight rests on the front wheels. Find the wheel weight percentage for the front and back wheels, and the percents for the weights carried by the front and back wheels.
   front wheel percentage 2,240 pounds, back wheel percentage 1,760 pounds; front wheel percent 56%, back wheel percent 44%

Express in your own words.

5. What do you like about the career of an automotive mechanic? **Answers may vary.**

6. Would you want to be an automotive mechanic? Why or why not?
   Answers may vary.

7. If you did not want to repair cars but wanted to be a part the automotive mechanic career, what else could you do? Explain why you would or wouldn't want to do this.
   sales and service representative, Answers may vary.

Time Students will need 1-2 days to complete pages 528-529. Students will need about one week to complete the project on page 530.

Teaching Notes Have students read pages 528-529 and complete the Exercises.

You may want to contact a 2-year college or an automotive shop to arrange for a field trip so that students can see the tools of the trade in action. Also, the automotive mechanic can explain what he/she is doing as he/she works. If this is not agreeable to the college or shop, see if one or two students could go for an interview. Students should have some questions ready ahead of time. Check to see if a video camera is allowed. Then have students report to the class. Or you may want to have an automotive mechanic speak to the class.

Have students discuss their responses to Exercises 5-7 on page 529 and continue to have students add to their list of interests, skills, likes, and dislikes.

Encourage students to use the resources listed at the end of the project. There are several automotive mechanics magazines at the library.

Project Description Students are going to write a presentation and make a graph to convince a board of superiors that consumers will pay for a car that gets 80 miles per gallon of gasoline. Some statistics are provided.

Project Notes Have students work in groups of three or four.

Groups should work on their projects at a regular pace rather than trying to finish at the end of the week. You may want to schedule class time for groups to work on their projects.

You may want to have students do research at several car dealerships to find the prices of various makes and models of cars and the miles per gallon that they get. They could use one of these cars as an example for their presentation.

As an extension, students might also want to do a survey of family members or people in the neighborhood to find out if they would pay a price increase between $1,000 and $2,000 for a car that gets 80 miles per gallon, pay less than a $1,000, or pay more than $2,000. Students could compare their findings with the statistics given.

PROJECT

When More is Less

Materials: poster board, various colored marking pens, straightedge, compass

Overview Suppose you work for a car manufacturer as a automotive mechanic. Your team of mechanics and engineers have just developed an engine that can travel 80 miles on one gallon of gasoline. The research department held focus group meetings and sent out questionnaires, and they have given you the statistical results of their surveys.

Procedure Work in groups. You are giving a presentation before a board of your superiors. You want to convince them that consumers are willing to pay a little more for the price of the car if the car is able to get 80 miles per gallon of gasoline. You also want to convince them of an increase in the price of the car that is agreeable with the majority of consumers. Plan what you are going to say and how you will present your statistics. Use poster board, colored marking pens, a straightedge, and a compass to make a very creative and colorful circle graph, line graph, or bar graph. It could look like a graph you have seen in a newspaper.

Use the following statistics to help make up a presentation.

- 43.2% of the consumers are willing to pay between $1,000 and $2,000 more.
- 42.1% are willing to pay less than $1,000 more.
- 14.7% of the consumers are willing to pay more than $2,000 more.

Give your presentation to the class. **See students' work.**

RESOURCES

For further information on a career as an automotive mechanic, write or call the following associations.

- National Automotive Technicians Education Foundation
  13505 Dulles Technology Drive
  Herndon, VA 22071-3415 (703) 713-0100

- Automotive Service Association
  1901 Airport Freeway, Suite 100
  P.O. Box 929
  Bedford, TX 76021-0929 (817) 283-6205

Check your local library for the following magazine.

- Automobile Magazine
  120 E. Liberty
  Ann Arbor, MI 48104 (313) 994-3500

CAREER DESCRIPTION

One of the main functions of a biomedical equipment specialist is to repair equipment that is not working properly. He or she determines the problem, determines the extent of the problem, and makes repairs. Another main function is to install equipment. The third area of responsibility is maintenance. He or she tries to find problems before they become serious problems.

In all three areas of responsibility, the specialist consults with physicians, administrators, engineers, and other related professionals.

The specialist must work well under pressure because health matters often demand quick decision-making and prompt repairs. She or he must be extremely precise and accurate in her or his work.

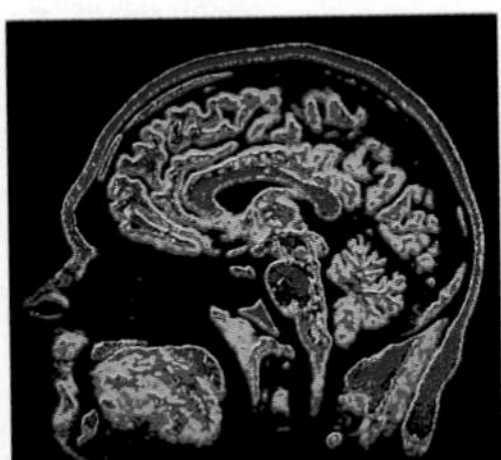

TOOLS OF THE TRADE

The tools of the trade may be as common as the screwdriver, hammer, level, wrench, vacuum sweeper or as precise as a micrometer, monitor, drill, or diagnostic equipment.

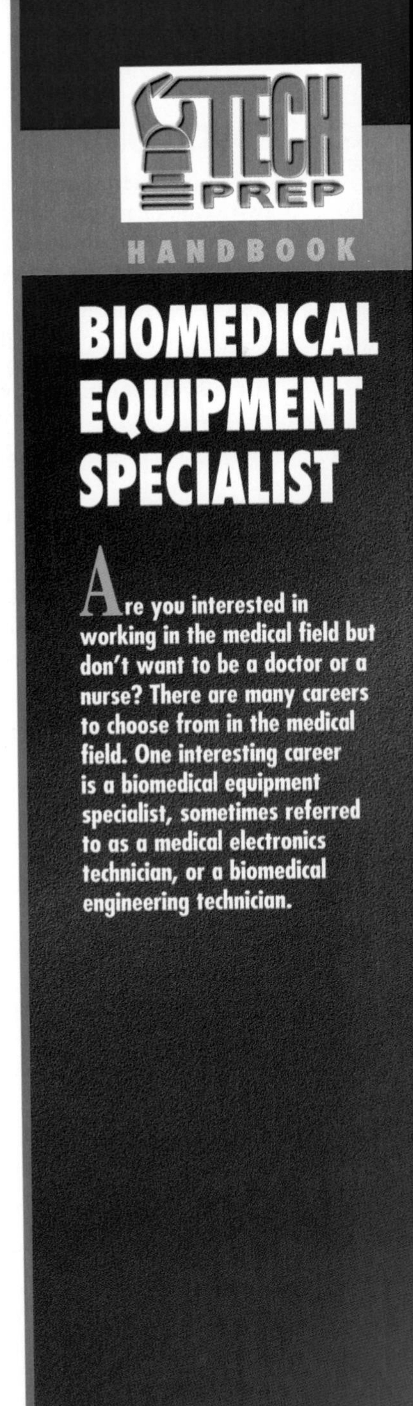

BIOMEDICAL EQUIPMENT SPECIALIST

Are you interested in working in the medical field but don't want to be a doctor or a nurse? There are many careers to choose from in the medical field. One interesting career is a biomedical equipment specialist, sometimes referred to as a medical electronics technician, or a biomedical engineering technician.

Applying Tech Prep to a Career

Objectives

- Familiarize students with the career description and education required for a career as a biomedical equipment specialist.
- Solve problems by using a graph.
- Determine interests and skills described in a career by expressing thoughts and feelings.
- Change electricity into sound by making a loudspeaker.

Mathematical Content Students should have completed Lesson 11-7 or know how to solve problems using graphs before attempting the Exercises.

Time Students will need 1 day to complete pages 531-532. Students will need 1-2 days to complete the project on page 533.

Teaching Notes Have students read pages 531-532 and complete the Exercises.

The following list names several pieces of equipment that a biomedical equipment specialist might work on. Students may need a medical dictionary from the library to find out what each piece of equipment does. You may want to assign 2-3 names to each group to research.

- ultrasound machine
- kidney machines
- x-ray units
- anesthesia apparatus
- sterilizers
- heart-lung machines
- defibrillators
- patient monitors
- blood-gas analyzer
- radiation monitors
- pacemakers
- CAT scan
- spectro-photometers
- spirometers

Have students discuss their responses to Exercises 5-8 on page 532 and continue to have students add to their list of interests, skills, likes, and dislikes.

Encourage students to use the resources listed at the end of the project. Students could also interview a biomedical equipment specialist or you may want to have a biomedical equipment specialist make a presentation to the class.

Project Description Students are going to make a loudspeaker from a portable radio, earphone plug, insulated wire, paper cup, glue, and a bar magnet. Students will demonstrate the loudspeaker and explain how they changed electricity into sound.

EDUCATION

If you are interested in being a biomedical equipment specialist, you should take at least two years of mathematics as well as biology, chemistry, and physics. Courses in English, electronics, and drafting are also helpful. Following high school, you will enroll in a 2-year program to earn an associate degree.

Besides course work, programs include practical experience in repairing and servicing equipment in a laboratory setting. You will learn about electrical components and circuits, the construction of common pieces of machinery, and computer technology.

EXERCISES

Solve. Use the graph on the right.

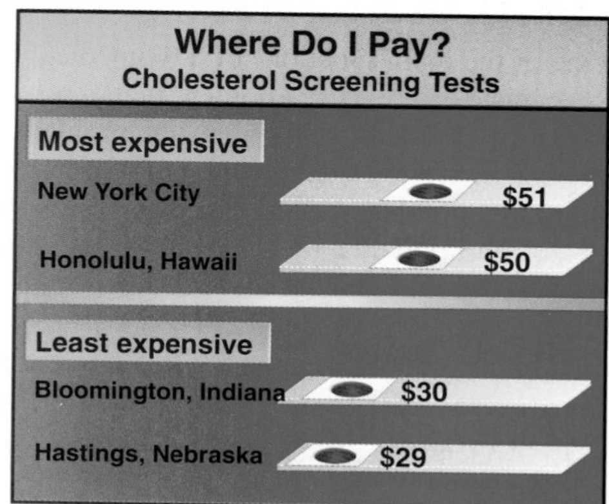

1. In what city is the cholesterol screening test the most expensive? **New York City**
2. In what city is the test the least expensive? **Hastings, NE**
3. How much more will it cost to be tested in New York City than in Hastings, Nebraska? **$22**
4. Why do think it is more expensive to have the same test using the same kind of equipment in New York City than in Hastings? **cost of living is higher, so hospital and testing costs and salaries are higher**

Express in your own words. **5. Answers may vary.**

5. What aspects of the biomedical equipment specialist's career do you like most?
6. What aspects of this career do you like least? **Answers may vary.**
7. Do you have some skills that you think would make you a good candidate for this career? **Answers may vary.**
8. Do you think you would be interested in being a biomedical equipment specialist? Explain. **Answers may vary.**

PROJECT

Changing Electricity into Sound

Materials: portable radio or portable CD player, earphone plug, 2 meters of thin insulated wire, paper cup, glue, bar magnet

Overview A biomedical equipment specialist might repair an ultrasound machine. Ultrasound is sound above human hearing. Some ultrasound machines change electric energy into ultrasonic (sound) waves. They have a disk that, when charged with electricity, vibrates so rapidly that ultrasonic waves are created. In medicine, this ultrasound machine can be used to destroy brain tumors and kidney stones.

You can change electric energy into sound energy by making a loudspeaker from a radio, earphone plug, insulated wire, paper cup, glue, and a bar magnet. When the magnet is near the coil, it causes the coil to vibrate. The motion of the coil causes the bottom of the cup to vibrate also, creating sound waves.

Procedure Follow the steps to build a loudspeaker.

1. Remove 1 centimeter of insulation from each end of a 2-meter length wire.
2. In the middle of the wire, coil the wire about ten times making turns that are 1 centimeter in diameter.
3. Glue the coil securely to the bottom of a paper cup. It may be necessary to allow the glue to dry overnight.
4. Attach one of the stripped ends of the wire to each wire of an earphone plug from a portable radio or CD player.
5. Turn on the radio, and insert the earphone plug into the earphone jack.
6. Hold the magnet very close to the coil.

Choose a team member to demonstrate your group's loudspeaker and to give a simple explanation in your group's own words how electric energy changes into sound energy. **See students' work.**

RESOURCES

For further information on a career as a biomedical equipment specialist, write or call the following associations.

- National Society of Biomedical Equipment Technicians
  3330 Washington Boulevard, Suite 400
  Arlington, VA 22201-4598 (703) 525-4890
- American Medical Technologists
  710 Higgins Road
  Park Ridge, IL 60068 (708) 823-5169

Project Description Students are going to make a loudspeaker from a portable radio, earphone plug, insulated wire, paper cup, glue, and a bar magnet. Students will demonstrate the loudspeaker and explain how they changed electricity into sound.

Project Notes Have students work in pairs or groups of three or four.

Groups should work on their projects at a regular pace rather than trying to finish at the end of the week. You may want to schedule class time for groups to work on their projects.

If a microphone and an oscilloscope are available, connect the microphone to the oscilloscope. Place the microphone in front of one of the group's loudspeaker. Have students observe the changing wave patterns of the sound on the screen. Have students note the change in the length of the waves and the number of waves for different sounds.

A microphone operates the opposite of a loudspeaker. It changes sound energy into electric energy. The system of loudspeaker and microphone make up medical ultrasonic devices.

Ultrasound is also used to monitor the growth of human fetuses. Electric energy is converted to ultrasonic waves to scan the fetuses. Echoes from the fetus' head, trunk, and limbs are received by a microphone and converted back into electric energy. Images of the fetus can be observed on a screen similar to an oscilloscope.

Applying Tech Prep to a Career

Objectives

- Familiarize students with the career description and education required for a career as a chemical assistant.
- Solve problems by using a graph.
- Determine interests and skills described in a career by expressing thoughts and feelings.
- Make an acid-base indicator.

Mathematical Content Students should have completed Lesson 13-5 or know that an equation can be used to plot points on the graph before attempting the Exercises.

HANDBOOK

CHEMICAL ASSISTANT

Do you ever stop to think where your vitamins, pain relievers, or medicines for a temporary illness come from? They are developed by chemists and doctors.

How do chemists in the field of medicine have time to conduct all the necessary tests and to do all the paperwork that is required for all the new and improved medicines being developed? How do chemists in any of the fields such as petroleum, aerospace, agriculture, and so on get all their work done? The answer is simple. Their help and support comes from chemical assistants.

CAREER DESCRIPTION

There are two types of chemical assistants. One is a chemical laboratory assistant; the other is a chemical engineering assistant.

Chemical laboratory assistants conduct tests to find the chemical content, strength, purity, and so on of a wide range of materials. These assistants test ores, foods, drugs, plastics, paints, petroleum and so on.

Chemical engineering technicians work closely with chemical engineers to develop and improve the equipment and processes used in chemical plants. They prepare tables, charts, diagrams, and flow charts that illustrate the results. These assistants also help build, install, and maintain chemical processing equipment.

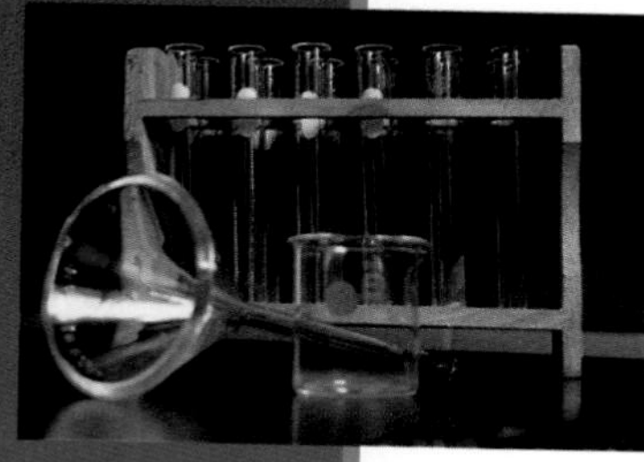

TOOLS OF THE TRADE

Some of the tools of the trade are gamma counters that test the amount of protein in cancer cells, laboratory equipment such as Bunsen burners, flasks and tubes, and large industrial machines. The tools are too numerous to mention all of them.

EDUCATION

A 2-year college program designed for training chemical assistants is required by most industry employers.

Most 2-year colleges require applicants to have algebra, geometry, trigonometry, chemistry, another physical science, and four years of English and language skills to be eligible for their chemical technology programs.

EXERCISES

The graph on the right shows that various kinds of salts, in varied amounts, will dissolve in water as the temperature of the solution increases.

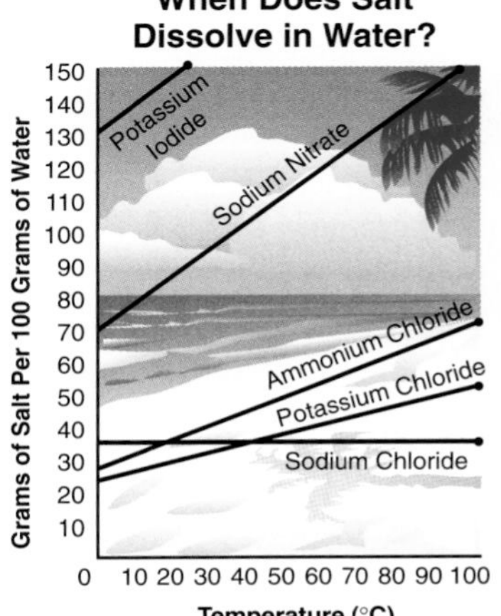

Find the number of grams of sodium nitrate that will dissolve in 100 grams of water for each temperature.

1. 20°C **85 g**
2. 30°C **95 g**
3. 85°C **140 g**
4. 55°C **115 g**

Find the number of grams of salt that will dissolve in 100 grams of water at 100°C for each kind of salt.

5. ammonium chloride **70 g**
6. potassium chloride **50 g**

Solve.

7. Which of the salts will dissolve the most at 25°C? **potassium iodide**
8. Which salt will dissolve the least at 10°C? **potassium chloride**

Express in your own words.

9. If you are interested in being a chemical assistant, are you willing to take the coursework? Explain. **Answers may vary.**
10. What skills and interests do you have that would make you consider being a chemical assistant? **Answers may vary.**
11. Would you like to be a chemical assistant? Explain your answer.
Answers may vary.

Time Students will need 1 day to complete pages 534-535. Students will need 2-3 days to complete the project on page 536.

Teaching Notes Have students read pages 534-535 and complete the Exercises.

You might want to display a periodic chart of the elements so students can see the chemical symbol for each chemical element.

Have students discuss their responses to Exercises 9-11 on page 535.

Continue to have students add to their list of interests, skills, likes, and dislikes. Once the lists are complete, discuss the following topics.

- Have students look over their lists to see if there is a pattern.
- What general interests or skills do they feel they have?
- From what they have listed, can they describe a career that might suit them?
- Are they willing to attend a 2-year college?
- If money is an object, where can they go for help? **Talk to the guidance counselor about scholarships and talk to the financial aid office at the college.**

Encourage students to use the resources listed at the end of the project. Students could also interview a chemical assistant or you may want to have a chemical assistant make a presentation to the class.

Project Description Students are going to make an acid-base indicator and test for the amount of vitamin C in certain fruits and vegetables. They will make a bar graph of the data from the test results.

Project Notes Have students work in groups of four.

Groups should work on their projects at a regular pace rather than trying to finish at the end of the week. You may want to schedule class time for groups to work on their projects.

Since students need to boil cabbage, you may want to check with the home-ec teacher about conducting Procedure One in the home-ec kitchen. Make sure students understand safety around the burners on the stove. Finish early so that there is time for students to clean up the kitchen before they leave.

You may want to conduct a test to determine how much vitamin C is contained in certain fruits and vegetables. Vitamin C is made up of absorbic acids so students can use their acid-base indicator.

Conduct the test using the following steps.

1. Place 50 milligrams of a crushed vitamin C tablet in 3 or 4 ounces of water.
2. Use the eyedropper to add the blue acid-base indicator drop by drop to the vitamin C. Count each drop. The solution should turn pink.
3. Continue adding drops of the blue indicator until a single drop retains its blue color. At this point, all the vitamin C has been used up.
4. Prepare test solutions for several fruit juices (3 or 4 ounces) that you think contain vitamin C. Then repeat steps 2-3.
5. Use the formula $C = \dfrac{Y \text{ drops} \times 50 \text{ mg vitamin C}}{X \text{ drops}}$, to calculate and record the milligrams of vitamin C for each test solution. X drops represent the number of drops of blue indicator used with the 50 mg of vitamin C solution and Y drops represent the number of drops used with each test solution. **Answers may vary.**

Have students make a bar graph to show the number of milligrams of vitamin C for each test solution. Each group should present their findings to the class and compare their results with other groups.

PROJECT

Making an Acid-Base Indicator

Materials: one head of red cabbage, hot plate, several clear glass containers or beakers that measure at least one cup, saucepan, eye droppers, pencils, tongs, graph paper, fruit juices, ammonia, 50-mg tablets of vitamin C

Overview One thing a chemical laboratory specialist might test a food product for is acid. To test for acid, they use an acid-base indicator. You will be making an acid-base indicator that indicates whether or not a substance contains acid. It will also tell you how strong the acid is in the substance.

Procedure **Work in groups as chemical specialists. Complete the following steps.**

1. Break up the red cabbage to fill at least 2 cups. Leaves with a deep purple color are best.
2. Put the cabbage leaves in a saucepan in just enough water to cover them.
3. Boil the cabbage until the color (dye) of the leaves becomes faded. Remove the faded cabbage leaves with tongs.
4. Add fresh leaves to deepen the color of the dye in the saucepan. Repeat step 3.
5. Remove the mixture from the heat and set it aside to cool.
6. Pour the cooled water into a clear glass.
7. Allow the purple water to sit overnight. It will change color from deep purple to deep blue. This is your acid-base indicator.

RESOURCES

For further information on a career as a chemical assistant, write or call the following associations.

- American Chemical Society
  Educational Activities Department
  1156 16th Street, NW
  Washington, DC 20036 (202) 872-4600
- Chemical Manufacturers Association
  2501 M Street, NW
  Washington, DC 20037 (202) 887-1100

ANSWERS

Chapter 4 Fractions: Adding and Subtracting

Page 137 Problem Solving

11. 19 ways

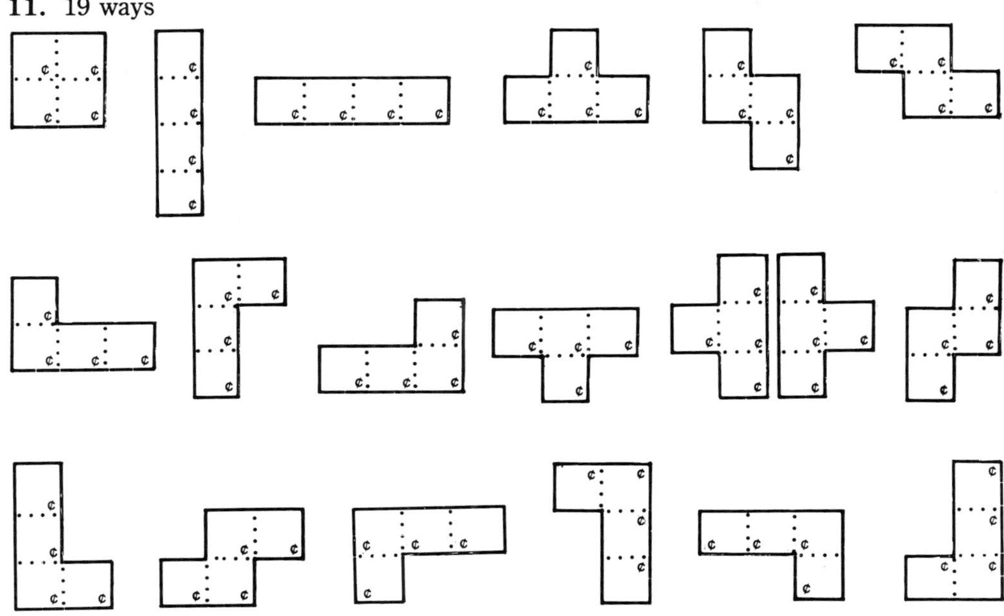

Chapter 5 Fractions: Multiplying and Dividing

Page 145 Activity

4. 8, $8\frac{1}{8}$

5. shade 5 parts, $\frac{5}{8}$

6. Separate a circle into the number of parts to match the denominator of the fraction and shade the number of parts that correspond to the whole number.

7. Fold the paper strip into fourths and shade one-fourth. Refold the strip into fourths. Then fold the strip in half lengthwise. Open the strip and shade $\frac{1}{2}$ of the shaded fourth. The part of the whole strip that is shaded twice is the product. $\frac{1}{8}$

8. Fold the strip into thirds and shade two parts. Refold the strip into thirds. Then fold the strip into fourths lengthwise. Open the strip and shade three of the fourths in each of the two shaded thirds. The part of a whole strip that is shaded twice is the product. $\frac{1}{2}$

9. Tape two strips together and fold each in half. Shade the part that represents $1\frac{1}{2}$ or $\frac{3}{2}$. Refold each strip into halves. Then fold the strip into fourths lengthwise and shade one fourth of each of the shaded halves. The three parts that are shaded twice represent the product. There are three eighths that are shaded twice. $\frac{3}{8}$

Chapter 7 Geometry

Page 213 Exercises

15. A B

16. K

17. M N

18. D E

19. A D

20. R S

23. A B

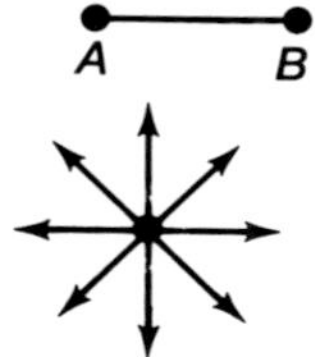

32.

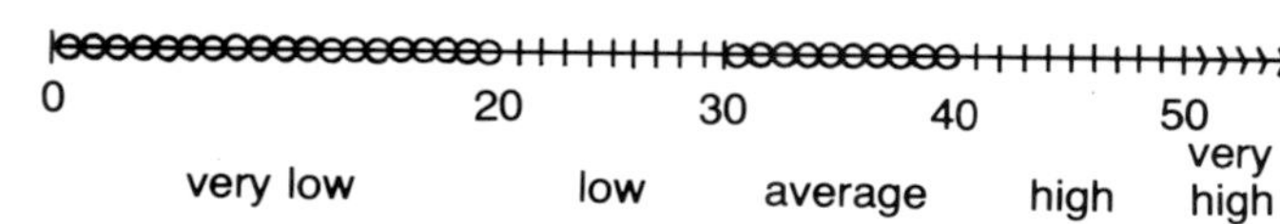

Page 221 Exercises

13. Construct a line perpendicular to a line at a point on the line.

Given: P

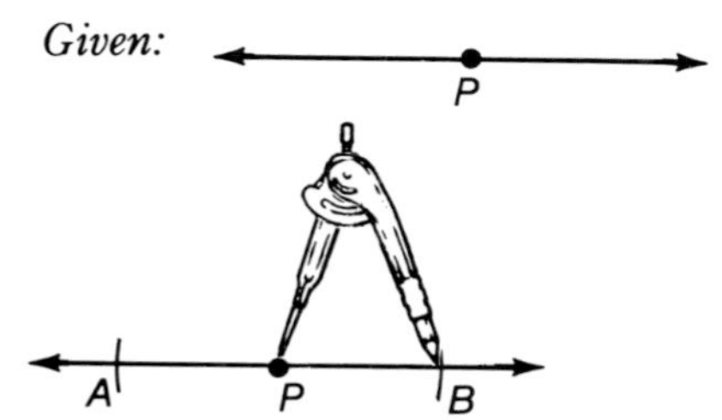

1. With the compass point at P, draw arcs intersecting the line at A and B.

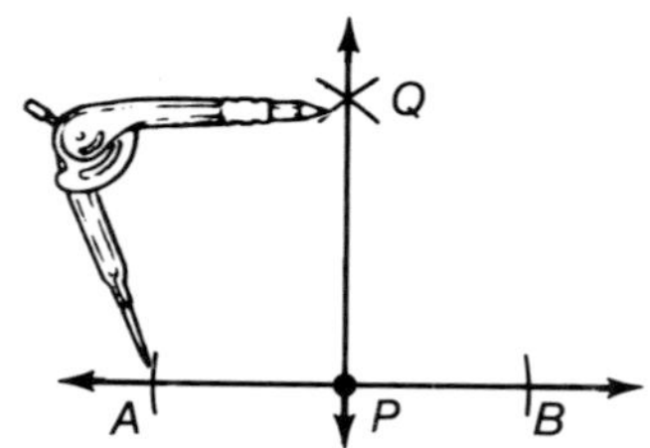

2. Use a larger compass setting to draw arcs from A and B which intersect at Q. Draw $\overleftrightarrow{PQ}$.

14. Construct parallel lines through two points on a line.

Given: A B

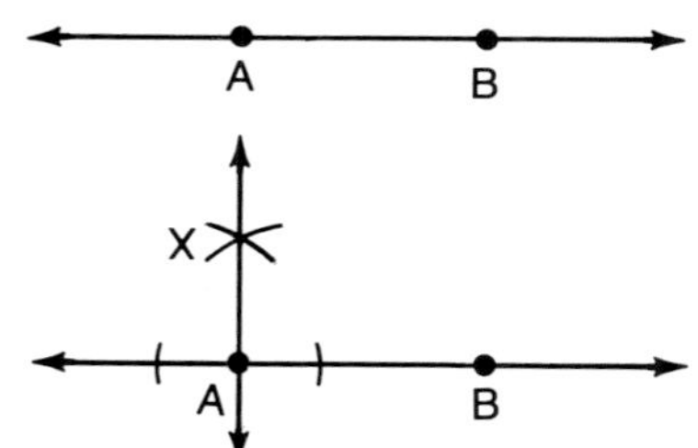

1. Construct a line perpendicular to $\overleftrightarrow{AB}$ at point A. Line AX is perpendicular to $\overleftrightarrow{AB}$.

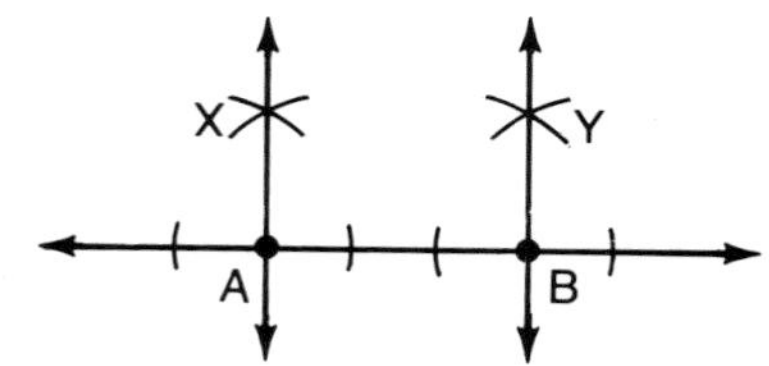

2. Construct $\overleftrightarrow{BY}$ perpendicular to $\overleftrightarrow{AB}$.

Page 225 Exercises

12.

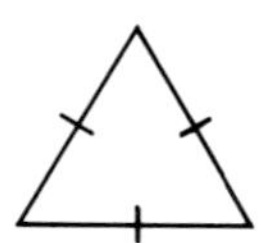

13.

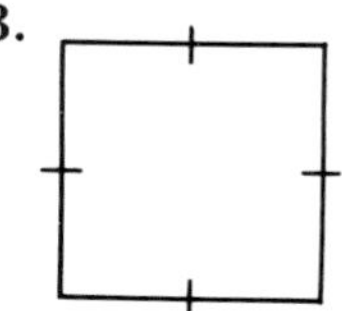

14.

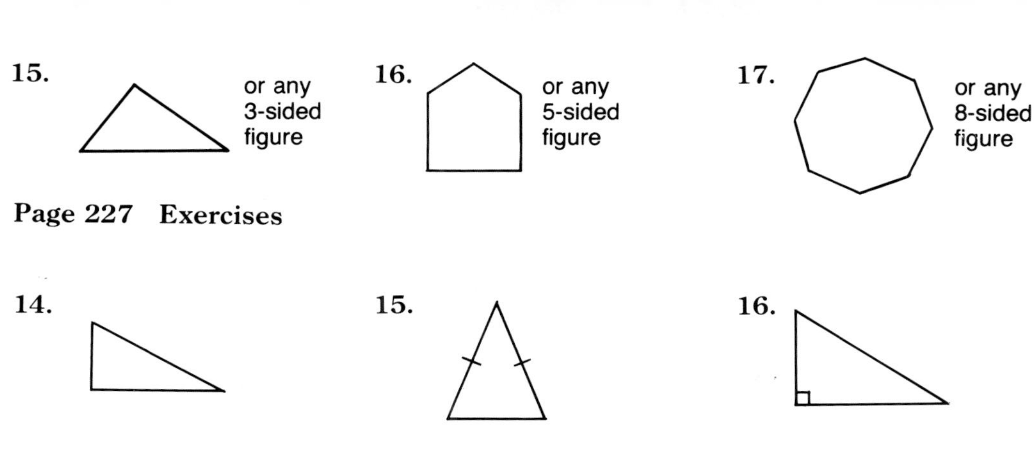

Page 227 Exercises

14. 15. 16.

17. 18.

19.

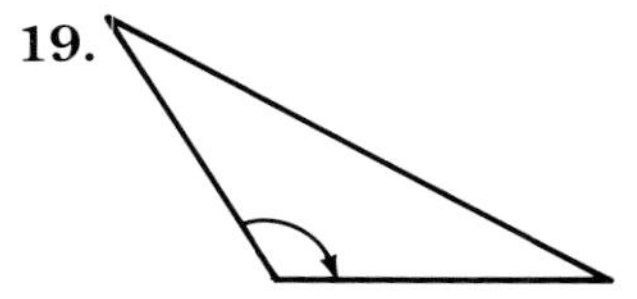

Page 234 Chapter 7 Review

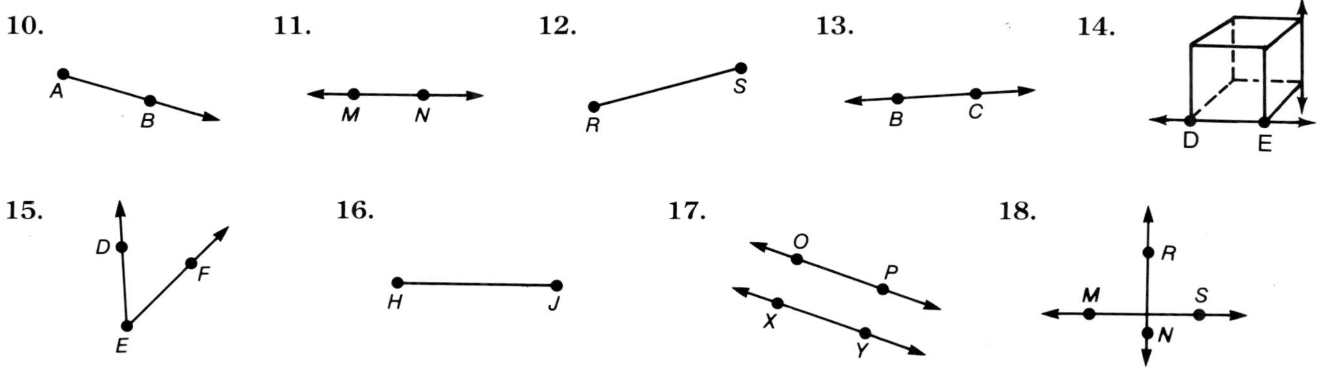

Chapter 10 Applying Percents

Page 318 Guided Practice

1. 8 families

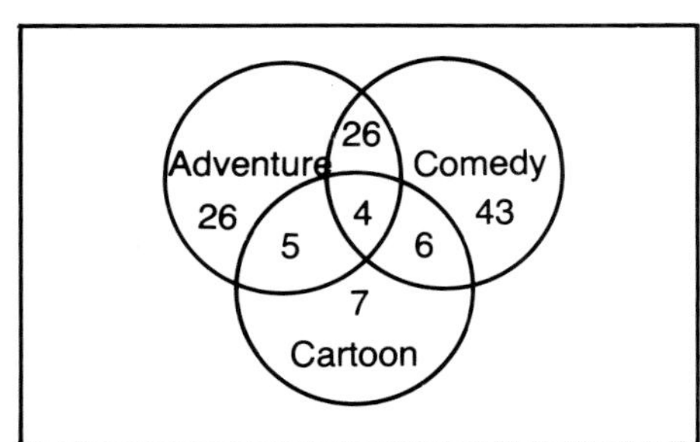

Page 318 Problem Solving

2. 62 students

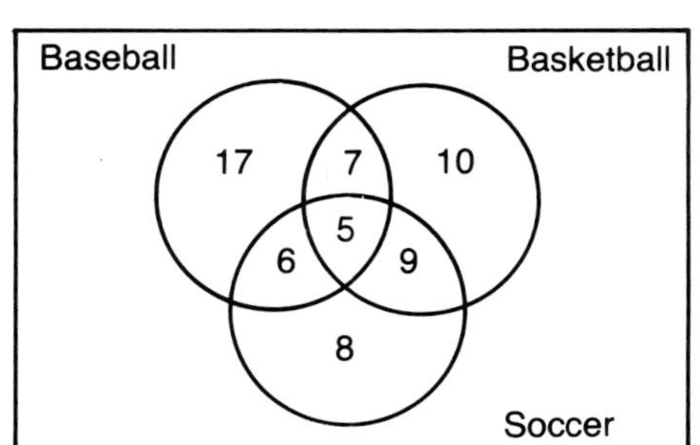

Chapter 11 Statistics and Graphs

Page 348 Guided Practice

1.

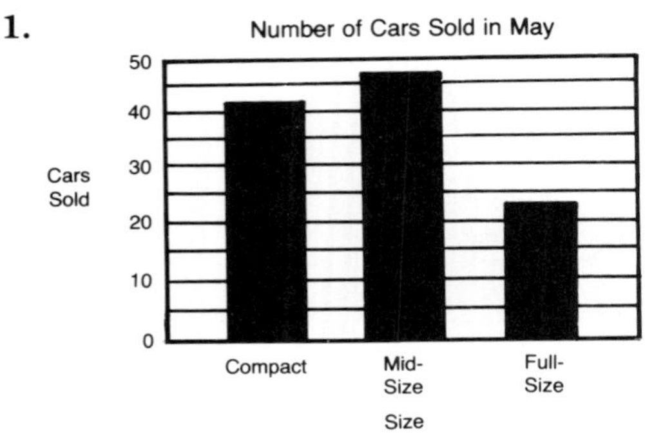

2.

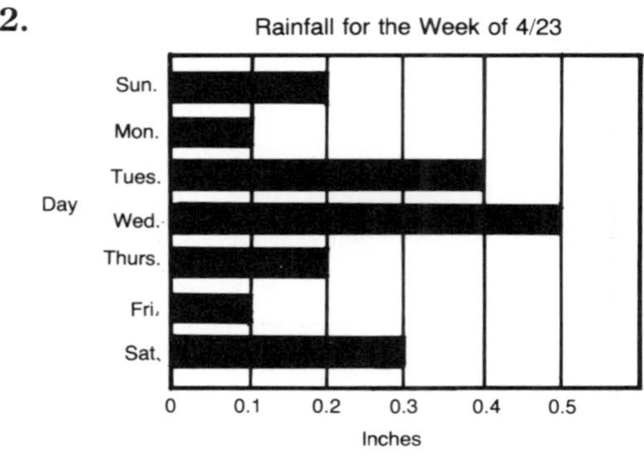

Page 349 Exercises

3.

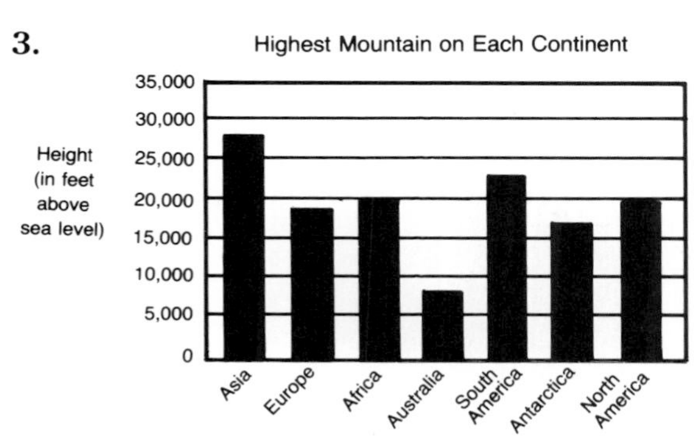

4.

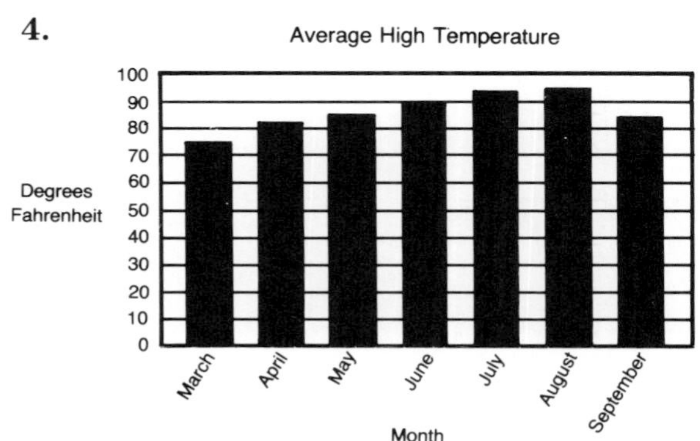

5.

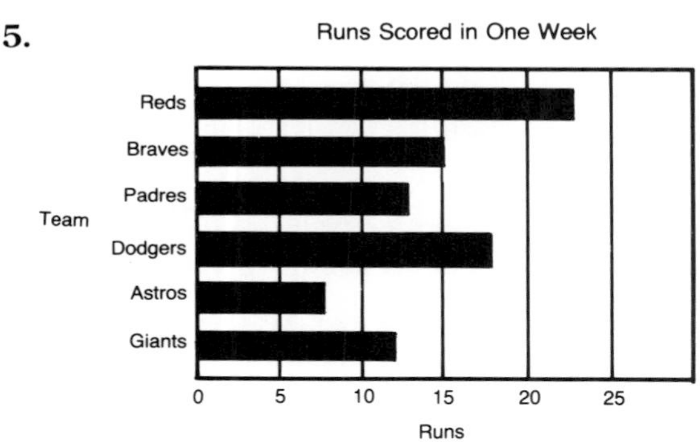

6.

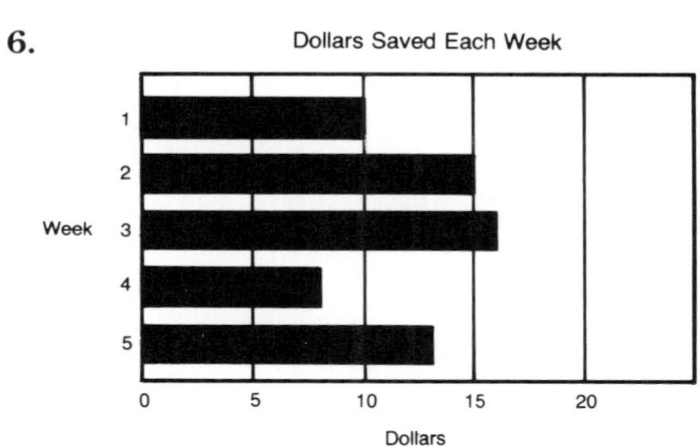

Page 350 Guided Practice

1.

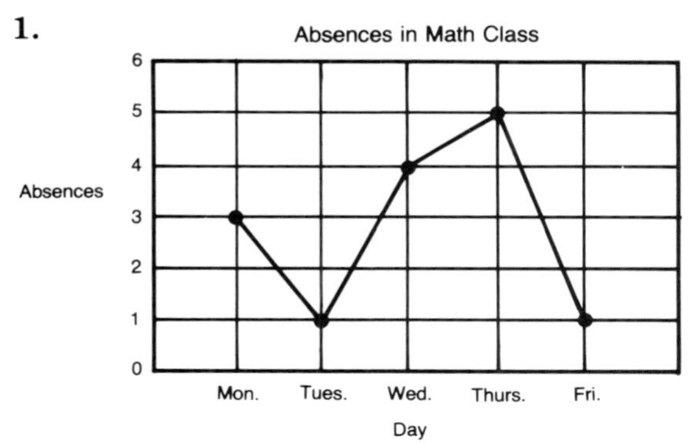

2.

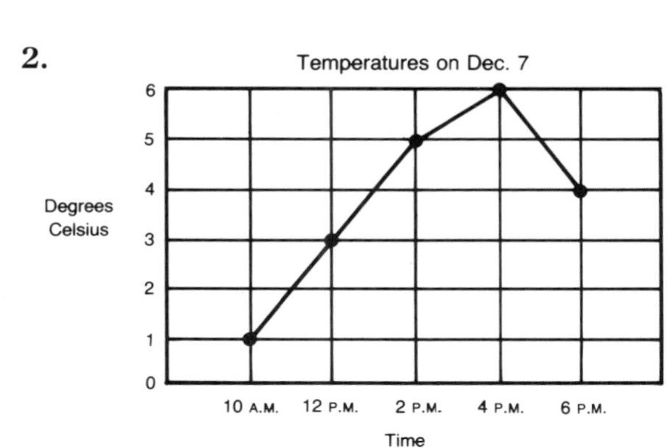

Page 351 Exercises

3.

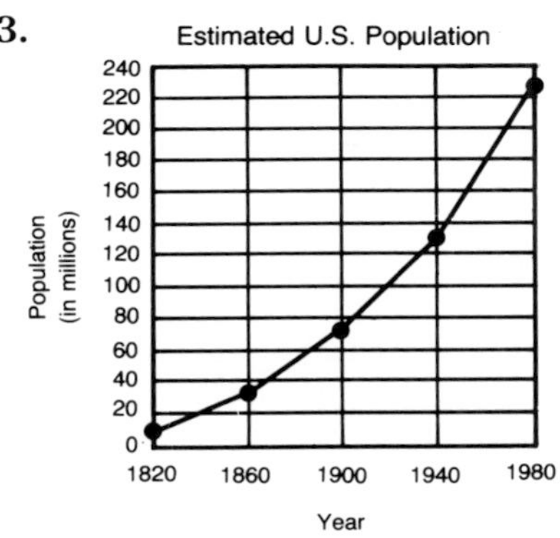

4.

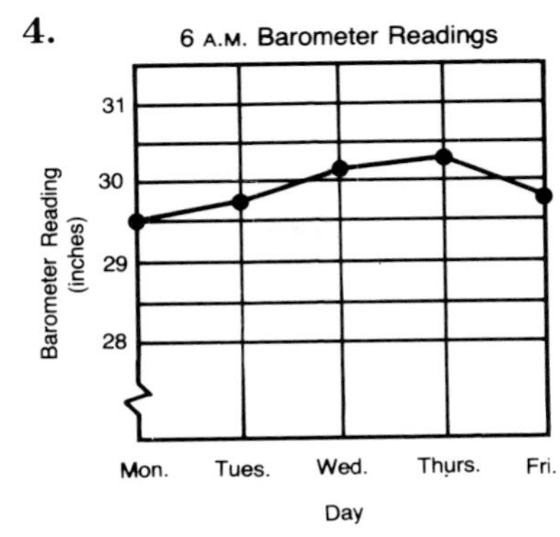

5.

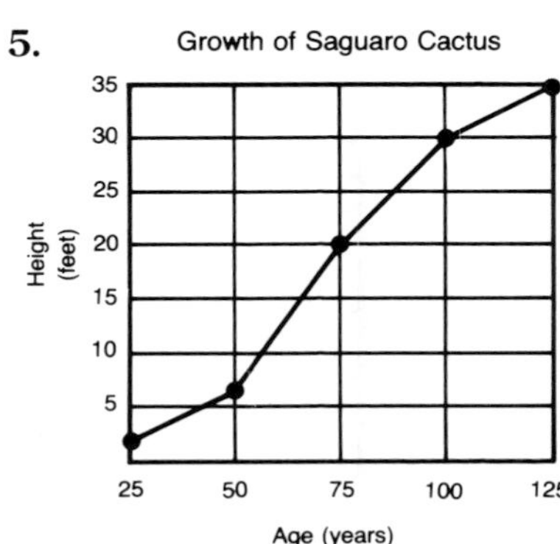

6.

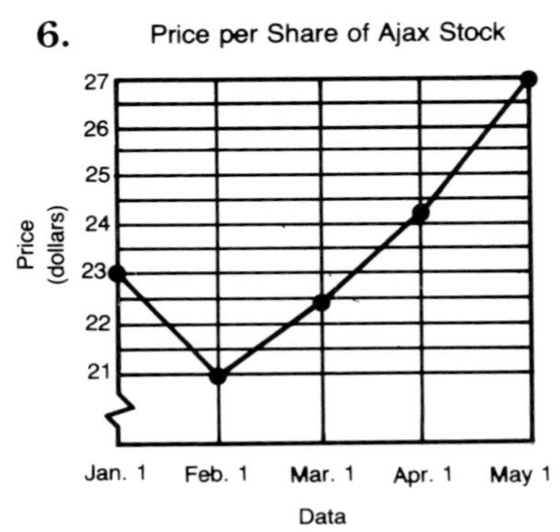

Page 352 Computer Application

1.

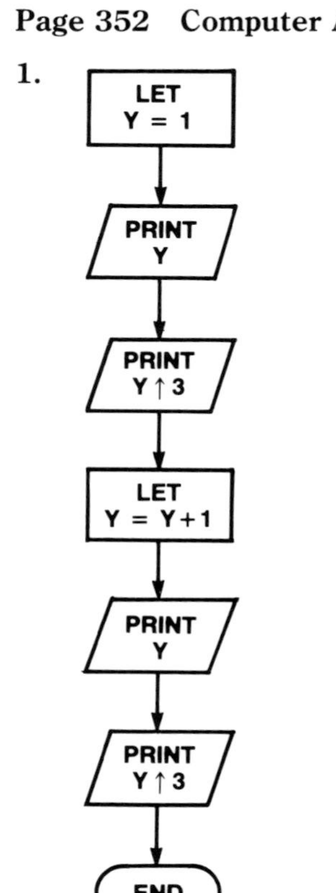

2.

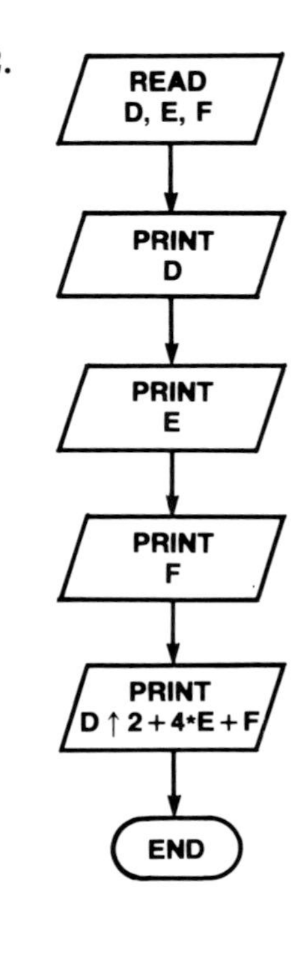

3.

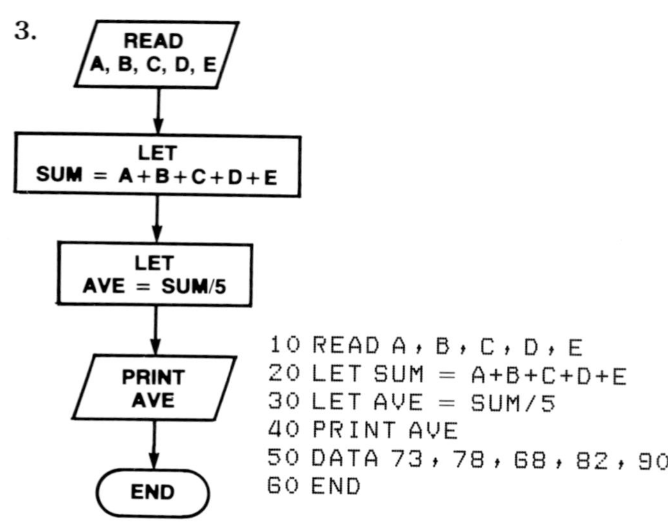


```
10 READ A, B, C, D, E
20 LET SUM = A+B+C+D+E
30 LET AVE = SUM/5
40 PRINT AVE
50 DATA 73, 78, 68, 82, 90
60 END
```

4.

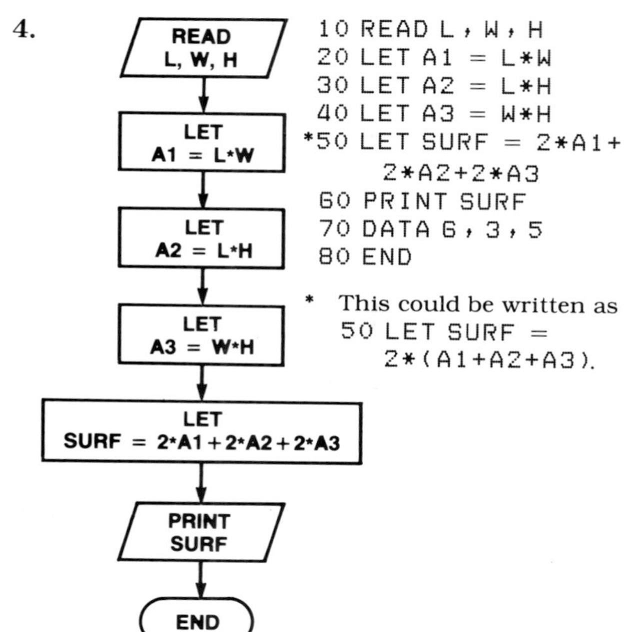


```
 10 READ L, W, H
 20 LET A1 = L*W
 30 LET A2 = L*H
 40 LET A3 = W*H
*50 LET SURF = 2*A1+
       2*A2+2*A3
 60 PRINT SURF
 70 DATA 6, 3, 5
 80 END
```

* This could be written as 50 LET SURF = 2*(A1+A2+A3).

Page 354 Guided Practice

5.

Apple Production in Four Orchards	
Adams	
Lynd	
Cooley	
Smith	
each = 50,000 apples	

6.

Boxes of Cereal Sold	
Corn Flakes	
Puffed Rice	
Bran Flakes	
Raisin Bran	
each = 100 boxes	

Page 355 Exercises

7.

Votes Received	
Johnson	✓✓✓✓✓✓✓✓
Kuhn	✓✓✓✓✓✓✓✓✓
Pitkin	✓✓✓✓✓✓✓✓✓✓✓✓
Van Dyke	✓✓✓✓✓✓
each ✓ = 100 votes	

8.

Number of Library Books Loaned	
Monday	
Tuesday	
Wednesday	
Thursday	
Friday	
each = 10 books	

Page 356 Guided Practice

5.

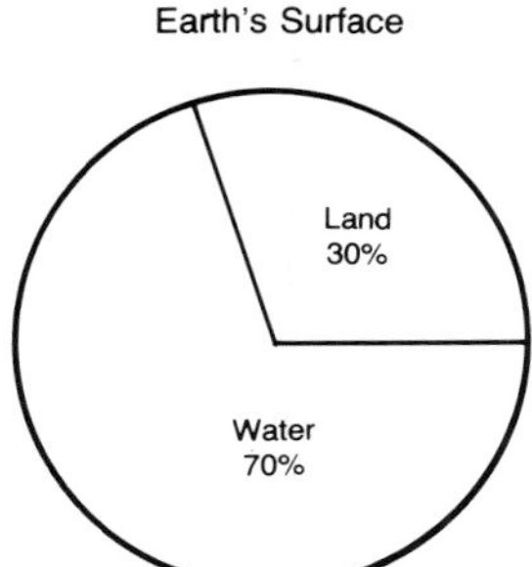

Page 357 Exercises

6.

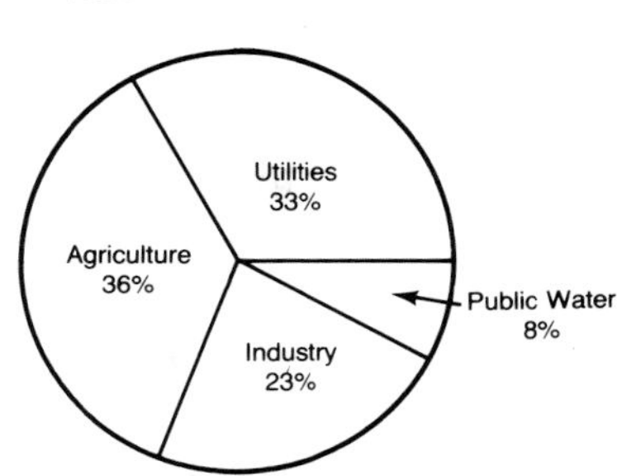

7.

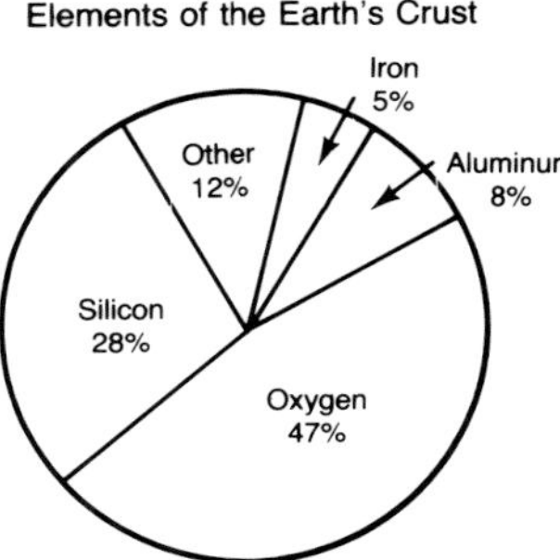

8.

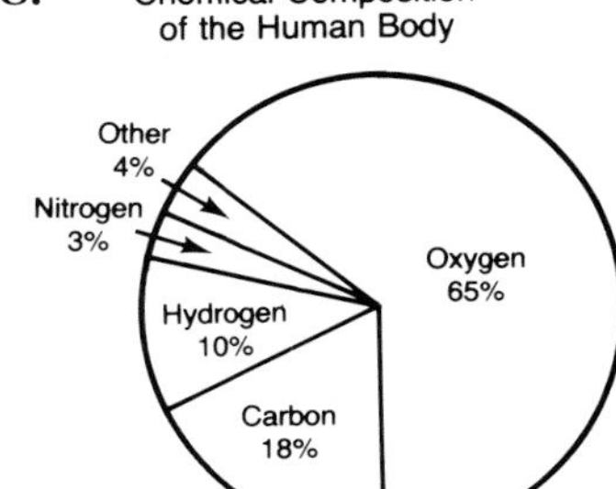

9.

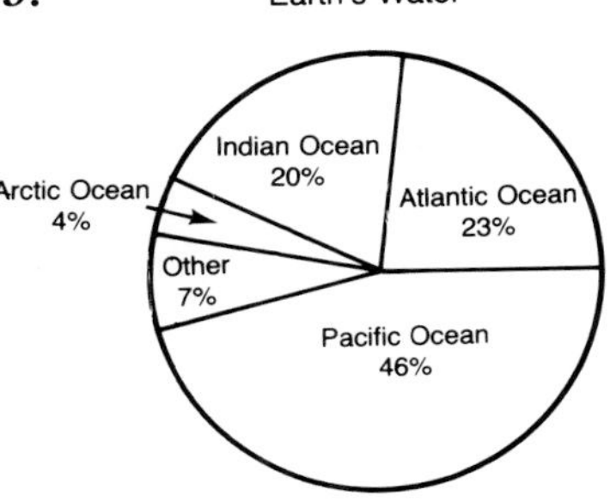

Page 367 Chapter 11 Review

20.

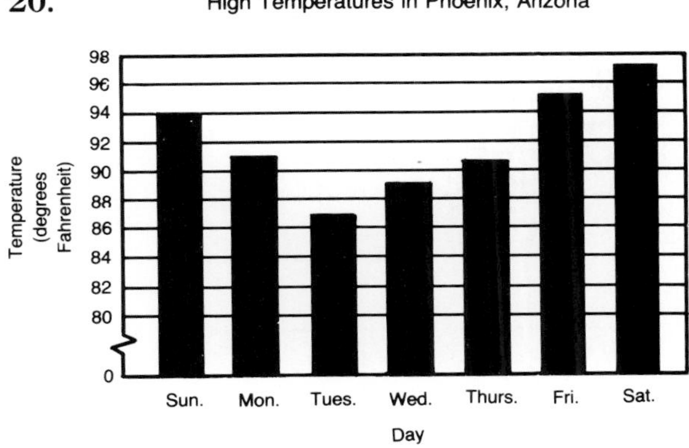

21.

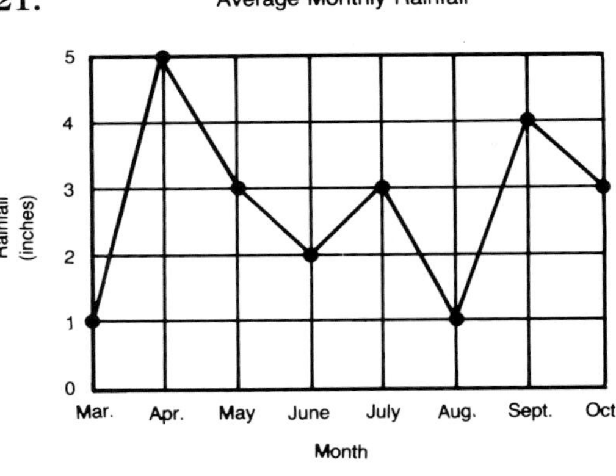

22.

Favorite Radio Station	
WXNY	
WNCI	
WQFM	
WRFD	
WBBY	
Each = 2 votes	

23.

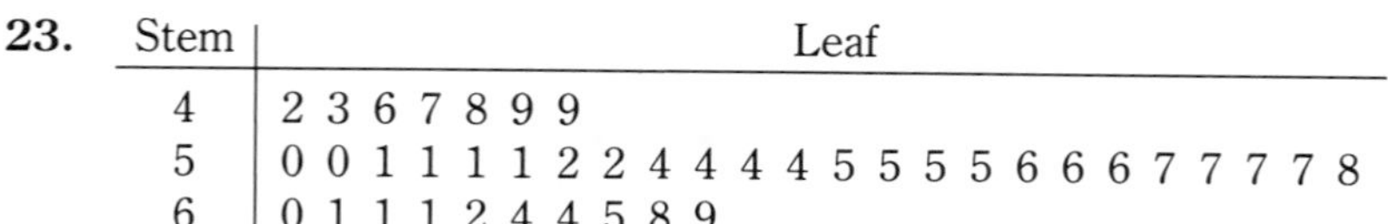

Stem	Leaf
4	2 3 6 7 8 9 9
5	0 0 1 1 1 1 2 2 4 4 4 4 5 5 5 5 6 6 6 7 7 7 7 8
6	0 1 1 1 2 4 4 5 8 9

24.

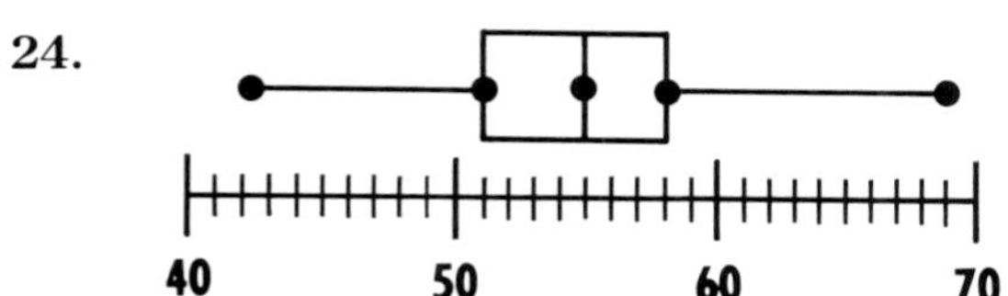

Page 368 Chapter 11 Test

7.

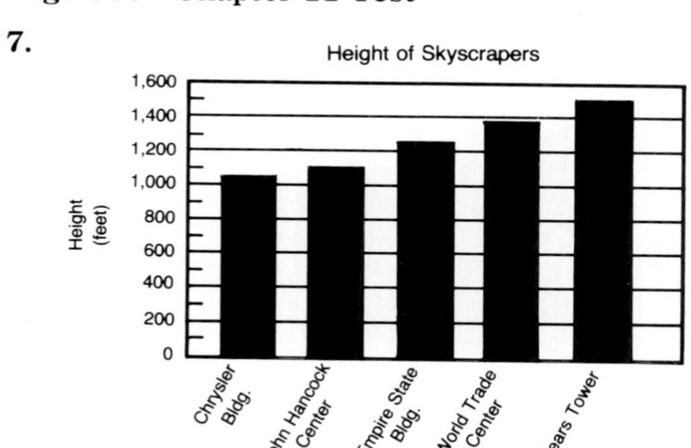

8.

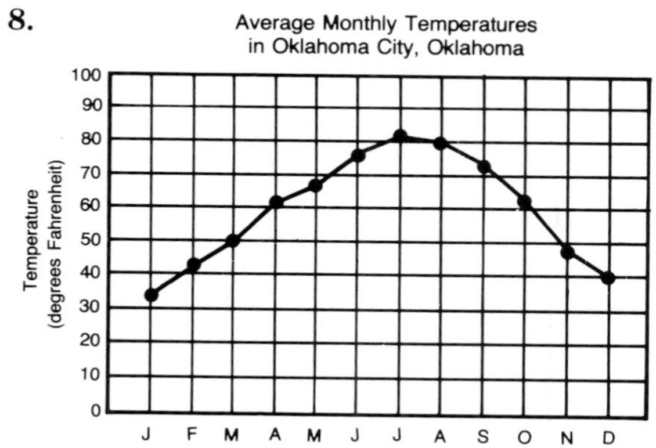

9.

Stem	Leaf
5	5 9 9
6	0 0 1 2 3 5 6 7 8
7	0 2 2 2 7 8

10.

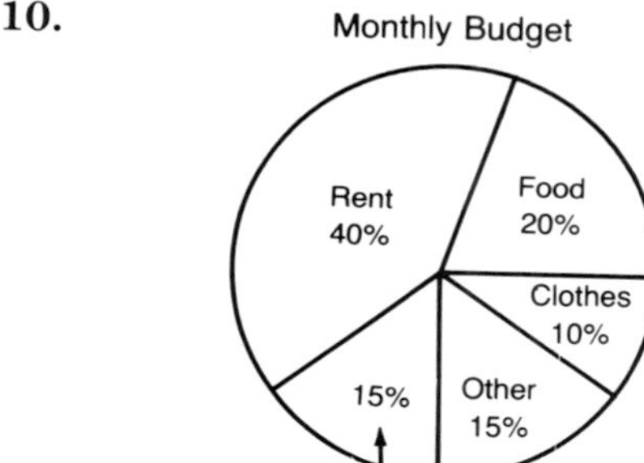

Chapter 12 Integers

Page 379 Guided Practice

3.

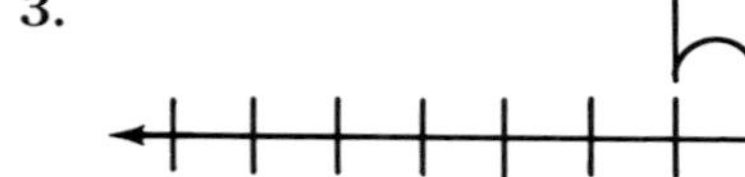

4.

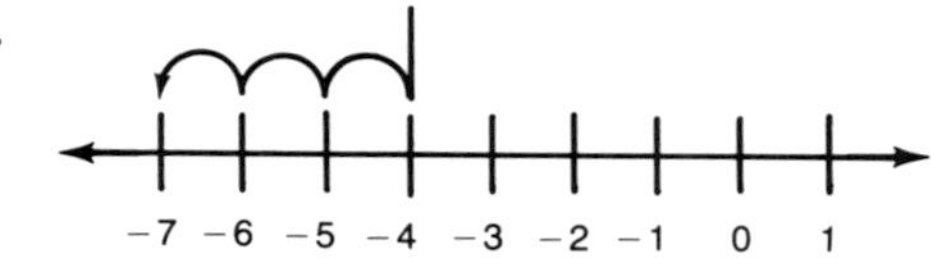

5.

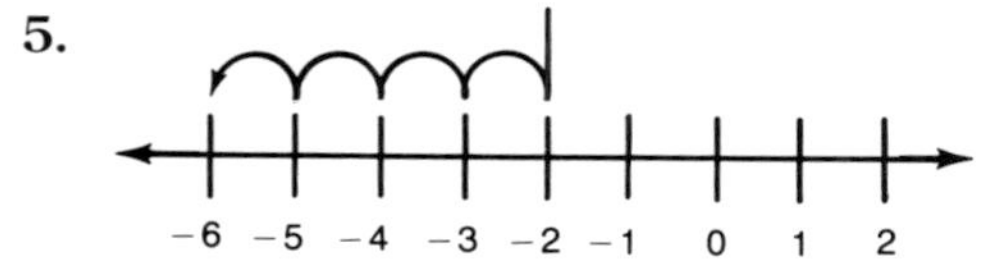

6.

7.

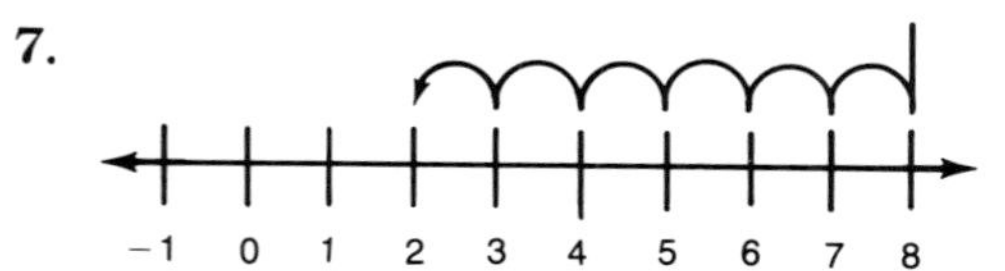

8.

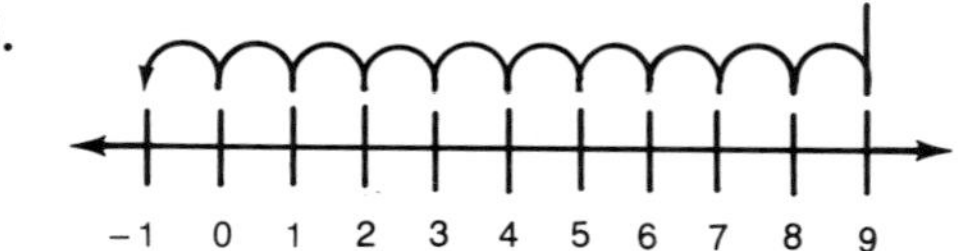

Chapter 13 Extending Algebra

Page 409 Guided Practice

7-15.

Page 409 Exercises

22-33.

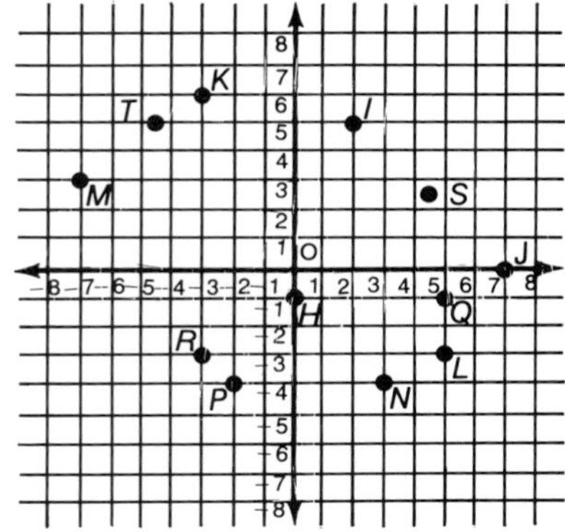

Page 411 Guided Practice

1.

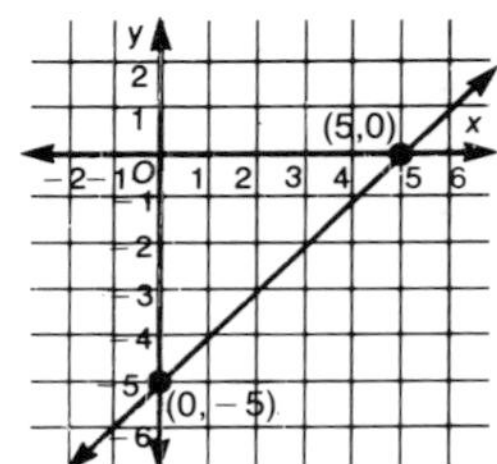

2.

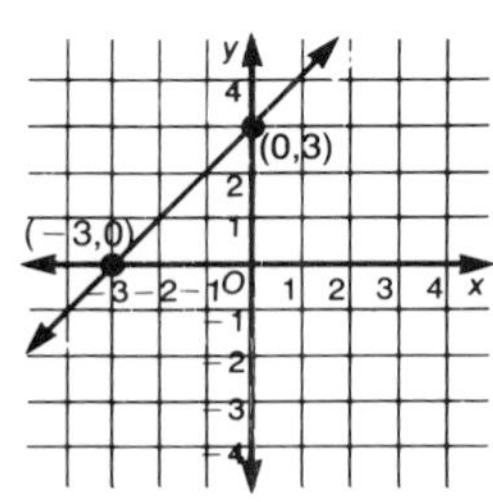

3.

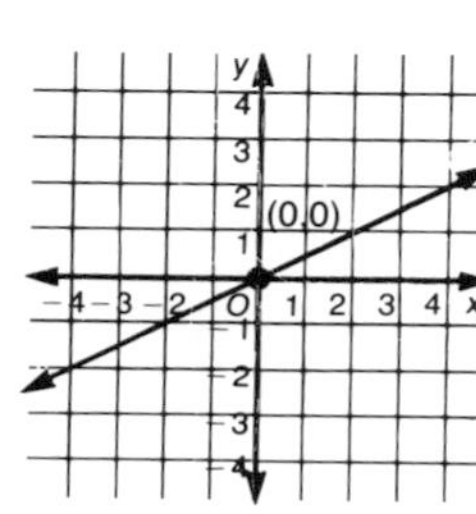

4.

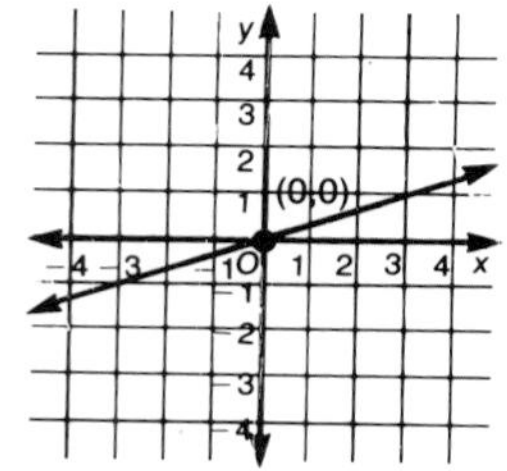

Page 411 Exercises

5.

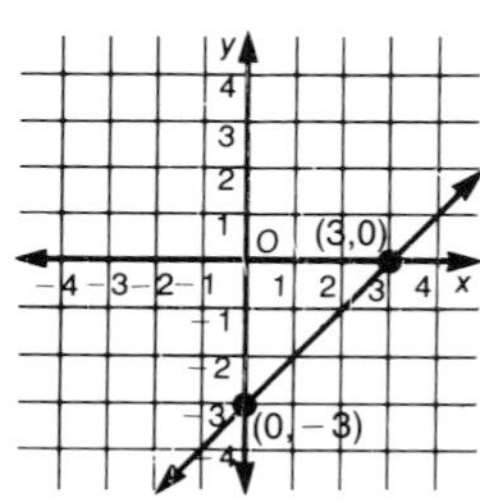

6.

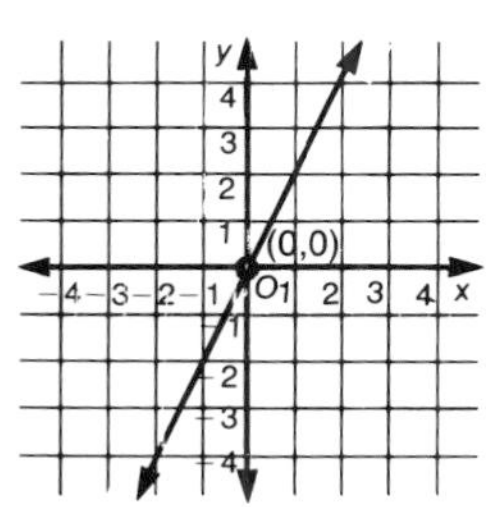

7.

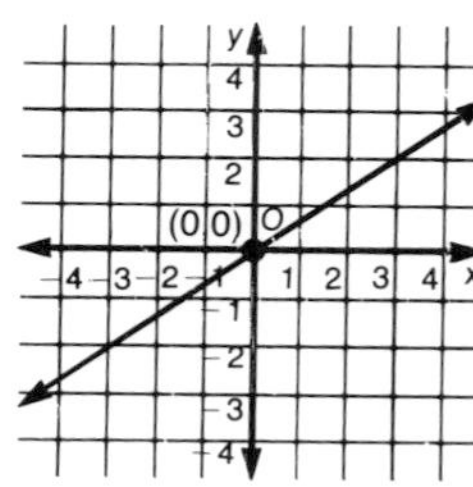

8.

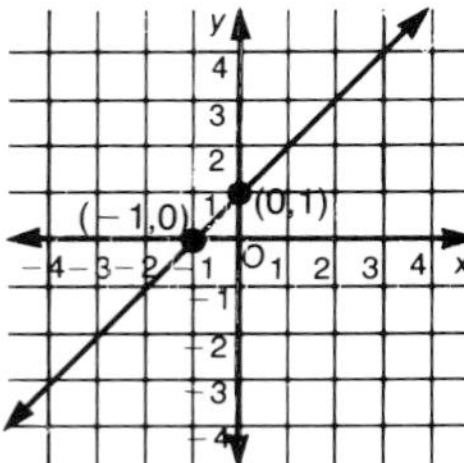

9.

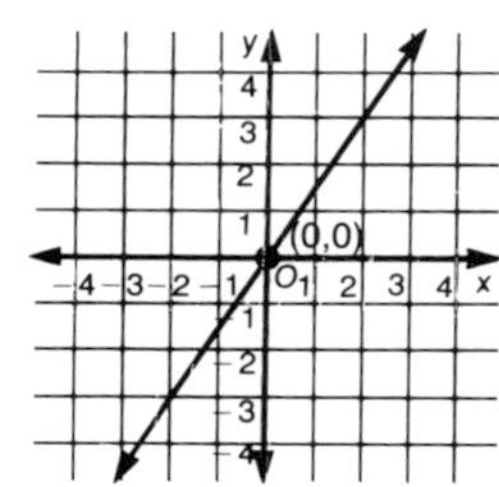

10.

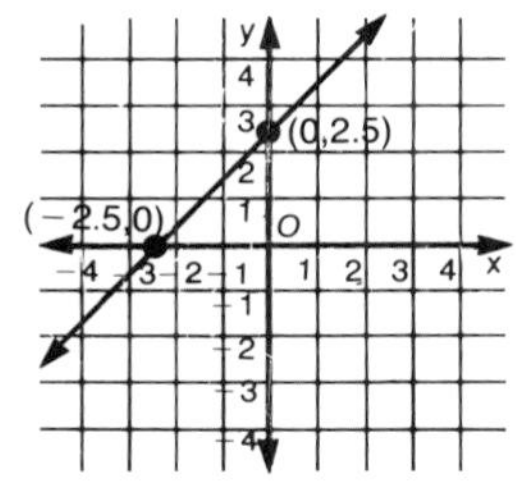

11.

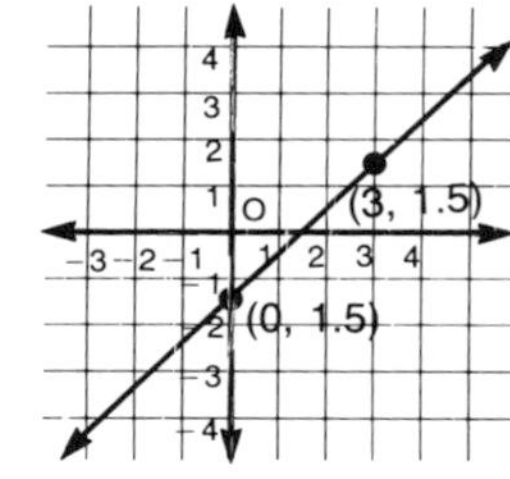

12.

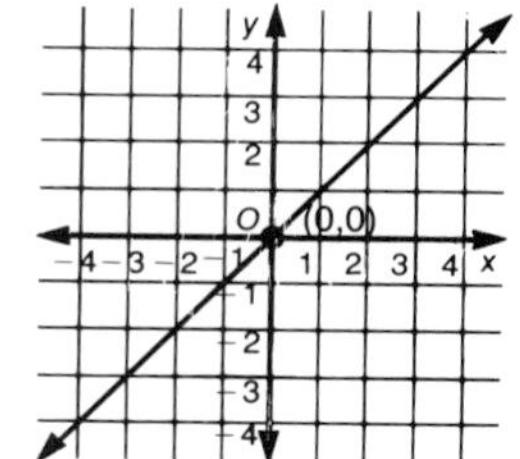

13.

(0,5)
(5,0)

14.

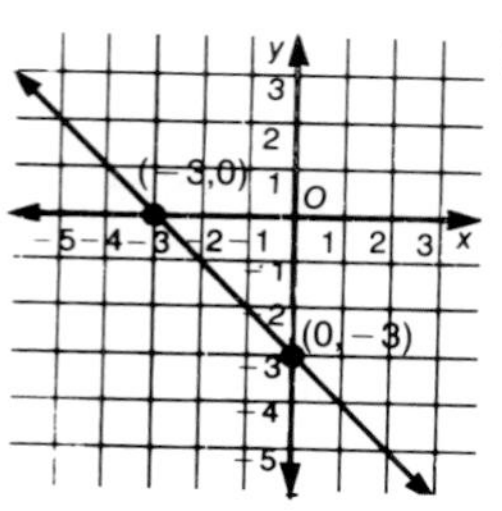

15.

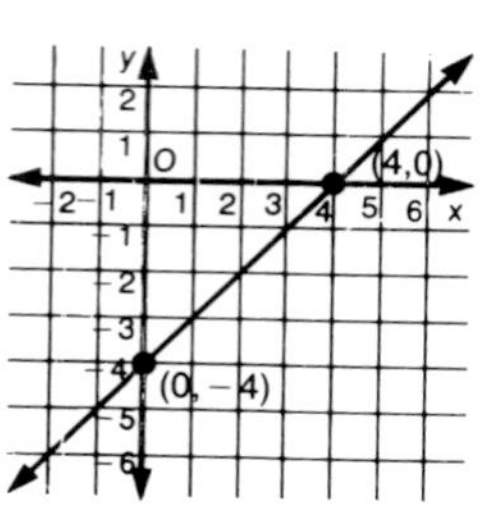

16.

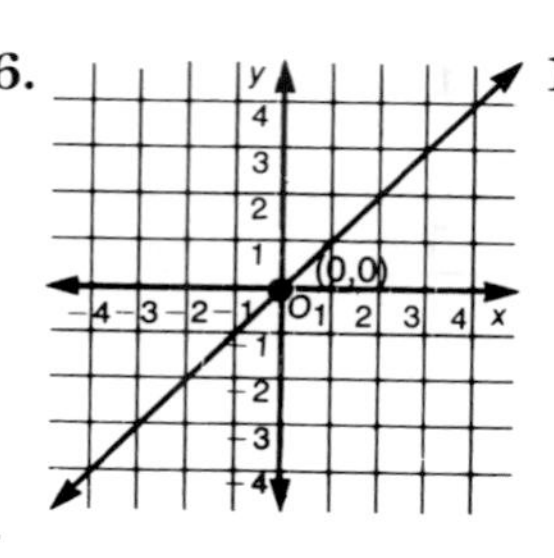

17.

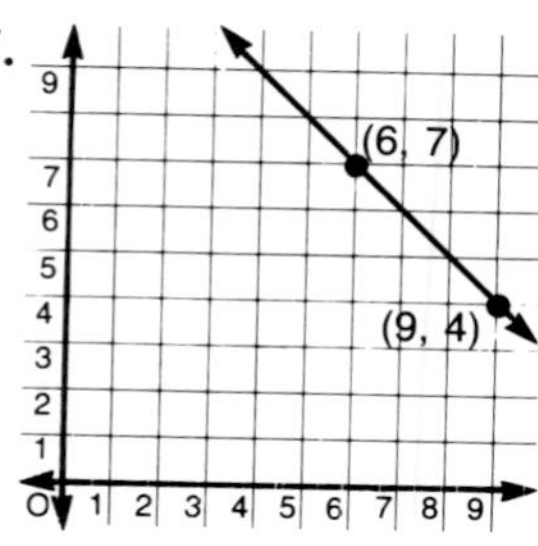

Page 417 Chapter 13 Review

30-35.

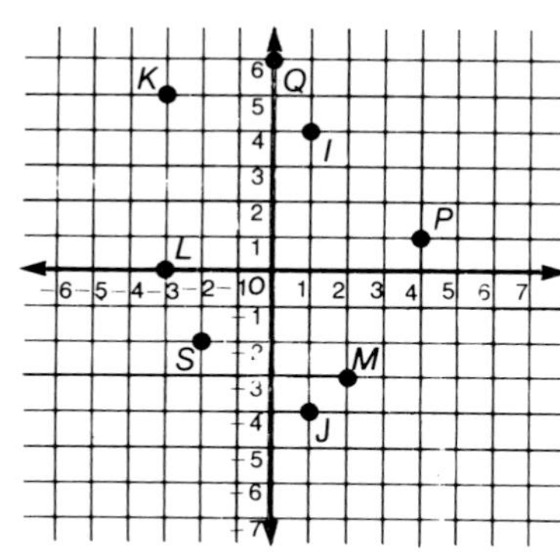

36.

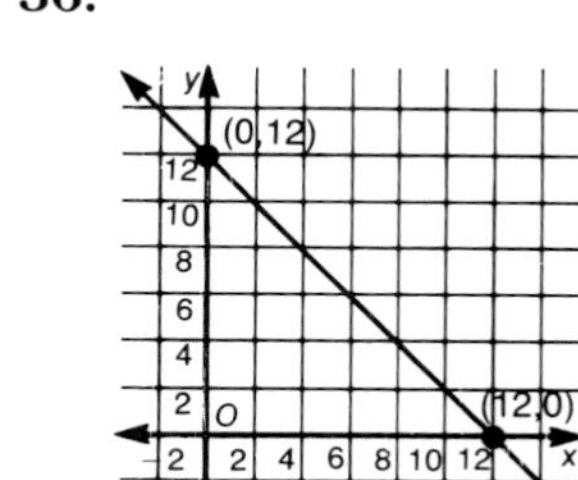

37.

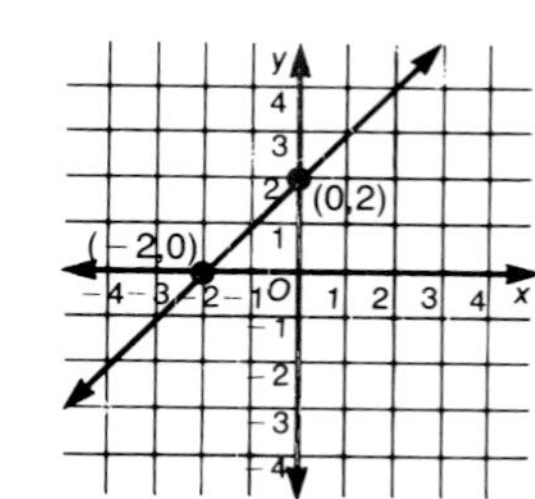

38.

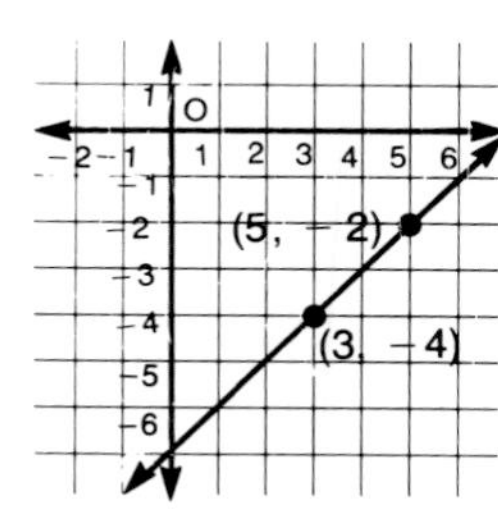

39.

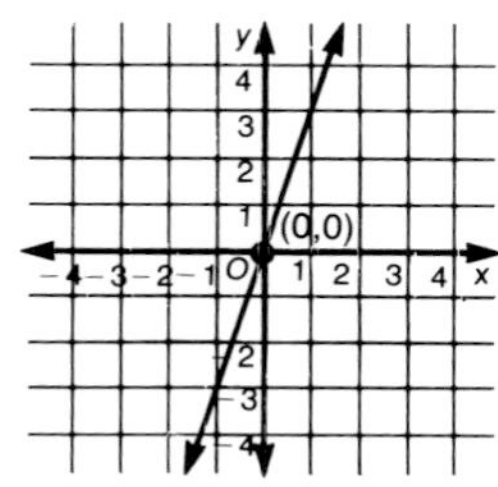

40.

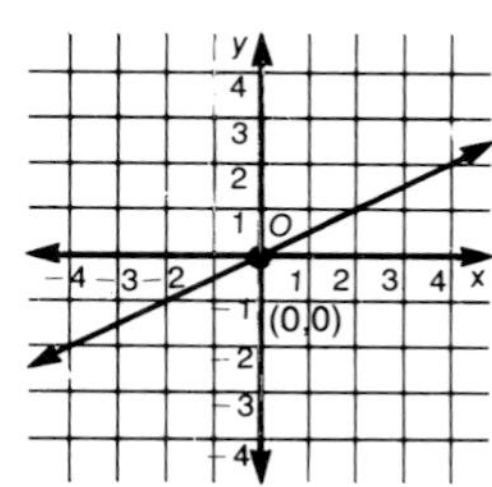

41.

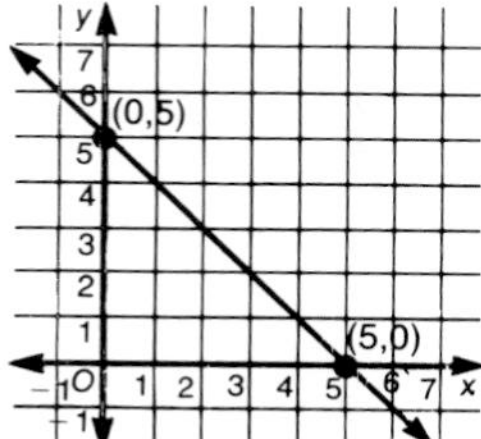

Page 418 Chapter 13 Test

14-17.

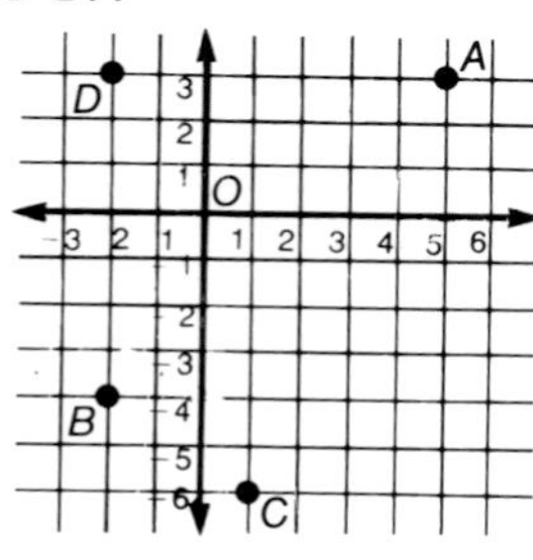

18.

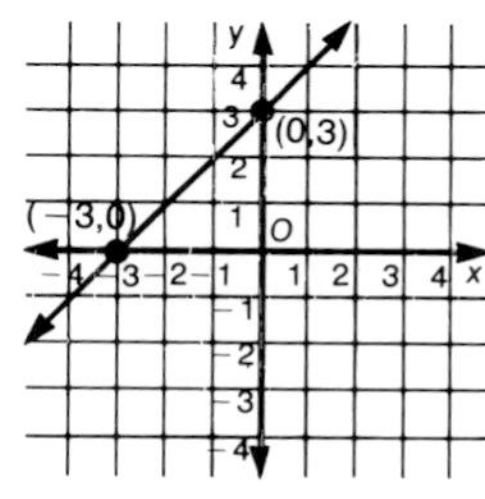

19.

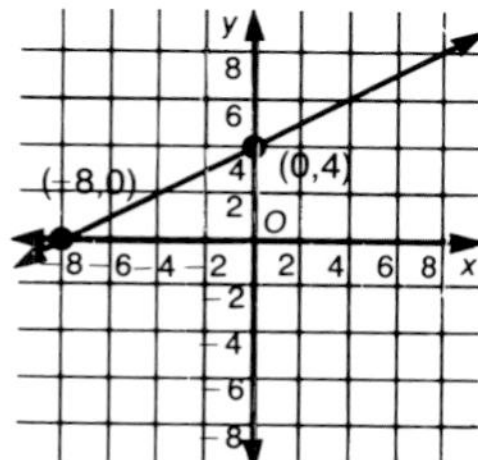

20.

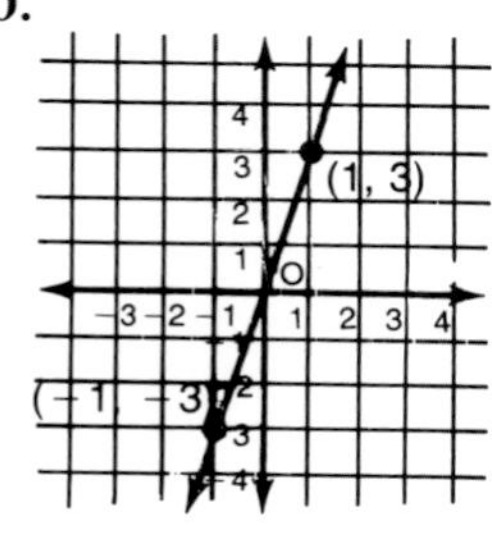

21.

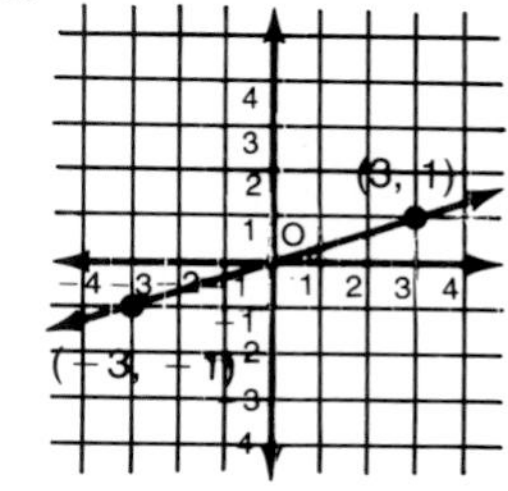

Page 420 Cumulative Review

17.

Ring Size	Tally	Frequency
5	||	2
6	~~||||~~	5
$6\frac{1}{2}$	|||	3
7	~~||||~~	5
$7\frac{1}{2}$	|||	3

18. 6.5, 6.5; two modes, 6, 7; $2\frac{1}{2}$

19.

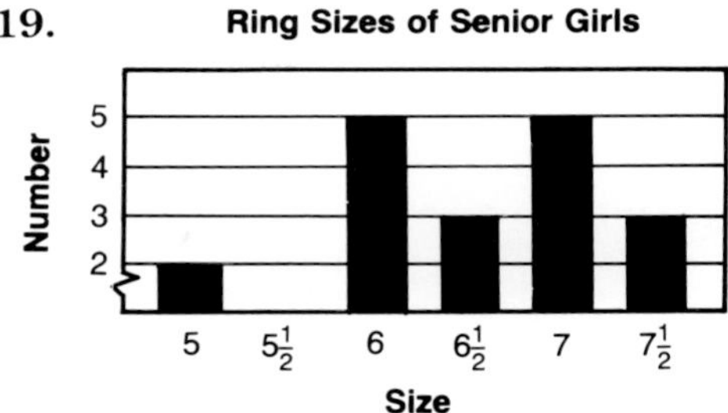

28.

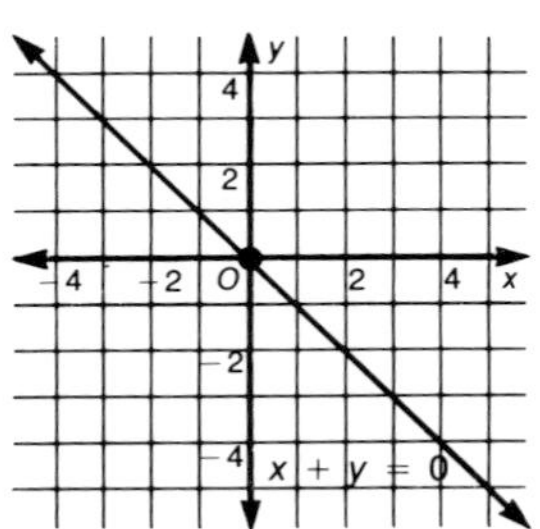

29.

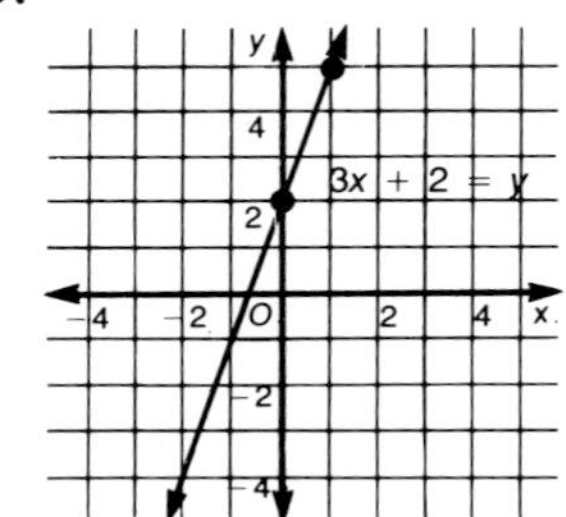

Chapter 14 Probability

Page 426 Guided Practice

1.

Dime	Spinner	Outcomes
Heads	Red	HR
	Yellow	HY
	Blue	HB
Tails	Red	TR
	Yellow	TY
	Blue	TB

2.

First Spinner	Second Spinner	Outcomes
Red	Red	RR
	Yellow	RY
	Blue	RB
	Green	RG
Yellow	Red	YR
	Yellow	YY
	Blue	YB
	Green	YG
Blue	Red	BR
	Yellow	BY
	Blue	BB
	Green	BG

Page 427 Exercises

5. $2 \times 2 \times 2 = 8$

Penny	Nickel	Dime	Outcomes
Heads	Heads	Heads	HHH
		Tails	HHT
	Tails	Heads	HTH
		Tails	HTT
Tails	Heads	Heads	THH
		Tails	THT
	Tails	Heads	TTH
		Tails	TTT

6. $3 \times 3 \times 2 = 18$

First Spinner	Second Spinner	Third Spinner	Outcomes
Red	Red	Black	RRB
		White	RRW
	Yellow	Black	RYB
		White	RYW
	Blue	Black	RBB
		White	RBW
Yellow	Red	Black	YRB
		White	YRW
	Yellow	Black	YYB
		White	YYW
	Blue	Black	YBB
		White	YBW
Blue	Red	Black	BRB
		White	BRW
	Yellow	Black	BYB
		White	BYW
	Blue	Black	BBB
		White	BBW

7. $2 \times 6 \times 2 = 24$

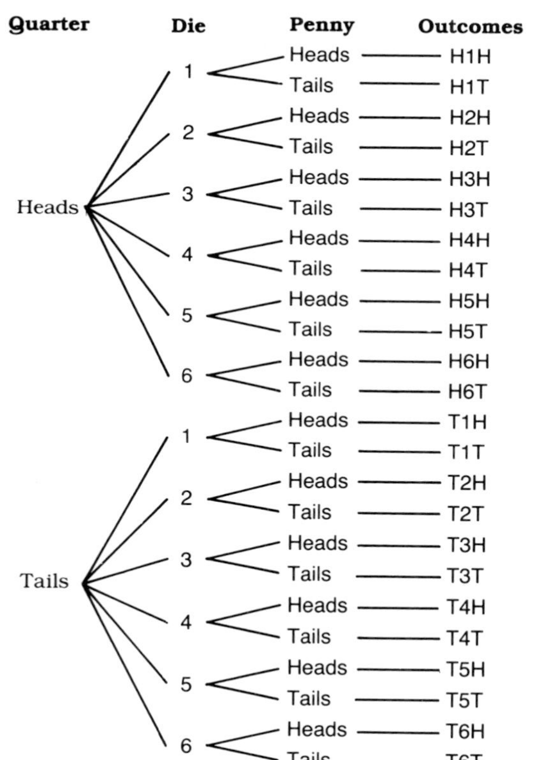

8. $2 \times 4 = 8$

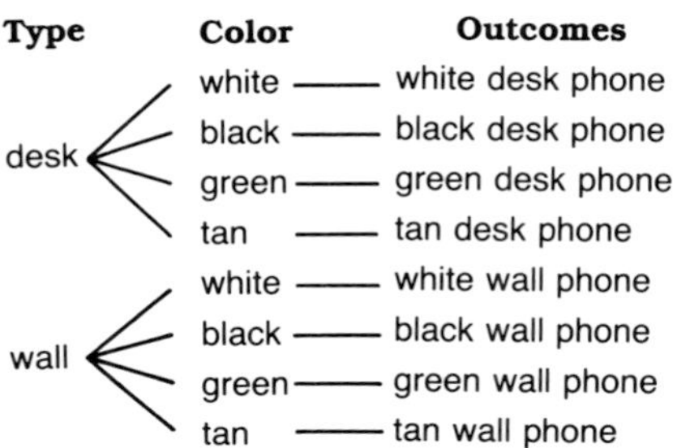

Page 442 Chapter 14 Review

8. $2 \times 3 = 6$

Type	Color	Outcomes
Probe	blue	blue Probe
	black	black Probe
	white	white Probe
Beretta	blue	blue Beretta
	black	black Beretta
	white	white Beretta

9. $3 \times 3 = 9$

Skirt	Blouse	Outcomes
green	white	green skirt, white blouse
	tan	green skirt, tan blouse
	yellow	green skirt, yellow blouse
brown	white	brown skirt, white blouse
	tan	brown skirt, tan blouse
	yellow	brown skirt, yellow blouse
black	white	black skirt, white blouse
	tan	black skirt, tan blouse
	yellow	black skirt, yellow blouse

Page 444 Chapter 14 Test

2.

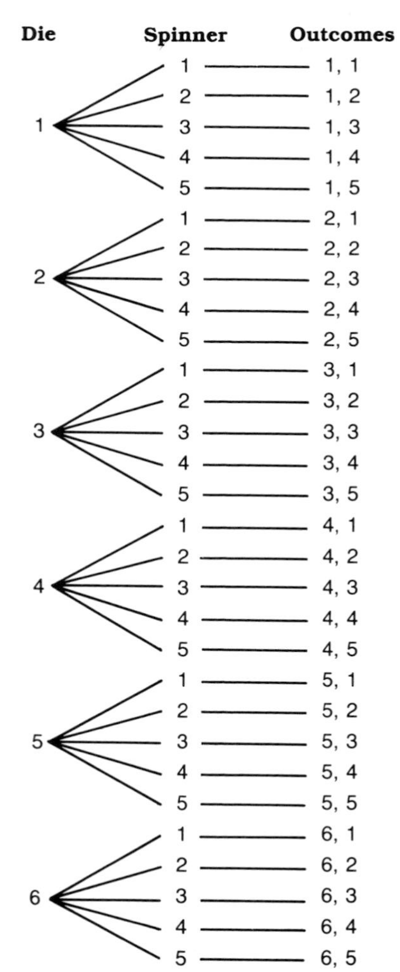

Extra Practice

Page 477 Lesson 11-4

1.

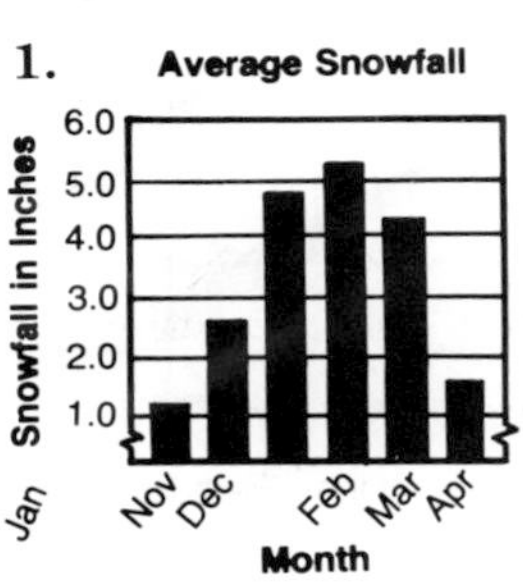

2.

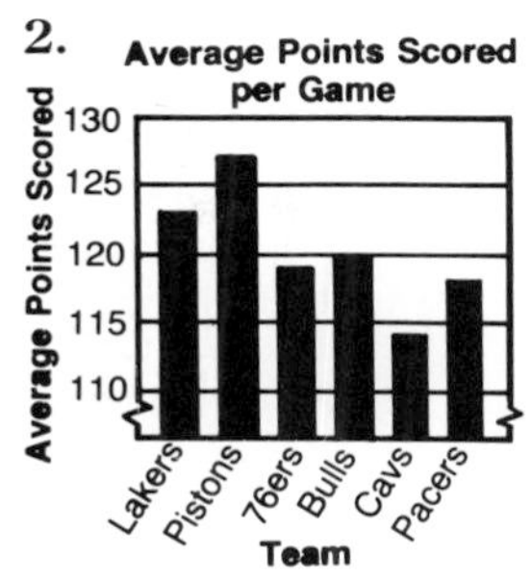

3.

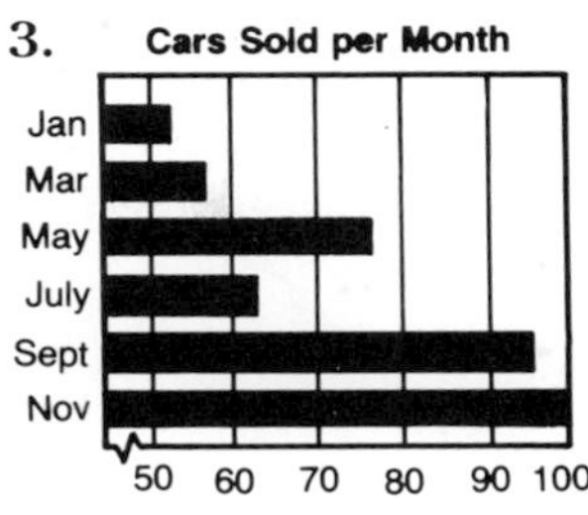

4.

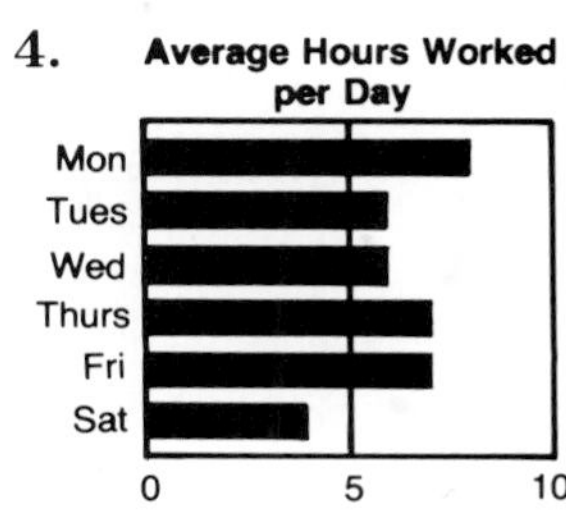

Page 478 Lesson 11-5

1.

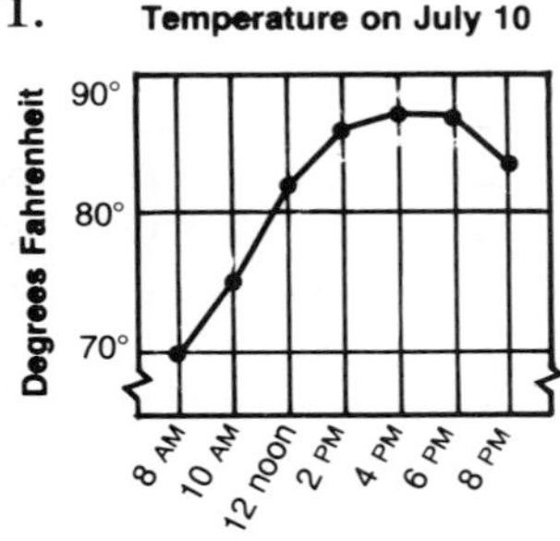

2.

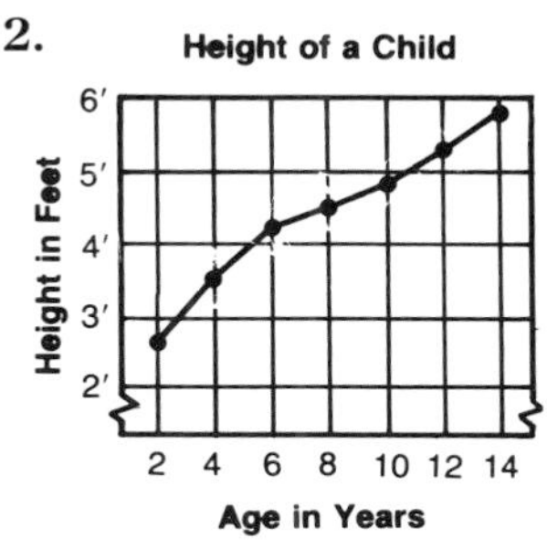

3.

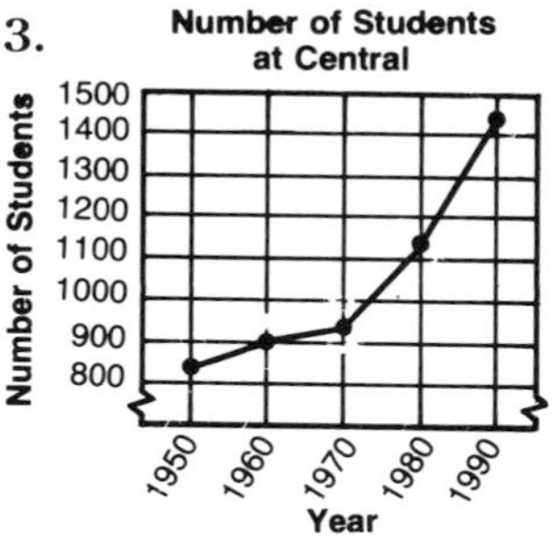

4.

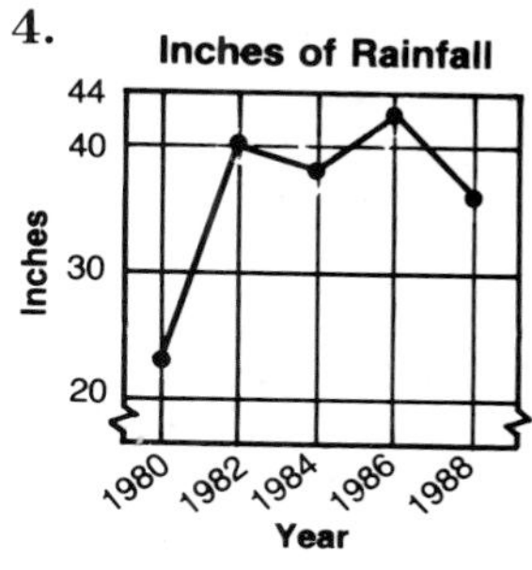

Page 478 Lesson 11-6

1.

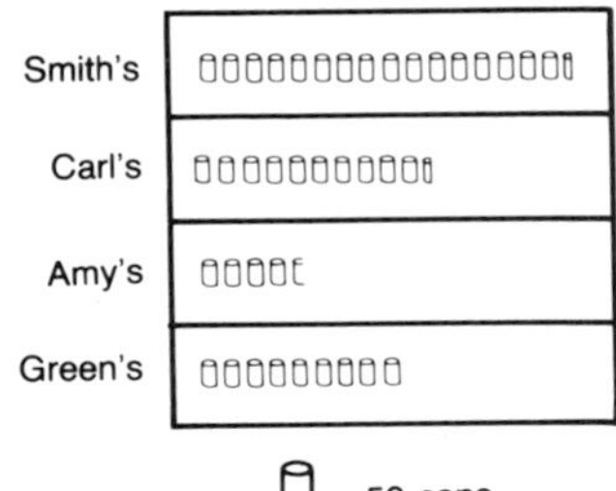

2.

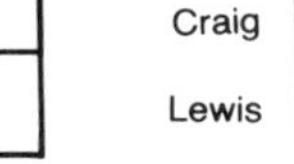

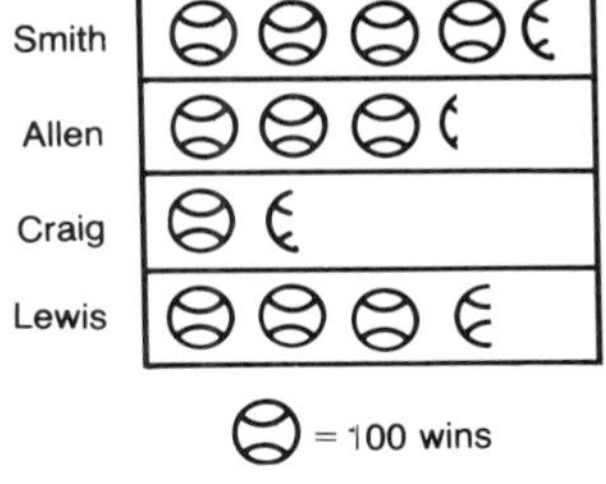

3.

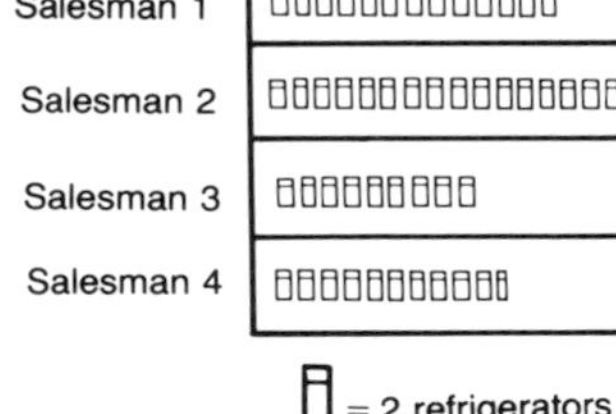

4.

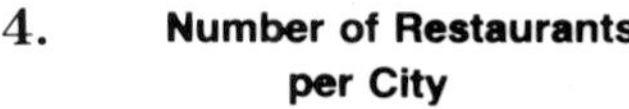

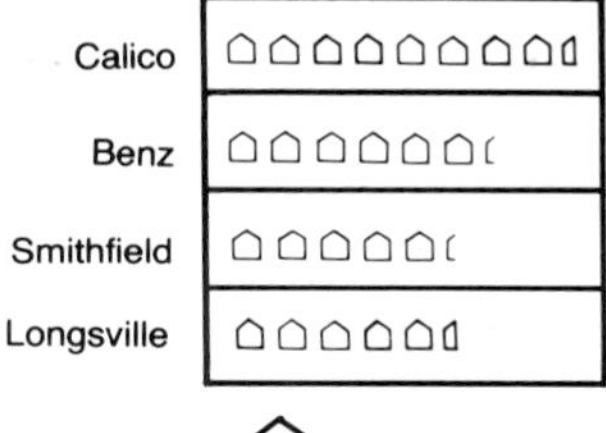

Page 478 Lesson 11-7

1. Household Income

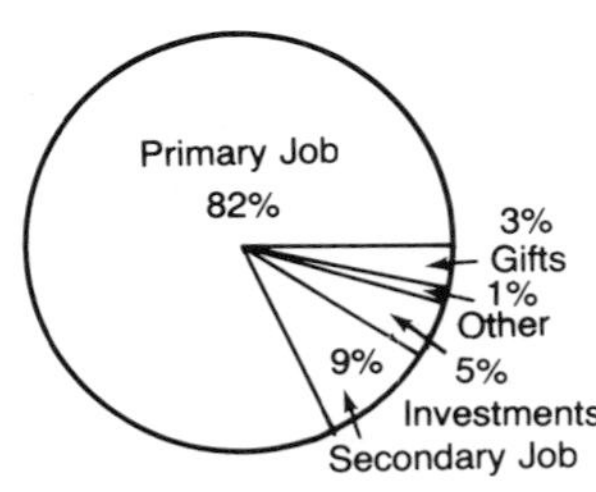

2. Sporting Goods Sales

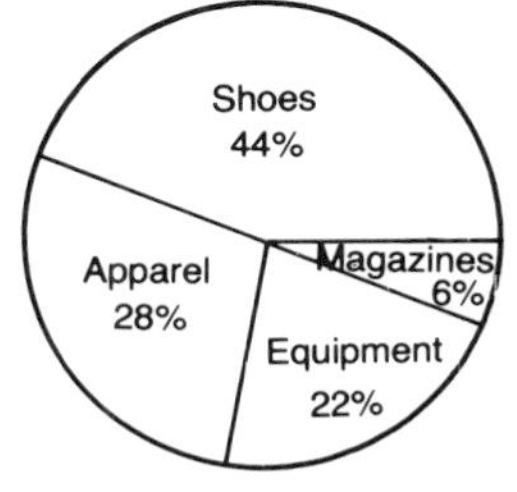

3. Energy Use in Home

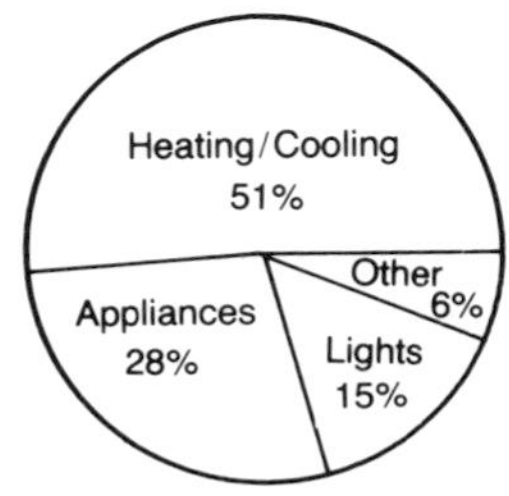

4. Students in North High School

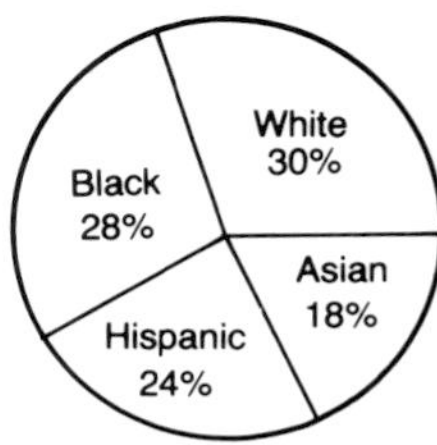

Page 482 Lesson 13-4

1-25.

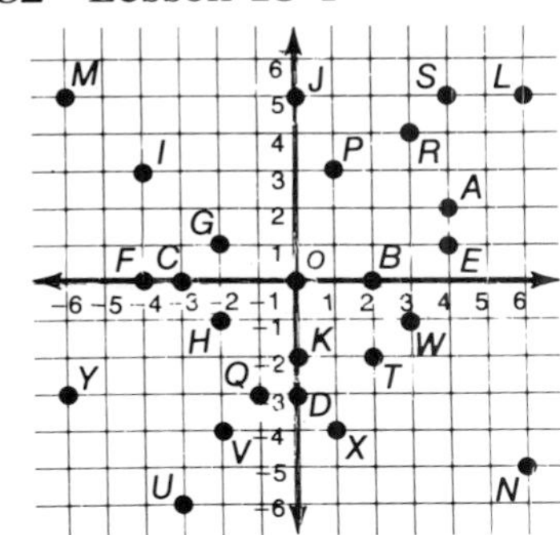

Page 483 Lesson 13-5

1. 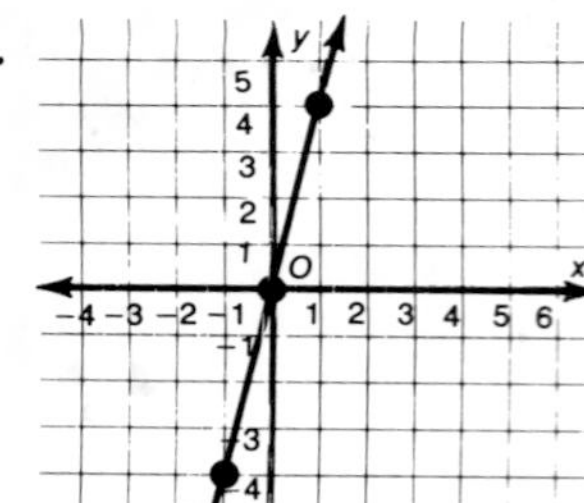

2. 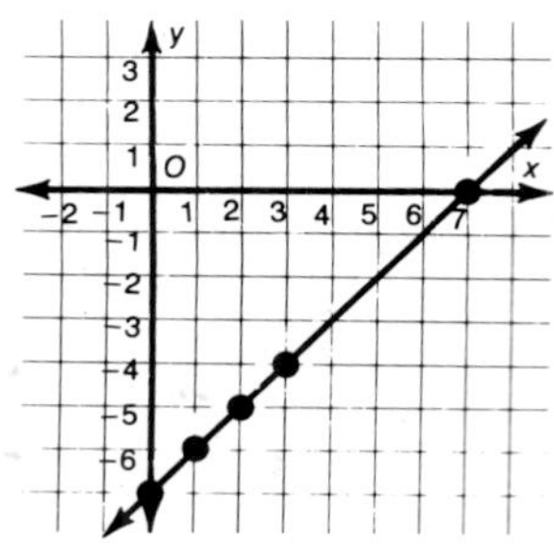

3. 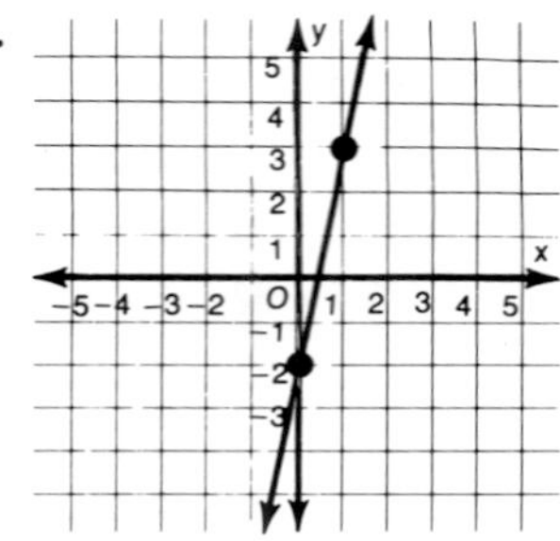

4. 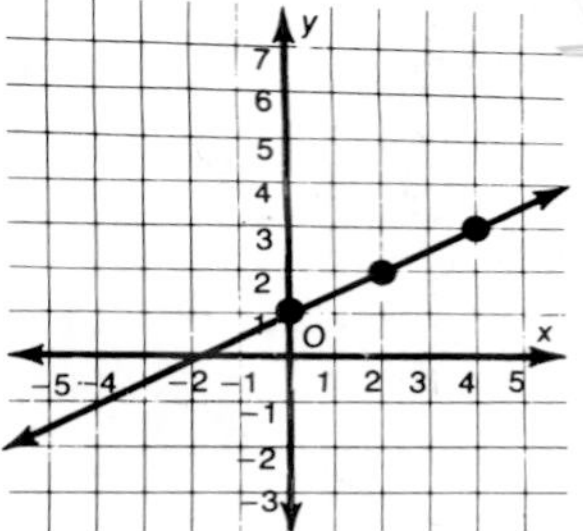

5. 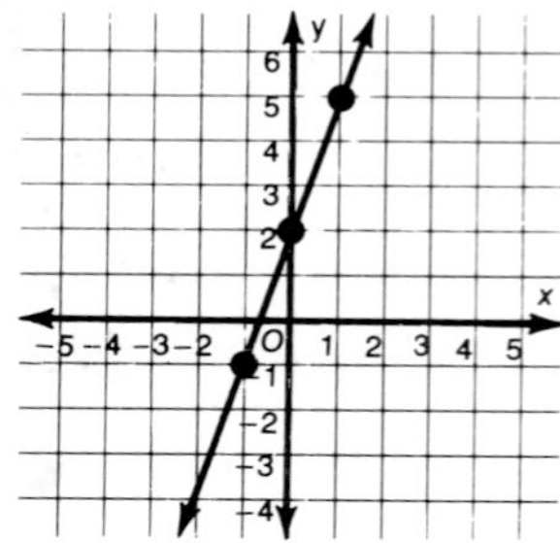

6. 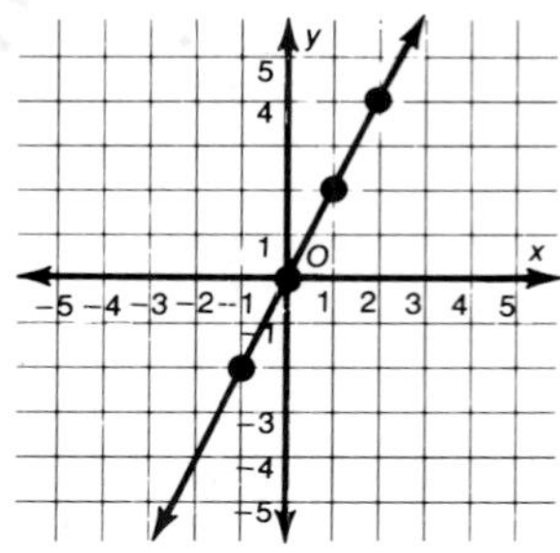

7. 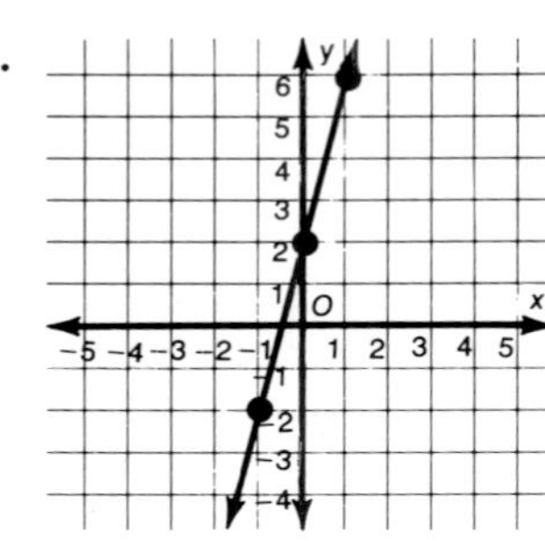

8. 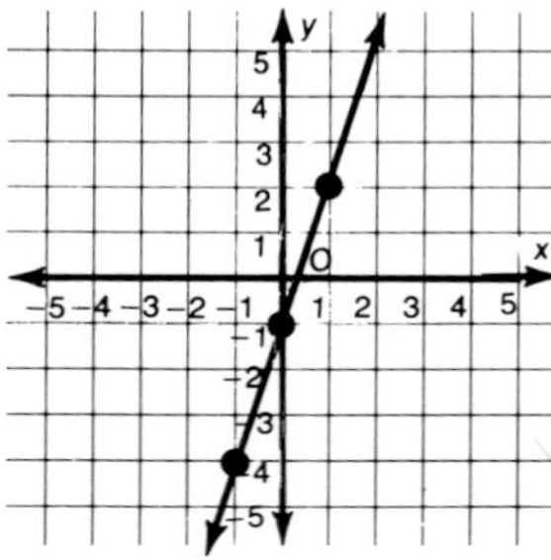

9. 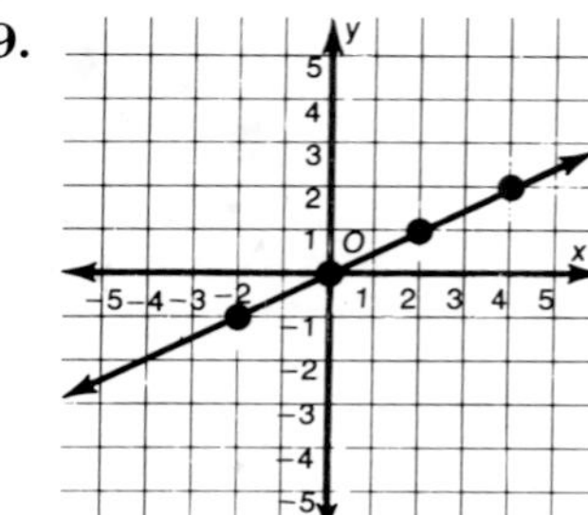

10. 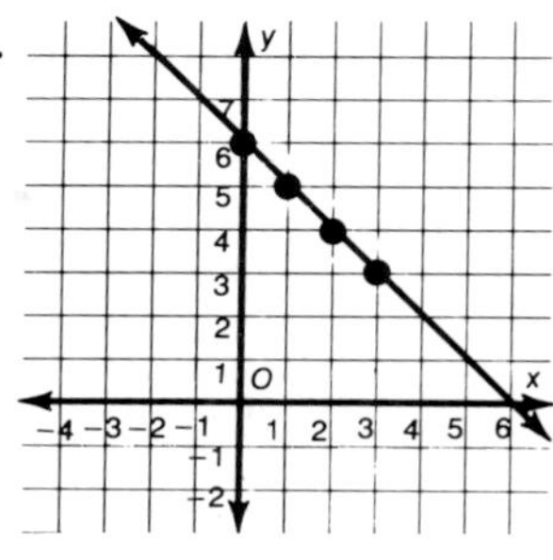

11. 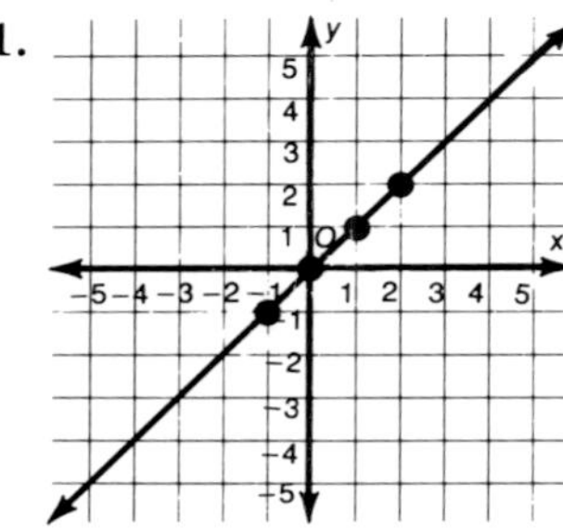

12. 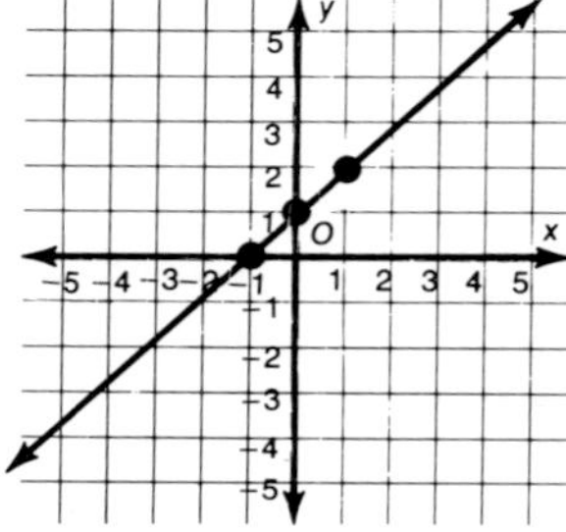

13. 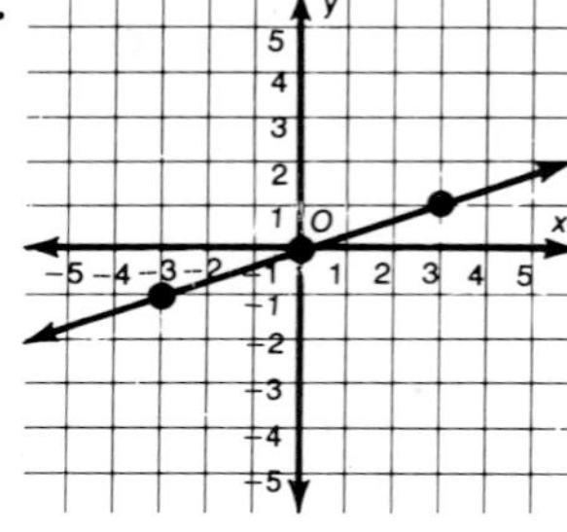

14. 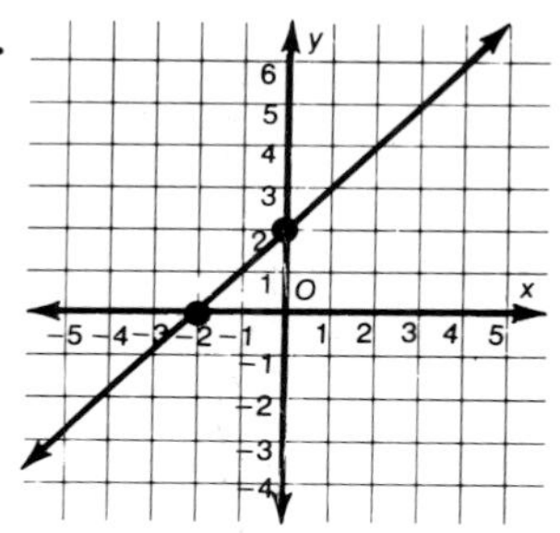

15. 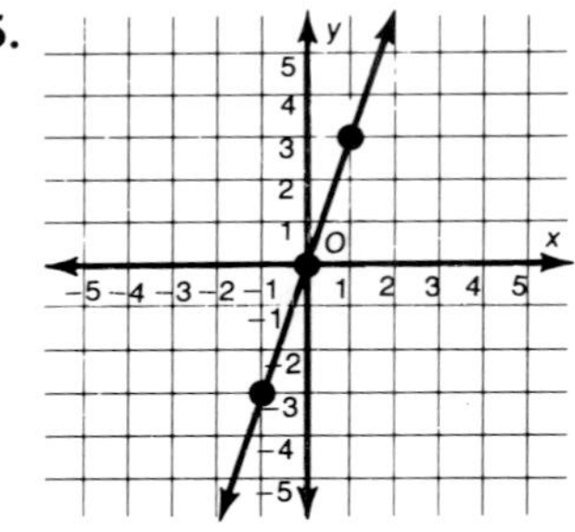

16. 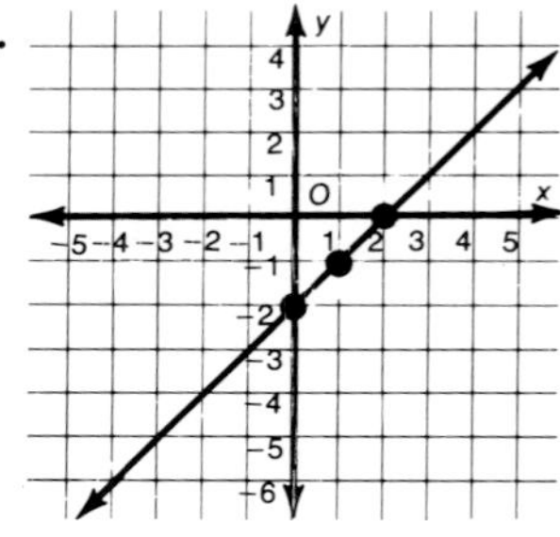

17. 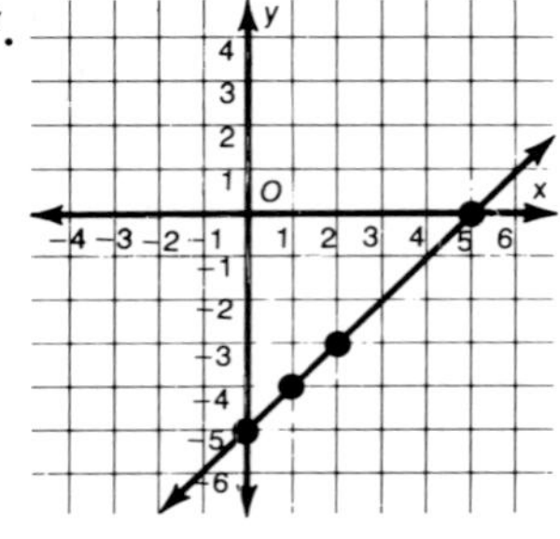

18. 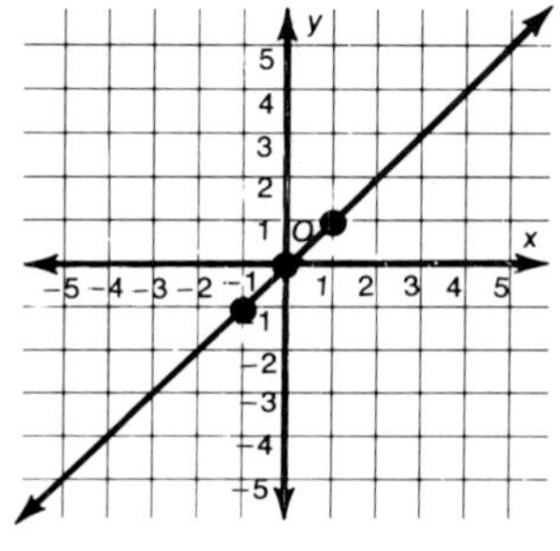

Page 483 Lesson 14-1

1. **Nickel Dime Outcomes**

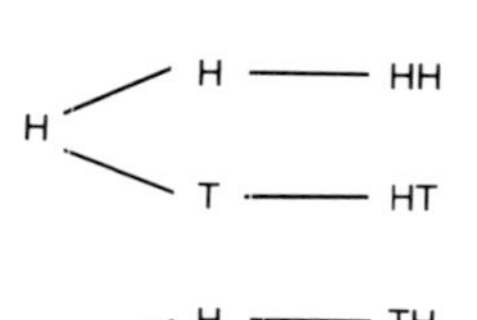

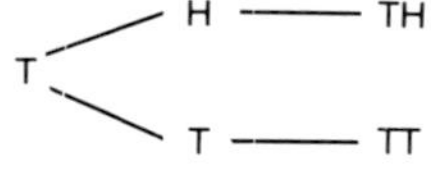

2. **Q D N Outcomes**

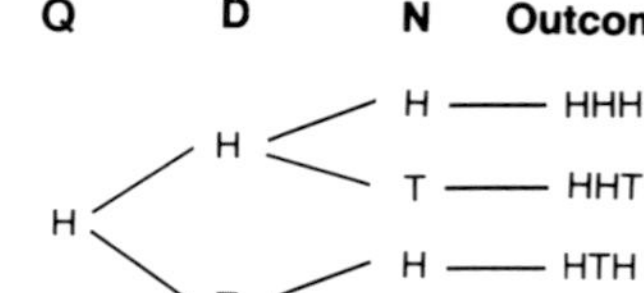

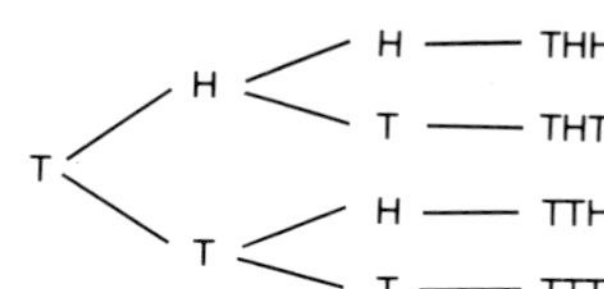